Unit Conversions (Equivalents)

D0086518

Length

1 in. = 2.54 cm (defined)
1 cm = 0.3937 in.
1 ft = 30.48 cm
1 m = 39.37 in. = 3.281 ft
1 mi = 5280 ft = 1.609 km
1 km = 0.6214 mi
1 nautical mile (U.S.) = 1.151 mi = 6076 ft = 1.852 km
1 fermi = 1 femtometer (fm) = 10^{-15} m
1 angstrom (Å) = 10^{-10} m = 0.1 nm
1 light-year (ly) = 9.461×10^{15} m
1 parsec = 3.26 ly = 3.09×10^{16} m

Volume

1 liter (L) = 1000 mL = 1000 cm^3 = 1.0×10^{-3} m^3 =
    1.057 qt (U.S.) = 61.02 $in.^3$
1 gal (U.S.) = 4 qt (U.S.) = 231 $in.^3$ = 3.785 L =
    0.8327 gal (British)
1 quart (U.S.) = 2 pints (U.S.) = 946 mL
1 pint (British) = 1.20 pints (U.S.) = 568 mL
1 m^3 = 35.31 ft^3

Speed

1 mi/h = 1.4667 ft/s = 1.6093 km/h = 0.4470 m/s
1 km/h = 0.2778 m/s = 0.6214 mi/h
1 ft/s = 0.3048 m/s (exact) = 0.6818 mi/h = 1.0973 km/h
1 m/s = 3.281 ft/s = 3.600 km/h = 2.237 mi/h
1 knot = 1.151 mi/h = 0.5144 m/s

Angle

1 radian (rad) = 57.30° = 57°18′
1° = 0.01745 rad
1 rev/min (rpm) = 0.1047 rad/s

Time

1 day = 8.640×10^4 s
1 year = 3.156×10^7 s

Mass

1 atomic mass unit (u) = 1.6605×10^{-27} kg
1 kg = 0.06852 slug
[1 kg has a weight of 2.20 lb where g = 9.80 m/s^2.]

Force

1 lb = 4.44822 N
1 N = 10^5 dyne = 0.2248 lb

Energy and Work

1 J = 10^7 ergs = 0.7376 ft·lb
1 ft·lb = 1.356 J = 1.29×10^{-3} Btu = 3.24×10^{-4} kcal
1 kcal = 4.19×10^3 J = 3.97 Btu
1 eV = 1.6022×10^{-19} J
1 kWh = 3.600×10^6 J = 860 kcal
1 Btu = 1.055×10^3 J

Power

1 W = 1 J/s = 0.7376 ft·lb/s = 3.41 Btu/h
1 hp = 550 ft·lb/s = 746 W

Pressure

1 atm = 1.01325 bar = 1.01325×10^5 N/m^2
    = 14.7 $lb/in.^2$ = 760 torr
1 $lb/in.^2$ = 6.895×10^3 N/m^2
1 Pa = 1 N/m^2 = 1.450×10^{-4} $lb/in.^2$

SI Derived Units and Their Abbreviations

Quantity	Unit	Abbreviation	In Terms of Base Units[†]
Force	newton	N	$kg·m/s^2$
Energy and work	joule	J	$kg·m^2/s^2$
Power	watt	W	$kg·m^2/s^3$
Pressure	pascal	Pa	$kg/(m·s^2)$
Frequency	hertz	Hz	s^{-1}
Electric charge	coulomb	C	A·s
Electric potential	volt	V	$kg·m^2/(A·s^3)$
Electric resistance	ohm	Ω	$kg·m^2/(A^2·s^3)$
Capacitance	farad	F	$A^2·s^4/(kg·m^2)$
Magnetic field	tesla	T	$kg/(A·s^2)$
Magnetic flux	weber	Wb	$kg·m^2/(A·s^2)$
Inductance	henry	H	$kg·m^2/(s^2·A^2)$

[†] kg = kilogram (mass), m = meter (length), s = second (time), A = ampere (electric current).

Metric (SI) Multipliers

Prefix	Abbreviation	Value
yotta	Y	10^{24}
zeta	Z	10^{21}
exa	E	10^{18}
peta	P	10^{15}
tera	T	10^{12}
giga	G	10^9
mega	M	10^6
kilo	k	10^3
hecto	h	10^2
deka	da	10^1
deci	d	10^{-1}
centi	c	10^{-2}
milli	m	10^{-3}
micro	μ	10^{-6}
nano	n	10^{-9}
pico	p	10^{-12}
femto	f	10^{-15}
atto	a	10^{-18}
zepto	z	10^{-21}
yocto	y	10^{-24}

FOURTH EDITION

PHYSICS

for
SCIENTISTS & ENGINEERS

DOUGLAS C. GIANCOLI

PEARSON

Prentice
Hall

Upper Saddle River, New Jersey 07458

Library of Congress Cataloging-in-Publication Data

Giancoli, Douglas C.
    Physics for scientists and engineers with modern physics / Douglas C.
Giancoli.—4th ed.
        p. cm.
    Includes bibliographical references and index.
    ISBN 0-13-227559-7
    1. Physics—Textbooks.   I. Title.
    QC21.3.G539 2008
    530—dc22

                                                    2006039431

President, ESM: Paul Corey
Sponsoring Editor: Christian Botting
Production Editor: Clare Romeo
Senior Managing Editor: Kathleen Schiaparelli
Art Director and Interior & Cover Designer: John Christiana
Manager, Art Production: Sean Hogan
Senior Development Editor: Karen Karlin
Copy Editor: Jocelyn Phillips
Proofreaders: Marne Evans, Jennie Kaufman, Karen Bosch, Colleen Brosnan,
    Gina Cheselka, and Traci Douglas
Senior Operations Specialist: Alan Fischer
Art Production Editor: Connie Long
Illustrators: Audrey Simonetti and Mark Landis
Photo Researchers: Mary Teresa Giancoli and Truitt & Marshall
Senior Administrative Coordinator: Trisha Tarricone
Composition: Emilcomp/Prepare Inc.;
                Pearson Education/Clara Bartunek, Lissette Quiñones
Photo credits appear on page A-62 which constitutes
    a continuation of the copyright page.

            Published by Pearson Education, Inc.
            Pearson Prentice Hall
            Pearson Education, Inc.
            Upper Saddle River, NJ 07458

Printed in the United States of America
10 9 8 7 6 5 4

ISBN   0-13-227559-7
ISBN 978-013-227559-0

Pearson Education LTD., *London*
Pearson Education Australia PTY, Limited, *Sydney*
Pearson Education Singapore, Pte. Ltd
Pearson Education North Asia Ltd, *Hong Kong*
Pearson Education Canada, Ltd., *Toronto*
Pearson Educación de Mexico, S.A. de C.V.
Pearson Education—Japan, *Tokyo*
Pearson Education Malaysia, Pte. Ltd

Brief Contents

1 Introduction, Measurement, Estimating 1

2 Describing Motion: Kinematics in One Dimension 18

3 Kinematics in Two or Three Dimensions; Vectors 51

4 Dynamics: Newton's Laws of Motion 83

5 Using Newton's Laws: Friction, Circular Motion, Drag Forces 112

6 Gravitation and Newton's Synthesis 139

7 Work and Energy 163

8 Conservation of Energy 183

9 Linear Momentum 214

10 Rotational Motion 248

11 Angular Momentum; General Rotation 284

12 Static Equilibrium; Elasticity and Fracture 311

13 Fluids 339

14 Oscillations 369

15 Wave Motion 395

16 Sound 424

17 Temperature, Thermal Expansion, and the Ideal Gas Law 454

18 Kinetic Theory of Gases 476

19 Heat and the First Law of Thermodynamics 496

20 Second Law of Thermodynamics 528

21 Electric Charge and Electric Field 559

22 Gauss's Law 591

23 Electric Potential 607

24 Capacitance, Dielectrics, Electric Energy Storage 628

25 Electric Currents and Resistance 651

26 DC Circuits 677

27 Magnetism 707

28 Sources of Magnetic Field 733

29 Electromagnetic Induction and Faraday's Law 758

30 Inductance, Electromagnetic Oscillations, and AC Circuits 785

31 Maxwell's Equations and Electromagnetic Waves 812

32 Light: Reflection and Refraction 837

33 Lenses and Optical Instruments 866

34 The Wave Nature of Light; Interference 900

35 Diffraction and Polarization 921

36 Special Theory of Relativity 951

37 Early Quantum Theory and Models of the Atom 987

Appendices

A Mathematical Formulas A–1
B Derivatives and Integrals A–6
C More on Dimensional Analysis A–8
D Gravitational Force due to a Spherical Mass Distribution A–9
E Differential Form of Maxwell's Equations A–12
F Selected Isotopes A–14

Contents

APPLICATIONS LIST   xii
PREFACE   xiv
TO STUDENTS   xviii
USE OF COLOR   xix

1 INTRODUCTION, MEASUREMENT, ESTIMATING   1

1–1   The Nature of Science   2
1–2   Models, Theories, and Laws   2
1–3   Measurement and Uncertainty; Significant Figures   3
1–4   Units, Standards, and the SI System   6
1–5   Converting Units   8
1–6   Order of Magnitude: Rapid Estimating   9
*1–7   Dimensions and Dimensional Analysis   12
   SUMMARY 14   QUESTIONS 14
   PROBLEMS 14   GENERAL PROBLEMS 16

2 DESCRIBING MOTION: KINEMATICS IN ONE DIMENSION   18

2–1   Reference Frames and Displacement   19
2–2   Average Velocity   20
2–3   Instantaneous Velocity   22
2–4   Acceleration   24
2–5   Motion at Constant Acceleration   28
2–6   Solving Problems   30
2–7   Freely Falling Objects   34
*2–8   Variable Acceleration; Integral Calculus   39
*2–9   Graphical Analysis and Numerical Integration   40
   SUMMARY 43   QUESTIONS 43
   PROBLEMS 44   GENERAL PROBLEMS 48

3 KINEMATICS IN TWO OR THREE DIMENSIONS; VECTORS   51

3–1   Vectors and Scalars   52
3–2   Addition of Vectors–Graphical Methods   52
3–3   Subtraction of Vectors, and Multiplication of a Vector by a Scalar   54
3–4   Adding Vectors by Components   55
3–5   Unit Vectors   59
3–6   Vector Kinematics   59
3–7   Projectile Motion   62
3–8   Solving Problems: Projectile Motion   64
3–9   Relative Velocity   71
   SUMMARY 74   QUESTIONS 75
   PROBLEMS 75   GENERAL PROBLEMS 80

4 DYNAMICS: NEWTON'S LAWS OF MOTION   83

4–1   Force   84
4–2   Newton's First Law of Motion   84
4–3   Mass   86
4–4   Newton's Second Law of Motion   86
4–5   Newton's Third Law of Motion   89
4–6   Weight—the Force of Gravity; the Normal Force   92
4–7   Solving Problems with Newton's Laws: Free-Body Diagrams   95
4–8   Problem Solving—A General Approach   102
   SUMMARY 102   QUESTIONS 103
   PROBLEMS 104   GENERAL PROBLEMS 109

5 USING NEWTON'S LAWS: FRICTION, CIRCULAR MOTION, DRAG FORCES   112

5–1   Applications of Newton's Laws Involving Friction   113
5–2   Uniform Circular Motion—Kinematics   119
5–3   Dynamics of Uniform Circular Motion   122
5–4   Highway Curves: Banked and Unbanked   126
*5–5   Nonuniform Circular Motion   128
*5–6   Velocity-Dependent Forces: Drag and Terminal Velocity   129
   SUMMARY 130   QUESTIONS 131
   PROBLEMS 132   GENERAL PROBLEMS 136

6 GRAVITATION AND NEWTON'S SYNTHESIS 139

6–1	Newton's Law of Universal Gravitation	140
6–2	Vector Form of Newton's Law of Universal Gravitation	143
6–3	Gravity Near the Earth's Surface; Geophysical Applications	143
6–4	Satellites and "Weightlessness"	146
6–5	Kepler's Laws and Newton's Synthesis	149
*6–6	Gravitational Field	154
6–7	Types of Forces in Nature	155
*6–8	Principle of Equivalence; Curvature of Space; Black Holes	155

SUMMARY 157   QUESTIONS 157
PROBLEMS 158   GENERAL PROBLEMS 160

Force   Displacement

7 WORK AND ENERGY 163

7–1	Work Done by a Constant Force	164
7–2	Scalar Product of Two Vectors	167
7–3	Work Done by a Varying Force	168
7–4	Kinetic Energy and the Work-Energy Principle	172

SUMMARY 176   QUESTIONS 177
PROBLEMS 177   GENERAL PROBLEMS 180

8 CONSERVATION OF ENERGY 183

8–1	Conservative and Nonconservative Forces	184
8–2	Potential Energy	186
8–3	Mechanical Energy and Its Conservation	189
8–4	Problem Solving Using Conservation of Mechanical Energy	190
8–5	The Law of Conservation of Energy	196
8–6	Energy Conservation with Dissipative Forces: Solving Problems	196
8–7	Gravitational Potential Energy and Escape Velocity	199
8–8	Power	201
*8–9	Potential Energy Diagrams; Stable and Unstable Equilibrium	204

SUMMARY 205   QUESTIONS 205
PROBLEMS 207   GENERAL PROBLEMS 211

9 LINEAR MOMENTUM 214

9–1	Momentum and Its Relation to Force	215
9–2	Conservation of Momentum	217
9–3	Collisions and Impulse	220
9–4	Conservation of Energy and Momentum in Collisions	222
9–5	Elastic Collisions in One Dimension	222
9–6	Inelastic Collisions	225
9–7	Collisions in Two or Three Dimensions	227
9–8	Center of Mass (CM)	230
9–9	Center of Mass and Translational Motion	234
*9–10	Systems of Variable Mass; Rocket Propulsion	236

SUMMARY 239   QUESTIONS 239
PROBLEMS 240   GENERAL PROBLEMS 245

10 ROTATIONAL MOTION 248

10–1	Angular Quantities	249
10–2	Vector Nature of Angular Quantities	254
10–3	Constant Angular Acceleration	255
10–4	Torque	256
10–5	Rotational Dynamics; Torque and Rotational Inertia	258
10–6	Solving Problems in Rotational Dynamics	260
10–7	Determining Moments of Inertia	263
10–8	Rotational Kinetic Energy	265
10–9	Rotational Plus Translational Motion; Rolling	267
*10–10	Why Does a Rolling Sphere Slow Down?	273

SUMMARY 274   QUESTIONS 275
PROBLEMS 276   GENERAL PROBLEMS 281

11 ANGULAR MOMENTUM; GENERAL ROTATION 284

11–1	Angular Momentum—Objects Rotating About a Fixed Axis	285
11–2	Vector Cross Product; Torque as a Vector	289
11–3	Angular Momentum of a Particle	291
11–4	Angular Momentum and Torque for a System of Particles; General Motion	292
11–5	Angular Momentum and Torque for a Rigid Object	294
11–6	Conservation of Angular Momentum	297
*11–7	The Spinning Top and Gyroscope	299
*11–8	Rotating Frames of Reference; Inertial Forces	300
*11–9	The Coriolis Effect	301

SUMMARY 302
QUESTIONS 303
PROBLEMS 303
GENERAL PROBLEMS 308

14 OSCILLATIONS 369

14–1 Oscillations of a Spring 370
14–2 Simple Harmonic Motion 372
14–3 Energy in the Simple Harmonic Oscillator 377
14–4 Simple Harmonic Motion Related to Uniform Circular Motion 379
14–5 The Simple Pendulum 379
*14–6 The Physical Pendulum and the Torsion Pendulum 381
14–7 Damped Harmonic Motion 382
14–8 Forced Oscillations; Resonance 385
SUMMARY 387 QUESTIONS 388
PROBLEMS 388 GENERAL PROBLEMS 392

12 STATIC EQUILIBRIUM; ELASTICITY AND FRACTURE 311

12–1 The Conditions for Equilibrium 312
12–2 Solving Statics Problems 313
12–3 Stability and Balance 317
12–4 Elasticity; Stress and Strain 318
12–5 Fracture 322
*12–6 Trusses and Bridges 324
*12–7 Arches and Domes 327
SUMMARY 329 QUESTIONS 329
PROBLEMS 330 GENERAL PROBLEMS 334

13 FLUIDS 339

13–1 Phases of Matter 340
13–2 Density and Specific Gravity 340
13–3 Pressure in Fluids 341
13–4 Atmospheric Pressure and Gauge Pressure 345
13–5 Pascal's Principle 346
13–6 Measurement of Pressure; Gauges and the Barometer 346
13–7 Buoyancy and Archimedes' Principle 348
13–8 Fluids in Motion; Flow Rate and the Equation of Continuity 352
13–9 Bernoulli's Equation 354
13–10 Applications of Bernoulli's Principle: Torricelli, Airplanes, Baseballs, TIA 356
*13–11 Viscosity 358
*13–12 Flow in Tubes: Poiseuille's Equation, Blood Flow 358
*13–13 Surface Tension and Capillarity 359
*13–14 Pumps, and the Heart 361
SUMMARY 361 QUESTIONS 362
PROBLEMS 363 GENERAL PROBLEMS 367

15 WAVE MOTION 395

15–1 Characteristics of Wave Motion 396
15–2 Types of Waves: Transverse and Longitudinal 398
15–3 Energy Transported by Waves 402
15–4 Mathematical Representation of a Traveling Wave 404
*15–5 The Wave Equation 406
15–6 The Principle of Superposition 408
15–7 Reflection and Transmission 409
15–8 Interference 410
15–9 Standing Waves; Resonance 412
*15–10 Refraction 415
*15–11 Diffraction 416
SUMMARY 417 QUESTIONS 417
PROBLEMS 418 GENERAL PROBLEMS 422

16 SOUND 424

16–1 Characteristics of Sound 425
16–2 Mathematical Representation of Longitudinal Waves 426
16–3 Intensity of Sound: Decibels 427
16–4 Sources of Sound: Vibrating Strings and Air Columns 431
*16–5 Quality of Sound, and Noise; Superposition 436
16–6 Interference of Sound Waves; Beats 437
16–7 Doppler Effect 439
*16–8 Shock Waves and the Sonic Boom 443
*16–9 Applications: Sonar, Ultrasound, and Medical Imaging 444
SUMMARY 446 QUESTIONS 447
PROBLEMS 448 GENERAL PROBLEMS 451

17 TEMPERATURE, THERMAL EXPANSION, AND THE IDEAL GAS LAW 454

17–1	Atomic Theory of Matter	455
17–2	Temperature and Thermometers	456
17–3	Thermal Equilibrium and the Zeroth Law of Thermodynamics	459
17–4	Thermal Expansion	459
*17–5	Thermal Stresses	463
17–6	The Gas Laws and Absolute Temperature	463
17–7	The Ideal Gas Law	465
17–8	Problem Solving with the Ideal Gas Law	466
17–9	Ideal Gas Law in Terms of Molecules: Avogadro's Number	468
*17–10	Ideal Gas Temperature Scale— a Standard	469

SUMMARY 470   QUESTIONS 471
PROBLEMS 471   GENERAL PROBLEMS 474

18 KINETIC THEORY OF GASES 476

18–1	The Ideal Gas Law and the Molecular Interpretation of Temperature	476
18–2	Distribution of Molecular Speeds	480
18–3	Real Gases and Changes of Phase	482
18–4	Vapor Pressure and Humidity	484
*18–5	Van der Waals Equation of State	486
*18–6	Mean Free Path	487
*18–7	Diffusion	489

SUMMARY 490   QUESTIONS 491
PROBLEMS 492   GENERAL PROBLEMS 494

19 HEAT AND THE FIRST LAW OF THERMODYNAMICS 496

19–1	Heat as Energy Transfer	497
19–2	Internal Energy	498
19–3	Specific Heat	499
19–4	Calorimetry—Solving Problems	500
19–5	Latent Heat	502
19–6	The First Law of Thermodynamics	505
19–7	The First Law of Thermodynamics Applied; Calculating the Work	507
19–8	Molar Specific Heats for Gases, and the Equipartition of Energy	511
19–9	Adiabatic Expansion of a Gas	514
19–10	Heat Transfer: Conduction, Convection, Radiation	515

SUMMARY 520   QUESTIONS 521
PROBLEMS 522   GENERAL PROBLEMS 526

20 SECOND LAW OF THERMODYNAMICS 528

20–1	The Second Law of Thermodynamics—Introduction	529
20–2	Heat Engines	530
20–3	Reversible and Irreversible Processes; the Carnot Engine	533
20–4	Refrigerators, Air Conditioners, and Heat Pumps	536
20–5	Entropy	539
20–6	Entropy and the Second Law of Thermodynamics	541
20–7	Order to Disorder	544
20–8	Unavailability of Energy; Heat Death	545
*20–9	Statistical Interpretation of Entropy and the Second Law	546
*20–10	Thermodynamic Temperature; Third Law of Thermodynamics	548
*20–11	Thermal Pollution, Global Warming, and Energy Resources	549

SUMMARY 551   QUESTIONS 552
PROBLEMS 552   GENERAL PROBLEMS 556

21 ELECTRIC CHARGE AND ELECTRIC FIELD 559

21–1	Static Electricity; Electric Charge and Its Conservation	560
21–2	Electric Charge in the Atom	561
21–3	Insulators and Conductors	561
21–4	Induced Charge; the Electroscope	562
21–5	Coulomb's Law	563
21–6	The Electric Field	568
21–7	Electric Field Calculations for Continuous Charge Distributions	572
21–8	Field Lines	575
21–9	Electric Fields and Conductors	577
21–10	Motion of a Charged Particle in an Electric Field	578
21–11	Electric Dipoles	579
*21–12	Electric Forces in Molecular Biology; DNA	581
*21–13	Photocopy Machines and Computer Printers Use Electrostatics	582

SUMMARY 584 QUESTIONS 584
PROBLEMS 585 GENERAL PROBLEMS 589

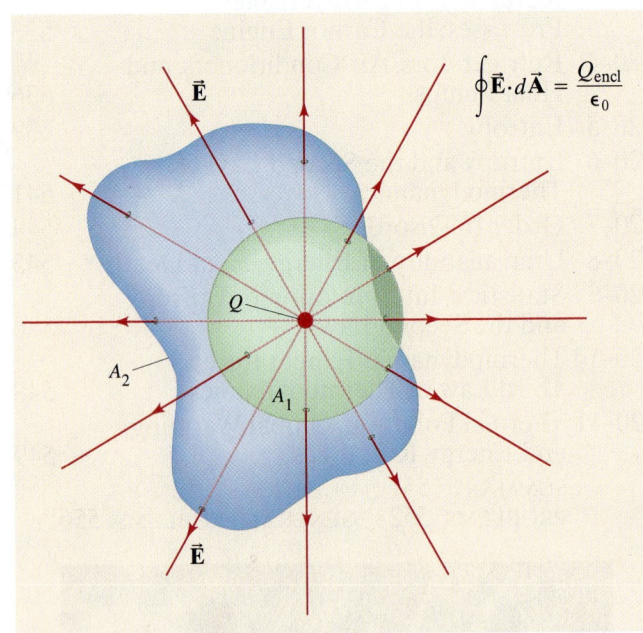

$$\oint \vec{\mathbf{E}} \cdot d\vec{\mathbf{A}} = \frac{Q_{encl}}{\epsilon_0}$$

22 GAUSS'S LAW 591

22–1	Electric Flux	592
22–2	Gauss's Law	593
22–3	Applications of Gauss's Law	595
*22–4	Experimental Basis of Gauss's and Coulomb's Laws	600

SUMMARY 601 QUESTIONS 601
PROBLEMS 601 GENERAL PROBLEMS 605

23 ELECTRIC POTENTIAL 607

23–1	Electric Potential Energy and Potential Difference	607
23–2	Relation between Electric Potential and Electric Field	610
23–3	Electric Potential Due to Point Charges	612
23–4	Potential Due to Any Charge Distribution	614
23–5	Equipotential Surfaces	616
23–6	Electric Dipole Potential	617
23–7	$\vec{\mathbf{E}}$ Determined from V	617
23–8	Electrostatic Potential Energy; the Electron Volt	619
*23–9	Cathode Ray Tube: TV and Computer Monitors, Oscilloscope	620

SUMMARY 622 QUESTIONS 622
PROBLEMS 623 GENERAL PROBLEMS 626

24 CAPACITANCE, DIELECTRICS, ELECTRIC ENERGY STORAGE 628

24–1	Capacitors	628
24–2	Determination of Capacitance	630
24–3	Capacitors in Series and Parallel	633
24–4	Electric Energy Storage	636
24–5	Dielectrics	638
*24–6	Molecular Description of Dielectrics	640

SUMMARY 643 QUESTIONS 643
PROBLEMS 644 GENERAL PROBLEMS 648

25 ELECTRIC CURRENTS AND RESISTANCE 651

25–1	The Electric Battery	652
25–2	Electric Current	654
25–3	Ohm's Law: Resistance and Resistors	655
25–4	Resistivity	658
25–5	Electric Power	660
25–6	Power in Household Circuits	662
25–7	Alternating Current	664
25–8	Microscopic View of Electric Current: Current Density and Drift Velocity	666
*25–9	Superconductivity	668
*25–10	Electrical Conduction in the Nervous System	669

SUMMARY 671 QUESTIONS 671
PROBLEMS 672 GENERAL PROBLEMS 675

26 DC CIRCUITS 677

26–1	EMF and Terminal Voltage	678
26–2	Resistors in Series and in Parallel	679
26–3	Kirchhoff's Rules	683
26–4	Series and Parallel EMFs; Battery Charging	686
26–5	Circuits Containing Resistor and Capacitor (RC Circuits)	687
26–6	Electric Hazards	692
*26–7	Ammeters and Voltmeters	695

SUMMARY 698 QUESTIONS 698
PROBLEMS 699 GENERAL PROBLEMS 704

I

27 MAGNETISM 707

27–1	Magnets and Magnetic Fields	707
27–2	Electric Currents Produce Magnetic Fields	710
27–3	Force on an Electric Current in a Magnetic Field; Definition of $\vec{B}$	710
27–4	Force on an Electric Charge Moving in a Magnetic Field	714
27–5	Torque on a Current Loop; Magnetic Dipole Moment	718
*27–6	Applications: Motors, Loudspeakers, Galvanometers	720
27–7	Discovery and Properties of the Electron	721
*27–8	The Hall Effect	723
*27–9	Mass Spectrometer	724
	SUMMARY 725 QUESTIONS 726	
	PROBLEMS 727 GENERAL PROBLEMS 730	

28 SOURCES OF MAGNETIC FIELD 733

28–1	Magnetic Field Due to a Straight Wire	734
28–2	Force between Two Parallel Wires	735
28–3	Definitions of the Ampere and the Coulomb	736
28–4	Ampère's Law	737
28–5	Magnetic Field of a Solenoid and a Toroid	741
28–6	Biot-Savart Law	743
28–7	Magnetic Materials—Ferromagnetism	746
*28–8	Electromagnets and Solenoids—Applications	747
*28–9	Magnetic Fields in Magnetic Materials; Hysteresis	748
*28–10	Paramagnetism and Diamagnetism	749
	SUMMARY 750 QUESTIONS 751	
	PROBLEMS 751 GENERAL PROBLEMS 755	

29 ELECTROMAGNETIC INDUCTION AND FARADAY'S LAW 758

29–1	Induced EMF	759
29–2	Faraday's Law of Induction; Lenz's Law	760
29–3	EMF Induced in a Moving Conductor	765
29–4	Electric Generators	766
*29–5	Back EMF and Counter Torque; Eddy Currents	768
29–6	Transformers and Transmission of Power	770
29–7	A Changing Magnetic Flux Produces an Electric Field	773
*29–8	Applications of Induction: Sound Systems, Computer Memory, Seismograph, GFCI	775
	SUMMARY 777 QUESTIONS 777	
	PROBLEMS 778 GENERAL PROBLEMS 782	

30 INDUCTANCE, ELECTROMAGNETIC OSCILLATIONS, AND AC CIRCUITS 785

30–1	Mutual Inductance	786
30–2	Self-Inductance	788
30–3	Energy Stored in a Magnetic Field	790
30–4	LR Circuits	790
30–5	LC Circuits and Electromagnetic Oscillations	793
30–6	LC Oscillations with Resistance (LRC Circuit)	795
30–7	AC Circuits with AC Source	796
30–8	LRC Series AC Circuit	799
30–9	Resonance in AC Circuits	802
*30–10	Impedance Matching	802
*30–11	Three-Phase AC	803
	SUMMARY 804 QUESTIONS 804	
	PROBLEMS 805 GENERAL PROBLEMS 809	

31 MAXWELL'S EQUATIONS AND ELECTROMAGNETIC WAVES 812

31–1	Changing Electric Fields Produce Magnetic Fields; Ampère's Law and Displacement Current	813
31–2	Gauss's Law for Magnetism	816
31–3	Maxwell's Equations	817
31–4	Production of Electromagnetic Waves	817
31–5	Electromagnetic Waves, and Their Speed, from Maxwell's Equations	819
31–6	Light as an Electromagnetic Wave and the Electromagnetic Spectrum	823
31–7	Measuring the Speed of Light	825
31–8	Energy in EM Waves; the Poynting Vector	826
31–9	Radiation Pressure	828
31–10	Radio and Television; Wireless Communication	829
	SUMMARY 832 QUESTIONS 832	
	PROBLEMS 833 GENERAL PROBLEMS 835	

32 LIGHT: REFLECTION AND REFRACTION 837

32–1	The Ray Model of Light	838
32–2	Reflection; Image Formation by a Plane Mirror	838
32–3	Formation of Images by Spherical Mirrors	842
32–4	Index of Refraction	850
32–5	Refraction: Snell's Law	850
32–6	Visible Spectrum and Dispersion	852
32–7	Total Internal Reflection; Fiber Optics	854
*32–8	Refraction at a Spherical Surface	856

SUMMARY 858   QUESTIONS 859
PROBLEMS 860   GENERAL PROBLEMS 864

33 LENSES AND OPTICAL INSTRUMENTS 866

33–1	Thin Lenses; Ray Tracing	867
33–2	The Thin Lens Equation; Magnification	870
33–3	Combinations of Lenses	874
*33–4	Lensmaker's Equation	876
33–5	Cameras: Film and Digital	878
33–6	The Human Eye; Corrective Lenses	882
33–7	Magnifying Glass	885
33–8	Telescopes	887
*33–9	Compound Microscope	890
*33–10	Aberrations of Lenses and Mirrors	891

SUMMARY 892   QUESTIONS 893
PROBLEMS 894   GENERAL PROBLEMS 897

34 THE WAVE NATURE OF LIGHT; INTERFERENCE 900

34–1	Waves Versus Particles; Huygens' Principle and Diffraction	901
34–2	Huygens' Principle and the Law of Refraction	902
34–3	Interference—Young's Double-Slit Experiment	903
*34–4	Intensity in the Double-Slit Interference Pattern	906
34–5	Interference in Thin Films	909
*34–6	Michelson Interferometer	914
*34–7	Luminous Intensity	915

SUMMARY 915   QUESTIONS 916
PROBLEMS 916   GENERAL PROBLEMS 918

35 DIFFRACTION AND POLARIZATION 921

35–1	Diffraction by a Single Slit or Disk	922
*35–2	Intensity in Single-Slit Diffraction Pattern	924
*35–3	Diffraction in the Double-Slit Experiment	927
35–4	Limits of Resolution; Circular Apertures	929
35–5	Resolution of Telescopes and Microscopes; the λ Limit	931
*35–6	Resolution of the Human Eye and Useful Magnification	932
35–7	Diffraction Grating	933
35–8	The Spectrometer and Spectroscopy	935
*35–9	Peak Widths and Resolving Power for a Diffraction Grating	937
35–10	X-Rays and X-Ray Diffraction	938
35–11	Polarization	940
*35–12	Liquid Crystal Displays (LCD)	943
*35–13	Scattering of Light by the Atmosphere	945

SUMMARY 945   QUESTIONS 946
PROBLEMS 946   GENERAL PROBLEMS 949

36 THE SPECIAL THEORY OF RELATIVITY   951

36–1	Galilean–Newtonian Relativity	952
*36–2	The Michelson–Morley Experiment	954
36–3	Postulates of the Special Theory of Relativity	957
36–4	Simultaneity	958
36–5	Time Dilation and the Twin Paradox	960
36–6	Length Contraction	964
36–7	Four-Dimensional Space–Time	967
36–8	Galilean and Lorentz Transformations	968
36–9	Relativistic Momentum	971
36–10	The Ultimate Speed	974
36–11	$E = mc^2$; Mass and Energy	974
36–12	Doppler Shift for Light	978
36–13	The Impact of Special Relativity	980

SUMMARY 981   QUESTIONS 981
PROBLEMS 982   GENERAL PROBLEMS 985

37 EARLY QUANTUM THEORY AND MODELS OF THE ATOM   987

37–1	Blackbody Radiation; Planck's Quantum Hypothesis	987
37–2	Photon Theory; Photoelectric Effect	989
37–3	Photon Energy, Mass, and Momentum	993
37–4	Compton Effect	994
37–5	Photon Interactions; Pair Production	996
37–6	Wave–Particle Duality; the Principle of Complementarity	997
37–7	Wave Nature of Matter	997
*37–8	Electron Microscopes	1000
37–9	Early Models of the Atom	1000
37–10	Atomic Spectra: Key to Atomic Structure	1001
37–11	The Bohr Model	1003
37–12	deBroglie's Hypothesis Applied to Atoms	1009

SUMMARY 1010 QUESTIONS 1011
PROBLEMS 1012 GENERAL PROBLEMS 1014

APPENDICES

A	MATHEMATICAL FORMULAS	A–1
B	DERIVATIVES AND INTEGRALS	A–6
C	MORE ON DIMENSIONAL ANALYSIS	A–8
D	GRAVITATIONAL FORCE DUE TO A SPHERICAL MASS DISTRIBUTION	A–9
E	DIFFERENTIAL FORM OF MAXWELL'S EQUATIONS	A–12
F	SELECTED ISOTOPES	A–14

ANSWERS TO ODD-NUMBERED PROBLEMS   A–18
INDEX   A–44
PHOTO CREDITS   A–62

APPLICATIONS (SELECTED)

Chapter 1
The 8000-m peaks	8
Estimating volume of a lake	10
Height by triangulation	11
Radius of the Earth	11
Heartbeats in a lifetime	12
Particulate pollution (Pr30)	15
Global positioning satellites (Pr39)	16
Lung capacity (Pr65)	17

Chapter 2
Airport runway design	29
Automobile air bags	31
Braking distances	32
CD error correction (Pr10)	44
CD playing time (Pr13)	45
Golfing uphill or down (Pr79)	48
Rapid transit (Pr83)	49

Chapter 3
Kicked football	66, 69
Ball sports (Problems)	77, 81, 82
Extreme sports (Pr41)	77

Chapter 4
Rocket acceleration	90
What force accelerates a car?	90
How we walk	90
Elevator and counterweight	99
Mechanical advantage of pulley	100
Bear sling (Q24)	104
High-speed elevators (Pr19)	105
Mountain climbing (Pr31, 82, 83)	106, 110
City planning, cars on hills (Pr71)	109
Bicyclists (Pr72, 73)	109
"Doomsday" asteroid (Pr84)	110

Chapter 5
Push or pull a sled?	116
Centrifugation	122
Not skidding on a curve	126–7
Banked highways	127
Simulated gravity (Q18, Pr48)	131, 134
"Rotor-ride" (Pr82)	136

Chapter 6
Oil/mineral exploration	144, 420
Artificial Earth satellites	146
Geosynchronous satellites	147
Weightlessness	148
Free fall in athletics	149
Planet discovery, extrasolar planets	152
Black holes	156
Asteroids (Pr44, 78)	159, 162
Navstar GPS (Pr58)	160
Black hole, galaxy center (Pr61, 64)	160, 161
Tides (Pr75)	162

Chapter 7
Car stopping distance of v^2	174
Lever (Pr6)	177
Spiderman (Pr54)	179
Bicycling on hills, gears (Pr85)	181
Child safety in car (Pr87)	181
Rock climber's rope (Pr90)	182

Chapter 8
Downhill ski runs	183
Roller coaster	191, 198
Pole vault	192–3
Toy dart gun	193
Escape velocity from Earth or Moon	201
Stair climbing power	202
Power needs of car	202–3
Cardiac treadmill (Pr104)	213

Chapter 9
Tennis serve	216
Rocket propulsion	219, 236–8
Rifle recoil	220
Karate blow	221
Billiards/bowling	223, 228
Nuclear collisions	225, 228
Ballistic pendulum	226
Conveyor belt	237
Gravitational slingshot (Pr105)	246
Crashworthiness (Pr109)	247
Asteroids, planets (Pr110, 112, 113)	247

Chapter 10
Hard drive and bit speed	253
Wrench/tire iron	256
Flywheel energy	266, 281
Yo-yo	271
Car braking forces	272–3
Bicycle odometer calibration (Q1)	275
Tightrope walker (Q11)	275
Triceps muscle and throwing (Pr38, 39)	278
CD speed (Pr84)	281
Bicycle gears (Pr89)	281

Chapter 11
Rotating skaters, divers	284, 286, 309
Neutron star collapse	287
Auto wheel balancing	296
Top and gyroscope	299–300
Coriolis effect	301–2
Hurricanes	302
SUV possible rollover (Pr67)	308
Triple axel jump (Pr79)	309
Bat's "sweet spot" (Pr82)	310

Chapter 12
Tragic collapse	311, 323
Lever's mechanical advantage	313
Cantilever	315
Biceps muscle force	315
Human balance with loads	318
Trusses and bridges	324–6, 335
Architecture: arches and domes	327–8
Forces on vertebrae (Pr87)	337

Chapter 13
Lifting water	345, 348
Hydraulic lift, brakes	346
Pressure gauges	346–7
Hydrometer	351
Helium balloon lift	352, 368
Blood flow	353, 357, 361
Airplane wings, lift	356
Sailing against the wind	357
Baseball curve	357
Blood to the brain, TIA	357
Blood flow and heart disease	359
Surface tension, capillarity	359–60
Walking on water	360
Pumps and the heart	361
Reynolds number (Pr69)	366

Chapter 14
Car shock absorbers	383
Resonance damage	386

Chapter 15
Echolocation by animals	400
Earthquake waves	401, 403, 416

Chapter 16
Distance from lightning	425
Autofocus camera	426
Wide range of human hearing	427–8, 431
Loudspeaker response	428
Stringed instruments	432–3
Wind instruments	433–6
Tuning with beats	439
Doppler blood flow meter	442, 453
Sonar: sonic boom	444
Ultrasound medical imaging	445–6
Motion sensor (Pr5)	448

Chapter 17
Hot air balloon	454
Expansion joints, highways	456, 460, 463
Gas tank overflow	462
Life under ice	462
Cold and hot tire pressure	468
Molecules in a breath	469
Thermostat (Q10)	471
Scuba/snorkeling (Pr38, 47, 82, 85)	473, 475

Chapter 18
Chemical reactions, temperature dependence	481
Superfluidity	483
Evaporation cools	484, 505
Humidity, weather	485–6
Chromatography	490
Pressure cooker (Pr35)	493

Chapter 19
Working off the calories	498
Cold floors	516
Heat loss through windows	516
How clothes insulate	516–7
R-values for thermal insulation	517
Convective house heating	517
Human radiative heat loss	518
Room comfort and metabolism	519
Radiation from Sun	519
Medical thermography	519
Astronomy—size of a star	520
Thermos bottle (Q30)	521
Weather, air parcel, adiabatic lapse rate (Pr56)	525

Chapter 20
Steam engine	530
Internal combustion engine	531, 535–6
Car efficiency	532
Refrigerators, air conditioners	537–8
Heat pump	538
Biological evolution, development	545
Thermal pollution, global warming	549–51
Energy resources	550
Diesel engine (Pr7)	553

Chapter 21
Static electricity   560, 589(Pr78)
Photocopiers   569, 582–3
Electric shielding, safety   577
DNA structure and
   replication   581–2
Biological cells: electric forces
   and kinetic theory   581–2, 617
Laser & inkjet printers   583

Chapter 23
Breakdown voltage   612
Lightning rods, corona   612
CRT, oscilloscopes,
   TV monitors   620–1, 723
Photocell (Pr75)   626
Geiger counter (Pr83)   627
Van de Graaff (Pr84)   627, 607

Chapter 24
Capacitor uses   628, 631
Very high capacitance   631
Computer key   631
Camera flash   636
Heart defibrillator   638
DRAM (Pr10, 57)   644, 647
Electrostatic air cleaner (Pr20)   645
CMOS circuits (Pr53)   647

Chapter 25
Light bulb   651, 653, 660
Battery construction   653
Loudspeaker wires   659
Resistance thermometer   660
Heating elements, bulb filament   660
Why bulbs burn out at turn on   661
Lightning bolt   662
Household circuits, shorts   662–3
Fuses, circuit breakers   662–3, 747, 776
Extension cord danger   663
Nervous system, conduction   669–70
Strain gauge (Pr 24)   673

Chapter 26
Car battery charging, jump start   686, 687
RC applications: flashers, wipers   691
Heart pacemaker   692, 787
Electric hazards   692–4
Proper grounding   693–4
Heart fibrillation   692
Meters, analog and digital   695–7
Potentiometers and bridges (Pr)   704, 705

Chapter 27
Compass and declination   709
Aurora borealis   717
Motors, loudspeakers,
   galvonometers   720–1
Mass spectrometer   724–5
Electromagnetic pumping (Q14)   726
Cyclotron (Pr66)   731
Beam steering (Pr67)   731

Chapter 28
Coaxial cable   740, 789
Solenoid switches: car starters,
   doorbell   747
Circuit breakers, magnetic   747, 776
Relay (Q16)   751
Atom trap (Pr73)   757

Chapter 29
Induction stove   762
EM blood-flow meter   765
Power plant generators   766–7
Car alternators   768
Motor overload   769
Airport metal detector   770
Eddy current damping   770
Transformers and uses, power   770–3
Car ignition, bulb ballast   772, 773
Microphone   775
Read/write on disks and tape   775
Digital coding   775
Credit card swipe   776
Ground fault circuit interrupter
   (GFCI)   776
Betatron (Pr55)   782
Search coil (Pr68)   783
Inductive battery charger (Pr81)   784

Chapter 30
Spark plug   785
Pacemaker   787
Surge protector   792
LC oscillators, resonance   794, 802
Capacitors as filters   799
Loudspeaker cross-over   799
Impedance matching   802–3
Three-phase AC   803
Q-value (Pr86, 87)   810

Chapter 31
Antennas   824, 831
Phone call lag time   825
Solar sail   829
Optical tweezers   829
Wireless: AM/FM, TV, tuning,
   cell phones, remotes   829–32

Chapter 32
How tall a mirror do you need   840–1
Close up and wide-view
   mirrors   842, 849, 859
Where you can *see* yourself in a
   concave mirror   848
Optical illusions   851, 903
Apparent depth in water   852
Rainbows   853
Colors underwater   854
Prism binoculars   855
Fiber optics in
   telecommunications   855–6, 865
Medical endoscopes   856
Highway reflectors (Pr86)   865

Chapter 33
Where you can *see* a lens image   869
Cameras, digital and film   878
Camera adjustments   879–80
Pixels and resolution   881
Human eye   882–5, 892
Corrective lenses   883–5
Contact lenses   885
Seeing under water   885
Magnifying glass   885–7
Telescopes   887–9, 931–2
Microscopes   890–1, 931, 933

Chapter 34
Bubbles, reflected color   900, 912–13
Mirages   903
Colors in thin soap film, details   912–13

Lens coatings   913–14
Multiple coating (Pr52)   919

Chapter 35
Lens and mirror resolution   929–30
Hubble Space Telescope   930
Eye resolution,
   useful magnification   930, 932–3
Radiotelescopes   931
Telescope resolution, λ rule   931
Spectroscopy   935–6
X-ray diffraction in biology   939
Polarized sunglasses   942
LCDs—liquid crystal displays   943–4
Sky color   945

Chapter 36
Space travel   963
Global positioning system (GPS)   964

Chapter 37
Photocells   992
Photodiodes   992
Photosynthesis   993
Measuring bone density   995
Electron microscopes   1000

Preface

I was motivated from the beginning to write a textbook different from others that present physics as a sequence of facts, like a Sears catalog: "here are the facts and you better learn them." Instead of that approach in which topics are begun formally and dogmatically, I have sought to begin each topic with concrete observations and experiences students can relate to: start with specifics and only then go to the great generalizations and the more formal aspects of a topic, showing *why* we believe what we believe. This approach reflects how science is actually practiced.

Why a Fourth Edition?

Two recent trends in physics texbooks are disturbing: (1) their revision cycles have become short—they are being revised every 3 or 4 years; (2) the books are getting larger, some over 1500 pages. I don't see how either trend can be of benefit to students. My response: (1) It has been 8 years since the previous edition of this book. (2) This book makes use of physics education research, although it avoids the detail a Professor may need to say in class but in a book shuts down the reader. And this book still remains among the shortest.

This new edition introduces some important new pedagogic tools. It contains new physics (such as in cosmology) and many new appealing applications (list on previous page). Pages and page breaks have been carefully formatted to make the physics easier to follow: no turning a page in the middle of a derivation or Example. Great efforts were made to make the book attractive so students will want to *read* it.

Some of the new features are listed below.

What's New

Chapter-Opening Questions: Each Chapter begins with a multiple-choice question, whose responses include common misconceptions. Students are asked to answer before starting the Chapter, to get them involved in the material and to get any preconceived notions out on the table. The issues reappear later in the Chapter, usually as Exercises, after the material has been covered. The Chapter-Opening Questions also show students the power and usefulness of Physics.

APPROACH paragraph in worked-out numerical Examples: A short introductory paragraph before the Solution, outlining an approach and the steps we can take to get started. Brief NOTES after the Solution may remark on the Solution, may give an alternate approach, or mention an application.

Step-by-Step Examples: After many Problem Solving Strategies (more than 20 in the book), the next Example is done step-by-step following precisely the steps just seen.

Exercises within the text, after an Example or derivation, give students a chance to see if they have understood enough to answer a simple question or do a simple calculation. Many are multiple choice.

Greater clarity: No topic, no paragraph in this book was overlooked in the search to improve the clarity and conciseness of the presentation. Phrases and sentences that may slow down the principal argument have been eliminated: keep to the essentials at first, give the elaborations later.

$\vec{F}, \vec{v}, \vec{B}$ *Vector notation, arrows*: The symbols for vector quantities in the text and Figures now have a tiny arrow over them, so they are similar to what we write by hand.

Cosmological Revolution: With generous help from top experts in the field, readers have the latest results.

Page layout: more than in the previous edition, serious attention has been paid to how each page is formatted. Examples and all important derivations and arguments are on facing pages. Students then don't have to turn back and forth. Throughout, readers see, on two facing pages, an important slice of physics.

New Applications: LCDs, digital cameras and electronic sensors (CCD, CMOS), electric hazards, GFCIs, photocopiers, inkjet and laser printers, metal detectors, underwater vision, curve balls, airplane wings, DNA, how we actually *see* images. (Turn back a page to see a longer list.)

Examples modified: more math steps are spelled out, and many new Examples added. About 10% of all Examples are Estimation Examples.

This Book is Shorter than other complete full-service books at this level. Shorter explanations are easier to understand and more likely to be read.

Content and Organizational Changes

- **Rotational Motion**: Chapters 10 and 11 have been reorganized. All of angular momentum is now in Chapter 11.
- **First law of thermodynamics**, in Chapter 19, has been rewritten and extended. The full form is given: $\Delta K + \Delta U + \Delta E_{int} = Q - W$, where internal energy is E_{int}, and U is potential energy; the form $Q - W$ is kept so that $dW = P\,dV$.
- Kinematics and Dynamics of Circular Motion are now treated together in Chapter 5.
- Work and Energy, Chapters 7 and 8, have been carefully revised.
- Work done by friction is discussed now with energy conservation (energy terms due to friction).
- Chapters on Inductance and AC Circuits have been combined into one: Chapter 30.
- Graphical Analysis and Numerical Integration is a new optional Section 2–9. Problems requiring a computer or graphing calculator are found at the end of most Chapters.
- Length of an object is a script ℓ rather than normal l, which looks like 1 or I (moment of inertia, current), as in $F = I\ell B$. Capital L is for angular momentum, latent heat, inductance, dimensions of length $[L]$.
- Newton's law of gravitation remains in Chapter 6. Why? Because the $1/r^2$ law is too important to relegate to a late chapter that might not be covered at all late in the semester; furthermore, it is one of the basic forces in nature. In Chapter 8 we can treat real gravitational potential energy and have a fine instance of using $U = -\int \vec{\mathbf{F}} \cdot d\vec{\boldsymbol{\ell}}$.
- New Appendices include the differential form of Maxwell's equations and more on dimensional analysis.
- Problem Solving Strategies are found on pages 30, 58, 64, 96, 102, 125, 166, 198, 229, 261, 314, 504, 551, 571, 600, 685, 716, 740, 763, 849, 871, and 913.

Organization

Some instructors may find that this book contains more material than can be covered in their courses. The text offers great flexibility. Sections marked with a star * are considered optional. These contain slightly more advanced physics material, or material not usually covered in typical courses and/or interesting applications; they contain no material needed in later Chapters (except perhaps in later optional Sections). For a brief course, all optional material could be dropped as well as major parts of Chapters 1, 13, 16, 26, 30, and 35, and selected parts of Chapters 9, 12, 19, 20, 33, and the modern physics Chapters. Topics not covered in class can be a valuable resource for later study by students. Indeed, this text can serve as a useful reference for years because of its wide range of coverage.

Versions of this Book

Complete version: 44 Chapters including 9 Chapters of modern physics.

Classic version: 37 Chapters including one each on relativity and quantum theory.

3 Volume version: Available separately or packaged together (Vols. 1 & 2 or all 3 Volumes):

Volume 1: Chapters 1–20 on mechanics, including fluids, oscillations, waves, plus heat and thermodynamics.

Volume 2: Chapters 21–35 on electricity and magnetism, plus light and optics.

Volume 3: Chapters 36–44 on modern physics: relativity, quantum theory, atomic physics, condensed matter, nuclear physics, elementary particles, cosmology and astrophysics.

Thanks

Many physics professors provided input or direct feedback on every aspect of this textbook. They are listed below, and I owe each a debt of gratitude.

Mario Affatigato, Coe College
Lorraine Allen, United States Coast Guard Academy
Zaven Altounian, McGill University
Bruce Barnett, Johns Hopkins University
Michael Barnett, Lawrence Berkeley Lab
Anand Batra, Howard University
Cornelius Bennhold, George Washington University
Bruce Birkett, University of California Berkeley
Dr. Robert Boivin, Auburn University
Subir Bose, University of Central Florida
David Branning, Trinity College
Meade Brooks, Collin County Community College
Bruce Bunker, University of Notre Dame
Grant Bunker, Illinois Institute of Technology
Wayne Carr, Stevens Institute of Technology
Charles Chiu, University of Texas Austin
Robert Coakley, University of Southern Maine
David Curott, University of North Alabama
Biman Das, SUNY Potsdam
Bob Davis, Taylor University
Kaushik De, University of Texas Arlington
Michael Dennin, University of California Irvine
Kathy Dimiduk, University of New Mexico
John DiNardo, Drexel University
Scott Dudley, United States Air Force Academy
John Essick, Reed College
Cassandra Fesen, Dartmouth College
Alex Filippenko, University of California Berkeley
Richard Firestone, Lawrence Berkeley Lab
Mike Fortner, Northern Illinois University
Tom Furtak, Colorado School of Mines
Edward Gibson, California State University Sacramento
John Hardy, Texas A&M
J. Erik Hendrickson, University of Wisconsin Eau Claire
Laurent Hodges, Iowa State University
David Hogg, New York University
Mark Hollabaugh, Normandale Community College
Andy Hollerman, University of Louisiana at Lafayette
Bob Jacobsen, University of California Berkeley
Teruki Kamon, Texas A&M
Daryao Khatri, University of the District of Columbia
Jay Kunze, Idaho State University

Jim LaBelle, Dartmouth College
M.A.K. Lodhi, Texas Tech
Bruce Mason, University of Oklahoma
Dan Mazilu, Virginia Tech
Linda McDonald, North Park College
Bill McNairy, Duke University
Raj Mohanty, Boston University
Giuseppe Molesini, Istituto Nazionale di Ottica Florence
Lisa K. Morris, Washington State University
Blaine Norum, University of Virginia
Alexandria Oakes, Eastern Michigan University
Michael Ottinger, Missouri Western State University
Lyman Page, Princeton and WMAP
Bruce Partridge, Haverford College
R. Daryl Pedigo, University of Washington
Robert Pelcovitz, Brown University
Vahe Peroomian, UCLA
James Rabchuk, Western Illinois University
Michele Rallis, Ohio State University
Paul Richards, University of California Berkeley
Peter Riley, University of Texas Austin
Larry Rowan, University of North Carolina Chapel Hill
Cindy Schwarz, Vassar College
Peter Sheldon, Randolph-Macon Woman's College
Natalia A. Sidorovskaia, University of Louisiana at Lafayette
George Smoot, University of California Berkeley
Mark Sprague, East Carolina University
Michael Strauss, University of Oklahoma
Laszlo Takac, University of Maryland Baltimore Co.
Franklin D. Trumpy, Des Moines Area Community College
Ray Turner, Clemson University
Som Tyagi, Drexel University
John Vasut, Baylor University
Robert Webb, Texas A&M
Robert Weidman, Michigan Technological University
Edward A. Whittaker, Stevens Institute of Technology
John Wolbeck, Orange County Community College
Stanley George Wojcicki, Stanford University
Edward Wright, UCLA
Todd Young, Wayne State College
William Younger, College of the Albemarle
Hsiao-Ling Zhou, Georgia State University

I owe special thanks to Prof. Bob Davis for much valuable input, and especially for working out all the Problems and producing the Solutions Manual for all Problems, as well as for providing the answers to odd-numbered Problems at the end of this book. Many thanks also to J. Erik Hendrickson who collaborated with Bob Davis on the solutions, and to the team they managed (Profs. Anand Batra, Meade Brooks, David Currott, Blaine Norum, Michael Ottinger, Larry Rowan, Ray Turner, John Vasut, William Younger). Many thanks to Katherine Whatley and Judith Beck who provided answers to the conceptual Questions at the end of each Chapter. I am grateful to Profs. John Essick, Bruce Barnett, Robert Coakley, Biman Das, Michael Dennin, Kathy Dimiduk, John DiNardo, Scott Dudley, David Hogg, Cindy Schwarz, Ray Turner, and Som Tyagi, who inspired many of the Examples, Questions, Problems, and significant clarifications.

Crucial for rooting out errors, as well as providing excellent suggestions, were Profs. Kathy Dimiduk, Ray Turner, and Lorraine Allen. A huge thank you to them and to Prof. Giuseppe Molesini for his suggestions and his exceptional photographs for optics.

For Chapter 44 on Cosmology and Astrophysics, I was fortunate to receive generous input from some of the top experts in the field, to whom I owe a debt of gratitude: George Smoot, Paul Richards, and Alex Filippenko (UC Berkeley), Lyman Page (Princeton and WMAP), Edward Wright (UCLA and WMAP), and Michael Strauss (University of Oklahoma).

I especially wish to thank Profs. Howard Shugart, Chair Marjorie Shapiro, and many others at the University of California, Berkeley, Physics Department for helpful discussions, and for hospitality. Thanks also to Prof. Tito Arecchi and others at the Istituto Nazionale di Ottica, Florence, Italy.

Finally, I am grateful to the many people at Prentice Hall with whom I worked on this project, especially Paul Corey, Christian Botting, Sean Hogan, Clare Romeo, Frank Weihenig, John Christiana, and Karen Karlin.

The final responsibility for all errors lies with me. I welcome comments, corrections, and suggestions as soon as possible to benefit students for the next reprint.

D.C.G.

email: Paul_Corey@Prenhall.com

Post: Paul Corey
     One Lake Street
     Upper Saddle River, NJ 07458

About the Author

Douglas C. Giancoli obtained his BA in physics (summa cum laude) from the University of California, Berkeley, his MS in physics at the Massachusetts Institute of Technology, and his PhD in elementary particle physics at the University of California, Berkeley. He spent 2 years as a post-doctoral fellow at UC Berkeley's Virus lab developing skills in molecular biology and biophysics. His mentors include Nobel winners Emilio Segré and Donald Glaser.

He has taught a wide range of undergraduate courses, traditional as well as innovative ones, and continues to update his textbooks meticulously, seeking ways to better provide an understanding of physics for students.

Doug's favorite spare-time activity is the outdoors, especially climbing peaks (here on the summit of one of Colorado's fourteeners (>14,000 ft)). He says climbing peaks is like learning physics: it takes effort and the rewards are great.

Online Supplements (partial list)

MasteringPhysics™ (www.masteringphysics.com)
is a sophisticated online tutoring and homework system developed specially for courses using calculus-based physics. Originally developed by David Pritchard and collaborators at MIT, MasteringPhysics provides **students** with individualized online tutoring by responding to their wrong answers and providing hints for solving multi-step problems when they get stuck. It gives them immediate and up-to-date assessment of their progress, and shows where they need to practice more. MasteringPhysics provides **instructors** with a fast and effective way to assign tried-and-tested online homework assignments that comprise a range of problem types. The powerful post-assignment diagnostics allow instructors to assess the progress of their class as a whole as well as individual students, and quickly identify areas of difficulty.

WebAssign (www.webassign.com)

CAPA and LON-CAPA (www.lon-capa.org/)

Student Supplements (partial list)

Student Study Guide & Selected Solutions Manual (Volume I: 0-13-2273241, Volumes II & III: 0-13-227325X) by Frank Wolfs

Student Pocket Companion (0-13-2273268) by Biman Das

Tutorials in Introductory Physics (0-13-097069-7)
by Lillian C. McDermott, Peter S. Schaffer, and the Physics Education Group at the University of Washington

Physlet® Physics (0-13-101969-4)
by Wolfgang Christian and Mario Belloni

Ranking Task Exercises in Physics, Student Edition (0-13-144851-X) by Thomas L. O'Kuma, David P. Maloney, and Curtis J. Hieggelke

E&M TIPERs: Electricity & Magnetism Tasks Inspired by Physics Education Research (0-13-185499-2) by Curtis J. Hieggelke, David P. Maloney, Stephen E. Kanim, and Thomas L. O'Kuma

Mathematics for Physics with Calculus (0-13-1913360) by Biman Das

To Students

HOW TO STUDY

1. Read the Chapter. Learn new vocabulary and notation. Try to respond to questions and exercises as they occur.

2. Attend all class meetings. Listen. Take notes, especially about aspects you do not remember seeing in the book. Ask questions (everyone else wants to, but maybe you will have the courage). You will get more out of class if you read the Chapter first.

3. Read the Chapter again, paying attention to details. Follow derivations and worked-out Examples. Absorb their logic. Answer Exercises and as many of the end of Chapter Questions as you can.

4. Solve 10 to 20 end of Chapter Problems (or more), especially those assigned. In doing Problems you find out what you learned and what you didn't. Discuss them with other students. Problem solving is one of the great learning tools. Don't just look for a formula—it won't cut it.

NOTES ON THE FORMAT AND PROBLEM SOLVING

1. Sections marked with a star (*) are considered **optional**. They can be omitted without interrupting the main flow of topics. No later material depends on them except possibly later starred Sections. They may be fun to read, though.

2. The customary **conventions** are used: symbols for quantities (such as m for mass) are italicized, whereas units (such as m for meter) are not italicized. Symbols for vectors are shown in boldface with a small arrow above: $\vec{\mathbf{F}}$.

3. Few equations are valid in all situations. Where practical, the **limitations** of important equations are stated in square brackets next to the equation. The equations that represent the great laws of physics are displayed with a tan background, as are a few other indispensable equations.

4. At the end of each Chapter is a set of **Problems** which are ranked as Level I, II, or III, according to estimated difficulty. Level I Problems are easiest, Level II are standard Problems, and Level III are "challenge problems." These ranked Problems are arranged by Section, but Problems for a given Section may depend on earlier material too. There follows a group of General Problems, which are not arranged by Section nor ranked as to difficulty. Problems that relate to optional Sections are starred (*). Most Chapters have 1 or 2 Computer/Numerical Problems at the end, requiring a computer or graphing calculator. Answers to odd-numbered Problems are given at the end of the book.

5. Being able to solve **Problems** is a crucial part of learning physics, and provides a powerful means for understanding the concepts and principles. This book contains many aids to problem solving: (a) worked-out **Examples** and their solutions in the text, which should be studied as an integral part of the text; (b) some of the worked-out Examples are **Estimation Examples**, which show how rough or approximate results can be obtained even if the given data are sparse (see Section 1–6); (c) special **Problem Solving Strategies** placed throughout the text to suggest a step-by-step approach to problem solving for a particular topic—but remember that the basics remain the same; most of these "Strategies" are followed by an Example that is solved by explicitly following the suggested steps; (d) special problem-solving Sections; (e) "Problem Solving" marginal notes which refer to hints within the text for solving Problems; (f) **Exercises** within the text that you should work out immediately, and then check your response against the answer given at the bottom of the last page of that Chapter; (g) the Problems themselves at the end of each Chapter (point 4 above).

6. **Conceptual Examples** pose a question which hopefully starts you to think and come up with a response. Give yourself a little time to come up with your own response before reading the Response given.

7. **Math** review, plus some additional topics, are found in Appendices. Useful data, conversion factors, and math formulas are found inside the front and back covers.

USE OF COLOR

Vectors

A general vector

    resultant vector (sum) is slightly thicker

    components of any vector are dashed

Displacement ($\vec{\mathbf{D}}, \vec{\mathbf{r}}$)

Velocity ($\vec{\mathbf{v}}$)

Acceleration ($\vec{\mathbf{a}}$)

Force ($\vec{\mathbf{F}}$)

    Force on second or

    third object in same figure

Momentum ($\vec{\mathbf{p}}$ or $m\vec{\mathbf{v}}$)

Angular momentum ($\vec{\mathbf{L}}$)

Angular velocity ($\vec{\boldsymbol{\omega}}$)

Torque ($\vec{\boldsymbol{\tau}}$)

Electric field ($\vec{\mathbf{E}}$)

Magnetic field ($\vec{\mathbf{B}}$)

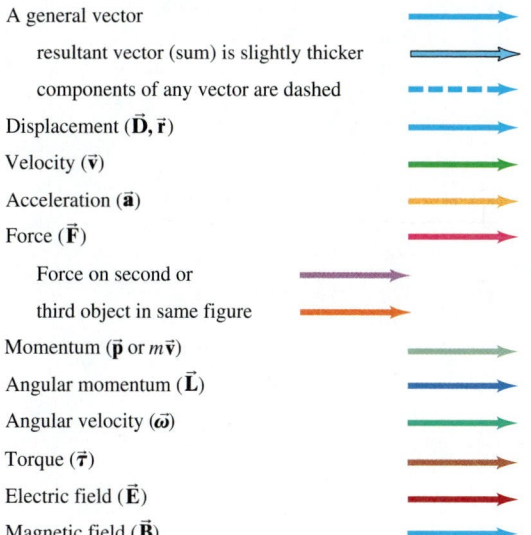

Electricity and magnetism

Electric field lines

Equipotential lines

Magnetic field lines

Electric charge (+)   + or ● +

Electric charge (−)   − or ● −

Electric circuit symbols

Wire, with switch S

Resistor

Capacitor

Inductor

Battery

Ground

Optics

Light rays

Object

Real image (dashed)

Virtual image (dashed and paler)

Other

Energy level (atom, etc.)

Measurement lines |←1.0 m→|

Path of a moving object

Direction of motion or current

Image of the Earth from a NASA satellite. The sky appears black from out in space because there are so few molecules to reflect light. (Why the sky appears blue to us on Earth has to do with scattering of light by molecules of the atmosphere, as discussed in Chapter 35.) Note the storm off the coast of Mexico.

Introduction, Measurement, Estimating

CHAPTER-OPENING QUESTION—Guess now!

Suppose you wanted to actually measure the radius of the Earth, at least roughly, rather than taking other people's word for what it is. Which response below describes the best approach?

(a) Give up; it is impossible using ordinary means.
(b) Use an extremely long measuring tape.
(c) It is only possible by flying high enough to see the actual curvature of the Earth.
(d) Use a standard measuring tape, a step ladder, and a large smooth lake.
(e) Use a laser and a mirror on the Moon or on a satellite.

[*We start each Chapter with a Question, like the one above. Try to answer it right away. Don't worry about getting the right answer now—the idea is to get your preconceived notions out on the table. If they are misconceptions, we expect them to be cleared up as you read the Chapter. You will usually get another chance at the Question later in the Chapter when the appropriate material has been covered. These Chapter-Opening Questions will also help you to see the power and usefulness of physics.*]

C H A P T E R

1

CONTENTS

1–1 The Nature of Science

1–2 Models, Theories, and Laws

1–3 Measurement and Uncertainty; Significant Figures

1–4 Units, Standards, and the SI System

1–5 Converting Units

1–6 Order of Magnitude: Rapid Estimating

*1–7 Dimensions and Dimensional Analysis

1

(a)

(b)

FIGURE 1–1 (a) This Roman aqueduct was built 2000 years ago and still stands. (b) The Hartford Civic Center collapsed in 1978, just two years after it was built.

Physics is the most basic of the sciences. It deals with the behavior and structure of matter. The field of physics is usually divided into *classical physics* which includes motion, fluids, heat, sound, light, electricity and magnetism; and *modern physics* which includes the topics of relativity, atomic structure, condensed matter, nuclear physics, elementary particles, and cosmology and astrophysics. We will cover all these topics in this book, beginning with motion (or mechanics, as it is often called) and ending with the most recent results in our study of the cosmos.

An understanding of physics is crucial for anyone making a career in science or technology. Engineers, for example, must know how to calculate the forces within a structure to design it so that it remains standing (Fig. 1–1a). Indeed, in Chapter 12 we will see a worked-out Example of how a simple physics calculation—or even intuition based on understanding the physics of forces—would have saved hundreds of lives (Fig. 1–1b). We will see many examples in this book of how physics is useful in many fields, and in everyday life.

1–1 The Nature of Science

The principal aim of all sciences, including physics, is generally considered to be the search for order in our observations of the world around us. Many people think that science is a mechanical process of collecting facts and devising theories. But it is not so simple. Science is a creative activity that in many respects resembles other creative activities of the human mind.

One important aspect of science is **observation** of events, which includes the design and carrying out of experiments. But observation and experiment require imagination, for scientists can never include everything in a description of what they observe. Hence, scientists must make judgments about what is relevant in their observations and experiments.

Consider, for example, how two great minds, Aristotle (384–322 B.C.) and Galileo (1564–1642), interpreted motion along a horizontal surface. Aristotle noted that objects given an initial push along the ground (or on a tabletop) always slow down and stop. Consequently, Aristotle argued that the natural state of an object is to be at rest. Galileo, in his reexamination of horizontal motion in the 1600s, imagined that if friction could be eliminated, an object given an initial push along a horizontal surface would continue to move indefinitely without stopping. He concluded that for an object to be in motion was just as natural as for it to be at rest. By inventing a new approach, Galileo founded our modern view of motion (Chapters 2, 3, and 4), and he did so with a leap of the imagination. Galileo made this leap conceptually, without actually eliminating friction.

Observation, with careful experimentation and measurement, is one side of the scientific process. The other side is the invention or creation of theories to explain and order the observations. Theories are never derived directly from observations. Observations may help inspire a theory, and theories are accepted or rejected based on the results of observation and experiment.

The great theories of science may be compared, as creative achievements, with great works of art or literature. But how does science differ from these other creative activities? One important difference is that science requires **testing** of its ideas or theories to see if their predictions are borne out by experiment.

Although the testing of theories distinguishes science from other creative fields, it should not be assumed that a theory is "proved" by testing. First of all, no measuring instrument is perfect, so exact confirmation is not possible. Furthermore, it is not possible to test a theory in every single possible circumstance. Hence a theory cannot be absolutely verified. Indeed, the history of science tells us that long-held theories can be replaced by new ones.

1–2 Models, Theories, and Laws

When scientists are trying to understand a particular set of phenomena, they often make use of a **model**. A model, in the scientist's sense, is a kind of analogy or mental image of the phenomena in terms of something we are familiar with. One example is the wave model of light. We cannot see waves of light as we can water waves. But it is valuable to think of light as made up of waves because experiments indicate that light behaves in many respects as water waves do.

The purpose of a model is to give us an approximate mental or visual picture—something to hold on to—when we cannot see what actually is happening. Models often give us a deeper understanding: the analogy to a known system (for instance, water waves in the above example) can suggest new experiments to perform and can provide ideas about what other related phenomena might occur.

You may wonder what the difference is between a theory and a model. Usually a model is relatively simple and provides a structural similarity to the phenomena being studied. A **theory** is broader, more detailed, and can give quantitatively testable predictions, often with great precision.

It is important, however, not to confuse a model or a theory with the real system or the phenomena themselves.

Scientists give the title **law** to certain concise but general statements about how nature behaves (that energy is conserved, for example). Sometimes the statement takes the form of a relationship or equation between quantities (such as Newton's second law, $F = ma$).

To be called a law, a statement must be found experimentally valid over a wide range of observed phenomena. For less general statements, the term **principle** is often used (such as Archimedes' principle).

Scientific laws are different from political laws in that the latter are *prescriptive*: they tell us how we ought to behave. Scientific laws are *descriptive*: they do not say how nature *should* behave, but rather are meant to describe how nature *does* behave. As with theories, laws cannot be tested in the infinite variety of cases possible. So we cannot be sure that any law is absolutely true. We use the term "law" when its validity has been tested over a wide range of cases, and when any limitations and the range of validity are clearly understood.

Scientists normally do their research as if the accepted laws and theories were true. But they are obliged to keep an open mind in case new information should alter the validity of any given law or theory.

1–3 Measurement and Uncertainty; Significant Figures

In the quest to understand the world around us, scientists seek to find relationships among physical quantities that can be measured.

Uncertainty

Reliable measurements are an important part of physics. But no measurement is absolutely precise. There is an uncertainty associated with every measurement. Among the most important sources of uncertainty, other than blunders, are the limited accuracy of every measuring instrument and the inability to read an instrument beyond some fraction of the smallest division shown. For example, if you were to use a centimeter ruler to measure the width of a board (Fig. 1–2), the result could be claimed to be precise to about 0.1 cm (1 mm), the smallest division on the ruler, although half of this value might be a valid claim as well. The reason is that it is difficult for the observer to estimate (or interpolate) between the smallest divisions. Furthermore, the ruler itself may not have been manufactured to an accuracy very much better than this.

When giving the result of a measurement, it is important to state the **estimated uncertainty** in the measurement. For example, the width of a board might be written as 8.8 ± 0.1 cm. The ± 0.1 cm ("plus or minus 0.1 cm") represents the estimated uncertainty in the measurement, so that the actual width most likely lies between 8.7 and 8.9 cm. The **percent uncertainty** is the ratio of the uncertainty to the measured value, multiplied by 100. For example, if the measurement is 8.8 and the uncertainty about 0.1 cm, the percent uncertainty is

$$\frac{0.1}{8.8} \times 100\% \approx 1\%,$$

where $\approx$ means "is approximately equal to."

FIGURE 1–2 Measuring the width of a board with a centimeter ruler. The uncertainty is about ± 1 mm.

Often the uncertainty in a measured value is not specified explicitly. In such cases, the uncertainty is generally assumed to be one or a few units in the last digit specified. For example, if a length is given as 8.8 cm, the uncertainty is assumed to be about 0.1 cm or 0.2 cm. It is important in this case that you do not write 8.80 cm, for this implies an uncertainty on the order of 0.01 cm; it assumes that the length is probably between 8.79 cm and 8.81 cm, when actually you believe it is between 8.7 and 8.9 cm.

Significant Figures

The number of reliably known digits in a number is called the number of **significant figures**. Thus there are four significant figures in the number 23.21 cm and two in the number 0.062 cm (the zeros in the latter are merely place holders that show where the decimal point goes). The number of significant figures may not always be clear. Take, for example, the number 80. Are there one or two significant figures? We need words here: If we say it is *roughly* 80 km between two cities, there is only one significant figure (the 8) since the zero is merely a place holder. If there is no suggestion that the 80 is a rough approximation, then we can often assume (as we will in this book) that it is 80 km within an accuracy of about 1 or 2 km, and then the 80 has two significant figures. If it is precisely 80 km, to within ± 0.1 km, then we write 80.0 km (three significant figures).

When making measurements, or when doing calculations, you should avoid the temptation to keep more digits in the final answer than is justified. For example, to calculate the area of a rectangle 11.3 cm by 6.8 cm, the result of multiplication would be 76.84 cm². But this answer is clearly not accurate to 0.01 cm², since (using the outer limits of the assumed uncertainty for each measurement) the result could be between 11.2 cm × 6.7 cm = 75.04 cm² and 11.4 cm × 6.9 cm = 78.66 cm². At best, we can quote the answer as 77 cm², which implies an uncertainty of about 1 or 2 cm². The other two digits (in the number 76.84 cm²) must be dropped because they are not significant. As a rough general rule (i.e., in the absence of a detailed consideration of uncertainties), we can say that *the final result of a multiplication or division should have only as many digits as the number with the least number of significant figures used in the calculation.* In our example, 6.8 cm has the least number of significant figures, namely two. Thus the result 76.84 cm² needs to be rounded off to 77 cm².

EXERCISE A The area of a rectangle 4.5 cm by 3.25 cm is correctly given by (*a*) 14.625 cm²; (*b*) 14.63 cm²; (*c*) 14.6 cm²; (*d*) 15 cm².

When adding or subtracting numbers, the final result is no more precise than the least precise number used. For example, the result of subtracting 0.57 from 3.6 is 3.0 (and not 3.03).

Keep in mind when you use a calculator that all the digits it produces may not be significant. When you divide 2.0 by 3.0, the proper answer is 0.67, and not some such thing as 0.666666666. Digits should not be quoted in a result, unless they are truly significant figures. However, to obtain the most accurate result, you should normally *keep one or more extra significant figures throughout a calculation, and round off only in the final result.* (With a calculator, you can keep all its digits in intermediate results.) Note also that calculators sometimes give too few significant figures. For example, when you multiply 2.5 × 3.2, a calculator may give the answer as simply 8. But the answer is accurate to two significant figures, so the proper answer is 8.0. See Fig. 1–3.

(a)   (b)

FIGURE 1–3 These two calculators show the wrong number of significant figures. In (a), 2.0 was divided by 3.0. The correct final result would be 0.67. In (b), 2.5 was multiplied by 3.2. The correct result is 8.0.

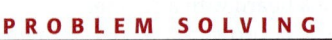
FIGURE 1–4 Example 1–1. A protractor used to measure an angle.

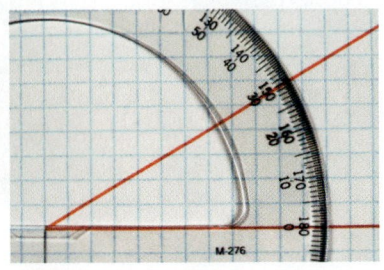

CONCEPTUAL EXAMPLE 1–1 **Significant figures.** Using a protractor (Fig. 1–4), you measure an angle to be 30°. (*a*) How many significant figures should you quote in this measurement? (*b*) Use a calculator to find the cosine of the angle you measured.

RESPONSE (*a*) If you look at a protractor, you will see that the precision with which you can measure an angle is about one degree (certainly not 0.1°). So you can quote two significant figures, namely, 30° (not 30.0°). (*b*) If you enter cos 30° in your calculator, you will get a number like 0.866025403. However, the angle you entered is known only to two significant figures, so its cosine is correctly given by 0.87; you must round your answer to two significant figures.

NOTE Cosine and other trigonometric functions are reviewed in Appendix A.

EXERCISE B Do 0.00324 and 0.00056 have the same number of significant figures?

Be careful not to confuse significant figures with the number of decimal places.

EXERCISE C For each of the following numbers, state the number of significant figures and the number of decimal places: (a) 1.23; (b) 0.123; (c) 0.0123.

Scientific Notation

We commonly write numbers in "powers of ten," or "scientific" notation—for instance 36,900 as 3.69×10^4, or 0.0021 as 2.1×10^{-3}. One advantage of scientific notation is that it allows the number of significant figures to be clearly expressed. For example, it is not clear whether 36,900 has three, four, or five significant figures. With powers of ten notation the ambiguity can be avoided: if the number is known to three significant figures, we write 3.69×10^4, but if it is known to four, we write 3.690×10^4.

EXERCISE D Write each of the following in scientific notation and state the number of significant figures for each: (a) 0.0258, (b) 42,300, (c) 344.50.

Percent Uncertainty versus Significant Figures

The significant figures rule is only approximate, and in some cases may underestimate the accuracy (or uncertainty) of the answer. Suppose for example we divide 97 by 92:

$$\frac{97}{92} = 1.05 \approx 1.1.$$

Both 97 and 92 have two significant figures, so the rule says to give the answer as 1.1. Yet the numbers 97 and 92 both imply an uncertainty of ± 1 if no other uncertainty is stated. Now 92 ± 1 and 97 ± 1 both imply an uncertainty of about 1% ($1/92 \approx 0.01 = 1\%$). But the final result to two significant figures is 1.1, with an implied uncertainty of ± 0.1, which is an uncertainty of $0.1/1.1 \approx 0.1 \approx 10\%$. In this case it is better to give the answer as 1.05 (which is three significant figures). Why? Because 1.05 implies an uncertainty of ± 0.01 which is $0.01/1.05 \approx 0.01 \approx 1\%$, just like the uncertainty in the original numbers 92 and 97.

SUGGESTION: Use the significant figures rule, but consider the % uncertainty too, and add an extra digit if it gives a more realistic estimate of uncertainty.

Approximations

Much of physics involves approximations, often because we do not have the means to solve a problem precisely. For example, we may choose to ignore air resistance or friction in doing a Problem even though they are present in the real world, and then our calculation is only an approximation. In doing Problems, we should be aware of what approximations we are making, and be aware that the precision of our answer may not be nearly as good as the number of significant figures given in the result.

Accuracy versus Precision

There is a technical difference between "precision" and "accuracy." **Precision** in a strict sense refers to the repeatability of the measurement using a given instrument. For example, if you measure the width of a board many times, getting results like 8.81 cm, 8.85 cm, 8.78 cm, 8.82 cm (interpolating between the 0.1 cm marks as best as possible each time), you could say the measurements give a *precision* a bit better than 0.1 cm. **Accuracy** refers to how close a measurement is to the true value. For example, if the ruler shown in Fig. 1–2 was manufactured with a 2% error, the accuracy of its measurement of the board's width (about 8.8 cm) would be about 2% of 8.8 cm or about ± 0.2 cm. Estimated uncertainty is meant to take both accuracy and precision into account.

1–4 Units, Standards, and the SI System

The measurement of any quantity is made relative to a particular standard or **unit**, and this unit must be specified along with the numerical value of the quantity. For example, we can measure length in British units such as inches, feet, or miles, or in the metric system in centimeters, meters, or kilometers. To specify that the length of a particular object is 18.6 is meaningless. The unit *must* be given; for clearly, 18.6 meters is very different from 18.6 inches or 18.6 millimeters.

For any unit we use, such as the meter for distance or the second for time, we need to define a **standard** which defines exactly how long one meter or one second is. It is important that standards be chosen that are readily reproducible so that anyone needing to make a very accurate measurement can refer to the standard in the laboratory.

Length

The first truly international standard was the **meter** (abbreviated m) established as the standard of **length** by the French Academy of Sciences in the 1790s. The standard meter was originally chosen to be one ten-millionth of the distance from the Earth's equator to either pole,[†] and a platinum rod to represent this length was made. (One meter is, very roughly, the distance from the tip of your nose to the tip of your finger, with arm and hand stretched out to the side.) In 1889, the meter was defined more precisely as the distance between two finely engraved marks on a particular bar of platinum–iridium alloy. In 1960, to provide greater precision and reproducibility, the meter was redefined as 1,650,763.73 wavelengths of a particular orange light emitted by the gas krypton-86. In 1983 the meter was again redefined, this time in terms of the speed of light (whose best measured value in terms of the older definition of the meter was 299,792,458 m/s, with an uncertainty of 1 m/s). The new definition reads: "The meter is the length of path traveled by light in vacuum during a time interval of 1/299,792,458 of a second."[‡]

British units of length (inch, foot, mile) are now defined in terms of the meter. The inch (in.) is defined as precisely 2.54 centimeters (cm; 1 cm = 0.01 m). Other conversion factors are given in the Table on the inside of the front cover of this book. Table 1–1 presents some typical lengths, from very small to very large, rounded off to the nearest power of ten. See also Fig. 1–5. [Note that the abbreviation for inches (in.) is the only one with a period, to distinguish it from the word "in".]

Time

The standard unit of **time** is the **second** (s). For many years, the second was defined as 1/86,400 of a mean solar day (24 h/day × 60 min/h × 60 s/min = 86,400 s/day). The standard second is now defined more precisely in terms of the frequency of radiation emitted by cesium atoms when they pass between two particular states. [Specifically, one second is defined as the time required for 9,192,631,770 periods of this radiation.] There are, by definition, 60 s in one minute (min) and 60 minutes in one hour (h). Table 1–2 presents a range of measured time intervals, rounded off to the nearest power of ten.

Mass

The standard unit of **mass** is the **kilogram** (kg). The standard mass is a particular platinum–iridium cylinder, kept at the International Bureau of Weights and Measures near Paris, France, whose mass is defined as exactly 1 kg. A range of masses is presented in Table 1–3. [For practical purposes, 1 kg weighs about 2.2 pounds on Earth.]

[†]Modern measurements of the Earth's circumference reveal that the intended length is off by about one-fiftieth of 1%. Not bad!

[‡]The new definition of the meter has the effect of giving the speed of light the exact value of 299,792,458 m/s.

TABLE 1–1 Some Typical Lengths or Distances
(order of magnitude)

Length (or Distance)	Meters (approximate)
Neutron or proton (diameter)	10^{-15} m
Atom (diameter)	10^{-10} m
Virus [see Fig. 1–5a]	10^{-7} m
Sheet of paper (thickness)	10^{-4} m
Finger width	10^{-2} m
Football field length	10^{2} m
Height of Mt. Everest [see Fig. 1–5b]	10^{4} m
Earth diameter	10^{7} m
Earth to Sun	10^{11} m
Earth to nearest star	10^{16} m
Earth to nearest galaxy	10^{22} m
Earth to farthest galaxy visible	10^{26} m

FIGURE 1–5 Some lengths: (a) viruses (about 10^{-7} m long) attacking a cell; (b) Mt. Everest's height is on the order of 10^{4} m (8850 m, to be precise).

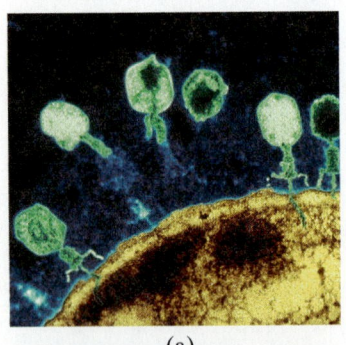

(a)

(b)

TABLE 1–2 Some Typical Time Intervals

Time Interval	Seconds (approximate)
Lifetime of very unstable subatomic particle	10^{-23} s
Lifetime of radioactive elements	10^{-22} s to 10^{28} s
Lifetime of muon	10^{-6} s
Time between human heartbeats	10^{0} s (= 1 s)
One day	10^{5} s
One year	3×10^{7} s
Human life span	2×10^{9} s
Length of recorded history	10^{11} s
Humans on Earth	10^{14} s
Life on Earth	10^{17} s
Age of Universe	10^{18} s

TABLE 1–3 Some Masses

Object	Kilograms (approximate)
Electron	10^{-30} kg
Proton, neutron	10^{-27} kg
DNA molecule	10^{-17} kg
Bacterium	10^{-15} kg
Mosquito	10^{-5} kg
Plum	10^{-1} kg
Human	10^{2} kg
Ship	10^{8} kg
Earth	6×10^{24} kg
Sun	2×10^{30} kg
Galaxy	10^{41} kg

When dealing with atoms and molecules, we usually use the **unified atomic mass unit** (u). In terms of the kilogram,

$$1\,\text{u} = 1.6605 \times 10^{-27}\,\text{kg}.$$

The definitions of other standard units for other quantities will be given as we encounter them in later Chapters. (Precise values of this and other numbers are given inside the front cover.)

Unit Prefixes

In the metric system, the larger and smaller units are defined in multiples of 10 from the standard unit, and this makes calculation particularly easy. Thus 1 kilometer (km) is 1000 m, 1 centimeter is $\frac{1}{100}$ m, 1 millimeter (mm) is $\frac{1}{1000}$ m or $\frac{1}{10}$ cm, and so on. The prefixes "centi-," "kilo-," and others are listed in Table 1–4 and can be applied not only to units of length but to units of volume, mass, or any other metric unit. For example, a centiliter (cL) is $\frac{1}{100}$ liter (L), and a kilogram (kg) is 1000 grams (g).

Systems of Units

When dealing with the laws and equations of physics it is very important to use a consistent set of units. Several systems of units have been in use over the years. Today the most important is the **Système International** (French for International System), which is abbreviated SI. In SI units, the standard of length is the meter, the standard for time is the second, and the standard for mass is the kilogram. This system used to be called the MKS (meter-kilogram-second) system.

A second metric system is the **cgs system**, in which the centimeter, gram, and second are the standard units of length, mass, and time, as abbreviated in the title. The **British engineering system** has as its standards the foot for length, the pound for force, and the second for time.

We use SI units almost exclusively in this book.

Base versus Derived Quantities

Physical quantities can be divided into two categories: *base quantities* and *derived quantities*. The corresponding units for these quantities are called *base units* and *derived units*. A **base quantity** must be defined in terms of a standard. Scientists, in the interest of simplicity, want the smallest number of base quantities possible consistent with a full description of the physical world. This number turns out to be seven, and those used in the SI are given in Table 1–5. All other quantities can be defined in terms of these seven base quantities,[†] and hence are referred to as **derived quantities**. An example of a derived quantity is speed, which is defined as distance divided by the time it takes to travel that distance. A Table inside the front cover lists many derived quantities and their units in terms of base units. To define any quantity, whether base or derived, we can specify a rule or procedure, and this is called an **operational definition**.

[†]The only exceptions are for angle (radians—see Chapter 8) and solid angle (steradian). No general agreement has been reached as to whether these are base or derived quantities.

TABLE 1–4 Metric (SI) Prefixes

Prefix	Abbreviation	Value
yotta	Y	10^{24}
zetta	Z	10^{21}
exa	E	10^{18}
peta	P	10^{15}
tera	T	10^{12}
giga	G	10^{9}
mega	M	10^{6}
kilo	k	10^{3}
hecto	h	10^{2}
deka	da	10^{1}
deci	d	10^{-1}
centi	c	10^{-2}
milli	m	10^{-3}
micro[†]	μ	10^{-6}
nano	n	10^{-9}
pico	p	10^{-12}
femto	f	10^{-15}
atto	a	10^{-18}
zepto	z	10^{-21}
yocto	y	10^{-24}

[†]μ is the Greek letter "mu."

TABLE 1–5 SI Base Quantities and Units

Quantity	Unit	Unit Abbreviation
Length	meter	m
Time	second	s
Mass	kilogram	kg
Electric current	ampere	A
Temperature	kelvin	K
Amount of substance	mole	mol
Luminous intensity	candela	cd

1–5 Converting Units

Any quantity we measure, such as a length, a speed, or an electric current, consists of a number *and* a unit. Often we are given a quantity in one set of units, but we want it expressed in another set of units. For example, suppose we measure that a table is 21.5 inches wide, and we want to express this in centimeters. We must use a **conversion factor**, which in this case is (by definition) exactly

$$1 \text{ in.} = 2.54 \text{ cm}$$

or, written another way,

$$1 = 2.54 \text{ cm/in.}$$

Since multiplying by one does not change anything, the width of our table, in cm, is

$$21.5 \text{ inches} = (21.5 \text{ in.}) \times \left(2.54 \frac{\text{cm}}{\text{in.}}\right) = 54.6 \text{ cm.}$$

Note how the units (inches in this case) cancelled out. A Table containing many unit conversions is found inside the front cover of this book. Let's consider some Examples.

FIGURE 1–6 The world's second highest peak, K2, whose summit is considered the most difficult of the "8000-ers." K2 is seen here from the north (China).

TABLE 1–6
The 8000-m Peaks

Peak	Height (m)
Mt. Everest	8850
K2	8611
Kangchenjunga	8586
Lhotse	8516
Makalu	8462
Cho Oyu	8201
Dhaulagiri	8167
Manaslu	8156
Nanga Parbat	8125
Annapurna	8091
Gasherbrum I	8068
Broad Peak	8047
Gasherbrum II	8035
Shisha Pangma	8013

EXAMPLE 1–2 **The 8000-m peaks.** The fourteen tallest peaks in the world (Fig. 1–6 and Table 1–6) are referred to as "eight-thousanders," meaning their summits are over 8000 m above sea level. What is the elevation, in feet, of an elevation of 8000 m?

APPROACH We need simply to convert meters to feet, and we can start with the conversion factor 1 in. = 2.54 cm, which is exact. That is, 1 in. = 2.5400 cm to any number of significant figures, because it is *defined* to be.

SOLUTION One foot is 12 in., so we can write

$$1 \text{ ft} = (12 \text{ in.})\left(2.54 \frac{\text{cm}}{\text{in.}}\right) = 30.48 \text{ cm} = 0.3048 \text{ m,}$$

which is exact. Note how the units cancel (colored slashes). We can rewrite this equation to find the number of feet in 1 meter:

$$1 \text{ m} = \frac{1 \text{ ft}}{0.3048} = 3.28084 \text{ ft.}$$

We multiply this equation by 8000.0 (to have five significant figures):

$$8000.0 \text{ m} = (8000.0 \text{ m})\left(3.28084 \frac{\text{ft}}{\text{m}}\right) = 26,247 \text{ ft.}$$

An elevation of 8000 m is 26,247 ft above sea level.

NOTE We could have done the conversion all in one line:

$$8000.0 \text{ m} = (8000.0 \text{ m})\left(\frac{100 \text{ cm}}{1 \text{ m}}\right)\left(\frac{1 \text{ in.}}{2.54 \text{ cm}}\right)\left(\frac{1 \text{ ft}}{12 \text{ in.}}\right) = 26,247 \text{ ft.}$$

The key is to multiply conversion factors, each equal to one (= 1.0000), and to make sure the units cancel.

EXERCISE E There are only 14 eight-thousand-meter peaks in the world (see Example 1–2), and their names and elevations are given in Table 1–6. They are all in the Himalaya mountain range in India, Pakistan, Tibet, and China. Determine the elevation of the world's three highest peaks in feet.

EXAMPLE 1–3 **Apartment area.** You have seen a nice apartment whose floor area is 880 square feet (ft^2). What is its area in square meters?

APPROACH We use the same conversion factor, 1 in. = 2.54 cm, but this time we have to use it twice.

SOLUTION Because 1 in. = 2.54 cm = 0.0254 m, then $1\,\text{ft}^2 = (12\,\text{in.})^2(0.0254\,\text{m/in.})^2 = 0.0929\,\text{m}^2$. So $880\,\text{ft}^2 = (880\,\text{ft}^2)(0.0929\,\text{m}^2/\text{ft}^2) \approx 82\,\text{m}^2$.

NOTE As a rule of thumb, an area given in ft^2 is roughly 10 times the number of square meters (more precisely, about $10.8\times$).

EXAMPLE 1–4 **Speeds.** Where the posted speed limit is 55 miles per hour (mi/h or mph), what is this speed (a) in meters per second (m/s) and (b) in kilometers per hour (km/h)?

APPROACH We again use the conversion factor 1 in. = 2.54 cm, and we recall that there are 5280 ft in a mile and 12 inches in a foot; also, one hour contains $(60\,\text{min/h}) \times (60\,\text{s/min}) = 3600\,\text{s/h}$.

SOLUTION (a) We can write 1 mile as

$$1\,\text{mi} = (5280\,\cancel{\text{ft}})\left(12\,\frac{\cancel{\text{in.}}}{\cancel{\text{ft}}}\right)\left(2.54\,\frac{\cancel{\text{cm}}}{\cancel{\text{in.}}}\right)\left(\frac{1\,\text{m}}{100\,\cancel{\text{cm}}}\right) = 1609\,\text{m}.$$

We also know that 1 hour contains 3600 s, so

$$55\,\frac{\text{mi}}{\text{h}} = \left(55\,\frac{\cancel{\text{mi}}}{\cancel{\text{h}}}\right)\left(1609\,\frac{\text{m}}{\cancel{\text{mi}}}\right)\left(\frac{1\,\cancel{\text{h}}}{3600\,\text{s}}\right) = 25\,\frac{\text{m}}{\text{s}},$$

where we rounded off to two significant figures.
(b) Now we use 1 mi = 1609 m = 1.609 km; then

$$55\,\frac{\text{mi}}{\text{h}} = \left(55\,\frac{\cancel{\text{mi}}}{\text{h}}\right)\left(1.609\,\frac{\text{km}}{\cancel{\text{mi}}}\right) = 88\,\frac{\text{km}}{\text{h}}.$$

NOTE Each conversion factor is equal to one. You can look up most conversion factors in the Table inside the front cover.

PROBLEM SOLVING
Conversion factors = 1

EXERCISE F Would a driver traveling at 15 m/s in a 35 mi/h zone be exceeding the speed limit?

When changing units, you can avoid making an error in the use of conversion factors by checking that units cancel out properly. For example, in our conversion of 1 mi to 1609 m in Example 1–4(a), if we had incorrectly used the factor $\left(\frac{100\,\text{cm}}{1\,\text{m}}\right)$ instead of $\left(\frac{1\,\text{m}}{100\,\text{cm}}\right)$, the centimeter units would not have cancelled out; we would not have ended up with meters.

PROBLEM SOLVING
Unit conversion is wrong if units do not cancel

1–6 Order of Magnitude: Rapid Estimating

We are sometimes interested only in an approximate value for a quantity. This might be because an accurate calculation would take more time than it is worth or would require additional data that are not available. In other cases, we may want to make a rough estimate in order to check an accurate calculation made on a calculator, to make sure that no blunders were made when the numbers were entered.

A rough estimate is made by rounding off all numbers to one significant figure and its power of 10, and after the calculation is made, again only one significant figure is kept. Such an estimate is called an **order-of-magnitude estimate** and can be accurate within a factor of 10, and often better. In fact, the phrase "order of magnitude" is sometimes used to refer simply to the power of 10.

PROBLEM SOLVING
How to make a rough estimate

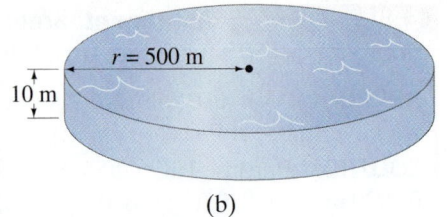

(b)

(a)

FIGURE 1–7 Example 1–5. (a) How much water is in this lake? (Photo is of one of the Rae Lakes in the Sierra Nevada of California.) (b) Model of the lake as a cylinder. [We could go one step further and estimate the mass or weight of this lake. We will see later that water has a density of 1000 kg/m³, so this lake has a mass of about $(10^3 \, \text{kg/m}^3)(10^7 \, \text{m}^3) \approx 10^{10}$ kg, which is about 10 billion kg or 10 million metric tons. (A metric ton is 1000 kg, about 2200 lbs, slightly larger than a British ton, 2000 lbs.)]

PHYSICS APPLIED

Estimating the volume (or mass) of a lake; see also Fig. 1–7

EXAMPLE 1–5 **ESTIMATE** **Volume of a lake.** Estimate how much water there is in a particular lake, Fig. 1–7a, which is roughly circular, about 1 km across, and you guess it has an average depth of about 10 m.

APPROACH No lake is a perfect circle, nor can lakes be expected to have a perfectly flat bottom. We are only estimating here. To estimate the volume, we can use a simple model of the lake as a cylinder: we multiply the average depth of the lake times its roughly circular surface area, as if the lake were a cylinder (Fig. 1–7b).

SOLUTION The volume V of a cylinder is the product of its height h times the area of its base: $V = h\pi r^2$, where r is the radius of the circular base.[†] The radius r is $\frac{1}{2}$ km = 500 m, so the volume is approximately

$$V = h\pi r^2 \approx (10 \, \text{m}) \times (3) \times (5 \times 10^2 \, \text{m})^2 \approx 8 \times 10^6 \, \text{m}^3 \approx 10^7 \, \text{m}^3,$$

where π was rounded off to 3. So the volume is on the order of $10^7 \, \text{m}^3$, ten million cubic meters. Because of all the estimates that went into this calculation, the order-of-magnitude estimate $(10^7 \, \text{m}^3)$ is probably better to quote than the $8 \times 10^6 \, \text{m}^3$ figure.

NOTE To express our result in U.S. gallons, we see in the Table on the inside front cover that 1 liter = $10^{-3} \, \text{m}^3 \approx \frac{1}{4}$ gallon. Hence, the lake contains $(8 \times 10^6 \, \text{m}^3)(1 \, \text{gallon}/4 \times 10^{-3} \, \text{m}^3) \approx 2 \times 10^9$ gallons of water.

EXAMPLE 1–6 **ESTIMATE** **Thickness of a page.** Estimate the thickness of a page of this book.

APPROACH At first you might think that a special measuring device, a micrometer (Fig. 1–8), is needed to measure the thickness of one page since an ordinary ruler clearly won't do. But we can use a trick or, to put it in physics terms, make use of a *symmetry*: we can make the reasonable assumption that all the pages of this book are equal in thickness.

PROBLEM SOLVING

Use symmetry when possible

SOLUTION We can use a ruler to measure hundreds of pages at once. If you measure the thickness of the first 500 pages of this book (page 1 to page 500), you might get something like 1.5 cm. Note that 500 numbered pages,

---

[†]Formulas like this for volume, area, etc., are found inside the back cover of this book.

counted front and back, is 250 separate sheets of paper. So one page must have a thickness of about

$$\frac{1.5 \text{ cm}}{250 \text{ pages}} \approx 6 \times 10^{-3} \text{ cm} = 6 \times 10^{-2} \text{ mm},$$

or less than a tenth of a millimeter (0.1 mm).

EXAMPLE 1–7 ESTIMATE Height by triangulation. Estimate the height of the building shown in Fig. 1–9, by "triangulation," with the help of a bus-stop pole and a friend.

APPROACH By standing your friend next to the pole, you estimate the height of the pole to be 3 m. You next step away from the pole until the top of the pole is in line with the top of the building, Fig. 1–9a. You are 5 ft 6 in. tall, so your eyes are about 1.5 m above the ground. Your friend is taller, and when she stretches out her arms, one hand touches you, and the other touches the pole, so you estimate that distance as 2 m (Fig. 1–9a). You then pace off the distance from the pole to the base of the building with big, 1-m-long steps, and you get a total of 16 steps or 16 m.

SOLUTION Now you draw, to scale, the diagram shown in Fig. 1–9b using these measurements. You can measure, right on the diagram, the last side of the triangle to be about $x = 13$ m. Alternatively, you can use similar triangles to obtain the height x:

$$\frac{1.5 \text{ m}}{2 \text{ m}} = \frac{x}{18 \text{ m}}, \quad \text{so} \quad x \approx 13\tfrac{1}{2} \text{ m}.$$

Finally you add in your eye height of 1.5 m above the ground to get your final result: the building is about 15 m tall.

EXAMPLE 1–8 ESTIMATE Estimating the radius of Earth. Believe it or not, you can estimate the radius of the Earth without having to go into space (see the photograph on page 1). If you have ever been on the shore of a large lake, you may have noticed that you cannot see the beaches, piers, or rocks at water level across the lake on the opposite shore. The lake seems to bulge out between you and the opposite shore—a good clue that the Earth is round. Suppose you climb a stepladder and discover that when your eyes are 10 ft (3.0 m) above the water, you can just see the rocks at water level on the opposite shore. From a map, you estimate the distance to the opposite shore as $d \approx 6.1$ km. Use Fig. 1–10 with $h = 3.0$ m to estimate the radius R of the Earth.

APPROACH We use simple geometry, including the theorem of Pythagoras, $c^2 = a^2 + b^2$, where c is the length of the hypotenuse of any right triangle, and a and b are the lengths of the other two sides.

SOLUTION For the right triangle of Fig. 1–10, the two sides are the radius of the Earth R and the distance $d = 6.1$ km $= 6100$ m. The hypotenuse is approximately the length $R + h$, where $h = 3.0$ m. By the Pythagorean theorem,

$$R^2 + d^2 \approx (R + h)^2$$
$$\approx R^2 + 2hR + h^2.$$

We solve algebraically for R, after cancelling R^2 on both sides:

$$R \approx \frac{d^2 - h^2}{2h} = \frac{(6100 \text{ m})^2 - (3.0 \text{ m})^2}{6.0 \text{ m}} = 6.2 \times 10^6 \text{ m} = 6200 \text{ km}.$$

NOTE Precise measurements give 6380 km. But look at your achievement! With a few simple rough measurements and simple geometry, you made a good estimate of the Earth's radius. You did not need to go out in space, nor did you need a very long measuring tape. Now you know the answer to the Chapter-Opening Question on p. 1.

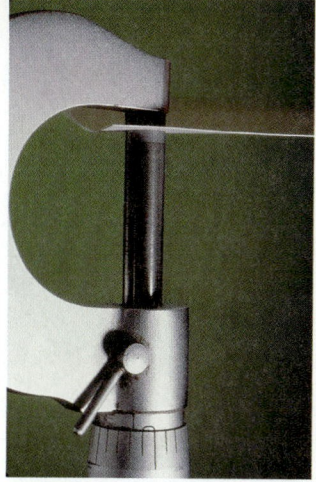

FIGURE 1–8 Example 1–6. Micrometer used for measuring small thicknesses.

FIGURE 1–9 Example 1–7. Diagrams are really useful!

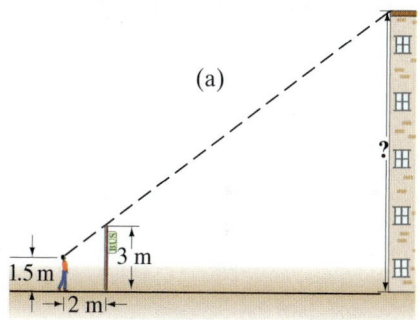

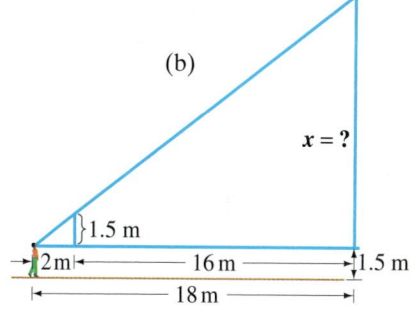

(a)

(b)

$x = ?$

FIGURE 1–10 Example 1–8, but not to scale. You can see small rocks at water level on the opposite shore of a lake 6.1 km wide if you stand on a stepladder.

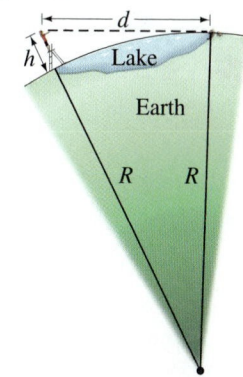

EXAMPLE 1–9 ESTIMATE **Total number of heartbeats.** Estimate the total number of beats a typical human heart makes in a lifetime.

APPROACH A typical resting heart rate is 70 beats/min. But during exercise it can be a lot higher. A reasonable average might be 80 beats/min.

SOLUTION One year in terms of seconds is $(24\,\text{h})(3600\,\text{s/h})(365\,\text{d}) \approx 3 \times 10^7\,\text{s}$. If an average person lives 70 years $= (70\,\text{yr})(3 \times 10^7\,\text{s/yr}) \approx 2 \times 10^9\,\text{s}$, then the total number of heartbeats would be about

$$\left(80\,\frac{\text{beats}}{\text{min}}\right)\left(\frac{1\,\text{min}}{60\,\text{s}}\right)(2 \times 10^9\,\text{s}) \approx 3 \times 10^9,$$

or 3 trillion.

Another technique for estimating, this one made famous by Enrico Fermi to his physics students, is to estimate the number of piano tuners in a city, say, Chicago or San Francisco. To get a rough order-of-magnitude estimate of the number of piano tuners today in San Francisco, a city of about 700,000 inhabitants, we can proceed by estimating the number of functioning pianos, how often each piano is tuned, and how many pianos each tuner can tune. To estimate the number of pianos in San Francisco, we note that certainly not everyone has a piano. A guess of 1 family in 3 having a piano would correspond to 1 piano per 12 persons, assuming an average family of 4 persons. As an order of magnitude, let's say 1 piano per 10 people. This is certainly more reasonable than 1 per 100 people, or 1 per every person, so let's proceed with the estimate that 1 person in 10 has a piano, or about 70,000 pianos in San Francisco. Now a piano tuner needs an hour or two to tune a piano. So let's estimate that a tuner can tune 4 or 5 pianos a day. A piano ought to be tuned every 6 months or a year—let's say once each year. A piano tuner tuning 4 pianos a day, 5 days a week, 50 weeks a year can tune about 1000 pianos a year. So San Francisco, with its (very) roughly 70,000 pianos, needs about 70 piano tuners. This is, of course, only a rough estimate.[†] It tells us that there must be many more than 10 piano tuners, and surely not as many as 1000.

PROBLEM SOLVING
Estimating how many piano tuners there are in a city

*1–7 Dimensions and Dimensional Analysis

When we speak of the **dimensions** of a quantity, we are referring to the type of base units or base quantities that make it up. The dimensions of area, for example, are always length squared, abbreviated $[L^2]$, using square brackets; the units can be square meters, square feet, cm², and so on. Velocity, on the other hand, can be measured in units of km/h, m/s, or mi/h, but the dimensions are always a length $[L]$ divided by a time $[T]$: that is, $[L/T]$.

The formula for a quantity may be different in different cases, but the dimensions remain the same. For example, the area of a triangle of base b and height h is $A = \frac{1}{2}bh$, whereas the area of a circle of radius r is $A = \pi r^2$. The formulas are different in the two cases, but the dimensions of area are always $[L^2]$.

Dimensions can be used as a help in working out relationships, a procedure referred to as **dimensional analysis**. One useful technique is the use of dimensions to check if a relationship is *incorrect*. Note that we add or subtract quantities only if they have the same dimensions (we don't add centimeters and hours); and the quantities on each side of an equals sign must have the same dimensions. (In numerical calculations, the units must also be the same on both sides of an equation.)

For example, suppose you derived the equation $v = v_0 + \frac{1}{2}at^2$, where v is the speed of an object after a time t, v_0 is the object's initial speed, and the object undergoes an acceleration a. Let's do a dimensional check to see if this equation

---

[†] A check of the San Francisco Yellow Pages (done after this calculation) reveals about 50 listings. Each of these listings may employ more than one tuner, but on the other hand, each may also do repairs as well as tuning. In any case, our estimate is reasonable.

[*] Some Sections of this book, such as this one, may be considered *optional* at the discretion of the instructor, and they are marked with an asterisk (*). See the Preface for more details.

could be correct or is surely incorrect. Note that numerical factors, like the $\frac{1}{2}$ here, do not affect dimensional checks. We write a dimensional equation as follows, remembering that the dimensions of speed are $[L/T]$ and (as we shall see in Chapter 2) the dimensions of acceleration are $[L/T^2]$:

$$\left[\frac{L}{T}\right] \overset{?}{=} \left[\frac{L}{T}\right] + \left[\frac{L}{T^2}\right][T^2] = \left[\frac{L}{T}\right] + [L].$$

The dimensions are incorrect: on the right side, we have the sum of quantities whose dimensions are not the same. Thus we conclude that an error was made in the derivation of the original equation.

A dimensional check can only tell you when a relationship is wrong. It can't tell you if it is completely right. For example, a dimensionless numerical factor (such as $\frac{1}{2}$ or 2π) could be missing.

Dimensional analysis can also be used as a quick check on an equation you are not sure about. For example, suppose that you can't remember whether the equation for the period of a simple pendulum T (the time to make one back-and-forth swing) of length ℓ is $T = 2\pi\sqrt{\ell/g}$ or $T = 2\pi\sqrt{g/\ell}$, where g is the acceleration due to gravity and, like all accelerations, has dimensions $[L/T^2]$. (Do not worry about these formulas—the correct one will be derived in Chapter 14; what we are concerned about here is a person's recalling whether it contains ℓ/g or g/ℓ.) A dimensional check shows that the former (ℓ/g) is correct:

$$[T] = \sqrt{\frac{[L]}{[L/T^2]}} = \sqrt{[T^2]} = [T],$$

whereas the latter (g/ℓ) is not:

$$[T] \neq \sqrt{\frac{[L/T^2]}{[L]}} = \sqrt{\frac{1}{[T^2]}} = \frac{1}{[T]}.$$

Note that the constant 2π has no dimensions and so can't be checked using dimensions.

Further uses of dimensional analysis are found in Appendix C.

EXAMPLE 1–10 Planck length. The smallest meaningful measure of length is called the "Planck length," and is defined in terms of three fundamental constants in nature, the speed of light $c = 3.00 \times 10^8$ m/s, the gravitational constant $G = 6.67 \times 10^{-11}$ m^3/kg·s^2, and Planck's constant $h = 6.63 \times 10^{-34}$ kg·m^2/s. The Planck length λ_P (λ is the Greek letter "lambda") is given by the following combination of these three constants:

$$\lambda_P = \sqrt{\frac{Gh}{c^3}}.$$

Show that the dimensions of λ_P are length $[L]$, and find the order of magnitude of λ_P.

APPROACH We rewrite the above equation in terms of dimensions. The dimensions of c are $[L/T]$, of G are $[L^3/MT^2]$, and of h are $[ML^2/T]$.

SOLUTION The dimensions of λ_P are

$$\sqrt{\frac{[L^3/MT^2][ML^2/T]}{[L^3/T^3]}} = \sqrt{[L^2]} = [L]$$

which is a length. The value of the Planck length is

$$\lambda_P = \sqrt{\frac{Gh}{c^3}} = \sqrt{\frac{(6.67 \times 10^{-11}\,\text{m}^3/\text{kg·s}^2)(6.63 \times 10^{-34}\,\text{kg·m}^2/\text{s})}{(3.0 \times 10^8\,\text{m/s})^3}} \approx 4 \times 10^{-35}\,\text{m},$$

which is on the order of 10^{-34} or 10^{-35} m.

NOTE Some recent theories (Chapters 43 and 44) suggest that the smallest particles (quarks, leptons) have sizes on the order of the Planck length, 10^{-35} m. These theories also suggest that the "Big Bang," with which the Universe is believed to have begun, started from an initial size on the order of the Planck length.

Summary

[The Summary that appears at the end of each Chapter in this book gives a brief overview of the main ideas of the Chapter. The Summary *cannot* serve to give an understanding of the material, which can be accomplished only by a detailed reading of the Chapter.]

Physics, like other sciences, is a creative endeavor. It is not simply a collection of facts. Important **theories** are created with the idea of explaining **observations**. To be accepted, theories are **tested** by comparing their predictions with the results of actual experiments. Note that, in general, a theory cannot be "proved" in an absolute sense.

Scientists often devise models of physical phenomena. A **model** is a kind of picture or analogy that helps to describe the phenomena in terms of something we already know. A **theory**, often developed from a model, is usually deeper and more complex than a simple model.

A scientific **law** is a concise statement, often expressed in the form of an equation, which quantitatively describes a wide range of phenomena.

Measurements play a crucial role in physics, but can never be perfectly precise. It is important to specify the **uncertainty** of a measurement either by stating it directly using the ± notation, and/or by keeping only the correct number of **significant figures**.

Physical quantities are always specified relative to a particular standard or **unit**, and the unit used should always be stated. The commonly accepted set of units today is the **Système International** (SI), in which the standard units of length, mass, and time are the **meter**, **kilogram**, and **second**.

When converting units, check all **conversion factors** for correct cancellation of units.

Making rough, **order-of-magnitude estimates** is a very useful technique in science as well as in everyday life.

[*The **dimensions** of a quantity refer to the combination of base quantities that comprise it. Velocity, for example, has dimensions of [length/time] or $[L/T]$. **Dimensional analysis** can be used to check a relationship for correct form.]

Questions

1. What are the merits and drawbacks of using a person's foot as a standard? Consider both (a) a particular person's foot, and (b) any person's foot. Keep in mind that it is advantageous that fundamental standards be accessible (easy to compare to), invariable (do not change), indestructible, and reproducible.

2. Why is it incorrect to think that the more digits you represent in your answer, the more accurate it is?

3. When traveling a highway in the mountains, you may see elevation signs that read "914 m (3000 ft)." Critics of the metric system claim that such numbers show the metric system is more complicated. How would you alter such signs to be more consistent with a switch to the metric system?

4. What is wrong with this road sign:
   Memphis 7 mi (11.263 km)?

5. For an answer to be complete, the units need to be specified. Why?

6. Discuss how the notion of symmetry could be used to estimate the number of marbles in a 1-liter jar.

7. You measure the radius of a wheel to be 4.16 cm. If you multiply by 2 to get the diameter, should you write the result as 8 cm or as 8.32 cm? Justify your answer.

8. Express the sine of 30.0° with the correct number of significant figures.

9. A recipe for a soufflé specifies that the measured ingredients must be exact, or the soufflé will not rise. The recipe calls for 6 large eggs. The size of "large" eggs can vary by 10%, according to the USDA specifications. What does this tell you about how exactly you need to measure the other ingredients?

10. List assumptions useful to estimate the number of car mechanics in (a) San Francisco, (b) your hometown, and then make the estimates.

11. Suggest a way to measure the distance from Earth to the Sun.

*12. Can you set up a complete set of base quantities, as in Table 1–5, that does not include length as one of them?

Problems

[The Problems at the end of each Chapter are ranked I, II, or III according to estimated difficulty, with (I) Problems being easiest. Level (III) Problems are meant mainly as a challenge for the best students, for "extra credit." The Problems are arranged by Sections, meaning that the reader should have read up to and including that Section, but not only that Section—Problems often depend on earlier material. Each Chapter also has a group of General Problems that are not arranged by Section and not ranked.]

1–3 Measurement, Uncertainty, Significant Figures

(*Note:* In Problems, assume a number like 6.4 is accurate to ± 0.1; and 950 is ± 10 unless 950 is said to be "precisely" or "very nearly" 950, in which case assume 950 ± 1.)

1. (I) The age of the universe is thought to be about 14 billion years. Assuming two significant figures, write this in powers of ten in (a) years, (b) seconds.

2. (I) How many significant figures do each of the following numbers have: (a) 214, (b) 81.60, (c) 7.03, (d) 0.03, (e) 0.0086, (f) 3236, and (g) 8700?

3. (I) Write the following numbers in powers of ten notation: (a) 1.156, (b) 21.8, (c) 0.0068, (d) 328.65, (e) 0.219, and (f) 444.

4. (I) Write out the following numbers in full with the correct number of zeros: (a) 8.69×10^4, (b) 9.1×10^3, (c) 8.8×10^{-1}, (d) 4.76×10^2, and (e) 3.62×10^{-5}.

5. (II) What is the percent uncertainty in the measurement 5.48 ± 0.25 m?

6. (II) Time intervals measured with a stopwatch typically have an uncertainty of about 0.2 s, due to human reaction time at the start and stop moments. What is the percent uncertainty of a hand-timed measurement of (a) 5 s, (b) 50 s, (c) 5 min?

7. (II) Add $(9.2 \times 10^3 \text{ s}) + (8.3 \times 10^4 \text{ s}) + (0.008 \times 10^6 \text{ s})$.

8. (II) Multiply 2.079×10^2 m by 0.082×10^{-1}, taking into account significant figures.

9. (III) For small angles θ, the numerical value of $\sin\theta$ is approximately the same as the numerical value of $\tan\theta$. Find the largest angle for which sine and tangent agree to within two significant figures.

10. (III) What, roughly, is the percent uncertainty in the volume of a spherical beach ball whose radius is $r = 0.84 \pm 0.04$ m?

1–4 and 1–5 Units, Standards, SI, Converting Units

11. (I) Write the following as full (decimal) numbers with standard units: (a) 286.6 mm, (b) 85 μV, (c) 760 mg, (d) 60.0 ps, (e) 22.5 fm, (f) 2.50 gigavolts.

12. (I) Express the following using the prefixes of Table 1–4: (a) 1×10^6 volts, (b) 2×10^{-6} meters, (c) 6×10^3 days, (d) 18×10^2 bucks, and (e) 8×10^{-8} seconds.

13. (I) Determine your own height in meters, and your mass in kg.

14. (I) The Sun, on average, is 93 million miles from Earth. How many meters is this? Express (a) using powers of ten, and (b) using a metric prefix.

15. (II) What is the conversion factor between (a) ft^2 and yd^2, (b) m^2 and ft^2?

16. (II) An airplane travels at 950 km/h. How long does it take to travel 1.00 km?

17. (II) A typical atom has a diameter of about 1.0×10^{-10} m. (a) What is this in inches? (b) Approximately how many atoms are there along a 1.0-cm line?

18. (II) Express the following sum with the correct number of significant figures: 1.80 m $+ 142.5$ cm $+ 5.34 \times 10^5 \mu$m.

19. (II) Determine the conversion factor between (a) km/h and mi/h, (b) m/s and ft/s, and (c) km/h and m/s.

20. (II) How much longer (percentage) is a one-mile race than a 1500-m race ("the metric mile")?

21. (II) A *light-year* is the distance light travels in one year (at speed $= 2.998 \times 10^8$ m/s). (a) How many meters are there in 1.00 light-year? (b) An astronomical unit (AU) is the average distance from the Sun to Earth, 1.50×10^8 km. How many AU are there in 1.00 light-year? (c) What is the speed of light in AU/h?

22. (II) If you used only a keyboard to enter data, how many years would it take to fill up the hard drive in your computer that can store 82 gigabytes (82×10^9 bytes) of data? Assume "normal" eight-hour working days, and that one byte is required to store one keyboard character, and that you can type 180 characters per minute.

23. (III) The diameter of the Moon is 3480 km. (a) What is the surface area of the Moon? (b) How many times larger is the surface area of the Earth?

1–6 Order-of-Magnitude Estimating

(*Note*: Remember that for rough estimates, only round numbers are needed both as input to calculations and as final results.)

24. (I) Estimate the order of magnitude (power of ten) of: (a) 2800, (b) 86.30×10^2, (c) 0.0076, and (d) 15.0×10^8.

25. (II) Estimate how many books can be shelved in a college library with 3500 m^2 of floor space. Assume 8 shelves high, having books on both sides, with corridors 1.5 m wide. Assume books are about the size of this one, on average.

26. (II) Estimate how many hours it would take a runner to run (at 10 km/h) across the United States from New York to California.

27. (II) Estimate the number of liters of water a human drinks in a lifetime.

28. (II) Estimate how long it would take one person to mow a football field using an ordinary home lawn mower (Fig. 1–11). Assume the mower moves with a 1-km/h speed, and has a 0.5-m width.

FIGURE 1–11
Problem 28.

29. (II) Estimate the number of dentists (a) in San Francisco and (b) in your town or city.

30. (III) The rubber worn from tires mostly enters the atmosphere as *particulate pollution*. Estimate how much rubber (in kg) is put into the air in the United States every year. To get started, a good estimate for a tire tread's depth is 1 cm when new, and rubber has a mass of about 1200 kg per m^3 of volume.

31. (III) You are in a hot air balloon, 200 m above the flat Texas plains. You look out toward the horizon. How far out can you see—that is, how far is your horizon? The Earth's radius is about 6400 km.

32. (III) I agree to hire you for 30 days and you can decide between two possible methods of payment: either (1) $1000 a day, or (2) one penny on the first day, two pennies on the second day and continue to double your daily pay each day up to day 30. Use quick estimation to make your decision, and justify it.

33. (III) Many sailboats are moored at a marina 4.4 km away on the opposite side of a lake. You stare at one of the sailboats because, when you are lying flat at the water's edge, you can just see its deck but none of the side of the sailboat. You then go to that sailboat on the other side of the lake and measure that the deck is 1.5 m above the level of the water. Using Fig. 1–12, where $h = 1.5$ m, estimate the radius R of the Earth.

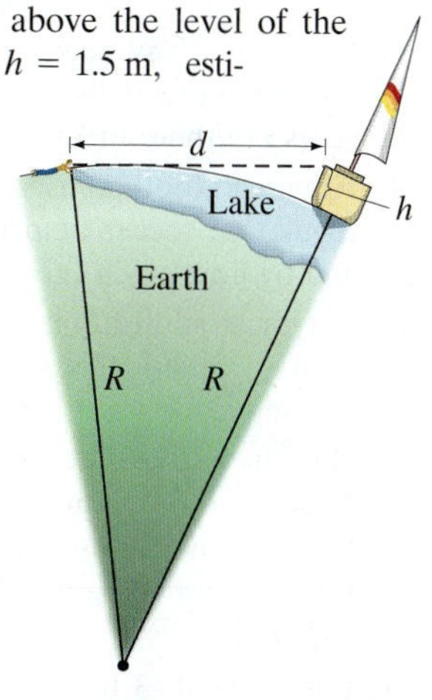

FIGURE 1–12 Problem 33.
You see a sailboat across a lake (not to scale). R is the radius of the Earth. You are a distance $d = 4.4$ km from the sailboat when you can see only its deck and not its side. Because of the curvature of the Earth, the water "bulges out" between you and the boat.

34. (III) Another experiment you can do also uses the radius of the Earth. The Sun sets, fully disappearing over the horizon as you lie on the beach, your eyes 20 cm above the sand. You immediately jump up, your eyes now 150 cm above the sand, and you can again see the top of the Sun. If you count the number of seconds ($= t$) until the Sun fully disappears again, you can estimate the radius of the Earth. But for this Problem, use the known radius of the Earth and calculate the time t.

*35. (I) What are the dimensions of density, which is mass per volume?

*36. (II) The speed v of an object is given by the equation $v = At^3 - Bt$, where t refers to time. (a) What are the dimensions of A and B? (b) What are the SI units for the constants A and B?

*37. (II) Three students derive the following equations in which x refers to distance traveled, v the speed, a the acceleration (m/s^2), t the time, and the subscript zero $(_0)$ means a quantity at time $t = 0$: (a) $x = vt^2 + 2at$, (b) $x = v_0 t + \frac{1}{2}at^2$, and (c) $x = v_0 t + 2at^2$. Which of these could possibly be correct according to a dimensional check?

*38. (II) Show that the following combination of the three fundamental constants of nature that we used in Example 1–10 (that is G, c, and h) forms a quantity with the dimensions of time:

$$t_P = \sqrt{\frac{Gh}{c^5}}.$$

This quantity, t_P, is called the *Planck time* and is thought to be the earliest time, after the creation of the Universe, at which the currently known laws of physics can be applied.

General Problems

39. *Global positioning satellites* (GPS) can be used to determine positions with great accuracy. If one of the satellites is at a distance of 20,000 km from you, what percent uncertainty in the distance does a 2-m uncertainty represent? How many significant figures are needed in the distance?

40. *Computer chips* (Fig. 1–13) etched on circular silicon wafers of thickness 0.300 mm are sliced from a solid cylindrical silicon crystal of length 25 cm. If each wafer can hold 100 chips, what is the maximum number of chips that can be produced from one entire cylinder?

FIGURE 1–13 Problem 40. The wafer held by the hand (above) is shown below, enlarged and illuminated by colored light. Visible are rows of integrated circuits (chips).

41. (a) How many seconds are there in 1.00 year? (b) How many nanoseconds are there in 1.00 year? (c) How many years are there in 1.00 second?

42. American football uses a field that is 100 yd long, whereas a regulation soccer field is 100 m long. Which field is longer, and by how much (give yards, meters, and percent)?

43. A typical adult human lung contains about 300 million tiny cavities called alveoli. Estimate the average diameter of a single alveolus.

44. One hectare is defined as $1.000 \times 10^4 \, m^2$. One acre is $4.356 \times 10^4 \, ft^2$. How many acres are in one hectare?

45. Estimate the number of gallons of gasoline consumed by the total of all automobile drivers in the United States, per year.

46. Use Table 1–3 to estimate the total number of protons or neutrons in (a) a bacterium, (b) a DNA molecule, (c) the human body, (d) our Galaxy.

47. An average family of four uses roughly 1200 L (about 300 gallons) of water per day $(1 \, L = 1000 \, cm^3)$. How much depth would a lake lose per year if it uniformly covered an area of 50 km^2 and supplied a local town with a population of 40,000 people? Consider only population uses, and neglect evaporation and so on.

48. Estimate the number of gumballs in the machine of Fig. 1–14.

FIGURE 1–14 Problem 48. Estimate the number of gumballs in the machine.

49. Estimate how many kilograms of laundry soap are used in the U.S. in one year (and therefore pumped out of washing machines with the dirty water). Assume each load of laundry takes 0.1 kg of soap.

50. How big is a ton? That is, what is the volume of something that weighs a ton? To be specific, estimate the diameter of a 1-ton rock, but first make a wild guess: will it be 1 ft across, 3 ft, or the size of a car? [Hint: Rock has mass per volume about 3 times that of water, which is 1 kg per liter $(10^3 \, cm^3)$ or 62 lb per cubic foot.]

51. A certain audio compact disc (CD) contains 783.216 megabytes of digital information. Each byte consists of exactly 8 bits. When played, a CD player reads the CD's digital information at a constant rate of 1.4 megabits per second. How many minutes does it take the player to read the entire CD?

52. Hold a pencil in front of your eye at a position where its blunt end just blocks out the Moon (Fig. 1–15). Make appropriate measurements to estimate the diameter of the Moon, given that the Earth–Moon distance is 3.8×10^5 km.

FIGURE 1–15 Problem 52. How big is the Moon?

53. A heavy rainstorm dumps 1.0 cm of rain on a city 5 km wide and 8 km long in a 2-h period. How many metric tons (1 metric ton = 10^3 kg) of water fell on the city? (1 cm^3 of water has a mass of 1 g = 10^{-3} kg.) How many gallons of water was this?

54. Noah's ark was ordered to be 300 cubits long, 50 cubits wide, and 30 cubits high. The cubit was a unit of measure equal to the length of a human forearm, elbow to the tip of the longest finger. Express the dimensions of Noah's ark in meters, and estimate its volume (m^3).

55. Estimate how many days it would take to walk around the world, assuming 10 h walking per day at 4 km/h.

56. One liter $(1000 \, cm^3)$ of oil is spilled onto a smooth lake. If the oil spreads out uniformly until it makes an oil slick just one molecule thick, with adjacent molecules just touching, estimate the diameter of the oil slick. Assume the oil molecules have a diameter of 2×10^{-10} m.

57. Jean camps beside a wide river and wonders how wide it is. She spots a large rock on the bank directly across from her. She then walks upstream until she judges that the angle between her and the rock, which she can still see clearly, is now at an angle of 30° downstream (Fig. 1–16). Jean measures her stride to be about 1 yard long. The distance back to her camp is 120 strides. About how far across, both in yards and in meters, is the river?

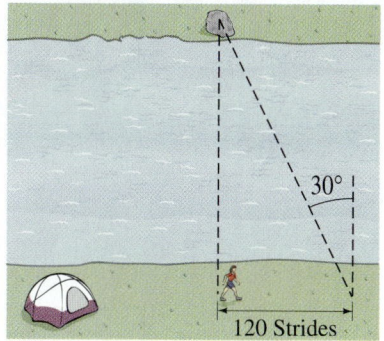

FIGURE 1–16
Problem 57.

120 Strides

30°

58. A watch manufacturer claims that its watches gain or lose no more than 8 seconds in a year. How accurate is this watch, expressed as a percentage?

59. An angstrom (symbol Å) is a unit of length, defined as 10^{-10} m, which is on the order of the diameter of an atom. (a) How many nanometers are in 1.0 angstrom? (b) How many femtometers or fermis (the common unit of length in nuclear physics) are in 1.0 angstrom? (c) How many angstroms are in 1.0 m? (d) How many angstroms are in 1.0 light-year (see Problem 21)?

60. The diameter of the Moon is 3480 km. What is the volume of the Moon? How many Moons would be needed to create a volume equal to that of Earth?

61. Determine the percent uncertainty in θ, and in $\sin\theta$, when (a) $\theta = 15.0° \pm 0.5°$, (b) $\theta = 75.0° \pm 0.5°$.

62. If you began walking along one of Earth's lines of longitude and walked north until you had changed latitude by 1 minute of arc (there are 60 minutes per degree), how far would you have walked (in miles)? This distance is called a "nautical mile."

63. Make a rough estimate of the volume of your body (in m^3).

64. Estimate the number of bus drivers (a) in Washington, D.C., and (b) in your town.

65. The American Lung Association gives the following formula for an average person's expected lung capacity V (in liters, where $1 \, L = 10^3 \, cm^3$):

$$V = 4.1H - 0.018A - 2.69,$$

where H and A are the person's height (in meters), and age (in years), respectively. In this formula, what are the units of the numbers 4.1, 0.018, and 2.69?

66. The density of an object is defined as its mass divided by its volume. Suppose the mass and volume of a rock are measured to be 8 g and 2.8325 cm^3. To the correct number of significant figures, determine the rock's density.

67. To the correct number of significant figures, use the information inside the front cover of this book to determine the ratio of (a) the surface area of Earth compared to the surface area of the Moon; (b) the volume of Earth compared to the volume of the Moon.

68. One mole of atoms consists of 6.02×10^{23} individual atoms. If a mole of atoms were spread uniformly over the surface of the Earth, how many atoms would there be per square meter?

69. Recent findings in astrophysics suggest that the observable Universe can be modeled as a sphere of radius $R = 13.7 \times 10^9$ light-years with an average mass density of about $1 \times 10^{-26} \, kg/m^3$, where only about 4% of the Universe's total mass is due to "ordinary" matter (such as protons, neutrons, and electrons). Use this information to estimate the total mass of ordinary matter in the observable Universe. (1 light-year = 9.46×10^{15} m.)

Answers to Exercises

A: (d).

B: No: they have 3 and 2, respectively.

C: All three have three significant figures, although the number of decimal places is (a) 2, (b) 3, (c) 4.

D: (a) 2.58×10^{-2}, 3; (b) 4.23×10^4, 3 (probably); (c) 3.4450×10^2, 5.

E: Mt. Everest, 29,035 ft; K2, 28,251 ft; Kangchenjunga, 28,169 ft.

F: No: 15 m/s $\approx$ 34 mi/h.

A high-speed car has released a parachute to reduce its speed quickly. The directions of the car's velocity and acceleration are shown by the green ($\vec{v}$) and gold ($\vec{a}$) arrows.

Motion is described using the concepts of velocity and acceleration. In the case shown here, the acceleration $\vec{a}$ is in the opposite direction from the velocity $\vec{v}$, which means the object is slowing down. We examine in detail motion with constant acceleration, including the vertical motion of objects falling under gravity.

C H A P T E R

2

Describing Motion: Kinematics in One Dimension

CONTENTS

2–1 Reference Frames and Displacement

2–2 Average Velocity

2–3 Instantaneous Velocity

2–4 Acceleration

2–5 Motion at Constant Acceleration

2–6 Solving Problems

2–7 Freely Falling Objects

*2–8 Variable Acceleration; Integral Calculus

*2–9 Graphical Analysis and Numerical Integration

CHAPTER-OPENING QUESTION—Guess now!

[*Don't worry about getting the right answer now—you will get another chance later in the Chapter. See also p. 1 of Chapter 1 for more explanation.*]

Two small heavy balls have the same diameter but one weighs twice as much as the other. The balls are dropped from a second-story balcony at the exact same time. The time to reach the ground below will be:

(a) twice as long for the lighter ball as for the heavier one.

(b) longer for the lighter ball, but not twice as long.

(c) twice as long for the heavier ball as for the lighter one.

(d) longer for the heavier ball, but not twice as long.

(e) nearly the same for both balls.

The motion of objects—baseballs, automobiles, joggers, and even the Sun and Moon—is an obvious part of everyday life. It was not until the sixteenth and seventeenth centuries that our modern understanding of motion was established. Many individuals contributed to this understanding, particularly Galileo Galilei (1564–1642) and Isaac Newton (1642–1727).

18

The study of the motion of objects, and the related concepts of force and energy, form the field called **mechanics**. Mechanics is customarily divided into two parts: **kinematics**, which is the description of how objects move, and **dynamics**, which deals with force and why objects move as they do. This Chapter and the next deal with kinematics.

For now we only discuss objects that move without rotating (Fig. 2–1a). Such motion is called **translational motion**. In this Chapter we will be concerned with describing an object that moves along a straight-line path, which is one-dimensional translational motion. In Chapter 3 we will describe translational motion in two (or three) dimensions along paths that are not straight.

We will often use the concept, or *model*, of an idealized **particle** which is considered to be a mathematical **point** with no spatial extent (no size). A point particle can undergo only translational motion. The particle model is useful in many real situations where we are interested only in translational motion and the object's size is not significant. For example, we might consider a billiard ball, or even a spacecraft traveling toward the Moon, as a particle for many purposes.

2–1 Reference Frames and Displacement

Any measurement of position, distance, or speed must be made with respect to a **reference frame**, or **frame of reference**. For example, while you are on a train traveling at 80 km/h, suppose a person walks past you toward the front of the train at a speed of, say, 5 km/h (Fig. 2–2). This 5 km/h is the person's speed with respect to the train as frame of reference. With respect to the ground, that person is moving at a speed of 80 km/h + 5 km/h = 85 km/h. It is always important to specify the frame of reference when stating a speed. In everyday life, we usually mean "with respect to the Earth" without even thinking about it, but the reference frame must be specified whenever there might be confusion.

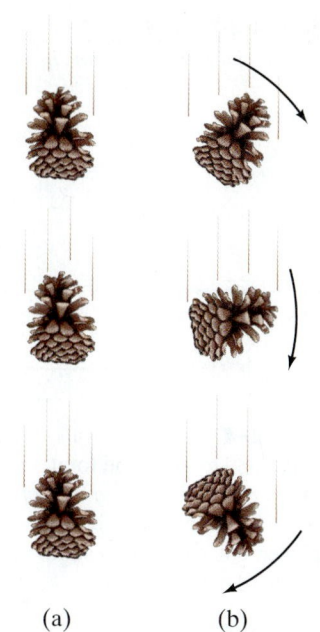

FIGURE 2–1 The pinecone in (a) undergoes pure translation as it falls, whereas in (b) it is rotating as well as translating.

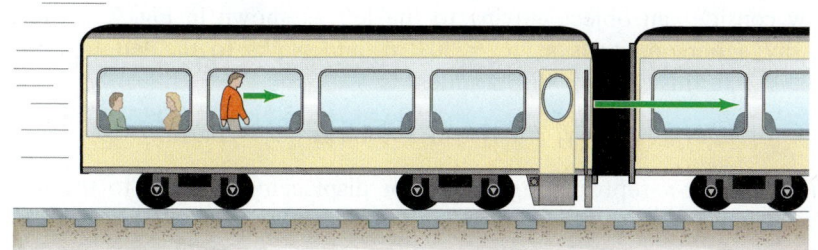

FIGURE 2–2 A person walks toward the front of a train at 5 km/h. The train is moving 80 km/h with respect to the ground, so the walking person's speed, relative to the ground, is 85 km/h.

When specifying the motion of an object, it is important to specify not only the speed but also the direction of motion. Often we can specify a direction by using the cardinal points, north, east, south, and west, and by "up" and "down." In physics, we often draw a set of **coordinate axes**, as shown in Fig. 2–3, to represent a frame of reference. We can always place the origin 0, and the directions of the x and y axes, as we like for convenience. The x and y axes are always perpendicular to each other. Objects positioned to the right of the origin of coordinates (0) on the x axis have an x coordinate which we usually choose to be positive; then points to the left of 0 have a negative x coordinate. The position along the y axis is usually considered positive when above 0, and negative when below 0, although the reverse convention can be used if convenient. Any point on the plane can be specified by giving its x and y coordinates. In three dimensions, a z axis perpendicular to the x and y axes is added.

For one-dimensional motion, we often choose the x axis as the line along which the motion takes place. Then the **position** of an object at any moment is given by its x coordinate. If the motion is vertical, as for a dropped object, we usually use the y axis.

FIGURE 2–3 Standard set of xy coordinate axes.

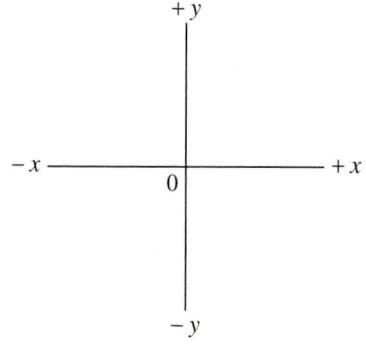

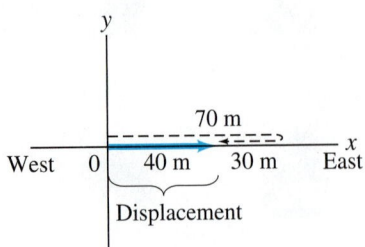

FIGURE 2–4 A person walks 70 m east, then 30 m west. The total distance traveled is 100 m (path is shown dashed in black); but the displacement, shown as a solid blue arrow, is 40 m to the east.

FIGURE 2–5 The arrow represents the displacement $x_2 - x_1$. Distances are in meters.

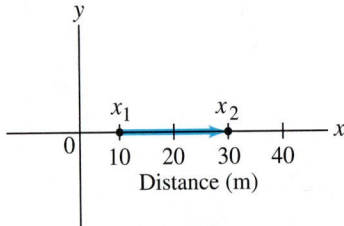

FIGURE 2–6 For the displacement $\Delta x = x_2 - x_1 = 10.0\,\text{m} - 30.0\,\text{m}$, the displacement vector points to the left.

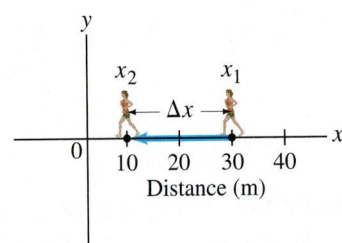

We need to make a distinction between the *distance* an object has traveled and its **displacement**, which is defined as the *change in position* of the object. That is, *displacement is how far the object is from its starting point*. To see the distinction between total distance and displacement, imagine a person walking 70 m to the east and then turning around and walking back (west) a distance of 30 m (see Fig. 2–4). The total *distance* traveled is 100 m, but the *displacement* is only 40 m since the person is now only 40 m from the starting point.

Displacement is a quantity that has both magnitude and direction. Such quantities are called **vectors**, and are represented by arrows in diagrams. For example, in Fig. 2–4, the blue arrow represents the displacement whose magnitude is 40 m and whose direction is to the right (east).

We will deal with vectors more fully in Chapter 3. For now, we deal only with motion in one dimension, along a line. In this case, vectors which point in one direction will have a positive sign, whereas vectors that point in the opposite direction will have a negative sign, along with their magnitude.

Consider the motion of an object over a particular time interval. Suppose that at some initial time, call it t_1, the object is on the x axis at the position x_1 in the coordinate system shown in Fig. 2–5. At some later time, t_2, suppose the object has moved to position x_2. The displacement of our object is $x_2 - x_1$, and is represented by the arrow pointing to the right in Fig. 2–5. It is convenient to write

$$\Delta x = x_2 - x_1,$$

where the symbol Δ (Greek letter delta) means "change in." Then Δx means "the change in x," or "change in position," which is the displacement. Note that the "change in" any quantity means the final value of that quantity, minus the initial value.

Suppose $x_1 = 10.0\,\text{m}$ and $x_2 = 30.0\,\text{m}$. Then

$$\Delta x = x_2 - x_1 = 30.0\,\text{m} - 10.0\,\text{m} = 20.0\,\text{m},$$

so the displacement is 20.0 m in the positive direction, Fig. 2–5.

Now consider an object moving to the left as shown in Fig. 2–6. Here the object, say, a person, starts at $x_1 = 30.0\,\text{m}$ and walks to the left to the point $x_2 = 10.0\,\text{m}$. In this case her displacement is

$$\Delta x = x_2 - x_1 = 10.0\,\text{m} - 30.0\,\text{m} = -20.0\,\text{m},$$

and the blue arrow representing the vector displacement points to the left. For one-dimensional motion along the x axis, a vector pointing to the right has a positive sign, whereas a vector pointing to the left has a negative sign.

EXERCISE A An ant starts at $x = 20\,\text{cm}$ on a piece of graph paper and walks along the x axis to $x = -20\,\text{cm}$. It then turns around and walks back to $x = -10\,\text{cm}$. What is the ant's displacement and total distance traveled?

2–2 Average Velocity

The most obvious aspect of the motion of a moving object is how fast it is moving—its speed or velocity.

The term "speed" refers to how far an object travels in a given time interval, regardless of direction. If a car travels 240 kilometers (km) in 3 hours (h), we say its average speed was 80 km/h. In general, the **average speed** of an object is defined as *the total distance traveled along its path divided by the time it takes to travel this distance*:

$$\text{average speed} = \frac{\text{distance traveled}}{\text{time elapsed}}. \tag{2–1}$$

The terms "velocity" and "speed" are often used interchangeably in ordinary language. But in physics we make a distinction between the two. Speed is simply a positive number, with units. **Velocity**, on the other hand, is used to signify both the *magnitude* (numerical value) of how fast an object is moving and also the *direction* in which it is moving. (Velocity is therefore a vector.) There is a second difference between speed and velocity: namely, the **average velocity** is defined in terms of *displacement*, rather than total distance traveled:

$$\text{average velocity} = \frac{\text{displacement}}{\text{time elapsed}} = \frac{\text{final position} - \text{initial position}}{\text{time elapsed}}.$$

Average speed and average velocity have the same magnitude when the motion is all in one direction. In other cases, they may differ: recall the walk we described earlier, in Fig. 2–4, where a person walked 70 m east and then 30 m west. The total distance traveled was $70\,\text{m} + 30\,\text{m} = 100\,\text{m}$, but the displacement was 40 m. Suppose this walk took 70 s to complete. Then the average speed was:

$$\frac{\text{distance}}{\text{time elapsed}} = \frac{100\,\text{m}}{70\,\text{s}} = 1.4\,\text{m/s}.$$

The magnitude of the average velocity, on the other hand, was:

$$\frac{\text{displacement}}{\text{time elapsed}} = \frac{40\,\text{m}}{70\,\text{s}} = 0.57\,\text{m/s}.$$

This difference between the speed and the magnitude of the velocity can occur when we calculate *average* values.

To discuss one-dimensional motion of an object in general, suppose that at some moment in time, call it t_1, the object is on the x axis at position x_1 in a coordinate system, and at some later time, t_2, suppose it is at position x_2. The **elapsed time** is $\Delta t = t_2 - t_1$; during this time interval the displacement of our object is $\Delta x = x_2 - x_1$. Then the average velocity, defined as *the displacement divided by the elapsed time*, can be written

$$\bar{v} = \frac{x_2 - x_1}{t_2 - t_1} = \frac{\Delta x}{\Delta t}, \qquad \textbf{(2–2)}$$

where v stands for velocity and the bar ($^-$) over the v is a standard symbol meaning "average."

For the usual case of the $+x$ axis to the right, note that if x_2 is less than x_1, the object is moving to the left, and then $\Delta x = x_2 - x_1$ is less than zero. The sign of the displacement, and thus of the average velocity, indicates the direction: the average velocity is positive for an object moving to the right along the $+x$ axis and negative when the object moves to the left. The direction of the average velocity is always the same as the direction of the displacement.

Note that it is always important to choose (and state) the *elapsed time*, or *time interval*, $t_2 - t_1$, the time that passes during our chosen period of observation.

EXAMPLE 2–1 **Runner's average velocity.** The position of a runner as a function of time is plotted as moving along the x axis of a coordinate system. During a 3.00-s time interval, the runner's position changes from $x_1 = 50.0\,\text{m}$ to $x_2 = 30.5\,\text{m}$, as shown in Fig. 2–7. What was the runner's average velocity?

APPROACH We want to find the average velocity, which is the displacement divided by the elapsed time.

SOLUTION The displacement is $\Delta x = x_2 - x_1 = 30.5\,\text{m} - 50.0\,\text{m} = -19.5\,\text{m}$. The elapsed time, or time interval, is $\Delta t = 3.00\,\text{s}$. The average velocity is

$$\bar{v} = \frac{\Delta x}{\Delta t} = \frac{-19.5\,\text{m}}{3.00\,\text{s}} = -6.50\,\text{m/s}.$$

The displacement and average velocity are negative, which tells us that the runner is moving to the left along the x axis, as indicated by the arrow in Fig. 2–7. Thus we can say that the runner's average velocity is 6.50 m/s to the left.

⚠ **CAUTION**
Average speed is not necessarily equal to the magnitude of the average velocity

PROBLEM SOLVING
$+$ or $-$ sign can signify the direction for linear motion

FIGURE 2–7 Example 2–1. A person runs from $x_1 = 50.0\,\text{m}$ to $x_2 = 30.5\,\text{m}$. The displacement is $-19.5\,\text{m}$.

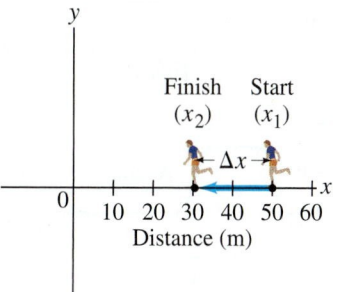

EXAMPLE 2–2 **Distance a cyclist travels.** How far can a cyclist travel in 2.5 h along a straight road if her average velocity is 18 km/h?

APPROACH We want to find the distance traveled, so we solve Eq. 2–2 for Δx.

SOLUTION We rewrite Eq. 2–2 as $\Delta x = \bar{v} \Delta t$, and find

$$\Delta x = \bar{v} \Delta t = (18 \text{ km/h})(2.5 \text{ h}) = 45 \text{ km}.$$

EXERCISE B A car travels at a constant 50 km/h for 100 km. It then speeds up to 100 km/h and is driven another 100 km. What is the car's average speed for the 200 km trip? (*a*) 67 km/h; (*b*) 75 km/h; (*c*) 81 km/h; (*d*) 50 km/h.

2–3 Instantaneous Velocity

If you drive a car along a straight road for 150 km in 2.0 h, the magnitude of your average velocity is 75 km/h. It is unlikely, though, that you were moving at precisely 75 km/h at every instant. To describe this situation we need the concept of *instantaneous velocity*, which is the velocity at any instant of time. (Its magnitude is the number, with units, indicated by a speedometer, Fig. 2–8.) More precisely, the **instantaneous velocity** at any moment is defined as *the average velocity over an infinitesimally short time interval*. That is, Eq. 2–2 is to be evaluated in the limit of Δt becoming extremely small, approaching zero. We can write the definition of instantaneous velocity, v, for one-dimensional motion as

FIGURE 2–8 Car speedometer showing mi/h in white, and km/h in orange.

$$v = \lim_{\Delta t \to 0} \frac{\Delta x}{\Delta t}. \tag{2–3}$$

The notation $\lim_{\Delta t \to 0}$ means the ratio $\Delta x / \Delta t$ is to be evaluated in the limit of Δt approaching zero. But we do not simply set $\Delta t = 0$ in this definition, for then Δx would also be zero, and we would have an undefined number. Rather, we are considering the *ratio* $\Delta x / \Delta t$, as a whole. As we let Δt approach zero, Δx approaches zero as well. But the ratio $\Delta x / \Delta t$ approaches some definite value, which is the instantaneous velocity at a given instant.

In Eq. 2–3, the limit as $\Delta t \to 0$ is written in calculus notation as dx/dt and is called the *derivative* of x with respect to t:

$$v = \lim_{\Delta t \to 0} \frac{\Delta x}{\Delta t} = \frac{dx}{dt}. \tag{2–4}$$

FIGURE 2–9 Velocity of a car as a function of time: (a) at constant velocity; (b) with varying velocity.

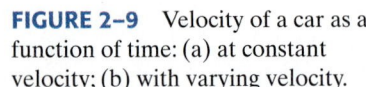
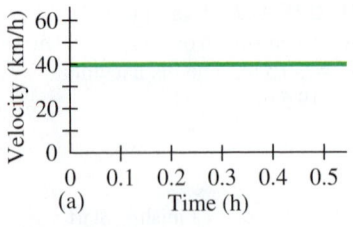

This equation is the definition of instantaneous velocity for one-dimensional motion.

For instantaneous velocity we use the symbol v, whereas for average velocity we use $\bar{v}$, with a bar above. In the rest of this book, when we use the term "velocity" it will refer to instantaneous velocity. When we want to speak of the average velocity, we will make this clear by including the word "average."

Note that the *instantaneous* speed always equals the magnitude of the instantaneous velocity. Why? Because distance traveled and the magnitude of the displacement become the same when they become infinitesimally small.

If an object moves at a uniform (that is, constant) velocity during a particular time interval, then its instantaneous velocity at any instant is the same as its average velocity (see Fig. 2–9a). But in many situations this is not the case. For example, a car may start from rest, speed up to 50 km/h, remain at that velocity for a time, then slow down to 20 km/h in a traffic jam, and finally stop at its destination after traveling a total of 15 km in 30 min. This trip is plotted on the graph of Fig. 2–9b. Also shown on the graph is the average velocity (dashed line), which is $\bar{v} = \Delta x / \Delta t = 15 \text{ km}/0.50 \text{ h} = 30 \text{ km/h}.$

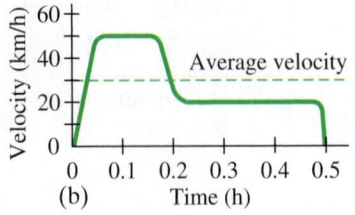

To better understand instantaneous velocity, let us consider a graph of the position of a particular particle versus time (x vs. t), as shown in Fig. 2–10. (Note that this is different from showing the "path" of a particle on an x vs. y plot.) The particle is at position x_1 at a time t_1, and at position x_2 at time t_2. P_1 and P_2 represent these two points on the graph. A straight line drawn from point $P_1(x_1, t_1)$ to point $P_2(x_2, t_2)$ forms the hypotenuse of a right triangle whose sides are Δx and Δt. The ratio $\Delta x/\Delta t$ is the **slope** of the straight line $P_1 P_2$. But $\Delta x/\Delta t$ is also the average velocity of the particle during the time interval $\Delta t = t_2 - t_1$. Therefore, we conclude that the average velocity of a particle during any time interval $\Delta t = t_2 - t_1$ is equal to the slope of the straight line (or *chord*) connecting the two points (x_1, t_1) and (x_2, t_2) on an x vs. t graph.

Consider now a time t_i, intermediate between t_1 and t_2, at which time the particle is at x_i (Fig. 2–11). The slope of the straight line $P_1 P_i$ is less than the slope of $P_1 P_2$ in this case. Thus the average velocity during the time interval $t_i - t_1$ is less than during the time interval $t_2 - t_1$.

Now let us imagine that we take the point P_i in Fig. 2–11 to be closer and closer to point P_1. That is, we let the interval $t_i - t_1$, which we now call Δt, to become smaller and smaller. The slope of the line connecting the two points becomes closer and closer to the slope of a line tangent to the curve at point P_1. The average velocity (equal to the slope of the chord) thus approaches the slope of the tangent at point P_1. The definition of the instantaneous velocity (Eq. 2–3) is the limiting value of the average velocity as Δt approaches zero. Thus the *instantaneous velocity equals the slope of the tangent to the curve* at that point (which we can simply call "the slope of the curve" at that point).

Because the velocity at any instant equals the slope of the tangent to the x vs. t graph at that instant, we can obtain the velocity at any instant from such a graph. For example, in Fig. 2–12 (which shows the same curve as in Figs. 2–10 and 2–11), as our object moves from x_1 to x_2, the slope continually increases, so the velocity is increasing. For times after t_2, however, the slope begins to decrease and in fact reaches zero (so $v = 0$) where x has its maximum value, at point P_3 in Fig. 2–12. Beyond this point, the slope is negative, as for point P_4. The velocity is therefore negative, which makes sense since x is now decreasing—the particle is moving toward decreasing values of x, to the left on a standard xy plot.

If an object moves with constant velocity over a particular time interval, its instantaneous velocity is equal to its average velocity. The graph of x vs. t in this case will be a straight line whose slope equals the velocity. The curve of Fig. 2–10 has no straight sections, so there are no time intervals when the velocity is constant.

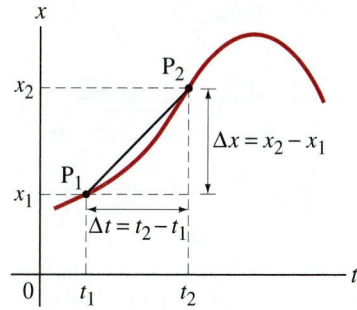

FIGURE 2–10 Graph of a particle's position x vs. time t. The slope of the straight line $P_1 P_2$ represents the average velocity of the particle during the time interval $\Delta t = t_2 - t_1$.

FIGURE 2–11 Same position vs. time curve as in Fig. 2–10, but note that the average velocity over the time interval $t_i - t_1$ (which is the slope of $P_1 P_i$) is less than the average velocity over the time interval $t_2 - t_1$. The slope of the thin line tangent to the curve at point P_1 equals the instantaneous velocity at time t_1.

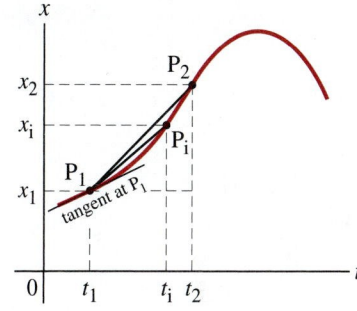

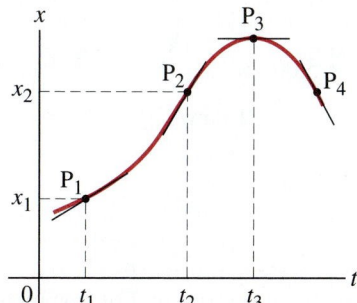

FIGURE 2–12 Same x vs. t curve as in Figs. 2–10 and 2–11, but here showing the slope at four different points: At P_3, the slope is zero, so $v = 0$. At P_4 the slope is negative, so $v < 0$.

EXERCISE C What is your speed at the instant you turn around to move in the opposite direction? (*a*) Depends on how quickly you turn around; (*b*) always zero; (*c*) always negative; (*d*) none of the above.

The derivatives of various functions are studied in calculus courses, and this book gives a summary in Appendix B. The derivatives of polynomial functions (which we use a lot) are:

$$\frac{d}{dt}(Ct^n) = nCt^{n-1} \quad \text{and} \quad \frac{dC}{dt} = 0,$$

where C is any constant.

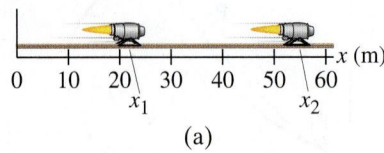

(a)

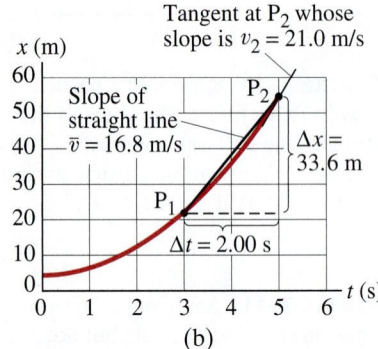

(b)

FIGURE 2–13 Example 2–3.
(a) Engine traveling on a straight track.
(b) Graph of x vs. t: $x = At^2 + B$.

EXAMPLE 2–3 **Given x as a function of t.** A jet engine moves along an experimental track (which we call the x axis) as shown in Fig. 2–13a. We will treat the engine as if it were a particle. Its position as a function of time is given by the equation $x = At^2 + B$, where $A = 2.10 \text{ m/s}^2$ and $B = 2.80 \text{ m}$, and this equation is plotted in Fig. 2–13b. (a) Determine the displacement of the engine during the time interval from $t_1 = 3.00 \text{ s}$ to $t_2 = 5.00 \text{ s}$. (b) Determine the average velocity during this time interval. (c) Determine the magnitude of the instantaneous velocity at $t = 5.00 \text{ s}$.

APPROACH We substitute values for t_1 and t_2 in the given equation for x to obtain x_1 and x_2. The average velocity can be found from Eq. 2–2. We take the derivative of the given x equation with respect to t to find the instantaneous velocity, using the formulas just given.

SOLUTION (a) At $t_1 = 3.00 \text{ s}$, the position (point P_1 in Fig. 2–13b) is

$$x_1 = At_1^2 + B = (2.10 \text{ m/s}^2)(3.00 \text{ s})^2 + 2.80 \text{ m} = 21.7 \text{ m}.$$

At $t_2 = 5.00 \text{ s}$, the position (P_2 in Fig. 2–13b) is

$$x_2 = (2.10 \text{ m/s}^2)(5.00 \text{ s})^2 + 2.80 \text{ m} = 55.3 \text{ m}.$$

The displacement is thus

$$x_2 - x_1 = 55.3 \text{ m} - 21.7 \text{ m} = 33.6 \text{ m}.$$

(b) The magnitude of the average velocity can then be calculated as

$$\bar{v} = \frac{\Delta x}{\Delta t} = \frac{x_2 - x_1}{t_2 - t_1} = \frac{33.6 \text{ m}}{2.00 \text{ s}} = 16.8 \text{ m/s}.$$

This equals the slope of the straight line joining points P_1 and P_2 shown in Fig. 2–13b.

(c) The instantaneous velocity at $t = t_2 = 5.00 \text{ s}$ equals the slope of the tangent to the curve at point P_2 shown in Fig. 2–13b. We could measure this slope off the graph to obtain v_2. But we can calculate v more precisely for any time t, using the given formula

$$x = At^2 + B,$$

which is the engine's position x as a function of time t. We take the derivative of x with respect to time (see formulas at bottom of previous page):

$$v = \frac{dx}{dt} = \frac{d}{dt}\left(At^2 + B\right) = 2At.$$

We are given $A = 2.10 \text{ m/s}^2$, so for $t = t_2 = 5.00 \text{ s}$,

$$v_2 = 2At = 2(2.10 \text{ m/s}^2)(5.00 \text{ s}) = 21.0 \text{ m/s}.$$

2–4 Acceleration

An object whose velocity is changing is said to be accelerating. For instance, a car whose velocity increases in magnitude from zero to 80 km/h is accelerating. Acceleration specifies how rapidly the velocity of an object is changing.

Average Acceleration

Average acceleration is defined as the change in velocity divided by the time taken to make this change:

$$\text{average acceleration} = \frac{\text{change of velocity}}{\text{time elapsed}}.$$

In symbols, the average acceleration over a time interval $\Delta t = t_2 - t_1$ during which the velocity changes by $\Delta v = v_2 - v_1$, is defined as

$$\bar{a} = \frac{v_2 - v_1}{t_2 - t_1} = \frac{\Delta v}{\Delta t}. \qquad \text{(2-5)}$$

Because velocity is a vector, acceleration is a vector too. But for one-dimensional motion, we need only use a plus or minus sign to indicate acceleration direction relative to a chosen coordinate axis.

EXAMPLE 2–4 **Average acceleration.** A car accelerates along a straight road from rest to 90 km/h in 5.0 s, Fig. 2–14. What is the magnitude of its average acceleration?

APPROACH Average acceleration is the change in velocity divided by the elapsed time, 5.0 s. The car starts from rest, so $v_1 = 0$. The final velocity is $v_2 = 90$ km/h $= 90 \times 10^3$ m/3600 s $= 25$ m/s.

SOLUTION From Eq. 2–5, the average acceleration is

$$\bar{a} = \frac{v_2 - v_1}{t_2 - t_1} = \frac{25 \text{ m/s} - 0 \text{ m/s}}{5.0 \text{ s}} = 5.0 \frac{\text{m/s}}{\text{s}}.$$

This is read as "five meters per second per second" and means that, on average, the velocity changed by 5.0 m/s during each second. That is, assuming the acceleration was constant, during the first second the car's velocity increased from zero to 5.0 m/s. During the next second its velocity increased by another 5.0 m/s, reaching a velocity of 10.0 m/s at $t = 2.0$ s, and so on. See Fig. 2–14.

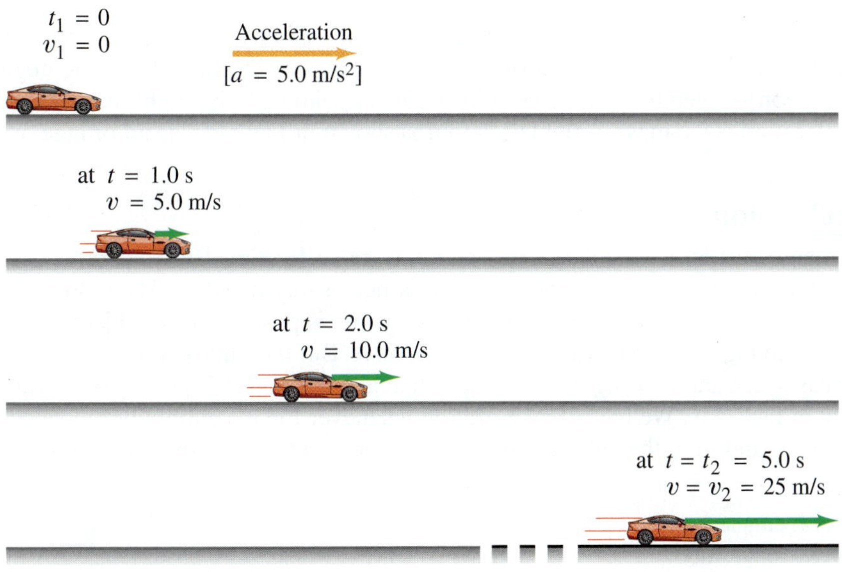

FIGURE 2–14 Example 2–4. The car is shown at the start with $v_1 = 0$ at $t_1 = 0$. The car is shown three more times, at $t = 1.0$ s, $t = 2.0$ s, and at the end of our time interval, $t_2 = 5.0$ s. We assume the acceleration is constant and equals 5.0 m/s^2. The green arrows represent the velocity vectors; the length of each arrow represents the magnitude of the velocity at that moment. The acceleration vector is the orange arrow. Distances are not to scale.

We almost always write the units for acceleration as m/s^2 (meters per second squared) instead of m/s/s. This is possible because:

$$\frac{\text{m/s}}{\text{s}} = \frac{\text{m}}{\text{s} \cdot \text{s}} = \frac{\text{m}}{\text{s}^2}.$$

According to the calculation in Example 2–4, the velocity changed on average by 5.0 m/s during each second, for a total change of 25 m/s over the 5.0 s; the average acceleration was 5.0 m/s^2.

Note that *acceleration tells us how quickly the velocity changes*, whereas *velocity tells us how quickly the position changes.*

CONCEPTUAL EXAMPLE 2–5 | **Velocity and acceleration.** (*a*) If the velocity of an object is zero, does it mean that the acceleration is zero? (*b*) If the acceleration is zero, does it mean that the velocity is zero? Think of some examples.

RESPONSE A zero velocity does not necessarily mean that the acceleration is zero, nor does a zero acceleration mean that the velocity is zero. (*a*) For example, when you put your foot on the gas pedal of your car which is at rest, the velocity starts from zero but the acceleration is not zero since the velocity of the car changes. (How else could your car start forward if its velocity weren't changing—that is, accelerating?) (*b*) As you cruise along a straight highway at a constant velocity of 100 km/h, your acceleration is zero: $a = 0$, $v \neq 0$.

EXERCISE D A powerful car is advertised to go from zero to 60 mi/h in 6.0 s. What does this say about the car: (*a*) it is fast (high speed); or (*b*) it accelerates well?

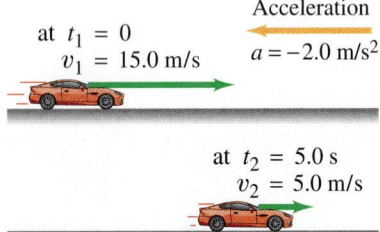

at $t_1 = 0$
$v_1 = 15.0$ m/s

Acceleration
$a = -2.0$ m/s^2

at $t_2 = 5.0$ s
$v_2 = 5.0$ m/s

FIGURE 2–15 Example 2–6, showing the position of the car at times t_1 and t_2, as well as the car's velocity represented by the green arrows. The acceleration vector (orange) points to the left as the car slows down while moving to the right.

EXAMPLE 2–6 **Car slowing down.** An automobile is moving to the right along a straight highway, which we choose to be the positive *x* axis (Fig. 2–15). Then the driver puts on the brakes. If the initial velocity (when the driver hits the brakes) is $v_1 = 15.0$ m/s, and it takes 5.0 s to slow down to $v_2 = 5.0$ m/s, what was the car's average acceleration?

APPROACH We put the given initial and final velocities, and the elapsed time, into Eq. 2–5 for $\bar{a}$.

SOLUTION In Eq. 2–5, we call the initial time $t_1 = 0$, and set $t_2 = 5.0$ s. (Note that our choice of $t_1 = 0$ doesn't affect the calculation of $\bar{a}$ because only $\Delta t = t_2 - t_1$ appears in Eq. 2–5.) Then

$$\bar{a} = \frac{5.0 \text{ m/s} - 15.0 \text{ m/s}}{5.0 \text{ s}} = -2.0 \text{ m/s}^2.$$

The negative sign appears because the final velocity is less than the initial velocity. In this case the direction of the acceleration is to the left (in the negative *x* direction)—even though the velocity is always pointing to the right. We say that the acceleration is 2.0 m/s^2 to the left, and it is shown in Fig. 2–15 as an orange arrow.

Deceleration

⚠ **CAUTION**

Deceleration means the magnitude of the velocity is decreasing; a is not *necessarily negative*

When an object is slowing down, we can say it is **decelerating**. But be careful: deceleration does *not* mean that the acceleration is necessarily negative. The velocity of an object moving to the right along the positive *x* axis is positive; if the object is slowing down (as in Fig. 2–15), the acceleration *is* negative. But the same car moving to the left (decreasing *x*), and slowing down, has positive acceleration that points to the right, as shown in Fig. 2–16. We have a deceleration whenever the magnitude of the velocity is decreasing, and then the velocity and acceleration point in opposite directions.

FIGURE 2–16 The car of Example 2–6, now moving to the *left* and decelerating. The acceleration is

$$
\begin{aligned}
a &= \frac{v_2 - v_1}{\Delta t} \\
&= \frac{(-5.0 \text{ m/s}) - (-15.0 \text{ m/s})}{5.0 \text{ s}} \\
&= \frac{-5.0 \text{ m/s} + 15.0 \text{ m/s}}{5.0 \text{ s}} = +2.0 \text{ m/s}.
\end{aligned}
$$

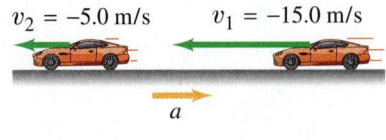

$v_2 = -5.0$ m/s $v_1 = -15.0$ m/s

a

EXERCISE E A car moves along the *x* axis. What is the sign of the car's acceleration if it is moving in the positive *x* direction with (*a*) increasing speed or (*b*) decreasing speed? What is the sign of the acceleration if the car moves in the negative direction with (*c*) increasing speed or (*d*) decreasing speed?

Instantaneous Acceleration

The **instantaneous acceleration**, a, is defined as the *limiting value of the average acceleration as we let Δt approach zero*:

$$a = \lim_{\Delta t \to 0} \frac{\Delta v}{\Delta t} = \frac{dv}{dt}. \qquad (2\text{–}6)$$

This limit, dv/dt, is the derivative of v with respect to t. We will use the term "acceleration" to refer to the instantaneous value. If we want to discuss the average acceleration, we will always include the word "average."

If we draw a graph of the velocity, v, vs. time, t, as shown in Fig. 2–17, then the average acceleration over a time interval $\Delta t = t_2 - t_1$ is represented by the slope of the straight line connecting the two points P_1 and P_2 as shown. [Compare this to the position vs. time graph of Fig. 2–10 for which the slope of the straight line represents the average velocity.] The instantaneous acceleration at any time, say t_1, is the slope of the tangent to the v vs. t curve at that time, which is also shown in Fig. 2–17. Let us use this fact for the situation graphed in Fig. 2–17; as we go from time t_1 to time t_2 the velocity continually increases, but the acceleration (the rate at which the velocity changes) is decreasing since the slope of the curve is decreasing.

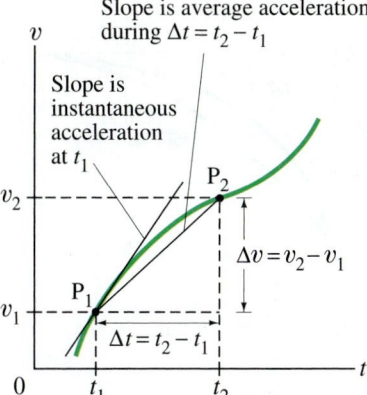

FIGURE 2–17 A graph of velocity v vs. time t. The average acceleration over a time interval $\Delta t = t_2 - t_1$ is the slope of the straight line $P_1 P_2$: $\bar{a} = \Delta v / \Delta t$. The instantaneous acceleration at time t_1 is the slope of the v vs. t curve at that instant.

FIGURE 2–18 Example 2–7. Graphs of (a) x vs. t, (b) v vs. t, and (c) a vs. t for the motion $x = At^2 + B$. Note that v increases linearly with t and that the acceleration a is constant. Also, v is the slope of the x vs. t curve, whereas a is the slope of the v vs. t curve.

EXAMPLE 2–7 **Acceleration given $x(t)$.** A particle is moving in a straight line so that its position is given by the relation $x = (2.10 \text{ m/s}^2)t^2 + (2.80 \text{ m})$, as in Example 2–3. Calculate (a) its average acceleration during the time interval from $t_1 = 3.00$ s to $t_2 = 5.00$ s, and (b) its instantaneous acceleration as a function of time.

APPROACH To determine acceleration, we first must find the velocity at t_1 and t_2 by differentiating x: $v = dx/dt$. Then we use Eq. 2–5 to find the average acceleration, and Eq. 2–6 to find the instantaneous acceleration.

SOLUTION (a) The velocity at any time t is

$$v = \frac{dx}{dt} = \frac{d}{dt}\left[(2.10 \text{ m/s}^2)t^2 + 2.80 \text{ m}\right] = (4.20 \text{ m/s}^2)t,$$

as we saw in Example 2–3c. Therefore, at $t_1 = 3.00$ s, $v_1 = (4.20 \text{ m/s}^2)(3.00 \text{ s}) = 12.6$ m/s and at $t_2 = 5.00$ s, $v_2 = 21.0$ m/s. Therefore,

$$\bar{a} = \frac{\Delta v}{\Delta t} = \frac{21.0 \text{ m/s} - 12.6 \text{ m/s}}{5.00 \text{ s} - 3.00 \text{ s}} = 4.20 \text{ m/s}^2.$$

(b) With $v = (4.20 \text{ m/s}^2)t$, the instantaneous acceleration at any time is

$$a = \frac{dv}{dt} = \frac{d}{dt}\left[(4.20 \text{ m/s}^2)t\right] = 4.20 \text{ m/s}^2.$$

The acceleration in this case is constant; it does not depend on time. Figure 2–18 shows graphs of (a) x vs. t (the same as Fig. 2–13b), (b) v vs. t, which is linearly increasing as calculated above, and (c) a vs. t, which is a horizontal straight line because $a =$ constant.

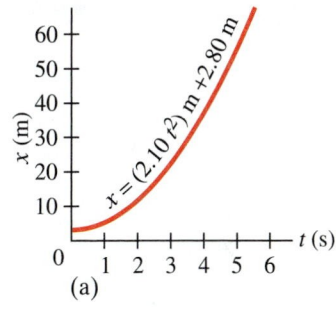

(b)

Like velocity, acceleration is a rate. The velocity of an object is the rate at which its displacement changes with time; its acceleration, on the other hand, is the rate at which its velocity changes with time. In a sense, acceleration is a "rate of a rate." This can be expressed in equation form as follows: since $a = dv/dt$ and $v = dx/dt$, then

$$a = \frac{dv}{dt} = \frac{d}{dt}\left(\frac{dx}{dt}\right) = \frac{d^2x}{dt^2}.$$

Here d^2x/dt^2 is the *second derivative* of x with respect to time: we first take the derivative of x with respect to time (dx/dt), and then we again take the derivative with respect to time, $(d/dt)(dx/dt)$, to get the acceleration.

EXERCISE F The position of a particle is given by the following equation:

$$x = (2.00 \text{ m/s}^3)t^3 + (2.50 \text{ m/s})t.$$

What is the acceleration of the particle at $t = 2.00$ s? (a) 13.0 m/s²; (b) 22.5 m/s²; (c) 24.0 m/s²; (d) 2.00 m/s².

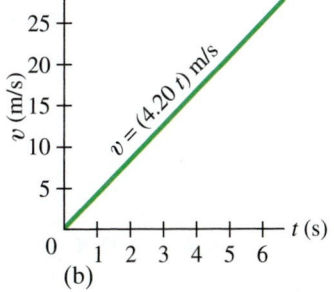

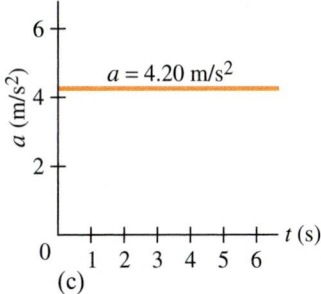

(c)

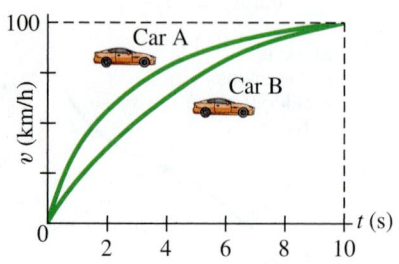

FIGURE 2–19 Example 2–8.

CONCEPTUAL EXAMPLE 2–8 | **Analyzing with graphs.** Figure 2–19 shows the velocity as a function of time for two cars accelerating from 0 to 100 km/h in a time of 10.0 s. Compare (a) the average acceleration; (b) instantaneous acceleration; and (c) total distance traveled for the two cars.

RESPONSE (a) Average acceleration is $\Delta v / \Delta t$. Both cars have the same Δv (100 km/h) and the same Δt (10.0 s), so the average acceleration is the same for both cars. (b) Instantaneous acceleration is the slope of the tangent to the v vs. t curve. For about the first 4 s, the top curve is steeper than the bottom curve, so car A has a greater acceleration during this interval. The bottom curve is steeper during the last 6 s, so car B has the larger acceleration for this period. (c) Except at $t = 0$ and $t = 10.0$ s, car A is always going faster than car B. Since it is going faster, it will go farther in the same time.

2–5 Motion at Constant Acceleration

We now examine the situation when the magnitude of the acceleration is constant and the motion is in a straight line. In this case, the instantaneous and average accelerations are equal. We use the definitions of average velocity and acceleration to derive a set of valuable equations that relate x, v, a, and t when a is constant, allowing us to determine any one of these variables if we know the others.

To simplify our notation, let us take the initial time in any discussion to be zero, and we call it t_0: $t_1 = t_0 = 0$. (This is effectively starting a stopwatch at t_0.) We can then let $t_2 = t$ be the elapsed time. The initial position (x_1) and the initial velocity (v_1) of an object will now be represented by x_0 and v_0, since they represent x and v at $t = 0$. At time t the position and velocity will be called x and v (rather than x_2 and v_2). The average velocity during the time interval $t - t_0$ will be (Eq. 2–2)

$$\bar{v} = \frac{\Delta x}{\Delta t} = \frac{x - x_0}{t - t_0} = \frac{x - x_0}{t}$$

since we chose $t_0 = 0$. The acceleration, assumed constant in time, is (Eq. 2–5)

$$a = \frac{v - v_0}{t}.$$

A common problem is to determine the velocity of an object after any elapsed time t, when we are given the object's constant acceleration. We can solve such problems by solving for v in the last equation to obtain:

$$v = v_0 + at. \qquad \text{[constant acceleration]} \quad \textbf{(2–7)}$$

If an object starts from rest ($v_0 = 0$) and accelerates at 4.0 m/s², after an elapsed time $t = 6.0$ s its velocity will be $v = at = (4.0 \text{ m/s}^2)(6.0 \text{ s}) = 24$ m/s.

Next, let us see how to calculate the position x of an object after a time t when it undergoes constant acceleration. The definition of average velocity (Eq. 2–2) is $\bar{v} = (x - x_0)/t$, which we can rewrite as

$$x = x_0 + \bar{v}t. \qquad \qquad \textbf{(2–8)}$$

Because the velocity increases at a uniform rate, the average velocity, $\bar{v}$, will be midway between the initial and final velocities:

$$\bar{v} = \frac{v_0 + v}{2}. \qquad \text{[constant acceleration]} \quad \textbf{(2–9)}$$

> **⚠ CAUTION**
> *Average velocity, but only if*
> *a = constant*

(Careful: Equation 2–9 is not necessarily valid if the acceleration is not constant.) We combine the last two Equations with Eq. 2–7 and find

$$x = x_0 + \bar{v}t$$
$$= x_0 + \left(\frac{v_0 + v}{2}\right)t$$
$$= x_0 + \left(\frac{v_0 + v_0 + at}{2}\right)t$$

or

$$x = x_0 + v_0 t + \tfrac{1}{2}at^2. \qquad \text{[constant acceleration]} \quad \textbf{(2–10)}$$

Equations 2–7, 2–9, and 2–10 are three of the four most useful equations for motion at constant acceleration. We now derive the fourth equation, which is useful in situations where the time t is not known. We substitute Eq. 2–9 into Eq. 2–8:

$$x = x_0 + \bar{v}t = x_0 + \left(\frac{v + v_0}{2}\right)t.$$

Next we solve Eq. 2–7 for t, obtaining

$$t = \frac{v - v_0}{a},$$

and substituting this into the previous equation we have

$$x = x_0 + \left(\frac{v + v_0}{2}\right)\left(\frac{v - v_0}{a}\right) = x_0 + \frac{v^2 - v_0^2}{2a}.$$

We solve this for v^2 and obtain

$$v^2 = v_0^2 + 2a(x - x_0), \qquad \text{[constant acceleration]} \quad \textbf{(2–11)}$$

which is the useful equation we sought.

We now have four equations relating position, velocity, acceleration, and time, when the acceleration a is constant. We collect these kinematic equations here in one place for future reference (the tan background screen emphasizes their usefulness):

$$v = v_0 + at \qquad\qquad\qquad [a = \text{constant}] \quad \textbf{(2–12a)}$$
$$x = x_0 + v_0 t + \tfrac{1}{2}at^2 \qquad [a = \text{constant}] \quad \textbf{(2–12b)}$$
$$v^2 = v_0^2 + 2a(x - x_0) \qquad [a = \text{constant}] \quad \textbf{(2–12c)}$$
$$\bar{v} = \frac{v + v_0}{2}. \qquad\qquad\qquad [a = \text{constant}] \quad \textbf{(2–12d)}$$

Kinematic equations
for constant acceleration
(we'll use them a lot)

These useful equations are not valid unless a is a constant. In many cases we can set $x_0 = 0$, and this simplifies the above equations a bit. Note that x represents position, not distance, that $x - x_0$ is the displacement, and that t is the elapsed time.

EXAMPLE 2–9 **Runway design.** You are designing an airport for small planes. One kind of airplane that might use this airfield must reach a speed before takeoff of at least 27.8 m/s (100 km/h), and can accelerate at 2.00 m/s². (a) If the runway is 150 m long, can this airplane reach the required speed for takeoff? (b) If not, what minimum length must the runway have?

PHYSICS APPLIED
Airport design

APPROACH The plane's acceleration is constant, so we can use the kinematic equations for constant acceleration. In (a), we want to find v, and we are given:

Known	Wanted
$x_0 = 0$	v
$v_0 = 0$	
$x = 150$ m	
$a = 2.00$ m/s²	

SOLUTION (a) Of the above four equations, Eq. 2–12c will give us v when we know v_0, a, x, and x_0:

$$v^2 = v_0^2 + 2a(x - x_0)$$
$$= 0 + 2(2.00 \text{ m/s}^2)(150 \text{ m}) = 600 \text{ m}^2/\text{s}^2$$
$$v = \sqrt{600 \text{ m}^2/\text{s}^2} = 24.5 \text{ m/s}.$$

PROBLEM SOLVING
Equations 2–12 are valid only when the acceleration is constant, which we assume in this Example

This runway length is *not* sufficient.

(b) Now we want to find the minimum length of runway, $x - x_0$, given $v = 27.8$ m/s and $a = 2.00$ m/s². So we again use Eq. 2–12c, but rewritten as

$$(x - x_0) = \frac{v^2 - v_0^2}{2a} = \frac{(27.8 \text{ m/s})^2 - 0}{2(2.00 \text{ m/s}^2)} = 193 \text{ m}.$$

A 200-m runway is more appropriate for this plane.

NOTE We did this Example as if the plane were a particle, so we round off our answer to 200 m.

EXERCISE G A car starts from rest and accelerates at a constant 10 m/s² during a $\tfrac{1}{4}$ mile (402 m) race. How fast is the car going at the finish line? (a) 8090 m/s; (b) 90 m/s; (c) 81 m/s; (d) 809 m/s.

2–6 Solving Problems

Before doing more worked-out Examples, let us look at how to approach problem solving. First, it is important to note that physics is *not* a collection of equations to be memorized. Simply searching for an equation that might work can lead you to a wrong result and will surely not help you understand physics. A better approach is to use the following (rough) procedure, which we put in a special "Problem Solving Strategy." (Other such Problem Solving Strategies, as an aid, will be found throughout the book.)

PROBLEM SOLVING

1. Read and **reread** the whole problem carefully before trying to solve it.

2. Decide what **object** (or objects) you are going to study, and for what **time interval**. You can often choose the initial time to be $t = 0$.

3. **Draw** a **diagram** or picture of the situation, with coordinate axes wherever applicable. [You can place the origin of coordinates and the axes wherever you like to make your calculations easier.]

4. Write down what quantities are "**known**" or "given," and then what you *want* to know. Consider quantities both at the beginning and at the end of the chosen time interval.

5. Think about which **principles of physics** apply in this problem. Use common sense and your own experiences. Then plan an approach.

6. Consider which **equations** (and/or definitions) relate the quantities involved. Before using them, be sure their **range of validity** includes your problem (for example, Eqs. 2–12 are valid only when the acceleration is constant). If you find an applicable equation that involves only known quantities and one desired unknown, **solve** the equation algebraically for the

unknown. Sometimes several sequential calculations, or a combination of equations, may be needed. It is often preferable to solve algebraically for the desired unknown before putting in numerical values.

7. Carry out the **calculation** if it is a numerical problem. Keep one or two extra digits during the calculations, but round off the final answer(s) to the correct number of significant figures (Section 1–3).

8. Think carefully about the result you obtain: Is it **reasonable**? Does it make sense according to your own intuition and experience? A good check is to do a rough **estimate** using only powers of ten, as discussed in Section 1–6. Often it is preferable to do a rough estimate at the *start* of a numerical problem because it can help you focus your attention on finding a path toward a solution.

9. A very important aspect of doing problems is keeping track of **units**. An equals sign implies the units on each side must be the same, just as the numbers must. If the units do not balance, a mistake has no doubt been made. This can serve as a **check** on your solution (but it only tells you if you're wrong, not if you're right). Always use a consistent set of units.

EXAMPLE 2–10 **Acceleration of a car.** How long does it take a car to cross a 30.0-m-wide intersection after the light turns green, if the car accelerates from rest at a constant $2.00 \, \text{m/s}^2$?

APPROACH We follow the Problem Solving Strategy above, step by step.
SOLUTION

1. **Reread** the problem. Be sure you understand what it asks for (here, a time interval).

2. The **object** under study is the car. We choose the **time interval**: $t = 0$, the initial time, is the moment the car starts to accelerate from rest $(v_0 = 0)$; the time t is the instant the car has traveled the full 30.0-m width of the intersection.

3. **Draw** a **diagram**: the situation is shown in Fig. 2–20, where the car is shown moving along the positive x axis. We choose $x_0 = 0$ at the front bumper of the car before it starts to move.

FIGURE 2–20 Example 2–10.

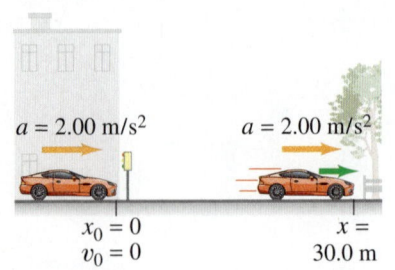

$a = 2.00 \, \text{m/s}^2$ $a = 2.00 \, \text{m/s}^2$

$x_0 = 0$ $x =$
$v_0 = 0$ $30.0 \, \text{m}$

4. The "**knowns**" and the "**wanted**" are shown in the Table in the margin, and we choose $x_0 = 0$. Note that "starting from rest" means $v = 0$ at $t = 0$; that is, $v_0 = 0$.

Known	Wanted
$x_0 = 0$	t
$x = 30.0$ m	
$a = 2.00$ m/s^2	
$v_0 = 0$	

5. The **physics**: the motion takes place at constant acceleration, so we can use the kinematic equations, Eqs. 2–12.

6. **Equations**: we want to find the time, given the distance and acceleration; Eq. 2–12b is perfect since the only unknown quantity is t. Setting $v_0 = 0$ and $x_0 = 0$ in Eq. 2–12b $\left(x = x_0 + v_0 t + \frac{1}{2}at^2\right)$, we can solve for t:

$$x = \tfrac{1}{2}at^2,$$

$$t^2 = \frac{2x}{a},$$

so

$$t = \sqrt{\frac{2x}{a}}.$$

7. The **calculation**:

$$t = \sqrt{\frac{2x}{a}} = \sqrt{\frac{2(30.0 \text{ m})}{2.00 \text{ m/s}^2}} = 5.48 \text{ s}.$$

This is our answer. Note that the units come out correctly.

8. We can check the **reasonableness** of the answer by calculating the final velocity $v = at = (2.00 \text{ m/s}^2)(5.48 \text{ s}) = 10.96 \text{ m/s}$, and then finding $x = x_0 + \bar{v}t = 0 + \frac{1}{2}(10.96 \text{ m/s} + 0)(5.48 \text{ s}) = 30.0 \text{ m}$, which is our given distance.

9. We checked the **units**, and they came out perfectly (seconds).

NOTE In steps 6 and 7, when we took the square root, we should have written $t = \pm\sqrt{2x/a} = \pm 5.48 \text{ s}$. Mathematically there are two solutions. But the second solution, $t = -5.48 \text{ s}$, is a time *before* our chosen time interval and makes no sense physically. We say it is "unphysical" and ignore it.

We explicitly followed the steps of the Problem Solving Strategy for Example 2–10. In upcoming Examples, we will use our usual "Approach" and "Solution" to avoid being wordy.

EXAMPLE 2–11 **ESTIMATE** **Air bags.** Suppose you want to design an air-bag system that can protect the driver at a speed of 100 km/h (60 mph) if the car hits a brick wall. Estimate how fast the air bag must inflate (Fig. 2–21) to effectively protect the driver. How does the use of a seat belt help the driver?

APPROACH We assume the acceleration is roughly constant, so we can use Eqs. 2–12. Both Eqs. 2–12a and 2–12b contain t, our desired unknown. They both contain a, so we must first find a, which we can do using Eq. 2–12c if we know the distance x over which the car crumples. A rough estimate might be about 1 meter. We choose the time interval to start at the instant of impact with the car moving at $v_0 = 100$ km/h, and to end when the car comes to rest $(v = 0)$ after traveling 1 m.

SOLUTION We convert the given initial speed to SI units: $100 \text{ km/h} = 100 \times 10^3 \text{ m}/3600 \text{ s} = 28 \text{ m/s}$. We then find the acceleration from Eq. 2–12c:

$$a = -\frac{v_0^2}{2x} = -\frac{(28 \text{ m/s})^2}{2.0 \text{ m}} = -390 \text{ m/s}^2.$$

This enormous acceleration takes place in a time given by (Eq. 2–12a):

$$t = \frac{v - v_0}{a} = \frac{0 - 28 \text{ m/s}}{-390 \text{ m/s}^2} = 0.07 \text{ s}.$$

To be effective, the air bag would need to inflate faster than this.

What does the air bag do? It spreads the force over a large area of the chest (to avoid puncture of the chest by the steering wheel). The seat belt keeps the person in a stable position against the expanding air bag.

PHYSICS APPLIED
Car safety—air bags

FIGURE 2–21 Example 2–11. An air bag deploying on impact.

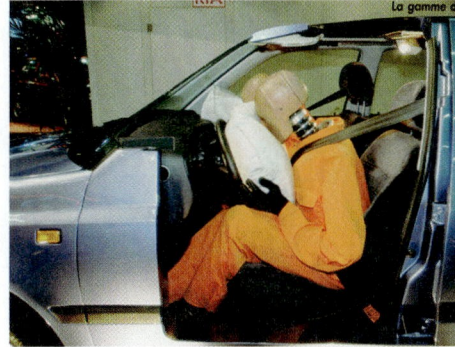

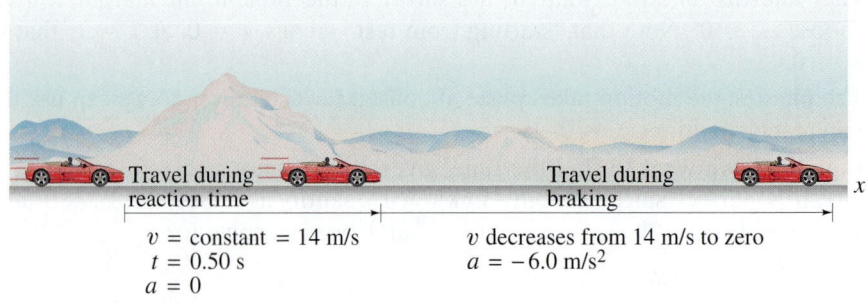

FIGURE 2–22 Example 2–12: stopping distance for a braking car.

Travel during reaction time
$v = $ constant $= 14$ m/s
$t = 0.50$ s
$a = 0$

Travel during braking
v decreases from 14 m/s to zero
$a = -6.0$ m/s²

Part 1: Reaction time

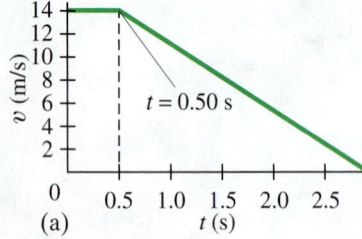

Known	Wanted
$t = 0.50$ s	x
$v_0 = 14$ m/s	
$v = 14$ m/s	
$a = 0$	
$x_0 = 0$	

Part 2: Braking

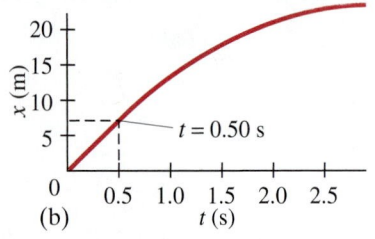

Known	Wanted
$x_0 = 7.0$ m	x
$v_0 = 14$ m/s	
$v = 0$	
$a = -6.0$ m/s²	

FIGURE 2–23 Example 2–12. Graphs of (a) v vs. t and (b) x vs. t.

(a) [graph: v (m/s) vs t (s), $t = 0.50$ s]

(b) [graph: x (m) vs t (s), $t = 0.50$ s]

EXAMPLE 2–12 **ESTIMATE** **Braking distances.** Estimate the minimum stopping distance for a car, which is important for traffic safety and traffic design. The problem is best dealt with in two parts, two separate time intervals. (1) The first time interval begins when the driver decides to hit the brakes, and ends when the foot touches the brake pedal. This is the "reaction time" during which the speed is constant, so $a = 0$. (2) The second time interval is the actual braking period when the vehicle slows down ($a \neq 0$) and comes to a stop. The stopping distance depends on the reaction time of the driver, the initial speed of the car (the final speed is zero), and the acceleration of the car. For a dry road and good tires, good brakes can decelerate a car at a rate of about 5 m/s² to 8 m/s². Calculate the total stopping distance for an initial velocity of 50 km/h ($= 14$ m/s ≈ 31 mi/h) and assume the acceleration of the car is -6.0 m/s² (the minus sign appears because the velocity is taken to be in the positive x direction and its magnitude is decreasing). Reaction time for normal drivers varies from perhaps 0.3 s to about 1.0 s; take it to be 0.50 s.

APPROACH During the "reaction time," part (1), the car moves at constant speed of 14 m/s, so $a = 0$. Once the brakes are applied, part (2), the acceleration is $a = -6.0$ m/s² and is constant over this time interval. For both parts a is constant, so we can use Eqs. 2–12.

SOLUTION Part (1). We take $x_0 = 0$ for the first time interval, when the driver is reacting (0.50 s): the car travels at a constant speed of 14 m/s so $a = 0$. See Fig. 2–22 and the Table in the margin. To find x, the position of the car at $t = 0.50$ s (when the brakes are applied), we cannot use Eq. 2–12c because x is multiplied by a, which is zero. But Eq. 2–12b works:

$$x = v_0 t + 0 = (14 \text{ m/s})(0.50 \text{ s}) = 7.0 \text{ m}.$$

Thus the car travels 7.0 m during the driver's reaction time, until the instant the brakes are applied. We will use this result as input to part (2).

Part (2). During the second time interval, the brakes are applied and the car is brought to rest. The initial position is $x_0 = 7.0$ m (result of part (1)), and other variables are shown in the second Table in the margin. Equation 2–12a doesn't contain x; Eq. 2–12b contains x but also the unknown t. Equation 2–12c, $v^2 - v_0^2 = 2a(x - x_0)$, is what we want; after setting $x_0 = 7.0$ m, we solve for x, the final position of the car (when it stops):

$$x = x_0 + \frac{v^2 - v_0^2}{2a}$$

$$= 7.0 \text{ m} + \frac{0 - (14 \text{ m/s})^2}{2(-6.0 \text{ m/s}^2)} = 7.0 \text{ m} + \frac{-196 \text{ m}^2/\text{s}^2}{-12 \text{ m/s}^2}$$

$$= 7.0 \text{ m} + 16 \text{ m} = 23 \text{ m}.$$

The car traveled 7.0 m while the driver was reacting and another 16 m during the braking period before coming to a stop, for a total distance traveled of 23 m. Figure 2–23 shows graphs of (a) v vs. t and (b) x vs. t.

NOTE From the equation above for x, we see that the stopping distance after the driver hit the brakes ($= x - x_0$) increases with the *square* of the initial speed, not just linearly with speed. If you are traveling twice as fast, it takes four times the distance to stop.

EXAMPLE 2-13 **ESTIMATE** **Two Moving Objects: Police and Speeder.**
A car speeding at 150 km/h passes a still police car which immediately takes off in hot pursuit. Using simple assumptions, such as that the speeder continues at constant speed, estimate how long it takes the police car to overtake the speeder. Then estimate the police car's speed at that moment and decide if the assumptions were reasonable.

APPROACH When the police car takes off, it accelerates, and the simplest assumption is that its acceleration is constant. This may not be reasonable, but let's see what happens. We can estimate the acceleration if we have noticed automobile ads, which claim cars can accelerate from rest to 100 km/h in 5.0 s. So the average acceleration of the police car could be approximately

$$a_P = \frac{100 \text{ km/h}}{5.0 \text{ s}} = 20 \frac{\text{km/h}}{\text{s}} \left(\frac{1000 \text{ m}}{1 \text{ km}}\right)\left(\frac{1 \text{ h}}{3600 \text{ s}}\right) = 5.6 \text{ m/s}^2.$$

SOLUTION We need to set up the kinematic equations to determine the unknown quantities, and since there are two moving objects, we need two separate sets of equations. We denote the speeding car's position by x_S and the police car's position by x_P. Because we are interested in solving for the time when the two vehicles arrive at the same position on the road, we use Eq. 2-12b for each car:

$$x_S = v_{0S}t + \tfrac{1}{2}a_S t^2 = (150 \text{ km/h})t = (42 \text{ m/s})t$$
$$x_P = v_{0P}t + \tfrac{1}{2}a_P t^2 = \tfrac{1}{2}(5.6 \text{ m/s}^2)t^2,$$

where we have set $v_{0P} = 0$ and $a_S = 0$ (speeder assumed to move at constant speed). We want the time when the cars meet, so we set $x_S = x_P$ and solve for t:

$$(42 \text{ m/s})t = (2.8 \text{ m/s}^2)t^2.$$

The solutions are

$$t = 0 \quad \text{and} \quad t = \frac{42 \text{ m/s}}{2.8 \text{ m/s}^2} = 15 \text{ s}.$$

The first solution corresponds to the instant the speeder passed the police car. The second solution tells us when the police car catches up to the speeder, 15 s later. This is our answer, but is it reasonable? The police car's speed at $t = 15$ s is

$$v_P = v_{0P} + a_P t = 0 + (5.6 \text{ m/s}^2)(15 \text{ s}) = 84 \text{ m/s}$$

or 300 km/h ($\approx$ 190 mi/h). Not reasonable, and highly dangerous.

NOTE More reasonable is to give up the assumption of constant acceleration. The police car surely cannot maintain constant acceleration at those speeds. Also, the speeder, if a reasonable person, would slow down upon hearing the police siren. Figure 2-24 shows (a) x vs. t and (b) v vs. t graphs, based on the original assumption of $a_P = $ constant, whereas (c) shows v vs. t for more reasonable assumptions.

⚠ **CAUTION**
Initial assumptions need to be checked out for reasonableness

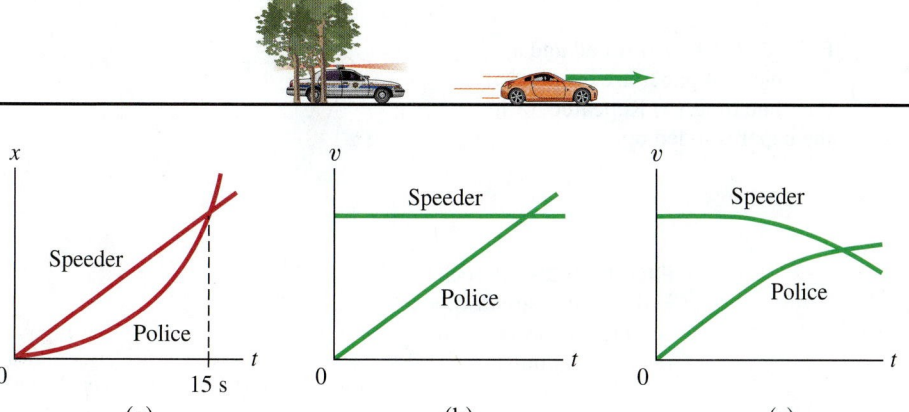

FIGURE 2-24 Example 2-13.

FIGURE 2–25 Galileo Galilei (1564–1642).

FIGURE 2–26 Multiflash photograph of a falling apple, at equal time intervals. The apple falls farther during each successive interval, which means it is accelerating.

Acceleration due to gravity

2–7 Freely Falling Objects

One of the most common examples of uniformly accelerated motion is that of an object allowed to fall freely near the Earth's surface. That a falling object is accelerating may not be obvious at first. And beware of thinking, as was widely believed before the time of Galileo (Fig. 2–25), that heavier objects fall faster than lighter objects and that the speed of fall is proportional to how heavy the object is.

Galileo made use of his new technique of imagining what would happen in idealized (simplified) cases. For free fall, he postulated that all objects would fall with the *same constant acceleration* in the absence of air or other resistance. He showed that this postulate predicts that for an object falling from rest, the distance traveled will be proportional to the square of the time (Fig. 2–26); that is, $d \propto t^2$. We can see this from Eq. 2–12b; but Galileo was the first to derive this mathematical relation.

To support his claim that falling objects increase in speed as they fall, Galileo made use of a clever argument: a heavy stone dropped from a height of 2 m will drive a stake into the ground much further than will the same stone dropped from a height of only 0.2 m. Clearly, the stone must be moving faster in the former case.

Galileo claimed that *all* objects, light or heavy, fall with the *same* acceleration, at least in the absence of air. If you hold a piece of paper horizontally in one hand and a heavier object—say, a baseball—in the other, and release them at the same time as in Fig. 2–27a, the heavier object will reach the ground first. But if you repeat the experiment, this time crumpling the paper into a small wad (see Fig. 2–27b), you will find that the two objects reach the floor at nearly the same time.

Galileo was sure that air acts as a resistance to very light objects that have a large surface area. But in many ordinary circumstances this air resistance is negligible. In a chamber from which the air has been removed, even light objects like a feather or a horizontally held piece of paper will fall with the same acceleration as any other object (see Fig. 2–28). Such a demonstration in vacuum was not possible in Galileo's time, which makes Galileo's achievement all the greater. Galileo is often called the "father of modern science," not only for the content of his science (astronomical discoveries, inertia, free fall) but also for his approach to science (idealization and simplification, mathematization of theory, theories that have testable consequences, experiments to test theoretical predictions).

Galileo's specific contribution to our understanding of the motion of falling objects can be summarized as follows:

at a given location on the Earth and in the absence of air resistance, all objects fall with the same constant acceleration.

We call this acceleration the **acceleration due to gravity** on the surface of the Earth, and we give it the symbol g. Its magnitude is approximately

$$g = 9.80 \text{ m/s}^2. \qquad \text{[at surface of Earth]}$$

In British units g is about 32 ft/s². Actually, g varies slightly according to latitude and elevation, but these variations are so small that we will ignore them for most

(a)   (b)

FIGURE 2–27 (a) A ball and a light piece of paper are dropped at the same time. (b) Repeated, with the paper wadded up.

FIGURE 2–28 A rock and a feather are dropped simultaneously (a) in air, (b) in a vacuum.

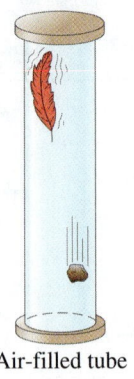

Air-filled tube   Evacuated tube
(a)   (b)

purposes. The effects of air resistance are often small, and we will neglect them for the most part. However, air resistance will be noticeable even on a reasonably heavy object if the velocity becomes large.[†] Acceleration due to gravity is a vector as is any acceleration, and its direction is downward, toward the center of the Earth.

When dealing with freely falling objects we can make use of Eqs. 2–12, where for a we use the value of g given above. Also, since the motion is vertical we will substitute y in place of x, and y_0 in place of x_0. We take $y_0 = 0$ unless otherwise specified. *It is arbitrary whether we choose y to be positive in the upward direction or in the downward direction; but we must be consistent about it throughout a problem's solution.*

EXERCISE H Return to the Chapter-Opening Question, page 18, and answer it again now. Try to explain why you may have answered differently the first time.

EXAMPLE 2–14 **Falling from a tower.** Suppose that a ball is dropped $(v_0 = 0)$ from a tower 70.0 m high. How far will it have fallen after a time $t_1 = 1.00$ s, $t_2 = 2.00$ s, and $t_3 = 3.00$ s? Ignore air resistance.

APPROACH Let us take y as positive downward, so the acceleration is $a = g = +9.80$ m/s². We set $v_0 = 0$ and $y_0 = 0$. We want to find the position y of the ball after three different time intervals. Equation 2–12b, with x replaced by y, relates the given quantities (t, a, and v_0) to the unknown y.

SOLUTION We set $t = t_1 = 1.00$ s in Eq. 2–12b:
$$y_1 = v_0 t_1 + \tfrac{1}{2} a t_1^2 = 0 + \tfrac{1}{2} a t_1^2 = \tfrac{1}{2}(9.80 \text{ m/s}^2)(1.00 \text{ s})^2 = 4.90 \text{ m}.$$
The ball has fallen a distance of 4.90 m during the time interval $t = 0$ to $t_1 = 1.00$ s. Similarly, after 2.00 s ($= t_2$), the ball's position is
$$y_2 = \tfrac{1}{2} a t_2^2 = \tfrac{1}{2}(9.80 \text{ m/s}^2)(2.00 \text{ s})^2 = 19.6 \text{ m}.$$
Finally, after 3.00 s ($= t_3$), the ball's position is (see Fig. 2–29)
$$y_3 = \tfrac{1}{2} a t_3^2 = \tfrac{1}{2}(9.80 \text{ m/s}^2)(3.00 \text{ s})^2 = 44.1 \text{ m}.$$

EXAMPLE 2–15 **Thrown down from a tower.** Suppose the ball in Example 2–14 is *thrown* downward with an initial velocity of 3.00 m/s, instead of being dropped. (*a*) What then would be its position after 1.00 s and 2.00 s? (*b*) What would its speed be after 1.00 s and 2.00 s? Compare with the speeds of a dropped ball.

APPROACH Again we use Eq. 2–12b, but now v_0 is not zero, it is $v_0 = 3.00$ m/s.

SOLUTION (*a*) At $t = 1.00$ s, the position of the ball as given by Eq. 2–12b is
$$y = v_0 t + \tfrac{1}{2} a t^2 = (3.00 \text{ m/s})(1.00 \text{ s}) + \tfrac{1}{2}(9.80 \text{ m/s}^2)(1.00 \text{ s})^2 = 7.90 \text{ m}.$$
At $t = 2.00$ s, (time interval $t = 0$ to $t = 2.00$ s), the position is
$$y = v_0 t + \tfrac{1}{2} a t^2 = (3.00 \text{ m/s})(2.00 \text{ s}) + \tfrac{1}{2}(9.80 \text{ m/s}^2)(2.00 \text{ s})^2 = 25.6 \text{ m}.$$
As expected, the ball falls farther each second than if it were dropped with $v_0 = 0$. (*b*) The velocity is obtained from Eq. 2–12a:
$$v = v_0 + at$$
$$= 3.00 \text{ m/s} + (9.80 \text{ m/s}^2)(1.00 \text{ s}) = 12.8 \text{ m/s} \quad [\text{at } t_1 = 1.00 \text{ s}]$$
$$= 3.00 \text{ m/s} + (9.80 \text{ m/s}^2)(2.00 \text{ s}) = 22.6 \text{ m/s}. \quad [\text{at } t_2 = 2.00 \text{ s}]$$
In Example 2–14, when the ball was dropped ($v_0 = 0$), the first term (v_0) in these equations was zero, so
$$v = 0 + at$$
$$= (9.80 \text{ m/s}^2)(1.00 \text{ s}) = 9.80 \text{ m/s} \quad [\text{at } t_1 = 1.00 \text{ s}]$$
$$= (9.80 \text{ m/s}^2)(2.00 \text{ s}) = 19.6 \text{ m/s}. \quad [\text{at } t_2 = 2.00 \text{ s}]$$

NOTE For both Examples 2–14 and 2–15, the speed increases linearly in time by 9.80 m/s during each second. But the speed of the downwardly thrown ball at any instant is always 3.00 m/s (its initial speed) higher than that of a dropped ball.

PROBLEM SOLVING
You can choose y to be positive either up or down

FIGURE 2–29 Example 2–14. (a) An object dropped from a tower falls with progressively greater speed and covers greater distance with each successive second. (See also Fig. 2–26.) (b) Graph of y vs. t.

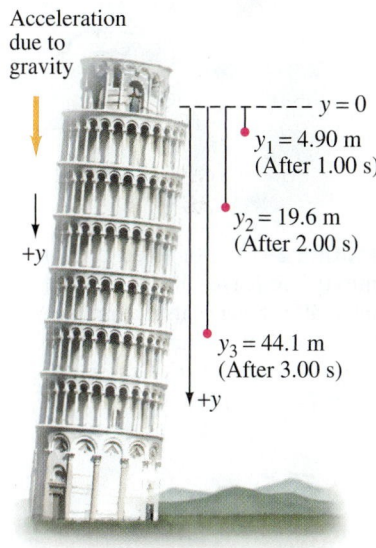

(a)

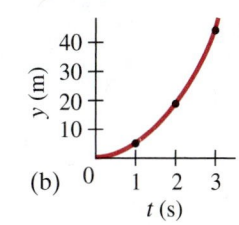

(b)

[†]The speed of an object falling in air (or other fluid) does not increase indefinitely. If the object falls far enough, it will reach a maximum velocity called the **terminal velocity** due to air resistance.

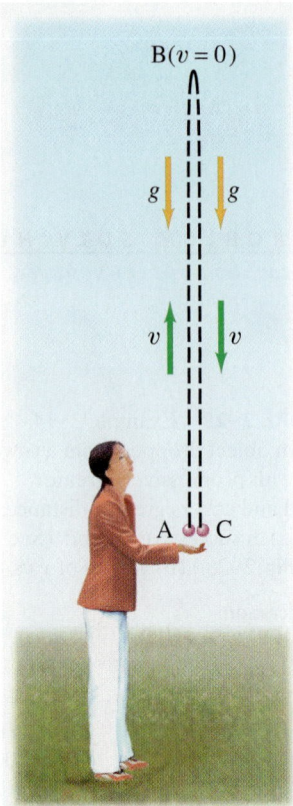

B(v = 0)

g ↓   ↓ g

v ↑   ↓ v

A ●● C

FIGURE 2–30 An object thrown into the air leaves the thrower's hand at A, reaches its maximum height at B, and returns to the original position at C. Examples 2–16, 2–17, 2–18, and 2–19.

EXAMPLE 2–16 **Ball thrown upward, I.** A person throws a ball *upward* into the air with an initial velocity of 15.0 m/s. Calculate (a) how high it goes, and (b) how long the ball is in the air before it comes back to the hand. Ignore air resistance.

APPROACH We are not concerned here with the throwing action, but only with the motion of the ball *after* it leaves the thrower's hand (Fig. 2–30) and until it comes back to the hand again. Let us choose y to be positive in the upward direction and negative in the downward direction. (This is a different convention from that used in Examples 2–14 and 2–15, and so illustrates our options.) The acceleration due to gravity is downward and so will have a negative sign, $a = -g = -9.80 \text{ m/s}^2$. As the ball rises, its speed decreases until it reaches the highest point (B in Fig. 2–30), where its speed is zero for an instant; then it descends, with increasing speed.

SOLUTION (a) We consider the time interval from when the ball leaves the thrower's hand until the ball reaches the highest point. To determine the maximum height, we calculate the position of the ball when its velocity equals zero ($v = 0$ at the highest point). At $t = 0$ (point A in Fig. 2–30) we have $y_0 = 0$, $v_0 = 15.0 \text{ m/s}$, and $a = -9.80 \text{ m/s}^2$. At time t (maximum height), $v = 0$, $a = -9.80 \text{ m/s}^2$, and we wish to find y. We use Eq. 2–12c, replacing x with y: $v^2 = v_0^2 + 2ay$. We solve this equation for y:

$$y = \frac{v^2 - v_0^2}{2a} = \frac{0 - (15.0 \text{ m/s})^2}{2(-9.80 \text{ m/s}^2)} = 11.5 \text{ m}.$$

The ball reaches a height of 11.5 m above the hand.

(b) Now we need to choose a different time interval to calculate how long the ball is in the air before it returns to the hand. We could do this calculation in two parts by first determining the time required for the ball to reach its highest point, and then determining the time it takes to fall back down. However, it is simpler to consider the time interval for the entire motion from A to B to C (Fig. 2–30) in one step and use Eq. 2–12b. We can do this because y represents position or displacement, and not the total distance traveled. Thus, at both points A and C, $y = 0$. We use Eq. 2–12b with $a = -9.80 \text{ m/s}^2$ and find

$$y = y_0 + v_0 t + \tfrac{1}{2}at^2$$
$$0 = 0 + (15.0 \text{ m/s})t + \tfrac{1}{2}(-9.80 \text{ m/s}^2)t^2.$$

This equation is readily factored (we factor out one t):

$$(15.0 \text{ m/s} - 4.90 \text{ m/s}^2\, t)t = 0.$$

There are two solutions:

$$t = 0 \quad \text{and} \quad t = \frac{15.0 \text{ m/s}}{4.90 \text{ m/s}^2} = 3.06 \text{ s}.$$

The first solution ($t = 0$) corresponds to the initial point (A) in Fig. 2–30, when the ball was first thrown from $y = 0$. The second solution, $t = 3.06$ s, corresponds to point C, when the ball has returned to $y = 0$. Thus the ball is in the air 3.06 s.

NOTE We have ignored air resistance, which could be significant, so our result is only an approximation to a real, practical situation.

We did not consider the throwing action in this Example. Why? Because during the throw, the thrower's hand is touching the ball and accelerating the ball at a rate unknown to us—the acceleration is *not* g. We consider only the time when the ball is in the air and the acceleration is equal to g.

Every quadratic equation (where the variable is squared) mathematically produces two solutions. In physics, sometimes only one solution corresponds to the real situation, as in Example 2–10, in which case we ignore the "unphysical" solution. But in Example 2–16, both solutions to our equation in t^2 are physically meaningful: $t = 0$ and $t = 3.06$ s.

⚠ **CAUTION**

Quadratic equations have two solutions. Sometimes only one corresponds to reality, sometimes both

CONCEPTUAL EXAMPLE 2-17 | **Two possible misconceptions.** Give examples to show the error in these two common misconceptions: (1) that acceleration and velocity are always in the same direction, and (2) that an object thrown upward has zero acceleration at the highest point (B in Fig. 2–30).

RESPONSE Both are wrong. (1) Velocity and acceleration are *not* necessarily in the same direction. When the ball in Example 2–16 is moving upward, its velocity is positive (upward), whereas the acceleration is negative (downward). (2) At the highest point (B in Fig. 2–30), the ball has zero velocity for an instant. Is the acceleration also zero at this point? No. The velocity near the top of the arc points upward, then becomes zero (for zero time) at the highest point, and then points downward. Gravity does not stop acting, so $a = -g = -9.80 \text{ m/s}^2$ even there. Thinking that $a = 0$ at point B would lead to the conclusion that upon reaching point B, the ball would stay there: if the acceleration (= rate of change of velocity) were zero, the velocity would stay zero at the highest point, and the ball would stay up there without falling. In sum, the acceleration of gravity always points down toward the Earth, even when the object is moving up.

> ⚠️ **CAUTION**
> *(1) Velocity and acceleration are not always in the same direction; the acceleration (of gravity) always points down*
> *(2) $a \neq 0$ even at the highest point of a trajectory*

EXAMPLE 2-18 | **Ball thrown upward, II.** Let us consider again the ball thrown upward of Example 2–16, and make more calculations. Calculate (*a*) how much time it takes for the ball to reach the maximum height (point B in Fig. 2–30), and (*b*) the velocity of the ball when it returns to the thrower's hand (point C).

APPROACH Again we assume the acceleration is constant, so we can use Eqs. 2–12. We have the height of 11.5 m from Example 2–16. Again we take *y* as positive upward.

SOLUTION (*a*) We consider the time interval between the throw $(t = 0, v_0 = 15.0 \text{ m/s})$ and the top of the path $(y = +11.5 \text{ m}, v = 0)$, and we want to find *t*. The acceleration is constant at $a = -g = -9.80 \text{ m/s}^2$. Both Eqs. 2–12a and 2–12b contain the time *t* with other quantities known. Let us use Eq. 2–12a with $a = -9.80 \text{ m/s}^2$, $v_0 = 15.0 \text{ m/s}$, and $v = 0$:

$$v = v_0 + at;$$

setting $v = 0$ and solving for *t* gives

$$t = -\frac{v_0}{a} = -\frac{15.0 \text{ m/s}}{-9.80 \text{ m/s}^2} = 1.53 \text{ s}.$$

This is just half the time it takes the ball to go up and fall back to its original position [3.06 s, calculated in part (*b*) of Example 2–16]. Thus it takes the same time to reach the maximum height as to fall back to the starting point.

(*b*) Now we consider the time interval from the throw $(t = 0, v_0 = 15.0 \text{ m/s})$ until the ball's return to the hand, which occurs at $t = 3.06 \text{ s}$ (as calculated in Example 2–16), and we want to find *v* when $t = 3.06 \text{ s}$:

$$v = v_0 + at = 15.0 \text{ m/s} - (9.80 \text{ m/s}^2)(3.06 \text{ s}) = -15.0 \text{ m/s}.$$

NOTE The ball has the same speed (magnitude of velocity) when it returns to the starting point as it did initially, but in the opposite direction (this is the meaning of the negative sign). And, as we saw in part (*a*), the time is the same up as down. Thus the motion is *symmetrical* about the maximum height.

The acceleration of objects such as rockets and fast airplanes is often given as a multiple of $g = 9.80 \text{ m/s}^2$. For example, a plane pulling out of a dive and undergoing 3.00 *g*'s would have an acceleration of $(3.00)(9.80 \text{ m/s}^2) = 29.4 \text{ m/s}^2$.

| **EXERCISE I** If a car is said to accelerate at 0.50 *g*, what is its acceleration in m/s^2?

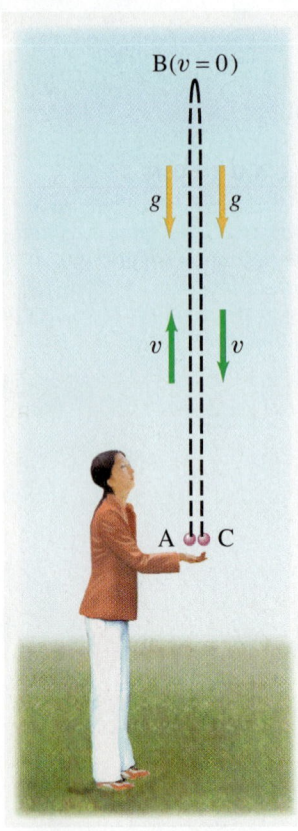

FIGURE 2–30
(Repeated for Example 2–19)

EXAMPLE 2–19 **Ball thrown upward, III; the quadratic formula.** For the ball in Example 2–18, calculate at what time t the ball passes a point 8.00 m above the person's hand. (See repeated Fig. 2–30 here).

APPROACH We choose the time interval from the throw $(t = 0, v_0 = 15.0 \text{ m/s})$ until the time t (to be determined) when the ball is at position $y = 8.00$ m, using Eq. 2–12b.

SOLUTION We want to find t, given $y = 8.00$ m, $y_0 = 0$, $v_0 = 15.0$ m/s, and $a = -9.80 \text{ m/s}^2$. We use Eq. 2–12b:

$$y = y_0 + v_0 t + \tfrac{1}{2} a t^2$$

$$8.00 \text{ m} = 0 + (15.0 \text{ m/s})t + \tfrac{1}{2}(-9.80 \text{ m/s}^2)t^2.$$

To solve any quadratic equation of the form $at^2 + bt + c = 0$, where a, b, and c are constants (a is *not* acceleration here), we use the **quadratic formula**:

$$t = \frac{-b \pm \sqrt{b^2 - 4ac}}{2a}.$$

We rewrite our y equation just above in standard form, $at^2 + bt + c = 0$:

$$(4.90 \text{ m/s}^2)t^2 - (15.0 \text{ m/s})t + (8.00 \text{ m}) = 0.$$

So the coefficient a is 4.90 m/s^2, b is -15.0 m/s, and c is 8.00 m. Putting these into the quadratic formula, we obtain

$$t = \frac{15.0 \text{ m/s} \pm \sqrt{(15.0 \text{ m/s})^2 - 4(4.90 \text{ m/s}^2)(8.00 \text{ m})}}{2(4.90 \text{ m/s}^2)},$$

which gives us $t = 0.69$ s and $t = 2.37$ s. Are both solutions valid? Yes, because the ball passes $y = 8.00$ m when it goes up $(t = 0.69 \text{ s})$ and again when it comes down $(t = 2.37 \text{ s})$.

NOTE Figure 2–31 shows graphs of (*a*) y vs. t and (*b*) v vs. t for the ball thrown upward in Fig. 2–30, incorporating the results of Examples 2–16, 2–18, and 2–19.

FIGURE 2–31 Graphs of (*a*) y vs. t, (*b*) v vs. t for a ball thrown upward, Examples 2–16, 2–18, and 2–19.

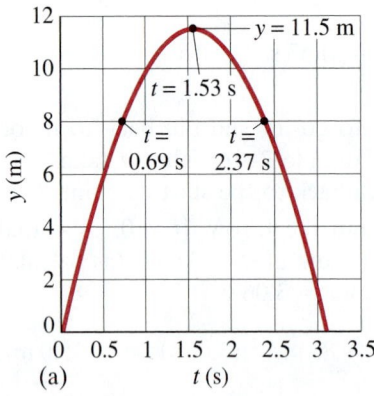
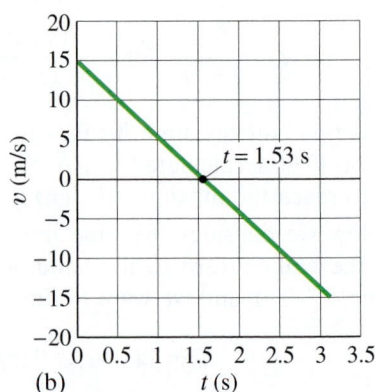

EXAMPLE 2–20 **Ball thrown upward at edge of cliff.** Suppose that the person of Examples 2–16, 2–18, and 2–19 is standing on the edge of a cliff, so that the ball can fall to the base of the cliff 50.0 m below as in Fig. 2–32. (*a*) How long does it take the ball to reach the base of the cliff? (*b*) What is the total distance traveled by the ball? Ignore air resistance (likely to be significant, so our result is an approximation).

APPROACH We again use Eq. 2–12b, but this time we set $y = -50.0$ m, the bottom of the cliff, which is 50.0 m below the initial position $(y_0 = 0)$.

SOLUTION (*a*) We use Eq. 2–12b with $a = -9.80 \text{ m/s}^2$, $v_0 = 15.0 \text{ m/s}$, $y_0 = 0$, and $y = -50.0 \text{ m}$:

$$y = y_0 + v_0 t + \tfrac{1}{2}at^2$$
$$-50.0 \text{ m} = 0 + (15.0 \text{ m/s})t - \tfrac{1}{2}(9.80 \text{ m/s}^2)t^2.$$

Rewriting in the standard form we have

$$(4.90 \text{ m/s}^2)t^2 - (15.0 \text{ m/s})t - (50.0 \text{ m}) = 0.$$

Using the quadratic formula, we find as solutions $t = 5.07 \text{ s}$ and $t = -2.01 \text{ s}$. The first solution, $t = 5.07 \text{ s}$, is the answer we are seeking: the time it takes the ball to rise to its highest point and then fall to the base of the cliff. To rise and fall back to the top of the cliff took 3.06 s (Example 2–16); so it took an additional 2.01 s to fall to the base. But what is the meaning of the other solution, $t = -2.01 \text{ s}$? This is a time before the throw, when our calculation begins, so it isn't relevant here.[†]

(*b*) From Example 2–16, the ball moves up 11.5 m, falls 11.5 m back down to the top of the cliff, and then down another 50.0 m to the base of the cliff, for a total distance traveled of 73.0 m. Note that the *displacement*, however, was -50.0 m. Figure 2–33 shows the *y* vs. *t* graph for this situation.

EXERCISE J Two balls are thrown from a cliff. One is thrown directly up, the other directly down, each with the same initial speed, and both hit the ground below the cliff. Which ball hits the ground at the greater speed: (*a*) the ball thrown upward, (*b*) the ball thrown downward, or (*c*) both the same? Ignore air resistance.

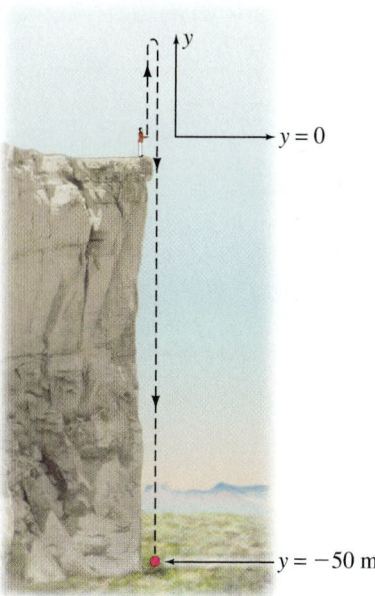

FIGURE 2–32 Example 2–20. The person in Fig. 2–30 stands on the edge of a cliff. The ball falls to the base of the cliff, 50.0 m below.

FIGURE 2–33 Example 2–20, the *y* vs. *t* graph.

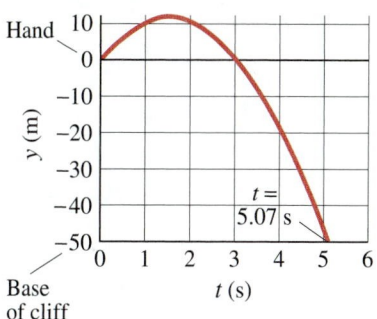

*2–8 Variable Acceleration; Integral Calculus

In this brief optional Section we use integral calculus to derive the kinematic equations for constant acceleration, Eqs. 2–12a and b. We also show how calculus can be used when the acceleration is not constant. If you have not yet studied simple integration in your calculus course, you may want to postpone reading this Section until you have. We discuss integration in more detail in Section 7–3, where we begin to use it in the physics.

First we derive Eq. 2–12a, assuming as we did in Section 2–5 that an object has velocity v_0 at $t = 0$ and a constant acceleration *a*. We start with the definition of instantaneous acceleration, $a = dv/dt$, which we rewrite as

$$dv = a \, dt.$$

We take the definite integral of both sides of this equation, using the same notation we did in Section 2–5:

$$\int_{v=v_0}^{v} dv = \int_{t=0}^{t} a \, dt$$

which gives, since $a =$ constant,

$$v - v_0 = at.$$

This is Eq. 2–12a, $v = v_0 + at$.

Next we derive Eq. 2–12b starting with the definition of instantaneous velocity, Eq. 2–4, $v = dx/dt$. We rewrite this as

$$dx = v \, dt$$

or

$$dx = (v_0 + at) \, dt$$

where we substituted in Eq. 2–12a.

[†]The solution $t = -2.01 \text{ s}$ could be meaningful in a different physical situation. Suppose that a person standing on top of a 50.0-m-high cliff sees a rock pass by him at $t = 0$ moving upward at 15.0 m/s; at what time did the rock leave the base of the cliff, and when did it arrive back at the base of the cliff? The equations will be precisely the same as for our original Example, and the answers $t = -2.01 \text{ s}$ and $t = 5.07 \text{ s}$ will be the correct answers. Note that we cannot put all the information for a problem into the mathematics, so we have to use common sense in interpreting results.

Now we integrate:

$$\int_{x=x_0}^{x} dx = \int_{t=0}^{t} (v_0 + at)\, dt$$

$$x - x_0 = \int_{t=0}^{t} v_0\, dt + \int_{t=0}^{t} at\, dt$$

$$x - x_0 = v_0 t + \tfrac{1}{2}at^2$$

since v_0 and a are constants. This result is just Eq. 2–12b, $x = x_0 + v_0 t + \tfrac{1}{2}at^2$.

Finally let us use calculus to find velocity and displacement, given an acceleration that is not constant but varies in time.

EXAMPLE 2–21 **Integrating a time-varying acceleration.** An experimental vehicle starts from rest $(v_0 = 0)$ at $t = 0$ and accelerates at a rate given by $a = (7.00 \text{ m/s}^3)t$. What is ($a$) its velocity and ($b$) its displacement 2.00 s later?

APPROACH We cannot use Eqs. 2–12 because a is not constant. We integrate the acceleration $a = dv/dt$ over time to find v as a function of time; and then integrate $v = dx/dt$ to get the displacement.

SOLUTION From the definition of acceleration, $a = dv/dt$, we have

$$dv = a\, dt.$$

We take the integral of both sides from $v = 0$ at $t = 0$ to velocity v at an arbitrary time t:

$$\int_{0}^{v} dv = \int_{0}^{t} a\, dt$$

$$v = \int_{0}^{t} (7.00 \text{ m/s}^3)t\, dt$$

$$= (7.00 \text{ m/s}^3)\left(\frac{t^2}{2}\right)\Big|_{0}^{t} = (7.00 \text{ m/s}^3)\left(\frac{t^2}{2} - 0\right) = (3.50 \text{ m/s}^3)t^2.$$

At $t = 2.00$ s, $v = (3.50 \text{ m/s}^3)(2.00 \text{ s})^2 = 14.0$ m/s.

(b) To get the displacement, we assume $x_0 = 0$ and start with $v = dx/dt$ which we rewrite as $dx = v\, dt$. Then we integrate from $x = 0$ at $t = 0$ to position x at time t:

$$\int_{0}^{x} dx = \int_{0}^{t} v\, dt$$

$$x = \int_{0}^{2.00 \text{ s}} (3.50 \text{ m/s}^3)t^2\, dt = (3.50 \text{ m/s}^3)\frac{t^3}{3}\Big|_{0}^{2.00 \text{ s}} = 9.33 \text{ m}.$$

In sum, at $t = 2.00$ s, $v = 14.0$ m/s and $x = 9.33$ m.

---

*2–9 Graphical Analysis and Numerical Integration

This Section is optional. It discusses how to solve certain Problems numerically, often needing a computer to do the sums. Some of this material is also covered in Chapter 7, Section 7–3.

If we are given the velocity v of an object as a function of time t, we can obtain the displacement, x. Suppose the velocity as a function of time, $v(t)$, is given as a graph (rather than as an equation that could be integrated as discussed in Section 2–8), as shown in Fig 2–34a. If we are interested in the time interval from t_1 to t_2, as shown, we divide the time axis into many small subintervals, $\Delta t_1, \Delta t_2, \Delta t_3, \ldots$, which are indicated by the dashed vertical lines. For each subinterval, a horizontal dashed line is drawn to indicate the average velocity during that time interval. The displacement during any subinterval is given by Δx_i, where the subscript i represents the particular subinterval

$(i = 1, 2, 3, \ldots)$. From the definition of average velocity (Eq. 2–2) we have

$$\Delta x_i = \bar{v}_i \, \Delta t_i.$$

Thus the displacement during each subinterval equals the product of $\bar{v}_i$ and Δt_i, and equals the area of the dark rectangle in Fig. 2–34a for that subinterval. The total displacement between times t_1 and t_2 is the sum of the displacements over all the subintervals:

$$x_2 - x_1 = \sum_{t_1}^{t_2} \bar{v}_i \, \Delta t_i, \tag{2–13a}$$

where x_1 is the position at t_1 and x_2 is the position at t_2. This sum equals the area of all the rectangles shown.

It is often difficult to estimate $\bar{v}_i$ with precision for each subinterval from the graph. We can get greater accuracy in our calculation of $x_2 - x_1$ by breaking the interval $t_2 - t_1$ into more, but narrower, subintervals. Ideally, we can let each Δt_i approach zero, so we approach (in principle) an infinite number of subintervals. In this limit the area of all these infinitesimally thin rectangles becomes exactly equal to the area under the curve (Fig. 2–34b). Thus *the total displacement between any two times is equal to the area between the velocity curve and the t axis between the two times t_1 and t_2.* This limit can be written

$$x_2 - x_1 = \lim_{\Delta t \to 0} \sum_{t_1}^{t_2} \bar{v}_i \, \Delta t_i$$

or, using standard calculus notation,

$$x_2 - x_1 = \int_{t_1}^{t_2} v(t) \, dt. \tag{2–13b}$$

We have let $\Delta t \to 0$ and renamed it dt to indicate that it is now infinitesimally small. The average velocity, $\bar{v}$, over an infinitesimal time dt is the instantaneous velocity at that instant, which we have written $v(t)$ to remind us that v is a function of t. The symbol $\int$ is an elongated S and indicates a sum over an infinite number of infinitesimal subintervals. We say that we are taking the *integral* of $v(t)$ over dt from time t_1 to time t_2, and this is equal to the area between the $v(t)$ curve and the t axis between the times t_1 and t_2 (Fig. 2–34b). The integral in Eq. 2–13b is a *definite integral*, since the limits t_1 and t_2 are specified.

Similarly, if we know the acceleration as a function of time, we can obtain the velocity by the same process. We use the definition of average acceleration (Eq. 2–5) and solve for Δv:

$$\Delta v = \bar{a} \, \Delta t.$$

If a is known as a function of t over some time interval t_1 to t_2, we can subdivide this time interval into many subintervals, Δt_i, just as we did in Fig. 2–34a. The change in velocity during each subinterval is $\Delta v_i = \bar{a}_i \, \Delta t_i$. The total change in velocity from time t_1 until time t_2 is

$$v_2 - v_1 = \sum_{t_1}^{t_2} \bar{a}_i \, \Delta t_i, \tag{2–14a}$$

where v_2 represents the velocity at t_2 and v_1 the velocity at t_1. This relation can be written as an integral by letting $\Delta t \to 0$ (the number of intervals then approaches infinity)

$$v_2 - v_1 = \lim_{\Delta t \to 0} \sum_{t_1}^{t_2} \bar{a}_i \, \Delta t_i$$

or

$$v_2 - v_1 = \int_{t_1}^{t_2} a(t) \, dt. \tag{2–14b}$$

Equations 2–14 will allow us to determine the velocity v_2 at some time t_2 if the velocity is known at t_1 and a is known as a function of time.

If the acceleration or velocity is known at discrete intervals of time, we can use the summation forms of the above equations, Eqs. 2–13a and 2–14a, to estimate velocity or displacement. This technique is known as **numerical integration**. We now take an Example that can also be evaluated analytically, so we can compare the results.

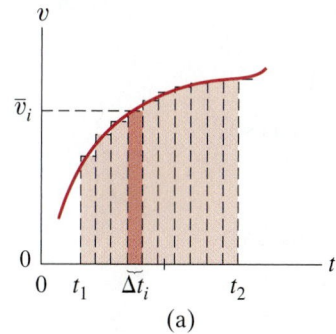

(a)

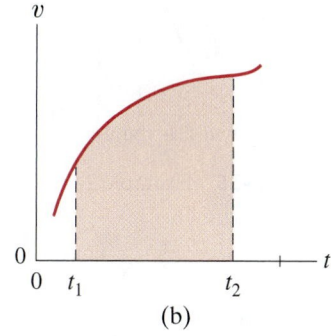

(b)

FIGURE 2–34 Graph of v vs. t for the motion of a particle. In (a), the time axis is broken into subintervals of width Δt_i, the average velocity during each Δt_i is $\bar{v}_i$, and the area of all the rectangles, $\Sigma \bar{v}_i \Delta t_i$, is numerically equal to the total displacement $(x_2 - x_1)$ during the total time $(t_2 - t_1)$. In (b), $\Delta t_i \to 0$ and the area under the curve is equal to $(x_2 - x_1)$.

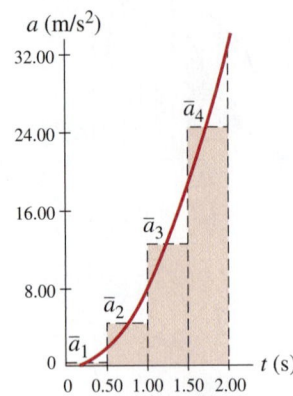

a (m/s^2)

32.00

24.00 — $\bar{a}_4$

16.00 — $\bar{a}_3$

8.00 — $\bar{a}_2$

$\bar{a}_1$

0 — t (s)

0   0.50 1.00 1.50 2.00

FIGURE 2–35   Example 2–22.

EXAMPLE 2–22   **Numerical integration.** An object starts from rest at $t = 0$ and accelerates at a rate $a(t) = (8.00 \text{ m/s}^4)t^2$. Determine its velocity after 2.00 s using numerical methods.

APPROACH   Let us first divide up the interval $t = 0.00$ s to $t = 2.00$ s into four subintervals each of duration $\Delta t_i = 0.50$ s (Fig. 2–35). We use Eq. 2–14a with $v_2 = v$, $v_1 = 0$, $t_2 = 2.00$ s, and $t_1 = 0$. For each of the subintervals we need to estimate $\bar{a}_i$. There are various ways to do this and we use the simple method of choosing $\bar{a}_i$ to be the acceleration $a(t)$ at the midpoint of each interval (an even simpler but usually less accurate procedure would be to use the value of a at the start of the subinterval). That is, we evaluate $a(t) = (8.00 \text{ m/s}^4)t^2$ at $t = 0.25$ s (which is midway between 0.00 s and 0.50 s), 0.75 s, 1.25 s, and 1.75 s.

SOLUTION   The results are as follows:

i	1	2	3	4
$\bar{a}_i (\text{m/s}^2)$	0.50	4.50	12.50	24.50

Now we use Eq. 2–14a, and note that all Δt_i equal 0.50 s (so they can be factored out):

$$v(t = 2.00 \text{ s}) = \sum_{t=0}^{t=2.00 \text{ s}} \bar{a}_i \Delta t_i$$
$$= (0.50 \text{ m/s}^2 + 4.50 \text{ m/s}^2 + 12.50 \text{ m/s}^2 + 24.50 \text{ m/s}^2)(0.50 \text{ s})$$
$$= 21.0 \text{ m/s}.$$

We can compare this result to the analytic solution given by Eq. 2–14b since the functional form for a is integrable analytically:

$$v = \int_0^{2.00 \text{ s}} (8.00 \text{ m/s}^4)t^2 \, dt = \frac{8.00 \text{ m/s}^4}{3} t^3 \Big|_0^{2.00 \text{ s}}$$
$$= \frac{8.00 \text{ m/s}^4}{3} \left[(2.00 \text{ s})^3 - (0)^3 \right] = 21.33 \text{ m/s}$$

or 21.3 m/s to the proper number of significant figures. This analytic solution is precise, and we see that our numerical estimate is not far off even though we only used four Δt intervals. It may not be close enough for purposes requiring high accuracy. If we use more and smaller subintervals, we will get a more accurate result. If we use 10 subintervals, each with $\Delta t = 2.00 \text{ s}/10 = 0.20$ s, we have to evaluate $a(t)$ at $t = 0.10$ s, 0.30 s, ..., 1.90 s to get the $\bar{a}_i$, and these are as follows:

i	1	2	3	4	5	6	7	8	9	10
$\bar{a}_i (\text{m/s}^2)$	0.08	0.72	2.00	3.92	6.48	9.68	13.52	18.00	23.12	28.88

Then, from Eq. 2–14a we obtain

$$v(t = 2.00 \text{ s}) = \sum \bar{a}_i \Delta t_i = \left(\sum \bar{a}_i \right)(0.200 \text{ s})$$
$$= (106.4 \text{ m/s}^2)(0.200 \text{ s}) = 21.28 \text{ m/s},$$

where we have kept an extra significant figure to show that this result is much closer to the (precise) analytic one but still is not quite identical to it. The percentage difference has dropped from 1.4% $(0.3 \text{ m/s}^2/21.3 \text{ m/s}^2)$ for the four-subinterval computation to only 0.2% $(0.05/21.3)$ for the 10-subinterval one.

In the Example above we were given an analytic function that was integrable, so we could compare the accuracy of the numerical calculation to the known precise one. But what do we do if the function is not integrable, so we can't compare our numerical result to an analytic one? That is, how do we know if we've taken enough subintervals so that we can trust our calculated estimate to be accurate to within some desired uncertainty, say 1 percent? What we can do is compare two successive numerical calculations: the first done with n subintervals and the second with, say, twice as many subintervals ($2n$). If the two results are within the desired uncertainty (say 1 percent), we can usually assume that the calculation with more subintervals is within the desired uncertainty of the true value. If the two calculations are not that close, then a third calculation, with more subintervals (maybe double, maybe 10 times as many, depending on how good the previous approximation was) must be done, and compared to the previous one. The procedure is easy to automate using a computer spreadsheet application.

If we wanted to also obtain the displacement x at some time, we would have to do a second numerical integration over v, which means we would first need to calculate v for many different times. Programmable calculators and computers are very helpful for doing the long sums.

Problems that use these numerical techniques are found at the end of many Chapters of this book; they are labeled Numerical/Computer and are given an asterisk to indicate that they are optional.

Summary

[The Summary that appears at the end of each Chapter in this book gives a brief overview of the main ideas of the Chapter. The Summary *cannot* serve to give an understanding of the material, which can be accomplished only by a detailed reading of the Chapter.]

Kinematics deals with the description of how objects move. The description of the motion of any object must always be given relative to some particular **reference frame**.

The **displacement** of an object is the change in position of the object.

Average speed is the distance traveled divided by the elapsed time or time interval, Δt, the time period over which we choose to make our observations. An object's **average velocity** over a particular time interval Δt is its displacement Δx during that time interval, divided by Δt:

$$\bar{v} = \frac{\Delta x}{\Delta t}. \qquad (2\text{–}2)$$

The **instantaneous velocity**, whose magnitude is the same as the *instantaneous speed*, is defined as the average velocity taken over an infinitesimally short time interval ($\Delta t \to 0$):

$$v = \lim_{\Delta t \to 0} \frac{\Delta x}{\Delta t} = \frac{dx}{dt}, \qquad (2\text{–}4)$$

where dx/dt is the derivative of x with respect to t.

On a graph of position vs. time, the *slope* is equal to the instantaneous velocity.

Acceleration is the change of velocity per unit time. An object's **average acceleration** over a time interval Δt is

$$\bar{a} = \frac{\Delta v}{\Delta t}, \qquad (2\text{–}5)$$

where Δv is the change of velocity during the time interval Δt.

Instantaneous acceleration is the average acceleration taken over an infinitesimally short time interval:

$$a = \lim_{\Delta t \to 0} \frac{\Delta v}{\Delta t} = \frac{dv}{dt}. \qquad (2\text{–}6)$$

If an object moves in a straight line with *constant acceleration*, the velocity v and position x are related to the acceleration a, the elapsed time t, the initial position x_0, and the initial velocity v_0 by Eqs. 2–12:

$$v = v_0 + at, \qquad x = x_0 + v_0 t + \tfrac{1}{2}at^2,$$
$$v^2 = v_0^2 + 2a(x - x_0), \qquad \bar{v} = \frac{v + v_0}{2}. \qquad (2\text{–}12)$$

Objects that move vertically near the surface of the Earth, either falling or having been projected vertically up or down, move with the constant downward **acceleration due to gravity**, whose magnitude is $g = 9.80 \text{ m/s}^2$ if air resistance can be ignored.

[*The kinematic Equations 2–12 can be derived using integral calculus.]

Questions

1. Does a car speedometer measure speed, velocity, or both?
2. Can an object have a varying speed if its velocity is constant? Can it have varying velocity if its speed is constant? If yes, give examples in each case.
3. When an object moves with constant velocity, does its average velocity during any time interval differ from its instantaneous velocity at any instant?
4. If one object has a greater speed than a second object, does the first necessarily have a greater acceleration? Explain, using examples.
5. Compare the acceleration of a motorcycle that accelerates from 80 km/h to 90 km/h with the acceleration of a bicycle that accelerates from rest to 10 km/h in the same time.
6. Can an object have a northward velocity and a southward acceleration? Explain.
7. Can the velocity of an object be negative when its acceleration is positive? What about vice versa?
8. Give an example where both the velocity and acceleration are negative.
9. Two cars emerge side by side from a tunnel. Car A is traveling with a speed of 60 km/h and has an acceleration of 40 km/h/min. Car B has a speed of 40 km/h and has an acceleration of 60 km/h/min. Which car is passing the other as they come out of the tunnel? Explain your reasoning.
10. Can an object be increasing in speed as its acceleration decreases? If so, give an example. If not, explain.
11. A baseball player hits a ball straight up into the air. It leaves the bat with a speed of 120 km/h. In the absence of air resistance, how fast would the ball be traveling when the catcher catches it?
12. As a freely falling object speeds up, what is happening to its acceleration—does it increase, decrease, or stay the same? (*a*) Ignore air resistance. (*b*) Consider air resistance.
13. You travel from point A to point B in a car moving at a constant speed of 70 km/h. Then you travel the same distance from point B to another point C, moving at a constant speed of 90 km/h. Is your average speed for the entire trip from A to C 80 km/h? Explain why or why not.
14. Can an object have zero velocity and nonzero acceleration at the same time? Give examples.
15. Can an object have zero acceleration and nonzero velocity at the same time? Give examples.
16. Which of these motions is *not* at constant acceleration: a rock falling from a cliff, an elevator moving from the second floor to the fifth floor making stops along the way, a dish resting on a table?
17. In a lecture demonstration, a 3.0-m-long vertical string with ten bolts tied to it at equal intervals is dropped from the ceiling of the lecture hall. The string falls on a tin plate, and the class hears the clink of each bolt as it hits the plate. The sounds will not occur at equal time intervals. Why? Will the time between clinks increase or decrease near the end of the fall? How could the bolts be tied so that the clinks occur at equal intervals?

18. Describe in words the motion plotted in Fig. 2–36 in terms of v, a, etc. [*Hint:* First try to duplicate the motion plotted by walking or moving your hand.]

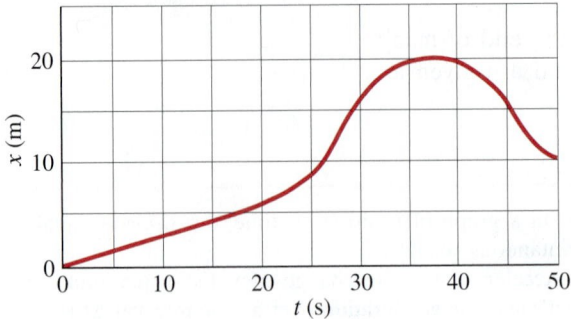

FIGURE 2–36 Question 18, Problems 9 and 86.

19. Describe in words the motion of the object graphed in Fig. 2–37.

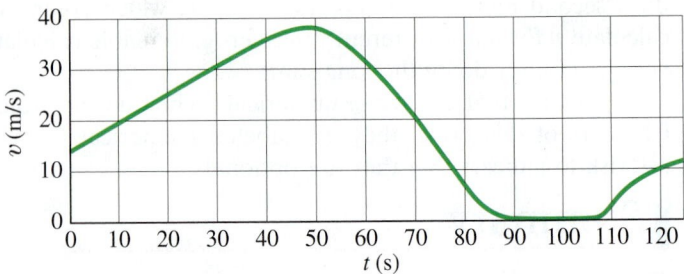

FIGURE 2–37 Question 19, Problem 23.

Problems

[The Problems at the end of each Chapter are ranked I, II, or III according to estimated difficulty, with (I) Problems being easiest. Level III are meant as challenges for the best students. The Problems are arranged by Section, meaning that the reader should have read up to and including that Section, but not only that Section—Problems often depend on earlier material. Finally, there is a set of unranked "General Problems" not arranged by Section number.]

2–1 to 2–3 Speed and Velocity

1. (I) If you are driving 110 km/h along a straight road and you look to the side for 2.0 s, how far do you travel during this inattentive period?

2. (I) What must your car's average speed be in order to travel 235 km in 3.25 h?

3. (I) A particle at $t_1 = -2.0$ s is at $x_1 = 4.3$ cm and at $t_2 = 4.5$ s is at $x_2 = 8.5$ cm. What is its average velocity? Can you calculate its average speed from these data?

4. (I) A rolling ball moves from $x_1 = 3.4$ cm to $x_2 = -4.2$ cm during the time from $t_1 = 3.0$ s to $t_2 = 5.1$ s. What is its average velocity?

5. (II) According to a rule-of-thumb, every five seconds between a lightning flash and the following thunder gives the distance to the flash in miles. Assuming that the flash of light arrives in essentially no time at all, estimate the speed of sound in m/s from this rule. What would be the rule for kilometers?

6. (II) You are driving home from school steadily at 95 km/h for 130 km. It then begins to rain and you slow to 65 km/h. You arrive home after driving 3 hours and 20 minutes. (*a*) How far is your hometown from school? (*b*) What was your average speed?

7. (II) A horse canters away from its trainer in a straight line, moving 116 m away in 14.0 s. It then turns abruptly and gallops halfway back in 4.8 s. Calculate (*a*) its average speed and (*b*) its average velocity for the entire trip, using "away from the trainer" as the positive direction.

8. (II) The position of a small object is given by $x = 34 + 10t - 2t^3$, where t is in seconds and x in meters. (*a*) Plot x as a function of t from $t = 0$ to $t = 3.0$ s. (*b*) Find the average velocity of the object between 0 and 3.0 s. (*c*) At what time between 0 and 3.0 s is the instantaneous velocity zero?

9. (II) The position of a rabbit along a straight tunnel as a function of time is plotted in Fig. 2–36. What is its instantaneous velocity (*a*) at $t = 10.0$ s and (*b*) at $t = 30.0$ s? What is its average velocity (*c*) between $t = 0$ and $t = 5.0$ s, (*d*) between $t = 25.0$ s and $t = 30.0$ s, and (*e*) between $t = 40.0$ s and $t = 50.0$ s?

10. (II) On an audio compact disc (CD), digital bits of information are encoded sequentially along a spiral path. Each bit occupies about 0.28 μm. A CD player's readout laser scans along the spiral's sequence of bits at a constant speed of about 1.2 m/s as the CD spins. (*a*) Determine the number N of digital bits that a CD player reads every second. (*b*) The audio information is sent to each of the two loudspeakers 44,100 times per second. Each of these samplings requires 16 bits and so one would (at first glance) think the required bit rate for a CD player is

$$N_0 = 2\left(44{,}100 \frac{\text{samplings}}{\text{second}}\right)\left(16 \frac{\text{bits}}{\text{sampling}}\right) = 1.4 \times 10^6 \frac{\text{bits}}{\text{second}},$$

where the 2 is for the 2 loudspeakers (the 2 stereo channels). Note that N_0 is less than the number N of bits actually read per second by a CD player. The excess number of bits $(= N - N_0)$ is needed for encoding and error-correction. What percentage of the bits on a CD are dedicated to encoding and error-correction?

11. (II) A car traveling 95 km/h is 110 m behind a truck traveling 75 km/h. How long will it take the car to reach the truck?

12. (II) Two locomotives approach each other on parallel tracks. Each has a speed of 95 km/h with respect to the ground. If they are initially 8.5 km apart, how long will it be before they reach each other? (See Fig. 2–38).

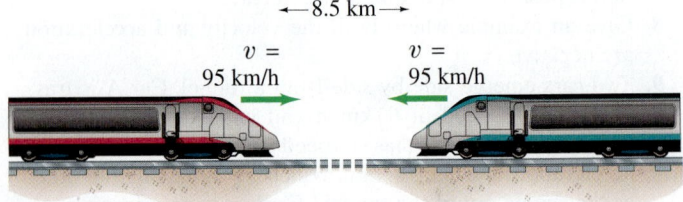

FIGURE 2–38 Problem 12.

13. (II) Digital bits on a 12.0-cm diameter audio CD are encoded along an outward spiraling path that starts at radius $R_1 = 2.5$ cm and finishes at radius $R_2 = 5.8$ cm. The distance between the centers of neighboring spiral-windings is $1.6 \, \mu\text{m} \, (= 1.6 \times 10^{-6} \, \text{m})$. (a) Determine the total length of the spiraling path. [*Hint*: Imagine "unwinding" the spiral into a straight path of width $1.6 \, \mu\text{m}$, and note that the original spiral and the straight path both occupy the same area.] (b) To read information, a CD player adjusts the rotation of the CD so that the player's readout laser moves along the spiral path at a constant speed of 1.25 m/s. Estimate the maximum playing time of such a CD.

14. (II) An airplane travels 3100 km at a speed of 720 km/h, and then encounters a tailwind that boosts its speed to 990 km/h for the next 2800 km. What was the total time for the trip? What was the average speed of the plane for this trip? [*Hint*: Does Eq. 2–12d apply, or not?]

15. (II) Calculate the average speed and average velocity of a complete round trip in which the outgoing 250 km is covered at 95 km/h, followed by a 1.0-h lunch break, and the return 250 km is covered at 55 km/h.

16. (II) The position of a ball rolling in a straight line is given by $x = 2.0 - 3.6 \, t + 1.1 \, t^2$, where x is in meters and t in seconds. (a) Determine the position of the ball at $t = 1.0$ s, 2.0 s, and 3.0 s. (b) What is the average velocity over the interval $t = 1.0$ s to $t = 3.0$ s? (c) What is its instantaneous velocity at $t = 2.0$ s and at $t = 3.0$ s?

17. (II) A dog runs 120 m away from its master in a straight line in 8.4 s, and then runs halfway back in one-third the time. Calculate (a) its average speed and (b) its average velocity.

18. (III) An automobile traveling 95 km/h overtakes a 1.10-km-long train traveling in the same direction on a track parallel to the road. If the train's speed is 75 km/h, how long does it take the car to pass it, and how far will the car have traveled in this time? See Fig. 2–39. What are the results if the car and train are traveling in opposite directions?

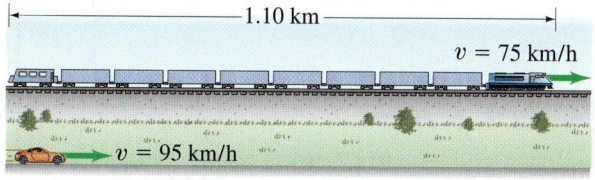

FIGURE 2–39 Problem 18.

19. (III) A bowling ball traveling with constant speed hits the pins at the end of a bowling lane 16.5 m long. The bowler hears the sound of the ball hitting the pins 2.50 s after the ball is released from his hands. What is the speed of the ball, assuming the speed of sound is 340 m/s?

2–4 Acceleration

20. (I) A sports car accelerates from rest to 95 km/h in 4.5 s. What is its average acceleration in m/s²?

21. (I) At highway speeds, a particular automobile is capable of an acceleration of about 1.8 m/s². At this rate, how long does it take to accelerate from 80 km/h to 110 km/h?

22. (I) A sprinter accelerates from rest to 9.00 m/s in 1.28 s. What is her acceleration in (a) m/s²; (b) km/h²?

23. (I) Figure 2–37 shows the velocity of a train as a function of time. (a) At what time was its velocity greatest? (b) During what periods, if any, was the velocity constant? (c) During what periods, if any, was the acceleration constant? (d) When was the magnitude of the acceleration greatest?

24. (II) A sports car moving at constant speed travels 110 m in 5.0 s. If it then brakes and comes to a stop in 4.0 s, what is the magnitude of its acceleration in m/s², and in g's $(g = 9.80 \, \text{m/s}^2)$?

25. (II) A car moving in a straight line starts at $x = 0$ at $t = 0$. It passes the point $x = 25.0$ m with a speed of 11.0 m/s at $t = 3.00$ s. It passes the point $x = 385$ m with a speed of 45.0 m/s at $t = 20.0$ s. Find (a) the average velocity and (b) the average acceleration between $t = 3.00$ s and $t = 20.0$ s.

26. (II) A particular automobile can accelerate approximately as shown in the velocity vs. time graph of Fig. 2–40. (The short flat spots in the curve represent shifting of the gears.) Estimate the average acceleration of the car in (a) second gear; and (b) fourth gear. (c) What is its average acceleration through the first four gears?

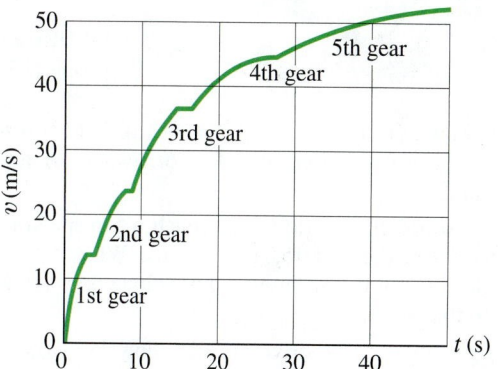

FIGURE 2–40 Problem 26. The velocity of a high-performance automobile as a function of time, starting from a dead stop. The flat spots in the curve represent gear shifts.

27. (II) A particle moves along the x axis. Its position as a function of time is given by $x = 6.8 \, t + 8.5 \, t^2$, where t is in seconds and x is in meters. What is the acceleration as a function of time?

28. (II) The position of a racing car, which starts from rest at $t = 0$ and moves in a straight line, is given as a function of time in the following Table. Estimate (a) its velocity and (b) its acceleration as a function of time. Display each in a Table and on a graph.

t(s)	0	0.25	0.50	0.75	1.00	1.50	2.00	2.50
x(m)	0	0.11	0.46	1.06	1.94	4.62	8.55	13.79

t(s)	3.00	3.50	4.00	4.50	5.00	5.50	6.00
x(m)	20.36	28.31	37.65	48.37	60.30	73.26	87.16

29. (II) The position of an object is given by $x = At + Bt^2$, where x is in meters and t is in seconds. (a) What are the units of A and B? (b) What is the acceleration as a function of time? (c) What are the velocity and acceleration at $t = 5.0$ s? (d) What is the velocity as a function of time if $x = At + Bt^{-3}$?

2–5 and 2–6 Motion at Constant Acceleration

30. (I) A car slows down from 25 m/s to rest in a distance of 85 m. What was its acceleration, assumed constant?

31. (I) A car accelerates from 12 m/s to 21 m/s in 6.0 s. What was its acceleration? How far did it travel in this time? Assume constant acceleration.

32. (I) A light plane must reach a speed of 32 m/s for takeoff. How long a runway is needed if the (constant) acceleration is 3.0 m/s^2?

33. (II) A baseball pitcher throws a baseball with a speed of 41 m/s. Estimate the average acceleration of the ball during the throwing motion. In throwing the baseball, the pitcher accelerates the ball through a displacement of about 3.5 m, from behind the body to the point where it is released (Fig. 2–41).

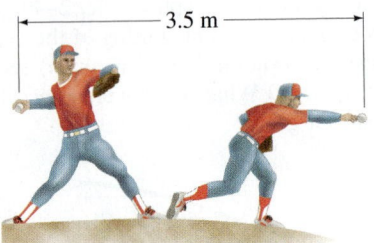

FIGURE 2–41
Problem 33.

34. (II) Show that $\bar{v} = (v + v_0)/2$ (see Eq. 2–12d) is not valid when the acceleration $a = A + Bt$, where A and B are constants.

35. (II) A world-class sprinter can reach a top speed (of about 11.5 m/s) in the first 15.0 m of a race. What is the average acceleration of this sprinter and how long does it take her to reach that speed?

36. (II) An inattentive driver is traveling 18.0 m/s when he notices a red light ahead. His car is capable of decelerating at a rate of 3.65 m/s^2. If it takes him 0.200 s to get the brakes on and he is 20.0 m from the intersection when he sees the light, will he be able to stop in time?

37. (II) A car slows down uniformly from a speed of 18.0 m/s to rest in 5.00 s. How far did it travel in that time?

38. (II) In coming to a stop, a car leaves skid marks 85 m long on the highway. Assuming a deceleration of 4.00 m/s^2, estimate the speed of the car just before braking.

39. (II) A car traveling 85 km/h slows down at a constant 0.50 m/s^2 just by "letting up on the gas." Calculate (a) the distance the car coasts before it stops, (b) the time it takes to stop, and (c) the distance it travels during the first and fifth seconds.

40. (II) A car traveling at 105 km/h strikes a tree. The front end of the car compresses and the driver comes to rest after traveling 0.80 m. What was the magnitude of the average acceleration of the driver during the collision? Express the answer in terms of "g's," where $1.00 \text{ g} = 9.80 \text{ m/s}^2$.

41. (II) Determine the stopping distances for an automobile with an initial speed of 95 km/h and human reaction time of 1.0 s: (a) for an acceleration $a = -5.0 \text{ m/s}^2$; (b) for $a = -7.0 \text{ m/s}^2$.

42. (II) A space vehicle accelerates uniformly from 65 m/s at $t = 0$ to 162 m/s at $t = 10.0$ s. How far did it move between $t = 2.0$ s and $t = 6.0$ s?

43. (II) A 75-m-long train begins uniform acceleration from rest. The front of the train has a speed of 23 m/s when it passes a railway worker who is standing 180 m from where the front of the train started. What will be the speed of the last car as it passes the worker? (See Fig. 2–42.)

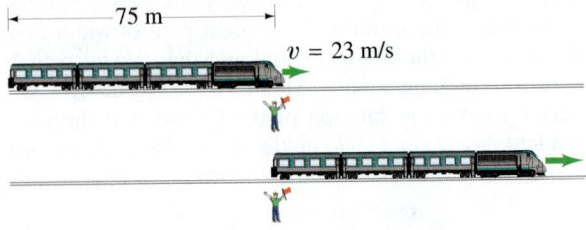

FIGURE 2–42 Problem 43.

44. (II) An unmarked police car traveling a constant 95 km/h is passed by a speeder traveling 135 km/h. Precisely 1.00 s after the speeder passes, the police officer steps on the accelerator; if the police car's acceleration is 2.00 m/s^2, how much time passes before the police car overtakes the speeder (assumed moving at constant speed)?

45. (III) Assume in Problem 44 that the speeder's speed is not known. If the police car accelerates uniformly as given above and overtakes the speeder after accelerating for 7.00 s, what was the speeder's speed?

46. (III) A runner hopes to complete the 10,000-m run in less than 30.0 min. After running at constant speed for exactly 27.0 min, there are still 1100 m to go. The runner must then accelerate at 0.20 m/s^2 for how many seconds in order to achieve the desired time?

47. (III) Mary and Sally are in a foot race (Fig. 2–43). When Mary is 22 m from the finish line, she has a speed of 4.0 m/s and is 5.0 m behind Sally, who has a speed of 5.0 m/s. Sally thinks she has an easy win and so, during the remaining portion of the race, decelerates at a constant rate of 0.50 m/s^2 to the finish line. What constant acceleration does Mary now need during the remaining portion of the race, if she wishes to cross the finish line side-by-side with Sally?

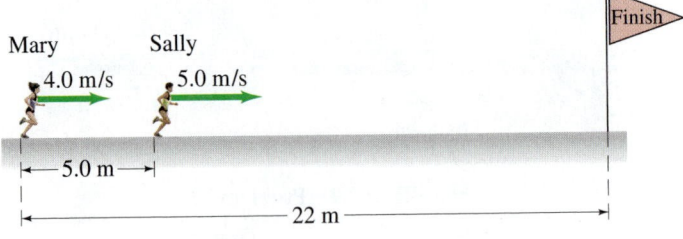

FIGURE 2–43 Problem 47.

2–7 Freely Falling Objects

[Neglect air resistance.]

48. (I) A stone is dropped from the top of a cliff. It is seen to hit the ground below after 3.75 s. How high is the cliff?

49. (I) If a car rolls gently ($v_0 = 0$) off a vertical cliff, how long does it take it to reach 55 km/h?

50. (I) Estimate (a) how long it took King Kong to fall straight down from the top of the Empire State Building (380 m high), and (b) his velocity just before "landing."

51. (II) A baseball is hit almost straight up into the air with a speed of about 20 m/s. (a) How high does it go? (b) How long is it in the air?

52. (II) A ball player catches a ball 3.2 s after throwing it vertically upward. With what speed did he throw it, and what height did it reach?

53. (II) A kangaroo jumps to a vertical height of 1.65 m. How long was it in the air before returning to Earth?

54. (II) The best rebounders in basketball have a vertical leap (that is, the vertical movement of a fixed point on their body) of about 120 cm. (a) What is their initial "launch" speed off the ground? (b) How long are they in the air?

55. (II) A helicopter is ascending vertically with a speed of 5.10 m/s. At a height of 105 m above the Earth, a package is dropped from a window. How much time does it take for the package to reach the ground? [*Hint*: v_0 for the package equals the speed of the helicopter.]

56. (II) For an object falling freely from rest, show that the distance traveled during each successive second increases in the ratio of successive odd integers (1, 3, 5, etc.). (This was first shown by Galileo.) See Figs. 2–26 and 2–29.

57. (II) A baseball is seen to pass upward by a window 23 m above the street with a vertical speed of 14 m/s. If the ball was thrown from the street, (a) what was its initial speed, (b) what altitude does it reach, (c) when was it thrown, and (d) when does it reach the street again?

58. (II) A rocket rises vertically, from rest, with an acceleration of 3.2 m/s² until it runs out of fuel at an altitude of 950 m. After this point, its acceleration is that of gravity, downward. (a) What is the velocity of the rocket when it runs out of fuel? (b) How long does it take to reach this point? (c) What maximum altitude does the rocket reach? (d) How much time (total) does it take to reach maximum altitude? (e) With what velocity does it strike the Earth? (f) How long (total) is it in the air?

59. (II) Roger sees water balloons fall past his window. He notices that each balloon strikes the sidewalk 0.83 s after passing his window. Roger's room is on the third floor, 15 m above the sidewalk. (a) How fast are the balloons traveling when they pass Roger's window? (b) Assuming the balloons are being released from rest, from what floor are they being released? Each floor of the dorm is 5.0 m high.

60. (II) A stone is thrown vertically upward with a speed of 24.0 m/s. (a) How fast is it moving when it reaches a height of 13.0 m? (b) How much time is required to reach this height? (c) Why are there two answers to (b)?

61. (II) A falling stone takes 0.33 s to travel past a window 2.2 m tall (Fig. 2–44). From what height above the top of the window did the stone fall?

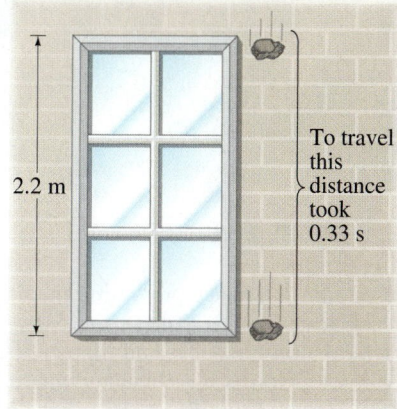

FIGURE 2–44   Problem 61.

2.2 m

To travel this distance took 0.33 s

62. (II) Suppose you adjust your garden hose nozzle for a hard stream of water. You point the nozzle vertically upward at a height of 1.5 m above the ground (Fig. 2–45). When you quickly turn off the nozzle, you hear the water striking the ground next to you for another 2.0 s. What is the water speed as it leaves the nozzle?

1.5 m

FIGURE 2–45
Problem 62.

63. (III) A toy rocket moving vertically upward passes by a 2.0-m-high window whose sill is 8.0 m above the ground. The rocket takes 0.15 s to travel the 2.0 m height of the window. What was the launch speed of the rocket, and how high will it go? Assume the propellant is burned very quickly at blastoff.

64. (III) A ball is dropped from the top of a 50.0-m-high cliff. At the same time, a carefully aimed stone is thrown straight up from the bottom of the cliff with a speed of 24.0 m/s. The stone and ball collide part way up. How far above the base of the cliff does this happen?

65. (III) A rock is dropped from a sea cliff and the sound of it striking the ocean is heard 3.4 s later. If the speed of sound is 340 m/s, how high is the cliff?

66. (III) A rock is thrown vertically upward with a speed of 12.0 m/s. Exactly 1.00 s later, a ball is thrown up vertically along the same path with a speed of 18.0 m/s. (a) At what time will they strike each other? (b) At what height will the collision occur? (c) Answer (a) and (b) assuming that the order is reversed: the ball is thrown 1.00 s before the rock.

*2–8 Variable Acceleration; Calculus

*67. (II) Given $v(t) = 25 + 18t$, where v is in m/s and t is in s, use calculus to determine the total displacement from $t_1 = 1.5$ s to $t_2 = 3.1$ s.

*68. (III) The acceleration of a particle is given by $a = A\sqrt{t}$ where $A = 2.0$ m/s$^{5/2}$. At $t = 0$, $v = 7.5$ m/s and $x = 0$. (a) What is the speed as a function of time? (b) What is the displacement as a function of time? (c) What are the acceleration, speed and displacement at $t = 5.0$ s?

*69. (III) Air resistance acting on a falling body can be taken into account by the approximate relation for the acceleration:

$$a = \frac{dv}{dt} = g - kv,$$

where k is a constant. (a) Derive a formula for the velocity of the body as a function of time assuming it starts from rest ($v = 0$ at $t = 0$). [*Hint*: Change variables by setting $u = g - kv$.] (b) Determine an expression for the terminal velocity, which is the maximum value the velocity reaches.

*2–9 Graphical Analysis and Numerical Integration

[See Problems 95–97 at the end of this Chapter.]

General Problems

70. A fugitive tries to hop on a freight train traveling at a constant speed of 5.0 m/s. Just as an empty box car passes him, the fugitive starts from rest and accelerates at $a = 1.2$ m/s^2 to his maximum speed of 6.0 m/s. (*a*) How long does it take him to catch up to the empty box car? (*b*) What is the distance traveled to reach the box car?

71. The acceleration due to gravity on the Moon is about one-sixth what it is on Earth. If an object is thrown vertically upward on the Moon, how many times higher will it go than it would on Earth, assuming the same initial velocity?

72. A person jumps from a fourth-story window 15.0 m above a firefighter's safety net. The survivor stretches the net 1.0 m before coming to rest, Fig. 2–46. (*a*) What was the average deceleration experienced by the survivor when she was slowed to rest by the net? (*b*) What would you do to make it "safer" (that is, to generate a smaller deceleration): would you stiffen or loosen the net? Explain.

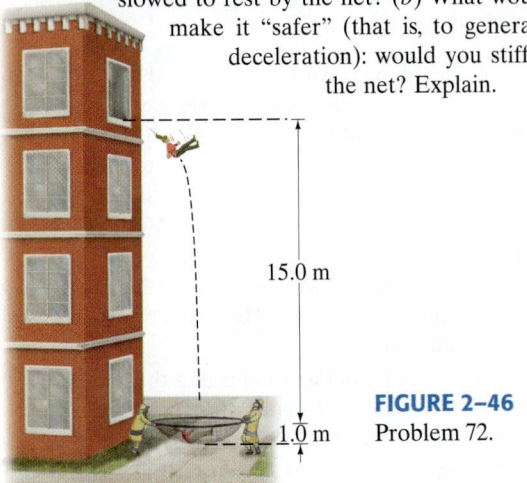

15.0 m

1.0 m

FIGURE 2–46 Problem 72.

73. A person who is properly restrained by an over-the-shoulder seat belt has a good chance of surviving a car collision if the deceleration does not exceed 30 "*g*'s" $(1.00\ g = 9.80$ m/s$^2)$. Assuming uniform deceleration of this value, calculate the distance over which the front end of the car must be designed to collapse if a crash brings the car to rest from 100 km/h.

74. Pelicans tuck their wings and free-fall straight down when diving for fish. Suppose a pelican starts its dive from a height of 16.0 m and cannot change its path once committed. If it takes a fish 0.20 s to perform evasive action, at what minimum height must it spot the pelican to escape? Assume the fish is at the surface of the water.

75. Suppose a car manufacturer tested its cars for front-end collisions by hauling them up on a crane and dropping them from a certain height. (*a*) Show that the speed just before a car hits the ground, after falling from rest a vertical distance H, is given by $\sqrt{2gH}$. What height corresponds to a collision at (*b*) 50 km/h? (*c*) 100 km/h?

76. A stone is dropped from the roof of a high building. A second stone is dropped 1.50 s later. How far apart are the stones when the second one has reached a speed of 12.0 m/s?

77. A bicyclist in the Tour de France crests a mountain pass as he moves at 15 km/h. At the bottom, 4.0 km farther, his speed is 75 km/h. What was his average acceleration (in m/s^2) while riding down the mountain?

78. Consider the street pattern shown in Fig. 2–47. Each intersection has a traffic signal, and the speed limit is 50 km/h. Suppose you are driving from the west at the speed limit. When you are 10.0 m from the first intersection, all the lights turn green. The lights are green for 13.0 s each. (*a*) Calculate the time needed to reach the third stoplight. Can you make it through all three lights without stopping? (*b*) Another car was stopped at the first light when all the lights turned green. It can accelerate at the rate of 2.00 m/s^2 to the speed limit. Can the second car make it through all three lights without stopping? By how many seconds would it make it or not?

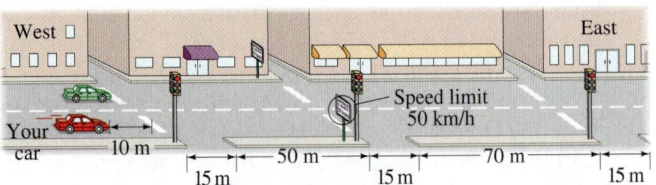

West / East / Speed limit 50 km/h / Your car / 10 m / 50 m / 70 m / 15 m / 15 m / 15 m

FIGURE 2–47 Problem 78.

79. In putting, the force with which a golfer strikes a ball is planned so that the ball will stop within some small distance of the cup, say 1.0 m long or short, in case the putt is missed. Accomplishing this from an uphill lie (that is, putting the ball downhill, see Fig. 2–48) is more difficult than from a downhill lie. To see why, assume that on a particular green the ball decelerates constantly at 1.8 m/s^2 going downhill, and constantly at 2.8 m/s^2 going uphill. Suppose we have an uphill lie 7.0 m from the cup. Calculate the allowable range of initial velocities we may impart to the ball so that it stops in the range 1.0 m short to 1.0 m long of the cup. Do the same for a downhill lie 7.0 m from the cup. What in your results suggests that the downhill putt is more difficult?

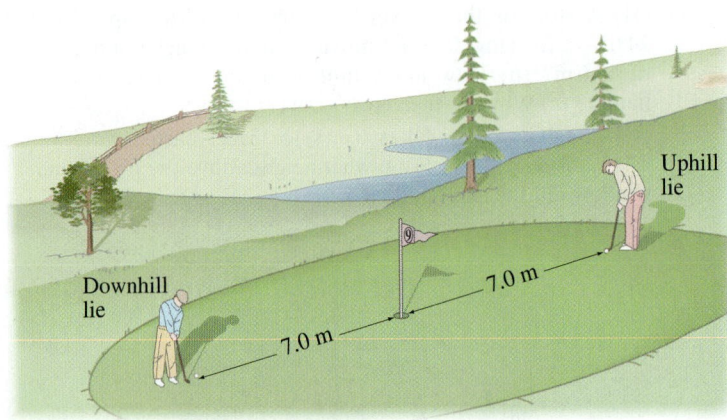

Uphill lie

Downhill lie

7.0 m

7.0 m

FIGURE 2–48 Problem 79.

80. A robot used in a pharmacy picks up a medicine bottle at $t = 0$. It accelerates at 0.20 m/s^2 for 5.0 s, then travels without acceleration for 68 s and finally decelerates at -0.40 m/s^2 for 2.5 s to reach the counter where the pharmacist will take the medicine from the robot. From how far away did the robot fetch the medicine?

81. A stone is thrown vertically upward with a speed of 12.5 m/s from the edge of a cliff 75.0 m high (Fig. 2–49). (*a*) How much later does it reach the bottom of the cliff? (*b*) What is its speed just before hitting? (*c*) What total distance did it travel?

$y = -75$ m **FIGURE 2–49**
Problem 81.

82. Figure 2–50 is a position versus time graph for the motion of an object along the *x* axis. Consider the time interval from A to B. (*a*) Is the object moving in the positive or negative direction? (*b*) Is the object speeding up or slowing down? (*c*) Is the acceleration of the object positive or negative? Next, consider the time interval from D to E. (*d*) Is the object moving in the positive or negative direction? (*e*) Is the object speeding up or slowing down? (*f*) Is the acceleration of the object positive or negative? (*g*) Finally, answer these same three questions for the time interval from C to D.

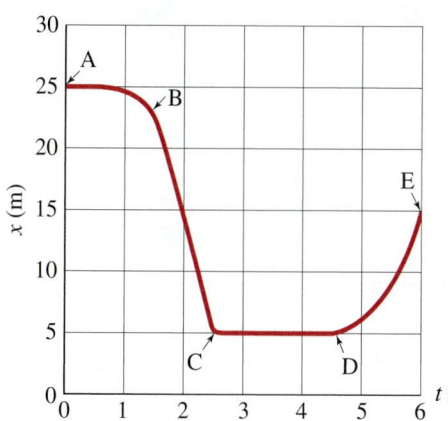

FIGURE 2–50
Problem 82.

83. In the design of a *rapid transit system*, it is necessary to balance the average speed of a train against the distance between stops. The more stops there are, the slower the train's average speed. To get an idea of this problem, calculate the time it takes a train to make a 9.0-km trip in two situations: (*a*) the stations at which the trains must stop are 1.8 km apart (a total of 6 stations, including those at the ends); and (*b*) the stations are 3.0 km apart (4 stations total). Assume that at each station the train accelerates at a rate of 1.1 m/s² until it reaches 95 km/h, then stays at this speed until its brakes are applied for arrival at the next station, at which time it decelerates at −2.0 m/s². Assume it stops at each intermediate station for 22 s.

84. A person jumps off a diving board 4.0 m above the water's surface into a deep pool. The person's downward motion stops 2.0 m below the surface of the water. Estimate the average deceleration of the person while under the water.

85. Bill can throw a ball vertically at a speed 1.5 times faster than Joe can. How many times higher will Bill's ball go than Joe's?

86. Sketch the *v* vs. *t* graph for the object whose displacement as a function of time is given by Fig. 2–36.

87. A person driving her car at 45 km/h approaches an intersection just as the traffic light turns yellow. She knows that the yellow light lasts only 2.0 s before turning to red, and she is 28 m away from the near side of the intersection (Fig. 2–51). Should she try to stop, or should she speed up to cross the intersection before the light turns red? The intersection is 15 m wide. Her car's maximum deceleration is −5.8 m/s², whereas it can accelerate from 45 km/h to 65 km/h in 6.0 s. Ignore the length of her car and her reaction time.

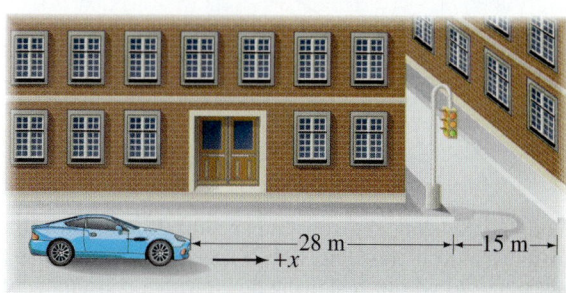

FIGURE 2–51 Problem 87.

88. A car is behind a truck going 25 m/s on the highway. The driver looks for an opportunity to pass, guessing that his car can accelerate at 1.0 m/s², and he gauges that he has to cover the 20-m length of the truck, plus 10-m clear room at the rear of the truck and 10 m more at the front of it. In the oncoming lane, he sees a car approaching, probably also traveling at 25 m/s. He estimates that the car is about 400 m away. Should he attempt the pass? Give details.

89. Agent Bond is standing on a bridge, 13 m above the road below, and his pursuers are getting too close for comfort. He spots a flatbed truck approaching at 25 m/s, which he measures by knowing that the telephone poles the truck is passing are 25 m apart in this country. The bed of the truck is 1.5 m above the road, and Bond quickly calculates how many poles away the truck should be when he jumps down from the bridge onto the truck, making his getaway. How many poles is it?

90. A police car at rest, passed by a speeder traveling at a constant 130 km/h, takes off in hot pursuit. The police officer catches up to the speeder in 750 m, maintaining a constant acceleration. (*a*) Qualitatively plot the position vs. time graph for both cars from the police car's start to the catch-up point. Calculate (*b*) how long it took the police officer to overtake the speeder, (*c*) the required police car acceleration, and (*d*) the speed of the police car at the overtaking point.

91. A fast-food restaurant uses a conveyor belt to send the burgers through a grilling machine. If the grilling machine is 1.1 m long and the burgers require 2.5 min to cook, how fast must the conveyor belt travel? If the burgers are spaced 15 cm apart, what is the rate of burger production (in burgers/min)?

92. Two students are asked to find the height of a particular building using a barometer. Instead of using the barometer as an altitude-measuring device, they take it to the roof of the building and drop it off, timing its fall. One student reports a fall time of 2.0 s, and the other, 2.3 s. What % difference does the 0.3 s make for the estimates of the building's height?

General Problems   **49**

93. Figure 2–52 shows the position vs. time graph for two bicycles, A and B. (*a*) Is there any instant at which the two bicycles have the same velocity? (*b*) Which bicycle has the larger acceleration? (*c*) At which instant(s) are the bicycles passing each other? Which bicycle is passing the other? (*d*) Which bicycle has the highest instantaneous velocity? (*e*) Which bicycle has the higher average velocity?

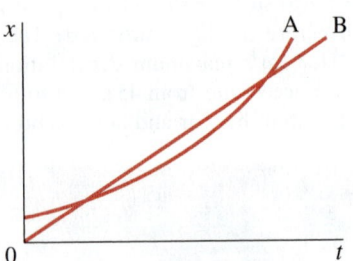

FIGURE 2–52 Problem 93.

94. You are traveling at a constant speed v_M, and there is a car in front of you traveling with a speed v_A. You notice that $v_M > v_A$, so you start slowing down with a constant acceleration a when the distance between you and the other car is x. What relationship between a and x determines whether or not you run into the car in front of you?

*Numerical/Computer

*95. (II) The Table below gives the speed of a particular drag racer as a function of time. (*a*) Calculate the average acceleration (m/s^2) during each time interval. (*b*) Using numerical integration (see Section 2–9) estimate the total distance traveled (m) as a function of time. [*Hint*: for $\bar{v}$ in each interval sum the velocities at the beginning and end of the interval and divide by 2; for example, in the second interval use $\bar{v} = (6.0 + 13.2)/2 = 9.6$] (*c*) Graph each of these.

$t(\text{s})$	0	0.50	1.00	1.50	2.00	2.50	3.00	3.50	4.00	4.50	5.00
$v(\text{km/h})$	0.0	6.0	13.2	22.3	32.2	43.0	53.5	62.6	70.6	78.4	85.1

*96. (III) The acceleration of an object (in m/s^2) is measured at 1.00-s intervals starting at $t = 0$ to be as follows: 1.25, 1.58, 1.96, 2.40, 2.66, 2.70, 2.74, 2.72, 2.60, 2.30, 2.04, 1.76, 1.41, 1.09, 0.86, 0.51, 0.28, 0.10. Use numerical integration (see Section 2–9) to estimate (*a*) the velocity (assume that $v = 0$ at $t = 0$) and (*b*) the displacement at $t = 17.00\,\text{s}$.

*97. (III) A lifeguard standing at the side of a swimming pool spots a child in distress, Fig. 2–53. The lifeguard runs with average speed v_R along the pool's edge for a distance x, then jumps into the pool and swims with average speed v_S on a straight path to the child. (*a*) Show that the total time t it takes the lifeguard to get to the child is given by

$$t = \frac{x}{v_R} + \frac{\sqrt{D^2 + (d - x)^2}}{v_S}.$$

(*b*) Assume $v_R = 4.0\,\text{m/s}$ and $v_S = 1.5\,\text{m/s}$. Use a graphing calculator or computer to plot t vs. x in part (*a*), and from this plot determine the optimal distance x the lifeguard should run before jumping into the pool (that is, find the value of x that minimizes the time t to get to the child).

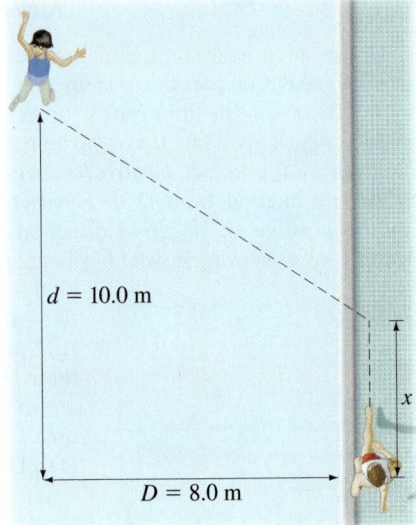

FIGURE 2–53 Problem 97.

Answers to Exercises

A: −30 cm; 50 cm.

B: (*a*).

C: (*b*).

D: (*b*).

E: (*a*) +; (*b*) −; (*c*) −; (*d*) +.

F: (*c*).

G: (*b*).

H: (*e*).

I: $4.9\,\text{m/s}^2$.

J: (*c*).

$\vec{g}$

This snowboarder flying through the air shows an example of motion in two dimensions. In the absence of air resistance, the path would be a perfect parabola. The gold arrow represents the downward acceleration of gravity, $\vec{g}$. Galileo analyzed the motion of objects in 2 dimensions under the action of gravity near the Earth's surface (now called "projectile motion") into its horizontal and vertical components.

We will discuss how to manipulate vectors and how to add them. Besides analyzing projectile motion, we will also see how to work with relative velocity.

3

Kinematics in Two or Three Dimensions; Vectors

CHAPTER-OPENING QUESTION—Guess now!

[*Don't worry about getting the right answer now—you will get another chance later in the Chapter. See also p. 1 of Chapter 1 for more explanation.*]

A small heavy box of emergency supplies is dropped from a moving helicopter at point A as it flies along in a horizontal direction. Which path in the drawing below best describes the path of the box (neglecting air resistance) as seen by a person standing on the ground?

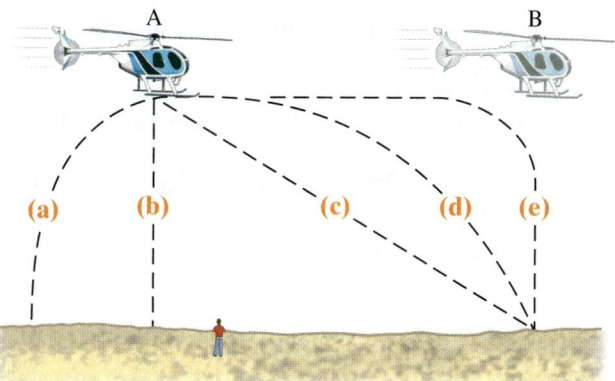

A   B

(a)   (b)   (c)   (d)   (e)

I n Chapter 2 we dealt with motion along a straight line. We now consider the description of the motion of objects that move in paths in two (or three) dimensions. To do so, we first need to discuss vectors and how they are added. We will examine the description of motion in general, followed by an interesting special case, the motion of projectiles near the Earth's surface. We also discuss how to determine the relative velocity of an object as measured in different reference frames.

CONTENTS

3–1 Vectors and Scalars

3–2 Addition of Vectors— Graphical Methods

3–3 Subtraction of Vectors, and Multiplication of a Vector by a Scalar

3–4 Adding Vectors by Components

3–5 Unit Vectors

3–6 Vector Kinematics

3–7 Projectile Motion

3–8 Solving Problems Involving Projectile Motion

3–9 Relative Velocity

3–1 Vectors and Scalars

We mentioned in Chapter 2 that the term *velocity* refers not only to how fast an object is moving but also to its direction. A quantity such as velocity, which has *direction* as well as *magnitude*, is a **vector** quantity. Other quantities that are also vectors are displacement, force, and momentum. However, many quantities have no direction associated with them, such as mass, time, and temperature. They are specified completely by a number and units. Such quantities are called **scalar** quantities.

Drawing a diagram of a particular physical situation is always helpful in physics, and this is especially true when dealing with vectors. On a diagram, each vector is represented by an arrow. The arrow is always drawn so that it points in the direction of the vector quantity it represents. The length of the arrow is drawn proportional to the magnitude of the vector quantity. For example, in Fig. 3–1, green arrows have been drawn representing the velocity of a car at various places as it rounds a curve. The magnitude of the velocity at each point can be read off Fig. 3–1 by measuring the length of the corresponding arrow and using the scale shown (1 cm = 90 km/h).

When we write the symbol for a vector, we will always use boldface type, with a tiny arrow over the symbol. Thus for velocity we write $\vec{v}$. If we are concerned only with the magnitude of the vector, we will write simply v, in italics, as we do for other symbols.

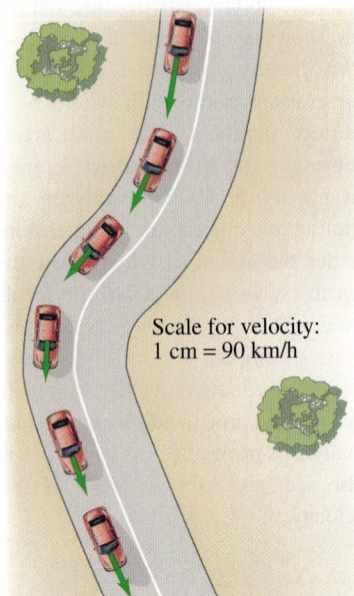

FIGURE 3–1 Car traveling on a road, slowing down to round the curve. The green arrows represent the velocity vector at each position.

Scale for velocity:
1 cm = 90 km/h

3–2 Addition of Vectors—Graphical Methods

Because vectors are quantities that have direction as well as magnitude, they must be added in a special way. In this Chapter, we will deal mainly with displacement vectors, for which we now use the symbol $\vec{D}$, and velocity vectors, $\vec{v}$. But the results will apply for other vectors we encounter later.

We use simple arithmetic for adding scalars. Simple arithmetic can also be used for adding vectors if they are in the same direction. For example, if a person walks 8 km east one day, and 6 km east the next day, the person will be 8 km + 6 km = 14 km east of the point of origin. We say that the *net* or *resultant* displacement is 14 km to the east (Fig. 3–2a). If, on the other hand, the person walks 8 km east on the first day, and 6 km west (in the reverse direction) on the second day, then the person will end up 2 km from the origin (Fig. 3–2b), so the resultant displacement is 2 km to the east. In this case, the resultant displacement is obtained by subtraction: 8 km − 6 km = 2 km.

But simple arithmetic cannot be used if the two vectors are not along the same line. For example, suppose a person walks 10.0 km east and then walks 5.0 km north. These displacements can be represented on a graph in which the positive y axis points north and the positive x axis points east, Fig. 3–3. On this graph, we draw an arrow, labeled $\vec{D}_1$, to represent the 10.0-km displacement to the east. Then we draw a second arrow, $\vec{D}_2$, to represent the 5.0-km displacement to the north. Both vectors are drawn to scale, as in Fig. 3–3.

FIGURE 3–2 Combining vectors in one dimension.

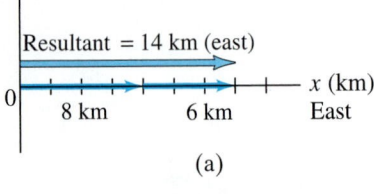

(a)

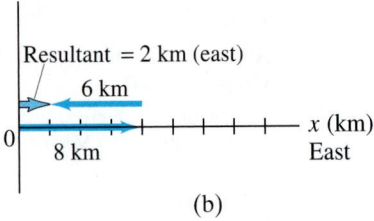

(b)

FIGURE 3–3 A person walks 10.0 km east and then 5.0 km north. These two displacements are represented by the vectors $\vec{D}_1$ and $\vec{D}_2$, which are shown as arrows. The resultant displacement vector, $\vec{D}_R$, which is the vector sum of $\vec{D}_1$ and $\vec{D}_2$, is also shown. Measurement on the graph with ruler and protractor shows that $\vec{D}_R$ has a magnitude of 11.2 km and points at an angle $\theta = 27°$ north of east.

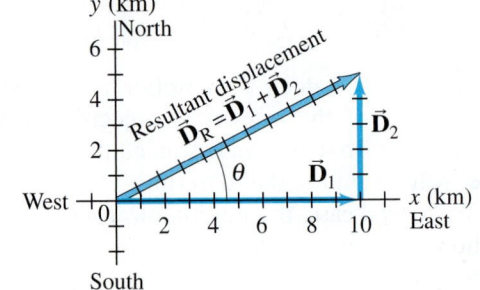

After taking this walk, the person is now 10.0 km east and 5.0 km north of the point of origin. The **resultant displacement** is represented by the arrow labeled $\vec{D}_R$ in Fig. 3–3. Using a ruler and a protractor, you can measure on this diagram that the person is 11.2 km from the origin at an angle $\theta = 27°$ north of east. In other words, the resultant displacement vector has a magnitude of 11.2 km and makes an angle $\theta = 27°$ with the positive x axis. The magnitude (length) of $\vec{D}_R$ can also be obtained using the theorem of Pythagoras in this case, since D_1, D_2, and D_R form a right triangle with D_R as the hypotenuse. Thus

$$D_R = \sqrt{D_1^2 + D_2^2} = \sqrt{(10.0\,\text{km})^2 + (5.0\,\text{km})^2}$$
$$= \sqrt{125\,\text{km}^2} = 11.2\,\text{km}.$$

You can use the Pythagorean theorem, of course, only when the vectors are *perpendicular* to each other.

The resultant displacement vector, $\vec{D}_R$, is the sum of the vectors $\vec{D}_1$ and $\vec{D}_2$. That is,

$$\vec{D}_R = \vec{D}_1 + \vec{D}_2.$$

This is a *vector* equation. An important feature of adding two vectors that are not along the same line is that the magnitude of the resultant vector is not equal to the sum of the magnitudes of the two separate vectors, but is smaller than their sum. That is,

$$D_R \leq D_1 + D_2,$$

where the equals sign applies only if the two vectors point in the same direction. In our example (Fig. 3–3), $D_R = 11.2\,\text{km}$, whereas $D_1 + D_2$ equals 15 km, which is the total distance traveled. Note also that we cannot set $\vec{D}_R$ equal to 11.2 km, because we have a vector equation and 11.2 km is only a part of the resultant vector, its magnitude. We could write something like this, though: $\vec{D}_R = \vec{D}_1 + \vec{D}_2 = (11.2\,\text{km}, 27°\,\text{N of E})$.

| **EXERCISE A** Under what conditions can the magnitude of the resultant vector above be $D_R = D_1 + D_2$?

Figure 3–3 illustrates the general rules for graphically adding two vectors together, no matter what angles they make, to get their sum. The rules are as follows:

1. On a diagram, draw one of the vectors—call it $\vec{D}_1$—to scale.
2. Next draw the second vector, $\vec{D}_2$, to scale, placing its tail at the tip of the first vector and being sure its direction is correct.
3. The arrow drawn from the tail of the first vector to the tip of the second vector represents the *sum*, or **resultant**, of the two vectors.

The length of the resultant vector represents its magnitude. Note that vectors can be translated parallel to themselves (maintaining the same length and angle) to accomplish these manipulations. The length of the resultant can be measured with a ruler and compared to the scale. Angles can be measured with a protractor. This method is known as the **tail-to-tip method of adding vectors**.

The resultant is not affected by the order in which the vectors are added. For example, a displacement of 5.0 km north, to which is added a displacement of 10.0 km east, yields a resultant of 11.2 km and angle $\theta = 27°$ (see Fig. 3–4), the same as when they were added in reverse order (Fig. 3–3). That is, now using $\vec{V}$ to represent any type of vector,

$$\vec{V}_1 + \vec{V}_2 = \vec{V}_2 + \vec{V}_1, \qquad \text{[commutative property]} \quad \textbf{(3–1a)}$$

which is known as the *commutative* property of vector addition.

FIGURE 3–4 If the vectors are added in reverse order, the resultant is the same. (Compare to Fig. 3–3.)

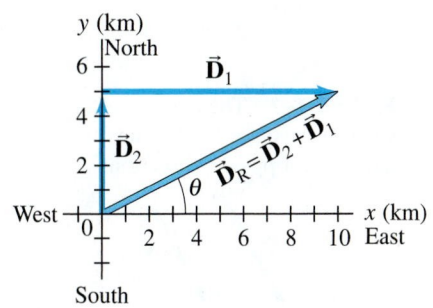

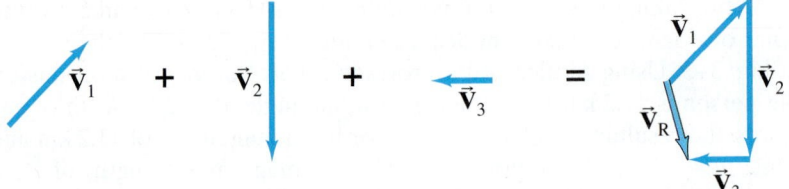

FIGURE 3–5 The resultant of three vectors: $\vec{V}_R = \vec{V}_1 + \vec{V}_2 + \vec{V}_3$.

The tail-to-tip method of adding vectors can be extended to three or more vectors. The resultant is drawn from the tail of the first vector to the tip of the last one added. An example is shown in Fig. 3–5; the three vectors could represent displacements (northeast, south, west) or perhaps three forces. Check for yourself that you get the same resultant no matter in which order you add the three vectors; that is,

$$(\vec{V}_1 + \vec{V}_2) + \vec{V}_3 = \vec{V}_1 + (\vec{V}_2 + \vec{V}_3), \quad \text{[associative property]} \quad \textbf{(3–1b)}$$

which is known as the *associative* property of vector addition.

A second way to add two vectors is the **parallelogram method**. It is fully equivalent to the tail-to-tip method. In this method, the two vectors are drawn starting from a common origin, and a parallelogram is constructed using these two vectors as adjacent sides as shown in Fig. 3–6b. The resultant is the diagonal drawn from the common origin. In Fig. 3–6a, the tail-to-tip method is shown, and it is clear that both methods yield the same result.

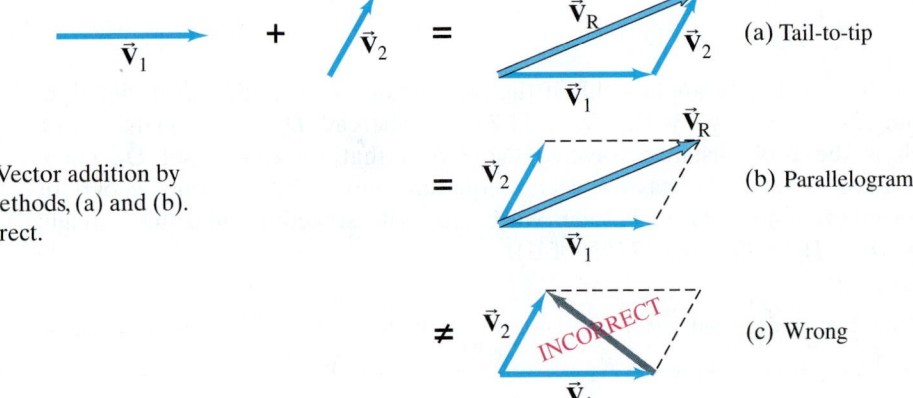

FIGURE 3–6 Vector addition by two different methods, (a) and (b). Part (c) is incorrect.

(a) Tail-to-tip

(b) Parallelogram

(c) Wrong

⚠ CAUTION

Be sure to use the correct diagonal on parallelogram to get the resultant

It is a common error to draw the sum vector as the diagonal running between the tips of the two vectors, as in Fig. 3–6c. *This is incorrect*: it does not represent the sum of the two vectors. (In fact, it represents their difference, $\vec{V}_2 - \vec{V}_1$, as we will see in the next Section.)

CONCEPTUAL EXAMPLE 3–1 | **Range of vector lengths.** Suppose two vectors each have length 3.0 units. What is the range of possible lengths for the vector representing the sum of the two?

RESPONSE The sum can take on any value from 6.0 (= 3.0 + 3.0) where the vectors point in the same direction, to 0 (= 3.0 − 3.0) when the vectors are antiparallel.

EXERCISE B If the two vectors of Example 3–1 are perpendicular to each other, what is the resultant vector length?

3–3 Subtraction of Vectors, and Multiplication of a Vector by a Scalar

FIGURE 3–7 The negative of a vector is a vector having the same length but opposite direction.

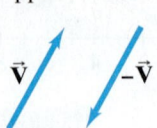

Given a vector $\vec{V}$, we define the *negative* of this vector $(-\vec{V})$ to be a vector with the same magnitude as $\vec{V}$ but opposite in direction, Fig. 3–7. Note, however, that no vector is ever negative in the sense of its magnitude: the magnitude of every vector is positive. Rather, a minus sign tells us about its direction.

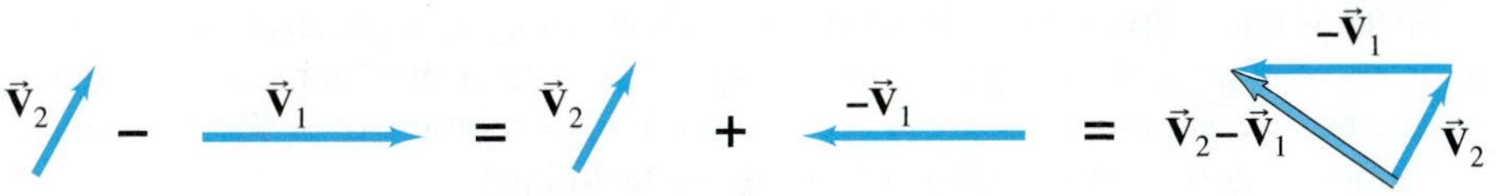

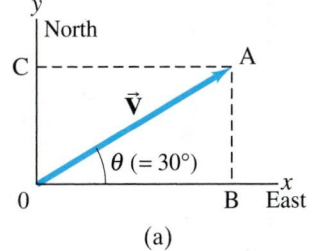

FIGURE 3–8 Subtracting two vectors: $\vec{V}_2 - \vec{V}_1$.

$$\vec{V}_2 - \vec{V}_1 \quad = \quad \vec{V}_2 + (-\vec{V}_1) \quad = \quad \vec{V}_2 - \vec{V}_1$$

We can now define the subtraction of one vector from another: the difference between two vectors $\vec{V}_2 - \vec{V}_1$ is defined as

$$\vec{V}_2 - \vec{V}_1 = \vec{V}_2 + (-\vec{V}_1).$$

That is, the difference between two vectors is equal to the sum of the first plus the negative of the second. Thus our rules for addition of vectors can be applied as shown in Fig. 3–8 using the tail-to-tip method.

A vector $\vec{V}$ can be multiplied by a scalar c. We define their product so that $c\vec{V}$ has the same direction as $\vec{V}$ and has magnitude cV. That is, multiplication of a vector by a positive scalar c changes the magnitude of the vector by a factor c but doesn't alter the direction. If c is a negative scalar, the magnitude of the product $c\vec{V}$ is still $|c|V$ (where $|c|$ means the magnitude of c), but the direction is precisely opposite to that of $\vec{V}$. See Fig. 3–9.

FIGURE 3–9 Multiplying a vector $\vec{V}$ by a scalar c gives a vector whose magnitude is c times greater and in the same direction as $\vec{V}$ (or opposite direction if c is negative).

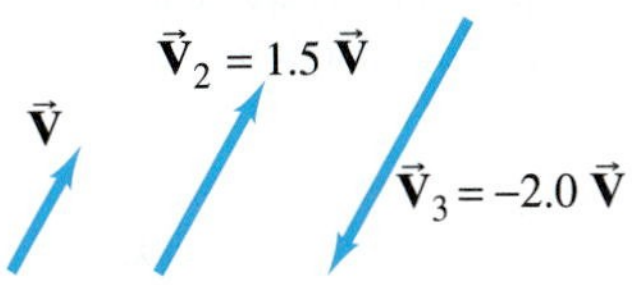

$$\vec{V}_2 = 1.5\,\vec{V}$$
$$\vec{V} \qquad \vec{V}_3 = -2.0\,\vec{V}$$

EXERCISE C What does the "incorrect" vector in Fig. 3–6c represent? (a) $\vec{V}_2 - \vec{V}_1$, (b) $\vec{V}_1 - \vec{V}_2$, (c) something else (specify).

3–4 Adding Vectors by Components

Adding vectors graphically using a ruler and protractor is often not sufficiently accurate and is not useful for vectors in three dimensions. We discuss now a more powerful and precise method for adding vectors. But do not forget graphical methods—they are useful for visualizing, for checking your math, and thus for getting the correct result.

Consider first a vector $\vec{V}$ that lies in a particular plane. It can be expressed as the sum of two other vectors, called the **components** of the original vector. The components are usually chosen to be along two perpendicular directions, such as the x and y axes. The process of finding the components is known as **resolving the vector into its components**. An example is shown in Fig. 3–10; the vector $\vec{V}$ could be a displacement vector that points at an angle $\theta = 30°$ north of east, where we have chosen the positive x axis to be to the east and the positive y axis north. This vector $\vec{V}$ is resolved into its x and y components by drawing dashed lines out from the tip (A) of the vector (lines AB and AC) making them perpendicular to the x and y axes. Then the lines OB and OC represent the x and y components of $\vec{V}$, respectively, as shown in Fig. 3–10b. These *vector components* are written $\vec{V}_x$ and $\vec{V}_y$. We generally show vector components as arrows, like vectors, but dashed. The *scalar components*, V_x and V_y, are the magnitudes of the vector components, with units, accompanied by a positive or negative sign depending on whether they point along the positive or negative x or y axis. As can be seen in Fig. 3–10, $\vec{V}_x + \vec{V}_y = \vec{V}$ by the parallelogram method of adding vectors.

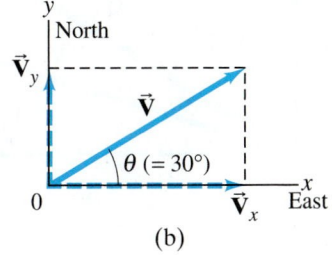

(a)   (b)

FIGURE 3–10 Resolving a vector $\vec{V}$ into its components along an arbitrarily chosen set of x and y axes. The components, once found, themselves represent the vector. That is, the components contain as much information as the vector itself.

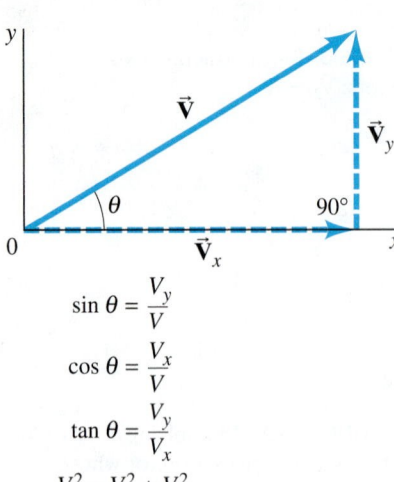

$$\sin \theta = \frac{V_y}{V}$$

$$\cos \theta = \frac{V_x}{V}$$

$$\tan \theta = \frac{V_y}{V_x}$$

$$V^2 = V_x^2 + V_y^2$$

FIGURE 3–11 Finding the components of a vector using trigonometric functions.

Space is made up of three dimensions, and sometimes it is necessary to resolve a vector into components along three mutually perpendicular directions. In rectangular coordinates the components are $\vec{V}_x$, $\vec{V}_y$, and $\vec{V}_z$. Resolution of a vector in three dimensions is merely an extension of the above technique.

The use of trigonometric functions for finding the components of a vector is illustrated in Fig. 3–11, where a vector and its two components are thought of as making up a right triangle. (See also Appendix A for other details on trigonometric functions and identities.) We then see that the sine, cosine, and tangent are as given in Fig. 3–11. If we multiply the definition of $\sin \theta = V_y/V$ by V on both sides, we get

$$V_y = V \sin \theta. \tag{3–2a}$$

Similarly, from the definition of $\cos \theta$, we obtain

$$V_x = V \cos \theta. \tag{3–2b}$$

Note that θ is chosen (by convention) to be the angle that the vector makes with the positive x axis, measured positive counterclockwise.

The components of a given vector will be different for different choices of coordinate axes. It is therefore crucial to specify the choice of coordinate system when giving the components.

There are two ways to specify a vector in a given coordinate system:

1. We can give its components, V_x and V_y.
2. We can give its magnitude V and the angle θ it makes with the positive x axis.

We can shift from one description to the other using Eqs. 3–2, and, for the reverse, by using the theorem of Pythagoras[†] and the definition of tangent:

$$V = \sqrt{V_x^2 + V_y^2} \tag{3–3a}$$

$$\tan \theta = \frac{V_y}{V_x} \tag{3–3b}$$

as can be seen in Fig. 3–11.

We can now discuss how to add vectors using components. The first step is to resolve each vector into its components. Next we can see, using Fig. 3–12, that the addition of any two vectors $\vec{V}_1$ and $\vec{V}_2$ to give a resultant, $\vec{V} = \vec{V}_1 + \vec{V}_2$, implies that

$$V_x = V_{1x} + V_{2x}$$
$$V_y = V_{1y} + V_{2y}. \tag{3–4}$$

That is, the sum of the x components equals the x component of the resultant, and the sum of the y components equals the y component of the resultant, as can be verified by a careful examination of Fig. 3–12. Note that we do *not* add x components to y components.

[†]In three dimensions, the theorem of Pythagoras becomes $V = \sqrt{V_x^2 + V_y^2 + V_z^2}$, where V_z is the component along the third, or z, axis.

FIGURE 3–12 The components of $\vec{V} = \vec{V}_1 + \vec{V}_2$ are
$$V_x = V_{1x} + V_{2x}$$
$$V_y = V_{1y} + V_{2y}.$$

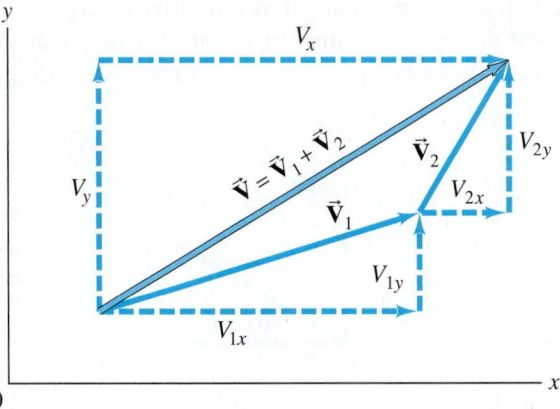

If the magnitude and direction of the resultant vector are desired, they can be obtained using Eqs. 3–3.

The components of a given vector depend on the choice of coordinate axes. You can often reduce the work involved in adding vectors by a good choice of axes—for example, by choosing one of the axes to be in the same direction as one of the vectors. Then that vector will have only one nonzero component.

EXAMPLE 3–2 **Mail carrier's displacement.** A rural mail carrier leaves the post office and drives 22.0 km in a northerly direction. She then drives in a direction 60.0° south of east for 47.0 km (Fig. 3–13a). What is her displacement from the post office?

APPROACH We choose the positive x axis to be east and the positive y axis to be north, since those are the compass directions used on most maps. The origin of the xy coordinate system is at the post office. We resolve each vector into its x and y components. We add the x components together, and then the y components together, giving us the x and y components of the resultant.

SOLUTION Resolve each displacement vector into its components, as shown in Fig. 3–13b. Since $\vec{D}_1$ has magnitude 22.0 km and points north, it has only a y component:

$$D_{1x} = 0, \qquad D_{1y} = 22.0 \text{ km}.$$

$\vec{D}_2$ has both x and y components:

$$D_{2x} = +(47.0 \text{ km})(\cos 60°) = +(47.0 \text{ km})(0.500) = +23.5 \text{ km}$$
$$D_{2y} = -(47.0 \text{ km})(\sin 60°) = -(47.0 \text{ km})(0.866) = -40.7 \text{ km}.$$

Notice that D_{2y} is negative because this vector component points along the negative y axis. The resultant vector, $\vec{D}$, has components:

$$D_x = D_{1x} + D_{2x} = 0 \text{ km} + 23.5 \text{ km} = +23.5 \text{ km}$$
$$D_y = D_{1y} + D_{2y} = 22.0 \text{ km} + (-40.7 \text{ km}) = -18.7 \text{ km}.$$

This specifies the resultant vector completely:

$$D_x = 23.5 \text{ km}, \qquad D_y = -18.7 \text{ km}.$$

We can also specify the resultant vector by giving its magnitude and angle using Eqs. 3–3:

$$D = \sqrt{D_x^2 + D_y^2} = \sqrt{(23.5 \text{ km})^2 + (-18.7 \text{ km})^2} = 30.0 \text{ km}$$
$$\tan \theta = \frac{D_y}{D_x} = \frac{-18.7 \text{ km}}{23.5 \text{ km}} = -0.796.$$

A calculator with an INV TAN, an ARC TAN, or a TAN^{-1} key gives $\theta = \tan^{-1}(-0.796) = -38.5°$. The negative sign means $\theta = 38.5°$ below the x axis, Fig. 3–13c. So, the resultant displacement is 30.0 km directed at 38.5° in a southeasterly direction.

NOTE Always be attentive about the quadrant in which the resultant vector lies. An electronic calculator does not fully give this information, but a good diagram does.

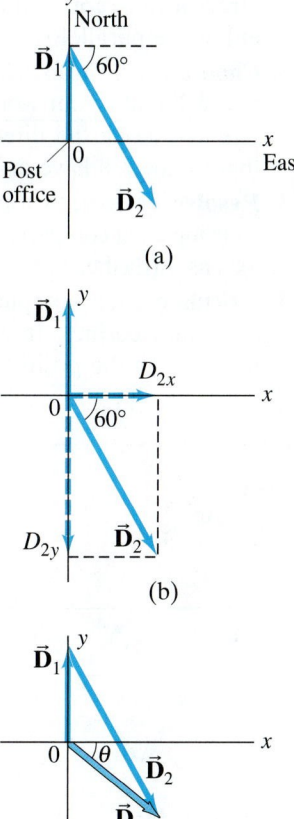

FIGURE 3–13 Example 3–2. (a) The two displacement vectors, $\vec{D}_1$ and $\vec{D}_2$. (b) $\vec{D}_2$ is resolved into its components. (c) $\vec{D}_1$ and $\vec{D}_2$ are added graphically to obtain the resultant $\vec{D}$. The component method of adding the vectors is explained in the Example.

The signs of trigonometric functions depend on which "quadrant" the angle falls in: for example, the tangent is positive in the first and third quadrants (from 0° to 90°, and 180° to 270°), but negative in the second and fourth quadrants; see Appendix A. The best way to keep track of angles, and to check any vector result, is always to draw a vector diagram. A vector diagram gives you something tangible to look at when analyzing a problem, and provides a check on the results.

The following Problem Solving Strategy should not be considered a prescription. Rather it is a summary of things to do to get you thinking and involved in the problem at hand.

PROBLEM SOLVING

Identify the correct quadrant by drawing a careful diagram

PROBLEM SOLVING

Adding Vectors

Here is a brief summary of how to add two or more vectors using components:

1. **Draw a diagram**, adding the vectors graphically by either the parallelogram or tail-to-tip method.
2. **Choose x and y axes.** Choose them in a way, if possible, that will make your work easier. (For example, choose one axis along the direction of one of the vectors so that vector will have only one component.)
3. **Resolve** each vector into its x and y **components**, showing each component along its appropriate (x or y) axis as a (dashed) arrow.
4. **Calculate each component** (when not given) using sines and cosines. If θ_1 is the angle that vector $\vec{V}_1$ makes with the positive x axis, then:
$$V_{1x} = V_1 \cos\theta_1, \qquad V_{1y} = V_1 \sin\theta_1.$$

Pay careful attention to **signs**: any component that points along the negative x or y axis gets a minus sign.

5. **Add** the x **components** together to get the x component of the resultant. Ditto for y:
$$V_x = V_{1x} + V_{2x} + \text{any others}$$
$$V_y = V_{1y} + V_{2y} + \text{any others}.$$
This is the answer: the components of the resultant vector. Check signs to see if they fit the quadrant shown in your diagram (point 1 above).

6. If you want to know the **magnitude and direction** of the resultant vector, use Eqs. 3–3:
$$V = \sqrt{V_x^2 + V_y^2}, \qquad \tan\theta = \frac{V_y}{V_x}.$$
The vector diagram you already drew helps to obtain the correct position (quadrant) of the angle θ.

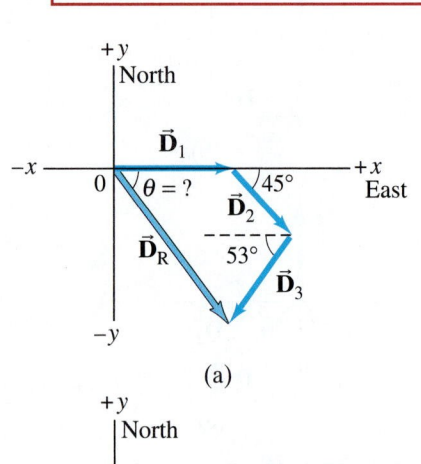

(a)

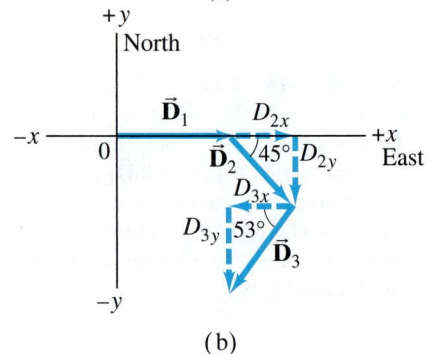

(b)

FIGURE 3–14 Example 3–3.

Vector	Components	
	x (km)	y (km)
$\vec{D}_1$	620	0
$\vec{D}_2$	311	−311
$\vec{D}_3$	−331	−439
$\vec{D}_R$	600	−750

EXAMPLE 3–3 **Three short trips.** An airplane trip involves three legs, with two stopovers, as shown in Fig. 3–14a. The first leg is due east for 620 km; the second leg is southeast (45°) for 440 km; and the third leg is at 53° south of west, for 550 km, as shown. What is the plane's total displacement?

APPROACH We follow the steps in the Problem Solving Strategy above.
SOLUTION
1. **Draw a diagram** such as Fig. 3–14a, where $\vec{D}_1$, $\vec{D}_2$, and $\vec{D}_3$ represent the three legs of the trip, and $\vec{D}_R$ is the plane's total displacement.
2. **Choose axes**: Axes are also shown in Fig. 3–14a: x is east, y north.
3. **Resolve components**: It is imperative to draw a good diagram. The components are drawn in Fig. 3–14b. Instead of drawing all the vectors starting from a common origin, as we did in Fig. 3–13b, here we draw them "tail-to-tip" style, which is just as valid and may make it easier to see.
4. **Calculate the components**:
$$\vec{D}_1: D_{1x} = +D_1 \cos 0° = D_1 = 620\,\text{km}$$
$$D_{1y} = +D_1 \sin 0° = 0\,\text{km}$$
$$\vec{D}_2: D_{2x} = +D_2 \cos 45° = +(440\,\text{km})(0.707) = +311\,\text{km}$$
$$D_{2y} = -D_2 \sin 45° = -(440\,\text{km})(0.707) = -311\,\text{km}$$
$$\vec{D}_3: D_{3x} = -D_3 \cos 53° = -(550\,\text{km})(0.602) = -331\,\text{km}$$
$$D_{3y} = -D_3 \sin 53° = -(550\,\text{km})(0.799) = -439\,\text{km}.$$
We have given a minus sign to each component that in Fig. 3–14b points in the $-x$ or $-y$ direction. The components are shown in the Table in the margin.
5. **Add the components**: We add the x components together, and we add the y components together to obtain the x and y components of the resultant:
$$D_x = D_{1x} + D_{2x} + D_{3x} = 620\,\text{km} + 311\,\text{km} - 331\,\text{km} = 600\,\text{km}$$
$$D_y = D_{1y} + D_{2y} + D_{3y} = 0\,\text{km} - 311\,\text{km} - 439\,\text{km} = -750\,\text{km}.$$
The x and y components are 600 km and −750 km, and point respectively to the east and south. This is one way to give the answer.
6. **Magnitude and direction**: We can also give the answer as
$$D_R = \sqrt{D_x^2 + D_y^2} = \sqrt{(600)^2 + (-750)^2}\,\text{km} = 960\,\text{km}$$
$$\tan\theta = \frac{D_y}{D_x} = \frac{-750\,\text{km}}{600\,\text{km}} = -1.25, \qquad \text{so } \theta = -51°.$$
Thus, the total displacement has magnitude 960 km and points 51° below the x axis (south of east), as was shown in our original sketch, Fig. 3–14a.

3–5 Unit Vectors

Vectors can be conveniently written in terms of *unit vectors*. A **unit vector** is defined to have a magnitude exactly equal to one (1). It is useful to define unit vectors that point along coordinate axes, and in an x, y, z rectangular coordinate system these unit vectors are called $\hat{\mathbf{i}}, \hat{\mathbf{j}}$, and $\hat{\mathbf{k}}$. They point, respectively, along the positive x, y, and z axes as shown in Fig. 3–15. Like other vectors, $\hat{\mathbf{i}}, \hat{\mathbf{j}}$, and $\hat{\mathbf{k}}$ do not have to be placed at the origin, but can be placed elsewhere as long as the direction and unit length remain unchanged. It is common to write unit vectors with a "hat": $\hat{\mathbf{i}}, \hat{\mathbf{j}}, \hat{\mathbf{k}}$ (and we will do so in this book) as a reminder that each is a unit vector.

Because of the definition of multiplication of a vector by a scalar (Section 3–3), the components of a vector $\vec{\mathbf{V}}$ can be written $\vec{\mathbf{V}}_x = V_x\hat{\mathbf{i}}$, $\vec{\mathbf{V}}_y = V_y\hat{\mathbf{j}}$, and $\vec{\mathbf{V}}_z = V_z\hat{\mathbf{k}}$. Hence any vector $\vec{\mathbf{V}}$ can be written in terms of its components as

$$\vec{\mathbf{V}} = V_x\hat{\mathbf{i}} + V_y\hat{\mathbf{j}} + V_z\hat{\mathbf{k}}. \qquad (3\text{–}5)$$

Unit vectors are helpful when adding vectors analytically by components. For example, Eq. 3–4 can be seen to be true by using unit vector notation for each vector (which we write for the two-dimensional case, with the extension to three dimensions being straightforward):

$$\begin{aligned}\vec{\mathbf{V}} &= (V_x)\hat{\mathbf{i}} + (V_y)\hat{\mathbf{j}} = \vec{\mathbf{V}}_1 + \vec{\mathbf{V}}_2 \\ &= (V_{1x}\hat{\mathbf{i}} + V_{1y}\hat{\mathbf{j}}) + (V_{2x}\hat{\mathbf{i}} + V_{2y}\hat{\mathbf{j}}) \\ &= (V_{1x} + V_{2x})\hat{\mathbf{i}} + (V_{1y} + V_{2y})\hat{\mathbf{j}}.\end{aligned}$$

Comparing the first line to the third line, we get Eq. 3–4.

EXAMPLE 3–4 Using unit vectors. Write the vectors of Example 3–2 in unit vector notation, and perform the addition.

APPROACH We use the components we found in Example 3–2,

$$D_{1x} = 0, \quad D_{1y} = 22.0\,\text{km}, \quad \text{and} \quad D_{2x} = 23.5\,\text{km}, \quad D_{2y} = -40.7\,\text{km},$$

and we now write them in the form of Eq. 3–5.

SOLUTION We have

$$\vec{\mathbf{D}}_1 = 0\hat{\mathbf{i}} + 22.0\,\text{km}\,\hat{\mathbf{j}}$$
$$\vec{\mathbf{D}}_2 = 23.5\,\text{km}\,\hat{\mathbf{i}} - 40.7\,\text{km}\,\hat{\mathbf{j}}.$$

Then

$$\begin{aligned}\vec{\mathbf{D}} = \vec{\mathbf{D}}_1 + \vec{\mathbf{D}}_2 &= (0 + 23.5)\,\text{km}\,\hat{\mathbf{i}} + (22.0 - 40.7)\,\text{km}\,\hat{\mathbf{j}} \\ &= 23.5\,\text{km}\,\hat{\mathbf{i}} - 18.7\,\text{km}\,\hat{\mathbf{j}}.\end{aligned}$$

The components of the resultant displacement, $\vec{\mathbf{D}}$, are $D_x = 23.5\,\text{km}$ and $D_y = -18.7\,\text{km}$. The magnitude of $\vec{\mathbf{D}}$ is $D = \sqrt{(23.5\,\text{km})^2 + (18.7\,\text{km})^2} = 30.0\,\text{km}$, just as in Example 3–2.

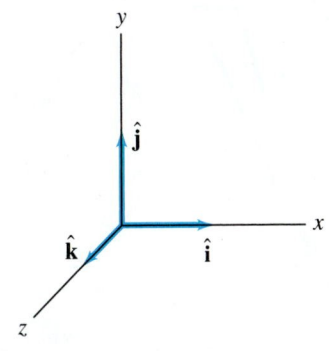

FIGURE 3–15 Unit vectors $\hat{\mathbf{i}}, \hat{\mathbf{j}}$, and $\hat{\mathbf{k}}$ along the x, y, and z axes.

3–6 Vector Kinematics

We can now extend our definitions of velocity and acceleration in a formal way to two- and three-dimensional motion. Suppose a particle follows a path in the xy plane as shown in Fig. 3–16. At time t_1, the particle is at point P_1, and at time t_2, it is at point P_2. The vector $\vec{\mathbf{r}}_1$ is the position vector of the particle at time t_1 (it represents the displacement of the particle from the origin of the coordinate system). And $\vec{\mathbf{r}}_2$ is the position vector at time t_2.

In one dimension, we defined displacement as the *change in position* of the particle. In the more general case of two or three dimensions, the **displacement vector** is defined as the vector representing change in position. We call it $\Delta\vec{\mathbf{r}}$,[†] where

$$\Delta\vec{\mathbf{r}} = \vec{\mathbf{r}}_2 - \vec{\mathbf{r}}_1.$$

This represents the displacement during the time interval $\Delta t = t_2 - t_1$.

FIGURE 3–16 Path of a particle in the xy plane. At time t_1 the particle is at point P_1 given by the position vector $\vec{\mathbf{r}}_1$; at t_2 the particle is at point P_2 given by the position vector $\vec{\mathbf{r}}_2$. The displacement vector for the time interval $t_2 - t_1$ is $\Delta\vec{\mathbf{r}} = \vec{\mathbf{r}}_2 - \vec{\mathbf{r}}_1$.

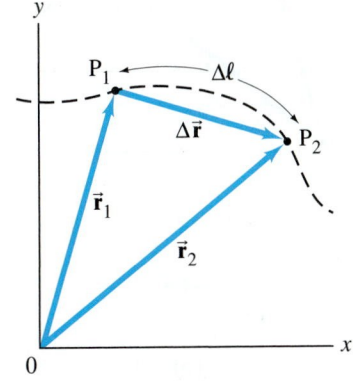

[†]We used $\vec{\mathbf{D}}$ for the displacement vector earlier in the Chapter for illustrating vector addition. The new notation here, $\Delta\vec{\mathbf{r}}$, emphasizes that it is the difference between two position vectors.

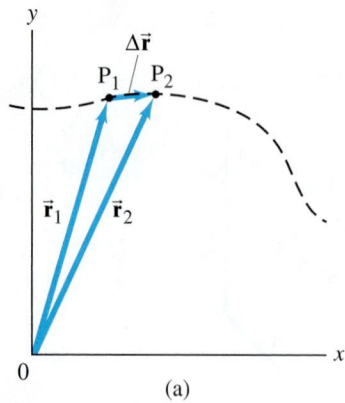

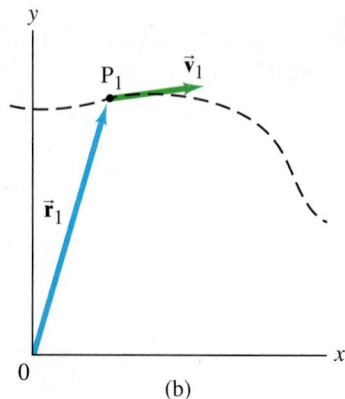

FIGURE 3–17 (a) As we take Δt and $\Delta \vec{\mathbf{r}}$ smaller and smaller [compare to Fig. 3–16] we see that the direction of $\Delta \vec{\mathbf{r}}$ and of the instantaneous velocity ($\Delta \vec{\mathbf{r}}/\Delta t$, where $\Delta t \rightarrow 0$) is (b) tangent to the curve at P_1.

FIGURE 3–18 (a) Velocity vectors $\vec{\mathbf{v}}_1$ and $\vec{\mathbf{v}}_2$ at instants t_1 and t_2 for a particle at points P_1 and P_2, as in Fig. 3–16. (b) The direction of the average acceleration is in the direction of $\Delta \vec{\mathbf{v}} = \vec{\mathbf{v}}_2 - \vec{\mathbf{v}}_1$.

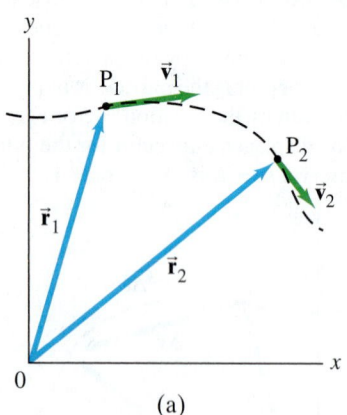

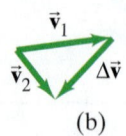

In unit vector notation, we can write

$$\vec{\mathbf{r}}_1 = x_1 \hat{\mathbf{i}} + y_1 \hat{\mathbf{j}} + z_1 \hat{\mathbf{k}}, \tag{3–6a}$$

where x_1, y_1, and z_1 are the coordinates of point P_1. Similarly,

$$\vec{\mathbf{r}}_2 = x_2 \hat{\mathbf{i}} + y_2 \hat{\mathbf{j}} + z_2 \hat{\mathbf{k}}.$$

Hence

$$\Delta \vec{\mathbf{r}} = (x_2 - x_1)\hat{\mathbf{i}} + (y_2 - y_1)\hat{\mathbf{j}} + (z_2 - z_1)\hat{\mathbf{k}}. \tag{3–6b}$$

If the motion is along the x axis only, then $y_2 - y_1 = 0$, $z_2 - z_1 = 0$, and the magnitude of the displacement is $\Delta r = x_2 - x_1$, which is consistent with our earlier one-dimensional equation (Section 2–1). Even in one dimension, displacement is a vector, as are velocity and acceleration.

The **average velocity vector** over the time interval $\Delta t = t_2 - t_1$ is defined as

$$\text{average velocity} = \frac{\Delta \vec{\mathbf{r}}}{\Delta t}. \tag{3–7}$$

Now let us consider shorter and shorter time intervals—that is, we let Δt approach zero so that the distance between points P_2 and P_1 also approaches zero, Fig. 3–17. We define the **instantaneous velocity vector** as the limit of the average velocity as Δt approaches zero:

$$\vec{\mathbf{v}} = \lim_{\Delta t \to 0} \frac{\Delta \vec{\mathbf{r}}}{\Delta t} = \frac{d\vec{\mathbf{r}}}{dt}. \tag{3–8}$$

The direction of $\vec{\mathbf{v}}$ at any moment is along the line tangent to the path at that moment (Fig. 3–17).

Note that the magnitude of the average velocity in Fig. 3–16 is not equal to the average speed, which is the actual distance traveled along the path, $\Delta \ell$, divided by Δt. In some special cases, the average speed and average velocity are equal (such as motion along a straight line in one direction), but in general they are not. However, in the limit $\Delta t \to 0$, Δr always approaches $\Delta \ell$, so the instantaneous speed *always* equals the magnitude of the instantaneous velocity at any time.

The instantaneous velocity (Eq. 3–8) is equal to the derivative of the position vector with respect to time. Equation 3–8 can be written in terms of components starting with Eq. 3–6a as:

$$\vec{\mathbf{v}} = \frac{d\vec{\mathbf{r}}}{dt} = \frac{dx}{dt}\hat{\mathbf{i}} + \frac{dy}{dt}\hat{\mathbf{j}} + \frac{dz}{dt}\hat{\mathbf{k}} = v_x\hat{\mathbf{i}} + v_y\hat{\mathbf{j}} + v_z\hat{\mathbf{k}}, \tag{3–9}$$

where $v_x = dx/dt$, $v_y = dy/dt$, $v_z = dz/dt$ are the x, y, and z components of the velocity. Note that $d\hat{\mathbf{i}}/dt = d\hat{\mathbf{j}}/dt = d\hat{\mathbf{k}}/dt = 0$ since these unit vectors are constant in both magnitude and direction.

Acceleration in two or three dimensions is treated in a similar way. The **average acceleration vector**, over a time interval $\Delta t = t_2 - t_1$ is defined as

$$\text{average acceleration} = \frac{\Delta \vec{\mathbf{v}}}{\Delta t} = \frac{\vec{\mathbf{v}}_2 - \vec{\mathbf{v}}_1}{t_2 - t_1}, \tag{3–10}$$

where $\Delta \vec{\mathbf{v}}$ is the change in the instantaneous velocity vector during that time interval: $\Delta \vec{\mathbf{v}} = \vec{\mathbf{v}}_2 - \vec{\mathbf{v}}_1$. Note that $\vec{\mathbf{v}}_2$ in many cases, such as in Fig. 3–18a, may not be in the same direction as $\vec{\mathbf{v}}_1$. Hence the average acceleration vector may be in a different direction from either $\vec{\mathbf{v}}_1$ or $\vec{\mathbf{v}}_2$ (Fig. 3–18b). Furthermore, $\vec{\mathbf{v}}_2$ and $\vec{\mathbf{v}}_1$ may have the same magnitude but different directions, and the difference of two such vectors will not be zero. Hence acceleration can result from either a change in the magnitude of the velocity, or from a change in direction of the velocity, or a change in both.

The **instantaneous acceleration vector** is defined as the limit of the average acceleration vector as the time interval Δt is allowed to approach zero:

$$\vec{\mathbf{a}} = \lim_{\Delta \to 0} \frac{\Delta \vec{\mathbf{v}}}{\Delta t} = \frac{d\vec{\mathbf{v}}}{dt}, \tag{3–11}$$

and is thus the derivative of $\vec{\mathbf{v}}$ with respect to t.

We can write $\vec{a}$ using components:

$$\vec{a} = \frac{d\vec{v}}{dt} = \frac{dv_x}{dt}\hat{i} + \frac{dv_y}{dt}\hat{j} + \frac{dv_z}{dt}\hat{k}$$
$$= a_x\hat{i} + a_y\hat{j} + a_z\hat{k}, \qquad\qquad (3\text{--}12)$$

where $a_x = dv_x/dt$, etc. Because $v_x = dx/dt$, then $a_x = dv_x/dt = d^2x/dt^2$, as we saw in Section 2–4. Thus we can also write the acceleration as

$$\vec{a} = \frac{d^2x}{dt^2}\hat{i} + \frac{d^2y}{dt^2}\hat{j} + \frac{d^2z}{dt^2}\hat{k}. \qquad\qquad (3\text{--}12c)$$

The instantaneous acceleration will be nonzero not only when the magnitude of the velocity changes but also if its direction changes. For example, a person riding in a car traveling at constant speed around a curve, or a child riding on a merry-go-round, will both experience an acceleration because of a change in the direction of the velocity, even though the speed may be constant. (More on this in Chapter 5.)

In general, we will use the terms "velocity" and "acceleration" to mean the instantaneous values. If we want to discuss average values, we will use the word "average."

EXAMPLE 3–5 **Position given as a function of time.** The position of a particle as a function of time is given by

$$\vec{r} = [(5.0\,\text{m/s})t + (6.0\,\text{m/s}^2)t^2]\hat{i} + [(7.0\,\text{m}) - (3.0\,\text{m/s}^3)t^3]\hat{j},$$

where r is in meters and t is in seconds. (a) What is the particle's displacement between $t_1 = 2.0\,\text{s}$ and $t_2 = 3.0\,\text{s}$? (b) Determine the particle's instantaneous velocity and acceleration as a function of time. (c) Evaluate $\vec{v}$ and $\vec{a}$ at $t = 3.0\,\text{s}$.

APPROACH For (a), we find $\Delta\vec{r} = \vec{r}_2 - \vec{r}_1$, inserting $t_1 = 2.0\,\text{s}$ for finding $\vec{r}_1$, and $t_2 = 3.0\,\text{s}$ for $\vec{r}_2$. For (b), we take derivatives (Eqs. 3–9 and 3–11), and for (c) we substitute $t = 3.0\,\text{s}$ into our results in (b).

SOLUTION (a) At $t_1 = 2.0\,\text{s}$,

$$\vec{r}_1 = [(5.0\,\text{m/s})(2.0\,\text{s}) + (6.0\,\text{m/s}^2)(2.0\,\text{s})^2]\hat{i} + [(7.0\,\text{m}) - (3.0\,\text{m/s}^3)(2.0\,\text{s})^3]\hat{j}$$
$$= (34\,\text{m})\hat{i} - (17\,\text{m})\hat{j}.$$

Similarly, at $t_2 = 3.0\,\text{s}$,

$$\vec{r}_2 = (15\,\text{m} + 54\,\text{m})\hat{i} + (7.0\,\text{m} - 81\,\text{m})\hat{j} = (69\,\text{m})\hat{i} - (74\,\text{m})\hat{j}.$$

Thus

$$\Delta\vec{r} = \vec{r}_2 - \vec{r}_1 = (69\,\text{m} - 34\,\text{m})\hat{i} + (-74\,\text{m} + 17\,\text{m})\hat{j} = (35\,\text{m})\hat{i} - (57\,\text{m})\hat{j}.$$

That is, $\Delta x = 35\,\text{m}$, and $\Delta y = -57\,\text{m}$.

(b) To find velocity, we take the derivative of the given $\vec{r}$ with respect to time, noting (Appendix B–2) that $d(t^2)/dt = 2t$, and $d(t^3)/dt = 3t^2$:

$$\vec{v} = \frac{d\vec{r}}{dt} = [5.0\,\text{m/s} + (12\,\text{m/s}^2)t]\hat{i} + [0 - (9.0\,\text{m/s}^3)t^2]\hat{j}.$$

The acceleration is (keeping only two significant figures):

$$\vec{a} = \frac{d\vec{v}}{dt} = (12\,\text{m/s}^2)\hat{i} - (18\,\text{m/s}^3)t\hat{j}.$$

Thus $a_x = 12\,\text{m/s}^2$ is constant; but $a_y = -(18\,\text{m/s}^3)t$ depends linearly on time, increasing in magnitude with time in the negative y direction.

(c) We substitute $t = 3.0\,\text{s}$ into the equations we just derived for $\vec{v}$ and $\vec{a}$:

$$\vec{v} = (5.0\,\text{m/s} + 36\,\text{m/s})\hat{i} - (81\,\text{m/s})\hat{j} = (41\,\text{m/s})\hat{i} - (81\,\text{m/s})\hat{j}$$
$$\vec{a} = (12\,\text{m/s}^2)\hat{i} - (54\,\text{m/s}^2)\hat{j}.$$

Their magnitudes at $t = 3.0\,\text{s}$ are $v = \sqrt{(41\,\text{m/s})^2 + (81\,\text{m/s})^2} = 91\,\text{m/s}$, and $a = \sqrt{(12\,\text{m/s}^2)^2 + (54\,\text{m/s}^2)^2} = 55\,\text{m/s}^2$.

Constant Acceleration

In Chapter 2 we studied the important case of one-dimensional motion for which the acceleration is constant. In two or three dimensions, if the acceleration vector, $\vec{a}$, is constant in magnitude and direction, then $a_x = $ constant, $a_y = $ constant, $a_z = $ constant. The average acceleration in this case is equal to the instantaneous acceleration at any moment. The equations we derived in Chapter 2 for one dimension, Eqs. 2–12a, b, and c, apply separately to each perpendicular component of two- or three-dimensional motion. In two dimensions we let $\vec{v}_0 = v_{x0}\hat{\mathbf{i}} + v_{y0}\hat{\mathbf{j}}$ be the initial velocity, and we apply Eqs. 3–6a, 3–9, and 3–12b for the position vector, $\vec{r}$, velocity, $\vec{v}$, and acceleration, $\vec{a}$. We can then write Eqs. 2–12a, b, and c, for two dimensions as shown in Table 3–1.

TABLE 3–1 Kinematic Equations for Constant Acceleration in 2 Dimensions

x component (horizontal)		y component (vertical)
$v_x = v_{x0} + a_x t$	(Eq. 2–12a)	$v_y = v_{y0} + a_y t$
$x = x_0 + v_{x0}t + \frac{1}{2}a_x t^2$	(Eq. 2–12b)	$y = y_0 + v_{y0}t + \frac{1}{2}a_y t^2$
$v_x^2 = v_{x0}^2 + 2a_x(x - x_0)$	(Eq. 2–12c)	$v_y^2 = v_{y0}^2 + 2a_y(y - y_0)$

The first two of the equations in Table 3–1 can be written more formally in vector notation.

$$\vec{v} = \vec{v}_0 + \vec{a}t \qquad [\vec{a} = \text{constant}] \quad (3\text{–}13a)$$
$$\vec{r} = \vec{r}_0 + \vec{v}_0 t + \frac{1}{2}\vec{a}t^2. \qquad [\vec{a} = \text{constant}] \quad (3\text{–}13b)$$

Here, $\vec{r}$ is the position vector at any time, and $\vec{r}_0$ is the position vector at $t = 0$. These equations are the vector equivalent of Eqs. 2–12a and b. In practical situations, we usually use the component form given in Table 3–1.

3–7 Projectile Motion

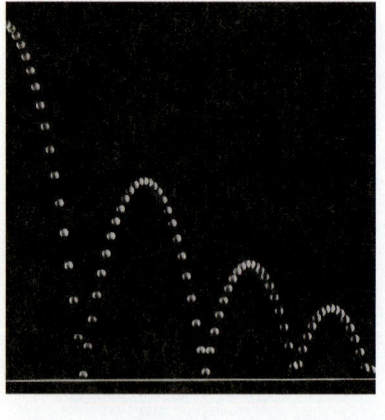

FIGURE 3–19 This strobe photograph of a ball making a series of bounces shows the characteristic "parabolic" path of projectile motion.

In Chapter 2, we studied one-dimensional motion of an object in terms of displacement, velocity, and acceleration, including purely vertical motion of a falling object undergoing acceleration due to gravity. Now we examine the more general translational motion of objects moving through the air in two dimensions near the Earth's surface, such as a golf ball, a thrown or batted baseball, kicked footballs, and speeding bullets. These are all examples of **projectile motion** (see Fig. 3–19), which we can describe as taking place in two dimensions.

Although air resistance is often important, in many cases its effect can be ignored, and we will ignore it in the following analysis. We will not be concerned now with the process by which the object is thrown or projected. We consider only its motion *after* it has been projected, and *before* it lands or is caught—that is, we analyze our projected object only when it is moving freely through the air under the action of gravity alone. Then the acceleration of the object is that due to gravity, which acts downward with magnitude $g = 9.80\,\text{m/s}^2$, and we assume it is constant.[†]

Galileo was the first to describe projectile motion accurately. He showed that it could be understood by analyzing the horizontal and vertical components of the motion separately. For convenience, we assume that the motion begins at time $t = 0$ at the origin of an xy coordinate system (so $x_0 = y_0 = 0$).

Let us look at a (tiny) ball rolling off the end of a horizontal table with an initial velocity in the horizontal (x) direction, v_{x0}. See Fig. 3–20, where an object falling vertically is also shown for comparison. The velocity vector $\vec{v}$ at each instant points in the direction of the ball's motion at that instant and is always tangent to the path. Following Galileo's ideas, we treat the horizontal and vertical components of the velocity, v_x and v_y, separately, and we can apply the kinematic equations (Eqs. 2–12a through 2–12c) to the x and y components of the motion.

First we examine the vertical (y) component of the motion. At the instant the ball leaves the table's top ($t = 0$), it has only an x component of velocity. Once the

---

[†]This restricts us to objects whose distance traveled and maximum height above the Earth are small compared to the Earth's radius (6400 km).

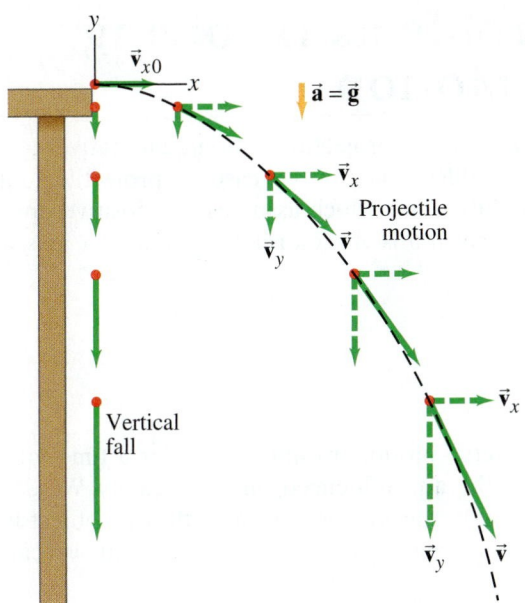

FIGURE 3–20 Projectile motion of a small ball projected horizontally. The dashed black line represents the path of the object. The velocity vector $\vec{v}$ at each point is in the direction of motion and thus is tangent to the path. The velocity vectors are green arrows, and velocity components are dashed. (A vertically falling object starting at the same point is shown at the left for comparison; v_y is the same for the falling object and the projectile.)

ball leaves the table (at $t = 0$), it experiences a vertically downward acceleration g, the acceleration due to gravity. Thus v_y is initially zero $(v_{y0} = 0)$ but increases continually in the downward direction (until the ball hits the ground). Let us take y to be positive upward. Then $a_y = -g$, and from Eq. 2–12a we can write $v_y = -gt$ since we set $v_{y0} = 0$. The vertical displacement is given by $y = -\frac{1}{2}gt^2$.

In the horizontal direction, on the other hand, the acceleration is zero (we are ignoring air resistance). With $a_x = 0$, the horizontal component of velocity, v_x, remains constant, equal to its initial value, v_{x0}, and thus has the same magnitude at each point on the path. The horizontal displacement is then given by $x = v_{x0}t$. The two vector components, $\vec{v}_x$ and $\vec{v}_y$, can be added vectorially at any instant to obtain the velocity $\vec{v}$ at that time (that is, for each point on the path), as shown in Fig. 3–20.

One result of this analysis, which Galileo himself predicted, is that *an object projected horizontally will reach the ground in the same time as an object dropped vertically.* This is because the vertical motions are the same in both cases, as shown in Fig. 3–20. Figure 3–21 is a multiple-exposure photograph of an experiment that confirms this.

| **EXERCISE D** Return to the Chapter-Opening Question, page 51, and answer it again now. Try to explain why you may have answered differently the first time.

If an object is projected at an upward angle, as in Fig. 3–22, the analysis is similar, except that now there is an initial vertical component of velocity, v_{y0}. Because of the downward acceleration of gravity, the upward component of velocity v_y gradually decreases with time until the object reaches the highest point on its path, at which point $v_y = 0$. Subsequently the object moves downward (Fig. 3–22) and v_y increases in the downward direction, as shown (that is, becoming more negative). As before, v_x remains constant.

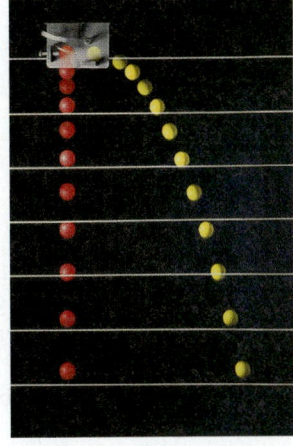

FIGURE 3–21 Multiple-exposure photograph showing positions of two balls at equal time intervals. One ball was dropped from rest at the same time the other was projected horizontally outward. The vertical position of each ball is seen to be the same at each instant.

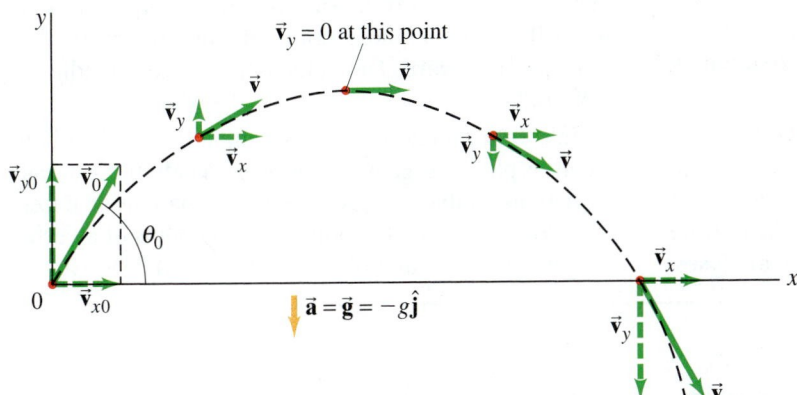

FIGURE 3–22 Path of a projectile fired with initial velocity $\vec{v}_0$ at angle θ_0 to the horizontal. Path is shown dashed in black, the velocity vectors are green arrows, and velocity components are dashed. The acceleration $\vec{a} = d\vec{v}/dt$ is downward. That is, $\vec{a} = \vec{g} = -g\hat{j}$ where $\hat{j}$ is the unit vector in the positive y direction.

3–8 Solving Problems Involving Projectile Motion

We now work through several Examples of projectile motion quantitatively.

We can simplify Eqs. 2–12 (Table 3–1) for the case of projectile motion because we can set $a_x = 0$. See Table 3–2, which assumes y is positive upward, so $a_y = -g = -9.80 \text{ m/s}^2$. Note that if θ is chosen relative to the $+x$ axis, as in Fig. 3–22, then

$$v_{x0} = v_0 \cos \theta_0,$$

$$v_{y0} = v_0 \sin \theta_0.$$

In doing Problems involving projectile motion, we must consider a time interval for which our chosen object is in the air, influenced only by gravity. We do not consider the throwing (or projecting) process, nor the time after the object lands or is caught, because then other influences act on the object, and we can no longer set $\vec{a} = \vec{g}$.

TABLE 3–2 Kinematic Equations for Projectile Motion
(*y* positive upward; $a_x = 0$, $a_y = -g = -9.80 \text{ m/s}^2$)

Horizontal Motion $\left(a_x = 0, v_x = \text{constant}\right)$		Vertical Motion[†] $\left(a_y = -g = \text{constant}\right)$
$v_x = v_{x0}$	(Eq. 2–12a)	$v_y = v_{y0} - gt$
$x = x_0 + v_{x0}t$	(Eq. 2–12b)	$y = y_0 + v_{y0}t - \frac{1}{2}gt^2$
	(Eq. 2–12c)	$v_y^2 = v_{y0}^2 - 2g(y - y_0)$

[†] If *y* is taken positive downward, the minus ($-$) signs in front of *g* become plus ($+$) signs.

PROBLEM SOLVING

Projectile Motion

Our approach to solving problems in Section 2–6 also applies here. Solving problems involving projectile motion can require creativity, and cannot be done just by following some rules. Certainly you must avoid just plugging numbers into equations that seem to "work."

1. As always, **read** carefully; **choose** the **object** (or objects) you are going to analyze.

2. **Draw** a careful **diagram** showing what is happening to the object.

3. **Choose** an origin and an *xy* **coordinate system**.

4. Decide on the **time interval**, which for projectile motion can only include motion under the effect of gravity alone, not throwing or landing. The time interval must be the same for the *x* and *y* analyses.

The *x* and *y* motions are connected by the common time.

5. **Examine** the horizontal (*x*) and vertical (*y*) **motions** separately. If you are given the initial velocity, you may want to resolve it into its *x* and *y* components.

6. List the **known** and **unknown** quantities, choosing $a_x = 0$ and $a_y = -g$ or $+g$, where $g = 9.80 \text{ m/s}^2$, and using the $+$ or $-$ sign, depending on whether you choose *y* positive down or up. Remember that v_x never changes throughout the trajectory, and that $v_y = 0$ at the highest point of any trajectory that returns downward. The velocity just before landing is generally not zero.

7. Think for a minute before jumping into the equations. A little planning goes a long way. **Apply** the **relevant equations** (Table 3–2), combining equations if necessary. You may need to combine components of a vector to get magnitude and direction (Eqs. 3–3).

EXAMPLE 3–6 **Driving off a cliff.** A movie stunt driver on a motorcycle speeds horizontally off a 50.0-m-high cliff. How fast must the motorcycle leave the cliff top to land on level ground below, 90.0 m from the base of the cliff where the cameras are? Ignore air resistance.

APPROACH We explicitly follow the steps of the Problem Solving Strategy above.

SOLUTION

1. and 2. **Read, choose the object, and draw a diagram**. Our object is the motorcycle and driver, taken as a single unit. The diagram is shown in Fig. 3–23.

3. **Choose a coordinate system**. We choose the y direction to be positive upward, with the top of the cliff as $y_0 = 0$. The x direction is horizontal with $x_0 = 0$ at the point where the motorcycle leaves the cliff.

4. **Choose a time interval**. We choose our time interval to begin $(t = 0)$ just as the motorcycle leaves the cliff top at position $x_0 = 0$, $y_0 = 0$; our time interval ends just before the motorcycle hits the ground below.

5. **Examine x and y motions**. In the horizontal (x) direction, the acceleration $a_x = 0$, so the velocity is constant. The value of x when the motorcycle reaches the ground is $x = +90.0$ m. In the vertical direction, the acceleration is the acceleration due to gravity, $a_y = -g = -9.80 \text{ m/s}^2$. The value of y when the motorcycle reaches the ground is $y = -50.0$ m. The initial velocity is horizontal and is our unknown, v_{x0}; the initial vertical velocity is zero, $v_{y0} = 0$.

6. **List knowns and unknowns**. See the Table in the margin. Note that in addition to not knowing the initial horizontal velocity v_{x0} (which stays constant until landing), we also do not know the time t when the motorcycle reaches the ground.

7. **Apply relevant equations**. The motorcycle maintains constant v_x as long as it is in the air. The time it stays in the air is determined by the y motion—when it hits the ground. So we first find the time using the y motion, and then use this time value in the x equations. To find out how long it takes the motorcycle to reach the ground below, we use Eq. 2–12b (Table 3–2) for the vertical (y) direction with $y_0 = 0$ and $v_{y0} = 0$:

$$y = y_0 + v_{y0}t + \tfrac{1}{2}a_y t^2$$
$$= 0 + 0 + \tfrac{1}{2}(-g)t^2$$

or

$$y = -\tfrac{1}{2}gt^2.$$

We solve for t and set $y = -50.0$ m:

$$t = \sqrt{\frac{2y}{-g}} = \sqrt{\frac{2(-50.0 \text{ m})}{-9.80 \text{ m/s}^2}} = 3.19 \text{ s}.$$

To calculate the initial velocity, v_{x0}, we again use Eq. 2–12b, but this time for the horizontal (x) direction, with $a_x = 0$ and $x_0 = 0$:

$$x = x_0 + v_{x0}t + \tfrac{1}{2}a_x t^2$$
$$= 0 + v_{x0}t + 0$$

or

$$x = v_{x0}t.$$

Then

$$v_{x0} = \frac{x}{t} = \frac{90.0 \text{ m}}{3.19 \text{ s}} = 28.2 \text{ m/s},$$

which is about 100 km/h (roughly 60 mi/h).

NOTE In the time interval of the projectile motion, the only acceleration is g in the negative y direction. The acceleration in the x direction is zero.

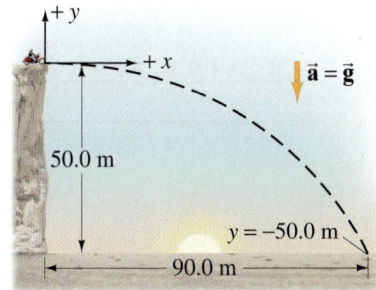

FIGURE 3–23 Example 3–6.

Known	Unknown
$x_0 = y_0 = 0$	v_{x0}
$x = 90.0$ m	t
$y = -50.0$ m	
$a_x = 0$	
$a_y = -g = -9.80 \text{ m/s}^2$	
$v_{y0} = 0$	

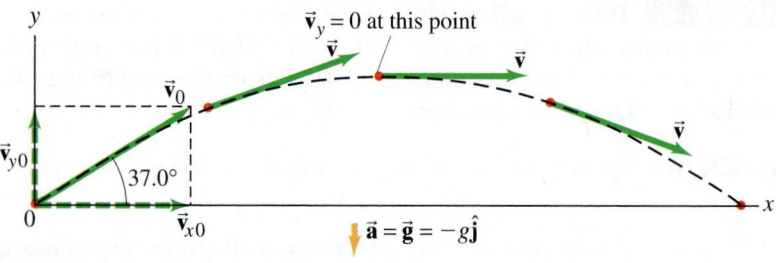

FIGURE 3–24 Example 3–7.

$\vec{v}_y = 0$ at this point

$\vec{a} = \vec{g} = -g\hat{j}$

EXAMPLE 3–7 **A kicked football.** A football is kicked at an angle $\theta_0 = 37.0°$ with a velocity of 20.0 m/s, as shown in Fig. 3–24. Calculate (*a*) the maximum height, (*b*) the time of travel before the football hits the ground, (*c*) how far away it hits the ground, (*d*) the velocity vector at the maximum height, and (*e*) the acceleration vector at maximum height. Assume the ball leaves the foot at ground level, and ignore air resistance and rotation of the ball.

APPROACH This may seem difficult at first because there are so many questions. But we can deal with them one at a time. We take the y direction as positive upward, and treat the x and y motions separately. The total time in the air is again determined by the y motion. The x motion occurs at constant velocity. The y component of velocity varies, being positive (upward) initially, decreasing to zero at the highest point, and then becoming negative as the football falls.

SOLUTION We resolve the initial velocity into its components (Fig. 3–24):

$$v_{x0} = v_0 \cos 37.0° = (20.0 \text{ m/s})(0.799) = 16.0 \text{ m/s}$$
$$v_{y0} = v_0 \sin 37.0° = (20.0 \text{ m/s})(0.602) = 12.0 \text{ m/s}.$$

(*a*) We consider a time interval that begins just after the football loses contact with the foot until it reaches its maximum height. During this time interval, the acceleration is g downward. At the maximum height, the velocity is horizontal (Fig. 3–24), so $v_y = 0$; and this occurs at a time given by $v_y = v_{y0} - gt$ with $v_y = 0$ (see Eq. 2–12a in Table 3–2). Thus

$$t = \frac{v_{y0}}{g} = \frac{(12.0 \text{ m/s})}{(9.80 \text{ m/s}^2)} = 1.224 \text{ s} \approx 1.22 \text{ s}.$$

From Eq. 2–12b, with $y_0 = 0$, we have

$$y = v_{y0}t - \tfrac{1}{2}gt^2$$
$$= (12.0 \text{ m/s})(1.224 \text{ s}) - \tfrac{1}{2}(9.80 \text{ m/s}^2)(1.224 \text{ s})^2 = 7.35 \text{ m}.$$

Alternatively, we could have used Eq. 2–12c, solved for y, and found

$$y = \frac{v_{y0}^2 - v_y^2}{2g} = \frac{(12.0 \text{ m/s})^2 - (0 \text{ m/s})^2}{2(9.80 \text{ m/s}^2)} = 7.35 \text{ m}.$$

The maximum height is 7.35 m.

(*b*) To find the time it takes for the ball to return to the ground, we consider a different time interval, starting at the moment the ball leaves the foot ($t = 0$, $y_0 = 0$) and ending just before the ball touches the ground ($y = 0$ again). We can use Eq. 2–12b with $y_0 = 0$ and also set $y = 0$ (ground level):

$$y = y_0 + v_{y0}t - \tfrac{1}{2}gt^2$$
$$0 = 0 + v_{y0}t - \tfrac{1}{2}gt^2.$$

This equation can be easily factored:

$$t\left(\tfrac{1}{2}gt - v_{y0}\right) = 0.$$

There are two solutions, $t = 0$ (which corresponds to the initial point, y_0), and

$$t = \frac{2v_{y0}}{g} = \frac{2(12.0 \text{ m/s})}{(9.80 \text{ m/s}^2)} = 2.45 \text{ s},$$

which is the total travel time of the football.

NOTE The time needed for the whole trip, $t = 2v_{y0}/g = 2.45$ s, is double the time to reach the highest point, calculated in (a). That is, the time to go up equals the time to come back down to the same level (ignoring air resistance).

(c) The total distance traveled in the x direction is found by applying Eq. 2–12b with $x_0 = 0$, $a_x = 0$, $v_{x0} = 16.0$ m/s:

$$x = v_{x0}t = (16.0 \text{ m/s})(2.45 \text{ s}) = 39.2 \text{ m}.$$

(d) At the highest point, there is no vertical component to the velocity. There is only the horizontal component (which remains constant throughout the flight), so $v = v_{x0} = v_0 \cos 37.0° = 16.0$ m/s.

(e) The acceleration vector is the same at the highest point as it is throughout the flight, which is 9.80 m/s² downward.

NOTE We treated the football as if it were a particle, ignoring its rotation. We also ignored air resistance. Because air resistance is significant on a football, our results are only estimates.

EXERCISE E Two balls are thrown in the air at different angles, but each reaches the same height. Which ball remains in the air longer: the one thrown at the steeper angle or the one thrown at a shallower angle?

CONCEPTUAL EXAMPLE 3–8 **Where does the apple land?** A child sits upright in a wagon which is moving to the right at constant speed as shown in Fig. 3–25. The child extends her hand and throws an apple straight upward (from her own point of view, Fig. 3–25a), while the wagon continues to travel forward at constant speed. If air resistance is neglected, will the apple land (a) behind the wagon, (b) in the wagon, or (c) in front of the wagon?

RESPONSE The child throws the apple straight up from her own reference frame with initial velocity $\vec{v}_{y0}$ (Fig. 3–25a). But when viewed by someone on the ground, the apple also has an initial horizontal component of velocity equal to the speed of the wagon, $\vec{v}_{x0}$. Thus, to a person on the ground, the apple will follow the path of a projectile as shown in Fig. 3–25b. The apple experiences no horizontal acceleration, so $\vec{v}_{x0}$ will stay constant and equal to the speed of the wagon. As the apple follows its arc, the wagon will be directly under the apple at all times because they have the same horizontal velocity. When the apple comes down, it will drop right into the outstretched hand of the child. The answer is (b).

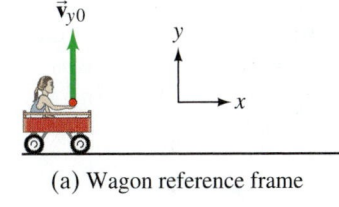
(a) Wagon reference frame

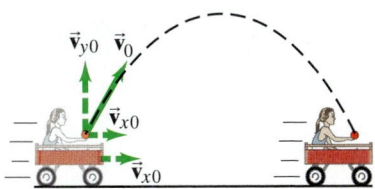
(b) Ground reference frame

FIGURE 3–25 Example 3–8.

CONCEPTUAL EXAMPLE 3–9 **The wrong strategy.** A boy on a small hill aims his water-balloon slingshot horizontally, straight at a second boy hanging from a tree branch a distance d away, Fig. 3–26. At the instant the water balloon is released, the second boy lets go and falls from the tree, hoping to avoid being hit. Show that he made the wrong move. (He hadn't studied physics yet.) Ignore air resistance.

RESPONSE Both the water balloon and the boy in the tree start falling at the same instant, and in a time t they each fall the same vertical distance $y = \frac{1}{2}gt^2$, much like Fig. 3–21. In the time it takes the water balloon to travel the horizontal distance d, the balloon will have the same y position as the falling boy. Splat. If the boy had stayed in the tree, he would have avoided the humiliation.

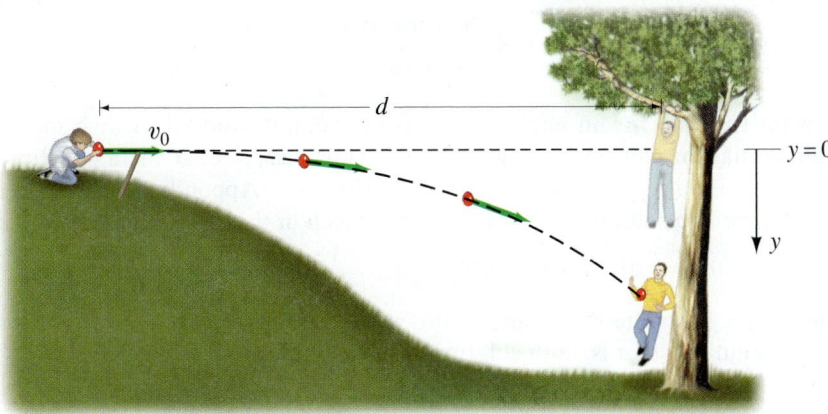

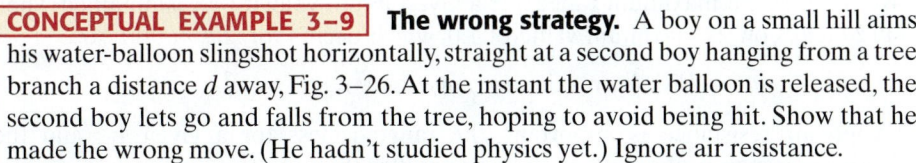

FIGURE 3–26 Example 3–9.

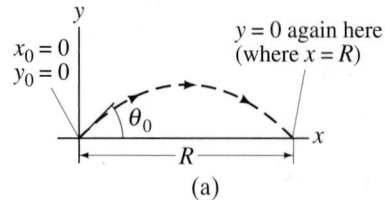

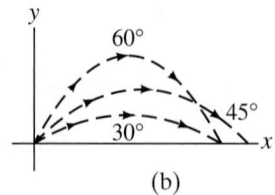

FIGURE 3–27 Example 3–10.
(a) The range R of a projectile;
(b) there are generally two angles θ_0 that will give the same range. Can you show that if one angle is θ_{01}, the other is $\theta_{02} = 90° - \theta_{01}$?

EXAMPLE 3–10 **Level horizontal range.** (a) Derive a formula for the horizontal range R of a projectile in terms of its initial speed v_0 and angle θ_0. The horizontal *range* is defined as the horizontal distance the projectile travels before returning to its original height (which is typically the ground); that is, y (final) $= y_0$. See Fig. 3–27a. (b) Suppose one of Napoleon's cannons had a muzzle speed, v_0, of 60.0 m/s. At what angle should it have been aimed (ignore air resistance) to strike a target 320 m away?

APPROACH The situation is the same as in Example 3–7, except we are now not given numbers in (a). We will algebraically manipulate equations to obtain our result.

SOLUTION (a) We set $x_0 = 0$ and $y_0 = 0$ at $t = 0$. After the projectile travels a horizontal distance R, it returns to the same level, $y = 0$, the final point. We choose our time interval to start $(t = 0)$ just after the projectile is fired and to end when it returns to the same vertical height. To find a general expression for R, we set both $y = 0$ and $y_0 = 0$ in Eq. 2–12b for the vertical motion, and obtain

$$y = y_0 + v_{y0}t + \tfrac{1}{2}a_y t^2$$

so

$$0 = 0 + v_{y0}t - \tfrac{1}{2}gt^2.$$

We solve for t, which gives two solutions: $t = 0$ and $t = 2v_{y0}/g$. The first solution corresponds to the initial instant of projection and the second is the time when the projectile returns to $y = 0$. Then the range, R, will be equal to x at the moment t has this value, which we put into Eq. 2–12b for the *horizontal* motion $(x = v_{x0}t$, with $x_0 = 0)$. Thus we have:

$$R = v_{x0}t = v_{x0}\left(\frac{2v_{y0}}{g}\right) = \frac{2v_{x0}v_{y0}}{g} = \frac{2v_0^2 \sin\theta_0 \cos\theta_0}{g}, \qquad [y = y_0]$$

where we have written $v_{x0} = v_0 \cos\theta_0$ and $v_{y0} = v_0 \sin\theta_0$. This is the result we sought. It can be rewritten, using the trigonometric identity $2\sin\theta\cos\theta = \sin 2\theta$ (Appendix A or inside the rear cover):

$$R = \frac{v_0^2 \sin 2\theta_0}{g}. \qquad [\text{only if } y \text{ (final)} = y_0]$$

We see that the maximum range, for a given initial velocity v_0, is obtained when $\sin 2\theta$ takes on its maximum value of 1.0, which occurs for $2\theta_0 = 90°$; so

$$\theta_0 = 45° \text{ for maximum range, and } R_{\text{max}} = v_0^2/g.$$

[When air resistance is important, the range is less for a given v_0, and the maximum range is obtained at an angle smaller than 45°.]

NOTE The maximum range increases by the square of v_0, so doubling the muzzle velocity of a cannon increases its maximum range by a factor of 4.

(b) We put $R = 320$ m into the equation we just derived, and (assuming, unrealistically, no air resistance) we solve it to find

$$\sin 2\theta_0 = \frac{Rg}{v_0^2} = \frac{(320 \text{ m})(9.80 \text{ m/s}^2)}{(60.0 \text{ m/s})^2} = 0.871.$$

We want to solve for an angle θ_0 that is between 0° and 90°, which means $2\theta_0$ in this equation can be as large as 180°. Thus, $2\theta_0 = 60.6°$ is a solution, but $2\theta_0 = 180° - 60.6° = 119.4°$ is also a solution (see Appendix A–9). In general we will have two solutions (see Fig. 3–27b), which in the present case are given by

$$\theta_0 = 30.3° \quad \text{or} \quad 59.7°.$$

Either angle gives the same range. Only when $\sin 2\theta_0 = 1$ (so $\theta_0 = 45°$) is there a single solution (that is, both solutions are the same).

EXERCISE F The maximum range of a projectile is found to be 100 m. If the projectile strikes the ground a distance of 82 m away, what was the angle of launch? (*a*) 35° or 55°; (*b*) 30° or 60°; (*c*) 27.5° or 62.5°; (*d*) 13.75° or 76.25°.

The level range formula derived in Example 3–10 applies only if takeoff and landing are at the same height $(y = y_0)$. Example 3–11 below considers a case where they are not equal heights $(y \neq y_0)$.

EXAMPLE 3–11 **A punt.** Suppose the football in Example 3–7 was punted and left the punter's foot at a height of 1.00 m above the ground. How far did the football travel before hitting the ground? Set $x_0 = 0$, $y_0 = 0$.

APPROACH The x and y motions are again treated separately. But we cannot use the range formula from Example 3–10 because it is valid only if y (final) $= y_0$, which is not the case here. Now we have $y_0 = 0$, and the football hits the ground where $y = -1.00$ m (see Fig. 3–28). We choose our time interval to start when the ball leaves his foot $(t = 0, y_0 = 0, x_0 = 0)$ and end just before the ball hits the ground $(y = -1.00$ m$)$. We can get x from Eq. 2–12b, $x = v_{x0}t$, since we know that $v_{x0} = 16.0$ m/s from Example 3–7. But first we must find t, the time at which the ball hits the ground, which we obtain from the y motion.

PHYSICS APPLIED
Sports

PROBLEM SOLVING
Do not use any formula unless you are sure its range of validity fits the problem; the range formula does not apply here because $y \neq y_0$

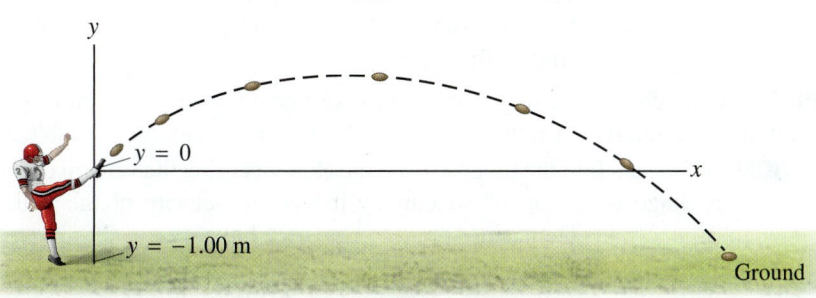

FIGURE 3–28 Example 3–11: the football leaves the punter's foot at $y = 0$, and reaches the ground where $y = -1.00$ m.

SOLUTION With $y = -1.00$ m and $v_{y0} = 12.0$ m/s (see Example 3–7), we use the equation

$$y = y_0 + v_{y0}t - \tfrac{1}{2}gt^2,$$

and obtain

$$-1.00 \text{ m} = 0 + (12.0 \text{ m/s})t - (4.90 \text{ m/s}^2)t^2.$$

We rearrange this equation into standard form $(ax^2 + bx + c = 0)$ so we can use the quadratic formula:

$$(4.90 \text{ m/s}^2)t^2 - (12.0 \text{ m/s})t - (1.00 \text{ m}) = 0.$$

The quadratic formula (Appendix A–1) gives

$$t = \frac{12.0 \text{ m/s} \pm \sqrt{(-12.0 \text{ m/s})^2 - 4(4.90 \text{ m/s}^2)(-1.00 \text{ m})}}{2(4.90 \text{ m/s}^2)}$$

$$= 2.53 \text{ s} \quad \text{or} \quad -0.081 \text{ s}.$$

The second solution would correspond to a time prior to our chosen time interval that begins at the kick, so it doesn't apply. With $t = 2.53$ s for the time at which the ball touches the ground, the horizontal distance the ball traveled is (using $v_{x0} = 16.0$ m/s from Example 3–7):

$$x = v_{x0}t = (16.0 \text{ m/s})(2.53 \text{ s}) = 40.5 \text{ m}.$$

Our assumption in Example 3–7 that the ball leaves the foot at ground level would result in an underestimate of about 1.3 m in the distance our punt traveled.

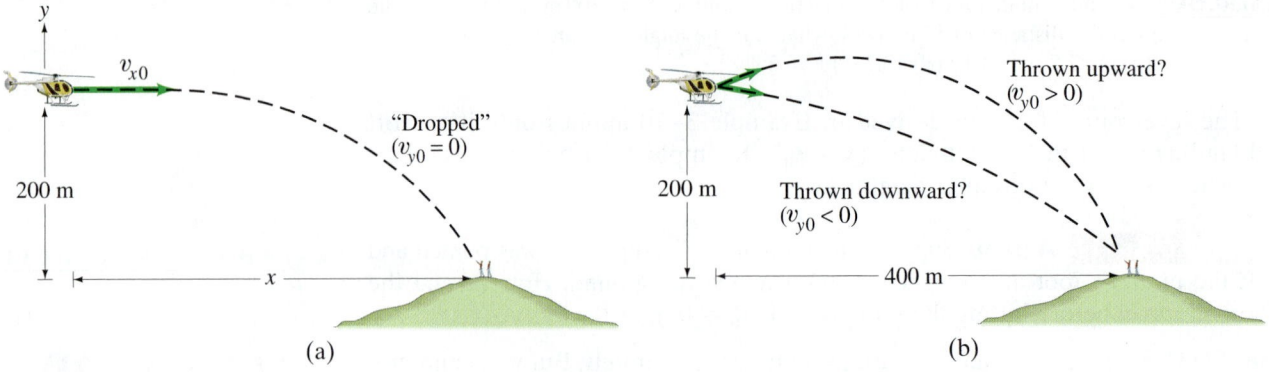

FIGURE 3–29 Example 3–12.

EXAMPLE 3–12 **Rescue helicopter drops supplies.** A rescue helicopter wants to drop a package of supplies to isolated mountain climbers on a rocky ridge 200 m below. If the helicopter is traveling horizontally with a speed of 70 m/s (250 km/h), (*a*) how far in advance of the recipients (horizontal distance) must the package be dropped (Fig. 3–29a)? (*b*) Suppose, instead, that the helicopter releases the package a horizontal distance of 400 m in advance of the mountain climbers. What vertical velocity should the package be given (up or down) so that it arrives precisely at the climbers' position (Fig. 3–29b)? (*c*) With what speed does the package land in the latter case?

APPROACH We choose the origin of our xy coordinate system at the initial position of the helicopter, taking $+y$ upward, and use the kinematic equations (Table 3–2).

SOLUTION (*a*) We can find the time to reach the climbers using the vertical distance of 200 m. The package is "dropped" so initially it has the velocity of the helicopter, $v_{x0} = 70$ m/s, $v_{y0} = 0$. Then, since $y = -\frac{1}{2}gt^2$, we have

$$t = \sqrt{\frac{-2y}{g}} = \sqrt{\frac{-2(-200 \text{ m})}{9.80 \text{ m/s}^2}} = 6.39 \text{ s.}$$

The horizontal motion of the falling package is at constant speed of 70 m/s. So

$$x = v_{x0}t = (70 \text{ m/s})(6.39 \text{ s}) = 447 \text{ m} \approx 450 \text{ m,}$$

assuming the given numbers were good to two significant figures.

(*b*) We are given $x = 400$ m, $v_{x0} = 70$ m/s, $y = -200$ m, and we want to find v_{y0} (see Fig. 3–29b). Like most problems, this one can be approached in various ways. Instead of searching for a formula or two, let's try to reason it out in a simple way, based on what we did in part (*a*). If we know t, perhaps we can get v_{y0}. Since the horizontal motion of the package is at constant speed (once it is released we don't care what the helicopter does), we have $x = v_{x0}t$, so

$$t = \frac{x}{v_{x0}} = \frac{400 \text{ m}}{70 \text{ m/s}} = 5.71 \text{ s.}$$

Now let's try to use the vertical motion to get v_{y0}: $y = y_0 + v_{y0}t - \frac{1}{2}gt^2$. Since $y_0 = 0$ and $y = -200$ m, we can solve for v_{y0}:

$$v_{y0} = \frac{y + \frac{1}{2}gt^2}{t} = \frac{-200 \text{ m} + \frac{1}{2}(9.80 \text{ m/s}^2)(5.71 \text{ s})^2}{5.71 \text{ s}} = -7.0 \text{ m/s.}$$

Thus, in order to arrive at precisely the mountain climbers' position, the package must be thrown *downward* from the helicopter with a speed of 7.0 m/s.

(*c*) We want to know v of the package at $t = 5.71$ s. The components are:

$$v_x = v_{x0} = 70 \text{ m/s}$$

$$v_y = v_{y0} - gt = -7.0 \text{ m/s} - (9.80 \text{ m/s}^2)(5.71 \text{ s}) = -63 \text{ m/s.}$$

So $v = \sqrt{(70 \text{ m/s})^2 + (-63 \text{ m/s})^2} = 94$ m/s. (Better not to release the package from such an altitude, or use a parachute.)

Projectile Motion Is Parabolic

We now show that the path followed by any projectile is a *parabola*, if we can ignore air resistance and can assume that $\vec{\mathbf{g}}$ is constant. To do so, we need to find y as a function of x by eliminating t between the two equations for horizontal and vertical motion (Eq. 2–12b in Table 3–2), and for simplicity we set $x_0 = y_0 = 0$:

$$x = v_{x0}t$$

$$y = v_{y0}t - \tfrac{1}{2}gt^2$$

From the first equation, we have $t = x/v_{x0}$, and we substitute this into the second one to obtain

$$y = \left(\frac{v_{y0}}{v_{x0}}\right)x - \left(\frac{g}{2v_{x0}^2}\right)x^2. \tag{3–14}$$

We see that y as a function of x has the form

$$y = Ax - Bx^2,$$

where A and B are constants for any specific projectile motion. This is the well-known equation for a parabola. See Figs. 3–19 and 3–30.

The idea that projectile motion is parabolic was, in Galileo's day, at the forefront of physics research. Today we discuss it in Chapter 3 of introductory physics!

FIGURE 3–30 Examples of projectile motion—sparks (small hot glowing pieces of metal), water, and fireworks. The parabolic path characteristic of projectile motion is affected by air resistance.

3–9 Relative Velocity

We now consider how observations made in different frames of reference are related to each other. For example, consider two trains approaching one another, each with a speed of 80 km/h with respect to the Earth. Observers on the Earth beside the train tracks will measure 80 km/hr for the speed of each of the trains. Observers on either one of the trains (a different frame of reference) will measure a speed of 160 km/h for the train approaching them.

Similarly, when one car traveling 90 km/h passes a second car traveling in the same direction at 75 km/h, the first car has a speed relative to the second car of 90 km/h − 75 km/h = 15 km/h.

When the velocities are along the same line, simple addition or subtraction is sufficient to obtain the relative velocity. But if they are not along the same line, we must make use of vector addition. We emphasize, as mentioned in Section 2–1, that when specifying a velocity, it is important to specify what the reference frame is.

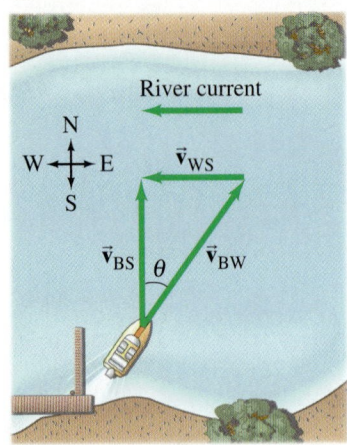

FIGURE 3–31 To move directly across the river, the boat must head upstream at an angle θ. Velocity vectors are shown as green arrows:

$\vec{\mathbf{v}}_{BS}$ = velocity of **B**oat with respect to the **S**hore,

$\vec{\mathbf{v}}_{BW}$ = velocity of **B**oat with respect to the **W**ater,

$\vec{\mathbf{v}}_{WS}$ = velocity of the **W**ater with respect to the **S**hore (river current).

When determining relative velocity, it is easy to make a mistake by adding or subtracting the wrong velocities. It is important, therefore, to draw a diagram and use a careful labeling process. Each velocity is labeled by *two subscripts: the first refers to the object, the second to the reference frame in which it has this velocity.* For example, suppose a boat is to cross a river to the opposite side, as shown in Fig. 3–31. We let $\vec{\mathbf{v}}_{BW}$ be the velocity of the **B**oat with respect to the **W**ater. (This is also what the boat's velocity would be relative to the shore if the water were still.) Similarly, $\vec{\mathbf{v}}_{BS}$ is the velocity of the **B**oat with respect to the **S**hore, and $\vec{\mathbf{v}}_{WS}$ is the velocity of the **W**ater with respect to the **S**hore (this is the river current). Note that $\vec{\mathbf{v}}_{BW}$ is what the boat's motor produces (against the water), whereas $\vec{\mathbf{v}}_{BS}$ is equal to $\vec{\mathbf{v}}_{BW}$ plus the effect of the current, $\vec{\mathbf{v}}_{WS}$. Therefore, the velocity of the boat relative to the shore is (see vector diagram, Fig. 3–31)

$$\vec{\mathbf{v}}_{BS} = \vec{\mathbf{v}}_{BW} + \vec{\mathbf{v}}_{WS}. \tag{3-15}$$

By writing the subscripts using this convention, we see that the inner subscripts (the two W's) on the right-hand side of Eq. 3–15 are the same, whereas the outer subscripts on the right of Eq. 3–15 (the B and the S) are the same as the two subscripts for the sum vector on the left, $\vec{\mathbf{v}}_{BS}$. By following this convention (first subscript for the object, second for the reference frame), you can write down the correct equation relating velocities in different reference frames.[†] Figure 3–32 gives a derivation of Eq. 3–15.

Equation 3–15 is valid in general and can be extended to three or more velocities. For example, if a fisherman on the boat walks with a velocity $\vec{\mathbf{v}}_{FB}$ relative to the boat, his velocity relative to the shore is $\vec{\mathbf{v}}_{FS} = \vec{\mathbf{v}}_{FB} + \vec{\mathbf{v}}_{BW} + \vec{\mathbf{v}}_{WS}$. The equations involving relative velocity will be correct when adjacent inner subscripts are identical and when the outermost ones correspond exactly to the two on the velocity on the left of the equation. But this works only with plus signs (on the right), not minus signs.

It is often useful to remember that for any two objects or reference frames, A and B, the velocity of A relative to B has the same magnitude, but opposite direction, as the velocity of B relative to A:

$$\vec{\mathbf{v}}_{BA} = -\vec{\mathbf{v}}_{AB}. \tag{3-16}$$

For example, if a train is traveling 100 km/h relative to the Earth in a certain direction, objects on the Earth (such as trees) appear to an observer on the train to be traveling 100 km/h in the opposite direction.

[†]We thus would know by inspection that (for example) the equation $\vec{\mathbf{V}}_{BW} = \vec{\mathbf{V}}_{BS} + \vec{\mathbf{V}}_{WS}$ is wrong.

FIGURE 3–32 Derivation of relative velocity equation (Eq. 3–15), in this case for a person walking along the corridor in a train. We are looking down on the train and two reference frames are shown: xy on the Earth and $x'y'$ fixed on the train. We have:

$\vec{\mathbf{r}}_{PT}$ = position vector of person (P) relative to train (T),

$\vec{\mathbf{r}}_{PE}$ = position vector of person (P) relative to Earth (E),

$\vec{\mathbf{r}}_{TE}$ = position vector of train's coordinate system (T) relative to Earth (E).

From the diagram we see that

$$\vec{\mathbf{r}}_{PE} = \vec{\mathbf{r}}_{PT} + \vec{\mathbf{r}}_{TE}.$$

We take the derivative with respect to time to obtain

$$\frac{d}{dt}(\vec{\mathbf{r}}_{PE}) = \frac{d}{dt}(\vec{\mathbf{r}}_{PT}) + \frac{d}{dt}(\vec{\mathbf{r}}_{TE}).$$

or, since $d\vec{\mathbf{r}}/dt = \vec{\mathbf{v}}$,

$$\vec{\mathbf{v}}_{PE} = \vec{\mathbf{v}}_{PT} + \vec{\mathbf{v}}_{TE}.$$

This is the equivalent of Eq. 3–15 for the present situation (check the subscripts!).

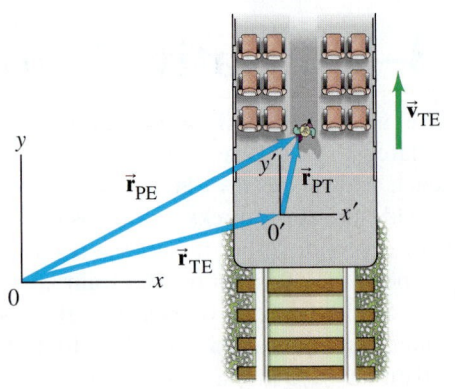

CONCEPTUAL EXAMPLE 3–13 **Crossing a river.** A woman in a small motor boat is trying to cross a river that flows due west with a strong current. The woman starts on the south bank and is trying to reach the north bank directly north from her starting point. Should she (*a*) head due north, (*b*) head due west, (*c*) head in a north-westerly direction, (*d*) head in a northeasterly direction?

RESPONSE If the woman heads straight across the river, the current will drag the boat downstream (westward). To overcome the river's westward current, the boat must acquire an eastward component of velocity as well as a northward compo-nent. Thus the boat must (*d*) head in a northeasterly direction (see Fig. 3–33). The actual angle depends on the strength of the current and how fast the boat moves relative to the water. If the current is weak and the motor is strong, then the boat can head almost, but not quite, due north.

EXAMPLE 3–14 **Heading upstream.** A boat's speed in still water is $v_{BW} = 1.85$ m/s. If the boat is to travel directly across a river whose current has speed $v_{WS} = 1.20$ m/s, at what upstream angle must the boat head? (See Fig. 3–33.)

APPROACH We reason as in Example 3–13, and use subscripts as in Eq. 3–15. Figure 3–33 has been drawn with $\vec{v}_{BS}$, the velocity of the **B**oat relative to the **S**hore, pointing directly across the river because this is how the boat is supposed to move. (Note that $\vec{v}_{BS} = \vec{v}_{BW} + \vec{v}_{WS}$.) To accomplish this, the boat needs to head upstream to offset the current pulling it downstream.

SOLUTION Vector $\vec{v}_{BW}$ points upstream at an angle θ as shown. From the diagram,

$$\sin \theta = \frac{v_{WS}}{v_{BW}} = \frac{1.20 \text{ m/s}}{1.85 \text{ m/s}} = 0.6486.$$

Thus $\theta = 40.4°$, so the boat must head upstream at a 40.4° angle.

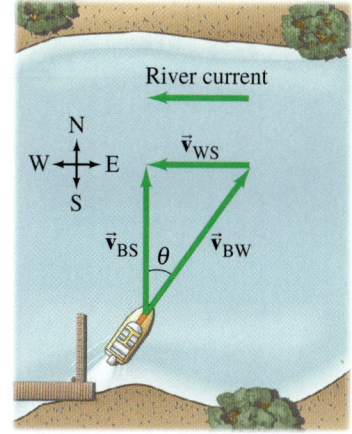

FIGURE 3–33 Examples 3–13 and 3–14.

EXAMPLE 3–15 **Heading across the river.** The same boat ($v_{BW} = 1.85$ m/s) now heads directly across the river whose current is still 1.20 m/s. (*a*) What is the velocity (magnitude and direction) of the boat relative to the shore? (*b*) If the river is 110 m wide, how long will it take to cross and how far downstream will the boat be then?

APPROACH The boat now heads directly across the river and is pulled down-stream by the current, as shown in Fig. 3–34. The boat's velocity with respect to the shore, $\vec{v}_{BS}$, is the sum of its velocity with respect to the water, $\vec{v}_{BW}$, plus the velocity of the water with respect to the shore, $\vec{v}_{WS}$:

$$\vec{v}_{BS} = \vec{v}_{BW} + \vec{v}_{WS},$$

just as before.

SOLUTION (*a*) Since $\vec{v}_{BW}$ is perpendicular to $\vec{v}_{WS}$, we can get v_{BS} using the theorem of Pythagoras:

$$v_{BS} = \sqrt{v_{BW}^2 + v_{WS}^2} = \sqrt{(1.85 \text{ m/s})^2 + (1.20 \text{ m/s})^2} = 2.21 \text{ m/s}.$$

We can obtain the angle (note how θ is defined in the diagram) from:

$$\tan \theta = v_{WS}/v_{BW} = (1.20 \text{ m/s})/(1.85 \text{ m/s}) = 0.6486.$$

Thus $\theta = \tan^{-1}(0.6486) = 33.0°$. Note that this angle is not equal to the angle calculated in Example 3–14.

(*b*) The travel time for the boat is determined by the time it takes to cross the river. Given the river's width $D = 110$ m, we can use the velocity component in the direction of D, $v_{BW} = D/t$. Solving for t, we get $t = 110$ m/1.85 m/s = 59.5 s. The boat will have been carried downstream, in this time, a distance

$$d = v_{WS}t = (1.20 \text{ m/s})(59.5 \text{ s}) = 71.4 \text{ m} \approx 71 \text{ m}.$$

NOTE There is no acceleration in this Example, so the motion involves only constant velocities (of the boat or of the river).

FIGURE 3–34 Example 3–15. A boat heading directly across a river whose current moves at 1.20 m/s.

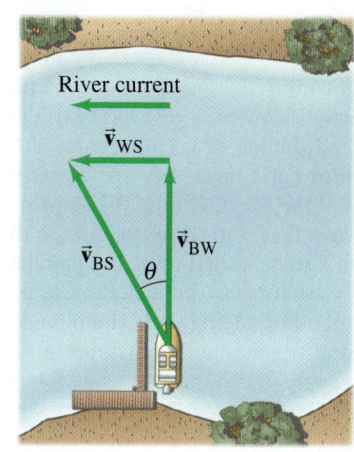

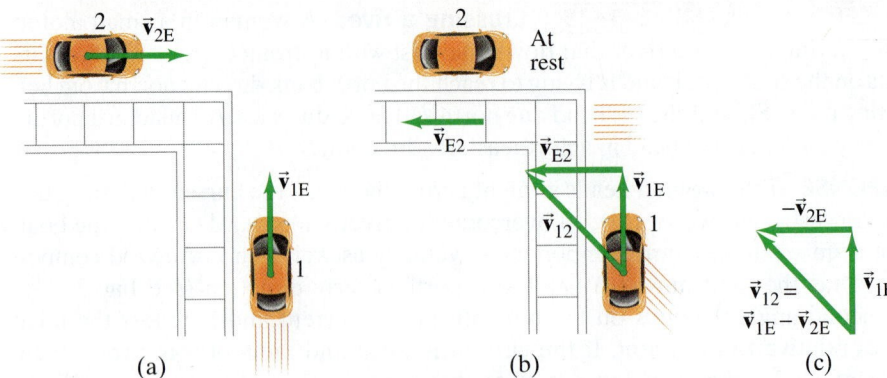

FIGURE 3–35 Example 3–16.

(a)   (b)   (c)

EXAMPLE 3–16 **Car velocities at 90°.** Two automobiles approach a street corner at right angles to each other with the same speed of 40.0 km/h (= 11.11 m/s), as shown in Fig. 3–35a. What is the relative velocity of one car with respect to the other? That is, determine the velocity of car 1 as seen by car 2.

APPROACH Figure 3–35a shows the situation in a reference frame fixed to the Earth. But we want to view the situation from a reference frame in which car 2 is at rest, and this is shown in Fig. 3–35b. In this reference frame (the world as seen by the driver of car 2), the Earth moves toward car 2 with velocity $\vec{v}_{E2}$ (speed of 40.0 km/h), which is of course equal and opposite to $\vec{v}_{2E}$, the velocity of car 2 with respect to the Earth (Eq. 3–16):

$$\vec{v}_{2E} = -\vec{v}_{E2}.$$

Then the velocity of car 1 as seen by car 2 is (see Eq. 3–15)

$$\vec{v}_{12} = \vec{v}_{1E} + \vec{v}_{E2}$$

SOLUTION Because $\vec{v}_{E2} = -\vec{v}_{2E}$, then

$$\vec{v}_{12} = \vec{v}_{1E} - \vec{v}_{2E}.$$

That is, the velocity of car 1 as seen by car 2 is the difference of their velocities, $\vec{v}_{1E} - \vec{v}_{2E}$, both measured relative to the Earth (see Fig. 3–35c). Since the magnitudes of $\vec{v}_{1E}$, $\vec{v}_{2E}$, and $\vec{v}_{E2}$ are equal (40.0 km/h = 11.11 m/s), we see (Fig. 3–35b) that $\vec{v}_{12}$ points at a 45° angle toward car 2; the speed is

$$v_{12} = \sqrt{(11.11 \text{ m/s})^2 + (11.11 \text{ m/s})^2} = 15.7 \text{ m/s} \ (= 56.6 \text{ km/h}).$$

Summary

A quantity that has both a magnitude and a direction is called a **vector**. A quantity that has only a magnitude is called a **scalar**.

Addition of vectors can be done graphically by placing the tail of each successive arrow (representing each vector) at the tip of the previous one. The sum, or **resultant vector**, is the arrow drawn from the tail of the first to the tip of the last. Two vectors can also be added using the parallelogram method.

Vectors can be added more accurately using the analytical method of adding their **components** along chosen axes with the aid of trigonometric functions. A vector of magnitude V making an angle θ with the x axis has components

$$V_x = V \cos \theta \qquad V_y = V \sin \theta. \qquad (3–2)$$

Given the components, we can find the magnitude and direction from

$$V = \sqrt{V_x^2 + V_y^2}, \qquad \tan \theta = \frac{V_y}{V_x}. \qquad (3–3)$$

It is often helpful to express a vector in terms of its components along chosen axes using **unit vectors**, which are vectors of unit length along the chosen coordinate axes; for Cartesian coordinates the unit vectors along the x, y, and z axes are called $\hat{\mathbf{i}}$, $\hat{\mathbf{j}}$, and $\hat{\mathbf{k}}$.

The general definitions for the **instantaneous velocity**, $\vec{\mathbf{v}}$, and **acceleration**, $\vec{\mathbf{a}}$, of a particle (in one, two, or three dimensions) are

$$\vec{\mathbf{v}} = \frac{d\vec{\mathbf{r}}}{dt} \qquad (3–8)$$

$$\vec{\mathbf{a}} = \frac{d\vec{\mathbf{v}}}{dt}, \qquad (3–11)$$

where $\vec{\mathbf{r}}$ is the position vector of the particle. The kinematic equations for motion with constant acceleration can be written for each of the x, y, and z components of the motion and have the same form as for one-dimensional motion (Eqs. 2–12). Or they can be written in the more general vector form:

$$\vec{\mathbf{v}} = \vec{\mathbf{v}}_0 + \vec{\mathbf{a}}t$$

$$\vec{\mathbf{r}} = \vec{\mathbf{r}}_0 + \vec{\mathbf{v}}_0 t + \tfrac{1}{2}\vec{\mathbf{a}}t^2 \qquad (3–13)$$

Projectile motion of an object moving in the air near the Earth's surface can be analyzed as two separate motions if air resistance can be ignored. The horizontal component of the motion is at constant velocity, whereas the vertical component is at constant acceleration, g, just as for an object falling vertically under the action of gravity.

The velocity of an object relative to one frame of reference can be found by vector addition if its velocity relative to a second frame of reference, and the **relative velocity** of the two reference frames, are known.

Questions

1. One car travels due east at 40 km/h, and a second car travels north at 40 km/h. Are their velocities equal? Explain.

2. Can you conclude that a car is not accelerating if its speedometer indicates a steady 60 km/h?

3. Can you give several examples of an object's motion in which a great distance is traveled but the displacement is zero?

4. Can the displacement vector for a particle moving in two dimensions ever be longer than the length of path traveled by the particle over the same time interval? Can it ever be less? Discuss.

5. During baseball practice, a batter hits a very high fly ball and then runs in a straight line and catches it. Which had the greater displacement, the player or the ball?

6. If $\vec{V} = \vec{V}_1 + \vec{V}_2$, is V necessarily greater than V_1 and/or V_2? Discuss.

7. Two vectors have length $V_1 = 3.5$ km and $V_2 = 4.0$ km. What are the maximum and minimum magnitudes of their vector sum?

8. Can two vectors, of unequal magnitude, add up to give the zero vector? Can *three* unequal vectors? Under what conditions?

9. Can the magnitude of a vector ever (*a*) equal, or (*b*) be less than, one of its components?

10. Can a particle with constant speed be accelerating? What if it has constant velocity?

11. Does the odometer of a car measure a scalar or a vector quantity? What about the speedometer?

12. A child wishes to determine the speed a slingshot imparts to a rock. How can this be done using only a meter stick, a rock, and the slingshot?

13. In archery, should the arrow be aimed directly at the target? How should your angle of aim depend on the distance to the target?

14. A projectile is launched at an upward angle of 30° to the horizontal with a speed of 30 m/s. How does the horizontal component of its velocity 1.0 s after launch compare with its horizontal component of velocity 2.0 s after launch, ignoring air resistance?

15. A projectile has the least speed at what point in its path?

16. It was reported in World War I that a pilot flying at an altitude of 2 km caught in his bare hands a bullet fired at the plane! Using the fact that a bullet slows down considerably due to air resistance, explain how this incident occurred.

17. Two cannonballs, A and B, are fired from the ground with identical initial speeds, but with θ_A larger than θ_B. (*a*) Which cannonball reaches a higher elevation? (*b*) Which stays longer in the air? (*c*) Which travels farther?

18. A person sitting in an enclosed train car, moving at constant velocity, throws a ball straight up into the air in her reference frame. (*a*) Where does the ball land? What is your answer if the car (*b*) accelerates, (*c*) decelerates, (*d*) rounds a curve, (*e*) moves with constant velocity but is open to the air?

19. If you are riding on a train that speeds past another train moving in the same direction on an adjacent track, it appears that the other train is moving backward. Why?

20. Two rowers, who can row at the same speed in still water, set off across a river at the same time. One heads straight across and is pulled downstream somewhat by the current. The other one heads upstream at an angle so as to arrive at a point opposite the starting point. Which rower reaches the opposite side first?

21. If you stand motionless under an umbrella in a rainstorm where the drops fall vertically you remain relatively dry. However, if you start running, the rain begins to hit your legs even if they remain under the umbrella. Why?

Problems

3–2 to 3–5 Vector Addition; Unit Vectors

1. (I) A car is driven 225 km west and then 78 km southwest (45°). What is the displacement of the car from the point of origin (magnitude and direction)? Draw a diagram.

2. (I) A delivery truck travels 28 blocks north, 16 blocks east, and 26 blocks south. What is its final displacement from the origin? Assume the blocks are equal length.

3. (I) If $V_x = 7.80$ units and $V_y = -6.40$ units, determine the magnitude and direction of $\vec{V}$.

4. (II) Graphically determine the resultant of the following three vector displacements: (1) 24 m, 36° north of east; (2) 18 m, 37° east of north; and (3) 26 m, 33° west of south.

5. (II) $\vec{V}$ is a vector 24.8 units in magnitude and points at an angle of 23.4° above the negative x axis. (*a*) Sketch this vector. (*b*) Calculate V_x and V_y. (*c*) Use V_x and V_y to obtain (again) the magnitude and direction of $\vec{V}$. [*Note*: Part (*c*) is a good way to check if you've resolved your vector correctly.]

6. (II) Figure 3–36 shows two vectors, $\vec{A}$ and $\vec{B}$, whose magnitudes are $A = 6.8$ units and $B = 5.5$ units. Determine $\vec{C}$ if (*a*) $\vec{C} = \vec{A} + \vec{B}$, (*b*) $\vec{C} = \vec{A} - \vec{B}$, (*c*) $\vec{C} = \vec{B} - \vec{A}$. Give the magnitude and direction for each.

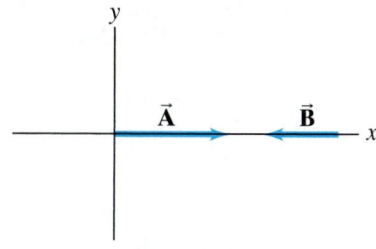

FIGURE 3–36 Problem 6.

7. (II) An airplane is traveling 835 km/h in a direction 41.5° west of north (Fig. 3–37). (*a*) Find the components of the velocity vector in the northerly and westerly directions. (*b*) How far north and how far west has the plane traveled after 2.50 h?

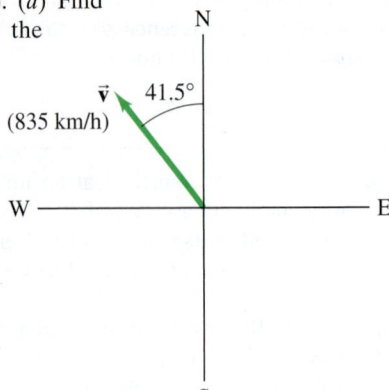

FIGURE 3–37
Problem 7.

8. (II) Let $\vec{V}_1 = -6.0\hat{i} + 8.0\hat{j}$ and $\vec{V}_2 = 4.5\hat{i} - 5.0\hat{j}$. Determine the magnitude and direction of (*a*) $\vec{V}_1$, (*b*) $\vec{V}_2$, (*c*) $\vec{V}_1 + \vec{V}_2$ and (*d*) $\vec{V}_2 - \vec{V}_1$.

9. (II) (*a*) Determine the magnitude and direction of the sum of the three vectors $\vec{V}_1 = 4.0\hat{i} - 8.0\hat{j}$, $\vec{V}_2 = \hat{i} + \hat{j}$, and $\vec{V}_3 = -2.0\hat{i} + 4.0\hat{j}$. (*b*) Determine $\vec{V}_1 - \vec{V}_2 + \vec{V}_3$.

10. (II) Three vectors are shown in Fig. 3–38. Their magnitudes are given in arbitrary units. Determine the sum of the three vectors. Give the resultant in terms of (*a*) components, (*b*) magnitude and angle with *x* axis.

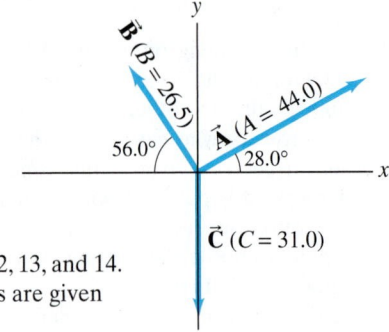

FIGURE 3–38
Problems 10, 11, 12, 13, and 14.
Vector magnitudes are given in arbitrary units.

11. (II) (*a*) Given the vectors $\vec{A}$ and $\vec{B}$ shown in Fig. 3–38, determine $\vec{B} - \vec{A}$. (*b*) Determine $\vec{A} - \vec{B}$ without using your answer in (*a*). Then compare your results and see if they are opposite.

12. (II) Determine the vector $\vec{A} - \vec{C}$, given the vectors $\vec{A}$ and $\vec{C}$ in Fig. 3–38.

13. (II) For the vectors shown in Fig. 3–38, determine (*a*) $\vec{B} - 2\vec{A}$, (*b*) $2\vec{A} - 3\vec{B} + 2\vec{C}$.

14. (II) For the vectors given in Fig. 3–38, determine (*a*) $\vec{A} - \vec{B} + \vec{C}$, (*b*) $\vec{A} + \vec{B} - \vec{C}$, and (*c*) $\vec{C} - \vec{A} - \vec{B}$.

15. (II) The summit of a mountain, 2450 m above base camp, is measured on a map to be 4580 m horizontally from the camp in a direction 32.4° west of north. What are the components of the displacement vector from camp to summit? What is its magnitude? Choose the *x* axis east, *y* axis north, and *z* axis up.

16. (III) You are given a vector in the *xy* plane that has a magnitude of 90.0 units and a *y* component of −55.0 units. (*a*) What are the two possibilities for its *x* component? (*b*) Assuming the *x* component is known to be positive, specify the vector which, if you add it to the original one, would give a resultant vector that is 80.0 units long and points entirely in the −*x* direction.

3–6 Vector Kinematics

17. (I) The position of a particular particle as a function of time is given by $\vec{r} = (9.60t\,\hat{i} + 8.85\hat{j} - 1.00t^2\hat{k})$ m. Determine the particle's velocity and acceleration as a function of time.

18. (I) What was the average velocity of the particle in Problem 17 between $t = 1.00$ s and $t = 3.00$ s? What is the magnitude of the instantaneous velocity at $t = 2.00$ s?

19. (II) What is the shape of the path of the particle of Problem 17?

20. (II) A car is moving with speed 18.0 m/s due south at one moment and 27.5 m/s due east 8.00 s later. Over this time interval, determine the magnitude and direction of (*a*) its average velocity, (*b*) its average acceleration. (*c*) What is its average speed. [*Hint:* Can you determine all these from the information given?]

21. (II) At $t = 0$, a particle starts from rest at $x = 0$, $y = 0$, and moves in the *xy* plane with an acceleration $\vec{a} = (4.0\hat{i} + 3.0\hat{j})$ m/s². Determine (*a*) the *x* and *y* components of velocity, (*b*) the speed of the particle, and (*c*) the position of the particle, all as a function of time. (*d*) Evaluate all the above at $t = 2.0$ s.

22. (II) (*a*) A skier is accelerating down a 30.0° hill at 1.80 m/s² (Fig. 3–39). What is the vertical component of her acceleration? (*b*) How long will it take her to reach the bottom of the hill, assuming she starts from rest and accelerates uniformly, if the elevation change is 325 m?

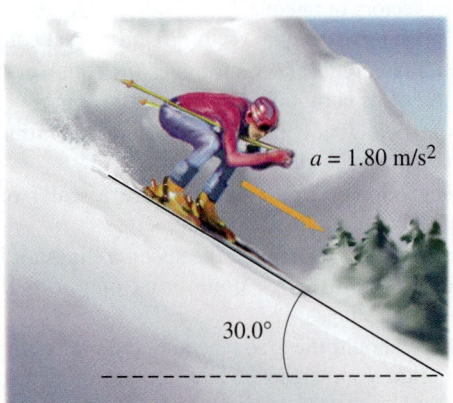

FIGURE 3–39 Problem 22.

23. (II) An ant walks on a piece of graph paper straight along the *x* axis a distance of 10.0 cm in 2.00 s. It then turns left 30.0° and walks in a straight line another 10.0 cm in 1.80 s. Finally, it turns another 70.0° to the left and walks another 10.0 cm in 1.55 s. Determine (*a*) the *x* and *y* components of the ant's average velocity, and (*b*) its magnitude and direction.

24. (II) A particle starts from the origin at $t = 0$ with an initial velocity of 5.0 m/s along the positive *x* axis. If the acceleration is $(-3.0\hat{i} + 4.5\hat{j})$ m/s², determine the velocity and position of the particle at the moment it reaches its maximum *x* coordinate.

25. (II) Suppose the position of an object is given by $\vec{r} = (3.0t^2\hat{i} - 6.0t^3\hat{j})$ m. (*a*) Determine its velocity $\vec{v}$ and acceleration $\vec{a}$, as a function of time. (*b*) Determine $\vec{r}$ and $\vec{v}$ at time $t = 2.5$ s.

26. (II) An object, which is at the origin at time $t = 0$, has initial velocity $\vec{v}_0 = (-14.0\hat{i} - 7.0\hat{j})$ m/s and constant acceleration $\vec{a} = (6.0\hat{i} + 3.0\hat{j})$ m/s². Find the position $\vec{r}$ where the object comes to rest (momentarily).

27. (II) A particle's position as a function of time t is given by $\vec{\mathbf{r}} = (5.0t + 6.0t^2)\,\text{m}\,\hat{\mathbf{i}} + (7.0 - 3.0t^3)\,\text{m}\,\hat{\mathbf{j}}$. At $t = 5.0\,\text{s}$, find the magnitude and direction of the particle's displacement vector $\Delta\vec{\mathbf{r}}$ relative to the point $\vec{\mathbf{r}}_0 = (0.0\hat{\mathbf{i}} + 7.0\hat{\mathbf{j}})\,\text{m}$.

3–7 and 3–8 Projectile Motion (neglect air resistance)

28. (I) A tiger leaps horizontally from a 7.5-m-high rock with a speed of 3.2 m/s. How far from the base of the rock will she land?

29. (I) A diver running 2.3 m/s dives out horizontally from the edge of a vertical cliff and 3.0 s later reaches the water below. How high was the cliff and how far from its base did the diver hit the water?

30. (II) Estimate how much farther a person can jump on the Moon as compared to the Earth if the takeoff speed and angle are the same. The acceleration due to gravity on the Moon is one-sixth what it is on Earth.

31. (II) A fire hose held near the ground shoots water at a speed of 6.5 m/s. At what angle(s) should the nozzle point in order that the water land 2.5 m away (Fig. 3–40)? Why are there two different angles? Sketch the two trajectories.

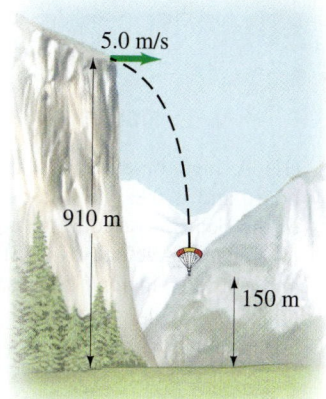

FIGURE 3–40
Problem 31.

θ_0

$\longleftarrow$ 2.5 m $\longrightarrow$

32. (II) A ball is thrown horizontally from the roof of a building 9.0 m tall and lands 9.5 m from the base. What was the ball's initial speed?

33. (II) A football is kicked at ground level with a speed of 18.0 m/s at an angle of 38.0° to the horizontal. How much later does it hit the ground?

34. (II) A ball thrown horizontally at 23.7 m/s from the roof of a building lands 31.0 m from the base of the building. How high is the building?

35. (II) A shot-putter throws the shot (mass = 7.3 kg) with an initial speed of 14.4 m/s at a 34.0° angle to the horizontal. Calculate the horizontal distance traveled by the shot if it leaves the athlete's hand at a height of 2.10 m above the ground.

36. (II) Show that the time required for a projectile to reach its highest point is equal to the time for it to return to its original height if air resistance is neglible.

37. (II) You buy a plastic dart gun, and being a clever physics student you decide to do a quick calculation to find its maximum horizontal range. You shoot the gun straight up, and it takes 4.0 s for the dart to land back at the barrel. What is the maximum horizontal range of your gun?

38. (II) A baseball is hit with a speed of 27.0 m/s at an angle of 45.0°. It lands on the flat roof of a 13.0-m-tall nearby building. If the ball was hit when it was 1.0 m above the ground, what horizontal distance does it travel before it lands on the building?

39. (II) In Example 3–11 we chose the x axis to the right and y axis up. Redo this problem by defining the x axis to the left and y axis down, and show that the conclusion remains the same—the football lands on the ground 40.5 m to the right of where it departed the punter's foot.

40. (II) A grasshopper hops down a level road. On each hop, the grasshopper launches itself at angle $\theta_0 = 45°$ and achieves a range $R = 1.0\,\text{m}$. What is the average horizontal speed of the grasshopper as it progresses down the road? Assume that the time spent on the ground between hops is negligible.

41. (II) Extreme-sports enthusiasts have been known to jump off the top of El Capitan, a sheer granite cliff of height 910 m in Yosemite National Park. Assume a jumper runs horizontally off the top of El Capitan with speed 5.0 m/s and enjoys a freefall until she is 150 m above the valley floor, at which time she opens her parachute (Fig. 3–41). (a) How long is the jumper in freefall? Ignore air resistance. (b) It is important to be as far away from the cliff as possible before opening the parachute. How far from the cliff is this jumper when she opens her chute?

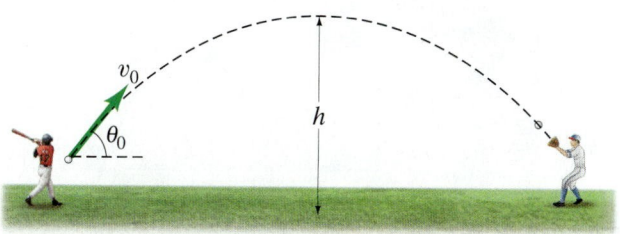

5.0 m/s

910 m

150 m

FIGURE 3–41
Problem 41.

42. (II) Here is something to try at a sporting event. Show that the maximum height h attained by an object projected into the air, such as a baseball, football, or soccer ball, is approximately given by

$$h \approx 1.2\,t^2\,\text{m},$$

where t is the total time of flight for the object in seconds. Assume that the object returns to the same level as that from which it was launched, as in Fig. 3–42. For example, if you count to find that a baseball was in the air for $t = 5.0\,\text{s}$, the maximum height attained was $h = 1.2 \times (5.0)^2 = 30\,\text{m}$. The beauty of this relation is that h can be determined without knowledge of the launch speed v_0 or launch angle θ_0.

v_0

θ_0

h

FIGURE 3–42 Problem 42.

43. (II) The pilot of an airplane traveling 170 km/h wants to drop supplies to flood victims isolated on a patch of land 150 m below. The supplies should be dropped how many seconds before the plane is directly overhead?

44. (II) (a) A long jumper leaves the ground at 45° above the horizontal and lands 8.0 m away. What is her "takeoff" speed v_0? (b) Now she is out on a hike and comes to the left bank of a river. There is no bridge and the right bank is 10.0 m away horizontally and 2.5 m, vertically below. If she long jumps from the edge of the left bank at 45° with the speed calculated in (a), how long, or short, of the opposite bank will she land (Fig. 3–43)?

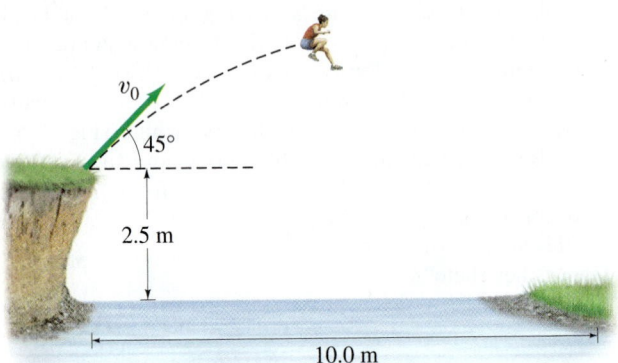

FIGURE 3–43 Problem 44.

45. (II) A high diver leaves the end of a 5.0-m-high diving board and strikes the water 1.3 s later, 3.0 m beyond the end of the board. Considering the diver as a particle, determine (a) her initial velocity, $\vec{v}_0$; (b) the maximum height reached; and (c) the velocity $\vec{v}_f$ with which she enters the water.

46. (II) A projectile is shot from the edge of a cliff 115 m above ground level with an initial speed of 65.0 m/s at an angle of 35.0° with the horizontal, as shown in Fig. 3–44. (a) Determine the time taken by the projectile to hit point P at ground level. (b) Determine the distance X of point P from the base of the vertical cliff. At the instant just before the projectile hits point P, find (c) the horizontal and the vertical components of its velocity, (d) the magnitude of the velocity, and (e) the angle made by the velocity vector with the horizontal. (f) Find the maximum height above the cliff top reached by the projectile.

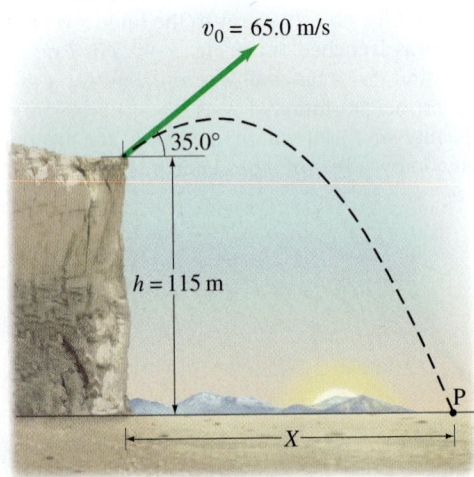

FIGURE 3–44 Problem 46.

47. (II) Suppose the kick in Example 3–7 is attempted 36.0 m from the goalposts, whose crossbar is 3.00 m above the ground. If the football is directed perfectly between the goalposts, will it pass over the bar and be a field goal? Show why or why not. If not, from what horizontal distance must this kick be made if it is to score?

48. (II) Exactly 3.0 s after a projectile is fired into the air from the ground, it is observed to have a velocity $\vec{v} = (8.6\hat{i} + 4.8\hat{j})$ m/s, where the x axis is horizontal and the y axis is positive upward. Determine (a) the horizontal range of the projectile, (b) its maximum height above the ground, and (c) its speed and angle of motion just before it strikes the ground.

49. (II) Revisit Example 3–9, and assume that the boy with the slingshot is *below* the boy in the tree (Fig. 3–45) and so aims *upward*, directly at the boy in the tree. Show that again the boy in the tree makes the wrong move by letting go at the moment the water balloon is shot.

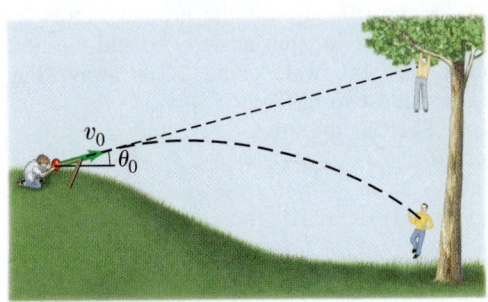

FIGURE 3–45 Problem 49.

50. (II) A stunt driver wants to make his car jump over 8 cars parked side by side below a horizontal ramp (Fig. 3–46). (a) With what minimum speed must he drive off the horizontal ramp? The vertical height of the ramp is 1.5 m above the cars and the horizontal distance he must clear is 22 m. (b) If the ramp is now tilted upward, so that "takeoff angle" is 7.0° above the horizontal, what is the new minimum speed?

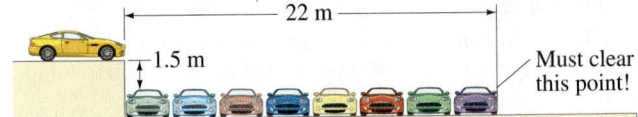

FIGURE 3–46 Problem 50.

51. (II) A ball is thrown horizontally from the top of a cliff with initial speed v_0 (at $t = 0$). At any moment, its direction of motion makes an angle θ to the horizontal (Fig. 3–47). Derive a formula for θ as a function of time, t, as the ball follows a projectile's path.

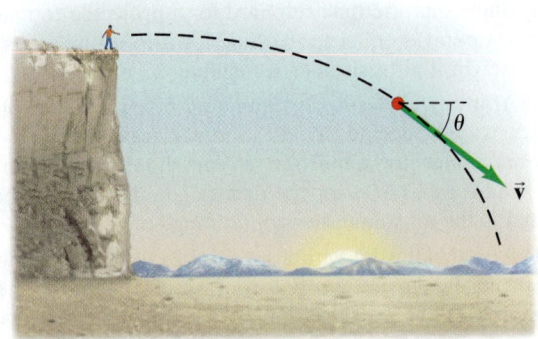

FIGURE 3–47 Problem 51.

52. (II) At what projection angle will the range of a projectile equal its maximum height?

53. (II) A projectile is fired with an initial speed of 46.6 m/s at an angle of 42.2° above the horizontal on a long flat firing range. Determine (a) the maximum height reached by the projectile, (b) the total time in the air, (c) the total horizontal distance covered (that is, the range), and (d) the velocity of the projectile 1.50 s after firing.

54. (II) An athlete executing a long jump leaves the ground at a 27.0° angle and lands 7.80 m away. (a) What was the takeoff speed? (b) If this speed were increased by just 5.0%, how much longer would the jump be?

55. (III) A person stands at the base of a hill that is a straight incline making an angle ϕ with the horizontal (Fig. 3–48). For a given initial speed v_0, at what angle θ (to the horizontal) should objects be thrown so that the distance d they land up the hill is as large as possible?

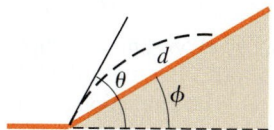

FIGURE 3–48 Problem 55. Given ϕ and v_0, determine θ to make d maximum.

56. (III) Derive a formula for the horizontal range R, of a projectile when it lands at a height h above its initial point. (For $h < 0$, it lands a distance $-h$ below the starting point.) Assume it is projected at an angle θ_0 with initial speed v_0.

3–9 Relative Velocity

57. (I) A person going for a morning jog on the deck of a cruise ship is running toward the bow (front) of the ship at 2.0 m/s while the ship is moving ahead at 8.5 m/s. What is the velocity of the jogger relative to the water? Later, the jogger is moving toward the stern (rear) of the ship. What is the jogger's velocity relative to the water now?

58. (I) Huck Finn walks at a speed of 0.70 m/s across his raft (that is, he walks perpendicular to the raft's motion relative to the shore). The raft is traveling down the Mississippi River at a speed of 1.50 m/s relative to the river bank (Fig. 3–49). What is Huck's velocity (speed and direction) relative to the river bank?

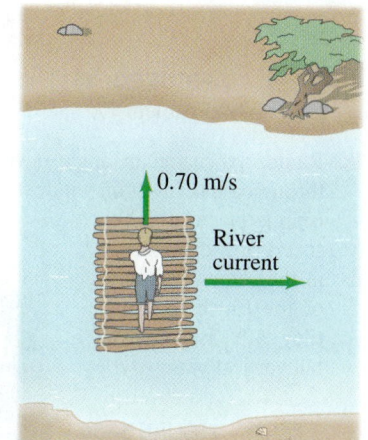

FIGURE 3–49
Problem 58.

59. (II) Determine the speed of the boat with respect to the shore in Example 3–14.

60. (II) Two planes approach each other head-on. Each has a speed of 780 km/h, and they spot each other when they are initially 12.0 km apart. How much time do the pilots have to take evasive action?

61. (II) A child, who is 45 m from the bank of a river, is being carried helplessly downstream by the river's swift current of 1.0 m/s. As the child passes a lifeguard on the river's bank, the lifeguard starts swimming in a straight line until she reaches the child at a point downstream (Fig. 3–50). If the lifeguard can swim at a speed of 2.0 m/s relative to the water, how long does it take her to reach the child? How far downstream does the lifeguard intercept the child?

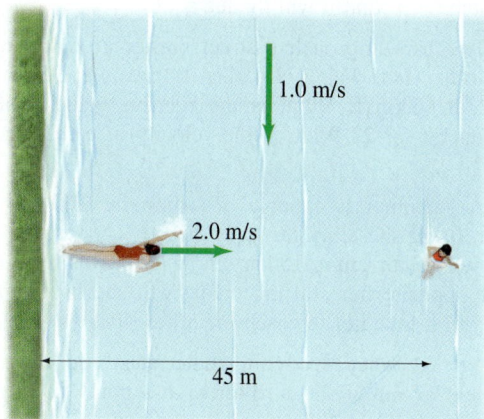

FIGURE 3–50 Problem 61.

62. (II) A passenger on a boat moving at 1.70 m/s on a still lake walks up a flight of stairs at a speed of 0.60 m/s, Fig. 3–51. The stairs are angled at 45° pointing in the direction of motion as shown. Write the vector velocity of the passenger relative to the water.

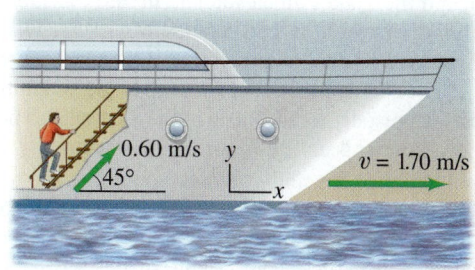

FIGURE 3–51 Problem 62.

63. (II) A person in the passenger basket of a hot-air balloon throws a ball horizontally outward from the basket with speed 10.0 m/s (Fig. 3–52). What initial velocity (magnitude and direction) does the ball have relative to a person standing on the ground (a) if the hot-air balloon is rising at 5.0 m/s relative to the ground during this throw, (b) if the hot-air balloon is descending at 5.0 m/s relative to the ground.

FIGURE 3–52
Problem 63.

64. (II) An airplane is heading due south at a speed of 580 km/h. If a wind begins blowing from the southwest at a speed of 90.0 km/h (average), calculate (*a*) the velocity (magnitude and direction) of the plane, relative to the ground, and (*b*) how far from its intended position it will be after 11.0 min if the pilot takes no corrective action. [*Hint*: First draw a diagram.]

65. (II) In what direction should the pilot aim the plane in Problem 64 so that it will fly due south?

66. (II) Two cars approach a street corner at right angles to each other (see Fig. 3–35). Car 1 travels at 35 km/h and car 2 at 45 km/h. What is the relative velocity of car 1 as seen by car 2? What is the velocity of car 2 relative to car 1?

67. (II) A swimmer is capable of swimming 0.60 m/s in still water. (*a*) If she aims her body directly across a 55-m-wide river whose current is 0.50 m/s, how far downstream (from a point opposite her starting point) will she land? (*b*) How long will it take her to reach the other side?

68. (II) (*a*) At what upstream angle must the swimmer in Problem 67 aim, if she is to arrive at a point directly across the stream? (*b*) How long will it take her?

69. (II) A motorboat whose speed in still water is 3.40 m/s must aim upstream at an angle of 19.5° (with respect to a line perpendicular to the shore) in order to travel directly across the stream. (*a*) What is the speed of the current? (*b*) What is the resultant speed of the boat with respect to the shore? (See Fig. 3–31.)

70. (II) A boat, whose speed in still water is 2.70 m/s, must cross a 280-m-wide river and arrive at a point 120 m upstream from where it starts (Fig. 3–53). To do so, the pilot must head the boat at a 45.0° upstream angle. What is the speed of the river's current?

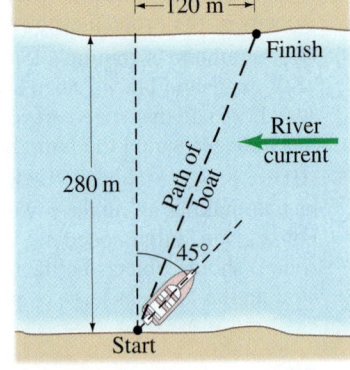

FIGURE 3–53
Problem 70.

71. (III) An airplane, whose air speed is 580 km/h, is supposed to fly in a straight path 38.0° N of E. But a steady 72 km/h wind is blowing from the north. In what direction should the plane head?

General Problems

72. Two vectors, $\vec{V}_1$ and $\vec{V}_2$, add to a resultant $\vec{V} = \vec{V}_1 + \vec{V}_2$. Describe $\vec{V}_1$ and $\vec{V}_2$ if (*a*) $V = V_1 + V_2$, (*b*) $V^2 = V_1^2 + V_2^2$, (*c*) $V_1 + V_2 = V_1 - V_2$.

73. A plumber steps out of his truck, walks 66 m east and 35 m south, and then takes an elevator 12 m into the subbasement of a building where a bad leak is occurring. What is the displacement of the plumber relative to his truck? Give your answer in components; also give the magnitude and angles, with respect to the *x* axis, in the vertical and horizontal plane. Assume *x* is east, *y* is north, and *z* is up.

74. On mountainous downhill roads, escape routes are sometimes placed to the side of the road for trucks whose brakes might fail. Assuming a constant upward slope of 26°, calculate the horizontal and vertical components of the acceleration of a truck that slowed from 110 km/h to rest in 7.0 s. See Fig. 3–54.

FIGURE 3–54
Problem 74.

75. A light plane is headed due south with a speed relative to still air of 185 km/h. After 1.00 h, the pilot notices that they have covered only 135 km and their direction is not south but southeast (45.0°). What is the wind velocity?

76. An Olympic long jumper is capable of jumping 8.0 m. Assuming his horizontal speed is 9.1 m/s as he leaves the ground, how long is he in the air and how high does he go? Assume that he lands standing upright—that is, the same way he left the ground.

77. Romeo is chucking pebbles gently up to Juliet's window, and he wants the pebbles to hit the window with only a horizontal component of velocity. He is standing at the edge of a rose garden 8.0 m below her window and 9.0 m from the base of the wall (Fig. 3–55). How fast are the pebbles going when they hit her window?

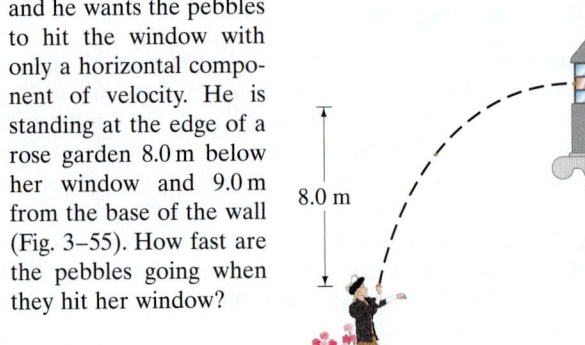

FIGURE 3–55
Problem 77.

78. Raindrops make an angle θ with the vertical when viewed through a moving train window (Fig. 3–56). If the speed of the train is v_T, what is the speed of the raindrops in the reference frame of the Earth in which they are assumed to fall vertically?

FIGURE 3–56
Problem 78.

79. Apollo astronauts took a "nine iron" to the Moon and hit a golf ball about 180 m. Assuming that the swing, launch angle, and so on, were the same as on Earth where the same astronaut could hit it only 32 m, estimate the acceleration due to gravity on the surface of the Moon. (We neglect air resistance in both cases, but on the Moon there is none.)

80. A hunter aims directly at a target (on the same level) 68.0 m away. (*a*) If the bullet leaves the gun at a speed of 175 m/s, by how much will it miss the target? (*b*) At what angle should the gun be aimed so the target will be hit?

81. The cliff divers of Acapulco push off horizontally from rock platforms about 35 m above the water, but they must clear rocky outcrops at water level that extend out into the water 5.0 m from the base of the cliff directly under their launch point. See Fig. 3–57. What minimum pushoff speed is necessary to clear the rocks? How long are they in the air?

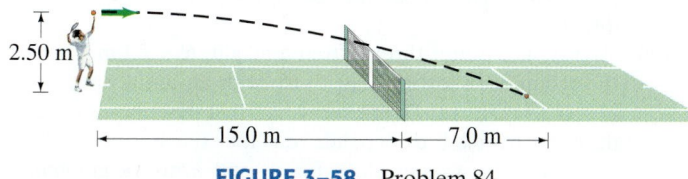

FIGURE 3–57
Problem 81.

82. When Babe Ruth hit a homer over the 8.0-m-high right-field fence 98 m from home plate, roughly what was the minimum speed of the ball when it left the bat? Assume the ball was hit 1.0 m above the ground and its path initially made a 36° angle with the ground.

83. The speed of a boat in still water is v. The boat is to make a round trip in a river whose current travels at speed u. Derive a formula for the time needed to make a round trip of total distance D if the boat makes the round trip by moving (*a*) upstream and back downstream, and (*b*) directly across the river and back. We must assume $u < v$; why?

84. At serve, a tennis player aims to hit the ball horizontally. What minimum speed is required for the ball to clear the 0.90-m-high net about 15.0 m from the server if the ball is "launched" from a height of 2.50 m? Where will the ball land if it just clears the net (and will it be "good" in the sense that it lands within 7.0 m of the net)? How long will it be in the air? See Fig. 3–58.

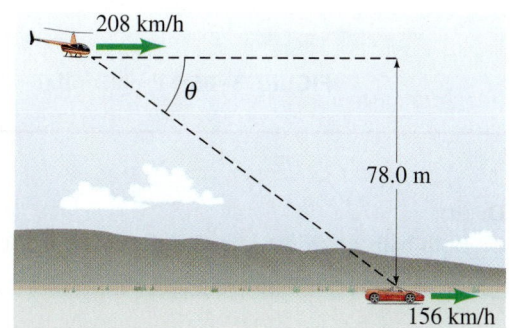

FIGURE 3–58 Problem 84.

85. Spymaster Chris, flying a constant 208 km/h horizontally in a low-flying helicopter, wants to drop secret documents into her contact's open car which is traveling 156 km/h on a level highway 78.0 m below. At what angle (with the horizontal) should the car be in her sights when the packet is released (Fig. 3–59)?

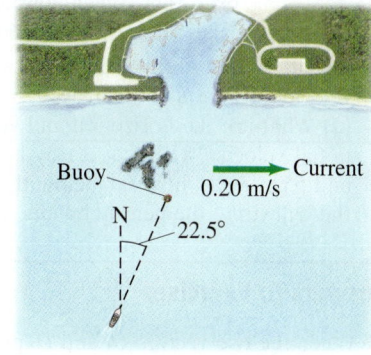

FIGURE 3–59
Problem 85.

86. A basketball leaves a player's hands at a height of 2.10 m above the floor. The basket is 3.05 m above the floor. The player likes to shoot the ball at a 38.0° angle. If the shot is made from a horizontal distance of 11.00 m and must be accurate to ±0.22 m (horizontally), what is the range of initial speeds allowed to make the basket?

87. A particle has a velocity of $\vec{v} = (-2.0\hat{i} + 3.5t\hat{j})$ m/s. The particle starts at $\vec{r} = (1.5\hat{i} - 3.1\hat{j})$ m at $t = 0$. Give the position and acceleration as a function of time. What is the shape of the resulting path?

88. A projectile is launched from ground level to the top of a cliff which is 195 m away and 135 m high (see Fig. 3–60). If the projectile lands on top of the cliff 6.6 s after it is fired, find the initial velocity of the projectile (magnitude and direction). Neglect air resistance.

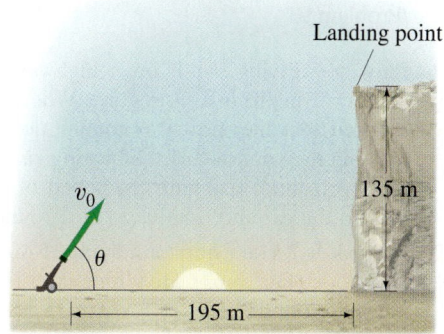

FIGURE 3–60
Problem 88.

89. In hot pursuit, Agent Logan of the FBI must get directly across a 1200-m-wide river in minimum time. The river's current is 0.80 m/s, he can row a boat at 1.60 m/s, and he can run 3.00 m/s. Describe the path he should take (rowing plus running along the shore) for the minimum crossing time, and determine the minimum time.

90. A boat can travel 2.20 m/s in still water. (*a*) If the boat points its prow directly across a stream whose current is 1.30 m/s, what is the velocity (magnitude and direction) of the boat relative to the shore? (*b*) What will be the position of the boat, relative to its point of origin, after 3.00 s?

91. A boat is traveling where there is a current of 0.20 m/s east (Fig. 3–61). To avoid some offshore rocks, the boat must clear a buoy that is NNE (22.5°) and 3.0 km away. The boat's speed through still water is 2.1 m/s. If the boat wants to pass the buoy 0.15 km on its right, at what angle should the boat head?

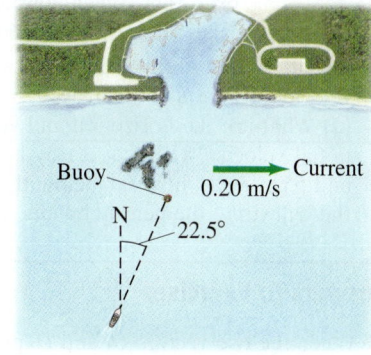

FIGURE 3–61
Problem 91.

92. A child runs down a 12° hill and then suddenly jumps upward at a 15° angle above horizontal and lands 1.4 m down the hill as measured along the hill. What was the child's initial speed?

93. A basketball is shot from an initial height of 2.4 m (Fig. 3–62) with an initial speed $v_0 = 12$ m/s directed at an angle $\theta_0 = 35°$ above the horizontal. (*a*) How far from the basket was the player if he made a basket? (*b*) At what angle to the horizontal did the ball enter the basket?

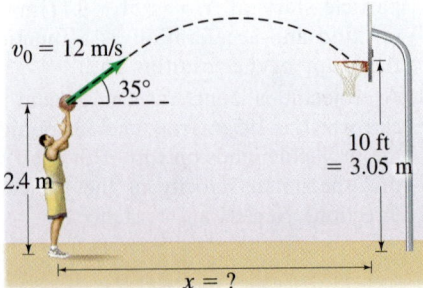

FIGURE 3–62
Problem 93.

94. You are driving south on a highway at 25 m/s (approximately 55 mi/h) in a snowstorm. When you last stopped, you noticed that the snow was coming down vertically, but it is passing the windows of the moving car at an angle of 37° to the horizontal. Estimate the speed of the snowflakes relative to the car and relative to the ground.

95. A rock is kicked horizontally at 15 m/s from a hill with a 45° slope (Fig. 3–63). How long does it take for the rock to hit the ground?

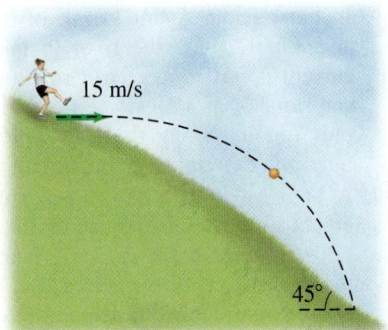

FIGURE 3–63
Problem 95.

96. A batter hits a fly ball which leaves the bat 0.90 m above the ground at an angle of 61° with an initial speed of 28 m/s heading toward centerfield. Ignore air resistance. (*a*) How far from home plate would the ball land if not caught? (*b*) The ball is caught by the centerfielder who, starting at a distance of 105 m from home plate, runs straight toward home plate at a constant speed and makes the catch at ground level. Find his speed.

97. A ball is shot from the top of a building with an initial velocity of 18 m/s at an angle $\theta = 42°$ above the horizontal. (*a*) What are the horizontal and vertical components of the initial velocity? (*b*) If a nearby building is the same height and 55 m away, how far below the top of the building will the ball strike the nearby building?

98. At $t = 0$ a batter hits a baseball with an initial speed of 28 m/s at a 55° angle to the horizontal. An outfielder is 85 m from the batter at $t = 0$ and, as seen from home plate, the line of sight to the outfielder makes a horizontal angle of 22° with the plane in which the ball moves (see Fig. 3–64). What speed and direction must the fielder take to catch the ball at the same height from which it was struck? Give the angle with respect to the outfielder's line of sight to home plate.

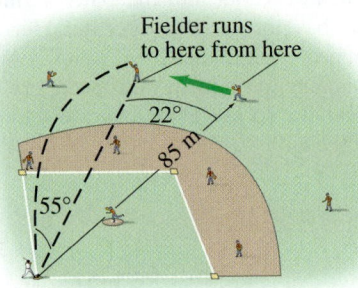

FIGURE 3–64
Problem 98.

*Numerical/Computer

***99.** (II) Students shoot a plastic ball horizontally from a projectile launcher. They measure the distance x the ball travels horizontally, the distance y the ball falls vertically, and the total time t the ball is in the air for six different heights of the projectile launcher. Here is their data.

Time, t (s)	Horizontal distance, x (m)	Vertical distance, y (m)
0.217	0.642	0.260
0.376	1.115	0.685
0.398	1.140	0.800
0.431	1.300	0.915
0.478	1.420	1.150
0.491	1.480	1.200

(*a*) Determine the best-fit straight line that represents x as a function of t. What is the initial speed of the ball obtained from the best-fit straight line? (*b*) Determine the best-fit quadratic equation that represents y as a function of t. What is the acceleration of the ball in the vertical direction?

***100.** (III) A shot-putter throws from a height $h = 2.1$ m above the ground as shown in Fig. 3–65, with an initial speed of $v_0 = 13.5$ m/s. (*a*) Derive a relation that describes how the distance traveled d depends on the release angle θ_0. (*b*) Using the given values for v_0 and h, use a graphing calculator or computer to plot d vs. θ_0. According to your plot, what value for θ_0 maximizes d?

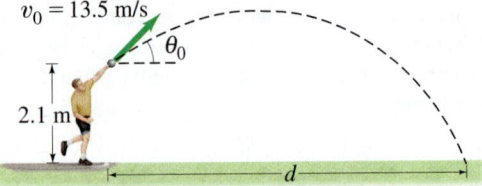

FIGURE 3–65 Problem 100.

Answers to Exercises

A: When the two vectors D_1 and D_2 point in the same direction.

B: $3\sqrt{2} = 4.24$.

C: (*a*).

D: (*d*).

E: Both balls reach the same height, so are in the air for the same length of time.

F: (*c*).

The space shuttle Discovery is carried out into space by powerful rockets. They are accelerating, increasing in speed rapidly. To do so, a force must be exerted on them according to Newton's second law, $\Sigma\vec{F} = m\vec{a}$. What exerts this force? The rocket engines exert a force on the gases they push out (expel) from the rear of the rockets (labeled $\vec{F}_{GR}$). According to Newton's third law, these ejected gases exert an equal and opposite force on the rockets in the forward direction. It is this "reaction" force exerted on the rockets by the gases, labeled $\vec{F}_{RG}$, that accelerates the rockets forward.

Dynamics:
Newton's Laws of Motion

CHAPTER-OPENING QUESTIONS—Guess now!

A 150-kg football player collides head-on with a 75-kg running back. During the collision, the heavier player exerts a force of magnitude F_A on the smaller player. If the smaller player exerts a force F_B back on the heavier player, which response is most accurate?

- **(a)** $F_B = F_A$.
- **(b)** $F_B < F_A$.
- **(c)** $F_B > F_A$.
- **(d)** $F_B = 0$.
- **(e)** We need more information.

Second Question:

A line by the poet T. S. Eliot (from *Murder in the Cathedral*) has the women of Canterbury say "the earth presses up against our feet." What force is this?

- **(a)** Gravity.
- **(b)** The normal force.
- **(c)** A friction force.
- **(d)** Centrifugal force.
- **(e)** No force—they are being poetic.

CONTENTS

4–1 Force

4–2 Newton's First Law of Motion

4–3 Mass

4–4 Newton's Second Law of Motion

4–5 Newton's Third Law of Motion

4–6 Weight—the Force of Gravity; and the Normal Force

4–7 Solving Problems with Newton's Laws: Free-Body Diagrams

4–8 Problem Solving—A General Approach

e have discussed how motion is described in terms of velocity and acceleration. Now we deal with the question of *why* objects move as they do: What makes an object at rest begin to move? What causes an object to accelerate or decelerate? What is involved when an object moves in a curved path? We can answer in each case that a force is required. In this Chapter[†], we will investigate the connection between force and motion, which is the subject called **dynamics**.

4–1 Force

Intuitively, we experience **force** as any kind of a push or a pull on an object. When you push a stalled car or a grocery cart (Fig. 4–1), you are exerting a force on it. When a motor lifts an elevator, or a hammer hits a nail, or the wind blows the leaves of a tree, a force is being exerted. We often call these *contact forces* because the force is exerted when one object comes in contact with another object. On the other hand, we say that an object falls because of the *force of gravity*.

If an object is at rest, to start it moving requires force—that is, a force is needed to accelerate an object from zero velocity to a nonzero velocity. For an object already moving, if you want to change its velocity—either in direction or in magnitude—a force is required. In other words, to accelerate an object, a force is always required. In Section 4–4 we discuss the precise relation between acceleration and net force, which is Newton's second law.

One way to measure the magnitude (or strength) of a force is to use a spring scale (Fig. 4–2). Normally, such a spring scale is used to find the weight of an object; by weight we mean the force of gravity acting on the object (Section 4–6). The spring scale, once calibrated, can be used to measure other kinds of forces as well, such as the pulling force shown in Fig. 4–2.

A force exerted in a different direction has a different effect. Force has direction as well as magnitude, and is indeed a vector that follows the rules of vector addition discussed in Chapter 3. We can represent any force on a diagram by an arrow, just as we did with velocity. The direction of the arrow is the direction of the push or pull, and its length is drawn proportional to the magnitude of the force.

FIGURE 4–1 A force exerted on a grocery cart—in this case exerted by a person.

FIGURE 4–2 A spring scale used to measure a force.

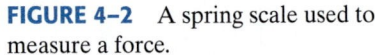
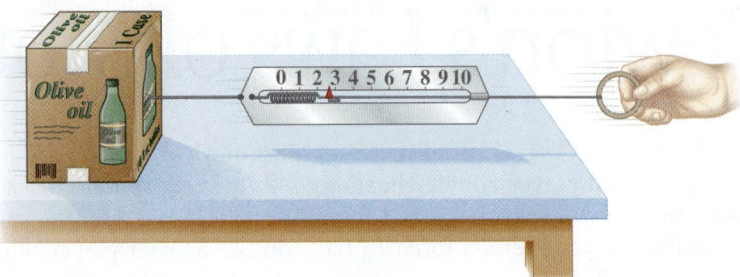

4–2 Newton's First Law of Motion

What is the relationship between force and motion? Aristotle (384–322 B.C.) believed that a force was required to keep an object moving along a horizontal plane. To Aristotle, the natural state of an object was at rest, and a force was believed necessary to keep an object in motion. Furthermore, Aristotle argued, the greater the force on the object, the greater its speed.

Some 2000 years later, Galileo disagreed: he maintained that it is just as natural for an object to be in motion with a constant velocity as it is for it to be at rest.

To understand Galileo's idea, consider the following observations involving motion along a horizontal plane. To push an object with a rough surface along a

[†]We treat everyday objects in motion here; the treatment of the submicroscopic world of atoms and molecules, and when velocities are extremely high, close to the speed of light $(3.0 \times 10^8 \, \text{m/s})$, are treated using quantum theory (Chapter 37 ff), and the theory of relativity (Chapter 36).

tabletop at constant speed requires a certain amount of force. To push an equally heavy object with a very smooth surface across the table at the same speed will require less force. If a layer of oil or other lubricant is placed between the surface of the object and the table, then almost no force is required to keep the object moving. Notice that in each successive step, less force is required. As the next step, we imagine that the object does not rub against the table at all—or there is a perfect lubricant between the object and the table—and theorize that once started, the object would move across the table at constant speed with *no* force applied. A steel ball bearing rolling on a hard horizontal surface approaches this situation. So does a puck on an air table, in which a thin layer of air reduces friction almost to zero.

It was Galileo's genius to imagine such an idealized world—in this case, one where there is no friction—and to see that it could lead to a more accurate and richer understanding of the real world. This idealization led him to his remarkable conclusion that if no force is applied to a moving object, it will continue to move with constant speed in a straight line. An object slows down only if a force is exerted on it. Galileo thus interpreted friction as a force akin to ordinary pushes and pulls.

To push an object across a table at constant speed requires a force from your hand that can balance out the force of friction (Fig. 4–3). When the object moves at constant speed, your pushing force is equal in magnitude to the friction force, but these two forces are in opposite directions, so the *net* force on the object (the vector sum of the two forces) is zero. This is consistent with Galileo's viewpoint, for the object moves with constant speed when no net force is exerted on it.

Upon this foundation laid by Galileo, Isaac Newton (Fig. 4–4) built his great theory of motion. Newton's analysis of motion is summarized in his famous "three laws of motion." In his great work, the *Principia* (published in 1687), Newton readily acknowledged his debt to Galileo. In fact, **Newton's first law of motion** is close to Galileo's conclusions. It states that

> **Every object continues in its state of rest, or of uniform velocity in a straight line, as long as no net force acts on it.**

The tendency of an object to maintain its state of rest or of uniform velocity in a straight line is called **inertia**. As a result, Newton's first law is often called the **law of inertia**.

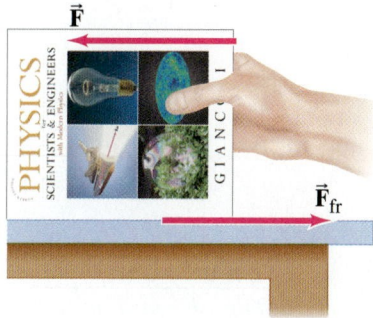

FIGURE 4–3 $\vec{F}$ represents the force applied by the person and $\vec{F}_{fr}$ represents the force of friction.

NEWTON'S FIRST LAW OF MOTION

FIGURE 4–4
Isaac Newton (1642–1727).

| **CONCEPTUAL EXAMPLE 4–1** | **Newton's first law.** A school bus comes to a sudden stop, and all of the backpacks on the floor start to slide forward. What force causes them to do that? |

RESPONSE It isn't "force" that does it. By Newton's first law, the backpacks continue their state of motion, maintaining their velocity. The backpacks slow down if a force is applied, such as friction with the floor.

Inertial Reference Frames

Newton's first law does not hold in every reference frame. For example, if your reference frame is fixed in an accelerating car, an object such as a cup resting on the dashboard may begin to move toward you (it stayed at rest as long as the car's velocity remained constant). The cup accelerated toward you, but neither you nor anything else exerted a force on it in that direction. Similarly, in the reference frame of the decelerating bus in Example 4–1, there was no force pushing the backpacks forward. In accelerating reference frames, Newton's first law does not hold. Reference frames in which Newton's first law does hold are called **inertial reference frames** (the law of inertia is valid in them). For most purposes, we usually make the approximation that a reference frame fixed on the Earth is an inertial frame. This is not precisely true, due to the Earth's rotation, but usually it is close enough.

Any reference frame that moves with constant velocity (say, a car or an airplane) relative to an inertial frame is also an inertial reference frame. Reference frames where the law of inertia does *not* hold, such as the accelerating reference frames discussed above, are called **noninertial** reference frames. How can we be sure a reference frame is inertial or not? By checking to see if Newton's first law holds. Thus Newton's first law serves as the definition of inertial reference frames.

4–3 Mass

Newton's second law, which we come to in the next Section, makes use of the concept of mass. Newton used the term *mass* as a synonym for *quantity of matter*. This intuitive notion of the mass of an object is not very precise because the concept "quantity of matter" is not very well defined. More precisely, we can say that **mass** is a *measure of the inertia* of an object. The more mass an object has, the greater the force needed to give it a particular acceleration. It is harder to start it moving from rest, or to stop it when it is moving, or to change its velocity sideways out of a straight-line path. A truck has much more inertia than a baseball moving at the same speed, and a much greater force is needed to change the truck's velocity at the same rate as the ball's. The truck therefore has much more mass.

To quantify the concept of mass, we must define a standard. In SI units, the unit of mass is the **kilogram** (kg) as we discussed in Chapter 1, Section 1–4.

The terms *mass* and *weight* are often confused with one another, but it is important to distinguish between them. Mass is a property of an object itself (a measure of an object's inertia, or its "quantity of matter"). Weight, on the other hand, is a force, the pull of gravity acting on an object. To see the difference, suppose we take an object to the Moon. The object will weigh only about one-sixth as much as it did on Earth, since the force of gravity is weaker. But its mass will be the same. It will have the same amount of matter as on Earth, and will have just as much inertia—for in the absence of friction, it will be just as hard to start it moving on the Moon as on Earth, or to stop it once it is moving. (More on weight in Section 4–6.)

⚠ CAUTION

Distinguish mass from weight

4–4 Newton's Second Law of Motion

Newton's first law states that if no net force is acting on an object at rest, the object remains at rest; or if the object is moving, it continues moving with constant speed in a straight line. But what happens if a net force *is* exerted on an object? Newton perceived that the object's velocity will change (Fig. 4–5). A net force exerted on an object may make its velocity increase. Or, if the net force is in a direction opposite to the motion, the force will reduce the object's velocity. If the net force acts sideways on a moving object, the *direction* of the object's velocity changes (and the magnitude may as well). Since a change in velocity is an acceleration (Section 2–4), we can say that *a net force causes acceleration*.

What precisely is the relationship between acceleration and force? Everyday experience can suggest an answer. Consider the force required to push a cart when friction is small enough to ignore. (If there is friction, consider the *net* force, which is the force you exert minus the force of friction.) If you push the cart with a gentle but constant force for a certain period of time, you will make the cart accelerate from rest up to some speed, say 3 km/h. If you push with twice the force, the cart will reach 3 km/h in half the time. The acceleration will be twice as great. If you triple the force, the acceleration is tripled, and so on. Thus, the acceleration of an object is directly proportional to the net applied force. But the acceleration depends on the mass of the object as well. If you push an empty grocery cart with the same force as you push one that is filled with groceries, you will find that the full cart accelerates more slowly. The greater the mass, the less the acceleration for the same net force. The mathematical relation, as Newton argued, is that the acceleration of an object is inversely proportional to its mass. These relationships are found to hold in general and can be summarized as follows:

> **The acceleration of an object is directly proportional to the net force acting on it, and is inversely proportional to the object's mass. The direction of the acceleration is in the direction of the net force acting on the object.**

This is **Newton's second law of motion**.

FIGURE 4–5 The bobsled accelerates because the team exerts a force.

NEWTON'S SECOND LAW
OF MOTION

Newton's second law can be written as an equation:

$$\vec{\mathbf{a}} = \frac{\Sigma \vec{\mathbf{F}}}{m},$$

where $\vec{\mathbf{a}}$ stands for acceleration, m for the mass, and $\Sigma \vec{\mathbf{F}}$ for the *net force* on the object. The symbol Σ (Greek "sigma") stands for "sum of"; $\vec{\mathbf{F}}$ stands for force, so $\Sigma \vec{\mathbf{F}}$ means the *vector sum of all forces* acting on the object, which we define as the **net force**.

We rearrange this equation to obtain the familiar statement of Newton's second law:

$$\Sigma \vec{\mathbf{F}} = m\vec{\mathbf{a}}. \qquad\qquad \textbf{(4–1a)}$$

NEWTON'S SECOND LAW OF MOTION

Newton's second law relates the description of motion (acceleration) to the cause of motion (force). It is one of the most fundamental relationships in physics. From Newton's second law we can make a more precise definition of **force** as *an action capable of accelerating an object.*

Every force $\vec{\mathbf{F}}$ is a vector, with magnitude and direction. Equation 4–1a is a vector equation valid in any inertial reference frame. It can be written in component form in rectangular coordinates as

$$\Sigma F_x = ma_x, \qquad \Sigma F_y = ma_y, \qquad \Sigma F_z = ma_z, \qquad \textbf{(4–1b)}$$

where

$$\vec{\mathbf{F}} = F_x\hat{\mathbf{i}} + F_y\hat{\mathbf{j}} + F_z\hat{\mathbf{k}}.$$

The component of acceleration in each direction is affected only by the component of the net force in that direction.

In SI units, with the mass in kilograms, the unit of force is called the **newton** (N). One newton, then, is the force required to impart an acceleration of 1 m/s^2 to a mass of 1 kg. Thus $1 \text{ N} = 1 \text{ kg} \cdot \text{m/s}^2$.

In cgs units, the unit of mass is the gram (g) as mentioned earlier.[†] The unit of force is the *dyne*, which is defined as the net force needed to impart an acceleration of 1 cm/s^2 to a mass of 1 g. Thus $1 \text{ dyne} = 1 \text{ g} \cdot \text{cm/s}^2$. It is easy to show that $1 \text{ dyne} = 10^{-5} \text{ N}$.

In the British system, the unit of force is the *pound* (abbreviated lb), where $1 \text{ lb} = 4.448222 \text{ N} \approx 4.45 \text{ N}$. The unit of mass is the *slug*, which is defined as that mass which will undergo an acceleration of 1 ft/s^2 when a force of 1 lb is applied to it. Thus $1 \text{ lb} = 1 \text{ slug} \cdot \text{ft/s}^2$. Table 4–1 summarizes the units in the different systems.

It is very important that only one set of units be used in a given calculation or problem, with the SI being preferred. If the force is given in, say, newtons, and the mass in grams, then before attempting to solve for the acceleration in SI units, we must change the mass to kilograms. For example, if the force is given as 2.0 N along the x axis and the mass is 500 g, we change the latter to 0.50 kg, and the acceleration will then automatically come out in m/s^2 when Newton's second law is used:

$$a_x = \frac{\Sigma F_x}{m} = \frac{2.0 \text{ N}}{0.50 \text{ kg}} = \frac{2.0 \text{ kg} \cdot \text{m/s}^2}{0.50 \text{ kg}} = 4.0 \text{ m/s}^2.$$

TABLE 4–1
Units for Mass and Force

System	Mass	Force
SI	kilogram (kg)	newton (N) $(= \text{kg} \cdot \text{m/s}^2)$
cgs	gram (g)	dyne $(= \text{g} \cdot \text{cm/s}^2)$
British	slug	pound (lb)

Conversion factors: $1 \text{ dyne} = 10^{-5} \text{ N}$; $1 \text{ lb} \approx 4.45 \text{ N}$.

PROBLEM SOLVING
Use a consistent set of units

EXAMPLE 4–2 ESTIMATE | **Force to accelerate a fast car.** Estimate the net force needed to accelerate (*a*) a 1000-kg car at $\frac{1}{2}g$; (*b*) a 200-g apple at the same rate.

APPROACH We use Newton's second law to find the net force needed for each object. This is an estimate (the $\frac{1}{2}$ is not said to be precise) so we round off to one significant figure.

SOLUTION (*a*) The car's acceleration is $a = \frac{1}{2}g = \frac{1}{2}(9.8 \text{ m/s}^2) \approx 5 \text{ m/s}^2$. We use Newton's second law to get the net force needed to achieve this acceleration:

$$\Sigma F = ma \approx (1000 \text{ kg})(5 \text{ m/s}^2) = 5000 \text{ N}.$$

(If you are used to British units, to get an idea of what a 5000-N force is, you can divide by 4.45 N/lb and get a force of about 1000 lb.)

(*b*) For the apple, $m = 200 \text{ g} = 0.2 \text{ kg}$, so

$$\Sigma F = ma \approx (0.2 \text{ kg})(5 \text{ m/s}^2) = 1 \text{ N}.$$

[†]Be careful not to confuse g for gram with g for the acceleration due to gravity. The latter is always italicized (or boldface when a vector).

EXAMPLE 4–3 **Force to stop a car.** What average net force is required to bring a 1500-kg car to rest from a speed of 100 km/h within a distance of 55 m?

APPROACH We use Newton's second law, $\Sigma F = ma$, to determine the force, but first we need to calculate the acceleration a. We assume the acceleration is constant, so we can use the kinematic equations, Eqs. 2–12, to calculate it.

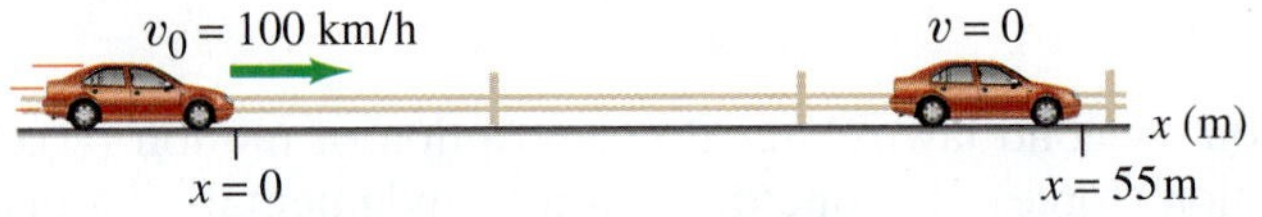

FIGURE 4–6
Example 4–3.

$v_0 = 100$ km/h $\qquad\qquad\qquad v = 0$

x (m)

$x = 0 \qquad\qquad\qquad\qquad\qquad x = 55\,\text{m}$

SOLUTION We assume the motion is along the $+x$ axis (Fig. 4–6). We are given the initial velocity $v_0 = 100\,\text{km/h} = 27.8\,\text{m/s}$ (Section 1–5), the final velocity $v = 0$, and the distance traveled $x - x_0 = 55\,\text{m}$. From Eq. 2–12c, we have

$$v^2 = v_0^2 + 2a(x - x_0),$$

so

$$a = \frac{v^2 - v_0^2}{2(x - x_0)} = \frac{0 - (27.8\,\text{m/s})^2}{2(55\,\text{m})} = -7.0\,\text{m/s}^2.$$

The net force required is then

$$\Sigma F = ma = (1500\,\text{kg})(-7.0\,\text{m/s}^2) = -1.1 \times 10^4\,\text{N}.$$

The force must be exerted in the direction *opposite* to the initial velocity, which is what the negative sign means.

NOTE If the acceleration is not precisely constant, then we are determining an "average" acceleration and we obtain an "average" net force.

Newton's second law, like the first law, is valid only in inertial reference frames (Section 4–2). In the noninertial reference frame of an accelerating car, for example, a cup on the dashboard starts sliding—it accelerates—even though the net force on it is zero; thus $\Sigma \vec{F} = m\vec{a}$ doesn't work in such an accelerating reference frame ($\Sigma \vec{F} = 0$, but $\vec{a} \neq 0$ in this noninertial frame).

EXERCISE A Suppose you watch a cup slide on the (smooth) dashboard of an accelerating car as we just discussed, but this time from an inertial reference frame outside the car, on the street. From your inertial frame, Newton's laws are valid. What force pushes the cup off the dashboard?

Precise Definition of Mass

As mentioned in Section 4–3, we can quantify the concept of mass using its definition as a measure of inertia. How to do this is evident from Eq. 4–1a, where we see that the acceleration of an object is inversely proportional to its mass. If the same net force ΣF acts to accelerate each of two masses, m_1 and m_2, then the ratio of their masses can be defined as the inverse ratio of their accelerations:

$$\frac{m_2}{m_1} = \frac{a_1}{a_2}.$$

If one of the masses is known (it could be the standard kilogram) and the two accelerations are precisely measured, then the unknown mass is obtained from this definition. For example, if $m_1 = 1.00\,\text{kg}$, and for a particular force $a_1 = 3.00\,\text{m/s}^2$ and $a_2 = 2.00\,\text{m/s}^2$, then $m_2 = 1.50\,\text{kg}$.

4–5 Newton's Third Law of Motion

Newton's second law of motion describes quantitatively how forces affect motion. But where, we may ask, do forces come from? Observations suggest that a force exerted on any object is always exerted *by another object.* A horse pulls a wagon, a person pushes a grocery cart, a hammer pushes on a nail, a magnet attracts a paper clip. In each of these examples, a force is exerted *on* one object, and that force is exerted *by* another object. For example, the force exerted *on* the nail is exerted *by* the hammer.

But Newton realized that things are not so one-sided. True, the hammer exerts a force on the nail (Fig. 4–7). But the nail evidently exerts a force back on the hammer as well, for the hammer's speed is rapidly reduced to zero upon contact. Only a strong force could cause such a rapid deceleration of the hammer. Thus, said Newton, the two objects must be treated on an equal basis. The hammer exerts a force on the nail, and the nail exerts a force back on the hammer. This is the essence of **Newton's third law of motion**:

> **Whenever one object exerts a force on a second object, the second exerts an equal force in the opposite direction on the first.**

This law is sometimes paraphrased as "to every action there is an equal and opposite reaction." This is perfectly valid. But to avoid confusion, it is very important to remember that the "action" force and the "reaction" force are acting on *different* objects.

As evidence for the validity of Newton's third law, look at your hand when you push against the edge of a desk, Fig. 4–8. Your hand's shape is distorted, clear evidence that a force is being exerted on it. You can *see* the edge of the desk pressing into your hand. You can even *feel* the desk exerting a force on your hand; it hurts! The harder you push against the desk, the harder the desk pushes back on your hand. (You only feel forces exerted *on* you; when you exert a force on another object, what you feel is that object pushing back on you.)

FIGURE 4–7 A hammer striking a nail. The hammer exerts a force on the nail and the nail exerts a force back on the hammer. The latter force decelerates the hammer and brings it to rest.

NEWTON'S THIRD LAW OF MOTION

⚠ **CAUTION**

Action and reaction forces act on different objects

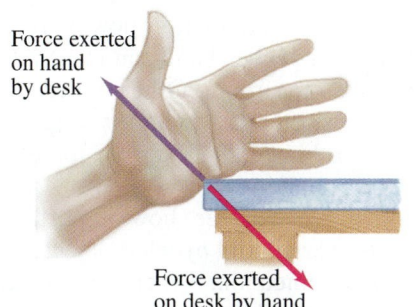

Force exerted on hand by desk

Force exerted on desk by hand

FIGURE 4–8 If your hand pushes against the edge of a desk (the force vector is shown in red), the desk pushes back against your hand (this force vector is shown in a different color, violet, to remind us that this force acts on a different object).

The force the desk exerts on your hand has the same magnitude as the force your hand exerts on the desk. This is true not only if the desk is at rest but is true even if the desk is accelerating due to the force your hand exerts.

As another demonstration of Newton's third law, consider the ice skater in Fig. 4–9. There is very little friction between her skates and the ice, so she will move freely if a force is exerted on her. She pushes against the wall; and then *she* starts moving backward. The force she exerts on the wall cannot make *her* start moving, for that force acts on the wall. Something had to exert a force *on her* to start her moving, and that force could only have been exerted by the wall. The force with which the wall pushes on her is, by Newton's third law, equal and opposite to the force she exerts on the wall.

When a person throws a package out of a small boat (initially at rest), the boat starts moving in the opposite direction. The person exerts a force on the package. The package exerts an equal and opposite force back on the person, and this force propels the person (and the boat) backward slightly.

FIGURE 4–9 An example of Newton's third law: when an ice skater pushes against the wall, the wall pushes back and this force causes her to accelerate away.

Force on skater   Force on wall

FIGURE 4–10 Another example of Newton's third law: the launch of a rocket. The rocket engine pushes the gases downward, and the gases exert an equal and opposite force upward on the rocket, accelerating it upward. (A rocket does *not* accelerate as a result of its propelling gases pushing against the ground.)

FIGURE 4–11 We can walk forward because, when one foot pushes backward against the ground, the ground pushes forward on that foot (Newton's third law). The two forces shown *act on different objects*.

Horizontal force exerted on the **g**round by **p**erson's foot

$\vec{\mathbf{F}}_{GP}$

Horizontal force exerted on the **p**erson's foot by the **g**round

$\vec{\mathbf{F}}_{PG}$

NEWTON'S THIRD LAW OF MOTION

Rocket propulsion also is explained using Newton's third law (Fig. 4–10). A common misconception is that rockets accelerate because the gases rushing out the back of the engine push against the ground or the atmosphere. Not true. What happens, instead, is that a rocket exerts a strong force on the gases, expelling them; and the gases exert an equal and opposite force *on the rocket*. It is this latter force that propels the rocket forward—the force exerted *on* the rocket *by* the gases (see Chapter-Opening photo, page 83). Thus, a space vehicle is maneuvered in empty space by firing its rockets in the direction opposite to that in which it needs to accelerate. When the rocket pushes on the gases in one direction, the gases push back on the rocket in the opposite direction. Jet aircraft too accelerate because the gases they thrust out backwards exert a forward force on the engines (Newton's third law).

Consider how we walk. A person begins walking by pushing with the foot backward against the ground. The ground then exerts an equal and opposite force forward on the person (Fig. 4–11), and it is this force, *on the person*, that moves the person forward. (If you doubt this, try walking normally where there is no friction, such as on very smooth slippery ice.) In a similar way, a bird flies forward by exerting a backward force on the air, but it is the air pushing forward (Newton's third law) on the bird's wings that propels the bird forward.

| **CONCEPTUAL EXAMPLE 4–4** | **What exerts the force to move a car?** What makes a car go forward? |

RESPONSE A common answer is that the engine makes the car move forward. But it is not so simple. The engine makes the wheels go around. But if the tires are on slick ice or deep mud, they just spin. Friction is needed. On firm ground, the tires push backward against the ground because of friction. By Newton's third law, the ground pushes on the tires in the opposite direction, accelerating the car forward.

We tend to associate forces with active objects such as humans, animals, engines, or a moving object like a hammer. It is often difficult to see how an inanimate object at rest, such as a wall or a desk, or the wall of an ice rink (Fig. 4–9), can exert a force. The explanation is that every material, no matter how hard, is elastic (springy) at least to some degree. A stretched rubber band can exert a force on a wad of paper and accelerate it to fly across the room. Other materials may not stretch as readily as rubber, but they do stretch or compress when a force is applied to them. And just as a stretched rubber band exerts a force, so does a stretched (or compressed) wall, desk, or car fender.

From the examples discussed above, we can see how important it is to remember *on* what object a given force is exerted and *by* what object that force is exerted. A force influences the motion of an object only when it is applied *on* that object. A force exerted *by* an object does not influence that same object; it only influences the other object *on* which it is exerted. Thus, to avoid confusion, the two prepositions *on* and *by* must always be used—and used with care.

One way to keep clear which force acts on which object is to use double subscripts. For example, the force exerted on the **P**erson by the **G**round as the person walks in Fig. 4–11 can be labeled $\vec{\mathbf{F}}_{PG}$. And the force exerted on the ground by the person is $\vec{\mathbf{F}}_{GP}$. By Newton's third law

$$\vec{\mathbf{F}}_{GP} = -\vec{\mathbf{F}}_{PG}. \tag{4–2}$$

$\vec{\mathbf{F}}_{GP}$ and $\vec{\mathbf{F}}_{PG}$ have the same magnitude (Newton's third law), and the minus sign reminds us that these two forces are in opposite directions.

Note carefully that the two forces shown in Fig. 4–11 act on different objects—hence we used slightly different colors for the vector arrows representing these forces. These two forces would never appear together in a sum of forces in Newton's second law, $\Sigma\vec{\mathbf{F}} = m\vec{\mathbf{a}}$. Why not? Because they act on different objects: $\vec{\mathbf{a}}$ is the acceleration of one particular object, and $\Sigma\vec{\mathbf{F}}$ must include *only* the forces on that *one* object.

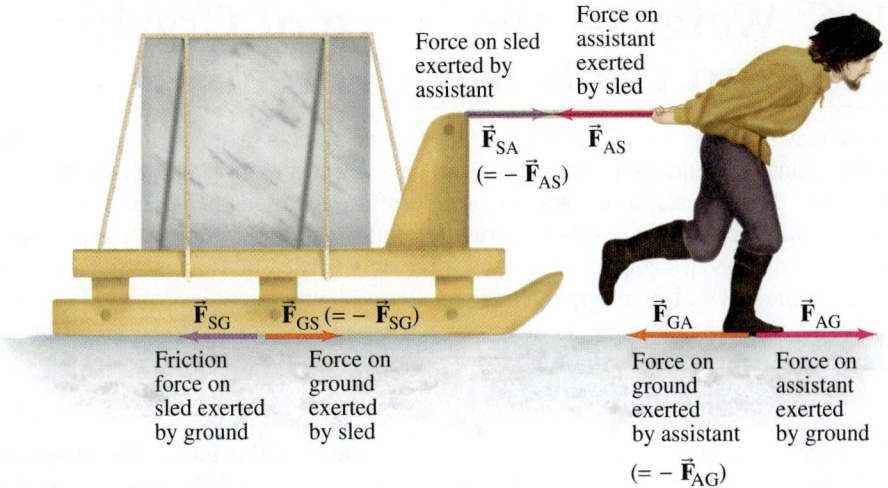

Force on sled exerted by assistant

Force on assistant exerted by sled

$\vec{\mathbf{F}}_{SA}$
$(= -\vec{\mathbf{F}}_{AS})$

$\vec{\mathbf{F}}_{AS}$

$\vec{\mathbf{F}}_{SG}$

$\vec{\mathbf{F}}_{GS} (= -\vec{\mathbf{F}}_{SG})$

$\vec{\mathbf{F}}_{GA}$

$\vec{\mathbf{F}}_{AG}$

Friction force on sled exerted by ground

Force on ground exerted by sled

Force on ground exerted by assistant
$(= -\vec{\mathbf{F}}_{AG})$

Force on assistant exerted by ground

FIGURE 4–12 Example 4–5, showing only horizontal forces. Michelangelo has selected a fine block of marble for his next sculpture. Shown here is his assistant pulling it on a sled away from the quarry. Forces on the assistant are shown as red (magenta) arrows. Forces on the sled are purple arrows. Forces acting on the ground are orange arrows. Action–reaction forces that are equal and opposite are labeled by the same subscripts but reversed (such as $\vec{\mathbf{F}}_{GA}$ and $\vec{\mathbf{F}}_{AG}$) and are of different colors because they act on different objects.

CONCEPTUAL EXAMPLE 4–5 **Third law clarification.** Michelangelo's assistant has been assigned the task of moving a block of marble using a sled (Fig. 4–12). He says to his boss, "When I exert a forward force on the sled, the sled exerts an equal and opposite force backward. So how can I ever start it moving? No matter how hard I pull, the backward reaction force always equals my forward force, so the net force must be zero. I'll never be able to move this load." Is he correct?

RESPONSE No. Although it is true that the action and reaction forces are equal in magnitude, the assistant has forgotten that they are exerted on different objects. The forward ("action") force is exerted by the assistant on the sled (Fig. 4–12), whereas the backward "reaction" force is exerted by the sled on the assistant. To determine if the *assistant* moves or not, we must consider only the forces *on the assistant* and then apply $\Sigma\vec{\mathbf{F}} = m\vec{\mathbf{a}}$, where $\Sigma\vec{\mathbf{F}}$ is the net force *on the assistant*, $\vec{\mathbf{a}}$ is the acceleration of the assistant, and m is the assistant's mass. There are two forces on the assistant that affect his forward motion; they are shown as bright red (magenta) arrows in Figs. 4–12 and 4–13: they are (1) the horizontal force $\vec{\mathbf{F}}_{AG}$ exerted on the assistant by the ground (the harder he pushes backward against the ground, the harder the ground pushes forward on him—Newton's third law), and (2) the force $\vec{\mathbf{F}}_{AS}$ exerted on the assistant by the sled, pulling backward on him; see Fig. 4–13. If he pushes hard enough on the ground, the force on him exerted by the ground, $\vec{\mathbf{F}}_{AG}$, will be larger than the sled pulling back, $\vec{\mathbf{F}}_{AS}$, and the assistant accelerates forward (Newton's second law). The sled, on the other hand, accelerates forward when the force on *it* exerted by the assistant is greater than the frictional force exerted backward on it by the ground (that is, when $\vec{\mathbf{F}}_{SA}$ has greater magnitude than $\vec{\mathbf{F}}_{SG}$ in Fig. 4–12).

PROBLEM SOLVING

A study of Newton's second and third laws

Force on assistant exerted by sled

$\vec{\mathbf{F}}_{AS}$

$\vec{\mathbf{F}}_{AG}$

Force on assistant exerted by ground

FIGURE 4–13 Example 4–5. The horizontal forces on the assistant.

Using double subscripts to clarify Newton's third law can become cumbersome, and we won't usually use them in this way. We will usually use a single subscript referring to what exerts the force on the object being discussed. Nevertheless, if there is any confusion in your mind about a given force, go ahead and use two subscripts to identify *on* what object and *by* what object the force is exerted.

EXERCISE B Return to the first Chapter-Opening Question, page 83, and answer it again now. Try to explain why you may have answered differently the first time.

EXERCISE C A massive truck collides head-on with a small sports car. (*a*) Which vehicle experiences the greater force of impact? (*b*) Which experiences the greater acceleration during the impact? (*c*) Which of Newton's laws are useful to obtain the correct answers?

EXERCISE D If you push on a heavy desk, does it always push back on you? (*a*) Not unless someone else also pushes on it. (*b*) Yes, if it is out in space. (*c*) A desk never pushes to start with. (*d*) No. (*e*) Yes.

4–6 Weight—the Force of Gravity; and the Normal Force

As we saw in Chapter 2, Galileo claimed that all objects dropped near the surface of the Earth would fall with the same acceleration, $\vec{\mathbf{g}}$, if air resistance was negligible. The force that causes this acceleration is called the *force of gravity* or *gravitational force*. What exerts the gravitational force on an object? It is the Earth, as we will discuss in Chapter 6, and the force acts vertically[†] downward, toward the center of the Earth. Let us apply Newton's second law to an object of mass m falling freely due to gravity. For the acceleration, $\vec{\mathbf{a}}$, we use the downward acceleration due to gravity, $\vec{\mathbf{g}}$. Thus, the **gravitational force** on an object, $\vec{\mathbf{F}}_G$, can be written as

$$\vec{\mathbf{F}}_G = m\vec{\mathbf{g}}. \tag{4–3}$$

The direction of this force is down toward the center of the Earth. The magnitude of the force of gravity on an object, mg, is commonly called the object's **weight**.

In SI units, $g = 9.80 \text{ m/s}^2 = 9.80 \text{ N/kg}$,[‡] so the weight of a 1.00-kg mass on Earth is $1.00 \text{ kg} \times 9.80 \text{ m/s}^2 = 9.80 \text{ N}$. We will mainly be concerned with the weight of objects on Earth, but we note that on the Moon, on other planets, or in space, the weight of a given mass will be different than it is on Earth. For example, on the Moon the acceleration due to gravity is about one-sixth what it is on Earth, and a 1.0-kg mass weighs only 1.6 N. Although we will not use British units, we note that for practical purposes on the Earth, a mass of 1 kg weighs about 2.2 lb. (On the Moon, 1 kg weighs only about 0.4 lb.)

The force of gravity acts on an object when it is falling. When an object is at rest on the Earth, the gravitational force on it does not disappear, as we know if we weigh it on a spring scale. The same force, given by Eq. 4–3, continues to act. Why, then, doesn't the object move? From Newton's second law, the net force on an object that remains at rest is zero. There must be another force on the object to balance the gravitational force. For an object resting on a table, the table exerts this upward force; see Fig. 4–14a. The table is compressed slightly beneath the object, and due to its elasticity, it pushes up on the object as shown. The force exerted by the table is often called a **contact force**, since it occurs when two objects are in contact. (The force of your hand pushing on a cart is also a contact force.) When a contact force acts *perpendicular* to the common surface of contact, it is referred to as the **normal force** ("normal" means perpendicular); hence it is labeled $\vec{\mathbf{F}}_N$ in Fig. 4–14a.

The two forces shown in Fig. 4–14a are both acting on the statue, which remains at rest, so the vector sum of these two forces must be zero (Newton's second law). Hence $\vec{\mathbf{F}}_G$ and $\vec{\mathbf{F}}_N$ must be of equal magnitude and in opposite directions. But they are *not* the equal and opposite forces spoken of in Newton's third law. The action and reaction forces of Newton's third law act on *different objects*, whereas the two forces shown in Fig. 4–14a act on the *same* object. For each of the forces shown in Fig. 4–14a, we can ask, "What is the reaction force?" The upward force, $\vec{\mathbf{F}}_N$, on the statue is exerted by the table. The reaction to this force is a force exerted by the statue downward on the table. It is shown in Fig. 4–14b, where it is labeled $\vec{\mathbf{F}}_N'$. This force, $\vec{\mathbf{F}}_N'$, exerted on the table by the statue, is the reaction force to $\vec{\mathbf{F}}_N$ in accord with Newton's third law. What about the other force on the statue, the force of gravity $\vec{\mathbf{F}}_G$ exerted by the Earth? Can you guess what the reaction is to this force? We will see in Chapter 6 that the reaction force is also a gravitational force, exerted on the Earth by the statue.

EXERCISE E Return to the second Chapter-Opening Question, page 83, and answer it again now. Try to explain why you may have answered differently the first time.

[†]The concept of "vertical" is tied to gravity. The best definition of *vertical* is that it is the direction in which objects fall. A surface that is "horizontal," on the other hand, is a surface on which a round object won't start rolling: gravity has no effect. Horizontal is perpendicular to vertical.

[‡]Since $1 \text{ N} = 1 \text{ kg} \cdot \text{m/s}^2$ (Section 4–4), then $1 \text{ m/s}^2 = 1 \text{ N/kg}$.

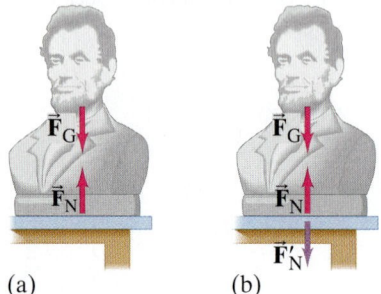

FIGURE 4–14 (a) The net force on an object at rest is zero according to Newton's second law. Therefore the downward force of gravity ($\vec{\mathbf{F}}_G$) on an object at rest must be balanced by an upward force (the normal force $\vec{\mathbf{F}}_N$) exerted by the table in this case. (b) $\vec{\mathbf{F}}_N'$ is the force exerted on the table by the statue and is the reaction force to $\vec{\mathbf{F}}_N$ by Newton's third law. ($\vec{\mathbf{F}}_N'$ is shown in a different color to remind us it acts on a different object.) The reaction force to $\vec{\mathbf{F}}_G$ is not shown.

⚠ **CAUTION**

*Weight and normal force are **not** action–reaction pairs*

EXAMPLE 4–6 **Weight, normal force, and a box.** A friend has given you a special gift, a box of mass 10.0 kg with a mystery surprise inside. The box is resting on the smooth (frictionless) horizontal surface of a table (Fig. 4–15a). (a) Determine the weight of the box and the normal force exerted on it by the table. (b) Now your friend pushes down on the box with a force of 40.0 N, as in Fig. 4–15b. Again determine the normal force exerted on the box by the table. (c) If your friend pulls upward on the box with a force of 40.0 N (Fig. 4–15c), what now is the normal force exerted on the box by the table?

APPROACH The box is at rest on the table, so the net force on the box in each case is zero (Newton's second law). The weight of the box has magnitude mg in all three cases.

SOLUTION (a) The weight of the box is $mg = (10.0 \text{ kg})(9.80 \text{ m/s}^2) = 98.0 \text{ N}$, and this force acts downward. The only other force on the box is the normal force exerted upward on it by the table, as shown in Fig. 4–15a. We chose the upward direction as the positive y direction; then the net force ΣF_y on the box is $\Sigma F_y = F_N - mg$; the minus sign means mg acts in the negative y direction (m and g are magnitudes). The box is at rest, so the net force on it must be zero (Newton's second law, $\Sigma F_y = ma_y$, and $a_y = 0$). Thus

$$\Sigma F_y = ma_y$$
$$F_N - mg = 0,$$

so we have

$$F_N = mg.$$

The normal force on the box, exerted by the table, is 98.0 N upward, and has magnitude equal to the box's weight.

(b) Your friend is pushing down on the box with a force of 40.0 N. So instead of only two forces acting on the box, now there are three forces acting on the box, as shown in Fig. 4–15b. The weight of the box is still $mg = 98.0 \text{ N}$. The net force is $\Sigma F_y = F_N - mg - 40.0 \text{ N}$, and is equal to zero because the box remains at rest ($a = 0$). Newton's second law gives

$$\Sigma F_y = F_N - mg - 40.0 \text{ N} = 0.$$

We solve this equation for the normal force:

$$F_N = mg + 40.0 \text{ N} = 98.0 \text{ N} + 40.0 \text{ N} = 138.0 \text{ N},$$

which is greater than in (a). The table pushes back with more force when a person pushes down on the box. The normal force is not always equal to the weight!

(c) The box's weight is still 98.0 N and acts downward. The force exerted by your friend and the normal force both act upward (positive direction), as shown in Fig. 4–15c. The box doesn't move since your friend's upward force is less than the weight. The net force, again set to zero in Newton's second law because $a = 0$, is

$$\Sigma F_y = F_N - mg + 40.0 \text{ N} = 0,$$

so

$$F_N = mg - 40.0 \text{ N} = 98.0 \text{ N} - 40.0 \text{ N} = 58.0 \text{ N}.$$

The table does not push against the full weight of the box because of the upward pull exerted by your friend.

NOTE The weight of the box ($= mg$) does not change as a result of your friend's push or pull. Only the normal force is affected.

Recall that the normal force is elastic in origin (the table in Fig. 4–15 sags slightly under the weight of the box). The normal force in Example 4–6 is vertical, perpendicular to the horizontal table. The normal force is not always vertical, however. When you push against a wall, for example, the normal force with which the wall pushes back on you is horizontal (Fig. 4–9). For an object on a plane inclined at an angle to the horizontal, such as a skier or car on a hill, the normal force acts perpendicular to the plane and so is not vertical.

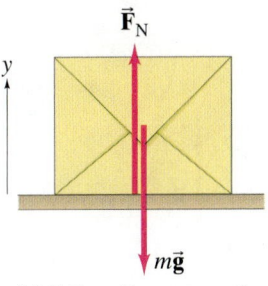

(a) $\Sigma F_y = F_N - mg = 0$

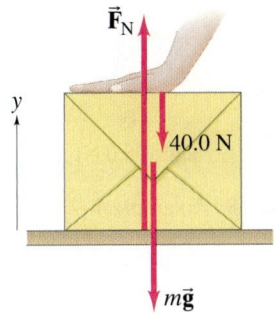

(b) $\Sigma F_y = F_N - mg - 40.0 \text{ N} = 0$

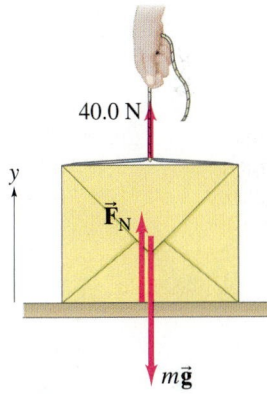

(c) $\Sigma F_y = F_N - mg + 40.0 \text{ N} = 0$

FIGURE 4–15 Example 4–6. (a) A 10-kg gift box is at rest on a table. (b) A person pushes down on the box with a force of 40.0 N. (c) A person pulls upward on the box with a force of 40.0 N. The forces are all assumed to act along a line; they are shown slightly displaced in order to be distinguishable. Only forces acting on the box are shown.

⚠ **CAUTION**
The normal force is not always equal to the weight

⚠ **CAUTION**
The normal force, $\vec{F}_N$, is not necessarily vertical

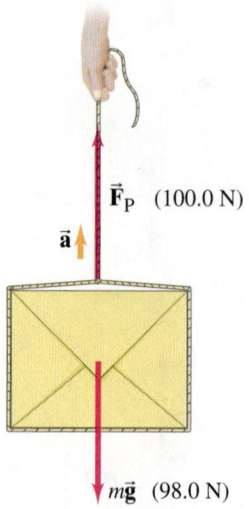

$\vec{F}_P$ (100.0 N)

$\vec{a}$

$m\vec{g}$ (98.0 N)

FIGURE 4–16 Example 4–7. The box accelerates upward because $F_P > mg$.

EXAMPLE 4–7 **Accelerating the box.** What happens when a person pulls upward on the box in Example 4–6c with a force equal to, or greater than, the box's weight? For example, let $F_P = 100.0\,\text{N}$ (Fig. 4–16) rather than the 40.0 N shown in Fig. 4–15c.

APPROACH We can start just as in Example 4–6, but be ready for a surprise.

SOLUTION The net force on the box is

$$\Sigma F_y = F_N - mg + F_P$$
$$= F_N - 98.0\,\text{N} + 100.0\,\text{N},$$

and if we set this equal to zero (thinking the acceleration might be zero), we would get $F_N = -2.0\,\text{N}$. This is nonsense, since the negative sign implies F_N points downward, and the table surely cannot *pull* down on the box (unless there's glue on the table). The least F_N can be is zero, which it will be in this case. What really happens here is that the box accelerates upward because the net force is not zero. The net force (setting the normal force $F_N = 0$) is

$$\Sigma F_y = F_P - mg = 100.0\,\text{N} - 98.0\,\text{N}$$
$$= 2.0\,\text{N}$$

upward. See Fig. 4–16. We apply Newton's second law and see that the box moves upward with an acceleration

$$a_y = \frac{\Sigma F_y}{m} = \frac{2.0\,\text{N}}{10.0\,\text{kg}}$$
$$= 0.20\,\text{m/s}^2.$$

FIGURE 4–17 Example 4–8. The acceleration vector is shown in gold to distinguish it from the red force vectors.

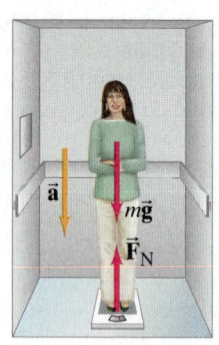

EXAMPLE 4–8 **Apparent weight loss.** A 65-kg woman descends in an elevator that briefly accelerates at 0.20g downward. She stands on a scale that reads in kg. (*a*) During this acceleration, what is her weight and what does the scale read? (*b*) What does the scale read when the elevator descends at a constant speed of 2.0 m/s?

APPROACH Figure 4–17 shows all the forces that act on the woman (and *only* those that act on her). The direction of the acceleration is downward, so we choose the positive direction as down (this is the opposite choice from Examples 4–6 and 4–7).

SOLUTION (*a*) From Newton's second law,

$$\Sigma F = ma$$
$$mg - F_N = m(0.20g).$$

We solve for F_N:

$$F_N = mg - 0.20mg = 0.80mg,$$

and it acts upward. The normal force $\vec{F}_N$ is the force the scale exerts on the person, and is equal and opposite to the force she exerts on the scale: $F_N' = 0.80mg$ downward. Her weight (force of gravity on her) is still $mg = (65\,\text{kg})(9.8\,\text{m/s}^2) = 640\,\text{N}$. But the scale, needing to exert a force of only $0.80mg$, will give a reading of $0.80m = 52\,\text{kg}$.

(*b*) Now there is no acceleration, $a = 0$, so by Newton's second law, $mg - F_N = 0$ and $F_N = mg$. The scale reads her true mass of 65 kg.

NOTE The scale in (*a*) may give a reading of 52 kg (as an "apparent mass"), but her mass doesn't change as a result of the acceleration: it stays at 65 kg.

4–7 Solving Problems with Newton's Laws: Free-Body Diagrams

Newton's second law tells us that the acceleration of an object is proportional to the *net force* acting on the object. The **net force**, as mentioned earlier, is the *vector sum* of all forces acting on the object. Indeed, extensive experiments have shown that forces do add together as vectors precisely according to the rules we developed in Chapter 3. For example, in Fig. 4–18, two forces of equal magnitude (100 N each) are shown acting on an object at right angles to each other. Intuitively, we can see that the object will start moving at a 45° angle and thus the net force acts at a 45° angle. This is just what the rules of vector addition give. From the theorem of Pythagoras, the magnitude of the resultant force is $F_R = \sqrt{(100\,\text{N})^2 + (100\,\text{N})^2} = 141\,\text{N}$.

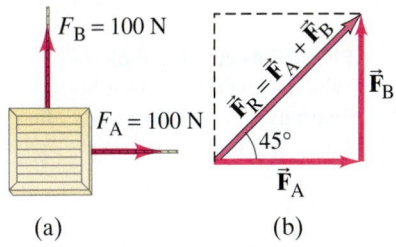

FIGURE 4–18 (a) Two forces, $\vec{\mathbf{F}}_A$ and $\vec{\mathbf{F}}_B$, exerted by workers A and B, act on a crate. (b) The sum, or resultant, of $\vec{\mathbf{F}}_A$ and $\vec{\mathbf{F}}_B$ is $\vec{\mathbf{F}}_R$.

EXAMPLE 4–9 **Adding force vectors.** Calculate the sum of the two forces exerted on the boat by workers A and B in Fig. 4–19a.

APPROACH We add force vectors like any other vectors as described in Chapter 3. The first step is to choose an *xy* coordinate system (see Fig. 4–19a), and then resolve vectors into their components.

SOLUTION The two force vectors are shown resolved into components in Fig. 4–19b. We add the forces using the method of components. The components of $\vec{\mathbf{F}}_A$ are

$$F_{Ax} = F_A \cos 45.0° = (40.0\,\text{N})(0.707) = 28.3\,\text{N},$$
$$F_{Ay} = F_A \sin 45.0° = (40.0\,\text{N})(0.707) = 28.3\,\text{N}.$$

The components of $\vec{\mathbf{F}}_B$ are

$$F_{Bx} = +F_B \cos 37.0° = +(30.0\,\text{N})(0.799) = +24.0\,\text{N},$$
$$F_{By} = -F_B \sin 37.0° = -(30.0\,\text{N})(0.602) = -18.1\,\text{N}.$$

F_{By} is negative because it points along the negative *y* axis. The components of the resultant force are (see Fig. 4–19c)

$$F_{Rx} = F_{Ax} + F_{Bx} = 28.3\,\text{N} + 24.0\,\text{N} = 52.3\,\text{N},$$
$$F_{Ry} = F_{Ay} + F_{By} = 28.3\,\text{N} - 18.1\,\text{N} = 10.2\,\text{N}.$$

To find the magnitude of the resultant force, we use the Pythagorean theorem

$$F_R = \sqrt{F_{Rx}^2 + F_{Ry}^2} = \sqrt{(52.3)^2 + (10.2)^2}\,\text{N} = 53.3\,\text{N}.$$

The only remaining question is the angle θ that the net force $\vec{\mathbf{F}}_R$ makes with the x axis. We use:

$$\tan \theta = \frac{F_{Ry}}{F_{Rx}} = \frac{10.2\,\text{N}}{52.3\,\text{N}} = 0.195,$$

and $\tan^{-1}(0.195) = 11.0°$. The net force on the boat has magnitude 53.3 N and acts at an 11.0° angle to the x axis.

FIGURE 4–19 Example 4–9: Two force vectors act on a boat.

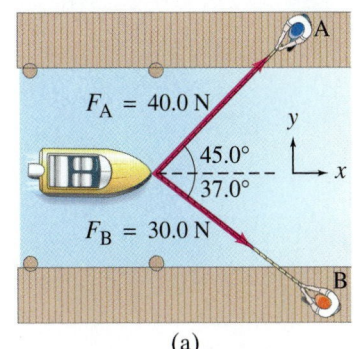

(a)

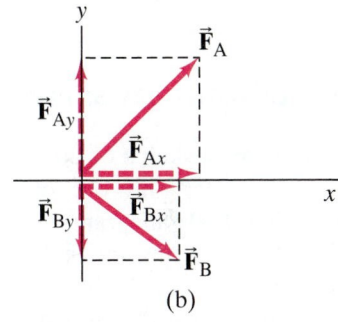

(b)

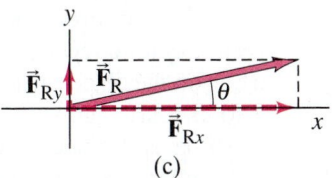

(c)

When solving problems involving Newton's laws and force, it is very important to draw a diagram showing all the forces acting *on* each object involved. Such a diagram is called a **free-body diagram**, or **force diagram**: choose one object, and draw an arrow to represent each force acting on it. Include *every* force acting on that object. Do not show forces that the chosen object exerts on *other* objects. To help you identify each and every force that is exerted on your chosen object, ask yourself what other objects could exert a force on it. If your problem involves more than one object, a separate free-body diagram is needed for each object. For now, the likely forces that could be acting are *gravity* and *contact forces* (one object pushing or pulling another, normal force, friction). Later we will consider air resistance, drag, buoyancy, pressure, as well as electric and magnetic forces.

PROBLEM SOLVING
Free-body diagram

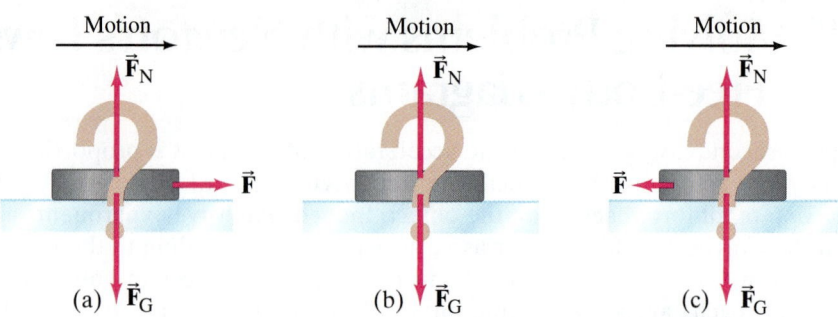

FIGURE 4–20 Example 4–10. Which is the correct free-body diagram for a hockey puck sliding across frictionless ice?

Motion

(a) $\vec{\mathbf{F}}_G$ (b) $\vec{\mathbf{F}}_G$ (c) $\vec{\mathbf{F}}_G$

CONCEPTUAL EXAMPLE 4–10 | **The hockey puck.** A hockey puck is sliding at constant velocity across a flat horizontal ice surface that is assumed to be frictionless. Which of the sketches in Fig. 4–20 is the correct free-body diagram for this puck? What would your answer be if the puck slowed down?

RESPONSE Did you choose (*a*)? If so, can you answer the question: what exerts the horizontal force labeled $\vec{\mathbf{F}}$ on the puck? If you say that it is the force needed to maintain the motion, ask yourself: what exerts this force? Remember that another object must exert any force—and there simply isn't any possibility here. Therefore, (*a*) is wrong. Besides, the force $\vec{\mathbf{F}}$ in Fig. 4–20a would give rise to an acceleration by Newton's second law. It is (*b*) that is correct. No net force acts on the puck, and the puck slides at constant velocity across the ice.

In the real world, where even smooth ice exerts at least a tiny friction force, then (*c*) is the correct answer. The tiny friction force is in the direction opposite to the motion, and the puck's velocity decreases, even if very slowly.

Here now is a brief summary of how to approach solving problems involving Newton's laws.

Newton's Laws; Free-Body Diagrams

1. **Draw a sketch** of the situation.
2. Consider only one object (at a time), and draw a **free-body diagram** for that object, showing *all* the forces acting *on* that object. Include any unknown forces that you have to solve for. Do not show any forces that the chosen object exerts on other objects.

   Draw the arrow for each force vector reasonably accurately for direction and magnitude. Label each force acting on the object, including forces you must solve for, as to its source (gravity, person, friction, and so on).

   If several objects are involved, draw a free-body diagram for each object *separately*, showing all the forces acting *on that object* (and *only* forces acting on that object). For each (and every) force, you must be clear about: *on* what object that force acts, and *by* what object that force is exerted. Only forces acting *on* a given object can be included in $\Sigma\vec{\mathbf{F}} = m\vec{\mathbf{a}}$ for that object.

3. Newton's second law involves vectors, and it is usually important to **resolve vectors** into components. **Choose** *x* and *y* axes in a way that simplifies the calculation. For example, it often saves work if you choose one coordinate axis to be in the direction of the acceleration.

4. For each object, **apply Newton's second law** to the *x* and *y* components separately. That is, the *x* component of the net force on that object is related to the *x* component of that object's acceleration: $\Sigma F_x = ma_x$, and similarly for the *y* direction.

5. **Solve** the equation or equations for the unknown(s).

This Problem Solving Strategy should not be considered a prescription. Rather it is a summary of things to do that will start you thinking and getting involved in the problem at hand.

When we are concerned only about translational motion, all the forces on a given object can be drawn as acting at the center of the object, thus treating the object as a *point particle*. However, for problems involving rotation or statics, the place *where* each force acts is also important, as we shall see in Chapters 10, 11, and 12.

CAUTION

Treating an object as a particle

In the Examples that follow, we assume that all surfaces are very smooth so that friction can be ignored. (Friction, and Examples using it, are discussed in Chapter 5).

EXAMPLE 4–11 **Pulling the mystery box.** Suppose a friend asks to examine the 10.0-kg box you were given (Example 4–6, Fig. 4–15), hoping to guess what is inside; and you respond, "Sure, pull the box over to you." She then pulls the box by the attached cord, as shown in Fig. 4–21a, along the smooth surface of the table. The magnitude of the force exerted by the person is $F_P = 40.0$ N, and it is exerted at a 30.0° angle as shown. Calculate (*a*) the acceleration of the box, and (*b*) the magnitude of the upward force F_N exerted by the table on the box. Assume that friction can be neglected.

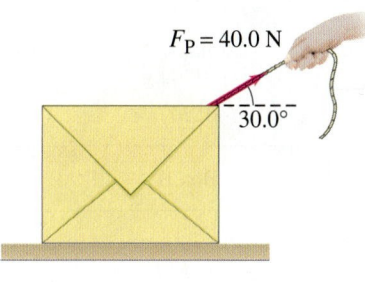

(a)

APPROACH We follow the Problem Solving Strategy on the previous page.

SOLUTION

1. **Draw a sketch**: The situation is shown in Fig. 4–21a; it shows the box and the force applied by the person, F_P.

2. **Free-body diagram**: Figure 4–21b shows the free-body diagram of the box. To draw it correctly, we show *all* the forces acting on the box and *only* the forces acting on the box. They are: the force of gravity $m\vec{g}$; the normal force exerted by the table $\vec{F}_N$; and the force exerted by the person $\vec{F}_P$. We are interested only in translational motion, so we can show the three forces acting at a point, Fig. 4–21c.

3. **Choose axes and resolve vectors**: We expect the motion to be horizontal, so we choose the *x* axis horizontal and the *y* axis vertical. The pull of 40.0 N has components

$$F_{Px} = (40.0 \text{ N})(\cos 30.0°) = (40.0 \text{ N})(0.866) = 34.6 \text{ N},$$
$$F_{Py} = (40.0 \text{ N})(\sin 30.0°) = (40.0 \text{ N})(0.500) = 20.0 \text{ N}.$$

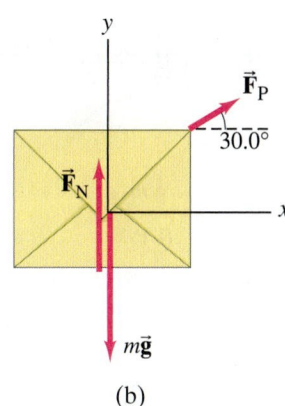

(b)

In the horizontal (*x*) direction, $\vec{F}_N$ and $m\vec{g}$ have zero components. Thus the horizontal component of the net force is F_{Px}.

4. (*a*) **Apply Newton's second law** to determine the *x* component of the acceleration:

$$F_{Px} = ma_x.$$

5. (*a*) **Solve**:

$$a_x = \frac{F_{Px}}{m} = \frac{(34.6 \text{ N})}{(10.0 \text{ kg})} = 3.46 \text{ m/s}^2.$$

The acceleration of the box is 3.46 m/s² to the right.

(*b*) Next we want to find F_N.

4′. (*b*) **Apply Newton's second law** to the vertical (*y*) direction, with upward as positive:

$$\Sigma F_y = ma_y$$
$$F_N - mg + F_{Py} = ma_y.$$

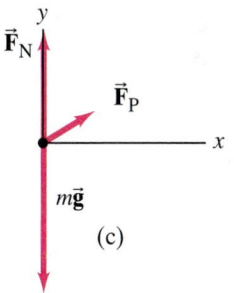

(c)

5′. (*b*) **Solve**: We have $mg = (10.0 \text{ kg})(9.80 \text{ m/s}^2) = 98.0$ N and, from point 3 above, $F_{Py} = 20.0$ N. Furthermore, since $F_{Py} < mg$, the box does not move vertically, so $a_y = 0$. Thus

$$F_N - 98.0 \text{ N} + 20.0 \text{ N} = 0,$$

so

$$F_N = 78.0 \text{ N}.$$

FIGURE 4–21 (a) Pulling the box, Example 4–11; (b) is the free-body diagram for the box, and (c) is the free-body diagram considering all the forces to act at a point (translational motion only, which is what we have here).

NOTE F_N is less than mg: the table does not push against the full weight of the box because part of the pull exerted by the person is in the upward direction.

EXERCISE F A 10.0-kg box is dragged on a horizontal frictionless surface by a horizontal force of 10.0 N. If the applied force is doubled, the normal force on the box will (a) increase; (b) remain the same; (c) decrease.

Tension in a Flexible Cord

When a flexible cord pulls on an object, the cord is said to be under **tension**, and the force it exerts on the object is the tension F_T. If the cord has negligible mass, the force exerted at one end is transmitted undiminished to each adjacent piece of cord along the entire length to the other end. Why? Because $\Sigma\vec{F} = m\vec{a} = 0$ for the cord if the cord's mass m is zero (or negligible) no matter what $\vec{a}$ is. Hence the forces pulling on the cord at its two ends must add up to zero (F_T and $-F_T$). Note that flexible cords and strings can only pull. They can't push because they bend.

PROBLEM SOLVING
Cords can pull but can't push;
tension exists throughout a cord

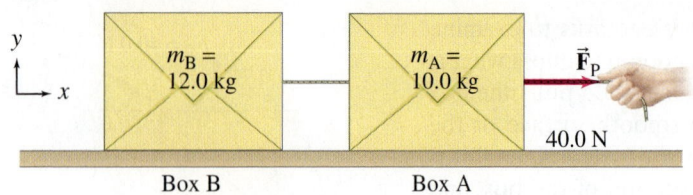

FIGURE 4–22 Example 4–12. (a) Two boxes, A and B, are connected by a cord. A person pulls horizontally on box A with force $F_P = 40.0$ N. (b) Free-body diagram for box A. (c) Free-body diagram for box B.

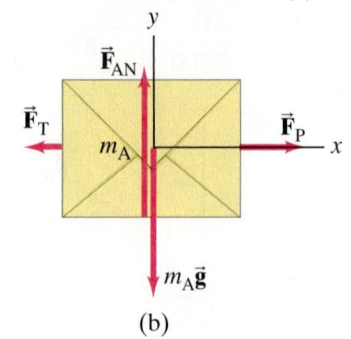

(b)

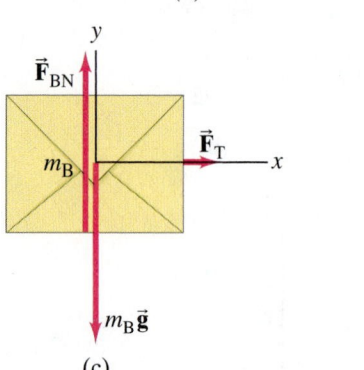

(c)

Our next Example involves two boxes connected by a cord. We can refer to this group of objects as a system. A *system* is any group of one or more objects we choose to consider and study.

EXAMPLE 4–12 **Two boxes connected by a cord.** Two boxes, A and B, are connected by a lightweight cord and are resting on a smooth (frictionless) table. The boxes have masses of 12.0 kg and 10.0 kg. A horizontal force F_P of 40.0 N is applied to the 10.0-kg box, as shown in Fig. 4–22a. Find (a) the acceleration of each box, and (b) the tension in the cord connecting the boxes.

APPROACH We streamline our approach by not listing each step. We have two boxes so we draw a free-body diagram for each. To draw them correctly, we must consider the forces on *each* box by itself, so that Newton's second law can be applied to each. The person exerts a force F_P on box A. Box A exerts a force F_T on the connecting cord, and the cord exerts an opposite but equal magnitude force F_T back on box A (Newton's third law). These two horizontal forces on box A are shown in Fig. 4–22b, along with the force of gravity $m_A \vec{g}$ downward and the normal force $\vec{F}_{AN}$ exerted upward by the table. The cord is light, so we neglect its mass. The tension at each end of the cord is thus the same. Hence the cord exerts a force F_T on the second box. Figure 4–22c shows the forces on box B, which are $\vec{F}_T$, $m_B \vec{g}$, and the normal force $\vec{F}_{BN}$. There will be only horizontal motion. We take the positive x axis to the right.

SOLUTION (a) We apply $\Sigma F_x = ma_x$ to box A:

$$\Sigma F_x = F_P - F_T = m_A a_A. \qquad \text{[box A]}$$

For box B, the only horizontal force is F_T, so

$$\Sigma F_x = F_T = m_B a_B. \qquad \text{[box B]}$$

The boxes are connected, and if the cord remains taut and doesn't stretch, then the two boxes will have the same acceleration a. Thus $a_A = a_B = a$. We are given $m_A = 10.0$ kg and $m_B = 12.0$ kg. We can add the two equations above to eliminate an unknown (F_T) and obtain

$$(m_A + m_B)a = F_P - F_T + F_T = F_P$$

or

$$a = \frac{F_P}{m_A + m_B} = \frac{40.0 \text{ N}}{22.0 \text{ kg}} = 1.82 \text{ m/s}^2.$$

This is what we sought.

Alternate Solution We would have obtained the same result had we considered a single system, of mass $m_A + m_B$, acted on by a net horizontal force equal to F_P. (The tension forces F_T would then be considered internal to the system as a whole, and summed together would make zero contribution to the net force on the *whole* system.)

(b) From the equation above for box B $(F_T = m_B a_B)$, the tension in the cord is

$$F_T = m_B a = (12.0 \text{ kg})(1.82 \text{ m/s}^2) = 21.8 \text{ N}.$$

Thus, F_T is less than F_P ($= 40.0$ N), as we expect, since F_T acts to accelerate only m_B.

NOTE It might be tempting to say that the force the person exerts, F_P, acts not only on box A but also box B. It doesn't. F_P acts only on box A. It affects box B via the tension in the cord, F_T, which acts on box B and accelerates it.

⚠️ **CAUTION**

For any object, use only the forces on that object in calculating $\Sigma F = ma$

EXAMPLE 4–13 Elevator and counterweight (Atwood's machine). A system of two objects suspended over a pulley by a flexible cable, as shown in Fig. 4–23a, is sometimes referred to as an *Atwood's machine*. Consider the real-life application of an elevator (m_E) and its counterweight (m_C). To minimize the work done by the motor to raise and lower the elevator safely, m_E and m_C are made similar in mass. We leave the motor out of the system for this calculation, and assume that the cable's mass is negligible and that the mass of the pulley, as well as any friction, is small and ignorable. These assumptions ensure that the tension F_T in the cable has the same magnitude on both sides of the pulley. Let the mass of the counterweight be $m_C = 1000$ kg. Assume the mass of the empty elevator is 850 kg, and its mass when carrying four passengers is $m_E = 1150$ kg. For the latter case $(m_E = 1150$ kg$)$, calculate (a) the acceleration of the elevator and (b) the tension in the cable.

APPROACH Again we have two objects, and we will need to apply Newton's second law to each of them separately. Each mass has two forces acting on it: gravity downward and the cable tension pulling upward, $\vec{F}_T$. Figures 4–23b and c show the free-body diagrams for the elevator (m_E) and for the counterweight (m_C). The elevator, being the heavier, will accelerate downward, whereas the counterweight will accelerate upward. The magnitudes of their accelerations will be equal (we assume the cable doesn't stretch). For the counterweight, $m_C g = (1000$ kg$)(9.80$ m/s$^2) = 9800$ N, so F_T must be greater than 9800 N (in order that m_C will accelerate upward). For the elevator, $m_E g = (1150$ kg$)(9.80$ m/s$^2) = 11{,}300$ N, which must have greater magnitude than F_T so that m_E accelerates downward. Thus our calculation must give F_T between 9800 N and 11,300 N.

SOLUTION (a) To find F_T as well as the acceleration a, we apply Newton's second law, $\Sigma F = ma$, to each object. We take upward as the positive y direction for both objects. With this choice of axes, $a_C = a$ because m_C accelerates upward, and $a_E = -a$ because m_E accelerates downward. Thus

$$F_T - m_E g = m_E a_E = -m_E a$$
$$F_T - m_C g = m_C a_C = +m_C a.$$

We can subtract the first equation from the second to get

$$(m_E - m_C)g = (m_E + m_C)a,$$

where a is now the only unknown. We solve this for a:

$$a = \frac{m_E - m_C}{m_E + m_C} g = \frac{1150 \text{ kg} - 1000 \text{ kg}}{1150 \text{ kg} + 1000 \text{ kg}} g = 0.070g = 0.68 \text{ m/s}^2.$$

The elevator (m_E) accelerates downward (and the counterweight m_C upward) at $a = 0.070g = 0.68$ m/s^2.

(b) The tension in the cable F_T can be obtained from either of the two $\Sigma F = ma$ equations, setting $a = 0.070g = 0.68$ m/s^2:

$$F_T = m_E g - m_E a = m_E(g - a)$$
$$= 1150 \text{ kg} (9.80 \text{ m/s}^2 - 0.68 \text{ m/s}^2) = 10{,}500 \text{ N},$$

or

$$F_T = m_C g + m_C a = m_C(g + a)$$
$$= 1000 \text{ kg} (9.80 \text{ m/s}^2 + 0.68 \text{ m/s}^2) = 10{,}500 \text{ N},$$

which are consistent. As predicted, our result lies between 9800 N and 11,300 N.

NOTE We can check our equation for the acceleration a in this Example by noting that if the masses were equal $(m_E = m_C)$, then our equation above for a would give $a = 0$, as we should expect. Also, if one of the masses is zero (say, $m_C = 0$), then the other mass $(m_E \neq 0)$ would be predicted by our equation to accelerate at $a = g$, again as expected.

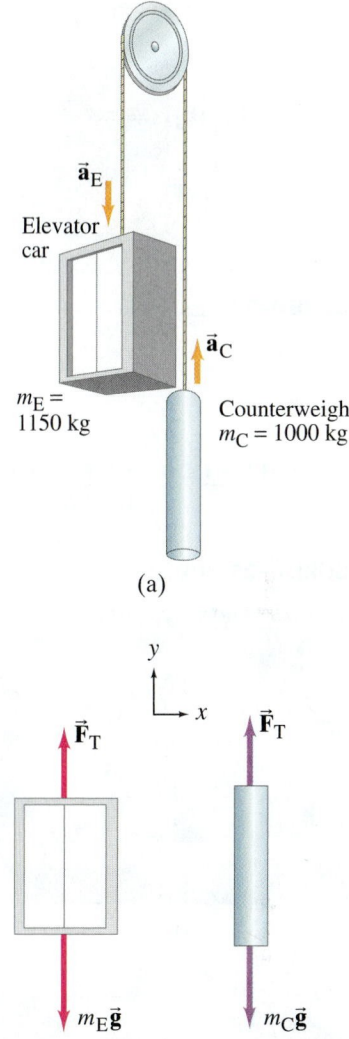

(a)

(b)   (c)

FIGURE 4–23 Example 4–13. (a) Atwood's machine in the form of an elevator–counterweight system. (b) and (c) Free-body diagrams for the two objects.

PROBLEM SOLVING
Check your result by seeing if it works in situations where the answer is easily guessed

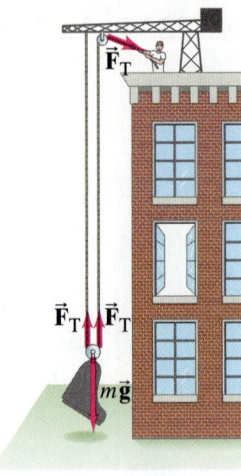

FIGURE 4–24 Example 4–14.

CONCEPTUAL EXAMPLE 4–14
The advantage of a pulley. A mover is trying to lift a piano (slowly) up to a second-story apartment (Fig. 4–24). He is using a rope looped over two pulleys as shown. What force must he exert on the rope to slowly lift the piano's 2000-N weight?

RESPONSE The magnitude of the tension force F_T within the rope is the same at any point along the rope if we assume we can ignore its mass. First notice the forces acting on the lower pulley at the piano. The weight of the piano pulls down on the pulley via a short cable. The tension in the rope, looped through this pulley, pulls up *twice*, once on each side of the pulley. Let us apply Newton's second law to the pulley–piano combination (of mass m), choosing the upward direction as positive:

$$2F_T - mg = ma.$$

To move the piano with constant speed (set $a = 0$ in this equation) thus requires a tension in the rope, and hence a pull on the rope, of $F_T = mg/2$. The mover can exert a force equal to half the piano's weight. We say the pulley has given a **mechanical advantage** of 2, since without the pulley the mover would have to exert twice the force.

PHYSICS APPLIED
Accelerometer

FIGURE 4–25 Example 4–15.

(a)

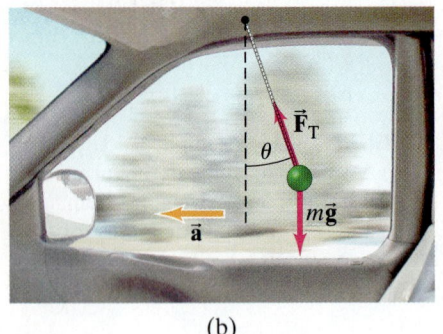

(b)

EXAMPLE 4–15 **Accelerometer.** A small mass m hangs from a thin string and can swing like a pendulum. You attach it above the window of your car as shown in Fig. 4–25a. When the car is at rest, the string hangs vertically. What angle θ does the string make (a) when the car accelerates at a constant $a = 1.20 \, \text{m/s}^2$, and (b) when the car moves at constant velocity, $v = 90 \, \text{km/h}$?

APPROACH The free-body diagram of Fig. 4–25b shows the pendulum at some angle θ and the forces on it: $m\vec{g}$ downward, and the tension $\vec{F}_T$ in the cord. These forces do not add up to zero if $\theta \neq 0$, and since we have an acceleration a, we therefore expect $\theta \neq 0$. Note that θ is the angle relative to the vertical.

SOLUTION (a) The acceleration $a = 1.20 \, \text{m/s}^2$ is horizontal, so from Newton's second law,

$$ma = F_T \sin \theta$$

for the horizontal component, whereas the vertical component gives

$$0 = F_T \cos \theta - mg.$$

Dividing these two equations, we obtain

$$\tan \theta = \frac{F_T \sin \theta}{F_T \cos \theta} = \frac{ma}{mg} = \frac{a}{g}$$

or

$$\tan \theta = \frac{1.20 \, \text{m/s}^2}{9.80 \, \text{m/s}^2}$$

$$= 0.122,$$

so

$$\theta = 7.0°.$$

(b) The velocity is constant, so $a = 0$ and $\tan \theta = 0$. Hence the pendulum hangs vertically $(\theta = 0°)$.

NOTE This simple device is an **accelerometer**—it can be used to measure acceleration.

Inclines

Now we consider what happens when an object slides down an incline, such as a hill or ramp. Such problems are interesting because gravity is the accelerating force, yet the acceleration is not vertical. Solving such problems is usually easier if we choose the xy coordinate system so that one axis points in the direction of the acceleration. Thus we often take the x axis to point along the incline and the y axis perpendicular to the incline, as shown in Fig. 4–26a. Note also that the normal force is not vertical, but is perpendicular to the plane, Fig. 4–26b.

PROBLEM SOLVING

Good choice of coordinate system simplifies the calculation

EXAMPLE 4–16 **Box slides down an incline.** A box of mass m is placed on a smooth (frictionless) incline that makes an angle θ with the horizontal, as shown in Fig. 4–26a. (a) Determine the normal force on the box. (b) Determine the box's acceleration. (c) Evaluate for a mass $m = 10\,\text{kg}$ and an incline of $\theta = 30°$.

APPROACH We expect the motion to be along the incline, so we choose the x axis along the slope, positive down the slope (the direction of motion). The y axis is perpendicular to the incline, upward. The free-body diagram is shown in Fig. 4–26b. The forces on the box are its weight mg vertically downward, which is shown resolved into its components parallel and perpendicular to the incline, and the normal force F_N. The incline acts as a constraint, allowing motion along its surface. The "constraining" force is the normal force.

SOLUTION (a) There is no motion in the y direction, so $a_y = 0$. Applying Newton's second law we have

$$F_y = ma_y$$

$$F_N - mg \cos\theta = 0,$$

where F_N and the y component of gravity ($mg \cos\theta$) are all the forces acting on the box in the y direction. Thus the normal force is given by

$$F_N = mg \cos\theta.$$

Note carefully that unless $\theta = 0°$, F_N has magnitude less than the weight mg.

(b) In the x direction the only force acting is the x component of $m\vec{\mathbf{g}}$, which we see from the diagram is $mg \sin\theta$. The acceleration a is in the x direction so

$$F_x = ma_x$$

$$mg \sin\theta = ma,$$

and we see that the acceleration down the plane is

$$a = g \sin\theta.$$

Thus the acceleration along an incline is always less than g, except at $\theta = 90°$, for which $\sin\theta = 1$ and $a = g$. This makes sense since $\theta = 90°$ is pure vertical fall. For $\theta = 0°$, $a = 0$, which makes sense because $\theta = 0°$ means the plane is horizontal so gravity causes no acceleration. Note too that the acceleration does not depend on the mass m.

(c) For $\theta = 30°$, $\cos\theta = 0.866$ and $\sin\theta = 0.500$, so

$$F_N = 0.866mg = 85\,\text{N},$$

and

$$a = 0.500g = 4.9\,\text{m/s}^2.$$

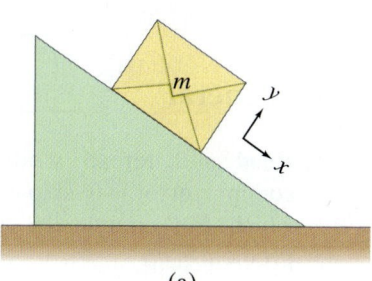

(a)

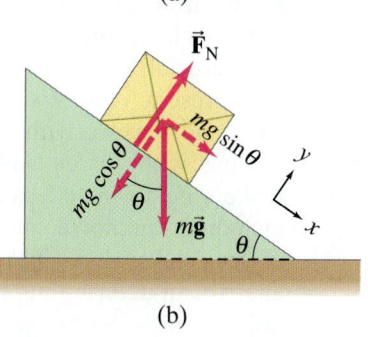

(b)

FIGURE 4–26 Example 4–16. (a) Box sliding on inclined plane. (b) Free-body diagram of box.

We will discuss more Examples of motion on an incline in the next Chapter, where friction will be included.

4–8 Problem Solving—A General Approach

A basic part of a physics course is solving problems effectively. The approach discussed here, though emphasizing Newton's laws, can be applied generally for other topics discussed throughout this book.

In General

1. **Read** and reread written problems carefully. A common error is to skip a word or two when reading, which can completely change the meaning of a problem.

2. **Draw** an accurate picture or diagram of the situation. (This is probably the most overlooked, yet most crucial, part of solving a problem.) Use arrows to represent vectors such as velocity or force, and label the vectors with appropriate symbols. When dealing with forces and applying Newton's laws, make sure to include all forces on a given object, including unknown ones, and make clear what forces act on what object (otherwise you may make an error in determining the *net force* on a particular object).

3. A separate **free-body diagram** needs to be drawn for each object involved, and it must show *all* the forces acting on a given object (and only on that object). Do not show forces that act on other objects.

4. Choose a convenient *xy* **coordinate system** (one that makes your calculations easier, such as one axis in the direction of the acceleration). Vectors are to be resolved into components along the coordinate axes. When using Newton's second law, apply $\Sigma \vec{F} = m\vec{a}$ separately to *x* and *y* components, remembering that *x* direction forces are related to a_x, and similarly for *y*. If more than one object is involved, you can choose different (convenient) coordinate systems for each.

5. List the knowns and the unknowns (what you are trying to determine), and decide what you need in order to find the unknowns. For problems in the present Chapter, we use Newton's laws. More generally, it may help to see if one or more **relationships** (or **equations**) relate the unknowns to the knowns.

But be sure each relationship is applicable in the given case. It is very important to know the limitations of each formula or relationship—when it is valid and when not. In this book, the more general equations have been given numbers, but even these can have a limited range of validity (often stated in brackets to the right of the equation).

6. Try to solve the problem approximately, to see if it is doable (to check if enough information has been given) and reasonable. Use your intuition, and make **rough calculations**—see "Order of Magnitude Estimating" in Section 1–6. A rough calculation, or a reasonable guess about what the range of final answers might be, is very useful. And a rough calculation can be checked against the final answer to catch errors in calculation, such as in a decimal point or the powers of 10.

7. **Solve** the problem, which may include algebraic manipulation of equations and/or numerical calculations. Recall the mathematical rule that you need as many independent equations as you have unknowns; if you have three unknowns, for example, then you need three independent equations. It is usually best to work out the algebra symbolically before putting in the numbers. Why? Because (*a*) you can then solve a whole class of similar problems with different numerical values; (*b*) you can check your result for cases already understood (say, $\theta = 0°$ or 90°); (*c*) there may be cancellations or other simplifications; (*d*) there is usually less chance for numerical error; and (*e*) you may gain better insight into the problem.

8. Be sure to keep track of **units**, for they can serve as a check (they must balance on both sides of any equation).

9. Again consider if your answer is **reasonable**. The use of dimensional analysis, described in Section 1–7, can also serve as a check for many problems.

Summary

Newton's three laws of motion are the basic classical laws describing motion.

Newton's first law (the **law of inertia**) states that if the net force on an object is zero, an object originally at rest remains at rest, and an object in motion remains in motion in a straight line with constant velocity.

Newton's second law states that the acceleration of an object is directly proportional to the net force acting on it, and inversely proportional to its mass:

$$\Sigma \vec{F} = m\vec{a}. \tag{4–1a}$$

Newton's second law is one of the most important and fundamental laws in classical physics.

Newton's third law states that whenever one object exerts a force on a second object, the second object always exerts a force on the first object which is equal in magnitude but opposite in direction:

$$\vec{F}_{AB} = -\vec{F}_{BA}, \qquad (4\text{--}2)$$

where $\vec{F}_{BA}$ is the force on object B exerted by object A. This is true even if objects are moving and accelerating, and/or have different masses.

The tendency of an object to resist a change in its motion is called **inertia**. **Mass** is a measure of the inertia of an object.

Weight refers to the **gravitational force** on an object, and is equal to the product of the object's mass m and the acceleration of gravity $\vec{g}$:

$$\vec{F}_G = m\vec{g}. \qquad (4\text{--}3)$$

Force, which is a vector, can be considered as a push or pull; or, from Newton's second law, force can be defined as an action capable of giving rise to acceleration. The **net force** on an object is the vector sum of all forces acting on that object.

For solving problems involving the forces on one or more objects, it is essential to draw a **free-body diagram** for each object, showing all the forces acting on only that object. Newton's second law can be applied to the vector components for each object.

Questions

1. Why does a child in a wagon seem to fall backward when you give the wagon a sharp pull forward?

2. A box rests on the (frictionless) bed of a truck. The truck driver starts the truck and accelerates forward. The box immediately starts to slide toward the rear of the truck bed. Discuss the motion of the box, in terms of Newton's laws, as seen (*a*) by Andrea standing on the ground beside the truck, and (*b*) by Jim who is riding on the truck (Fig. 4–27).

FIGURE 4–27 Question 2.

3. If the acceleration of an object is zero, are no forces acting on it? Explain.

4. If an object is moving, is it possible for the net force acting on it to be zero?

5. Only one force acts on an object. Can the object have zero acceleration? Can it have zero velocity? Explain.

6. When a golf ball is dropped to the pavement, it bounces back up. (*a*) Is a force needed to make it bounce back up? (*b*) If so, what exerts the force?

7. If you walk along a log floating on a lake, why does the log move in the opposite direction?

8. Why might your foot hurt if you kick a heavy desk or a wall?

9. When you are running and want to stop quickly, you must decelerate quickly. (*a*) What is the origin of the force that causes you to stop? (*b*) Estimate (using your own experience) the maximum rate of deceleration of a person running at top speed to come to rest.

10. (*a*) Why do you push down harder on the pedals of a bicycle when first starting out than when moving at constant speed? (*b*) Why do you need to pedal at all when cycling at constant speed?

11. A father and his young daughter are ice skating. They face each other at rest and push each other, moving in opposite directions. Which one has the greater final speed?

12. Suppose that you are standing on a cardboard carton that just barely supports you. What would happen to it if you jumped up into the air? It would (*a*) collapse; (*b*) be unaffected; (*c*) spring upward a bit; (*d*) move sideways.

13. A stone hangs by a fine thread from the ceiling, and a section of the same thread dangles from the bottom of the stone (Fig. 4–28). If a person gives a sharp pull on the dangling thread, where is the thread likely to break: below the stone or above it? What if the person gives a slow and steady pull? Explain your answers.

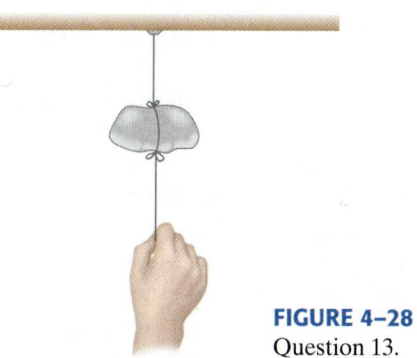

FIGURE 4–28
Question 13.

14. The force of gravity on a 2-kg rock is twice as great as that on a 1-kg rock. Why then doesn't the heavier rock fall faster?

15. Would a spring scale carried to the Moon give accurate results if the scale had been calibrated on Earth, (*a*) in pounds, or (*b*) in kilograms?

16. You pull a box with a constant force across a frictionless table using an attached rope held horizontally. If you now pull the rope with the same force at an angle to the horizontal (with the box remaining flat on the table), does the acceleration of the box (*a*) remain the same, (*b*) increase, or (*c*) decrease? Explain.

17. When an object falls freely under the influence of gravity there is a net force mg exerted on it by the Earth. Yet by Newton's third law the object exerts an equal and opposite force on the Earth. Does the Earth move?

18. Compare the effort (or force) needed to lift a 10-kg object when you are on the Moon with the force needed to lift it on Earth. Compare the force needed to throw a 2-kg object horizontally with a given speed on the Moon and on Earth.

19. Which of the following objects weighs about 1 N: (*a*) an apple, (*b*) a mosquito, (*c*) this book, (*d*) you?

20. According to Newton's third law, each team in a tug of war (Fig. 4–29) pulls with equal force on the other team. What, then, determines which team will win?

FIGURE 4–29 Question 20. A tug of war. Describe the forces on each of the teams and on the rope.

21. When you stand still on the ground, how large a force does the ground exert on you? Why doesn't this force make you rise up into the air?

22. Whiplash sometimes results from an automobile accident when the victim's car is struck violently from the rear. Explain why the head of the victim seems to be thrown backward in this situation. Is it really?

23. Mary exerts an upward force of 40 N to hold a bag of groceries. Describe the "reaction" force (Newton's third law) by stating (*a*) its magnitude, (*b*) its direction, (*c*) *on* what object it is exerted, and (*d*) *by* what object it is exerted.

24. A bear sling, Fig. 4–30, is used in some national parks for placing backpackers' food out of the reach of bears. Explain why the force needed to pull the backpack up increases as the backpack gets higher and higher. Is it possible to pull the rope hard enough so that it doesn't sag at all?

FIGURE 4–30 Question 24.

Problems

4–4 to 4–6 Newton's Laws, Gravitational Force, Normal Force

1. (I) What force is needed to accelerate a child on a sled (total mass = 55 kg) at $1.4 \, \text{m/s}^2$?

2. (I) A net force of 265 N accelerates a bike and rider at $2.30 \, \text{m/s}^2$. What is the mass of the bike and rider together?

3. (I) What is the weight of a 68-kg astronaut (*a*) on Earth, (*b*) on the Moon ($g = 1.7 \, \text{m/s}^2$), (*c*) on Mars ($g = 3.7 \, \text{m/s}^2$), (*d*) in outer space traveling with constant velocity?

4. (I) How much tension must a rope withstand if it is used to accelerate a 1210-kg car horizontally along a frictionless surface at $1.20 \, \text{m/s}^2$?

5. (II) Superman must stop a 120-km/h train in 150 m to keep it from hitting a stalled car on the tracks. If the train's mass is $3.6 \times 10^5 \, \text{kg}$, how much force must he exert? Compare to the weight of the train (give as %). How much force does the train exert on Superman?

6. (II) What average force is required to stop a 950-kg car in 8.0 s if the car is traveling at 95 km/h?

7. (II) Estimate the average force exerted by a shot-putter on a 7.0-kg shot if the shot is moved through a distance of 2.8 m and is released with a speed of 13 m/s.

8. (II) A 0.140-kg baseball traveling 35.0 m/s strikes the catcher's mitt, which, in bringing the ball to rest, recoils backward 11.0 cm. What was the average force applied by the ball on the glove?

9. (II) A fisherman yanks a fish vertically out of the water with an acceleration of $2.5 \, \text{m/s}^2$ using very light fishing line that has a breaking strength of $18 \, \text{N} (\approx 4 \, \text{lb})$. The fisherman unfortunately loses the fish as the line snaps. What can you say about the mass of the fish?

10. (II) A 20.0-kg box rests on a table. (*a*) What is the weight of the box and the normal force acting on it? (*b*) A 10.0-kg box is placed on top of the 20.0-kg box, as shown in Fig. 4–31. Determine the normal force that the table exerts on the 20.0-kg box and the normal force that the 20.0-kg box exerts on the 10.0-kg box.

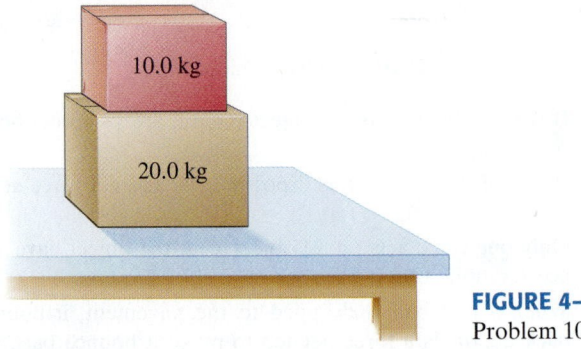

FIGURE 4–31
Problem 10.

11. (II) What average force is needed to accelerate a 9.20-gram pellet from rest to 125 m/s over a distance of 0.800 m along the barrel of a rifle?

12. (II) How much tension must a cable withstand if it is used to accelerate a 1200-kg car vertically upward at $0.70 \, \text{m/s}^2$?

13. (II) A 14.0-kg bucket is lowered vertically by a rope in which there is 163 N of tension at a given instant. What is the acceleration of the bucket? Is it up or down?

14. (II) A particular race car can cover a quarter-mile track (402 m) in 6.40 s starting from a standstill. Assuming the acceleration is constant, how many "g's" does the driver experience? If the combined mass of the driver and race car is 535 kg, what horizontal force must the road exert on the tires?

15. (II) A 75-kg petty thief wants to escape from a third-story jail window. Unfortunately, a makeshift rope made of sheets tied together can support a mass of only 58 kg. How might the thief use this "rope" to escape? Give a quantitative answer.

16. (II) An elevator (mass 4850 kg) is to be designed so that the maximum acceleration is 0.0680g. What are the maximum and minimum forces the motor should exert on the supporting cable?

17. (II) Can cars "stop on a dime"? Calculate the acceleration of a 1400-kg car if it can stop from 35 km/h on a dime (diameter = 1.7 cm.) How many g's is this? What is the force felt by the 68-kg occupant of the car?

18. (II) A person stands on a bathroom scale in a motionless elevator. When the elevator begins to move, the scale briefly reads only 0.75 of the person's regular weight. Calculate the acceleration of the elevator, and find the direction of acceleration.

19. (II) High-speed elevators function under two limitations: (1) the maximum magnitude of vertical acceleration that a typical human body can experience without discomfort is about 1.2 m/s², and (2) the typical maximum speed attainable is about 9.0 m/s. You board an elevator on a skyscraper's ground floor and are transported 180 m above the ground level in three steps: acceleration of magnitude 1.2 m/s² from rest to 9.0 m/s, followed by constant upward velocity of 9.0 m/s, then deceleration of magnitude 1.2 m/s² from 9.0 m/s to rest. (a) Determine the elapsed time for each of these 3 stages. (b) Determine the change in the magnitude of the normal force, expressed as a % of your normal weight during each stage. (c) What fraction of the total transport time does the normal force not equal the person's weight?

20. (II) Using focused laser light, *optical tweezers* can apply a force of about 10 pN to a 1.0-μm diameter polystyrene bead, which has a density about equal to that of water: a volume of 1.0 cm³ has a mass of about 1.0 g. Estimate the bead's acceleration in g's.

21. (II) A rocket with a mass of 2.75×10^6 kg exerts a vertical force of 3.55×10^7 N on the gases it expels. Determine (a) the acceleration of the rocket, (b) its velocity after 8.0 s, and (c) how long it takes to reach an altitude of 9500 m. Assume g remains constant, and ignore the mass of gas expelled (not realistic).

22. (II) (a) What is the acceleration of two falling sky divers (mass = 132 kg including parachute) when the upward force of air resistance is equal to one-fourth of their weight? (b) After popping open the parachute, the divers descend leisurely to the ground at constant speed. What now is the force of air resistance on the sky divers and their parachute? See Fig. 4–32.

FIGURE 4–32   Problem 22.

23. (II) An exceptional standing jump would raise a person 0.80 m off the ground. To do this, what force must a 68-kg person exert against the ground? Assume the person crouches a distance of 0.20 m prior to jumping, and thus the upward force has this distance to act over before he leaves the ground.

24. (II) The cable supporting a 2125-kg elevator has a maximum strength of 21,750 N. What maximum upward acceleration can it give the elevator without breaking?

25. (III) The 100-m dash can be run by the best sprinters in 10.0 s. A 66-kg sprinter accelerates uniformly for the first 45 m to reach top speed, which he maintains for the remaining 55 m. (a) What is the average horizontal component of force exerted on his feet by the ground during acceleration? (b) What is the speed of the sprinter over the last 55 m of the race (i.e., his top speed)?

26. (III) A person jumps from the roof of a house 3.9-m high. When he strikes the ground below, he bends his knees so that his torso decelerates over an approximate distance of 0.70 m. If the mass of his torso (excluding legs) is 42 kg, find (a) his velocity just before his feet strike the ground, and (b) the average force exerted on his torso by his legs during deceleration.

4–7 Using Newton's Laws

27. (I) A box weighing 77.0 N rests on a table. A rope tied to the box runs vertically upward over a pulley and a weight is hung from the other end (Fig. 4–33). Determine the force that the table exerts on the box if the weight hanging on the other side of the pulley weighs (a) 30.0 N, (b) 60.0 N, and (c) 90.0 N.

FIGURE 4–33
Problem 27.

28. (I) Draw the free-body diagram for a basketball player (a) just before leaving the ground on a jump, and (b) while in the air. See Fig. 4–34.

FIGURE 4–34
Problem 28.

29. (I) Sketch the free-body diagram of a baseball (a) at the moment it is hit by the bat, and again (b) after it has left the bat and is flying toward the outfield.

30. (I) A 650-N force acts in a northwesterly direction. A second 650-N force must be exerted in what direction so that the resultant of the two forces points westward? Illustrate your answer with a vector diagram.

31. (II) Christian is making a Tyrolean traverse as shown in Fig. 4–35. That is, he traverses a chasm by stringing a rope between a tree on one side of the chasm and a tree on the opposite side, 25 m away. The rope must sag sufficiently so it won't break. Assume the rope can provide a tension force of up to 29 kN before breaking, and use a "safety factor" of 10 (that is, the rope should only be required to undergo a tension force of 2.9 kN) at the center of the Tyrolean traverse. (a) Determine the distance x that the rope must sag if it is to be within its recommended safety range and Christian's mass is 72.0 kg. (b) If the Tyrolean traverse is incorrectly set up so that the rope sags by only one-fourth the distance found in (a), determine the tension force in the rope. Will the rope break?

FIGURE 4–35 Problem 31.

32. (II) A window washer pulls herself upward using the bucket–pulley apparatus shown in Fig. 4–36. (a) How hard must she pull downward to raise herself slowly at constant speed? (b) If she increases this force by 15%, what will her acceleration be? The mass of the person plus the bucket is 72 kg.

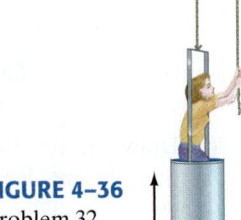

FIGURE 4–36
Problem 32.

33. (II) One 3.2-kg paint bucket is hanging by a massless cord from another 3.2-kg paint bucket, also hanging by a massless cord, as shown in Fig. 4–37. (a) If the buckets are at rest, what is the tension in each cord? (b) If the two buckets are pulled upward with an acceleration of 1.25 m/s^2 by the upper cord, calculate the tension in each cord.

FIGURE 4–37
Problems 33 and 34.

34. (II) The cords accelerating the buckets in Problem 33b, Fig. 4–37, each has a weight of 2.0 N. Determine the tension in each cord at the three points of attachment.

35. (II) Two snowcats in Antarctica are towing a housing unit to a new location, as shown in Fig. 4–38. The sum of the forces $\vec{F}_A$ and $\vec{F}_B$ exerted on the unit by the horizontal cables is parallel to the line L, and $F_A = 4500 \text{ N}$. Determine F_B and the magnitude of $\vec{F}_A + \vec{F}_B$.

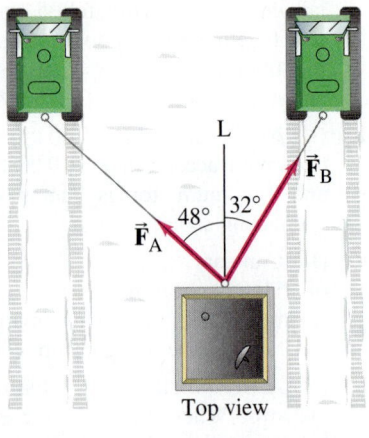

FIGURE 4–38
Problem 35.

Top view

36. (II) A train locomotive is pulling two cars of the same mass behind it, Fig. 4–39. Determine the ratio of the tension in the coupling (think of it as a cord) between the locomotive and the first car (F_{T1}), to that between the first car and the second car (F_{T2}), for any nonzero acceleration of the train.

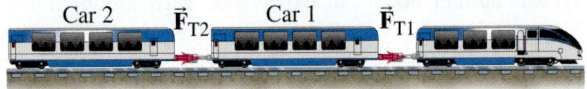

FIGURE 4–39 Problem 36.

37. (II) The two forces $\vec{F}_1$ and $\vec{F}_2$ shown in Fig. 4–40a and b (looking down) act on a 18.5-kg object on a frictionless tabletop. If $F_1 = 10.2 \text{ N}$ and $F_2 = 16.0 \text{ N}$, find the net force on the object and its acceleration for (a) and (b).

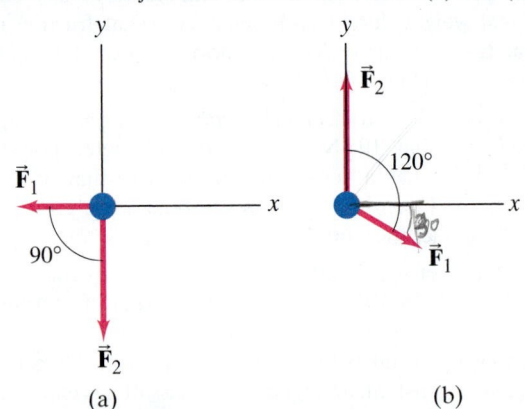

FIGURE 4–40 Problem 37.

38. (II) At the instant a race began, a 65-kg sprinter exerted a force of 720 N on the starting block at a 22° angle with respect to the ground. (a) What was the horizontal acceleration of the sprinter? (b) If the force was exerted for 0.32 s, with what speed did the sprinter leave the starting block?

39. (II) A mass m is at rest on a horizontal frictionless surface at $t = 0$. Then a constant force F_0 acts on it for a time t_0. Suddenly the force doubles to $2F_0$ and remains constant until $t = 2t_0$. Determine the total distance traveled from $t = 0$ to $t = 2t_0$.

40. (II) A 3.0-kg object has the following two forces acting on it:

$$\vec{F}_1 = (16\hat{i} + 12\hat{j}) \text{ N}$$
$$\vec{F}_2 = (-10\hat{i} + 22\hat{j}) \text{ N}$$

If the object is initially at rest, determine its velocity $\vec{v}$ at $t = 3.0 \text{ s}$.

41. (II) Uphill escape ramps are sometimes provided to the side of steep downhill highways for trucks with overheated brakes. For a simple 11° upward ramp, what length would be needed for a runaway truck traveling 140 km/h? Note the large size of your calculated length. (If sand is used for the bed of the ramp, its length can be reduced by a factor of about 2.)

42. (II) A child on a sled reaches the bottom of a hill with a velocity of 10.0 m/s and travels 25.0 m along a horizontal straightaway to a stop. If the child and sled together have a mass of 60.0 kg, what is the average retarding force on the sled on the horizontal straightaway?

43. (II) A skateboarder, with an initial speed of 2.0 m/s, rolls virtually friction free down a straight incline of length 18 m in 3.3 s. At what angle θ is the incline oriented above the horizontal?

44. (II) As shown in Fig. 4–41, five balls (masses 2.00, 2.05, 2.10, 2.15, 2.20 kg) hang from a crossbar. Each mass is supported by "5-lb test" fishing line which will break when its tension force exceeds 22.2 N (= 5 lb). When this device is placed in an elevator, which accelerates upward, only the lines attached to the 2.05 and 2.00 kg masses do not break. Within what range is the elevator's acceleration?

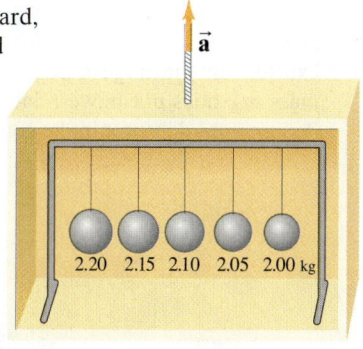

FIGURE 4–41
Problem 44.

45. (II) A 27-kg chandelier hangs from a ceiling on a vertical 4.0-m-long wire. (a) What horizontal force would be necessary to displace its position 0.15 m to one side? (b) What will be the tension in the wire?

46. (II) Three blocks on a frictionless horizontal surface are in contact with each other as shown in Fig. 4–42. A force $\vec{F}$ is applied to block A (mass m_A). (a) Draw a free-body diagram for each block. Determine (b) the acceleration of the system (in terms of m_A, m_B, and m_C), (c) the net force on each block, and (d) the force of contact that each block exerts on its neighbor. (e) If $m_A = m_B = m_C = 10.0$ kg and $F = 96.0$ N, give numerical answers to (b), (c), and (d). Explain how your answers make sense intuitively.

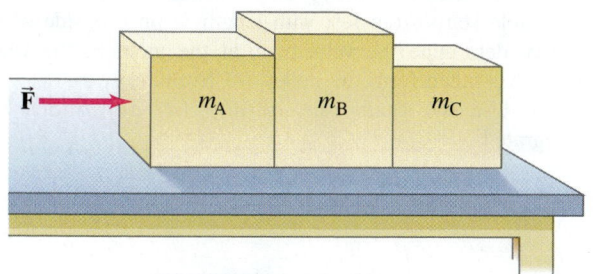

FIGURE 4–42   Problem 46.

47. (II) Redo Example 4–13 but (a) set up the equations so that the direction of the acceleration $\vec{a}$ of each object is in the direction of motion of that object. (In Example 4–13, we took $\vec{a}$ as positive upward for both masses.) (b) Solve the equations to obtain the same answers as in Example 4–13.

48. (II) The block shown in Fig. 4–43 has mass $m = 7.0$ kg and lies on a fixed smooth frictionless plane tilted at an angle $\theta = 22.0°$ to the horizontal. (a) Determine the acceleration of the block as it slides down the plane. (b) If the block starts from rest 12.0 m up the plane from its base, what will be the block's speed when it reaches the bottom of the incline?

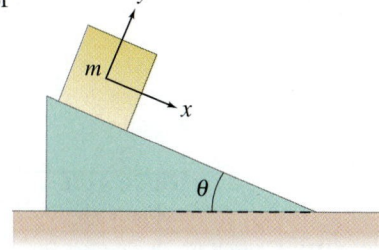

FIGURE 4–43
Block on inclined plane. Problems 48 and 49.

49. (II) A block is given an initial speed of 4.5 m/s up the 22° plane shown in Fig. 4–43. (a) How far up the plane will it go? (b) How much time elapses before it returns to its starting point? Ignore friction.

50. (II) An object is hanging by a string from your rearview mirror. While you are accelerating at a constant rate from rest to 28 m/s in 6.0 s, what angle θ does the string make with the vertical? See Fig. 4–44.

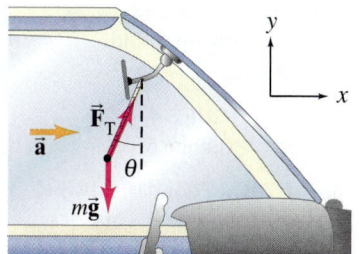

FIGURE 4–44
Problem 50.

51. (II) Figure 4–45 shows a block (mass m_A) on a smooth horizontal surface, connected by a thin cord that passes over a pulley to a second block (m_B), which hangs vertically. (a) Draw a free-body diagram for each block, showing the force of gravity on each, the force (tension) exerted by the cord, and any normal force. (b) Apply Newton's second law to find formulas for the acceleration of the system and for the tension in the cord. Ignore friction and the masses of the pulley and cord.

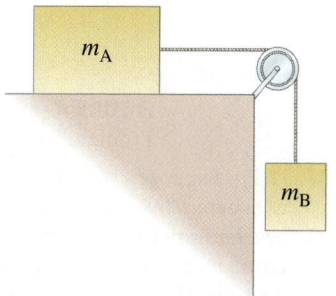

FIGURE 4–45
Problems 51, 52, and 53. Mass m_A rests on a smooth horizontal surface, m_B hangs vertically.

52. (II) (a) If $m_A = 13.0$ kg and $m_B = 5.0$ kg in Fig. 4–45, determine the acceleration of each block. (b) If initially m_A is at rest 1.250 m from the edge of the table, how long does it take to reach the edge of the table if the system is allowed to move freely? (c) If $m_B = 1.0$ kg, how large must m_A be if the acceleration of the system is to be kept at $\frac{1}{100}g$?

53. (III) Determine a formula for the acceleration of the system shown in Fig. 4–45 (see Problem 51) if the cord has a non-negligible mass m_C. Specify in terms of ℓ_A and ℓ_B, the lengths of cord from the respective masses to the pulley. (The total cord length is $\ell = \ell_A + \ell_B$.)

54. (III) Suppose the pulley in Fig. 4–46 is suspended by a cord C. Determine the tension in this cord after the masses are released and before one hits the ground. Ignore the mass of the pulley and cords.

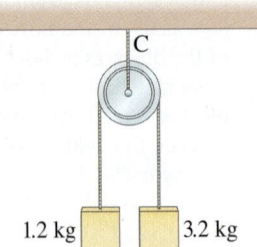

FIGURE 4–46
Problem 54.

55. (III) A small block of mass m rests on the sloping side of a triangular block of mass M which itself rests on a horizontal table as shown in Fig. 4–47. Assuming all surfaces are frictionless, determine the magnitude of the force $\vec{F}$ that must be applied to M so that m remains in a fixed position relative to M (that is, m doesn't move on the incline). [*Hint*: Take x and y axes horizontal and vertical.]

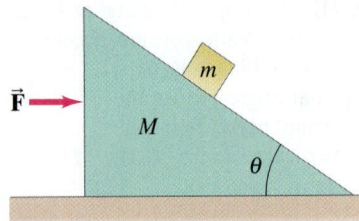

FIGURE 4–47
Problem 55.

56. (III) The double Atwood machine shown in Fig. 4–48 has frictionless, massless pulleys and cords. Determine (*a*) the acceleration of masses m_A, m_B, and m_C, and (*b*) the tensions F_{TA} and F_{TC} in the cords.

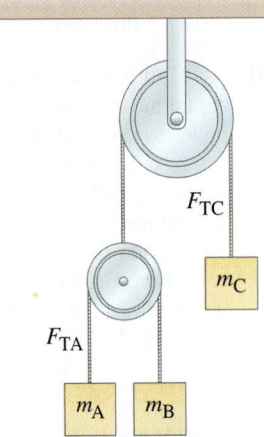

FIGURE 4–48
Problem 56.

57. (III) Suppose two boxes on a frictionless table are connected by a heavy cord of mass 1.0 kg. Calculate the acceleration of each box and the tension at each end of the cord, using the free-body diagrams shown in Fig. 4–49. Assume $F_P = 35.0$ N, and ignore sagging of the cord. Compare your results to Example 4–12 and Fig. 4–22.

58. (III) The two masses shown in Fig. 4–50 are each initially 1.8 m above the ground, and the massless frictionless pulley is 4.8 m above the ground. What maximum height does the lighter object reach after the system is released? [*Hint*: First determine the acceleration of the lighter mass and then its velocity at the moment the heavier one hits the ground. This is its "launch" speed. Assume the mass doesn't hit the pulley. Ignore the mass of the cord.]

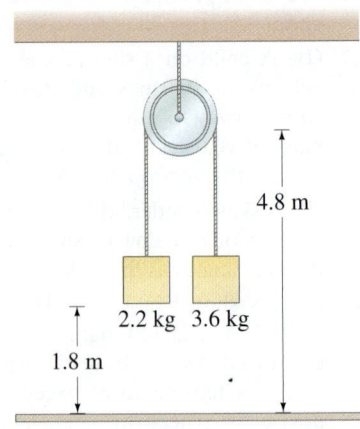

FIGURE 4–50
Problem 58.

59. (III) Determine a formula for the magnitude of the force $\vec{F}$ exerted on the large block (m_C) in Fig. 4–51 so that the mass m_A does not move relative to m_C. Ignore all friction. Assume m_B does not make contact with m_C.

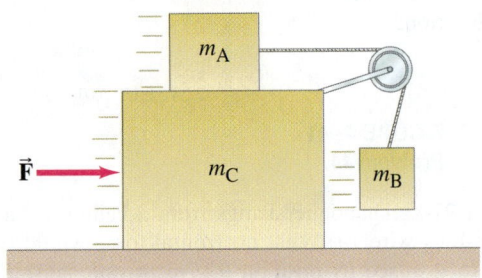

FIGURE 4–51   Problem 59.

60. (III) A particle of mass m, initially at rest at $x = 0$, is accelerated by a force that increases in time as $F = Ct^2$. Determine its velocity v and position x as a function of time.

61. (III) A heavy steel cable of length ℓ and mass M passes over a small massless, frictionless pulley. (*a*) If a length y hangs on one side of the pulley (so $\ell - y$ hangs on the other side), calculate the acceleration of the cable as a function of y. (*b*) Assuming the cable starts from rest with length y_0 on one side of the pulley, determine the velocity v_f at the moment the whole cable has fallen from the pulley. (*c*) Evaluate v_f for $y_0 = \frac{2}{3}\ell$. [*Hint*: Use the chain rule, $dv/dt = (dv/dy)(dy/dt)$, and integrate.]

FIGURE 4–49   Problem 57. Free-body diagrams for each of the objects of the system shown in Fig. 4–22a. Vertical forces, $\vec{F}_N$ and $\vec{F}_G$, are not shown.

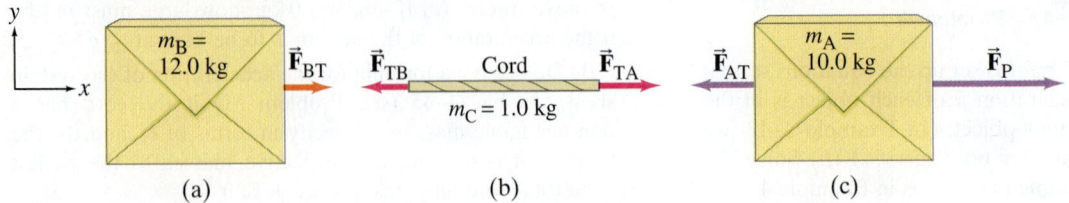

General Problems

62. A person has a reasonable chance of surviving an automobile crash if the deceleration is no more than 30 g's. Calculate the force on a 65-kg person accelerating at this rate. What distance is traveled if brought to rest at this rate from 95 km/h?

63. A 2.0-kg purse is dropped 58 m from the top of the Leaning Tower of Pisa and falls 55 m before reaching the ground with a speed of 27 m/s. What was the average force of air resistance?

64. Tom's hang glider supports his weight using the six ropes shown in Fig. 4–52. Each rope is designed to support an equal fraction of Tom's weight. Tom's mass is 74.0 kg. What is the tension in each of the support ropes?

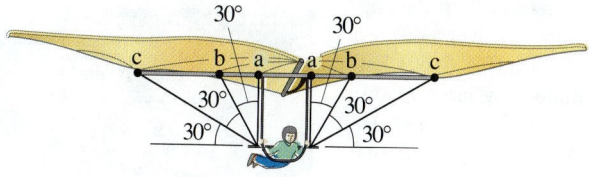

FIGURE 4–52 Problem 64.

65. A wet bar of soap ($m = 150$ g) slides freely down a ramp 3.0 m long inclined at 8.5°. How long does it take to reach the bottom? How would this change if the soap's mass were 300 g?

66. A crane's trolley at point P in Fig. 4–53 moves for a few seconds to the right with constant acceleration, and the 870-kg load hangs at a 5.0° angle to the vertical as shown. What is the acceleration of the trolley and load?

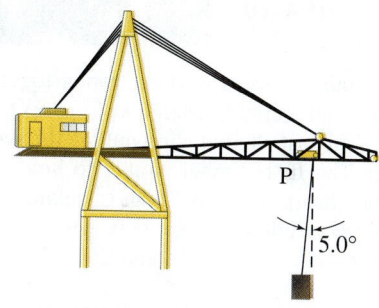

FIGURE 4–53 Problem 66.

67. A block (mass m_A) lying on a fixed frictionless inclined plane is connected to a mass m_B by a cord passing over a pulley, as shown in Fig. 4–54. (*a*) Determine a formula for the acceleration of the system in terms of m_A, m_B, θ, and g. (*b*) What conditions apply to masses m_A and m_B for the acceleration to be in one direction (say, m_A down the plane), or in the opposite direction? Ignore the mass of the cord and pulley.

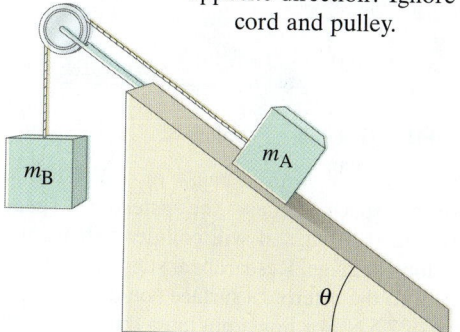

FIGURE 4–54 Problems 67 and 68.

68. (*a*) In Fig. 4–54, if $m_A = m_B = 1.00$ kg and $\theta = 33.0°$, what will be the acceleration of the system? (*b*) If $m_A = 1.00$ kg and the system remains at rest, what must the mass m_B be? (*c*) Calculate the tension in the cord for (*a*) and (*b*).

69. The masses m_A and m_B slide on the smooth (frictionless) inclines fixed as shown in Fig. 4–55. (*a*) Determine a formula for the acceleration of the system in terms of m_A, m_B, θ_A, θ_B, and g. (*b*) If $\theta_A = 32°$, $\theta_B = 23°$, and $m_A = 5.0$ kg, what value of m_B would keep the system at rest? What would be the tension in the cord (negligible mass) in this case? (*c*) What ratio, m_A/m_B, would allow the masses to move at constant speed along their ramps in either direction?

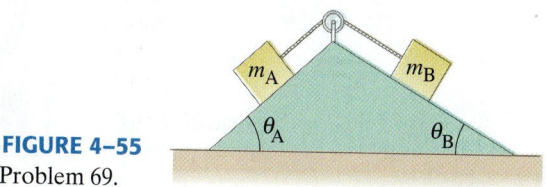

FIGURE 4–55 Problem 69.

70. A 75.0-kg person stands on a scale in an elevator. What does the scale read (in N and in kg) when (*a*) the elevator is at rest, (*b*) the elevator is climbing at a constant speed of 3.0 m/s, (*c*) the elevator is descending at 3.0 m/s, (*d*) the elevator is accelerating upward at 3.0 m/s², (*e*) the elevator is accelerating downward at 3.0 m/s²?

71. A city planner is working on the redesign of a hilly portion of a city. An important consideration is how steep the roads can be so that even low-powered cars can get up the hills without slowing down. A particular small car, with a mass of 920 kg, can accelerate on a level road from rest to 21 m/s (75 km/h) in 12.5 s. Using these data, calculate the maximum steepness of a hill.

72. If a bicyclist of mass 65 kg (including the bicycle) can coast down a 6.5° hill at a steady speed of 6.0 km/h because of air resistance, how much force must be applied to climb the hill at the same speed (and the same air resistance)?

73. A bicyclist can coast down a 5.0° hill at a constant speed of 6.0 km/h. If the force of air resistance is proportional to the speed v so that $F_{air} = cv$, calculate (*a*) the value of the constant c, and (*b*) the average force that must be applied in order to descend the hill at 18.0 km/h. The mass of the cyclist plus bicycle is 80.0 kg.

74. Francesca dangles her watch from a thin piece of string while the jetliner she is in accelerates for takeoff, which takes about 16 s. Estimate the takeoff speed of the aircraft if the string makes an angle of 25° with respect to the vertical, Fig. 4–56.

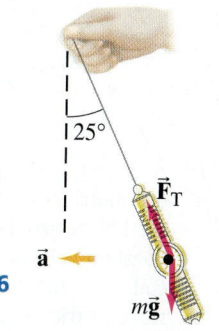

FIGURE 4–56 Problem 74.

75. (*a*) What minimum force F is needed to lift the piano (mass M) using the pulley apparatus shown in Fig. 4–57? (*b*) Determine the tension in each section of rope: F_{T1}, F_{T2}, F_{T3}, and F_{T4}.

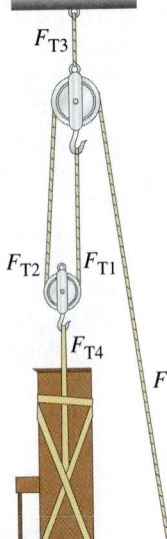

FIGURE 4–57
Problem 75.

76. In the design of a supermarket, there are to be several ramps connecting different parts of the store. Customers will have to push grocery carts up the ramps and it is obviously desirable that this not be too difficult. The engineer has done a survey and found that almost no one complains if the force required is no more than 18 N. Ignoring friction, at what maximum angle θ should the ramps be built, assuming a full 25-kg grocery cart?

77. A jet aircraft is accelerating at 3.8 m/s^2 as it climbs at an angle of 18° above the horizontal (Fig. 4–58). What is the total force that the cockpit seat exerts on the 75-kg pilot?

FIGURE 4–58
Problem 77.

78. A 7650-kg helicopter accelerates upward at 0.80 m/s^2 while lifting a 1250-kg frame at a construction site, Fig. 4–59. (*a*) What is the lift force exerted by the air on the helicopter rotors? (*b*) What is the tension in the cable (ignore its mass) that connects the frame to the helicopter? (*c*) What force does the cable exert on the helicopter?

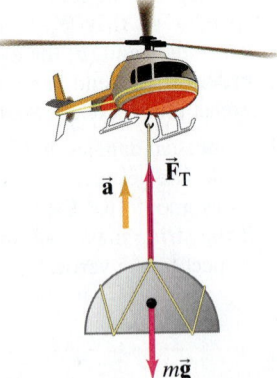

FIGURE 4–59
Problem 78.

79. A super high-speed 14-car Italian train has a mass of 640 metric tons (640,000 kg). It can exert a maximum force of 400 kN horizontally against the tracks, whereas at maximum constant velocity (300 km/h), it exerts a force of about 150 kN. Calculate (*a*) its maximum acceleration, and (*b*) estimate the force of friction and air resistance at top speed.

80. A fisherman in a boat is using a "10-lb test" fishing line. This means that the line can exert a force of 45 N without breaking (1 lb = 4.45 N). (*a*) How heavy a fish can the fisherman land if he pulls the fish up vertically at constant speed? (*b*) If he accelerates the fish upward at 2.0 m/s^2, what maximum weight fish can he land? (*c*) Is it possible to land a 15-lb trout on 10-lb test line? Why or why not?

81. An elevator in a tall building is allowed to reach a maximum speed of 3.5 m/s going down. What must the tension be in the cable to stop this elevator over a distance of 2.6 m if the elevator has a mass of 1450 kg including occupants?

82. Two rock climbers, Bill and Karen, use safety ropes of similar length. Karen's rope is more elastic, called a *dynamic rope* by climbers. Bill has a *static rope*, not recommended for safety purposes in pro climbing. (*a*) Karen falls freely about 2.0 m and then the rope stops her over a distance of 1.0 m (Fig. 4–60). Estimate how large a force (assume constant) she will feel from the rope. (Express the result in multiples of her weight.) (*b*) In a similar fall, Bill's rope stretches by only 30 cm. How many times his weight will the rope pull on him? Which climber is more likely to be hurt?

FIGURE 4–60
Problem 82.

83. Three mountain climbers who are roped together in a line are ascending an icefield inclined at 31.0° to the horizontal (Fig. 4–61). The last climber slips, pulling the second climber off his feet. The first climber is able to hold them both. If each climber has a mass of 75 kg, calculate the tension in each of the two sections of rope between the three climbers. Ignore friction between the ice and the fallen climbers.

FIGURE 4–61   Problem 83.

84. A "doomsday" asteroid with a mass of 1.0×10^{10} kg is hurtling through space. Unless the asteroid's speed is changed by about 0.20 cm/s, it will collide with Earth and cause tremendous damage. Researchers suggest that a small "space tug" sent to the asteroid's surface could exert a gentle constant force of 2.5 N. For how long must this force act?

85. A 450-kg piano is being unloaded from a truck by rolling it down a ramp inclined at 22°. There is negligible friction and the ramp is 11.5 m long. Two workers slow the rate at which the piano moves by pushing with a combined force of 1420 N parallel to the ramp. If the piano starts from rest, how fast is it moving at the bottom?

86. Consider the system shown in Fig. 4–62 with $m_A = 9.5$ kg and $m_B = 11.5$ kg. The angles $\theta_A = 59°$ and $\theta_B = 32°$. (a) In the absence of friction, what force $\vec{F}$ would be required to pull the masses at a constant velocity up the fixed inclines? (b) The force $\vec{F}$ is now removed. What is the magnitude and direction of the acceleration of the two blocks? (c) In the absence of $\vec{F}$, what is the tension in the string?

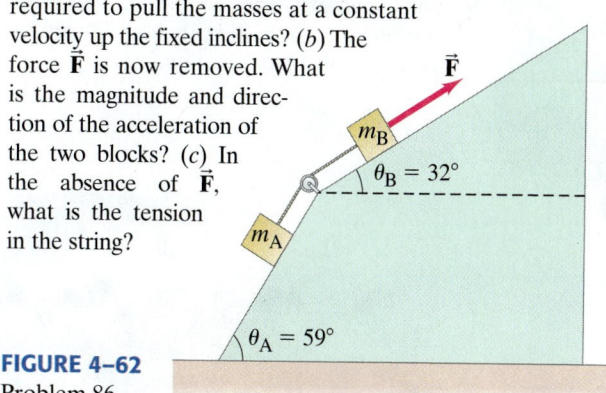

FIGURE 4–62
Problem 86.

87. A 1.5-kg block rests on top of a 7.5-kg block (Fig. 4–63). The cord and pulley have negligible mass, and there is no significant friction anywhere. (a) What force F must be applied to the bottom block so the top block accelerates to the right at 2.5 m/s²? (b) What is the tension in the connecting cord?

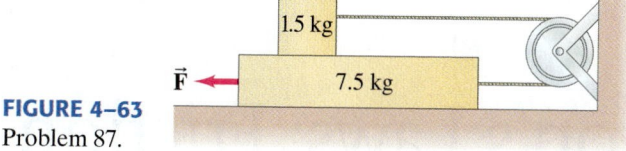

FIGURE 4–63
Problem 87.

88. You are driving home in your 750-kg car at 15 m/s. At a point 45 m from the beginning of an intersection, you see a green traffic light change to yellow, which you expect will last 4.0 s, and the distance to the far side of the intersection is 65 m (Fig. 4–64). (a) If you choose to accelerate, your car's engine will furnish a forward force of 1200 N. Will you make it completely through the intersection before the light turns red? (b) If you decide to panic stop, your brakes will provide a force of 1800 N. Will you stop before entering the intersection?

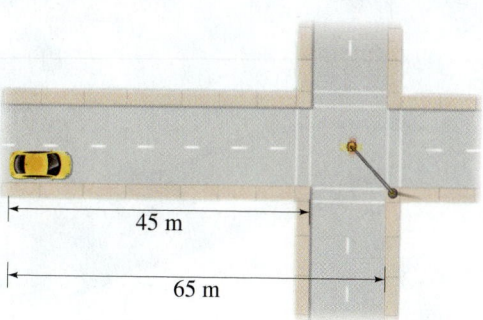

FIGURE 4–64 Problem 88.

Numerical/Computer

*89. (II) A large crate of mass 1500 kg starts sliding from rest along a frictionless ramp, whose length is ℓ and whose inclination with the horizontal is θ. (a) Determine as a function of θ: (i) the acceleration a of the crate as it goes downhill, (ii) the time t to reach the bottom of the incline, (iii) the final velocity v of the crate when it reaches the bottom of the ramp, and (iv) the normal force F_N on the crate. (b) Now assume $\ell = 100$ m. Use a spreadsheet to calculate and graph a, t, v, and F_N as functions of θ from $\theta = 0°$ to 90° in 1° steps. Are your results consistent with the known result for the limiting cases $\theta = 0°$ and $\theta = 90°$?

Answers to Exercises

A: No force is needed. The car accelerates out from under the cup. Think of Newton's first law (see Example 4–1).

B: (a).

C: (a) The same; (b) the sports car; (c) third law for part (a), second law for part (b).

D: (e).

E: (b).

F: (b).

Newton's laws are fundamental in physics. These photos show two situations of using Newton's laws which involve some new elements in addition to those discussed in the previous Chapter. The downhill skier illustrates *friction* on an incline, although at this moment she is not touching the snow, and so is retarded only by air resistance which is a velocity-dependent force (an optional topic in this Chapter). The people on the rotating amusement park ride below illustrate the dynamics of circular motion.

5 Using Newton's Laws: Friction, Circular Motion, Drag Forces

C H A P T E R

CONTENTS

5–1 Applications of Newton's Laws Involving Friction

5–2 Uniform Circular Motion—Kinematics

5–3 Dynamics of Uniform Circular Motion

5–4 Highway Curves: Banked and Unbanked

*5–5 Nonuniform Circular Motion

*5–6 Velocity-Dependent Forces: Drag and Terminal Velocity

CHAPTER-OPENING QUESTION—Guess now!
You revolve a ball around you in a horizontal circle at constant speed on a string, as shown here from above. Which path will the ball follow if you let go of the string at point P?

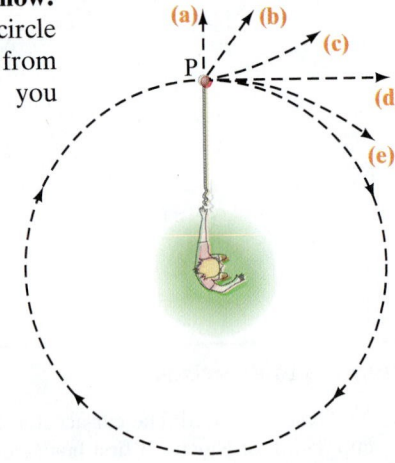

This chapter continues our study of Newton's laws and emphasizes their fundamental importance in physics. We cover some important applications of Newton's laws, including friction and circular motion. Although some material in this Chapter may seem to repeat topics covered in Chapter 4, in fact, new elements are involved.

5–1 Applications of Newton's Laws Involving Friction

Until now we have ignored friction, but it must be taken into account in most practical situations. Friction exists between two solid surfaces because even the smoothest looking surface is quite rough on a microscopic scale, Fig. 5–1. When we try to slide an object across another surface, these microscopic bumps impede the motion. Exactly what is happening at the microscopic level is not yet fully understood. It is thought that the atoms on a bump of one surface may come so close to the atoms of the other surface that attractive electric forces between the atoms could "bond" as a tiny weld between the two surfaces. Sliding an object across a surface is often jerky, perhaps due to the making and breaking of these bonds. Even when a round object rolls across a surface, there is still some friction, called *rolling friction*, although it is generally much less than when objects slide across a surface. We focus our attention now on sliding friction, which is usually called **kinetic friction** (*kinetic* is from the Greek for "moving").

When an object slides along a rough surface, the force of kinetic friction acts opposite to the direction of the object's velocity. The magnitude of the force of kinetic friction depends on the nature of the two sliding surfaces. For given surfaces, experiment shows that the friction force is approximately proportional to the *normal force* between the two surfaces, which is the force that either object exerts on the other and is perpendicular to their common surface of contact (see Fig. 5–2). The force of friction between hard surfaces in many cases depends very little on the total surface area of contact; that is, the friction force on this book is roughly the same whether it is being slid on its wide face or on its spine, assuming the surfaces have the same smoothness. We consider a simple model of friction in which we make this assumption that the friction force is independent of area. Then we write the proportionality between the magnitudes of the friction force F_{fr} and the normal force F_N as an equation by inserting a constant of proportionality, μ_k:

$$F_{fr} = \mu_k F_N. \qquad \text{[kinetic friction]}$$

This relation is not a fundamental law; it is an experimental relation between the magnitude of the friction force F_{fr}, which acts parallel to the two surfaces, and the magnitude of the normal force F_N, which acts perpendicular to the surfaces. It is *not* a vector equation since the two forces have directions perpendicular to one another. The term μ_k is called the *coefficient of kinetic friction*, and its value depends on the nature of the two surfaces. Measured values for a variety of surfaces are given in Table 5–1. These are only approximate, however, since μ depends on whether the surfaces are wet or dry, on how much they have been sanded or rubbed, if any burrs remain, and other such factors. But μ_k is roughly independent of the sliding speed, as well as the area in contact.

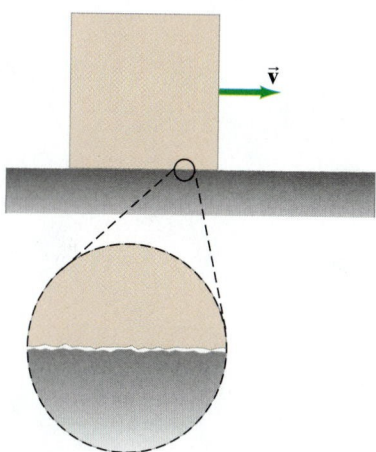

FIGURE 5–1 An object moving to the right on a table or floor. The two surfaces in contact are rough, at least on a microscopic scale.

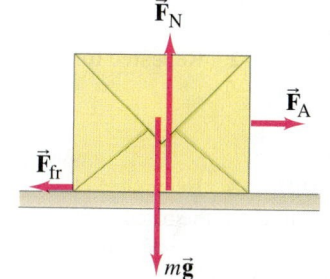

FIGURE 5–2 When an object is pulled along a surface by an applied force ($\vec{F}_A$), the force of friction $\vec{F}_{fr}$ opposes the motion. The magnitude of $\vec{F}_{fr}$ is proportional to the magnitude of the normal force (F_N).

TABLE 5–1 Coefficients of Friction†

Surfaces	Coefficient of Static Friction, μ_s	Coefficient of Kinetic Friction, μ_k
Wood on wood	0.4	0.2
Ice on ice	0.1	0.03
Metal on metal (lubricated)	0.15	0.07
Steel on steel (unlubricated)	0.7	0.6
Rubber on dry concrete	1.0	0.8
Rubber on wet concrete	0.7	0.5
Rubber on other solid surfaces	1–4	1
Teflon® on Teflon in air	0.04	0.04
Teflon on steel in air	0.04	0.04
Lubricated ball bearings	<0.01	<0.01
Synovial joints (in human limbs)	0.01	0.01

†Values are approximate and intended only as a guide.

What we have been discussing up to now is *kinetic friction*, when one object slides over another. There is also **static friction**, which refers to a force parallel to the two surfaces that can arise even when they are not sliding. Suppose an object such as a desk is resting on a horizontal floor. If no horizontal force is exerted on the desk, there also is no friction force. But now suppose you try to push the desk, and it doesn't move. You are exerting a horizontal force, but the desk isn't moving, so there must be another force on the desk keeping it from moving (the net force is zero on an object at rest). This is the force of *static friction* exerted by the floor on the desk. If you push with a greater force without moving the desk, the force of static friction also has increased. If you push hard enough, the desk will eventually start to move, and kinetic friction takes over. At this point, you have exceeded the maximum force of static friction, which is given by $(F_{fr})_{max} = \mu_s F_N$, where μ_s is the *coefficient of static friction* (Table 5–1). Because the force of static friction can vary from zero to this maximum value, we write

$$F_{fr} \leq \mu_s F_N. \qquad \text{[static friction]}$$

You may have noticed that it is often easier to keep a heavy object sliding than it is to start it sliding in the first place. This is consistent with μ_s generally being greater than μ_k (see Table 5–1).

EXAMPLE 5–1 **Friction: static and kinetic.** Our 10.0-kg mystery box rests on a horizontal floor. The coefficient of static friction is $\mu_s = 0.40$ and the coefficient of kinetic friction is $\mu_k = 0.30$. Determine the force of friction, F_{fr}, acting on the box if a horizontal external applied force F_A is exerted on it of magnitude: (a) 0, (b) 10 N, (c) 20 N, (d) 38 N, and (e) 40 N.

APPROACH We don't know, right off, if we are dealing with static friction or kinetic friction, nor if the box remains at rest or accelerates. We need to draw a free-body diagram, and then determine in each case whether or not the box will move: the box starts moving if F_A is greater than the maximum static friction force (Newton's second law). The forces on the box are gravity $m\vec{g}$, the normal force exerted by the floor $\vec{F}_N$, the horizontal applied force $\vec{F}_A$, and the friction force $\vec{F}_{fr}$, as shown in Fig. 5–2.

SOLUTION The free-body diagram of the box is shown in Fig. 5–2. In the vertical direction there is no motion, so Newton's second law in the vertical direction gives $\Sigma F_y = ma_y = 0$, which tells us $F_N - mg = 0$. Hence the normal force is

$$F_N = mg = (10.0 \text{ kg})(9.80 \text{ m/s}^2) = 98.0 \text{ N}.$$

(a) Because $F_A = 0$ in this first case, the box doesn't move, and $F_{fr} = 0$.
(b) The force of static friction will oppose any applied force up to a maximum of

$$\mu_s F_N = (0.40)(98.0 \text{ N}) = 39 \text{ N}.$$

When the applied force is $F_A = 10 \text{ N}$, the box will not move. Newton's second law gives $\Sigma F_x = F_A - F_{fr} = 0$, so $F_{fr} = 10 \text{ N}$.
(c) An applied force of 20 N is also not sufficient to move the box. Thus $F_{fr} = 20 \text{ N}$ to balance the applied force.
(d) The applied force of 38 N is still not quite large enough to move the box; so the friction force has now increased to 38 N to keep the box at rest.
(e) A force of 40 N will start the box moving since it exceeds the maximum force of static friction, $\mu_s F_N = (0.40)(98 \text{ N}) = 39 \text{ N}$. Instead of static friction, we now have kinetic friction, and its magnitude is

$$F_{fr} = \mu_k F_N = (0.30)(98.0 \text{ N}) = 29 \text{ N}.$$

There is now a net (horizontal) force on the box of magnitude $F = 40 \text{ N} - 29 \text{ N} = 11 \text{ N}$, so the box will accelerate at a rate

$$a_x = \frac{\Sigma F}{m} = \frac{11 \text{ N}}{10.0 \text{ kg}} = 1.1 \text{ m/s}^2$$

as long as the applied force is 40 N. Figure 5–3 shows a graph that summarizes this Example.

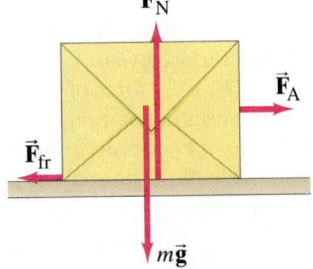

FIGURE 5–2 Repeated for Example 5–1.

FIGURE 5–3 Example 5–1. Magnitude of the force of friction as a function of the external force applied to an object initially at rest. As the applied force is increased in magnitude, the force of static friction increases in proportion until the applied force equals $\mu_s F_N$. If the applied force increases further, the object will begin to move, and the friction force drops to a roughly constant value characteristic of kinetic friction.

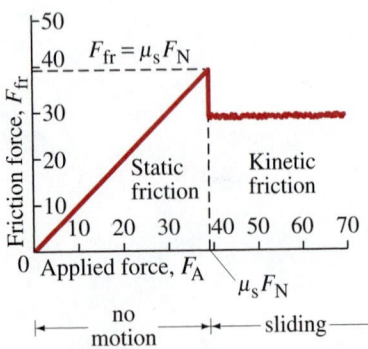

Friction can be a hindrance. It slows down moving objects and causes heating and binding of moving parts in machinery. Friction can be reduced by using lubricants such as oil. More effective in reducing friction between two surfaces is to maintain a layer of air or other gas between them. Devices using this concept, which is not practical for most situations, include air tracks and air tables in which the layer of air is maintained by forcing air through many tiny holes. Another technique to maintain the air layer is to suspend objects in air using magnetic fields ("magnetic levitation"). On the other hand, friction can be helpful. Our ability to walk depends on friction between the soles of our shoes (or feet) and the ground. (Walking involves static friction, not kinetic friction. Why?) The movement of a car, and also its stability, depend on friction. When friction is low, such as on ice, safe walking or driving becomes difficult.

CONCEPTUAL EXAMPLE 5–2 | **A box against a wall.** You can hold a box against a rough wall (Fig. 5–4) and prevent it from slipping down by pressing hard horizontally. How does the application of a horizontal force keep an object from moving vertically?

RESPONSE This won't work well if the wall is slippery. You need friction. Even then, if you don't press hard enough, the box will slip. The horizontal force you apply produces a normal force on the box exerted by the wall (net force horizontally is zero since box doesn't move horizontally.) The force of gravity mg, acting downward on the box, can now be balanced by an upward static friction force whose maximum magnitude is proportional to the normal force. The harder you push, the greater F_N is and the greater F_{fr} can be. If you don't press hard enough, then $mg > \mu_s F_N$ and the box begins to slide down.

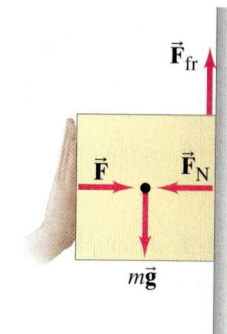

FIGURE 5–4 Example 5–2.

EXERCISE A If $\mu_s = 0.40$ and $mg = 20\,\text{N}$, what minimum force F will keep the box from falling: (a) 100 N; (b) 80 N; (c) 50 N; (d) 20 N; (e) 8 N?

EXAMPLE 5–3 **Pulling against friction.** A 10.0-kg box is pulled along a horizontal surface by a force F_P of 40.0 N applied at a 30.0° angle above horizontal. This is like Example 4–11 except now there is friction, and we assume a coefficient of kinetic friction of 0.30. Calculate the acceleration.

APPROACH The free-body diagram is shown in Fig. 5–5. It is much like that in Fig. 4–21, but with one more force, that of friction.

FIGURE 5–5 Example 5–3.

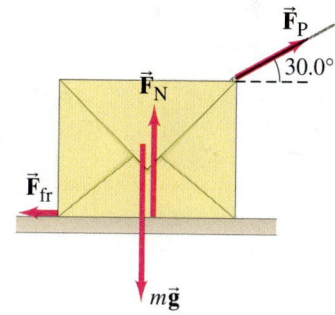

SOLUTION The calculation for the vertical (y) direction is just the same as in Example 4–11, $mg = (10.0\,\text{kg})(9.80\,\text{m/s}^2) = 98.0\,\text{N}$ and $F_{Py} = (40.0\,\text{N})(\sin 30.0°) = 20.0\,\text{N}$. With y positive upward and $a_y = 0$, we have

$$F_N - mg + F_{Py} = ma_y$$
$$F_N - 98.0\,\text{N} + 20.0\,\text{N} = 0,$$

so the normal force is $F_N = 78.0\,\text{N}$. Now we apply Newton's second law for the horizontal (x) direction (positive to the right), and include the friction force:

$$F_{Px} - F_{fr} = ma_x.$$

The friction force is kinetic as long as $F_{fr} = \mu_k F_N$ is less than $F_{Px} = (40.0\,\text{N})\cos 30.0° = 34.6\,\text{N}$, which it is:

$$F_{fr} = \mu_k F_N = (0.30)(78.0\,\text{N}) = 23.4\,\text{N}.$$

Hence the box does accelerate:

$$a_x = \frac{F_{Px} - F_{fr}}{m} = \frac{34.6\,\text{N} - 23.4\,\text{N}}{10.0\,\text{kg}} = 1.1\,\text{m/s}^2.$$

In the absence of friction, as we saw in Example 4–11, the acceleration would be much greater than this.

NOTE Our final answer has only two significant figures because our least significant input value $(\mu_k = 0.30)$ has two.

EXERCISE B If $\mu_k F_N$ were greater than F_{Px}, what would you conclude?

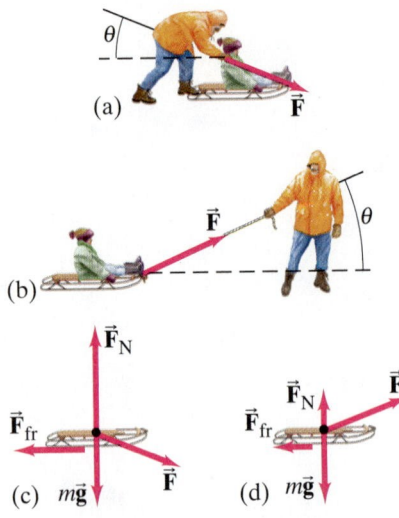

(a)

(b)

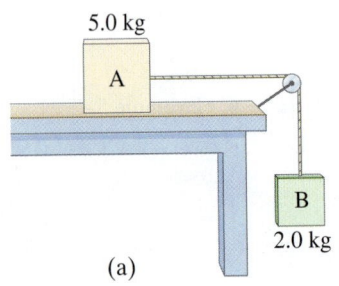

(c) $m\vec{\mathbf{g}}$ (d) $m\vec{\mathbf{g}}$

FIGURE 5–6 Example 5–4.

FIGURE 5–7 Example 5–5.

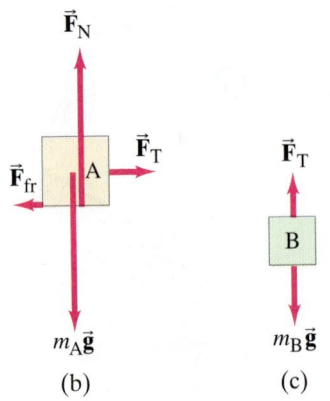

5.0 kg

A

B

2.0 kg

(a)

$\vec{\mathbf{F}}_N$

$\vec{\mathbf{F}}_{fr}$ A $\vec{\mathbf{F}}_T$

$m_A\vec{\mathbf{g}}$

(b)

$\vec{\mathbf{F}}_T$

B

$m_B\vec{\mathbf{g}}$

(c)

CONCEPTUAL EXAMPLE 5–4 **To push or to pull a sled?** Your little sister wants a ride on her sled. If you are on flat ground, will you exert less force if you push her or pull her? See Figs. 5–6a and b. Assume the same angle θ in each case.

RESPONSE Let us draw free-body diagrams for the sled–sister combination, as shown in Figs. 5–6c and d. They show, for the two cases, the forces exerted by you, $\vec{\mathbf{F}}$ (an unknown), by the snow, $\vec{\mathbf{F}}_N$ and $\vec{\mathbf{F}}_{fr}$, and gravity $m\vec{\mathbf{g}}$. (a) If you push her, and $\theta > 0$, there is a vertically downward component to your force. Hence the normal force upward exerted by the ground (Fig. 5–6c) will be larger than mg (where m is the mass of sister plus sled). (b) If you pull her, your force has a vertically upward component, so the normal force F_N will be less than mg, Fig. 5–6d. Because the friction force is proportional to the normal force, F_{fr} will be less if you pull her. So you exert less force if you pull her.

EXAMPLE 5–5 **Two boxes and a pulley.** In Fig. 5–7a, two boxes are connected by a cord running over a pulley. The coefficient of kinetic friction between box A and the table is 0.20. We ignore the mass of the cord and pulley and any friction in the pulley, which means we can assume that a force applied to one end of the cord will have the same magnitude at the other end. We wish to find the acceleration, a, of the system, which will have the same magnitude for both boxes assuming the cord doesn't stretch. As box B moves down, box A moves to the right.

APPROACH The free-body diagrams for each box are shown in Figs. 5–7b and c. The forces on box A are the pulling force of the cord F_T, gravity $m_A g$, the normal force exerted by the table F_N, and a friction force exerted by the table F_{fr}; the forces on box B are gravity $m_B g$, and the cord pulling up, F_T.

SOLUTION Box A does not move vertically, so Newton's second law tells us the normal force just balances the weight,

$$F_N = m_A g = (5.0\,\text{kg})(9.8\,\text{m/s}^2) = 49\,\text{N}.$$

In the horizontal direction, there are two forces on box A (Fig. 5–7b): F_T, the tension in the cord (whose value we don't know), and the force of friction

$$F_{fr} = \mu_k F_N = (0.20)(49\,\text{N}) = 9.8\,\text{N}.$$

The horizontal acceleration is what we wish to find; we use Newton's second law in the x direction, $\Sigma F_{Ax} = m_A a_x$, which becomes (taking the positive direction to the right and setting $a_{Ax} = a$):

$$\Sigma F_{Ax} = F_T - F_{fr} = m_A a. \qquad \text{[box A]}$$

Next consider box B. The force of gravity $m_B g = (2.0\,\text{kg})(9.8\,\text{m/s}^2) = 19.6\,\text{N}$ pulls downward; and the cord pulls upward with a force F_T. So we can write Newton's second law for box B (taking the downward direction as positive):

$$\Sigma F_{By} = m_B g - F_T = m_B a. \qquad \text{[box B]}$$

[Notice that if $a \neq 0$, then F_T is not equal to $m_B g$.]

We have two unknowns, a and F_T, and we also have two equations. We solve the box A equation for F_T:

$$F_T = F_{fr} + m_A a,$$

and substitute this into the box B equation:

$$m_B g - F_{fr} - m_A a = m_B a.$$

Now we solve for a and put in numerical values:

$$a = \frac{m_B g - F_{fr}}{m_A + m_B} = \frac{19.6\,\text{N} - 9.8\,\text{N}}{5.0\,\text{kg} + 2.0\,\text{kg}} = 1.4\,\text{m/s}^2,$$

which is the acceleration of box A to the right, and of box B down.

If we wish, we can calculate F_T using the third equation up from here:

$$F_T = F_{fr} + m_A a = 9.8\,\text{N} + (5.0\,\text{kg})(1.4\,\text{m/s}^2) = 17\,\text{N}.$$

NOTE Box B is not in free fall. It does not fall at $a = g$ because an additional force, F_T, is acting upward on it.

In Chapter 4 we examined motion on ramps and inclines, and saw that it is usually an advantage to choose the x axis along the plane, in the direction of acceleration. There we ignored friction, but now we take it into account.

EXAMPLE 5–6 **The skier.** The skier in Fig. 5–8a is descending a 30° slope, at constant speed. What can you say about the coefficient of kinetic friction μ_k?

APPROACH We choose the x axis along the slope, positive pointing downslope in the direction of the skier's motion. The y axis is perpendicular to the surface as shown in Fig. 5–8b, which is the free-body diagram for our system which we choose as the skier and her skis (total mass m). The forces acting are gravity, $\vec{F}_G = m\vec{g}$, which points vertically downward (*not* perpendicular to the slope), and the two forces exerted on her skis by the snow—the normal force perpendicular to the snowy slope (*not* vertical), and the friction force parallel to the surface. These three forces are shown acting at one point in Fig. 5–8b, for convenience.

SOLUTION We have to resolve only one vector into components, the weight $\vec{F}_G$, and its components are shown as dashed lines in Fig. 5–8c:

$$F_{Gx} = mg \sin\theta,$$
$$F_{Gy} = -mg \cos\theta,$$

where we have stayed general by using θ rather than 30° for now. There is no acceleration, so Newton's second law applied to the x and y components gives

$$\Sigma F_y = F_N - mg \cos\theta = ma_y = 0$$
$$\Sigma F_x = mg \sin\theta - \mu_k F_N = ma_x = 0.$$

From the first equation, we have $F_N = mg \cos\theta$. We substitute this into the second equation:

$$mg \sin\theta - \mu_k (mg \cos\theta) = 0.$$

Now we solve for μ_k:

$$\mu_k = \frac{mg \sin\theta}{mg \cos\theta} = \frac{\sin\theta}{\cos\theta} = \tan\theta$$

which for $\theta = 30°$ is

$$\mu_k = \tan\theta = \tan 30° = 0.58.$$

Notice that we could use the equation

$$\mu_k = \tan\theta$$

to determine μ_k under a variety of conditions. All we need to do is observe at what slope angle the skier descends at constant speed. Here is another reason why it is often useful to plug in numbers only at the end: we obtained a general result useful for other situations as well.

In problems involving a slope or "inclined plane," avoid making errors in the directions of the normal force and gravity. The normal force is *not* vertical: it is perpendicular to the slope or plane. And gravity is *not* perpendicular to the slope—gravity acts vertically downward toward the center of the Earth.

PHYSICS APPLIED
Skiing

FIGURE 5–8 Example 5–6. A skier descending a slope; $\vec{F}_G = m\vec{g}$ is the force of gravity (weight) on the skier.

(a)

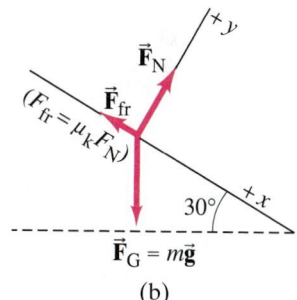

(b)

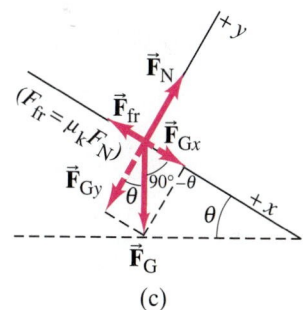

(c)

CAUTION
Directions of gravity and the normal force

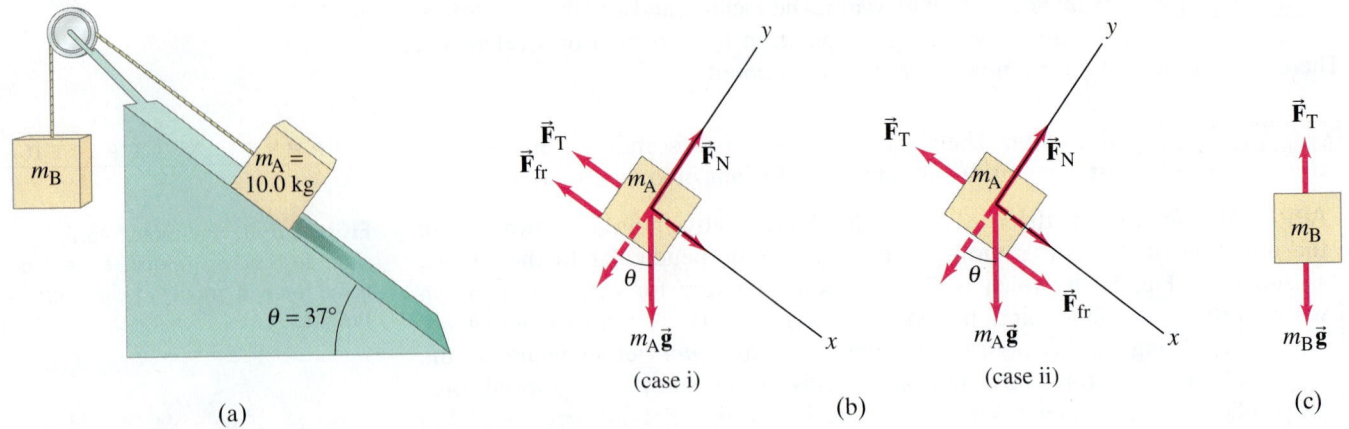

(a)

(case i)   (case ii)
(b)   (c)

FIGURE 5–9 Example 5–7. Note choice of x and y axes.

EXAMPLE 5–7 **A ramp, a pulley, and two boxes.** A box of mass $m_A = 10.0\,\text{kg}$ rests on a surface inclined at $\theta = 37°$ to the horizontal. It is connected by a light-weight cord, which passes over a massless and frictionless pulley, to a second box of mass m_B, which hangs freely as shown in Fig. 5–9a. (a) If the coefficient of static friction is $\mu_s = 0.40$, determine what range of values for mass m_B will keep the system at rest. (b) If the coefficient of kinetic friction is $\mu_k = 0.30$, and $m_B = 10.0\,\text{kg}$, determine the acceleration of the system.

APPROACH Figure 5–9b shows two free-body diagrams for box m_A because the force of friction can be either up or down the slope, depending on which direction the box slides: (i) if $m_B = 0$ or is sufficiently small, m_A would tend to slide down the incline, so $\vec{F}_{fr}$ would be directed up the incline; (ii) if m_B is large enough, m_A will tend to be pulled up the plane, so $\vec{F}_{fr}$ would point down the plane. The tension force exerted by the cord is labeled $\vec{F}_T$.

SOLUTION (a) For both cases (i) and (ii), Newton's second law for the y direction (perpendicular to the plane) is the same:

$$F_N - m_A g \cos\theta = m_A a_y = 0$$

since there is no y motion. So

$$F_N = m_A g \cos\theta.$$

Now for the x motion. We consider case (i) first for which $\Sigma F = ma$ gives

$$m_A g \sin\theta - F_T - F_{fr} = m_A a_x.$$

We want $a_x = 0$ and we solve for F_T since F_T is related to m_B (whose value we are seeking) by $F_T = m_B g$ (see Fig. 5–9c). Thus

$$m_A g \sin\theta - F_{fr} = F_T = m_B g.$$

We solve this for m_B and set F_{fr} at its maximum value $\mu_s F_N = \mu_s m_A g \cos\theta$ to find the minimum value that m_B can have to prevent motion $(a_x = 0)$:

$$m_B = m_A \sin\theta - \mu_s m_A \cos\theta$$

$$= (10.0\,\text{kg})(\sin 37° - 0.40 \cos 37°) = 2.8\,\text{kg}.$$

Thus if $m_B < 2.8\,\text{kg}$, then box A will slide down the incline.

Now for case (ii) in Fig. 5–9b, box A being pulled *up* the incline. Newton's second law is

$$m_A g \sin\theta + F_{fr} - F_T = m_A a_x = 0.$$

Then the maximum value m_B can have without causing acceleration is given by

$$F_T = m_B g = m_A g \sin\theta + \mu_s m_A g \cos\theta$$

or

$$m_B = m_A \sin\theta + \mu_s m_A \cos\theta$$

$$= (10.0\,\text{kg})(\sin 37° + 0.40 \cos 37°) = 9.2\,\text{kg}.$$

Thus, to prevent motion, we have the condition

$$2.8\,\text{kg} < m_B < 9.2\,\text{kg}.$$

(b) If $m_B = 10.0\,\text{kg}$ and $\mu_k = 0.30$, then m_B will fall and m_A will rise up the plane (case ii). To find their acceleration a, we use $\Sigma F = ma$ for box A:

$$m_A a = F_T - m_A g \sin\theta - \mu_k F_N.$$

Since m_B accelerates downward, Newton's second law for box B (Fig. 5–9c) tells us $m_B a = m_B g - F_T$, or $F_T = m_B g - m_B a$, and we substitute this into the equation above:

$$m_A a = m_B g - m_B a - m_A g \sin\theta - \mu_k F_N.$$

We solve for the acceleration a and substitute $F_N = m_A g \cos\theta$, and then $m_A = m_B = 10.0\,\text{kg}$, to find

$$a = \frac{m_B g - m_A g \sin\theta - \mu_k m_A g \cos\theta}{m_A + m_B}$$

$$= \frac{(10.0\,\text{kg})(9.80\,\text{m/s}^2)(1 - \sin 37° - 0.30 \cos 37°)}{20.0\,\text{kg}}$$

$$= 0.079g = 0.78\,\text{m/s}^2.$$

NOTE It is worth comparing this equation for acceleration a with that obtained in Example 5–5: if here we let $\theta = 0$, the plane is horizontal as in Example 5–5, and we obtain $a = (m_B g - \mu_k m_A g)/(m_A + m_B)$ just as in Example 5–5.

5–2 Uniform Circular Motion—Kinematics

An object moves in a straight line if the net force on it acts in the direction of motion, or the net force is zero. If the net force acts at an angle to the direction of motion at any moment, then the object moves in a curved path. An example of the latter is projectile motion, which we discussed in Chapter 3. Another important case is that of an object moving in a circle, such as a ball at the end of a string revolving around one's head, or the nearly circular motion of the Moon about the Earth.

An object that moves in a circle at constant speed v is said to experience **uniform circular motion**. The *magnitude* of the velocity remains constant in this case, but the *direction* of the velocity continuously changes as the object moves around the circle (Fig. 5–10). Because acceleration is defined as the rate of change of velocity, a change in direction of velocity constitutes an acceleration, just as a change in magnitude of velocity does. Thus, an object revolving in a circle is continuously accelerating, even when the speed remains constant $(v_1 = v_2 = v)$. We now investigate this acceleration quantitatively.

FIGURE 5–10 A small object moving in a circle, showing how the velocity changes. At each point, the instantaneous velocity is in a direction tangent to the circular path.

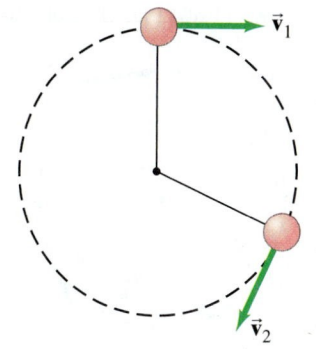

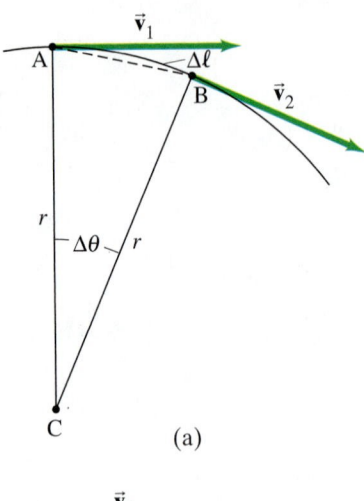

(a)

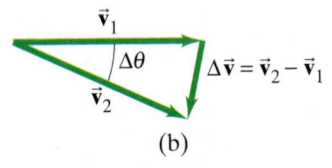

$\Delta\vec{v} = \vec{v}_2 - \vec{v}_1$

(b)

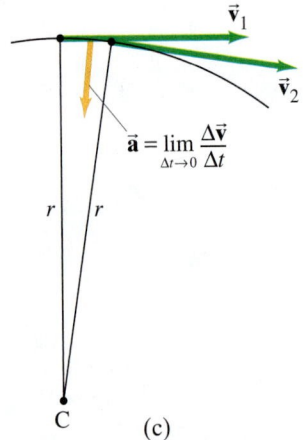

$\vec{a} = \lim\limits_{\Delta t \to 0} \dfrac{\Delta\vec{v}}{\Delta t}$

(c)

FIGURE 5–11 Determining the change in velocity, $\Delta\vec{v}$, for a particle moving in a circle. The length $\Delta\ell$ is the distance along the arc, from A to B.

⚠ **CAUTION**

In uniform circular motion, the speed is constant, but the acceleration is not zero

FIGURE 5–12 For uniform circular motion, $\vec{a}$ is always perpendicular to $\vec{v}$.

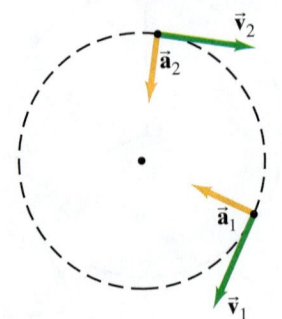

Acceleration is defined as

$$\vec{a} = \lim_{\Delta t \to 0} \frac{\Delta\vec{v}}{\Delta t} = \frac{d\vec{v}}{dt},$$

where $\Delta\vec{v}$ is the change in velocity during the short time interval Δt. We will eventually consider the situation in which Δt approaches zero and thus obtain the instantaneous acceleration. But for purposes of making a clear drawing (Fig. 5–11), we consider a nonzero time interval. During the time interval Δt, the particle in Fig. 5–11a moves from point A to point B, covering a distance $\Delta\ell$ *along the arc* which subtends an angle $\Delta\theta$. The change in the velocity vector is $\vec{v}_2 - \vec{v}_1 = \Delta\vec{v}$, and is shown in Fig. 5–11b.

Now we let Δt be very small, approaching zero. Then $\Delta\ell$ and $\Delta\theta$ are also very small, and $\vec{v}_2$ will be almost parallel to $\vec{v}_1$ (Fig. 5–11c); $\Delta\vec{v}$ will be essentially perpendicular to them. Thus $\Delta\vec{v}$ points toward the center of the circle. Since $\vec{a}$, by definition, is in the same direction as $\Delta\vec{v}$, it too must point toward the center of the circle. Therefore, this acceleration is called **centripetal acceleration** ("center-pointing" acceleration) or **radial acceleration** (since it is directed along the radius, toward the center of the circle), and we denote it by $\vec{a}_R$.

We next determine the magnitude of the radial (centripetal) acceleration, a_R. Because CA in Fig. 5–11a is perpendicular to $\vec{v}_1$, and CB is perpendicular to $\vec{v}_2$, it follows that the angle $\Delta\theta$, defined as the angle between CA and CB, is also the angle between $\vec{v}_1$ and $\vec{v}_2$. Hence the vectors $\vec{v}_1, \vec{v}_2$, and $\Delta\vec{v}$ in Fig. 5–11b form a triangle that is geometrically similar[†] to triangle CAB in Fig. 5–11a. If we take $\Delta\theta$ to be very small (letting Δt be very small) and setting $v = v_1 = v_2$ because the magnitude of the velocity is assumed not to change, we can write

$$\frac{\Delta v}{v} \approx \frac{\Delta\ell}{r},$$

or

$$\Delta v \approx \frac{v}{r}\Delta\ell.$$

This is an exact equality when Δt approaches zero, for then the arc length $\Delta\ell$ equals the chord length AB. We want to find the instantaneous acceleration, a_R, so we use the expression above to write

$$a_R = \lim_{\Delta t \to 0} \frac{\Delta v}{\Delta t} = \lim_{\Delta t \to 0} \frac{v}{r}\frac{\Delta\ell}{\Delta t}.$$

Then, because

$$\lim_{\Delta t \to 0} \frac{\Delta\ell}{\Delta t}$$

is just the linear speed, v, of the object, we have for the centripetal (radial) acceleration

$$a_R = \frac{v^2}{r}. \qquad \text{[centripetal (radial) acceleration]} \quad \textbf{(5–1)}$$

Equation 5–1 is valid even when v is not constant.

To summarize, *an object moving in a circle of radius r at constant speed v has an acceleration whose direction is toward the center of the circle and whose magnitude is* $a_R = v^2/r$. It is not surprising that this acceleration depends on v and r. The greater the speed v, the faster the velocity changes direction; and the larger the radius, the less rapidly the velocity changes direction.

The acceleration vector points toward the center of the circle. But the velocity vector always points in the direction of motion, which is tangential to the circle. Thus the velocity and acceleration vectors are perpendicular to each other at every point in the path for uniform circular motion (Fig. 5–12). This is another example that illustrates the error in thinking that acceleration and velocity are always in the same direction. For an object falling vertically, $\vec{a}$ and $\vec{v}$ are indeed parallel. But in circular motion, $\vec{a}$ and $\vec{v}$ are perpendicular, not parallel (nor were they parallel in projectile motion, Section 3–7).

EXERCISE C Can Equations 2–12, the kinematic equations for constant acceleration, be used for uniform circular motion? For example, could Eq. 2–12b be used to calculate the time for the revolving ball in Fig. 5–12 to make one revolution?

[†] Appendix A contains a review of geometry.

Circular motion is often described in terms of the **frequency** f, the number of revolutions per second. The **period** T of an object revolving in a circle is the time required for one complete revolution. Period and frequency are related by

$$T = \frac{1}{f}. \tag{5-2}$$

For example, if an object revolves at a frequency of 3 rev/s, then each revolution takes $\frac{1}{3}$ s. For an object revolving in a circle (of circumference $2\pi r$) at constant speed v, we can write

$$v = \frac{2\pi r}{T},$$

since in one revolution the object travels one circumference.

EXAMPLE 5-8 **Acceleration of a revolving ball.** A 150-g ball at the end of a string is revolving uniformly in a horizontal circle of radius 0.600 m, as in Fig. 5–10 or 5–12. The ball makes 2.00 revolutions in a second. What is its centripetal acceleration?

APPROACH The centripetal acceleration is $a_R = v^2/r$. We are given r, and we can find the speed of the ball, v, from the given radius and frequency.

SOLUTION If the ball makes two complete revolutions per second, then the ball travels in a complete circle in a time interval equal to 0.500 s, which is its period T. The distance traveled in this time is the circumference of the circle, $2\pi r$, where r is the radius of the circle. Therefore, the ball has speed

$$v = \frac{2\pi r}{T} = \frac{2\pi(0.600\,\text{m})}{(0.500\,\text{s})} = 7.54\,\text{m/s}.$$

The centripetal acceleration[†] is

$$a_R = \frac{v^2}{r} = \frac{(7.54\,\text{m/s})^2}{(0.600\,\text{m})} = 94.7\,\text{m/s}^2.$$

EXERCISE D If the radius is doubled to 1.20 m but the period stays the same, by what factor will the centripetal acceleration change? (a) 2, (b) 4, (c) $\frac{1}{2}$, (d) $\frac{1}{4}$, (e) none of these.

EXAMPLE 5-9 **Moon's centripetal acceleration.** The Moon's nearly circular orbit about the Earth has a radius of about 384,000 km and a period T of 27.3 days. Determine the acceleration of the Moon toward the Earth.

APPROACH Again we need to find the velocity v in order to find a_R. We will need to convert to SI units to get v in m/s.

SOLUTION In one orbit around the Earth, the Moon travels a distance $2\pi r$, where $r = 3.84 \times 10^8$ m is the radius of its circular path. The time required for one complete orbit is the Moon's period of 27.3 d. The speed of the Moon in its orbit about the Earth is $v = 2\pi r/T$. The period T in seconds is $T = (27.3\,\text{d})(24.0\,\text{h/d})(3600\,\text{s/h}) = 2.36 \times 10^6$ s. Therefore,

$$a_R = \frac{v^2}{r} = \frac{(2\pi r)^2}{T^2 r} = \frac{4\pi^2 r}{T^2} = \frac{4\pi^2(3.84 \times 10^8\,\text{m})}{(2.36 \times 10^6\,\text{s})^2}$$

$$= 0.00272\,\text{m/s}^2 = 2.72 \times 10^{-3}\,\text{m/s}^2.$$

We can write this acceleration in terms of $g = 9.80\,\text{m/s}^2$ (the acceleration of gravity at the Earth's surface) as

$$a = 2.72 \times 10^{-3}\,\text{m/s}^2\left(\frac{g}{9.80\,\text{m/s}^2}\right) = 2.78 \times 10^{-4}g.$$

NOTE The centripetal acceleration of the Moon, $a = 2.78 \times 10^{-4}\,g$, is *not* the acceleration of gravity for objects at the Moon's surface due to the Moon's gravity. Rather, it is the acceleration due to the *Earth's* gravity for any object (such as the Moon) that is 384,000 km from the Earth. Notice how small this acceleration is compared to the acceleration of objects near the Earth's surface.

⚠ **CAUTION**

Distinguish the Moon's gravity on objects at its surface, from the Earth's gravity acting on the Moon (this Example)

[†]Differences in the final digit can depend on whether you keep all digits in your calculator for v (which gives $a_R = 94.7\,\text{m/s}^2$), or if you use $v = 7.54\,\text{m/s}$ in which case you get $a_R = 94.8\,\text{m/s}^2$. Both results are valid since our assumed accuracy is about $\pm 0.1\,\text{m/s}$ (see Section 1–3).

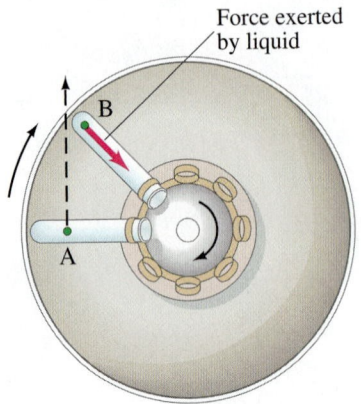

Force exerted
by liquid

B

A

FIGURE 5–13 Two positions of a rotating test tube in a centrifuge (top view). At A, the green dot represents a macromolecule or other particle being sedimented. It would tend to follow the dashed line, heading toward the bottom of the tube, but the fluid resists this motion by exerting a force on the particle as shown at point B.

*Centrifugation

Centrifuges and very high speed ultracentrifuges, are used to sediment materials quickly or to separate materials. Test tubes held in the centrifuge rotor are accelerated to very high rotational speeds: see Fig. 5–13, where one test tube is shown in two positions as the rotor turns. The small green dot represents a small particle, perhaps a macromolecule, in a fluid-filled test tube. At position A the particle has a tendency to move in a straight line, but the fluid resists the motion of the particles, exerting a centripetal force that keeps the particles moving nearly in a circle. The resistive force exerted by the fluid (liquid, gas, or gel, depending on the application) usually does not quite equal mv^2/r, and the particles move slowly toward the bottom of the tube. A centrifuge provides an "effective gravity" much larger than normal gravity because of the high rotational speeds, thus causing more rapid sedimentation.

EXAMPLE 5–10 **Ultracentrifuge.** The rotor of an ultracentrifuge rotates at 50,000 rpm (revolutions per minute). A particle at the top of a test tube (Fig. 5–13) is 6.00 cm from the rotation axis. Calculate its centripetal acceleration, in "g's."

APPROACH We calculate the centripetal acceleration from $a_R = v^2/r$.

SOLUTION The test tube makes 5.00×10^4 revolutions each minute, or, dividing by 60 s/min, 833 rev/s. The time to make one revolution, the period T, is

$$T = \frac{1}{(833 \text{ rev/s})} = 1.20 \times 10^{-3} \text{ s/rev}.$$

At the top of the tube, a particle revolves in a circle of circumference $2\pi r = (2\pi)(0.0600 \text{ m}) = 0.377$ m per revolution. The speed of the particle is then

$$v = \frac{2\pi r}{T} = \left(\frac{0.377 \text{ m/rev}}{1.20 \times 10^{-3} \text{ s/rev}} \right) = 3.14 \times 10^2 \text{ m/s}.$$

The centripetal acceleration is

$$a_R = \frac{v^2}{r} = \frac{(3.14 \times 10^2 \text{ m/s})^2}{0.0600 \text{ m}} = 1.64 \times 10^6 \text{ m/s}^2,$$

which, dividing by $g = 9.80 \text{ m/s}^2$, is 1.67×10^5 g's = 167,000 g's.

FIGURE 5–14 A force is required to keep an object moving in a circle. If the speed is constant, the force is directed toward the circle's center.

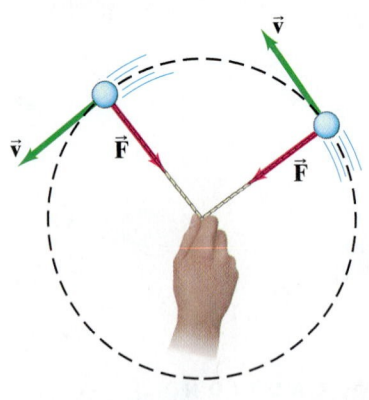

$\vec{v}$

$\vec{v}$

$\vec{F}$

$\vec{F}$

⚠ **CAUTION**

Centripetal force is not a new kind of force (Every force must be exerted by an object)

5–3 Dynamics of Uniform Circular Motion

According to Newton's second law $(\Sigma\vec{F} = m\vec{a})$, an object that is accelerating must have a net force acting on it. An object moving in a circle, such as a ball on the end of a string, must therefore have a force applied to it to keep it moving in that circle. That is, a net force is necessary to give it centripetal acceleration. The magnitude of the required force can be calculated using Newton's second law for the radial component, $\Sigma F_R = ma_R$, where a_R is the centripetal acceleration, $a_R = v^2/r$, and ΣF_R is the total (or net) force in the radial direction:

$$\Sigma F_R = ma_R = m\frac{v^2}{r}. \qquad \text{[circular motion]} \quad \textbf{(5–3)}$$

For uniform circular motion $(v = \text{constant})$, the acceleration is a_R, which is directed toward the center of the circle at any moment. Thus the *net force too must be directed toward the center of the circle*, Fig. 5–14. A net force is necessary because if no net force were exerted on the object, it would not move in a circle but in a straight line, as Newton's first law tells us. The direction of the net force is continually changing so that it is always directed toward the center of the circle. This force is sometimes called a centripetal ("pointing toward the center") force. But be aware that "centripetal force" does not indicate some new kind of force. The term merely describes the *direction* of the net force needed to provide a circular path: the net force is directed toward the circle's center. The force *must be applied by other objects*. For example, to swing a ball in a circle on the end of a string, you pull on the string and the string exerts the force on the ball. (Try it.)

There is a common misconception that an object moving in a circle has an outward force acting on it, a so-called centrifugal ("center-fleeing") force. This is incorrect: *there is no outward force* on the revolving object. Consider, for example, a person swinging a ball on the end of a string around her head (Fig. 5–15). If you have ever done this yourself, you know that you feel a force pulling outward on your hand. The misconception arises when this pull is interpreted as an outward "centrifugal" force pulling on the ball that is transmitted along the string to your hand. This is not what is happening at all. To keep the ball moving in a circle, you pull *inwardly* on the string, and the string exerts this force on the ball. The ball exerts an equal and opposite force on the string (Newton's third law), and *this* is the outward force your hand feels (see Fig. 5–15).

The force *on the ball* is the one exerted *inwardly* on it by you, via the string. To see even more convincing evidence that a "centrifugal force" does not act on the ball, consider what happens when you let go of the string. If a centrifugal force were acting, the ball would fly outward, as shown in Fig. 5–16a. But it doesn't; the ball flies off tangentially (Fig. 5–16b), in the direction of the velocity it had at the moment it was released, because the inward force no longer acts. Try it and see!

EXERCISE E Return to the Chapter-Opening Question, page 112, and answer it again now. Try to explain why you may have answered differently the first time.

⚠ **CAUTION**
There is no real "centrifugal force"

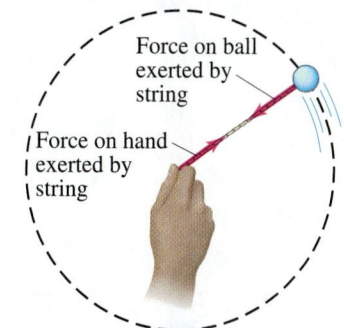

FIGURE 5–15 Swinging a ball on the end of a string.

FIGURE 5–16 If centrifugal force existed, the revolving ball would fly outward as in (a) when released. In fact, it flies off tangentially as in (b). For example, in (c) sparks fly in straight lines tangentially from the edge of a rotating grinding wheel.

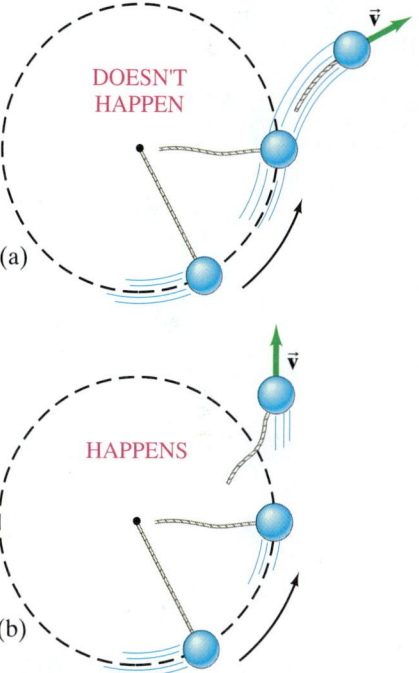

EXAMPLE 5–11 **ESTIMATE** **Force on revolving ball (horizontal).** Estimate the force a person must exert on a string attached to a 0.150-kg ball to make the ball revolve in a horizontal circle of radius 0.600 m. The ball makes 2.00 revolutions per second $(T = 0.500\,\text{s})$, as in Example 5–8. Ignore the string's mass.

APPROACH First we need to draw the free-body diagram for the ball. The forces acting on the ball are the force of gravity, $m\vec{g}$ downward, and the tension force $\vec{F}_T$ that the string exerts toward the hand at the center (which occurs because the person exerts that same force on the string). The free-body diagram for the ball is as shown in Fig. 5–17. The ball's weight complicates matters and makes it impossible to revolve a ball with the cord perfectly horizontal. We assume the weight is small, and put $\phi \approx 0$ in Fig. 5–17. Thus $\vec{F}_T$ will act nearly horizontally and, in any case, provides the force necessary to give the ball its centripetal acceleration.

SOLUTION We apply Newton's second law to the radial direction, which we assume is horizontal:

$$(\Sigma F)_R = ma_R,$$

where $a_R = v^2/r$ and $v = 2\pi r/T = 2\pi(0.600\,\text{m})/(0.500\,\text{s}) = 7.54\,\text{m/s}$. Thus

$$F_T = m\frac{v^2}{r} = (0.150\,\text{kg})\frac{(7.54\,\text{m/s})^2}{(0.600\,\text{m})} \approx 14\,\text{N}.$$

NOTE We keep only two significant figures in the answer because we ignored the ball's weight; it is $mg = (0.150\,\text{kg})(9.80\,\text{m/s}^2) = 1.5\,\text{N}$, about $\frac{1}{10}$ of our result, which is small but not so small as to justify stating a more precise answer for F_T.

NOTE To include the effect of $m\vec{g}$, resolve $\vec{F}_T$ in Fig. 5–17 into components, and set the horizontal component of $\vec{F}_T$ equal to mv^2/r and its vertical component equal to mg.

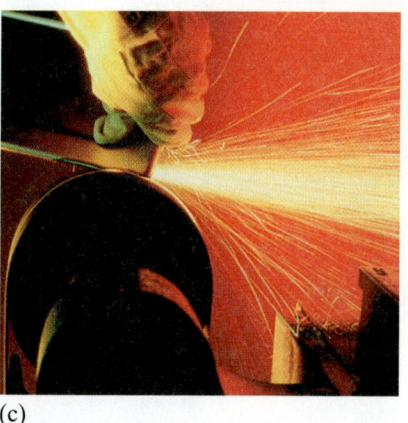

(c)

FIGURE 5–17 Example 5–11.

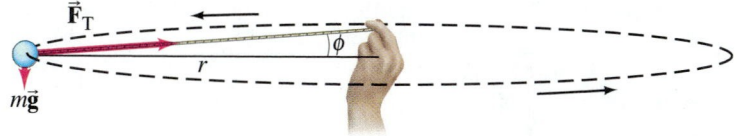

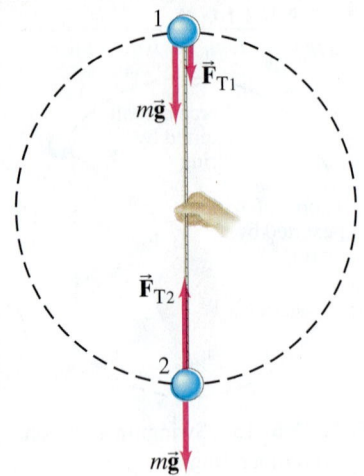

FIGURE 5–18 Example 5–12. Free-body diagrams for positions 1 and 2.

⚠ **CAUTION**

Circular motion only if cord is under tension

EXAMPLE 5–12 Revolving ball (vertical circle). A 0.150-kg ball on the end of a 1.10-m-long cord (negligible mass) is swung in a *vertical* circle. (a) Determine the minimum speed the ball must have at the top of its arc so that the ball continues moving in a circle. (b) Calculate the tension in the cord at the bottom of the arc, assuming the ball is moving at twice the speed of part (a).

APPROACH The ball moves in a vertical circle and is *not* undergoing uniform circular motion. The radius is assumed constant, but the speed v changes because of gravity. Nonetheless, Eq. 5–1 is valid at each point along the circle, and we use it at the top and bottom points. The free-body diagram is shown in Fig. 5–18 for both positions.

SOLUTION (a) At the top (point 1), two forces act on the ball: $m\vec{g}$, the force of gravity, and $\vec{F}_{T1}$, the tension force the cord exerts at point 1. Both act downward, and their vector sum acts to give the ball its centripetal acceleration a_R. We apply Newton's second law, for the vertical direction, choosing downward as positive since the acceleration is downward (toward the center):

$$(\Sigma F)_R = ma_R$$

$$F_{T1} + mg = m\frac{v_1^2}{r}. \qquad \text{[at top]}$$

From this equation we can see that the tension force F_{T1} at point 1 will get larger if v_1 (ball's speed at top of circle) is made larger, as expected. But we are asked for the *minimum* speed to keep the ball moving in a circle. The cord will remain taut as long as there is tension in it. But if the tension disappears (because v_1 is too small) the cord can go limp, and the ball will fall out of its circular path. Thus, the minimum speed will occur if $F_{T1} = 0$, for which we have

$$mg = m\frac{v_1^2}{r}. \qquad \text{[minimum speed at top]}$$

We solve for v_1, keeping an extra digit for use in (b):

$$v_1 = \sqrt{gr} = \sqrt{(9.80\,\text{m/s}^2)(1.10\,\text{m})} = 3.283\,\text{m/s}.$$

This is the minimum speed at the top of the circle if the ball is to continue moving in a circular path.

(b) When the ball is at the bottom of the circle (point 2 in Fig. 5–18), the cord exerts its tension force F_{T2} upward, whereas the force of gravity, $m\vec{g}$, still acts downward. Choosing *upward* as positive, Newton's second law gives:

$$(\Sigma F)_R = ma_R$$

$$F_{T2} - mg = m\frac{v_2^2}{r}. \qquad \text{[at bottom]}$$

The speed v_2 is given as twice that in (a), namely 6.566 m/s. We solve for F_{T2}:

$$F_{T2} = m\frac{v_2^2}{r} + mg$$

$$= (0.150\,\text{kg})\frac{(6.566\,\text{m/s})^2}{(1.10\,\text{m})} + (0.150\,\text{kg})(9.80\,\text{m/s}^2) = 7.35\,\text{N}.$$

EXERCISE F A rider on a Ferris wheel moves in a vertical circle of radius r at constant speed v (Fig. 5–19). Is the normal force that the seat exerts on the rider at the top of the wheel (a) less than, (b) more than, or (c) the same as, the force the seat exerts at the bottom of the wheel?

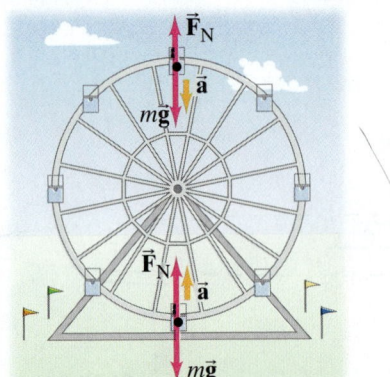

FIGURE 5–19 Exercise F.

EXAMPLE 5–13 **Conical pendulum.** A small ball of mass m, suspended by a cord of length ℓ, revolves in a circle of radius $r = \ell \sin\theta$, where θ is the angle the string makes with the vertical (Fig. 5–20). (a) In what direction is the acceleration of the ball, and what causes the acceleration? (b) Calculate the speed and period (time required for one revolution) of the ball in terms of ℓ, θ, g, and m.

APPROACH We can answer (a) by looking at Fig. 5–20, which shows the forces on the revolving ball at one instant: the acceleration points horizontally toward the center of the ball's circular path (not along the cord). The force responsible for the acceleration is the *net* force which here is the vector sum of the forces acting on the mass m: its weight $\vec{\mathbf{F}}_G$ (of magnitude $F_G = mg$) and the force exerted by the tension in the cord, $\vec{\mathbf{F}}_T$. The latter has horizontal and vertical components of magnitude $F_T \sin\theta$ and $F_T \cos\theta$, respectively.

SOLUTION (b) We apply Newton's second law to the horizontal and vertical directions. In the vertical direction, there is no motion, so the acceleration is zero and the net force in the vertical direction is zero:

$$F_T \cos\theta - mg = 0.$$

In the horizontal direction there is only one force, of magnitude $F_T \sin\theta$, that acts toward the center of the circle and gives rise to the acceleration v^2/r. Newton's second law tells us:

$$F_T \sin\theta = m\frac{v^2}{r}.$$

We solve the second equation for v, and substitute for F_T from the first equation (and use $r = \ell \sin\theta$):

$$v = \sqrt{\frac{rF_T \sin\theta}{m}} = \sqrt{\frac{r}{m}\left(\frac{mg}{\cos\theta}\right)\sin\theta}$$

$$= \sqrt{\frac{\ell g \sin^2\theta}{\cos\theta}}.$$

The period T is the time required to make one revolution, a distance of $2\pi r = 2\pi\ell \sin\theta$. The speed v can thus be written $v = 2\pi\ell \sin\theta/T$; then

$$T = \frac{2\pi\ell \sin\theta}{v} = \frac{2\pi\ell \sin\theta}{\sqrt{\dfrac{\ell g \sin^2\theta}{\cos\theta}}}$$

$$= 2\pi\sqrt{\frac{\ell \cos\theta}{g}}.$$

NOTE The speed and period do not depend on the mass m of the ball. They do depend on ℓ and θ.

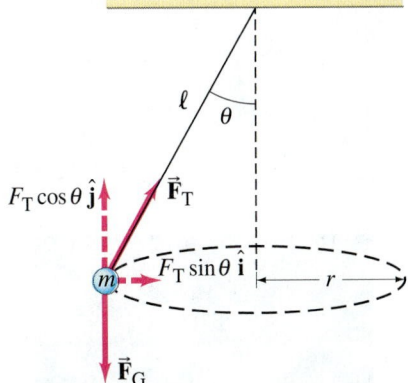

FIGURE 5–20 Example 5–13. Conical pendulum.

PROBLEM SOLVING

Uniform Circular Motion

1. **Draw a free-body diagram**, showing all the forces acting on each object under consideration. Be sure you can identify the source of each force (tension in a cord, Earth's gravity, friction, normal force, and so on). Don't put in something that doesn't belong (like a centrifugal force).

2. **Determine** which of the forces, or which of their components, act to provide the centripetal acceleration—that is, all the **forces or components that act radially**, toward or away from the center of the circular path. The sum of these forces (or components) provides the centripetal acceleration, $a_R = v^2/r$.

3. **Choose a convenient coordinate system**, preferably with one axis along the acceleration direction.

4. **Apply Newton's second law** to the radial component:

$$(\Sigma F)_R = ma_R = m\frac{v^2}{r}. \qquad \text{[radial direction]}$$

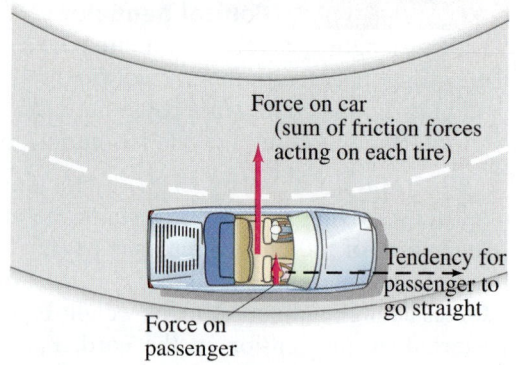

FIGURE 5–21 The road exerts an inward force on a car (friction against the tires) to make it move in a circle. The car exerts an inward force on the passenger.

Force on car (sum of friction forces acting on each tire)

Tendency for passenger to go straight

Force on passenger

FIGURE 5–22 Race car heading into a curve. From the tire marks we see that most cars experienced a sufficient friction force to give them the needed centripetal acceleration for rounding the curve safely. But, we also see tire tracks of cars on which there was not sufficient force—and which unfortunately followed more nearly straight-line paths.

FIGURE 5–23 Example 5–14. Forces on a car rounding a curve on a flat road. (a) Front view, (b) top view.

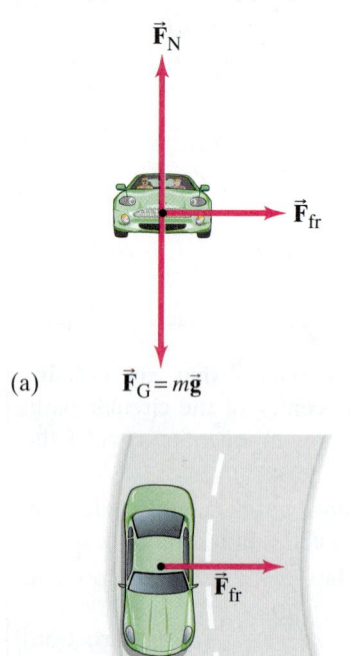

$\vec{\mathbf{F}}_N$

$\vec{\mathbf{F}}_{fr}$

(a)

$\vec{\mathbf{F}}_G = m\vec{\mathbf{g}}$

$\vec{\mathbf{F}}_{fr}$

(b)

5–4 Highway Curves: Banked and Unbanked

An example of circular dynamics occurs when an automobile rounds a curve, say to the left. In such a situation, you may feel that you are thrust outward toward the right side door. But there is no mysterious centrifugal force pulling on you. What is happening is that you tend to move in a straight line, whereas the car has begun to follow a curved path. To make you go in the curved path, the seat (friction) or the door of the car (direct contact) exerts a force on you (Fig. 5–21). The car also must have a force exerted on it toward the center of the curve if it is to move in that curve. On a flat road, this force is supplied by friction between the tires and the pavement.

If the wheels and tires of the car are rolling normally without slipping or sliding, the bottom of the tire is at rest against the road at each instant; so the friction force the road exerts on the tires is static friction. But if the static friction force is not great enough, as under icy conditions or high speed, sufficient friction force cannot be applied and the car will skid out of a circular path into a more nearly straight path. See Fig. 5–22. Once a car skids or slides, the friction force becomes kinetic friction, which is less than static friction.

EXAMPLE 5–14 **Skidding on a curve.** A 1000-kg car rounds a curve on a flat road of radius 50 m at a speed of 15 m/s (54 km/h). Will the car follow the curve, or will it skid? Assume: (a) the pavement is dry and the coefficient of static friction is $\mu_s = 0.60$; (b) the pavement is icy and $\mu_s = 0.25$.

APPROACH The forces on the car are gravity mg downward, the normal force F_N exerted upward by the road, and a horizontal friction force due to the road. They are shown in Fig. 5–23, which is the free-body diagram for the car. The car will follow the curve if the maximum static friction force is greater than the mass times the centripetal acceleration.

SOLUTION In the vertical direction there is no acceleration. Newton's second law tells us that the normal force F_N on the car is equal to the weight mg:

$$F_N = mg = (1000\,\text{kg})(9.80\,\text{m/s}^2) = 9800\,\text{N}.$$

In the horizontal direction the only force is friction, and we must compare it to the force needed to produce the centripetal acceleration to see if it is sufficient. The net horizontal force required to keep the car moving in a circle around the curve is

$$(\Sigma F)_R = ma_R = m\frac{v^2}{r} = (1000\,\text{kg})\frac{(15\,\text{m/s})^2}{(50\,\text{m})} = 4500\,\text{N}.$$

Now we compute the maximum total static friction force (the sum of the friction forces acting on each of the four tires) to see if it can be large enough to provide a safe centripetal acceleration. For (a), $\mu_s = 0.60$, and the maximum friction force attainable (recall from Section 5–1 that $F_{fr} \le \mu_s F_N$) is

$$(F_{fr})_{max} = \mu_s F_N = (0.60)(9800\,\text{N}) = 5880\,\text{N}.$$

Since a force of only 4500 N is needed, and that is, in fact, how much will be exerted by the road as a static friction force, the car can follow the curve. But in

(b) the maximum static friction force possible is

$$(F_{fr})_{max} = \mu_s F_N = (0.25)(9800 \text{ N}) = 2450 \text{ N}.$$

The car will skid because the ground cannot exert sufficient force (4500 N is needed) to keep it moving in a curve of radius 50 m at a speed of 54 km/h.

The banking of curves can reduce the chance of skidding. The normal force exerted by a banked road, acting perpendicular to the road, will have a component toward the center of the circle (Fig. 5–24), thus reducing the reliance on friction. For a given banking angle θ, there will be one speed for which no friction at all is required. This will be the case when the horizontal component of the normal force toward the center of the curve, $F_N \sin \theta$ (see Fig. 5–24), is just equal to the force required to give a vehicle its centripetal acceleration—that is, when

$$F_N \sin \theta = m \frac{v^2}{r}. \qquad \text{[no friction required]}$$

The banking angle of a road, θ, is chosen so that this condition holds for a particular speed, called the "design speed."

EXAMPLE 5–15 **Banking angle.** (a) For a car traveling with speed v around a curve of radius r, determine a formula for the angle at which a road should be banked so that no friction is required. (b) What is this angle for an expressway off-ramp curve of radius 50 m at a design speed of 50 km/h?

APPROACH Even though the road is banked, the car is still moving along a horizontal circle, so the centripetal acceleration needs to be horizontal. We choose our x and y axes as horizontal and vertical so that a_R, which is horizontal, is along the x axis. The forces on the car are the Earth's gravity mg downward, and the normal force F_N exerted by the road perpendicular to its surface. See Fig. 5–24, where the components of F_N are also shown. We don't need to consider the friction of the road because we are designing a road to be banked so as to eliminate dependence on friction.

SOLUTION (a) Since there is no vertical motion, $\Sigma F_y = ma_y$ gives us

$$F_N \cos \theta - mg = 0.$$

Thus,

$$F_N = \frac{mg}{\cos \theta}.$$

[Note in this case that $F_N \geq mg$ since $\cos \theta \leq 1$.]
We substitute this relation for F_N into the equation for the horizontal motion,

$$F_N \sin \theta = m \frac{v^2}{r},$$

and obtain

$$\frac{mg}{\cos \theta} \sin \theta = m \frac{v^2}{r}$$

or

$$\tan \theta = \frac{v^2}{rg}.$$

This is the formula for the banking angle θ: no friction needed at speed v.
(b) For $r = 50$ m and $v = 50$ km/h (or 14 m/s),

$$\tan \theta = \frac{(14 \text{ m/s})^2}{(50 \text{ m})(9.8 \text{ m/s}^2)} = 0.40,$$

so $\theta = 22°$.

EXERCISE G The banking angle of a curve for a design speed v is θ_1. What banking angle θ_2 is needed for a design speed of $2v$? (a) $\theta_2 = 4\theta_1$; (b) $\theta_2 = 2\theta_1$; (c) $\tan \theta_2 = 4 \tan \theta_1$; (d) $\tan \theta_2 = 2 \tan \theta_1$.

EXERCISE H Can a heavy truck and a small car travel safely at the same speed around an icy banked-curve road?

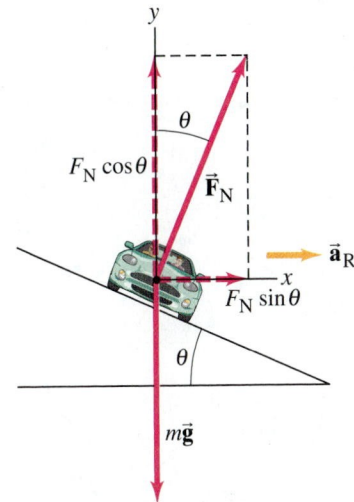

FIGURE 5–24 Normal force on a car rounding a banked curve, resolved into its horizontal and vertical components. The centripetal acceleration is horizontal (*not* parallel to the sloping road). The friction force on the tires, not shown, could point up or down along the slope, depending on the car's speed. The friction force will be zero for one particular speed.

CAUTION
F_N is not always equal to mg

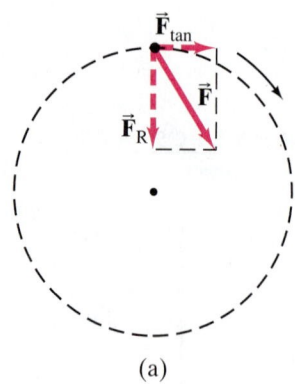

(a)

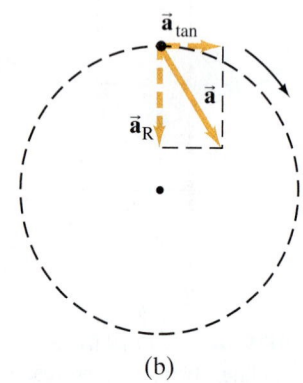

(b)

FIGURE 5–25 The speed of an object moving in a circle changes if the force on it has a tangential component, F_{tan}. Part (a) shows the force $\vec{F}$ and its vector components; part (b) shows the acceleration vector and its vector components.

*5–5 Nonuniform Circular Motion

Circular motion at constant speed occurs when the net force on an object is exerted toward the center of the circle. If the net force is not directed toward the center but is at an angle, as shown in Fig. 5–25a, the force has two components. The component directed toward the center of the circle, $\vec{F}_R$, gives rise to the centripetal acceleration, $\vec{a}_R$, and keeps the object moving in a circle. The component tangent to the circle, $\vec{F}_{tan}$, acts to increase (or decrease) the speed, and thus gives rise to a component of the acceleration tangent to the circle, $\vec{a}_{tan}$. When the speed of the object is changing, a tangential component of force is acting.

When you first start revolving a ball on the end of a string around your head, you must give it tangential acceleration. You do this by pulling on the string with your hand displaced from the center of the circle. In athletics, a hammer thrower accelerates the hammer tangentially in a similar way so that it reaches a high speed before release.

The tangential component of the acceleration, a_{tan}, has magnitude equal to the rate of change of the *magnitude* of the object's velocity:

$$a_{tan} = \frac{dv}{dt}. \qquad (5\text{–}4)$$

The radial (centripetal) acceleration arises from the change in *direction* of the velocity and, as we have seen, has magnitude

$$a_R = \frac{v^2}{r}.$$

The tangential acceleration always points in a direction tangent to the circle, and is in the direction of motion (parallel to $\vec{v}$, which is always tangent to the circle) if the speed is increasing, as shown in Fig. 5–25b. If the speed is decreasing, $\vec{a}_{tan}$ points antiparallel to $\vec{v}$. In either case, $\vec{a}_{tan}$ and $\vec{a}_R$ are always perpendicular to each other; and *their directions change* continually as the object moves along its circular path. The total vector acceleration $\vec{a}$ is the sum of the two components:

$$\vec{a} = \vec{a}_{tan} + \vec{a}_R. \qquad (5\text{–}5)$$

Since $\vec{a}_R$ and $\vec{a}_{tan}$ are always perpendicular to each other, the magnitude of $\vec{a}$ at any moment is

$$a = \sqrt{a_{tan}^2 + a_R^2}.$$

EXAMPLE 5–16 Two components of acceleration. A race car starts from rest in the pit area and accelerates at a uniform rate to a speed of 35 m/s in 11 s, moving on a circular track of radius 500 m. Assuming constant tangential acceleration, find (a) the tangential acceleration, and (b) the radial acceleration, at the instant when the speed is $v = 15$ m/s.

APPROACH The tangential acceleration relates to the change in speed of the car, and can be calculated as $a_{tan} = \Delta v/\Delta t$. The centripetal acceleration relates to the change in the *direction* of the velocity vector and is calculated using $a_R = v^2/r$.

SOLUTION (a) During the 11-s time interval, we assume the tangential acceleration a_{tan} is constant. Its magnitude is

$$a_{tan} = \frac{\Delta v}{\Delta t} = \frac{(35\ \text{m/s} - 0\ \text{m/s})}{11\ \text{s}} = 3.2\ \text{m/s}^2.$$

(b) When $v = 15$ m/s, the centripetal acceleration is

$$a_R = \frac{v^2}{r} = \frac{(15\ \text{m/s})^2}{(500\ \text{m})} = 0.45\ \text{m/s}^2.$$

NOTE The radial acceleration increases continually, whereas the tangential acceleration stays constant.

EXERCISE I When the speed of the race car in Example 5–16 is 30 m/s, how are (a) a_{tan} and (b) a_R changed?

These concepts can be used for an object moving along any curved path, such as that shown in Fig. 5–26. We can treat any portion of the curve as an arc of a circle with a **radius of curvature** r. The velocity at any point is always tangent to the path. The acceleration can be written, in general, as a vector sum of two components: the tangential component $a_{tan} = dv/dt$, and the radial (centripetal) component $a_R = v^2/r$.

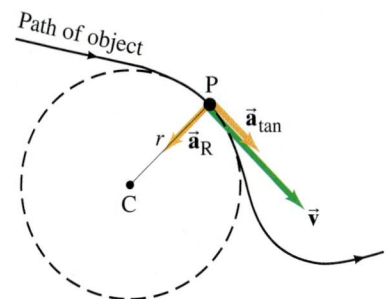

*5–6 Velocity-Dependent Forces: Drag and Terminal Velocity

FIGURE 5–26 Object following a curved path (solid line). At point P the path has a radius of curvature r. The object has velocity $\vec{v}$, tangential acceleration $\vec{a}_{tan}$ (the object is here increasing in speed), and radial (centripetal) acceleration $\vec{a}_R$ (magnitude $a_R = v^2/r$) which points toward the center of curvature C.

When an object slides along a surface, the force of friction acting on the object is nearly independent of how fast the object is moving. But other types of resistive forces do depend on the object's velocity. The most important example is for an object moving through a liquid or gas, such as air. The fluid offers resistance to the motion of the object, and this resistive force, or **drag force**, depends on the velocity of the object.[†]

The way the drag force varies with velocity is complicated in general. But for small objects at very low speeds, a good approximation can often be made by assuming that the drag force, F_D, is directly proportional to the magnitude of the velocity, v:

$$F_D = -bv. \qquad (5\text{–}6)$$

The minus sign is necessary because the drag force opposes the motion. Here b is a constant (approximately) that depends on the viscosity of the fluid and on the size and shape of the object. Equation 5–6 works well for small objects moving at low speed in a viscous liquid. It also works for very small objects moving in air at very low speeds, such as dust particles. For objects moving at high speeds, such as an airplane, a sky diver, a baseball, or an automobile, the force of air resistance can be better approximated as being proportional to v^2:

$$F_D \propto v^2.$$

For accurate calculations, however, more complicated forms and numerical integration generally need to be used. For objects moving through liquids, Eq. 5–6 works well for everyday objects at normal speeds (e.g., a boat in water).

Let us consider an object that falls from rest, through air or other fluid, under the action of gravity and a resistive force proportional to v. The forces acting on the object are the force of gravity, mg, acting downward, and the drag force, $-bv$, acting upward (Fig. 5–27a). Since the velocity $\vec{v}$ points downward, let us take the positive direction as downward. Then the net force on the object can be written

$$\Sigma F = mg - bv.$$

From Newton's second law $\Sigma F = ma$, we have

$$mg - bv = m\frac{dv}{dt}, \qquad (5\text{–}7)$$

where we have written the acceleration according to its definition as rate of change of velocity, $a = dv/dt$. At $t = 0$, we set $v = 0$ and the acceleration $dv/dt = g$. As the object falls and increases in speed, the resistive force increases, and this reduces the acceleration, dv/dt (see Fig. 5–27b). The velocity continues to increase, but at a slower rate. Eventually, the velocity becomes so large that the magnitude of the resistive force, bv, approaches that of the gravitational force, mg; when the two are equal, we have

$$mg - bv = 0. \qquad (5\text{–}8)$$

At this point $dv/dt = 0$ and the object no longer increases in speed. It has reached its **terminal velocity** and continues to fall at this constant velocity until it hits the ground. This sequence of events is shown in the graph of Fig. 5–27b. The value of the terminal velocity v_T can be obtained from Eq. 5–8.

$$v_T = \frac{mg}{b}. \qquad (5\text{–}9)$$

If the resistive force is assumed proportional to v^2, or an even higher power of v, the sequence of events is similar and a terminal velocity reached, although it will not be given by Eq. 5–9.

FIGURE 5–27 (a) Forces acting on an object falling downward. (b) Graph of the velocity of an object falling due to gravity when the air resistance drag force is $F_D = -bv$. Initially, $v = 0$ and $dv/dt = g$, but as time goes on dv/dt (= slope of curve) decreases because of F_D. Eventually, v approaches a maximum value, v_T, the terminal velocity, which occurs when F_D has magnitude equal to mg.

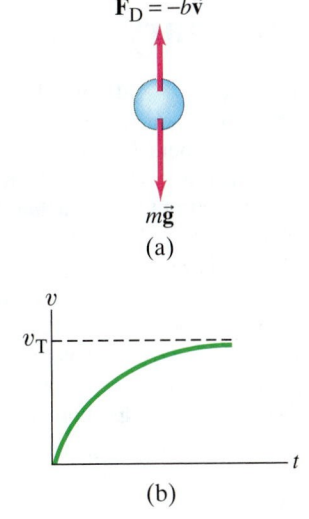

[†]Any buoyant force (Chapter 13) is ignored in this Section.

EXAMPLE 5–17 **Force proportional to velocity.** Determine the velocity as a function of time for an object falling vertically from rest when there is a resistive force linearly proportional to v.

APPROACH This is a derivation and we start with Eq. 5–7, which we rewrite as

$$\frac{dv}{dt} = g - \frac{b}{m}v.$$

SOLUTION In this equation there are two variables, v and t. We collect variables of the same type on one or the other side of the equation:

$$\frac{dv}{g - \frac{b}{m}v} = dt \qquad \text{or} \qquad \frac{dv}{v - \frac{mg}{b}} = -\frac{b}{m}dt.$$

Now we can integrate, remembering $v = 0$ at $t = 0$:

$$\int_0^v \frac{dv}{v - \frac{mg}{b}} = -\frac{b}{m}\int_0^t dt$$

which gives

$$\ln\left(v - \frac{mg}{b}\right) - \ln\left(-\frac{mg}{b}\right) = -\frac{b}{m}t$$

or

$$\ln\frac{v - mg/b}{-mg/b} = -\frac{b}{m}t.$$

We raise each side to the exponential [note that the natural log and the exponential are inverse operations of each other: $e^{\ln x} = x$, or $\ln(e^x) = x$] and obtain

$$v - \frac{mg}{b} = -\frac{mg}{b}e^{-\frac{b}{m}t}$$

so

$$v = \frac{mg}{b}\left(1 - e^{-\frac{b}{m}t}\right).$$

This relation gives the velocity v as a function of time and corresponds to the graph of Fig. 5–27b. As a check, note that at $t = 0$, and $v = 0$

$$a(t = 0) = \frac{dv}{dt} = \frac{mg}{b}\frac{d}{dt}\left(1 - e^{-\frac{b}{m}t}\right) = \frac{mg}{b}\left(\frac{b}{m}\right) = g,$$

as expected (see also Eq. 5–7). At large t, $e^{-\frac{b}{m}t}$ approaches zero, so v approaches mg/b, which is the terminal velocity, v_T, as we saw earlier. If we set $\tau = m/b$, then $v = v_T(1 - e^{-t/\tau})$. So $\tau = m/b$ is the time required for the velocity to reach 63% of the terminal velocity (since $e^{-1} = 0.37$). Figure 5–27b shows a plot of speed v vs. time t, where the terminal velocity $v_T = mg/b$.

Summary

When two objects slide over one another, the force of **friction** that each exerts on the other can be written approximately as $F_{fr} = \mu_k F_N$, where F_N is the **normal force** (the force each object exerts on the other perpendicular to their contact surfaces), and μ_k is the coefficient of **kinetic friction**. If the objects are at rest relative to each other, even though forces act, then F_{fr} is just large enough to hold them at rest and satisfies the inequality $F_{fr} \leq \mu_s F_N$, where μ_s is the coefficient of **static friction**.

An object moving in a circle of radius r with constant speed v is said to be in **uniform circular motion**. It has a **radial acceleration** a_R that is directed radially toward the center of the circle (also called **centripetal acceleration**), and has magnitude

$$a_R = \frac{v^2}{r}. \tag{5–1}$$

The direction of the velocity vector and that of the accelera-

tion $\vec{a}_R$ are continually changing in direction, but are perpendicular to each other at each moment.

A force is needed to keep an object revolving uniformly in a circle, and the direction of this force is toward the center of the circle. This force may be gravity (as for the Moon), or tension in a cord, or a component of the normal force, or another type of force or a combination of forces.

[*When the speed of circular motion is not constant, the acceleration has two components, tangential as well as radial. The force too has tangential and radial components.]

[*A **drag force** acts on an object moving through a fluid, such as air or water. The drag force F_D can often be approximated by $F_D = -bv$ or $F_D \propto v^2$, where v is the speed of the object relative to the fluid.]

Questions

1. A heavy crate rests on the bed of a flatbed truck. When the truck accelerates, the crate remains where it is on the truck, so it, too, accelerates. What force causes the crate to accelerate?

2. A block is given a push so that it slides up a ramp. After the block reaches its highest point, it slides back down, but the magnitude of its acceleration is less on the descent than on the ascent. Why?

3. Why is the stopping distance of a truck much shorter than for a train going the same speed?

4. Can a coefficient of friction exceed 1.0?

5. Cross-country skiers prefer their skis to have a large coefficient of static friction but a small coefficient of kinetic friction. Explain why. [*Hint*: Think of uphill and downhill.]

6. When you must brake your car very quickly, why is it safer if the wheels don't lock? When driving on slick roads, why is it advisable to apply the brakes slowly?

7. When attempting to stop a car quickly on dry pavement, which of the following methods will stop the car in the least time? (*a*) Slam on the brakes as hard as possible, locking the wheels and *skidding* to a stop. (*b*) Press the brakes as hard as possible without locking the wheels and *rolling* to a stop. Explain.

8. You are trying to push your stalled car. Although you apply a horizontal force of 400 N to the car, it doesn't budge, and neither do you. Which force(s) must also have a magnitude of 400 N: (*a*) the force exerted by the car on you; (*b*) the friction force exerted by the car on the road; (*c*) the normal force exerted by the road on you; (*d*) the friction force exerted by the road on you?

9. It is not easy to walk on an icy sidewalk without slipping. Even your gait looks different than on dry pavement. Describe what you need to do differently on the icy surface and why.

10. A car rounds a curve at a steady 50 km/h. If it rounds the same curve at a steady 70 km/h, will its acceleration be any different? Explain.

11. Will the acceleration of a car be the same when a car travels around a sharp curve at a constant 60 km/h as when it travels around a gentle curve at the same speed? Explain.

12. Describe all the forces acting on a child riding a horse on a merry-go-round. Which of these forces provides the centripetal acceleration of the child?

13. A child on a sled comes flying over the crest of a small hill, as shown in Fig. 5–28. His sled does not leave the ground, but he feels the normal force between his chest and the sled decrease as he goes over the hill. Explain this decrease using Newton's second law.

FIGURE 5–28
Question 13.

14. Sometimes it is said that water is removed from clothes in a spin dryer by centrifugal force throwing the water outward. Is this correct? Discuss.

15. Technical reports often specify only the rpm for centrifuge experiments. Why is this inadequate?

16. A girl is whirling a ball on a string around her head in a horizontal plane. She wants to let go at precisely the right time so that the ball will hit a target on the other side of the yard. When should she let go of the string?

17. The game of tetherball is played with a ball tied to a pole with a string. When the ball is struck, it whirls around the pole as shown in Fig. 5–29. In what direction is the acceleration of the ball, and what causes the acceleration?

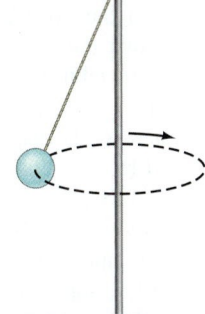

FIGURE 5–29
Problem 17.

18. Astronauts who spend long periods in outer space could be adversely affected by weightlessness. One way to simulate gravity is to shape the spaceship like a cylindrical shell that rotates, with the astronauts walking on the inside surface (Fig. 5–30). Explain how this simulates gravity. Consider (*a*) how objects fall, (*b*) the force we feel on our feet, and (*c*) any other aspects of gravity you can think of.

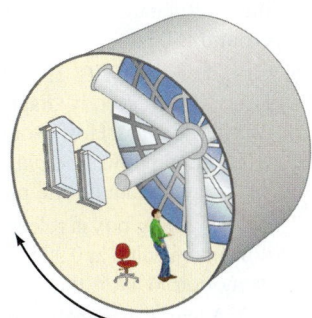

FIGURE 5–30
Question 18.

19. A bucket of water can be whirled in a vertical circle without the water spilling out, even at the top of the circle when the bucket is upside down. Explain.

20. A car maintains a constant speed v as it traverses the hill and valley shown in Fig. 5–31. Both the hill and valley have a radius of curvature R. At which point, A, B, or C, is the normal force acting on the car (*a*) the largest, (*b*) the smallest? Explain. (*c*) Where would the driver feel heaviest and (*d*) lightest? Explain. (*e*) How fast can the car go without losing contact with the road at A?

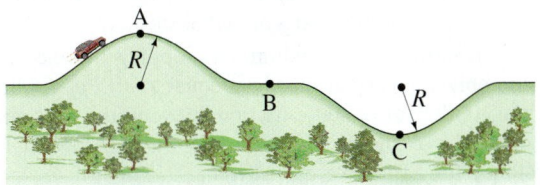

FIGURE 5–31 Question 20.

21. Why do bicycle riders lean in when rounding a curve at high speed?

22. Why do airplanes bank when they turn? How would you compute the banking angle given the airspeed and radius of the turn? [*Hint*: Assume an aerodynamic "lift" force acts perpendicular to the wings.]

*23. For a drag force of the form $F = -bv$, what are the units of b?

*24. Suppose two forces act on an object, one force proportional to v and the other proportional to v^2. Which force dominates at high speed?

Problems

5-1 Friction and Newton's Laws

1. (I) If the coefficient of kinetic friction between a 22-kg crate and the floor is 0.30, what horizontal force is required to move the crate at a steady speed across the floor? What horizontal force is required if μ_k is zero?

2. (I) A force of 35.0 N is required to start a 6.0-kg box moving across a horizontal concrete floor. (a) What is the coefficient of static friction between the box and the floor? (b) If the 35.0-N force continues, the box accelerates at 0.60 m/s^2. What is the coefficient of kinetic friction?

3. (I) Suppose you are standing on a train accelerating at $0.20 \, g$. What minimum coefficient of static friction must exist between your feet and the floor if you are not to slide?

4. (I) The coefficient of static friction between hard rubber and normal street pavement is about 0.90. On how steep a hill (maximum angle) can you leave a car parked?

5. (I) What is the maximum acceleration a car can undergo if the coefficient of static friction between the tires and the ground is 0.90?

6. (II) (a) A box sits at rest on a rough 33° inclined plane. Draw the free-body diagram, showing all the forces acting on the box. (b) How would the diagram change if the box were sliding down the plane? (c) How would it change if the box were sliding up the plane after an initial shove?

7. (II) A 25.0-kg box is released on a 27° incline and accelerates down the incline at 0.30 m/s^2. Find the friction force impeding its motion. What is the coefficient of kinetic friction?

8. (II) A car can decelerate at -3.80 m/s^2 without skidding when coming to rest on a level road. What would its deceleration be if the road is inclined at 9.3° and the car moves uphill? Assume the same static friction coefficient.

9. (II) A skier moves down a 27° slope at constant speed. What can you say about the coefficient of friction, μ_k? Assume the speed is low enough that air resistance can be ignored.

10. (II) A wet bar of soap slides freely down a ramp 9.0 m long inclined at 8.0°. How long does it take to reach the bottom? Assume $\mu_k = 0.060$.

11. (II) A box is given a push so that it slides across the floor. How far will it go, given that the coefficient of kinetic friction is 0.15 and the push imparts an initial speed of 3.5 m/s?

12. (II) (a) Show that the minimum stopping distance for an automobile traveling at speed v is equal to $v^2/2 \, \mu_s g$, where μ_s is the coefficient of static friction between the tires and the road, and g is the acceleration of gravity. (b) What is this distance for a 1200-kg car traveling 95 km/h if $\mu_s = 0.65$? (c) What would it be if the car were on the Moon (the acceleration of gravity on the Moon is about $g/6$) but all else stayed the same?

13. (II) A 1280-kg car pulls a 350-kg trailer. The car exerts a horizontal force of 3.6×10^3 N against the ground in order to accelerate. What force does the car exert on the trailer? Assume an effective friction coefficient of 0.15 for the trailer.

14. (II) Police investigators, examining the scene of an accident involving two cars, measure 72-m-long skid marks of one of the cars, which nearly came to a stop before colliding. The coefficient of kinetic friction between rubber and the pavement is about 0.80. Estimate the initial speed of that car assuming a level road.

15. (II) Piles of snow on slippery roofs can become dangerous projectiles as they melt. Consider a chunk of snow at the ridge of a roof with a slope of 34°. (a) What is the minimum value of the coefficient of static friction that will keep the snow from sliding down? (b) As the snow begins to melt the coefficient of static friction decreases and the snow finally slips. Assuming that the distance from the chunk to the edge of the roof is 6.0 m and the coefficient of kinetic friction is 0.20, calculate the speed of the snow chunk when it slides off the roof. (c) If the edge of the roof is 10.0 m above ground, estimate the speed of the snow when it hits the ground.

16. (II) A small box is held in place against a rough vertical wall by someone pushing on it with a force directed upward at 28° above the horizontal. The coefficients of static and kinetic friction between the box and wall are 0.40 and 0.30, respectively. The box slides down unless the applied force has magnitude 23 N. What is the mass of the box?

17. (II) Two crates, of mass 65 kg and 125 kg, are in contact and at rest on a horizontal surface (Fig. 5–32). A 650-N force is exerted on the 65-kg crate. If the coefficient of kinetic friction is 0.18, calculate (a) the acceleration of the system, and (b) the force that each crate exerts on the other. (c) Repeat with the crates reversed.

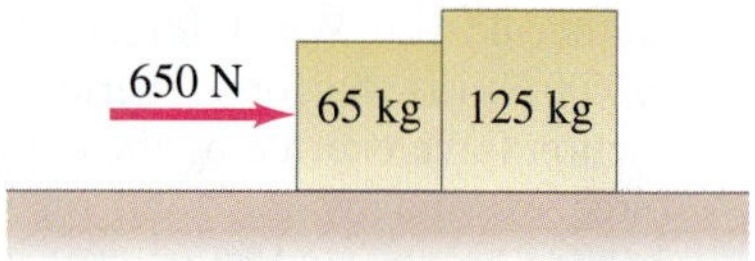

FIGURE 5–32
Problem 17.

18. (II) The crate shown in Fig. 5–33 lies on a plane tilted at an angle $\theta = 25.0°$ to the horizontal, with $\mu_k = 0.19$. (a) Determine the acceleration of the crate as it slides down the plane. (b) If the crate starts from rest 8.15 m up the plane from its base, what will be the crate's speed when it reaches the bottom of the incline?

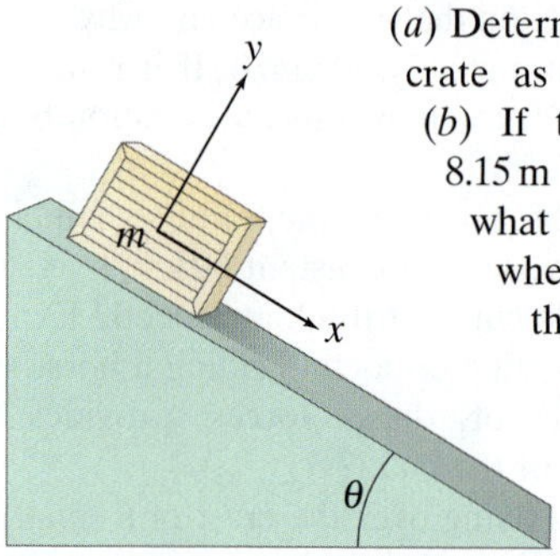

FIGURE 5–33
Crate on inclined plane.
Problems 18 and 19.

19. (II) A crate is given an initial speed of 3.0 m/s up the 25.0° plane shown in Fig. 5–33. (a) How far up the plane will it go? (b) How much time elapses before it returns to its starting point? Assume $\mu_k = 0.17$.

20. (II) Two blocks made of different materials connected together by a thin cord, slide down a plane ramp inclined at an angle θ to the horizontal as shown in Fig. 5–34 (block B is above block A). The masses of the blocks are m_A and m_B, and the coefficients of friction are μ_A and μ_B. If $m_A = m_B = 5.0$ kg, and $\mu_A = 0.20$ and $\mu_B = 0.30$, determine (a) the acceleration of the blocks and (b) the tension in the cord, for an angle $\theta = 32°$.

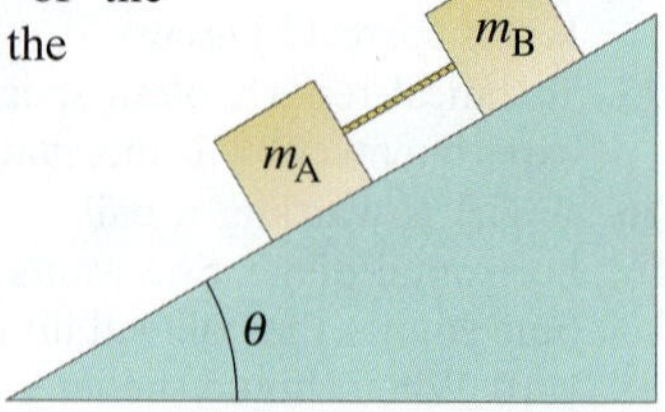

FIGURE 5–34
Problems 20 and 21.

21. (II) For two blocks, connected by a cord and sliding down the incline shown in Fig. 5–34 (see Problem 20), describe the motion (*a*) if $\mu_A < \mu_B$, and (*b*) if $\mu_A > \mu_B$. (*c*) Determine a formula for the acceleration of each block and the tension F_T in the cord in terms of m_A, m_B, and θ; interpret your results in light of your answers to (*a*) and (*b*).

22. (II) A flatbed truck is carrying a heavy crate. The coefficient of static friction between the crate and the bed of the truck is 0.75. What is the maximum rate at which the driver can decelerate and still avoid having the crate slide against the cab of the truck?

23. (II) In Fig. 5–35 the coefficient of static friction between mass m_A and the table is 0.40, whereas the coefficient of kinetic friction is 0.30 (*a*) What minimum value of m_A will keep the system from starting to move? (*b*) What value(s) of m_A will keep the system moving at constant speed?

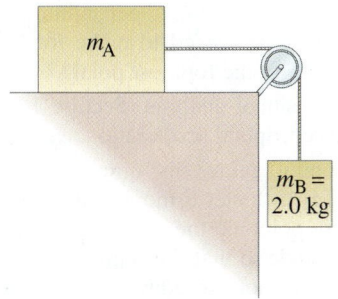

FIGURE 5–35 Problems 23 and 24.

24. (II) Determine a formula for the acceleration of the system shown in Fig. 5–35 in terms of m_A, m_B, and the mass of the cord, m_C. Define any other variables needed.

25. (II) A small block of mass m is given an initial speed v_0 up a ramp inclined at angle θ to the horizontal. It travels a distance d up the ramp and comes to rest. (*a*) Determine a formula for the coefficient of kinetic friction between block and ramp. (*b*) What can you say about the value of the coefficient of static friction?

26. (II) A 75-kg snowboarder has an initial velocity of 5.0 m/s at the top of a 28° incline (Fig. 5–36). After sliding down the 110-m long incline (on which the coefficient of kinetic friction is $\mu_k = 0.18$), the snowboarder has attained a velocity v. The snowboarder then slides along a flat surface (on which $\mu_k = 0.15$) and comes to rest after a distance x. Use Newton's second law to find the snowboarder's acceleration while on the incline and while on the flat surface. Then use these accelerations to determine x.

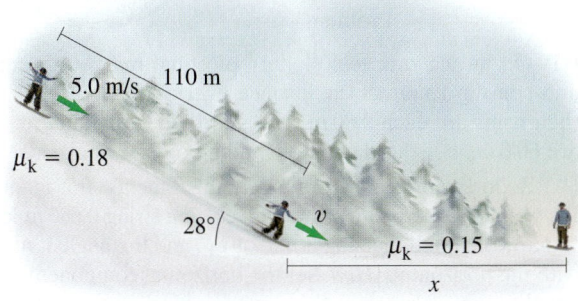

FIGURE 5–36 Problem 26.

27. (II) A package of mass m is dropped vertically onto a horizontal conveyor belt whose speed is $v = 1.5$ m/s, and the coefficient of kinetic friction between the package and the belt is $\mu_k = 0.70$. (*a*) For how much time does the package slide on the belt (until it is at rest relative to the belt)? (*b*) How far does the package move during this time?

28. (II) Two masses $m_A = 2.0$ kg and $m_B = 5.0$ kg are on inclines and are connected together by a string as shown in Fig. 5–37. The coefficient of kinetic friction between each mass and its incline is $\mu_k = 0.30$. If m_A moves up, and m_B moves down, determine their acceleration.

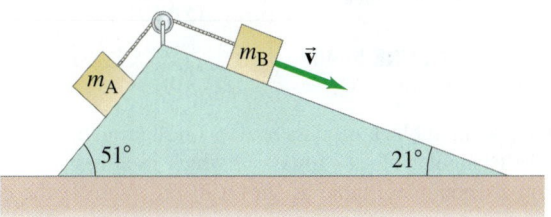

FIGURE 5–37 Problem 28.

29. (II) A child slides down a slide with a 34° incline, and at the bottom her speed is precisely half what it would have been if the slide had been frictionless. Calculate the coefficient of kinetic friction between the slide and the child.

30. (II) (*a*) Suppose the coefficient of kinetic friction between m_A and the plane in Fig. 5–38 is $\mu_k = 0.15$, and that $m_A = m_B = 2.7$ kg. As m_B moves down, determine the magnitude of the acceleration of m_A and m_B, given $\theta = 34°$. (*b*) What smallest value of μ_k will keep the system from accelerating?

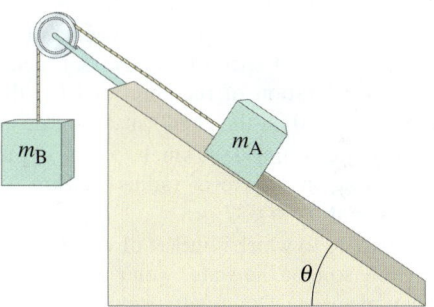

FIGURE 5–38 Problem 30.

31. (III) A 3.0-kg block sits on top of a 5.0-kg block which is on a horizontal surface. The 5.0-kg block is pulled to the right with a force $\vec{F}$ as shown in Fig. 5–39. The coefficient of static friction between all surfaces is 0.60 and the kinetic coefficient is 0.40. (*a*) What is the minimum value of F needed to move the two blocks? (*b*) If the force is 10% greater than your answer for (*a*), what is the acceleration of each block?

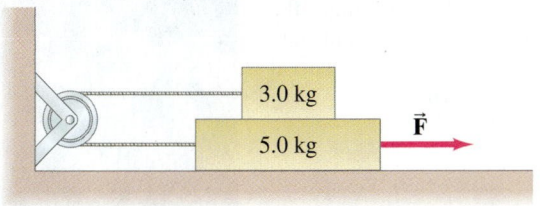

FIGURE 5–39 Problem 31.

32. (III) A 4.0-kg block is stacked on top of a 12.0-kg block, which is accelerating along a horizontal table at $a = 5.2\,\text{m/s}^2$ (Fig. 5–40). Let $\mu_k = \mu_s = \mu$. (*a*) What minimum coefficient of friction μ between the two blocks will prevent the 4.0-kg block from sliding off? (*b*) If μ is only half this minimum value, what is the acceleration of the 4.0-kg block with respect to the table, and (*c*) with respect to the 12.0-kg block? (*d*) What is the force that must be applied to the 12.0-kg block in (*a*) and in (*b*), assuming that the table is frictionless?

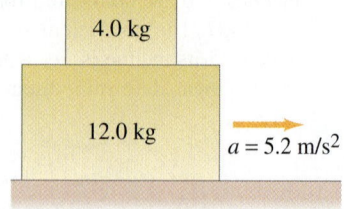

FIGURE 5–40
Problem 32.

33. (III) A small block of mass m rests on the rough, sloping side of a triangular block of mass M which itself rests on a horizontal frictionless table as shown in Fig. 5–41. If the coefficient of static friction is μ, determine the minimum horizontal force F applied to M that will cause the small block m to start moving up the incline.

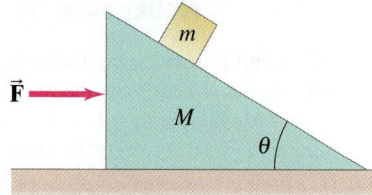

FIGURE 5–41
Problem 33.

5–2 to 5–4 Uniform Circular Motion

34. (I) What is the maximum speed with which a 1200-kg car can round a turn of radius 80.0 m on a flat road if the coefficient of friction between tires and road is 0.65? Is this result independent of the mass of the car?

35. (I) A child sitting 1.20 m from the center of a merry-go-around moves with a speed of 1.30 m/s. Calculate (*a*) the centripetal acceleration of the child and (*b*) the net horizontal force exerted on the child (mass = 22.5 kg).

36. (I) A jet plane traveling 1890 km/h (525 m/s) pulls out of a dive by moving in an arc of radius 4.80 km. What is the plane's acceleration in g's?

37. (II) Is it possible to whirl a bucket of water fast enough in a vertical circle so that the water won't fall out? If so, what is the minimum speed? Define all quantities needed.

38. (II) How fast (in rpm) must a centrifuge rotate if a particle 8.00 cm from the axis of rotation is to experience an acceleration of 125,000 g's?

39. (II) Highway curves are marked with a suggested speed. If this speed is based on what would be safe in wet weather, estimate the radius of curvature for a curve marked 50 km/h. Use Table 5–1.

40. (II) At what minimum speed must a roller coaster be traveling when upside down at the top of a circle (Fig. 5–42) so that the passengers do not fall out? Assume a radius of curvature of 7.6 m.

FIGURE 5–42
Problem 40.

41. (II) A sports car crosses the bottom of a valley with a radius of curvature equal to 95 m. At the very bottom, the normal force on the driver is twice his weight. At what speed was the car traveling?

42. (II) How large must the coefficient of static friction be between the tires and the road if a car is to round a level curve of radius 85 m at a speed of 95 km/h?

43. (II) Suppose the space shuttle is in orbit 400 km from the Earth's surface, and circles the Earth about once every 90 min. Find the centripetal acceleration of the space shuttle in its orbit. Express your answer in terms of g, the gravitational acceleration at the Earth's surface.

44. (II) A bucket of mass 2.00 kg is whirled in a vertical circle of radius 1.10 m. At the lowest point of its motion the tension in the rope supporting the bucket is 25.0 N. (*a*) Find the speed of the bucket. (*b*) How fast must the bucket move at the top of the circle so that the rope does not go slack?

45. (II) How many revolutions per minute would a 22-m-diameter Ferris wheel need to make for the passengers to feel "weightless" at the topmost point?

46. (II) Use dimensional analysis (Section 1–7) to obtain the form for the centripetal acceleration, $a_R = v^2/r$.

47. (II) A jet pilot takes his aircraft in a vertical loop (Fig. 5–43). (*a*) If the jet is moving at a speed of 1200 km/h at the lowest point of the loop, determine the minimum radius of the circle so that the centripetal acceleration at the lowest point does not exceed 6.0 g's. (*b*) Calculate the 78-kg pilot's effective weight (the force with which the seat pushes up on him) at the bottom of the circle, and (*c*) at the top of the circle (assume the same speed).

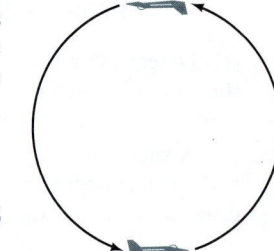

FIGURE 5–43
Problem 47.

48. (II) A proposed space station consists of a circular tube that will rotate about its center (like a tubular bicycle tire), Fig. 5–44. The circle formed by the tube has a diameter of about 1.1 km. What must be the rotation speed (revolutions per day) if an effect equal to gravity at the surface of the Earth (1.0 g) is to be felt?

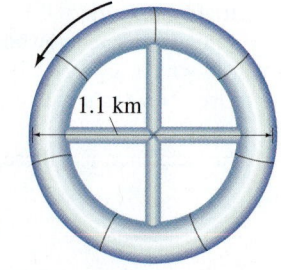

FIGURE 5–44
Problem 48.

49. (II) On an ice rink two skaters of equal mass grab hands and spin in a mutual circle once every 2.5 s. If we assume their arms are each 0.80 m long and their individual masses are 60.0 kg, how hard are they pulling on one another?

50. (II) Redo Example 5–11, precisely this time, by not ignoring the weight of the ball which revolves on a string 0.600 m long. In particular, find the magnitude of $\vec{\mathbf{F}}_T$, and the angle it makes with the horizontal. [*Hint:* Set the horizontal component of $\vec{\mathbf{F}}_T$ equal to ma_R; also, since there is no vertical motion, what can you say about the vertical component of $\vec{\mathbf{F}}_T$?]

51. (II) A coin is placed 12.0 cm from the axis of a rotating turntable of variable speed. When the speed of the turntable is slowly increased, the coin remains fixed on the turntable until a rate of 35.0 rpm (revolutions per minute) is reached, at which point the coin slides off. What is the coefficient of static friction between the coin and the turntable?

52. (II) The design of a new road includes a straight stretch that is horizontal and flat but that suddenly dips down a steep hill at 22°. The transition should be rounded with what minimum radius so that cars traveling 95 km/h will not leave the road (Fig. 5–45)?

FIGURE 5–45
Problem 52.

53. (II) A 975-kg sports car (including driver) crosses the rounded top of a hill (radius = 88.0 m) at 12.0 m/s. Determine (a) the normal force exerted by the road on the car, (b) the normal force exerted by the car on the 72.0-kg driver, and (c) the car speed at which the normal force on the driver equals zero.

54. (II) Two blocks, with masses m_A and m_B, are connected to each other and to a central post by cords as shown in Fig. 5–46. They rotate about the post at frequency f (revolutions per second) on a frictionless horizontal surface at distances r_A and r_B from the post. Derive an algebraic expression for the tension in each segment of the cord (assumed massless).

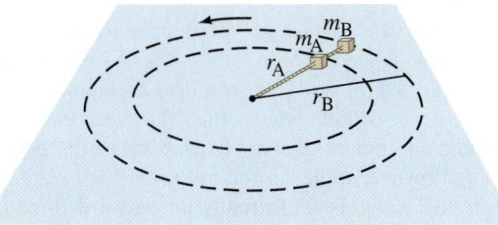

FIGURE 5–46   Problem 54.

55. (II) Tarzan plans to cross a gorge by swinging in an arc from a hanging vine (Fig. 5–47). If his arms are capable of exerting a force of 1350 N on the rope, what is the maximum speed he can tolerate at the lowest point of his swing? His mass is 78 kg and the vine is 5.2 m long.

FIGURE 5–47
Problem 55.

56. (II) A pilot performs an evasive maneuver by diving vertically at 310 m/s. If he can withstand an acceleration of 9.0 g's without blacking out, at what altitude must he begin to pull out of the dive to avoid crashing into the sea?

57. (III) The position of a particle moving in the xy plane is given by $\vec{r} = 2.0 \cos(3.0 \text{ rad/s } t)\hat{i} + 2.0 \sin(3.0 \text{ rad/s } t)\hat{j}$, where r is in meters and t is in seconds. (a) Show that this represents circular motion of radius 2.0 m centered at the origin. (b) Determine the velocity and acceleration vectors as functions of time. (c) Determine the speed and magnitude of the acceleration. (d) Show that $a = v^2/r$. (e) Show that the acceleration vector always points toward the center of the circle.

58. (III) If a curve with a radius of 85 m is properly banked for a car traveling 65 km/h, what must be the coefficient of static friction for a car not to skid when traveling at 95 km/h?

59. (III) A curve of radius 68 m is banked for a design speed of 85 km/h. If the coefficient of static friction is 0.30 (wet pavement), at what range of speeds can a car safely make the curve? [*Hint*: Consider the direction of the friction force when the car goes too slow or too fast.]

*5–5 Nonuniform Circular Motion

*60.** (II) A particle starting from rest revolves with uniformly increasing speed in a clockwise circle in the xy plane. The center of the circle is at the origin of an xy coordinate system. At $t = 0$, the particle is at $x = 0.0$, $y = 2.0$ m. At $t = 2.0$ s, it has made one-quarter of a revolution and is at $x = 2.0$ m, $y = 0.0$. Determine (a) its speed at $t = 2.0$ s, (b) the average velocity vector, and (c) the average acceleration vector during this interval.

*61.** (II) In Problem 60 assume the tangential acceleration is constant and determine the components of the instantaneous acceleration at (a) $t = 0.0$, (b) $t = 1.0$ s, and (c) $t = 2.0$ s.

*62.** (II) An object moves in a circle of radius 22 m with its speed given by $v = 3.6 + 1.5t^2$, with v in meters per second and t in seconds. At $t = 3.0$ s, find (a) the tangential acceleration and (b) the radial acceleration.

*63.** (III) A particle rotates in a circle of radius 3.80 m. At a particular instant its acceleration is 1.15 m/s² in a direction that makes an angle of 38.0° to its direction of motion. Determine its speed (a) at this moment and (b) 2.00 s later, assuming constant tangential acceleration.

*64.** (III) An object of mass m is constrained to move in a circle of radius r. Its tangential acceleration as a function of time is given by $a_{tan} = b + ct^2$, where b and c are constants. If $v = v_0$ at $t = 0$, determine the tangential and radial components of the force, F_{tan} and F_R, acting on the object at any time $t > 0$.

*5–6 Velocity-Dependent Forces

*65.** (I) Use dimensional analysis (Section 1–7) in Example 5–17 to determine if the time constant τ is $\tau = m/b$ or $\tau = b/m$.

*66.** (II) The terminal velocity of a 3×10^{-5} kg raindrop is about 9 m/s. Assuming a drag force $F_D = -bv$, determine (a) the value of the constant b and (b) the time required for such a drop, starting from rest, to reach 63% of terminal velocity.

*67.** (II) An object moving vertically has $\vec{v} = \vec{v}_0$ at $t = 0$. Determine a formula for its velocity as a function of time assuming a resistive force $F = -bv$ as well as gravity for two cases: (a) $\vec{v}_0$ is downward and (b) $\vec{v}_0$ is upward.

*68. (III) The drag force on large objects such as cars, planes, and sky divers moving through air is more nearly $F_D = -bv^2$. (a) For this quadratic dependence on v, determine a formula for the terminal velocity v_T, of a vertically falling object. (b) A 75-kg sky diver has a terminal velocity of about 60 m/s; determine the value of the constant b. (c) Sketch a curve like that of Fig. 5–27b for this case of $F_D \propto v^2$. For the same terminal velocity, would this curve lie above or below that in Fig. 5–27? Explain why.

*69. (III) A bicyclist can coast down a 7.0° hill at a steady 9.5 km/h. If the drag force is proportional to the square of the speed v, so that $F_D = -cv^2$, calculate (a) the value of the constant c and (b) the average force that must be applied in order to descend the hill at 25 km/h. The mass of the cyclist plus bicycle is 80.0 kg. Ignore other types of friction.

*70. (III) Two drag forces act on a bicycle and rider: F_{D1} due to rolling resistance, which is essentially velocity independent; and F_{D2} due to air resistance, which is proportional to v^2. For a specific bike plus rider of total mass 78 kg, $F_{D1} \approx 4.0$ N; and for a speed of 2.2 m/s, $F_{D2} \approx 1.0$ N. (a) Show that the total drag force is

$$F_D = 4.0 + 0.21v^2,$$

where v is in m/s, and F_D is in N and opposes the motion. (b) Determine at what slope angle θ the bike and rider can coast downhill at a constant speed of 8.0 m/s.

*71. (III) Determine a formula for the position and acceleration of a falling object as a function of time if the object starts from rest at $t = 0$ and undergoes a resistive force $F = -bv$, as in Example 5–17.

*72. (III) A block of mass m slides along a horizontal surface lubricated with a thick oil which provides a drag force proportional to the square root of velocity:

$$F_D = -bv^{\frac{1}{2}}.$$

If $v = v_0$ at $t = 0$, determine v and x as functions of time.

*73. (III) Show that the maximum distance the block in Problem 72 can travel is $2m\, v_0^{3/2}/3b$.

*74. (III) You dive straight down into a pool of water. You hit the water with a speed of 5.0 m/s, and your mass is 75 kg. Assuming a drag force of the form $F_D = -(1.00 \times 10^4 \text{ kg/s})\, v$, how long does it take you to reach 2% of your original speed? (Ignore any effects of buoyancy.)

*75. (III) A motorboat traveling at a speed of 2.4 m/s shuts off its engines at $t = 0$. How far does it travel before coming to rest if it is noted that after 3.0 s its speed has dropped to half its original value? Assume that the drag force of the water is proportional to v.

General Problems

76. A coffee cup on the horizontal dashboard of a car slides forward when the driver decelerates from 45 km/h to rest in 3.5 s or less, but not if she decelerates in a longer time. What is the coefficient of static friction between the cup and the dash? Assume the road and the dashboard are level (horizontal).

77. A 2.0-kg silverware drawer does not slide readily. The owner gradually pulls with more and more force, and when the applied force reaches 9.0 N, the drawer suddenly opens, throwing all the utensils to the floor. What is the coefficient of static friction between the drawer and the cabinet?

78. A roller coaster reaches the top of the steepest hill with a speed of 6.0 km/h. It then descends the hill, which is at an average angle of 45° and is 45.0 m long. What will its speed be when it reaches the bottom? Assume $\mu_k = 0.12$.

79. An 18.0-kg box is released on a 37.0° incline and accelerates down the incline at 0.220 m/s². Find the friction force impeding its motion. How large is the coefficient of friction?

80. A flat puck (mass M) is revolved in a circle on a frictionless air hockey table top, and is held in this orbit by a light cord which is connected to a dangling mass (mass m) through a central hole as shown in Fig. 5–48. Show that the speed of the puck is given by $v = \sqrt{mgR/M}$.

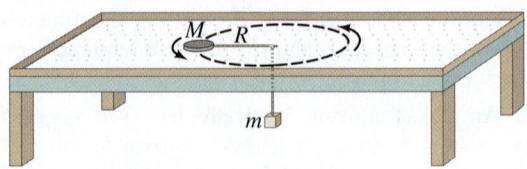

FIGURE 5–48 Problem 80.

81. A motorcyclist is coasting with the engine off at a steady speed of 20.0 m/s but enters a sandy stretch where the coefficient of kinetic friction is 0.70. Will the cyclist emerge from the sandy stretch without having to start the engine if the sand lasts for 15 m? If so, what will be the speed upon emerging?

82. In a "Rotor-ride" at a carnival, people rotate in a vertical cylindrically walled "room." (See Fig. 5–49). If the room radius was 5.5 m, and the rotation frequency 0.50 revolutions per second when the floor drops out, what minimum coefficient of static friction keeps the people from slipping down? People on this ride said they were "pressed against the wall." Is there really an outward force pressing them against the wall? If so, what is its source? If not, what is the proper description of their situation (besides nausea)? [Hint: Draw a free-body diagram for a person.]

FIGURE 5–49 Problem 82.

83. A device for training astronauts and jet fighter pilots is designed to rotate the trainee in a horizontal circle of radius 11.0 m. If the force felt by the trainee is 7.45 times her own weight, how fast is she rotating? Express your answer in both m/s and rev/s.

84. A 1250-kg car rounds a curve of radius 72 m banked at an angle of 14°. If the car is traveling at 85 km/h, will a friction force be required? If so, how much and in what direction?

85. Determine the tangential and centripetal components of the net force exerted on a car (by the ground) when its speed is 27 m/s, and it has accelerated to this speed from rest in 9.0 s on a curve of radius 450 m. The car's mass is 1150 kg.

86. The 70.0-kg climber in Fig. 5–50 is supported in the "chimney" by the friction forces exerted on his shoes and back. The static coefficients of friction between his shoes and the wall, and between his back and the wall, are 0.80 and 0.60, respectively. What is the minimum normal force he must exert? Assume the walls are vertical and that the static friction forces are both at their maximum. Ignore his grip on the rope.

FIGURE 5–50
Problem 86.

87. A small mass m is set on the surface of a sphere, Fig. 5–51. If the coefficient of static friction is $\mu_s = 0.70$, at what angle ϕ would the mass start sliding?

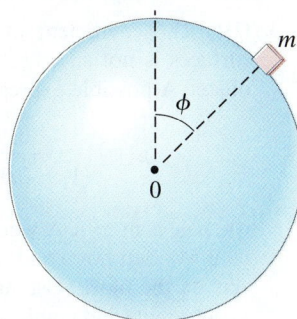

FIGURE 5–51
Problem 87.

88. A 28.0-kg block is connected to an empty 2.00-kg bucket by a cord running over a frictionless pulley (Fig. 5–52). The coefficient of static friction between the table and the block is 0.45 and the coefficient of kinetic friction between the table and the block is 0.32. Sand is gradually added to the bucket until the system just begins to move. (*a*) Calculate the mass of sand added to the bucket. (*b*) Calculate the acceleration of the system.

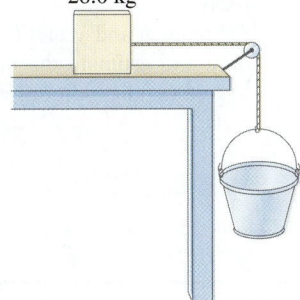

FIGURE 5–52
Problem 88.

89. A car is heading down a slippery road at a speed of 95 km/h. The minimum distance within which it can stop without skidding is 66 m. What is the sharpest curve the car can negotiate on the icy surface at the same speed without skidding?

90. What is the acceleration experienced by the tip of the 1.5-cm-long sweep second hand on your wrist watch?

91. An airplane traveling at 480 km/h needs to reverse its course. The pilot decides to accomplish this by banking the wings at an angle of 38°. (*a*) Find the time needed to reverse course. (*b*) Describe any additional force the passengers experience during the turn. [*Hint:* Assume an aerodynamic "lift" force that acts perpendicularly to the flat wings; see Fig. 5–53.]

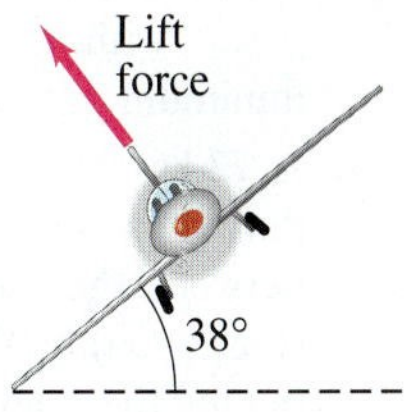

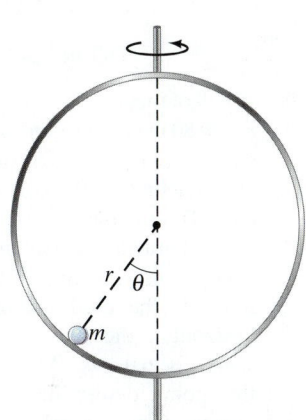

FIGURE 5–53
Problem 91.

92. A banked curve of radius R in a new highway is designed so that a car traveling at speed v_0 can negotiate the turn safely on glare ice (zero friction). If a car travels too slowly, then it will slip toward the center of the circle. If it travels too fast, it will slip away from the center of the circle. If the coefficient of static friction increases, it becomes possible for a car to stay on the road while traveling at a speed within a range from $v_{\min}$ to $v_{\max}$. Derive formulas for $v_{\min}$ and $v_{\max}$ as functions of μ_s, v_0, and R.

93. A small bead of mass m is constrained to slide without friction inside a circular vertical hoop of radius r which rotates about a vertical axis (Fig. 5–54) at a frequency f. (*a*) Determine the angle θ where the bead will be in equilibrium—that is, where it will have no tendency to move up or down along the hoop. (*b*) If $f = 2.00$ rev/s and $r = 22.0$ cm, what is θ? (*c*) Can the bead ride as high as the center of the circle ($\theta = 90°$)? Explain.

FIGURE 5–54
Problem 93.

94. *Earth is not quite an inertial frame.* We often make measurements in a reference frame fixed on the Earth, assuming Earth is an inertial reference frame. But the Earth rotates, so this assumption is not quite valid. Show that this assumption is off by 3 parts in 1000 by calculating the acceleration of an object at Earth's equator due to Earth's daily rotation, and compare to $g = 9.80$ m/s², the acceleration due to gravity.

95. While fishing, you get bored and start to swing a sinker weight around in a circle below you on a 0.45-m piece of fishing line. The weight makes a complete circle every 0.50 s. What is the angle that the fishing line makes with the vertical? [*Hint*: See Fig. 5–20.]

96. Consider a train that rounds a curve with a radius of 570 m at a speed of 160 km/h (approximately 100 mi/h). (*a*) Calculate the friction force needed on a train passenger of mass 75 kg if the track is not banked and the train does not tilt. (*b*) Calculate the friction force on the passenger if the train tilts at an angle of 8.0° toward the center of the curve.

97. A car starts rolling down a 1-in-4 hill (1-in-4 means that for each 4 m traveled along the road, the elevation change is 1 m). How fast is it going when it reaches the bottom after traveling 55 m? (*a*) Ignore friction. (*b*) Assume an effective coefficient of friction equal to 0.10.

98. The sides of a cone make an angle ϕ with the vertical. A small mass m is placed on the inside of the cone and the cone, with its point down, is revolved at a frequency f (revolutions per second) about its symmetry axis. If the coefficient of static friction is μ_s, at what positions on the cone can the mass be placed without sliding on the cone? (Give the maximum and minimum distances, r, from the axis).

99. A 72-kg water skier is being accelerated by a ski boat on a flat ("glassy") lake. The coefficient of kinetic friction between the skier's skis and the water surface is $\mu_k = 0.25$ (Fig. 5–55). (a) What is the skier's acceleration if the rope pulling the skier behind the boat applies a horizontal tension force of magnitude $F_T = 240 \text{ N}$ to the skier ($\theta = 0°$)? (b) What is the skier's horizontal acceleration if the rope pulling the skier exerts a force of $F_T = 240 \text{ N}$ on the skier at an upward angle $\theta = 12°$? (c) Explain why the skier's acceleration in part (b) is greater than that in part (a).

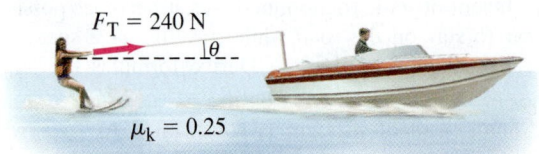

$F_T = 240 \text{ N}$

θ

$\mu_k = 0.25$

FIGURE 5–55 Problem 99.

100. A ball of mass $m = 1.0 \text{ kg}$ at the end of a thin cord of length $r = 0.80 \text{ m}$ revolves in a vertical circle about point O, as shown in Fig. 5–56. During the time we observe it, the only forces acting on the ball are gravity and the tension in the cord. The motion is circular but not uniform because of the force of gravity. The ball increases in speed as it descends and decelerates as it rises on the other side of the circle. At the moment the cord makes an angle $\theta = 30°$ below the horizontal, the ball's speed is 6.0 m/s. At this point, determine the tangential acceleration, the radial acceleration, and the tension in the cord, F_T. Take θ increasing downward as shown.

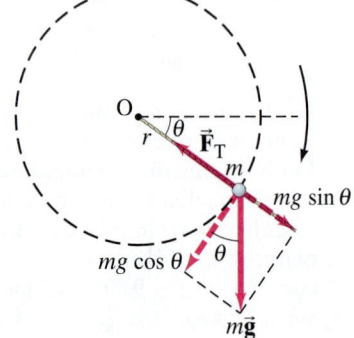

FIGURE 5–56
Problem 100.

101. A car drives at a constant speed around a banked circular track with a diameter of 127 m. The motion of the car can be described in a coordinate system with its origin at the center of the circle. At a particular instant the car's acceleration in the horizontal plane is given by

$$\vec{a} = (-15.7\hat{i} - 23.2\hat{j}) \text{ m/s}^2.$$

(a) What is the car's speed? (b) Where (x and y) is the car at this instant?

*Numerical/Computer

* **102.** (III) The force of air resistance (drag force) on a rapidly falling body such as a skydiver has the form $F_D = -kv^2$, so that Newton's second law applied to such an object is

$$m\frac{dv}{dt} = mg - kv^2,$$

where the downward direction is taken to be positive. (a) Use numerical integration [Section 2–9] to estimate (within 2%) the position, speed, and acceleraton, from $t = 0$ up to $t = 15.0 \text{ s}$, for a 75-kg skydiver who starts from rest, assuming $k = 0.22 \text{ kg/m}$. (b) Show that the diver eventually reaches a steady speed, the *terminal speed*, and explain why this happens. (c) How long does it take for the skydiver to reach 99.5% of the terminal speed?

* **103.** (III) The coefficient of kinetic friction μ_k between two surfaces is not strictly independent of the velocity of the object. A possible expression for μ_k for wood on wood is

$$\mu_k = \frac{0.20}{(1 + 0.0020v^2)^2},$$

where v is in m/s. A wooden block of mass 8.0 kg is at rest on a wooden floor, and a constant horizontal force of 41 N acts on the block. Use numerical integration [Section 2–9] to determine and graph (a) the speed of the block, and (b) its position, as a function of time from 0 to 5.0 s. (c) Determine the percent difference for the speed and position at 5.0 s if μ_k is constant and equal to 0.20.

* **104.** (III) Assume a net force $F = -mg - kv^2$ acts during the upward vertical motion of a 250-kg rocket, starting at the moment ($t = 0$) when the fuel has burned out and the rocket has an upward speed of 120 m/s. Let $k = 0.65 \text{ kg/m}$. Estimate v and y at 1.0-s intervals for the upward motion only, and estimate the maximum height reached. Compare to free-flight conditions without air resistance ($k = 0$).

Answers to Exercises

A: (c).

B: F_{Px} is insufficient to keep the box moving for long.

C: No—the acceleration is not constant (in direction).

D: (a), it doubles.

E: (d).

F: (a).

G: (c).

H: Yes.

I: (a) No change; (b) 4 times larger.

The astronauts in the upper left of this photo are working on the Space Shuttle. As they orbit the Earth—at a rather high speed—they experience apparent weightlessness. The Moon, in the background, also is orbiting the Earth at high speed. What keeps the Moon and the space shuttle (and its astronauts) from moving off in a straight line away from Earth? It is the force of gravity. Newton's law of universal gravitation states that all objects attract all other objects with a force proportional to their masses and inversely proportional to the square of the distance between them.

Gravitation and Newton's Synthesis

C H A P T E R

6

CHAPTER-OPENING QUESTION—Guess now!

A space station revolves around the Earth as a satellite, 100 km above Earth's surface. What is the net force on an astronaut at rest inside the space station?

(a) Equal to her weight on Earth.

(b) A little less than her weight on Earth.

(c) Less than half her weight on Earth.

(d) Zero (she is weightless).

(e) Somewhat larger than her weight on Earth.

Sir Isaac Newton not only put forth the three great laws of motion that serve as the foundation for the study of dynamics. He also conceived of another great law to describe one of the basic forces in nature, gravitation, and he applied it to understand the motion of the planets. This new law, published in 1687 in his book *Philosophiae Naturalis Principia Mathematica* (the *Principia*, for short), is called Newton's law of universal gravitation. It was the capstone of Newton's analysis of the physical world. Indeed, Newtonian mechanics, with its three laws of motion and the law of universal gravitation, was accepted for centuries as a mechanical basis for the way the universe works.

CONTENTS

6–1 Newton's Law of Universal Gravitation

6–2 Vector Form of Newton's Law of Universal Gravitation

6–3 Gravity Near the Earth's Surface; Geophysical Applications

6–4 Satellites and "Weightlessness"

6–5 Kepler's Laws and Newton's Synthesis

*6–6 Gravitational Field

6–7 Types of Forces in Nature

*6–8 Principle of Equivalence; Curvature of Space; Black Holes

6–1 Newton's Law of Universal Gravitation

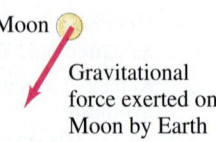

FIGURE 6–1 Anywhere on Earth, whether in Alaska, Australia, or Peru, the force of gravity acts downward toward the center of the Earth.

Among his many great accomplishments, Sir Isaac Newton examined the motion of the heavenly bodies—the planets and the Moon. In particular, he wondered about the nature of the force that must act to keep the Moon in its nearly circular orbit around the Earth.

Newton was also thinking about the problem of gravity. Since falling objects accelerate, Newton had concluded that they must have a force exerted on them, a force we call the force of gravity. Whenever an object has a force exerted *on* it, that force is exerted *by* some other object. But what exerts the force of gravity? Every object on the surface of the Earth feels the force of gravity, and no matter where the object is, the force is directed toward the center of the Earth (Fig. 6–1). Newton concluded that it must be the Earth itself that exerts the gravitational force on objects at its surface.

According to legend, Newton noticed an apple drop from a tree. He is said to have been struck with a sudden inspiration: If gravity acts at the tops of trees, and even at the tops of mountains, then perhaps it acts all the way to the Moon! With this idea that it is Earth's gravity that holds the Moon in its orbit, Newton developed his great theory of gravitation. But there was controversy at the time. Many thinkers had trouble accepting the idea of a force "acting at a distance." Typical forces act through contact—your hand pushes a cart and pulls a wagon, a bat hits a ball, and so on. But gravity acts without contact, said Newton: the Earth exerts a force on a falling apple and on the Moon, even though there is no contact, and the two objects may even be very far apart.

Newton set about determining the magnitude of the gravitational force that the Earth exerts on the Moon as compared to the gravitational force on objects at the Earth's surface. At the surface of the Earth, the force of gravity accelerates objects at 9.80 m/s^2. The centripetal acceleration of the Moon is calculated from $a_R = v^2/r$ (see Example 5–9) and gives $a_R = 0.00272 \text{ m/s}^2$. In terms of the acceleration of gravity at the Earth's surface, g, this is equivalent to

$$a_R = \frac{0.00272 \text{ m/s}^2}{9.80 \text{ m/s}^2} g \approx \frac{1}{3600} g.$$

That is, the acceleration of the Moon toward the Earth is about $\frac{1}{3600}$ as great as the acceleration of objects at the Earth's surface. The Moon is 384,000 km from the Earth, which is about 60 times the Earth's radius of 6380 km. That is, the Moon is 60 times farther from the Earth's center than are objects at the Earth's surface. But $60 \times 60 = 60^2 = 3600$. Again that number 3600! Newton concluded that the gravitational force F exerted by the Earth on any object decreases with the square of its distance, r, from the Earth's center:

$$F \propto \frac{1}{r^2}.$$

The Moon is 60 Earth radii away, so it feels a gravitational force only $\frac{1}{60^2} = \frac{1}{3600}$ times as strong as it would if it were a point at the Earth's surface.

Newton realized that the force of gravity on an object depends not only on distance but also on the object's mass. In fact, it is directly proportional to its mass, as we have seen. According to Newton's third law, when the Earth exerts its gravitational force on any object, such as the Moon, that object exerts an equal and opposite force on the Earth (Fig. 6–2). Because of this *symmetry*, Newton reasoned, the magnitude of the force of gravity must be proportional to *both* the masses. Thus

$$F \propto \frac{m_E m_B}{r^2},$$

where m_E is the mass of the Earth, m_B the mass of the other object, and r the distance from the Earth's center to the center of the other object.

FIGURE 6–2 The gravitational force one object exerts on a second object is directed toward the first object, and is equal and opposite to the force exerted by the second object on the first.

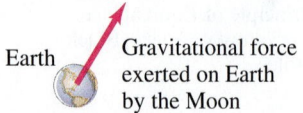

Newton went a step further in his analysis of gravity. In his examination of the orbits of the planets, he concluded that the force required to hold the different planets in their orbits around the Sun seems to diminish as the inverse square of their distance from the Sun. This led him to believe that it is also the gravitational force that acts between the Sun and each of the planets to keep them in their orbits. And if gravity acts between these objects, why not between all objects? Thus he proposed his **law of universal gravitation**, which we can state as follows:

> **Every particle in the universe attracts every other particle with a force that is proportional to the product of their masses and inversely proportional to the square of the distance between them. This force acts along the line joining the two particles.**

The magnitude of the gravitational force can be written as

$$F = G\frac{m_1 m_2}{r^2}, \tag{6–1}$$

NEWTON'S

LAW

OF

UNIVERSAL

GRAVITATION

where m_1 and m_2 are the masses of the two particles, r is the distance between them, and G is a universal constant which must be measured experimentally and has the same numerical value for all objects.

The value of G must be very small, since we are not aware of any force of attraction between ordinary-sized objects, such as between two baseballs. The force between two ordinary objects was first measured by Henry Cavendish in 1798, over 100 years after Newton published his law. To detect and measure the incredibly small force between ordinary objects, he used an apparatus like that shown in Fig. 6–3. Cavendish confirmed Newton's hypothesis that two objects attract one another and that Eq. 6–1 accurately describes this force. In addition, because Cavendish could measure F, m_1, m_2, and r accurately, he was able to determine the value of the constant G as well. The accepted value today is

$$G = 6.67 \times 10^{-11}\,\text{N}\cdot\text{m}^2/\text{kg}^2.$$

(See Table inside front cover for values of all constants to highest known precision.)

Strictly speaking, Eq. 6–1 gives the magnitude of the gravitational force that one particle exerts on a second particle that is a distance r away. For an extended object (that is, not a point), we must consider how to measure the distance r. You might think that r would be the distance between the centers of the objects. This is true for two spheres, and is often a good approximation for other objects. A correct calculation treats each extended body as a collection of particles, and the total force is the sum of the forces due to all the particles. The sum over all these particles is often best done using integral calculus, which Newton himself invented. When extended bodies are small compared to the distance between them (as for the Earth–Sun system), little inaccuracy results from considering them as point particles.

Newton was able to show (see derivation in Appendix D) that the *gravitational force exerted on a particle outside a sphere, with a spherically symmetric mass distribution, is the same as if the entire mass of the sphere was concentrated at its center.* Thus Eq. 6–1 gives the correct force between two uniform spheres where r is the distance between their centers.

FIGURE 6–3 Schematic diagram of Cavendish's apparatus. Two spheres are attached to a light horizontal rod, which is suspended at its center by a thin fiber. When a third sphere (labeled A) is brought close to one of the suspended spheres, the gravitational force causes the latter to move, and this twists the fiber slightly. The tiny movement is magnified by the use of a narrow light beam directed at a mirror mounted on the fiber. The beam reflects onto a scale. Previous determination of how large a force will twist the fiber a given amount then allows the experimenter to determine the magnitude of the gravitational force between two objects.

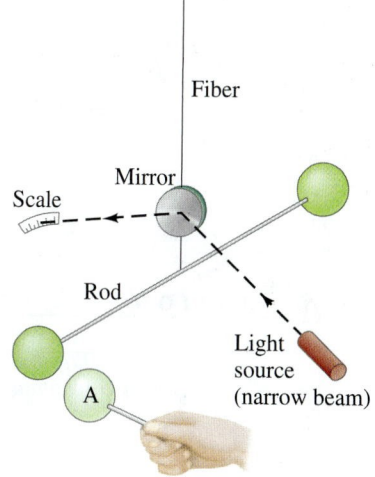

EXAMPLE 6–1 ESTIMATE Can you attract another person gravitationally?
A 50-kg person and a 70-kg person are sitting on a bench close to each other. Estimate the magnitude of the gravitational force each exerts on the other.

APPROACH This is an estimate: we let the distance between the centers of the two people be $\frac{1}{2}$ m (about as close as you can get).

SOLUTION We use Eq. 6–1, which gives

$$F = \frac{(6.67 \times 10^{-11}\,\text{N}\cdot\text{m}^2/\text{kg}^2)(50\,\text{kg})(70\,\text{kg})}{(0.5\,\text{m})^2} \approx 10^{-6}\,\text{N},$$

rounded off to an order of magnitude. Such a force is unnoticeably small unless extremely sensitive instruments are used.

NOTE As a fraction of their weight, this force is $(10^{-6}\,\text{N})/(70\,\text{kg})(9.8\,\text{m/s}^2) \approx 10^{-9}$.

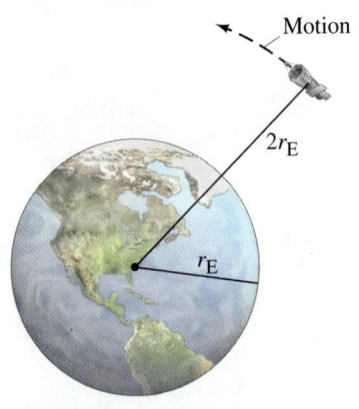

Motion

$2r_E$

r_E

FIGURE 6–4 Example 6–2.

FIGURE 6–5 Example 6–3. Orientation of Sun (S), Earth (E), and Moon (M) at right angles to each other (not to scale).

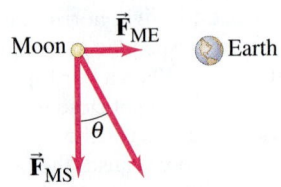

Moon $\vec{F}_{ME}$ Earth

θ

$\vec{F}_{MS}$

Sun

EXAMPLE 6–2 **Spacecraft at $2r_E$.** What is the force of gravity acting on a 2000-kg spacecraft when it orbits two Earth radii from the Earth's center (that is, a distance $r_E = 6380$ km above the Earth's surface, Fig. 6–4)? The mass of the Earth is $m_E = 5.98 \times 10^{24}$ kg.

APPROACH We could plug all the numbers into Eq. 6–1, but there is a simpler approach. The spacecraft is twice as far from the Earth's center as when it is at the surface of the Earth. Therefore, since the force of gravity decreases as the square of the distance (and $\frac{1}{2^2} = \frac{1}{4}$), the force of gravity on the satellite will be only one-fourth its weight at the Earth's surface.

SOLUTION At the surface of the Earth, $F_G = mg$. At a distance from the Earth's center of $2r_E$, F_G is $\frac{1}{4}$ as great:

$$F_G = \tfrac{1}{4}mg = \tfrac{1}{4}(2000\,\text{kg})(9.80\,\text{m/s}^2) = 4900\,\text{N}.$$

EXAMPLE 6–3 **Force on the Moon.** Find the net force on the Moon $(m_M = 7.35 \times 10^{22}\,\text{kg})$ due to the gravitational attraction of both the Earth $(m_E = 5.98 \times 10^{24}\,\text{kg})$ and the Sun $(m_S = 1.99 \times 10^{30}\,\text{kg})$, assuming they are at right angles to each other as in Fig. 6–5.

APPROACH The forces on our object, the Moon, are the gravitational force exerted on the Moon by the Earth F_{ME} and the force exerted by the Sun F_{MS}, as shown in the free-body diagram of Fig. 6–5. We use the law of universal gravitation to find the magnitude of each force, and then add the two forces as vectors.

SOLUTION The Earth is 3.84×10^5 km $= 3.84 \times 10^8$ m from the Moon, so F_{ME} (the gravitational force on the Moon due to the Earth) is

$$F_{ME} = \frac{(6.67 \times 10^{-11}\,\text{N}\cdot\text{m}^2/\text{kg}^2)(7.35 \times 10^{22}\,\text{kg})(5.98 \times 10^{24}\,\text{kg})}{(3.84 \times 10^8\,\text{m})^2} = 1.99 \times 10^{20}\,\text{N}.$$

The Sun is 1.50×10^8 km from the Earth and the Moon, so F_{MS} (the gravitational force on the Moon due to the Sun) is

$$F_{MS} = \frac{(6.67 \times 10^{-11}\,\text{N}\cdot\text{m}^2/\text{kg}^2)(7.35 \times 10^{22}\,\text{kg})(1.99 \times 10^{30}\,\text{kg})}{(1.50 \times 10^{11}\,\text{m})^2} = 4.34 \times 10^{20}\,\text{N}.$$

The two forces act at right angles in the case we are considering (Fig. 6–5), so we can apply the Pythagorean theorem to find the magnitude of the total force:

$$F = \sqrt{(1.99 \times 10^{20}\,\text{N})^2 + (4.34 \times 10^{20}\,\text{N})^2} = 4.77 \times 10^{20}\,\text{N}.$$

The force acts at an angle θ (Fig. 6–5) given by $\theta = \tan^{-1}(1.99/4.34) = 24.6°$.

NOTE The two forces, F_{ME} and F_{MS}, have the same order of magnitude $(10^{20}\,\text{N})$. This may be surprising. Is it reasonable? The Sun is much farther from Earth than the Moon (a factor of $10^{11}\,\text{m}/10^8\,\text{m} \approx 10^3$), but the Sun is also much more massive (a factor of $10^{30}\,\text{kg}/10^{23}\,\text{kg} \approx 10^7$). Mass divided by distance squared $(10^7/10^6)$ comes out within an order of magnitude, and we have ignored factors of 3 or more. Yes, it is reasonable.

Note carefully that the law of universal gravitation describes a *particular* force (gravity), whereas Newton's second law of motion ($F = ma$) tells how an object accelerates due to *any* type of force.

*Spherical Shells

Newton was able to show, using the calculus he invented for the purpose, that a thin uniform spherical shell exerts a force on a particle *outside* it as if all the shell's mass were at its center; and that such a thin uniform shell exerts *zero* force on a particle *inside* the shell. (The derivation is given in Appendix D.) The Earth can be modelled as a series of concentric shells starting at its center, each shell uniform but perhaps having a different density to take into account Earth's varying density in various layers. As a simple example, suppose the Earth were uniform throughout; what is the gravitational force on a particle placed exactly halfway from Earth's center to its surface? Only the mass *inside* this radius $r = \frac{1}{2} r_E$ would exert a net force on this particle. The mass of a sphere is proportional to its volume $V = \frac{4}{3} \pi r^3$, so the mass m inside $r = \frac{1}{2} r_E$ is $\left(\frac{1}{2}\right)^3 = \frac{1}{8}$ the mass of the entire Earth. The gravitational force on the particle at $r = \frac{1}{2} r_E$, which is proportional to m/r^2 (Eq. 6–1), is reduced to $\left(\frac{1}{8}\right)/\left(\frac{1}{2}\right)^2 = \frac{1}{2}$ the gravitational force it would experience at Earth's surface.

6–2 Vector Form of Newton's Law of Universal Gravitation

We can write Newton's law of universal gravitation in vector form as

$$\vec{F}_{12} = -G \frac{m_1 m_2}{r_{21}^2} \hat{r}_{21}, \tag{6–2}$$

where $\vec{F}_{12}$ is the vector force on particle 1 (of mass m_1) exerted by particle 2 (of mass m_2), which is a distance r_{21} away; $\hat{r}_{21}$ is a unit vector that points from particle 2 toward particle 1 along the line joining them so that $\hat{r}_{21} = \vec{r}_{21}/r_{21}$, where $\vec{r}_{21}$ is the displacement vector as shown in Fig. 6–6. The minus sign in Eq. 6–2 is necessary because the force on particle 1 due to particle 2 points toward m_2, in the direction opposite to $\hat{r}_{21}$. The displacement vector $\vec{r}_{12}$ is a vector of the same magnitude as $\vec{r}_{21}$, but it points in the opposite direction so that

$$\vec{r}_{12} = -\vec{r}_{21}.$$

By Newton's third law, the force $\vec{F}_{21}$ acting on m_2 exerted by m_1 must have the same magnitude as $\vec{F}_{12}$ but acts in the opposite direction (Fig. 6–7), so that

$$\vec{F}_{21} = -\vec{F}_{12} = G \frac{m_1 m_2}{r_{21}^2} \hat{r}_{21}$$

$$= -G \frac{m_2 m_1}{r_{12}^2} \hat{r}_{12}.$$

The force of gravity exerted on one particle by a second particle is always directed toward the second particle, as in Fig. 6–6. When many particles interact, the total gravitational force on a given particle is the vector sum of the forces exerted by each of the others. For example, the total force on particle number 1 is

$$\vec{F}_1 = \vec{F}_{12} + \vec{F}_{13} + \vec{F}_{14} + \cdots + \vec{F}_{1n} = \sum_{i=2}^{n} \vec{F}_{1i} \tag{6–3}$$

where $\vec{F}_{1i}$ means the force on particle 1 exerted by particle i, and n is the total number of particles.

This vector notation can be very helpful, especially when sums over many particles are needed. However, in many cases we do not need to be so formal and we can deal with directions by making careful diagrams.

6–3 Gravity Near the Earth's Surface; Geophysical Applications

When Eq. 6–1 is applied to the gravitational force between the Earth and an object at its surface, m_1 becomes the mass of the Earth m_E, m_2 becomes the mass of the object m, and r becomes the distance of the object from the Earth's center, which is the radius of the Earth r_E. This force of gravity due to the Earth is the weight of the object, which we have been writing as mg. Thus,

$$mg = G \frac{m m_E}{r_E^2}.$$

We can solve this for g, the acceleration of gravity at the Earth's surface:

$$g = G \frac{m_E}{r_E^2}. \tag{6–4}$$

Thus, the acceleration of gravity at the surface of the Earth, g, is determined by m_E and r_E. (Don't confuse G with g; they are very different quantities, but are related by Eq. 6–4.)

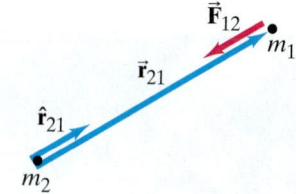

FIGURE 6–6 The displacement vector $\vec{r}_{21}$ points from particle of mass m_2 to particle of mass m_1. The unit vector shown, $\hat{r}_{21}$ is in the same direction as $\vec{r}_{21}$, but is defined as having length one.

FIGURE 6–7 By Newton's third law, the gravitational force on particle 1 exerted by particle 2, $\vec{F}_{12}$, is equal and opposite to that on particle 2 exerted by particle 1, $\vec{F}_{21}$; that is $\vec{F}_{21} = -\vec{F}_{12}$.

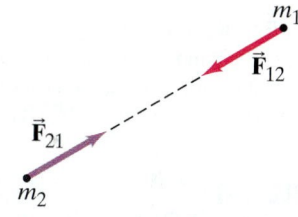

⚠ **CAUTION**

Distinguish G from g

Until G was measured, the mass of the Earth was not known. But once G was measured, Eq. 6–4 could be used to calculate the Earth's mass, and Cavendish was the first to do so. Since $g = 9.80 \text{ m/s}^2$ and the radius of the Earth is $r_E = 6.38 \times 10^6 \text{ m}$, then, from Eq. 6–4, we obtain

$$m_E = \frac{gr_E^2}{G} = \frac{(9.80 \text{ m/s}^2)(6.38 \times 10^6 \text{ m})^2}{6.67 \times 10^{-11} \text{ N}\cdot\text{m}^2/\text{kg}^2}$$

$$= 5.98 \times 10^{24} \text{ kg}$$

for the mass of the Earth.

Equation 6–4 can be applied to other planets, where g, m, and r would refer to that planet.

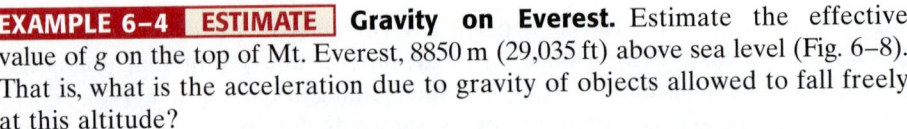

EXAMPLE 6–4 **ESTIMATE** **Gravity on Everest.** Estimate the effective value of g on the top of Mt. Everest, 8850 m (29,035 ft) above sea level (Fig. 6–8). That is, what is the acceleration due to gravity of objects allowed to fall freely at this altitude?

APPROACH The force of gravity (and the acceleration due to gravity g) depends on the distance from the center of the Earth, so there will be an effective value g' on top of Mt. Everest which will be smaller than g at sea level. We assume the Earth is a uniform sphere (a reasonable "estimate").

SOLUTION We use Eq. 6–4, with r_E replaced by $r = 6380 \text{ km} + 8.9 \text{ km} = 6389 \text{ km} = 6.389 \times 10^6 \text{ m}$:

$$g = G\frac{m_E}{r^2} = \frac{(6.67 \times 10^{-11} \text{ N}\cdot\text{m}^2/\text{kg}^2)(5.98 \times 10^{24} \text{ kg})}{(6.389 \times 10^6 \text{ m})^2}$$

$$= 9.77 \text{ m/s}^2,$$

which is a reduction of about 3 parts in a thousand (0.3%).

NOTE This is an estimate because, among other things, we ignored the mass accumulated under the mountaintop.

FIGURE 6–8 Example 6–4. Mount Everest, 8850 m (29,035 ft) above sea level; in the foreground, the author with sherpas at 5500 m (18,000 ft).

TABLE 6–1
Acceleration Due to Gravity at Various Locations on Earth

Location	Elevation (m)	g (m/s²)
New York	0	9.803
San Francisco	0	9.800
Denver	1650	9.796
Pikes Peak	4300	9.789
Sydney, Australia	0	9.798
Equator	0	9.780
North Pole (calculated)	0	9.832

🚶 PHYSICS APPLIED
Geology—mineral and oil exploration

Note that Eq. 6–4 does not give precise values for g at different locations because the Earth is not a perfect sphere. The Earth not only has mountains and valleys, and bulges at the equator, but also its mass is not distributed precisely uniformly (see Table 6–1). The Earth's rotation also affects the value of g (see Example 6–5). However, for most practical purposes when an object is near the Earth's surface, we will simply use $g = 9.80 \text{ m/s}^2$ and write the weight of an object as mg.

EXERCISE A Suppose you could double the mass of a planet but kept its volume the same. How would the acceleration of gravity, g, at the surface change?

The value of g can vary locally on the Earth's surface because of the presence of irregularities and rocks of different densities. Such variations in g, known as "gravity anomalies," are very small—on the order of 1 part per 10^6 or 10^7 in the value of g. But they can be measured ("gravimeters" today can detect variations in g to 1 part in 10^9). Geophysicists use such measurements as part of their investigations into the structure of the Earth's crust, and in mineral and oil exploration. Mineral deposits, for example, often have a greater density than surrounding material; because of the greater mass in a given volume, g can have a slightly greater value on top of such a deposit than at its flanks. "Salt domes," under which petroleum is often found, have a lower than average density and searches for a slight reduction in the value of g in certain locales have led to the discovery of oil.

EXAMPLE 6–5 **Effect of Earth's rotation on g.** Assuming the Earth is a perfect sphere, determine how the Earth's rotation affects the value of g at the equator compared to its value at the poles.

APPROACH Figure 6–9 shows a person of mass m standing on a doctor's scale at two places on the Earth. At the North Pole there are two forces acting on the mass m: the force of gravity, $\vec{F}_G = m\vec{g}$, and the force with which the scale pushes up on the mass, $\vec{w}$. We call this latter force w because it is what the scale reads as the weight of the object, and by Newton's third law it equals the force with which the mass pushes down on the scale. Since the mass is not accelerating, Newton's second law tells us

$$mg - w = 0,$$

so $w = mg$. Thus the weight w that the spring registers equals mg, which is no surprise. Next, at the equator, there *is* an acceleration because the Earth is rotating. The same magnitude of the force of gravity $F_G = mg$ acts downward (we are letting g represent the acceleration of gravity in the absence of rotation and we ignore the slight bulging of the equator). The scale pushes upward with a force w'; w' is also the force with which the person pushes down on the scale (Newton's third law) and hence is the weight registered on the scale. From Newton's second law we now have (see Fig. 6–9)

$$mg - w' = m\frac{v^2}{r_E},$$

because the person of mass m now has a centripetal acceleration due to Earth's rotation; $r_E = 6.38 \times 10^6$ m is the Earth's radius and v is the speed of m due to the Earth's daily rotation.

SOLUTION First we determine the speed v of an object at rest on the Earth's equator, rembering that Earth makes one rotation (distance = circumference of Earth = $2\pi r_E$) in 1 day = $(24\text{ h})(60\text{ min/h})(60\text{ s/min}) = 8.64 \times 10^4$ s:

$$v = \frac{2\pi r_E}{1\text{ day}} = \frac{(6.283)(6.38 \times 10^6\text{ m})}{(8.64 \times 10^4\text{ s})}$$
$$= 4.640 \times 10^2\text{ m/s}.$$

The effective weight is $w' = mg'$ where g' is the effective value of g, and so $g' = w'/m$. Solving the equation above for w', we have

$$w' = m\left(g - \frac{v^2}{r_E}\right),$$

so

$$g' = \frac{w'}{m} = g - \frac{v^2}{r_E}.$$

Hence

$$\Delta g = g - g' = \frac{v^2}{r_E} = \frac{(4.640 \times 10^2\text{ m/s})^2}{(6.38 \times 10^6\text{ m})}$$
$$= 0.0337\text{ m/s}^2,$$

which is about $\Delta g \approx 0.003g$, a difference of 0.3%.

NOTE In Table 6–1 we see that the difference in g at the pole and equator is actually greater than this: $(9.832 - 9.780)\text{ m/s}^2 = 0.052\text{ m/s}^2$. This discrepancy is due mainly to the Earth being slightly fatter at the equator (by 21 km) than at the poles.

NOTE The calculation of the effective value of g at latitudes other than at the poles or equator is a two-dimensional problem because $\vec{F}_G$ acts radially toward the Earth's center whereas the centripetal acceleration is directed perpendicular to the axis of rotation, parallel to the equator and that means that a plumb line (the effective direction of g) is not precisely vertical except at the equator and the poles.

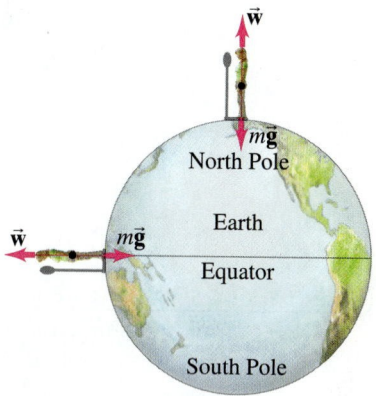

FIGURE 6–9 Example 6–5.

Earth as Inertial Reference Frame

We often make the assumption that reference frames fixed on the Earth are inertial reference frames. Our calculation in Example 6–5 above shows that this assumption can result in errors no larger than about 0.3% in the use of Newton's second law, for example. We discuss the effects of Earth's rotation and reference frames in more detail in Chapter 11, including the Coriolis effect.

6–4 Satellites and "Weightlessness"

Satellite Motion

Artificial satellites circling the Earth are now commonplace (Fig. 6–10). A satellite is put into orbit by accelerating it to a sufficiently high tangential speed with the use of rockets, as shown in Fig. 6–11. If the speed is too high, the spacecraft will not be confined by the Earth's gravity and will escape, never to return. If the speed is too low, it will return to Earth. Satellites are usually put into circular (or nearly circular) orbits, because such orbits require the least takeoff speed.

FIGURE 6–10 A satellite, the International Space Station, circling the Earth.

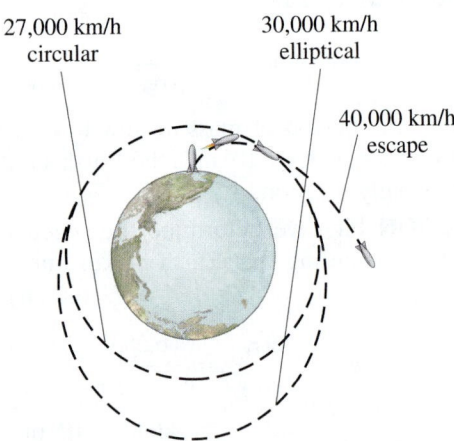

FIGURE 6–11 Artificial satellites launched at different speeds.

It is sometimes asked: "What keeps a satellite up?" The answer is: its high speed. If a satellite stopped moving, it would fall directly to Earth. But at the very high speed a satellite has, it would quickly fly out into space (Fig. 6–12) if it weren't for the gravitational force of the Earth pulling it into orbit. In fact, a satellite *is* falling (accelerating toward Earth), but its high tangential speed keeps it from hitting Earth.

For satellites that move in a circle (at least approximately), the needed acceleration is centripetal and equals v^2/r. The force that gives a satellite this acceleration is the force of gravity exerted by the Earth, and since a satellite may be at a considerable distance from the Earth, we must use Newton's law of universal gravitation (Eq. 6–1) for the force acting on it. When we apply Newton's second law, $\Sigma F_R = ma_R$ in the radial direction, we find

$$G\frac{mm_E}{r^2} = m\frac{v^2}{r},$$

(6–5)

where m is the mass of the satellite. This equation relates the distance of the satellite from the Earth's center, r, to its speed, v, in a circular orbit. Note that only one force—gravity—is acting on the satellite, and that r is the sum of the Earth's radius r_E plus the satellite's height h above the Earth: $r = r_E + h$.

FIGURE 6–12 A moving satellite "falls" out of a straight-line path toward the Earth.

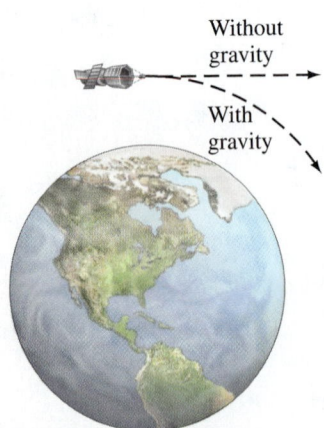

Without gravity

With gravity

EXAMPLE 6–6 **Geosynchronous satellite.** A *geosynchronous* satellite is one that stays above the same point on the Earth, which is possible only if it is above a point on the equator. Such satellites are used for TV and radio transmission, for weather forecasting, and as communication relays. Determine (a) the height above the Earth's surface such a satellite must orbit, and (b) such a satellite's speed. (c) Compare to the speed of a satellite orbiting 200 km above Earth's surface.

PHYSICS APPLIED
Geosynchronous satellites

APPROACH To remain above the same point on Earth as the Earth rotates, the satellite must have a period of 24 hours. We can apply Newton's second law, $F = ma$, where $a = v^2/r$ if we assume the orbit is circular.

SOLUTION (a) The only force on the satellite is the force of universal gravitation due to the Earth. (We can ignore the gravitational force exerted by the Sun. Why?) We apply Eq. 6–5, assuming the satellite moves in a circle:

$$G\frac{m_{\text{Sat}}\, m_{\text{E}}}{r^2} = m_{\text{Sat}}\frac{v^2}{r}.$$

This equation has two unknowns, r and v. But the satellite revolves around the Earth with the same period that the Earth rotates on its axis, namely once in 24 hours. Thus the speed of the satellite must be

$$v = \frac{2\pi r}{T},$$

where $T = 1\,\text{day} = (24\,\text{h})(3600\,\text{s/h}) = 86{,}400\,\text{s}$. We substitute this into the "satellite equation" above and obtain (after canceling m_{Sat} on both sides)

$$G\frac{m_{\text{E}}}{r^2} = \frac{(2\pi r)^2}{rT^2}.$$

After cancelling an r, we can solve for r^3:

$$r^3 = \frac{Gm_{\text{E}}T^2}{4\pi^2} = \frac{(6.67 \times 10^{-11}\,\text{N·m}^2/\text{kg}^2)(5.98 \times 10^{24}\,\text{kg})(86{,}400\,\text{s})^2}{4\pi^2}$$
$$= 7.54 \times 10^{22}\,\text{m}^3.$$

We take the cube root and find

$$r = 4.23 \times 10^7\,\text{m},$$

or 42,300 km from the Earth's center. We subtract the Earth's radius of 6380 km to find that a geosynchronous satellite must orbit about 36,000 km (about $6\,r_{\text{E}}$) above the Earth's surface.

(b) We solve for v in the satellite equation, Eq. 6–5:

$$v = \sqrt{\frac{Gm_{\text{E}}}{r}} = \sqrt{\frac{(6.67 \times 10^{-11}\,\text{N·m}^2/\text{kg}^2)(5.98 \times 10^{24}\,\text{kg})}{(4.23 \times 10^7\,\text{m})}} = 3070\,\text{m/s}.$$

We get the same result if we use $v = 2\pi r/T$.

(c) The equation in part (b) for v shows $v \propto \sqrt{1/r}$. So for $r = r_{\text{E}} + h = 6380\,\text{km} + 200\,\text{km} = 6580\,\text{km}$, we get

$$v' = v\sqrt{\frac{r}{r'}} = (3070\,\text{m/s})\sqrt{\frac{(42{,}300\,\text{km})}{(6580\,\text{km})}} = 7780\,\text{m/s}.$$

NOTE The center of a satellite orbit is always at the center of the Earth; so it is not possible to have a satellite orbiting above a fixed point on the Earth at any latitude other than $0°$.

CONCEPTUAL EXAMPLE 6–7 **Catching a satellite.** You are an astronaut in the space shuttle pursuing a satellite in need of repair. You find yourself in a circular orbit of the same radius as the satellite, but 30 km behind it. How will you catch up with it?

RESPONSE We saw in Example 6–6 (or see Eq. 6–5) that the velocity is proportional to $1/\sqrt{r}$. Thus you need to aim for a smaller orbit in order to increase your speed. Note that you cannot just increase your speed without changing your orbit. After passing the satellite, you will need to slow down and rise upward again.

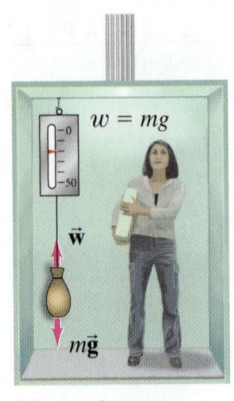

(a) $\vec{\mathbf{a}} = 0$

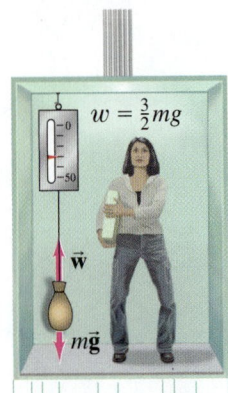

(b) $\uparrow$ $\vec{\mathbf{a}} = -\frac{1}{2}\vec{\mathbf{g}}$ (up)

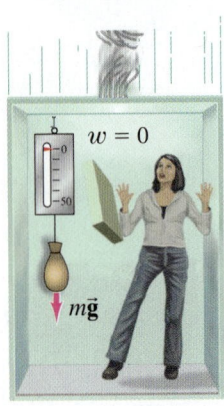

(c) $\downarrow$ $\vec{\mathbf{a}} = \vec{\mathbf{g}}$ (down)

FIGURE 6–13 (a) An object in an elevator at rest exerts a force on a spring scale equal to its weight. (b) In an elevator accelerating upward at $\frac{1}{2}g$, the object's apparent weight is $1\frac{1}{2}$ times larger than its true weight. (c) In a freely falling elevator, the object experiences "weightlessness": the scale reads zero.

EXERCISE B Two satellites orbit the Earth in circular orbits of the same radius. One satellite is twice as massive as the other. Which of the following statements is true about the speeds of these satellites? (a) The heavier satellite moves twice as fast as the lighter one. (b) The two satellites have the same speed. (c) The lighter satellite moves twice as fast as the heavier one. (d) The heavier satellite moves four times as fast as the lighter one.

Weightlessness

People and other objects in a satellite circling the Earth are said to experience apparent weightlessness. Let us first look at a simpler case, that of a falling elevator. In Fig. 6–13a, an elevator is at rest with a bag hanging from a spring scale. The scale reading indicates the downward force exerted on it by the bag. This force, exerted *on* the scale, is equal and opposite to the force exerted *by* the scale upward on the bag, and we call its magnitude w. Two forces act on the bag: the downward gravitational force and the upward force exerted by the scale equal to w. Because the bag is not accelerating ($a = 0$) when we apply $\Sigma F = ma$ to the bag in Fig. 6–13a we obtain

$$w - mg = 0,$$

where mg is the weight of the bag. Thus, $w = mg$, and since the scale indicates the force w exerted on it by the bag, it registers a force equal to the weight of the bag, as we expect.

Now let the elevator have an acceleration, a. Applying Newton's second law, $\Sigma F = ma$, to the bag as seen from an inertial reference frame (the elevator itself is not an inertial frame) we have

$$w - mg = ma.$$

Solving for w, we have

$$w = mg + ma. \qquad\qquad [a \text{ is } + \text{ upward}]$$

We have chosen the positive direction up. Thus, if the acceleration a is up, a is positive; and the scale, which measures w, will read more than mg. We call w the *apparent weight* of the bag, which in this case would be greater than its actual weight (mg). If the elevator accelerates downward, a will be negative and w, the apparent weight, will be less than mg. The direction of the velocity $\vec{\mathbf{v}}$ doesn't matter. Only the direction of the acceleration $\vec{\mathbf{a}}$ (and its magnitude) influences the scale reading.

Suppose, for example, the elevator's acceleration is $\frac{1}{2}g$ upward; then we find

$$w = mg + m\left(\tfrac{1}{2}g\right) = \tfrac{3}{2}mg.$$

That is, the scale reads $1\frac{1}{2}$ times the actual weight of the bag (Fig. 6–13b). The apparent weight of the bag is $1\frac{1}{2}$ times its real weight. The same is true of the person: her apparent weight (equal to the normal force exerted on her by the elevator floor) is $1\frac{1}{2}$ times her real weight. We can say that she is experiencing $1\frac{1}{2}g$'s, just as astronauts experience so many g's at a rocket's launch.

If, instead, the elevator's acceleration is $a = -\frac{1}{2}g$ (downward), then $w = mg - \frac{1}{2}mg = \frac{1}{2}mg$. That is, the scale reads half the actual weight. If the elevator is in *free fall* (for example, if the cables break), then $a = -g$ and $w = mg - mg = 0$. The scale reads zero. See Fig. 6–13c. The bag appears weightless. If the person in the elevator accelerating at $-g$ let go of a pencil, say, it would not fall to the floor. True, the pencil would be falling with acceleration g. But so would the floor of the elevator and the person. The pencil would hover right in front of the person. This phenomenon is called *apparent weightlessness* because in the reference frame of the person, objects don't fall or seem to have weight—yet gravity does not disappear. Gravity is still acting on each object, whose weight is still mg. The person and other objects seem weightless only because the elevator is accelerating in free fall, and there is no contact force on the person to make her feel the weight.

The "weightlessness" experienced by people in a satellite orbit close to the Earth (Fig. 6–14) is the same apparent weightlessness experienced in a freely falling elevator. It may seem strange, at first, to think of a satellite as freely falling. But a satellite is indeed falling toward the Earth, as was shown in Fig. 6–12. The force of gravity causes it to "fall" out of its natural straight-line path. The acceleration of the satellite must be the acceleration due to gravity at that point, since the only force acting on it is gravity. (We used this to obtain Eq. 6–5.) Thus, although the force of gravity acts on objects within the satellite, the objects experience an apparent weightlessness because they, and the satellite, are accelerating together as in free fall.

EXERCISE C Return to the Chapter-Opening Question, page 139, and answer it again now. Try to explain why you may have answered differently the first time.

Figure 6–15 shows some examples of "free fall," or apparent weightlessness, experienced by people on Earth for brief moments.

A completely different situation occurs if a spacecraft is out in space far from the Earth, the Moon, and other attracting bodies. The force of gravity due to the Earth and other heavenly bodies will then be quite small because of the distances involved, and persons in such a spacecraft would experience real weightlessness.

EXERCISE D Could astronauts in a spacecraft far out in space easily play catch with a bowling ball ($m = 7\,\text{kg}$)?

FIGURE 6–14 This astronaut is moving outside the International Space Station. He must feel very free because he is experiencing apparent weightlessness.

FIGURE 6–15 Experiencing "weightlessness" on Earth.

(a)

(b)

(c)

6–5 Kepler's Laws and Newton's Synthesis

More than a half century before Newton proposed his three laws of motion and his law of universal gravitation, the German astronomer Johannes Kepler (1571–1630) had worked out a detailed description of the motion of the planets about the Sun. Kepler's work resulted in part from the many years he spent examining data collected by Tycho Brahe (1546–1601) on the positions of the planets in their motion through the heavens.

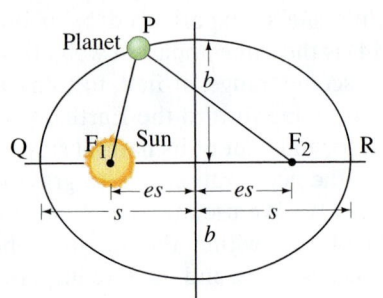

FIGURE 6–16 *Kepler's first law.* An ellipse is a closed curve such that the sum of the distances from any point P on the curve to two fixed points (called the foci, F_1 and F_2) remains constant. That is, the sum of the distances, $F_1P + F_2P$, is the same for all points on the curve. A circle is a special case of an ellipse in which the two foci coincide, at the center of the circle. The semimajor axis is s (that is, the long axis is $2s$) and the semiminor axis is b, as shown. The *eccentricity, e,* is defined as the ratio of the distance from either focus to the center divided by the semimajor axis s. Thus es is the distance from the center to either focus, as shown. For a circle, $e = 0$. The Earth and most of the other planets have nearly circular orbits. For Earth $e = 0.017$.

Among Kepler's writings were three empirical findings that we now refer to as **Kepler's laws of planetary motion**. These are summarized as follows, with additional explanation in Figs. 6–16 and 6–17.

Kepler's first law: The path of each planet about the Sun is an ellipse with the Sun at one focus (Fig. 6–16).

Kepler's second law: Each planet moves so that an imaginary line drawn from the Sun to the planet sweeps out equal areas in equal periods of time (Fig. 6–17).

Kepler's third law: The ratio of the squares of the periods of any two planets revolving about the Sun is equal to the ratio of the cubes of their semimajor axes. [The semimajor axis is half the long (major) axis of the orbit, as shown in Fig. 6–16, and represents the planet's mean distance from the Sun.[†]] That is, if T_1 and T_2 represent the periods (the time needed for one revolution about the Sun) for any two planets, and s_1 and s_2 represent their semimajor axes, then

$$\left(\frac{T_1}{T_2}\right)^2 = \left(\frac{s_1}{s_2}\right)^3.$$

We can rewrite this as

$$\frac{s_1^3}{T_1^2} = \frac{s_2^3}{T_2^2},$$

meaning that s^3/T^2 should be the same for each planet. Present-day data are given in Table 6–2; see the last column.

Kepler arrived at his laws through careful analysis of experimental data. Fifty years later, Newton was able to show that Kepler's laws could be derived mathematically from the law of universal gravitation and the laws of motion. He also showed that for any reasonable form for the gravitational force law, only one that depends on the inverse square of the distance is fully consistent with Kepler's laws. He thus used Kepler's laws as evidence in favor of his law of universal gravitation, Eq. 6–1.

We will derive Kepler's second law later, in Chapter 11. Here we derive Kepler's third law, and we do it for the special case of a circular orbit, in which case the semimajor axis is the radius r of the circle. (Most planetary orbits are close to a circle.) First, we write Newton's second law of motion, $\Sigma F = ma$. For F we use the law of universal gravitation (Eq. 6–1) for the force between the Sun and a planet of mass m_1, and for a the centripetal acceleration, v^2/r. We assume the mass of the Sun M_S is much greater than the mass of its planets, so we ignore the effects of the planets on each other. Then

$$\Sigma F = ma$$

$$G\frac{m_1 M_S}{r_1^2} = m_1\frac{v_1^2}{r_1}.$$

FIGURE 6–17 *Kepler's second law.* The two shaded regions have equal areas. The planet moves from point 1 to point 2 in the same time as it takes to move from point 3 to point 4. Planets move fastest in that part of their orbit where they are closest to the Sun. Exaggerated scale.

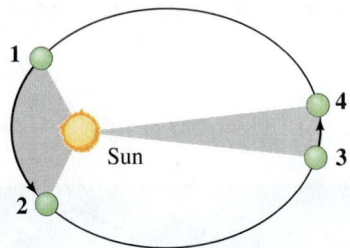

TABLE 6–2 Planetary Data Applied to Kepler's Third Law

Planet	Mean Distance from Sun, s (10^6 km)	Period, T (Earth yr)	s^3/T^2 $\left(10^{24}\frac{km^3}{yr^2}\right)$
Mercury	57.9	0.241	3.34
Venus	108.2	0.615	3.35
Earth	149.6	1.0	3.35
Mars	227.9	1.88	3.35
Jupiter	778.3	11.86	3.35
Saturn	1427	29.5	3.34
Uranus	2870	84.0	3.35
Neptune	4497	165	3.34
Pluto	5900	248	3.34

[†]The semimajor axis is equal to the planet's mean distance from the Sun in the sense that it equals half the sum of the planet's nearest and farthest distances from the Sun (points Q and R in Fig. 6–16). Most planetary orbits are close to circles, and for a circle the semimajor axis is the radius of the circle.

Here m_1 is the mass of a particular planet, r_1 its distance from the Sun, and v_1 its average speed in orbit; M_S is the mass of the Sun, since it is the gravitational attraction of the Sun that keeps each planet in its orbit. The period T_1 of the planet is the time required for one complete orbit, which is a distance equal to $2\pi r_1$, the circumference of a circle. Thus

$$v_1 = \frac{2\pi r_1}{T_1}.$$

We substitute this formula for v_1 into the equation above:

$$G\frac{m_1 M_S}{r_1^2} = m_1 \frac{4\pi^2 r_1}{T_1^2}.$$

We rearrange this to get

$$\frac{T_1^2}{r_1^3} = \frac{4\pi^2}{GM_S}. \tag{6-6}$$

We derived this for planet 1 (say, Mars). The same derivation would apply for a second planet (say, Saturn) orbiting the Sun,

$$\frac{T_2^2}{r_2^3} = \frac{4\pi^2}{GM_S},$$

where T_2 and r_2 are the period and orbit radius, respectively, for the second planet. Since the right sides of the two previous equations are equal, we have $T_1^2/r_1^3 = T_2^2/r_2^3$ or, rearranging,

$$\left(\frac{T_1}{T_2}\right)^2 = \left(\frac{r_1}{r_2}\right)^3, \tag{6-7}$$

which is Kepler's third law. Equations 6–6 and 6–7 are valid also for elliptical orbits if we replace r with the semimajor axis s.

The derivations of Eqs. 6–6 and 6–7 (Kepler's third law) compared two planets revolving around the Sun. But they are general enough to be applied to other systems. For example, we could apply Eq. 6–6 to our Moon revolving around Earth (then M_S would be replaced by M_E, the mass of the Earth). Or we could apply Eq. 6–7 to compare two moons revolving around Jupiter. But Kepler's third law, Eq. 6–7, applies only to objects orbiting the same attracting center. Do not use Eq. 6–7 to compare, say, the Moon's orbit around the Earth to the orbit of Mars around the Sun because they depend on different attracting centers.

In the following Examples, we assume the orbits are circles, although it is not quite true in general.

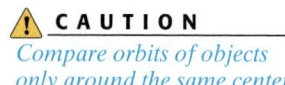

CAUTION

Compare orbits of objects only around the same center

EXAMPLE 6–8 **Where is Mars?** Mars' period (its "year") was first noted by Kepler to be about 687 days (Earth-days), which is $(687\,\text{d}/365\,\text{d}) = 1.88\,\text{yr}$ (Earth years). Determine the mean distance of Mars from the Sun using the Earth as a reference.

APPROACH We are given the ratio of the periods of Mars and Earth. We can find the distance from Mars to the Sun using Kepler's third law, given the Earth–Sun distance as $1.50 \times 10^{11}\,\text{m}$ (Table 6–2; also Table inside front cover).

SOLUTION Let the distance of Mars from the Sun be r_{MS}, and the Earth–Sun distance be $r_{ES} = 1.50 \times 10^{11}\,\text{m}$. From Kepler's third law (Eq. 6–7):

$$\frac{r_{MS}}{r_{ES}} = \left(\frac{T_M}{T_E}\right)^{\frac{2}{3}} = \left(\frac{1.88\,\text{yr}}{1\,\text{yr}}\right)^{\frac{2}{3}} = 1.52.$$

So Mars is 1.52 times the Earth's distance from the Sun, or $2.28 \times 10^{11}\,\text{m}$.

EXAMPLE 6–9 **The Sun's mass determined.** Determine the mass of the Sun given the Earth's distance from the Sun as $r_{ES} = 1.5 \times 10^{11}$ m.

APPROACH Equation 6–6 relates the mass of the Sun M_S to the period and distance of any planet. We use the Earth.

SOLUTION The Earth's period is $T_E = 1$ yr $= (365\frac{1}{4}$ d$)(24$ h/d$)(3600$ s/h$) = 3.16 \times 10^7$ s. We solve Eq. 6–6 for M_S:

$$M_S = \frac{4\pi^2 r_{ES}^3}{GT_E^2} = \frac{4\pi^2 (1.5 \times 10^{11} \text{ m})^3}{(6.67 \times 10^{-11} \text{ N} \cdot \text{m}^2/\text{kg}^2)(3.16 \times 10^7 \text{ s})^2} = 2.0 \times 10^{30} \text{ kg}.$$

EXERCISE E Suppose there were a planet in circular orbit exactly halfway between the orbits of Mars and Jupiter. What would its period be in Earth-years? Use Table 6–2.

Accurate measurements on the orbits of the planets indicated that they did not precisely follow Kepler's laws. For example, slight deviations from perfectly elliptical orbits were observed. Newton was aware that this was to be expected because any planet would be attracted gravitationally not only by the Sun but also (to a much lesser extent) by the other planets. Such deviations, or **perturbations**, in the orbit of Saturn were a hint that helped Newton formulate the law of universal gravitation, that all objects attract gravitationally. Observation of other perturbations later led to the discovery of Neptune and Pluto. Deviations in the orbit of Uranus, for example, could not all be accounted for by perturbations due to the other known planets. Careful calculation in the nineteenth century indicated that these deviations could be accounted for if another planet existed farther out in the solar system. The position of this planet was predicted from the deviations in the orbit of Uranus, and telescopes focused on that region of the sky quickly found it; the new planet was called Neptune. Similar but much smaller perturbations of Neptune's orbit led to the discovery of Pluto in 1930.

Starting in the mid 1990s, planets revolving about distant stars (Fig. 6–18) were inferred from the regular "wobble" of each star due to the gravitational attraction of the revolving planet(s). Many such "extrasolar" planets are now known.

The development by Newton of the law of universal gravitation and the three laws of motion was a major intellectual achievement: with these laws, he was able to describe the motion of objects on Earth and in the heavens. The motions of heavenly bodies and objects on Earth were seen to follow the same laws (not recognized previously). For this reason, and also because Newton integrated the results of earlier scientists into his system, we sometimes speak of **Newton's synthesis**.

The laws formulated by Newton are referred to as **causal laws**. By **causality** we mean the idea that one occurrence can cause another. When a rock strikes a window, we infer that the rock *causes* the window to break. This idea of "cause and effect" relates to Newton's laws: the acceleration of an object was seen to be *caused* by the net force acting on it.

As a result of Newton's theories the universe came to be viewed by many scientists and philosophers as a big machine whose parts move in a *deterministic* way. This deterministic view of the universe, however, had to be modified by scientists in the twentieth century (Chapter 38).

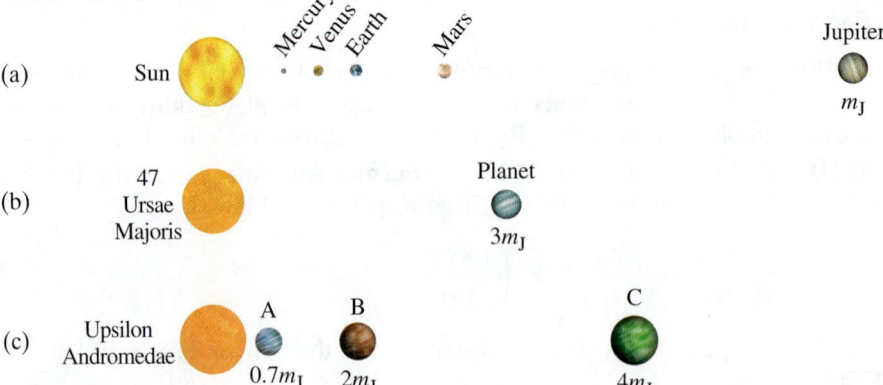

FIGURE 6–18 Our solar system (a) is compared to recently discovered planets orbiting (b) the star 47 Ursae Majoris and (c) the star Upsilon Andromedae with at least three planets. m_J is the mass of Jupiter. (Sizes not to scale.)

EXAMPLE 6–10 **Lagrange Point.** The mathematician Joseph-Louis Lagrange discovered five special points in the vicinity of the Earth's orbit about the Sun where a small satellite (mass m) can orbit the Sun with the same period T as Earth's ($= 1$ year). One of these "Lagrange Points," called L1, lies between the Earth (mass M_E) and Sun (mass M_S), on the line connecting them (Fig. 6–19). That is, the Earth and the satellite are always separated by a distance d. If the Earth's orbital radius is R_{ES}, then the satellite's orbital radius is $(R_{ES} - d)$. Determine d.

APPROACH We use Newton's law of universal gravitation and set it equal to the mass times the centripetal acceleration. But how could an object with a smaller orbit than Earth's have the same period as Earth? Kepler's third law clearly tells us a smaller orbit around the Sun results in a smaller period. But that law depends on only the Sun's gravitational attraction. Our mass m is pulled by both the Sun and the Earth.

SOLUTION Because the satellite is assumed to have negligible mass in comparison to the masses of the Earth and Sun, to an excellent approximation the Earth's orbit will be determined solely by the Sun. Applying Newton's second law to the Earth gives

$$\frac{GM_E M_S}{R_{ES}^2} = M_E \frac{v^2}{R_{ES}} = \frac{M_E}{R_{ES}} \frac{(2\pi R_{ES})^2}{T^2}$$

or

$$\frac{GM_S}{R_{ES}^2} = \frac{4\pi^2 R_{ES}}{T^2}. \qquad \textbf{(i)}$$

Next we apply Newton's second law to the satellite m (which has the same period T as Earth), including the pull of both Sun and Earth (see simplified form, Eq. (i))

$$\frac{GM_S}{(R_{ES} - d)^2} - \frac{GM_E}{d^2} = \frac{4\pi^2 (R_{ES} - d)}{T^2},$$

which we rewrite as

$$\frac{GM_S}{R_{ES}^2}\left(1 - \frac{d}{R_{ES}}\right)^{-2} - \frac{GM_E}{d^2} = \frac{4\pi^2 R_{ES}}{T^2}\left(1 - \frac{d}{R_{ES}}\right).$$

We now use the binomial expansion $(1 + x)^n \approx 1 + nx$, if $x \ll 1$. Setting $x = d/R_{ES}$ and assuming $d \ll R_{ES}$, we have

$$\frac{GM_S}{R_{ES}^2}\left(1 + 2\frac{d}{R_{ES}}\right) - \frac{GM_E}{d^2} = \frac{4\pi^2 R_{ES}}{T^2}\left(1 - \frac{d}{R_{ES}}\right). \qquad \textbf{(ii)}$$

Substituting GM_S/R_{ES}^2 from Eq. (i) into Eq. (ii) we find

$$\frac{GM_S}{R_{ES}^2}\left(1 + 2\frac{d}{R_{ES}}\right) - \frac{GM_E}{d^2} = \frac{GM_S}{R_{ES}^2}\left(1 - \frac{d}{R_{ES}}\right).$$

Simplifying, we have

$$\frac{GM_S}{R_{ES}^2}\left(3\frac{d}{R_{ES}}\right) = \frac{GM_E}{d^2}.$$

We solve for d to find

$$d = \left(\frac{M_E}{3M_S}\right)^{\frac{1}{3}} R_{ES}.$$

Substituting in values we find

$$d = 1.0 \times 10^{-2} R_{ES} = 1.5 \times 10^6 \text{ km}.$$

NOTE Since $d/R_{ES} = 10^{-2}$, we were justified in using the binomial expansion.

NOTE Placing a satellite at L1 has two advantages: the satellite's view of the Sun is never eclipsed by the Earth, and it is always close enough to Earth to transmit data easily. The L1 point of the Earth–Sun system is currently home to the Solar and Heliospheric Observatory (SOHO) satellite, Fig. 6–20.

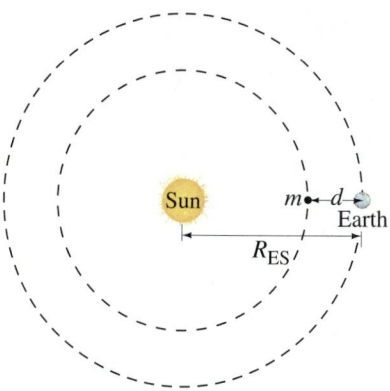

FIGURE 6–19 Finding the position of the Lagrange Point L1 for a satellite that can remain along the revolving line between the Sun and Earth, at distance d from the Earth. Thus a mass m at L1 has the same period around the Sun as the Earth has. (Not to scale.)

FIGURE 6–20 Artist's rendition of the Solar and Heliospheric Observatory (SOHO) satellite in orbit.

*6–6 Gravitational Field

Most of the forces we meet in everyday life are contact forces: you push or pull on a lawn mower, a tennis racket exerts a force on a tennis ball when they make contact, or a ball exerts a force on a window when they make contact. But the gravitational force acts over a distance: there is a force even when the two objects are not in contact. The Earth, for example, exerts a force on a falling apple. It also exerts a force on the Moon, 384,000 km away. And the Sun exerts a gravitational force on the Earth. The idea of a force *acting at a distance* was a difficult one for early thinkers. Newton himself felt uneasy with this concept when he published his law of universal gravitation.

Another point of view that helps with these conceptual difficulties is the concept of the **field**, developed in the nineteenth century by Michael Faraday (1791–1867) to aid understanding of electric and magnetic forces which also act over a distance. Only later was it applied to gravity. According to the field concept, a **gravitational field** surrounds every object that has mass, and this field permeates all of space. A second object at a particular location near the first object experiences a force because of the gravitational field that exists there. Because the gravitational field at the location of the second mass is considered to act directly on this mass, we are a little closer to the idea of a contact force.

To be quantitative, we can define the **gravitational field** as the gravitational force per unit mass at any point in space. If we want to measure the gravitational field at any point, we place a small "test" mass m at that point and measure the force $\vec{F}$ exerted on it (making sure only gravitational forces are acting). Then the gravitational field, $\vec{g}$, at that point is defined as

$$\vec{g} = \frac{\vec{F}}{m}. \qquad \text{[gravitational field]} \quad \textbf{(6–8)}$$

The units of $\vec{g}$ are N/kg.

From Eq. 6–8 we see that the gravitational field an object experiences has magnitude equal to the acceleration due to gravity at that point. (When we speak of acceleration, however, we use units m/s^2, which is equivalent to N/kg, since $1\,\text{N} = 1\,\text{kg} \cdot \text{m/s}^2$.)

If the gravitational field is due to a single spherically symmetric (or small) object of mass M, such as when m is near the Earth's surface, then the gravitational field at a distance r from M has magnitude

$$g = \frac{1}{m} G \frac{mM}{r^2} = G \frac{M}{r^2}.$$

In vector notation we write

$$\vec{g} = -\frac{GM}{r^2} \hat{\mathbf{r}}, \qquad \begin{bmatrix} \text{due to a} \\ \text{single mass } M \end{bmatrix}$$

where $\hat{\mathbf{r}}$ is a unit vector pointing radially outward from mass M, and the minus sign reminds us that the field points toward mass M (see Eqs. 6–1, 6–2, and 6–4). If several different bodies contribute significantly to the gravitational field, then we write the gravitational field $\vec{g}$ as the vector sum of all these contributions. In interplanetary space, for example, $\vec{g}$ at any point in space is the vector sum of terms due to the Earth, Sun, Moon, and other bodies that contribute. The gravitational field $\vec{g}$ at any point in space does not depend on the value of our test mass, m, placed at that point; $\vec{g}$ depends only on the masses (and locations) of the bodies that create the field there.

6–7 Types of Forces in Nature

We have already discussed that Newton's law of universal gravitation, Eq. 6–1, describes how a particular type of force—gravity—depends on the distance between, and masses of, the objects involved. Newton's second law, $\Sigma\vec{\mathbf{F}} = m\vec{\mathbf{a}}$, on the other hand, tells how an object will accelerate due to *any* type of force. But what are the types of forces that occur in nature besides gravity?

In the twentieth century, physicists came to recognize four different fundamental forces in nature: (1) the gravitational force; (2) the electromagnetic force (we shall see later that electric and magnetic forces are intimately related); (3) the strong nuclear force; and (4) the weak nuclear force. In this Chapter, we discussed the gravitational force in detail. The nature of the electromagnetic force will be discussed in detail in Chapters 21 to 31. The strong and weak nuclear forces operate at the level of the atomic nucleus; although they manifest themselves in such phenomena as radioactivity and nuclear energy (Chapters 41 to 43), they are much less obvious in our daily lives.

Physicists have been working on theories that would unify these four forces—that is, to consider some or all of these forces as different manifestations of the same basic force. So far, the electromagnetic and weak nuclear forces have been theoretically united to form *electroweak* theory, in which the electromagnetic and weak forces are seen as two different manifestations of a single *electroweak force*. Attempts to further unify the forces, such as in *grand unified theories* (GUT), are hot research topics today.

But where do everyday forces fit into this scheme? Ordinary forces, other than gravity, such as pushes, pulls, and other contact forces like the normal force and friction, are today considered to be due to the electromagnetic force acting at the atomic level. For example, the force your fingers exert on a pencil is the result of electrical repulsion between the outer electrons of the atoms of your finger and those of the pencil.

*6–8 Principle of Equivalence; Curvature of Space; Black Holes

We have dealt with two aspects of mass. In Chapter 4, we defined mass as a measure of the inertia of a body. Newton's second law relates the force acting on a body to its acceleration and its **inertial mass**, as we call it. We might say that inertial mass represents a resistance to any force. In this Chapter we have dealt with mass as a property related to the gravitational force—that is, mass as a quantity that determines the strength of the gravitational force between two bodies. This we call the **gravitational mass**.

It is not obvious that the inertial mass of a body should be equal to its gravitational mass. The force of gravity might have depended on a different property of a body, just as the electrical force depends on a property called electric charge. Newton's and Cavendish's experiments indicated that the two types of mass are equal for a body, and modern experiments confirm it to a precision of about 1 part in 10^{12}.

Albert Einstein (1879–1955) called this equivalence between gravitational and inertial masses the **principle of equivalence**, and he used it as a foundation for his *general theory of relativity* (c. 1916). The principle of equivalence can be stated in another way: there is no experiment observers can perform to distinguish if an acceleration arises because of a gravitational force or because their reference frame is accelerating. If you were far out in space and an apple fell to the floor of your spacecraft, you might assume a gravitational force was acting on the apple. But it would also be possible that the apple fell because your spacecraft accelerated upward (relative to an inertial system). The effects would be indistinguishable, according to the principle of equivalence, because the apple's inertial and gravitational masses—that determine how a body "reacts" to outside influences—are indistinguishable.

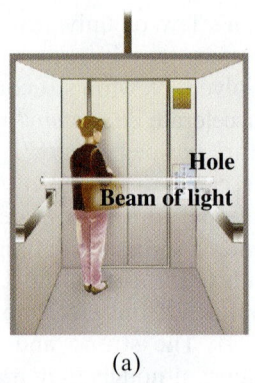
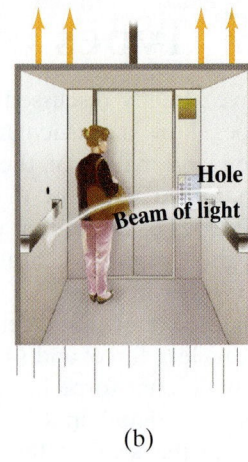

FIGURE 6–21 (a) Light beam goes straight across an elevator that is not accelerating. (b) The light beam bends (exaggerated) in an elevator accelerating in an upward direction.

(a)   (b)

The principle of equivalence can be used to show that light ought to be deflected due to the gravitational force of a massive object. Let us consider a thought experiment in an elevator in free space where virtually no gravity acts. If a light beam enters a hole in the side of the elevator, the beam travels straight across the elevator and makes a spot on the opposite side if the elevator is at rest (Fig. 6–21a). If the elevator is accelerating upward as in Fig. 6–21b, the light beam still travels straight as observed in the original reference frame at rest. In the upwardly accelerating elevator, however, the beam is observed to curve downward. Why? Because during the time the light travels from one side of the elevator to the other, the elevator is moving upward at ever-increasing speed.

According to the equivalence principle, an upwardly accelerating reference frame is equivalent to a downward gravitational field. Hence, we can picture the curved light path in Fig. 6–21b as being the effect of a gravitational field. Thus we expect gravity to exert a force on a beam of light and to bend it out of a straight-line path!

Einstein's general theory of relativity predicts that light should be affected by gravity. It was calculated that light from a distant star would be deflected by 1.75″ of arc (tiny but detectable) as it passed near the Sun, as shown in Fig. 6–22. Such a deflection was measured and confirmed in 1919 during an eclipse of the Sun. (The eclipse reduced the brightness of the Sun so that the stars in line with its edge at that moment would be visible.)

That a light beam can follow a curved path suggests that *space itself is curved* and that it is gravitational mass that causes the curvature. The curvature is greatest near very massive objects. To visualize this curvature of space, we might think of space as being like a thin rubber sheet; if a heavy weight is hung from it, it curves as shown in Fig. 6–23. The weight corresponds to a huge mass that causes space (space itself!) to curve.

The extreme curvature of space-time shown in Fig. 6–23 could be produced by a **black hole**, a star that becomes so dense and massive that gravity would be so strong that even light could not escape it. Light would be pulled back in by the force of gravity. Since no light could escape from such a massive star, we could not see it—it would be black. An object might pass by it and be deflected by its gravitational field, but if the object came too close it would be swallowed up, never to escape. Hence the name black holes. Experimentally there is good evidence for their existence. One likely possibility is a giant black hole at the center of our Galaxy and probably at the center of other galaxies.

FIGURE 6–22 (a) Three stars in the sky. (b) If the light from one of these stars passes very near the Sun, whose gravity bends the light beam, the star will appear higher than it actually is.

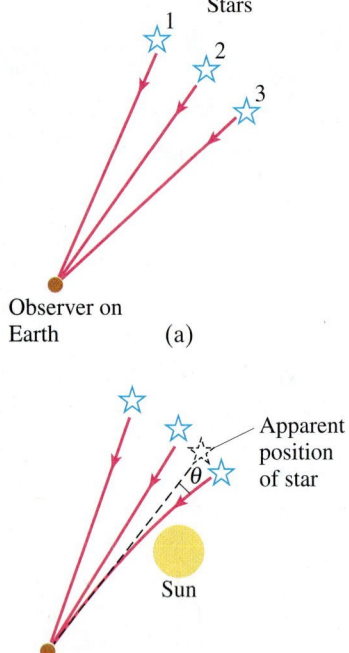

FIGURE 6–23 Rubber-sheet analogy for space (technically space-time) curved by matter.

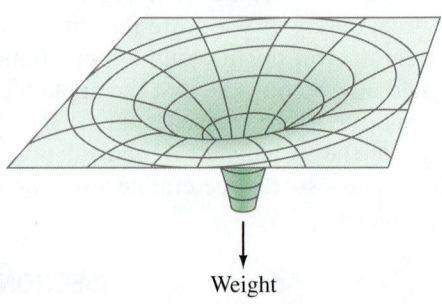

Weight

Summary

Newton's **law of universal gravitation** states that every particle in the universe attracts every other particle with a force proportional to the product of their masses and inversely proportional to the square of the distance between them:

$$F = G \frac{m_1 m_2}{r^2}. \tag{6-1}$$

The direction of this force is along the line joining the two particles, and is always attractive. It is this gravitational force that keeps the Moon revolving around the Earth and the planets revolving around the Sun.

The total gravitational force on any object is the vector sum of the forces exerted by all other objects; frequently the effects of all but one or two objects can be ignored.

Satellites revolving around the Earth are acted on by gravity, but "stay up" because of their high tangential speed.

Newton's three laws of motion, plus his law of universal gravitation, constituted a wide-ranging theory of the universe. With them, motion of objects on Earth and in the heavens could be accurately described. And they provided a theoretical base for **Kepler's laws** of planetary motion.

[*According to the **field** concept, a **gravitational field** surrounds every object that has mass, and it permeates all of space. The gravitational field at any point in space is the vector sum of the fields due to all massive objects and can be defined as

$$\vec{\mathbf{g}} = \frac{\vec{\mathbf{F}}}{m} \tag{6-8}$$

where $\vec{\mathbf{F}}$ is the force acting on a small "test" mass m placed at that point.]

The four fundamental forces in nature are (1) the gravitational force, (2) electromagnetic force, (3) strong nuclear force, and (4) weak nuclear force. The first two fundamental forces are responsible for nearly all "everyday" forces.

Questions

1. Does an apple exert a gravitational force on the Earth? If so, how large a force? Consider an apple (*a*) attached to a tree and (*b*) falling.

2. The Sun's gravitational pull on the Earth is much larger than the Moon's. Yet the Moon's is mainly responsible for the tides. Explain. [*Hint*: Consider the difference in gravitational pull from one side of the Earth to the other.]

3. Will an object weigh more at the equator or at the poles? What two effects are at work? Do they oppose each other?

4. Why is more fuel required for a spacecraft to travel from the Earth to the Moon than it does to return from the Moon to the Earth?

5. The gravitational force on the Moon due to the Earth is only about half the force on the Moon due to the Sun (see Example 6–3). Why isn't the Moon pulled away from the Earth?

6. How did the scientists of Newton's era determine the distance from the Earth to the Moon, despite not knowing about spaceflight or the speed of light? [*Hint*: Think about why two eyes are useful for depth perception.]

7. If it were possible to drill a hole all the way through the Earth along a diameter, then it would be possible to drop a ball through the hole. When the ball was right at the center of the Earth, what would be the total gravitational force exerted on it by the Earth?

8. Why is it not possible to put a satellite in geosynchronous orbit above the North Pole?

9. Which pulls harder gravitationally, the Earth on the Moon, or the Moon on the Earth? Which accelerates more?

10. Would it require less speed to launch a satellite (*a*) toward the east or (*b*) toward the west? Consider the Earth's rotation direction.

11. An antenna loosens and becomes detached from a satellite in a circular orbit around the Earth. Describe the antenna's motion subsequently. If it will land on the Earth, describe where; if not, describe how it could be made to land on the Earth.

12. Describe how careful measurements of the variation in g in the vicinity of an ore deposit might be used to estimate the amount of ore present.

13. The Sun is below us at midnight, nearly in line with the Earth's center. Are we then heavier at midnight, due to the Sun's gravitational force on us, than we are at noon? Explain.

14. When will your apparent weight be the greatest, as measured by a scale in a moving elevator: when the elevator (*a*) accelerates downward, (*b*) accelerates upward, (*c*) is in free fall, or (*d*) moves upward at constant speed? In which case would your apparent weight be the least? When would it be the same as when you are on the ground?

15. If the Earth's mass were double what it actually is, in what ways would the Moon's orbit be different?

16. The source of the Mississippi River is closer to the center of the Earth than is its outlet in Louisiana (since the Earth is fatter at the equator than at the poles). Explain how the Mississippi can flow "uphill."

17. People sometimes ask, "What keeps a satellite up in its orbit around the Earth?" How would you respond?

18. Explain how a runner experiences "free fall" or "apparent weightlessness" between steps.

19. If you were in a satellite orbiting the Earth, how might you cope with walking, drinking, or putting a pair of scissors on a table?

20. Is the centripetal acceleration of Mars in its orbit around the Sun larger or smaller than the centripetal acceleration of the Earth?

21. The mass of the planet Pluto was not known until it was discovered to have a moon. Explain how this enabled an estimate of Pluto's mass.

22. The Earth moves faster in its orbit around the Sun in January than in July. Is the Earth closer to the Sun in January, or in July? Explain. [*Note*: This is not much of a factor in producing the seasons—the main factor is the tilt of the Earth's axis relative to the plane of its orbit.]

23. Kepler's laws tell us that a planet moves faster when it is closer to the Sun than when it is farther from the Sun. What causes this change in speed of the planet?

*24. Does your body directly sense a gravitational field? (Compare to what you would feel in free fall.)

*25. Discuss the conceptual differences between $\vec{\mathbf{g}}$ as acceleration due to gravity and $\vec{\mathbf{g}}$ as gravitational field.

Problems

6–1 to 6–3 Law of Universal Gravitation

1. (I) Calculate the force of Earth's gravity on a spacecraft 2.00 Earth radii above the Earth's surface if its mass is 1480 kg.

2. (I) Calculate the acceleration due to gravity on the Moon. The Moon's radius is 1.74×10^6 m and its mass is 7.35×10^{22} kg.

3. (I) A hypothetical planet has a radius 2.3 times that of Earth, but has the same mass. What is the acceleration due to gravity near its surface?

4. (I) A hypothetical planet has a mass 1.80 times that of Earth, but the same radius. What is g near its surface?

5. (I) If you doubled the mass and tripled the radius of a planet, by what factor would g at its surface change?

6. (II) Calculate the effective value of g, the acceleration of gravity, at (a) 6400 m, and (b) 6400 km, above the Earth's surface.

7. (II) You are explaining to friends why astronauts feel weightless orbiting in the space shuttle, and they respond that they thought gravity was just a lot weaker up there. Convince them and yourself that it isn't so by calculating how much weaker gravity is 300 km above the Earth's surface.

8. (II) Every few hundred years most of the planets line up on the same side of the Sun. Calculate the total force on the Earth due to Venus, Jupiter, and Saturn, assuming all four planets are in a line, Fig. 6–24. The masses are $M_V = 0.815\,M_E$, $M_J = 318\,M_E$, $M_{Sat} = 95.1\,M_E$, and the mean distances of the four planets from the Sun are 108, 150, 778, and 1430 million km. What fraction of the Sun's force on the Earth is this?

FIGURE 6–24   Problem 8 (not to scale).

9. (II) Four 8.5-kg spheres are located at the corners of a square of side 0.80 m. Calculate the magnitude and direction of the gravitational force exerted on one sphere by the other three.

10. (II) Two objects attract each other gravitationally with a force of 2.5×10^{-10} N when they are 0.25 m apart. Their total mass is 4.00 kg. Find their individual masses.

11. (II) Four masses are arranged as shown in Fig. 6–25. Determine the x and y components of the gravitational force on the mass at the origin (m). Write the force in vector notation $(\hat{\mathbf{i}}, \hat{\mathbf{j}})$.

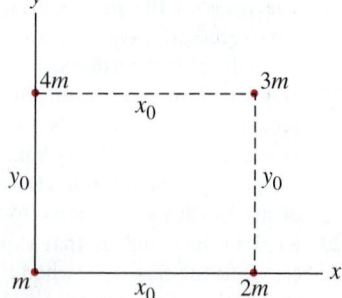

FIGURE 6–25
Problem 11.

12. (II) Estimate the acceleration due to gravity at the surface of Europa (one of the moons of Jupiter) given that its mass is 4.9×10^{22} kg and making the assumption that its density is the same as Earth's.

13. (II) Suppose the mass of the Earth were doubled, but it kept the same density and spherical shape. How would the weight of objects at the Earth's surface change?

14. (II) Given that the acceleration of gravity at the surface of Mars is 0.38 of what it is on Earth, and that Mars' radius is 3400 km, determine the mass of Mars.

15. (II) At what distance from the Earth will a spacecraft traveling directly from the Earth to the Moon experience zero net force because the Earth and Moon pull with equal and opposite forces?

16. (II) Determine the mass of the Sun using the known value for the period of the Earth and its distance from the Sun. [Hint: The force on the Earth due to the Sun is related to the centripetal acceleration of the Earth.] Compare your answer to that obtained using Kepler's laws, Example 6–9.

17. (II) Two identical point masses, each of mass M, always remain separated by a distance of $2R$. A third mass m is then placed a distance x along the perpendicular bisector of the original two masses, as shown in Fig. 6–26. Show that the gravitational force on the third mass is directed inward along the perpendicular bisector and has a magnitude of

$$F = \frac{2GMmx}{\left(x^2 + R^2\right)^{\frac{3}{2}}}.$$

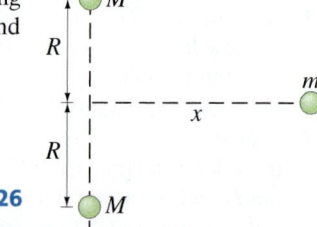

FIGURE 6–26
Problem 17.

18. (II) A mass M is ring shaped with radius r. A small mass m is placed at a distance x along the ring's axis as shown in Fig. 6–27. Show that the gravitational force on the mass m due to the ring is directed inward along the axis and has magnitude

$$F = \frac{GMmx}{\left(x^2 + r^2\right)^{\frac{3}{2}}}.$$

[Hint: Think of the ring as made up of many small point masses dM; sum over the forces due to each dM, and use symmetry.]

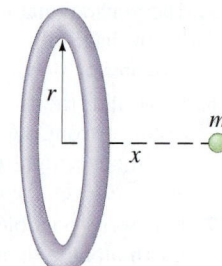

FIGURE 6–27
Problem 18.

19. (III) (a) Use the binomial expansion

$$(1 \pm x)^n = 1 \pm nx + \frac{n(n-1)}{2}x^2 \pm \cdots$$

to show that the value of g is altered by approximately

$$\Delta g \approx -2g\frac{\Delta r}{r_E}$$

at a height Δr above the Earth's surface, where r_E is the radius of the Earth, as long as $\Delta r \ll r_E$. (b) What is the meaning of the minus sign in this relation? (c) Use this result to compute the effective value of g at 125 km above the Earth's surface. Compare to a direct use of Eq. 6–1.

20. (III) The center of a 1.00 km diameter spherical pocket of oil is 1.00 km beneath the Earth's surface. Estimate by what percentage g directly above the pocket of oil would differ from the expected value of g for a uniform Earth? Assume the density of oil is 8.0×10^2 kg/m³.

21. (III) Determine the magnitude and direction of the effective value of $\vec{g}$ at a latitude of 45° on the Earth. Assume the Earth is a rotating sphere.

*22. (III) It can be shown (Appendix D) that for a uniform sphere the force of gravity at a point inside the sphere depends only on the mass closer to the center than that point. The net force of gravity due to points outside the radius of the point cancels. How far would you have to drill into the Earth, to reach a point where your weight is reduced by 5.0%? Approximate the Earth as a uniform sphere.

6–4 Satellites and Weightlessness

23. (I) The space shuttle releases a satellite into a circular orbit 680 km above the Earth. How fast must the shuttle be moving (relative to Earth's center) when the release occurs?

24. (I) Calculate the speed of a satellite moving in a stable circular orbit about the Earth at a height of 5800 km.

25. (II) You know your mass is 65 kg, but when you stand on a bathroom scale in an elevator, it says your mass is 76 kg. What is the acceleration of the elevator, and in which direction?

26. (II) A 13.0-kg monkey hangs from a cord suspended from the ceiling of an elevator. The cord can withstand a tension of 185 N and breaks as the elevator accelerates. What was the elevator's minimum acceleration (magnitude and direction)?

27. (II) Calculate the period of a satellite orbiting the Moon, 120 km above the Moon's surface. Ignore effects of the Earth. The radius of the Moon is 1740 km.

28. (II) Two satellites orbit Earth at altitudes of 5000 km and 15,000 km. Which satellite is faster, and by what factor?

29. (II) What will a spring scale read for the weight of a 53-kg woman in an elevator that moves (a) upward with constant speed 5.0 m/s, (b) downward with constant speed 5.0 m/s, (c) upward with acceleration 0.33 g, (d) downward with acceleration 0.33 g, and (e) in free fall?

30. (II) Determine the time it takes for a satellite to orbit the Earth in a circular "near-Earth" orbit. A "near-Earth" orbit is at a height above the surface of the Earth that is very small compared to the radius of the Earth. [*Hint*: You may take the acceleration due to gravity as essentially the same as that on the surface.] Does your result depend on the mass of the satellite?

31. (II) What is the apparent weight of a 75-kg astronaut 2500 km from the center of the Earth's Moon in a space vehicle (a) moving at constant velocity and (b) accelerating toward the Moon at 2.3 m/s²? State "direction" in each case.

32. (II) A Ferris wheel 22.0 m in diameter rotates once every 12.5 s (see Fig. 5–19). What is the ratio of a person's apparent weight to her real weight (a) at the top, and (b) at the bottom?

33. (II) Two equal-mass stars maintain a constant distance apart of 8.0×10^{11} m and rotate about a point midway between them at a rate of one revolution every 12.6 yr. (a) Why don't the two stars crash into one another due to the gravitational force between them? (b) What must be the mass of each star?

34. (III) (a) Show that if a satellite orbits very near the surface of a planet with period T, the density (= mass per unit volume) of the planet is $\rho = m/V = 3\pi/GT^2$. (b) Estimate the density of the Earth, given that a satellite near the surface orbits with a period of 85 min. Approximate the Earth as a uniform sphere.

35. (III) Three bodies of identical mass M form the vertices of an equilateral triangle of side ℓ and rotate in circular orbits about the center of the triangle. They are held in place by their mutual gravitation. What is the speed of each?

36. (III) An inclined plane, fixed to the inside of an elevator, makes a 32° angle with the floor. A mass m slides on the plane without friction. What is its acceleration relative to the plane if the elevator (a) accelerates upward at 0.50 g, (b) accelerates downward at 0.50 g, (c) falls freely, and (d) moves upward at constant speed?

6–5 Kepler's Laws

37. (I) Use Kepler's laws and the period of the Moon (27.4 d) to determine the period of an artificial satellite orbiting very near the Earth's surface.

38. (I) Determine the mass of the Earth from the known period and distance of the Moon.

39. (I) Neptune is an average distance of 4.5×10^9 km from the Sun. Estimate the length of the Neptunian year using the fact that the Earth is 1.50×10^8 km from the Sun on the average.

40. (II) Planet A and planet B are in circular orbits around a distant star. Planet A is 9.0 times farther from the star than is planet B. What is the ratio of their speeds v_A/v_B?

41. (II) Our Sun rotates about the center of our Galaxy $(m_G \approx 4 \times 10^{41} \text{ kg})$ at a distance of about 3×10^4 light-years $[1 \text{ly} = (3.00 \times 10^8 \text{ m/s}) \cdot (3.16 \times 10^7 \text{ s/yr}) \cdot (1.00 \text{ yr})]$. What is the period of the Sun's orbital motion about the center of the Galaxy?

42. (II) Table 6–3 gives the mean distance, period, and mass for the four largest moons of Jupiter (those discovered by Galileo in 1609). (a) Determine the mass of Jupiter using the data for Io. (b) Determine the mass of Jupiter using data for each of the other three moons. Are the results consistent?

TABLE 6–3 Principal Moons of Jupiter
(Problems 42, 43, and 47)

Moon	Mass (kg)	Period (Earth days)	Mean distance from Jupiter (km)
Io	8.9×10^{22}	1.77	422×10^3
Europa	4.9×10^{22}	3.55	671×10^3
Ganymede	15×10^{22}	7.16	1070×10^3
Callisto	11×10^{22}	16.7	1883×10^3

43. (II) Determine the mean distance from Jupiter for each of Jupiter's moons, using Kepler's third law. Use the distance of Io and the periods given in Table 6–3. Compare your results to the values in the Table.

44. (II) The asteroid belt between Mars and Jupiter consists of many fragments (which some space scientists think came from a planet that once orbited the Sun but was destroyed). (a) If the mean orbital radius of the asteroid belt (where the planet would have been) is about three times farther from the Sun than the Earth is, how long would it have taken this hypothetical planet to orbit the Sun? (b) Can we use these data to deduce the mass of this planet?

45. (III) The comet Hale–Bopp has a period of 2400 years. (a) What is its mean distance from the Sun? (b) At its closest approach, the comet is about 1.0 AU from the Sun (1 AU = distance from Earth to the Sun). What is the farthest distance? (c) What is the ratio of the speed at the closest point to the speed at the farthest point?

46. (III) (a) Use Kepler's second law to show that the ratio of the speeds of a planet at its nearest and farthest points from the Sun is equal to the inverse ratio of the near and far distances: $v_N/v_F = d_F/d_N$. (b) Given that the Earth's distance from the Sun varies from 1.47 to 1.52×10^{11} m, determine the minimum and maximum velocities of the Earth in its orbit around the Sun.

47. (III) The orbital periods T and mean orbital distances r for Jupiter's four largest moons are given in Table 6–3, on the previous page. (a) Starting with Kepler's third law in the form

$$T^2 = \left(\frac{4\pi^2}{Gm_J}\right)r^3,$$

where m_J is the mass of Jupiter, show that this relation implies that a plot of $\log(T)$ vs. $\log(r)$ will yield a straight line. Explain what Kepler's third law predicts about the slope and y-intercept of this straight-line plot. (b) Using the data for Jupiter's four moons, plot $\log(T)$ vs. $\log(r)$ and show that you get a straight line. Determine the slope of this plot and compare it to the value you expect if the data are consistent with Kepler's third law. Determine the y-intercept of the plot and use it to compute the mass of Jupiter.

General Problems

51. How far above the Earth's surface will the acceleration of gravity be half what it is at the surface?

52. At the surface of a certain planet, the gravitational acceleration g has a magnitude of 12.0 m/s^2. A 13.0-kg brass ball is transported to this planet. What is (a) the mass of the brass ball on the Earth and on the planet, and (b) the weight of the brass ball on the Earth and on the planet?

53. A certain white dwarf star was once an average star like our Sun. But now it is in the last stage of its evolution and is the size of our Moon but has the mass of our Sun. (a) Estimate gravity on the surface on this star. (b) How much would a 65-kg person weigh on this star? (c) What would be the speed of a baseball dropped from a height of 1.0 m when it hit the surface?

54. What is the distance from the Earth's center to a point outside the Earth where the gravitational acceleration due to the Earth is $\frac{1}{10}$ of its value at the Earth's surface?

55. The rings of Saturn are composed of chunks of ice that orbit the planet. The inner radius of the rings is 73,000 km, while the outer radius is 170,000 km. Find the period of an orbiting chunk of ice at the inner radius and the period of a chunk at the outer radius. Compare your numbers with Saturn's mean rotation period of 10 hours and 39 minutes. The mass of Saturn is $5.7 \times 10^{26} \text{ kg}$.

56. During an *Apollo* lunar landing mission, the command module continued to orbit the Moon at an altitude of about 100 km. How long did it take to go around the Moon once?

57. Halley's comet orbits the Sun roughly once every 76 years. It comes very close to the surface of the Sun on its closest approach (Fig. 6–28). Estimate the greatest distance of the comet from the Sun. Is it still "in" the solar system? What planet's orbit is nearest when it is out there?

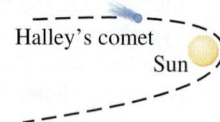

FIGURE 6–28
Problem 57.

58. The Navstar Global Positioning System (GPS) utilizes a group of 24 satellites orbiting the Earth. Using "triangulation" and signals transmitted by these satellites, the position of a receiver on the Earth can be determined to within an accuracy of a few centimeters. The satellite orbits are distributed evenly around the Earth, with four satellites in each of six orbits, allowing continuous navigational "fixes." The satellites orbit at an altitude of approximately 11,000 nautical miles [1 nautical mile = 1.852 km = 6076 ft]. (a) Determine the speed of each satellite. (b) Determine the period of each satellite.

*6–6 Gravitational Field

*48. (II) What is the magnitude and direction of the gravitational field midway between the Earth and Moon? Ignore effects of the Sun.

*49. (II) (a) What is the gravitational field at the surface of the Earth due to the Sun? (b) Will this affect your weight significantly?

*50. (III) Two identical particles, each of mass m, are located on the x axis at $x = +x_0$ and $x = -x_0$. (a) Determine a formula for the gravitational field due to these two particles for points on the y axis; that is, write $\vec{g}$ as a function of y, m, x_0, and so on. (b) At what point (or points) on the y axis is the magnitude of $\vec{g}$ a maximum value, and what is its value there? [*Hint*: Take the derivative $d\vec{g}/dy$.]

59. Jupiter is about 320 times as massive as the Earth. Thus, it has been claimed that a person would be crushed by the force of gravity on a planet the size of Jupiter since people can't survive more than a few g's. Calculate the number of g's a person would experience at the equator of such a planet. Use the following data for Jupiter: mass $= 1.9 \times 10^{27} \text{ kg}$, equatorial radius $= 7.1 \times 10^4 \text{ km}$, rotation period $= 9 \text{ hr } 55 \text{ min}$. Take the centripetal acceleration into account.

60. The Sun rotates about the center of the Milky Way Galaxy (Fig. 6–29) at a distance of about 30,000 light-years from the center $(1 \text{ ly} = 9.5 \times 10^{15} \text{ m})$. If it takes about 200 million years to make one rotation, estimate the mass of our Galaxy. Assume that the mass distribution of our Galaxy is concentrated mostly in a central uniform sphere. If all the stars had about the mass of our Sun $(2 \times 10^{30} \text{ kg})$, how many stars would there be in our Galaxy?

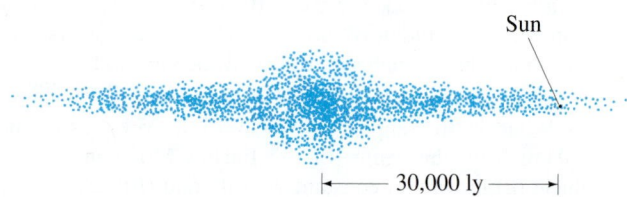

FIGURE 6–29 Edge-on view of our galaxy. Problem 60.

61. Astronomers have observed an otherwise normal star, called S2, closely orbiting an extremely massive but small object at the center of the Milky Way Galaxy called SgrA. S2 moves in an elliptical orbit around SgrA with a period of 15.2 yr and an eccentricity $e = 0.87$ (Fig. 6–16). In 2002, S2 reached its closest approach to SgrA, a distance of only 123 AU $(1 \text{ AU} = 1.50 \times 10^{11} \text{ m}$ is the mean Earth–Sun distance). Determine the mass M of SgrA, the massive compact object (believed to be a supermassive black hole) at the center of our Galaxy. State M in kg and in terms of the mass of our Sun.

62. A satellite of mass 5500 kg orbits the Earth and has a period of 6200 s. Determine (a) the radius of its circular orbit, (b) the magnitude of the Earth's gravitational force on the satellite, and (c) the altitude of the satellite.

63. Show that the rate of change of your weight is

$$-2G\frac{m_{\rm E}\,m}{r^3}\,v$$

if you are traveling directly away from Earth at constant speed v. Your mass is m, and r is your distance from the center of the Earth at any moment.

64. Astronomers using the Hubble Space Telescope deduced the presence of an extremely massive core in the distant galaxy M87, so dense that it could be a black hole (from which no light escapes). They did this by measuring the speed of gas clouds orbiting the core to be 780 km/s at a distance of 60 light-years $(5.7 \times 10^{17}\,{\rm m})$ from the core. Deduce the mass of the core, and compare it to the mass of our Sun.

65. Suppose all the mass of the Earth were compacted into a small spherical ball. What radius must the sphere have so that the acceleration due to gravity at the Earth's new surface was equal to the acceleration due to gravity at the surface of the Sun?

66. A plumb bob (a mass m hanging on a string) is deflected from the vertical by an angle θ due to a massive mountain nearby (Fig. 6–30). (a) Find an approximate formula for θ in terms of the mass of the mountain, $m_{\rm M}$, the distance to its center, $D_{\rm M}$, and the radius and mass of the Earth. (b) Make a rough estimate of the mass of Mt. Everest, assuming it has the shape of a cone 4000 m high and base of diameter 4000 m. Assume its mass per unit volume is 3000 kg per m^3. (c) Estimate the angle θ of the plumb bob if it is 5 km from the center of Mt. Everest.

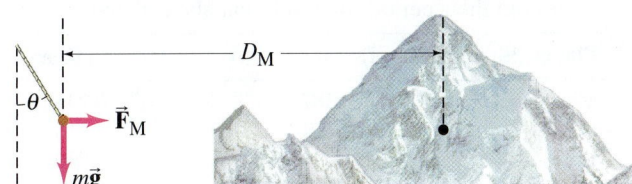

FIGURE 6–30 Problem 66.

67. A geologist searching for oil finds that the gravity at a certain location is 2 parts in 10^7 smaller than average. Assume that a deposit of oil is located 2000 m directly below. Estimate the size of the deposit, assumed spherical. Take the density (mass per unit volume) of rock to be 3000 kg/m^3 and that of oil to be 800 kg/m^3.

68. You are an astronaut in the space shuttle pursuing a satellite in need of repair. You are in a circular orbit of the same radius as the satellite (400 km above the Earth), but 25 km behind it. (a) How long will it take to overtake the satellite if you reduce your orbital radius by 1.0 km? (b) By how much must you reduce your orbital radius to catch up in 7.0 h?

69. A science-fiction tale describes an artificial "planet" in the form of a band completely encircling a sun (Fig. 6–31). The inhabitants live on the inside surface (where it is always noon). Imagine that this sun is exactly like our own, that the distance to the band is the same as the Earth–Sun distance (to make the climate temperate), and that the ring rotates quickly enough to produce an apparent gravity of g as on Earth. What will be the period of revolution, this planet's year, in Earth days?

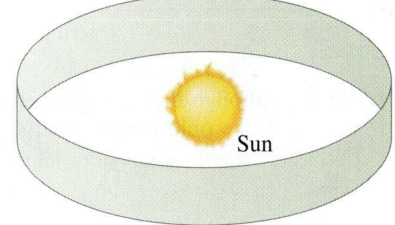

FIGURE 6–31 Problem 69.

70. How long would a day be if the Earth were rotating so fast that objects at the equator were apparently weightless?

71. An asteroid of mass m is in a circular orbit of radius r around the Sun with a speed v. It has an impact with another asteroid of mass M and is kicked into a new circular orbit with a speed of $1.5\,v$. What is the radius of the new orbit in terms of r?

72. Newton had the data listed in Table 6–4, plus the relative sizes of these objects: in terms of the Sun's radius R, the radii of Jupiter and Earth were $0.0997R$ and $0.0109R$. Newton used this information to determine that the average density $\rho\,(=$ mass/volume) of Jupiter is slightly less than that of the Sun, while the average density of the Earth is four times that of the Sun. Thus, without leaving his home planet, Newton was able to predict that the composition of the Sun and Jupiter is markedly different than that of Earth. Reproduce Newton's calculation and find his values for the ratios $\rho_{\rm J}/\rho_{\rm Sun}$ and $\rho_{\rm E}/\rho_{\rm Sun}$ (the modern values for these ratios are 0.93 and 3.91, respectively).

TABLE 6–4 Problem 72

	Orbital Radius, R (in AU = 1.50×10^{11} m)	Orbital Period, T (Earth days)
Venus about Sun	0.724	224.70
Callisto about Jupiter	0.01253	16.69
Moon about Earth	0.003069	27.32

73. A satellite circles a spherical planet of unknown mass in a circular orbit of radius 2.0×10^7 m. The magnitude of the gravitational force exerted on the satellite by the planet is 120 N. (a) What would be the magnitude of the gravitational force exerted on the satellite by the planet if the radius of the orbit were increased to 3.0×10^7 m? (b) If the satellite circles the planet once every 2.0 h in the larger orbit, what is the mass of the planet?

74. A uniform sphere has mass M and radius r. A spherical cavity (no mass) of radius $r/2$ is then carved within this sphere as shown in Fig. 6–32 (the cavity's surface passes through the sphere's center and just touches the sphere's outer surface). The centers of the original sphere and the cavity lie on a straight line, which defines the x axis. With what gravitational force will the hollowed-out sphere attract a point mass m which lies on the x axis a distance d from the sphere's center? [Hint: Subtract the effect of the "small" sphere (the cavity) from that of the larger entire sphere.]

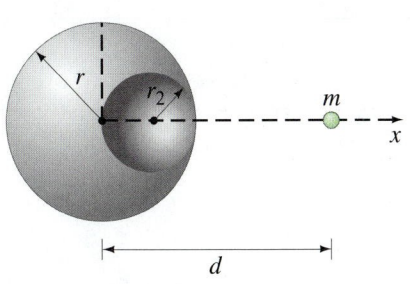

FIGURE 6–32 Problem 74.

75. The gravitational force at different places on Earth due to the Sun and the Moon depends on each point's distance from the Sun or Moon, and this variation is what causes the **tides**. Use the values inside the front cover of this book for the Earth–Moon distance R_{EM}, the Earth–Sun distance R_{ES}, the Moon's mass M_M, the Sun's mass, M_S, and the Earth's radius R_E. (*a*) First consider two small pieces of the Earth, each of mass m, one on the side of the Earth nearest the Moon, the other on the side farthest from the Moon. Show that the ratio of the Moon's gravitational forces on these two masses is

$$\left(\frac{F_{near}}{F_{far}}\right)_M = 1.0687.$$

(*b*) Next consider two small pieces of the Earth, each of mass m, one on the nearest point of Earth to the Sun, the other at the farthest point from the Sun. Show that the ratio of the Sun's gravitational forces on these two masses is

$$\left(\frac{F_{near}}{F_{far}}\right)_S = 1.000171.$$

(*c*) Show that the ratio of the Sun's average gravitational force on the Earth compared to that of the Moon's is

$$\left(\frac{F_S}{F_M}\right)_{avg} = 178.$$

Note that the Moon's smaller force varies much more across the Earth's diameter than the Sun's larger force. (*d*) Estimate the resulting "force difference" (the cause of the tides)

$$\Delta F = F_{near} - F_{far} = F_{far}\left(\frac{F_{near}}{F_{far}} - 1\right) \approx F_{avg}\left(\frac{F_{near}}{F_{far}} - 1\right)$$

for the Moon and for the Sun. Show that the ratio of the tide-causing force differences due to the Moon compared to the Sun is

$$\frac{\Delta F_M}{\Delta F_S} \approx 2.3.$$

Thus the Moon's influence on tide production is over two times as great as the Sun's.

*76. A particle is released at a height r_E (radius of Earth) above the Earth's surface. Determine its velocity when it hits the Earth. Ignore air resistance. [*Hint*: Use Newton's second law, the law of universal gravitation, the chain rule, and integrate.]

77. Estimate the value of the gravitational constant G in Newton's law of universal gravitation using the following data: the acceleration due to gravity at the Earth's surface is about $10\ \text{m/s}^2$; the Earth has a circumference of about $40 \times 10^6\ \text{m}$; rocks found on the Earth's surface typically have densities of about $3000\ \text{kg/m}^3$ and assume this density is constant throughout (even though you suspect it is not true).

78. Between the orbits of Mars and Jupiter, several thousand small objects called asteroids move in nearly circular orbits around the Sun. Consider an asteroid that is spherically shaped with radius r and density $2700\ \text{kg/m}^3$. (*a*) You find yourself on the surface of this asteroid and throw a baseball at a speed of $22\ \text{m/s}$ (about $50\ \text{mi/h}$). If the baseball is to travel around the asteroid in a circular orbit, what is the largest radius asteroid on which you are capable of accomplishing this feat? (*b*) After you throw the baseball, you turn around and face the opposite direction and catch the baseball. How much time T elapses between your throw and your catch?

*Numerical/Computer

*79. (II) The accompanying table shows the data for the mean distances of planets (except Pluto) from the Sun in our solar system, and their periods of revolution about the Sun.

Planet	Mean Distance (AU)	Period (Years)
Mercury	0.387	0.241
Venus	0.723	0.615
Earth	1.000	1.000
Mars	1.524	1.881
Jupiter	5.203	11.88
Saturn	9.539	29.46
Uranus	19.18	84.01
Neptune	30.06	164.8

(*a*) Graph the square of the periods as a function of the cube of the average distances, and find the best-fit straight line. (*b*) If the period of Pluto is 247.7 years, estimate the mean distance of Pluto from the Sun from the best-fit line.

Answers to Exercises

A: g would double.

B: (*b*).

C: (*b*).

D: No; even though they are experiencing weightlessness, the massive ball would require a large force to throw and to decelerate when caught (inertial mass, Newton's second law).

E: 6.17 yr.

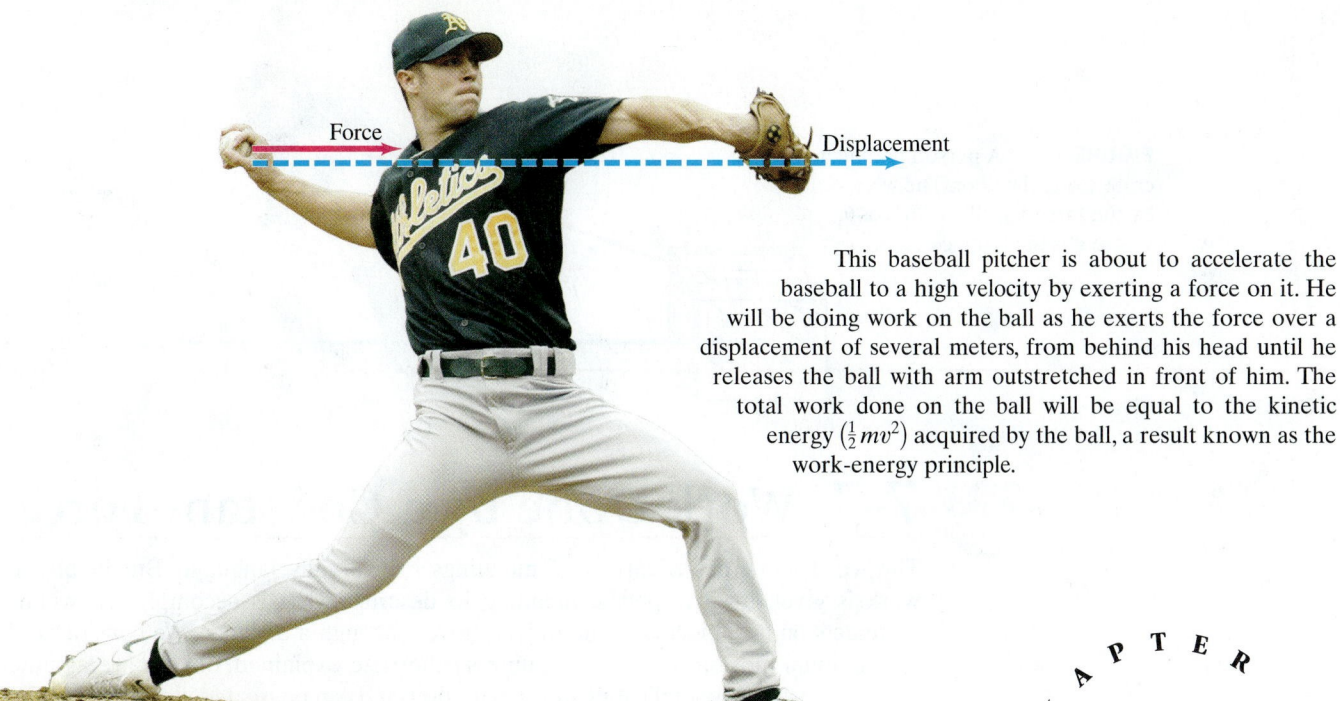

Force

Displacement

This baseball pitcher is about to accelerate the baseball to a high velocity by exerting a force on it. He will be doing work on the ball as he exerts the force over a displacement of several meters, from behind his head until he releases the ball with arm outstretched in front of him. The total work done on the ball will be equal to the kinetic energy $\left(\frac{1}{2}mv^2\right)$ acquired by the ball, a result known as the work-energy principle.

C H A P T E R

7

Work and Energy

CHAPTER-OPENING QUESTION—Guess now!

You push very hard on a heavy desk, trying to move it. You do work on the desk:

(a) Whether or not it moves, as long as you are exerting a force.

(b) Only if it starts moving.

(c) Only if it doesn't move.

(d) Never—it does work on you.

(e) None of the above.

U ntil now we have been studying the translational motion of an object in terms of Newton's three laws of motion. In that analysis, *force* has played a central role as the quantity determining the motion. In this Chapter and the two that follow, we discuss an alternative analysis of the translational motion of objects in terms of the quantities *energy* and *momentum*. The significance of energy and momentum is that they are *conserved*. In quite general circumstances they remain constant. That conserved quantities exist gives us not only a deeper insight into the nature of the world but also gives us another way to approach solving practical problems.

The conservation laws of energy and momentum are especially valuable in dealing with systems of many objects, in which a detailed consideration of the forces involved would be difficult or impossible. These laws are applicable to a wide range of phenomena, including the atomic and subatomic worlds, where Newton's laws cannot be applied.

This Chapter is devoted to the very important concept of *energy* and the closely related concept of *work*. These two quantities are scalars and so have no direction associated with them, which often makes them easier to work with than vector quantities such as acceleration and force.

CONTENTS

7–1 Work Done by a Constant Force

7–2 Scalar Product of Two Vectors

7–3 Work Done by a Varying Force

7–4 Kinetic Energy and the Work-Energy Principle

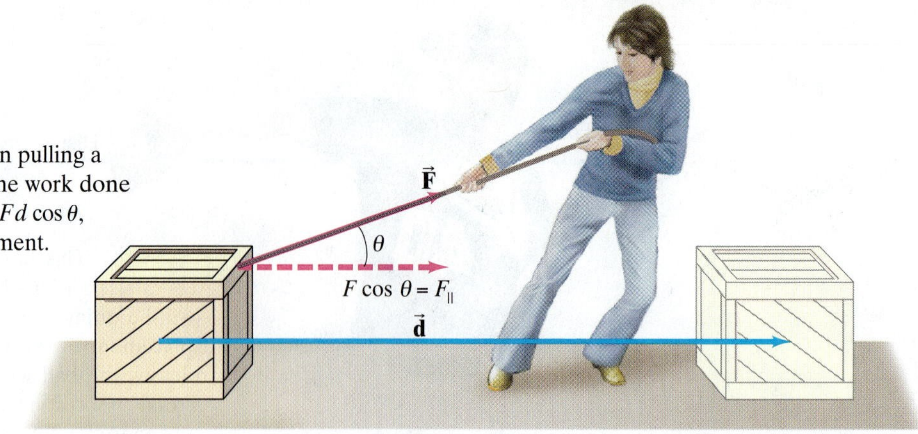

FIGURE 7–1 A person pulling a crate along the floor. The work done by the force $\vec{F}$ is $W = Fd\cos\theta$, where $\vec{d}$ is the displacement.

$$\vec{F}$$

$$\theta$$

$$F\cos\theta = F_{\parallel}$$

$$\vec{d}$$

7–1 Work Done by a Constant Force

The word *work* has a variety of meanings in everyday language. But in physics, work is given a very specific meaning to describe what is accomplished when a force acts on an object, and the object moves through a distance. We consider only translational motion for now and, unless otherwise explained, objects are assumed to be rigid with no complicating internal motion, and can be treated like particles. Then the **work** done on an object by a constant force (constant in both magnitude and direction) is defined to be *the product of the magnitude of the displacement times the component of the force parallel to the displacement*. In equation form, we can write

$$W = F_{\parallel}d,$$

where $F_{\parallel}$ is the component of the constant force $\vec{F}$ parallel to the displacement $\vec{d}$. We can also write

$$W = Fd\cos\theta, \tag{7–1}$$

where F is the magnitude of the constant force, d is the magnitude of the displacement of the object, and θ is the angle between the directions of the force and the displacement (Fig. 7–1). The $\cos\theta$ factor appears in Eq. 7–1 because $F\cos\theta\,(= F_{\parallel})$ is the component of $\vec{F}$ that is parallel to $\vec{d}$. Work is a scalar quantity—it has only magnitude, which can be positive or negative.

Let us consider the case in which the motion and the force are in the same direction, so $\theta = 0$ and $\cos\theta = 1$; in this case, $W = Fd$. For example, if you push a loaded grocery cart a distance of 50 m by exerting a horizontal force of 30 N on the cart, you do $30\,\text{N} \times 50\,\text{m} = 1500\,\text{N}\cdot\text{m}$ of work on the cart.

As this example shows, in SI units work is measured in newton-meters ($\text{N}\cdot\text{m}$). A special name is given to this unit, the **joule** (J): $1\,\text{J} = 1\,\text{N}\cdot\text{m}$.

[In the cgs system, the unit of work is called the *erg* and is defined as $1\,\text{erg} = 1\,\text{dyne}\cdot\text{cm}$. In British units, work is measured in foot-pounds. It is easy to show that $1\,\text{J} = 10^7\,\text{erg} = 0.7376\,\text{ft}\cdot\text{lb}$.]

A force can be exerted on an object and yet do no work. If you hold a heavy bag of groceries in your hands at rest, you do no work on it. You do exert a force on the bag, but the displacement of the bag is zero, so the work done by you on the bag is $W = 0$. You need both a force and a displacement to do work. You also do no work on the bag of groceries if you carry it as you walk horizontally across the floor at constant velocity, as shown in Fig. 7–2. No horizontal force is required to move the bag at a constant velocity. The person shown in Fig. 7–2 does exert an upward force $\vec{F}_P$ on the bag equal to its weight. But this upward force is perpendicular to the horizontal displacement of the bag and thus is doing no work. This conclusion comes from our definition of work, Eq. 7–1: $W = 0$, because $\theta = 90°$ and $\cos 90° = 0$. Thus, when a particular force is perpendicular to the displacement, no work is done by that force. When you start or stop walking, there is a horizontal acceleration and you do briefly exert a horizontal force, and thus do work on the bag.

FIGURE 7–2 The person does no work on the bag of groceries since $\vec{F}_P$ is perpendicular to the displacement $\vec{d}$.

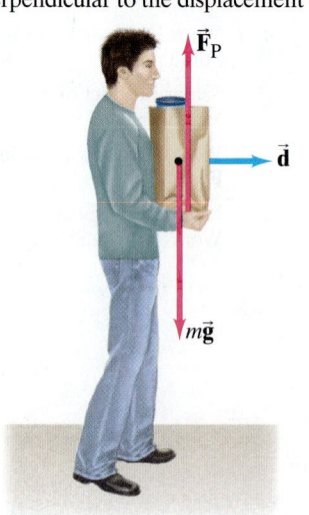

$$\vec{F}_P$$

$$\vec{d}$$

$$m\vec{g}$$

When we deal with work, as with force, it is necessary to specify whether you are talking about work done *by* a specific object or done *on* a specific object. It is also important to specify whether the work done is due to one particular force (and which one), or the total (net) work done by the *net force* on the object.

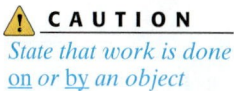
EXAMPLE 7–1 **Work done on a crate.** A person pulls a 50-kg crate 40 m along a horizontal floor by a constant force $F_P = 100$ N, which acts at a 37° angle as shown in Fig. 7–3. The floor is rough and exerts a friction force, $F_{fr} = 50$ N. Determine (*a*) the work done by each force acting on the crate, and (*b*) the net work done on the crate.

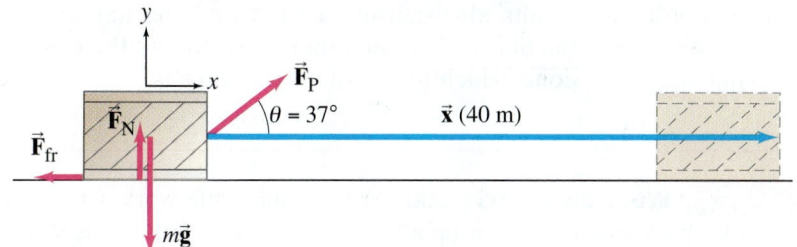

FIGURE 7–3 Example 7–1. A 50-kg crate is pulled along a floor.

APPROACH We choose our coordinate system so that $\vec{x}$ can be the vector that represents the 40-m displacement (that is, along the x axis). Four forces act on the crate, as shown in Fig. 7–3: the force exerted by the person $\vec{F}_P$; the friction force $\vec{F}_{fr}$; the gravitational force exerted by the Earth, $m\vec{g}$; and the normal force $\vec{F}_N$ exerted upward by the floor. The net force on the crate is the vector sum of these four forces.

SOLUTION (*a*) The work done by the gravitational and normal forces is zero, since they are perpendicular to the displacement $\vec{x}$ ($\theta = 90°$ in Eq. 7–1):

$$W_G = mgx \cos 90° = 0 \qquad \text{and} \qquad W_N = F_N x \cos 90° = 0.$$

The work done by $\vec{F}_P$ is

$$W_P = F_P x \cos \theta = (100\,\text{N})(40\,\text{m}) \cos 37° = 3200\,\text{J}.$$

The work done by the friction force is

$$W_{fr} = F_{fr} x \cos 180° = (50\,\text{N})(40\,\text{m})(-1) = -2000\,\text{J}.$$

The angle between the displacement $\vec{x}$ and $\vec{F}_{fr}$ is 180° because they point in opposite directions. Since the force of friction is opposing the motion (and $\cos 180° = -1$), it does *negative* work on the crate.

(*b*) The net work can be calculated in two equivalent ways:

(1) The net work done on an object is the algebraic sum of the work done by each force, since work is a scalar:

$$W_{net} = W_G + W_N + W_P + W_{fr} = 0 + 0 + 3200\,\text{J} - 2000\,\text{J} = 1200\,\text{J}.$$

(2) The net work can also be calculated by first determining the net force on the object and then taking its component along the displacement: $(F_{net})_x = F_P \cos \theta - F_{fr}$. Then the net work is

$$W_{net} = (F_{net})_x x = (F_P \cos \theta - F_{fr})x$$
$$= (100\,\text{N} \cos 37° - 50\,\text{N})(40\,\text{m}) = 1200\,\text{J}.$$

In the vertical (*y*) direction, there is no displacement and no work done.

In Example 7–1 we saw that friction did negative work. In general, the work done by a force is negative whenever the force (or the component of the force, $F_∥$) acts in the direction opposite to the direction of motion.

EXERCISE A A box is dragged a distance d across a floor by a force $\vec{F}_P$ which makes an angle θ with the horizontal as in Fig. 7–1 or 7–3. If the magnitude of $\vec{F}_P$ is held constant but the angle θ is increased, the work done by $\vec{F}_P$ (*a*) remains the same; (*b*) increases; (*c*) decreases; (*d*) first increases, then decreases.

EXERCISE B Return to the Chapter-Opening Question, page 163, and answer it again now. Try to explain why you may have answered differently the first time.

Work

1. **Draw a free-body diagram** showing all the forces acting on the object you choose to study.
2. **Choose** an *xy* coordinate system. If the object is in motion, it may be convenient to choose one of the coordinate directions as the direction of one of the forces, or as the direction of motion. [Thus, for an object on an incline, you might choose one coordinate axis to be parallel to the incline.]
3. **Apply Newton's laws** to determine any unknown forces.
4. Find the **work done** *by a specific force* on the object by using $W = Fd \cos \theta$ for a constant force. Note that the work done is negative when a force tends to oppose the displacement.
5. To find the **net work** done on the object, either (*a*) find the work done by each force and add the results algebraically; or (*b*) find the net force on the object, F_{net}, and then use it to find the net work done, which for constant net force is:
$$W_{net} = F_{net}\, d \cos \theta.$$

FIGURE 7–4 Example 7–2.

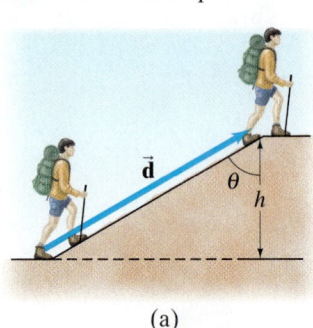

(a)

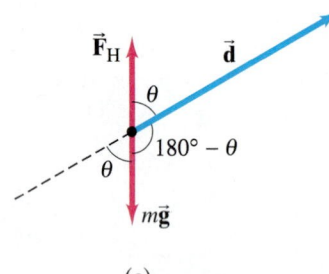

(b)

(c)

EXAMPLE 7–2 **Work on a backpack.** (*a*) Determine the work a hiker must do on a 15.0-kg backpack to carry it up a hill of height $h = 10.0\,\text{m}$, as shown in Fig. 7–4a. Determine also (*b*) the work done by gravity on the backpack, and (*c*) the net work done on the backpack. For simplicity, assume the motion is smooth and at constant velocity (i.e., acceleration is zero).

APPROACH We explicitly follow the steps of the Problem Solving Strategy above.

SOLUTION
1. **Draw a free-body diagram.** The forces on the backpack are shown in Fig. 7–4b: the force of gravity, $m\vec{\mathbf{g}}$, acting downward; and $\vec{\mathbf{F}}_H$, the force the hiker must exert upward to support the backpack. The acceleration is zero, so horizontal forces on the backpack are negligible.
2. **Choose a coordinate system.** We are interested in the vertical motion of the backpack, so we choose the *y* coordinate as positive vertically upward.
3. **Apply Newton's laws.** Newton's second law applied in the vertical direction to the backpack gives
$$\Sigma F_y = ma_y$$
$$F_H - mg = 0$$
since $a_y = 0$. Hence,
$$F_H = mg = (15.0\,\text{kg})(9.80\,\text{m/s}^2) = 147\,\text{N}.$$
4. **Work done by a specific force.** (*a*) To calculate the work done by the hiker on the backpack, we write Eq. 7–1 as
$$W_H = F_H(d \cos \theta),$$
and we note from Fig. 7–4a that $d \cos \theta = h$. So the work done by the hiker is
$$W_H = F_H(d \cos \theta) = F_H h = mgh$$
$$= (147\,\text{N})(10.0\,\text{m}) = 1470\,\text{J}.$$
Note that the work done depends only on the change in elevation and not on the angle of the hill, θ. The hiker would do the same work to lift the pack vertically the same height h.
(*b*) The work done by gravity on the backpack is (from Eq. 7–1 and Fig. 7–4c)
$$W_G = F_G d \cos(180° - \theta).$$
Since $\cos(180° - \theta) = -\cos \theta$, we have
$$W_G = F_G d(-\cos \theta) = mg(-d \cos \theta)$$
$$= -mgh$$
$$= -(15.0\,\text{kg})(9.80\,\text{m/s}^2)(10.0\,\text{m}) = -1470\,\text{J}.$$

NOTE The work done by gravity (which is negative here) doesn't depend on the angle of the incline, only on the vertical height h of the hill. This is because gravity acts vertically, so only the vertical component of displacement contributes to work done.
5. **Net work done.** (*c*) The *net* work done on the backpack is $W_{net} = 0$, since the net force on the backpack is zero (it is assumed not to accelerate significantly). We can also determine the net work done by adding the work done by each force:
$$W_{net} = W_G + W_H = -1470\,\text{J} + 1470\,\text{J} = 0.$$

NOTE Even though the *net* work done by all the forces on the backpack is zero, the hiker *does* do work on the backpack equal to 1470 J.

CONCEPTUAL EXAMPLE 7–3 **Does the Earth do work on the Moon?** The Moon revolves around the Earth in a nearly circular orbit, with approximately constant tangential speed, kept there by the gravitational force exerted by the Earth. Does gravity do (*a*) positive work, (*b*) negative work, or (*c*) no work at all on the Moon?

RESPONSE The gravitational force $\vec{\mathbf{F}}_G$ on the Moon (Fig. 7–5) acts toward the Earth and provides its centripetal force, inward along the radius of the Moon's orbit. The Moon's displacement at any moment is tangent to the circle, in the direction of its velocity, perpendicular to the radius and perpendicular to the force of gravity. Hence the angle θ between the force $\vec{\mathbf{F}}_G$ and the instantaneous displacement of the Moon is 90°, and the work done by gravity is therefore zero (cos 90° = 0). This is why the Moon, as well as artificial satellites, can stay in orbit without expenditure of fuel: no work needs to be done against the force of gravity.

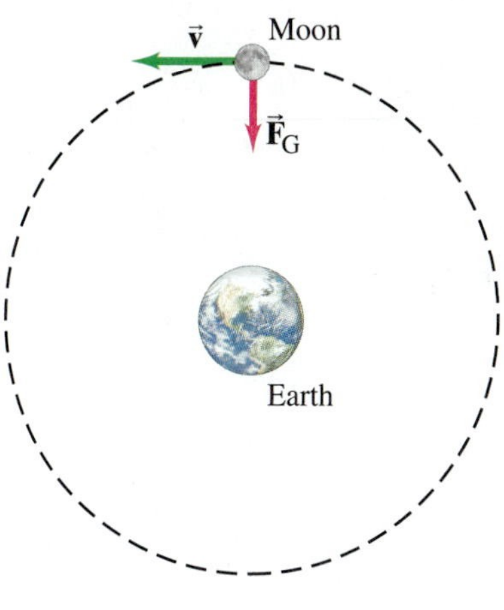

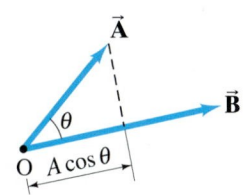

FIGURE 7–5 Example 7–3.

7–2 Scalar Product of Two Vectors

Although work is a scalar, it involves the product of two quantities, force and displacement, both of which are vectors. Therefore, we now investigate the multiplication of vectors, which will be useful throughout the book, and apply it to work.

Because vectors have direction as well as magnitude, they cannot be multiplied in the same way that scalars are. Instead we must *define* what the operation of vector multiplication means. Among the possible ways to define how to multiply vectors, there are three ways that we find useful in physics: (1) multiplication of a vector by a scalar, which was discussed in Section 3–3; (2) multiplication of one vector by a second vector to produce a scalar; (3) multiplication of one vector by a second vector to produce another vector. The third type, called the *vector product*, will be discussed later, in Section 11–2.

We now discuss the second type, called the *scalar product*, or *dot product* (because a dot is used to indicate the multiplication). If we have two vectors, $\vec{\mathbf{A}}$ and $\vec{\mathbf{B}}$, then their **scalar** (or **dot**) **product** is defined to be

$$\vec{\mathbf{A}} \cdot \vec{\mathbf{B}} = AB \cos \theta, \qquad (7\text{–}2)$$

where A and B are the magnitudes of the vectors and θ is the angle ($< 180°$) between them when their tails touch, Fig. 7–6. Since A, B, and $\cos \theta$ are scalars, then so is the scalar product $\vec{\mathbf{A}} \cdot \vec{\mathbf{B}}$ (read "A dot B").

This definition, Eq. 7–2, fits perfectly with our definition of the work done by a constant force, Eq. 7–1. That is, we can write the work done by a constant force as the scalar product of force and displacement:

$$W = \vec{\mathbf{F}} \cdot \vec{\mathbf{d}} = Fd \cos \theta. \qquad (7\text{–}3)$$

Indeed, the definition of scalar product, Eq. 7–2, is so chosen because many physically important quantities, such as work (and others we will meet later), can be described as the scalar product of two vectors.

An equivalent definition of the scalar product is that it is the product of the magnitude of one vector (say B) and the component (or projection) of the other vector along the direction of the first ($A \cos \theta$). See Fig. 7–6.

Since A, B, and $\cos \theta$ are scalars, it doesn't matter in what order they are multiplied. Hence the scalar product is **commutative**:

$$\vec{\mathbf{A}} \cdot \vec{\mathbf{B}} = \vec{\mathbf{B}} \cdot \vec{\mathbf{A}}. \qquad \text{[commutative property]}$$

It is also easy to show that it is **distributive** (see Problem 33 for the proof):

$$\vec{\mathbf{A}} \cdot (\vec{\mathbf{B}} + \vec{\mathbf{C}}) = \vec{\mathbf{A}} \cdot \vec{\mathbf{B}} + \vec{\mathbf{A}} \cdot \vec{\mathbf{C}}. \qquad \text{[distributive property]}$$

FIGURE 7–6 The scalar product, or dot product, of two vectors $\vec{\mathbf{A}}$ and $\vec{\mathbf{B}}$ is $\vec{\mathbf{A}} \cdot \vec{\mathbf{B}} = AB \cos \theta$. The scalar product can be interpreted as the magnitude of one vector (B in this case) times the projection of the other vector, $A \cos \theta$, onto $\vec{\mathbf{B}}$.

Let us write our vectors $\vec{A}$ and $\vec{B}$ in terms of their rectangular components using unit vectors (Section 3–5, Eq. 3–5) as

$$\vec{A} = A_x\hat{i} + A_y\hat{j} + A_z\hat{k}$$
$$\vec{B} = B_x\hat{i} + B_y\hat{j} + B_z\hat{k}.$$

We will take the scalar product, $\vec{A} \cdot \vec{B}$, of these two vectors, remembering that the unit vectors, $\hat{i}$, $\hat{j}$, and $\hat{k}$, are perpendicular to each other

$$\hat{i} \cdot \hat{i} = \hat{j} \cdot \hat{j} = \hat{k} \cdot \hat{k} = 1$$
$$\hat{i} \cdot \hat{j} = \hat{i} \cdot \hat{k} = \hat{j} \cdot \hat{k} = 0.$$

Thus the scalar product equals

$$\vec{A} \cdot \vec{B} = \left(A_x\hat{i} + A_y\hat{j} + A_z\hat{k}\right) \cdot \left(B_x\hat{i} + B_y\hat{j} + B_z\hat{k}\right)$$
$$= A_x B_x + A_y B_y + A_z B_z. \tag{7–4}$$

Equation 7–4 is very useful.

If $\vec{A}$ is perpendicular to $\vec{B}$, then Eq. 7–2 tells us $\vec{A} \cdot \vec{B} = AB \cos 90° = 0$. But the converse, given that $\vec{A} \cdot \vec{B} = 0$, can come about in three different ways: $\vec{A} = 0$, $\vec{B} = 0$, or $\vec{A} \perp \vec{B}$.

FIGURE 7–7 Example 7–4. Work done by a force $\vec{F}_P$ acting at an angle θ to the ground is $W = \vec{F}_P \cdot \vec{d}$.

EXAMPLE 7–4 **Using the dot product.** The force shown in Fig. 7–7 has magnitude $F_P = 20\,\text{N}$ and makes an angle of 30° to the ground. Calculate the work done by this force using Eq. 7–4 when the wagon is dragged 100 m along the ground.

APPROACH We choose the x axis horizontal to the right and the y axis vertically upward, and write $\vec{F}_P$ and $\vec{d}$ in terms of unit vectors.

SOLUTION

$$\vec{F}_P = F_x\hat{i} + F_y\hat{j} = \left(F_P \cos 30°\right)\hat{i} + \left(F_P \sin 30°\right)\hat{j} = (17\,\text{N})\hat{i} + (10\,\text{N})\hat{j},$$

whereas $\vec{d} = (100\,\text{m})\hat{i}$. Then, using Eq. 7–4,

$$W = \vec{F}_P \cdot \vec{d} = (17\,\text{N})(100\,\text{m}) + (10\,\text{N})(0) + (0)(0) = 1700\,\text{J}.$$

Note that by choosing the x axis along $\vec{d}$ we simplified the calculation because $\vec{d}$ then has only one component.

7–3 Work Done by a Varying Force

If the force acting on an object is constant, the work done by that force can be calculated using Eq. 7–1. In many cases, however, the force varies in magnitude or direction during a process. For example, as a rocket moves away from Earth, work is done to overcome the force of gravity, which varies as the inverse square of the distance from the Earth's center. Other examples are the force exerted by a spring, which increases with the amount of stretch, or the work done by a varying force exerted to pull a box or cart up an uneven hill.

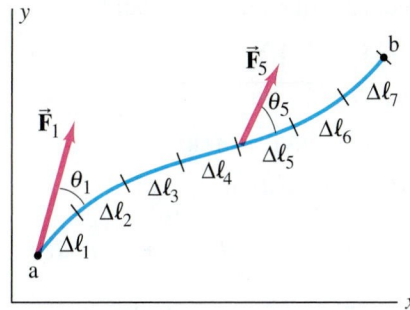

FIGURE 7–8 A particle acted on by a variable force, $\vec{\mathbf{F}}$, moves along the path shown from point a to point b.

Figure 7–8 shows the path of an object in the xy plane as it moves from point a to point b. The path has been divided into short intervals each of length $\Delta\ell_1, \Delta\ell_2, \dots, \Delta\ell_7$. A force $\vec{\mathbf{F}}$ acts at each point on the path, and is indicated at two points as $\vec{\mathbf{F}}_1$ and $\vec{\mathbf{F}}_5$. During each small interval $\Delta\ell$, the force is approximately constant. For the first interval, the force does work ΔW of approximately (see Eq. 7–1)

$$\Delta W \approx F_1 \cos\theta_1 \Delta\ell_1.$$

In the second interval the work done is approximately $F_2 \cos\theta_2 \Delta\ell_2$, and so on. The total work done in moving the particle the total distance $\ell = \Delta\ell_1 + \Delta\ell_2 + \dots + \Delta\ell_7$ is the sum of all these terms:

$$W \approx \sum_{i=1}^{7} F_i \cos\theta_i \Delta\ell_i. \tag{7–5}$$

We can examine this graphically by plotting $F\cos\theta$ versus the distance ℓ along the path as shown in Fig. 7–9a. The distance ℓ has been subdivided into the same seven intervals (see the vertical dashed lines). The value of $F\cos\theta$ at the center of each interval is indicated by the *horizontal* dashed lines. Each of the shaded rectangles has an area $(F_i\cos\theta)(\Delta\ell_i)$, which is a good estimate of the work done during the interval. The estimate of the work done along the entire path given by Eq. 7–5, equals the sum of the areas of all the rectangles. If we subdivide the distance into a greater number of intervals, so that each $\Delta\ell_i$ is smaller, the estimate of the work done becomes more accurate (the assumption that F is constant over each interval is more accurate). Letting each $\Delta\ell_i$ approach zero (so we approach an infinite number of intervals), we obtain an exact result for the work done:

$$W = \lim_{\Delta\ell_i \to 0} \Sigma F_i \cos\theta_i \Delta\ell_i = \int_a^b F\cos\theta \, d\ell. \tag{7–6}$$

This limit as $\Delta\ell_i \to 0$ is the *integral* of $(F\cos\theta \, d\ell)$ from point a to point b. The symbol for the integral, $\int$, is an elongated S to indicate an infinite sum; and $\Delta\ell$ has been replaced by $d\ell$, meaning an infinitesimal distance. [We also discussed this in the optional Section 2–9.]

In this limit as $\Delta\ell$ approaches zero, the total area of the rectangles (Fig. 7–9a) approaches the area between the $(F\cos\theta)$ curve and the ℓ axis from a to b as shown shaded in Fig. 7–9b. That is, *the work done by a variable force in moving an object between two points is equal to the area under the $(F\cos\theta)$ versus (ℓ) curve between those two points.*

In the limit as $\Delta\ell$ approaches zero, the infinitesimal distance $d\ell$ equals the magnitude of the infinitesimal displacement vector $d\vec{\boldsymbol{\ell}}$. The direction of the vector $d\vec{\boldsymbol{\ell}}$ is along the tangent to the path at that point, so θ is the angle between $\vec{\mathbf{F}}$ and $d\vec{\boldsymbol{\ell}}$ at any point. Thus we can rewrite Eq. 7–6, using dot-product notation:

$$W = \int_a^b \vec{\mathbf{F}} \cdot d\vec{\boldsymbol{\ell}}. \tag{7–7}$$

This is a *general definition of work.* In this equation, a and b represent two points in space, (x_a, y_a, z_a) and (x_b, y_b, z_b). The integral in Eq. 7–7 is called a *line integral* since it is the integral of $F\cos\theta$ along the line that represents the path of the object. (Equation 7–1 for a constant force is a special case of Eq. 7–7.)

FIGURE 7–9 Work done by a force F is (a) approximately equal to the sum of the areas of the rectangles, (b) exactly equal to the area under the curve of $F\cos\theta$ vs. ℓ.

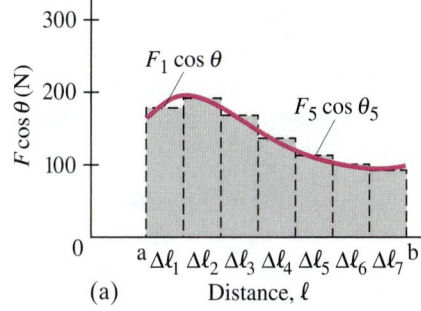

(a)

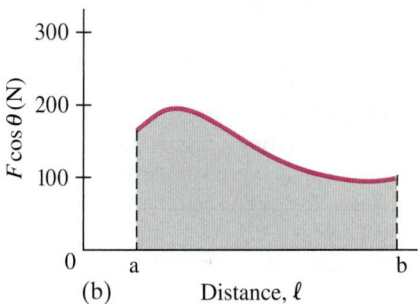

(b)

In rectangular coordinates, any force can be written

$$\vec{F} = F_x\hat{i} + F_y\hat{j} + F_z\hat{k}$$

and the displacement $d\vec{\ell}$ is

$$d\vec{\ell} = dx\hat{i} + dy\hat{j} + dz\hat{k}.$$

Then the work done can be written

$$W = \int_{x_a}^{x_b} F_x\, dx + \int_{y_a}^{y_b} F_y\, dy + \int_{z_a}^{z_b} F_z\, dz.$$

To actually use Eq. 7–6 or 7–7 to calculate the work, there are several options: (1) If $F \cos\theta$ is known as a function of position, a graph like that of Fig. 7–9b can be made and the area determined graphically. (2) Another possibility is to use numerical integration (numerical summing), perhaps with the aid of a computer or calculator. (3) A third possibility is to use the analytical methods of integral calculus, when it is doable. To do so, we must be able to write $\vec{F}$ as a function of position, $F(x, y, z)$, and we must know the path. Let's look at some specific examples.

Work Done by a Spring Force

Let us determine the work needed to stretch or compress a coiled spring, such as that shown in Fig. 7–10. For a person to hold a spring either stretched or compressed an amount x from its normal (relaxed) length requires a force F_P that is directly proportional to x. That is,

$$F_P = kx,$$

where k is a constant, called the *spring constant* (or *spring stiffness constant*), and is a measure of the stiffness of the particular spring. The spring itself exerts a force in the opposite direction (Fig. 7–10b or c):

$$F_S = -kx. \tag{7–8}$$

This force is sometimes called a "restoring force" because the spring exerts its force in the direction opposite the displacement (hence the minus sign), and thus acts to return the spring to its normal length. Equation 7–8 is known as the **spring equation** or **Hooke's law**, and is accurate for springs as long as x is not too great (see Section 12–4) and no permanent deformation occurs.

Let us calculate the work a person does to stretch (or compress) a spring from its normal (unstretched) length, $x_a = 0$, to an extra length, $x_b = x$. We assume the stretching is done slowly, so that the acceleration is essentially zero. The force $\vec{F}_P$ is exerted parallel to the axis of the spring, along the x axis, so $\vec{F}_P$ and $d\vec{\ell}$ are parallel. Hence, since $d\vec{\ell} = dx\hat{i}$ in this case, the work done by the person is[†]

$$W_P = \int_{x_a=0}^{x_b=x} [F_P(x)\hat{i}] \cdot [dx\hat{i}] = \int_0^x F_P(x)\, dx = \int_0^x kx\, dx = \tfrac{1}{2}kx^2\Big|_0^x = \tfrac{1}{2}kx^2.$$

(As is frequently done, we have used x to represent both the variable of integration, and the particular value of x at the end of the interval $x_a = 0$ to $x_b = x$.) Thus we see that the work needed is proportional to the square of the distance stretched (or compressed), x.

This same result can be obtained by computing the area under the graph of F vs. x (with $\cos\theta = 1$ in this case) as shown in Fig. 7–11. Since the area is a triangle of altitude kx and base x, the work a person does to stretch or compress a spring an amount x is

$$W = \tfrac{1}{2}(x)(kx) = \tfrac{1}{2}kx^2,$$

which is the same result as before. Because $W \propto x^2$, it takes the same amount of work to stretch a spring or compress it the same amount x.

[†]See the Table of Integrals, Appendix B.

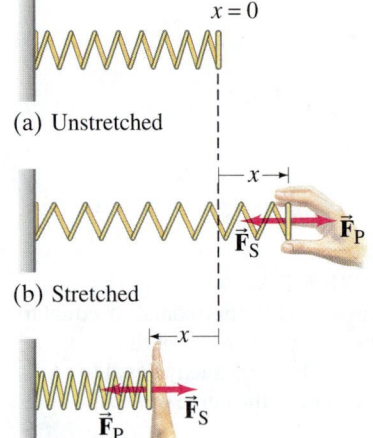

(a) Unstretched

(b) Stretched

(c) Compressed

FIGURE 7–10 (a) Spring in normal (unstretched) position. (b) Spring is stretched by a person exerting a force $\vec{F}_P$ to the right (positive direction). The spring pulls back with a force $\vec{F}_S$ where $F_S = -kx$. (c) Person compresses the spring ($x < 0$), and the spring pushes back with a force $F_S = -kx$ where $F_S > 0$ because $x < 0$.

FIGURE 7–11 Work done to stretch a spring a distance x equals the triangular area under the curve $F = kx$. The area of a triangle is $\tfrac{1}{2} \times$ base $\times$ altitude, so $W = \tfrac{1}{2}(x)(kx) = \tfrac{1}{2}kx^2$.

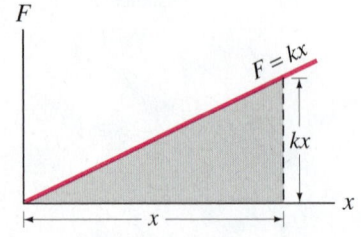

EXAMPLE 7–5 **Work done on a spring.** (a) A person pulls on the spring in Fig. 7–10, stretching it 3.0 cm, which requires a maximum force of 75 N. How much work does the person do? (b) If, instead, the person compresses the spring 3.0 cm, how much work does the person do?

APPROACH The force $F = kx$ holds at each point, including x_{max}. Hence F_{max} occurs at $x = x_{max}$.

SOLUTION (a) First we need to calculate the spring constant k:

$$k = \frac{F_{max}}{x_{max}} = \frac{75\,\text{N}}{0.030\,\text{m}} = 2.5 \times 10^3\,\text{N/m}.$$

Then the work done by the person on the spring is

$$W = \tfrac{1}{2}kx_{max}^2 = \tfrac{1}{2}(2.5 \times 10^3\,\text{N/m})(0.030\,\text{m})^2 = 1.1\,\text{J}.$$

(b) The force that the person exerts is still $F_P = kx$, though now both x and F_P are negative (x is positive to the right). The work done is

$$W_P = \int_{x=0}^{x=-0.030\,\text{m}} F_P(x)\,dx = \int_0^{x=-0.030\,\text{m}} kx\,dx = \tfrac{1}{2}kx^2 \Big|_0^{-0.030\,\text{m}}$$

$$= \tfrac{1}{2}(2.5 \times 10^3\,\text{N/m})(-0.030\,\text{m})^2 = 1.1\,\text{J},$$

which is the same as for stretching it.

NOTE We cannot use $W = Fd$ (Eq. 7–1) for a spring because the force is not constant.

A More Complex Force Law—Robot Arm

EXAMPLE 7–6 **Force as function of x.** A robot arm that controls the position of a video camera (Fig. 7–12) in an automated surveillance system is manipulated by a motor that exerts a force on the arm. The force is given by

$$F(x) = F_0\left(1 + \frac{1}{6}\frac{x^2}{x_0^2}\right),$$

where $F_0 = 2.0\,\text{N}$, $x_0 = 0.0070\,\text{m}$, and x is the position of the end of the arm. If the arm moves from $x_1 = 0.010\,\text{m}$ to $x_2 = 0.050\,\text{m}$, how much work did the motor do?

APPROACH The force applied by the motor is not a linear function of x. We can determine the integral $\int F(x)\,dx$, or the area under the $F(x)$ curve (shown in Fig. 7–13).

SOLUTION We integrate to find the work done by the motor:

$$W_M = F_0\int_{x_1}^{x_2}\left(1 + \frac{x^2}{6x_0^2}\right)dx = F_0\int_{x_1}^{x_2}dx + \frac{F_0}{6x_0^2}\int_{x_1}^{x_2}x^2\,dx$$

$$= F_0\left(x + \frac{1}{6x_0^2}\frac{x^3}{3}\right)\Big|_{x_1}^{x_2}.$$

We put in the values given and obtain

$$W_M = 2.0\,\text{N}\left[(0.050\,\text{m} - 0.010\,\text{m}) + \frac{(0.050\,\text{m})^3 - (0.010\,\text{m})^3}{(3)(6)(0.0070\,\text{m})^2}\right] = 0.36\,\text{J}.$$

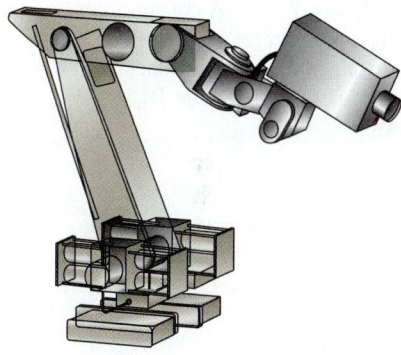

FIGURE 7–12 Robot arm positions a video camera.

FIGURE 7–13 Example 7–6.

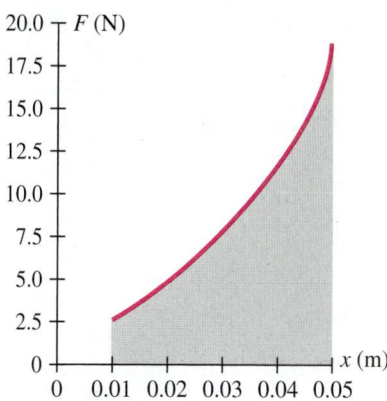

7–4 Kinetic Energy and the Work-Energy Principle

Energy is one of the most important concepts in science. Yet we cannot give a simple general definition of energy in only a few words. Nonetheless, each specific type of energy can be defined fairly simply. In this Chapter we define translational kinetic energy; in the next Chapter, we take up potential energy. In later Chapters we will examine other types of energy, such as that related to heat (Chapters 19 and 20). The crucial aspect of all the types of energy is that the sum of all types, the *total energy*, is the same after any process as it was before: that is, energy is a conserved quantity.

For the purposes of this Chapter, we can define energy in the traditional way as "the ability to do work." This simple definition is not very precise, nor is it really valid for all types of energy.[†] It works, however, for mechanical energy which we discuss in this Chapter and the next. We now define and discuss one of the basic types of energy, kinetic energy.

A moving object can do work on another object it strikes. A flying cannonball does work on a brick wall it knocks down; a moving hammer does work on a nail it drives into wood. In either case, a moving object exerts a force on a second object which undergoes a displacement. An object in motion has the ability to do work and thus can be said to have energy. The energy of motion is called **kinetic energy**, from the Greek word *kinetikos*, meaning "motion."

To obtain a quantitative definition for kinetic energy, let us consider a simple rigid object of mass m (treated as a particle) that is moving in a straight line with an initial speed v_1. To accelerate it uniformly to a speed v_2, a constant net force F_{net} is exerted on it parallel to its motion over a displacement d, Fig. 7–14.

FIGURE 7–14 A constant net force F_{net} accelerates a car from speed v_1 to speed v_2 over a displacement d. The net work done is $W_{net} = F_{net} d$.

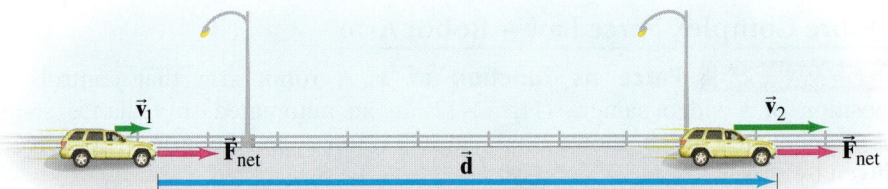

Then the net work done on the object is $W_{net} = F_{net} d$. We apply Newton's second law, $F_{net} = ma$, and use Eq. 2–12c $(v_2^2 = v_1^2 + 2ad)$, which we rewrite as

$$a = \frac{v_2^2 - v_1^2}{2d},$$

where v_1 is the initial speed and v_2 the final speed. Substituting this into $F_{net} = ma$, we determine the work done:

$$W_{net} = F_{net} d = mad = m\left(\frac{v_2^2 - v_1^2}{2d}\right)d = m\left(\frac{v_2^2 - v_1^2}{2}\right)$$

or

$$W_{net} = \tfrac{1}{2}mv_2^2 - \tfrac{1}{2}mv_1^2. \tag{7–9}$$

We *define* the quantity $\tfrac{1}{2}mv^2$ to be the **translational kinetic energy**, K, of the object:

Kinetic energy (defined)

$$K = \tfrac{1}{2}mv^2. \tag{7–10}$$

(We call this "translational" kinetic energy to distinguish it from rotational kinetic energy, which we discuss in Chapter 10.) Equation 7–9, derived here for one-dimensional motion with a constant force, is valid in general for translational motion of an object in three dimensions and even if the force varies, as we will show at the end of this Section.

[†]Energy associated with heat is often not available to do work, as we will discuss in Chapter 20.

We can rewrite Eq. 7–9 as:

$$W_{\text{net}} = K_2 - K_1$$

or

$$W_{\text{net}} = \Delta K = \tfrac{1}{2}mv_2^2 - \tfrac{1}{2}mv_1^2. \qquad \textbf{(7–11)}$$

Equation 7–11 (or Eq. 7–9) is a useful result known as the **work-energy principle**. It can be stated in words:

The net work done on an object is equal to the change in the object's kinetic energy.

Notice that we made use of Newton's second law, $F_{\text{net}} = ma$, where F_{net} is the *net* force—the sum of all forces acting on the object. Thus, the work-energy principle is valid only if W is the *net work* done on the object—that is, the work done by all forces acting on the object.

The work-energy principle is a very useful reformulation of Newton's laws. It tells us that if (positive) net work W is done on an object, the object's kinetic energy increases by an amount W. The principle also holds true for the reverse situation: if the net work W done on an object is negative, the object's kinetic energy decreases by an amount W. That is, a net force exerted on an object opposite to the object's direction of motion decreases its speed and its kinetic energy. An example is a moving hammer (Fig. 7–15) striking a nail. The net force on the hammer ($-\vec{F}$ in Fig. 7–15, where $\vec{F}$ is assumed constant for simplicity) acts toward the left, whereas the displacement $\vec{d}$ of the hammer is toward the right. So the net work done on the hammer, $W_h = (F)(d)(\cos 180°) = -Fd$, is negative and the hammer's kinetic energy decreases (usually to zero).

Figure 7–15 also illustrates how energy can be considered the ability to do work. The hammer, as it slows down, does positive work on the nail: $W_n = (+F)(+d)(\cos 0°) = Fd$ and is positive. The decrease in kinetic energy of the hammer ($= Fd$ by Eq. 7–11) is equal to the work the hammer can do on another object, the nail in this case.

The translational kinetic energy ($= \tfrac{1}{2}mv^2$) is directly proportional to the mass of the object, and it is also proportional to the *square* of the speed. Thus, if the mass is doubled, the kinetic energy is doubled. But if the speed is doubled, the object has four times as much kinetic energy and is therefore capable of doing four times as much work.

Because of the direct connection between work and kinetic energy, energy is measured in the same units as work: joules in SI units. [The energy unit is ergs in the cgs, and foot-pounds in the British system.] Like work, kinetic energy is a scalar quantity. The kinetic energy of a group of objects is the sum of the kinetic energies of the individual objects.

The work-energy principle can be applied to a particle, and also to an object that can be approximated as a particle, such as an object that is rigid or whose internal motions are insignificant. It is very useful in simple situations, as we will see in the Examples below. The work-energy principle is not as powerful and encompassing as the law of conservation of energy which we treat in the next Chapter, and should not itself be considered a statement of energy conservation.

⚠ **CAUTION**
Work-energy valid only for net work

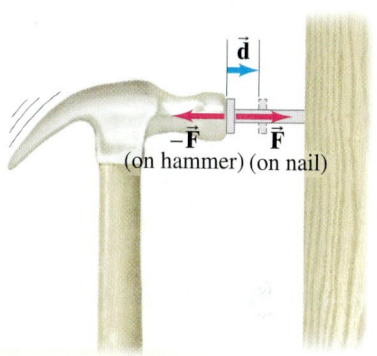

FIGURE 7–15 A moving hammer strikes a nail and comes to rest. The hammer exerts a force F on the nail; the nail exerts a force $-F$ on the hammer (Newton's third law). The work done on the nail by the hammer is positive ($W_n = Fd > 0$). The work done on the hammer by the nail is negative ($W_h = -Fd$).

EXAMPLE 7–7 **Kinetic energy and work done on a baseball.** A 145-g baseball is thrown so that it acquires a speed of 25 m/s. (*a*) What is its kinetic energy? (*b*) What was the net work done on the ball to make it reach this speed, if it started from rest?

APPROACH We use $K = \tfrac{1}{2}mv^2$, and the work-energy principle, Eq. 7–11.

SOLUTION (*a*) The kinetic energy of the ball after the throw is

$$K = \tfrac{1}{2}mv^2 = \tfrac{1}{2}(0.145\,\text{kg})(25\,\text{m/s})^2 = 45\,\text{J}.$$

(*b*) Since the initial kinetic energy was zero, the net work done is just equal to the final kinetic energy, 45 J.

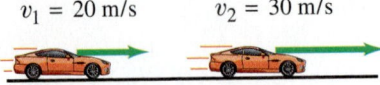

$v_1 = 20$ m/s $\qquad$ $v_2 = 30$ m/s

FIGURE 7–16 Example 7–8.

EXAMPLE 7–8 **ESTIMATE** **Work on a car, to increase its kinetic energy.**
How much net work is required to accelerate a 1000-kg car from 20 m/s to 30 m/s (Fig. 7–16)?

APPROACH A car is a complex system. The engine turns the wheels and tires which push against the ground, and the ground pushes back (see Example 4–4). We aren't interested right now in those complications. Instead, we can get a useful result using the work-energy principle, but only if we model the car as a particle or simple rigid object.

SOLUTION The net work needed is equal to the increase in kinetic energy:

$$W = K_2 - K_1 = \tfrac{1}{2}mv_2^2 - \tfrac{1}{2}mv_1^2$$
$$= \tfrac{1}{2}(1000\,\text{kg})(30\,\text{m/s})^2 - \tfrac{1}{2}(1000\,\text{kg})(20\,\text{m/s})^2 = 2.5 \times 10^5\,\text{J}.$$

EXERCISE C (a) Make a guess: will the work needed to accelerate the car in Example 7–8 from rest to 20 m/s be more than, less than, or equal to the work already calculated to accelerate it from 20 m/s to 30 m/s? (b) Make the calculation.

FIGURE 7–17 Example 7–9.

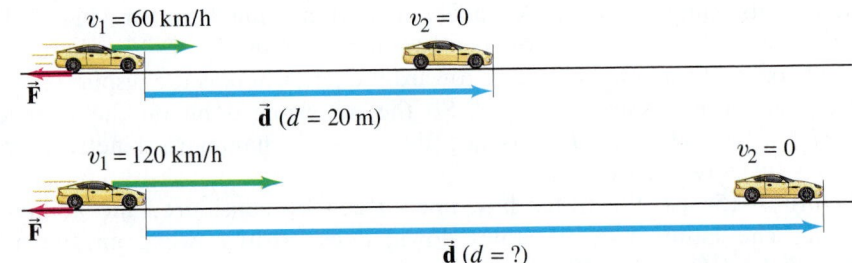

(a) $v_1 = 60$ km/h $\qquad$ $v_2 = 0$ $\qquad$ $\vec{\mathbf{F}}$ $\qquad$ $\vec{\mathbf{d}}$ $(d = 20$ m$)$

(b) $v_1 = 120$ km/h $\qquad$ $v_2 = 0$ $\qquad$ $\vec{\mathbf{F}}$ $\qquad$ $\vec{\mathbf{d}}$ $(d = ?)$

CONCEPTUAL EXAMPLE 7–9 **Work to stop a car.** A car traveling 60 km/h can brake to a stop within a distance d of 20 m (Fig. 7–17a). If the car is going twice as fast, 120 km/h, what is its stopping distance (Fig. 7–17b)? Assume the maximum braking force is approximately independent of speed.

RESPONSE Again we model the car as if it were a particle. Because the net stopping force F is approximately constant, the work needed to stop the car, Fd, is proportional to the distance traveled. We apply the work-energy principle, noting that $\vec{\mathbf{F}}$ and $\vec{\mathbf{d}}$ are in opposite directions and that the final speed of the car is zero:

$$W_{\text{net}} = Fd\cos 180° = -Fd.$$

Then

$$-Fd = \Delta K = \tfrac{1}{2}mv_2^2 - \tfrac{1}{2}mv_1^2$$
$$= 0 - \tfrac{1}{2}mv_1^2.$$

Thus, since the force and mass are constant, we see that the stopping distance, d, increases with the square of the speed:

$$d \propto v^2.$$

If the car's initial speed is doubled, the stopping distance is $(2)^2 = 4$ times as great, or 80 m.

PHYSICS APPLIED
Car's stopping distance $\propto$ initial speed squared

EXERCISE D Can kinetic energy ever be negative?

EXERCISE E (a) If the kinetic energy of an arrow is doubled, by what factor has its speed increased? (b) If its speed is doubled, by what factor does its kinetic energy increase?

EXAMPLE 7–10 **A compressed spring.** A horizontal spring has spring constant $k = 360$ N/m. (a) How much work is required to compress it from its uncompressed length ($x = 0$) to $x = 11.0$ cm? (b) If a 1.85-kg block is placed against the spring and the spring is released, what will be the speed of the block when it separates from the spring at $x = 0$? Ignore friction. (c) Repeat part (b) but assume that the block is moving on a table as in Fig. 7–18 and that some kind of constant drag force $F_D = 7.0$ N is acting to slow it down, such as friction (or perhaps your finger).

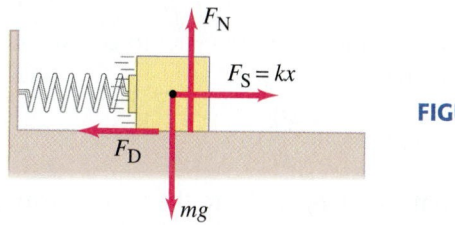

FIGURE 7–18 Example 7–10.

APPROACH We use our result from Section 7–3 that the net work, W, needed to stretch or compress a spring by a distance x is $W = \frac{1}{2}kx^2$. In (b) and (c) we use the work-energy principle.

SOLUTION (a) The work needed to compress the spring a distance $x = 0.110$ m is

$$W = \frac{1}{2}(360 \text{ N/m})(0.110 \text{ m})^2 = 2.18 \text{ J},$$

where we have converted all units to SI.

(b) In returning to its uncompressed length, the spring does 2.18 J of work on the block (same calculation as in part (a), only in reverse). According to the work-energy principle, the block acquires kinetic energy of 2.18 J. Since $K = \frac{1}{2}mv^2$, the block's speed must be

$$v = \sqrt{\frac{2K}{m}}$$

$$= \sqrt{\frac{2(2.18 \text{ J})}{1.85 \text{ kg}}} = 1.54 \text{ m/s}.$$

(c) There are two forces on the block: that exerted by the spring and that exerted by the drag force, $\vec{\mathbf{F}}_D$. Work done by a force such as friction is complicated. For one thing, heat (or, rather, "thermal energy") is produced—try rubbing your hands together. Nonetheless, the product $\vec{\mathbf{F}}_D \cdot \vec{\mathbf{d}}$ for the drag force, even when it is friction, can be used in the work-energy principle to give correct results for a particle-like object. The spring does 2.18 J of work on the block. The work done by the friction or drag force on the block, in the negative x direction, is

$$W_D = -F_D x = -(7.0 \text{ N})(0.110 \text{ m}) = -0.77 \text{ J}.$$

This work is negative because the drag force acts in the direction opposite to the displacement x. The net work done on the block is $W_{net} = 2.18 \text{ J} - 0.77 \text{ J} = 1.41 \text{ J}$. From the work-energy principle, Eq. 7–11 (with $v_2 = v$ and $v_1 = 0$), we have

$$v = \sqrt{\frac{2W_{net}}{m}}$$

$$= \sqrt{\frac{2(1.41 \text{ J})}{1.85 \text{ kg}}} = 1.23 \text{ m/s}$$

for the block's speed at the moment it separates from the spring ($x = 0$).

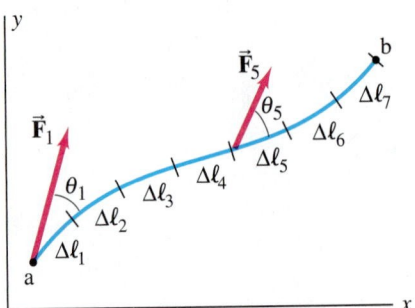

FIGURE 7–8 (repeated)
A particle acted on by a variable force $\vec{F}$, moves along the path shown from point a to point b.

General Derivation of the Work-Energy Principle

We derived the work-energy principle, Eq. 7–11, for motion in one dimension with a constant force. It is valid even if the force is variable and the motion is in two or three dimensions, as we now show. Suppose the net force $\vec{F}_{net}$ on a particle varies in both magnitude and direction, and the path of the particle is a curve as in Fig. 7–8. The net force may be considered to be a function of ℓ, the distance along the curve. The net work done is (Eq. 7–6):

$$W_{net} = \int \vec{F}_{net} \cdot d\vec{\ell} = \int F_{net} \cos\theta \, d\ell = \int F_{\parallel} \, d\ell,$$

where $F_{\parallel}$ represents the component of the net force parallel to the curve at any point. By Newton's second law,

$$F_{\parallel} = ma_{\parallel} = m\frac{dv}{dt},$$

where $a_{\parallel}$, the component of a parallel to the curve at any point, is equal to the rate of change of speed, dv/dt. We can think of v as a function of ℓ, and using the chain rule for derivatives, we have

$$\frac{dv}{dt} = \frac{dv}{d\ell}\frac{d\ell}{dt} = \frac{dv}{d\ell}v,$$

since $d\ell/dt$ is the speed v. Thus (letting 1 and 2 refer to the initial and final quantities, respectively):

$$W_{net} = \int_1^2 F_{\parallel} \, d\ell = \int_1^2 m\frac{dv}{dt} \, d\ell = \int_1^2 mv\frac{dv}{d\ell} \, d\ell = \int_1^2 mv \, dv,$$

which integrates to

$$W_{net} = \tfrac{1}{2}mv_2^2 - \tfrac{1}{2}mv_1^2 = \Delta K.$$

This is again the work-energy principle, which we have now derived for motion in three dimensions with a variable net force, using the definitions of work and kinetic energy plus Newton's second law.

Notice in this derivation that only the component of $\vec{F}_{net}$ parallel to the motion, $F_{\parallel}$, contributes to the work. Indeed, a force (or component of a force) acting perpendicular to the velocity vector does no work. Such a force changes only the direction of the velocity. It does not affect the magnitude of the velocity. One example of this is uniform circular motion in which an object moving with constant speed in a circle has a ("centripetal") force acting on it toward the center of the circle. This force does no work on the object, because (as we saw in Example 7–3) it is always perpendicular to the object's displacement $d\vec{\ell}$.

Summary

Work is done on an object by a force when the object moves through a distance, d. The **work** W done by a constant force $\vec{F}$ on an object whose position changes by a displacement $\vec{d}$ is given by

$$W = Fd\cos\theta = \vec{F} \cdot \vec{d}, \qquad (7\text{-}1, 7\text{-}3)$$

where θ is the angle between $\vec{F}$ and $\vec{d}$.

The last expression is called the scalar product of $\vec{F}$ and $\vec{d}$. In general, the **scalar product** of any two vectors $\vec{A}$ and $\vec{B}$ is defined as

$$\vec{A} \cdot \vec{B} = AB\cos\theta \qquad (7\text{-}2)$$

where θ is the angle between $\vec{A}$ and $\vec{B}$. In rectangular coordinates we can also write

$$\vec{A} \cdot \vec{B} = A_x B_x + A_y B_y + A_z B_z. \qquad (7\text{-}4)$$

The work W done by a variable force $\vec{F}$ on an object that moves from point a to point b is

$$W = \int_a^b \vec{F} \cdot d\vec{\ell} = \int_a^b F\cos\theta \, d\ell, \qquad (7\text{-}7)$$

where $d\vec{\ell}$ represents an infinitesimal displacement along the path of the object and θ is the angle between $d\vec{\ell}$ and $\vec{F}$ at each point of the object's path.

The translational **kinetic energy**, K, of an object of mass m moving with speed v is defined to be

$$K = \tfrac{1}{2}mv^2. \qquad (7\text{-}10)$$

The **work-energy principle** states that the net work done on an object by the net resultant force is equal to the change in kinetic energy of the object:

$$W_{net} = \Delta K = \tfrac{1}{2}mv_2^2 - \tfrac{1}{2}mv_1^2. \qquad (7\text{-}11)$$

Questions

1. In what ways is the word "work" as used in everyday language the same as defined in physics? In what ways is it different? Give examples of both.

2. A woman swimming upstream is not moving with respect to the shore. Is she doing any work? If she stops swimming and merely floats, is work done on her?

3. Can a centripetal force ever do work on an object? Explain.

4. Why is it tiring to push hard against a solid wall even though you are doing no work?

5. Does the scalar product of two vectors depend on the choice of coordinate system?

6. Can a dot product ever be negative? If yes, under what conditions?

7. If $\vec{A} \cdot \vec{C} = \vec{B} \cdot \vec{C}$, is it necessarily true that $\vec{A} = \vec{B}$?

8. Does the dot product of two vectors have direction as well as magnitude?

9. Can the normal force on an object ever do work? Explain.

10. You have two springs that are identical except that spring 1 is stiffer than spring 2 $(k_1 > k_2)$. On which spring is more work done: (a) if they are stretched using the same force; (b) if they are stretched the same distance?

11. If the speed of a particle triples, by what factor does its kinetic energy increase?

12. In Example 7–10, it was stated that the block separates from the compressed spring when the spring reached its equilibrium length $(x = 0)$. Explain why separation doesn't take place before (or after) this point.

13. Two bullets are fired at the same time with the same kinetic energy. If one bullet has twice the mass of the other, which has the greater speed and by what factor? Which can do the most work?

14. Does the net work done on a particle depend on the choice of reference frame? How does this affect the work-energy principle?

15. A hand exerts a constant horizontal force on a block that is free to slide on a frictionless surface (Fig. 7–19). The block starts from rest at point A, and by the time it has traveled a distance d to point B it is traveling with speed v_B. When the block has traveled another distance d to point C, will its speed be greater than, less than, or equal to $2v_B$? Explain your reasoning.

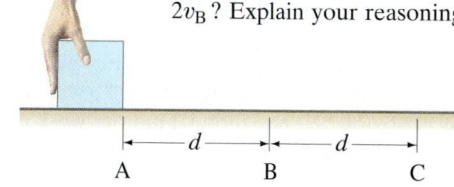

FIGURE 7–19
Question 15.

Problems

7–1 Work, Constant Force

1. (I) How much work is done by the gravitational force when a 280-kg pile driver falls 2.80 m?

2. (I) How high will a 1.85-kg rock go if thrown straight up by someone who does 80.0 J of work on it? Neglect air resistance.

3. (I) A 75.0-kg firefighter climbs a flight of stairs 20.0 m high. How much work is required?

4. (I) A hammerhead with a mass of 2.0 kg is allowed to fall onto a nail from a height of 0.50 m. What is the maximum amount of work it could do on the nail? Why do people not just "let it fall" but add their own force to the hammer as it falls?

5. (II) Estimate the work you do to mow a lawn 10 m by 20 m with a 50-cm wide mower. Assume you push with a force of about 15 N.

6. (II) A lever such as that shown in Fig. 7–20 can be used to lift objects we might not otherwise be able to lift. Show that the ratio of output force, F_O, to input force, F_I, is related to the lengths ℓ_I and ℓ_O from the pivot by $F_O/F_I = \ell_I/\ell_O$. Ignore friction and the mass of the lever, and assume the work output equals work input.

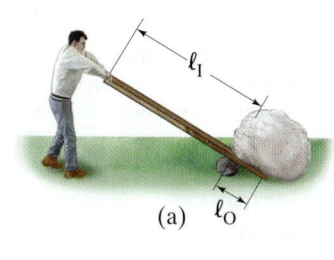
(a)
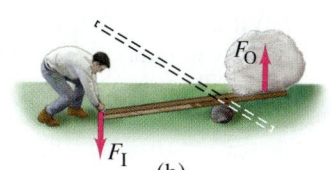
(b)

FIGURE 7–20
A lever. Problem 6.

7. (II) What is the minimum work needed to push a 950-kg car 310 m up along a 9.0° incline? Ignore friction.

8. (II) Eight books, each 4.0 cm thick with mass 1.8 kg, lie flat on a table. How much work is required to stack them one on top of another?

9. (II) A box of mass 6.0 kg is accelerated from rest by a force across a floor at a rate of 2.0 m/s² for 7.0 s. Find the net work done on the box.

10. (II) (a) What magnitude force is required to give a helicopter of mass M an acceleration of $0.10\,g$ upward? (b) What work is done by this force as the helicopter moves a distance h upward?

11. (II) A 380-kg piano slides 3.9 m down a 27° incline and is kept from accelerating by a man who is pushing back on it *parallel to the incline* (Fig. 7–21). Determine: (a) the force exerted by the man, (b) the work done by the man on the piano, (c) the work done by the force of gravity, and (d) the net work done on the piano. Ignore friction.

FIGURE 7–21
Problem 11.

12. (II) A gondola can carry 20 skiers, with a total mass of up to 2250 kg. The gondola ascends at a constant speed from the base of a mountain, at 2150 m, to the summit at 3345 m. (a) How much work does the motor do in moving a full gondola up the mountain? (b) How much work does gravity do on the gondola? (c) If the motor is capable of generating 10% more work than found in (a), what is the acceleration of the gondola?

13. (II) A 17,000-kg jet takes off from an aircraft carrier via a catapult (Fig. 7–22a). The gases thrust out from the jet's engines exert a constant force of 130 kN on the jet; the force exerted on the jet by the catapult is plotted in Fig. 7–22b. Determine: (a) the work done on the jet by the gases expelled by its engines during launch of the jet; and (b) the work done on the jet by the catapult during launch of the jet.

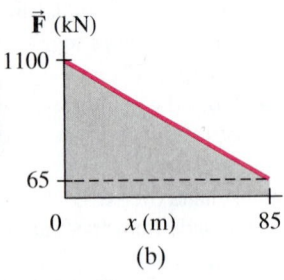

(a)   (b)

FIGURE 7–22 Problem 13.

14. (II) A 2200-N crate rests on the floor. How much work is required to move it at constant speed (a) 4.0 m along the floor against a drag force of 230 N, and (b) 4.0 m vertically?

15. (II) A grocery cart with mass of 16 kg is being pushed at constant speed up a flat 12° ramp by a force F_P which acts at an angle of 17° below the horizontal. Find the work done by each of the forces $(m\vec{\mathbf{g}}, \vec{\mathbf{F}}_N, \vec{\mathbf{F}}_P)$ on the cart if the ramp is 15 m long.

7–2 Scalar Product

16. (I) What is the dot product of $\vec{\mathbf{A}} = 2.0x^2\hat{\mathbf{i}} - 4.0x\hat{\mathbf{j}} + 5.0\hat{\mathbf{k}}$ and $\vec{\mathbf{B}} = 11.0\hat{\mathbf{i}} + 2.5x\hat{\mathbf{j}}$?

17. (I) For any vector $\vec{\mathbf{V}} = V_x\hat{\mathbf{i}} + V_y\hat{\mathbf{j}} + V_z\hat{\mathbf{k}}$ show that
$$V_x = \hat{\mathbf{i}} \cdot \vec{\mathbf{V}}, \qquad V_y = \hat{\mathbf{j}} \cdot \vec{\mathbf{V}}, \qquad V_z = \hat{\mathbf{k}} \cdot \vec{\mathbf{V}}.$$

18. (I) Calculate the angle between the vectors:
$\vec{\mathbf{A}} = 6.8\hat{\mathbf{i}} - 3.4\hat{\mathbf{j}} - 6.2\hat{\mathbf{k}}$ and $\vec{\mathbf{B}} = 8.2\hat{\mathbf{i}} + 2.3\hat{\mathbf{j}} - 7.0\hat{\mathbf{k}}$.

19. (I) Show that $\vec{\mathbf{A}} \cdot (-\vec{\mathbf{B}}) = -\vec{\mathbf{A}} \cdot \vec{\mathbf{B}}$.

20. (I) Vector $\vec{\mathbf{V}}_1$ points along the z axis and has magnitude $V_1 = 75$. Vector $\vec{\mathbf{V}}_2$ lies in the xz plane, has magnitude $V_2 = 58$, and makes a $-48°$ angle with the x axis (points below x axis). What is the scalar product $\vec{\mathbf{V}}_1 \cdot \vec{\mathbf{V}}_2$?

21. (II) Given the vector $\vec{\mathbf{A}} = 3.0\hat{\mathbf{i}} + 1.5\hat{\mathbf{j}}$, find a vector $\vec{\mathbf{B}}$ that is perpendicular to $\vec{\mathbf{A}}$.

22. (II) A constant force $\vec{\mathbf{F}} = (2.0\hat{\mathbf{i}} + 4.0\hat{\mathbf{j}})$ N acts on an object as it moves along a straight-line path. If the object's displacement is $\vec{\mathbf{d}} = (1.0\hat{\mathbf{i}} + 5.0\hat{\mathbf{j}})$ m, calculate the work done by $\vec{\mathbf{F}}$ using these alternate ways of writing the dot product: (a) $W = Fd \cos\theta$; (b) $W = F_x d_x + F_y d_y$.

23. (II) If $\vec{\mathbf{A}} = 9.0\hat{\mathbf{i}} - 8.5\hat{\mathbf{j}}$, $\vec{\mathbf{B}} = -8.0\hat{\mathbf{i}} + 7.1\hat{\mathbf{j}} + 4.2\hat{\mathbf{k}}$, and $\vec{\mathbf{C}} = 6.8\hat{\mathbf{i}} - 9.2\hat{\mathbf{j}}$, determine (a) $\vec{\mathbf{A}} \cdot (\vec{\mathbf{B}} + \vec{\mathbf{C}})$; (b) $(\vec{\mathbf{A}} + \vec{\mathbf{C}}) \cdot \vec{\mathbf{B}}$; (c) $(\vec{\mathbf{B}} + \vec{\mathbf{A}}) \cdot \vec{\mathbf{C}}$.

24. (II) Prove that $\vec{\mathbf{A}} \cdot \vec{\mathbf{B}} = A_x B_x + A_y B_y + A_z B_z$, starting from Eq. 7–2 and using the distributive property (p. 167, proved in Problem 33).

25. (II) Given vectors $\vec{\mathbf{A}} = -4.8\hat{\mathbf{i}} + 6.8\hat{\mathbf{j}}$ and $\vec{\mathbf{B}} = 9.6\hat{\mathbf{i}} + 6.7\hat{\mathbf{j}}$, determine the vector $\vec{\mathbf{C}}$ that lies in the xy plane, is perpendicular to $\vec{\mathbf{B}}$, and whose dot product with $\vec{\mathbf{A}}$ is 20.0.

26. (II) Show that if two nonparallel vectors have the same magnitude, their sum must be perpendicular to their difference.

27. (II) Let $\vec{\mathbf{V}} = 20.0\hat{\mathbf{i}} + 22.0\hat{\mathbf{j}} - 14.0\hat{\mathbf{k}}$. What angles does this vector make with the x, y, and z axes?

28. (II) Use the scalar product to prove the *law of cosines* for a triangle:
$$c^2 = a^2 + b^2 - 2ab \cos\theta,$$
where a, b, and c are the lengths of the sides of a triangle and θ is the angle opposite side c.

29. (II) Vectors $\vec{\mathbf{A}}$ and $\vec{\mathbf{B}}$ are in the xy plane and their scalar product is 20.0 units. If $\vec{\mathbf{A}}$ makes a 27.4° angle with the x axis and has magnitude $A = 12.0$ units, and $\vec{\mathbf{B}}$ has magnitude $B = 24.0$ units, what can you say about the direction of $\vec{\mathbf{B}}$?

30. (II) $\vec{\mathbf{A}}$ and $\vec{\mathbf{B}}$ are two vectors in the xy plane that make angles α and β with the x axis respectively. Evaluate the scalar product of $\vec{\mathbf{A}}$ and $\vec{\mathbf{B}}$ and deduce the following trigonometric identity: $\cos(\alpha - \beta) = \cos\alpha \cos\beta + \sin\alpha \sin\beta$.

31. (II) Suppose $\vec{\mathbf{A}} = 1.0\hat{\mathbf{i}} + 1.0\hat{\mathbf{j}} - 2.0\hat{\mathbf{k}}$ and $\vec{\mathbf{B}} = -1.0\hat{\mathbf{i}} + 1.0\hat{\mathbf{j}} + 2.0\hat{\mathbf{k}}$, (a) what is the angle between these two vectors? (b) Explain the significance of the sign in part (a).

32. (II) Find a vector of unit length in the xy plane that is perpendicular to $3.0\hat{\mathbf{i}} + 4.0\hat{\mathbf{j}}$.

33. (III) Show that the scalar product of two vectors is distributive: $\vec{\mathbf{A}} \cdot (\vec{\mathbf{B}} + \vec{\mathbf{C}}) = \vec{\mathbf{A}} \cdot \vec{\mathbf{B}} + \vec{\mathbf{A}} \cdot \vec{\mathbf{C}}$. [*Hint*: Use a diagram showing all three vectors in a plane and indicate dot products on the diagram.]

7–3 Work, Varying Force

34. (I) In pedaling a bicycle uphill, a cyclist exerts a downward force of 450 N during each stroke. If the diameter of the circle traced by each pedal is 36 cm, calculate how much work is done in each stroke.

35. (II) A spring has $k = 65$ N/m. Draw a graph like that in Fig. 7–11 and use it to determine the work needed to stretch the spring from $x = 3.0$ cm to $x = 6.5$ cm, where $x = 0$ refers to the spring's unstretched length.

36. (II) If the hill in Example 7–2 (Fig. 7–4) was not an even slope but rather an irregular curve as in Fig. 7–23, show that the same result would be obtained as in Example 7–2: namely, that the work done by gravity depends only on the height of the hill and not on its shape or the path taken.

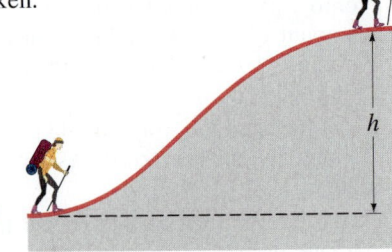

FIGURE 7–23
Problem 36.

37. (II) The net force exerted on a particle acts in the positive x direction. Its magnitude increases linearly from zero at $x = 0$, to 380 N at $x = 3.0$ m. It remains constant at 380 N from $x = 3.0$ m to $x = 7.0$ m, and then decreases linearly to zero at $x = 12.0$ m. Determine the work done to move the particle from $x = 0$ to $x = 12.0$ m graphically, by determining the area under the F_x versus x graph.

38. (II) If it requires 5.0 J of work to stretch a particular spring by 2.0 cm from its equilibrium length, how much more work will be required to stretch it an additional 4.0 cm?

39. (II) In Fig. 7–9 assume the distance axis is the x axis and that a = 10.0 m and b = 30.0 m. Estimate the work done by this force in moving a 3.50-kg object from a to b.

40. (II) The force on a particle, acting along the x axis, varies as shown in Fig. 7–24. Determine the work done by this force to move the particle along the x axis: (*a*) from $x = 0.0$ to $x = 10.0$ m; (*b*) from $x = 0.0$ to $x = 15.0$ m.

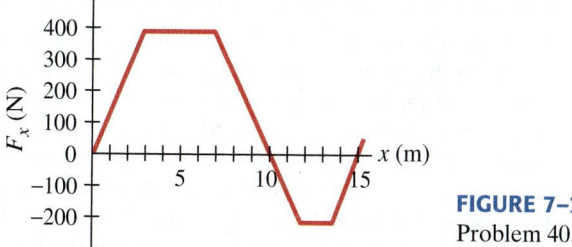

FIGURE 7–24
Problem 40.

41. (II) A child is pulling a wagon down the sidewalk. For 9.0 m the wagon stays on the sidewalk and the child pulls with a horizontal force of 22 N. Then one wheel of the wagon goes off on the grass so the child has to pull with a force of 38 N at an angle of 12° to the side for the next 5.0 m. Finally the wagon gets back on the sidewalk so the child makes the rest of the trip, 13.0 m, with a force of 22 N. How much total work did the child do on the wagon?

42. (II) The resistance of a packing material to a sharp object penetrating it is a force proportional to the fourth power of the penetration depth x; that is, $\vec{F} = -kx^4\hat{i}$. Calculate the work done to force a sharp object a distance d into the material.

43. (II) The force needed to hold a particular spring compressed an amount x from its normal length is given by $F = kx + ax^3 + bx^4$. How much work must be done to compress it by an amount X, starting from $x = 0$?

44. (II) At the top of a pole vault, an athlete actually can do work pushing on the pole before releasing it. Suppose the pushing force that the pole exerts back on the athlete is given by $F(x) = (1.5 \times 10^2 \, \text{N/m})x - (1.9 \times 10^2 \, \text{N/m}^2)x^2$ acting over a distance of 0.20 m. How much work is done on the athlete?

45. (II) Consider a force $F_1 = A/\sqrt{x}$ which acts on an object during its journey along the x axis from $x = 0.0$ to $x = 1.0$ m, where $A = 2.0 \, \text{N} \cdot \text{m}^{1/2}$. Show that during this journey, even though F_1 is infinite at $x = 0.0$, the work done on the object by this force is finite.

46. (II) Assume that a force acting on an object is given by $\vec{F} = ax\hat{i} + by\hat{j}$, where the constants $a = 3.0 \, \text{N} \cdot \text{m}^{-1}$ and $b = 4.0 \, \text{N} \cdot \text{m}^{-1}$. Determine the work done on the object by this force as it moves in a straight line from the origin to $\vec{r} = (10.0\hat{i} + 20.0\hat{j})$ m.

47. (II) An object, moving along the circumference of a circle with radius R, is acted upon by a force of constant magnitude F. The force is directed at all times at a 30° angle with respect to the tangent to the circle as shown in Fig. 7–25. Determine the work done by this force when the object moves along the half circle from A to B.

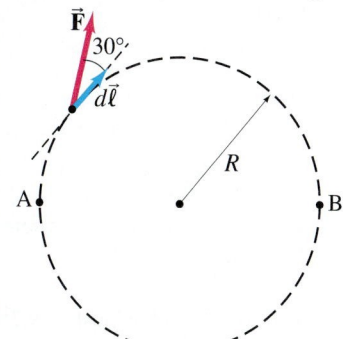

FIGURE 7–25
Problem 47.

48. (III) A 2800-kg space vehicle, initially at rest, falls vertically from a height of 3300 km above the Earth's surface. Determine how much work is done by the force of gravity in bringing the vehicle to the Earth's surface.

49. (III) A 3.0-m-long steel chain is stretched out along the top level of a horizontal scaffold at a construction site, in such a way that 2.0 m of the chain remains on the top level and 1.0 m hangs vertically, Fig. 7–26. At this point, the force on the hanging segment is sufficient to pull the entire chain over the edge. Once the chain is moving, the kinetic friction is so small that it can be neglected. How much work is performed on the chain by the force of gravity as the chain falls from the point where 2.0 m remains on the scaffold to the point where the entire chain has left the scaffold? (Assume that the chain has a linear weight density of 18 N/m.)

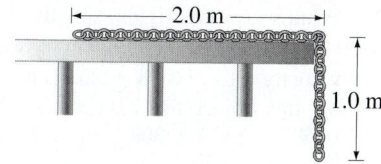

FIGURE 7–26
Problem 49.

7–4 Kinetic Energy; Work-Energy Principle

50. (I) At room temperature, an oxygen molecule, with mass of 5.31×10^{-26} kg, typically has a kinetic energy of about 6.21×10^{-21} J. How fast is it moving?

51. (I) (*a*) If the kinetic energy of a particle is tripled, by what factor has its speed increased? (*b*) If the speed of a particle is halved, by what factor does its kinetic energy change?

52. (I) How much work is required to stop an electron $(m = 9.11 \times 10^{-31} \, \text{kg})$ which is moving with a speed of 1.40×10^6 m/s?

53. (I) How much work must be done to stop a 1300-kg car traveling at 95 km/h?

54. (II) Spiderman uses his spider webs to save a runaway train, Fig. 7–27. His web stretches a few city blocks before the 10^4-kg train comes to a stop. Assuming the web acts like a spring, estimate the spring constant.

FIGURE 7–27
Problem 54.

55. (II) A baseball $(m = 145 \, \text{g})$ traveling 32 m/s moves a fielder's glove backward 25 cm when the ball is caught. What was the average force exerted by the ball on the glove?

56. (II) An 85-g arrow is fired from a bow whose string exerts an average force of 105 N on the arrow over a distance of 75 cm. What is the speed of the arrow as it leaves the bow?

57. (II) A mass m is attached to a spring which is held stretched a distance x by a force F (Fig. 7–28), and then released. The spring compresses, pulling the mass. Assuming there is no friction, determine the speed of the mass m when the spring returns: (*a*) to its normal length ($x = 0$); (*b*) to half its original extension ($x/2$).

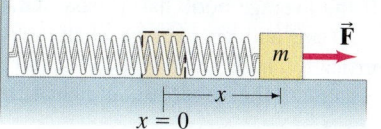

FIGURE 7–28
Problem 57.

58. (II) If the speed of a car is increased by 50%, by what factor will its minimum braking distance be increased, assuming all else is the same? Ignore the driver's reaction time.

59. (II) A 1200-kg car rolling on a horizontal surface has speed $v = 66$ km/h when it strikes a horizontal coiled spring and is brought to rest in a distance of 2.2 m. What is the spring constant of the spring?

60. (II) One car has twice the mass of a second car, but only half as much kinetic energy. When both cars increase their speed by 7.0 m/s, they then have the same kinetic energy. What were the original speeds of the two cars?

61. (II) A 4.5-kg object moving in two dimensions initially has a velocity $\vec{v}_1 = (10.0\hat{i} + 20.0\hat{j})$ m/s. A net force $\vec{F}$ then acts on the object for 2.0 s, after which the object's velocity is $\vec{v}_2 = (15.0\hat{i} + 30.0\hat{j})$ m/s. Determine the work done by $\vec{F}$ on the object.

62. (II) A 265-kg load is lifted 23.0 m vertically with an acceleration $a = 0.150\,g$ by a single cable. Determine (a) the tension in the cable; (b) the net work done on the load; (c) the work done by the cable on the load; (d) the work done by gravity on the load; (e) the final speed of the load assuming it started from rest.

63. (II) (a) How much work is done by the horizontal force $F_P = 150$ N on the 18-kg block of Fig. 7–29 when the force pushes the block 5.0 m up along the 32° frictionless incline? (b) How much work is done by the gravitational force on the block during this displacement? (c) How much work is done by the normal force? (d) What is the speed of the block (assume that it is zero initially) after this displacement? [*Hint*: Work-energy involves *net* work done.]

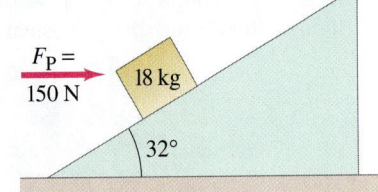

FIGURE 7–29
Problems 63 and 64.

64. (II) Repeat Problem 63 assuming a coefficient of friction $\mu_k = 0.10$.

65. (II) At an accident scene on a level road, investigators measure a car's skid mark to be 98 m long. It was a rainy day and the coefficient of friction was estimated to be 0.38. Use these data to determine the speed of the car when the driver slammed on (and locked) the brakes. (Why does the car's mass not matter?)

66. (II) A 46.0-kg crate, starting from rest, is pulled across a floor with a constant horizontal force of 225 N. For the first 11.0 m the floor is frictionless, and for the next 10.0 m the coefficient of friction is 0.20. What is the final speed of the crate after being pulled these 21.0 m?

67. (II) A train is moving along a track with constant speed v_1 relative to the ground. A person on the train holds a ball of mass m and throws it toward the front of the train with a speed v_2 relative to the train. Calculate the change in kinetic energy of the ball (a) in the Earth frame of reference, and (b) in the train frame of reference. (c) Relative to each frame of reference, how much work was done on the ball? (d) Explain why the results in part (c) are not the same for the two frames—after all, it's the same ball.

68. (III) We usually neglect the mass of a spring if it is small compared to the mass attached to it. But in some applications, the mass of the spring must be taken into account. Consider a spring of unstretched length ℓ and mass M_S uniformly distributed along the length of the spring. A mass m is attached to the end of the spring. One end of the spring is fixed and the mass m is allowed to vibrate horizontally without friction (Fig. 7–30). Each point on the spring moves with a velocity proportional to the distance from that point to the fixed end. For example, if the mass on the end moves with speed v_0, the midpoint of the spring moves with speed $v_0/2$. Show that the kinetic energy of the mass plus spring when the mass is moving with velocity v is

$$K = \tfrac{1}{2}Mv^2$$

where $M = m + \tfrac{1}{3}M_S$ is the "effective mass" of the system. [*Hint*: Let D be the total length of the stretched spring. Then the velocity of a mass dm of a spring of length dx located at x is $v(x) = v_0(x/D)$. Note also that $dm = dx(M_S/D)$.]

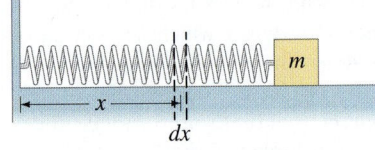

FIGURE 7–30
Problem 68.

69. (III) An elevator cable breaks when a 925-kg elevator is 22.5 m above the top of a huge spring $(k = 8.00 \times 10^4$ N/m$)$ at the bottom of the shaft. Calculate (a) the work done by gravity on the elevator before it hits the spring; (b) the speed of the elevator just before striking the spring; (c) the amount the spring compresses (note that here work is done by both the spring and gravity).

General Problems

70. (a) A 3.0-g locust reaches a speed of 3.0 m/s during its jump. What is its kinetic energy at this speed? (b) If the locust transforms energy with 35% efficiency, how much energy is required for the jump?

71. In a certain library the first shelf is 12.0 cm off the ground, and the remaining 4 shelves are each spaced 33.0 cm above the previous one. If the average book has a mass of 1.40 kg with a height of 22.0 cm, and an average shelf holds 28 books (standing vertically), how much work is required to fill all the shelves, assuming the books are all laying flat on the floor to start?

72. A 75-kg meteorite buries itself 5.0 m into soft mud. The force between the meteorite and the mud is given by $F(x) = (640\,\text{N/m}^3)x^3$, where x is the depth in the mud. What was the speed of the meteorite when it initially impacted the mud?

73. A 6.10-kg block is pushed 9.25 m up a smooth 37.0° inclined plane by a horizontal force of 75.0 N. If the initial speed of the block is 3.25 m/s up the plane, calculate (a) the initial kinetic energy of the block; (b) the work done by the 75.0-N force; (c) the work done by gravity; (d) the work done by the normal force; (e) the final kinetic energy of the block.

74. The arrangement of atoms in zinc is an example of "hexagonal close-packed" structure. Three of the nearest neighbors are found at the following (x, y, z) coordinates, given in nanometers (10^{-9} m): atom 1 is at $(0, 0, 0)$; atom 2 is at $(0.230, 0.133, 0)$; atom 3 is at $(0.077, 0.133, 0.247)$. Find the angle between two vectors: one that connects atom 1 with atom 2 and another that connects atom 1 with atom 3.

75. Two forces, $\vec{F}_1 = (1.50\hat{i} - 0.80\hat{j} + 0.70\hat{k})$ N and $\vec{F}_2 = (-0.70\hat{i} + 1.20\hat{j})$ N, are applied on a moving object of mass 0.20 kg. The displacement vector produced by the two forces is $\vec{d} = (8.0\hat{i} + 6.0\hat{j} + 5.0\hat{k})$ m. What is the work done by the two forces?

76. The barrels of the 16-in. guns (bore diameter = 16 in. = 41 cm) on the World War II battleship *U.S.S. Massachusetts* were each 15 m long. The shells each had a mass of 1250 kg and were fired with sufficient explosive force to provide them with a muzzle velocity of 750 m/s. Use the work-energy principle to determine the explosive force (assumed to be a constant) that was applied to the shell within the barrel of the gun. Express your answer in both newtons and in pounds.

77. A varying force is given by $F = Ae^{-kx}$, where x is the position; A and k are constants that have units of N and m^{-1}, respectively. What is the work done when x goes from 0.10 m to infinity?

78. The force required to compress an imperfect horizontal spring an amount x is given by $F = 150x + 12x^3$, where x is in meters and F in newtons. If the spring is compressed 2.0 m, what speed will it give to a 3.0-kg ball held against it and then released?

79. A force $\vec{F} = (10.0\hat{i} + 9.0\hat{j} + 12.0\hat{k})$ kN acts on a small object of mass 95 g. If the displacement of the object is $\vec{d} = (5.0\hat{i} + 4.0\hat{j})$ m, find the work done by the force. What is the angle between $\vec{F}$ and $\vec{d}$?

80. In the game of paintball, players use guns powered by pressurized gas to propel 33-g gel capsules filled with paint at the opposing team. Game rules dictate that a paintball cannot leave the barrel of a gun with a speed greater than 85 m/s. Model the shot by assuming the pressurized gas applies a constant force F to a 33-g capsule over the length of the 32-cm barrel. Determine F (a) using the work-energy principle, and (b) using the kinematic equations (Eqs. 2–12) and Newton's second law.

81. A softball having a mass of 0.25 kg is pitched horizontally at 110 km/h. By the time it reaches the plate, it may have slowed by 10%. Neglecting gravity, estimate the average force of air resistance during a pitch, if the distance between the plate and the pitcher is about 15 m.

82. An airplane pilot fell 370 m after jumping from an aircraft without his parachute opening. He landed in a snowbank, creating a crater 1.1 m deep, but survived with only minor injuries. Assuming the pilot's mass was 88 kg and his terminal velocity was 45 m/s, estimate: (a) the work done by the snow in bringing him to rest; (b) the average force exerted on him by the snow to stop him; and (c) the work done on him by air resistance as he fell. Model him as a particle.

83. Many cars have "5 mi/h (8 km/h) bumpers" that are designed to compress and rebound elastically without any physical damage at speeds below 8 km/h. If the material of the bumpers permanently deforms after a compression of 1.5 cm, but remains like an elastic spring up to that point, what must be the effective spring constant of the bumper material, assuming the car has a mass of 1050 kg and is tested by ramming into a solid wall?

84. What should be the spring constant k of a spring designed to bring a 1300-kg car to rest from a speed of 90 km/h so that the occupants undergo a maximum acceleration of $5.0\,g$?

85. Assume a cyclist of weight mg can exert a force on the pedals equal to 0.90 mg on the average. If the pedals rotate in a circle of radius 18 cm, the wheels have a radius of 34 cm, and the front and back sprockets on which the chain runs have 42 and 19 teeth respectively (Fig. 7–31), determine the maximum steepness of hill the cyclist can climb at constant speed. Assume the mass of the bike is 12 kg and that of the rider is 65 kg. Ignore friction. Assume the cyclist's average force is always: (a) downward; (b) tangential to pedal motion.

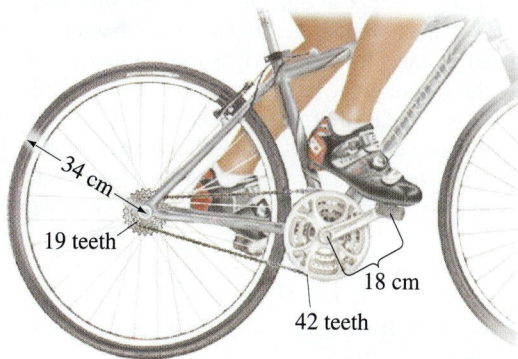

FIGURE 7–31 Problem 85.

86. A simple pendulum consists of a small object of mass m (the "bob") suspended by a cord of length ℓ (Fig. 7–32) of negligible mass. A force $\vec{F}$ is applied in the horizontal direction (so $\vec{F} = F\hat{i}$), moving the bob very slowly so the acceleration is essentially zero. (Note that the magnitude of $\vec{F}$ will need to vary with the angle θ that the cord makes with the vertical at any moment.) (a) Determine the work done by this force, $\vec{F}$, to move the pendulum from $\theta = 0$ to $\theta = \theta_0$. (b) Determine the work done by the gravitational force on the bob, $\vec{F}_G = m\vec{g}$, and the work done by the force $\vec{F}_T$ that the cord exerts on the bob.

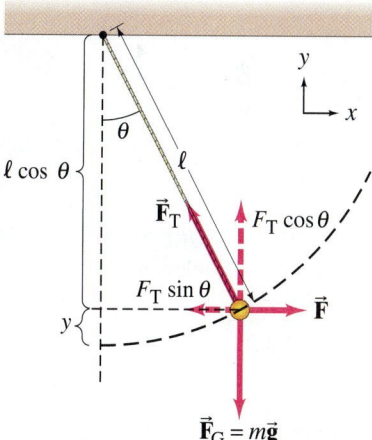

FIGURE 7–32
Problem 86.

87. A car passenger buckles himself in with a seat belt and holds his 18-kg toddler on his lap. Use the work-energy principle to answer the following questions. (a) While traveling 25 m/s, the driver has to make an emergency stop over a distance of 45 m. Assuming constant deceleration, how much force will the arms of the parent need to exert on the child during this deceleration period? Is this force achievable by an average parent? (b) Now assume that the car $(v = 25\text{ m/s})$ is in an accident and is brought to stop over a distance of 12 m. Assuming constant deceleration, how much force will the parent need to exert on the child? Is this force achievable by an average parent?

88. As an object moves along the x axis from $x = 0.0$ m to $x = 20.0$ m it is acted upon by a force given by $F = (100 - (x - 10)^2)$ N. Determine the work done by the force on the object: (a) by first sketching the F vs. x graph and estimating the area under this curve; (b) by evaluating the integral $\int_{x=0.0\,m}^{x=20\,m} F\,dx$.

89. A cyclist starts from rest and coasts down a 4.0° hill. The mass of the cyclist plus bicycle is 85 kg. After the cyclist has traveled 250 m, (a) what was the net work done by gravity on the cyclist? (b) How fast is the cyclist going? Ignore air resistance.

90. Stretchable ropes are used to safely arrest the fall of rock climbers. Suppose one end of a rope with unstretched length ℓ is anchored to a cliff and a climber of mass m is attached to the other end. When the climber is a height ℓ above the anchor point, he slips and falls under the influence of gravity for a distance 2ℓ, after which the rope becomes taut and stretches a distance x as it stops the climber (see Fig. 7–33). Assume a stretchy rope behaves as a spring with spring constant k. (a) Applying the work-energy principle, show that

$$x = \frac{mg}{k}\left[1 + \sqrt{1 + \frac{4k\ell}{mg}}\right].$$

(b) Assuming $m = 85$ kg, $\ell = 8.0$ m and $k = 850$ N/m, determine x/ℓ (the fractional stretch of the rope) and kx/mg (the force that the rope exerts on the climber compared to his own weight) at the moment the climber's fall has been stopped.

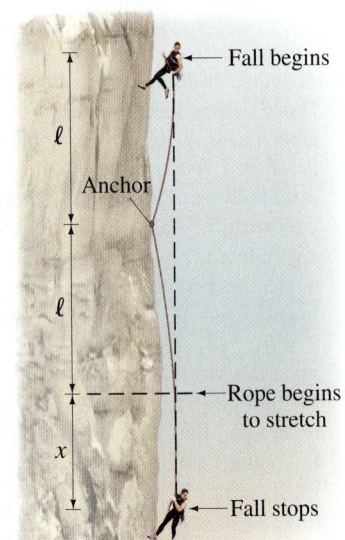

Fall begins

ℓ

Anchor

ℓ

Rope begins to stretch

x

Fall stops

FIGURE 7–33
Problem 90.

91. A small mass m hangs at rest from a vertical rope of length ℓ that is fixed to the ceiling. A force $\vec{F}$ then pushes on the mass, perpendicular to the taut rope at all times, until the rope is oriented at an angle $\theta = \theta_0$ and the mass has been raised by a vertical distance h (Fig. 7–34). Assume the force's magnitude F is adjusted so that the mass moves at constant speed along its curved trajectory. Show that the work done by $\vec{F}$ during this process equals mgh, which is equivalent to the amount of work it takes to slowly lift a mass m straight up by a height h. [Hint: When the angle is increased by $d\theta$ (in radians), the mass moves along an arc length $ds = \ell\,d\theta$.]

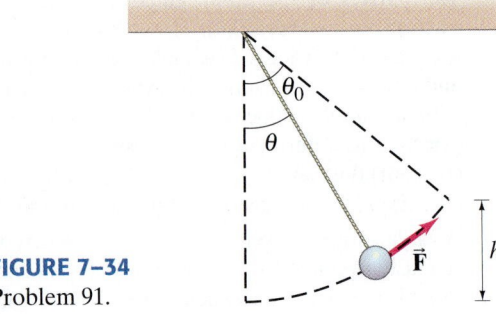

FIGURE 7–34
Problem 91.

*Numerical/Computer

*92. (II) The net force along the linear path of a particle of mass 480 g has been measured at 10.0-cm intervals, starting at $x = 0.0$, to be 26.0, 28.5, 28.8, 29.6, 32.8, 40.1, 46.6, 42.2, 48.8, 52.6, 55.8, 60.2, 60.6, 58.2, 53.7, 50.3, 45.6, 45.2, 43.2, 38.9, 35.1, 30.8, 27.2, 21.0, 22.2, and 18.6, all in newtons. Determine the total work done on the particle over this entire range.

*93. (II) When different masses are suspended from a spring, the spring stretches by different amounts as shown in the Table below. Masses are ±1.0 gram.

Mass (g)	0	50	100	150	200	250	300	350	400
Stretch (cm)	0	5.0	9.8	14.8	19.4	24.5	29.6	34.1	39.2

(a) Graph the applied force (in Newtons) versus the stretch (in meters) of the spring, and determine the best-fit straight line. (b) Determine the spring constant (N/m) of the spring from the slope of the best-fit line. (c) If the spring is stretched by 20.0 cm, estimate the force acting on the spring using the best-fit line.

A polevaulter running toward the high bar has kinetic energy. When he plants the pole and puts his weight on it, his kinetic energy gets transformed: first into elastic potential energy of the bent pole and then into gravitational potential energy as his body rises. As he crosses the bar, the pole is straight and has given up all its elastic potential energy to the athlete's gravitational potential energy. Nearly all his kinetic energy has disappeared, also becoming gravitational potential energy of his body at the great height of the bar (world record over 6 m), which is exactly what he wants. In these, and all other energy transformations that continually take place in the world, the total energy is always conserved. Indeed, the conservation of energy is one of the greatest laws of physics, and finds applications in a wide range of other fields.

C H A P T E R

8

Conservation of Energy

CHAPTER-OPENING QUESTION—Guess now!

A skier starts at the top of a hill. On which run does her gravitational potential energy change the most: (a), (b), (c), or (d); or are they (e) all the same? On which run would her speed at the bottom be the fastest if the runs are icy and we assume no friction? Recognizing that there is always some friction, answer the above two questions again. List your four answers now.

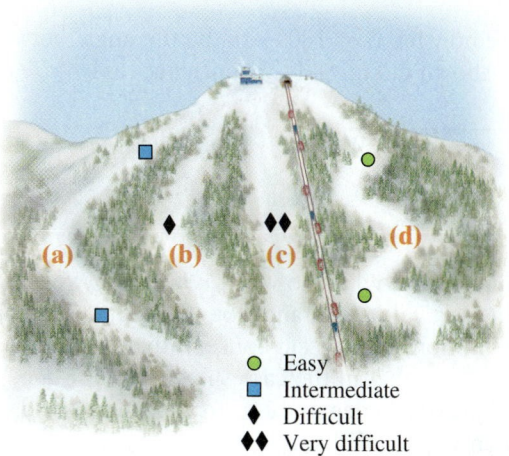

● Easy
■ Intermediate
◆ Difficult
◆◆ Very difficult

CONTENTS

8–1 Conservative and Nonconservative Forces

8–2 Potential Energy

8–3 Mechanical Energy and Its Conservation

8–4 Problem Solving Using Conservation of Mechanical Energy

8–5 The Law of Conservation of Energy

8–6 Energy Conservation with Dissipative Forces: Solving Problems

8–7 Gravitational Potential Energy and Escape Velocity

8–8 Power

*8–9 Potential Energy Diagrams; Stable and Unstable Equilibrium

This chapter continues the discussion of the concepts of work and energy begun in Chapter 7 and introduces additional types of energy, in particular potential energy. Now we will see why the concept of energy is so important. The reason, ultimately, is that energy is conserved—the total energy *always* remains constant in any process. That a quantity can be defined which remains constant, as far as our best experiments can tell, is a remarkable statement about nature. The law of conservation of energy is, in fact, one of the great unifying principles of science.

The law of conservation of energy also gives us another tool, another approach, to solving problems. There are many situations for which an analysis based on Newton's laws would be difficult or impossible—the forces may not be known or accessible to measurement. But often these situations can be dealt with using the law of conservation of energy.

In this Chapter we will mainly treat objects as if they were particles or rigid objects that undergo only translational motion, with no internal or rotational motion.

8–1 Conservative and Nonconservative Forces

We will find it important to categorize forces into two types: conservative and nonconservative. By definition, we call any force a **conservative force** if

> **the work done by the force on an object moving from one point to another depends only on the initial and final positions of the object, and is independent of the particular path taken.**

A conservative force can be a function *only of position*, and cannot depend on other variables like time or velocity.

We can readily show that the force of gravity is a conservative force. The gravitational force on an object of mass m near the Earth's surface is $\vec{\mathbf{F}} = m\vec{\mathbf{g}}$, where $\vec{\mathbf{g}}$ is a constant. The work done by this gravitational force on an object that falls a vertical distance h is $W_G = Fd = mgh$ (see Fig. 8–1a). Now suppose that instead of moving vertically downward or upward, an object follows some arbitrary path in the xy plane, as shown in Fig. 8–1b. The object starts at a vertical height y_1 and reaches a height y_2, where $y_2 - y_1 = h$. To calculate the work done by gravity, W_G, we use Eq. 7–7:

$$W_G = \int_1^2 \vec{\mathbf{F}}_G \cdot d\vec{\boldsymbol{\ell}}$$

$$= \int_1^2 mg \cos \theta \, d\ell.$$

We now let $\phi = 180° - \theta$ be the angle between $d\vec{\boldsymbol{\ell}}$ and its vertical component dy, as shown in Fig. 8–1b. Then, since $\cos \theta = -\cos \phi$ and $dy = d\ell \cos \phi$, we have

$$W_G = -\int_{y_1}^{y_2} mg \, dy$$

$$= -mg(y_2 - y_1). \tag{8–1}$$

Since $(y_2 - y_1)$ is the vertical height h, we see that the work done depends only on the vertical height and does *not* depend on the particular path taken! Hence, by definition, gravity is a conservative force.

Note that in the case shown in Fig. 8–1b, $y_2 > y_1$ and therefore the work done by gravity is negative. If on the other hand $y_2 < y_1$, so that the object is falling, then W_G is positive.

FIGURE 8–1 Object of mass m: (a) falls a height h vertically; (b) is raised along an arbitrary two-dimensional path.

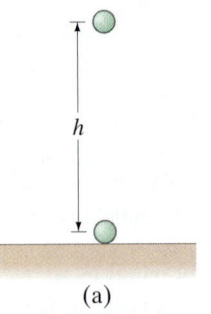

(a)

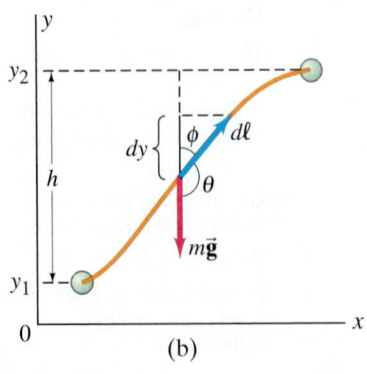

(b)

We can give the definition of a conservative force in another, completely equivalent way:

a force is conservative if the net work done by the force on an object moving around any closed path is zero.

To see why this is equivalent to our earlier definition, consider a small object that moves from point 1 to point 2 via either of two paths labeled A and B in Fig. 8–2a. If we assume a conservative force acts on the object, the work done by this force is the same whether the object takes path A or path B, by our first definition. This work to get from point 1 to point 2 we will call W. Now consider the round trip shown in Fig. 8–2b. The object moves from 1 to 2 via path A and our force does work W. Our object then returns to point 1 via path B. How much work is done during the return? In going from 1 to 2 via path B the work done is W, which by definition equals $\int_1^2 \vec{F} \cdot d\vec{l}$. In doing the reverse, going from 2 to 1, the force $\vec{F}$ at each point is the same, but $d\vec{l}$ is directed in precisely the opposite direction. Consequently $\vec{F} \cdot d\vec{l}$ has the opposite sign at each point so the total work done in making the return trip from 2 to 1 must be $-W$. Hence the total work done in going from 1 to 2 and back to 1 is $W + (-W) = 0$, which proves the equivalence of the two above definitions for a conservative force.

The second definition of a conservative force illuminates an important aspect of such a force: the *work done by a conservative force is recoverable* in the sense that if positive work is done *by* an object (on something else) on one part of a closed path, an equivalent amount of negative work will be done by the object on its return.

As we saw above, the force of gravity is conservative, and it is easy to show that the elastic force $(F = -kx)$ is also conservative.

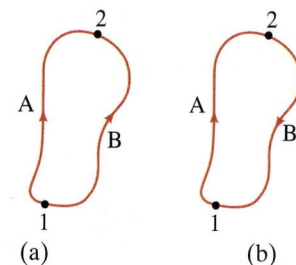

FIGURE 8–2 (a) A tiny object moves between points 1 and 2 via two different paths, A and B. (b) The object makes a round trip, via path A from point 1 to point 2 and via path B back to point 1.

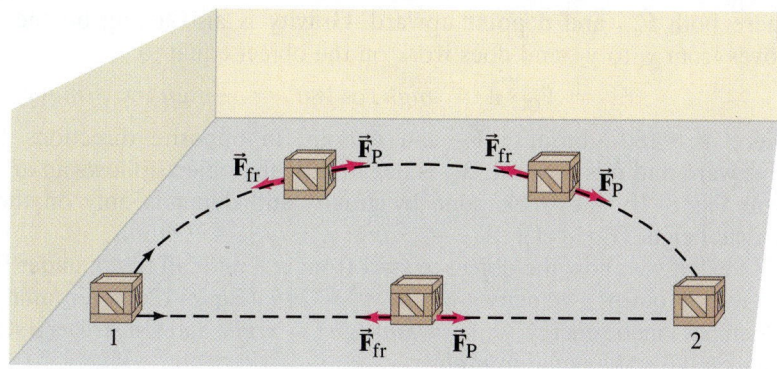

FIGURE 8–3   A crate is pushed slowly at constant speed across a rough floor from position 1 to position 2 via two paths: one straight and one curved. The pushing force $\vec{F}_P$ is in the direction of motion at each point. (The friction force opposes the motion.) Hence for a constant magnitude pushing force, the work it does is $W = F_P d$, so if d is greater (as for the curved path), then W is greater. The work done does not depend only on points 1 and 2; it also depends on the path taken.

Many forces, such as friction and a push or pull exerted by a person, are **nonconservative forces** since any work they do depends on the path. For example, if you push a crate across a floor from one point to another, the work you do depends on whether the path taken is straight, or is curved. As shown in Fig. 8–3, if a crate is pushed slowly from point 1 to point 2 along the longer semicircular path, you do more work against friction than if you push it along the straight path. This is because the distance is greater and, unlike the gravitational force, the pushing force $\vec{F}_P$ is in the direction of motion at each point. Thus the work done by the person in Fig. 8–3 does not depend *only* on points 1 and 2; it depends also on the path taken. The force of kinetic friction, also shown in Fig. 8–3, always opposes the motion; it too is a nonconservative force, and we discuss how to treat it later in this Chapter (Section 8–6). Table 8–1 lists a few conservative and nonconservative forces.

TABLE 8–1 Conservative and Nonconservative Forces

Conservative Forces	Nonconservative Forces
Gravitational	Friction
Elastic	Air resistance
Electric	Tension in cord
	Motor or rocket propulsion
	Push or pull by a person

8–2 Potential Energy

In Chapter 7 we discussed the energy associated with a moving object, which is its kinetic energy $K = \frac{1}{2}mv^2$. Now we introduce **potential energy**, which is the energy associated with forces that depend on the position or configuration of objects relative to the surroundings. Various types of potential energy can be defined, and each type is associated with a particular conservative force.

The wound-up spring of a toy is an example of potential energy. The spring acquired its potential energy because work was done *on* it by the person winding the toy. As the spring unwinds, it exerts a force and does work to make the toy move.

Gravitational Potential Energy

Perhaps the most common example of potential energy is *gravitational potential energy*. A heavy brick held above the ground has potential energy because of its position relative to the Earth. The raised brick has the ability to do work, for if it is released, it will fall to the ground due to the gravitational force, and can do work on a stake, driving it into the ground. Let us seek the form for the gravitational potential energy of an object near the surface of the Earth. For an object of mass m to be lifted vertically, an upward force at least equal to its weight, mg, must be exerted on it, say by a person's hand. To lift it without acceleration a vertical displacement of height h, from position y_1 to y_2 in Fig. 8–4 (upward direction chosen positive), a person must do work equal to the product of the "external" force she exerts, $F_{ext} = mg$ upward, times the vertical displacement h. That is,

$$W_{ext} = \vec{\mathbf{F}}_{ext} \cdot \vec{\mathbf{d}} = mgh \cos 0° = mgh = mg(y_2 - y_1)$$

where both $\vec{\mathbf{F}}_{ext}$ and $\vec{\mathbf{d}}$ point upward. Gravity is also acting on the object as it moves from y_1 to y_2, and does work on the object equal to

$$W_G = \vec{\mathbf{F}}_G \cdot \vec{\mathbf{d}} = mgh \cos 180° = -mgh = -mg(y_2 - y_1),$$

where $\theta = 180°$ because $\vec{\mathbf{F}}_G$ and $\vec{\mathbf{d}}$ point in opposite directions. Since $\vec{\mathbf{F}}_G$ is downward and $\vec{\mathbf{d}}$ is upward, W_G is negative. If the object follows an arbitrary path, as in Fig. 8–1b, the work done by gravity still depends only on the change in vertical height (Eq. 8–1): $W_G = -mg(y_2 - y_1) = -mgh$.

Next, if we allow the object to start from rest and fall freely under the action of gravity, it acquires a velocity given by $v^2 = 2gh$ (Eq. 2–12c) after falling a height h. It then has kinetic energy $\frac{1}{2}mv^2 = \frac{1}{2}m(2gh) = mgh$, and if it strikes a stake it can do work on the stake equal to mgh.

To summarize, raising an object of mass m to a height h *requires* an amount of work equal to mgh. And once at height h, the object has the *ability* to do an amount of work equal to mgh. Thus we can say that the work done in lifting the object has been stored as gravitational potential energy.

Indeed, we can define the *change in gravitational potential energy U*, when an object moves from a height y_1 to a height y_2, as equal to the work done by a net external force to accomplish this without acceleration:

$$\Delta U = U_2 - U_1 = W_{ext} = mg(y_2 - y_1).$$

Equivalently, we can define the change in gravitational potential energy as equal to the negative of the work done by gravity itself in the process:

$$\Delta U = U_2 - U_1 = -W_G = mg(y_2 - y_1). \tag{8–2}$$

Equation 8–2 defines the change in gravitational potential energy when an object of mass m moves between two points near the surface of the Earth.[†] The gravitational potential energy, U, at any point a vertical height y above some reference point (the origin of the coordinate system) can be defined as

$$U_{grav} = mgy. \qquad \text{[gravity only]} \tag{8–3}$$

Note that the potential energy is associated with the force of gravity between the Earth and the mass m. Hence U_{grav} represents the gravitational potential energy, not simply of the mass m alone, but of the mass–Earth system.

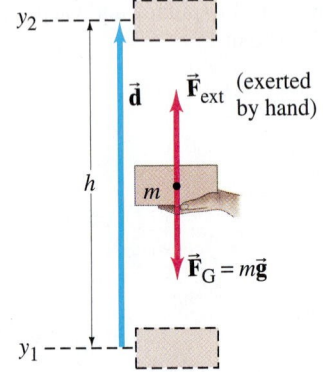

FIGURE 8–4 A person exerts an upward force $F_{ext} = mg$ to lift a brick from y_1 to y_2.

[†]Section 8–7 deals with the $1/r^2$ dependence of Newton's law of universal gravitation.

Gravitational potential energy depends on the *vertical height* of the object *above some reference level*, $U = mgy$. Sometimes you may wonder from what point to measure y. The gravitational potential energy of a book held high above a table, for example, depends on whether we measure y from the top of the table, from the floor, or from some other reference point. What is physically important in any situation is the *change* in potential energy, ΔU, because that is what is related to the work done, and it is ΔU that can be measured. We can thus choose to measure y from any reference point that is convenient, but we must choose the reference point at the start and be consistent throughout any given calculation. The *change* in potential energy between any two points does not depend on this choice.

Potential energy belongs to a system, and not to a single object alone. Potential energy is associated with a force, and a force on one object is always exerted by some other object. Thus potential energy is a property of the system as a whole. For an object raised to a height y above the Earth's surface, the change in gravitational potential energy is mgy. The system here is the object plus the Earth, and properties of both are involved: object (m) and Earth (g). In general, a *system* is one or more objects that we choose to study. The choice of what makes up a system is always ours, and we often try to choose a simple system. Below, when we deal with the potential energy of an object in contact with a spring, our system will be the object and the spring.

EXERCISE A Return to the Chapter-Opening Question, page 183, and answer it again now. Try to explain why you may have answered differently the first time.

EXAMPLE 8–1 **Potential energy changes for a roller coaster.** A 1000-kg roller-coaster car moves from point 1, Fig. 8–5, to point 2 and then to point 3. (*a*) What is the gravitational potential energy at points 2 and 3 relative to point 1? That is, take $y = 0$ at point 1. (*b*) What is the change in potential energy when the car goes from point 2 to point 3? (*c*) Repeat parts (*a*) and (*b*), but take the reference point ($y = 0$) to be at point 3.

APPROACH We are interested in the potential energy of the car–Earth system. We take upward as the positive y direction, and use the definition of gravitational potential energy to calculate the potential energy.

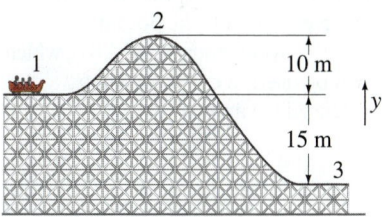

FIGURE 8–5 Example 8–1.

SOLUTION (*a*) We measure heights from point 1 ($y_1 = 0$), which means initially that the gravitational potential energy is zero. At point 2, where $y_2 = 10\,\text{m}$,

$$U_2 = mgy_2 = (1000\,\text{kg})(9.8\,\text{m/s}^2)(10\,\text{m}) = 9.8 \times 10^4\,\text{J}.$$

At point 3, $y_3 = -15\,\text{m}$, since point 3 is below point 1. Therefore,

$$U_3 = mgy_3 = (1000\,\text{kg})(9.8\,\text{m/s}^2)(-15\,\text{m}) = -1.5 \times 10^5\,\text{J}.$$

(*b*) In going from point 2 to point 3, the potential energy change ($U_{\text{final}} - U_{\text{initial}}$) is

$$U_3 - U_2 = (-1.5 \times 10^5\,\text{J}) - (9.8 \times 10^4\,\text{J}) = -2.5 \times 10^5\,\text{J}.$$

The gravitational potential energy decreases by $2.5 \times 10^5\,\text{J}$.

(*c*) Now we set $y_3 = 0$. Then $y_1 = +15\,\text{m}$ at point 1, so the potential energy initially (at point 1) is

$$U_1 = (1000\,\text{kg})(9.8\,\text{m/s}^2)(15\,\text{m}) = 1.5 \times 10^5\,\text{J}.$$

At point 2, $y_2 = 25\,\text{m}$, so the potential energy is

$$U_2 = 2.5 \times 10^5\,\text{J}.$$

At point 3, $y_3 = 0$, so the potential energy is zero. The change in potential energy going from point 2 to point 3 is

$$U_3 - U_2 = 0 - 2.5 \times 10^5\,\text{J} = -2.5 \times 10^5\,\text{J},$$

which is the same as in part (*b*).

NOTE Work done by gravity depends only on the vertical height, so changes in gravitational potential energy do not depend on the path taken.

EXERCISE B By how much does the potential energy change when a 1200-kg car climbs to the top of a 300-m-tall hill? (*a*) $3.6 \times 10^5\,\text{J}$; (*b*) $3.5 \times 10^6\,\text{J}$; (*c*) $4\,\text{J}$; (*d*) $40\,\text{J}$; (*e*) $39.2\,\text{J}$.

Potential Energy in General

We have defined the change in gravitational potential energy (Eq. 8–2) to be equal to the negative of the work done by gravity when the object moves from height y_1 to y_2, which we now write as

$$\Delta U = -W_G = -\int_1^2 \vec{\mathbf{F}}_G \cdot d\vec{\boldsymbol{\ell}}.$$

There are other types of potential energy besides gravitational. In general, we define the *change in potential energy associated with a particular conservative force* $\vec{\mathbf{F}}$ *as the negative of the work done by that force*:

$$\Delta U = U_2 - U_1 = -\int_1^2 \vec{\mathbf{F}} \cdot d\vec{\boldsymbol{\ell}} = -W. \qquad \textbf{(8–4)}$$

⚠ CAUTION

Potential energy can be defined only for conservative forces

However, we cannot use this definition to define a potential energy for all possible forces. It makes sense only for conservative forces such as gravity, for which the integral depends only on the end points and not on the path taken. It does not apply to nonconservative forces like friction, because the integral in Eq. 8–4 would *not* have a unique value depending on the end points 1 and 2. Thus the concept of potential energy cannot be defined and is meaningless for a nonconservative force.

Elastic Potential Energy

We consider now potential energy associated with elastic materials, which includes a great variety of practical applications.

Consider a simple coil spring as shown in Fig. 8–6, whose mass is so small that we can ignore it. When the spring is compressed and then released, it can do work on a ball (mass m). Thus the spring–ball system has potential energy when compressed (or stretched). Like other elastic materials, a spring is described by Hooke's law (see Section 7–3) as long as the displacement x is not too great. Let us take our coordinate system so the end of the uncompressed spring is at $x = 0$ (Fig. 8–6a) and x is positive to the right. To hold the spring compressed (or stretched) a distance x from its natural (unstretched) length requires the person's hand to exert a force $F_P = kx$ on the spring (Fig. 8–6b), where k is the spring stiffness constant. The spring pushes back with a force (Newton's third law),

$$F_S = -kx,$$

FIGURE 8–6 A spring (a) can store energy (elastic potential energy) when compressed (b), which can be used to do work when released (c) and (d).

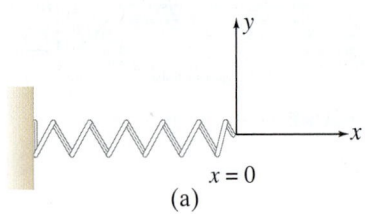

(a)

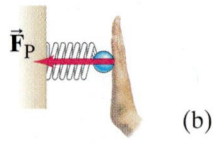

(b)

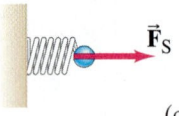

(c)

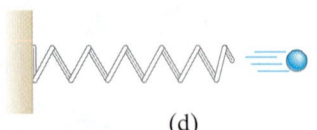

(d)

Fig. 8–6c. The negative sign appears because the force $\vec{\mathbf{F}}_S$ is in the direction opposite to the displacement x. From Eq. 8–4, the change in potential energy when the spring is compressed or stretched from $x_1 = 0$ (its uncompressed position) to $x_2 = x$ (where x can be + or −) is

$$\Delta U = U(x) - U(0) = -\int_1^2 \vec{\mathbf{F}}_S \cdot d\vec{\boldsymbol{\ell}} = -\int_0^x (-kx)\,dx = \tfrac{1}{2}kx^2.$$

Here, $U(x)$ means the potential energy at x, and $U(0)$ means U at $x = 0$. It is usually convenient to choose the potential energy at $x = 0$ to be zero: $U(0) = 0$, so the potential energy of a spring compressed or stretched an amount x from equilibrium is

$$U_{el}(x) = \tfrac{1}{2}kx^2. \qquad \text{[elastic spring]} \quad \textbf{(8–5)}$$

Potential Energy Related to Force (1–D)

In the one-dimensional case, where a conservative force can be written as a function of x, say, the potential energy can be written as an indefinite integral

$$U(x) = -\int F(x)\,dx + C, \qquad \textbf{(8–6)}$$

where the constant C represents the value of U at $x = 0$; we can sometimes choose $C = 0$. Equation 8–6 tells us how to obtain $U(x)$ when given $F(x)$. If, instead, we are given $U(x)$, we can obtain $F(x)$ by inverting the above equation: that is, we take the derivative of both sides, remembering that integration and differentiation are inverse operations:

$$\frac{d}{dx} \int F(x) \, dx = F(x).$$

Thus

$$F(x) = -\frac{dU(x)}{dx}. \qquad \textbf{(8–7)}$$

EXAMPLE 8–2 **Determine F from U.** Suppose $U(x) = -ax/(b^2 + x^2)$, where a and b are constants. What is F as a function of x?

APPROACH Since $U(x)$ depends only on x, this is a one-dimensional problem.

SOLUTION Equation 8–7 gives

$$F(x) = -\frac{dU}{dx} = -\frac{d}{dx}\left[-\frac{ax}{b^2 + x^2}\right] = \frac{a}{b^2 + x^2} - \frac{ax}{(b^2 + x^2)^2} 2x = \frac{a(b^2 - x^2)}{(b^2 + x^2)^2}.$$

---

*Potential Energy in Three Dimensions

In three dimensions, we can write the relation between $\vec{F}(x, y, z)$ and U as:

$$F_x = -\frac{\partial U}{\partial x}, \qquad F_y = -\frac{\partial U}{\partial y}, \qquad F_z = -\frac{\partial U}{\partial z},$$

or

$$\vec{F}(x, y, z) = -\hat{i}\frac{\partial U}{\partial x} - \hat{j}\frac{\partial U}{\partial y} - \hat{k}\frac{\partial U}{\partial z}.$$

Here, $\partial/\partial x$, $\partial/\partial y$ and $\partial/\partial z$ are called partial derivatives; $\partial/\partial x$, for example, means that although U may be a function of x, y, and z, written $U(x, y, z)$, we take the derivative only with respect to x with the other variables held constant.

8–3 Mechanical Energy and Its Conservation

Let us consider a conservative system (meaning only conservative forces do work) in which energy is transformed from kinetic to potential or vice versa. Again, we must consider a system because potential energy does not exist for an isolated object. Our system might be a mass m oscillating on the end of a spring or moving in the Earth's gravitational field.

According to the work-energy principle (Eq. 7–11), the net work W_{net} done on an object is equal to its change in kinetic energy:

$$W_{net} = \Delta K.$$

(If more than one object of our system has work done on it, then W_{net} and ΔK can represent the sum for all of them.) Since we assume a conservative system, we can write the net work done on an object or objects in terms of the change in total potential energy (see Eq. 8–4) between points 1 and 2:

$$\Delta U_{total} = -\int_1^2 \vec{F}_{net} \cdot d\vec{\ell} = -W_{net}. \qquad \textbf{(8–8)}$$

We combine the previous two equations, letting U be the total potential energy:

$$\Delta K + \Delta U = 0 \qquad \text{[conservative forces only]} \quad \textbf{(8–9a)}$$

or

$$(K_2 - K_1) + (U_2 - U_1) = 0. \qquad \text{[conservative forces only]} \quad \textbf{(8–9b)}$$

We now define a quantity E, called the **total mechanical energy** of our system, as the sum of the kinetic energy plus the potential energy of the system at any moment

$$E = K + U.$$

We can rewrite Eq. 8–9b as

$$K_2 + U_2 = K_1 + U_1 \qquad \text{[conservative forces only]} \quad \textbf{(8–10a)}$$

or

$$E_2 = E_1 = \text{constant.} \qquad \text{[conservative forces only]} \quad \textbf{(8–10b)}$$

Equations 8–10 express a useful and profound principle regarding the total mechanical energy—it is a **conserved quantity**, as long as no nonconservative forces do work; that is, the quantity $E = K + U$ at some initial time 1 is equal to $K + U$ at any later time 2.

To say it another way, consider Eq. 8–9a which tells us $\Delta U = -\Delta K$; that is, if the kinetic energy K increases, then the potential energy U must decrease by an equivalent amount to compensate. Thus the total, $K + U$, remains constant. This is called the **principle of conservation of mechanical energy** for conservative forces:

> **If only conservative forces are doing work, the total mechanical energy of a system neither increases nor decreases in any process. It stays constant—it is conserved.**

We now see the reason for the term "conservative force"—because for such forces, mechanical energy is conserved.

If only one object of a system[†] has significant kinetic energy, then Eqs. 8–10 become

$$E = \tfrac{1}{2}mv^2 + U = \text{constant.} \quad \text{[conservative forces only]} \quad \textbf{(8–11a)}$$

If we let v_1 and U_1 represent the velocity and potential energy at one instant, and v_2 and U_2 represent them at a second instant, then we can rewrite this as

$$\tfrac{1}{2}mv_1^2 + U_1 = \tfrac{1}{2}mv_2^2 + U_2. \qquad \text{[conservative system]} \quad \textbf{(8–11b)}$$

From this equation we can see again that it doesn't make any difference where we choose the potential energy to be zero: adding a constant to U merely adds a constant to both sides of Eq. 8–11b, and these cancel. A constant also doesn't affect the force obtained using Eq. 8–7, $F = -dU/dx$, since the derivative of a constant is zero. Only changes in the potential energy matter.

8–4 Problem Solving Using Conservation of Mechanical Energy

A simple example of the conservation of mechanical energy (neglecting air resistance) is a rock allowed to fall due to Earth's gravity from a height h above the ground, as shown in Fig. 8–7. If the rock starts from rest, all of the initial energy is potential energy. As the rock falls, the potential energy mgy decreases (because y decreases), but the rock's kinetic energy increases to compensate, so that the sum of the two remains constant. At any point along the path, the total mechanical energy is given by

$$E = K + U = \tfrac{1}{2}mv^2 + mgy$$

where y is the rock's height above the ground at a given instant and v is its speed at that point. If we let the subscript 1 represent the rock at one point along its path (for example, the initial point), and the subscript 2 represent it at some other point, then we can write

total mechanical energy at point 1 = total mechanical energy at point 2

or (see also Eq. 8–11b)

$$\tfrac{1}{2}mv_1^2 + mgy_1 = \tfrac{1}{2}mv_2^2 + mgy_2. \qquad \text{[gravity only]} \quad \textbf{(8–12)}$$

Just before the rock hits the ground, where we chose $y = 0$, all of the initial potential energy will have been transformed into kinetic energy.

FIGURE 8–7 The rock's potential energy changes to kinetic energy as it falls. Note bar graphs representing potential energy U and kinetic energy K for the three different positions.

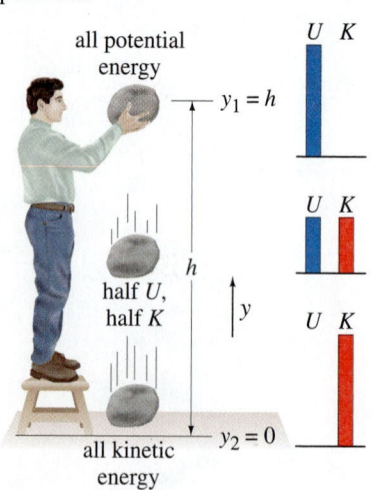

[†]For an object moving under the influence of Earth's gravity, the kinetic energy of the Earth can usually be ignored. For a mass oscillating at the end of a spring, the mass of the spring, and hence its kinetic energy, can often be ignored.

EXAMPLE 8–3 **Falling rock.** If the original height of the rock in Fig. 8–7 is $y_1 = h = 3.0\,\text{m}$, calculate the rock's speed when it has fallen to 1.0 m above the ground.

APPROACH We apply the principle of conservation of mechanical energy, Eq. 8–12, with only gravity acting on the rock. We choose the ground as our reference level $(y = 0)$.

SOLUTION At the moment of release (point 1) the rock's position is $y_1 = 3.0\,\text{m}$ and it is at rest: $v_1 = 0$. We want to find v_2 when the rock is at position $y_2 = 1.0\,\text{m}$. Equation 8–12 gives

$$\tfrac{1}{2}mv_1^2 + mgy_1 = \tfrac{1}{2}mv_2^2 + mgy_2.$$

The m's cancel out; setting $v_1 = 0$ and solving for v_2 we find

$$v_2 = \sqrt{2g(y_1 - y_2)} = \sqrt{2(9.8\,\text{m/s}^2)\big[(3.0\,\text{m}) - (1.0\,\text{m})\big]} = 6.3\,\text{m/s}.$$

The rock's speed 1.0 m above the ground is 6.3 m/s downward.

NOTE The velocity of the rock is independent of the rock's mass.

EXERCISE C In Example 8–3, what is the rock's speed just before it hits the ground? (a) 6.5 m/s; (b) 7.0 m/s; (c) 7.7 m/s; (d) 8.3 m/s; (e) 9.8 m/s.

Equation 8–12 can be applied to any object moving without friction under the action of gravity. For example, Fig. 8–8 shows a roller-coaster car starting from rest at the top of a hill, and coasting without friction to the bottom and up the hill on the other side. True, there is another force besides gravity acting on the car, the normal force exerted by the tracks. But this "constraint" force acts perpendicular to the direction of motion at each point and so does zero work. We ignore rotational motion of the car's wheels and treat the car as a particle undergoing simple translation. Initially, the car has only potential energy. As it coasts down the hill, it loses potential energy and gains in kinetic energy, but the sum of the two remains constant. At the bottom of the hill it has its maximum kinetic energy, and as it climbs up the other side the kinetic energy changes back to potential energy. When the car comes to rest again at the same height from which it started, all of its energy will be potential energy. Given that the gravitational potential energy is proportional to the vertical height, energy conservation tells us that (in the absence of friction) the car comes to rest at a height equal to its original height. If the two hills are the same height, the car will just barely reach the top of the second hill when it stops. If the second hill is lower than the first, not all of the car's kinetic energy will be transformed to potential energy and the car can continue over the top and down the other side. If the second hill is higher, the car will only reach a height on it equal to its original height on the first hill. This is true (in the absence of friction) no matter how steep the hill is, since potential energy depends only on the vertical height.

FIGURE 8–8 A roller-coaster car moving without friction illustrates the conservation of mechanical energy.

EXAMPLE 8–4 **Roller-coaster car speed using energy conservation.** Assuming the height of the hill in Fig. 8–8 is 40 m, and the roller-coaster car starts from rest at the top, calculate (a) the speed of the roller-coaster car at the bottom of the hill, and (b) at what height it will have half this speed. Take $y = 0$ at the bottom of the hill.

APPROACH We choose point 1 to be where the car starts from rest $(v_1 = 0)$ at the top of the hill $(y_1 = 40\,\text{m})$. Point 2 is the bottom of the hill, which we choose as our reference level, so $y_2 = 0$. We use conservation of mechanical energy.

SOLUTION (a) We use Eq. 8–12 with $v_1 = 0$ and $y_2 = 0$, which gives

$$mgy_1 = \tfrac{1}{2}mv_2^2$$

or

$$v_2 = \sqrt{2gy_1} = \sqrt{2(9.8\,\text{m/s}^2)(40\,\text{m})} = 28\,\text{m/s}.$$

(b) We again use conservation of energy,

$$\tfrac{1}{2}mv_1^2 + mgy_1 = \tfrac{1}{2}mv_2^2 + mgy_2,$$

but now $v_2 = \tfrac{1}{2}(28\,\text{m/s}) = 14\,\text{m/s}$, $v_1 = 0$, and y_2 is the unknown. Thus

$$y_2 = y_1 - \frac{v_2^2}{2g} = 30\,\text{m}.$$

That is, the car has a speed of 14 m/s when it is 30 *vertical* meters above the lowest point, both when descending the left-hand hill and when ascending the right-hand hill.

The mathematics of the roller-coaster Example 8–4 is almost the same as in Example 8–3. But there is an important difference between them. In Example 8–3 the motion is all vertical and could have been solved using force, acceleration, and the kinematic equations (Eqs. 2–12). But for the roller coaster, where the motion is not vertical, we could not have used Eqs. 2–12 because a is not constant on the curved track; but energy conservation readily gives us the answer.

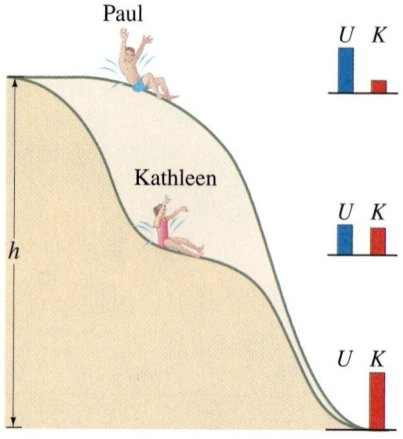

FIGURE 8–9 Example 8–5.

CONCEPTUAL EXAMPLE 8–5 | **Speeds on two water slides.** Two water slides at a pool are shaped differently, but start at the same height h (Fig. 8–9). Two riders, Paul and Kathleen, start from rest at the same time on different slides. (*a*) Which rider, Paul or Kathleen, is traveling faster at the bottom? (*b*) Which rider makes it to the bottom first? Ignore friction and assume both slides have the same path length.

RESPONSE (*a*) Each rider's initial potential energy mgh gets transformed to kinetic energy, so the speed v at the bottom is obtained from $\frac{1}{2}mv^2 = mgh$. The mass cancels and so the speed will be the same, regardless of the mass of the rider. Since they descend the same vertical height, they will finish with the same speed. (*b*) Note that Kathleen is consistently at a lower elevation than Paul at any instant, until the end. This means she has converted her potential energy to kinetic energy earlier. Consequently, she is traveling faster than Paul for the whole trip, and because the distance is the same, Kathleen gets to the bottom first.

EXERCISE D Two balls are released from the same height above the floor. Ball A falls freely through the air, whereas ball B slides on a curved frictionless track to the floor. How do the speeds of the balls compare when they reach the floor?

PROBLEM SOLVING
Use energy, or Newton's laws?

PHYSICS APPLIED
Sports

FIGURE 8–10 Transformation of energy during a pole vault.

You may wonder sometimes whether to approach a problem using work and energy, or instead to use Newton's laws. As a rough guideline, if the force(s) involved are constant, either approach may succeed. If the forces are not constant, and/or the path is not simple, energy is probably the better approach.

There are many interesting examples of the conservation of energy in sports, such as the pole vault illustrated in Fig. 8–10. We often have to make approximations, but the sequence of events in broad outline for the pole vault is as follows. The initial kinetic energy of the running athlete is transformed into elastic potential energy of the bending pole and, as the athlete leaves the ground, into gravitational potential energy. When the vaulter reaches the top and the pole has straightened out again, the energy has all been transformed into gravitational potential energy (if we ignore the vaulter's low horizontal speed over the bar). The pole does not supply any energy, but it acts as a device to *store* energy and thus aid in the transformation of kinetic energy into gravitational potential energy, which is the net result. The energy required to pass over the bar depends on how high the center of mass (CM) of the vaulter must be raised. By bending their bodies, pole vaulters keep their CM so low that it can actually pass slightly beneath the bar (Fig. 8–11), thus enabling them to cross over a higher bar than would otherwise be possible. (Center of mass is covered in Chapter 9.)

FIGURE 8–11 By bending their bodies, pole vaulters can keep their center of mass so low that it may even pass below the bar. By changing their kinetic energy (of running) into gravitational potential energy $(= mgy)$ in this way, vaulters can cross over a higher bar than if the change in potential energy were accomplished without carefully bending the body.

EXAMPLE 8–6 **ESTIMATE** **Pole vault.** Estimate the kinetic energy and the speed required for a 70-kg pole vaulter to just pass over a bar 5.0 m high. Assume the vaulter's center of mass is initially 0.90 m off the ground and reaches its maximum height at the level of the bar itself.

APPROACH We equate the total energy just before the vaulter places the end of the pole onto the ground (and the pole begins to bend and store potential energy) with the vaulter's total energy when passing over the bar (we ignore the small amount of kinetic energy at this point). We choose the initial position of the vaulter's center of mass to be $y_1 = 0$. The vaulter's body must then be raised to a height $y_2 = 5.0\,\text{m} - 0.9\,\text{m} = 4.1\,\text{m}$.

SOLUTION We use Eq. 8–12,

$$\tfrac{1}{2}mv_1^2 + 0 = 0 + mgy_2$$

so

$$K_1 = \tfrac{1}{2}mv_1^2 = mgy_2 = (70\,\text{kg})(9.8\,\text{m/s}^2)(4.1\,\text{m}) = 2.8 \times 10^3\,\text{J}.$$

The speed is

$$v_1 = \sqrt{\frac{2K_1}{m}} = \sqrt{\frac{2(2800\,\text{J})}{70\,\text{kg}}} = 8.9\,\text{m/s} \approx 9\,\text{m/s}.$$

NOTE This is an approximation because we have ignored such things as the vaulter's speed while crossing over the bar, mechanical energy transformed when the pole is planted in the ground, and work done by the vaulter on the pole. All would increase the needed initial kinetic energy.

As another example of the conservation of mechanical energy, let us consider an object of mass m connected to a horizontal spring (Fig. 8–6) whose own mass can be neglected and whose spring stiffness constant is k. The mass m has speed v at any moment. The potential energy of the system (object plus spring) is $\tfrac{1}{2}kx^2$, where x is the displacement of the spring from its unstretched length. If neither friction nor any other force is acting, conservation of mechanical energy tells us that

$$\tfrac{1}{2}mv_1^2 + \tfrac{1}{2}kx_1^2 = \tfrac{1}{2}mv_2^2 + \tfrac{1}{2}kx_2^2, \qquad \text{[elastic PE only]} \quad \textbf{(8–13)}$$

where the subscripts 1 and 2 refer to the velocity and displacement at two different moments.

EXAMPLE 8–7 **Toy dart gun.** A dart of mass 0.100 kg is pressed against the spring of a toy dart gun as shown in Fig. 8–12a. The spring (with spring stiffness constant $k = 250\,\text{N/m}$ and ignorable mass) is compressed 6.0 cm and released. If the dart detaches from the spring when the spring reaches its natural length ($x = 0$), what speed does the dart acquire?

APPROACH The dart is initially at rest (point 1), so $K_1 = 0$. We ignore friction and use conservation of mechanical energy; the only potential energy is elastic.

SOLUTION We use Eq. 8–13 with point 1 being at the maximum compression of the spring, so $v_1 = 0$ (dart not yet released) and $x_1 = -0.060\,\text{m}$. Point 2 we choose to be the instant the dart flies off the end of the spring (Fig. 8–12b), so $x_2 = 0$ and we want to find v_2. Thus Eq. 8–13 can be written

$$0 + \tfrac{1}{2}kx_1^2 = \tfrac{1}{2}mv_2^2 + 0.$$

Then

$$v_2^2 = \frac{kx_1^2}{m}$$

and

$$v_2 = \sqrt{\frac{(250\,\text{N/m})(-0.060\,\text{m})^2}{(0.100\,\text{kg})}} = 3.0\,\text{m/s}.$$

NOTE In the horizontal direction, the only force on the dart (neglecting friction) was the force exerted by the spring. Vertically, gravity was counterbalanced by the normal force exerted on the dart by the gun barrel. After it leaves the barrel, the dart will follow a projectile's path under gravity.

FIGURE 8–12 Example 8–7. (a) A dart is pushed against a spring, compressing it 6.0 cm. The dart is then released, and in (b) it leaves the spring at velocity v_2.

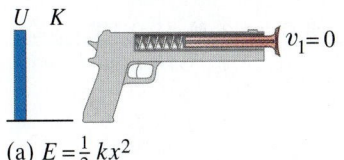

(a) $E = \tfrac{1}{2}kx^2$

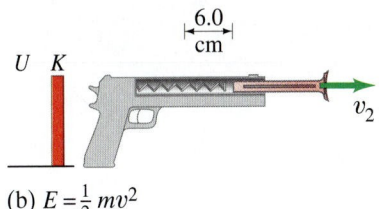

(b) $E = \tfrac{1}{2}mv^2$

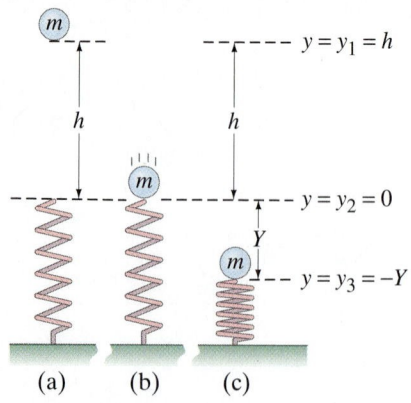

(a)   (b)   (c)

FIGURE 8–13 Example 8–8.

EXAMPLE 8–8 **Two kinds of potential energy.** A ball of mass $m = 2.60\,\text{kg}$, starting from rest, falls a vertical distance $h = 55.0\,\text{cm}$ before striking a vertical coiled spring, which it compresses an amount $Y = 15.0\,\text{cm}$ (Fig. 8–13). Determine the spring stiffness constant of the spring. Assume the spring has negligible mass, and ignore air resistance. Measure all distances from the point where the ball first touches the uncompressed spring ($y = 0$ at this point).

APPROACH The forces acting on the ball are the gravitational pull of the Earth and the elastic force exerted by the spring. Both forces are conservative, so we can use conservation of mechanical energy, including both types of potential energy. We must be careful, however: gravity acts throughout the fall (Fig. 8–13), whereas the elastic force does not act until the ball touches the spring (Fig. 8–13b). We choose y positive upward, and $y = 0$ at the end of the spring in its natural (uncompressed) state.

SOLUTION We divide this solution into two parts. (An alternate solution follows.)
Part 1: Let us first consider the energy changes as the ball falls from a height $y_1 = h = 0.55\,\text{m}$, Fig. 8–13a, to $y_2 = 0$, just as it touches the spring, Fig. 8–13b. Our system is the ball acted on by gravity plus the spring (which up to this point doesn't do anything). Thus

$$\tfrac{1}{2}mv_1^2 + mgy_1 = \tfrac{1}{2}mv_2^2 + mgy_2$$
$$0 + mgh = \tfrac{1}{2}mv_2^2 + 0.$$

We solve for $v_2 = \sqrt{2gh} = \sqrt{2(9.80\,\text{m/s}^2)(0.550\,\text{m})} = 3.283\,\text{m/s} \approx 3.28\,\text{m/s}$. This is the speed of the ball just as it touches the top of the spring, Fig. 8–13b.
Part 2: As the ball compresses the spring, Figs. 8–13b to c, there are two conservative forces on the ball—gravity and the spring force. So our conservation of energy equation is

$$E_2 \text{ (ball touches spring)} = E_3 \text{ (spring compressed)}$$
$$\tfrac{1}{2}mv_2^2 + mgy_2 + \tfrac{1}{2}ky_2^2 = \tfrac{1}{2}mv_3^2 + mgy_3 + \tfrac{1}{2}ky_3^2.$$

Substituting $y_2 = 0$, $v_2 = 3.283\,\text{m/s}$, $v_3 = 0$ (the ball comes to rest for an instant), and $y_3 = -Y = -0.150\,\text{m}$, we have

$$\tfrac{1}{2}mv_2^2 + 0 + 0 = 0 - mgY + \tfrac{1}{2}k(-Y)^2.$$

We know m, v_2, and Y, so we can solve for k:

$$k = \frac{2}{Y^2}\left[\tfrac{1}{2}mv_2^2 + mgY\right] = \frac{m}{Y^2}\left[v_2^2 + 2gY\right]$$
$$= \frac{(2.60\,\text{kg})}{(0.150\,\text{m})^2}\left[(3.283\,\text{m/s})^2 + 2(9.80\,\text{m/s}^2)(0.150\,\text{m})\right] = 1590\,\text{N/m}.$$

PROBLEM SOLVING
Alternate Solution

Alternate Solution Instead of dividing the solution into two parts, we can do it all at once. After all, we get to choose what two points are used on the left and right of the energy equation. Let us write the energy equation for points 1 and 3 in Fig. 8–13. Point 1 is the initial point just before the ball starts to fall (Fig. 8–13a), so $v_1 = 0$, and $y_1 = h = 0.550\,\text{m}$. Point 3 is when the spring is fully compressed (Fig. 8–13c), so $v_3 = 0$, $y_3 = -Y = -0.150\,\text{m}$. The forces on the ball in this process are gravity and (at least part of the time) the spring. So conservation of energy tells us

$$\tfrac{1}{2}mv_1^2 + mgy_1 + \tfrac{1}{2}k(0)^2 = \tfrac{1}{2}mv_3^2 + mgy_3 + \tfrac{1}{2}ky_3^2$$
$$0 + mgh + 0 = 0 - mgY + \tfrac{1}{2}kY^2$$

where we have set $y = 0$ for the spring at point 1 because it is not acting and is not compressed or stretched. We solve for k:

$$k = \frac{2mg(h + y)}{Y^2} = \frac{2(2.60\,\text{kg})(9.80\,\text{m/s}^2)(0.550\,\text{m} + 0.150\,\text{m})}{(0.150\,\text{m})^2} = 1590\,\text{N/m}$$

just as in our first method of solution.

EXAMPLE 8–9 **A swinging pendulum.** The simple pendulum shown in Fig. 8–14 consists of a small bob of mass m suspended by a massless cord of length ℓ. The bob is released (without a push) at $t = 0$, where the cord makes an angle $\theta = \theta_0$ to the vertical. (a) Describe the motion of the bob in terms of kinetic energy and potential energy. Then determine the speed of the bob (b) as a function of position θ as it swings back and forth, and (c) at the lowest point of the swing. (d) Find the tension in the cord, $\vec{\mathbf{F}}_T$. Ignore friction and air resistance.

APPROACH We use the law of conservation of mechanical energy (only the conservative force of gravity does work), except in (d) where we use Newton's second law.

SOLUTION (a) At the moment of release, the bob is at rest, so its kinetic energy $K = 0$. As the bob moves down, it loses potential energy and gains kinetic energy. At the lowest point its kinetic energy is a maximum and the potential energy is a minimum. The bob continues its swing until it reaches an equal height and angle (θ_0) on the opposite side, at which point the potential energy is a maximum and $K = 0$. It continues the swinging motion as $U \rightarrow K \rightarrow U$ and so on, but it can never go higher than $\theta = \pm \theta_0$ (conservation of mechanical energy).

(b) The cord is assumed to be massless, so we need to consider only the bob's kinetic energy, and the gravitational potential energy. The bob has two forces acting on it at any moment: gravity, mg, and the force the cord exerts on it, $\vec{\mathbf{F}}_T$. The latter (a constraint force) always acts perpendicular to the motion, so it does no work. We need be concerned only with gravity, for which we can write the potential energy. The mechanical energy of the system is

$$E = \tfrac{1}{2}mv^2 + mgy,$$

where y is the vertical height of the bob at any moment. We take $y = 0$ at the lowest point of the bob's swing. Hence at $t = 0$,

$$y = y_0 = \ell - \ell \cos \theta_0 = \ell(1 - \cos \theta_0)$$

as can be seen from the diagram. At the moment of release

$$E = mgy_0,$$

since $v = v_0 = 0$. At any other point along the swing

$$E = \tfrac{1}{2}mv^2 + mgy = mgy_0.$$

We solve this for v:

$$v = \sqrt{2g(y_0 - y)}.$$

In terms of the angle θ of the cord, we can write

$$(y_0 - y) = (\ell - \ell \cos \theta_0) - (\ell - \ell \cos \theta) = \ell(\cos \theta - \cos \theta_0)$$

so

$$v = \sqrt{2g\ell(\cos \theta - \cos \theta_0)}.$$

(c) At the lowest point, $y = 0$, so

$$v = \sqrt{2gy_0}$$

or

$$v = \sqrt{2g\ell(1 - \cos \theta_0)}.$$

(d) The tension in the cord is the force $\vec{\mathbf{F}}_T$ that the cord exerts on the bob. As we've seen, there is no work done by this force, but we can calculate the force simply by using Newton's second law $\Sigma \vec{\mathbf{F}} = m\vec{\mathbf{a}}$ and by noting that at any point the acceleration of the bob in the inward radial direction is v^2/ℓ, since the bob is constrained to move in an arc of a circle of radius ℓ. In the radial direction, $\vec{\mathbf{F}}_T$ acts inward, and a component of gravity equal to $mg \cos \theta$ acts outward. Hence

$$m\frac{v^2}{\ell} = F_T - mg \cos \theta.$$

We solve for F_T and use the result of part (b) for v^2:

$$F_T = m\left(\frac{v^2}{\ell} + g \cos \theta\right) = 2mg(\cos \theta - \cos \theta_0) + mg \cos \theta$$

$$= (3 \cos \theta - 2 \cos \theta_0)mg.$$

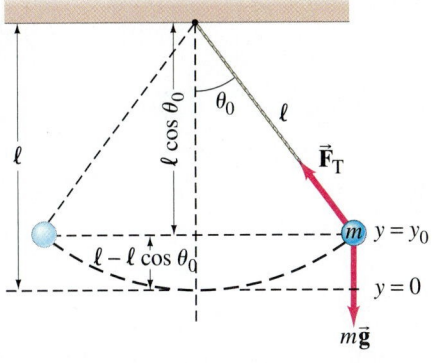

FIGURE 8–14 Example 8–9: a simple pendulum; y is measured positive upward.

8–5 The Law of Conservation of Energy

We now take into account nonconservative forces such as friction, since they are important in real situations. For example, consider again the roller-coaster car in Fig. 8–8, but this time let us include friction. The car will not in this case reach the same height on the second hill as it had on the first hill because of friction.

In this, and in other natural processes, the mechanical energy (sum of the kinetic and potential energies) does not remain constant but decreases. Because frictional forces reduce the mechanical energy (but *not* the total energy), they are called **dissipative forces**. Historically, the presence of dissipative forces hindered the formulation of a comprehensive conservation of energy law until well into the nineteenth century. It was not until then that heat, which is always produced when there is friction (try rubbing your hands together), was interpreted in terms of energy. Quantitative studies in the nineteenth-century (Chapter 19) demonstrated that if heat is considered as a transfer of energy (sometimes called **thermal energy**), then the total energy is conserved in any process. A block sliding freely across a table, for example, comes to rest because of friction. Its initial kinetic energy is all transformed into thermal energy. The block and table are a little warmer as a result of this process.

According to the atomic theory, thermal energy represents kinetic energy of rapidly moving molecules. We shall see in Chapter 18 that a rise in temperature corresponds to an increase in the average kinetic energy of the molecules. Because thermal energy represents the energy of atoms and molecules that make up an object, it is often called **internal energy**. Internal energy, from the atomic point of view, includes kinetic energy of molecules as well as potential energy (usually electrical in nature) due to the relative positions of atoms within molecules. For example the energy stored in food, or in a fuel such as gasoline, is regarded as electric potential energy (chemical bonds). The energy is released through chemical reactions (Fig. 8–15). This is analogous to a compressed spring which, when released, can do work.

To establish the more general law of conservation of energy, nineteenth-century physicists had to recognize electrical, chemical, and other forms of energy in addition to heat, and to explore if in fact they could fit into a conservation law. For each type of force, conservative or nonconservative, it has always been possible to define a type of energy that corresponds to the work done by such a force. And it has been found experimentally that the total energy E always remains constant. That is, the change in total energy, kinetic plus potential plus all other forms of energy, equals zero:

$$\Delta K + \Delta U + [\text{change in all other forms of energy}] = 0. \quad \textbf{(8–14)}$$

This is one of the most important principles in physics. It is called the **law of conservation of energy** and can be stated as follows:

> **The total energy is neither increased nor decreased in any process. Energy can be transformed from one form to another, and transferred from one object to another, but the total amount remains constant.**

For conservative mechanical systems, this law can be derived from Newton's laws (Section 8–3) and thus is equivalent to them. But in its full generality, the validity of the law of conservation of energy rests on experimental observation.

Even though Newton's laws have been found to fail in the submicroscopic world of the atom, the law of conservation of energy has been found to hold there and in every experimental situation so far tested.

8–6 Energy Conservation with Dissipative Forces: Solving Problems

In Section 8–4 we discussed several Examples of the law of conservation of energy for conservative systems. Now let us consider in detail some examples that involve nonconservative forces.

FIGURE 8–15 The burning of fuel (a chemical reaction) releases energy to boil water in this steam engine. The steam produced expands against a piston to do work in turning the wheels.

Suppose, for example, that the roller-coaster car rolling on the hills of Fig. 8–8 is subject to frictional forces. In going from some point 1 to a second point 2, the energy dissipated by the friction force $\vec{F}_{fr}$ acting on the car (treating it as a particle) is $\int_1^2 \vec{F}_{fr} \cdot d\vec{\ell}$. If $\vec{F}_{fr}$ is constant in magnitude, $W_{fr} = -F_{fr}\ell$, where ℓ is the actual distance along the path traveled by the car from point 1 to point 2. (The minus sign appears because $\vec{F}_{fr}$ and $d\vec{\ell}$ are in opposite directions.) From the work-energy principle (Eq. 7–11), the net work W_{net} done on an object is equal to the change in its kinetic energy:

$$\Delta K = W_{net}.$$

We can separate W_{net} into two parts:

$$W_{net} = W_C + W_{NC},$$

where W_C is the work done by conservative forces (gravity for our car) and W_{NC} is the work done by nonconservative forces (friction for our car). We saw in Eq. 8–4 that the work done by a conservative force can be written in terms of potential energy:

$$W_C = \int_1^2 \vec{F} \cdot d\vec{\ell} = -\Delta U.$$

Thus $\Delta K = W_{net}$ means $\Delta K = W_C + W_{NC} = -\Delta U + W_{NC}$. Hence

$$\Delta K + \Delta U = W_{NC}. \tag{8–15a}$$

This equation represents the general form of the work-energy principle. It also represents the conservation of energy. For our car, W_{NC} is the work done by the friction force and represents production of thermal energy. Equation 8–15a is valid in general. Based on our derivation from the work-energy principle, W_{NC} on the right side of Eq. 8–15a must be the total work done by all forces that are not included in the potential energy term, ΔU, on the left side.

Let us rewrite Eq. 8–15a for our roller-coaster car example of Fig. 8–8, setting $W_{NC} = -F_{fr}\ell$ as we discussed above:

$$W_{NC} = -F_{fr}\ell = \Delta K + \Delta U = \left(\tfrac{1}{2}mv_2^2 - \tfrac{1}{2}mv_1^2\right) + \left(mgy_2 - mgy_1\right)$$

or

$$\tfrac{1}{2}mv_1^2 + mgy_1 = \tfrac{1}{2}mv_2^2 + mgy_2 + F_{fr}\ell. \quad \left[\begin{array}{c}\text{gravity and} \\ \text{friction acting}\end{array}\right] \tag{8–15b}$$

On the left we have the mechanical energy of the system initially. It equals the mechanical energy at any subsequent point along the path plus the amount of thermal (or internal) energy produced in the process. Total energy is conserved.

Other nonconservative forces can be treated similarly. If you are not sure about the sign of the extra term $\left(\int \vec{F} \cdot d\vec{\ell}\right)$, use your intuition: is the mechanical energy increased or decreased in the process.

Work-Energy versus Energy Conservation

The law of conservation of energy is more general and more powerful than the work-energy principle. Indeed, the work-energy principle should *not* be viewed as a statement of conservation of energy. It is nonetheless useful for some mechanical problems; and whether you use it, or use the more powerful conservation of energy, can depend on your *choice of the system* under study. If you choose as your system a particle or rigid object on which external forces do work, then you can use the work-energy principle: the work done by the external forces on your object equals the change in its kinetic energy.

On the other hand, if you choose a system on which no external forces do work, then you need to apply conservation of energy to that system directly.

Consider, for example, a spring connected to a block on a frictionless table (Fig. 8–16). If you choose the block as your system, then the work done on the block by the spring equals the change in kinetic energy of the block: the work-energy principle. (Energy conservation does not apply to this system—the block's energy changes.) If instead you choose the block plus the spring as your system, no external forces do work (since the spring is part of the chosen system). To this system you need to apply conservation of energy: if you compress the spring and then release it, the spring still exerts a force on the block, but the subsequent motion can be discussed in terms of kinetic energy $\left(\tfrac{1}{2}mv^2\right)$ plus potential energy $\left(\tfrac{1}{2}kx^2\right)$, whose total remains constant.

FIGURE 8–16 A spring connected to a block on a frictionless table. If you choose your system to be the block plus spring, then

$$E = \tfrac{1}{2}mv^2 + \tfrac{1}{2}kx^2$$

is conserved.

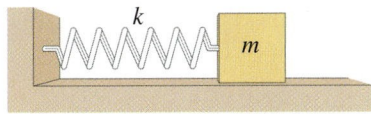

Problem solving is not a process that can be done by simply following a set of rules. The following Problem Solving Strategy, like all others, is thus *not* a prescription, but is a summary to help you get started solving problems involving energy.

Conservation of Energy

1. **Draw a picture** of the physical situation.
2. Determine **the system** for which you will apply energy conservation: the object or objects and the forces acting.
3. Ask yourself what quantity you are looking for, and **choose initial** (point 1) **and final** (point 2) **positions**.
4. If the object under investigation changes its height during the problem, then **choose a reference frame** with a convenient $y = 0$ level for gravitational potential energy; the lowest point in the problem is often a good choice.

   If springs are involved, choose the unstretched spring position to be x (or y) = 0.

5. **Is mechanical energy conserved?** If no friction or other nonconservative forces act, then conservation of mechanical energy holds:

$$K_1 + U_1 = K_2 + U_2.$$

6. **Apply conservation of energy.** If friction (or other nonconservative forces) are present, then an additional term of the form $\int \vec{F} \cdot d\vec{\ell}$ will be needed. For a constant friction force acting over a distance ℓ

$$K_1 + U_1 = K_2 + U_2 + F_{fr}\ell.$$

For other nonconservative forces use your intuition for the sign of $\int \vec{F} \cdot d\vec{\ell}$: is the total mechanical energy increased or decreased in the process?

7. Use the equation(s) you develop to **solve** for the unknown quantity.

FIGURE 8–17 Example 8–10. Because of friction, a roller-coaster car does not reach the original height on the second hill. (Not to scale)

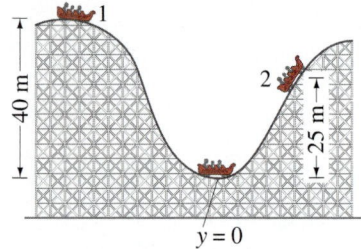

EXAMPLE 8–10 ESTIMATE **Friction on the roller-coaster car.** The roller-coaster car in Example 8–4 reaches a vertical height of only 25 m on the second hill before coming to a momentary stop (Fig. 8–17). It traveled a total distance of 400 m. Determine the thermal energy produced and estimate the average friction force (assume it is roughly constant) on the car, whose mass is 1000 kg.

APPROACH We explicitly follow the Problem Solving Strategy above.

SOLUTION

1. **Draw a picture.** See Fig. 8–17.
2. **The system.** The system is the roller-coaster car and the Earth (which exerts the gravitational force). The forces acting on the car are gravity and friction. (The normal force also acts on the car, but does no work, so it does not affect the energy.)
3. **Choose initial and final positions.** We take point 1 to be the instant when the car started coasting (at the top of the first hill), and point 2 to be the instant it stopped 25 m up the second hill.
4. **Choose a reference frame.** We choose the lowest point in the motion to be $y = 0$ for the gravitational potential energy.
5. **Is mechanical energy conserved?** No. Friction is present.
6. **Apply conservation of energy.** There is friction acting on the car, so we use conservation of energy in the form of Eq. 8–15, with $v_1 = 0$, $y_1 = 40$ m, $v_2 = 0$, $y_2 = 25$ m, and $\ell = 400$ m. Thus

$$0 + (1000 \text{ kg})(9.8 \text{ m/s}^2)(40 \text{ m}) = 0 + (1000 \text{ kg})(9.8 \text{ m/s}^2)(25 \text{ m}) + F_{fr}\ell.$$

7. **Solve.** We solve the above equation for $F_{fr}\ell$, the energy dissipated to thermal energy: $F_{fr}\ell = (1000 \text{ kg})(9.8 \text{ m/s}^2)(40 \text{ m} - 25 \text{ m}) = 147{,}000 \text{ J}$. The average force of friction was $F_{fr} = (1.47 \times 10^5 \text{ J})/400 \text{ m} = 370 \text{ N}$. [This result is only a rough average: the friction force at various points depends on the normal force, which varies with slope.]

EXAMPLE 8–11 **Friction with a spring.** A block of mass m sliding along a rough horizontal surface is traveling at a speed v_0 when it strikes a massless spring head-on (see Fig. 8–18) and compresses the spring a maximum distance X. If the spring has stiffness constant k, determine the coefficient of kinetic friction between block and surface.

APPROACH At the moment of collision, the block has $K = \frac{1}{2}mv_0^2$ and the spring is presumably uncompressed, so $U = 0$. Initially the mechanical energy of the system is $\frac{1}{2}mv_0^2$. By the time the spring reaches maximum compression, $K = 0$ and $U = \frac{1}{2}kX^2$. In the meantime, the friction force $(= \mu_k F_N = \mu_k mg)$ has transformed energy $F_{fr}X = \mu_k mgX$ into thermal energy.

SOLUTION Conservation of energy allows us to write

energy (initial) = energy (final)

$$\tfrac{1}{2}mv_0^2 = \tfrac{1}{2}kX^2 + \mu_k mgX.$$

We solve for μ_k and find

$$\mu_k = \frac{v_0^2}{2gX} - \frac{kX}{2mg}.$$

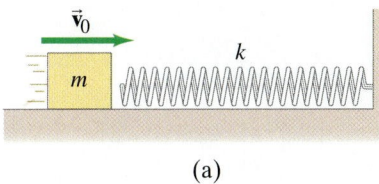

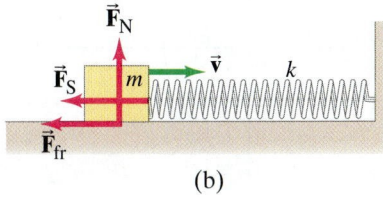

FIGURE 8–18 Example 8–11.

8–7 Gravitational Potential Energy and Escape Velocity

We have been dealing with gravitational potential energy so far in this Chapter assuming the force of gravity is constant, $\vec{F} = m\vec{g}$. This is an accurate assumption for ordinary objects near the surface of the Earth. But to deal with gravity more generally, for points not close to the Earth's surface, we must consider that the gravitational force exerted by the Earth on a particle of mass m decreases inversely as the square of the distance r from the Earth's center. The precise relationship is given by Newton's law of universal gravitation (Sections 6–1 and 6–2):

$$\vec{F} = -G\frac{mM_E}{r^2}\hat{r} \qquad\qquad [r > r_E]$$

where M_E is the mass of the Earth and $\hat{r}$ is a unit vector (at the position of m) directed radially away from the Earth's center. The minus sign indicates that the force on m is directed toward the Earth's center, in the direction opposite to $\hat{r}$. This equation can also be used to describe the gravitational force on a mass m in the vicinity of other heavenly bodies, such as the Moon, a planet, or the Sun, in which case M_E must be replaced by that body's mass.

Suppose an object of mass m moves from one position to another along an arbitrary path (Fig. 8–19) so that its distance from the Earth's center changes from r_1 to r_2. The work done by the gravitational force is

$$W = \int_1^2 \vec{F} \cdot d\vec{\ell} = -GmM_E\int_1^2 \frac{\hat{r} \cdot d\vec{\ell}}{r^2},$$

where $d\vec{\ell}$ represents an infinitesimal displacement. Since $\hat{r} \cdot d\vec{\ell} = dr$ is the component of $d\vec{\ell}$ along $\hat{r}$ (see Fig. 8–19), then

$$W = -GmM_E\int_{r_1}^{r_2} \frac{dr}{r^2} = GmM_E\left(\frac{1}{r_2} - \frac{1}{r_1}\right)$$

or

$$W = \frac{GmM_E}{r_2} - \frac{GmM_E}{r_1}.$$

Because the value of the integral depends only on the position of the end points (r_1 and r_2) and not on the path taken, the gravitational force is a conservative force.

FIGURE 8–19 Arbitrary path of particle of mass m moving from point 1 to point 2.

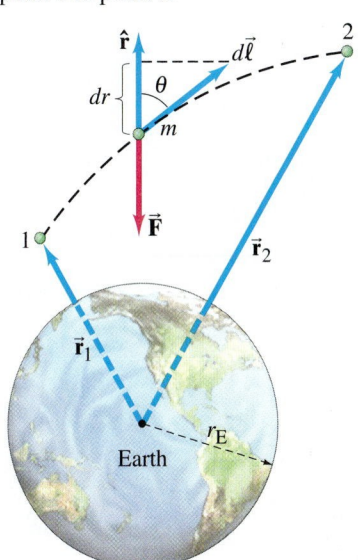

We can therefore use the concept of potential energy for the gravitational force. Since the change in potential energy is always defined (Section 8–2) as the negative of the work done by the force, we have

$$\Delta U = U_2 - U_1 = -\frac{GmM_E}{r_2} + \frac{GmM_E}{r_1}. \qquad (8\text{–}16)$$

From Eq. 8–16 the potential energy at any distance r from the Earth's center can be written:

$$U(r) = -\frac{GmM_E}{r} + C,$$

where C is a constant. It is usual to choose $C = 0$, so that

$$U(r) = -\frac{GmM_E}{r}. \qquad \left[\begin{matrix} \text{gravity} \\ (r > r_E) \end{matrix}\right] \quad (8\text{–}17)$$

With this choice for C, $U = 0$ at $r = \infty$. As an object approaches the Earth, its potential energy decreases and is always negative (Fig. 8–20).

Equation 8–16 reduces to Eq. 8–2, $\Delta U = mg(y_2 - y_1)$, for objects near the surface of the Earth (see Problem 48).

For a particle of mass m, which experiences only the force of the Earth's gravity, the total energy is conserved because gravity is a conservative force. Therefore we can write

$$\tfrac{1}{2}mv_1^2 - G\frac{mM_E}{r_1} = \tfrac{1}{2}mv_2^2 - G\frac{mM_E}{r_2} = \text{constant}. \qquad \left[\begin{matrix} \text{gravity} \\ \text{only} \end{matrix}\right] \quad (8\text{–}18)$$

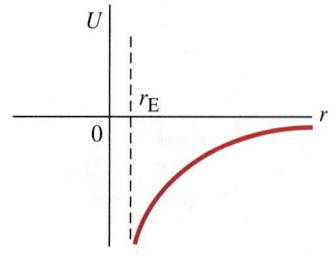

FIGURE 8–20 Gravitational potential energy plotted as a function of r, the distance from Earth's center. Valid only for points $r > r_E$, the radius of the Earth.

EXAMPLE 8–12 **Package dropped from high-speed rocket.** A box of empty film canisters is allowed to fall from a rocket traveling outward from Earth at a speed of 1800 m/s when 1600 km above the Earth's surface. The package eventually falls to the Earth. Estimate its speed just before impact. Ignore air resistance.

APPROACH We use conservation of energy. The package initially has a speed relative to Earth equal to the speed of the rocket from which it falls.

SOLUTION Conservation of energy in this case is expressed by Eq. 8–18:

$$\tfrac{1}{2}mv_1^2 - G\frac{mM_E}{r_1} = \tfrac{1}{2}mv_2^2 - G\frac{mM_E}{r_2}$$

where $v_1 = 1.80 \times 10^3$ m/s, $r_1 = (1.60 \times 10^6 \text{ m}) + (6.38 \times 10^6 \text{ m}) = 7.98 \times 10^6$ m, and $r_2 = 6.38 \times 10^6$ m (the radius of the Earth). We solve for v_2:

$$v_2 = \sqrt{v_1^2 - 2GM_E\left(\frac{1}{r_1} - \frac{1}{r_2}\right)}$$

$$= \sqrt{\begin{matrix}(1.80 \times 10^3 \text{ m/s})^2 - 2(6.67 \times 10^{-11} \text{ N·m}^2/\text{kg}^2)(5.98 \times 10^{24} \text{ kg}) \\ \times \left(\dfrac{1}{7.98 \times 10^6 \text{ m}} - \dfrac{1}{6.38 \times 10^6 \text{ m}}\right)\end{matrix}}$$

$$= 5320 \text{ m/s}.$$

NOTE In reality, the speed will be considerably less than this because of air resistance. Note, incidentally, that the direction of the velocity never entered into this calculation, and this is one of the advantages of the energy method. The rocket could have been heading away from the Earth, or toward it, or at some other angle, and the result would be the same.

Escape Velocity

When an object is projected into the air from the Earth, it will return to Earth unless its speed is very high. But if the speed *is* high enough, it will continue out into space never to return to Earth (barring other forces or collisions). The minimum initial velocity needed to prevent an object from returning to the Earth is called the **escape velocity** from Earth, v_{esc}. To determine v_{esc} from the Earth's surface (ignoring air resistance), we use Eq. 8–18 with $v_1 = v_{esc}$ and $r_1 = r_E = 6.38 \times 10^6$ m, the radius of the Earth. Since we want the minimum speed for escape, we need the object to reach $r_2 = \infty$ with merely zero speed, $v_2 = 0$. Applying Eq. 8–18 we have

$$\tfrac{1}{2} m v_{esc}^2 - G \frac{m M_E}{r_E} = 0 + 0$$

or

$$v_{esc} = \sqrt{2 G M_E / r_E} = 1.12 \times 10^4 \, \text{m/s} \qquad \textbf{(8–19)}$$

or 11.2 km/s. It is important to note that although a mass can escape from the Earth (or solar system) never to return, the force on it due to the Earth's gravitational field is never actually zero for a finite value of r.

EXAMPLE 8–13 **Escaping the Earth or the Moon.** (*a*) Compare the escape velocities of a rocket from the Earth and from the Moon. (*b*) Compare the energies required to launch the rockets. For the Moon, $M_M = 7.35 \times 10^{22}$ kg and $r_M = 1.74 \times 10^6$ m, and for Earth, $M_E = 5.98 \times 10^{24}$ kg and $r_E = 6.38 \times 10^6$ m.

APPROACH We use Eq. 8–19, replacing M_E and r_E with M_M and r_M for finding v_{esc} from the Moon.

SOLUTION (*a*) Using Eq. 8–19, the ratio of the escape velocities is

$$\frac{v_{esc}(\text{Earth})}{v_{esc}(\text{Moon})} = \sqrt{\frac{M_E}{M_M} \frac{r_M}{r_E}} = 4.7.$$

To escape Earth requires a speed 4.7 times that required to escape the Moon.
(*b*) The fuel that must be burned provides energy proportional to v^2 ($K = \tfrac{1}{2} m v^2$); so to launch a rocket to escape Earth requires $(4.7)^2 = 22$ times as much energy as to escape from the Moon.

8–8 Power

Power is defined as the *rate at which work is done*. The *average power*, $\overline{P}$, equals the work W done divided by the time t it takes to do it:

$$\overline{P} = \frac{W}{t}. \qquad \textbf{(8–20a)}$$

Since the work done in a process involves the transformation of energy from one type (or object) to another, power can also be defined as the *rate at which energy is transformed*:

$$\overline{P} = \frac{W}{t} = \frac{\text{energy transformed}}{\text{time}}.$$

The *instantaneous power*, P, is

$$P = \frac{dW}{dt}. \qquad \textbf{(8–20b)}$$

The work done in a process is equal to the energy transferred from one object to another. For example, as the potential energy stored in the spring of Fig. 8–6c is transformed to kinetic energy of the ball, the spring is doing work on the ball. Similarly, when you throw a ball or push a grocery cart, *whenever work is done, energy is being transferred from one body to another*. Hence we can also say that power is the *rate at which energy is transformed*:

$$P = \frac{dE}{dt}. \qquad \textbf{(8–20c)}$$

The power of a horse refers to how much work it can do per unit of time.

The power rating of an engine refers to how much chemical or electrical energy can be transformed into mechanical energy per unit of time. In SI units, power is measured in joules per second, and this unit is given a special name, the **watt** (W): $1 \, \text{W} = 1 \, \text{J/s}$. We are most familiar with the watt for electrical devices: the rate at which an electric lightbulb or heater changes electric energy into light or thermal energy. But the watt is used for other types of energy transformation as well. In the British system, the unit of power is the foot-pound per second (ft·lb/s). For practical purposes a larger unit is often used, the **horsepower**. One horsepower[†] (hp) is defined as 550 ft·lb/s, which equals 746 watts. An engine's power is usually specified in hp or in kW $\left(1 \, \text{kW} \approx 1\frac{1}{3} \, \text{hp}\right)$.

To see the distinction between energy and power, consider the following example. A person is limited in the work he or she can do, not only by the total energy required, but also by how fast this energy is transformed: that is, by power. For example, a person may be able to walk a long distance or climb many flights of stairs before having to stop because so much energy has been expended. On the other hand, a person who runs very quickly up stairs may feel exhausted after only a flight or two. He or she is limited in this case by power, the rate at which his or her body can transform chemical energy into mechanical energy.

FIGURE 8–21 Example 8–14.

EXAMPLE 8–14 **Stair-climbing power.** A 60-kg jogger runs up a long flight of stairs in 4.0 s (Fig. 8–21). The vertical height of the stairs is 4.5 m. (*a*) Estimate the jogger's power output in watts and horsepower. (*b*) How much energy did this require?

APPROACH The work done by the jogger is against gravity, and equals $W = mgy$. To find her average output, we divide W by the time it took.

SOLUTION (*a*) The average power output was

$$\overline{P} = \frac{W}{t} = \frac{mgy}{t} = \frac{(60 \, \text{kg})(9.8 \, \text{m/s}^2)(4.5 \, \text{m})}{4.0 \, \text{s}} = 660 \, \text{W}.$$

Since there are 746 W in 1 hp, the jogger is doing work at a rate of just under 1 hp. A human cannot do work at this rate for very long.

(*b*) The energy required is $E = \overline{P}t = (660 \, \text{J/s})(4.0 \, \text{s}) = 2600 \, \text{J}$. This result equals $W = mgy$.

NOTE The person had to transform more energy than this 2600 J. The total energy transformed by a person or an engine always includes some thermal energy (recall how hot you get running up stairs).

Automobiles do work to overcome the force of friction (and air resistance), to climb hills, and to accelerate. A car is limited by the rate it can do work, which is why automobile engines are rated in horsepower. A car needs power most when it is climbing hills and when accelerating. In the next Example, we will calculate how much power is needed in these situations for a car of reasonable size. Even when a car travels on a level road at constant speed, it needs some power just to do work to overcome the retarding forces of internal friction and air resistance. These forces depend on the conditions and speed of the car, but are typically in the range 400 N to 1000 N.

It is often convenient to write the power in terms of the net force $\vec{\mathbf{F}}$ applied to an object and its velocity $\vec{\mathbf{v}}$. Since $P = dW/dt$ and $dW = \vec{\mathbf{F}} \cdot d\vec{\boldsymbol{\ell}}$ (Eq. 7–7), then

$$P = \frac{dW}{dt} = \vec{\mathbf{F}} \cdot \frac{d\vec{\boldsymbol{\ell}}}{dt} = \vec{\mathbf{F}} \cdot \vec{\mathbf{v}}. \qquad (8\text{–}21)$$

[†]The unit was chosen by James Watt (1736–1819), who needed a way to specify the power of his newly developed steam engines. He found by experiment that a good horse can work all day at an average rate of about 360 ft·lb/s. So as not to be accused of exaggeration in the sale of his steam engines, he multiplied this by roughly $1\frac{1}{2}$ when he defined the hp.

EXAMPLE 8–15 **Power needs of a car.** Calculate the power required of a 1400-kg car under the following circumstances: (a) the car climbs a 10° hill (a fairly steep hill) at a steady 80 km/h; and (b) the car accelerates along a level road from 90 to 110 km/h in 6.0 s to pass another car. Assume that the average retarding force on the car is $F_R = 700\ N$ throughout. See Fig. 8–22.

APPROACH First we must be careful not to confuse $\vec{F}_R$, which is due to air resistance and friction that retards the motion, with the force $\vec{F}$ needed to accelerate the car, which is the frictional force exerted by the road on the tires—the reaction to the motor-driven tires pushing against the road. We must determine the latter force F before calculating the power.

SOLUTION (a) To move at a steady speed up the hill, the car must, by Newton's second law, exert a force F equal to the sum of the retarding force, 700 N, and the component of gravity parallel to the hill, $mg \sin 10°$. Thus

$$F = 700\ N + mg \sin 10°$$
$$= 700\ N + (1400\ kg)(9.80\ m/s^2)(0.174) = 3100\ N.$$

Since $\bar{v} = 80$ km/h = 22 m/s and is parallel to $\vec{F}$, then (Eq. 8–21) the power is

$$\bar{P} = F\bar{v} = (3100\ N)(22\ m/s) = 6.80 \times 10^4\ W = 68.0\ kW = 91\ hp.$$

(b) The car accelerates from 25.0 m/s to 30.6 m/s (90 to 110 km/h). Thus the car must exert a force that overcomes the 700-N retarding force plus that required to give it the acceleration

$$\bar{a}_x = \frac{(30.6\ m/s - 25.0\ m/s)}{6.0\ s} = 0.93\ m/s^2.$$

We apply Newton's second law with x being the direction of motion:

$$ma_x = \Sigma F_x = F - F_R.$$

We solve for the force required, F:

$$F = ma_x + F_R$$
$$= (1400\ kg)(0.93\ m/s^2) + 700\ N = 1300\ N + 700\ N = 2000\ N.$$

Since $P = \vec{F} \cdot \vec{v}$, the required power increases with speed and the motor must be able to provide a maximum power output of

$$\bar{P} = (2000\ N)(30.6\ m/s) = 6.12 \times 10^4\ W = 61.2\ kW = 82\ hp.$$

NOTE Even taking into account the fact that only 60 to 80% of the engine's power output reaches the wheels, it is clear from these calculations that an engine of 75 to 100 kW (100 to 130 hp) is adequate from a practical point of view.

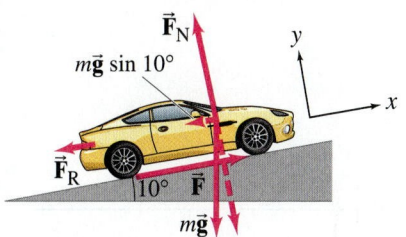

FIGURE 8–22 Example 8–15: Calculation of power needed for a car to climb a hill.

We mentioned in the Example above that only part of the energy output of a car engine reaches the wheels. Not only is some energy wasted in getting from the engine to the wheels, in the engine itself much of the input energy (from the gasoline) does not end up doing useful work. An important characteristic of all engines is their overall efficiency e, defined as the ratio of the useful power output of the engine, P_{out}, to the power input, P_{in}:

$$e = \frac{P_{out}}{P_{in}}.$$

The efficiency is always less than 1.0 because no engine can create energy, and in fact, cannot even transform energy from one form to another without some going to friction, thermal energy, and other nonuseful forms of energy. For example, an automobile engine converts chemical energy released in the burning of gasoline into mechanical energy that moves the pistons and eventually the wheels. But nearly 85% of the input energy is "wasted" as thermal energy that goes into the cooling system or out the exhaust pipe, plus friction in the moving parts. Thus car engines are roughly only about 15% efficient. We discuss efficiency in detail in Chapter 20.

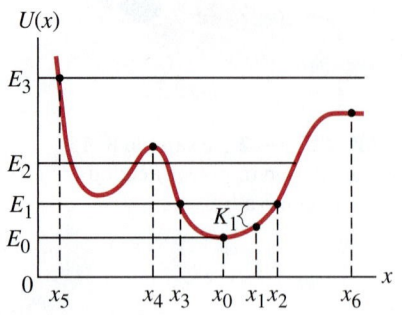

FIGURE 8–23 A potential energy diagram.

*8–9 Potential Energy Diagrams; Stable and Unstable Equilibrium

If only conservative forces do work on an object, we can learn a great deal about its motion simply by examining a potential energy diagram—the graph of $U(x)$ versus x. An example of a potential energy diagram is shown in Fig. 8–23. The rather complex curve represents some complicated potential energy $U(x)$. The total energy $E = K + U$ is constant and can be represented as a horizontal line on this graph. Four different possible values for E are shown, labeled E_0, E_1, E_2, and E_3. What the actual value of E will be for a given system depends on the initial conditions. (For example, the total energy E of a mass oscillating on the end of a spring depends on the amount the spring is initially compressed or stretched.) Kinetic energy $K = \frac{1}{2}mv^2$ cannot be less than zero (v would be imaginary), and because $E = U + K = $ constant, then $U(x)$ must be less than or equal to E for all situations: $U(x) \leq E$. Thus the minimum value which the total energy can take for the potential energy shown in Fig. 8–23 is that labeled E_0. For this value of E, the mass can only be at rest at $x = x_0$. The system has potential energy but no kinetic energy at this position.

If the system's total energy E is greater than E_0, say it is E_1 on our plot, the system can have both kinetic and potential energy. Because energy is conserved,

$$K = E - U(x).$$

Since the curve represents $U(x)$ at each x, the kinetic energy at any value of x is represented by the distance between the E line and the curve $U(x)$ at that value of x. In the diagram, the kinetic energy for an object at x_1, when its total energy is E_1, is indicated by the notation K_1.

An object with energy E_1 can oscillate only between the points x_2 and x_3. This is because if $x > x_2$ or $x < x_3$, the potential energy would be greater than E, meaning $K = \frac{1}{2}mv^2 < 0$ and v would be imaginary, and so impossible. At x_2 and x_3 the velocity is zero, since $E = U$ at these points. Hence x_2 and x_3 are called the **turning points** of the motion. If the object is at x_0, say, moving to the right, its kinetic energy (and speed) decreases until it reaches zero at $x = x_2$. The object then reverses direction, proceeding to the left and increasing in speed until it passes x_0 again. It continues to move, decreasing in speed until it reaches $x = x_3$, where again $v = 0$, and the object again reverses direction.

If the object has energy $E = E_2$ in Fig. 8–23, there are four turning points. The object can move in only one of the two potential energy "valleys," depending on where it is initially. It cannot get from one valley to the other because of the barrier between them—for example at a point such as x_4, $U > E_2$, which means v would be imaginary.[†] For energy E_3, there is only one turning point since $U(x) < E_3$ for all $x > x_5$. Thus our object, if moving initially to the left, varies in speed as it passes the potential valleys but eventually stops and turns around at $x = x_5$. It then proceeds to the right indefinitely, never to return.

How do we know the object reverses direction at the turning points? Because of the force exerted on it. The force F is related to the potential energy U by Eq. 8–7, $F = -dU/dx$. The force F is equal to the negative of the slope of the U-versus-x curve at any point x. At $x = x_2$, for example, the slope is positive so the force is negative, which means it acts to the left (toward decreasing values of x).

At $x = x_0$ the slope is zero, so $F = 0$. At such a point the particle is said to be in **equilibrium**. This term means simply that the net force on the object is zero. Hence, its acceleration is zero, and so if it is initially at rest, it remains at rest. If the object at rest at $x = x_0$ were moved slightly to the left or right, a nonzero force would act on it in the direction to move it back toward x_0. An object that returns

---

[†]Although this is true according to Newtonian physics, modern quantum mechanics predicts that objects can "tunnel" through such a barrier, and such processes have been observed at the atomic and subatomic level.

toward its equilibrium point when displaced slightly is said to be at a point of **stable equilibrium**. Any *minimum* in the potential energy curve represents a point of stable equilibrium.

An object at $x = x_4$ would also be in equilibrium, since $F = -dU/dx = 0$. If the object were displaced a bit to either side of x_4, a force would act to pull the object *away* from the equilibrium point. Points like x_4, where the potential energy curve has a maximum, are points of **unstable equilibrium**. The object will *not* return to equilibrium if displaced slightly, but instead will move farther away.

When an object is in a region over which U is constant, such as near $x = x_6$ in Fig. 8–23, the force is zero over some distance. The object is in equilibrium and if displaced slightly to one side the force is still zero. The object is said to be in **neutral equilibrium** in this region.

Summary

A **conservative force** is one for which the work done by the force in moving an object from one position to another depends only on the two positions and not on the path taken. The work done by a conservative force is recoverable, which is not true for nonconservative forces, such as friction.

Potential energy, U, is energy associated with conservative forces that depend on the position or configuration of objects. Gravitational potential energy is

$$U_{grav} = mgy, \tag{8–3}$$

where the mass m is near the Earth's surface, a height y above some reference point. Elastic potential energy is given by

$$U_{el} = \tfrac{1}{2}kx^2 \tag{8–5}$$

for a spring with stiffness constant k stretched or compressed a displacement x from equilibrium. Other potential energies include chemical, electrical, and nuclear energy.

Potential energy is always associated with a conservative force, and the change in potential energy, ΔU, between two points under the action of a conservative force $\vec{F}$ is defined as the negative of the work done by the force:

$$\Delta U = U_2 - U_1 = -\int_1^2 \vec{F} \cdot d\vec{\ell}. \tag{8–4}$$

Inversely, we can write, for the one-dimensional case,

$$F = -\frac{dU(x)}{dx}. \tag{8–7}$$

Only *changes* in potential energy are physically meaningful, so the position where $U = 0$ can be chosen for convenience.

Potential energy is not a property of an object but is associated with the interaction of two or more objects.

When only conservative forces act, the total **mechanical energy**, E, defined as the sum of kinetic and potential energies, is conserved:

$$E = K + U = \text{constant.} \tag{8–10}$$

If nonconservative forces also act, additional types of energy are involved, such as thermal energy. It has been found experimentally that, when all forms of energy are included, the total energy is conserved. This is the **law of conservation of energy**:

$$\Delta K + \Delta U + \Delta(\text{other energy types}) = 0. \tag{8–14}$$

The gravitational force as described by Newton's law of universal gravitation is a conservative force. The potential energy of an object of mass m due to the gravitational force exerted on it by the Earth is given by

$$U(r) = -\frac{GmM_E}{r}, \tag{8–17}$$

where M_E is the mass of the Earth and r is the distance of the object from the Earth's center ($r \geq$ radius of Earth).

Power is defined as the rate at which work is done or the rate at which energy is transformed from one form to another:

$$P = \frac{dW}{dt} = \frac{dE}{dt}, \tag{8–20}$$

or

$$P = \vec{F} \cdot \vec{v}. \tag{8–21}$$

Questions

1. List some everyday forces that are not conservative, and explain why they aren't.

2. You lift a heavy book from a table to a high shelf. List the forces on the book during this process, and state whether each is conservative or nonconservative.

3. The net force acting on a particle is conservative and increases the kinetic energy by 300 J. What is the change in (*a*) the potential energy, and (*b*) the total energy, of the particle?

4. When a "superball" is dropped, can it rebound to a greater height than its original height?

5. A hill has a height h. A child on a sled (total mass m) slides down starting from rest at the top. Does the velocity at the bottom depend on the angle of the hill if (*a*) it is icy and there is no friction, and (*b*) there is friction (deep snow)?

6. Why is it tiring to push hard against a solid wall even though no work is done?

7. Analyze the motion of a simple swinging pendulum in terms of energy, (*a*) ignoring friction, and (*b*) taking friction into account. Explain why a grandfather clock has to be wound up.

8. Describe precisely what is "wrong" physically in the famous Escher drawing shown in Fig. 8–24.

FIGURE 8–24
Question 8.

9. In Fig. 8–25, water balloons are tossed from the roof of a building, all with the same speed but with different launch angles. Which one has the highest speed when it hits the ground? Ignore air resistance.

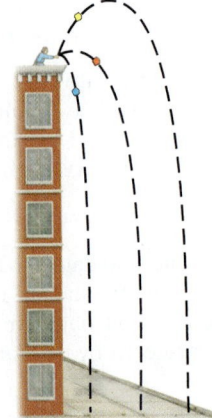

FIGURE 8–25
Question 9.

10. A coil spring of mass m rests upright on a table. If you compress the spring by pressing down with your hand and then release it, can the spring leave the table? Explain using the law of conservation of energy.

11. What happens to the gravitational potential energy when water at the top of a waterfall falls to the pool below?

12. Experienced hikers prefer to step over a fallen log in their path rather than stepping on top and jumping down on the other side. Explain.

13. (a) Where does the kinetic energy come from when a car accelerates uniformly starting from rest? (b) How is the increase in kinetic energy related to the friction force the road exerts on the tires?

14. The Earth is closest to the Sun in winter (Northern Hemisphere). When is the gravitational potential energy the greatest?

15. Can the total mechanical energy $E = K + U$ ever be negative? Explain.

16. Suppose that you wish to launch a rocket from the surface of the Earth so that it escapes the Earth's gravitational field. You wish to use minimum fuel in doing this. From what point on the surface of the Earth should you make the launch and in what direction? Do the launch location and direction matter? Explain.

17. Recall from Chapter 4, Example 4–14, that you can use a pulley and ropes to decrease the force needed to raise a heavy load (see Fig. 8–26). But for every meter the load is raised, how much rope must be pulled up? Account for this, using energy concepts.

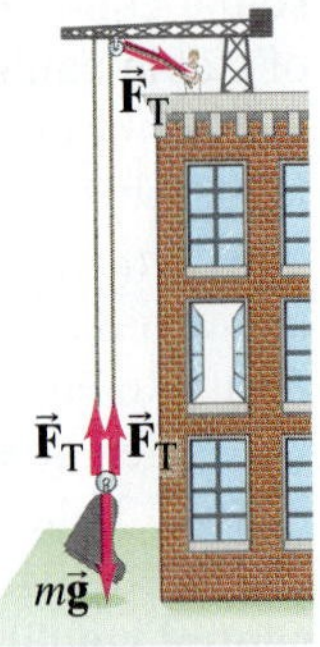

FIGURE 8–26
Question 17.

18. Two identical arrows, one with twice the speed of the other, are fired into a bale of hay. Assuming the hay exerts a constant "frictional" force on the arrows, the faster arrow will penetrate how much farther than the slower arrow? Explain.

19. A bowling ball is hung from the ceiling by a steel wire (Fig. 8–27). The instructor pulls the ball back and stands against the wall with the ball against his nose. To avoid injury the instructor is supposed to release the ball without pushing it. Why?

FIGURE 8–27
Question 19.

20. A pendulum is launched from a point that is a height h above its lowest point in two different ways (Fig. 8–28). During both launches, the pendulum is given an initial speed of 3.0 m/s. On the first launch, the initial velocity of the pendulum is directed upward along the trajectory, and on the second launch it is directed downward along the trajectory. Which launch will cause the highest speed when the pendulum bob passes the lowest point of its swing? Explain.

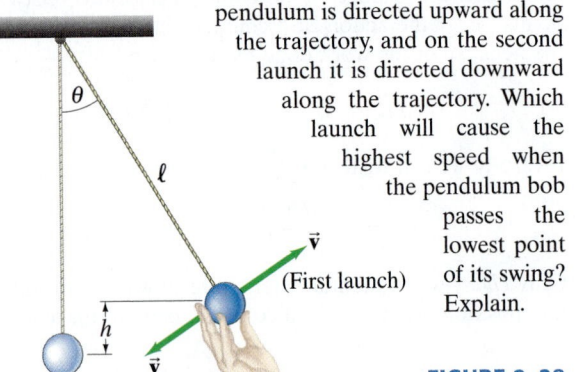

FIGURE 8–28
Question 20.

21. Describe the energy transformations when a child hops around on a pogo stick.

22. Describe the energy transformations that take place when a skier starts skiing down a hill, but after a time is brought to rest by striking a snowdrift.

23. Suppose you lift a suitcase from the floor to a table. The work you do on the suitcase depends on which of the following: (a) whether you lift it straight up or along a more complicated path, (b) the time the lifting takes, (c) the height of the table, and (d) the weight of the suitcase?

24. Repeat Question 23 for the *power* needed instead of the work.

25. Why is it easier to climb a mountain via a zigzag trail rather than to climb straight up?

***26.** Figure 8–29 shows a potential energy curve, $U(x)$. (a) At which point does the force have greatest magnitude? (b) For each labeled point, state whether the force acts to the left or to the right, or is zero. (c) Where is there equilibrium and of what type is it?

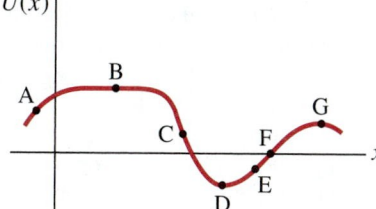

FIGURE 8–29
Question 26.

***27.** (a) Describe in detail the velocity changes of a particle that has energy E_3 in Fig. 8–23 as it moves from x_6 to x_5 and back to x_6. (b) Where is its kinetic energy the greatest and the least?

***28.** Name the type of equilibrium for each position of the balls in Fig. 8–30.

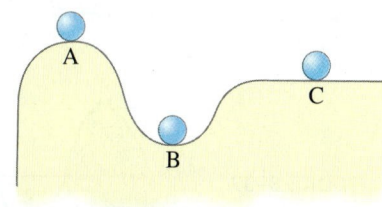

FIGURE 8–30
Question 28.

Problems

8–1 and 8–2 Conservative Forces and Potential Energy

1. (I) A spring has a spring constant k of 82.0 N/m. How much must this spring be compressed to store 35.0 J of potential energy?

2. (I) A 6.0-kg monkey swings from one branch to another 1.3 m higher. What is the change in gravitational potential energy?

3. (II) A spring with $k = 63$ N/m hangs vertically next to a ruler. The end of the spring is next to the 15-cm mark on the ruler. If a 2.5-kg mass is now attached to the end of the spring, where will the end of the spring line up with the ruler marks?

4. (II) A 56.5-kg hiker starts at an elevation of 1270 m and climbs to the top of a 2660-m peak. (a) What is the hiker's change in potential energy? (b) What is the minimum work required of the hiker? (c) Can the actual work done be greater than this? Explain.

5. (II) A 1.60-m tall person lifts a 1.95-kg book off the ground so it is 2.20 m above the ground. What is the potential energy of the book relative to (a) the ground, and (b) the top of the person's head? (c) How is the work done by the person related to the answers in parts (a) and (b)?

6. (II) A 1200-kg car rolling on a horizontal surface has speed $v = 75$ km/h when it strikes a horizontal coiled spring and is brought to rest in a distance of 2.2 m. What is the spring stiffness constant of the spring?

7. (II) A particular spring obeys the force law $\vec{F} = \left(-kx + ax^3 + bx^4\right)\hat{i}$. (a) Is this force conservative? Explain why or why not. (b) If it is conservative, determine the form of the potential energy function.

8. (II) If $U = 3x^2 + 2xy + 4y^2z$, what is the force, $\vec{F}$?

9. (II) A particle is constrained to move in one dimension along the x axis and is acted upon by a force given by

$$\vec{F}(x) = -\frac{k}{x^3}\hat{i}$$

where k is a constant with units appropriate to the SI system. Find the potential energy function $U(x)$, if U is arbitrarily defined to be zero at $x = 2.0$ m, so that $U(2.0\text{ m}) = 0$.

10. (II) A particle constrained to move in one dimension is subject to a force $F(x)$ that varies with position x as

$$\vec{F}(x) = A\sin(kx)\hat{i}$$

where A and k are constants. What is the potential energy function $U(x)$, if we take $U = 0$ at the point $x = 0$?

8–3 and 8–4 Conservation of Mechanical Energy

11. (I) A novice skier, starting from rest, slides down a frictionless 13.0° incline whose vertical height is 125 m. How fast is she going when she reaches the bottom?

12. (I) Jane, looking for Tarzan, is running at top speed (5.0 m/s) and grabs a vine hanging vertically from a tall tree in the jungle. How high can she swing upward? Does the length of the vine affect your answer?

13. (II) In the high jump, the kinetic energy of an athlete is transformed into gravitational potential energy without the aid of a pole. With what minimum speed must the athlete leave the ground in order to lift his center of mass 2.10 m and cross the bar with a speed of 0.70 m/s?

14. (II) A sled is initially given a shove up a frictionless 23.0° incline. It reaches a maximum vertical height 1.12 m higher than where it started. What was its initial speed?

15. (II) A 55-kg bungee jumper leaps from a bridge. She is tied to a bungee cord that is 12 m long when unstretched, and falls a total of 31 m. (a) Calculate the spring constant k of the bungee cord assuming Hooke's law applies. (b) Calculate the maximum acceleration she experiences.

16. (II) A 72-kg trampoline artist jumps vertically upward from the top of a platform with a speed of 4.5 m/s. (a) How fast is he going as he lands on the trampoline, 2.0 m below (Fig. 8–31)? (b) If the trampoline behaves like a spring of spring constant 5.8×10^4 N/m, how far does he depress it?

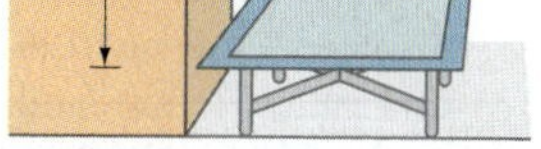

2.0 m

FIGURE 8–31
Problem 16.

17. (II) The total energy E of an object of mass m that moves in one dimension under the influence of only conservative forces can be written as

$$E = \frac{1}{2}mv^2 + U.$$

Use conservation of energy, $dE/dt = 0$, to predict Newton's second law.

18. (II) A 0.40-kg ball is thrown with a speed of 8.5 m/s at an upward angle of 36°. (a) What is its speed at its highest point, and (b) how high does it go? (Use conservation of energy.)

19. (II) A vertical spring (ignore its mass), whose spring constant is 875 N/m, is attached to a table and is compressed down by 0.160 m. (a) What upward speed can it give to a 0.380-kg ball when released? (b) How high above its original position (spring compressed) will the ball fly?

20. (II) A roller-coaster car shown in Fig. 8–32 is pulled up to point 1 where it is released from rest. Assuming no friction, calculate the speed at points 2, 3, and 4.

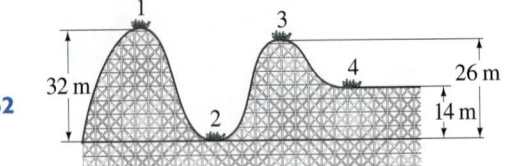

FIGURE 8–32
Problems 20 and 34.

21. (II) When a mass m sits at rest on a spring, the spring is compressed by a distance d from its undeformed length (Fig. 8–33a). Suppose instead that the mass is released from rest when it barely touches the undeformed spring (Fig. 8–33b). Find the distance D that the spring is compressed before it is able to stop the mass. Does $D = d$? If not, why not?

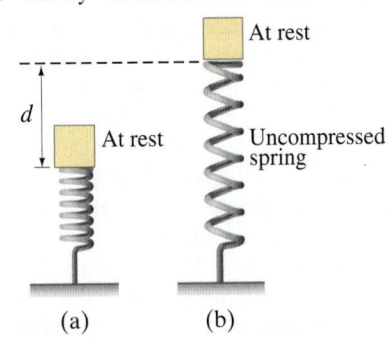

FIGURE 8–33
Problem 21.

(a)   (b)

22. (II) Two masses are connected by a string as shown in Fig. 8–34. Mass $m_A = 4.0\,kg$ rests on a frictionless inclined plane, while $m_B = 5.0\,kg$ is initially held at a height of $h = 0.75\,m$ above the floor. (a) If m_B is allowed to fall, what will be the resulting acceleration of the masses? (b) If the masses were initially at rest, use the kinematic equations (Eqs. 2–12) to find their velocity just before m_B hits the floor. (c) Use conservation of energy to find the velocity of the masses just before m_B hits the floor. You should get the same answer as in part (b).

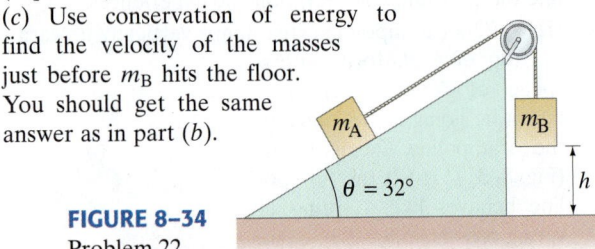

FIGURE 8–34
Problem 22.

23. (II) A block of mass m is attached to the end of a spring (spring stiffness constant k), Fig. 8–35. The mass is given an initial displacement x_0 from equilibrium, and an initial speed v_0. Ignoring friction and the mass of the spring, use energy methods to find (a) its maximum speed, and (b) its maximum stretch from equilibrium, in terms of the given quantities.

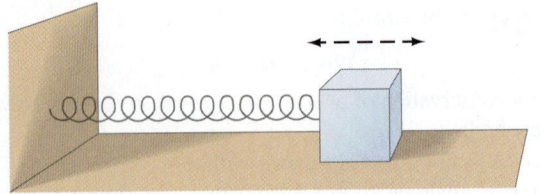

FIGURE 8–35 Problems 23, 37, and 38.

24. (II) A cyclist intends to cycle up a 9.50° hill whose vertical height is 125 m. The pedals turn in a circle of diameter 36.0 cm. Assuming the mass of bicycle plus person is 75.0 kg, (a) calculate how much work must be done against gravity. (b) If each complete revolution of the pedals moves the bike 5.10 m along its path, calculate the average force that must be exerted on the pedals tangent to their circular path. Neglect work done by friction and other losses.

25. (II) A pendulum 2.00 m long is released (from rest) at an angle $\theta_0 = 30.0°$ (Fig. 8–14). Determine the speed of the 70.0-g bob: (a) at the lowest point $(\theta = 0)$; (b) at $\theta = 15.0°$, (c) at $\theta = -15.0°$ (i.e., on the opposite side). (d) Determine the tension in the cord at each of these three points. (e) If the bob is given an initial speed $v_0 = 1.20\,m/s$ when released at $\theta = 30.0°$, recalculate the speeds for parts (a), (b), and (c).

26. (II) What should be the spring constant k of a spring designed to bring a 1200-kg car to rest from a speed of 95 km/h so that the occupants undergo a maximum acceleration of 5.0 g?

27. (III) An engineer is designing a spring to be placed at the bottom of an elevator shaft. If the elevator cable breaks when the elevator is at a height h above the top of the spring, calculate the value that the spring constant k should have so that passengers undergo an acceleration of no more than 5.0 g when brought to rest. Let M be the total mass of the elevator and passengers.

28. (III) A skier of mass m starts from rest at the top of a solid sphere of radius r and slides down its frictionless surface. (a) At what angle θ (Fig. 8–36) will the skier leave the sphere? (b) If friction were present, would the skier fly off at a greater or lesser angle?

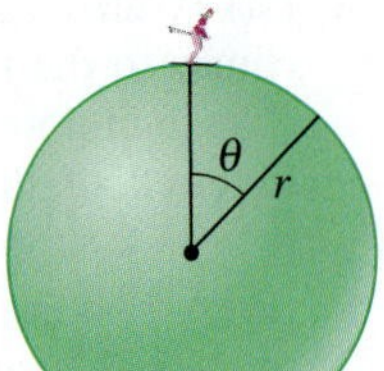

FIGURE 8–36
Problem 28.

8–5 and 8–6 Law of Conservation of Energy

29. (I) Two railroad cars, each of mass 56,000 kg, are traveling 95 km/h toward each other. They collide head-on and come to rest. How much thermal energy is produced in this collision?

30. (I) A 16.0-kg child descends a slide 2.20 m high and reaches the bottom with a speed of 1.25 m/s. How much thermal energy due to friction was generated in this process?

31. (II) A ski starts from rest and slides down a 28° incline 85 m long. (a) If the coefficient of friction is 0.090, what is the ski's speed at the base of the incline? (b) If the snow is level at the foot of the incline and has the same coefficient of friction, how far will the ski travel along the level? Use energy methods.

32. (II) A 145-g baseball is dropped from a tree 14.0 m above the ground. (a) With what speed would it hit the ground if air resistance could be ignored? (b) If it actually hits the ground with a speed of 8.00 m/s, what is the average force of air resistance exerted on it?

33. (II) A 96-kg crate, starting from rest, is pulled across a floor with a constant horizontal force of 350 N. For the first 15 m the floor is frictionless, and for the next 15 m the coefficient of friction is 0.25. What is the final speed of the crate?

34. (II) Suppose the roller-coaster car in Fig. 8–32 passes point 1 with a speed of 1.70 m/s. If the average force of friction is equal to 0.23 of its weight, with what speed will it reach point 2? The distance traveled is 45.0 m.

35. (II) A skier traveling 9.0 m/s reaches the foot of a steady upward 19° incline and glides 12 m up along this slope before coming to rest. What was the average coefficient of friction?

36. (II) Consider the track shown in Fig. 8–37. The section AB is one quadrant of a circle of radius 2.0 m and is frictionless. B to C is a horizontal span 3.0 m long with a coefficient of kinetic friction $\mu_k = 0.25$. The section CD under the spring is frictionless. A block of mass 1.0 kg is released from rest at A. After sliding on the track, it compresses the spring by 0.20 m. Determine: (a) the velocity of the block at point B; (b) the thermal energy produced as the block slides from B to C; (c) the velocity of the block at point C; (d) the stiffness constant k for the spring.

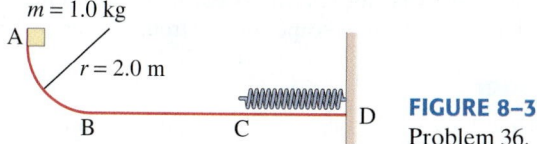

$m = 1.0$ kg

A

$r = 2.0$ m

B   C   D

FIGURE 8–37
Problem 36.

37. (II) A 0.620-kg wood block is firmly attached to a very light horizontal spring $(k = 180 \text{ N/m})$ as shown in Fig. 8–35. This block–spring system, when compressed 5.0 cm and released, stretches out 2.3 cm beyond the equilibrium position before stopping and turning back. What is the coefficient of kinetic friction between the block and the table?

38. (II) A 180-g wood block is firmly attached to a very light horizontal spring, Fig. 8–35. The block can slide along a table where the coefficient of friction is 0.30. A force of 25 N compresses the spring 18 cm. If the spring is released from this position, how far beyond its equilibrium position will it stretch on its first cycle?

39. (II) You drop a ball from a height of 2.0 m, and it bounces back to a height of 1.5 m. (a) What fraction of its initial energy is lost during the bounce? (b) What is the ball's speed just before and just after the bounce? (c) Where did the energy go?

40. (II) A 56-kg skier starts from rest at the top of a 1200-m-long trail which drops a total of 230 m from top to bottom. At the bottom, the skier is moving 11.0 m/s. How much energy was dissipated by friction?

41. (II) How much does your gravitational energy change when you jump as high as you can (say, 1.0 m)?

42. (III) A spring $(k = 75 \text{ N/m})$ has an equilibrium length of 1.00 m. The spring is compressed to a length of 0.50 m and a mass of 2.0 kg is placed at its free end on a frictionless slope which makes an angle of 41° with respect to the horizontal (Fig. 8–38). The spring is then released. (a) If the mass is *not* attached to the spring, how far up the slope will the mass move before coming to rest? (b) If the mass *is* attached to the spring, how far up the slope will the mass move before coming to rest? (c) Now the incline has a coefficient of kinetic friction μ_k. If the block, attached to the spring, is observed to stop just as it reaches the spring's equilibrium position, what is the coefficient of friction μ_k?

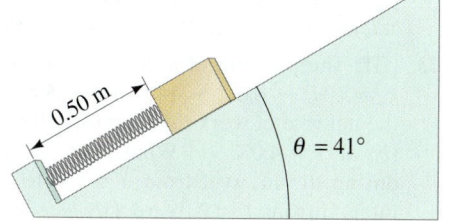

0.50 m

$\theta = 41°$

FIGURE 8–38
Problem 42.

43. (III) A 2.0-kg block slides along a horizontal surface with a coefficient of kinetic friction $\mu_k = 0.30$. The block has a speed $v = 1.3$ m/s when it strikes a massless spring head-on (as in Fig. 8–18). (a) If the spring has force constant $k = 120$ N/m, how far is the spring compressed? (b) What minimum value of the coefficient of static friction, μ_S, will assure that the spring remains compressed at the maximum compressed position? (c) If μ_S is less than this, what is the speed of the block when it detaches from the decompressing spring? [*Hint*: Detachment occurs when the spring reaches its natural length $(x = 0)$; explain why.]

44. (III) Early test flights for the space shuttle used a "glider" (mass of 980 kg including pilot). After a horizontal launch at 480 km/h at a height of 3500 m, the glider eventually landed at a speed of 210 km/h. (a) What would its landing speed have been in the absence of air resistance? (b) What was the average force of air resistance exerted on it if it came in at a constant glide angle of 12° to the Earth's surface?

8–7 Gravitational Potential Energy

45. (I) For a satellite of mass m_S in a circular orbit of radius r_S around the Earth, determine (a) its kinetic energy K, (b) its potential energy U $(U = 0$ at infinity), and (c) the ratio K/U.

46. (I) Jill and her friends have built a small rocket that soon after lift-off reaches a speed of 850 m/s. How high above the Earth can it rise? Ignore air friction.

47. (I) The escape velocity from planet A is double that for planet B. The two planets have the same mass. What is the ratio of their radii, r_A/r_B?

48. (II) Show that Eq. 8–16 for gravitational potential energy reduces to Eq. 8–2, $\Delta U = mg(y_2 - y_1)$, for objects near the surface of the Earth.

49. (II) Determine the escape velocity from the Sun for an object (a) at the Sun's surface $(r = 7.0 \times 10^5 \text{ km},$ $M = 2.0 \times 10^{30} \text{ kg})$, and (b) at the average distance of the Earth $(1.50 \times 10^8 \text{ km})$. Compare to the speed of the Earth in its orbit.

50. (II) Two Earth satellites, A and B, each of mass $m = 950$ kg, are launched into circular orbits around the Earth's center. Satellite A orbits at an altitude of 4200 km, and satellite B orbits at an altitude of 12,600 km. (a) What are the potential energies of the two satellites? (b) What are the kinetic energies of the two satellites? (c) How much work would it require to change the orbit of satellite A to match that of satellite B?

51. (II) Show that the escape velocity for any satellite in a circular orbit is $\sqrt{2}$ times its velocity.

52. (II) (a) Show that the total mechanical energy of a satellite (mass m) orbiting at a distance r from the center of the Earth (mass M_E) is

$$E = -\frac{1}{2}\frac{GmM_E}{r},$$

if $U = 0$ at $r = \infty$. (b) Show that although friction causes the value of E to decrease slowly, kinetic energy must actually increase if the orbit remains a circle.

53. (II) Take into account the Earth's rotational speed (1 rev/day) and determine the necessary speed, with respect to Earth, for a rocket to escape if fired from the Earth at the equator in a direction (a) eastward; (b) westward; (c) vertically upward.

54. (II) (a) Determine a formula for the maximum height h that a rocket will reach if launched vertically from the Earth's surface with speed v_0 ($< v_{esc}$). Express in terms of v_0, r_E, M_E, and G. (b) How high does a rocket go if $v_0 = 8.35$ km/s? Ignore air resistance and the Earth's rotation.

55. (II) (a) Determine the rate at which the escape velocity from the Earth changes with distance from the center of the Earth, dv_{esc}/dr. (b) Use the approximation $\Delta v \approx (dv/dr)\,\Delta r$ to determine the escape velocity for a spacecraft orbiting the Earth at a height of 320 km.

56. (II) A meteorite has a speed of 90.0 m/s when 850 km above the Earth. It is falling vertically (ignore air resistance) and strikes a bed of sand in which it is brought to rest in 3.25 m. (a) What is its speed just before striking the sand? (b) How much work does the sand do to stop the meteorite (mass = 575 kg)? (c) What is the average force exerted by the sand on the meteorite? (d) How much thermal energy is produced?

57. (II) How much work would be required to move a satellite of mass m from a circular orbit of radius $r_1 = 2r_E$ about the Earth to another circular orbit of radius $r_2 = 3r_E$? (r_E is the radius of the Earth.)

58. (II) (a) Suppose we have three masses, m_1, m_2, and m_3, that initially are infinitely far apart from each other. Show that the work needed to bring them to the positions shown in Fig. 8–39 is

$$W = -G\left(\frac{m_1 m_2}{r_{12}} + \frac{m_1 m_3}{r_{13}} + \frac{m_2 m_3}{r_{23}}\right).$$

(b) Can we say that this formula also gives the potential energy of the system, or the potential energy of one or two of the objects? (c) Is W equal to the binding energy of the system—that is, equal to the energy required to separate the components by an infinite distance? Explain.

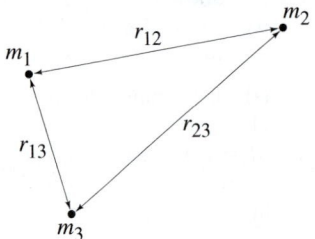

FIGURE 8–39
Problem 58.

59. (II) A NASA satellite has just observed an asteroid that is on a collision course with the Earth. The asteroid has an estimated mass, based on its size, of 5×10^9 kg. It is approaching the Earth on a head-on course with a velocity of 660 m/s relative to the Earth and is now 5.0×10^6 km away. With what speed will it hit the Earth's surface, neglecting friction with the atmosphere?

60. (II) A sphere of radius r_1 has a concentric spherical cavity of radius r_2 (Fig. 8–40). Assume this spherical shell of thickness $r_1 - r_2$ is uniform and has a total mass M. Show that the gravitational potential energy of a mass m at a distance r from the center of the shell ($r > r_1$) is given by

$$U = -\frac{GmM}{r}.$$

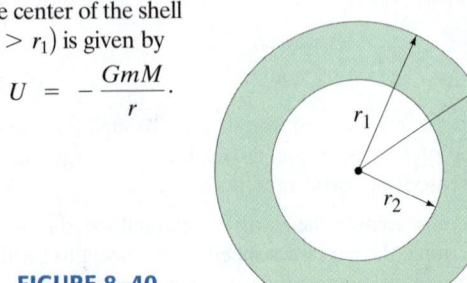

FIGURE 8–40
Problem 60.

61. (III) To escape the solar system, an interstellar spacecraft must overcome the gravitational attraction of both the Earth and Sun. Ignore the effects of other bodies in the solar system. (a) Show that the escape velocity is

$$v = \sqrt{v_E^2 + (v_S - v_0)^2} = 16.7 \text{ km/s},$$

where: v_E is the escape velocity from the Earth (Eq. 8–19); $v_S = \sqrt{2GM_S/r_{SE}}$ is the escape velocity from the gravitational field of the Sun at the orbit of the Earth but far from the Earth's influence (r_{SE} is the Sun–Earth distance); and v_0 is the Earth's orbital velocity about the Sun. (b) Show that the energy required is 1.40×10^8 J per kilogram of spacecraft mass. [*Hint*: Write the energy equation for escape from Earth with v' as the velocity, relative to Earth, but far from Earth; then let $v' + v_0$ be the escape velocity from the Sun.]

8–8 Power

62. (I) How long will it take a 1750-W motor to lift a 335-kg piano to a sixth-story window 16.0 m above?

63. (I) If a car generates 18 hp when traveling at a steady 95 km/h, what must be the average force exerted on the car due to friction and air resistance?

64. (I) An 85-kg football player traveling 5.0 m/s is stopped in 1.0 s by a tackler. (a) What is the original kinetic energy of the player? (b) What average power is required to stop him?

65. (II) A driver notices that her 1080-kg car slows down from 95 km/h to 65 km/h in about 7.0 s on the level when it is in neutral. Approximately what power (watts and hp) is needed to keep the car traveling at a constant 80 km/h?

66. (II) How much work can a 3.0-hp motor do in 1.0 h?

67. (II) An outboard motor for a boat is rated at 55 hp. If it can move a particular boat at a steady speed of 35 km/h, what is the total force resisting the motion of the boat?

68. (II) A 1400-kg sports car accelerates from rest to 95 km/h in 7.4 s. What is the average power delivered by the engine?

69. (II) During a workout, football players ran up the stadium stairs in 75 s. The stairs are 78 m long and inclined at an angle of 33°. If a player has a mass of 92 kg, estimate his average power output on the way up. Ignore friction and air resistance.

70. (II) A pump lifts 21.0 kg of water per minute through a height of 3.50 m. What minimum output rating (watts) must the pump motor have?

71. (II) A ski area claims that its lifts can move 47,000 people per hour. If the average lift carries people about 200 m (vertically) higher, estimate the maximum total power needed.

72. (II) A 75-kg skier grips a moving rope that is powered by an engine and is pulled at constant speed to the top of a 23° hill. The skier is pulled a distance $x = 220$ m along the incline and it takes 2.0 min to reach the top of the hill. If the coefficient of kinetic friction between the snow and skis is $\mu_k = 0.10$, what horsepower engine is required if 30 such skiers (max) are on the rope at one time?

73. (III) The position of a 280-g object is given (in meters) by $x = 5.0t^3 - 8.0t^2 - 44t$, where t is in seconds. Determine the net rate of work done on this object (a) at $t = 2.0$ s and (b) at $t = 4.0$ s. (c) What is the average net power input during the interval from $t = 0$ s to $t = 2.0$ s, and in the interval from $t = 2.0$ s to 4.0 s?

74. (III) A bicyclist coasts down a 6.0° hill at a steady speed of 4.0 m/s. Assuming a total mass of 75 kg (bicycle plus rider), what must be the cyclist's power output to climb the same hill at the same speed?

*8–9 Potential Energy Diagrams

***75. (II)** Draw a potential energy diagram, U vs. x, and analyze the motion of a mass m resting on a frictionless horizontal table and connected to a horizontal spring with stiffness constant k. The mass is pulled a distance to the right so that the spring is stretched a distance x_0 initially, and then the mass is released from rest.

***76. (II)** The spring of Problem 75 has a stiffness constant $k = 160$ N/m. The mass $m = 5.0$ kg is released from rest when the spring is stretched $x_0 = 1.0$ m from equilibrium. Determine (a) the total energy of the system; (b) the kinetic energy when $x = \frac{1}{2}x_0$; (c) the maximum kinetic energy; (d) the maximum speed and at what positions it occurs; (e) the maximum acceleration and where it occurs.

***77. (III)** The potential energy of the two atoms in a diatomic (two-atom) molecule can be written

$$U(r) = -\frac{a}{r^6} + \frac{b}{r^{12}},$$

where r is the distance between the two atoms and a and b are positive constants. (a) At what values of r is $U(r)$ a minimum? A maximum? (b) At what values of r is $U(r) = 0$? (c) Plot $U(r)$ as a function of r from $r = 0$ to r at a value large enough for all the features in (a) and (b) to show. (d) Describe the motion of one atom with respect to the second atom when $E < 0$, and when $E > 0$. (e) Let F be the force one atom exerts on the other. For what values of r is $F > 0, F < 0, F = 0$? (f) Determine F as a function of r.

***78. (III)** The *binding energy* of a two-particle system is defined as the energy required to separate the two particles from their state of lowest energy to $r = \infty$. Determine the binding energy for the molecule discussed in Problem 77.

General Problems

79. What is the average power output of an elevator that lifts 885 kg a vertical height of 32.0 m in 11.0 s?

80. A projectile is fired at an upward angle of 48.0° from the top of a 135-m-high cliff with a speed of 165 m/s. What will be its speed when it strikes the ground below? (Use conservation of energy.)

81. Water flows over a dam at the rate of 580 kg/s and falls vertically 88 m before striking the turbine blades. Calculate (a) the speed of the water just before striking the turbine blades (neglect air resistance), and (b) the rate at which mechanical energy is transferred to the turbine blades, assuming 55% efficiency.

82. A bicyclist of mass 75 kg (including the bicycle) can coast down a 4.0° hill at a steady speed of 12 km/h. Pumping hard, the cyclist can descend the hill at a speed of 32 km/h. Using the same power, at what speed can the cyclist climb the same hill? Assume the force of friction is proportional to the square of the speed v; that is, $F_{fr} = bv^2$, where b is a constant.

83. A 62-kg skier starts from rest at the top of a ski jump, point A in Fig. 8–41, and travels down the ramp. If friction and air resistance can be neglected, (a) determine her speed v_B when she reaches the horizontal end of the ramp at B. (b) Determine the distance s to where she strikes the ground at C.

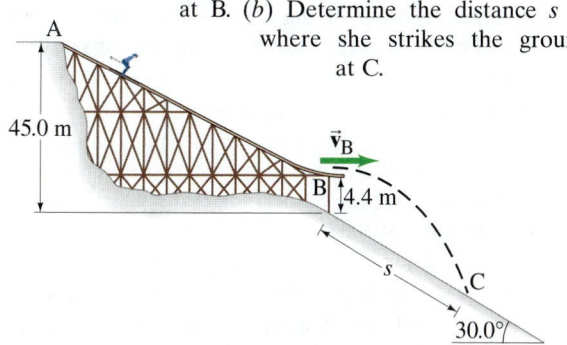

FIGURE 8–41 Problems 83 and 84.

84. Repeat Problem 83, but now assume the ski jump turns upward at point B and gives her a vertical component of velocity (at B) of 3.0 m/s.

85. A ball is attached to a horizontal cord of length ℓ whose other end is fixed, Fig. 8–42. (a) If the ball is released, what will be its speed at the lowest point of its path? (b) A peg is located a distance h directly below the point of attachment of the cord. If $h = 0.80\ell$, what will be the speed of the ball when it reaches the top of its circular path about the peg?

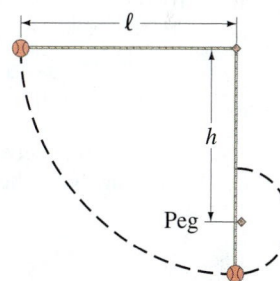

FIGURE 8–42
Problems 85 and 86.

86. Show that h must be greater than 0.60ℓ if the ball in Fig. 8–42 is to make a complete circle about the peg.

87. Show that on a roller coaster with a circular vertical loop (Fig. 8–43), the difference in your apparent weight at the top of the loop and the bottom of the loop is 6 g's—that is, six times your weight. Ignore friction. Show also that as long as your speed is above the minimum needed, this answer doesn't depend on the size of the loop or how fast you go through it.

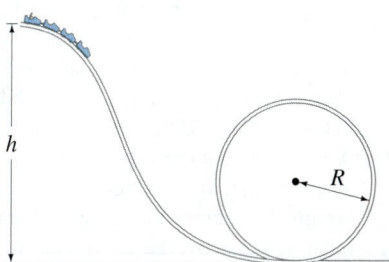

FIGURE 8–43
Problem 87.

88. If you stand on a bathroom scale, the spring inside the scale compresses 0.50 mm, and it tells you your weight is 760 N. Now if you jump on the scale from a height of 1.0 m, what does the scale read at its peak?

89. A 65-kg hiker climbs to the top of a 4200-m-high mountain. The climb is made in 5.0 h starting at an elevation of 2800 m. Calculate (a) the work done by the hiker against gravity, (b) the average power output in watts and in horsepower, and (c) assuming the body is 15% efficient, what rate of energy input was required.

90. The small mass m sliding without friction along the looped track shown in Fig. 8–44 is to remain on the track at all times, even at the very top of the loop of radius r. (a) In terms of the given quantities, determine the minimum release height h. Next, if the actual release height is $2h$, calculate the normal force exerted (b) by the track at the bottom of the loop, (c) by the track at the top of the loop, and (d) by the track after the block exits the loop onto the flat section.

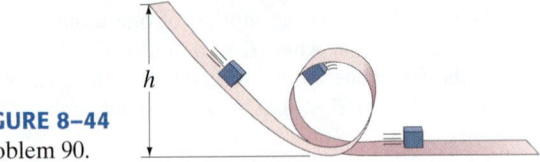

FIGURE 8–44
Problem 90.

91. A 56-kg student runs at 5.0 m/s, grabs a hanging rope, and swings out over a lake (Fig. 8–45). He releases the rope when his velocity is zero. (a) What is the angle θ when he releases the rope? (b) What is the tension in the rope just before he releases it? (c) What is the maximum tension in the rope?

FIGURE 8–45
Problem 91.

92. The nuclear force between two neutrons in a nucleus is described roughly by the Yukawa potential

$$U(r) = -U_0 \frac{r_0}{r} e^{-r/r_0},$$

where r is the distance between the neutrons and U_0 and $r_0 (\approx 10^{-15}\, \text{m})$ are constants. (a) Determine the force $F(r)$. (b) What is the ratio $F(3r_0)/F(r_0)$? (c) Calculate this same ratio for the force between two electrically charged particles where $U(r) = -C/r$, with C a constant. Why is the Yukawa force referred to as a "short-range" force?

93. A fire hose for use in urban areas must be able to shoot a stream of water to a maximum height of 33 m. The water leaves the hose at ground level in a circular stream 3.0 cm in diameter. What minimum power is required to create such a stream of water? Every cubic meter of water has a mass of 1.00×10^3 kg.

94. A 16-kg sled starts up a 28° incline with a speed of 2.4 m/s. The coefficient of kinetic friction is $\mu_k = 0.25$. (a) How far up the incline does the sled travel? (b) What condition must you put on the coefficient of static friction if the sled is not to get stuck at the point determined in part (a)? (c) If the sled slides back down, what is its speed when it returns to its starting point?

95. The Lunar Module could make a safe landing if its vertical velocity at impact is 3.0 m/s or less. Suppose that you want to determine the greatest height h at which the pilot could shut off the engine if the velocity of the lander relative to the surface is (a) zero; (b) 2.0 m/s downward; (c) 2.0 m/s upward. Use conservation of energy to determine h in each case. The acceleration due to gravity at the surface of the Moon is 1.62 m/s².

96. Proper design of automobile braking systems must account for heat buildup under heavy braking. Calculate the thermal energy dissipated from brakes in a 1500-kg car that descends a 17° hill. The car begins braking when its speed is 95 km/h and slows to a speed of 35 km/h in a distance of 0.30 km measured along the road.

97. Some electric power companies use water to store energy. Water is pumped by reversible turbine pumps from a low reservoir to a high reservoir. To store the energy produced in 1.0 hour by a 180-MW electric power plant, how many cubic meters of water will have to be pumped from the lower to the upper reservoir? Assume the upper reservoir is 380 m above the lower one, and we can neglect the small change in depths of each. Water has a mass of 1.00×10^3 kg for every 1.0 m³.

98. Estimate the energy required from fuel to launch a 1465-kg satellite into orbit 1375 km above the Earth's surface. Consider two cases: (a) the satellite is launched into an equatorial orbit from a point on the Earth's equator, and (b) it is launched from the North Pole into a polar orbit.

99. A satellite is in an elliptic orbit around the Earth (Fig. 8–46). Its speed at the perigee A is 8650 m/s. (a) Use conservation of energy to determine its speed at B. The radius of the Earth is 6380 km. (b) Use conservation of energy to determine the speed at the apogee C.

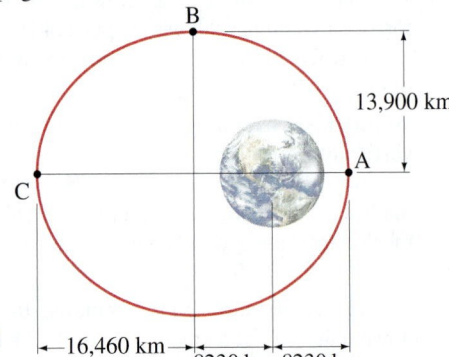

FIGURE 8–46
Problem 99.

—16,460 km—
8230 km 8230 km

B
13,900 km
A
C

100. Suppose the gravitational potential energy of an object of mass m at a distance r from the center of the Earth is given by

$$U(r) = -\frac{GMm}{r} e^{-\alpha r}$$

where α is a positive constant and e is the exponential function. (Newton's law of universal gravitation has $\alpha = 0$). (a) What would be the force on the object as a function of r? (b) What would be the object's escape velocity in terms of the Earth's radius R_E?

101. (a) If the human body could convert a candy bar directly into work, how high could a 76-kg man climb a ladder if he were fueled by one bar $(= 1100\, \text{kJ})$? (b) If the man then jumped off the ladder, what will be his speed when he reaches the bottom?

102. Electric energy units are often expressed in the form of "kilowatt-hours." (a) Show that one kilowatt-hour (kWh) is equal to 3.6×10^6 J. (b) If a typical family of four uses electric energy at an average rate of 580 W, how many kWh would their electric bill show for one month, and (c) how many joules would this be? (d) At a cost of $0.12 per kWh, what would their monthly bill be in dollars? Does the monthly bill depend on the *rate* at which they use the electric energy?

103. Chris jumps off a bridge with a bungee cord (a heavy stretchable cord) tied around his ankle, Fig. 8–47. He falls for 15 m before the bungee cord begins to stretch. Chris's mass is 75 kg and we assume the cord obeys Hooke's law, $F = -kx$, with $k = 50$ N/m. If we neglect air resistance, estimate how far below the bridge Chris's foot will be before coming to a stop. Ignore the mass of the cord (not realistic, however) and treat Chris as a particle.

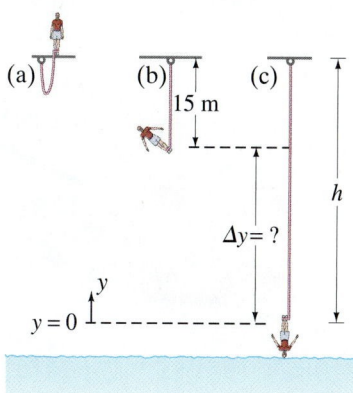

FIGURE 8–47 Problem 103. (a) Bungee jumper about to jump. (b) Bungee cord at its unstretched length. (c) Maximum stretch of cord.

104. In a common test for cardiac function (the "stress test"), the patient walks on an inclined treadmill (Fig. 8–48). Estimate the power required from a 75-kg patient when the treadmill is sloping at an angle of 12° and the velocity is 3.3 km/h. (How does this power compare to the power rating of a lightbulb?)

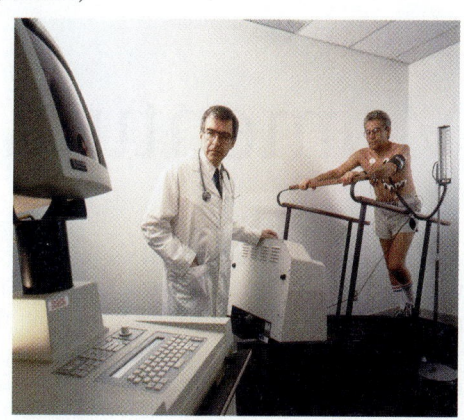

FIGURE 8–48
Problem 104.

105. (a) If a volcano spews a 450-kg rock vertically upward a distance of 320 m, what was its velocity when it left the volcano? (b) If the volcano spews 1000 rocks of this size every minute, estimate its power output.

106. A film of Jesse Owens's famous long jump (Fig. 8–49) in the 1936 Olympics shows that his center of mass rose 1.1 m from launch point to the top of the arc. What minimum speed did he need at launch if he was traveling at 6.5 m/s at the top of the arc?

FIGURE 8–49
Problem 106.

107. An elevator cable breaks when a 920-kg elevator is 24 m above a huge spring $(k = 2.2 \times 10^5$ N/m$)$ at the bottom of the shaft. Calculate (a) the work done by gravity on the elevator before it hits the spring, (b) the speed of the elevator just before striking the spring, and (c) the amount the spring compresses (note that work is done by both the spring and gravity in this part).

108. A particle moves where its potential energy is given by $U(r) = U_0\left[(2/r^2) - (1/r)\right]$. (a) Plot $U(r)$ versus r. Where does the curve cross the $U(r) = 0$ axis? At what value of r does the minimum value of $U(r)$ occur? (b) Suppose that the particle has an energy of $E = -0.050U_0$. Sketch in the approximate turning points of the motion of the particle on your diagram. What is the maximum kinetic energy of the particle, and for what value of r does this occur?

109. A particle of mass m moves under the influence of a potential energy

$$U(x) = \frac{a}{x} + bx$$

where a and b are positive constants and the particle is restricted to the region $x > 0$. Find a point of equilibrium for the particle and demonstrate that it is stable.

*Numerical/Computer

* **110.** (III) The two atoms in a diatomic molecule exert an attractive force on each other at large distances and a repulsive force at short distances. The magnitude of the force between two atoms in a diatomic molecule can be approximated by the Lennard-Jones force, or $F(r) = F_0\left[2(\sigma/r)^{13} - (\sigma/r)^7\right]$, where r is the separation between the two atoms, and σ and F_0 are constant. For an oxygen molecule (which is diatomic) $F_0 = 9.60 \times 10^{-11}$ N and $\sigma = 3.50 \times 10^{-11}$ m. (a) Integrate the equation for $F(r)$ to determine the potential energy $U(r)$ of the oxygen molecule. (b) Find the equilibrium distance r_0 between the two atoms. (c) Graph $F(r)$ and $U(r)$ between $0.9 r_0$ and $2.5 r_0$.

Answers to Exercises

A: (e), (e); (e), (c).

B: (b).

C: (c).

D: Equal speeds.

Conservation of linear momentum is another great conservation law of physics. Collisions, such as between billiard or pool balls, illustrate this vector law very nicely: the total vector momentum just before the collision equals the total vector momentum just after the collision. In this photo, the moving cue ball strikes the 11 ball at rest. Both balls move after the collision, at angles, but the sum of their vector momenta equals the initial vector momentum of the incoming cue ball.

We will consider both elastic collisions (where kinetic energy is also conserved) and inelastic collisions. We also examine the concept of center of mass, and how it helps us in the study of complex motion.

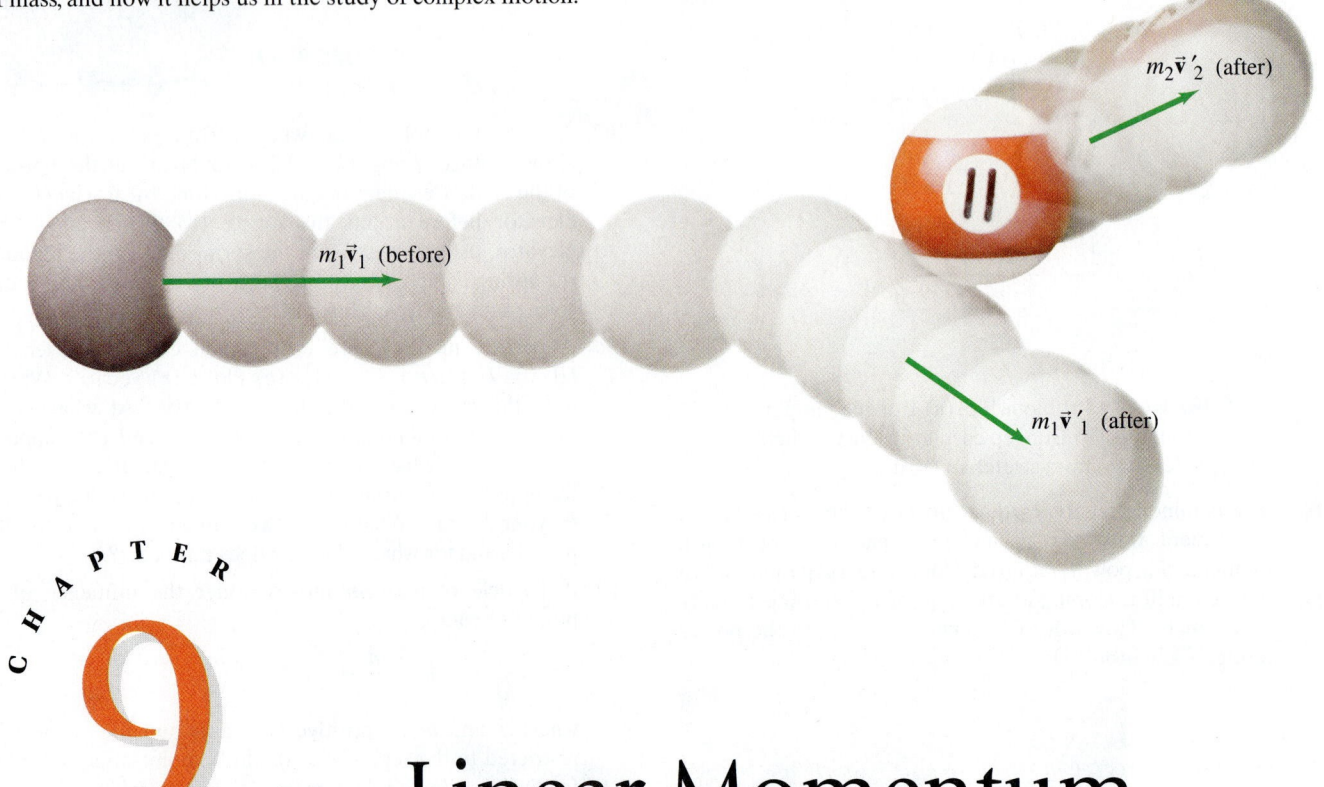

$m_2\vec{\mathbf{v}}'_2$ (after)

$m_1\vec{\mathbf{v}}_1$ (before)

$m_1\vec{\mathbf{v}}'_1$ (after)

CHAPTER

9

Linear Momentum

CONTENTS

9–1 Momentum and Its Relation to Force

9–2 Conservation of Momentum

9–3 Collisions and Impulse

9–4 Conservation of Energy and Momentum in Collisions

9–5 Elastic Collisions in One Dimension

9–6 Inelastic Collisions

9–7 Collisions in Two or Three Dimensions

9–8 Center of Mass (CM)

9–9 Center of Mass and Translational Motion

*9–10 Systems of Variable Mass; Rocket Propulsion

CHAPTER-OPENING QUESTIONS—Guess now!

1. A railroad car loaded with rocks coasts on a level track without friction. A worker on board starts throwing the rocks horizontally backward from the car. Then what happens?

(a) The car slows down.
(b) The car speeds up.
(c) First the car speeds up and then it slows down.
(d) The car's speed remains constant.
(e) None of these.

2. Which answer would you choose if the rocks fall out through a hole in the floor of the car, one at a time?

The law of conservation of energy, which we discussed in the previous Chapter, is one of several great conservation laws in physics. Among the other quantities found to be conserved are linear momentum, angular momentum, and electric charge. We will eventually discuss all of these because the conservation laws are among the most important ideas in science. In this Chapter, we discuss linear momentum, and its conservation. The law of conservation of momentum is essentially a reworking of Newton's laws that gives us tremendous physical insight and problem-solving power.

We make use of the laws of conservation of linear momentum and of energy to analyze collisions. Indeed, the law of conservation of momentum is particularly useful when dealing with a system of two or more objects that interact with each other, such as in collisions of ordinary objects or nuclear particles.

Our focus up to now has been mainly on the motion of a single object, often thought of as a "particle" in the sense that we have ignored any rotation or internal motion. In this Chapter we will deal with systems of two or more objects, and toward the end of the Chapter, the concept of center of mass.

9–1 Momentum and Its Relation to Force

The **linear momentum** (or "momentum" for short) of an object is defined as the product of its mass and its velocity. Momentum (plural is *momenta*) is represented by the symbol $\vec{p}$. If we let m represent the mass of an object and $\vec{v}$ represent its velocity, then its momentum $\vec{p}$ is defined as

$$\vec{p} = m\vec{v}. \tag{9–1}$$

Velocity is a vector, so momentum too is a vector. The direction of the momentum is the direction of the velocity, and the magnitude of the momentum is $p = mv$. Because velocity depends on the reference frame, so does momentum; thus the reference frame must be specified. The unit of momentum is that of mass × velocity, which in SI units is kg·m/s. There is no special name for this unit.

Everyday usage of the term *momentum* is in accord with the definition above. According to Eq. 9–1, a fast-moving car has more momentum than a slow-moving car of the same mass; a heavy truck has more momentum than a small car moving with the same speed. The more momentum an object has, the harder it is to stop it, and the greater effect it will have on another object if it is brought to rest by striking that object. A football player is more likely to be stunned if tackled by a heavy opponent running at top speed than by a lighter or slower-moving tackler. A heavy, fast-moving truck can do more damage than a slow-moving motorcycle.

A force is required to change the momentum of an object, whether it is to increase the momentum, to decrease it, or to change its direction. Newton originally stated his second law in terms of momentum (although he called the product mv the "quantity of motion"). Newton's statement of the **second law of motion**, translated into modern language, is as follows:

The rate of change of momentum of an object is equal to the net force applied to it.

We can write this as an equation,

$$\Sigma\vec{F} = \frac{d\vec{p}}{dt}, \tag{9–2}$$

NEWTON'S SECOND LAW

where $\Sigma\vec{F}$ is the net force applied to the object (the vector sum of all forces acting on it). We can readily derive the familiar form of the second law, $\Sigma\vec{F} = m\vec{a}$, from Eq. 9–2 for the case of constant mass. If $\vec{v}$ is the velocity of an object at any moment, then Eq. 9–2 gives

⚠ **CAUTION**

The change in the momentum vector is in the direction of the net force

$$\Sigma\vec{F} = \frac{d\vec{p}}{dt} = \frac{d(m\vec{v})}{dt} = m\frac{d\vec{v}}{dt} = m\vec{a} \qquad \text{[constant mass]}$$

because, by definition, $\vec{a} = d\vec{v}/dt$ and we assume $m = $ constant. Newton's statement, Eq. 9–2, is actually more general than the more familiar one because it includes the situation in which the mass may change. This is important in certain circumstances, such as for rockets which lose mass as they burn fuel (Section 9–10) and in relativity theory (Chapter 36).

EXERCISE A Light carries momentum, so if a light beam strikes a surface, it will exert a force on that surface. If the light is reflected rather than absorbed, the force will be (*a*) the same, (*b*) less, (*c*) greater, (*d*) impossible to tell, (*e*) none of these.

FIGURE 9–1 Example 9–1.

EXAMPLE 9–1 **ESTIMATE** **Force of a tennis serve.** For a top player, a tennis ball may leave the racket on the serve with a speed of 55 m/s (about 120 mi/h), Fig. 9–1. If the ball has a mass of 0.060 kg and is in contact with the racket for about 4 ms $(4 \times 10^{-3}\,\text{s})$, estimate the average force on the ball. Would this force be large enough to lift a 60-kg person?

APPROACH We write Newton's second law, Eq. 9–2, for the average force as

$$F_{\text{avg}} = \frac{\Delta p}{\Delta t} = \frac{mv_2 - mv_1}{\Delta t},$$

where mv_1 and mv_2 are the initial and final momenta. The tennis ball is hit when its initial velocity v_1 is very nearly zero at the top of the throw, so we set $v_1 = 0$, whereas $v_2 = 55$ m/s is in the horizontal direction. We ignore all other forces on the ball, such as gravity, in comparison to the force exerted by the tennis racket.

SOLUTION The force exerted on the ball by the racket is

$$F_{\text{avg}} = \frac{\Delta p}{\Delta t} = \frac{mv_2 - mv_1}{\Delta t} = \frac{(0.060\,\text{kg})(55\,\text{m/s}) - 0}{0.004\,\text{s}}$$
$$\approx 800\,\text{N}.$$

This is a large force, larger than the weight of a 60-kg person, which would require a force $mg = (60\,\text{kg})(9.8\,\text{m/s}^2) \approx 600\,\text{N}$ to lift.

NOTE The force of gravity acting on the tennis ball is $mg = (0.060\,\text{kg})(9.8\,\text{m/s}^2) = 0.59\,\text{N}$, which justifies our ignoring it compared to the enormous force the racket exerts.

NOTE High-speed photography and radar can give us an estimate of the contact time and the velocity of the ball leaving the racket. But a direct measurement of the force is not practical. Our calculation shows a handy technique for determining an unknown force in the real world.

FIGURE 9–2 Example 9–2.

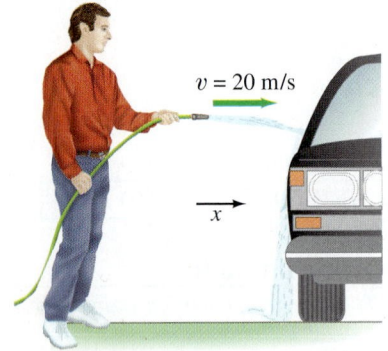

$v = 20$ m/s

x

EXAMPLE 9–2 **Washing a car: momentum change and force.** Water leaves a hose at a rate of 1.5 kg/s with a speed of 20 m/s and is aimed at the side of a car, which stops it, Fig. 9–2. (That is, we ignore any splashing back.) What is the force exerted by the water on the car?

APPROACH The water leaving the hose has mass and velocity, so it has a momentum p_{initial} in the horizontal (x) direction, and we assume gravity doesn't pull the water down significantly. When the water hits the car, the water loses this momentum $(p_{\text{final}} = 0)$. We use Newton's second law in the momentum form to find the force that the car exerts on the water to stop it. By Newton's third law, the force exerted by the water on the car is equal and opposite. We have a continuing process: 1.5 kg of water leaves the hose in each 1.0-s time interval. So let us write $F = \Delta p/\Delta t$ where $\Delta t = 1.0\,\text{s}$, and $mv_{\text{initial}} = (1.5\,\text{kg})(20\,\text{m/s})$.

SOLUTION The force (assumed constant) that the car must exert to change the momentum of the water is

$$F = \frac{\Delta p}{\Delta t} = \frac{p_{\text{final}} - p_{\text{initial}}}{\Delta t} = \frac{0 - 30\,\text{kg}\cdot\text{m/s}}{1.0\,\text{s}} = -30\,\text{N}.$$

The minus sign indicates that the force exerted by the car on the water is opposite to the water's original velocity. The car exerts a force of 30 N to the left to stop the water, so by Newton's third law, the water exerts a force of 30 N to the right on the car.

NOTE Keep track of signs, although common sense helps too. The water is moving to the right, so common sense tells us the force on the car must be to the right.

EXERCISE B If the water splashes back from the car in Example 9–2, would the force on the car be larger or smaller?

9–2 Conservation of Momentum

The concept of momentum is particularly important because, if no net external force acts on a system, the total momentum of the system is a conserved quantity. Consider, for example, the head-on collision of two billiard balls, as shown in Fig. 9–3. We assume the net external force on this system of two balls is zero—that is, the only significant forces during the collision are the forces that each ball exerts on the other. Although the momentum of each of the two balls changes as a result of the collision, the *sum* of their momenta is found to be the same before as after the collision. If $m_A \vec{\mathbf{v}}_A$ is the momentum of ball A and $m_B \vec{\mathbf{v}}_B$ the momentum of ball B, both measured just before the collision, then the total momentum of the two balls before the collision is the vector sum $m_A \vec{\mathbf{v}}_A + m_B \vec{\mathbf{v}}_B$. Immediately after the collision, the balls each have a different velocity and momentum, which we designate by a "prime" on the velocity: $m_A \vec{\mathbf{v}}'_A$ and $m_B \vec{\mathbf{v}}'_B$. The total momentum after the collision is the vector sum $m_A \vec{\mathbf{v}}'_A + m_B \vec{\mathbf{v}}'_B$. No matter what the velocities and masses are, experiments show that the total momentum before the collision is the same as afterward, whether the collision is head-on or not, as long as no net external force acts:

$$\text{momentum before} \;=\; \text{momentum after}$$

$$m_A \vec{\mathbf{v}}_A + m_B \vec{\mathbf{v}}_B \;=\; m_A \vec{\mathbf{v}}'_A + m_B \vec{\mathbf{v}}'_B. \qquad \left[\Sigma \vec{\mathbf{F}}_{\text{ext}} = 0\right] \quad \textbf{(9–3)}$$

That is, the total vector momentum of the system of two colliding balls is conserved: it stays constant.

Although the law of conservation of momentum was discovered experimentally, it can be derived from Newton's laws of motion, which we now show. Let us consider two objects of mass m_A and m_B that have momenta $\vec{\mathbf{p}}_A$ and $\vec{\mathbf{p}}_B$ before they collide and $\vec{\mathbf{p}}'_A$ and $\vec{\mathbf{p}}'_B$ after they collide, as in Fig. 9–4. During the collision, suppose that the force exerted by object A on object B at any instant is $\vec{\mathbf{F}}$. Then, by Newton's third law, the force exerted by object B on object A is $-\vec{\mathbf{F}}$. During the brief collision time, we assume no other (external) forces are acting (or that $\vec{\mathbf{F}}$ is much greater than any other external forces acting). Thus we have

$$\vec{\mathbf{F}} = \frac{d\vec{\mathbf{p}}_B}{dt}$$

and

$$-\vec{\mathbf{F}} = \frac{d\vec{\mathbf{p}}_A}{dt}.$$

We add these two equations together and find

$$0 = \frac{d(\vec{\mathbf{p}}_A + \vec{\mathbf{p}}_B)}{dt}$$

which tells us that

$$\vec{\mathbf{p}}_A + \vec{\mathbf{p}}_B = \text{constant}.$$

The total momentum thus is conserved.

We have put this derivation in the context of a collision. As long as no external forces act, it is valid over any time interval, and conservation of momentum is always valid as long as no external forces act. In the real world, external forces do act: friction on billiard balls, gravity acting on a tennis ball, and so on. So it may seem that conservation of momentum cannot be applied. Or can it? In a collision, the force each object exerts on the other acts only over a very brief time interval, and is very strong relative to the other forces. If we measure the momenta immediately before and after the collision, momentum will be very nearly conserved. We cannot wait for the external forces to produce their effect before measuring $\vec{\mathbf{p}}'_A$ and $\vec{\mathbf{p}}'_B$.

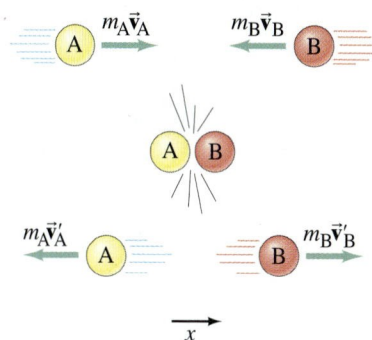

FIGURE 9–3 Momentum is conserved in a collision of two balls, labeled A and B.

> *CONSERVATION OF MOMENTUM*
> *(two objects colliding)*

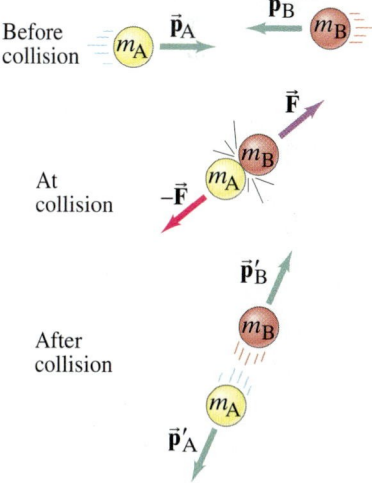

FIGURE 9–4 Collision of two objects. Their momenta before collision are $\vec{\mathbf{p}}_A$ and $\vec{\mathbf{p}}_B$, and after collision are $\vec{\mathbf{p}}'_A$ and $\vec{\mathbf{p}}'_B$. At any moment during the collision each exerts a force on the other of equal magnitude but opposite direction.

For example, when a racket hits a tennis ball or a bat hits a baseball, both before and after the "collision" the ball moves as a projectile under the action of gravity and air resistance. However, when the bat or racket hits the ball, during this brief time of the collision those external forces are insignificant compared to the collision force the bat or racket exerts on the ball. Momentum is conserved (or very nearly so) as long as we measure $\vec{p}_A$ and $\vec{p}_B$ just before the collision and $\vec{p}'_A$ and $\vec{p}'_B$ immediately after the collision (Eq. 9–3).

Our derivation of the conservation of momentum can be extended to include any number of interacting objects. Let $\vec{P}$ represent the total momentum of a system of n interacting objects which we number from 1 to n:

$$\vec{P} = m_1 \vec{v}_1 + m_2 \vec{v}_2 + \cdots + m_n \vec{v}_n = \Sigma \vec{p}_i.$$

We differentiate with respect to time:

$$\frac{d\vec{P}}{dt} = \Sigma \frac{d\vec{p}_i}{dt} = \Sigma \vec{F}_i \qquad\qquad (9\text{–}4)$$

where $\vec{F}_i$ represents the *net* force on the ith object. The forces can be of two types: (1) *external forces* on objects of the system, exerted by objects outside the system, and (2) *internal forces* that objects within the system exert on other objects in the system. By Newton's third law, the internal forces occur in pairs: if one object exerts a force on a second object, the second exerts an equal and opposite force on the first object. Thus, in the sum over all the forces in Eq. 9–4, all the internal forces cancel each other in pairs. Thus we have

> *NEWTON'S SECOND LAW*
> *(for a system of objects)*

$$\frac{d\vec{P}}{dt} = \Sigma \vec{F}_{ext}, \qquad\qquad (9\text{–}5)$$

where $\Sigma \vec{F}_{ext}$ is the sum of all external forces acting on our system. If the net external force is zero, then $d\vec{P}/dt = 0$, so $\Delta\vec{P} = 0$ or $\vec{P} = $ constant. Thus we see that

> **when the net external force on a system of objects is zero, the total momentum of the system remains constant.**

> *LAW OF*
> *CONSERVATION*
> *OF LINEAR*
> *MOMENTUM*

This is the **law of conservation of momentum**. It can also be stated as

> **the total momentum of an isolated system of objects remains constant.**

By an **isolated system**, we mean one on which no external forces act—the only forces acting are those between objects of the system.

If a net external force acts on a system, then the law of conservation of momentum will not apply. However, if the "system" can be redefined so as to include the other objects exerting these forces, then the conservation of momentum principle can apply. For example, if we take as our system a falling rock, it does not conserve momentum since an external force, the force of gravity exerted by the Earth, is acting on it and its momentum changes. However, if we include the Earth in the system, the total momentum of rock plus Earth is conserved. (This means that the Earth comes up to meet the rock. Since the Earth's mass is so great, its upward velocity is very tiny.)

Although the law of conservation of momentum follows from Newton's second law, as we have seen, it is in fact more general than Newton's laws. In the tiny world of the atom, Newton's laws fail, but the great conservation laws—those of energy, momentum, angular momentum, and electric charge—have been found to hold in every experimental situation tested. It is for this reason that the conservation laws are considered more basic than Newton's laws.

EXAMPLE 9–3 **Railroad cars collide: momentum conserved.** A 10,000-kg railroad car, A, traveling at a speed of 24.0 m/s strikes an identical car, B, at rest. If the cars lock together as a result of the collision, what is their common speed immediately after the collision? See Fig. 9–5.

APPROACH We choose our system to be the two railroad cars. We consider a very brief time interval, from just before the collision until just after, so that external

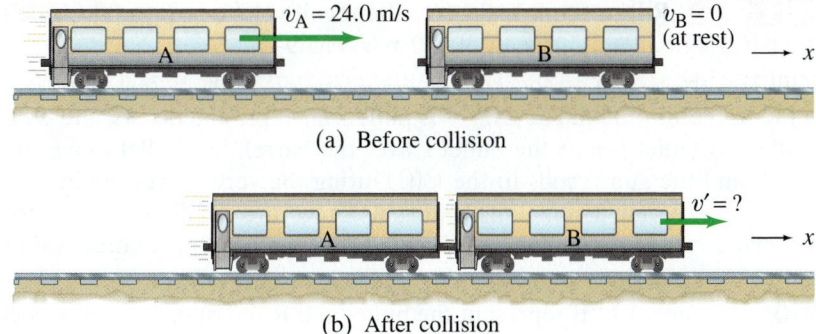

(a) Before collision

$v_A = 24.0 \text{ m/s}$

$v_B = 0$ (at rest)

(b) After collision

$v' = ?$

FIGURE 9–5 Example 9–3.

forces such as friction can be ignored. Then we apply conservation of momentum:

$$P_{\text{initial}} = P_{\text{final}}.$$

SOLUTION The initial total momentum is

$$P_{\text{initial}} = m_A v_A + m_B v_B = m_A v_A$$

because car B is at rest initially $(v_B = 0)$. The direction is to the right in the $+x$ direction. After the collision, the two cars become attached, so they will have the same speed, call it v'. Then the total momentum after the collision is

$$P_{\text{final}} = (m_A + m_B)v'.$$

We have assumed there are no external forces, so momentum is conserved:

$$P_{\text{initial}} = P_{\text{final}}$$
$$m_A v_A = (m_A + m_B)v'.$$

Solving for v', we obtain

$$v' = \frac{m_A}{m_A + m_B} v_A = \left(\frac{10,000 \text{ kg}}{10,000 \text{ kg} + 10,000 \text{ kg}} \right)(24.0 \text{ m/s}) = 12.0 \text{ m/s},$$

to the right. Their mutual speed after collision is half the initial speed of car A because their masses are equal.

NOTE We kept symbols until the very end, so we have an equation we can use in other (related) situations.

NOTE We haven't mentioned friction here. Why? Because we are examining speeds just before and just after the very brief time interval of the collision, and during that brief time friction can't do much—it is ignorable (but not for long: the cars will slow down because of friction).

EXERCISE C A 50-kg child runs off a dock at 2.0 m/s (horizontally) and lands in a waiting rowboat of mass 150 kg. At what speed does the rowboat move away from the dock?

EXERCISE D In Example 9–3, what result would you get if (a) $m_B = 3m_A$, (b) m_B is much larger than m_A $(m_B \gg m_A)$, and (c) $m_B \ll m_A$?

The law of conservation of momentum is particularly useful when we are dealing with fairly simple systems such as colliding objects and certain types of "explosions." For example, *rocket propulsion*, which we saw in Chapter 4 can be understood on the basis of action and reaction, can also be explained on the basis of the conservation of momentum. We can consider the rocket and fuel as an isolated system if it is far out in space (no external forces). In the reference frame of the rocket before any fuel is ejected, the total momentum of rocket plus fuel is zero. When the fuel burns, the total momentum remains unchanged: the backward momentum of the expelled gases is just balanced by the forward momentum gained by the rocket itself (see Fig. 9–6). Thus, a rocket can accelerate in empty space. There is no need for the expelled gases to push against the Earth or the air (as is sometimes erroneously thought). Similar examples of (nearly) isolated systems where momentum is conserved are the recoil of a gun when a bullet is fired, and the movement of a rowboat just after a package is thrown from it.

FIGURE 9–6 (a) A rocket, containing fuel, at rest in some reference frame. (b) In the same reference frame, the rocket fires, and gases are expelled at high speed out the rear. The total vector momentum, $\vec{P} = \vec{P}_{\text{gas}} + \vec{P}_{\text{rocket}}$, remains zero.

(a)

$$\vec{P} = 0$$

(b)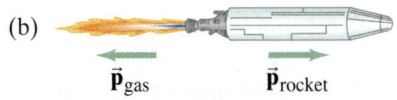

$\vec{P}_{\text{gas}}$ $\vec{P}_{\text{rocket}}$

PHYSICS APPLIED

Rocket propulsion

⚠ **CAUTION**

A rocket pushes on the gases released by the fuel, not on the Earth or other objects

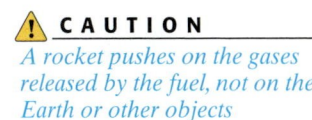

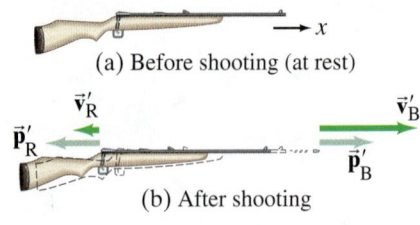

(a) Before shooting (at rest)

$\vec{\mathbf{v}}'_R$ $\vec{\mathbf{v}}'_B$

$\vec{\mathbf{p}}'_R$ $\vec{\mathbf{p}}'_B$

(b) After shooting

FIGURE 9–7 Example 9–4.

EXAMPLE 9–4 **Rifle recoil.** Calculate the recoil velocity of a 5.0-kg rifle that shoots a 0.020-kg bullet at a speed of 620 m/s, Fig. 9–7.

APPROACH Our system is the rifle and the bullet, both at rest initially, just before the trigger is pulled. The trigger is pulled, an explosion occurs, and we look at the rifle and bullet just as the bullet leaves the barrel. The bullet moves to the right ($+x$), and the gun recoils to the left. During the very short time interval of the explosion, we can assume the external forces are small compared to the forces exerted by the exploding gunpowder. Thus we can apply conservation of momentum, at least approximately.

SOLUTION Let subscript B represent the bullet and R the rifle; the final velocities are indicated by primes. Then momentum conservation in the x direction gives

$$\text{momentum before} = \text{momentum after}$$
$$m_B v_B + m_R v_R = m_B v'_B + m_R v'_R$$
$$0 + 0 = m_B v'_B + m_R v'_R$$

so

$$v'_R = -\frac{m_B v'_B}{m_R} = -\frac{(0.020\,\text{kg})(620\,\text{m/s})}{(5.0\,\text{kg})} = -2.5\,\text{m/s}.$$

Since the rifle has a much larger mass, its (recoil) velocity is much less than that of the bullet. The minus sign indicates that the velocity (and momentum) of the rifle is in the negative x direction, opposite to that of the bullet.

CONCEPTUAL EXAMPLE 9–5 **Falling on or off a sled.** (a) An empty sled is sliding on frictionless ice when Susan drops vertically from a tree above onto the sled. When she lands, does the sled speed up, slow down, or keep the same speed? (b) Later: Susan falls sideways off the sled. When she drops off, does the sled speed up, slow down, or keep the same speed?

RESPONSE (a) Because Susan falls vertically onto the sled, she has no initial horizontal momentum. Thus the total horizontal momentum afterward equals the momentum of the sled initially. Since the mass of the system (sled + person) has increased, the speed must decrease. (b) At the instant Susan falls off, she is moving with the same horizontal speed as she was while on the sled. At the moment she leaves the sled, she has the same momentum she had an instant before. Because momentum is conserved, the sled keeps the same speed.

EXERCISE E Return to the Chapter-Opening Questions, page 214, and answer them again now. Try to explain why you may have answered differently the first time.

9–3 Collisions and Impulse

Conservation of momentum is a very useful tool for dealing with everyday collision processes, such as a tennis racket or a baseball bat striking a ball, two billiard balls colliding, a hammer hitting a nail. At the subatomic level, scientists learn about the structure of nuclei and their constituents, and about the nature of the forces involved, by careful study of collisions between nuclei and/or elementary particles.

During a collision of two ordinary objects, both objects are deformed, often considerably, because of the large forces involved (Fig. 9–8). When the collision occurs, the force each exerts on the other usually jumps from zero at the moment of contact to a very large value within a very short time, and then abruptly returns to zero again. A graph of the magnitude of the force one object exerts on the other during a collision, as a function of time, is something like the red curve in Fig. 9–9. The time interval Δt is usually very distinct and usually very small.

From Newton's second law, Eq. 9–2, the *net* force on an object is equal to the rate of change of its momentum:

$$\vec{\mathbf{F}} = \frac{d\vec{\mathbf{p}}}{dt}.$$

(We have written $\vec{\mathbf{F}}$ instead of $\Sigma\vec{\mathbf{F}}$ for the net force, which we assume is entirely due to the brief but large force that acts during the collision.) This equation applies

FIGURE 9–8 Tennis racket striking a ball. Both the ball and the racket strings are deformed due to the large force each exerts on the other.

FIGURE 9–9 Force as a function of time during a typical collision: F can become very large; Δt is typically milliseconds for macroscopic collisions.

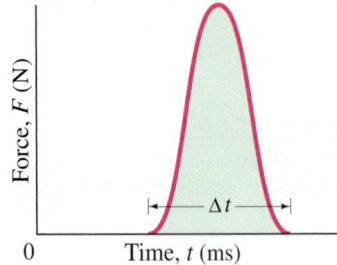

to *each* of the objects in a collision. During the infinitesimal time interval dt, the momentum changes by

$$d\vec{\mathbf{p}} = \vec{\mathbf{F}} \, dt.$$

If we integrate this over the duration of a collision, we have

$$\int_i^f d\vec{\mathbf{p}} = \vec{\mathbf{p}}_f - \vec{\mathbf{p}}_i = \int_{t_i}^{t_f} \vec{\mathbf{F}} \, dt,$$

where $\vec{\mathbf{p}}_i$ and $\vec{\mathbf{p}}_f$ are the initial and final momenta of the object, just before and just after the collision. The integral of the net force over the time interval during which it acts is called the **impulse, $\vec{\mathbf{J}}$**:

$$\vec{\mathbf{J}} = \int_{t_i}^{t_f} \vec{\mathbf{F}} \, dt.$$

Thus the change in momentum of an object, $\Delta\vec{\mathbf{p}} = \vec{\mathbf{p}}_f - \vec{\mathbf{p}}_i$, is equal to the impulse acting on it:

$$\Delta\vec{\mathbf{p}} = \vec{\mathbf{p}}_f - \vec{\mathbf{p}}_i = \int_{t_i}^{t_f} \vec{\mathbf{F}} \, dt = \vec{\mathbf{J}}. \qquad \textbf{(9–6)}$$

The units for impulse are the same as for momentum, $kg \cdot m/s$ (or $N \cdot s$) in SI. Since $\vec{\mathbf{J}} = \int \vec{\mathbf{F}} \, dt$, we can state that the impulse $\vec{\mathbf{J}}$ of a force is equal to the area under the F versus t curve, as indicated by the shading in Fig. 9–9.

Equation 9–6 is true only if $\vec{\mathbf{F}}$ is the *net* force on the object. It is valid for *any* net force $\vec{\mathbf{F}}$ where $\vec{\mathbf{p}}_i$ and $\vec{\mathbf{p}}_f$ correspond precisely to the times t_i and t_f. But the impulse concept is really most useful for so-called *impulsive forces*—that is, for a force like that shown in Fig. 9–9, which has a very large magnitude over a very short time interval and is essentially zero outside this time interval. For most collision processes, the impulsive force is much larger than any other force acting, and the others can be neglected. For such an impulsive force, the time interval over which we take the integral in Eq. 9–6 is not critical as long as we start before t_i and end after t_f, since $\vec{\mathbf{F}}$ is essentially zero outside this time interval. (Of course, if the chosen time interval is too large, the effect of the other forces does become significant—such as the flight of a tennis ball which, after the impulsive force administered by the racket, begins to fall under gravity.)

It is sometimes useful to speak of the average force, $\vec{\mathbf{F}}_{avg}$, during a collision, defined as that constant force which, if acting over the same time interval $\Delta t = t_f - t_i$ as the actual force, would produce the same impulse and change in momentum. Thus

$$\vec{\mathbf{F}}_{avg} \, \Delta t = \int_{t_i}^{t_f} \vec{\mathbf{F}} \, dt.$$

Figure 9–10 shows the magnitude of the average force, F_{avg}, for the impulsive force of Fig. 9–9. The rectangular area $F_{avg} \, \Delta t$ equals the area under the impulsive force curve.

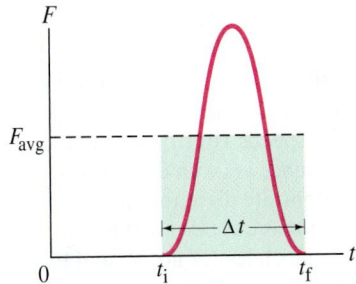

FIGURE 9–10 The average force F_{avg} acting over a very brief time interval Δt gives the same impulse $(F_{avg} \, \Delta t)$ as the actual force.

EXAMPLE 9–6 ESTIMATE Karate blow. Estimate the impulse and the average force delivered by a karate blow (Fig. 9–11) that breaks a board a few cm thick. Assume the hand moves at roughly 10 m/s when it hits the board.

APPROACH We use the momentum-impulse relation, Eq. 9–6. The hand's speed changes from 10 m/s to zero over a distance of perhaps one cm (roughly how much your hand and the board compress before your hand comes to a stop, or nearly so, and the board begins to give way). The hand's mass should probably include part of the arm, and we take it to be roughly $m \approx 1 \, kg$.

SOLUTION The impulse J equals the change in momentum

$$J = \Delta p = (1 \, kg)(10 \, m/s - 0) = 10 \, kg \cdot m/s.$$

We obtain the force from the definition of impulse $F_{avg} = J/\Delta t$; but what is Δt? The hand is brought to rest over the distance of roughly a centimeter: $\Delta x \approx 1 \, cm$. The average speed during the impact is $\bar{v} = (10 \, m/s + 0)/2 = 5 \, m/s$ and equals $\Delta x/\Delta t$. Thus $\Delta t = \Delta x/\bar{v} \approx (10^{-2} \, m)/(5 \, m/s) = 2 \times 10^{-3} \, s$ or about 2 ms. The force is thus (Eq. 9–6) about

$$F_{avg} = \frac{J}{\Delta t} = \frac{10 \, kg \cdot m/s}{2 \times 10^{-3} \, s} \approx 5000 \, N = 5 \, kN.$$

FIGURE 9–11 Example 9–6.

9–4 Conservation of Energy and Momentum in Collisions

During most collisions, we usually don't know how the collision force varies over time, and so analysis using Newton's second law becomes difficult or impossible. But by making use of the conservation laws for momentum and energy, we can still determine a lot about the motion after a collision, given the motion before the collision. We saw in Section 9–2 that in the collision of two objects such as billiard balls, the total momentum is conserved. If the two objects are very hard and no heat or other form of energy is produced in the collision, then the kinetic energy of the two objects is the same after the collision as before. For the brief moment during which the two objects are in contact, some (or all) of the energy is stored momentarily in the form of elastic potential energy. But if we compare the total kinetic energy just before the collision with the total kinetic energy just after the collision, and they are found to be the same, then we say that the total kinetic energy is conserved. Such a collision is called an **elastic collision**. If we use the subscripts A and B to represent the two objects, we can write the equation for conservation of total kinetic energy as

total kinetic energy before = total kinetic energy after

$$\tfrac{1}{2} m_A v_A^2 + \tfrac{1}{2} m_B v_B^2 = \tfrac{1}{2} m_A v_A'^2 + \tfrac{1}{2} m_B v_B'^2. \qquad \text{[elastic collision]} \quad \textbf{(9–7)}$$

Primed quantities ($'$) mean after the collision, and unprimed mean before the collision, just as in Eq. 9–3 for conservation of momentum.

At the atomic level the collisions of atoms and molecules are often elastic. But in the "macroscopic" world of ordinary objects, an elastic collision is an ideal that is never quite reached, since at least a little thermal energy (and perhaps sound and other forms of energy) is always produced during a collision. The collision of two hard elastic balls, such as billiard balls, however, is very close to being perfectly elastic, and we often treat it as such.

We do need to remember that even when the kinetic energy is not conserved, the *total* energy is always conserved.

Collisions in which kinetic energy is not conserved are said to be **inelastic collisions**. The kinetic energy that is lost is changed into other forms of energy, often thermal energy, so that the total energy (as always) is conserved. In this case,

$$K_A + K_B = K_A' + K_B' + \text{thermal and other forms of energy.}$$

See Fig. 9–12, and the details in its caption. We discuss inelastic collisions in Section 9–6.

(a) Approach

(b) Collision

(c) If elastic

(d) If inelastic

FIGURE 9–12 Two equal-mass objects (a) approach each other with equal speeds, (b) collide, and then (c) bounce off with equal speeds in the opposite directions if the collision is elastic, or (d) bounce back much less or not at all if the collision is inelastic.

FIGURE 9–13 Two small objects of masses m_A and m_B, (a) before the collision and (b) after the collision.

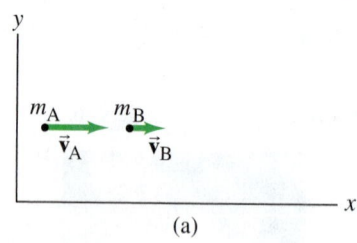

(a)

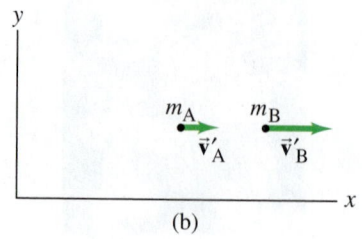

(b)

9–5 Elastic Collisions in One Dimension

We now apply the conservation laws for momentum and kinetic energy to an elastic collision between two small objects that collide head-on, so all the motion is along a line. Let us assume that the two objects are moving with velocities v_A and v_B along the x axis before the collision, Fig. 9–13a. After the collision, their velocities are v_A' and v_B', Fig. 9–13b. For any $v > 0$, the object is moving to the right (increasing x), whereas for $v < 0$, the object is moving to the left (toward decreasing values of x).

From conservation of momentum, we have

$$m_A v_A + m_B v_B = m_A v_A' + m_B v_B'.$$

Because the collision is assumed to be elastic, kinetic energy is also conserved:

$$\tfrac{1}{2} m_A v_A^2 + \tfrac{1}{2} m_B v_B^2 = \tfrac{1}{2} m_A v_A'^2 + \tfrac{1}{2} m_B v_B'^2.$$

We have two equations, so we can solve for two unknowns. If we know the masses and velocities before the collision, then we can solve these two equations for the velocities after the collision, v'_A and v'_B. We derive a helpful result by rewriting the momentum equation as

$$m_A(v_A - v'_A) = m_B(v'_B - v_B), \qquad \text{(i)}$$

and we rewrite the kinetic energy equation as

$$m_A(v_A^2 - v_A'^2) = m_B(v_B'^2 - v_B^2).$$

Noting that algebraically $(a - b)(a + b) = a^2 - b^2$, we write this last equation as

$$m_A(v_A - v'_A)(v_A + v'_A) = m_B(v'_B - v_B)(v'_B + v_B). \qquad \text{(ii)}$$

We divide Eq. (ii) by Eq. (i), and (assuming $v_A \neq v'_A$ and $v_B \neq v'_B$)[†] obtain

$$v_A + v'_A = v'_B + v_B.$$

We can rewrite this equation as

$$v_A - v_B = v'_B - v'_A$$

or

$$v_A - v_B = -(v'_A - v'_B). \qquad \text{[head-on (1-D) elastic collision]} \quad \text{(9–8)}$$

This is an interesting result: it tells us that for any elastic head-on collision, the relative speed of the two objects after the collision has the same magnitude (but opposite direction) as before the collision, no matter what the masses are.

Equation 9–8 was derived from conservation of kinetic energy for elastic collisions, and can be used in place of it. Because the v's are not squared in Eq. 9–8, it is simpler to use in calculations than the conservation of kinetic energy equation (Eq. 9–7) directly.

EXAMPLE 9–7 Equal masses. Billiard ball A of mass m moving with speed v_A collides head-on with ball B of equal mass. What are the speeds of the two balls after the collision, assuming it is elastic? Assume (*a*) both balls are moving initially (v_A and v_B), (*b*) ball B is initially at rest ($v_B = 0$).

APPROACH There are two unknowns, v'_A and v'_B, so we need two independent equations. We focus on the time interval from just before the collision until just after. No net external force acts on our system of two balls (mg and the normal force cancel), so momentum is conserved. Conservation of kinetic energy applies as well because we are told the collision is elastic.

SOLUTION (*a*) The masses are equal ($m_A = m_B = m$) so conservation of momentum gives

$$v_A + v_B = v'_A + v'_B.$$

We need a second equation, because there are two unknowns. We could use the conservation of kinetic energy equation, or the simpler Eq. 9–8 derived from it:

$$v_A - v_B = v'_B - v'_A.$$

We add these two equations and obtain

$$v'_B = v_A$$

and then subtract the two equations to obtain

$$v'_A = v_B.$$

That is, the balls exchange velocities as a result of the collision: ball B acquires the velocity that ball A had before the collision, and vice versa.

(*b*) If ball B is at rest initially, so that $v_B = 0$, we have

$$v'_B = v_A \quad \text{and} \quad v'_A = 0.$$

That is, ball A is brought to rest by the collision, whereas ball B acquires the original velocity of ball A. This result is often observed by billiard and pool players, and is valid only if the two balls have equal masses (and no spin is given to the balls). See Fig. 9–14.

FIGURE 9–14 In this multi-flash photo of a head-on collision between two balls of equal mass, the white cue ball is accelerated from rest by the cue stick and then strikes the red ball, initially at rest. The white ball stops in its tracks and the (equal mass) red ball moves off with the same speed as the white ball had before the collision. See Example 9–7.

[†]Note that Eqs. (i) and (ii), which are the conservation laws for momentum and kinetic energy, are both satisfied by the solution $v'_A = v_A$ and $v'_B = v_B$. This is a valid solution, but not very interesting. It corresponds to no collision at all—when the two objects miss each other.

EXAMPLE 9–8 **Unequal masses, target at rest.** A very common practical situation is for a moving object (m_A) to strike a second object (m_B, the "target") at rest ($v_B = 0$). Assume the objects have unequal masses, and that the collision is elastic and occurs along a line (head-on). (a) Derive equations for v_B' and v_A' in terms of the initial velocity v_A of mass m_A and the masses m_A and m_B. (b) Determine the final velocities if the moving object is much more massive than the target ($m_A \gg m_B$). (c) Determine the final velocities if the moving object is much less massive than the target ($m_A \ll m_B$).

APPROACH The momentum equation (with $v_B = 0$) is

$$m_B v_B' = m_A(v_A - v_A').$$

Kinetic energy is also conserved, and to use it we use Eq. 9–8 and rewrite it as

$$v_A' = v_B' - v_A.$$

SOLUTION (a) We substitute the above v_A' equation into the momentum equation and rearrange to find

$$v_B' = v_A\left(\frac{2m_A}{m_A + m_B}\right).$$

We substitute this value for v_B' back into the equation $v_A' = v_B' - v_A$ to obtain

$$v_A' = v_A\left(\frac{m_A - m_B}{m_A + m_B}\right).$$

To check these two equations we have derived, we let $m_A = m_B$, and we obtain

$$v_B' = v_A \quad \text{and} \quad v_A' = 0.$$

This is the same case treated in Example 9–7, and we get the same result: for objects of equal mass, one of which is initially at rest, the velocity of the one moving initially is completely transferred to the object originally at rest.

(b) We are given $v_B = 0$ and $m_A \gg m_B$. A very heavy moving object strikes a light object at rest, and we have, using the relations for v_B' and v_A' above,

$$v_B' \approx 2v_A$$
$$v_A' \approx v_A.$$

Thus the velocity of the heavy incoming object is practically unchanged, whereas the light object, originally at rest, takes off with twice the velocity of the heavy one. The velocity of a heavy bowling ball, for example, is hardly affected by striking a much lighter bowling pin.

(c) This time we have $v_B = 0$ and $m_A \ll m_B$. A moving light object strikes a very massive object at rest. In this case, using the equations in part (a)

$$v_B' \approx 0$$
$$v_A' \approx -v_A.$$

The massive object remains essentially at rest and the very light incoming object rebounds with essentially its same speed but in the opposite direction. For example, a tennis ball colliding head-on with a stationary bowling ball will hardly affect the bowling ball, but will rebound with nearly the same speed it had initially, just as if it had struck a hard wall.

It can readily be shown (it is given as Problem 40) for any elastic head-on collision that

$$v_B' = v_A\left(\frac{2m_A}{m_A + m_B}\right) + v_B\left(\frac{m_B - m_A}{m_A + m_B}\right)$$

and

$$v_A' = v_A\left(\frac{m_A - m_B}{m_A + m_B}\right) + v_B\left(\frac{2m_B}{m_A + m_B}\right).$$

These general equations, however, should not be memorized. They can always be derived quickly from the conservation laws. For many problems, it is simplest just to start from scratch, as we did in the special cases above and as shown in the next Example.

EXAMPLE 9–9 **A nuclear collision.** A proton (p) of mass 1.01 u (unified atomic mass units) traveling with a speed of 3.60×10^4 m/s has an elastic head-on collision with a helium (He) nucleus $(m_{He} = 4.00\ u)$ initially at rest. What are the velocities of the proton and helium nucleus after the collision? (As mentioned in Chapter 1, $1\ u = 1.66 \times 10^{-27}$ kg, but we won't need this fact.) Assume the collision takes place in nearly empty space.

APPROACH This is an elastic head-on collision. The only external force is Earth's gravity, but it is insignificant compared to the strong force during the collision. So again we use the conservation laws of momentum and of kinetic energy, and apply them to our system of two particles.

SOLUTION Let the proton (p) be particle A and the helium nucleus (He) be particle B. We have $v_B = v_{He} = 0$ and $v_A = v_p = 3.60 \times 10^4$ m/s. We want to find the velocities v'_p and v'_{He} after the collision. From conservation of momentum,

$$m_p v_p + 0 = m_p v'_p + m_{He} v'_{He}.$$

Because the collision is elastic, the kinetic energy of our system of two particles is conserved and we can use Eq. 9–8, which becomes

$$v_p - 0 = v'_{He} - v'_p.$$

Thus

$$v'_p = v'_{He} - v_p,$$

and substituting this into our momentum equation displayed above, we get

$$m_p v_p = m_p v'_{He} - m_p v_p + m_{He} v'_{He}.$$

Solving for v'_{He}, we obtain

$$v'_{He} = \frac{2 m_p v_p}{m_p + m_{He}} = \frac{2(1.01\ u)(3.60 \times 10^4\ m/s)}{5.01\ u} = 1.45 \times 10^4\ m/s.$$

The other unknown is v'_p, which we can now obtain from

$$v'_p = v'_{He} - v_p = (1.45 \times 10^4\ m/s) - (3.60 \times 10^4\ m/s) = -2.15 \times 10^4\ m/s.$$

The minus sign for v'_p tells us that the proton reverses direction upon collision, and we see that its speed is less than its initial speed (see Fig. 9–15).

NOTE This result makes sense: the lighter proton would be expected to "bounce back" from the more massive helium nucleus, but not with its full original velocity as from a rigid wall (which corresponds to extremely large, or infinite, mass).

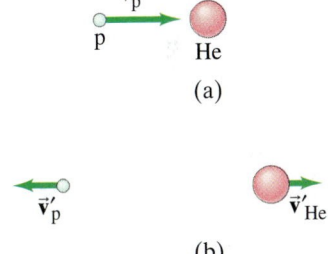

FIGURE 9–15 Example 9–9: (a) before collision, (b) after collision.

9–6 Inelastic Collisions

Collisions in which kinetic energy is not conserved are called **inelastic collisions**. Some of the initial kinetic energy is transformed into other types of energy, such as thermal or potential energy, so the total kinetic energy after the collision is less than the total kinetic energy before the collision. The inverse can also happen when potential energy (such as chemical or nuclear) is released, in which case the total kinetic energy after the interaction can be greater than the initial kinetic energy. Explosions are examples of this type.

Typical macroscopic collisions are inelastic, at least to some extent, and often to a large extent. If two objects stick together as a result of a collision, the collision is said to be **completely inelastic**. Two colliding balls of putty that stick together or two railroad cars that couple together when they collide are examples of completely inelastic collisions. The kinetic energy in some cases is all transformed to other forms of energy in an inelastic collision, but in other cases only part of it is. In Example 9–3, for instance, we saw that when a traveling railroad car collided with a stationary one, the coupled cars traveled off with some kinetic energy. In a completely inelastic collision, the maximum amount of kinetic energy is transformed to other forms consistent with conservation of momentum. Even though kinetic energy is not conserved in inelastic collisions, the total energy is always conserved, and the total vector momentum is also conserved.

EXAMPLE 9–10 **Railroad cars again.** For the completely inelastic collision of two railroad cars that we considered in Example 9–3, calculate how much of the initial kinetic energy is transformed to thermal or other forms of energy.

APPROACH The railroad cars stick together after the collision, so this is a completely inelastic collision. By subtracting the total kinetic energy after the collision from the total initial kinetic energy, we can find how much energy is transformed to other types of energy.

SOLUTION Before the collision, only car A is moving, so the total initial kinetic energy is $\frac{1}{2}m_A v_A^2 = \frac{1}{2}(10{,}000\,\text{kg})(24.0\,\text{m/s})^2 = 2.88 \times 10^6\,\text{J}$. After the collision, both cars are moving with a speed of 12.0 m/s, by conservation of momentum (Example 9–3). So the total kinetic energy afterward is $\frac{1}{2}(20{,}000\,\text{kg})(12.0\,\text{m/s})^2 = 1.44 \times 10^6\,\text{J}$. Hence the energy transformed to other forms is

$$(2.88 \times 10^6\,\text{J}) - (1.44 \times 10^6\,\text{J}) = 1.44 \times 10^6\,\text{J},$$

which is half the original kinetic energy.

EXAMPLE 9–11 **Ballistic pendulum.** The *ballistic pendulum* is a device used to measure the speed of a projectile, such as a bullet. The projectile, of mass m, is fired into a large block (of wood or other material) of mass M, which is suspended like a pendulum. (Usually, M is somewhat greater than m.) As a result of the collision, the pendulum and projectile together swing up to a maximum height h, Fig. 9–16. Determine the relationship between the initial horizontal speed of the projectile, v, and the maximum height h.

APPROACH We can analyze the process by dividing it into two parts or two time intervals: (1) the time interval from just before to just after the collision itself, and (2) the subsequent time interval in which the pendulum moves from the vertical hanging position to the maximum height h.

In part (1), Fig. 9–16a, we assume the collision time is very short, so that the projectile comes to rest in the block before the block has moved significantly from its rest position directly below its support. Thus there is effectively no net external force, and we can apply conservation of momentum to this completely inelastic collision. In part (2), Fig. 9–16b, the pendulum begins to move, subject to a net external force (gravity, tending to pull it back to the vertical position); so for part (2), we cannot use conservation of momentum. But we can use conservation of mechanical energy because gravity is a conservative force (Chapter 8). The kinetic energy immediately after the collision is changed entirely to gravitational potential energy when the pendulum reaches its maximum height, h.

SOLUTION In part (1) momentum is conserved:

$$\text{total } P \text{ before} = \text{total } P \text{ after}$$
$$mv = (m + M)v', \tag{i}$$

where v' is the speed of the block and embedded projectile just after the collision, before they have moved significantly.

In part (2), mechanical energy is conserved. We choose $y = 0$ when the pendulum hangs vertically, and then $y = h$ when the pendulum–projectile system reaches its maximum height. Thus we write

$$(K + U) \text{ just after collision} = (K + U) \text{ at pendulum's maximum height}$$

or

$$\tfrac{1}{2}(m + M)v'^2 + 0 = 0 + (m + M)gh. \tag{ii}$$

We solve for v':

$$v' = \sqrt{2gh}.$$

Inserting this result for v' into Eq. (i) above, and solving for v, gives

$$v = \frac{m + M}{m}v' = \frac{m + M}{m}\sqrt{2gh},$$

which is our final result.

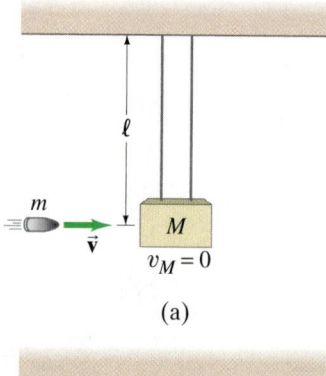

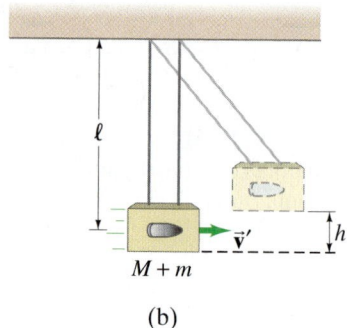

FIGURE 9–16 Ballistic pendulum. Example 9–11.

NOTE The separation of the process into two parts was crucial. Such an analysis is a powerful problem-solving tool. But how do you decide how to make such a division? Think about the conservation laws. They are your *tools*. Start a problem by asking yourself whether the conservation laws apply in the given situation. Here, we determined that momentum is conserved only during the brief collision, which we called part (1). But in part (1), because the collision is inelastic, the conservation of mechanical energy is not valid. Then in part (2), conservation of mechanical energy is valid, but not conservation of momentum.

Note, however, that if there had been significant motion of the pendulum during the deceleration of the projectile in the block, then there *would* have been an external force (gravity) during the collision, so conservation of momentum would not have been valid in part (1).

9–7 Collisions in Two or Three Dimensions

Conservation of momentum and energy can also be applied to collisions in two or three dimensions, where the vector nature of momentum is especially important. One common type of non-head-on collision is that in which a moving object (called the "projectile") strikes a second object initially at rest (the "target"). This is the common situation in games such as billiards and pool, and for experiments in atomic and nuclear physics (the projectiles, from radioactive decay or a high-energy accelerator, strike a stationary target nucleus; Fig. 9–17).

Figure 9–18 shows the incoming projectile, m_A, heading along the x axis toward the target object, m_B, which is initially at rest. If these are billiard balls, m_A strikes m_B not quite head-on and they go off at the angles θ'_A and θ'_B, respectively, which are measured relative to m_A's initial direction (the x axis).[†]

Let us apply the law of conservation of momentum to a collision like that of Fig. 9–18. We choose the xy plane to be the plane in which the initial and final momenta lie. Momentum is a vector, and because the total momentum is conserved, its components in the x and y directions also are conserved. The x component of momentum conservation gives

$$p_{Ax} + p_{Bx} = p'_{Ax} + p'_{Bx}$$

or, with $p_{Bx} = m_B v_{Bx} = 0$,

$$m_A v_A = m_A v'_A \cos \theta'_A + m_B v'_B \cos \theta'_B, \qquad \textbf{(9–9a)} \qquad p_x \text{ conserved}$$

where the primes (') refer to quantities *after* the collision. Because there is no motion in the y direction initially, the y component of the total momentum is zero before the collision. The y component equation of momentum conservation is then

$$p_{Ay} + p_{By} = p'_{Ay} + p'_{By}$$

or

$$0 = m_A v'_A \sin \theta'_A + m_B v'_B \sin \theta'_B. \qquad \textbf{(9–9b)} \qquad p_y \text{ conserved}$$

When we have two independent equations, we can solve for two unknowns, at most.

FIGURE 9–17 A recent color-enhanced version of a cloud-chamber photograph made in the early days (1920s) of nuclear physics. Green lines are paths of helium nuclei (He) coming from the left. One He, highlighted in yellow, strikes a proton of the hydrogen gas in the chamber, and both scatter at an angle; the scattered proton's path is shown in red.

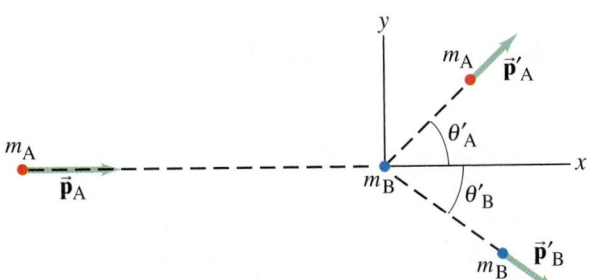

FIGURE 9–18 Object A, the projectile, collides with object B, the target. After the collision, they move off with momenta $\vec{p}'_A$ and $\vec{p}'_B$ at angles θ'_A and θ'_B. The objects are shown here as particles, as we would visualize them in atomic or nuclear physics. But they could also be macroscopic pool balls.

[†]The objects may begin to deflect even before they touch if electric, magnetic, or nuclear forces act between them. You might think, for example, of two magnets oriented so that they repel each other: when one moves toward the other, the second moves away before the first one touches it.

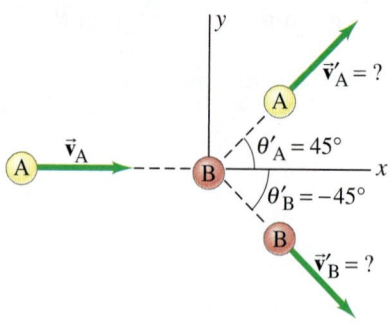

FIGURE 9–19 Example 9–12.

EXAMPLE 9–12 **Billiard ball collision in 2-D.** Billiard ball A moving with speed $v_A = 3.0\,\text{m/s}$ in the $+x$ direction (Fig. 9–19) strikes an equal-mass ball B initially at rest. The two balls are observed to move off at 45° to the x axis, ball A above the x axis and ball B below. That is, $\theta'_A = 45°$ and $\theta'_B = -45°$ in Fig. 9–19. What are the speeds of the two balls after the collision?

APPROACH There is no net external force on our system of two balls, assuming the table is level (the normal force balances gravity). Thus momentum conservation applies, and we apply it to both the x and y components using the xy coordinate system shown in Fig. 9–19. We get two equations, and we have two unknowns, v'_A and v'_B. From *symmetry* we might guess that the two balls have the same speed. But let us not assume that now. Even though we are not told whether the collision is elastic or inelastic, we can still use conservation of momentum.

SOLUTION We apply conservation of momentum for the x and y components, Eqs. 9–9a and b, and we solve for v'_A and v'_B. We are given $m_A = m_B(= m)$, so

$$\text{(for } x\text{)} \qquad mv_A = mv'_A \cos(45°) + mv'_B \cos(-45°)$$

and

$$\text{(for } y\text{)} \qquad 0 = mv'_A \sin(45°) + mv'_B \sin(-45°).$$

The m's cancel out in both equations (the masses are equal). The second equation yields [recall that $\sin(-\theta) = -\sin\theta$]:

$$v'_B = -v'_A \frac{\sin(45°)}{\sin(-45°)} = -v'_A \left(\frac{\sin 45°}{-\sin 45°} \right) = v'_A.$$

So they do have equal speeds as we guessed at first. The x component equation gives [recall that $\cos(-\theta) = \cos\theta$]:

$$v_A = v'_A \cos(45°) + v'_B \cos(45°) = 2v'_A \cos(45°),$$

so

$$v'_A = v'_B = \frac{v_A}{2\cos(45°)} = \frac{3.0\,\text{m/s}}{2(0.707)} = 2.1\,\text{m/s}.$$

If we know that a collision is elastic, we can also apply conservation of kinetic energy and obtain a third equation in addition to Eqs. 9–9a and b:

$$K_A + K_B = K'_A + K'_B$$

or, for the collision shown in Fig. 9–18 or 9–19,

$$\tfrac{1}{2}m_A v_A^2 = \tfrac{1}{2}m_A v_A'^2 + \tfrac{1}{2}m_B v_B'^2. \qquad \text{[elastic collision]} \quad \textbf{(9–9c)}$$

If the collision is elastic, we have three independent equations and can solve for three unknowns. If we are given m_A, m_B, v_A (and v_B, if it is not zero), we cannot, for example, predict the final variables, v'_A, v'_B, θ'_A, and θ'_B, because there are four of them. However, if we measure one of these variables, say θ'_A, then the other three variables (v'_A, v'_B, and θ'_B) are uniquely determined, and we can determine them using Eqs. 9–9a, b, and c.

A note of caution: Eq. 9–8 does *not* apply for two-dimensional collisions. It works only when a collision occurs along a line.

⚠ **CAUTION**

Equation 9–8 applies only in 1-D

EXAMPLE 9–13 **Proton–proton collision.** A proton traveling with speed $8.2 \times 10^5\,\text{m/s}$ collides elastically with a stationary proton in a hydrogen target as in Fig. 9–18. One of the protons is observed to be scattered at a 60° angle. At what angle will the second proton be observed, and what will be the velocities of the two protons after the collision?

APPROACH We saw a two-dimensional collision in Example 9–12, where we needed to use only conservation of momentum. Now we are given less information: we have three unknowns instead of two. Because the collision is elastic, we can use the kinetic energy equation as well as the two momentum equations.

SOLUTION Since $m_A = m_B$, Eqs. 9–9a, b, and c become

$$v_A = v'_A \cos \theta'_A + v'_B \cos \theta'_B \qquad \text{(i)}$$
$$0 = v'_A \sin \theta'_A + v'_B \sin \theta'_B \qquad \text{(ii)}$$
$$v_A^2 = v'^2_A + v'^2_B, \qquad \text{(iii)}$$

where $v_A = 8.2 \times 10^5$ m/s and $\theta'_A = 60°$ are given. In the first and second equations, we move the v'_A terms to the left side and square both sides of the equations:

$$v_A^2 - 2v_A v'_A \cos \theta'_A + v'^2_A \cos^2 \theta'_A = v'^2_B \cos^2 \theta'_B$$
$$v'^2_A \sin^2 \theta'_A = v'^2_B \sin^2 \theta'_B.$$

We add these two equations and use $\sin^2 \theta + \cos^2 \theta = 1$ to get:

$$v_A^2 - 2v_A v'_A \cos \theta'_A + v'^2_A = v'^2_B.$$

Into this equation we substitute $v'^2_B = v_A^2 - v'^2_A$, from equation (iii) above, and get

$$2v'^2_A = 2v_A v'_A \cos \theta'_A$$

or

$$v'_A = v_A \cos \theta'_A = (8.2 \times 10^5 \text{ m/s})(\cos 60°) = 4.1 \times 10^5 \text{ m/s}.$$

To obtain v'_B, we use equation (iii) above (conservation of kinetic energy):

$$v'_B = \sqrt{v_A^2 - v'^2_A} = 7.1 \times 10^5 \text{ m/s}.$$

Finally, from equation (ii), we have

$$\sin \theta'_B = -\frac{v'_A}{v'_B} \sin \theta'_A = -\left(\frac{4.1 \times 10^5 \text{ m/s}}{7.1 \times 10^5 \text{ m/s}}\right)(0.866) = -0.50,$$

so $\theta'_B = -30°$. (The minus sign means particle B moves at an angle below the x axis if particle A is above the axis, as in Fig. 9–19.) An example of such a collision is shown in the bubble chamber photo of Fig. 9–20. Notice that the two trajectories are at right angles to each other after the collision. This can be shown to be true in general for non-head-on elastic collisions of two particles of equal mass, one of which was at rest initially (see Problem 61).

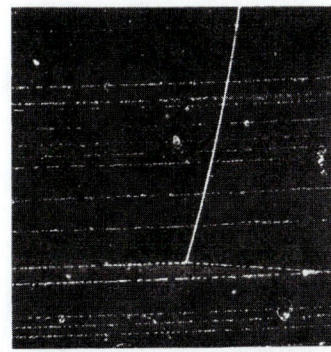

FIGURE 9–20 Photo of a proton–proton collision in a hydrogen bubble chamber (a device that makes visible the paths of elementary particles). The many lines represent incoming protons which can strike the protons of the hydrogen in the chamber.

Momentum Conservation and Collisions

1. Choose your **system**. If the situation is complex, think about how you might break it up into separate parts when one or more conservation laws apply.

2. If a significant **net external force** acts on your chosen system, be sure the time interval Δt is so short that the effect on momentum is negligible. That is, the forces that act between the interacting objects must be the only significant ones if momentum conservation is to be used. [Note: If this is valid for a portion of the problem, you can use momentum conservation only for that portion.]

3. Draw a **diagram** of the initial situation, just before the interaction (collision, explosion) takes place, and represent the momentum of each object with an arrow and a label. Do the same for the final situation, just after the interaction.

4. Choose a **coordinate system** and "+" and "−" directions. (For a head-on collision, you will need only an x axis.) It is often convenient to choose the $+x$ axis in the direction of one object's initial velocity.

5. Apply the **momentum conservation** equation(s):

   total initial momentum = total final momentum.

   You have one equation for each component (x, y, z): only one equation for a head-on collision.

6. If the collision is elastic, you can also write down a **conservation of kinetic energy** equation:

   total initial kinetic energy = total final kinetic energy.

   [Alternately, you could use Eq. 9–8: $v_A - v_B = v'_B - v'_A$, if the collision is one dimensional (head-on).]

7. Solve for the **unknown(s)**.

8. **Check** your work, check the units, and ask yourself whether the results are reasonable.

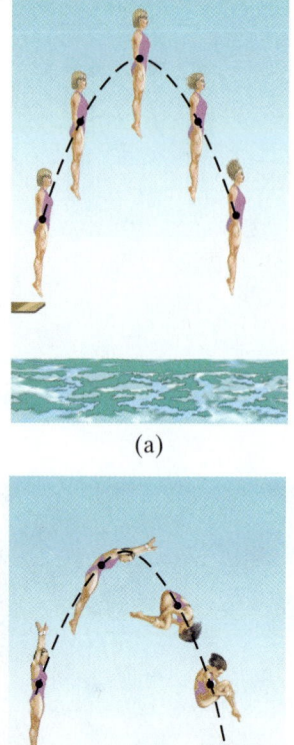

(a)

(b)

FIGURE 9–21 The motion of the diver is pure translation in (a), but is translation plus rotation in (b). The black dot represents the diver's CM at each moment.

9–8 Center of Mass (CM)

Momentum is a powerful concept not only for analyzing collisions but also for analyzing the translational motion of real extended objects. Until now, whenever we have dealt with the motion of an extended object (that is, an object that has size), we have assumed that it could be approximated as a point particle or that it undergoes only translational motion. Real extended objects, however, can undergo rotational and other types of motion as well. For example, the diver in Fig. 9–21a undergoes only translational motion (all parts of the object follow the same path), whereas the diver in Fig. 9–21b undergoes both translational and rotational motion. We will refer to motion that is not pure translation as *general motion*.

Observations indicate that even if an object rotates, or several parts of a system of objects move relative to one another, there is one point that moves in the same path that a particle would move if subjected to the same net force. This point is called the **center of mass** (abbreviated CM). The general motion of an extended object (or system of objects) can be considered as *the sum of the translational motion of the CM, plus rotational, vibrational, or other types of motion about the CM*.

As an example, consider the motion of the center of mass of the diver in Fig. 9–21; the CM follows a parabolic path even when the diver rotates, as shown in Fig. 9–21b. This is the same parabolic path that a projected particle follows when acted on only by the force of gravity (projectile motion, Section 3–7). Other points in the rotating diver's body, such as her feet or head, follow more complicated paths.

Figure 9–22 shows a wrench acted on by zero net force, translating and rotating along a horizontal surface. Note that its CM, marked by a red cross, moves in a straight line, as shown by the dashed white line.

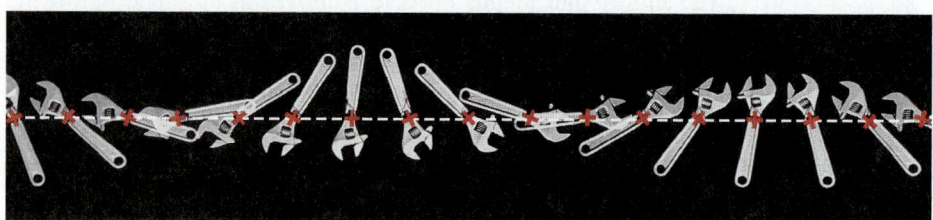

FIGURE 9–22 Translation plus rotation: a wrench moving over a horizontal surface. The CM, marked with a red cross, moves in a straight line.

We will show in Section 9–9 that the important properties of the CM follow from Newton's laws if the CM is defined in the following way. We can consider any extended object as being made up of many tiny particles. But first we consider a system made up of only two particles (or small objects), of masses m_A and m_B. We choose a coordinate system so that both particles lie on the x axis at positions x_A and x_B, Fig. 9–23. The center of mass of this system is defined to be at the position x_{CM}, given by

$$x_{CM} = \frac{m_A x_A + m_B x_B}{m_A + m_B} = \frac{m_A x_A + m_B x_B}{M},$$

where $M = m_A + m_B$ is the total mass of the system. The center of mass lies on the line joining m_A and m_B. If the two masses are equal $(m_A = m_B = m)$, then x_{CM} is midway between them, since in this case

$$x_{CM} = \frac{m(x_A + x_B)}{2m} = \frac{(x_A + x_B)}{2}.$$

If one mass is greater than the other, say, $m_A > m_B$, then the CM is closer to the larger mass. If all the mass is concentrated at x_B, so $m_A = 0$, then $x_{CM} = (0 x_A + m_B x_B)/(0 + m_B) = x_B$, as we would expect.

FIGURE 9–23 The center of mass of a two-particle system lies on the line joining the two masses. Here $m_A > m_B$, so the CM is closer to m_A than to m_B.

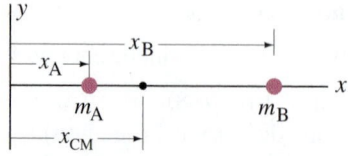

Now let us consider a system consisting of n particles, where n could be very large. This system could be an extended object which we consider as being made up of n tiny particles. If these n particles are all along a straight line (call it the x axis), we define the CM of the system to be located at

$$x_{CM} = \frac{m_1 x_1 + m_2 x_2 + \cdots + m_n x_n}{m_1 + m_2 + \cdots + m_n} = \frac{\sum_{i=1}^{n} m_i x_i}{M}, \qquad (9\text{–}10)$$

where $m_1, m_2, \ldots m_n$ are the masses of each particle and $x_1, x_2, \ldots x_n$ are their positions. The symbol $\sum_{i=1}^{n}$ is the summation sign meaning to sum over all the particles, where i takes on integer values from 1 to n. (Often we simply write $\sum m_i x_i$, leaving out the $i = 1$ to n.) The total mass of the system is $M = \sum m_i$.

EXAMPLE 9–14 **CM of three guys on a raft.** Three people of roughly equal masses m on a lightweight (air-filled) banana boat sit along the x axis at positions $x_A = 1.0$ m, $x_B = 5.0$ m, and $x_C = 6.0$ m, measured from the left-hand end as shown in Fig. 9–24. Find the position of the CM. Ignore the boat's mass.

APPROACH We are given the mass and location of the three people, so we use three terms in Eq. 9–10. We approximate each person as a point particle. Equivalently, the location of each person is the position of that person's own CM.

SOLUTION We use Eq. 9–10 with three terms:

$$x_{CM} = \frac{m x_A + m x_B + m x_C}{m + m + m} = \frac{m(x_A + x_B + x_C)}{3m}$$
$$= \frac{(1.0\,\text{m} + 5.0\,\text{m} + 6.0\,\text{m})}{3} = \frac{12.0\,\text{m}}{3} = 4.0\,\text{m}.$$

The CM is 4.0 m from the left-hand end of the boat. This makes sense—it should be closer to the two people in front than the one at the rear.

FIGURE 9–24 Example 9–14.

Note that the coordinates of the CM depend on the reference frame or coordinate system chosen. But the physical location of the CM is independent of that choice.

EXERCISE F Calculate the CM of the three people in Example 9–14 taking the origin at the driver $(x_C = 0)$ on the right. Is the physical location of the CM the same?

If the particles are spread out in two or three dimensions, as for a typical extended object, then we define the coordinates of the CM as

$$x_{CM} = \frac{\sum m_i x_i}{M}, \qquad y_{CM} = \frac{\sum m_i y_i}{M}, \qquad z_{CM} = \frac{\sum m_i z_i}{M}, \qquad (9\text{–}11)$$

where x_i, y_i, z_i are the coordinates of the particle of mass m_i and again $M = \sum m_i$ is the total mass.

Although from a practical point of view we usually calculate the components of the CM (Eq. 9–11), it is sometimes convenient (for example, for derivations) to write Eq. 9–11 in vector form. If $\vec{r}_i = x_i \hat{i} + y_i \hat{j} + z_i \hat{k}$ is the position vector of the ith particle, and $\vec{r}_{CM} = x_{CM} \hat{i} + y_{CM} \hat{j} + z_{CM} \hat{k}$ is the position vector of the center of mass, then

$$\vec{r}_{CM} = \frac{\sum m_i \vec{r}_i}{M}. \qquad (9\text{–}12)$$

EXAMPLE 9–15 **Three particles in 2-D.** Three particles, each of mass 2.50 kg, are located at the corners of a right triangle whose sides are 2.00 m and 1.50 m long, as shown in Fig. 9–25. Locate the center of mass.

APPROACH We choose our coordinate system as shown (to simplify calculations) with m_A at the origin and m_B on the x axis. Then m_A has coordinates $x_A = y_A = 0$; m_B has coordinates $x_B = 2.0$ m, $y_B = 0$; and m_C has coordinates $x_C = 2.0$ m, $y_C = 1.5$ m.

SOLUTION From Eqs. 9–11,

$$x_{CM} = \frac{(2.50\,\text{kg})(0) + (2.50\,\text{kg})(2.00\,\text{m}) + (2.50\,\text{kg})(2.00\,\text{m})}{3(2.50\,\text{kg})} = 1.33\,\text{m}$$

$$y_{CM} = \frac{(2.50\,\text{kg})(0) + (2.50\,\text{kg})(0) + (2.50\,\text{kg})(1.50\,\text{m})}{7.50\,\text{kg}} = 0.50\,\text{m}.$$

The CM and the position vector $\vec{r}_{CM}$ are shown in Fig. 9–25, inside the "triangle" as we should expect.

FIGURE 9–25 Example 9–15.

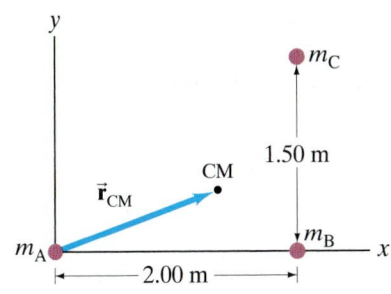

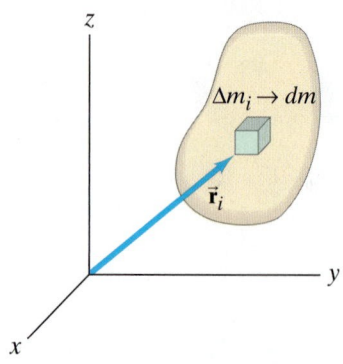

FIGURE 9–26 An extended object, here shown in only two dimensions, can be considered to be made up of many tiny particles (n), each having a mass Δm_i. One such particle is shown located at a point $\vec{r}_i = x_i\hat{i} + y_i\hat{j} + z_i\hat{k}$. We take the limit of $n \to \infty$ so Δm_i becomes the infinitesimal dm.

FIGURE 9–27 Example 9–16.

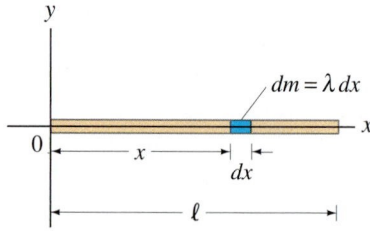

EXERCISE G A diver does a high dive involving a flip and a half-pike (legs and arms straight, but body bent in half). What can you say about the diver's center of mass? (a) It accelerates with a magnitude of 9.8 m/s^2 (ignoring air friction). (b) It moves in a circular path because of the rotation of the diver. (c) It must always be roughly located inside the diver's body, somewhere in the geometric center. (d) All of the above are true.

It is often convenient to think of an extended object as made up of a continuous distribution of matter. In other words, we consider the object to be made up of n particles, each of mass Δm_i in a tiny volume around a point x_i, y_i, z_i, and we take the limit of n approaching infinity (Fig. 9–26). Then Δm_i becomes the infinitesimal mass dm at points x, y, z. The summations in Eqs. 9–11 and 9–12 become integrals:

$$x_{CM} = \frac{1}{M}\int x\,dm, \quad y_{CM} = \frac{1}{M}\int y\,dm, \quad z_{CM} = \frac{1}{M}\int z\,dm, \quad \textbf{(9–13)}$$

where the sum over all the mass elements is $\int dm = M$, the total mass of the object. In vector notation, this becomes

$$\vec{r}_{CM} = \frac{1}{M}\int \vec{r}\,dm. \qquad \textbf{(9–14)}$$

A concept similar to *center of mass* is **center of gravity** (CG). The CG of an object is that point at which the force of gravity can be considered to act. The force of gravity actually acts on *all* the different parts or particles of an object, but for purposes of determining the translational motion of an object as a whole, we can assume that the entire weight of the object (which is the sum of the weights of all its parts) acts at the CG. There is a conceptual difference between the center of gravity and the center of mass, but for nearly all practical purposes, they are at the same point.[†]

EXAMPLE 9–16 **CM of a thin rod.** (a) Show that the CM of a uniform thin rod of length ℓ and mass M is at its center. (b) Determine the CM of the rod assuming its linear mass density λ (its mass per unit length) varies linearly from $\lambda = \lambda_0$ at the left end to double that value, $\lambda = 2\lambda_0$, at the right end.

APPROACH We choose a coordinate system so that the rod lies on the x axis with the left end at $x = 0$, Fig. 9–27. Then $y_{CM} = 0$ and $z_{CM} = 0$.

SOLUTION (a) The rod is uniform, so its mass per unit length (linear mass density λ) is constant and we write it as $\lambda = M/\ell$. We now imagine the rod as divided into infinitesimal elements of length dx, each of which has mass $dm = \lambda\,dx$. We use Eq. 9–13:

$$x_{CM} = \frac{1}{M}\int_{x=0}^{\ell} x\,dm = \frac{1}{M}\int_0^{\ell}\lambda x\,dx = \frac{\lambda}{M}\frac{x^2}{2}\bigg|_0^{\ell} = \frac{\lambda\ell^2}{2M} = \frac{\ell}{2}$$

where we used $\lambda = M/\ell$. This result, x_{CM} at the center, is what we expected.
(b) Now we have $\lambda = \lambda_0$ at $x = 0$ and we are told that λ increases linearly to $\lambda = 2\lambda_0$ at $x = \ell$. So we write

$$\lambda = \lambda_0(1 + \alpha x)$$

which satisfies $\lambda = \lambda_0$ at $x = 0$, increases linearly, and gives $\lambda = 2\lambda_0$ at $x = \ell$ if $(1 + \alpha\ell) = 2$. In other words, $\alpha = 1/\ell$. Again we use Eq. 9–13, with $\lambda = \lambda_0(1 + x/\ell)$:

$$x_{CM} = \frac{1}{M}\int_{x=0}^{\ell}\lambda x\,dx = \frac{1}{M}\lambda_0\int_0^{\ell}\left(1 + \frac{x}{\ell}\right)x\,dx = \frac{\lambda_0}{M}\left(\frac{x^2}{2} + \frac{x^3}{3\ell}\right)\bigg|_0^{\ell} = \frac{5}{6}\frac{\lambda_0}{M}\ell^2.$$

Now let us write M in terms of λ_0 and ℓ. We can write

$$M = \int_{x=0}^{\ell} dm = \int_0^{\ell}\lambda\,dx = \lambda_0\int_0^{\ell}\left(1 + \frac{x}{\ell}\right)dx = \lambda_0\left(x + \frac{x^2}{2\ell}\right)\bigg|_0^{\ell} = \frac{3}{2}\lambda_0\ell.$$

Then

$$x_{CM} = \frac{5}{6}\frac{\lambda_0}{M}\ell^2 = \frac{5}{9}\ell,$$

which is more than halfway along the rod, as we would expect since there is more mass to the right.

[†]There would be a difference between the CM and CG only in the unusual case of an object so large that the acceleration due to gravity, g, was different at different parts of the object.

For symmetrically shaped objects of uniform composition, such as spheres, cylinders, and rectangular solids, the CM is located at the geometric center of the object. Consider a uniform circular cylinder, such as a solid circular disk. We expect the CM to be at the center of the circle. To show that it is, we first choose a coordinate system whose origin is at the center of the circle with the z axis perpendicular to the disk (Fig. 9–28). When we take the sum $\Sigma m_i x_i$ in Eqs. 9–11, there is as much mass at any $+x_i$ as there is at $-x_i$. So all terms cancel out in pairs and $x_{CM} = 0$. The same is true for y_{CM}. In the vertical (z) direction, the CM must lie halfway between the circular faces: if we choose our origin of coordinates at that point, there is as much mass at any $+z_i$ as at $-z_i$, so $z_{CM} = 0$. For other uniform, symmetrically shaped objects, we can make similar arguments to show that the CM must lie on a line of symmetry. If a symmetric body is *not* uniform, then these arguments do not hold. For example, the CM of a wheel or disk weighted on one side is not at the geometric center but closer to the weighted side.

To locate the center of mass of a group of extended objects, we can use Eqs. 9–11, where the m_i are the masses of these objects and x_i, y_i, and z_i are the coordinates of the CM of each of the objects.

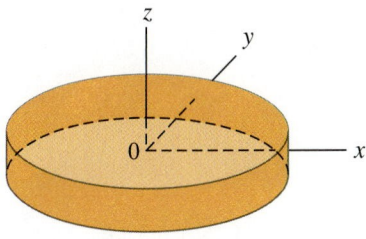

FIGURE 9–28 Cylindrical disk with origin of coordinates at geometric center.

FIGURE 9–29 Example 9–17. This L-shaped object has thickness t (not shown on diagram).

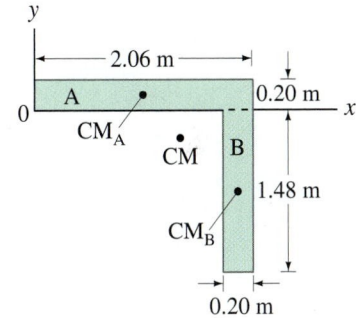

EXAMPLE 9–17 CM of L-shaped flat object. Determine the CM of the uniform thin L-shaped construction brace shown in Fig. 9–29.

APPROACH We can consider the object as two rectangles: rectangle A, which is 2.06 m × 0.20 m, and rectangle B, which is 1.48 m × 0.20 m. We choose the origin at 0 as shown. We assume a uniform thickness t.

SOLUTION The CM of rectangle A is at

$$x_A = 1.03 \text{ m}, \quad y_A = 0.10 \text{ m}.$$

The CM of B is at

$$x_B = 1.96 \text{ m}, \quad y_B = -0.74 \text{ m}.$$

The mass of A, whose thickness is t, is

$$M_A = (2.06 \text{ m})(0.20 \text{ m})(t)(\rho) = (0.412 \text{ m}^2)(\rho t),$$

where ρ is the density (mass per unit volume). The mass of B is

$$M_B = (1.48 \text{ m})(0.20 \text{ m})(\rho t) = (0.296 \text{ m}^2)(\rho t),$$

and the total mass is $M = (0.708 \text{ m}^2)(\rho t)$. Thus

$$x_{CM} = \frac{M_A x_A + M_B x_B}{M} = \frac{(0.412 \text{ m}^2)(1.03 \text{ m}) + (0.296 \text{ m}^2)(1.96 \text{ m})}{(0.708 \text{ m}^2)} = 1.42 \text{ m},$$

where ρt was canceled out in numerator and denominator. Similarly,

$$y_{CM} = \frac{(0.412 \text{ m}^2)(0.10 \text{ m}) + (0.296 \text{ m}^2)(-0.74 \text{ m})}{(0.708 \text{ m}^2)} = -0.25 \text{ m},$$

which puts the CM approximately at the point so labeled in Fig. 9–29. In thickness, $z_{CM} = t/2$, since the object is assumed to be uniform.

Note in this last Example that the CM can actually lie *outside* the object. Another example is a doughnut whose CM is at the center of the hole.

It is often easier to determine the CM or CG of an extended object experimentally rather than analytically. If an object is suspended from any point, it will swing (Fig. 9–30) due to the force of gravity on it, unless it is placed so its CG lies on a vertical line directly below the point from which it is suspended. If the object is two-dimensional, or has a plane of *symmetry*, it need only be hung from two different pivot points and the respective vertical (plumb) lines drawn. Then the CG will be at the intersection of the two lines, as in Fig. 9–31. If the object doesn't have a plane of symmetry, the CG with respect to the third dimension is found by suspending the object from at least three points whose plumb lines do not lie in the same plane. For symmetrically shaped objects, the CM is located at the geometric center of the object.

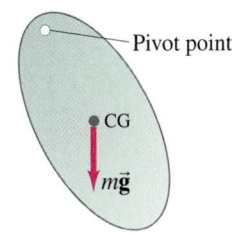

FIGURE 9–30 Determining the CM of a flat uniform body.

FIGURE 9–31 Finding the CG.

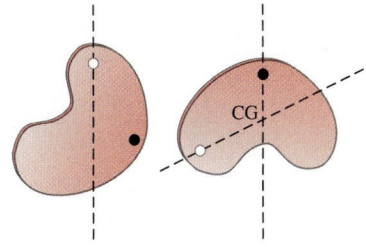

9–9 Center of Mass and Translational Motion

As mentioned in Section 9–8, a major reason for the importance of the concept of center of mass is that the translational motion of the CM for a system of particles (or an extended object) is directly related to the net force acting on the system as a whole. We now show this, by examining the motion of a system of n particles of total mass M, and we assume all the masses remain constant. We begin by rewriting Eq. 9–12 as

$$M\vec{\mathbf{r}}_{CM} = \Sigma m_i \vec{\mathbf{r}}_i.$$

We differentiate this equation with respect to time:

$$M\frac{d\vec{\mathbf{r}}_{CM}}{dt} = \Sigma m_i \frac{d\vec{\mathbf{r}}_i}{dt}$$

or

$$M\vec{\mathbf{v}}_{CM} = \Sigma m_i \vec{\mathbf{v}}_i, \qquad\qquad (9\text{–}15)$$

where $\vec{\mathbf{v}}_i = d\vec{\mathbf{r}}_i/dt$ is the velocity of the ith particle of mass m_i, and $\vec{\mathbf{v}}_{CM}$ is the velocity of the CM. We take the derivative with respect to time again and obtain

$$M\frac{d\vec{\mathbf{v}}_{CM}}{dt} = \Sigma m_i \vec{\mathbf{a}}_i,$$

where $\vec{\mathbf{a}}_i = d\vec{\mathbf{v}}_i/dt$ is the acceleration of the ith particle. Now $d\vec{\mathbf{v}}_{CM}/dt$ is the acceleration of the CM, $\vec{\mathbf{a}}_{CM}$. By Newton's second law, $m_i \vec{\mathbf{a}}_i = \vec{\mathbf{F}}_i$ where $\vec{\mathbf{F}}_i$ is the net force on the ith particle. Therefore

$$M\vec{\mathbf{a}}_{CM} = \vec{\mathbf{F}}_1 + \vec{\mathbf{F}}_2 + \cdots + \vec{\mathbf{F}}_n = \Sigma \vec{\mathbf{F}}_i. \qquad (9\text{–}16)$$

That is, the vector sum of all the forces acting on the system is equal to the total mass of the system times the acceleration of its center of mass. Note that our system of n particles could be the n particles that make up one or more extended objects.

The forces $\vec{\mathbf{F}}_i$ exerted on the particles of the system can be divided into two types: (1) *external forces* exerted by objects outside the system and (2) *internal forces* that particles within the system exert on one another. By Newton's third law, the internal forces occur in pairs: if one particle exerts a force on a second particle in our system, the second must exert an equal and opposite force on the first. Thus, in the sum over all the forces in Eq. 9–16, these internal forces cancel each other in pairs. We are left, then, with only the external forces on the right side of Eq. 9–16:

NEWTON'S SECOND LAW
(for a system)

$$M\vec{\mathbf{a}}_{CM} = \Sigma \vec{\mathbf{F}}_{ext}, \qquad\qquad [\text{constant } M] \quad (9\text{–}17)$$

where $\Sigma \vec{\mathbf{F}}_{ext}$ is the sum of all the external forces acting on our system, which is the *net force* acting on the system. Thus

the sum of all the forces acting on the system is equal to the total mass of the system times the acceleration of its center of mass.

This is **Newton's second law** for a system of particles. It also applies to an extended object (which can be thought of as a collection of particles), and to a system of objects. Thus we conclude that

TRANSLATIONAL MOTION OF CM

the center of mass of a system of particles (or objects) with total mass M moves like a single particle of mass M acted upon by the same net external force.

That is, the system translates as if all its mass were concentrated at the CM and all the external forces acted at that point. We can thus treat the *translational motion* of any object or system of objects as the motion of a particle (see Figs. 9–21 and 9–22).

This result clearly simplifies our analysis of the motion of complex systems and extended objects. Although the motion of various parts of the system may be complicated, we may often be satisfied with knowing the motion of the center of mass. This result also allows us to solve certain types of problems very easily, as illustrated by the following Example.

CONCEPTUAL EXAMPLE 9–18 | **A two-stage rocket.** A rocket is shot into the air as shown in Fig. 9–32. At the moment it reaches its highest point, a horizontal distance d from its starting point, a prearranged explosion separates it into two parts of equal mass. Part I is stopped in midair by the explosion and falls vertically to Earth. Where does part II land? Assume $\vec{\mathbf{g}}$ = constant.

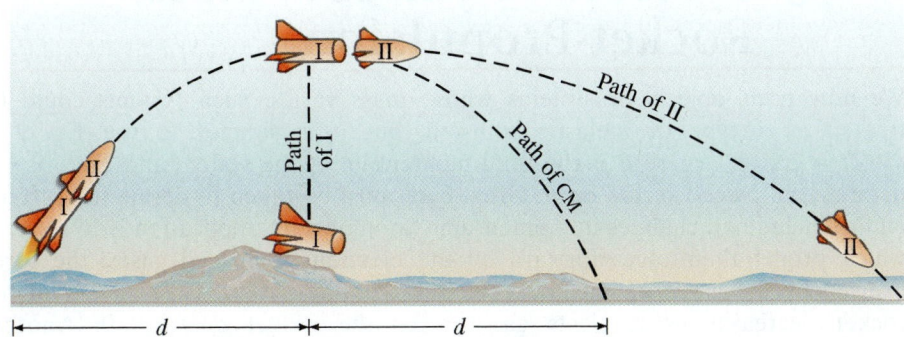

FIGURE 9–32 Example 9–18.

RESPONSE After the rocket is fired, the path of the CM of the system continues to follow the parabolic trajectory of a projectile acted on only by a constant gravitational force. The CM will thus arrive at a point $2d$ from the starting point. Since the masses of I and II are equal, the CM must be midway between them. Therefore, part II lands a distance $3d$ from the starting point.

NOTE If part I had been given a kick up or down, instead of merely falling, the solution would have been somewhat more complicated.

EXERCISE H A woman stands up in a rowboat and walks from one end of the boat to the other. How does the boat move, as seen from the shore?

We can write Eq. 9–17, $M\vec{\mathbf{a}}_{CM} = \Sigma\vec{\mathbf{F}}_{ext}$, in terms of the total momentum $\vec{\mathbf{P}}$ of a system of particles. $\vec{\mathbf{P}}$ is defined, as we saw in Section 9–2 as

$$\vec{\mathbf{P}} = m_1\vec{\mathbf{v}}_1 + m_2\vec{\mathbf{v}}_2 + \cdots + m_n\vec{\mathbf{v}}_n = \Sigma\vec{\mathbf{p}}_i.$$

From Eq. 9–15 $\left(M\vec{\mathbf{v}}_{CM} = \Sigma m_i\vec{\mathbf{v}}_i\right)$ we have

$$\vec{\mathbf{P}} = M\vec{\mathbf{v}}_{CM}. \tag{9–18}$$

Thus, *the total linear momentum of a system of particles is equal to the product of the total mass M and the velocity of the center of mass of the system.* Or, *the linear momentum of an extended object is the product of the object's mass and the velocity of its CM.*

If we differentiate Eq. 9–18 with respect to time, we obtain (assuming the total mass M is constant)

$$\frac{d\vec{\mathbf{P}}}{dt} = M\frac{d\vec{\mathbf{v}}_{CM}}{dt} = M\vec{\mathbf{a}}_{CM}.$$

From Eq. 9–17, we see that

$$\frac{d\vec{\mathbf{P}}}{dt} = \Sigma\vec{\mathbf{F}}_{ext}, \qquad \text{[same as Eq. 9–5]}$$

NEWTON'S SECOND LAW (for a system)

where $\Sigma\vec{\mathbf{F}}_{ext}$ is the net external force on the system. This is just Eq. 9–5 obtained earlier: **Newton's second law for a system of objects**. It is valid for any definite fixed system of particles or objects. If we know $\Sigma\vec{\mathbf{F}}_{ext}$, we can determine how the total momentum changes.

An interesting application is the discovery of nearby stars (see Section 6–5) that seem to "wobble." What could cause such a wobble? It could be that a planet orbits the star, and each exerts a gravitational force on the other. The planets are too small and too far away to have been observed directly by existing telescopes. But the slight wobble in the motion of the star suggests that both the planet and the star (its sun) orbit about their mutual center of mass, and hence the star appears to have a wobble. Irregularities in the star's motion can be obtained to high accuracy, and from the data the size of the planets' orbits can be obtained as well as their masses. See Fig. 6–18 in Chapter 6.

*9–10 Systems of Variable Mass; Rocket Propulsion

We now treat objects or systems whose mass varies. Such systems could be treated as a type of inelastic collision, but it is simpler to use Eq. 9–5, $d\vec{P}/dt = \Sigma\vec{F}_{ext}$, where $\vec{P}$ is the total momentum of the system and $\Sigma\vec{F}_{ext}$ is the net external force exerted on it. Great care must be taken to define the system, and to include all changes in momentum. An important application is to rockets, which propel themselves forward by the ejection of burned gases: the force exerted by the gases on the rocket accelerates the rocket. The mass M of the rocket decreases as it ejects gas, so for the rocket $dM/dt < 0$. Another application is the dropping of material (gravel, packaged goods) onto a conveyor belt. In this situation, the mass M of the loaded conveyor belt increases and $dM/dt > 0$.

To treat the general case of variable mass, let us consider the system shown in Fig. 9–33. At some time t, we have a system of mass M and momentum $M\vec{v}$. We also have a tiny (infinitesimal) mass dM traveling with velocity $\vec{u}$ which is about to enter our system. An infinitesimal time dt later, the mass dM combines with the system. For simplicity we will refer to this as a "collision." So our system has changed in mass from M to $M + dM$ in the time dt. Note that dM can be less than zero, as for a rocket propelled by ejected gases whose mass M thus decreases.

In order to apply Eq. 9–5, $d\vec{P}/dt = \Sigma\vec{F}_{ext}$, we must consider a definite fixed system of particles. That is, in considering the change in momentum, $d\vec{P}$, we must consider the momentum of the same particles initially and finally. We will define our *total system* as including M plus dM. Then initially, at time t, the total momentum is $M\vec{v} + \vec{u}\,dM$ (Fig. 9–33). At time $t + dt$, after dM has combined with M, the velocity of the whole is now $\vec{v} + d\vec{v}$ and the total momentum is $(M + dM)(\vec{v} + d\vec{v})$. So the change in momentum $d\vec{P}$ is

$$d\vec{P} = (M + dM)(\vec{v} + d\vec{v}) - (M\vec{v} + \vec{u}\,dM)$$
$$= M\,d\vec{v} + \vec{v}\,dM + dM\,d\vec{v} - \vec{u}\,dM.$$

The term $dM\,d\vec{v}$ is the product of two differentials and is zero even after we "divide by dt," which we do, and apply Eq. 9–5 to obtain

$$\Sigma\vec{F}_{ext} = \frac{d\vec{P}}{dt} = \frac{M\,d\vec{v} + \vec{v}\,dM - \vec{u}\,dM}{dt}.$$

Thus we get

$$\Sigma\vec{F}_{ext} = M\frac{d\vec{v}}{dt} - (\vec{u} - \vec{v})\frac{dM}{dt}. \qquad \textbf{(9–19a)}$$

Note that the quantity $(\vec{u} - \vec{v})$ is the relative velocity, $\vec{v}_{rel}$, of dM with respect to M.

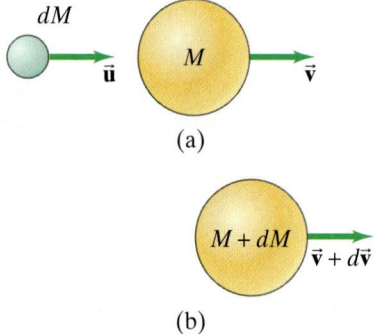

FIGURE 9–33 (a) At time t, a mass dM is about to be added to our system M. (b) At time $t + dt$, the mass dM has been added to our system.

That is,

$$\vec{v}_{rel} = \vec{u} - \vec{v}$$

is the velocity of the entering mass dM as seen by an observer on M. We can rearrange Eq. 9–19a:

$$M\frac{d\vec{v}}{dt} = \Sigma\vec{F}_{ext} + \vec{v}_{rel}\frac{dM}{dt}. \qquad (9\text{–}19b)$$

We can interpret this equation as follows. $M\,d\vec{v}/dt$ is the mass times the acceleration of M. The first term on the right, $\Sigma\vec{F}_{ext}$, refers to the external force on the mass M (for a rocket, it would include the force of gravity and air resistance). It does *not* include the force that dM exerts on M as a result of their collision. This is taken care of by the second term on the right, $\vec{v}_{rel}(dM/dt)$, which represents the rate at which momentum is being transferred into (or out of) the mass M because of the mass that is added to (or leaves) it. It can thus be interpreted as the force exerted on the mass M due to the addition (or ejection) of mass. For a rocket this term is called the *thrust*, since it represents the force exerted on the rocket by the expelled gases. For a rocket ejecting burned fuel, $dM/dt < 0$, but so is $\vec{v}_{rel}$ (gases are forced out the back), so the second term in Eq. 9–19b acts to increase $\vec{v}$.

EXAMPLE 9–19 Conveyor belt. You are designing a conveyor system for a gravel yard. A hopper drops gravel at a rate of 75.0 kg/s onto a conveyor belt that moves at a constant speed $v = 2.20$ m/s (Fig. 9–34). (*a*) Determine the additional force (over and above internal friction) needed to keep the conveyor belt moving as gravel falls on it. (*b*) What power output would be needed from the motor that drives the conveyor belt?

APPROACH We assume that the hopper is at rest so $u = 0$, and that the hopper has just begun dropping gravel so $dM/dt = 75.0$ kg/s.

SOLUTION (*a*) The belt needs to move at a constant speed $(dv/dt = 0)$, so Eq. 9–19 as written for one dimension, gives:

$$\begin{aligned}
F_{ext} &= M\frac{dv}{dt} - (u - v)\frac{dM}{dt} \\
&= 0 - (0 - v)\frac{dM}{dt} \\
&= v\frac{dM}{dt} = (2.20\text{ m/s})(75.0\text{ kg/s}) = 165\text{ N}.
\end{aligned}$$

(*b*) This force does work at the rate (Eq. 8–21)

$$\begin{aligned}
\frac{dW}{dt} &= \vec{F}_{ext} \cdot \vec{v} = v^2\frac{dM}{dt} \\
&= 363\text{ W},
\end{aligned}$$

which is the power output required of the motor.

NOTE This work does not all go into kinetic energy of the gravel, since

$$\frac{dK}{dt} = \frac{d}{dt}\left(\frac{1}{2}Mv^2\right) = \frac{1}{2}\frac{dM}{dt}v^2,$$

which is only half the work done by $\vec{F}_{ext}$. The other half of the external work done goes into thermal energy produced by friction between the gravel and the belt (the same friction force that accelerates the gravel).

PHYSICS APPLIED
Moving conveyor belt

FIGURE 9–34 Example 9–19. Gravel dropped from hopper onto conveyor belt.

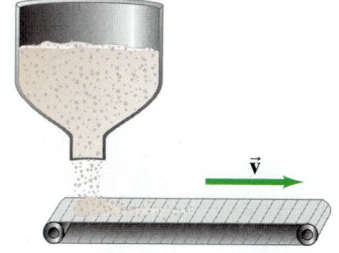

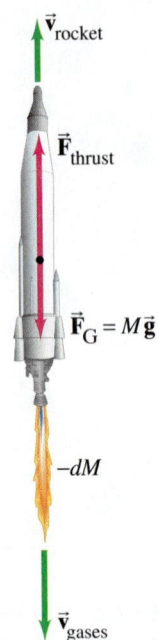

FIGURE 9–35 Example 9–20;
$\vec{v}_{rel} = \vec{v}_{gases} - \vec{v}_{rocket}$. M is the mass
of the rocket at any instant and is
decreasing until burnout.

EXAMPLE 9–20 **Rocket propulsion.** A fully fueled rocket has a mass of 21,000 kg, of which 15,000 kg is fuel. The burned fuel is spewed out the rear at a rate of 190 kg/s with a speed of 2800 m/s relative to the rocket. If the rocket is fired vertically upward (Fig. 9–35) calculate: (*a*) the thrust of the rocket; (*b*) the net force on the rocket at blastoff, and just before burnout (when all the fuel has been used up); (*c*) the rocket's velocity as a function of time, and (*d*) its final velocity at burnout. Ignore air resistance and assume the acceleration due to gravity is constant at $g = 9.80 \text{ m/s}^2$.

APPROACH To begin, the thrust is defined (see discussion after Eq. 9–19b) as the last term in Eq. 9–19b, $v_{rel}(dM/dt)$. The net force [for (*b*)] is the vector sum of the thrust and gravity. The velocity is found from Eq. 9–19b.

SOLUTION (*a*) The thrust is:

$$F_{thrust} = v_{rel}\frac{dM}{dt} = (-2800 \text{ m/s})(-190 \text{ kg/s}) = 5.3 \times 10^5 \text{ N},$$

where we have taken upward as positive so v_{rel} is negative because it is downward, and dM/dt is negative because the rocket's mass is diminishing.

(*b*) $F_{ext} = Mg = (2.1 \times 10^4 \text{ kg})(9.80 \text{ m/s}^2) = 2.1 \times 10^5 \text{ N}$ initially, and at burnout $F_{ext} = (6.0 \times 10^3 \text{ kg})(9.80 \text{ m/s}^2) = 5.9 \times 10^4 \text{ N}$. Hence, the net force on the rocket at blastoff is

$$F_{net} = 5.3 \times 10^5 \text{ N} - 2.1 \times 10^5 \text{ N} = 3.2 \times 10^5 \text{ N}, \quad [\text{blastoff}]$$

and just before burnout it is

$$F_{net} = 5.3 \times 10^5 \text{ N} - 5.9 \times 10^4 \text{ N} = 4.7 \times 10^5 \text{ N}. \quad [\text{burnout}]$$

After burnout, of course, the net force is that of gravity, $-5.9 \times 10^4 \text{ N}$.

(*c*) From Eq. 9–19b we have

$$dv = \frac{F_{ext}}{M}dt + v_{rel}\frac{dM}{M},$$

where $F_{ext} = -Mg$, and M is the mass of the rocket and is a function of time. Since v_{rel} is constant, we can integrate this easily:

$$\int_{v_0}^{v} dv = -\int_0^t g\,dt + v_{rel}\int_{M_0}^{M}\frac{dM}{M}$$

or

$$v(t) = v_0 - gt + v_{rel}\ln\frac{M}{M_0},$$

where $v(t)$ is the rocket's velocity and M its mass at any time t. Note that v_{rel} is negative (-2800 m/s in our case) because it is opposite to the motion, and that $\ln(M/M_0)$ is also negative because $M_0 > M$. Hence, the last term—which represents the thrust—is positive and acts to increase the velocity.

(*d*) The time required to reach burnout is the time needed to use up all the fuel (15,000 kg) at a rate of 190 kg/s; so at burnout,

$$t = \frac{1.50 \times 10^4 \text{ kg}}{190 \text{ kg/s}} = 79 \text{ s}.$$

If we take $v_0 = 0$, then using the result of part (*c*):

$$v = -(9.80 \text{ m/s}^2)(79 \text{ s}) + (-2800 \text{ m/s})\left(\ln\frac{6000 \text{ kg}}{21{,}000 \text{ kg}}\right) = 2700 \text{ m/s}.$$

Summary

The **linear momentum**, $\vec{p}$, of an object is defined as the product of its mass times its velocity,

$$\vec{p} = m\vec{v}. \tag{9-1}$$

In terms of momentum, **Newton's second law** can be written as

$$\Sigma\vec{F} = \frac{d\vec{p}}{dt}. \tag{9-2}$$

That is, the rate of change of momentum of an object equals the net force exerted on it.

When the net external force on a system of objects is zero, the total momentum remains constant. This is the **law of conservation of momentum**. Stated another way, the total momentum of an isolated system of objects remains constant.

The law of conservation of momentum is very useful in dealing with the class of events known as **collisions**. In a collision, two (or more) objects interact with each other for a very short time, and the force each exerts on the other during this time interval is very large compared to any other forces acting. The **impulse** of such a force on an object is defined as

$$\vec{J} = \int \vec{F}\, dt$$

and is equal to the change in momentum of the object as long as $\vec{F}$ is the net force on the object:

$$\Delta\vec{p} = \vec{p}_f - \vec{p}_i = \int_{t_i}^{t_f}\vec{F}\, dt = \vec{J}. \tag{9-6}$$

Total momentum is conserved in any collision:

$$\vec{p}_A + \vec{p}_B = \vec{p}'_A + \vec{p}'_B.$$

The total energy is also conserved; but this may not be useful unless kinetic energy is conserved, in which case the collision is called an **elastic collision**:

$$\tfrac{1}{2}m_A v_A^2 + \tfrac{1}{2}m_B v_B^2 = \tfrac{1}{2}m_A v_A'^2 + \tfrac{1}{2}m_B v_B'^2. \tag{9-7}$$

If kinetic energy is not conserved, the collision is called **inelastic**.

If two colliding objects stick together as the result of a collision, the collision is said to be **completely inelastic**.

For a system of particles, or for an extended object that can be considered as having a continuous distribution of matter, the **center of mass** (CM) is defined as

$$x_{CM} = \frac{\Sigma m_i x_i}{M}, \qquad y_{CM} = \frac{\Sigma m_i y_i}{M}, \qquad z_{CM} = \frac{\Sigma m_i z_i}{M} \tag{9-11}$$

or

$$x_{CM} = \frac{1}{M}\int x\, dm, \qquad y_{CM} = \frac{1}{M}\int y\, dm, \qquad z_{CM} = \frac{1}{M}\int z\, dm, \tag{9-13}$$

where M is the total mass of the system.

The center of mass of a system is important because this point moves like a single particle of mass M acted on by the same net external force, $\Sigma\vec{F}_{ext}$. In equation form, this is just Newton's second law for a system of particles (or extended objects):

$$M\vec{a}_{CM} = \Sigma\vec{F}_{ext}, \tag{9-17}$$

where M is the total mass of the system, $\vec{a}_{CM}$ is the acceleration of the CM of the system, and $\Sigma\vec{F}_{ext}$ is the total (net) external force acting on all parts of the system.

For a system of particles of total linear momentum $\vec{P} = \Sigma m_i\vec{v}_i = M\vec{v}_{CM}$, Newton's second law is

$$\frac{d\vec{P}}{dt} = \Sigma\vec{F}_{ext}. \tag{9-5}$$

[*If the mass M of an object is not constant, then

$$M\frac{d\vec{v}}{dt} = \Sigma\vec{F}_{ext} + \vec{v}_{rel}\frac{dM}{dt} \tag{9-19b}$$

where $\vec{v}$ is the velocity of the object at any instant and $\vec{v}_{rel}$ is the relative velocity at which mass enters (or leaves) the object.]

Questions

1. We claim that momentum is conserved. Yet most moving objects eventually slow down and stop. Explain.

2. Two blocks of mass m_1 and m_2 rest on a frictionless table and are connected by a spring. The blocks are pulled apart, stretching the spring, and then released. Describe the subsequent motion of the two blocks.

3. A light object and a heavy object have the same kinetic energy. Which has the greater momentum? Explain.

4. When a person jumps from a tree to the ground, what happens to the momentum of the person upon striking the ground?

5. Explain, on the basis of conservation of momentum, how a fish propels itself forward by swishing its tail back and forth.

6. Two children float motionlessly in a space station. The 20-kg girl pushes on the 40-kg boy and he sails away at 1.0 m/s. The girl (a) remains motionless; (b) moves in the same direction at 1.0 m/s; (c) moves in the opposite direction at 1.0 m/s; (d) moves in the opposite direction at 2.0 m/s; (e) none of these.

7. A truck going 15 km/h has a head-on collision with a small car going 30 km/h. Which statement best describes the situation? (a) The truck has the greater change of momentum because it has the greater mass. (b) The car has the greater change of momentum because it has the greater speed. (c) Neither the car nor the truck changes its momentum in the collision because momentum is conserved. (d) They both have the same change in magnitude of momentum because momentum is conserved. (e) None of the above is necessarily true.

8. If a falling ball were to make a perfectly elastic collision with the floor, would it rebound to its original height? Explain.

9. A boy stands on the back of a rowboat and dives into the water. What happens to the rowboat as the boy leaves it? Explain.

10. It is said that in ancient times a rich man with a bag of gold coins was stranded on the surface of a frozen lake. Because the ice was frictionless, he could not push himself to shore and froze to death. What could he have done to save himself had he not been so miserly?

11. The speed of a tennis ball on the return of a serve can be just as fast as the serve, even though the racket isn't swung very fast. How can this be?

12. Is it possible for an object to receive a larger impulse from a small force than from a large force? Explain.

13. How could a force give zero impulse over a nonzero time interval even though the force is not zero for at least a part of that time interval?

14. In a collision between two cars, which would you expect to be more damaging to the occupants: if the cars collide and remain together, or if the two cars collide and rebound backward? Explain.

15. A superball is dropped from a height h onto a hard steel plate (fixed to the Earth), from which it rebounds at very nearly its original speed. (a) Is the momentum of the ball conserved during any part of this process? (b) If we consider the ball and the Earth as our system, during what parts of the process is momentum conserved? (c) Answer part (b) for a piece of putty that falls and sticks to the steel plate.

16. Cars used to be built as rigid as possible to withstand collisions. Today, though, cars are designed to have "crumple zones" that collapse upon impact. What is the advantage of this new design?

17. At a hydroelectric power plant, water is directed at high speed against turbine blades on an axle that turns an electric generator. For maximum power generation, should the turbine blades be designed so that the water is brought to a dead stop, or so that the water rebounds?

18. A squash ball hits a wall at a 45° angle as shown in Fig. 9–36. What is the direction (a) of the change in momentum of the ball, (b) of the force on the wall?

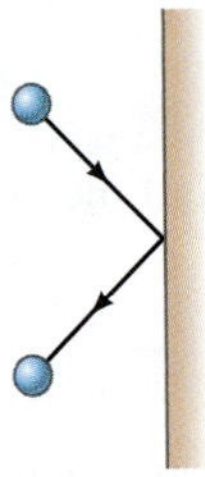

FIGURE 9–36
Question 18.

19. Why can a batter hit a pitched baseball farther than a ball he himself has tossed up in the air?

20. Describe a collision in which all kinetic energy is lost.

21. Inelastic and elastic collisions are similar in that (a) momentum and kinetic energy are conserved in both; (b) momentum is conserved in both; (c) momentum and potential energy are conserved in both; (d) kinetic energy is conserved in both.

22. If a 20-passenger plane is not full, sometimes passengers are told they must sit in certain seats and may not move to empty seats. Why might this be?

23. Why do you tend to lean backward when carrying a heavy load in your arms?

24. Why is the CM of a 1-m length of pipe at its midpoint, whereas this is not true for your arm or leg?

25. Show on a diagram how your CM shifts when you move from a lying position to a sitting position.

26. Describe an analytic way of determining the CM of any thin, triangular-shaped, uniform plate.

27. Place yourself facing the edge of an open door. Position your feet astride the door with your nose and abdomen touching the door's edge. Try to rise on your tiptoes. Why can't this be done?

28. If only an external force can change the momentum of the center of mass of an object, how can the internal force of the engine accelerate a car?

29. A rocket following a parabolic path through the air suddenly explodes into many pieces. What can you say about the motion of this system of pieces?

30. How can a rocket change direction when it is far out in space and essentially in a vacuum?

31. In observations of nuclear β-decay, the electron and recoil nucleus often do not separate along the same line. Use conservation of momentum in two dimensions to explain why this implies the emission of at least one other particle in the disintegration.

32. Bob and Jim decide to play tug-of-war on a frictionless (icy) surface. Jim is considerably stronger than Bob, but Bob weighs 160 lbs while Jim weighs 145 lbs. Who loses by crossing over the midline first?

33. At a carnival game you try to knock over a heavy cylinder by throwing a small ball at it. You have a choice of throwing either a ball that will stick to the cylinder, or a second ball of equal mass and speed that will bounce backward off the cylinder. Which ball is more likely to make the cylinder move?

Problems

9–1 Momentum

1. (I) Calculate the force exerted on a rocket when the propelling gases are being expelled at a rate of 1300 kg/s with a speed of 4.5×10^4 m/s.

2. (I) A constant friction force of 25 N acts on a 65-kg skier for 15 s. What is the skier's change in velocity?

3. (II) The momentum of a particle, in SI units, is given by $\vec{p} = 4.8\,t^2\hat{i} - 8.0\hat{j} - 8.9\,t\hat{k}$. What is the force as a function of time?

4. (II) The force on a particle of mass m is given by $\vec{F} = 26\hat{i} - 12\,t^2\hat{j}$ where F is in N and t in s. What will be the change in the particle's momentum between $t = 1.0$ s and $t = 2.0$ s?

5. (II) A 145-g baseball, moving along the x axis with speed 30.0 m/s, strikes a fence at a 45° angle and rebounds along the y axis with unchanged speed. Give its change in momentum using unit vector notation.

6. (II) A 0.145-kg baseball pitched horizontally at 32.0 m/s strikes a bat and is popped straight up to a height of 36.5 m. If the contact time between bat and ball is 2.5 ms, calculate the average force between the ball and bat during contact.

7. (II) A rocket of total mass 3180 kg is traveling in outer space with a velocity of 115 m/s. To alter its course by 35.0°, its rockets can be fired briefly in a direction perpendicular to its original motion. If the rocket gases are expelled at a speed of 1750 m/s, how much mass must be expelled?

8. (III) Air in a 120-km/h wind strikes head-on the face of a building 45 m wide by 65 m high and is brought to rest. If air has a mass of 1.3 kg per cubic meter, determine the average force of the wind on the building.

9–2 Conservation of Momentum

9. (I) A 7700-kg boxcar traveling 18 m/s strikes a second car. The two stick together and move off with a speed of 5.0 m/s. What is the mass of the second car?

10. (I) A 9150-kg railroad car travels alone on a level frictionless track with a constant speed of 15.0 m/s. A 4350-kg load, initially at rest, is dropped onto the car. What will be the car's new speed?

11. (I) An atomic nucleus at rest decays radioactively into an alpha particle and a smaller nucleus. What will be the speed of this recoiling nucleus if the speed of the alpha particle is 2.8×10^5 m/s? Assume the recoiling nucleus has a mass 57 times greater than that of the alpha particle.

12. (I) A 130-kg tackler moving at 2.5 m/s meets head-on (and tackles) an 82-kg halfback moving at 5.0 m/s. What will be their mutual speed immediately after the collision?

13. (II) A child in a boat throws a 5.70-kg package out horizontally with a speed of 10.0 m/s, Fig. 9–37. Calculate the velocity of the boat immediately after, assuming it was initially at rest. The mass of the child is 24.0 kg and that of the boat is 35.0 kg.

FIGURE 9–37
Problem 13.

14. (II) An atomic nucleus initially moving at 420 m/s emits an alpha particle in the direction of its velocity, and the remaining nucleus slows to 350 m/s. If the alpha particle has a mass of 4.0 u and the original nucleus has a mass of 222 u, what speed does the alpha particle have when it is emitted?

15. (II) An object at rest is suddenly broken apart into two fragments by an explosion. One fragment acquires twice the kinetic energy of the other. What is the ratio of their masses?

16. (II) A 22-g bullet traveling 210 m/s penetrates a 2.0-kg block of wood and emerges going 150 m/s. If the block is stationary on a frictionless surface when hit, how fast does it move after the bullet emerges?

17. (II) A rocket of mass m traveling with speed v_0 along the x axis suddenly shoots out fuel equal to one-third its mass, perpendicular to the x axis (along the y axis) with speed $2v_0$. Express the final velocity of the rocket in $\hat{\mathbf{i}}, \hat{\mathbf{j}}, \hat{\mathbf{k}}$ notation.

18. (II) The decay of a neutron into a proton, an electron, and a neutrino is an example of a three-particle decay process. Use the vector nature of momentum to show that if the neutron is initially at rest, the velocity vectors of the three must be coplanar (that is, all in the same plane). The result is not true for numbers greater than three.

19. (II) A mass $m_A = 2.0$ kg, moving with velocity $\vec{\mathbf{v}}_A = (4.0\hat{\mathbf{i}} + 5.0\hat{\mathbf{j}} - 2.0\hat{\mathbf{k}})$ m/s, collides with mass $m_B = 3.0$ kg, which is initially at rest. Immediately after the collision, mass m_A is observed traveling at velocity $\vec{\mathbf{v}}'_A = (-2.0\hat{\mathbf{i}} + 3.0\hat{\mathbf{k}})$ m/s. Find the velocity of mass m_B after the collision. Assume no outside force acts on the two masses during the collision.

20. (II) A 925-kg two-stage rocket is traveling at a speed of 6.60×10^3 m/s away from Earth when a predesigned explosion separates the rocket into two sections of equal mass that then move with a speed of 2.80×10^3 m/s relative to each other along the original line of motion. (a) What is the speed and direction of each section (relative to Earth) after the explosion? (b) How much energy was supplied by the explosion? [Hint: What is the change in kinetic energy as a result of the explosion?]

21. (III) A 224-kg projectile, fired with a speed of 116 m/s at a 60.0° angle, breaks into three pieces of equal mass at the highest point of its arc (where its velocity is horizontal). Two of the fragments move with the same speed right after the explosion as the entire projectile had just before the explosion; one of these moves vertically downward and the other horizontally. Determine (a) the velocity of the third fragment immediately after the explosion and (b) the energy released in the explosion.

9–3 Collisions and Impulse

22. (I) A 0.145-kg baseball pitched at 35.0 m/s is hit on a horizontal line drive straight back at the pitcher at 56.0 m/s. If the contact time between bat and ball is 5.00×10^{-3} s, calculate the force (assumed to be constant) between the ball and bat.

23. (II) A golf ball of mass 0.045 kg is hit off the tee at a speed of 45 m/s. The golf club was in contact with the ball for 3.5×10^{-3} s. Find (a) the impulse imparted to the golf ball, and (b) the average force exerted on the ball by the golf club.

24. (II) A 12-kg hammer strikes a nail at a velocity of 8.5 m/s and comes to rest in a time interval of 8.0 ms. (a) What is the impulse given to the nail? (b) What is the average force acting on the nail?

25. (II) A tennis ball of mass $m = 0.060$ kg and speed $v = 25$ m/s strikes a wall at a 45° angle and rebounds with the same speed at 45° (Fig. 9–38). What is the impulse (magnitude and direction) given to the ball?

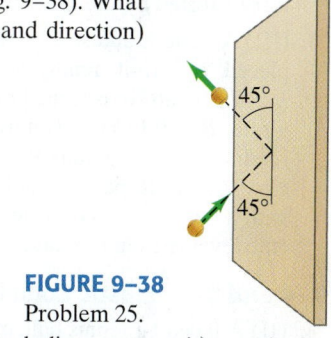

FIGURE 9–38
Problem 25.

26. (II) A 130-kg astronaut (including space suit) acquires a speed of 2.50 m/s by pushing off with his legs from a 1700-kg space capsule. (a) What is the change in speed of the space capsule? (b) If the push lasts 0.500 s, what is the average force exerted by each on the other? As the reference frame, use the position of the capsule before the push. (c) What is the kinetic energy of each after the push?

27. (II) Rain is falling at the rate of 5.0 cm/h and accumulates in a pan. If the raindrops hit at 8.0 m/s, estimate the force on the bottom of a 1.0 m² pan due to the impacting rain which does not rebound. Water has a mass of 1.00×10^3 kg per m³.

28. (II) Suppose the force acting on a tennis ball (mass 0.060 kg) points in the $+x$ direction and is given by the graph of Fig. 9–39 as a function of time. Use graphical methods to estimate (a) the total impulse given the ball, and (b) the velocity of the ball after being struck, assuming the ball is being served so it is nearly at rest initially.

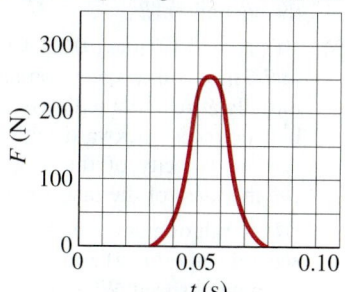

FIGURE 9–39
Problem 28.

29. (II) With what impulse does a 0.50-kg newspaper have to be thrown to give it a velocity of 3.0 m/s?

30. (II) The force on a bullet is given by the formula $F = [740 - (2.3 \times 10^5 \, \text{s}^{-1})t] \, \text{N}$ over the time interval $t = 0$ to $t = 3.0 \times 10^{-3}$ s. (a) Plot a graph of F versus t for $t = 0$ to $t = 3.0$ ms. (b) Use the graph to estimate the impulse given the bullet. (c) Determine the impulse by integration. (d) If the bullet achieves a speed of 260 m/s as a result of this impulse, given to it in the barrel of a gun, what must the bullet's mass be? (e) What is the recoil speed of the 4.5-kg gun?

31. (II) (a) A molecule of mass m and speed v strikes a wall at right angles and rebounds back with the same speed. If the collision time is Δt, what is the average force on the wall during the collision? (b) If molecules, all of this type, strike the wall at intervals a time t apart (on the average) what is the average force on the wall averaged over a long time?

32. (III) (a) Calculate the impulse experienced when a 65-kg person lands on firm ground after jumping from a height of 3.0 m. (b) Estimate the average force exerted on the person's feet by the ground if the landing is stiff-legged, and again (c) with bent legs. With stiff legs, assume the body moves 1.0 cm during impact, and when the legs are bent, about 50 cm. [Hint: The average net force on her which is related to impulse, is the vector sum of gravity and the force exerted by the ground.]

33. (III) A scale is adjusted so that when a large, shallow pan is placed on it, it reads zero. A water faucet at height $h = 2.5$ m above is turned on and water falls into the pan at a rate $R = 0.14$ kg/s. Determine (a) a formula for the scale reading as a function of time t and (b) the reading for $t = 9.0$ s. (c) Repeat (a) and (b), but replace the shallow pan with a tall, narrow cylindrical container of area $A = 20$ cm^2 (the level rises in this case).

9–4 and 9–5 Elastic Collisions

34. (II) A 0.060-kg tennis ball, moving with a speed of 4.50 m/s, has a head-on collision with a 0.090-kg ball initially moving in the same direction at a speed of 3.00 m/s. Assuming a perfectly elastic collision, determine the speed and direction of each ball after the collision.

35. (II) A 0.450-kg hockey puck, moving east with a speed of 4.80 m/s, has a head-on collision with a 0.900-kg puck initially at rest. Assuming a perfectly elastic collision, what will be the speed and direction of each object after the collision?

36. (II) A 0.280-kg croquet ball makes an elastic head-on collision with a second ball initially at rest. The second ball moves off with half the original speed of the first ball. (a) What is the mass of the second ball? (b) What fraction of the original kinetic energy ($\Delta K/K$) gets transferred to the second ball?

37. (II) A ball of mass 0.220 kg that is moving with a speed of 7.5 m/s collides head-on and elastically with another ball initially at rest. Immediately after the collision, the incoming ball bounces backward with a speed of 3.8 m/s. Calculate (a) the velocity of the target ball after the collision, and (b) the mass of the target ball.

38. (II) A ball of mass m makes a head-on elastic collision with a second ball (at rest) and rebounds with a speed equal to 0.350 its original speed. What is the mass of the second ball?

39. (II) Determine the fraction of kinetic energy lost by a neutron ($m_1 = 1.01$ u) when it collides head-on and elastically with a target particle at rest which is (a) $^1_1$H ($m = 1.01$ u); (b) $^2_1$H (heavy hydrogen, $m = 2.01$ u); (c) $^{12}_6$C ($m = 12.00$ u); (d) $^{208}_{82}$Pb (lead, $m = 208$ u).

40. (II) Show that, in general, for any head-on one-dimensional elastic collision, the speeds after collision are

$$v'_B = v_A \left(\frac{2m_A}{m_A + m_B} \right) + v_B \left(\frac{m_B - m_A}{m_A + m_B} \right)$$

and

$$v'_A = v_A \left(\frac{m_A - m_B}{m_A + m_B} \right) + v_B \left(\frac{2m_B}{m_A + m_B} \right),$$

where v_A and v_B are the initial speeds of the two objects of mass m_A and m_B.

41. (III) A 3.0-kg block slides along a frictionless tabletop at 8.0 m/s toward a second block (at rest) of mass 4.5 kg. A coil spring, which obeys Hooke's law and has spring constant $k = 850$ N/m, is attached to the second block in such a way that it will be compressed when struck by the moving block, Fig. 9–40. (a) What will be the maximum compression of the spring? (b) What will be the final velocities of the blocks after the collision? (c) Is the collision elastic? Ignore the mass of the spring.

FIGURE 9–40 Problem 41.

9–6 Inelastic Collisions

42. (I) In a ballistic pendulum experiment, projectile 1 results in a maximum height h of the pendulum equal to 2.6 cm. A second projectile (of the same mass) causes the pendulum to swing twice as high, $h_2 = 5.2$ cm. The second projectile was how many times faster than the first?

43. (II) (a) Derive a formula for the fraction of kinetic energy lost, $\Delta K/K$, in terms of m and M for the ballistic pendulum collision of Example 9–11. (b) Evaluate for $m = 16.0$ g and $M = 380$ g.

44. (II) A 28-g rifle bullet traveling 210 m/s buries itself in a 3.6-kg pendulum hanging on a 2.8-m-long string, which makes the pendulum swing upward in an arc. Determine the vertical and horizontal components of the pendulum's maximum displacement.

45. (II) An internal explosion breaks an object, initially at rest, into two pieces, one of which has 1.5 times the mass of the other. If 7500 J is released in the explosion, how much kinetic energy does each piece acquire?

46. (II) A 920-kg sports car collides into the rear end of a 2300-kg SUV stopped at a red light. The bumpers lock, the brakes are locked, and the two cars skid forward 2.8 m before stopping. The police officer, estimating the coefficient of kinetic friction between tires and road to be 0.80, calculates the speed of the sports car at impact. What was that speed?

47. (II) You drop a 12-g ball from a height of 1.5 m and it only bounces back to a height of 0.75 m. What was the total impulse on the ball when it hit the floor? (Ignore air resistance).

48. (II) Car A hits car B (initially at rest and of equal mass) from behind while going 35 m/s. Immediately after the collision, car B moves forward at 25 m/s and car A is at rest. What fraction of the initial kinetic energy is lost in the collision?

49. (II) A measure of inelasticity in a head-on collision of two objects is the *coefficient of restitution, e,* defined as

$$e = \frac{v'_A - v'_B}{v_B - v_A},$$

where $v'_A - v'_B$ is the relative velocity of the two objects after the collision and $v_B - v_A$ is their relative velocity before it. (a) Show that $e = 1$ for a perfectly elastic collision, and $e = 0$ for a completely inelastic collision. (b) A simple method for measuring the coefficient of restitution for an object colliding with a very hard surface like steel is to drop the object onto a heavy steel plate, as shown in Fig. 9–41. Determine a formula for e in terms of the original height h and the maximum height h' reached after collision.

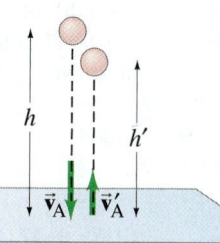

FIGURE 9–41 Problem 49. Measurement of coefficient of restitution.

50. (II) A pendulum consists of a mass M hanging at the bottom end of a massless rod of length ℓ, which has a frictionless pivot at its top end. A mass m, moving as shown in Fig. 9–42 with velocity v, impacts M and becomes embedded. What is the smallest value of v sufficient to cause the pendulum (with embedded mass m) to swing clear over the top of its arc?

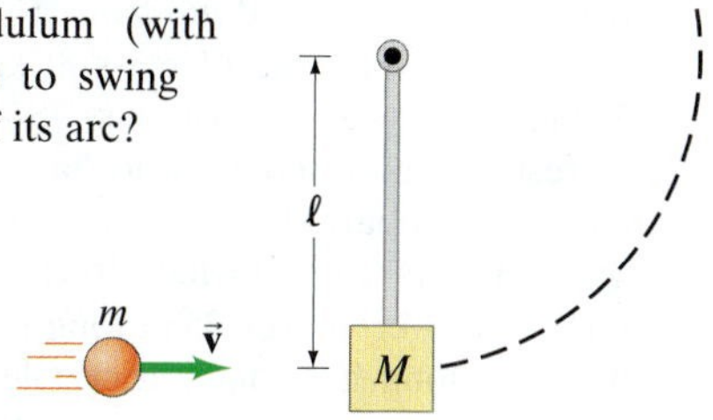

FIGURE 9–42 Problem 50.

51. (II) A bullet of mass $m = 0.0010\,\text{kg}$ embeds itself in a wooden block with mass $M = 0.999\,\text{kg}$, which then compresses a spring $(k = 120\,\text{N/m})$ by a distance $x = 0.050\,\text{m}$ before coming to rest. The coefficient of kinetic friction between the block and table is $\mu = 0.50$. (a) What is the initial speed of the bullet? (b) What fraction of the bullet's initial kinetic energy is dissipated (in damage to the wooden block, rising temperature, etc.) in the collision between the bullet and the block?

52. (II) A 144-g baseball moving $28.0\,\text{m/s}$ strikes a stationary 5.25-kg brick resting on small rollers so it moves without significant friction. After hitting the brick, the baseball bounces straight back, and the brick moves forward at $1.10\,\text{m/s}$. (a) What is the baseball's speed after the collision? (b) Find the total kinetic energy before and after the collision.

53. (II) A 6.0-kg object moving in the $+x$ direction at $5.5\,\text{m/s}$ collides head-on with an 8.0-kg object moving in the $-x$ direction at $4.0\,\text{m/s}$. Find the final velocity of each mass if: (a) the objects stick together; (b) the collision is elastic; (c) the 6.0-kg object is at rest after the collision; (d) the 8.0-kg object is at rest after the collision; (e) the 6.0-kg object has a velocity of $4.0\,\text{m/s}$ in the $-x$ direction after the collision. Are the results in (c), (d), and (e) "reasonable"? Explain.

9–7 Collisions in Two Dimensions

54. (II) Billiard ball A of mass $m_A = 0.120\,\text{kg}$ moving with speed $v_A = 2.80\,\text{m/s}$ strikes ball B, initially at rest, of mass $m_B = 0.140\,\text{kg}$. As a result of the collision, ball A is deflected off at an angle of $30.0°$ with a speed $v'_A = 2.10\,\text{m/s}$. (a) Taking the x axis to be the original direction of motion of ball A, write down the equations expressing the conservation of momentum for the components in the x and y directions separately. (b) Solve these equations for the speed, v'_B, and angle, θ'_B, of ball B. Do not assume the collision is elastic.

55. (II) A radioactive nucleus at rest decays into a second nucleus, an electron, and a neutrino. The electron and neutrino are emitted at right angles and have momenta of $9.6 \times 10^{-23}\,\text{kg·m/s}$ and $6.2 \times 10^{-23}\,\text{kg·m/s}$, respectively. Determine the magnitude and the direction of the momentum of the second (recoiling) nucleus.

56. (II) Two billiard balls of equal mass move at right angles and meet at the origin of an xy coordinate system. Initially ball A is moving along the y axis at $+2.0\,\text{m/s}$, and ball B is moving to the right along the x axis with speed $+3.7\,\text{m/s}$. After the collision (assumed elastic), the second ball is moving along the positive y axis (Fig. 9–43). What is the final direction of ball A, and what are the speeds of the two balls?

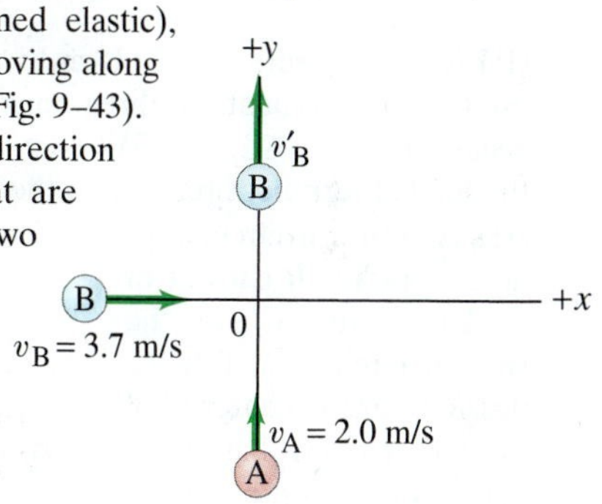

FIGURE 9–43 Problem 56. (Ball A after the collision is not shown.)

57. (II) An atomic nucleus of mass m traveling with speed v collides elastically with a target particle of mass $2m$ (initially at rest) and is scattered at $90°$. (a) At what angle does the target particle move after the collision? (b) What are the final speeds of the two particles? (c) What fraction of the initial kinetic energy is transferred to the target particle?

58. (II) A neutron collides elastically with a helium nucleus (at rest initially) whose mass is four times that of the neutron. The helium nucleus is observed to move off at an angle $\theta'_{He} = 45°$. Determine the angle of the neutron, θ'_n, and the speeds of the two particles, v'_n and v'_{He}, after the collision. The neutron's initial speed is $6.2 \times 10^5\,\text{m/s}$.

59. (III) A neon atom ($m = 20.0\,\text{u}$) makes a perfectly elastic collision with another atom at rest. After the impact, the neon atom travels away at a $55.6°$ angle from its original direction and the unknown atom travels away at a $-50.0°$ angle. What is the mass (in u) of the unknown atom? [*Hint:* You could use the law of sines.]

60. (III) For an elastic collision between a projectile particle of mass m_A and a target particle (at rest) of mass m_B, show that the scattering angle, θ'_A, of the projectile (a) can take any value, 0 to $180°$, for $m_A < m_B$, but (b) has a maximum angle ϕ given by $\cos^2 \phi = 1 - (m_B/m_A)^2$ for $m_A > m_B$.

61. (III) Prove that in the elastic collision of two objects of identical mass, with one being a target initially at rest, the angle between their final velocity vectors is always $90°$.

9–8 Center of Mass (CM)

62. (I) The CM of an empty 1250-kg car is 2.50 m behind the front of the car. How far from the front of the car will the CM be when two people sit in the front seat 2.80 m from the front of the car, and three people sit in the back seat 3.90 m from the front? Assume that each person has a mass of 70.0 kg.

63. (I) The distance between a carbon atom ($m = 12\,u$) and an oxygen atom ($m = 16\,u$) in the CO molecule is 1.13×10^{-10} m. How far from the carbon atom is the center of mass of the molecule?

64. (II) Three cubes, of side ℓ_0, $2\ell_0$, and $3\ell_0$, are placed next to one another (in contact) with their centers along a straight line as shown in Fig. 9–44. What is the position, along this line, of the CM of this system? Assume the cubes are made of the same uniform material.

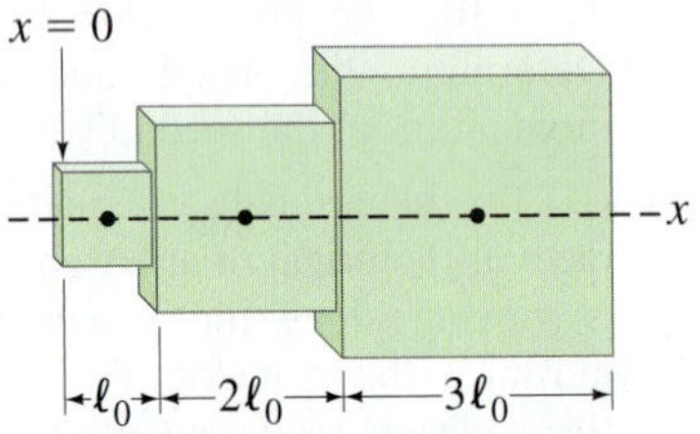

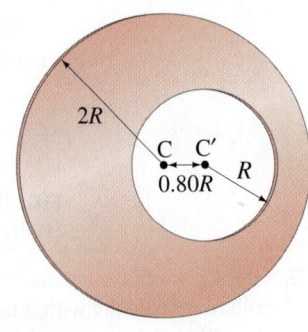

FIGURE 9–44
Problem 64.

65. (II) A square uniform raft, 18 m by 18 m, of mass 6200 kg, is used as a ferryboat. If three cars, each of mass 1350 kg, occupy the NE, SE, and SW corners, determine the CM of the loaded ferryboat relative to the center of the raft.

66. (II) A uniform circular plate of radius $2R$ has a circular hole of radius R cut out of it. The center C' of the smaller circle is a distance $0.80R$ from the center C of the larger circle, Fig. 9–45. What is the position of the center of mass of the plate? [*Hint:* Try subtraction.]

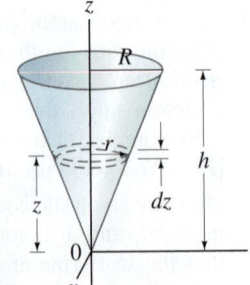

FIGURE 9–45
Problem 66.

67. (II) A uniform thin wire is bent into a semicircle of radius r. Determine the coordinates of its center of mass with respect to an origin of coordinates at the center of the "full" circle.

68. (II) Find the center of mass of the ammonia molecule. The chemical formula is NH_3. The hydrogens are at the corners of an equilateral triangle (with sides 0.16 nm) that forms the base of a pyramid, with nitrogen at the apex (0.037 nm vertically above the plane of the triangle).

69. (III) Determine the CM of a machine part that is a uniform cone of height h and radius R, Fig. 9–46. [*Hint:* Divide the cone into an infinite number of disks of thickness dz, one of which is shown.]

FIGURE 9–46
Problem 69.

70. (III) Determine the CM of a uniform pyramid that has four triangular faces and a square base with equal sides all of length s. [*Hint:* See Problem 69.]

71. (III) Determine the CM of a thin, uniform, semicircular plate.

9–9 CM and Translational Motion

72. (II) Mass $M_A = 35$ kg and mass $M_B = 25$ kg. They have velocities (in m/s) $\vec{v}_A = 12\hat{i} - 16\hat{j}$ and $\vec{v}_B = -20\hat{i} + 14\hat{j}$. Determine the velocity of the center of mass of the system.

73. (II) The masses of the Earth and Moon are 5.98×10^{24} kg and 7.35×10^{22} kg, respectively, and their centers are separated by 3.84×10^8 m. (a) Where is the CM of this system located? (b) What can you say about the motion of the Earth–Moon system about the Sun, and of the Earth and Moon separately about the Sun?

74. (II) A mallet consists of a uniform cylindrical head of mass 2.80 kg and a diameter 0.0800 m mounted on a uniform cylindrical handle of mass 0.500 kg and length 0.240 m, as shown in Fig. 9–47. If this mallet is tossed, spinning, into the air, how far above the bottom of the handle is the point that will follow a parabolic trajectory?

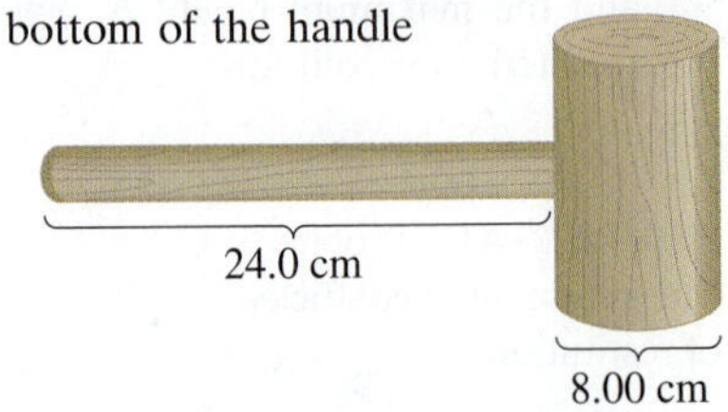

FIGURE 9–47
Problem 74.

24.0 cm

8.00 cm

75. (II) A 55-kg woman and a 72-kg man stand 10.0 m apart on frictionless ice. (a) How far from the woman is their CM? (b) If each holds one end of a rope, and the man pulls on the rope so that he moves 2.5 m, how far from the woman will he be now? (c) How far will the man have moved when he collides with the woman?

76. (II) Suppose that in Example 9–18 (Fig. 9–32), $m_{II} = 3m_I$. (a) Where then would m_{II} land? (b) What if $m_I = 3m_{II}$?

77. (II) Two people, one of mass 85 kg and the other of mass 55 kg, sit in a rowboat of mass 78 kg. With the boat initially at rest, the two people, who have been sitting at opposite ends of the boat, 3.0 m apart from each other, now exchange seats. How far and in what direction will the boat move?

78. (III) A 280-kg flatcar 25 m long is moving with a speed of 6.0 m/s along horizontal frictionless rails. A 95-kg worker starts walking from one end of the car to the other in the direction of motion, with speed 2.0 m/s with respect to the car. In the time it takes for him to reach the other end, how far has the flatcar moved?

79. (III) A huge balloon and its gondola, of mass M, are in the air and stationary with respect to the ground. A passenger, of mass m, then climbs out and slides down a rope with speed v, measured with respect to the balloon. With what speed and direction (relative to Earth) does the balloon then move? What happens if the passenger stops?

*9–10 Variable Mass

***80.** (II) A 3500-kg rocket is to be accelerated at $3.0\,g$ at take-off from the Earth. If the gases can be ejected at a rate of 27 kg/s, what must be their exhaust speed?

***81.** (II) Suppose the conveyor belt of Example 9–19 is retarded by a friction force of 150 N. Determine the required output power (hp) of the motor as a function of time from the moment gravel first starts falling ($t = 0$) until 3.0 s after the gravel begins to be dumped off the end of the 22-m-long conveyor belt.

***82.** (II) The jet engine of an airplane takes in 120 kg of air per second, which is burned with 4.2 kg of fuel per second. The burned gases leave the plane at a speed of 550 m/s (relative to the plane). If the plane is traveling 270 m/s (600 mi/h), determine: (a) the thrust due to ejected fuel; (b) the thrust due to accelerated air passing through the engine; and (c) the power (hp) delivered.

*83. (II) A rocket traveling 1850 m/s away from the Earth at an altitude of 6400 km fires its rockets, which eject gas at a speed of 1300 m/s (relative to the rocket). If the mass of the rocket at this moment is 25,000 kg and an acceleration of 1.5 m/s^2 is desired, at what rate must the gases be ejected?

*84. (III) A sled filled with sand slides without friction down a 32° slope. Sand leaks out a hole in the sled at a rate of 2.0 kg/s. If the sled starts from rest with an initial total mass of 40.0 kg, how long does it take the sled to travel 120 m along the slope?

General Problems

85. A novice pool player is faced with the corner pocket shot shown in Fig. 9–48. Relative dimensions are also shown. Should the player worry that this might be a "scratch shot," in which the cue ball will also fall into a pocket? Give details. Assume equal mass balls and an elastic collision.

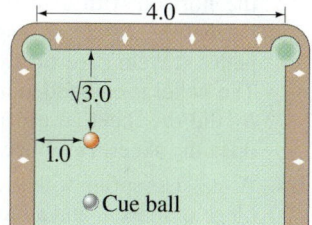

FIGURE 9–48
Problem 85.

86. During a Chicago storm, winds can whip horizontally at speeds of 120 km/h. If the air strikes a person at the rate of 45 kg/s per square meter and is brought to rest, calculate the force of the wind on a person. Assume the person is 1.60 m high and 0.50 m wide. Compare to the typical maximum force of friction ($\mu \approx 1.0$) between the person and the ground, if the person has a mass of 75 kg.

87. A ball is dropped from a height of 1.50 m and rebounds to a height of 1.20 m. Approximately how many rebounds will the ball make before losing 90% of its energy?

88. In order to convert a tough split in bowling, it is necessary to strike the pin a glancing blow as shown in Fig. 9–49. Assume that the bowling ball, initially traveling at 13.0 m/s, has five times the mass of a pin and that the pin goes off at 75° from the original direction of the ball. Calculate the speed (a) of the pin and (b) of the ball just after collision, and (c) calculate the angle through which the ball was deflected. Assume the collision is elastic and ignore any spin of the ball.

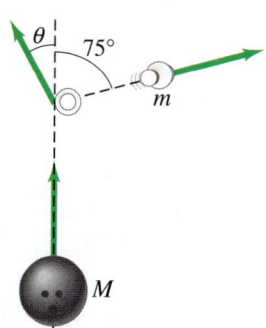

FIGURE 9–49
Problem 88.

89. A gun fires a bullet vertically into a 1.40-kg block of wood at rest on a thin horizontal sheet, Fig. 9–50. If the bullet has a mass of 24.0 g and a speed of 310 m/s, how high will the block rise into the air after the bullet becomes embedded in it?

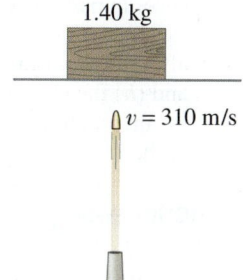

FIGURE 9–50
Problem 89.

90. A hockey puck of mass $4m$ has been rigged to explode, as part of a practical joke. Initially the puck is at rest on a frictionless ice rink. Then it bursts into three pieces. One chunk, of mass m, slides across the ice at velocity $v\hat{\mathbf{i}}$. Another chunk, of mass $2m$, slides across the ice at velocity $2v\hat{\mathbf{j}}$. Determine the velocity of the third chunk.

91. For the completely inelastic collision of two railroad cars that we considered in Example 9–3, calculate how much of the initial kinetic energy is transformed to thermal or other forms of energy.

92. A 4800-kg open railroad car coasts along with a constant speed of 8.60 m/s on a level track. Snow begins to fall vertically and fills the car at a rate of 3.80 kg/min. Ignoring friction with the tracks, what is the speed of the car after 60.0 min? (See Section 9–2.)

*93. Consider the railroad car of Problem 92, which is slowly filling with snow. (a) Determine the speed of the car as a function of time using Eqs. 9–19. (b) What is the speed of the car after 60.0 min? Does this agree with the simpler calculation (Problem 92)?

94. Two blocks of mass m_A and m_B, resting on a frictionless table, are connected by a stretched spring and then released (Fig. 9–51). (a) Is there a net external force on the system? (b) Determine the ratio of their speeds, v_A/v_B. (c) What is the ratio of their kinetic energies? (d) Describe the motion of the CM of this system. (e) How would the presence of friction alter the above results?

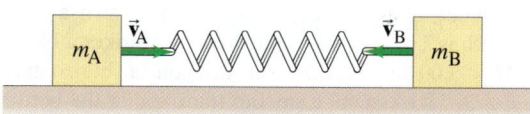

FIGURE 9–51   Problem 94.

95. You have been hired as an expert witness in a court case involving an automobile accident. The accident involved car A of mass 1500 kg which crashed into stationary car B of mass 1100 kg. The driver of car A applied his brakes 15 m before he skidded and crashed into car B. After the collision, car A slid 18 m while car B slid 30 m. The coefficient of kinetic friction between the locked wheels and the road was measured to be 0.60. Show that the driver of car A was exceeding the 55-mi/h (90 km/h) speed limit before applying the brakes.

96. A meteor whose mass was about 2.0×10^8 kg struck the Earth ($m_E = 6.0 \times 10^{24}$ kg) with a speed of about 25 km/s and came to rest in the Earth. (a) What was the Earth's recoil speed (relative to Earth at rest before the collision)? (b) What fraction of the meteor's kinetic energy was transformed to kinetic energy of the Earth? (c) By how much did the Earth's kinetic energy change as a result of this collision?

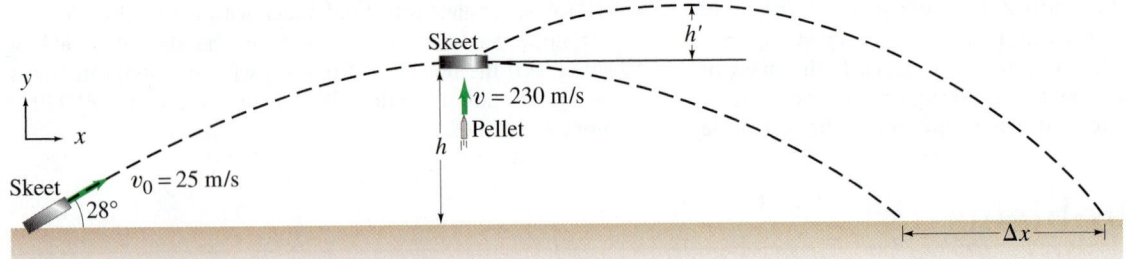

FIGURE 9–54
Problem 103.

97. Two astronauts, one of mass 65 kg and the other 85 kg, are initially at rest in outer space. They then push each other apart. How far apart are they when the lighter astronaut has moved 12 m?

98. A 22-g bullet strikes and becomes embedded in a 1.35-kg block of wood placed on a horizontal surface just in front of the gun. If the coefficient of kinetic friction between the block and the surface is 0.28, and the impact drives the block a distance of 8.5 m before it comes to rest, what was the muzzle speed of the bullet?

99. Two balls, of masses $m_A = 45$ g and $m_B = 65$ g, are suspended as shown in Fig. 9–52. The lighter ball is pulled away to a 66° angle with the vertical and released. (a) What is the velocity of the lighter ball before impact? (b) What is the velocity of each ball after the elastic collision? (c) What will be the maximum height of each ball after the elastic collision?

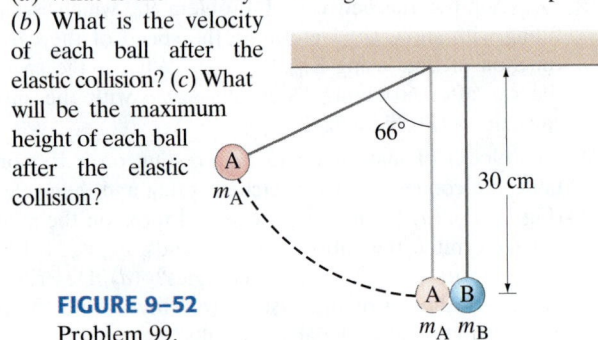

FIGURE 9–52
Problem 99.

100. A block of mass $m = 2.20$ kg slides down a 30.0° incline which is 3.60 m high. At the bottom, it strikes a block of mass $M = 7.00$ kg which is at rest on a horizontal surface, Fig. 9–53. (Assume a smooth transition at the bottom of the incline.) If the collision is elastic, and friction can be ignored, determine (a) the speeds of the two blocks after the collision, and (b) how far back up the incline the smaller mass will go.

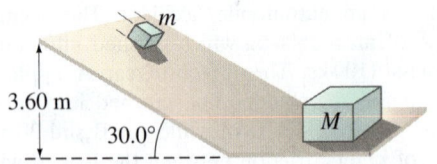

FIGURE 9–53   Problems 100 and 101.

101. In Problem 100 (Fig. 9–53), what is the upper limit on mass m if it is to rebound from M, slide up the incline, stop, slide down the incline, and collide with M again?

102. After a completely inelastic collision between two objects of equal mass, each having initial speed, v, the two move off together with speed $v/3$. What was the angle between their initial directions?

103. A 0.25-kg skeet (clay target) is fired at an angle of 28° to the horizon with a speed of 25 m/s (Fig. 9–54). When it reaches the maximum height, h, it is hit from below by a 15-g pellet traveling vertically upward at a speed of 230 m/s. The pellet is embedded in the skeet. (a) How much higher, h', did the skeet go up? (b) How much extra distance, Δx, does the skeet travel because of the collision?

104. A massless spring with spring constant k is placed between a block of mass m and a block of mass $3\,m$. Initially the blocks are at rest on a frictionless surface and they are held together so that the spring between them is compressed by an amount D from its equilibrium length. The blocks are then released and the spring pushes them off in opposite directions. Find the speeds of the two blocks when they detach from the spring.

105. *The gravitational slingshot effect.* Figure 9–55 shows the planet Saturn moving in the negative x direction at its orbital speed (with respect to the Sun) of 9.6 km/s. The mass of Saturn is 5.69×10^{26} kg. A spacecraft with mass 825 kg approaches Saturn. When far from Saturn, it moves in the $+x$ direction at 10.4 km/s. The gravitational attraction of Saturn (a conservative force) acting on the spacecraft causes it to swing around the planet (orbit shown as dashed line) and head off in the opposite direction. Estimate the final speed of the spacecraft after it is far enough away to be considered free of Saturn's gravitational pull.

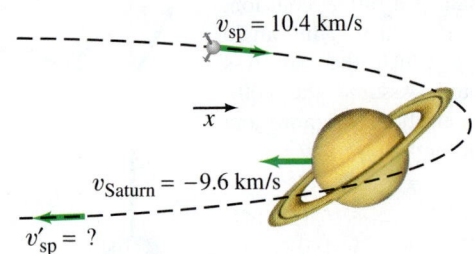

FIGURE 9–55   Problem 105.

106. Two bumper cars in an amusement park ride collide elastically as one approaches the other directly from the rear (Fig. 9–56). Car A has a mass of 450 kg and car B 490 kg, owing to differences in passenger mass. If car A approaches at 4.50 m/s and car B is moving at 3.70 m/s, calculate (a) their velocities after the collision, and (b) the change in momentum of each.

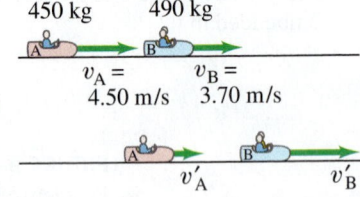

FIGURE 9–56
Problem 106:
(a) before collision,
(b) after collision.

107. In a physics lab, a cube slides down a frictionless incline as shown in Fig. 9–57 and elastically strikes another cube at the bottom that is only one-half its mass. If the incline is 35 cm high and the table is 95 cm off the floor, where does each cube land? [*Hint*: Both leave the incline moving horizontally.]

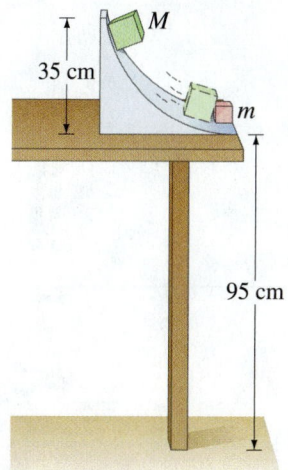

35 cm

95 cm

M

m

FIGURE 9–57
Problem 107.

108. The space shuttle launches an 850-kg satellite by ejecting it from the cargo bay. The ejection mechanism is activated and is in contact with the satellite for 4.0 s to give it a velocity of 0.30 m/s in the z-direction relative to the shuttle. The mass of the shuttle is 92,000 kg. (*a*) Determine the component of velocity v_f of the shuttle in the minus z-direction resulting from the ejection. (*b*) Find the average force that the shuttle exerts on the satellite during the ejection.

109. You are the design engineer in charge of the crashworthiness of new automobile models. Cars are tested by smashing them into fixed, massive barriers at 45 km/h. A new model of mass 1500 kg takes 0.15 s from the time of impact until it is brought to rest. (*a*) Calculate the average force exerted on the car by the barrier. (*b*) Calculate the average deceleration of the car.

110. Astronomers estimate that a 2.0-km-wide asteroid collides with the Earth once every million years. The collision could pose a threat to life on Earth. (*a*) Assume a spherical asteroid has a mass of 3200 kg for each cubic meter of volume and moves toward the Earth at 15 km/s. How much destructive energy could be released when it embeds itself in the Earth? (*b*) For comparison, a nuclear bomb could release about 4.0×10^{16} J. How many such bombs would have to explode simultaneously to release the destructive energy of the asteroid collision with the Earth?

111. An astronaut of mass 210 kg including his suit and jet pack wants to acquire a velocity of 2.0 m/s to move back toward his space shuttle. Assuming the jet pack can eject gas with a velocity of 35 m/s, what mass of gas will need to be ejected?

112. An extrasolar planet can be detected by observing the wobble it produces on the star around which it revolves. Suppose an extrasolar planet of mass m_B revolves around its star of mass m_A. If no external force acts on this simple two-object system, then its CM is stationary. Assume m_A and m_B are in circular orbits with radii r_A and r_B about the system's CM. (*a*) Show that

$$r_A = \frac{m_B}{m_A} r_B.$$

(*b*) Now consider a Sun-like star and a single planet with the same characteristics as Jupiter. That is, $m_B = 1.0 \times 10^{-3} m_A$ and the planet has an orbital radius of 8.0×10^{11} m. Determine the radius r_A of the star's orbit about the system's CM. (*c*) When viewed from Earth, the distant system appears to wobble over a distance of $2 r_A$. If astronomers are able to detect angular displacements θ of about 1 milliarcsec (1 arcsec = $\frac{1}{3600}$ of a degree), from what distance d (in light-years) can the star's wobble be detected $(1 \text{ ly} = 9.46 \times 10^{15} \text{ m})$? (*d*) The star nearest to our Sun is about 4 ly away. Assuming stars are uniformly distributed throughout our region of the Milky Way Galaxy, about how many stars can this technique be applied to in the search for extrasolar planetary systems?

113. Suppose two asteroids strike head on. Asteroid A $(m_A = 7.5 \times 10^{12} \text{ kg})$ has velocity 3.3 km/s before the collision, and asteroid B $(m_B = 1.45 \times 10^{13} \text{ kg})$ has velocity 1.4 km/s before the collision in the opposite direction. If the asteroids stick together, what is the velocity (magnitude and direction) of the new asteroid after the collision?

*Numerical/Computer

*114. (III) A particle of mass m_A traveling with speed v_A collides elastically head-on with a stationary particle of smaller mass m_B. (*a*) Show that the speed of m_B after the collision is

$$v'_B = \frac{2 v_A}{1 + m_B/m_A}.$$

(*b*) Consider now a third particle of mass m_C at rest between m_A and m_B so that m_A first collides head on with m_C and then m_C collides head on with m_B. Both collisions are elastic. Show that in this case,

$$v'_B = 4 v_A \frac{m_C m_A}{(m_C + m_A)(m_B + m_C)}.$$

(*c*) From the result of part (*b*), show that for maximum v'_B, $m_C = \sqrt{m_A m_B}$. (*d*) Assume $m_B = 2.0$ kg, $m_A = 18.0$ kg and $v_A = 2.0$ m/s. Use a spreadsheet to calculate and graph the values of v'_B from $m_C = 0.0$ kg to $m_C = 50.0$ kg in steps of 1.0 kg. For what value of m_C is the value of v'_B maximum? Does your numerical result agree with your result in part (*c*)?

Answers to Exercises

A: (*c*) because the momentum change is greater.

B: Larger (Δp is greater).

C: 0.50 m/s.

D: (*a*) 6.0 m/s; (*b*) almost zero; (*c*) almost 24.0 m/s.

E: (*b*); (*d*).

F: $x_{CM} = -2.0$ m; yes.

G: (*a*).

H: The boat moves in the opposite direction.

You too can experience rapid rotation—if your stomach can take the high angular velocity and centripetal acceleration of some of the faster amusement park rides. If not, try the slower merry-go-round or Ferris wheel. Rotating carnival rides have rotational kinetic energy as well as angular momentum. Angular acceleration is produced by a net torque, and rotating objects have rotational kinetic energy.

C H A P T E R

10

Rotational Motion

CONTENTS

10–1 Angular Quantities

10–2 Vector Nature of Angular Quantities

10–3 Constant Angular Acceleration

10–4 Torque

10–5 Rotational Dynamics; Torque and Rotational Inertia

10–6 Solving Problems in Rotational Dynamics

10–7 Determining Moments of Inertia

10–8 Rotational Kinetic Energy

10–9 Rotational Plus Translational Motion; Rolling

*10–10 Why Does a Rolling Sphere Slow Down?

CHAPTER-OPENING QUESTION—Guess Now!

A solid ball and a solid cylinder roll down a ramp. They both start from rest at the same time. Which gets to the bottom first?

(a) They get there at the same time.

(b) They get there at almost exactly the same time except for frictional differences.

(c) The ball gets there first.

(d) The cylinder gets there first.

(e) Can't tell without knowing the mass and radius of each.

U ntil now, we have been concerned mainly with translational motion. We discussed the kinematics and dynamics of translational motion (the role of force), and the energy and momentum associated with it. In this Chapter and the next we will deal with rotational motion. We will discuss the kinematics of rotational motion and then its dynamics (involving torque), as well as rotational kinetic energy and angular momentum (the rotational analog of linear momentum). Our understanding of the world around us will be increased significantly—from rotating bicycle wheels and compact discs to amusement park rides, a spinning skater, the rotating Earth, and a centrifuge—and there may be a few surprises.

We will consider mainly the rotation of rigid objects. A **rigid object** is an object with a definite shape that doesn't change, so that the particles composing it stay in fixed positions relative to one another. Any real object is capable of vibrating or deforming when a force is exerted on it. But these effects are often very small, so the concept of an ideal rigid object is very useful as a good approximation.

Our development of rotational motion will parallel our discussion of translational motion: rotational position, angular velocity, angular acceleration, rotational inertia, and the rotational analog of force, "torque."

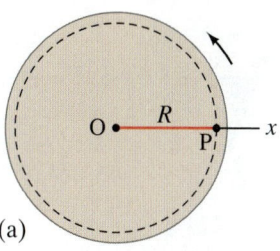

(a)

10–1 Angular Quantities

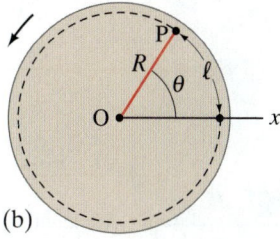

(b)

The motion of a rigid object can be analyzed as the translational motion of its center of mass plus rotational motion *about* its center of mass (Sections 9–8 and 9–9). We have already discussed translational motion in detail, so now we focus our attention on purely rotational motion. By *purely rotational motion* of an object about a fixed axis, we mean that all points in the object move in circles, such as the point P on the rotating wheel of Fig. 10–1, and that the centers of these circles all lie on a line called the **axis of rotation**. In Fig. 10–1 the axis of rotation is perpendicular to the page and passes through point O. We assume the axis is fixed in an inertial reference frame, but we will not always insist that the axis pass through the center of mass.

For a three-dimensional rigid object rotating about a fixed axis, we will use the symbol R to represent the perpendicular distance of a point or particle from the axis of rotation. We do this to distinguish R from r, which will continue to represent the position of a particle with reference to the origin (point) of some coordinate system. This distinction is illustrated in Fig. 10–2. This distinction may seem like a small point, but not being fully aware of it can cause huge errors when working with rotational motion. For a flat, very thin object, like a wheel, with the origin in the plane of the object (at the center of a wheel, for example), R and r will be nearly the same.

Every point in an object rotating about a fixed axis moves in a circle (shown dashed in Fig. 10–1 for point P) whose center is on the axis of rotation and whose radius is R, the distance of that point from the axis of rotation. A straight line drawn from the axis to any point in the object sweeps out the same angle θ in the same time interval.

To indicate the angular position of the object, or how far it has rotated, we specify the angle θ of some particular line in the object (red in Fig. 10–1) with respect to some reference line, such as the x axis in Fig. 10–1. A point in the object, such as P in Fig. 10–1b, moves through an angle θ when it travels the distance ℓ measured along the circumference of its circular path. Angles are commonly stated in degrees, but the mathematics of circular motion is much simpler if we use the *radian* for angular measure. One **radian** (abbreviated rad) is defined as the angle subtended by an arc whose length is equal to the radius. For example, in Fig. 10–1, point P is a distance R from the axis of rotation, and it has moved a distance ℓ along the arc of a circle. The arc length ℓ is said to "subtend" the angle θ. In general, any angle θ is given by

$$\theta = \frac{\ell}{R}, \qquad \text{[θ in radians]} \quad \textbf{(10–1a)}$$

where R is the radius of the circle and ℓ is the arc length subtended by the angle θ, which is specified in radians. If $\ell = R$, then $\theta = 1$ rad.

The radian, being the ratio of two lengths, is dimensionless. We thus do not have to mention it in calculations, although it is usually best to include it to remind us the angle is in radians and not degrees. We can rewrite Eq. 10–1a in terms of arc length ℓ:

$$\ell = R\theta. \qquad \textbf{(10–1b)}$$

Radians can be related to degrees in the following way. In a complete circle there are 360°, which must correspond to an arc length equal to the circumference of the circle, $\ell = 2\pi R$. Thus $\theta = \ell/R = 2\pi R/R = 2\pi$ rad in a complete circle, so

$$360° = 2\pi \text{ rad}.$$

One radian is therefore $360°/2\pi \approx 360°/6.28 \approx 57.3°$. An object that makes one complete revolution (rev) has rotated through 360°, or 2π radians:

$$1 \text{ rev} = 360° = 2\pi \text{ rad}.$$

FIGURE 10–1 Looking at a wheel that is rotating counterclockwise about an axis through the wheel's center at O (axis perpendicular to the page). Each point, such as point P, moves in a circular path; ℓ is the distance P travels as the wheel rotates through the angle θ.

FIGURE 10–2 Showing the distinction between $\vec{r}$ (the position vector) and R (the distance from the rotation axis) for a point P on the edge of a cylinder rotating about the z axis.

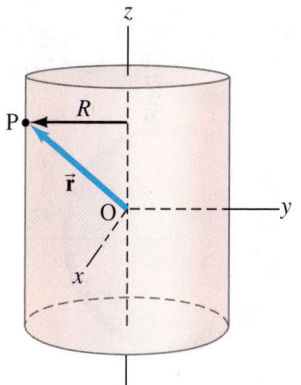

⚠ **C A U T I O N**

Use radians in calculating, not degrees

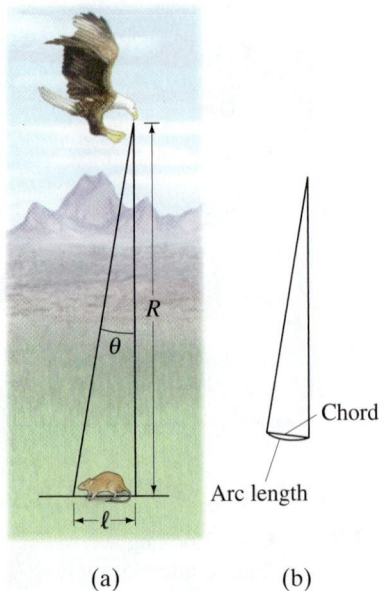

(a)   (b)

FIGURE 10–3 (a) Example 10–1.
(b) For small angles, arc length and
the chord length (straight line) are
nearly equal. For an angle as large as
15°, the error in making this estimate
is only 1%. For larger angles the
error increases rapidly.

FIGURE 10–4 A wheel rotates
from (a) initial position θ_1 to
(b) final position θ_2. The angular
displacement is $\Delta\theta = \theta_2 - \theta_1$.

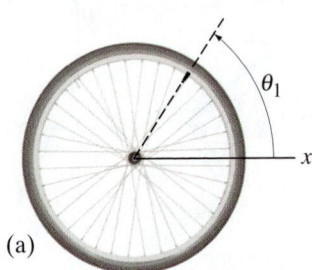

(a)

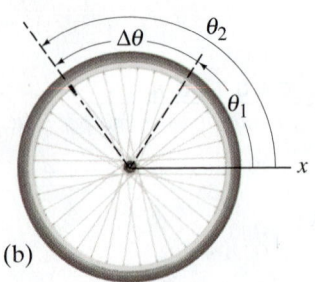

(b)

EXAMPLE 10–1 **Birds of prey—in radians.** A particular bird's eye can just distinguish objects that subtend an angle no smaller than about 3×10^{-4} rad. (a) How many degrees is this? (b) How small an object can the bird just distinguish when flying at a height of 100 m (Fig. 10–3a)?

APPROACH For (a) we use the relation $360° = 2\pi$ rad. For (b) we use Eq. 10–1b, $\ell = R\theta$, to find the arc length.

SOLUTION (a) We convert 3×10^{-4} rad to degrees:

$$(3 \times 10^{-4}\,\text{rad})\left(\frac{360°}{2\pi\,\text{rad}}\right) = 0.017°,$$

or about 0.02°.

(b) We use Eq. 10–1b, $\ell = R\theta$. For small angles, the arc length ℓ and the chord length are approximately the same (Fig. 10–3b). Since $R = 100$ m and $\theta = 3 \times 10^{-4}$ rad, we find

$$\ell = (100\,\text{m})(3 \times 10^{-4}\,\text{rad}) = 3 \times 10^{-2}\,\text{m} = 3\,\text{cm}.$$

A bird can distinguish a small mouse (about 3 cm long) from a height of 100 m. That is good eyesight.

NOTE Had the angle been given in degrees, we would first have had to convert it to radians to make this calculation. Equation 10–1 is valid *only* if the angle is specified in radians. Degrees (or revolutions) won't work.

To describe rotational motion, we make use of angular quantities, such as angular velocity and angular acceleration. These are defined in analogy to the corresponding quantities in linear motion, and are chosen to describe the rotating object as a whole, so they have the same value for each point in the rotating object. Each point in a rotating object also has translational velocity and acceleration, but they have different values for different points in the object.

When an object, such as the bicycle wheel in Fig. 10–4, rotates from some initial position, specified by θ_1, to some final position, θ_2, its **angular displacement** is

$$\Delta\theta = \theta_2 - \theta_1.$$

The *angular velocity* (denoted by ω, the Greek lowercase letter omega) is defined in analogy with linear (translational) velocity that was discussed in Chapter 2. Instead of linear displacement, we use the angular displacement. Thus the **average angular velocity** of an object rotating about a fixed axis is defined as the time rate of change of angular position:

$$\overline{\omega} = \frac{\Delta\theta}{\Delta t}, \tag{10–2a}$$

where $\Delta\theta$ is the angle through which the object has rotated in the time interval Δt. The **instantaneous angular velocity** is the limit of this ratio as Δt approaches zero:

$$\omega = \lim_{\Delta t \to 0} \frac{\Delta\theta}{\Delta t} = \frac{d\theta}{dt}. \tag{10–2b}$$

Angular velocity has units of radians per second (rad/s). Note that *all points in a rigid object rotate with the same angular velocity*, since every position in the object moves through the same angle in the same time interval.

An object such as the wheel in Fig. 10–4 can rotate about a fixed axis either clockwise or counterclockwise. The direction can be specified with a + or − sign, just as we did in Chapter 2 for linear motion toward the +x or −x direction. The usual convention is to choose the angular displacement $\Delta\theta$ and angular velocity ω as positive when the wheel rotates counterclockwise. If the rotation is clockwise, then θ would decrease, so $\Delta\theta$ and ω would be negative.

Angular acceleration (denoted by α, the Greek lowercase letter alpha), in analogy to linear acceleration, is defined as the change in angular velocity divided by the time required to make this change. The **average angular acceleration** is defined as

$$\bar{\alpha} = \frac{\omega_2 - \omega_1}{\Delta t} = \frac{\Delta \omega}{\Delta t}, \qquad (10\text{–}3a)$$

where ω_1 is the angular velocity initially, and ω_2 is the angular velocity after a time interval Δt. **Instantaneous angular acceleration** is defined as the limit of this ratio as Δt approaches zero:

$$\alpha = \lim_{\Delta t \to 0} \frac{\Delta \omega}{\Delta t} = \frac{d\omega}{dt}. \qquad (10\text{–}3b)$$

Since ω is the same for all points of a rotating object, Eq. 10–3 tells us that α also will be the same for all points. Thus, ω and α are properties of the rotating object as a whole. With ω measured in radians per second and t in seconds, α has units of radians per second squared (rad/s^2).

Each point or particle of a rotating rigid object has, at any instant, a linear velocity v and a linear acceleration a. We can relate the linear quantities at each point, v and a, to the angular quantities, ω and α, of the rotating object. Consider a point P located a distance R from the axis of rotation, as in Fig. 10–5. If the object rotates with angular velocity ω, any point will have a linear velocity whose direction is tangent to its circular path. The magnitude of that point's linear velocity is $v = d\ell/dt$. From Eq. 10–1b, a change in rotation angle $d\theta$ (in radians) is related to the linear distance traveled by $d\ell = R\,d\theta$. Hence

$$v = \frac{d\ell}{dt} = R\frac{d\theta}{dt}$$

or

$$v = R\omega, \qquad (10\text{–}4)$$

where R is a fixed distance from the rotation axis and ω is given in rad/s. Thus, although ω is the same for every point in the rotating object at any instant, the linear velocity v is greater for points farther from the axis (Fig. 10–6). Note that Eq. 10–4 is valid both instantaneously and on the average.

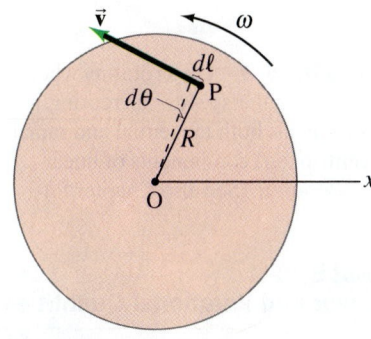

FIGURE 10–5 A point P on a rotating wheel has a linear velocity $\vec{v}$ at any moment.

FIGURE 10–6 A wheel rotating uniformly counterclockwise. Two points on the wheel, at distances R_A and R_B from the center, have the same angular velocity ω because they travel through the same angle θ in the same time interval. But the two points have different linear velocities because they travel different distances in the same time interval. Since $R_B > R_A$, then $v_B > v_A$ (because $v = R\omega$).

CONCEPTUAL EXAMPLE 10–2 | **Is the lion faster than the horse?** On a rotating carousel or merry-go-round, one child sits on a horse near the outer edge and another child sits on a lion halfway out from the center. (a) Which child has the greater linear velocity? (b) Which child has the greater angular velocity?

RESPONSE (a) The *linear* velocity is the distance traveled divided by the time interval. In one rotation the child on the outer edge travels a longer distance than the child near the center, but the time interval is the same for both. Thus the child at the outer edge, on the horse, has the greater linear velocity. (b) The *angular* velocity is the angle of rotation divided by the time interval. In one rotation both children rotate through the same angle $(360° = 2\pi \text{ rad})$. The two children have the same angular velocity.

If the angular velocity of a rotating object changes, the object as a whole—and each point in it—has an angular acceleration. Each point also has a linear acceleration whose direction is tangent to that point's circular path. We use Eq. 10–4 $(v = R\omega)$ to show that the angular acceleration α is related to the tangential linear acceleration a_{tan} of a point in the rotating object by

$$a_{\text{tan}} = \frac{dv}{dt} = R\frac{d\omega}{dt}$$

or

$$a_{\text{tan}} = R\alpha. \qquad (10\text{–}5)$$

In this equation, R is the radius of the circle in which the particle is moving, and the subscript "tan" in a_{tan} stands for "tangential."

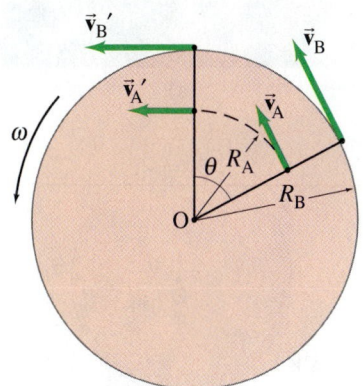

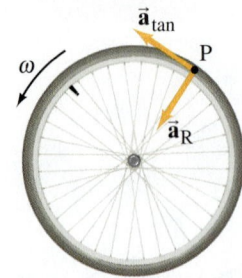

FIGURE 10–7 On a rotating wheel whose angular speed is increasing, a point P has both tangential and radial (centripetal) components of linear acceleration. (See also Chapter 5.)

TABLE 10–1
Linear and Rotational Quantities

Linear	Type	Rotational	Relation (θ in radians)
x	displacement	θ	$x = R\theta$
v	velocity	ω	$v = R\omega$
a_{tan}	acceleration	α	$a_{\text{tan}} = R\alpha$

FIGURE 10–8 Example 10–3. The total acceleration vector $\vec{\mathbf{a}} = \vec{\mathbf{a}}_{\text{tan}} + \vec{\mathbf{a}}_R$, at $t = 8.0$ s.

(a)

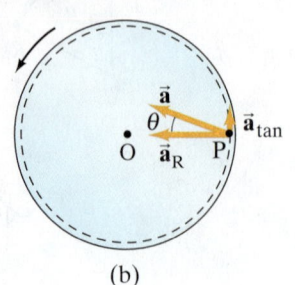

(b)

The total linear acceleration of a point at any instant is the vector sum of two components:

$$\vec{\mathbf{a}} = \vec{\mathbf{a}}_{\text{tan}} + \vec{\mathbf{a}}_R,$$

where the radial component, $\vec{\mathbf{a}}_R$, is the radial or "centripetal" acceleration and its direction is toward the center of the point's circular path; see Fig. 10–7. We saw in Chapter 5 (Eq. 5–1) that a particle moving in a circle of radius R with linear speed v has a radial acceleration $a_R = v^2/R$; we can rewrite this in terms of ω using Eq. 10–4:

$$a_R = \frac{v^2}{R} = \frac{(R\omega)^2}{R} = \omega^2 R. \tag{10–6}$$

Equation 10–6 applies to any particle of a rotating object. Thus the centripetal acceleration is greater the farther you are from the axis of rotation: the children farthest out on a carousel experience the greatest acceleration.

Table 10–1 summarizes the relationships between the angular quantities describing the rotation of an object to the linear quantities for each point of the object.

EXAMPLE 10–3 Angular and linear velocities and accelerations. A carousel is initially at rest. At $t = 0$ it is given a constant angular acceleration $\alpha = 0.060$ rad/s^2, which increases its angular velocity for 8.0 s. At $t = 8.0$ s, determine the magnitude of the following quantities: (a) the angular velocity of the carousel; (b) the linear velocity of a child (Fig. 10–8a) located 2.5 m from the center, point P in Fig. 10–8b; (c) the tangential (linear) acceleration of that child; (d) the centripetal acceleration of the child; and (e) the total linear acceleration of the child.

APPROACH The angular acceleration α is constant, so we can use $\alpha = \Delta\omega/\Delta t$ to solve for ω after a time $t = 8.0$ s. With this ω and the given α, we determine the other quantities using the relations we just developed, Eqs. 10–4, 10–5, and 10–6.

SOLUTION (a) In Eq. 10–3a, $\bar{\alpha} = (\omega_2 - \omega_1)/\Delta t$, we put $\Delta t = 8.0$ s, $\bar{\alpha} = 0.060$ rad/s^2, and $\omega_1 = 0$. Solving for ω_2, we get

$$\omega_2 = \omega_1 + \bar{\alpha}\,\Delta t = 0 + (0.060\text{ rad/s}^2)(8.0\text{ s}) = 0.48\text{ rad/s}.$$

During the 8.0-s interval, the carousel has accelerated from $\omega_1 = 0$ (rest) to $\omega_2 = 0.48$ rad/s.

(b) The linear velocity of the child, with $R = 2.5$ m at time $t = 8.0$ s, is found using Eq. 10–4:

$$v = R\omega = (2.5\text{ m})(0.48\text{ rad/s}) = 1.2\text{ m/s}.$$

Note that the "rad" has been dropped here because it is dimensionless (and only a reminder)—it is a ratio of two distances, Eq. 10–1a.

(c) The child's tangential acceleration is given by Eq. 10–5:

$$a_{\text{tan}} = R\alpha = (2.5\text{ m})(0.060\text{ rad/s}^2) = 0.15\text{ m/s}^2,$$

and it is the same throughout the 8.0-s acceleration interval.

(d) The child's centripetal acceleration at $t = 8.0$ s is given by Eq. 10–6:

$$a_R = \frac{v^2}{R} = \frac{(1.2\text{ m/s})^2}{(2.5\text{ m})} = 0.58\text{ m/s}^2.$$

(e) The two components of linear acceleration calculated in parts (c) and (d) are perpendicular to each other. Thus the total linear acceleration at $t = 8.0$ s has magnitude

$$a = \sqrt{a_{\text{tan}}^2 + a_R^2} = \sqrt{(0.15\text{ m/s}^2)^2 + (0.58\text{ m/s}^2)^2} = 0.60\text{ m/s}^2.$$

Its direction (Fig. 10–8b) is

$$\theta = \tan^{-1}\left(\frac{a_{\text{tan}}}{a_R}\right) = \tan^{-1}\left(\frac{0.15\text{ m/s}^2}{0.58\text{ m/s}^2}\right) = 0.25\text{ rad},$$

so $\theta \approx 15°$.

NOTE The linear acceleration at this chosen instant is mostly centripetal, keeping the child moving in a circle with the carousel. The tangential component that speeds up the motion is smaller.

We can relate the angular velocity ω to the frequency of rotation, f. The **frequency** is the number of complete revolutions (rev) per second, as we saw in Chapter 5. One revolution (of a wheel, say) corresponds to an angle of 2π radians, and thus $1 \text{ rev/s} = 2\pi \text{ rad/s}$. Hence, in general, the frequency f is related to the angular velocity ω by

$$f = \frac{\omega}{2\pi}$$

or

$$\omega = 2\pi f. \tag{10–7}$$

The unit for frequency, revolutions per second (rev/s), is given the special name the hertz (Hz). That is

$$1 \text{ Hz} = 1 \text{ rev/s}.$$

Note that "revolution" is not really a unit, so we can also write $1 \text{ Hz} = 1 \text{ s}^{-1}$.

The time required for one complete revolution is the **period** T, and it is related to the frequency by

$$T = \frac{1}{f}. \tag{10–8}$$

If a particle rotates at a frequency of three revolutions per second, then the period of each revolution is $\frac{1}{3}$ s.

EXERCISE A In Example 10–3, we found that the carousel, after 8.0 s, rotates at an angular velocity $\omega = 0.48 \text{ rad/s}$, and continues to do so after $t = 8.0 \text{ s}$ because the acceleration ceased. Determine the frequency and period of the carousel after it has reached a constant angular velocity.

EXAMPLE 10–4 **Hard drive.** The platter of the hard drive of a computer rotates at 7200 rpm (rpm = revolutions per minute = rev/min). (a) What is the angular velocity (rad/s) of the platter? (b) If the reading head of the drive is located 3.00 cm from the rotation axis, what is the linear speed of the point on the platter just below it? (c) If a single bit requires $0.50 \, \mu\text{m}$ of length along the direction of motion, how many bits per second can the writing head write when it is 3.00 cm from the axis?

PHYSICS APPLIED
Hard drive and bit speed

APPROACH We use the given frequency f to find the angular velocity ω of the platter and then the linear speed of a point on the platter $(v = R\omega)$. The bit rate is found by dividing the linear speed by the length of one bit $(v = \text{distance}/\text{time})$.

SOLUTION (a) First we find the frequency in rev/s, given $f = 7200 \text{ rev/min}$:

$$f = \frac{(7200 \text{ rev/min})}{(60 \text{ s/min})} = 120 \text{ rev/s} = 120 \text{ Hz}.$$

Then the angular velocity is

$$\omega = 2\pi f = 754 \text{ rad/s}.$$

(b) The linear speed of a point 3.00 cm out from the axis is given by Eq. 10–4:

$$v = R\omega = (3.00 \times 10^{-2} \text{ m})(754 \text{ rad/s}) = 22.6 \text{ m/s}.$$

(c) Each bit requires 0.50×10^{-6} m, so at a speed of 22.6 m/s, the number of bits passing the head per second is

$$\frac{22.6 \text{ m/s}}{0.50 \times 10^{-6} \text{ m/bit}} = 45 \times 10^6 \text{ bits per second},$$

or 45 megabits/s (Mbps).

EXAMPLE 10–5 **Given ω as function of time.** A disk of radius $R = 3.0$ m rotates at an angular velocity $\omega = (1.6 + 1.2\,t)$ rad/s, where t is in seconds. At the instant $t = 2.0$ s, determine (a) the angular acceleration, and (b) the speed v and the components of the acceleration a of a point on the edge of the disk.

APPROACH We use $\alpha = d\omega/dt$, $v = R\omega$, $a_{tan} = R\alpha$, and $a_R = \omega^2 R$, which are Eqs. 10–3b, 10–4, 10–5 and 10–6. We can write ω explicitly showing units of the constants (in case we want to check later): $\omega = [1.6\,\mathrm{s}^{-1} + (1.2\,\mathrm{s}^{-2})t]$ which will give us $\mathrm{s}^{-1}(= \mathrm{rad/s})$ for each term.

SOLUTION (a) The angular acceleration is

$$\alpha = \frac{d\omega}{dt} = \frac{d}{dt}(1.6 + 1.2\,t)\mathrm{s}^{-1} = 1.2\,\mathrm{rad/s^2}.$$

(b) The speed v of a point 3.0 m from the center of the rotating disk at $t = 2.0$ s is, using Eq. 10–4,

$$v = R\omega = (3.0\,\mathrm{m})(1.6 + 1.2\,t)\mathrm{s}^{-1} = (3.0\,\mathrm{m})(4.0\,\mathrm{s}^{-1}) = 12.0\,\mathrm{m/s}.$$

The components of the linear acceleration of this point at $t = 2.0$ s are

$$a_{tan} = R\alpha = (3.0\,\mathrm{m})(1.2\,\mathrm{rad/s^2}) = 3.6\,\mathrm{m/s^2}$$

$$a_R = \omega^2 R = [(1.6 + 1.2\,t)\mathrm{s}^{-1}]^2 (3.0\,\mathrm{m}) = (4.0\,\mathrm{s}^{-1})^2(3.0\,\mathrm{m}) = 48\,\mathrm{m/s^2}.$$

10–2 Vector Nature of Angular Quantities

Both $\vec{\omega}$ and $\vec{\alpha}$ can be treated as vectors, and we define their directions in the following way. Consider the rotating wheel shown in Fig. 10–9a. The linear velocities of different particles of the wheel point in all different directions. The only unique direction in space associated with the rotation is along the axis of rotation, perpendicular to the actual motion. We therefore choose the axis of rotation to be the direction of the angular velocity vector, $\vec{\omega}$. Actually, there is still an ambiguity since $\vec{\omega}$ could point in either direction along the axis of rotation (up or down in Fig. 10–9a). The convention we use, called the **right-hand rule**, is this: when the fingers of the right hand are curled around the rotation axis and point in the direction of the rotation, then the thumb points in the direction of $\vec{\omega}$. This is shown in Fig. 10–9b. Note that $\vec{\omega}$ points in the direction a right-handed screw would move when turned in the direction of rotation. Thus, if the rotation of the wheel in Fig. 10–9a is counterclockwise, the direction of $\vec{\omega}$ is upward as shown in Fig. 10–9b. If the wheel rotates clockwise, then $\vec{\omega}$ points in the opposite direction, downward.[†] Note that no part of the rotating object moves in the direction of $\vec{\omega}$.

If the axis of rotation is fixed in direction, then $\vec{\omega}$ can change only in magnitude. Thus $\vec{\alpha} = d\vec{\omega}/dt$ must also point along the axis of rotation. If the rotation is counterclockwise as in Fig. 10–9a and the magnitude of ω is increasing, then $\vec{\alpha}$ points upward; but if ω is decreasing (the wheel is slowing down), $\vec{\alpha}$ points downward. If the rotation is clockwise, $\vec{\alpha}$ points downward if ω is increasing, and $\vec{\alpha}$ points upward if ω is decreasing.

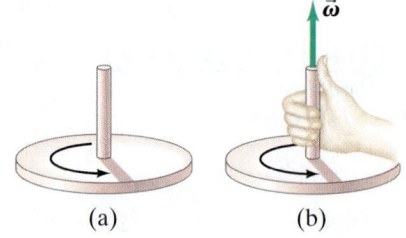

FIGURE 10–9 (a) Rotating wheel. (b) Right-hand rule for obtaining direction of $\vec{\omega}$.

FIGURE 10–10 (a) Velocity is a true vector. The reflection of $\vec{v}$ points in the same direction. (b) Angular velocity is a pseudovector since it does not follow this rule. As can be seen, the reflection of the wheel rotates in the opposite direction, so the direction of $\vec{\omega}$ is opposite for the reflection.

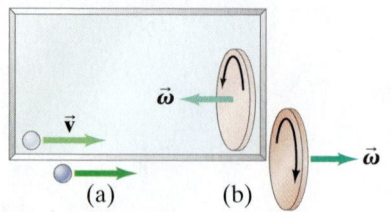

[†]Strictly speaking, $\vec{\omega}$ and $\vec{\alpha}$ are not quite vectors. The problem is that they do not behave like vectors under reflection. Suppose, as we are looking directly into a mirror, a particle moving with velocity $\vec{v}$ to the right passes in front of and parallel to the mirror. In the reflection of the mirror, $\vec{v}$ still points to the right, Fig. 10–10a. Thus a true vector, like velocity, when pointing parallel to the face of the mirror has the same direction in the reflection as in actuality. Now consider a wheel rotating in front of the mirror, so $\vec{\omega}$ points to the right. (We will be looking at the edge of the wheel.) As viewed in the mirror, Fig. 10–10b, the wheel will be rotating in the opposite direction. So $\vec{\omega}$ will point in the opposite direction (to the left) in the mirror. Because $\vec{\omega}$ is different under reflection than a true vector, $\vec{\omega}$ is called a *pseudovector* or *axial vector*. The angular acceleration $\vec{\alpha}$ is also a pseudovector, as are all cross products of true vectors (Section 11–2). The difference between true vectors and pseudovectors is important in elementary particle physics, but will not concern us in this book.

10–3 Constant Angular Acceleration

In Chapter 2, we derived the useful kinematic equations (Eqs. 2–12) that relate acceleration, velocity, distance, and time for the special case of uniform linear acceleration. Those equations were derived from the definitions of linear velocity and acceleration, assuming constant acceleration. The definitions of angular velocity and angular acceleration are the same as those for their linear counterparts, except that θ has replaced the linear displacement x, ω has replaced v, and α has replaced a. Therefore, the angular equations for **constant angular acceleration** will be analogous to Eqs. 2–12 with x replaced by θ, v by ω, and a by α, and they can be derived in exactly the same way. We summarize them here, opposite their linear equivalents (we have chosen $x_0 = 0$ and $\theta_0 = 0$ at the initial time $t = 0$):

Angular	Linear		
$\omega = \omega_0 + \alpha t$	$v = v_0 + at$	[constant α, a]	**(10–9a)**
$\theta = \omega_0 t + \frac{1}{2}\alpha t^2$	$x = v_0 t + \frac{1}{2}at^2$	[constant α, a]	**(10–9b)**
$\omega^2 = \omega_0^2 + 2\alpha\theta$	$v^2 = v_0^2 + 2ax$	[constant α, a]	**(10–9c)**
$\bar{\omega} = \dfrac{\omega + \omega_0}{2}$	$\bar{v} = \dfrac{v + v_0}{2}$	[constant α, a]	**(10–9d)**

Kinematic equations
for constant
angular acceleration
$(x_0 = 0,\ \theta_0 = 0)$

Note that ω_0 represents the angular velocity at $t = 0$, whereas θ and ω represent the angular position and velocity, respectively, at time t. Since the angular acceleration is constant, $\alpha = \bar{\alpha}$.

EXAMPLE 10–6 **Centrifuge acceleration.** A centrifuge rotor is accelerated from rest to 20,000 rpm in 30 s. (a) What is its average angular acceleration? (b) Through how many revolutions has the centrifuge rotor turned during its acceleration period, assuming constant angular acceleration?

APPROACH To determine $\bar{\alpha} = \Delta\omega/\Delta t$, we need the initial and final angular velocities. For (b), we use Eqs. 10–9 (recall that one revolution corresponds to $\theta = 2\pi$ rad).

SOLUTION (a) The initial angular velocity is $\omega = 0$. The final angular velocity is

$$\omega = 2\pi f = (2\pi\ \text{rad/rev})\frac{(20{,}000\ \text{rev/min})}{(60\ \text{s/min})} = 2100\ \text{rad/s}.$$

Then, since $\bar{\alpha} = \Delta\omega/\Delta t$ and $\Delta t = 30$ s, we have

$$\bar{\alpha} = \frac{\omega - \omega_0}{\Delta t} = \frac{2100\ \text{rad/s} - 0}{30\ \text{s}} = 70\ \text{rad/s}^2.$$

That is, every second the rotor's angular velocity increases by 70 rad/s, or by $(70/2\pi) = 11$ revolutions per second.

(b) To find θ we could use either Eq. 10–9b or 10–9c, or both to check our answer. The former gives

$$\theta = 0 + \tfrac{1}{2}(70\ \text{rad/s}^2)(30\ \text{s})^2 = 3.15 \times 10^4\ \text{rad},$$

where we have kept an extra digit because this is an intermediate result. To find the total number of revolutions, we divide by 2π rad/rev and obtain

$$\frac{3.15 \times 10^4\ \text{rad}}{2\pi\ \text{rad/rev}} = 5.0 \times 10^3\ \text{rev}.$$

NOTE Let us calculate θ using Eq. 10–9c:

$$\theta = \frac{\omega^2 - \omega_0^2}{2\alpha} = \frac{(2100\ \text{rad/s})^2 - 0}{2(70\ \text{rad/s}^2)} = 3.15 \times 10^4\ \text{rad}$$

which checks our answer using Eq. 10–9b perfectly.

10–4 Torque

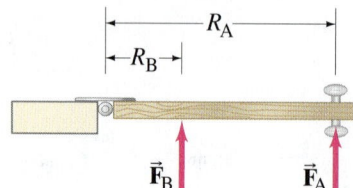

FIGURE 10–11 Top view of a door. Applying the same force with different lever arms, R_A and R_B. If $R_A = 3R_B$, then to create the same effect (angular acceleration), F_B needs to be three times F_A, or $F_A = \frac{1}{3}F_B$.

FIGURE 10–12 (a) A tire iron too can have a long lever arm. (b) A plumber can exert greater torque using a wrench with a long lever arm.

Axis of rotation   Axis of rotation

      (a)   (b)

FIGURE 10–13 (a) Forces acting at different angles at the doorknob. (b) The lever arm is defined as the perpendicular distance from the axis of rotation (the hinge) to the line of action of the force.

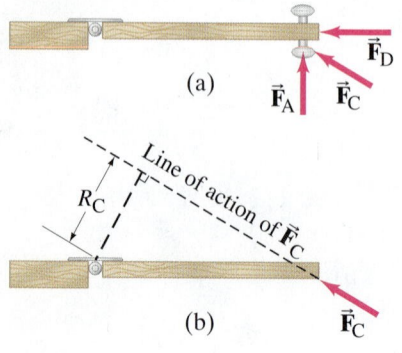

We have so far discussed rotational kinematics—the description of rotational motion in terms of angular position, angular velocity, and angular acceleration. Now we discuss the dynamics, or causes, of rotational motion. Just as we found analogies between linear and rotational motion for the description of motion, so rotational equivalents for dynamics exist as well.

To make an object start rotating about an axis clearly requires a force. But the direction of this force, and where it is applied, are also important. Take, for example, an ordinary situation such as the overhead view of the door in Fig. 10–11. If you apply a force $\vec{F}_A$ to the door as shown, you will find that the greater the magnitude, F_A, the more quickly the door opens. But now if you apply the same magnitude force at a point closer to the hinge—say, $\vec{F}_B$ in Fig. 10–11—the door will not open so quickly. The effect of the force is less: *where* the force acts, as well as its magnitude and direction, affects how quickly the door opens. Indeed, if only this one force acts, the angular acceleration of the door is proportional not only to the magnitude of the force, but is also directly proportional to *the perpendicular distance from the axis of rotation to the line along which the force acts*. This distance is called the **lever arm**, or **moment arm**, of the force, and is labeled R_A and R_B for the two forces in Fig. 10–11. Thus, if R_A in Fig. 10–11 is three times larger than R_B, then the angular acceleration of the door will be three times as great, assuming that the magnitudes of the forces are the same. To say it another way, if $R_A = 3R_B$, then F_B must be three times as large as F_A to give the same angular acceleration. (Figure 10–12 shows two examples of tools whose long lever arms are very effective.)

The angular acceleration, then, is proportional to the product of the *force times the lever arm*. This product is called the *moment of the force* about the axis, or, more commonly, it is called the **torque**, and is represented by τ (Greek lowercase letter tau). Thus, the angular acceleration α of an object is directly proportional to the net applied torque τ:

$$\alpha \propto \tau,$$

and we see that it is torque that gives rise to angular acceleration. This is the rotational analog of Newton's second law for linear motion, $a \propto F$.

We defined the lever arm as the *perpendicular* distance from the axis of rotation to the line of action of the force—that is, the distance which is perpendicular both to the axis of rotation and to an imaginary line drawn along the direction of the force. We do this to take into account the effect of forces acting at an angle. It is clear that a force applied at an angle, such as $\vec{F}_C$ in Fig. 10–13, will be less effective than the same magnitude force applied perpendicular to the door, such as $\vec{F}_A$ (Fig. 10–13a). And if you push on the end of the door so that the force is directed at the hinge (the axis of rotation), as indicated by $\vec{F}_D$, the door will not rotate at all.

The lever arm for a force such as $\vec{F}_C$ is found by drawing a line along the direction of $\vec{F}_C$ (this is the "line of action" of $\vec{F}_C$). Then we draw another line, perpendicular to this line of action, that goes to the axis of rotation and is perpendicular also to it. The length of this second line is the lever arm for $\vec{F}_C$ and is labeled R_C in Fig. 10–13b. The lever arm for $\vec{F}_A$ is the full distance from the hinge to the door knob, R_A; thus R_C is much smaller than R_A.

The magnitude of the torque associated with $\vec{F}_C$ is then $R_C F_C$. This short lever arm R_C and the corresponding smaller torque associated with $\vec{F}_C$ is consistent with the observation that $\vec{F}_C$ is less effective in accelerating the door than is $\vec{F}_A$. When the lever arm is defined in this way, experiment shows that the relation $\alpha \propto \tau$ is valid in general. Notice in Fig. 10–13 that the line of action of the force $\vec{F}_D$ passes through the hinge, and hence its lever arm is zero. Consequently, zero torque is associated with $\vec{F}_D$ and it gives rise to no angular acceleration, in accord with everyday experience.

In general, then, we can write the magnitude of the torque about a given axis as

$$\tau = R_\perp F, \qquad (10\text{--}10a)$$

where $R_\perp$ is the lever arm, and the perpendicular symbol ($\perp$) reminds us that we must use the distance from the axis of rotation that is perpendicular to the line of action of the force (Fig. 10–14a).

An equivalent way of determining the torque associated with a force is to resolve the force into components parallel and perpendicular to the line that connects the axis to the point of application of the force, as shown in Fig. 10–14b. The component $F_\parallel$ exerts no torque since it is directed at the rotation axis (its moment arm is zero). Hence the torque will be equal to $F_\perp$ times the distance R from the axis to the point of application of the force:

$$\tau = RF_\perp. \qquad (10\text{--}10b)$$

This gives the same result as Eq. 10–10a because $F_\perp = F \sin\theta$ and $R_\perp = R \sin\theta$. So

$$\tau = RF \sin\theta \qquad (10\text{--}10c)$$

in either case. [Note that θ is the angle between the directions of $\vec{F}$ and R (radial line from the axis to the point where $\vec{F}$ acts)]. We can use any of Eqs. 10–10 to calculate the torque, whichever is easiest.

Because torque is a distance times a force, it is measured in units of $m \cdot N$ in SI units,[†] $cm \cdot dyne$ in the cgs system, and $ft \cdot lb$ in the English system.

When more than one torque acts on an object, the angular acceleration α is found to be proportional to the *net* torque. If all the torques acting on an object tend to rotate it about a fixed axis of rotation in the same direction, the net torque is the sum of the torques. But if, say, one torque acts to rotate an object in one direction, and a second torque acts to rotate the object in the opposite direction (as in Fig. 10–15), the net torque is the difference of the two torques. We normally assign a positive sign to torques that act to rotate the object counterclockwise (just as θ is usually positive counterclockwise), and a negative sign to torques that act to rotate the object clockwise, when the rotation axis is fixed.

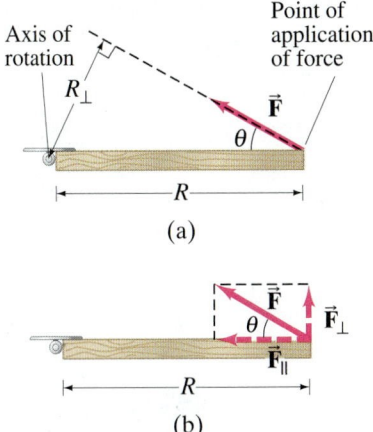

FIGURE 10–14 Torque $= R_\perp F = RF_\perp$.

EXAMPLE 10–7 **Torque on a compound wheel.** Two thin disk-shaped wheels, of radii $R_A = 30\,cm$ and $R_B = 50\,cm$, are attached to each other on an axle that passes through the center of each, as shown in Fig. 10–15. Calculate the net torque on this compound wheel due to the two forces shown, each of magnitude 50 N.

APPROACH The force $\vec{F}_A$ acts to rotate the system counterclockwise, whereas $\vec{F}_B$ acts to rotate it clockwise. So the two forces act in opposition to each other. We must choose one direction of rotation to be positive—say, counterclockwise. Then $\vec{F}_A$ exerts a positive torque, $\tau_A = R_A F_A$, since the lever arm is R_A. On the other hand, $\vec{F}_B$ produces a negative (clockwise) torque and does not act perpendicular to R_B, so we must use its perpendicular component to calculate the torque it produces: $\tau_B = -R_B F_{B\perp} = -R_B F_B \sin\theta$, where $\theta = 60°$. (Note that θ must be the angle between $\vec{F}_B$ and a radial line from the axis.)

SOLUTION The net torque is

$$\tau = R_A F_A - R_B F_B \sin 60°$$
$$= (0.30\,m)(50\,N) - (0.50\,m)(50\,N)(0.866) = -6.7\,m\cdot N.$$

This net torque acts to accelerate the rotation of the wheel in the clockwise direction.

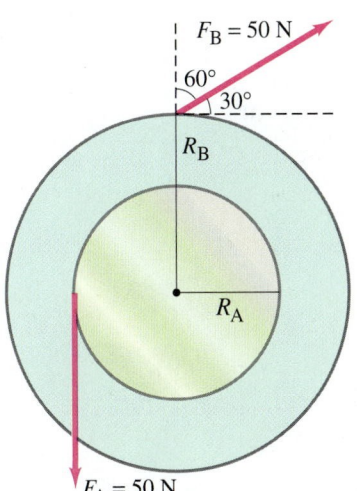

FIGURE 10–15 Example 10–7. The torque due to $\vec{F}_A$ tends to accelerate the wheel counterclockwise, whereas the torque due to $\vec{F}_B$ tends to accelerate the wheel clockwise.

EXERCISE B Two forces ($F_B = 20\,N$ and $F_A = 30\,N$) are applied to a meter stick which can rotate about its left end, Fig. 10–16. Force $\vec{F}_B$ is applied perpendicularly at the midpoint. Which force exerts the greater torque: F_A, F_B, or both the same?

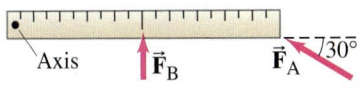

FIGURE 10–16 Exercise B.

[†]Note that the units for torque are the same as those for energy. We write the unit for torque here as $m \cdot N$ (in SI) to distinguish it from energy ($N \cdot m$) because the two quantities are very different. An obvious difference is that energy is a scalar, whereas torque has a direction and is a vector (as we will see in Chapter 11). The special name *joule* ($1\,J = 1\,N \cdot m$) is used only for energy (and for work), *never* for torque.

10–5 Rotational Dynamics; Torque and Rotational Inertia

We discussed in Section 10–4 that the angular acceleration α of a rotating object is proportional to the net torque τ applied to it:

$$\alpha \propto \Sigma\tau,$$

where we write $\Sigma\tau$ to remind us that it is the *net* torque (sum of all torques acting on the object) that is proportional to α. This corresponds to Newton's second law for translational motion, $a \propto \Sigma F$, but here torque has taken the place of force, and, correspondingly, the angular acceleration α takes the place of the linear acceleration a. In the linear case, the acceleration is not only proportional to the net force, but it is also inversely proportional to the inertia of the object, which we call its mass, m. Thus we could write $a = \Sigma F/m$. But what plays the role of mass for the rotational case? That is what we now set out to determine. At the same time, we will see that the relation $\alpha \propto \Sigma\tau$ follows directly from Newton's second law, $\Sigma F = ma$.

We first consider a very simple case: a particle of mass m rotating in a circle of radius R at the end of a string or rod whose mass we can ignore compared to m (Fig. 10–17), and we assume that a single force F acts on m tangent to the circle as shown. The torque that gives rise to the angular acceleration is $\tau = RF$. If we use Newton's second law for linear quantities, $\Sigma F = ma$, and Eq. 10–5 relating the angular acceleration to the tangential linear acceleration, $a_{\text{tan}} = R\alpha$, then we have

$$F = ma$$
$$= mR\alpha,$$

where α is given in rad/s². When we multiply both sides of this equation by R, we find that the torque $\tau = RF = R(mR\alpha)$, or

$$\tau = mR^2\alpha. \qquad \text{[single particle]} \quad \textbf{(10–11)}$$

Here at last we have a direct relation between the angular acceleration and the applied torque τ. The quantity mR^2 represents the *rotational inertia* of the particle and is called its *moment of inertia*.

Now let us consider a rotating rigid object, such as a wheel rotating about a fixed axis through its center, such as an axle. We can think of the wheel as consisting of many particles located at various distances from the axis of rotation. We can apply Eq. 10–11 to each particle of the object; that is, we write $\tau_i = m_i R_i^2 \alpha$ for the ith particle of the object. Then we sum over all the particles. The sum of the various torques is just the total torque, $\Sigma\tau$, so we obtain:

$$\Sigma\tau_i = (\Sigma m_i R_i^2)\alpha \qquad \text{[axis fixed]} \quad \textbf{(10–12)}$$

where we factored out the α since it is the same for all the particles of a rigid object. The resultant torque, $\Sigma\tau$, represents the sum of all internal torques that each particle exerts on another, plus all external torques applied from the outside: $\Sigma\tau = \Sigma\tau_{\text{ext}} + \Sigma\tau_{\text{int}}$. The sum of the internal torques is zero from Newton's third law. Hence $\Sigma\tau$ represents the resultant *external* torque.

The sum $\Sigma m_i R_i^2$ in Eq. 10–12 represents the sum of the masses of each particle in the object multiplied by the square of the distance of that particle from the axis of rotation. If we give each particle a number $(1, 2, 3, \ldots)$, then

$$\Sigma m_i R_i^2 = m_1 R_1^2 + m_2 R_2^2 + m_3 R_3^2 + \cdots.$$

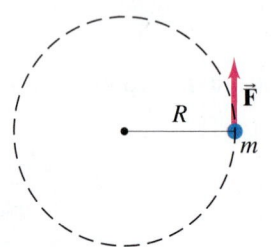

FIGURE 10–17 A mass m rotating in a circle of radius R about a fixed point.

This summation is called the **moment of inertia** (or *rotational inertia*) I of the object:

$$I = \Sigma m_i R_i^2 = m_1 R_1^2 + m_2 R_2^2 + \cdots. \qquad \textbf{(10-13)}$$

Combining Eqs. 10–12 and 10–13, we can write

$$\Sigma\tau = I\alpha. \qquad \left[\begin{array}{c}\text{axis fixed in}\\\text{inertial reference frame}\end{array}\right] \textbf{(10-14)}$$

This is the rotational equivalent of Newton's second law. It is valid for the rotation of a rigid object about a fixed axis.[†] It can be shown (see Chapter 11) that Eq. 10–14 is valid even when the object is translating with acceleration, as long as I and α are calculated about the center of mass of the object, and the rotation axis through the CM doesn't change direction. (A ball rolling down a ramp is an example.) Then

$$(\Sigma\tau)_{\text{CM}} = I_{\text{CM}}\alpha_{\text{CM}}, \qquad \left[\begin{array}{c}\text{axis fixed in direction,}\\\text{but may accelerate}\end{array}\right] \textbf{(10-15)}$$

where the subscript CM means "calculated about the center of mass."

We see that the moment of inertia, I, which is a measure of the rotational inertia of an object, plays the same role for rotational motion that mass does for translational motion. As can be seen from Eq. 10–13, the rotational inertia of an object depends not only on its mass, but also on how that mass is distributed with respect to the axis. For example, a large-diameter cylinder will have greater rotational inertia than one of equal mass but smaller diameter (and therefore greater length), Fig. 10–18. The former will be harder to start rotating, and harder to stop. When the mass is concentrated farther from the axis of rotation, the rotational inertia is greater. For rotational motion, the mass of an object *cannot* be considered as concentrated at its center of mass.

NEWTON'S SECOND LAW FOR ROTATION

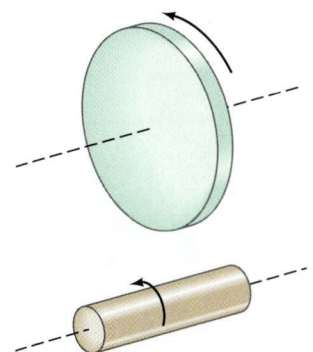

FIGURE 10–18 A large-diameter cylinder has greater rotational inertia than one of equal mass but smaller diameter.

⚠️ **CAUTION**

Mass can not be considered concentrated at CM for rotational motion

EXAMPLE 10–8 **Two weights on a bar: different axis, different I.** Two small "weights," of mass 5.0 kg and 7.0 kg, are mounted 4.0 m apart on a light rod (whose mass can be ignored), as shown in Fig. 10–19. Calculate the moment of inertia of the system (*a*) when rotated about an axis halfway between the weights, Fig. 10–19a, and (*b*) when rotated about an axis 0.50 m to the left of the 5.0-kg mass (Fig. 10–19b).

APPROACH In each case, the moment of inertia of the system is found by summing over the two parts using Eq. 10–13.

SOLUTION (*a*) Both weights are the same distance, 2.0 m, from the axis of rotation. Thus

$$\begin{aligned}I &= \Sigma mR^2 = (5.0\,\text{kg})(2.0\,\text{m})^2 + (7.0\,\text{kg})(2.0\,\text{m})^2\\&= 20\,\text{kg}\cdot\text{m}^2 + 28\,\text{kg}\cdot\text{m}^2 = 48\,\text{kg}\cdot\text{m}^2.\end{aligned}$$

(*b*) The 5.0-kg mass is now 0.50 m from the axis, and the 7.0-kg mass is 4.50 m from the axis. Then

$$\begin{aligned}I &= \Sigma mR^2 = (5.0\,\text{kg})(0.50\,\text{m})^2 + (7.0\,\text{kg})(4.5\,\text{m})^2\\&= 1.3\,\text{kg}\cdot\text{m}^2 + 142\,\text{kg}\cdot\text{m}^2 = 143\,\text{kg}\cdot\text{m}^2.\end{aligned}$$

NOTE This Example illustrates two important points. First, the moment of inertia of a given system is different for different axes of rotation. Second, we see in part (*b*) that mass close to the axis of rotation contributes little to the total moment of inertia; here, the 5.0-kg object contributed less than 1% to the total.

FIGURE 10–19 Example 10–8. Calculating the moment of inertia.

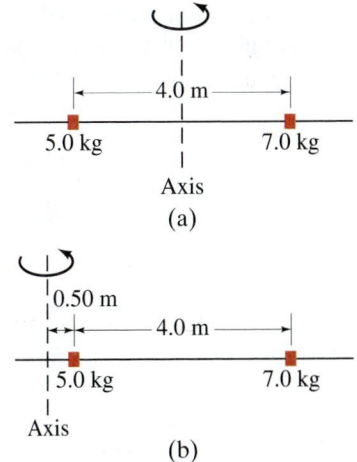

⚠️ **CAUTION**

I depends on axis of rotation and on distribution of mass

[†]That is, the axis is fixed relative to the object and is fixed in an inertial reference frame. This includes an axis moving at uniform velocity in an inertial frame, since the axis can be considered fixed in a second inertial frame that moves with respect to the first.

Object	Location of axis		Moment of inertia
(a) **Thin hoop,** radius R_0	Through center		MR_0^2
(b) **Thin hoop,** radius R_0 width w	Through central diameter		$\frac{1}{2}MR_0^2 + \frac{1}{12}Mw^2$
(c) **Solid cylinder,** radius R_0	Through center		$\frac{1}{2}MR_0^2$
(d) **Hollow cylinder,** inner radius R_1 outer radius R_2	Through center		$\frac{1}{2}M(R_1^2 + R_2^2)$
(e) **Uniform sphere,** radius r_0	Through center		$\frac{2}{5}Mr_0^2$
(f) **Long uniform rod,** length ℓ	Through center		$\frac{1}{12}M\ell^2$
(g) **Long uniform rod,** length ℓ	Through end		$\frac{1}{3}M\ell^2$
(h) **Rectangular thin plate,** length ℓ, width w	Through center		$\frac{1}{12}M(\ell^2 + w^2)$

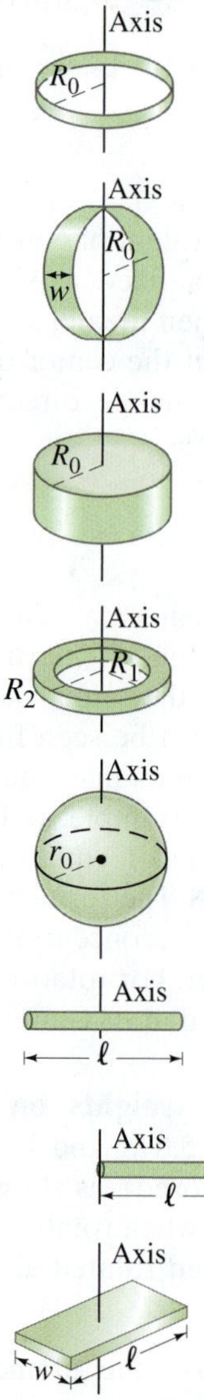

FIGURE 10–20 Moments of inertia for various objects of uniform composition. [We use R for radial distance from an axis, and r for distance from a point (only in e, the sphere), as discussed in Fig. 10–2.]

For most ordinary objects, the mass is distributed continuously, and the calculation of the moment of inertia, ΣmR^2, can be difficult. Expressions can, however, be worked out (using calculus) for the moments of inertia of regularly shaped objects in terms of the dimensions of the objects, as we will discuss in Section 10–7. Figure 10–20 gives these expressions for a number of solids rotated about the axes specified. The only one for which the result is obvious is that for the thin hoop or ring rotated about an axis passing through its center perpendicular to the plane of the hoop (Fig. 10–20a). For this hoop, all the mass is concentrated at the same distance from the axis, R_0. Thus $\Sigma mR^2 = (\Sigma m)R_0^2 = MR_0^2$, where M is the total mass of the hoop.

When calculation is difficult, I can be determined experimentally by measuring the angular acceleration α about a fixed axis due to a known net torque, $\Sigma\tau$, and applying Newton's second law, $I = \Sigma\tau/\alpha$, Eq. 10–14.

10–6 Solving Problems in Rotational Dynamics

When working with torque and angular acceleration (Eq. 10–14), it is important to use a consistent set of units, which in SI is: α in rad/s^2; τ in m·N; and the moment of inertia, I, in kg·m^2.

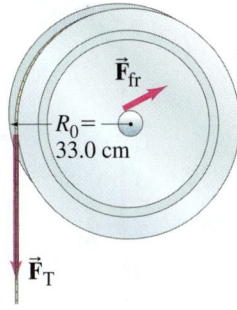

Rotational Motion

1. As always, draw a clear and complete **diagram**.
2. Choose the object or objects that will be the **system** to be studied.
3. Draw a **free-body diagram** for the object under consideration (or for each object, if more than one), showing all (and only) the forces acting on that object and exactly where they act, so you can determine the torque due to each. Gravity acts at the CG of the object (Section 9–8).
4. Identify the axis of rotation and determine the **torques** about it. Choose positive and negative directions of

rotation (counterclockwise and clockwise), and assign the correct sign to each torque.

5. Apply **Newton's second law for rotation**, $\Sigma\tau = I\alpha$. If the moment of inertia is not given, and it is not the unknown sought, you need to determine it first. Use consistent units, which in SI are: α in rad/s²; τ in m·N; and I in kg·m².
6. Also apply **Newton's second law for translation**, $\Sigma\vec{F} = m\vec{a}$, and **other** laws or principles as needed.
7. **Solve** the resulting equation(s) for the unknown(s).
8. Do a rough **estimate** to determine if your answer is reasonable.

EXAMPLE 10–9 **A heavy pulley.** A 15.0-N force (represented by $\vec{F}_T$) is applied to a cord wrapped around a pulley of mass $M = 4.00$ kg and radius $R_0 = 33.0$ cm, Fig. 10–21. The pulley accelerates uniformly from rest to an angular speed of 30.0 rad/s in 3.00 s. If there is a frictional torque $\tau_{fr} = 1.10$ m·N at the axle, determine the moment of inertia of the pulley. The pulley rotates about its center.

APPROACH We follow the steps of the Problem Solving Strategy above.

SOLUTION

1. **Draw a diagram.** The pulley and the attached cord are shown in Fig. 10–21.
2. **Choose the system:** the pulley.
3. **Draw a free-body diagram.** The cord exerts a force F_T on the pulley as shown in Fig. 10–21. The friction force retards the motion and acts all around the axle in a clockwise direction, as suggested by the arrow $\vec{F}_{fr}$ in Fig. 10–21; we are given its torque, which is all we need. Two other forces should be included in the diagram: the force of gravity mg down, and whatever force holds the axle in place. They do not contribute to the torque (their lever arms are zero) and so we omit them for convenience (or tidiness).
4. **Determine the torques.** The torque exerted by the cord equals $R_0 F_T$ and is counterclockwise, which we choose to be positive. The frictional torque is given as $\tau_{fr} = 1.10$ m·N; it opposes the motion and is negative.
5. **Apply Newton's second law for rotation.** The net torque is
$$\Sigma\tau = R_0 F_T - \tau_{fr} = (0.330\,\text{m})(15.0\,\text{N}) - 1.10\,\text{m·N} = 3.85\,\text{m·N}.$$
The angular acceleration α is found from the given data that it takes 3.0 s to accelerate the pulley from rest to $\omega = 30.0$ rad/s:
$$\alpha = \frac{\Delta\omega}{\Delta t} = \frac{30.0\,\text{rad/s} - 0}{3.00\,\text{s}} = 10.0\,\text{rad/s}^2.$$
We can now solve for I in Newton's second law (see step 7).
6. **Other calculations:** None needed.
7. **Solve for unknowns.** We solve for I in Newton's second law for rotation, $\Sigma\tau = I\alpha$, and insert our values for $\Sigma\tau$ and α:
$$I = \frac{\Sigma\tau}{\alpha} = \frac{3.85\,\text{m·N}}{10.0\,\text{rad/s}^2} = 0.385\,\text{kg·m}^2.$$
8. **Do a rough estimate.** We can do a rough estimate of the moment of inertia by assuming the pulley is a uniform cylinder and using Fig. 10–20c:
$$I \approx \tfrac{1}{2}MR_0^2 = \tfrac{1}{2}(4.00\,\text{kg})(0.330\,\text{m})^2 = 0.218\,\text{kg·m}^2.$$

This is the same order of magnitude as our result, but numerically somewhat less. This makes sense, though, because a pulley is not usually a uniform cylinder but instead has more of its mass concentrated toward the outside edge. Such a pulley would be expected to have a greater moment of inertia than a solid cylinder of equal mass. A thin hoop, Fig. 10–20a, ought to have a greater I than our pulley, and indeed it does: $I = MR_0^2 = 0.436\,\text{kg·m}^2$.

FIGURE 10–21 Example 10–9.

PROBLEM SOLVING
Usefulness and power of rough estimates

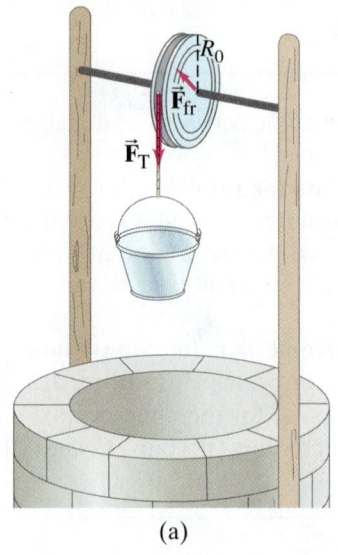

(a)

(b)

FIGURE 10–22 Example 10–10. (a) Pulley and falling bucket of mass m. (b) Free-body diagram for the bucket.

EXAMPLE 10–10 **Pulley and bucket.** Consider again the pulley in Fig. 10–21 and Example 10–9 with the same friction. But this time, instead of a constant 15.0-N force being exerted on the cord, we now have a bucket of weight $w = 15.0\,\text{N}$ (mass $m = w/g = 1.53\,\text{kg}$) hanging from the cord. See Fig. 10–22a. We assume the cord has negligible mass and does not stretch or slip on the pulley. (a) Calculate the angular acceleration α of the pulley and the linear acceleration a of the bucket. (b) Determine the angular velocity ω of the pulley and the linear velocity v of the bucket at $t = 3.00\,\text{s}$ if the pulley (and bucket) start from rest at $t = 0$.

APPROACH This situation looks a lot like Example 10–9, Fig. 10–21. But there is a big difference: the tension in the cord is now an unknown, and it is no longer equal to the weight of the bucket if the bucket accelerates. Our system has two parts: the bucket, which can undergo translational motion (Fig. 10–22b is its free-body diagram); and the pulley. The pulley does not translate, but it can rotate. We apply the rotational version of Newton's second law to the pulley, $\Sigma\tau = I\alpha$, and the linear version to the bucket, $\Sigma F = ma$.

SOLUTION (a) Let F_T be the tension in the cord. Then a force F_T acts at the edge of the pulley, and we apply Newton's second law, Eq. 10–14, for the rotation of the pulley:

$$I\alpha = \Sigma\tau = R_0 F_T - \tau_{\text{fr}}. \qquad \text{[pulley]}$$

Next we look at the (linear) motion of the bucket of mass m. Figure 10–22b, the free-body diagram for the bucket, shows that two forces act on the bucket: the force of gravity mg acts downward, and the tension of the cord F_T pulls upward. Applying Newton's second law, $\Sigma F = ma$, for the bucket, we have (taking downward as positive):

$$mg - F_T = ma. \qquad \text{[bucket]}$$

Note that the tension F_T, which is the force exerted on the edge of the pulley, is *not* equal to the weight of the bucket $(= mg = 15.0\,\text{N})$. There must be a net force on the bucket if it is accelerating, so $F_T < mg$. Indeed from the last equation above, $F_T = mg - ma$.

To obtain α, we note that the tangential acceleration of a point on the edge of the pulley is the same as the acceleration of the bucket if the cord doesn't stretch or slip. Hence we can use Eq. 10–5, $a_{\text{tan}} = a = R_0\alpha$. Substituting $F_T = mg - ma = mg - mR_0\alpha$ into the first equation above (Newton's second law for rotation of the pulley), we obtain

$$I\alpha = \Sigma\tau = R_0 F_T - \tau_{\text{fr}} = R_0(mg - mR_0\alpha) - \tau_{\text{fr}} = mgR_0 - mR_0^2\alpha - \tau_{\text{fr}}.$$

The variable α appears on the left and in the second term on the right, so we bring that term to the left side and solve for α:

$$\alpha = \frac{mgR_0 - \tau_{\text{fr}}}{I + mR_0^2}.$$

The numerator $(mgR_0 - \tau_{\text{fr}})$ is the net torque, and the denominator $(I + mR_0^2)$ is the total rotational inertia of the system. Then, since $I = 0.385\,\text{kg}\cdot\text{m}^2$, $m = 1.53\,\text{kg}$, and $\tau_{\text{fr}} = 1.10\,\text{m}\cdot\text{N}$ (from Example 10–9),

$$\alpha = \frac{(15.0\,\text{N})(0.330\,\text{m}) - 1.10\,\text{m}\cdot\text{N}}{0.385\,\text{kg}\cdot\text{m}^2 + (1.53\,\text{kg})(0.330\,\text{m})^2} = 6.98\,\text{rad/s}^2.$$

The angular acceleration is somewhat less in this case than the $10.0\,\text{rad/s}^2$ of Example 10–9. Why? Because $F_T\,(= mg - ma)$ is less than the 15.0-N weight of the bucket, mg. The linear acceleration of the bucket is

$$a = R_0\alpha = (0.330\,\text{m})(6.98\,\text{rad/s}^2) = 2.30\,\text{m/s}^2.$$

NOTE The tension in the cord F_T is less than mg because the bucket accelerates.

(b) Since the angular acceleration is constant, after 3.00 s

$$\omega = \omega_0 + \alpha t = 0 + (6.98 \text{ rad/s}^2)(3.00 \text{ s}) = 20.9 \text{ rad/s}.$$

The velocity of the bucket is the same as that of a point on the wheel's edge:

$$v = R_0 \omega = (0.330 \text{ m})(20.9 \text{ rad/s}) = 6.91 \text{ m/s}.$$

The same result can also be obtained by using the linear equation $v = v_0 + at = 0 + (2.30 \text{ m/s}^2)(3.00 \text{ s}) = 6.90 \text{ m/s}$. (The difference is due to rounding off.)

EXAMPLE 10–11 **Rotating rod.** A uniform rod of mass M and length ℓ can pivot freely (i.e., we ignore friction) about a hinge or pin attached to the case of a large machine, as in Fig. 10–23. The rod is held horizontally and then released. At the moment of release (when you are no longer exerting a force holding it up), determine (a) the angular acceleration of the rod and (b) the linear acceleration of the tip of the rod. Assume the force of gravity acts at the center of mass of the rod, as shown.

APPROACH (a) The only torque on the rod about the hinge is that due to gravity, which acts with a force $F = Mg$ downward with a lever arm $\ell/2$ at the moment of release (the CM is at the center of a uniform rod). There is also a force on the rod at the hinge, but with the hinge as axis of rotation, the lever arm of this force is zero. The moment of inertia of a uniform rod pivoted about its end is (Fig. 10–20g) $I = \frac{1}{3} M\ell^2$. In part (b) we use $a_{\tan} = R\alpha$.

SOLUTION We use Eq. 10–14, solving for α to obtain the initial angular acceleration of the rod:

$$\alpha = \frac{\tau}{I} = \frac{Mg\dfrac{\ell}{2}}{\frac{1}{3}M\ell^2} = \frac{3}{2}\frac{g}{\ell}.$$

As the rod descends, the force of gravity on it is constant but the torque due to this force is not constant since the lever arm changes. Hence the rod's angular acceleration is not constant.

(b) The linear acceleration of the tip of the rod is found from the relation $a_{\tan} = R\alpha$ (Eq. 10–5) with $R = \ell$:

$$a_{\tan} = \ell\alpha = \frac{3}{2}g.$$

NOTE The tip of the rod falls with an acceleration greater than g! A small object balanced on the tip of the rod would be left behind when the rod is released. In contrast, the CM of the rod, at a distance $\ell/2$ from the pivot, has acceleration $a_{\tan} = (\ell/2)\alpha = \frac{3}{4}g$.

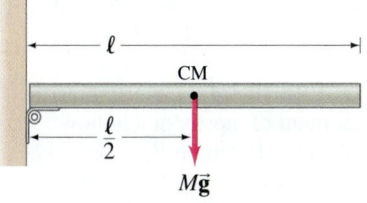

FIGURE 10–23 Example 10–11.

10–7 Determining Moments of Inertia

By Experiment

The moment of inertia of any object about any axis can be determined experimentally, such as by measuring the net torque $\Sigma\tau$ required to give the object an angular acceleration α. Then, from Eq. 10–14, $I = \Sigma\tau/\alpha$. See Example 10–9.

Using Calculus

For simple systems of masses or particles, the moment of inertia can be calculated directly, as in Example 10–8. Many objects can be considered as a continuous distribution of mass. In this case, Eq. 10–13 defining moment of inertia becomes

$$I = \int R^2 \, dm, \tag{10-16}$$

where dm represents the mass of any infinitesimal particle of the object and R is the perpendicular distance of this particle from the axis of rotation. The integral is taken over the whole object. This is easily done only for objects of simple geometric shape.

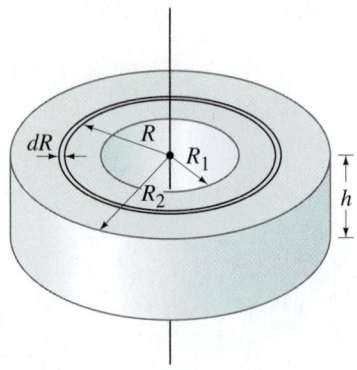

FIGURE 10–24 Determining the moment of inertia of a hollow cylinder (Example 10–12).

EXAMPLE 10–12 **Cylinder, solid or hollow.** (*a*) Show that the moment of inertia of a uniform hollow cylinder of inner radius R_1, outer radius R_2, and mass M, is $I = \frac{1}{2}M(R_1^2 + R_2^2)$, as stated in Fig. 10–20d, if the rotation axis is through the center along the axis of symmetry. (*b*) Obtain the moment of inertia for a solid cylinder.

APPROACH We know that the moment of inertia of a thin ring of radius R is mR^2. So we divide the cylinder into thin concentric cylindrical rings or hoops of thickness dR, one of which is indicated in Fig. 10–24. If the density (mass per unit volume) is ρ, then

$$dm = \rho \, dV,$$

where dV is the volume of the thin ring of radius R, thickness dR, and height h. Since $dV = (2\pi R)(dR)(h)$, we have

$$dm = 2\pi \rho h R \, dR.$$

SOLUTION (*a*) The moment of inertia is obtained by integrating (summing) over all these rings:

$$I = \int R^2 \, dm = \int_{R_1}^{R_2} 2\pi \rho h R^3 \, dR = 2\pi \rho h \left[\frac{R_2^4 - R_1^4}{4} \right] = \frac{\pi \rho h}{2}(R_2^4 - R_1^4),$$

where we are given that the cylinder has uniform density, $\rho = $ constant. (If this were not so, we would have to know ρ as a function of R before the integration could be carried out.) The volume V of this hollow cylinder is $V = (\pi R_2^2 - \pi R_1^2)h$, so its mass M is

$$M = \rho V = \rho \pi (R_2^2 - R_1^2)h.$$

Since $(R_2^4 - R_1^4) = (R_2^2 - R_1^2)(R_2^2 + R_1^2)$, we have

$$I = \frac{\pi \rho h}{2}(R_2^2 - R_1^2)(R_2^2 + R_1^2) = \frac{1}{2}M(R_1^2 + R_2^2),$$

as stated in Fig. 10–20d.

(*b*) For a solid cylinder, $R_1 = 0$ and if we set $R_2 = R_0$, then

$$I = \frac{1}{2}MR_0^2,$$

which is that given in Fig. 10–20c for a solid cylinder of mass M and radius R_0.

The Parallel-Axis Theorem

There are two simple theorems that are helpful in obtaining moments of inertia. The first is called the **parallel-axis theorem**. It relates the moment of inertia I of an object of total mass M about any axis, and its moment of inertia I_{CM} about an axis passing through the center of mass and parallel to the first axis. If the two axes are a distance h apart, then

$$I = I_{CM} + Mh^2. \qquad \text{[parallel axis]} \quad \textbf{(10–17)}$$

Thus, for example, if the moment of inertia about an axis through the CM is known, the moment of inertia about any axis parallel to this axis is easily obtained.

FIGURE 10–25 Example 10–13.

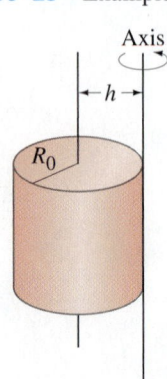

EXAMPLE 10–13 **Parallel axis.** Determine the moment of inertia of a solid cylinder of radius R_0 and mass M about an axis tangent to its edge and parallel to its symmetry axis, Fig. 10–25.

APPROACH We use the parallel-axis theorem with $I_{CM} = \frac{1}{2}MR_0^2$ (Fig. 10–20c).

SOLUTION Since $h = R_0$, Eq. 10–17 gives

$$I = I_{CM} + Mh^2 = \frac{3}{2}MR_0^2.$$

EXERCISE C In Figs. 10–20f and g, the moments of inertia for a thin rod about two different axes are given. Are they related by the parallel-axis theorem? Please show how.

*Proof of the Parallel-Axis Theorem

The proof of the parallel-axis theorem is as follows. We choose our coordinate system so the origin is at the CM, and I_{CM} is the moment of inertia about the z axis. Figure 10–26 shows a cross section of an object of arbitrary shape in the xy plane. We let I represent the moment of inertia of the object about an axis parallel to the z axis that passes through the point A in Fig. 10–26 where the point A has coordinates x_A and y_A. Let x_i, y_i, and m_i represent the coordinates and mass of an arbitrary particle of the object. The square of the distance from this point to A is $[(x_i - x_A)^2 + (y_i - y_A)^2]$. So the moment of inertia, I, about the axis through A is

$$I = \Sigma m_i[(x_i - x_A)^2 + (y_i - y_A)^2]$$
$$= \Sigma m_i(x_i^2 + y_i^2) - 2x_A \Sigma m_i x_i - 2y_A \Sigma m_i y_i + (\Sigma m_i)(x_A^2 + y_A^2).$$

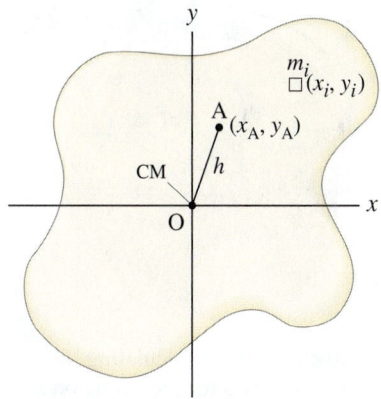

FIGURE 10–26 Derivation of the parallel-axis theorem.

The first term on the right is just $I_{CM} = \Sigma m_i(x_i^2 + y_i^2)$ since the CM is at the origin. The second and third terms are zero since, by definition of the CM, $\Sigma m_i x_i = \Sigma m_i y_i = 0$ because $x_{CM} = y_{CM} = 0$. The last term is Mh^2 since $\Sigma m_i = M$ and $(x_A^2 + y_A^2) = h^2$ where h is the distance of A from the CM. Thus we have proved $I = I_{CM} + Mh^2$, which is Eq. 10–17.

*The Perpendicular-Axis Theorem

The parallel-axis theorem can be applied to any object. The second theorem, the **perpendicular-axis theorem**, can be applied only to plane (flat) objects—that is, to two-dimensional objects, or objects of uniform thickness whose thickness can be neglected compared to the other dimensions. This theorem states that the sum of the moments of inertia of a plane object about any two perpendicular axes in the plane of the object, is equal to the moment of inertia about an axis through their point of intersection perpendicular to the plane of the object. That is, if the object is in the xy plane (Fig. 10–27),

$$I_z = I_x + I_y. \qquad \text{[object in } xy \text{ plane]} \quad \textbf{(10–18)}$$

FIGURE 10–27 The perpendicular-axis theorem.

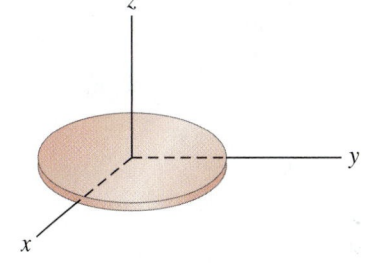

Here I_z, I_x, I_y are moments of inertia about the z, x, and y axes. The proof is simple: since $I_x = \Sigma m_i y_i^2$, $I_y = \Sigma m_i x_i^2$, and $I_z = \Sigma m_i(x_i^2 + y_i^2)$, Eq. 10–18 follows directly.

10–8 Rotational Kinetic Energy

The quantity $\frac{1}{2}mv^2$ is the kinetic energy of an object undergoing translational motion. An object rotating about an axis is said to have **rotational kinetic energy**. By analogy with translational kinetic energy, we would expect this to be given by the expression $\frac{1}{2}I\omega^2$ where I is the moment of inertia of the object and ω is its angular velocity. We can indeed show that this is true.

Consider any rigid rotating object as made up of many tiny particles, each of mass m_i. If we let R_i represent the distance of any one particle from the axis of rotation, then its linear velocity is $v_i = R_i\omega$. The total kinetic energy of the whole object will be the sum of the kinetic energies of all its particles:

$$K = \Sigma(\tfrac{1}{2}m_i v_i^2) = \Sigma(\tfrac{1}{2}m_i R_i^2 \omega^2)$$
$$= \tfrac{1}{2}\Sigma(m_i R_i^2)\omega^2.$$

We have factored out the $\frac{1}{2}$ and the ω^2 since they are the same for every particle of a rigid object. Since $\Sigma m_i R_i^2 = I$, the moment of inertia, we see that the kinetic energy, K, of an object rotating about a fixed axis is, as expected,

$$K = \tfrac{1}{2}I\omega^2. \qquad \text{[rotation about a fixed axis]} \quad \textbf{(10–19)}$$

If the axis is not fixed in space, the rotational kinetic energy can take on a more complicated form.

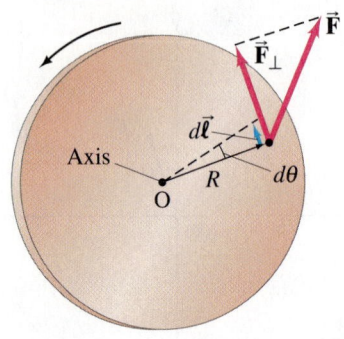

FIGURE 10–28 Calculating the work done by a torque acting on a rigid object rotating about a fixed axis.

The work done on an object rotating about a fixed axis can be written in terms of angular quantities. Suppose a force $\vec{F}$ is exerted at a point whose distance from the axis of rotation is R, as in Fig. 10–28. The work done by this force is

$$W = \int \vec{F} \cdot d\vec{\ell} = \int F_\perp R \, d\theta,$$

where $d\vec{\ell}$ is an infinitesimal distance perpendicular to R with magnitude $d\ell = R \, d\theta$, and $F_\perp$ is the component of $\vec{F}$ perpendicular to R and parallel to $d\vec{\ell}$ (Fig. 10–28). But $F_\perp R$ is the torque about the axis, so

$$W = \int_{\theta_1}^{\theta_2} \tau \, d\theta \tag{10–20}$$

is the work done by a torque τ to rotate an object through the angle $\theta_2 - \theta_1$. The rate of work done, or power P, at any instant is

$$P = \frac{dW}{dt} = \tau \frac{d\theta}{dt} = \tau\omega. \tag{10–21}$$

The work-energy principle holds for rotation of a rigid object about a fixed axis. From Eq. 10–14 we have

$$\tau = I\alpha = I\frac{d\omega}{dt} = I\frac{d\omega}{d\theta}\frac{d\theta}{dt} = I\omega\frac{d\omega}{d\theta},$$

where we used the chain rule and $\omega = d\theta/dt$. Then $\tau \, d\theta = I\omega \, d\omega$ and

$$W = \int_{\theta_1}^{\theta_2} \tau \, d\theta = \int_{\omega_1}^{\omega_2} I\omega \, d\omega = \tfrac{1}{2}I\omega_2^2 - \tfrac{1}{2}I\omega_1^2. \tag{10–22}$$

This is the work-energy principle for a rigid object rotating about a fixed axis. It states that the work done in rotating an object through an angle $\theta_2 - \theta_1$ is equal to the change in rotational kinetic energy of the object.

PHYSICS APPLIED

Energy from a flywheel

EXAMPLE 10–14 ESTIMATE Flywheel. Flywheels, which are simply large rotating disks, have been suggested as a means of storing energy for solar-powered generating systems. Estimate the kinetic energy that can be stored in an 80,000-kg (80-ton) flywheel with a diameter of 10 m (a three-story building). Assume it could hold together (without flying apart due to internal stresses) at 100 rpm.

APPROACH We use Eq. 10–19, $K = \tfrac{1}{2}I\omega^2$, but only after changing 100 rpm to ω in rad/s.

SOLUTION We are given

$$\omega = 100 \text{ rpm} = \left(100 \frac{\text{rev}}{\text{min}}\right)\left(\frac{1 \text{ min}}{60 \text{ sec}}\right)\left(\frac{2\pi \text{ rad}}{\text{rev}}\right) = 10.5 \text{ rad/s}.$$

The kinetic energy stored in the disk (for which $I = \tfrac{1}{2}MR_0^2$) is

$$K = \tfrac{1}{2}I\omega^2 = \tfrac{1}{2}\left(\tfrac{1}{2}MR_0^2\right)\omega^2$$
$$= \tfrac{1}{4}(8.0 \times 10^4 \text{ kg})(5 \text{ m})^2(10.5 \text{ rad/s})^2 = 5.5 \times 10^7 \text{ J}.$$

NOTE In terms of kilowatt-hours $[1 \text{ kWh} = (1000 \text{ J/s})(3600 \text{ s/h})(1 \text{ h}) = 3.6 \times 10^6 \text{ J}]$, this energy is only about 15 kWh, which is not a lot of energy (one 3-kW oven would use it all in 5 h). Thus flywheels seem unlikely for this application.

EXAMPLE 10–15 **Rotating rod.** A rod of mass M is pivoted on a frictionless hinge at one end, as shown in Fig. 10–29. The rod is held at rest horizontally and then released. Determine the angular velocity of the rod when it reaches the vertical position, and the speed of the rod's tip at this moment.

APPROACH We can use the work-energy principle here. The work done is due to gravity, and is equal to the change in gravitational potential energy of the rod.

SOLUTION Since the CM of the rod drops a vertical distance $\ell/2$, the work done by gravity is

$$W = Mg\frac{\ell}{2}.$$

The initial kinetic energy is zero. Hence, from the work-energy principle,

$$\tfrac{1}{2}I\omega^2 = Mg\frac{\ell}{2}.$$

Since $I = \tfrac{1}{3}M\ell^2$ for a rod pivoted about its end (Fig. 10–20g), we can solve for ω:

$$\omega = \sqrt{\frac{3g}{\ell}}.$$

The tip of the rod will have a linear speed (see Eq. 10–4)

$$v = \ell\omega = \sqrt{3g\ell}.$$

NOTE By comparison, an object that falls vertically a height ℓ has a speed $v = \sqrt{2g\ell}$.

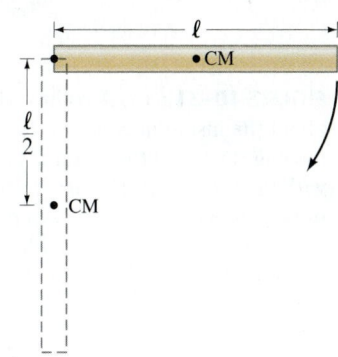

FIGURE 10–29 Example 10–15.

EXERCISE D Estimate the energy stored in the rotational motion of a hurricane. Model the hurricane as a uniform cylinder 300 km in diameter and 5 km high, made of air whose mass is 1.3 kg per m³. Estimate the outer edge of the hurricane to move at a speed of 200 km/h.

10–9 Rotational Plus Translational Motion; Rolling

Rolling Without Slipping

The rolling motion of a ball or wheel is familiar in everyday life: a ball rolling across the floor, or the wheels and tires of a car or bicycle rolling along the pavement. Rolling *without slipping* depends on static friction between the rolling object and the ground. The friction is static because the rolling object's point of contact with the ground is at rest at each moment.

Rolling without slipping involves both rotation and translation. There is a simple relation between the linear speed v of the axle and the angular velocity ω of the rotating wheel or sphere: namely, $v = R\omega$, where R is the radius, as we now show. Figure 10–30a shows a wheel rolling to the right without slipping. At the instant shown, point P on the wheel is in contact with the ground and is momentarily at rest. The velocity of the axle at the wheel's center C is $\vec{v}$. In Fig. 10–30b we have put ourselves in the reference frame of the wheel—that is, we are moving to the right with velocity $\vec{v}$ relative to the ground. In this reference frame the axle C is at rest, whereas the ground and point P are moving to the left with velocity $-\vec{v}$ as shown. Here we are seeing pure rotation. We can then use Eq. 10–4 to obtain $v = R\omega$, where R is the radius of the wheel. This is the same v as in Fig. 10–30a, so we see that the linear speed v of the axle relative to the ground is related to the angular velocity ω by

$$v = R\omega. \qquad \text{[rolling without slipping]}$$

This is valid only if there is no slipping.

FIGURE 10–30 (a) A wheel rolling to the right. Its center C moves with velocity $\vec{v}$. Point P is at rest at this instant. (b) The same wheel as seen from a reference frame in which the axle of the wheel C is at rest—that is, we are moving to the right with velocity $\vec{v}$ relative to the ground. Point P, which was at rest in (a), here in (b) is moving to the left with velocity $-\vec{v}$ as shown. (See also Section 3–9 on relative velocity.)

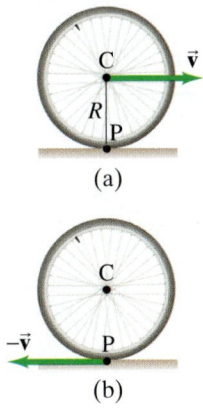

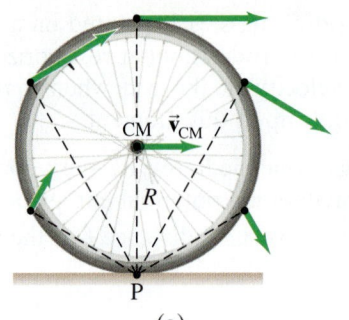

FIGURE 10–31 (a) A rolling wheel rotates about the instantaneous axis (perpendicular to the page) passing through the point of contact with the ground, P. The arrows represent the instantaneous velocity of each point. (b) Photograph of a rolling wheel. The spokes are more blurred where the speed is greater.

(a)   (b)

Instantaneous Axis

When a wheel rolls without slipping, the point of contact of the wheel with the ground is instantaneously at rest. It is sometimes useful to think of the motion of the wheel as pure rotation about this "instantaneous axis" passing through that point P (Fig. 10–31a). Points close to the ground have a small linear speed, as they are close to this instantaneous axis, whereas points farther away have a greater linear speed. This can be seen in a photograph of a real rolling wheel (Fig. 10–31b): spokes near the top of the wheel appear more blurry because they are moving faster than those near the bottom of the wheel.

Total Kinetic Energy = $K_{CM} + K_{rot}$

An object that rotates while its center of mass (CM) undergoes translational motion will have both translational and rotational kinetic energy. Equation 10–19, $K = \frac{1}{2}I\omega^2$, gives the rotational kinetic energy if the rotation axis is fixed. If the object is moving, such as a wheel rolling along the ground, Fig. 10–32, this equation is still valid as long as the rotation axis is fixed in direction. To obtain the total kinetic energy, we note that the rolling wheel undergoes pure rotation about its instantaneous point of contact P, Fig. 10–31. As we saw above relative to the discussion of Fig. 10–30, the speed v of the CM relative to the ground equals the speed of a point on the edge of the wheel relative to its center. Both of these speeds are related to the radius R by $v = \omega R$. Thus the angular velocity ω about point P is the same ω for the wheel about its center, and the total kinetic energy is

$$K_{tot} = \frac{1}{2}I_P\omega^2,$$

where I_P is the rolling object's moment of inertia about the instantaneous axis at P. We can write K_{tot} in terms of the center of mass using the parallel-axis theorem: $I_P = I_{CM} + MR^2$, where we have substituted $h = R$ in Eq. 10–17. Thus

$$K_{tot} = \frac{1}{2}I_{CM}\omega^2 + \frac{1}{2}MR^2\omega^2.$$

But $R\omega = v_{CM}$, the speed of the center of mass. So the total kinetic energy of a rolling object is

$$K_{tot} = \frac{1}{2}I_{CM}\omega^2 + \frac{1}{2}Mv_{CM}^2, \tag{10–23}$$

where v_{CM} is the linear velocity of the CM, I_{CM} is the moment of inertia about an axis through the CM, ω is the angular velocity about this axis, and M is the total mass of the object.

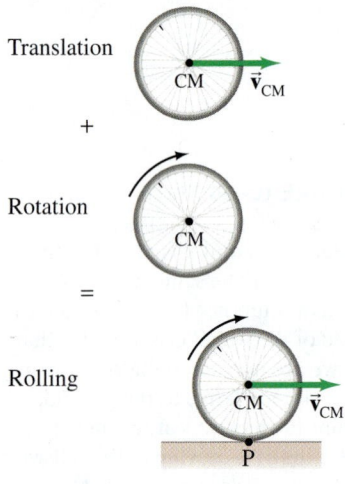

FIGURE 10–32 A wheel rolling without slipping can be considered as translation of the wheel as a whole with velocity $\vec{v}_{CM}$ plus rotation about the CM.

Translation

+

Rotation

=

Rolling

EXAMPLE 10–16 **Sphere rolling down an incline.** What will be the speed of a solid sphere of mass M and radius r_0 when it reaches the bottom of an incline if it starts from rest at a vertical height H and rolls without slipping? See Fig. 10–33. (Assume no slipping occurs because of static friction, which does no work.) Compare your result to that for an object *sliding* down a frictionless incline.

APPROACH We use the law of conservation of energy with gravitational potential energy, now including rotational as well as translational kinetic energy.

SOLUTION The total energy at any point a vertical distance y above the base of the incline is

$$\tfrac{1}{2}Mv^2 + \tfrac{1}{2}I_{CM}\omega^2 + Mgy,$$

where v is the speed of the center of mass, and Mgy is the gravitational potential energy. Applying conservation of energy, we equate the total energy at the top ($y = H$, $v = 0$, $\omega = 0$) to the total energy at the bottom ($y = 0$):

$$0 + 0 + MgH = \tfrac{1}{2}Mv^2 + \tfrac{1}{2}I_{CM}\omega^2 + 0.$$

The moment of inertia of a solid sphere about an axis through its center of mass is $I_{CM} = \tfrac{2}{5}Mr_0^2$, Fig. 10–20e. Since the sphere rolls without slipping, we have $\omega = v/r_0$ (recall Fig. 10–30). Hence

$$MgH = \tfrac{1}{2}Mv^2 + \tfrac{1}{2}\left(\tfrac{2}{5}Mr_0^2\right)\left(\frac{v^2}{r_0^2}\right).$$

Canceling the M's and r_0's we obtain

$$\left(\tfrac{1}{2} + \tfrac{1}{5}\right)v^2 = gH$$

or

$$v = \sqrt{\tfrac{10}{7}gH}.$$

We can compare this result for the speed of a rolling sphere to that for an object sliding down a plane without rotating and without friction, $\tfrac{1}{2}mv^2 = mgH$ (see our energy equation above, removing the rotational term). For the sliding object, $v = \sqrt{2gH}$, which is greater than for a rolling sphere. An object sliding without friction or rotation transforms its initial potential energy entirely into translational kinetic energy (none into rotational kinetic energy), so the speed of its center of mass is greater.

NOTE Our result for the rolling sphere shows (perhaps surprisingly) that v is independent of both the mass M and the radius r_0 of the sphere.

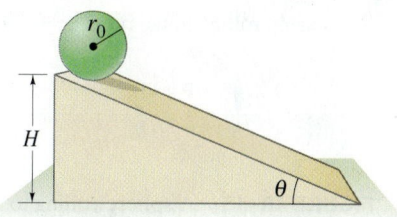

FIGURE 10–33 A sphere rolling down a hill has both translational and rotational kinetic energy. Example 10–16.

⚠ **CAUTION**

Rolling objects go slower than sliding objects because of rotational kinetic energy

CONCEPTUAL EXAMPLE 10–17 | **Which is fastest?** Several objects roll without slipping down an incline of vertical height H, all starting from rest at the same moment. The objects are a thin hoop (or a plain wedding band), a spherical marble, a solid cylinder (a D-cell battery), and an empty soup can. In what order do they reach the bottom of the incline? Compare also to a greased box that slides down an incline at the same angle, ignoring sliding friction.

RESPONSE We use conservation of energy with gravitational potential energy plus rotational and translational kinetic energy. The sliding box would be fastest because the potential energy loss (MgH) is transformed completely into translational kinetic energy of the box, whereas for rolling objects the initial potential energy is shared between translational and rotational kinetic energies, and so the speed of the CM is less. For each of the rolling objects we can state that the loss in potential energy equals the increase in kinetic energy:

$$MgH = \tfrac{1}{2}Mv^2 + \tfrac{1}{2}I_{CM}\omega^2.$$

For all our rolling objects, the moment of inertia I_{CM} is a numerical factor times the mass M and the radius R^2 (Fig. 10–20). The mass M is in each term, so the translational speed v_{CM} doesn't depend on M; nor does it depend on the radius R since $\omega = v/R$, so R^2 cancels out for all the rolling objects. Thus the speed v at the bottom depends only on that numerical factor in I_{CM} which expresses how the mass is distributed. The hoop, with all its mass concentrated at radius R ($I_{CM} = MR^2$), has the largest moment of inertia; hence it will have the lowest v_{CM} and will arrive at the bottom behind the D-cell ($I_{CM} = \tfrac{1}{2}MR^2$), which in turn will be behind the marble ($I_{CM} = \tfrac{2}{5}MR^2$). The empty can, which is mainly a hoop plus a small disk, has most of its mass concentrated at R; so it will be a bit faster than the pure hoop but slower than the D-cell. See Fig. 10–34.

NOTE The objects do not have to have the same radius: the speed at the bottom does not depend on the object's mass M or radius R, but only on the shape (and the height of the hill H).

FIGURE 10–34 Example 10–17.

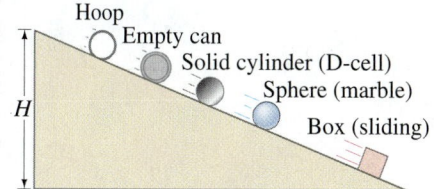

Hoop
Empty can
Solid cylinder (D-cell)
Sphere (marble)
Box (sliding)
H

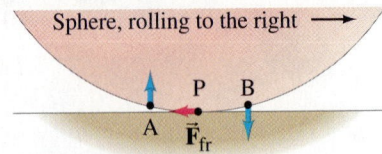

Sphere, rolling to the right →

FIGURE 10–35 A sphere rolling to the right on a plane surface. The point in contact with the ground at any moment, point P, is momentarily at rest. Point A to the left of P is moving nearly vertically upward at the instant shown, and point B to the right is moving nearly vertically downward. An instant later, point B will touch the plane and be at rest momentarily. Thus no work is done by the force of static friction.

⚠ **CAUTION**
When is $\Sigma\tau = I\alpha$ valid?

FIGURE 10–36 Example 10–18.

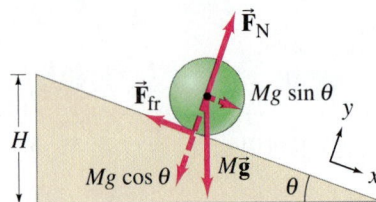

If there had been little or no static friction between the rolling objects and the plane in these Examples, the round objects would have slid rather than rolled, or a combination of both. Static friction must be present to make a round object roll. We did not need to take friction into account in the energy equation for the rolling objects because it is *static* friction and does no work—the point of contact of a sphere at each instant does not slide, but moves perpendicular to the plane (first down and then up as shown in Fig. 10–35) as it rolls. Thus, no work is done by the static friction force because the force and the motion (displacement) are perpendicular. The reason the rolling objects in Examples 10–16 and 10–17 move down the slope more slowly than if they were sliding is *not* because friction slows them down. Rather, it is because some of the gravitational potential energy is converted to rotational kinetic energy, leaving less for the translational kinetic energy.

EXERCISE E Return to the Chapter-Opening Question, p. 248, and answer it again now. Try to explain why you may have answered differently the first time.

Using $\Sigma\tau_{CM} = I_{CM}\alpha_{CM}$

We can examine objects rolling down a plane not only from the point of view of kinetic energy, as we did in Examples 10–16 and 10–17, but also in terms of forces and torques. If we calculate torques about an axis fixed in direction (even if the axis is accelerating) which passes through the center of mass of the rolling sphere, then

$$\Sigma\tau_{CM} = I_{CM}\alpha_{CM}$$

is valid, as we discussed in Section 10–5. See Eq. 10–15, whose validity we will show in Chapter 11. Be careful, however: Do not assume $\Sigma\tau = I\alpha$ is always valid. You cannot just calculate τ, I, and α about any axis unless the axis is (1) fixed in an inertial reference frame or (2) fixed in direction but passes through the CM of the object.

EXAMPLE 10–18 **Analysis of a sphere on an incline using forces.** Analyze the rolling sphere of Example 10–16, Fig. 10–33, in terms of forces and torques. In particular, find the velocity v and the magnitude of the friction force, F_{fr}, Fig. 10–36.

APPROACH We analyze the motion as translation of the CM plus rotation about the CM. F_{fr} is due to static friction and we cannot assume $F_{fr} = \mu_s F_N$, only $F_{fr} \leq \mu_s F_N$.

SOLUTION For translation in the x direction we have from $\Sigma F = ma$,

$$Mg\sin\theta - F_{fr} = Ma,$$

and in the y direction

$$F_N - Mg\cos\theta = 0$$

since there is no acceleration perpendicular to the plane. This last equation merely tells us the magnitude of the normal force,

$$F_N = Mg\cos\theta.$$

For the rotational motion about the CM, we use Newton's second law for rotation $\Sigma\tau_{CM} = I_{CM}\alpha_{CM}$ (Eq. 10–15), calculating about an axis passing through the CM but fixed in direction:

$$F_{fr}r_0 = \left(\tfrac{2}{5}Mr_0^2\right)\alpha.$$

The other forces, $\vec{F}_N$ and $M\vec{g}$, point through the axis of rotation (CM), so have lever arms equal to zero and do not appear here. As we saw in Example 10–16 and Fig. 10–30, $\omega = v/r_0$ where v is the speed of the CM. Taking derivatives of $\omega = v/r_0$ with respect to time we have $\alpha = a/r_0$; substituting into the last equation we find

$$F_{fr} = \tfrac{2}{5}Ma.$$

When we substitute this into the top equation, we get

$$Mg\sin\theta - \tfrac{2}{5}Ma = Ma,$$

or

$$a = \tfrac{5}{7}g\sin\theta.$$

We thus see that the acceleration of the CM of a rolling sphere is less than that for an object sliding without friction ($a = g\sin\theta$). The sphere started from rest at the top of the incline (height H). To find the speed v at the bottom we use Eq. 2–12c where the total distance traveled along the plane is $x = H/\sin\theta$ (see Fig. 10–36). Thus

$$v = \sqrt{2ax} = \sqrt{2\left(\tfrac{5}{7}g\sin\theta\right)\left(\frac{H}{\sin\theta}\right)} = \sqrt{\frac{10}{7}gH}.$$

This is the same result obtained in Example 10–16 although less effort was needed there. To get the magnitude of the force of friction, we use the equations obtained above:

$$F_{\text{fr}} = \tfrac{2}{5}Ma = \tfrac{2}{5}M\left(\tfrac{5}{7}g\sin\theta\right) = \tfrac{2}{7}Mg\sin\theta.$$

NOTE If the coefficient of static friction is sufficiently small, or θ sufficiently large so that $F_{\text{fr}} > \mu_s F_N$ (that is, if[†] $\tan\theta > \tfrac{7}{2}\mu_s$), the sphere will not simply roll but will slip as it moves down the plane.

*More Advanced Examples

Here we do three more Examples, all of them fun and interesting. When they use $\Sigma\tau = I\alpha$, we must remember that this equation is valid only if τ, α, and I are calculated about an axis that either (1) is fixed in an inertial reference frame, or (2) passes through the CM of the object and remains fixed in direction.

EXAMPLE 10–19 **A falling yo-yo.** String is wrapped around a uniform solid cylinder (something like a yo-yo) of mass M and radius R, and the cylinder starts falling from rest, Fig. 10–37a. As the cylinder falls, find (a) its acceleration and (b) the tension in the string.

APPROACH As always we begin with a free-body diagram, Fig. 10–37b, which shows the weight of the cylinder acting at the CM and the tension of the string $\vec{F}_T$ acting at the edge of the cylinder. We write Newton's second law for the linear motion (down is positive)

$$Ma = \Sigma F$$
$$= Mg - F_T.$$

Since we do not know the tension in the string, we cannot immediately solve for a. So we try Newton's second law for the rotational motion, calculated about the center of mass:

$$\Sigma\tau_{\text{CM}} = I_{\text{CM}}\alpha_{\text{CM}}$$
$$F_T R = \tfrac{1}{2}MR^2\alpha.$$

Because the cylinder "rolls without slipping" down the string, we have the additional relation that $a = \alpha R$ (Eq. 10–5).

SOLUTION The torque equation becomes

$$F_T R = \tfrac{1}{2}MR^2\left(\frac{a}{R}\right) = \tfrac{1}{2}MRa$$

so

$$F_T = \tfrac{1}{2}Ma.$$

Substituting this into the force equation, we obtain

$$Ma = Mg - F_T$$
$$= Mg - \tfrac{1}{2}Ma.$$

Solving for a, we find that $a = \tfrac{2}{3}g$. That is, the linear acceleration is less than what it would be if the cylinder were simply dropped. This makes sense since gravity is not the only vertical force acting; the tension in the string is acting as well. (b) Since $a = \tfrac{2}{3}g$, $F_T = \tfrac{1}{2}Ma = \tfrac{1}{3}Mg$.

EXERCISE F Find the acceleration a of a yo-yo whose spindle has radius $\tfrac{1}{2}R$. Assume the moment of inertia is still $\tfrac{1}{2}MR^2$ (ignore the mass of the spindle).

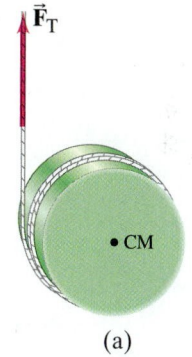

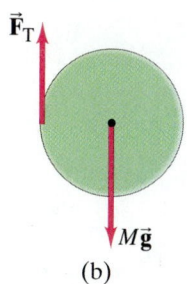

FIGURE 10–37 Example 10–19.

[†]$F_{\text{fr}} > \mu_s F_N$ is equivalent to $\tan\theta > \tfrac{7}{2}\mu_s$ because $F_{\text{fr}} = \tfrac{2}{7}Mg\sin\theta$ and $\mu_s F_N = \mu_s Mg\cos\theta$.

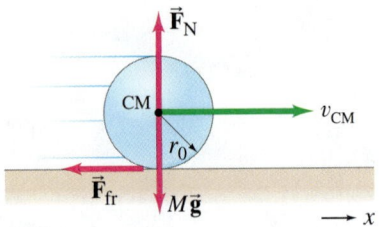

FIGURE 10–38 Example 10–20.

EXAMPLE 10–20 **What if a rolling ball slips?** A bowling ball of mass M and radius r_0 is thrown along a level surface so that initially $(t = 0)$ it slides with a linear speed v_0 but does not rotate. As it slides, it begins to spin, and eventually rolls without slipping. How long does it take to begin rolling without slipping?

APPROACH The free-body diagram is shown in Fig. 10–38, with the ball moving to the right. The friction force does two things: it acts to slow down the translational motion of the CM; and it immediately acts to start the ball rotating clockwise.

SOLUTION Newton's second law for translation gives

$$Ma_x = \Sigma F_x = -F_{fr} = -\mu_k F_N = -\mu_k Mg,$$

where μ_k is the coefficient of kinetic friction because the ball is sliding. Thus $a_x = -\mu_k g$. The velocity of the CM is

$$v_{CM} = v_0 + a_x t = v_0 - \mu_k gt.$$

Next we apply Newton's second law for rotation about the CM, $I_{CM}\alpha_{CM} = \Sigma\tau_{CM}$:

$$\tfrac{2}{5}Mr_0^2\alpha_{CM} = F_{fr}r_0$$
$$= \mu_k Mgr_0.$$

The angular acceleration is thus $\alpha_{CM} = 5\mu_k g/2r_0$, which is constant. Then the angular velocity of the ball is (Eq. 10–9a)

$$\omega_{CM} = \omega_0 + \alpha_{CM}t = 0 + \frac{5\mu_k gt}{2r_0}.$$

The ball starts rolling immediately after it touches the ground, but it rolls and slips at the same time to begin with. It eventually stops slipping, and then rolls without slipping. The condition for rolling without slipping is that

$$v_{CM} = \omega_{CM}r_0,$$

which is Eq. 10–4, and is *not* valid if there is slipping. This rolling without slipping begins at a time $t = t_1$ given by $v_{CM} = \omega_{CM}r_0$ and we apply the equations for v_{CM} and ω_{CM} above:

$$v_0 - \mu_k gt_1 = \frac{5\mu_k gt_1}{2r_0}r_0$$

so

$$t_1 = \frac{2v_0}{7\mu_k g}.$$

Braking distribution of a car

FIGURE 10–39 Forces on a braking car (Example 10–21).

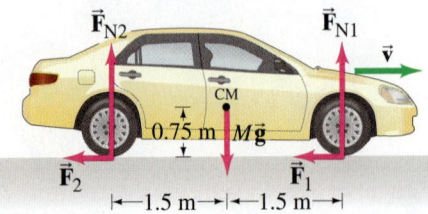

EXAMPLE 10–21 **ESTIMATE** **Braking a car.** When the brakes of a car are applied, the front of the car dips down a bit; and the force on the front tires is greater than on the rear tires. To see why, estimate the magnitude of the normal forces, F_{N1} and F_{N2}, on the front and rear tires of the car shown in Fig. 10–39 when the car brakes and decelerates at a rate $a = 0.50\,g$. The car has mass $M = 1200\,\text{kg}$, the distance between the front and rear axles is 3.0 m, and its CM (where the force of gravity acts) is midway between the axles 75 cm above the ground.

APPROACH Figure 10–39 is the free-body diagram showing all the forces on the car. F_1 and F_2 are the frictional forces that decelerate the car. We let F_1 be the sum of the forces on both front tires, and F_2 likewise for the two rear tires. F_{N1} and F_{N2} are the normal forces the road exerts on the tires and, for our estimate, we assume the static friction force acts the same for all the tires, so that F_1 and F_2 are proportional respectively to F_{N1} and F_{N2}:

$$F_1 = \mu F_{N1} \quad \text{and} \quad F_2 = \mu F_{N2}.$$

SOLUTION The friction forces F_1 and F_2 decelerate the car, so Newton's second law gives

$$F_1 + F_2 = Ma$$
$$= (1200\,\text{kg})(0.50)(9.8\,\text{m/s}^2) = 5900\,\text{N}. \qquad \textbf{(i)}$$

While the car is braking its motion is only translational, so the net torque on the car is zero. If we calculate the torques about the CM as axis, the forces F_1, F_2, and F_{N2} all act to rotate the car clockwise, and only F_{N1} acts to rotate it counterclockwise; so F_{N1} must balance the other three. Hence, F_{N1} must be significantly greater than F_{N2}. Mathematically, we have for the torques calculated about the CM:

$$(1.5\,\text{m})F_{N1} - (1.5\,\text{m})F_{N2} - (0.75\,\text{m})F_1 - (0.75\,\text{m})F_2 = 0.$$

Since F_1 and F_2 are proportional[†] to F_{N1} and F_{N2} $(F_1 = \mu F_{N1},\ F_2 = \mu F_{N2})$, we can write this as

$$(1.5\,\text{m})(F_{N1} - F_{N2}) - (0.75\,\text{m})(\mu)(F_{N1} + F_{N2}) = 0. \qquad \textbf{(ii)}$$

Also, since the car does not accelerate vertically, we have

$$Mg = F_{N1} + F_{N2} = \frac{F_1 + F_2}{\mu}. \qquad \textbf{(iii)}$$

Comparing (iii) to (i), we see that $\mu = a/g = 0.50$. Now we solve (ii) for F_{N1} and use $\mu = 0.50$ to obtain

$$F_{N1} = F_{N2}\left(\frac{2 + \mu}{2 - \mu}\right) = \frac{5}{3}F_{N2}.$$

Thus F_{N1} is $1\frac{2}{3}$ times greater than F_{N2}. Actual magnitudes are determined from (iii) and (i): $F_{N1} + F_{N2} = (5900\,\text{N})/(0.50) = 11{,}800\,\text{N}$ which equals $F_{N2}(1 + \frac{5}{3})$; so $F_{N2} = 4400\,\text{N}$ and $F_{N1} = 7400\,\text{N}$.

NOTE Because the force on the front tires is generally greater than on the rear tires, cars are often designed with larger brake pads on the front wheels than on the rear. Or, to say it another way, if the brake pads are equal, the front ones wear out a lot faster.

*10–10 Why Does a Rolling Sphere Slow Down?

A sphere of mass M and radius r_0 rolling on a horizontal flat surface eventually comes to rest. What force causes it to come to rest? You might think it is friction, but when you examine the problem from a simple straightforward point of view, a paradox seems to arise.

Suppose a sphere is rolling to the right as shown in Fig. 10–40, and is slowing down. By Newton's second law, $\Sigma\vec{F} = M\vec{a}$, there must be a force $\vec{F}$ (presumably frictional) acting to the left as shown, so that the acceleration $\vec{a}$ will also point to the left and v will decrease. Curiously enough, though, if we now look at the torque equation (calculated about the center of mass), $\Sigma\tau_{CM} = I_{CM}\alpha$, we see that the force $\vec{F}$ acts to increase the angular acceleration α, and thus to *increase* the velocity of the sphere. Thus the paradox. The force $\vec{F}$ acts to decelerate the sphere if we look at the translational motion, but speeds it up if we look at the rotational motion.

FIGURE 10–40 Sphere rolling to the right.

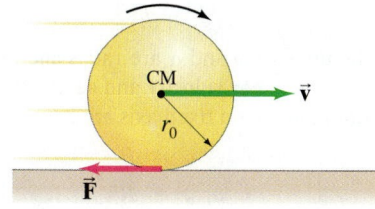

[†]Our proportionality constant μ is not equal to μ_s, the static coefficient of friction $(F_{fr} \le \mu_s F_N)$, unless the car is just about to skid.

The resolution of this apparent paradox is that some other force must be acting. The only other forces acting are gravity, $M\vec{g}$, and the normal force $\vec{F}_N (= -M\vec{g})$. These act vertically and hence do not affect the horizontal translational motion. If we assume the sphere and plane are rigid, so the sphere is in contact at only one point, these forces give rise to no torques about the CM either, since they act through the CM.

The only recourse we have to resolve the paradox is to give up our idealization that the objects are rigid. In fact, all objects are deformable to some extent. Our sphere flattens slightly and the level surface also acquires a slight depression where the two are in contact. There is an *area* of contact, not a point. Hence there can be a torque at this area of contact which acts in the opposite direction to the torque associated with $\vec{F}$, and thus acts to slow down the rotation of the sphere. This torque is associated with the normal force $\vec{F}_N$ that the table exerts on the sphere over the whole area of contact. The net effect is that we can consider $\vec{F}_N$ acting vertically a distance ℓ in front of the CM as shown in Fig. 10–41 (where the deformation is greatly exaggerated).

Is it reasonable that the normal force $\vec{F}_N$ should effectively act in *front* of the CM as shown in Fig. 10–41? Yes. The sphere is rolling, and the leading edge strikes the surface with a slight impulse. The table therefore pushes upward a bit more strongly on the front part of the sphere than it would if the sphere were at rest. At the back part of the area of contact, the sphere is starting to move upward and so the table pushes upward on it less strongly than when the sphere is at rest. The table pushing up more strongly on the front part of the area of contact gives rise to the necessary torque and justifies the effective acting point of $\vec{F}_N$ being in front of the CM.

When other forces are present, the tiny torque τ_N due to $\vec{F}_N$ can usually be ignored. For example, when a sphere or cylinder rolls down an incline, the force of gravity has far more influence than τ_N, so the latter can be ignored. For many purposes (but not all), we can assume a hard sphere is in contact with a hard surface at essentially one point.

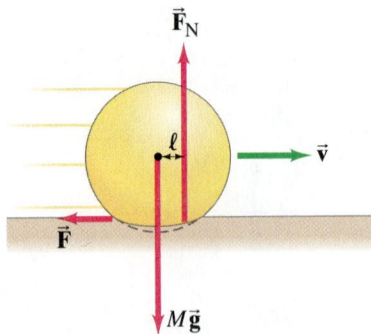

FIGURE 10–41 The normal force, $\vec{F}_N$, exerts a torque that slows down the sphere. The deformation of the sphere and the surface it moves on has been exaggerated for detail.

Summary

When a rigid object rotates about a fixed axis, each point of the object moves in a circular path. Lines drawn perpendicularly from the rotation axis to different points in the object all sweep out the same angle θ in any given time interval.

Angles are conveniently measured in **radians**. One radian is the angle subtended by an arc whose length is equal to the radius, or

$$2\pi \text{ rad} = 360° \quad \text{so} \quad 1 \text{ rad} \approx 57.3°.$$

All parts of a rigid object rotating about a fixed axis have the same **angular velocity** ω and the same **angular acceleration** α at any instant, where

$$\omega = \frac{d\theta}{dt} \tag{10-2b}$$

and

$$\alpha = \frac{d\omega}{dt}. \tag{10-3b}$$

The units of ω and α are rad/s and rad/s^2.

The linear velocity and acceleration of any point in an object rotating about a fixed axis are related to the angular quantities by

$$v = R\omega \tag{10-4}$$

$$a_{\text{tan}} = R\alpha \tag{10-5}$$

$$a_R = \omega^2 R \tag{10-6}$$

where R is the perpendicular distance of the point from the rotation axis, and a_{tan} and a_R are the tangential and radial components of the linear acceleration. The frequency f and period T are related to ω (rad/s) by

$$\omega = 2\pi f \tag{10-7}$$

$$T = 1/f. \tag{10-8}$$

Angular velocity and angular acceleration are vectors. For a rigid object rotating about a fixed axis, both $\vec{\omega}$ and $\vec{\alpha}$ point along the rotation axis. The direction of $\vec{\omega}$ is given by the **right-hand rule**.

If a rigid object undergoes uniformly accelerated rotational motion (α = constant), equations analogous to those for linear motion are valid:

$$\omega = \omega_0 + \alpha t; \qquad \theta = \omega_0 t + \tfrac{1}{2}\alpha t^2;$$
$$\omega^2 = \omega_0^2 + 2\alpha\theta; \qquad \bar{\omega} = \frac{\omega + \omega_0}{2}. \tag{10-9}$$

The **torque** due to a force $\vec{F}$ exerted on a rigid object is equal to

$$\tau = R_\perp F = RF_\perp = RF\sin\theta, \tag{10-10}$$

where $R_\perp$, called the **lever arm**, is the perpendicular distance from the axis of rotation to the line along which the force acts, and θ is the angle between $\vec{F}$ and R.

The rotational equivalent of Newton's second law is

$$\Sigma\tau = I\alpha, \tag{10-14}$$

where $I = \Sigma m_i R_i^2$ is the **moment of inertia** of the object about the axis of rotation. This relation is valid for a rigid object rotating about an axis fixed in an inertial reference frame, or when τ, I, and α are calculated about the center of mass of an object even if the CM is moving.

The **rotational kinetic energy** of an object rotating about a fixed axis with angular velocity ω is

$$K = \tfrac{1}{2}I\omega^2. \qquad (10\text{--}19)$$

For an object undergoing both translational and rotational motion, the total kinetic energy is the sum of the translational kinetic energy of the object's CM plus the rotational kinetic energy of the object about its CM:

$$K_{\text{tot}} = \tfrac{1}{2}Mv_{\text{CM}}^2 + \tfrac{1}{2}I_{\text{CM}}\omega^2 \qquad (10\text{--}23)$$

as long as the rotation axis is fixed in direction.

The following Table summarizes angular (or rotational) quantities, comparing them to their translational analogs.

Translation	Rotation	Connection
x	θ	$x = R\theta$
v	ω	$v = R\omega$
a	α	$a = R\alpha$
m	I	$I = \Sigma mR^2$
F	τ	$\tau = RF\sin\theta$
$K = \tfrac{1}{2}mv^2$	$\tfrac{1}{2}I\omega^2$	
$W = Fd$	$W = \tau\theta$	
$\Sigma F = ma$	$\Sigma\tau = I\alpha$	

Questions

1. A bicycle odometer (which counts revolutions and is calibrated to report distance traveled) is attached near the wheel hub and is calibrated for 27-inch wheels. What happens if you use it on a bicycle with 24-inch wheels?

2. Suppose a disk rotates at constant angular velocity. Does a point on the rim have radial and/or tangential acceleration? If the disk's angular velocity increases uniformly, does the point have radial and/or tangential acceleration? For which cases would the magnitude of either component of linear acceleration change?

3. Could a nonrigid object be described by a single value of the angular velocity ω? Explain.

4. Can a small force ever exert a greater torque than a larger force? Explain.

5. Why is it more difficult to do a sit-up with your hands behind your head than when your arms are stretched out in front of you? A diagram may help you to answer this.

6. Mammals that depend on being able to run fast have slender lower legs with flesh and muscle concentrated high, close to the body (Fig. 10–42). On the basis of rotational dynamics, explain why this distribution of mass is advantageous.

FIGURE 10–42
Question 6.
A gazelle.

7. If the net force on a system is zero, is the net torque also zero? If the net torque on a system is zero, is the net force zero?

8. Two inclines have the same height but make different angles with the horizontal. The same steel ball is rolled down each incline. On which incline will the speed of the ball at the bottom be greater? Explain.

9. Two spheres look identical and have the same mass. However, one is hollow and the other is solid. Describe an experiment to determine which is which.

10. Two solid spheres simultaneously start rolling (from rest) down an incline. One sphere has twice the radius and twice the mass of the other. Which reaches the bottom of the incline first? Which has the greater speed there? Which has the greater total kinetic energy at the bottom?

11. Why do tightrope walkers (Fig. 10–43) carry a long, narrow beam?

FIGURE 10–43   Question 11.

12. A sphere and a cylinder have the same radius and the same mass. They start from rest at the top of an incline. Which reaches the bottom first? Which has the greater speed at the bottom? Which has the greater total kinetic energy at the bottom? Which has the greater rotational kinetic energy?

13. The moment of inertia of this textbook would be the least about which symmetry axis through its center?

14. The moment of inertia of a rotating solid disk about an axis through its CM is $\tfrac{1}{2}MR^2$ (Fig. 10–20c). Suppose instead that a parallel axis of rotation passes through a point on the edge of the disk. Will the moment of inertia be the same, larger, or smaller?

15. The angular velocity of a wheel rotating on a horizontal axle points west. In what direction is the linear velocity of a point on the top of the wheel? If the angular acceleration points east, describe the tangential linear acceleration of this point at the top of the wheel. Is the angular speed increasing or decreasing?

Problems

10–1 Angular Quantities

1. (I) Express the following angles in radians: (*a*) 45.0°, (*b*) 60.0°, (*c*) 90.0°, (*d*) 360.0°, and (*e*) 445°. Give as numerical values and as fractions of π.

2. (I) The Sun subtends an angle of about 0.5° to us on Earth, 150 million km away. Estimate the radius of the Sun.

3. (I) A laser beam is directed at the Moon, 380,000 km from Earth. The beam diverges at an angle θ (Fig. 10–44) of 1.4×10^{-5} rad. What diameter spot will it make on the Moon?

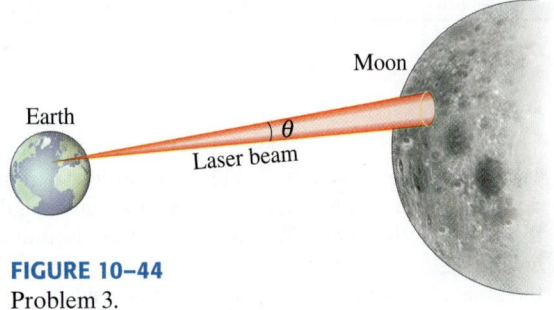

FIGURE 10–44
Problem 3.

4. (I) The blades in a blender rotate at a rate of 6500 rpm. When the motor is turned off during operation, the blades slow to rest in 4.0 s. What is the angular acceleration as the blades slow down?

5. (II) (*a*) A grinding wheel 0.35 m in diameter rotates at 2500 rpm. Calculate its angular velocity in rad/s. (*b*) What are the linear speed and acceleration of a point on the edge of the grinding wheel?

6. (II) A bicycle with tires 68 cm in diameter travels 7.2 km. How many revolutions do the wheels make?

7. (II) Calculate the angular velocity of (*a*) the second hand, (*b*) the minute hand, and (*c*) the hour hand, of a clock. State in rad/s. (*d*) What is the angular acceleration in each case?

8. (II) A rotating merry-go-round makes one complete revolution in 4.0 s (Fig. 10–45). (*a*) What is the linear speed of a child seated 1.2 m from the center? (*b*) What is her acceleration (give components)?

FIGURE 10–45
Problem 8.

9. (II) What is the linear speed of a point (*a*) on the equator, (*b*) on the Arctic Circle (latitude 66.5° N), and (*c*) at a latitude of 45.0° N, due to the Earth's rotation?

10. (II) Calculate the angular velocity of the Earth (*a*) in its orbit around the Sun, and (*b*) about its axis.

11. (II) How fast (in rpm) must a centrifuge rotate if a particle 7.0 cm from the axis of rotation is to experience an acceleration of 100,000 *g*'s?

12. (II) A 64-cm-diameter wheel accelerates uniformly about its center from 130 rpm to 280 rpm in 4.0 s. Determine (*a*) its angular acceleration, and (*b*) the radial and tangential components of the linear acceleration of a point on the edge of the wheel 2.0 s after it has started accelerating.

13. (II) In traveling to the Moon, astronauts aboard the *Apollo* spacecraft put themselves into a slow rotation to distribute the Sun's energy evenly. At the start of their trip, they accelerated from no rotation to 1.0 revolution every minute during a 12-min time interval. The spacecraft can be thought of as a cylinder with a diameter of 8.5 m. Determine (*a*) the angular acceleration, and (*b*) the radial and tangential components of the linear acceleration of a point on the skin of the ship 7.0 min after it started this acceleration.

14. (II) A turntable of radius R_1 is turned by a circular rubber roller of radius R_2 in contact with it at their outer edges. What is the ratio of their angular velocities, ω_1/ω_2?

10–2 Vector Nature of $\vec{\omega}$ and $\vec{\alpha}$

15. (II) The axle of a wheel is mounted on supports that rest on a rotating turntable as shown in Fig. 10–46. The wheel has angular velocity $\omega_1 = 44.0$ rad/s about its axle, and the turntable has angular velocity $\omega_2 = 35.0$ rad/s about a vertical axis. (Note arrows showing these motions in the figure.) (*a*) What are the directions of $\vec{\omega}_1$ and $\vec{\omega}_2$ at the instant shown? (*b*) What is the resultant angular velocity of the wheel, as seen by an outside observer, at the instant shown? Give the magnitude and direction. (*c*) What is the magnitude and direction of the angular acceleration of the wheel at the instant shown? Take the *z* axis vertically upward and the direction of the axle at the moment shown to be the *x* axis pointing to the right.

FIGURE 10–46
Problem 15.

10–3 Constant Angular Acceleration

16. (I) An automobile engine slows down from 3500 rpm to 1200 rpm in 2.5 s. Calculate (*a*) its angular acceleration, assumed constant, and (*b*) the total number of revolutions the engine makes in this time.

17. (I) A centrifuge accelerates uniformly from rest to 15,000 rpm in 220 s. Through how many revolutions did it turn in this time?

18. (I) Pilots can be tested for the stresses of flying high-speed jets in a whirling "human centrifuge," which takes 1.0 min to turn through 20 complete revolutions before reaching its final speed. (*a*) What was its angular acceleration (assumed constant), and (*b*) what was its final angular speed in rpm?

19. (II) A cooling fan is turned off when it is running at 850 rev/min. It turns 1350 revolutions before it comes to a stop. (*a*) What was the fan's angular acceleration, assumed constant? (*b*) How long did it take the fan to come to a complete stop?

20. (II) Using calculus, derive the angular kinematic equations 10–9a and 10–9b for constant angular acceleration. Start with $\alpha = d\omega/dt$.

21. (II) A small rubber wheel is used to drive a large pottery wheel. The two wheels are mounted so that their circular edges touch. The small wheel has a radius of 2.0 cm and accelerates at the rate of 7.2 rad/s², and it is in contact with the pottery wheel (radius 21.0 cm) without slipping. Calculate (a) the angular acceleration of the pottery wheel, and (b) the time it takes the pottery wheel to reach its required speed of 65 rpm.

22. (II) The angle through which a rotating wheel has turned in time t is given by $\theta = 8.5t - 15.0t^2 + 1.6t^4$, where θ is in radians and t in seconds. Determine an expression (a) for the instantaneous angular velocity ω and (b) for the instantaneous angular acceleration α. (c) Evaluate ω and α at $t = 3.0$ s. (d) What is the average angular velocity, and (e) the average angular acceleration between $t = 2.0$ s and $t = 3.0$ s?

23. (II) The angular acceleration of a wheel, as a function of time, is $\alpha = 5.0t^2 - 8.5t$, where α is in rad/s² and t in seconds. If the wheel starts from rest ($\theta = 0$, $\omega = 0$, at $t = 0$), determine a formula for (a) the angular velocity ω and (b) the angular position θ, both as a function of time. (c) Evaluate ω and θ at $t = 2.0$ s.

10–4 Torque

24. (I) A 62-kg person riding a bike puts all her weight on each pedal when climbing a hill. The pedals rotate in a circle of radius 17 cm. (a) What is the maximum torque she exerts? (b) How could she exert more torque?

25. (I) Calculate the net torque about the axle of the wheel shown in Fig. 10–47. Assume that a friction torque of 0.40 m·N opposes the motion.

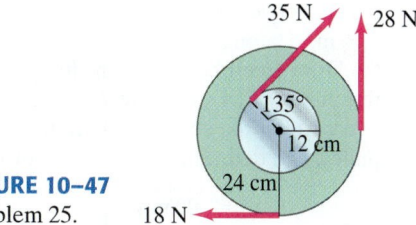

FIGURE 10–47
Problem 25.

26. (II) A person exerts a horizontal force of 32 N on the end of a door 96 cm wide. What is the magnitude of the torque if the force is exerted (a) perpendicular to the door and (b) at a 60.0° angle to the face of the door?

27. (II) Two blocks, each of mass m, are attached to the ends of a massless rod which pivots as shown in Fig. 10–48. Initially the rod is held in the horizontal position and then released. Calculate the magnitude and direction of the net torque on this system when it is first released.

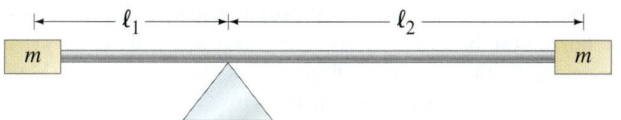

FIGURE 10–48   Problem 27.

28. (II) A wheel of diameter 27.0 cm is constrained to rotate in the xy plane, about the z axis, which passes through its center. A force $\vec{F} = (-31.0\hat{i} + 43.4\hat{j})$ N acts at a point on the edge of the wheel that lies exactly on the x axis at a particular instant. What is the torque about the rotation axis at this instant?

29. (II) The bolts on the cylinder head of an engine require tightening to a torque of 75 m·N. If a wrench is 28 cm long, what force perpendicular to the wrench must the mechanic exert at its end? If the six-sided bolt head is 15 mm across (Fig. 10–49), estimate the force applied near each of the six points by a socket wrench.

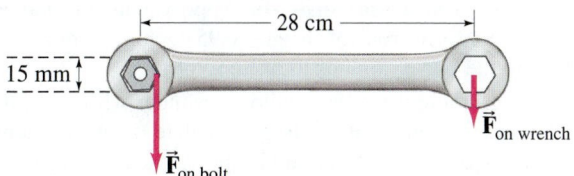

FIGURE 10–49   Problem 29.

30. (II) Determine the net torque on the 2.0-m-long uniform beam shown in Fig. 10–50. Calculate about (a) point C, the CM, and (b) point P at one end.

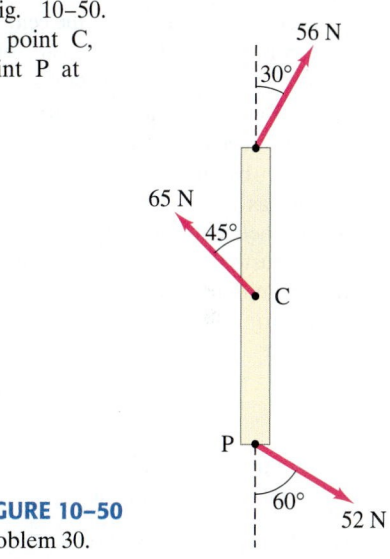

FIGURE 10–50
Problem 30.

10–5 and 10–6 Rotational Dynamics

31. (I) Determine the moment of inertia of a 10.8-kg sphere of radius 0.648 m when the axis of rotation is through its center.

32. (I) Estimate the moment of inertia of a bicycle wheel 67 cm in diameter. The rim and tire have a combined mass of 1.1 kg. The mass of the hub can be ignored (why?).

33. (II) A potter is shaping a bowl on a potter's wheel rotating at constant angular speed (Fig. 10–51). The friction force between her hands and the clay is 1.5 N total. (a) How large is her torque on the wheel, if the diameter of the bowl is 12 cm? (b) How long would it take for the potter's wheel to stop if the only torque acting on it is due to the potter's hand? The initial angular velocity of the wheel is 1.6 rev/s, and the moment of inertia of the wheel and the bowl is 0.11 kg·m².

FIGURE 10–51
Problem 33.

34. (II) An oxygen molecule consists of two oxygen atoms whose total mass is 5.3×10^{-26} kg and whose moment of inertia about an axis perpendicular to the line joining the two atoms, midway between them, is 1.9×10^{-46} kg·m². From these data, estimate the effective distance between the atoms.

35. (II) A softball player swings a bat, accelerating it from rest to 2.7 rev/s in a time of 0.20 s. Approximate the bat as a 2.2-kg uniform rod of length 0.95 m, and compute the torque the player applies to one end of it.

36. (II) A grinding wheel is a uniform cylinder with a radius of 8.50 cm and a mass of 0.380 kg. Calculate (a) its moment of inertia about its center, and (b) the applied torque needed to accelerate it from rest to 1750 rpm in 5.00 s if it is known to slow down from 1500 rpm to rest in 55.0 s.

37. (II) A small 650-g ball on the end of a thin, light rod is rotated in a horizontal circle of radius 1.2 m. Calculate (a) the moment of inertia of the ball about the center of the circle, and (b) the torque needed to keep the ball rotating at constant angular velocity if air resistance exerts a force of 0.020 N on the ball. Ignore the rod's moment of inertia and air resistance.

38. (II) The forearm in Fig. 10–52 accelerates a 3.6-kg ball at 7.0 m/s² by means of the triceps muscle, as shown. Calculate (a) the torque needed, and (b) the force that must be exerted by the triceps muscle. Ignore the mass of the arm.

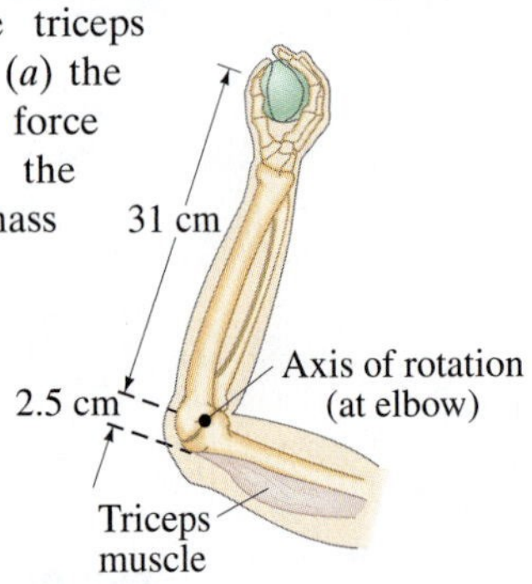

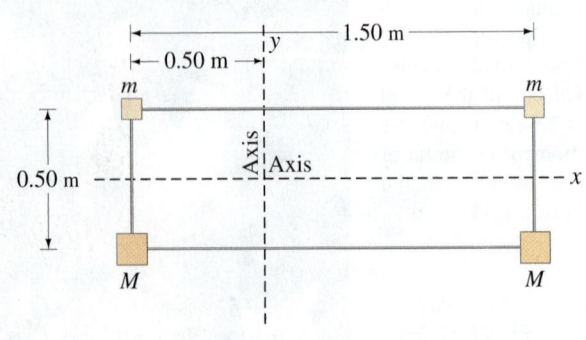

FIGURE 10–52
Problems 38 and 39.

39. (II) Assume that a 1.00-kg ball is thrown solely by the action of the forearm, which rotates about the elbow joint under the action of the triceps muscle, Fig. 10–52. The ball is accelerated uniformly from rest to 8.5 m/s in 0.35 s, at which point it is released. Calculate (a) the angular acceleration of the arm, and (b) the force required of the triceps muscle. Assume that the forearm has a mass of 3.7 kg and rotates like a uniform rod about an axis at its end.

40. (II) Calculate the moment of inertia of the array of point objects shown in Fig. 10–53 about (a) the vertical axis, and (b) the horizontal axis. Assume $m = 2.2$ kg, $M = 3.1$ kg, and the objects are wired together by very light, rigid pieces of wire. The array is rectangular and is split through the middle by the horizontal axis. (c) About which axis would it be harder to accelerate this array?

FIGURE 10–53 Problem 40.

41. (II) A merry-go-round accelerates from rest to 0.68 rad/s in 24 s. Assuming the merry-go-round is a uniform disk of radius 7.0 m and mass 31,000 kg, calculate the net torque required to accelerate it.

42. (II) A 0.72-m-diameter solid sphere can be rotated about an axis through its center by a torque of 10.8 m·N which accelerates it uniformly from rest through a total of 180 revolutions in 15.0 s. What is the mass of the sphere?

43. (II) Suppose the force F_T in the cord hanging from the pulley of Example 10–9, Fig. 10–21, is given by the relation $F_T = 3.00 t - 0.20 t^2$ (newtons) where t is in seconds. If the pulley starts from rest, what is the linear speed of a point on its rim 8.0 s later? Ignore friction.

44. (II) A dad pushes tangentially on a small hand-driven merry-go-round and is able to accelerate it from rest to a frequency of 15 rpm in 10.0 s. Assume the merry-go-round is a uniform disk of radius 2.5 m and has a mass of 760 kg, and two children (each with a mass of 25 kg) sit opposite each other on the edge. Calculate the torque required to produce the acceleration, neglecting frictional torque. What force is required at the edge?

45. (II) Four equal masses M are spaced at equal intervals, ℓ, along a horizontal straight rod whose mass can be ignored. The system is to be rotated about a vertical axis passing through the mass at the left end of the rod and perpendicular to it. (a) What is the moment of inertia of the system about this axis? (b) What minimum force, applied to the farthest mass, will impart an angular acceleration α? (c) What is the direction of this force?

46. (II) Two blocks are connected by a light string passing over a pulley of radius 0.15 m and moment of inertia I. The blocks move (towards the right) with an acceleration of 1.00 m/s² along their frictionless inclines (see Fig. 10–54). (a) Draw free-body diagrams for each of the two blocks and the pulley. (b) Determine F_{TA} and F_{TB}, the tensions in the two parts of the string. (c) Find the net torque acting on the pulley, and determine its moment of inertia, I.

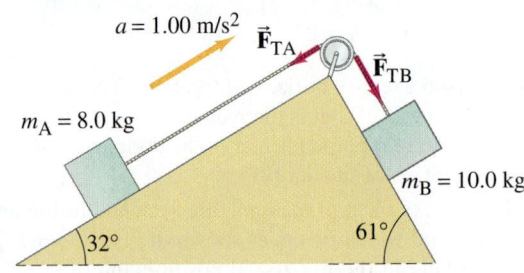

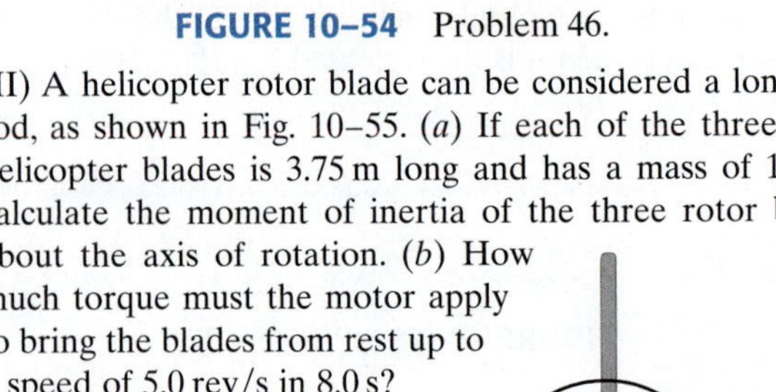

FIGURE 10–54 Problem 46.

47. (II) A helicopter rotor blade can be considered a long thin rod, as shown in Fig. 10–55. (a) If each of the three rotor helicopter blades is 3.75 m long and has a mass of 135 kg, calculate the moment of inertia of the three rotor blades about the axis of rotation. (b) How much torque must the motor apply to bring the blades from rest up to a speed of 5.0 rev/s in 8.0 s?

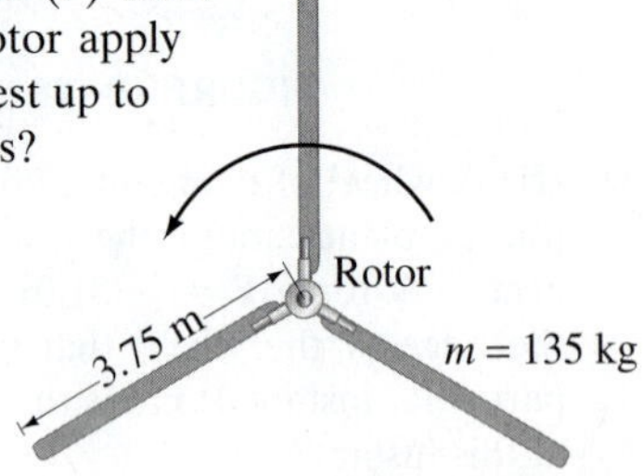

FIGURE 10–55
Problem 47.

48. (II) A centrifuge rotor rotating at 10,300 rpm is shut off and is eventually brought uniformly to rest by a frictional torque of 1.20 m·N. If the mass of the rotor is 3.80 kg and it can be approximated as a solid cylinder of radius 0.0710 m, through how many revolutions will the rotor turn before coming to rest, and how long will it take?

49. (II) When discussing moments of inertia, especially for unusual or irregularly shaped objects, it is sometimes convenient to work with the **radius of gyration**, k. This radius is defined so that if all the mass of the object were concentrated at this distance from the axis, the moment of inertia would be the same as that of the original object. Thus, the moment of inertia of any object can be written in terms of its mass M and the radius of gyration as $I = Mk^2$. Determine the radius of gyration for each of the objects (hoop, cylinder, sphere, etc.) shown in Fig. 10–20.

50. (II) To get a flat, uniform cylindrical satellite spinning at the correct rate, engineers fire four tangential rockets as shown in Fig. 10–56. If the satellite has a mass of 3600 kg, a radius of 4.0 m, and the rockets each add a mass of 250 kg, what is the required steady force of each rocket if the satellite is to reach 32 rpm in 5.0 min, starting from rest?

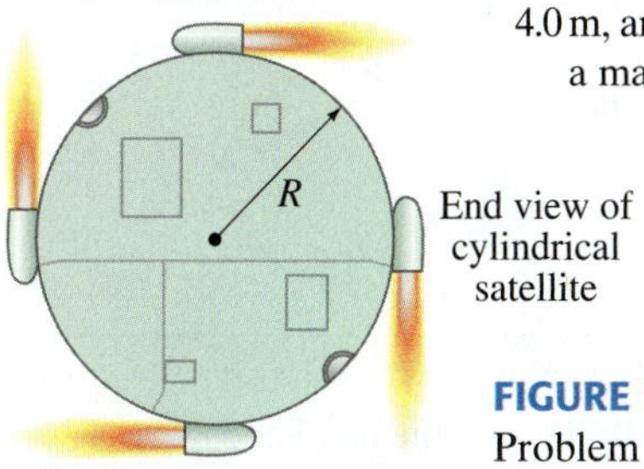

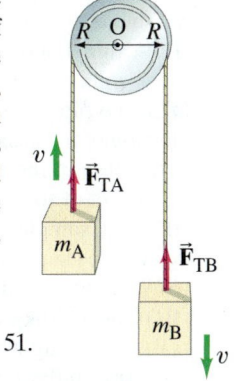

FIGURE 10–56
Problem 50.

End view of cylindrical satellite

51. (III) An *Atwood's machine* consists of two masses, m_A and m_B, which are connected by a massless inelastic cord that passes over a pulley, Fig. 10–57. If the pulley has radius R and moment of inertia I about its axle, determine the acceleration of the masses m_A and m_B, and compare to the situation in which the moment of inertia of the pulley is ignored. [*Hint*: The tensions F_{TA} and F_{TB} are not equal. We discussed the Atwood machine in Example 4–13, assuming $I = 0$ for the pulley.]

FIGURE 10–57   Problem 51.
Atwood's machine.

52. (III) A string passing over a pulley has a 3.80-kg mass hanging from one end and a 3.15-kg mass hanging from the other end. The pulley is a uniform solid cylinder of radius 4.0 cm and mass 0.80 kg. (*a*) If the bearings of the pulley were frictionless, what would be the acceleration of the two masses? (*b*) In fact, it is found that if the heavier mass is given a downward speed of 0.20 m/s, it comes to rest in 6.2 s. What is the average frictional torque acting on the pulley?

53. (III) A hammer thrower accelerates the hammer (mass = 7.30 kg) from rest within four full turns (revolutions) and releases it at a speed of 26.5 m/s. Assuming a uniform rate of increase in angular velocity and a horizontal circular path of radius 1.20 m, calculate (*a*) the angular acceleration, (*b*) the (linear) tangential acceleration, (*c*) the centripetal acceleration just before release, (*d*) the net force being exerted on the hammer by the athlete just before release, and (*e*) the angle of this force with respect to the radius of the circular motion. Ignore gravity.

54. (III) A thin rod of length ℓ stands vertically on a table. The rod begins to fall, but its lower end does not slide. (*a*) Determine the angular velocity of the rod as a function of the angle ϕ it makes with the tabletop. (*b*) What is the speed of the tip of the rod just before it strikes the table?

10–7 Moment of Inertia

55. (I) Use the parallel-axis theorem to show that the moment of inertia of a thin rod about an axis perpendicular to the rod at one end is $I = \frac{1}{3}M\ell^2$, given that if the axis passes through the center, $I = \frac{1}{12}M\ell^2$ (Fig. 10–20f and g).

56. (II) Determine the moment of inertia of a 19-kg door that is 2.5 m high and 1.0 m wide and is hinged along one side. Ignore the thickness of the door.

57. (II) Two uniform solid spheres of mass M and radius r_0 are connected by a thin (massless) rod of length r_0 so that the centers are $3r_0$ apart. (*a*) Determine the moment of inertia of this system about an axis perpendicular to the rod at its center. (*b*) What would be the percentage error if the masses of each sphere were assumed to be concentrated at their centers and a very simple calculation made?

58. (II) A ball of mass M and radius r_1 on the end of a thin massless rod is rotated in a horizontal circle of radius R_0 about an axis of rotation AB, as shown in Fig. 10–58. (*a*) Considering the mass of the ball to be concentrated at its center of mass, calculate its moment of inertia about AB. (*b*) Using the parallel-axis theorem and considering the finite radius of the ball, calculate the moment of inertia of the ball about AB. (*c*) Calculate the percentage error introduced by the point mass approximation for $r_1 = 9.0$ cm and $R_0 = 1.0$ m.

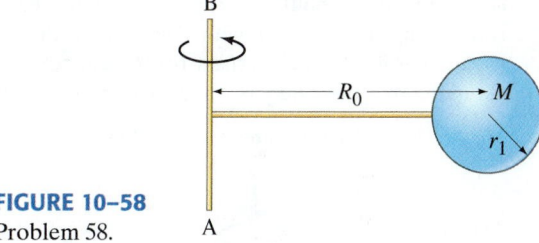

FIGURE 10–58
Problem 58.

59. (II) A thin 7.0-kg wheel of radius 32 cm is weighted to one side by a 1.50-kg weight, small in size, placed 22 cm from the center of the wheel. Calculate (*a*) the position of the center of mass of the weighted wheel and (*b*) the moment of inertia about an axis through its CM, perpendicular to its face.

60. (III) Derive the formula for the moment of inertia of a uniform thin rod of length ℓ about an axis through its center, perpendicular to the rod (see Fig. 10–20f).

61. (III) (*a*) Derive the formula given in Fig. 10–20h for the moment of inertia of a uniform, flat, rectangular plate of dimensions $\ell \times w$ about an axis through its center, perpendicular to the plate. (*b*) What is the moment of inertia about each of the axes through the center that are parallel to the edges of the plate?

10–8 Rotational Kinetic Energy

62. (I) An automobile engine develops a torque of 255 m·N at 3750 rpm. What is the horsepower of the engine?

63. (I) A centrifuge rotor has a moment of inertia of 4.25×10^{-2} kg·m². How much energy is required to bring it from rest to 9750 rpm?

64. (II) A rotating uniform cylindrical platform of mass 220 kg and radius 5.5 m slows down from 3.8 rev/s to rest in 16 s when the driving motor is disconnected. Estimate the power output of the motor (hp) required to maintain a steady speed of 3.8 rev/s.

65. (II) A merry-go-round has a mass of 1640 kg and a radius of 7.50 m. How much net work is required to accelerate it from rest to a rotation rate of 1.00 revolution per 8.00 s? Assume it is a solid cylinder.

66. (II) A uniform thin rod of length ℓ and mass M is suspended freely from one end. It is pulled to the side an angle θ and released. If friction can be ignored, what is its angular velocity, and the speed of its free end, at the lowest point?

67. (II) Two masses, $m_A = 35.0$ kg and $m_B = 38.0$ kg, are connected by a rope that hangs over a pulley (as in Fig. 10–59). The pulley is a uniform cylinder of radius 0.381 m and mass 3.1 kg. Initially m_A is on the ground and m_B rests 2.5 m above the ground. If the system is released, use conservation of energy to determine the speed of m_B just before it strikes the ground. Assume the pulley bearing is frictionless.

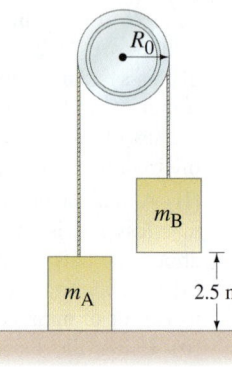

FIGURE 10–59
Problem 67.

68. (III) A 4.00-kg mass and a 3.00-kg mass are attached to opposite ends of a thin 42.0-cm-long horizontal rod (Fig. 10–60). The system is rotating at angular speed $\omega = 5.60$ rad/s about a vertical axle at the center of the rod. Determine (a) the kinetic energy K of the system, and (b) the net force on each mass. (c) Repeat parts (a) and (b) assuming that the axle passes through the CM of the system.

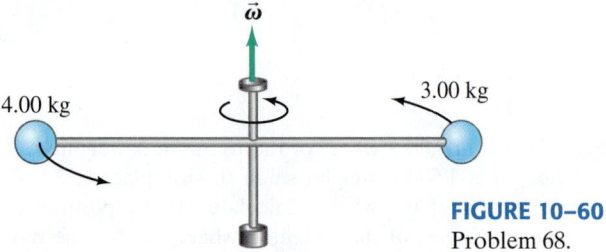

FIGURE 10–60
Problem 68.

69. (III) A 2.30-m-long pole is balanced vertically on its tip. It starts to fall and its lower end does not slip. What will be the speed of the upper end of the pole just before it hits the ground? [*Hint*: Use conservation of energy.]

10–9 Rotational Plus Translational Motion

70. (I) Calculate the translational speed of a cylinder when it reaches the foot of an incline 7.20 m high. Assume it starts from rest and rolls without slipping.

71. (I) A bowling ball of mass 7.3 kg and radius 9.0 cm rolls without slipping down a lane at 3.7 m/s. Calculate its total kinetic energy.

72. (I) Estimate the kinetic energy of the Earth with respect to the Sun as the sum of two terms, (a) that due to its daily rotation about its axis, and (b) that due to its yearly revolution about the Sun. [Assume the Earth is a uniform sphere with mass $= 6.0 \times 10^{24}$ kg, radius $= 6.4 \times 10^6$ m, and is 1.5×10^8 km from the Sun.]

73. (II) A sphere of radius $r_0 = 24.5$ cm and mass $m = 1.20$ kg starts from rest and rolls without slipping down a 30.0° incline that is 10.0 m long. (a) Calculate its translational and rotational speeds when it reaches the bottom. (b) What is the ratio of translational to rotational kinetic energy at the bottom? Avoid putting in numbers until the end so you can answer: (c) do your answers in (a) and (b) depend on the radius of the sphere or its mass?

74. (II) A narrow but solid spool of thread has radius R and mass M. If you pull up on the thread so that the CM of the spool remains suspended in the air at the same place as it unwinds, (a) what force must you exert on the thread? (b) How much work have you done by the time the spool turns with angular velocity ω?

75. (II) A ball of radius r_0 rolls on the inside of a track of radius R_0 (see Fig. 10–61). If the ball starts from rest at the vertical edge of the track, what will be its speed when it reaches the lowest point of the track, rolling without slipping?

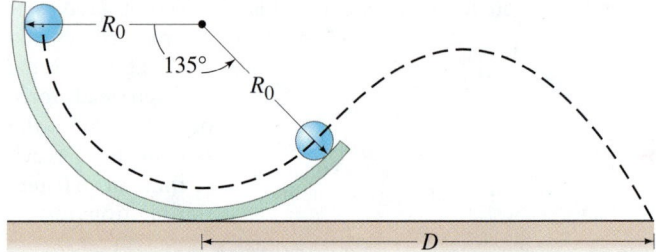

FIGURE 10–61 Problems 75 and 81.

76. (II) A solid rubber ball rests on the floor of a railroad car when the car begins moving with acceleration a. Assuming the ball rolls without slipping, what is its acceleration relative to (a) the car and (b) the ground?

*77. (II) A thin, hollow 0.545-kg section of pipe of radius 10.0 cm starts rolling (from rest) down a 17.5° incline 5.60 m long. (a) If the pipe rolls without slipping, what will be its speed at the base of the incline? (b) What will be its total kinetic energy at the base of the incline? (c) What minimum value must the coefficient of static friction have if the pipe is not to slip?

*78. (II) In Example 10–20, (a) how far has the ball moved down the lane when it starts rolling without slipping? (b) What are its final linear and rotational speeds?

79. (III) The 1100-kg mass of a car includes four tires, each of mass (including wheels) 35 kg and diameter 0.80 m. Assume each tire and wheel combination acts as a solid cylinder. Determine (a) the total kinetic energy of the car when traveling 95 km/h and (b) the fraction of the kinetic energy in the tires and wheels. (c) If the car is initially at rest and is then pulled by a tow truck with a force of 1500 N, what is the acceleration of the car? Ignore frictional losses. (d) What percent error would you make in part (c) if you ignored the rotational inertia of the tires and wheels?

*80. (III) A wheel with rotational inertia $I = \frac{1}{2}MR^2$ about its horizontal central axle is set spinning with initial angular speed ω_0. It is then lowered, and at the instant its edge touches the ground the speed of the axle (and CM) is zero. Initially the wheel slips when it touches the ground, but then begins to move forward and eventually rolls without slipping. (a) In what direction does friction act on the slipping wheel? (b) How long does the wheel slip before it begins to roll without slipping? (c) What is the wheel's final translational speed, v_{CM}? [*Hint*: Use $\Sigma\vec{F} = m\vec{a}$, $\Sigma\tau_{CM} = I_{CM}\alpha_{CM}$, and recall that only when there is rolling without slipping is $v_{CM} = \omega R$.]

81. (III) A small sphere of radius $r_0 = 1.5$ cm rolls without slipping on the track shown in Fig. 10–61 whose radius is $R_0 = 26.0$ cm. The sphere starts rolling at a height R_0 above the bottom of the track. When it leaves the track after passing through an angle of 135° as shown, (a) what will be its speed, and (b) at what distance D from the base of the track will the sphere hit the ground?

*10–10 Rolling Sphere Slows Down

*82. (I) A rolling ball slows down because the normal force does not pass exactly through the CM of the ball, but passes in front of the CM. Using Fig. 10–41, show that the torque resulting from the normal force ($\tau_N = \ell F_N$ in Fig. 10–41) is $\frac{7}{5}$ of that due to the frictional force, $\tau_{fr} = r_0 F$ where r_0 is the ball's radius; that is, show that $\tau_N = \frac{7}{5}\tau_{fr}$.

General Problems

83. A large spool of rope rolls on the ground with the end of the rope lying on the top edge of the spool. A person grabs the end of the rope and walks a distance ℓ, holding onto it, Fig. 10–62. The spool rolls behind the person without slipping. What length of rope unwinds from the spool? How far does the spool's center of mass move?

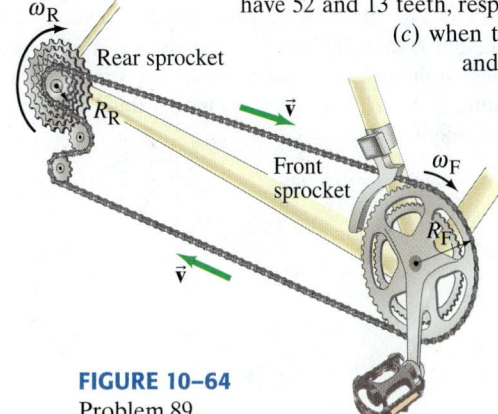

FIGURE 10–62
Problem 83.

84. On a 12.0-cm-diameter audio compact disc (CD), digital bits of information are encoded sequentially along an outward spiraling path. The spiral starts at radius $R_1 = 2.5$ cm and winds its way out to radius $R_2 = 5.8$ cm. To read the digital information, a CD player rotates the CD so that the player's readout laser scans along the spiral's sequence of bits at a constant linear speed of 1.25 m/s. Thus the player must accurately adjust the rotational frequency f of the CD as the laser moves outward. Determine the values for f (in units of rpm) when the laser is located at R_1 and when it is at R_2.

85. (a) A yo-yo is made of two solid cylindrical disks, each of mass 0.050 kg and diameter 0.075 m, joined by a (concentric) thin solid cylindrical hub of mass 0.0050 kg and diameter 0.010 m. Use conservation of energy to calculate the linear speed of the yo-yo just before it reaches the end of its 1.0-m-long string, if it is released from rest. (b) What fraction of its kinetic energy is rotational?

86. A cyclist accelerates from rest at a rate of 1.00 m/s². How fast will a point at the top of the rim of the tire (diameter = 68 cm) be moving after 2.5 s? [*Hint*: At any moment, the lowest point on the tire is in contact with the ground and is at rest—see Fig. 10–63.]

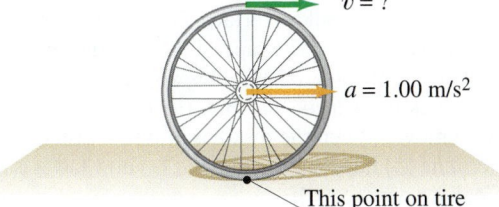

$v = ?$

$a = 1.00$ m/s²

This point on tire at rest momentarily

FIGURE 10–63   Problem 86.

87. Suppose David puts a 0.50-kg rock into a sling of length 1.5 m and begins whirling the rock in a nearly horizontal circle, accelerating it from rest to a rate of 85 rpm after 5.0 s. What is the torque required to achieve this feat, and where does the torque come from?

88. A 1.4-kg grindstone in the shape of a uniform cylinder of radius 0.20 m acquires a rotational rate of 18 rev/s from rest over a 6.0-s interval at constant angular acceleration. Calculate the torque delivered by the motor.

89. Bicycle gears: (a) How is the angular velocity ω_R of the rear wheel of a bicycle related to the angular velocity ω_F of the front sprocket and pedals? Let N_F and N_R be the number of teeth on the front and rear sprockets, respectively, Fig. 10–64. The teeth are spaced the same on both sprockets and the rear sprocket is firmly attached to the rear wheel. (b) Evaluate the ratio ω_R/ω_F when the front and rear sprockets have 52 and 13 teeth, respectively, and (c) when they have 42 and 28 teeth.

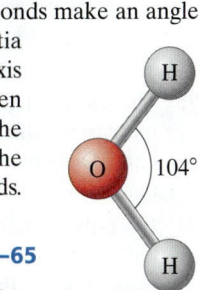

FIGURE 10–64
Problem 89.

90. Figure 10–65 illustrates an H_2O molecule. The O—H bond length is 0.096 nm and the H—O—H bonds make an angle of 104°. Calculate the moment of inertia for the H_2O molecule about an axis passing through the center of the oxygen atom (a) perpendicular to the plane of the molecule, and (b) in the plane of the molecule, bisecting the H—O—H bonds.

FIGURE 10–65
Problem 90.

91. One possibility for a low-pollution automobile is for it to use energy stored in a heavy rotating flywheel. Suppose such a car has a total mass of 1100 kg, uses a uniform cylindrical flywheel of diameter 1.50 m and mass 240 kg, and should be able to travel 350 km without needing a flywheel "spinup." (a) Make reasonable assumptions (average frictional retarding force = 450 N, twenty acceleration periods from rest to 95 km/h, equal uphill and downhill, and that energy can be put back into the flywheel as the car goes downhill), and estimate what total energy needs to be stored in the flywheel. (b) What is the angular velocity of the flywheel when it has a full "energy charge"? (c) About how long would it take a 150-hp motor to give the flywheel a full energy charge before a trip?

92. A hollow cylinder (hoop) is rolling on a horizontal surface at speed $v = 3.3 \, \text{m/s}$ when it reaches a 15° incline. (a) How far up the incline will it go? (b) How long will it be on the incline before it arrives back at the bottom?

93. A wheel of mass M has radius R. It is standing vertically on the floor, and we want to exert a horizontal force F at its axle so that it will climb a step against which it rests (Fig. 10–66). The step has height h, where $h < R$. What minimum force F is needed?

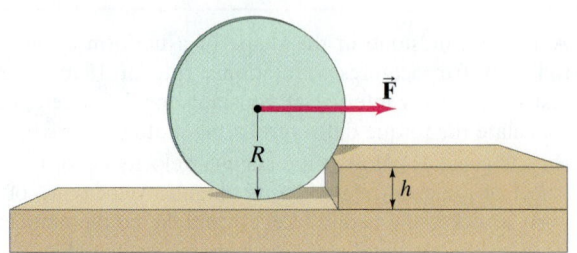

FIGURE 10–66 Problem 93.

94. A marble of mass m and radius r rolls along the looped rough track of Fig. 10–67. What is the minimum value of the vertical height h that the marble must drop if it is to reach the highest point of the loop without leaving the track? (a) Assume $r \ll R$; (b) do not make this assumption. Ignore frictional losses.

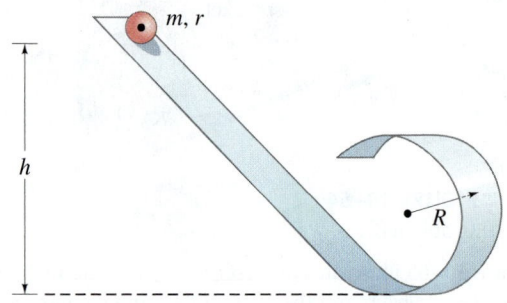

FIGURE 10–67 Problem 94.

95. The density (mass per unit length) of a thin rod of length ℓ increases uniformly from λ_0 at one end to $3\lambda_0$ at the other end. Determine the moment of inertia about an axis perpendicular to the rod through its geometric center.

96. If a billiard ball is hit in just the right way by a cue stick, the ball will roll without slipping immediately after losing contact with the stick. Consider a billiard ball (radius r, mass M) at rest on a horizontal pool table. A cue stick exerts a constant horizontal force F on the ball for a time t at a point that is a height h above the table's surface (see Fig. 10–68). Assume that the coefficient of kinetic friction between the ball and table is μ_k. Determine the value for h so that the ball will roll without slipping immediately after losing contact with the stick.

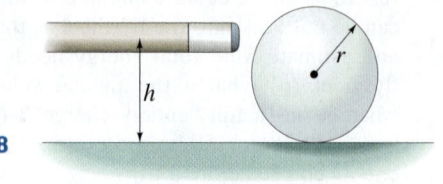

FIGURE 10–68
Problem 96.

97. If the coefficient of static friction between tires and pavement is 0.65, calculate the minimum torque that must be applied to the 66-cm-diameter tire of a 950-kg automobile in order to "lay rubber" (make the wheels spin, slipping as the car accelerates). Assume each wheel supports an equal share of the weight.

98. A cord connected at one end to a block which can slide on an inclined plane has its other end wrapped around a cylinder resting in a depression at the top of the plane as shown in Fig. 10–69. Determine the speed of the block after it has traveled 1.80 m along the plane, starting from rest. Assume (a) there is no friction, (b) the coefficient of friction between all surfaces is $\mu = 0.035$. [*Hint:* In part (b) first determine the normal force on the cylinder, and make any reasonable assumptions needed.]

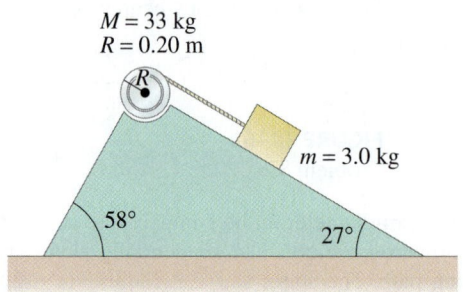

$M = 33 \, \text{kg}$
$R = 0.20 \, \text{m}$

$m = 3.0 \, \text{kg}$

58°   27°

FIGURE 10–69 Problem 98.

99. The radius of the roll of paper shown in Fig. 10–70 is 7.6 cm and its moment of inertia is $I = 3.3 \times 10^{-3} \, \text{kg} \cdot \text{m}^2$. A force of 2.5 N is exerted on the end of the roll for 1.3 s, but the paper does not tear so it begins to unroll. A constant friction torque of 0.11 m·N is exerted on the roll which gradually brings it to a stop. Assuming that the paper's thickness is negligible, calculate (a) the length of paper that unrolls during the time that the force is applied (1.3 s) and (b) the length of paper that unrolls from the time the force ends to the time when the roll has stopped moving.

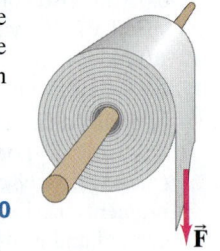

FIGURE 10–70
Problem 99.

100. A solid uniform disk of mass 21.0 kg and radius 85.0 cm is at rest flat on a frictionless surface. Figure 10–71 shows a view from above. A string is wrapped around the rim of the disk and a constant force of 35.0 N is applied to the string. The string does not slip on the rim. (a) In what direction does the CM move? When the disk has moved a distance of 5.5 m, determine (b) how fast it is moving, (c) how fast it is spinning (in radians per second), and (d) how much string has unwrapped from around the rim.

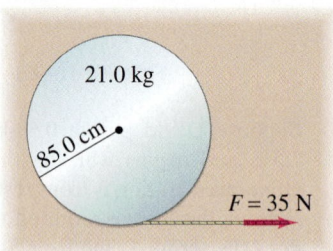

21.0 kg

85.0 cm

$F = 35 \, \text{N}$

FIGURE 10–71
Problem 100, looking
down on the disk.

101. When bicycle and motorcycle riders "pop a wheelie," a large acceleration causes the bike's front wheel to leave the ground. Let M be the total mass of the bike-plus-rider system; let x and y be the horizontal and vertical distance of this system's CM from the rear wheel's point of contact with the ground (Fig. 10–72). (a) Determine the horizontal acceleration a required to barely lift the bike's front wheel off of the ground. (b) To minimize the acceleration necessary to pop a wheelie, should x be made as small or as large as possible? How about y? How should a rider position his or her body on the bike in order to achieve these optimal values for x and y? (c) If $x = 35$ cm and $y = 95$ cm, find a.

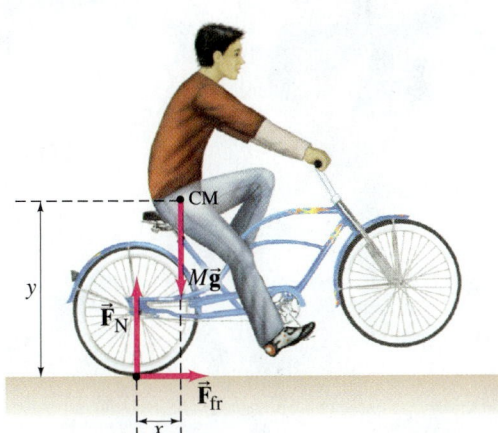

FIGURE 10–72 Problem 101.

102. A crucial part of a piece of machinery starts as a flat uniform cylindrical disk of radius R_0 and mass M. It then has a circular hole of radius R_1 drilled into it (Fig. 10–73). The hole's center is a distance h from the center of the disk. Find the moment of inertia of this disk (with off-center hole) when rotated about its center, C. [Hint: Consider a solid disk and "subtract" the hole; use the parallel-axis theorem.]

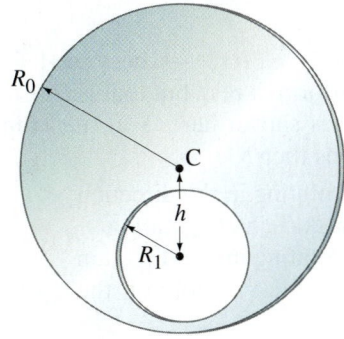

FIGURE 10–73
Problem 102.

103. A thin uniform stick of mass M and length ℓ is positioned vertically, with its tip on a frictionless table. It is released and allowed to fall. Determine the speed of its CM just before it hits the table (Fig. 10–74).

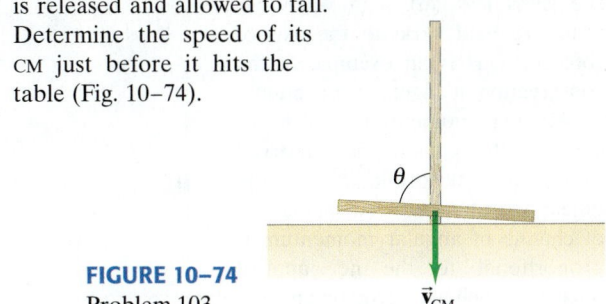

FIGURE 10–74
Problem 103.

* **104.** (a) For the yo-yo-like cylinder of Example 10–19, we saw that the downward acceleration of its CM was $a = \frac{2}{3} g$. If it starts from rest, what will be the CM velocity after it has fallen a distance h? (b) Now use conservation of energy to determine the cylinder's CM velocity after it has fallen a distance h, starting from rest.

*Numerical/Computer

* **105.** (II) Determine the torque produced about the support A of the rigid structure, shown in Fig. 10–75, as a function of the leg angle θ if a force $F = 500$ N is applied at the point P perpendicular to the leg end. Graph the values of the torque τ as a function of θ from $\theta = 0°$ to $90°$, in $1°$ increments.

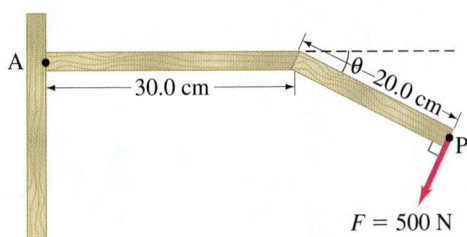

FIGURE 10–75 Problem 105.

* **106.** (II) Use the expression that was derived in Problem 51 for the acceleration of masses on an Atwood's machine to investigate at what point the moment of inertia of the pulley becomes negligible. Assume $m_A = 0.150$ kg, $m_B = 0.350$ kg, and $R = 0.040$ m. (a) Graph the acceleration as a function of the moment of inertia. (b) Find the acceleration of the masses when the moment of inertia goes to zero. (c) Using your graph to guide you, at what minimum value of I does the calculated acceleration deviate by 2.0% from the acceleration found in part (b)? (d) If the pulley could be thought of as a uniform disk, find the mass of the pulley using the I found in part (c).

Answers to Exercises

A: $f = 0.076$ Hz; $T = 13$ s.

B: $\vec{F}_A$.

C: Yes; $\frac{1}{12} M\ell^2 + M(\frac{1}{2}\ell)^2 = \frac{1}{3} M\ell^2$.

D: 4×10^{17} J.

E: (c).

F: $a = \frac{1}{3} g$.

This skater is doing a spin. When her arms are spread outward horizontally, she spins less fast than when her arms are held close to the axis of rotation. This is an example of the conservation of angular momentum.

Angular momentum, which we study in this Chapter, is conserved only if no net torque acts on the object or system. Otherwise, the rate of change of angular momentum is proportional to the net applied torque—which, if zero, means the angular momentum is *conserved*. In this Chapter we also examine more complicated aspects of rotational motion.

C H A P T E R
11
Angular Momentum; General Rotation

CONTENTS

11–1 Angular Momentum—Objects Rotating About a Fixed Axis

11–2 Vector Cross Product; Torque as a Vector

11–3 Angular Momentum of a Particle

11–4 Angular Momentum and Torque for a System of Particles; General Motion

11–5 Angular Momentum and Torque for a Rigid Object

11–6 Conservation of Angular Momentum

*11–7 The Spinning Top and Gyroscope

*11–8 Rotating Frames of Reference; Inertial Forces

*11–9 The Coriolis Effect

CHAPTER-OPENING QUESTION—Guess now!

You are standing on a platform at rest, but that is free to rotate. You hold a spinning bicycle wheel by its axle as shown here. You then flip the wheel over so its axle points down. What happens then?

(a) The platform starts rotating in the direction the bicycle wheel was originally rotating.

(b) The platform starts rotating in the direction opposite to the original rotation of the bicycle wheel.

(c) The platform stays at rest.

(d) The platform turns only while you are flipping the wheel.

(e) None of these is correct.

I n Chapter 10 we dealt with the kinematics and dynamics of the rotation of a rigid object about an axis whose direction is fixed in an inertial reference frame. We analyzed the motion in terms of the rotational equivalent of Newton's laws (torque plays the role that force does for translational motion), as well as rotational kinetic energy.

To keep the axis of a rotating object fixed, the object must usually be constrained by external supports (such as bearings at the end of an axle). The motion of objects that are not constrained to move about a fixed axis is more difficult to describe and analyze. Indeed, the complete analysis of the general rotational motion of an object (or system of objects) is very complicated, and we will only look at some aspects of general rotational motion in this Chapter.

We start this Chapter by introducing the concept of *angular momentum*, which is the rotational analog of linear momentum. We first treat angular momentum and its conservation for an object rotating about a fixed axis. After that, we examine the vector nature of torque and angular momentum. We will derive some general theorems and apply them to some interesting types of motion.

11–1 Angular Momentum—Objects Rotating About a Fixed Axis

In Chapter 10 we saw that if we use the appropriate angular variables, the kinematic and dynamic equations for rotational motion are analogous to those for ordinary linear motion. In like manner, the linear momentum, $p = mv$, has a rotational analog. It is called **angular momentum**, L, and for an object rotating about a fixed axis with angular velocity ω, it is defined as

$$L = I\omega, \tag{11–1}$$

where I is the moment of inertia. The SI units for L are $kg \cdot m^2/s$; there is no special name for this unit.

We saw in Chapter 9 (Section 9–1) that Newton's second law can be written not only as $\Sigma F = ma$, but also more generally in terms of momentum (Eq. 9–2), $\Sigma F = dp/dt$. In a similar way, the rotational equivalent of Newton's second law, which we saw in Eqs. 10–14 and 10–15 can be written as $\Sigma\tau = I\alpha$, can also be written in terms of angular momentum: since the angular acceleration $\alpha = d\omega/dt$ (Eq. 10–3), then $I\alpha = I(d\omega/dt) = d(I\omega)/dt = dL/dt$, so

$$\Sigma\tau = \frac{dL}{dt}. \tag{11–2}$$

NEWTON'S SECOND LAW FOR ROTATION

This derivation assumes that the moment of inertia, I, remains constant. However, Eq. 11–2 is valid even if the moment of inertia changes, and applies also to a system of objects rotating about a fixed axis where $\Sigma\tau$ is the net external torque (discussed in Section 11–4). Equation 11–2 is Newton's second law for rotational motion about a fixed axis, and is also valid for a moving object if its rotation is about an axis passing through its center of mass (as for Eq. 10–15).

Conservation of Angular Momentum

Angular momentum is an important concept in physics because, under certain conditions, it is a conserved quantity. What are the conditions for which it is conserved? From Eq. 11–2 we see immediately that if the net external torque $\Sigma\tau$ on an object (or system of objects) is zero, then

$$\frac{dL}{dt} = 0 \quad \text{and} \quad L = I\omega = \text{constant.} \qquad [\Sigma\tau = 0]$$

This, then, is the **law of conservation of angular momentum** for a rotating object:

The total angular momentum of a rotating object remains constant if the net external torque acting on it is zero.

CONSERVATION OF ANGULAR MOMENTUM

The law of conservation of angular momentum is one of the great conservation laws of physics, along with those for energy and linear momentum.

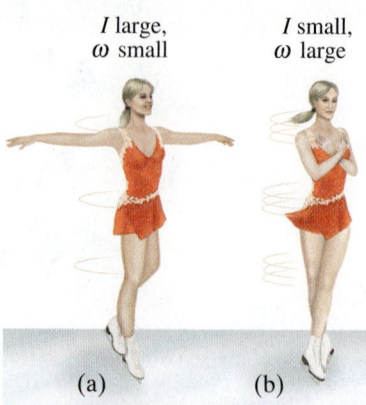

I large,
ω small

I small,
ω large

(a)   (b)

FIGURE 11–1 A skater doing a spin on ice, illustrating conservation of angular momentum: (a) *I* is large and ω is small; (b) *I* is smaller so ω is larger.

FIGURE 11–2 A diver rotates faster when arms and legs are tucked in than when they are outstretched. Angular momentum is conserved.

When there is zero net torque acting on an object, and the object is rotating about a fixed axis or about an axis through its center of mass whose direction doesn't change, we can write

$$I\omega = I_0\omega_0 = \text{constant.}$$

I_0 and ω_0 are the moment of inertia and angular velocity, respectively, about the axis at some initial time ($t = 0$), and I and ω are their values at some other time. The parts of the object may alter their positions relative to one another, so that I changes. But then ω changes as well and the product $I\omega$ remains constant.

Many interesting phenomena can be understood on the basis of conservation of angular momentum. Consider a skater doing a spin on the tips of her skates, Fig. 11–1. She rotates at a relatively low speed when her arms are outstretched, but when she brings her arms in close to her body, she suddenly spins much faster. From the definition of moment of inertia, $I = \Sigma mR^2$, it is clear that when she pulls her arms in closer to the axis of rotation, R is reduced for the arms so her moment of inertia is reduced. Since the angular momentum $I\omega$ remains constant (we ignore the small torque due to friction), if I decreases, then the angular velocity ω must increase. If the skater reduces her moment of inertia by a factor of 2, she will then rotate with twice the angular velocity.

A similar example is the diver shown in Fig. 11–2. The push as she leaves the board gives her an initial angular momentum about her center of mass. When she curls herself into the tuck position, she rotates quickly one or more times. She then stretches out again, increasing her moment of inertia which reduces the angular velocity to a small value, and then she enters the water. The change in moment of inertia from the straight position to the tuck position can be a factor of as much as $3\frac{1}{2}$.

Note that for angular momentum to be conserved, the net torque must be zero, but the net force does not necessarily have to be zero. The net force on the diver in Fig. 11–2, for example, is not zero (gravity is acting), but the net torque about her CM is zero because the force of gravity acts at her center of mass.

EXAMPLE 11–1 **Object rotating on a string of changing length.** A small mass m attached to the end of a string revolves in a circle on a frictionless tabletop. The other end of the string passes through a hole in the table (Fig. 11–3). Initially, the mass revolves with a speed $v_1 = 2.4\,\text{m/s}$ in a circle of radius $R_1 = 0.80\,\text{m}$. The string is then pulled slowly through the hole so that the radius is reduced to $R_2 = 0.48\,\text{m}$. What is the speed, v_2, of the mass now?

APPROACH There is no net torque on the mass m because the force exerted by the string to keep it moving in a circle is exerted toward the axis; hence the lever arm is zero. We can thus apply conservation of angular momentum.

SOLUTION Conservation of angular momentum gives

$$I_1\omega_1 = I_2\omega_2.$$

Our small mass is essentially a particle whose moment of inertia about the hole is $I = mR^2$ (Eq. 10–11), so we have

$$mR_1^2\omega_1 = mR_2^2\omega_2,$$

or

$$\omega_2 = \omega_1\left(\frac{R_1^2}{R_2^2}\right).$$

Then, since $v = R\omega$, we can write

$$v_2 = R_2\omega_2 = R_2\omega_1\left(\frac{R_1^2}{R_2^2}\right) = R_2\frac{v_1}{R_1}\left(\frac{R_1^2}{R_2^2}\right) = v_1\frac{R_1}{R_2}$$

$$= (2.4\,\text{m/s})\left(\frac{0.80\,\text{m}}{0.48\,\text{m}}\right) = 4.0\,\text{m/s}.$$

The speed increases as the radius decreases.

FIGURE 11–3 Example 11–1.

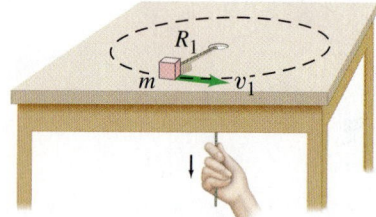

EXAMPLE 11–2 **Clutch.** A simple clutch consists of two cylindrical plates that can be pressed together to connect two sections of an axle, as needed, in a piece of machinery. The two plates have masses $M_A = 6.0$ kg and $M_B = 9.0$ kg, with equal radii $R_0 = 0.60$ m. They are initially separated (Fig. 11–4). Plate M_A is accelerated from rest to an angular velocity $\omega_1 = 7.2$ rad/s in time $\Delta t = 2.0$ s. Calculate (a) the angular momentum of M_A, and (b) the torque required to have accelerated M_A from rest to ω_1. (c) Next, plate M_B, initially at rest but free to rotate without friction, is placed in firm contact with freely rotating plate M_A, and the two plates both rotate at a constant angular velocity ω_2, which is considerably less than ω_1. Why does this happen, and what is ω_2?

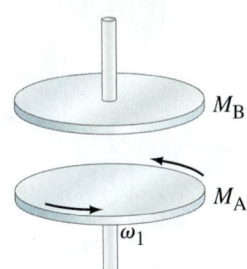

FIGURE 11–4 Example 11–2.

APPROACH We use the definition of angular momentum $L = I\omega$ (Eq. 11–1) plus Newton's second law for rotation, Eq. 11–2.

SOLUTION (a) The angular momentum of M_A will be

$$L_A = I_A \omega_1 = \tfrac{1}{2} M_A R_0^2 \omega_1 = \tfrac{1}{2}(6.0 \text{ kg})(0.60 \text{ m})^2 (7.2 \text{ rad/s}) = 7.8 \text{ kg} \cdot \text{m}^2/\text{s}.$$

(b) The plate started from rest so the torque, assumed constant, was

$$\tau = \frac{\Delta L}{\Delta t} = \frac{7.8 \text{ kg} \cdot \text{m}^2/\text{s} - 0}{2.0 \text{ s}} = 3.9 \text{ m} \cdot \text{N}.$$

(c) Initially, M_A is rotating at constant ω_1 (we ignore friction). When plate B comes in contact, why is their joint rotation speed less? You might think in terms of the torque each exerts on the other upon contact. But quantitatively, it's easier to use conservation of angular momentum, since no external torques are assumed to act. Thus

$$\text{angular momentum before} = \text{angular momentum after}$$
$$I_A \omega_1 = (I_A + I_B)\omega_2.$$

Solving for ω_2 we find

$$\omega_2 = \left(\frac{I_A}{I_A + I_B}\right)\omega_1 = \left(\frac{M_A}{M_A + M_B}\right)\omega_1 = \left(\frac{6.0 \text{ kg}}{15.0 \text{ kg}}\right)(7.2 \text{ rad/s}) = 2.9 \text{ rad/s}.$$

EXAMPLE 11–3 **ESTIMATE** **Neutron star.** Astronomers detect stars that are rotating extremely rapidly, known as neutron stars. A neutron star is believed to form from the inner core of a larger star that collapsed, under its own gravitation, to a star of very small radius and very high density. Before collapse, suppose the core of such a star is the size of our Sun ($r \approx 7 \times 10^5$ km) with mass 2.0 times as great as the Sun, and is rotating at a frequency of 1.0 revolution every 100 days. If it were to undergo gravitational collapse to a neutron star of radius 10 km, what would its rotation frequency be? Assume the star is a uniform sphere at all times, and loses no mass.

PHYSICS APPLIED
Neutron star

APPROACH We assume the star is isolated (no external forces), so we can use conservation of angular momentum for this process. We use r for the radius of a sphere, as compared to R used for distance from an axis of rotation or cylindrical symmetry: see Fig. 10–2.

SOLUTION From conservation of angular momentum,

$$I_1 \omega_1 = I_2 \omega_2,$$

where the subscripts 1 and 2 refer to initial (normal star) and final (neutron star), respectively. Then, assuming no mass is lost in the process,

$$\omega_2 = \left(\frac{I_1}{I_2}\right)\omega_1 = \left(\frac{\tfrac{2}{5} m_1 r_1^2}{\tfrac{2}{5} m_2 r_2^2}\right)\omega_1 = \frac{r_1^2}{r_2^2}\omega_1.$$

The frequency $f = \omega/2\pi$, so

$$f_2 = \frac{\omega_2}{2\pi} = \frac{r_1^2}{r_2^2} f_1$$
$$= \left(\frac{7 \times 10^5 \text{ km}}{10 \text{ km}}\right)^2 \left(\frac{1.0 \text{ rev}}{100 \text{ d}(24 \text{ h/d})(3600 \text{ s/h})}\right) \approx 6 \times 10^2 \text{ rev/s}.$$

Directional Nature of Angular Momentum

Angular momentum is a vector, as we shall discuss later in this Chapter. For now we consider the simple case of an object rotating about a fixed axis, and the direction of $\vec{L}$ is specified by a plus or minus sign, just as we did for one-dimensional linear motion in Chapter 2.

For a symmetrical object rotating about a symmetry axis (such as a cylinder or wheel), the direction of the angular momentum[†] can be taken as the direction of the angular velocity $\vec{\omega}$. That is,

$$\vec{L} = I\vec{\omega}.$$

As a simple example, consider a person standing at rest on a circular platform capable of rotating friction-free about an axis through its center (that is, a simplified merry-go-round). If the person now starts to walk along the edge of the platform, Fig. 11–5a, the platform starts rotating in the opposite direction. Why? One explanation is that the person's foot exerts a force on the platform. Another explanation (and this is the most useful analysis here) is as an example of the conservation of angular momentum. If the person starts walking counterclockwise, the person's angular momentum will be pointed upward along the axis of rotation (remember how we defined the direction of $\vec{\omega}$ using the right-hand rule in Section 10–2). The magnitude of the person's angular momentum will be $L = I\omega = (mR^2)(v/R)$, where v is the person's speed (relative to the Earth, not the platform), R is his distance from the rotation axis, m is his mass, and mR^2 is his moment of inertia if we consider him a particle (mass concentrated at one point). The platform rotates in the opposite direction, so its angular momentum points downward. If the initial total angular momentum was zero (person and platform at rest), it will remain zero after the person starts walking. That is, the upward angular momentum of the person just balances the oppositely directed downward angular momentum of the platform (Fig. 11–5b), so the total vector angular momentum remains zero. Even though the person exerts a force (and torque) on the platform, the platform exerts an equal and opposite torque on the person. So the net torque on the *system* of person plus platform is zero (ignoring friction) and the total angular momentum remains constant.

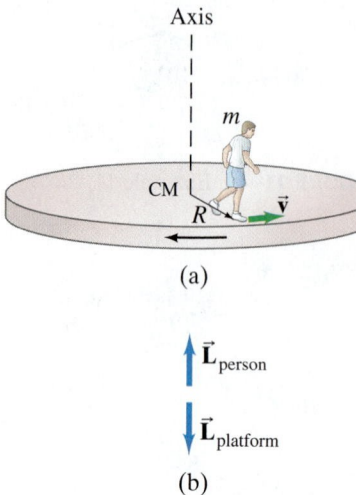

FIGURE 11–5 (a) A person on a circular platform, both initially at rest, begins walking along the edge at speed v. The platform, assumed to be mounted on friction-free bearings, begins rotating in the opposite direction, so that the total angular momentum remains zero, as shown in (b).

EXAMPLE 11–4 **Running on a circular platform.** Suppose a 60-kg person stands at the edge of a 6.0-m-diameter circular platform, which is mounted on frictionless bearings and has a moment of inertia of 1800 kg·m². The platform is at rest initially, but when the person begins running at a speed of 4.2 m/s (with respect to the Earth) around its edge, the platform begins to rotate in the opposite direction as in Fig. 11–5. Calculate the angular velocity of the platform.

APPROACH We use conservation of angular momentum. The total angular momentum is zero initially. Since there is no net torque, $\vec{L}$ is conserved and will remain zero, as in Fig. 11–5. The person's angular momentum is $L_{per} = (mR^2)(v/R)$, and we take this as positive. The angular momentum of the platform is $L_{plat} = -I\omega$.

SOLUTION Conservation of angular momentum gives

$$L = L_{per} + L_{plat}$$

$$0 = mR^2\left(\frac{v}{R}\right) - I\omega.$$

So

$$\omega = \frac{mRv}{I} = \frac{(60\text{ kg})(3.0\text{ m})(4.2\text{ m/s})}{1800\text{ kg·m}^2} = 0.42\text{ rad/s}.$$

NOTE The frequency of rotation is $f = \omega/2\pi = 0.067$ rev/s and the period $T = 1/f = 15$ s per revolution.

[†]For more complicated situations of objects rotating about a fixed axis, there will be a component of $\vec{L}$ along the direction of $\vec{\omega}$ and its magnitude will be equal to $I\vec{\omega}$, but there could be other components as well. If the total angular momentum is conserved, then the component $I\vec{\omega}$ will also be conserved. So our results here can be applied to any rotation about a fixed axis.

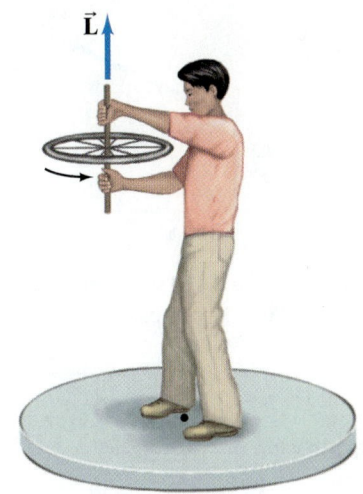

CONCEPTUAL EXAMPLE 11–5 **Spinning bicycle wheel.** Your physics teacher is holding a spinning bicycle wheel while he stands on a stationary frictionless turntable (Fig. 11–6). What will happen if the teacher suddenly flips the bicycle wheel over so that it is spinning in the opposite direction?

RESPONSE We consider the system of turntable, teacher, and bicycle wheel. The total angular momentum initially is $\vec{\mathbf{L}}$ vertically upward. That is also what the system's angular momentum must be afterward, since $\vec{\mathbf{L}}$ is conserved when there is no net torque. Thus, if the wheel's angular momentum after being flipped over is $-\vec{\mathbf{L}}$ downward, then the angular momentum of teacher plus turntable will have to be $+2\vec{\mathbf{L}}$ upward. We can safely predict that the teacher will begin spinning around in the same direction the wheel was spinning originally.

EXERCISE A In Example 11–5, what if he moves the axis only 90° so it is horizontal? (*a*) The same direction and speed as above; (*b*) the same as above, but slower; (*c*) the opposite result.

EXERCISE B Return to the Chapter-Opening Question, page 284, and answer it again now. Try to explain why you may have answered differently the first time.

EXERCISE C Suppose you are standing on the edge of a large freely rotating turntable. If you walk toward the center, (*a*) the turntable slows down; (*b*) the turntable speeds up; (*c*) its rotation speed is unchanged; (*d*) you need to know the walking speed to answer.

FIGURE 11–6 Example 11–5.

11–2 Vector Cross Product; Torque as a Vector

Vector Cross Product

To deal with the vector nature of angular momentum and torque in general, we will need the concept of the *vector cross product* (often called simply the *vector product* or *cross product*). In general, the **vector** or **cross product** of two vectors $\vec{\mathbf{A}}$ and $\vec{\mathbf{B}}$ is defined as another vector $\vec{\mathbf{C}} = \vec{\mathbf{A}} \times \vec{\mathbf{B}}$ *whose magnitude* is

$$C = |\vec{\mathbf{A}} \times \vec{\mathbf{B}}| = AB \sin \theta, \tag{11–3a}$$

where θ is the angle ($< 180°$) *between $\vec{\mathbf{A}}$ and $\vec{\mathbf{B}}$, and whose direction is perpendicular to both $\vec{\mathbf{A}}$ and $\vec{\mathbf{B}}$ in the sense of the right-hand rule*, Fig. 11–7. The angle θ is measured between $\vec{\mathbf{A}}$ and $\vec{\mathbf{B}}$ when their tails are together at the same point. According to the right-hand rule, as shown in Fig. 11–7, you orient your right hand so your fingers point along $\vec{\mathbf{A}}$, and when you bend your fingers they point along $\vec{\mathbf{B}}$. When your hand is correctly oriented in this way, your thumb will point along the direction of $\vec{\mathbf{C}} = \vec{\mathbf{A}} \times \vec{\mathbf{B}}$.

The cross product of two vectors, $\vec{\mathbf{A}} = A_x\hat{\mathbf{i}} + A_y\hat{\mathbf{j}} + A_z\hat{\mathbf{k}}$, and $\vec{\mathbf{B}} = B_x\hat{\mathbf{i}} + B_y\hat{\mathbf{j}} + B_z\hat{\mathbf{k}}$, can be written in component form (see Problem 26) as

FIGURE 11–7 The vector $\vec{\mathbf{C}} = \vec{\mathbf{A}} \times \vec{\mathbf{B}}$ is perpendicular to the plane containing $\vec{\mathbf{A}}$ and $\vec{\mathbf{B}}$; its direction is given by the right-hand rule.

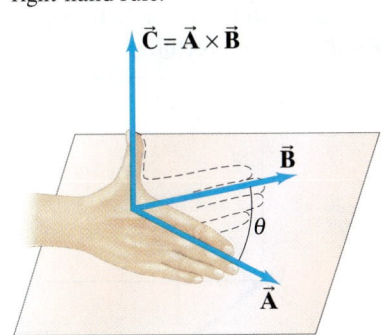

$$\vec{\mathbf{A}} \times \vec{\mathbf{B}} = \begin{vmatrix} \hat{\mathbf{i}} & \hat{\mathbf{j}} & \hat{\mathbf{k}} \\ A_x & A_y & A_z \\ B_x & B_y & B_z \end{vmatrix} \tag{11–3b}$$

$$= (A_yB_z - A_zB_y)\hat{\mathbf{i}} + (A_zB_x - A_xB_z)\hat{\mathbf{j}} + (A_xB_y - A_yB_x)\hat{\mathbf{k}}. \tag{11–3c}$$

Equation 11–3b is meant to be evaluated using the rules of determinants (and obtain Eq. 11–3c).

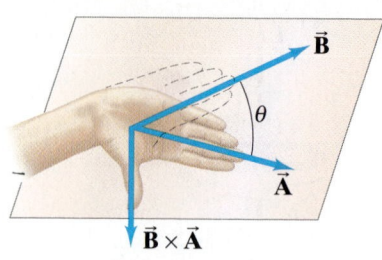

FIGURE 11–8 The vector $\vec{\mathbf{B}} \times \vec{\mathbf{A}}$ equals $-\vec{\mathbf{A}} \times \vec{\mathbf{B}}$; compare to Fig. 11–7.

FIGURE 11–9 Exercise D.

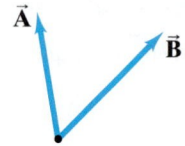

FIGURE 11–10 The torque due to the force $\vec{\mathbf{F}}$ (in the plane of the wheel) starts the wheel rotating counterclockwise so $\vec{\boldsymbol{\omega}}$ and $\vec{\boldsymbol{\alpha}}$ point out of the page.

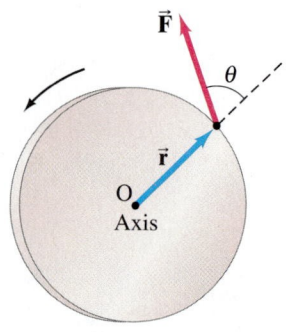

FIGURE 11–11 $\vec{\boldsymbol{\tau}} = \vec{\mathbf{r}} \times \vec{\mathbf{F}}$, where $\vec{\mathbf{r}}$ is the position vector.

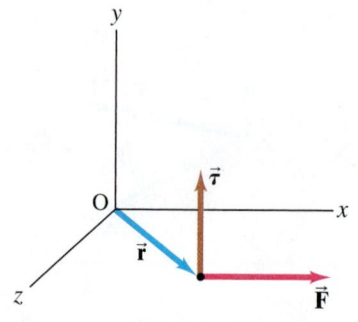

Some properties of the cross product are the following:

$$\vec{\mathbf{A}} \times \vec{\mathbf{A}} = 0 \qquad \text{(11–4a)}$$

$$\vec{\mathbf{A}} \times \vec{\mathbf{B}} = -\vec{\mathbf{B}} \times \vec{\mathbf{A}} \qquad \text{(11–4b)}$$

$$\vec{\mathbf{A}} \times (\vec{\mathbf{B}} + \vec{\mathbf{C}}) = (\vec{\mathbf{A}} \times \vec{\mathbf{B}}) + (\vec{\mathbf{A}} \times \vec{\mathbf{C}}) \qquad \text{[distributive law]} \quad \text{(11–4c)}$$

$$\frac{d}{dt}(\vec{\mathbf{A}} \times \vec{\mathbf{B}}) = \frac{d\vec{\mathbf{A}}}{dt} \times \vec{\mathbf{B}} + \vec{\mathbf{A}} \times \frac{d\vec{\mathbf{B}}}{dt}. \qquad \text{(11–4d)}$$

Equation 11–4a follows from Eqs. 11–3 (since $\theta = 0$). So does Eq. 11–4b, since the magnitude of $\vec{\mathbf{B}} \times \vec{\mathbf{A}}$ is the same as that for $\vec{\mathbf{A}} \times \vec{\mathbf{B}}$, but by the right-hand rule the direction is opposite (see Fig. 11–8). Thus the order of the two vectors is crucial. If you change the order, you change the result. That is, the commutative law does *not* hold for the cross product $(\vec{\mathbf{A}} \times \vec{\mathbf{B}} \neq \vec{\mathbf{B}} \times \vec{\mathbf{A}})$, although it does hold for the dot product of two vectors and for the product of scalars. Note in Eq. 11–4d that the order of quantities in the two products on the right must not be changed (because of Eq. 11–4b).

EXERCISE D For the vectors $\vec{\mathbf{A}}$ and $\vec{\mathbf{B}}$ in the plane of the page as shown in Fig. 11–9, in what direction is (i) $\vec{\mathbf{A}} \cdot \vec{\mathbf{B}}$, (ii) $\vec{\mathbf{A}} \times \vec{\mathbf{B}}$, (iii) $\vec{\mathbf{B}} \times \vec{\mathbf{A}}$? (*a*) Into the page; (*b*) out of the page; (*c*) between $\vec{\mathbf{A}}$ and $\vec{\mathbf{B}}$; (*d*) it is a scalar and has no direction; (*e*) it is zero and has no direction.

The Torque Vector

Torque is an example of a quantity that can be expressed as a cross product. To see this, let us take a simple example: the thin wheel shown in Fig. 11–10 which is free to rotate about an axis through its center at point O. A force $\vec{\mathbf{F}}$ acts at the edge of the wheel, at a point whose position relative to the center O is given by the position vector $\vec{\mathbf{r}}$ as shown. The force $\vec{\mathbf{F}}$ tends to rotate the wheel (assumed initially at rest) counterclockwise, so the angular velocity $\vec{\boldsymbol{\omega}}$ will point out of the page toward the viewer (remember the right-hand rule from Section 10–2). The torque due to $\vec{\mathbf{F}}$ will tend to increase $\vec{\boldsymbol{\omega}}$ so $\vec{\boldsymbol{\alpha}}$ also points outward along the rotation axis. The relation between angular acceleration and torque that we developed in Chapter 10 for an object rotating about a fixed axis is

$$\Sigma\tau = I\alpha,$$

(Eq. 10–14) where I is the moment of inertia. This scalar equation is the rotational equivalent of $\Sigma F = ma$, and we would like to make it a vector equation just as $\Sigma\vec{\mathbf{F}} = m\vec{\mathbf{a}}$ is a vector equation. To do so in the case of Fig. 11–10 we must have the direction of $\vec{\boldsymbol{\tau}}$ point outward along the rotation axis, since $\vec{\boldsymbol{\alpha}}$ ($= d\vec{\boldsymbol{\omega}}/dt$) has that direction; and the magnitude of the torque must be (see Eqs. 10–10 and Fig. 11–10) $\tau = rF_\perp = rF\sin\theta$. We can achieve this by defining the **torque vector** to be the cross product of $\vec{\mathbf{r}}$ and $\vec{\mathbf{F}}$:

$$\vec{\boldsymbol{\tau}} = \vec{\mathbf{r}} \times \vec{\mathbf{F}}. \qquad \text{(11–5)}$$

From the definition of the cross product above (Eq. 11–3a) the magnitude of $\vec{\boldsymbol{\tau}}$ will be $rF\sin\theta$ and the direction will be along the axis, as required for this special case.

We will see in Sections 11–3 through 11–5 that if we take Eq. 11–5 as the *general definition of torque*, then the vector relation $\Sigma\vec{\boldsymbol{\tau}} = I\vec{\boldsymbol{\alpha}}$ will hold in general. Thus we state now that Eq. 11–5 is the general definition of torque. It contains both magnitude and direction information. Note that this definition involves the position vector $\vec{\mathbf{r}}$ and thus the torque is being calculated about a point. We can choose that point O as we wish.

For a particle of mass m on which a force $\vec{\mathbf{F}}$ is applied, we define the torque about a point O as

$$\vec{\boldsymbol{\tau}} = \vec{\mathbf{r}} \times \vec{\mathbf{F}}$$

where $\vec{\mathbf{r}}$ is the position vector of the particle relative to O (Fig. 11–11). If we have a system of particles (which could be the particles making up a rigid object) the total torque $\vec{\boldsymbol{\tau}}$ on the system will be the sum of the torques on the individual particles:

$$\vec{\boldsymbol{\tau}} = \Sigma(\vec{\mathbf{r}}_i \times \vec{\mathbf{F}}_i),$$

where $\vec{\mathbf{r}}_i$ is the position vector of the ith particle and $\vec{\mathbf{F}}_i$ is the net force on the ith particle.

EXAMPLE 11–6 Torque Vector. Suppose the vector $\vec{r}$ is in the xz plane, as in Fig. 11–11, and is given by $\vec{r} = (1.2\,\text{m})\hat{i} + (1.2\,\text{m})\hat{k}$. Calculate the torque vector $\vec{\tau}$ if $\vec{F} = (150\,\text{N})\hat{i}$.

APPROACH We use the determinant form, Eq. 11–3b.

SOLUTION
$$\vec{\tau} = \vec{r} \times \vec{F} = \begin{vmatrix} \hat{i} & \hat{j} & \hat{k} \\ 1.2\,\text{m} & 0 & 1.2\,\text{m} \\ 150\,\text{N} & 0 & 0 \end{vmatrix} = 0\hat{i} + (180\,\text{m}\cdot\text{N})\hat{j} + 0\hat{k}.$$

So τ has magnitude 180 m·N and points along the positive y axis.

EXERCISE E If $\vec{F} = 5.0\,\text{N}\,\hat{i}$ and $\vec{r} = 2.0\,\text{m}\,\hat{j}$, what is $\vec{\tau}$? (*a*) 10 mN, (*b*) −10 mN, (*c*) 10 mN $\hat{k}$, (*d*) −10 mN $\hat{j}$, (*e*) −10 mN $\hat{k}$.

11–3 Angular Momentum of a Particle

The most general way of writing Newton's second law for the translational motion of a particle (or system of particles) is in terms of the linear momentum $\vec{p} = m\vec{v}$ as given by Eq. 9–2 (or 9–5):

$$\Sigma\vec{F} = \frac{d\vec{p}}{dt}.$$

The rotational analog of linear momentum is *angular momentum*. Just as the rate of change of $\vec{p}$ is related to the net force $\Sigma\vec{F}$, so we might expect the rate of change of angular momentum to be related to the net torque. Indeed, we saw this was true in Section 11–1 for the special case of a rigid object rotating about a fixed axis. Now we will see it is true in general. We first treat a single particle.

Suppose a particle of mass m has momentum $\vec{p}$ and position vector $\vec{r}$ with respect to the origin O in some chosen inertial reference frame. Then the general definition of the **angular momentum**, $\vec{L}$, of the particle about point O is the vector cross product of $\vec{r}$ and $\vec{p}$:

$$\vec{L} = \vec{r} \times \vec{p}. \qquad \text{[particle]} \quad \textbf{(11–6)}$$

Angular momentum is a vector.[†] Its direction is perpendicular to both $\vec{r}$ and $\vec{p}$ as given by the right-hand rule (Fig. 11–12). Its magnitude is given by

$$L = rp\sin\theta$$

or

$$L = rp_\perp = r_\perp p$$

where θ is the angle between $\vec{r}$ and $\vec{p}$ and $p_\perp (= p\sin\theta)$ and $r_\perp (= r\sin\theta)$ are the components of $\vec{p}$ and $\vec{r}$ perpendicular to $\vec{r}$ and $\vec{p}$, respectively.

Now let us find the relation between angular momentum and torque for a particle. If we take the derivative of $\vec{L}$ with respect to time we have

$$\frac{d\vec{L}}{dt} = \frac{d}{dt}(\vec{r} \times \vec{p}) = \frac{d\vec{r}}{dt} \times \vec{p} + \vec{r} \times \frac{d\vec{p}}{dt}.$$

But

$$\frac{d\vec{r}}{dt} \times \vec{p} = \vec{v} \times m\vec{v} = m(\vec{v} \times \vec{v}) = 0,$$

since $\sin\theta = 0$ for this case. Thus

$$\frac{d\vec{L}}{dt} = \vec{r} \times \frac{d\vec{p}}{dt}.$$

If we let $\Sigma\vec{F}$ represent the resultant force on the particle, then in an inertial reference frame, $\Sigma\vec{F} = d\vec{p}/dt$ and

$$\vec{r} \times \Sigma\vec{F} = \vec{r} \times \frac{d\vec{p}}{dt} = \frac{d\vec{L}}{dt}.$$

But $\vec{r} \times \Sigma\vec{F} = \Sigma\vec{\tau}$ is the net torque on our particle. Hence

$$\Sigma\vec{\tau} = \frac{d\vec{L}}{dt}. \qquad \text{[particle, inertial frame]} \quad \textbf{(11–7)}$$

The time rate of change of angular momentum of a particle is equal to the net torque applied to it. Equation 11–7 is the rotational equivalent of Newton's

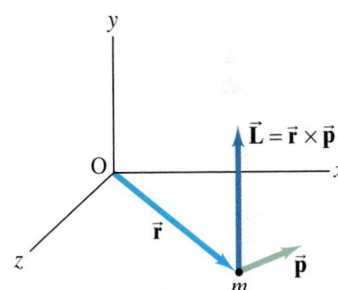

FIGURE 11–12 The angular momentum of a particle of mass m is given by $\vec{L} = \vec{r} \times \vec{p} = \vec{r} \times m\vec{v}$.

[†]Actually a pseudovector; see footnote in Section 10–2.

second law for a particle, written in its most general form. Equation 11–7 is valid only in an inertial frame since only then is it true that $\Sigma\vec{F} = d\vec{p}/dt$, which was used in the proof.

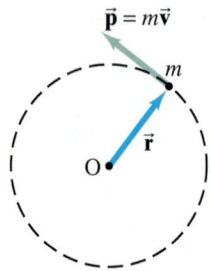

FIGURE 11–13 The angular momentum of a particle of mass m rotating in a circle of radius $\vec{r}$ with velocity $\vec{v}$ is $\vec{L} = \vec{r} \times m\vec{v}$ (Example 11–7).

CONCEPTUAL EXAMPLE 11–7 **A particle's angular momentum.** What is the angular momentum of a particle of mass m moving with speed v in a circle of radius r in a counterclockwise direction?

RESPONSE The value of the angular momentum depends on the choice of the point O. Let us calculate $\vec{L}$ with respect to the center of the circle, Fig. 11–13. Then $\vec{r}$ is perpendicular to $\vec{p}$ so $L = |\vec{r} \times \vec{p}| = rmv$. By the right-hand rule, the direction of $\vec{L}$ is perpendicular to the plane of the circle, outward toward the viewer. Since $v = \omega r$ and $I = mr^2$ for a single particle rotating about an axis a distance r away, we can write

$$L = mvr = mr^2\omega = I\omega.$$

11–4 Angular Momentum and Torque for a System of Particles; General Motion

Relation Between Angular Momentum and Torque

Consider a system of n particles which have angular momenta $\vec{L}_1, \vec{L}_2, \ldots, \vec{L}_n$. The system could be anything from a rigid object to a loose assembly of particles whose positions are not fixed relative to each other. The total angular momentum $\vec{L}$ of the system is defined as the vector sum of the angular momenta of all the particles in the system:

$$\vec{L} = \sum_{i=1}^{n} \vec{L}_i. \tag{11–8}$$

The resultant torque acting on the system is the sum of the net torques acting on all the particles:

$$\vec{\tau}_{net} = \sum \vec{\tau}_i.$$

This sum includes (1) internal torques due to internal forces that particles of the system exert on other particles of the system, and (2) external torques due to forces exerted by objects outside our system. By Newton's third law, the force each particle exerts on another is equal and opposite (and acts along the same line as) the force that the second particle exerts on the first. Hence the sum of all internal torques adds to zero, and

$$\vec{\tau}_{net} = \sum_i \vec{\tau}_i = \sum \vec{\tau}_{ext}.$$

Now we take the time derivative of Eq. 11–8 and use Eq. 11–7 for each particle to obtain

$$\frac{d\vec{L}}{dt} = \sum_i \frac{d\vec{L}_i}{dt} = \sum \vec{\tau}_{ext}$$

or

NEWTON'S SECOND LAW
(rotation, system of particles)

$$\frac{d\vec{L}}{dt} = \sum \vec{\tau}_{ext}. \qquad \text{[inertial reference frame]} \quad \textbf{(11–9a)}$$

This fundamental result states that the time rate of change of the total angular momentum of a system of particles (or a rigid object) equals the resultant external torque on the system. It is the rotational equivalent of Eq. 9–5, $d\vec{P}/dt = \Sigma\vec{F}_{ext}$ for translational motion. Note that $\vec{L}$ and $\Sigma\vec{\tau}$ must be calculated about the same origin O.

Equation 11–9a is valid when $\vec{L}$ and $\vec{\tau}_{ext}$ are calculated with reference to a point fixed in an inertial reference frame. (In the derivation, we used Eq. 11–7 which is valid only in this case.) It is also valid when $\vec{\tau}_{\text{ext}}$ and $\vec{L}$ are calculated about a point which is moving uniformly in an inertial reference frame since such a point can be considered the origin of a second inertial reference frame. It is *not* valid in general when $\vec{\tau}_{\text{ext}}$ and $\vec{L}$ are calculated about a point that is *accelerating*, except for one special (and very important) case—when that point is the center of mass (CM) of the system:

$$\frac{d\vec{L}_{\text{CM}}}{dt} = \sum \vec{\tau}_{\text{CM}}. \qquad \text{[even if accelerating]} \quad \textbf{(11–9b)}$$

NEWTON'S SECOND LAW
(for CM *even if accelerating*)

Equation 11–9b is valid no matter how the CM moves, and $\Sigma\tau_{\text{CM}}$ is the net external torque calculated about the center of mass. The derivation is in the optional subsection below.

It is because of the validity of Eq. 11–9b that we are justified in describing the general motion of a system of particles, as we did in Chapter 10, as *translational* motion of the center of mass plus *rotation* about the center of mass. Equations 11–9b plus 9–5 $\left(d\vec{P}_{\text{CM}}/dt = \Sigma\vec{F}_{\text{ext}}\right)$ provide the more general statement of this principle. (See also Section 9–8.)

*Derivation of $d\vec{L}_{\text{CM}}/dt = \Sigma\vec{\tau}_{\text{CM}}$

The proof of Eq. 11–9b is as follows. Let $\vec{r}_i$ be the position vector of the ith particle in an inertial reference frame, and $\vec{r}_{\text{CM}}$ be the position vector of the center of mass of the system in this reference frame. The position of the ith particle with respect to the CM is $\vec{r}_i^*$ where (see Fig. 11–14)

$$\vec{r}_i = \vec{r}_{\text{CM}} + \vec{r}_i^*.$$

If we multiply each term by m_i and take the derivative of this equation, we can write

$$\vec{p}_i = m_i\frac{d\vec{r}_i}{dt} = m_i\frac{d}{dt}(\vec{r}_i^* + \vec{r}_{\text{CM}}) = m_i\vec{v}_i^* + m_i\vec{v}_{\text{CM}} = \vec{p}_i^* + m_i\vec{v}_{\text{CM}}.$$

The angular momentum with respect to the CM is

$$\vec{L}_{\text{CM}} = \sum_i(\vec{r}_i^* \times \vec{p}_i^*) = \sum_i \vec{r}_i^* \times (\vec{p}_i - m_i\vec{v}_{\text{CM}}).$$

Then, taking the time derivative, we have

$$\frac{d\vec{L}_{\text{CM}}}{dt} = \sum_i\left(\frac{d\vec{r}_i^*}{dt} \times \vec{p}_i^*\right) + \sum_i\left(\vec{r}_i^* \times \frac{d\vec{p}_i^*}{dt}\right).$$

The first term on the right is $\vec{v}_i^* \times m\vec{v}_i^*$ and equals zero because $\vec{v}_i^*$ is parallel to itself ($\sin\theta = 0$). Thus

$$\frac{d\vec{L}_{\text{CM}}}{dt} = \sum_i\vec{r}_i^* \times \frac{d}{dt}(\vec{p}_i - m_i\vec{v}_{\text{CM}})$$

$$= \sum_i\vec{r}_i^* \times \frac{d\vec{p}_i}{dt} - \left(\sum_i m_i\vec{r}_i^*\right) \times \frac{d\vec{v}_{\text{CM}}}{dt}.$$

The second term on the right is zero since, by Eq. 9–12, $\Sigma m_i\vec{r}_i^* = M\vec{r}_{\text{CM}}^*$, and $\vec{r}_{\text{CM}}^* = 0$ by definition (the position of the CM is at the origin of the CM reference frame). Furthermore, by Newton's second law we have

$$\frac{d\vec{p}_i}{dt} = \vec{F}_i,$$

where $\vec{F}_i$ is the net force on m_i. (Note that $d\vec{p}_i^*/dt \neq \vec{F}_i$ because the CM may be accelerating and Newton's second law does not hold in a noninertial reference frame.) Consequently

$$\frac{d\vec{L}_{\text{CM}}}{dt} = \sum_i\vec{r}_i^* \times \vec{F}_i = \sum_i(\vec{\tau}_i)_{\text{CM}} = \Sigma\vec{\tau}_{\text{CM}},$$

where $\Sigma\vec{\tau}_{\text{CM}}$ is the resultant external torque on the entire system calculated about the CM. (By Newton's third law, the sum over all the $\vec{\tau}_i$ eliminates the net torque due to internal forces, as we saw on p. 292.) This last equation is Eq. 11–9b, and this concludes its proof.

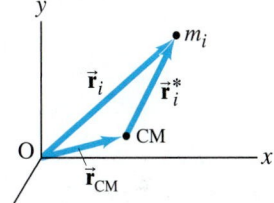

FIGURE 11–14 The position of m_i in the inertial frame is $\vec{r}_i$; with regard to the CM (which could be accelerating) it is $\vec{r}_i^*$, where $\vec{r}_i = \vec{r}_i^* + \vec{r}_{\text{CM}}$ and $\vec{r}_{\text{CM}}$ is the position of the CM in the inertial frame.

To summarize, the relation

$$\sum \vec{\tau}_{ext} = \frac{d\vec{L}}{dt}$$

is valid *only* when $\vec{\tau}_{ext}$ and $\vec{L}$ are calculated with respect to either (1) the origin of an inertial reference frame or (2) the center of mass of a system of particles (or of a rigid object).

11–5 Angular Momentum and Torque for a Rigid Object

Let us now consider the rotation of a rigid object about an axis that has a fixed direction in space, using the general principles just developed.

Let us calculate the component of angular momentum along the rotation axis of the rotating object. We will call this component L_ω since the angular velocity $\vec{\omega}$ points along the rotation axis. For each particle of the object,

$$\vec{L}_i = \vec{r}_i \times \vec{p}_i.$$

Let ϕ be the angle between $\vec{L}_i$ and the rotation axis. (See Fig. 11–15; ϕ is *not* the angle between $\vec{r}_i$ and $\vec{p}_i$, which is 90°). Then the component of $\vec{L}_i$ along the rotation axis is

$$L_{i\omega} = r_i p_i \cos\phi = m_i v_i r_i \cos\phi,$$

where m_i is the mass and v_i the velocity of the ith particle. Now $v_i = R_i \omega$ where ω is the angular velocity of the object and R_i is the perpendicular distance of m_i from the axis of rotation. Furthermore, $R_i = r_i \cos\phi$, as can be seen in Fig. 11–15, so

$$L_{i\omega} = m_i v_i (r_i \cos\phi) = m_i R_i^2 \omega.$$

We sum over all the particles to obtain

$$L_\omega = \sum_i L_{i\omega} = \left(\sum_i m_i R_i^2 \right)\omega.$$

But $\sum m_i R_i^2$ is the moment of inertia I of the object about the axis of rotation. Therefore the component of the total angular momentum along the rotation axis is given by

$$L_\omega = I\omega. \qquad (11\text{--}10)$$

Note that we would obtain Eq. 11–10 no matter where we choose the point O (for measuring $\vec{r}_i$) as long as it is on the axis of rotation. Equation 11–10 is the same as Eq. 11–1, which we have now proved from the general definition of angular momentum.

If the object rotates about a symmetry axis through the center of mass, then L_ω is the only component of $\vec{L}$, as we now show. For each point on one side of the axis there will be a corresponding point on the opposite side. We can see from Fig. 11–15 that each $\vec{L}_i$ has a component parallel to the axis $(L_{i\omega})$ and a component perpendicular to the axis. The components parallel to the axis add together for each pair of opposite points, but the components perpendicular to the axis for opposite points will have the same magnitude but opposite direction and so will cancel. Hence, for an object rotating about a symmetry axis, the angular momentum vector is parallel to the axis and we can write

$$\vec{L} = I\vec{\omega}, \qquad [\text{rotation axis = symmetry axis, through CM}] \quad (11\text{--}11)$$

where $\vec{L}$ is measured relative to the center of mass.

The general relation between angular momentum and torque is Eq. 11–9:

$$\sum \vec{\tau} = \frac{d\vec{L}}{dt}.$$

where $\sum \vec{\tau}$ and $\vec{L}$ are calculated either about (1) the origin of an inertial reference frame, or (2) the center of mass of the system. This is a vector relation, and must

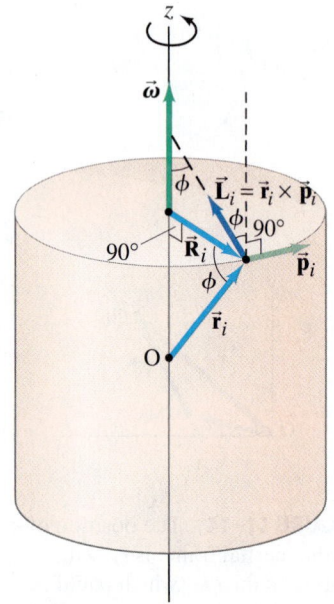

FIGURE 11–15
Calculating $L_\omega = L_z = \Sigma L_{iz}$. Note that $\vec{L}_i$ is perpendicular to $\vec{r}_i$, and $\vec{R}_i$ is perpendicular to the z axis, so the three angles marked ϕ are equal.

therefore be valid for each component. Hence, for a rigid object, the component along the rotation axis is

$$\Sigma\tau_{\text{axis}} = \frac{dL_\omega}{dt} = \frac{d}{dt}(I\omega) = I\frac{d\omega}{dt} = I\alpha,$$

which is valid for a rigid object rotating about an axis fixed relative to the object; also this axis must be either (1) fixed in an inertial system or (2) passing through the CM of the object. This is equivalent to Eqs. 10–14 and 10–15, which we now see are special cases of Eq. 11–9, $\Sigma\vec{\tau} = d\vec{L}/dt$.

EXAMPLE 11–8 **Atwood's machine.** An *Atwood machine* consists of two masses, m_A and m_B, which are connected by an inelastic cord of negligible mass that passes over a pulley, Fig. 11–16. If the pulley has radius R_0 and moment of inertia I about its axle, determine the acceleration of the masses m_A and m_B, and compare to the situation where the moment of inertia of the pulley is ignored.

APPROACH We first determine the angular momentum of the system, and then apply Newton's second law, $\tau = dL/dt$.

SOLUTION The angular momentum is calculated about an axis along the axle through the center O of the pulley. The pulley has angular momentum $I\omega$, where $\omega = v/R_0$ and v is the velocity of m_A and m_B at any instant. The angular momentum of m_A is $R_0 m_A v$ and that of m_B is $R_0 m_B v$. The total angular momentum is

$$L = (m_A + m_B)vR_0 + I\frac{v}{R_0}.$$

The external torque on the system, calculated about the axis O (taking clockwise as positive), is

$$\tau = m_B g R_0 - m_A g R_0.$$

(The force on the pulley exerted by the support on its axle gives rise to no torque because the lever arm is zero.) We apply Eq. 11–9a:

$$\tau = \frac{dL}{dt}$$

$$(m_B - m_A)gR_0 = (m_A + m_B)R_0\frac{dv}{dt} + \frac{I}{R_0}\frac{dv}{dt}.$$

Solving for $a = dv/dt$, we get

$$a = \frac{dv}{dt} = \frac{(m_B - m_A)g}{(m_A + m_B) + I/R_0^2}.$$

If we were to ignore I, $a = (m_B - m_A)g/(m_B + m_A)$ and we see that the effect of the moment of inertia of the pulley is to slow down the system. This is just what we would expect.

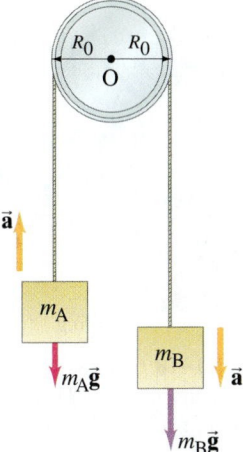

FIGURE 11–16 Atwood's machine, Example 11–8. We also discussed this in Example 4–13.

CONCEPTUAL EXAMPLE 11–9 **Bicycle wheel.** Suppose you are holding a bicycle wheel by a handle connected to its axle as in Fig. 11–17a. The wheel is spinning rapidly so its angular momentum $\vec{L}$ points horizontally as shown. Now you suddenly try to tilt the axle upward as shown by the dashed line in Fig. 11–17a (so the CM moves vertically). You expect the wheel to go up (and it would if it weren't rotating), but it unexpectedly swerves to the right! Explain.

RESPONSE To explain this seemingly odd behavior—you may need to do it to believe it—we only need to use the relation $\vec{\tau}_{\text{net}} = d\vec{L}/dt$. In the short time Δt, you exert a net torque (about an axis through your wrist) that points along the x axis perpendicular to $\vec{L}$. Thus the change in $\vec{L}$ is

$$\Delta\vec{L} \approx \vec{\tau}_{\text{net}}\Delta t;$$

so $\Delta\vec{L}$ must also point (approximately) along the x axis, since $\vec{\tau}_{\text{net}}$ does (Fig. 11–17b). Thus the new angular momentum, $\vec{L} + \Delta\vec{L}$, points to the right, looking along the axis of the wheel, as shown in Fig. 11–17b. Since the angular momentum is directed along the axle of the wheel, we see that the axle, which now is along $\vec{L} + \Delta\vec{L}$, must move sideways to the right, which is what we observe.

FIGURE 11–17 When you try to tilt a rotating bicycle wheel vertically upward, it swerves to the side instead.

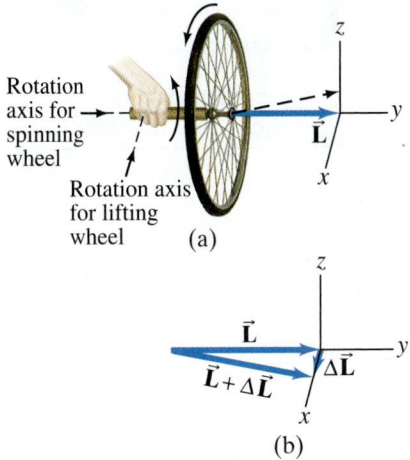

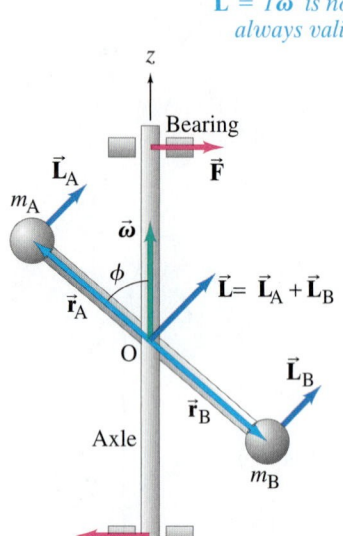

FIGURE 11–18 In this system $\vec{\mathbf{L}}$ and $\vec{\boldsymbol{\omega}}$ are not parallel. This is an example of rotational imbalance.

Ⓧ **PHYSICS APPLIED**

Automobile wheel balancing

FIGURE 11–19 Unbalanced automobile wheel.

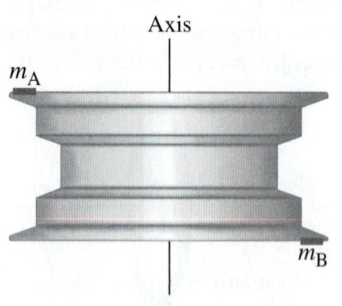

Although Eq. 11–11, $\vec{\mathbf{L}} = I\vec{\boldsymbol{\omega}}$, is often very useful, it is not valid in general unless the rotation axis is along a *symmetry* axis through the center of mass. Nonetheless, it can be shown that every rigid object, no matter what its shape, has three "principal axes" about which Eq. 11–11 is valid (we will not go into the details here). As an example of a case where Eq. 11–11 is not valid, consider the nonsymmetrical object shown in Fig. 11–18. It consists of two equal masses, m_A and m_B, attached to the ends of a rigid (massless) rod which makes an angle ϕ with the axis of rotation. We calculate the angular momentum about the CM at point O. At the moment shown, m_A is coming toward the viewer, and m_B is moving away, so $\vec{\mathbf{L}}_A = \vec{\mathbf{r}}_A \times \vec{\mathbf{p}}_A$ and $\vec{\mathbf{L}}_B = \vec{\mathbf{r}}_B \times \vec{\mathbf{p}}_B$ are as shown. The total angular momentum is $\vec{\mathbf{L}} = \vec{\mathbf{L}}_A + \vec{\mathbf{L}}_B$, which is clearly *not* along $\vec{\boldsymbol{\omega}}$ if $\phi \neq 90°$.

*Rotational Imbalance

Let us go one step further with the system shown in Fig. 11–18, since it is a fine illustration of $\Sigma\vec{\boldsymbol{\tau}} = d\vec{\mathbf{L}}/dt$. If the system rotates with constant angular velocity, ω, the magnitude of $\vec{\mathbf{L}}$ will not change, but its direction will. As the rod and two masses rotate about the z axis, $\vec{\mathbf{L}}$ also rotates about the axis. At the moment shown in Fig. 11–18, $\vec{\mathbf{L}}$ is in the plane of the paper. A time dt later, when the rod has rotated through an angle $d\theta = \omega\, dt$, $\vec{\mathbf{L}}$ will also have rotated through an angle $d\theta$ (it remains perpendicular to the rod). $\vec{\mathbf{L}}$ will then have a component pointing into the page. Thus $d\vec{\mathbf{L}}$ points into the page and so must $d\vec{\mathbf{L}}/dt$. Because

$$\Sigma\vec{\boldsymbol{\tau}} = \frac{d\vec{\mathbf{L}}}{dt},$$

we see that a net torque, directed into the page at the moment shown, must be applied to the axle on which the rod is mounted. The torque is supplied by bearings (or other constraint) at the ends of the axle. The forces $\vec{\mathbf{F}}$ exerted by the bearings on the axle are shown in Fig. 11–18. The direction of each force $\vec{\mathbf{F}}$ rotates as the system does, always being in the plane of $\vec{\mathbf{L}}$ and $\vec{\boldsymbol{\omega}}$ for this system. If the torque due to these forces were not present, the system would not rotate about the fixed axis as desired.

The axle tends to move in the direction of $\vec{\mathbf{F}}$ and thus tends to wobble as it rotates. This has many practical applications, such as the vibrations felt in a car whose wheels are not balanced. Consider an automobile wheel that is symmetrical except for an extra mass m_A on one rim and an equal mass m_B opposite it on the other rim, as shown in Fig. 11–19. Because of the nonsymmetry of m_A and m_B, the wheel bearings would have to exert a force perpendicular to the axle at all times simply to keep the wheel rotating, just as in Fig. 11–18. The bearings would wear excessively and the wobble of the wheel would be felt by occupants of the car. When the wheels are balanced, they rotate smoothly without wobble. This is why "dynamic balancing" of automobile wheels and tires is important. The wheel of Fig. 11–19 would balance *statically* just fine. If equal masses m_C and m_D are added symmetrically, below m_A and above m_B, the wheel will be balanced dynamically as well ($\vec{\mathbf{L}}$ will be parallel to $\vec{\boldsymbol{\omega}}$, and $\vec{\boldsymbol{\tau}}_{\text{ext}} = 0$).

EXAMPLE 11–10 **Torque on imbalanced system.** Determine the magnitude of the net torque τ_{net} needed to keep the system turning in Fig. 11–18.

APPROACH Figure 11–20 is a view of the angular momentum vector, looking down the rotation axis (z axis) of the object depicted in Fig. 11–18, as it rotates. $L\cos\phi$ is the component of $\vec{\mathbf{L}}$ perpendicular to the axle (points to the right in Fig. 11–18). We find dL from Fig. 11–20 and use $\tau_{\text{net}} = dL/dt$.

SOLUTION In a time dt, $\vec{\mathbf{L}}$ changes by an amount (Fig. 11–20 and Eq. 10–2b)

$$dL = (L\cos\phi)\,d\theta = L\cos\phi\,\omega\,dt,$$

where $\omega = d\theta/dt$. Hence

$$\tau_{\text{net}} = \frac{dL}{dt} = \omega L \cos\phi.$$

Now $L = L_A + L_B = r_A m_A v_A + r_B m_B v_B = r_A m_A (\omega r_A \sin\phi) + r_B m_B (\omega r_B \sin\phi) = (m_A r_A^2 + m_B r_B^2)\omega \sin\phi$. Since $I = (m_A r_A^2 + m_B r_B^2)\sin^2\phi$ is the moment of inertia about the axis of rotation, then $L = I\omega/\sin\phi$. So

$$\tau_{\text{net}} = \omega L \cos\phi = (m_A r_A^2 + m_B r_B^2)\omega^2 \sin\phi \cos\phi = I\omega^2/\tan\phi.$$

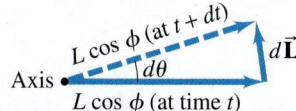

FIGURE 11–20 Angular momentum vector looking down along the rotation axis of the system of Fig. 11–18 as it rotates during a time dt.

The situation of Fig. 11–18 illustrates the usefulness of the vector nature of torque and angular momentum. If we had considered only the components of angular momentum and torque along the rotation axis, we could not have calculated the torque due to the bearings (since the forces $\vec{\mathbf{F}}$ act at the axle and hence produce no torque along that axis). By using the concept of vector angular momentum we have a far more powerful technique for understanding and for attacking problems.

11–6 Conservation of Angular Momentum

In Chapter 9 we saw that the most general form of Newton's second law for the translational motion of a particle or system of particles is

$$\sum \vec{\mathbf{F}}_{\text{ext}} = \frac{d\vec{\mathbf{P}}}{dt},$$

where $\vec{\mathbf{P}}$ is the (linear) momentum, defined as $m\vec{\mathbf{v}}$ for a particle, or $M\vec{\mathbf{v}}_{\text{CM}}$ for a system of particles of total mass M whose CM moves with velocity $\vec{\mathbf{v}}_{\text{CM}}$, and $\sum\vec{\mathbf{F}}_{\text{ext}}$ is the net external force acting on the particle or system. This relation is valid only in an inertial reference frame.

In this Chapter, we have found a similar relation to describe the general rotation of a system of particles (including rigid objects):

$$\sum \vec{\boldsymbol{\tau}} = \frac{d\vec{\mathbf{L}}}{dt},$$

where $\sum\vec{\boldsymbol{\tau}}$ is the net external torque acting on the system, and $\vec{\mathbf{L}}$ is the total angular momentum. This relation is valid when $\sum\vec{\boldsymbol{\tau}}$ and $\vec{\mathbf{L}}$ are calculated about a point fixed in an inertial reference frame, or about the CM of the system.

For translational motion, if the net force on the system is zero, $d\vec{\mathbf{P}}/dt = 0$, so the total linear momentum of the system remains constant. This is the law of conservation of linear momentum. For rotational motion, if the net torque on the system is zero, then

$$\frac{d\vec{\mathbf{L}}}{dt} = 0 \quad \text{and} \quad \vec{\mathbf{L}} = \text{constant}. \qquad \left[\sum\vec{\boldsymbol{\tau}} = 0\right] \quad \textbf{(11–12)}$$

In words:

The total angular momentum of a system remains constant if the net external torque acting on the system is zero.

CONSERVATION OF ANGULAR MOMENTUM

This is the **law of conservation of angular momentum** in full vector form. It ranks with the laws of conservation of energy and linear momentum (and others to be discussed later) as one of the great laws of physics. In Section 11–1 we saw some Examples of this important law applied to the special case of a rigid object rotating about a fixed axis. Here we have it in general form. We use it now in interesting Examples.

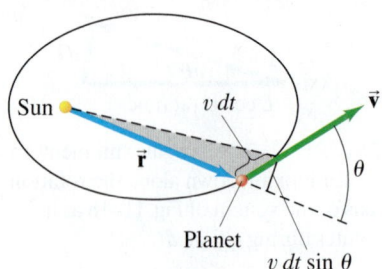

FIGURE 11–21 Kepler's second law of planetary motion (Example 11–11).

EXAMPLE 11–11 **Kepler's second law derived.** Kepler's second law states that each planet moves so that a line from the Sun to the planet sweeps out equal areas in equal times (Section 6–5). Use conservation of angular momentum to show this.

APPROACH We determine the angular momentum of a planet in terms of the area swept out with the help of Fig. 11–21.

SOLUTION The planet moves in an ellipse as shown in Fig. 11–21. In a time dt, the planet moves a distance $v\,dt$ and sweeps out an area dA equal to the area of a triangle of base r and height $v\,dt\sin\theta$ (shown exaggerated in Fig. 11–21). Hence

$$dA = \tfrac{1}{2}(r)(v\,dt\sin\theta)$$

and

$$\frac{dA}{dt} = \tfrac{1}{2}rv\sin\theta.$$

The magnitude of the angular momentum $\vec{L}$ about the Sun is

$$L = |\vec{r} \times m\vec{v}| = mrv\sin\theta,$$

so

$$\frac{dA}{dt} = \frac{1}{2m}L.$$

But $L = $ constant, since the gravitational force $\vec{F}$ is directed toward the Sun so the torque it produces is zero (we ignore the pull of the other planets). Hence $dA/dt = $ constant, which is what we set out to prove.

FIGURE 11–22 Bullet strikes and becomes embedded in cylinder at its edge (Example 11–12).

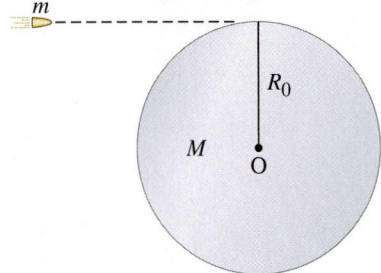

EXAMPLE 11–12 **Bullet strikes cylinder edge.** A bullet of mass m moving with velocity v strikes and becomes embedded at the edge of a cylinder of mass M and radius R_0, as shown in Fig. 11–22. The cylinder, initially at rest, begins to rotate about its symmetry axis, which remains fixed in position. Assuming no frictional torque, what is the angular velocity of the cylinder after this collision? Is kinetic energy conserved?

APPROACH We take as our system the bullet and cylinder, on which there is no net external torque. Thus we can use conservation of angular momentum, and we calculate all angular momenta about the center O of the cylinder.

SOLUTION Initially, because the cylinder is at rest, the total angular momentum about O is solely that of the bullet:

$$L = |\vec{r} \times \vec{p}| = R_0 mv,$$

since R_0 is the perpendicular distance of $\vec{p}$ from O. After the collision, the cylinder $\left(I_{\text{cyl}} = \tfrac{1}{2}MR_0^2\right)$ rotates with the bullet $\left(I_b = mR_0^2\right)$ embedded in it at angular velocity ω:

$$L = I\omega = \left(I_{\text{cyl}} + mR_0^2\right)\omega = \left(\tfrac{1}{2}M + m\right)R_0^2\omega.$$

Hence, because angular momentum is conserved, we find that ω is

$$\omega = \frac{L}{\left(\tfrac{1}{2}M + m\right)R_0^2} = \frac{mvR_0}{\left(\tfrac{1}{2}M + m\right)R_0^2} = \frac{mv}{\left(\tfrac{1}{2}M + m\right)R_0}.$$

Angular momentum is conserved in this collision, but kinetic energy is not:

$$
\begin{aligned}
K_f - K_i &= \tfrac{1}{2}I_{\text{cyl}}\omega^2 + \tfrac{1}{2}\left(mR_0^2\right)\omega^2 - \tfrac{1}{2}mv^2 \\
&= \tfrac{1}{2}\left(\tfrac{1}{2}MR_0^2\right)\omega^2 + \tfrac{1}{2}\left(mR_0^2\right)\omega^2 - \tfrac{1}{2}mv^2 \\
&= \tfrac{1}{2}\left(\tfrac{1}{2}M + m\right)\left(\frac{mv}{\tfrac{1}{2}M + m}\right)^2 - \tfrac{1}{2}mv^2 \\
&= -\frac{mM}{2M + 4m}v^2,
\end{aligned}
$$

which is less than zero. Hence $K_f < K_i$. This energy is transformed to thermal energy as a result of the inelastic collision.

*11–7 The Spinning Top and Gyroscope

The motion of a rapidly spinning top, or a gyroscope, is an interesting example of rotational motion and of the use of the vector equation

$$\sum \vec{\tau} = \frac{d\vec{L}}{dt}.$$

Consider a symmetrical top of mass M spinning rapidly about its symmetry axis, as in Fig. 11–23. The top is balanced on its tip at point O in an inertial reference frame. If the axis of the top makes an angle ϕ to the vertical (z axis), when the top is carefully released its axis will move, sweeping out a cone about the vertical as shown by the dashed lines in Fig. 11–23. This type of motion, in which a torque produces a change in the direction of the rotation axis, is called **precession**. The rate at which the rotation axis moves about the vertical (z) axis is called the angular velocity of precession, Ω (capital Greek omega). Let us now try to understand the reasons for this motion, and calculate Ω.

If the top were not spinning, it would immediately fall to the ground when released due to the pull of gravity. The apparent mystery of a top is that when it is spinning, it does not immediately fall to the ground but instead precesses—it moves slowly sideways. But this is not really so mysterious if we examine it from the point of view of angular momentum and torque, which we calculate about the point O. When the top is spinning with angular velocity ω about its symmetry axis, it has an angular momentum $\vec{L}$ directed along its axis, as shown in Fig. 11–23. (There is also angular momentum due to the precessional motion, so that the total $\vec{L}$ is not exactly along the axis of the top; but if $\Omega \ll \omega$, which is usually the case, we can ignore this.) To change the angular momentum, a torque is required. If no torque were applied to the top, $\vec{L}$ would remain constant in magnitude and direction; the top would neither fall nor precess. But the slightest tip to the side results in a net torque about O, equal to $\vec{\tau}_{net} = \vec{r} \times M\vec{g}$, where $\vec{r}$ is the position vector of the top's center of mass with respect to O, and M is the mass of the top. The direction of $\vec{\tau}_{net}$ is perpendicular to both $\vec{r}$ and $M\vec{g}$ and by the right-hand rule is, as shown in Fig. 11–23, in the horizontal (xy) plane. The change in $\vec{L}$ in a time dt is

$$d\vec{L} = \vec{\tau}_{net} \, dt,$$

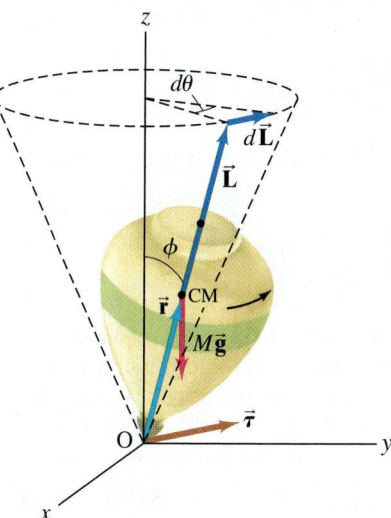

FIGURE 11–23 Spinning top.

which is perpendicular to $\vec{L}$ and horizontal (parallel to $\vec{\tau}_{net}$), as shown in Fig. 11–23. Since $d\vec{L}$ is perpendicular to $\vec{L}$, the magnitude of $\vec{L}$ does not change. Only the direction of $\vec{L}$ changes. Since $\vec{L}$ points along the axis of the top, we see that this axis moves to the right in Fig. 11–23. That is, the upper end of the top's axis moves in a horizontal direction perpendicular to $\vec{L}$. This explains why the top precesses rather than falls. The vector $\vec{L}$ and the top's axis move together in a horizontal circle. As they do so, $\vec{\tau}_{net}$ and $d\vec{L}$ rotate as well so as to be horizontal and perpendicular to $\vec{L}$.

To determine Ω, we see from Fig. 11–23 that the angle $d\theta$ (which is in a horizontal plane) is related to dL by

$$dL = L \sin\phi \, d\theta,$$

since $\vec{L}$ makes an angle ϕ to the z axis. The angular velocity of precession is $\Omega = d\theta/dt$, which becomes (since $d\theta = dL/L\sin\phi$)

$$\Omega = \frac{1}{L\sin\phi} \frac{dL}{dt} = \frac{\tau}{L\sin\phi}. \qquad \text{[spinning top]} \quad \textbf{(11–13a)}$$

But $\tau_{net} = |\vec{r} \times M\vec{g}| = rMg\sin\phi$ [because $\sin(\pi - \phi) = \sin\phi$] so we can also write

$$\Omega = \frac{Mgr}{L}. \qquad \text{[spinning top]} \quad \textbf{(11–13b)}$$

Thus the rate of precession does not depend on the angle ϕ; but it is inversely proportional to the top's angular momentum. The faster the top spins, the greater L is and the slower the top precesses.

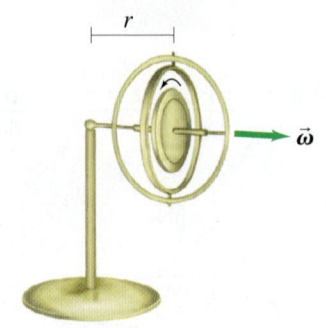

FIGURE 11–24 A toy gyroscope.

From Eq. 11–1 (or Eq. 11–11) we can write $L = I\omega$, where I and ω are the moment of inertia and angular velocity of the spinning top about its spin axis. Then Eq. 11–13b for the top's precession angular velocity becomes

$$\Omega = \frac{Mgr}{I\omega}. \tag{11–13c}$$

Equations 11–13 apply also to a toy gyroscope, which consists of a rapidly spinning wheel mounted on an axle (Fig. 11–24). One end of the axle rests on a support. The other end of the axle is free and will precess like a top if its "spin" angular velocity ω is large compared to the precession rate ($\omega \gg \Omega$). As ω decreases due to friction and air resistance, the gyroscope will begin to fall, just as does a top.

*11–8 Rotating Frames of Reference; Inertial Forces

Inertial and Noninertial Reference Frames

Up to now, we have examined the motion of objects, including circular and rotational motion, from the outside, as observers fixed on the Earth. Sometimes it is convenient to place ourselves (in theory, if not physically) into a reference frame that is rotating. Let us examine the motion of objects from the point of view, or frame of reference, of persons seated on a rotating platform such as a merry-go-round. It looks to them as if the rest of the world is going around *them*. But let us focus attention on what they observe when they place a tennis ball on the floor of the rotating platform, which we assume is frictionless. If they put the ball down gently, without giving it any push, they will observe that it accelerates from rest and moves outward as shown in Fig. 11–25a. According to Newton's first law, an object initially at rest should stay at rest if no net force acts on it. But, according to the observers on the rotating platform, the ball starts moving even though there is no net force acting on it. To observers on the ground this is all very clear: the ball has an initial velocity when it is released (because the platform is moving), and it simply continues moving in a straight-line path as shown in Fig. 11–25b, in accordance with Newton's first law.

But what shall we do about the frame of reference of the observers on the rotating platform? Since the ball starts moving without any net force on it, Newton's first law, the law of inertia, does not hold in this rotating frame of reference. For this reason, such a frame is called a **noninertial reference frame**. An **inertial reference frame** (as we discussed in Chapter 4) is one in which the law of inertia—Newton's first law—does hold, and so do Newton's second and third laws. In a noninertial reference frame, such as our rotating platform, Newton's second law also does not hold. For instance in the situation described above, there is no net force on the ball; yet, with respect to the rotating platform, the ball accelerates.

Fictitious (Inertial) Forces

Because Newton's laws do not hold when observations are made with respect to a rotating frame of reference, calculation of motion can be complicated. However, we can still make use of Newton's laws in such a reference frame if we make use of a trick. The ball on the rotating platform of Fig. 11–25a flies outward when released (even though no force is actually acting on it). So the trick we use is to write down the equation $\Sigma F = ma$ as if a force equal to mv^2/r (or $m\omega^2 r$) were acting radially outward on the object in addition to any other forces that may be acting. This extra force, which might be designated as "centrifugal force" since it *seems* to act outward, is called a **fictitious force** or **pseudoforce**. It is a pseudoforce ("pseudo" means "false") because there is no object that exerts this force. Furthermore, when viewed from an inertial reference frame, the effect doesn't exist at all. We have made up this pseudoforce so that we can make calculations in a noninertial frame using Newton's second law, $\Sigma F = ma$. Thus the observer in the noninertial frame of Fig. 11–25a uses Newton's second law for the ball's outward motion by assuming that a force equal to mv^2/r acts on it. Such pseudoforces are also called **inertial forces** since they arise only because the reference frame is not an inertial one.

FIGURE 11–25 Path of a ball released on a rotating merry-go-round (a) in the reference frame of the merry-go-round, and (b) in a reference frame fixed on the ground.

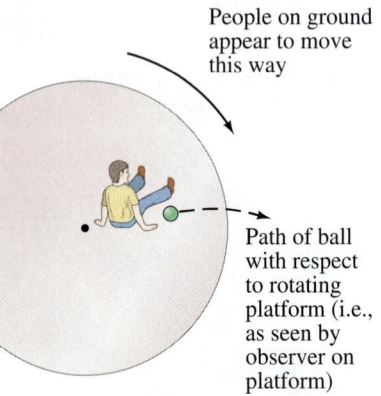

People on ground appear to move this way

Path of ball with respect to rotating platform (i.e., as seen by observer on platform)

(a) Rotating reference frame

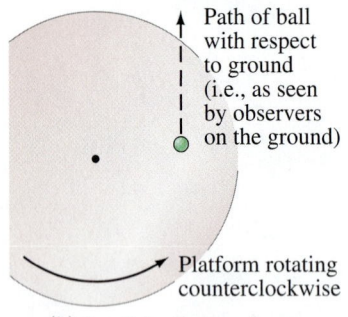

Path of ball with respect to ground (i.e., as seen by observers on the ground)

Platform rotating counterclockwise

(b) Inertial reference frame

The Earth itself is rotating on its axis. Thus, strictly speaking, Newton's laws are not valid on the Earth. However, the effect of the Earth's rotation is usually so small that it can be ignored, although it does influence the movement of large air masses and ocean currents. Because of the Earth's rotation, the material of the Earth is concentrated slightly more at the equator. The Earth is thus not a perfect sphere but is slightly fatter at the equator than at the poles.

*11–9 The Coriolis Effect

In a reference frame that rotates at a constant angular speed ω (relative to an inertial frame), there exists another pseudoforce known as the *Coriolis force*. It appears to act on an object in a rotating reference frame only if the object is moving relative to that rotating reference frame, and it acts to deflect the object sideways. It, too, is an effect of the rotating reference frame being noninertial and hence is referred to as an *inertial force*. It too affects the weather.

To see how the Coriolis force arises, consider two people, A and B, at rest on a platform rotating with angular speed ω, as shown in Fig. 11–26a. They are situated at distances r_A and r_B from the axis of rotation (at O). The woman at A throws a ball with a horizontal velocity $\bar{\mathbf{v}}$ (in her reference frame) radially outward toward the man at B on the outer edge of the platform. In Fig. 11–26a, we view the situation from an inertial reference frame. The ball initially has not only the velocity $\bar{\mathbf{v}}$ radially outward, but also a tangential velocity $\bar{\mathbf{v}}_A$ due to the rotation of the platform. Now Eq. 10–4 tells us that $v_A = r_A\omega$, where r_A is the woman's radial distance from the axis of rotation at O. If the man at B had this same velocity v_A, the ball would reach him perfectly. But his speed is $v_B = r_B\omega$, which is greater than v_A because $r_B > r_A$. Thus, when the ball reaches the outer edge of the platform, it passes a point that the man at B has already gone by because his speed in that direction is greater than the ball's. So the ball passes behind him.

Figure 11–26b shows the situation as seen from the rotating platform as frame of reference. Both A and B are at rest, and the ball is thrown with velocity $\bar{\mathbf{v}}$ toward B, but the ball deflects to the right as shown and passes behind B as previously described. This is not a centrifugal-force effect, for the latter acts radially outward. Instead, this effect acts sideways, perpendicular to $\bar{\mathbf{v}}$, and is called a **Coriolis acceleration**; it is said to be due to the Coriolis force, which is a fictitious inertial force. Its explanation as seen from an inertial system was given above: it is an effect of being in a rotating system, wherein a point farther from the rotation axis has a higher linear speed. On the other hand, when viewed from the rotating system, we can describe the motion using Newton's second law, $\Sigma\vec{\mathbf{F}} = m\vec{\mathbf{a}}$, if we add a "pseudoforce" term corresponding to this Coriolis effect.

Let us determine the magnitude of the Coriolis acceleration for the simple case described above. (We assume v is large and distances short, so we can ignore gravity.) We do the calculation from the inertial reference frame (Fig. 11–26a). The ball moves radially outward a distance $r_B - r_A$ at speed v in a short time t given by

$$r_B - r_A = vt.$$

During this time, the ball moves to the side a distance s_A given by

$$s_A = v_A t.$$

The man at B, in this time t, moves a distance

$$s_B = v_B t.$$

The ball therefore passes behind him a distance s (Fig. 11–26a) given by

$$s = s_B - s_A = (v_B - v_A)t.$$

We saw earlier that $v_A = r_A\omega$ and $v_B = r_B\omega$, so

$$s = (r_B - r_A)\omega t.$$

We substitute $r_B - r_A = vt$ (see above) and get

$$s = \omega v t^2. \tag{11–14}$$

This same s equals the sideways displacement as seen from the noninertial rotating system (Fig. 11–26b).

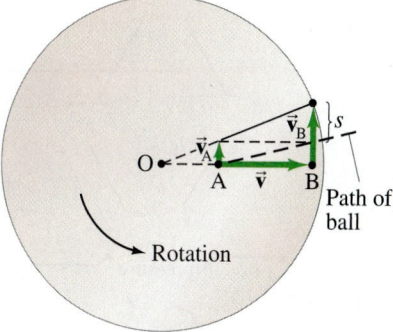

(a) Inertial reference frame

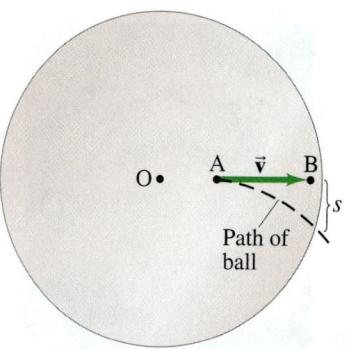

(b) Rotating reference frame

FIGURE 11–26 The origin of the Coriolis effect. Looking down on a rotating platform, (a) as seen from a nonrotating inertial reference frame, and (b) as seen from the rotating platform as frame of reference.

FIGURE 11–27 (a) Winds (moving air masses) would flow directly toward a low-pressure area if the Earth did not rotate. (b) and (c): Because of the Earth's rotation, the winds are deflected to the right in the Northern Hemisphere (as in Fig. 11–26) as if a fictitious (Coriolis) force were acting.

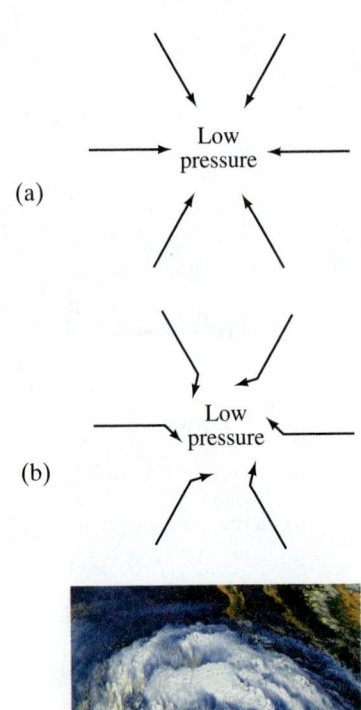

(a)

(b)

(c)

We see immediately that Eq. 11–14 corresponds to motion at constant acceleration. For as we saw in Chapter 2 (Eq. 2–12b), $y = \frac{1}{2}at^2$ for a constant acceleration (with zero initial velocity in the y direction). Thus, if we write Eq. 11–14 in the form $s = \frac{1}{2}a_{Cor}t^2$, we see that the Coriolis acceleration a_{Cor} is

$$a_{Cor} = 2\omega v. \tag{11–15}$$

This relation is valid for any velocity in the plane of rotation perpendicular to the axis of rotation† (in Fig. 11–26, the axis through point O perpendicular to the page).

Because the Earth rotates, the Coriolis effect has some interesting manifestations on the Earth. It affects the movement of air masses and thus has an influence on weather. In the absence of the Coriolis effect, air would rush directly into a region of low pressure, as shown in Fig. 11–27a. But because of the Coriolis effect, the winds are deflected to the right in the Northern Hemisphere (Fig. 11–27b), since the Earth rotates from west to east. So there tends to be a counterclockwise wind pattern around a low-pressure area. The reverse is true in the Southern Hemisphere. Thus cyclones rotate counterclockwise in the Northern Hemisphere and clockwise in the Southern Hemisphere. The same effect explains the easterly trade winds near the equator: any winds heading south toward the equator will be deflected toward the west (that is, as if coming from the east).

The Coriolis effect also acts on a falling object. An object released from the top of a high tower will not hit the ground directly below the release point, but will be deflected slightly to the east. Viewed from an inertial frame, this happens because the top of the tower revolves with the Earth at a slightly higher speed than the bottom of the tower.

---

†The Coriolis acceleration can be written in general in terms of the vector cross product as $\vec{a}_{Cor} = -2\vec{\omega} \times \vec{v}$ where $\vec{\omega}$ has direction along the rotation axis; its magnitude is $a_{Cor} = 2wv_\perp$ where $v_\perp$ is the component of velocity perpendicular to the rotation axis.

Summary

The **angular momentum** $\vec{L}$ of a rigid object rotating about a fixed axis is given by

$$L = I\omega. \tag{11–1}$$

Newton's second law, in terms of angular momentum, is

$$\Sigma\tau = \frac{dL}{dt}. \tag{11–2}$$

If the net torque on an object is zero, $dL/dt = 0$, so $L =$ constant. This is the **law of conservation of angular momentum**.

The **vector product** or **cross product** of two vectors $\vec{A}$ and $\vec{B}$ is another vector $\vec{C} = \vec{A} \times \vec{B}$ whose magnitude is $AB \sin\theta$ and whose direction is perpendicular to both $\vec{A}$ and $\vec{B}$ in the sense of the right-hand rule.

The **torque** $\vec{\tau}$ due to a force $\vec{F}$ is a vector quantity and is always calculated about some point O (the origin of a coordinate system) as follows:

$$\vec{\tau} = \vec{r} \times \vec{F}, \tag{11–5}$$

where $\vec{r}$ is the position vector of the point at which the force $\vec{F}$ acts.

Angular momentum is also a vector. For a particle having momentum $\vec{p} = m\vec{v}$, the angular momentum $\vec{L}$ about some point O is

$$\vec{L} = \vec{r} \times \vec{p}, \tag{11–6}$$

where $\vec{r}$ is the position vector of the particle relative to the point O

at any instant. The net torque $\Sigma\vec{\tau}$ on a particle is related to its angular momentum by

$$\Sigma\vec{\tau} = \frac{d\vec{L}}{dt}. \tag{11–7}$$

For a system of particles, the total angular momentum $\vec{L} = \Sigma\vec{L}_i$. The total angular momentum of the system is related to the total net torque $\Sigma\vec{\tau}$ on the system by

$$\Sigma\vec{\tau} = \frac{d\vec{L}}{dt}. \tag{11–9}$$

This last relation is the vector rotational equivalent of Newton's second law. It is valid when $\vec{L}$ and $\Sigma\vec{\tau}$ are calculated about an origin (1) fixed in an inertial reference system or (2) situated at the CM of the system. For a rigid object rotating about a fixed axis, the component of angular momentum about the rotation axis is given by $L_\omega = I\omega$. If an object rotates about an axis of symmetry, then the vector relation $\vec{L} = I\vec{\omega}$ holds, but this is not true in general.

If the total net torque on a system is zero, then the total vector angular momentum $\vec{L}$ remains constant. This is the important **law of conservation of angular momentum**. It applies to the vector $\vec{L}$ and therefore also to each of its components.

Questions

1. If there were a great migration of people toward the Earth's equator, would the length of the day (a) get longer because of conservation of angular momentum; (b) get shorter because of conservation of angular momentum; (c) get shorter because of conservation of energy; (d) get longer because of conservation of energy; or (e) remain unaffected?

2. Can the diver of Fig. 11–2 do a somersault without having any initial rotation when she leaves the board?

3. Suppose you are sitting on a rotating stool holding a 2-kg mass in each outstretched hand. If you suddenly drop the masses, will your angular velocity increase, decrease, or stay the same? Explain.

4. When a motorcyclist leaves the ground on a jump and leaves the throttle on (so the rear wheel spins), why does the front of the cycle rise up?

5. Suppose you are standing on the edge of a large freely rotating turntable. What happens if you walk toward the center?

6. A shortstop may leap into the air to catch a ball and throw it quickly. As he throws the ball, the upper part of his body rotates. If you look quickly you will notice that his hips and legs rotate in the opposite direction (Fig. 11–28). Explain.

FIGURE 11–28
Question 6. A shortstop in the air, throwing the ball.

7. If all the components of the vectors $\vec{V}_1$ and $\vec{V}_2$ were reversed in direction, how would this alter $\vec{V}_1 \times \vec{V}_2$?

8. Name the four different conditions that could make $\vec{V}_1 \times \vec{V}_2 = 0$.

9. A force $\vec{F} = F\hat{j}$ is applied to an object at a position $\vec{r} = x\hat{i} + y\hat{j} + z\hat{k}$ where the origin is at the CM. Does the torque about the CM depend on x? On y? On z?

10. A particle moves with constant speed along a straight line. How does its angular momentum, calculated about any point not on its path, change in time?

11. If the net force on a system is zero, is the net torque also zero? If the net torque on a system is zero, is the net force zero? Give examples.

12. Explain how a child "pumps" on a swing to make it go higher.

13. Describe the torque needed if the person in Fig. 11–17 is to tilt the axle of the rotating wheel directly upward without it swerving to the side.

14. An astronaut floats freely in a weightless environment. Describe how the astronaut can move her limbs so as to (a) turn her body upside down and (b) turn her body about-face.

15. On the basis of the law of conservation of angular momentum, discuss why a helicopter must have more than one rotor (or propeller). Discuss one or more ways the second propeller can operate in order to keep the helicopter stable.

16. A wheel is rotating freely about a vertical axis with constant angular velocity. Small parts of the wheel come loose and fly off. How does this affect the rotational speed of the wheel? Is angular momentum conserved? Is kinetic energy conserved? Explain.

17. Consider the following vector quantities: displacement, velocity, acceleration, momentum, angular momentum, torque. (a) Which of these are independent of the choice of origin of coordinates? (Consider different points as origin which are at rest with respect to each other.) (b) Which are independent of the velocity of the coordinate system?

18. How does a car make a right turn? Where does the torque come from that is needed to change the angular momentum?

*19. The axis of the Earth precesses with a period of about 25,000 years. This is much like the precession of a top. Explain how the Earth's equatorial bulge gives rise to a torque exerted by the Sun and Moon on the Earth; see Fig. 11–29, which is drawn for the winter solstice (December 21). About what axis would you expect the Earth's rotation axis to precess as a result of the torque due to the Sun? Does the torque exist three months later? Explain.

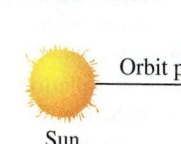

FIGURE 11–29
Question 19.
(Not to scale.)

*20. Why is it that at most locations on the Earth, a plumb bob does not hang precisely in the direction of the Earth's center?

*21. In a rotating frame of reference, Newton's first and second laws remain useful if we assume that a pseudoforce equal to $m\omega^2 r$ is acting. What effect does this assumption have on the validity of Newton's third law?

*22. In the battle of the Falkland Islands in 1914, the shots of British gunners initially fell wide of their marks because their calculations were based on naval battles fought in the Northern Hemisphere. The Falklands are in the Southern Hemisphere. Explain the origin of their problem.

Problems

11–1 Angular Momentum

1. (I) What is the angular momentum of a 0.210-kg ball rotating on the end of a thin string in a circle of radius 1.35 m at an angular speed of 10.4 rad/s?

2. (I) (a) What is the angular momentum of a 2.8-kg uniform cylindrical grinding wheel of radius 18 cm when rotating at 1300 rpm? (b) How much torque is required to stop it in 6.0 s?

3. (II) A person stands, hands at his side, on a platform that is rotating at a rate of 0.90 rev/s. If he raises his arms to a horizontal position, Fig. 11–30, the speed of rotation decreases to 0.70 rev/s. (a) Why? (b) By what factor has his moment of inertia changed?

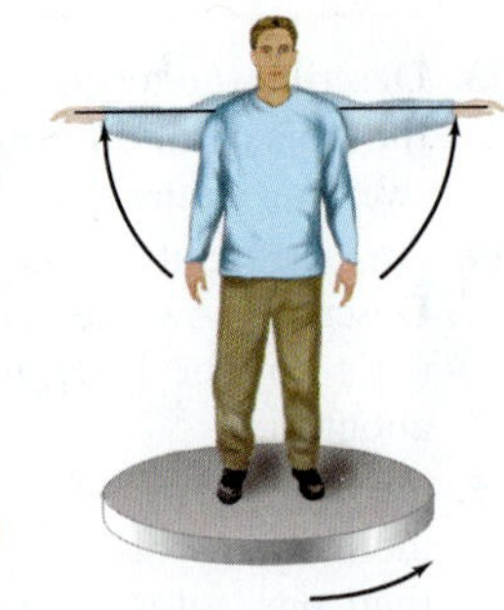

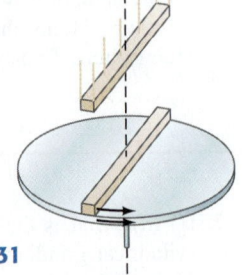

FIGURE 11–30
Problem 3.

4. (II) A figure skater can increase her spin rotation rate from an initial rate of 1.0 rev every 1.5 s to a final rate of 2.5 rev/s. If her initial moment of inertia was $4.6 \text{ kg} \cdot \text{m}^2$, what is her final moment of inertia? How does she physically accomplish this change?

5. (II) A diver (such as the one shown in Fig. 11–2) can reduce her moment of inertia by a factor of about 3.5 when changing from the straight position to the tuck position. If she makes 2.0 rotations in 1.5 s when in the tuck position, what is her angular speed (rev/s) when in the straight position?

6. (II) A uniform horizontal rod of mass M and length ℓ rotates with angular velocity ω about a vertical axis through its center. Attached to each end of the rod is a small mass m. Determine the angular momentum of the system about the axis.

7. (II) Determine the angular momentum of the Earth (a) about its rotation axis (assume the Earth is a uniform sphere), and (b) in its orbit around the Sun (treat the Earth as a particle orbiting the Sun). The Earth has mass $= 6.0 \times 10^{24} \text{ kg}$ and radius $= 6.4 \times 10^6 \text{ m}$, and is $1.5 \times 10^8 \text{ km}$ from the Sun.

8. (II) (a) What is the angular momentum of a figure skater spinning at 2.8 rev/s with arms in close to her body, assuming her to be a uniform cylinder with a height of 1.5 m, a radius of 15 cm, and a mass of 48 kg? (b) How much torque is required to slow her to a stop in 5.0 s, assuming she does not move her arms?

9. (II) A person stands on a platform, initially at rest, that can rotate freely without friction. The moment of inertia of the person plus the platform is I_P. The person holds a spinning bicycle wheel with its axis horizontal. The wheel has moment of inertia I_W and angular velocity ω_W. What will be the angular velocity ω_P of the platform if the person moves the axis of the wheel so that it points (a) vertically upward, (b) at a 60° angle to the vertical, (c) vertically downward? (d) What will ω_P be if the person reaches up and stops the wheel in part (a)?

10. (II) A uniform disk turns at 3.7 rev/s around a frictionless spindle. A nonrotating rod, of the same mass as the disk and length equal to the disk's diameter, is dropped onto the freely spinning disk, Fig. 11–31. They then turn together around the spindle with their centers superposed. What is the angular frequency in rev/s of the combination?

FIGURE 11–31
Problem 10.

11. (II) A person of mass 75 kg stands at the center of a rotating merry-go-round platform of radius 3.0 m and moment of inertia $920 \text{ kg} \cdot \text{m}^2$. The platform rotates without friction with angular velocity 0.95 rad/s. The person walks radially to the edge of the platform. (a) Calculate the angular velocity when the person reaches the edge. (b) Calculate the rotational kinetic energy of the system of platform plus person before and after the person's walk.

12. (II) A potter's wheel is rotating around a vertical axis through its center at a frequency of 1.5 rev/s. The wheel can be considered a uniform disk of mass 5.0 kg and diameter 0.40 m. The potter then throws a 2.6-kg chunk of clay, approximately shaped as a flat disk of radius 8.0 cm, onto the center of the rotating wheel. What is the frequency of the wheel after the clay sticks to it?

13. (II) A 4.2-m-diameter merry-go-round is rotating freely with an angular velocity of 0.80 rad/s. Its total moment of inertia is $1760 \text{ kg} \cdot \text{m}^2$. Four people standing on the ground, each of mass 65 kg, suddenly step onto the edge of the merry-go-round. What is the angular velocity of the merry-go-round now? What if the people were on it initially and then jumped off in a radial direction (relative to the merry-go-round)?

14. (II) A woman of mass m stands at the edge of a solid cylindrical platform of mass M and radius R. At $t = 0$, the platform is rotating with negligible friction at angular velocity ω_0 about a vertical axis through its center, and the woman begins walking with speed v (relative to the platform) toward the center of the platform. (a) Determine the angular velocity of the system as a function of time. (b) What will be the angular velocity when the woman reaches the center?

15. (II) A nonrotating cylindrical disk of moment of inertia I is dropped onto an identical disk rotating at angular speed ω. Assuming no external torques, what is the final common angular speed of the two disks?

16. (II) Suppose our Sun eventually collapses into a white dwarf, losing about half its mass in the process, and winding up with a radius 1.0% of its existing radius. Assuming the lost mass carries away no angular momentum, what would the Sun's new rotation rate be? (Take the Sun's current period to be about 30 days.) What would be its final kinetic energy in terms of its initial kinetic energy of today?

17. (III) Hurricanes can involve winds in excess of 120 km/h at the outer edge. Make a crude estimate of (a) the energy, and (b) the angular momentum, of such a hurricane, approximating it as a rigidly rotating uniform cylinder of air (density 1.3 kg/m^3) of radius 85 km and height 4.5 km.

18. (III) An asteroid of mass $1.0 \times 10^5 \text{ kg}$, traveling at a speed of 35 km/s relative to the Earth, hits the Earth at the equator tangentially, in the direction of Earth's rotation, and is embedded there. Use angular momentum to estimate the percent change in the angular speed of the Earth as a result of the collision.

19. (III) Suppose a 65-kg person stands at the edge of a 6.5-m diameter merry-go-round turntable that is mounted on frictionless bearings and has a moment of inertia of $1850 \text{ kg} \cdot \text{m}^2$. The turntable is at rest initially, but when the person begins running at a speed of 3.8 m/s (with respect to the turntable) around its edge, the turntable begins to rotate in the opposite direction. Calculate the angular velocity of the turntable.

11–2 Vector Cross Product and Torque

20. (I) If vector $\vec{\mathbf{A}}$ points along the negative x axis and vector $\vec{\mathbf{B}}$ along the positive z axis, what is the direction of (a) $\vec{\mathbf{A}} \times \vec{\mathbf{B}}$ and (b) $\vec{\mathbf{B}} \times \vec{\mathbf{A}}$? (c) What is the magnitude of $\vec{\mathbf{A}} \times \vec{\mathbf{B}}$ and $\vec{\mathbf{B}} \times \vec{\mathbf{A}}$?

21. (I) Show that (a) $\hat{\mathbf{i}} \times \hat{\mathbf{i}} = \hat{\mathbf{j}} \times \hat{\mathbf{j}} = \hat{\mathbf{k}} \times \hat{\mathbf{k}} = 0$, (b) $\hat{\mathbf{i}} \times \hat{\mathbf{j}} = \hat{\mathbf{k}}$, $\hat{\mathbf{i}} \times \hat{\mathbf{k}} = -\hat{\mathbf{j}}$, and $\hat{\mathbf{j}} \times \hat{\mathbf{k}} = \hat{\mathbf{i}}$.

22. (I) The directions of vectors $\vec{\mathbf{A}}$ and $\vec{\mathbf{B}}$ are given below for several cases. For each case, state the direction of $\vec{\mathbf{A}} \times \vec{\mathbf{B}}$. (a) $\vec{\mathbf{A}}$ points east, $\vec{\mathbf{B}}$ points south. (b) $\vec{\mathbf{A}}$ points east, $\vec{\mathbf{B}}$ points straight down. (c) $\vec{\mathbf{A}}$ points straight up, $\vec{\mathbf{B}}$ points north. (d) $\vec{\mathbf{A}}$ points straight up, $\vec{\mathbf{B}}$ points straight down.

23. (II) What is the angle θ between two vectors $\vec{\mathbf{A}}$ and $\vec{\mathbf{B}}$, if $|\vec{\mathbf{A}} \times \vec{\mathbf{B}}| = \vec{\mathbf{A}} \cdot \vec{\mathbf{B}}$?

24. (II) A particle is located at $\vec{\mathbf{r}} = (4.0\hat{\mathbf{i}} + 3.5\hat{\mathbf{j}} + 6.0\hat{\mathbf{k}})$ m. A force $\vec{\mathbf{F}} = (9.0\hat{\mathbf{j}} - 4.0\hat{\mathbf{k}})$ N acts on it. What is the torque, calculated about the origin?

25. (II) Consider a particle of a rigid object rotating about a fixed axis. Show that the tangential and radial vector components of the linear acceleration are:

$$\vec{\mathbf{a}}_{\text{tan}} = \vec{\boldsymbol{\alpha}} \times \vec{\mathbf{r}} \quad \text{and} \quad \vec{\mathbf{a}}_{\text{R}} = \vec{\boldsymbol{\omega}} \times \vec{\mathbf{v}}.$$

26. (II) (a) Show that the cross product of two vectors, $\vec{\mathbf{A}} = A_x\hat{\mathbf{i}} + A_y\hat{\mathbf{j}} + A_z\hat{\mathbf{k}}$, and $\vec{\mathbf{B}} = B_x\hat{\mathbf{i}} + B_y\hat{\mathbf{j}} + B_z\hat{\mathbf{k}}$ is

$$\vec{\mathbf{A}} \times \vec{\mathbf{B}} = (A_y B_z - A_z B_y)\hat{\mathbf{i}} + (A_z B_x - A_x B_z)\hat{\mathbf{j}}$$
$$+ (A_x B_y - A_y B_x)\hat{\mathbf{k}}.$$

(b) Then show that the cross product can be written

$$\vec{\mathbf{A}} \times \vec{\mathbf{B}} = \begin{vmatrix} \hat{\mathbf{i}} & \hat{\mathbf{j}} & \hat{\mathbf{k}} \\ A_x & A_y & A_z \\ B_x & B_y & B_z \end{vmatrix},$$

where we use the rules for evaluating a determinant. (Note, however, that this is not really a determinant, but a memory aid.)

27. (II) An engineer estimates that under the most adverse expected weather conditions, the total force on the highway sign in Fig. 11–32 will be $\vec{\mathbf{F}} = (\pm 2.4\hat{\mathbf{i}} - 4.1\hat{\mathbf{j}})$ kN, acting at the CM. What torque does this force exert about the base O?

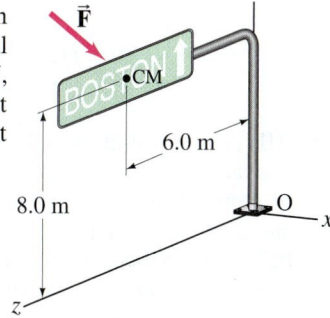

FIGURE 11–32
Problem 27.

28. (II) The origin of a coordinate system is at the center of a wheel which rotates in the xy plane about its axle which is the z axis. A force $F = 215$ N acts in the xy plane, at a $+33.0°$ angle to the x axis, at the point $x = 28.0$ cm, $y = 33.5$ cm. Determine the magnitude and direction of the torque produced by this force about the axis.

29. (II) Use the result of Problem 26 to determine (a) the vector product $\vec{\mathbf{A}} \times \vec{\mathbf{B}}$ and (b) the angle between $\vec{\mathbf{A}}$ and $\vec{\mathbf{B}}$ if $\vec{\mathbf{A}} = 5.4\hat{\mathbf{i}} - 3.5\hat{\mathbf{j}}$ and $\vec{\mathbf{B}} = -8.5\hat{\mathbf{i}} + 5.6\hat{\mathbf{j}} + 2.0\hat{\mathbf{k}}$.

30. (III) Show that the velocity $\vec{\mathbf{v}}$ of any point in an object rotating with angular velocity $\vec{\boldsymbol{\omega}}$ about a fixed axis can be written

$$\vec{\mathbf{v}} = \vec{\boldsymbol{\omega}} \times \vec{\mathbf{r}}$$

where $\vec{\mathbf{r}}$ is the position vector of the point relative to an origin O located on the axis of rotation. Can O be anywhere on the rotation axis? Will $\vec{\mathbf{v}} = \vec{\boldsymbol{\omega}} \times \vec{\mathbf{r}}$ if O is located at a point not on the axis of rotation?

31. (III) Let $\vec{\mathbf{A}}, \vec{\mathbf{B}}$, and $\vec{\mathbf{C}}$ be three vectors, which for generality we assume do not all lie in the same plane. Show that $\vec{\mathbf{A}} \cdot (\vec{\mathbf{B}} \times \vec{\mathbf{C}}) = \vec{\mathbf{B}} \cdot (\vec{\mathbf{C}} \times \vec{\mathbf{A}}) = \vec{\mathbf{C}} \cdot (\vec{\mathbf{A}} \times \vec{\mathbf{B}})$.

11–3 Angular Momentum of a Particle

32. (I) What are the x, y, and z components of the angular momentum of a particle located at $\vec{\mathbf{r}} = x\hat{\mathbf{i}} + y\hat{\mathbf{j}} + z\hat{\mathbf{k}}$ which has momentum $\vec{\mathbf{p}} = p_x\hat{\mathbf{i}} + p_y\hat{\mathbf{j}} + p_z\hat{\mathbf{k}}$?

33. (I) Show that the kinetic energy K of a particle of mass m, moving in a circular path, is $K = L^2/2I$, where L is its angular momentum and I is its moment of inertia about the center of the circle.

34. (I) Calculate the angular momentum of a particle of mass m moving with constant velocity v for two cases (see Fig. 11–33): (a) about origin O, and (b) about O'.

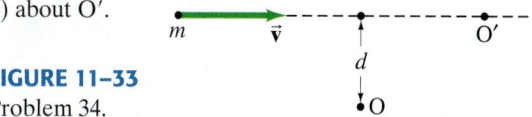

FIGURE 11–33
Problem 34.

35. (II) Two identical particles have equal but opposite momenta, $\vec{\mathbf{p}}$ and $-\vec{\mathbf{p}}$, but they are not traveling along the same line. Show that the total angular momentum of this system does not depend on the choice of origin.

36. (II) Determine the angular momentum of a 75-g particle about the origin of coordinates when the particle is at $x = 4.4$ m, $y = -6.0$ m, and it has velocity $v = (3.2\hat{\mathbf{i}} - 8.0\hat{\mathbf{k}})$ m/s.

37. (II) A particle is at the position $(x, y, z) = (1.0, 2.0, 3.0)$ m. It is traveling with a vector velocity $(-5.0, +2.8, -3.1)$ m/s. Its mass is 3.8 kg. What is its vector angular momentum about the origin?

11–4 and 11–5 Angular Momentum and Torque: General Motion; Rigid Objects

38. (II) An Atwood machine (Fig. 11–16) consists of two masses, $m_A = 7.0$ kg and $m_B = 8.2$ kg, connected by a cord that passes over a pulley free to rotate about a fixed axis. The pulley is a solid cylinder of radius $R_0 = 0.40$ m and mass 0.80 kg. (a) Determine the acceleration a of each mass. (b) What percentage of error in a would be made if the moment of inertia of the pulley were ignored? Ignore friction in the pulley bearings.

39. (II) Four identical particles of mass m are mounted at equal intervals on a thin rod of length ℓ and mass M, with one mass at each end of the rod. If the system is rotated with angular velocity ω about an axis perpendicular to the rod through one of the end masses, determine (a) the kinetic energy and (b) the angular momentum of the system.

40. (II) Two lightweight rods 24 cm in length are mounted perpendicular to an axle and at 180° to each other (Fig. 11–34). At the end of each rod is a 480-g mass. The rods are spaced 42 cm apart along the axle. The axle rotates at 4.5 rad/s. (a) What is the component of the total angular momentum along the axle? (b) What angle does the vector angular momentum make with the axle? [*Hint*: Remember that the vector angular momentum must be calculated about the *same* point for *both* masses, which could be the CM.]

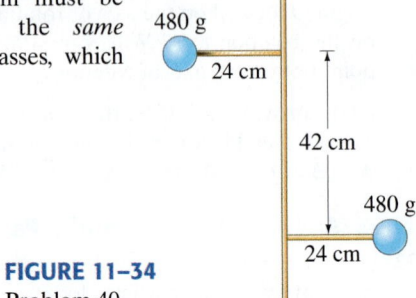

FIGURE 11–34
Problem 40.

41. (II) Figure 11–35 shows two masses connected by a cord passing over a pulley of radius R_0 and moment of inertia I. Mass M_A slides on a frictionless surface, and M_B hangs freely. Determine a formula for (a) the angular momentum of the system about the pulley axis, as a function of the speed v of mass M_A or M_B, and (b) the acceleration of the masses.

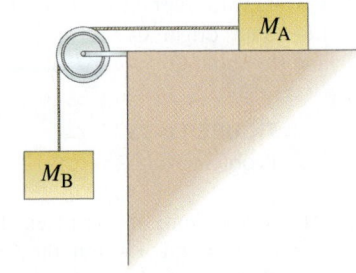

FIGURE 11–35
Problem 41.

42. (III) A thin rod of length ℓ and mass M rotates about a vertical axis through its center with angular velocity ω. The rod makes an angle ϕ with the rotation axis. Determine the magnitude and direction of $\vec{L}$.

43. (III) Show that the total angular momentum $\vec{L} = \Sigma \vec{r}_i \times \vec{p}_i$ of a system of particles about the origin of an inertial reference frame can be written as the sum of the angular momentum about the CM, $\vec{L}^*$ (spin angular momentum), plus the angular momentum of the CM about the origin (orbital angular momentum): $\vec{L} = \vec{L}^* + \vec{r}_{CM} \times M\vec{v}_{CM}$. [*Hint*: See the derivation of Eq. 11–9b.]

*44. (III) What is the magnitude of the force $\vec{F}$ exerted by each bearing in Fig. 11–18 (Example 11–10)? The bearings are a distance d from point O. Ignore the effects of gravity.

*45. (III) Suppose in Fig. 11–18 that $m_B = 0$; that is, only one mass, m_A, is actually present. If the bearings are each a distance d from O, determine the forces F_A and F_B at the upper and lower bearings respectively. [*Hint*: Choose an origin—different than O in Fig. 11–18—such that $\vec{L}$ is parallel to $\vec{\omega}$. Ignore effects of gravity.]

*46. (III) Suppose in Fig. 11–18 that $m_A = m_B = 0.60$ kg, $r_A = r_B = 0.30$ m, and the distance between the bearings is 0.23 m. Evaluate the force that each bearing must exert on the axle if $\phi = 34.0°$, $\omega = 11.0$ rad/s.

11–6 Angular Momentum Conservation

47. (II) A thin rod of mass M and length ℓ is suspended vertically from a frictionless pivot at its upper end. A mass m of putty traveling horizontally with a speed v strikes the rod at its CM and sticks there. How high does the bottom of the rod swing?

48. (II) A uniform stick 1.0 m long with a total mass of 270 g is pivoted at its center. A 3.0-g bullet is shot through the stick midway between the pivot and one end (Fig. 11–36). The bullet approaches at 250 m/s and leaves at 140 m/s. With what angular speed is the stick spinning after the collision?

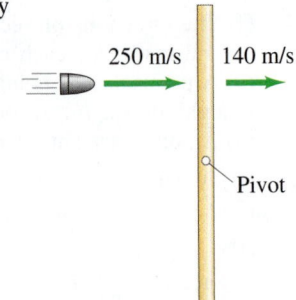

FIGURE 11–36
Problems 48 and 83.

49. (II) Suppose a 5.8×10^{10} kg meteorite struck the Earth at the equator with a speed $v = 2.2 \times 10^4$ m/s, as shown in Fig. 11–37 and remained stuck. By what factor would this affect the rotational frequency of the Earth (1 rev/day)?

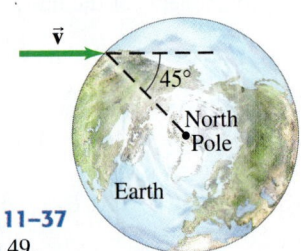

FIGURE 11–37
Problem 49.

50. (III) A 230-kg beam 2.7 m in length slides broadside down the ice with a speed of 18 m/s (Fig. 11–38). A 65-kg man at rest grabs one end as it goes past and hangs on as both he and the beam go spinning down the ice. Assume frictionless motion. (a) How fast does the center of mass of the system move after the collision? (b) With what angular velocity does the system rotate about its CM?

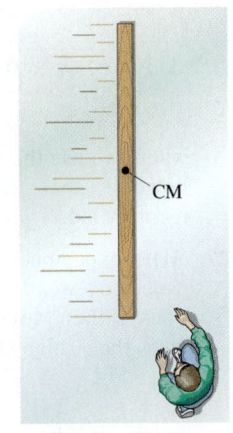

FIGURE 11–38
Problem 50.

51. (III) A thin rod of mass M and length ℓ rests on a frictionless table and is struck at a point $\ell/4$ from its CM by a clay ball of mass m moving at speed v (Fig. 11–39). The ball sticks to the rod. Determine the translational and rotational motion of the rod after the collision.

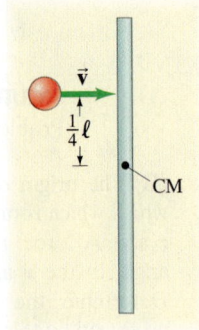

FIGURE 11–39
Problems 51 and 84.

52. (III) On a level billiards table a cue ball, initially at rest at point O on the table, is struck so that it leaves the cue stick with a center-of-mass speed v_0 and a "reverse" spin of angular speed ω_0 (see Fig. 11–40). A kinetic friction force acts on the ball as it initially skids across the table. (a) Explain why the ball's angular momentum is conserved about point O. (b) Using conservation of angular momentum, find the critical angular speed ω_C such that, if $\omega_0 = \omega_C$, kinetic friction will bring the ball to a complete (as opposed to momentary) stop. (c) If ω_0 is 10% smaller than ω_C, i.e., $\omega_0 = 0.90\,\omega_C$, determine the ball's CM velocity v_{CM} when it starts to roll without slipping. (d) If ω_0 is 10% larger than ω_C, i.e., $\omega_0 = 1.10\,\omega_C$, determine the ball's CM velocity v_{CM} when it starts to roll without slipping. [*Hint:* The ball possesses two types of angular momentum, the first due to the linear speed v_{CM} of its CM relative to point O, the second due to the spin at angular velocity ω about its own CM. The ball's total L about O is the sum of these two angular momenta.]

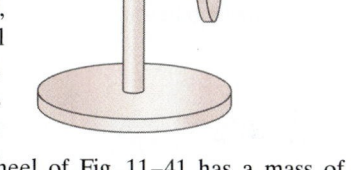

FIGURE 11–40
Problem 52.

*11–7 Spinning Top

*53. **(II)** A 220-g top spinning at 15 rev/s makes an angle of 25° to the vertical and precesses at a rate of 1.00 rev per 6.5 s. If its CM is 3.5 cm from its tip along its symmetry axis, what is the moment of inertia of the top?

*54. **(II)** A toy gyroscope consists of a 170-g disk with a radius of 5.5 cm mounted at the center of a thin axle 21 cm long (Fig. 11–41). The gyroscope spins at 45 rev/s. One end of its axle rests on a stand and the other end precesses horizontally about the stand. (a) How long does it take the gyroscope to precess once around? (b) If all the dimensions of the gyroscope were doubled (radius = 11 cm, axle = 42 cm), how long would it take to precess once?

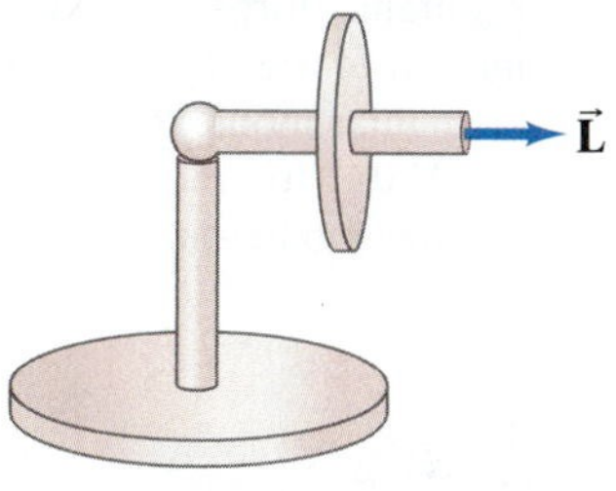

FIGURE 11–41 A wheel, rotating about a horizontal axle supported at one end, precesses. Problems 54, 55, and 56.

*55. **(II)** Suppose the solid wheel of Fig. 11–41 has a mass of 300 g and rotates at 85 rad/s; it has radius 6.0 cm and is mounted at the center of a horizontal thin axle 25 cm long. At what rate does the axle precess?

*56. **(II)** If a mass equal to half the mass of the wheel in Problem 55 is placed at the free end of the axle, what will be the precession rate now? Treat the extra mass as insignificant in size.

*57. **(II)** A bicycle wheel of diameter 65 cm and mass m rotates on its axle; two 20-cm-long wooden handles, one on each side of the wheel, act as the axle. You tie a rope to a small hook on the end of one of the handles, and then spin the bicycle wheel with a flick of the hand. When you release the spinning wheel, it precesses about the vertical axis defined by the rope, instead of falling to the ground (as it would if it were not spinning). Estimate the rate and direction of precession if the wheel rotates counterclockwise at 2.0 rev/s and its axle remains horizontal.

*11–8 Rotating Reference Frames

*58. **(II)** If a plant is allowed to grow from seed on a rotating platform, it will grow at an angle, pointing inward. Calculate what this angle will be (put yourself in the rotating frame) in terms of g, r, and ω. Why does it grow inward rather than outward?

*59. **(III)** Let $\vec{g}'$ be the effective acceleration of gravity at a point on the rotating Earth, equal to the vector sum of the "true" value $\vec{g}$ plus the effect of the rotating reference frame ($m\omega^2 r$ term). See Fig. 11–42. Determine the magnitude and direction of $\vec{g}'$ relative to a radial line from the center of the Earth (a) at the North Pole, (b) at a latitude of 45.0° north, and (c) at the equator. Assume that g (if ω were zero) is a constant 9.80 m/s².

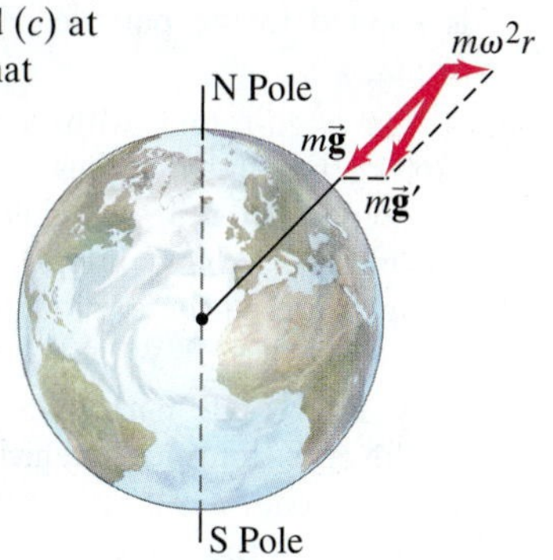

FIGURE 11–42
Problem 59.

*11–9 Coriolis Effect

*60. **(II)** Suppose the man at B in Fig. 11–26 throws the ball toward the woman at A. (a) In what direction is the ball deflected as seen in the noninertial system? (b) Determine a formula for the amount of deflection and for the (Coriolis) acceleration in this case.

*61. **(II)** For what directions of velocity would the Coriolis effect on an object moving at the Earth's equator be zero?

*62. **(III)** We can alter Eqs. 11–14 and 11–15 for use on Earth by considering only the component of $\vec{v}$ perpendicular to the axis of rotation. From Fig. 11–43, we see that this is $v\cos\lambda$ for a vertically falling object, where λ is the latitude of the place on the Earth. If a lead ball is dropped vertically from a 110-m-high tower in Florence, Italy (latitude = 44°), how far from the base of the tower is it deflected by the Coriolis force?

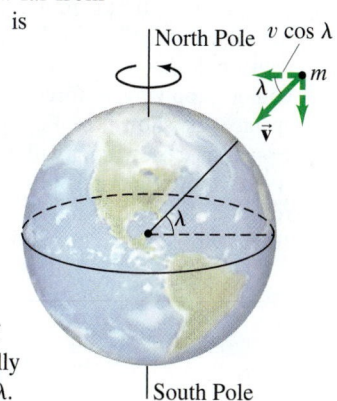

FIGURE 11–43
Problem 62. Object of mass m falling vertically to Earth at a latitude λ.

*63. **(III)** An ant crawls with constant speed outward along a radial spoke of a wheel rotating at constant angular velocity ω about a vertical axis. Write a vector equation for all the forces (including inertial forces) acting on the ant. Take the x axis along the spoke, y perpendicular to the spoke pointing to the ant's left, and the z axis vertically upward. The wheel rotates counterclockwise as seen from above.

General Problems

64. A thin string is wrapped around a cylindrical hoop of radius R and mass M. One end of the string is fixed, and the hoop is allowed to fall vertically, starting from rest, as the string unwinds. (a) Determine the angular momentum of the hoop about its CM as a function of time. (b) What is the tension in the string as function of time?

65. A particle of mass $1.00 \, \text{kg}$ is moving with velocity $\vec{v} = (7.0\hat{i} + 6.0\hat{j}) \, \text{m/s}$. (a) Find the angular momentum $\vec{L}$ relative to the origin when the particle is at $\vec{r} = (2.0\hat{j} + 4.0\hat{k}) \, \text{m}$. (b) At position $\vec{r}$ a force of $\vec{F} = 4.0 \, \text{N}\hat{i}$ is applied to the particle. Find the torque relative to the origin.

66. A merry-go-round with a moment of inertia equal to $1260 \, \text{kg} \cdot \text{m}^2$ and a radius of $2.5 \, \text{m}$ rotates with negligible friction at $1.70 \, \text{rad/s}$. A child initially standing still next to the merry-go-round jumps onto the edge of the platform straight toward the axis of rotation causing the platform to slow to $1.25 \, \text{rad/s}$. What is her mass?

67. Why might tall narrow SUVs and buses be prone to "rollover"? Consider a vehicle rounding a curve of radius R on a flat road. When just on the verge of rollover, its tires on the inside of the curve are about to leave the ground, so the friction and normal force on these two tires are zero. The total normal force on the outside tires is F_N and the total friction force is F_{fr}. Assume that the vehicle is not skidding. (a) Analysts define a static stability factor SSF = $w/2h$, where a vehicle's "track width" w is the distance between tires on the same axle, and h is the height of the CM above the ground. Show that the critical rollover speed is

$$v_C = \sqrt{Rg\left(\frac{w}{2h}\right)}.$$

[Hint: Take torques about an axis through the center of mass of the SUV, parallel to its direction of motion.] (b) Determine the ratio of highway curve radii (minimum possible) for a typical passenger car with SSF = 1.40 and an SUV with SSF = 1.05 at a speed of 90 km/h.

68. A spherical asteroid with radius $r = 123 \, \text{m}$ and mass $M = 2.25 \times 10^{10} \, \text{kg}$ rotates about an axis at four revolutions per day. A "tug" spaceship attaches itself to the asteroid's south pole (as defined by the axis of rotation) and fires its engine, applying a force F tangentially to the asteroid's surface as shown in Fig. 11–44. If $F = 265 \, \text{N}$, how long will it take the tug to rotate the asteroid's axis of rotation through an angle of $10.0°$ by this method?

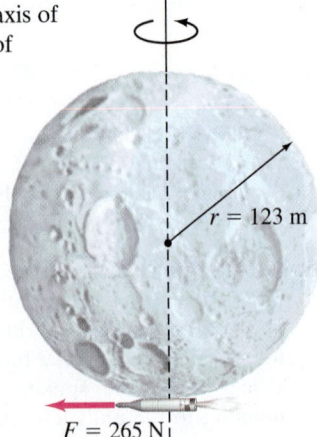

FIGURE 11–44
Problem 68.

$F = 265 \, \text{N}$

69. The time-dependent position of a point object which moves counterclockwise along the circumference of a circle (radius R) in the xy plane with constant speed v is given by

$$\vec{r} = \hat{i} R \cos \omega t + \hat{j} R \sin \omega t$$

where the constant $\omega = v/R$. Determine the velocity $\vec{v}$ and angular velocity $\vec{\omega}$ of this object and then show that these three vectors obey the relation $\vec{v} = \vec{\omega} \times \vec{r}$.

70. The position of a particle with mass m traveling on a helical path (see Fig. 11–45) is given by

$$\vec{r} = R \cos\left(\frac{2\pi z}{d}\right)\hat{i} + R \sin\left(\frac{2\pi z}{d}\right)\hat{j} + z\hat{k}$$

where R and d are the radius and pitch of the helix, respectively, and z has time dependence $z = v_z t$ where v_z is the (constant) component of velocity in the z direction. Determine the time-dependent angular momentum $\vec{L}$ of the particle about the origin.

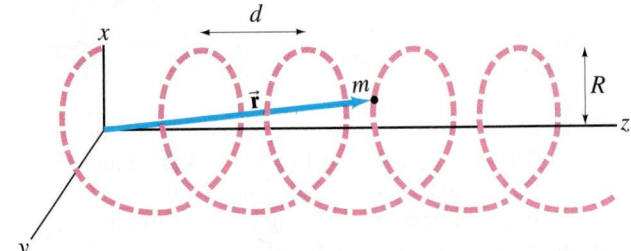

FIGURE 11–45 Problem 70.

71. A boy rolls a tire along a straight level street. The tire has mass 8.0 kg, radius 0.32 m and moment of inertia about its central axis of symmetry of $0.83 \, \text{kg} \cdot \text{m}^2$. The boy pushes the tire forward away from him at a speed of 2.1 m/s and sees that the tire leans 12° to the right (Fig. 11–46). (a) How will the resultant torque affect the subsequent motion of the tire? (b) Compare the change in angular momentum caused by this torque in 0.20 s to the original magnitude of angular momentum.

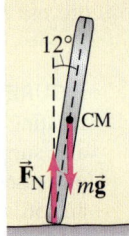

FIGURE 11–46
Problem 71.

72. A 70-kg person stands on a tiny rotating platform with arms outstretched. (a) Estimate the moment of inertia of the person using the following approximations: the body (including head and legs) is a 60-kg cylinder, 12 cm in radius and 1.70 m high; and each arm is a 5.0-kg thin rod, 60 cm long, attached to the cylinder. (b) Using the same approximations, estimate the moment of inertia when the arms are at the person's sides. (c) If one rotation takes 1.5 s when the person's arms are outstretched, what is the time for each rotation with arms at the sides? Ignore the moment of inertia of the lightweight platform. (d) Determine the change in kinetic energy when the arms are lifted from the sides to the horizontal position. (e) From your answer to part (d), would you expect it to be harder or easier to lift your arms when rotating or when at rest?

73. Water drives a waterwheel (or turbine) of radius $R = 3.0\,\text{m}$ as shown in Fig. 11–47. The water enters at a speed $v_1 = 7.0\,\text{m/s}$ and exits from the waterwheel at a speed $v_2 = 3.8\,\text{m/s}$. (a) If 85 kg of water passes through per second, what is the rate at which the water delivers angular momentum to the waterwheel? (b) What is the torque the water applies to the waterwheel? (c) If the water causes the waterwheel to make one revolution every 5.5 s, how much power is delivered to the wheel?

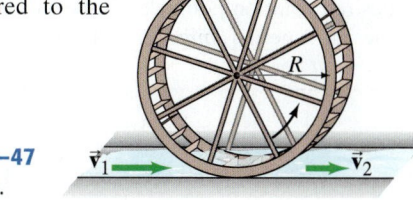

FIGURE 11–47
Problem 73.

74. The Moon orbits the Earth such that the same side always faces the Earth. Determine the ratio of the Moon's spin angular momentum (about its own axis) to its orbital angular momentum. (In the latter case, treat the Moon as a particle orbiting the Earth.)

75. A particle of mass m uniformly accelerates as it moves counterclockwise along the circumference of a circle of radius R:

$$\vec{r} = \hat{i}\,R\cos\theta + \hat{j}\,R\sin\theta$$

with $\theta = \omega_0 t + \frac{1}{2}\alpha t^2$, where the constants ω_0 and α are the initial angular velocity and angular acceleration, respectively. Determine the object's tangential acceleration $\vec{a}_{\text{tan}}$ and determine the torque acting on the object using (a) $\vec{\tau} = \vec{r} \times \vec{F}$, (b) $\vec{\tau} = I\vec{\alpha}$.

76. A projectile with mass m is launched from the ground and follows a trajectory given by

$$\vec{r} = \left(v_{x0}t\right)\hat{i} + \left(v_{y0}t - \frac{1}{2}gt^2\right)\hat{j}$$

where v_{x0} and v_{y0} are the initial velocities in the x and y direction, respectively, and g is the acceleration due to gravity. The launch position is defined to be the origin. Determine the torque acting on the projectile about the origin using (a) $\vec{\tau} = \vec{r} \times \vec{F}$, (b) $\vec{\tau} = d\vec{L}/dt$.

77. Most of our Solar System's mass is contained in the Sun, and the planets possess almost all of the Solar System's angular momentum. This observation plays a key role in theories attempting to explain the formation of our Solar System. Estimate the fraction of the Solar System's total angular momentum that is possessed by planets using a simplified model which includes only the large outer planets with the most angular momentum. The central Sun (mass $1.99 \times 10^{30}\,\text{kg}$, radius $6.96 \times 10^8\,\text{m}$) spins about its axis once every 25 days and the planets Jupiter, Saturn, Uranus, and Neptune move in nearly circular orbits around the Sun with orbital data given in the Table below. Ignore each planet's spin about its own axis.

Planet	Mean Distance from Sun ($\times 10^6$ km)	Orbital Period (Earth Years)	Mass ($\times 10^{25}$ kg)
Jupiter	778	11.9	190
Saturn	1427	29.5	56.8
Uranus	2870	84.0	8.68
Neptune	4500	165	10.2

78. A bicyclist traveling with speed $v = 9.2\,\text{m/s}$ on a flat road is making a turn with a radius $r = 12\,\text{m}$. The forces acting on the cyclist and cycle are the normal force ($\vec{F}_N$) and friction force ($\vec{F}_{\text{fr}}$) exerted by the road on the tires and $m\vec{g}$, the total weight of the cyclist and cycle. Ignore the small mass of the wheels. (a) Explain carefully why the angle θ the bicycle makes with the vertical (Fig. 11–48) must be given by $\tan\theta = F_{\text{fr}}/F_N$ if the cyclist is to maintain balance. (b) Calculate θ for the values given. [Hint: Consider the "circular" translational motion of the bicycle and rider.] (c) If the coefficient of static friction between tires and road is $\mu_s = 0.65$, what is the minimum turning radius?

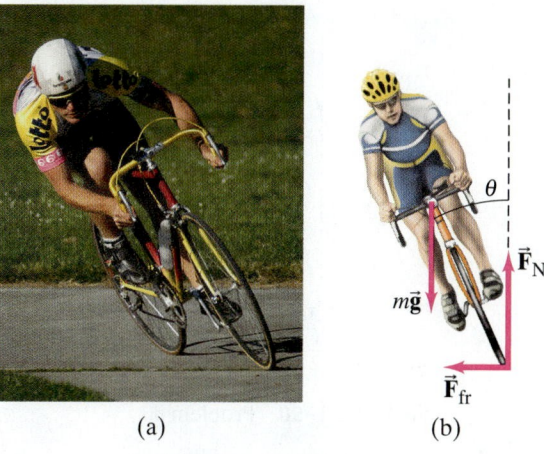

(a) (b)

FIGURE 11–48 Problem 78.

79. Competitive ice skaters commonly perform single, double, and triple axel jumps in which they rotate $1\frac{1}{2}$, $2\frac{1}{2}$, and $3\frac{1}{2}$ revolutions, respectively, about a vertical axis while airborne. For all these jumps, a typical skater remains airborne for about 0.70 s. Suppose a skater leaves the ground in an "open" position (e.g., arms outstretched) with moment of inertia I_0 and rotational frequency $f_0 = 1.2\,\text{rev/s}$, maintaining this position for 0.10 s. The skater then assumes a "closed" position (arms brought closer) with moment of inertia I, acquiring a rotational frequency f, which is maintained for 0.50 s. Finally, the skater immediately returns to the "open" position for 0.10 s until landing (see Fig. 11–49). (a) Why is angular momentum conserved during the skater's jump? Neglect air resistance. (b) Determine the minimum rotational frequency f during the flight's middle section for the skater to successfully complete a single and a triple axel. (c) Show that, according to this model, a skater must be able to reduce his or her moment of inertia in midflight by a factor of about 2 and 5 in order to complete a single and triple axel, respectively.

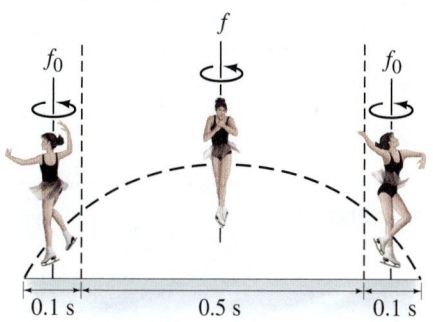

FIGURE 11–49 Problem 79.

80. A radio transmission tower has a mass of 80 kg and is 12 m high. The tower is anchored to the ground by a flexible joint at its base, but it is secured by three cables 120° apart (Fig. 11–50). In an analysis of a potential failure, a mechanical engineer needs to determine the behavior of the tower if one of the cables broke. The tower would fall away from the broken cable, rotating about its base. Determine the speed of the top of the tower as a function of the rotation angle θ. Start your analysis with the rotational dynamics equation of motion $d\vec{L}/dt = \vec{\tau}_{net}$. Approximate the tower as a tall thin rod.

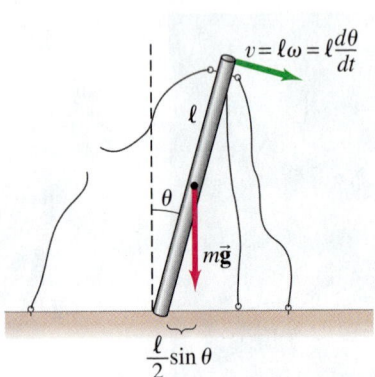

FIGURE 11–50   Problem 80.

81. Suppose a star the size of our Sun, but with mass 8.0 times as great, were rotating at a speed of 1.0 revolution every 9.0 days. If it were to undergo gravitational collapse to a neutron star of radius 12 km, losing $\frac{3}{4}$ of its mass in the process, what would its rotation speed be? Assume the star is a uniform sphere at all times. Assume also that the thrown-off mass carries off either (a) no angular momentum, or (b) its proportional share $\left(\frac{3}{4}\right)$ of the initial angular momentum.

82. A baseball bat has a "sweet spot" where a ball can be hit with almost effortless transmission of energy. A careful analysis of baseball dynamics shows that this special spot is located at the point where an applied force would result in pure rotation of the bat about the handle grip. Determine the location of the sweet spot of the bat shown in Fig. 11–51. The linear mass density of the bat is given roughly by $(0.61 + 3.3x^2)$ kg/m, where x is in meters measured from the end of the handle. The entire bat is 0.84 m long. The desired rotation point should be 5.0 cm from the end where the bat is held. [*Hint:* Where is the CM of the bat?]

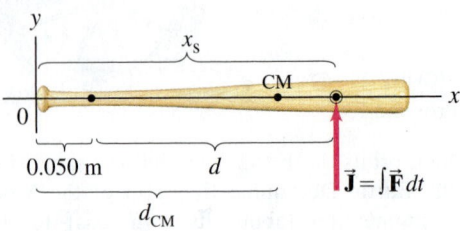

FIGURE 11–51   Problem 82.

*Numerical/Computer

*83. (II) A uniform stick 1.00 m long with a total mass of 330 g is pivoted at its center. A 3.0-g bullet is shot through the stick a distance x from the pivot. The bullet approaches at 250 m/s and leaves at 140 m/s (Fig. 11–36). (a) Determine a formula for the angular speed of the spinning stick after the collision as a function of x. (b) Graph the angular speed as a function of x, from $x = 0$ to $x = 0.50$ m.

*84. (III) Figure 11–39 shows a thin rod of mass M and length ℓ resting on a frictionless table. The rod is struck at a distance x from its CM by a clay ball of mass m moving at speed v. The ball sticks to the rod. (a) Determine a formula for the rotational motion of the system after the collision. (b) Graph the rotational motion of the system as a function of x, from $x = 0$ to $x = \ell/2$, with values of $M = 450$ g, $m = 15$ g, $\ell = 1.20$ m, and $v = 12$ m/s. (c) Does the translational motion depend on x? Explain.

Answers to Exercises

A: (b).

B: (a).

C: (b).

D: (i) (d); (ii) (a); (iii) (b).

E: (e).

Our whole built environment, from modern bridges to skyscrapers, has required architects and engineers to determine the forces and stresses within these structures. The object is to keep these structures static—that is, not in motion, especially not falling down.

CONTENTS

12–1 The Conditions for Equilibrium
12–2 Solving Statics Problems
12–3 Stability and Balance
12–4 Elasticity; Stress and Strain
12–5 Fracture
*12–6 Trusses and Bridges
*12–7 Arches and Domes

C H A P T E R

12

Static Equilibrium; Elasticity and Fracture

CHAPTER-OPENING QUESTION—Guess Now!

The diving board shown here is held by two supports at A and B. Which statement is true about the forces exerted *on* the diving board at A and B?

(a) $\vec{F}_A$ is down, $\vec{F}_B$ is up, and F_B is larger than F_A.
(b) Both forces are up and F_B is larger than F_A.
(c) $\vec{F}_A$ is down, $\vec{F}_B$ is up, and F_A is larger than F_B.
(d) Both forces are down and approximately equal.
(e) $\vec{F}_B$ is down, $\vec{F}_A$ is up, and they are equal.

W e now study a special case in mechanics—when the net force and the net torque on an object, or system of objects, are both zero. In this case both the linear acceleration and the angular acceleration of the object or system are zero. The object is either at rest, or its center of mass is moving at constant velocity. We will be concerned mainly with the first situation, in which the object or objects are at rest.

We will see how to determine the forces (and torques) that act within a structure. Just how and where these forces act can be very important for buildings, bridges, and other structures, and in the human body.

Statics is concerned with the calculation of the forces acting on and within structures that are in *equilibrium*. Determination of these forces, which occupies us in the first part of this Chapter, then allows a determination of whether the structures can sustain the forces without significant deformation or fracture, since any material can break or buckle if too much force is applied (Fig. 12–1).

FIGURE 12–1 Elevated walkway collapse in a Kansas City hotel in 1981. How a simple physics calculation could have prevented the tragic loss of over 100 lives is considered in Example 12–9.

311

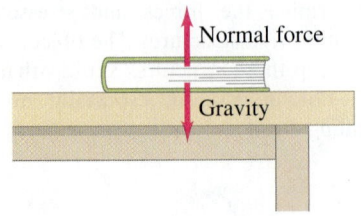

FIGURE 12-2 The book is in equilibrium; the net force on it is zero.

12-1 The Conditions for Equilibrium

Objects in daily life have at least one force acting on them (gravity). If they are at rest, then there must be other forces acting on them as well so that the net force is zero. A book at rest on a table, for example, has two forces acting on it, the downward force of gravity and the normal force the table exerts upward on it (Fig. 12-2). Because the book is at rest, Newton's second law tells us the net force on it is zero. Thus the upward force exerted by the table on the book must be equal in magnitude to the force of gravity acting downward on the book. Such an object is said to be in **equilibrium** (Latin for "equal forces" or "balance") under the action of these two forces.

Do not confuse the two forces in Fig. 12-2 with the equal and opposite forces of Newton's third law, which act on different objects. Here, both forces act on the same object; and they add up to zero.

The First Condition for Equilibrium

For an object to be at rest, Newton's second law tells us that the sum of the forces acting on it must add up to zero. Since force is a vector, the components of the net force must each be zero. Hence, a condition for equilibrium is that

$$\Sigma F_x = 0, \qquad \Sigma F_y = 0, \qquad \Sigma F_z = 0. \qquad \textbf{(12-1)}$$

We will mainly be dealing with forces that act in a plane, so we usually need only the x and y components. We must remember that if a particular force component points along the negative x or y axis, it must have a negative sign. Equations 12-1 are called the **first condition for equilibrium**.

FIGURE 12-3 Example 12-1.

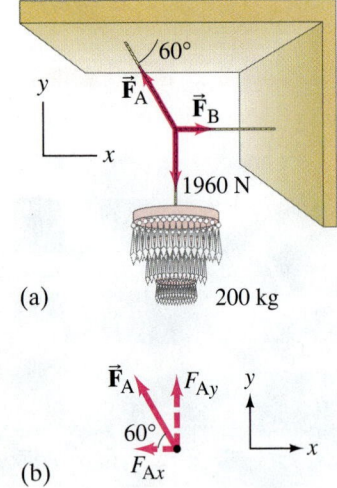

(a)

(b)

EXAMPLE 12-1 **Chandelier cord tension.** Calculate the tensions $\vec{F}_A$ and $\vec{F}_B$ in the two cords that are connected to the vertical cord supporting the 200-kg chandelier in Fig. 12-3. Ignore the mass of the cords.

APPROACH We need a free-body diagram, but for which object? If we choose the chandelier, the cord supporting it must exert a force equal to the chandelier's weight $mg = (200 \text{ kg})(9.8 \text{ m/s}^2) = 1960 \text{ N}$. But the forces $\vec{F}_A$ and $\vec{F}_B$ don't get involved. Instead, let us choose as our object the point where the three cords join (it could be a knot). The free-body diagram is then as shown in Fig. 12-3a. The three forces—$\vec{F}_A$, $\vec{F}_B$, and the tension in the vertical cord equal to the weight of the 200-kg chandelier—act at this point where the three cords join. For this junction point we write $\Sigma F_x = 0$ and $\Sigma F_y = 0$, since the problem is laid out in two dimensions. The directions of $\vec{F}_A$ and $\vec{F}_B$ are known, since tension in a cord can only be along the cord—any other direction would cause the cord to bend, as already pointed out in Chapter 4. Thus, our unknowns are the magnitudes F_A and F_B.

SOLUTION We first resolve $\vec{F}_A$ into its horizontal (x) and vertical (y) components. Although we don't know the value of F_A, we can write (see Fig. 12-3b) $F_{Ax} = -F_A \cos 60°$ and $F_{Ay} = F_A \sin 60°$. $\vec{F}_B$ has only an x component. In the vertical direction, we have the downward force exerted by the vertical cord equal to the weight of the chandelier $= (200 \text{ kg})(g)$, and the vertical component of $\vec{F}_A$ upward:

$$\Sigma F_y = 0$$
$$F_A \sin 60° - (200 \text{ kg})(g) = 0$$

so

$$F_A = \frac{(200 \text{ kg})g}{\sin 60°} = (231 \text{ kg})g = 2260 \text{ N}.$$

In the horizontal direction, with $\Sigma F_x = 0$,

$$\Sigma F_x = F_B - F_A \cos 60° = 0.$$

Thus

$$F_B = F_A \cos 60° = (231 \text{ kg})(g)(0.500) = (115 \text{ kg})g = 1130 \text{ N}.$$

The magnitudes of $\vec{F}_A$ and $\vec{F}_B$ determine the strength of cord or wire that must be used. In this case, the cord must be able to hold more than 230 kg.

| **EXERCISE A** In Example 12-1, F_A has to be greater than the chandelier's weight, mg. Why?

The Second Condition for Equilibrium

Although Eqs. 12–1 are a necessary condition for an object to be in equilibrium, they are not always a sufficient condition. Figure 12–4 shows an object on which the net force is zero. Although the two forces labeled $\vec{F}$ add up to give zero net force on the object, they do give rise to a net torque that will rotate the object. Referring to Eq. 10–14, $\Sigma\tau = I\alpha$, we see that if an object is to remain at rest, the net torque applied to it (calculated about *any* axis) must be zero. Thus we have the **second condition for equilibrium**: that the sum of the torques acting on an object, as calculated about any axis, must be zero:

$$\Sigma\tau = 0. \tag{12–2}$$

This condition will ensure that the angular acceleration, α, about any axis will be zero. If the object is not rotating initially ($\omega = 0$), it will not start rotating. Equations 12–1 and 12–2 are the only requirements for an object to be in equilibrium.

We will mainly consider cases in which the forces all act in a plane (we call it the *xy* plane). In such cases the torque is calculated about an axis that is perpendicular to the *xy* plane. *The choice of this axis is arbitrary.* If the object is at rest, then $\Sigma\tau = 0$ about any axis whatever. Therefore we can choose any axis that makes our calculation easier. Once the axis is chosen, all torques must be calculated about that axis.

CONCEPTUAL EXAMPLE 12–2 | A lever. The bar in Fig. 12–5 is being used as a lever to pry up a large rock. The small rock acts as a fulcrum (pivot point). The force F_P required at the long end of the bar can be quite a bit smaller than the rock's weight mg, since it is the *torques* that balance in the rotation about the fulcrum. If, however, the leverage isn't sufficient, and the large rock isn't budged, what are two ways to increase the leverage?

RESPONSE One way is to increase the lever arm of the force F_P by slipping a pipe over the end of the bar and thereby pushing with a longer lever arm. A second way is to move the fulcrum closer to the large rock. This may change the long lever arm R only a little, but it changes the short lever arm r by a substantial fraction and therefore changes the ratio of R/r dramatically. In order to pry the rock, the torque due to F_P must at least balance the torque due to mg; that is, $mgr = F_P R$ and

$$\frac{r}{R} = \frac{F_P}{mg}.$$

With r smaller, the weight mg can be balanced with less force F_P. The ratio of the load force to your applied force ($= mg/F_P$ here) is the **mechanical advantage** of the system, and here equals R/r. A lever is a "simple machine." We discussed another simple machine, the pulley, in Chapter 4, Example 4–14.

EXERCISE B For simplicity, we wrote the equation in Example 12–2 as if the lever were perpendicular to the forces. Would the equation be valid even for a lever at an angle as shown in Fig. 12–5?

12–2 Solving Statics Problems

The subject of statics is important because it allows us to calculate certain forces on (or within) a structure when some of the forces on it are already known. We will mainly consider situations in which all the forces act in a plane, so we can have two force equations (*x* and *y* components) and one torque equation, for a total of three equations. Of course, you do not have to use all three equations if they are not needed. When using a torque equation, a torque that tends to rotate the object counterclockwise is usually considered positive, whereas a torque that tends to rotate it clockwise is considered negative. (But the opposite convention would be okay too.)

One of the forces that acts on objects is the force of gravity. As we discussed in Section 9–8, we can consider the force of gravity on an object as acting at its center of gravity (CG) or center of mass (CM), which for practical purposes are the same point. For uniform symmetrically shaped objects, the CG is at the geometric center. For more complicated objects, the CG can be determined as discussed in Section 9–8.

There is no single technique for attacking statics problems, but the following procedure may be helpful.

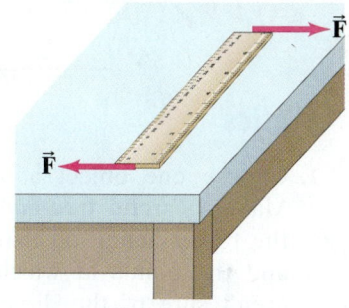

FIGURE 12–4 Although the net force on it is zero, the ruler will move (rotate). A pair of equal forces acting in opposite directions but at different points on an object (as shown here) is referred to as a *couple*.

⚠ CAUTION

Axis choice for $\Sigma\tau = 0$ is arbitrary. All torques must be calculated about the same axis.

🏃 PHYSICS APPLIED

The lever

FIGURE 12–5 Example 12–2. A lever can "multiply" your force.

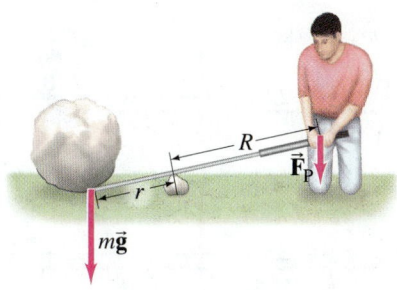

PROBLEM SOLVING

$\tau > 0$ counterclockwise
$\tau < 0$ clockwise

Statics

1. Choose one object at a time for consideration. Make a careful **free-body diagram** by showing all the forces acting on that object, including gravity, and the points at which these forces act. If you aren't sure of the direction of a force, choose a direction; if the actual direction is opposite, your eventual calculation will give a result with a minus sign.

2. Choose a convenient **coordinate system**, and resolve the forces into their components.

3. Using letters to represent unknowns, write down the **equilibrium equations** for the **forces**:

$$\Sigma F_x = 0 \quad \text{and} \quad \Sigma F_y = 0,$$

assuming all the forces act in a plane.

4. For the **torque equation**,

$$\Sigma \tau = 0,$$

choose any axis perpendicular to the xy plane that might make the calculation easier. (For example, you can reduce the number of unknowns in the resulting equation by choosing the axis so that one of the unknown forces acts through that axis; then this force will have zero lever arm and produce zero torque, and so won't appear in the torque equation.) Pay careful attention to determining the lever arm for each force correctly. Give each torque a + or − sign to indicate torque direction. For example, if torques tending to rotate the object counterclockwise are positive, then those tending to rotate it clockwise are negative.

5. **Solve** these equations for the unknowns. Three equations allow a maximum of three unknowns to be solved for. They can be forces, distances, or even angles.

PHYSICS APPLIED
Balancing a seesaw

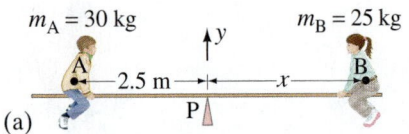

$m_A = 30$ kg $\qquad m_B = 25$ kg

(a)

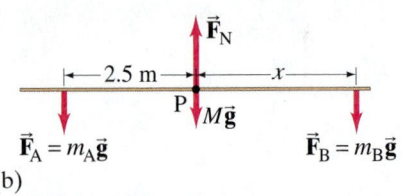

(b)

FIGURE 12–6 (a) Two children on a seesaw, Example 12–3. (b) Free-body diagram of the board.

EXAMPLE 12–3 **Balancing a seesaw.** A board of mass $M = 2.0$ kg serves as a seesaw for two children, as shown in Fig. 12–6a. Child A has a mass of 30 kg and sits 2.5 m from the pivot point, P (his center of gravity is 2.5 m from the pivot). At what distance x from the pivot must child B, of mass 25 kg, place herself to balance the seesaw? Assume the board is uniform and centered over the pivot.

APPROACH We follow the steps of the Problem Solving Strategy above.

SOLUTION

1. **Free-body diagram.** We choose the board as our object, and assume it is horizontal. Its free-body diagram is shown in Fig. 12–6b. The forces acting on the board are the forces exerted downward on it by each child, $\vec{F}_A$ and $\vec{F}_B$, the upward force exerted by the pivot $\vec{F}_N$, and the force of gravity on the board $(= M\vec{g})$ which acts at the center of the uniform board.

2. **Coordinate system.** We choose y to be vertical, with positive upward, and x horizontal to the right, with origin at the pivot.

3. **Force equation.** All the forces are in the y (vertical) direction, so

$$\Sigma F_y = 0$$
$$F_N - m_A g - m_B g - Mg = 0,$$

where $F_A = m_A g$ and $F_B = m_B g$ because each child is in equilibrium when the seesaw is balanced.

4. **Torque equation.** Let us calculate the torque about an axis through the board at the pivot point, P. Then the lever arms for F_N and for the weight of the board are zero, and they will contribute zero torque about point P. Thus the torque equation will involve only the forces $\vec{F}_A$ and $\vec{F}_B$, which are equal to the weights of the children. The torque exerted by each child will be mg times the appropriate lever arm, which here is the distance of each child from the pivot point. Hence the torque equation is

$$\Sigma \tau = 0$$
$$m_A g(2.5 \text{ m}) - m_B g x + Mg(0 \text{ m}) + F_N(0 \text{ m}) = 0$$

or

$$m_A g(2.5 \text{ m}) - m_B g x = 0,$$

where two terms were dropped because their lever arms were zero.

5. Solve. We solve the torque equation for x and find

$$x = \frac{m_A}{m_B}(2.5\,\text{m}) = \frac{30\,\text{kg}}{25\,\text{kg}}(2.5\,\text{m}) = 3.0\,\text{m}.$$

To balance the seesaw, child B must sit so that her CM is 3.0 m from the pivot point. This makes sense: since she is lighter, she must sit farther from the pivot than the heavier child in order to provide equal torque.

EXERCISE C We did not need to use the force equation to solve Example 12–3 because of our choice of the axis. Use the force equation to find the force exerted by the pivot.

Figure 12–7 shows a uniform beam that extends beyond its support like a diving board. Such a beam is called a **cantilever**. The forces acting on the beam in Fig. 12–7 are those due to the supports, $\vec{F}_A$ and $\vec{F}_B$, and the force of gravity which acts at the CG, 5.0 m to the right of the right-hand support. If you follow the procedure of the last Example and calculate F_A and F_B, assuming they point upward as shown in Fig. 12–7, you will find that F_A comes out negative. If the beam has a mass of 1200 kg and a weight $mg = 12,000\,\text{N}$, then $F_B = 15,000\,\text{N}$ and $F_A = -3000\,\text{N}$ (see Problem 9). Whenever an unknown force comes out negative, it merely means that the force actually points in the opposite direction from what you assumed. Thus in Fig. 12–7, $\vec{F}_A$ actually points downward. With a little reflection it should become clear that the left-hand support must indeed pull downward on the beam (by means of bolts, screws, fasteners and/or glue) if the beam is to be in equilibrium; otherwise the sum of the torques about the CG (or about the point where $\vec{F}_B$ acts) could not be zero.

EXERCISE D Return to the Chapter-Opening Question, p. 311, and answer it again now. Try to explain why you may have answered differently the first time.

EXAMPLE 12–4 **Force exerted by biceps muscle.** How much force must the biceps muscle exert when a 5.0-kg ball is held in the hand (a) with the arm horizontal as in Fig. 12–8a, and (b) when the arm is at a 45° angle as in Fig. 12–8b? The biceps muscle is connected to the forearm by a tendon attached 5.0 cm from the elbow joint. Assume that the mass of forearm and hand together is 2.0 kg and their CG is as shown.

APPROACH The free-body diagram for the forearm is shown in Fig. 12–8; the forces are the weights of the arm and ball, the upward force $\vec{F}_M$ exerted by the muscle, and a force $\vec{F}_J$ exerted at the joint by the bone in the upper arm (all assumed to act vertically). We wish to find the magnitude of $\vec{F}_M$, which is done most easily by using the torque equation and by choosing our axis through the joint so that $\vec{F}_J$ contributes zero torque.

SOLUTION (a) We calculate torques about the point where $\vec{F}_J$ acts in Fig. 12–8a. The $\Sigma\tau = 0$ equation gives

$$(0.050\,\text{m})F_M - (0.15\,\text{m})(2.0\,\text{kg})g - (0.35\,\text{m})(5.0\,\text{kg})g = 0.$$

We solve for F_M:

$$F_M = \frac{(0.15\,\text{m})(2.0\,\text{kg})g + (0.35\,\text{m})(5.0\,\text{kg})g}{0.050\,\text{m}} = (41\,\text{kg})g = 400\,\text{N}.$$

(b) The lever arm, as calculated about the joint, is reduced by the factor cos 45° for all three forces. Our torque equation will look like the one just above, except that each term will have its lever arm reduced by the same factor, which will cancel out. The same result is obtained, $F_M = 400\,\text{N}$.

NOTE The force required of the muscle (400 N) is quite large compared to the weight of the object lifted ($= mg = 49\,\text{N}$). Indeed, the muscles and joints of the body are generally subjected to quite large forces.

EXERCISE E How much mass could the person in Example 12–4 hold in the hand with a biceps force of 450 N if the tendon was attached 6.0 cm from the elbow instead of 5.0 cm?

PHYSICS APPLIED
Cantilever

PROBLEM SOLVING
If a force comes out negative

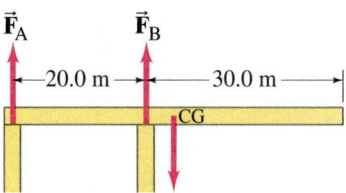
FIGURE 12–7 A cantilever. The force vectors shown are hypothetical—one may even have a different direction.

PHYSICS APPLIED
Forces in muscles and joints

FIGURE 12–8 Example 12–4.

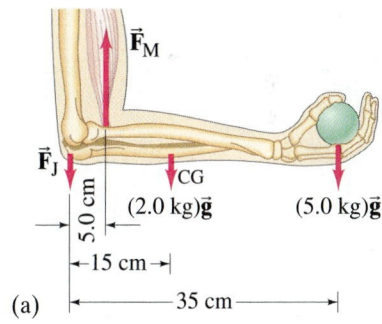
(a)

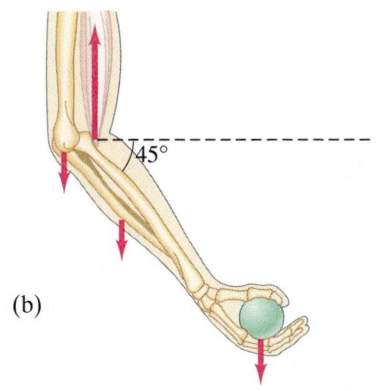
(b)

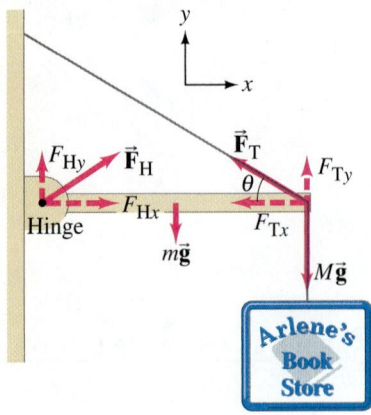

FIGURE 12–9 Example 12–5.

Our next Example involves a beam that is attached to a wall by a hinge and is supported by a cable or cord (Fig. 12–9). It is important to remember that a flexible cable can support a force only along its length. (If there were a component of force perpendicular to the cable, it would bend because it is flexible.) But for a rigid device, such as the hinge in Fig. 12–9, the force can be in any direction and we can know the direction only after solving the problem. The hinge is assumed small and smooth, so it can exert no internal torque (about its center) on the beam.

EXAMPLE 12–5 **Hinged beam and cable.** A uniform beam, 2.20 m long with mass $m = 25.0\,\text{kg}$, is mounted by a small hinge on a wall as shown in Fig. 12–9. The beam is held in a horizontal position by a cable that makes an angle $\theta = 30.0°$. The beam supports a sign of mass $M = 28.0\,\text{kg}$ suspended from its end. Determine the components of the force $\vec{F}_H$ that the (smooth) hinge exerts on the beam, and the tension F_T in the supporting cable.

APPROACH Figure 12–9 is the free-body diagram for the beam, showing all the forces acting on the beam. It also shows the components of $\vec{F}_T$ and a guess for $\vec{F}_H$. We have three unknowns, F_{Hx}, F_{Hy}, and F_T (we are given θ), so we will need all three equations, $\Sigma F_x = 0$, $\Sigma F_y = 0$, $\Sigma\tau = 0$.

SOLUTION The sum of the forces in the vertical (y) direction is
$$\Sigma F_y = 0$$
$$F_{Hy} + F_{Ty} - mg - Mg = 0. \tag{i}$$
In the horizontal (x) direction, the sum of the forces is
$$\Sigma F_x = 0$$
$$F_{Hx} - F_{Tx} = 0. \tag{ii}$$
For the torque equation, we choose the axis at the point where $\vec{F}_T$ and $M\vec{g}$ act (so our equation then contains only one unknown, F_{Hy}). We choose torques that tend to rotate the beam counterclockwise as positive. The weight mg of the (uniform) beam acts at its center, so we have
$$\Sigma\tau = 0$$
$$-(F_{Hy})(2.20\,\text{m}) + mg(1.10\,\text{m}) = 0.$$
We solve for F_{Hy}:
$$F_{Hy} = \left(\frac{1.10\,\text{m}}{2.20\,\text{m}}\right)mg = (0.500)(25.0\,\text{kg})(9.80\,\text{m/s}^2) = 123\,\text{N}. \tag{iii}$$

Next, since the tension $\vec{F}_T$ in the cable acts along the cable ($\theta = 30.0°$), we see from Fig. 12–9 that $\tan\theta = F_{Ty}/F_{Tx}$, or
$$F_{Ty} = F_{Tx}\tan\theta = F_{Tx}(\tan 30.0°) = 0.577\,F_{Tx}. \tag{iv}$$
Equation (i) above gives
$$F_{Ty} = (m + M)g - F_{Hy} = (53.0\,\text{kg})(9.80\,\text{m/s}^2) - 123\,\text{N} = 396\,\text{N};$$
Equations (iv) and (ii) give
$$F_{Tx} = F_{Ty}/0.577 = 687\,\text{N};$$
$$F_{Hx} = F_{Tx} = 687\,\text{N}.$$
The components of $\vec{F}_H$ are $F_{Hy} = 123\,\text{N}$ and $F_{Hx} = 687\,\text{N}$. The tension in the wire is $F_T = \sqrt{F_{Tx}^2 + F_{Ty}^2} = 793\,\text{N}$.

Alternate Solution Let us see the effect of choosing a different axis for calculating torques, such as an axis through the hinge. Then the lever arm for F_H is zero, and the torque equation ($\Sigma\tau = 0$) becomes
$$-mg(1.10\,\text{m}) - Mg(2.20\,\text{m}) + F_{Ty}(2.20\,\text{m}) = 0.$$
We solve this for F_{Ty} and find
$$F_{Ty} = \frac{m}{2}g + Mg = (12.5\,\text{kg} + 28.0\,\text{kg})(9.80\,\text{m/s}^2) = 397\,\text{N}.$$

We get the same result, within the precision of our significant figures.

NOTE It doesn't matter which axis we choose for $\Sigma\tau = 0$. Using a second axis can serve as a check.

EXAMPLE 12–6 **Ladder.** A 5.0-m-long ladder leans against a smooth wall at a point 4.0 m above a cement floor as shown in Fig. 12–10. The ladder is uniform and has mass $m = 12.0$ kg. Assuming the wall is frictionless (but the floor is not), determine the forces exerted on the ladder by the floor and by the wall.

APPROACH Figure 12–10 is the free-body diagram for the ladder, showing all the forces acting on the ladder. The wall, since it is frictionless, can exert a force only perpendicular to the wall, and we label that force $\vec{F}_W$. The cement floor exerts a force $\vec{F}_C$ which has both horizontal and vertical force components: F_{Cx} is frictional and F_{Cy} is the normal force. Finally, gravity exerts a force $mg = (12.0\,\text{kg})(9.80\,\text{m/s}^2) = 118$ N on the ladder at its midpoint, since the ladder is uniform.

SOLUTION Again we use the equilibrium conditions, $\Sigma F_x = 0$, $\Sigma F_y = 0$, $\Sigma \tau = 0$. We will need all three since there are three unknowns: F_W, F_{Cx}, and F_{Cy}. The y component of the force equation is

$$\Sigma F_y = F_{Cy} - mg = 0,$$

so immediately we have

$$F_{Cy} = mg = 118\,\text{N}.$$

The x component of the force equation is

$$\Sigma F_x = F_{Cx} - F_W = 0.$$

To determine both F_{Cx} and F_W, we need a torque equation. If we choose to calculate torques about an axis through the point where the ladder touches the cement floor, then $\vec{F}_C$, which acts at this point, will have a lever arm of zero and so won't enter the equation. The ladder touches the floor a distance $x_0 = \sqrt{(5.0\,\text{m})^2 - (4.0\,\text{m})^2} = 3.0$ m from the wall (right triangle, $c^2 = a^2 + b^2$). The lever arm for mg is half this, or 1.5 m, and the lever arm for F_W is 4.0 m, Fig. 12–10. We get

$$\Sigma \tau = (4.0\,\text{m})F_W - (1.5\,\text{m})mg = 0.$$

Thus

$$F_W = \frac{(1.5\,\text{m})(12.0\,\text{kg})(9.8\,\text{m/s}^2)}{4.0\,\text{m}} = 44\,\text{N}.$$

Then, from the x component of the force equation,

$$F_{Cx} = F_W = 44\,\text{N}.$$

Since the components of $\vec{F}_C$ are $F_{Cx} = 44$ N and $F_{Cy} = 118$ N, then

$$F_C = \sqrt{(44\,\text{N})^2 + (118\,\text{N})^2} = 126\,\text{N} \approx 130\,\text{N}$$

(rounded off to two significant figures), and it acts at an angle to the floor of

$$\theta = \tan^{-1}(118\,\text{N}/44\,\text{N}) = 70°.$$

NOTE The force $\vec{F}_C$ does *not* have to act along the ladder's direction because the ladder is rigid and not flexible like a cord or cable.

EXERCISE F Why is it reasonable to ignore friction along the wall, but not reasonable to ignore it along the floor?

12–3 Stability and Balance

An object in static equilibrium, if left undisturbed, will undergo no translational or rotational acceleration since the sum of all the forces and the sum of all the torques acting on it are zero. However, if the object is displaced slightly, three outcomes are possible: (1) the object returns to its original position, in which case it is said to be in **stable equilibrium**; (2) the object moves even farther from its original position, and it is said to be in **unstable equilibrium**; or (3) the object remains in its new position, and it is said to be in **neutral equilibrium**.

Consider the following examples. A ball suspended freely from a string is in stable equilibrium, for if it is displaced to one side, it will return to its original position (Fig. 12–11a) due to the net force and torque exerted on it. On the other hand, a pencil standing on its point is in unstable equilibrium. If its center of gravity is directly over its tip (Fig. 12–11b), the net force and net torque on it will be zero. But if it is displaced ever so slightly as shown—say, by a slight vibration or tiny air current—there will be a torque on it, and this torque acts to make the pencil continue to fall in the direction of the original displacement. Finally, an example of an object in neutral equilibrium is a sphere resting on a horizontal tabletop. If it is placed slightly to one side, it will remain in its new position—no net torque acts on it.

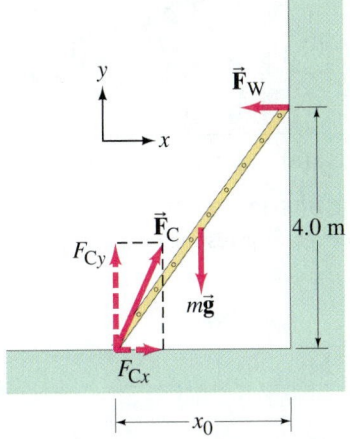

FIGURE 12–10 A ladder leaning against a wall. Example 12–6.

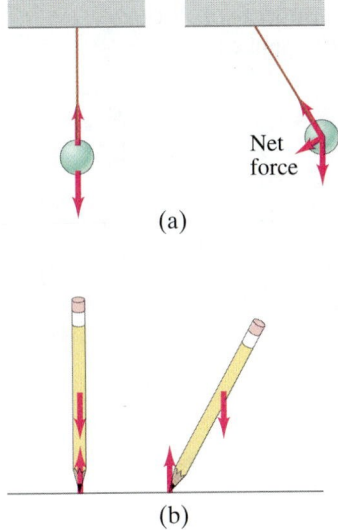

FIGURE 12–11 (a) Stable equilibrium, and (b) unstable equilibrium.

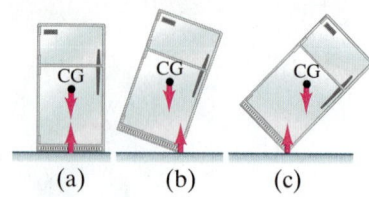

FIGURE 12–12 Equilibrium of a refrigerator resting on a flat floor.

FIGURE 12–13 Humans adjust their posture to achieve stability when carrying loads.

Total CG

In most situations, such as in the design of structures and in working with the human body, we are interested in maintaining stable equilibrium, or *balance*, as we sometimes say. In general, an object whose center of gravity (CG) is below its point of support, such as a ball on a string, will be in stable equilibrium. If the CG is above the base of support, we have a more complicated situation. Consider a standing refrigerator (Fig. 12–12a). If it is tipped slightly, it will return to its original position due to the torque on it as shown in Fig. 12–12b. But if it is tipped too far, Fig. 12–12c, it will fall over. The critical point is reached when the CG shifts from one side of the pivot point to the other. When the CG is on one side, the torque pulls the object back onto its original base of support, Fig. 12–12b. If the object is tipped further, the CG goes past the pivot point and the torque causes the object to topple, Fig. 12–12c. In general, *an object whose center of gravity is above its base of support will be stable if a vertical line projected downward from the CG falls within the base of support*. This is because the normal force upward on the object (which balances out gravity) can be exerted only within the area of contact, so if the force of gravity acts beyond this area, a net torque will act to topple the object.

Stability, then, can be relative. A brick lying on its widest face is more stable than a brick standing on its end, for it will take more of an effort to tip it over. In the extreme case of the pencil in Fig. 12–11b, the base is practically a point and the slightest disturbance will topple it. In general, the larger the base and the lower the CG, the more stable the object.

In this sense, humans are less stable than four-legged mammals, which have a larger base of support because of their four legs, and most also have a lower center of gravity. When walking and performing other kinds of movement, a person continually shifts the body so that its CG is over the feet, although in the normal adult this requires no conscious thought. Even as simple a movement as bending over requires moving the hips backward so that the CG remains over the feet, and you do this repositioning without thinking about it. To see this, position yourself with your heels and back to a wall and try to touch your toes. You won't be able to do it without falling. People carrying heavy loads automatically adjust their posture so that the CG of the total mass is over their feet, Fig. 12–13.

12–4 Elasticity; Stress and Strain

In the first part of this Chapter we studied how to calculate the forces on objects in equilibrium. In this Section we study the effects of these forces: any object changes shape under the action of applied forces. If the forces are great enough, the object will break, or *fracture*, as we will discuss in Section 12–5.

Elasticity and Hooke's Law

If a force is exerted on an object, such as the vertically suspended metal rod shown in Fig. 12–14, the length of the object changes. If the amount of elongation, $\Delta\ell$, is small compared to the length of the object, experiment shows that $\Delta\ell$ is proportional to the force exerted on the object. This proportionality, as we saw in Section 7–3, can be written as an equation:

$$F = k\,\Delta\ell. \tag{12–3}$$

Here F represents the force pulling on the object, $\Delta\ell$ is the change in length, and k is a proportionality constant. Equation 12–3, which is sometimes called **Hooke's law**[†] after Robert Hooke (1635–1703), who first noted it, is found to be valid for almost any solid material from iron to bone—but it is valid only up to a point. For if the force is too great, the object stretches excessively and eventually breaks.

Figure 12–15 shows a typical graph of applied force versus elongation. Up to a point called the **proportional limit**, Eq. 12–3 is a good approximation for many

FIGURE 12–14 Hooke's law: $\Delta\ell \propto$ applied force.

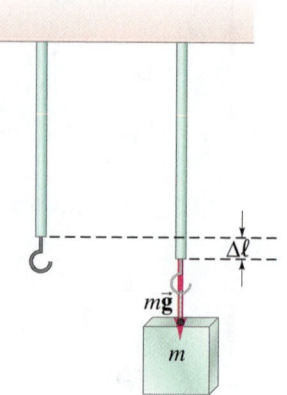

$m\vec{g}$

m

$\Delta\ell$

[†]The term "law" applied to this relation is not really appropriate, since first of all, it is only an approximation, and second, it refers only to a limited set of phenomena. Most physicists prefer to reserve the word "law" for those relations that are deeper and more encompassing and precise, such as Newton's laws of motion or the law of conservation of energy.

common materials, and the curve is a straight line. Beyond this point, the graph deviates from a straight line, and no simple relationship exists between F and $\Delta\ell$. Nonetheless, up to a point farther along the curve called the **elastic limit**, the object will return to its original length if the applied force is removed. The region from the origin to the elastic limit is called the *elastic region*. If the object is stretched beyond the elastic limit, it enters the *plastic region*: it does not return to the original length upon removal of the external force, but remains permanently deformed (such as a bent paper clip). The maximum elongation is reached at the *breaking point*. The maximum force that can be applied without breaking is called the **ultimate strength** of the material (actually, force per unit area, as we discuss in Section 12–5).

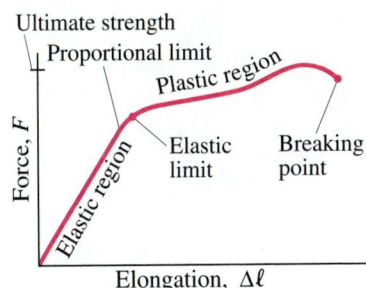

FIGURE 12–15 Applied force vs. elongation for a typical metal under tension.

Young's Modulus

The amount of elongation of an object, such as the rod shown in Fig. 12–14, depends not only on the force applied to it, but also on the material of which it is made and on its dimensions. That is, the constant k in Eq. 12–3 can be written in terms of these factors.

If we compare rods made of the same material but of different lengths and cross-sectional areas, it is found that for the same applied force, the amount of stretch (again assumed small compared to the total length) is proportional to the original length and inversely proportional to the cross-sectional area. That is, the longer the object, the more it elongates for a given force; and the thicker it is, the less it elongates. These findings can be combined with Eq. 12–3 to yield

$$\Delta\ell = \frac{1}{E}\frac{F}{A}\ell_0, \tag{12–4}$$

where ℓ_0 is the original length of the object, A is the cross-sectional area, and $\Delta\ell$ is the change in length due to the applied force F. E is a constant of proportionality[†] known as the **elastic modulus**, or **Young's modulus**; its value depends only on the material. The value of Young's modulus for various materials is given in Table 12–1 (the shear modulus and bulk modulus in this Table are discussed later in this Section). Because E is a property only of the material and is independent of the object's size or shape, Eq. 12–4 is far more useful for practical calculation than Eq. 12–3.

[†]The fact that E is in the denominator, so $1/E$ is the actual proportionality constant, is merely a convention. When we rewrite Eq. 12–4 to get Eq. 12–5, E is found in the numerator.

TABLE 12–1 Elastic Moduli

Material	Young's Modulus, E (N/m²)	Shear Modulus, G (N/m²)	Bulk Modulus, B (N/m²)
Solids			
Iron, cast	100×10^9	40×10^9	90×10^9
Steel	200×10^9	80×10^9	140×10^9
Brass	100×10^9	35×10^9	80×10^9
Aluminum	70×10^9	25×10^9	70×10^9
Concrete	20×10^9		
Brick	14×10^9		
Marble	50×10^9		70×10^9
Granite	45×10^9		45×10^9
Wood (pine) (parallel to grain)	10×10^9		
(perpendicular to grain)	1×10^9		
Nylon	5×10^9		
Bone (limb)	15×10^9	80×10^9	
Liquids			
Water			2.0×10^9
Alcohol (ethyl)			1.0×10^9
Mercury			2.5×10^9
Gases[†]			
Air, H_2, He, CO_2			1.01×10^5

[†]At normal atmospheric pressure; no variation in temperature during process.

EXAMPLE 12–7 **Tension in piano wire.** A 1.60-m-long steel piano wire has a diameter of 0.20 cm. How great is the tension in the wire if it stretches 0.25 cm when tightened?

APPROACH We assume Hooke's law holds, and use it in the form of Eq. 12–4, finding E for steel in Table 12–1.

SOLUTION We solve for F in Eq. 12–4 and note that the area of the wire is $A = \pi r^2 = (3.14)(0.0010\,\text{m})^2 = 3.14 \times 10^{-6}\,\text{m}^2$. Then

$$F = E\frac{\Delta\ell}{\ell_0}A = (2.0 \times 10^{11}\,\text{N/m}^2)\left(\frac{0.0025\,\text{m}}{1.60\,\text{m}}\right)(3.14 \times 10^{-6}\,\text{m}^2) = 980\,\text{N}.$$

NOTE The large tension in all the wires in a piano must be supported by a strong frame.

EXERCISE G Two steel wires have the same length and are under the same tension. But wire A has twice the diameter of wire B. Which of the following is true? (*a*) Wire B stretches twice as much as wire A. (*b*) Wire B stretches four times as much as wire A. (*c*) Wire A stretches twice as much as wire B. (*d*) Wire A stretches four times as much as wire B. (*e*) Both wires stretch the same amount.

Stress and Strain

From Eq. 12–4, we see that the change in length of an object is directly proportional to the product of the object's length ℓ_0 and the force per unit area F/A applied to it. It is general practice to define the force per unit area as the **stress**:

$$\text{stress} = \frac{\text{force}}{\text{area}} = \frac{F}{A},$$

which has SI units of N/m². Also, the **strain** is defined to be the ratio of the change in length to the original length:

$$\text{strain} = \frac{\text{change in length}}{\text{original length}} = \frac{\Delta\ell}{\ell_0},$$

and is dimensionless (no units). Strain is thus the fractional change in length of the object, and is a measure of how much the rod has been deformed. Stress is applied to the material by external agents, whereas strain is the material's response to the stress. Equation 12–4 can be rewritten as

$$\frac{F}{A} = E\frac{\Delta\ell}{\ell_0} \qquad\qquad (12\text{–}5)$$

or

$$E = \frac{F/A}{\Delta\ell/\ell_0} = \frac{\text{stress}}{\text{strain}}.$$

Thus we see that the strain is directly proportional to the stress, in the linear (elastic) region of Fig. 12–15.

Tension, Compression, and Shear Stress

The rod shown in Fig. 12–16a is said to be under *tension* or **tensile stress**. Not only is there a force pulling down on the rod at its lower end, but since the rod is in equilibrium, we know that the support at the top is exerting an equal[†] upward force on the rod at its upper end, Fig. 12–16a. In fact, this tensile stress exists throughout the material. Consider, for example, the lower half of a suspended rod as shown in Fig. 12–16b. This lower half is in equilibrium, so there must be an upward force on it to balance the downward force at its lower end. What exerts this upward force? It must be the upper part of the rod. Thus we see that external forces applied to an object give rise to internal forces, or stress, within the material itself.

[†]Or a greater force if the weight of the rod cannot be ignored compared to F.

FIGURE 12–16 Stress exists *within* the material.

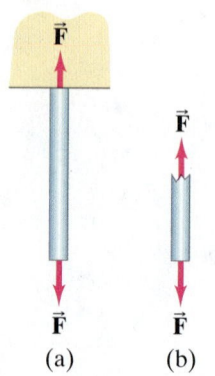

(a)   (b)

Strain or deformation due to tensile stress is but one type of stress to which materials can be subjected. There are two other common types of stress: compressive and shear. **Compressive stress** is the exact opposite of tensile stress. Instead of being stretched, the material is compressed: the forces act inwardly on the object. Columns that support a weight, such as the columns of a Greek temple (Fig. 12–17), are subjected to compressive stress. Equations 12–4 and 12–5 apply equally well to compression and tension, and the values for the modulus E are usually the same.

FIGURE 12–17 This Greek temple, in Agrigento, Sicily, built 2500 years ago, shows the post-and-beam construction. The columns are under compression.

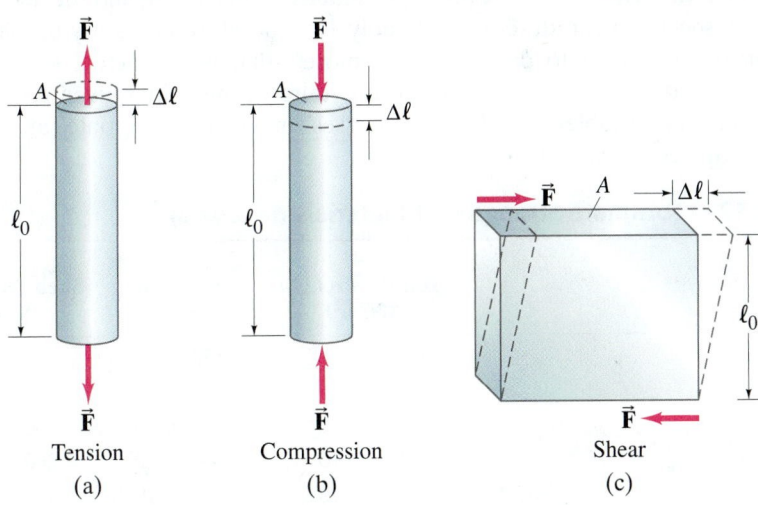

Tension
(a)

Compression
(b)

Shear
(c)

FIGURE 12–18 The three types of stress for rigid objects.

Figure 12–18 compares tensile and compressive stresses as well as the third type, shear stress. An object under **shear stress** has equal and opposite forces applied *across* its opposite faces. A simple example is a book or brick firmly attached to a tabletop, on which a force is exerted parallel to the top surface. The table exerts an equal and opposite force along the bottom surface. Although the dimensions of the object do not change significantly, the shape of the object does change, Fig. 12–18c. An equation similar to Eq. 12–4 can be applied to calculate shear strain:

$$\Delta \ell = \frac{1}{G} \frac{F}{A} \ell_0, \tag{12–6}$$

but $\Delta \ell$, ℓ_0, and A must be reinterpreted as indicated in Fig. 12–18c. Note that A is the area of the surface *parallel* to the applied force (and not perpendicular as for tension and compression), and $\Delta \ell$ is *perpendicular* to ℓ_0. The constant of proportionality G is called the **shear modulus** and is generally one-half to one-third the value of Young's modulus E (see Table 12–1). Figure 12–19 suggests why $\Delta \ell \propto \ell_0$: the fatter book shifts more for the same shearing force.

FIGURE 12–19 The fatter book (a) shifts more than the thinner book (b) with the same applied shear force.

(a)

(b)

Volume Change—Bulk Modulus

If an object is subjected to inward forces from all sides, its volume will decrease. A common situation is an object submerged in a fluid; in this case, the fluid exerts a pressure on the object in all directions, as we shall see in Chapter 13. *Pressure* is defined as force per unit area, and thus is the equivalent of stress. For this situation the change in volume, ΔV, is proportional to the original volume, V_0, and to the change in the pressure, ΔP. We thus obtain a relation of the same form as Eq. 12–4 but with a proportionality constant called the **bulk modulus** B:

$$\frac{\Delta V}{V_0} = -\frac{1}{B} \Delta P \tag{12–7}$$

or

$$B = -\frac{\Delta P}{\Delta V/V_0}.$$

The minus sign means the volume *decreases* with an increase in pressure.

Values for the bulk modulus are given in Table 12–1. Since liquids and gases do not have a fixed shape, only the bulk modulus (not the Young's or shear moduli) applies to them.

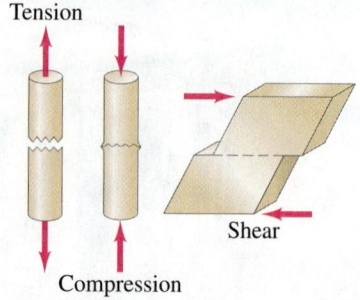

Tension

Compression

Shear

FIGURE 12–20 Fracture as a result of the three types of stress.

12–5 Fracture

If the stress on a solid object is too great, the object fractures, or breaks (Fig. 12–20). Table 12–2 lists the ultimate strengths for tension, compression, and shear for a variety of materials. These values give the maximum force per unit area, or stress, that an object can withstand under each of these three types of stress for various types of material. They are, however, representative values only, and the actual value for a given specimen can differ considerably. It is therefore necessary to maintain a *safety factor* of from 3 to perhaps 10 or more—that is, the actual stresses on a structure should not exceed one-tenth to one-third of the values given in the Table. You may encounter tables of "allowable stresses" in which appropriate safety factors have already been included.

TABLE 12–2 Ultimate Strengths of Materials (force/area)

Material	Tensile Strength (N/m²)	Compressive Strength (N/m²)	Shear Strength (N/m²)
Iron, cast	170×10^6	550×10^6	170×10^6
Steel	500×10^6	500×10^6	250×10^6
Brass	250×10^6	250×10^6	200×10^6
Aluminum	200×10^6	200×10^6	200×10^6
Concrete	2×10^6	20×10^6	2×10^6
Brick		35×10^6	
Marble		80×10^6	
Granite		170×10^6	
Wood (pine) (parallel to grain) (perpendicular to grain)	40×10^6	35×10^6 10×10^6	5×10^6
Nylon	500×10^6		
Bone (limb)	130×10^6	170×10^6	

EXAMPLE 12–8 ESTIMATE **Breaking the piano wire.** The steel piano wire we discussed in Example 12–7 was 1.60 m long with a diameter of 0.20 cm. Approximately what tension force would break it?

APPROACH We set the tensile stress F/A equal to the tensile strength of steel given in Table 12–2.

SOLUTION The area of the wire is $A = \pi r^2$, where $r = 0.10$ cm $= 1.0 \times 10^{-3}$ m. Table 12–2 tells us

$$\frac{F}{A} = 500 \times 10^6 \, \text{N/m}^2,$$

so the wire would likely break if the force exceeded

$$F = (500 \times 10^6 \, \text{N/m}^2)(\pi)(1.0 \times 10^{-3} \, \text{m})^2 = 1600 \, \text{N}.$$

As can be seen in Table 12–2, concrete (like stone and brick) is reasonably strong under compression but extremely weak under tension. Thus concrete can be used as vertical columns placed under compression, but is of little value as a beam because it cannot withstand the tensile forces that result from the inevitable sagging of the lower edge of a beam (see Fig. 12–21).

FIGURE 12–21 A beam sags, at least a little (but is exaggerated here), even under its own weight. The beam thus changes shape: the upper edge is compressed, and the lower edge is under tension (elongated). Shearing stress also occurs within the beam.

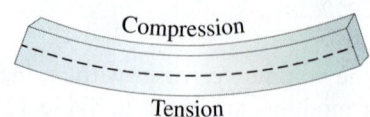

Compression

Tension

Reinforced concrete, in which iron rods are embedded in the concrete (Fig. 12–22), is much stronger. Stronger still is *prestressed concrete*, which also contains iron rods or a wire mesh, but during the pouring of the concrete, the rods or wire are held under tension. After the concrete dries, the tension on the iron is released, putting the concrete under compression. The amount of compressive stress is carefully predetermined so that when loads are applied to the beam, they reduce the compression on the lower edge, but never put the concrete into tension.

FIGURE 12–22 Steel rods around which concrete is poured for strength.

| **CONCEPTUAL EXAMPLE 12–9** | **A tragic substitution.** Two walkways, one |

above the other, are suspended from vertical rods attached to the ceiling of a high hotel lobby, Fig. 12–23a. The original design called for single rods 14 m long, but when such long rods proved to be unwieldy to install, it was decided to replace each long rod with two shorter ones as shown schematically in Fig. 12–23b. Determine the net force exerted by the rods on the supporting pin A (assumed to be the same size) for each design. Assume each vertical rod supports a mass m of each bridge.

RESPONSE The single long vertical rod in Fig. 12–23a exerts an upward force equal to mg on pin A to support the mass m of the upper bridge. Why? Because the pin is in equilibrium, and the other force that balances this is the downward force mg exerted on it by the upper bridge (Fig. 12–23c). There is thus a shear stress on the pin because the rod pulls up on one side of the pin, and the bridge pulls down on the other side. The situation when two shorter rods support the bridges (Fig. 12–23b) is shown in Fig. 12–23d, in which only the connections at the upper bridge are shown. The lower rod exerts a force mg downward on the lower of the two pins because it supports the lower bridge. The upper rod exerts a force $2mg$ on the upper pin (labelled A) because the upper rod supports both bridges. Thus we see that when the builders substituted two shorter rods for each single long one, the stress in the supporting pin A was *doubled*. What perhaps seemed like a simple substitution did, in fact, lead to a tragic collapse in 1981 with a loss of life of over 100 people (see Fig. 12–1). Having a feel for physics, and being able to make simple calculations based on physics, can have a great effect, literally, on people's lives.

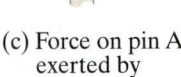
FIGURE 12–23 Example 12–9.

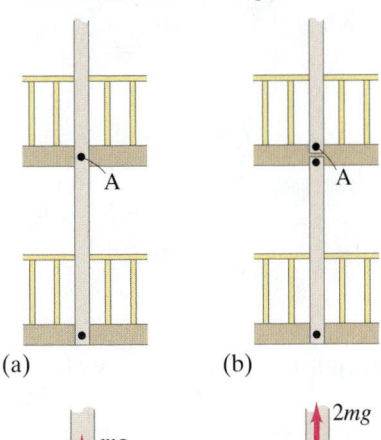

(a)   (b)

(c) Force on pin A exerted by vertical rod

(d) Forces on pins at A exerted by vertical rods

| **EXAMPLE 12–10** | **Shear on a beam.** A uniform pine beam, 3.6 m long and |

9.5 cm × 14 cm in cross section, rests on two supports near its ends, as shown in Fig. 12–24. The beam's mass is 25 kg and two vertical roof supports rest on it, each one-third of the way from the ends. What maximum load force F_L can each of the roof supports exert without shearing the pine beam at its supports? Use a safety factor of 5.0.

APPROACH The *symmetry* present simplifies our calculation. We first find the shear strength of pine in Table 12–2 and use the safety factor of 5.0 to get F from $F/A \leq \frac{1}{5}$(shear strength). Then we use $\Sigma\tau = 0$ to find F_L.

SOLUTION Each support exerts an upward force F (there is symmetry) that can be at most (see Table 12–2)

$$F = \frac{1}{5} A(5 \times 10^6 \, \text{N/m}^2) = \frac{1}{5}(0.095 \, \text{m})(0.14 \, \text{m})(5 \times 10^6 \, \text{N/m}^2) = 13,000 \, \text{N}.$$

To determine the maximum load force F_L, we calculate the torque about the left end of the beam (counterclockwise positive):

$$\Sigma\tau = -F_L(1.2 \, \text{m}) - (25 \, \text{kg})(9.8 \, \text{m/s}^2)(1.8 \, \text{m}) - F_L(2.4 \, \text{m}) + F(3.6 \, \text{m}) = 0$$

so each of the two roof supports can exert

$$F_L = \frac{(13,000 \, \text{N})(3.6 \, \text{m}) - (250 \, \text{N})(1.8 \, \text{m})}{(1.2 + 2.4)} = 13,000 \, \text{N}.$$

The total mass of roof the beam can support is $(2)(13,000 \, \text{N})/(9.8 \, \text{m/s}^2) = 2600 \, \text{kg}$.

FIGURE 12–24 Example 12–10.

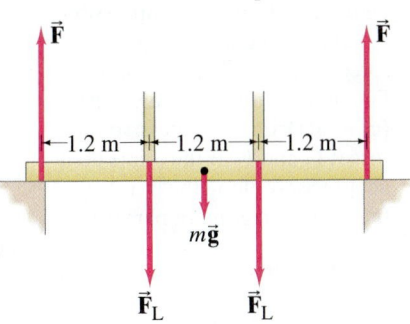

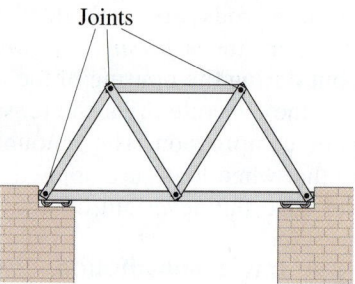

FIGURE 12–25 A truss bridge.

Joints

*12-6 Trusses and Bridges

A beam used to span a wide space, as for a bridge, is subject to strong stresses of all three types as we saw in Fig. 12–21: compression, tension and shear. A basic engineering device to support large spans is the *truss*, an example of which is shown in Fig. 12–25. Wooden truss bridges were first designed by the great architect Andrea Palladio (1518–1580), famous for his design of public buildings and villas. With the introduction of steel in the nineteenth century, much stronger steel trusses came into use, although wood trusses are still used to support the roofs of houses and mountain lodges (Fig. 12–26).

Basically, a **truss** is a framework of rods or struts joined together at their ends by pins or rivets, always arranged as triangles. (Triangles are relatively stable, as compared to a rectangle, which easily becomes a parallelogram under sideways forces and then collapses.) The place where the struts are joined by a pin is called a **joint**.

It is commonly assumed that the struts of a truss are under pure compression or pure tension—that is, the forces act along the length of each strut, Fig. 12–27a. This is an ideal, valid only if a strut has no mass and supports no weight along its length, in which case a strut has only two forces on it, at the ends, as shown in Fig. 12–27a. If the strut is in equilibrium, these two forces must be equal and opposite in direction $(\Sigma\vec{F} = 0)$. But couldn't they be at an angle, as in Fig. 12–27b? No, because then $\Sigma\vec{\tau}$ would not be zero. The two forces *must* act along the strut if the strut is in equilibrium. But in a real case of a strut with mass, there are three forces on the strut, as shown in Fig. 12–27c, and $\vec{F}_1$ and $\vec{F}_2$ do not act along the strut; the vector diagram in Fig. 12–27d shows $\Sigma\vec{F} = \vec{F}_1 + \vec{F}_2 + m\vec{g} = 0$. Can you see why $\vec{F}_1$ and $\vec{F}_2$ both point *above* the strut? (Do $\Sigma\tau$ about each end.)

FIGURE 12–26 A roof truss.

FIGURE 12–27 (a) Each massless strut (or rod) of a truss is assumed to be under tension or compression. (b) The two equal and opposite forces must be along the same line or a net torque would exist. (c) Real struts have mass, so the forces $\vec{F}_1$ and $\vec{F}_2$ at the joints do not act precisely along the strut. (d) Vector diagram of part (c).

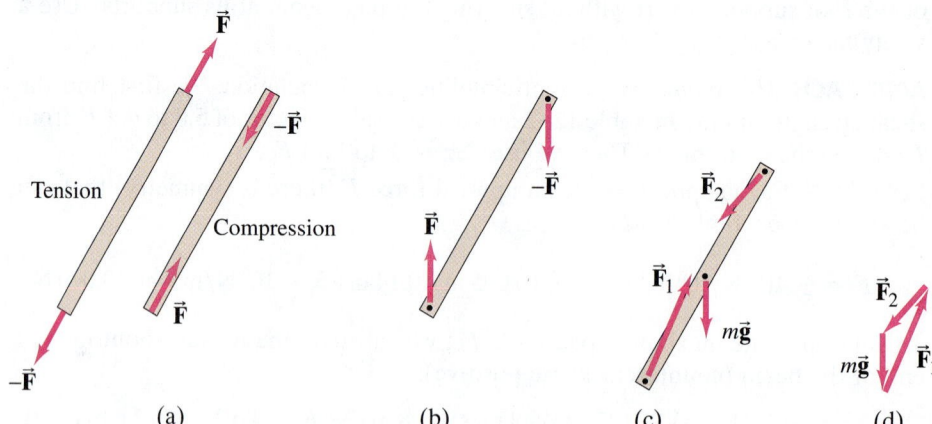

Consider again the simple beam in Example 12–5, Fig. 12–9. The force $\vec{F}_H$ at the pin is *not* along the beam, but acts at an upward angle. If that beam were massless, we see from Eq. (iii) in Example 12–5 with $m = 0$, that $F_{Hy} = 0$, and $\vec{F}_H$ would be along the beam.

The assumption that the forces in each strut of a truss are purely along the strut is still very useful whenever the loads act only at the joints and are much greater than the weight of the struts themselves.

EXAMPLE 12–11 **A truss bridge.** Determine the tension or compression in each of the struts of the truss bridge shown in Fig. 12–28a. The bridge is 64 m long and supports a uniform level concrete roadway whose total mass is 1.40×10^6 kg. Use the **method of joints**, which involves (1) drawing a free-body diagram of the truss as a whole, and (2) drawing a free-body diagram for each of the pins (joints), one by one, and setting $\Sigma \vec{F} = 0$ for each pin. Ignore the mass of the struts. Assume all triangles are equilateral.

APPROACH Any bridge has two trusses, one on each side of the roadway. Consider only one truss, Fig. 12–28a, and it will support half the weight of the roadway. That is, our truss supports a total mass $M = 7.0 \times 10^5$ kg. First we draw a free-body diagram for the entire truss as a single unit, which we assume rests on supports at either end that exert upward forces $\vec{F}_1$ and $\vec{F}_2$, Fig. 12–28b. We assume the mass of the roadway acts entirely at the center, on pin C, as shown. From *symmetry* we can see that each of the end supports carries half the weight [or do a torque equation about, say, point A: $(F_2)(\ell) - Mg(\ell/2) = 0$], so

$$F_1 = F_2 = \tfrac{1}{2} Mg.$$

SOLUTION We look at pin A and apply $\Sigma \vec{F} = 0$ to it. We label the forces on pin A due to each strut with two subscripts: $\vec{F}_{AB}$ means the force exerted by the strut AB and $\vec{F}_{AC}$ is the force exerted by strut AC. $\vec{F}_{AB}$ and $\vec{F}_{AC}$ act along their respective struts; but not knowing whether each is compressive or tensile, we could draw four different free-body diagrams, as shown in Fig. 12–28c. Only the one on the left could provide $\Sigma \vec{F} = 0$, so we immediately know the directions of $\vec{F}_{AB}$ and $\vec{F}_{AC}$.[†] These forces act on the pin. The force that pin A exerts on strut AB is opposite in direction to $\vec{F}_{AB}$ (Newton's third law), so strut AB is under compression and strut AC is under tension. Now let's calculate the magnitudes of $\vec{F}_{AB}$ and $\vec{F}_{AC}$. At pin A:

$$\Sigma F_x = F_{AC} - F_{AB} \cos 60° = 0$$
$$\Sigma F_y = F_1 - F_{AB} \sin 60° = 0.$$

Thus

$$F_{AB} = \frac{F_1}{\sin 60°} = \frac{\tfrac{1}{2} Mg}{\tfrac{1}{2} \sqrt{3}} = \frac{1}{\sqrt{3}} Mg,$$

which equals $(7.0 \times 10^5 \text{ kg})(9.8 \text{ m/s}^2)/\sqrt{3} = 4.0 \times 10^6$ N; and

$$F_{AC} = F_{AB} \cos 60° = \frac{1}{2\sqrt{3}} Mg.$$

Next we look at pin B, and Fig. 12–28d is the free-body diagram. [Convince yourself that if $\vec{F}_{BD}$ or $\vec{F}_{BC}$ were in the opposite direction, $\Sigma \vec{F}$ could not be zero; note that $\vec{F}_{BA} = -\vec{F}_{AB}$ (and $F_{BA} = F_{AB}$) because now we are at the opposite end of strut AB.] We see that BC is under tension and BD compression. (Recall that the forces on the struts are opposite to the forces shown which are on the pin.) We set $\Sigma \vec{F} = 0$:

$$\Sigma F_x = F_{BA} \cos 60° + F_{BC} \cos 60° - F_{BD} = 0$$
$$\Sigma F_y = F_{BA} \sin 60° - F_{BC} \sin 60° = 0.$$

Then, because $F_{BA} = F_{AB}$, we have

$$F_{BC} = F_{AB} = \frac{1}{\sqrt{3}} Mg,$$

and

$$F_{BD} = F_{AB} \cos 60° + F_{BC} \cos 60° = \frac{1}{\sqrt{3}} Mg \left(\tfrac{1}{2}\right) + \frac{1}{\sqrt{3}} Mg \left(\tfrac{1}{2}\right) = \frac{1}{\sqrt{3}} Mg.$$

The solution is complete. By symmetry, $F_{DE} = F_{AB}$, $F_{CE} = F_{AC}$, and $F_{CD} = F_{BC}$.

NOTE As a check, calculate ΣF_x and ΣF_y for pin C and see if they equal zero. Figure 12–28e shows the free-body diagram.

[†]If we were to choose the direction of a force on a diagram opposite to what it really is, we would get a minus sign.

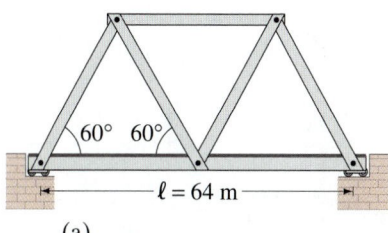

(a)

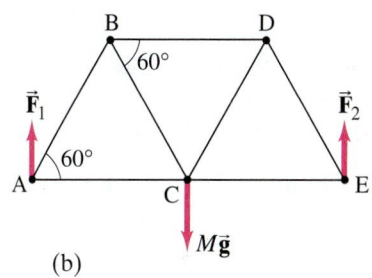

(b)

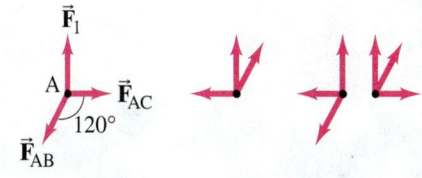

(c) Pin A (different guesses)

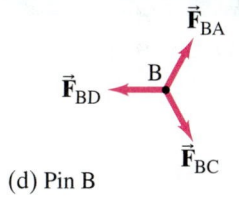

(d) Pin B

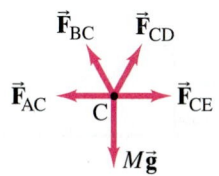

(e) Pin C

FIGURE 12–28
Example 12–11. (a) A truss bridge. Free-body diagrams:
(b) for the entire truss,
(c) for pin A (different guesses),
(d) for pin B and (e) for pin C.

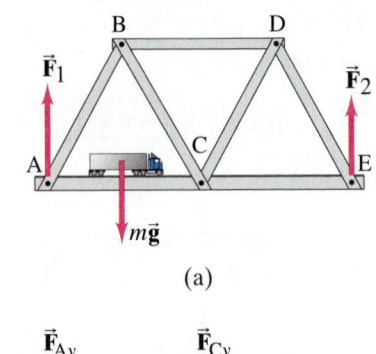

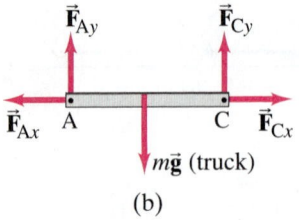

FIGURE 12-29 (a) Truss with truck of mass m at center of strut AC. (b) Forces on strut AC.

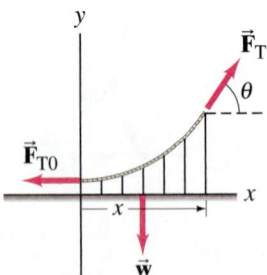

FIGURE 12-30 Suspension bridges (Brooklyn and Manhattan bridges, NY).

FIGURE 12-31 Example 12-12.

Example 12–11 put the roadway load at the center, C. Now consider a heavy load, such as a heavy truck, supported by strut AC at its middle, as shown in Fig. 12–29a. The strut AC sags under this load, telling us there is shear stress in strut AC. Figure 12–29b shows the forces exerted on strut AC: the weight of the truck $m\vec{\mathbf{g}}$, and the forces $\vec{\mathbf{F}}_A$ and $\vec{\mathbf{F}}_C$ that pins A and C exert on the strut. [Note that $\vec{\mathbf{F}}_1$ does not appear because it is a force (exerted by external supports) that acts on pin A, not on strut AC.] The forces that pins A and C exert on strut AC will act not only along the strut, but will have vertical components too, perpendicular to the strut to balance the weight of the truck, $m\vec{\mathbf{g}}$, creating shear stress. The other struts, not bearing weight, remain under pure tension or compression. Problems 53 and 54 deal with this situation, and an early step in their solution is to calculate the forces $\vec{\mathbf{F}}_A$ and $\vec{\mathbf{F}}_C$ by using torque equations for the strut.

For very large bridges, truss structures are too heavy. One solution is to build suspension bridges, with the load being carried by relatively light suspension cables under tension, supporting the roadway by means of closely spaced vertical wires, as shown in Fig. 12–30, and in the photo on the first page of this Chapter.

EXAMPLE 12–12 Suspension bridge. Determine the shape of the cable between the two towers of a suspension bridge (as in Fig. 12–30), assuming the weight of the roadway is supported uniformly along its length. Ignore the weight of the cable.

APPROACH We take $x = 0$, $y = 0$ at the center of the span, as shown in Fig. 12–31. Let $\vec{\mathbf{F}}_{T0}$ be the tension in the cable at $x = 0$; it acts horizontally as shown. Let F_T be the tension in the cable at some other place where the horizontal coordinate is x, as shown. This section of cable supports a portion of the roadway whose weight w is proportional to the distance x, since the roadway is assumed uniform; that is,

$$w = \lambda x$$

where λ is the weight per unit length.

SOLUTION We set $\Sigma \vec{\mathbf{F}} = 0$:

$$\Sigma F_x = F_T \cos\theta - F_{T0} = 0$$
$$\Sigma F_y = F_T \sin\theta - w = 0.$$

We divide these two equations,

$$\tan\theta = \frac{w}{F_{T0}} = \frac{\lambda x}{F_{T0}}.$$

The slope of our curve (the cable) at any point is

$$\frac{dy}{dx} = \tan\theta$$

or

$$\frac{dy}{dx} = \frac{\lambda}{F_{T0}} x.$$

We integrate this:

$$\int dy = \frac{\lambda}{F_{T0}} \int x \, dx$$
$$y = Ax^2 + B$$

where we set $A = \lambda/F_{T0}$ and B is a constant of integration. This is just the equation of a parabola.

NOTE Real bridges have cables that do have mass, so the cables hang only approximately as a parabola, although often it is quite close.

*12–7 Arches and Domes

There are various ways that engineers and architects can span a space, such as beams, trusses, and suspension bridges. In this Section we discuss arches and domes.

FIGURE 12–32 Round arches in the Roman Forum. The one in the background is the Arch of Titus.

FIGURE 12–33 An arch is used here to good effect in spanning a chasm on the California coast.

The semicircular **arch** (Figs. 12–32 and 12–33) was introduced by the ancient Romans 2000 years ago. Aside from its aesthetic appeal, it was a tremendous technological innovation. The advantage of the "true" or semicircular arch is that, if well designed, its wedge-shaped stones experience stress which is mainly compressive even when supporting a large load such as the wall and roof of a cathedral. Because the stones are forced to squeeze against one another, they are mainly under compression (see Fig. 12–34). Note, however, that the arch transfers horizontal as well as vertical forces to the supports. A round arch consisting of many well-shaped stones could span a very wide space. However, considerable buttressing on the sides is needed to support the horizontal components of the forces.

PHYSICS APPLIED

Architecture: Beams, arches and domes

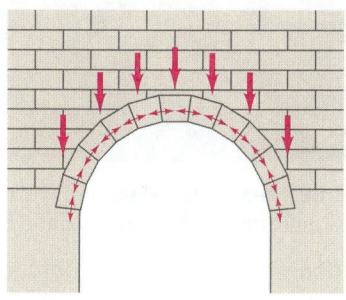

FIGURE 12–34 Stones in a round arch (see Fig. 12–32) are mainly under compression.

The pointed arch came into use about A.D. 1100 and became the hallmark of the great Gothic cathedrals. It too was an important technical innovation, and was first used to support heavy loads such as the tower and arch of a cathedral. Apparently the builders realized that, because of the steepness of the pointed arch, the forces due to the weight above could be brought down more nearly vertically, so less horizontal buttressing would be needed. The pointed arch reduced the load on the walls, so there could be more openness and light. The smaller buttressing needed was provided on the outside by graceful flying buttresses (Fig. 12–35).

FIGURE 12–35 Flying buttresses (on the cathedral of Notre Dame, in Paris).

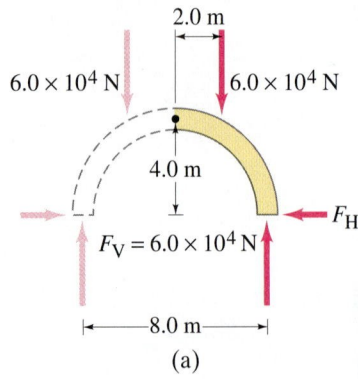

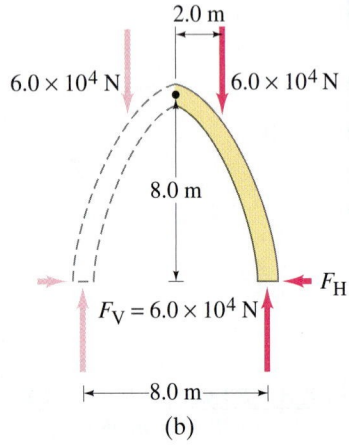

FIGURE 12–36 (a) Forces in a round arch, compared (b) with those in a pointed arch.

To make an accurate analysis of a stone arch is quite difficult in practice. But if we make some simplifying assumptions, we can show why the horizontal component of the force at the base is less for a pointed arch than for a round one. Figure 12–36 shows a round arch and a pointed arch, each with an 8.0-m span. The height of the round arch is thus 4.0 m, whereas that of the pointed arch is larger and has been chosen to be 8.0 m. Each arch supports a weight of 12.0×10^4 N $(= 12,000$ kg $\times g)$ which, for simplicity, we have divided into two parts (each 6.0×10^4 N) acting on the two halves of each arch as shown. To be in equilibrium, each of the supports must exert an upward force of 6.0×10^4 N. For rotational equilibrium, each support also exerts a horizontal force, F_H, at the base of the arch, and it is this we want to calculate. We focus only on the right half of each arch. We set equal to zero the total torque calculated about the apex of the arch due to the forces exerted on that half arch. For the round arch, the torque equation $(\Sigma\tau = 0)$ is (see Fig. 12–36a)

$$(4.0\,\text{m})(6.0 \times 10^4\,\text{N}) - (2.0\,\text{m})(6.0 \times 10^4\,\text{N}) - (4.0\,\text{m})(F_H) = 0.$$

Thus $F_H = 3.0 \times 10^4$ N for the round arch. For the pointed arch, the torque equation is (see Fig. 12–36b)

$$(4.0\,\text{m})(6.0 \times 10^4\,\text{N}) - (2.0\,\text{m})(6.0 \times 10^4\,\text{N}) - (8.0\,\text{m})(F_H) = 0.$$

Solving, we find that $F_H = 1.5 \times 10^4$ N—only half as much as for the round arch! From this calculation we can see that the horizontal buttressing force required for a pointed arch is less because the arch is higher, and there is therefore a longer lever arm for this force. Indeed, the steeper the arch, the less the horizontal component of the force needs to be, and hence the more nearly vertical is the force exerted at the base of the arch.

Whereas an arch spans a two-dimensional space, a **dome**—which is basically an arch rotated about a vertical axis—spans a three-dimensional space. The Romans built the first large domes. Their shape was hemispherical and some still stand, such as that of the Pantheon in Rome (Fig. 12–37), built 2000 years ago.

FIGURE 12–37 Interior of the Pantheon in Rome, built almost 2000 years ago. This view, showing the great dome and its central opening for light, was painted about 1740 by Panini. Photographs do not capture its grandeur as well as this painting does.

FIGURE 12–38 The skyline of Florence, showing Brunelleschi's dome on the cathedral.

Fourteen centuries later, a new cathedral was being built in Florence. It was to have a dome 43 m in diameter to rival that of the Pantheon, whose construction has remained a mystery. The new dome was to rest on a "drum" with no external abutments. Filippo Brunelleschi (1377–1446) designed a pointed dome (Fig. 12–38), since a pointed dome, like a pointed arch, exerts a smaller side thrust against its base. A dome, like an arch, is not stable until all the stones are in place. To support smaller domes during construction, wooden frameworks were used. But no trees big enough or strong enough could be found to span the 43-m space required. Brunelleschi decided to try to build the dome in horizontal layers, each bonded to the previous one, holding it in place until the last stone of the circle was placed. Each closed ring was then strong enough to support the next layer. It was an amazing feat. Only in the twentieth century were larger domes built, the largest being that of the Superdome in New Orleans, completed in 1975. Its 200-m diameter dome is made of steel trusses and concrete.

Summary

An object at rest is said to be in **equilibrium**. The subject concerned with the determination of the forces within a structure at rest is called **statics**.

The two necessary conditions for an object to be in equilibrium are that (1) the vector sum of all the forces on it must be zero, and (2) the sum of all the torques (calculated about any arbitrary axis) must also be zero. For a 2-dimensional problem we can write

$$\Sigma F_x = 0, \qquad \Sigma F_y = 0, \qquad \Sigma \tau = 0. \qquad \textbf{(12–1, 12–2)}$$

It is important when doing statics problems to apply the equilibrium conditions to only one object at a time.

An object in static equilibrium is said to be in (a) **stable**, (b) **unstable**, or (c) **neutral equilibrium**, depending on whether a slight displacement leads to (a) a return to the original position, (b) further movement away from the original position, or (c) rest in the new position. An object in stable equilibrium is also said to be in **balance**.

Hooke's law applies to many elastic solids, and states that the change in length of an object is proportional to the applied force:

$$F = k \, \Delta \ell. \qquad \textbf{(12–3)}$$

If the force is too great, the object will exceed its **elastic limit**, which means it will no longer return to its original shape when the distorting force is removed. If the force is even greater, the **ultimate strength** of the material can be exceeded, and the object will **fracture**. The force per unit area acting on an object is the **stress**, and the resulting fractional change in length is the **strain**. The stress on an object is present within the object and can be of three types: **compression**, **tension**, or **shear**. The ratio of stress to strain is called the **elastic modulus** of the material. **Young's modulus** applies for compression and tension, and the **shear modulus** for shear. **Bulk modulus** applies to an object whose volume changes as a result of pressure on all sides. All three moduli are constants for a given material when distorted within the elastic region.

Questions

1. Describe several situations in which an object is not in equilibrium, even though the net force on it is zero.

2. A bungee jumper momentarily comes to rest at the bottom of the dive before he springs back upward. At that moment, is the bungee jumper in equilibrium? Explain.

3. You can find the center of gravity of a meter stick by resting it horizontally on your two index fingers, and then slowly drawing your fingers together. First the meter stick will slip on one finger, and then on the other, but eventually the fingers meet at the CG. Why does this work?

4. Your doctor's scale has arms on which weights slide to counter your weight, Fig. 12–39. These weights are much lighter than you are. How does this work?

Weights

FIGURE 12–39
Question 4.

5. A ground retaining wall is shown in Fig. 12–40a. The ground, particularly when wet, can exert a significant force F on the wall. (a) What force produces the torque to keep the wall upright? (b) Explain why the retaining wall in Fig. 12–40b would be much less likely to overturn than that in Fig. 12–40a.

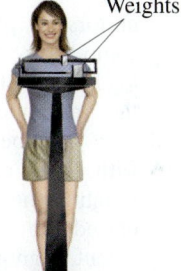

(a)   (b)

FIGURE 12–40 Question 5.

6. Can the sum of the torques on an object be zero while the net force on the object is nonzero? Explain.

7. A ladder, leaning against a wall, makes a 60° angle with the ground. When is it more likely to slip: when a person stands on the ladder near the top or near the bottom? Explain.

8. A uniform meter stick supported at the 25-cm mark is in equilibrium when a 1-kg rock is suspended at the 0-cm end (as shown in Fig. 12–41). Is the mass of the meter stick greater than, equal to, or less than the mass of the rock? Explain your reasoning.

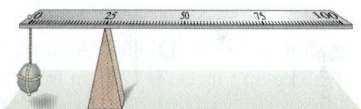

FIGURE 12–41 Question 8.

9. Why do you tend to lean backward when carrying a heavy load in your arms?

10. Figure 12–42 shows a cone. Explain how to lay it on a flat table so that it is in (a) stable equilibrium, (b) unstable equilibrium, (c) neutral equilibrium.

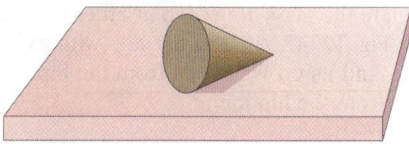

FIGURE 12–42 Question 10.

11. Place yourself facing the edge of an open door. Position your feet astride the door with your nose and abdomen touching the door's edge. Try to rise on your tiptoes. Why can't this be done?

12. Why is it not possible to sit upright in a chair and rise to your feet without first leaning forward?

13. Why is it more difficult to do sit-ups when your knees are bent than when your legs are stretched out?

14. Which of the configurations of brick, (*a*) or (*b*) of Fig. 12–43, is the more likely to be stable? Why?

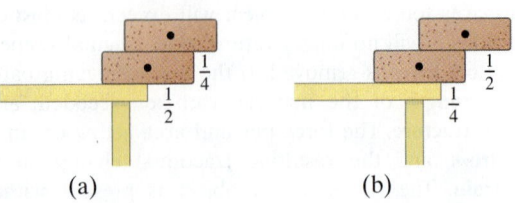

(a) (b)

FIGURE 12–43 Question 14. The dots indicate the CG of each brick. The fractions $\frac{1}{4}$ and $\frac{1}{2}$ indicate what portion of each brick is hanging beyond its support.

15. Name the type of equilibrium for each position of the ball in Fig. 12–44.

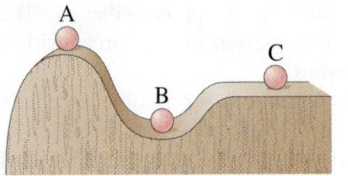

FIGURE 12–44
Question 15.

16. Is the Young's modulus for a bungee cord smaller or larger than that for an ordinary rope?

17. Examine how a pair of scissors or shears cuts through a piece of cardboard. Is the name "shears" justified? Explain.

18. Materials such as ordinary concrete and stone are very weak under tension or shear. Would it be wise to use such a material for either of the supports of the cantilever shown in Fig. 12–7? If so, which one(s)? Explain.

Problems

12–1 and 12–2 Equilibrium

1. (I) Three forces are applied to a tree sapling, as shown in Fig. 12–45, to stabilize it. If $\vec{F}_A = 385\ \text{N}$ and $\vec{F}_B = 475\ \text{N}$, find $\vec{F}_C$ in magnitude and direction.

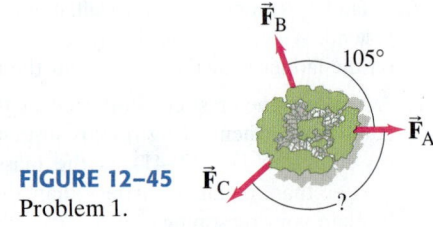

FIGURE 12–45
Problem 1.

2. (I) Approximately what magnitude force, F_M, must the extensor muscle in the upper arm exert on the lower arm to hold a 7.3-kg shot put (Fig. 12–46)? Assume the lower arm has a mass of 2.3 kg and its CG is 12.0 cm from the elbow-joint pivot.

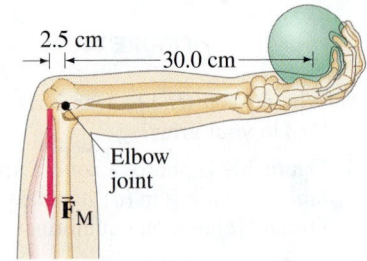

FIGURE 12–46
Problem 2.

3. (I) Calculate the mass *m* needed in order to suspend the leg shown in Fig. 12–47. Assume the leg (with cast) has a mass of 15.0 kg, and its CG is 35.0 cm from the hip joint; the sling is 78.0 cm from the hip joint.

FIGURE 12–47 Problem 3.

4. (I) A tower crane (Fig. 12–48a) must always be carefully balanced so that there is no net torque tending to tip it. A particular crane at a building site is about to lift a 2800-kg air-conditioning unit. The crane's dimensions are shown in Fig. 12–48b. (*a*) Where must the crane's 9500-kg counterweight be placed when the load is lifted from the ground? (Note that the counterweight is usually moved automatically via sensors and motors to precisely compensate for the load.) (*b*) Determine the maximum load that can be lifted with this counterweight when it is placed at its full extent. Ignore the mass of the beam.

(a)

FIGURE 12–48
Problem 4.

(b)

5. (II) Calculate the forces F_A and F_B that the supports exert on the diving board of Fig. 12–49 when a 52-kg person stands at its tip. (*a*) Ignore the weight of the board. (*b*) Take into account the board's mass of 28 kg. Assume the board's CG is at its center.

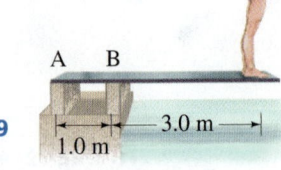

FIGURE 12–49
Problem 5.

6. (II) Two cords support a chandelier in the manner shown in Fig 12–3 except that the upper cord makes an angle of 45° with the ceiling. If the cords can sustain a force of 1660 N without breaking, what is the maximum chandelier weight that can be supported?

7. (II) The two trees in Fig. 12–50 are 6.6 m apart. A back-packer is trying to lift his pack out of the reach of bears. Calculate the magnitude of the force $\vec{F}$ that he must exert downward to hold a 19-kg backpack so that the rope sags at its midpoint by (a) 1.5 m, (b) 0.15 m.

FIGURE 12–50
Problems 7 and 83.

8. (II) A 110-kg horizontal beam is supported at each end. A 320-kg piano rests a quarter of the way from one end. What is the vertical force on each of the supports?

9. (II) Calculate F_A and F_B for the uniform cantilever shown in Fig. 12–7 whose mass is 1200 kg.

10. (II) A 75-kg adult sits at one end of a 9.0-m-long board. His 25-kg child sits on the other end. (a) Where should the pivot be placed so that the board is balanced, ignoring the board's mass? (b) Find the pivot point if the board is uniform and has a mass of 15 kg.

11. (II) Find the tension in the two cords shown in Fig. 12–51. Neglect the mass of the cords, and assume that the angle θ is 33° and the mass m is 190 kg.

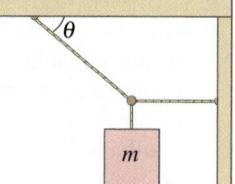

FIGURE 12–51
Problem 11.

12. (II) Find the tension in the two wires supporting the traffic light shown in Fig. 12–52.

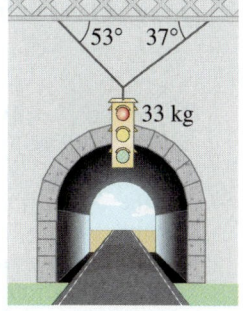

FIGURE 12–52
Problem 12.

13. (II) How close to the edge of the 24.0-kg table shown in Fig. 12–53 can a 66.0-kg person sit without tipping it over?

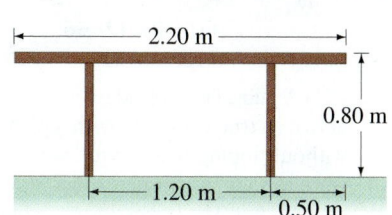

FIGURE 12–53
Problem 13.

14. (II) The force required to pull the cork out of the top of a wine bottle is in the range of 200 to 400 N. A common bottle opener is shown in Fig. 12–54. What range of forces F is required to open a wine bottle with this device?

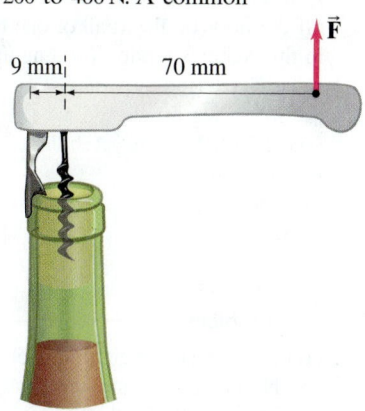

FIGURE 12–54
Problem 14.

15. (II) Calculate F_A and F_B for the beam shown in Fig. 12–55. The downward forces represent the weights of machinery on the beam. Assume the beam is uniform and has a mass of 280 kg.

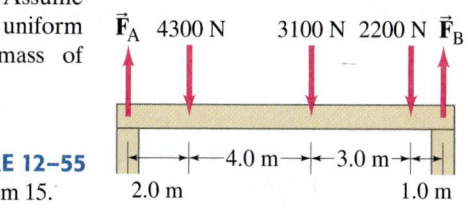

FIGURE 12–55
Problem 15.

16. (II) (a) Calculate the magnitude of the force, F_M, required of the "deltoid" muscle to hold up the outstretched arm shown in Fig. 12–56. The total mass of the arm is 3.3 kg. (b) Calculate the magnitude of the force F_J exerted by the shoulder joint on the upper arm and the angle (to the horizontal) at which it acts.

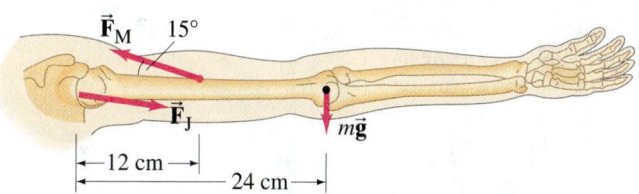

FIGURE 12–56 Problems 16 and 17.

17. (II) Suppose the hand in Problem 16 holds an 8.5-kg mass. What force, F_M, is required of the deltoid muscle, assuming the mass is 52 cm from the shoulder joint?

18. (II) Three children are trying to balance on a seesaw, which includes a fulcrum rock acting as a pivot at the center, and a very light board 3.2 m long (Fig. 12–57). Two play-mates are already on either end. Boy A has a mass of 45 kg, and boy B a mass of 35 kg. Where should girl C, whose mass is 25 kg, place herself so as to balance the seesaw?

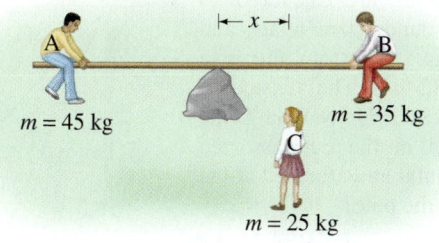

FIGURE 12–57 Problem 18.

19. (II) The Achilles tendon is attached to the rear of the foot as shown in Fig. 12–58. When a person elevates himself just barely off the floor on the "ball of one foot," estimate the tension F_T in the Achilles tendon (pulling upward), and the (downward) force F_B exerted by the lower leg bone on the foot. Assume the person has a mass of 72 kg and D is twice as long as d.

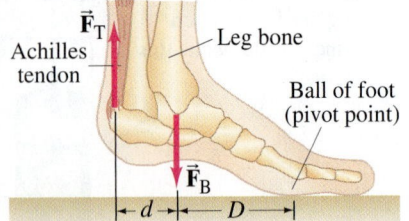

FIGURE 12–58
Problem 19.

20. (II) A shop sign weighing 215 N is supported by a uniform 155-N beam as shown in Fig. 12–59. Find the tension in the guy wire and the horizontal and vertical forces exerted by the hinge on the beam.

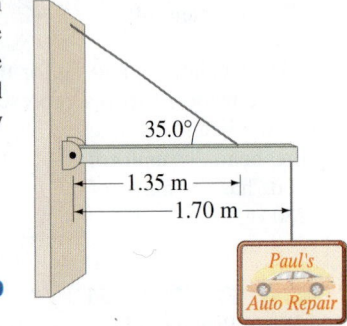

FIGURE 12–59
Problem 20.

21. (II) A traffic light hangs from a pole as shown in Fig. 12–60. The uniform aluminum pole AB is 7.20 m long and has a mass of 12.0 kg. The mass of the traffic light is 21.5 kg. Determine (a) the tension in the horizontal massless cable CD, and (b) the vertical and horizontal components of the force exerted by the pivot A on the aluminum pole.

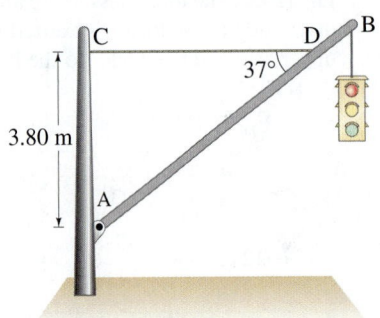

FIGURE 12–60
Problem 21.

22. (II) A uniform steel beam has a mass of 940 kg. On it is resting half of an identical beam, as shown in Fig. 12–61. What is the vertical support force at each end?

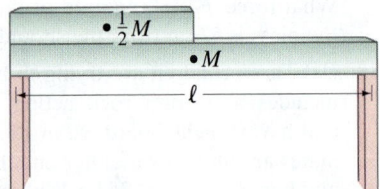

FIGURE 12–61
Problem 22.

23. (II) Two wires run from the top of a pole 2.6 m tall that supports a volleyball net. The two wires are anchored to the ground 2.0 m apart, and each is 2.0 m from the pole (Fig. 12–62). The tension in each wire is 115 N. What is the tension in the net, assumed horizontal and attached at the top of the pole?

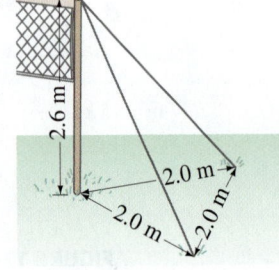

FIGURE 12–62
Problem 23.

24. (II) A large 62.0-kg board is propped at a 45° angle against the edge of a barn door that is 2.6 m wide. How great a horizontal force must a person behind the door exert (at the edge) in order to open it? Assume that there is negligible friction between the door and the board but that the board is firmly set against the ground.

25. (II) Repeat Problem 24 assuming the coefficient of friction between the board and the door is 0.45.

26. (II) A 0.75-kg sheet hangs from a massless clothesline as shown in Fig. 12–63. The clothesline on either side of the sheet makes an angle of 3.5° with the horizontal. Calculate the tension in the clothesline on either side of the sheet. Why is the tension so much greater than the weight of the sheet?

FIGURE 12–63
Problem 26.

27. (II) A uniform rod AB of length 5.0 m and mass $M = 3.8$ kg is hinged at A and held in equilibrium by a light cord, as shown in Fig. 12–64. A load $W = 22$ N hangs from the rod at a distance x so that the tension in the cord is 85 N. (a) Draw a free-body diagram for the rod. (b) Determine the vertical and horizontal forces on the rod exerted by the hinge. (c) Determine d from the appropriate torque equation.

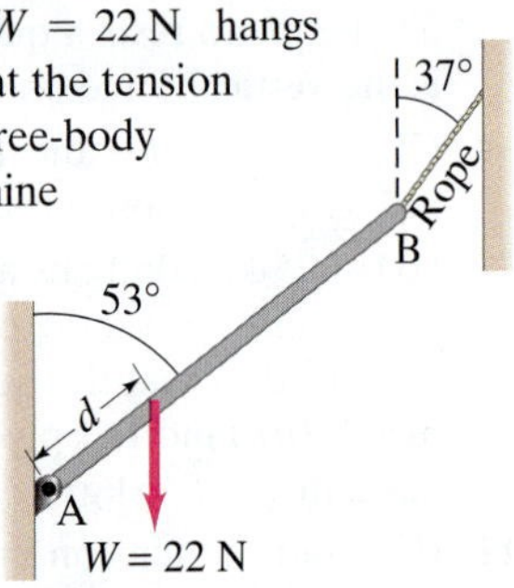

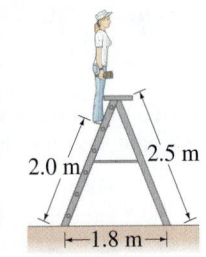

FIGURE 12–64
Problem 27.

28. (III) A 56.0-kg person stands 2.0 m from the bottom of the stepladder shown in Fig. 12–65. Determine (a) the tension in the horizontal tie rod, which is halfway up the ladder, (b) the normal force the ground exerts on each side of the ladder, and (c) the force (magnitude and direction) that the left side of the ladder exerts on the right side at the hinge on the top. Ignore the mass of the ladder and assume the ground is frictionless. [*Hint:* Consider free-body diagrams for each section of the ladder.]

FIGURE 12–65
Problem 28.

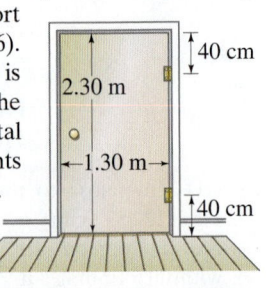

29. (III) A door 2.30 m high and 1.30 m wide has a mass of 13.0 kg. A hinge 0.40 m from the top and another hinge 0.40 m from the bottom each support half the door's weight (Fig. 12–66). Assume that the center of gravity is at the geometrical center of the door, and determine the horizontal and vertical force components exerted by each hinge on the door.

FIGURE 12–66
Problem 29.

30. (III) A cubic crate of side $s = 2.0$ m is top-heavy: its CG is 18 cm above its true center. How steep an incline can the crate rest on without tipping over? What would your answer be if the crate were to slide at constant speed down the plane without tipping over? [*Hint:* The normal force would act at the lowest corner.]

31. (III) A refrigerator is approximately a uniform rectangular solid 1.9 m tall, 1.0 m wide, and 0.75 m deep. If it sits upright on a truck with its 1.0-m dimension in the direction of travel, and if the refrigerator cannot slide on the truck, how rapidly can the truck accelerate without tipping the refrigerator over? [*Hint*: The normal force would act at one corner.]

32. (III) A uniform ladder of mass m and length ℓ leans at an angle θ against a frictionless wall, Fig. 12–67. If the coefficient of static friction between the ladder and the ground is μ_s, determine a formula for the minimum angle at which the ladder will not slip.

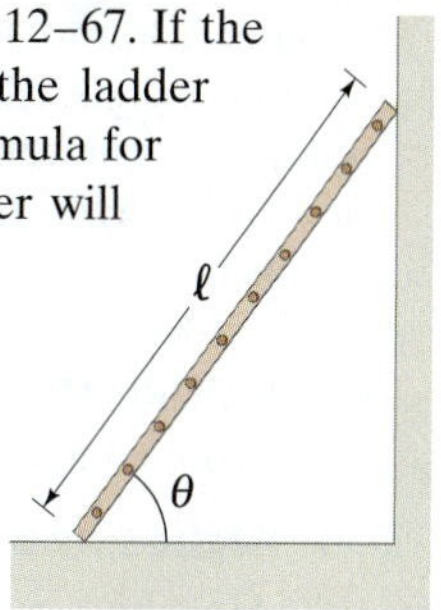

FIGURE 12–67
Problem 32.

12–3 Stability and Balance

33. (II) The Leaning Tower of Pisa is 55 m tall and about 7.0 m in diameter. The top is 4.5 m off center. Is the tower in stable equilibrium? If so, how much farther can it lean before it becomes unstable? Assume the tower is of uniform composition.

12–4 Elasticity; Stress and Strain

34. (I) A nylon string on a tennis racket is under a tension of 275 N. If its diameter is 1.00 mm, by how much is it lengthened from its untensioned length of 30.0 cm?

35. (I) A marble column of cross-sectional area 1.4 m² supports a mass of 25,000 kg. (a) What is the stress within the column? (b) What is the strain?

36. (I) By how much is the column in Problem 35 shortened if it is 8.6 m high?

37. (I) A sign (mass 1700 kg) hangs from the end of a vertical steel girder with a cross-sectional area of 0.012 m². (a) What is the stress within the girder? (b) What is the strain on the girder? (c) If the girder is 9.50 m long, how much is it lengthened? (Ignore the mass of the girder itself.)

38. (II) How much pressure is needed to compress the volume of an iron block by 0.10%? Express your answer in N/m², and compare it to atmospheric pressure $(1.0 \times 10^5 \text{ N/m}^2)$.

39. (II) A 15-cm-long tendon was found to stretch 3.7 mm by a force of 13.4 N. The tendon was approximately round with an average diameter of 8.5 mm. Calculate Young's modulus of this tendon.

40. (II) At depths of 2000 m in the sea, the pressure is about 200 times atmospheric pressure $(1 \text{ atm} = 1.0 \times 10^5 \text{ N/m}^2)$. By what percentage does the interior space of an iron bathysphere's volume change at this depth?

41. (III) A pole projects horizontally from the front wall of a shop. A 6.1-kg sign hangs from the pole at a point 2.2 m from the wall (Fig. 12–68). (a) What is the torque due to this sign calculated about the point where the pole meets the wall? (b) If the pole is not to fall off, there must be another torque exerted to balance it. What exerts this torque? Use a diagram to show how this torque must act. (c) Discuss whether compression, tension, and/or shear play a role in part (b).

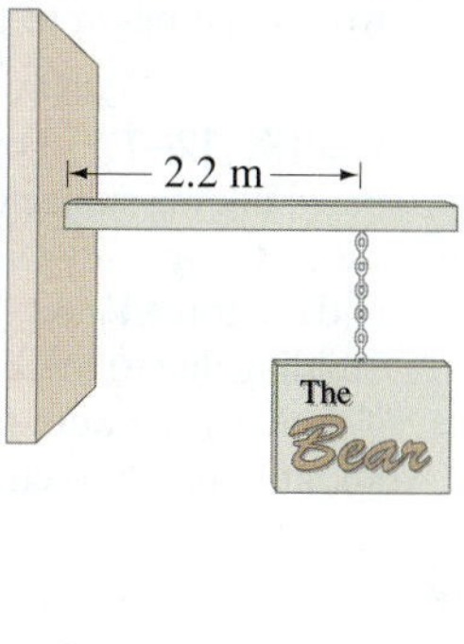

FIGURE 12–68
Problem 41.

12–5 Fracture

42. (I) The femur bone in the human leg has a minimum effective cross section of about 3.0 cm² $(= 3.0 \times 10^{-4} \text{ m}^2)$. How much compressive force can it withstand before breaking?

43. (II) (a) What is the maximum tension possible in a 1.00-mm-diameter nylon tennis racket string? (b) If you want tighter strings, what do you do to prevent breakage: use thinner or thicker strings? Why? What causes strings to break when they are hit by the ball?

44. (II) If a compressive force of 3.3 × 10⁴ N is exerted on the end of a 22-cm-long bone of cross-sectional area 3.6 cm², (a) will the bone break, and (b) if not, by how much does it shorten?

45. (II) (a) What is the minimum cross-sectional area required of a vertical steel cable from which is suspended a 270-kg chandelier? Assume a safety factor of 7.0. (b) If the cable is 7.5 m long, how much does it elongate?

46. (II) Assume the supports of the uniform cantilever shown in Fig. 12–69 $(m = 2900 \text{ kg})$ are made of wood. Calculate the minimum cross-sectional area required of each, assuming a safety factor of 9.0.

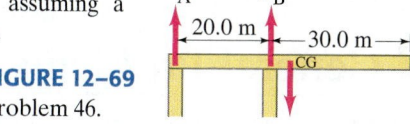

FIGURE 12–69
Problem 46.

47. (II) An iron bolt is used to connect two iron plates together. The bolt must withstand shear forces up to about 3300 N. Calculate the minimum diameter for the bolt, based on a safety factor of 7.0.

48. (III) A steel cable is to support an elevator whose total (loaded) mass is not to exceed 3100 kg. If the maximum acceleration of the elevator is 1.2 m/s², calculate the diameter of cable required. Assume a safety factor of 8.0.

*12–6 Trusses and Bridges

*49. (II) A heavy load $Mg = 66.0$ kN hangs at point E of the single cantilever truss shown in Fig. 12–70. (a) Use a torque equation for the truss as a whole to determine the tension F_T in the support cable, and then determine the force $\vec{F}_A$ on the truss at pin A. (b) Determine the force in each member of the truss. Neglect the weight of the trusses, which is small compared to the load.

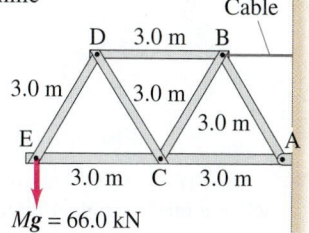

FIGURE 12–70
Problem 49. $Mg = 66.0$ kN

*50. (II) Figure 12–71 shows a simple truss that carries a load at the center (C) of 1.35 × 10⁴ N. (a) Calculate the force on each strut at the pins, A, B, C, D, and (b) determine which struts (ignore their masses) are under tension and which under compression.

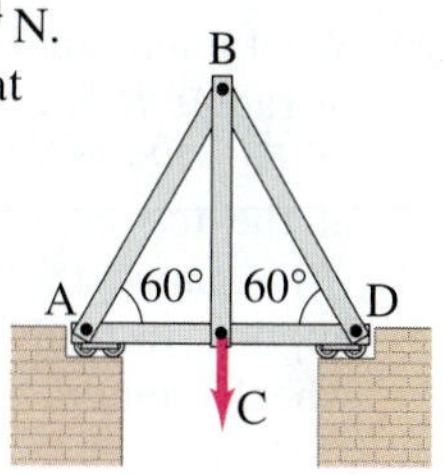

FIGURE 12–71
Problem 50.

*51. (II) (a) What minimum cross-sectional area must the trusses have in Example 12–11 if they are of steel (and all the same size for looks), using a safety factor of 7.0? (b) If at any time the bridge may carry as many as 60 trucks with an average mass of 1.3 × 10⁴ kg, estimate again the area needed for the truss members.

*52. (II) Consider again Example 12–11 but this time assume the roadway is supported uniformly so that $\frac{1}{2}$ its mass M $(= 7.0 \times 10^5 \text{ kg})$ acts at the center and $\frac{1}{4} M$ at each end support (think of the bridge as two spans, AC and CE, so the center pin supports two span ends). Calculate the magnitude of the force in each truss member and compare to Example 12–11.

*53. (III) The truss shown in Fig. 12–72 supports a railway bridge. Determine the compressive or tension force in each strut if a 53-ton $(1 \text{ ton} = 10^3 \text{ kg})$ train locomotive is stopped at the midpoint between the center and one end. Ignore the masses of the rails and truss, and use only $\frac{1}{2}$ the mass of train because there are two trusses (one on each side of the train). Assume all triangles are equilateral. [*Hint*: See Fig. 12–29.]

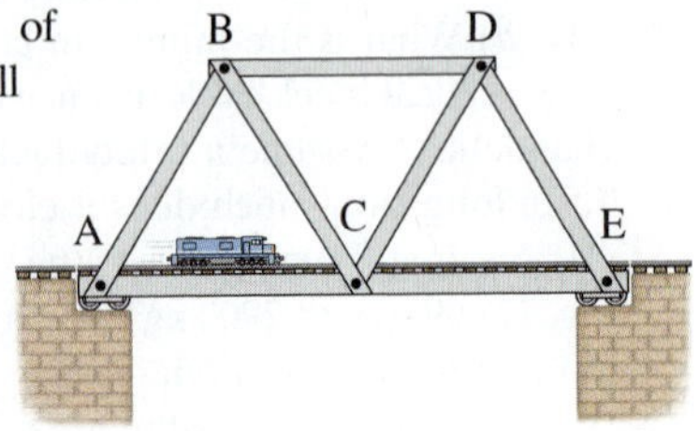

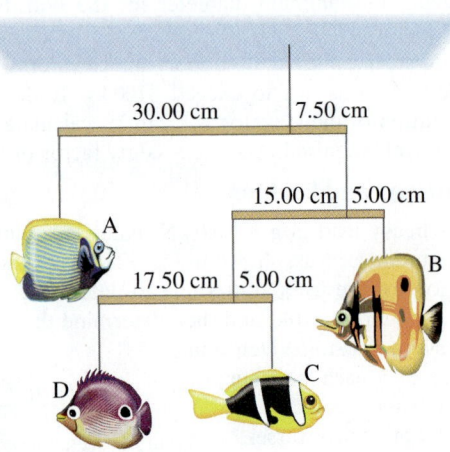

FIGURE 12–72
Problem 53.

*54. (III) Suppose in Example 12–11, a 23-ton truck $(m = 23 \times 10^3 \text{ kg})$ has its CM located 22 m from the left end of the bridge (point A). Determine the magnitude of the force and type of stress in each strut. [*Hint*: See Fig. 12–29.]

*55. (III) For the "Pratt truss" shown in Fig. 12–73, determine the force on each member and whether it is tensile or compressive. Assume the truss is loaded as shown, and give results in terms of F. The vertical height is a and each of the four lower horizontal spans has length a.

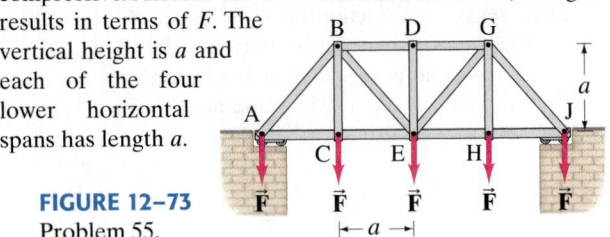

FIGURE 12–73
Problem 55.

*12–7 Arches and Domes

*56. (II) How high must a pointed arch be if it is to span a space 8.0 m wide and exert one-third the horizontal force at its base that a round arch would?

General Problems

57. The mobile in Fig. 12–74 is in equilibrium. Object B has mass of 0.748 kg. Determine the masses of objects A, C, and D. (Neglect the weights of the crossbars.)

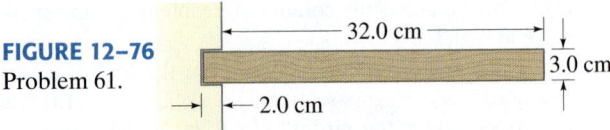

30.00 cm   7.50 cm

15.00 cm   5.00 cm

A

17.50 cm   5.00 cm

B

D

C

FIGURE 12–74 Problem 57.

58. A tightly stretched "high wire" is 36 m long. It sags 2.1 m when a 60.0-kg tightrope walker stands at its center. What is the tension in the wire? Is it possible to increase the tension in the wire so that there is no sag?

59. What minimum horizontal force F is needed to pull a wheel of radius R and mass M over a step of height h as shown in Fig. 12–75 $(R > h)$? (a) Assume the force is applied at the top edge as shown. (b) Assume the force is applied instead at the wheel's center.

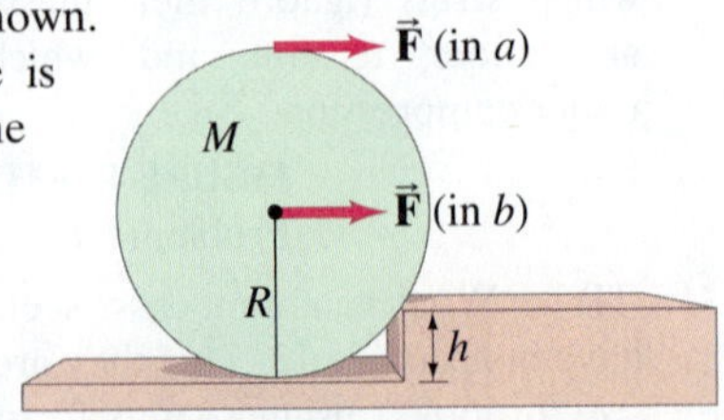

$\vec{F}$ (in *a*)

M

$\vec{F}$ (in *b*)

R

h

FIGURE 12–75
Problem 59.

60. A 28-kg round table is supported by three legs equal distances apart on the edge. What minimum mass, placed on the table's edge, will cause the table to overturn?

61. When a wood shelf of mass 6.6 kg is fastened inside a slot in a vertical support as shown in Fig. 12–76, the support exerts a torque on the shelf. (a) Draw a free-body diagram for the shelf, assuming three vertical forces (two exerted by the support slot—explain why). Then calculate (b) the magnitudes of the three forces and (c) the torque exerted by the support (about the left end of the shelf).

FIGURE 12–76
Problem 61.

32.0 cm

3.0 cm

2.0 cm

62. A 50-story building is being planned. It is to be 180.0 m high with a base 46.0 m by 76.0 m. Its total mass will be about 1.8×10^7 kg, and its weight therefore about 1.8×10^8 N. Suppose a 200-km/h wind exerts a force of 950 N/m² over the 76.0-m-wide face (Fig. 12–77). Calculate the torque about the potential pivot point, the rear edge of the building (where $\vec{F}_E$ acts in Fig. 12–77), and determine whether the building will topple. Assume the total force of the wind acts at the midpoint of the building's face, and that the building is not anchored in bedrock. [*Hint*: $\vec{F}_E$ in Fig. 12–77 represents the force that the Earth would exert on the building in the case where the building would just begin to tip.]

FIGURE 12–77 Forces on a building subjected to wind ($\vec{F}_A$), gravity ($m\vec{g}$), and the force $\vec{F}_E$ on the building due to the Earth if the building were just about to tip. Problem 62.

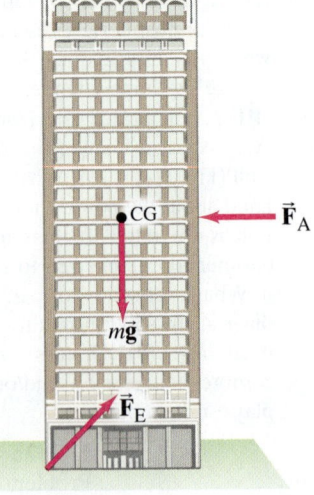

CG $\vec{F}_A$

$m\vec{g}$

$\vec{F}_E$

63. The center of gravity of a loaded truck depends on how the truck is packed. If it is 4.0 m high and 2.4 m wide, and its CG is 2.2 m above the ground, how steep a slope can the truck be parked on without tipping over (Fig. 12–78)?

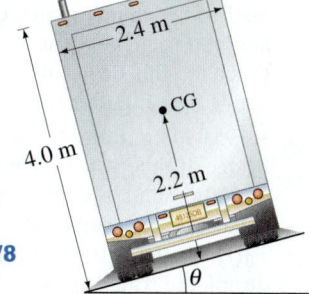

FIGURE 12–78
Problem 63.

64. In Fig. 12–79, consider the right-hand (northernmost) section of the Golden Gate Bridge, which has a length $d_1 = 343$ m. Assume the CG of this span is halfway between the tower and anchor. Determine F_{T1} and F_{T2} (which act on the northernmost cable) in terms of mg, the weight of the northernmost span, and calculate the tower height h needed for equilibrium. Assume the roadway is supported only by the suspension cables, and neglect the mass of the cables and vertical wires. [*Hint*: F_{T3} does not act on this section.]

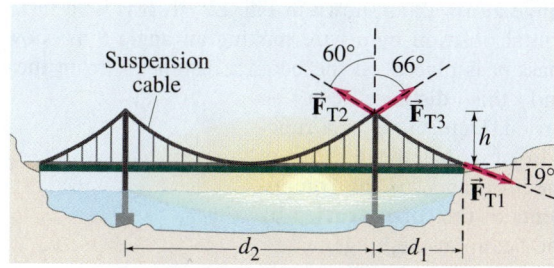

FIGURE 12–79 Problems 64 and 65.

65. Assume that a single-span suspension bridge such as the Golden Gate Bridge has the symmetrical configuration indicated in Fig. 12–79. Assume that the roadway is uniform over the length of the bridge and that each segment of the suspension cable provides the sole support for the roadway directly below it. The ends of the cable are anchored to the ground only, not to the roadway. What must the ratio of d_2 to d_1 be so that the suspension cable exerts no net horizontal force on the towers? Neglect the mass of the cables and the fact that the roadway isn't precisely horizontal.

66. When a mass of 25 kg is hung from the middle of a fixed straight aluminum wire, the wire sags to make an angle of 12° with the horizontal as shown in Fig. 12–80. (*a*) Determine the radius of the wire. (*b*) Would the Al wire break under these conditions? If so, what other material (see Table 12–2) might work?

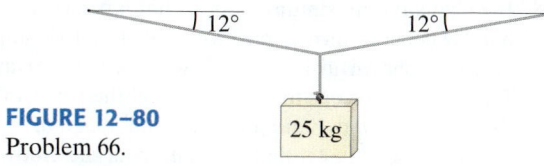

FIGURE 12–80
Problem 66.

67. The forces acting on a 77,000-kg aircraft flying at constant velocity are shown in Fig. 12–81. The engine thrust, $F_T = 5.0 \times 10^5$ N, acts on a line 1.6 m below the CM. Determine the drag force F_D and the distance above the CM that it acts. Assume $\vec{F}_D$ and $\vec{F}_T$ are horizontal. ($\vec{F}_L$ is the "lift" force on the wing.)

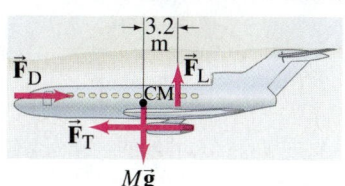

FIGURE 12–81
Problem 67.

68. A uniform flexible steel cable of weight mg is suspended between two points at the same elevation as shown in Fig. 12–82, where $\theta = 56°$. Determine the tension in the cable (*a*) at its lowest point, and (*b*) at the points of attachment. (*c*) What is the direction of the tension force in each case?

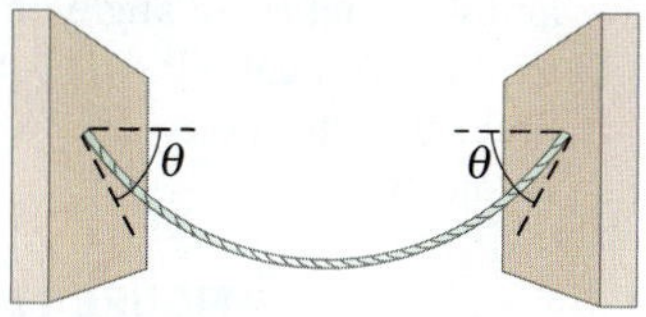

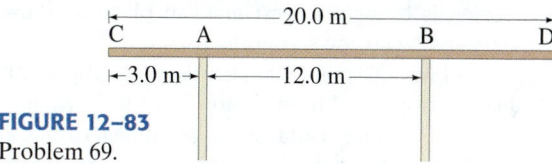

FIGURE 12–82
Problem 68.

69. A 20.0-m-long uniform beam weighing 650 N rests on walls A and B, as shown in Fig. 12–83. (*a*) Find the maximum weight of a person who can walk to the extreme end D without tipping the beam. Find the forces that the walls A and B exert on the beam when the person is standing: (*b*) at D; (*c*) at a point 2.0 m to the right of B; (*d*) 2.0 m to the right of A.

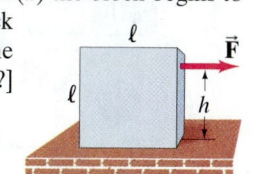

FIGURE 12–83
Problem 69.

70. A cube of side ℓ rests on a rough floor. It is subjected to a steady horizontal pull F, exerted a distance h above the floor as shown in Fig. 12–84. As F is increased, the block will either begin to slide, or begin to tip over. Determine the coefficient of static friction μ_s so that (*a*) the block begins to slide rather than tip; (*b*) the block begins to tip. [*Hint*: Where will the normal force on the block act if it tips?]

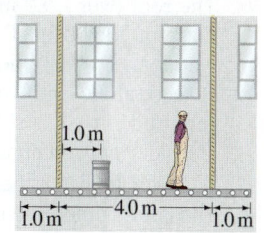

FIGURE 12–84
Problem 70.

71. A 65.0-kg painter is on a uniform 25-kg scaffold supported from above by ropes (Fig. 12–85). There is a 4.0-kg pail of paint to one side, as shown. Can the painter walk safely to both ends of the scaffold? If not, which end(s) is dangerous, and how close to the end can he approach safely?

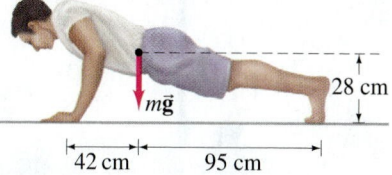

FIGURE 12–85
Problem 71.

72. A man doing push-ups pauses in the position shown in Fig. 12–86. His mass $m = 68$ kg. Determine the normal force exerted by the floor (*a*) on each hand; (*b*) on each foot.

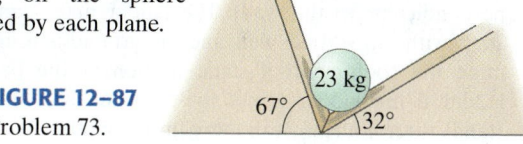

FIGURE 12–86
Problem 72.

73. A 23-kg sphere rests between two smooth planes as shown in Fig. 12–87. Determine the magnitude of the force acting on the sphere exerted by each plane.

FIGURE 12–87
Problem 73.

74. A 15.0-kg ball is supported from the ceiling by rope A. Rope B pulls downward and to the side on the ball. If the angle of A to the vertical is 22° and if B makes an angle of 53° to the vertical (Fig. 12–88), find the tensions in ropes A and B.

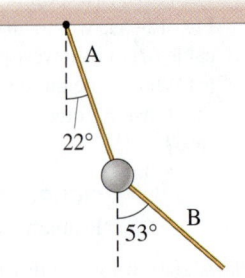

FIGURE 12–88
Problem 74.

75. Parachutists whose chutes have failed to open have been known to survive if they land in deep snow. Assume that a 75-kg parachutist hits the ground with an area of impact of $0.30 \, m^2$ at a velocity of 55 m/s, and that the ultimate strength of body tissue is $5 \times 10^5 \, N/m^2$. Assume that the person is brought to rest in 1.0 m of snow. Show that the person may escape serious injury.

76. A steel wire 2.3 mm in diameter stretches by 0.030% when a mass is suspended from it. How large is the mass?

77. A 2500-kg trailer is attached to a stationary truck at point B, Fig. 12–89. Determine the normal force exerted by the road on the rear tires at A, and the vertical force exerted on the trailer by the support B.

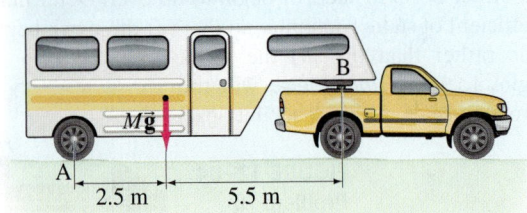

FIGURE 12–89 Problem 77.

78. The roof over a 9.0-m × 10.0-m room in a school has a total mass of 13,600 kg. The roof is to be supported by vertical wooden "2 × 4s" (actually about 4.0 cm × 9.0 cm) equally spaced along the 10.0-m sides. How many supports are required on each side, and how far apart must they be? Consider only compression, and assume a safety factor of 12.

79. A 25-kg object is being lifted by pulling on the ends of a 1.15-mm-diameter nylon cord that goes over two 3.00-m-high poles that are 4.0 m apart, as shown in Fig. 12–90. How high above the floor will the object be when the cord breaks?

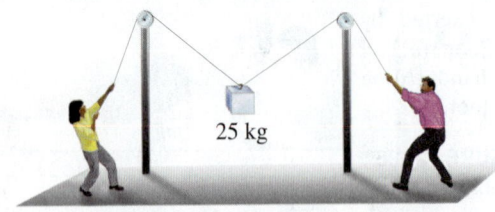

FIGURE 12–90 Problem 79.

80. A uniform 6.0-m-long ladder of mass 16.0 kg leans against a smooth wall (so the force exerted by the wall, $\vec{F}_W$, is perpendicular to the wall). The ladder makes an angle of 20.0° with the vertical wall, and the ground is rough. Determine the coefficient of static friction at the base of the ladder if the ladder is not to slip when a 76.0-kg person stands three-fourths of the way up the ladder.

81. There is a maximum height of a uniform vertical column made of any material that can support itself without buckling, and it is independent of the cross-sectional area (why?). Calculate this height for (a) steel (density $7.8 \times 10^3 \, kg/m^3$), and (b) granite (density $2.7 \times 10^3 \, kg/m^3$).

82. A 95,000-kg train locomotive starts across a 280-m-long bridge at time $t = 0$. The bridge is a uniform beam of mass 23,000 kg and the train travels at a constant 80.0 km/h. What are the magnitudes of the vertical forces, $F_A(t)$ and $F_B(t)$, on the two end supports, written as a function of time during the train's passage?

83. A 23.0-kg backpack is suspended midway between two trees by a light cord as in Fig. 12–50. A bear grabs the backpack and pulls vertically downward with a constant force, so that each section of cord makes an angle of 27° below the horizontal. Initially, without the bear pulling, the angle was 15°; the tension in the cord with the bear pulling is double what it was when he was not. Calculate the force the bear is exerting on the backpack.

84. A uniform beam of mass M and length ℓ is mounted on a hinge at a wall as shown in Fig. 12–91. It is held in a horizontal position by a wire making an angle θ as shown. A mass m is placed on the beam a distance x from the wall, and this distance can be varied. Determine, as a function of x, (a) the tension in the wire and (b) the components of the force exerted by the beam on the hinge.

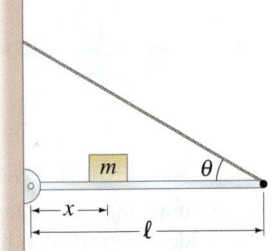

FIGURE 12–91
Problem 84.

85. Two identical, uniform beams are symmetrically set up against each other (Fig. 12–92) on a floor with which they have a coefficient of friction $\mu_s = 0.50$. What is the minimum angle the beams can make with the floor and still not fall?

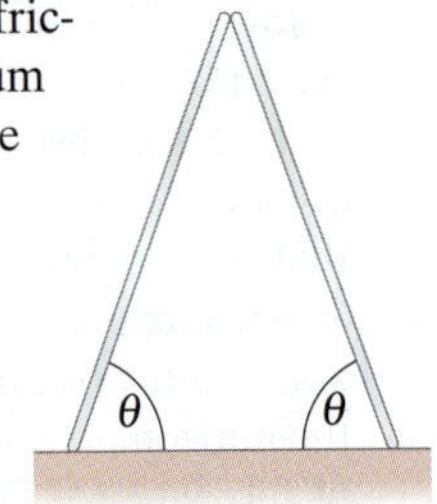

FIGURE 12–92
Problem 85.

86. If 35 kg is the maximum mass m that a person can hold in a hand when the arm is positioned with a 105° angle at the elbow as shown in Fig. 12–93, what is the maximum force F_{max} that the biceps muscle exerts on the forearm? Assume the forearm and hand have a total mass of 2.0 kg with a CG that is 15 cm from the elbow, and that the biceps muscle attaches 5.0 cm from the elbow.

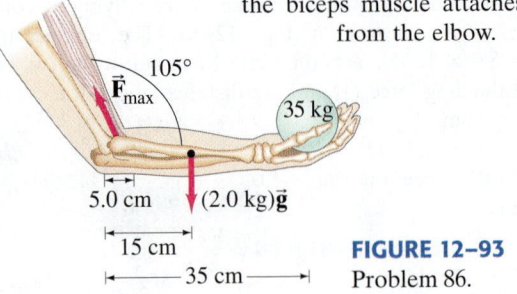

FIGURE 12–93
Problem 86.

87. (*a*) Estimate the magnitude of the force $\vec{\mathbf{F}}_M$ the muscles exert on the back to support the upper body when a person bends forward. Use the model shown in Fig. 12–94b. (*b*) Estimate the magnitude and direction of the force $\vec{\mathbf{F}}_V$ acting on the fifth lumbar vertebra (exerted by the spine below).

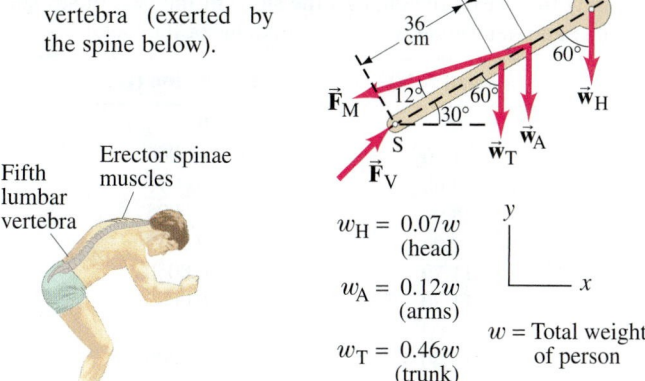

$w_H = 0.07w$
(head)

$w_A = 0.12w$
(arms)

$w_T = 0.46w$
(trunk)

w = Total weight of person

(a)　　　　(b)

FIGURE 12–94 Problem 87.

88. One rod of the square frame shown in Fig. 12–95 contains a turnbuckle which, when turned, can put the rod under tension or compression. If the turnbuckle puts rod AB under a compressive force F, determine the forces produced in the other rods. Ignore the mass of the rods and assume the diagonal rods cross each other freely at the center without friction. [*Hint:* Use the symmetry of the situation.]

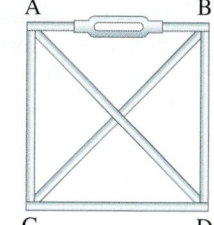

FIGURE 12–95
Problem 88.

89. A steel rod of radius $R = 15\ \text{cm}$ and length ℓ_0 stands upright on a firm surface. A 65-kg man climbs atop the rod. (*a*) Determine the percent decrease in the rod's length. (*b*) When a metal is compressed, each atom throughout its bulk moves closer to its neighboring atom by exactly the same fractional amount. If iron atoms in steel are normally $2.0 \times 10^{-10}\ \text{m}$ apart, by what distance did this interatomic spacing have to change in order to produce the normal force required to support the man? [*Note:* Neighboring atoms repel each other, and this repulsion accounts for the observed normal force.]

90. A home mechanic wants to raise the 280-kg engine out of a car. The plan is to stretch a rope vertically from the engine to a branch of a tree 6.0 m above, and back to the bumper (Fig. 12–96). When the mechanic climbs up a stepladder and pulls horizontally on the rope at its midpoint, the engine rises out of the car. (*a*) How much force must the mechanic exert to hold the engine 0.50 m above its normal position? (*b*) What is the system's mechanical advantage?

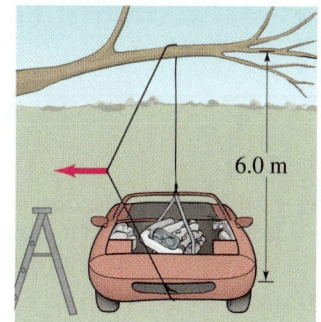

FIGURE 12–96
Problem 90.

91. A 2.0-m-high box with a 1.0-m-square base is moved across a rough floor as in Fig. 12–97. The uniform box weighs 250 N and has a coefficient of static friction with the floor of 0.60. What minimum force must be exerted on the box to make it slide? What is the maximum height h above the floor that this force can be applied without tipping the box over? Note that as the box tips, the normal force and the friction force will act at the lowest corner.

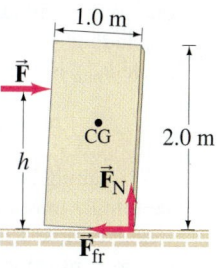

FIGURE 12–97
Problem 91.

92. You are on a pirate ship and being forced to walk the plank (Fig. 12–98). You are standing at the point marked C. The plank is nailed onto the deck at point A, and rests on the support 0.75 m away from A. The center of mass of the uniform plank is located at point B. Your mass is 65 kg and the mass of the plank is 45 kg. What is the minimum downward force the nails must exert on the plank to hold it in place?

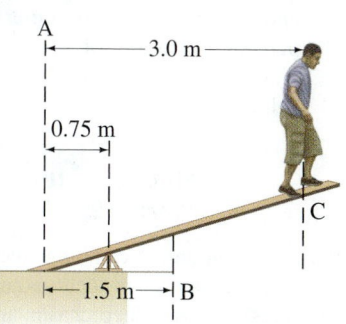

FIGURE 12–98
Problem 92.

93. A uniform sphere of weight mg and radius r_0 is tethered to a wall by a rope of length ℓ. The rope is tied to the wall a distance h above the contact point of the sphere, as shown in Fig. 12–99. The rope makes an angle θ with respect to the wall and is not in line with the ball's center. The coefficient of static friction between the wall and sphere is μ. (*a*) Determine the value of the frictional force on the sphere due to the wall. [*Hint:* A wise choice of axis will make this calculation easy.] (*b*) Suppose the sphere is just on the verge of slipping. Derive an expression for μ in terms of h and θ.

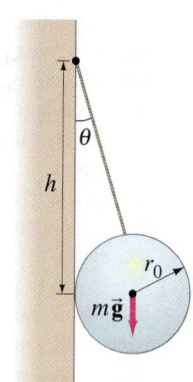

FIGURE 12–99
Problem 93.

*94. Use the method of joints to determine the force in each member of the truss shown in Fig. 12–100. State whether each member is in tension or compression.

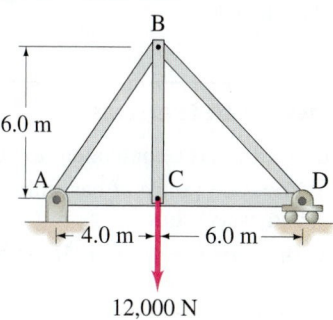

FIGURE 12–100
Problem 94.

95. A uniform ladder of mass m and length ℓ leans at an angle θ against a wall, Fig. 12–101. The coefficients of static friction between ladder–ground and ladder–wall are μ_G and μ_W, respectively. The ladder will be on the verge of slipping when both the static friction forces due to the ground and due to the wall take on their maximum values. (a) Show that the ladder will be stable if $\theta \geq \theta_{min}$, where the minimum angle θ_{min} is given by

$$\tan \theta_{min} = \frac{1}{2\mu_G}(1 - \mu_G \mu_W).$$

(b) "Leaning ladder problems" are often analyzed under the seemingly unrealistic assumption that the wall is frictionless (see Example 12–6). You wish to investigate the magnitude of error introduced by modeling the wall as frictionless, if in reality it is frictional. Using the relation found in part (a), calculate the true value of θ_{min} for a frictional wall, taking $\mu_G = \mu_W = 0.40$. Then, determine the approximate value of θ_{min} for the "frictionless wall" model by taking $\mu_G = 0.40$ and $\mu_W = 0$. Finally, determine the percent deviation of the approximate value of θ_{min} from its true value.

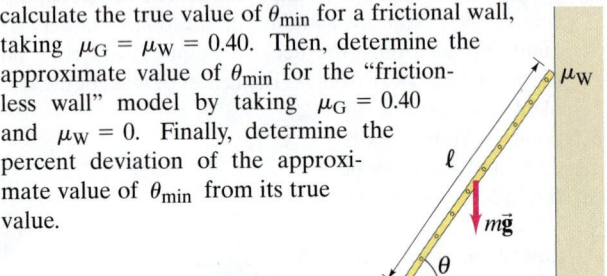

FIGURE 12–101
Problem 95.

96. In a mountain-climbing technique called the "Tyrolean traverse," a rope is anchored on both ends (to rocks or strong trees) across a deep chasm, and then a climber traverses the rope while attached by a sling as in Fig. 12–102. This technique generates tremendous forces in the rope and anchors, so a basic understanding of physics is crucial for safety. A typical climbing rope can undergo a tension force of perhaps 29 kN before breaking, and a "safety factor" of 10 is usually recommended. The length of rope used in the Tyrolean traverse must allow for some "sag" to remain in the recommended safety range. Consider a 75-kg climber at the center of a Tyrolean traverse, spanning a 25-m chasm. (a) To be within its recommended safety range, what minimum distance x must the rope sag? (b) If the Tyrolean traverse is set up incorrectly so that the rope sags by only one-fourth the distance found in (a), determine the tension in the rope. Will the rope break?

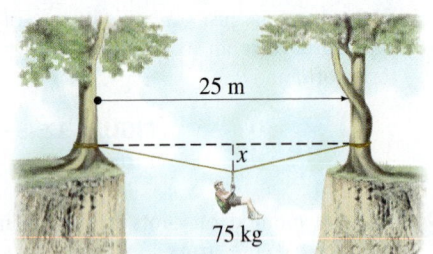

FIGURE 12–102
Problem 96.

***97.** (III) A metal cylinder has an original diameter of 1.00 cm and a length of 5.00 cm. A tension test was performed on the specimen and the data are listed in the Table. (a) Graph the stress on the specimen vs. the strain. (b) Considering only the elastic region, find the slope of the best-fit straight line and determine the elastic modulus of the metal.

Load (kN)	Elongation (cm)
0	0
1.50	0.0005
4.60	0.0015
8.00	0.0025
11.00	0.0035
11.70	0.0050
11.80	0.0080
12.00	0.0200
16.60	0.0400
20.00	0.1000
21.50	0.2800
19.50	0.4000
18.50	0.4600

***98.** (III) Two springs, attached by a rope, are connected as shown in Fig. 12–103. The length AB is 4.0 m and $AC = BC$. The spring constant of each spring is $k = 20.0$ N/m. A force F acts downward at C on the rope. Graph θ as a function of F from $\theta = 0$ to 75°, assuming the springs are unstretched at $\theta = 0$.

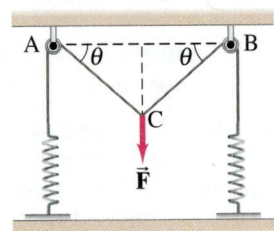

FIGURE 12–103   Problem 98.

Answers to Exercises

A: F_A also has a component to balance the sideways force F_B.

B: Yes: $\cos \theta$ (angle of bar with ground) appears on both sides and cancels out.

C: $F_N = m_A g + m_B g + Mg = 560$ N.

D: (a).

E: 7.0 kg.

F: Static friction at the cement floor $(= F_{Cx})$ is crucial, or else the ladder would slip. At the top, the ladder can move and adjust, so we wouldn't need or expect a strong static friction force there.

G: (b).

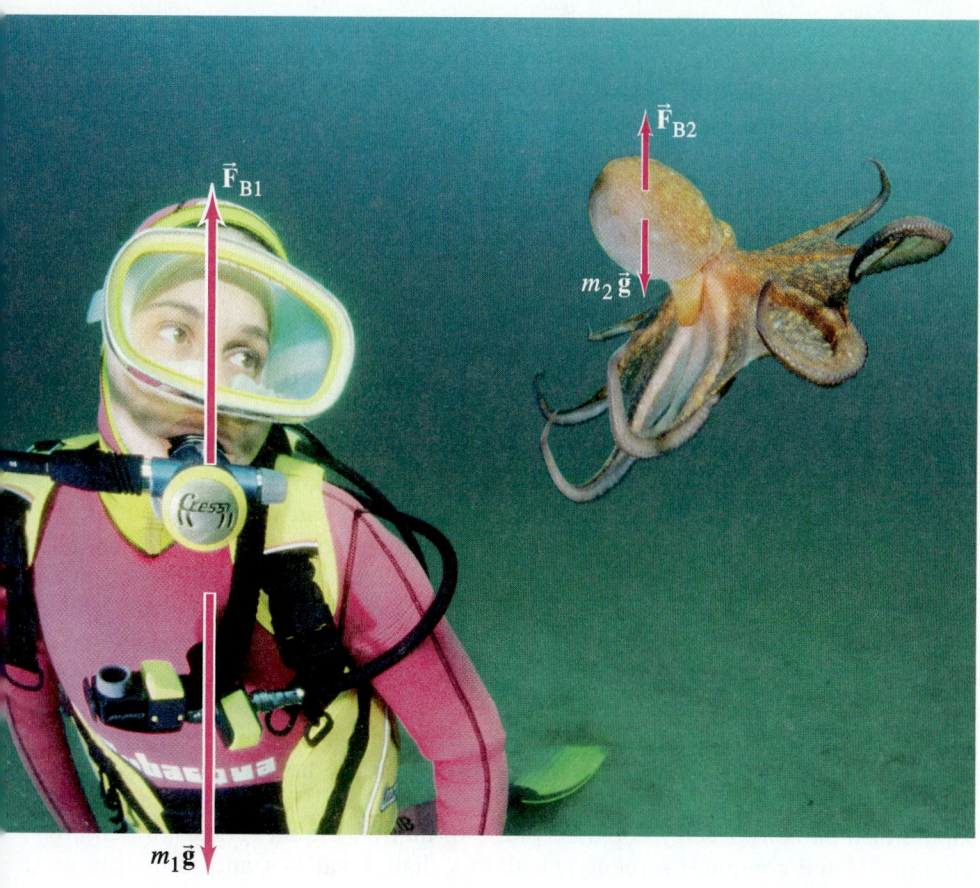

Underwater divers and sea creatures experience a buoyant force ($\vec{F}_B$) that closely balances their weight $m\vec{g}$. The buoyant force is equal to the weight of the volume of fluid displaced (Archimedes' principle) and arises because the pressure increases with depth in the fluid. Sea creatures have a density very close to that of water, so their weight very nearly equals the buoyant force. Humans have a density slightly less than water, so they can float.

When fluids flow, interesting effects occur because the pressure in the fluid is lower where the fluid velocity is higher (Bernoulli's principle).

C H A P T E R

13

Fluids

CHAPTER-OPENING QUESTIONS—Guess now!

1. Which container has the largest pressure at the bottom? Assume each container holds the same volume of water.

(a)   (b)   (c)   (d)   (e)

The pressures are equal.

2. Two balloons are tied and hang with their nearest edges about 3 cm apart. If you blow between the balloons (not *at* the balloons, but at the opening between them), what will happen?
(a) Nothing.
(b) The balloons will move closer together.
(c) The balloons will move farther apart.

CONTENTS

13–1 Phases of Matter

13–2 Density and Specific Gravity

13–3 Pressure in Fluids

13–4 Atmospheric Pressure and Gauge Pressure

13–5 Pascal's Principle

13–6 Measurement of Pressure; Gauges and the Barometer

13–7 Buoyancy and Archimedes' Principle

13–8 Fluids in Motion; Flow Rate and the Equation of Continuity

13–9 Bernoulli's Equation

13–10 Applications of Bernoulli's Principle: Torricelli, Airplanes, Baseballs, TIA

*13–11 Viscosity

*13–12 Flow in Tubes: Poiseuille's Equation, Blood Flow

*13–13 Surface Tension and Capillarity

*13–14 Pumps, and the Heart

339

TABLE 13–1
Densities of Substances†

Substance	Density, ρ (kg/m³)
Solids	
Aluminum	2.70×10^3
Iron and steel	7.8×10^3
Copper	8.9×10^3
Lead	11.3×10^3
Gold	19.3×10^3
Concrete	2.3×10^3
Granite	2.7×10^3
Wood (typical)	$0.3–0.9 \times 10^3$
Glass, common	$2.4–2.8 \times 10^3$
Ice (H_2O)	0.917×10^3
Bone	$1.7–2.0 \times 10^3$
Liquids	
Water (4°C)	1.00×10^3
Blood, plasma	1.03×10^3
Blood, whole	1.05×10^3
Sea water	1.025×10^3
Mercury	13.6×10^3
Alcohol, ethyl	0.79×10^3
Gasoline	0.68×10^3
Gases	
Air	1.29
Helium	0.179
Carbon dioxide	1.98
Steam (water, 100°C)	0.598

†Densities are given at 0°C and 1 atm pressure unless otherwise specified.

I n previous Chapters we considered objects that were solid and assumed to maintain their shape except for a small amount of elastic deformation. We sometimes treated objects as point particles. Now we are going to shift our attention to materials that are very deformable and can flow. Such "fluids" include liquids and gases. We will examine fluids both at rest (fluid statics) and in motion (fluid dynamics).

13–1 Phases of Matter

The three common **phases**, or **states**, of matter are solid, liquid, and gas. We can distinguish these three phases as follows. A **solid** maintains a fixed shape and a fixed size; even if a large force is applied to a solid, it does not readily change in shape or volume. A **liquid** does not maintain a fixed shape—it takes on the shape of its container—but like a solid it is not readily compressible, and its volume can be changed significantly only by a very large force. A **gas** has neither a fixed shape nor a fixed volume—it will expand to fill its container. For example, when air is pumped into an automobile tire, the air does not all run to the bottom of the tire as a liquid would; it spreads out to fill the whole volume of the tire. Since liquids and gases do not maintain a fixed shape, they both have ability to flow; they are thus often referred to collectively as **fluids**.

The division of matter into three phases is not always simple. How, for example, should butter be classified? Furthermore, a fourth phase of matter can be distinguished, the **plasma** phase, which occurs only at very high temperatures and consists of ionized atoms (electrons separated from the nuclei). Some scientists believe that so-called colloids (suspensions of tiny particles in a liquid) should also be considered a separate phase of matter. **Liquid crystals**, which are used in TV and computer screens, calculators, digital watches, and so on, can be considered a phase of matter intermediate between solids and liquids. However, for our present purposes we will mainly be interested in the three ordinary phases of matter.

13–2 Density and Specific Gravity

It is sometimes said that iron is "heavier" than wood. This cannot really be true since a large log clearly weighs more than an iron nail. What we should say is that iron is more *dense* than wood.

The **density**, ρ, of a substance (ρ is the lowercase Greek letter rho) is defined as its mass per unit volume:

$$\rho = \frac{m}{V}, \tag{13–1}$$

where m is the mass of a sample of the substance and V is its volume. Density is a characteristic property of any pure substance. Objects made of a particular pure substance, such as pure gold, can have any size or mass, but the density will be the same for each.

We will sometimes use the concept of density, Eq. 13–1, to write the mass of an object as

$$m = \rho V,$$

and the weight of an object as

$$mg = \rho V g.$$

The SI unit for density is kg/m³. Sometimes densities are given in g/cm³. Note that since $1 \text{ kg/m}^3 = 1000 \text{ g}/(100 \text{ cm})^3 = 10^3 \text{ g}/10^6 \text{ cm}^3 = 10^{-3} \text{ g/cm}^3$, then a density given in g/cm³ must be multiplied by 1000 to give the result in kg/m³. Thus the density of aluminum is $\rho = 2.70 \text{ g/cm}^3$, which is equal to 2700 kg/m³. The densities of a variety of substances are given in Table 13–1. The Table specifies temperature and atmospheric pressure because they affect the density of substances (although the effect is slight for liquids and solids). Note that air is roughly 1000 times less dense than water.

EXAMPLE 13–1 **Mass, given volume and density.** What is the mass of a solid iron wrecking ball of radius 18 cm?

APPROACH First we use the standard formula $V = \frac{4}{3}\pi r^3$ (see inside rear cover) to obtain the volume of the sphere. Then Eq. 13–1 and Table 13–1 give us the mass m.

SOLUTION The volume of the sphere is
$$V = \frac{4}{3}\pi r^3 = \frac{4}{3}(3.14)(0.18\,\text{m})^3 = 0.024\,\text{m}^3.$$
From Table 13–1, the density of iron is $\rho = 7800\,\text{kg/m}^3$, so Eq. 13–1 gives
$$m = \rho V = (7800\,\text{kg/m}^3)(0.024\,\text{m}^3) = 190\,\text{kg}.$$

The **specific gravity** of a substance is defined as the ratio of the density of that substance to the density of water at 4.0°C. Because specific gravity (abbreviated SG) is a ratio, it is a simple number without dimensions or units. The density of water is $1.00\,\text{g/cm}^3 = 1.00 \times 10^3\,\text{kg/m}^3$, so the specific gravity of any substance will be equal numerically to its density specified in g/cm^3, or 10^{-3} times its density specified in kg/m^3. For example (see Table 13–1), the specific gravity of lead is 11.3, and that of alcohol is 0.79.

The concepts of density and specific gravity are especially helpful in the study of fluids because we are not always dealing with a fixed volume or mass.

13–3 Pressure in Fluids

Pressure and force are related, but they are not the same thing. **Pressure** is defined as force per unit area, where the force F is understood to be the magnitude of the force acting perpendicular to the surface area A:

$$\text{pressure} = P = \frac{F}{A}. \tag{13–2}$$

Although force is a vector, pressure is a scalar. Pressure has magnitude only. The SI unit of pressure is N/m^2. This unit has the official name **pascal** (Pa), in honor of Blaise Pascal (see Section 13–5); that is, $1\,\text{Pa} = 1\,\text{N/m}^2$. However, for simplicity, we will often use N/m^2. Other units sometimes used are dynes/cm^2, and lb/in.^2 (abbreviated "psi"). Several other units for pressure are discussed, along with conversions between them, in Section 13–6 (see also the Table inside the front cover).

⚠ **CAUTION**
Pressure is a scalar, not a vector

EXAMPLE 13–2 **Calculating pressure.** The two feet of a 60-kg person cover an area of $500\,\text{cm}^2$. (*a*) Determine the pressure exerted by the two feet on the ground. (*b*) If the person stands on one foot, what will the pressure be under that foot?

APPROACH Assume the person is at rest. Then the ground pushes up on her with a force equal to her weight mg, and she exerts a force mg on the ground where her feet (or foot) contact it. Because $1\,\text{cm}^2 = (10^{-2}\,\text{m})^2 = 10^{-4}\,\text{m}^2$, then $500\,\text{cm}^2 = 0.050\,\text{m}^2$.

SOLUTION (*a*) The pressure on the ground exerted by the two feet is
$$P = \frac{F}{A} = \frac{mg}{A} = \frac{(60\,\text{kg})(9.8\,\text{m/s}^2)}{(0.050\,\text{m}^2)} = 12 \times 10^3\,\text{N/m}^2.$$
(*b*) If the person stands on one foot, the force is still equal to the person's weight, but the area will be half as much, so the pressure will be twice as much: $24 \times 10^3\,\text{N/m}^2$.

FIGURE 13–1 Pressure is the same in every direction in a nonmoving fluid at a given depth. If this weren't true, the fluid would be in motion.

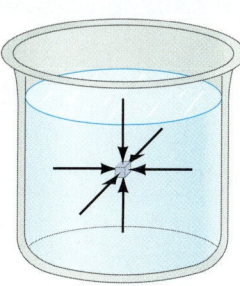

Pressure is particularly useful for dealing with fluids. It is an experimental observation that *a fluid exerts pressure in any direction*. This is well known to swimmers and divers who feel the water pressure on all parts of their bodies. At any depth in a fluid at rest, the pressure is the same in all directions at a given depth. To see why, consider a tiny cube of the fluid (Fig. 13–1) which is so small that we can consider it a point and can ignore the force of gravity on it. The pressure on one side of it must equal the pressure on the opposite side. If this weren't true, there would be a net force on the cube and it would start moving. If the fluid is not flowing, then the pressures must be equal.

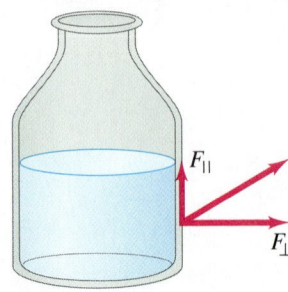

FIGURE 13–2 If there were a component of force parallel to the solid surface of the container, the liquid would move in response to it. For a liquid at rest, $F_\parallel = 0$.

FIGURE 13–3 Calculating the pressure at a depth h in a liquid.

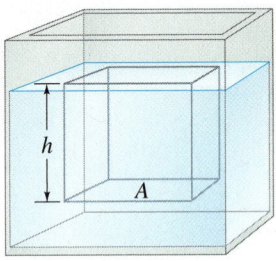

FIGURE 13–4 Forces on a flat, slablike volume of fluid for determining the pressure P at a height y in the fluid.

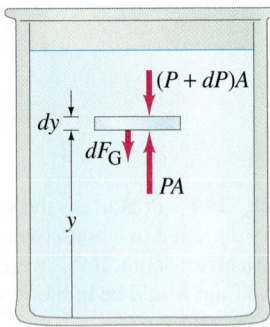

For a fluid at rest, the force due to fluid pressure always acts *perpendicular* to any solid surface it touches. If there were a component of the force parallel to the surface, as shown in Fig. 13–2, then according to Newton's third law the solid surface would exert a force back on the fluid that also would have a component parallel to the surface. Such a component would cause the fluid to flow, in contradiction to our assumption that the fluid is at rest. Thus the force due to the pressure in a fluid at rest is always perpendicular to the surface.

Let us now calculate quantitatively how the pressure in a liquid of uniform density varies with depth. Consider a point at a depth h below the surface of the liquid, as shown in Fig. 13–3 (that is, the surface is a height h above this point). The pressure due to the liquid at this depth h is due to the weight of the column of liquid above it. Thus the force due to the weight of liquid acting on the area A is $F = mg = (\rho V)g = \rho Ahg$, where Ah is the volume of the column of liquid, ρ is the density of the liquid (assumed to be constant), and g is the acceleration of gravity. The pressure P due to the weight of liquid is then

$$P = \frac{F}{A} = \frac{\rho Ahg}{A}$$

$$P = \rho gh. \qquad \text{[liquid]} \quad \textbf{(13–3)}$$

Note that the area A doesn't affect the pressure at a given depth. The fluid pressure is directly proportional to the density of the liquid and to the depth within the liquid. In general, the pressure at equal depths within a uniform liquid is the same.

| **EXERCISE A** Return to Chapter-Opening Question 1, page 339, and answer it again now. Try to explain why you may have answered differently the first time.

Equation 13–3 tells us what the pressure is at a depth h in the liquid, due to the liquid itself. But what if there is additional pressure exerted at the surface of the liquid, such as the pressure of the atmosphere or a piston pushing down? And what if the density of the fluid is not constant? Gases are quite compressible and hence their density can vary significantly with depth. Liquids, too, can be compressed, although we can often ignore the variation in density. (One exception is in the depths of the ocean where the great weight of water above significantly compresses the water and increases its density.) To cover these, and other cases, we now treat the general case of determining how the pressure in a fluid varies with depth.

As shown in Fig. 13–4, let us determine the pressure at any height y above some reference point[†] (such as the ocean floor or the bottom of a tank or swimming pool). Within this fluid, at the height y, we consider a tiny, flat, slablike volume of the fluid whose area is A and whose (infinitesimal) thickness is dy, as shown. Let the pressure acting upward on its lower surface (at height y) be P. The pressure acting downward on the top surface of our tiny slab (at height $y + dy$) is designated $P + dP$. The fluid pressure acting on our slab thus exerts a force equal to PA upward on our slab and a force equal to $(P + dP)A$ downward on it. The only other force acting vertically on the slab is the (infinitesimal) force of gravity dF_G, which on our slab of mass dm is

$$dF_G = (dm)g = \rho g\, dV = \rho g A\, dy,$$

where ρ is the density of the fluid at the height y. Since the fluid is assumed to be at rest, our slab is in equilibrium so the net force on it must be zero. Therefore we have

$$PA - (P + dP)A - \rho g A\, dy = 0,$$

which when simplified becomes

$$\frac{dP}{dy} = -\rho g. \qquad \textbf{(13–4)}$$

This relation tells us how the pressure within the fluid varies with height above any reference point. The minus sign indicates that the pressure decreases with an increase in height; or that the pressure increases with depth (reduced height).

[†]Now we are measuring y positive upwards, the reverse of what we did to get Eq. 13–3 where we measured the depth (i.e. downward as positive).

If the pressure at a height y_1 in the fluid is P_1, and at height y_2 it is P_2, then we can integrate Eq. 13–4 to obtain

$$\int_{P_1}^{P_2} dP = -\int_{y_1}^{y_2} \rho g \, dy$$

$$P_2 - P_1 = -\int_{y_1}^{y_2} \rho g \, dy, \qquad\qquad (13-5)$$

where we assume ρ is a function of height y: $\rho = \rho(y)$. This is a general relation, and we apply it now to two special cases: (1) pressure in liquids of uniform density and (2) pressure variations in the Earth's atmosphere.

For liquids in which any variation in density can be ignored, $\rho = $ constant and Eq. 13–5 is readily integrated:

$$P_2 - P_1 = -\rho g(y_2 - y_1). \qquad\qquad (13-6a)$$

For the everyday situation of a liquid in an open container—such as water in a glass, a swimming pool, a lake, or the ocean—there is a free surface at the top exposed to the atmosphere. It is convenient to measure distances from this top surface. That is, we let h be the *depth* in the liquid where $h = y_2 - y_1$ as shown in Fig. 13–5. If we let y_2 be the position of the top surface, then P_2 represents the atmospheric pressure, P_0, at the top surface. Then, from Eq. 13–6a, the pressure $P (= P_1)$ at a depth h in the fluid is

$$P = P_0 + \rho g h. \qquad \text{[h is depth in liquid]} \quad (13-6b)$$

Note that Eq. 13–6b is simply the liquid pressure (Eq. 13–3) plus the pressure P_0 due to the atmosphere above.

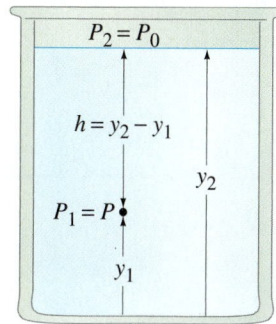

FIGURE 13–5 Pressure at a depth $h = (y_2 - y_1)$ in a liquid of density ρ is $P = P_0 + \rho g h$, where P_0 is the external pressure at the liquid's top surface.

EXAMPLE 13–3 **Pressure at a faucet.** The surface of the water in a storage tank is 30 m above a water faucet in the kitchen of a house, Fig. 13–6. Calculate the difference in water pressure between the faucet and the surface of the water in the tank.

APPROACH Water is practically incompressible, so ρ is constant even for $h = 30$ m when used in Eq. 13–6b. Only h matters; we can ignore the "route" of the pipe and its bends.

SOLUTION We assume the atmospheric pressure at the surface of the water in the storage tank is the same as at the faucet. So, the water pressure difference between the faucet and the surface of the water in the tank is

$$\Delta P = \rho g h = (1.0 \times 10^3 \text{ kg/m}^3)(9.8 \text{ m/s}^2)(30 \text{ m}) = 2.9 \times 10^5 \text{ N/m}^2.$$

NOTE The height h is sometimes called the **pressure head**. In this Example, the head of water is 30 m at the faucet. The very different diameters of the tank and faucet don't affect the result—only pressure does.

FIGURE 13–6 Example 13–3.

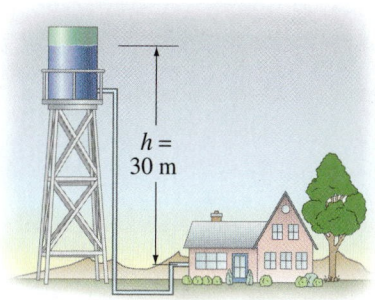

EXAMPLE 13–4 **Force on aquarium window.** Calculate the force due to water pressure exerted on a 1.0 m × 3.0 m aquarium viewing window whose top edge is 1.0 m below the water surface, Fig. 13–7.

APPROACH At a depth h, the pressure due to the water is given by Eq. 13–6b. Divide the window up into thin horizontal strips of length $\ell = 3.0$ m and thickness dy, as shown in Fig. 13–7. We choose a coordinate system with $y = 0$ at the surface of the water and y is positive downward. (With this choice, the minus sign in Eq. 13–6a becomes plus, or we use Eq. 13–6b with $y = h$.) The force due to water pressure on each strip is $dF = P \, dA = \rho g y \ell \, dy$.

SOLUTION The total force on the window is given by the integral:

$$\int_{y_1=1.0\,\text{m}}^{y_2=2.0\,\text{m}} \rho g y \ell \, dy = \tfrac{1}{2}\rho g \ell (y_2^2 - y_1^2)$$

$$= \tfrac{1}{2}(1000 \text{ kg/m}^3)(9.8 \text{ m/s}^2)(3.0 \text{ m})[(2.0 \text{ m})^2 - (1.0 \text{ m})^2] = 44,000 \text{ N}.$$

NOTE To check our answer, we can do an estimate: multiply the area of the window (3.0 m^2) times the pressure at the middle of the window $(h = 1.5 \text{ m})$ using Eq. 13–3, $P = \rho g h = (1000 \text{ kg/m}^3)(9.8 \text{ m/s}^2)(1.5 \text{ m}) \approx 1.5 \times 10^4 \text{ N/m}^2$. So $F = PA \approx (1.5 \times 10^4 \text{ N/m}^2)(3.0 \text{ m})(1.0 \text{ m}) \approx 4.5 \times 10^4 \text{ N}$. Good!

FIGURE 13–7 Example 13–4.

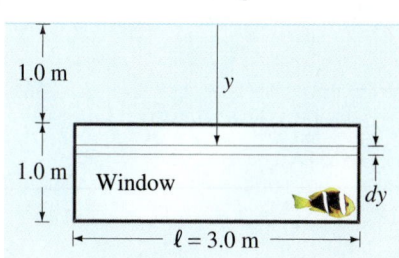

EXERCISE B A dam holds back a lake that is 85 m deep at the dam. If the lake is 20 km long, how much thicker should the dam be than if the lake were smaller, only 1.0 km long?

Now let us apply Eq. 13–4 or 13–5 to gases. The density of gases is normally quite small, so the difference in pressure at different heights can usually be ignored if $y_2 - y_1$ is not large (which is why, in Example 13–3, we could ignore the difference in air pressure between the faucet and the top of the storage tank). Indeed, for most ordinary containers of gas, we can assume that the pressure is the same throughout. However, if $y_2 - y_1$ is very large, we cannot make this assumption. An interesting example is the air of Earth's atmosphere, whose pressure at sea level is about $1.013 \times 10^5 \, \text{N/m}^2$ and decreases slowly with altitude.

EXAMPLE 13–5 **Elevation effect on atmospheric pressure.** (*a*) Determine the variation in pressure in the Earth's atmosphere as a function of height y above sea level, assuming g is constant and that the density of the air is proportional to the pressure. (This last assumption is not terribly accurate, in part because temperature and other weather effects are important.) (*b*) At what elevation is the air pressure equal to half the pressure at sea level?

APPROACH We start with Eq. 13–4 and integrate it from the surface of the Earth where $y = 0$ and $P = P_0$, up to height y at pressure P. In (*b*) we choose $P = \frac{1}{2}P_0$.

SOLUTION (*a*) We are assuming that ρ is proportional to P, so we can write

$$\frac{\rho}{\rho_0} = \frac{P}{P_0},$$

where $P_0 = 1.013 \times 10^5 \, \text{N/m}^2$ is atmospheric pressure at sea level and $\rho_0 = 1.29 \, \text{kg/m}^3$ is the density of air at sea level at 0°C (Table 13–1). From the differential change in pressure with height, Eq. 13–4, we have

$$\frac{dP}{dy} = -\rho g = -P\left(\frac{\rho_0}{P_0}\right)g,$$

so

$$\frac{dP}{P} = -\frac{\rho_0}{P_0} g \, dy.$$

We integrate this from $y = 0$ (Earth's surface) and $P = P_0$, to the height y where the pressure is P:

$$\int_{P_0}^{P} \frac{dP}{P} = -\frac{\rho_0}{P_0} g \int_0^y dy$$

$$\ln \frac{P}{P_0} = -\frac{\rho_0}{P_0} gy,$$

since $\ln P - \ln P_0 = \ln(P/P_0)$. Then

$$P = P_0 e^{-(\rho_0 g/P_0)y}.$$

So, based on our assumptions, we find that the air pressure in our atmosphere decreases approximately exponentially with height.

NOTE The atmosphere does not have a distinct top surface, so there is no natural point from which to measure depth in the atmosphere, as we can do for a liquid.

(*b*) The constant $(\rho_0 g/P_0)$ has the value

$$\frac{\rho_0 g}{P_0} = \frac{(1.29 \, \text{kg/m}^3)(9.80 \, \text{m/s}^2)}{(1.013 \times 10^5 \, \text{N/m}^2)} = 1.25 \times 10^{-4} \, \text{m}^{-1}.$$

Then, when we set $P = \frac{1}{2}P_0$ in our expression derived in (*a*), we obtain

$$\tfrac{1}{2} = e^{-(1.25 \times 10^{-4} \, \text{m}^{-1})y}$$

or, taking natural logarithms of both sides,

$$\ln \tfrac{1}{2} = (-1.25 \times 10^{-4} \, \text{m}^{-1})y$$

so (recall $\ln \frac{1}{2} = -\ln 2$, Appendix A–7, Eq. ii)

$$y = (\ln 2.00)/(1.25 \times 10^{-4} \, \text{m}^{-1}) = 5550 \, \text{m}.$$

Thus, at an elevation of about 5500 m (about 18,000 ft), atmospheric pressure drops to half what it is at sea level. It is not surprising that mountain climbers often use oxygen tanks at very high altitudes.

13–4 Atmospheric Pressure and Gauge Pressure

Atmospheric Pressure

The pressure of the air at a given place on Earth varies slightly according to the weather. At sea level, the pressure of the atmosphere on average is $1.013 \times 10^5 \, \text{N/m}^2$ (or $14.7 \, \text{lb/in.}^2$). This value lets us define a commonly used unit of pressure, the **atmosphere** (abbreviated atm):

$$1 \, \text{atm} = 1.013 \times 10^5 \, \text{N/m}^2 = 101.3 \, \text{kPa}.$$

Another unit of pressure sometimes used (in meteorology and on weather maps) is the **bar**, which is defined as

$$1 \, \text{bar} = 1.000 \times 10^5 \, \text{N/m}^2.$$

Thus standard atmospheric pressure is slightly more than 1 bar.

The pressure due to the weight of the atmosphere is exerted on all objects immersed in this great sea of air, including our bodies. How does a human body withstand the enormous pressure on its surface? The answer is that living cells maintain an internal pressure that closely equals the external pressure, just as the pressure inside a balloon closely matches the outside pressure of the atmosphere. An automobile tire, because of its rigidity, can maintain internal pressures much greater than the external pressure.

PHYSICS APPLIED
Pressure on living cells

CONCEPTUAL EXAMPLE 13–6 | **Finger holds water in a straw.** You insert a straw of length ℓ into a tall glass of water. You place your finger over the top of the straw, capturing some air above the water but preventing any additional air from getting in or out, and then you lift the straw from the water. You find that the straw retains most of the water (see Fig. 13–8a). Does the air in the space between your finger and the top of the water have a pressure P that is greater than, equal to, or less than the atmospheric pressure P_0 outside the straw?

RESPONSE Consider the forces on the column of water (Fig. 13–8b). Atmospheric pressure outside the straw pushes upward on the water at the bottom of the straw, gravity pulls the water downward, and the air pressure inside the top of the straw pushes downward on the water. Since the water is in equilibrium, the upward force due to atmospheric pressure P_0 must balance the two downward forces. The only way this is possible is for the air pressure inside the straw to be *less than* the atmosphere pressure outside the straw. (When you initially remove the straw from the glass of water, a little water may leave the bottom of the straw, thus increasing the volume of trapped air and reducing its density and pressure.)

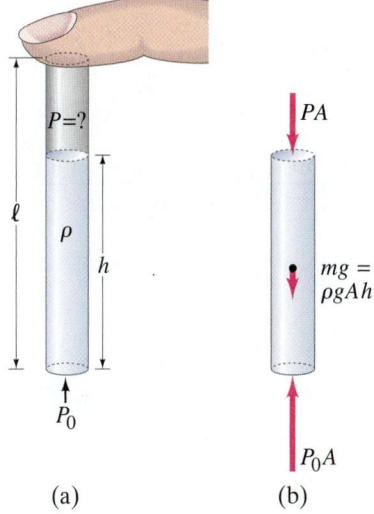

FIGURE 13–8 Example 13–6.

Gauge Pressure

It is important to note that tire gauges, and most other pressure gauges, register the pressure above and beyond atmospheric pressure. This is called **gauge pressure**. Thus, to get the **absolute pressure**, P, we must add the atmospheric pressure, P_0, to the gauge pressure, P_G:

$$P = P_0 + P_G.$$

If a tire gauge registers $220 \, \text{kPa}$, the absolute pressure within the tire is $220 \, \text{kPa} + 101 \, \text{kPa} = 321 \, \text{kPa}$, equivalent to about $3.2 \, \text{atm}$ ($2.2 \, \text{atm}$ gauge pressure).

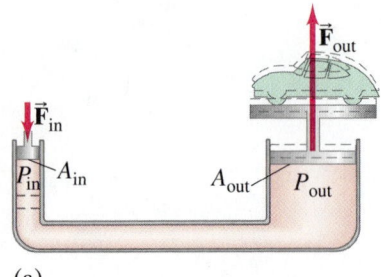

(a)

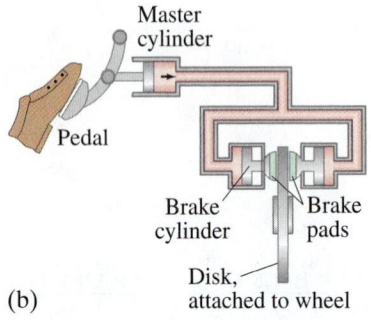

Master cylinder

Pedal

Brake cylinder

Brake pads

Disk, attached to wheel

(b)

FIGURE 13–9 Applications of Pascal's principle: (a) hydraulic lift; (b) hydraulic brakes in a car.

13–5 Pascal's Principle

The Earth's atmosphere exerts a pressure on all objects with which it is in contact, including other fluids. External pressure acting on a fluid is transmitted throughout that fluid. For instance, according to Eq. 13–3, the pressure due to the water at a depth of 100 m below the surface of a lake is $P = \rho g h = (1000 \text{ kg/m}^3)(9.8 \text{ m/s}^2)(100 \text{ m}) = 9.8 \times 10^5 \text{ N/m}^2$, or 9.7 atm. However, the total pressure at this point is due to the pressure of water plus the pressure of the air above it. Hence the total pressure (if the lake is near sea level) is 9.7 atm + 1.0 atm = 10.7 atm. This is just one example of a general principle attributed to the French philosopher and scientist Blaise Pascal (1623–1662). **Pascal's principle** states that *if an external pressure is applied to a confined fluid, the pressure at every point within the fluid increases by that amount.*

A number of practical devices make use of Pascal's principle. One example is the hydraulic lift, illustrated in Fig. 13–9a, in which a small input force is used to exert a large output force by making the area of the output piston larger than the area of the input piston. To see how this works, we assume the input and output pistons are at the same height (at least approximately). Then the external input force F_{in}, by Pascal's principle, increases the pressure equally throughout. Therefore, at the same level (see Fig. 13–9a),

$$P_{\text{out}} = P_{\text{in}}$$

where the input quantities are represented by the subscript "in" and the output by "out." Since $P = F/A$, we write the above equality as

$$\frac{F_{\text{out}}}{A_{\text{out}}} = \frac{F_{\text{in}}}{A_{\text{in}}},$$

or

$$\frac{F_{\text{out}}}{F_{\text{in}}} = \frac{A_{\text{out}}}{A_{\text{in}}}.$$

The quantity $F_{\text{out}}/F_{\text{in}}$ is called the **mechanical advantage** of the hydraulic lift, and it is equal to the ratio of the areas. For example, if the area of the output piston is 20 times that of the input cylinder, the force is multiplied by a factor of 20. Thus a force of 200 lb could lift a 4000-lb car.

Figure 13–9b illustrates the brake system of a car. When the driver presses the brake pedal, the pressure in the master cylinder increases. This pressure increase occurs throughout the brake fluid, thus pushing the brake pads against the disk attached to the car's wheel.

13–6 Measurement of Pressure; Gauges and the Barometer

Many devices have been invented to measure pressure, some of which are shown in Fig. 13–10. The simplest is the open-tube *manometer* (Fig 13–10a) which is a U-shaped tube partially filled with a liquid, usually mercury or water. The pressure P being measured is related to the difference in height Δh of the two levels of the liquid by the relation

$$P = P_0 + \rho g \, \Delta h,$$

where P_0 is atmospheric pressure (acting on the top of the liquid in the left-hand tube), and ρ is the density of the liquid. Note that the quantity $\rho g \, \Delta h$ is the gauge pressure—the amount by which P exceeds atmospheric pressure P_0. If the liquid in the left-hand column were lower than that in the right-hand column, P would have to be less than atmospheric pressure (and Δh would be negative).

Instead of calculating the product $\rho g \, \Delta h$, sometimes only the change in height Δh is specified. In fact, pressures are sometimes specified as so many "millimeters of mercury" (mm-Hg) or "mm of water" (mm-H$_2$O). The unit mm-Hg is equivalent to a pressure of 133 N/m^2, since $\rho g \, \Delta h$ for 1 mm = 1.0×10^{-3} m of mercury gives

$$\rho g \, \Delta h = (13.6 \times 10^3 \text{ kg/m}^3)(9.80 \text{ m/s}^2)(1.00 \times 10^{-3} \text{ m}) = 1.33 \times 10^2 \text{ N/m}^2.$$

The unit mm-Hg is also called the **torr** in honor of Evangelista Torricelli (1608–1647), a student of Galileo's who invented the barometer (see next page).

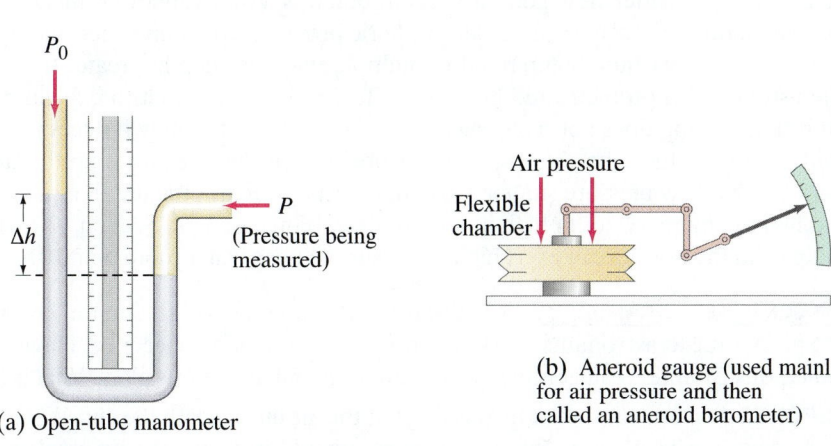

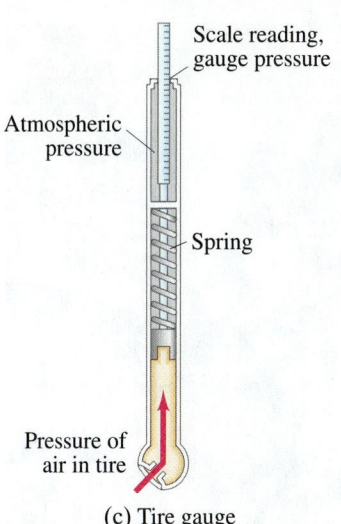

(b) Aneroid gauge (used mainly for air pressure and then called an aneroid barometer)

(a) Open-tube manometer

(c) Tire gauge

FIGURE 13–10 Pressure gauges: (a) open-tube manometer, (b) aneroid gauge, and (c) common tire-pressure gauge.

Conversion factors among the various units of pressure (an incredible nuisance!) are given in Table 13–2. It is important that only $N/m^2 = Pa$, the proper SI unit, be used in calculations involving other quantities specified in SI units.

PROBLEM SOLVING

In calculations, use SI units:
$1 \, Pa = 1 \, N/m^2$

TABLE 13–2 Conversion Factors Between Different Units of Pressure

In Terms of $1 \, Pa = 1 \, N/m^2$	1 atm in Different Units
$1 \, atm = 1.013 \times 10^5 \, N/m^2$	$1 \, atm = 1.013 \times 10^5 \, N/m^2$
$= 1.013 \times 10^5 \, Pa = 101.3 \, kPa$	
$1 \, bar = 1.000 \times 10^5 \, N/m^2$	$1 \, atm = 1.013 \, bar$
$1 \, dyne/cm^2 = 0.1 \, N/m^2$	$1 \, atm = 1.013 \times 10^6 \, dyne/cm^2$
$1 \, lb/in.^2 = 6.90 \times 10^3 \, N/m^2$	$1 \, atm = 14.7 \, lb/in.^2$
$1 \, lb/ft^2 = 47.9 \, N/m^2$	$1 \, atm = 2.12 \times 10^3 \, lb/ft^2$
$1 \, cm\text{-}Hg = 1.33 \times 10^3 \, N/m^2$	$1 \, atm = 76.0 \, cm\text{-}Hg$
$1 \, mm\text{-}Hg = 133 \, N/m^2$	$1 \, atm = 760 \, mm\text{-}Hg$
$1 \, torr = 133 \, N/m^2$	$1 \, atm = 760 \, torr$
$1 \, mm\text{-}H_2O \, (4°C) = 9.80 \, N/m^2$	$1 \, atm = 1.03 \times 10^4 \, mm\text{-}H_2O \, (4°C)$

Another type of pressure gauge is the aneroid gauge (Fig. 13–10b) in which the pointer is linked to the flexible ends of an evacuated thin metal chamber. In an electronic gauge, the pressure may be applied to a thin metal diaphragm whose resulting distortion is translated into an electrical signal by a transducer. A common tire gauge is shown in Fig. 13–10c.

Atmospheric pressure can be measured by a modified kind of mercury manometer with one end closed, called a mercury **barometer** (Fig. 13–11). The glass tube is completely filled with mercury and then inverted into the bowl of mercury. If the tube is long enough, the level of the mercury will drop, leaving a vacuum at the top of the tube, since atmospheric pressure can support a column of mercury only about 76 cm high (exactly 76.0 cm at standard atmospheric pressure). That is, a column of mercury 76 cm high exerts the same pressure as the atmosphere[†]:

$$P = \rho g \, \Delta h$$
$$= (13.6 \times 10^3 \, kg/m^3)(9.80 \, m/s^2)(0.760 \, m) = 1.013 \times 10^5 \, N/m^2 = 1.00 \, atm.$$

[†]This calculation confirms the entry in Table 13–2, $1 \, atm = 76.0 \, cm\text{-}Hg$.

FIGURE 13–11 A mercury barometer, invented by Torricelli, is shown here when the air pressure is standard atmospheric, 76.0 cm-Hg.

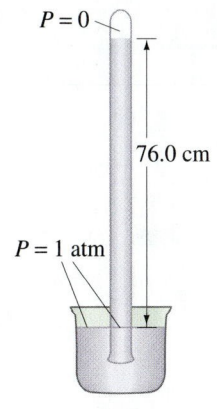

A calculation similar to what we just did will show that atmospheric pressure can maintain a column of water 10.3 m high in a tube whose top is under vacuum (Fig. 13–12). No matter how good a vacuum pump is, water cannot be made to rise more than about 10 m using normal atmospheric pressure. To pump water out of deep mine shafts with a vacuum pump requires multiple stages for depths greater than 10 m. Galileo studied this problem, and his student Torricelli was the first to explain it. The point is that a pump does not really suck water up a tube—it merely reduces the pressure at the top of the tube. Atmospheric air pressure *pushes* the water up the tube if the top end is at low pressure (under a vacuum), just as it is air pressure that pushes (or maintains) the mercury 76 cm high in a barometer. [Force pumps (Section 13–14) that push up from the bottom can exert higher pressure to push water more than 10 m high.]

CONCEPTUAL EXAMPLE 13–7 | **Suction.** A student suggests suction-cup shoes for Space Shuttle astronauts working on the exterior of a spacecraft. Having just studied this Chapter, you gently remind him of the fallacy of this plan. What is it?

RESPONSE Suction cups work by pushing out the air underneath the cup. What holds the suction cup in place is the air pressure outside it. (This can be a substantial force when on Earth. For example, a 10-cm-diameter suction cup has an area of $7.9 \times 10^{-3} \, \text{m}^2$. The force of the atmosphere on it is $(7.9 \times 10^{-3} \, \text{m}^2)(1.0 \times 10^5 \, \text{N/m}^2) \approx 800 \, \text{N}$, about 180 lbs!) But in outer space, there is no air pressure to push the suction cup onto the spacecraft.

We sometimes mistakenly think of suction as something we actively do. For example, we intuitively think that we pull the soda up through a straw. Instead, what we do is lower the pressure at the top of the straw, and the atmosphere *pushes* the soda up the straw.

13–7 Buoyancy and Archimedes' Principle

Objects submerged in a fluid appear to weigh less than they do when outside the fluid. For example, a large rock that you would have difficulty lifting off the ground can often be easily lifted from the bottom of a stream. When the rock breaks through the surface of the water, it suddenly seems to be much heavier. Many objects, such as wood, float on the surface of water. These are two examples of *buoyancy*. In each example, the force of gravity is acting downward. But in addition, an upward *buoyant force* is exerted by the liquid. The buoyant force on fish and underwater divers (as in the Chapter-Opening photo) almost exactly balances the force of gravity downward, and allows them to "hover" in equilibrium.

The buoyant force occurs because the pressure in a fluid increases with depth. Thus the upward pressure on the bottom surface of a submerged object is greater than the downward pressure on its top surface. To see this effect, consider a cylinder of height Δh whose top and bottom ends have an area A and which is completely submerged in a fluid of density ρ_F, as shown in Fig. 13–13. The fluid exerts a pressure $P_1 = \rho_F g h_1$ at the top surface of the cylinder (Eq. 13–3). The force due to this pressure on top of the cylinder is $F_1 = P_1 A = \rho_F g h_1 A$, and it is directed downward. Similarly, the fluid exerts an upward force on the bottom of the cylinder equal to $F_2 = P_2 A = \rho_F g h_2 A$. The net force on the cylinder exerted by the fluid pressure, which is the **buoyant force**, $\vec{F}_B$, acts upward and has the magnitude

$$
\begin{aligned}
F_B = F_2 - F_1 &= \rho_F g A (h_2 - h_1) \\
&= \rho_F g A \, \Delta h \\
&= \rho_F V g \\
&= m_F g,
\end{aligned}
$$

where $V = A \, \Delta h$ is the volume of the cylinder, the product $\rho_F V$ is the mass of the fluid displaced, and $\rho_F V g = m_F g$ is the weight of fluid which takes up a volume equal to the volume of the cylinder. Thus the buoyant force on the cylinder is equal to the weight of fluid displaced by the cylinder.

FIGURE 13–12 A water barometer: a full tube of water is inserted into a tub of water, keeping the tube's spigot at the top closed. When the bottom end of the tube is unplugged, some water flows out of the tube into the tub, leaving a vacuum between the water's upper surface and the spigot. Why? Because air pressure can not support a column of water more than 10 m high.

FIGURE 13–13 Determination of the buoyant force.

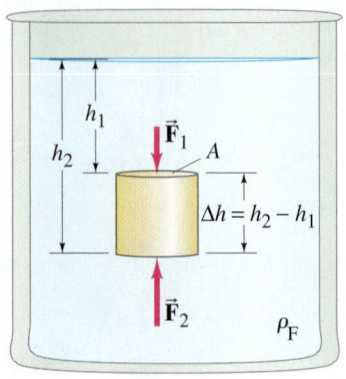

This result is valid no matter what the shape of the object. Its discovery is credited to Archimedes (287?–212 B.C.), and it is called **Archimedes' principle**: *the buoyant force on an object immersed in a fluid is equal to the weight of the fluid displaced by that object.*

By "fluid displaced," we mean a volume of fluid equal to the submerged volume of the object (or that part of the object that is submerged). If the object is placed in a glass or tub initially filled to the brim with water, the water that flows over the top represents the water displaced by the object.

We can derive Archimedes' principle in general by the following simple but elegant argument. The irregularly shaped object D shown in Fig. 13–14a is acted on by the force of gravity (its weight, $m\vec{g}$, downward) and the buoyant force, $\vec{F}_B$, upward. We wish to determine F_B. To do so, we next consider a body (D′ in Fig. 13–14b), this time made of the fluid itself, with the same shape and size as the original object, and located at the same depth. You might think of this body of fluid as being separated from the rest of the fluid by an imaginary membrane. The buoyant force F_B on this body of fluid will be exactly the same as that on the original object since the surrounding fluid, which exerts F_B, is in exactly the same configuration. This body of fluid D′ is in equilibrium (the fluid as a whole is at rest). Therefore, $F_B = m'g$, where $m'g$ is the weight of the body of fluid. Hence the buoyant force F_B is equal to the weight of the body of fluid whose volume equals the volume of the original submerged object, which is Archimedes' principle.

Archimedes' discovery was made by experiment. What we have done in the last two paragraphs is to show that Archimedes' principle can be derived from Newton's laws.

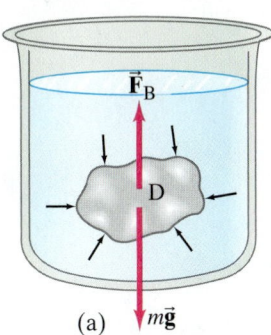

(a)

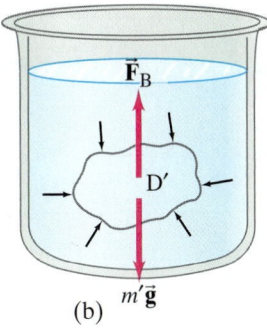

(b)

FIGURE 13–14
Archimedes' principle.

CONCEPTUAL EXAMPLE 13–8 | **Two pails of water.** Consider two identical pails of water filled to the brim. One pail contains only water, the other has a piece of wood floating in it. Which pail has the greater weight?

RESPONSE Both pails weigh the same. Recall Archimedes' principle: the wood displaces a volume of water with weight equal to the weight of the wood. Some water will overflow the pail, but Archimedes' principle tells us the spilled water has weight equal to the weight of the wood object; so the pails have the same weight.

EXAMPLE 13–9 **Recovering a submerged statue.** A 70-kg ancient statue lies at the bottom of the sea. Its volume is $3.0 \times 10^4 \, \text{cm}^3$. How much force is needed to lift it?

APPROACH The force F needed to lift the statue is equal to the statue's weight mg minus the buoyant force F_B. Figure 13–15 is the free-body diagram.

SOLUTION The buoyant force on the statue due to the water is equal to the weight of $3.0 \times 10^4 \, \text{cm}^3 = 3.0 \times 10^{-2} \, \text{m}^3$ of water (for seawater, $\rho = 1.025 \times 10^3 \, \text{kg/m}^3$):

$$F_B = m_{H_2O}g = \rho_{H_2O}Vg$$

$$= (1.025 \times 10^3 \, \text{kg/m}^3)(3.0 \times 10^{-2} \, \text{m}^3)(9.8 \, \text{m/s}^2)$$

$$= 3.0 \times 10^2 \, \text{N}.$$

The weight of the statue is $mg = (70 \, \text{kg})(9.8 \, \text{m/s}^2) = 6.9 \times 10^2 \, \text{N}$. Hence the force F needed to lift it is $690 \, \text{N} - 300 \, \text{N} = 390 \, \text{N}$. It is as if the statue had a mass of only $(390 \, \text{N})/(9.8 \, \text{m/s}^2) = 40 \, \text{kg}$.

NOTE Here $F = 390 \, \text{N}$ is the force needed to lift the statue without acceleration when it is under water. As the statue comes *out* of the water, the force F increases, reaching $690 \, \text{N}$ when the statue is fully out of the water.

FIGURE 13–15 Example 13–9. The force needed to lift the statue is $\vec{F}$.

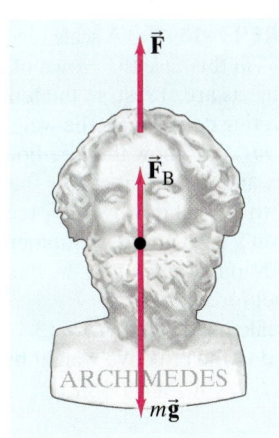

Archimedes is said to have discovered his principle in his bath while thinking how he might determine whether the king's new crown was pure gold or a fake. Gold has a specific gravity of 19.3, somewhat higher than that of most metals, but a determination of specific gravity or density is not readily done directly because, even if the mass is known, the volume of an irregularly shaped object is not easily calculated. However, if the object is weighed in air ($= w$) and also "weighed" while it is under water ($= w'$), the density can be determined using Archimedes' principle, as the following Example shows. The quantity w' is called the *apparent weight* in water, and is what a scale reads when the object is submerged in water (see Fig. 13–16); w' equals the true weight ($w = mg$) minus the buoyant force.

EXAMPLE 13–10 **Archimedes: Is the crown gold?** When a crown of mass 14.7 kg is submerged in water, an accurate scale reads only 13.4 kg. Is the crown made of gold?

APPROACH If the crown is gold, its density and specific gravity must be very high, SG = 19.3 (see Section 13–2 and Table 13–1). We determine the specific gravity using Archimedes' principle and the two free-body diagrams shown in Fig. 13–16.

SOLUTION The apparent weight of the submerged object (the crown) is w' (what the scale reads), and is the force pulling down on the scale hook. By Newton's third law, w' equals the force F'_T that the scale exerts on the crown in Fig. 13–16b. The sum of the forces on the crown is zero, so w' equals the actual weight w ($= mg$) minus the buoyant force F_B:

$$w' = F'_T = w - F_B$$

so

$$w - w' = F_B.$$

Let V be the volume of the completely submerged object and ρ_O its density (so $\rho_O V$ is its mass), and let ρ_F be the density of the fluid (water). Then $(\rho_F V)g$ is the weight of fluid displaced ($= F_B$). Now we can write

$$w = mg = \rho_O V g$$
$$w - w' = F_B = \rho_F V g.$$

We divide these two equations and obtain

$$\frac{w}{w - w'} = \frac{\rho_O V g}{\rho_F V g} = \frac{\rho_O}{\rho_F}.$$

We see that $w/(w - w')$ is equal to the specific gravity of the object if the fluid in which it is submerged is water ($\rho_F = 1.00 \times 10^3 \, \text{kg/m}^3$). Thus

$$\frac{\rho_O}{\rho_{H_2O}} = \frac{w}{w - w'} = \frac{(14.7 \, \text{kg})g}{(14.7 \, \text{kg} - 13.4 \, \text{kg})g} = \frac{14.7 \, \text{kg}}{1.3 \, \text{kg}} = 11.3.$$

This corresponds to a density of 11,300 kg/m³. The crown is not gold, but seems to be made of lead (see Table 13–1).

FIGURE 13–16 (a) A scale reads the mass of an object in air—in this case the crown of Example 13–10. All objects are at rest, so the tension F_T in the connecting cord equals the weight w of the object: $F_T = mg$. We show the free-body diagram of the crown, and F_T is what causes the scale reading (it is equal to the net downward force on the scale, by Newton's third law). (b) Submerged, the crown has an additional force on it, the buoyant force F_B. The net force is zero, so $F'_T + F_B = mg$ ($= w$). The scale now reads $m' = 13.4$ kg, where m' is related to the effective weight by $w' = m'g$. Thus $F'_T = w' = w - F_B$.

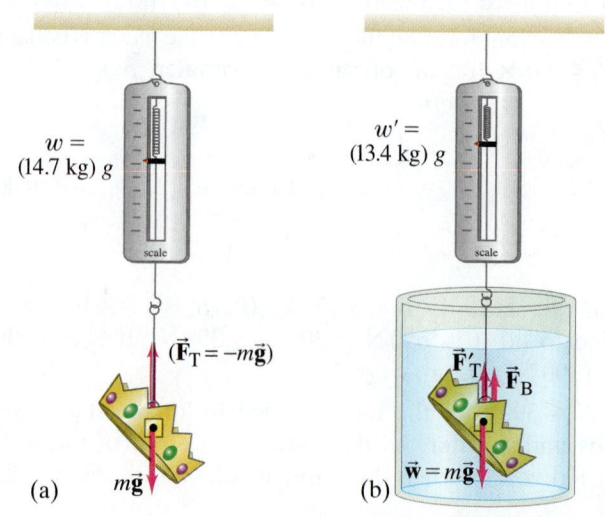

$w = (14.7 \, \text{kg}) \, g$

$w' = (13.4 \, \text{kg}) \, g$

$(\vec{F}_T = -m\vec{g})$

$\vec{F}'_T$ $\vec{F}_B$

(a) $m\vec{g}$

(b) $\vec{w} = m\vec{g}$

Archimedes' principle applies equally well to objects that float, such as wood. In general, *an object floats on a fluid if its density* (ρ_O) *is less than that of the fluid* (ρ_F). This is readily seen from Fig. 13–17a, where a submerged log will experience a net upward force and float to the surface if $F_B > mg$; that is, if $\rho_F V g > \rho_O V g$ or $\rho_F > \rho_O$. At equilibrium—that is, when floating—the buoyant force on an object has magnitude equal to the weight of the object. For example, a log whose specific gravity is 0.60 and whose volume is 2.0 m³ has a mass $m = \rho_O V = (0.60 \times 10^3 \text{ kg/m}^3)(2.0 \text{ m}^3) = 1200$ kg. If the log is fully submerged, it will displace a mass of water $m_F = \rho_F V = (1000 \text{ kg/m}^3)(2.0 \text{ m}^3) = 2000$ kg. Hence the buoyant force on the log will be greater than its weight, and it will float upward to the surface (Fig. 13–17). The log will come to equilibrium when it displaces 1200 kg of water, which means that 1.2 m³ of its volume will be submerged. This 1.2 m³ corresponds to 60% of the volume of the log (1.2/2.0 = 0.60), so 60% of the log is submerged.

In general when an object floats, we have $F_B = mg$, which we can write as (see Fig. 13–18)

$$F_B = mg$$
$$\rho_F V_{\text{displ}} g = \rho_O V_O g,$$

where V_O is the full volume of the object and V_{displ} is the volume of fluid it displaces (= volume submerged). Thus

$$\frac{V_{\text{displ}}}{V_O} = \frac{\rho_O}{\rho_F}.$$

That is, the fraction of the object submerged is given by the ratio of the object's density to that of the fluid. If the fluid is water, this fraction equals the specific gravity of the object.

EXAMPLE 13–11 **Hydrometer calibration.** A **hydrometer** is a simple instrument used to measure the specific gravity of a liquid by indicating how deeply the instrument sinks in the liquid. A particular hydrometer (Fig. 13–19) consists of a glass tube, weighted at the bottom, which is 25.0 cm long and 2.00 cm² in cross-sectional area, and has a mass of 45.0 g. How far from the end should the 1.000 mark be placed?

APPROACH The hydrometer will float in water if its density ρ is less than $\rho_w = 1.000 \text{ g/cm}^3$, the density of water. The fraction of the hydrometer submerged ($V_{\text{displaced}}/V_{\text{total}}$) is equal to the density ratio ρ/ρ_w.

SOLUTION The hydrometer has an overall density

$$\rho = \frac{m}{V} = \frac{45.0 \text{ g}}{(2.00 \text{ cm}^2)(25.0 \text{ cm})} = 0.900 \text{ g/cm}^3.$$

Thus, when placed in water, it will come to equilibrium when 0.900 of its volume is submerged. Since it is of uniform cross section, $(0.900)(25.0 \text{ cm}) = 22.5$ cm of its length will be submerged. The specific gravity of water is defined to be 1.000, so the mark should be placed 22.5 cm from the weighted end.

EXERCISE C On the hydrometer of Example 13–11, will the marks above the 1.000 mark represent higher or lower values of density of the liquid in which it is submerged?

Archimedes' principle is also useful in geology. According to the theories of plate tectonics and continental drift, the continents float on a fluid "sea" of slightly deformable rock (mantle rock). Some interesting calculations can be done using very simple models, which we consider in the Problems at the end of the Chapter.

Air is a fluid, and it too exerts a buoyant force. Ordinary objects weigh less in air than they do if weighed in a vacuum. Because the density of air is so small, the effect for ordinary solids is slight. There are objects, however, that *float* in air—helium-filled balloons, for example, because the density of helium is less than the density of air.

EXERCISE D Which of the following objects, submerged in water, experiences the largest magnitude of the buoyant force? (*a*) A 1-kg helium balloon; (*b*) 1 kg of wood; (*c*) 1 kg of ice; (*d*) 1 kg of iron; (*e*) all the same.

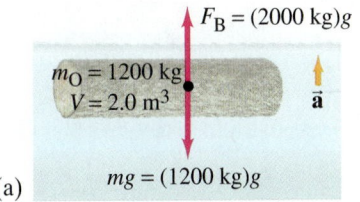

(a)

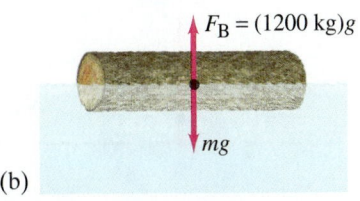

(b)

FIGURE 13–17 (a) The fully submerged log accelerates upward because $F_B > mg$. It comes to equilibrium (b) when $\Sigma F = 0$, so $F_B = mg = (1200 \text{ kg})g$. Thus 1200 kg, or 1.2 m³, of water is displaced.

FIGURE 13–18 An object floating in equilibrium: $F_B = mg$.

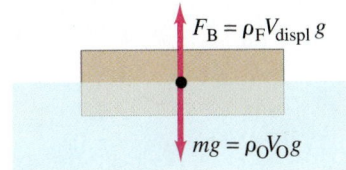

FIGURE 13–19 A hydrometer. Example 13–11.

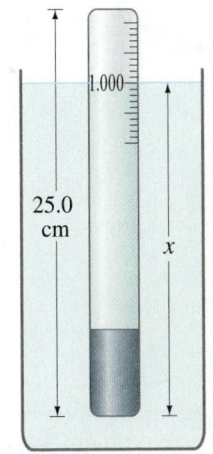

PHYSICS APPLIED
Continental drift—plate tectonics

EXAMPLE 13–12 **Helium balloon.** What volume V of helium is needed if a balloon is to lift a load of 180 kg (including the weight of the empty balloon)?

APPROACH The buoyant force on the helium balloon, F_B, which is equal to the weight of displaced air, must be at least equal to the weight of the helium plus the weight of the balloon and load (Fig. 13–20). Table 13–1 gives the density of helium as 0.179 kg/m^3.

SOLUTION The buoyant force must have a minimum value of

$$F_B = (m_{He} + 180 \text{ kg})g.$$

This equation can be written in terms of density using Archimedes' principle:

$$\rho_{air} V g = (\rho_{He} V + 180 \text{ kg})g.$$

Solving now for V, we find

$$V = \frac{180 \text{ kg}}{\rho_{air} - \rho_{He}} = \frac{180 \text{ kg}}{(1.29 \text{ kg/m}^3 - 0.179 \text{ kg/m}^3)} = 160 \text{ m}^3.$$

NOTE This is the minimum volume needed near the Earth's surface, where $\rho_{air} = 1.29 \text{ kg/m}^3$. To reach a high altitude, a greater volume would be needed since the density of air decreases with altitude.

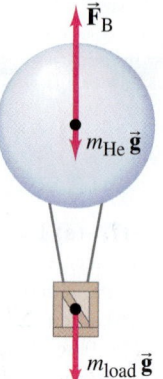

FIGURE 13–20 Example 13–12.

13–8 Fluids in Motion; Flow Rate and the Equation of Continuity

We now turn to the subject of fluids in motion, which is called **fluid dynamics**, or (especially if the fluid is water) **hydrodynamics**.

We can distinguish two main types of fluid flow. If the flow is smooth, such that neighboring layers of the fluid slide by each other smoothly, the flow is said to be **streamline** or **laminar flow**.[†] In streamline flow, each particle of the fluid follows a smooth path, called a **streamline**, and these paths do not cross one another (Fig. 13–21a). Above a certain speed, the flow becomes turbulent. **Turbulent flow** is characterized by erratic, small, whirlpool-like circles called *eddy currents* or *eddies* (Fig. 13–21b). Eddies absorb a great deal of energy, and although a certain amount of internal friction called **viscosity** is present even during streamline flow, it is much greater when the flow is turbulent. A few tiny drops of ink or food coloring dropped into a moving liquid can quickly reveal whether the flow is streamline or turbulent.

[†]The word *laminar* means "in layers."

FIGURE 13–21 (a) Streamline, or laminar, flow; (b) turbulent flow. The photos show airflow around an airfoil or airplane wing (more in Section 13–10).

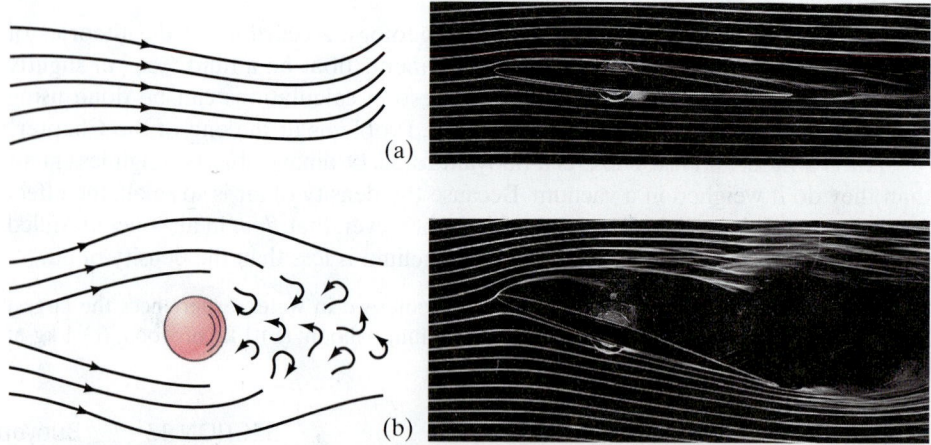

(a)

(b)

Let us consider the steady laminar flow of a fluid through an enclosed tube or pipe as shown in Fig. 13–22. First we determine how the speed of the fluid changes when the size of the tube changes. The mass **flow rate** is defined as the mass Δm of fluid that passes a given point per unit time Δt:

$$\text{mass flow rate} = \frac{\Delta m}{\Delta t}.$$

In Fig. 13–22, the volume of fluid passing point 1 (that is, through area A_1) in a time Δt is $A_1 \Delta \ell_1$, where $\Delta \ell_1$ is the distance the fluid moves in time Δt. Since the velocity† of fluid passing point 1 is $v_1 = \Delta \ell_1 / \Delta t$, the mass flow rate through area A_1 is

$$\frac{\Delta m_1}{\Delta t} = \frac{\rho_1 \Delta V_1}{\Delta t} = \frac{\rho_1 A_1 \Delta \ell_1}{\Delta t} = \rho_1 A_1 v_1,$$

where $\Delta V_1 = A_1 \Delta \ell_1$ is the volume of mass Δm_1, and ρ_1 is the fluid density. Similarly, at point 2 (through area A_2), the flow rate is $\rho_2 A_2 v_2$. Since no fluid flows in or out the sides, the flow rates through A_1 and A_2 must be equal. Thus, since

$$\frac{\Delta m_1}{\Delta t} = \frac{\Delta m_2}{\Delta t},$$

then

$$\rho_1 A_1 v_1 = \rho_2 A_2 v_2. \tag{13–7a}$$

This is called the **equation of continuity**.

If the fluid is incompressible (ρ doesn't change with pressure), which is an excellent approximation for liquids under most circumstances (and sometimes for gases as well), then $\rho_1 = \rho_2$, and the equation of continuity becomes

$$A_1 v_1 = A_2 v_2. \qquad [\rho = \text{constant}] \tag{13–7b}$$

The product Av represents the *volume rate of flow* (volume of fluid passing a given point per second), since $\Delta V / \Delta t = A \Delta \ell / \Delta t = Av$, which in SI units is m^3/s. Equation 13–7b tells us that where the cross-sectional area is large, the velocity is small, and where the area is small, the velocity is large. That this is reasonable can be seen by looking at a river. A river flows slowly through a meadow where it is broad, but speeds up to torrential speed when passing through a narrow gorge.

EXAMPLE 13–13 **ESTIMATE** **Blood flow.** In humans, blood flows from the heart into the aorta, from which it passes into the major arteries. These branch into the small arteries (arterioles), which in turn branch into myriads of tiny capillaries, Fig. 13–23. The blood returns to the heart via the veins. The radius of the aorta is about 1.2 cm, and the blood passing through it has a speed of about 40 cm/s. A typical capillary has a radius of about 4×10^{-4} cm, and blood flows through it at a speed of about 5×10^{-4} m/s. Estimate the number of capillaries that are in the body.

APPROACH We assume the density of blood doesn't vary significantly from the aorta to the capillaries. By the equation of continuity, the volume flow rate in the aorta must equal the volume flow rate through *all* the capillaries. The total area of all the capillaries is given by the area of one capillary multiplied by the total number N of capillaries.

SOLUTION Let A_1 be the area of the aorta and A_2 be the area of *all* the capillaries through which blood flows. Then $A_2 = N \pi r_{cap}^2$, where $r_{cap} \approx 4 \times 10^{-4}$ cm is the estimated average radius of one capillary. From the equation of continuity (Eq. 13–7b), we have

$$v_2 A_2 = v_1 A_1$$
$$v_2 N \pi r_{cap}^2 = v_1 \pi r_{aorta}^2$$

so

$$N = \frac{v_1}{v_2} \frac{r_{aorta}^2}{r_{cap}^2} = \left(\frac{0.40 \text{ m/s}}{5 \times 10^{-4} \text{ m/s}}\right)\left(\frac{1.2 \times 10^{-2} \text{ m}}{4 \times 10^{-6} \text{ m}}\right)^2 \approx 7 \times 10^9,$$

or on the order of 10 billion capillaries.

---

†If there were no viscosity, the velocity would be the same across a cross section of the tube. Real fluids have viscosity, and this internal friction causes different layers of the fluid to flow at different speeds. In this case v_1 and v_2 represent the average speeds at each cross section.

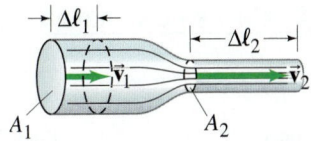

FIGURE 13–22 Fluid flow through a pipe of varying diameter.

FIGURE 13–23
Human circulatory system.

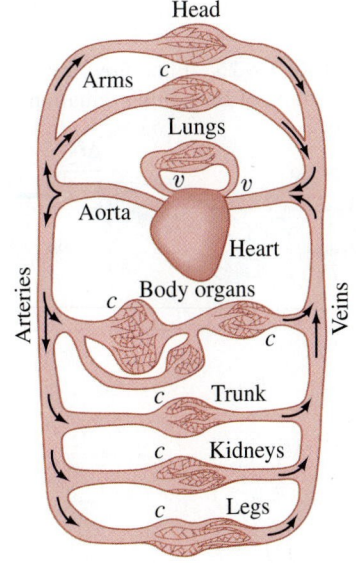

v = valves
c = capillaries

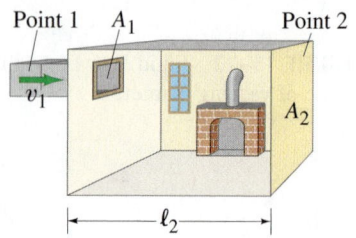

Point 1 A_1 Point 2

v_1

A_2

ℓ_2

FIGURE 13–24 Example 13–14.

EXAMPLE 13–14 **Heating duct to a room.** What area must a heating duct have if air moving 3.0 m/s along it can replenish the air every 15 minutes in a room of volume 300 m³? Assume the air's density remains constant.

APPROACH We apply the equation of continuity at constant density, Eq. 13–7b, to the air that flows through the duct (point 1 in Fig. 13–24) and then into the room (point 2). The volume flow rate in the room equals the volume of the room divided by the 15-min replenishing time.

SOLUTION Consider the room as a large section of the duct, Fig. 13–24, and think of air equal to the volume of the room as passing by point 2 in $t = 15$ min $= 900$ s. Reasoning in the same way we did to obtain Eq. 13–7a (changing Δt to t), we write $v_2 = \ell_2/t$ so $A_2 v_2 = A_2 \ell_2/t = V_2/t$, where V_2 is the volume of the room. Then the equation of continuity becomes $A_1 v_1 = A_2 v_2 = V_2/t$ and

$$A_1 = \frac{V_2}{v_1 t} = \frac{300 \text{ m}^3}{(3.0 \text{ m/s})(900 \text{ s})} = 0.11 \text{ m}^2.$$

If the duct is square, then each side has length $\ell = \sqrt{A} = 0.33$ m, or 33 cm. A rectangular duct 20 cm × 55 cm will also do.

13–9 Bernoulli's Equation

Have you ever wondered why an airplane can fly, or how a sailboat can move against the wind? These are examples of a principle worked out by Daniel Bernoulli (1700–1782) concerning fluids in motion. In essence, **Bernoulli's principle** states that *where the velocity of a fluid is high, the pressure is low, and where the velocity is low, the pressure is high.* For example, if the pressures in the fluid at points 1 and 2 of Fig. 13–22 are measured, it will be found that the pressure is lower at point 2, where the velocity is greater, than it is at point 1, where the velocity is smaller. At first glance, this might seem strange; you might expect that the greater speed at point 2 would imply a higher pressure. But this cannot be the case. For if the pressure in the fluid at point 2 were higher than at point 1, this higher pressure would slow the fluid down, whereas in fact it has sped up in going from point 1 to point 2. Thus the pressure at point 2 must be less than at point 1, to be consistent with the fact that the fluid accelerates.

To help clarify any misconceptions, a faster fluid *would* exert a greater force on an obstacle placed in its path. But that is not what we mean by the pressure in a fluid, and besides we are not considering obstacles that interrupt the flow. We are examining smooth streamline flow. The fluid pressure is exerted on the walls of a tube or pipe, or on the surface of any material the fluid passes over.

Bernoulli developed an equation that expresses this principle quantitatively. To derive Bernoulli's equation, we assume the flow is steady and laminar, the fluid is incompressible, and the viscosity is small enough to be ignored. To be general, we assume the fluid is flowing in a tube of nonuniform cross section that varies in height above some reference level, Fig. 13–25. We will consider the volume of fluid shown in color and calculate the work done to move it from the position shown in Fig. 13–25a to that shown in Fig. 13–25b. In this process, fluid entering area A_1 flows a distance $\Delta \ell_1$ and forces the fluid at area A_2 to move a distance $\Delta \ell_2$. The fluid to the left of area A_1 exerts a pressure P_1 on our section of fluid and does an amount of work

$$W_1 = F_1 \Delta \ell_1 = P_1 A_1 \Delta \ell_1.$$

At area A_2, the work done on our cross section of fluid is

$$W_2 = -P_2 A_2 \Delta \ell_2.$$

The negative sign is present because the force exerted on the fluid is opposite to the motion (thus the fluid shown in color does work on the fluid to the right of point 2). Work is also done on the fluid by the force of gravity. The net effect of the process shown in Fig. 13–25 is to move a mass m of volume $A_1 \Delta \ell_1$ (= $A_2 \Delta \ell_2$, since the

FIGURE 13–25 Fluid flow: for derivation of Bernoulli's equation.

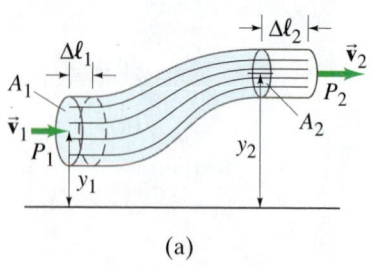

$\Delta \ell_1$

A_1

$\vec{v}_1$

P_1

y_1

$\Delta \ell_2$

$\vec{v}_2$

P_2

A_2

y_2

(a)

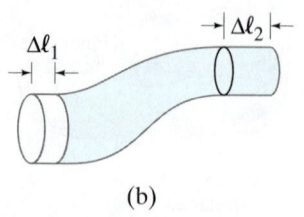

$\Delta \ell_1$

$\Delta \ell_2$

(b)

fluid is incompressible) from point 1 to point 2, so the work done by gravity is

$$W_3 = -mg(y_2 - y_1),$$

where y_1 and y_2 are heights of the center of the tube above some (arbitrary) reference level. In the case shown in Fig. 13–25, this term is negative since the motion is uphill against the force of gravity. The net work W done on the fluid is thus

$$W = W_1 + W_2 + W_3$$
$$W = P_1 A_1 \Delta\ell_1 - P_2 A_2 \Delta\ell_2 - mgy_2 + mgy_1.$$

According to the work-energy principle (Section 7–4), the net work done on a system is equal to its change in kinetic energy. Hence

$$\tfrac{1}{2}mv_2^2 - \tfrac{1}{2}mv_1^2 = P_1 A_1 \Delta\ell_1 - P_2 A_2 \Delta\ell_2 - mgy_2 + mgy_1.$$

The mass m has volume $A_1 \Delta\ell_1 = A_2 \Delta\ell_2$ for an incompressible fluid. Thus we can substitute $m = \rho A_1 \Delta\ell_1 = \rho A_2 \Delta\ell_2$, and then divide through by $A_1 \Delta\ell_1 = A_2 \Delta\ell_2$, to obtain

$$\tfrac{1}{2}\rho v_2^2 - \tfrac{1}{2}\rho v_1^2 = P_1 - P_2 - \rho g y_2 + \rho g y_1,$$

which we rearrange to get

$$P_1 + \tfrac{1}{2}\rho v_1^2 + \rho g y_1 = P_2 + \tfrac{1}{2}\rho v_2^2 + \rho g y_2. \qquad \textbf{(13–8)} \qquad \textit{Bernoulli's equation}$$

This is **Bernoulli's equation**. Since points 1 and 2 can be any two points along a tube of flow, Bernoulli's equation can be written as

$$P + \tfrac{1}{2}\rho v^2 + \rho g y = \text{constant}$$

at every point in the fluid, where y is the height of the center of the tube above a fixed reference level. [Note that if there is no flow $(v_1 = v_2 = 0)$, then Eq. 13–8 reduces to the hydrostatic equation, Eq. 13–6a: $P_2 - P_1 = -\rho g(y_2 - y_1)$.]

Bernoulli's equation is an expression of the law of energy conservation, since we derived it from the work-energy principle.

EXERCISE F As water in a level pipe passes from a narrow cross section of pipe to a wider cross section, how does the pressure against the walls change?

EXAMPLE 13–15 **Flow and pressure in a hot-water heating system.** Water circulates throughout a house in a hot-water heating system. If the water is pumped at a speed of 0.50 m/s through a 4.0-cm-diameter pipe in the basement under a pressure of 3.0 atm, what will be the flow speed and pressure in a 2.6-cm-diameter pipe on the second floor 5.0 m above? Assume the pipes do not divide into branches.

APPROACH We use the equation of continuity at constant density to determine the flow speed on the second floor, and then Bernoulli's equation to find the pressure.

SOLUTION We take v_2 in the equation of continuity, Eq. 13–7, as the flow speed on the second floor, and v_1 as the flow speed in the basement. Noting that the areas are proportional to the radii squared $(A = \pi r^2)$, we obtain

$$v_2 = \frac{v_1 A_1}{A_2} = \frac{v_1 \pi r_1^2}{\pi r_2^2} = (0.50 \text{ m/s}) \frac{(0.020 \text{ m})^2}{(0.013 \text{ m})^2} = 1.2 \text{ m/s}.$$

To find the pressure on the second floor, we use Bernoulli's equation (Eq. 13–8):

$$P_2 = P_1 + \rho g(y_1 - y_2) + \tfrac{1}{2}\rho(v_1^2 - v_2^2)$$
$$= (3.0 \times 10^5 \text{ N/m}^2) + (1.0 \times 10^3 \text{ kg/m}^3)(9.8 \text{ m/s}^2)(-5.0 \text{ m})$$
$$+ \tfrac{1}{2}(1.0 \times 10^3 \text{ kg/m}^3)[(0.50 \text{ m/s})^2 - (1.2 \text{ m/s})^2]$$
$$= (3.0 \times 10^5 \text{ N/m}^2) - (4.9 \times 10^4 \text{ N/m}^2) - (6.0 \times 10^2 \text{ N/m}^2)$$
$$= 2.5 \times 10^5 \text{ N/m}^2 = 2.5 \text{ atm}.$$

NOTE The velocity term contributes very little in this case.

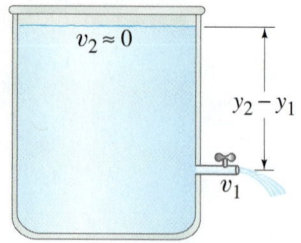

FIGURE 13–26 Torricelli's theorem: $v_1 = \sqrt{2g(y_2 - y_1)}$.

13–10 Applications of Bernoulli's Principle: Torricelli, Airplanes, Baseballs, TIA

Bernoulli's equation can be applied to many situations. One example is to calculate the velocity, v_1, of a liquid flowing out of a spigot at the bottom of a reservoir, Fig. 13–26. We choose point 2 in Eq. 13–8 to be the top surface of the liquid. Assuming the diameter of the reservoir is large compared to that of the spigot, v_2 will be almost zero. Points 1 (the spigot) and 2 (top surface) are open to the atmosphere, so the pressure at both points is equal to atmospheric pressure: $P_1 = P_2$. Then Bernoulli's equation becomes

$$\tfrac{1}{2}\rho v_1^2 + \rho g y_1 = \rho g y_2$$

or

$$v_1 = \sqrt{2g(y_2 - y_1)}. \tag{13–9}$$

This result is called **Torricelli's theorem**. Although it is seen to be a special case of Bernoulli's equation, it was discovered a century earlier by Evangelista Torricelli. Equation 13–9 tells us that the liquid leaves the spigot with the same speed that a freely falling object would attain if falling from the same height. This should not be too surprising since Bernoulli's equation relies on the conservation of energy.

Another special case of Bernoulli's equation arises when a fluid is flowing horizontally with no significant change in height; that is, $y_1 = y_2$. Then Eq. 13–8 becomes

$$P_1 + \tfrac{1}{2}\rho v_1^2 = P_2 + \tfrac{1}{2}\rho v_2^2, \tag{13–10}$$

which tells us quantitatively that the speed is high where the pressure is low, and vice versa. It explains many common phenomena, some of which are illustrated in Figs. 13–27 to 13–32. The pressure in the air blown at high speed across the top of the vertical tube of a perfume atomizer (Fig. 13–27a) is less than the normal air pressure acting on the surface of the liquid in the bowl. Thus atmospheric pressure in the bowl pushes the perfume up the tube because of the lower pressure at the top. A Ping-Pong ball can be made to float above a blowing jet of air (some vacuum cleaners can blow air), Fig. 13–27b; if the ball begins to leave the jet of air, the higher pressure in the still air outside the jet pushes the ball back in.

FIGURE 13–27 Examples of Bernoulli's principle: (a) atomizer, (b) Ping-Pong ball in jet of air.

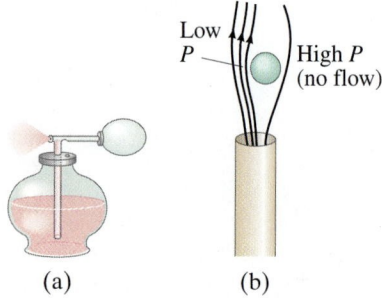

Low P — High P (no flow)

(a) (b)

EXERCISE G Return to Chapter-Opening Question 2, page 339, and answer it again now. Try to explain why you may have answered differently the first time. Try it and see.

Airplane Wings and Dynamic Lift

Airplanes experience a "lift" force on their wings, keeping them up in the air, if they are moving at a sufficiently high speed relative to the air and the wing is tilted upward at a small angle (the "attack angle"), as in Fig. 13–28, where streamlines of air are shown rushing by the wing. (We are in the reference frame of the wing, as if sitting on the wing.) The upward tilt, as well as the rounded upper surface of the wing, causes the streamlines to be forced upward and to be crowded together above the wing. The area for air flow between any two streamlines is reduced as the streamlines get closer together, so from the equation of continuity $(A_1 v_1 = A_2 v_2)$, the air speed increases above the wing where the streamlines are squished together. (Recall also how the crowded streamlines in a pipe constriction, Fig. 13–22, indicate the velocity is higher in the constriction.) Because the air speed is greater above the wing than below it, the pressure above the wing is less than the pressure below the wing (Bernoulli's principle). Hence there is a net upward force on the wing called **dynamic lift**. Experiments show that the speed of air above the wing can even be double the speed of the air below it. (Friction between the air and wing exerts a *drag force*, toward the rear, which must be overcome by the plane's engines.)

A flat wing, or one with symmetric cross section, will experience lift as long as the front of the wing is tilted upward (attack angle). The wing shown in Fig. 13–28 can experience lift even if the attack angle is zero, because the rounded upper surface deflects air up, squeezing the streamlines together. Airplanes can fly upside down, experiencing lift, if the attack angle is sufficient to deflect streamlines up and closer together.

FIGURE 13–28 Lift on an airplane wing. We are in the reference frame of the wing, seeing the air flow by.

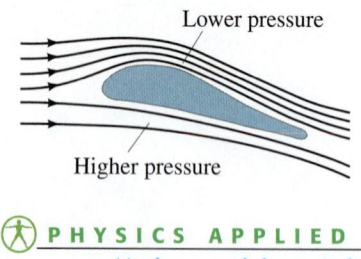

Lower pressure

Higher pressure

⊛ PHYSICS APPLIED
Airplanes and dynamic lift

Our picture considers streamlines; but if the attack angle is larger than about 15°, turbulence sets in (Fig. 13–21b) leading to greater drag and less lift, causing the wing to "stall" and the plane to drop.

From another point of view, the upward tilt of a wing means the air moving horizontally in front of the wing is deflected downward; the change in momentum of the rebounding air molecules results in an upward force on the wing (Newton's third law).

Sailboats

A sailboat can move *against* the wind, with the aid of the Bernoulli effect, by setting the sails at an angle, as shown in Fig. 13–29. The air travels rapidly over the bulging front surface of the sail, and the relatively still air filling the sail exerts a greater pressure behind the sail, resulting in a net force on the sail, $\vec{\mathbf{F}}_{wind}$. This force would tend to make the boat move sideways if it weren't for the keel that extends vertically downward beneath the water: the water exerts a force $(\vec{\mathbf{F}}_{water})$ on the keel nearly perpendicular to the keel. The resultant of these two forces $(\vec{\mathbf{F}}_R)$ is almost directly forward as shown.

Baseball Curve

Why a spinning pitched baseball (or tennis ball) curves can also be explained using Bernoulli's principle. It is simplest if we put ourselves in the reference frame of the ball, with the air rushing by, just as we did for the airplane wing. Suppose the ball is rotating counterclockwise as seen from above, Fig. 13–30. A thin layer of air ("boundary layer") is being dragged around by the ball. We are looking down on the ball, and at point A in Fig. 13–30, this boundary layer tends to slow down the oncoming air. At point B, the air rotating with the ball adds its speed to that of the oncoming air, so the air speed is higher at B than at A. The higher speed at B means the pressure is lower at B than at A, resulting in a net force toward B. The ball's path curves toward the left (as seen by the pitcher).

Lack of Blood to the Brain—TIA

In medicine, one of many applications of Bernoulli's principle is to explain a TIA, a transient ischemic attack (meaning a temporary lack of blood supply to the brain). A person suffering a TIA may experience symptoms such as dizziness, double vision, headache, and weakness of the limbs. A TIA can occur as follows. Blood normally flows up to the brain at the back of the head via the two vertebral arteries—one going up each side of the neck—which meet to form the basilar artery just below the brain, as shown in Fig. 13–31. The vertebral arteries issue from the subclavian arteries, as shown, before the latter pass to the arms. When an arm is exercised vigorously, blood flow increases to meet the needs of the arm's muscles. If the subclavian artery on one side of the body is partially blocked, however, as in arteriosclerosis (hardening of the arteries), the blood velocity will have to be higher on that side to supply the needed blood. (Recall the equation of continuity: smaller area means larger velocity for the same flow rate, Eqs. 13–7.) The increased blood velocity past the opening to the vertebral artery results in lower pressure (Bernoulli's principle). Thus blood rising in the vertebral artery on the "good" side at normal pressure can be *diverted down* into the other vertebral artery because of the low pressure on that side, instead of passing upward to the brain. Hence the blood supply to the brain is reduced.

Other Applications

A **venturi tube** is essentially a pipe with a narrow constriction (the throat). The flowing fluid speeds up as it passes through this constriction, so the pressure is lower in the throat. A *venturi meter*, Fig. 13–32, is used to measure the flow speed of gases and liquids, including blood velocity in arteries.

Why does smoke go up a chimney? It's partly because hot air rises (it's less dense and therefore buoyant). But Bernoulli's principle also plays a role. When wind blows across the top of a chimney, the pressure is less there than inside the house. Hence, air and smoke are pushed up the chimney by the higher indoor pressure. Even on an apparently still night there is usually enough ambient air flow at the top of a chimney to assist upward flow of smoke.

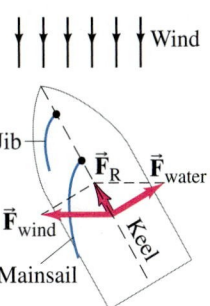

FIGURE 13–29 Sailboat sailing against the wind.

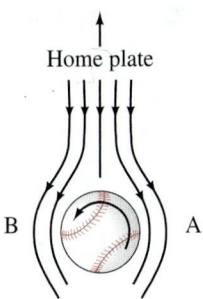

FIGURE 13–30 Looking down on a pitched baseball heading toward home plate. We are in the reference frame of the baseball, with the air flowing by.

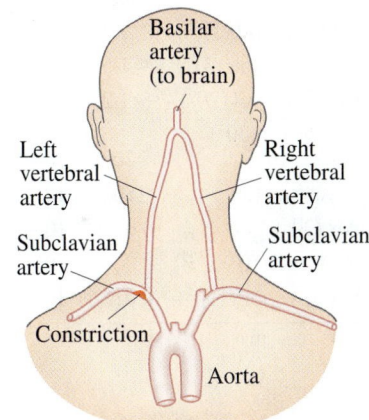

FIGURE 13–31 Rear of the head and shoulders showing arteries leading to the brain and to the arms. High blood velocity past the constriction in the left subclavian artery causes low pressure in the left vertebral artery, in which a reverse (downward) blood flow can then occur, resulting in a TIA, a loss of blood to the brain.

FIGURE 13–32 Venturi meter.

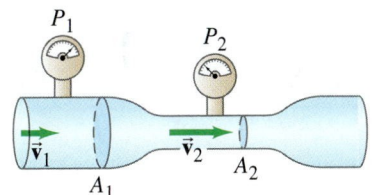

Bernoulli's equation ignores the effects of friction (viscosity) and the compressibility of the fluid. The energy that is transformed to internal (or potential) energy due to compression and to thermal energy by friction can be taken into account by adding terms to Eq. 13–8. These terms are difficult to calculate theoretically and are normally determined empirically. They do not significantly alter the explanations for the phenomena described above.

*13–11 Viscosity

Real fluids have a certain amount of internal friction called **viscosity**, as mentioned in Section 13–8. Viscosity exists in both liquids and gases, and is essentially a frictional force between adjacent layers of fluid as the layers move past one another. In liquids, viscosity is due to the electrical cohesive forces between the molecules. In gases, it arises from collisions between the molecules.

The viscosity of different fluids can be expressed quantitatively by a *coefficient of viscosity*, η (the Greek lowercase letter eta), which is defined in the following way. A thin layer of fluid is placed between two flat plates. One plate is stationary and the other is made to move, Fig. 13–33. The fluid directly in contact with each plate is held to the surface by the adhesive force between the molecules of the liquid and those of the plate. Thus the upper surface of the fluid moves with the same speed v as the upper plate, whereas the fluid in contact with the stationary plate remains stationary. The stationary layer of fluid retards the flow of the layer just above it, which in turn retards the flow of the next layer, and so on. Thus the velocity varies continuously from 0 to v, as shown. The increase in velocity divided by the distance over which this change is made—equal to v/ℓ—is called the *velocity gradient*. To move the upper plate requires a force, which you can verify by moving a flat plate across a puddle of syrup on a table. For a given fluid, it is found that the force required, F, is proportional to the area of fluid in contact with each plate, A, and to the speed, v, and is inversely proportional to the separation, ℓ, of the plates: $F \propto vA/\ell$. For different fluids, the more viscous the fluid, the greater is the required force. Hence the proportionality constant for this equation is defined as the coefficient of viscosity, η:

$$F = \eta A \frac{v}{\ell}. \tag{13–11}$$

Solving for η, we find $\eta = F\ell/vA$. The SI unit for η is $N \cdot s/m^2 = Pa \cdot s$ (pascal·second). In the cgs system, the unit is dyne·s/cm², which is called a *poise* (P). Viscosities are often given in centipoise ($1\,cP = 10^{-2}\,P$). Table 13–3 lists the coefficient of viscosity for various fluids. The temperature is also specified, since it has a strong effect; the viscosity of liquids such as motor oil, for example, decreases rapidly as temperature increases.[†]

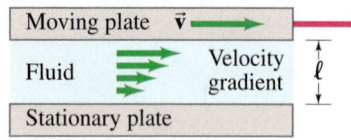

FIGURE 13–33
Determination of viscosity.

TABLE 13–3
Coefficients of Viscosity

Fluid (temperature in °C)	Coefficient of Viscosity, η (Pa·s)[†]
Water (0°)	1.8×10^{-3}
(20°)	1.0×10^{-3}
(100°)	0.3×10^{-3}
Whole blood (37°)	$\approx 4 \times 10^{-3}$
Blood plasma (37°)	$\approx 1.5 \times 10^{-3}$
Ethyl alcohol (20°)	1.2×10^{-3}
Engine oil (30°) (SAE 10)	200×10^{-3}
Glycerine (20°)	1500×10^{-3}
Air (20°)	0.018×10^{-3}
Hydrogen (0°)	0.009×10^{-3}
Water vapor (100°)	0.013×10^{-3}

[†] $1\,Pa \cdot s = 10\,P = 1000\,cP$.

*13–12 Flow in Tubes: Poiseuille's Equation, Blood Flow

If a fluid had no viscosity, it could flow through a level tube or pipe without a force being applied. Viscosity acts like a sort of friction (between fluid layers moving at slightly different speeds), so a pressure difference between the ends of a level tube is necessary for the steady flow of any real fluid, be it water or oil in a pipe, or blood in the circulatory system of a human.

The French scientist J. L. Poiseuille (1799–1869), who was interested in the physics of blood circulation (and after whom the poise is named), determined how the variables affect the flow rate of an incompressible fluid undergoing laminar flow in a cylindrical tube. His result, known as *Poiseuille's equation*, is:

$$Q = \frac{\pi R^4 (P_1 - P_2)}{8\eta\ell}, \tag{13–12}$$

where R is the inside radius of the tube, ℓ is the tube length, $P_1 - P_2$ is the pressure

[†]The Society of Automotive Engineers assigns numbers to represent the viscosity of oils: 30 weight (SAE 30) is more viscous than 10 weight. Multigrade oils, such as 20–50, are designed to maintain viscosity as temperature increases; 20–50 means the oil is 20 wt when cool but is like a 50-wt pure oil when it is hot (engine running temperature).

difference between the ends, η is the coefficient of viscosity, and Q is the volume rate of flow (volume of fluid flowing past a given point per unit time which in SI has units of m^3/s). Equation 13–12 applies only to laminar flow.

Poiseuille's equation tells us that the flow rate Q is directly proportional to the "pressure gradient," $(P_1 - P_2)/\ell$, and it is inversely proportional to the viscosity of the fluid. This is just what we might expect. It may be surprising, however, that Q also depends on the *fourth* power of the tube's radius. This means that for the same pressure gradient, if the tube radius is halved, the flow rate is decreased by a factor of 16! Thus the rate of flow, or alternately the pressure required to maintain a given flow rate, is greatly affected by only a small change in tube radius.

An interesting example of this R^4 dependence is *blood flow* in the human body. Poiseuille's equation is valid only for the streamline flow of an incompressible fluid. So it cannot be precisely accurate for blood whose flow is not without turbulence and that contains blood cells (whose diameter is almost equal to that of a capillary). Nonetheless, Poiseuille's equation does give a reasonable first approximation. Because the radius of arteries is reduced as a result of arteriosclerosis (thickening and hardening of artery walls) and by cholesterol buildup, the pressure gradient must be increased to maintain the same flow rate. If the radius is reduced by half, the heart would have to increase the pressure by a factor of about $2^4 = 16$ in order to maintain the same blood-flow rate. The heart must work much harder under these conditions, but usually cannot maintain the original flow rate. Thus, high blood pressure is an indication both that the heart is working harder and that the blood-flow rate is reduced.

*13–13 Surface Tension and Capillarity

The *surface* of a liquid at rest behaves in an interesting way, almost as if it were a stretched membrane under tension. For example, a drop of water on the end of a dripping faucet, or hanging from a thin branch in the early morning dew (Fig. 13–34), forms into a nearly spherical shape as if it were a tiny balloon filled with water. A steel needle can be made to float on the surface of water even though it is denser than the water. The surface of a liquid acts like it is under tension, and this tension, acting along the surface, arises from the attractive forces between the molecules. This effect is called **surface tension**. More specifically, a quantity called the *surface tension*, γ (the Greek letter gamma), is defined as the force F per unit length ℓ that acts perpendicular to any line or cut in a liquid surface, tending to pull the surface closed:

$$\gamma = \frac{F}{\ell}. \qquad\qquad (13-13)$$

To understand this, consider the U-shaped apparatus shown in Fig. 13–35 which encloses a thin film of liquid. Because of surface tension, a force F is required to pull the movable wire and thus increase the surface area of the liquid. The liquid contained by the wire apparatus is a thin film having both a top and a bottom surface. Hence the total length of the surface being increased is 2ℓ, and the surface tension is $\gamma = F/2\ell$. A delicate apparatus of this type can be used to measure the surface tension of various liquids. The surface tension of water is $0.072\,N/m$ at 20°C. Table 13–4 gives the values for several substances. Note that temperature has a considerable effect on the surface tension.

FIGURE 13–34 Spherical water droplets, dew on a blade of grass.

TABLE 13–4
Surface Tension of Some Substances

Substance (temperature in °C)	Surface Tension (N/m)
Mercury (20°)	0.44
Blood, whole (37°)	0.058
Blood, plasma (37°)	0.073
Alcohol, ethyl (20°)	0.023
Water (0°)	0.076
(20°)	0.072
(100°)	0.059
Benzene (20°)	0.029
Soap solution (20°)	≈0.025
Oxygen (−193°)	0.016

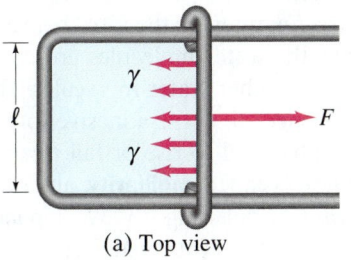

(a) Top view

FIGURE 13–35 U-shaped wire apparatus holding a film of liquid to measure surface tension ($\gamma = F/2\ell$).

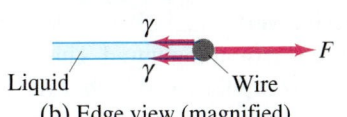

Liquid   Wire
(b) Edge view (magnified)

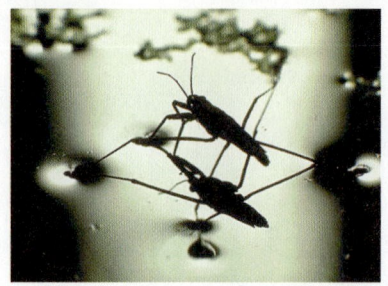

FIGURE 13–36 A water strider.

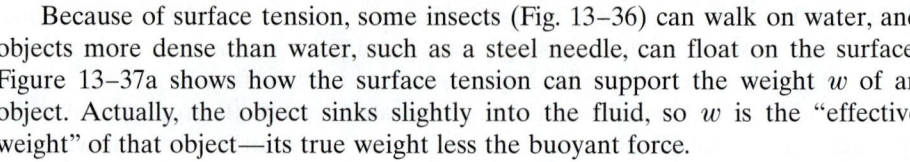

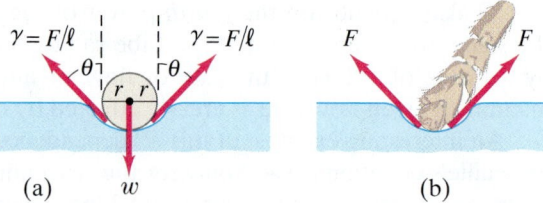

FIGURE 13–37 Surface tension acting on (a) a sphere, and (b) an insect leg. Example 13–16.

(a) w (b)

Because of surface tension, some insects (Fig. 13–36) can walk on water, and objects more dense than water, such as a steel needle, can float on the surface. Figure 13–37a shows how the surface tension can support the weight w of an object. Actually, the object sinks slightly into the fluid, so w is the "effective weight" of that object—its true weight less the buoyant force.

EXAMPLE 13–16 **ESTIMATE** **Insect walks on water.** The base of an insect's leg is approximately spherical in shape, with a radius of about 2.0×10^{-5} m. The 0.0030-g mass of the insect is supported equally by its six legs. Estimate the angle θ (see Fig. 13–37) for an insect on the surface of water. Assume the water temperature is 20°C.

APPROACH Since the insect is in equilibrium, the upward surface tension force is equal to the pull of gravity downward on each leg. We ignore the buoyant force for this estimate.

SOLUTION For each leg, we assume the surface tension force acts all around a circle of radius r, at an angle θ, as shown in Fig. 13–37a. Only the vertical component, $\gamma \cos \theta$, acts to balance the weight mg. So we set the length ℓ in Eq. 13–13 equal to the circumference of the circle, $\ell \approx 2\pi r$. Then the net upward force due to surface tension is $F_y \approx (\gamma \cos \theta)\ell \approx 2\pi r \gamma \cos \theta$. We set this surface tension force equal to one-sixth the weight of the insect since it has six legs:

$$2\pi r \gamma \cos \theta \approx \tfrac{1}{6}mg$$
$$(6.28)(2.0 \times 10^{-5}\,\text{m})(0.072\,\text{N/m}) \cos \theta \approx \tfrac{1}{6}(3.0 \times 10^{-6}\,\text{kg})(9.8\,\text{m/s}^2)$$
$$\cos \theta \approx \frac{0.49}{0.90} = 0.54.$$

So $\theta \approx 57°$. If $\cos \theta$ had come out greater than 1, the surface tension would not be great enough to support the insect's weight.

NOTE Our estimate ignored the buoyant force and ignored any difference between the radius of the insect's "foot" and the radius of the surface depression.

Soaps and detergents lower the surface tension of water. This is desirable for washing and cleaning since the high surface tension of pure water prevents it from penetrating easily between the fibers of material and into tiny crevices. Substances that reduce the surface tension of a liquid are called *surfactants*.

Surface tension plays a role in another interesting phenomenon, capillarity. It is a common observation that water in a glass container rises up slightly where it touches the glass, Fig. 13–38a. The water is said to "wet" the glass. Mercury, on the other hand, is depressed when it touches the glass, Fig. 13–38b; the mercury does not wet the glass. Whether a liquid wets a solid surface is determined by the relative strength of the cohesive forces between the molecules of the liquid compared to the adhesive forces between the molecules of the liquid and those of the container. *Cohesion* refers to the force between molecules of the same type, whereas *adhesion* refers to the force between molecules of different types. Water wets glass because the water molecules are more strongly attracted to the glass molecules than they are to other water molecules. The opposite is true for mercury: the cohesive forces are stronger than the adhesive forces.

In tubes having very small diameters, liquids are observed to rise or fall relative to the level of the surrounding liquid. This phenomenon is called **capillarity**, and such thin tubes are called **capillaries**. Whether the liquid rises or falls (Fig. 13–39) depends on the relative strengths of the adhesive and cohesive forces. Thus water rises in a glass tube, whereas mercury falls. The actual amount of rise (or fall) depends on the surface tension—which is what keeps the liquid surface from breaking apart.

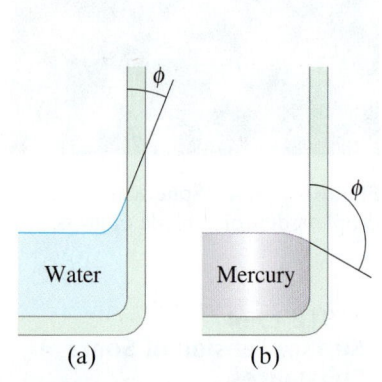

Water   Mercury

(a)   (b)

FIGURE 13–38 (a) Water "wets" the surface of glass, whereas (b) mercury does not "wet" the glass.

FIGURE 13–39 Capillarity.

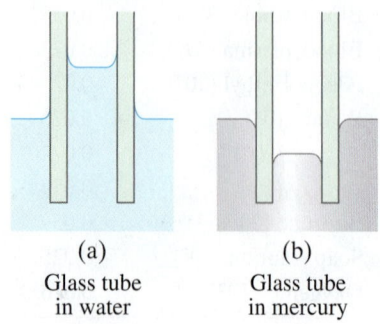

(a)   (b)
Glass tube   Glass tube
in water   in mercury

*13–14 Pumps, and the Heart

We conclude this Chapter with a brief discussion of pumps, including the heart. Pumps can be classified into categories according to their function. A *vacuum pump* is designed to reduce the pressure (usually of air) in a given vessel. A *force pump*, on the other hand, is a pump that is intended to increase the pressure—for example, to lift a liquid (such as water from a well) or to push a fluid through a pipe. Figure 13–40 illustrates the principle behind a simple reciprocating pump. It could be a vacuum pump, in which case the intake is connected to the vessel to be evacuated. A similar mechanism is used in some force pumps, and in this case the fluid is forced under increased pressure through the outlet.

A centrifugal pump (Fig. 13–41), or any force pump, can be used as a *circulating pump*—that is, to circulate a fluid around a closed path, such as the cooling water or lubricating oil in an automobile.

The heart of a human (and of other animals as well) is essentially a circulating pump. The action of a human heart is shown in Fig. 13–42. There are actually two separate paths for blood flow. The longer path takes blood to the parts of the body, via the arteries, bringing oxygen to body tissues and picking up carbon dioxide, which it carries back to the heart via veins. This blood is then pumped to the lungs (the second path), where the carbon dioxide is released and oxygen is taken up. The oxygen-laden blood is returned to the heart, where it is again pumped to the tissues of the body.

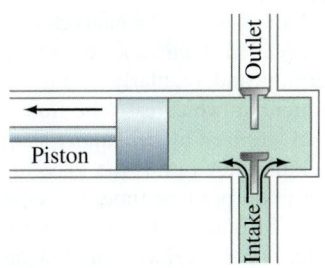

FIGURE 13–40 One kind of pump: the intake valve opens and air (or fluid that is being pumped) fills the empty space when the piston moves to the left. When the piston moves to the right (not shown), the outlet valve opens and fluid is forced out.

FIGURE 13–41 Centrifugal pump: the rotating blades force fluid through the outlet pipe; this kind of pump is used in vacuum cleaners and as a water pump in automobiles.

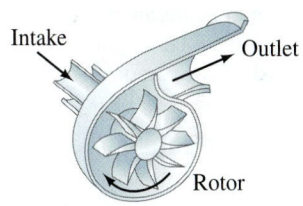

FIGURE 13–42 (a) In the diastole phase, the heart relaxes between beats. Blood moves into the heart; both atria fill rapidly. (b) When the atria contract, the systole or pumping phase begins. The contraction pushes the blood through the mitral and tricuspid valves into the ventricles. (c) The contraction of the ventricles forces the blood through the semilunar valves into the pulmonary artery, which leads to the lungs, and to the aorta (the body's largest artery), which leads to the arteries serving all the body. (d) When the heart relaxes, the semilunar valves close; blood fills the atria, beginning the cycle again.

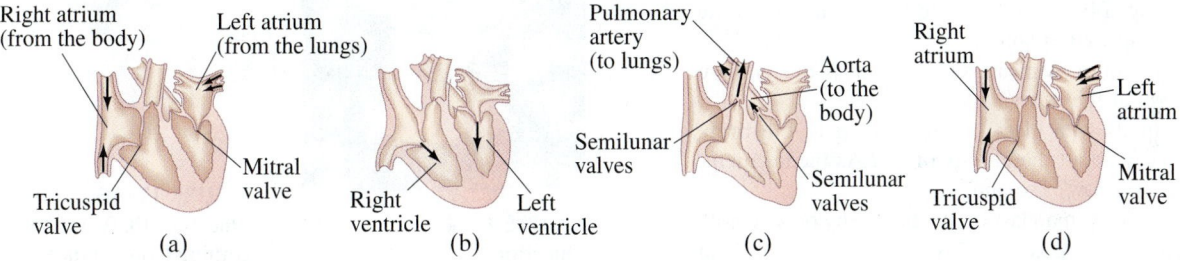

Summary

The three common phases of matter are **solid**, **liquid**, and **gas**. Liquids and gases are collectively called **fluids**, meaning they have the ability to flow. The **density** of a material is defined as its mass per unit volume:

$$\rho = \frac{m}{V}. \tag{13–1}$$

Specific gravity is the ratio of the density of the material to the density of water (at 4°C).

Pressure is defined as force per unit area:

$$P = \frac{F}{A}. \tag{13–2}$$

The pressure P at a depth h in a liquid is given by

$$P = \rho g h, \tag{13–3}$$

where ρ is the density of the liquid and g is the acceleration due to gravity. If the density of a fluid is not uniform, the pressure P varies with height y as

$$\frac{dP}{dy} = -\rho g. \tag{13–4}$$

Pascal's principle says that an external pressure applied to a confined fluid is transmitted throughout the fluid.

Pressure is measured using a manometer or other type of gauge. A **barometer** is used to measure atmospheric pressure. Standard **atmospheric pressure** (average at sea level) is $1.013 \times 10^5 \ \text{N/m}^2$. **Gauge pressure** is the total (absolute) pressure less atmospheric pressure.

Archimedes' principle states that an object submerged wholly or partially in a fluid is buoyed up by a force equal to the weight of fluid it displaces ($F_B = m_F g = \rho_F V_{\text{displ}} g$).

Fluid flow can be characterized either as **streamline** (sometimes called **laminar**), in which the layers of fluid move smoothly and regularly along paths called **streamlines**, or as **turbulent**, in which case the flow is not smooth and regular but is characterized by irregularly shaped whirlpools.

Fluid flow rate is the mass or volume of fluid that passes a given point per unit time. The **equation of continuity** states that for an incompressible fluid flowing in an enclosed tube, the product of the velocity of flow and the cross-sectional area of the tube remains constant:

$$Av = \text{constant.} \qquad \textbf{(13–7b)}$$

Bernoulli's principle tells us that where the velocity of a fluid is high, the pressure in it is low, and where the velocity is low, the pressure is high. For steady laminar flow of an incompressible and nonviscous fluid, **Bernoulli's equation**, which is based on the law of conservation of energy, is

$$P_1 + \tfrac{1}{2}\rho v_1^2 + \rho g y_1 = P_2 + \tfrac{1}{2}\rho v_2^2 + \rho g y_2, \qquad \textbf{(13–8)}$$

for two points along the flow.

[***Viscosity** refers to friction within a fluid and is essentially a frictional force between adjacent layers of fluid as they move past one another.]

[*Liquid surfaces hold together as if under tension (**surface tension**), allowing drops to form and objects like needles and insects to stay on the surface.]

Questions

1. If one material has a higher density than another, must the molecules of the first be heavier than those of the second? Explain.
2. Airplane travelers sometimes note that their cosmetics bottles and other containers have leaked during a flight. What might cause this?
3. The three containers in Fig. 13–43 are filled with water to the same height and have the same surface area at the base; hence the water pressure, and the total force on the base of each, is the same. Yet the total weight of water is different for each. Explain this "hydrostatic paradox."

FIGURE 13–43
Question 3.

4. Consider what happens when you push both a pin and the blunt end of a pen against your skin with the same force. Decide what determines whether your skin is cut—the net force applied to it or the pressure.
5. A small amount of water is boiled in a 1-gallon metal can. The can is removed from the heat and the lid put on. As the can cools, it collapses. Explain.
6. When blood pressure is measured, why must the cuff be held at the level of the heart?
7. An ice cube floats in a glass of water filled to the brim. What can you say about the density of ice? As the ice melts, will the water overflow? Explain.
8. Will an ice cube float in a glass of alcohol? Why or why not?
9. A submerged can of Coke® will sink, but a can of Diet Coke® will float. (Try it!) Explain.
10. Why don't ships made of iron sink?
11. Explain how the tube in Fig. 13–44, known as a **siphon**, can transfer liquid from one container to a lower one even though the liquid must flow uphill for part of its journey. (Note that the tube must be filled with liquid to start with.)

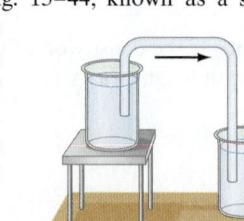

FIGURE 13–44
Question 11. A siphon.

12. A barge filled high with sand approaches a low bridge over the river and cannot quite pass under it. Should sand be added to, or removed from, the barge? [*Hint*: Consider Archimedes' principle.]
13. Explain why helium weather balloons, which are used to measure atmospheric conditions at high altitudes, are normally released while filled to only 10–20% of their maximum volume.

14. A row boat floats in a swimming pool, and the level of the water at the edge of the pool is marked. Consider the following situations and explain whether the level of the water will rise, fall, or stay the same. (*a*) The boat is removed from the water. (*b*) The boat in the water holds an iron anchor which is removed from the boat and placed on the shore. (*c*) The iron anchor is removed from the boat and dropped in the pool.
15. Will an empty balloon have precisely the same apparent weight on a scale as a balloon filled with air? Explain.
16. Why do you float higher in salt water than in fresh water?
17. If you dangle two pieces of paper vertically, a few inches apart (Fig. 13–45), and blow between them, how do you think the papers will move? Try it and see. Explain.

FIGURE 13–45
Question 17.

FIGURE 13–46
Question 18. Water coming from a faucet.

18. Why does the stream of water from a faucet become narrower as it falls (Fig. 13–46)?
19. Children are told to avoid standing too close to a rapidly moving train because they might get sucked under it. Is this possible? Explain.
20. A tall Styrofoam cup is filled with water. Two holes are punched in the cup near the bottom, and water begins rushing out. If the cup is dropped so it falls freely, will the water continue to flow from the holes? Explain.
21. Why do airplanes normally take off into the wind?
22. Two ships moving in parallel paths close to one another risk colliding. Why?
23. Why does the canvas top of a convertible bulge out when the car is traveling at high speed? [*Hint*: The windshield deflects air upward, pushing streamlines closer together.]
24. Roofs of houses are sometimes "blown" off (or are they pushed off?) during a tornado or hurricane. Explain using Bernoulli's principle.

Problems

13–2 Density and Specific Gravity

1. (I) The approximate volume of the granite monolith known as El Capitan in Yosemite National Park (Fig. 13–47) is about 10^8 m^3. What is its approximate mass?

FIGURE 13–47 Problem 1.

2. (I) What is the approximate mass of air in a living room $5.6 \text{ m} \times 3.8 \text{ m} \times 2.8 \text{ m}$?

3. (I) If you tried to smuggle gold bricks by filling your backpack, whose dimensions are $56 \text{ cm} \times 28 \text{ cm} \times 22 \text{ cm}$, what would its mass be?

4. (I) State your mass and then estimate your volume. [*Hint*: Because you can swim on or just under the surface of the water in a swimming pool, you have a pretty good idea of your density.]

5. (II) A bottle has a mass of 35.00 g when empty and 98.44 g when filled with water. When filled with another fluid, the mass is 89.22 g. What is the specific gravity of this other fluid?

6. (II) If 5.0 L of antifreeze solution (specific gravity = 0.80) is added to 4.0 L of water to make a 9.0-L mixture, what is the specific gravity of the mixture?

7. (III) The Earth is not a uniform sphere, but has regions of varying density. Consider a simple model of the Earth divided into three regions—inner core, outer core, and mantle. Each region is taken to have a unique constant density (the average density of that region in the real Earth):

Region	Radius (km)	Density (kg/m³)
Inner Core	0–1220	13,000
Outer Core	1220–3480	11,100
Mantle	3480–6371	4,400

(*a*) Use this model to predict the average density of the entire Earth. (*b*) The measured radius of the Earth is 6371 km and its mass is 5.98×10^{24} kg. Use these data to determine the actual average density of the Earth and compare it (as a percent difference) with the one you determined in (*a*).

13–3 to 13–6 Pressure; Pascal's Principle

8. (I) Estimate the pressure needed to raise a column of water to the same height as a 35-m-tall oak tree.

9. (I) Estimate the pressure exerted on a floor by (*a*) one pointed chair leg (66 kg on all four legs) of area $= 0.020 \text{ cm}^2$, and (*b*) a 1300-kg elephant standing on one foot (area $= 800 \text{ cm}^2$).

10. (I) What is the difference in blood pressure (mm-Hg) between the top of the head and bottom of the feet of a 1.70-m-tall person standing vertically?

11. (II) How high would the level be in an alcohol barometer at normal atmospheric pressure?

12. (II) In a movie, Tarzan evades his captors by hiding underwater for many minutes while breathing through a long, thin reed. Assuming the maximum pressure difference his lungs can manage and still breathe is -85 mm-Hg, calculate the deepest he could have been.

13. (II) The maximum gauge pressure in a hydraulic lift is 17.0 atm. What is the largest-size vehicle (kg) it can lift if the diameter of the output line is 22.5 cm?

14. (II) The gauge pressure in each of the four tires of an automobile is 240 kPa. If each tire has a "footprint" of 220 cm^2, estimate the mass of the car.

15. (II) (*a*) Determine the total force and the absolute pressure on the bottom of a swimming pool 28.0 m by 8.5 m whose uniform depth is 1.8 m. (*b*) What will be the pressure against the *side* of the pool near the bottom?

16. (II) A house at the bottom of a hill is fed by a full tank of water 5.0 m deep and connected to the house by a pipe that is 110 m long at an angle of 58° from the horizontal (Fig. 13–48). (*a*) Determine the water gauge pressure at the house. (*b*) How high could the water shoot if it came vertically out of a broken pipe in front of the house?

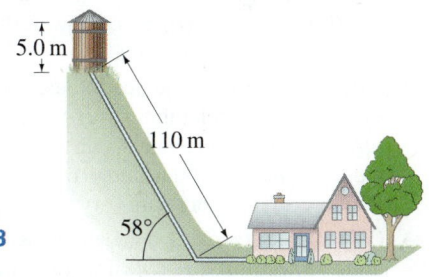

FIGURE 13–48
Problem 16.

17. (II) Water and then oil (which don't mix) are poured into a U-shaped tube, open at both ends. They come to equilibrium as shown in Fig. 13–49. What is the density of the oil? [*Hint*: Pressures at points a and b are equal. Why?]

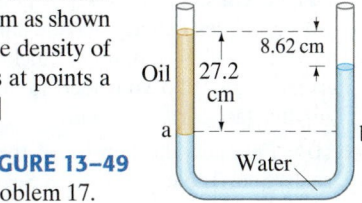

FIGURE 13–49
Problem 17.

18. (II) In working out his principle, Pascal showed dramatically how force can be multiplied with fluid pressure. He placed a long, thin tube of radius $r = 0.30$ cm vertically into a wine barrel of radius $R = 21$ cm, Fig. 13–50. He found that when the barrel was filled with water and the tube filled to a height of 12 m, the barrel burst. Calculate (*a*) the mass of water in the tube, and (*b*) the net force exerted by the water in the barrel on the lid just before rupture.

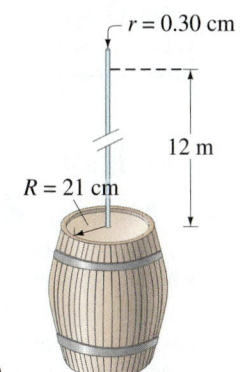

FIGURE 13–50
Problem 18 (not to scale).

19. (II) What is the normal pressure of the atmosphere at the summit of Mt. Everest, 8850 m above sea level?

20. (II) A hydraulic press for compacting powdered samples has a large cylinder which is 10.0 cm in diameter, and a small cylinder with a diameter of 2.0 cm (Fig. 13–51). A lever is attached to the small cylinder as shown. The sample, which is placed on the large cylinder, has an area of 4.0 cm². What is the pressure on the sample if 350 N is applied to the lever?

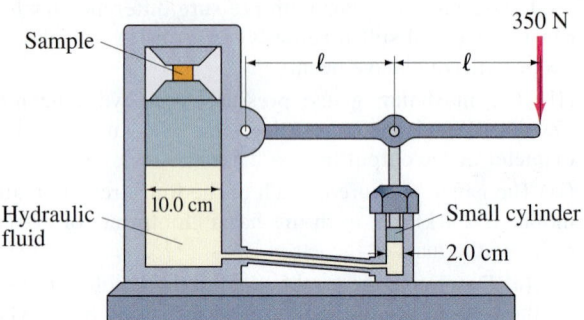

FIGURE 13–51 Problem 20.

21. (II) An open-tube mercury manometer is used to measure the pressure in an oxygen tank. When the atmospheric pressure is 1040 mbar, what is the absolute pressure (in Pa) in the tank if the height of the mercury in the open tube is (*a*) 21.0 cm higher, (*b*) 5.2 cm lower, than the mercury in the tube connected to the tank?

22. (III) A beaker of liquid accelerates from rest, on a horizontal surface, with acceleration *a* to the right. (*a*) Show that the surface of the liquid makes an angle $\theta = \tan^{-1}(a/g)$ with the horizontal. (*b*) Which edge of the water surface is higher? (*c*) How does the pressure vary with depth below the surface?

23. (III) Water stands at a height *h* behind a vertical dam of uniform width *b*. (*a*) Use integration to show that the total force of the water on the dam is $F = \frac{1}{2}\rho g h^2 b$. (*b*) Show that the torque about the base of the dam due to this force can be considered to act with a lever arm equal to *h*/3. (*c*) For a freestanding concrete dam of uniform thickness *t* and height *h*, what minimum thickness is needed to prevent overturning? Do you need to add in atmospheric pressure for this last part? Explain.

24. (III) Estimate the density of the water 5.4 km deep in the sea. (See Table 12–1 and Section 12–4 regarding bulk modulus.) By what fraction does it differ from the density at the surface?

25. (III) A cylindrical bucket of liquid (density ρ) is rotated about its symmetry axis, which is vertical. If the angular velocity is ω, show that the pressure at a distance *r* from the rotation axis is

$$P = P_0 + \frac{1}{2}\rho\omega^2 r^2,$$

where P_0 is the pressure at $r = 0$.

13–7 Buoyancy and Archimedes' Principle

26. (I) What fraction of a piece of iron will be submerged when it floats in mercury?

27. (I) A geologist finds that a Moon rock whose mass is 9.28 kg has an apparent mass of 6.18 kg when submerged in water. What is the density of the rock?

28. (II) A crane lifts the 16,000-kg steel hull of a sunken ship out of the water. Determine (*a*) the tension in the crane's cable when the hull is fully submerged in the water, and (*b*) the tension when the hull is completely out of the water.

29. (II) A spherical balloon has a radius of 7.35 m and is filled with helium. How large a cargo can it lift, assuming that the skin and structure of the balloon have a mass of 930 kg? Neglect the buoyant force on the cargo volume itself.

30. (II) A 74-kg person has an apparent mass of 54 kg (because of buoyancy) when standing in water that comes up to the hips. Estimate the mass of each leg. Assume the body has SG = 1.00.

31. (II) What is the likely identity of a metal (see Table 13–1) if a sample has a mass of 63.5 g when measured in air and an apparent mass of 55.4 g when submerged in water?

32. (II) Calculate the true mass (in vacuum) of a piece of aluminum whose apparent mass is 3.0000 kg when weighed in air.

33. (II) Because gasoline is less dense than water, drums containing gasoline will float in water. Suppose a 230-L steel drum is completely full of gasoline. What total volume of steel can be used in making the drum if the gasoline-filled drum is to float in fresh water?

34. (II) A scuba diver and her gear displace a volume of 65.0 L and have a total mass of 68.0 kg. (*a*) What is the buoyant force on the diver in seawater? (*b*) Will the diver sink or float?

35. (II) The specific gravity of ice is 0.917, whereas that of seawater is 1.025. What percent of an iceberg is above the surface of the water?

36. (II) Archimedes' principle can be used not only to determine the specific gravity of a solid using a known liquid (Example 13–10); the reverse can be done as well. (*a*) As an example, a 3.80-kg aluminum ball has an apparent mass of 2.10 kg when submerged in a particular liquid: calculate the density of the liquid. (*b*) Derive a formula for determining the density of a liquid using this procedure.

37. (II) (*a*) Show that the buoyant force F_B on a partially submerged object such as a ship acts at the center of gravity of the fluid before it is displaced. This point is called the **center of buoyancy**. (*b*) To ensure that a ship is in stable equilibrium, would it be better if its center of buoyancy was above, below, or at the same point as, its center of gravity? Explain. (See Fig. 13–52.)

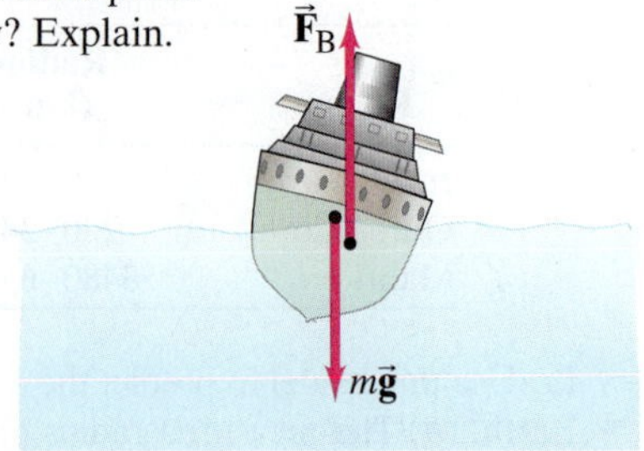

FIGURE 13–52
Problem 37.

38. (II) A cube of side length 10.0 cm and made of unknown material floats at the surface between water and oil. The oil has a density of 810 kg/m³. If the cube floats so that it is 72% in the water and 28% in the oil, what is the mass of the cube and what is the buoyant force on the cube?

39. (II) How many helium-filled balloons would it take to lift a person? Assume the person has a mass of 75 kg and that each helium-filled balloon is spherical with a diameter of 33 cm.

40. (II) A scuba tank, when fully submerged, displaces 15.7 L of seawater. The tank itself has a mass of 14.0 kg and, when "full," contains 3.00 kg of air. Assuming only a weight and buoyant force act, determine the net force (magnitude and direction) on the fully submerged tank at the beginning of a dive (when it is full of air) and at the end of a dive (when it no longer contains any air).

41. (III) If an object floats in water, its density can be determined by tying a sinker to it so that both the object and the sinker are submerged. Show that the specific gravity is given by $w/(w_1 - w_2)$, where w is the weight of the object alone in air, w_1 is the apparent weight when a sinker is tied to it and the sinker only is submerged, and w_2 is the apparent weight when both the object and the sinker are submerged.

42. (III) A 3.25-kg piece of wood (SG = 0.50) floats on water. What minimum mass of lead, hung from the wood by a string, will cause it to sink?

13–8 to 13–10 Fluid Flow, Bernoulli's Equation

43. (I) A 15-cm-radius air duct is used to replenish the air of a room $8.2 \text{ m} \times 5.0 \text{ m} \times 3.5 \text{ m}$ every 12 min. How fast does the air flow in the duct?

44. (I) Using the data of Example 13–13, calculate the average speed of blood flow in the major arteries of the body which have a total cross-sectional area of about 2.0 cm^2.

45. (I) How fast does water flow from a hole at the bottom of a very wide, 5.3-m-deep storage tank filled with water? Ignore viscosity.

46. (II) A fish tank has dimensions 36 cm wide by 1.0 m long by 0.60 m high. If the filter should process all the water in the tank once every 4.0 h, what should the flow speed be in the 3.0-cm-diameter input tube for the filter?

47. (II) What gauge pressure in the water mains is necessary if a firehose is to spray water to a height of 18 m?

48. (II) A $\frac{5}{8}$-in. (inside) diameter garden hose is used to fill a round swimming pool 6.1 m in diameter. How long will it take to fill the pool to a depth of 1.2 m if water flows from the hose at a speed of 0.40 m/s?

49. (II) A 180-km/h wind blowing over the flat roof of a house causes the roof to lift off the house. If the house is $6.2 \text{ m} \times 12.4 \text{ m}$ in size, estimate the weight of the roof. Assume the roof is not nailed down.

50. (II) A 6.0-cm-diameter horizontal pipe gradually narrows to 4.5 cm. When water flows through this pipe at a certain rate, the gauge pressure in these two sections is 32.0 kPa and 24.0 kPa, respectively. What is the volume rate of flow?

51. (II) Estimate the air pressure inside a category 5 hurricane, where the wind speed is 300 km/h (Fig. 13–53).

FIGURE 13–53   Problem 51.

52. (II) What is the lift (in newtons) due to Bernoulli's principle on a wing of area 88 m^2 if the air passes over the top and bottom surfaces at speeds of 280 m/s and 150 m/s, respectively?

53. (II) Show that the power needed to drive a fluid through a pipe with uniform cross-section is equal to the volume rate of flow, Q, times the pressure difference, $P_1 - P_2$.

54. (II) Water at a gauge pressure of 3.8 atm at street level flows into an office building at a speed of 0.68 m/s through a pipe 5.0 cm in diameter. The pipe tapers down to 2.8 cm in diameter by the top floor, 18 m above (Fig. 13–54), where the faucet has been left open. Calculate the flow velocity and the gauge pressure in the pipe on the top floor. Assume no branch pipes and ignore viscosity.

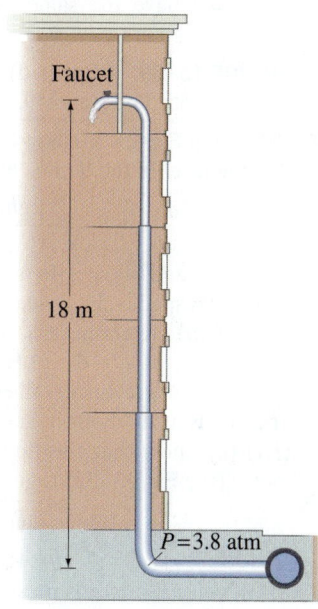

FIGURE 13–54
Problem 54.

55. (II) In Fig. 13–55, take into account the speed of the top surface of the tank and show that the speed of fluid leaving the opening at the bottom is

$$v_1 = \sqrt{\frac{2gh}{(1 - A_1^2/A_2^2)}},$$

where $h = y_2 - y_1$, and A_1 and A_2 are the areas of the opening and of the top surface, respectively. Assume $A_1 \ll A_2$ so that the flow remains nearly steady and laminar.

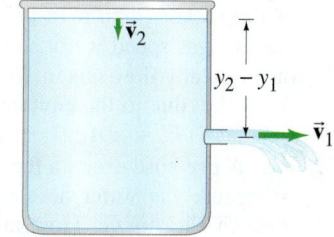

FIGURE 13–55
Problems 55, 56, 58, and 59.

56. (II) Suppose the top surface of the vessel in Fig. 13–55 is subjected to an external gauge pressure P_2. (a) Derive a formula for the speed, v_1, at which the liquid flows from the opening at the bottom into atmospheric pressure, P_0. Assume the velocity of the liquid surface, v_2, is approximately zero. (b) If $P_2 = 0.85 \text{ atm}$ and $y_2 - y_1 = 2.4 \text{ m}$, determine v_1 for water.

57. (II) You are watering your lawn with a hose when you put your finger over the hose opening to increase the distance the water reaches. If you are pointing the hose at the same angle, and the distance the water reaches increases by a factor of 4, what fraction of the hose opening did you block?

58. (III) Suppose the opening in the tank of Fig. 13–55 is a height h_1 above the base and the liquid surface is a height h_2 above the base. The tank rests on level ground. (a) At what horizontal distance from the base of the tank will the fluid strike the ground? (b) At what other height, h_1', can a hole be placed so that the emerging liquid will have the same "range"? Assume $v_2 \approx 0$.

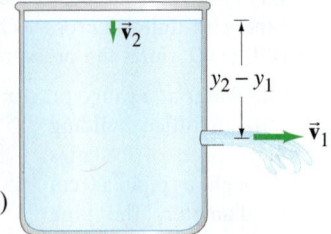

FIGURE 13–55 (repeated) Problems 55, 56, 58, and 59.

59. (III) (a) In Fig. 13–55, show that Bernoulli's principle predicts that the level of the liquid, $h = y_2 - y_1$, drops at a rate

$$\frac{dh}{dt} = -\sqrt{\frac{2ghA_1^2}{A_2^2 - A_1^2}},$$

where A_1 and A_2 are the areas of the opening and the top surface, respectively, assuming $A_1 \ll A_2$, and viscosity is ignored. (b) Determine h as a function of time by integrating. Let $h = h_0$ at $t = 0$. (c) How long would it take to empty a 10.6-cm-tall cylinder filled with 1.3 L of water if the opening is at the bottom and has a 0.50-cm diameter?

60. (III) (a) Show that the flow speed measured by a venturi meter (see Fig. 13–32) is given by the relation

$$v_1 = A_2 \sqrt{\frac{2(P_1 - P_2)}{\rho(A_1^2 - A_2^2)}}.$$

(b) A venturi meter is measuring the flow of water; it has a main diameter of 3.0 cm tapering down to a throat diameter of 1.0 cm. If the pressure difference is measured to be 18 mm-Hg, what is the speed of the water entering the venturi throat?

61. (III) *Thrust of a rocket.* (a) Use Bernoulli's equation and the equation of continuity to show that the emission speed of the propelling gases of a rocket is

$$v = \sqrt{2(P - P_0)/\rho},$$

where ρ is the density of the gas, P is the pressure of the gas inside the rocket, and P_0 is atmospheric pressure just outside the exit orifice. Assume that the gas density stays approximately constant, and that the area of the exit orifice, A_0, is much smaller than the cross-sectional area, A, of the inside of the rocket (take it to be a large cylinder). Assume also that the gas speed is not so high that significant turbulence or nonsteady flow sets in. (b) Show that the thrust force on the rocket due to the emitted gases is

$$F = 2A_0(P - P_0).$$

62. (III) A fire hose exerts a force on the person holding it. This is because the water accelerates as it goes from the hose through the nozzle. How much force is required to hold a 7.0-cm-diameter hose delivering 450 L/min through a 0.75-cm-diameter nozzle?

* 13–11 Viscosity

*63. (II) A viscometer consists of two concentric cylinders, 10.20 cm and 10.60 cm in diameter. A liquid fills the space between them to a depth of 12.0 cm. The outer cylinder is fixed, and a torque of 0.024 m·N keeps the inner cylinder turning at a steady rotational speed of 57 rev/min. What is the viscosity of the liquid?

*64. (III) A long vertical hollow tube with an inner diameter of 1.00 cm is filled with SAE 10 motor oil. A 0.900-cm-diameter, 30.0-cm-long 150-g rod is dropped vertically through the oil in the tube. What is the maximum speed attained by the rod as it falls?

* 13–12 Flow in Tubes; Poiseuille's Equation

*65. (I) Engine oil (assume SAE 10, Table 13–3) passes through a fine 1.80-mm-diameter tube that is 8.6 cm long. What pressure difference is needed to maintain a flow rate of 6.2 mL/min?

*66. (I) A gardener feels it is taking too long to water a garden with a $\frac{3}{8}$-in.-diameter hose. By what factor will the time be cut using a $\frac{5}{8}$-in.-diameter hose instead? Assume nothing else is changed.

*67. (II) What diameter must a 15.5-m-long air duct have if the ventilation and heating system is to replenish the air in a room 8.0 m × 14.0 m × 4.0 m every 12.0 min? Assume the pump can exert a gauge pressure of 0.710×10^{-3} atm.

*68. (II) What must be the pressure difference between the two ends of a 1.9-km section of pipe, 29 cm in diameter, if it is to transport oil ($\rho = 950$ kg/m^3, $\eta = 0.20$ Pa·s) at a rate of 650 cm^3/s?

*69. (II) Poiseuille's equation does not hold if the flow velocity is high enough that turbulence sets in. The onset of turbulence occurs when the **Reynolds number**, Re, exceeds approximately 2000. Re is defined as

$$Re = \frac{2\bar{v}r\rho}{\eta},$$

where $\bar{v}$ is the average speed of the fluid, ρ is its density, η is its viscosity, and r is the radius of the tube in which the fluid is flowing. (a) Determine if blood flow through the aorta is laminar or turbulent when the average speed of blood in the aorta ($r = 0.80$ cm) during the resting part of the heart's cycle is about 35 cm/s. (b) During exercise, the blood-flow speed approximately doubles. Calculate the Reynolds number in this case, and determine if the flow is laminar or turbulent.

*70. (II) Assuming a constant pressure gradient, if blood flow is reduced by 85%, by what factor is the radius of a blood vessel decreased?

*71. (III) A patient is to be given a blood transfusion. The blood is to flow through a tube from a raised bottle to a needle inserted in the vein (Fig. 13–56). The inside diameter of the 25-mm-long needle is 0.80 mm, and the required flow rate is 2.0 cm^3 of blood per minute. How high h should the bottle be placed above the needle? Obtain ρ and η from the Tables. Assume the blood pressure is 78 torr above atmospheric pressure.

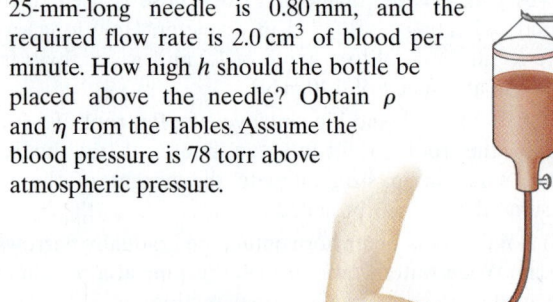

FIGURE 13–56 Problems 71 and 79.

* 13–13 Surface Tension and Capillarity

*72. (I) If the force F needed to move the wire in Fig. 13–35 is 3.4×10^{-3} N, calculate the surface tension γ of the enclosed fluid. Assume $\ell = 0.070$ m.

*73. (I) Calculate the force needed to move the wire in Fig. 13–35 if it is immersed in a soapy solution and the wire is 24.5 cm long.

*74. (II) The surface tension of a liquid can be determined by measuring the force F needed to just lift a circular platinum ring of radius r from the surface of the liquid. (a) Find a formula for γ in terms of F and r. (b) At 30°C, if $F = 5.80 \times 10^{-3}$ N and $r = 2.8$ cm, calculate γ for the tested liquid.

*75. (III) Estimate the diameter of a steel needle that can just "float" on water due to surface tension.

*76. (III) Show that inside a soap bubble, there must be a pressure ΔP in excess of that outside equal to $\Delta P = 4\gamma/r$, where r is the radius of the bubble and γ is the surface tension. [*Hint*: Think of the bubble as two hemispheres in contact with each other; and remember that there are two surfaces to the bubble. Note that this result applies to any kind of membrane, where 2γ is the tension per unit length in that membrane.]

*77. (III) A common effect of surface tension is the ability of a liquid to rise up a narrow tube due to what is called capillary action. Show that for a narrow tube of radius r placed in a liquid of density ρ and surface tension γ, the liquid in the tube will reach a height $h = 2\gamma/\rho g r$ above the level of the liquid outside the tube, where g is the gravitational acceleration. Assume that the liquid "wets" the capillary (the liquid surface is vertical at the contact with the inside of the tube).

General Problems

78. A 2.8-N force is applied to the plunger of a hypodermic needle. If the diameter of the plunger is 1.3 cm and that of the needle 0.20 mm, (*a*) with what force does the fluid leave the needle? (*b*) What force on the plunger would be needed to push fluid into a vein where the gauge pressure is 75 mm-Hg? Answer for the instant just before the fluid starts to move.

79. Intravenous infusions are often made under gravity, as shown in Fig. 13–56. Assuming the fluid has a density of $1.00\,g/cm^3$, at what height h should the bottle be placed so the liquid pressure is (*a*) 55 mm-Hg, and (*b*) 650 mm-H_2O? (*c*) If the blood pressure is 78 mm-Hg above atmospheric pressure, how high should the bottle be placed so that the fluid just barely enters the vein?

80. A beaker of water rests on an electronic balance that reads 998.0 g. A 2.6-cm-diameter solid copper ball attached to a string is submerged in the water, but does not touch the bottom. What are the tension in the string and the new balance reading?

81. Estimate the difference in air pressure between the top and the bottom of the Empire State building in New York City. It is 380 m tall and is located at sea level. Express as a fraction of atmospheric pressure at sea level.

82. A hydraulic lift is used to jack a 920-kg car 42 cm off the floor. The diameter of the output piston is 18 cm, and the input force is 350 N. (*a*) What is the area of the input piston? (*b*) What is the work done in lifting the car 42 cm? (*c*) If the input piston moves 13 cm in each stroke, how high does the car move up for each stroke? (*d*) How many strokes are required to jack the car up 42 cm? (*e*) Show that energy is conserved.

83. When you ascend or descend a great deal when driving in a car, your ears "pop," which means that the pressure behind the eardrum is being equalized to that outside. If this did not happen, what would be the approximate force on an eardrum of area $0.20\,cm^2$ if a change in altitude of 950 m takes place?

84. Giraffes are a wonder of cardiovascular engineering. Calculate the difference in pressure (in atmospheres) that the blood vessels in a giraffe's head must accommodate as the head is lowered from a full upright position to ground level for a drink. The height of an average giraffe is about 6 m.

85. Suppose a person can reduce the pressure in his lungs to -75 mm-Hg gauge pressure. How high can water then be "sucked" up a straw?

86. Airlines are allowed to maintain a minimum air pressure within the passenger cabin equivalent to that at an altitude of 8000 ft (2400 m) to avoid adverse health effects among passengers due to oxygen deprivation. Estimate this minimum pressure (in atm).

87. A simple model (Fig. 13–57) considers a continent as a block (density $\approx 2800\,kg/m^3$) floating in the mantle rock around it (density $\approx 3300\,kg/m^3$). Assuming the continent is 35 km thick (the average thickness of the Earth's continental crust), estimate the height of the continent above the surrounding rock.

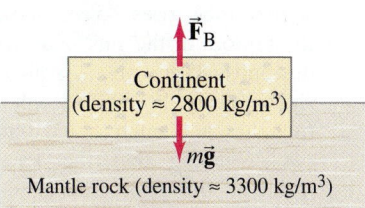

FIGURE 13–57
Problem 87.

88. A ship, carrying fresh water to a desert island in the Caribbean, has a horizontal cross-sectional area of $2240\,m^2$ at the waterline. When unloaded, the ship rises 8.50 m higher in the sea. How many cubic meters of water was delivered?

89. During ascent, and especially during descent, volume changes of trapped air in the middle ear can cause ear discomfort until the middle-ear pressure and exterior pressure are equalized. (*a*) If a rapid descent at a rate of 7.0 m/s or faster commonly causes ear discomfort, what is the maximum rate of increase in atmospheric pressure (that is, dP/dt) tolerable to most people? (*b*) In a 350-m-tall building, what will be the fastest possible descent time for an elevator traveling from the top to ground floor, assuming the elevator is properly designed to account for human physiology?

90. A raft is made of 12 logs lashed together. Each is 45 cm in diameter and has a length of 6.1 m. How many people can the raft hold before they start getting their feet wet, assuming the average person has a mass of 68 kg? Do *not* neglect the weight of the logs. Assume the specific gravity of wood is 0.60.

91. Estimate the total mass of the Earth's atmosphere, using the known value of atmospheric pressure at sea level.

92. During each heartbeat, approximately $70\,cm^3$ of blood is pushed from the heart at an average pressure of 105 mm-Hg. Calculate the power output of the heart, in watts, assuming 70 beats per minute.

93. Four lawn sprinkler heads are fed by a 1.9-cm-diameter pipe. The water comes out of the heads at an angle of 35° to the horizontal and covers a radius of 7.0 m. (*a*) What is the velocity of the water coming out of each sprinkler head? (Assume zero air resistance.) (*b*) If the output diameter of each head is 3.0 mm, how many liters of water do the four heads deliver per second? (*c*) How fast is the water flowing inside the 1.9-cm-diameter pipe?

94. A bucket of water is accelerated upward at $1.8\,g$. What is the buoyant force on a 3.0-kg granite rock $(SG = 2.7)$ submerged in the water? Will the rock float? Why or why not?

95. The stream of water from a faucet decreases in diameter as it falls (Fig. 13–58). Derive an equation for the diameter of the stream as a function of the distance y below the faucet, given that the water has speed v_0 when it leaves the faucet, whose diameter is d.

FIGURE 13–58 Problem 95. Water coming from a faucet.

96. You need to siphon water from a clogged sink. The sink has an area of $0.38\,m^2$ and is filled to a height of 4.0 cm. Your siphon tube rises 45 cm above the bottom of the sink and then descends 85 cm to a pail as shown in Fig. 13–59. The siphon tube has a diameter of 2.0 cm. (a) Assuming that the water level in the sink has almost zero velocity, estimate the water velocity when it enters the pail. (b) Estimate how long it will take to empty the sink.

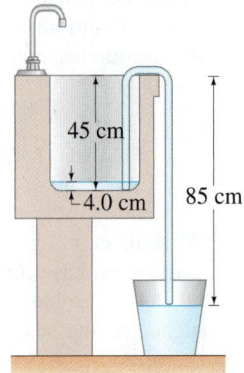

FIGURE 13–59 Problem 96.

97. An airplane has a mass of $1.7 \times 10^6\,kg$, and the air flows past the lower surface of the wings at 95 m/s. If the wings have a surface area of $1200\,m^2$, how fast must the air flow over the upper surface of the wing if the plane is to stay in the air?

98. A drinking fountain shoots water about 14 cm up in the air from a nozzle of diameter 0.60 cm. The pump at the base of the unit (1.1 m below the nozzle) pushes water into a 1.2-cm-diameter supply pipe that goes up to the nozzle. What gauge pressure does the pump have to provide? Ignore the viscosity; your answer will therefore be an underestimate.

99. A hurricane-force wind of 200 km/h blows across the face of a storefront window. Estimate the force on the $2.0\,m \times 3.0\,m$ window due to the difference in air pressure inside and outside the window. Assume the store is airtight so the inside pressure remains at 1.0 atm. (This is why you should not tightly seal a building in preparation for a hurricane).

100. Blood from an animal is placed in a bottle 1.30 m above a 3.8-cm-long needle, of inside diameter 0.40 mm, from which it flows at a rate of $4.1\,cm^3/min$. What is the viscosity of this blood?

101. Three forces act significantly on a freely floating helium-filled balloon: gravity, air resistance (or drag force), and a buoyant force. Consider a spherical helium-filled balloon of radius $r = 15\,cm$ rising upward through $0°C$ air, and $m = 2.8\,g$ is the mass of the (deflated) balloon itself. For all speeds v, except the very slowest ones, the flow of air past a rising balloon is turbulent, and the drag force F_D is given by the relation

$$F_D = \tfrac{1}{2}C_D \rho_{air}\pi r^2 v^2$$

where the constant $C_D = 0.47$ is the "drag coefficient" for a smooth sphere of radius r. If this balloon is released from rest, it will accelerate very quickly (in a few tenths of a second) to its terminal velocity v_T, where the buoyant force is cancelled by the drag force and the balloon's total weight. Assuming the balloon's acceleration takes place over a negligible time and distance, how long does it take the released balloon to rise a distance $h = 12\,m$?

*102. If cholesterol buildup reduces the diameter of an artery by 15%, by what % will the blood flow rate be reduced, assuming the same pressure difference?

103. A two-component model used to determine percent body fat in a human body assumes that a fraction $f\,(< 1)$ of the body's total mass m is composed of fat with a density of $0.90\,g/cm^3$, and that the remaining mass of the body is composed of fat-free tissue with a density of $1.10\,g/cm^3$. If the specific gravity of the entire body's density is X, show that the percent body fat $(= f \times 100)$ is given by

$$\% \text{ Body fat} = \frac{495}{X} - 450.$$

*Numerical/Computer

*104. (III) Air pressure decreases with altitude. The following data show the air pressure at different altitudes.

Altitude (m)	Pressure (kPa)
0	101.3
1000	89.88
2000	79.50
3000	70.12
4000	61.66
5000	54.05
6000	47.22
7000	41.11
8000	35.65
9000	30.80
10,000	26.50

(a) Determine the best-fit quadratic equation that shows how the air pressure changes with altitude. (b) Determine the best-fit exponential equation that describes the change of air pressure with altitude. (c) Use each fit to find the air pressure at the summit of the mountain K2 at 8611 m, and give the % difference.

Answers to Exercises

A: (d).

B: The same. Pressure depends on depth, not on length.

C: Lower.

D: (a).

E: (e).

F: Increases.

G: (b).

An object attached to a coil spring can exhibit oscillatory motion. Many kinds of oscillatory motion are sinusoidal in time, or nearly so, and are referred to as being simple harmonic motion. Real systems generally have at least some friction, causing the motion to be damped. The automobile spring shown here has a shock absorber (yellow) that purposefully dampens the oscillation to make for a smooth ride. When an external sinusoidal force is exerted on a system able to oscillate, resonance occurs if the driving force is at or near the natural frequency of oscillation.

C H A P T E R

14

Oscillations

CHAPTER-OPENING QUESTION—Guess now!

A simple pendulum consists of a mass m (the "bob") hanging on the end of a thin string of length ℓ and negligible mass. The bob is pulled sideways so the string makes a 5.0° angle to the vertical; when released, it oscillates back and forth at a frequency f. If the pendulum was raised to a 10.0° angle instead, its frequency would be

(a) twice as great.
(b) half as great.
(c) the same, or very close to it.
(d) not quite twice as great.
(e) a bit more than half as great.

CONTENTS

14–1 Oscillations of a Spring

14–2 Simple Harmonic Motion

14–3 Energy in the Simple Harmonic Oscillator

14–4 Simple Harmonic Motion Related to Uniform Circular Motion

14–5 The Simple Pendulum

*14–6 The Physical Pendulum and the Torsion Pendulum

14–7 Damped Harmonic Motion

14–8 Forced Oscillations; Resonance

Many objects vibrate or oscillate—an object on the end of a spring, a tuning fork, the balance wheel of an old watch, a pendulum, a plastic ruler held firmly over the edge of a table and gently struck, the strings of a guitar or piano. Spiders detect prey by the vibrations of their webs; cars oscillate up and down when they hit a bump; buildings and bridges vibrate when heavy trucks pass or the wind is fierce. Indeed, because most solids are elastic (see Chapter 12), they vibrate (at least briefly) when given an impulse. Electrical oscillations are necessary in radio and television sets. At the atomic level, atoms vibrate within a molecule, and the atoms of a solid vibrate about their relatively fixed positions. Because it is so common in everyday life and occurs in so many areas of physics, oscillatory motion is of great importance. Mechanical oscillations are fully described on the basis of Newtonian mechanics.

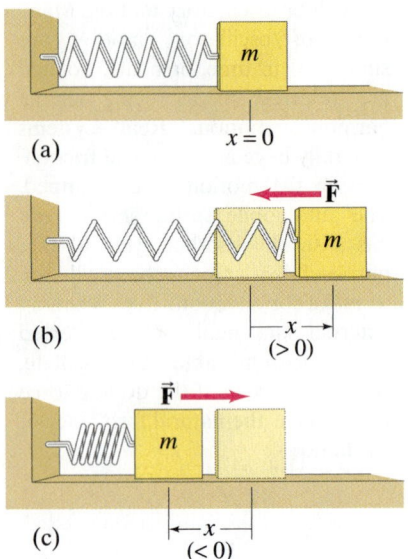

(a) $x = 0$

(b) $x \rightarrow$
       (> 0)

(c) x
       (< 0)

FIGURE 14–1 A mass oscillating at the end of a uniform spring.

FIGURE 14–2 Force on, and velocity of, a mass at different positions of its oscillation cycle on a frictionless surface.

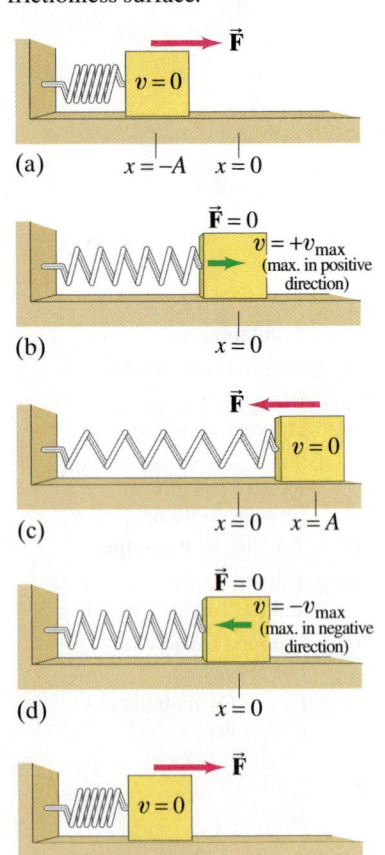

(a) $x = -A$ $x = 0$

(b) $\vec{F} = 0$
       $v = +v_{max}$
       (max. in positive direction)
       $x = 0$

(c) $x = 0$ $x = A$

(d) $\vec{F} = 0$
       $v = -v_{max}$
       (max. in negative direction)
       $x = 0$

(e) $x = -A$ $x = 0$

14–1 Oscillations of a Spring

When an object **vibrates** or **oscillates** back and forth, over the same path, each oscillation taking the same amount of time, the motion is **periodic**. The simplest form of periodic motion is represented by an object oscillating on the end of a uniform coil spring. Because many other types of oscillatory motion closely resemble this system, we will look at it in detail. We assume that the mass of the spring can be ignored, and that the spring is mounted horizontally, as shown in Fig. 14–1a, so that the object of mass m slides without friction on the horizontal surface. Any spring has a natural length at which it exerts no force on the mass m. The position of the mass at this point is called the **equilibrium position**. If the mass is moved either to the left, which compresses the spring, or to the right, which stretches it, the spring exerts a force on the mass that acts in the direction of returning the mass to the equilibrium position; hence it is called a *restoring force*. We consider the common situation where we can assume the restoring force F is directly proportional to the displacement x the spring has been stretched (Fig. 14–1b) or compressed (Fig. 14–1c) from the equilibrium position:

$$F = -kx. \qquad \text{[force exerted by spring]} \quad \textbf{(14–1)}$$

Note that the equilibrium position has been chosen at $x = 0$ and the minus sign in Eq. 14–1 indicates that the restoring force is always in the direction opposite to the displacement x. For example, if we choose the positive direction to the right in Fig. 14–1, x is positive when the spring is stretched (Fig. 14–1b), but the direction of the restoring force is to the left (negative direction). If the spring is compressed, x is negative (to the left) but the force F acts toward the right (Fig. 14–1c).

Equation 14–1 is often referred to as Hooke's law (Sections 7–3, 8–2 and 12–4), and is accurate only if the spring is not compressed to where the coils are close to touching, or stretched beyond the elastic region (see Fig. 12–15). Hooke's law works not only for springs but for other oscillating solids as well; it thus has wide applicability, even though it is valid only over a certain range of F and x values.

The proportionality constant k in Eq. 14–1 is called the *spring constant* for that particular spring, or its *spring stiffness constant*. To stretch the spring a distance x, one has to exert an (external) force on the free end of the spring with a magnitude at least equal to

$$F_{ext} = +kx. \qquad \text{[external force on spring]}$$

The greater the value of k, the greater the force needed to stretch a spring a given distance. That is, the stiffer the spring, the greater the spring constant k.

Note that the force F in Eq. 14–1 is *not* a constant, but varies with position. Therefore the acceleration of the mass m is not constant, so we *cannot* use the equations for constant acceleration developed in Chapter 2.

Let us examine what happens when our uniform spring is initially compressed a distance $x = -A$, as shown in Fig. 14–2a, and then released on the frictionless surface. The spring exerts a force on the mass that pushes it toward the equilibrium position. But because the mass has inertia, it passes the equilibrium position with considerable speed. Indeed, as the mass reaches the equilibrium position, the force on it decreases to zero, but its speed at this point is a maximum, v_{max} (Fig. 14–2b). As the mass moves farther to the right, the force on it acts to slow it down, and it stops for an instant at $x = A$ (Fig. 14–2c). It then begins moving back in the opposite direction, accelerating until it passes the equilibrium point (Fig. 14–2d), and then slows down until it reaches zero speed at the original starting point, $x = -A$ (Fig. 14–2e). It then repeats the motion, moving back and forth symmetrically between $x = A$ and $x = -A$.

EXERCISE A An object is oscillating back and forth. Which of the following statements are true at some time during the course of the motion? (*a*) The object can have zero velocity and, simultaneously, nonzero acceleration. (*b*) The object can have zero velocity and, simultaneously, zero acceleration. (*c*) The object can have zero acceleration and, simultaneously, nonzero velocity. (*d*) The object can have nonzero velocity and nonzero acceleration simultaneously.

EXERCISE B A mass is oscillating on a frictionless surface at the end of a horizontal spring. Where, if anywhere, is the acceleration of the mass zero (see Fig. 14–2)? (*a*) At $x = -A$; (*b*) at $x = 0$; (*c*) at $x = +A$; (*d*) at both $x = -A$ and $x = +A$; (*e*) nowhere.

To discuss oscillatory motion, we need to define a few terms. The distance x of the mass from the equilibrium point at any moment is called the **displacement**. The maximum displacement—the greatest distance from the equilibrium point—is called the **amplitude**, A. One **cycle** refers to the complete to-and-fro motion from some initial point back to that same point—say, from $x = -A$ to $x = A$ and back to $x = -A$. The **period**, T, is defined as the time required to complete one cycle. Finally, the **frequency**, f, is the number of complete cycles per second. Frequency is generally specified in hertz (Hz), where $1\ \text{Hz} = 1$ cycle per second (s^{-1}). It is easy to see, from their definitions, that frequency and period are inversely related, as we saw earlier (Eqs. 5–2 and 10–8):

$$f = \frac{1}{T} \quad \text{and} \quad T = \frac{1}{f}; \qquad \textbf{(14–2)}$$

for example, if the frequency is 5 cycles per second, then each cycle takes $\frac{1}{5}\,\text{s}$.

The oscillation of a spring hung vertically is essentially the same as that of a horizontal spring. Because of gravity, the length of a vertical spring with a mass m on the end will be longer at equilibrium than when that same spring is horizontal, as shown in Fig. 14–3. The spring is in equilibrium when $\Sigma F = 0 = mg - kx_0$, so the spring stretches an extra amount $x_0 = mg/k$ to be in equilibrium. If x is measured from this new equilibrium position, Eq. 14–1 can be used directly with the same value of k.

EXERCISE C If an oscillating mass has a frequency of 1.25 Hz, it makes 100 oscillations in (a) 12.5 s, (b) 125 s, (c) 80 s, (d) 8.0 s.

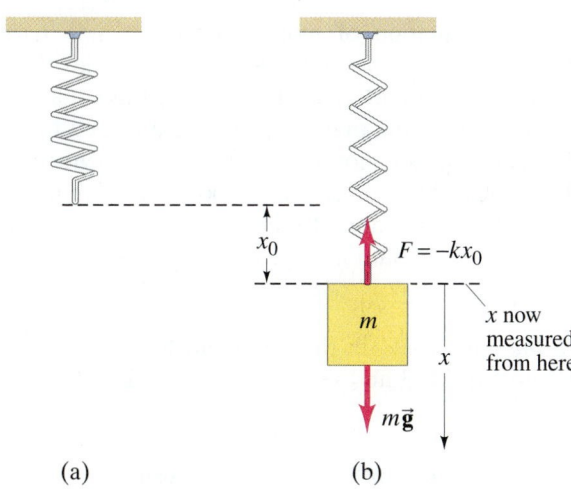

(a) (b)

FIGURE 14–3
(a) Free spring, hung vertically.
(b) Mass m attached to spring in new equilibrium position, which occurs when $\Sigma F = 0 = mg - kx_0$.

EXAMPLE 14–1 Car springs. When a family of four with a total mass of 200 kg step into their 1200-kg car, the car's springs compress 3.0 cm. (a) What is the spring constant of the car's springs (Fig. 14–4), assuming they act as a single spring? (b) How far will the car lower if loaded with 300 kg rather than 200 kg?

APPROACH We use Hooke's law: the weight of the people, mg, causes a 3.0-cm displacement.

SOLUTION (a) The added force of $(200\ \text{kg})(9.8\ \text{m/s}^2) = 1960\ \text{N}$ causes the springs to compress $3.0 \times 10^{-2}\ \text{m}$. Therefore (Eq. 14–1), the spring constant is

$$k = \frac{F}{x} = \frac{1960\ \text{N}}{3.0 \times 10^{-2}\ \text{m}} = 6.5 \times 10^4\ \text{N/m}.$$

(b) If the car is loaded with 300 kg, Hooke's law gives

$$x = \frac{F}{k} = \frac{(300\ \text{kg})(9.8\ \text{m/s}^2)}{(6.5 \times 10^4\ \text{N/m})} = 4.5 \times 10^{-2}\ \text{m},$$

or 4.5 cm.

NOTE In (b), we could have obtained x without solving for k: since x is proportional to F, if 200 kg compresses the spring 3.0 cm, then 1.5 times the force will compress the spring 1.5 times as much, or 4.5 cm.

FIGURE 14–4 Photo of a car's spring. (Also visible is the shock absorber, in blue—see Section 14–7.)

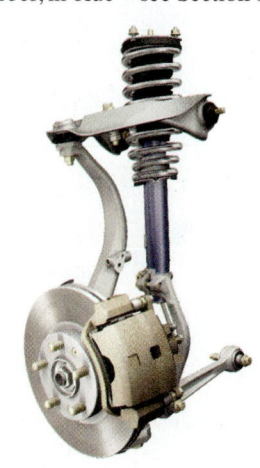

14–2 Simple Harmonic Motion

Any oscillating system for which the net restoring force is directly proportional to the negative of the displacement (as in Eq. 14–1, $F = -kx$) is said to exhibit **simple harmonic motion** (SHM). Such a system is often called a **simple harmonic oscillator** (SHO). We saw in Chapter 12 (Section 12–4) that most solid materials stretch or compress according to Eq. 14–1 as long as the displacement is not too great. Because of this, many natural oscillations are simple harmonic or close to it.

EXERCISE D Which of the following represents a simple harmonic oscillator: (a) $F = -0.5x^2$, (b) $F = -2.3y$, (c) $F = 8.6x$, (d) $F = -4\theta$?

Let us now determine the position x as a function of time for a mass attached to the end of a simple spring with spring constant k. To do so, we make use of Newton's second law, $F = ma$. Since the acceleration $a = d^2x/dt^2$, we have

$$ma = \Sigma F$$
$$m\frac{d^2x}{dt^2} = -kx,$$

where m is the mass[†] which is oscillating. We rearrange this to obtain

$$\frac{d^2x}{dt^2} + \frac{k}{m}x = 0, \qquad \text{[SHM]} \quad \textbf{(14–3)}$$

which is known as the **equation of motion** for the simple harmonic oscillator. Mathematically it is called a *differential equation*, since it involves derivatives. We want to determine what function of time, $x(t)$, satisfies this equation. We might guess the form of the solution by noting that if a pen were attached to an oscillating mass (Fig. 14–5) and a sheet of paper moved at a steady rate beneath it, the pen would trace the curve shown. The shape of this curve looks a lot like it might be **sinusoidal** (such as cosine or sine) as a function of time, and its height is the amplitude A. Let us then guess that the general solution to Eq. 14–3 can be written in a form such as

$$x = A\cos(\omega t + \phi), \qquad \textbf{(14–4)}$$

where we include the constant ϕ in the argument to be general.[‡] Let us now put this trial solution into Eq. 14–3 and see if it really works. We need to differentiate the $x = x(t)$ twice:

$$\frac{dx}{dt} = \frac{d}{dt}[A\cos(\omega t + \phi)] = -\omega A\sin(\omega t + \phi)$$

$$\frac{d^2x}{dt^2} = -\omega^2 A\cos(\omega t + \phi).$$

We now put the latter into Eq. 14–3, along with Eq. 14–4 for x:

$$\frac{d^2x}{dt^2} + \frac{k}{m}x = 0$$

$$-\omega^2 A\cos(\omega t + \phi) + \frac{k}{m}A\cos(\omega t + \phi) = 0$$

or

$$\left(\frac{k}{m} - \omega^2\right)A\cos(\omega t + \phi) = 0.$$

Our solution, Eq. 14–4, does indeed satisfy the equation of motion (Eq. 14–3) for

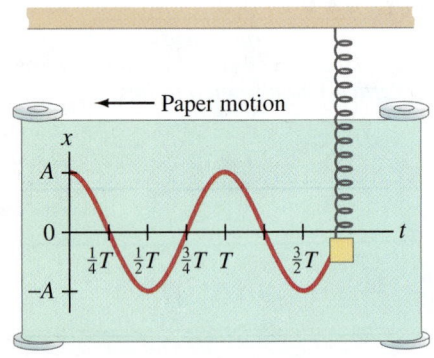

FIGURE 14–5 Sinusoidal nature of SHM as a function of time. In this case, $x = A\cos(2\pi t/T)$.

[†]In the case of a mass, m', on the end of a spring, the spring itself also oscillates and at least a part of its mass must be included. It can be shown—see the Problems—that approximately one-third the mass of the spring, m_s, must be included, so then $m = m' + \frac{1}{3}m_s$ in our equation. Often m_s is small enough to be ignored.

[‡]Another possible way to write the solution is the combination $x = a\cos\omega t + b\sin\omega t$, where a and b are constants. This is equivalent to Eq. 14–4 as can be seen using the trigonometric identity $\cos(A \pm B) = \cos A\cos B \mp \sin A\sin B$.

any time t, but only if $(k/m - \omega^2) = 0$. Hence

$$\omega^2 = \frac{k}{m}. \qquad \text{(14–5)}$$

Equation 14–4 is the general solution to Eq. 14–3, and it contains two arbitrary constants A and ϕ, which we should expect because the second derivative in Eq. 14–3 implies that two integrations are needed, each yielding a constant. They are "arbitrary" only in a calculus sense, in that they can be anything and still satisfy the differential equation, Eq. 14–3. In real physical situations, however, A and ϕ are determined by the **initial conditions**. Suppose, for example, that the mass is started at its maximum displacement and is released from rest. This is, in fact, what is shown in Fig. 14–5, and for this case $x = A \cos \omega t$. Let us confirm it: we are given $v = 0$ at $t = 0$, where

$$v = \frac{dx}{dt} = \frac{d}{dt}[A \cos(\omega t + \phi)] = -\omega A \sin(\omega t + \phi) = 0. \qquad [\text{at } t = 0]$$

For v to be zero at $t = 0$, then $\sin(\omega t + \phi) = \sin(0 + \phi)$ is zero if $\phi = 0$ (ϕ could also be $\pi, 2\pi$, etc.), and when $\phi = 0$, then

$$x = A \cos \omega t,$$

as we expected. We see immediately that A is the amplitude of the motion, and it is determined initially by how far you pulled the mass m from equilibrium before releasing it.

Consider another interesting case: at $t = 0$, the mass m is at $x = 0$ and is struck, giving it an initial velocity toward increasing values of x. Then at $t = 0$, $x = 0$, so we can write $x = A \cos(\omega t + \phi) = A \cos \phi = 0$, which can happen only if $\phi = \pm \pi/2$ (or $\pm 90°$). Whether $\phi = +\pi/2$ or $-\pi/2$ depends on $v = dx/dt = -\omega A \sin(\omega t + \phi) = -\omega A \sin \phi$ at $t = 0$, which we are given as positive ($v > 0$ at $t = 0$); hence $\phi = -\pi/2$ because $\sin(-90°) = -1$. Thus our solution for this case is

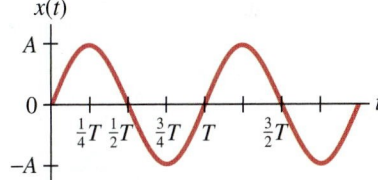

$x(t)$

FIGURE 14–6 Special case of SHM where the mass m starts, at $t = 0$, at the equilibrium position $x = 0$ and has initial velocity toward positive values of x ($v > 0$ at $t = 0$).

$$x = A \cos\left(\omega t - \frac{\pi}{2}\right)$$

$$= A \sin \omega t,$$

where we used $\cos(\theta - \pi/2) = \sin \theta$. The solution in this case is a pure sine wave, Fig. 14–6, where A is still the amplitude.

Many other situations are possible, such as that shown in Fig. 14–7. The constant ϕ is called the **phase angle**, and it tells us how long after (or before) $t = 0$ the peak at $x = A$ is reached. Notice that the value of ϕ does not affect the shape of the $x(t)$ curve, but only affects the displacement at some arbitrary time, $t = 0$. Simple harmonic motion is thus always *sinusoidal*. Indeed, simple harmonic motion is *defined* as motion that is purely sinusoidal.

Since our oscillating mass repeats its motion after a time equal to its period T, it must be at the same position and moving in the same direction at $t = T$ as it was at $t = 0$. Since a sine or cosine function repeats itself after every 2π radians, then from Eq. 14–4, we must have

$$\omega T = 2\pi.$$

Hence

$$\omega = \frac{2\pi}{T} = 2\pi f,$$

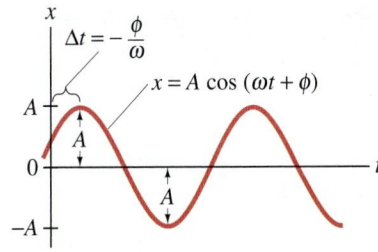

FIGURE 14–7 A plot of $x = A \cos(\omega t + \phi)$ when $\phi < 0$.

where f is the frequency of the motion. Strictly speaking, we call ω the **angular frequency** (units are rad/s) to distinguish it from the frequency f (units are $\text{s}^{-1} = \text{Hz}$); sometimes the word "angular" is dropped, so the symbol ω or f needs to be specified. Because $\omega = 2\pi f = 2\pi/T$, we can write Eq. 14–4 as

$$x = A \cos\left(\frac{2\pi t}{T} + \phi\right) \qquad \text{(14–6a)}$$

or

$$x = A \cos(2\pi f t + \phi). \qquad \text{(14–6b)}$$

Because $\omega = 2\pi f = \sqrt{k/m}$ (Eq. 14–5), then

$$f = \frac{1}{2\pi}\sqrt{\frac{k}{m}}, \qquad \text{(14–7a)}$$

$$T = 2\pi\sqrt{\frac{m}{k}}. \qquad \text{(14–7b)}$$

Note that the *frequency and period do not depend on the amplitude*. Changing the amplitude of a simple harmonic oscillator does not affect its frequency. Equation 14–7a tells us that the greater the mass, the lower the frequency; and the stiffer the spring, the higher the frequency. This makes sense since a greater mass means more inertia and therefore a slower response (or acceleration); and larger k means greater force and therefore quicker response. The frequency f (Eq. 14–7a) at which a SHO oscillates naturally is called its **natural frequency** (to distinguish it from a frequency at which it might be forced to oscillate by an outside force, as discussed in Section 14–8).

The simple harmonic oscillator is important in physics because whenever we have a net restoring force proportional to the displacement $(F = -kx)$, which is at least a good approximation for a variety of systems, then the motion is simple harmonic—that is, sinusoidal.

⊕ **PHYSICS APPLIED**
Car springs

EXAMPLE 14–2 **Car springs again.** Determine the period and frequency of the car in Example 14–1a after hitting a bump. Assume the shock absorbers are poor, so the car really oscillates up and down.

APPROACH We put $m = 1400$ kg and $k = 6.5 \times 10^4$ N/m from Example 14–1a into Eqs. 14–7.

SOLUTION From Eq. 14–7b,

$$T = 2\pi\sqrt{\frac{m}{k}} = 2\pi\sqrt{\frac{1400 \text{ kg}}{6.5 \times 10^4 \text{ N/m}}} = 0.92 \text{ s},$$

or slightly less than a second. The frequency $f = 1/T = 1.09$ Hz.

EXERCISE E By how much should the mass on the end of a spring be changed to halve the frequency of its oscillations? (*a*) No change; (*b*) doubled; (*c*) quadrupled; (*d*) halved; (*e*) quartered.

EXERCISE F The position of a SHO is given by $x = (0.80 \text{ m}) \cos(3.14t - 0.25)$. The frequency is (*a*) 3.14 Hz, (*b*) 1.0 Hz, (*c*) 0.50 Hz, (*d*) 9.88 Hz, (*e*) 19.8 Hz.

Let us continue our analysis of a simple harmonic oscillator. The velocity and acceleration of the oscillating mass can be obtained by differentiation of Eq. 14–4, $x = A\cos(\omega t + \phi)$:

$$v = \frac{dx}{dt} = -\omega A \sin(\omega t + \phi) \qquad \text{(14–8a)}$$

$$a = \frac{d^2x}{dt^2} = \frac{dv}{dt} = -\omega^2 A \cos(\omega t + \phi). \qquad \text{(14–8b)}$$

The velocity and acceleration of a SHO also vary sinusoidally. In Fig. 14–8 we plot the displacement, velocity, and acceleration of a SHO as a function of time for the case when $\phi = 0$. As can be seen, the speed reaches its maximum

$$v_{\max} = \omega A = \sqrt{\frac{k}{m}}\, A \qquad \text{(14–9a)}$$

when the oscillating object is passing through its equilibrium point, $x = 0$. And the speed is zero at points of maximum displacement, $x = \pm A$. This is in accord with our discussion of Fig. 14–2. Similarly, the acceleration has its maximum value

$$a_{\max} = \omega^2 A = \frac{k}{m}\, A \qquad \text{(14–9b)}$$

which occurs where $x = \pm A$; and a is zero at $x = 0$, as we expect, since $ma = F = -kx$.

FIGURE 14–8 Displacement, x, velocity, dx/dt, and acceleration, d^2x/dt^2, of a simple harmonic oscillator when $\phi = 0$.

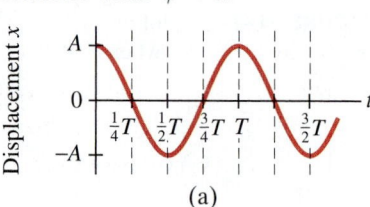

(a)

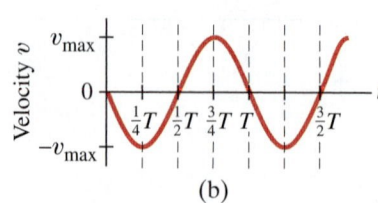

(b)

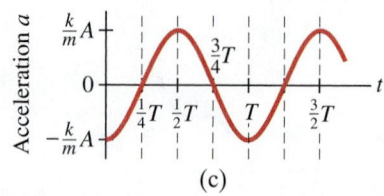

(c)

For the general case when $\phi \neq 0$, we can relate the constants A and ϕ to the initial values of x, v, and a by setting $t = 0$ in Eqs. 14–4, 14–8, and 14–9:

$$x_0 = x(0) = A \cos \phi$$

$$v_0 = v(0) = -\omega A \sin \phi = -v_{max} \sin \phi$$

$$a_0 = a(0) = -\omega^2 A \cos \phi = -a_{max} \cos \phi.$$

EXAMPLE 14–3 ESTIMATE A vibrating floor. A large motor in a factory causes the floor to vibrate at a frequency of 10 Hz. The amplitude of the floor's motion near the motor is about 3.0 mm. Estimate the maximum acceleration of the floor near the motor.

APPROACH Assuming the motion of the floor is roughly SHM we can make an estimate for the maximum acceleration using Eq. 14–9b.

SOLUTION Given $\omega = 2\pi f = (2\pi)(10 \, \text{s}^{-1}) = 62.8 \, \text{rad/s}$, then Eq. 14–9b gives

$$a_{max} = \omega^2 A = (62.8 \, \text{rad/s})^2 (0.0030 \, \text{m}) = 12 \, \text{m/s}^2.$$

NOTE The maximum acceleration is a little over g, so when the floor accelerates down, objects sitting on the floor will actually lose contact momentarily, which will cause noise and serious wear.

EXAMPLE 14–4 Loudspeaker. The cone of a loudspeaker (Fig. 14–9) oscillates in SHM at a frequency of 262 Hz ("middle C"). The amplitude at the center of the cone is $A = 1.5 \times 10^{-4} \, \text{m}$, and at $t = 0, x = A$. (a) What equation describes the motion of the center of the cone? (b) What are the velocity and acceleration as a function of time? (c) What is the position of the cone at $t = 1.00 \, \text{ms} \, (= 1.00 \times 10^{-3} \, \text{s})$?

APPROACH The motion begins $(t = 0)$ with the cone at its maximum displacement $(x = A$ at $t = 0)$. So we use the cosine function, $x = A \cos \omega t$, with $\phi = 0$.

SOLUTION (a) The amplitude $A = 1.5 \times 10^{-4} \, \text{m}$ and

$$\omega = 2\pi f = (6.28 \, \text{rad})(262 \, \text{s}^{-1}) = 1650 \, \text{rad/s}.$$

The motion is described as

$$x = A \cos \omega t = (1.5 \times 10^{-4} \, \text{m}) \cos(1650t),$$

where t is in seconds.
(b) The maximum velocity, from Eq. 14–9a, is

$$v_{max} = \omega A = (1650 \, \text{rad/s})(1.5 \times 10^{-4} \, \text{m}) = 0.25 \, \text{m/s},$$

so

$$v = -(0.25 \, \text{m/s}) \sin(1650t).$$

From Eq. 14–9b the maximum acceleration is $a_{max} = \omega^2 A = (1650 \, \text{rad/s})^2 (1.5 \times 10^{-4} \, \text{m}) = 410 \, \text{m/s}^2$, which is more than 40 g's. Then

$$a = -(410 \, \text{m/s}^2) \cos(1650t).$$

(c) At $t = 1.00 \times 10^{-3} \, \text{s}$,

$$x = A \cos \omega t = (1.5 \times 10^{-4} \, \text{m}) \cos[(1650 \, \text{rad/s})(1.00 \times 10^{-3} \, \text{s})]$$

$$= (1.5 \times 10^{-4} \, \text{m}) \cos(1.65 \, \text{rad}) = -1.2 \times 10^{-5} \, \text{m}.$$

NOTE Be sure your calculator is set in RAD mode, not DEG mode, for these $\cos \omega t$ calculations.

FIGURE 14–9 Example 14–4. A loudspeaker cone.

CAUTION
Always be sure your calculator is in the correct mode for angles

EXAMPLE 14–5 **Spring calculations.** A spring stretches 0.150 m when a 0.300-kg mass is gently attached to it as in Fig. 14–3b. The spring is then set up horizontally with the 0.300-kg mass resting on a frictionless table as in Fig. 14–2. The mass is pushed so that the spring is compressed 0.100 m from the equilibrium point, and released from rest. Determine: (a) the spring stiffness constant k and angular frequency ω; (b) the amplitude of the horizontal oscillation A; (c) the magnitude of the maximum velocity v_{max}; (d) the magnitude of the maximum acceleration a_{max} of the mass; (e) the period T and frequency f; (f) the displacement x as a function of time; and (g) the velocity at $t = 0.150$ s.

APPROACH When the 0.300-kg mass hangs at rest from the spring as in Fig. 14–3b, we apply Newton's second law for the vertical forces: $\Sigma F = 0 = mg - kx_0$, so $k = mg/x_0$. For the horizontal oscillations, the amplitude is given, and the other quantities can be found from Eqs. 14–4, 14–5, 14–7, and 14–9. We choose x positive to the right.

SOLUTION (a) The spring stretches 0.150 m due to the 0.300-kg load, so

$$k = \frac{F}{x_0} = \frac{mg}{x_0} = \frac{(0.300 \text{ kg})(9.80 \text{ m/s}^2)}{0.150 \text{ m}} = 19.6 \text{ N/m}.$$

From Eq. 14–5,

$$\omega = \sqrt{\frac{k}{m}} = \sqrt{\frac{19.6 \text{ N/m}}{0.300 \text{ kg}}} = 8.08 \text{ s}^{-1}.$$

(b) The spring is now horizontal (on a table). It is compressed 0.100 m from equilibrium and is given no initial speed, so $A = 0.100$ m.

(c) From Eq. 14–9a, the maximum velocity has magnitude

$$v_{max} = \omega A = (8.08 \text{ s}^{-1})(0.100 \text{ m}) = 0.808 \text{ m/s}.$$

(d) Since $F = ma$, the maximum acceleration occurs where the force is greatest—that is, when $x = \pm A = \pm 0.100$ m. Thus its magnitude is

$$a_{max} = \frac{F}{m} = \frac{kA}{m} = \frac{(19.6 \text{ N/m})(0.100 \text{ m})}{0.300 \text{ kg}} = 6.53 \text{ m/s}^2.$$

[This result could also have been obtained directly from Eq. 14–9b, but it is often useful to go back to basics as we did here.]

(e) Equations 14–7b and 14–2 give

$$T = 2\pi \sqrt{\frac{m}{k}} = 2\pi \sqrt{\frac{0.300 \text{ kg}}{19.6 \text{ N/m}}} = 0.777 \text{ s}$$

$$f = \frac{1}{T} = 1.29 \text{ Hz}.$$

(f) The motion begins at a point of maximum compression. If we take x positive to the right in Fig. 14–2, then at $t = 0$, $x = -A = -0.100$ m. So we need a sinusoidal curve that has its maximum negative value at $t = 0$; this is just a negative cosine:

$$x = -A \cos \omega t.$$

To write this in the form of Eq. 14–4 (no minus sign), recall that $\cos \theta = -\cos(\theta - \pi)$. Then, putting in numbers, and recalling $-\cos \theta = \cos(\pi - \theta) = \cos(\theta - \pi)$, we have

$$x = -(0.100 \text{ m}) \cos 8.08t$$
$$= (0.100 \text{ m}) \cos(8.08t - \pi),$$

where t is in seconds and x is in meters. Note that the phase angle (Eq. 14–4) is $\phi = -\pi$ or $-180°$.

(g) The velocity at any time t is dx/dt (see also part c):

$$v = \frac{dx}{dt} = \omega A \sin \omega t = (0.808 \text{ m/s}) \sin 8.08t.$$

At $t = 0.150$ s, $v = (0.808 \text{ m/s}) \sin(1.21 \text{ rad}) = 0.756 \text{ m/s}$, and is to the right (+).

Spring is started with a push. Suppose the spring of Example 14–5 is compressed 0.100 m from equilibrium $(x_0 = -0.100\text{ m})$ but is given a shove to create a velocity in the $+x$ direction of $v_0 = 0.400\text{ m/s}$. Determine (a) the phase angle ϕ, (b) the amplitude A, and (c) the displacement x as a function of time, $x(t)$.

APPROACH We use Eq. 14–8a, at $t = 0$, to write $v_0 = -\omega A \sin\phi$, and Eq. 14–4 to write $x_0 = A\cos\phi$. Combining these, we can obtain ϕ. We obtain A by using Eq. 14–4 again at $t = 0$. From Example 14–5, $\omega = 8.08\text{ s}^{-1}$.

SOLUTION (a) We combine Eqs. 14–8a and 14–4 at $t = 0$ and solve for the tangent:

$$\tan\phi = \frac{\sin\phi}{\cos\phi} = \frac{(v_0/-\omega A)}{(x_0/A)} = -\frac{v_0}{\omega x_0} = -\frac{0.400\text{ m/s}}{(8.08\text{ s}^{-1})(-0.100\text{ m})} = 0.495.$$

A calculator gives the angle as 26.3°, but we note from this equation that both the sine and cosine are negative, so our angle is in the third quadrant. Hence

$$\phi = 26.3° + 180° = 206.3° = 3.60\text{ rad}.$$

(b) Again using Eq. 14–4 at $t = 0$, as given in the Approach above,

$$A = \frac{x_0}{\cos\phi} = \frac{(-0.100\text{ m})}{\cos(3.60\text{ rad})} = 0.112\text{ m}.$$

(c) $x = A\cos(\omega t + \phi) = (0.112\text{ m})\cos(8.08t + 3.60).$

14–3 Energy in the Simple Harmonic Oscillator

When forces are not constant, as is the case here with simple harmonic motion, it is often convenient and useful to use the energy approach, as we saw in Chapters 7 and 8.

For a simple harmonic oscillator, such as a mass m oscillating on the end of a massless spring, the restoring force is given by

$$F = -kx.$$

The potential energy function, as we saw in Chapter 8, is given by

$$U = -\int F\,dx = \tfrac{1}{2}kx^2,$$

where we set the constant of integration equal to zero so $U = 0$ at $x = 0$ (the equilibrium position).

The total mechanical energy is the sum of the kinetic and potential energies,

$$E = \tfrac{1}{2}mv^2 + \tfrac{1}{2}kx^2,$$

where v is the velocity of the mass m when it is a distance x from the equilibrium position. SHM can occur only if there is no friction, so the total mechanical energy E remains constant. As the mass oscillates back and forth, the energy continuously changes from potential energy to kinetic energy, and back again (Fig. 14–10). At the extreme points, $x = A$ and $x = -A$, all the energy is stored in the spring as potential energy (and is the same whether the spring is compressed or stretched to the full amplitude). At these extreme points, the mass stops for an instant as it changes direction, so $v = 0$ and:

$$E = \tfrac{1}{2}m(0)^2 + \tfrac{1}{2}kA^2 = \tfrac{1}{2}kA^2. \tag{14–10a}$$

Thus, the *total mechanical energy of a simple harmonic oscillator is proportional to the square of the amplitude.* At the equilibrium point, $x = 0$, all the energy is kinetic:

$$E = \tfrac{1}{2}mv^2 + \tfrac{1}{2}k(0)^2 = \tfrac{1}{2}mv_{max}^2, \tag{14–10b}$$

where v_{max} is the maximum velocity during the motion. At intermediate points the energy is part kinetic and part potential, and because energy is conserved

$$E = \tfrac{1}{2}mv^2 + \tfrac{1}{2}kx^2. \tag{14–10c}$$

We can confirm Eqs. 14–10a and b explicitly by inserting Eqs. 14–4 and 14–8a into this last relation:

$$E = \tfrac{1}{2}m\omega^2 A^2 \sin^2(\omega t + \phi) + \tfrac{1}{2}kA^2\cos^2(\omega t + \phi).$$

Substituting with $\omega^2 = k/m$, or $kA^2 = m\omega^2 A^2 = mv_{max}^2$, and recalling the important trigonometric identity $\sin^2(\omega t + \phi) + \cos^2(\omega t + \phi) = 1$, we obtain Eqs. 14–10a and b:

$$E = \tfrac{1}{2}kA^2 = \tfrac{1}{2}mv_{max}^2.$$

[Note that we can check for consistency by solving this equation for v_{max}, and we do get Eq. 14–9a.]

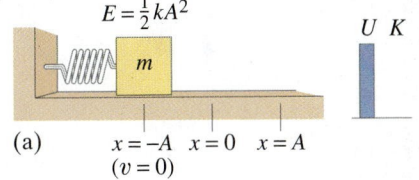

(a) $x = -A$ $x = 0$ $x = A$
$(v = 0)$

$E = \tfrac{1}{2}kA^2$

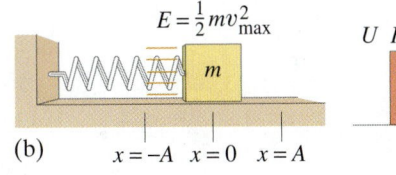

(b) $x = -A$ $x = 0$ $x = A$

$E = \tfrac{1}{2}mv_{max}^2$

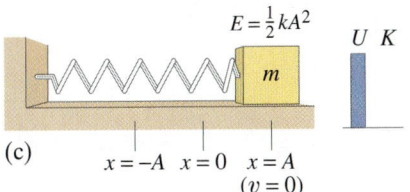

(c) $x = -A$ $x = 0$ $x = A$
$(v = 0)$

$E = \tfrac{1}{2}kA^2$

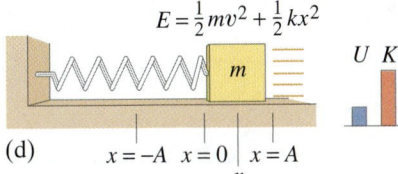

(d) $x = -A$ $x = 0$ $x = A$
x

$E = \tfrac{1}{2}mv^2 + \tfrac{1}{2}kx^2$

FIGURE 14–10 Energy changes from potential energy to kinetic energy and back again as the spring oscillates. Energy bar graphs (on the right) are described in Section 8–4.

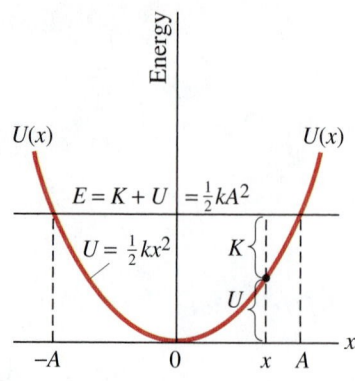

FIGURE 14–11 Graph of potential energy, $U = \frac{1}{2}kx^2$. $K + U = E =$ constant for any point x where $-A \leq x \leq A$. Values of K and U are indicated for an arbitrary position x.

We can now obtain an equation for the velocity v as a function of x by solving for v in Eq. 14–10c:

$$v = \pm\sqrt{\frac{k}{m}(A^2 - x^2)} \qquad (14\text{–}11\text{a})$$

or, since $v_{max} = A\sqrt{k/m}$,

$$v = \pm v_{max}\sqrt{1 - \frac{x^2}{A^2}}. \qquad (14\text{–}11\text{b})$$

Again we see that v is a maximum at $x = 0$, and is zero at $x = \pm A$.

The potential energy, $U = \frac{1}{2}kx^2$, is plotted in Fig. 14–11 (see also Section 8–9). The upper horizontal line represents a particular value of the total energy $E = \frac{1}{2}kA^2$. The distance between the E line and the U curve represents the kinetic energy, K, and the motion is restricted to x values between $-A$ and $+A$. These results are, of course, consistent with our full solution of the previous Section.

Energy conservation is a convenient way to obtain v, for example, if x is given (or vice versa), without having to deal with time t.

EXAMPLE 14–7 **Energy calculations.** For the simple harmonic oscillation of Example 14–5, determine (a) the total energy, (b) the kinetic and potential energies as a function of time, (c) the velocity when the mass is 0.050 m from equilibrium, (d) the kinetic and potential energies at half amplitude ($x = \pm A/2$).

APPROACH We use conservation of energy for a spring–mass system, Eqs. 14–10 and 14–11.

SOLUTION (a) From Example 14–5, $k = 19.6\,\text{N/m}$ and $A = 0.100\,\text{m}$, so the total energy E from Eq. 14–10a is

$$E = \tfrac{1}{2}kA^2 = \tfrac{1}{2}(19.6\,\text{N/m})(0.100\,\text{m})^2 = 9.80 \times 10^{-2}\,\text{J}.$$

(b) We have, from parts (f) and (g) of Example 14–5, $x = -(0.100\,\text{m})\cos 8.08t$ and $v = (0.808\,\text{m/s})\sin 8.08t$, so

$$U = \tfrac{1}{2}kx^2 = \tfrac{1}{2}(19.6\,\text{N/m})(0.100\,\text{m})^2\cos^2 8.08t = (9.80 \times 10^{-2}\,\text{J})\cos^2 8.08t$$

$$K = \tfrac{1}{2}mv^2 = \tfrac{1}{2}(0.300\,\text{kg})(0.808\,\text{m/s})^2\sin^2 8.08t = (9.80 \times 10^{-2}\,\text{J})\sin^2 8.08t.$$

(c) We use Eq. 14–11b and find

$$v = v_{max}\sqrt{1 - x^2/A^2} = (0.808\,\text{m/s})\sqrt{1 - (\tfrac{1}{2})^2} = 0.70\,\text{m/s}.$$

(d) At $x = A/2 = 0.050\,\text{m}$, we have

$$U = \tfrac{1}{2}kx^2 = \tfrac{1}{2}(19.6\,\text{N/m})(0.050\,\text{m})^2 = 2.5 \times 10^{-2}\,\text{J}$$

$$K = E - U = 7.3 \times 10^{-2}\,\text{J}.$$

CONCEPTUAL EXAMPLE 14–8 **Doubling the amplitude.** Suppose the spring in Fig. 14–10 is stretched twice as far (to $x = 2A$). What happens to (a) the energy of the system, (b) the maximum velocity of the oscillating mass, (c) the maximum acceleration of the mass?

RESPONSE (a) From Eq. 14–10a, the total energy is proportional to the square of the amplitude A, so stretching it twice as far quadruples the energy ($2^2 = 4$). You may protest, "I did work stretching the spring from $x = 0$ to $x = A$. Don't I do the same work stretching it from A to $2A$?" No. The force you exert is proportional to the displacement x, so for the second displacement, from $x = A$ to $2A$, you do more work than for the first displacement ($x = 0$ to A). (b) From Eq. 14–10b, we can see that when the energy is quadrupled, the maximum velocity must be doubled. $[v_{max} \propto \sqrt{E} \propto A.]$ (c) Since the force is twice as great when we stretch the spring twice as far, the acceleration is also twice as great: $a \propto F \propto x$.

EXERCISE G Suppose the spring in Fig. 14–10 is compressed to $x = -A$, but is given a push to the right so that the initial speed of the mass m is v_0. What effect does this push have on (a) the energy of the system, (b) the maximum velocity, (c) the maximum acceleration?

14–4 Simple Harmonic Motion Related to Uniform Circular Motion

Simple harmonic motion has a simple relationship to a particle rotating in a circle with uniform speed. Consider a mass m rotating in a circle of radius A with speed v_M on top of a table as shown in Fig. 14–12. As viewed from above, the motion is a circle. But a person who looks at the motion from the edge of the table sees an oscillatory motion back and forth, and this corresponds precisely to SHM as we shall now see. What the person sees, and what we are interested in, is the projection of the circular motion onto the x axis, Fig. 14–12. To see that this motion is analogous to SHM, let us calculate the x component of the velocity v_M which is labeled v in Fig. 14–12. The two right triangles indicated in Fig. 14–12 are similar, so

$$\frac{v}{v_M} = \frac{\sqrt{A^2 - x^2}}{A}$$

or

$$v = v_M \sqrt{1 - \frac{x^2}{A^2}}.$$

This is exactly the equation for the speed of a mass oscillating with SHM, Eq. 14–11b, where $v_M = v_{max}$. Furthermore, we can see from Fig. 14–12 that if the angular displacement at $t = 0$ is ϕ, then after a time t the particle will have rotated through an angle $\theta = \omega t$, and so

$$x = A \cos(\theta + \phi) = A \cos(\omega t + \phi).$$

But what is ω here? The linear velocity v_M of our particle undergoing rotational motion is related to ω by $v_M = \omega A$, where A is the radius of the circle (see Eq. 10–4, $v = R\omega$). To make one revolution requires a time T, so we also have $v_M = 2\pi A/T$ where $2\pi A$ is the circle's circumference. Hence

$$\omega = \frac{v_M}{A} = \frac{2\pi A/T}{A} = 2\pi/T = 2\pi f$$

where T is the time required for one rotation and f is the frequency. This corresponds precisely to the back-and-forth motion of a simple harmonic oscillator. Thus, the projection on the x axis of a particle rotating in a circle has the same motion as a mass undergoing SHM. Indeed, we can say that the projection of circular motion onto a straight line is simple harmonic motion.

The projection of uniform circular motion onto the y axis is also simple harmonic. Thus uniform circular motion can be thought of as two simple harmonic motions operating at right angles.

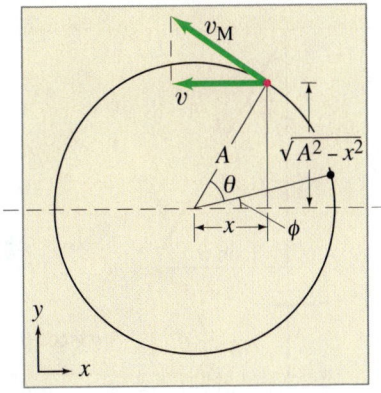

(a)

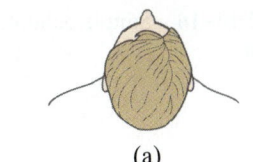

(b)

FIGURE 14–12 Analysis of simple harmonic motion as a side view (b) of circular motion (a).

FIGURE 14–13 Strobe-light photo of an oscillating pendulum photographed at equal time intervals.

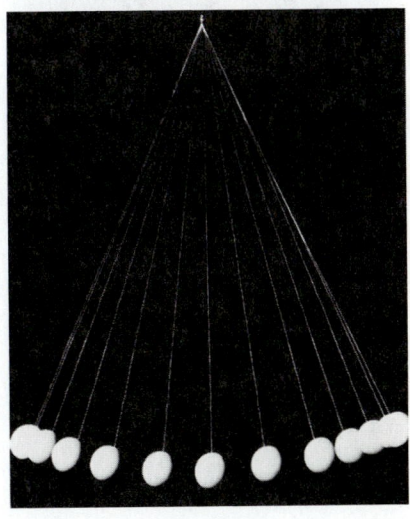

14–5 The Simple Pendulum

A **simple pendulum** consists of a small object (the pendulum bob) suspended from the end of a lightweight cord, Fig. 14–13. We assume that the cord does not stretch and that its mass can be ignored relative to that of the bob. The motion of a simple pendulum moving back and forth with negligible friction resembles simple harmonic motion: the pendulum oscillates along the arc of a circle with equal amplitude on either side of its equilibrium point and as it passes through the equilibrium point (where it would hang vertically) it has its maximum speed. But is it really undergoing SHM? That is, is the restoring force proportional to its displacement? Let us find out.

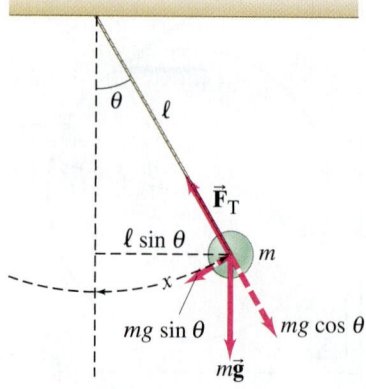

FIGURE 14–14 Simple pendulum.

FIGURE 14–15 The swinging motion of this lamp, hanging by a very long cord from the ceiling of the cathedral at Pisa, is said to have been observed by Galileo and to have inspired him to the conclusion that the period of a pendulum does not depend on amplitude.

The displacement of the pendulum along the arc is given by $x = \ell\theta$, where θ is the angle (in radians) that the cord makes with the vertical and ℓ is the length of the cord (Fig. 14–14). If the restoring force is proportional to x or to θ, the motion will be simple harmonic. The restoring force is the net force on the bob, equal to the component of the weight, mg, tangent to the arc:

$$F = -mg \sin\theta,$$

where g is the acceleration of gravity. The minus sign here, as in Eq. 14–1, means that the force is in the direction opposite to the angular displacement θ. Since F is proportional to the sine of θ and not to θ itself, the motion is *not* SHM. However, if θ is small, then $\sin\theta$ is very nearly equal to θ when the latter is specified in radians. This can be seen by looking at the series expansion[†] of $\sin\theta$ (or by looking at the trigonometry Table in Appendix A), or simply by noting in Fig. 14–14 that the arc length x ($=\ell\theta$) is nearly the same length as the chord ($=\ell\sin\theta$) indicated by the straight dashed line, *if θ is small*. For angles less than 15°, the difference between θ (in radians) and $\sin\theta$ is less than 1%. Thus, to a very good approximation for small angles,

$$F = -mg \sin\theta \approx -mg\theta.$$

Substituting $x = \ell\theta$ or $\theta = x/\ell$, we have

$$F \approx -\frac{mg}{\ell} x.$$

Thus, for small displacements, the motion is essentially simple harmonic, since this equation fits Hooke's law, $F = -kx$. The effective force constant is $k = mg/\ell$. Thus we can write

$$\theta = \theta_{max} \cos(\omega t + \phi)$$

where θ_{max} is the maximum angular displacement and $\omega = 2\pi f = 2\pi/T$. To obtain ω we use Eq. 14–5, where for k we substitute mg/ℓ: that is,[‡] $\omega = \sqrt{k/m} = \sqrt{(mg/\ell)/m}$, or

$$\omega = \sqrt{\frac{g}{\ell}}. \qquad [\theta \text{ small}] \quad \textbf{(14–12a)}$$

Then the frequency f is

$$f = \frac{\omega}{2\pi} = \frac{1}{2\pi}\sqrt{\frac{g}{\ell}}, \qquad [\theta \text{ small}] \quad \textbf{(14–12b)}$$

and the period T is

$$T = \frac{1}{f} = 2\pi\sqrt{\frac{\ell}{g}}. \qquad [\theta \text{ small}] \quad \textbf{(14–12c)}$$

The mass m of the pendulum bob does not appear in these formulas for T and f. Thus we have the surprising result that the period and frequency of a simple pendulum do not depend on the mass of the pendulum bob. You may have noticed this if you pushed a small child and then a large one on the same swing.

We also see from Eq. 14–12c that the period of a pendulum does not depend on the amplitude (like any SHM, Section 14–2), as long as the amplitude θ is small. Galileo is said to have first noted this fact while watching a swinging lamp in the cathedral at Pisa (Fig. 14–15). This discovery led to the invention of the pendulum clock, the first really precise timepiece, which became the standard for centuries.

Because a pendulum does not undergo *precisely* SHM, the period does depend slightly on the amplitude, the more so for large amplitudes. The accuracy of a pendulum clock would be affected, after many swings, by the decrease in amplitude due to friction. But the mainspring in a pendulum clock (or the falling weight in a grandfather clock) supplies energy to compensate for the friction and to maintain the amplitude constant, so that the timing remains precise.

[†]$\sin\theta = \theta - \dfrac{\theta^3}{3!} + \dfrac{\theta^5}{5!} - \dfrac{\theta^7}{7!} + \cdots.$

[‡]Be careful not to think that $\omega = d\theta/dt$ as in rotational motion. Here θ is the angle of the pendulum at any instant (Fig. 14–14), but we use ω now to represent *not* the rate this angle θ changes, but rather a constant related to the period, $\omega = 2\pi f = \sqrt{g/\ell}$.

EXERCISE H If a simple pendulum is taken from sea level to the top of a high mountain and started at the same angle of 5°, it would oscillate at the top of the mountain (*a*) slightly slower, (*b*) slightly faster, (*c*) at exactly the same frequency, (*d*) not at all—it would stop, (*e*) none of these.

EXERCISE I Return to the Chapter-Opening Question, p. 369, and answer it again now. Try to explain why you may have answered differently the first time.

EXAMPLE 14–9 **Measuring g.** A geologist uses a simple pendulum that has a length of 37.10 cm and a frequency of 0.8190 Hz at a particular location on the Earth. What is the acceleration of gravity at this location?

APPROACH We can use the length ℓ and frequency f of the pendulum in Eq. 14-12b, which contains our unknown, g.

SOLUTION We solve Eq. 14–12b for g and obtain

$$g = (2\pi f)^2 \ell = (6.283 \times 0.8190\,\text{s}^{-1})^2 (0.3710\,\text{m}) = 9.824\,\text{m/s}^2.$$

EXERCISE J (*a*) Estimate the length of a simple pendulum that makes one swing back and forth per second. (*b*) What would be the period of a 1.0-m-long pendulum?

*14–6 The Physical Pendulum and the Torsion Pendulum

Physical Pendulum

The term physical pendulum refers to any real extended object which oscillates back and forth, in contrast to the rather idealized simple pendulum where all the mass is assumed concentrated in the tiny pendulum bob. An example of a physical pendulum is a baseball bat suspended from the point O, as shown in Fig. 14–16. The force of gravity acts at the center of gravity (CG) of the object located a distance h from the pivot point O. The physical pendulum is best analyzed using the equations of rotational motion. The torque on a physical pendulum, calculated about point O, is

$$\tau = -mgh\sin\theta.$$

Newton's second law for rotational motion, Eq. 10–14, states that

$$\Sigma\tau = I\alpha = I\frac{d^2\theta}{dt^2},$$

where I is the moment of inertia of the object about the pivot point and $\alpha = d^2\theta/dt^2$ is the angular acceleration. Thus we have

$$I\frac{d^2\theta}{dt^2} = -mgh\sin\theta$$

or

$$\frac{d^2\theta}{dt^2} + \frac{mgh}{I}\sin\theta = 0,$$

where I is calculated about an axis through point O. For small angular amplitude, $\sin\theta \approx \theta$, so we have

$$\frac{d^2\theta}{dt^2} + \left(\frac{mgh}{I}\right)\theta = 0. \qquad \text{[small angular displacement]} \quad \textbf{(14–13)}$$

This is just the equation for SHM, Eq. 14–3, except that θ replaces x and mgh/I replaces k/m. Thus, for small angular displacements, a physical pendulum undergoes SHM, given by

$$\theta = \theta_{\text{max}}\cos(\omega t + \phi),$$

where θ_{max} is the maximum angular displacement and $\omega = 2\pi/T$. The period, T, is (see Eq. 14–7b, replacing m/k with I/mgh):

$$T = 2\pi\sqrt{\frac{I}{mgh}}. \qquad \text{[small angular displacement]} \quad \textbf{(14–14)}$$

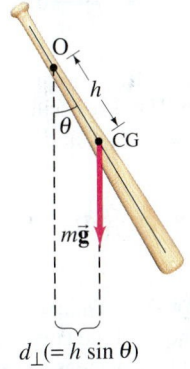

FIGURE 14–16 A physical pendulum suspended from point O.

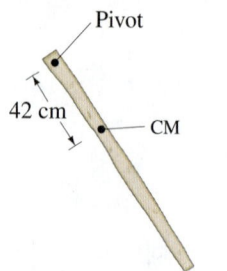

FIGURE 14–17 Example 14–10.

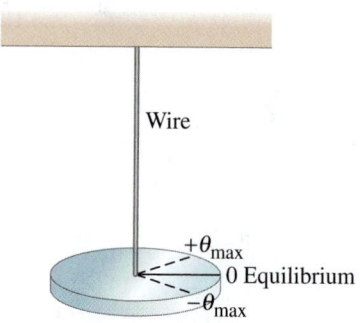

FIGURE 14–18 A torsion pendulum. The disc oscillates in SHM between θ_{max} and $-\theta_{max}$.

FIGURE 14–19 Damped harmonic motion. The solid red curve represents a cosine times a decreasing exponential (the dashed curves).

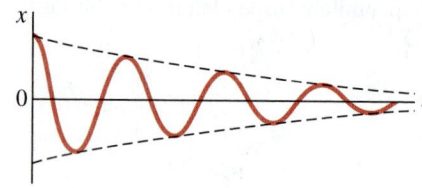

EXAMPLE 14–10 **Moment of inertia measurement.** An easy way to measure the moment of inertia of an object about any axis is to measure the period of oscillation about that axis. (a) Suppose a nonuniform 1.0-kg stick can be balanced at a point 42 cm from one end. If it is pivoted about that end (Fig. 14–17), it oscillates with a period of 1.6 s. What is its moment of inertia about this end? (b) What is its moment of inertia about an axis perpendicular to the stick through its center of mass?

APPROACH We put the given values into Eq. 14–14 and solve for I. For (b) we use the parallel-axis theorem (Section 10–7).

SOLUTION (a) Given $T = 1.6$ s, and $h = 0.42$ m, Eq. 14–14 gives

$$I = mghT^2/4\pi^2 = 0.27 \text{ kg} \cdot \text{m}^2.$$

(b) We use the parallel-axis theorem, Eq. 10–17. The CM is where the stick balanced, 42 cm from the end, so

$$I_{CM} = I - mh^2 = 0.27 \text{ kg} \cdot \text{m}^2 - (1.0 \text{ kg})(0.42 \text{ m})^2 = 0.09 \text{ kg} \cdot \text{m}^2.$$

NOTE Since an object does not oscillate about its CM, we cannot measure I_{CM} directly, but the parallel-axis theorem provides a convenient method to determine I_{CM}.

Torsion Pendulum

Another type of oscillatory motion is a **torsion pendulum**, in which a disc (Fig. 14–18) or a bar (as in Cavendish's apparatus, Fig. 6–3) is suspended from a wire. The twisting (torsion) of the wire serves as the elastic force. The motion here will be SHM since the restoring torque is very closely proportional to the negative of the angular displacement,

$$\tau = -K\theta,$$

where K is a constant that depends on the wire stiffness. Then

$$\omega = \sqrt{K/I}.$$

There is no small angle restriction here, as there is for the physical pendulum (where gravity acts), as long as the wire responds linearly in accordance with Hooke's law.

14–7 Damped Harmonic Motion

The amplitude of any real oscillating spring or swinging pendulum slowly decreases in time until the oscillations stop altogether. Figure 14–19 shows a typical graph of the displacement as a function of time. This is called **damped harmonic motion**. The damping[†] is generally due to the resistance of air and to internal friction within the oscillating system. The energy that is dissipated to thermal energy is reflected in a decreased amplitude of oscillation.

Since natural oscillating systems are damped in general, why do we even talk about (undamped) simple harmonic motion? The answer is that SHM is much easier to deal with mathematically. And if the damping is not large, the oscillations can be thought of as simple harmonic motion on which the damping is superposed, as represented by the dashed curves in Fig. 14–19. Although damping does alter the frequency of vibration, the effect is usually small if the damping is small. Let us look at this in more detail.

The damping force depends on the speed of the oscillating object, and opposes the motion. In some simple cases the damping force can be approximated as being directly proportional to the speed:

$$F_{damping} = -bv,$$

where b is a constant.[‡] For a mass oscillating on the end of a spring, the restoring force of the spring is $F = -kx$; so Newton's second law ($ma = \Sigma F$) becomes

$$ma = -kx - bv.$$

We bring all terms to the left side of the equation and substitute $v = dx/dt$ and

---

[†]To "damp" means to diminish, restrain, or extinguish, as to "dampen one's spirits."
[‡]Such velocity-dependent forces were discussed in Section 5–6.

$a = d^2x/dt^2$ to obtain

$$m\frac{d^2x}{dt^2} + b\frac{dx}{dt} + kx = 0, \tag{14–15}$$

which is the equation of motion. To solve this equation, we guess at a solution and then check to see if it works. If the damping constant b is small, x as a function of t is as plotted in Fig. 14–19, which looks like a cosine function times a factor (represented by the dashed lines) that decreases in time. A simple function that does this is the exponential, $e^{-\gamma t}$, and the solution that satisfies Eq. 14–15 is

$$x = Ae^{-\gamma t}\cos\omega' t, \tag{14–16}$$

where A, γ, and ω' are assumed to be constants, and $x = A$ at $t = 0$. We have called the angular frequency ω' (and not ω) because it is not the same as the ω for SHM without damping $(\omega = \sqrt{k/m})$.

If we substitute Eq. 14–16 into Eq. 14–15 (we do this in the optional subsection below), we find that Eq. 14–16 is indeed a solution if γ and ω' have the values

$$\gamma = \frac{b}{2m} \tag{14–17}$$

$$\omega' = \sqrt{\frac{k}{m} - \frac{b^2}{4m^2}}. \tag{14–18}$$

Thus x as a function of time t for a (lightly) damped harmonic oscillator is

$$x = Ae^{(-b/2m)t}\cos\omega' t. \tag{14–19}$$

Of course a phase constant, ϕ, can be added to the argument of the cosine in Eq. 14–19. As it stands with $\phi = 0$, it is clear that the constant A in Eq. 14–19 is simply the initial displacement, $x = A$ at $t = 0$. The frequency f' is

$$f' = \frac{\omega'}{2\pi} = \frac{1}{2\pi}\sqrt{\frac{k}{m} - \frac{b^2}{4m^2}}. \tag{14–20}$$

The frequency is lower, and the period longer, than for undamped SHM. (In many practical cases of light damping, however, ω' differs only slightly from $\omega = \sqrt{k/m}$.) This makes sense since we expect damping to slow down the motion. Equation 14–20 reduces to Eq. 14–7a, as it should, when there is no damping $(b = 0)$. The constant $\gamma = b/2m$ is a measure of how quickly the oscillations decrease toward zero (Fig. 14–19). The time $t_L = 2m/b$ is the time taken for the oscillations to drop to $1/e$ of the original amplitude; t_L is called the "mean lifetime" of the oscillations. Note that the larger b is, the more quickly the oscillations die away.

The solution, Eq. 14–19, is not valid if b is so large that

$$b^2 > 4mk$$

since then ω' (Eq. 14–18) would become imaginary. In this case the system does not oscillate at all but returns directly to its equilibrium position, as we now discuss.

Three common cases of *heavily damped* systems are shown in Fig. 14–20. Curve C represents the situation when the damping is so large $(b^2 \gg 4mk)$ that it takes a long time to reach equilibrium; the system is **overdamped**. Curve A represents an **underdamped** situation in which the system makes several swings before coming to rest $(b^2 < 4mk)$ and corresponds to a more heavily damped version of Eq. 14–19. Curve B represents **critical damping**: $b^2 = 4mk$; in this case equilibrium is reached in the shortest time. These terms all derive from the use of practical damped systems such as door-closing mechanisms and **shock absorbers** in a car (Fig. 14–21), which are usually designed to give critical damping. But as they wear out, underdamping occurs: a door slams and a car bounces up and down several times whenever it hits a bump.

In many systems, the oscillatory motion is what counts, as in clocks and watches, and damping needs to be minimized. In other systems, oscillations are the problem, such as a car's springs, so a proper amount of damping (i.e., critical) is desired. Well-designed damping is needed for all kinds of applications. Large buildings, especially in California, are now built (or retrofitted) with huge dampers to reduce earthquake damage.

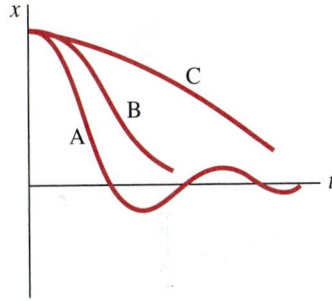

FIGURE 14–20 Underdamped (A), critically damped (B), and overdamped (C) motion.

FIGURE 14–21 Automobile spring and shock absorber provide damping so that a car won't bounce up and down endlessly.

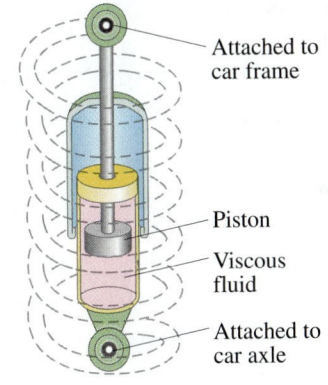

Attached to car frame

Piston

Viscous fluid

Attached to car axle

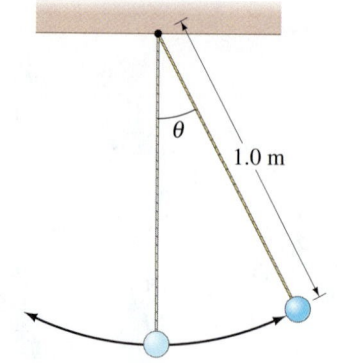

FIGURE 14–22 Example 14–11.

EXAMPLE 14–11 **Simple pendulum with damping.** A simple pendulum has a length of 1.0 m (Fig. 14–22). It is set swinging with small-amplitude oscillations. After 5.0 minutes, the amplitude is only 50% of what it was initially. (a) What is the value of γ for the motion? (b) By what factor does the frequency, f', differ from f, the undamped frequency?

APPROACH We assume the damping force is proportional to angular speed, $d\theta/dt$. The equation of motion for damped harmonic motion is

$$x = Ae^{-\gamma t}\cos\omega' t, \quad \text{where} \quad \gamma = \frac{b}{2m} \quad \text{and} \quad \omega' = \sqrt{\frac{k}{m} - \frac{b^2}{4m^2}},$$

for motion of a mass on the end of a spring. For the simple pendulum without damping, we saw in Section 14–5 that

$$F = -mg\theta$$

for small θ. Since $F = ma$, where a can be written in terms of the angular acceleration $\alpha = d^2\theta/dt^2$ as $a = \ell\alpha = \ell\, d^2\theta/dt^2$, then $F = m\ell\, d^2\theta/dt^2$, and

$$\ell\frac{d^2\theta}{dt^2} + g\theta = 0.$$

Introducing a damping term, $b(d\theta/dt)$, we have

$$\ell\frac{d^2\theta}{dt^2} + b\frac{d\theta}{dt} + g\theta = 0,$$

which is the same as Eq. 14–15 with θ replacing x, and ℓ and g replacing m and k.

SOLUTION (a) We compare Eq. 14–15 with our equation just above and see that our equation $x = Ae^{-\gamma t}\cos\omega' t$ becomes an equation for θ with

$$\gamma = \frac{b}{2\ell} \quad \text{and} \quad \omega' = \sqrt{\frac{g}{\ell} - \frac{b^2}{4\ell^2}}.$$

At $t = 0$, we rewrite Eq. 14–16 with θ replacing x as

$$\theta_0 = Ae^{-\gamma \cdot 0}\cos\omega' \cdot 0 = A.$$

Then at $t = 5.0\,\text{min} = 300\,\text{s}$, the amplitude given by Eq. 14–16 has fallen to $0.50\,A$, so

$$0.50A = Ae^{-\gamma(300\,\text{s})}.$$

We solve this for γ and obtain $\gamma = \ln 2.0/(300\,\text{s}) = 2.3 \times 10^{-3}\,\text{s}^{-1}$.

(b) We have $\ell = 1.0\,\text{m}$, so $b = 2\gamma\ell = 2(2.3 \times 10^{-3}\,\text{s}^{-1})(1.0\,\text{m}) = 4.6 \times 10^{-3}\,\text{m/s}$. Thus $(b^2/4\ell^2)$ is very much less than $g/\ell\,(= 9.8\,\text{s}^{-2})$, and the angular frequency of the motion remains almost the same as that of the undamped motion. Specifically (see Eq. 14–20),

$$f' = \frac{1}{2\pi}\sqrt{\frac{g}{\ell}}\left[1 - \frac{\ell}{g}\left(\frac{b^2}{4\ell^2}\right)\right]^{\frac{1}{2}} \approx \frac{1}{2\pi}\sqrt{\frac{g}{\ell}}\left[1 - \frac{1}{2}\frac{\ell}{g}\left(\frac{b^2}{4\ell^2}\right)\right]$$

where we used the binomial expansion. Then, with $f = (1/2\pi)\sqrt{g/\ell}$ (Eq. 14–12b),

$$\frac{f - f'}{f} \approx \frac{1}{2}\frac{\ell}{g}\left(\frac{b^2}{4\ell^2}\right) = 2.7 \times 10^{-7}.$$

So f' differs from f by less than one part in a million.

*Showing $x = Ae^{-\gamma t}\cos\omega' t$ is a Solution

We start with Eq. 14–16, to see if it is a solution to Eq. 14–15. First we take the first and second derivatives

$$\frac{dx}{dt} = -\gamma Ae^{-\gamma t}\cos\omega' t - \omega' Ae^{-\gamma t}\sin\omega' t$$

$$\frac{d^2x}{dt^2} = \gamma^2 Ae^{-\gamma t}\cos\omega' t + \gamma A\omega' e^{-\gamma t}\sin\omega' t + \omega'\gamma Ae^{-\gamma t}\sin\omega' t - \omega'^2 Ae^{-\gamma t}\cos\omega' t.$$

We next substitute these relations back into Eq. 14–15 and reorganize to obtain

$$Ae^{-\gamma t}\left[(m\gamma^2 - m\omega'^2 - b\gamma + k)\cos\omega't + (2\omega'\gamma m - b\omega')\sin\omega't\right] = 0. \quad \textbf{(i)}$$

The left side of this equation must equal zero for all times t, but this can only be so for certain values of γ and ω'. To determine γ and ω', we choose two values of t that will make their evaluation easy. At $t = 0$, $\sin\omega't = 0$, so the above relation reduces to $A(m\gamma^2 - m\omega'^2 - b\gamma + k) = 0$, which means[†] that

$$m\gamma^2 - m\omega'^2 - b\gamma + k = 0. \quad \textbf{(ii)}$$

Then at $t = \pi/2\omega'$, $\cos\omega't = 0$ so Eq. (i) can be valid only if

$$2\gamma m - b = 0. \quad \textbf{(iii)}$$

From Eq. (iii) we have

$$\gamma = \frac{b}{2m}$$

and from Eq. (ii)

$$\omega' = \sqrt{\gamma^2 - \frac{b\gamma}{m} + \frac{k}{m}} = \sqrt{\frac{k}{m} - \frac{b^2}{4m^2}}.$$

Thus we see that Eq. 14–16 is a solution to the equation of motion for the damped harmonic oscillator as long as γ and ω' have these specific values, as already given in Eqs. 14–17 and 14–18.

14–8 Forced Oscillations; Resonance

When an oscillating system is set into motion, it oscillates at its natural frequency (Eqs. 14–7a and 14–12b). However, a system may have an external force applied to it that has its own particular frequency and then we have a **forced oscillation**.

For example, we might pull the mass on the spring of Fig. 14–1 back and forth at a frequency f. The mass then oscillates at the frequency f of the external force, even if this frequency is different from the **natural frequency** of the spring, which we will now denote by f_0 where (see Eqs. 14–5 and 14–7a)

$$\omega_0 = 2\pi f_0 = \sqrt{\frac{k}{m}}.$$

In a forced oscillation the amplitude of oscillation, and hence the energy transferred to the oscillating system, is found to depend on the difference between f and f_0 as well as on the amount of damping, reaching a maximum when the frequency of the external force equals the natural frequency of the system—that is, when $f = f_0$. The amplitude is plotted in Fig. 14–23 as a function of the external frequency f. Curve A represents light damping and curve B heavy damping. The amplitude can become large when the driving frequency f is near the natural frequency, $f \approx f_0$, as long as the damping is not too large. When the damping is small, the increase in amplitude near $f = f_0$ is very large (and often dramatic). This effect is known as **resonance**. The natural frequency f_0 of a system is called its **resonant frequency**.

A simple illustration of resonance is pushing a child on a swing. A swing, like any pendulum, has a natural frequency of oscillation that depends on its length ℓ. If you push on the swing at a random frequency, the swing bounces around and reaches no great amplitude. But if you push with a frequency equal to the natural frequency of the swing, the amplitude increases greatly. At resonance, relatively little effort is required to obtain a large amplitude.

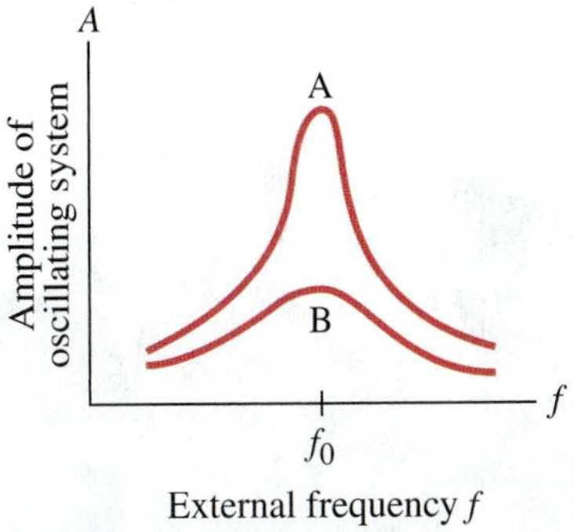

FIGURE 14–23 Resonance for lightly damped (A) and heavily damped (B) systems. (See Fig. 14–26 for a more detailed graph.)

[†]It would also be satisfied by $A = 0$, but this gives the trivial and uninteresting solution $x = 0$ for all t—that is, no oscillation.

FIGURE 14–24 This goblet breaks as it vibrates in resonance to a trumpet call.

FIGURE 14–25 (a) Large-amplitude oscillations of the Tacoma Narrows Bridge, due to gusty winds, led to its collapse (1940). (b) Collapse of a freeway in California, due to the 1989 earthquake.

(a)

(b)

The great tenor Enrico Caruso was said to be able to shatter a crystal goblet by singing a note of just the right frequency at full voice. This is an example of resonance, for the sound waves emitted by the voice act as a forced oscillation on the glass. At resonance, the resulting oscillation of the goblet may be large enough in amplitude that the glass exceeds its elastic limit and breaks (Fig. 14–24).

Since material objects are, in general, elastic, resonance is an important phenomenon in a variety of situations. It is particularly important in structural engineering, although the effects are not always foreseen. For example, it has been reported that a railway bridge collapsed because a nick in one of the wheels of a crossing train set up a resonant oscillation in the bridge. Indeed, marching soldiers break step when crossing a bridge to avoid the possibility that their normal rhythmic march might match a resonant frequency of the bridge. The famous collapse of the Tacoma Narrows Bridge (Fig. 14–25a) in 1940 occurred as a result of strong gusting winds driving the span into large-amplitude oscillatory motion. The Oakland freeway collapse in the 1989 California earthquake (Fig. 14–25b) involved resonant oscillation of a section built on mudfill that readily transmitted that frequency.

We will meet important examples of resonance later. We will also see that vibrating objects often have not one, but many resonant frequencies.

*Equation of Motion and Its Solution

We now look at the equation of motion for a forced oscillation and its solution. Suppose the external force is sinusoidal and can be represented by

$$F_{\text{ext}} = F_0 \cos \omega t,$$

where $\omega = 2\pi f$ is the angular frequency applied externally to the oscillator. Then the equation of motion (with damping) is

$$ma = -kx - bv + F_0 \cos \omega t.$$

This can be written as

$$m \frac{d^2 x}{dt^2} + b \frac{dx}{dt} + kx = F_0 \cos \omega t. \qquad (14\text{–}21)$$

The external force, on the right of the equation, is the only term that does not involve x or one of its derivatives. Problem 68 asks you to show that

$$x = A_0 \sin(\omega t + \phi_0) \qquad (14\text{–}22)$$

is a solution to Eq. 14–21, by direct substitution, where

$$A_0 = \frac{F_0}{m\sqrt{(\omega^2 - \omega_0^2)^2 + b^2\omega^2/m^2}} \qquad (14\text{–}23)$$

and

$$\phi_0 = \tan^{-1} \frac{\omega_0^2 - \omega^2}{\omega(b/m)}. \qquad (14\text{–}24)$$

Actually, the general solution to Eq. 14–21 is Eq. 14–22 plus another term of the form of Eq. 14–19 for the natural damped motion of the oscillator; this second term approaches zero in time, so in many cases we need to be concerned only with Eq. 14–22.

The amplitude of forced harmonic motion, A_0, depends strongly on the difference between the applied and the natural frequency. A plot of A_0 (Eq. 14–23) as a function of the applied frequency, ω, is shown in Fig. 14–26 (a more detailed version of Fig. 14–23) for three specific values of the damping constant b. Curve A $\left(b = \frac{1}{6}m\omega_0\right)$ represents light damping, curve B $\left(b = \frac{1}{2}m\omega_0\right)$ fairly heavy damping, and curve C $\left(b = \sqrt{2}m\omega_0\right)$ overdamped motion. The amplitude can become large when the driving frequency is near the natural frequency, $\omega \approx \omega_0$, as long as the damping is not too large. When the damping is small, the increase in amplitude near $\omega = \omega_0$ is very large and, as we saw, is known as *resonance*.

The natural oscillating frequency $f_0 \,(= \omega_0/2\pi)$ of a system is its *resonant frequency*.[†] If $b = 0$, resonance occurs at $\omega = \omega_0$ and the resonant peak (of A_0) becomes infinite; in such a case, energy is being continuously transferred into the system and none is dissipated. For real systems, b is never precisely zero, and the resonant peak is finite. The peak does not occur precisely at $\omega = \omega_0$ (because of the term $b^2\omega^2/m^2$ in the denominator of Eq. 14–23), although it is quite close to ω_0 unless the damping is very large. If the damping is large, there is little or no peak (curve C in Fig. 14–26).

*Q value

The height and narrowness of a resonant peak is often specified by its **quality factor** or **Q value**, defined as

$$Q = \frac{m\omega_0}{b}. \qquad (14\text{–}25)$$

In Fig. 14–26, curve A has $Q = 6$, curve B has $Q = 2$, and curve C has $Q = 1/\sqrt{2}$. The smaller the damping constant b, the larger the Q value becomes, and the higher the resonance peak. The Q value is also a measure of the width of the peak. To see why, let ω_1 and ω_2 be the frequencies where the square of the amplitude A_0 has half its maximum value (we use the square because the power transferred to the system is proportional to A_0^2); then $\Delta\omega = \omega_1 - \omega_2$, which is called the *width* of the resonance peak, is related to Q by

$$\frac{\Delta\omega}{\omega_0} = \frac{1}{Q}. \qquad (14\text{–}26)$$

This relation is accurate only for weak damping. The larger the Q value, the narrower will be the resonance peak relative to its height. Thus a large Q value, representing a system of high quality, has a high, narrow resonance peak.

[†]Sometimes the resonant frequency is defined as the actual value of ω at which the amplitude has its maximum value, and this depends somewhat on the damping constant. Except for very heavy damping, this value is quite close to ω_0.

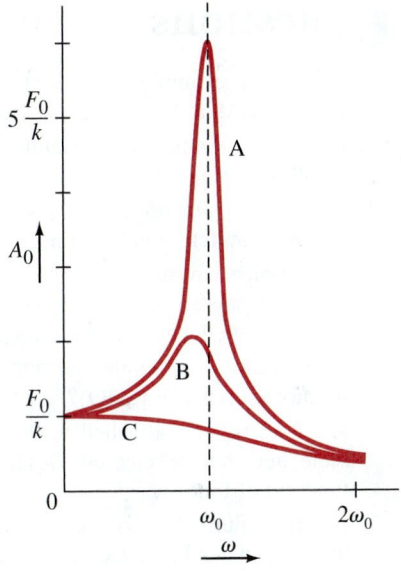

FIGURE 14–26 Amplitude of a forced harmonic oscillator as a function of ω. Curves A, B, and C correspond to light, heavy, and overdamped systems, respectively ($Q = m\omega_0/b = 6, 2, 0.71$).

Summary

An oscillating object undergoes **simple harmonic motion** (SHM) if the restoring force is proportional to the displacement,

$$F = -kx. \qquad (14\text{–}1)$$

The maximum displacement from equilibrium is called the **amplitude**.

The **period**, T, is the time required for one complete cycle (back and forth), and the **frequency**, f, is the number of cycles per second; they are related by

$$f = \frac{1}{T}. \qquad (14\text{–}2)$$

The period of oscillation for a mass m on the end of an ideal massless spring is given by

$$T = 2\pi\sqrt{\frac{m}{k}}. \qquad (14\text{–}7b)$$

SHM is **sinusoidal**, which means that the displacement as a function of time follows a sine or cosine curve. The general solution can be written

$$x = A\cos(\omega t + \phi) \qquad (14\text{–}4)$$

where A is the amplitude, ϕ is the **phase angle**, and

$$\omega = 2\pi f = \sqrt{\frac{k}{m}}. \qquad (14\text{–}5)$$

The values of A and ϕ depend on the **initial conditions** (x and v at $t = 0$).

During SHM, the total energy $E = \frac{1}{2}mv^2 + \frac{1}{2}kx^2$ is continually changing from potential to kinetic and back again.

A **simple pendulum** of length ℓ approximates SHM if its amplitude is small and friction can be ignored. For small amplitudes, its period is given by

$$T = 2\pi\sqrt{\frac{\ell}{g}}, \qquad (14\text{–}12c)$$

where g is the acceleration of gravity.

When friction is present (for all real springs and pendulums), the motion is said to be **damped**. The maximum displacement decreases in time, and the mechanical energy is eventually all transformed to thermal energy. If the friction is very large, so no oscillations occur, the system is said to be **overdamped**. If the friction is small enough that oscillations occur, the system is **underdamped**, and the displacement is given by

$$x = Ae^{-\gamma t}\cos\omega' t, \qquad (14\text{–}16)$$

where γ and ω' are constants. For a **critically damped** system, no oscillations occur and equilibrium is reached in the shortest time.

If an oscillating force is applied to a system capable of vibrating, the amplitude of vibration can be very large if the frequency of the applied force is near the **natural** (or **resonant**) **frequency** of the oscillator; this is called **resonance**.

Questions

1. Give some examples of everyday vibrating objects. Which exhibit SHM, at least approximately?

2. Is the acceleration of a simple harmonic oscillator ever zero? If so, where?

3. Explain why the motion of a piston in an automobile engine is approximately simple harmonic.

4. Real springs have mass. Will the true period and frequency be larger or smaller than given by the equations for a mass oscillating on the end of an idealized massless spring? Explain.

5. How could you double the maximum speed of a simple harmonic oscillator (SHO)?

6. A 5.0-kg trout is attached to the hook of a vertical spring scale, and then is released. Describe the scale reading as a function of time.

7. If a pendulum clock is accurate at sea level, will it gain or lose time when taken to high altitude? Why?

8. A tire swing hanging from a branch reaches nearly to the ground (Fig. 14–27). How could you estimate the height of the branch using only a stopwatch?

FIGURE 14–27 Question 8.

9. For a simple harmonic oscillator, when (if ever) are the displacement and velocity vectors in the same direction? When are the displacement and acceleration vectors in the same direction?

10. A 100-g mass hangs from a long cord forming a pendulum. The mass is pulled a short distance to one side and released from rest. The time to swing over and back is carefully measured to be 2.0 s. If the 100-g mass is replaced by a 200-g mass, which is then pulled over the same distance and released from rest, the time will be (a) 1.0 s, (b) 1.41 s, (c) 2.0 s, (d) 2.82 s, (e) 4.0 s.

11. Two equal masses are attached to separate identical springs next to one another. One mass is pulled so its spring stretches 20 cm and the other is pulled so its spring stretches only 10 cm. The masses are released simultaneously. Which mass reaches the equilibrium point first?

12. Does a car bounce on its springs faster when it is empty or when it is fully loaded?

13. What is the approximate period of your walking step?

14. What happens to the period of a playground swing if you rise up from sitting to a standing position?

*15. A thin uniform rod of mass m is suspended from one end and oscillates with a frequency f. If a small sphere of mass $2m$ is attached to the other end, does the frequency increase or decrease? Explain.

16. A tuning fork of natural frequency 264 Hz sits on a table at the front of a room. At the back of the room, two tuning forks, one of natural frequency 260 Hz and one of 420 Hz are initially silent, but when the tuning fork at the front of the room is set into vibration, the 260-Hz fork spontaneously begins to vibrate but the 420-Hz fork does not. Explain.

17. Why can you make water slosh back and forth in a pan only if you shake the pan at a certain frequency?

18. Give several everyday examples of resonance.

19. Is a rattle in a car ever a resonance phenomenon? Explain.

20. Over the years, buildings have been able to be built out of lighter and lighter materials. How has this affected the natural oscillation frequencies of buildings and the problems of resonance due to passing trucks, airplanes, or by wind and other natural sources of vibration?

Problems

14–1 and 14–2 Simple Harmonic Motion

1. (I) If a particle undergoes SHM with amplitude 0.18 m, what is the total distance it travels in one period?

2. (I) An elastic cord is 65 cm long when a weight of 75 N hangs from it but is 85 cm long when a weight of 180 N hangs from it. What is the "spring" constant k of this elastic cord?

3. (I) The springs of a 1500-kg car compress 5.0 mm when its 68-kg driver gets into the driver's seat. If the car goes over a bump, what will be the frequency of oscillations? Ignore damping.

4. (I) (a) What is the equation describing the motion of a mass on the end of a spring which is stretched 8.8 cm from equilibrium and then released from rest, and whose period is 0.66 s? (b) What will be its displacement after 1.8 s?

5. (II) Estimate the stiffness of the spring in a child's pogo stick if the child has a mass of 35 kg and bounces once every 2.0 seconds.

6. (II) A fisherman's scale stretches 3.6 cm when a 2.4-kg fish hangs from it. (a) What is the spring stiffness constant and (b) what will be the amplitude and frequency of oscillation if the fish is pulled down 2.5 cm more and released so that it oscillates up and down?

7. (II) Tall buildings are designed to sway in the wind. In a 100-km/h wind, for example, the top of the 110-story Sears Tower oscillates horizontally with an amplitude of 15 cm. The building oscillates at its natural frequency, which has a period of 7.0 s. Assuming SHM, find the maximum horizontal velocity and acceleration experienced by a Sears employee as she sits working at her desk located on the top floor. Compare the maximum acceleration (as a percentage) with the acceleration due to gravity.

8. (II) Construct a Table indicating the position x of the mass in Fig. 14–2 at times $t = 0, \frac{1}{4}T, \frac{1}{2}T, \frac{3}{4}T, T$, and $\frac{5}{4}T$, where T is the period of oscillation. On a graph of x vs. t, plot these six points. Now connect these points with a smooth curve. Based on these simple considerations, does your curve resemble that of a cosine or sine wave?

9. (II) A small fly of mass 0.25 g is caught in a spider's web. The web oscillates predominately with a frequency of 4.0 Hz. (a) What is the value of the effective spring stiffness constant k for the web? (b) At what frequency would you expect the web to oscillate if an insect of mass 0.50 g were trapped?

10. (II) A mass m at the end of a spring oscillates with a frequency of 0.83 Hz. When an additional 680-g mass is added to m, the frequency is 0.60 Hz. What is the value of m?

11. (II) A uniform meter stick of mass M is pivoted on a hinge at one end and held horizontal by a spring with spring constant k attached at the other end (Fig. 14–28). If the stick oscillates up and down slightly, what is its frequency? [Hint: Write a torque equation about the hinge.]

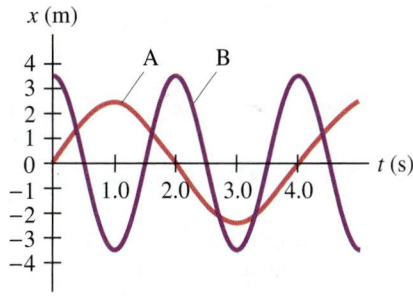

FIGURE 14–28
Problem 11.

12. (II) A balsa wood block of mass 55 g floats on a lake, bobbing up and down at a frequency of 3.0 Hz. (a) What is the value of the effective spring constant of the water? (b) A partially filled water bottle of mass 0.25 kg and almost the same size and shape of the balsa block is tossed into the water. At what frequency would you expect the bottle to bob up and down? Assume SHM.

13. (II) Figure 14–29 shows two examples of SHM, labeled A and B. For each, what is (a) the amplitude, (b) the frequency, and (c) the period? (d) Write the equations for both A and B in the form of a sine or cosine.

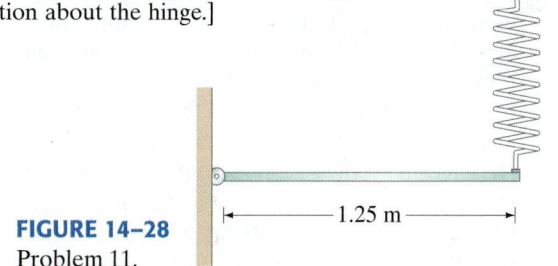

FIGURE 14–29   Problem 13.

14. (II) Determine the phase constant ϕ in Eq. 14–4 if, at $t = 0$, the oscillating mass is at (a) $x = -A$, (b) $x = 0$, (c) $x = A$, (d) $x = \frac{1}{2}A$, (e) $x = -\frac{1}{2}A$, (f) $x = A/\sqrt{2}$.

15. (II) A vertical spring with spring stiffness constant 305 N/m oscillates with an amplitude of 28.0 cm when 0.260 kg hangs from it. The mass passes through the equilibrium point $(y = 0)$ with positive velocity at $t = 0$. (a) What equation describes this motion as a function of time? (b) At what times will the spring be longest and shortest?

16. (II) The graph of displacement vs. time for a small mass m at the end of a spring is shown in Fig. 14–30. At $t = 0$, $x = 0.43$ cm. (a) If $m = 9.5$ g, find the spring constant, k. (b) Write the equation for displacement x as a function of time.

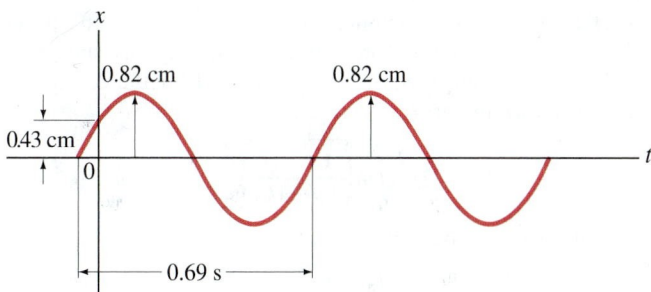

FIGURE 14–30   Problem 16.

17. (II) The position of a SHO as a function of time is given by $x = 3.8 \cos(5\pi t/4 + \pi/6)$ where t is in seconds and x in meters. Find (a) the period and frequency, (b) the position and velocity at $t = 0$, and (c) the velocity and acceleration at $t = 2.0$ s.

18. (II) A tuning fork oscillates at a frequency of 441 Hz and the tip of each prong moves 1.5 mm to either side of center. Calculate (a) the maximum speed and (b) the maximum acceleration of the tip of a prong.

19. (II) An object of unknown mass m is hung from a vertical spring of unknown spring constant k, and the object is observed to be at rest when the spring has extended by 14 cm. The object is then given a slight push and executes SHM. Determine the period T of this oscillation.

20. (II) A 1.25-kg mass stretches a vertical spring 0.215 m. If the spring is stretched an additional 0.130 m and released, how long does it take to reach the (new) equilibrium position again?

21. (II) Consider two objects, A and B, both undergoing SHM, but with different frequencies, as described by the equations $x_A = (2.0 \text{ m}) \sin(2.0 t)$ and $x_B = (5.0 \text{ m}) \sin(3.0 t)$, where t is in seconds. After $t = 0$, find the next three times t at which both objects simultaneously pass through the origin.

22. (II) A 1.60-kg object oscillates from a vertically hanging light spring once every 0.55 s. (a) Write down the equation giving its position y (+ upward) as a function of time t, assuming it started by being compressed 16 cm from the equilibrium position (where $y = 0$), and released. (b) How long will it take to get to the equilibrium position for the first time? (c) What will be its maximum speed? (d) What will be its maximum acceleration, and where will it first be attained?

23. (II) A bungee jumper with mass 65.0 kg jumps from a high bridge. After reaching his lowest point, he oscillates up and down, hitting a low point eight more times in 43.0 s. He finally comes to rest 25.0 m below the level of the bridge. Estimate the spring stiffness constant and the unstretched length of the bungee cord assuming SHM.

24. (II) A block of mass m is supported by two identical parallel vertical springs, each with spring stiffness constant k (Fig. 14–31). What will be the frequency of vertical oscillation?

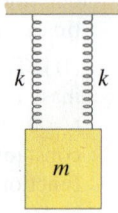

FIGURE 14–31
Problem 24.

25. (III) A mass m is connected to two springs, with spring constants k_1 and k_2, in two different ways as shown in Fig. 14–32a and b. Show that the period for the configuration shown in part (a) is given by

$$T = 2\pi \sqrt{m\left(\frac{1}{k_1} + \frac{1}{k_2}\right)}$$

and for that in part (b) is given by

$$T = 2\pi \sqrt{\frac{m}{k_1 + k_2}}.$$

Ignore friction.

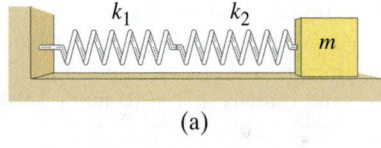

(a)

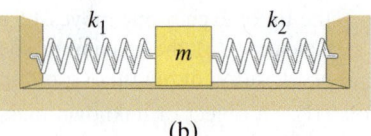

FIGURE 14–32
Problem 25.

(b)

26. (III) A mass m is at rest on the end of a spring of spring constant k. At $t = 0$ it is given an impulse J by a hammer. Write the formula for the subsequent motion in terms of m, k, J, and t.

14–3 Energy in SHM

27. (I) A 1.15-kg mass oscillates according to the equation $x = 0.650 \cos 7.40t$ where x is in meters and t in seconds. Determine (a) the amplitude, (b) the frequency, (c) the total energy, and (d) the kinetic energy and potential energy when $x = 0.260$ m.

28. (I) (a) At what displacement of a SHO is the energy half kinetic and half potential? (b) What fraction of the total energy of a SHO is kinetic and what fraction potential when the displacement is one third the amplitude?

29. (II) Draw a graph like Fig. 14–11 for a horizontal spring whose spring constant is 95 N/m and which has a mass of 55 g on the end of it. Assume the spring was started with an initial amplitude of 2.0 cm. Neglect the mass of the spring and any friction with the horizontal surface. Use your graph to estimate (a) the potential energy, (b) the kinetic energy, and (c) the speed of the mass, for $x = 1.5$ cm.

30. (II) A 0.35-kg mass at the end of a spring oscillates 2.5 times per second with an amplitude of 0.15 m. Determine (a) the velocity when it passes the equilibrium point, (b) the velocity when it is 0.10 m from equilibrium, (c) the total energy of the system, and (d) the equation describing the motion of the mass, assuming that at $t = 0$, x was a maximum.

31. (II) It takes a force of 95.0 N to compress the spring of a toy popgun 0.175 m to "load" a 0.160-kg ball. With what speed will the ball leave the gun if fired horizontally?

32. (II) A 0.0125-kg bullet strikes a 0.240-kg block attached to a fixed horizontal spring whose spring constant is 2.25×10^3 N/m and sets it into oscillation with an amplitude of 12.4 cm. What was the initial speed of the bullet if the two objects move together after impact?

33. (II) If one oscillation has 5.0 times the energy of a second one of equal frequency and mass, what is the ratio of their amplitudes?

34. (II) A mass of 240 g oscillates on a horizontal frictionless surface at a frequency of 3.0 Hz and with amplitude of 4.5 cm. (a) What is the effective spring constant for this motion? (b) How much energy is involved in this motion?

35. (II) A mass resting on a horizontal, frictionless surface is attached to one end of a spring; the other end is fixed to a wall. It takes 3.6 J of work to compress the spring by 0.13 m. If the spring is compressed, and the mass is released from rest, it experiences a maximum acceleration of 15 m/s². Find the value of (a) the spring constant and (b) the mass.

36. (II) An object with mass 2.7 kg is executing simple harmonic motion, attached to a spring with spring constant $k = 280$ N/m. When the object is 0.020 m from its equilibrium position, it is moving with a speed of 0.55 m/s. (a) Calculate the amplitude of the motion. (b) Calculate the maximum speed attained by the object.

37. (II) Agent Arlene devised the following method of measuring the muzzle velocity of a rifle (Fig. 14–33). She fires a bullet into a 4.648-kg wooden block resting on a smooth surface, and attached to a spring of spring constant $k = 142.7$ N/m. The bullet, whose mass is 7.870 g, remains embedded in the wooden block. She measures the maximum distance that the block compresses the spring to be 9.460 cm. What is the speed v of the bullet?

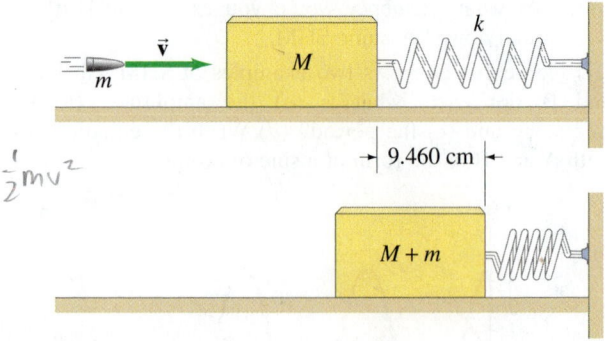

FIGURE 14–33 Problem 37.

38. (II) Obtain the displacement x as a function of time for the simple harmonic oscillator using the conservation of energy, Eqs. 14–10. [*Hint*: Integrate Eq. 14–11a with $v = dx/dt$.]

39. (II) At $t = 0$, a 785-g mass at rest on the end of a horizontal spring ($k = 184$ N/m) is struck by a hammer which gives it an initial speed of 2.26 m/s. Determine (a) the period and frequency of the motion, (b) the amplitude, (c) the maximum acceleration, (d) the position as a function of time, (e) the total energy, and (f) the kinetic energy when $x = 0.40A$ where A is the amplitude.

40. (II) A pinball machine uses a spring launcher that is compressed 6.0 cm to launch a ball up a 15° ramp. Assume that the pinball is a solid uniform sphere of radius $r = 1.0$ cm and mass $m = 25$ g. If it is rolling without slipping at a speed of 3.0 m/s when it leaves the launcher, what is the spring constant of the spring launcher?

14–5 Simple Pendulum

41. (I) A pendulum has a period of 1.35 s on Earth. What is its period on Mars, where the acceleration of gravity is about 0.37 that on Earth?

42. (I) A pendulum makes 32 oscillations in exactly 50 s. What is its (a) period and (b) frequency?

43. (II) A simple pendulum is 0.30 m long. At $t = 0$ it is released from rest starting at an angle of 13°. Ignoring friction, what will be the angular position of the pendulum at (a) $t = 0.35$ s, (b) $t = 3.45$ s, and (c) $t = 6.00$ s?

44. (II) What is the period of a simple pendulum 53 cm long (a) on the Earth, and (b) when it is in a freely falling elevator?

45. (II) A simple pendulum oscillates with an amplitude of 10.0°. What fraction of the time does it spend between +5.0° and −5.0°? Assume SHM.

46. (II) Your grandfather clock's pendulum has a length of 0.9930 m. If the clock loses 26 s per day, how should you adjust the length of the pendulum?

47. (II) Derive a formula for the maximum speed v_{max} of a simple pendulum bob in terms of g, the length ℓ, and the maximum angle of swing θ_{max}.

*14–6 Physical Pendulum and Torsion Pendulum

*48. (II) A pendulum consists of a tiny bob of mass M and a uniform cord of mass m and length ℓ. (a) Determine a formula for the period using the small angle approximation. (b) What would be the fractional error if you use the formula for a simple pendulum, Eq. 14–12c?

*49. (II) The balance wheel of a watch is a thin ring of radius 0.95 cm and oscillates with a frequency of 3.10 Hz. If a torque of 1.1×10^{-5} m·N causes the wheel to rotate 45°, calculate the mass of the balance wheel.

*50. (II) The human leg can be compared to a physical pendulum, with a "natural" swinging period at which walking is easiest. Consider the leg as two rods joined rigidly together at the knee; the axis for the leg is the hip joint. The length of each rod is about the same, 55 cm. The upper rod has a mass of 7.0 kg and the lower rod has a mass of 4.0 kg. (a) Calculate the natural swinging period of the system. (b) Check your answer by standing on a chair and measuring the time for one or more complete back-and-forth swings. The effect of a shorter leg is a shorter swinging period, enabling a faster "natural" stride.

*51. (II) (a) Determine the equation of motion (for θ as a function of time) for a torsion pendulum, Fig. 14–18, and show that the motion is simple harmonic. (b) Show that the period T is $T = 2\pi\sqrt{I/K}$. [The balance wheel of a mechanical watch is an example of a torsion pendulum in which the restoring torque is applied by a coil spring.]

*52. (II) A student wants to use a meter stick as a pendulum. She plans to drill a small hole through the meter stick and suspend it from a smooth pin attached to the wall (Fig. 14–34). Where in the meter stick should she drill the hole to obtain the shortest possible period? How short an oscillation period can she obtain with a meter stick in this way?

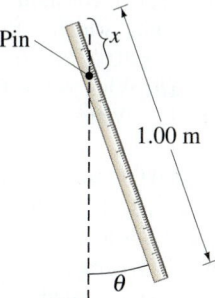

FIGURE 14–34
Problem 52.

*53. (II) A meter stick is hung at its center from a thin wire (Fig. 14–35a). It is twisted and oscillates with a period of 5.0 s. The meter stick is sawed off to a length of 70.0 cm. This piece is again balanced at its center and set in oscillation (Fig. 14–35b). With what period does it oscillate?

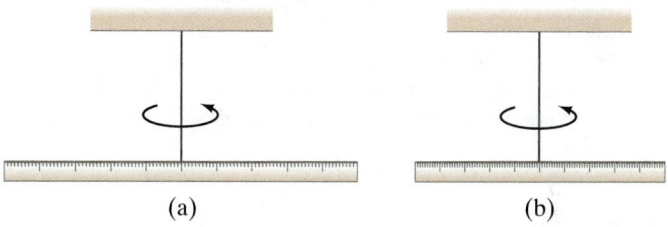

(a)   (b)

FIGURE 14–35 Problem 53.

*54. (II) An aluminum disk, 12.5 cm in diameter and 375 g in mass, is mounted on a vertical shaft with very low friction (Fig. 14–36). One end of a flat coil spring is attached to the disk, the other end to the base of the apparatus. The disk is set into rotational oscillation and the frequency is 0.331 Hz. What is the torsional spring constant K ($\tau = -K\theta$)?

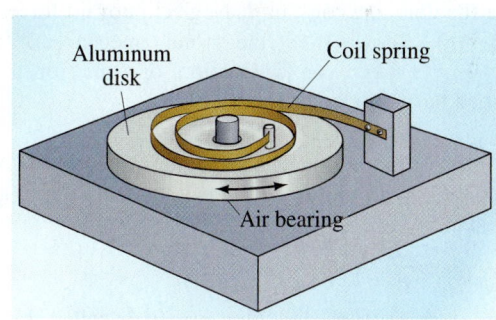

FIGURE 14–36 Problem 54.

*55. (II) A plywood disk of radius 20.0 cm and mass 2.20 kg has a small hole drilled through it, 2.00 cm from its edge (Fig. 14–37). The disk is hung from the wall by means of a metal pin through the hole, and is used as a pendulum. What is the period of this pendulum for small oscillations?

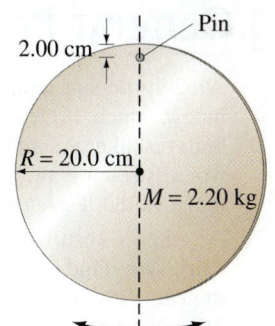

FIGURE 14–37
Problem 55.

14–7 Damping

56. (II) A 0.835-kg block oscillates on the end of a spring whose spring constant is $k = 41.0$ N/m. The mass moves in a fluid which offers a resistive force $F = -bv$, where $b = 0.662$ N·s/m. (a) What is the period of the motion? (b) What is the fractional decrease in amplitude per cycle? (c) Write the displacement as a function of time if at $t = 0$, $x = 0$, and at $t = 1.00$ s, $x = 0.120$ m.

57. (II) Estimate how the damping constant changes when a car's shock absorbers get old and the car bounces three times after going over a speed bump.

58. (II) A physical pendulum consists of an 85-cm-long, 240-g-mass, uniform wooden rod hung from a nail near one end (Fig. 14–38). The motion is damped because of friction in the pivot; the damping force is approximately proportional to $d\theta/dt$. The rod is set in oscillation by displacing it 15° from its equilibrium position and releasing it. After 8.0 s, the amplitude of the oscillation has been reduced to 5.5°. If the angular displacement can be written as $\theta = Ae^{-\gamma t}\cos\omega't$, find (a) γ, (b) the approximate period of the motion, and (c) how long it takes for the amplitude to be reduced to $\frac{1}{2}$ of its original value.

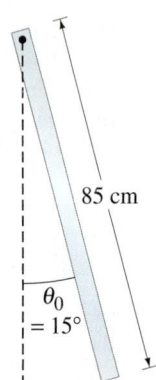

85 cm

$\theta_0 = 15°$

FIGURE 14–38
Problem 58.

59. (II) A damped harmonic oscillator loses 6.0% of its mechanical energy per cycle. (a) By what percentage does its frequency differ from the natural frequency $f_0 = (1/2\pi)\sqrt{k/m}$? (b) After how many periods will the amplitude have decreased to $1/e$ of its original value?

60. (II) A vertical spring of spring constant 115 N/m supports a mass of 75 g. The mass oscillates in a tube of liquid. If the mass is initially given an amplitude of 5.0 cm, the mass is observed to have an amplitude of 2.0 cm after 3.5 s. Estimate the damping constant b. Neglect buoyant forces.

61. (III) (a) Show that the total mechanical energy, $E = \frac{1}{2}mv^2 + \frac{1}{2}kx^2$, as a function of time for a lightly damped harmonic oscillator is

$$E = \tfrac{1}{2}kA^2e^{-(b/m)t} = E_0e^{-(b/m)t},$$

where E_0 is the total mechanical energy at $t = 0$. (Assume $\omega' \gg b/2m$.) (b) Show that the fractional energy lost per period is

$$\frac{\Delta E}{E} = \frac{2\pi b}{m\omega_0} = \frac{2\pi}{Q},$$

where $\omega_0 = \sqrt{k/m}$ and $Q = m\omega_0/b$ is called the **quality factor** or **Q value** of the system. A larger Q value means the system can undergo oscillations for a longer time.

62. (III) A glider on an air track is connected by springs to either end of the track (Fig. 14–39). Both springs have the same spring constant, k, and the glider has mass M. (a) Determine the frequency of the oscillation, assuming no damping, if $k = 125$ N/m and $M = 215$ g. (b) It is observed that after 55 oscillations, the amplitude of the oscillation has dropped to one-half of its initial value. Estimate the value of γ, using Eq. 14–16. (c) How long does it take the amplitude to decrease to one-quarter of its initial value?

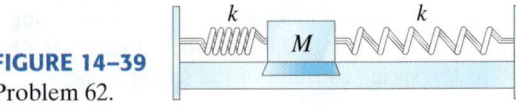

k k

M

FIGURE 14–39
Problem 62.

14–8 Forced Oscillations; Resonance

63. (II) (a) For a forced oscillation at resonance $(\omega = \omega_0)$, what is the value of the phase angle ϕ_0 in Eq. 14–22? (b) What, then, is the displacement at a time when the driving force F_{ext} is a maximum, and at a time when $F_{ext} = 0$? (c) What is the phase difference (in degrees) between the driving force and the displacement in this case?

64. (II) Differentiate Eq. 14–23 to show that the resonant amplitude peaks at

$$\omega = \sqrt{\omega_0^2 - \frac{b^2}{2m^2}}.$$

65. (II) An 1150 kg automobile has springs with $k = 16{,}000$ N/m. One of the tires is not properly balanced; it has a little extra mass on one side compared to the other, causing the car to shake at certain speeds. If the tire radius is 42 cm, at what speed will the wheel shake most?

*66. (II) Construct an accurate resonance curve, from $\omega = 0$ to $\omega = 2\omega_0$, for $Q = 6.0$.

*67. (II) The amplitude of a driven harmonic oscillator reaches a value of $23.7F_0/k$ at a resonant frequency of 382 Hz. What is the Q value of this system?

68. (III) By direct substitution, show that Eq. 14–22, with Eqs. 14–23 and 14–24, is a solution of the equation of motion (Eq. 14–21) for the forced oscillator. [*Hint*: To find $\sin\phi_0$ and $\cos\phi_0$ from $\tan\phi_0$, draw a right triangle.]

*69. (III) Consider a simple pendulum (point mass bob) 0.50 m long with a Q of 350. (a) How long does it take for the amplitude (assumed small) to decrease by two-thirds? (b) If the amplitude is 2.0 cm and the bob has mass 0.27 kg, what is the initial energy loss rate of the pendulum in watts? (c) If we are to stimulate resonance with a sinusoidal driving force, how close must the driving frequency be to the natural frequency of the pendulum (give $\Delta f = f - f_0$)?

General Problems

70. A 62-kg person jumps from a window to a fire net 20.0 m below, which stretches the net 1.1 m. Assume that the net behaves like a simple spring. (a) Calculate how much it would stretch if the same person were lying in it. (b) How much would it stretch if the person jumped from 38 m?

71. An energy-absorbing car bumper has a spring constant of 430 kN/m. Find the maximum compression of the bumper if the car, with mass 1300 kg, collides with a wall at a speed of 2.0 m/s (approximately 5 mi/h).

72. The length of a simple pendulum is 0.63 m, the pendulum bob has a mass of 295 g, and it is released at an angle of 15° to the vertical. (a) With what frequency does it oscillate? (b) What is the pendulum bob's speed when it passes through the lowest point of the swing? Assume SHM. (c) What is the total energy stored in this oscillation assuming no losses?

73. A simple pendulum oscillates with frequency f. What is its frequency if the entire pendulum accelerates at $0.50g$ (a) upward, and (b) downward?

74. A 0.650-kg mass oscillates according to the equation $x = 0.25 \sin(5.50t)$ where x is in meters and t is in seconds. Determine (a) the amplitude, (b) the frequency, (c) the period, (d) the total energy, and (e) the kinetic energy and potential energy when x is 15 cm.

75. (a) A crane has hoisted a 1350-kg car at the junkyard. The crane's steel cable is 20.0 m long and has a diameter of 6.4 mm. If the car starts bouncing at the end of the cable, what is the period of the bouncing? [*Hint*: Refer to Table 12–1.] (b) What amplitude of bouncing will likely cause the cable to snap? (See Table 12–2, and assume Hooke's law holds all the way up to the breaking point.)

76. An oxygen atom at a particular site within a DNA molecule can be made to execute simple harmonic motion when illuminated by infrared light. The oxygen atom is bound with a spring-like chemical bond to a phosphorus atom, which is rigidly attached to the DNA backbone. The oscillation of the oxygen atom occurs with frequency $f = 3.7 \times 10^{13}$ Hz. If the oxygen atom at this site is chemically replaced with a sulfur atom, the spring constant of the bond is unchanged (sulfur is just below oxygen in the Periodic Table). Predict the frequency for a DNA molecule after the sulfur substitution.

77. A "seconds" pendulum has a period of exactly 2.000 s. That is, each one-way swing takes 1.000 s. What is the length of a seconds pendulum in Austin, Texas, where $g = 9.793$ m/s^2? If the pendulum is moved to Paris, where $g = 9.809$ m/s^2, by how many millimeters must we lengthen the pendulum? What is the length of a seconds pendulum on the Moon, where $g = 1.62$ m/s^2?

78. A 320-kg wooden raft floats on a lake. When a 75-kg man stands on the raft, it sinks 3.5 cm deeper into the water. When he steps off, the raft oscillates for a while. (a) What is the frequency of oscillation? (b) What is the total energy of oscillation (ignoring damping)?

79. At what displacement from equilibrium is the speed of a SHO half the maximum value?

80. A diving board oscillates with simple harmonic motion of frequency 2.5 cycles per second. What is the maximum amplitude with which the end of the board can oscillate in order that a pebble placed there (Fig. 14–40) does not lose contact with the board during the oscillation?

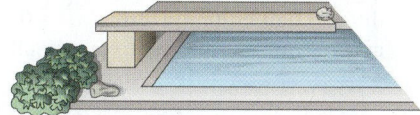

FIGURE 14–40
Problem 80.

81. A rectangular block of wood floats in a calm lake. Show that, if friction is ignored, when the block is pushed gently down into the water and then released, it will then oscillate with SHM. Also, determine an equation for the force constant.

82. A 950-kg car strikes a huge spring at a speed of 25 m/s (Fig. 14–41), compressing the spring 5.0 m. (a) What is the spring stiffness constant of the spring? (b) How long is the car in contact with the spring before it bounces off in the opposite direction?

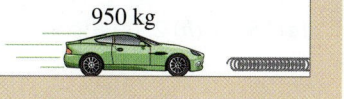
950 kg

FIGURE 14–41
Problem 82.

83. A 1.60-kg table is supported on four springs. A 0.80-kg chunk of modeling clay is held above the table and dropped so that it hits the table with a speed of 1.65 m/s (Fig. 14–42). The clay makes an inelastic collision with the table, and the table and clay oscillate up and down. After a long time the table comes to rest 6.0 cm below its original position. (a) What is the effective spring constant of all four springs taken together? (b) With what maximum amplitude does the platform oscillate?

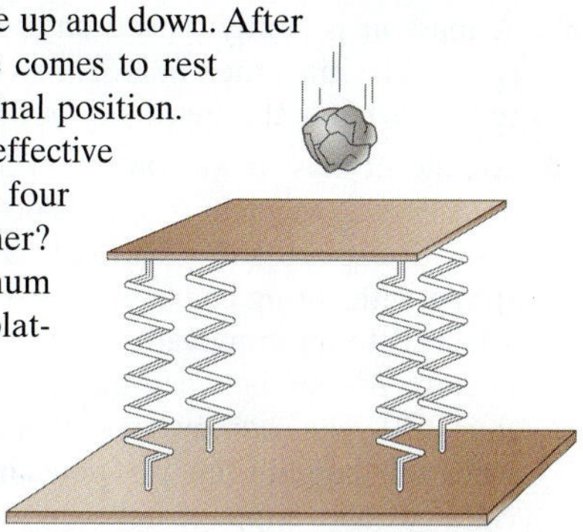

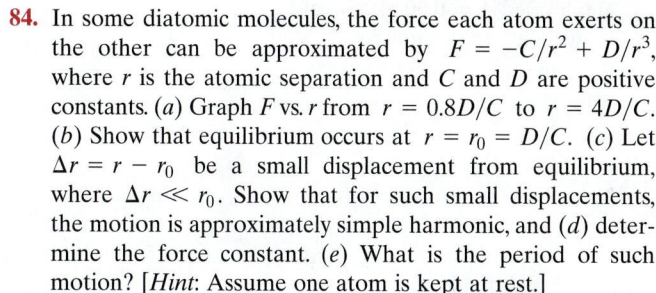
FIGURE 14–42
Problem 83.

84. In some diatomic molecules, the force each atom exerts on the other can be approximated by $F = -C/r^2 + D/r^3$, where r is the atomic separation and C and D are positive constants. (a) Graph F vs. r from $r = 0.8D/C$ to $r = 4D/C$. (b) Show that equilibrium occurs at $r = r_0 = D/C$. (c) Let $\Delta r = r - r_0$ be a small displacement from equilibrium, where $\Delta r \ll r_0$. Show that for such small displacements, the motion is approximately simple harmonic, and (d) determine the force constant. (e) What is the period of such motion? [*Hint*: Assume one atom is kept at rest.]

85. A mass attached to the end of a spring is stretched a distance x_0 from equilibrium and released. At what distance from equilibrium will it have (a) velocity equal to half its maximum velocity, and (b) acceleration equal to half its maximum acceleration?

86. Carbon dioxide is a linear molecule. The carbon–oxygen bonds in this molecule act very much like springs. Figure 14–43 shows one possible way the oxygen atoms in this molecule can oscillate: the oxygen atoms oscillate symmetrically in and out, while the central carbon atom remains at rest. Hence each oxygen atom acts like a simple harmonic oscillator with a mass equal to the mass of an oxygen atom. It is observed that this oscillation occurs with a frequency of $f = 2.83 \times 10^{13}$ Hz. What is the spring constant of the C—O bond?

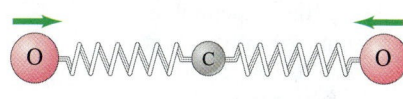

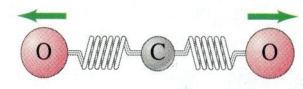
FIGURE 14–43
Problem 86, the CO_2 molecule.

87. Imagine that a 10-cm-diameter circular hole was drilled all the way through the center of the Earth (Fig. 14–44). At one end of the hole, you drop an apple into the hole. Show that, if you assume that the Earth has a constant density, the subsequent motion of the apple is simple harmonic. How long will the apple take to return? Assume that we can ignore all frictional effects. [*Hint*: See Appendix D.]

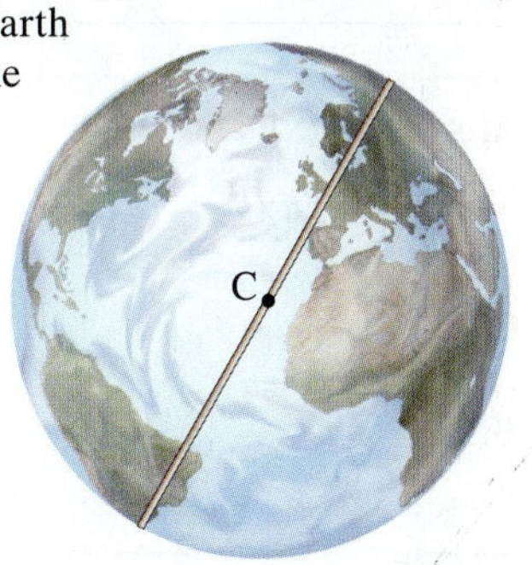

FIGURE 14–44
Problem 87.

88. A thin, straight, uniform rod of length $\ell = 1.00\,\text{m}$ and mass $m = 215\,\text{g}$ hangs from a pivot at one end. (a) What is its period for small-amplitude oscillations? (b) What is the length of a simple pendulum that will have the same period?

89. A mass m is gently placed on the end of a freely hanging spring. The mass then falls 32.0 cm before it stops and begins to rise. What is the frequency of the oscillation?

90. A child of mass m sits on top of a rectangular slab of mass $M = 35\,\text{kg}$, which in turn rests on the frictionless horizontal floor at a pizza shop. The slab is attached to a horizontal spring with spring constant $k = 430\,\text{N/m}$ (the other end is attached to an immovable wall, Fig. 14–45). The coefficient of static friction between the child and the top of the slab is $\mu = 0.40$. The shop owner's intention is that, when displaced from the equilibrium position and released, the slab and child (with no slippage between the two) execute SHM with amplitude $A = 0.50\,\text{m}$. Should there be a weight restriction for this ride? If so, what is it?

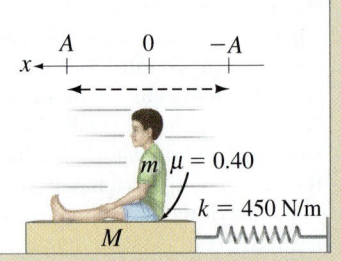

FIGURE 14–45
Problem 90.

91. Estimate the effective spring constant of a trampoline.

92. In Section 14–5, the oscillation of a simple pendulum (Fig. 14–46) is viewed as linear motion along the arc length x and analyzed via $F = ma$. Alternatively, the pendulum's movement can be regarded as rotational motion about its point of support and analyzed using $\tau = I\alpha$. Carry out this alternative analysis and show that

$$\theta(t) = \theta_{\text{max}} \cos\left(\sqrt{\frac{g}{\ell}}\,t + \phi\right),$$

where $\theta(t)$ is the angular displacement of the pendulum from the vertical at time t, as long as its maximum value is less than about 15°.

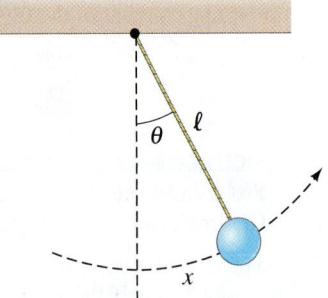

FIGURE 14–46
Problem 92.

*93. (II) A mass m on a frictionless surface is attached to a spring with spring constant k as shown in Fig. 14–47. This mass–spring system is then observed to execute simple harmonic motion with a period T. The mass m is changed several times and the associated period T is measured in each case, generating the following data Table:

Mass m (kg)	Period T (s)
0.5	0.445
1.0	0.520
2.0	0.630
3.0	0.723
4.5	0.844

(a) Starting with Eq. 14–7b, show why a graph of T^2 vs. m is expected to yield a straight line. How can k be determined from the straight line's slope? What is the line's y-intercept expected to be? (b) Using the data in the Table, plot T^2 vs. m and show that this graph yields a straight line. Determine the slope and (nonzero) y-intercept. (c) Show that a nonzero y-intercept can be expected in our plot theoretically if, rather than simply using m for the mass in Eq. 14–7b, we use $m + m_0$, where m_0 is a constant. That is, repeat part (a) using $m + m_0$ for the mass in Eq. 14–7b. Then use the result of this analysis to determine k and m_0 from your graph's slope and y-intercept. (d) Offer a physical interpretation for m_0, a mass that appears to be oscillating in addition to the attached mass m.

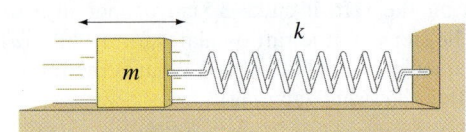

FIGURE 14–47   Problem 93.

*94. (III) *Damping proportional to* v^2. Suppose the oscillator of Example 14–5 is damped by a force proportional to the square of the velocity, $F_{\text{damping}} = -cv^2$, where $c = 0.275\,\text{kg/m}$ is a constant. Numerically integrate[†] the differential equation from $t = 0$ to $t = 2.00\,\text{s}$ to an accuracy of 2%, and plot your results.

[†]See Section 2–9.

Answers to Exercises

A: (a), (c), (d).

B: (b).

C: (c).

D: (b), (d).

E: (c).

F: (c).

G: All are larger.

H: (a).

I: (c).

J: (a) 25 cm; (b) 2.0 s.

Waves—such as these water waves—spread outward from a source. The source in this case is a small spot of water oscillating up and down briefly where a rock was thrown in (left photo). Other kinds of waves include waves on a cord or string, which also are produced by a vibration. Waves move away from their source, but we also study waves that seem to stand still ("standing waves"). Waves reflect, and they can interfere with each other when they pass through any point at the same time.

Wave Motion

C H A P T E R

15

CHAPTER-OPENING QUESTION—Guess now!

You drop a rock into a pond, and water waves spread out in circles.

- **(a)** The waves carry water outward, away from where the rock hit. That moving water carries energy outward.
- **(b)** The waves only make the water move up and down. No energy is carried outward from where the rock hit.
- **(c)** The waves only make the water move up and down, but the waves do carry energy outward, away from where the rock hit.

When you throw a stone into a lake or pool of water, circular waves form and move outward, as shown in the photos above. Waves will also travel along a cord that is stretched out flat on a table if you vibrate one end back and forth as shown in Fig. 15–1. Water waves and waves on a cord are two common examples of **mechanical waves**, which propagate as oscillations of matter. We will discuss other kinds of waves in later Chapters, including electromagnetic waves and light.

CONTENTS

15–1 Characteristics of Wave Motion

15–2 Types of Waves: Transverse and Longitudinal

15–3 Energy Transported by Waves

15–4 Mathematical Representation of a Traveling Wave

*15–5 The Wave Equation

15–6 The Principle of Superposition

15–7 Reflection and Transmission

15–8 Interference

15–9 Standing Waves; Resonance

*15–10 Refraction

*15–11 Diffraction

FIGURE 15–1 Wave traveling on a cord. The wave travels to the right along the cord. Particles of the cord oscillate back and forth on the tabletop.

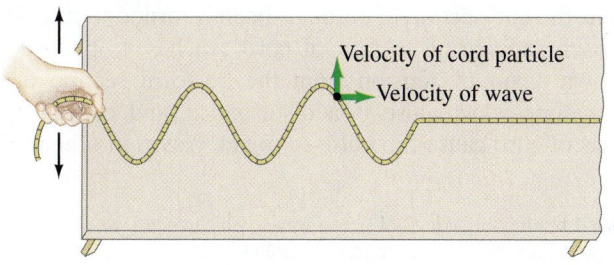

Velocity of cord particle
Velocity of wave

If you have ever watched ocean waves moving toward shore before they break, you may have wondered if the waves were carrying water from far out at sea onto the beach. They don't.[†] Water waves move with a recognizable velocity. But each particle (or molecule) of the water itself merely oscillates about an equilibrium point. This is clearly demonstrated by observing leaves on a pond as waves move by. The leaves (or a cork) are not carried forward by the waves, but simply oscillate about an equilibrium point because this is the motion of the water itself.

CONCEPTUAL EXAMPLE 15–1 **Wave vs. particle velocity.** Is the velocity of a wave moving along a cord the same as the velocity of a particle of the cord? See Fig. 15–1.

RESPONSE No. The two velocities are different, both in magnitude and direction. The wave on the cord of Fig. 15–1 moves to the right along the tabletop, but each piece of the cord only vibrates to and fro. (The cord clearly does not travel in the direction that the wave on it does.)

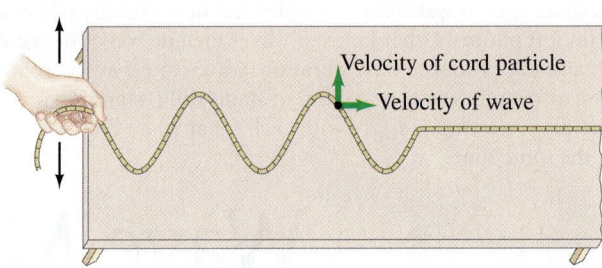

FIGURE 15–1 (repeated) Wave traveling on a cord. The wave travels to the right along the cord. Particles of the cord oscillate back and forth on the tabletop.

Waves can move over large distances, but the medium (the water or the cord) itself has only a limited movement, oscillating about an equilibrium point as in simple harmonic motion. Thus, although a wave is not matter, the wave pattern can travel in matter. A wave consists of oscillations that move without carrying matter with them.

Waves carry energy from one place to another. Energy is given to a water wave, for example, by a rock thrown into the water, or by wind far out at sea. The energy is transported by waves to the shore. The oscillating hand in Fig. 15–1 transfers energy to the cord, and that energy is transported down the cord and can be transferred to an object at the other end. All forms of traveling waves transport energy.

EXERCISE A Return to the Chapter-Opening Question, page 395, and answer it again now. Try to explain why you may have answered differently the first time.

15–1 Characteristics of Wave Motion

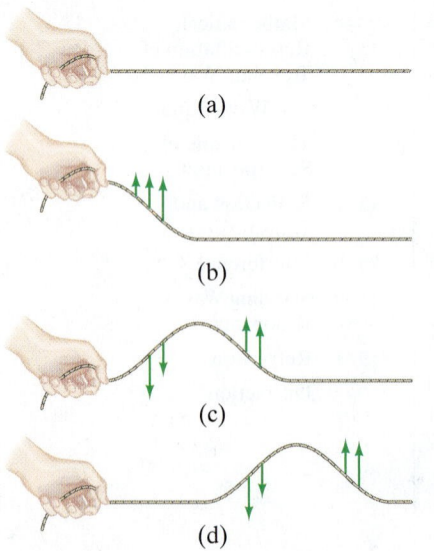

FIGURE 15–2 Motion of a wave pulse to the right. Arrows indicate velocity of cord particles.

(a)

(b)

(c)

(d)

Let us look a little more closely at how a wave is formed and how it comes to "travel." We first look at a single wave bump, or **pulse**. A single pulse can be formed on a cord by a quick up-and-down motion of the hand, Fig. 15–2. The hand pulls up on one end of the cord. Because the end section is attached to adjacent sections, these also feel an upward force and they too begin to move upward. As each succeeding section of cord moves upward, the wave crest moves outward along the cord. Meanwhile, the end section of cord has been returned to its original position by the hand. As each succeeding section of cord reaches its peak position, it too is pulled back down again by tension from the adjacent section of cord. Thus the source of a traveling wave pulse is a disturbance, and cohesive forces between adjacent sections of cord cause the pulse to travel. Waves in other media are created

[†]Do not be confused by the "breaking" of ocean waves, which occurs when a wave interacts with the ground in shallow water and hence is no longer a simple wave.

396 CHAPTER 15 Wave Motion

and propagate outward in a similar fashion. A dramatic example of a wave pulse is a tsunami or tidal wave that is created by an earthquake in the Earth's crust under the ocean. The bang you hear when a door slams is a sound wave pulse.

A **continuous** or **periodic wave**, such as that shown in Fig. 15–1, has as its source a disturbance that is continuous and oscillating; that is, the source is a *vibration* or *oscillation*. In Fig. 15–1, a hand oscillates one end of the cord. Water waves may be produced by any vibrating object at the surface, such as your hand; or the water itself is made to vibrate when wind blows across it or a rock is thrown into it. A vibrating tuning fork or drum membrane gives rise to sound waves in air. And we will see later that oscillating electric charges give rise to light waves. Indeed, almost any vibrating object sends out waves.

The source of any wave, then, is a vibration. And it is a *vibration* that propagates outward and thus constitutes the wave. If the source vibrates sinusoidally in SHM, then the wave itself—if the medium is perfectly elastic—will have a sinusoidal shape both in space and in time. (1) In space: if you take a picture of the wave in space at a given instant of time, the wave will have the shape of a sine or cosine as a function of position. (2) In time: if you look at the motion of the medium at one place over a long period of time—for example, if you look between two closely spaced posts of a pier or out of a ship's porthole as water waves pass by—the up-and-down motion of that small segment of water will be simple harmonic motion. The water moves up and down sinusoidally in time.

Some of the important quantities used to describe a periodic sinusoidal wave are shown in Fig. 15–3. The high points on a wave are called *crests*; the low points, *troughs*. The **amplitude**, A, is the maximum height of a crest, or depth of a trough, relative to the normal (or equilibrium) level. The total swing from a crest to a trough is twice the amplitude. The distance between two successive crests is called the **wavelength**, λ (the Greek letter lambda). The wavelength is also equal to the distance between *any* two successive identical points on the wave. The **frequency**, f, is the number of crests—or complete cycles—that pass a given point per unit time. The **period**, T, equals $1/f$ and is the time elapsed between two successive crests passing by the same point in space.

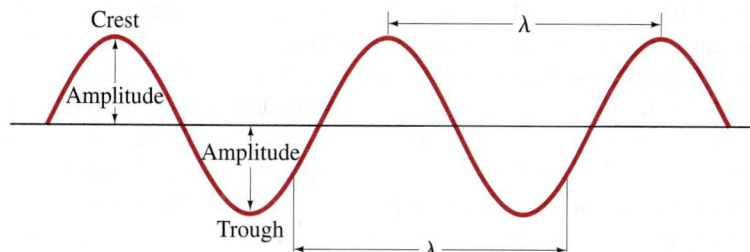

FIGURE 15–3 Characteristics of a single-frequency continuous wave moving through space.

The **wave velocity**, v, is the velocity at which wave crests (or any other part of the waveform) move forward. The wave velocity must be distinguished from the velocity of a particle of the medium itself as we saw in Example 15–1.

A wave crest travels a distance of one wavelength, λ, in a time equal to one period, T. Thus the wave velocity is $v = \lambda/T$. Then, since $1/T = f$,

$$v = \lambda f. \tag{15–1}$$

For example, suppose a wave has a wavelength of 5 m and a frequency of 3 Hz. Since three crests pass a given point per second, and the crests are 5 m apart, the first crest (or any other part of the wave) must travel a distance of 15 m during the 1 s. So the wave velocity is 15 m/s.

EXERCISE B You notice a water wave pass by the end of a pier with about 0.5 s between crests. Therefore (a) the frequency is 0.5 Hz; (b) the velocity is 0.5 m/s; (c) the wavelength is 0.5 m; (d) the period is 0.5 s.

15–2 Types of Waves: Transverse and Longitudinal

When a wave travels down a cord—say, from left to right as in Fig. 15–1—the particles of the cord vibrate up and down in a direction transverse (that is, perpendicular) to the motion of the wave itself. Such a wave is called a **transverse wave** (Fig. 15–4a). There exists another type of wave known as a **longitudinal wave**. In a longitudinal wave, the vibration of the particles of the medium is *along* the direction of the wave's motion. Longitudinal waves are readily formed on a stretched spring or Slinky by alternately compressing and expanding one end. This is shown in Fig. 15–4b, and can be compared to the transverse wave in Fig. 15–4a.

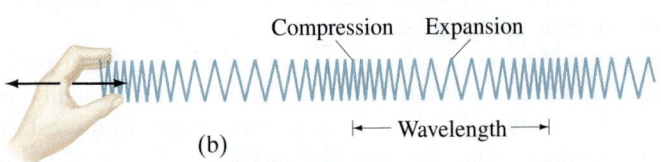

FIGURE 15–4 (a) Transverse wave; (b) longitudinal wave.

A series of compressions and expansions propagate along the spring. The *compressions* are those areas where the coils are momentarily close together. *Expansions* (sometimes called *rarefactions*) are regions where the coils are momentarily far apart. Compressions and expansions correspond to the crests and troughs of a transverse wave.

An important example of a longitudinal wave is a sound wave in air. A vibrating drumhead, for instance, alternately compresses and rarefies the air in contact with it, producing a longitudinal wave that travels outward in the air, as shown in Fig. 15–5.

As in the case of transverse waves, each section of the medium in which a longitudinal wave passes oscillates over a very small distance, whereas the wave itself can travel large distances. Wavelength, frequency, and wave velocity all have meaning for a longitudinal wave. The wavelength is the distance between successive compressions (or between successive expansions), and frequency is the number of compressions that pass a given point per second. The wave velocity is the velocity with which each compression appears to move; it is equal to the product of wavelength and frequency, $v = \lambda f$ (Eq. 15–1).

A longitudinal wave can be represented graphically by plotting the density of air molecules (or coils of a Slinky) versus position at a given instant, as shown in Fig. 15–6. Such a graphical representation makes it easy to illustrate what is happening. Note that the graph looks much like a transverse wave.

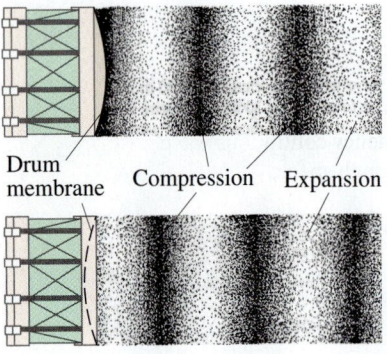

FIGURE 15–5 Production of a sound wave, which is longitudinal, shown at two moments in time about a half period ($\frac{1}{2}T$) apart.

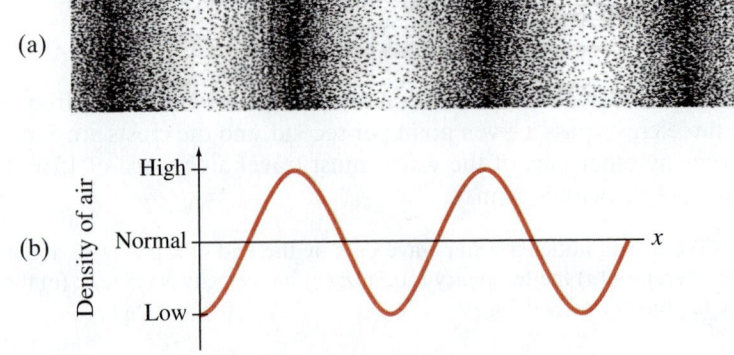

FIGURE 15–6 (a) A longitudinal wave with (b) its graphical representation at a particular instant in time.

Velocity of Transverse Waves

The velocity of a wave depends on the properties of the medium in which it travels. The velocity of a transverse wave on a stretched string or cord, for example, depends on the tension in the cord, F_T, and on the mass per unit length of the cord, μ (the Greek letter mu, where here $\mu = m/\ell$). For waves of small amplitude, the relationship is

$$v = \sqrt{\frac{F_T}{\mu}}.$$ $\qquad \left[\begin{array}{c}\text{transverse} \\ \text{wave on a cord}\end{array}\right]$ **(15–2)**

Before giving a derivation of this formula, it is worth noting that at least qualitatively it makes sense on the basis of Newtonian mechanics. That is, we do expect the tension to be in the numerator and the mass per unit length in the denominator. Why? Because when the tension is greater, we expect the velocity to be greater since each segment of cord is in tighter contact with its neighbor. And, the greater the mass per unit length, the more inertia the cord has and the more slowly the wave would be expected to propagate.

EXERCISE C A wave starts at the left end of a long cord (see Fig. 15–1) when someone shakes the cord back and forth at the rate of 2.0 Hz. The wave is observed to move to the right at 4.0 m/s. If the frequency is increased from 2.0 to 3.0 Hz, the new speed of the wave is (a) 1.0 m/s, (b) 2.0 m/s, (c) 4.0 m/s, (d) 8.0 m/s, (e) 16.0 m/s.

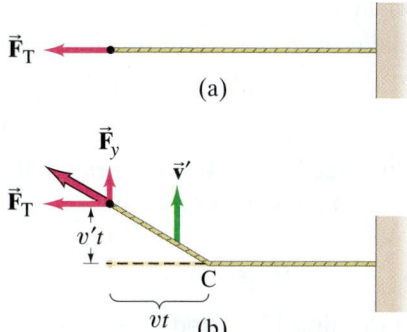

FIGURE 15–7 Diagram of simple wave pulse on a cord for derivation of Eq. 15–2. The vector shown in (b) as the resultant of $\vec{F}_T + \vec{F}_y$ has to be directed along the cord because the cord is flexible. (Diagram is not to scale: we assume $v' \ll v$; the upward angle of the cord is exaggerated for visibility.)

We can make a simple derivation of Eq. 15–2 using a simple model of a cord under a tension F_T as shown in Fig. 15–7a. The cord is pulled upward at a speed v' by the force F_y. As shown in Fig. 15–7b all points of the cord to the left of point C move upward at the speed v', and those to the right are still at rest. The speed of propagation, v, of this wave pulse is the speed of point C, the leading edge of the pulse. Point C moves to the right a distance vt in a time t, whereas the end of the cord moves upward a distance $v't$. By similar triangles we have the approximate relation

$$\frac{F_T}{F_y} = \frac{vt}{v't} = \frac{v}{v'},$$

which is accurate for small displacements ($v't \ll vt$) so that F_T does not change appreciably. As we saw in Chapter 9, the impulse given to an object is equal to its change in momentum. During the time t the total upward impulse is $F_y t = (v'/v)F_T t$. The change in momentum of the cord, Δp, is the mass of cord moving upward times its velocity. Since the upward moving segment of cord has mass equal to the mass per unit length μ times its length vt we have

$$F_y t = \Delta p$$

$$\frac{v'}{v} F_T t = (\mu vt)v'.$$

Solving for v we find $v = \sqrt{F_T/\mu}$ which is Eq. 15–2. Although it was derived for a special case, it is valid for any wave shape since other shapes can be considered to be made up of many tiny such lengths. But it is valid only for small displacements (as was our derivation). Experiment is in accord with this result derived from Newtonian mechanics.

EXAMPLE 15–2 **Pulse on a wire.** An 80.0-m-long, 2.10-mm-diameter copper wire is stretched between two poles. A bird lands at the center point of the wire, sending a small wave pulse out in both directions. The pulses reflect at the ends and arrive back at the bird's location 0.750 seconds after it landed. Determine the tension in the wire.

APPROACH From Eq. 15–2, the tension is given by $F_T = \mu v^2$. The speed v is distance divided by the time. The mass per unit length μ is calculated from the density of copper and the dimensions of the wire.

SOLUTION Each wave pulse travels 40.0 m to the pole and back again ($= 80.0$ m) in 0.750 s. Thus their speed is $v = (80.0 \text{ m})/(0.750 \text{ s}) = 107$ m/s. We take (Table 13–1) the density of copper as 8.90×10^3 kg/m³. The volume of copper in the wire is the cross-sectional area (πr^2) times the length ℓ, and the mass of the wire is the volume times the density: $m = \rho(\pi r^2)\ell$ for a wire of radius r and length ℓ. Then $\mu = m/\ell$ is

$$\mu = \rho\pi r^2\ell/\ell = \rho\pi r^2 = (8.90 \times 10^3 \text{ kg/m}^3)\pi(1.05 \times 10^{-3} \text{ m})^2 = 0.0308 \text{ kg/m}.$$

Thus, the tension is $F_T = \mu v^2 = (0.0308 \text{ kg/m})(107 \text{ m/s})^2 = 353$ N.

Velocity of Longitudinal Waves

The velocity of a longitudinal wave has a form similar to that for a transverse wave on a cord (Eq. 15–2); that is,

$$v = \sqrt{\frac{\text{elastic force factor}}{\text{inertia factor}}}.$$

In particular, for a longitudinal wave traveling down a long solid rod,

$$v = \sqrt{\frac{E}{\rho}}, \qquad \left[\begin{array}{l}\text{longitudinal} \\ \text{wave in a long rod}\end{array}\right] \quad \textbf{(15–3)}$$

where E is the elastic modulus (Section 12–4) of the material and ρ is its density. For a longitudinal wave traveling in a liquid or gas,

$$v = \sqrt{\frac{B}{\rho}}, \qquad \left[\begin{array}{l}\text{longitudinal wave} \\ \text{in a fluid}\end{array}\right] \quad \textbf{(15–4)}$$

where B is the bulk modulus (Section 12–4) and ρ again is the density.

PHYSICS APPLIED
Space perception
by animals, using sound waves

EXAMPLE 15–3 **Echolocation.** Echolocation is a form of sensory perception used by animals such as bats, toothed whales, and dolphins. The animal emits a pulse of sound (a longitudinal wave) which, after reflection from objects, returns and is detected by the animal. Echolocation waves can have frequencies of about 100,000 Hz. (*a*) Estimate the wavelength of a sea animal's echolocation wave. (*b*) If an obstacle is 100 m from the animal, how long after the animal emits a wave is its reflection detected?

APPROACH We first compute the speed of longitudinal (sound) waves in sea water, using Eq. 15–4 and Tables 12–1 and 13–1. The wavelength is $\lambda = v/f$.

SOLUTION (*a*) The speed of longitudinal waves in sea water, which is slightly more dense than pure water, is

$$v = \sqrt{\frac{B}{\rho}} = \sqrt{\frac{2.0 \times 10^9 \text{ N/m}^2}{1.025 \times 10^3 \text{ kg/m}^3}} = 1.4 \times 10^3 \text{ m/s}.$$

Then, using Eq. 15–1, we find

$$\lambda = \frac{v}{f} = \frac{(1.4 \times 10^3 \text{ m/s})}{(1.0 \times 10^5 \text{ Hz})} = 14 \text{ mm}.$$

(*b*) The time required for the round-trip between the animal and the object is

$$t = \frac{\text{distance}}{\text{speed}} = \frac{2(100 \text{ m})}{1.4 \times 10^3 \text{ m/s}} = 0.14 \text{ s}.$$

NOTE We shall see later that waves can be used to "resolve" (or detect) objects only if the wavelength is comparable to or smaller than the object. Thus, a dolphin can resolve objects on the order of a centimeter or larger in size.

We now derive Eq. 15–4. Consider a wave pulse traveling in a fluid in a long tube, so that the wave motion is one dimensional. The tube is fitted with a piston at the end and is filled with a fluid which, at $t = 0$, is of uniform density ρ and at uniform pressure P_0, Fig. 15–8a. At this moment the piston is abruptly made to start moving to the right with speed v', compressing the fluid in front of it. In the (short) time t the piston moves a distance $v't$. The compressed fluid itself also moves with speed v', but the leading edge of the compressed region moves to the right at the characteristic speed v of compression waves in that fluid; we assume the wave speed v is much larger than the piston speed v'. The leading edge of the compression (which at $t = 0$ was at the piston face) thus moves a distance vt in time t as shown in Fig. 15–8b. Let the pressure in the compression be $P_0 + \Delta P$, which is ΔP higher than in the uncompressed fluid. To move the piston to the right requires an external force $(P_0 + \Delta P)S$ acting to the right, where S is the cross-sectional area of the tube. (S for "surface area"; we save A for amplitude.) The *net* force on the compressed region of the fluid is

$$F_{\text{net}} = (P_0 + \Delta P)S - P_0S = S\,\Delta P$$

since the uncompressed fluid exerts a force $P_0 S$ to the left at the leading edge. Hence the impulse given to the compressed fluid, which equals its change in momentum, is

$$F_{\text{net}}\,t = \Delta m v'$$
$$S\,\Delta P\,t = (\rho S v t)v',$$

where $(\rho S v t)$ represents the mass of fluid which is given the speed v' (the compressed fluid of area S moves a distance vt, Fig. 15–8, so the volume moved is Svt). Hence we have

$$\Delta P = \rho v v'.$$

From the definition of the bulk modulus, B (Eq. 12–7):

$$B = -\frac{\Delta P}{\Delta V/V_0} = -\frac{\rho v v'}{\Delta V/V_0},$$

where $\Delta V/V_0$ is the fractional change in volume due to compression. The original volume of the compressed fluid is $V_0 = Svt$ (see Fig. 15–8), and it has been compressed by an amount $\Delta V = -Sv't$ (Fig. 15–8b). Thus

$$B = -\frac{\rho v v'}{\Delta V/V_0} = -\rho v v'\left(\frac{Svt}{-Sv't}\right) = \rho v^2,$$

and so

$$v = \sqrt{\frac{B}{\rho}},$$

which is what we set out to show, Eq. 15–4.

The derivation of Eq. 15–3 follows similar lines, but takes into account the expansion of the sides of a rod when the end of a rod is compressed.

Other Waves

Both transverse and longitudinal waves are produced when an **earthquake** occurs. The transverse waves that travel through the body of the Earth are called S waves (S for shear), and the longitudinal waves are called P waves (P for pressure) or *compression* waves. Both longitudinal and transverse waves can travel through a solid since the atoms or molecules can vibrate about their relatively fixed positions in any direction. But only longitudinal waves can propagate through a fluid, because any transverse motion would not experience any restoring force since a fluid is readily deformable. This fact was used by geophysicists to infer that a portion of the Earth's core must be liquid: after an earthquake, longitudinal waves are detected diametrically across the Earth, but not transverse waves.

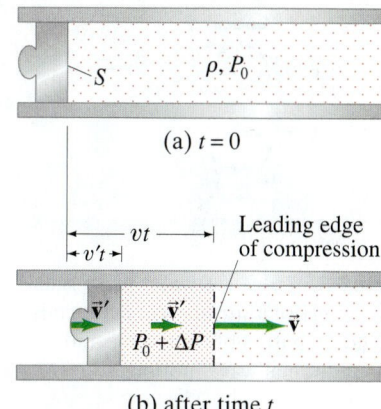

FIGURE 15–8 Determining the speed of a one-dimensional longitudinal wave in a fluid contained in a long narrow tube.

(a) $t = 0$

(b) after time t

⊛ **PHYSICS APPLIED**
Earthquake waves

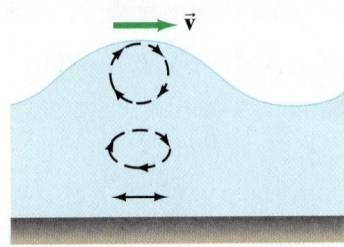

FIGURE 15–9 A water wave is an example of a *surface wave*, which is a combination of transverse and longitudinal wave motions.

FIGURE 15–10 How a wave breaks. The green arrows represent the local velocity of water molecules.

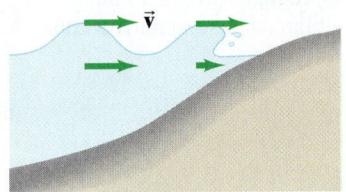

Besides these two types of waves that can pass through the body of the Earth (or other substance), there can also be *surface waves* that travel along the boundary between two materials. A wave on water is actually a surface wave that moves on the boundary between water and air. The motion of each particle of water at the surface is circular or elliptical (Fig. 15–9), so it is a combination of transverse and longitudinal motions. Below the surface, there is also transverse plus longitudinal wave motion, as shown. At the bottom, the motion is only longitudinal. (When a wave approaches shore, the water drags at the bottom and is slowed down, while the crests move ahead at higher speed (Fig. 15–10) and "spill" over the top.)

Surface waves are also set up on the Earth when an earthquake occurs. The waves that travel along the surface are mainly responsible for the damage caused by earthquakes.

Waves which travel along a line in one dimension, such as transverse waves on a stretched string, or longitudinal waves in a rod or fluid-filled tube, are *linear* or *one-dimensional waves*. Surface waves, such as the water waves pictured at the start of this Chapter, are *two-dimensional waves*. Finally, waves that move out from a source in all directions, such as sound from a loudspeaker or earthquake waves through the Earth, are *three-dimensional waves*.

15–3 Energy Transported by Waves

Waves transport energy from one place to another. As waves travel through a medium, the energy is transferred as vibrational energy from particle to particle of the medium. For a sinusoidal wave of frequency f, the particles move in simple harmonic motion (Chapter 14) as a wave passes, and each particle has energy $E = \frac{1}{2}kA^2$ where A is the maximum displacement (amplitude) of its motion, either transversely or longitudinally (Eq. 14–10a). Using Eq. 14–7a we can write $k = 4\pi^2 mf^2$, where m is the mass of a particle (or small volume) of the medium. Then in terms of the frequency f and amplitude A,

$$E = \tfrac{1}{2}kA^2 = 2\pi^2 mf^2 A^2.$$

FIGURE 15–11 Calculating the energy carried by a wave moving with velocity v.

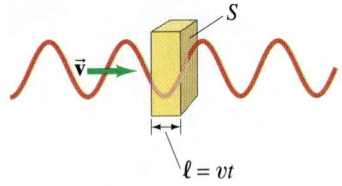

For three-dimensional waves traveling in an elastic medium, the mass $m = \rho V$, where ρ is the density of the medium and V is the volume of a small slice of the medium. The volume $V = S\ell$ where S is the cross-sectional area through which the wave travels (Fig. 15–11), and we can write ℓ as the distance the wave travels in a time t as $\ell = vt$, where v is the speed of the wave. Thus $m = \rho V = \rho S\ell = \rho Svt$ and

$$E = 2\pi^2 \rho Svt f^2 A^2. \tag{15–5}$$

From this equation we have the important result that *the energy transported by a wave is proportional to the square of the amplitude, and to the square of the frequency.* The average *rate* of energy transferred is the average power P:

$$\overline{P} = \frac{E}{t} = 2\pi^2 \rho Sv f^2 A^2. \tag{15–6}$$

Finally, the **intensity**, I, of a wave is defined as the average power transferred across unit area perpendicular to the direction of energy flow:

$$I = \frac{\overline{P}}{S} = 2\pi^2 v\rho f^2 A^2. \tag{15–7}$$

If a wave flows out from the source in all directions, it is a three-dimensional wave. Examples are sound traveling in the open air, earthquake waves, and light waves.

If the medium is isotropic (same in all directions), the wave from a point source is a *spherical wave* (Fig. 15–12). As the wave moves outward, the energy it carries is spread over a larger and larger area since the surface area of a sphere of radius r is $4\pi r^2$. Thus the intensity of a wave is

$$I = \frac{\overline{P}}{S} = \frac{\overline{P}}{4\pi r^2}.$$

If the power output $\overline{P}$ is constant, then the intensity decreases as the inverse square of the distance from the source:

$$I \propto \frac{1}{r^2}. \qquad\qquad \text{[spherical wave]} \quad \textbf{(15–8a)}$$

If we consider two points at distances r_1 and r_2 from the source, as in Fig. 15–12, then $I_1 = \overline{P}/4\pi r_1^2$ and $I_2 = \overline{P}/4\pi r_2^2$, so

$$\frac{I_2}{I_1} = \frac{\overline{P}/4\pi r_2^2}{\overline{P}/4\pi r_1^2} = \frac{r_1^2}{r_2^2}. \qquad\qquad \textbf{(15–8b)}$$

Thus, for example, when the distance doubles $(r_2/r_1 = 2)$, then the intensity is reduced to $\frac{1}{4}$ of its earlier value: $I_2/I_1 = \left(\frac{1}{2}\right)^2 = \frac{1}{4}$.

The amplitude of a wave also decreases with distance. Since the intensity is proportional to the square of the amplitude (Eq. 15–7), $I \propto A^2$, the amplitude A must decrease as $1/r$, so that I can be proportional to $1/r^2$ (Eq. 15–8a). Hence

$$A \propto \frac{1}{r}.$$

To see this directly from Eq. 15–6, consider again two different distances from the source, r_1 and r_2. For constant power output, $S_1 A_1^2 = S_2 A_2^2$ where A_1 and A_2 are the amplitudes of the wave at r_1 and r_2, respectively. Since $S_1 = 4\pi r_1^2$ and $S_2 = 4\pi r_2^2$, we have $\left(A_1^2 r_1^2\right) = \left(A_2^2 r_2^2\right)$, or

$$\frac{A_2}{A_1} = \frac{r_1}{r_2}.$$

When the wave is twice as far from the source, the amplitude is half as large, and so on (ignoring damping due to friction).

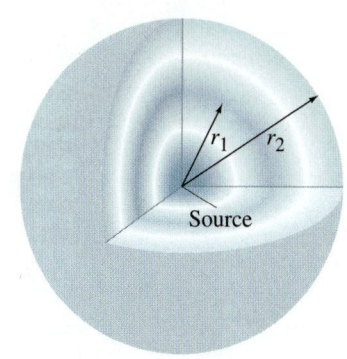

FIGURE 15–12 Wave traveling outward from a point source has spherical shape. Two different crests (or compressions) are shown, of radius r_1 and r_2.

EXAMPLE 15–4 **Earthquake intensity.** The intensity of an earthquake P wave traveling through the Earth and detected 100 km from the source is 1.0×10^6 W/m². What is the intensity of that wave if detected 400 km from the source?

APPROACH We assume the wave is spherical, so the intensity decreases as the square of the distance from the source.

SOLUTION At 400 km the distance is 4 times greater than at 100 km, so the intensity will be $\left(\frac{1}{4}\right)^2 = \frac{1}{16}$ of its value at 100 km, or $(1.0 \times 10^6 \text{ W/m}^2)/16 = 6.3 \times 10^4$ W/m².

NOTE Using Eq. 15–8b directly gives

$$I_2 = I_1 r_1^2/r_2^2 = \left(1.0 \times 10^6 \text{ W/m}^2\right)(100 \text{ km})^2/(400 \text{ km})^2 = 6.3 \times 10^4 \text{ W/m}^2.$$

The situation is different for a one-dimensional wave, such as a transverse wave on a string or a longitudinal wave pulse traveling down a thin uniform metal rod. The area remains constant, so the amplitude A also remains constant (ignoring friction). Thus the amplitude and the intensity do not decrease with distance.

In practice, frictional damping is generally present, and some of the energy is transformed into thermal energy. Thus the amplitude and intensity of a one-dimensional wave decrease with distance from the source. For a three-dimensional wave, the decrease will be greater than that discussed above, although the effect may often be small.

15–4 Mathematical Representation of a Traveling Wave

Let us now consider a one-dimensional wave traveling along the x axis. It could be, for example, a transverse wave on a cord or a longitudinal wave traveling in a rod or in a fluid-filled tube. Let us assume the wave shape is sinusoidal and has a particular wavelength λ and frequency f. At $t = 0$, suppose the wave shape is given by

$$D(x) = A \sin \frac{2\pi}{\lambda} x, \tag{15–9}$$

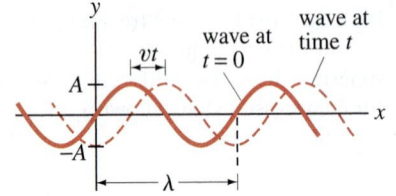

FIGURE 15–13 A traveling wave. In time t, the wave moves a distance vt.

as shown by the solid curve in Fig. 15–13: $D(x)$ is the **displacement**[†] of the wave (be it a longitudinal or transverse wave) at position x, and A is the **amplitude** (maximum displacement) of the wave. This relation gives a shape that repeats itself every wavelength, which is needed so that the displacement is the same at $x = 0$, $x = \lambda$, $x = 2\lambda$, and so on (since $\sin 4\pi = \sin 2\pi = \sin 0$).

Now suppose the wave is moving to the right with velocity v. Then, after a time t, each part of the wave (indeed, the whole wave "shape") has moved to the right a distance vt; see the dashed curve in Fig. 15–13. Consider any point on the wave at $t = 0$: say, a crest which is at some position x. After a time t, that crest will have traveled a distance vt so its new position is a distance vt greater than its old position. To describe this same point on the wave shape, the argument of the sine function must be the same, so we replace x in Eq. 15–9 by $(x - vt)$:

1-D wave moving in positive x direction

$$D(x,t) = A \sin \left[\frac{2\pi}{\lambda} (x - vt) \right]. \tag{15–10a}$$

Said another way, if you are riding on a crest, the argument of the sine function, $(2\pi/\lambda)(x - vt)$, remains the same ($= \pi/2, 5\pi/2$, and so on); as t increases, x must increase at the same rate so that $(x - vt)$ remains constant.

Equation 15–10a is the mathematical representation of a sinusoidal wave traveling along the x axis to the right (increasing x). It gives the displacement $D(x,t)$ of the wave at any chosen point x at any time t. The function $D(x,t)$ describes a curve that represents the actual shape of the wave in space at time t. Since $v = \lambda f$ (Eq. 15–1) we can write Eq. 15–10a in other ways that are often convenient:

1-D wave

moving in

positive x

direction

$$D(x,t) = A \sin \left(\frac{2\pi x}{\lambda} - \frac{2\pi t}{T} \right), \tag{15–10b}$$

where $T = 1/f = \lambda/v$ is the period; and

$$D(x,t) = A \sin(kx - \omega t), \tag{15–10c}$$

where $\omega = 2\pi f = 2\pi/T$ is the angular frequency and

$$k = \frac{2\pi}{\lambda} \tag{15–11}$$

CAUTION

Don't confuse wave number k with spring constant k

is called the **wave number**. (Do not confuse the wave number k with the spring constant k; they are very different quantities.) All three forms, Eqs. 15–10a, b, and c, are equivalent; Eq. 15–10c is the simplest to write and is perhaps the most common. The quantity $(kx - \omega t)$, and its equivalent in the other two equations, is called the **phase** of the wave. The velocity v of the wave is often called the **phase velocity**, since it describes the velocity of the phase (or shape) of the wave and it can be written in terms of ω and k:

$$v = \lambda f = \left(\frac{2\pi}{k} \right) \left(\frac{\omega}{2\pi} \right) = \frac{\omega}{k}. \tag{15–12}$$

[†]Some books use $y(x)$ in place of $D(x)$. To avoid confusion, we reserve y (and z) for the coordinate positions of waves in two or three dimensions. Our $D(x)$ can stand for pressure (in longitudinal waves), position displacement (transverse mechanical waves), or—as we will see later—electric or magnetic fields (for electromagnetic waves).

For a wave traveling along the x axis to the left (decreasing values of x), we start again with Eq. 15–9 and note that the velocity is now $-v$. A particular point on the wave changes position by $-vt$ in a time t, so x in Eq. 15–9 must be replaced by $(x + vt)$. Thus, for a wave traveling to the left with velocity v,

$$D(x, t) = A \sin\left[\frac{2\pi}{\lambda}(x + vt)\right] \qquad \textbf{(15–13a)}$$

1-D wave moving in negative x direction

$$= A \sin\left(\frac{2\pi x}{\lambda} + \frac{2\pi t}{T}\right) \qquad \textbf{(15–13b)}$$

$$= A \sin(kx + \omega t). \qquad \textbf{(15–13c)}$$

In other words, we simply replace v in Eqs. 15–10 by $-v$.

Let us look at Eq. 15–13c (or, just as well, at Eq. 15–10c). At $t = 0$ we have

$$D(x, 0) = A \sin kx,$$

which is what we started with, a sinusoidal wave shape. If we look at the wave shape in space at a particular later time t_1, then we have

$$D(x, t_1) = A \sin(kx + \omega t_1).$$

That is, if we took a picture of the wave at $t = t_1$, we would see a sine wave with a phase constant ωt_1. Thus, for fixed $t = t_1$, the wave has a sinusoidal shape in space. On the other hand, if we consider a fixed point in space, say $x = 0$, we can see how the wave varies in time:

$$D(0, t) = A \sin \omega t$$

where we used Eq. 15–13c. This is just the equation for simple harmonic motion (Section 14–2). For any other fixed value of x, say $x = x_1$, $D = A \sin(\omega t + kx_1)$ which differs only by a phase constant kx_1. Thus, at any fixed point in space, the displacement undergoes the oscillations of simple harmonic motion in time. Equations 15–10 and 15–13 combine both these aspects to give us the representation for a **traveling sinusoidal wave** (also called a **harmonic wave**).

The argument of the sine in Eqs. 15–10 and 15–13 can in general contain a phase angle ϕ, which for Eq. 15–10c is

$$D(x, t) = A \sin(kx - \omega t + \phi),$$

to adjust for the position of the wave at $t = 0$, $x = 0$, just as in Section 14–2 (see Fig. 14–7). If the displacement is zero at $t = 0$, $x = 0$, as in Fig. 14–6 (or Fig. 15–13), then $\phi = 0$.

Now let us consider a general wave (or wave pulse) of any shape. If frictional losses are small, experiment shows that the wave maintains its shape as it travels. Thus we can make the same arguments as we did right after Eq. 15–9. Suppose our wave has some shape at $t = 0$, given by

$$D(x, 0) = D(x)$$

where $D(x)$ is the displacement of the wave at x and is not necessarily sinusoidal. At some later time, if the wave is traveling to the right along the x axis, the wave will have the same shape but all parts will have moved a distance vt where v is the phase velocity of the wave. Hence we must replace x by $x - vt$ to obtain the amplitude at time t:

$$D(x, t) = D(x - vt). \qquad \textbf{(15–14)}$$

Similarly, if the wave moves to the left, we must replace x by $x + vt$, so

$$D(x, t) = D(x + vt). \qquad \textbf{(15–15)}$$

Thus, any wave traveling along the x axis must have the form of Eq. 15–14 or 15–15.

EXERCISE D A wave is given by $D(x, t) = (5.0 \text{ mm}) \sin(2.0x - 20.0t)$ where x is in meters and t is in seconds. What is the speed of the wave? (a) 10 m/s, (b) 0.10 m/s, (c) 40 m/s, (d) 0.005 m/s, (e) $2.5 \times 10^{-4} \text{ m/s}$.

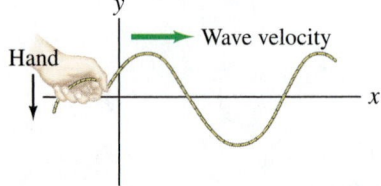

Wave velocity

Hand

FIGURE 15–14 Example 15–5. The wave at $t = 0$ (the hand is falling). Not to scale.

EXAMPLE 15–5 **A traveling wave.** The left-hand end of a long horizontal stretched cord oscillates transversely in SHM with frequency $f = 250 \, \text{Hz}$ and amplitude 2.6 cm. The cord is under a tension of 140 N and has a linear density $\mu = 0.12 \, \text{kg/m}$. At $t = 0$, the end of the cord has an upward displacement of 1.6 cm and is falling (Fig. 15–14). Determine (a) the wavelength of waves produced and (b) the equation for the traveling wave.

APPROACH We first find the phase velocity of the transverse wave from Eq. 15–2; then $\lambda = v/f$. In (b), we need to find the phase ϕ using the initial conditions.

SOLUTION (a) The wave velocity is

$$v = \sqrt{\frac{F_T}{\mu}} = \sqrt{\frac{140 \, \text{N}}{0.12 \, \text{kg/m}}} = 34 \, \text{m/s}.$$

Then

$$\lambda = \frac{v}{f} = \frac{34 \, \text{m/s}}{250 \, \text{Hz}} = 0.14 \, \text{m} \quad \text{or} \quad 14 \, \text{cm}.$$

(b) Let $x = 0$ at the left-hand end of the cord. The phase of the wave at $t = 0$ is not zero in general as was assumed in Eqs. 15–9, 10, and 13. The general form for a wave traveling to the right is

$$D(x, t) = A \sin(kx - \omega t + \phi),$$

where ϕ is the phase angle. In our case, the amplitude $A = 2.6 \, \text{cm}$; and at $t = 0$, $x = 0$, we are given $D = 1.6 \, \text{cm}$. Thus

$$1.6 = 2.6 \sin \phi,$$

so $\phi = \sin^{-1}(1.6/2.6) = 38° = 0.66 \, \text{rad}$. We also have $\omega = 2\pi f = 1570 \, \text{s}^{-1}$ and $k = 2\pi/\lambda = 2\pi/0.14 \, \text{m} = 45 \, \text{m}^{-1}$. Hence

$$D = (0.026 \, \text{m}) \sin\big[(45 \, \text{m}^{-1})x - (1570 \, \text{s})t + 0.66\big]$$

which we can write more simply as

$$D = 0.026 \sin(45x - 1570t + 0.66),$$

and we specify clearly that D and x are in meters and t in seconds.

*15–5 The Wave Equation

Many types of waves satisfy an important general equation that is the equivalent of Newton's second law of motion for particles. This "equation of motion for a wave" is called the **wave equation**, and we derive it now for waves traveling on a stretched horizontal string.

We assume the amplitude of the wave is small compared to the wavelength so that each point on the string can be assumed to move only vertically and the tension in the string, F_T, does not vary during a vibration. We apply Newton's second law, $\Sigma F = ma$, to the vertical motion of a tiny section of the string as shown in Fig. 15–15. The amplitude of the wave is small, so the angles θ_1 and θ_2 that the string makes with the horizontal are small. The length of this section is then approximately Δx, and its mass is $\mu \, \Delta x$, where μ is the mass per unit length of the string. The net vertical force on this section of string is $F_T \sin \theta_2 - F_T \sin \theta_1$. So Newton's second law applied to the vertical (y) direction gives

$$\Sigma F_y = ma_y$$

$$F_T \sin \theta_2 - F_T \sin \theta_1 = (\mu \, \Delta x) \frac{\partial^2 D}{\partial t^2}. \tag{i}$$

FIGURE 15–15 Deriving the wave equation from Newton's second law: a segment of string under tension F_T.

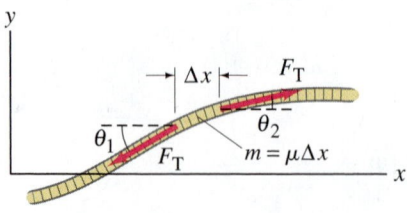

We have written the acceleration as $a_y = \partial^2 D/\partial t^2$ since the motion is only vertical, and we use the partial derivative notation because the displacement D is a function of both x and t.

Because the angles θ_1 and θ_2 are assumed small, $\sin \theta \approx \tan \theta$ and $\tan \theta$ is equal to the slope s of the string at each point:

$$\sin \theta \approx \tan \theta = \frac{\partial D}{\partial x} = s.$$

Thus our equation (i) at the bottom of the previous page becomes

$$F_T(s_2 - s_1) = \mu\, \Delta x \frac{\partial^2 D}{\partial t^2}$$

or

$$F_T \frac{\Delta s}{\Delta x} = \mu \frac{\partial^2 D}{\partial t^2}, \qquad \text{(ii)}$$

where $\Delta s = s_2 - s_1$ is the difference in the slope between the two ends of our tiny section. Now we take the limit of $\Delta x \to 0$, so that

$$F_T \lim_{\Delta x \to 0} \frac{\Delta s}{\Delta x} = F_T \frac{\partial s}{\partial x}$$

$$= F_T \frac{\partial}{\partial x}\left(\frac{\partial D}{\partial x}\right) = F_T \frac{\partial^2 D}{\partial x^2}$$

since the slope $s = \partial D / \partial x$, as we wrote above. Substituting this into the equation labeled (ii) above gives

$$F_T \frac{\partial^2 D}{\partial x^2} = \mu \frac{\partial^2 D}{\partial t^2}$$

or

$$\frac{\partial^2 D}{\partial x^2} = \frac{\mu}{F_T} \frac{\partial^2 D}{\partial t^2}.$$

We saw earlier in this Chapter (Eq. 15–2) that the velocity of waves on a string is given by $v = \sqrt{F_T/\mu}$, so we can write this last equation as

$$\frac{\partial^2 D}{\partial x^2} = \frac{1}{v^2} \frac{\partial^2 D}{\partial t^2}. \qquad \text{(15–16)}$$

This is the **one-dimensional wave equation**, and it can describe not only small amplitude waves on a stretched string, but also small amplitude longitudinal waves (such as sound waves) in gases, liquids, and elastic solids, in which case D can refer to the pressure variations. In this case, the wave equation is a direct consequence of Newton's second law applied to a continuous elastic medium. The wave equation also describes electromagnetic waves for which D refers to the electric or magnetic field, as we shall see in Chapter 31. Equation 15–16 applies to waves traveling in one dimension only. For waves spreading out in three dimensions, the wave equation is the same, with the addition of $\partial^2 D/\partial y^2$ and $\partial^2 D/\partial z^2$ to the left side of Eq. 15–16.

The wave equation is a *linear* equation: the displacement D appears singly in each term. There are no terms that contain D^2, or $D(\partial D/\partial x)$, or the like in which D appears more than once. Thus, if $D_1(x, t)$ and $D_2(x, t)$ are two different solutions of the wave equation, then the linear combination

$$D_3(x, t) = aD_1(x, t) + bD_2(x, t),$$

where a and b are constants, is also a solution. This is readily seen by direct substitution into the wave equation. This is the essence of the *superposition principle*, which we discuss in the next Section. Basically it says that if two waves pass through the same region of space at the same time, the actual displacement is the sum of the separate displacements. For waves on a string, or for sound waves, this is valid only for small-amplitude waves. If the amplitude is not small enough, the equations for wave propagation may become nonlinear and the principle of superposition would not hold and more complicated effects may occur.

EXAMPLE 15–6 **Wave equation solution.** Verify that the sinusoidal wave of Eq. 15–10c, $D(x, t) = A \sin(kx - \omega t)$, satisfies the wave equation.

APPROACH We substitute Eq. 15–10c into the wave equation, Eq. 15–16.

SOLUTION We take the derivative of Eq. 15–10c twice with respect to t:

$$\frac{\partial D}{\partial t} = -\omega A \cos(kx - \omega t)$$

$$\frac{\partial^2 D}{\partial t^2} = -\omega^2 A \sin(kx - \omega t).$$

With respect to x, the derivatives are

$$\frac{\partial D}{\partial x} = kA \cos(kx - \omega t)$$

$$\frac{\partial^2 D}{\partial x^2} = -k^2 A \sin(kx - \omega t).$$

If we now divide the second derivatives we get

$$\frac{\partial^2 D/\partial t^2}{\partial^2 D/\partial x^2} = \frac{-\omega^2 A \sin(kx - \omega t)}{-k^2 A \sin(kx - \omega t)} = \frac{\omega^2}{k^2}.$$

From Eq. 15–12 we have $\omega^2/k^2 = v^2$, so we see that Eq. 15–10 does satisfy the wave equation (Eq. 15–16).

15–6 The Principle of Superposition

When two or more waves pass through the same region of space at the same time, it is found that for many waves *the actual displacement is the vector* (or *algebraic*) *sum of the separate displacements*. This is called the **principle of superposition**. It is valid for mechanical waves as long as the displacements are not too large and there is a linear relationship between the displacement and the restoring force of the oscillating medium.[†] If the amplitude of a mechanical wave, for example, is so large that it goes beyond the elastic region of the medium, and Hooke's law is no longer operative, the superposition principle is no longer accurate.[‡] For the most part, we will consider systems for which the superposition principle can be assumed to hold.

One result of the superposition principle is that if two waves pass through the same region of space, they continue to move independently of one another. You may have noticed, for example, that the ripples on the surface of water (two-dimensional waves) that form from two rocks striking the water at different places will pass through each other.

Figure 15–16 shows an example of the superposition principle. In this case there are three waves present on a stretched string, each of different amplitude and frequency. At any time, such as at the instant shown, the actual displacement at any position x is the algebraic sum of the displacements of the three waves at that position. The actual wave is not a simple sinusoidal wave and is called a *composite* (or *complex*) *wave*. (Amplitudes are exaggerated in Fig. 15–16.)

It can be shown that any complex wave can be considered as being composed of many simple sinusoidal waves of different amplitudes, wavelengths, and frequencies. This is known as *Fourier's theorem*. A complex periodic wave of period T can be represented as a sum of pure sinusoidal terms whose frequencies are integral multiples of $f = 1/T$. If the wave is not periodic, the sum becomes an integral (called a *Fourier integral*). Although we will not go into the details here, we see the importance of considering sinusoidal waves (and simple harmonic motion): because any other wave shape can be considered a sum of such pure sinusoidal waves.

[†]For electromagnetic waves in vacuum, Chapter 31, the superposition principle always holds.

[‡]Intermodulation distortion in high-fidelity equipment is an example of the superposition principle not holding when two frequencies do not combine linearly in the electronics.

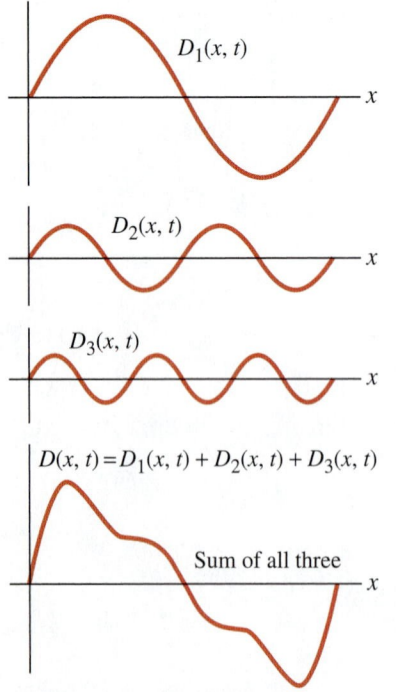

FIGURE 15–16 The superposition principle for one-dimensional waves. The composite wave formed from three sinusoidal waves of different amplitudes and frequencies $(f_0, 2f_0, 3f_0)$ is shown at the bottom at a certain instant in time. The displacement of the composite wave at each point in space, at any time, is the algebraic sum of the displacements of the component waves. Amplitudes are shown exaggerated; for the superposition principle to hold, they must be small compared to the wavelengths.

CONCEPTUAL EXAMPLE 15–7 | **Making a square wave.** At $t = 0$, three waves are given by $D_1 = A \cos kx$, $D_2 = -\frac{1}{3} A \cos 3kx$, and $D_3 = \frac{1}{5} A \cos 5kx$, where $A = 1.0 \, \text{m}$ and $k = 10 \, \text{m}^{-1}$. Plot the sum of the three waves from $x = -0.4 \, \text{m}$ to $+0.4 \, \text{m}$. (These three waves are the first three Fourier components of a "square wave.")

RESPONSE The first wave, D_1, has amplitude of 1.0 m and wavelength $\lambda = 2\pi/k = (2\pi/10) \, \text{m} = 0.628 \, \text{m}$. The second wave, D_2, has amplitude of 0.33 m and wavelength $\lambda = 2\pi/3k = (2\pi/30) \, \text{m} = 0.209 \, \text{m}$. The third wave, D_3, has amplitude of 0.20 m and wavelength $\lambda = 2\pi/5k = (2\pi/50) \, \text{m} = 0.126 \, \text{m}$. Each wave is plotted in Fig. 15–17a. The sum of the three waves is shown in Fig. 15–17b. The sum begins to resemble a "square wave," shown in blue in Fig. 15–17b.

When the restoring force is not precisely proportional to the displacement for mechanical waves in some continuous medium, the speed of sinusoidal waves depends on the frequency. The variation of speed with frequency is called **dispersion**. The different sinusoidal waves that compose a complex wave will travel with slightly different speeds in such a case. Consequently, a complex wave will change shape as it travels if the medium is "dispersive." A pure sine wave will not change shape under these conditions, however, except by the influence of friction or dissipative forces. If there is no dispersion (or friction), even a complex linear wave does not change shape.

15–7 Reflection and Transmission

When a wave strikes an obstacle, or comes to the end of the medium in which it is traveling, at least a part of the wave is reflected. You have probably seen water waves reflect off a rock or the side of a swimming pool. And you may have heard a shout reflected from a distant cliff—which we call an "echo."

A wave pulse traveling down a cord is reflected as shown in Fig. 15–18. The reflected pulse returns inverted as in Fig. 15–18a if the end of the cord is fixed; it returns right side up if the end is free as in Fig. 15–18b. When the end is fixed to a support, as in Fig. 15–18a, the pulse reaching that fixed end exerts a force (upward) on the support. The support exerts an equal but opposite force downward on the cord (Newton's third law). This downward force on the cord is what "generates" the inverted reflected pulse.

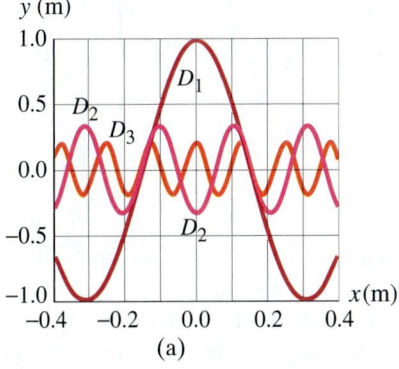
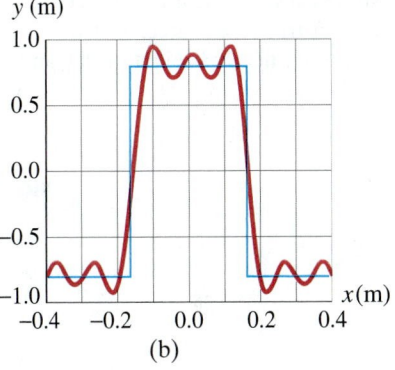

FIGURE 15–17 Example 15–7. Making a square wave.

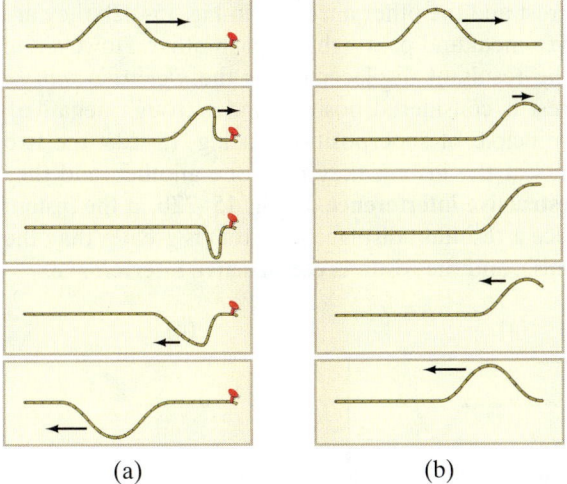

(a) (b)

FIGURE 15–18 Reflection of a wave pulse on a cord lying on a table. (a) The end of the cord is fixed to a peg. (b) The end of the cord is free to move.

Consider next a pulse that travels down a cord which consists of a light section and a heavy section, as shown in Fig. 15–19. When the wave pulse reaches the boundary between the two sections, part of the pulse is reflected and part is transmitted, as shown. The heavier the second section of the cord, the less the energy that is transmitted. (When the second section is a wall or rigid support, very little is transmitted and most is reflected, as in Fig. 15–18a.) For a periodic wave, the frequency of the transmitted wave does not change across the boundary since the boundary point oscillates at that frequency. Thus if the transmitted wave has a lower speed, its wavelength is also less $(\lambda = v/f)$.

FIGURE 15–19 When a wave pulse traveling to the right along a thin cord (a) reaches a discontinuity where the cord becomes thicker and heavier, then part is reflected and part is transmitted (b).

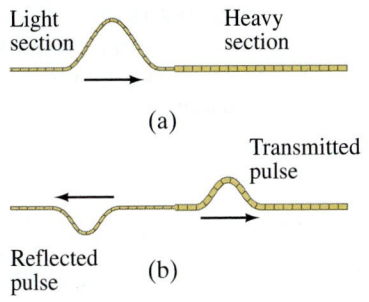

Light section Heavy section

(a)

Transmitted pulse

Reflected pulse (b)

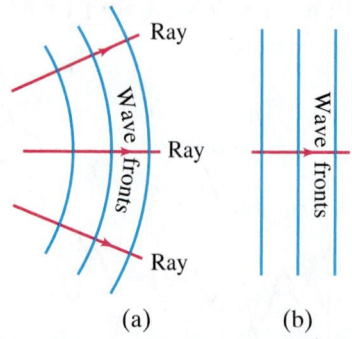

FIGURE 15–20 Rays, signifying the direction of motion, are always perpendicular to the wave fronts (wave crests). (a) Circular or spherical waves near the source. (b) Far from the source, the wave fronts are nearly straight or flat, and are called plane waves.

For a two- or three-dimensional wave, such as a water wave, we are concerned with **wave fronts**, by which we mean all the points along the wave forming the wave crest (what we usually refer to simply as a "wave" at the seashore). A line drawn in the direction of motion, perpendicular to the wave front, is called a **ray**, as shown in Fig. 15–20. Wave fronts far from the source have lost almost all their curvature (Fig. 15–20b) and are nearly straight, as ocean waves often are; they are then called **plane waves**.

For reflection of a two- or three-dimensional plane wave, as shown in Fig. 15–21, the angle that the incoming or *incident wave* makes with the reflecting surface is equal to the angle made by the reflected wave. This is the **law of reflection**:

the angle of reflection equals the angle of incidence.

The "angle of incidence" is defined as the angle (θ_i) the incident ray makes with the perpendicular to the reflecting surface (or the wave front makes with a tangent to the surface), and the "angle of reflection" is the corresponding angle (θ_r) for the reflected wave.

FIGURE 15–21
Law of reflection: $\theta_r = \theta_i$.

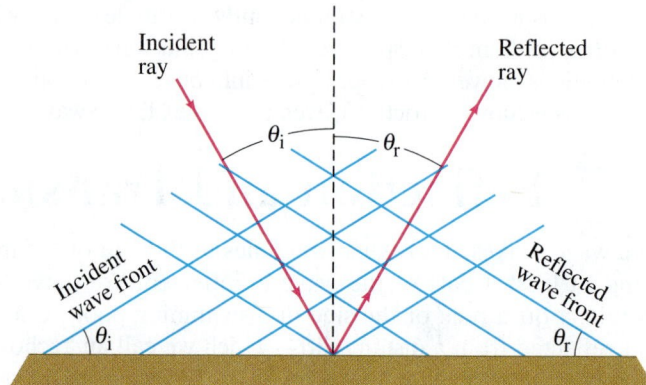

15–8 Interference

Interference refers to what happens when two waves pass through the same region of space at the same time. Consider, for example, the two wave pulses on a cord traveling toward each other as shown in Fig. 15–22. In Fig. 15–22a the two pulses have the same amplitude, but one is a crest and the other a trough; in Fig. 15–22b they are both crests. In both cases, the waves meet and pass right by each other. However, in the region where they overlap, the resultant displacement is the *algebraic sum of their separate displacements* (a crest is considered positive and a trough negative). This is another example of the principle of superposition. In Fig. 15–22a, the two waves have opposite displacements at the instant they pass one another, and they add to zero. The result is called **destructive interference**. In Fig. 15–22b, at the instant the two pulses overlap, they produce a resultant displacement that is greater than the displacement of either separate pulse, and the result is **constructive interference**.

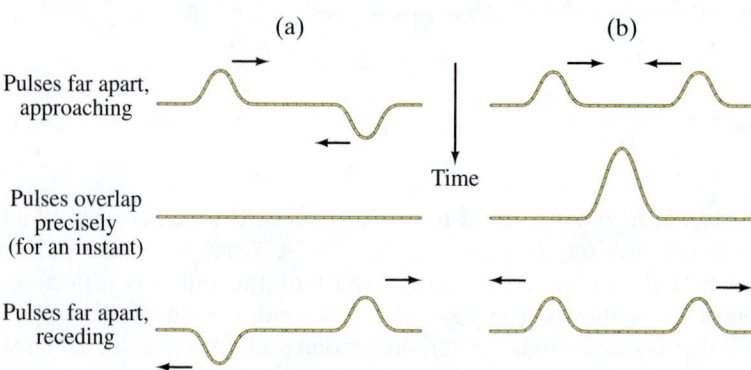

FIGURE 15–22 Two wave pulses pass each other. Where they overlap, interference occurs: (a) destructive, and (b) constructive.

[You may wonder where the energy is at the moment of destructive interference in Fig. 15–22a; the cord may be straight at this instant, but the central parts of it are still moving up or down.]

(a)   (b)

FIGURE 15–23 (a) Interference of water waves. (b) Constructive interference occurs where one wave's maximum (a crest) meets the other's maximum. Destructive interference ("flat water") occurs where one wave's maximum (a crest) meets the other's miminum (a trough).

When two rocks are thrown into a pond simultaneously, the two sets of circular waves interfere with one another as shown in Fig. 15–23a. In some areas of overlap, crests of one wave repeatedly meet crests of the other (and troughs meet troughs), Fig. 15–23b. Constructive interference is occurring at these points, and the water continuously oscillates up and down with greater amplitude than either wave separately. In other areas, destructive interference occurs where the water does not move up and down at all over time. This is where crests of one wave meet troughs of the other, and vice versa. Figure 15–24a shows the displacement of two identical waves graphically as a function of time, as well as their sum, for the case of constructive interference. For constructive interference (Fig. 15–24a), the two waves are **in phase**. At points where destructive interference occurs (Fig. 15–24b) crests of one wave repeatedly meet troughs of the other wave and the two waves are **out of phase** by one-half wavelength or 180°. The crests of one wave occur a half wavelength behind the crests of the other wave. The relative phase of the two water waves in Fig. 15–23 in most areas is intermediate between these two extremes, resulting in *partially* destructive interference, as illustrated in Fig. 15–24c. If the amplitudes of two interfering waves are not equal, fully destructive interference (as in Fig. 15–24b) does not occur.

FIGURE 15–24 Graphs showing two identical waves, and their sum, as a function of time at three locations. In (a) the two waves interfere constructively, in (b) destructively, and in (c) partially destructively.

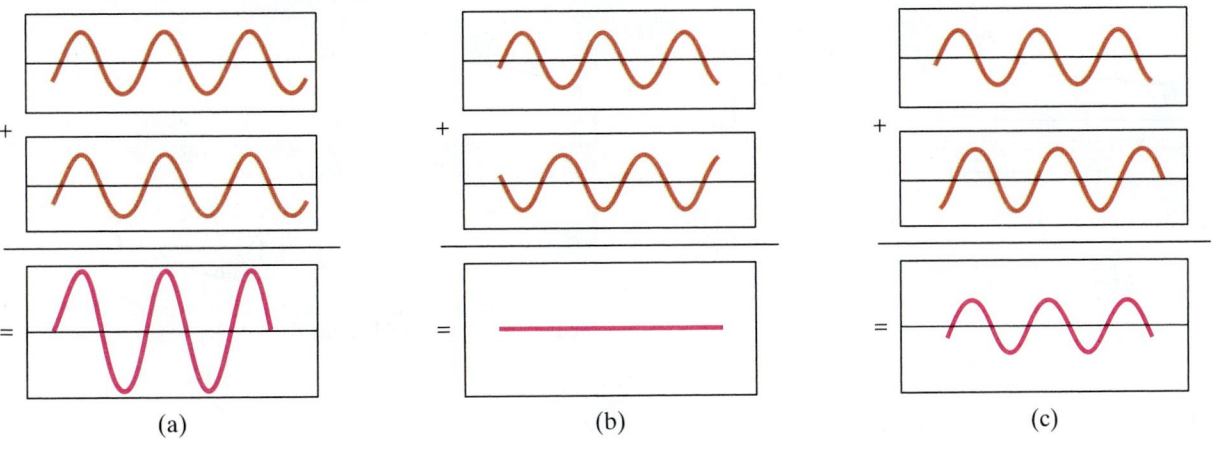

(a)   (b)   (c)

15–9 Standing Waves; Resonance

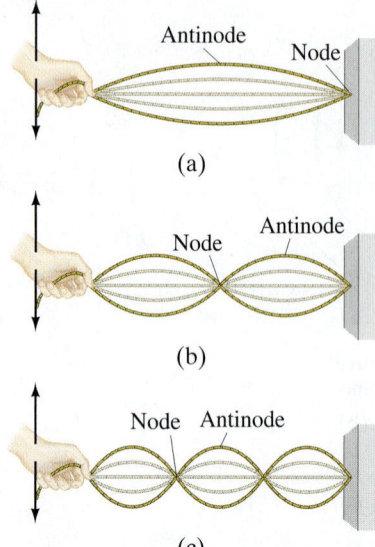

Antinode   Node

(a)

Node   Antinode

(b)

Node   Antinode

(c)

FIGURE 15–25 Standing waves corresponding to three resonant frequencies.

If you shake one end of a cord and the other end is kept fixed, a continuous wave will travel down to the fixed end and be reflected back, inverted, as we saw in Fig. 15–18a. As you continue to vibrate the cord, waves will travel in both directions, and the wave traveling along the cord, away from your hand, will interfere with the reflected wave coming back. Usually there will be quite a jumble. But if you vibrate the cord at just the right frequency, the two traveling waves will interfere in such a way that a large-amplitude **standing wave** will be produced, Fig. 15–25. It is called a "standing wave" because it does not appear to be traveling. The cord simply appears to have segments that oscillate up and down in a fixed pattern. The points of destructive interference, where the cord remains still at all times, are called **nodes**. Points of constructive interference, where the cord oscillates with maximum amplitude, are called **antinodes**. The nodes and antinodes remain in fixed positions for a particular frequency.

Standing waves can occur at more than one frequency. The lowest frequency of vibration that produces a standing wave gives rise to the pattern shown in Fig. 15–25a. The standing waves shown in Figs. 15–25b and 15–25c are produced at precisely twice and three times the lowest frequency, respectively, assuming the tension in the cord is the same. The cord can also vibrate with four loops (four antinodes) at four times the lowest frequency, and so on.

The frequencies at which standing waves are produced are the **natural frequencies** or **resonant frequencies** of the cord, and the different standing wave patterns shown in Fig. 15–25 are different "resonant modes of vibration." A standing wave on a cord is the result of the interference of two waves traveling in opposite directions. A standing wave can also be considered a vibrating object at resonance. Standing waves represent the same phenomenon as the resonance of a vibrating spring or pendulum, which we discussed in Chapter 14. However, a spring or pendulum has only one resonant frequency, whereas the cord has an infinite number of resonant frequencies, each of which is a whole-number multiple of the lowest resonant frequency.

Consider a string stretched between two supports that is plucked like a guitar or violin string, Fig. 15–26a. Waves of a great variety of frequencies will travel in both directions along the string, will be reflected at the ends, and will travel back in the opposite direction. Most of these waves interfere with each other and quickly die out. However, those waves that correspond to the resonant frequencies of the string will persist. The ends of the string, since they are fixed, will be nodes. There may be other nodes as well. Some of the possible resonant modes of vibration (standing waves) are shown in Fig. 15–26b. Generally, the motion will be a combination of these different resonant modes, but only those frequencies that correspond to a resonant frequency will be present.

FIGURE 15–26 (a) A string is plucked. (b) Only standing waves corresponding to resonant frequencies persist for long.

(a)

(b)

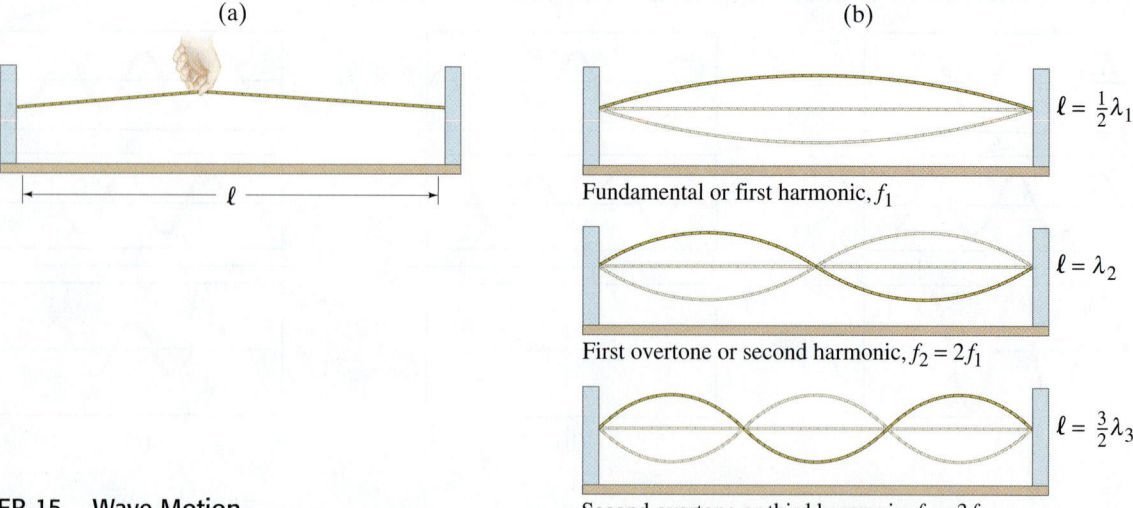

ℓ

$\ell = \frac{1}{2}\lambda_1$

Fundamental or first harmonic, f_1

$\ell = \lambda_2$

First overtone or second harmonic, $f_2 = 2f_1$

$\ell = \frac{3}{2}\lambda_3$

Second overtone or third harmonic, $f_3 = 3f_1$

To determine the resonant frequencies, we first note that the wavelengths of the standing waves bear a simple relationship to the length ℓ of the string. The lowest frequency, called the **fundamental frequency**, corresponds to one antinode (or loop). And as can be seen in Fig. 15–26b, the whole length corresponds to one-half wavelength. Thus $\ell = \frac{1}{2}\lambda_1$, where λ_1 stands for the wavelength of the fundamental frequency. The other natural frequencies are called **overtones**; for a vibrating string they are whole-number (integral) multiples of the fundamental, and then are also called **harmonics**, with the fundamental being referred to as the **first harmonic**.[†] The next mode of vibration after the fundamental has two loops and is called the **second harmonic** (or first overtone), Fig. 15–26b. The length of the string ℓ at the second harmonic corresponds to one complete wavelength: $\ell = \lambda_2$. For the third and fourth harmonics, $\ell = \frac{3}{2}\lambda_3$, and $\ell = 2\lambda_4$, respectively, and so on. In general, we can write

$$\ell = \frac{n\lambda_n}{2}, \qquad \text{where } n = 1, 2, 3, \cdots.$$

The integer n labels the number of the harmonic: $n = 1$ for the fundamental, $n = 2$ for the second harmonic, and so on. We solve for λ_n and find

$$\lambda_n = \frac{2\ell}{n}, \qquad n = 1, 2, 3, \cdots. \qquad \left[\begin{array}{l}\text{string fixed} \\ \text{at both ends}\end{array}\right] \quad \textbf{(15–17a)}$$

To find the frequency f of each vibration we use Eq. 15–1, $f = v/\lambda$, and we see that

$$f_n = \frac{v}{\lambda_n} = n\frac{v}{2\ell} = nf_1, \qquad n = 1, 2, 3, \cdots, \qquad \textbf{(15–17b)}$$

where $f_1 = v/\lambda_1 = v/2\ell$ is the fundamental frequency. We see that each resonant frequency is an integer multiple of the fundamental frequency.

Because a standing wave is equivalent to two traveling waves moving in opposite directions, the concept of wave velocity still makes sense and is given by Eq. 15–2 in terms of the tension F_T in the string and its mass per unit length $(\mu = m/\ell)$. That is, $v = \sqrt{F_T/\mu}$ for waves traveling in both directions.

EXAMPLE 15–8 **Piano string.** A piano string is 1.10 m long and has a mass of 9.00 g. (a) How much tension must the string be under if it is to vibrate at a fundamental frequency of 131 Hz? (b) What are the frequencies of the first four harmonics?

APPROACH To determine the tension, we need to find the wave speed using Eq. 15–1 $(v = \lambda f)$, and then use Eq. 15–2, solving it for F_T.

SOLUTION (a) The wavelength of the fundamental is $\lambda = 2\ell = 2.20$ m (Eq. 15–17a with $n = 1$). The speed of the wave on the string is $v = \lambda f = (2.20\ \text{m})(131\ \text{s}^{-1}) = 288$ m/s. Then we have (Eq. 15–2)

$$F_T = \mu v^2 = \frac{m}{\ell}v^2 = \left(\frac{9.00 \times 10^{-3}\ \text{kg}}{1.10\ \text{m}}\right)(288\ \text{m/s})^2 = 679\ \text{N}.$$

(b) The frequencies of the second, third, and fourth harmonics are two, three, and four times the fundamental frequency: 262, 393, and 524 Hz, respectively.

NOTE The speed of the wave on the string is *not* the same as the speed of the sound wave that the piano string produces in the air (as we shall see in Chapter 16).

A standing wave does appear to be standing in place (and a traveling wave appears to move). The term "standing" wave is also meaningful from the point of view of energy. Since the string is at rest at the nodes, no energy flows past these points. Hence the energy is not transmitted down the string but "stands" in place in the string.

Standing waves are produced not only on strings, but on any object that is struck, such as a drum membrane or an object made of metal or wood. The resonant frequencies depend on the dimensions of the object, just as for a string they depend on its length. Large objects have lower resonant frequencies than small objects. All musical instruments, from stringed instruments to wind instruments (in which a column of air vibrates as a standing wave) to drums and other percussion instruments, depend on standing waves to produce their particular musical sounds, as we shall see in Chapter 16.

[†]The term "harmonic" comes from music, because such integral multiples of frequencies "harmonize."

Mathematical Representation of a Standing Wave

As we saw, a standing wave can be considered to consist of two traveling waves that move in opposite directions. These can be written (see Eqs. 15–10c and 15–13c)

$$D_1(x, t) = A \sin(kx - \omega t) \quad \text{and} \quad D_2(x, t) = A \sin(kx + \omega t)$$

since, assuming no damping, the amplitudes are equal as are the frequencies and wavelengths. The sum of these two traveling waves produces a standing wave which can be written mathematically as

$$D = D_1 + D_2 = A[\sin(kx - \omega t) + \sin(kx + \omega t)].$$

From the trigonometric identity $\sin \theta_1 + \sin \theta_2 = 2 \sin \frac{1}{2}(\theta_1 + \theta_2) \cos \frac{1}{2}(\theta_1 - \theta_2)$, we can rewrite this as

$$D = 2A \sin kx \cos \omega t. \tag{15–18}$$

If we let $x = 0$ at the left-hand end of the string, then the right-hand end is at $x = \ell$ where ℓ is the length of the string. Since the string is fixed at its two ends (Fig. 15–26), $D(x, t)$ must be zero at $x = 0$ and at $x = \ell$. Equation 15–18 already satisfies the first condition ($D = 0$ at $x = 0$) and satisfies the second condition if $\sin k\ell = 0$ which means

$$k\ell = \pi, 2\pi, 3\pi, \cdots, n\pi, \cdots$$

where $n =$ an integer. Since $k = 2\pi/\lambda$, then $\lambda = 2\ell/n$, which is just Eq. 15–17a.

Equation 15–18, with the condition $\lambda = 2\ell/n$, is the mathematical representation of a standing wave. We see that a particle at any position x vibrates in simple harmonic motion (because of the factor $\cos \omega t$). All particles of the string vibrate with the same frequency $f = \omega/2\pi$, but the amplitude depends on x and equals $2A \sin kx$. (Compare this to a traveling wave for which all particles vibrate with the same amplitude.) The amplitude has a maximum, equal to $2A$, when $kx = \pi/2, 3\pi/2, 5\pi/2$, and so on—that is, at

$$x = \frac{\lambda}{4}, \frac{3\lambda}{4}, \frac{5\lambda}{4}, \cdots.$$

These are, of course, the positions of the antinodes (see Fig. 15–26).

EXAMPLE 15–9 **Wave forms.** Two waves traveling in opposite directions on a string fixed at $x = 0$ are described by the functions

$$D_1 = (0.20 \text{ m}) \sin(2.0x - 4.0t) \quad \text{and} \quad D_2 = (0.20 \text{ m}) \sin(2.0x + 4.0t)$$

(where x is in m, t is in s), and they produce a standing wave pattern. Determine (a) the function for the standing wave, (b) the maximum amplitude at $x = 0.45$ m, (c) where the other end is fixed ($x > 0$), (d) the maximum amplitude, and where it occurs.

APPROACH We use the principle of superposition to add the two waves. The given waves have the form we used to obtain Eq. 15–18, which we thus can use.

SOLUTION (a) The two waves are of the form $D = A \sin(kx \pm \omega t)$, so

$$k = 2.0 \text{ m}^{-1} \quad \text{and} \quad \omega = 4.0 \text{ s}^{-1}.$$

These combine to form a standing wave of the form of Eq. 15–18:

$$D = 2A \sin kx \cos \omega t = (0.40 \text{ m}) \sin(2.0x) \cos(4.0t),$$

where x is in meters and t in seconds.

(b) At $x = 0.45$ m,

$$D = (0.40 \text{ m}) \sin(0.90) \cos(4.0t) = (0.31 \text{ m}) \cos(4.0t).$$

The maximum amplitude at this point is $D = 0.31$ m and occurs when $\cos(4.0t) = 1$.

(c) These waves make a standing wave pattern, so both ends of the string must be nodes. Nodes occur every half wavelength, which for our string is

$$\frac{\lambda}{2} = \frac{1}{2} \frac{2\pi}{k} = \frac{\pi}{2.0} \text{ m} = 1.57 \text{ m}.$$

If the string includes only one loop, its length is $\ell = 1.57$ m. But without more information, it could be twice as long, $\ell = 3.14$ m, or any integral number times 1.57 m, and still provide a standing wave pattern, Fig. 15–27.

FIGURE 15–27 Example 15–9: possible lengths for the string.

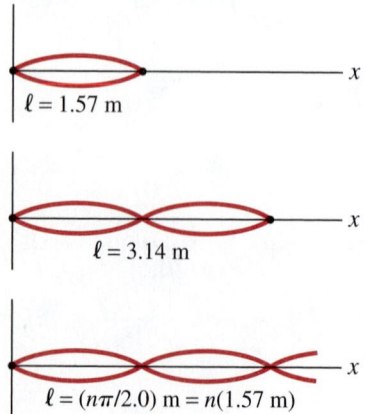

$\ell = 1.57$ m

$\ell = 3.14$ m

$\ell = (n\pi/2.0) \text{ m} = n(1.57 \text{ m})$

(d) The nodes occur at $x = 0$, $x = 1.57\,\text{m}$, and, if the string is longer than $\ell = 1.57\,\text{m}$, at $x = 3.14\,\text{m}$, $4.71\,\text{m}$, and so on. The maximum amplitude (antinode) is 0.40 m [from part (b) above] and occurs midway between the nodes. For $\ell = 1.57\,\text{m}$, there is only one antinode, at $x = 0.79\,\text{m}$.

*15–10 Refraction[†]

When any wave strikes a boundary, some of the energy is reflected and some is transmitted or absorbed. When a two- or three-dimensional wave traveling in one medium crosses a boundary into a medium where its speed is different, the transmitted wave may move in a different direction than the incident wave, as shown in Fig. 15–28. This phenomenon is known as **refraction**. One example is a water wave; the velocity decreases in shallow water and the waves refract, as shown in Fig. 15–29 below. [When the wave velocity changes gradually, as in Fig. 15–29, without a sharp boundary, the waves change direction (refract) gradually.]

In Fig. 15–28, the velocity of the wave in medium 2 is less than in medium 1. In this case, the wave front bends so that it travels more nearly parallel to the boundary. That is, the *angle of refraction*, θ_r, is less than the *angle of incidence*, θ_i. To see why this is so, and to help us get a quantitative relation between θ_r and θ_i, let us think of each wave front as a row of soldiers. The soldiers are marching from firm ground (medium 1) into mud (medium 2) and hence are slowed down after the boundary. The soldiers that reach the mud first are slowed down first, and the row bends as shown in Fig. 15–30a. Let us consider the wave front (or row of soldiers) labeled A in Fig. 15–30b. In the same time t that A_1 moves a distance $\ell_1 = v_1 t$, we see that A_2 moves a distance $\ell_2 = v_2 t$. The two right triangles in Fig. 15–30b, shaded yellow and green, have the side labeled a in common. Thus

$$\sin \theta_1 = \frac{\ell_1}{a} = \frac{v_1 t}{a}$$

since a is the hypotenuse, and

$$\sin \theta_2 = \frac{\ell_2}{a} = \frac{v_2 t}{a}.$$

Dividing these two equations, we obtain the **law of refraction**:

$$\frac{\sin \theta_2}{\sin \theta_1} = \frac{v_2}{v_1}. \qquad \textbf{(15–19)}$$

Since θ_1 is the angle of incidence (θ_i), and θ_2 is the angle of refraction (θ_r), Eq. 15–19 gives the quantitative relation between the two. If the wave were going in the opposite direction, the geometry would not change; only θ_1 and θ_2 would change roles: θ_1 would be the angle of refraction and θ_2 the angle of incidence. Clearly then, if the wave travels into a medium where it can move faster, it will bend the opposite way, $\theta_r > \theta_i$. We see from Eq. 15–19 that if the velocity increases, the angle increases, and vice versa.

Earthquake waves refract within the Earth as they travel through rock layers of different densities (and therefore the velocity is different) just as water waves do. Light waves refract as well, and when we discuss light, we shall find Eq. 15–19 very useful.

[†]This Section and the next are covered in more detail in Chapters 32 to 35 on optics.

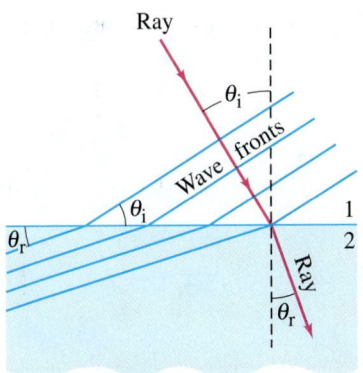

FIGURE 15–28 Refraction of waves passing a boundary.

FIGURE 15–29 Water waves refract gradually as they approach the shore, as their velocity decreases. There is no distinct boundary, as in Fig. 15–28, because the wave velocity changes gradually.

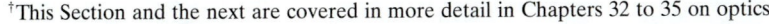

(a)

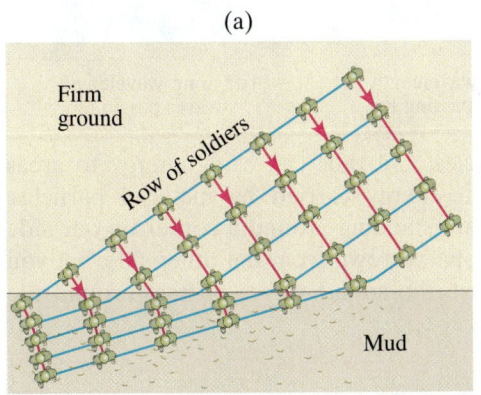

(b)

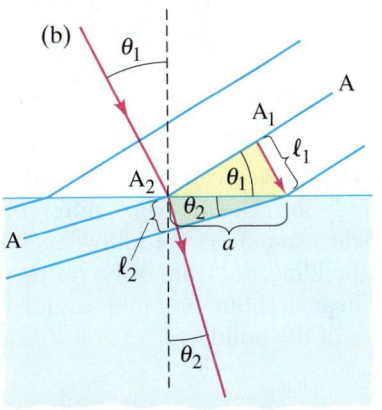

FIGURE 15–30 (a) Soldier analogy to derive (b) law of refraction for waves.

EXAMPLE 15–10 **Refraction of an earthquake wave.** An earthquake P wave passes across a boundary in rock where its velocity increases from 6.5 km/s to 8.0 km/s. If it strikes this boundary at 30°, what is the angle of refraction?

APPROACH We apply the law of refraction, Eq. 15–19, $\sin\theta_2/\sin\theta_1 = v_2/v_1$.

SOLUTION Since $\sin 30° = 0.50$, Eq. 15–19 yields

$$\sin\theta_2 = \frac{(8.0 \text{ m/s})}{(6.5 \text{ m/s})}(0.50) = 0.62.$$

So $\theta_2 = \sin^{-1}(0.62) = 38°$.

NOTE Be careful with angles of incidence and refraction. As we discussed in Section 15–7 (Fig. 15–21), these angles are between the wave front and the boundary line, or—equivalently—between the ray (direction of wave motion) and the line perpendicular to the boundary. Inspect Fig. 15–30b carefully.

*15–11 Diffraction

Waves spread as they travel. When they encounter an obstacle, they bend around it somewhat and pass into the region behind it, as shown in Fig. 15–31 for water waves. This phenomenon is called **diffraction**.

The amount of diffraction depends on the wavelength of the wave and on the size of the obstacle, as shown in Fig. 15–32. If the wavelength is much larger than the object, as with the grass blades of Fig. 15–32a, the wave bends around them almost as if they are not there. For larger objects, parts (b) and (c), there is more of a "shadow" region behind the obstacle where we might not expect the waves to penetrate—but they do, at least a little. Then notice in part (d), where the obstacle is the same as in part (c) but the wavelength is longer, that there is more diffraction into the shadow region. As a rule of thumb, *only if the wavelength is smaller than the size of the object will there be a significant shadow region*. This rule applies to *reflection* from an obstacle as well. Very little of a wave is reflected unless the wavelength is smaller than the size of the obstacle.

A rough guide to the amount of diffraction is

$$\theta(\text{radians}) \approx \frac{\lambda}{\ell},$$

where θ is roughly the angular spread of waves after they have passed through an opening of width ℓ or around an obstacle of width ℓ.

FIGURE 15–31 Wave diffraction. The waves are coming from the upper left. Note how the waves, as they pass the obstacle, bend around it, into the "shadow region" behind the obstacle.

FIGURE 15–32 Water waves passing objects of various sizes. Note that the longer the wavelength compared to the size of the object, the more diffraction there is into the "shadow region."

(a) Water waves passing blades of grass

(b) Stick in water

(c) Short-wavelength waves passing log

(d) Long-wavelength waves passing log

That waves can bend around obstacles, and thus can carry energy to areas behind obstacles, is very different from energy carried by material particles. A clear example is the following: if you are standing around a corner on one side of a building, you cannot be hit by a baseball thrown from the other side, but you can hear a shout or other sound because the sound waves diffract around the edges of the building.

Summary

Vibrating objects act as sources of **waves** that travel outward from the source. Waves on water and on a string are examples. The wave may be a **pulse** (a single crest) or it may be continuous (many crests and troughs).

The **wavelength** of a continuous wave is the distance between two successive crests (or any two identical points on the wave shape).

The **frequency** is the number of full wavelengths (or crests) that pass a given point per unit time.

The **wave velocity** (how fast a crest moves) is equal to the product of wavelength and frequency,

$$v = \lambda f. \qquad (15-1)$$

The **amplitude** of a wave is the maximum height of a crest, or depth of a trough, relative to the normal (or equilibrium) level.

In a **transverse wave**, the oscillations are perpendicular to the direction in which the wave travels. An example is a wave on a string.

In a **longitudinal wave**, the oscillations are along (parallel to) the line of travel; sound is an example.

The velocity of both longitudinal and transverse waves in matter is proportional to the square root of an elastic force factor divided by an inertia factor (or density).

Waves carry energy from place to place without matter being carried. The **intensity** of a wave (energy transported across unit area per unit time) is proportional to the square of the amplitude of the wave.

For a wave traveling outward in three dimensions from a point source, the intensity (ignoring damping) decreases with the square of the distance from the source,

$$I \propto \frac{1}{r^2}. \qquad (15-8a)$$

The amplitude decreases linearly with distance from the source.

A one-dimensional transverse wave traveling in a medium to the right along the x axis (x increasing) can be represented by a formula for the displacement of the medium from equilibrium at any point x as a function of time as

$$D(x, t) = A \sin\left[\left(\frac{2\pi}{\lambda}\right)(x - vt)\right] \qquad (15-10a)$$

$$= A \sin(kx - \omega t) \qquad (15-10c)$$

where

$$k = \frac{2\pi}{\lambda} \qquad (15-11)$$

and

$$\omega = 2\pi f.$$

If a wave is traveling toward decreasing values of x,

$$D(x, t) = A \sin(kx + \omega t). \qquad (15-13c)$$

[*Waves can be described by the **wave equation**, which in one dimension is $\partial^2 D/\partial x^2 = (1/v^2)\,\partial^2 D/\partial t^2$, Eq. 15–16.]

When two or more waves pass through the same region of space at the same time, the displacement at any given point will be the vector sum of the displacements of the separate waves. This is the **principle of superposition**. It is valid for mechanical waves if the amplitudes are small enough that the restoring force of the medium is proportional to displacement.

Waves reflect off objects in their path. When the **wave front** of a two- or three-dimensional wave strikes an object, the *angle* of *reflection* equals the *angle* of *incidence*, which is the **law of reflection**. When a wave strikes a boundary between two materials in which it can travel, part of the wave is reflected and part is transmitted.

When two waves pass through the same region of space at the same time, they **interfere**. From the superposition principle, the resultant displacement at any point and time is the sum of their separate displacements. This can result in **constructive interference**, **destructive interference**, or something in between depending on the amplitudes and relative phases of the waves.

Waves traveling on a cord of fixed length interfere with waves that have reflected off the end and are traveling back in the opposite direction. At certain frequencies, **standing waves** can be produced in which the waves seem to be standing still rather than traveling. The cord (or other medium) is vibrating as a whole. This is a resonance phenomenon, and the frequencies at which standing waves occur are called **resonant frequencies**. The points of destructive interference (no vibration) are called **nodes**. Points of constructive interference (maximum amplitude of vibration) are called **antinodes**. On a cord of length ℓ fixed at both ends, the wavelengths of standing waves are given by

$$\lambda_n = 2\ell/n \qquad (15-17a)$$

where n is an integer.

[*Waves change direction, or **refract**, when traveling from one medium into a second medium where their speed is different. Waves spread, or **diffract**, as they travel and encounter obstacles. A rough guide to the amount of diffraction is $\theta \approx \lambda/\ell$, where λ is the wavelength and ℓ the width of an opening or obstacle. There is a significant "shadow region" only if the wavelength λ is smaller than the size of the obstacle.]

Questions

1. Is the frequency of a simple periodic wave equal to the frequency of its source? Why or why not?

2. Explain the difference between the speed of a transverse wave traveling down a cord and the speed of a tiny piece of the cord.

3. You are finding it a challenge to climb from one boat up onto a higher boat in heavy waves. If the climb varies from 2.5 m to 4.3 m, what is the amplitude of the wave? Assume the centers of the two boats are a half wavelength apart.

4. What kind of waves do you think will travel down a horizontal metal rod if you strike its end (*a*) vertically from above and (*b*) horizontally parallel to its length?

5. Since the density of air decreases with an increase in temperature, but the bulk modulus B is nearly independent of temperature, how would you expect the speed of sound waves in air to vary with temperature?

6. Describe how you could estimate the speed of water waves across the surface of a pond.

7. The speed of sound in most solids is somewhat greater than in air, yet the density of solids is much greater (10^3 to 10^4 times). Explain.

8. Give two reasons why circular water waves decrease in amplitude as they travel away from the source.

9. Two linear waves have the same amplitude and speed, and otherwise are identical, except one has half the wavelength of the other. Which transmits more energy? By what factor?

10. Will any function of $(x - vt)$—see Eq. 15–14—represent a wave motion? Why or why not? If not, give an example.

11. When a sinusoidal wave crosses the boundary between two sections of cord as in Fig. 15–19, the frequency does not change (although the wavelength and velocity do change). Explain why.

12. If a sinusoidal wave on a two-section cord (Fig. 15–19) is inverted upon reflection, does the transmitted wave have a longer or shorter wavelength?

13. Is energy always conserved when two waves interfere? Explain.

14. If a string is vibrating as a standing wave in three segments, are there any places you could touch it with a knife blade without disturbing the motion?

15. When a standing wave exists on a string, the vibrations of incident and reflected waves cancel at the nodes. Does this mean that energy was destroyed? Explain.

16. Can the amplitude of the standing waves in Fig. 15–25 be greater than the amplitude of the vibrations that cause them (up and down motion of the hand)?

17. When a cord is vibrated as in Fig. 15–25 by hand or by a mechanical oscillator, the "nodes" are not quite true nodes (at rest). Explain. [Hint: Consider damping and energy flow from hand or oscillator.]

*18. AM radio signals can usually be heard behind a hill, but FM often cannot. That is, AM signals bend more than FM. Explain. (Radio signals, as we shall see, are carried by electromagnetic waves whose wavelength for AM is typically 200 to 600 m and for FM about 3 m.)

*19. If we knew that energy was being transmitted from one place to another, how might we determine whether the energy was being carried by particles (material objects) or by waves?

Problems

15–1 and 15–2 Characteristics of Waves

1. (I) A fisherman notices that wave crests pass the bow of his anchored boat every 3.0 s. He measures the distance between two crests to be 8.0 m. How fast are the waves traveling?

2. (I) A sound wave in air has a frequency of 262 Hz and travels with a speed of 343 m/s. How far apart are the wave crests (compressions)?

3. (I) Calculate the speed of longitudinal waves in (a) water, (b) granite, and (c) steel.

4. (I) AM radio signals have frequencies between 550 kHz and 1600 kHz (kilohertz) and travel with a speed of 3.0×10^8 m/s. What are the wavelengths of these signals? On FM the frequencies range from 88 MHz to 108 MHz (megahertz) and travel at the same speed. What are their wavelengths?

5. (I) Determine the wavelength of a 5800-Hz sound wave traveling along an iron rod.

6. (II) A cord of mass 0.65 kg is stretched between two supports 8.0 m apart. If the tension in the cord is 140 N, how long will it take a pulse to travel from one support to the other?

7. (II) A 0.40-kg cord is stretched between two supports, 7.8 m apart. When one support is struck by a hammer, a transverse wave travels down the cord and reaches the other support in 0.85 s. What is the tension in the cord?

8. (II) A sailor strikes the side of his ship just below the surface of the sea. He hears the echo of the wave reflected from the ocean floor directly below 2.8 s later. How deep is the ocean at this point?

9. (II) A ski gondola is connected to the top of a hill by a steel cable of length 660 m and diameter 1.5 cm. As the gondola comes to the end of its run, it bumps into the terminal and sends a wave pulse along the cable. It is observed that it took 17 s for the pulse to return. (a) What is the speed of the pulse? (b) What is the tension in the cable?

10. (II) P and S waves from an earthquake travel at different speeds, and this difference helps locate the earthquake "epicenter" (where the disturbance took place). (a) Assuming typical speeds of 8.5 km/s and 5.5 km/s for P and S waves, respectively, how far away did the earthquake occur if a particular seismic station detects the arrival of these two types of waves 1.7 min apart? (b) Is one seismic station sufficient to determine the position of the epicenter? Explain.

11. (II) The wave on a string shown in Fig. 15–33 is moving to the right with a speed of 1.10 m/s. (a) Draw the shape of the string 1.00 s later and indicate which parts of the string are moving up and which down at that instant. (b) Estimate the vertical speed of point A on the string at the instant shown in the Figure.

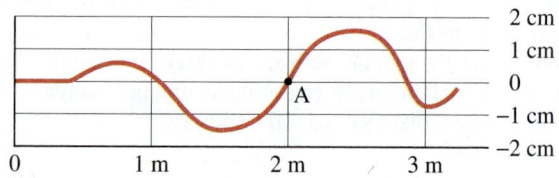

FIGURE 15–33 Problem 11.

12. (II) A 5.0 kg ball hangs from a steel wire 1.00 mm in diameter and 5.00 m long. What would be the speed of a wave in the steel wire?

13. (II) Two children are sending signals along a cord of total mass 0.50 kg tied between tin cans with a tension of 35 N. It takes the vibrations in the string 0.50 s to go from one child to the other. How far apart are the children?

*14. (II) **Dimensional analysis.** Waves on the surface of the ocean do not depend significantly on the properties of water such as density or surface tension. The primary "return force" for water piled up in the wave crests is due to the gravitational attraction of the Earth. Thus the speed v (m/s) of ocean waves depends on the acceleration due to gravity g. It is reasonable to expect that v might also depend on water depth h and the wave's wavelength λ. Assume the wave speed is given by the functional form $v = Cg^\alpha h^\beta \lambda^\gamma$, where α, β, γ, and C are numbers without dimension. (a) In deep water, the water deep below does not affect the motion of waves at the surface. Thus v should be independent of depth h (i.e., $\beta = 0$). Using only dimensional analysis (Section 1–7), determine the formula for the speed of surface waves in deep water. (b) In shallow water, the speed of surface waves is found experimentally to be independent of the wavelength (i.e., $\gamma = 0$). Using only dimensional analysis, determine the formula for the speed of waves in shallow water.

15–3 Energy Transported by Waves

15. (I) Two earthquake waves of the same frequency travel through the same portion of the Earth, but one is carrying 3.0 times the energy. What is the ratio of the amplitudes of the two waves?

16. (II) What is the ratio of (a) the intensities, and (b) the amplitudes, of an earthquake P wave passing through the Earth and detected at two points 15 km and 45 km from the source?

17. (II) Show that if damping is ignored, the amplitude A of circular water waves decreases as the square root of the distance r from the source: $A \propto 1/\sqrt{r}$.

18. (II) The intensity of an earthquake wave passing through the Earth is measured to be $3.0 \times 10^6 \, \text{J/m}^2 \cdot \text{s}$ at a distance of 48 km from the source. (a) What was its intensity when it passed a point only 1.0 km from the source? (b) At what rate did energy pass through an area of $2.0 \, \text{m}^2$ at 1.0 km?

19. (II) A small steel wire of diameter 1.0 mm is connected to an oscillator and is under a tension of 7.5 N. The frequency of the oscillator is 60.0 Hz and it is observed that the amplitude of the wave on the steel wire is 0.50 cm. (a) What is the power output of the oscillator, assuming that the wave is not reflected back? (b) If the power output stays constant but the frequency is doubled, what is the amplitude of the wave?

20. (II) Show that the intensity of a wave is equal to the energy density (energy per unit volume) in the wave times the wave speed.

21. (II) (a) Show that the average rate with which energy is transported along a cord by a mechanical wave of frequency f and amplitude A is
$$\bar{P} = 2\pi^2 \mu v f^2 A^2,$$
where v is the speed of the wave and μ is the mass per unit length of the cord. (b) If the cord is under a tension $F_T = 135 \, \text{N}$ and has mass per unit length 0.10 kg/m, what power is required to transmit 120-Hz transverse waves of amplitude 2.0 cm?

15–4 Mathematical Representation of Traveling Wave

22. (I) A transverse wave on a wire is given by $D(x, t) = 0.015 \sin(25x - 1200t)$ where D and x are in meters and t is in seconds. (a) Write an expression for a wave with the same amplitude, wavelength, and frequency but traveling in the opposite direction. (b) What is the speed of either wave?

23. (I) Suppose at $t = 0$, a wave shape is represented by $D = A \sin(2\pi x/\lambda + \phi)$; that is, it differs from Eq. 15–9 by a constant phase factor ϕ. What then will be the equation for a wave traveling to the left along the x axis as a function of x and t?

24. (II) A transverse traveling wave on a cord is represented by $D = 0.22 \sin(5.6x + 34t)$ where D and x are in meters and t is in seconds. For this wave determine (a) the wavelength, (b) frequency, (c) velocity (magnitude and direction), (d) amplitude, and (e) maximum and minimum speeds of particles of the cord.

25. (II) Consider the point $x = 1.00 \, \text{m}$ on the cord of Example 15–5. Determine (a) the maximum velocity of this point, and (b) its maximum acceleration. (c) What is its velocity and acceleration at $t = 2.50 \, \text{s}$?

26. (II) A transverse wave on a cord is given by $D(x, t) = 0.12 \sin(3.0x - 15.0t)$, where D and x are in m and t is in s. At $t = 0.20 \, \text{s}$, what are the displacement and velocity of the point on the cord where $x = 0.60 \, \text{m}$?

27. (II) A transverse wave pulse travels to the right along a string with a speed $v = 2.0 \, \text{m/s}$. At $t = 0$ the shape of the pulse is given by the function
$$D = 0.45 \cos(2.6x + 1.2),$$
where D and x are in meters. (a) Plot D vs. x at $t = 0$. (b) Determine a formula for the wave pulse at any time t assuming there are no frictional losses. (c) Plot $D(x, t)$ vs. x at $t = 1.0 \, \text{s}$. (d) Repeat parts (b) and (c) assuming the pulse is traveling to the left. Plot all 3 graphs on the same axes for easy comparison.

28. (II) A 524-Hz longitudinal wave in air has a speed of 345 m/s. (a) What is the wavelength? (b) How much time is required for the phase to change by 90° at a given point in space? (c) At a particular instant, what is the phase difference (in degrees) between two points 4.4 cm apart?

29. (II) Write the equation for the wave in Problem 28 traveling to the right, if its amplitude is 0.020 cm, and $D = -0.020 \, \text{cm}$, at $t = 0$ and $x = 0$.

30. (II) A sinusoidal wave traveling on a string in the negative x direction has amplitude 1.00 cm, wavelength 3.00 cm, and frequency 245 Hz. At $t = 0$, the particle of string at $x = 0$ is displaced a distance $D = 0.80 \, \text{cm}$ above the origin and is moving upward. (a) Sketch the shape of the wave at $t = 0$ and (b) determine the function of x and t that describes the wave.

*15–5 The Wave Equation

***31.** (II) Determine if the function $D = A \sin kx \cos \omega t$ is a solution of the wave equation.

***32.** (II) Show by direct substitution that the following functions satisfy the wave equation: (a) $D(x, t) = A \ln(x + vt)$; (b) $D(x, t) = (x - vt)^4$.

***33.** (II) Show that the wave forms of Eqs. 15–13 and 15–15 satisfy the wave equation, Eq. 15–16.

***34.** (II) Let two linear waves be represented by $D_1 = f_1(x, t)$ and $D_2 = f_2(x, t)$. If both these waves satisfy the wave equation (Eq. 15–16), show that any combination $D = C_1 D_1 + C_2 D_2$ does as well, where C_1 and C_2 are constants.

***35.** (II) Does the function $D(x, t) = e^{-(kx - \omega t)^2}$ satisfy the wave equation? Why or why not?

***36.** (II) In deriving Eq. 15–2, $v = \sqrt{F_T/\mu}$, for the speed of a transverse wave on a string, it was assumed that the wave's amplitude A is much less than its wavelength λ. Assuming a sinusoidal wave shape $D = A \sin(kx - \omega t)$, show via the partial derivative $v' = \partial D/\partial t$ that the assumption $A \ll \lambda$ implies that the maximum transverse speed v'_{max} of the string itself is much less than the wave velocity. If $A = \lambda/100$ determine the ratio v'_{max}/v.

15–7 Reflection and Transmission

37. (II) A cord has two sections with linear densities of 0.10 kg/m and 0.20 kg/m, Fig. 15–34. An incident wave, given by $D = (0.050 \, \text{m}) \sin(7.5x - 12.0t)$, where x is in meters and t in seconds, travels along the lighter cord. (a) What is the wavelength on the lighter section of the cord? (b) What is the tension in the cord? (c) What is the wavelength when the wave travels on the heavier section?

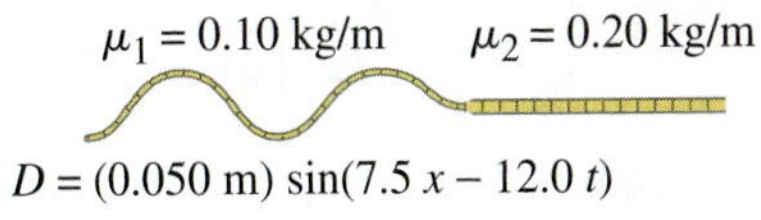

$\mu_1 = 0.10 \, \text{kg/m}$ $\quad$ $\mu_2 = 0.20 \, \text{kg/m}$

FIGURE 15–34
Problem 37. $\qquad D = (0.050 \, \text{m}) \sin(7.5 \, x - 12.0 \, t)$

38. (II) Consider a sine wave traveling down the stretched two-part cord of Fig. 15–19. Determine a formula (a) for the ratio of the speeds of the wave in the two sections, v_H/v_L, and (b) for the ratio of the wavelengths in the two sections. (The frequency is the same in both sections. Why?) (c) Is the wavelength larger in the heavier cord or the lighter?

39. (II) Seismic reflection prospecting is commonly used to map deeply buried formations containing oil. In this technique, a seismic wave generated on the Earth's surface (for example, by an explosion or falling weight) reflects from the subsurface formation and is detected upon its return to ground level. By placing ground-level detectors at a variety of locations relative to the source, and observing the variation in the source-to-detector travel times, the depth of the subsurface formation can be determined. (a) Assume a ground-level detector is placed a distance x away from a seismic-wave source and that a horizontal boundary between overlying rock and a subsurface formation exists at depth D (Fig. 15–35a). Determine an expression for the time t taken by the reflected wave to travel from source to detector, assuming the seismic wave propagates at constant speed v. (b) Suppose several detectors are placed along a line at different distances x from the source as in Fig. 15–35b. Then, when a seismic wave is generated, the different travel times t for each detector are measured. Starting with your result from part (a), explain how a graph of t^2 vs. x^2 can be used to determine D.

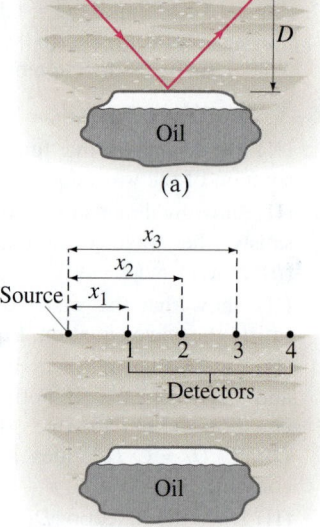

FIGURE 15–35
Problem 39.

(a)

(b)

40. (III) A cord stretched to a tension F_T consists of two sections (as in Fig. 15–19) whose linear densities are μ_1 and μ_2. Take $x = 0$ to be the point (a knot) where they are joined, with μ_1 referring to that section of cord to the left and μ_2 that to the right. A sinusoidal wave, $D = A \sin[k_1(x - v_1 t)]$, starts at the left end of the cord. When it reaches the knot, part of it is reflected and part is transmitted. Let the equation of the reflected wave be $D_R = A_R \sin[k_1(x + v_1 t)]$ and that for the transmitted wave be $D_T = A_T \sin[k_2(x - v_2 t)]$. Since the frequency must be the same in both sections, we have $\omega_1 = \omega_2$ or $k_1 v_1 = k_2 v_2$. (a) Because the cord is continuous, a point an infinitesimal distance to the left of the knot has the same displacement at any moment (due to incident plus reflected waves) as a point just to the right of the knot (due to the transmitted wave). Thus show that $A = A_T + A_R$. (b) Assuming that the slope $(\partial D/\partial x)$ of the cord just to the left of the knot is the same as the slope just to the right of the knot, show that the amplitude of the reflected wave is given by

$$A_R = \left(\frac{v_1 - v_2}{v_1 + v_2}\right) A = \left(\frac{k_2 - k_1}{k_2 + k_1}\right) A.$$

(c) What is A_T in terms of A?

15–8 Interference

41. (I) The two pulses shown in Fig. 15–36 are moving toward each other. (a) Sketch the shape of the string at the moment they directly overlap. (b) Sketch the shape of the string a few moments later. (c) In Fig. 15–22a, at the moment the pulses pass each other, the string is straight. What has happened to the energy at this moment?

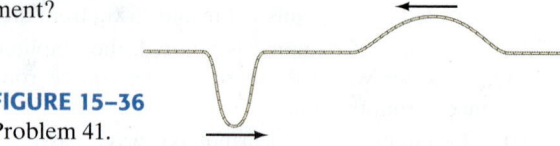

FIGURE 15–36
Problem 41.

42. (II) Suppose two linear waves of equal amplitude and frequency have a phase difference ϕ as they travel in the same medium. They can be represented by

$$D_1 = A \sin(kx - \omega t)$$
$$D_2 = A \sin(kx - \omega t + \phi).$$

(a) Use the trigonometric identity $\sin\theta_1 + \sin\theta_2 = 2\sin\frac{1}{2}(\theta_1 + \theta_2)\cos\frac{1}{2}(\theta_1 - \theta_2)$ to show that the resultant wave is given by

$$D = \left(2A\cos\frac{\phi}{2}\right)\sin\left(kx - \omega t + \frac{\phi}{2}\right).$$

(b) What is the amplitude of this resultant wave? Is the wave purely sinusoidal, or not? (c) Show that constructive interference occurs if $\phi = 0, 2\pi, 4\pi$, and so on, and destructive interference occurs if $\phi = \pi, 3\pi, 5\pi$, etc. (d) Describe the resultant wave, by equation and in words, if $\phi = \pi/2$.

15–9 Standing Waves; Resonance

43. (I) A violin string vibrates at 441 Hz when unfingered. At what frequency will it vibrate if it is fingered one-third of the way down from the end? (That is, only two-thirds of the string vibrates as a standing wave.)

44. (I) If a violin string vibrates at 294 Hz as its fundamental frequency, what are the frequencies of the first four harmonics?

45. (I) In an earthquake, it is noted that a footbridge oscillated up and down in a one-loop (fundamental standing wave) pattern once every 1.5 s. What other possible resonant periods of motion are there for this bridge? What frequencies do they correspond to?

46. (I) A particular string resonates in four loops at a frequency of 280 Hz. Name at least three other frequencies at which it will resonate.

47. (II) A cord of length 1.0 m has two equal-length sections with linear densities of 0.50 kg/m and 1.00 kg/m. The tension in the entire cord is constant. The ends of the cord are oscillated so that a standing wave is set up in the cord with a single node where the two sections meet. What is the ratio of the oscillatory frequencies?

48. (II) The velocity of waves on a string is 96 m/s. If the frequency of standing waves is 445 Hz, how far apart are the two adjacent nodes?

49. (II) If two successive harmonics of a vibrating string are 240 Hz and 320 Hz, what is the frequency of the fundamental?

50. (II) A guitar string is 90.0 cm long and has a mass of 3.16 g. From the bridge to the support post ($= \ell$) is 60.0 cm and the string is under a tension of 520 N. What are the frequencies of the fundamental and first two overtones?

51. (II) Show that the frequency of standing waves on a cord of length ℓ and linear density μ, which is stretched to a tension F_T, is given by

$$f = \frac{n}{2\ell}\sqrt{\frac{F_T}{\mu}}$$

where n is an integer.

52. (II) One end of a horizontal string of linear density 6.6×10^{-4} kg/m is attached to a small-amplitude mechanical 120-Hz oscillator. The string passes over a pulley, a distance $\ell = 1.50$ m away, and weights are hung from this end, Fig. 15–37. What mass m must be hung from this end of the string to produce (a) one loop, (b) two loops, and (c) five loops of a standing wave? Assume the string at the oscillator is a node, which is nearly true.

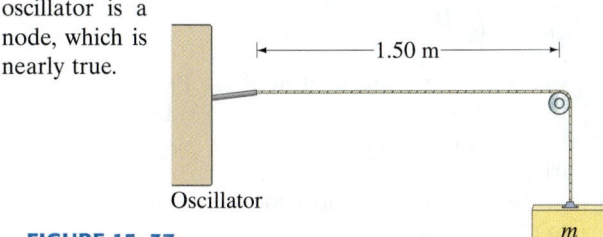

FIGURE 15–37
Problems 52 and 53.

53. (II) In Problem 52, Fig. 15–37, the length of the string may be adjusted by moving the pulley. If the hanging mass m is fixed at 0.070 kg, how many different standing wave patterns may be achieved by varying ℓ between 10 cm and 1.5 m?

54. (II) The displacement of a standing wave on a string is given by $D = 2.4\sin(0.60x)\cos(42t)$, where x and D are in centimeters and t is in seconds. (a) What is the distance (cm) between nodes? (b) Give the amplitude, frequency, and speed of each of the component waves. (c) Find the speed of a particle of the string at $x = 3.20$ cm when $t = 2.5$ s.

55. (II) The displacement of a transverse wave traveling on a string is represented by $D_1 = 4.2\sin(0.84x - 47t + 2.1)$, where D_1 and x are in cm and t in s. (a) Find an equation that represents a wave which, when traveling in the opposite direction, will produce a standing wave when added to this one. (b) What is the equation describing the standing wave?

56. (II) When you slosh the water back and forth in a tub at just the right frequency, the water alternately rises and falls at each end, remaining relatively calm at the center. Suppose the frequency to produce such a standing wave in a 45-cm-wide tub is 0.85 Hz. What is the speed of the water wave?

57. (II) A particular violin string plays at a frequency of 294 Hz. If the tension is increased 15%, what will the new frequency be?

58. (II) Two traveling waves are described by the functions

$$D_1 = A\sin(kx - \omega t)$$
$$D_2 = A\sin(kx + \omega t),$$

where $A = 0.15$ m, $k = 3.5$ m^{-1}, and $\omega = 1.8$ s^{-1}. (a) Plot these two waves, from $x = 0$ to a point $x(>0)$ that includes one full wavelength. Choose $t = 1.0$ s. (b) Plot the sum of the two waves and identify the nodes and antinodes in the plot, and compare to the analytic (mathematical) representation.

59. (II) Plot the two waves given in Problem 58 and their sum, as a function of time from $t = 0$ to $t = T$ (one period). Choose (a) $x = 0$ and (b) $x = \lambda/4$. Interpret your results.

60. (II) A standing wave on a 1.64-m-long horizontal string displays three loops when the string vibrates at 120 Hz. The maximum swing of the string (top to bottom) at the center of each loop is 8.00 cm. (a) What is the function describing the standing wave? (b) What are the functions describing the two equal-amplitude waves traveling in opposite directions that make up the standing wave?

61. (II) On an electric guitar, a "pickup" under each string transforms the string's vibrations directly into an electrical signal. If a pickup is placed 16.25 cm from one of the fixed ends of a 65.00-cm-long string, which of the harmonics from $n = 1$ to $n = 12$ will not be "picked up" by this pickup?

62. (II) A 65-cm guitar string is fixed at both ends. In the frequency range between 1.0 and 2.0 kHz, the string is found to resonate only at frequencies 1.2, 1.5, and 1.8 kHz. What is the speed of traveling waves on this string?

63. (II) Two oppositely directed traveling waves given by $D_1 = (5.0 \text{ mm})\cos[(2.0\text{ m}^{-1})x - (3.0\text{ rad/s})t]$ and $D_2 = (5.0\text{ mm})\cos[(2.0\text{ m}^{-1})x + (3.0\text{ rad/s})t]$ form a standing wave. Determine the position of nodes along the x axis.

64. (II) A wire is composed of aluminum with length $\ell_1 = 0.600$ m and mass per unit length $\mu_1 = 2.70$ g/m joined to a steel section with length $\ell_2 = 0.882$ m and mass per unit length $\mu_2 = 7.80$ g/m. This composite wire is fixed at both ends and held at a uniform tension of 135 N. Find the lowest frequency standing wave that can exist on this wire, assuming there is a node at the joint between aluminum and steel. How many nodes (including the two at the ends) does this standing wave possess?

*15–10 Refraction

*65. (I) An earthquake P wave traveling 8.0 km/s strikes a boundary within the Earth between two kinds of material. If it approaches the boundary at an incident angle of 52° and the angle of refraction is 31°, what is the speed in the second medium?

*66. (I) Water waves approach an underwater "shelf" where the velocity changes from 2.8 m/s to 2.5 m/s. If the incident wave crests make a 35° angle with the shelf, what will be the angle of refraction?

*67. (II) A sound wave is traveling in warm air (25°C) when it hits a layer of cold (-15°C) denser air. If the sound wave hits the cold air interface at an angle of 33°, what is the angle of refraction? The speed of sound as a function of temperature can be approximated by $v = (331 + 0.60\,T)$ m/s, where T is in °C.

*68. (II) Any type of wave that reaches a boundary beyond which its speed is increased, there is a maximum incident angle if there is to be a transmitted refracted wave. This maximum incident angle θ_{iM} corresponds to an angle of refraction equal to 90°. If $\theta_i > \theta_{iM}$, all the wave is reflected at the boundary and none is refracted, because this would correspond to $\sin\theta_r > 1$ (where θ_r is the angle of refraction), which is impossible. This phenomenon is referred to as *total internal reflection*. (a) Find a formula for θ_{iM} using the law of refraction, Eq. 15–19. (b) How far from the bank should a trout fisherman stand (Fig. 15–38) so trout won't be frightened by his voice (1.8 m above the ground)? The speed of sound is about 343 m/s in air and 1440 m/s in water.

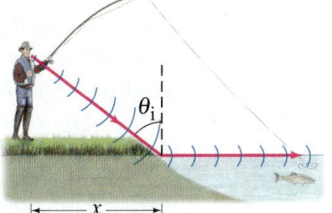

FIGURE 15–38
Problem 68b.

A longitudinal earthquake wave strikes a boundary between two types of rock at a 38° angle. As the wave crosses the boundary, the specific gravity of the rock changes from 3.6 to 2.8. Assuming that the elastic modulus is the same for both types of rock, determine the angle of refraction.

* **70.** (II) A satellite dish is about 0.5 m in diameter. According to the user's manual, the dish has to be pointed in the direction of the satellite, but an error of about 2° to either side is allowed without loss of reception. Estimate the wavelength of the electromagnetic waves (speed = 3×10^8 m/s) received by the dish.

General Problems

71. A sinusoidal traveling wave has frequency 880 Hz and phase velocity 440 m/s. (*a*) At a given time, find the distance between any two locations that correspond to a difference in phase of $\pi/6$ rad. (*b*) At a fixed location, by how much does the phase change during a time interval of 1.0×10^{-4} s?

72. When you walk with a cup of coffee (diameter 8 cm) at just the right pace of about one step per second, the coffee sloshes higher and higher in your cup until eventually it starts to spill over the top, Fig 15–39. Estimate the speed of the waves in the coffee.

FIGURE 15–39 Problem 72.

73. Two solid rods have the same bulk modulus but one is 2.5 times as dense as the other. In which rod will the speed of longitudinal waves be greater, and by what factor?

74. Two waves traveling along a stretched string have the same frequency, but one transports 2.5 times the power of the other. What is the ratio of the amplitudes of the two waves?

75. A bug on the surface of a pond is observed to move up and down a total vertical distance of 0.10 m, lowest to highest point, as a wave passes. (*a*) What is the amplitude of the wave? (*b*) If the amplitude increases to 0.15 m, by what factor does the bug's maximum kinetic energy change?

76. A guitar string is supposed to vibrate at 247 Hz, but is measured to actually vibrate at 255 Hz. By what percentage should the tension in the string be changed to get the frequency to the correct value?

77. An earthquake-produced surface wave can be approximated by a sinusoidal transverse wave. Assuming a frequency of 0.60 Hz (typical of earthquakes, which actually include a mixture of frequencies), what amplitude is needed so that objects begin to leave contact with the ground? [*Hint*: Set the acceleration $a > g$.]

78. A uniform cord of length ℓ and mass m is hung vertically from a support. (*a*) Show that the speed of transverse waves in this cord is $\sqrt{gh}$, where h is the height above the lower end. (*b*) How long does it take for a pulse to travel upward from one end to the other?

79. A transverse wave pulse travels to the right along a string with a speed $v = 2.4$ m/s. At $t = 0$ the shape of the pulse is given by the function

$$D = \frac{4.0 \text{ m}^3}{x^2 + 2.0 \text{ m}^2},$$

where D and x are in meters. (*a*) Plot D vs. x at $t = 0$ from $x = -10$ m to $x = +10$ m. (*b*) Determine a formula for the wave pulse at any time t assuming there are no frictional losses. (*c*) Plot $D(x, t)$ vs. x at $t = 1.00$ s. (*d*) Repeat parts (*b*) and (*c*) assuming the pulse is traveling to the left.

80. (*a*) Show that if the tension in a stretched string is changed by a small amount ΔF_T, the frequency of the fundamental is changed by an amount $\Delta f = \frac{1}{2}(\Delta F_T/F_T)f$. (*b*) By what percent must the tension in a piano string be increased or decreased to raise the frequency from 436 Hz to 442 Hz. (*c*) Does the formula in part (*a*) apply to the overtones as well?

81. Two strings on a musical instrument are tuned to play at 392 Hz (G) and 494 Hz (B). (*a*) What are the frequencies of the first two overtones for each string? (*b*) If the two strings have the same length and are under the same tension, what must be the ratio of their masses (m_G/m_B)? (*c*) If the strings, instead, have the same mass per unit length and are under the same tension, what is the ratio of their lengths (ℓ_G/ℓ_B)? (*d*) If their masses and lengths are the same, what must be the ratio of the tensions in the two strings?

82. The ripples in a certain groove 10.8 cm from the center of a 33-rpm phonograph record have a wavelength of 1.55 mm. What will be the frequency of the sound emitted?

83. A 10.0-m-long wire of mass 152 g is stretched under a tension of 255 N. A pulse is generated at one end, and 20.0 ms later a second pulse is generated at the opposite end. Where will the two pulses first meet?

84. A wave with a frequency of 220 Hz and a wavelength of 10.0 cm is traveling along a cord. The maximum speed of particles on the cord is the same as the wave speed. What is the amplitude of the wave?

85. A string can have a "free" end if that end is attached to a ring that can slide without friction on a vertical pole (Fig. 15–40). Determine the wavelengths of the resonant vibrations of such a string with one end fixed and the other free.

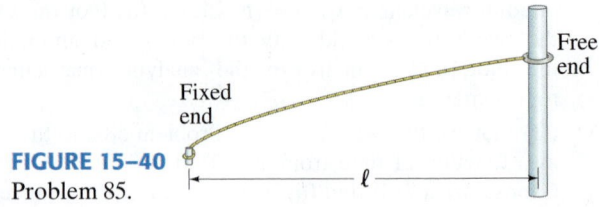

FIGURE 15–40
Problem 85.

86. A highway overpass was observed to resonate as one full loop $(\frac{1}{2}\lambda)$ when a small earthquake shook the ground vertically at 3.0 Hz. The highway department put a support at the center of the overpass, anchoring it to the ground as shown in Fig. 15–41. What resonant frequency would you now expect for the overpass? It is noted that earthquakes rarely do significant shaking above 5 or 6 Hz. Did the modifications do any good? Explain.

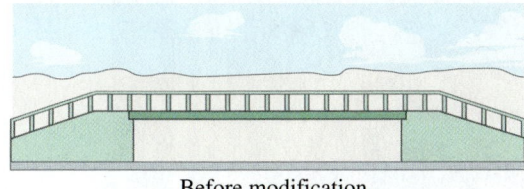

Before modification

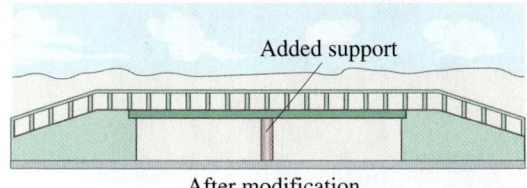

Added support

After modification

FIGURE 15–41 Problem 86.

87. Figure 15–42 shows the wave shape at two instants of time for a sinusoidal wave traveling to the right. What is the mathematical representation of this wave?

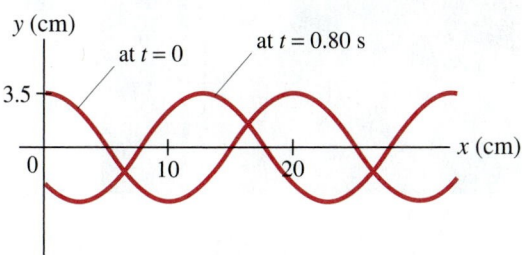

FIGURE 15–42 Problem 87.

88. Estimate the average power of a water wave when it hits the chest of an adult standing in the water at the seashore. Assume that the amplitude of the wave is 0.50 m, the wavelength is 2.5 m, and the period is 4.0 s.

89. A tsunami of wavelength 215 km and velocity 550 km/h travels across the Pacific Ocean. As it approaches Hawaii, people observe an unusual decrease of sea level in the harbors. Approximately how much time do they have to run to safety? (In the absence of knowledge and warning, people have died during tsunamis, some of them attracted to the shore to see stranded fishes and boats.)

90. Two wave pulses are traveling in opposite directions with the same speed of 7.0 cm/s as shown in Fig. 15–43. At $t = 0$, the leading edges of the two pulses are 15 cm apart. Sketch the wave pulses at $t = 1.0, 2.0$ and 3.0 s.

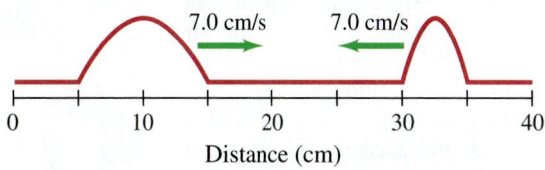

FIGURE 15–43 Problem 90.

91. For a spherical wave traveling uniformly away from a point source, show that the displacement can be represented by

$$D = \left(\frac{A}{r}\right)\sin(kr - \omega t),$$

where r is the radial distance from the source and A is a constant.

92. What frequency of sound would have a wavelength the same size as a 1.0-m-wide window? (The speed of sound is 344 m/s at 20°C.) What frequencies would diffract through the window?

*Numerical/Computer

*93. (II) Consider a wave generated by the periodic vibration of a source and given by the expression $D(x, t) = A\sin^2 k(x - ct)$, where x represents position (in meters), t represents time (in seconds), and c is a positive constant. We choose $A = 5.0$ m, $k = 1.0$ m^{-1}, and $c = 0.50$ m/s. Use a spreadsheet to make a graph with three curves of $D(x, t)$ from $x = -5.0$ m to $+5.0$ m in steps of 0.050 m at times $t = 0.0$, 1.0, and 2.0 s. Determine the speed, direction of motion, period, and wavelength of the wave.

*94. (II) The displacement of a bell-shaped wave pulse is described by a relation that involves the exponential function:

$$D(x, t) = Ae^{-\alpha(x - vt)^2}$$

where the constants $A = 10.0$ m, $\alpha = 2.0$ m^{-2}, and $v = 3.0$ m/s. (a) Over the range -10.0 m $\leq x \leq +10.0$ m, use a graphing calculator or computer program to plot $D(x, t)$ at each of the three times $t = 0$, $t = 1.0$, and $t = 2.0$ s. Do these three plots demonstrate the wave-pulse shape shifting along the x axis by the expected amount over the span of each 1.0-s interval? (b) Repeat part (a) but assume $D(x, t) = Ae^{-\alpha(x + vt)^2}$.

Answers to Exercises

A: (c).

B: (d).

C: (c).

D: (a).

"If music be the food of physics, play on." [See Shakespeare, *Twelfth Night*, line 1.]

Stringed instruments depend on transverse standing waves on strings to produce their harmonious sounds. The sound of wind instruments originates in longitudinal standing waves of an air column. Percussion instruments create more complicated standing waves.

Besides examining sources of sound, we also study the decibel scale of sound level, sound wave interference and beats, the Doppler effect, shock waves and sonic booms, and ultrasound imaging.

C H A P T E R
16

Sound

CONTENTS

16–1 Characteristics of Sound

16–2 Mathematical Representation of Longitudinal Waves

16–3 Intensity of Sound: Decibels

16–4 Sources of Sound: Vibrating Strings and Air Columns

*16–5 Quality of Sound, and Noise; Superposition

16–6 Interference of Sound Waves; Beats

16–7 Doppler Effect

*16–8 Shock Waves and the Sonic Boom

*16–9 Applications: Sonar, Ultrasound, and Medical Imaging

424

CHAPTER-OPENING QUESTION—Guess now!

A pianist plays the note "middle C." The sound is made by the vibration of the piano string and is propagated outward as a vibration of the air (which can reach your ear). Comparing the vibration on the string to the vibration in the air, which of the following is true?

(a) The vibration on the string and the vibration in the air have the same wavelength.

(b) They have the same frequency.

(c) They have the same speed.

(d) Neither wavelength, frequency, nor speed are the same in the air as on the string.

S ound is associated with our sense of hearing and, therefore, with the physiology of our ears and the psychology of our brain, which interprets the sensations that reach our ears. The term *sound* also refers to the physical sensation that stimulates our ears: namely, longitudinal waves.

We can distinguish three aspects of any sound. First, there must be a *source* for a sound; as with any mechanical wave, the source of a sound wave is a vibrating object. Second, the energy is transferred from the source in the form of longitudinal sound *waves*. And third, the sound is *detected* by an ear or by a microphone. We start this Chapter by looking at some aspects of sound waves themselves.

16–1 Characteristics of Sound

We saw in Chapter 15, Fig. 15–5, how a vibrating drumhead produces a sound wave in air. Indeed, we usually think of sound waves traveling in the air, for normally it is the vibrations of the air that force our eardrums to vibrate. But sound waves can also travel in other materials.

Two stones struck together under water can be heard by a swimmer beneath the surface, for the vibrations are carried to the ear by the water. When you put your ear flat against the ground, you can hear an approaching train or truck. In this case the ground does not actually touch your eardrum, but the longitudinal wave transmitted by the ground is called a sound wave just the same, for its vibrations cause the outer ear and the air within it to vibrate. Sound cannot travel in the absence of matter. For example, a bell ringing inside an evacuated jar cannot be heard, nor does sound travel through the empty reaches of outer space.

The **speed of sound** is different in different materials. In air at 0°C and 1 atm, sound travels at a speed of 331 m/s. We saw in Eq. 15–4 $\left(v = \sqrt{B/\rho}\right)$ that the speed depends on the elastic modulus, B, and the density, ρ, of the material. Thus for helium, whose density is much less than that of air but whose elastic modulus is not greatly different, the speed is about three times as great as in air. In liquids and solids, which are much less compressible and therefore have much greater elastic moduli, the speed is larger still. The speed of sound in various materials is given in Table 16–1. The values depend somewhat on temperature, but this is significant mainly for gases. For example, in air at normal (ambient) temperatures, the speed increases approximately 0.60 m/s for each Celsius degree increase in temperature:

$$v \approx (331 + 0.60\,T) \text{ m/s}, \qquad \text{[speed of sound in air]}$$

where T is the temperature in °C. Unless stated otherwise, we will assume in this Chapter that $T = 20$°C, so that[†] $v = \left[331 + (0.60)(20)\right] \text{ m/s} = 343 \text{ m/s}$.

CONCEPTUAL EXAMPLE 16–1 | **Distance from a lightning strike.** A rule of thumb that tells how close lightning has struck is, "one mile for every five seconds before the thunder is heard." Explain why this works, noting that the speed of light is so high (3×10^8 m/s, almost a million times faster than sound) that the time for light to travel to us is negligible compared to the time for the sound.

RESPONSE The speed of sound in air is about 340 m/s, so to travel $1 \text{ km} = 1000 \text{ m}$ takes about 3 seconds. One mile is about 1.6 kilometers, so the time for the thunder to travel a mile is about $(1.6)(3) \approx 5$ seconds.

EXERCISE A What would be the rule used in Example 16–1 in terms of kilometers?

Two aspects of any sound are immediately evident to a human listener: "loudness" and "pitch." Each refers to a sensation in the consciousness of the listener. But to each of these subjective sensations there corresponds a physically measurable quantity. **Loudness** is related to the intensity (energy per unit time crossing unit area) in the sound wave, and we shall discuss it in Section 16–3.

The **pitch** of a sound refers to whether it is high, like the sound of a piccolo or violin, or low, like the sound of a bass drum or string bass. The physical quantity that determines pitch is the frequency, as was first noted by Galileo. The lower the frequency, the lower the pitch; the higher the frequency, the higher the pitch.[‡] The best human ears can respond to frequencies from about 20 Hz to almost 20,000 Hz. (Recall that 1 Hz is 1 cycle per second.) This frequency range is called the **audible range**. These limits vary somewhat from one individual to another. One general trend is that as people age, they are less able to hear high frequencies, so the high-frequency limit may be 10,000 Hz or less.

TABLE 16–1 Speed of Sound in Various Materials (20°C and 1 atm)

Material	Speed (m/s)
Air	343
Air (0°C)	331
Helium	1005
Hydrogen	1300
Water	1440
Sea water	1560
Iron and steel	≈5000
Glass	≈4500
Aluminum	≈5100
Hardwood	≈4000
Concrete	≈3000

PHYSICS APPLIED
How far away is the lightning?

[†]We treat the 20°C ("room temperature") as accurate to 2 significant figures.

[‡]Although pitch is determined mainly by frequency, it also depends to a slight extent on loudness. For example, a very loud sound may seem slightly lower in pitch than a quiet sound of the same frequency.

Sound waves whose frequencies are outside the audible range may reach the ear, but we are not generally aware of them. Frequencies above 20,000 Hz are called **ultrasonic** (do not confuse with *supersonic*, which is used for an object moving with a speed faster than the speed of sound). Many animals can hear ultrasonic frequencies; dogs, for example, can hear sounds as high as 50,000 Hz, and bats can detect frequencies as high as 100,000 Hz. Ultrasonic waves have many useful applications in medicine and other fields, which we discuss later in this Chapter.

PHYSICS APPLIED

Autofocusing camera

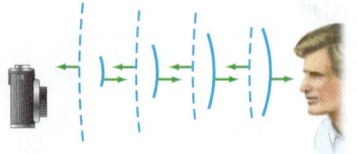

FIGURE 16–1 Example 16–2. Autofocusing camera emits an ultrasonic pulse. Solid lines represent the wave front of the outgoing wave pulse moving to the right; dashed lines represent the wave front of the pulse reflected off the person's face, returning to the camera. The time information allows the camera mechanism to adjust the lens to focus at the proper distance.

EXAMPLE 16–2 **Autofocusing with sound waves.** Older autofocusing cameras determine the distance by emitting a pulse of very high frequency (ultrasonic) sound that travels to the object being photographed, and include a sensor that detects the returning reflected sound, as shown in Fig. 16–1. To get an idea of the time sensitivity of the detector, calculate the travel time of the pulse for an object (*a*) 1.0 m away, and (*b*) 20 m away.

APPROACH If we assume the temperature is about 20°C, then the speed of sound is 343 m/s. Using this speed v and the total distance d back and forth in each case, we can obtain the time ($v = d/t$).

SOLUTION (*a*) The pulse travels 1.0 m to the object and 1.0 m back, for a total of 2.0 m. We solve for t in $v = d/t$:

$$t = \frac{d}{v} = \frac{2.0 \text{ m}}{343 \text{ m/s}} = 0.0058 \text{ s} = 5.8 \text{ ms}.$$

(*b*) The total distance now is $2 \times 20 \text{ m} = 40 \text{ m}$, so

$$t = \frac{40 \text{ m}}{343 \text{ m/s}} = 0.12 \text{ s} = 120 \text{ ms}.$$

NOTE Newer autofocus cameras use infrared light ($v = 3 \times 10^8$ m/s) instead of ultrasound, and/or a digital sensor array that detects light intensity differences between adjacent receptors as the lens is automatically moved back and forth, choosing the lens position that provides maximum intensity differences (sharpest focus).

Sound waves whose frequencies are below the audible range (that is, less than 20 Hz) are called **infrasonic**. Sources of infrasonic waves include earthquakes, thunder, volcanoes, and waves produced by vibrating heavy machinery. This last source can be particularly troublesome to workers, for infrasonic waves—even though inaudible—can cause damage to the human body. These low-frequency waves act in a resonant fashion, causing motion and irritation of the body's organs.

16–2 Mathematical Representation of Longitudinal Waves

In Section 15–4, we saw that a one-dimensional sinusoidal wave traveling along the x axis can be represented by the relation (Eq. 15–10c)

$$D = A \sin(kx - \omega t), \tag{16–1}$$

where D is the displacement of the wave at position x and time t, and A is its *amplitude* (maximum value). The wave number k is related to the wavelength λ by $k = 2\pi/\lambda$, and $\omega = 2\pi f$ where f is the frequency. For a transverse wave—such as a wave on a string—the displacement D is perpendicular to the direction of wave propagation along the x axis. But for a longitudinal wave the displacement D is *along the direction of wave propagation*. That is, D is parallel to x and represents the displacement of a tiny volume element of the medium from its equilibrium position.

Longitudinal (sound) waves can also be considered from the point of view of variations in pressure rather than displacement. Indeed, longitudinal waves are often called **pressure waves**. The pressure variation is usually easier to measure than the displacement (see Example 16–7). As can be seen in Fig. 16–2, in a wave "compression" (where molecules are closest together), the pressure is higher than normal, whereas in an expansion (or rarefaction) the pressure is less than normal.

FIGURE 16–2 Longitudinal sound wave traveling to the right, and its graphical representation in terms of pressure.

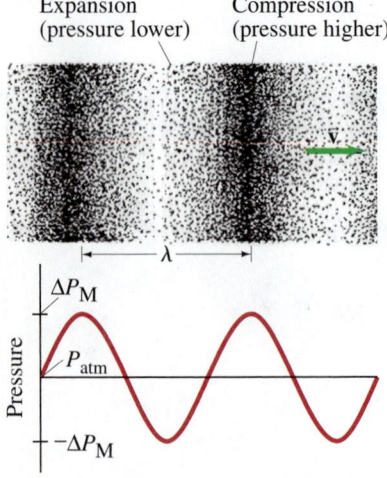

Figure 16–3 shows a graphical representation of a sound wave in air in terms of (a) displacement and (b) pressure. Note that the displacement wave is a quarter wavelength, or 90° ($\pi/2$ rad), out of phase with the pressure wave: where the pressure is a maximum or minimum, the displacement from equilibrium is zero; and where the pressure variation is zero, the displacement is a maximum or minimum.

Pressure Wave Derivation

Let us now derive the mathematical representation of the pressure variation in a traveling longitudinal wave. From the definition of the bulk modulus, B (Eq. 12–7),

$$\Delta P = -B(\Delta V/V),$$

where ΔP represents the pressure difference from the normal pressure P_0 (no wave present) and $\Delta V/V$ is the fractional change in volume of the medium due to the pressure change ΔP. The negative sign reflects the fact that the volume decreases ($\Delta V < 0$) if the pressure is increased. Consider now a layer of fluid through which the longitudinal wave is passing (Fig. 16–4). If this layer has thickness Δx and area S, then its volume is $V = S\,\Delta x$. As a result of pressure variation in the wave, the volume will change by an amount $\Delta V = S\,\Delta D$, where ΔD is the change in thickness of this layer as it compresses or expands. (Remember that D represents the displacement of the medium.) Thus we have

$$\Delta P = -B\frac{S\,\Delta D}{S\,\Delta x}.$$

To be precise, we take the limit of $\Delta x \to 0$, so we obtain

$$\Delta P = -B\frac{\partial D}{\partial x}, \qquad \textbf{(16–2)}$$

where we use the partial derivative notation since D is a function of both x and t. If the displacement D is sinusoidal as given by Eq. 16–1, then we have from Eq. 16–2 that

$$\Delta P = -(BAk)\cos(kx - \omega t). \qquad \textbf{(16–3)}$$

(Here A is the displacement amplitude, not area which is S.) Thus the pressure varies sinusoidally as well, but is out of phase from the displacement by 90° or a quarter wavelength, as in Fig. 16–3. The quantity BAk is called the **pressure amplitude**, ΔP_M. It represents the maximum and minimum amounts by which the pressure varies from the normal ambient pressure. We can thus write

$$\Delta P = -\Delta P_M \cos(kx - \omega t), \qquad \textbf{(16–4)}$$

where, using $v = \sqrt{B/\rho}$ (Eq. 15–4), and $k = \omega/v = 2\pi f/v$ (Eq. 15–12), then

$$\begin{aligned}\Delta P_M &= BAk \\ &= \rho v^2\, Ak \\ &= 2\pi\rho v\, Af. \qquad \textbf{(16–5)}\end{aligned}$$

16–3 Intensity of Sound: Decibels

Loudness is a sensation in the consciousness of a human being and is related to a physically measurable quantity, the **intensity** of the wave. Intensity is defined as the energy transported by a wave per unit time across a unit area perpendicular to the energy flow. As we saw in Chapter 15, intensity is proportional to the square of the wave amplitude. Intensity has units of power per unit area, or watts/meter^2 (W/m^2).

The human ear can detect sounds with an intensity as low as 10^{-12} W/m^2 and as high as 1 W/m^2 (and even higher, although above this it is painful). This is an incredibly wide range of intensity, spanning a factor of 10^{12} from lowest to highest. Presumably because of this wide range, what we perceive as loudness is not directly proportional to the intensity. To produce a sound that sounds about twice as loud requires a sound wave that has about 10 times the intensity. This is roughly valid at any sound level for frequencies near the middle of the audible range. For example, a sound wave of intensity 10^{-2} W/m^2 sounds to an average human being like it is about twice as loud as one whose intensity is 10^{-3} W/m^2, and four times as loud as 10^{-4} W/m^2.

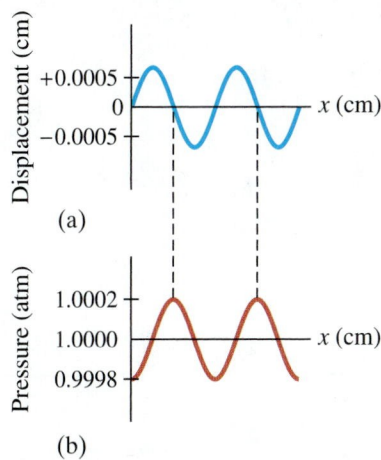

FIGURE 16–3 Representation of a sound wave in space at a given instant in terms of (a) displacement, and (b) pressure.

FIGURE 16–4 Longitudinal wave in a fluid moves to the right. A thin layer of fluid, in a thin cylinder of area S and thickness Δx, changes in volume as a result of pressure variation as the wave passes. At the moment shown, the pressure will increase as the wave moves to the right, so the thickness of our layer will decrease, by an amount ΔD.

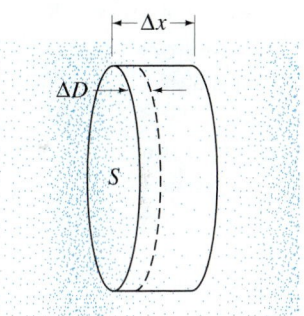

PHYSICS APPLIED
Wide range of human hearing

Sound Level

Because of this relationship between the subjective sensation of loudness and the physically measurable quantity "intensity," sound intensity levels are usually specified on a logarithmic scale. The unit on this scale is a **bel**, after the inventor Alexander Graham Bell, or much more commonly, the **decibel** (dB), which is $\frac{1}{10}$ bel (10 dB = 1 bel). The **sound level**, β, of any sound is defined in terms of its intensity, I, as

$$\beta \text{ (in dB)} = 10 \log \frac{I}{I_0}, \tag{16-6}$$

where I_0 is the intensity of a chosen reference level, and the logarithm is to the base 10. I_0 is usually taken as the minimum intensity audible to a good ear—the "threshold of hearing," which is $I_0 = 1.0 \times 10^{-12} \text{ W/m}^2$. Thus, for example, the sound level of a sound whose intensity $I = 1.0 \times 10^{-10} \text{ W/m}^2$ will be

$$\beta = 10 \log \left(\frac{1.0 \times 10^{-10} \text{ W/m}^2}{1.0 \times 10^{-12} \text{ W/m}^2} \right) = 10 \log 100 = 20 \text{ dB},$$

since log 100 is equal to 2.0. (Appendix A has a brief review of logarithms.) Notice that the sound level at the threshold of hearing is 0 dB. That is, $\beta = 10 \log 10^{-12}/10^{-12} = 10 \log 1 = 0$ since $\log 1 = 0$. Notice too that an increase in intensity by a factor of 10 corresponds to a sound level increase of 10 dB. An increase in intensity by a factor of 100 corresponds to a sound level increase of 20 dB. Thus a 50-dB sound is 100 times more intense than a 30-dB sound, and so on.

Intensities and sound levels for a number of common sounds are listed in Table 16–2.

> ⚠️ **CAUTION**
> *0 dB does not mean zero intensity*

TABLE 16–2
Intensity of Various Sounds

Source of the Sound	Sound Level (dB)	Intensity (W/m²)
Jet plane at 30 m	140	100
Threshold of pain	120	1
Loud rock concert	120	1
Siren at 30 m	100	1×10^{-2}
Truck traffic	90	1×10^{-3}
Busy street traffic	80	1×10^{-4}
Noisy restaurant	70	1×10^{-5}
Talk, at 50 cm	65	3×10^{-6}
Quiet radio	40	1×10^{-8}
Whisper	30	1×10^{-9}
Rustle of leaves	10	1×10^{-11}
Threshold of hearing	0	1×10^{-12}

EXAMPLE 16–3 **Sound intensity on the street.** At a busy street corner, the sound level is 75 dB. What is the intensity of sound there?

APPROACH We have to solve Eq. 16–6 for intensity I, remembering that $I_0 = 1.0 \times 10^{-12} \text{ W/m}^2$.

SOLUTION From Eq. 16–6

$$\log \frac{I}{I_0} = \frac{\beta}{10},$$

so

$$\frac{I}{I_0} = 10^{\beta/10}.$$

With $\beta = 75$, then

$$I = I_0 10^{\beta/10} = (1.0 \times 10^{-12} \text{ W/m}^2)(10^{7.5}) = 3.2 \times 10^{-5} \text{ W/m}^2.$$

NOTE Recall that $x = \log y$ is the same as $y = 10^x$ (Appendix A).

🚶 **PHYSICS APPLIED**
Loudspeaker response ($\pm 3 \, dB$)

EXAMPLE 16–4 **Loudspeaker response.** A high-quality loudspeaker is advertised to reproduce, at full volume, frequencies from 30 Hz to 18,000 Hz with uniform sound level ± 3 dB. That is, over this frequency range, the sound level output does not vary by more than 3 dB for a given input level. By what factor does the intensity change for the maximum change of 3 dB in output sound level?

APPROACH Let us call the average intensity I_1 and the average sound level β_1. Then the maximum intensity, I_2, corresponds to a level $\beta_2 = \beta_1 + 3$ dB. We then use the relation between intensity and sound level, Eq. 16–6.

SOLUTION Equation 16–6 gives

$$\beta_2 - \beta_1 = 10 \log \frac{I_2}{I_0} - 10 \log \frac{I_1}{I_0}$$

$$3\,\text{dB} = 10 \left(\log \frac{I_2}{I_0} - \log \frac{I_1}{I_0} \right)$$

$$= 10 \log \frac{I_2}{I_1}$$

because $(\log a - \log b) = \log a/b$ (see Appendix A). This last equation gives

$$\log \frac{I_2}{I_1} = 0.30,$$

or

$$\frac{I_2}{I_1} = 10^{0.30} = 2.0.$$

So $\pm 3\,\text{dB}$ corresponds to a doubling or halving of the intensity.

It is worth noting that a sound-level difference of 3 dB (which corresponds to a doubled intensity, as we just saw) corresponds to only a very small change in the subjective sensation of apparent loudness. Indeed, the average human can distinguish a difference in sound level of only about 1 or 2 dB.

EXERCISE B If an increase of 3 dB means "twice as intense," what does an increase of 6 dB mean?

CONCEPTUAL EXAMPLE 16–5 **Trumpet players.** A trumpeter plays at a sound level of 75 dB. Three equally loud trumpet players join in. What is the new sound level?

RESPONSE The intensity of four trumpets is four times the intensity of one trumpet $(= I_1)$ or $4I_1$. The sound level of the four trumpets would be

$$\beta = 10 \log \frac{4I_1}{I_0} = 10 \log 4 + 10 \log \frac{I_1}{I_0}$$

$$= 6.0\,\text{dB} + 75\,\text{dB} = 81\,\text{dB}.$$

EXERCISE C From Table 16–2, we see that ordinary conversation corresponds to a sound level of about 65 dB. If two people are talking at once, the sound level is (a) 65 dB, (b) 68 dB, (c) 75 dB, (d) 130 dB, (e) 62 dB.

Normally, the loudness or intensity of a sound decreases as you get farther from the source of the sound. In interior rooms, this effect is altered because of reflections from the walls. However, if a source is in the open so that sound can radiate out freely in all directions, the intensity decreases as the inverse square of the distance,

$$I \propto \frac{1}{r^2},$$

as we saw in Section 15–3. Over large distances, the intensity decreases faster than $1/r^2$ because some of the energy is transferred into irregular motion of air molecules. This loss happens more for higher frequencies, so any sound of mixed frequencies will be less "bright" at a distance.

EXAMPLE 16–6 **Airplane roar.** The sound level measured 30 m from a jet plane is 140 dB. What is the sound level at 300 m? (Ignore reflections from the ground.)

APPROACH Given the sound level, we can determine the intensity at 30 m using Eq. 16–6. Because intensity decreases as the square of the distance, ignoring reflections, we can find I at 300 m and again apply Eq. 16–6 to obtain the sound level.

SOLUTION The intensity I at 30 m is

$$140 \text{ dB} = 10 \log\left(\frac{I}{10^{-12} \text{ W/m}^2}\right)$$

or

$$14 = \log\left(\frac{I}{10^{-12} \text{ W/m}^2}\right).$$

We raise both sides of this equation to the power 10 (recall $10^{\log x} = x$) and have

$$10^{14} = \frac{I}{10^{-12} \text{ W/m}^2},$$

so $I = (10^{14})(10^{-12} \text{ W/m}^2) = 10^2 \text{ W/m}^2$. At 300 m, 10 times as far, the intensity will be $\left(\frac{1}{10}\right)^2 = 1/100$ as much, or 1 W/m². Hence, the sound level is

$$\beta = 10 \log\left(\frac{1 \text{ W/m}^2}{10^{-12} \text{ W/m}^2}\right) = 120 \text{ dB}.$$

Even at 300 m, the sound is at the threshold of pain. This is why workers at airports wear ear covers to protect their ears from damage (Fig. 16–5).

NOTE Here is a simpler approach that avoids Eq. 16–6. Because the intensity decreases as the square of the distance, at 10 times the distance the intensity decreases by $\left(\frac{1}{10}\right)^2 = \frac{1}{100}$. We can use the result that 10 dB corresponds to an intensity change by a factor of 10 (see text just before Example 16–3). Then an intensity change by a factor of 100 corresponds to a sound-level change of $(2)(10 \text{ dB}) = 20 \text{ dB}$. This confirms our result above: $140 \text{ dB} - 20 \text{ dB} = 120 \text{ dB}$.

FIGURE 16–5 Example 16–6. Airport worker with sound-intensity-reducing ear covers (headphones).

Intensity Related to Amplitude

The intensity I of a wave is proportional to the square of the wave amplitude, as we saw in Chapter 15. We can therefore relate the amplitude quantitatively to the intensity I or level β, as the following Example shows.

EXAMPLE 16–7 **How tiny the displacement is.** (*a*) Calculate the displacement of air molecules for a sound having a frequency of 1000 Hz at the threshold of hearing. (*b*) Determine the maximum pressure variation in such a sound wave.

APPROACH In Section 15–3 we found a relation between intensity I and displacement amplitude A of a wave, Eq. 15–7. The amplitude of oscillation of air molecules is what we want to solve for, given the intensity. The pressure is found from Eq. 16–5.

SOLUTION (*a*) At the threshold of hearing, $I = 1.0 \times 10^{-12} \text{ W/m}^2$ (Table 16–2). We solve for the amplitude A in Eq. 15–7:

$$
\begin{aligned}
A &= \frac{1}{\pi f}\sqrt{\frac{I}{2\rho v}} \\
&= \frac{1}{(3.14)(1.0 \times 10^3 \text{ s}^{-1})}\sqrt{\frac{1.0 \times 10^{-12} \text{ W/m}^2}{(2)(1.29 \text{ kg/m}^3)(343 \text{ m/s})}} \\
&= 1.1 \times 10^{-11} \text{ m},
\end{aligned}
$$

where we have taken the density of air to be 1.29 kg/m³ and the speed of sound in air (assumed 20°C) as 343 m/s.

NOTE We see how incredibly sensitive the human ear is: it can detect displacements of air molecules which are actually less than the diameter of atoms (about 10^{-10} m).

(*b*) Now we are dealing with sound as a pressure wave (Section 16–2). From Eq. 16–5,

$$
\begin{aligned}
\Delta P_\text{M} &= 2\pi\rho v A f \\
&= 2\pi(1.29 \text{ kg/m}^3)(343 \text{ m/s})(1.1 \times 10^{-11} \text{ m})(1.0 \times 10^3 \text{ s}^{-1}) = 3.1 \times 10^{-5} \text{ Pa}
\end{aligned}
$$

or 3.1×10^{-10} atm. Again we see that the human ear is incredibly sensitive.

By combining Eqs. 15–7 and 16–5, we can write the intensity in terms of the pressure amplitude, ΔP_M:

$$I = 2\pi^2 v \rho f^2 A^2 = 2\pi^2 v \rho f^2 \left(\frac{\Delta P_M}{2\pi \rho v f} \right)^2$$

$$I = \frac{(\Delta P_M)^2}{2 v \rho}. \qquad (16\text{–}7)$$

The intensity, when given in terms of pressure amplitude, thus does not depend on frequency.

The Ear's Response

The ear is not equally sensitive to all frequencies. To hear the same loudness for sounds of different frequencies requires different intensities. Studies averaged over large numbers of people have produced the curves shown in Fig. 16–6. On this graph, each curve represents sounds that seemed to be equally loud. The number labeling each curve represents the **loudness level** (the units are called *phons*), which is numerically equal to the sound level in dB at 1000 Hz. For example, the curve labeled 40 represents sounds that are heard by an average person to have the same loudness as a 1000-Hz sound with a sound level of 40 dB. From this 40-phon curve, we see that a 100-Hz tone must be at a level of about 62 dB to be perceived as loud as a 1000-Hz tone of only 40 dB.

The lowest curve in Fig. 16–6 (labeled 0) represents the sound level, as a function of frequency, for the *threshold of hearing*, the softest sound that is just audible by a very good ear. Note that the ear is most sensitive to sounds of frequency between 2000 and 4000 Hz, which are common in speech and music. Note too that whereas a 1000-Hz sound is audible at a level of 0 dB, a 100-Hz sound must be nearly 40 dB to be heard. The top curve in Fig. 16–6, labeled 120 phons, represents the *threshold of pain*. Sounds above this level can actually be felt and cause pain.

Figure 16–6 shows that at lower sound levels, our ears are less sensitive to the high and low frequencies relative to middle frequencies. The "loudness" control on some stereo systems is intended to compensate for this low-volume insensitivity. As the volume is turned down, the loudness control boosts the high and low frequencies relative to the middle frequencies so that the sound will have a more "normal-sounding" frequency balance. Many listeners, however, find the sound more pleasing or natural without the loudness control.

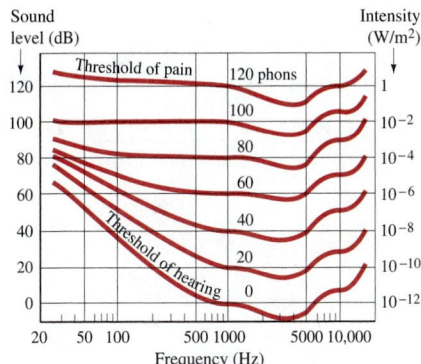

FIGURE 16–6 Sensitivity of the human ear as a function of frequency (see text). Note that the frequency scale is "logarithmic" in order to cover a wide range of frequencies.

16–4 Sources of Sound: Vibrating Strings and Air Columns

The source of any sound is a vibrating object. Almost any object can vibrate and hence be a source of sound. We now discuss some simple sources of sound, particularly musical instruments. In musical instruments, the source is set into vibration by striking, plucking, bowing, or blowing. Standing waves are produced and the source vibrates at its natural resonant frequencies. The vibrating source is in contact with the air (or other medium) and pushes on it to produce sound waves that travel outward. The frequencies of the waves are the same as those of the source, but the speed and wavelengths can be different. A drum has a stretched membrane that vibrates. Xylophones and marimbas have metal or wood bars that can be set into vibration. Bells, cymbals, and gongs also make use of a vibrating metal. Many instruments make use of vibrating strings, such as the violin, guitar, and piano, or make use of vibrating columns of air, such as the flute, trumpet, and pipe organ. We have already seen that the pitch of a pure sound is determined by the frequency. Typical frequencies for musical notes on the "equally tempered chromatic scale" are given in Table 16–3 for the octave beginning with middle C. Note that one octave corresponds to a doubling of frequency. For example, middle C has frequency of 262 Hz whereas C′ (C above middle C) has twice that frequency, 524 Hz. [Middle C is the C or "do" note at the middle of a piano keyboard.]

TABLE 16–3 Equally Tempered Chromatic Scale[†]

Note	Frequency (Hz)
C	262
C# or D♭	277
D	294
D# or E♭	311
E	330
F	349
F# or G♭	370
G	392
G# or A♭	415
A	440
A# or B♭	466
B	494
C′	524

[†]Only one octave is included.

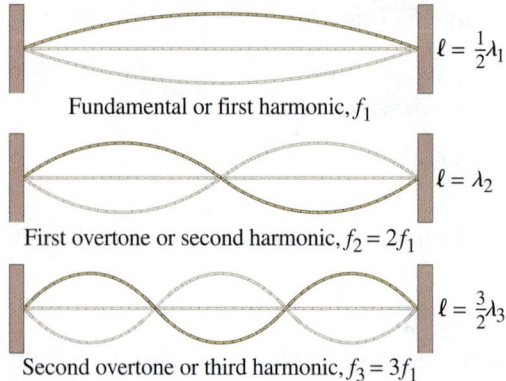

$\ell = \frac{1}{2}\lambda_1$

Fundamental or first harmonic, f_1

$\ell = \lambda_2$

First overtone or second harmonic, $f_2 = 2f_1$

$\ell = \frac{3}{2}\lambda_3$

Second overtone or third harmonic, $f_3 = 3f_1$

FIGURE 16–7 Standing waves on a string—only the lowest three frequencies are shown.

Stringed Instruments

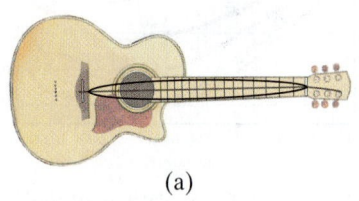

(a)

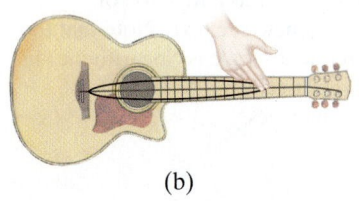

(b)

FIGURE 16–8 The wavelength of (a) an unfingered string is longer than that of (b) a fingered string. Hence, the frequency of the fingered string is higher. Only one string is shown on this guitar, and only the simplest standing wave, the fundamental, is shown.

We saw in Chapter 15, Fig. 15–26b, how standing waves are established on a string, and we show this again here in Fig. 16–7. Such standing waves are the basis for all stringed instruments. The pitch is normally determined by the lowest resonant frequency, the **fundamental**, which corresponds to nodes occurring only at the ends. The string vibrating up and down as a whole corresponds to a half wavelength as shown at the top of Fig. 16–7; so the wavelength of the fundamental on the string is equal to twice the length of the string. Therefore, the fundamental frequency is $f_1 = v/\lambda = v/2\ell$, where v is the velocity of the wave on the string (*not* in the air). The possible frequencies for standing waves on a stretched string are whole-number multiples of the fundamental frequency:

$$f_n = nf_1 = n\frac{v}{2\ell}, \qquad n = 1, 2, 3, \cdots$$

where $n = 1$ refers to the fundamental and $n = 2, 3, \cdots$ are the overtones. All of the standing waves, $n = 1, 2, 3, \cdots$, are called harmonics,[†] as we saw in Section 15–9.

When a finger is placed on the string of a guitar or violin, the effective length of the string is shortened. So its fundamental frequency, and pitch, is higher since the wavelength of the fundamental is shorter (Fig. 16–8). The strings on a guitar or violin are all the same length. They sound at a different pitch because the strings have different mass per unit length, μ, which affects the velocity on the string, Eq. 15–2,

$$v = \sqrt{F_T/\mu}. \qquad \text{[stretched string]}$$

Thus the velocity on a heavier string is lower and the frequency will be lower for the same wavelength. The tension F_T may also be different. Adjusting the tension is the means for tuning the pitch of each string. In pianos and harps the strings are of different lengths. For the lower notes the strings are not only longer, but heavier as well, and the reason is illustrated in the following Example.

EXAMPLE 16–8 **Piano strings.** The highest key on a piano corresponds to a frequency about 150 times that of the lowest key. If the string for the highest note is 5.0 cm long, how long would the string for the lowest note have to be if it had the same mass per unit length and was under the same tension?

APPROACH Since $v = \sqrt{F_T/\mu}$, the velocity would be the same on each string. So the frequency is inversely proportional to the length ℓ of the string ($f = v/\lambda = v/2\ell$).

SOLUTION We can write, for the fundamental frequencies of each string, the ratio

$$\frac{\ell_L}{\ell_H} = \frac{f_H}{f_L},$$

where the subscripts L and H refer to the lowest and highest notes, respectively. Thus $\ell_L = \ell_H(f_H/f_L) = (5.0\text{ cm})(150) = 750\text{ cm},$ or 7.5 m. This would be ridiculously long (≈ 25 ft) for a piano.

NOTE The longer strings of lower frequency are made heavier, of higher mass per unit length, so even on grand pianos the strings are less than 3 m long.

[†]When the resonant frequencies above the fundamental (that is, the overtones) are integral multiples of the fundamental, as here, they are called harmonics. But if the overtones are not integral multiples of the fundamental, as is the case for a vibrating drumhead, for example, they are not harmonics.

EXERCISE D Two strings have the same length and tension, but one is more massive than the other. Which plays the higher note?

EXAMPLE 16–9 **Frequencies and wavelengths in the violin.** A 0.32-m-long violin string is tuned to play A above middle C at 440 Hz. (*a*) What is the wavelength of the fundamental string vibration, and (*b*) what are the frequency and wavelength of the sound wave produced? (*c*) Why is there a difference?

APPROACH The wavelength of the fundamental string vibration equals twice the length of the string (Fig. 16–7). As the string vibrates, it pushes on the air, which is thus forced to oscillate at the same frequency as the string.

SOLUTION (*a*) From Fig. 16–7 the wavelength of the fundamental is

$$\lambda = 2\ell = 2(0.32 \text{ m}) = 0.64 \text{ m} = 64 \text{ cm}.$$

This is the wavelength of the standing wave on the string.

(*b*) The sound wave that travels outward in the air (to reach our ears) has the same frequency, 440 Hz. Its wavelength is

$$\lambda = \frac{v}{f} = \frac{343 \text{ m/s}}{440 \text{ Hz}} = 0.78 \text{ m} = 78 \text{ cm},$$

where *v* is the speed of sound in air (assumed at 20°C), Section 16–1.

(*c*) The wavelength of the sound wave is different from that of the standing wave on the string because the speed of sound in air (343 m/s at 20°C) is different from the speed of the wave on the string ($= f\lambda = 440 \text{ Hz} \times 0.64 \text{ m} = 280 \text{ m/s}$) which depends on the tension in the string and its mass per unit length.

NOTE The frequencies on the string and in the air are the same: the string and air are in contact, and the string "forces" the air to vibrate at the same frequency. But the wavelengths are different because the wave speed on the string is different than that in air.

(a)

(b)

FIGURE 16–9 (a) Piano, showing sounding board to which the strings are attached; (b) sounding box (guitar).

Stringed instruments would not be very loud if they relied on their vibrating strings to produce the sound waves since the strings are too thin to compress and expand much air. Stringed instruments therefore make use of a kind of mechanical amplifier known as a *sounding board* (piano) or *sounding box* (guitar, violin), which acts to amplify the sound by putting a greater surface area in contact with the air (Fig. 16–9). When the strings are set into vibration, the sounding board or box is set into vibration as well. Since it has much greater area in contact with the air, it can produce a more intense sound wave. On an electric guitar, the sounding box is not so important since the vibrations of the strings are amplified electronically.

Wind Instruments

Instruments such as woodwinds, the brasses, and the pipe organ produce sound from the vibrations of standing waves in a column of air within a tube (Fig. 16–10). Standing waves can occur in the air of any cavity, but the frequencies present are complicated for any but very simple shapes such as the uniform, narrow tube of a flute or an organ pipe. In some instruments, a vibrating reed or the vibrating lip of the player helps to set up vibrations of the air column. In others, a stream of air is directed against one edge of the opening or mouthpiece, leading to turbulence which sets up the vibrations. Because of the disturbance, whatever its source, the air within the tube vibrates with a variety of frequencies, but only frequencies that correspond to standing waves will persist.

For a string fixed at both ends, Fig. 16–7, we saw that the standing waves have nodes (no movement) at the two ends, and one or more antinodes (large amplitude of vibration) in between. A node separates successive antinodes. The lowest-frequency standing wave, the *fundamental*, corresponds to a single antinode. The higher-frequency standing waves are called **overtones** or **harmonics**, as we saw in Section 15–9. Specifically, the first harmonic is the fundamental, the second harmonic (= first overtone) has twice the frequency of the fundamental, and so on.

FIGURE 16–10 Wind instruments: flute (left) and clarinet.

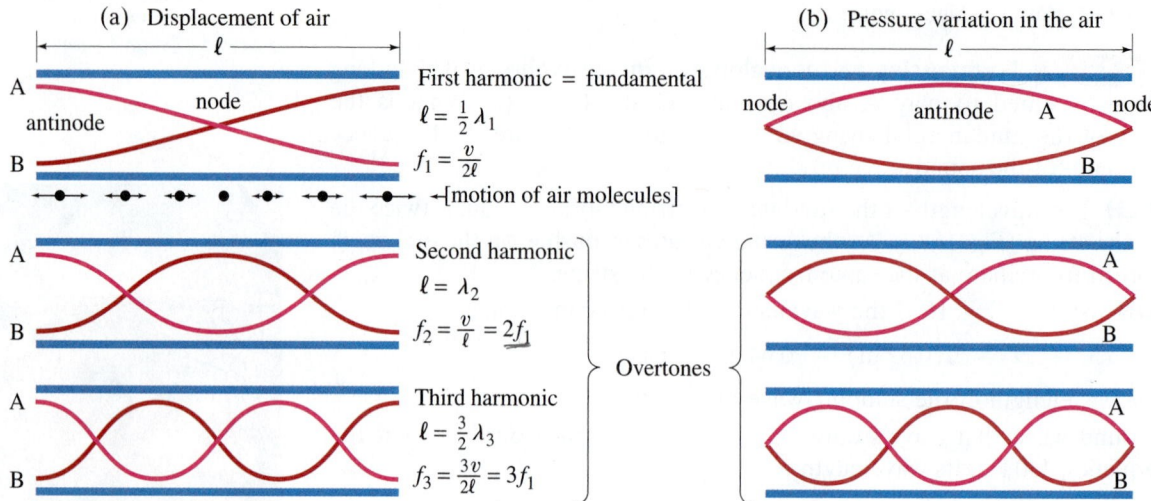

(a) Displacement of air

A
antinode   node
B
←—[motion of air molecules]

First harmonic = fundamental
$\ell = \frac{1}{2}\lambda_1$
$f_1 = \frac{v}{2\ell}$

A
B
Second harmonic
$\ell = \lambda_2$
$f_2 = \frac{v}{\ell} = 2f_1$

A
B
Third harmonic
$\ell = \frac{3}{2}\lambda_3$
$f_3 = \frac{3v}{2\ell} = 3f_1$

Overtones

(b) Pressure variation in the air

node   antinode   A   node
B

A
B

A
B

FIGURE 16–11 Graphs of the three simplest modes of vibration (standing waves) for a uniform tube open at both ends ("open tube"). These simplest modes of vibration are graphed in (a), on the left, in terms of the motion of the air (displacement), and in (b), on the right, in terms of air pressure. Each graph shows the wave format at two times, A and B, a half period apart. The actual motion of molecules for one case, the fundamental, is shown just below the tube at top left.

The situation is similar for a column of air in a tube of uniform diameter, but we must remember that it is now air itself that is vibrating. We can describe the waves either in terms of the flow of the air—that is, in terms of the *displacement* of air—or in terms of the *pressure* in the air (see Figs. 16–2 and 16–3). In terms of displacement, the air at the closed end of a tube is a displacement node since the air is not free to move there, whereas near the open end of a tube there will be an antinode because the air can move freely in and out. The air within the tube vibrates in the form of longitudinal standing waves. The possible modes of vibration for a tube open at both ends (called an **open tube**) are shown graphically in Fig. 16–11. They are shown for a tube that is open at one end but closed at the other (called a **closed tube**) in Fig. 16–12. [A tube closed at *both* ends, having no connection to the outside air, would be useless as an instrument.] The graphs in part (a) of each Figure (left-hand sides) represent the displacement amplitude of the vibrating air in the tube. Note that these are graphs, and that the air molecules themselves oscillate *horizontally*, parallel to the tube length, as shown by the small arrows in the top diagram of Fig. 16–11a (on the left). The exact position of the antinode near the open end of a tube depends on the diameter of the tube, but if the diameter is small compared to the length, which is the usual case, the antinode occurs very close to the end as shown. We assume this is the case in what follows. (The position of the antinode may also depend slightly on the wavelength and other factors.)

Let us look in detail at the open tube, in Fig. 16–11a, which might be an organ pipe or a flute. An open tube has displacement antinodes at both ends since the air is free to move at open ends. There must be at least one node within an open tube if there is to be a standing wave at all. A single node corresponds to the *fundamental frequency* of the tube. Since the distance between two successive nodes, or between two successive antinodes, is $\frac{1}{2}\lambda$, there is one-half of a wavelength within the length of the tube for the simplest case of the fundamental (top diagram in Fig. 16–11a): $\ell = \frac{1}{2}\lambda$, or $\lambda = 2\ell$. So the fundamental frequency is $f_1 = v/\lambda = v/2\ell$, where v is the velocity of sound in air (the air in the tube). The standing wave with two nodes is the *first overtone* or *second harmonic* and has half the wavelength $(\ell = \lambda)$ and twice the frequency of the fundamental. Indeed, in a uniform tube open at both ends, the frequency of each overtone is an integral multiple of the fundamental frequency, as shown in Fig. 16–11a. This is just what is found for a string.

For a closed tube, shown in Fig. 16–12a, which could be an organ pipe, there is always a displacement node at the closed end (because the air is not free to move) and an antinode at the open end (where the air can move freely). Since the distance between a node and the nearest antinode is $\frac{1}{4}\lambda$, we see that the fundamental in a closed tube corresponds to only one-fourth of a wavelength within the length of the tube: $\ell = \lambda/4$, and $\lambda = 4\ell$. The fundamental frequency is thus $f_1 = v/4\ell$, or half that for an open pipe of the same length. There is another

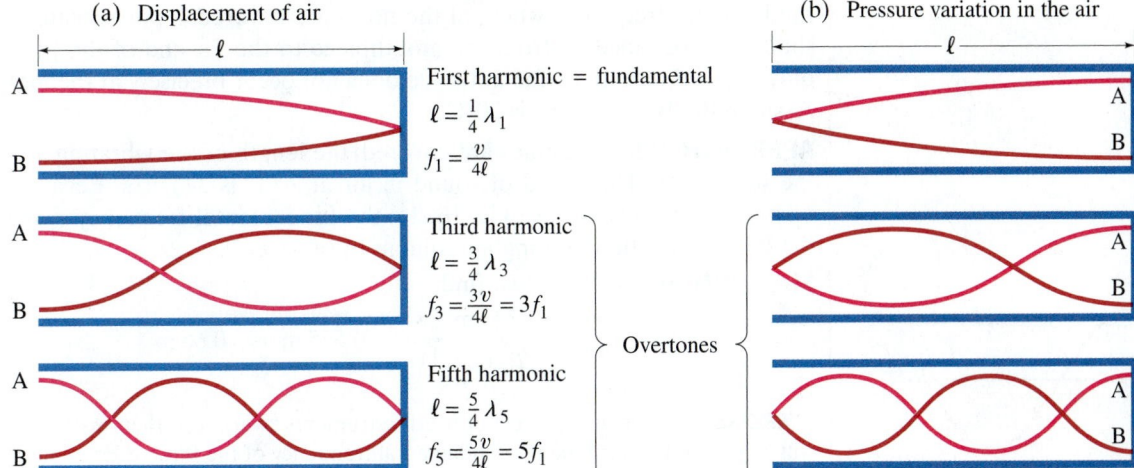

(a) Displacement of air

A

B

First harmonic = fundamental
$\ell = \frac{1}{4}\lambda_1$
$f_1 = \frac{v}{4\ell}$

A

B

Third harmonic
$\ell = \frac{3}{4}\lambda_3$
$f_3 = \frac{3v}{4\ell} = 3f_1$

Overtones

A

B

Fifth harmonic
$\ell = \frac{5}{4}\lambda_5$
$f_5 = \frac{5v}{4\ell} = 5f_1$

(b) Pressure variation in the air

A

B

A

B

A

B

FIGURE 16–12 Modes of vibration (standing waves) for a tube closed at one end ("closed tube"). See caption for Fig. 16–11.

difference, for as we can see from Fig. 16–12a, only the odd harmonics are present in a closed tube: the overtones have frequencies equal to $3, 5, 7, \dots$ times the fundamental frequency. There is no way for waves with $2, 4, 6, \dots$ times the fundamental frequency to have a node at one end and an antinode at the other, and thus they cannot exist as standing waves in a closed tube.

Another way to analyze the vibrations in a uniform tube is to consider a description in terms of the *pressure* in the air, shown in part (b) of Figs. 16–11 and 16–12 (right-hand sides). Where the air in a wave is compressed, the pressure is higher, whereas in a wave expansion (or rarefaction), the pressure is less than normal. The open end of a tube is open to the atmosphere. Hence the pressure variation at an open end must be a *node*: the pressure does not alternate, but remains at the outside atmospheric pressure. If a tube has a closed end, the pressure at that closed end can readily alternate to be above or below atmospheric pressure. Hence there is a pressure *antinode* at a closed end of a tube. There can be pressure nodes and antinodes within the tube. Some of the possible vibrational modes in terms of pressure for an open tube are shown in Fig. 16–11b, and for a closed tube are shown in Fig. 16–12b.

EXAMPLE 16–10 **Organ pipes.** What will be the fundamental frequency and first three overtones for a 26-cm-long organ pipe at 20°C if it is (*a*) open and (*b*) closed?

APPROACH All our calculations can be based on Figs. 16–11a and 16–12a.

SOLUTION (*a*) For the open pipe, Fig. 16–11a, the fundamental frequency is

$$f_1 = \frac{v}{2\ell} = \frac{343 \text{ m/s}}{2(0.26 \text{ m})} = 660 \text{ Hz}.$$

The speed v is the speed of sound in air (the air vibrating in the pipe). The overtones include all harmonics: 1320 Hz, 1980 Hz, 2640 Hz, and so on.

(*b*) For a closed pipe, Fig. 16–12a, the fundamental frequency is

$$f_1 = \frac{v}{4\ell} = \frac{343 \text{ m/s}}{4(0.26 \text{ m})} = 330 \text{ Hz}.$$

Only odd harmonics are present: the first three overtones are 990 Hz, 1650 Hz, and 2310 Hz.

NOTE The closed pipe plays 330 Hz, which, from Table 16–3, is E above middle C, whereas the open pipe of the same length plays 660 Hz, an octave higher.

Pipe organs use both open and closed pipes, with lengths from a few centimeters to 5 m or more. A flute acts as an open tube, for it is open not only where you blow into it, but also at the opposite end. The different notes on a flute are obtained by shortening the length of the vibrating air column, by uncovering holes along the tube (so a displacement antinode can occur at the hole). The shorter the length of the vibrating air column, the higher the fundamental frequency.

EXAMPLE 16–11 Flute. A flute is designed to play middle C (262 Hz) as the fundamental frequency when all the holes are covered. Approximately how long should the distance be from the mouthpiece to the far end of the flute? (This is only approximate since the antinode does not occur precisely at the mouthpiece.) Assume the temperature is 20°C.

APPROACH When all holes are covered, the length of the vibrating air column is the full length. The speed of sound in air at 20°C is 343 m/s. Because a flute is open at both ends, we use Fig. 16–11: the fundamental frequency f_1 is related to the length ℓ of the vibrating air column by $f = v/2\ell$.

SOLUTION Solving for ℓ, we find

$$\ell = \frac{v}{2f} = \frac{343 \text{ m/s}}{2(262 \text{ s}^{-1})} = 0.655 \text{ m} \approx 0.66 \text{ m}.$$

EXERCISE E To see why players of wind instruments "warm up" their instruments (so they will be in tune), determine the fundamental frequency of the flute of Example 16–11 when all holes are covered and the temperature is 10°C instead of 20°C.

EXERCISE F Return to the Chapter-Opening Question, page 424, and answer it again now. Try to explain why you may have answered differently the first time.

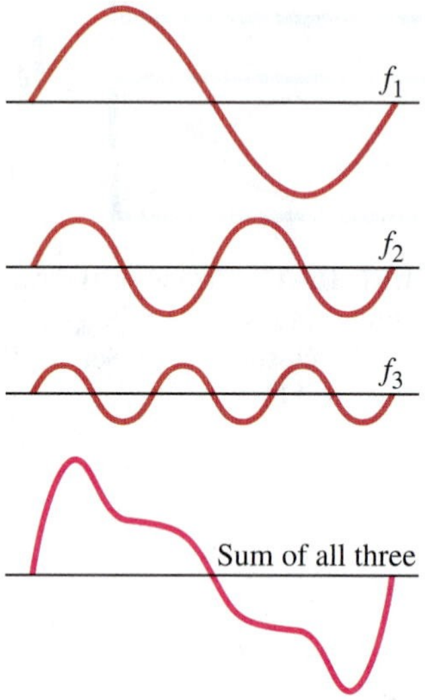

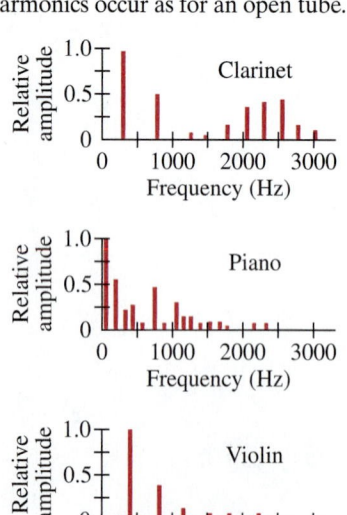

FIGURE 16–13 The displacements of the fundamental and first two overtones are added at each point to get the "sum," or composite waveform.

FIGURE 16–14 Sound spectra for different instruments. The spectra change when the instruments play different notes. The clarinet is a bit complicated: it acts like a closed tube at lower frequencies, having only odd harmonics, but at higher frequencies all harmonics occur as for an open tube.

*16–5 Quality of Sound, and Noise; Superposition

Whenever we hear a sound, particularly a musical sound, we are aware of its loudness, its pitch, and also of a third aspect called its *timbre* or "quality." For example, when a piano and then a flute play a note of the same loudness and pitch (say, middle C), there is a clear difference in the overall sound. We would never mistake a piano for a flute. This is what is meant by the timbre or *quality* of a sound. For musical instruments, the term *tone color* is also used.

Just as loudness and pitch can be related to physically measurable quantities, so too can quality. The quality of a sound depends on the presence of overtones—their number and their relative amplitudes. Generally, when a note is played on a musical instrument, the fundamental as well as overtones are present simultaneously. Figure 16–13 illustrates how the *principle of superposition* (Section 15–6) applies to three wave forms, in this case the fundamental and first two overtones (with particular amplitudes): they add together at each point to give a composite **waveform** (by which we mean the displacement as a function of position at a given time). Normally, more than two overtones are present. [Any complex wave can be analyzed into a superposition of sinusoidal waves of appropriate amplitudes, wavelengths, and frequencies—see Section 15–6. Such an analysis is called a *Fourier analysis*.]

The relative amplitudes of the overtones for a given note are different for different musical instruments, which is what gives each instrument its characteristic quality or timbre. A bar graph showing the relative amplitudes of the harmonics for a given note produced by an instrument is called a *sound spectrum*. Several typical examples for different musical instruments are shown in Fig. 16–14. The fundamental usually has the greatest amplitude, and its frequency is what is heard as the pitch.

The manner in which an instrument is played strongly influences the sound quality. Plucking a violin string, for example, makes a very different sound than pulling a bow across it. The sound spectrum at the very start (or end) of a note (as when a hammer strikes a piano string) can be very different from the subsequent sustained tone. This too affects the subjective tone quality of an instrument.

An ordinary sound, like that made by striking two stones together, is a noise that has a certain quality, but a clear pitch is not discernible. Such a noise is a mixture of many frequencies which bear little relation to one another. A sound spectrum made of that noise would not show discrete lines like those of Fig. 16–14. Instead it would show a continuous, or nearly continuous, spectrum of frequencies. Such a sound we call "noise" in comparison with the more harmonious sounds which contain frequencies that are simple multiples of the fundamental.

16–6 Interference of Sound Waves; Beats

Interference in Space

We saw in Section 15–8 that when two waves simultaneously pass through the same region of space, they interfere with one another. Interference also occurs with sound waves.

Consider two large loudspeakers, A and B, a distance d apart on the stage of an auditorium as shown in Fig. 16–15. Let us assume the two speakers are emitting sound waves of the same single frequency and that they are in phase: that is, when one speaker is forming a compression, so is the other. (We ignore reflections from walls, floor, etc.) The curved lines in the diagram represent the crests of sound waves from each speaker at one instant in time. We must remember that for a sound wave, a crest is a compression in the air whereas a trough—which falls between two crests—is a rarefaction. A human ear or detector at a point such as C, which is the same distance from each speaker, will experience a loud sound because the interference will be constructive—two crests reach it at one moment, two troughs reach it a moment later. On the other hand, at a point such as D in the diagram, little if any sound will be heard because destructive interference occurs—compressions of one wave meet rarefactions of the other and vice versa (see Fig. 15–24 and the related discussion on water waves in Section 15–8).

An analysis of this situation is perhaps clearer if we graphically represent the waveforms as in Fig. 16–16. In Fig. 16–16a, it can be seen that at point C, constructive interference occurs since both waves simultaneously have crests or simultaneously have troughs when they arrive at C. In Fig. 16–16b we see that, to reach point D, the wave from speaker B must travel a greater distance than the wave from A. Thus the wave from B lags behind that from A. In this diagram, point E is chosen so that the distance ED is equal to AD. Thus we see that if the distance BE is equal to precisely one-half the wavelength of the sound, the two waves will be exactly out of phase when they reach D, and destructive interference occurs. This then is the criterion for determining at what points destructive interference occurs: destructive interference occurs at any point whose distance from one speaker is one-half wavelength greater than its distance from the other speaker. Notice that if this extra distance (BE in Fig. 16–16b) is equal to a whole wavelength (or 2, 3, ... wavelengths), then the two waves will be in phase and *constructive interference* occurs. If the distance BE equals $\frac{1}{2}, 1\frac{1}{2}, 2\frac{1}{2}, \dots$ wavelengths, *destructive interference* occurs.

It is important to realize that a person at point D in Fig. 16–15 or 16–16 hears nothing at all (or nearly so), yet sound is coming from both speakers. Indeed, if one of the speakers is turned off, the sound from the other speaker will be clearly heard.

If a loudspeaker emits a whole range of frequencies, only specific wavelengths will destructively interfere completely at a given point.

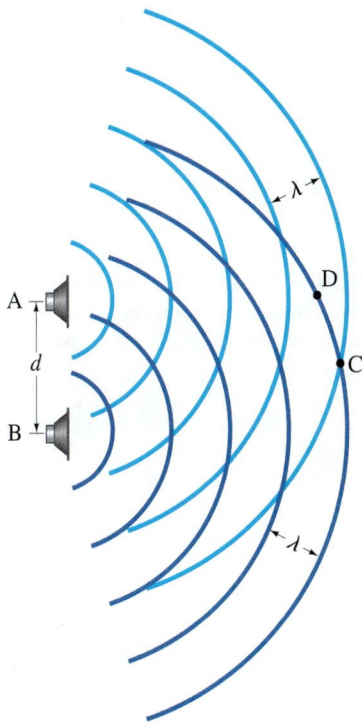

FIGURE 16–15 Sound waves from two loudspeakers interfere.

FIGURE 16–16 Sound waves of a single frequency from loudspeakers A and B (see Fig. 16–15) constructively interfere at C and destructively interfere at D. [Shown here are graphical representations, not the actual longitudinal sound waves.]

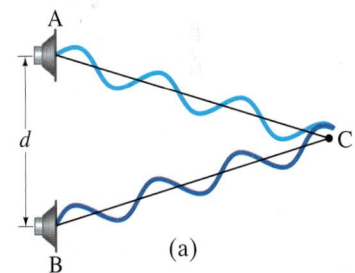

(a)

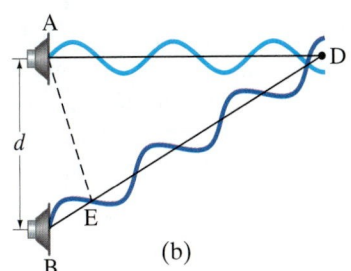

(b)

EXAMPLE 16–12 **Loudspeakers' interference.** Two loudspeakers are 1.00 m apart. A person stands 4.00 m from one speaker. How far must this person be from the second speaker to detect destructive interference when the speakers emit an 1150-Hz sound? Assume the temperature is 20°C.

APPROACH To sense destructive interference, the person must be one-half wavelength closer to or farther from one speaker than from the other—that is, at a distance = 4.00 m $\pm \lambda/2$. We can determine λ since we know f and v.

SOLUTION The speed of sound at 20°C is 343 m/s, so the wavelength of this sound is (Eq. 15–1)

$$\lambda = \frac{v}{f} = \frac{343 \text{ m/s}}{1150 \text{ Hz}} = 0.30 \text{ m}.$$

For destructive interference to occur, the person must be one-half wavelength farther from one loudspeaker than from the other, or 0.15 m. Thus the person must be 3.85 m or 4.15 m from the second speaker.

NOTE If the speakers are less than 0.15 m apart, there will be no point that is 0.15 m farther from one speaker than the other, and there will be no point where destructive interference could occur.

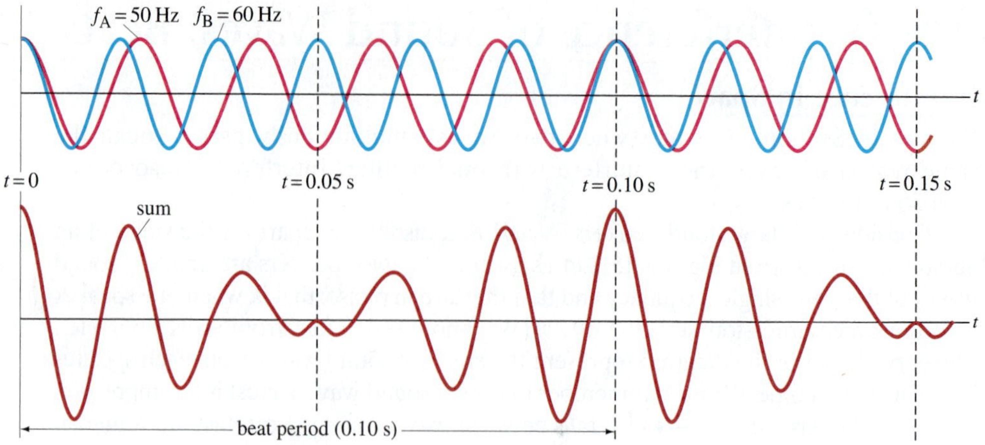

FIGURE 16–17 Beats occur as a result of the superposition of two sound waves of slightly different frequency.

Beats—Interference in Time

We have been discussing interference of sound waves that takes place in space. An interesting and important example of interference that occurs in time is the phenomenon known as **beats**: If two sources of sound—say, two tuning forks—are close in frequency but not exactly the same, sound waves from the two sources interfere with each other. The sound level at a given position alternately rises and falls in time, because the two waves are sometimes in phase and sometimes out of phase due to their different wavelengths. The regularly spaced intensity changes are called beats.

To see how beats arise, consider two equal-amplitude sound waves of frequency $f_A = 50\,\text{Hz}$ and $f_B = 60\,\text{Hz}$, respectively. In 1.00 s, the first source makes 50 vibrations whereas the second makes 60. We now examine the waves at one point in space equidistant from the two sources. The waveforms for each wave as a function of time, at a fixed position, are shown on the top graph of Fig. 16–17; the red line represents the 50-Hz wave, and the blue line represents the 60-Hz wave. The lower graph in Fig. 16–17 shows the sum of the two waves as a function of time. At time $t = 0$ the two waves are shown to be in phase and interfere constructively. Because the two waves vibrate at different rates, at time $t = 0.05\,\text{s}$ they are completely out of phase and interfere destructively. At $t = 0.10\,\text{s}$, they are again in phase and the resultant amplitude again is large. Thus the resultant amplitude is large every 0.10 s and drops drastically in between. This rising and falling of the intensity is what is heard as beats.[†] In this case the beats are 0.10 s apart. That is, the **beat frequency** is ten per second, or 10 Hz. This result, that the beat frequency equals the difference in frequency of the two waves, is valid in general, as we now show.

Let the two waves, of frequencies f_1 and f_2, be represented at a fixed point in space by

$$D_1 = A \sin 2\pi f_1 t$$

and

$$D_2 = A \sin 2\pi f_2 t.$$

The resultant displacement, by the principle of superposition, is

$$D = D_1 + D_2 = A\left(\sin 2\pi f_1 t + \sin 2\pi f_2 t\right).$$

Using the trigonometric identity $\sin\theta_1 + \sin\theta_2 = 2\sin\frac{1}{2}(\theta_1 + \theta_2)\cos\frac{1}{2}(\theta_1 - \theta_2)$, we have

$$D = \left[2A\cos 2\pi\left(\frac{f_1 - f_2}{2}\right)t\right]\sin 2\pi\left(\frac{f_1 + f_2}{2}\right)t. \qquad \textbf{(16–8)}$$

We can interpret Eq. 16–8 as follows. The superposition of the two waves results in a wave that vibrates at the average frequency of the two components, $(f_1 + f_2)/2$. This vibration has an amplitude given by the expression in brackets, and this amplitude varies in time, from zero to a maximum of $2A$ (the sum of the separate amplitudes), with a frequency of $(f_1 - f_2)/2$. A beat occurs whenever $\cos 2\pi[(f_1 - f_2)/2]t$ equals $+1$ or -1 (see Fig. 16–17); that is, two beats occur per cycle, so the beat frequency is twice $(f_1 - f_2)/2$ which is just $f_1 - f_2$, the difference in frequency of the component waves.

[†]Beats will be heard even if the amplitudes are not equal, as long as the difference in amplitude is not great.

The phenomenon of beats can occur with any kind of wave and is a very sensitive method for comparing frequencies. For example, to tune a piano, a piano tuner listens for beats produced between his standard tuning fork and that of a particular string on the piano, and knows it is in tune when the beats disappear. The members of an orchestra tune up by listening for beats between their instruments and that of a standard tone (usually A above middle C at 440 Hz) produced by a piano or an oboe. A beat frequency is perceived as an intensity modulation (a wavering between loud and soft) for beat frequencies below 20 Hz or so, and as a separate low tone for higher beat frequencies (audible if the tones are strong enough).

PHYSICS APPLIED

Tuning a piano

EXAMPLE 16–13 **Beats.** A tuning fork produces a steady 400-Hz tone. When this tuning fork is struck and held near a vibrating guitar string, twenty beats are counted in five seconds. What are the possible frequencies produced by the guitar string?

APPROACH For beats to occur, the string must vibrate at a frequency different from 400 Hz by whatever the beat frequency is.

SOLUTION The beat frequency is

$$f_{\text{beat}} = 20 \, \text{vibrations}/5 \, \text{s} = 4 \, \text{Hz}.$$

This is the difference of the frequencies of the two waves. Because one wave is known to be 400 Hz, the other must be either 404 Hz or 396 Hz.

16–7 Doppler Effect

You may have noticed that you hear the pitch of the siren on a speeding fire truck drop abruptly as it passes you. Or you may have noticed the change in pitch of a blaring horn on a fast-moving car as it passes by you. The pitch of the engine noise of a racecar changes as the car passes an observer. When a source of sound is moving toward an observer, the pitch the observer hears is higher than when the source is at rest; and when the source is traveling away from the observer, the pitch is lower. This phenomenon is known as the **Doppler effect**[†] and occurs for all types of waves. Let us now see why it occurs, and calculate the difference between the perceived and source frequencies when there is relative motion between source and observer.

Consider the siren of a fire truck at rest, which is emitting sound of a particular frequency in all directions as shown in Fig. 16–18a. The sound waves are moving at the speed of sound in air, v_{snd}, which is independent of the velocity of the source or observer. If our source, the fire truck, is moving, the siren emits sound at the same frequency as it does at rest. But the sound wavefronts it emits forward, in front of it, are closer together than when the fire truck is at rest, as shown in Fig. 16–18b. This is because the fire truck, as it moves, is "chasing" the previously emitted wavefronts, and emits each crest closer to the previous one. Thus an observer on the sidewalk in front of the truck will detect more wave crests passing per second, so the frequency heard is higher. The wavefronts emitted behind the truck, on the other hand, are farther apart than when the truck is at rest because the truck is speeding away from them. Hence, fewer wave crests per second pass by an observer behind the moving truck (Fig. 16–18b) and the perceived pitch is lower.

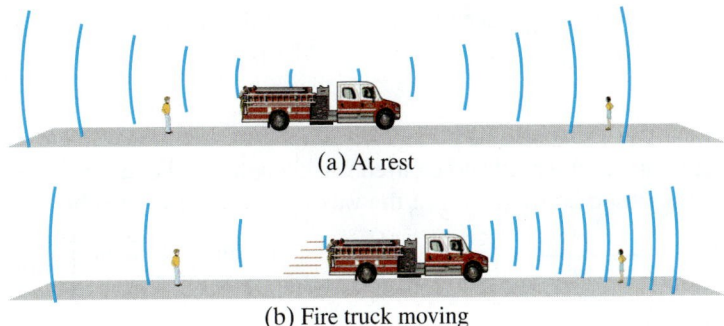

(a) At rest

(b) Fire truck moving

FIGURE 16–18 (a) Both observers on the sidewalk hear the same frequency from a fire truck at rest. (b) Doppler effect: observer toward whom the fire truck moves hears a higher-frequency sound, and observer behind the fire truck hears a lower-frequency sound.

[†]After J. C. Doppler (1803–1853).

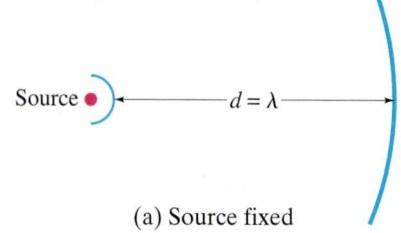

(a) Source fixed

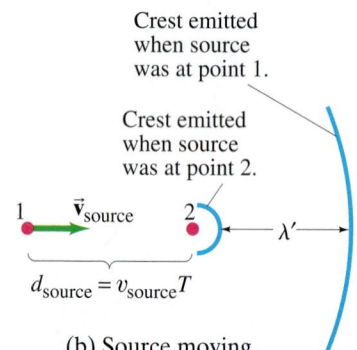

Crest emitted when source was at point 1.

Crest emitted when source was at point 2.

(b) Source moving

FIGURE 16–19 Determination of the frequency shift in the Doppler effect (see text). The red dot is the source.

We can calculate the frequency shift perceived by making use of Fig. 16–19, and we assume the air (or other medium) is at rest in our reference frame. (The stationary observer is off to the right.) In Fig. 16–19a, the source of the sound is shown as a red dot, and is at rest. Two successive wave crests are shown, the second of which has just been emitted and so is still near the source. The distance between these crests is λ, the wavelength. If the frequency of the source is f, then the time between emissions of wave crests is

$$ T = \frac{1}{f} = \frac{\lambda}{v_{snd}}. $$

In Fig. 16–19b, the source is moving with a velocity v_{source} toward the observer. In a time T (as just defined), the first wave crest has moved a distance $d = v_{snd} T = \lambda$, where v_{snd} is the velocity of the sound wave in air (which is the same whether the source is moving or not). In this same time, the source has moved a distance $d_{source} = v_{source} T$. Then the distance between successive wave crests, which is the wavelength λ' the observer will perceive, is

$$ \lambda' = d - d_{source} $$
$$ = \lambda - v_{source} T $$
$$ = \lambda - v_{source} \frac{\lambda}{v_{snd}} $$
$$ = \lambda \left(1 - \frac{v_{source}}{v_{snd}} \right). $$

We subtract λ from both sides of this equation and find that the shift in wavelength, $\Delta\lambda$, is

$$ \Delta\lambda = \lambda' - \lambda = -\lambda \frac{v_{source}}{v_{snd}}. $$

So the shift in wavelength is directly proportional to the source speed v_{source}. The frequency f' that will be perceived by our stationary observer on the ground is given by

$$ f' = \frac{v_{snd}}{\lambda'} = \frac{v_{snd}}{\lambda \left(1 - \frac{v_{source}}{v_{snd}} \right)}. $$

Since $v_{snd}/\lambda = f$, then

$$ f' = \frac{f}{\left(1 - \frac{v_{source}}{v_{snd}} \right)}. \qquad \begin{bmatrix} \text{source moving toward} \\ \text{stationary observer} \end{bmatrix} \quad \textbf{(16–9a)} $$

Because the denominator is less than 1, the observed frequency f' is greater than the source frequency f. That is, $f' > f$. For example, if a source emits a sound of frequency 400 Hz when at rest, then when the source moves toward a fixed observer with a speed of 30 m/s, the observer hears a frequency (at 20°C) of

$$ f' = \frac{400 \text{ Hz}}{1 - \frac{30 \text{ m/s}}{343 \text{ m/s}}} = 438 \text{ Hz}. $$

Now consider a source moving *away* from the stationary observer at a speed v_{source}. Using the same arguments as above, the wavelength λ' perceived by our observer will have the minus sign on d_{source} (second equation on this page) changed to plus:

$$ \lambda' = d + d_{source} $$
$$ = \lambda \left(1 + \frac{v_{source}}{v_{snd}} \right). $$

The difference between the observed and emitted wavelengths will be $\Delta\lambda = \lambda' - \lambda = +\lambda(v_{source}/v_{snd})$. The observed frequency of the wave, $f' = v_{snd}/\lambda'$, will be

$$ f' = \frac{f}{\left(1 + \frac{v_{source}}{v_{snd}} \right)}. \qquad \begin{bmatrix} \text{source moving away from} \\ \text{stationary observer} \end{bmatrix} \quad \textbf{(16–9b)} $$

If a source emitting at 400 Hz is moving away from a fixed observer at 30 m/s, the observer hears a frequency $f' = (400 \text{ Hz})/[1 + (30 \text{ m/s})/(343 \text{ m/s})] = 368 \text{ Hz}$.

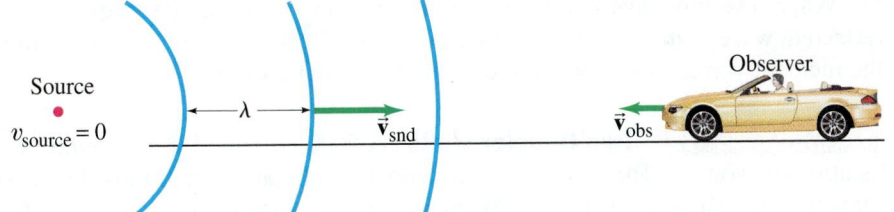

The Doppler effect also occurs when the source is at rest and the observer is in motion. If the observer is traveling *toward* the source, the pitch heard is higher than that of the emitted source frequency. If the observer is traveling *away* from the source, the pitch heard is lower. Quantitatively the change in frequency is different than for the case of a moving source. With a fixed source and a moving observer, the distance between wave crests, the wavelength λ, is not changed. But the velocity of the crests with respect to the observer *is* changed. If the observer is moving toward the source, Fig. 16–20, the speed v' of the waves relative to the observer is a simple addition of velocities: $v' = v_{snd} + v_{obs}$, where v_{snd} is the velocity of sound in air (we assume the air is still) and v_{obs} is the velocity of the observer. Hence, the frequency heard is

$$f' = \frac{v'}{\lambda} = \frac{v_{snd} + v_{obs}}{\lambda}.$$

Because $\lambda = v_{snd}/f$, then

$$f' = \frac{(v_{snd} + v_{obs})f}{v_{snd}},$$

or

$$f' = \left(1 + \frac{v_{obs}}{v_{snd}}\right)f. \qquad \left[\begin{array}{l} \text{observer moving toward} \\ \text{stationary source} \end{array}\right] \quad \textbf{(16–10a)}$$

If the observer is moving away from the source, the relative velocity is $v' = v_{snd} - v_{obs}$, so

$$f' = \left(1 - \frac{v_{obs}}{v_{snd}}\right)f. \qquad \left[\begin{array}{l} \text{observer moving away} \\ \text{from stationary source} \end{array}\right] \quad \textbf{(16–10b)}$$

EXAMPLE 16–14 **A moving siren.** The siren of a police car at rest emits at a predominant frequency of 1600 Hz. What frequency will you hear if you are at rest and the police car moves at 25.0 m/s (*a*) toward you, and (*b*) away from you?

APPROACH The observer is fixed, and the source moves, so we use Eqs. 16–9. The frequency you (the observer) hear is the emitted frequency f divided by the factor $(1 \pm v_{source}/v_{snd})$ where v_{source} is the speed of the police car. Use the minus sign when the car moves toward you (giving a higher frequency); use the plus sign when the car moves away from you (lower frequency).

SOLUTION (*a*) The car is moving toward you, so (Eq. 16–9a)

$$f' = \frac{f}{\left(1 - \dfrac{v_{source}}{v_{snd}}\right)} = \frac{1600\ \text{Hz}}{\left(1 - \dfrac{25.0\ \text{m/s}}{343\ \text{m/s}}\right)} = 1726\ \text{Hz} \approx 1730\ \text{Hz}.$$

(*b*) The car is moving away from you, so (Eq. 16–9b)

$$f' = \frac{f}{\left(1 + \dfrac{v_{source}}{v_{snd}}\right)} = \frac{1600\ \text{Hz}}{\left(1 + \dfrac{25.0\ \text{m/s}}{343\ \text{m/s}}\right)} = 1491\ \text{Hz} \approx 1490\ \text{Hz}.$$

EXERCISE G Suppose the police car of Example 16–14 is at rest and emits at 1600 Hz. What frequency would you hear if you were moving at 25.0 m/s (*a*) toward it, and (*b*) away from it?

When a sound wave is reflected from a moving obstacle, the frequency of the reflected wave will, because of the Doppler effect, be different from that of the incident wave. This is illustrated in the following Example.

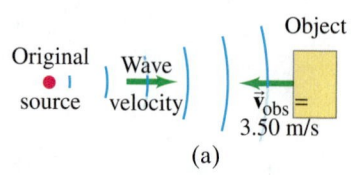

(a)

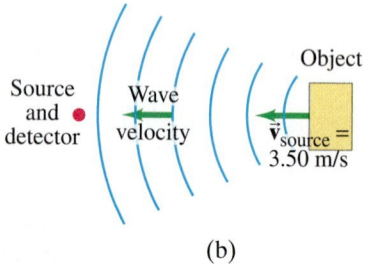

(b)

FIGURE 16–21 Example 16–15.

EXAMPLE 16–15 **Two Doppler shifts.** A 5000-Hz sound wave is emitted by a stationary source. This sound wave reflects from an object moving 3.50 m/s toward the source (Fig. 16–21). What is the frequency of the wave reflected by the moving object as detected by a detector at rest near the source?

APPROACH There are actually two Doppler shifts in this situation. First, the moving object acts like an observer moving toward the source with speed $v_{obs} = 3.50$ m/s (Fig. 16–21a) and so "detects" a sound wave of frequency (Eq. 16–10a) $f' = f[1 + (v_{obs}/v_{snd})]$. Second, reflection of the wave from the moving object is equivalent to the object reemitting the wave, acting effectively as a moving source with speed $v_{source} = 3.50$ m/s (Fig. 16–21b). The final frequency detected, f'', is given by $f'' = f'/[1 - v_{source}/v_{snd}]$, Eq. 16–9a.

SOLUTION The frequency f' that is "detected" by the moving object is (Eq. 16–10a):

$$f' = \left(1 + \frac{v_{obs}}{v_{snd}}\right)f = \left(1 + \frac{3.50 \text{ m/s}}{343 \text{ m/s}}\right)(5000 \text{ Hz}) = 5051 \text{ Hz}.$$

The moving object now "emits" (reflects) a sound of frequency (Eq. 16–9a)

$$f'' = \frac{f'}{\left(1 - \frac{v_{source}}{v_{snd}}\right)} = \frac{5051 \text{ Hz}}{\left(1 - \frac{3.50 \text{ m/s}}{343 \text{ m/s}}\right)} = 5103 \text{ Hz}.$$

Thus the frequency shifts by 103 Hz.

NOTE Bats use this technique to be aware of their surroundings. This is also the principle behind Doppler radar as speed-measuring devices for vehicles and other objects.

The incident wave and the reflected wave in Example 16–15, when mixed together (say, electronically), interfere with one another and beats are produced. The beat frequency is equal to the difference in the two frequencies, 103 Hz. This Doppler technique is used in a variety of medical applications, usually with ultrasonic waves in the megahertz frequency range. For example, ultrasonic waves reflected from red blood cells can be used to determine the velocity of blood flow. Similarly, the technique can be used to detect the movement of the chest of a young fetus and to monitor its heartbeat.

For convenience, we can write Eqs. 16–9 and 16–10 as a single equation that covers all cases of both source and observer in motion:

$$f' = f\left(\frac{v_{snd} \pm v_{obs}}{v_{snd} \mp v_{source}}\right). \qquad \begin{bmatrix} \text{source and} \\ \text{observer moving} \end{bmatrix} \quad \textbf{(16–11)}$$

PHYSICS APPLIED
Doppler blood-flow meter and other medical uses

PROBLEM SOLVING
Getting the signs right

To get the signs right, recall from your own experience that the frequency is higher when observer and source approach each other, and lower when they move apart. Thus the upper signs in numerator and denominator apply if source and/or observer move toward each other; the lower signs apply if they are moving apart.

EXERCISE H How fast would a source have to approach an observer for the observed frequency to be one octave above (twice) the produced frequency? (*a*) $\frac{1}{2}v_{snd}$, (*b*) v_{snd}, (*c*) $2v_{snd}$, (*d*) $4v_{snd}$.

Doppler Effect for Light

The Doppler effect occurs for other types of waves as well. Light and other types of electromagnetic waves (such as radar) exhibit the Doppler effect: although the formulas for the frequency shift are not identical to Eqs. 16–9 and 16–10, as discussed in Chapter 44, the effect is similar. One important application is for weather forecasting using radar. The time delay between the emission of radar pulses and their reception after being reflected off raindrops gives the position of precipitation. Measuring the Doppler shift in frequency (as in Example 16–15) tells how fast the storm is moving and in which direction.

Another important application is to astronomy, where the velocities of distant galaxies can be determined from the Doppler shift. Light from distant galaxies is shifted toward lower frequencies, indicating that the galaxies are moving away from us. This is called the **redshift** since red has the lowest frequency of visible light. The greater the frequency shift, the greater the velocity of recession. It is found that the farther the galaxies are from us, the faster they move away. This observation is the basis for the idea that the universe is expanding, and is one basis for the idea that the universe began as a great explosion, affectionately called the "Big Bang" (Chapter 44).

*16–8 Shock Waves and the Sonic Boom

An object such as an airplane traveling faster than the speed of sound is said to have a **supersonic speed**. Such a speed is often given as a **Mach**[†] number, which is defined as the ratio of the speed of the object to the speed of sound in the surrounding medium. For example, a plane traveling 600 m/s high in the atmosphere, where the speed of sound is only 300 m/s, has a speed of Mach 2.

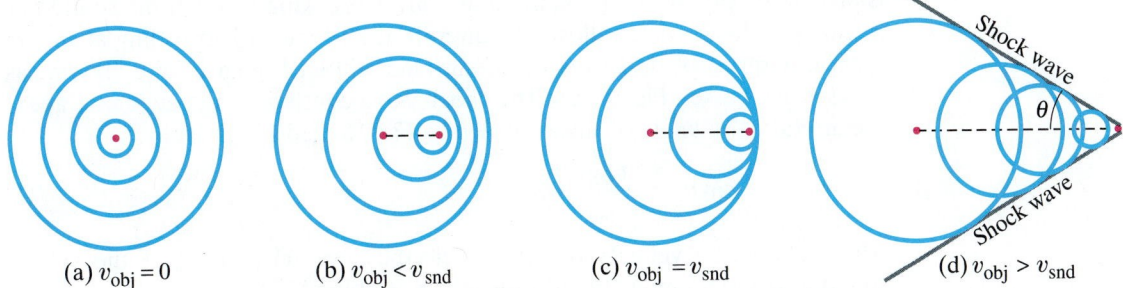

(a) $v_{obj} = 0$ (b) $v_{obj} < v_{snd}$ (c) $v_{obj} = v_{snd}$ (d) $v_{obj} > v_{snd}$

FIGURE 16–22 Sound waves emitted by an object (a) at rest or (b, c, and d) moving. (b) If the object's velocity is less than the velocity of sound, the Doppler effect occurs; (d) if its velocity is greater than the velocity of sound, a shock wave is produced.

When a source of sound moves at subsonic speeds (less than the speed of sound), the pitch of the sound is altered as we have seen (the Doppler effect); see also Fig. 16–22a and b. But if a source of sound moves faster than the speed of sound, a more dramatic effect known as a **shock wave** occurs. In this case, the source is actually "outrunning" the waves it produces. As shown in Fig. 16–22c, when the source is traveling *at* the speed of sound, the wave fronts it emits in the forward direction "pile up" directly in front of it. When the object moves faster, at a supersonic speed, the wave fronts pile up on one another along the sides, as shown in Fig. 16–22d. The different wave crests overlap one another and form a single very large crest which is the shock wave. Behind this very large crest there is usually a very large trough. A shock wave is essentially the result of constructive interference of a large number of wave fronts. A shock wave in air is analogous to the bow wave of a boat traveling faster than the speed of the water waves it produces, Fig. 16–23.

FIGURE 16–23 Bow waves produced by a boat.

[†]After the Austrian physicist Ernst Mach (1838–1916).

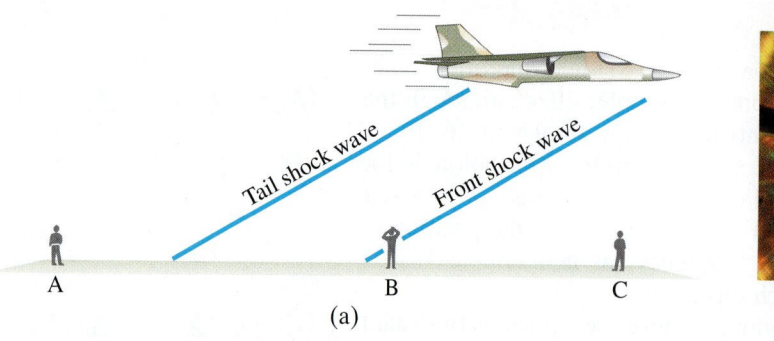
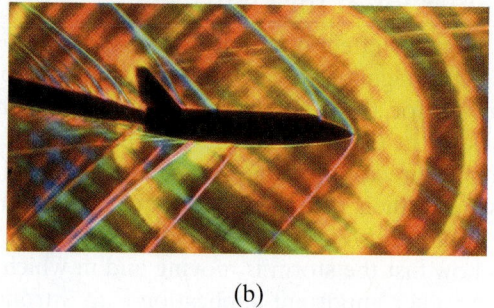

(a)   (b)

FIGURE 16–24 (a) The (double) sonic boom has already been heard by person A on the left. The front shock wave is just being heard by person B in the center. And it will shortly be heard by person C on the right. (b) Special photo of supersonic aircraft showing shock waves produced in the air. (Several closely spaced shock waves are produced by different parts of the aircraft.)

When an airplane travels at supersonic speeds, the noise it makes and its disturbance of the air form into a shock wave containing a tremendous amount of sound energy. When the shock wave passes a listener, it is heard as a loud *sonic boom*. A sonic boom lasts only a fraction of a second, but the energy it contains is often sufficient to break windows and cause other damage. Actually, a sonic boom is made up of two or more booms since major shock waves can form at the front and the rear of the aircraft, as well as at the wings, etc. (Fig. 16–24). Bow waves of a boat are also multiple, as can be seen in Fig. 16–23.

When an aircraft approaches the speed of sound, it encounters a barrier of sound waves in front of it (see Fig. 16–22c). To exceed the speed of sound, the aircraft needs extra thrust to pass through this "sound barrier." This is called "breaking the sound barrier." Once a supersonic speed is attained, this barrier no longer impedes the motion. It is sometimes erroneously thought that a sonic boom is produced only at the moment an aircraft is breaking through the sound barrier. Actually, a shock wave follows the aircraft at all times it is traveling at supersonic speeds. A series of observers on the ground will each hear a loud "boom" as the shock wave passes, Fig. 16–24. The shock wave consists of a cone whose apex is at the aircraft. The angle of this cone, θ (see Fig. 16–22d), is given by

$$\sin \theta = \frac{v_{\text{snd}}}{v_{\text{obj}}}, \tag{16–12}$$

where v_{obj} is the velocity of the object (the aircraft) and v_{snd} is the velocity of sound in the medium. (The proof is left as Problem 75.)

PHYSICS APPLIED
Sonic boom

*16–9 Applications: Sonar, Ultrasound, and Medical Imaging

*Sonar

PHYSICS APPLIED
Sonar: depth finding, Earth soundings

The reflection of sound is used in many applications to determine distance. The **sonar**[†] or pulse-echo technique is used to locate underwater objects. A transmitter sends out a sound pulse through the water, and a detector receives its reflection, or echo, a short time later. This time interval is carefully measured, and from it the distance to the reflecting object can be determined since the speed of sound in water is known. The depth of the sea and the location of reefs, sunken ships, submarines, or schools of fish can be determined in this way. The interior structure of the Earth is studied in a similar way by detecting reflections of waves traveling through the Earth whose source was a deliberate explosion (called "soundings"). An analysis of waves reflected from various structures and boundaries within the Earth reveals characteristic patterns that are also useful in the exploration for oil and minerals.

[†] Sonar stands for "*so*und *na*vigation *r*anging."

Sonar generally makes use of **ultrasonic** frequencies: that is, waves whose frequencies are above 20 kHz, beyond the range of human detection. For sonar, the frequencies are typically in the range 20 kHz to 100 kHz. One reason for using ultrasound waves, other than the fact that they are inaudible, is that for shorter wavelengths there is less diffraction (Section 15–11) so the beam spreads less and smaller objects can be detected.

*Ultrasound Medical Imaging

The diagnostic use of ultrasound in medicine, in the form of images (sometimes called *sonograms*), is an important and interesting application of physical principles. A **pulse-echo technique** is used, much like sonar, except that the frequencies used are in the range of 1 to 10 MHz $(1\ \text{MHz} = 10^6\ \text{Hz})$. A high-frequency sound pulse is directed into the body, and its reflections from boundaries or interfaces between organs and other structures and lesions in the body are then detected. Tumors and other abnormal growths, or pockets of fluid, can be distinguished; the action of heart valves and the development of a fetus can be examined; and information about various organs of the body, such as the brain, heart, liver, and kidneys, can be obtained. Although ultrasound does not replace X-rays, for certain kinds of diagnosis it is more helpful. Some kinds of tissue or fluid are not detected in X-ray photographs, but ultrasound waves are reflected from their boundaries. "Real-time" ultrasound images are like a movie of a section of the interior of the body.

The pulse-echo technique for medical imaging works as follows. A brief pulse of ultrasound is emitted by a transducer that transforms an electrical pulse into a sound-wave pulse. Part of the pulse is reflected as echoes at each interface in the body, and most of the pulse (usually) continues on, Fig. 16–25a. The detection of reflected pulses by the same transducer can then be displayed on the screen of a display terminal or monitor. The time elapsed from when the pulse is emitted to when each reflection (echo) is received is proportional to the distance to the reflecting surface. For example, if the distance from transducer to the vertebra is 25 cm, the pulse travels a round-trip distance of $2 \times 25\ \text{cm} = 0.50\ \text{m}$. The speed of sound in human tissue is about 1540 m/s (close to that of sea water), so the time taken is

$$t = \frac{d}{v} = \frac{(0.50\ \text{m})}{(1540\ \text{m/s})} = 320\ \mu s.$$

The *strength* of a reflected pulse depends mainly on the difference in density of the two materials on either side of the interface and can be displayed as a pulse or as a dot (Figs. 16–25b and c). Each echo dot (Fig. 16–25c) can be represented as a point whose position is given by the time delay and whose brightness depends on the strength of the echo.

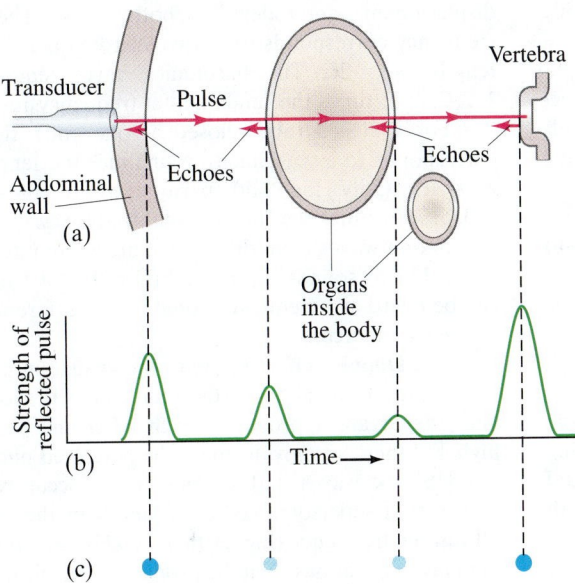

FIGURE 16–25 (a) Ultrasound pulse passes through the abdomen, reflecting from surfaces in its path. (b) Reflected pulses plotted as a function of time when received by transducer. The vertical dashed lines point out which reflected pulse goes with which surface. (c) Dot display for the same echoes: brightness of each dot is related to signal strength.

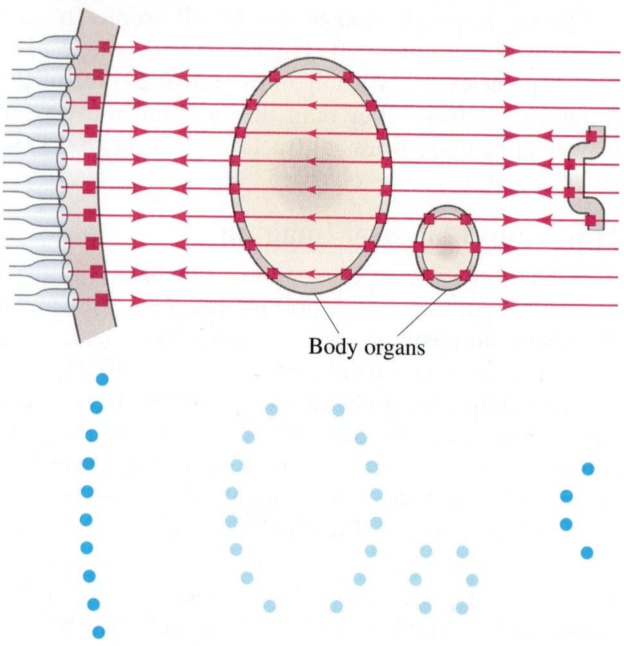

FIGURE 16–26 (a) Ten traces are made across the abdomen by moving the transducer, or by using an array of transducers. (b) The echoes are plotted as dots to produce the image. More closely spaced traces would give a more detailed image.

(a)

Body organs

(b)

FIGURE 16–27 Ultrasound image of a human fetus within the uterus.

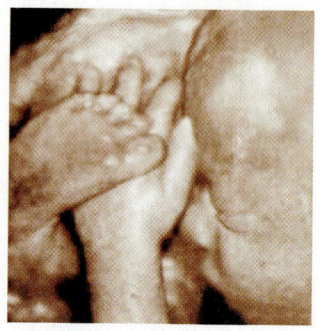

A two-dimensional image can then be formed out of these dots from a series of scans. The transducer is moved, or an array of transducers is used, each of which sends out a pulse at each position and receives echoes as shown in Fig. 16–26a. Each trace can be plotted, spaced appropriately one below the other, to form an image on a monitor screen as shown in Fig. 16–26b. Only 10 lines are shown in Fig. 16–26, so the image is crude. More lines give a more precise image.[†] An ultrasound image is shown in Fig. 16–27.

[†]*Radar* used for aircraft involves a similar pulse-echo technique except that it uses electromagnetic (EM) waves, which, like light, travel with a speed of 3×10^8 m/s.

Summary

Sound travels as a longitudinal wave in air and other materials. In air, the speed of sound increases with temperature; at 20°C, it is about 343 m/s.

The **pitch** of a sound is determined by the frequency; the higher the frequency, the higher the pitch.

The **audible range** of frequencies for humans is roughly 20 Hz to 20,000 Hz (1 Hz = 1 cycle per second).

The **loudness** or **intensity** of a sound is related to the amplitude squared of the wave. Because the human ear can detect sound intensities from 10^{-12} W/m² to over 1 W/m², sound levels are specified on a logarithmic scale. The **sound level** β, specified in decibels, is defined in terms of intensity I as

$$\beta(\text{in dB}) = 10\log\left(\frac{I}{I_0}\right), \qquad \textbf{(16–6)}$$

where the reference intensity I_0 is usually taken to be 10^{-12} W/m².

Musical instruments are simple sources of sound in which *standing waves* are produced.

The strings of a stringed instrument may vibrate as a whole with nodes only at the ends; the frequency at which this standing wave occurs is called the **fundamental**. The fundamental frequency corresponds to a wavelength equal to twice the length of the string, $\lambda_1 = 2\ell$. The string can also vibrate at higher frequencies, called **overtones** or **harmonics**, in which there are one or more additional nodes. The frequency of each harmonic is a whole-number multiple of the fundamental.

In wind instruments, standing waves are set up in the column of air within the tube.

The vibrating air in an **open tube** (open at both ends) has displacement antinodes at both ends. The fundamental frequency corresponds to a wavelength equal to twice the tube length: $\lambda_1 = 2\ell$. The harmonics have frequencies that are 1, 2, 3, 4, ... times the fundamental frequency, just as for strings.

For a **closed tube** (closed at one end), the fundamental corresponds to a wavelength four times the length of the tube: $\lambda_1 = 4\ell$. Only the odd harmonics are present, equal to 1, 3, 5, 7, ... times the fundamental frequency.

Sound waves from different sources can interfere with each other. If two sounds are at slightly different frequencies, **beats** can be heard at a frequency equal to the difference in frequency of the two sources.

The **Doppler effect** refers to the change in pitch of a sound due to the motion either of the source or of the listener. If source and listener are approaching each other, the perceived pitch is higher; if they are moving apart, the perceived pitch is lower.

[*Shock waves and a sonic boom occur when an object moves at a supersonic speed—faster than the speed of sound. Ultrasonic-frequency (higher than 20 kHz) sound waves are used in many applications, including sonar and medical imaging.]

Questions

1. What is the evidence that sound travels as a wave?

2. What is the evidence that sound is a form of energy?

3. Children sometimes play with a homemade "telephone" by attaching a string to the bottoms of two paper cups. When the string is stretched and a child speaks into one cup, the sound can be heard at the other cup (Fig. 16–28). Explain clearly how the sound wave travels from one cup to the other.

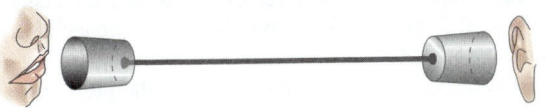

FIGURE 16–28   Question 3.

4. When a sound wave passes from air into water, do you expect the frequency or wavelength to change?

5. What evidence can you give that the speed of sound in air does not depend significantly on frequency?

6. The voice of a person who has inhaled helium sounds very high-pitched. Why?

7. What is the main reason the speed of sound in hydrogen is greater than the speed of sound in air?

8. Two tuning forks oscillate with the same amplitude, but one has twice the frequency. Which (if either) produces the more intense sound?

9. How will the air temperature in a room affect the pitch of organ pipes?

10. Explain how a tube might be used as a filter to reduce the amplitude of sounds in various frequency ranges. (An example is a car muffler.)

11. Why are the frets on a guitar (Fig. 16–29) spaced closer together as you move up the fingerboard toward the bridge?

FIGURE 16–29
Question 11.

12. A noisy truck approaches you from behind a building. Initially you hear it but cannot see it. When it emerges and you do see it, its sound is suddenly "brighter"—you hear more of the high-frequency noise. Explain. [*Hint*: See Section 15–11 on diffraction.]

13. Standing waves can be said to be due to "interference in space," whereas beats can be said to be due to "interference in time." Explain.

14. In Fig. 16–15, if the frequency of the speakers is lowered, would the points D and C (where destructive and constructive interference occur) move farther apart or closer together?

15. Traditional methods of protecting the hearing of people who work in areas with very high noise levels have consisted mainly of efforts to block or reduce noise levels. With a relatively new technology, headphones are worn that do not block the ambient noise. Instead, a device is used which detects the noise, inverts it electronically, then feeds it to the headphones *in addition to* the ambient noise. How could adding *more* noise reduce the sound levels reaching the ears?

16. Consider the two waves shown in Fig. 16–30. Each wave can be thought of as a superposition of two sound waves with slightly different frequencies, as in Fig. 16–17. In which of the waves, (a) or (b), are the two component frequencies farther apart? Explain.

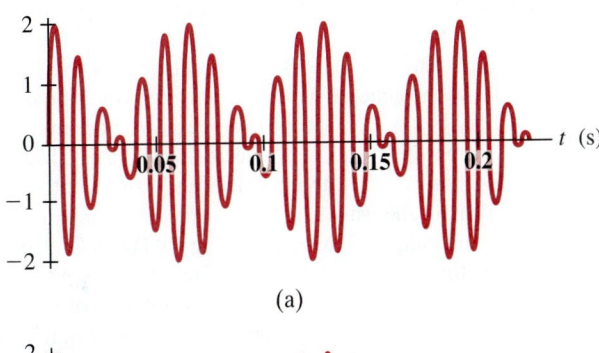

(a)

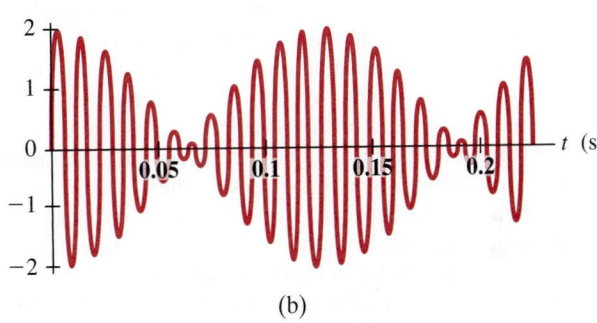

(b)

FIGURE 16–30   Question 16.

17. Is there a Doppler shift if the source and observer move in the same direction, with the same velocity? Explain.

18. If a wind is blowing, will this alter the frequency of the sound heard by a person at rest with respect to the source? Is the wavelength or velocity changed?

19. Figure 16–31 shows various positions of a child on a swing moving toward a person on the ground who is blowing a whistle. At which position, A through E, will the child hear the highest frequency for the sound of the whistle? Explain your reasoning.

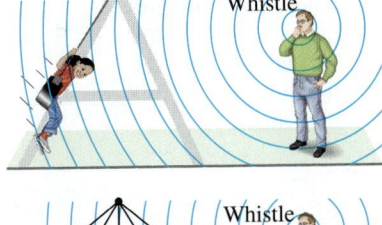

FIGURE 16–31
Question 19.

20. Approximately how many octaves are there in the human audible range?

21. At a race track, you can estimate the speed of cars just by listening to the difference in pitch of the engine noise between approaching and receding cars. Suppose the sound of a certain car drops by a full octave (frequency halved) as it goes by on the straightaway. How fast is it going?

Problems

[Unless stated otherwise, assume $T = 20°C$ and $v_{sound} = 343$ m/s in air.]

16–1 Characteristics of Sound

1. (I) A hiker determines the length of a lake by listening for the echo of her shout reflected by a cliff at the far end of the lake. She hears the echo 2.0 s after shouting. Estimate the length of the lake.

2. (I) A sailor strikes the side of his ship just below the waterline. He hears the echo of the sound reflected from the ocean floor directly below 2.5 s later. How deep is the ocean at this point? Assume the speed of sound in sea water is 1560 m/s (Table 16–1) and does not vary significantly with depth.

3. (I) (a) Calculate the wavelengths in air at 20°C for sounds in the maximum range of human hearing, 20 Hz to 20,000 Hz. (b) What is the wavelength of a 15-MHz ultrasonic wave?

4. (I) On a warm summer day (27°C), it takes 4.70 s for an echo to return from a cliff across a lake. On a winter day, it takes 5.20 s. What is the temperature on the winter day?

5. (II) A *motion sensor* can accurately measure the distance d to an object repeatedly via the sonar technique used in Example 16–2. A short ultrasonic pulse is emitted and reflects from any objects it encounters, creating echo pulses upon their arrival back at the sensor. The sensor measures the time interval t between the emission of the original pulse and the arrival of the first echo. (a) The smallest time interval t that can be measured with high precision is 1.0 ms. What is the smallest distance (at 20°C) that can be measured with the motion sensor? (b) If the motion sensor makes 15 distance measurements every second (that is, it emits 15 sound pulses per second at evenly spaced time intervals), the measurement of t must be completed within the time interval between the emissions of successive pulses. What is the largest distance (at 20°C) that can be measured with the motion sensor? (c) Assume that during a lab period the room's temperature increases from 20°C to 23°C. What percent error will this introduce into the motion sensor's distance measurements?

6. (II) An ocean fishing boat is drifting just above a school of tuna on a foggy day. Without warning, an engine backfire occurs on another boat 1.35 km away (Fig. 16–32). How much time elapses before the backfire is heard (a) by the fish, and (b) by the fishermen?

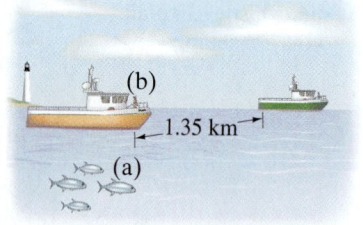

FIGURE 16–32 Problem 6.

7. (II) A stone is dropped from the top of a cliff. The splash it makes when striking the water below is heard 3.0 s later. How high is the cliff?

8. (II) A person, with his ear to the ground, sees a huge stone strike the concrete pavement. A moment later two sounds are heard from the impact: one travels in the air and the other in the concrete, and they are 0.75 s apart. How far away did the impact occur? See Table 16–1.

9. (II) Calculate the percent error made over one mile of distance by the "5-second rule" for estimating the distance from a lightning strike if the temperature is (a) 30°C, and (b) 10°C.

16–2 Mathematical Representation of Waves

10. (I) The pressure amplitude of a sound wave in air ($\rho = 1.29$ kg/m³) at 0°C is 3.0×10^{-3} Pa. What is the displacement amplitude if the frequency is (a) 150 Hz and (b) 15 kHz?

11. (I) What must be the pressure amplitude in a sound wave in air (0°C) if the air molecules undergo a maximum displacement equal to the diameter of an oxygen molecule, about 3.0×10^{-10} m? Assume a sound-wave frequency of (a) 55 Hz and (b) 5.5 kHz.

12. (II) Write an expression that describes the pressure variation as a function of x and t for the waves described in Problem 11.

13. (II) The pressure variation in a sound wave is given by

$$\Delta P = 0.0035 \sin(0.38\pi x - 1350\pi t),$$

where ΔP is in pascals, x in meters, and t in seconds. Determine (a) the wavelength, (b) the frequency, (c) the speed, and (d) the displacement amplitude of the wave. Assume the density of the medium to be $\rho = 2.3 \times 10^3$ kg/m³.

16–3 Intensity of Sound; Decibels

14. (I) What is the intensity of a sound at the pain level of 120 dB? Compare it to that of a whisper at 20 dB.

15. (I) What is the sound level of a sound whose intensity is 2.0×10^{-6} W/m²?

16. (I) What are the lowest and highest frequencies that an ear can detect when the sound level is 40 dB? (See Fig. 16–6.)

17. (II) Your auditory system can accommodate a huge range of sound levels. What is the ratio of highest to lowest intensity at (a) 100 Hz, (b) 5000 Hz? (See Fig. 16–6.)

18. (II) You are trying to decide between two new stereo amplifiers. One is rated at 100 W per channel and the other is rated at 150 W per channel. In terms of dB, how much louder will the more powerful amplifier be when both are producing sound at their maximum levels?

19. (II) At a painfully loud concert, a 120-dB sound wave travels away from a loudspeaker at 343 m/s. How much sound wave energy is contained in each 1.0-cm³ volume of air in the region near this loudspeaker?

20. (II) If two firecrackers produce a sound level of 95 dB when fired simultaneously at a certain place, what will be the sound level if only one is exploded?

21. (II) A person standing a certain distance from an airplane with four equally noisy jet engines is experiencing a sound level of 130 dB. What sound level would this person experience if the captain shut down all but one engine?

22. (II) A cassette player is said to have a signal-to-noise ratio of 62 dB, whereas for a CD player it is 98 dB. What is the ratio of intensities of the signal and the background noise for each device?

23. (II) (a) Estimate the power output of sound from a person speaking in normal conversation. Use Table 16–2. Assume the sound spreads roughly uniformly over a sphere centered on the mouth. (b) How many people would it take to produce a total sound output of 75 W of ordinary conversation? [*Hint*: Add intensities, not dBs.]

24. (II) A 50-dB sound wave strikes an eardrum whose area is 5.0×10^{-5} m². (a) How much energy is received by the eardrum per second? (b) At this rate, how long would it take your eardrum to receive a total energy of 1.0 J?

25. (II) Expensive amplifier A is rated at 250 W, while the more modest amplifier B is rated at 45 W. (a) Estimate the sound level in decibels you would expect at a point 3.5 m from a loudspeaker connected in turn to each amp. (b) Will the expensive amp sound twice as loud as the cheaper one?

26. (II) At a rock concert, a dB meter registered 130 dB when placed 2.2 m in front of a loudspeaker on the stage. (a) What was the power output of the speaker, assuming uniform spherical spreading of the sound and neglecting absorption in the air? (b) How far away would the sound level be a somewhat reasonable 85 dB?

27. (II) A fireworks shell explodes 100 m above the ground, creating a colorful display of sparks. How much greater is the sound level of the explosion for a person standing at a point directly below the explosion than for a person a horizontal distance of 200 m away (Fig. 16–33)?

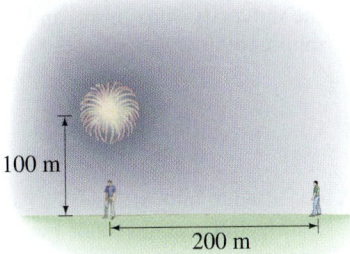

FIGURE 16–33
Problem 27.

28. (II) If the amplitude of a sound wave is made 2.5 times greater, (a) by what factor will the intensity increase? (b) By how many dB will the sound level increase?

29. (II) Two sound waves have equal displacement amplitudes, but one has 2.6 times the frequency of the other. (a) Which has the greater pressure amplitude and by what factor is it greater? (b) What is the ratio of their intensities?

30. (II) What would be the sound level (in dB) of a sound wave in air that corresponds to a displacement amplitude of vibrating air molecules of 0.13 mm at 380 Hz?

31. (II) (a) Calculate the maximum displacement of air molecules when a 330-Hz sound wave passes whose intensity is at the threshold of pain (120 dB). (b) What is the pressure amplitude in this wave?

32. (II) A jet plane emits 5.0×10^5 J of sound energy per second. (a) What is the sound level 25 m away? Air absorbs sound at a rate of about 7.0 dB/km; calculate what the sound level will be (b) 1.00 km and (c) 7.50 km away from this jet plane, taking into account air absorption.

16–4 Sources of Sound: Strings and Air Columns

33. (I) What would you estimate for the length of a bass clarinet, assuming that it is modeled as a closed tube and that the lowest note that it can play is a D♭ whose frequency is 69.3 Hz?

34. (I) The A string on a violin has a fundamental frequency of 440 Hz. The length of the vibrating portion is 32 cm, and it has a mass of 0.35 g. Under what tension must the string be placed?

35. (I) An organ pipe is 124 cm long. Determine the fundamental and first three audible overtones if the pipe is (a) closed at one end, and (b) open at both ends.

36. (I) (a) What resonant frequency would you expect from blowing across the top of an empty soda bottle that is 21 cm deep, if you assumed it was a closed tube? (b) How would that change if it was one-third full of soda?

37. (I) If you were to build a pipe organ with open-tube pipes spanning the range of human hearing (20 Hz to 20 kHz), what would be the range of the lengths of pipes required?

38. (II) Estimate the frequency of the "sound of the ocean" when you put your ear very near a 20-cm-diameter seashell (Fig. 16–34).

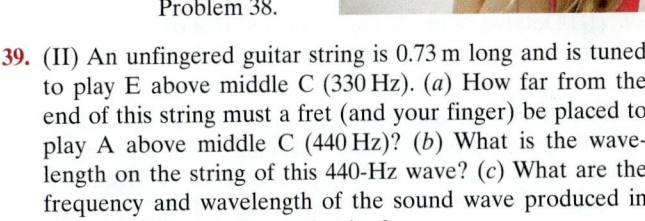

FIGURE 16–34
Problem 38.

39. (II) An unfingered guitar string is 0.73 m long and is tuned to play E above middle C (330 Hz). (a) How far from the end of this string must a fret (and your finger) be placed to play A above middle C (440 Hz)? (b) What is the wavelength on the string of this 440-Hz wave? (c) What are the frequency and wavelength of the sound wave produced in air at 25°C by this fingered string?

40. (II) (a) Determine the length of an open organ pipe that emits middle C (262 Hz) when the temperature is 15°C. (b) What are the wavelength and frequency of the fundamental standing wave in the tube? (c) What are λ and f in the traveling sound wave produced in the outside air?

41. (II) An organ is in tune at 22.0°C. By what percent will the frequency be off at 5.0°C?

42. (II) How far from the mouthpiece of the flute in Example 16–11 should the hole be that must be uncovered to play F above middle C at 349 Hz?

43. (II) A bugle is simply a tube of fixed length that behaves as if it is open at both ends. A bugler, by adjusting his lips correctly and blowing with proper air pressure, can cause a harmonic (usually other than the fundamental) of the air column within the tube to sound loudly. Standard military tunes like *Taps* and *Reveille* require only four musical notes: G4 (392 Hz), C5 (523 Hz), E5 (659 Hz), and G5 (784 Hz). (a) For a certain length ℓ, a bugle will have a sequence of four consecutive harmonics whose frequencies very nearly equal those associated with the notes G4, C5, E5, and G5. Determine this ℓ. (b) Which harmonic is each of the (approximate) notes G4, C5, E5, and G5 for the bugle?

44. (II) A particular organ pipe can resonate at 264 Hz, 440 Hz, and 616 Hz, but not at any other frequencies in between. (a) Show why this is an open or a closed pipe. (b) What is the fundamental frequency of this pipe?

45. (II) When a player's finger presses a guitar string down onto a fret, the length of the vibrating portion of the string is shortened, thereby increasing the string's fundamental frequency (see Fig. 16–35). The string's tension and mass per unit length remain unchanged. If the unfingered length of the string is ℓ = 65.0 cm, determine the positions x of the first six frets, if each fret raises the pitch of the fundamental by one musical note in comparison to the neighboring fret. On the equally tempered chromatic scale, the ratio of frequencies of neighboring notes is $2^{1/12}$.

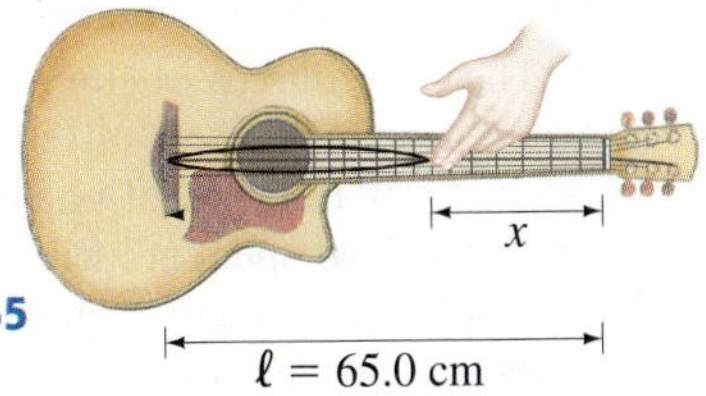

FIGURE 16–35
Problem 45.

46. (II) A uniform narrow tube 1.80 m long is open at both ends. It resonates at two successive harmonics of frequencies 275 Hz and 330 Hz. What is (a) the fundamental frequency, and (b) the speed of sound in the gas in the tube?

47. (II) A pipe in air at 23.0°C is to be designed to produce two successive harmonics at 240 Hz and 280 Hz. How long must the pipe be, and is it open or closed?

48. (II) How many overtones are present within the audible range for a 2.48-m-long organ pipe at 20°C (a) if it is open, and (b) if it is closed?

49. (II) Determine the fundamental and first overtone frequencies for an 8.0-m-long hallway with all doors closed. Model the hallway as a tube closed at both ends.

50. (II) In a *quartz oscillator*, used as a stable clock in electronic devices, a transverse (shear) standing sound wave is excited across the thickness d of a quartz disk and its frequency f is detected electronically. The parallel faces of the disk are unsupported and so behave as "free ends" when the sound wave reflects from them (see Fig. 16–36). If the oscillator is designed to operate with the first harmonic, determine the required disk thickness if $f = 12.0$ MHz. The density and shear modulus of quartz are $\rho = 2650$ kg/m^3 and $G = 2.95 \times 10^{10}$ N/m^2.

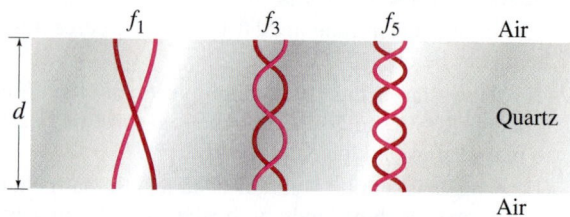

FIGURE 16–36 Problem 50.

51. (III) The human ear canal is approximately 2.5 cm long. It is open to the outside and is closed at the other end by the eardrum. Estimate the frequencies (in the audible range) of the standing waves in the ear canal. What is the relationship of your answer to the information in the graph of Fig. 16–6?

*16–5 Quality of Sound, Superposition

*52. (II) Approximately what are the intensities of the first two overtones of a violin compared to the fundamental? How many decibels softer than the fundamental are the first and second overtones? (See Fig. 16–14.)

16–6 Interference; Beats

53. (I) A piano tuner hears one beat every 2.0 s when trying to adjust two strings, one of which is sounding 370 Hz. How far off in frequency is the other string?

54. (I) What is the beat frequency if middle C (262 Hz) and C$^\sharp$ (277 Hz) are played together? What if each is played two octaves lower (each frequency reduced by a factor of 4)?

55. (II) A guitar string produces 4 beats/s when sounded with a 350-Hz tuning fork and 9 beats/s when sounded with a 355-Hz tuning fork. What is the vibrational frequency of the string? Explain your reasoning.

56. (II) The two sources of sound in Fig. 16–15 face each other and emit sounds of equal amplitude and equal frequency (294 Hz) but 180° out of phase. For what minimum separation of the two speakers will there be some point at which (a) complete constructive interference occurs and (b) complete destructive interference occurs. (Assume $T = 20$°C.)

57. (II) How many beats will be heard if two identical flutes, each 0.66 m long, try to play middle C (262 Hz), but one is at 5.0°C and the other at 28°C?

58. (II) Two loudspeakers are placed 3.00 m apart, as shown in Fig. 16–37. They emit 494-Hz sounds, in phase. A microphone is placed 3.20 m distant from a point midway between the two speakers, where an intensity maximum is recorded. (a) How far must the microphone be moved to the right to find the first intensity minimum? (b) Suppose the speakers are reconnected so that the 494-Hz sounds they emit are exactly out of phase. At what positions are the intensity maximum and minimum now?

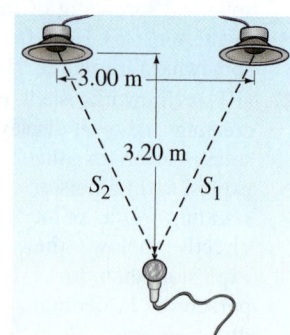

FIGURE 16–37
Problem 58.

59. (II) Two piano strings are supposed to be vibrating at 220 Hz, but a piano tuner hears three beats every 2.0 s when they are played together. (a) If one is vibrating at 220.0 Hz, what must be the frequency of the other (is there only one answer)? (b) By how much (in percent) must the tension be increased or decreased to bring them in tune?

60. (II) A source emits sound of wavelengths 2.64 m and 2.72 m in air. (a) How many beats per second will be heard? (Assume $T = 20$°C.) (b) How far apart in space are the regions of maximum intensity?

16–7 Doppler Effect

61. (I) The predominant frequency of a certain fire truck's siren is 1350 Hz when at rest. What frequency do you detect if you move with a speed of 30.0 m/s (a) toward the fire truck, and (b) away from it?

62. (I) A bat at rest sends out ultrasonic sound waves at 50.0 kHz and receives them returned from an object moving directly away from it at 30.0 m/s. What is the received sound frequency?

63. (II) (a) Compare the shift in frequency if a 2300-Hz source is moving toward you at 18 m/s, versus you moving toward it at 18 m/s. Are the two frequencies exactly the same? Are they close? (b) Repeat the calculation for 160 m/s and then again (c) for 320 m/s. What can you conclude about the asymmetry of the Doppler formulas? (d) Show that at low speeds (relative to the speed of sound), the two formulas—source approaching and detector approaching—yield the same result.

64. (II) Two automobiles are equipped with the same single-frequency horn. When one is at rest and the other is moving toward the first at 15 m/s, the driver at rest hears a beat frequency of 4.5 Hz. What is the frequency the horns emit? Assume $T = 20$°C.

65. (II) A police car sounding a siren with a frequency of 1280 Hz is traveling at 120.0 km/h. (a) What frequencies does an observer standing next to the road hear as the car approaches and as it recedes? (b) What frequencies are heard in a car traveling at 90.0 km/h in the opposite direction before and after passing the police car? (c) The police car passes a car traveling in the same direction at 80.0 km/h. What two frequencies are heard in this car?

66. (II) A bat flies toward a wall at a speed of 7.0 m/s. As it flies, the bat emits an ultrasonic sound wave with frequency 30.0 kHz. What frequency does the bat hear in the reflected wave?

67. (II) In one of the original Doppler experiments, a tuba was played on a moving flat train car at a frequency of 75 Hz, and a second identical tuba played the same tone while at rest in the railway station. What beat frequency was heard in the station if the train car approached the station at a speed of 12.0 m/s?

68. (II) If a speaker mounted on an automobile broadcasts a song, with what speed (km/h) does the automobile have to move toward a stationary listener so that the listener hears the song with each musical note shifted up by one note in comparison to the song heard by the automobile's driver? On the equally tempered chromatic scale, the ratio of frequencies of neighboring notes is $2^{1/12}$.

69. (II) A wave on the surface of the ocean with wavelength 44 m is traveling east at a speed of 18 m/s relative to the ocean floor. If, on this stretch of ocean surface, a powerboat is moving at 15 m/s (relative to the ocean floor), how often does the boat encounter a wave crest, if the boat is traveling (a) west, and (b) east?

70. (III) A factory whistle emits sound of frequency 720 Hz. When the wind velocity is 15.0 m/s from the north, what frequency will observers hear who are located, at rest, (a) due north, (b) due south, (c) due east, and (d) due west, of the whistle? What frequency is heard by a cyclist heading (e) north or (f) west, toward the whistle at 12.0 m/s? Assume $T = 20°C$.

71. (III) The Doppler effect using ultrasonic waves of frequency 2.25×10^6 Hz is used to monitor the heartbeat of a fetus. A (maximum) beat frequency of 260 Hz is observed. Assuming that the speed of sound in tissue is 1.54×10^3 m/s, calculate the maximum velocity of the surface of the beating heart.

*16–8 Shock Waves; Sonic Boom

*72. (II) An airplane travels at Mach 2.0 where the speed of sound is 310 m/s. (a) What is the angle the shock wave makes with the direction of the airplane's motion? (b) If the plane is flying at a height of 6500 m, how long after it is directly overhead will a person on the ground hear the shock wave?

*73. (II) A space probe enters the thin atmosphere of a planet where the speed of sound is only about 45 m/s. (a) What is the probe's Mach number if its initial speed is 15,000 km/h? (b) What is the angle of the shock wave relative to the direction of motion?

*74. (II) A meteorite traveling 8800 m/s strikes the ocean. Determine the shock wave angle it produces (a) in the air just before entering the ocean, and (b) in the water just after entering. Assume $T = 20°C$.

*75. (II) Show that the angle θ a sonic boom makes with the path of a supersonic object is given by Eq. 16–12.

*76. (II) You look directly overhead and see a plane exactly 1.25 km above the ground flying faster than the speed of sound. By the time you hear the sonic boom, the plane has traveled a horizontal distance of 2.0 km. See Fig. 16–38. Determine (a) the angle of the shock cone, θ, and (b) the speed of the plane (the Mach number). Assume the speed of sound is 330 m/s.

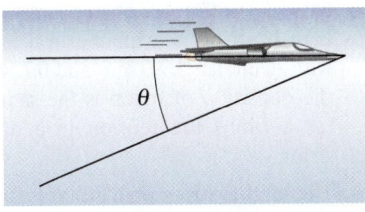

FIGURE 16–38
Problem 76.

*77. (II) A supersonic jet traveling at Mach 2.2 at an altitude of 9500 m passes directly over an observer on the ground. Where will the plane be relative to the observer when the latter hears the sonic boom? (See Fig. 16–39.)

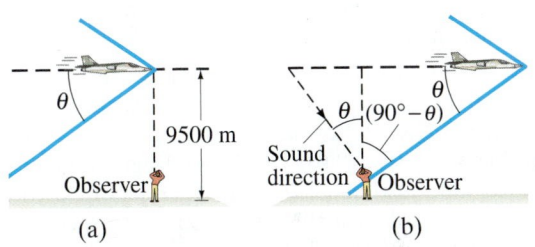

FIGURE 16–39 Problem 77.

General Problems

78. A fish finder uses a sonar device that sends 20,000-Hz sound pulses downward from the bottom of the boat, and then detects echoes. If the maximum depth for which it is designed to work is 75 m, what is the minimum time between pulses (in fresh water)?

79. A science museum has a display called a sewer pipe symphony. It consists of many plastic pipes of various lengths, which are open on both ends. (a) If the pipes have lengths of 3.0 m, 2.5 m, 2.0 m, 1.5 m and 1.0 m, what frequencies will be heard by a visitor's ear placed near the ends of the pipes? (b) Why does this display work better on a noisy day than on a quiet day?

80. A single mosquito 5.0 m from a person makes a sound close to the threshold of human hearing (0 dB). What will be the sound level of 100 such mosquitoes?

81. What is the resultant sound level when an 82-dB sound and an 89-dB sound are heard simultaneously?

82. The sound level 9.00 m from a loudspeaker, placed in the open, is 115 dB. What is the acoustic power output (W) of the speaker, assuming it radiates equally in all directions?

83. A stereo amplifier is rated at 175 W output at 1000 Hz. The power output drops by 12 dB at 15 kHz. What is the power output in watts at 15 kHz?

84. Workers around jet aircraft typically wear protective devices over their ears. Assume that the sound level of a jet airplane engine, at a distance of 30 m, is 130 dB, and that the average human ear has an effective radius of 2.0 cm. What would be the power intercepted by an unprotected ear at a distance of 30 m from a jet airplane engine?

85. In audio and communications systems, the *gain*, β, in decibels is defined as

$$\beta = 10 \log\left(\frac{P_{out}}{P_{in}}\right),$$

where P_{in} is the power input to the system and P_{out} is the power output. A particular stereo amplifier puts out 125 W of power for an input of 1.0 mW. What is its gain in dB?

86. For large concerts, loudspeakers are sometimes used to amplify a singer's sound. The human brain interprets sounds that arrive within 50 ms of the original sound as if they came from the same source. Thus if the sound from a loudspeaker reaches a listener first, it would sound as if the loudspeaker is the source of the sound. Conversely, if the singer is heard first and the loudspeaker adds to the sound within 50 ms, the sound would seem to come from the singer, who would now seem to be singing louder. The second situation is desired. Because the signal to the loudspeaker travels at the speed of light $(3 \times 10^8 \text{ m/s})$, which is much faster than the speed of sound, a delay is added to the signal sent to the loudspeaker. How much delay must be added if the loudspeaker is 3.0 m behind the singer and we want its sound to arrive 30 ms after the singer's?

87. Manufacturers typically offer a particular guitar string in a choice of diameters so that players can tune their instruments with a preferred string tension. For example, a nylon high-E string is available in a low- and high-tension model with diameter 0.699 mm and 0.724 mm, respectively. Assuming the density ρ of nylon is the same for each model, compare (as a ratio) the tension in a tuned high- and low-tension string.

88. The high-E string on a guitar is fixed at both ends with length $\ell = 65.0$ cm and fundamental frequency $f_1 = 329.6$ Hz. On an acoustic guitar, this string typically has a diameter of 0.33 mm and is commonly made of brass (7760 kg/m^3), while on an electric guitar it has a diameter of 0.25 mm and is made of nickel-coated steel (7990 kg/m^3). Compare (as a ratio) the high-E string tension on an acoustic versus an electric guitar.

89. The A string of a violin is 32 cm long between fixed points with a fundamental frequency of 440 Hz and a mass per unit length of 7.2×10^{-4} kg/m. (a) What are the wave speed and tension in the string? (b) What is the length of the tube of a simple wind instrument (say, an organ pipe) closed at one end whose fundamental is also 440 Hz if the speed of sound is 343 m/s in air? (c) What is the frequency of the first overtone of each instrument?

90. A tuning fork is set into vibration above a vertical open tube filled with water (Fig. 16–40). The water level is allowed to drop slowly. As it does so, the air in the tube above the water level is heard to resonate with the tuning fork when the distance from the tube opening to the water level is 0.125 m and again at 0.395 m. What is the frequency of the tuning fork?

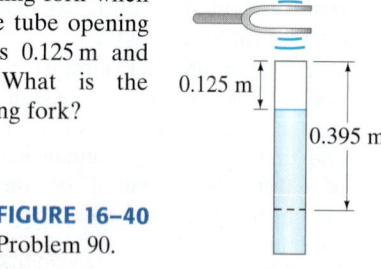

FIGURE 16–40
Problem 90.

0.125 m

0.395 m

91. Two identical tubes, each closed at one end, have a fundamental frequency of 349 Hz at 25.0°C. The air temperature is increased to 30.0°C in one tube. If the two pipes are sounded together now, what beat frequency results?

92. Each string on a violin is tuned to a frequency $1\frac{1}{2}$ times that of its neighbor. The four equal-length strings are to be placed under the same tension; what must be the mass per unit length of each string relative to that of the lowest string?

93. The diameter D of a tube does affect the node at the open end of a tube. The end correction can be roughly approximated as adding $D/3$ to the effective length of the tube. For a closed tube of length 0.60 m and diameter 3.0 cm, what are the first four harmonics, taking the end correction into consideration?

94. A person hears a pure tone in the 500 to 1000-Hz range coming from two sources. The sound is loudest at points equidistant from the two sources. To determine exactly what the frequency is, the person moves about and finds that the sound level is minimal at a point 0.28 m farther from one source than the other. What is the frequency of the sound?

95. The frequency of a steam train whistle as it approaches you is 552 Hz. After it passes you, its frequency is measured as 486 Hz. How fast was the train moving (assume constant velocity)?

96. Two trains emit 516-Hz whistles. One train is stationary. The conductor on the stationary train hears a 3.5-Hz beat frequency when the other train approaches. What is the speed of the moving train?

97. Two loudspeakers are at opposite ends of a railroad car as it moves past a stationary observer at 10.0 m/s, as shown in Fig. 16–41. If the speakers have identical sound frequencies of 348 Hz, what is the beat frequency heard by the observer when (a) he listens from the position A, in front of the car, (b) he is between the speakers, at B, and (c) he hears the speakers after they have passed him, at C?

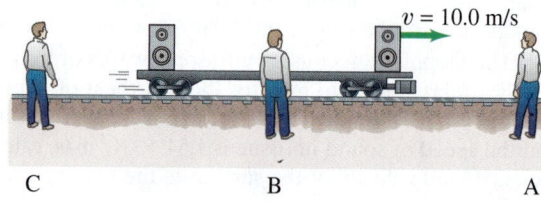

$v = 10.0$ m/s

C   B   A

FIGURE 16–41 Problem 97.

98. Two open organ pipes, sounding together, produce a beat frequency of 8.0 Hz. The shorter one is 2.40 m long. How long is the other?

99. A bat flies toward a moth at speed 7.5 m/s while the moth is flying toward the bat at speed 5.0 m/s. The bat emits a sound wave of 51.35 kHz. What is the frequency of the wave detected by the bat after that wave reflects off the moth?

100. If the velocity of blood flow in the aorta is normally about 0.32 m/s, what beat frequency would you expect if 3.80-MHz ultrasound waves were directed along the flow and reflected from the red blood cells? Assume that the waves travel with a speed of 1.54×10^3 m/s.

101. A bat emits a series of high-frequency sound pulses as it approaches a moth. The pulses are approximately 70.0 ms apart, and each is about 3.0 ms long. How far away can the moth be detected by the bat so that the echo from one pulse returns before the next pulse is emitted?

102. (a) Use the binomial expansion to show that Eqs. 16–9a and 16–10a become essentially the same for small relative velocity between source and observer. (b) What percent error would result if Eq. 16–10a were used instead of Eq. 16–9a for a relative velocity of 18.0 m/s?

103. Two loudspeakers face each other at opposite ends of a long corridor. They are connected to the same source which produces a pure tone of 282 Hz. A person walks from one speaker toward the other at a speed of 1.4 m/s. What "beat" frequency does the person hear?

104. A *Doppler flow meter* is used to measure the speed of blood flow. Transmitting and receiving elements are placed on the skin, as shown in Fig. 16–42. Typical sound-wave frequencies of about 5.0 MHz are used, which have a reasonable chance of being reflected from red blood cells. By measuring the frequency of the reflected waves, which are Doppler-shifted because the red blood cells are moving, the speed of the blood flow can be deduced. "Normal" blood flow speed is about 0.1 m/s. Suppose that an artery is partly constricted, so that the speed of the blood flow is increased, and the flow meter measures a Doppler shift of 780 Hz. What is the speed of blood flow in the constricted region? The effective angle between the sound waves (both transmitted and reflected) and the direction of blood flow is 45°. Assume the velocity of sound in tissue is 1540 m/s.

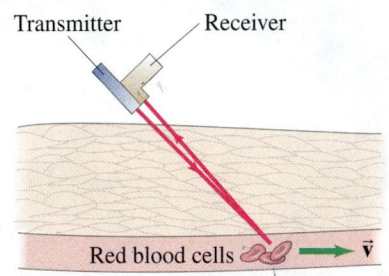

FIGURE 16–42
Problem 104.

Red blood cells $\vec{v}$

105. The wake of a speedboat is 15° in a lake where the speed of the water wave is 2.2 km/h. What is the speed of the boat?

106. A source of sound waves (wavelength λ) is a distance ℓ from a detector. Sound reaches the detector directly, and also by reflecting off an obstacle, as shown in Fig. 16–43. The obstacle is equidistant from source and detector. When the obstacle is a distance d to the right of the line of sight between source and detector, as shown, the two waves arrive in phase. How much farther to the right must the obstacle be moved if the two waves are to be out of phase by $\frac{1}{2}$ wavelength, so destructive interference occurs? (Assume $\lambda \ll \ell, d$.)

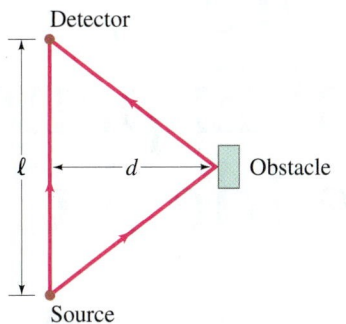

FIGURE 16–43
Problem 106.

Detector

ℓ — d — Obstacle

Source

107. A dramatic demonstration, called "singing rods," involves a long, slender aluminum rod held in the hand near the rod's midpoint. The rod is stroked with the other hand. With a little practice, the rod can be made to "sing," or emit a clear, loud, ringing sound. For a 75-cm-long rod, (*a*) what is the fundamental frequency of the sound? (*b*) What is its wavelength in the rod, and (*c*) what is the wavelength of the sound in air at 20°C?

108. Assuming that the maximum displacement of the air molecules in a sound wave is about the same as that of the speaker cone that produces the sound (Fig. 16–44), estimate by how much a loudspeaker cone moves for a fairly loud (105 dB) sound of (*a*) 8.0 kHz, and (*b*) 35 Hz.

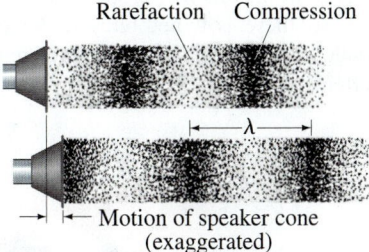

Rarefaction   Compression

λ

FIGURE 16–44
Problem 108.

Motion of speaker cone
(exaggerated)

***Numerical/Computer**

* **109.** (III) The manner in which a string is plucked determines the mixture of harmonic amplitudes in the resulting wave. Consider a string exactly $\frac{1}{2}$-m long that is fixed at both its ends located at $x = 0.0$ and $x = \frac{1}{2}$ m. The first five harmonics of this string have wavelengths $\lambda_1 = 1.0$ m, $\lambda_2 = \frac{1}{2}$ m, $\lambda_3 = \frac{1}{3}$ m, $\lambda_4 = \frac{1}{4}$ m, and $\lambda_5 = \frac{1}{5}$ m. According to Fourier's theorem, any shape of this string can be formed by a sum of its harmonics, with each harmonic having its own unique amplitude A. We limit the sum to the first five harmonics in the expression

$$D(x) = A_1 \sin\left(\frac{2\pi}{\lambda_1} x\right) + A_2 \sin\left(\frac{2\pi}{\lambda_2} x\right)$$
$$+ A_3 \sin\left(\frac{2\pi}{\lambda_3} x\right) + A_4 \sin\left(\frac{2\pi}{\lambda_4} x\right) + A_5 \sin\left(\frac{2\pi}{\lambda_5} x\right),$$

and D is the displacement of the string at a time $t = 0$. Imagine plucking this string at its midpoint (Fig. 16–45a) or at a point two-thirds from the left end (Fig. 16–45b). Using a graphing calculator or computer program, show that the above expression can fairly accurately represent the shape in: (*a*) Fig. 16–45a, if $A_1 = 1.00$, $A_2 = 0.00$, $A_3 = -0.11$, $A_4 = 0.00$, and $A_5 = 0.040$; and in (*b*) Fig. 16–45b, if $A_1 = 0.87$, $A_2 = -0.22$, $A_3 = 0.00$, $A_4 = 0.054$, and $A_5 = -0.035$.

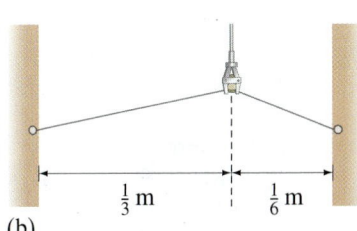

(a) $\frac{1}{4}$ m $\frac{1}{4}$ m

$\frac{1}{3}$ m $\frac{1}{6}$ m

FIGURE 16–45
Problem 109.   (b)

Answers to Exercises

A: 1 km for every 3 s before the thunder is heard.

B: 4 times as intense.

C: (*b*).

D: The less-massive one.

E: 257 Hz.

F: (*b*).

G: (*a*) 1717 Hz, (*b*) 1483 Hz.

H: (*a*).

Heating the air inside a "hot-air" balloon raises the air's temperature, causing it to expand, and forces air out the opening at the bottom. The reduced amount of air inside means its density is lower than the outside air, so there is a net buoyant force upward on the balloon. In this Chapter we study temperature and its effects on matter: thermal expansion and the gas laws.

17

Temperature, Thermal Expansion, and the Ideal Gas Law

CONTENTS

17–1 Atomic Theory of Matter

17–2 Temperature and Thermometers

17–3 Thermal Equilibrium and the Zeroth Law of Thermodynamics

17–4 Thermal Expansion

*17–5 Thermal Stresses

17–6 The Gas Laws and Absolute Temperature

17–7 The Ideal Gas Law

17–8 Problem Solving with the Ideal Gas Law

17–9 Ideal Gas Law in Terms of Molecules: Avogadro's Number

*17–10 Ideal Gas Temperature Scale—a Standard

CHAPTER-OPENING QUESTION—Guess now!

A hot-air balloon, open at one end (see photos above), rises when the air inside is heated by a flame. For the following properties, is the air inside the balloon higher, lower, or the same as for the air outside the balloon?

(i) Temperature,
(ii) Pressure,
(iii) Density.

In the next four Chapters, Chapters 17 through 20, we study temperature, heat and thermodynamics, and the kinetic theory of gases.

We will often consider a particular **system**, by which we mean a particular object or set of objects; everything else in the universe is called the "environment." We can describe the **state** (or condition) of a particular system—such as a gas in a container—from either a microscopic or macroscopic point of view. A **microscopic** description would involve details of the motion of all the atoms or molecules making up the system, which could be very complicated. A **macroscopic** description is given in terms of quantities that are detectable directly by our senses and instruments, such as volume, mass, pressure, and temperature.

The description of processes in terms of macroscopic quantities is the field of **thermodynamics**. Quantities that can be used to describe the state of a system are called **state variables**. To describe the state of a pure gas in a container, for example, requires only three state variables, which are typically the volume, the pressure, and the temperature. More complex systems require more than three state variables to describe them.

The emphasis in this Chapter is on the concept of temperature. We begin, however, with a brief discussion of the theory that matter is made up of atoms and that these atoms are in continual random motion. This theory is called *kinetic theory* ("kinetic," you may recall, is Greek for "moving"), and we discuss it in more detail in Chapter 18.

17–1 Atomic Theory of Matter

The idea that matter is made up of atoms dates back to the ancient Greeks. According to the Greek philosopher Democritus, if a pure substance—say, a piece of iron—were cut into smaller and smaller bits, eventually a smallest piece of that substance would be obtained which could not be divided further. This smallest piece was called an **atom**, which in Greek means "indivisible."[†]

Today the atomic theory is universally accepted. The experimental evidence in its favor, however, came mainly in the eighteenth, nineteenth, and twentieth centuries, and much of it was obtained from the analysis of chemical reactions.

We will often speak of the relative masses of individual atoms and molecules—what we call the **atomic mass** or **molecular mass**, respectively.[‡] These are based on arbitrarily assigning the abundant carbon atom, ^{12}C, the atomic mass of exactly 12.0000 **unified atomic mass units** (u). In terms of kilograms,

$$1\,u = 1.6605 \times 10^{-27}\,kg.$$

The atomic mass of hydrogen is then 1.0078 u, and the values for other atoms are as listed in the Periodic Table inside the back cover of this book, and also in Appendix F. The molecular mass of a compound is the sum of atomic masses of the atoms making up the molecules of that compound.[§]

An important piece of evidence for the atomic theory is called **Brownian motion**, named after the biologist Robert Brown, who is credited with its discovery in 1827. While he was observing tiny pollen grains suspended in water under his microscope, Brown noticed that the tiny grains moved about in tortuous paths (Fig. 17–1), even though the water appeared to be perfectly still. The atomic theory easily explains Brownian motion if the further reasonable assumption is made that the atoms of any substance are continually in motion. Then Brown's tiny pollen grains are jostled about by the vigorous barrage of rapidly moving molecules of water.

In 1905, Albert Einstein examined Brownian motion from a theoretical point of view and was able to calculate from the experimental data the approximate size and mass of atoms and molecules. His calculations showed that the diameter of a typical atom is about 10^{-10} m.

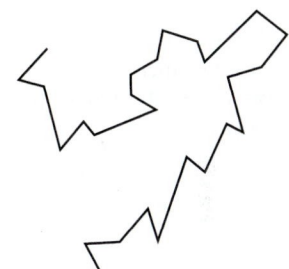

FIGURE 17–1 Path of a tiny particle (pollen grain, for example) suspended in water. The straight lines connect observed positions of the particle at equal time intervals.

[†]Today we do not consider the atom as indivisible, but rather as consisting of a nucleus (containing protons and neutrons) and electrons.

[‡]The terms *atomic weight* and *molecular weight* are sometimes used for these quantities, but properly speaking we are comparing masses.

[§]An *element* is a substance, such as gold, iron, or copper, that cannot be broken down into simpler substances by chemical means. *Compounds* are substances made up of elements, and can be broken down into them; examples are carbon dioxide and water. The smallest piece of an element is an atom; the smallest piece of a compound is a molecule. Molecules are made up of atoms; a molecule of water, for example, is made up of two atoms of hydrogen and one of oxygen; its chemical formula is H_2O.

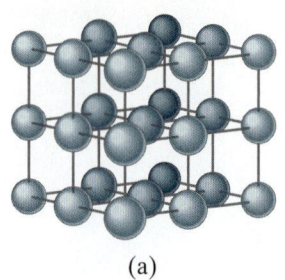

(a)

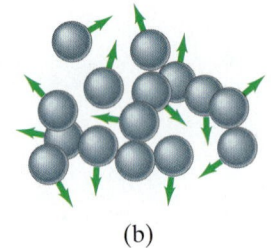

(b)

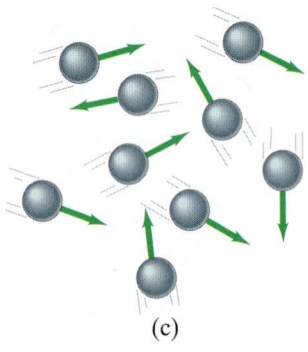

(c)

FIGURE 17–2
Atomic arrangements in
(a) a crystalline solid, (b) a liquid,
and (c) a gas.

At the start of Chapter 13, we distinguished the three common phases (or states) of matter—solid, liquid, gas—based on **macroscopic**, or "large-scale," properties. Now let us see how these three phases of matter differ, from the atomic or **microscopic** point of view. Clearly, atoms and molecules must exert attractive forces on each other. For how else could a brick or a block of aluminum hold together in one piece? The attractive forces between molecules are of an electrical nature (more on this in later Chapters). When molecules come too close together, the force between them must become repulsive (electric repulsion between their outer electrons), for how else could matter take up space? Thus molecules maintain a minimum distance from each other. In a solid material, the attractive forces are strong enough that the atoms or molecules move only slightly (oscillate) about relatively fixed positions, often in an array known as a crystal lattice, as shown in Fig. 17–2a. In a liquid, the atoms or molecules are moving more rapidly, or the forces between them are weaker, so that they are sufficiently free to pass around one another, as in Fig. 17–2b. In a gas, the forces are so weak, or the speeds so high, that the molecules do not even stay close together. They move rapidly every which way, Fig. 17–2c, filling any container and occasionally colliding with one another. On average, the speeds are sufficiently high in a gas that when two molecules collide, the force of attraction is not strong enough to keep them close together and they fly off in new directions.

EXAMPLE 17–1 ESTIMATE **Distance between atoms.** The density of copper is $8.9 \times 10^3 \, \text{kg/m}^3$, and each copper atom has a mass of 63 u. Estimate the average distance between the centers of neighboring copper atoms.

APPROACH We consider a cube of copper 1 m on a side. From the given density ρ we can calculate the mass m of a cube of volume $V = 1 \, \text{m}^3$ $(m = \rho V)$. We divide this by the mass of one atom (63 u) to obtain the number of atoms in $1 \, \text{m}^3$. We assume the atoms are in a uniform array, and we let N be the number of atoms in a 1-m length; then $(N)(N)(N) = N^3$ equals the total number of atoms in $1 \, \text{m}^3$.

SOLUTION The mass of 1 copper atom is $63 \, \text{u} = 63 \times 1.66 \times 10^{-27} \, \text{kg} = 1.05 \times 10^{-25} \, \text{kg}$. This means that in a cube of copper 1 m on a side $(\text{volume} = 1 \, \text{m}^3)$, there are

$$\frac{8.9 \times 10^3 \, \text{kg/m}^3}{1.05 \times 10^{-25} \, \text{kg/atom}} = 8.5 \times 10^{28} \, \text{atoms/m}^3.$$

The volume of a cube of side ℓ is $V = \ell^3$, so on one edge of the 1-m-long cube there are $(8.5 \times 10^{28})^{\frac{1}{3}}$ atoms $= 4.4 \times 10^9$ atoms. Hence the distance between neighboring atoms is

$$\frac{1 \, \text{m}}{4.4 \times 10^9 \, \text{atoms}} = 2.3 \times 10^{-10} \, \text{m}.$$

NOTE Watch out for units. Even though "atoms" is not a unit, it is helpful to include it to make sure you calculate correctly.

FIGURE 17–3 Expansion joint on a bridge.

17–2 Temperature and Thermometers

In everyday life, **temperature** is a measure of how hot or cold something is. A hot oven is said to have a high temperature, whereas the ice of a frozen lake is said to have a low temperature.

Many properties of matter change with temperature. For example, most materials expand when their temperature is increased.[†] An iron beam is longer when hot than when cold. Concrete roads and sidewalks expand and contract slightly according to temperature, which is why compressible spacers or expansion joints (Fig. 17–3) are placed at regular intervals. The electrical resistance of matter changes with temperature (Chapter 25). So too does the color radiated by objects, at least at high temperatures: you may have noticed that the heating element of an electric stove glows with a red color when hot.

[†]Most materials expand when their temperature is raised, but not all. Water, for example, in the range 0°C to 4°C contracts with an increase in temperature (see Section 17–4).

At higher temperatures, solids such as iron glow orange or even white. The white light from an ordinary incandescent lightbulb comes from an extremely hot tungsten wire. The surface temperatures of the Sun and other stars can be measured by the predominant color (more precisely, wavelengths) of light they emit.

Instruments designed to measure temperature are called **thermometers**. There are many kinds of thermometers, but their operation always depends on some property of matter that changes with temperature. Many common thermometers rely on the expansion of a material with an increase in temperature. The first idea for a thermometer, by Galileo, made use of the expansion of a gas. Common thermometers today consist of a hollow glass tube filled with mercury or with alcohol colored with a red dye, as were the earliest usable thermometers (Fig. 17–4).

Inside a common liquid-in-glass thermometer, the liquid expands more than the glass when the temperature is increased, so the liquid level rises in the tube (Fig. 17–5a). Although metals also expand with temperature, the change in length of a metal rod, say, is generally too small to measure accurately for ordinary changes in temperature. However, a useful thermometer can be made by bonding together two dissimilar metals whose rates of expansion are different (Fig. 17–5b). When the temperature is increased, the different amounts of expansion cause the bimetallic strip to bend. Often the bimetallic strip is in the form of a coil, one end of which is fixed while the other is attached to a pointer, Fig. 17–6. This kind of thermometer is used as ordinary air thermometers, oven thermometers, automatic off switches in electric coffeepots, and in room thermostats for determining when the heater or air conditioner should go on or off. Very precise thermometers make use of electrical properties (Chapter 25), such as resistance thermometers, thermocouples, and thermistors, often with a digital readout.

Temperature Scales

In order to measure temperature quantitatively, some sort of numerical scale must be defined. The most common scale today is the **Celsius** scale, sometimes called the **centigrade** scale. In the United States, the **Fahrenheit** scale is also common. The most important scale in scientific work is the absolute, or Kelvin, scale, and it will be discussed later in this Chapter.

One way to define a temperature scale is to assign arbitrary values to two readily reproducible temperatures. For both the Celsius and Fahrenheit scales these two fixed points are chosen to be the freezing point and the boiling point[†] of water, both taken at standard atmospheric pressure. On the Celsius scale, the freezing point of water is chosen to be 0°C ("zero degrees Celsius") and the boiling point 100°C. On the Fahrenheit scale, the freezing point is defined as 32°F and the boiling point 212°F. A practical thermometer is calibrated by placing it in carefully prepared environments at each of the two temperatures and marking the position of the liquid or pointer. For a Celsius scale, the distance between the two marks is divided into one hundred equal intervals representing each degree between 0°C and 100°C (hence the name "centigrade scale" meaning "hundred steps"). For a Fahrenheit scale, the two points are labeled 32°F and 212°F and the distance between them is divided into 180 equal intervals. For temperatures below the freezing point of water and above the boiling point of water, the scales may be extended using the same equally spaced intervals. However, thermometers can be used only over a limited temperature range because of their own limitations—for example, the liquid mercury in a mercury-in-glass thermometer solidifies at some point, below which the thermometer will be useless. It is also rendered useless above temperatures where the fluid, such as alcohol, vaporizes. For very low or very high temperatures, specialized thermometers are required, some of which we will mention later.

[†]The freezing point of a substance is defined as that temperature at which the solid and liquid phases coexist in equilibrium—that is, without any net liquid changing into the solid or vice versa. Experimentally, this is found to occur at only one definite temperature, for a given pressure. Similarly, the boiling point is defined as that temperature at which the liquid and gas coexist in equilibrium. Since these points vary with pressure, the pressure must be specified (usually it is 1 atm).

FIGURE 17–4 Thermometers built by the Accademia del Cimento (1657–1667) in Florence, Italy, are among the earliest known. These sensitive and exquisite instruments contained alcohol, sometimes colored, like many thermometers today.

FIGURE 17–5 (a) Mercury- or alcohol-in-glass thermometer; (b) bimetallic strip.

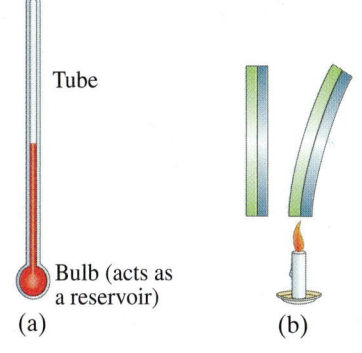

Tube

Bulb (acts as a reservoir)

(a)   (b)

FIGURE 17–6 Photograph of a thermometer using a coiled bimetallic strip.

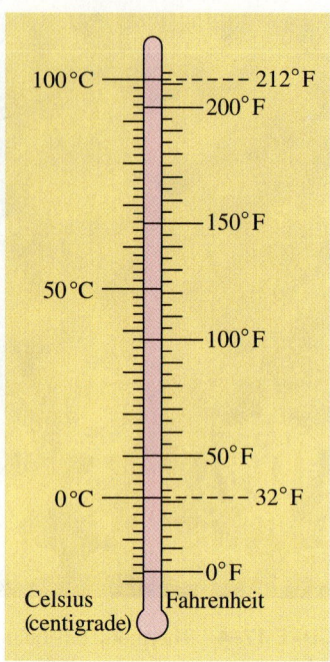

FIGURE 17–7 Celsius and Fahrenheit scales compared.

⚠️ **C A U T I O N**

Convert temperature by remembering 0°C = 32°F and a change of 5 C° = 9 F°

Every temperature on the Celsius scale corresponds to a particular temperature on the Fahrenheit scale, Fig. 17–7. It is easy to convert from one to the other if you remember that 0°C corresponds to 32°F and that a range of 100° on the Celsius scale corresponds to a range of 180° on the Fahrenheit scale. Thus, one Fahrenheit degree (1 F°) corresponds to $100/180 = \frac{5}{9}$ of a Celsius degree (1 C°). That is, $1\,F° = \frac{5}{9}C°$. (Notice that when we refer to a specific temperature, we say "degrees Celsius," as in 20°C; but when we refer to a *change* in temperature or a temperature interval, we say "Celsius degrees," as in "2 C°.") The conversion between the two temperature scales can be written

$$T(°C) = \tfrac{5}{9}\big[T(°F) - 32\big]$$

or

$$T(°F) = \tfrac{9}{5}T(°C) + 32.$$

Rather than memorizing these relations (it would be easy to confuse them), it is usually easier simply to remember that 0°C = 32°F and that a change of 5 C° = a change of 9 F°.

EXAMPLE 17–2 Taking your temperature. Normal body temperature is 98.6°F. What is this on the Celsius scale?

APPROACH We recall that 0°C = 32°F and 5 C° = 9 F°.

SOLUTION First we relate the given temperature to the freezing point of water (0°C). That is, 98.6°F is $98.6 - 32.0 = 66.6\,F°$ above the freezing point of water. Since each F° is equal to $\frac{5}{9}C°$, this corresponds to $66.6 \times \frac{5}{9} = 37.0$ Celsius degrees above the freezing point. The freezing point is 0°C, so the temperature is 37.0°C.

EXERCISE A Determine the temperature at which both scales give the same numerical reading $(T_C = T_F)$.

Different materials do not expand in quite the same way over a wide temperature range. Consequently, if we calibrate different kinds of thermometers exactly as described above, they will not usually agree precisely. Because of how we calibrate them, they will agree at 0°C and at 100°C. But because of different expansion properties, they may not agree precisely at intermediate temperatures (remember we arbitrarily divided the thermometer scale into 100 equal divisions between 0°C and 100°C). Thus a carefully calibrated mercury-in-glass thermometer might register 52.0°C, whereas a carefully calibrated thermometer of another type might read 52.6°C. Discrepancies below 0°C and above 100°C can also be significant.

Because of such discrepancies, some standard kind of thermometer must be chosen so that all temperatures can be precisely defined. The chosen standard for this purpose is the **constant-volume gas thermometer**. As shown in the simplified diagram of Fig. 17–8, this thermometer consists of a bulb filled with a dilute gas connected by a thin tube to a mercury manometer (Section 13–6). The volume of the gas is kept constant by raising or lowering the right-hand tube of the manometer so that the mercury in the left-hand tube coincides with the reference mark. An increase in temperature causes a proportional increase in pressure in the bulb. Thus the tube must be lifted higher to keep the gas volume constant. The height of the mercury in the right-hand column is then a measure of the temperature. This thermometer gives the same results for all gases in the limit of reducing the gas pressure in the bulb toward zero. The resulting scale serves as a basis for the standard temperature scale (Section 17–10).

FIGURE 17–8 Constant-volume gas thermometer.

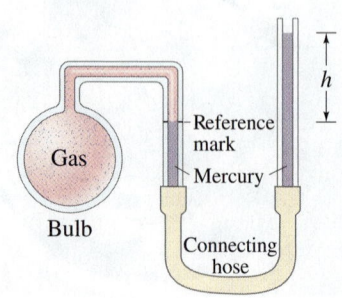

17–3 Thermal Equilibrium and the Zeroth Law of Thermodynamics

We are all familiar with the fact that if two objects at different temperatures are placed in thermal contact (meaning thermal energy can transfer from one to the other), the two objects will eventually reach the same temperature. They are then said to be in **thermal equilibrium**. For example, you leave a fever thermometer in your mouth until it comes into thermal equilibrium with that environment, and then you read it. Two objects are defined to be in thermal equilibrium if, when placed in thermal contact, no net energy flows from one to the other, and their temperatures don't change. Experiments indicate that

> **if two systems are in thermal equilibrium with a third system, then they are in thermal equilibrium with each other.**

This postulate is called the **zeroth law of thermodynamics**. It has this unusual name because it was not until after the great first and second laws of thermodynamics (Chapters 19 and 20) were worked out that scientists realized that this apparently obvious postulate needed to be stated first.

Temperature is a property of a system that determines whether the system will be in thermal equilibrium with other systems. When two systems are in thermal equilibrium, their temperatures are, by definition, equal, and no net thermal energy will be exchanged between them. This is consistent with our everyday notion of temperature, since when a hot object and a cold one are put into contact, they eventually come to the same temperature. Thus the importance of the zeroth law is that it allows a useful definition of temperature.

17–4 Thermal Expansion

Most substances expand when heated and contract when cooled. However, the amount of expansion or contraction varies, depending on the material.

Linear Expansion

Experiments indicate that the change in length $\Delta \ell$ of almost all solids is, to a good approximation, directly proportional to the change in temperature ΔT, as long as ΔT is not too large. The change in length is also proportional to the original length of the object, ℓ_0. That is, for the same temperature increase, a 4-m-long iron rod will increase in length twice as much as a 2-m-long iron rod. We can write this proportionality as an equation:

$$\Delta \ell = \alpha \ell_0 \Delta T, \tag{17–1a}$$

where α, the proportionality constant, is called the *coefficient of linear expansion* for the particular material and has units of $(C°)^{-1}$. We write $\ell = \ell_0 + \Delta \ell$, Fig. 17–9, and rewrite this equation as $\ell = \ell_0 + \Delta \ell = \ell_0 + \alpha \ell_0 \Delta T$, or

$$\ell = \ell_0(1 + \alpha \Delta T), \tag{17–1b}$$

where ℓ_0 is the length initially, at temperature T_0, and ℓ is the length after heating or cooling to a temperature T. If the temperature change $\Delta T = T - T_0$ is negative, then $\Delta \ell = \ell - \ell_0$ is also negative; the length shortens as the temperature decreases.

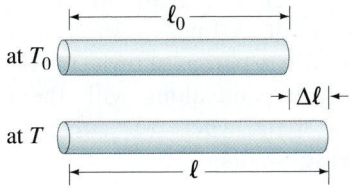

at T_0

ℓ_0

$\Delta \ell$

at T

ℓ

FIGURE 17–9 A thin rod of length ℓ_0 at temperature T_0 is heated to a new uniform temperature T and acquires length ℓ, where $\ell = \ell_0 + \Delta \ell$.

TABLE 17–1 Coefficients of Expansion, near 20°C

Material	Coefficient of Linear Expansion, α (C°)$^{-1}$	Coefficient of Volume Expansion, β (C°)$^{-1}$
Solids		
Aluminum	25×10^{-6}	75×10^{-6}
Brass	19×10^{-6}	56×10^{-6}
Copper	17×10^{-6}	50×10^{-6}
Gold	14×10^{-6}	42×10^{-6}
Iron or steel	12×10^{-6}	35×10^{-6}
Lead	29×10^{-6}	87×10^{-6}
Glass (Pyrex®)	3×10^{-6}	9×10^{-6}
Glass (ordinary)	9×10^{-6}	27×10^{-6}
Quartz	0.4×10^{-6}	1×10^{-6}
Concrete and brick	$\approx 12 \times 10^{-6}$	$\approx 36 \times 10^{-6}$
Marble	$1.4–3.5 \times 10^{-6}$	$4–10 \times 10^{-6}$
Liquids		
Gasoline		950×10^{-6}
Mercury		180×10^{-6}
Ethyl alcohol		1100×10^{-6}
Glycerin		500×10^{-6}
Water		210×10^{-6}
Gases		
Air (and most other gases at atmospheric pressure)		3400×10^{-6}

The values of α for various materials at 20°C are listed in Table 17–1. Actually, α does vary slightly with temperature (which is why thermometers made of different materials do not agree precisely). However, if the temperature range is not too great, the variation can usually be ignored.

PHYSICS APPLIED

Expansion in structures

EXAMPLE 17–3 **Bridge expansion.** The steel bed of a suspension bridge is 200 m long at 20°C. If the extremes of temperature to which it might be exposed are $-30°C$ to $+40°C$, how much will it contract and expand?

APPROACH We assume the bridge bed will expand and contract linearly with temperature, as given by Eq. 17–1a.

SOLUTION From Table 17–1, we find that $\alpha = 12 \times 10^{-6}(C°)^{-1}$ for steel. The increase in length when it is at 40°C will be

$$\Delta\ell = \alpha\ell_0\Delta T = (12 \times 10^{-6}/C°)(200\,\text{m})(40°C - 20°C) = 4.8 \times 10^{-2}\,\text{m},$$

or 4.8 cm. When the temperature decreases to $-30°C$, $\Delta T = -50\,C°$. Then

$$\Delta\ell = (12 \times 10^{-6}/C°)(200\,\text{m})(-50\,C°) = -12.0 \times 10^{-2}\,\text{m},$$

or a decrease in length of 12 cm. The total range the expansion joints must accommodate is $12\,\text{cm} + 4.8\,\text{cm} \approx 17\,\text{cm}$ (Fig. 17–3).

CONCEPTUAL EXAMPLE 17–4 **Do holes expand or contract?** If you heat a thin, circular ring (Fig. 17–10a) in the oven, does the ring's hole get larger or smaller?

RESPONSE You might guess that the metal expands into the hole, making the hole smaller. But it is not so. Imagine the ring is solid, like a coin (Fig. 17–10b). Draw a circle on it with a pen as shown. When the metal expands, the material inside the circle will expand along with the rest of the metal; so the circle expands. Cutting the metal where the circle is makes clear to us that the hole in Fig. 17–10a increases in diameter.

FIGURE 17–10 Example 17–4.

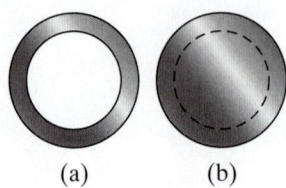

(a)   (b)

EXAMPLE 17–5 **Ring on a rod.** An iron ring is to fit snugly on a cylindrical iron rod. At 20°C, the diameter of the rod is 6.445 cm and the inside diameter of the ring is 6.420 cm. To slip over the rod, the ring must be slightly larger than the rod diameter by about 0.008 cm. To what temperature must the ring be brought if its hole is to be large enough so it will slip over the rod?

APPROACH The hole in the ring must be increased from a diameter of 6.420 cm to 6.445 cm + 0.008 cm = 6.453 cm. The ring must be heated since the hole diameter will increase linearly with temperature (Example 17–4).

SOLUTION We solve for ΔT in Eq. 17–1a and find

$$\Delta T = \frac{\Delta \ell}{\alpha \ell_0} = \frac{6.453 \text{ cm} - 6.420 \text{ cm}}{(12 \times 10^{-6}/\text{C}°)(6.420 \text{ cm})} = 430 \text{ C}°.$$

So it must be raised at least to $T = (20°C + 430 \text{ C}°) = 450°C$.

NOTE In doing Problems, do not forget the last step, adding in the initial temperature (20°C here).

CONCEPTUAL EXAMPLE 17–6 **Opening a tight jar lid.** When the lid of a glass jar is tight, holding the lid under hot water for a short time will often make it easier to open (Fig. 17–11). Why?

RESPONSE The lid may be struck by the hot water more directly than the glass and so expand sooner. But even if not, metals generally expand more than glass for the same temperature change (α is greater—see Table 17–1).

NOTE If you put a hard-boiled egg in cold water immediately after cooking it, it is easier to peel: the different thermal expansions of the shell and egg cause the egg to separate from the shell.

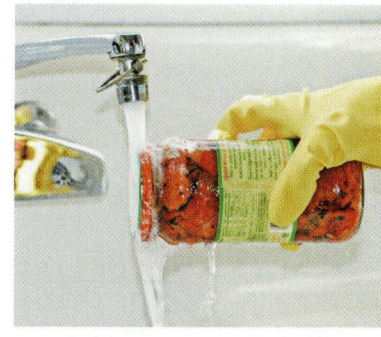

FIGURE 17–11 Example 17–6.

Volume Expansion

The change in *volume* of a material which undergoes a temperature change is given by a relation similar to Eq. 17–1a, namely,

$$\Delta V = \beta V_0 \Delta T, \qquad (17-2)$$

where ΔT is the change in temperature, V_0 is the original volume, ΔV is the change in volume, and β is the *coefficient of volume expansion*. The units of β are $(\text{C}°)^{-1}$.

Values of β for various materials are given in Table 17–1. Notice that for solids, β is normally equal to approximately 3α. To see why, consider a rectangular solid of length ℓ_0, width W_0, and height H_0. When its temperature is changed by ΔT, its volume changes from $V_0 = \ell_0 W_0 H_0$ to

$$V = \ell_0(1 + \alpha \Delta T)W_0(1 + \alpha \Delta T)H_0(1 + \alpha \Delta T),$$

using Eq. 17–1b and assuming α is the same in all directions. Thus,

$$\Delta V = V - V_0 = V_0(1 + \alpha \Delta T)^3 - V_0 = V_0[3\alpha \Delta T + 3(\alpha \Delta T)^2 + (\alpha \Delta T)^3].$$

If the amount of expansion is much smaller than the original size of the object, then $\alpha \Delta T \ll 1$ and we can ignore all but the first term and obtain

$$\Delta V \approx (3\alpha)V_0 \Delta T.$$

This is Eq. 17–2 with $\beta \approx 3\alpha$. For solids that are not isotropic (having the same properties in all directions), however, the relation $\beta \approx 3\alpha$ is not valid. Note also that linear expansion has no meaning for liquids and gases since they do not have fixed shapes.

EXERCISE B A long thin bar of aluminum at 0°C is 1.0 m long and has a volume of $1.0000 \times 10^{-3} \text{ m}^3$. When heated to 100°C, the length of the bar becomes 1.0025 m. What is the approximate volume of the bar at 100°C? (a) $1.0000 \times 10^{-3} \text{ m}^3$; (b) $1.0025 \times 10^{-3} \text{ m}^3$; (c) $1.0050 \times 10^{-3} \text{ m}^3$; (d) $1.0075 \times 10^{-3} \text{ m}^3$; (e) $2.5625 \times 10^{-3} \text{ m}^3$.

Equations 17–1 and 17–2 are accurate only if $\Delta \ell$ (or ΔV) is small compared to ℓ_0 (or V_0). This is of particular concern for liquids and even more so for gases because of the large values of β. Furthermore, β itself varies substantially with temperature for gases. Therefore, a more convenient way of dealing with gases is needed, and will be discussed starting in Section 17–6.

EXAMPLE 17–7 **Gas tank in the Sun.** The 70-liter (L) steel gas tank of a car is filled to the top with gasoline at 20°C. The car sits in the Sun and the tank reaches a temperature of 40°C (104°F). How much gasoline do you expect to overflow from the tank?

APPROACH Both the gasoline and the tank expand as the temperature increases, and we assume they do so linearly as described by Eq. 17–2. The volume of overflowing gasoline equals the volume increase of the gasoline minus the increase in volume of the tank.

SOLUTION The gasoline expands by

$$\Delta V = \beta V_0 \Delta T = (950 \times 10^{-6}/C°)(70\,L)(40°C - 20°C) = 1.3\,L.$$

The tank also expands. We can think of it as a steel shell that undergoes volume expansion $(\beta \approx 3\alpha = 36 \times 10^{-6}/C°)$. If the tank were solid, the surface layer (the shell) would expand just the same. Thus the tank increases in volume by

$$\Delta V = (36 \times 10^{-6}/C°)(70\,L)(40°C - 20°C) = 0.050\,L,$$

so the tank expansion has little effect. More than a liter of gas could spill out.

NOTE Want to save a few pennies? You pay for gas by volume, so fill your gas tank when it is cool and the gas is denser—more molecules for the same price. But don't fill the tank quite all the way.

Anomalous Behavior of Water Below 4°C

Most substances expand more or less uniformly with an increase in temperature, as long as no phase change occurs. Water, however, does not follow the usual pattern. If water at 0°C is heated, it actually *decreases* in volume until it reaches 4°C. Above 4°C water behaves normally and expands in volume as the temperature is increased, Fig. 17–12. Water thus has its greatest density at 4°C. This anomalous behavior of water is of great importance for the survival of aquatic life during cold winters. When the water in a lake or river is above 4°C and begins to cool by contact with cold air, the water at the surface sinks because of its greater density. It is replaced by warmer water from below. This mixing continues until the temperature reaches 4°C. As the surface water cools further, it remains on the surface because it is less dense than the 4°C water below. Water then freezes first at the surface, and the ice remains on the surface since ice (specific gravity = 0.917) is less dense than water. The water at the bottom remains liquid unless it is so cold that the whole body of water freezes. If water were like most substances, becoming more dense as it cools, the water at the bottom of a lake would be frozen first. Lakes would freeze solid more easily since circulation would bring the warmer water to the surface to be efficiently cooled. The complete freezing of a lake would cause severe damage to its plant and animal life. Because of the unusual behavior of water below 4°C, it is rare for any large body of water to freeze completely, and this is helped by the layer of ice on the surface which acts as an insulator to reduce the flow of heat out of the water into the cold air above. Without this peculiar but wonderful property of water, life on this planet as we know it might not have been possible.

Not only does water expand as it cools from 4°C to 0°C, it expands even more as it freezes to ice. This is why ice cubes float in water and pipes break when water inside them freezes.

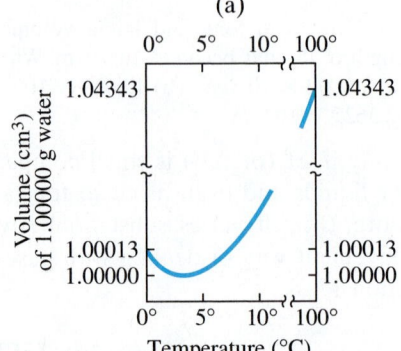 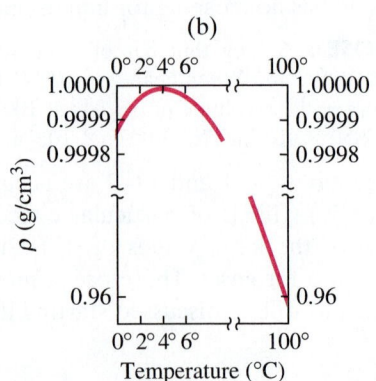

FIGURE 17–12 Behavior of water as a function of temperature near 4°C. (a) Volume of 1.00000 g of water, as a function of temperature. (b) Density vs. temperature. [Note the break in each axis.]

*17–5 Thermal Stresses

In many situations, such as in buildings and roads, the ends of a beam or slab of material are rigidly fixed, which greatly limits expansion or contraction. If the temperature should change, large compressive or tensile stresses, called *thermal stresses*, will occur. The magnitude of such stresses can be calculated using the concept of elastic modulus developed in Chapter 12. To calculate the internal stress, we can think of this process as occurring in two steps: (1) the beam tries to expand (or contract) by an amount $\Delta\ell$ given by Eq. 17–1; (2) the solid in contact with the beam exerts a force to compress (or expand) it, keeping it at its original length. The force F required is given by Eq. 12–4:

$$\Delta\ell = \frac{1}{E}\frac{F}{A}\ell_0,$$

where E is Young's modulus for the material. To calculate the internal stress, F/A, we then set $\Delta\ell$ in Eq. 17–1a equal to $\Delta\ell$ in the equation above and find

$$\alpha\ell_0\Delta T = \frac{1}{E}\frac{F}{A}\ell_0.$$

Hence, the stress

$$\frac{F}{A} = \alpha E\,\Delta T.$$

EXAMPLE 17–8 **Stress in concrete on a hot day.** A highway is to be made of blocks of concrete 10 m long placed end to end with no space between them to allow for expansion. If the blocks were placed at a temperature of 10°C, what compressive stress would occur if the temperature reached 40°C? The contact area between each block is 0.20 m². Will fracture occur?

PHYSICS APPLIED
Highway buckling

APPROACH We use the expression for the stress F/A we just derived, and find the value of E from Table 12–1. To see if fracture occurs, we compare this stress to the ultimate strength of concrete in Table 12–2.

SOLUTION

$$\frac{F}{A} = \alpha E\,\Delta T = (12\times 10^{-6}/\text{C}°)(20\times 10^9\,\text{N/m}^2)(30\,\text{C}°) = 7.2\times 10^6\,\text{N/m}^2.$$

This stress is not far from the ultimate strength of concrete under compression (Table 12–2) and exceeds it for tension and shear. If the concrete is not perfectly aligned, part of the force will act in shear, and fracture is likely. This is why soft spacers or expansion joints (Fig. 17–3) are used in concrete sidewalks, highways, and bridges.

EXERCISE C How much space would you allow between the 10-m-long concrete blocks if you expected a temperature range of 0°F to 110°F?

17–6 The Gas Laws and Absolute Temperature

Equation 17–2 is not very useful for describing the expansion of a gas, partly because the expansion can be so great, and partly because gases generally expand to fill whatever container they are in. Indeed, Eq. 17–2 is meaningful only if the pressure is kept constant. The volume of a gas depends very much on the pressure as well as on the temperature. It is therefore valuable to determine a relation between the volume, the pressure, the temperature, and the mass of a gas. Such a relation is called an **equation of state**. (By the word *state*, we mean the physical condition of the system.)

If the state of a system is changed, we will always wait until the pressure and temperature have reached the same values throughout. We thus consider only **equilibrium states** of a system—when the variables that describe it (such as temperature and pressure) are the same throughout the system and are not changing in time. We also note that the results of this Section are accurate only for gases that are not too dense (the pressure is not too high, on the order of an atmosphere or less) and not close to the liquefaction (boiling) point.

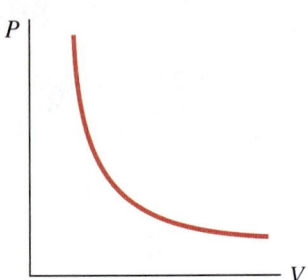

FIGURE 17–13 Pressure vs. volume of a fixed amount of gas at a constant temperature, showing the inverse relationship as given by Boyle's law: as the pressure decreases, the volume increases.

FIGURE 17–14 Volume of a fixed amount of gas as a function of (a) Celsius temperature, and (b) Kelvin temperature, when the pressure is kept constant.

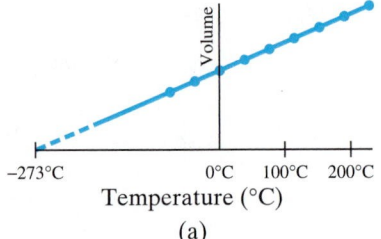

(a)

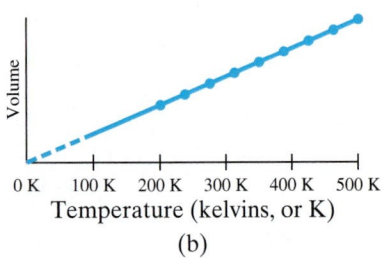

(b)

For a given quantity of gas it is found experimentally that, to a good approximation, *the volume of a gas is inversely proportional to the absolute pressure applied to it when the temperature is kept constant*. That is,

$$V \propto \frac{1}{P},$$
  [constant T]

where P is the absolute pressure (*not* "gauge pressure"—see Section 13–4). For example, if the pressure on a gas is doubled, the volume is reduced to half its original volume. This relation is known as **Boyle's law**, after Robert Boyle (1627–1691), who first stated it on the basis of his own experiments. A graph of P vs. V for a fixed temperature is shown in Fig. 17–13. Boyle's law can also be written

$$PV = \text{constant}.$$
  [constant T]

That is, at constant temperature, if either the pressure or volume of a fixed amount of gas is allowed to vary, the other variable also changes so that the product PV remains constant.

Temperature also affects the volume of a gas, but a quantitative relationship between V and T was not found until more than a century after Boyle's work. The Frenchman Jacques Charles (1746–1823) found that when the pressure is not too high and is kept constant, the volume of a gas increases with temperature at a nearly linear rate, as shown in Fig. 17–14a. However, all gases liquefy at low temperatures (for example, oxygen liquefies at −183°C), so the graph cannot be extended below the liquefaction point. Nonetheless, the graph is essentially a straight line and if projected to lower temperatures, as shown by the dashed line, it crosses the axis at about −273°C.

Such a graph can be drawn for any gas, and the straight line always projects back to −273°C at zero volume. This seems to imply that if a gas could be cooled to −273°C, it would have zero volume, and at lower temperatures a negative volume, which makes no sense. It could be argued that −273°C is the lowest temperature possible; indeed, many other more recent experiments indicate that this is so. This temperature is called the **absolute zero** of temperature. Its value has been determined to be −273.15°C.

Absolute zero forms the basis of a temperature scale known as the **absolute scale** or **Kelvin scale**, and it is used extensively in scientific work. On this scale the temperature is specified as degrees Kelvin or, preferably, simply as *kelvins* (K) without the degree sign. The intervals are the same as for the Celsius scale, but the zero on this scale (0 K) is chosen as absolute zero. Thus the freezing point of water (0°C) is 273.15 K, and the boiling point of water is 373.15 K. Indeed, any temperature on the Celsius scale can be changed to kelvins by adding 273.15 to it:

$$T(\text{K}) = T(°\text{C}) + 273.15.$$

Now let us look at Fig. 17–14b, where the graph of the volume of a gas versus absolute temperature is a straight line that passes through the origin. Thus, to a good approximation, *the volume of a given amount of gas is directly proportional to the absolute temperature when the pressure is kept constant*. This is known as **Charles's law**, and is written

$$V \propto T.$$
  [constant P]

A third gas law, known as **Gay-Lussac's law**, after Joseph Gay-Lussac (1778–1850), states that *at constant volume, the absolute pressure of a gas is directly proportional to the absolute temperature*:

$$P \propto T.$$
  [constant V]

The laws of Boyle, Charles, and Gay-Lussac are not really laws in the sense that we use this term today (precise, deep, wide-ranging validity). They are really only approximations that are accurate for real gases only as long as the pressure and density of the gas are not too high, and the gas is not too close to liquefaction (condensation). The term *law* applied to these three relationships has become traditional, however, so we have stuck with that usage.

CONCEPTUAL EXAMPLE 17–9 | **Why you should not throw a closed glass jar into a campfire.** What can happen if you did throw an empty glass jar, with the lid on tight, into a fire, and why?

RESPONSE The inside of the jar is not empty. It is filled with air. As the fire heats the air inside, its temperature rises. The volume of the glass jar changes only slightly due to the heating. According to Gay-Lussac's law the pressure P of the air inside the jar can increase dramatically, enough to cause the jar to explode, throwing glass pieces outward.

17–7 The Ideal Gas Law

The gas laws of Boyle, Charles, and Gay-Lussac were obtained by means of a very useful scientific technique: namely, holding one or more variables constant to see clearly the effects on one variable due to changing one other variable. These laws can now be combined into a single more general relation between the absolute pressure, volume, and absolute temperature of a fixed quantity of gas:

$$PV \propto T.$$

This relation indicates how any of the quantities P, V, or T will vary when the other two quantities change. This relation reduces to Boyle's, Charles's, or Gay-Lussac's law when either T, P, or V, respectively, is held constant.

Finally, we must incorporate the effect of the amount of gas present. Anyone who has blown up a balloon knows that the more air forced into the balloon, the bigger it gets (Fig. 17–15). Indeed, careful experiments show that at constant temperature and pressure, the volume V of an enclosed gas increases in direct proportion to the mass m of gas present. Hence we write

$$PV \propto mT.$$

This proportion can be made into an equation by inserting a constant of proportionality. Experiment shows that this constant has a different value for different gases. However, the constant of proportionality turns out to be the same for all gases if, instead of the mass m, we use the number of *moles*.

One **mole** (abbreviated mol) is defined as the amount of substance that contains as many atoms or molecules as there are in precisely 12 grams of carbon 12 (whose atomic mass is exactly 12 u). A simpler but equivalent definition is this: 1 mol is that quantity of substance whose mass in grams is numerically equal to the molecular mass (Section 17–1) of the substance. For example, the molecular mass of hydrogen gas (H_2) is 2.0 u (since each molecule contains two atoms of hydrogen and each atom has an atomic mass of 1.0 u). Thus 1 mol of H_2 has a mass of 2.0 g. Similarly, 1 mol of neon gas has a mass of 20 g, and 1 mol of CO_2 has a mass of $[12 + (2 \times 16)] = 44$ g since oxygen has atomic mass of 16 (see Periodic Table inside the rear cover). The mole is the official unit of amount of substance in the SI system. In general, the number of moles, n, in a given sample of a pure substance is equal to the mass of the sample in grams divided by the molecular mass specified as grams per mole:

$$n \, (\text{mole}) = \frac{\text{mass (grams)}}{\text{molecular mass (g/mol)}}.$$

For example, the number of moles in 132 g of CO_2 (molecular mass 44 u) is

$$n = \frac{132 \, \text{g}}{44 \, \text{g/mol}} = 3.0 \, \text{mole}.$$

FIGURE 17–15 Blowing up a balloon means putting more air (more air molecules) into the balloon, which increases its volume. The pressure is nearly constant (atmospheric) except for the small effect of the balloon's elasticity.

We can now write the proportion discussed above ($PV \propto mT$) as an equation:

IDEAL GAS LAW

$$PV = nRT, \qquad \textbf{(17–3)}$$

where n represents the number of moles and R is the constant of proportionality. R is called the **universal gas constant** because its value is found experimentally to be the same for all gases. The value of R, in several sets of units (only the first is the proper SI unit), is

$$\begin{aligned} R &= 8.314 \, \text{J/(mol} \cdot \text{K)} \qquad\qquad \text{[SI units]} \\ &= 0.0821 \, \text{(L} \cdot \text{atm)/(mol} \cdot \text{K)} \\ &= 1.99 \, \text{calories/(mol} \cdot \text{K).}^{\dagger} \end{aligned}$$

Equation 17–3 is called the **ideal gas law**, or the **equation of state for an ideal gas**. We use the term "ideal" because real gases do not follow Eq. 17–3 precisely, particularly at high pressure (and density) or when the gas is near the liquefaction point (= boiling point). However, at pressures less than an atmosphere or so, and when T is not close to the liquefaction point of the gas, Eq. 17–3 is quite accurate and useful for real gases.

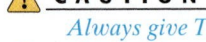 **CAUTION**

Always give T in kelvins and P as absolute (not gauge) pressure

Always remember, when using the ideal gas law, that temperatures must be given in kelvins (K) and that the pressure P must always be *absolute* pressure, not gauge pressure (Section 13–4).

EXERCISE D Return to the Chapter-Opening Question, page 454, and answer it again now. Try to explain why you may have answered differently the first time.

EXERCISE E An ideal gas is contained in a steel sphere at 27.0°C and 1.00 atm absolute pressure. If no gas is allowed to escape and the temperature is raised to 127°C, what will be the new pressure? (*a*) 1.33 atm; (*b*) 0.75 atm; (*c*) 4.7 atm; (*d*) 0.21 atm; (*e*) 1.00 atm.

17–8 Problem Solving with the Ideal Gas Law

The ideal gas law is an extremely useful tool, and we now consider some Examples. We will often refer to "standard conditions" or **standard temperature and pressure** (**STP**), which means:

STP

$$T = 273 \, \text{K} \, (0°\text{C}) \quad \text{and} \quad P = 1.00 \, \text{atm} = 1.013 \times 10^{5} \, \text{N/m}^{2} = 101.3 \, \text{kPa.}$$

EXAMPLE 17–10 **Volume of one mole at STP.** Determine the volume of 1.00 mol of any gas, assuming it behaves like an ideal gas, at STP.

APPROACH We use the ideal gas law, solving for V.

SOLUTION We solve for V in Eq. 17–3:

$$V = \frac{nRT}{P} = \frac{(1.00 \, \text{mol})(8.314 \, \text{J/mol} \cdot \text{K})(273 \, \text{K})}{(1.013 \times 10^{5} \, \text{N/m}^{2})} = 22.4 \times 10^{-3} \, \text{m}^{3}.$$

 **PROBLEM SOLVING**

1 mol of gas at STP has V = 22.4 L

Since 1 liter (L) is $1000 \, \text{cm}^{3} = 1.00 \times 10^{-3} \, \text{m}^{3}$, 1.00 mol of any (ideal) gas has volume $V = 22.4 \, \text{L}$ at STP.

The value of 22.4 L for the volume of 1 mol of an ideal gas at STP is worth remembering, for it sometimes makes calculation simpler.

EXERCISE F What is the volume of 1.00 mol of ideal gas at 546 K ($= 2 \times 273 \, \text{K}$) and 2.0 atm absolute pressure? (*a*) 11.2 L, (*b*) 22.4 L, (*c*) 44.8 L, (*d*) 67.2 L, (*e*) 89.6 L.

---

†Calories will be defined in Section 19–1; sometimes it is useful to use R as given in terms of calories.

EXAMPLE 17–11 **Helium balloon.** A helium party balloon, assumed to be a perfect sphere, has a radius of 18.0 cm. At room temperature (20°C), its internal pressure is 1.05 atm. Find the number of moles of helium in the balloon and the mass of helium needed to inflate the balloon to these values.

APPROACH We can use the ideal gas law to find n, since we are given P and T, and can find V from the given radius.

SOLUTION We get the volume V from the formula for a sphere:

$$V = \tfrac{4}{3}\pi r^3$$
$$= \tfrac{4}{3}\pi (0.180\,\text{m})^3 = 0.0244\,\text{m}^3.$$

The pressure is given as $1.05\,\text{atm} = 1.064 \times 10^5\,\text{N/m}^2$. The temperature must be expressed in kelvins, so we change 20°C to $(20 + 273)\text{K} = 293\,\text{K}$. Finally, we use the value $R = 8.314\,\text{J/(mol·K)}$ because we are using SI units. Thus

$$n = \frac{PV}{RT} = \frac{(1.064 \times 10^5\,\text{N/m}^2)(0.0244\,\text{m}^3)}{(8.314\,\text{J/mol·K})(293\,\text{K})} = 1.066\,\text{mol}.$$

The mass of helium (atomic mass = 4.00 g/mol as given in the Periodic Table or Appendix F) can be obtained from

$$\text{mass} = n \times \text{molecular mass} = (1.066\,\text{mol})(4.00\,\text{g/mol}) = 4.26\,\text{g}$$

or $4.26 \times 10^{-3}\,\text{kg}$.

EXAMPLE 17–12 **ESTIMATE** **Mass of air in a room.** Estimate the mass of air in a room whose dimensions are $5\,\text{m} \times 3\,\text{m} \times 2.5\,\text{m}$ high, at STP.

APPROACH First we determine the number of moles n using the given volume. Then we can multiply by the mass of one mole to get the total mass.

SOLUTION Example 17–10 told us that 1 mol of a gas at 0°C has a volume of 22.4 L. The room's volume is $5\,\text{m} \times 3\,\text{m} \times 2.5\,\text{m}$, so

$$n = \frac{(5\,\text{m})(3\,\text{m})(2.5\,\text{m})}{22.4 \times 10^{-3}\,\text{m}^3} \approx 1700\,\text{mol}.$$

Air is a mixture of about 20% oxygen (O_2) and 80% nitrogen (N_2). The molecular masses are $2 \times 16\,\text{u} = 32\,\text{u}$ and $2 \times 14\,\text{u} = 28\,\text{u}$, respectively, for an average of about 29 u. Thus, 1 mol of air has a mass of about $29\,\text{g} = 0.029\,\text{kg}$, so our room has a mass of air

$$m \approx (1700\,\text{mol})(0.029\,\text{kg/mol}) \approx 50\,\text{kg}.$$

NOTE That is roughly 100 lb of air!

PHYSICS APPLIED
Mass (and weight)
of the air in a room

EXERCISE G At 20°C, would there be (a) more, (b) less, or (c) the same air mass in a room than at 0°C?

Frequently, volume is specified in liters and pressure in atmospheres. Rather than convert these to SI units, we can instead use the value of R given in Section 17–7 as $0.0821\,\text{L·atm/mol·K}$.

In many situations it is not necessary to use the value of R at all. For example, many problems involve a change in the pressure, temperature, and volume of a fixed amount of gas. In this case, $PV/T = nR = $ constant, since n and R remain constant. If we now let P_1, V_1, and T_1 represent the appropriate variables initially, and P_2, V_2, T_2 represent the variables after the change is made, then we can write

PROBLEM SOLVING
Using the ideal gas law as a ratio

$$\frac{P_1 V_1}{T_1} = \frac{P_2 V_2}{T_2}.$$

If we know any five of the quantities in this equation, we can solve for the sixth. Or, if one of the three variables is constant ($V_1 = V_2$, or $P_1 = P_2$, or $T_1 = T_2$) then we can use this equation to solve for one unknown when given the other three quantities.

FIGURE 17–16 Example 17–13.

EXAMPLE 17–13 **Check tires cold.** An automobile tire is filled (Fig. 17–16) to a gauge pressure of 200 kPa at 10°C. After a drive of 100 km, the temperature within the tire rises to 40°C. What is the pressure within the tire now?

APPROACH We do not know the number of moles of gas, or the volume of the tire, but we assume they are constant. We use the ratio form of the ideal gas law.

SOLUTION Since $V_1 = V_2$, then

$$\frac{P_1}{T_1} = \frac{P_2}{T_2}.$$

This is, incidentally, a statement of Gay-Lussac's law. Since the pressure given is the gauge pressure (Section 13–4), we must add atmospheric pressure ($= 101\,\text{kPa}$) to get the absolute pressure $P_1 = (200\,\text{kPa} + 101\,\text{kPa}) = 301\,\text{kPa}$. We convert temperatures to kelvins by adding 273 and solve for P_2:

$$P_2 = P_1\left(\frac{T_2}{T_1}\right) = (3.01 \times 10^5\,\text{Pa})\left(\frac{313\,\text{K}}{283\,\text{K}}\right) = 333\,\text{kPa}.$$

Subtracting atmospheric pressure, we find the resulting gauge pressure to be 232 kPa, which is a 16% increase. This Example shows why car manuals suggest checking tire pressure when the tires are cold.

17–9 Ideal Gas Law in Terms of Molecules: Avogadro's Number

The fact that the gas constant, R, has the same value for all gases is a remarkable reflection of simplicity in nature. It was first recognized, although in a slightly different form, by the Italian scientist Amedeo Avogadro (1776–1856). Avogadro stated that *equal volumes of gas at the same pressure and temperature contain equal numbers of molecules*. This is sometimes called **Avogadro's hypothesis**. That this is consistent with R being the same for all gases can be seen as follows. From Eq. 17–3, $PV = nRT$, we see that for the same number of moles, n, and the same pressure and temperature, the volume will be the same for all gases as long as R is the same. Second, the number of molecules in 1 mole is the same for all gases.[†] Thus Avogadro's hypothesis is equivalent to R being the same for all gases.

The number of molecules in one mole of any pure substance is known as **Avogadro's number**, N_A. Although Avogadro conceived the notion, he was not able to actually determine the value of N_A. Indeed, precise measurements were not done until the twentieth century.

A number of methods have been devised to measure N_A, and the accepted value today is

$$N_A = 6.02 \times 10^{23}.$$ [molecules/mole]

Since the total number of molecules, N, in a gas is equal to the number per mole times the number of moles $(N = nN_A)$, the ideal gas law, Eq. 17–3, can be written in terms of the number of molecules present:

$$PV = nRT = \frac{N}{N_A}RT,$$

or

IDEAL GAS LAW
(in terms of molecules)

$$PV = NkT,$$ **(17–4)**

where $k = R/N_A$ is called the **Boltzmann constant** and has the value

$$k = \frac{R}{N_A} = \frac{8.314\,\text{J/mol·K}}{6.02 \times 10^{23}/\text{mol}} = 1.38 \times 10^{-23}\,\text{J/K}.$$

[†]For example, the molecular mass of H_2 gas is 2.0 atomic mass units (u), whereas that of O_2 gas is 32.0 u. Thus 1 mol of H_2 has a mass of 0.0020 kg and 1 mol of O_2 gas, 0.0320 kg. The number of molecules in a mole is equal to the total mass M of a mole divided by the mass m of one molecule; since this ratio (M/m) is the same for all gases by definition of the mole, a mole of any gas must contain the same number of molecules.

EXAMPLE 17–14 Hydrogen atom mass. Use Avogadro's number to determine the mass of a hydrogen atom.

APPROACH The mass of one atom equals the mass of 1 mol divided by the number of atoms in 1 mol, N_A.

SOLUTION One mole of hydrogen atoms (atomic mass = 1.008 u, Section 17–1 or Appendix F) has a mass of 1.008×10^{-3} kg and contains 6.02×10^{23} atoms. Thus one atom has a mass

$$m = \frac{1.008 \times 10^{-3} \text{ kg}}{6.02 \times 10^{23}}$$

$$= 1.67 \times 10^{-27} \text{ kg.}$$

EXAMPLE 17–15 ESTIMATE How many molecules in one breath? Estimate how many molecules you breathe in with a 1.0-L breath of air.

APPROACH We determine what fraction of a mole 1.0 L is by using the result of Example 17–10 that 1 mole has a volume of 22.4 L at STP, and then multiply that by N_A to get the number of molecules in this number of moles.

SOLUTION One mole corresponds to 22.4 L at STP, so 1.0 L of air is $(1.0 \text{ L})/(22.4 \text{ L/mol}) = 0.045$ mol. Then 1.0 L of air contains

$$(0.045 \text{ mol})(6.02 \times 10^{23} \text{ molecules/mol}) \approx 3 \times 10^{22} \text{ molecules.}$$

*17–10 Ideal Gas Temperature Scale — a Standard

It is important to have a very precisely defined temperature scale so that measurements of temperature made at different laboratories around the world can be accurately compared. We now discuss such a scale that has been accepted by the general scientific community.

The standard thermometer for this scale is the constant-volume gas thermometer discussed in Section 17–2. The scale itself is called the **ideal gas temperature scale**, since it is based on the property of an ideal gas that the pressure is directly proportional to the absolute temperature (Gay-Lussac's law). A real gas, which would need to be used in any real constant-volume gas thermometer, approaches this ideal at low density. In other words, the temperature at any point in space is *defined* as being proportional to the pressure in the (nearly) ideal gas used in the thermometer. To set up a scale we need two fixed points. One fixed point will be $P = 0$ at $T = 0$ K. The second fixed point is chosen to be the **triple point** of water, which is that point where water in the solid, liquid, and gas states can coexist in equilibrium. This occurs only at a unique temperature and pressure,[†] and can be reproduced at different laboratories with great precision. The pressure at the triple point of water is 4.58 torr and the temperature is 0.01°C. This temperature corresponds to 273.16 K, since absolute zero is about −273.15°C. In fact, the triple point is now *defined* to be exactly 273.16 K.

[†]Liquid water and steam can coexist (the boiling point) at a range of temperatures depending on the pressure. Water boils at a lower temperature when the pressure is less, such as high in the mountains. The triple point represents a more precisely reproducible fixed point than does either the freezing point or boiling point of water at, say, 1 atm. See Section 18–3 for further discussion.

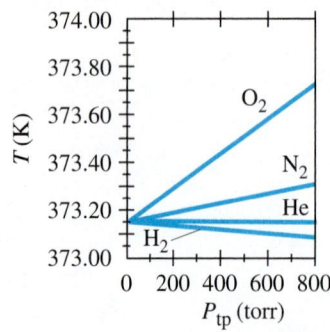

FIGURE 17–17 Temperature readings of a constant-volume gas thermometer for the boiling point of water at 1.00 atm are plotted, for different gases, as a function of the gas pressure in the thermometer at the triple point (P_{tp}). Note that as the amount of gas in the thermometer is reduced, so that $P_{tp} \to 0$, all gases give the same reading, 373.15 K. For pressure less than 0.10 atm (76 torr), the variation shown is less than 0.07 K.

The absolute or Kelvin temperature T at any point is then defined, using a constant-volume gas thermometer for an ideal gas, as

$$T = (273.16 \text{ K})\left(\frac{P}{P_{tp}}\right). \quad \text{[ideal gas; constant volume]} \quad \textbf{(17–5a)}$$

In this relation, P_{tp} is the pressure of the gas in the thermometer at the triple point temperature of water, and P is the pressure in the thermometer when it is at the point where T is being determined. Note that if we let $P = P_{tp}$ in this relation, then $T = 273.16 \text{ K}$, as it must.

The definition of temperature, Eq. 17–5a, with a constant-volume gas thermometer filled with a real gas is only approximate because we find that we get different results for the temperature depending on the type of gas that is used in the thermometer. Temperatures determined in this way also vary depending on the amount of gas in the bulb of the thermometer: for example, the boiling point of water at 1.00 atm is found from Eq. 17–5a to be 373.87 K when the gas is O_2 and $P_{tp} = 1000$ torr. If the amount of O_2 in the bulb is reduced so that at the triple point $P_{tp} = 500$ torr, the boiling point of water from Eq. 17–5a is then found to be 373.51 K. If H_2 gas is used instead, the corresponding values are 373.07 K and 373.11 K (see Fig. 17–17). But now suppose we use a particular real gas and make a series of measurements in which the amount of gas in the thermometer bulb is reduced to smaller and smaller amounts, so that P_{tp} becomes smaller and smaller. It is found experimentally that an extrapolation of such data to $P_{tp} = 0$ always gives the *same value* for the temperature of a given system (such as $T = 373.15$ K for the boiling point of water at 1.00 atm) as shown in Fig. 17–17. Thus the temperature T at any point in space, determined using a constant-volume gas thermometer containing a real gas, is defined using this limiting process:

$$T = (273.16 \text{ K})\lim_{P_{tp} \to 0}\left(\frac{P}{P_{tp}}\right). \quad \text{[constant volume]} \quad \textbf{(17–5b)}$$

This defines the **ideal gas temperature scale**. One of the great advantages of this scale is that the value for T does not depend on the kind of gas used. But the scale does depend on the properties of gases in general. Helium has the lowest condensation point of all gases; at very low pressures it liquefies at about 1 K, so temperatures below this cannot be defined on this scale.

Summary

The atomic theory of matter postulates that all matter is made up of tiny entities called **atoms**, which are typically 10^{-10} m in diameter.

Atomic and **molecular masses** are specified on a scale where ordinary carbon (^{12}C) is arbitrarily given the value 12.0000 u (atomic mass units).

The distinction between solids, liquids, and gases can be attributed to the strength of the attractive forces between the atoms or molecules and to their average speed.

Temperature is a measure of how hot or cold something is. **Thermometers** are used to measure temperature on the **Celsius** (°C), **Fahrenheit** (°F), and **Kelvin** (K) scales. Two standard points on each scale are the freezing point of water (0°C, 32°F, 273.15 K) and the boiling point of water (100°C, 212°F, 373.15 K). A one-kelvin change in temperature equals a change of one Celsius degree or $\frac{9}{5}$ Fahrenheit degrees. Kelvins are related to °C by $T(\text{K}) = T(°\text{C}) + 273.15$.

The change in length, $\Delta\ell$, of a solid, when its temperature changes by an amount ΔT, is directly proportional to the temperature change and to its original length ℓ_0. That is,

$$\Delta\ell = \alpha\ell_0\Delta T, \quad \textbf{(17–1a)}$$

where α is the *coefficient of linear expansion*.

The change in volume of most solids, liquids, and gases is proportional to the temperature change and to the original volume V_0:

$$\Delta V = \beta V_0 \Delta T. \quad \textbf{(17–2)}$$

The *coefficient of volume expansion*, β, is approximately equal to 3α for uniform solids.

Water is unusual because, unlike most materials whose volume increases with temperature, its volume actually decreases as the temperature increases in the range from 0°C to 4°C.

The **ideal gas law**, or **equation of state for an ideal gas**, relates the pressure P, volume V, and temperature T (in kelvins) of n moles of gas by the equation

$$PV = nRT, \quad \textbf{(17–3)}$$

where $R = 8.314 \text{ J/mol·K}$ for all gases. Real gases obey the ideal gas law quite accurately if they are not at too high a pressure or near their liquefaction point.

One **mole** is that amount of a substance whose mass in grams is numerically equal to the atomic or molecular mass of that substance.

Avogadro's number, $N_A = 6.02 \times 10^{23}$, is the number of atoms or molecules in 1 mol of any pure substance.

The ideal gas law can be written in terms of the number of molecules N in the gas as

$$PV = NkT, \quad \textbf{(17–4)}$$

where $k = R/N_A = 1.38 \times 10^{-23} \text{ J/K}$ is Boltzmann's constant.

Questions

1. Which has more atoms: 1 kg of iron or 1 kg of aluminum? See the Periodic Table or Appendix F.
2. Name several properties of materials that could be exploited to make a thermometer.
3. Which is larger, $1C°$ or $1F°$?
4. If system A is in equilibrium with system B, but B is not in equilibrium with system C, what can you say about the temperatures of A, B, and C?
5. Suppose system C is not in equilibrium with system A nor in equilibrium with system B. Does this imply that A and B are not in equilibrium? What can you infer regarding the temperatures of A, B, and C?
6. In the relation $\Delta \ell = \alpha \ell_0 \Delta T$, should ℓ_0 be the initial length, the final length, or does it matter?
7. A flat bimetallic strip consists of a strip of aluminum riveted to a strip of iron. When heated, the strip will bend. Which metal will be on the outside of the curve? Why?
8. Long steam pipes that are fixed at the ends often have a section in the shape of a ∪. Why?
9. A flat uniform cylinder of lead floats in mercury at 0°C. Will the lead float higher or lower if the temperature is raised?
10. Figure 17–18 shows a diagram of a simple **thermostat** used to control a furnace (or other heating or cooling system). The bimetallic strip consists of two strips of different metals bonded together. The electric switch (attached to the bimetallic strip) is a glass vessel containing liquid mercury that conducts electricity when it can flow to touch both contact wires. Explain how this device controls the furnace and how it can be set at different temperatures.

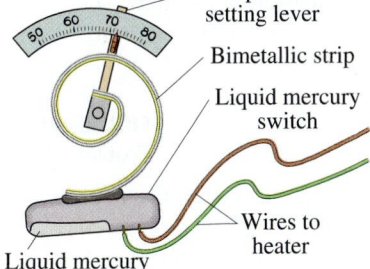

Temperature setting lever
Bimetallic strip
Liquid mercury switch
Wires to heater
Liquid mercury

FIGURE 17–18 A thermostat (Question 10).

11. Explain why it is advisable to add water to an overheated automobile engine only slowly, and only with the engine running.
12. The units for the coefficients of expansion α are $(C°)^{-1}$, and there is no mention of a length unit such as meters. Would the expansion coefficient change if we used feet or millimeters instead of meters?
13. When a cold mercury-in-glass thermometer is first placed in a hot tub of water, the mercury initially descends a bit and then rises. Explain.
14. The principal virtue of Pyrex glass is that its coefficient of linear expansion is much smaller than that for ordinary glass (Table 17–1). Explain why this gives rise to the higher resistance to heat of Pyrex.
15. Will a grandfather clock, accurate at 20°C, run fast or slow on a hot day (30°C)? The clock uses a pendulum supported on a long thin brass rod.
16. Freezing a can of soda will cause its bottom and top to bulge so badly the can will not stand up. What has happened?
17. Why might you expect an alcohol-in-glass thermometer to be more precise than a mercury-in-glass thermometer?
18. Will the buoyant force on an aluminum sphere submerged in water increase, decrease, or remain the same, if the temperature is increased from 20°C to 40°C?
19. If an atom is measured to have a mass of 6.7×10^{-27} kg, what atom do you think it is?
20. From a practical point of view, does it really matter what gas is used in a constant-volume gas thermometer? If so, explain. [*Hint*: See Fig. 17–17.]
21. A ship loaded in sea water at 4°C later sailed up a river into fresh water where it sank in a storm. Explain why a ship might be more likely to sink in fresh water than on the open sea. [*Hint*: Consider the buoyant force due to water.]

Problems

17–1 Atomic Theory

1. (I) How does the number of atoms in a 21.5-g gold ring compare to the number in a silver ring of the same mass?
2. (I) How many atoms are there in a 3.4-g copper penny?

17–2 Temperature and Thermometers

3. (I) (*a*) "Room temperature" is often taken to be 68°F. What is this on the Celsius scale? (*b*) The temperature of the filament in a lightbulb is about 1900°C. What is this on the Fahrenheit scale?
4. (I) Among the highest and lowest natural air temperatures recorded are 136°F in the Libyan desert and −129°F in Antarctica. What are these temperatures on the Celsius scale?
5. (I) A thermometer tells you that you have a fever of 39.4°C. What is this in Fahrenheit?
6. (II) In an alcohol-in-glass thermometer, the alcohol column has length 11.82 cm at 0.0°C and length 21.85 cm at 100.0°C. What is the temperature if the column has length (*a*) 18.70 cm, and (*b*) 14.60 cm?

17–4 Thermal Expansion

7. (I) The Eiffel Tower (Fig. 17–19) is built of wrought iron approximately 300 m tall. Estimate how much its height changes between January (average temperature of 2°C) and July (average temperature of 25°C). Ignore the angles of the iron beams and treat the tower as a vertical beam.

FIGURE 17–19 Problem 7. The Eiffel Tower in Paris.

8. (I) A concrete highway is built of slabs 12 m long (15°C). How wide should the expansion cracks between the slabs be (at 15°C) to prevent buckling if the range of temperature is −30°C to +50°C?

9. (I) Super Invar™, an alloy of iron and nickel, is a strong material with a very low coefficient of thermal expansion $(0.20 \times 10^{-6}/C°)$. A 1.6-m-long tabletop made of this alloy is used for sensitive laser measurements where extremely high tolerances are required. How much will this alloy table expand along its length if the temperature increases 5.0 C°? Compare to tabletops made of steel.

10. (II) To what temperature would you have to heat a brass rod for it to be 1.0% longer than it is at 25°C?

11. (II) The density of water at 4°C is $1.00 \times 10^3 \text{ kg/m}^3$. What is water's density at 94°C? Assume a constant coefficient of volume expansion.

12. (II) At a given latitude, ocean water in the so-called "mixed layer" (from the surface to a depth of about 50 m) is at approximately the same temperature due to the mixing action of waves. Assume that because of global warming, the temperature of the mixed layer is everywhere increased by 0.5°C, while the temperature of the deeper portions of the ocean remains unchanged. Estimate the resulting rise in sea level. The ocean covers about 70% of the Earth's surface.

13. (II) To make a secure fit, rivets that are larger than the rivet hole are often used and the rivet is cooled (usually in dry ice) before it is placed in the hole. A steel rivet 1.872 cm in diameter is to be placed in a hole 1.870 cm in diameter in a metal at 20°C. To what temperature must the rivet be cooled if it is to fit in the hole?

14. (II) A uniform rectangular plate of length ℓ and width w has a coefficient of linear expansion α. Show that, if we neglect very small quantities, the change in area of the plate due to a temperature change ΔT is $\Delta A = 2\alpha \ell w \, \Delta T$. See Fig. 17–20.

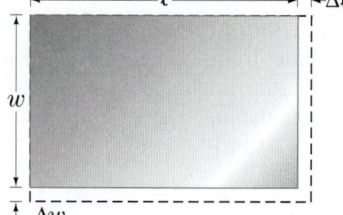

FIGURE 17–20
Problem 14.
A rectangular plate is heated.

15. (II) An aluminum sphere is 8.75 cm in diameter. What will be its change in volume if it is heated from 30°C to 180°C?

16. (II) A typical car has 17 L of liquid coolant circulating at a temperature of 93°C through the engine's cooling system. Assume that, in this normal condition, the coolant completely fills the 3.5-L volume of the aluminum radiator and the 13.5-L internal cavities within the steel engine. When a car overheats, the radiator, engine, and coolant expand and a small reservoir connected to the radiator catches any resultant coolant overflow. Estimate how much coolant overflows to the reservoir if the system is heated from 93°C to 105°C. Model the radiator and engine as hollow shells of aluminum and steel, respectively. The coefficient of volume expansion for coolant is $\beta = 410 \times 10^{-6}/C°$.

17. (II) It is observed that 55.50 mL of water at 20°C completely fills a container to the brim. When the container and the water are heated to 60°C, 0.35 g of water is lost. (a) What is the coefficient of volume expansion of the container? (b) What is the most likely material of the container? Density of water at 60°C is 0.98324 g/mL.

18. (II) (a) A brass plug is to be placed in a ring made of iron. At 15°C, the diameter of the plug is 8.753 cm and that of the inside of the ring is 8.743 cm. They must both be brought to what common temperature in order to fit? (b) What if the plug were iron and the ring brass?

19. (II) If a fluid is contained in a long narrow vessel so it can expand in essentially one direction only, show that the effective coefficient of linear expansion α is approximately equal to the coefficient of volume expansion β.

20. (II) (a) Show that the change in the density ρ of a substance, when the temperature changes by ΔT, is given by $\Delta \rho = -\beta \rho \, \Delta T$. (b) What is the fractional change in density of a lead sphere whose temperature decreases from 25°C to −55°C?

21. (II) Wine bottles are never completely filled: a small volume of air is left in the glass bottle's cylindrically shaped neck (inner diameter $d = 18.5 \text{ mm}$) to allow for wine's fairly large coefficient of thermal expansion. The distance H between the surface of the liquid contents and the bottom of the cork is called the "headspace height" (Fig. 17–21), and is typically $H = 1.5 \text{ cm}$ for a 750-mL bottle filled at 20°C. Due to its alcoholic content, wine's coefficient of volume expansion is about double that of water; in comparison, the thermal expansion of glass can be neglected. Estimate H if the bottle is kept (a) at 10°C, (b) at 30°C.

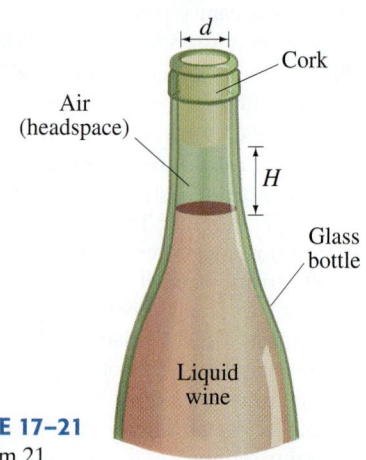

FIGURE 17–21
Problem 21.

22. (III) (a) Determine a formula for the change in surface area of a uniform solid sphere of radius r if its coefficient of linear expansion is α (assumed constant) and its temperature is changed by ΔT. (b) What is the increase in area of a solid iron sphere of radius 60.0 cm if its temperature is raised from 15°C to 275°C?

23. (III) The pendulum in a grandfather clock is made of brass and keeps perfect time at 17°C. How much time is gained or lost in a year if the clock is kept at 28°C? (Assume the frequency dependence on length for a simple pendulum applies.)

24. (III) A 28.4-kg solid aluminum cylindrical wheel of radius 0.41 m is rotating about its axle in frictionless bearings with angular velocity $\omega = 32.8 \text{ rad/s}$. If its temperature is then raised from 20.0°C to 95.0°C, what is the fractional change in ω?

*17–5 Thermal Stresses

*25. (I) An aluminum bar has the desired length when at 18°C. How much stress is required to keep it at this length if the temperature increases to 35°C?

*26. (II) (a) A horizontal steel I-beam of cross-sectional area 0.041 m² is rigidly connected to two vertical steel girders. If the beam was installed when the temperature was 25°C, what stress is developed in the beam when the temperature drops to −25°C? (b) Is the ultimate strength of the steel exceeded? (c) What stress is developed if the beam is concrete and has a cross-sectional area of 0.13 m²? Will it fracture?

*27. (III) A barrel of diameter 134.122 cm at 20°C is to be enclosed by an iron band. The circular band has an inside diameter of 134.110 cm at 20°C. It is 9.4 cm wide and 0.65 cm thick. (a) To what temperature must the band be heated so that it will fit over the barrel? (b) What will be the tension in the band when it cools to 20°C?

17–6 Gas Laws; Absolute Temperature

28. (I) What are the following temperatures on the Kelvin scale: (a) 66°C, (b) 92°F, (c) −55°C, (d) 5500°C?

29. (I) Absolute zero is what temperature on the Fahrenheit scale?

30. (II) Typical temperatures in the interior of the Earth and Sun are about 4000°C and 15×10^6 °C, respectively. (a) What are these temperatures in kelvins? (b) What percent error is made in each case if a person forgets to change °C to K?

17–7 and 17–8 Ideal Gas Law

31. (I) If 3.80 m^3 of a gas initially at STP is placed under a pressure of 3.20 atm, the temperature of the gas rises to 38.0°C. What is the volume?

32. (I) In an internal combustion engine, air at atmospheric pressure and a temperature of about 20°C is compressed in the cylinder by a piston to $\frac{1}{8}$ of its original volume (compression ratio = 8.0). Estimate the temperature of the compressed air, assuming the pressure reaches 40 atm.

33. (II) Calculate the density of nitrogen at STP using the ideal gas law.

34. (II) If 14.00 mol of helium gas is at 10.0°C and a gauge pressure of 0.350 atm, calculate (a) the volume of the helium gas under these conditions, and (b) the temperature if the gas is compressed to precisely half the volume at a gauge pressure of 1.00 atm.

35. (II) A stoppered test tube traps 25.0 cm^3 of air at a pressure of 1.00 atm and temperature of 18°C. The cylindrically shaped stopper at the test tube's mouth has a diameter of 1.50 cm and will "pop off" the test tube if a net upward force of 10.0 N is applied to it. To what temperature would one have to heat the trapped air in order to "pop off" the stopper? Assume the air surrounding the test tube is always at a pressure of 1.00 atm.

36. (II) A storage tank contains 21.6 kg of nitrogen (N_2) at an absolute pressure of 3.85 atm. What will the pressure be if the nitrogen is replaced by an equal mass of CO_2 at the same temperature?

37. (II) A storage tank at STP contains 28.5 kg of nitrogen (N_2). (a) What is the volume of the tank? (b) What is the pressure if an additional 25.0 kg of nitrogen is added without changing the temperature?

38. (II) A scuba tank is filled with air to a pressure of 204 atm when the air temperature is 29°C. A diver then jumps into the ocean and, after a short time treading water on the ocean surface, checks the tank's pressure and finds that it is only 194 atm. Assuming the diver has inhaled a negligible amount of air from the tank, what is the temperature of the ocean water?

39. (II) What is the pressure inside a 38.0-L container holding 105.0 kg of argon gas at 20.0°C?

40. (II) A tank contains 30.0 kg of O_2 gas at a gauge pressure of 8.20 atm. If the oxygen is replaced by helium at the same temperature, how many kilograms of the latter will be needed to produce a gauge pressure of 7.00 atm?

41. (II) A sealed metal container contains a gas at 20.0°C and 1.00 atm. To what temperature must the gas be heated for the pressure to double to 2.00 atm? (Ignore expansion of the container.)

42. (II) A tire is filled with air at 15°C to a gauge pressure of 250 kPa. If the tire reaches a temperature of 38°C, what fraction of the original air must be removed if the original pressure of 250 kPa is to be maintained?

43. (II) If 61.5 L of oxygen at 18.0°C and an absolute pressure of 2.45 atm are compressed to 48.8 L and at the same time the temperature is raised to 56.0°C, what will the new pressure be?

44. (II) A helium-filled balloon escapes a child's hand at sea level and 20.0°C. When it reaches an altitude of 3600 m, where the temperature is 5.0°C and the pressure only 0.68 atm, how will its volume compare to that at sea level?

45. (II) A sealed metal container can withstand a pressure difference of 0.50 atm. The container initially is filled with an ideal gas at 18°C and 1.0 atm. To what temperature can you cool the container before it collapses? (Ignore any changes in the container's volume due to thermal expansion.)

46. (II) You buy an "airtight" bag of potato chips packaged at sea level, and take the chips on an airplane flight. When you take the potato chips out of your luggage, you notice it has noticeably "puffed up." Airplane cabins are typically pressurized at 0.75 atm, and assuming the temperature inside an airplane is about the same as inside a potato chip processing plant, by what percentage has the bag "puffed up" in comparison to when it was packaged?

47. (II) A typical scuba tank, when fully charged, contains 12 L of air at 204 atm. Assume an "empty" tank contains air at 34 atm and is connected to an air compressor at sea level. The air compressor intakes air from the atmosphere, compresses it to high pressure, and then inputs this high-pressure air into the scuba tank. If the (average) flow rate of air from the atmosphere into the intake port of the air compressor is 290 L/min, how long will it take to fully charge the scuba tank? Assume the tank remains at the same temperature as the surrounding air during the filling process.

48. (III) A sealed container containing 4.0 mol of gas is squeezed, changing its volume from 0.020 m^3 to 0.018 m^3. During this process, the temperature decreases by 9.0 K while the pressure increases by 450 Pa. What was the original pressure and temperature of the gas in the container?

49. (III) Compare the value for the density of water vapor at exactly 100°C and 1 atm (Table 13–1) with the value predicted from the ideal gas law. Why would you expect a difference?

50. (III) An air bubble at the bottom of a lake 37.0 m deep has a volume of 1.00 cm^3. If the temperature at the bottom is 5.5°C and at the top 18.5°C, what is the volume of the bubble just before it reaches the surface?

17–9 Ideal Gas Law in Terms of Molecules; Avogadro's Number

51. (I) Calculate the number of molecules/m^3 in an ideal gas at STP.

52. (I) How many moles of water are there in 1.000 L at STP? How many molecules?

53. (II) What is the pressure in a region of outer space where there is 1 molecule/cm^3 and the temperature is 3 K?

54. (II) Estimate the number of (a) moles and (b) molecules of water in all the Earth's oceans. Assume water covers 75% of the Earth to an average depth of 3 km.

55. (II) The lowest pressure attainable using the best available vacuum techniques is about $10^{-12}\,\text{N/m}^2$. At such a pressure, how many molecules are there per cm^3 at 0°C?

56. (II) Is a gas mostly empty space? Check by assuming that the spatial extent of common gas molecules is about $\ell_0 = 0.3\,\text{nm}$ so one gas molecule occupies an approximate volume equal to ℓ_0^3. Assume STP.

57. (III) Estimate how many molecules of air are in each 2.0-L breath you inhale that were also in the last breath Galileo took. [*Hint*: Assume the atmosphere is about 10 km high and of constant density.]

*17–10 Ideal Gas Temperature Scale

*58. (I) In a constant-volume gas thermometer, what is the limiting ratio of the pressure at the boiling point of water at 1 atm to that at the triple point? (Keep five significant figures.)

*59. (I) At the boiling point of sulfur (444.6°C) the pressure in a constant-volume gas thermometer is 187 torr. Estimate (a) the pressure at the triple point of water, (b) the temperature when the pressure in the thermometer is 118 torr.

*60. (II) Use Fig. 17–17 to determine the inaccuracy of a constant-volume gas thermometer using oxygen if it reads a pressure $P = 268$ torr at the boiling point of water at 1 atm. Express answer (a) in kelvins and (b) as a percentage.

*61. (III) A constant-volume gas thermometer is being used to determine the temperature of the melting point of a substance. The pressure in the thermometer at this temperature is 218 torr; at the triple point of water, the pressure is 286 torr. Some gas is now released from the thermometer bulb so that the pressure at the triple point of water becomes 163 torr. At the temperature of the melting substance, the pressure is 128 torr. Estimate, as accurately as possible, the melting-point temperature of the substance.

General Problems

62. A Pyrex measuring cup was calibrated at normal room temperature. How much error will be made in a recipe calling for 350 mL of cool water, if the water and the cup are hot, at 95°C, instead of at room temperature? Neglect the glass expansion.

63. A precise steel tape measure has been calibrated at 15°C. At 36°C, (a) will it read high or low, and (b) what will be the percentage error?

64. A cubic box of volume $6.15 \times 10^{-2}\,\text{m}^3$ is filled with air at atmospheric pressure at 15°C. The box is closed and heated to 185°C. What is the net force on each side of the box?

65. The gauge pressure in a helium gas cylinder is initially 32 atm. After many balloons have been blown up, the gauge pressure has decreased to 5 atm. What fraction of the original gas remains in the cylinder?

66. If a rod of original length ℓ_1 has its temperature changed from T_1 to T_2, determine a formula for its new length ℓ_2 in terms of T_1, T_2, and α. Assume (a) $\alpha = $ constant, (b) $\alpha = \alpha(T)$ is some function of temperature, and (c) $\alpha = \alpha_0 + bT$ where α_0 and b are constants.

67. If a scuba diver fills his lungs to full capacity of 5.5 L when 8.0 m below the surface, to what volume would his lungs expand if he quickly rose to the surface? Is this advisable?

68. (a) Use the ideal gas law to show that, for an ideal gas at constant pressure, the coefficient of volume expansion is equal to $\beta = 1/T$, where T is the kelvin temperature. Compare to Table 17–1 for gases at $T = 293\,\text{K}$. (b) Show that the bulk modulus (Section 12–4) for an ideal gas held at constant temperature is $B = P$, where P is the pressure.

69. A house has a volume of $870\,\text{m}^3$. (a) What is the total mass of air inside the house at 15°C? (b) If the temperature drops to −15°C, what mass of air enters or leaves the house?

70. Assume that in an alternate universe, the laws of physics are very different from ours and that "ideal" gases behave as follows: (i) At constant temperature, pressure is inversely proportional to the square of the volume. (ii) At constant pressure, the volume varies directly with the $\frac{2}{3}$ power of the temperature. (iii) At 273.15 K and 1.00 atm pressure, 1.00 mole of an ideal gas is found to occupy 22.4 L. Obtain the form of the ideal gas law in this alternate universe, including the value of the gas constant R.

71. An iron cube floats in a bowl of liquid mercury at 0°C. (a) If the temperature is raised to 25°C, will the cube float higher or lower in the mercury? (b) By what percent will the fraction of volume submerged change?

72. (a) The tube of a mercury thermometer has an inside diameter of 0.140 mm. The bulb has a volume of $0.275\,\text{cm}^3$. How far will the thread of mercury move when the temperature changes from 10.5°C to 33.0°C? Take into account expansion of the Pyrex glass. (b) Determine a formula for the change in length of the mercury column in terms of relevant variables. Ignore tube volume compared to bulb volume.

73. From the known value of atmospheric pressure at the surface of the Earth, estimate the total number of air molecules in the Earth's atmosphere.

74. Estimate the percent difference in the density of iron at STP, and when it is a solid deep in the Earth where the temperature is 2000°C and under 5000 atm of pressure. Assume the bulk modulus $(90 \times 10^9\,\text{N/m}^2)$ and the coefficient of volume expansion do not vary with temperature and are the same as at STP.

75. What is the average distance between nitrogen molecules at STP?

76. A helium balloon, assumed to be a perfect sphere, has a radius of 22.0 cm. At room temperature (20°C), its internal pressure is 1.06 atm. Determine the number of moles of helium in the balloon, and the mass of helium needed to inflate the balloon to these values.

77. A standard cylinder of oxygen used in a hospital has gauge pressure $= 2000\,\text{psi}$ (13,800 kPa) and volume $= 14\,\text{L}\,(0.014\,\text{m}^3)$ at $T = 295\,\text{K}$. How long will the cylinder last if the flow rate, measured at atmospheric pressure, is constant at 2.4 L/min?

78. A brass lid screws tightly onto a glass jar at 15°C. To help open the jar, it can be placed into a bath of hot water. After this treatment, the temperatures of the lid and the jar are both 75°C. The inside diameter of the lid is 8.0 cm. Find the size of the gap (difference in radius) that develops by this procedure.

79. The density of gasoline at 0°C is $0.68 \times 10^3\,\text{kg/m}^3$. (a) What is the density on a hot day, when the temperature is 35°C? (b) What is the percent change in density?

80. A helium balloon has volume V_0 and temperature T_0 at sea level where the pressure is P_0 and the air density is ρ_0. The balloon is allowed to float up in the air to altitude y where the temperature is T_1. (a) Show that the volume occupied by the balloon is then $V = V_0(T_1/T_0)e^{+cy}$ where $c = \rho_0 g/P_0 = 1.25 \times 10^{-4}\,\mathrm{m}^{-1}$. (b) Show that the buoyant force does not depend on altitude y. Assume that the skin of the balloon maintains the helium pressure at a constant factor of 1.05 times greater than the outside pressure. [Hint: Assume that the pressure change with altitude is $P = P_0 e^{-cy}$, as in Example 13–5, Chapter 13.]

81. The first real length standard, adopted more than 200 years ago, was a platinum bar with two very fine marks separated by what was defined to be exactly one meter. If this standard bar was to be accurate to within $\pm 1.0\,\mu\mathrm{m}$, how carefully would the trustees have needed to control the temperature? The coefficient of linear expansion is $9 \times 10^{-6}/\mathrm{C}°$.

82. A scuba tank when fully charged has a pressure of 180 atm at 20°C. The volume of the tank is 11.3 L. (a) What would the volume of the air be at 1.00 atm and at the same temperature? (b) Before entering the water, a person consumes 2.0 L of air in each breath, and breathes 12 times a minute. At this rate, how long would the tank last? (c) At a depth of 20.0 m in sea water at a temperature of 10°C, how long would the same tank last assuming the breathing rate does not change?

83. A temperature controller, designed to work in a steam environment, involves a bimetallic strip constructed of brass and steel, connected at their ends by rivets. Each of the metals is 2.0 mm thick. At 20°C, the strip is 10.0 cm long and straight. Find the radius of curvature r of the assembly at 100°C. See Fig. 17–22.

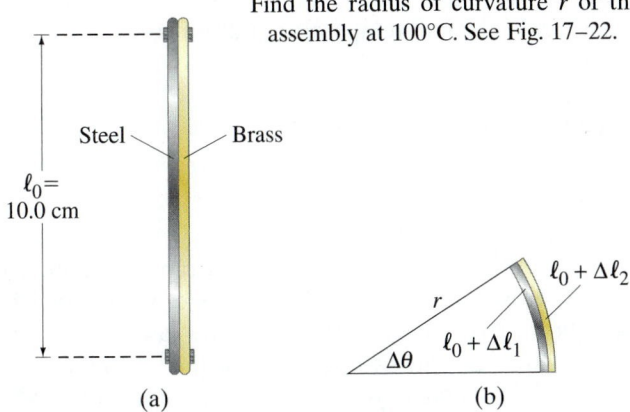

FIGURE 17–22 Problem 83.

84. A copper wire sags 50.0 cm between two utility poles 30.0 m apart when the temperature is −15°C. Estimate the amount of sag when the temperature is +35°C. [Hint: An estimate can be made by assuming the shape of the wire is approximately an arc of a circle; hard equations can sometimes be solved by guessing values.]

85. Snorkelers breathe through short tubular "snorkels" while swimming under water very near the surface. One end of the snorkel attaches to the snorkeler's mouth while the other end protrudes above the water's surface. Unfortunately, snorkels cannot support breathing to any great depth: it is said that a typical snorkeler below a water depth of only about 30 cm cannot draw a breath through a snorkel. Based on this claim, what is the approximate fractional change in a typical person's lung volume when drawing a breath? Assume, in equilibrium, the air pressure in a snorkeler's lungs matches that of the surrounding water pressure.

*Numerical/Computer

*86. (II) A thermocouple consists of a junction of two different types of materials that produces a voltage depending on its temperature. A thermocouple's voltages were recorded when at different temperatures as follows:

Temperature (°C)	50	100	200	300
Voltage (mV)	1.41	2.96	5.90	8.92

Use a spreadsheet to fit these data to a cubic equation and determine the temperature when the thermocouple produces 3.21 mV. Get a second value of the temperature by fitting the data to a quadratic equation.

*87. (III) You have a vial of an unknown liquid which might be octane (gasoline), water, glycerin, or ethyl alcohol. You are trying to determine its identity by studying how its volume changes with temperature changes. You fill a Pyrex graduated cylinder to 100.00 mL with the liquid when the liquid and the cylinder are at 0.000°C. You raise the temperature in five-degree increments, allowing the graduated cylinder and liquid to come to equilibrium at each temperature. You read the volumes listed below off the graduated cylinder at each temperature. Take into account the expansion of the Pyrex glass cylinder. Graph the data, possibly using a spreadsheet program, and determine the slope of the line to find the effective (combined) coefficient of volume expansion β. Then determine β for the liquid and which liquid is in the vial.

Temperature (°C)	Volume Reading (apparent mL)
0.000	100.00
5.000	100.24
10.000	100.50
15.000	100.72
20.000	100.96
25.000	101.26
30.000	101.48
35.000	101.71
40.000	101.97
45.000	102.20
50.000	102.46

Answers to Exercises

A: −40°.

B: (d).

C: 8 mm.

D: (i) Higher, (ii) same, (iii) lower.

E: (a).

F: (b).

G: (b) Less.

In this winter scene in Yellowstone Park, we recognize the three states of matter for water: as a liquid, as a solid (snow and ice), and as a gas (steam). In this Chapter we examine the microscopic theory of matter as atoms or molecules that are always in motion, which we call kinetic theory. We will see that the temperature of a gas is directly related to the average translational kinetic energy of its molecules. We will consider ideal gases, but we will also look at real gases and how they change phase, including evaporation, vapor pressure, and humidity.

C H A P T E R

18

Kinetic Theory of Gases

CHAPTER-OPENING QUESTION—Guess now!
The typical speed of an air molecule at room temperature (20°C) is

(a) nearly at rest (<10 km/h).
(b) on the order of 10 km/h.
(c) on the order of 100 km/h.
(d) on the order of 1000 km/h.
(e) nearly the speed of light.

CONTENTS

18–1 The Ideal Gas Law and the Molecular Interpretation of Temperature

18–2 Distribution of Molecular Speeds

18–3 Real Gases and Changes of Phase

18–4 Vapor Pressure and Humidity

*18–5 Van der Waals Equation of State

*18–6 Mean Free Path

*18–7 Diffusion

T he analysis of matter in terms of atoms in continuous random motion is called **kinetic theory**. We now investigate the properties of a gas from the point of view of kinetic theory, which is based on the laws of classical mechanics. But to apply Newton's laws to each one of the vast number of molecules in a gas ($>10^{25}/m^3$ at STP) is far beyond the capability of any present computer. Instead we take a statistical approach and determine averages of certain quantities, and these averages correspond to macroscopic variables. We will, of course, demand that our microscopic description correspond to the macroscopic properties of gases; otherwise our theory would be of little value. Most importantly, we will arrive at an important relation between the average kinetic energy of molecules in a gas and the absolute temperature.

18–1 The Ideal Gas Law and the Molecular Interpretation of Temperature

We make the following assumptions about the molecules in a gas. These assumptions reflect a simple view of a gas, but nonetheless the results they predict correspond well to the essential features of real gases that are at low pressures and far from the liquefaction point. Under these conditions real gases follow the ideal gas law quite closely, and indeed the gas we now describe is referred to as an **ideal gas**.

The assumptions, which represent the basic postulates of the kinetic theory for an ideal gas, are

1. There are a large number of molecules, N, each of mass m, moving in random directions with a variety of speeds. This assumption is in accord with our observation that a gas fills its container and, in the case of air on Earth, is kept from escaping only by the force of gravity.

2. The molecules are, on average, far apart from one another. That is, their average separation is much greater than the diameter of each molecule.

3. The molecules are assumed to obey the laws of classical mechanics, and are assumed to interact with one another only when they collide. Although molecules exert weak attractive forces on each other between collisions, the potential energy associated with these forces is small compared to the kinetic energy, and we ignore it for now.

4. Collisions with another molecule or the wall of the vessel are assumed to be perfectly elastic, like the collisions of perfectly elastic billiard balls (Chapter 9). We assume the collisions are of very short duration compared to the time between collisions. Then we can ignore the potential energy associated with collisions in comparison to the kinetic energy between collisions.

We can see immediately how this kinetic view of a gas can explain Boyle's law (Section 17–6). The pressure exerted on a wall of a container of gas is due to the constant bombardment of molecules. If the volume is reduced by (say) half, the molecules are closer together and twice as many will be striking a given area of the wall per second. Hence we expect the pressure to be twice as great, in agreement with Boyle's law.

Now let us calculate quantitatively the pressure a gas exerts on its container as based on kinetic theory. We imagine that the molecules are inside a rectangular container (at rest) whose ends have area A and whose length is ℓ, as shown in Fig. 18–1a. The pressure exerted by the gas on the walls of its container is, according to our model, due to the collisions of the molecules with the walls. Let us focus our attention on the wall, of area A, at the left end of the container and examine what happens when one molecule strikes this wall, as shown in Fig. 18–1b. This molecule exerts a force on the wall, and according to Newton's third law the wall exerts an equal and opposite force back on the molecule. The magnitude of this force on the molecule, according to Newton's second law, is equal to the molecule's rate of change of momentum, $F = dp/dt$ (Eq. 9–2). Assuming the collision is elastic, only the x component of the molecule's momentum changes, and it changes from $-mv_x$ (it is moving in the negative x direction) to $+mv_x$. Thus the change in the molecule's momentum, $\Delta(mv)$, which is the final momentum minus the initial momentum, is

$$\Delta(mv) = mv_x - (-mv_x) = 2mv_x$$

for one collision. This molecule will make many collisions with the wall, each separated by a time Δt, which is the time it takes the molecule to travel across the container and back again, a distance (x component) equal to 2ℓ. Thus $2\ell = v_x \Delta t$, or

$$\Delta t = \frac{2\ell}{v_x}.$$

The time Δt between collisions is very small, so the number of collisions per second is very large. Thus the average force—averaged over many collisions—will be equal to the momentum change during one collision divided by the time between collisions (Newton's second law):

$$F = \frac{\Delta(mv)}{\Delta t} = \frac{2mv_x}{2\ell/v_x} = \frac{mv_x^2}{\ell}. \qquad \text{[due to one molecule]}$$

During its passage back and forth across the container, the molecule may collide with the tops and sides of the container, but this does not alter its x component of momentum and thus does not alter our result. It may also collide with other molecules, which may change its v_x. However, any loss (or gain) of momentum is acquired by other molecules, and because we will eventually sum over all the molecules, this effect will be included. So our result above is not altered.

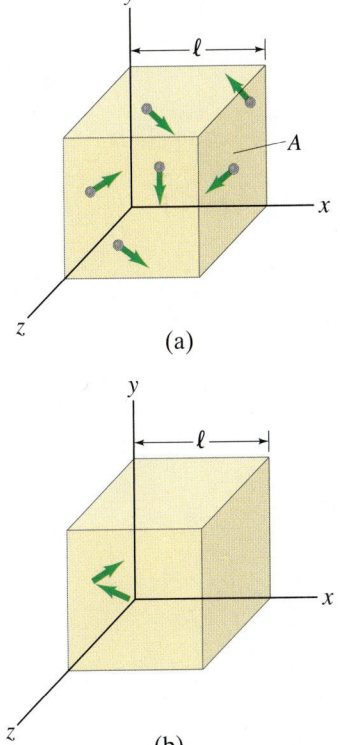

FIGURE 18–1 (a) Molecules of a gas moving about in a rectangular container. (b) Arrows indicate the momentum of one molecule as it rebounds from the end wall.

The actual force due to one molecule is intermittent, but because a huge number of molecules are striking the wall per second, the force is, on average, nearly constant. To calculate the force due to *all* the molecules in the container, we have to add the contributions of each. Thus the net force on the wall is

$$F = \frac{m}{\ell}(v_{x1}^2 + v_{x2}^2 + \cdots + v_{xN}^2),$$

where v_{x1} means v_x for molecule number 1 (we arbitrarily assign each molecule a number) and the sum extends over the total number of molecules N in the container. The average value of the square of the x component of velocity is

$$\overline{v_x^2} = \frac{v_{x1}^2 + v_{x2}^2 + \cdots + v_{xN}^2}{N}, \tag{18-1}$$

where the overbar ($^-$) means "average." Thus we can write the force as

$$F = \frac{m}{\ell}N\overline{v_x^2}.$$

We know that the square of any vector is equal to the sum of the squares of its components (theorem of Pythagoras). Thus $v^2 = v_x^2 + v_y^2 + v_z^2$ for any velocity v. Taking averages, we obtain

$$\overline{v^2} = \overline{v_x^2} + \overline{v_y^2} + \overline{v_z^2}.$$

Since the velocities of the molecules in our gas are assumed to be random, there is no preference to one direction or another. Hence

$$\overline{v_x^2} = \overline{v_y^2} = \overline{v_z^2}.$$

Combining this relation with the one just above, we get

$$\overline{v^2} = 3\overline{v_x^2}.$$

We substitute this into the equation for net force F:

$$F = \frac{m}{\ell}N\frac{\overline{v^2}}{3}.$$

The pressure on the wall is then

$$P = \frac{F}{A} = \frac{1}{3}\frac{Nm\overline{v^2}}{A\ell}$$

or

$$P = \frac{1}{3}\frac{Nm\overline{v^2}}{V}, \tag{18-2}$$

where $V = \ell A$ is the volume of the container. This is the result we were seeking, the pressure exerted by a gas on its container expressed in terms of molecular properties.

Equation 18–2 can be rewritten in a clearer form by multiplying both sides by V and rearranging the right-hand side:

$$PV = \tfrac{2}{3}N(\tfrac{1}{2}m\overline{v^2}). \tag{18-3}$$

The quantity $\frac{1}{2}m\overline{v^2}$ is the average translational kinetic energy $\overline{K}$ of the molecules in the gas. If we compare Eq. 18–3 with Eq. 17–4, the ideal gas law $PV = NkT$, we see that the two agree if

$$\tfrac{2}{3}(\tfrac{1}{2}m\overline{v^2}) = kT,$$

TEMPERATURE RELATED TO AVERAGE KINETIC ENERGY OF MOLECULES

or

$$\overline{K} = \tfrac{1}{2}m\overline{v^2} = \tfrac{3}{2}kT. \qquad\text{[ideal gas]} \quad (18-4)$$

This equation tells us that

> **the average translational kinetic energy of molecules in random motion in an ideal gas is directly proportional to the absolute temperature of the gas.**

The higher the temperature, according to kinetic theory, the faster the molecules are moving on the average. This relation is one of the triumphs of the kinetic theory.

EXAMPLE 18–1 **Molecular kinetic energy.** What is the average translational kinetic energy of molecules in an ideal gas at 37°C?

APPROACH We use the absolute temperature in Eq. 18–4.

SOLUTION We change 37°C to 310 K and insert into Eq. 18–4:

$$\overline{K} = \tfrac{3}{2}kT = \tfrac{3}{2}(1.38 \times 10^{-23}\,\text{J/K})(310\,\text{K}) = 6.42 \times 10^{-21}\,\text{J}.$$

NOTE A mole of molecules would have a total translational kinetic energy equal to $(6.42 \times 10^{-21}\,\text{J})(6.02 \times 10^{23}) = 3860\,\text{J}$, which equals the kinetic energy of a 1-kg stone traveling almost 90 m/s.

EXERCISE A In a mixture of the gases oxygen and helium, which statement is valid: (a) the helium molecules will be moving faster than the oxygen molecules, on average; (b) both kinds of molecules will be moving at the same speed; (c) the oxygen molecules will, on average, be moving more rapidly than the helium molecules; (d) the kinetic energy of the helium will exceed that of the oxygen; (e) none of the above.

Equation 18–4 holds not only for gases, but also applies reasonably accurately to liquids and solids. Thus the result of Example 18–1 would apply to molecules within living cells at body temperature (37°C).

We can use Eq. 18–4 to calculate how fast molecules are moving on the average. Notice that the average in Eqs. 18–1 through 18–4 is over the *square* of the speed. The square root of $\overline{v^2}$ is called the **root-mean-square** speed, v_{rms} (since we are taking the square *root* of the *mean* of the *square* of the speed):

$$v_{\text{rms}} = \sqrt{\overline{v^2}} = \sqrt{\frac{3kT}{m}}. \qquad (18\text{–}5)$$

EXAMPLE 18–2 **Speeds of air molecules.** What is the rms speed of air molecules (O_2 and N_2) at room temperature (20°C)?

APPROACH To obtain v_{rms}, we need the masses of O_2 and N_2 molecules and then apply Eq. 18–5 to oxygen and nitrogen separately, since they have different masses.

SOLUTION The masses of one molecule of O_2 (molecular mass = 32 u) and N_2 (molecular mass = 28 u) are (where 1 u = 1.66×10^{-27} kg)

$$m(O_2) = (32)(1.66 \times 10^{-27}\,\text{kg}) = 5.3 \times 10^{-26}\,\text{kg},$$
$$m(N_2) = (28)(1.66 \times 10^{-27}\,\text{kg}) = 4.6 \times 10^{-26}\,\text{kg}.$$

Thus, for oxygen

$$v_{\text{rms}} = \sqrt{\frac{3kT}{m}} = \sqrt{\frac{(3)(1.38 \times 10^{-23}\,\text{J/K})(293\,\text{K})}{(5.3 \times 10^{-26}\,\text{kg})}} = 480\,\text{m/s},$$

and for nitrogen the result is $v_{\text{rms}} = 510$ m/s. These speeds are more than 1700 km/h or 1000 mi/h, and are greater than the speed of sound ≈ 340 m/s at 20°C (Chapter 16).

NOTE The speed v_{rms} is a magnitude only. The *velocity* of molecules averages to zero: the velocity has direction, and as many molecules move to the right as to the left, as many up as down, as many inward as outward.

EXERCISE B Now you can return to the Chapter-Opening Question, page 476, and answer it correctly. Try to explain why you may have answered differently the first time.

EXERCISE C If you double the volume of a gas while keeping the pressure and number of moles constant, the average (rms) speed of the molecules (a) doubles, (b) quadruples, (c) increases by $\sqrt{2}$, (d) is half, (e) is $\tfrac{1}{4}$.

EXERCISE D By what factor must the absolute temperature change to double v_{rms}? (a) $\sqrt{2}$; (b) 2; (c) $2\sqrt{2}$; (d) 4; (e) 16.

CONCEPTUAL EXAMPLE 18–3 | **Less gas in the tank.** A tank of helium is used to fill balloons. As each balloon is filled, the number of helium atoms remaining in the tank decreases. How does this affect the rms speed of molecules remaining in the tank?

RESPONSE The rms speed is given by Eq. 18–5: $v_{rms} = \sqrt{3kT/m}$. So only the temperature matters, not pressure P or number of moles n. If the tank remains at a constant (ambient) temperature, then the rms speed remains constant even though the pressure of helium in the tank decreases.

In a collection of molecules, the **average speed**, $\bar{v}$, is the average of the magnitudes of the speeds themselves; $\bar{v}$ is generally not equal to v_{rms}. To see the difference between the average speed and the rms speed, consider the following Example.

EXAMPLE 18–4 | **Average speed and rms speed.** Eight particles have the following speeds, given in m/s: 1.0, 6.0, 4.0, 2.0, 6.0, 3.0, 2.0, 5.0. Calculate (a) the average speed and (b) the rms speed.

APPROACH In (a) we sum the speeds and divide by $N = 8$. In (b) we square each speed, sum the squares, divide by $N = 8$, and take the square root.

SOLUTION (a) The average speed is

$$\bar{v} = \frac{1.0 + 6.0 + 4.0 + 2.0 + 6.0 + 3.0 + 2.0 + 5.0}{8} = 3.6 \text{ m/s}.$$

(b) The rms speed is (Eq. 18–1):

$$v_{rms} = \sqrt{\frac{(1.0)^2 + (6.0)^2 + (4.0)^2 + (2.0)^2 + (6.0)^2 + (3.0)^2 + (2.0)^2 + (5.0)^2}{8}} \text{ m/s}$$

$$= 4.0 \text{ m/s}.$$

We see in this Example that $\bar{v}$ and v_{rms} are not necessarily equal. In fact, for an ideal gas they differ by about 8%. We will see in the next Section how to calculate $\bar{v}$ for an ideal gas. We already have the tool to calculate v_{rms} (Eq. 18–5).

*Kinetic Energy Near Absolute Zero

Equation 18–4, $\bar{K} = \frac{3}{2}kT$, implies that as the temperature approaches absolute zero, the kinetic energy of molecules approaches zero. Modern quantum theory, however, tells us this is not quite so. Instead, as absolute zero is approached, the kinetic energy approaches a very small nonzero minimum value. Even though all real gases become liquid or solid near 0 K, molecular motion does not cease, even at absolute zero.

18–2 Distribution of Molecular Speeds

The Maxwell Distribution

The molecules in a gas are assumed to be in random motion, which means that many molecules have speeds less than the average speed and others have speeds greater than the average. In 1859, James Clerk Maxwell (1831–1879) worked out a formula for the most probable distribution of speeds in a gas containing N molecules. We will not give a derivation here but merely quote his result:

$$f(v) = 4\pi N \left(\frac{m}{2\pi kT}\right)^{\frac{3}{2}} v^2 e^{-\frac{1}{2}\frac{mv^2}{kT}} \qquad \text{(18–6)}$$

where $f(v)$ is called the **Maxwell distribution of speeds**, and is plotted in Fig. 18–2. The quantity $f(v)\,dv$ represents the number of molecules that have speed between v and $v + dv$. Notice that $f(v)$ does not give the number of molecules with speed v; $f(v)$ must be multiplied by dv to give the number of molecules (the number of molecules depends on the "width" or "range" of velocities included, dv). In the formula for $f(v)$, m is the mass of a single molecule, T is the absolute temperature, and k is the Boltzmann constant. Since N is the total number of molecules in the gas,

FIGURE 18–2 Distribution of speeds of molecules in an ideal gas. Note that $\bar{v}$ and v_{rms} are not at the peak of the curve. This is because the curve is skewed to the right: it is not symmetrical. The speed at the peak of the curve is the "most probable speed," v_p.

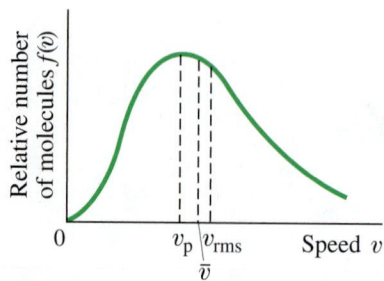

when we sum over all the molecules in the gas we must get N; thus we must have

$$\int_0^\infty f(v)\,dv \;=\; N.$$

(Problem 22 is an exercise to show that this is true.)

Experiments to determine the distribution of speeds in real gases, starting in the 1920s, confirmed with considerable accuracy the Maxwell distribution (for gases at not too high a pressure) and the direct proportion between average kinetic energy and absolute temperature, Eq. 18–4.

The Maxwell distribution for a given gas depends only on the absolute temperature. Figure 18–3 shows the distributions for two different temperatures. Just as v_{rms} increases with temperature, so the whole distribution curve shifts to the right at higher temperatures.

Figure 18–3 illustrates how kinetic theory can be used to explain why many chemical reactions, including those in biological cells, take place more rapidly as the temperature increases. Most chemical reactions take place in a liquid solution, and the molecules in a liquid have a distribution of speeds close to the Maxwell distribution. Two molecules may chemically react only if their kinetic energy is great enough so that when they collide, they partially penetrate into each other. The minimum energy required is called the *activation energy*, E_A, and it has a specific value for each chemical reaction. The molecular speed corresponding to a kinetic energy of E_A for a particular reaction is indicated in Fig. 18–3. The relative number of molecules with energy greater than this value is given by the area under the curve to the right of $v(E_A)$, shown in Fig. 18–3 by the two different shadings. We see that the number of molecules that have kinetic energies in excess of E_A increases greatly for only a small increase in temperature. The rate at which a chemical reaction occurs is proportional to the number of molecules with energy greater than E_A, and thus we see why reaction rates increase rapidly with increased temperature.

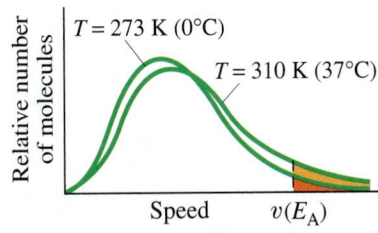

FIGURE 18–3 Distribution of molecular speeds for two different temperatures.

PHYSICS APPLIED

How chemical reactions depend on temperature

*Calculations Using the Maxwell Distribution

Let us see how the Maxwell distribution can be used to obtain some interesting results.

EXAMPLE 18–5 **Determining $\bar{v}$ and v_p.** Determine formulas for (*a*) the average speed, $\bar{v}$, and (*b*) the most probable speed, v_p, of molecules in an ideal gas at temperature T.

APPROACH (*a*) The average value of any quantity is found by multiplying each possible value of the quantity (here, speed) by the number of molecules that have that value, and then summing all these numbers and dividing by N (the total number). For (*b*), we want to find where the curve of Fig. 18–2 has zero slope; so we set $df/dv = 0$.

SOLUTION (*a*) We are given a continuous distribution of speeds (Eq. 18–6), so the sum over the speeds becomes an integral over the product of v and the number $f(v)\,dv$ that have speed v:

$$\bar{v} \;=\; \frac{\displaystyle\int_0^\infty v\,f(v)\,dv}{N} \;=\; 4\pi\left(\frac{m}{2\pi kT}\right)^{\frac{3}{2}}\int_0^\infty v^3 e^{-\frac{1}{2}\frac{mv^2}{kT}}\,dv.$$

We can integrate by parts or look up the definite integral in a Table, and obtain

$$\bar{v} \;=\; 4\pi\left(\frac{m}{2\pi kT}\right)^{\frac{3}{2}}\left(\frac{2k^2T^2}{m^2}\right) \;=\; \sqrt{\frac{8}{\pi}\frac{kT}{m}} \;\approx\; 1.60\sqrt{\frac{kT}{m}}\,.$$

(*b*) The *most probable speed* is that speed which occurs more than any others, and thus is that speed where $f(v)$ has its maximum value. At the maximum of the curve, the slope is zero: $df(v)/dv = 0$. Taking the derivative of Eq. 18–6 gives

$$\frac{df(v)}{dv} \;=\; 4\pi N\left(\frac{m}{2\pi kT}\right)^{\frac{3}{2}}\left(2v\,e^{-\frac{mv^2}{2kT}} - \frac{2mv^3}{2kT}e^{-\frac{mv^2}{2kT}}\right) \;=\; 0.$$

Solving for v, we find

$$v_p \;=\; \sqrt{\frac{2kT}{m}} \;\approx\; 1.41\sqrt{\frac{kT}{m}}\,.$$

(Another solution is $v = 0$, but this corresponds to a minimum, not a maximum.)

In summary,

Most probable speed, v_p

$$v_p = \sqrt{2\frac{kT}{m}} \approx 1.41\sqrt{\frac{kT}{m}}$$ **(18–7a)**

Average speed, $\bar{v}$

$$\bar{v} = \sqrt{\frac{8}{\pi}\frac{kT}{m}} \approx 1.60\sqrt{\frac{kT}{m}}$$ **(18–7b)**

and from Eq. 18–5

rms speed, v_{rms}

$$v_{rms} = \sqrt{3\frac{kT}{m}} \approx 1.73\sqrt{\frac{kT}{m}}.$$

These are all indicated in Fig. 18–2. From Eq. 18–6 and Fig. 18–2, it is clear that the speeds of molecules in a gas vary from zero up to many times the average speed, but as can be seen from the graph, most molecules have speeds that are not far from the average. Less than 1% of the molecules exceed four times v_{rms}.

18–3 Real Gases and Changes of Phase

The ideal gas law

$$PV = NkT$$

is an accurate description of the behavior of a real gas as long as the pressure is not too high and as long as the temperature is far from the liquefaction point. But what happens when these two criteria are not satisfied? First we discuss real gas behavior, and then we examine how kinetic theory can help us understand this behavior.

Let us look at a graph of pressure plotted against volume for a given amount of gas. On such a "PV diagram," Fig. 18–4, each point represents an equilibrium state of the given substance. The various curves (labeled A, B, C, and D) show how the pressure varies, as the volume is changed at constant temperature, for several different values of the temperature. The dashed curve A′ represents the behavior of a gas as predicted by the ideal gas law; that is, PV = constant. The solid curve A represents the behavior of a real gas at the same temperature. Notice that at high pressure, the volume of a real gas is less than that predicted by the ideal gas law. The curves B and C in Fig. 18–4 represent the gas at successively lower temperatures, and we see that the behavior deviates even more from the curves predicted by the ideal gas law (for example, B′), and the deviation is greater the closer the gas is to liquefying.

To explain this, we note that at higher pressure we expect the molecules to be closer together. And, particularly at lower temperatures, the potential energy associated with the attractive forces between the molecules (which we ignored before) is no longer negligible compared to the now reduced kinetic energy of the molecules. These attractive forces tend to pull the molecules closer together so that at a given pressure, the volume is less than expected from the ideal gas law, as in Fig. 18–4. At still lower temperatures, these forces cause liquefaction, and the molecules become very close together. Section 18–5 discusses in more detail the effect of these attractive molecular forces, as well as the effect of the volume which the molecules themselves occupy.

Curve D represents the situation when liquefaction occurs. At low pressure on curve D (on the right in Fig. 18–4), the substance is a gas and occupies a large volume. As the pressure is increased, the volume decreases until point b is reached. Beyond b, the volume decreases with no change in pressure; the substance is gradually changing from the gas to the liquid phase. At point a, all of the substance has changed to

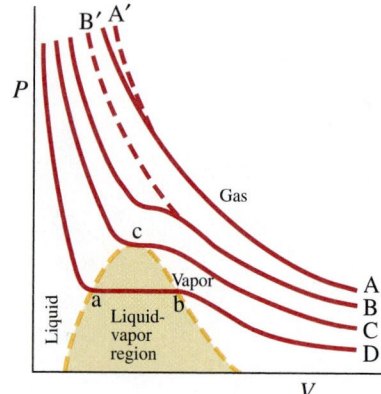

FIGURE 18–4 *PV* diagram for a real substance. Curves A, B, C, and D represent the same substance at different temperatures $(T_A > T_B > T_C > T_D)$.

liquid. Further increase in pressure reduces the volume only slightly—liquids are nearly incompressible—so on the left the curve is very steep as shown. The colored area under the dashed line represents the region where the gas and liquid phases exist together in equilibrium.

Curve C in Fig. 18–4 represents the behavior of the substance at its **critical temperature**; the point c (the one point where curve C is horizontal) is called the **critical point**. At temperatures less than the critical temperature (and this is the definition of the term), a gas will change to the liquid phase if sufficient pressure is applied. Above the critical temperature, no amount of pressure can cause a gas to change phase and become a liquid. The critical temperatures for various gases are given in Table 18–1. Scientists tried for many years to liquefy oxygen without success. Only after the discovery of the behavior of substances associated with the critical point was it realized that oxygen can be liquefied only if first cooled below its critical temperature of $-118°C$.

Often a distinction is made between the terms "gas" and "vapor": a substance below its critical temperature in the gaseous state is called a **vapor**; above the critical temperature, it is called a **gas**.

The behavior of a substance can be diagrammed not only on a PV diagram but also on a PT diagram. A PT diagram, often called a **phase diagram**, is particularly convenient for comparing the different phases of a substance. Figure 18–5 is the phase diagram for water. The curve labeled ℓ-v represents those points where the liquid and vapor phases are in equilibrium—it is thus a graph of the boiling point versus pressure. Note that the curve correctly shows that at a pressure of 1 atm the boiling point is 100°C and that the boiling point is lowered for a decreased pressure. The curve s-ℓ represents points where solid and liquid exist in equilibrium and thus is a graph of the freezing point versus pressure. At 1 atm, the freezing point of water is 0°C, as shown. Notice also in Fig. 18–5 that at a pressure of 1 atm, the substance is in the liquid phase if the temperature is between 0°C and 100°C, but is in the solid or vapor phase if the temperature is below 0°C or above 100°C. The curve labeled s-v is the *sublimation point* versus pressure curve. **Sublimation** refers to the process whereby at low pressures a solid changes directly into the vapor phase without passing through the liquid phase. For water, sublimation occurs if the pressure of the water vapor is less than 0.0060 atm. Carbon dioxide, which in the solid phase is called dry ice, sublimates even at atmospheric pressure (Fig. 18–6).

The intersection of the three curves (in Fig. 18–5) is the **triple point**. For water this occurs at $T = 273.16\,\text{K}$ and $P = 6.03 \times 10^{-3}\,\text{atm}$. It is only at the triple point that the three phases can exist together in equilibrium. Because the triple point corresponds to a unique value of temperature and pressure, it is precisely reproducible and is often used as a point of reference. For example, the standard of temperature is usually specified as exactly 273.16 K at the triple point of water, rather than 273.15 K at the freezing point of water at 1 atm.

Notice that the solid liquid (s-ℓ) curve for water slopes upward to the left. This is true only of substances that *expand* upon freezing: at a higher pressure, a lower temperature is needed to cause the liquid to freeze. More commonly, substances contract upon freezing and the s-ℓ curve slopes upward to the right, as shown for carbon dioxide (CO_2) in Fig. 18–6.

The phase transitions we have been discussing are the common ones. Some substances, however, can exist in several forms in the solid phase. A transition from one phase to another occurs at a particular temperature and pressure, just like ordinary phase changes. For example, ice has been observed in at least eight forms at very high pressure. Ordinary helium has two distinct liquid phases, called helium I and II. They exist only at temperatures within a few degrees of absolute zero. Helium II exhibits very unusual properties referred to as **superfluidity**. It has essentially zero viscosity and exhibits strange properties such as climbing up the sides of an open container. Also interesting are **liquid crystals** (used for TV and computer monitors, Section 35–11) which can be considered to be in a phase between liquid and solid.

TABLE 18–1 Critical Temperatures and Pressures

Substance	Critical Temperature °C	K	Critical Pressure (atm)
Water	374	647	218
CO_2	31	304	72.8
Oxygen	−118	155	50
Nitrogen	−147	126	33.5
Hydrogen	−239.9	33.3	12.8
Helium	−267.9	5.3	2.3

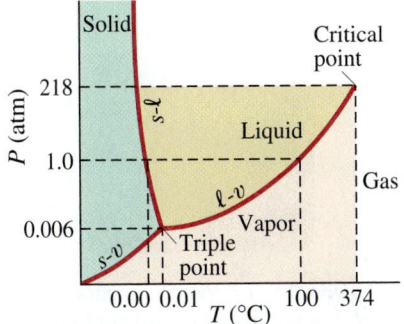

FIGURE 18–5 Phase diagram for water (note that the scales are not linear).

FIGURE 18–6 Phase diagram for carbon dioxide.

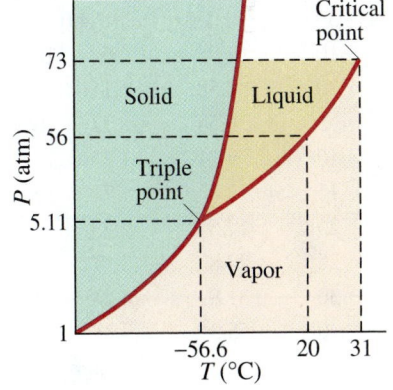

PHYSICS APPLIED
Liquid crystals

SECTION 18–3 Real Gases and Changes of Phase **483**

18–4 Vapor Pressure and Humidity

Evaporation

If a glass of water is left out overnight, the water level will have dropped by morning. We say the water has evaporated, meaning that some of the water has changed to the vapor or gas phase.

This process of **evaporation** can be explained on the basis of kinetic theory. The molecules in a liquid move past one another with a variety of speeds that follow, approximately, the Maxwell distribution. There are strong attractive forces between these molecules, which is what keeps them close together in the liquid phase. A molecule near the surface of the liquid may, because of its speed, leave the liquid momentarily. But just as a rock thrown into the air returns to the Earth, so the attractive forces of the other molecules can pull the vagabond molecule back to the liquid surface—that is, if its velocity is not too large. A molecule with a high enough velocity, however, will escape the liquid entirely (like an object leaving Earth with a high enough speed, Section 8–7), and become part of the gas phase. Only those molecules that have kinetic energy above a particular value can escape to the gas phase. We have already seen that kinetic theory predicts that the relative number of molecules with kinetic energy above a particular value (such as E_A in Fig. 18–3) increases with temperature. This is in accord with the well-known observation that the evaporation rate is greater at higher temperatures.

Because it is the fastest molecules that escape from the surface, the average speed of those remaining is less. When the average speed is less, the absolute temperature is less. Thus kinetic theory predicts that *evaporation is a cooling process.* You have no doubt noticed this effect when you stepped out of a warm shower and felt cold as the water on your body began to evaporate; and after working up a sweat on a hot day, even a slight breeze makes you feel cool through evaporation.

Vapor Pressure

Air normally contains water vapor (water in the gas phase), and it comes mainly from evaporation. To look at this process in a little more detail, consider a closed container that is partially filled with water (or another liquid) and from which the air has been removed (Fig. 18–7). The fastest moving molecules quickly evaporate into the empty space above the liquid's surface. As they move about, some of these molecules strike the liquid surface and again become part of the liquid phase: this is called **condensation**. The number of molecules in the vapor increases until a point is reached when the number of molecules returning to the liquid equals the number leaving in the same time interval. Equilibrium then exists, and the space above the liquid surface is said to be *saturated*. The pressure of the vapor when it is saturated is called the **saturated vapor pressure** (or sometimes simply the vapor pressure).

The saturated vapor pressure does not depend on the volume of the container. If the volume above the liquid were reduced suddenly, the density of molecules in the vapor phase would be increased temporarily. More molecules would then be striking the liquid surface per second. There would be a net flow of molecules back to the liquid phase until equilibrium was again reached, and this would occur at the same value of the saturated vapor pressure, as long as the temperature had not changed.

The saturated vapor pressure of any substance depends on the temperature. At higher temperatures, more molecules have sufficient kinetic energy to break from the liquid surface into the vapor phase. Hence equilibrium will be reached at a higher pressure. The saturated vapor pressure of water at various temperatures is given in Table 18–2. Notice that even solids—for example, ice—have a measurable saturated vapor pressure.

In everyday situations, evaporation from a liquid takes place into the air above it rather than into a vacuum. This does not materially alter the discussion above relating to Fig. 18–7. Equilibrium will still be reached when there are sufficient molecules in the gas phase that the number reentering the liquid equals the number leaving. The concentration of particular molecules (such as water) in the gas phase

FIGURE 18–7 Vapor appears above a liquid in a closed container.

TABLE 18–2 Saturated Vapor Pressure of Water

Temp- erature (°C)	Saturated Vapor Pressure	
	torr (= mm-Hg)	Pa (= N/m²)
−50	0.030	4.0
−10	1.95	2.60×10^2
0	4.58	6.11×10^2
5	6.54	8.72×10^2
10	9.21	1.23×10^3
15	12.8	1.71×10^3
20	17.5	2.33×10^3
25	23.8	3.17×10^3
30	31.8	4.24×10^3
40	55.3	7.37×10^3
50	92.5	1.23×10^4
60	149	1.99×10^4
70[†]	234	3.12×10^4
80	355	4.73×10^4
90	526	7.01×10^4
100[‡]	760	1.01×10^5
120	1489	1.99×10^5
150	3570	4.76×10^5

[†]Boiling point on summit of Mt. Everest.
[‡]Boiling point at sea level.

is not affected by the presence of air, although collisions with air molecules may lengthen the time needed to reach equilibrium. Thus equilibrium occurs at the same value of the saturated vapor pressure as if air were not there.

If the container is large or is not closed, all the liquid may evaporate before saturation is reached. And if the container is not sealed—as, for example, a room in your house—it is not likely that the air will become saturated with water vapor (unless it is raining outside).

Boiling

The saturated vapor pressure of a liquid increases with temperature. When the temperature is raised to the point where the saturated vapor pressure at that temperature equals the external pressure, **boiling** occurs (Fig. 18–8). As the boiling point is approached, tiny bubbles tend to form in the liquid, which indicate a change from the liquid to the gas phase. However, if the vapor pressure inside the bubbles is less than the external pressure, the bubbles immediately are crushed. As the temperature is increased, the saturated vapor pressure inside a bubble eventually becomes equal to or exceeds the external pressure. The bubble will then not collapse but can rise to the surface. Boiling has then begun. *A liquid boils when its saturated vapor pressure equals the external pressure.* This occurs for water at a pressure of 1 atm (760 torr) at 100°C, as can be seen from Table 18–2.

The boiling point of a liquid clearly depends on the external pressure. At high elevations, the boiling point of water is somewhat less than at sea level since the air pressure is less up there. For example, on the summit of Mt. Everest (8850 m) the air pressure is about one-third of what it is at sea level, and from Table 18–2 we can see that water will boil at about 70°C. Cooking food by boiling takes longer at high elevations, since the temperature is less. Pressure cookers, however, reduce cooking time, because they build up a pressure as high as 2 atm, allowing higher boiling temperatures to be attained.

FIGURE 18–8 Boiling: bubbles of water vapor float upward from the bottom (where the temperature is highest).

Partial Pressure and Humidity

When we refer to the weather as being dry or humid, we are referring to the water vapor content of the air. In a gas such as air, which is a mixture of several types of gases, the total pressure is the sum of the *partial pressures* of each gas present.[†] By **partial pressure**, we mean the pressure each gas would exert if it alone were present. The partial pressure of water in the air can be as low as zero and can vary up to a maximum equal to the saturated vapor pressure of water at the given temperature. Thus, at 20°C, the partial pressure of water cannot exceed 17.5 torr (see Table 18–2). The **relative humidity** is defined as the ratio of the partial pressure of water vapor to the saturated vapor pressure at a given temperature. It is usually expressed as a percentage:

$$\text{Relative humidity} = \frac{\text{partial pressure of H}_2\text{O}}{\text{saturated vapor pressure of H}_2\text{O}} \times 100\%.$$

Thus, when the humidity is close to 100%, the air holds nearly all the water vapor it can.

EXAMPLE 18–6 **Relative humidity.** On a particular hot day, the temperature is 30°C and the partial pressure of water vapor in the air is 21.0 torr. What is the relative humidity?

APPROACH From Table 18–2, we see that the saturated vapor pressure of water at 30°C is 31.8 torr.

SOLUTION The relative humidity is thus

$$\frac{21.0 \text{ torr}}{31.8 \text{ torr}} \times 100\% = 66\%.$$

[†]For example, 78% (by volume) of air molecules are nitrogen and 21% oxygen, with much smaller amounts of water vapor, argon, and other gases. At an air pressure of 1 atm, oxygen exerts a partial pressure of 0.21 atm and nitrogen 0.78 atm.

Humans are sensitive to humidity. A relative humidity of 40–50% is generally optimum for both health *b* and comfort. High humidity, particularly on a hot day, reduces the evaporation of moisture from the skin, which is one of the body's vital mechanisms for regulating body temperature. Very low humidity, on the other hand, can dry the skin and mucous membranes.

Air is saturated with water vapor when the partial pressure of water in the air is equal to the saturated vapor pressure at that temperature. If the partial pressure of water exceeds the saturated vapor pressure, the air is said to be **supersaturated**. This situation can occur when a temperature decrease occurs. For example, suppose the temperature is 30°C and the partial pressure of water is 21 torr, which represents a humidity of 66% as we saw in Example 18–6. Suppose now that the temperature falls to, say, 20°C, as might happen at nightfall. From Table 18–2 we see that the saturated vapor pressure of water at 20°C is 17.5 torr. Hence the relative humidity would be greater than 100%, and the supersaturated air cannot hold this much water. The excess water may condense and appear as dew, or as fog or rain (Fig. 18–9).

When air containing a given amount of water is cooled, a temperature is reached where the partial pressure of water equals the saturated vapor pressure. This is called the **dew point**. Measurement of the dew point is the most accurate means of determining the relative humidity. One method uses a polished metal surface in contact with air, which is gradually cooled down. The temperature at which moisture begins to appear on the surface is the dew point, and the partial pressure of water can then be obtained from saturated vapor pressure Tables. If, for example, on a given day the temperature is 20°C and the dew point is 5°C, then the partial pressure of water (Table 18–2) in the 20°C air is 6.54 torr, whereas its saturated vapor pressure is 17.5 torr; hence the relative humidity is $6.54/17.5 = 37\%$.

FIGURE 18–9 Fog or mist settling around a castle where the temperature has dropped below the dew point.

EXERCISE E As the air warms up in the afternoon, how would the relative humidity change if there were no further evaporation? It would (*a*) increase, (*b*) decrease, (*c*) stay the same.

CONCEPTUAL EXAMPLE 18–7 | **Dryness in winter.** Why does the air inside heated buildings seem very dry on a cold winter day?

RESPONSE Suppose the relative humidity outside on a −10°C day is 50%. Table 18–2 tells us the partial pressure of water in the air is about 1.0 torr. If this air is brought indoors and heated to +20°C, the relative humidity is $(1.0 \, \text{torr})/(17.5 \, \text{torr}) = 5.7\%$. Even if the outside air were saturated at a partial pressure of 1.95 torr, the inside relative humidity would be at a low 11%.

*18–5 Van der Waals Equation of State

In Section 18–3, we discussed how real gases deviate from ideal gas behavior, particularly at high densities or when near condensing to a liquid. We would like to understand these deviations using a microscopic (molecular) point of view. J. D. van der Waals (1837–1923) analyzed this problem and in 1873 arrived at an equation of state which fits real gases more accurately than the ideal gas law. His analysis is based on kinetic theory but takes into account: (1) the finite size of molecules (we previously neglected the actual volume of the molecules themselves, compared to the total volume of the container, and this assumption becomes poorer as the density increases and molecules become closer together); (2) the range of the forces between molecules may be greater than the size of the molecules (we previously assumed that intermolecular forces act only during collisions, when the molecules are "in contact"). Let us now look at this analysis and derive the van der Waals equation of state.

Assume the molecules in a gas are spherical with radius *r*. If we assume these molecules behave like hard spheres, then two molecules collide and bounce off one another if the distance between their centers (Fig. 18–10) gets as small as $2r$. Thus the actual volume in which the molecules can move about is somewhat less than the volume *V* of the container holding the gas. The amount of "unavailable volume" depends on the number of molecules and on their size. Let *b* represent the "unavailable volume per mole" of gas.

FIGURE 18–10 Molecules, of radius *r*, colliding.

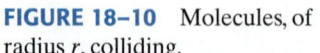
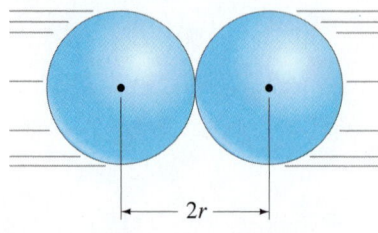

$2r$

Then in the ideal gas law we replace V by $(V - nb)$, where n is the number of moles, and we obtain

$$P(V - nb) = nRT.$$

If we divide through by n, we have

$$P\left(\frac{V}{n} - b\right) = RT. \qquad (18\text{–}8)$$

This relation (sometimes called the **Clausius equation of state**) predicts that for a given temperature T and volume V, the pressure P will be greater than for an ideal gas. This makes sense since the reduced "available" volume means the number of collisions with the walls is increased.

Next we consider the effects of attractive forces between molecules, which are responsible for holding molecules in the liquid and solid states at lower temperatures. These forces are electrical in nature and although they act even when molecules are not touching, we assume their range is small—that is, they act mainly between nearest neighbors. Molecules at the edge of the gas, headed toward a wall of the container, are slowed down by a net force pulling them back into the gas. Thus these molecules will exert less force and less pressure on the wall than if there were no attractive forces. The reduced pressure will be proportional to the density of molecules in the layer of gas at the surface, and also to the density in the next layer, which exerts the inward force.[†] Therefore we expect the pressure to be reduced by a factor proportional to the density squared $(n/V)^2$, here written as moles per volume. If the pressure P is given by Eq. 18–8, then we should reduce this by an amount $a(n/V)^2$ where a is a proportionality constant. Thus we have

$$P = \frac{RT}{(V/n) - b} - \frac{a}{(V/n)^2}$$

or

$$\left(P + \frac{a}{(V/n)^2}\right)\left(\frac{V}{n} - b\right) = RT, \qquad (18\text{–}9)$$

which is the **van der Waals equation of state**.

The constants a and b in the van der Waals equation are different for different gases and are determined by fitting to experimental data for each gas. For CO_2 gas, the best fit is obtained for $a = 0.36\ \mathrm{N \cdot m^4/mol^2}$ and $b = 4.3 \times 10^{-5}\ \mathrm{m^3/mol}$. Figure 18–11 shows a typical PV diagram for Eq. 18–9 (a "van der Waals gas") for four different temperatures, with detailed caption, and it should be compared to Fig. 18–4 for real gases.

Neither the van der Waals equation of state nor the many other equations of state that have been proposed are accurate for all gases under all conditions. Yet Eq. 18–9 is a very useful relation. And because it is quite accurate for many situations, its derivation gives us further insight into the nature of gases at the microscopic level. Note that at low densities, $a/(V/n)^2 \ll P$ and $b \ll V/n$, so that the van der Waals equation reduces to the equation of state for an ideal gas, $PV = nRT$.

*18–6 Mean Free Path

If gas molecules were truly point particles, they would have zero cross-section and never collide with one another. If you opened a perfume bottle, you would be able to smell it almost instantaneously across the room, since molecules travel hundreds of meters per second. In reality, it takes time before you detect an odor and, according to kinetic theory, this must be due to collisions between molecules of nonzero size.

If we were to follow the path of a particular molecule, we would expect to see it follow a zigzag path as shown in Fig. 18–12. Between each collision the molecule would move in a straight-line path. (Not quite true if we take account of the small intermolecular forces that act between collisions.) An important parameter for a given situation is the **mean free path**, which is defined as the average distance a molecule travels between collisions. We would expect that the greater the gas density, and the larger the molecules, the shorter the mean free path would be. We now determine the nature of this relationship for an ideal gas.

[†]This is similar to the gravitational force in which the force on mass m_1 due to mass m_2 is proportional to the product of their masses (Newton's law of universal gravitation, Chapter 6).

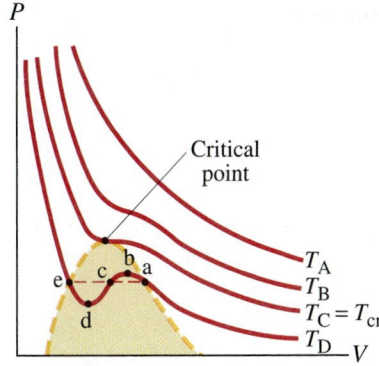

FIGURE 18–11 PV diagram for a van der Waals gas, shown for four different temperatures. For T_A, T_B, and T_C (T_C is chosen equal to the critical temperature), the curves fit experimental data very well for most gases. The curve labeled T_D, a temperature below the critical point, passes through the liquid–vapor region. The maximum (point b) and minimum (point d) would seem to be artifacts, since we usually see constant pressure, as indicated by the horizontal dashed line (and Fig. 18–4). However, for very pure supersaturated vapors or supercooled liquids, the sections ab and ed, respectively, have been observed. (The section bd would be unstable and has not been observed.)

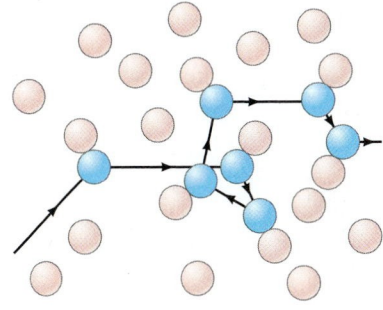

FIGURE 18–12 Zigzag path of a molecule colliding with other molecules.

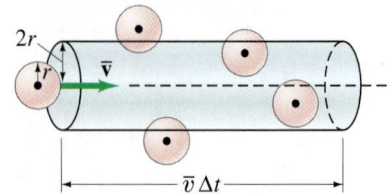

FIGURE 18–13 Molecule at left moves to the right with speed $\bar{v}$. It collides with any molecule whose center is within the cylinder of radius $2r$.

Suppose our gas is made up of molecules which are hard spheres of radius r. A collision will occur whenever the centers of two molecules come within a distance $2r$ of one another. Let us follow a molecule as it traces a straight-line path. In Fig. 18–13, the dashed line represents the path of our particle if it makes no collisions. Also shown is a cylinder of radius $2r$. If the center of another molecule lies within this cylinder, a collision will occur. (Of course, when a collision occurs the particle's path would change direction, as would our imagined cylinder, but our result won't be altered by unbending a zigzag cylinder into a straight one for purposes of calculation.) Assume our molecule is an average one, moving at the mean speed $\bar{v}$ in the gas. For the moment, let us assume that the other molecules are not moving, and that the concentration of molecules (number per unit volume) is N/V. Then the number of molecules whose centers lie within the cylinder of Fig. 18–13 is N/V times the volume of this cylinder, and this also represents the number of collisions that will occur. In a time Δt, our molecule travels a distance $\bar{v}\,\Delta t$, so the length of the cylinder is $\bar{v}\,\Delta t$ and its volume is $\pi(2r)^2\,\bar{v}\,\Delta t$. Hence the number of collisions that occur in a time Δt is $(N/V)\pi(2r)^2\,\bar{v}\,\Delta t$. We define the **mean free path**, ℓ_{M}, as the average distance between collisions. This distance is equal to the distance traveled ($\bar{v}\,\Delta t$) in a time Δt divided by the number of collisions made in time Δt:

$$\ell_{\mathrm{M}} = \frac{\bar{v}\,\Delta t}{(N/V)\pi(2r)^2\,\bar{v}\,\Delta t} = \frac{1}{4\pi r^2(N/V)}. \qquad \textbf{(18–10a)}$$

Thus we see that ℓ_{M} is inversely proportional to the cross-sectional area $(= \pi r^2)$ of the molecules and to their concentration (number/volume), N/V. However, Eq. 18–10a is not fully correct since we assumed the other molecules are all at rest. In fact, they are moving, and the number of collisions in a time Δt must depend on the *relative* speed of the colliding molecules, rather than on $\bar{v}$. Hence the number of collisions per second is $(N/V)\pi(2r)^2 v_{\mathrm{rel}}\,\Delta t$ (rather than $(N/V)\pi(2r)^2\,\bar{v}\,\Delta t$), where v_{rel} is the average relative speed of colliding molecules. A careful calculation shows that for a Maxwellian distribution of speeds $v_{\mathrm{rel}} = \sqrt{2}\,\bar{v}$. Hence the mean free path is

Mean free path

$$\ell_{\mathrm{M}} = \frac{1}{4\pi\sqrt{2}\,r^2(N/V)}. \qquad \textbf{(18–10b)}$$

EXAMPLE 18–8 **ESTIMATE** **Mean free path of air molecules at STP.** Estimate the mean free path of air molecules at STP, standard temperature and pressure (0°C, 1 atm). The diameter of O_2 and N_2 molecules is about 3×10^{-10} m.

APPROACH We saw in Example 17–10 that 1 mol of an ideal gas occupies a volume of $22.4 \times 10^{-3}\,\mathrm{m}^3$ at STP. We can thus determine N/V and apply Eq. 18–10b.

SOLUTION

$$\frac{N}{V} = \frac{6.02 \times 10^{23}\text{ molecules}}{22.4 \times 10^{-3}\,\mathrm{m}^3} = 2.69 \times 10^{25}\text{ molecules/m}^3.$$

Then

$$\ell_{\mathrm{M}} = \frac{1}{4\pi\sqrt{2}\,(1.5 \times 10^{-10}\,\mathrm{m})^2(2.7 \times 10^{25}\,\mathrm{m}^{-3})} \approx 9 \times 10^{-8}\,\mathrm{m}.$$

NOTE This is about 300 times the diameter of an air molecule.

At very low densities, such as in an evacuated vessel, the concept of mean free path loses meaning since collisions with the container walls may occur more frequently than collisions with other molecules. For example, in a cubical box that is (say) 20 cm on a side containing air at 10^{-7} torr $(\approx 10^{-10}\text{ atm})$, the mean free path would be about 900 m, which means many more collisions are made with the walls than with other molecules. (Note, nonetheless, that the box contains over 10^{12} molecules.) If the concept of mean free path included also collision with the walls, it would be closer to 0.2 m than to the 900 m calculated from Eq. 18–10b.

*18–7 Diffusion

If you carefully place a few drops of food coloring in a container of water as in Fig. 18–14, you will find that the color spreads throughout the water. The process may take some time (assuming you do not shake the glass), but eventually the color will become uniform. This mixing, known as **diffusion**, takes place because of the random movement of the molecules. Diffusion occurs in gases too. Common examples include perfume or smoke (or the odor of something cooking on the stove) diffusing in air, although convection (moving air currents) often plays a greater role in spreading odors than does diffusion. Diffusion depends on concentration, by which we mean the number of molecules or moles per unit volume. In general, *the diffusing substance moves from a region where its concentration is high to one where its concentration is low.*

(a)

(b)

(c)

FIGURE 18–14 A few drops of food coloring (a) dropped into water, (b) spreads slowly throughout the water, eventually (c) becoming uniform.

Diffusion can be readily understood on the basis of kinetic theory and the random motion of molecules. Consider a tube of cross-sectional area A containing molecules in a higher concentration on the left than on the right, Fig. 18–15. We assume the molecules are in random motion. Yet there will be a net flow of molecules to the right. To see why this is true, let us consider the small section of tube of length Δx as shown. Molecules from both regions 1 and 2 cross into this central section as a result of their random motion. The more molecules there are in a region, the more will strike a given area or cross a boundary. Since there is a greater concentration of molecules in region 1 than in region 2, more molecules cross into the central section from region 1 than from region 2. There is, then, a net flow of molecules from left to right, from high concentration toward low concentration. The net flow becomes zero only when the concentrations become equal.

You might expect that the greater the difference in concentration, the greater the flow rate. Indeed, the rate of diffusion, J (number of molecules or moles or kg per second), is directly proportional to the difference in concentration per unit distance, $(C_1 - C_2)/\Delta x$ (which is called the **concentration gradient**), and to the cross-sectional area A (see Fig. 18–15):

$$J = DA \frac{C_1 - C_2}{\Delta x},$$

or, in terms of derivatives,

$$J = DA \frac{dC}{dx}. \qquad (18\text{–}11)$$

D is a constant of proportionality called the **diffusion constant**. Equation 18–11 is known as the **diffusion equation**, or **Fick's law**. If the concentrations are given in mol/m^3, then J is the number of moles passing a given point per second. If the concentrations are given in kg/m^3, then J is the mass movement per second (kg/s). The length Δx is given in meters. The values of D for a variety of substances are given in Table 18–3.

FIGURE 18–15 Diffusion occurs from a region of high concentration to one of lower concentration (only one type of molecule is shown).

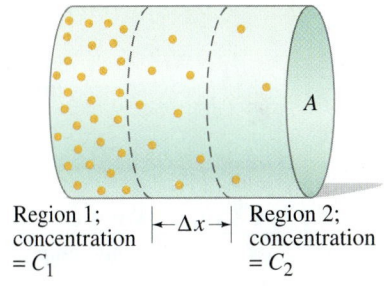

Region 1; concentration $= C_1$ $\;\;\leftarrow\!\Delta x\!\rightarrow\;$ Region 2; concentration $= C_2$

TABLE 18–3 Diffusion Constants, D (20°C, 1 atm)

Diffusing Molecules	Medium	D (m²/s)
H_2	Air	6.3×10^{-5}
O_2	Air	1.8×10^{-5}
O_2	Water	100×10^{-11}
Blood hemoglobin	Water	6.9×10^{-11}
Glycine (an amino acid)	Water	95×10^{-11}
DNA (mass 6×10^6 u)	Water	0.13×10^{-11}

EXAMPLE 18–9 ESTIMATE Diffusion of ammonia in air. To get an idea of the time required for diffusion, estimate how long it might take for ammonia (NH_3) to be detected 10 cm from a bottle after it is opened, assuming only diffusion is occurring.

APPROACH This will be an order-of-magnitude calculation. The rate of diffusion J can be set equal to the number of molecules N diffusing across area A in a time t: $J = N/t$. Then the time $t = N/J$, where J is given by Eq. 18–11. We will have to make some assumptions and rough approximations about concentrations to use Eq. 18–11.

SOLUTION Using Eq. 18–11, we find

$$t = \frac{N}{J} = \frac{N}{DA}\frac{\Delta x}{\Delta C}.$$

The average concentration (midway between bottle and nose) can be approximated by $\overline{C} \approx N/V$, where V is the volume over which the molecules move and is roughly of the order of $V \approx A\,\Delta x$, where Δx is 10 cm = 0.10 m. We substitute $N = \overline{C}V = \overline{C}A\,\Delta x$ into the above equation:

$$t \approx \frac{(\overline{C}A\,\Delta x)\,\Delta x}{DA\,\Delta C} = \frac{\overline{C}}{\Delta C}\frac{(\Delta x)^2}{D}.$$

The concentration of ammonia is high near the bottle (C) and low near the detecting nose (≈ 0), so $\overline{C} \approx C/2 \approx \Delta C/2$, or $(\overline{C}/\Delta C) \approx \frac{1}{2}$. Since NH_3 molecules have a size somewhere between H_2 and O_2, from Table 18–3 we can estimate $D \approx 4 \times 10^{-5}\,m^2/s$. Then

$$t \approx \frac{1}{2}\frac{(0.10\,m)^2}{(4 \times 10^{-5}\,m^2/s)} \approx 100\,s,$$

or about a minute or two.

NOTE This result seems rather long from experience, suggesting that air currents (convection) are more important than diffusion for transmitting odors.

CONCEPTUAL EXAMPLE 18–10 Colored rings on a paper towel. A child colors a small spot on a wet paper towel with a brown marker. Later, she discovers that instead of a brown spot, there are concentric colored rings around the marked spot. What happened?

RESPONSE The ink in a brown marker is composed of several different inks that mix to make brown. These inks each diffuse at different rates through the wet paper towel. After a period of time the inks have diffused far enough that the differences in distances traveled is sufficient to separate the different colors. Chemists and biochemists use a similar technique, called *chromatography*, to separate substances based on their diffusion rates through a medium.

PHYSICS APPLIED
Diffusion time

PHYSICS APPLIED
Chromatography

Summary

According to the **kinetic theory** of gases, which is based on the idea that a gas is made up of molecules that are moving rapidly and at random, the average translational kinetic energy of the molecules is proportional to the Kelvin temperature T:

$$\overline{K} = \tfrac{1}{2}m\overline{v^2} = \tfrac{3}{2}kT \qquad \textbf{(18–4)}$$

where k is Boltzmann's constant.

At any moment, there exists a wide distribution of molecular speeds within a gas. The **Maxwell distribution of speeds** is derived from simple kinetic theory assumptions, and is in good accord with experiment for gases at not too high a pressure.

The behavior of real gases at high pressure, and/or when near their liquefaction point, deviates from the ideal gas law due to the finite size of molecules and to the attractive forces between molecules.

Below the **critical temperature**, a gas can change to a liquid if sufficient pressure is applied; but if the temperature is higher than the critical temperature, no amount of pressure will cause a liquid surface to form.

The **triple point** of a substance is that unique temperature and pressure at which all three phases—solid, liquid, and gas—can coexist in equilibrium. Because of its precise reproducibility, the triple point of water is often taken as a standard reference point.

Evaporation of a liquid is the result of the fastest moving molecules escaping from the surface. Because the average molecular velocity is less after the fastest molecules escape, the temperature decreases when evaporation takes place.

Saturated vapor pressure refers to the pressure of the vapor above a liquid when the two phases are in equilibrium. The vapor pressure of a substance (such as water) depends strongly on temperature and is equal to atmospheric pressure at the boiling point.

Relative humidity of air at a given place is the ratio of the partial pressure of water vapor in the air to the saturated vapor pressure at that temperature; it is usually expressed as a percentage.

[*The **van der Waals equation of state** takes into account the finite volume of molecules, and the attractive forces between molecules, to better approximate the behavior of real gases.]

[*The **mean free path** is the average distance a molecule moves between collisions with other molecules.]

[*Diffusion** is the process whereby molecules of a substance move (on average) from one area to another because of a difference in that substance's concentration.]

Questions

1. Why doesn't the size of different molecules enter into the ideal gas law?

2. When a gas is rapidly compressed (say, by pushing down a piston) its temperature increases. When a gas expands against a piston, it cools. Explain these changes in temperature using the kinetic theory, in particular noting what happens to the momentum of molecules when they strike the moving piston.

3. In Section 18–1 we assumed the gas molecules made perfectly elastic collisions with the walls of the container. This assumption is not necessary as long as the walls are at the same temperature as the gas. Why?

4. Explain in words how Charles's law follows from kinetic theory and the relation between average kinetic energy and the absolute temperature.

5. Explain in words how Gay-Lussac's law follows from kinetic theory.

6. As you go higher in the Earth's atmosphere, the ratio of N_2 molecules to O_2 molecules increases. Why?

7. Can you determine the temperature of a vacuum?

8. Is temperature a macroscopic or microscopic variable?

9. Explain why the peak of the curve for 310 K in Fig. 18–3 is not as high as for 273 K. (Assume the total number of molecules is the same for both.)

10. Escape velocity for the Earth refers to the minimum speed an object must have to leave the Earth and never return. (a) The escape velocity for the Moon is about one-fifth what it is for the Earth due to the Moon's smaller mass; explain why the Moon has practically no atmosphere. (b) If hydrogen was once in the Earth's atmosphere, why would it have probably escaped?

11. If a container of gas is at rest, the average velocity of molecules must be zero. Yet the average speed is not zero. Explain.

12. If the pressure in a gas is doubled while its volume is held constant, by what factor do (a) v_{rms} and (b) $\bar{v}$ change?

13. What everyday observation would tell you that not all molecules in a material have the same speed?

14. We saw that the saturated vapor pressure of a liquid (say, water) does not depend on the external pressure. Yet the temperature of boiling does depend on the external pressure. Is there a contradiction? Explain.

15. Alcohol evaporates more quickly than water at room temperature. What can you infer about the molecular properties of one relative to the other?

16. Explain why a hot humid day is far more uncomfortable than a hot dry day at the same temperature.

17. Is it possible to boil water at room temperature (20°C) without heating it? Explain.

18. What exactly does it mean when we say that oxygen boils at −183°C?

19. A length of thin wire is placed over a block of ice (or an ice cube) at 0°C and weights are hung from the ends of the wire. It is found that the wire cuts its way through the ice cube, but leaves a solid block of ice behind it. This process is called *regelation*. Explain how this happens by inferring how the freezing point of water depends on pressure.

20. Consider two days when the air temperature is the same but the humidity is different. Which is more dense, the dry air or the humid air at the same *T*? Explain.

21. (a) Why does food cook faster in a pressure cooker? (b) Why does pasta or rice need to boil longer at high altitudes? (c) Is it harder to boil water at high altitudes?

22. How do a gas and a vapor differ?

23. (a) At suitable temperatures and pressures, can ice be melted by applying pressure? (b) At suitable temperatures and pressures, can carbon dioxide be melted by applying pressure?

24. Why does dry ice not last long at room temperature?

25. Under what conditions can liquid CO_2 exist? Be specific. Can it exist as a liquid at normal room temperature?

26. Why does exhaled air appear as a little white cloud in the winter (Fig. 18–16)?

FIGURE 18–16
Question 26.

*27. Discuss why sound waves can travel in a gas only if their wavelength is somewhat larger than the mean free path.

*28. Name several ways to reduce the mean free path in a gas.

Problems

18-1 Molecular Interpretation of Temperature

1. (I) (a) What is the average translational kinetic energy of an oxygen molecule at STP? (b) What is the total translational kinetic energy of 1.0 mol of O_2 molecules at 25°C?

2. (I) Calculate the rms speed of helium atoms near the surface of the Sun at a temperature of about 6000 K.

3. (I) By what factor will the rms speed of gas molecules increase if the temperature is increased from 0°C to 180°C?

4. (I) A gas is at 20°C. To what temperature must it be raised to triple the rms speed of its molecules?

5. (I) What speed would a 1.0-g paper clip have if it had the same kinetic energy as a molecule at 15°C?

6. (I) A 1.0-mol sample of helium gas has a temperature of 27°C. (a) What is the total kinetic energy of all the gas atoms in the sample? (b) How fast would a 65-kg person have to run to have the same kinetic energy?

7. (I) Twelve molecules have the following speeds, given in arbitrary units: 6.0, 2.0, 4.0, 6.0, 0.0, 4.0, 1.0, 8.0, 5.0, 3.0, 7.0, and 8.0. Calculate (a) the mean speed, and (b) the rms speed.

8. (II) The rms speed of molecules in a gas at 20.0°C is to be increased by 2.0%. To what temperature must it be raised?

9. (II) If the pressure in a gas is tripled while its volume is held constant, by what factor does v_{rms} change?

10. (II) Show that the rms speed of molecules in a gas is given by $v_{rms} = \sqrt{3P/\rho}$, where P is the pressure in the gas, and ρ is the gas density.

11. (II) Show that for a mixture of two gases at the same temperature, the ratio of their rms speeds is equal to the inverse ratio of the square roots of their molecular masses.

12. (II) What is the rms speed of nitrogen molecules contained in an 8.5-m³ volume at 3.1 atm if the total amount of nitrogen is 1800 mol?

13. (II) (a) For an ideal gas at temperature T show that

$$\frac{dv_{rms}}{dT} = \frac{1}{2}\frac{v_{rms}}{T},$$

and using the approximation $\Delta v_{rms} \approx \frac{dv_{rms}}{dT}\Delta T$, show that

$$\frac{\Delta v_{rms}}{v_{rms}} \approx \frac{1}{2}\frac{\Delta T}{T}.$$

(b) If the average air temperature changes from −5°C in winter to 25°C in summer, estimate the percent change in the rms speed of air molecules between these seasons.

14. (II) What is the average distance between oxygen molecules at STP?

15. (II) Two isotopes of uranium, ^{235}U and ^{238}U (the superscripts refer to their atomic masses), can be separated by a gas diffusion process by combining them with fluorine to make the gaseous compound UF_6. Calculate the ratio of the rms speeds of these molecules for the two isotopes, at constant T. Use Appendix F for masses.

16. (II) Can pockets of vacuum persist in an ideal gas? Assume that a room is filled with air at 20°C and that somehow a small spherical region of radius 1 cm within the room becomes devoid of air molecules. Estimate how long it will take for air to refill this region of vacuum. Assume the atomic mass of air is 29 u.

17. (II) Calculate (a) the rms speed of a nitrogen molecule at 0°C and (b) determine how many times per second it would move back and forth across a 5.0-m-long room on the average, assuming it made very few collisions with other molecules.

18. (III) Estimate how many air molecules rebound from a wall in a typical room per second, assuming an ideal gas of N molecules contained in a cubic room with sides of length ℓ at temperature T and pressure P. (a) Show that the frequency f with which gas molecules strike a wall is

$$f = \frac{\bar{v}_x}{2}\frac{P}{kT}\ell^2$$

where $\bar{v}_x$ is the average x component of the molecule's velocity. (b) Show that the equation can then be written as

$$f \approx \frac{P\ell^2}{\sqrt{4mkT}}$$

where m is the mass of a gas molecule. (c) Assume a cubic air-filled room is at sea level, has a temperature 20°C, and has sides of length $\ell = 3$ m. Determine f.

18-2 Distribution of Molecular Speeds

19. (I) If you double the mass of the molecules in a gas, is it possible to change the temperature to keep the velocity distribution from changing? If so, what do you need to do to the temperature?

20. (I) A group of 25 particles have the following speeds: two have speed 10 m/s, seven have 15 m/s, four have 20 m/s, three have 25 m/s, six have 30 m/s, one has 35 m/s, and two have 40 m/s. Determine (a) the average speed, (b) the rms speed, and (c) the most probable speed.

21. (II) A gas consisting of 15,200 molecules, each of mass 2.00×10^{-26} kg, has the following distribution of speeds, which crudely mimics the Maxwell distribution:

Number of Molecules	Speed (m/s)
1600	220
4100	440
4700	660
3100	880
1300	1100
400	1320

(a) Determine v_{rms} for this distribution of speeds. (b) Given your value for v_{rms}, what (effective) temperature would you assign to this gas? (c) Determine the mean speed $\bar{v}$ of this distribution and use this value to assign an (effective) temperature to the gas. Is the temperature you find here consistent with the one you determined in part (b)?

22. (III) Starting from the Maxwell distribution of speeds, Eq. 18-6, show (a) $\int_0^\infty f(v)\,dv = N$, and (b)

$$\int_0^\infty v^2 f(v)\,dv/N = 3kT/m.$$

18-3 Real Gases

23. (I) CO_2 exists in what phase when the pressure is 30 atm and the temperature is 30°C (Fig. 18-6)?

24. (I) (a) At atmospheric pressure, in what phases can CO_2 exist? (b) For what range of pressures and temperatures can CO_2 be a liquid? Refer to Fig. 18-6.

25. (I) Water is in which phase when the pressure is 0.01 atm and the temperature is (a) 90°C, (b) −20°C?

26. (II) You have a sample of water and are able to control temperature and pressure arbitrarily. (a) Using Fig. 18–5, describe the phase changes you would see if you started at a temperature of 85°C, a pressure of 180 atm, and decreased the pressure down to 0.004 atm while keeping the temperature fixed. (b) Repeat part (a) with the temperature at 0.0°C. Assume that you held the system at the starting conditions long enough for the system to stabilize before making further changes.

18–4 Vapor Pressure and Humidity

27. (I) What is the partial pressure of water vapor at 30°C if the humidity is 85%?

28. (I) What is the partial pressure of water on a day when the temperature is 25°C and the relative humidity is 55%?

29. (I) What is the air pressure at a place where water boils at 80°C?

30. (II) What is the dew point if the humidity is 75% on a day when the temperature is 25°C?

31. (II) If the air pressure at a particular place in the mountains is 0.75 atm, estimate the temperature at which water boils.

32. (II) What is the mass of water in a closed room 5.0 m × 6.0 m × 2.4 m when the temperature is 24.0°C and the relative humidity is 65%?

33. (II) What is the approximate pressure inside a pressure cooker if the water is boiling at a temperature of 120°C? Assume no air escaped during the heating process, which started at 12°C.

34. (II) If the humidity in a room of volume 440 m³ at 25°C is 65%, what mass of water can still evaporate from an open pan?

35. (II) A **pressure cooker** is a sealed pot designed to cook food with the steam produced by boiling water somewhat above 100°C. The pressure cooker in Fig. 18–17 uses a weight of mass m to allow steam to escape at a certain pressure through a small hole (diameter d) in the cooker's lid. If $d = 3.0$ mm, what should m be in order to cook food at 120°C? Assume that atmospheric pressure outside the cooker is 1.01×10^5 Pa.

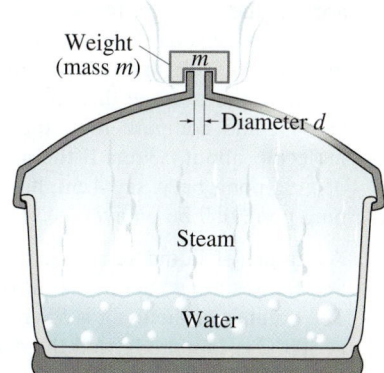

Weight (mass m)

Diameter d

Steam

Water

FIGURE 18–17
Problem 35.

36. (II) When using a mercury barometer (Section 13–6), the vapor pressure of mercury is usually assumed to be zero. At room temperature mercury's vapor pressure is about 0.0015 mm-Hg. At sea level, the height h of mercury in a barometer is about 760 mm. (a) If the vapor pressure of mercury is neglected, is the true atmospheric pressure greater or less than the value read from the barometer? (b) What is the percent error? (c) What is the percent error if you use a water barometer and ignore water's saturated vapor pressure at STP?

37. (II) If the humidity is 45% at 30.0°C, what is the dew point? Use linear interpolation to find the temperature of the dew point to the nearest degree.

38. (III) Air that is at its dew point of 5°C is drawn into a building where it is heated to 20°C. What will be the relative humidity at this temperature? Assume constant pressure of 1.0 atm. Take into account the expansion of the air.

39. (III) What is the mathematical relation between water's boiling temperature and atmospheric pressure? (a) Using the data from Table 18–2, in the temperature range from 50°C to 150°C, plot $\ln P$ versus $(1/T)$, where P is water's saturated vapor pressure (Pa) and T is temperature on the Kelvin scale. Show that a straight-line plot results and determine the slope and y-intercept of the line. (b) Show that your result implies

$$P = Be^{-A/T}$$

where A and B are constants. Use the slope and y-intercept from your plot to show that $A \approx 5000$ K and $B \approx 7 \times 10^{10}$ Pa.

*18–5 Van der Waals Equation of State

*40. (II) In the van der Waals equation of state, the constant b represents the amount of "unavailable volume" occupied by the molecules themselves. Thus V is replaced by $(V - nb)$, where n is the number of moles. For oxygen, b is about 3.2×10^{-5} m³/mol. Estimate the diameter of an oxygen molecule.

*41. (II) For oxygen gas, the van der Waals equation of state achieves its best fit for $a = 0.14$ N·m⁴/mol² and $b = 3.2 \times 10^{-5}$ m³/mol. Determine the pressure in 1.0 mol of the gas at 0°C if its volume is 0.70 L, calculated using (a) the van der Waals equation, (b) the ideal gas law.

*42. (III) A 0.5-mol sample of O_2 gas is in a large cylinder with a movable piston on one end so it can be compressed. The initial volume is large enough that there is not a significant difference between the pressure given by the ideal gas law and that given by the van der Waals equation. As the gas is slowly compressed at constant temperature (use 300 K), at what volume does the van der Waals equation give a pressure that is 5% different than the ideal gas law pressure? Let $a = 0.14$ N·m⁴/mol² and $b = 3.2 \times 10^{-5}$ m³/mol.

*43. (III) (a) From the van der Waals equation of state, show that the critical temperature and pressure are given by

$$T_{cr} = \frac{8a}{27bR}, \qquad P_{cr} = \frac{a}{27b^2}.$$

[Hint: Use the fact that the P versus V curve has an inflection point at the critical point so that the first and second derivatives are zero.] (b) Determine a and b for CO_2 from the measured values of $T_{cr} = 304$ K and $P_{cr} = 72.8$ atm.

*44. (III) How well does the ideal gas law describe the pressurized air in a scuba tank? (a) To fill a typical scuba tank, an air compressor intakes about 2300 L of air at 1.0 atm and compresses this gas into the tank's 12-L internal volume. If the filling process occurs at 20°C, show that a tank holds about 96 mol of air. (b) Assume the tank has 96 mol of air at 20°C. Use the ideal gas law to predict the air's pressure within the tank. (c) Use the van der Waals equation of state to predict the air's pressure within the tank. For air, the van der Waals constants are $a = 0.1373$ N·m⁴/mol² and $b = 3.72 \times 10^{-5}$ m³/mol. (d) Taking the van der Waals pressure as the true air pressure, show that the ideal gas law predicts a pressure that is in error by only about 3%.

* 45. (II) At about what pressure would the mean free path of air molecules be (a) 0.10 m and (b) equal to the diameter of air molecules, $\approx 3 \times 10^{-10}$ m? Assume $T = 20°C$.

* 46. (II) Below a certain threshold pressure, the air molecules (0.3-nm diameter) within a research vacuum chamber are in the "collision-free regime," meaning that a particular air molecule is as likely to cross the container and collide first with the opposite wall, as it is to collide with another air molecule. Estimate the threshold pressure for a vacuum chamber of side 1.0 m at 20°C.

* 47. (II) A very small amount of hydrogen gas is released into the air. If the air is at 1.0 atm and 15°C, estimate the mean free path for a H_2 molecule. What assumptions did you make?

* 48. (II) (a) The mean free path of CO_2 molecules at STP is measured to be about 5.6×10^{-8} m. Estimate the diameter of a CO_2 molecule. (b) Do the same for He gas for which $\ell_M \approx 25 \times 10^{-8}$ m at STP.

* 49. (II) (a) Show that the number of collisions a molecule makes per second, called the collision frequency, f, is given by $f = \bar{v}/\ell_M$, and thus $f = 4\sqrt{2}\,\pi r^2 \bar{v} N/V$. (b) What is the collision frequency for N_2 molecules in air at $T = 20°C$ and $P = 1.0 \times 10^{-2}$ atm?

* 50. (II) We saw in Example 18–8 that the mean free path of air molecules at STP, ℓ_M, is about 9×10^{-8} m. Estimate the collision frequency f, the number of collisions per unit time.

* 51. (II) A cubic box 1.80 m on a side is evacuated so the pressure of air inside is 10^{-6} torr. Estimate how many collisions molecules make with each other for each collision with a wall (0°C).

* 52. (III) Estimate the maximum allowable pressure in a 32-cm-long cathode ray tube if 98% of all electrons must hit the screen without first striking an air molecule.

* 53. (I) Approximately how long would it take for the ammonia of Example 18–9 to be detected 1.0 m from the bottle after it is opened? What does this suggest about the relative importance of diffusion and convection for carrying odors?

* 54. (II) Estimate the time needed for a glycine molecule (see Table 18–3) to diffuse a distance of 15 μm in water at 20°C if its concentration varies over that distance from 1.00 mol/m³ to 0.50 mol/m³? Compare this "speed" to its rms (thermal) speed. The molecular mass of glycine is about 75 u.

* 55. (II) Oxygen diffuses from the surface of insects to the interior through tiny tubes called tracheae. An average trachea is about 2 mm long and has cross-sectional area of 2×10^{-9} m². Assuming the concentration of oxygen inside is half what it is outside in the atmosphere, (a) show that the concentration of oxygen in the air (assume 21% is oxygen) at 20°C is about 8.7 mol/m³, then (b) calculate the diffusion rate J, and (c) estimate the average time for a molecule to diffuse in. Assume the diffusion constant is 1×10^{-5} m²/s.

General Problems

56. A sample of ideal gas must contain at least $N = 10^6$ molecules in order for the Maxwell distribution to be a valid description of the gas, and to assign it a meaningful temperature. For an ideal gas at STP, what is the smallest length scale ℓ (volume $V = \ell^3$) for which a valid temperature can be assigned?

57. In outer space the density of matter is about one atom per cm³, mainly hydrogen atoms, and the temperature is about 2.7 K. Calculate the rms speed of these hydrogen atoms, and the pressure (in atmospheres).

58. Calculate approximately the total translational kinetic energy of all the molecules in an E. coli bacterium of mass 2.0×10^{-15} kg at 37°C. Assume 70% of the cell, by weight, is water, and the other molecules have an average molecular mass on the order of 10^5 u.

59. (a) Estimate the rms speed of an amino acid, whose molecular mass is 89 u, in a living cell at 37°C. (b) What would be the rms speed of a protein of molecular mass 85,000 u at 37°C?

60. The escape speed from the Earth is 1.12×10^4 m/s, so that a gas molecule travelling away from Earth near the outer boundary of the Earth's atmosphere would, at this speed, be able to escape from the Earth's gravitational field and be lost to the atmosphere. At what temperature is the average speed of (a) oxygen molecules, and (b) helium atoms equal to 1.12×10^4 m/s? (c) Can you explain why our atmosphere contains oxygen but not helium?

61. The second postulate of kinetic theory is that the molecules are, on the average, far apart from one another. That is, their average separation is much greater than the diameter of each molecule. Is this assumption reasonable? To check, calculate the average distance between molecules of a gas at STP, and compare it to the diameter of a typical gas molecule, about 0.3 nm. If the molecules were the diameter of ping-pong balls, say 4 cm, how far away would the next ping-pong ball be on average?

62. A sample of liquid cesium is heated in an oven to 400°C and the resulting vapor is used to produce an atomic beam. The volume of the oven is 55 cm³, the vapor pressure of Cs at 400°C is 17 mm-Hg, and the diameter of cesium atoms in the vapor is 0.33 nm. (a) Calculate the mean speed of cesium atoms in the vapor. (b) Determine the number of collisions a single Cs atom undergoes with other cesium atoms per second. (c) Determine the total number of collisions per second between all of the cesium atoms in the vapor. Note that a collision involves two Cs atoms and assume the ideal gas law holds.

63. Consider a container of oxygen gas at a temperature of 20°C that is 1.00 m tall. Compare the gravitational potential energy of a molecule at the top of the container (assuming the potential energy is zero at the bottom) with the average kinetic energy of the molecules. Is it reasonable to neglect the potential energy?

64. In humid climates, people constantly *dehumidify* their cellars to prevent rot and mildew. If the cellar in a house (kept at 20°C) has 115 m^2 of floor space and a ceiling height of 2.8 m, what is the mass of water that must be removed from it in order to drop the humidity from 95% to a more reasonable 40%?

65. Assuming a typical nitrogen or oxygen molecule is about 0.3 nm in diameter, what percent of the room you are sitting in is taken up by the volume of the molecules themselves?

66. A scuba tank has a volume of 3100 cm^3. For very deep dives, the tank is filled with 50% (by volume) pure oxygen and 50% pure helium. (*a*) How many molecules are there of each type in the tank if at 20°C the gauge pressure reads 12 atm? (*b*) What is the ratio of the average kinetic energies of the two types of molecule? (*c*) What is the ratio of the rms speeds of the two types of molecule?

67. A space vehicle returning from the Moon enters the atmosphere at a speed of about 42,000 km/h. Molecules (assume nitrogen) striking the nose of the vehicle with this speed correspond to what temperature? (Because of this high temperature, the nose of a space vehicle must be made of special materials; indeed, part of it does vaporize, and this is seen as a bright blaze upon reentry.)

68. At room temperature, it takes approximately 2.45×10^3 J to evaporate 1.00 g of water. Estimate the average speed of evaporating molecules. What multiple of v_{rms} (at 20°C) for water molecules is this? (Assume Eq. 18–4 holds.)

69. Calculate the total water vapor pressure in the air on the following two days: (*a*) a hot summer day, with the temperature 30°C and the relative humidity at 65%; (*b*) a cold winter day, with the temperature 5°C and the relative humidity at 75%.

*70. At 300 K, an 8.50-mol sample of carbon dioxide occupies a volume of 0.220 m^3. Calculate the gas pressure, first by assuming the ideal gas law, and then by using the van der Waals equation of state. (The values for *a* and *b* are given in Section 18–5.) In this range of pressure and volume, the van der Waals equation is very accurate. What percent error did you make in assuming ideal-gas-law behavior?

*71. The density of atoms, mostly hydrogen, in interstellar space is about one per cubic centimeter. Estimate the mean free path of the hydrogen atoms, assuming an atomic diameter of 10^{-10} m.

*72. Using the ideal gas law, find an expression for the mean free path ℓ_M that involves pressure and temperature instead of (N/V). Use this expression to find the mean free path for nitrogen molecules at a pressure of 7.5 atm and 300 K.

73. A sauna has 8.5 m^3 of air volume, and the temperature is 90°C. The air is perfectly dry. How much water (in kg) should be evaporated if we want to increase the relative humidity from 0% to 10%? (See Table 18–2.)

74. A 0.50-kg trash-can lid is suspended against gravity by tennis balls thrown vertically upward at it. How many tennis balls per second must rebound from the lid elastically, assuming they have a mass of 0.060 kg and are thrown at 12 m/s?

*75. Sound waves in a gas can only propagate if the gas molecules collide with each other on the time scale of the sound wave's period. Thus the highest possible frequency f_{max} for a sound wave in a gas is approximately equal to the inverse of the average collision time between molecules. Assume a gas, composed of molecules with mass *m* and radius *r*, is at pressure *P* and temperature *T*. (*a*) Show that

$$f_{max} \approx 16Pr^2 \sqrt{\frac{\pi}{mkT}}.$$

(*b*) Determine f_{max} for 20°C air at sea level. How many times larger is f_{max} than the highest frequency in the human audio range (20 kHz)?

*Numerical/Computer

*76. (II) Use a spreadsheet to calculate and graph the fraction of molecules in each 50-m/s speed interval from 100 m/s to 5000 m/s if $T = 300$ K.

*77. (II) Use numerical integration [Section 2–9] to estimate (within 2%) the fraction of molecules in air at 1.00 atm and 20°C that have a speed greater than 1.5 times the most probable speed.

*78. (II) For oxygen gas the van der Waals constants are $a = 0.14$ N·m^4/mol^2 and $b = 3.2 \times 10^{-5}$ m^3/mol. Using these values, graph six curves of pressure vs. volume between $V = 2 \times 10^{-5}$ m^3 to 2.0×10^{-4} m^3, for 1 mol of oxygen gas at temperatures of 80 K, 100 K, 120 K, 130 K, 150 K, and 170 K. From the graphs determine approximately the critical temperature for oxygen.

Answers to Exercises

A: (*a*).

B: (*d*).

C: (*c*).

D: (*d*).

E: (*b*).

When it is cold, warm clothes act as insulators to reduce heat loss from the body to the environment by conduction and convection. Heat radiation from a campfire can warm you and your clothes. The fire can also transfer energy directly by heat convection and conduction to what you are cooking. Heat, like work, represents a transfer of energy. Heat is defined as a transfer of energy due to a difference of temperature. Work is a transfer of energy by mechanical means, not due to a temperature difference. The first law of thermodynamics links the two in a general statement of energy conservation: the heat Q added to a system minus the net work W done by the system equals the change in internal energy ΔE_{int} of the system: $\Delta E_{int} = Q - W$. Internal energy E_{int} is the sum total of all the energy of the molecules of the system.

C H A P T E R

19

Heat and the First Law of Thermodynamics

CONTENTS

19–1   Heat as Energy Transfer

19–2   Internal Energy

19–3   Specific Heat

19–4   Calorimetry—Solving Problems

19–5   Latent Heat

19–6   The First Law of Thermodynamics

19–7   The First Law of Thermodynamics Applied; Calculating the Work

19–8   Molar Specific Heats for Gases, and the Equipartition of Energy

19–9   Adiabatic Expansion of a Gas

19–10 Heat Transfer: Conduction, Convection, Radiation

CHAPTER-OPENING QUESTION—Guess now!

A 5-kg cube of warm iron (60°C) is put in thermal contact with a 10-kg cube of cold iron (15°C). Which statement is valid:

(a) Heat flows spontaneously from the warm cube to the cold cube until both cubes have the same heat content.

(b) Heat flows spontaneously from the warm cube to the cold cube until both cubes have the same temperature.

(c) Heat can flow spontaneously from the warm cube to the cold cube, but can also flow spontaneously from the cold cube to the warm cube.

(d) Heat can never flow from a cold object or area to a hot object or area.

(e) Heat flows from the larger cube to the smaller one because the larger one has more internal energy.

When a pot of cold water is placed on a hot burner of a stove, the temperature of the water increases. We say that heat "flows" from the hot burner to the cold water. When two objects at different temperatures are put in contact, heat spontaneously flows from the hotter one to the colder one. The spontaneous flow of heat is in the direction tending to equalize the temperature. If the two objects are kept in contact long enough for their temperatures to become equal, the objects are said to be in thermal equilibrium, and there is no further heat flow between them. For example, when a fever thermometer is first placed in your mouth, heat flows from your mouth to the thermometer. When the thermometer reaches the same temperature as the inside of your mouth, the thermometer and your mouth are then in equilibrium, and no more heat flows.

Heat and temperature are often confused. They are very different concepts and we will make the clear distinction between them. We start this Chapter by defining and using the concept of heat. We also begin our discussion of thermodynamics, which is the name we give to the study of processes in which energy is transferred as heat and as work.

19–1 Heat as Energy Transfer

We use the term "heat" in everyday life as if we knew what we meant. But the term is often used inconsistently, so it is important for us to define heat clearly, and to clarify the phenomena and concepts related to heat.

We commonly speak of the flow of heat—heat flows from a stove burner to a pot of soup, from the Sun to the Earth, from a person's mouth into a fever thermometer. Heat flows spontaneously from an object at higher temperature to one at lower temperature. Indeed, an eighteenth-century model of heat pictured heat flow as movement of a fluid substance called *caloric*. However, the caloric fluid could never be detected. In the nineteenth century, it was found that the various phenomena associated with heat could be described consistently using a new model that views heat as akin to work, as we will discuss in a moment. First we note that a common unit for heat, still in use today, is named after caloric. It is called the **calorie** (cal) and is defined as *the amount of heat necessary to raise the temperature of 1 gram of water by 1 Celsius degree.* [To be precise, the particular temperature range from 14.5°C to 15.5°C is specified because the heat required is very slightly different at different temperatures. The difference is less than 1% over the range 0 to 100°C, and we will ignore it for most purposes.] More often used than the calorie is the **kilocalorie** (kcal), which is 1000 calories. Thus *1 kcal is the heat needed to raise the temperature of 1 kg of water by 1 C°.* Often a kilocalorie is called a **Calorie** (with a capital C), and this Calorie (or the kJ) is used to specify the energy value of food. In the British system of units, heat is measured in British thermal units (Btu). One Btu is defined as the heat needed to raise the temperature of 1 lb of water by 1 F°. It can be shown (Problem 4) that 1 Btu = 0.252 kcal = 1056 J.

The idea that heat is related to energy transfer was pursued by a number of scientists in the 1800s, particularly by an English brewer, James Prescott Joule (1818–1889). One of Joule's experiments is shown (simplified) in Fig. 19–1. The falling weight causes the paddle wheel to turn. The friction between the water and the paddle wheel causes the temperature of the water to rise slightly (barely measurable, in fact, by Joule). In this and many other experiments (some involving electrical energy), Joule determined that a given amount of work done was always equivalent to a particular amount of heat input. Quantitatively, 4.186 joules (J) of work was found to be equivalent to 1 calorie (cal) of heat. This is known as the **mechanical equivalent of heat**:

$$4.186 \text{ J} = 1 \text{ cal};$$
$$4.186 \text{ kJ} = 1 \text{ kcal}.$$

As a result of these and other experiments, scientists came to interpret heat not as a substance, and not exactly as a form of energy. Rather, heat refers to a *transfer of energy*: when heat flows from a hot object to a cooler one, it is energy that is being transferred from the hot to the cold object. Thus, **heat** is *energy transferred from one object to another because of a difference in temperature.* In SI units, the unit for heat, as for any form of energy, is the joule. Nonetheless, calories and kcal are still sometimes used. Today the calorie is *defined* in terms of the joule (via the mechanical equivalent of heat, above), rather than in terms of the properties of water, as given previously. The latter is still handy to remember: 1 cal raises 1 g of water by 1 C°, or 1 kcal raises 1 kg of water by 1 C°.

Joule's result was crucial because it extended the work-energy principle to include processes involving heat. It also led to the establishment of the law of conservation of energy, which we shall discuss more fully later in this Chapter.

CAUTION

Heat is not a fluid

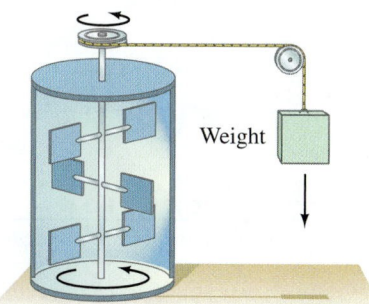

FIGURE 19–1 Joule's experiment on the mechanical equivalent of heat.

CAUTION

Heat is energy transferred because of a ΔT

Working off Calories

EXAMPLE 19–1 **ESTIMATE** **Working off the extra calories.** Suppose you throw caution to the wind and eat too much ice cream and cake on the order of 500 Calories. To compensate, you want to do an equivalent amount of work climbing stairs or a mountain. How much total height must you climb?

APPROACH The work W you need to do in climbing stairs equals the change in gravitational potential energy: $W = \Delta PE = mgh$, where h is the vertical height climbed. For this estimate, approximate your mass as $m \approx 60$ kg.

SOLUTION 500 food Calories are 500 kcal, which in joules is

$$(500 \text{ kcal})(4.186 \times 10^3 \text{ J/kcal}) = 2.1 \times 10^6 \text{ J}.$$

The work done to climb a vertical height h is $W = mgh$. We solve for h:

$$h = \frac{W}{mg} = \frac{2.1 \times 10^6 \text{ J}}{(60 \text{ kg})(9.80 \text{ m/s}^2)} = 3600 \text{ m}.$$

This is a huge elevation change (over 11,000 ft).

NOTE The human body does not transform food energy with 100% efficiency—it is more like 20% efficient. As we shall discuss in the next Chapter, some energy is always "wasted," so you would actually have to climb only about $(0.2)(3600 \text{ m}) \approx 700$ m, which is more reasonable (about 2300 ft of elevation gain).

19–2 Internal Energy

The sum total of all the energy of all the molecules in an object is called its **internal energy**. (Sometimes **thermal energy** is used to mean the same thing.) We introduce the concept of internal energy now since it will help clarify ideas about heat.

Distinguishing Temperature, Heat, and Internal Energy

Distinguish heat from internal energy and from temperature

Using the kinetic theory, we can make a clear distinction between temperature, heat, and internal energy. Temperature (in kelvins) is a measure of the *average* kinetic energy of individual molecules. Internal energy refers to the *total* energy of all the molecules within the object. (Thus two equal-mass hot ingots of iron may have the same temperature, but two of them have twice as much internal energy as one does.) Heat, finally, refers to a *transfer* of energy from one object to another because of a difference in temperature.

Direction of heat flow depends on temperature (not on amount of internal energy)

Notice that the direction of heat flow between two objects depends on their temperatures, not on how much internal energy each has. Thus, if 50 g of water at 30°C is placed in contact (or mixed) with 200 g of water at 25°C, heat flows *from* the water at 30°C *to* the water at 25°C even though the internal energy of the 25°C water is much greater because there is so much more of it.

EXERCISE A Return to the Chapter-Opening Question, page 496, and answer it again now. Try to explain why you may have answered differently the first time.

Internal Energy of an Ideal Gas

Let us calculate the internal energy of n moles of an ideal monatomic (one atom per molecule) gas. The internal energy, E_{int}, is the sum of the translational kinetic energies of all the atoms.[†] This sum is equal to the average kinetic energy per molecule times the total number of molecules, N:

$$E_{\text{int}} = N(\tfrac{1}{2}m\overline{v^2}).$$

Using Eq. 18–4, $\overline{K} = \tfrac{1}{2}m\overline{v^2} = \tfrac{3}{2}kT$, we can write this as

$$E_{\text{int}} = \tfrac{3}{2}NkT \tag{19–1a}$$

[†]The symbol U is used in some books for internal energy. The use of E_{int} avoids confusion with U which stands for potential energy (Chapter 8).

498 CHAPTER 19 Heat and the First Law of Thermodynamics

or (recall Section 17–9)

$$E_{\text{int}} = \tfrac{3}{2}nRT,$$ [ideal monatomic gas] **(19–1b)**

where n is the number of moles. Thus, the internal energy of an ideal gas depends only on temperature and the number of moles of gas.

If the gas molecules contain more than one atom, then the rotational and vibrational energy of the molecules (Fig. 19–2) must also be taken into account. The internal energy will be greater at a given temperature than for a monatomic gas, but it will still be a function only of temperature for an ideal gas.

The internal energy of real gases also depends mainly on temperature, but where real gases deviate from ideal gas behavior, their internal energy depends also somewhat on pressure and volume (due to atomic potential energy).

The internal energy of liquids and solids is quite complicated, for it includes electrical potential energy associated with the forces (or "chemical" bonds) between atoms and molecules.

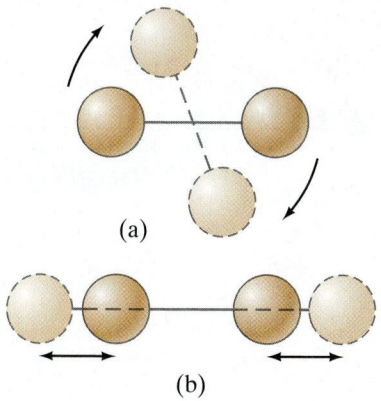

FIGURE 19–2 Besides translational kinetic energy, molecules can have (a) rotational kinetic energy, and (b) vibrational energy (both kinetic and potential).

19–3 Specific Heat

If heat flows into an object, the object's temperature rises (assuming no phase change). But how much does the temperature rise? That depends. As early as the eighteenth century, experimenters had recognized that the amount of heat Q required to change the temperature of a given material is proportional to the mass m of the material present and to the temperature change ΔT. This remarkable simplicity in nature can be expressed in the equation

$$Q = mc\,\Delta T,$$ **(19–2)**

where c is a quantity characteristic of the material called its **specific heat**. Because $c = Q/m\,\Delta T$, specific heat is specified in units[†] of $J/kg\cdot C°$ (the proper SI unit) or $kcal/kg\cdot C°$. For water at 15°C and a constant pressure of 1 atm, $c = 4.19 \times 10^3\,J/kg\cdot C°$ or $1.00\,kcal/kg\cdot C°$, since, by definition of the cal and the joule, it takes 1 kcal of heat to raise the temperature of 1 kg of water by 1 C°. Table 19–1 gives the values of specific heat for other solids and liquids at 20°C. The values of c for solids and liquids depend to some extent on temperature (as well as slightly on pressure), but for temperature changes that are not too great, c can often be considered constant.[‡] Gases are more complicated and are treated in Section 19–8.

EXAMPLE 19–2 How heat transferred depends on specific heat. (a) How much heat input is needed to raise the temperature of an empty 20-kg vat made of iron from 10°C to 90°C? (b) What if the vat is filled with 20 kg of water?

APPROACH We apply Eq. 19–2 to the different materials involved.

SOLUTION (a) Our system is the iron vat alone. From Table 19–1, the specific heat of iron is $450\,J/kg\cdot C°$. The change in temperature is $(90°C - 10°C) = 80\,C°$. Thus,

$$Q = mc\,\Delta T = (20\,kg)(450\,J/kg\cdot C°)(80\,C°) = 7.2 \times 10^5\,J = 720\,kJ.$$

(b) Our system is the vat plus the water. The water alone would require

$$Q = mc\,\Delta T = (20\,kg)(4186\,J/kg\cdot C°)(80\,C°) = 6.7 \times 10^6\,J = 6700\,kJ,$$

or almost 10 times what an equal mass of iron requires. The total, for the vat plus the water, is $720\,kJ + 6700\,kJ = 7400\,kJ$.

NOTE In (b), the iron vat and the water underwent the same temperature change, $\Delta T = 80\,C°$, but their specific heats are different.

[†]Note that $J/kg\cdot C°$ means $\dfrac{J}{kg\cdot C°}$ and *not* $(J/kg)\cdot C° = J\cdot C°/kg$ (otherwise we would have written it that way).

[‡]To take into account the dependence of c on T, we can write Eq. 19–2 in differential form: $dQ = mc(T)\,dT$, where $c(T)$ means c is a function of temperature T. Then the heat Q required to change the temperature from T_1 to T_2 is

$$Q = \int_{T_1}^{T_2} mc(T)\,dT.$$

TABLE 19–1 Specific Heats
(at 1 atm constant pressure and 20°C unless otherwise stated)

Substance	Specific Heat, c	
	kcal/kg · C° (= cal/g · C°)	J/kg · C°
Aluminum	0.22	900
Alcohol (ethyl)	0.58	2400
Copper	0.093	390
Glass	0.20	840
Iron or steel	0.11	450
Lead	0.031	130
Marble	0.21	860
Mercury	0.033	140
Silver	0.056	230
Wood	0.4	1700
Water		
Ice (−5°C)	0.50	2100
Liquid (15°C)	1.00	4186
Steam (110°C)	0.48	2010
Human body (average)	0.83	3470
Protein	0.4	1700

If the iron vat in part (*a*) of Example 19–2 had been *cooled* from 90°C to 10°C, 720 kJ of heat would have flowed *out* of the iron. In other words, Eq. 19–2 is valid for heat flow either in or out, with a corresponding increase or decrease in temperature. We saw in part (*b*) that water requires almost 10 times as much heat as an equal mass of iron to make the same temperature change. Water has one of the highest specific heats of all substances, which makes it an ideal substance for hot-water space-heating systems and other uses that require a minimal drop in temperature for a given amount of heat transfer. It is the water content, too, that causes the apples rather than the crust in hot apple pie to burn our tongues, through heat transfer.

PHYSICS APPLIED
Practical effects of water's high specific heat

19–4 Calorimetry—Solving Problems

In discussing heat and thermodynamics, we shall often refer to particular systems. As already mentioned in earlier Chapters, a **system** is any object or set of objects that we wish to consider. Everything else in the universe we will refer to as its "environment" or the "surroundings." There are several categories of systems. A **closed system** is one for which no mass enters or leaves (but energy may be exchanged with the environment). In an **open system**, mass may enter or leave (as may energy). Many (idealized) systems we study in physics are closed systems. But many systems, including plants and animals, are open systems since they exchange materials (food, oxygen, waste products) with the environment. A closed system is said to be **isolated** if no energy in any form passes across its boundaries; otherwise it is not isolated.

When different parts of an isolated system are at different temperatures, heat will flow (energy is transferred) from the part at higher temperature to the part at lower temperature—that is, within the system. If the system is truely isolated, no energy is transferred into or out of it. So the *conservation of energy* again plays an important role for us: the heat lost by one part of the system is equal to the heat gained by the other part:

$$\text{heat lost} \; = \; \text{heat gained}$$

or

$$\text{energy out of one part} \; = \; \text{energy into another part.}$$

These simple relations are very useful, but depend on the (often very good) approximation that the whole system is isolated (no other energy transfers occur). Let us take an Example.

EXAMPLE 19–3 **The cup cools the tea.** If 200 cm³ of tea at 95°C is poured into a 150-g glass cup initially at 25°C (Fig. 19–3), what will be the common final temperature T of the tea and cup when equilibrium is reached, assuming no heat flows to the surroundings?

APPROACH We apply conservation of energy to our system of tea plus cup, which we are assuming is isolated: all of the heat that leaves the tea flows into the cup. We use the specific heat equation, Eq. 19–2, to determine how the heat flow is related to the temperature changes.

SOLUTION Because tea is mainly water, its specific heat is 4186 J/kg·C° (Table 19–1), and its mass m is its density times its volume $(V = 200 \text{ cm}^3 = 200 \times 10^{-6} \text{ m}^3)$: $m = \rho V = (1.0 \times 10^3 \text{ kg/m}^3)(200 \times 10^{-6} \text{ m}^3) = 0.20 \text{ kg}$. We use Eq. 19–2, apply conservation of energy, and let T be the as yet unknown final temperature:

$$\text{heat lost by tea} \; = \; \text{heat gained by cup}$$
$$m_{\text{tea}} c_{\text{tea}}(95°C - T) \; = \; m_{\text{cup}} c_{\text{cup}}(T - 25°C).$$

Putting in numbers and using Table 19–1 ($c_{\text{cup}} = 840 \text{ J/kg·C°}$ for glass), we solve for T, and find

$$(0.20 \text{ kg})(4186 \text{ J/kg·C°})(95°C - T) \; = \; (0.15 \text{ kg})(840 \text{ J/kg·C°})(T - 25°C)$$
$$79{,}500 \text{ J} - (837 \text{ J/C°})T \; = \; (126 \text{ J/C°})T - 3150 \text{ J}$$
$$T \; = \; 86°C.$$

The tea drops in temperature by 9 C° by coming into equilibrium with the cup.

NOTE The cup increases in temperature by $86°C - 25°C = 61 \text{ C}°$. Its much greater change in temperature (compared with that of the tea water) is due to its much smaller specific heat compared to that of water.

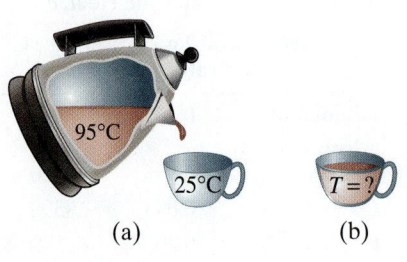

FIGURE 19–3 Example 19–3.

95°C

25°C $T = ?$

(a)   (b)

NOTE In this calculation, the ΔT (of Eq. 19–2, $Q = mc\,\Delta T$) is a positive quantity on both sides of our conservation of energy equation. On the left is "heat lost" and ΔT is the initial minus the final temperature ($95°C - T$), whereas on the right is "heat gained" and ΔT is the final minus the initial temperature. But consider the following alternate approach.

Alternate Solution We can set up this Example (and others) by a different approach. We can write that the total heat transferred into or out of the isolated system is zero:

$$\Sigma Q = 0.$$

Then each term is written as $Q = mc(T_f - T_i)$, and $\Delta T = T_f - T_i$ is always the final minus the initial temperature, and each ΔT can be positive or negative. In the present Example:

$$\Sigma Q = m_{cup}c_{cup}(T - 25°C) + m_{tea}c_{tea}(T - 95°C) = 0.$$

The second term is negative because T will be less than 95°C. Solving the algebra gives the same result.

! **CAUTION**

*When using
heat lost = heat gained,
ΔT is positive on both sides*

The exchange of energy, as exemplified in Example 19–3, is the basis for a technique known as **calorimetry**, which is the quantitative measurement of heat exchange. To make such measurements, a **calorimeter** is used; a simple water calorimeter is shown in Fig. 19–4. It is very important that the calorimeter be well insulated so that almost no heat is exchanged with the surroundings. One important use of the calorimeter is in the determination of specific heats of substances. In the technique known as the "method of mixtures," a sample of a substance is heated to a high temperature, which is accurately measured, and then quickly placed in the cool water of the calorimeter. The heat lost by the sample will be gained by the water and the calorimeter cup. By measuring the final temperature of the mixture, the specific heat can be calculated, as illustrated in the following Example.

PROBLEM SOLVING

Alternate approach: $\Sigma Q = 0$

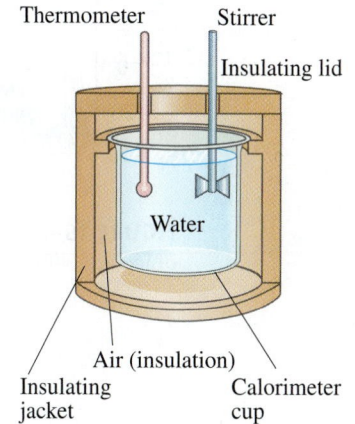

FIGURE 19–4 Simple water calorimeter.

EXAMPLE 19–4 Unknown specific heat determined by calorimetry. An engineer wishes to determine the specific heat of a new metal alloy. A 0.150-kg sample of the alloy is heated to 540°C. It is then quickly placed in 0.400 kg of water at 10.0°C, which is contained in a 0.200-kg aluminum calorimeter cup. (We do not need to know the mass of the insulating jacket since we assume the air space between it and the cup insulates it well, so that its temperature does not change significantly.) The final temperature of the system is 30.5°C. Calculate the specific heat of the alloy.

APPROACH We apply conservation of energy to our system, which we take to be the alloy sample, the water, and the calorimeter cup. We assume this system is isolated, so the energy lost by the hot alloy equals the energy gained by the water and calorimeter cup.

SOLUTION The heat lost equals the heat gained:

$$\begin{pmatrix} \text{heat lost} \\ \text{by alloy} \end{pmatrix} = \begin{pmatrix} \text{heat gained} \\ \text{by water} \end{pmatrix} + \begin{pmatrix} \text{heat gained by} \\ \text{calorimeter cup} \end{pmatrix}$$

$$m_a c_a \Delta T_a = m_w c_w \Delta T_w + m_{cal} c_{cal} \Delta T_{cal}$$

where the subscripts a, w, and cal refer to the alloy, water, and calorimeter, respectively, and each $\Delta T > 0$. When we put in values and use Table 19–1, this equation becomes

$$(0.150\,\text{kg})(c_a)(540°C - 30.5°C) = (0.400\,\text{kg})(4186\,\text{J/kg}\cdot\text{C}°)(30.5°C - 10.0°C)$$
$$+ (0.200\,\text{kg})(900\,\text{J/kg}\cdot\text{C}°)(30.5°C - 10.0°C)$$
$$(76.4\,\text{kg}\cdot\text{C}°)\,c_a = (34{,}300 + 3690)\,\text{J}$$
$$c_a = 497\,\text{J/kg}\cdot\text{C}°.$$

In making this calculation, we have ignored any heat transferred to the thermometer and the stirrer (which is used to quicken the heat transfer process and thus reduce heat loss to the outside). It can be taken into account by adding additional terms to the right side of the above equation and will result in a slight correction to the value of c_a.

In all Examples and Problems of this sort, be sure to include *all* objects that gain or lose heat (within reason). On the "heat loss" side here, it is only the hot metal alloy. On the "heat gain" side, it is both the water and the aluminum calorimeter cup. For simplicity, we have ignored very small masses, such as the thermometer and the stirrer, which will affect the energy balance only very slightly.

PROBLEM SOLVING

Be sure to consider all possible sources of energy transfer

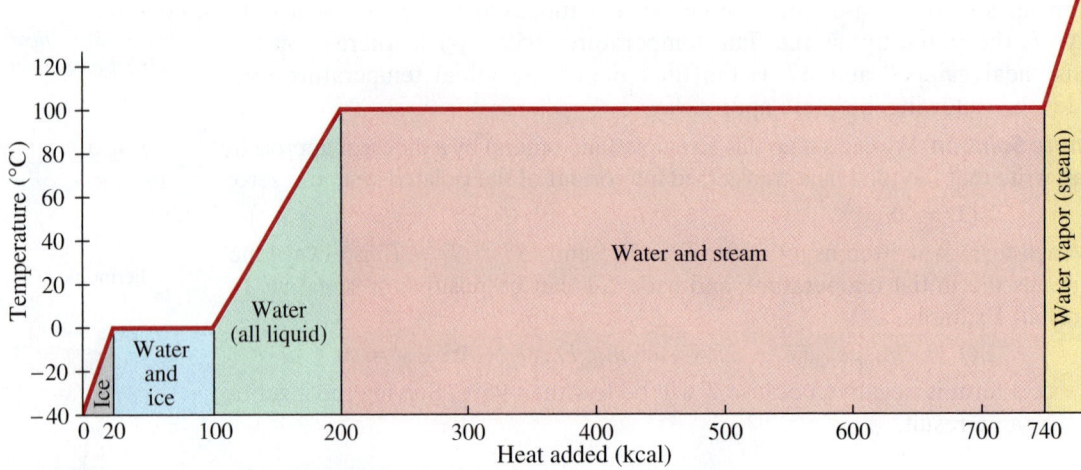

FIGURE 19–5 Temperature as a function of the heat added to bring 1.0 kg of ice at −40°C to steam above 100°C.

19–5 Latent Heat

When a material changes phase from solid to liquid, or from liquid to gas (see also Section 18–3), a certain amount of energy is involved in this **change of phase**. For example, let us trace what happens when a 1.0-kg block of ice at −40°C is heated at a slow steady rate until all the ice has changed to water, then the (liquid) water is heated to 100°C and changed to steam, and heated further above 100°C, all at 1 atm pressure. As shown in the graph of Fig. 19–5, as the ice is heated starting at −40°C, its temperature rises at a rate of about 2 C°/kcal of heat added (for ice, $c \approx 0.50$ kcal/kg·C°). However, when 0°C is reached, the temperature stops increasing even though heat is still being added. The ice gradually changes to water in the liquid state, with no change in temperature. After about 40 kcal has been added at 0°C, half the ice remains and half has changed to water. After about 80 kcal, or 330 kJ, has been added, all the ice has changed to water, still at 0°C. Continued addition of heat causes the water's temperature to again increase, now at a rate of 1 C°/kcal. When 100°C is reached, the temperature again remains constant as the heat added changes the liquid water to vapor (steam). About 540 kcal (2260 kJ) is required to change the 1.0 kg of water completely to steam, after which the graph rises again, indicating that the temperature of the steam rises as heat is added.

The heat required to change 1.0 kg of a substance from the solid to the liquid state is called the **heat of fusion**; it is denoted by L_F. The heat of fusion of water is 79.7 kcal/kg or, in proper SI units, 333 kJ/kg $(= 3.33 \times 10^5$ J/kg$)$. The heat required to change a substance from the liquid to the vapor phase is called the **heat of vaporization**, L_V. For water it is 539 kcal/kg or 2260 kJ/kg. Other substances follow graphs similar to Fig. 19–5, although the melting-point and boiling-point temperatures are different, as are the specific heats and heats of fusion and vaporization. Values for the heats of fusion and vaporization, which are also called the **latent heats**, are given in Table 19–2 for a number of substances.

The heats of vaporization and fusion also refer to the amount of heat *released* by a substance when it changes from a gas to a liquid, or from a liquid to a solid. Thus, steam releases 2260 kJ/kg when it changes to water, and water releases 333 kJ/kg when it becomes ice.

The heat involved in a change of phase depends not only on the latent heat but also on the total mass of the substance. That is,

$$Q = mL, \qquad \qquad \textbf{(19–3)}$$

where L is the latent heat of the particular process and substance, m is the mass of the substance, and Q is the heat added or released during the phase change. For example, when 5.00 kg of water freezes at 0°C, $(5.00$ kg$)(3.33 \times 10^5$ J/kg$) = 1.67 \times 10^6$ J of energy is released.

TABLE 19–2 Latent Heats (at 1 atm)

Substance	Melting Point (°C)	Heat of Fusion kcal/kg[†]	Heat of Fusion kJ/kg	Boiling Point (°C)	Heat of Vaporization kcal/kg[†]	Heat of Vaporization kJ/kg
Oxygen	−218.8	3.3	14	−183	51	210
Nitrogen	−210.0	6.1	26	−195.8	48	200
Ethyl alcohol	−114	25	104	78	204	850
Ammonia	−77.8	8.0	33	−33.4	33	137
Water	0	79.7	333	100	539	2260
Lead	327	5.9	25	1750	208	870
Silver	961	21	88	2193	558	2300
Iron	1808	69.1	289	3023	1520	6340
Tungsten	3410	44	184	5900	1150	4800

[†]Numerical values in kcal/kg are the same in cal/g.

EXERCISE B A pot of water is boiling on a gas stove, and then you turn up the heat. What happens? (*a*) The temperature of the water starts increasing. (*b*) There is a tiny decrease in the rate of water loss by evaporation. (*c*) The rate of water loss by boiling increases. (*d*) There is an appreciable increase in both the rate of boiling and the temperature of the water. (*e*) None of these.

Calorimetry sometimes involves a change of state, as the following Examples show. Indeed, latent heats are often measured using calorimetry.

EXAMPLE 19–5 Will all the ice melt? A 0.50-kg chunk of ice at −10°C is placed in 3.0 kg of "iced" tea at 20°C. At what temperature and in what phase will the final mixture be? The tea can be considered as water. Ignore any heat flow to the surroundings, including the container.

APPROACH Before we can write down an equation applying conservation of energy, we must first check to see if the final state will be all ice, a mixture of ice and water at 0°C, or all water. To bring the 3.0 kg of water at 20°C down to 0°C would require an energy release of (Eq. 19–2)

$$m_w c_w(20°C − 0°C) = (3.0\,\text{kg})(4186\,\text{J/kg·C°})(20\,\text{C°}) = 250\,\text{kJ}.$$

On the other hand, to raise the ice from −10°C to 0°C would require

$$m_{ice} c_{ice}[0°C − (−10°C)] = (0.50\,\text{kg})(2100\,\text{J/kg·C°})(10\,\text{C°}) = 10.5\,\text{kJ},$$

and to change the ice to water at 0°C would require (Eq. 19–3)

$$m_{ice} L_F = (0.50\,\text{kg})(333\,\text{kJ/kg}) = 167\,\text{kJ},$$

for a total of 10.5 kJ + 167 kJ = 177 kJ. This is not enough energy to bring the 3.0 kg of water at 20°C down to 0°C, so we see that the mixture must end up all water, somewhere between 0°C and 20°C.

SOLUTION To determine the final temperature *T*, we apply conservation of energy and write: heat gain = heat loss,

$$\begin{pmatrix}\text{heat to raise} \\ \text{0.50 kg of ice} \\ \text{from −10°C} \\ \text{to 0°C}\end{pmatrix} + \begin{pmatrix}\text{heat to change} \\ \text{0.50 kg} \\ \text{of ice} \\ \text{to water}\end{pmatrix} + \begin{pmatrix}\text{heat to raise} \\ \text{0.50 kg of water} \\ \text{from 0°C} \\ \text{to } T\end{pmatrix} = \begin{pmatrix}\text{heat lost by} \\ \text{3.0 kg of} \\ \text{water cooling} \\ \text{from 20°C to } T\end{pmatrix}.$$

Using some of the results from above, we obtain

$$10.5\,\text{kJ} + 167\,\text{kJ} + (0.50\,\text{kg})(4186\,\text{J/kg·C°})(T − 0°C)$$
$$= (3.0\,\text{kg})(4186\,\text{J/kg·C°})(20°C − T).$$

Solving for *T* we obtain

$$T = 5.0°C.$$

EXERCISE C How much more ice at −10°C would be needed in Example 19–5 to bring the tea down to 0°C, while just melting all the ice?

Calorimetry

1. Be sure you have sufficient information to apply energy conservation. Ask yourself: **is the system isolated** (or very nearly so, enough to get a good estimate)? Do we know or can we calculate all significant sources of energy transfer?

2. Apply **conservation of energy**:

   heat gained = heat lost.

   For each substance in the system, a heat (energy) term will appear on either the left or right side of this equation. [Alternatively, use $\Sigma Q = 0$.]

3. If **no phase changes** occur, each term in the energy conservation equation (above) will have the form

$$Q(\text{gain}) = mc(T_f - T_i)$$

   or

$$Q(\text{lost}) = mc(T_i - T_f)$$

   where T_i and T_f are the initial and final temperatures

of the substance, and m and c are its mass and specific heat, respectively.

4. If **phase changes** do or might occur, there may be terms in the energy conservation equation of the form $Q = mL$, where L is the latent heat. But *before* applying energy conservation, determine (or estimate) in which phase the final state will be, as we did in Example 19–5 by calculating the different contributing values for heat Q.

5. Be sure each term appears on the correct side of the **energy equation** (heat gained or heat lost) and that each ΔT is positive.

6. Note that when the system reaches thermal **equilibrium**, the final **temperature** of each substance will have the *same* value. There is only one T_f.

7. **Solve** your energy equation for the unknown.

EXAMPLE 19–6 **Determining a latent heat.** The specific heat of liquid mercury is $140\,\text{J/kg}\cdot\text{C}°$. When 1.0 kg of solid mercury at its melting point of $-39°\text{C}$ is placed in a 0.50-kg aluminum calorimeter filled with 1.2 kg of water at $20.0°\text{C}$, the mercury melts and the final temperature of the combination is found to be $16.5°\text{C}$. What is the heat of fusion of mercury in J/kg?

APPROACH We follow the Problem Solving Strategy above.

SOLUTION

1. **Is the system isolated?** The mercury is placed in a calorimeter, which we assume is well insulated. Our isolated system is the calorimeter, the water, and the mercury.

2. **Conservation of energy.** The heat gained by the mercury = the heat lost by the water and calorimeter.

3. and **4. Phase changes.** There is a phase change (of mercury), plus we use specific heat equations. The heat gained by the mercury (Hg) includes a term representing the melting of the Hg,

$$Q(\text{melt solid Hg}) = m_{Hg}L_{Hg},$$

   plus a term representing the heating of the liquid Hg from $-39°\text{C}$ to $+16.5°\text{C}$:

$$Q(\text{heat liquid Hg}) = m_{Hg}c_{Hg}[16.5°\text{C} - (-39°\text{C})]$$
$$= (1.0\,\text{kg})(140\,\text{J/kg}\cdot\text{C}°)(55.5\,\text{C}°) = 7770\,\text{J}.$$

   All of this heat gained by the mercury is obtained from the water and calorimeter, which cool down:

$$Q_{cal} + Q_w = m_{cal}c_{cal}(20.0°\text{C} - 16.5°\text{C}) + m_w c_w(20.0°\text{C} - 16.5°\text{C})$$
$$= (0.50\,\text{kg})(900\,\text{J/kg}\cdot\text{C}°)(3.5\,\text{C}°) + (1.2\,\text{kg})(4186\,\text{J/kg}\cdot\text{C}°)(3.5\,\text{C}°)$$
$$= 19{,}200\,\text{J}.$$

5. **Energy equation.** The conservation of energy tells us the heat lost by the water and calorimeter cup must equal the heat gained by the mercury:

$$Q_{cal} + Q_w = Q(\text{melt solid Hg}) + Q(\text{heat liquid Hg})$$

   or

$$19{,}200\,\text{J} = m_{Hg}L_{Hg} + 7770\,\text{J}.$$

6. **Equilibrium temperature.** It is given as $16.5°\text{C}$, and we already used it.

7. Solve. The only unknown in our energy equation (point 5) is L_{Hg}, the latent heat of fusion (or melting) of mercury. We solve for it, putting in $m_{Hg} = 1.0$ kg:

$$L_{Hg} = \frac{19{,}200\ \text{J}\ -\ 7770\ \text{J}}{1.0\ \text{kg}} = 11{,}400\ \text{J/kg} \approx 11\ \text{kJ/kg},$$

where we rounded off to 2 significant figures.

Evaporation

The latent heat to change a liquid to a gas is needed not only at the boiling point. Water can change from the liquid to the gas phase even at room temperature. This process is called **evaporation** (see also Section 18–4). The value of the heat of vaporization of water increases slightly with a decrease in temperature: at 20°C, for example, it is 2450 kJ/kg (585 kcal/kg) compared to 2260 kJ/kg (= 539 kcal/kg) at 100°C. When water evaporates, the remaining liquid cools, because the energy required (the latent heat of vaporization) comes from the water itself; so its internal energy, and therefore its temperature, must drop.[†]

Evaporation of water from the skin is one of the most important methods the body uses to control its temperature. When the temperature of the blood rises slightly above normal, the hypothalamus region of the brain detects this temperature increase and sends a signal to the sweat glands to increase their production. The energy (latent heat) required to vaporize this water comes from the body, and hence the body cools.

PHYSICS APPLIED
Body temperature

Kinetic Theory of Latent Heats

We can make use of kinetic theory to see why energy is needed to melt or vaporize a substance. At the melting point, the latent heat of fusion does not act to increase the average kinetic energy (and the temperature) of the molecules in the solid, but instead is used to overcome the potential energy associated with the forces between the molecules. That is, work must be done against these attractive forces to break the molecules loose from their relatively fixed positions in the solid so they can freely roll over one another in the liquid phase. Similarly, energy is required for molecules held close together in the liquid phase to escape into the gaseous phase. This process is a more violent reorganization of the molecules than is melting (the average distance between the molecules is greatly increased), and hence the heat of vaporization is generally much greater than the heat of fusion for a given substance.

19–6 The First Law of Thermodynamics

Up to now in this Chapter we have discussed internal energy and heat. But work too is often involved in thermodynamic processes.

In Chapter 8 we saw that work is done when energy is transferred from one object to another by mechanical means. In Section 19–1 we saw that heat is a transfer of energy from one object to a second one at a lower temperature. Thus, heat is much like work. To distinguish them, *heat* is defined as a *transfer of energy due to a difference in temperature*, whereas work is a transfer of energy that is not due to a temperature difference.

In Section 19–2, we defined the internal energy of a system as the sum total of all the energy of the molecules within the system. We would expect that the internal energy of a system would be increased if work was done on the system, or if heat were added to it. Similarly the internal energy would be decreased if heat flowed out of the system or if work were done by the system on something in the surroundings.

[†]According to kinetic theory, evaporation is a cooling process because it is the fastest-moving molecules that escape from the surface. Hence the average speed of the remaining molecules is less, so by Eq. 18–4 the temperature is less.

Thus it is reasonable to extend conservation of energy and propose an important law: the change in internal energy of a closed system, ΔE_{int}, will be equal to the energy added to the system by heating minus the work done by the system on the surroundings. In equation form we write

FIRST LAW OF
THERMODYNAMICS

$$\Delta E_{\text{int}} = Q - W \qquad\qquad \textbf{(19–4)}$$

⚠ CAUTION

Heat added is +
Heat lost is −
Work on system is −
Work by system is +

where Q is the net heat *added* to the system and W is the net work done *by* the system.[†] We must be careful and consistent in following the sign conventions for Q and W. Because W in Eq. 19–4 is the work done *by* the system, then if work is done *on* the system, W will be negative and E_{int} will increase. Similarly, Q is positive for heat added to the system, so if heat leaves the system, Q is negative.

Equation 19–4 is known as the **first law of thermodynamics**. It is one of the great laws of physics, and its validity rests on experiments (such as Joule's) to which no exceptions have been seen. Since Q and W represent energy transferred into or out of the system, the internal energy changes accordingly. Thus, the first law of thermodynamics is a great and broad statement of the *law of conservation of energy*.

It is worth noting that the conservation of energy law was not formulated until the nineteenth century, for it depended on the interpretation of heat as a transfer of energy.

Equation 19–4 applies to a closed system. It also applies to an open system (Section 19–4) if we take into account the change in internal energy due to the increase or decrease in the amount of matter. For an isolated system (p. 500), no work is done and no heat enters or leaves the system, so $W = Q = 0$, and hence $\Delta E_{\text{int}} = 0$.

A given system at any moment is in a particular state and can be said to have a certain amount of internal energy, E_{int}. But a system does not "have" a certain amount of heat or work. Rather, when work is done on a system (such as compressing a gas), or when heat is added or removed from a system, the state of the system *changes*. Thus, work and heat are involved in *thermodynamic processes* that can change the system from one state to another; they are not characteristic of the state itself. Quantities which describe the state of a system, such as internal energy E_{int}, pressure P, volume V, temperature T, and mass m or number of moles n, are called **state variables**. Q and W are *not* state variables.

Because E_{int} is a *state variable*, which depends only on the state of the system and not on how the system arrived in that state, we can write

$$\Delta E_{\text{int}} = E_{\text{int}, 2} - E_{\text{int}, 1} = Q - W$$

where $E_{\text{int}, 1}$ and $E_{\text{int}, 2}$ represent the internal energy of the system in states 1 and 2, and Q and W are the heat added to the system and work done by the system in going from state 1 to state 2.

It is sometimes useful to write the first law of thermodynamics in differential form:

$$dE_{\text{int}} = dQ - dW.$$

Here, dE_{int} represents an infinitesimal change in internal energy when an infinitesimal amount of heat dQ is added to the system, and the system does an infinitesimal amount of work dW.[‡]

[†]This convention relates historically to steam engines: the interest was in the heat *input* and the work *output*, both regarded as positive. In other books you may see the first law of thermodynamics written as $\Delta E_{\text{int}} = Q + W$, in which case W refers to the work done *on* the system.

[‡]The differential form of the first law is often written

$$dE_{\text{int}} = \bar{d}Q - \bar{d}W,$$

where the bars on the differential sign ($\bar{d}$) are used to remind us that W and Q are not functions of the state variables (such as P, V, T, n). Internal energy, E_{int}, is a function of the state variables, and so dE_{int} represents the differential (called an *exact differential*) of some function E_{int}. The differentials $\bar{d}W$ and $\bar{d}Q$ are not exact differentials (they are not the differential of some mathematical function); they thus only represent infinitesimal amounts. This issue won't really be of concern in this book.

EXAMPLE 19–7 **Using the first law.** 2500 J of heat is added to a system, and 1800 J of work is done on the system. What is the change in internal energy of the system?

APPROACH We apply the first law of thermodynamics, Eq. 19–4, to our system.

SOLUTION The heat added to the system is $Q = 2500$ J. The work W done *by* the system is -1800 J. Why the minus sign? Because 1800 J done *on* the system (as given) equals -1800 J done *by* the system, and it is the latter we need for the sign conventions we used in Eq. 19–4. Hence

$$\Delta E_{int} = 2500 \text{ J} - (-1800 \text{ J}) = 2500 \text{ J} + 1800 \text{ J} = 4300 \text{ J}.$$

You may have intuitively thought that the 2500 J and the 1800 J would need to be added together, since both refer to energy added to the system. You would have been right.

EXERCISE D What would be the internal energy change in Example 19–7 if 2500 J of heat is added to the system and 1800 J of work is done *by* the system (i.e., as output)?

*The First Law of Thermodynamics Extended

To write the first law of thermodynamics in its complete form, consider a system that has kinetic energy K (there is motion) as well as potential energy U. Then the first law of thermodynamics would have to include these terms and would be written as

$$\Delta K + \Delta U + \Delta E_{int} = Q - W. \tag{19–5}$$

EXAMPLE 19–8 **Kinetic energy transformed to thermal energy.** A 3.0-g bullet traveling at a speed of 400 m/s enters a tree and exits the other side with a speed of 200 m/s. Where did the bullet's lost kinetic energy go, and what was the energy transferred?

APPROACH Take the bullet and tree as our system. No potential energy is involved. No work is done on (or by) the system by outside forces, nor is any heat added because no energy was transferred to or from the system due to a temperature difference. Thus the kinetic energy gets transformed into internal energy of the bullet and tree.

SOLUTION From the first law of thermodynamics as given in Eq. 19–5, we are given $Q = W = \Delta U = 0$, so we have

$$\Delta K + \Delta E_{int} = 0$$

or, using subscripts i and f for initial and final velocities

$$\Delta E_{int} = -\Delta K = -(K_f - K_i) = \tfrac{1}{2}m(v_i^2 - v_f^2)$$
$$= \tfrac{1}{2}(3.0 \times 10^{-3} \text{ kg})[(400 \text{ m/s})^2 - (200 \text{ m/s})^2] = 180 \text{ J}.$$

NOTE The internal energy of the bullet and tree both increase, as both experience a rise in temperature. If we had chosen the bullet alone as our system, work would be done on it and heat transfer would occur.

19–7 The First Law of Thermodynamics Applied; Calculating the Work

Let us analyze some simple processes in the light of the first law of thermodynamics.

Isothermal Processes ($\Delta T = 0$)

First we consider an idealized process that is carried out at constant temperature. Such a process is called an **isothermal** process (from the Greek meaning "same temperature"). If the system is an ideal gas, then $PV = nRT$ (Eq. 17–3), so for a fixed amount of gas kept at constant temperature, $PV = $ constant. Thus the process follows a curve like AB on the PV diagram shown in Fig. 19–6, which is a curve for $PV = $ constant. Each point on the curve, such as point A, represents the state of the system at a given moment—that is, its pressure P and volume V. At a lower temperature, another isothermal process would be represented by a curve like A′B′ in Fig. 19–6 (the product $PV = nRT = $ constant is less when T is less). The curves shown in Fig. 19–6 are referred to as *isotherms*.

FIGURE 19–6 *PV* diagram for an ideal gas undergoing isothermal processes at two different temperatures.

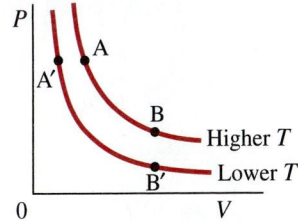

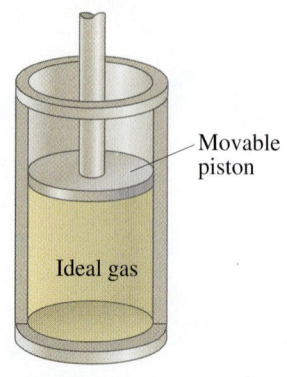

FIGURE 19–7 An ideal gas in a cylinder fitted with a movable piston.

Let us assume that the gas is enclosed in a container fitted with a movable piston, Fig. 19–7, and that the gas is in contact with a **heat reservoir** (a body whose mass is so large that, ideally, its temperature does not change significantly when heat is exchanged with our system). We also assume that the process of compression (volume decreases) or expansion (volume increases) is done **quasistatically** ("almost statically"), by which we mean extremely slowly, so that all of the gas moves between a series of equilibrium states each of which are at the same constant temperature. If an amount of heat Q is added to the system and temperature is to remain constant, the gas will expand and do an amount of work W on the environment (it exerts a force on the piston and moves it through a distance). The temperature and mass are kept constant so, from Eq. 19–1, the internal energy does not change: $\Delta E_{\text{int}} = \frac{3}{2} nR \Delta T = 0$. Hence, by the first law of thermodynamics, Eq. 19–4, $\Delta E_{\text{int}} = Q - W = 0$, so $W = Q$: the work done by the gas in an isothermal process equals the heat added to the gas.

Adiabatic Processes ($Q = 0$)

An **adiabatic** process is one in which no heat is allowed to flow into or out of the system: $Q = 0$. This situation can occur if the system is extremely well insulated, or the process happens so quickly that heat—which flows slowly—has no time to flow in or out. The very rapid expansion of gases in an internal combustion engine is one example of a process that is very nearly adiabatic. A slow adiabatic expansion of an ideal gas follows a curve like that labeled AC in Fig. 19–8. Since $Q = 0$, we have from Eq. 19–4 that $\Delta E_{\text{int}} = -W$. That is, the internal energy decreases if the gas expands; hence the temperature decreases as well (because $\Delta E_{\text{int}} = \frac{3}{2} nR \Delta T$). This is evident in Fig. 19–8 where the product PV ($= nRT$) is less at point C than at point B (curve AB is for an isothermal process, for which $\Delta E_{\text{int}} = 0$ and $\Delta T = 0$). In the reverse operation, an adiabatic compression (going from C to A, for example), work is done *on* the gas, and hence the internal energy increases and the temperature rises. In a diesel engine, the fuel–air mixture is rapidly compressed adiabatically by a factor of 15 or more; the temperature rise is so great that the mixture ignites spontaneously.

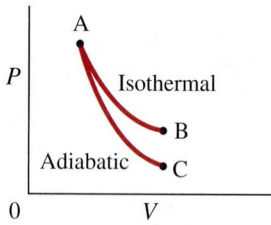

FIGURE 19–8 *PV* diagram for adiabatic (AC) and isothermal (AB) processes on an ideal gas.

Isobaric and Isovolumetric Processes

Isothermal and adiabatic processes are just two possible processes that can occur. Two other simple thermodynamic processes are illustrated on the *PV* diagrams of Fig. 19–9: (a) an **isobaric** process is one in which the pressure is kept constant, so the process is represented by a horizontal straight line on the *PV* diagram, Fig. 19–9a; (b) an **isovolumetric** (or *isochoric*) process is one in which the volume does not change (Fig. 19–9b). In these, and in all other processes, the first law of thermodynamics holds.

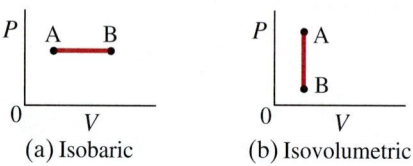

FIGURE 19–9 (a) Isobaric ("same pressure") process. (b) Isovolumetric ("same volume") process.

Work Done in Volume Changes

We often want to calculate the work done in a process. Suppose we have a gas confined to a cylindrical container fitted with a movable piston (Fig. 19–10). We must always be careful to define exactly what our system is. In this case we choose our system to be the gas; so the container's walls and the piston are parts of the environment. Now let us calculate the work done by the gas when it expands quasistatically, so that P and T are defined for the system at all instants.[†] The gas expands against the piston, whose area is A. The gas exerts a force $F = PA$ on the piston, where P is the pressure in the gas. The work done by the gas to move the piston an infinitesimal displacement $d\vec{\ell}$ is

$$dW = \vec{F} \cdot d\vec{\ell} = PA \, d\ell = P \, dV \tag{19–6}$$

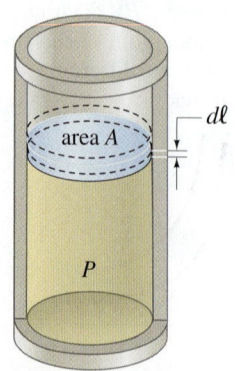

FIGURE 19–10 The work done by a gas when its volume increases by $dV = A \, d\ell$ is $dW = P \, dV$.

since the infinitesimal increase in volume is $dV = A \, d\ell$. If the gas was *compressed* so that $d\vec{\ell}$ pointed into the gas, the volume would decrease and $dV < 0$. The work done by the gas in this case would then be negative, which is equivalent to saying that positive work was done *on* the gas, not by it. For a finite change in volume

---

[†]If the gas expands or is compressed quickly, there would be turbulence and different parts would be at different pressure (and temperature).

from V_A to V_B, the work W done by the gas will be

$$W = \int dW = \int_{V_A}^{V_B} P\,dV. \qquad (19\text{--}7)$$

Equations 19–6 and 19–7 are valid for the work done in any volume change—by a gas, a liquid, or a solid—as long as it is done quasistatically.

In order to integrate Eq. 19–7, we need to know how the pressure varies during the process, and this depends on the type of process. Let us first consider a quasistatic isothermal expansion of an ideal gas. This process is represented by the curve between points A and B on the PV diagram of Fig. 19–11. The work done by the gas in this process, according to Eq. 19–7, is just the area between the PV curve and the V axis, and is shown shaded in Fig. 19–11. We can do the integral in Eq. 19–7 for an ideal gas by using the ideal gas law, $P = nRT/V$. The work done at constant T is

$$W = \int_{V_A}^{V_B} P\,dV = nRT \int_{V_A}^{V_B} \frac{dV}{V} = nRT \ln\frac{V_B}{V_A}. \quad \begin{bmatrix}\text{isothermal process;}\\ \text{ideal gas}\end{bmatrix} \quad (19\text{--}8)$$

Let us next consider a different way of taking an ideal gas between the same states A and B. This time, let us lower the pressure in the gas from P_A to P_B, as indicated by the line AD in Fig. 19–12. (In this *isovolumetric* process, heat must be allowed to flow out of the gas so its temperature drops.) Then let the gas expand from V_A to V_B at constant pressure $(= P_B)$, which is indicated by the line DB in Fig. 19–12. (In this *isobaric* process, heat is added to the gas to raise its temperature.) No work is done in the isovolumetric process AD, since $dV = 0$:

$$W = 0. \qquad \text{[isovolumetric process]}$$

In the isobaric process DB the pressure remains constant, so

$$W = \int_{V_A}^{V_B} P\,dV = P_B(V_B - V_A) = P\,\Delta V. \qquad \text{[isobaric process]} \quad (19\text{--}9a)$$

The work done is again represented on the PV diagram by the area between the curve (ADB) and the V axis, as indicated by the shading in Fig. 19–12. Using the ideal gas law, we can also write

$$W = P_B(V_B - V_A) = nRT_B\left(1 - \frac{V_A}{V_B}\right). \quad \begin{bmatrix}\text{isobaric process;}\\ \text{ideal gas}\end{bmatrix} \quad (19\text{--}9b)$$

As can be seen from the shaded areas in Figs. 19–11 and 19–12, or by putting in numbers in Eqs. 19–8 and 19–9 (try it for $V_B = 2V_A$), the work done in these two processes is different. This is a general result. *The work done in taking a system from one state to another depends not only on the initial and final states but also on the type of process (or "path").*

This result reemphasizes the fact that work cannot be considered a property of a system. The same is true of heat. The heat input required to change the gas from state A to state B depends on the process; for the isothermal process of Fig. 19–11, the heat input turns out to be greater than for the process ADB of Fig. 19–12. In general, *the amount of heat added or removed in taking a system from one state to another depends not only on the initial and final states but also on the path or process.*

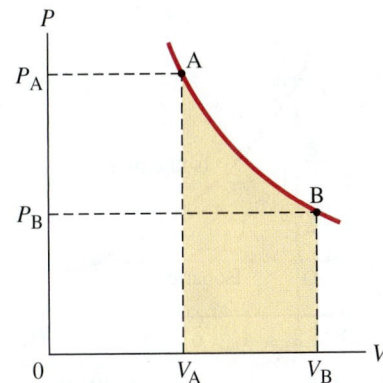

FIGURE 19–11 Work done by an ideal gas in an isothermal process equals the area under the PV curve. Shaded area equals the work done by the gas when it expands from V_A to V_B.

FIGURE 19–12 Process ADB consists of an isovolumetric (AD) and an isobaric (DB) process.

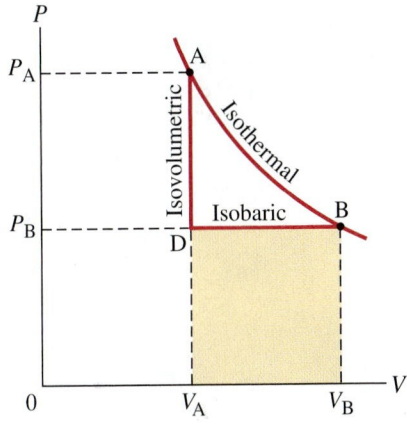

CONCEPTUAL EXAMPLE 19–9 | **Work in isothermal and adiabatic processes.**
In Fig. 19–8 we saw the PV diagrams for a gas expanding in two ways, isothermally and adiabatically. The initial volume V_A was the same in each case, and the final volumes were the same $(V_B = V_C)$. In which process was more work done by the gas?

RESPONSE Our system is the gas. More work was done by the gas in the isothermal process, which we can see in two simple ways by looking at Fig. 19–8. First, the "average" pressure was higher during the isothermal process AB, so $W = \overline{P}\,\Delta V$ was greater (ΔV is the same for both processes). Second, we can look at the area under each curve: the area under curve AB, which represents the work done, was greater (since curve AB is higher) than that under AC.

EXERCISE E Is the work done by the gas in process ADB of Fig. 19–12 greater than, less than, or equal to the work done in the isothermal process AB?

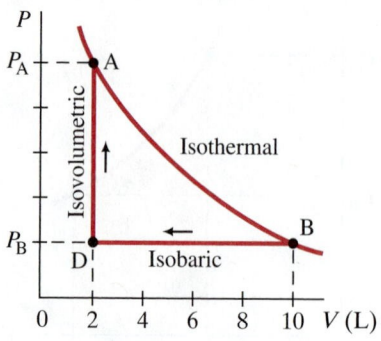

FIGURE 19–13 Example 19–10.

EXAMPLE 19–10 **First law in isobaric and isovolumetric processes.** An ideal gas is slowly compressed at a constant pressure of 2.0 atm from 10.0 L to 2.0 L. This process is represented in Fig. 19–13 as the path B to D. (In this process, some heat flows out of the gas and the temperature drops.) Heat is then added to the gas, holding the volume constant, and the pressure and temperature are allowed to rise (line DA) until the temperature reaches its original value $(T_A = T_B)$. Calculate (*a*) the total work done by the gas in the process BDA, and (*b*) the total heat flow into the gas.

APPROACH (*a*) Work is done only in the compression process BD. In process DA, the volume is constant so $\Delta V = 0$ and no work is done. (*b*) We use the first law of thermodynamics, Eq. 19–4.

SOLUTION (*a*) During the compression BD, the pressure is 2.0 atm = $2(1.01 \times 10^5 \,\text{N/m}^2)$ and the work done is (since $1\,\text{L} = 10^3\,\text{cm}^3 = 10^{-3}\,\text{m}^3$)

$$W = P\,\Delta V = (2.02 \times 10^5\,\text{N/m}^2)(2.0 \times 10^{-3}\,\text{m}^3 - 10.0 \times 10^{-3}\,\text{m}^3)$$
$$= -1.6 \times 10^3\,\text{J}.$$

The total work done *by* the gas is $-1.6 \times 10^3\,\text{J}$, where the minus sign means that $+1.6 \times 10^3\,\text{J}$ of work is done *on* the gas.

(*b*) Because the temperature at the beginning and at the end of process BDA is the same, there is no change in internal energy: $\Delta E_{int} = 0$. From the first law of thermodynamics we have

$$0 = \Delta E_{int} = Q - W$$

so $Q = W = -1.6 \times 10^3\,\text{J}$. Because Q is negative, 1600 J of heat flows out of the gas for the whole process, BDA.

EXERCISE F In Example 19–10, if the heat lost from the gas in the process BD is $8.4 \times 10^3\,\text{J}$, what is the change in internal energy of the gas during process BD?

EXAMPLE 19–11 **Work done in an engine.** In an engine, 0.25 mol of an ideal monatomic gas in the cylinder expands rapidly and adiabatically against the piston. In the process, the temperature of the gas drops from 1150 K to 400 K. How much work does the gas do?

APPROACH We take the gas as our system (the piston is part of the surroundings). The pressure is not constant, and its varying value is not given. Instead, we can use the first law of thermodynamics because we can determine ΔE_{int} given $Q = 0$ (the process is adiabatic).

SOLUTION We determine ΔE_{int} from Eq. 19–1 for the internal energy of an ideal monatomic gas, using subscripts f and i for final and initial states:

$$\Delta E_{int} = E_{int,f} - E_{int,i} = \tfrac{3}{2}nR(T_f - T_i)$$
$$= \tfrac{3}{2}(0.25\,\text{mol})(8.314\,\text{J/mol}\cdot\text{K})(400\,\text{K} - 1150\,\text{K})$$
$$= -2300\,\text{J}.$$

Then, from the first law of thermodynamics, Eq. 19–4,

$$W = Q - \Delta E_{int} = 0 - (-2300\,\text{J}) = 2300\,\text{J}.$$

Table 19–3 gives a brief summary of the processes we have discussed.

Free Expansion

One type of adiabatic process is a so-called **free expansion** in which a gas is allowed to expand in volume adiabatically without doing any work. The apparatus to

TABLE 19–3 Simple Thermodynamic Processes and the First Law

Process	What is constant:	The first law predicts:
Isothermal	T = constant	$\Delta T = 0$ makes $\Delta E_{int} = 0$, so $Q = W$
Isobaric	P = constant	$Q = \Delta E_{int} + W = \Delta E_{int} + P\,\Delta V$
Isovolumetric	V = constant	$\Delta V = 0$ makes $W = 0$, so $Q = \Delta E_{int}$
Adiabatic	$Q = 0$	$\Delta E_{int} = -W$

accomplish a free expansion is shown in Fig. 19–14. It consists of two well-insulated compartments (to ensure no heat flow in or out) connected by a valve or stopcock. One compartment is filled with gas, the other is empty. When the valve is opened, the gas expands to fill both containers. No heat flows in or out $(Q = 0)$, and no work is done because the gas does not move any other object. Thus $Q = W = 0$ and by the first law of thermodynamics, $\Delta E_{int} = 0$. *The internal energy of a gas does not change in a free expansion.* For an ideal gas, $\Delta T = 0$ also, since E_{int} depends only on T (Section 19–2). Experimentally, the free expansion has been used to determine if the internal energy of *real gases* depends only on T. The experiments are very difficult to do accurately, but it has been found that the temperature of a real gas drops very slightly in a free expansion. Thus the internal energy of real gases does depend, a little, on pressure or volume as well as on temperature.

A free expansion can not be plotted on a PV diagram, because the process is rapid, not quasistatic. The intermediate states are not equilibrium states, and hence the pressure (and even the volume at some instants) is not clearly defined.

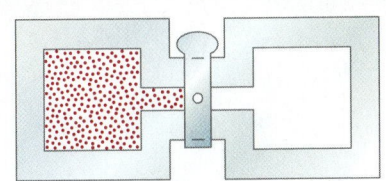

FIGURE 19–14 Free expansion.

19–8 Molar Specific Heats for Gases, and the Equipartition of Energy

In Section 19–3 we discussed the concept of specific heat and applied it to solids and liquids. Much more than for solids and liquids, the values of the specific heat for gases depends on how the process is carried out. Two important processes are those in which either the volume or the pressure is kept constant. Although for solids and liquids it matters little, Table 19–4 shows that the specific heats of gases at constant volume (c_V) and at constant pressure (c_P) are quite different.

Molar Specific Heats for Gases

The difference in specific heats for gases is nicely explained in terms of the first law of thermodynamics and kinetic theory. Our discussion is simplified if we use **molar specific heats**, C_V and C_P, which are defined as the heat required to raise 1 mol of the gas by 1 C° at constant volume and at constant pressure, respectively. That is, in analogy to Eq. 19–2, the heat Q needed to raise the temperature of n moles of gas by ΔT is

$$Q = nC_V \Delta T \qquad \text{[volume constant]} \quad \text{(19–10a)}$$

$$Q = nC_P \Delta T. \qquad \text{[pressure constant]} \quad \text{(19–10b)}$$

It is clear from the definition of molar specific heat (or by comparing Eqs. 19–2 and 19–10) that

$$C_V = Mc_V$$

$$C_P = Mc_P,$$

where M is the molecular mass of the gas $(M = m/n$ in grams/mol). The values for molar specific heats are included in Table 19–4, and we see that the values are nearly the same for different gases that have the same number of atoms per molecule.

TABLE 19–4 Specific Heats of Gases at 15°C

Gas	Specific heats (kcal/kg · K)		Molar specific heats (cal/mol · K)		$C_P - C_V$ (cal/mol · K)	$\gamma = \dfrac{C_P}{C_V}$
	c_V	c_P	C_V	C_P		
Monatomic						
He	0.75	1.15	2.98	4.97	1.99	1.67
Ne	0.148	0.246	2.98	4.97	1.99	1.67
Diatomic						
N_2	0.177	0.248	4.96	6.95	1.99	1.40
O_2	0.155	0.218	5.03	7.03	2.00	1.40
Triatomic						
CO_2	0.153	0.199	6.80	8.83	2.03	1.30
H_2O (100°C)	0.350	0.482	6.20	8.20	2.00	1.32

Now we use kinetic theory and imagine that an ideal gas is slowly heated via two different processes—first at constant volume, and then at constant pressure. In both of these processes, we let the temperature increase by the same amount, ΔT. In the process done at constant volume, no work is done since $\Delta V = 0$. Thus, according to the first law of thermodynamics, the heat added (which we denote by Q_V) all goes into increasing the internal energy of the gas:

$$Q_V = \Delta E_{int}.$$

In the process carried out at constant pressure, work is done, and hence the heat added, Q_P, must not only increase the internal energy but also is used to do the work $W = P\Delta V$. Thus, more heat must be added in this process at constant pressure than in the first process at constant volume. For the process at constant pressure, we have from the first law of thermodynamics

$$Q_P = \Delta E_{int} + P\Delta V.$$

Since ΔE_{int} is the same in the two processes (ΔT was chosen to be the same), we can combine the two above equations:

$$Q_P - Q_V = P\Delta V.$$

From the ideal gas law, $V = nRT/P$, so for a process at constant pressure we have $\Delta V = nR\,\Delta T/P$. Putting this into the above equation and using Eqs. 19–10, we find

$$nC_P\Delta T - nC_V\Delta T = P\left(\frac{nR\,\Delta T}{P}\right)$$

or, after cancellations,

$$C_P - C_V = R. \tag{19-11}$$

Since the gas constant $R = 8.314\ \text{J/mol}\cdot\text{K} = 1.99\ \text{cal/mol}\cdot\text{K}$, our prediction is that C_P will be larger than C_V by about 1.99 cal/mol·K. Indeed, this is very close to what is obtained experimentally, as shown in the next to last column of Table 19–4.

Now we calculate the molar specific heat of a monatomic gas using kinetic theory. In a process carried out at constant volume, no work is done; so the first law of thermodynamics tells us that if heat Q is added to the gas, the internal energy of the gas changes by

$$\Delta E_{int} = Q.$$

For an ideal monatomic gas, the internal energy E_{int} is the total kinetic energy of all the molecules,

$$E_{int} = N(\tfrac{1}{2}m\overline{v^2}) = \tfrac{3}{2}nRT$$

as we saw in Section 19–2. Then, using Eq. 19–10a, we can write $\Delta E_{int} = Q$ in the form

$$\Delta E_{int} = \tfrac{3}{2}nR\,\Delta T = nC_V\,\Delta T \tag{19-12}$$

or

$$C_V = \tfrac{3}{2}R. \tag{19-13}$$

Since $R = 8.314\ \text{J/mol}\cdot\text{K} = 1.99\ \text{cal/mol}\cdot\text{K}$, kinetic theory predicts that $C_V = 2.98\ \text{cal/mol}\cdot\text{K}$ for an ideal monatomic gas. This is very close to the experimental values for monatomic gases such as helium and neon (Table 19–4). From Eq. 19–11, C_P is predicted to be about 4.97 cal/mol·K, also in agreement with experiment.

Equipartition of Energy

The measured molar specific heats for more complex gases (Table 19–4), such as diatomic (two atoms) and triatomic (three atoms) gases, increase with the increased number of atoms per molecule. We can explain this by assuming that the internal energy includes not only translational kinetic energy but other forms of energy as well. Take, for example, a diatomic gas. As shown in Fig. 19–15 the two atoms can rotate about two different axes (but rotation about a third axis passing through the two atoms would give rise to very little energy since the moment of inertia is so small). The molecules can have rotational as well as translational kinetic energy. It is useful to introduce the idea of **degrees of freedom**, by which we mean the number of independent ways molecules can possess energy. For example, a monatomic gas is said to have three degrees of freedom, since an atom can have velocity along the x axis, the y axis, and the z axis. These are considered to be three independent motions because a change in any one of the components would not affect any of the others. A diatomic molecule has the same three degrees of freedom associated with translational kinetic

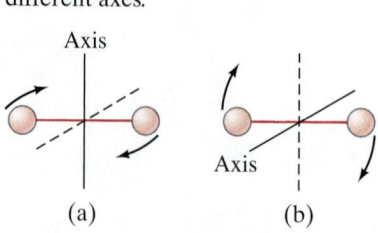

FIGURE 19–15 A diatomic molecule can rotate about two different axes.

Axis

Axis

(a)   (b)

energy plus two more degrees of freedom associated with rotational kinetic energy, for a total of five degrees of freedom. A quick look at Table 19–4 indicates that the C_V for diatomic gases is about $\frac{5}{3}$ times as great as for a monatomic gas—that is, in the same ratio as their degrees of freedom. This result led nineteenth-century physicists to an important idea, the **principle of equipartition of energy**. This principle states that energy is shared equally among the active degrees of freedom, and in particular each active degree of freedom of a molecule has on the average an energy equal to $\frac{1}{2}kT$. Thus, the average energy for a molecule of a monatomic gas would be $\frac{3}{2}kT$ (which we already knew) and of a diatomic gas $\frac{5}{2}kT$. Hence the internal energy of a diatomic gas would be $E_{\text{int}} = N\left(\frac{5}{2}kT\right) = \frac{5}{2}nRT$, where n is the number of moles. Using the same argument we did for monatomic gases, we see that for diatomic gases the molar specific heat at constant volume would be $\frac{5}{2}R = 4.97\,\text{cal/mol·K}$, in accordance with measured values. More complex molecules have even more degrees of freedom and thus greater molar specific heats.

The situation was complicated, however, by measurements that showed that for diatomic gases at very low temperatures, C_V has a value of only $\frac{3}{2}R$, as if it had only three degrees of freedom. And at very high temperatures, C_V was about $\frac{7}{2}R$, as if there were seven degrees of freedom. The explanation is that at low temperatures, nearly all molecules have only translational kinetic energy. That is, no energy goes into rotational energy, so only three degrees of freedom are "active." At very high temperatures, on the other hand, all five degrees of freedom are active plus two additional ones. We can interpret the two new degrees of freedom as being associated with the two atoms vibrating as if they were connected by a spring, as shown in Fig. 19–16. One degree of freedom comes from the kinetic energy of the vibrational motion, and the second comes from the potential energy of vibrational motion $\left(\frac{1}{2}kx^2\right)$. At room temperature, these two degrees of freedom are apparently not active. See Fig. 19–17.

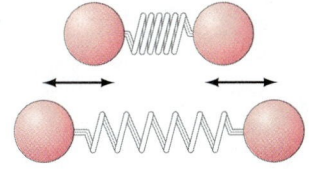

FIGURE 19–16 A diatomic molecule can vibrate, as if the two atoms were connected by a spring. Of course they are not connected by a spring; rather they exert forces on each other that are electrical in nature, but of a form that resembles a spring force.

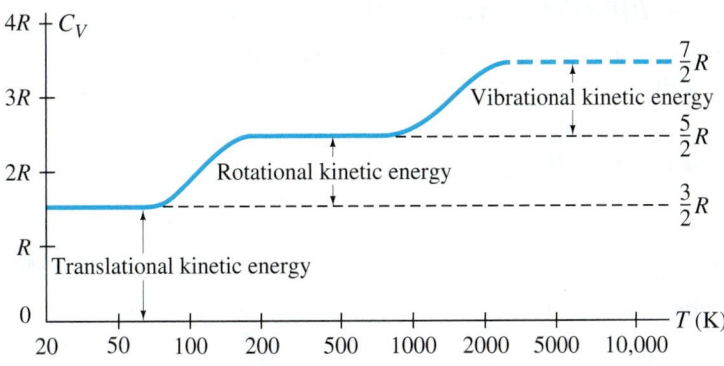

FIGURE 19–17 Molar specific heat C_V as a function of temperature for hydrogen molecules (H_2). As the temperature is increased, some of the translational kinetic energy can be transferred in collisions into rotational kinetic energy and, at still higher temperature, into vibrational kinetic energy. [Note: H_2 dissociates into two atoms at about 3200 K, so the last part of the curve is shown dashed.]

Just why fewer degrees of freedom are "active" at lower temperatures was eventually explained by Einstein using the quantum theory. [According to quantum theory, energy does not take on continuous values but is quantized—it can have only certain values, and there is a certain minimum energy. The minimum rotational and vibrational energies are higher than for simple translational kinetic energy, so at lower temperatures and lower translational kinetic energy, there is not enough energy to excite the rotational or vibrational kinetic energy.] Calculations based on kinetic theory and the principle of equipartition of energy (as modified by the quantum theory) give numerical results in accord with experiment.

*Solids

The principle of equipartition of energy can be applied to solids as well. The molar specific heat of any solid, at high temperature, is close to $3R$ ($6.0\,\text{cal/mol·K}$), Fig. 19–18. This is called the *Dulong and Petit* value after the scientists who first measured it in 1819. (Note that Table 19–1 gives the specific heats per kilogram, not per mole.) At high temperatures, each atom apparently has six degrees of freedom, although some are not active at low temperatures. Each atom in a crystalline solid can vibrate about its equilibrium position as if it were connected by springs to each of its neighbors. Thus it can have three degrees of freedom for kinetic energy and three more associated with potential energy of vibration in each of the x, y, and z directions, which is in accord with the measured values.

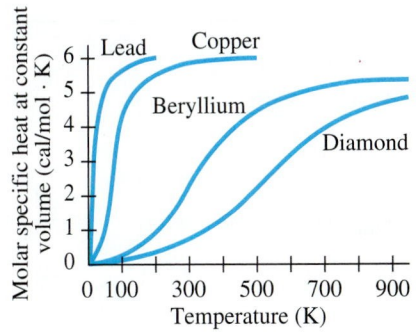

FIGURE 19–18 Molar specific heats of solids as a function of temperature.

19–9 Adiabatic Expansion of a Gas

The PV curve for the quasistatic (slow) adiabatic expansion $(Q = 0)$ of an ideal gas was shown in Fig. 19–8 (curve AC). It is somewhat steeper than for an isothermal process $(\Delta T = 0)$, which indicates that for the same change in volume the change in pressure will be greater. Hence the temperature of the gas must drop during an adiabatic expansion. Conversely, the temperature rises during an adiabatic compression.

We can derive the relation between the pressure P and the volume V of an ideal gas that is allowed to slowly expand adiabatically. We begin with the first law of thermodynamics, written in differential form:

$$dE_{int} = dQ - dW = -dW = -P \, dV,$$

since $dQ = 0$ for an adiabatic process. Equation 19–12 gives us a relation between ΔE_{int} and C_V, which is valid for any ideal gas process since E_{int} is a function only of T for an ideal gas. We write this in differential form:

$$dE_{int} = nC_V \, dT.$$

When we combine these last two equations, we obtain

$$nC_V \, dT + P \, dV = 0.$$

We next take the differential of the ideal gas law, $PV = nRT$, allowing P, V, and T to vary:

$$P \, dV + V \, dP = nR \, dT.$$

We solve for dT in this relation and substitute it into the previous relation and get

$$nC_V \left(\frac{P \, dV + V \, dP}{nR} \right) + P \, dV = 0$$

or, multiplying through by R and rearranging,

$$(C_V + R)P \, dV + C_V V \, dP = 0.$$

We note from Eq. 19–11 that $C_V + R = C_P$, so we have

$$C_P P \, dV + C_V V \, dP = 0,$$

or

$$\frac{C_P}{C_V} P \, dV + V \, dP = 0.$$

We define

$$\gamma = \frac{C_P}{C_V} \qquad\qquad\qquad \textbf{(19–14)}$$

so that our last equation becomes

$$\frac{dP}{P} + \gamma \frac{dV}{V} = 0.$$

This is integrated to become

$$\ln P + \gamma \ln V = \text{constant}.$$

This simplifies (using the rules for addition and multiplication of logarithms) to

$$PV^\gamma = \text{constant}. \qquad \left[\begin{array}{l}\text{quasistatic adiabatic} \\ \text{process; ideal gas}\end{array}\right] \quad \textbf{(19–15)}$$

This is the relation between P and V for a quasistatic adiabatic expansion or contraction. We will find it very useful when we discuss heat engines in the next Chapter. Table 19–4 (p. 511) gives values of γ for some real gases. Figure 19–8 compares an adiabatic expansion (Eq. 19–15) in curve AC to an isothermal expansion $(PV = \text{constant})$ in curve AB. It is important to remember that the ideal gas law, $PV = nRT$, continues to hold even for an adiabatic expansion $(PV^\gamma = \text{constant})$; clearly PV is not constant, meaning T is not constant.

EXAMPLE 19–12 **Compressing an ideal gas.** An ideal monatomic gas is compressed starting at point A in the PV diagram of Fig. 19–19, where $P_A = 100 \, \text{kPa}$, $V_A = 1.00 \, \text{m}^3$, and $T_A = 300 \, \text{K}$. The gas is first compressed adiabatically to state B $(P_B = 200 \, \text{kPa})$. The gas is then further compressed from point B to point C $(V_C = 0.50 \, \text{m}^3)$ in an isothermal process. (a) Determine V_B. (b) Calculate the work done on the gas for the whole process.

APPROACH Volume V_B is obtained using Eq. 19–15. The work done *by* a gas is given by Eq. 19–7, $W = \int P\,dV$. The work *on* the gas is the negative of this: $W_{on} = -\int P\,dV$.

SOLUTION In the adiabatic process, Eq. 19–15 tells us $PV^\gamma = $ constant. Therefore, $PV^\gamma = P_A V_A^\gamma = P_B V_B^\gamma$ where for a monatomic gas $\gamma = C_P/C_V = (5/2)/(3/2) = \frac{5}{3}$.

(*a*) Eq. 19–15 gives $V_B = V_A(P_A/P_B)^{\frac{1}{\gamma}} = (1.00\ \text{m}^3)(100\ \text{kPa}/200\ \text{kPa})^{\frac{3}{5}} = 0.66\ \text{m}^3$.

(*b*) The pressure P at any instant during the adiabatic process is given by $P = P_A V_A^\gamma V^{-\gamma}$. The work done on the gas in going from V_A to V_B is

$$W_{AB} = -\int_A^B P\,dV = -P_A V_A^\gamma \int_{V_A}^{V_B} V^{-\gamma}\,dV = -P_A V_A^\gamma \left(\frac{1}{-\gamma+1}\right)\left(V_B^{1-\gamma} - V_A^{1-\gamma}\right).$$

Since $\gamma = \frac{5}{3}$, then $-\gamma + 1 = 1 - \gamma = -\frac{2}{3}$, so

$$W_{AB} = -\left(P_A V_A^{\frac{5}{3}}\right)\left(-\frac{3}{2}\right)\left(V_A^{-\frac{2}{3}}\right)\left[\left(\frac{V_B}{V_A}\right)^{-\frac{2}{3}} - 1\right] = +\frac{3}{2}P_A V_A\left[\left(\frac{V_B}{V_A}\right)^{-\frac{2}{3}} - 1\right]$$

$$= +\frac{3}{2}(100\ \text{kPa})(1.00\ \text{m}^3)\left[(0.66)^{-\frac{2}{3}} - 1\right] = +48\ \text{kJ}.$$

For the isothermal process from B to C, the work is done at constant temperature, so the pressure at any instant during the process is $P = nRT_B/V$ and

$$W_{BC} = -\int_B^C P\,dV = -nRT_B \int_{V_B}^{V_C} \frac{dV}{V} = -nRT_B \ln \frac{V_C}{V_B} = -P_B V_B \ln \frac{V_C}{V_B} = +37\ \text{kJ}.$$

The total work done on the gas is $48\ \text{kJ} + 37\ \text{kJ} = 85\ \text{kJ}$.

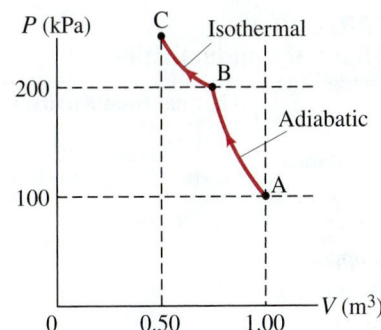

19–10 Heat Transfer: Conduction, Convection, Radiation

Heat is transferred from one place or body to another in three different ways: by *conduction*, *convection*, and *radiation*. We now discuss each of these in turn; but in practical situations, any two or all three may be operating at the same time. We start with conduction.

Conduction

When a metal poker is put in a hot fire, or a silver spoon is placed in a hot bowl of soup, the exposed end of the poker or spoon soon becomes hot as well, even though it is not directly in contact with the source of heat. We say that heat has been conducted from the hot end to the cold end.

Heat **conduction** in many materials can be visualized as being carried out via molecular collisions. As one end of an object is heated, the molecules there move faster and faster. As they collide with slower-moving neighbors, they transfer some of their kinetic energy to these other molecules, which in turn transfer energy by collision with molecules still farther along the object. In metals, collisions of free electrons are mainly responsible for conduction.

Heat conduction from one point to another takes place only if there is a difference in temperature between the two points. Indeed, it is found experimentally that the rate of heat flow through a substance is proportional to the difference in temperature between its ends. The rate of heat flow also depends on the size and shape of the object. To investigate this quantitatively, let us consider the heat flow through a uniform cylinder, as illustrated in Fig. 19–20. It is found experimentally that the heat flow ΔQ over a time interval Δt is given by the relation

$$\frac{\Delta Q}{\Delta t} = kA\frac{T_1 - T_2}{\ell} \tag{19–16a}$$

where A is the cross-sectional area of the object, ℓ is the distance between the two ends, which are at temperatures T_1 and T_2, and k is a proportionality constant called the **thermal conductivity** which is characteristic of the material. From Eq. 19–16a, we see that the rate of heat flow (units of J/s) is directly proportional to the cross-sectional area and to the temperature gradient $(T_1 - T_2)/\ell$.

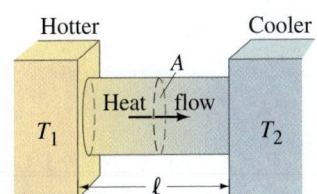

FIGURE 19–20 Heat conduction between areas at temperatures T_1 and T_2. If T_1 is greater than T_2, the heat flows to the right; the rate is given by Eq. 19–16a.

TABLE 19–5
Thermal Conductivities

Substance	Thermal conductivity, k	
	kcal $\dfrac{}{(\text{s} \cdot \text{m} \cdot \text{C}°)}$	J $\dfrac{}{(\text{s} \cdot \text{m} \cdot \text{C}°)}$
Silver	10×10^{-2}	420
Copper	9.2×10^{-2}	380
Aluminum	5.0×10^{-2}	200
Steel	1.1×10^{-2}	40
Ice	5×10^{-4}	2
Glass	2.0×10^{-4}	0.84
Brick	2.0×10^{-4}	0.84
Concrete	2.0×10^{-4}	0.84
Water	1.4×10^{-4}	0.56
Human tissue	0.5×10^{-4}	0.2
Wood	0.3×10^{-4}	0.1
Fiberglass	0.12×10^{-4}	0.048
Cork	0.1×10^{-4}	0.042
Wool	0.1×10^{-4}	0.040
Goose down	0.06×10^{-4}	0.025
Polyurethane	0.06×10^{-4}	0.024
Air	0.055×10^{-4}	0.023

FIGURE 19–21 Example 19–13.

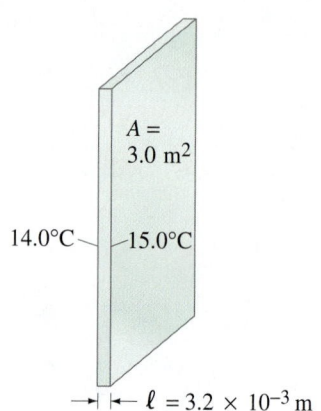

$A = 3.0 \text{ m}^2$

$14.0°C$ $15.0°C$

$\ell = 3.2 \times 10^{-3}$ m

PHYSICS APPLIED

Thermal windows

In some cases (such as when k or A cannot be considered constant) we need to consider the limit of an infinitesimally thin slab of thickness dx. Then Eq. 19–16a becomes

$$\frac{dQ}{dt} = -kA\frac{dT}{dx}, \qquad (19\text{–}16\text{b})$$

where dT/dx is the temperature gradient[†] and the negative sign is included since the heat flow is in the direction opposite to the temperature gradient.

The thermal conductivities, k, for a variety of substances are given in Table 19–5. Substances for which k is large conduct heat rapidly and are said to be good thermal **conductors**. Most metals fall in this category, although there is a wide range even among them, as you may observe by holding the ends of a silver spoon and a stainless-steel spoon immersed in the same hot cup of soup. Substances for which k is small, such as wool, fiberglass, polyurethane, and goose down, are poor conductors of heat and are therefore good thermal **insulators**.

The relative magnitudes of k can explain simple phenomena such as why a tile floor is much colder on the feet than a rug-covered floor at the same temperature. Tile is a better conductor of heat than the rug. Heat that flows from your foot to the rug is not conducted away rapidly, so the rug's surface quickly warms up to the temperature of your foot and feels good. But the tile conducts the heat away rapidly and thus can take more heat from your foot quickly, so your foot's surface temperature drops.

EXAMPLE 19–13 **Heat loss through windows.** A major source of heat loss from a house is through the windows. Calculate the rate of heat flow through a glass window 2.0 m × 1.5 m in area and 3.2 mm thick, if the temperatures at the inner and outer surfaces are 15.0°C and 14.0°C, respectively (Fig. 19–21).

APPROACH Heat flows by conduction through the glass from the higher inside temperature to the lower outside temperature. We use the heat conduction equation, Eq. 19–16a.

SOLUTION Here $A = (2.0 \text{ m})(1.5 \text{ m}) = 3.0 \text{ m}^2$ and $\ell = 3.2 \times 10^{-3}$ m. Using Table 19–5 to get k, we have

$$\frac{\Delta Q}{\Delta t} = kA\frac{T_1 - T_2}{\ell} = \frac{(0.84 \text{ J/s} \cdot \text{m} \cdot \text{C}°)(3.0 \text{ m}^2)(15.0°C - 14.0°C)}{(3.2 \times 10^{-3} \text{ m})}$$

$$= 790 \text{ J/s}.$$

NOTE This rate of heat flow is equivalent to $(790 \text{ J/s})/(4.19 \times 10^3 \text{ J/kcal}) = 0.19 \text{ kcal/s}$, or $(0.19 \text{ kcal/s}) \times (3600 \text{ s/h}) = 680 \text{ kcal/h}$.

You might notice in Example 19–13 that 15°C is not very warm for the living room of a house. The room itself may indeed be much warmer, and the outside might be colder than 14°C. But the temperatures of 15°C and 14°C were specified as those at the window surfaces, and there is usually a considerable drop in temperature of the air in the vicinity of the window both on the inside and the outside. That is, the layer of air on either side of the window acts as an insulator, and normally the major part of the temperature drop between the inside and outside of the house takes place across the air layer. If there is a heavy wind, the air outside a window will constantly be replaced with cold air; the temperature gradient across the glass will be greater and there will be a much greater rate of heat loss. Increasing the width of the air layer, such as using two panes of glass separated by an air gap, will reduce the heat loss more than simply increasing the glass thickness, since the thermal conductivity of air is much less than that for glass.

The insulating properties of clothing come from the insulating properties of air. Without clothes, our bodies in still air would heat the air in contact with the skin and would soon become reasonably comfortable because air is a very good insulator.

[†]Equations 19–16 are quite similar to the relations describing diffusion (Section 18–7) and the flow of fluids through a pipe (Section 13–12). In those cases, the flow of matter was found to be proportional to the concentration gradient dC/dx, or to the pressure gradient $(P_1 - P_2)/\ell$. This close similarity is one reason we speak of the "flow" of heat. Yet we must keep in mind that no substance is flowing in the case of heat—it is energy that is being transferred.

But since air moves—there are breezes and drafts, and people move about—the warm air would be replaced by cold air, thus increasing the temperature difference and the heat loss from the body. Clothes keep us warm by trapping air so it cannot move readily. It is not the cloth that insulates us, but the air that the cloth traps. Goose down is a very good insulator because even a small amount of it fluffs up and traps a great amount of air.

[For practical purposes the thermal properties of building materials, particularly when considered as insulation, are usually specified by R-values (or "thermal resistance"), defined for a given thickness ℓ of material as:

$$R = \ell/k.$$

The R-value of a given piece of material combines the thickness ℓ and the thermal conductivity k in one number. In the United States, R-values are given in British units as $\mathrm{ft^2 \cdot h \cdot F° / Btu}$ (for example, R-19 means $R = 19\ \mathrm{ft^2 \cdot h \cdot F° / Btu}$). Table 19–6 gives R-values for some common building materials. R-values increase directly with material thickness: for example, 2 inches of fiberglass is R-6, whereas 4 inches is R-12.]

Convection

Although liquids and gases are generally not very good conductors of heat, they can transfer heat quite rapidly by convection. **Convection** is the process whereby heat flows by the mass movement of molecules from one place to another. Whereas conduction involves molecules (and/or electrons) moving only over small distances and colliding, convection involves the movement of large numbers of molecules over large distances.

A forced-air furnace, in which air is heated and then blown by a fan into a room, is an example of *forced convection*. *Natural convection* occurs as well, and one familiar example is that hot air rises. For instance, the air above a radiator (or other type of heater) expands as it is heated (Chapter 17), and hence its density decreases. Because its density is less than that of the surrounding cooler air, it rises, just as a log submerged in water floats upward because its density is less than that of water. Warm or cold ocean currents, such as the balmy Gulf Stream, represent natural convection on a global scale. Wind is another example of convection, and weather in general is strongly influenced by convective air currents.

When a pot of water is heated (Fig. 19–22), convection currents are set up as the heated water at the bottom of the pot rises because of its reduced density. That heated water is replaced by cooler water from above. This principle is used in many heating systems, such as the hot-water radiator system shown in Fig. 19–23. Water is heated in the furnace, and as its temperature increases, it expands and rises as shown. This causes the water to circulate in the heating system. Hot water then enters the radiators, heat is transferred by conduction to the air, and the cooled water returns to the furnace. Thus, the water circulates because of convection; pumps are sometimes used to improve circulation. The air throughout the room also becomes heated as a result of convection. The air heated by the radiators rises and is replaced by cooler air, resulting in convective air currents, as shown by the green arrows in Fig. 19–23.

Other types of furnaces also depend on convection. Hot-air furnaces with registers (openings) near the floor often do not have fans but depend on natural convection, which can be appreciable. In other systems, a fan is used. In either case, it is important that cold air can return to the furnace so that convective currents circulate throughout the room if the room is to be uniformly heated. Convection is not always favorable. Much of the heat from a fireplace, for example, goes up the chimney and not out into the room.

Radiation

Convection and conduction require the presence of matter as a medium to carry the heat from the hotter to the colder region. But a third type of heat transfer occurs without any medium at all. All life on Earth depends on the transfer of energy from the Sun, and this energy is transferred to the Earth over empty (or nearly empty) space. This form of energy transfer is heat—since the Sun's surface temperature is much higher (6000 K) than Earth's—and is referred to as **radiation**. The warmth we receive from a fire is mainly radiant energy.

SECTION 19–10   **517**

PHYSICS APPLIED
Clothes insulate by trapping an air layer

PHYSICS APPLIED
R-values of thermal insulation

TABLE 19–6 R-values

Material	Thickness	R-value ($\mathrm{ft^2 \cdot h \cdot F° / Btu}$)
Glass	$\frac{1}{8}$ inch	1
Brick	$3\frac{1}{2}$ inches	0.6–1
Plywood	$\frac{1}{2}$ inch	0.6
Fiberglass insulation	4 inches	12

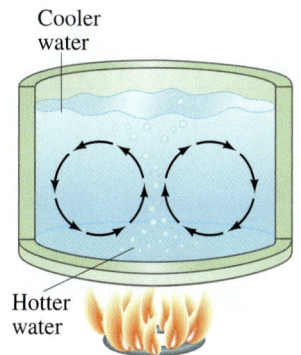

FIGURE 19–22 Convection currents in a pot of water being heated on a stove.

PHYSICS APPLIED
Convective home heating

FIGURE 19–23 Convection plays a role in heating a house. The circular arrows show the convective air currents in the rooms.

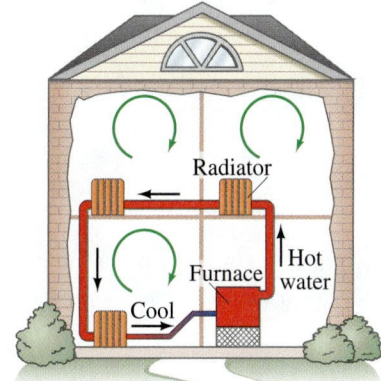

As we shall see in later Chapters, radiation consists essentially of electromagnetic waves. Suffice it to say for now that radiation from the Sun consists of visible light plus many other wavelengths that the eye is not sensitive to, including infrared (IR) radiation.

The rate at which an object radiates energy has been found to be proportional to the fourth power of the Kelvin temperature, T. That is, an object at 2000 K, as compared to one at 1000 K, radiates energy at a rate $2^4 = 16$ times as much. The rate of radiation is also proportional to the area A of the emitting object, so the rate at which energy leaves the object, $\Delta Q/\Delta t$, is

$$\frac{\Delta Q}{\Delta t} = \epsilon \sigma A T^4. \tag{19-17}$$

This is called the **Stefan-Boltzmann equation**, and σ is a universal constant called the **Stefan-Boltzmann constant** which has the value

$$\sigma = 5.67 \times 10^{-8} \, \text{W/m}^2 \cdot \text{K}^4.$$

The factor ϵ (Greek letter epsilon), called the **emissivity**, is a number between 0 and 1 that is characteristic of the surface of the radiating material. Very black surfaces, such as charcoal, have emissivity close to 1, whereas shiny metal surfaces have ϵ close to zero and thus emit correspondingly less radiation. The value depends somewhat on the temperature of the material.

Not only do shiny surfaces emit less radiation, but they absorb little of the radiation that falls upon them (most is reflected). Black and very dark objects are good emitters ($\epsilon \approx 1$), and they also absorb nearly all the radiation that falls on them—which is why light-colored clothing is usually preferable to dark clothing on a hot day. Thus, **a good absorber is also a good emitter**.

Any object not only emits energy by radiation but also absorbs energy radiated by other objects. If an object of emissivity ϵ and area A is at a temperature T_1, it radiates energy at a rate $\epsilon \sigma A T_1^4$. If the object is surrounded by an environment at temperature T_2, the rate at which the surroundings radiate energy is proportional to T_2^4, and the rate that energy is absorbed by the object is proportional to T_2^4. The *net* rate of radiant heat flow from the object is given by

$$\frac{\Delta Q}{\Delta t} = \epsilon \sigma A \left(T_1^4 - T_2^4 \right), \tag{19-18}$$

where A is the surface area of the object, T_1 its temperature and ϵ its emissivity (at temperature T_1), and T_2 is the temperature of the surroundings. This equation is consistent with the experimental fact that equilibrium between the object and its surroundings is reached when they come to the same temperature. That is, $\Delta Q/\Delta t$ must equal zero when $T_1 = T_2$, so ϵ must be the same for emission and absorption. This confirms the idea that a good emitter is a good absorber. Because both the object and its surroundings radiate energy, there is a net transfer of energy from one to the other unless everything is at the same temperature.

EXAMPLE 19–14 ESTIMATE Cooling by radiation. An athlete is sitting unclothed in a locker room whose dark walls are at a temperature of 15°C. Estimate his rate of heat loss by radiation, assuming a skin temperature of 34°C and $\epsilon = 0.70$. Take the surface area of the body not in contact with the chair to be 1.5 m^2.

APPROACH We use Eq. 19–18, with Kelvin temperatures.

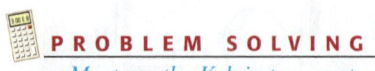

SOLUTION We have

$$\frac{\Delta Q}{\Delta t} = \epsilon \sigma A \left(T_1^4 - T_2^4 \right)$$
$$= (0.70)(5.67 \times 10^{-8} \, \text{W/m}^2 \cdot \text{K}^4)(1.5 \, \text{m}^2)\left[(307 \, \text{K})^4 - (288 \, \text{K})^4 \right] = 120 \, \text{W}.$$

NOTE The "output" of this resting person is a bit more than what a 100-W lightbulb uses.

A resting person naturally produces heat internally at a rate of about 100 W, which is less than the heat loss by radiation as calculated in Example 19–14. Hence, the person's temperature would drop, causing considerable discomfort. The body responds to excessive heat loss by increasing its metabolic rate, and shivering is one method by which the body increases its metabolism. Naturally, clothes help a lot. Example 19–14 illustrates that a person may be uncomfortable even if the temperature of the air is, say, 25°C, which is quite a warm room. If the walls or floor are cold, radiation to them occurs no matter how warm the air is. Indeed, it is estimated that radiation accounts for about 50% of the heat loss from a sedentary person in a normal room. Rooms are most comfortable when the walls and floor are warm and the air is not so warm. Floors and walls can be heated by means of hot-water conduits or electric heating elements. Such first-rate heating systems are becoming more common today, and it is interesting to note that 2000 years ago the Romans, even in houses in the remote province of Great Britain, made use of hot-water and steam conduits in the floor to heat their houses.

Heating of an object by radiation from the Sun cannot be calculated using Eq. 19–18 since this equation assumes a uniform temperature, T_2, of the environment surrounding the object, whereas the Sun is essentially a point source. Hence the Sun must be treated as a separate source of energy. Heating by the Sun is calculated using the fact that about 1350 J of energy strikes the atmosphere of the Earth from the Sun per second per square meter of area at right angles to the Sun's rays. This number, 1350 W/m², is called the **solar constant**. The atmosphere may absorb as much as 70% of this energy before it reaches the ground, depending on the cloud cover. On a clear day, about 1000 W/m² reaches the Earth's surface. An object of emissivity ε with area A facing the Sun absorbs energy from the Sun at a rate, in watts, of about

$$\frac{\Delta Q}{\Delta t} = (1000 \text{ W/m}^2)\,\epsilon\,A\cos\theta, \tag{19–19}$$

where θ is the angle between the Sun's rays and a line perpendicular to the area A (Fig. 19–24). That is, $A\cos\theta$ is the "effective" area, at right angles to the Sun's rays.

The explanation for the **seasons** and the polar ice caps (see Fig. 19–25) depends on this $\cos\theta$ factor in Eq. 19–19. The seasons are *not* a result of how close the Earth is to the Sun—in fact, in the Northern Hemisphere, summer occurs when the Earth is farthest from the Sun. It is the angle (i.e., $\cos\theta$) that really matters. Furthermore, the reason the Sun heats the Earth more at midday than at sunrise or sunset is also related to this $\cos\theta$ factor.

An interesting application of thermal radiation to diagnostic medicine is **thermography**. A special instrument, the thermograph, scans the body, measuring the intensity of radiation from many points and forming a picture that resembles an X-ray (Fig. 19–26). Areas where metabolic activity is high, such as in tumors, can often be detected on a thermogram as a result of their higher temperature and consequent increased radiation.

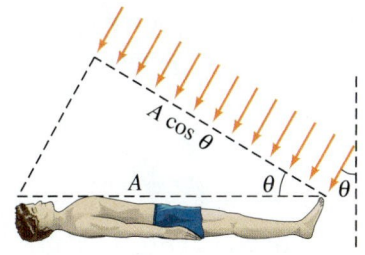

FIGURE 19–24 Radiant energy striking a body at an angle θ.

FIGURE 19–25 (a) Earth's seasons arise from the $23\frac{1}{2}°$ angle Earth's axis makes with its orbit around the Sun. (b) June sunlight makes an angle of about 23° with the equator. Thus θ in the southern United States (A) is near 0° (direct summer sunlight), whereas in the Southern Hemisphere (B), θ is 50° or 60°, and less heat can be absorbed—hence it is winter. Near the poles (C), there is never strong direct sunlight; $\cos\theta$ varies from about $\frac{1}{2}$ in summer to 0 in winter; so with little heating, ice can form.

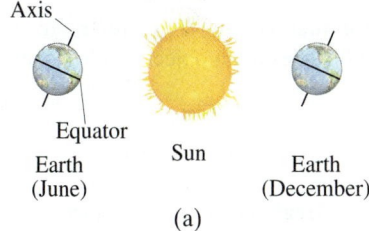

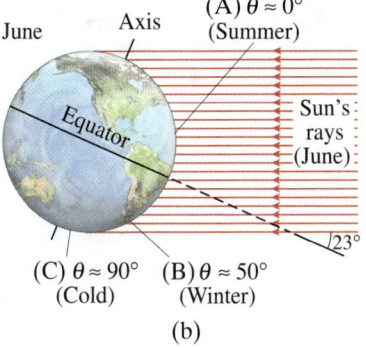

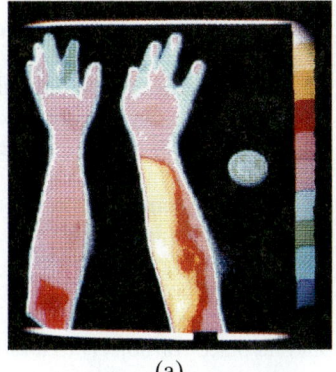

(a)

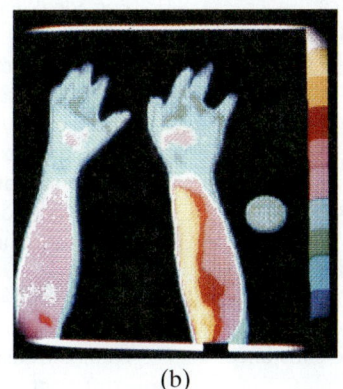

(b)

FIGURE 19–26 Thermograms of a healthy person's arms and hands (a) before and (b) after smoking a cigarette, showing a temperature decrease due to impaired blood circulation associated with smoking. The thermograms have been color-coded according to temperature; the scale on the right goes from blue (cold) to white (hot).

EXAMPLE 19–15 **ESTIMATE** **Star radius.** The giant star Betelgeuse emits radiant energy at a rate 10^4 times greater than our Sun, whereas its surface temperature is only half (2900 K) that of our Sun. Estimate the radius of Betelgeuse, assuming $\epsilon = 1$ for both. The Sun's radius is $r_S = 7 \times 10^8$ m.

APPROACH We assume both Betelgeuse and the Sun are spherical, with surface area $4\pi r^2$.

SOLUTION We solve Eq. 19–17 for A:

$$4\pi r^2 = A = \frac{(\Delta Q/\Delta t)}{\epsilon \sigma T^4}.$$

Then

$$\frac{r_B^2}{r_S^2} = \frac{(\Delta Q/\Delta t)_B}{(\Delta Q/\Delta t)_S} \cdot \frac{T_S^4}{T_B^4} = (10^4)(2^4) = 16 \times 10^4.$$

Hence $r_B = \sqrt{16 \times 10^4}\, r_S = (400)(7 \times 10^8 \text{ m}) \approx 3 \times 10^{11}$ m.

NOTE If Betelgeuse were our Sun, it would envelop us (Earth is 1.5×10^{11} m from the Sun).

EXERCISE G Fanning yourself on a hot day cools you by (*a*) increasing the radiation rate of the skin; (*b*) increasing conductivity; (*c*) decreasing the mean free path of air; (*d*) increasing the evaporation of perspiration; (*e*) none of these.

Summary

Internal energy, E_{int}, refers to the total energy of all the molecules in an object. For an ideal monatomic gas,

$$E_{int} = \tfrac{3}{2} NkT = \tfrac{3}{2} nRT \tag{19–1}$$

where N is the number of molecules or n is the number of moles.

Heat refers to the transfer of energy from one object to another because of a difference of temperature. Heat is thus measured in energy units, such as joules.

Heat and internal energy are also sometimes specified in calories or kilocalories (kcal), where

$$1 \text{ kcal} = 4.186 \text{ kJ}$$

is the amount of heat needed to raise the temperature of 1 kg of water by 1 C°.

The **specific heat**, c, of a substance is defined as the energy (or heat) required to change the temperature of unit mass of substance by 1 degree; as an equation,

$$Q = mc\,\Delta T, \tag{19–2}$$

where Q is the heat absorbed or given off, ΔT is the temperature increase or decrease, and m is the mass of the substance.

When heat flows between parts of an isolated system, conservation of energy tells us that the heat gained by one part of the system is equal to the heat lost by the other part of the system. This is the basis of **calorimetry**, which is the quantitative measurement of heat exchange.

Exchange of energy occurs, without a change in temperature, whenever a substance changes phase. The **heat of fusion** is the heat required to melt 1 kg of a solid into the liquid phase; it is also equal to the heat given off when the substance changes from liquid to solid. The **heat of vaporization** is the energy required to change 1 kg of a substance from the liquid to the vapor phase; it is also the energy given off when the substance changes from vapor to liquid.

The **first law of thermodynamics** states that the change in internal energy ΔE_{int} of a system is equal to the heat *added to* the system, Q, minus the work, W, done *by* the system:

$$\Delta E_{int} = Q - W. \tag{19–4}$$

This important law is a broad restatement of the conservation of energy and is found to hold for all processes.

Two simple thermodynamic processes are **isothermal**, which is a process carried out at constant temperature, and **adiabatic**, a process in which no heat is exchanged. Two more are **isobaric** (a process carried out at constant pressure) and **isovolumetric** (a process at constant volume).

The work done by (or on) a gas to change its volume by dV is $dW = P\,dV$, where P is the pressure.

Work and heat are not functions of the state of a system (as are P, V, T, n, and E_{int}) but depend on the type of process that takes a system from one state to another.

The **molar specific heat** of an ideal gas at constant volume, C_V, and at constant pressure, C_P, are related by

$$C_P - C_V = R, \tag{19–11}$$

where R is the gas constant. For a monatomic ideal gas, $C_V = \tfrac{3}{2} R$.

For ideal gases made up of diatomic or more complex molecules, C_V is equal to $\tfrac{1}{2} R$ times the number of **degrees of freedom** of the molecule. Unless the temperature is very high, some of the degrees of freedom may not be active and so do not contribute. According to the **principle of equipartition of energy**, energy is shared equally among the active degrees of freedom in an amount $\tfrac{1}{2} kT$ per molecule on average.

When an ideal gas expands (or contracts) adiabatically $(Q = 0)$, the relation $PV^\gamma = $ constant holds, where

$$\gamma = \frac{C_P}{C_V}. \tag{19–14}$$

Heat is transferred from one place (or object) to another in three different ways: conduction, convection, and radiation.

In **conduction**, energy is transferred by collisions between molecules or electrons with higher kinetic energy to slower-moving neighbors.

Convection is the transfer of energy by the mass movement of molecules over considerable distances.

Radiation, which does not require the presence of matter, is energy transfer by electromagnetic waves, such as from the Sun.

All objects radiate energy in an amount that is proportional to the fourth power of their Kelvin temperature (T^4) and to their surface area. The energy radiated (or absorbed) also depends on the nature of the surface, which is characterized by the emissivity, ϵ (dark surfaces absorb and radiate more than do bright shiny ones).

Radiation from the Sun arrives at the surface of the Earth on a clear day at a rate of about $1000 \, \text{W/m}^2$.

Questions

1. What happens to the work done on a jar of orange juice when it is vigorously shaken?

2. When a hot object warms a cooler object, does temperature flow between them? Are the temperature changes of the two objects equal? Explain.

3. (a) If two objects of different temperature are placed in contact, will heat naturally flow from the object with higher internal energy to the object with lower internal energy? (b) Is it possible for heat to flow even if the internal energies of the two objects are the same? Explain.

4. In warm regions where tropical plants grow but the temperature may drop below freezing a few times in the winter, the destruction of sensitive plants due to freezing can be reduced by watering them in the evening. Explain.

5. The specific heat of water is quite large. Explain why this fact makes water particularly good for heating systems (that is, hot-water radiators).

6. Why does water in a canteen stay cooler if the cloth jacket surrounding the canteen is kept moist?

7. Explain why burns caused by steam at 100°C on the skin are often more severe than burns caused by water at 100°C.

8. Explain why water cools (its temperature drops) when it evaporates, using the concepts of latent heat and internal energy.

9. Will potatoes cook faster if the water is boiling more vigorously?

10. Very high in the Earth's atmosphere the temperature can be 700°C. Yet an animal there would freeze to death rather than roast. Explain.

11. What happens to the internal energy of water vapor in the air that condenses on the outside of a cold glass of water? Is work done or heat exchanged? Explain.

12. Use the conservation of energy to explain why the temperature of a well-insulated gas increases when it is compressed—say, by pushing down on a piston—whereas the temperature decreases when the gas expands.

13. In an isothermal process, 3700 J of work is done by an ideal gas. Is this enough information to tell how much heat has been added to the system? If so, how much?

14. Explorers on failed Arctic expeditions have survived by covering themselves with snow. Why would they do that?

15. Why is wet sand at the beach cooler to walk on than dry sand?

16. When hot-air furnaces are used to heat a house, why is it important that there be a vent for air to return to the furnace? What happens if this vent is blocked by a bookcase?

17. Is it possible for the temperature of a system to remain constant even though heat flows into or out of it? If so, give examples.

18. Discuss how the first law of thermodynamics can apply to metabolism in humans. In particular, note that a person does work W, but very little heat Q is added to the body (rather, it tends to flow out). Why then doesn't the internal energy drop drastically in time?

19. Explain in words why C_P is greater than C_V.

20. Explain why the temperature of a gas increases when it is adiabatically compressed.

21. An ideal monatomic gas is allowed to expand slowly to twice its volume (1) isothermally; (2) adiabatically; (3) isobarically. Plot each on a PV diagram. In which process is ΔE_{int} the greatest, and in which is ΔE_{int} the least? In which is W the greatest and the least? In which is Q the greatest and the least?

22. Ceiling fans are sometimes reversible, so that they drive the air down in one season and pull it up in another season. Which way should you set the fan for summer? For winter?

23. Goose down sleeping bags and parkas are often specified as so many inches or centimeters of *loft*, the actual thickness of the garment when it is fluffed up. Explain.

24. Microprocessor chips nowadays have a "heat sink" glued on top that looks like a series of fins. Why is it shaped like that?

25. Sea breezes are often encountered on sunny days at the shore of a large body of water. Explain, assuming the temperature of the land rises more rapidly than that of the nearby water.

26. The Earth cools off at night much more quickly when the weather is clear than when cloudy. Why?

27. Explain why air-temperature readings are always taken with the thermometer in the shade.

28. A premature baby in an incubator can be dangerously cooled even when the air temperature in the incubator is warm. Explain.

29. The floor of a house on a foundation under which the air can flow is often cooler than a floor that rests directly on the ground (such as a concrete slab foundation). Explain.

30. Why is the liner of a **thermos bottle** silvered (Fig. 19–27), and why does it have a vacuum between its two walls?

FIGURE 19–27
Question 30.

31. A 22°C day is warm, while a swimming pool at 22°C feels cool. Why?

32. In the Northern Hemisphere the amount of heat required to heat a room where the windows face north is much higher than that required where the windows face south. Explain.

33. Heat loss occurs through windows by the following processes: (1) ventilation around edges; (2) through the frame, particularly if it is metal; (3) through the glass panes; and (4) radiation. (*a*) For the first three, what is (are) the mechanism(s): conduction, convection, or radiation? (*b*) Heavy curtains reduce which of these heat losses? Explain in detail.

34. Early in the day, after the Sun has reached the slope of a mountain, there tends to be a gentle upward movement of air. Later, after a slope goes into shadow, there is a gentle downdraft. Explain.

35. A piece of wood lying in the Sun absorbs more heat than a piece of shiny metal. Yet the metal feels hotter than the wood when you pick it up. Explain.

36. An "emergency blanket" is a thin shiny (metal-coated) plastic foil. Explain how it can help to keep an immobile person warm.

37. Explain why cities situated by the ocean tend to have less extreme temperatures than inland cities at the same latitude.

Problems

19–1 Heat as Energy Transfer

1. (I) To what temperature will 8700 J of heat raise 3.0 kg of water that is initially at 10.0°C?

2. (II) When a diver jumps into the ocean, water leaks into the gap region between the diver's skin and her wetsuit, forming a water layer about 0.5 mm thick. Assuming the total surface area of the wetsuit covering the diver is about 1.0 m², and that ocean water enters the suit at 10°C and is warmed by the diver to skin temperature of 35°C, estimate how much energy (in units of candy bars = 300 kcal) is required by this heating process.

3. (II) An average active person consumes about 2500 Cal a day. (*a*) What is this in joules? (*b*) What is this in kilowatt-hours? (*c*) If your power company charges about 10 ¢ per kilowatt-hour, how much would your energy cost per day if you bought it from the power company? Could you feed yourself on this much money per day?

4. (II) A British thermal unit (Btu) is a unit of heat in the British system of units. One Btu is defined as the heat needed to raise 1 lb of water by 1 F°. Show that

$$1 \text{ Btu} = 0.252 \text{ kcal} = 1056 \text{ J}.$$

5. (II) How many joules and kilocalories are generated when the brakes are used to bring a 1200-kg car to rest from a speed of 95 km/h?

6. (II) A small immersion heater is rated at 350 W. Estimate how long it will take to heat a cup of soup (assume this is 250 mL of water) from 15°C to 75°C.

19–3 and 19–4 Specific Heat; Calorimetry

7. (I) An automobile cooling system holds 18 L of water. How much heat does it absorb if its temperature rises from 15°C to 95°C?

8. (I) What is the specific heat of a metal substance if 135 kJ of heat is needed to raise 5.1 kg of the metal from 18.0°C to 37.2°C?

9. (II) (*a*) How much energy is required to bring a 1.0-L pot of water at 20°C to 100°C? (*b*) For how long could this amount of energy run a 100-W lightbulb?

10. (II) Samples of copper, aluminum, and water experience the same temperature rise when they absorb the same amount of heat. What is the ratio of their masses?

11. (II) How long does it take a 750-W coffeepot to bring to a boil 0.75 L of water initially at 8.0°C? Assume that the part of the pot which is heated with the water is made of 280 g of aluminum, and that no water boils away.

12. (II) A hot iron horseshoe (mass = 0.40 kg), just forged (Fig. 19–28), is dropped into 1.05 L of water in a 0.30-kg iron pot initially at 20.0°C. If the final equilibrium temperature is 25.0°C, estimate the initial temperature of the hot horseshoe.

FIGURE 19–28
Problem 12.

13. (II) A 31.5-g glass thermometer reads 23.6°C before it is placed in 135 mL of water. When the water and thermometer come to equilibrium, the thermometer reads 39.2°C. What was the original temperature of the water? [*Hint:* Ignore the mass of fluid inside the glass thermometer.]

14. (II) Estimate the Calorie content of 65 g of candy from the following measurements. A 15-g sample of the candy is placed in a small aluminum container of mass 0.325 kg filled with oxygen. This container is placed in 2.00 kg of water in an aluminum calorimeter cup of mass 0.624 kg at an initial temperature of 15.0°C. The oxygen-candy mixture in the small container is ignited, and the final temperature of the whole system is 53.5°C.

15. (II) When a 290-g piece of iron at 180°C is placed in a 95-g aluminum calorimeter cup containing 250 g of glycerin at 10°C, the final temperature is observed to be 38°C. Estimate the specific heat of glycerin.

16. (II) The *heat capacity, C*, of an object is defined as the amount of heat needed to raise its temperature by 1 C°. Thus, to raise the temperature by ΔT requires heat Q given by

$$Q = C \Delta T.$$

(*a*) Write the heat capacity C in terms of the specific heat, c, of the material. (*b*) What is the heat capacity of 1.0 kg of water? (*c*) Of 35 kg of water?

17. (II) The 1.20-kg head of a hammer has a speed of 7.5 m/s just before it strikes a nail (Fig. 19–29) and is brought to rest. Estimate the temperature rise of a 14-g iron nail generated by 10 such hammer blows done in quick succession. Assume the nail absorbs all the energy.

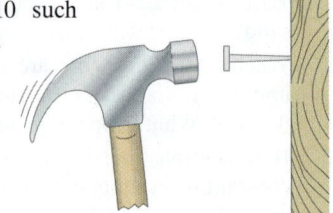

FIGURE 19–29
Problem 17.

19–5 Latent Heat

18. (I) How much heat is needed to melt 26.50 kg of silver that is initially at 25°C?

19. (I) During exercise, a person may give off 180 kcal of heat in 25 min by evaporation of water from the skin. How much water has been lost?

20. (II) A 35-g ice cube at its melting point is dropped into an insulated container of liquid nitrogen. How much nitrogen evaporates if it is at its boiling point of 77 K and has a latent heat of vaporization of 200 kJ/kg? Assume for simplicity that the specific heat of ice is a constant and is equal to its value near its melting point.

21. (II) High-altitude mountain climbers do not eat snow, but always melt it first with a stove. To see why, calculate the energy absorbed from your body if you (a) eat 1.0 kg of −10°C snow which your body warms to body temperature of 37°C. (b) You melt 1.0 kg of −10°C snow using a stove and drink the resulting 1.0 kg of water at 2°C, which your body has to warm to 37°C.

22. (II) An iron boiler of mass 180 kg contains 730 kg of water at 18°C. A heater supplies energy at the rate of 52,000 kJ/h. How long does it take for the water (a) to reach the boiling point, and (b) to all have changed to steam?

23. (II) In a hot day's race, a bicyclist consumes 8.0 L of water over the span of 3.5 hours. Making the approximation that all of the cyclist's energy goes into evaporating this water as sweat, how much energy in kcal did the rider use during the ride? (Since the efficiency of the rider is only about 20%, most of the energy consumed does go to heat, so our approximation is not far off.)

24. (II) The specific heat of mercury is 138 J/kg·C°. Determine the latent heat of fusion of mercury using the following calorimeter data: 1.00 kg of solid Hg at its melting point of −39.0°C is placed in a 0.620-kg aluminum calorimeter with 0.400 kg of water at 12.80°C; the resulting equilibrium temperature is 5.06°C.

25. (II) At a crime scene, the forensic investigator notes that the 7.2-g lead bullet that was stopped in a doorframe apparently melted completely on impact. Assuming the bullet was shot at room temperature (20°C), what does the investigator calculate as the minimum muzzle velocity of the gun?

26. (II) A 58-kg ice-skater moving at 7.5 m/s glides to a stop. Assuming the ice is at 0°C and that 50% of the heat generated by friction is absorbed by the ice, how much ice melts?

19–6 and 19–7 First Law of Thermodynamics

27. (I) Sketch a PV diagram of the following process: 2.0 L of ideal gas at atmospheric pressure are cooled at constant pressure to a volume of 1.0 L, and then expanded isothermally back to 2.0 L, whereupon the pressure is increased at constant volume until the original pressure is reached.

28. (I) A gas is enclosed in a cylinder fitted with a light frictionless piston and maintained at atmospheric pressure. When 1250 kcal of heat is added to the gas, the volume is observed to increase slowly from 12.0 m³ to 18.2 m³. Calculate (a) the work done by the gas and (b) the change in internal energy of the gas.

29. (II) The pressure in an ideal gas is cut in half slowly, while being kept in a container with rigid walls. In the process, 365 kJ of heat left the gas. (a) How much work was done during this process? (b) What was the change in internal energy of the gas during this process?

30. (II) A 1.0-L volume of air initially at 3.5 atm of (absolute) pressure is allowed to expand isothermally until the pressure is 1.0 atm. It is then compressed at constant pressure to its initial volume, and lastly is brought back to its original pressure by heating at constant volume. Draw the process on a PV diagram, including numbers and labels for the axes.

31. (II) Consider the following two-step process. Heat is allowed to flow out of an ideal gas at constant volume so that its pressure drops from 2.2 atm to 1.4 atm. Then the gas expands at constant pressure, from a volume of 5.9 L to 9.3 L, where the temperature reaches its original value. See Fig. 19–30. Calculate (a) the total work done by the gas in the process, (b) the change in internal energy of the gas in the process, and (c) the total heat flow into or out of the gas.

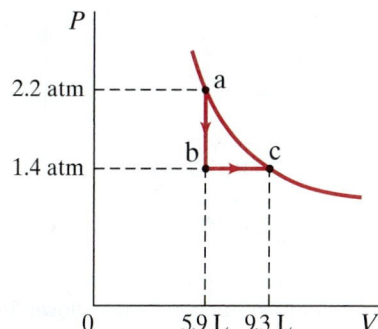

FIGURE 19–30
Problem 31.

32. (II) The PV diagram in Fig. 19–31 shows two possible states of a system containing 1.55 moles of a monatomic ideal gas. ($P_1 = P_2 = 455\ \text{N/m}^2$, $V_1 = 2.00\ \text{m}^3$, $V_2 = 8.00\ \text{m}^3$.) (a) Draw the process which depicts an isobaric expansion from state 1 to state 2, and label this process A. (b) Find the work done by the gas and the change in internal energy of the gas in process A. (c) Draw the two-step process which depicts an isothermal expansion from state 1 to the volume V_2, followed by an isovolumetric increase in temperature to state 2, and label this process B. (d) Find the change in internal energy of the gas for the two-step process B.

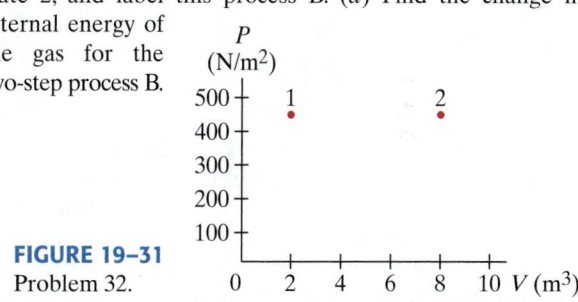

FIGURE 19–31
Problem 32.

33. (II) Suppose 2.60 mol of an ideal gas of volume $V_1 = 3.50\ \text{m}^3$ at $T_1 = 290\ \text{K}$ is allowed to expand isothermally to $V_2 = 7.00\ \text{m}^3$ at $T_2 = 290\ \text{K}$. Determine (a) the work done by the gas, (b) the heat added to the gas, and (c) the change in internal energy of the gas.

34. (II) In an engine, an almost ideal gas is compressed adiabatically to half its volume. In doing so, 2850 J of work is done on the gas. (a) How much heat flows into or out of the gas? (b) What is the change in internal energy of the gas? (c) Does its temperature rise or fall?

35. (II) One and one-half moles of an ideal monatomic gas expand adiabatically, performing 7500 J of work in the process. What is the change in temperature of the gas during this expansion?

36. (II) Determine (a) the work done and (b) the change in internal energy of 1.00 kg of water when it is all boiled to steam at 100°C. Assume a constant pressure of 1.00 atm.

37. (II) How much work is done by a pump to slowly compress, isothermally, 3.50 L of nitrogen at 0°C and 1.00 atm to 1.80 L at 0°C?

38. (II) When a gas is taken from a to c along the curved path in Fig. 19–32, the work done by the gas is $W = -35$ J and the heat added to the gas is $Q = -63$ J. Along path abc, the work done is $W = -54$ J. (a) What is Q for path abc? (b) If $P_c = \frac{1}{2}P_b$, what is W for path cda? (c) What is Q for path cda? (d) What is $E_{int, a} - E_{int, c}$? (e) If $E_{int, d} - E_{int, c} = 12$ J, what is Q for path da?

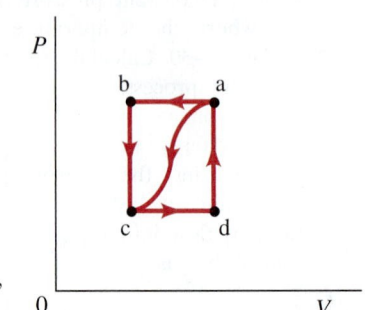

FIGURE 19–32
Problems 38, 39, and 40.

39. (III) In the process of taking a gas from state a to state c along the curved path shown in Fig. 19–32, 85 J of heat leaves the system and 55 J of work is done *on* the system. (a) Determine the change in internal energy, $E_{int, a} - E_{int, c}$. (b) When the gas is taken along the path cda, the work done by the gas is $W = 38$ J. How much heat Q is added to the gas in the process cda? (c) If $P_a = 2.2P_d$, how much work is done by the gas in the process abc? (d) What is Q for path abc? (e) If $E_{int, a} - E_{int, b} = 15$ J, what is Q for the process bc? Here is a summary of what is given:

$$Q_{a \to c} = -85 \text{ J}$$
$$W_{a \to c} = -55 \text{ J}$$
$$W_{cda} = 38 \text{ J}$$
$$E_{int, a} - E_{int, b} = 15 \text{ J}$$
$$P_a = 2.2P_d.$$

40. (III) Suppose a gas is taken clockwise around the rectangular cycle shown in Fig. 19–32, starting at b, then to a, to d, to c, and returning to b. Using the values given in Problem 39, (a) describe each leg of the process, and then calculate (b) the net work done during the cycle, (c) the total internal energy change during the cycle, and (d) the net heat flow during the cycle. (e) What percentage of the *intake* heat was turned into usable work: i.e., how efficient is this "rectangular" cycle (give as a percentage)?

*41. (III) Determine the work done by 1.00 mol of a van der Waals gas (Section 18–5) when it expands from volume V_1 to V_2 isothermally.

19–8 Molecular Specific Heat for Gases; Equipartition of Energy

42. (I) What is the internal energy of 4.50 mol of an ideal diatomic gas at 645 K, assuming all degrees of freedom are active?

43. (I) If a heater supplies 1.8×10^6 J/h to a room 3.5 m × 4.6 m × 3.0 m containing air at 20°C and 1.0 atm, by how much will the temperature rise in one hour, assuming no losses of heat or air mass to the outside? Assume air is an ideal diatomic gas with molecular mass 29.

44. (I) Show that if the molecules of a gas have n degrees of freedom, then theory predicts $C_V = \frac{1}{2}nR$ and $C_P = \frac{1}{2}(n + 2)R$.

45. (II) A certain monatomic gas has specific heat $c_V = 0.0356$ kcal/kg·C°, which changes little over a wide temperature range. What is the atomic mass of this gas? What gas is it?

46. (II) Show that the work done by n moles of an ideal gas when it expands adiabatically is $W = nC_V(T_1 - T_2)$, where T_1 and T_2 are the initial and final temperatures, and C_V is the molar specific heat at constant volume.

47. (II) An audience of 1800 fills a concert hall of volume 22,000 m³. If there were no ventilation, by how much would the temperature of the air rise over a period of 2.0 h due to the metabolism of the people (70 W/person)?

48. (II) The specific heat at constant volume of a particular gas is 0.182 kcal/kg·K at room temperature, and its molecular mass is 34. (a) What is its specific heat at constant pressure? (b) What do you think is the molecular structure of this gas?

49. (II) A 2.00 mole sample of N_2 gas at 0°C is heated to 150°C at constant pressure (1.00 atm). Determine (a) the change in internal energy, (b) the work the gas does, and (c) the heat added to it.

50. (III) A 1.00-mol sample of an ideal diatomic gas at a pressure of 1.00 atm and temperature of 420 K undergoes a process in which its pressure increases linearly with temperature. The final temperature and pressure are 720 K and 1.60 atm. Determine (a) the change in internal energy, (b) the work done by the gas, and (c) the heat added to the gas. (Assume five active degrees of freedom.)

19–9 Adiabatic Expansion of a Gas

51. (I) A 1.00-mol sample of an ideal diatomic gas, originally at 1.00 atm and 20°C, expands adiabatically to 1.75 times its initial volume. What are the final pressure and temperature for the gas? (Assume no molecular vibration.)

52. (II) Show, using Eqs. 19–6 and 19–15, that the work done by a gas that slowly expands adiabatically from pressure P_1 and volume V_1, to P_2 and V_2, is given by $W = (P_1V_1 - P_2V_2)/(\gamma - 1)$.

53. (II) A 3.65-mol sample of an ideal diatomic gas expands adiabatically from a volume of 0.1210 m³ to 0.750 m³. Initially the pressure was 1.00 atm. Determine: (a) the initial and final temperatures; (b) the change in internal energy; (c) the heat lost by the gas; (d) the work done on the gas. (Assume no molecular vibration.)

54. (II) An ideal monatomic gas, consisting of 2.8 mol of volume 0.086 m³, expands adiabatically. The initial and final temperatures are 25°C and −68°C. What is the final volume of the gas?

55. (III) A 1.00-mol sample of an ideal monatomic gas, originally at a pressure of 1.00 atm, undergoes a three-step process: (1) it is expanded adiabatically from $T_1 = 588$ K to $T_2 = 389$ K; (2) it is compressed at constant pressure until its temperature reaches T_3; (3) it then returns to its original pressure and temperature by a constant-volume process. (a) Plot these processes on a PV diagram. (b) Determine T_3. (c) Calculate the change in internal energy, the work done by the gas, and the heat added to the gas for each process, and (d) for the complete cycle.

56. (III) Consider a **parcel of air** moving to a different altitude y in the Earth's atmosphere (Fig. 19–33). As the parcel changes altitude it acquires the pressure P of the surrounding air. From Eq. 13–4 we have

$$\frac{dP}{dy} = -\rho g$$

where ρ is the parcel's altitude-dependent mass density.

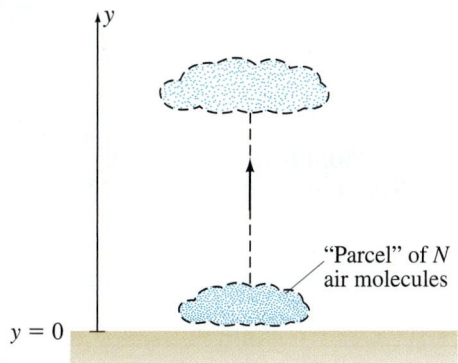

FIGURE 19–33
Problem 56.

During this motion, the parcel's volume will change and, because air is a poor heat conductor, we assume this expansion or contraction will take place adiabatically. (a) Starting with Eq. 19–15, $PV^{\gamma} = $ constant, show that for an ideal gas undergoing an adiabatic process, $P^{1-\gamma}T^{\gamma} = $ constant. Then show that the parcel's pressure and temperature are related by

$$(1 - \gamma)\frac{dP}{dy} + \gamma\frac{P}{T}\frac{dT}{dy} = 0$$

and thus

$$(1 - \gamma)(-\rho g) + \gamma\frac{P}{T}\frac{dT}{dy} = 0.$$

(b) Use the ideal gas law with the result from part (a) to show that the change in the parcel's temperature with change in altitude is given by

$$\frac{dT}{dy} = \frac{1 - \gamma}{\gamma}\frac{mg}{k}$$

where m is the average mass of an air molecule and k is the Boltzmann constant. (c) Given that air is a diatomic gas with an average molecular mass of 29, show that $dT/dy = -9.8\ \text{C}°/\text{km}$. This value is called the **adiabatic lapse rate** for dry air. (d) In California, the prevailing westerly winds descend from one of the highest elevations (the 4000-m Sierra Nevada mountains) to one of the lowest elevations (Death Valley, -100 m) in the continental United States. If a dry wind has a temperature of $-5°C$ at the top of the Sierra Nevada, what is the wind's temperature after it has descended to Death Valley?

19–10 Conduction, Convection, Radiation

57. (I) (a) How much power is radiated by a tungsten sphere (emissivity $\epsilon = 0.35$) of radius 16 cm at a temperature of 25°C? (b) If the sphere is enclosed in a room whose walls are kept at $-5°C$, what is the *net* flow rate of energy out of the sphere?

58. (I) One end of a 45-cm-long copper rod with a diameter of 2.0 cm is kept at 460°C, and the other is immersed in water at 22°C. Calculate the heat conduction rate along the rod.

59. (II) How long does it take the Sun to melt a block of ice at 0°C with a flat horizontal area $1.0\ \text{m}^2$ and thickness 1.0 cm? Assume that the Sun's rays make an angle of 35° with the vertical and that the emissivity of ice is 0.050.

60. (II) *Heat conduction to skin.* Suppose 150 W of heat flows by conduction from the blood capillaries beneath the skin to the body's surface area of $1.5\ \text{m}^2$. If the temperature difference is $0.50\ \text{C}°$, estimate the average distance of capillaries below the skin surface.

61. (II) A ceramic teapot ($\epsilon = 0.70$) and a shiny one ($\epsilon = 0.10$) each hold 0.55 L of tea at 95°C. (a) Estimate the rate of heat loss from each, and (b) estimate the temperature drop after 30 min for each. Consider only radiation, and assume the surroundings are at 20°C.

62. (II) A copper rod and an aluminum rod of the same length and cross-sectional area are attached end to end (Fig. 19–34). The copper end is placed in a furnace maintained at a constant temperature of 225°C. The aluminum end is placed in an ice bath held at constant temperature of 0.0°C. Calculate the temperature at the point where the two rods are joined.

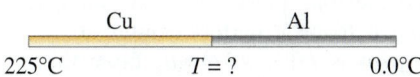

FIGURE 19–34   Problem 62.

63. (II) (a) Using the solar constant, estimate the rate at which the whole Earth receives energy from the Sun. (b) Assume the Earth radiates an equal amount back into space (that is, the Earth is in equilibrium). Then, assuming the Earth is a perfect emitter ($\epsilon = 1.0$), estimate its average surface temperature. [*Hint:* Use area $A = 4\pi r_{\text{E}}^2$, and state why.]

64. (II) A 100-W lightbulb generates 95 W of heat, which is dissipated through a glass bulb that has a radius of 3.0 cm and is 0.50 mm thick. What is the difference in temperature between the inner and outer surfaces of the glass?

65. (III) A house thermostat is normally set to 22°C, but at night it is turned down to 12°C for 9.0 h. Estimate how much more heat would be needed (state as a percentage of daily usage) if the thermostat were not turned down at night. Assume that the outside temperature averages 0°C for the 9.0 h at night and 8°C for the remainder of the day, and that the heat loss from the house is proportional to the difference in temperature inside and out. To obtain an estimate from the data, you will have to make other simplifying assumptions; state what these are.

66. (III) Approximately how long should it take 9.5 kg of ice at 0°C to melt when it is placed in a carefully sealed Styrofoam ice chest of dimensions 25 cm $\times$ 35 cm $\times$ 55 cm whose walls are 1.5 cm thick? Assume that the conductivity of Styrofoam is double that of air and that the outside temperature is 34°C.

67. (III) A cylindrical pipe has inner radius R_1 and outer radius R_2. The interior of the pipe carries hot water at temperature T_1. The temperature outside is T_2 $(< T_1)$. (a) Show that the rate of heat loss for a length ℓ of pipe is

$$\frac{dQ}{dt} = \frac{2\pi k(T_1 - T_2)\ell}{\ln(R_2/R_1)},$$

where k is the thermal conductivity of the pipe. (b) Suppose the pipe is steel with $R_1 = 3.3$ cm, $R_2 = 4.0$ cm, and $T_2 = 18°C$. If the pipe holds still water at $T_1 = 71°C$, what will be the initial rate of change of its temperature? (c) Suppose water at $71°C$ enters the pipe and moves at a speed of 8.0 cm/s. What will be its temperature drop per centimeter of travel?

68. (III) Suppose the insulating qualities of the wall of a house come mainly from a 4.0-in. layer of brick and an R-19 layer of insulation, as shown in Fig. 19–35. What is the total rate of heat loss through such a wall, if its total area is 195 ft^2 and the temperature difference across it is 12 F°?

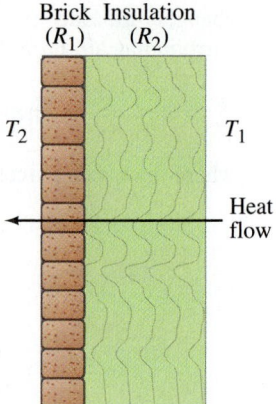

FIGURE 19–35 Problem 68. Two layers insulating a wall.

General Problems

69. A soft-drink can contains about 0.20 kg of liquid at 5°C. Drinking this liquid can actually consume some of the fat in the body, since energy is needed to warm the liquid to body temperature (37°C). How many food Calories should the drink have so that it is in perfect balance with the heat needed to warm the liquid (essentially water)?

70. (a) Find the total power radiated into space by the Sun, assuming it to be a perfect emitter at $T = 5500$ K. The Sun's radius is 7.0×10^8 m. (b) From this, determine the power per unit area arriving at the Earth, 1.5×10^{11} m away.

71. To get an idea of how much thermal energy is contained in the world's oceans, estimate the heat liberated when a cube of ocean water, 1 km on each side, is cooled by 1 K. (Approximate the ocean water as pure water for this estimate.)

72. A mountain climber wears a goose-down jacket 3.5 cm thick with total surface area 0.95 m^2. The temperature at the surface of the clothing is $-18°C$ and at the skin is 34°C. Determine the rate of heat flow by conduction through the jacket (a) assuming it is dry and the thermal conductivity k is that of goose down, and (b) assuming the jacket is wet, so k is that of water and the jacket has matted to 0.50 cm thickness.

73. During light activity, a 70-kg person may generate 200 kcal/h. Assuming that 20% of this goes into useful work and the other 80% is converted to heat, estimate the temperature rise of the body after 30 min if none of this heat is transferred to the environment.

74. Estimate the rate at which heat can be conducted from the interior of the body to the surface. Assume that the thickness of tissue is 4.0 cm, that the skin is at 34°C and the interior at 37°C, and that the surface area is 1.5 m^2. Compare this to the measured value of about 230 W that must be dissipated by a person working lightly. This clearly shows the necessity of convective cooling by the blood.

75. A marathon runner has an average metabolism rate of about 950 kcal/h during a race. If the runner has a mass of 55 kg, estimate how much water she would lose to evaporation from the skin for a race that lasts 2.2 h.

76. A house has well-insulated walls 19.5 cm thick (assume conductivity of air) and area 410 m^2, a roof of wood 5.5 cm thick and area 280 m^2, and uncovered windows 0.65 cm thick and total area 33 m^2. (a) Assuming that heat is lost only by conduction, calculate the rate at which heat must be supplied to this house to maintain its inside temperature at 23°C if the outside temperature is $-15°C$. (b) If the house is initially at 12°C, estimate how much heat must be supplied to raise the temperature to 23°C within 30 min. Assume that only the air needs to be heated and that its volume is 750 m^3. (c) If natural gas costs \$0.080 per kilogram and its heat of combustion is 5.4×10^7 J/kg, how much is the monthly cost to maintain the house as in part (a) for 24 h each day, assuming 90% of the heat produced is used to heat the house? Take the specific heat of air to be 0.24 kcal/kg·C°.

77. In a typical game of squash (Fig. 19–36), two people hit a soft rubber ball at a wall until they are about to drop due to dehydration and exhaustion. Assume that the ball hits the wall at a velocity of 22 m/s and bounces back with a velocity of 12 m/s, and that the kinetic energy lost in the process heats the ball. What will be the temperature increase of the ball after one bounce? (The specific heat of rubber is about 1200 J/kg·C°.)

FIGURE 19–36 Problem 77.

78. A bicycle pump is a cylinder 22 cm long and 3.0 cm in diameter. The pump contains air at 20.0°C and 1.0 atm. If the outlet at the base of the pump is blocked and the handle is pushed in very quickly, compressing the air to half its original volume, how hot does the air in the pump become?

79. A microwave oven is used to heat 250 g of water. On its maximum setting, the oven can raise the temperature of the liquid water from 20°C to 100°C in 1 min 45 s ($= 105$ s). (*a*) At what rate does the oven input energy to the liquid water? (*b*) If the power input from the oven to the water remains constant, determine how many grams of water will boil away if the oven is operated for 2 min (rather than just 1 min 45 s).

80. The temperature within the Earth's crust increases about 1.0 C° for each 30 m of depth. The thermal conductivity of the crust is 0.80 W/C°·m. (*a*) Determine the heat transferred from the interior to the surface for the entire Earth in 1.0 h. (*b*) Compare this heat to the amount of energy incident on the Earth in 1.0 h due to radiation from the Sun.

81. An ice sheet forms on a lake. The air above the sheet is at $-18°C$, whereas the water is at 0°C. Assume that the heat of fusion of the water freezing on the lower surface is conducted through the sheet to the air above. How much time will it take to form a sheet of ice 15 cm thick?

82. An iron meteorite melts when it enters the Earth's atmosphere. If its initial temperature was $-105°C$ outside of Earth's atmosphere, calculate the minimum velocity the meteorite must have had before it entered Earth's atmosphere.

83. A scuba diver releases a 3.60-cm-diameter (spherical) bubble of air from a depth of 14.0 m. Assume the temperature is constant at 298 K, and that the air behaves as an ideal gas. (*a*) How large is the bubble when it reaches the surface? (*b*) Sketch a *PV* diagram for the process. (*c*) Apply the first law of thermodynamics to the bubble, and find the work done by the air in rising to the surface, the change in its internal energy, and the heat added or removed from the air in the bubble as it rises. Take the density of water to be 1000 kg/m³.

84. A reciprocating compressor is a device that compresses air by a back-and-forth straight-line motion, like a piston in a cylinder. Consider a reciprocating compressor running at 150 rpm. During a compression stroke, 1.00 mol of air is compressed. The initial temperature of the air is 390 K, the engine of the compressor is supplying 7.5 kW of power to compress the air, and heat is being removed at the rate of 1.5 kW. Calculate the temperature change per compression stroke.

85. The temperature of the glass surface of a 75-W lightbulb is 75°C when the room temperature is 18°C. Estimate the temperature of a 150-W lightbulb with a glass bulb the same size. Consider only radiation, and assume that 90% of the energy is emitted as heat.

86. Suppose 3.0 mol of neon (an ideal monatomic gas) at STP are compressed slowly and isothermally to 0.22 the original volume. The gas is then allowed to expand quickly and adiabatically back to its original volume. Find the highest and lowest temperatures and pressures attained by the gas, and show on a *PV* diagram where these values occur.

87. At very low temperatures, the molar specific heat of many substances varies as the cube of the absolute temperature:

$$C = k\frac{T^3}{T_0^3},$$

which is sometimes called Debye's law. For rock salt, $T_0 = 281$ K and $k = 1940$ J/mol·K. Determine the heat needed to raise 2.75 mol of salt from 22.0 K to 48.0 K.

88. A diesel engine accomplishes ignition without a spark plug by an adiabatic compression of air to a temperature above the ignition temperature of the diesel fuel, which is injected into the cylinder at the peak of the compression. Suppose air is taken into the cylinder at 280 K and volume V_1 and is compressed adiabatically to 560°C ($\approx 1000°F$) and volume V_2. Assuming that the air behaves as an ideal gas whose ratio of C_P to C_V is 1.4, calculate the compression ratio V_1/V_2 of the engine.

89. When 6.30×10^5 J of heat is added to a gas enclosed in a cylinder fitted with a light frictionless piston maintained at atmospheric pressure, the volume is observed to increase from 2.2 m³ to 4.1 m³. Calculate (*a*) the work done by the gas, and (*b*) the change in internal energy of the gas. (*c*) Graph this process on a *PV* diagram.

90. In a cold environment, a person can lose heat by conduction and radiation at a rate of about 200 W. Estimate how long it would take for the body temperature to drop from 36.6°C to 35.6°C if metabolism were nearly to stop. Assume a mass of 70 kg. (See Table 19–1.)

*Numerical/Computer

***91.** (II) Suppose 1.0 mol of steam at 100°C of volume 0.50 m³ is expanded isothermally to volume 1.00 m³. Assume steam obeys the van der Waals equation $(P + n^2a/V^2)(V/n - b) = RT$, Eq. 18–9, with $a = 0.55$ N·m⁴/mol² and $b = 3.0 \times 10^{-5}$ m³/mol. Using the expression $dW = P\,dV$, determine numerically the total work done W. Your result should agree within 2% of the result obtained by integrating the expression for dW.

Answers to Exercises

A: (*b*).

B: (*c*).

C: 0.21 kg.

D: 700 J.

E: Less.

F: -6.8×10^3 J.

G: (*d*).

There are many uses for a heat engine, such as old steam trains and modern coal-burning power plants. Steam engines produce steam which does work: on turbines to generate electricity, and on a piston that moves linkage to turn locomotive wheels. The efficiency of any engine—no matter how carefully engineered—is limited by nature as described in the second law of thermodynamics. This great law is best stated in terms of a quantity called entropy, which is unlike any other. Entropy is *not* conserved, but instead is constrained always to increase in any real process. Entropy is a measure of disorder. The second law of thermodynamics tells us that as time moves forward, the disorder in the universe increases.

We discuss many practical matters including heat engines, heat pumps, and refrigeration.

20

Second Law of Thermodynamics

CONTENTS

20–1 The Second Law of Thermodynamics—Introduction

20–2 Heat Engines

20–3 Reversible and Irreversible Processes; the Carnot Engine

20–4 Refrigerators, Air Conditioners, and Heat Pumps

20–5 Entropy

20–6 Entropy and the Second Law of Thermodynamics

20–7 Order to Disorder

20–8 Unavailability of Energy; Heat Death

*20–9 Statistical Interpretation of Entropy and the Second Law

*20–10 Thermodynamic Temperature; Third Law of Thermodynamics

*20–11 Thermal Pollution, Global Warming, and Energy Resources

CHAPTER-OPENING QUESTION—Guess now!

Fossil-fuel electric generating plants produce "thermal pollution." Part of the heat produced by the burning fuel is not converted to electric energy. The reason for this waste is

(a) The efficiency is higher if some heat is allowed to escape.

(b) Engineering technology has not yet reached the point where 100% waste heat recovery is possible.

(c) Some waste heat *must* be produced: this is a fundamental property of nature when converting heat to useful work.

(d) The plants rely on fossil fuels, not nuclear fuel.

(e) None of the above.

In this final Chapter on heat and thermodynamics, we discuss the famous second law of thermodynamics, and the quantity "entropy" that arose from this fundamental law and is its quintessential expression. We also discuss heat engines—the engines that transform heat into work in power plants, trains, and motor vehicles—because they first showed us that a new law was needed. Finally, we briefly discuss the third law of thermodynamics.

20–1 The Second Law of Thermodynamics—Introduction

The first law of thermodynamics states that energy is conserved. There are, however, many processes we can imagine that conserve energy but are not observed to occur in nature. For example, when a hot object is placed in contact with a cold object, heat flows from the hotter one to the colder one, never spontaneously the reverse. If heat were to leave the colder object and pass to the hotter one, energy could still be conserved. Yet it does not happen spontaneously.[†] As a second example, consider what happens when you drop a rock and it hits the ground. The initial potential energy of the rock changes to kinetic energy as the rock falls. When the rock hits the ground, this energy in turn is transformed into internal energy of the rock and the ground in the vicinity of the impact; the molecules move faster and the temperature rises slightly. But have you seen the reverse happen—a rock at rest on the ground suddenly rise up in the air because the thermal energy of molecules is transformed into kinetic energy of the rock as a whole? Energy could be conserved in this process, yet we never see it happen.

There are many other examples of processes that occur in nature but whose reverse does not. Here are two more. (1) If you put a layer of salt in a jar and cover it with a layer of similar-sized grains of pepper, when you shake it you get a thorough mixture. But no matter how long you shake it, the mixture does not separate into two layers again. (2) Coffee cups and glasses break spontaneously if you drop them. But they do not go back together spontaneously (Fig. 20–1).

(a) Initial state.   (b) Later: cup reassembles   (c) Later still: cup lands on table.
and rises up.

FIGURE 20–1 Have you ever observed this process, a broken cup spontaneously reassembling and rising up onto a table? This process could conserve energy and other laws of mechanics.

The first law of thermodynamics (conservation of energy) would not be violated if any of these processes occurred in reverse. To explain this lack of reversibility, scientists in the latter half of the nineteenth century formulated a new principle known as the second law of thermodynamics.

The **second law of thermodynamics** is a statement about which processes occur in nature and which do not. It can be stated in a variety of ways, all of which are equivalent. One statement, due to R. J. E. Clausius (1822–1888), is that

> **heat can flow spontaneously from a hot object to a cold object; heat will not flow spontaneously from a cold object to a hot object.**

SECOND LAW OF THERMODYNAMICS (Clausius statement)

Since this statement applies to one particular process, it is not obvious how it applies to other processes. A more general statement is needed that will include other possible processes in a more obvious way.

The development of a general statement of the second law of thermodynamics was based partly on the study of heat engines. A **heat engine** is any device that changes thermal energy into mechanical work, such as a steam engine or automobile engine. We now examine heat engines, both from a practical point of view and to show their importance in developing the second law of thermodynamics.

[†]By spontaneously, we mean by itself without input of work of some sort. (A refrigerator does move heat from a cold environment to a warmer one, but only because its motor does work—Section 20–4.)

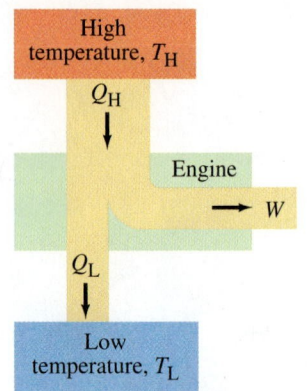

FIGURE 20–2 Schematic diagram of energy transfers for a heat engine.

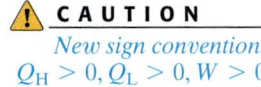
CAUTION

New sign convention:
$Q_H > 0, Q_L > 0, W > 0$

PHYSICS APPLIED

Engines

20–2 Heat Engines

It is easy to produce thermal energy by doing work—for example, by simply rubbing your hands together briskly, or indeed by any frictional process. But to get work from thermal energy is more difficult, and a practical device to do this was invented only about 1700 with the development of the steam engine.

The basic idea behind any heat engine is that mechanical energy can be obtained from thermal energy only when heat is allowed to flow from a high temperature to a low temperature. In the process, some of the heat can then be transformed to mechanical work, as diagrammed schematically in Fig. 20–2. We will be interested only in engines that run in a repeating *cycle* (that is, the system returns repeatedly to its starting point) and thus can run continuously. In each cycle the change in internal energy of the system is $\Delta E_{int} = 0$ because it returns to the starting state. Thus a heat input Q_H at a high temperature T_H is partly transformed into work W and partly exhausted as heat Q_L at a lower temperature T_L (Fig. 20–2). By conservation of energy, $Q_H = W + Q_L$. The high and low temperatures, T_H and T_L, are called the **operating temperatures** of the engine. Note carefully that we are now using a new sign convention: we take Q_H, Q_L, and W as always positive. The direction of each energy transfer is shown by the arrow on the applicable diagram, such as Fig. 20–2.

Steam Engine and Internal Combustion Engine

The operation of a steam engine is illustrated in Fig. 20–3. Steam engines are of two main types, each making use of steam heated by combustion of coal, oil or gas, or by nuclear energy. In a reciprocating engine, Fig. 20–3a, the heated steam passes through the intake valve and expands against a piston, forcing it to move. As the piston returns to its original position, it forces the gases out the exhaust valve. A steam turbine, Fig. 20–3b, is very similar except that the reciprocating piston is replaced by a rotating turbine that resembles a paddlewheel with many sets of blades. Most of our electricity today is generated using steam turbines.[†] The material that is heated and cooled, steam in this case, is called the **working substance**.

[†] Even nuclear power plants utilize steam turbines; the nuclear fuel—uranium—merely serves as fuel to heat the steam.

(a) Reciprocating type

(b) Turbine (boiler and condenser not shown)

High temperature

Boiler

Steam

Water

Heat input

Intake valve (open during expansion)

Exhaust valve (closed during expansion)

Piston

Pump

Condenser

Water

Low temperature

High-pressure steam, from boiler

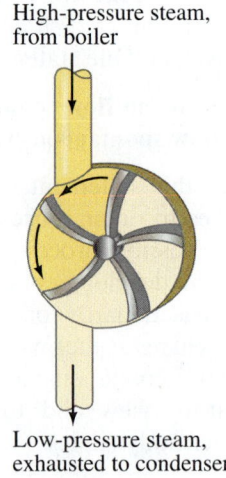

Low-pressure steam, exhausted to condenser

FIGURE 20–3 Steam engines.

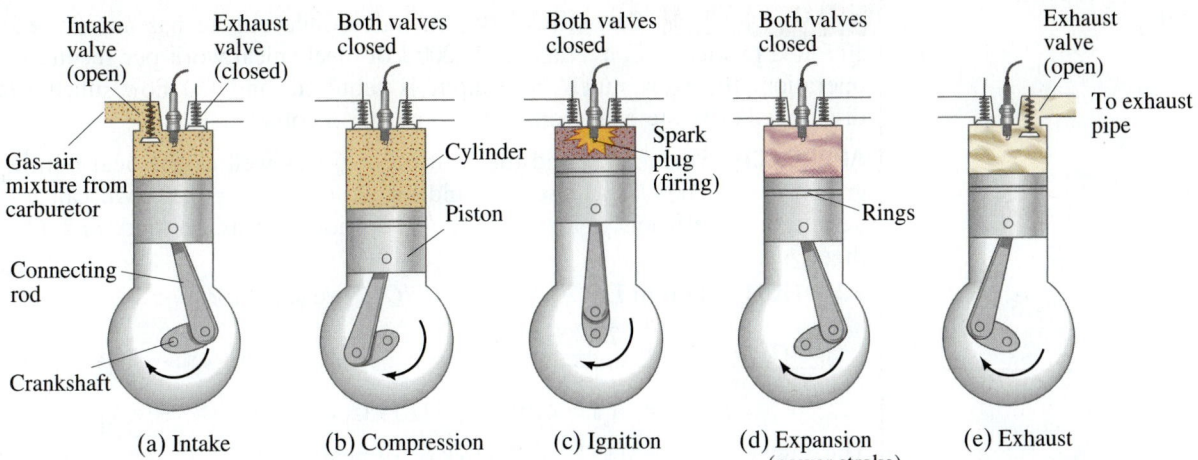

| (a) Intake | (b) Compression | (c) Ignition | (d) Expansion (power stroke) | (e) Exhaust |

In an internal combustion engine (used in most automobiles), the high temperature is achieved by burning the gasoline–air mixture in the cylinder itself (ignited by the spark plug), as described in Fig. 20–4.

Why a ΔT Is Needed to Drive a Heat Engine

To see why a *temperature difference* is required to run an engine, let us examine the steam engine. In the reciprocating engine, for example, suppose there were no condenser or pump (Fig. 20–3a), and that the steam was at the same temperature throughout the system. This would mean that the pressure of the gas being exhausted would be the same as that on intake. Thus, although work would be done by the gas *on* the piston when it expanded, an equal amount of work would have to be done *by* the piston to force the steam out the exhaust; hence, no net work would be done. In a real engine, the exhausted gas is cooled to a lower temperature and condensed so that the exhaust pressure is less than the intake pressure. Thus, although the piston must do work on the gas to expel it on the exhaust stroke, it is less than the work done by the gas on the piston during the intake. So a net amount of work can be obtained—but only if there is a difference of temperature. Similarly, in the gas turbine if the gas isn't cooled, the pressure on each side of the blades would be the same. By cooling the gas on the exhaust side, the pressure on the back side of the blade is less and hence the turbine turns.

Efficiency and the Second Law

The **efficiency**, e, of any heat engine can be defined as the ratio of the work it does, W, to the heat input at the high temperature, Q_H (Fig. 20–2):

$$e = \frac{W}{Q_H}.$$

This is a sensible definition since W is the output (what you get from the engine), whereas Q_H is what you put in and pay for in burned fuel. Since energy is conserved, the heat input Q_H must equal the work done plus the heat that flows out at the low temperature (Q_L):

$$Q_H = W + Q_L.$$

Thus $W = Q_H - Q_L$, and the efficiency of an engine is

$$e = \frac{W}{Q_H} \tag{20–1a}$$

$$= \frac{Q_H - Q_L}{Q_H} = 1 - \frac{Q_L}{Q_H}. \tag{20–1b}$$

To give the efficiency as a percent, we multiply Eqs. 20–1 by 100. Note that e could be 1.0 (or 100%) only if Q_L were zero—that is, only if no heat were exhausted to the environment.

FIGURE 20–4 Four-stroke-cycle internal combustion engine: (a) the gasoline–air mixture flows into the cylinder as the piston moves down; (b) the piston moves upward and compresses the gas; (c) the brief instant when firing of the spark plug ignites the highly compressed gasoline–air mixture, raising it to a high temperature; (d) the gases, now at high temperature and pressure, expand against the piston in this, the power stroke; (e) the burned gases are pushed out to the exhaust pipe; when the piston reaches the top, the exhaust valve closes and the intake valve opens, and the whole cycle repeats. (a), (b), (d), and (e) are the four strokes of the cycle.

EXAMPLE 20–1 **Car efficiency.** An automobile engine has an efficiency of 20% and produces an average of 23,000 J of mechanical work per second during operation. (*a*) How much heat input is required, and (*b*) how much heat is discharged as waste heat from this engine, per second?

APPROACH We want to find the heat input Q_H as well as the heat output Q_L, given $W = 23,000$ J each second and an efficiency $e = 0.20$. We can use the definition of efficiency, Eq. 20–1 in its various forms, to find first Q_H and then Q_L.

SOLUTION (*a*) From Eq. 20–1a, $e = W/Q_H$, we solve for Q_H:

$$Q_H = \frac{W}{e} = \frac{23,000 \text{ J}}{0.20}$$
$$= 1.15 \times 10^5 \text{ J} = 115 \text{ kJ}.$$

The engine requires 115 kJ/s = 115 kW of heat input.

(*b*) Now we use Eq. 20–1b $\left(e = 1 - Q_L/Q_H\right)$ and solve for Q_L:

$$Q_L = (1 - e)Q_H = (0.80)115 \text{ kJ} = 92 \text{ kJ}.$$

The engine discharges heat to the environment at a rate of 92 kJ/s = 92 kW.

NOTE Of the 115 kJ that enters the engine per second, only 23 kJ does useful work whereas 92 kJ is wasted as heat output.

NOTE The problem was stated in terms of energy per unit time. We could just as well have stated it in terms of power, since 1 J/s = 1 W.

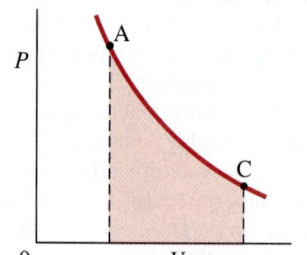

FIGURE 20–5 Adiabatic process, Exercise A.

EXERCISE A An adiabatic process is defined as one in which no heat flows in or out of the system. If an ideal gas expands as shown in Fig. 20–5 (see also Fig. 19–8), the work W done in this expansion equals the area under the graph, shown shaded. The efficiency of this process would be $e = W/Q$, much greater than 100% ($= \infty$ since $Q = 0$). Is this a violation of the second law?

It is clear from Eq. 20–1b, $e = 1 - Q_L/Q_H$, that the efficiency of an engine will be greater if Q_L can be made small. However, from experience with a wide variety of systems, it has not been found possible to reduce Q_L to zero. If Q_L could be reduced to zero we would have a 100% efficient engine, as diagrammed in Fig. 20–6.

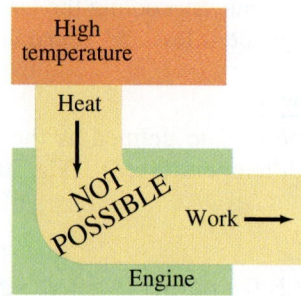

FIGURE 20–6 Schematic diagram of a hypothetical perfect heat engine in which all the heat input would be used to do work.

That such a perfect engine (running continuously in a cycle) is not possible is another way of expressing the second law of thermodynamics:

No device is possible whose sole effect is to transform a given amount of heat completely into work.

This is known as the **Kelvin-Planck statement of the second law of thermodynamics**. Said another way, *there can be no perfect (100% efficient) heat engine* such as that diagrammed in Fig. 20–6.

If the second law were not true, so that a perfect engine could be built, some rather remarkable things could happen. For example, if the engine of a ship did not need a low-temperature reservoir to exhaust heat into, the ship could sail across the ocean using the vast resources of the internal energy of the ocean water. Indeed, we would have no fuel problems at all!

20–3 Reversible and Irreversible Processes; the Carnot Engine

In the early nineteenth century, the French scientist N. L. Sadi Carnot (1796–1832) studied in detail the process of transforming heat into mechanical energy. His aim had been to determine how to increase the efficiency of heat engines, but his studies soon led him to investigate the foundations of thermodynamics itself. In 1824, Carnot invented (on paper) an idealized type of engine which we now call the *Carnot engine*. No Carnot engine actually exists, but as a theoretical idea it played an important role in the establishment and understanding of the second law of thermodynamics.

Reversible and Irreversible Processes

The Carnot engine involves *reversible processes*, so before we discuss it we must discuss what is meant by reversible and irreversible processes. A **reversible process** is one that is carried out infinitely slowly, so that the process can be considered as a series of equilibrium states, and the whole process could be done in reverse with no change in magnitude of the work done or heat exchanged. For example, a gas contained in a cylinder fitted with a tight, movable, but frictionless piston could be compressed isothermally in a reversible way if done infinitely slowly. Not all very slow (quasistatic) processes are reversible, however. If there is friction present, for example (as between the movable piston and cylinder just mentioned), the work done in one direction (going from some state A to state B) will not be the negative of the work done in the reverse direction (state B to state A). Such a process would not be considered reversible. A perfectly reversible process is not possible in reality because it would require an infinite time; reversible processes can be approached arbitrarily closely, however, and they are very important theoretically.

All real processes are **irreversible**: they are not done infinitely slowly. There could be turbulence in the gas, friction would be present, and so on. Any process could not be done precisely in reverse since the heat lost to friction would not reverse itself, the turbulence would be different, and so on. For any given volume there would not be a well-defined pressure P and temperature T since the system would not always be in an equilibrium state. Thus a real, irreversible, process cannot be plotted on a PV diagram, except insofar as it may approach an ideal reversible process. But a reversible process (since it is a quasistatic series of equilibrium states) always can be plotted on a PV diagram; and a reversible process, when done in reverse, retraces the same path on a PV diagram. Although all real processes are irreversible, reversible processes are conceptually important, just as the concept of an ideal gas is.

Carnot's Engine

Now let us look at Carnot's idealized engine. The **Carnot engine** makes use of a **reversible cycle**, by which we mean a series of reversible processes that take a given substance (the *working substance*) from an initial equilibrium state through many other equilibrium states and returns it again to the same initial state. In particular, the Carnot engine utilizes the **Carnot cycle**, which is illustrated in Fig. 20–7, with the working substance assumed to be an ideal gas. Let us take point a as the initial state. The gas is first expanded isothermally and reversibly, path ab, at temperature T_H, as heat Q_H is added to it. Next the gas is expanded adiabatically and reversibly, path bc; no heat is exchanged and the temperature of the gas is reduced to T_L. The third step is a reversible isothermal compression, path cd, during which heat Q_L flows out of the working substance. Finally, the gas is compressed adiabatically, path da, back to its original state. Thus a Carnot cycle consists of two isothermal and two adiabatic processes.

The net work done in one cycle by a Carnot engine (or any other type of engine using a reversible cycle) is equal to the area enclosed by the curve representing the cycle on the PV diagram, the curve abcd in Fig. 20–7. (See Section 19–7.)

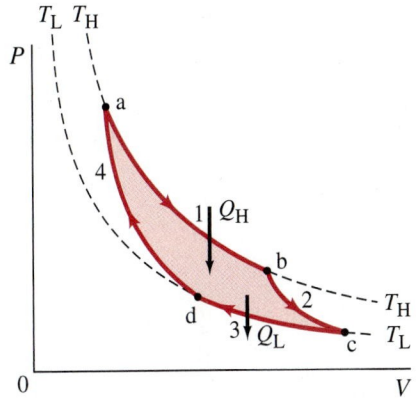

FIGURE 20–7 The Carnot cycle. Heat engines work in a cycle, and the cycle for the Carnot engine begins at point a on this PV diagram. (1) The gas is first expanded isothermally, with the addition of heat Q_H, along the path ab at temperature T_H. (2) Next the gas expands adiabatically from b to c—no heat is exchanged, but the temperature drops to T_L. (3) The gas is then compressed at constant temperature T_L, path cd, and heat Q_L flows out. (4) Finally, the gas is compressed adiabatically, path da, back to its original state. No Carnot engine actually exists, but as a theoretical idea it played an important role in the development of thermodynamics.

Carnot Efficiency and the Second Law of Thermodynamics

The efficiency of a Carnot engine, like any heat engine, is given by Eq. 20–1b:

$$e = 1 - \frac{Q_L}{Q_H}.$$

For a Carnot engine using an ideal gas, however, we can show that the efficiency depends only on the temperatures of the heat reservoirs, T_H and T_L. In the first isothermal process ab in Fig. 20–7, the work done by the gas is (see Eq. 19–8)

$$W_{ab} = nRT_H \ln \frac{V_b}{V_a},$$

where n is the number of moles of the ideal gas used as working substance. Because the internal energy of an ideal gas does not change when the temperature remains constant, the first law of thermodynamics tells us that the heat added to the gas equals the work done by the gas:

$$Q_H = nRT_H \ln \frac{V_b}{V_a}.$$

Similarly, the heat lost by the gas in the isothermal process cd is

$$Q_L = nRT_L \ln \frac{V_c}{V_d}.$$

The paths bc and da are adiabatic, so we have from Eq. 19–15:

$$P_b V_b^\gamma = P_c V_c^\gamma \quad \text{and} \quad P_d V_d^\gamma = P_a V_a^\gamma,$$

where $\gamma = C_P/C_V$ is the ratio of molar specific heats (Eq. 19–14). Also, from the ideal gas law,

$$\frac{P_b V_b}{T_H} = \frac{P_c V_c}{T_L} \quad \text{and} \quad \frac{P_d V_d}{T_L} = \frac{P_a V_a}{T_H}.$$

When we divide these last equations, term by term, into the corresponding set of equations on the line above, we obtain

$$T_H V_b^{\gamma-1} = T_L V_c^{\gamma-1} \quad \text{and} \quad T_L V_d^{\gamma-1} = T_H V_a^{\gamma-1}.$$

Next we divide the equation on the left by the one on the right and obtain

$$\left(\frac{V_b}{V_a} \right)^{\gamma-1} = \left(\frac{V_c}{V_d} \right)^{\gamma-1}$$

or

$$\frac{V_b}{V_a} = \frac{V_c}{V_d}.$$

Inserting this result in our equations for Q_H and Q_L above, we obtain

$$\frac{Q_L}{Q_H} = \frac{T_L}{T_H}. \qquad \text{[Carnot cycle]} \quad \textbf{(20–2)}$$

Hence the efficiency of a reversible Carnot engine can now be written

$$e_{ideal} = 1 - \frac{Q_L}{Q_H}$$

or

$$e_{ideal} = 1 - \frac{T_L}{T_H}. \qquad \left[\begin{array}{l} \text{Carnot efficiency;} \\ \text{Kelvin temperatures} \end{array} \right] \quad \textbf{(20–3)}$$

The temperatures T_L and T_H are the absolute or Kelvin temperatures as measured on the ideal gas temperature scale. Thus the efficiency of a Carnot engine depends only on the temperatures T_L and T_H.

We could imagine other possible reversible cycles that could be used for an ideal reversible engine. According to a theorem stated by Carnot:

> **All reversible engines operating between the same two constant temperatures T_H and T_L have the same efficiency. Any irreversible engine operating between the same two fixed temperatures will have an efficiency less than this.**

This is known as **Carnot's theorem.**[†] It tells us that Eq. 20–3, $e = 1 - (T_L/T_H)$, applies to any ideal reversible engine with fixed input and exhaust temperatures, T_H and T_L, and that this equation represents a maximum possible efficiency for a real (i.e., irreversible) engine.

In practice, the efficiency of real engines is always less than the Carnot efficiency. Well-designed engines reach perhaps 60% to 80% of Carnot efficiency.

EXAMPLE 20–2 A phony claim? An engine manufacturer makes the following claims: An engine's heat input per second is 9.0 kJ at 435 K. The heat output per second is 4.0 kJ at 285 K. Do you believe these claims?

APPROACH The engine's efficiency can be calculated from the definition, Eqs. 20–1. It must be less than the maximum possible, Eq. 20–3.

SOLUTION The claimed efficiency of the engine is (Eq. 20–1b)

$$e = 1 - \frac{Q_L}{Q_H} = 1 - \frac{4.0\,\text{kJ}}{9.0\,\text{kJ}} = 0.56,$$

or 56%. However, the maximum possible efficiency is given by the Carnot efficiency, Eq. 20–3:

$$e_{ideal} = 1 - \frac{T_L}{T_H} = 1 - \frac{285\,\text{K}}{435\,\text{K}} = 0.34,$$

or 34%. The manufacturer's claims violate the second law of thermodynamics and cannot be believed.

EXERCISE B A motor is running with an intake temperature $T_H = 400\,\text{K}$ and an exhaust temperature $T_L = 300\,\text{K}$. Which of the following is *not* a possible efficiency for the engine? (a) 0.10; (b) 0.16; (c) 0.24; (d) 0.30.

It is clear from Eq. 20–3 that a 100% efficient engine is not possible. Only if the exhaust temperature, T_L, were at absolute zero would 100% efficiency be obtainable. But reaching absolute zero is a practical (as well as theoretical) impossibility.[‡] Thus we can state, as we already did in Section 20–2, that **no device is possible whose sole effect is to transform a given amount of heat completely into work**. As we saw in Section 20–2, this is known as the *Kelvin-Planck statement of the second law of thermodynamics*. It tells us that there can be no perfect (100% efficient) heat engine such as the one diagrammed in Fig. 20–6.

EXERCISE C Return to the Chapter-Opening Question, page 528, and answer it again now. Try to explain why you may have answered differently the first time.

*The Otto Cycle

The operation of an automobile internal combustion engine (Fig. 20–4) can be approximated by a reversible cycle known as the *Otto cycle*, whose PV diagram is shown in Fig. 20–8. Unlike the Carnot cycle, the input and exhaust temperatures of the Otto cycle are *not* constant. Paths ab and cd are adiabatic, and paths bc and da are at constant volume. The gas (gasoline-air mixture) enters the cylinder at point a and is compressed adiabatically (compression stroke) to point b. At b ignition occurs (spark plug) and the burning of the gas adds heat Q_H to the system at constant volume (approximately in a real engine). The temperature and pressure rise, and then in the power stroke, cd, the gas expands adiabatically. In the exhaust stroke, da, heat Q_L is ejected to the environment (in a real engine, the gas leaves the engine and is replaced by a new mixture of air and fuel).

FIGURE 20–8 The Otto cycle.

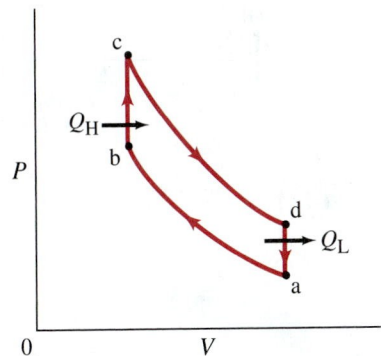

[†]Carnot's theorem can be shown to follow directly from either the Clausius or Kelvin-Planck statements of the second law of thermodynamics.

[‡]This result is known as the *third law of thermodynamics*, as discussed in Section 20–10.

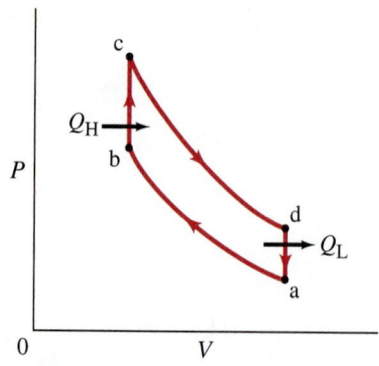

FIGURE 20–8 (repeated for Example 20–3) The Otto cycle.

EXAMPLE 20–3 The Otto cycle. (a) Show that for an ideal gas as working substance, the efficiency of an Otto cycle engine is

$$e = 1 - \left(\frac{V_a}{V_b}\right)^{1-\gamma}$$

where γ is the ratio of specific heats ($\gamma = C_P/C_V$, Eq. 19–14) and V_a/V_b is the *compression ratio*. (b) Calculate the efficiency for a compression ratio $V_a/V_b = 8.0$ assuming a diatomic gas like O_2 and N_2.

APPROACH We use the original definition of efficiency and the results from Chapter 19 for constant volume and adiabatic processes (Sections 19–8 and 19–9).

SOLUTION The heat exchanges take place at constant volume in the ideal Otto cycle, so from Eq. 19–10a:

$$Q_H = nC_V(T_c - T_b) \quad \text{and} \quad Q_L = nC_V(T_d - T_a).$$

Then from Eq. 20–1b,

$$e = 1 - \frac{Q_L}{Q_H} = 1 - \left[\frac{T_d - T_a}{T_c - T_b}\right].$$

To get this in terms of the compression ratio, V_a/V_b, we use the result from Section 19–9, Eq. 19–15, $PV^\gamma = $ constant during the adiabatic processes ab and cd. Thus

$$P_a V_a^\gamma = P_b V_b^\gamma \quad \text{and} \quad P_c V_c^\gamma = P_d V_d^\gamma.$$

We use the ideal gas law, $P = nRT/V$, and substitute P into these two equations

$$T_a V_a^{\gamma-1} = T_b V_b^{\gamma-1} \quad \text{and} \quad T_c V_c^{\gamma-1} = T_d V_d^{\gamma-1}.$$

Then the efficiency (see above) is

$$e = 1 - \left[\frac{T_d - T_a}{T_c - T_b}\right] = 1 - \left[\frac{T_c(V_c/V_d)^{\gamma-1} - T_b(V_b/V_a)^{\gamma-1}}{T_c - T_b}\right].$$

But processes bc and da are at constant volume, so $V_c = V_b$ and $V_d = V_a$. Hence $V_c/V_d = V_b/V_a$ and

$$e = 1 - \left[\frac{(V_b/V_a)^{\gamma-1}(T_c - T_b)}{T_c - T_b}\right] = 1 - \left(\frac{V_b}{V_a}\right)^{\gamma-1} = 1 - \left(\frac{V_a}{V_b}\right)^{1-\gamma}.$$

(b) For diatomic molecules (Section 19–8), $\gamma = C_P/C_V = 1.4$ so

$$e = 1 - (8.0)^{1-\gamma} = 1 - (8.0)^{-0.4} = 0.56.$$

Real engines do not reach this high efficiency because they do not follow perfectly the Otto cycle, plus there is friction, turbulence, heat loss and incomplete combustion of the gases.

FIGURE 20–9 Schematic diagram of energy transfers for a refrigerator or air conditioner.

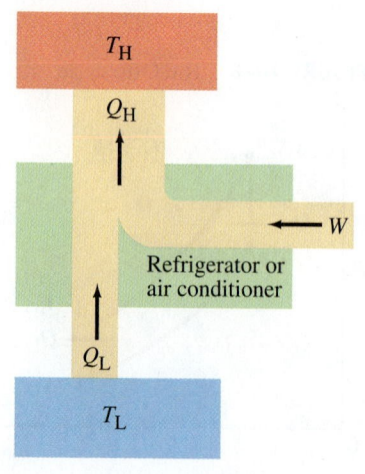

20–4 Refrigerators, Air Conditioners, and Heat Pumps

The operating principle of refrigerators, air conditioners, and heat pumps is just the reverse of a heat engine. Each operates to transfer heat out of a cool environment into a warm environment. As diagrammed in Fig. 20–9, by doing work W, heat is taken from a low-temperature region, T_L (such as inside a refrigerator), and a greater amount of heat is exhausted at a high temperature, T_H (the room). You can often feel this heated air blowing out beneath a refrigerator. The work W is usually done by an electric motor which compresses a fluid, as illustrated in Fig. 20–10.

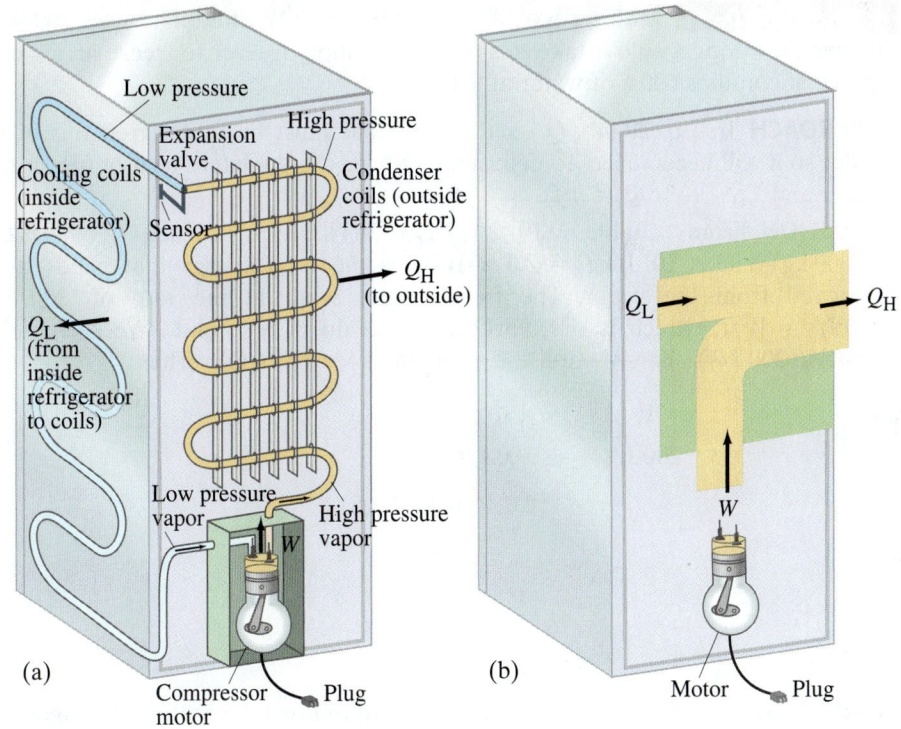

Low pressure

Expansion
valve   High pressure

Cooling coils
(inside
refrigerator)   Condenser
coils (outside
refrigerator)

Sensor

Q_H
(to outside)

Q_L
(from
inside
refrigerator
to coils)

Low pressure
vapor
High pressure
W vapor

(a)

Compressor   Plug
motor

Q_L Q_H

W

(b)

Motor   Plug

FIGURE 20–10 (a) Typical refrigerator system. The electric compressor motor forces a gas at high pressure through a heat exchanger (condenser) on the rear outside wall of the refrigerator where Q_H is given off and the gas cools to become liquid. The liquid passes from a high-pressure region, via a valve, to low-pressure tubes on the inside walls of the refrigerator; the liquid evaporates at this lower pressure and thus absorbs heat (Q_L) from the inside of the refrigerator. The fluid returns to the compressor where the cycle begins again. (b) Schematic diagram, like Fig. 20–9.

A perfect **refrigerator**—one in which no work is required to take heat from the low-temperature region to the high-temperature region—is not possible. This is the **Clausius statement of the second law of thermodynamics**, already mentioned in Section 20–1, which can be stated formally as

No device is possible whose sole effect is to transfer heat from one system at a temperature T_L into a second system at a higher temperature T_H.

To make heat flow from a low-temperature object (or system) to one at a higher temperature, work must be done. Thus, *there can be no perfect refrigerator.*

The **coefficient of performance** (COP) of a refrigerator is defined as the heat Q_L removed from the low-temperature area (inside a refrigerator) divided by the work W done to remove the heat (Fig. 20–9 or 20–10b):

⊛ PHYSICS APPLIED
Refrigerator

$$\text{COP} = \frac{Q_L}{W}. \qquad \begin{bmatrix} \text{refrigerator and} \\ \text{air conditioner} \end{bmatrix} \quad \textbf{(20–4a)}$$

This makes sense because the more heat Q_L that can be removed from the inside of the refrigerator for a given amount of work, the better (more efficient) the refrigerator is. Energy is conserved, so from the first law of thermodynamics we can write (see Fig. 20–9 or 20–10b) $Q_L + W = Q_H$, or $W = Q_H - Q_L$. Then Eq. 20–4a becomes

$$\text{COP} = \frac{Q_L}{W} = \frac{Q_L}{Q_H - Q_L}. \qquad \begin{bmatrix} \text{refrigerator} \\ \text{air conditioner} \end{bmatrix} \quad \textbf{(20–4b)}$$

For an ideal refrigerator (not a perfect one, which is impossible), the best we could do would be

$$\text{COP}_{\text{ideal}} = \frac{T_L}{T_H - T_L}, \qquad \begin{bmatrix} \text{refrigerator} \\ \text{air conditioner} \end{bmatrix} \quad \textbf{(20–4c)}$$

analagous to an ideal (Carnot) engine (Eqs. 20–2 and 20–3).

An **air conditioner** works very much like a refrigerator, although the actual construction details are different: an air conditioner takes heat Q_L from inside a room or building at a low temperature, and deposits heat Q_H outside to the environment at a higher temperature. Equations 20–4 also describe the coefficient of performance for an air conditioner.

⊛ PHYSICS APPLIED
Air conditioner

EXAMPLE 20–4 **Making ice.** A freezer has a COP of 3.8 and uses 200 W of power. How long would it take this otherwise empty freezer to freeze an ice-cube tray that contains 600 g of water at 0°C?

APPROACH In Eq. 20–4b, Q_L is the heat that must be transferred out of the water so it will become ice. To determine Q_L, we use the latent heat of fusion L of water and Eq. 19–3, $Q = mL$.

SOLUTION From Table 19–2, $L = 333\,\text{kJ/kg}$. Hence $Q = mL = (0.600\,\text{kg})(3.33 \times 10^5\,\text{J/kg}) = 2.0 \times 10^5\,\text{J}$ is the total energy that needs to be removed from the water. The freezer does work at the rate of $200\,\text{W} = 200\,\text{J/s} = W/t$, which is the work W it can do in t seconds. We solve for t: $t = W/(200\,\text{J/s})$. For W, we use Eq. 20–4a: $W = Q_L/\text{COP}$. Thus

$$t = \frac{W}{200\,\text{J/s}} = \frac{Q_L/\text{COP}}{200\,\text{J/s}}$$

$$= \frac{(2.0 \times 10^5\,\text{J})/(3.8)}{200\,\text{J/s}} = 260\,\text{s},$$

or about $4\frac{1}{2}\,\text{min}$.

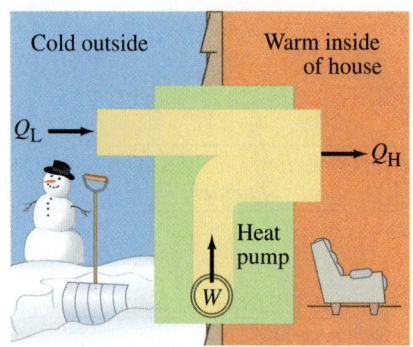

FIGURE 20–11 A heat pump uses an electric motor to "pump" heat from the cold outside to the warm inside of a house.

Heat naturally flows from high temperature to low temperature. Refrigerators and air conditioners do work to accomplish the opposite: to make heat flow from cold to hot. We might say they "pump" heat from cold areas to hotter areas, against the natural tendency of heat to flow from hot to cold, just as water can be pumped uphill, against the natural tendency to flow downhill. The term **heat pump** is usually reserved for a device that can heat a house in winter by using an electric motor that does work W to take heat Q_L from the outside at low temperature and delivers heat Q_H to the warmer inside of the house; see Fig. 20–11. As in a refrigerator, there is an indoor and an outdoor heat exchanger (coils of the refrigerator) and an electric compressor motor. The operating principle is like that for a refrigerator or air conditioner; but the objective of a heat pump is to heat (deliver Q_H), rather than to cool (remove Q_L). Thus, the coefficient of performance of a heat pump is defined differently than for an air conditioner because it is the heat Q_H delivered to the inside of the house that is important now:

$$\text{COP} = \frac{Q_H}{W}. \qquad \text{[heat pump]} \quad \textbf{(20–5)}$$

The COP is necessarily greater than 1. Most heat pumps can be "turned around" and used as air conditioners in the summer.

EXAMPLE 20–5 **Heat pump.** A heat pump has a coefficient of performance of 3.0 and is rated to do work at 1500 W. (*a*) How much heat can it add to a room per second? (*b*) If the heat pump were turned around to act as an air conditioner in the summer, what would you expect its coefficient of performance to be, assuming all else stays the same?

APPROACH We use the definitions of coefficient of performance, which are different for the two devices in (*a*) and (*b*).

SOLUTION (*a*) We use Eq. 20–5 for the heat pump, and, since our device does 1500 J of work per second, it can pour heat into the room at a rate of

$$Q_H = \text{COP} \times W = 3.0 \times 1500\,\text{J} = 4500\,\text{J}$$

per second, or at a rate of 4500 W.

(b) If our device is turned around in summer, it can take heat Q_L from inside the house, doing 1500 J of work per second to then dump $Q_H = 4500$ J per second to the hot outside. Energy is conserved, so $Q_L + W = Q_H$ (see Fig. 20–11, but reverse the inside and outside of the house). Then

$$Q_L = Q_H - W = 4500 \text{ J} - 1500 \text{ J} = 3000 \text{ J}.$$

The coefficient of performance as an air conditioner would thus be (Eq. 20–4a)

$$\text{COP} = \frac{Q_L}{W} = \frac{3000 \text{ J}}{1500 \text{ J}} = 2.0.$$

20–5 Entropy

Thus far we have stated the second law of thermodynamics for specific situations. What we really need is a general statement of the second law of thermodynamics that will cover all situations, including those discussed earlier in this Chapter that are not observed in nature even though they would not violate the first law of thermodynamics. It was not until the latter half of the nineteenth century that the second law of thermodynamics was finally stated in a general way—namely, in terms of a quantity called **entropy**, introduced by Clausius in the 1860s. In Section 20–7 we will see that entropy can be interpreted as a measure of the order or disorder of a system.

When we deal with entropy—as with potential energy—it is the *change* in entropy during a process that is important, not the absolute amount. According to Clausius, the change in entropy S of a system, when an amount of heat Q is *added* to it by a reversible process at constant temperature, is given by

$$\Delta S = \frac{Q}{T}, \tag{20–6}$$

where T is the kelvin temperature.

If the temperature is not constant, we define entropy S by the relation

$$dS = \frac{dQ}{T}. \tag{20–7}$$

Then the change in entropy of a system taken reversibly between two states a and b is given by[†]

$$\Delta S = S_b - S_a = \int_a^b dS = \int_a^b \frac{dQ}{T}. \quad \text{[reversible process]} \tag{20–8}$$

A careful analysis (see next page) shows that the change in entropy when a system moves by a reversible process from any state a to another state b does not depend on the process. That is, $\Delta S = S_b - S_a$ depends only on the states a and b of the system. Thus entropy (unlike heat) is a *state variable*. Any system in a given state has a temperature, a volume, a pressure, and also has a particular value of entropy.

It is easy to see why entropy is a state variable for a Carnot cycle. In Eq. 20–2 we saw that $Q_L/Q_H = T_L/T_H$, which we rewrite as

$$\frac{Q_L}{T_L} = \frac{Q_H}{T_H}.$$

In the *PV* diagram for a Carnot cycle, Fig. 20–7, the entropy change $\Delta S = Q/T$ in going from state a to state c along path abc $(= Q_H/T_H + 0)$ is thus the same as going along the path adc. That is, the change in entropy is path independent—it depends only on the initial and final states of the system.

[†]Equation 20–8 says nothing about the absolute value of S; it only gives the change in S. This is much like potential energy (Chapter 8). However, one form of the so-called *third law of thermodynamics* (see also Section 20–10) states that as $T \to 0$, $S \to 0$.

*Showing Entropy Is a State Variable

In our study of the Carnot cycle we found (Eq. 20–2) that $Q_L/Q_H = T_L/T_H$. We rewrite this as

$$\frac{Q_H}{T_H} = \frac{Q_L}{T_L}.$$

In this relation, both Q_H and Q_L are positive. But now let us recall our original convention as used in the first law (Section 19–6), that Q is positive when it represents a heat flow into the system (as Q_H) and negative for a heat flow out of the system (as $-Q_L$). Then this relation becomes

$$\frac{Q_H}{T_H} + \frac{Q_L}{T_L} = 0. \qquad \text{[Carnot cycle]} \quad \textbf{(20–9)}$$

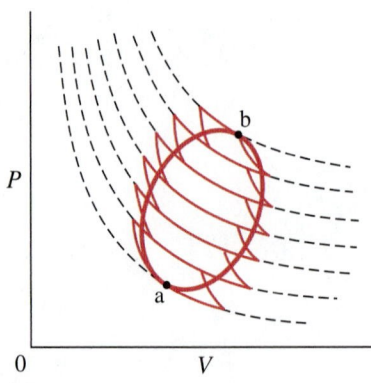

FIGURE 20–12 Any reversible cycle can be approximated as a series of Carnot cycles. (The dashed lines represent isotherms.)

Now consider *any* reversible cycle, as represented by the smooth (oval-shaped) curve in Fig. 20–12. Any reversible cycle can be approximated as a series of Carnot cycles. Figure 20–12 shows only six—the isotherms (dashed lines) are connected by adiabatic paths for each—and the approximation becomes better and better if we increase the number of Carnot cycles. Equation 20–9 is valid for each of these cycles, so we can write

$$\Sigma \frac{Q}{T} = 0 \qquad \text{[Carnot cycles]} \quad \textbf{(20–10)}$$

for the sum of all these cycles. But note that the heat output Q_L of one cycle crosses the boundary below it and is approximately equal to the negative of the heat input, Q_H, of the cycle below it (actual equality occurs in the limit of an infinite number of infinitely thin Carnot cycles). Hence the heat flows on the inner paths of all these Carnot cycles cancel out, so the net heat transferred, and the work done, is the same for the series of Carnot cycles as for the original cycle. Hence, in the limit of infinitely many Carnot cycles, Eq. 20–10 applies to any reversible cycle. In this case Eq. 20–10 becomes

$$\oint \frac{dQ}{T} = 0, \qquad \text{[reversible cycle]} \quad \textbf{(20–11)}$$

where dQ represents an infinitesimal heat flow.[†] The symbol $\oint$ means that the integral is taken around a closed path; the integral can be started at any point on the path such as at a or b in Fig. 20–12, and proceed in either direction. If we divide the cycle of Fig. 20–12 into two parts as indicated in Fig. 20–13, then

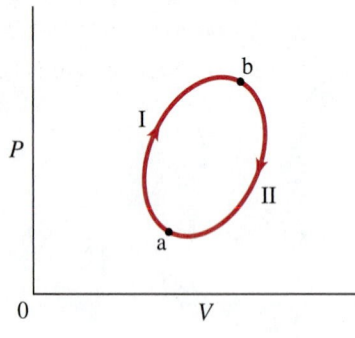

FIGURE 20–13 The integral, $\oint dS$, of the entropy for a reversible cycle is zero. Hence the difference in entropy between states a and b, $S_b - S_a = \int_a^b dS$, is the same for path I as for path II.

$$\underbrace{\int_a^b \frac{dQ}{T}}_{\text{I}} + \underbrace{\int_b^a \frac{dQ}{T}}_{\text{II}} = 0.$$

The first term is the integral from point a to point b along path I in Fig. 20–13, and the second term is the integral from b back to a along path II. If path II is taken in reverse, dQ at each point becomes $-dQ$, since the path is reversible. Therefore

$$\underbrace{\int_a^b \frac{dQ}{T}}_{\text{I}} = \underbrace{\int_a^b \frac{dQ}{T}}_{\text{II}}. \qquad \text{[reversible paths]} \quad \textbf{(20–12)}$$

The integral of dQ/T between any two equilibrium states, a and b, does not depend on the path of the process. By defining entropy as $dS = dQ/T$ (Eq. 20–7), we see from Eq. 20–12 that the change in entropy between any two states along a reversible path is *independent of the path between two points a and b*. Thus *entropy is a state variable*—its value depends only on the state of the system, and not on the process or the past history of how it got there.[‡] This is in clear distinction to Q and W which are *not* state variables; their values do depend on the processes undertaken.

[†]dQ is often written $đQ$: see footnote at the end of Section 19–6.

[‡]Real processes are irreversible. Because entropy is a state variable, the change in entropy ΔS for an irreversible process can be determined by calculating ΔS for a reversible process between the same two states.

20–6 Entropy and the Second Law of Thermodynamics

We have defined a new quantity, S, the entropy, which can be used to describe the state of the system, along with P, T, V, E_{int}, and n. But what does this rather abstract quantity have to do with the second law of thermodynamics? To answer this, let us take some examples in which we calculate the entropy changes during particular processes. But note first that Eq. 20–8 can be applied only to reversible processes. How then do we calculate $\Delta S = S_b - S_a$ for a real process that is irreversible? What we can do is this: we figure out some other *reversible* process that takes the system between the same two states, and calculate ΔS for this reversible process. This will equal ΔS for the irreversible process since ΔS depends only on the initial and final states of the system.

If the temperature varies during a process, a summation of the heat flow over the changing temperature can often be calculated using calculus or a computer. However, if the temperature change is not too great, a reasonable approximation can be made using the average value of the temperature, as indicated in the next Example.

EXAMPLE 20–6 **ESTIMATE** **Entropy change when mixing water.** A sample of 50.0 kg of water at 20.00°C is mixed with 50.0 kg of water at 24.00°C. Estimate the change in entropy.

APPROACH The final temperature of the mixture will be 22.00°C, since we started with equal amounts of water. We use the specific heat of water and the methods of calorimetry (Sections 19–3 and 19–4) to determine the heat transferred. Then we use the average temperature of each sample of water to estimate the entropy change ($\Delta Q/T$).

SOLUTION A quantity of heat,

$$Q = mc\,\Delta T = (50.0\,\text{kg})(4186\,\text{J/kg}\cdot\text{C}°)(2.00\,\text{C}°) = 4.186 \times 10^5\,\text{J},$$

flows out of the hot water as it cools down from 24°C to 22°C, and this heat flows into the cold water as it warms from 20°C to 22°C. The total change in entropy, ΔS, will be the sum of the changes in entropy of the hot water, ΔS_H, and that of the cold water, ΔS_C:

$$\Delta S = \Delta S_H + \Delta S_C.$$

We estimate entropy changes by writing $\Delta S = Q/\overline{T}$, where $\overline{T}$ is an "average" temperature for each process, which ought to give a reasonable estimate since the temperature change is small. For the hot water we use an average temperature of 23°C (296 K), and for the cold water an average temperature of 21°C (294 K). Thus

$$\Delta S_H \approx -\frac{4.186 \times 10^5\,\text{J}}{296\,\text{K}} = -1414\,\text{J/K}$$

which is negative because this heat flows out, whereas heat is added to the cold water:

$$\Delta S_C \approx \frac{4.186 \times 10^5\,\text{J}}{294\,\text{K}} = 1424\,\text{J/K}.$$

The entropy of the hot water (S_H) decreases since heat flows out of the hot water. But the entropy of the cold water (S_C) increases by a greater amount. The total change in entropy is

$$\Delta S = \Delta S_H + \Delta S_C \approx -1414\,\text{J/K} + 1424\,\text{J/K} \approx 10\,\text{J/K}.$$

We see that although the entropy of one part of the system decreased, the entropy of the other part increased by a greater amount so that the net change in entropy of the whole system is positive.

We can now show in general that for an isolated system of two objects, the flow of heat from the higher-temperature $(T_{\rm H})$ object to the lower-temperature $(T_{\rm L})$ object always results in an increase in the total entropy. The two objects eventually come to some intermediate temperature, $T_{\rm M}$. The heat lost by the hotter object $(Q_{\rm H} = -Q$, where Q is positive) is equal to the heat gained by the colder one $(Q_{\rm L} = Q)$, so the total change in entropy is

$$\Delta S = \Delta S_{\rm H} + \Delta S_{\rm L} = -\frac{Q}{T_{\rm HM}} + \frac{Q}{T_{\rm LM}},$$

where $T_{\rm HM}$ is some intermediate temperature between $T_{\rm H}$ and $T_{\rm M}$ for the hot object as it cools from $T_{\rm H}$ to $T_{\rm M}$, and $T_{\rm LM}$ is the counterpart for the cold object. Since the temperature of the hot object is, at all times during the process, greater than that of the cold object, then $T_{\rm HM} > T_{\rm LM}$. Hence

$$\Delta S = Q\left(\frac{1}{T_{\rm LM}} - \frac{1}{T_{\rm HM}}\right) > 0.$$

One object decreases in entropy, while the other gains in entropy, but the *total* change is positive.

EXAMPLE 20–7 **Entropy changes in a free expansion.** Consider the *adiabatic free expansion* of n moles of an ideal gas from volume V_1 to volume V_2, where $V_2 > V_1$ as was discussed in Section 19–7, Fig. 19–14. Calculate the change in entropy (a) of the gas and (b) of the surrounding environment. (c) Evaluate ΔS for 1.00 mole, with $V_2 = 2.00\,V_1$.

APPROACH We saw in Section 19–7 that the gas is initially in a closed container of volume V_1, and, with the opening of a valve, it expands adiabatically into a previously empty container. The total volume of the two containers is V_2. The whole apparatus is thermally insulated from the surroundings, so no heat flows into the gas, $Q = 0$. The gas does no work, $W = 0$, so there is no change in internal energy, $\Delta E_{\rm int} = 0$, and the temperature of the initial and final states is the same, $T_2 = T_1 = T$. The process takes place very quickly, and so is irreversible. Thus we cannot apply Eq. 20–8 to this process. Instead we must think of a reversible process that will take the gas from volume V_1 to V_2 at the same temperature, and use Eq. 20–8 on this reversible process to get ΔS. A reversible isothermal process will do the trick; in such a process, the internal energy does not change, so from the first law,

$$dQ = dW = P\, dV.$$

SOLUTION (a) For the gas,

$$\Delta S_{\rm gas} = \int \frac{dQ}{T} = \frac{1}{T}\int_{V_1}^{V_2} P\, dV.$$

The ideal gas law tells us $P = nRT/V$, so

$$\Delta S_{\rm gas} = \frac{nRT}{T}\int_{V_1}^{V_2}\frac{dV}{V} = nR\ln\frac{V_2}{V_1}.$$

Since $V_2 > V_1$, $\Delta S_{\rm gas} > 0$.

(b) Since no heat is transferred to the surrounding environment, there is no change of the state of the environment due to this process. Hence $\Delta S_{\rm env} = 0$. Note that the total change in entropy, $\Delta S_{\rm gas} + \Delta S_{\rm env}$, is greater than zero.

(c) Since $n = 1.00$ and $V_2 = 2.00\,V_1$, then $\Delta S_{\rm gas} = R\ln 2.00 = 5.76\,{\rm J/K}$.

EXAMPLE 20–8 **Heat transfer.** A red-hot 2.00-kg piece of iron at temperature $T_1 = 880\,\text{K}$ is thrown into a huge lake whose temperature is $T_2 = 280\,\text{K}$. Assume the lake is so large that its temperature rise is insignificant. Determine the change in entropy (a) of the iron and (b) of the surrounding environment (the lake).

APPROACH The process is irreversible, but the same entropy change will occur for a reversible process, and we use the concept of specific heat, Eq. 19–2.

SOLUTION (a) We assume the specific heat of the iron is constant at $c = 450\,\text{J/kg}\cdot\text{K}$. Then $dQ = mc\,dT$ and in a quasistatic reversible process

$$\Delta S_{\text{iron}} = \int \frac{dQ}{T} = mc \int_{T_1}^{T_2} \frac{dT}{T} = mc \ln \frac{T_2}{T_1} = -mc \ln \frac{T_1}{T_2}.$$

Putting in numbers, we find

$$\Delta S_{\text{iron}} = -(2.00\,\text{kg})(450\,\text{J/kg}\cdot\text{K}) \ln \frac{880\,\text{K}}{280\,\text{K}} = -1030\,\text{J/K}.$$

(b) The initial and final temperatures of the lake are the same, $T = 280\,\text{K}$. The lake receives from the iron an amount of heat

$$Q = mc(T_2 - T_1) = (2.00\,\text{kg})(450\,\text{J/kg}\cdot\text{K})(880\,\text{K} - 280\,\text{K}) = 540\,\text{kJ}.$$

Strictly speaking, this is an irreversible process (the lake heats up locally before equilibrium is reached), but is equivalent to a reversible isothermal transfer of heat $Q = 540\,\text{kJ}$ at $T = 280\,\text{K}$. Hence

$$\Delta S_{\text{env}} = \frac{540\,\text{kJ}}{280\,\text{K}} = 1930\,\text{J/K}.$$

Thus, although the entropy of the iron actually decreases, the *total* change in entropy of iron plus environment is positive: $1930\,\text{J/K} - 1030\,\text{J/K} = 900\,\text{J/K}$.

EXERCISE D A 1.00-kg piece of ice at 0°C melts very slowly to water at 0°C. Assume the ice is in contact with a heat reservoir whose temperature is only infinitesimally greater than 0°C. Determine the entropy change of (a) the ice cube and (b) the heat reservoir.

In each of these Examples, the entropy of our system plus that of the environment (or surroundings) either stayed constant or increased. For any *reversible* process, such as that in Exercise D, the total entropy change is zero. This can be seen in general as follows: any reversible process can be considered as a series of quasistatic isothermal transfers of heat ΔQ between a system and the environment, which differ in temperature only by an infinitesimal amount. Hence the change in entropy of either the system or environment is $\Delta Q/T$ and that of the other is $-\Delta Q/T$, so the total is

$$\Delta S = \Delta S_{\text{syst}} + \Delta S_{\text{env}} = 0. \qquad \text{[any reversible process]}$$

In Examples 20–6, 20–7, and 20–8, we found that the total entropy of system plus environment increases. Indeed, it has been found that for all real (irreversible) processes, the total entropy increases. No exceptions have been found. We can thus make the *general statement of the second law of thermodynamics* as follows:

The entropy of an isolated system never decreases. It either stays constant (reversible processes) or increases (irreversible processes).

Since all real processes are irreversible, we can equally well state the second law as:

The total entropy of any system plus that of its environment increases as a result of any natural process:

$$\Delta S = \Delta S_{\text{syst}} + \Delta S_{\text{env}} > 0. \qquad \textbf{(20–13)}$$

SECOND LAW OF THERMODYNAMICS (general statement)

Although the entropy of one part of the universe may decrease in any process (see the Examples above), the entropy of some other part of the universe always increases by a greater amount, so the total entropy always increases.

Now that we finally have a quantitative general statement of the second law of thermodynamics, we can see that it is an unusual law. It differs considerably from other laws of physics, which are typically equalities (such as $F = ma$) or conservation laws (such as for energy and momentum). The second law of thermodynamics introduces a new quantity, the entropy, S, but does not tell us it is conserved. Quite the opposite. Entropy is not conserved in natural processes. Entropy always increases in time.

"Time's Arrow"

The second law of thermodynamics summarizes which processes are observed in nature, and which are not. Or, said another way, it tells us about the *direction* processes go. For the reverse of any of the processes in the last few Examples, the entropy would decrease; and we never observe them. For example, we never observe heat flowing spontaneously from a cold object to a hot object, the reverse of Example 20–8. Nor do we ever observe a gas spontaneously compressing itself into a smaller volume, the reverse of Example 20–7 (gases always expand to fill their containers). Nor do we see thermal energy transform into kinetic energy of a rock so the rock rises spontaneously from the ground. Any of these processes would be consistent with the first law of thermodynamics (conservation of energy). But they are not consistent with the second law of thermodynamics, and this is why we need the second law. If you were to see a movie run backward, you would probably realize it immediately because you would see odd occurrences—such as rocks rising spontaneously from the ground, or air rushing in from the atmosphere to fill an empty balloon (the reverse of free expansion). When watching a movie or video, we are tipped off to a faked reversal of time by observing whether entropy is increasing or decreasing. Hence entropy has been called **time's arrow,** for it can tell us in which direction time is going.

20–7 Order to Disorder

The concept of entropy, as we have discussed it so far, may seem rather abstract. But we can relate it to the more ordinary concepts of *order* and *disorder*. In fact, the entropy of a system can be considered a *measure of the disorder of the system*. Then the second law of thermodynamics can be stated simply as:

Natural processes tend to move toward a state of greater disorder.

Exactly what we mean by disorder may not always be clear, so we now consider a few examples. Some of these will show us how this very general statement of the second law applies beyond what we usually consider as thermodynamics.

Let us look at the simple processes mentioned in Section 20–1. First, a jar containing separate layers of salt and pepper is more orderly than when the salt and pepper are all mixed up. Shaking a jar containing separate layers results in a mixture, and no amount of shaking brings the layers back again. The natural process is from a state of relative order (layers) to one of relative disorder (a mixture), not the reverse. That is, disorder increases. Next, a solid coffee cup is a more "orderly" and useful object than the pieces of a broken cup. Cups break when they fall, but they do not spontaneously mend themselves (as faked in Fig. 20–1). Again, the normal course of events is an increase of disorder.

Let us consider some processes for which we have actually calculated the entropy change, and see that an increase in entropy results in an increase in disorder (or vice versa). When ice melts to water at 0°C, the entropy of the water increases (Exercise D). Intuitively, we can think of solid water, ice, as being more ordered than the less orderly fluid state which can flow all over the place. This change from order to disorder can be seen more clearly from the molecular point of view: the orderly arrangement of water molecules in an ice crystal has changed to the disorderly and somewhat random motion of the molecules in the fluid state.

When a hot object is put in contact with a cold object, heat flows from the high temperature to the low until the two objects reach the same intermediate temperature. At the beginning of the process we can distinguish two classes of molecules: those with a high average kinetic energy (the hot object), and those with a low average kinetic energy (the cooler object). After the process in which heat flows, all the molecules are in one class with the same average kinetic energy. We no longer have the more orderly arrangement of molecules in two classes. Order has gone to disorder. Furthermore, the separate hot and cold objects could serve as the hot- and cold-temperature regions of a heat engine, and thus could be used to obtain useful work. But once the two objects are put in contact and reach the same temperature, no work can be obtained. Disorder has increased, since a system that has the ability to perform work must surely be considered to have a higher order than a system no longer able to do work.

When a stone falls to the ground, its macroscopic kinetic energy is transformed to thermal energy. Thermal energy is associated with the disorderly random motion of molecules, but the molecules in the falling stone all have the same velocity downward in addition to their own random velocities. Thus, the more orderly kinetic energy of the stone as a whole is changed to disordered thermal energy when the stone strikes the ground. Disorder increases in this process, as it does in all processes that occur in nature.

*Biological Evolution

An interesting example of the increase in entropy relates to biological evolution and to growth of organisms. Clearly, a human being is a highly ordered organism. The theory of evolution describes the process from the early macromolecules and simple forms of life to *Homo sapiens*, which is a process of increasing order. So, too, the development of an individual from a single cell to a grown person is a process of increasing order. Do these processes violate the second law of thermodynamics? No, they do not. In the processes of evolution and growth, and even during the mature life of an individual, waste products are eliminated. These small molecules that remain as a result of metabolism are simple molecules without much order. Thus they represent relatively higher disorder or entropy. Indeed, the total entropy of the molecules cast aside by organisms during the processes of evolution and growth is greater than the decrease in entropy associated with the order of the growing individual or evolving species.

20–8 Unavailability of Energy; Heat Death

In the process of heat conduction from a hot object to a cold one, we have seen that entropy increases and that order goes to disorder. The separate hot and cold objects could serve as the high- and low-temperature regions for a heat engine and thus could be used to obtain useful work. But after the two objects are put in contact with each other and reach the same uniform temperature, no work can be obtained from them. With regard to being able to do useful work, order has gone to disorder in this process.

The same can be said about a falling rock that comes to rest upon striking the ground. Before hitting the ground, all the kinetic energy of the rock could have been used to do useful work. But once the rock's mechanical kinetic energy becomes thermal energy, doing useful work is no longer possible.

Both these examples illustrate another important aspect of the second law of thermodynamics:

in any natural process, some energy becomes unavailable to do useful work.

In any process, no energy is ever lost (it is always conserved). Rather, energy becomes less useful—it can do less useful work. As time goes on, **energy is degraded**, in a sense; it goes from more orderly forms (such as mechanical) eventually to the least orderly form, internal, or thermal, energy. Entropy is a factor here because the amount of energy that becomes unavailable to do work is proportional to the change in entropy during any process.[†]

[†]It can be shown that the amount of energy that becomes unavailable to do useful work is equal to $T_L \Delta S$, where T_L is the lowest available temperature and ΔS is the total increase in entropy during the process.

A natural outcome of the degradation of energy is the prediction that as time goes on, the universe should approach a state of maximum disorder. Matter would become a uniform mixture, and heat will have flowed from high-temperature regions to low-temperature regions until the whole universe is at one temperature. No work could then be done. All the energy of the universe would have degraded to thermal energy. This prediction, called the **heat death** of the universe, has been much discussed, but would lie very far in the future. It is a complicated subject, and some scientists question whether thermodynamic modeling of the universe is possible or appropriate.

*20–9 Statistical Interpretation of Entropy and the Second Law

The ideas of entropy and disorder are made clearer with the use of a statistical or probabilistic analysis of the molecular state of a system. This statistical approach, which was first applied toward the end of the nineteenth century by Ludwig Boltzmann (1844–1906), makes a clear distinction between the "macrostate" and the "microstate" of a system. The **microstate** of a system would be specified in giving the position and velocity of every particle (or molecule). The **macrostate** of a system is specified by giving the macroscopic properties of the system—the temperature, pressure, number of moles, and so on. In reality, we can know only the macrostate of a system. We could not possibily know the velocity and position of every one of the huge number of molecules in a system at a given moment. Nonetheless, we can hypothesize a great many different microstates that can correspond to the *same* macrostate.

Let us take a very simple example. Suppose you repeatedly shake four coins in your hand and drop them on a table. Specifying the number of heads and the number of tails that appear on a given throw is the macrostate of this system. Specifying each coin as being a head or a tail is the microstate of the system. In the following Table we see how many microstates correspond to each macrostate:

Macrostate	Possible Microstates (H = heads, T = tails)	Number of Microstates
4 heads	H H H H	1
3 heads, 1 tail	H H H T, H H T H, H T H H, T H H H	4
2 heads, 2 tails	H H T T, H T H T, T H H T, H T T H, T H T H, T T H H	6
1 head, 3 tails	T T T H, T T H T, T H T T, H T T T	4
4 tails	T T T T	1

A basic assumption behind the statistical approach is that *each microstate is equally probable*. Thus the number of microstates that give the same macrostate corresponds to the relative probability of that macrostate occurring. The macrostate of two heads and two tails is the most probable one in our case of tossing four coins; out of the total of 16 possible microstates, six correspond to two heads and two tails, so the probability of throwing two heads and two tails is 6 out of 16, or 38%. The probability of throwing one head and three tails is 4 out of 16, or 25%. The probability of four heads is only 1 in 16, or 6%. If you threw the coins 16 times, you might not find that two heads and two tails appear exactly 6 times, or four tails exactly once. These are only probabilities or averages. But if you made 1600 throws, very nearly 38% of them would be two heads and two tails. The greater the number of tries, the closer the percentages are to the calculated probabilities.

EXERCISE E In the Table above, what is the probability that there will be at least two heads? (a) $\frac{1}{2}$; (b) $\frac{1}{16}$; (c) $\frac{1}{8}$; (d) $\frac{3}{8}$; (e) $\frac{11}{16}$.

If we toss more coins—say, 100 all at the same time—the relative probability of throwing all heads (or all tails) is greatly reduced. There is only one microstate corresponding to all heads. For 99 heads and 1 tail, there are 100 microstates since each of the coins could be the one tail. The relative probabilities for other macrostates are given in Table 20–1. About 1.3×10^{30} microstates are possible.[†] Thus the relative probability of finding all heads is 1 in 10^{30}, an incredibly unlikely event! The probability of obtaining 50 heads and 50 tails (see Table 20–1) is $(1.0 \times 10^{29})/1.3 \times 10^{30} = 0.08$ or 8%. The probability of obtaining anything between 45 and 55 heads is over 70%.

TABLE 20–1 Probabilities of Various Macrostates for 100 Coin Tosses

| Macrostate | | Number of | |
Heads	Tails	Microstates	Probability
100	0	1	7.9×10^{-31}
99	1	1.0×10^2	7.9×10^{-29}
90	10	1.7×10^{13}	1.4×10^{-17}
80	20	5.4×10^{20}	4.2×10^{-10}
60	40	1.4×10^{28}	0.01
55	45	6.1×10^{28}	0.05
50	50	1.0×10^{29}	0.08
45	55	6.1×10^{28}	0.05
40	60	1.4×10^{28}	0.01
20	80	5.4×10^{20}	4.2×10^{-10}
10	90	1.7×10^{13}	1.4×10^{-17}
1	99	1.0×10^2	7.9×10^{-29}
0	100	1	7.9×10^{-31}

Thus we see that as the number of coins increases, the probability of obtaining the most orderly arrangement (all heads or all tails) becomes extremely unlikely. The least orderly arrangement (half heads, half tails) is the most probable, and the probability of being within, say, 5% of the most probable arrangement greatly increases as the number of coins increases. These same ideas can be applied to the molecules of a system. For example, the most probable state of a gas (say, the air in a room) is one in which the molecules take up the whole space and move about randomly; this corresponds to the Maxwellian distribution, Fig. 20–14a (and see Section 18–2). On the other hand, the very orderly arrangement of all the molecules located in one corner of the room and all moving with the same velocity (Fig. 20–14b) is extremely unlikely.

From these examples, it is clear that probability is directly related to disorder and hence to entropy. That is, the most probable state is the one with greatest entropy or greatest disorder and randomness. Boltzmann showed, consistent with Clausius's definition $(dS = dQ/T)$, that the entropy of a system in a given (macro) state can be written

$$S = k \ln W, \tag{20–14}$$

where k is Boltzmann's constant $(k = R/N_A = 1.38 \times 10^{-23}\,\text{J/K})$ and W is the number of microstates corresponding to the given macrostate. That is, W is proportional to the probability of occurrence of that state. W is called the **thermodynamic probability**, or, sometimes, the disorder parameter.

[†]Each coin has two possibilities, heads or tails. Then the possible number of microstates is $2 \times 2 \times 2 \times \cdots = 2^{100} = 1.27 \times 10^{30}$ (using a calculator or logarithms).

FIGURE 20–14 (a) Most probable distribution of molecular speeds in a gas (Maxwellian, or random); (b) orderly, but highly unlikely, distribution of speeds in which all molecules have nearly the same speed.

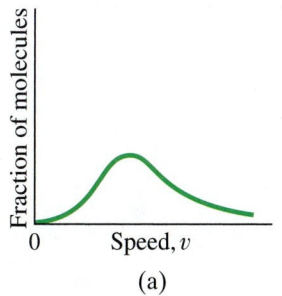

(a)

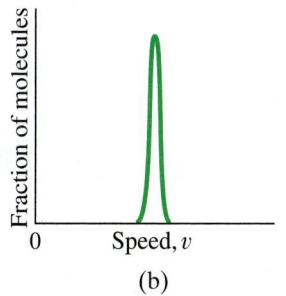

(b)

EXAMPLE 20-9 **Free expansion—statistical determination of entropy.**
Use Eq. 20–14 to determine the change in entropy for the adiabatic *free expansion*
of a gas, a calculation we did macroscopically in Example 20–7. Assume $\mathscr{W}$, the
number of microstates for each macrostate, is the number of possible positions.

APPROACH We assume the number of moles is $n = 1$, and then the number of
molecules is $N = nN_A = N_A$. We let the volume double, just as in Example 20–7.
Because the volume doubles, the number of possible positions for each molecule
doubles.

SOLUTION When the volume doubles, each molecule has two times as many
positions (microstates) available. For two molecules, the number of total
microstates increases by $2 \times 2 = 2^2$. For N_A molecules, the total number of
microstates increases by a factor of $2 \times 2 \times 2 \times \cdots = 2^{N_A}$. That is

$$\frac{\mathscr{W}_2}{\mathscr{W}_1} = 2^{N_A}.$$

The change in entropy is, from Eq. 20–14,

$$\Delta S = S_2 - S_1 = k\left(\ln \mathscr{W}_2 - \ln \mathscr{W}_1\right) = k \ln \frac{\mathscr{W}_2}{\mathscr{W}_1} = k \ln 2^{N_A} = k N_A \ln 2 = R \ln 2$$

which is the same result we obtained in Example 20–7.

In terms of probability, the second law of thermodynamics—which tells us
that entropy increases in any process—reduces to the statement that those
processes occur which are most probable. The second law thus becomes a trivial
statement. However there is an additional element now. The second law in
terms of probability does not *forbid* a decrease in entropy. Rather, it says the
probability is extremely low. It is not impossible that salt and pepper could
separate spontaneously into layers, or that a broken tea cup could mend
itself. It is even possible that a lake could freeze over on a hot summer day
(that is, heat flow out of the cold lake into the warmer surroundings). But
the probability for such events occurring is extremely small. In our coin
examples, we saw that increasing the number of coins from 4 to 100 drastically
reduced the probability of large deviations from the average, or most probable,
arrangement. In ordinary systems we are not dealing with 100 molecules, but
with incredibly large numbers of molecules: in 1 mol alone there are 6×10^{23}
molecules. Hence the probability of deviation far from the average is incredibly
tiny. For example, it has been calculated that the probability that a stone resting
on the ground should transform 1 cal of thermal energy into mechanical energy
and rise up into the air is much less likely than the probability that a group of
monkeys typing randomly would by chance produce the complete works of
Shakespeare.

*20–10 Thermodynamic Temperature; Third Law of Thermodynamics

In Section 20–3 we saw for a Carnot cycle that the ratio of the heat absorbed Q_H from
the high-temperature reservoir and the heat exhausted Q_L to the low-temperature
reservoir is directly related to the ratio of the temperatures of the two reservoirs
(Eq. 20–2):

$$\frac{Q_L}{Q_H} = \frac{T_L}{T_H}.$$

This result is valid for any reversible engine and does not depend on the working
substance. It can thus serve as the basis for the **Kelvin** or **thermodynamic
temperature scale**.

We use this relation and the ideal gas temperature scale (Section 17–10) to
complete the definition of the thermodynamic scale: we assign the value

$T_{tp} = 273.16 \text{ K}$ to the triple point of water so that

$$T = (273.16 \text{ K})\left(\frac{Q}{Q_{tp}}\right),$$

where Q and Q_{tp} are the magnitudes of the heats exchanged by a Carnot engine with reservoirs at temperatures T and T_{tp}. Then the thermodynamic scale is identical to the ideal gas scale over the latter's range of validity.

Very low temperatures are difficult to obtain experimentally. The closer the temperature is to absolute zero, the more difficult it is to reduce the temperature further, and it is generally accepted that *it is not possible to reach absolute zero in any finite number of processes*. This last statement is one way to state[†] the **third law of thermodynamics**. Since the maximum efficiency that any heat engine can have is the Carnot efficiency

$$e = 1 - \frac{T_L}{T_H},$$

and since T_L can never be zero, we see that a 100% efficient heat engine is not possible.

FIGURE 20–15 (a) An array of mirrors focuses sunlight on a boiler to produce steam at a solar energy installation. (b) A fossil-fuel steam plant (this one uses forest waste products, biomass). (c) Large cooling towers at an electric generating plant.

(a)   (b)   (c)

*20–11 Thermal Pollution, Global Warming, and Energy Resources

FIGURE 20–16 Mechanical energy is transformed to electric energy with a turbine and generator.

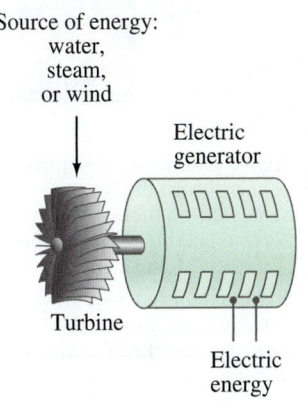

Much of the energy we utilize in everyday life—from motor vehicles to most of the electricity produced by power plants—makes use of a heat engine. Electricity produced by falling water at dams, by windmills, or by solar cells (Fig. 20–15a) does not involve a heat engine. But over 90% of the electric energy produced in the U.S. is generated at fossil-fuel steam plants (coal, oil, or gas—see Fig. 20–15b), and they make use of a heat engine (essentially steam engines). In electric power plants, the steam drives the turbines and generators (Fig. 20–16) whose output is electric energy. The various means to turn the turbine are discussed briefly in Table 20–2 (next page), along with some of the advantages and disadvantages of each. Even nuclear power plants use nuclear fuel to run a steam engine.

The heat output Q_L from every heat engine, from power plants to cars, is referred to as **thermal pollution** because this heat (Q_L) must be absorbed by the environment—such as by water from rivers or lakes, or by the air using large cooling towers (Fig. 20–15c). When water is the coolant, this heat raises the temperature of the water, altering the natural ecology of aquatic life (largely because warmer water holds less oxygen). In the case of air cooling towers, the output heat Q_L raises the temperature of the atmosphere, which affects the weather.

PHYSICS APPLIED
Heat engines and thermal pollution

[†]See also the statement in the footnote on p. 539.

TABLE 20–2 Electric Energy Resources

Form of Electric Energy Production	% of Production (approx.)		Advantages	Disadvantages
	U.S.	**World**		
Fossil-fuel steam plants: burn coal, oil, or natural gas to boil water, producing high-pressure steam that turns a turbine of a generator (Figs. 20–3b, 20–16); uses heat engine.	71	66	We know how to build them; for now relatively inexpensive.	Air pollution; thermal pollution; limited efficiency; land devastation from extraction of raw materials (mining); global warming; accidents such as oil spills at sea; limited fuel supply (estimates range from a couple of decades to a few centuries).
Nuclear energy:				
Fission: nuclei of uranium or plutonium atoms split ("fission") with release of energy (Chapter 42) that heats steam; uses heat engine.	20	16	Normally almost no air pollution; less contribution to global warming; relatively inexpensive.	Thermal pollution; accidents can release damaging radioactivity; difficult disposal of radioactive by-products; possible diversion of nuclear material by terrorists; limited fuel supply.
Fusion: energy released when isotopes of hydrogen (or other small nuclei) combine or "fuse" (Chapter 42).	0	0	Relatively "clean"; vast fuel supply (hydrogen in water molecules in oceans); less contribution to global warming.	Not yet workable.
Hydroelectric: falling water turns turbines at the base of a dam.	7	16	No heat engine needed; no air, water, or thermal pollution; relatively inexpensive; high efficiency; dams can control flooding.	Reservoirs behind dams inundate scenic or inhabited land; dams block upstream migration of salmon and other fish for reproduction; few locations remain for new dams; drought.
Geothermal: natural steam from inside the Earth comes to the surface (hot springs, geysers, steam vents); or cold water passed down into contact with hot, dry rock is heated to steam.	<1	<1	No heat engine needed; little air pollution; good efficiency; relatively inexpensive and "clean."	Few appropriate sites; small production; mineral content of spent hot water can pollute.
Wind power: 3-kW to 5-MW windmills (vanes up to 50 m wide) turn a generator.	<1	<1	No heat engine; no air, water, or thermal pollution; relatively inexpensive.	Large array of big windmills might affect weather and be eyesores; hazardous to migratory birds; winds not always strong.
Solar energy:	<1	<1		
Active solar heating: rooftop solar panels absorb the Sun's rays, which heat water in tubes for space heating and hot water supply.			No heat engine needed; no air or thermal pollution; unlimited fuel supply.	Space limitations; may require back-up; relatively expensive; less effective when cloudy.
Passive solar heating: architectural devices—windows along southern exposure, sunshade over windows to keep Sun's rays out in summer.			No heat engine needed; no air or thermal pollution; relatively inexpensive.	Almost none, but other methods needed too.
Solar cells (photovoltaic cells): convert sunlight directly into electricity without use of heat engine.			No heat engine; thermal, air, and water pollution very low; good efficiency (>30% and improving).	Expensive; chemical pollution at manufacture; large land area needed as Sun's energy not concentrated.

Air pollution—by which we mean the chemicals released in the burning of fossil fuels in cars, power plants, and industrial furnaces—gives rise to smog and other problems. Another much talked about issue is the buildup of CO_2 in the Earth's atmosphere due to the burning of fossil fuels. CO_2 absorbs some of the infrared radiation that the Earth naturally emits (Section 19–10) and thus can contribute to **global warming**. Limiting the burning of fossil fuels can help these problems.

Thermal pollution, however, is unavoidable. Engineers can try to design and build engines that are more efficient, but they cannot surpass the Carnot efficiency and must live with T_L being at best the ambient temperature of water or air. The second law of thermodynamics tells us the limit imposed by nature. What we can do, in the light of the second law of thermodynamics, is use less energy and conserve our fuel resources.

PROBLEM SOLVING

Thermodynamics

1. Define the **system** you are dealing with; distinguish the system under study from its surroundings.

2. Be careful of **signs** associated with **work** and **heat**. In the first law, work done *by* the system is positive; work done *on* the system is negative. Heat *added* to the system is positive, but heat *removed* from it is negative. With heat engines, we usually consider the heat intake, the heat exhausted, and the work done as positive.

3. Watch the **units** used for work and heat; work is most often expressed in joules, and heat can be in calories, kilocalories, or joules. Be consistent: choose only one unit for use throughout a given problem.

4. **Temperatures** must generally be expressed in kelvins; temperature *differences* may be expressed in C° or K.

5. **Efficiency** (or coefficient of performance) is a ratio of two energy transfers: useful output divided by required input. Efficiency (but *not* coefficient of performance) is always less than 1 in value, and hence is often stated as a percentage.

6. The **entropy** of a system increases when heat is added to the system, and decreases when heat is removed. If heat is transferred from system A to system B, the change in entropy of A is negative and the change in entropy of B is positive.

Summary

A **heat engine** is a device for changing thermal energy, by means of heat flow, into useful work.

The **efficiency** of a heat engine is defined as the ratio of the work W done by the engine to the heat input Q_H. Because of conservation of energy, the work output equals $Q_H - Q_L$, where Q_L is the heat exhausted to the environment; hence the efficiency

$$e = \frac{W}{Q_H} = 1 - \frac{Q_L}{Q_H}. \qquad (20\text{–}1)$$

Carnot's (idealized) engine consists of two isothermal and two adiabatic processes in a reversible cycle. For a **Carnot engine**, or any reversible engine operating between two temperatures, T_H and T_L (in kelvins), the efficiency is

$$e_{ideal} = 1 - \frac{T_L}{T_H}. \qquad (20\text{–}3)$$

Irreversible (real) engines always have an efficiency less than this.

The operation of **refrigerators** and **air conditioners** is the reverse of that of a heat engine: work is done to extract heat from a cool region and exhaust it to a region at a higher temperature. The coefficient of performance (COP) for either is

$$COP = \frac{Q_L}{W}, \qquad \begin{bmatrix} \text{refrigerator or} \\ \text{air conditioner} \end{bmatrix} \quad (20\text{–}4a)$$

where W is the work needed to remove heat Q_L from the area with the low temperature.

A **heat pump** does work W to bring heat Q_L from the cold outside and deliver heat Q_H to warm the interior. The coefficient of performance of a heat pump is

$$COP = \frac{Q_H}{W}. \qquad \text{[heat pump]} \quad (20\text{–}5)$$

The **second law of thermodynamics** can be stated in several equivalent ways:

(a) heat flows spontaneously from a hot object to a cold one, but not the reverse;

(b) there can be no 100% efficient heat engine—that is, one that can change a given amount of heat completely into work;

(c) natural processes tend to move toward a state of greater disorder or greater **entropy**.

Statement (c) is the most general statement of the second law of thermodynamics, and can be restated as: the total entropy, S, of any system plus that of its environment increases as a result of any natural process:

$$\Delta S > 0. \qquad (20\text{–}13)$$

Entropy, which is a state variable, is a quantitative measure of the disorder of a system. The change in entropy of a system during a reversible process is given by $\Delta S = \int dQ/T$.

The second law of thermodynamics tells us in which direction processes tend to proceed; hence entropy is called "time's arrow."

As time goes on, energy is degraded to less useful forms—that is, it is less available to do useful work.

[*All heat engines give rise to **thermal pollution** because they exhaust heat to the environment.]

Questions

1. Can mechanical energy ever be transformed completely into heat or internal energy? Can the reverse happen? In each case, if your answer is no, explain why not; if yes, give one or two examples.

2. Can you warm a kitchen in winter by leaving the oven door open? Can you cool the kitchen on a hot summer day by leaving the refrigerator door open? Explain.

3. Would a definition of heat engine efficiency as $e = W/Q_L$ be useful? Explain.

4. What plays the role of high-temperature and low-temperature areas in (a) an internal combustion engine, and (b) a steam engine? Are they, strictly speaking, heat reservoirs?

5. Which will give the greater improvement in the efficiency of a Carnot engine, a $10\,C°$ increase in the high-temperature reservoir, or a $10\,C°$ decrease in the low-temperature reservoir? Explain.

6. The oceans contain a tremendous amount of thermal (internal) energy. Why, in general, is it not possible to put this energy to useful work?

7. Discuss the factors that keep real engines from reaching Carnot efficiency.

8. The expansion valve in a refrigeration system, Fig. 20–10, is crucial for cooling the fluid. Explain how the cooling occurs.

9. Describe a process in nature that is nearly reversible.

10. (a) Describe how heat could be added to a system reversibly. (b) Could you use a stove burner to add heat to a system reversibly? Explain.

11. Suppose a gas expands to twice its original volume (a) adiabatically, (b) isothermally. Which process would result in a greater change in entropy? Explain.

12. Give three examples, other than those mentioned in this Chapter, of naturally occurring processes in which order goes to disorder. Discuss the observability of the reverse process.

13. Which do you think has the greater entropy, 1 kg of solid iron or 1 kg of liquid iron? Why?

14. (a) What happens if you remove the lid of a bottle containing chlorine gas? (b) Does the reverse process ever happen? Why or why not? (c) Can you think of two other examples of irreversibility?

15. You are asked to test a machine that the inventor calls an "in-room air conditioner": a big box, standing in the middle of the room, with a cable that plugs into a power outlet. When the machine is switched on, you feel a stream of cold air coming out of it. How do you know that this machine cannot cool the room?

16. Think up several processes (other than those already mentioned) that would obey the first law of thermodynamics, but, if they actually occurred, would violate the second law.

17. Suppose a lot of papers are strewn all over the floor; then you stack them neatly. Does this violate the second law of thermodynamics? Explain.

18. The first law of thermodynamics is sometimes whimsically stated as, "You can't get something for nothing," and the second law as, "You can't even break even." Explain how these statements could be equivalent to the formal statements.

19. Powdered milk is very slowly (quasistatically) added to water while being stirred. Is this a reversible process? Explain.

20. Two identical systems are taken from state a to state b by two different *irreversible* processes. Will the change in entropy for the system be the same for each process? For the environment? Answer carefully and completely.

21. It can be said that the *total change in entropy during a process is a measure of the irreversibility of the process.* Discuss why this is valid, starting with the fact that $\Delta S_{total} = \Delta S_{system} + \Delta S_{environment} = 0$ for a reversible process.

22. Use arguments, other than the principle of entropy increase, to show that for an adiabatic process, $\Delta S = 0$ if it is done reversibly and $\Delta S > 0$ if done irreversibly.

Problems

20–2 Heat Engines

1. (I) A heat engine exhausts 7800 J of heat while performing 2600 J of useful work. What is the efficiency of this engine?

2. (I) A certain power plant puts out 580 MW of electric power. Estimate the heat discharged per second, assuming that the plant has an efficiency of 35%.

3. (II) A typical compact car experiences a total drag force at 55 mi/h of about 350 N. If this car gets 35 miles per gallon of gasoline at this speed, and a liter of gasoline (1 gal = 3.8 L) releases about 3.2×10^7 J when burned, what is the car's efficiency?

4. (II) A four-cylinder gasoline engine has an efficiency of 0.22 and delivers 180 J of work per cycle per cylinder. The engine fires at 25 cycles per second. (a) Determine the work done per second. (b) What is the total heat input per second from the gasoline? (c) If the energy content of gasoline is 130 MJ per gallon, how long does one gallon last?

5. (II) The burning of gasoline in a car releases about 3.0×10^4 kcal/gal. If a car averages 38 km/gal when driving 95 km/h, which requires 25 hp, what is the efficiency of the engine under those conditions?

6. (II) Figure 20–17 is a *PV* diagram for a reversible heat engine in which 1.0 mol of argon, a nearly ideal monatomic gas, is initially at STP (point a). Points b and c are on an isotherm at $T = 423$ K. Process ab is at constant volume, process ac at constant pressure. (a) Is the path of the cycle carried out clockwise or counterclockwise? (b) What is the efficiency of this engine?

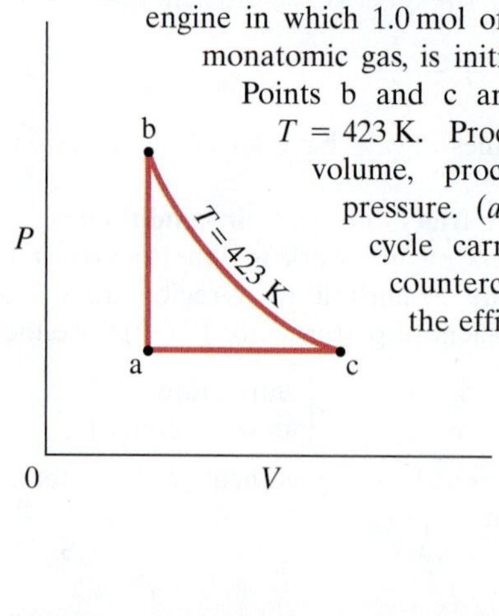

FIGURE 20–17
Problem 6.

7. (III) The operation of a *diesel engine* can be idealized by the cycle shown in Fig. 20–18. Air is drawn into the cylinder during the intake stroke (not part of the idealized cycle). The air is compressed adiabatically, path ab. At point b diesel fuel is injected into the cylinder which immediately burns since the temperature is very high. Combustion is slow, and during the first part of the power stroke, the gas expands at (nearly) constant pressure, path bc. After burning, the rest of the power stroke is adiabatic, path cd. Path da corresponds to the exhaust stroke. (a) Show that, for a quasistatic reversible engine undergoing this cycle using an ideal gas, the ideal efficiency is

$$e = 1 - \frac{(V_a/V_c)^{-\gamma} - (V_a/V_b)^{-\gamma}}{\gamma[(V_a/V_c)^{-1} - (V_a/V_b)^{-1}]},$$

where V_a/V_b is the "compression ratio", V_a/V_c is the "expansion ratio", and γ is defined by Eq. 19–14. (b) If $V_a/V_b = 16$ and $V_a/V_c = 4.5$, calculate the efficiency assuming the gas is diatomic (like N_2 and O_2) and ideal.

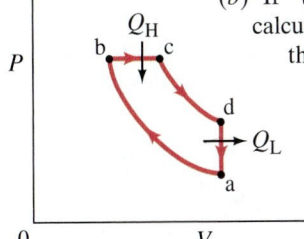

FIGURE 20–18
Problem 7.

20–3 Carnot Engine

8. (I) What is the maximum efficiency of a heat engine whose operating temperatures are 550°C and 365°C?

9. (I) It is not necessary that a heat engine's hot environment be hotter than ambient temperature. Liquid nitrogen (77 K) is about as cheap as bottled water. What would be the efficiency of an engine that made use of heat transferred from air at room temperature (293 K) to the liquid nitrogen "fuel" (Fig. 20–19)?

FIGURE 20–19
Problem 9.

10. (II) A heat engine exhausts its heat at 340°C and has a Carnot efficiency of 38%. What exhaust temperature would enable it to achieve a Carnot efficiency of 45%?

11. (II) (a) Show that the work done by a Carnot engine is equal to the area enclosed by the Carnot cycle on a PV diagram, Fig. 20–7. (See Section 19–7.) (b) Generalize this to any reversible cycle.

12. (II) A Carnot engine's operating temperatures are 210°C and 45°C. The engine's power output is 950 W. Calculate the rate of heat output.

13. (II) A nuclear power plant operates at 65% of its maximum theoretical (Carnot) efficiency between temperatures of 660°C and 330°C. If the plant produces electric energy at the rate of 1.2 GW, how much exhaust heat is discharged per hour?

14. (II) A Carnot engine performs work at the rate of 520 kW with an input of 950 kcal of heat per second. If the temperature of the heat source is 560°C, at what temperature is the waste heat exhausted?

15. (II) Assume that a 65 kg hiker needs 4.0×10^3 kcal of energy to supply a day's worth of metabolism. Estimate the maximum height the person can climb in one day, using only this amount of energy. As a rough prediction, treat the person as an isolated heat engine, operating between the internal temperature of 37°C (98.6°F) and the ambient air temperature of 20°C.

16. (II) A particular car does work at the rate of about 7.0 kJ/s when traveling at a steady 20.0 m/s along a level road. This is the work done against friction. The car can travel 17 km on 1 L of gasoline at this speed (about 40 mi/gal). What is the minimum value for T_H if T_L is 25°C? The energy available from 1 L of gas is 3.2×10^7 J.

17. (II) A heat engine utilizes a heat source at 580°C and has a Carnot efficiency of 32%. To increase the efficiency to 38%, what must be the temperature of the heat source?

18. (II) The working substance of a certain Carnot engine is 1.0 mol of an ideal monatomic gas. During the isothermal expansion portion of this engine's cycle, the volume of the gas doubles, while during the adiabatic expansion the volume increases by a factor of 5.7. The work output of the engine is 920 J in each cycle. Compute the temperatures of the two reservoirs between which this engine operates.

19. (III) A Carnot cycle, shown in Fig. 20–7, has the following conditions: $V_a = 7.5$ L, $V_b = 15.0$ L, $T_H = 470°C$, and $T_L = 260°C$. The gas used in the cycle is 0.50 mol of a diatomic gas, $\gamma = 1.4$. Calculate (a) the pressures at a and b; (b) the volumes at c and d. (c) What is the work done along process ab? (d) What is the heat lost along process cd? (e) Calculate the net work done for the whole cycle. (f) What is the efficiency of the cycle, using the definition $e = W/Q_H$? Show that this is the same as given by Eq. 20–3.

20. (III) One mole of monatomic gas undergoes a Carnot cycle with $T_H = 350°C$ and $T_L = 210°C$. The initial pressure is 8.8 atm. During the isothermal expansion, the volume doubles. (a) Find the values of the pressure and volume at the points a, b, c, and d (see Fig. 20–7). (b) Determine Q, W, and ΔE_{int} for each segment of the cycle. (c) Calculate the efficiency of the cycle using Eqs. 20–1 and 20–3.

*21. (III) In an engine that approximates the Otto cycle (Fig. 20–8), gasoline vapor must be ignited at the end of the cylinder's adiabatic compression by the spark from a spark plug. The ignition temperature of 87-octane gasoline vapor is about 430°C and, assuming that the working gas is diatomic and enters the cylinder at 25°C, determine the maximum compression ratio of the engine.

20–4 Refrigerators, Air Conditioners, Heat Pumps

22. (I) If an ideal refrigerator keeps its contents at 3.0°C when the house temperature is 22°C, what is its coefficient of performance?

23. (I) The low temperature of a freezer cooling coil is −15°C and the discharge temperature is 33°C. What is the maximum theoretical coefficient of performance?

24. (II) An ideal (Carnot) engine has an efficiency of 38%. If it were possible to run it backward as a heat pump, what would be its coefficient of performance?

25. (II) An ideal heat pump is used to maintain the inside temperature of a house at $T_{in} = 22°C$ when the outside temperature is T_{out}. Assume that when it is operating, the heat pump does work at a rate of 1500 W. Also assume that the house loses heat via conduction through its walls and other surfaces at a rate given by $(650 \text{ W/C°}) (T_{in} - T_{out})$. (a) For what outside temperature would the heat pump have to operate at all times in order to maintain the house at an inside temperature of 22°C? (b) If the outside temperature is 8°C, what percentage of the time does the heat pump have to operate in order to maintain the house at an inside temperature of 22°C?

26. (II) A restaurant refrigerator has a coefficient of performance of 5.0. If the temperature in the kitchen outside the refrigerator is 32°C, what is the lowest temperature that could be obtained inside the refrigerator if it were ideal?

27. (II) A heat pump is used to keep a house warm at 22°C. How much work is required of the pump to deliver 3100 J of heat into the house if the outdoor temperature is (a) 0°C, (b) −15°C? Assume ideal (Carnot) behavior.

28. (II) (a) Given that the coefficient of performance of a refrigerator is defined (Eq. 20–4a) as

$$COP = \frac{Q_L}{W},$$

show that for an ideal (Carnot) refrigerator,

$$COP_{ideal} = \frac{T_L}{T_H - T_L}.$$

(b) Write the COP in terms of the efficiency e of the reversible heat engine obtained by running the refrigerator backward. (c) What is the coefficient of performance for an ideal refrigerator that maintains a freezer compartment at −18°C when the condenser's temperature is 24°C?

29. (II) A "Carnot" refrigerator (reverse of a Carnot engine) absorbs heat from the freezer compartment at a temperature of −17°C and exhausts it into the room at 25°C. (a) How much work must be done by the refrigerator to change 0.40 kg of water at 25°C into ice at −17°C? (b) If the compressor output is 180 W, what minimum time is needed to take 0.40 kg of 25°C water and freeze it at 0°C?

30. (II) A central heat pump operating as an air conditioner draws 33,000 Btu per hour from a building and operates between the temperatures of 24°C and 38°C. (a) If its coefficient of performance is 0.20 that of a Carnot air conditioner, what is the effective coefficient of performance? (b) What is the power (kW) required of the compressor motor? (c) What is the power in terms of hp?

31. (II) What volume of water at 0°C can a freezer make into ice cubes in 1.0 h, if the coefficient of performance of the cooling unit is 7.0 and the power input is 1.2 kilowatt?

20–5 and 20–6 Entropy

32. (I) What is the change in entropy of 250 g of steam at 100°C when it is condensed to water at 100°C?

33. (I) A 7.5-kg box having an initial speed of 4.0 m/s slides along a rough table and comes to rest. Estimate the total change in entropy of the universe. Assume all objects are at room temperature (293 K).

34. (I) What is the change in entropy of 1.00 m^3 of water at 0°C when it is frozen to ice at 0°C?

35. (II) If 1.00 m^3 of water at 0°C is frozen and cooled to −10°C by being in contact with a great deal of ice at −10°C, estimate the total change in entropy of the process.

36. (II) If 0.45 kg of water at 100°C is changed by a reversible process to steam at 100°C, determine the change in entropy of (a) the water, (b) the surroundings, and (c) the universe as a whole. (d) How would your answers differ if the process were irreversible?

37. (II) An aluminum rod conducts 9.50 cal/s from a heat source maintained at 225°C to a large body of water at 22°C. Calculate the rate at which entropy increases in this process.

38. (II) A 2.8-kg piece of aluminum at 43.0°C is placed in 1.0 kg of water in a Styrofoam container at room temperature (20°C). Estimate the net change in entropy of the system.

39. (II) An ideal gas expands isothermally $(T = 410 \text{ K})$ from a volume of 2.50 L and a pressure of 7.5 atm to a pressure of 1.0 atm. What is the entropy change for this process?

40. (II) When 2.0 kg of water at 12.0°C is mixed with 3.0 kg of water at 38.0°C in a well-insulated container, what is the change in entropy of the system? (a) Make an estimate; (b) use the integral $\Delta S = \int dQ/T$.

41. (II) (a) An ice cube of mass m at 0°C is placed in a large 20°C room. Heat flows (from the room to the ice cube) such that the ice cube melts and the liquid water warms to 20°C. The room is so large that its temperature remains nearly 20°C at all times. Calculate the change in entropy for the (water + room) system due to this process. Will this process occur naturally? (b) A mass m of liquid water at 20°C is placed in a large 20°C room. Heat flows (from the water to the room) such that the liquid water cools to 0°C and then freezes into a 0°C ice cube. The room is so large that its temperature remains 20°C at all times. Calculate the change in entropy for the (water + room) system due to this process. Will this process occur naturally?

42. (II) The temperature of 2.0 mol of an ideal diatomic gas goes from 25°C to 55°C at a constant volume. What is the change in entropy? Use $\Delta S = \int dQ/T$.

43. (II) Calculate the change in entropy of 1.00 kg of water when it is heated from 0°C to 75°C. (a) Make an estimate; (b) use the integral $\Delta S = \int dQ/T$. (c) Does the entropy of the surroundings change? If so, by how much?

44. (II) An ideal gas of n moles undergoes the reversible process ab shown in the PV diagram of Fig. 20–20. The temperature T of the gas is the same at points a and b. Determine the change in entropy of the gas due to this process.

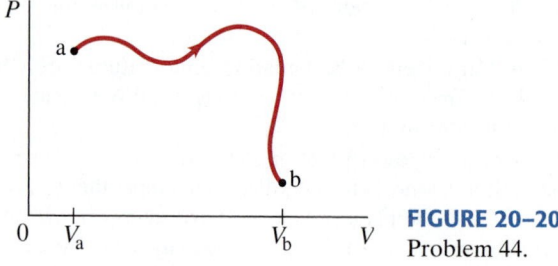

FIGURE 20–20
Problem 44.

45. (II) Two samples of an ideal gas are initially at the same temperature and pressure. They are each compressed reversibly from a volume V to volume $V/2$, one isothermally, the other adiabatically. (a) In which sample is the final pressure greater? (b) Determine the change in entropy of the gas for each process by integration. (c) What is the entropy change of the environment for each process?

46. (II) A 150-g insulated aluminum cup at 15°C is filled with 215 g of water at 100°C. Determine (a) the final temperature of the mixture, and (b) the total change in entropy as a result of the mixing process (use $\Delta S = \int dQ/T$).

47. (II) (a) Why would you expect the total entropy change in a Carnot cycle to be zero? (b) Do a calculation to show that it is zero.

48. (II) 1.00 mole of nitrogen (N_2) gas and 1.00 mole of argon (Ar) gas are in separate, equal-sized, insulated containers at the same temperature. The containers are then connected and the gases (assumed ideal) allowed to mix. What is the change in entropy (a) of the system and (b) of the environment? (c) Repeat part (a) but assume one container is twice as large as the other.

49. (II) Thermodynamic processes are sometimes represented on TS (temperature–entropy) diagrams, rather than PV diagrams. Determine the slope of a constant-volume process on a TS diagram when a system with n moles of an ideal gas with constant-volume molar specific heat C_V is at temperature T.

50. (III) The specific heat per mole of potassium at low temperatures is given by $C_V = aT + bT^3$, where $a = 2.08 \text{ mJ/mol·K}^2$ and $b = 2.57 \text{ mJ/mol·K}^4$. Determine (by integration) the entropy change of 0.15 mol of potassium when its temperature is lowered from 3.0 K to 1.0 K.

51. (III) Consider an ideal gas of n moles with molar specific heats C_V and C_P. (a) Starting with the first law, show that when the temperature and volume of this gas are changed by a reversible process, its change in entropy is given by

$$dS = nC_V \frac{dT}{T} + nR \frac{dV}{V}.$$

(b) Show that the expression in part (a) can be written as

$$dS = nC_V \frac{dP}{P} + nC_P \frac{dV}{V}.$$

(c) Using the expression from part (b), show that if $dS = 0$ for the reversible process (that is, the process is adiabatic), then $PV^\gamma = $ constant, where $\gamma = C_P/C_V$.

20–8 Unavailability of Energy

52. (III) A general theorem states that the amount of energy that becomes unavailable to do useful work in any process is equal to $T_L \Delta S$, where T_L is the lowest temperature available and ΔS is the total change in entropy during the process. Show that this is valid in the specific cases of (a) a falling rock that comes to rest when it hits the ground; (b) the free adiabatic expansion of an ideal gas; and (c) the conduction of heat, Q, from a high-temperature (T_H) reservoir to a low-temperature (T_L) reservoir. [Hint: In part (c) compare to a Carnot engine.]

53. (III) Determine the work available in a 3.5-kg block of copper at 490 K if the surroundings are at 290 K. Use results of Problem 52.

* 20–9 **Statistical Interpretation of Entropy**

*54. (I) Use Eq. 20–14 to determine the entropy of each of the five macrostates listed in the Table on page 546.

*55. (II) Suppose that you repeatedly shake six coins in your hand and drop them on the floor. Construct a table showing the number of microstates that correspond to each macrostate. What is the probability of obtaining (a) three heads and three tails and (b) six heads?

*56. (II) Calculate the relative probabilities, when you throw two dice, of obtaining (a) a 7, (b) an 11, (c) a 4.

*57. (II) (a) Suppose you have four coins, all with tails up. You now rearrange them so two heads and two tails are up. What was the change in entropy of the coins? (b) Suppose your system is the 100 coins of Table 20–1; what is the change in entropy of the coins if they are mixed randomly initially, 50 heads and 50 tails, and you arrange them so all 100 are heads? (c) Compare these entropy changes to ordinary thermodynamic entropy changes, such as Examples 20–6, 20–7, and 20–8.

*58. (III) Consider an isolated gas-like system consisting of a box that contains $N = 10$ distinguishable atoms, each moving at the same speed v. The number of unique ways that these atoms can be arranged so that N_L atoms are within the left-hand half of the box and N_R atoms are within the right-hand half of the box is given by $N!/N_L!N_R!$, where, for example, the factorial $4! = 4 \cdot 3 \cdot 2 \cdot 1$ (the only exception is that $0! = 1$). Define each unique arrangement of atoms within the box to be a microstate of this system. Now imagine the following two possible macrostates: state A where all of the atoms are within the left-hand half of the box and none are within the right-hand half; and state B where the distribution is uniform (that is, there is the same number in each half). See Fig. 20–21. (a) Assume the system is initially in state A and, at a later time, is found to be in state B. Determine the system's change in entropy. Can this process occur naturally? (b) Assume the system is initially in state B and, at a later time, is found to be in state A. Determine the system's change in entropy. Can this process occur naturally?

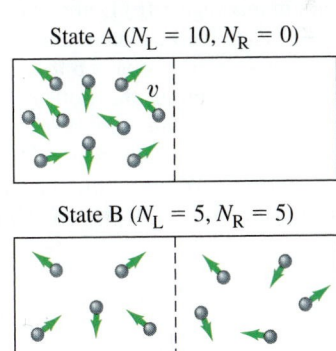

State A ($N_L = 10$, $N_R = 0$)

State B ($N_L = 5$, $N_R = 5$)

FIGURE 20–21 Problem 58.

* 20–11 **Energy Resources**

*59. (II) Energy may be stored for use during peak demand by pumping water to a high reservoir when demand is low and then releasing it to drive turbines when needed. Suppose water is pumped to a lake 135 m above the turbines at a rate of 1.35×10^5 kg/s for 10.0 h at night. (a) How much energy (kWh) is needed to do this each night? (b) If all this energy is released during a 14-h day, at 75% efficiency, what is the average power output?

*60. (II) If solar cells (Fig. 20–22) can produce about 40 W of electricity per square meter of surface area when directly facing the Sun, how large an area is required to supply the needs of a house that requires 22 kWh/day? Would this fit on the roof of an average house? (Assume the Sun shines about 9 h/day.)

FIGURE 20–22 Problem 60.

*61. (II) Water is stored in an artificial lake created by a dam (Fig. 20–23). The water depth is 38 m at the dam, and a steady flow rate of 32 m^3/s is maintained through hydroelectric turbines installed near the base of the dam. How much electrical power can be produced?

FIGURE 20–23 Problem 61.

General Problems

62. It has been suggested that a heat engine could be developed that made use of the temperature difference between water at the surface of the ocean and water several hundred meters deep. In the tropics, the temperatures may be 27°C and 4°C, respectively. (a) What is the maximum efficiency such an engine could have? (b) Why might such an engine be feasible in spite of the low efficiency? (c) Can you imagine any adverse environmental effects that might occur?

63. A heat engine takes a diatomic gas around the cycle shown in Fig. 20–24. (a) Using the ideal gas law, determine how many moles of gas are in this engine. (b) Determine the temperature at point c. (c) Calculate the heat input into the gas during the constant volume process from points b to c. (d) Calculate the work done by the gas during the isothermal process from points a to b. (e) Calculate the work done by the gas during the adiabatic process from points c to a. (f) Determine the engine's efficiency. (g) What is the maximum efficiency possible for an engine working between T_a and T_c?

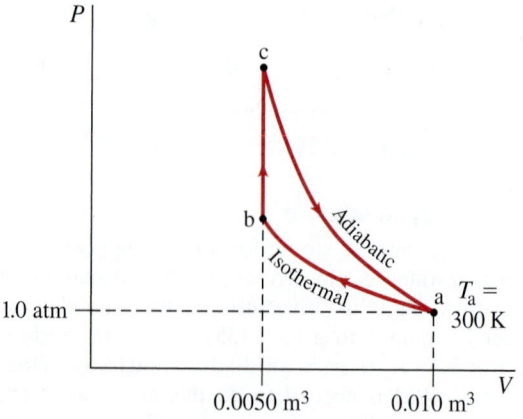

FIGURE 20–24 Problem 63.

64. A 126.5-g insulated aluminum cup at 18.00°C is filled with 132.5 g of water at 46.25°C. After a few minutes, equilibrium is reached. Determine (a) the final temperature, and (b) the total change in entropy.

65. (a) At a steam power plant, steam engines work in pairs, the heat output of the first one being the approximate heat input of the second. The operating temperatures of the first are 710°C and 430°C, and of the second 415°C and 270°C. If the heat of combustion of coal is 2.8 × 10^7 J/kg, at what rate must coal be burned if the plant is to put out 950 MW of power? Assume the efficiency of the engines is 65% of the ideal (Carnot) efficiency. (b) Water is used to cool the power plant. If the water temperature is allowed to increase by no more than 5.5 C°, estimate how much water must pass through the plant per hour.

66. Refrigeration units can be rated in "tons." A 1-ton air conditioning system can remove sufficient energy to freeze 1 British ton (2000 pounds = 909 kg) of 0°C water into 0°C ice in one 24-h day. If, on a 35°C day, the interior of a house is maintained at 22°C by the continuous operation of a 5-ton air conditioning system, how much does this cooling cost the homeowner per hour? Assume the work done by the refrigeration unit is powered by electricity that costs $0.10 per kWh and that the unit's coefficient of performance is 15% that of an ideal refrigerator. 1 kWh = 3.60 × 10^6 J.

67. A 35% efficient power plant puts out 920 MW of electrical power. Cooling towers are used to take away the exhaust heat. (a) If the air temperature (15°C) is allowed to rise 7.0 C°, estimate what volume of air (km^3) is heated per day. Will the local climate be heated significantly? (b) If the heated air were to form a layer 150 m thick, estimate how large an area it would cover for 24 h of operation. Assume the air has density 1.2 kg/m^3 and that its specific heat is about 1.0 kJ/kg·C° at constant pressure.

68. (a) What is the coefficient of performance of an ideal heat pump that extracts heat from 11°C air outside and deposits heat inside your house at 24°C? (b) If this heat pump operates on 1400 W of electrical power, what is the maximum heat it can deliver into your house each hour?

69. The operation of a certain heat engine takes an ideal monatomic gas through a cycle shown as the rectangle on the PV diagram of Fig. 20–25. (a) Determine the efficiency of this engine. Let Q_H and Q_L be the total heat input and total heat exhausted during one cycle of this engine. (b) Compare (as a ratio) the efficiency of this engine to that of a Carnot engine operating between T_H and T_L, where T_H and T_L are the highest and lowest temperatures achieved.

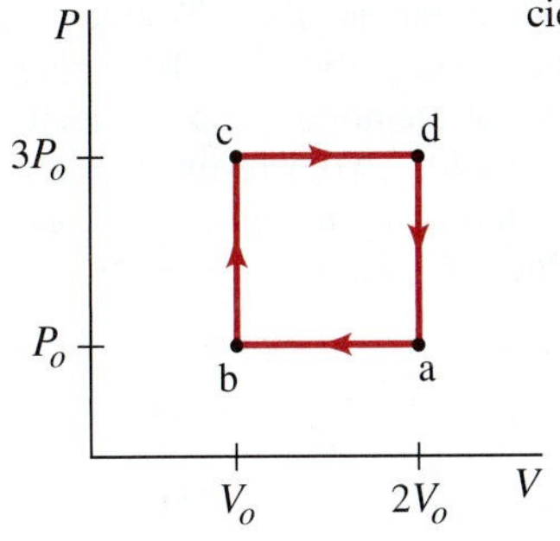

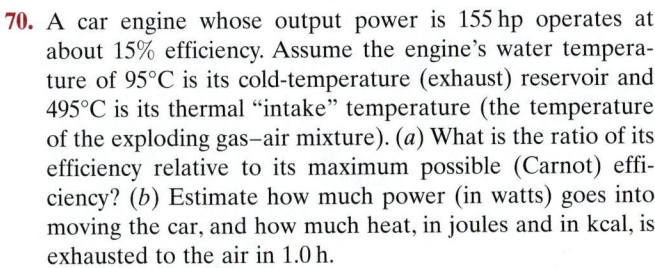

FIGURE 20–25
Problem 69.

70. A car engine whose output power is 155 hp operates at about 15% efficiency. Assume the engine's water temperature of 95°C is its cold-temperature (exhaust) reservoir and 495°C is its thermal "intake" temperature (the temperature of the exploding gas–air mixture). (a) What is the ratio of its efficiency relative to its maximum possible (Carnot) efficiency? (b) Estimate how much power (in watts) goes into moving the car, and how much heat, in joules and in kcal, is exhausted to the air in 1.0 h.

71. Suppose a power plant delivers energy at 850 MW using steam turbines. The steam goes into the turbines superheated at 625 K and deposits its unused heat in river water at 285 K. Assume that the turbine operates as an ideal Carnot engine. (a) If the river's flow rate is 34 m³/s, estimate the average temperature increase of the river water immediately downstream from the power plant. (b) What is the entropy increase per kilogram of the downstream river water in J/kg·K?

72. 1.00 mole of an ideal monatomic gas at STP first undergoes an isothermal expansion so that the volume at b is 2.5 times the volume at a (Fig. 20–26). Next, heat is extracted at a constant volume so that the pressure drops. The gas is then compressed adiabatically back to the original state. (a) Calculate the pressures at b and c. (b) Determine the temperature at c. (c) Determine the work done, heat input or extracted, and the change in entropy for each process. (d) What is the efficiency of this cycle?

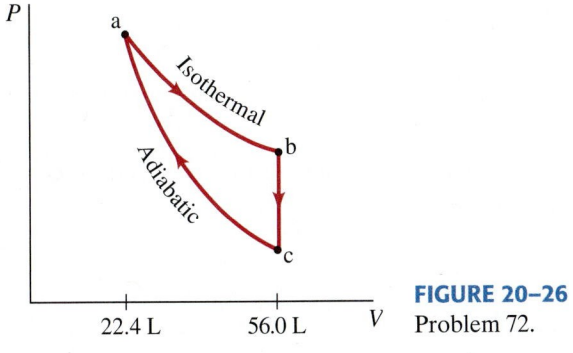

FIGURE 20–26
Problem 72.

73. Two 1100-kg cars are traveling 75 km/h in opposite directions when they collide and are brought to rest. Estimate the change in entropy of the universe as a result of this collision. Assume $T = 15°C$.

74. Metabolizing 1.0 kg of fat results in about 3.7×10^7 J of internal energy in the body. (a) In one day, how much fat does the body burn to maintain the body temperature of a person staying in bed and metabolizing at an average rate of 95 W? (b) How long would it take to burn 1.0-kg of fat this way assuming there is no food intake?

75. A cooling unit for a new freezer has an inner surface area of 6.0 m², and is bounded by walls 12 cm thick with a thermal conductivity of 0.050 W/m·K. The inside must be kept at −10°C in a room that is at 20°C. The motor for the cooling unit must run no more than 15% of the time. What is the minimum power requirement of the cooling motor?

76. An ideal air conditioner keeps the temperature inside a room at 21°C when the outside temperature is 32°C. If 3.3 kW of power enters a room through the windows in the form of direct radiation from the Sun, how much electrical power would be saved if the windows were shaded so only 500 W came through them?

77. The *Stirling cycle*, shown in Fig. 20–27, is useful to describe external combustion engines as well as solar-power systems. Find the efficiency of the cycle in terms of the parameters shown, assuming a monatomic gas as the working substance. The processes ab and cd are isothermal whereas bc and da are at constant volume. How does it compare to the Carnot efficiency?

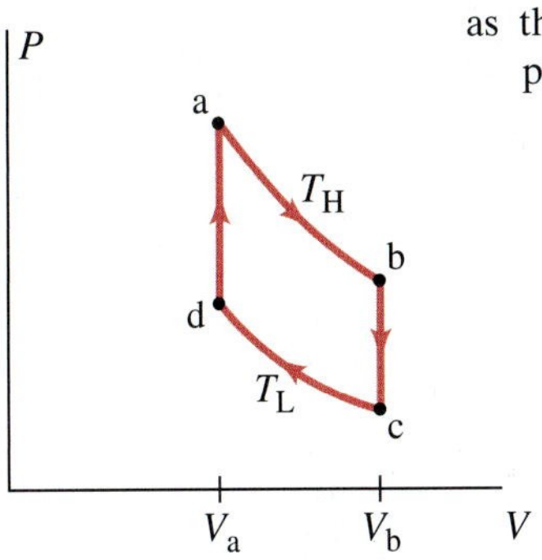

FIGURE 20–27
Problem 77.

78. A gas turbine operates under the *Brayton cycle*, which is depicted in the PV diagram of Fig. 20–28. In process ab the air–fuel mixture undergoes an adiabatic compression. This is followed, in process bc, with an isobaric (constant pressure) heating, by combustion. Process cd is an adiabatic expansion with expulsion of the products to the atmosphere. The return step, da, takes place at constant pressure. If the working gas behaves like an ideal gas, show that the efficiency of the Brayton cycle is

$$e = 1 - \left(\frac{P_b}{P_a}\right)^{\frac{1-\gamma}{\gamma}}.$$

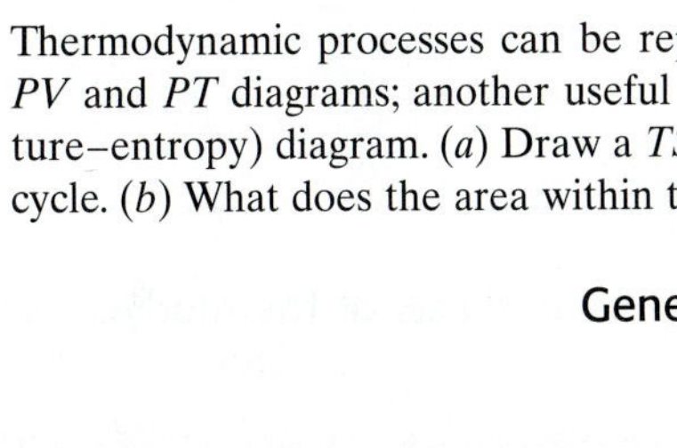

FIGURE 20–28
Problem 78.

79. Thermodynamic processes can be represented not only on PV and PT diagrams; another useful one is a TS (temperature–entropy) diagram. (a) Draw a TS diagram for a Carnot cycle. (b) What does the area within the curve represent?

80. An aluminum can, with negligible heat capacity, is filled with 450 g of water at 0°C and then is brought into thermal contact with a similar can filled with 450 g of water at 50°C. Find the change in entropy of the system if no heat is allowed to exchange with the surroundings. Use $\Delta S = \int dQ/T$.

81. A dehumidifier is essentially a "refrigerator with an open door." The humid air is pulled in by a fan and guided to a cold coil, whose temperature is less than the dew point, and some of the air's water condenses. After this water is extracted, the air is warmed back to its original temperature and sent into the room. In a well-designed dehumidifier, the heat is exchanged between the incoming and outgoing air. Thus the heat that is removed by the refrigerator coil mostly comes from the condensation of water vapor to liquid. Estimate how much water is removed in 1.0 h by an ideal dehumidifier, if the temperature of the room is 25°C, the water condenses at 8°C, and the dehumidifier does work at the rate of 650 W of electrical power.

*82. A bowl contains a large number of red, orange, and green jelly beans. You are to make a line of three jelly beans. (*a*) Construct a table showing the number of microstates that correspond to each macrostate. Then determine the probability of (*b*) all 3 beans red, and (*c*) 2 greens, 1 orange.

*Numerical/Computer

*83. (II) At low temperature the specific heat of diamond varies with absolute temperature T according to the Debye equation $C_V = 1.88 \times 10^3 (T/T_D)^3$ J·mol^{-1}·K^{-1} where the Debye temperature for diamond is $T_D = 2230$ K. Use a spreadsheet and numerical integration to determine the entropy change of 1.00 mol of diamond when it is heated at constant volume from 4 K to 40 K. Your result should agree within 2% of the result obtained by integrating the expression for dS. [*Hint*: $dS = nC_V \, dT/T$, where n is the number of moles.]

Answers to Exercises

A: No. Efficiency makes no sense for a single process. It is defined (Eqs. 20–1) only for cyclic processes that return to the initial state.

B: (*d*).

C: (*c*).

D: 1220 J/K; −1220 J/K. (Note that the *total* entropy change, $\Delta S_{ice} + \Delta S_{res}$, is zero.)

E: (*e*).

This comb has acquired a static electric charge, either from passing through hair, or being rubbed by a cloth or paper towel. The electrical charge on the comb induces a polarization (separation of charge) in scraps of paper, and thus attracts them.

Our introduction to electricity in this Chapter covers conductors and insulators, and Coulomb's law which relates the force between two point charges as a function of their distance apart. We also introduce the powerful concept of electric field.

Electric Charge and Electric Field

CHAPTER-OPENING QUESTION—Guess now!

Two identical tiny spheres have the same electric charge. If the electric charge on each of them is doubled, and their separation is also doubled, the force each exerts on the other will be

(a) half.
(b) double.
(c) four times larger.
(d) one-quarter as large.
(e) unchanged.

T he word "electricity" may evoke an image of complex modern technology: lights, motors, electronics, and computers. But the electric force plays an even deeper role in our lives. According to atomic theory, electric forces between atoms and molecules hold them together to form liquids and solids, and electric forces are also involved in the metabolic processes that occur within our bodies. Many of the forces we have dealt with so far, such as elastic forces, the normal force, and friction and other contact forces (pushes and pulls), are now considered to result from electric forces acting at the atomic level. Gravity, on the other hand, is a separate force.[†]

[†] As we discussed in Section 6–7, physicists in the twentieth century came to recognize four different fundamental forces in nature: (1) gravitational force, (2) electromagnetic force (we will see later that electric and magnetic forces are intimately related), (3) strong nuclear force, and (4) weak nuclear force. The last two forces operate at the level of the nucleus of an atom. Recent theory has combined the electromagnetic and weak nuclear forces so they are now considered to have a common origin known as the electroweak force. We will discuss these forces in later Chapters.

CONTENTS

21–1 Static Electricity; Electric Charge and Its Conservation

21–2 Electric Charge in the Atom

21–3 Insulators and Conductors

21–4 Induced Charge; the Electroscope

21–5 Coulomb's Law

21–6 The Electric Field

21–7 Electric Field Calculations for Continuous Charge Distributions

21–8 Field Lines

21–9 Electric Fields and Conductors

21–10 Motion of a Charged Particle in an Electric Field

21–11 Electric Dipoles

*21–12 Electric Forces in Molecular Biology; DNA

*21–13 Photocopy Machines and Computer Printers Use Electrostatics

559

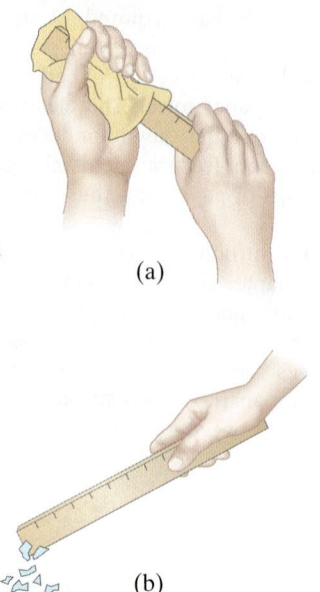

(a)

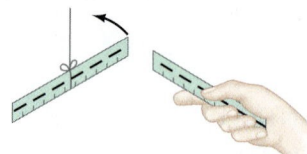

(b)

FIGURE 21–1 (a) Rub a plastic ruler and (b) bring it close to some tiny pieces of paper.

FIGURE 21–2 Like charges repel one another; unlike charges attract. (Note color coding: positive and negative charged objects are often colored pink and blue-green, respectively, when we want to emphasize them. We use these colors especially for point charges, but not often for real objects.)

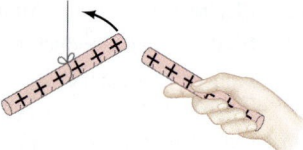

(a) Two charged plastic rulers repel

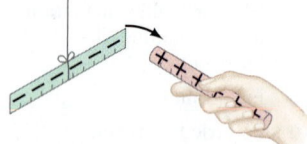

(b) Two charged glass rods repel

(c) Charged glass rod attracts charged plastic ruler

LAW OF CONSERVATION OF ELECTRIC CHARGE

The earliest studies on electricity date back to the ancients, but only in the past two centuries has electricity been studied in detail. We will discuss the development of ideas about electricity, including practical devices, as well as its relation to magnetism, in the next eleven Chapters.

21–1 Static Electricity; Electric Charge and Its Conservation

The word *electricity* comes from the Greek word *elektron*, which means "amber." Amber is petrified tree resin, and the ancients knew that if you rub a piece of amber with a cloth, the amber attracts small pieces of leaves or dust. A piece of hard rubber, a glass rod, or a plastic ruler rubbed with a cloth will also display this "amber effect," or **static electricity** as we call it today. You can readily pick up small pieces of paper with a plastic comb or ruler that you have just vigorously rubbed with even a paper towel. See the photo on the previous page and Fig. 21–1. You have probably experienced static electricity when combing your hair or when taking a synthetic blouse or shirt from a clothes dryer. And you may have felt a shock when you touched a metal doorknob after sliding across a car seat or walking across a nylon carpet. In each case, an object becomes "charged" as a result of rubbing, and is said to possess a net **electric charge**.

Is all electric charge the same, or is there more than one type? In fact, there are *two* types of electric charge, as the following simple experiments show. A plastic ruler suspended by a thread is vigorously rubbed with a cloth to charge it. When a second plastic ruler, which has been charged in the same way, is brought close to the first, it is found that one ruler *repels* the other. This is shown in Fig. 21–2a. Similarly, if a rubbed glass rod is brought close to a second charged glass rod, again a repulsive force is seen to act, Fig. 21–2b. However, if the charged glass rod is brought close to the charged plastic ruler, it is found that they *attract* each other, Fig. 21–2c. The charge on the glass must therefore be different from that on the plastic. Indeed, it is found experimentally that all charged objects fall into one of two categories. Either they are attracted to the plastic and repelled by the glass; or they are repelled by the plastic and attracted to the glass. Thus there seem to be two, and only two, types of electric charge. Each type of charge repels the same type but attracts the opposite type. That is: **unlike charges attract; like charges repel**.

The two types of electric charge were referred to as *positive* and *negative* by the American statesman, philosopher, and scientist Benjamin Franklin (1706–1790). The choice of which name went with which type of charge was arbitrary. Franklin's choice set the charge on the rubbed glass rod to be positive charge, so the charge on a rubbed plastic ruler (or amber) is called negative charge. We still follow this convention today.

Franklin argued that whenever a certain amount of charge is produced on one object, an equal amount of the opposite type of charge is produced on another object. The positive and negative are to be treated *algebraically*, so during any process, the net change in the amount of charge produced is zero. For example, when a plastic ruler is rubbed with a paper towel, the plastic acquires a negative charge and the towel acquires an equal amount of positive charge. The charges are separated, but the sum of the two is zero.

This is an example of a law that is now well established: the **law of conservation of electric charge**, which states that

the net amount of electric charge produced in any process is zero;

or, said another way,

no net electric charge can be created or destroyed.

If one object (or a region of space) acquires a positive charge, then an equal amount of negative charge will be found in neighboring areas or objects. No violations have ever been found, and this conservation law is as firmly established as those for energy and momentum.

21–2 Electric Charge in the Atom

Only within the past century has it become clear that an understanding of electricity originates inside the atom itself. In later Chapters we will discuss atomic structure and the ideas that led to our present view of the atom in more detail. But it will help our understanding of electricity if we discuss it briefly now.

A simplified model of an atom shows it as having a tiny but heavy, positively charged nucleus surrounded by one or more negatively charged electrons (Fig. 21–3). The nucleus contains protons, which are positively charged, and neutrons, which have no net electric charge. All protons and all electrons have exactly the same magnitude of electric charge; but their signs are opposite. Hence neutral atoms, having no net charge, contain equal numbers of protons and electrons. Sometimes an atom may lose one or more of its electrons, or may gain extra electrons, in which case it will have a net positive or negative charge and is called an **ion**.

In solid materials the nuclei tend to remain close to fixed positions, whereas some of the electrons may move quite freely. When an object is *neutral*, it contains equal amounts of positive and negative charge. The charging of a solid object by rubbing can be explained by the transfer of electrons from one object to the other. When a plastic ruler becomes negatively charged by rubbing with a paper towel, the transfer of electrons from the towel to the plastic leaves the towel with a positive charge equal in magnitude to the negative charge acquired by the plastic. In liquids and gases, nuclei or ions can move as well as electrons.

Normally when objects are charged by rubbing, they hold their charge only for a limited time and eventually return to the neutral state. Where does the charge go? Usually the charge "leaks off" onto water molecules in the air. This is because water molecules are **polar**—that is, even though they are neutral, their charge is not distributed uniformly, Fig. 21–4. Thus the extra electrons on, say, a charged plastic ruler can "leak off" into the air because they are attracted to the positive end of water molecules. A positively charged object, on the other hand, can be neutralized by transfer of loosely held electrons from water molecules in the air. On dry days, static electricity is much more noticeable since the air contains fewer water molecules to allow leakage. On humid or rainy days, it is difficult to make any object hold a net charge for long.

21–3 Insulators and Conductors

Suppose we have two metal spheres, one highly charged and the other electrically neutral (Fig. 21–5a). If we now place a metal object, such as a nail, so that it touches both spheres (Fig. 21–5b), the previously uncharged sphere quickly becomes charged. If, instead, we had connected the two spheres by a wooden rod or a piece of rubber (Fig. 21–5c), the uncharged ball would not become noticeably charged. Materials like the iron nail are said to be **conductors** of electricity, whereas wood and rubber are **nonconductors** or **insulators**.

Metals are generally good conductors, whereas most other materials are insulators (although even insulators conduct electricity very slightly). Nearly all natural materials fall into one or the other of these two very distinct categories. However, a few materials (notably silicon and germanium) fall into an intermediate category known as **semiconductors**.

From the atomic point of view, the electrons in an insulating material are bound very tightly to the nuclei. In a good conductor, on the other hand, some of the electrons are bound very loosely and can move about freely within the material (although they cannot *leave* the object easily) and are often referred to as *free electrons* or *conduction electrons*. When a positively charged object is brought close to or touches a conductor, the free electrons in the conductor are attracted by this positively charged object and move quickly toward it. On the other hand, the free electrons move swiftly away from a negatively charged object that is brought close to the conductor. In a semiconductor, there are many fewer free electrons, and in an insulator, almost none.

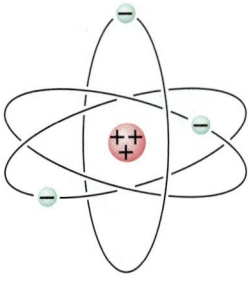

FIGURE 21–3 Simple model of the atom.

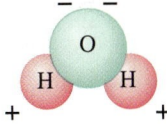

FIGURE 21–4 Diagram of a water molecule. Because it has opposite charges on different ends, it is called a "polar" molecule.

FIGURE 21–5 (a) A charged metal sphere and a neutral metal sphere. (b) The two spheres connected by a conductor (a metal nail), which conducts charge from one sphere to the other. (c) The original two spheres connected by an insulator (wood); almost no charge is conducted.

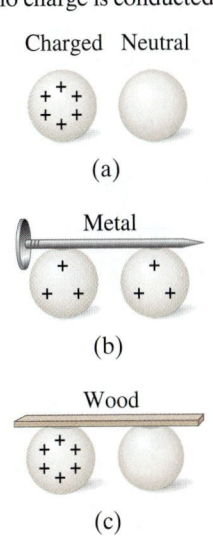

Charged Neutral

(a)

Metal

(b)

Wood

(c)

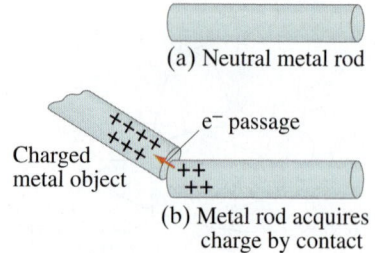

(a) Neutral metal rod

e⁻ passage

Charged
metal object

(b) Metal rod acquires
charge by contact

FIGURE 21–6 A neutral metal rod in (a) will acquire a positive charge if placed in contact (b) with a positively charged metal object. (Electrons move as shown by the orange arrow.) This is called charging by conduction.

21–4 Induced Charge; the Electroscope

Suppose a positively charged metal object is brought close to an uncharged metal object. If the two touch, the free electrons in the neutral one are attracted to the positively charged object and some will pass over to it, Fig. 21–6. Since the second object, originally neutral, is now missing some of its negative electrons, it will have a net positive charge. This process is called "charging by conduction," or "by contact," and the two objects end up with the same sign of charge.

Now suppose a positively charged object is brought close to a neutral metal rod, but does not touch it. Although the free electrons of the metal rod do not leave the rod, they still move within the metal toward the external positive charge, leaving a positive charge at the opposite end of the rod (Fig. 21–7). A charge is said to have been *induced* at the two ends of the metal rod. No net charge has been created in the rod: charges have merely been *separated*. The net charge on the metal rod is still zero. However, if the metal is separated into two pieces, we would have two charged objects: one charged positively and one charged negatively.

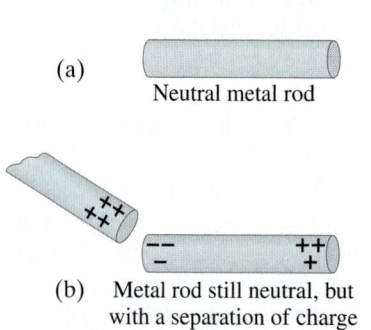

(a)   Neutral metal rod

(b)   Metal rod still neutral, but with a separation of charge

FIGURE 21–7 Charging by induction.

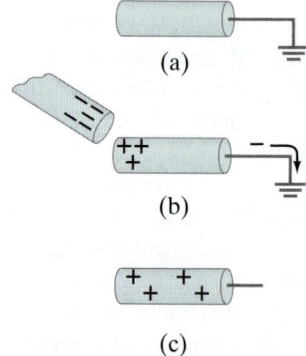

(a)

(b)

(c)

FIGURE 21–8 Inducing a charge on an object connected to ground.

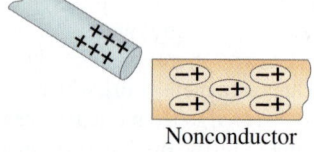

Nonconductor

FIGURE 21–9 A charged object brought near an insulator causes a charge separation within the insulator's molecules.

FIGURE 21–10 Electroscope.

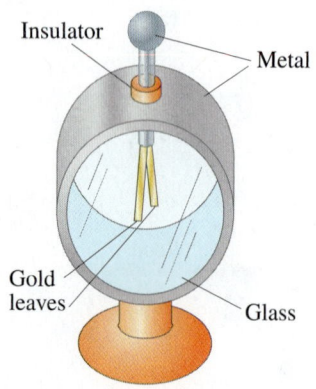

Insulator

Metal

Gold
leaves

Glass

Another way to induce a net charge on a metal object is to first connect it with a conducting wire to the ground (or a conducting pipe leading into the ground) as shown in Fig. 21–8a (the symbol ⏚ means connected to "ground"). The object is then said to be "grounded" or "earthed." The Earth, because it is so large and can conduct, easily accepts or gives up electrons; hence it acts like a reservoir for charge. If a charged object—say negative this time—is brought up close to the metal object, free electrons in the metal are repelled and many of them move down the wire into the Earth, Fig. 21–8b. This leaves the metal positively charged. If the wire is now cut, the metal object will have a positive induced charge on it (Fig. 21–8c). If the wire were cut after the negative object was moved away, the electrons would all have moved back into the metal object and it would be neutral.

Charge separation can also be done in nonconductors. If you bring a positively charged object close to a neutral nonconductor as shown in Fig. 21–9, almost no electrons can move about freely within the nonconductor. But they can move slightly within their own atoms and molecules. Each oval in Fig. 21–9 represents a molecule (not to scale); the negatively charged electrons, attracted to the external positive charge, tend to move in its direction within their molecules. Because the negative charges in the nonconductor are nearer to the external positive charge, the nonconductor as a whole is attracted to the external positive charge (see the Chapter-Opening Photo, page 559).

An **electroscope** is a device that can be used for detecting charge. As shown in Fig. 21–10, inside of a case are two movable metal leaves, often made of gold, connected to a metal knob on the outside. (Sometimes only one leaf is movable.)

If a positively charged object is brought close to the knob, a separation of charge is induced: electrons are attracted up into the knob, leaving the leaves positively charged, Fig. 21–11a. The two leaves repel each other as shown, because they are both positively charged. If, instead, the knob is charged by conduction, the whole apparatus acquires a net charge as shown in Fig. 21–11b. In either case, the greater the amount of charge, the greater the separation of the leaves.

Note that you cannot tell the sign of the charge in this way, since negative charge will cause the leaves to separate just as much as an equal amount of positive charge; in either case, the two leaves repel each other. An electroscope can, however, be used to determine the sign of the charge if it is first charged by conduction, say, negatively, as in Fig. 21–12a. Now if a negative object is brought close, as in Fig. 21–12b, more electrons are induced to move down into the leaves and they separate further. If a positive charge is brought close instead, the electrons are induced to flow upward, leaving the leaves less negative and their separation is reduced, Fig. 21–12c.

The electroscope was used in the early studies of electricity. The same principle, aided by some electronics, is used in much more sensitive modern **electrometers**.

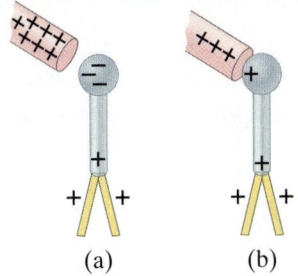

FIGURE 21–11 Electroscope charged (a) by induction, (b) by conduction.

FIGURE 21–12 A previously charged electroscope can be used to determine the sign of a charged object.

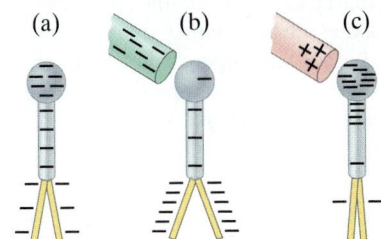

21–5 Coulomb's Law

We have seen that an electric charge exerts a force of attraction or repulsion on other electric charges. What factors affect the magnitude of this force? To find an answer, the French physicist Charles Coulomb (1736–1806) investigated electric forces in the 1780s using a torsion balance (Fig. 21–13) much like that used by Cavendish for his studies of the gravitational force (Chapter 6).

Precise instruments for the measurement of electric charge were not available in Coulomb's time. Nonetheless, Coulomb was able to prepare small spheres with different magnitudes of charge in which the *ratio* of the charges was known.[†] Although he had some difficulty with induced charges, Coulomb was able to argue that the force one tiny charged object exerts on a second tiny charged object is directly proportional to the charge on each of them. That is, if the charge on either one of the objects is doubled, the force is doubled; and if the charge on both of the objects is doubled, the force increases to four times the original value. This was the case when the distance between the two charges remained the same. If the distance between them was allowed to increase, he found that the force decreased with the *square of the distance* between them. That is, if the distance was doubled, the force fell to one-fourth of its original value. Thus, Coulomb concluded, the force one small charged object exerts on a second one is proportional to the product of the magnitude of the charge on one, Q_1, times the magnitude of the charge on the other, Q_2, and inversely proportional to the square of the distance r between them (Fig. 21–14). As an equation, we can write **Coulomb's law** as

$$F = k\frac{Q_1 Q_2}{r^2}, \qquad \text{[magnitudes]} \quad \textbf{(21–1)}$$

where k is a proportionality constant.[‡]

FIGURE 21–13 (below) Coulomb used a torsion balance to investigate how the electric force varies as a function of the magnitude of the charges and of the distance between them. When an external charged sphere is placed close to the charged one on the suspended bar, the bar rotates slightly. The suspending fiber resists the twisting motion, and the angle of twist is proportional to the electric force.

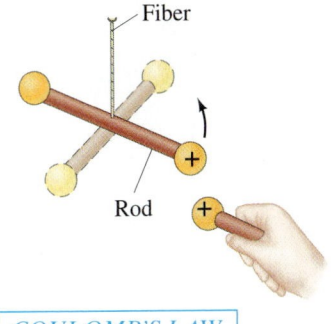

COULOMB'S LAW

FIGURE 21–14 Coulomb's law, Eq. 21–1, gives the force between two point charges, Q_1 and Q_2, a distance r apart.

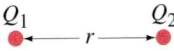

[†]Coulomb reasoned that if a charged conducting sphere is placed in contact with an identical uncharged sphere, the charge on the first would be shared equally by the two of them because of symmetry. He thus had a way to produce charges equal to $\frac{1}{2}, \frac{1}{4}$, and so on, of the original charge.

[‡]The validity of Coulomb's law today rests on precision measurements that are much more sophisticated than Coulomb's original experiment. The exponent 2 in Coulomb's law has been shown to be accurate to 1 part in 10^{16} [that is, $2 \pm (1 \times 10^{-16})$].

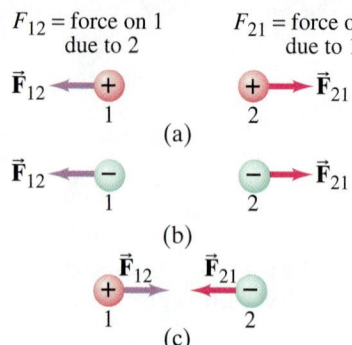

F_{12} = force on 1 due to 2 F_{21} = force on 2 due to 1

(a)

(b)

(c)

FIGURE 21–15 The direction of the static electric force one point charge exerts on another is always along the line joining the two charges, and depends on whether the charges have the same sign as in (a) and (b), or opposite signs (c).

As we just saw, Coulomb's law,

$$F = k\frac{Q_1 Q_2}{r^2},$$ [magnitudes] **(21–1)**

gives the *magnitude* of the electric force that either charge exerts on the other. The *direction* of the electric force *is always along the line joining the two charges.* If the two charges have the same sign, the force on either charge is directed away from the other (they repel each other). If the two charges have opposite signs, the force on one is directed toward the other (they attract). See Fig. 21–15. Notice that the force one charge exerts on the second is equal but opposite to that exerted by the second on the first, in accord with Newton's third law.

The SI unit of charge is the **coulomb** (C).[†] The precise definition of the coulomb today is in terms of electric current and magnetic field, and will be discussed later (Section 28–3). In SI units, the constant k in Coulomb's law has the value

$$k = 8.99 \times 10^9 \, \text{N} \cdot \text{m}^2/\text{C}^2$$

or, when we only need two significant figures,

$$k \approx 9.0 \times 10^9 \, \text{N} \cdot \text{m}^2/\text{C}^2.$$

Thus, 1 C is that amount of charge which, if placed on each of two point objects that are 1.0 m apart, will result in each object exerting a force of $(9.0 \times 10^9 \, \text{N} \cdot \text{m}^2/\text{C}^2)(1.0 \, \text{C})(1.0 \, \text{C})/(1.0 \, \text{m})^2 = 9.0 \times 10^9 \, \text{N}$ on the other. This would be an enormous force, equal to the weight of almost a million tons. We rarely encounter charges as large as a coulomb.

Charges produced by rubbing ordinary objects (such as a comb or plastic ruler) are typically around a microcoulomb $(1 \, \mu\text{C} = 10^{-6} \, \text{C})$ or less. Objects that carry a positive charge have a deficit of electrons, whereas negatively charged objects have an excess of electrons. The charge on one electron has been determined to have a magnitude of about $1.602 \times 10^{-19} \, \text{C}$, and is negative. This is the smallest charge found in nature,[‡] and because it is fundamental, it is given the symbol e and is often referred to as the *elementary charge*:

$$e = 1.602 \times 10^{-19} \, \text{C}.$$

Note that e is defined as a positive number, so the charge on the electron is $-e$. (The charge on a proton, on the other hand, is $+e$.) Since an object cannot gain or lose a fraction of an electron, the net charge on any object must be an integral multiple of this charge. Electric charge is thus said to be **quantized** (existing only in discrete amounts: $1e$, $2e$, $3e$, etc.). Because e is so small, however, we normally do not notice this discreteness in macroscopic charges ($1 \, \mu\text{C}$ requires about 10^{13} electrons), which thus seem continuous.

Coulomb's law looks a lot like the *law of universal gravitation*, $F = Gm_1 m_2/r^2$, which expresses the gravitational force a mass m_1 exerts on a mass m_2 (Eq. 6–1). Both are inverse square laws $(F \propto 1/r^2)$. Both also have a proportionality to a property of each object—mass for gravity, electric charge for electricity. And both act over a distance (that is, there is no need for contact). A major difference between the two laws is that gravity is always an attractive force, whereas the electric force can be either attractive or repulsive. Electric charge comes in two types, positive and negative; gravitational mass is only positive.

[†]In the once common cgs system of units, k is set equal to 1, and the unit of electric charge is called the *electrostatic unit* (esu) or the statcoulomb. One esu is defined as that charge, on each of two point objects 1 cm apart, that gives rise to a force of 1 dyne.

[‡]According to the standard model of elementary particle physics, subnuclear particles called quarks have a smaller charge than that on the electron, equal to $\frac{1}{3}e$ or $\frac{2}{3}e$. Quarks have not been detected directly as isolated objects, and theory indicates that free quarks may not be detectable.

The constant k in Eq. 21–1 is often written in terms of another constant, ϵ_0, called the **permittivity of free space**. It is related to k by $k = 1/4\pi\epsilon_0$. Coulomb's law can then be written

$$F = \frac{1}{4\pi\epsilon_0}\frac{Q_1 Q_2}{r^2},$$ (21–2)

where

$$\epsilon_0 = \frac{1}{4\pi k} = 8.85 \times 10^{-12}\,\mathrm{C^2/N\cdot m^2}.$$

COULOMB'S LAW
(in terms of ϵ_0)

Equation 21–2 looks more complicated than Eq. 21–1, but other fundamental equations we haven't seen yet are simpler in terms of ϵ_0 rather than k. It doesn't matter which form we use since Eqs. 21–1 and 21–2 are equivalent. (The latest precise values of e and ϵ_0 are given inside the front cover.)

[Our convention for units, such as $\mathrm{C^2/N\cdot m^2}$ for ϵ_0, means $\mathrm{m^2}$ is in the denominator. That is, $\mathrm{C^2/N\cdot m^2}$ does *not* mean $(\mathrm{C^2/N})\cdot \mathrm{m^2} = \mathrm{C^2\cdot m^2/N}$.]

Equations 21–1 and 21–2 apply to objects whose size is much smaller than the distance between them. Ideally, it is precise for **point charges** (spatial size negligible compared to other distances). For finite-sized objects, it is not always clear what value to use for r, particularly since the charge may not be distributed uniformly on the objects. If the two objects are spheres and the charge is known to be distributed uniformly on each, then r is the distance between their centers.

Coulomb's law describes the force between two charges when they are at rest. Additional forces come into play when charges are in motion, and will be discussed in later Chapters. In this Chapter we discuss only charges at rest, the study of which is called **electrostatics**, and Coulomb's law gives the **electrostatic force**.

When calculating with Coulomb's law, we usually ignore the signs of the charges and determine the direction of a force separately based on whether the force is attractive or repulsive.

PROBLEM SOLVING
Use magnitudes in Coulomb's law; find force direction from signs of charges

EXERCISE A Return to the Chapter-Opening Question, page 559, and answer it again now. Try to explain why you may have answered differently the first time.

CONCEPTUAL EXAMPLE 21–1 | **Which charge exerts the greater force?** Two positive point charges, $Q_1 = 50\,\mu\mathrm{C}$ and $Q_2 = 1\,\mu\mathrm{C}$, are separated by a distance ℓ, Fig. 21–16. Which is larger in magnitude, the force that Q_1 exerts on Q_2, or the force that Q_2 exerts on Q_1?

RESPONSE From Coulomb's law, the force on Q_1 exerted by Q_2 is

$$F_{12} = k\frac{Q_1 Q_2}{\ell^2}.$$

The force on Q_2 exerted by Q_1 is

$$F_{21} = k\frac{Q_2 Q_1}{\ell^2}$$

which is the same magnitude. The equation is symmetric with respect to the two charges, so $F_{21} = F_{12}$.

NOTE Newton's third law also tells us that these two forces must have equal magnitude.

FIGURE 21–16 Example 21–1.

EXERCISE B What is the magnitude of F_{12} (and F_{21}) in Example 21–1 if $\ell = 30\,\mathrm{cm}$?

Keep in mind that Eq. 21–2 (or 21–1) gives the force on a charge due to only *one* other charge. If several (or many) charges are present, the *net force on any one of them will be the vector sum of the forces due to each of the others*. This **principle of superposition** is based on experiment, and tells us that electric force vectors add like any other vector. For continuous distributions of charge, the sum becomes an integral.

EXAMPLE 21–2 **Three charges in a line.** Three charged particles are arranged in a line, as shown in Fig. 21–17. Calculate the net electrostatic force on particle 3 (the $-4.0\,\mu\text{C}$ on the right) due to the other two charges.

APPROACH The net force on particle 3 is the vector sum of the force $\vec{F}_{31}$ exerted on 3 by particle 1 and the force $\vec{F}_{32}$ exerted on 3 by particle 2: $\vec{F} = \vec{F}_{31} + \vec{F}_{32}$.

SOLUTION The magnitudes of these two forces are obtained using Coulomb's law, Eq. 21–1:

$$
\begin{aligned}
F_{31} &= k\frac{Q_3 Q_1}{r_{31}^2} \\
&= \frac{(9.0 \times 10^9\,\text{N}\cdot\text{m}^2/\text{C}^2)(4.0 \times 10^{-6}\,\text{C})(8.0 \times 10^{-6}\,\text{C})}{(0.50\,\text{m})^2} = 1.2\,\text{N},
\end{aligned}
$$

where $r_{31} = 0.50\,\text{m}$ is the distance from Q_3 to Q_1. Similarly,

$$
\begin{aligned}
F_{32} &= k\frac{Q_3 Q_2}{r_{32}^2} \\
&= \frac{(9.0 \times 10^9\,\text{N}\cdot\text{m}^2/\text{C}^2)(4.0 \times 10^{-6}\,\text{C})(3.0 \times 10^{-6}\,\text{C})}{(0.20\,\text{m})^2} = 2.7\,\text{N}.
\end{aligned}
$$

Since we were calculating the magnitudes of the forces, we omitted the signs of the charges. But we must be aware of them to get the direction of each force. Let the line joining the particles be the x axis, and we take it positive to the right. Then, because $\vec{F}_{31}$ is repulsive and $\vec{F}_{32}$ is attractive, the directions of the forces are as shown in Fig. 21–17b: F_{31} points in the positive x direction and F_{32} points in the negative x direction. The net force on particle 3 is then

$$
F = -F_{32} + F_{31} = -2.7\,\text{N} + 1.2\,\text{N} = -1.5\,\text{N}.
$$

The magnitude of the net force is 1.5 N, and it points to the left.

NOTE Charge Q_1 acts on charge Q_3 just as if Q_2 were not there (this is the principle of superposition). That is, the charge in the middle, Q_2, in no way blocks the effect of charge Q_1 acting on Q_3. Naturally, Q_2 exerts its own force on Q_3.

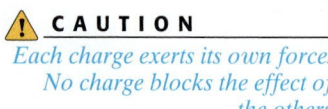

$$
\begin{aligned}
Q_1 &= &Q_2 &= &Q_3 &= \\
-8.0\,\mu\text{C} & &+3.0\,\mu\text{C} & &-4.0\,\mu\text{C}
\end{aligned}
$$

(a)

(b)

FIGURE 21–17 Example 21–2.

> ⚠ **CAUTION**
> *Each charge exerts its own force. No charge blocks the effect of the others*

| **EXERCISE C** Determine the magnitude and direction of the net force on Q_1 in Fig. 21–17a.

EXAMPLE 21–3 **Electric force using vector components.** Calculate the net electrostatic force on charge Q_3 shown in Fig. 21–18a due to the charges Q_1 and Q_2.

APPROACH We use Coulomb's law to find the magnitudes of the individual forces. The direction of each force will be along the line connecting Q_3 to Q_1 or Q_2. The forces $\vec{F}_{31}$ and $\vec{F}_{32}$ have the directions shown in Fig. 21–18a, since Q_1 exerts an attractive force on Q_3, and Q_2 exerts a repulsive force. The forces $\vec{F}_{31}$ and $\vec{F}_{32}$ are *not* along the same line, so to find the resultant force on Q_3 we resolve $\vec{F}_{31}$ and $\vec{F}_{32}$ into x and y components and perform the vector addition.

SOLUTION The magnitudes of $\vec{F}_{31}$ and $\vec{F}_{32}$ are (ignoring signs of the charges since we know the directions)

$$
F_{31} = k\frac{Q_3 Q_1}{r_{31}^2} = \frac{(9.0 \times 10^9\,\text{N}\cdot\text{m}^2/\text{C}^2)(6.5 \times 10^{-5}\,\text{C})(8.6 \times 10^{-5}\,\text{C})}{(0.60\,\text{m})^2} = 140\,\text{N},
$$

$$
F_{32} = k\frac{Q_3 Q_2}{r_{32}^2} = \frac{(9.0 \times 10^9\,\text{N}\cdot\text{m}^2/\text{C}^2)(6.5 \times 10^{-5}\,\text{C})(5.0 \times 10^{-5}\,\text{C})}{(0.30\,\text{m})^2} = 330\,\text{N}.
$$

We resolve $\vec{F}_{31}$ into its components along the x and y axes, as shown in Fig. 21–18a:

$$
F_{31x} = F_{31}\cos 30° = (140\,\text{N})\cos 30° = 120\,\text{N},
$$

$$
F_{31y} = -F_{31}\sin 30° = -(140\,\text{N})\sin 30° = -70\,\text{N}.
$$

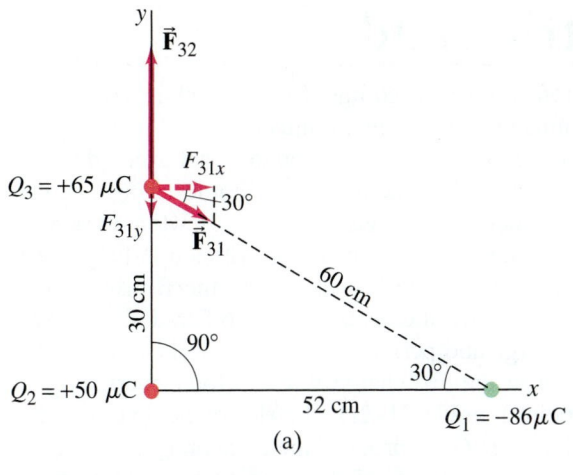

(a)

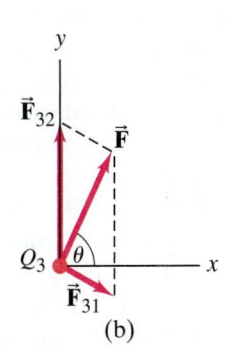

(b)

FIGURE 21–18 Determining the forces for Example 21–3. (a) The directions of the individual forces are as shown because $\vec{F}_{32}$ is repulsive (the force on Q_3 is in the direction away from Q_2 because Q_3 and Q_2 are both positive) whereas $\vec{F}_{31}$ is attractive (Q_3 and Q_1 have opposite signs), so $\vec{F}_{31}$ points toward Q_1. (b) Adding $\vec{F}_{32}$ to $\vec{F}_{31}$ to obtain the net force $\vec{F}$.

The force $\vec{F}_{32}$ has only a y component. So the net force $\vec{F}$ on Q_3 has components

$$F_x = F_{31x} = 120\,\text{N},$$
$$F_y = F_{32} + F_{31y} = 330\,\text{N} - 70\,\text{N} = 260\,\text{N}.$$

The magnitude of the net force is

$$F = \sqrt{F_x^2 + F_y^2} = \sqrt{(120\,\text{N})^2 + (260\,\text{N})^2} = 290\,\text{N};$$

and it acts at an angle θ (see Fig. 21–18b) given by

$$\tan\theta = \frac{F_y}{F_x} = \frac{260\,\text{N}}{120\,\text{N}} = 2.2,$$

so $\theta = \tan^{-1}(2.2) = 65°$.

NOTE Because $\vec{F}_{31}$ and $\vec{F}_{32}$ are not along the same line, the magnitude of $\vec{F}_3$ is not equal to the sum (or difference as in Example 21–2) of the separate magnitudes.

CONCEPTUAL EXAMPLE 21–4 **Make the force on Q_3 zero.** In Fig. 21–18, where could you place a fourth charge, $Q_4 = -50\,\mu\text{C}$, so that the net force on Q_3 would be zero?

RESPONSE By the principle of superposition, we need a force in exactly the opposite direction to the resultant $\vec{F}$ due to Q_2 and Q_1 that we calculated in Example 21–3, Fig. 21–18b. Our force must have magnitude 290 N, and must point down and to the left of Q_3 in Fig. 21–18b. So Q_4 must be along this line. See Fig. 21–19.

EXERCISE D (a) Consider two point charges of the same magnitude but opposite sign ($+Q$ and $-Q$), which are fixed a distance d apart. Can you find a location where a third positive charge Q could be placed so that the net electric force on this third charge is zero? (b) What if the first two charges were both $+Q$?

*Vector Form of Coulomb's Law

Coulomb's law can be written in vector form (as we did for Newton's law of universal gravitation in Chapter 6, Section 6–2) as

$$\vec{F}_{12} = k\frac{Q_1 Q_2}{r_{21}^2}\,\hat{r}_{21},$$

where $\vec{F}_{12}$ is the vector force on charge Q_1 due to Q_2 and $\hat{r}_{21}$ is the unit vector pointing from Q_2 toward Q_1. That is, $\hat{r}_{21}$ points from the "source" charge (Q_2) toward the charge on which we want to know the force (Q_1). See Fig. 21–20. The charges Q_1 and Q_2 can be either positive or negative, and this will affect the direction of the electric force. If Q_1 and Q_2 have the same sign, the product $Q_1 Q_2 > 0$ and the force on Q_1 points away from Q_2—that is, it is repulsive. If Q_1 and Q_2 have opposite signs, $Q_1 Q_2 < 0$ and $\vec{F}_{12}$ points toward Q_2—that is, it is attractive.

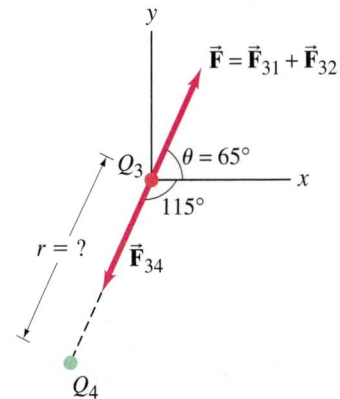

FIGURE 21–19 Example 21–4: Q_4 exerts force $(\vec{F}_{34})$ that makes the net force on Q_3 zero.

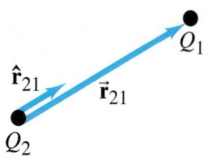

FIGURE 21–20 Determining the force on Q_1 due to Q_2, showing the direction of the unit vector $\hat{r}_{21}$.

21–6 The Electric Field

Many common forces might be referred to as "contact forces," such as your hands pushing or pulling a cart, or a tennis racket hitting a tennis ball.

In contrast, both the gravitational force and the electrical force act over a distance: there is a force between two objects even when the objects are not touching. The idea of a force *acting at a distance* was a difficult one for early thinkers. Newton himself felt uneasy with this idea when he published his law of universal gravitation. A helpful way to look at the situation uses the idea of the **field**, developed by the British scientist Michael Faraday (1791–1867). In the electrical case, according to Faraday, an *electric field* extends outward from every charge and permeates all of space (Fig. 21–21). If a second charge (call it Q_2) is placed near the first charge, it feels a force exerted by the electric field that is there (say, at point P in Fig. 21–21). The electric field at point P is considered to interact directly with charge Q_2 to produce the force on Q_2.

We can in principle investigate the electric field surrounding a charge or group of charges by measuring the force on a small positive **test charge** at rest. By a test charge we mean a charge so small that the force it exerts does not significantly affect the charges that create the field. If a tiny positive test charge q is placed at various locations in the vicinity of a single positive charge Q as shown in Fig. 21–22 (points A, B, C), the force exerted on q is as shown. The force at B is less than at A because B's distance from Q is greater (Coulomb's law); and the force at C is smaller still. In each case, the force on q is directed radially away from Q. The electric field is defined in terms of the force on such a positive test charge. In particular, the **electric field**, $\vec{E}$, at any point in space is defined as the force $\vec{F}$ exerted on a tiny positive test charge placed at that point divided by the magnitude of the test charge q:

$$\vec{E} = \frac{\vec{F}}{q}. \tag{21-3}$$

More precisely, $\vec{E}$ is defined as the limit of $\vec{F}/q$ as q is taken smaller and smaller, approaching zero. That is, q is so tiny that it exerts essentially no force on the other charges which created the field. From this definition (Eq. 21–3), we see that the electric field at any point in space is a vector whose direction is the direction of the force on a tiny positive test charge at that point, and whose magnitude is the *force per unit charge*. Thus $\vec{E}$ has SI units of newtons per coulomb (N/C).

The reason for defining $\vec{E}$ as $\vec{F}/q$ (with $q \rightarrow 0$) is so that $\vec{E}$ does not depend on the magnitude of the test charge q. This means that $\vec{E}$ describes only the effect of the charges creating the electric field at that point.

The electric field at any point in space can be measured, based on the definition, Eq. 21–3. For simple situations involving one or several point charges, we can calculate $\vec{E}$. For example, the electric field at a distance r from a single point charge Q would have magnitude

$$E = \frac{F}{q} = \frac{kqQ/r^2}{q}$$

$$E = k\frac{Q}{r^2}; \qquad \text{[single point charge]} \tag{21-4a}$$

or, in terms of ϵ_0 as in Eq. 21–2 $(k = 1/4\pi\epsilon_0)$:

$$E = \frac{1}{4\pi\epsilon_0}\frac{Q}{r^2}. \qquad \text{[single point charge]} \tag{21-4b}$$

Notice that E is independent of the test charge q—that is, E depends only on the charge Q which produces the field, and not on the value of the test charge q. Equations 21–4 are referred to as the electric field form of Coulomb's law.

If we are given the electric field $\vec{E}$ at a given point in space, then we can calculate the force $\vec{F}$ on any charge q placed at that point by writing (see Eq. 21–3):

$$\vec{F} = q\vec{E}. \tag{21-5}$$

This is valid even if q is not small as long as q does not cause the charges creating $\vec{E}$ to move. If q is positive, $\vec{F}$ and $\vec{E}$ point in the same direction. If q is negative, $\vec{F}$ and $\vec{E}$ point in opposite directions. See Fig. 21–23.

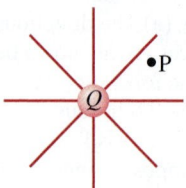

FIGURE 21–21 An electric field surrounds every charge. P is an arbitrary point.

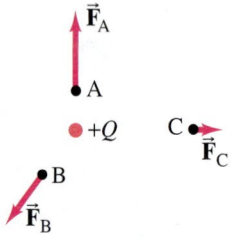

FIGURE 21–22 Force exerted by charge $+Q$ on a small test charge, q, placed at points A, B, and C.

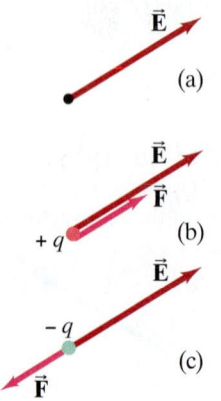

FIGURE 21–23 (a) Electric field at a given point in space. (b) Force on a positive charge at that point. (c) Force on a negative charge at that point.

EXAMPLE 21–5 **Photocopy machine.** A photocopy machine works by arranging positive charges (in the pattern to be copied) on the surface of a drum, then gently sprinkling negatively charged dry toner (ink) particles onto the drum. The toner particles temporarily stick to the pattern on the drum (Fig. 21–24) and are later transferred to paper and "melted" to produce the copy. Suppose each toner particle has a mass of 9.0×10^{-16} kg and carries an average of 20 extra electrons to provide an electric charge. Assuming that the electric force on a toner particle must exceed twice its weight in order to ensure sufficient attraction, compute the required electric field strength near the surface of the drum.

APPROACH The electric force on a toner particle of charge $q = 20e$ is $F = qE$, where E is the needed electric field. This force needs to be at least as great as twice the weight (mg) of the particle.

SOLUTION The minimum value of electric field satisfies the relation

$$qE = 2mg$$

where $q = 20e$. Hence

$$E = \frac{2mg}{q} = \frac{2(9.0 \times 10^{-16}\,\text{kg})(9.8\,\text{m/s}^2)}{20(1.6 \times 10^{-19}\,\text{C})} = 5.5 \times 10^3\,\text{N/C}.$$

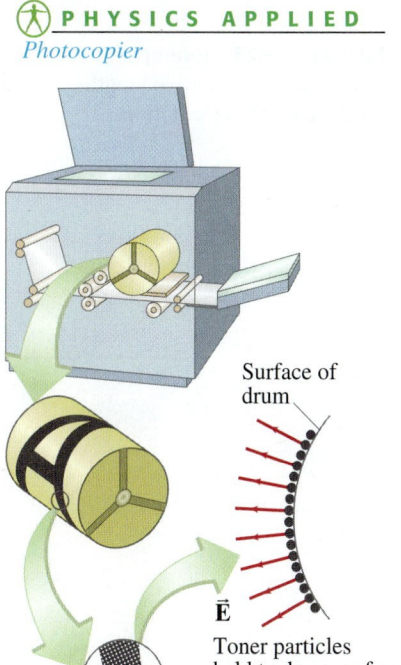

FIGURE 21–24 Example 21–5.

Surface of drum

$\vec{E}$

Toner particles held to drum surface by electric field $\vec{E}$

EXAMPLE 21–6 **Electric field of a single point charge.** Calculate the magnitude and direction of the electric field at a point P which is 30 cm to the right of a point charge $Q = -3.0 \times 10^{-6}$ C.

APPROACH The magnitude of the electric field due to a single point charge is given by Eq. 21–4. The direction is found using the sign of the charge Q.

SOLUTION The magnitude of the electric field is:

$$E = k\frac{Q}{r^2} = \frac{(9.0 \times 10^9\,\text{N} \cdot \text{m}^2/\text{C}^2)(3.0 \times 10^{-6}\,\text{C})}{(0.30\,\text{m})^2} = 3.0 \times 10^5\,\text{N/C}.$$

The direction of the electric field is *toward* the charge Q, to the left as shown in Fig. 21–25a, since we defined the direction as that of the force on a positive test charge which here would be attractive. If Q had been positive, the electric field would have pointed away, as in Fig. 21–25b.

NOTE There is no electric charge at point P. But there is an electric field there. The only real charge is Q.

FIGURE 21–25 Example 21–6. Electric field at point P (a) due to a negative charge Q, and (b) due to a positive charge Q, each 30 cm from P.

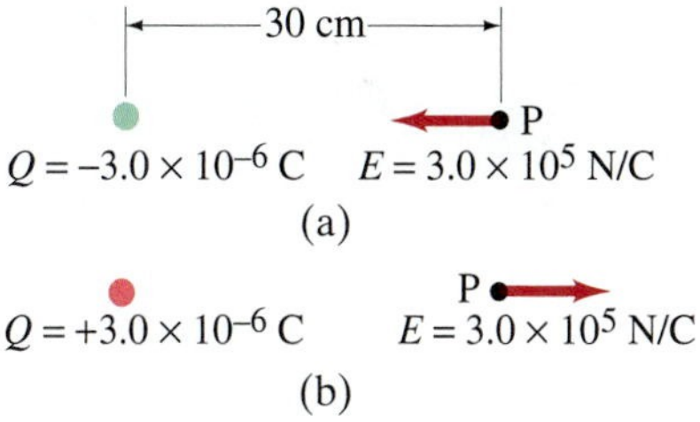

|←————30 cm————→|

$Q = -3.0 \times 10^{-6}$ C $E = 3.0 \times 10^5$ N/C P

(a)

$Q = +3.0 \times 10^{-6}$ C   P $E = 3.0 \times 10^5$ N/C

(b)

This Example illustrates a general result: The electric field $\vec{E}$ due to a positive charge points away from the charge, whereas $\vec{E}$ due to a negative charge points toward that charge.

EXERCISE E Four charges of equal magnitude, but possibly different sign, are placed on the corners of a square. What arrangement of charges will produce an electric field with the greatest magnitude at the center of the square? (*a*) All four positive charges; (*b*) all four negative charges; (*c*) three positive and one negative; (*d*) two positive and two negative; (*e*) three negative and one positive.

If the electric field at a given point in space is due to more than one charge, the individual fields (call them $\vec{E}_1$, $\vec{E}_2$, etc.) due to each charge are added vectorially to get the total field at that point:

$$\vec{E} = \vec{E}_1 + \vec{E}_2 + \cdots.$$

The validity of this **superposition principle** for electric fields is fully confirmed by experiment.

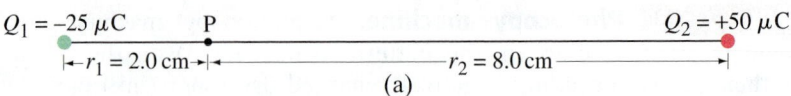

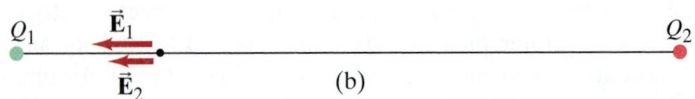

EXAMPLE 21–7 **E at a point between two charges.** Two point charges are separated by a distance of 10.0 cm. One has a charge of $-25\,\mu\text{C}$ and the other $+50\,\mu\text{C}$. (*a*) Determine the direction and magnitude of the electric field at a point P between the two charges that is 2.0 cm from the negative charge (Fig. 21–26a). (*b*) If an electron $(\text{mass} = 9.11 \times 10^{-31}\,\text{kg})$ is placed at rest at P and then released, what will be its initial acceleration (direction and magnitude)?

APPROACH The electric field at P will be the vector sum of the fields created separately by Q_1 and Q_2. The field due to the negative charge Q_1 points toward Q_1, and the field due to the positive charge Q_2 points away from Q_2. Thus both fields point to the left as shown in Fig. 21–26b and we can add the magnitudes of the two fields together algebraically, ignoring the signs of the charges. In (*b*) we use Newton's second law $(F = ma)$ to determine the acceleration, where $F = qE$ (Eq. 21–5).

SOLUTION (*a*) Each field is due to a point charge as given by Eq. 21–4, $E = kQ/r^2$. The total field is

$$E = k\frac{Q_1}{r_1^2} + k\frac{Q_2}{r_2^2} = k\left(\frac{Q_1}{r_1^2} + \frac{Q_2}{r_2^2}\right)$$

$$= \left(9.0 \times 10^9\,\text{N·m}^2/\text{C}^2\right)\left(\frac{25 \times 10^{-6}\,\text{C}}{\left(2.0 \times 10^{-2}\,\text{m}\right)^2} + \frac{50 \times 10^{-6}\,\text{C}}{\left(8.0 \times 10^{-2}\,\text{m}\right)^2}\right)$$

$$= 6.3 \times 10^8\,\text{N/C}.$$

(*b*) The electric field points to the left, so the electron will feel a force to the *right* since it is negatively charged. Therefore the acceleration $a = F/m$ (Newton's second law) will be to the right. The force on a charge q in an electric field E is $F = qE$ (Eq. 21–5). Hence the magnitude of the acceleration is

$$a = \frac{F}{m} = \frac{qE}{m} = \frac{\left(1.60 \times 10^{-19}\,\text{C}\right)\left(6.3 \times 10^8\,\text{N/C}\right)}{9.11 \times 10^{-31}\,\text{kg}} = 1.1 \times 10^{20}\,\text{m/s}^2.$$

NOTE By carefully considering the directions of *each* field ($\vec{E}_1$ and $\vec{E}_2$) before doing any calculations, we made sure our calculation could be done simply and correctly.

EXAMPLE 21–8 **$\vec{E}$ above two point charges.** Calculate the total electric field (*a*) at point A and (*b*) at point B in Fig. 21–27 due to both charges, Q_1 and Q_2.

APPROACH The calculation is much like that of Example 21–3, except now we are dealing with electric fields instead of force. The electric field at point A is the vector sum of the fields $\vec{E}_{A1}$ due to Q_1, and $\vec{E}_{A2}$ due to Q_2. We find the magnitude of the field produced by each point charge, then we add their components to find the total field at point A. We do the same for point B.

SOLUTION (*a*) The magnitude of the electric field produced at point A by each of the charges Q_1 and Q_2 is given by $E = kQ/r^2$, so

$$E_{A1} = \frac{\left(9.0 \times 10^9\,\text{N·m}^2/\text{C}^2\right)\left(50 \times 10^{-6}\,\text{C}\right)}{\left(0.60\,\text{m}\right)^2} = 1.25 \times 10^6\,\text{N/C},$$

$$E_{A2} = \frac{\left(9.0 \times 10^9\,\text{N·m}^2/\text{C}^2\right)\left(50 \times 10^{-6}\,\text{C}\right)}{\left(0.30\,\text{m}\right)^2} = 5.0 \times 10^6\,\text{N/C}.$$

The direction of E_{A1} points from A toward Q_1 (negative charge), whereas E_{A2} points

PROBLEM SOLVING

Ignore signs of charges and determine direction physically, showing directions on diagram

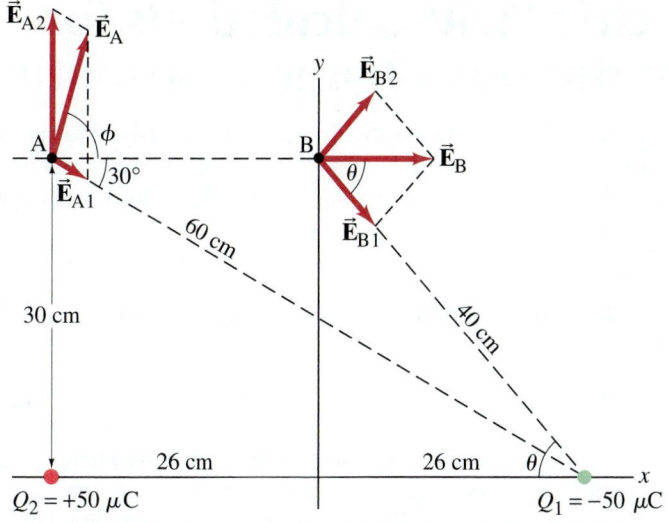

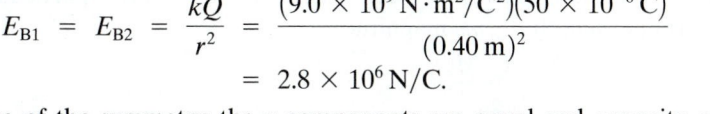

$Q_2 = +50\,\mu C$

$Q_1 = -50\,\mu C$

FIGURE 21–27 Calculation of the electric field at points A and B for Example 21–8.

from A away from Q_2, as shown; so the total electric field at A, $\vec{E}_A$, has components

$$E_{Ax} = E_{A1} \cos 30° = 1.1 \times 10^6\,\text{N/C},$$
$$E_{Ay} = E_{A2} - E_{A1} \sin 30° = 4.4 \times 10^6\,\text{N/C}.$$

Thus the magnitude of $\vec{E}_A$ is

$$E_A = \sqrt{(1.1)^2 + (4.4)^2} \times 10^6\,\text{N/C} = 4.5 \times 10^6\,\text{N/C},$$

and its direction is ϕ given by $\tan \phi = E_{Ay}/E_{Ax} = 4.4/1.1 = 4.0$, so $\phi = 76°$.

(b) Because B is equidistant from the two equal charges (40 cm by the Pythagorean theorem), the magnitudes of E_{B1} and E_{B2} are the same; that is,

$$E_{B1} = E_{B2} = \frac{kQ}{r^2} = \frac{(9.0 \times 10^9\,\text{N}\cdot\text{m}^2/\text{C}^2)(50 \times 10^{-6}\,\text{C})}{(0.40\,\text{m})^2}$$
$$= 2.8 \times 10^6\,\text{N/C}.$$

Also, because of the symmetry, the y components are equal and opposite, and so cancel out. Hence the total field E_B is horizontal and equals $E_{B1} \cos \theta + E_{B2} \cos \theta = 2E_{B1} \cos \theta$. From the diagram, $\cos \theta = 26\,\text{cm}/40\,\text{cm} = 0.65$. Then

$$E_B = 2E_{B1} \cos \theta = 2(2.8 \times 10^6\,\text{N/C})(0.65)$$
$$= 3.6 \times 10^6\,\text{N/C},$$

and the direction of $\vec{E}_B$ is along the $+x$ direction.

NOTE We could have done part (b) in the same way we did part (a). But symmetry allowed us to solve the problem with less effort.

PROBLEM SOLVING

Use symmetry to save work, when possible

Electrostatics: Electric Forces and Electric Fields

Solving electrostatics problems follows, to a large extent, the general problem-solving procedure discussed in Section 4–8. Whether you use electric field or electrostatic forces, the procedure is similar:

1. **Draw** a careful **diagram**—namely, a free-body diagram for each object, showing all the forces acting on that object, or showing the electric field at a point due to all significant charges present. Determine the **direction** of each force or electric field physically: like charges repel each other, unlike charges attract; fields point away from a + charge, and toward

a − charge. Show and label each vector force or field on your diagram.

2. **Apply Coulomb's law** to calculate the magnitude of the force that each contributing charge exerts on a charged object, or the magnitude of the electric field each charge produces at a given point. Deal only with magnitudes of charges (leaving out minus signs), and obtain the magnitude of each force or electric field.

3. **Add vectorially** all the forces on an object, or the contributing fields at a point, to get the resultant. Use **symmetry** (say, in the geometry) whenever possible.

4. **Check** your answer. Is it **reasonable**? If a function of distance, does it give reasonable results in limiting cases?

In many cases we can treat charge as being distributed continuously.[†] We can divide up a charge distribution into infinitesimal charges dQ, each of which will act as a tiny point charge. The contribution to the electric field at a distance r from each dQ is

$$dE = \frac{1}{4\pi\epsilon_0}\frac{dQ}{r^2}.$$ **(21–6a)**

Then the electric field, $\vec{\mathbf{E}}$, at any point is obtained by summing over all the infinitesimal contributions, which is the integral

$$\vec{\mathbf{E}} = \int d\vec{\mathbf{E}}.$$ **(21–6b)**

Note that $d\vec{\mathbf{E}}$ is a vector (Eq. 21–6a gives its magnitude). [In situations where Eq. 21–6b is difficult to evaluate, other techniques (discussed in the next two Chapters) can often be used instead to determine $\vec{\mathbf{E}}$. Numerical integration can also be used in many cases.]

EXAMPLE 21–9 **A ring of charge.** A thin, ring-shaped object of radius a holds a total charge $+Q$ distributed uniformly around it. Determine the electric field at a point P on its axis, a distance x from the center. See Fig. 21–28. Let λ be the charge per unit length (C/m).

APPROACH AND SOLUTION We explicitly follow the steps of the Problem Solving Strategy on page 571.
1. **Draw** a careful **diagram**. The **direction** of the electric field due to one infinitesimal length $d\ell$ of the charged ring is shown in Fig. 21–28.
2. **Apply Coulomb's law**. The electric field, $d\vec{\mathbf{E}}$, due to this particular segment of the ring of length $d\ell$ has magnitude

$$dE = \frac{1}{4\pi\epsilon_0}\frac{dQ}{r^2}.$$

The whole ring has length (circumference) of $2\pi a$, so the charge on a length $d\ell$ is

$$dQ = Q\left(\frac{d\ell}{2\pi a}\right) = \lambda\, d\ell$$

where $\lambda = Q/2\pi a$ is the charge per unit length. Now we write dE as

$$dE = \frac{1}{4\pi\epsilon_0}\frac{\lambda\, d\ell}{r^2}.$$

3. Add **vectorially** and use **symmetry**: The vector $d\vec{\mathbf{E}}$ has components dE_x along the x axis and $dE_\perp$ perpendicular to the x axis (Fig. 21–28). We are going to sum (integrate) around the entire ring. We note that an equal-length segment diametrically opposite the $d\ell$ shown will produce a $d\vec{\mathbf{E}}$ whose component perpendicular to the x axis will just cancel the $dE_\perp$ shown. This is true for all segments of the ring, so by symmetry $\vec{\mathbf{E}}$ will have zero y component, and so we need only sum the x components, dE_x. The total field is then

$$E = E_x = \int dE_x = \int dE\cos\theta = \frac{1}{4\pi\epsilon_0}\lambda\int\frac{d\ell}{r^2}\cos\theta.$$

Since $\cos\theta = x/r$, where $r = (x^2 + a^2)^{\frac{1}{2}}$, we have

$$E = \frac{\lambda}{(4\pi\epsilon_0)}\frac{x}{(x^2+a^2)^{\frac{3}{2}}}\int_0^{2\pi a}d\ell = \frac{1}{4\pi\epsilon_0}\frac{\lambda x(2\pi a)}{(x^2+a^2)^{\frac{3}{2}}} = \frac{1}{4\pi\epsilon_0}\frac{Qx}{(x^2+a^2)^{\frac{3}{2}}}.$$

4. To **check reasonableness**, note that at great distances, $x \gg a$, this result reduces to $E = Q/(4\pi\epsilon_0 x^2)$. We would expect this result because at great distances the ring would appear to be a point charge ($1/r^2$ dependence). Also note that our result gives $E = 0$ at $x = 0$, as we might expect because all components will cancel at the center of the circle.

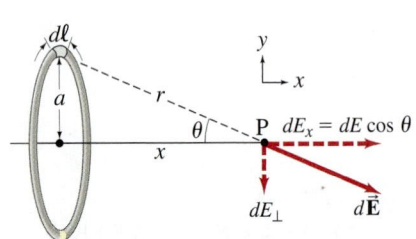

FIGURE 21–28 Example 21–9.

PROBLEM SOLVING

Use symmetry when possible

PROBLEM SOLVING

Check result by noting that at a great distance the ring looks like a point charge

[†]Because we believe there is a minimum charge (e), the treatment here is only for convenience; it is nonetheless useful and accurate since e is usually very much smaller than macroscopic charges.

Note in this Example three important problem-solving techniques that can be used elsewhere: (1) using symmetry to reduce the complexity of the problem; (2) expressing the charge dQ in terms of a charge density (here linear, $\lambda = Q/2\pi a$); and (3) checking the answer at the limit of large r, which serves as an indication (but not proof) of the correctness of the answer—if the result does not check at large r, your result has to be wrong.

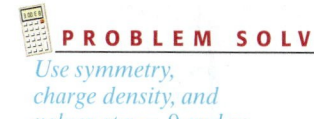
CONCEPTUAL EXAMPLE 21–10 | **Charge at the center of a ring.** Imagine a small positive charge placed at the center of a nonconducting ring carrying a uniformly distributed negative charge. Is the positive charge in equilibrium if it is displaced slightly from the center along the axis of the ring, and if so is it stable? What if the small charge is negative? Neglect gravity, as it is much smaller than the electrostatic forces.

RESPONSE The positive charge is in equilibrium because there is no net force on it, by *symmetry*. If the positive charge moves away from the center of the ring along the axis in either direction, the net force will be back towards the center of the ring and so the charge is in *stable* equilibrium. A negative charge at the center of the ring would feel no net force, but is in *unstable* equilibrium because if it moved along the ring's axis, the net force would be away from the ring and the charge would be pushed farther away.

EXAMPLE 21–11 | **Long line of charge.** Determine the magnitude of the electric field at any point P a distance x from the midpoint 0 of a very long line (a wire, say) of uniformly distributed positive charge, Fig. 21–29. Assume x is much smaller than the length of the wire, and let λ be the charge per unit length (C/m).

APPROACH We set up a coordinate system so the wire is on the y axis with origin 0 as shown. A segment of wire dy has charge $dQ = \lambda \, dy$. The field $d\vec{E}$ at point P due to this length dy of wire (at y) has magnitude

$$dE = \frac{1}{4\pi\epsilon_0}\frac{dQ}{r^2} = \frac{1}{4\pi\epsilon_0}\frac{\lambda \, dy}{(x^2 + y^2)},$$

where $r = (x^2 + y^2)^{\frac{1}{2}}$ as shown in Fig. 21–29. The vector $d\vec{E}$ has components dE_x and dE_y as shown where $dE_x = dE \cos\theta$ and $dE_y = dE \sin\theta$.

SOLUTION Because 0 is at the midpoint of the wire, the y component of $\vec{E}$ will be zero since there will be equal contributions to $E_y = \int dE_y$ from above and below point 0:

$$E_y = \int dE \sin\theta = 0.$$

Thus we have

$$E = E_x = \int dE \cos\theta = \frac{\lambda}{4\pi\epsilon_0}\int \frac{\cos\theta \, dy}{x^2 + y^2}.$$

The integration here is over y, along the wire, with x treated as constant. We must now write θ as a function of y, or y as a function of θ. We do the latter: since $y = x \tan\theta$, then $dy = x \, d\theta/\cos^2\theta$. Furthermore, because $\cos\theta = x/\sqrt{x^2 + y^2}$, then $1/(x^2 + y^2) = \cos^2\theta/x^2$ and our integrand above is $(\cos\theta)(x \, d\theta/\cos^2\theta)(\cos^2\theta/x^2) = \cos\theta \, d\theta/x$. Hence

$$E = \frac{\lambda}{4\pi\epsilon_0}\frac{1}{x}\int_{-\pi/2}^{\pi/2} \cos\theta \, d\theta = \frac{\lambda}{4\pi\epsilon_0 x}(\sin\theta)\Big|_{-\pi/2}^{\pi/2} = \frac{1}{2\pi\epsilon_0}\frac{\lambda}{x},$$

where we have assumed the wire is extremely long in both directions ($y \to \pm\infty$) which corresponds to the limits $\theta = \pm\pi/2$. Thus the field near a long straight wire of uniform charge decreases inversely as the first power of the distance from the wire.

NOTE This result, obtained for an infinite wire, is a good approximation for a wire of finite length as long as x is small compared to the distance of P from the ends of the wire.

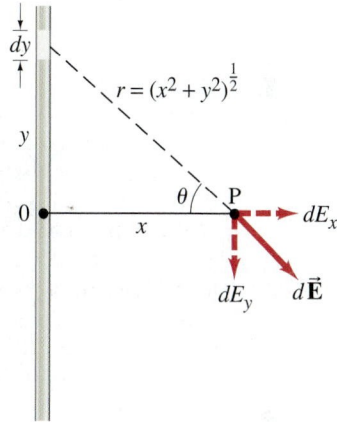

FIGURE 21–29 Example 21–11.

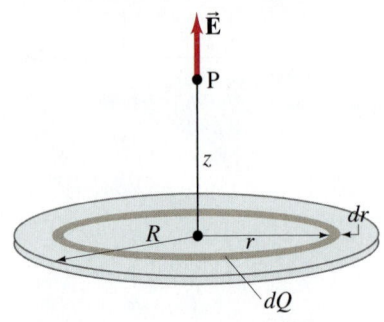

FIGURE 21–30 Example 21–12; a uniformly charged flat disk of radius R.

EXAMPLE 21–12 **Uniformly charged disk.** Charge is distributed uniformly over a thin circular disk of radius R. The charge per unit area (C/m^2) is σ. Calculate the electric field at a point P on the axis of the disk, a distance z above its center, Fig. 21–30.

APPROACH We can think of the disk as a set of concentric rings. We can then apply the result of Example 21–9 to each of these rings, and then sum over all the rings.

SOLUTION For the ring of radius r shown in Fig. 21–30, the electric field has magnitude (see result of Example 21–9)

$$ dE = \frac{1}{4\pi\epsilon_0} \frac{z\,dQ}{(z^2 + r^2)^{\frac{3}{2}}} $$

where we have written dE (instead of E) for this thin ring of total charge dQ. The ring has area $(dr)(2\pi r)$ and charge per unit area $\sigma = dQ/(2\pi r\,dr)$. We solve this for dQ $(= \sigma\,2\pi r\,dr)$ and insert it in the equation above for dE:

$$ dE = \frac{1}{4\pi\epsilon_0} \frac{z\sigma 2\pi r\,dr}{(z^2 + r^2)^{\frac{3}{2}}} = \frac{z\sigma r\,dr}{2\epsilon_0(z^2 + r^2)^{\frac{3}{2}}}. $$

Now we sum over all the rings, starting at $r = 0$ out to the largest with $r = R$:

$$ E = \frac{z\sigma}{2\epsilon_0} \int_0^R \frac{r\,dr}{(z^2 + r^2)^{\frac{3}{2}}} = \frac{z\sigma}{2\epsilon_0} \left[-\frac{1}{(z^2 + r^2)^{\frac{1}{2}}} \right]_0^R $$

$$ = \frac{\sigma}{2\epsilon_0} \left[1 - \frac{z}{(z^2 + R^2)^{\frac{1}{2}}} \right]. $$

This gives the magnitude of $\vec{E}$ at any point z along the axis of the disk. The direction of each $d\vec{E}$ due to each ring is along the z axis (as in Example 21–9), and therefore the direction of $\vec{E}$ is along z. If Q (and σ) are positive, $\vec{E}$ points away from the disk; if Q (and σ) are negative, $\vec{E}$ points toward the disk.

If the radius of the disk in Example 21–12 is much greater than the distance of our point P from the disk (i.e., $z \ll R$) then we can obtain a very useful result: the second term in the solution above becomes very small, so

$$ E = \frac{\sigma}{2\epsilon_0}. \qquad\qquad \text{[infinite plane]} \quad \textbf{(21–7)} $$

This result is valid for any point above (or below) an infinite plane of any shape holding a uniform charge density σ. It is also valid for points close to a finite plane, as long as the point is close to the plane compared to the distance to the edges of the plane. Thus the field near a large uniformly charged plane is uniform, and directed outward if the plane is positively charged.

It is interesting to compare here the distance dependence of the electric field due to a point charge $(E \sim 1/r^2)$, due to a very long uniform line of charge $(E \sim 1/r)$, and due to a very large uniform plane of charge $(E$ does not depend on $r)$.

EXAMPLE 21–13 **Two parallel plates.** Determine the electric field between two large parallel plates or sheets, which are very thin and are separated by a distance d which is small compared to their height and width. One plate carries a uniform surface charge density σ and the other carries a uniform surface charge density $-\sigma$, as shown in Fig. 21–31 (the plates extend upward and downward beyond the part shown).

APPROACH From Eq. 21–7, each plate sets up an electric field of magnitude $\sigma/2\epsilon_0$. The field due to the positive plate points away from that plate whereas the field due to the negative plate points toward that plate.

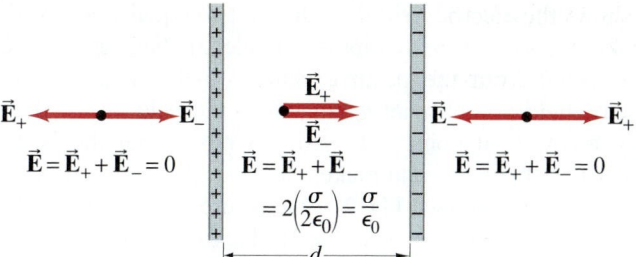

FIGURE 21–31 Example 21–13. (Only the center portion of these large plates is shown: their dimensions are large compared to their separation d.)

SOLUTION In the region between the plates, the fields add together as shown:

$$E = E_+ + E_- = \frac{\sigma}{2\epsilon_0} + \frac{\sigma}{2\epsilon_0} = \frac{\sigma}{\epsilon_0}.$$

The field is uniform, since the plates are very large compared to their separation, so this result is valid for any point, whether near one or the other of the plates, or midway between them as long as the point is far from the ends. Outside the plates, the fields cancel,

$$E = E_+ + E_- = \frac{\sigma}{2\epsilon_0} - \frac{\sigma}{2\epsilon_0} = 0,$$

as shown in Fig. 21–31. These results are valid ideally for infinitely large plates; they are a good approximation for finite plates if the separation is much less than the dimensions of the plate and for points not too close to the edge.

NOTE: These useful and extraordinary results illustrate the principle of superposition and its great power.

21–8 Field Lines

Since the electric field is a vector, it is sometimes referred to as a **vector field**. We could indicate the electric field with arrows at various points in a given situation, such as at A, B, and C in Fig. 21–32. The directions of $\vec{E}_A$, $\vec{E}_B$, and $\vec{E}_C$ are the same as that of the forces shown earlier in Fig. 21–22, but the magnitudes (arrow lengths) are different since we divide $\vec{F}$ in Fig. 21–22 by q to get $\vec{E}$. However, the relative lengths of $\vec{E}_A$, $\vec{E}_B$, and $\vec{E}_C$ are the same as for the forces since we divide by the same q each time. To indicate the electric field in such a way at *many* points, however, would result in many arrows, which might appear complicated or confusing. To avoid this, we use another technique, that of field lines.

To visualize the electric field, we draw a series of lines to indicate the direction of the electric field at various points in space. These **electric field lines** (sometimes called **lines of force**) are drawn so that they indicate the direction of the force due to the given field on a positive test charge. The lines of force due to a single isolated positive charge are shown in Fig. 21–33a, and for a single isolated negative charge in Fig. 21–33b. In part (a) the lines point radially outward from the charge, and in part (b) they point radially inward toward the charge because that is the direction the force would be on a positive test charge in each case (as in Fig. 21–25). Only a few representative lines are shown. We could just as well draw lines in between those shown since the electric field exists there as well. We can draw the lines so that the *number of lines starting on a positive charge, or ending on a negative charge, is proportional to the magnitude of the charge.* Notice that nearer the charge, where the electric field is greater $\left(F \propto 1/r^2\right)$, the lines are closer together. This is a general property of electric field lines: *the closer together the lines are, the stronger the electric field in that region.* In fact, field lines can be drawn so that the number of lines crossing unit area perpendicular to $\vec{E}$ is proportional to the magnitude of the electric field.

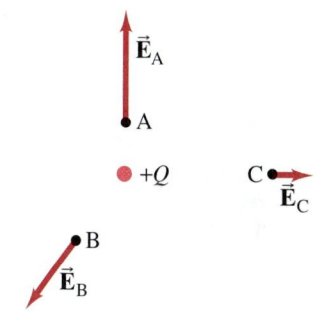

FIGURE 21–32 Electric field vector shown at three points, due to a single point charge Q. (Compare to Fig. 21–22.)

FIGURE 21–33 Electric field lines (a) near a single positive point charge, (b) near a single negative point charge.

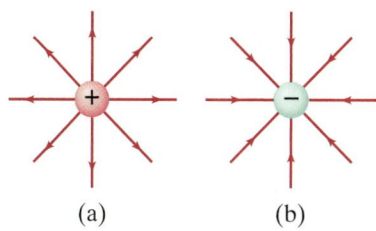

(a)   (b)

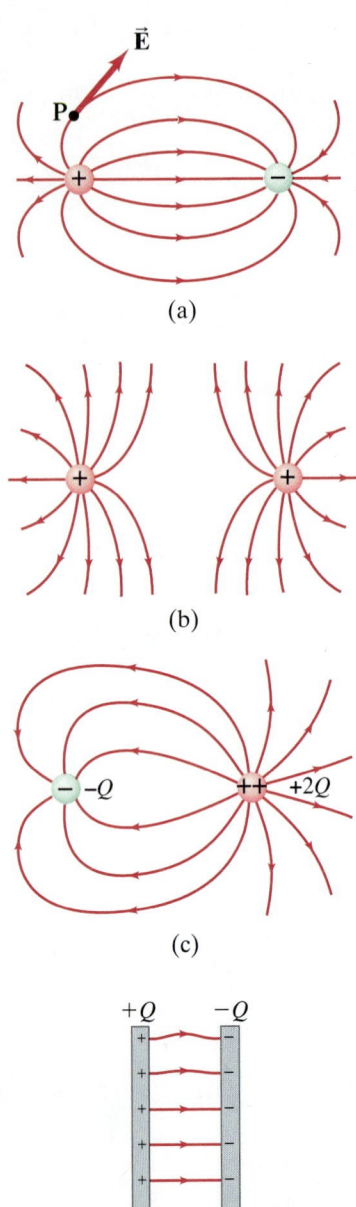

(a)

(b)

$-Q$ ++ $+2Q$

(c)

$+Q$ $-Q$

(d)

FIGURE 21–34 Electric field lines for four arrangements of charges.

Figure 21–34a shows the electric field lines due to two equal charges of opposite sign, a combination known as an **electric dipole**. The electric field lines are curved in this case and are directed from the positive charge to the negative charge. The direction of the electric field at any point is tangent to the field line at that point as shown by the vector arrow $\vec{E}$ at point P. To satisfy yourself that this is the correct pattern for the electric field lines, you can make a few calculations such as those done in Example 21–8 for just this case (see Fig. 21–27). Figure 21–34b shows the electric field lines for two equal positive charges, and Fig. 21–34c for unequal charges, $-Q$ and $+2Q$. Note that twice as many lines leave $+2Q$, as enter $-Q$ (number of lines is proportional to magnitude of Q). Finally, in Fig. 21–34d, we see the field lines between two parallel plates carrying equal but opposite charges. Notice that the electric field lines between the two plates start out perpendicular to the surface of the metal plates (we will see why this is true in the next Section) and go directly from one plate to the other, as we expect because a positive test charge placed between the plates would feel a strong repulsion from the positive plate and a strong attraction to the negative plate. The field lines between two close plates are parallel and equally spaced in the central region, but fringe outward near the edges. Thus, in the central region, the electric field has the same magnitude at all points, and we can write (see Example 21–13)

$$E = \text{constant} = \frac{\sigma}{\epsilon_0}. \quad \left[\begin{array}{l}\text{between two closely spaced,}\\ \text{oppositely charged, parallel plates}\end{array}\right] \quad \textbf{(21–8)}$$

The fringing of the field near the edges can often be ignored, particularly if the separation of the plates is small compared to their height and width.

We summarize the properties of field lines as follows:

1. Electric field lines indicate the direction of the electric field; the field points in the direction tangent to the field line at any point.
2. The lines are drawn so that the magnitude of the electric field, E, is proportional to the number of lines crossing unit area perpendicular to the lines. The closer together the lines, the stronger the field.
3. Electric field lines start on positive charges and end on negative charges; and the number starting or ending is proportional to the magnitude of the charge.

Also note that field lines never cross. Why not? Because the electric field can not have two directions at the same point, nor exert more than one force on a test charge.

Gravitational Field

The field concept can also be applied to the gravitational force as mentioned in Chapter 6. Thus we can say that a **gravitational field** exists for every object that has mass. One object attracts another by means of the gravitational field. The Earth, for example, can be said to possess a gravitational field (Fig. 21–35) which is responsible for the gravitational force on objects. The *gravitational field* is defined as the *force per unit mass*. The magnitude of the Earth's gravitational field at any point above the Earth's surface is thus (GM_E/r^2), where M_E is the mass of the Earth, r is the distance of the point from the Earth's center, and G is the gravitational constant (Chapter 6). At the Earth's surface, r is the radius of the Earth and the gravitational field is equal to g, the acceleration due to gravity. Beyond the Earth, the gravitational field can be calculated at any point as a sum of terms due to Earth, Sun, Moon, and other bodies that contribute significantly.

FIGURE 21–35 The Earth's gravitational field.

21–9 Electric Fields and Conductors

We now discuss some properties of conductors. First, *the electric field inside a conductor is zero in the static situation*—that is, when the charges are at rest. If there were an electric field within a conductor, there would be a force on the free electrons. The electrons would move until they reached positions where the electric field, and therefore the electric force on them, did become zero.

This reasoning has some interesting consequences. For one, *any net charge on a conductor distributes itself on the surface*. (If there *were* charges inside, there would be an electric field.) For a negatively charged conductor, you can imagine that the negative charges repel one another and race to the surface to get as far from one another as possible. Another consequence is the following. Suppose that a positive charge Q is surrounded by an isolated uncharged metal conductor whose shape is a spherical shell, Fig. 21–36. Because there can be no field within the metal, the lines leaving the central positive charge must end on negative charges on the inner surface of the metal. Thus an equal amount of negative charge, $-Q$, is induced on the inner surface of the spherical shell. Then, since the shell is neutral, a positive charge of the same magnitude, $+Q$, must exist on the outer surface of the shell. Thus, although no field exists in the metal itself, an electric field exists outside of it, as shown in Fig. 21–36, as if the metal were not even there.

A related property of static electric fields and conductors is that *the electric field is always perpendicular to the surface outside of a conductor*. If there were a component of $\vec{E}$ parallel to the surface (Fig. 21–37), it would exert a force on free electrons at the surface, causing the electrons to move along the surface until they reached positions where no net force was exerted on them parallel to the surface—that is, until the electric field was perpendicular to the surface.

These properties apply only to conductors. Inside a nonconductor, which does not have free electrons, a static electric field can exist as we will see in the next Chapter. Also, the electric field outside a nonconductor does not necessarily make an angle of 90° to the surface.

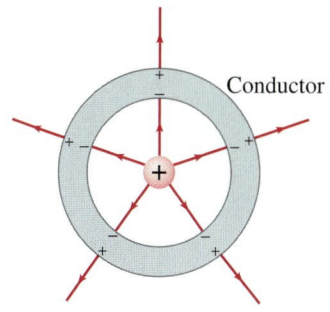

FIGURE 21–36 A charge inside a neutral spherical metal shell induces charge on its surfaces. The electric field exists even beyond the shell, but not within the conductor itself.

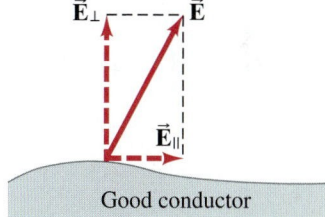

FIGURE 21–37 If the electric field $\vec{E}$ at the surface of a conductor had a component parallel to the surface, $\vec{E}_\parallel$, the latter would accelerate electrons into motion. In the static case, $\vec{E}_\parallel$ must be zero, and the electric field must be perpendicular to the conductor's surface: $\vec{E} = \vec{E}_\perp$.

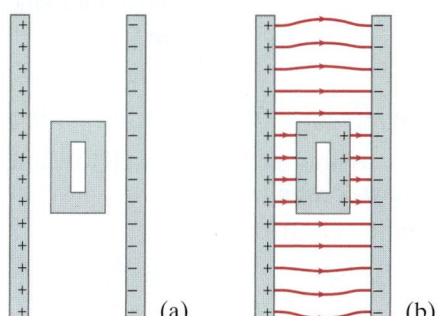

FIGURE 21–38 Example 21–14.

FIGURE 21–39 A strong electric field exists in the vicinity of this "Faraday cage," so strong that stray electrons in the atmosphere are accelerated to the kinetic energy needed to knock electrons out of air atoms, causing an avalanche of charge which flows to (or from) the metal cage. Yet the person inside the cage is not affected.

CONCEPTUAL EXAMPLE 21–14 **Shielding, and safety in a storm.** A neutral hollow metal box is placed between two parallel charged plates as shown in Fig. 21–38a. What is the field like inside the box?

RESPONSE If our metal box had been solid, and not hollow, free electrons in the box would have redistributed themselves along the surface until all their individual fields would have canceled each other inside the box. The net field inside the box would have been zero. For a hollow box, the external field is not changed since the electrons in the metal can move just as freely as before to the surface. Hence the field inside the hollow metal box is also zero, and the field lines are shown in Fig. 21–38b. A conducting box used in this way is an effective device for shielding delicate instruments and electronic circuits from unwanted external electric fields. We also can see that a relatively safe place to be during a lightning storm is inside a parked car, surrounded by metal. See also Fig. 21–39, where a person inside a porous "cage" is protected from a strong electric discharge.

Ⓟ **PHYSICS APPLIED**
Electrical shielding

21–10 Motion of a Charged Particle in an Electric Field

If an object having an electric charge q is at a point in space where the electric field is $\vec{E}$, the force on the object is given by

$$\vec{F} = q\vec{E}$$

(see Eq. 21–5). In the past few Sections we have seen how to determine $\vec{E}$ for some particular situations. Now let us suppose we know $\vec{E}$ and we want to find the force on a charged object and the object's subsequent motion. (We assume no other forces act.)

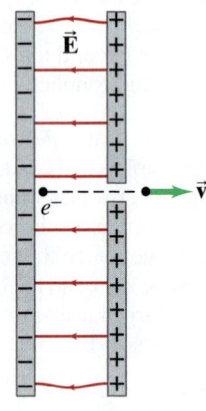

FIGURE 21–40 Example 21–15.

EXAMPLE 21–15 **Electron accelerated by electric field.** An electron (mass $m = 9.1 \times 10^{-31}$ kg) is accelerated in the uniform field $\vec{E}$ ($E = 2.0 \times 10^4$ N/C) between two parallel charged plates. The separation of the plates is 1.5 cm. The electron is accelerated from rest near the negative plate and passes through a tiny hole in the positive plate, Fig. 21–40. (a) With what speed does it leave the hole? (b) Show that the gravitational force can be ignored. Assume the hole is so small that it does not affect the uniform field between the plates.

APPROACH We can obtain the electron's velocity using the kinematic equations of Chapter 2, after first finding its acceleration from Newton's second law, $F = ma$. The magnitude of the force on the electron is $F = qE$ and is directed to the right.

SOLUTION (a) The magnitude of the electron's acceleration is

$$a = \frac{F}{m} = \frac{qE}{m}.$$

Between the plates $\vec{E}$ is uniform so the electron undergoes uniformly accelerated motion with acceleration

$$a = \frac{(1.6 \times 10^{-19}\,\text{C})(2.0 \times 10^4\,\text{N/C})}{(9.1 \times 10^{-31}\,\text{kg})} = 3.5 \times 10^{15}\,\text{m/s}^2.$$

It travels a distance $x = 1.5 \times 10^{-2}$ m before reaching the hole, and since its initial speed was zero, we can use the kinematic equation, $v^2 = v_0^2 + 2ax$ (Eq. 2–12c), with $v_0 = 0$:

$$v = \sqrt{2ax} = \sqrt{2(3.5 \times 10^{15}\,\text{m/s}^2)(1.5 \times 10^{-2}\,\text{m})} = 1.0 \times 10^7\,\text{m/s}.$$

There is no electric field outside the plates, so after passing through the hole, the electron moves with this speed, which is now constant.

(b) The magnitude of the electric force on the electron is

$$qE = (1.6 \times 10^{-19}\,\text{C})(2.0 \times 10^4\,\text{N/C}) = 3.2 \times 10^{-15}\,\text{N}.$$

The gravitational force is

$$mg = (9.1 \times 10^{-31}\,\text{kg})(9.8\,\text{m/s}^2) = 8.9 \times 10^{-30}\,\text{N},$$

which is 10^{14} times smaller! Note that the electric field due to the electron does not enter the problem (since a particle cannot exert a force on itself).

FIGURE 21–41 Example 21–16.

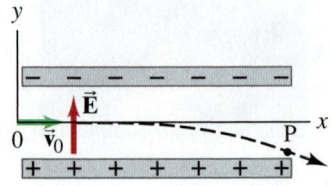

EXAMPLE 21–16 **Electron moving perpendicular to $\vec{E}$.** Suppose an electron traveling with speed v_0 enters a uniform electric field $\vec{E}$, which is at right angles to $\vec{v}_0$ as shown in Fig. 21–41. Describe its motion by giving the equation of its path while in the electric field. Ignore gravity.

APPROACH Again we use Newton's second law, with $F = qE$, and the kinematic equations from Chapter 2.

SOLUTION When the electron enters the electric field (at $x = y = 0$) it has velocity $\vec{v}_0 = v_0\hat{\mathbf{i}}$ in the x direction. The electric field $\vec{E}$, pointing vertically upward, imparts a uniform vertical acceleration to the electron of

$$a_y = \frac{F}{m} = \frac{qE}{m} = -\frac{eE}{m},$$

where we set $q = -e$ for the electron.

The electron's vertical position is given by Eq. 2–12b,

$$y = \frac{1}{2}a_y t^2 = -\frac{eE}{2m}t^2$$

since the motion is at constant acceleration. The horizontal position is given by

$$x = v_0 t$$

since $a_x = 0$. We eliminate t between these two equations and obtain

$$y = -\frac{eE}{2mv_0^2}x^2,$$

which is the equation of a parabola (just as in projectile motion, Section 3–7).

21–11 Electric Dipoles

The combination of two equal charges of opposite sign, $+Q$ and $-Q$, separated by a distance ℓ, is referred to as an **electric dipole**. The quantity $Q\ell$ is called the **dipole moment** and is represented[†] by the symbol p. The dipole moment can be considered to be a vector $\vec{\mathbf{p}}$, of magnitude $Q\ell$, that points from the negative to the positive charge as shown in Fig. 21–42. Many molecules, such as the diatomic molecule CO, have a dipole moment (C has a small positive charge and O a small negative charge of equal magnitude), and are referred to as **polar molecules**. Even though the molecule as a whole is neutral, there is a separation of charge that results from an uneven sharing of electrons by the two atoms.[‡] (Symmetric diatomic molecules, like O_2, have no dipole moment.) The water molecule, with its uneven sharing of electrons (O is negative, the two H are positive), also has a dipole moment—see Fig. 21–43.

Dipole in an External Field

First let us consider a dipole, of dipole moment $p = Q\ell$, that is placed in a uniform electric field $\vec{\mathbf{E}}$, as shown in Fig. 21–44. If the field is uniform, the force $Q\vec{\mathbf{E}}$ on the positive charge and the force $-Q\vec{\mathbf{E}}$ on the negative charge result in no net force on the dipole. There will, however, be a *torque* on the dipole (Fig. 21–44) which has magnitude (calculated about the center, 0, of the dipole)

$$\tau = QE\frac{\ell}{2}\sin\theta + QE\frac{\ell}{2}\sin\theta = pE\sin\theta. \qquad \textbf{(21–9a)}$$

This can be written in vector notation as

$$\vec{\boldsymbol{\tau}} = \vec{\mathbf{p}} \times \vec{\mathbf{E}}. \qquad \textbf{(21–9b)}$$

The effect of the torque is to try to turn the dipole so $\vec{\mathbf{p}}$ is parallel to $\vec{\mathbf{E}}$. The work done on the dipole by the electric field to change the angle θ from θ_1 to θ_2 is (see Eq. 10–22)

$$W = \int_{\theta_1}^{\theta_2} \tau \, d\theta.$$

We need to write the torque as $\tau = -pE\sin\theta$ because its direction is opposite to the direction of increasing θ (right-hand rule). Then

$$W = \int_{\theta_1}^{\theta_2} \tau \, d\theta = -pE\int_{\theta_1}^{\theta_2}\sin\theta \, d\theta = pE\cos\theta\Big|_{\theta_1}^{\theta_2} = pE(\cos\theta_2 - \cos\theta_1).$$

Positive work done by the field decreases the potential energy, U, of the dipole in this field. (Recall the relation between work and potential energy, Eq. 8–4, $\Delta U = -W$.) If we choose $U = 0$ when $\vec{\mathbf{p}}$ is perpendicular to $\vec{\mathbf{E}}$ (that is, choosing $\theta_1 = 90°$ so $\cos\theta_1 = 0$), and setting $\theta_2 = \theta$, then

$$U = -W = -pE\cos\theta = -\vec{\mathbf{p}} \cdot \vec{\mathbf{E}}. \qquad \textbf{(21–10)}$$

If the electric field is *not* uniform, the force on the $+Q$ of the dipole may not have the same magnitude as on the $-Q$, so there may be a net force as well as a torque.

[†]Be careful not to confuse this p for dipole moment with p for momentum.

[‡]The value of the separated charges may be a fraction of e (say $\pm 0.2e$ or $\pm 0.4e$) but note that such charges do not violate what we said about e being the smallest charge. These charges less than e cannot be isolated and merely represent how much time electrons spend around one atom or the other.

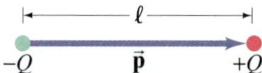

FIGURE 21–42 A dipole consists of equal but opposite charges, $+Q$ and $-Q$, separated by a distance ℓ. The dipole moment is $\vec{\mathbf{p}} = Q\vec{\boldsymbol{\ell}}$ and points from the negative to the positive charge.

FIGURE 21–43 In the water molecule (H_2O), the electrons spend more time around the oxygen atom than around the two hydrogen atoms. The net dipole moment $\vec{\mathbf{p}}$ can be considered as the vector sum of two dipole moments $\vec{\mathbf{p}}_1$ and $\vec{\mathbf{p}}_2$ that point from the O toward each H as shown: $\vec{\mathbf{p}} = \vec{\mathbf{p}}_1 + \vec{\mathbf{p}}_2$.

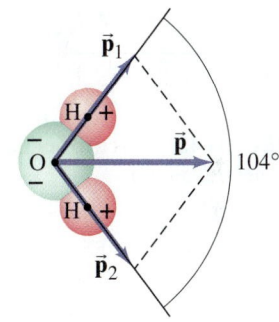

FIGURE 21–44 (below) An electric dipole in a uniform electric field.

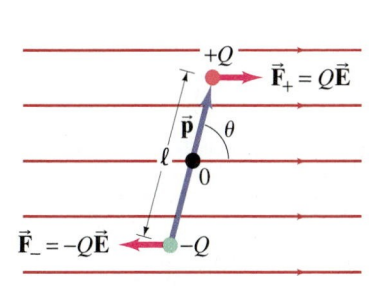

EXAMPLE 21–17 **Dipole in a field.** The dipole moment of a water molecule is $6.1 \times 10^{-30}\,\text{C·m}$. A water molecule is placed in a uniform electric field with magnitude $2.0 \times 10^5\,\text{N/C}$. (*a*) What is the magnitude of the maximum torque that the field can exert on the molecule? (*b*) What is the potential energy when the torque is at its maximum? (*c*) In what position will the potential energy take on its greatest value? Why is this different than the position where the torque is maximum?

APPROACH The torque is given by Eq. 21–9 and the potential energy by Eq. 21–10.

SOLUTION (*a*) From Eq. 21–9 we see that τ is maximized when θ is 90°. Then $\tau = pE = (6.1 \times 10^{-30}\,\text{C·m})(2.0 \times 10^5\,\text{N/C}) = 1.2 \times 10^{-24}\,\text{N·m}.$

(*b*) The potential energy for $\theta = 90°$ is zero (Eq. 21–10). Note that the potential energy is negative for smaller values of θ, so U is not a minimum for $\theta = 90°$.

(*c*) The potential energy U will be a maximum when $\cos\theta = -1$ in Eq. 21–10, so $\theta = 180°$, meaning $\vec{E}$ and $\vec{p}$ are antiparallel. The potential energy is maximized when the dipole is oriented so that it has to rotate through the largest angle, 180°, to reach the equilibrium position at $\theta = 0°$. The torque on the other hand is maximized when the electric forces are perpendicular to $\vec{p}$.

Electric Field Produced by a Dipole

We have just seen how an external electric field affects an electric dipole. Now let us suppose that there is no external field, and we want to determine the electric field produced *by* the dipole. For brevity, we restrict ourselves to points that are on the perpendicular bisector of the dipole, such as point P in Fig. 21–45 which is a distance r above the midpoint of the dipole. Note that r in Fig. 21–45 is not the distance from either charge to point P; the latter distance is $(r^2 + \ell^2/4)^{\frac{1}{2}}$ and this is what must be used in Eq. 21–4. The total field at P is

$$\vec{E} = \vec{E}_+ + \vec{E}_-,$$

where $\vec{E}_+$ and $\vec{E}_-$ are the fields due to the $+$ and $-$ charges respectively. The magnitudes E_+ and E_- are equal:

$$E_+ = E_- = \frac{1}{4\pi\epsilon_0}\frac{Q}{r^2 + \ell^2/4}.$$

Their y components cancel at point P (*symmetry* again), so the magnitude of the total field $\vec{E}$ is

$$E = 2E_+\cos\phi = \frac{1}{2\pi\epsilon_0}\left(\frac{Q}{r^2 + \ell^2/4}\right)\frac{\ell}{2(r^2 + \ell^2/4)^{\frac{1}{2}}}$$

or, setting $Q\ell = p$,

$$E = \frac{1}{4\pi\epsilon_0}\frac{p}{(r^2 + \ell^2/4)^{\frac{3}{2}}}. \qquad \left[\begin{matrix}\text{on perpendicular bisector} \\ \text{of dipole}\end{matrix}\right] \quad \textbf{(21–11)}$$

Far from the dipole, $r \gg \ell$, this reduces to

$$E = \frac{1}{4\pi\epsilon_0}\frac{p}{r^3}. \qquad \left[\begin{matrix}\text{on perpendicular bisector} \\ \text{of dipole; } r \gg \ell\end{matrix}\right] \quad \textbf{(21–12)}$$

So the field decreases more rapidly for a dipole than for a single point charge $(1/r^3$ versus $1/r^2)$, which we expect since at large distances the two opposite charges appear so close together as to neutralize each other. This $1/r^3$ dependence also applies for points not on the perpendicular bisector (see Problem 67).

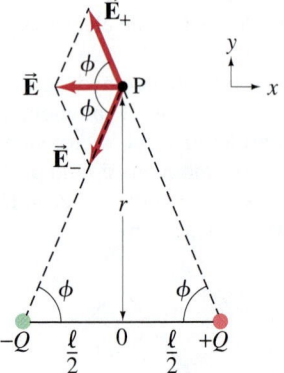

FIGURE 21–45 Electric field due to an electric dipole.

*21–12 Electric Forces in Molecular Biology; DNA

The interior of every biological cell is mainly water. We can imagine a cell as a vast sea of molecules continually in motion (kinetic theory, Chapter 18), colliding with one another with various amounts of kinetic energy. These molecules interact with one another because of *electrostatic attraction* between molecules.

Indeed, cellular processes are now considered to be the result of *random ("thermal") molecular motion plus the ordering effect of the electrostatic force*. As an example, we look at DNA structure and replication. The picture we present has not been seen "in action." Rather, it is a model of what happens based on physical theories and experiment.

The genetic information that is passed on from generation to generation in all living cells is contained in the chromosomes, which are made up of genes. Each gene contains the information needed to produce a particular type of protein molecule, and that information is built into the principal molecule of a chromosome, DNA (deoxyribonucleic acid), Fig. 21–46. DNA molecules are made up of many small molecules known as nucleotide bases which are each polar due to unequal sharing of electrons. There are four types of nucleotide bases in DNA: adenine (A), cytosine (C), guanine (G), and thymine (T).

The DNA of a chromosome generally consists of two long DNA strands wrapped about one another in the shape of a "double helix." The genetic information is contained in the specific order of the four bases (A, C, G, T) along the strand. As shown in Fig. 21–47, the two strands are attracted by electrostatic forces—that is, by the attraction of positive charges to negative charges that exist on parts of the molecules. We see in Fig. 21–47a that an A (adenine) on one strand is always opposite a T on the other strand; similarly, a G is always opposite a C. This important ordering effect occurs because the shapes of A, T, C, and G are such that a T fits closely only into an A, and a G into a C; and only in the case of this close proximity of the charged portions is the electrostatic force great enough to hold them together even for a short time (Fig. 21–47b), forming what are referred to as "weak bonds."

PHYSICS APPLIED

Inside a cell: kinetic theory plus electrostatic force

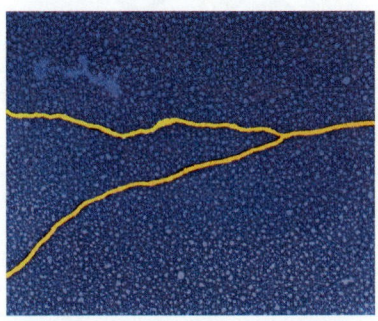

FIGURE 21–46 DNA replicating in a human HeLa cancer cell. This is a false-color image made by a transmission electron microscope (TEM; discussed in Chapter 37).

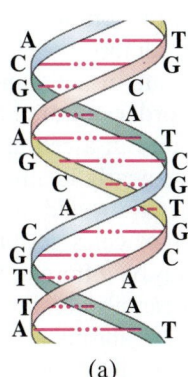

(a)

FIGURE 21–47 (a) Section of a DNA double helix. (b) "Close-up" view of the helix, showing how A and T attract each other and how G and C attract each other through electrostatic forces. The + and − signs represent net charges, usually a fraction of *e*, due to uneven sharing of electrons. The red dots indicate the electrostatic attraction (often called a "weak bond" or "hydrogen bond"—Section 40–3). Note that there are two weak bonds between A and T, and three between C and G.

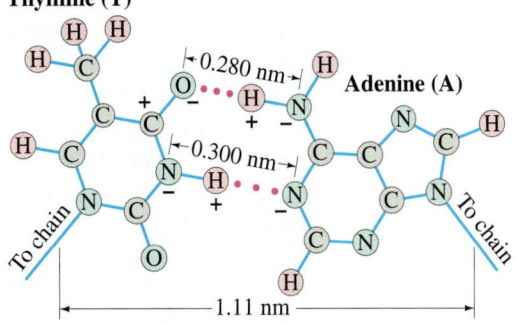

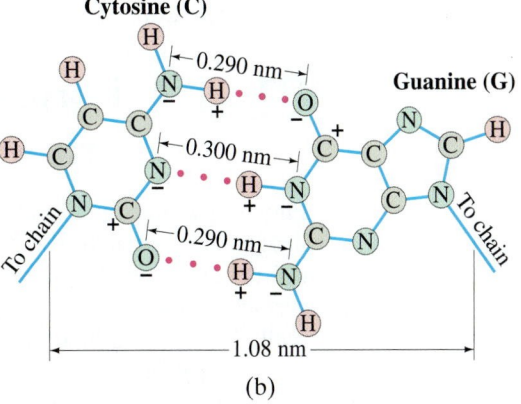

(b)

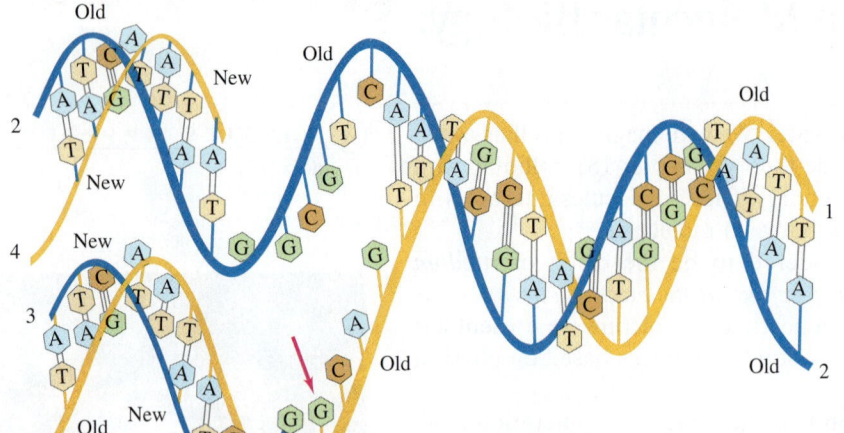

FIGURE 21–48 Replication of DNA.

When the DNA replicates (duplicates) itself just before cell division, the arrangement of A opposite T and G opposite C is crucial for ensuring that the genetic information is passed on accurately to the next generation, Fig. 21–48. The two strands of DNA separate (with the help of enzymes, which also operate via the electrostatic force), leaving the charged parts of the bases exposed. Once replication starts, let us see how the correct order of bases occurs by looking at the G molecule indicated by the red arrow in Fig. 21–48. Many unattached nucleotide bases of all four kinds are bouncing around in the cellular fluid, and the only type that will experience attraction to our G, if it bounces close to it, will be a C. The charges on the other three bases can not get close enough to those on the G to provide a significant attractive force—remember that the force decreases rapidly with distance $(\propto 1/r^2)$. Because the G does not attract an A, T, or G appreciably, an A, T, or G will be knocked away by collisions with other molecules before enzymes can attach it to the growing chain (number 3). But the electrostatic force will often hold a C opposite our G long enough so that an enzyme can attach the C to the growing end of the new chain. Thus we see that electrostatic forces are responsible for selecting the bases in the proper order during replication.

This process of DNA replication is often presented as if it occurred in clockwork fashion—as if each molecule knew its role and went to its assigned place. But this is not the case. The forces of attraction are rather weak, and if the molecular shapes are not just right, there is almost no electrostatic attraction, which is why there are few mistakes. Thus, out of the random motion of the molecules, the electrostatic force acts to bring order out of chaos.

The random (thermal) velocities of molecules in a cell affect *cloning*. When a bacterial cell divides, the two new bacteria have nearly identical DNA. Even if the DNA were perfectly identical, the two new bacteria would not end up behaving in the same way. Long protein, DNA, and RNA molecules get bumped into different shapes, and even the expression of genes can thus be different. Loosely held parts of large molecules such as a methyl group (CH_3) can also be knocked off by a strong collision with another molecule. Hence, cloned organisms are not identical, even if their DNA were identical. Indeed, there can not really be genetic determinism.

*21–13 Photocopy Machines and Computer Printers Use Electrostatics

Photocopy machines and laser printers use electrostatic attraction to print an image. They each use a different technique to project an image onto a special cylindrical drum. The drum is typically made of aluminum, a good conductor; its surface is coated with a thin layer of selenium, which has the interesting property (called "photoconductivity") of being an electrical nonconductor in the dark, but a conductor when exposed to light.

In a *photocopier*, lenses and mirrors focus an image of the original sheet of paper onto the drum, much like a camera lens focuses an image on film. Step 1 is

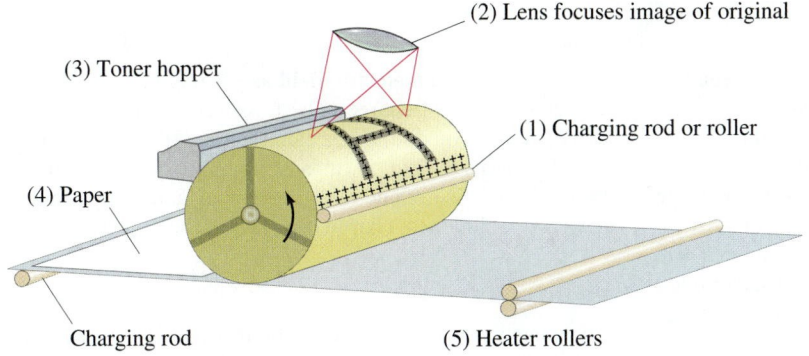

(2) Lens focuses image of original

(3) Toner hopper

(1) Charging rod or roller

(4) Paper

Charging rod

(5) Heater rollers

FIGURE 21–49 Inside a photocopy machine: (1) the selenium drum is given a + charge; (2) the lens focuses image on drum—only dark spots stay charged; (3) toner particles (negatively charged) are attracted to positive areas on drum; (4) the image is transferred to paper; (5) heat binds the image to the paper.

PHYSICS APPLIED
Photocopy machines

the placing of a uniform positive charge on the drum's selenium layer by a charged rod or roller, done in the dark. In step 2, the image to be copied or printed is projected onto the drum. For simplicity, let us assume the image is a dark letter A on a white background (as on the page of a book) as shown in Fig. 21–49. The letter A on the drum is dark, but all around it is light. At all these light places, the selenium becomes conducting and electrons flow in from the aluminum beneath, neutralizing those positive areas. In the dark areas of the letter A, the selenium is nonconducting and so retains a positive charge, Fig. 21–49. In step 3, a fine dark powder known as *toner* is given a negative charge, and brushed on the drum as it rotates. The negatively charged toner particles are attracted to the positive areas on the drum (the A in our case) and stick only there. In step 4, the rotating drum presses against a piece of paper which has been positively charged more strongly than the selenium, so the toner particles are transferred to the paper, forming the final image. Finally, step 5, the paper is heated to fix the toner particles firmly on the paper.

In a color copier (or printer), this process is repeated for each color—black, cyan (blue), magenta (red), and yellow. Combining these four colors in different proportions produces any desired color.

A *laser printer*, on the other hand, uses a computer output to program the intensity of a laser beam onto the selenium-coated drum. The thin beam of light from the laser is scanned (by a movable mirror) from side to side across the drum in a series of horizontal lines, each line just below the previous line. As the beam sweeps across the drum, the intensity of the beam is varied by the computer output, being strong for a point that is meant to be white or bright, and weak or zero for points that are meant to come out dark. After each sweep, the drum rotates very slightly for additional sweeps, Fig. 21–50, until a complete image is formed on it. The light parts of the selenium become conducting and lose their electric charge, and the toner sticks only to the dark, electrically charged areas. The drum then transfers the image to paper, as in a photocopier.

An *inkjet printer* does not use a drum. Instead nozzles spray tiny droplets of ink directly at the paper. The nozzles are swept across the paper, each sweep just above the previous one as the paper moves down. On each sweep, the ink makes dots on the paper, except for those points where no ink is desired, as directed by the computer. The image consists of a huge number of very tiny dots. The quality or resolution of a printer is usually specified in dots per inch (dpi) in each (linear) direction.

PHYSICS APPLIED
Laser printer

PHYSICS APPLIED
Inkjet printer

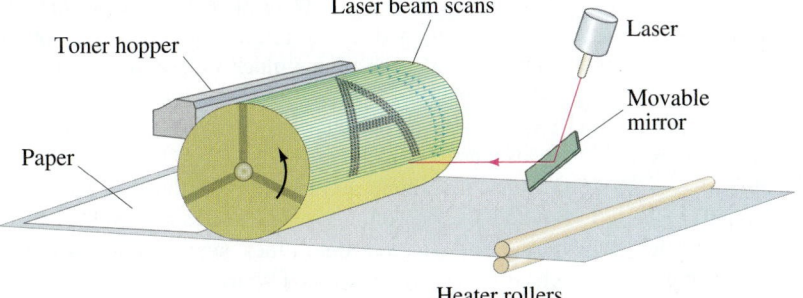

Toner hopper

Laser beam scans

Laser

Movable mirror

Paper

Heater rollers

FIGURE 21–50 Inside a laser printer: A movable mirror sweeps the laser beam in horizontal lines across the drum.

*SECTION 21–13 Photocopy Machines and Computer Printers Use Electrostatics **583**

Summary

There are two kinds of **electric charge**, positive and negative. These designations are to be taken algebraically—that is, any charge is plus or minus so many coulombs (C), in SI units.

Electric charge is **conserved**: if a certain amount of one type of charge is produced in a process, an equal amount of the opposite type is also produced; thus the *net* charge produced is zero.

According to atomic theory, electricity originates in the atom, each consisting of a positively charged nucleus surrounded by negatively charged electrons. Each electron has a charge $-e = -1.6 \times 10^{-19}\,\text{C}$.

Electric **conductors** are those materials in which many electrons are relatively free to move, whereas electric **insulators** are those in which very few electrons are free to move.

An object is negatively charged when it has an excess of electrons, and positively charged when it has less than its normal amount of electrons. The charge on any object is thus a whole number times $+e$ or $-e$. That is, charge is **quantized**.

An object can become charged by rubbing (in which electrons are transferred from one material to another), by conduction (which is transfer of charge from one charged object to another by touching), or by induction (the separation of charge within an object because of the close approach of another charged object but without touching).

Electric charges exert a force on each other. If two charges are of opposite types, one positive and one negative, they each exert an attractive force on the other. If the two charges are the same type, each repels the other.

The magnitude of the force one point charge exerts on another is proportional to the product of their charges, and inversely proportional to the square of the distance between them:

$$F = k\frac{Q_1 Q_2}{r^2} = \frac{1}{4\pi\epsilon_0}\frac{Q_1 Q_2}{r^2}; \quad \textbf{(21–1, 21–2)}$$

this is **Coulomb's law**. In SI units, k is often written as $1/4\pi\epsilon_0$.

We think of an **electric field** as existing in space around any charge or group of charges. The force on another charged object is then said to be due to the electric field present at its location.

The *electric field*, $\vec{E}$, at any point in space due to one or more charges, is defined as the force per unit charge that would act on a positive test charge q placed at that point:

$$\vec{E} = \frac{\vec{F}}{q}. \quad \textbf{(21–3)}$$

The magnitude of the electric field a distance r from a point charge Q is

$$E = k\frac{Q}{r^2}. \quad \textbf{(21–4a)}$$

The total electric field at a point in space is equal to the vector sum of the individual fields due to each contributing charge (**principle of superposition**).

Electric fields are represented by **electric field lines** that start on positive charges and end on negative charges. Their direction indicates the direction the force would be on a tiny positive test charge placed at each point. The lines can be drawn so that the number per unit area is proportional to the magnitude of E.

The static electric field inside a conductor is zero, and the electric field lines just outside a charged conductor are perpendicular to its surface.

An **electric dipole** is a combination of two equal but opposite charges, $+Q$ and $-Q$, separated by a distance ℓ. The **dipole moment** is $p = Q\ell$. A dipole placed in a uniform electric field feels no net force but does feel a net torque (unless $\vec{p}$ is parallel to $\vec{E}$). The electric field produced by a dipole decreases as the third power of the distance r from the dipole $\left(E \propto 1/r^3\right)$ for r large compared to ℓ.

[*In the replication of DNA, the electrostatic force plays a crucial role in selecting the proper molecules so the genetic information is passed on accurately from generation to generation.]

Questions

1. If you charge a pocket comb by rubbing it with a silk scarf, how can you determine if the comb is positively or negatively charged?

2. Why does a shirt or blouse taken from a clothes dryer sometimes cling to your body?

3. Explain why fog or rain droplets tend to form around ions or electrons in the air.

4. A positively charged rod is brought close to a neutral piece of paper, which it attracts. Draw a diagram showing the separation of charge in the paper, and explain why attraction occurs.

5. Why does a plastic ruler that has been rubbed with a cloth have the ability to pick up small pieces of paper? Why is this difficult to do on a humid day?

6. Contrast the *net charge* on a conductor to the "free charges" in the conductor.

7. Figures 21–7 and 21–8 show how a charged rod placed near an uncharged metal object can attract (or repel) electrons. There are a great many electrons in the metal, yet only some of them move as shown. Why not all of them?

8. When an electroscope is charged, the two leaves repel each other and remain at an angle. What balances the electric force of repulsion so that the leaves don't separate further?

9. The form of Coulomb's law is very similar to that for Newton's law of universal gravitation. What are the differences between these two laws? Compare also gravitational mass and electric charge.

10. We are not normally aware of the gravitational or electric force between two ordinary objects. What is the reason in each case? Give an example where we are aware of each one and why.

11. Is the electric force a conservative force? Why or why not? (See Chapter 8.)

12. What experimental observations mentioned in the text rule out the possibility that the numerator in Coulomb's law contains the sum $(Q_1 + Q_2)$ rather than the product $Q_1 Q_2$?

13. When a charged ruler attracts small pieces of paper, sometimes a piece jumps quickly away after touching the ruler. Explain.

14. Explain why the test charges we use when measuring electric fields must be small.

15. When determining an electric field, must we use a *positive* test charge, or would a negative one do as well? Explain.

16. Draw the electric field lines surrounding two negative electric charges a distance ℓ apart.

17. Assume that the two opposite charges in Fig. 21–34a are 12.0 cm apart. Consider the magnitude of the electric field 2.5 cm from the positive charge. On which side of this charge—top, bottom, left, or right—is the electric field the strongest? The weakest?

18. Consider the electric field at the three points indicated by the letters A, B, and C in Fig. 21–51. First draw an arrow at each point indicating the direction of the net force that a positive test charge would experience if placed at that point, then list the letters in order of *decreasing* field strength (strongest first).

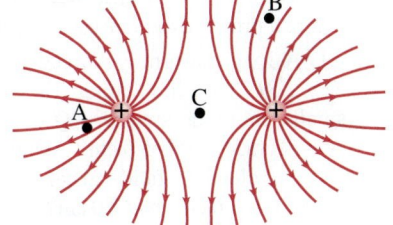

FIGURE 21–51
Question 18.

19. Why can electric field lines never cross?

20. Show, using the three rules for field lines given in Section 21–8, that the electric field lines starting or ending on a single point charge must be symmetrically spaced around the charge.

21. Given two point charges, Q and $2Q$, a distance ℓ apart, is there a point along the straight line that passes through them where $E = 0$ when their signs are (a) opposite, (b) the same? If yes, state roughly where this point will be.

22. Suppose the ring of Fig. 21–28 has a uniformly distributed negative charge Q. What is the magnitude and direction of $\vec{E}$ at point P?

23. Consider a small positive test charge located on an electric field line at some point, such as point P in Fig. 21–34a. Is the direction of the velocity and/or acceleration of the test charge along this line? Discuss.

24. We wish to determine the electric field at a point near a positively charged metal sphere (a good conductor). We do so by bringing a small test charge, q_0, to this point and measure the force F_0 on it. Will F_0/q_0 be greater than, less than, or equal to the electric field $\vec{E}$ as it was at that point before the test charge was present?

25. In what ways does the electron motion in Example 21–16 resemble projectile motion (Section 3–7)? In which ways not?

26. Describe the motion of the dipole shown in Fig. 21–44 if it is released from rest at the position shown.

27. Explain why there can be a net force on an electric dipole placed in a nonuniform electric field.

Problems

21–5 Coulomb's Law

$[1\ mC = 10^{-3}\ C,\ 1\ \mu C = 10^{-6}\ C,\ 1\ nC = 10^{-9}\ C.]$

1. (I) What is the magnitude of the electric force of attraction between an iron nucleus $(q = +26e)$ and its innermost electron if the distance between them is $1.5 \times 10^{-12}\ m$?

2. (I) How many electrons make up a charge of $-38.0\ \mu C$?

3. (I) What is the magnitude of the force a $+25\ \mu C$ charge exerts on a $+2.5\ mC$ charge 28 cm away?

4. (I) What is the repulsive electrical force between two protons $4.0 \times 10^{-15}\ m$ apart from each other in an atomic nucleus?

5. (II) When an object such as a plastic comb is charged by rubbing it with a cloth, the net charge is typically a few microcoulombs. If that charge is $3.0\ \mu C$, by what percentage does the mass of a 35-g comb change during charging?

6. (II) Two charged dust particles exert a force of $3.2 \times 10^{-2}\ N$ on each other. What will be the force if they are moved so they are only one-eighth as far apart?

7. (II) Two charged spheres are 8.45 cm apart. They are moved, and the force on each of them is found to have been tripled. How far apart are they now?

8. (II) A person scuffing her feet on a wool rug on a dry day accumulates a net charge of $-46\ \mu C$. How many excess electrons does she get, and by how much does her mass increase?

9. (II) What is the total charge of all the electrons in a 15-kg bar of gold? What is the net charge of the bar? (Gold has 79 electrons per atom and an atomic mass of 197 u.)

10. (II) Compare the electric force holding the electron in orbit $(r = 0.53 \times 10^{-10}\ m)$ around the proton nucleus of the hydrogen atom, with the gravitational force between the same electron and proton. What is the ratio of these two forces?

11. (II) Two positive point charges are a fixed distance apart. The sum of their charges is Q_T. What charge must each have in order to (a) maximize the electric force between them, and (b) minimize it?

12. (II) Particles of charge $+75$, $+48$, and $-85\ \mu C$ are placed in a line (Fig. 21–52). The center one is 0.35 m from each of the others. Calculate the net force on each charge due to the other two.

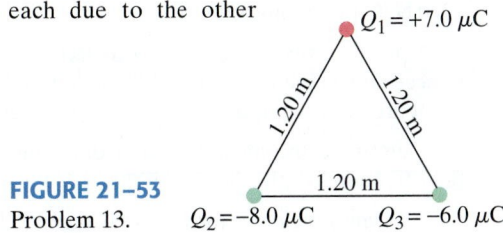

FIGURE 21–52
Problem 12.

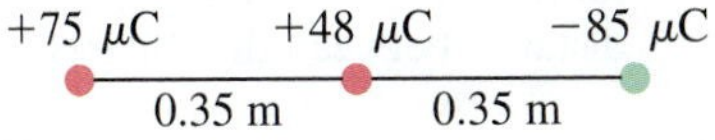

13. (II) Three charged particles are placed at the corners of an equilateral triangle of side 1.20 m (Fig. 21–53). The charges are $+7.0\ \mu C$, $-8.0\ \mu C$, and $-6.0\ \mu C$. Calculate the magnitude and direction of the net force on each due to the other two.

FIGURE 21–53
Problem 13.

14. (II) Two small nonconducting spheres have a total charge of $90.0\ \mu C$. (a) When placed 1.16 m apart, the force each exerts on the other is 12.0 N and is repulsive. What is the charge on each? (b) What if the force were attractive?

15. (II) A charge of 4.15 mC is placed at each corner of a square 0.100 m on a side. Determine the magnitude and direction of the force on each charge.

16. (II) Two negative and two positive point charges (magnitude $Q = 4.15$ mC) are placed on opposite corners of a square as shown in Fig. 21–54. Determine the magnitude and direction of the force on each charge.

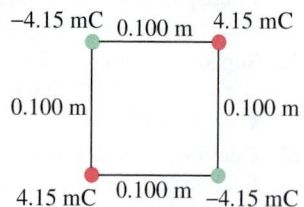

FIGURE 21–54
Problem 16.

17. (II) A charge Q is transferred from an initially uncharged plastic ball to an identical ball 12 cm away. The force of attraction is then 17 mN. How many electrons were transferred from one ball to the other?

18. (III) Two charges, $-Q_0$ and $-4Q_0$, are a distance ℓ apart. These two charges are free to move but do not because there is a third charge nearby. What must be the magnitude of the third charge and its placement in order for the first two to be in equilibrium?

19. (III) Two positive charges $+Q$ are affixed rigidly to the x axis, one at $x = +d$ and the other at $x = -d$. A third charge $+q$ of mass m, which is constrained to move only along the x axis, is displaced from the origin by a small distance $s \ll d$ and then released from rest. (a) Show that (to a good approximation) $+q$ will execute simple harmonic motion and determine an expression for its oscillation period T. (b) If these three charges are each singly ionized sodium atoms $(q = Q = +e)$ at the equilibrium spacing $d = 3 \times 10^{-10}$ m typical of the atomic spacing in a solid, find T in picoseconds.

20. (III) Two small charged spheres hang from cords of equal length ℓ as shown in Fig. 21–55 and make small angles θ_1 and θ_2 with the vertical. (a) If $Q_1 = Q$, $Q_2 = 2Q$, and $m_1 = m_2 = m$, determine the ratio θ_1/θ_2. (b) If $Q_1 = Q$, $Q_2 = 2Q$, $m_1 = m$, and $m_2 = 2m$, determine the ratio θ_1/θ_2. (c) Estimate the distance between the spheres for each case.

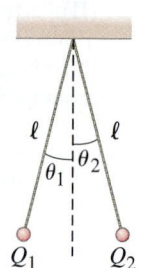

FIGURE 21–55
Problem 20.

21–6 to 21–8 Electric Field, Field Lines

21. (I) What are the magnitude and direction of the electric force on an electron in a uniform electric field of strength 1920 N/C that points due east?

22. (I) A proton is released in a uniform electric field, and it experiences an electric force of 2.18×10^{-14} N toward the south. What are the magnitude and direction of the electric field?

23. (I) Determine the magnitude and direction of the electric field 16.4 cm directly above an isolated 33.0×10^{-6} C charge.

24. (I) A downward electric force of 8.4 N is exerted on a $-8.8\,\mu$C charge. What are the magnitude and direction of the electric field at the position of this charge?

25. (I) The electric force on a $+4.20$-μC charge is $\vec{\mathbf{F}} = (7.22 \times 10^{-4}\,\text{N})\hat{\mathbf{j}}$. What is the electric field at the position of the charge?

26. (I) What is the electric field at a point when the force on a 1.25-μC charge placed at that point is $\vec{\mathbf{F}} = (3.0\hat{\mathbf{i}} - 3.9\hat{\mathbf{j}}) \times 10^{-3}$ N?

27. (II) Determine the magnitude of the acceleration experienced by an electron in an electric field of 576 N/C. How does the direction of the acceleration depend on the direction of the field at that point?

28. (II) Determine the magnitude and direction of the electric field at a point midway between a $-8.0\,\mu$C and a $+5.8\,\mu$C charge 8.0 cm apart. Assume no other charges are nearby.

29. (II) Draw, approximately, the electric field lines about two point charges, $+Q$ and $-3Q$, which are a distance ℓ apart.

30. (II) What is the electric field strength at a point in space where a proton experiences an acceleration of 1.8 million "g's"?

31. (II) A long uniformly charged thread (linear charge density $\lambda = 2.5$ C/m) lies along the x axis in Fig. 21–56. A small charged sphere $(Q = -2.0\,$C$)$ is at the point $x = 0$ cm, $y = -5.0$ cm. What is the electric field at the point $x = 7.0$ cm, $y = 7.0$ cm? $\vec{\mathbf{E}}_{\text{thread}}$ and $\vec{\mathbf{E}}_Q$ represent fields due to the long thread and the charge Q, respectively.

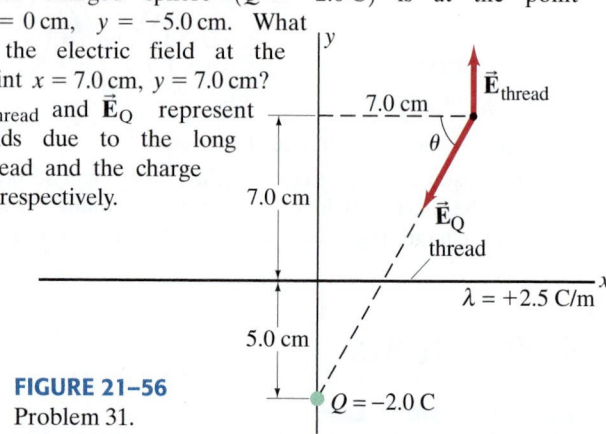

FIGURE 21–56
Problem 31.

32. (II) The electric field midway between two equal but opposite point charges is 586 N/C, and the distance between the charges is 16.0 cm. What is the magnitude of the charge on each?

33. (II) Calculate the electric field at one corner of a square 1.22 m on a side if the other three corners are occupied by 2.25×10^{-6} C charges.

34. (II) Calculate the electric field at the center of a square 52.5 cm on a side if one corner is occupied by a $-38.6\,\mu$C charge and the other three are occupied by $-27.0\,\mu$C charges.

35. (II) Determine the direction and magnitude of the electric field at the point P in Fig. 21–57. The charges are separated by a distance $2a$, and point P is a distance x from the midpoint between the two charges. Express your answer in terms of Q, x, a, and k.

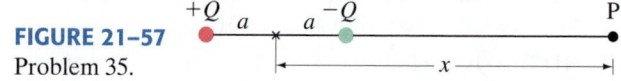

FIGURE 21–57
Problem 35.

36. (II) Two point charges, $Q_1 = -25\,\mu$C and $Q_2 = +45\,\mu$C, are separated by a distance of 12 cm. The electric field at the point P (see Fig. 21–58) is zero. How far from Q_1 is P?

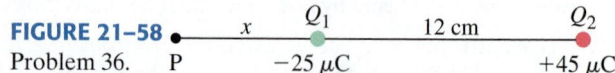

FIGURE 21–58
Problem 36.

37. (II) A very thin line of charge lies along the x axis from $x = -\infty$ to $x = +\infty$. Another similar line of charge lies along the y axis from $y = -\infty$ to $y = +\infty$. Both lines have a uniform charge per length λ. Determine the resulting electric field magnitude and direction (relative to the x axis) at a point (x, y) in the first quadrant of the xy plane.

38. (II) (*a*) Determine the electric field $\vec{\mathbf{E}}$ at the origin 0 in Fig. 21–59 due to the two charges at A and B. (*b*) Repeat, but let the charge at B be reversed in sign.

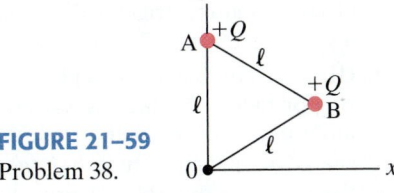

FIGURE 21–59
Problem 38.

39. (II) Draw, approximately, the electric field lines emanating from a uniformly charged straight wire whose length ℓ is not great. The spacing between lines near the wire should be much less than ℓ. [*Hint:* Also consider points very far from the wire.]

40. (II) Two parallel circular rings of radius R have their centers on the x axis separated by a distance ℓ as shown in Fig. 21–60. If each ring carries a uniformly distributed charge Q, find the electric field, $\vec{\mathbf{E}}(x)$, at points along the x axis.

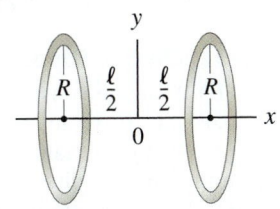

FIGURE 21–60
Problem 40.

41. (II) You are given two unknown point charges, Q_1 and Q_2. At a point on the line joining them, one-third of the way from Q_1 to Q_2, the electric field is zero (Fig. 21–61). What is the ratio Q_1/Q_2?

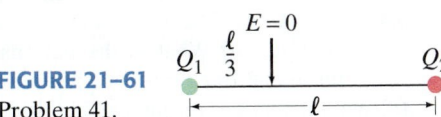

FIGURE 21–61
Problem 41.

42. (II) Use Coulomb's law to determine the magnitude and direction of the electric field at points A and B in Fig. 21–62 due to the two positive charges $(Q = 5.7\,\mu\text{C})$ shown. Are your results consistent with Fig. 21–34b?

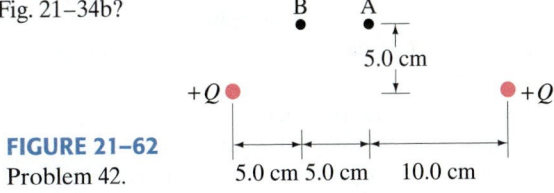

FIGURE 21–62
Problem 42.

43. (II) (*a*) Two equal charges Q are positioned at points $(x = \ell, y = 0)$ and $(x = -\ell, y = 0)$. Determine the electric field as a function of y for points along the y axis. (*b*) Show that the field is a maximum at $y = \pm \ell/\sqrt{2}$.

44. (II) At what position, $x = x_{\text{M}}$, is the magnitude of the electric field along the axis of the ring of Example 21–9 a maximum?

45. (II) Estimate the electric field at a point 2.40 cm perpendicular to the midpoint of a uniformly charged 2.00-m-long thin wire carrying a total charge of 4.75 μC.

46. (II) The uniformly charged straight wire in Fig. 21–29 has the length ℓ, where point 0 is at the midpoint. Show that the field at point P, a perpendicular distance x from 0, is given by

$$E = \frac{\lambda}{2\pi\epsilon_0} \frac{\ell}{x(\ell^2 + 4x^2)^{1/2}},$$

where λ is the charge per unit length.

47. (II) Use your result from Problem 46 to find the electric field (magnitude and direction) a distance z above the center of a square loop of wire, each of whose sides has length ℓ and uniform charge per length λ (Fig. 21–63).

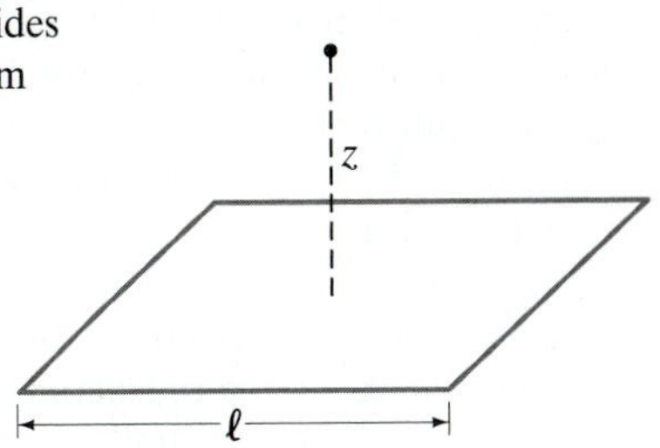

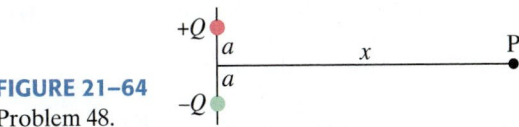

FIGURE 21–63
Problem 47.

48. (II) Determine the direction and magnitude of the electric field at the point P shown in Fig. 21–64. The two charges are separated by a distance of $2a$. Point P is on the perpendicular bisector of the line joining the charges, a distance x from the midpoint between them. Express your answers in terms of Q, x, a, and k.

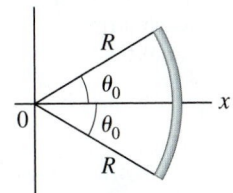

FIGURE 21–64
Problem 48.

49. (III) A thin rod bent into the shape of an arc of a circle of radius R carries a uniform charge per unit length λ. The arc subtends a total angle $2\theta_0$, symmetric about the x axis, as shown in Fig. 21–65. Determine the electric field $\vec{\mathbf{E}}$ at the origin 0.

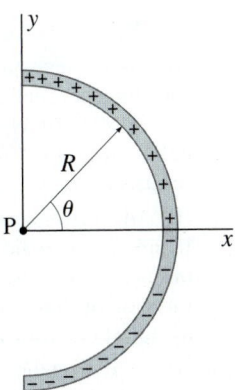

FIGURE 21–65
Problem 49.

50. (III) A thin glass rod is a semicircle of radius R, Fig. 21–66. A charge is nonuniformly distributed along the rod with a linear charge density given by $\lambda = \lambda_0 \sin \theta$, where λ_0 is a positive constant. Point P is at the center of the semicircle. (*a*) Find the electric field $\vec{\mathbf{E}}$ (magnitude and direction) at point P. [*Hint:* Remember $\sin(-\theta) = -\sin \theta$, so the two halves of the rod are oppositely charged.] (*b*) Determine the acceleration (magnitude and direction) of an electron placed at point P, assuming $R = 1.0$ cm and $\lambda_0 = 1.0\,\mu\text{C/m}$.

FIGURE 21–66
Problem 50.

51. (III) Suppose a uniformly charged wire starts at point 0 and rises vertically along the positive y axis to a length ℓ. (*a*) Determine the components of the electric field E_x and E_y at point $(x, 0)$. That is, calculate $\vec{\mathbf{E}}$ near one end of a long wire, in the plane perpendicular to the wire. (*b*) If the wire extends from $y = 0$ to $y = \infty$, so that $\ell = \infty$, show that $\vec{\mathbf{E}}$ makes a 45° angle to the horizontal for any x. [*Hint:* See Example 21–11 and Fig. 21–29.]

52. (III) Suppose in Example 21–11 that $x = 0.250\,\text{m}$, $Q = 3.15\,\mu\text{C}$, and that the uniformly charged wire is only 6.50 m long and extends along the y axis from $y = -4.00\,\text{m}$ to $y = +2.50\,\text{m}$. (*a*) Calculate E_x and E_y at point P. (*b*) Determine what the error would be if you simply used the result of Example 21–11, $E = \lambda/2\pi\epsilon_0 x$. Express this error as $(E_x - E)/E$ and E_y/E.

53. (III) A thin rod of length ℓ carries a total charge Q distributed uniformly along its length. See Fig. 21–67. Determine the electric field along the axis of the rod starting at one end—that is, find $E(x)$ for $x \geq 0$ in Fig. 21–67.

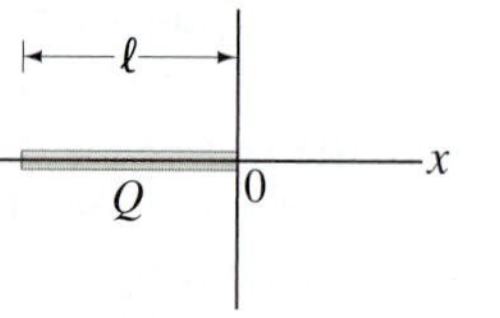

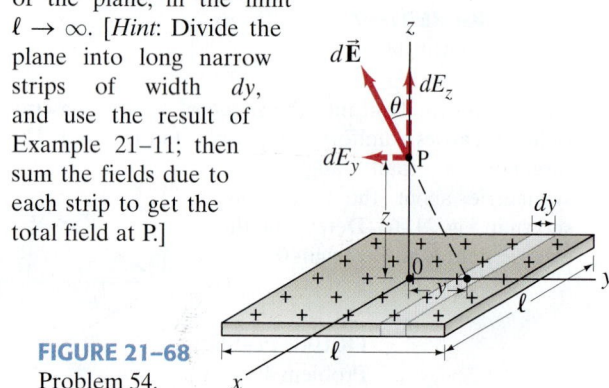

FIGURE 21–67
Problem 53.

54. (III) *Uniform plane of charge.* Charge is distributed uniformly over a large square plane of side ℓ, as shown in Fig. 21–68. The charge per unit area (C/m^2) is σ. Determine the electric field at a point P a distance z above the center of the plane, in the limit $\ell \to \infty$. [*Hint:* Divide the plane into long narrow strips of width dy, and use the result of Example 21–11; then sum the fields due to each strip to get the total field at P.]

FIGURE 21–68
Problem 54.

55. (III) Suppose the charge Q on the ring of Fig. 21–28 was all distributed uniformly on only the upper half of the ring, and no charge was on the lower half. Determine the electric field $\vec{\mathbf{E}}$ at P. (Take y vertically upward.)

21–10 Motion of Charges in an Electric Field

56. (II) An electron with speed $v_0 = 27.5 \times 10^6\,\text{m/s}$ is traveling parallel to a uniform electric field of magnitude $E = 11.4 \times 10^3\,\text{N/C}$. (*a*) How far will the electron travel before it stops? (*b*) How much time will elapse before it returns to its starting point?

57. (II) An electron has an initial velocity $\vec{\mathbf{v}}_0 = (8.0 \times 10^4\,\text{m/s})\hat{\mathbf{j}}$. It enters a region where $\vec{\mathbf{E}} = (2.0\hat{\mathbf{i}} + 8.0\hat{\mathbf{j}}) \times 10^4\,\text{N/C}$. (*a*) Determine the vector acceleration of the electron as a function of time. (*b*) At what angle θ is it moving (relative to its initial direction) at $t = 1.0\,\text{ns}$?

58. (II) An electron moving to the right at $7.5 \times 10^5\,\text{m/s}$ enters a uniform electric field parallel to its direction of motion. If the electron is to be brought to rest in the space of 4.0 cm, (*a*) what direction is required for the electric field, and (*b*) what is the strength of the field?

59. (II) At what angle will the electrons in Example 21–16 leave the uniform electric field at the end of the parallel plates (point P in Fig. 21–41)? Assume the plates are 4.9 cm long, $E = 5.0 \times 10^3\,\text{N/C}$, and $v_0 = 1.00 \times 10^7\,\text{m/s}$. Ignore fringing of the field.

60. (II) An electron is traveling through a uniform electric field. The field is constant and given by $\vec{\mathbf{E}} = (2.00 \times 10^{-11}\,\text{N/C})\hat{\mathbf{i}} - (1.20 \times 10^{-11}\,\text{N/C})\hat{\mathbf{j}}$. At $t = 0$, the electron is at the origin and traveling in the x direction with a speed of $1.90\,\text{m/s}$. What is its position 2.00 s later?

61. (II) A positive charge q is placed at the center of a circular ring of radius R. The ring carries a uniformly distributed negative charge of total magnitude $-Q$. (*a*) If the charge q is displaced from the center a small distance x as shown in Fig. 21–69, show that it will undergo simple harmonic motion when released. (*b*) If its mass is m, what is its period?

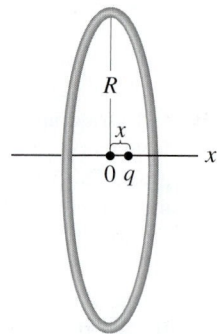

FIGURE 21–69
Problem 61.

21–11 Electric Dipoles

62. (II) A dipole consists of charges $+e$ and $-e$ separated by 0.68 nm. It is in an electric field $E = 2.2 \times 10^4\,\text{N/C}$. (*a*) What is the value of the dipole moment? (*b*) What is the torque on the dipole when it is perpendicular to the field? (*c*) What is the torque on the dipole when it is at an angle of 45° to the field? (*d*) What is the work required to rotate the dipole from being oriented parallel to the field to being antiparallel to the field?

63. (II) The HCl molecule has a dipole moment of about $3.4 \times 10^{-30}\,\text{C·m}$. The two atoms are separated by about $1.0 \times 10^{-10}\,\text{m}$. (*a*) What is the net charge on each atom? (*b*) Is this equal to an integral multiple of e? If not, explain. (*c*) What maximum torque would this dipole experience in a $2.5 \times 10^4\,\text{N/C}$ electric field? (*d*) How much energy would be needed to rotate one molecule 45° from its equilibrium position of lowest potential energy?

64. (II) Suppose both charges in Fig. 21–45 (for a dipole) were positive. (*a*) Show that the field on the perpendicular bisector, for $r \gg \ell$, is given by $(1/4\pi\epsilon_0)(2Q/r^2)$. (*b*) Explain why the field decreases as $1/r^2$ here whereas for a dipole it decreases as $1/r^3$.

65. (II) An electric dipole, of dipole moment p and moment of inertia I, is placed in a uniform electric field $\vec{\mathbf{E}}$. (*a*) If displaced by an angle θ as shown in Fig. 21–44 and released, under what conditions will it oscillate in simple harmonic motion? (*b*) What will be its frequency?

66. (III) Suppose a dipole $\vec{\mathbf{p}}$ is placed in a nonuniform electric field $\vec{\mathbf{E}} = E\hat{\mathbf{i}}$ that points along the x axis. If $\vec{\mathbf{E}}$ depends only on x, show that the net force on the dipole is

$$\vec{\mathbf{F}} = \left(\vec{\mathbf{p}} \cdot \frac{d\vec{\mathbf{E}}}{dx}\right)\hat{\mathbf{i}},$$

where $d\vec{\mathbf{E}}/dx$ is the gradient of the field in the x direction.

67. (III) (*a*) Show that at points along the axis of a dipole (along the same line that contains $+Q$ and $-Q$), the electric field has magnitude

$$E = \frac{1}{4\pi\epsilon_0}\frac{2p}{r^3}$$

for $r \gg \ell$ (Fig. 21–45), where r is the distance from a point to the center of the dipole. (*b*) In what direction does $\vec{\mathbf{E}}$ point?

General Problems

68. How close must two electrons be if the electric force between them is equal to the weight of either at the Earth's surface?

69. Given that the human body is mostly made of water, estimate the total amount of positive charge in a 65-kg person.

70. A 3.0-g copper penny has a positive charge of $38 \mu C$. What fraction of its electrons has it lost?

71. Measurements indicate that there is an electric field surrounding the Earth. Its magnitude is about 150 N/C at the Earth's surface and points inward toward the Earth's center. What is the magnitude of the electric charge on the Earth? Is it positive or negative? [*Hint:* The electric field outside a uniformly charged sphere is the same as if all the charge were concentrated at its center.]

72. (*a*) The electric field near the Earth's surface has magnitude of about 150 N/C. What is the acceleration experienced by an electron near the surface of the Earth? (*b*) What about a proton? (*c*) Calculate the ratio of each acceleration to $g = 9.8 \text{ m/s}^2$.

73. A water droplet of radius 0.018 mm remains stationary in the air. If the downward-directed electric field of the Earth is 150 N/C, how many excess electron charges must the water droplet have?

74. Estimate the net force between the CO group and the HN group shown in Fig. 21–70. The C and O have charges $\pm 0.40e$, and the H and N have charges $\pm 0.20e$, where $e = 1.6 \times 10^{-19} \text{ C}$. [*Hint:* Do not include the "internal" forces between C and O, or between H and N.]

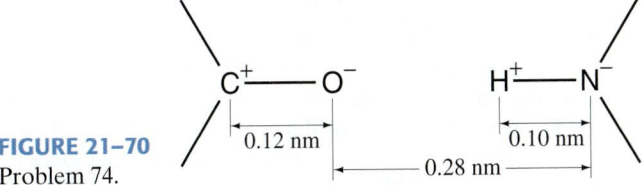

FIGURE 21–70
Problem 74.

75. Suppose that electrical attraction, rather than gravity, were responsible for holding the Moon in orbit around the Earth. If equal and opposite charges Q were placed on the Earth and the Moon, what should be the value of Q to maintain the present orbit? Use data given on the inside front cover of this book. Treat the Earth and Moon as point particles.

76. In a simple model of the hydrogen atom, the electron revolves in a circular orbit around the proton with a speed of $2.2 \times 10^6 \text{ m/s}$. Determine the radius of the electron's orbit. [*Hint:* See Chapter 5 on circular motion.]

77. A positive point charge $Q_1 = 2.5 \times 10^{-5} \text{ C}$ is fixed at the origin of coordinates, and a negative point charge $Q_2 = -5.0 \times 10^{-6} \text{ C}$ is fixed to the x axis at $x = +2.0 \text{ m}$. Find the location of the place(s) along the x axis where the electric field due to these two charges is zero.

78. When clothes are removed from a dryer, a 40-g sock is stuck to a sweater, even with the sock clinging to the sweater's underside. Estimate the minimum attractive force between the sock and the sweater. Then estimate the minimum charge on the sock and the sweater. Assume the charging came entirely from the sock rubbing against the sweater so that they have equal and opposite charges, and approximate the sweater as a flat sheet of uniform charge.

79. A small lead sphere is encased in insulating plastic and suspended vertically from an ideal spring (spring constant $k = 126 \text{ N/m}$) as in Fig. 21–71. The total mass of the coated sphere is 0.650 kg, and its center lies 15.0 cm above a tabletop when in equilibrium. The sphere is pulled down 5.00 cm below equilibrium, an electric charge $Q = -3.00 \times 10^{-6} \text{ C}$ is deposited on it, and then it is released. Using what you know about harmonic oscillation, write an expression for the electric field strength as a function of time that would be measured at the point on the tabletop (P) directly below the sphere.

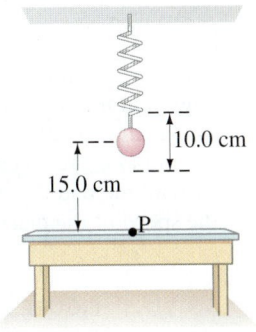

FIGURE 21–71
Problem 79.

80. A large electroscope is made with "leaves" that are 78-cm-long wires with tiny 24-g spheres at the ends. When charged, nearly all the charge resides on the spheres. If the wires each make a 26° angle with the vertical (Fig. 21–72), what total charge Q must have been applied to the electroscope? Ignore the mass of the wires.

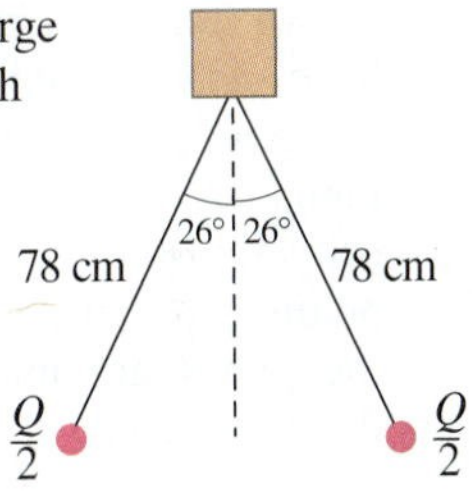

FIGURE 21–72
Problem 80.

81. Dry air will break down and generate a spark if the electric field exceeds about $3 \times 10^6 \text{ N/C}$. How much charge could be packed onto a green pea (diameter 0.75 cm) before the pea spontaneously discharges? [*Hint:* Eqs. 21–4 work outside a sphere if r is measured from its center.]

82. Two point charges, $Q_1 = -6.7 \mu C$ and $Q_2 = 1.8 \mu C$, are located between two oppositely charged parallel plates, as shown in Fig. 21–73. The two charges are separated by a distance of $x = 0.34 \text{ m}$. Assume that the electric field produced by the charged plates is uniform and equal to $E = 73,000 \text{ N/C}$. Calculate the net electrostatic force on Q_1 and give its direction.

FIGURE 21–73
Problem 82.

83. Packing material made of pieces of foamed polystyrene can easily become charged and stick to each other. Given that the density of this material is about 35 kg/m^3, estimate how much charge might be on a 2.0-cm-diameter foamed polystyrene sphere, assuming the electric force between two spheres stuck together is equal to the weight of one sphere.

84. One type of *electric quadrupole* consists of two dipoles placed end to end with their negative charges (say) overlapping; that is, in the center is $-2Q$ flanked (on a line) by a $+Q$ to either side (Fig. 21–74). Determine the electric field $\vec{E}$ at points along the perpendicular bisector and show that E decreases as $1/r^4$. Measure r from the $-2Q$ charge and assume $r \gg \ell$.

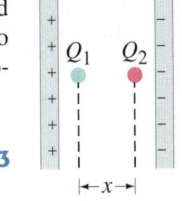

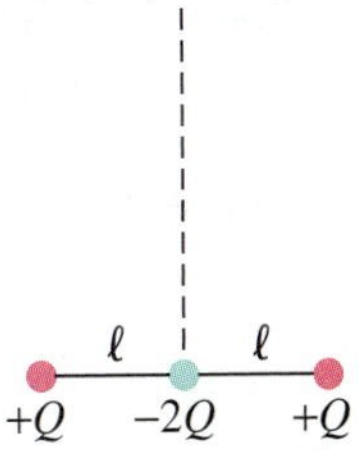

FIGURE 21–74
Problem 84.

85. Suppose electrons enter a uniform electric field midway between two plates at an angle θ_0 to the horizontal, as shown in Fig. 21–75. The path is symmetrical, so they leave at the same angle θ_0 and just barely miss the top plate. What is θ_0? Ignore fringing of the field.

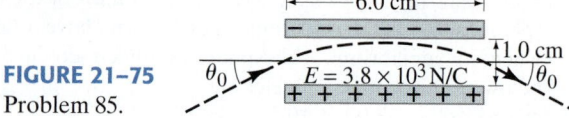

FIGURE 21–75
Problem 85.

86. An electron moves in a circle of radius r around a very long uniformly charged wire in a vacuum chamber, as shown in Fig. 21–76. The charge density on the wire is $\lambda = 0.14\ \mu C/m$. (*a*) What is the electric field at the electron (magnitude and direction in terms of r and λ)? (*b*) What is the speed of the electron?

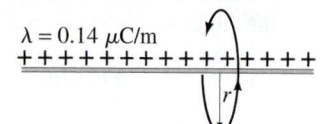

FIGURE 21–76
Problem 86.

87. Three very large square planes of charge are arranged as shown (on edge) in Fig. 21–77. From left to right, the planes have charge densities per unit area of $-0.50\ \mu C/m^2$, $+0.25\ \mu C/m^2$, and $-0.35\ \mu C/m^2$. Find the total electric field (direction and magnitude) at the points A, B, C, and D. Assume the plates are much larger than the distance AD.

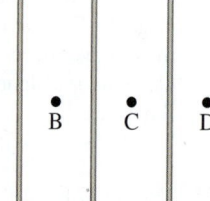

FIGURE 21–77
Problem 87.

88. A point charge ($m = 1.0$ g) at the end of an insulating cord of length 55 cm is observed to be in equilibrium in a uniform horizontal electric field of 15,000 N/C, when the pendulum's position is as shown in Fig. 21–78, with the charge 12 cm above the lowest (vertical) position. If the field points to the right in Fig. 21–78, determine the magnitude and sign of the point charge.

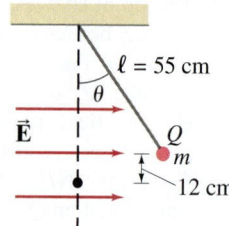

FIGURE 21–78
Problem 88.

89. Four equal positive point charges, each of charge $8.0\ \mu C$, are at the corners of a square of side 9.2 cm. What charge should be placed at the center of the square so that all charges are at equilibrium? Is this a stable or unstable equilibrium (Section 12–3) in the plane?

90. Two small, identical conducting spheres A and B are a distance R apart; each carries the same charge Q. (*a*) What is the force sphere B exerts on sphere A? (*b*) An identical sphere with zero charge, sphere C, makes contact with sphere B and is then moved very far away. What is the net force now acting on sphere A? (*c*) Sphere C is brought back and now makes contact with sphere A and is then moved far away. What is the force on sphere A in this third case?

91. A point charge of mass 0.210 kg, and net charge $+0.340\ \mu C$, hangs at rest at the end of an insulating cord above a large sheet of charge. The horizontal sheet of fixed uniform charge creates a uniform vertical electric field in the vicinity of the point charge. The tension in the cord is measured to be 5.18 N. (*a*) Calculate the magnitude and direction of the electric field due to the sheet of charge (Fig. 21–79). (*b*) What is the surface charge density $\sigma\ (C/m^2)$ on the sheet?

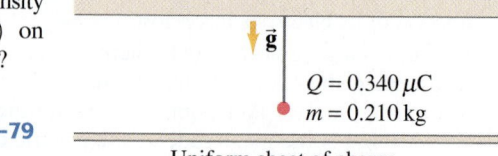

FIGURE 21–79
Problem 91.

Uniform sheet of charge

92. A one-dimensional row of positive ions, each with charge $+Q$ and separated from its neighbors by a distance d, occupies the right-hand half of the x axis. That is, there is a $+Q$ charge at $x = 0$, $x = +d$, $x = +2d$, $x = +3d$, and so on out to ∞. (*a*) If an electron is placed at the position $x = -d$, determine F, the magnitude of force that this row of charges exerts on the electron. (*b*) If the electron is instead placed at $x = -3d$, what is the value of F? [*Hint:* The infinite sum $\sum_{n=1}^{n=\infty} \dfrac{1}{n^2} = \dfrac{\pi^2}{6}$, where n is a positive integer.]

*Numerical/Computer

*93. (III) A thin ring-shaped object of radius a contains a total charge Q uniformly distributed over its length. The electric field at a point on its axis a distance x from its center is given in Example 21–9 as

$$E = \frac{1}{4\pi\epsilon_0} \frac{Qx}{(x^2 + a^2)^{\frac{3}{2}}}.$$

(*a*) Take the derivative to find where on the x axis ($x > 0$) E_x is a maximum. Assume $Q = 6.00\ \mu C$ and $a = 10.0$ cm. (*b*) Calculate the electric field for $x = 0$ to $x = +12.0$ cm in steps of 0.1 cm, and make a graph of the electric field. Does the maximum of the graph coincide with the maximum of the electric field you obtained analytically? Also, calculate and graph the electric field (*c*) due to the ring, and (*d*) due to a point charge $Q = 6.00\ \mu C$ at the center of the ring. Make a single graph, from $x = 0$ (or $x = 1.0$ cm) out to $x = 50.0$ cm in 1.0 cm steps, with two curves of the electric fields, and show that both fields converge at large distances from the center. (*e*) At what distance does the electric field of the ring differ from that of the point charge by 10%?

*94. (III) An 8.00 μC charge is on the x axis of a coordinate system at $x = +5.00$ cm. A $-2.00\ \mu C$ charge is at $x = -5.00$ cm. (*a*) Plot the x component of the electric field for points on the x axis from $x = -30.0$ cm to $x = +30.0$ cm. The sign of E_x is positive when $\vec{\mathbf{E}}$ points to the right and negative when it points to the left. (*b*) Make a plot of E_x and E_y for points on the y axis from $y = -30.0$ to $+30.0$ cm.

Answers to Exercises

A: (*e*).

B: 5 N.

C: 1.2 N, to the right.

D: (*a*) No; (*b*) yes, midway between them.

E: (*d*), if the two + charges are not at opposite corners (use symmetry).

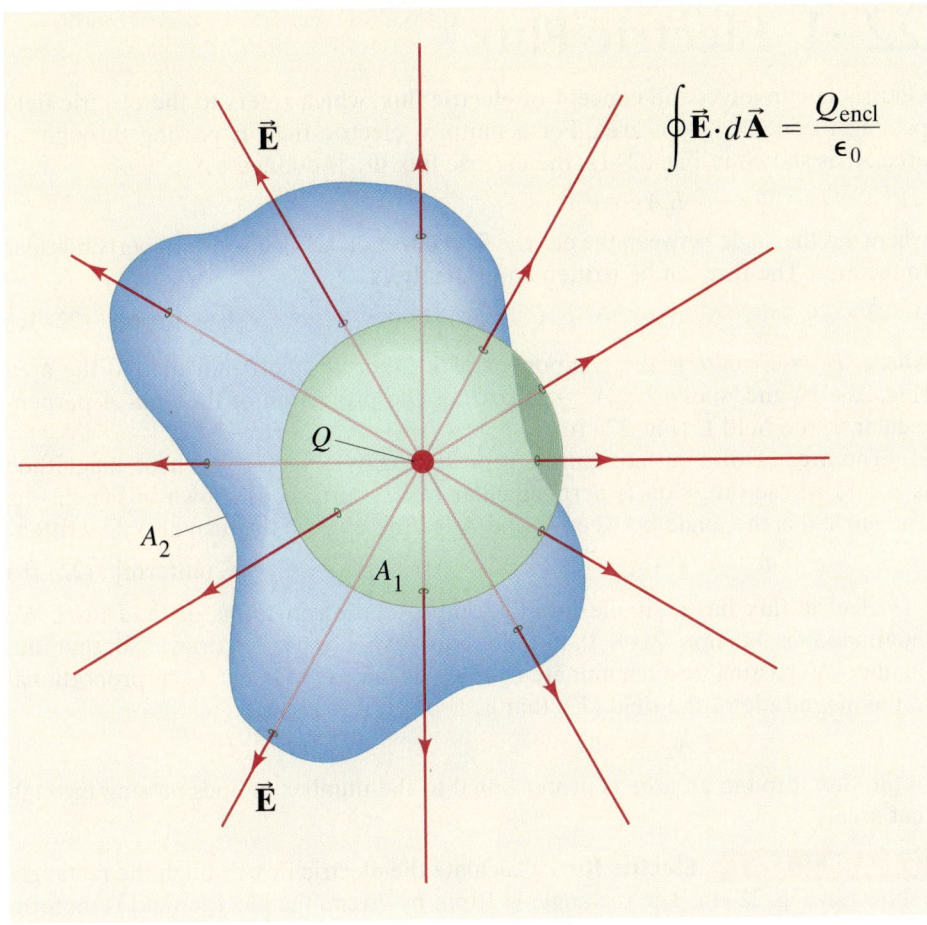

$$\oint \vec{E} \cdot d\vec{A} = \frac{Q_{encl}}{\epsilon_0}$$

Gauss's law is an elegant relation between electric charge and electric field. It is more general than Coulomb's law. Gauss's law involves an integral of the electric field $\vec{E}$ at each point on a closed surface. The surface is only imaginary, and we choose the shape and placement of the surface so that we can evaluate the integral. In this drawing, two different 3-D surfaces are shown (one green, one blue), both enclosing a point charge Q. Gauss's law states that the product $\vec{E} \cdot d\vec{A}$, where $d\vec{A}$ is an infinitesimal area of the surface, integrated over the entire surface, equals the charge enclosed by the surface Q_{encl} divided by ϵ_0. Both surfaces here enclose the same charge Q. Hence $\oint \vec{E} \cdot d\vec{A}$ will give the same result for both surfaces.

22

Gauss's Law

CHAPTER-OPENING QUESTION—Guess now!

A nonconducting sphere has a uniform charge density throughout. How does the magnitude of the electric field vary inside with distance from the center?

(a) The electric field is zero throughout.
(b) The electric field is constant but nonzero throughout.
(c) The electric field is linearly increasing from the center to the outer edge.
(d) The electric field is exponentially increasing from the center to the outer edge.
(e) The electric field increases quadratically from the center to the outer edge.

The great mathematician Karl Friedrich Gauss (1777–1855) developed an important relation, now known as Gauss's law, which we develop and discuss in this Chapter. It is a statement of the relation between electric charge and electric field and is a more general and elegant form of Coulomb's law.

We can, in principle, determine the electric field due to any given distribution of electric charge using Coulomb's law. The total electric field at any point will be the vector sum (or integral) of contributions from all charges present (see Eq. 21–6). Except for some simple cases, the sum or integral can be quite complicated to evaluate. For situations in which an analytic solution (such as we carried out in the Examples of Sections 21–6 and 21–7) is not possible, a computer can be used.

In some cases, however, the electric field due to a given charge distribution can be calculated more easily or more elegantly using Gauss's law, as we shall see in this Chapter. But the major importance of Gauss's law is that it gives us additional insight into the nature of electrostatic fields, and a more general relationship between charge and field.

Before discussing Gauss's law itself, we first discuss the concept of *flux*.

CONTENTS

22–1 Electric Flux

22–2 Gauss's Law

22–3 Applications of Gauss's Law

*22–4 Experimental Basis of Gauss's and Coulomb's Laws

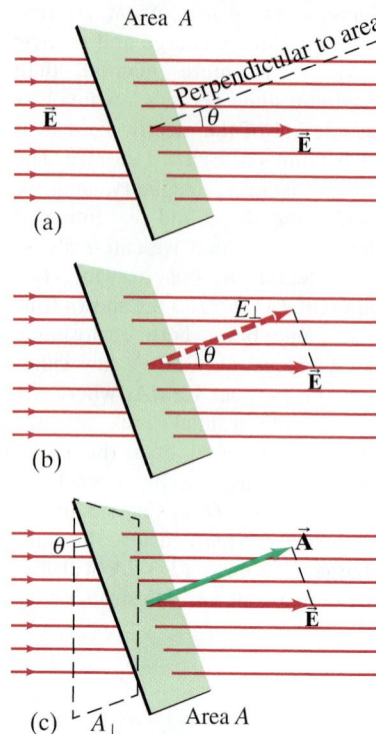

Area A

Perpendicular to area

$\vec{E}$

θ

$\vec{E}$

(a)

$E_\perp$

θ

$\vec{E}$

(b)

θ

$\vec{A}$

$\vec{E}$

(c) $A_\perp$ Area A

FIGURE 22–1 (a) A uniform electric field $\vec{E}$ passing through a flat area A. (b) $E_\perp = E\cos\theta$ is the component of $\vec{E}$ perpendicular to the plane of area A. (c) $A_\perp = A\cos\theta$ is the projection (dashed) of the area A perpendicular to the field $\vec{E}$.

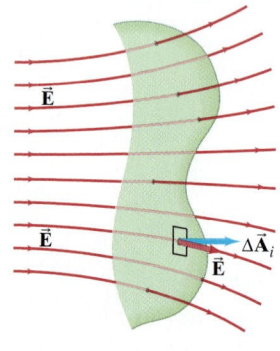

FIGURE 22–2 Electric flux through a curved surface. One small area of the surface, $\Delta\vec{A}_i$, is indicated.

$\vec{E}$

$\vec{E}$

$\Delta\vec{A}_i$

$\vec{E}$

FIGURE 22–3 Electric flux through a closed surface.

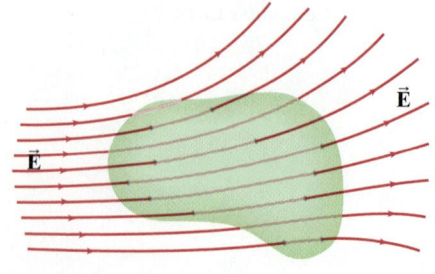

$\vec{E}$

$\vec{E}$

22–1 Electric Flux

Gauss's law involves the concept of **electric flux**, which refers to the electric field passing through a given area. For a uniform electric field $\vec{E}$ passing through an area A, as shown in Fig. 22–1a, the electric flux Φ_E is defined as

$$\Phi_E = EA\cos\theta,$$

where θ is the angle between the electric field direction and a line drawn perpendicular to the area. The flux can be written equivalently as

$$\Phi_E = E_\perp A = EA_\perp = EA\cos\theta, \qquad [\vec{E}\text{ uniform}] \quad \textbf{(22–1a)}$$

where $E_\perp = E\cos\theta$ is the component of $\vec{E}$ along the perpendicular to the area (Fig. 22–1b) and, similarly, $A_\perp = A\cos\theta$ is the projection of the area A perpendicular to the field $\vec{E}$ (Fig. 22–1c).

The area A of a surface can be represented by a vector $\vec{A}$ whose magnitude is A and whose direction is perpendicular to the surface, as shown in Fig. 22–1c. The angle θ is the angle between $\vec{E}$ and $\vec{A}$, so the electric flux can also be written

$$\Phi_E = \vec{E}\cdot\vec{A}. \qquad [\vec{E}\text{ uniform}] \quad \textbf{(22–1b)}$$

Electric flux has a simple intuitive interpretation in terms of field lines. We mentioned in Section 21–8 that field lines can always be drawn so that the number (N) passing through unit area perpendicular to the field ($A_\perp$) is proportional to the magnitude of the field (E): that is, $E \propto N/A_\perp$. Hence,

$$N \propto EA_\perp = \Phi_E,$$

so the flux through an area is proportional to the number of lines passing through that area.

EXAMPLE 22–1 **Electric flux.** Calculate the electric flux through the rectangle shown in Fig. 22–1a. The rectangle is 10 cm by 20 cm, the electric field is uniform at 200 N/C, and the angle θ is 30°.

APPROACH We use the definition of flux, $\Phi_E = \vec{E}\cdot\vec{A} = EA\cos\theta$.

SOLUTION The electric flux is

$$\Phi_E = (200\,\text{N/C})(0.10\,\text{m}\times0.20\,\text{m})\cos30° = 3.5\,\text{N}\cdot\text{m}^2/\text{C}.$$

EXERCISE A Which of the following would cause a change in the electric flux through a circle lying in the xz plane where the electric field is $(10\,\text{N})\hat{j}$? (*a*) Changing the magnitude of the electric field. (*b*) Changing the size of the circle. (*c*) Tipping the circle so that it is lying in the xy plane. (*d*) All of the above. (*e*) None of the above.

In the more general case, when the electric field $\vec{E}$ is not uniform and the surface is not flat, Fig. 22–2, we divide up the chosen surface into n small elements of surface whose areas are $\Delta A_1, \Delta A_2, \cdots, \Delta A_n$. We choose the division so that each ΔA_i is small enough that (1) it can be considered flat, and (2) the electric field varies so little over this small area that it can be considered uniform. Then the electric flux through the entire surface is approximately

$$\Phi_E \approx \sum_{i=1}^{n} \vec{E}_i\cdot\Delta\vec{A}_i,$$

where $\vec{E}_i$ is the field passing through $\Delta\vec{A}_i$. In the limit as we let $\Delta\vec{A}_i \to 0$, the sum becomes an integral over the entire surface and the relation becomes mathematically exact:

$$\Phi_E = \int \vec{E}\cdot d\vec{A}. \qquad \textbf{(22–2)}$$

Gauss's law involves the *total* flux through a *closed* surface—a surface of any shape that completely encloses a volume of space, such as that shown in Fig. 22–3. In this case, the net flux through the enclosing surface is given by

$$\Phi_E = \oint \vec{E}\cdot d\vec{A}, \qquad \textbf{(22–3)}$$

where the integral sign is written $\oint$ to indicate that the integral is over the value of $\vec{E}$ at every point on an enclosing surface.

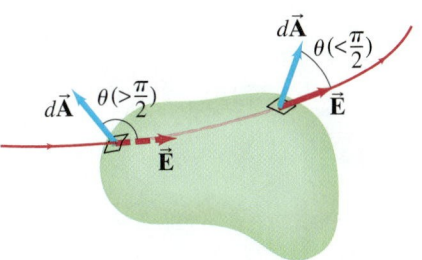

Up to now we have not been concerned with an ambiguity in the direction of the vector $\vec{A}$ or $d\vec{A}$ that represents a surface. For example, in Fig. 22–1c, the vector $\vec{A}$ could point upward and to the right (as shown) or downward to the left and still be perpendicular to the surface. For a closed surface, we define (arbitrarily) the direction of $\vec{A}$, or of $d\vec{A}$, to point *outward* from the enclosed volume, Fig. 22–4. For an electric field line leaving the enclosed volume (on the right in Fig. 22–4), the angle θ between $\vec{E}$ and $d\vec{A}$ must be less than $\pi/2$ ($= 90°$); hence $\cos\theta > 0$. For a line entering the volume (on the left in Fig. 22–4) $\theta > \pi/2$; hence $\cos\theta < 0$. Hence, *flux entering the enclosed volume is negative* $\left(\int E \cos\theta\, dA < 0\right)$, whereas *flux leaving the volume is positive*. Consequently, Eq. 22–3 gives the net flux *out of* the volume. If Φ_E is negative, there is a net flux *into* the volume.

In Figs. 22–3 and 22–4, each field line that enters the volume also leaves the volume. Hence $\Phi_E = \oint \vec{E} \cdot d\vec{A} = 0$. There is no net flux into or out of this enclosed surface. The flux, $\oint \vec{E} \cdot d\vec{A}$, will be nonzero only if one or more lines start or end within the surface. Since electric field lines start and stop only on electric charges, the flux will be nonzero only if the surface encloses a net charge. For example, the surface labeled A_1 in Fig. 22–5 encloses a positive charge and there is a net outward flux through this surface $\left(\Phi_E > 0\right)$. The surface A_2 encloses an equal magnitude negative charge and there is a net inward flux $\left(\Phi_E < 0\right)$. For the configuration shown in Fig. 22–6, the flux through the surface shown is negative (count the lines). The value of Φ_E depends on the charge enclosed by the surface, and this is what Gauss's law is all about.

FIGURE 22–4 The direction of an element of area $d\vec{A}$ is taken to point outward from an enclosed surface.

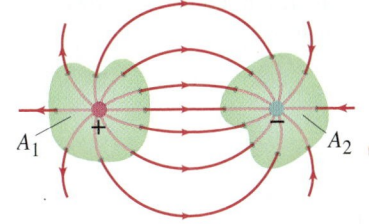

FIGURE 22–5 An electric dipole. Flux through surface A_1 is positive. Flux through A_2 is negative.

22–2 Gauss's Law

FIGURE 22–6 Net flux through surface A is negative.

The precise relation between the electric flux through a closed surface and the net charge Q_{encl} enclosed within that surface is given by **Gauss's law**:

$$\oint \vec{E} \cdot d\vec{A} = \frac{Q_{encl}}{\epsilon_0}, \qquad (22–4)$$

where ϵ_0 is the same constant (permittivity of free space) that appears in Coulomb's law. The integral on the left is over the value of $\vec{E}$ on any closed surface, and we choose that surface for our convenience in any given situation. The charge Q_{encl} is the net charge *enclosed* by that surface. It doesn't matter where or how the charge is distributed within the surface. Any charge outside this surface must not be included. A charge outside the chosen surface may affect the position of the electric field lines, but will not affect the net number of lines entering or leaving the surface. For example, Q_{encl} for the gaussian surface A_1 in Fig. 22–5 would be the positive charge enclosed by A_1; the negative charge does contribute to the electric field at A_1 but it is *not* enclosed by surface A_1 and so is not included in Q_{encl}.

Now let us see how Gauss's law is related to Coulomb's law. First, we show that Coulomb's law follows from Gauss's law. In Fig. 22–7 we have a single isolated charge Q. For our "gaussian surface," we choose an imaginary sphere of radius r centered on the charge. Because Gauss's law is supposed to be valid for any surface, we have chosen one that will make our calculation easy. Because of the *symmetry* of this (imaginary) sphere about the charge at its center, we know that $\vec{E}$ must have the same magnitude at any point on the surface, and that $\vec{E}$ points radially outward (inward for a negative charge) parallel to $d\vec{A}$, an element of the surface area. Hence, we write the integral in Gauss's law as

FIGURE 22–7 A single point charge Q at the center of an imaginary sphere of radius r (our "gaussian surface"—that is, the closed surface we choose to use for applying Gauss's law in this case).

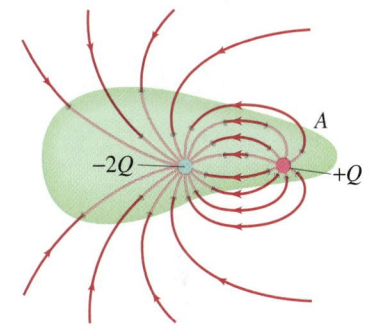

$$\oint \vec{E} \cdot d\vec{A} = \oint E\, dA = E \oint dA = E(4\pi r^2)$$

since the surface area of a sphere of radius r is $4\pi r^2$, and the magnitude of $\vec{E}$ is the same at all points on this spherical surface. Then Gauss's law becomes, with $Q_{encl} = Q$,

$$\frac{Q}{\epsilon_0} = \oint \vec{E} \cdot d\vec{A} = E(4\pi r^2)$$

because $\vec{E}$ and $d\vec{A}$ are both perpendicular to the surface at each point, and $\cos\theta = 1$. Solving for E we obtain

$$E = \frac{Q}{4\pi\epsilon_0 r^2},$$

which is the electric field form of Coulomb's law, Eq. 21–4b.

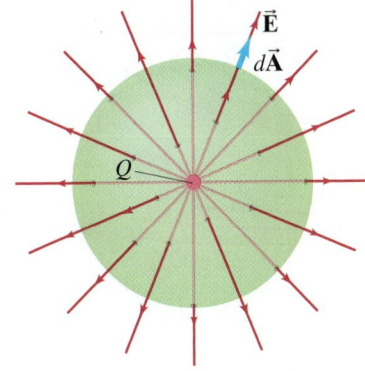

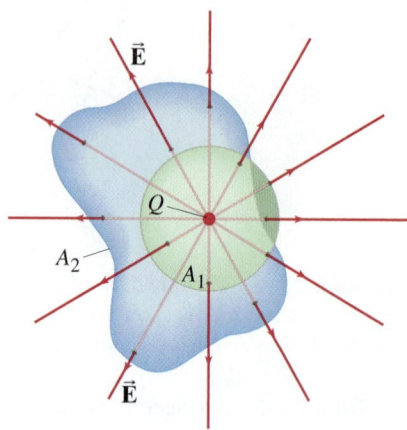

$\vec{E}$

Q

A_2

A_1

$\vec{E}$

FIGURE 22–8 A single point charge surrounded by a spherical surface, A_1, and an irregular surface, A_2.

Now let us do the reverse, and derive Gauss's law from Coulomb's law for static electric charges[†]. First we consider a single point charge Q surrounded by an imaginary spherical surface as in Fig. 22–7(and shown again, green, in Fig. 22–8). Coulomb's law tells us that the electric field at the spherical surface is $E = (1/4\pi\epsilon_0)(Q/r^2)$. Reversing the argument we just used, we have

$$\oint \vec{E} \cdot d\vec{A} = \oint \frac{1}{4\pi\epsilon_0} \frac{Q}{r^2} dA = \frac{Q}{4\pi\epsilon_0 r^2}(4\pi r^2) = \frac{Q}{\epsilon_0}.$$

This is Gauss's law, with $Q_{encl} = Q$, and we derived it for the special case of a spherical surface enclosing a point charge at its center. But what about some other surface, such as the irregular surface labeled A_2 in Fig. 22–8? The same number of field lines (due to our charge Q) pass through surface A_2, as pass through the spherical surface, A_1. Therefore, because the flux through a surface is proportional to the number of lines through it as we saw in Section 22–1, the flux through A_2 is the same as through A_1:

$$\oint_{A_2} \vec{E} \cdot d\vec{A} = \oint_{A_1} \vec{E} \cdot d\vec{A} = \frac{Q}{\epsilon_0}.$$

Hence, we can expect that

$$\oint \vec{E} \cdot d\vec{A} = \frac{Q}{\epsilon_0}$$

would be valid for *any* surface surrounding a single point charge Q.

Finally, let us look at the case of more than one charge. For each charge, Q_i, enclosed by the chosen surface,

$$\oint \vec{E}_i \cdot d\vec{A} = \frac{Q_i}{\epsilon_0},$$

where $\vec{E}_i$ refers to the electric field produced by Q_i alone. By the superposition principle for electric fields (Section 21–6), the total field $\vec{E}$ is equal to the sum of the fields due to each separate charge, $\vec{E} = \Sigma\vec{E}_i$. Hence

$$\oint \vec{E} \cdot d\vec{A} = \oint (\Sigma\vec{E}_i) \cdot d\vec{A} = \Sigma \frac{Q_i}{\epsilon_0} = \frac{Q_{encl}}{\epsilon_0},$$

where $Q_{encl} = \Sigma Q_i$ is the total net charge enclosed within the surface. Thus we see, based on this simple argument, that Gauss's law follows from Coulomb's law for any distribution of static electric charge enclosed within a closed surface of any shape.

The derivation of Gauss's law from Coulomb's law is valid for electric fields produced by static electric charges. We will see later that electric fields can also be produced by changing magnetic fields. Coulomb's law cannot be used to describe such electric fields. But Gauss's law *is* found to hold also for electric fields produced in any of these ways. Hence *Gauss's law is a more general law than Coulomb's law*. It holds for any electric field whatsoever.

Even for the case of static electric fields that we are considering in this Chapter, it is important to recognize that $\vec{E}$ on the left side of Gauss's law is not necessarily due only to the charge Q_{encl} that appears on the right. For example, in Fig. 22–9 there is an electric field $\vec{E}$ at all points on the imaginary gaussian surface, but it is not due to the charge enclosed by the surface (which is $Q_{encl} = 0$ in this case). The electric field $\vec{E}$ which appears on the left side of Gauss's law is the *total* electric field at each point, on the gaussian surface chosen, not just that due to the charge Q_{encl}, which appears on the right side. Gauss's law has been found to be valid for the total field at any surface. It tells us that any *difference* between the input and output flux of the electric field over any surface is due to charge within that surface.

FIGURE 22–9 Electric flux through a closed surface. (Same as Fig. 22–3.) No electric charge is enclosed by this surface ($Q_{encl} = 0$).

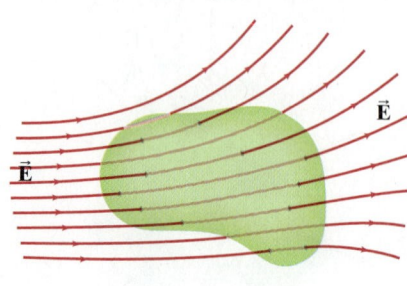

$\vec{E}$

$\vec{E}$

[†]Note that Gauss's law would look more complicated in terms of the constant $k = 1/4\pi\epsilon_0$ that we originally used in Coulomb's law (Eqs. 21–1 or 21–4a):

Coulomb's law	*Gauss's law*
$E = k\dfrac{Q}{r^2}$	$\oint \vec{E} \cdot d\vec{A} = 4\pi kQ$
$E = \dfrac{1}{4\pi\epsilon_0}\dfrac{Q}{r^2}$	$\oint \vec{E} \cdot d\vec{A} = \dfrac{Q}{\epsilon_0}.$

Gauss's law has a simpler form using ϵ_0; Coulomb's law is simpler using k. The normal convention is to use ϵ_0 rather than k because Gauss's law is considered more general and therefore it is preferable to have it in simpler form.

CONCEPTUAL EXAMPLE 22–2 **Flux from Gauss's law.** Consider the two gaussian surfaces, A_1 and A_2, shown in Fig. 22–10. The only charge present is the charge Q at the center of surface A_1. What is the net flux through each surface, A_1 and A_2?

RESPONSE The surface A_1 encloses the charge $+Q$. By Gauss's law, the net flux through A_1 is then Q/ϵ_0. For surface A_2, the charge $+Q$ is outside the surface. Surface A_2 encloses zero net charge, so the net electric flux through A_2 is zero, by Gauss's law. Note that all field lines that enter the volume enclosed by surface A_2 also leave it.

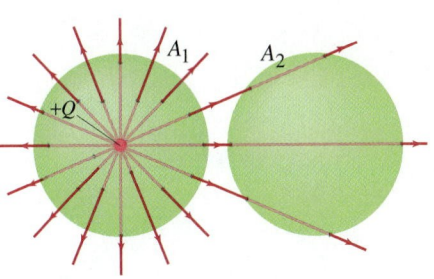

FIGURE 22–10 Example 22–2. Two gaussian surfaces.

EXERCISE B A point charge Q is at the center of a spherical gaussian surface A. When a second charge Q is placed just outside A, the total flux through this spherical surface A is (a) unchanged, (b) doubled, (c) halved, (d) none of these.

EXERCISE C Three $2.95 \, \mu C$ charges are in a small box. What is the net flux leaving the box? (a) $3.3 \times 10^{12} \, \text{N} \cdot \text{m}^2/\text{C}$, (b) $3.3 \times 10^5 \, \text{N} \cdot \text{m}^2/\text{C}$, (c) $1.0 \times 10^{12} \, \text{N} \cdot \text{m}^2/\text{C}$, (d) $1.0 \times 10^6 \, \text{N} \cdot \text{m}^2/\text{C}$, (e) $6.7 \times 10^6 \, \text{N} \cdot \text{m}^2/\text{C}$.

We note that the integral in Gauss's law is often rather difficult to carry out in practice. We rarely need to do it except for some fairly simple situations that we now discuss.

22–3 Applications of Gauss's Law

Gauss's law is a very compact and elegant way to write the relation between electric charge and electric field. It also offers a simple way to determine the electric field when the charge distribution is simple and/or possesses a high degree of *symmetry*. In order to apply Gauss's law, however, we must choose the "gaussian" surface very carefully (for the integral on the left side of Gauss's law) so we can determine $\vec{\mathbf{E}}$. We normally try to think of a surface that has just the symmetry needed so that E will be constant on all or on parts of its surface. Sometimes we choose a surface so the flux through part of the surface is zero.

EXAMPLE 22–3 **Spherical conductor.** A thin spherical shell of radius r_0 possesses a total net charge Q that is uniformly distributed on it (Fig. 22–11). Determine the electric field at points (a) outside the shell, and (b) inside the shell. (c) What if the conductor were a solid sphere?

APPROACH Because the charge is distributed symmetrically, the electric field must also be *symmetric*. Thus the field outside the sphere must be directed radially outward (inward if $Q < 0$) and must depend only on r, not on angle (spherical coordinates).

SOLUTION (a) The electric field will have the same magnitude at all points on an imaginary gaussian surface, if we choose that surface as a sphere of radius r ($r > r_0$) concentric with the shell, and shown in Fig. 22–11 as the dashed circle A_1. Because $\vec{\mathbf{E}}$ is perpendicular to this surface, the cosine of the angle between $\vec{\mathbf{E}}$ and $d\vec{\mathbf{A}}$ is always 1. Gauss's law then gives (with $Q_{\text{encl}} = Q$ in Eq. 22–4)

$$\oint \vec{\mathbf{E}} \cdot d\vec{\mathbf{A}} = E(4\pi r^2) = \frac{Q}{\epsilon_0},$$

where $4\pi r^2$ is the surface area of our sphere (gaussian surface) of radius r. Thus

$$E = \frac{1}{4\pi\epsilon_0} \frac{Q}{r^2}. \qquad [r > r_0]$$

Thus the field outside a uniformly charged spherical shell is the same as if all the charge were concentrated at the center as a point charge.

(b) Inside the shell, the electric field must also be symmetric. So E must again have the same value at all points on a spherical gaussian surface (A_2 in Fig. 22–11) concentric with the shell. Thus E can be factored out of the integral and, with $Q_{\text{encl}} = 0$ because the charge enclosed within the sphere A_2 is zero, we have

$$\oint \vec{\mathbf{E}} \cdot d\vec{\mathbf{A}} = E(4\pi r^2) = 0.$$

Hence

$$E = 0 \qquad [r < r_0]$$

inside a uniform spherical shell of charge.

(c) These same results also apply to a uniformly charged solid spherical conductor, since all the charge would lie in a thin layer at the surface.

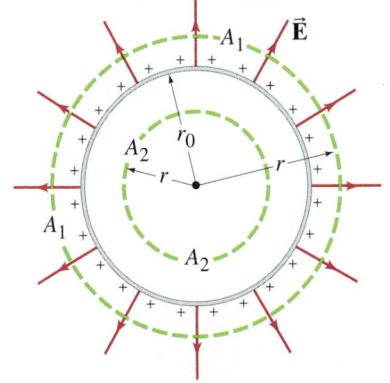

FIGURE 22–11 Cross-sectional drawing of a thin spherical shell of radius r_0, carrying a net charge Q uniformly distributed. A_1 and A_2 represent two gaussian surfaces we use to determine $\vec{\mathbf{E}}$. Example 22–3.

EXERCISE D A charge Q is placed on a hollow metal ball. We saw in Chapter 21 that the charge is all on the surface of the ball because metal is a conductor. How does the charge distribute itself on the ball? (*a*) Half on the inside surface and half on the outside surface. (*b*) Part on each surface in inverse proportion to the two radii. (*c*) Part on each surface but with a more complicated dependence on the radii than in answer (*b*). (*d*) All on the inside surface. (*e*) All on the outside surface.

EXAMPLE 22–4 **Solid sphere of charge.** An electric charge Q is distributed uniformly throughout a nonconducting sphere of radius r_0, Fig. 22–12. Determine the electric field (*a*) outside the sphere $(r > r_0)$ and (*b*) inside the sphere $(r < r_0)$.

APPROACH Since the charge is distributed symmetrically in the sphere, the electric field at all points must again be *symmetric*. $\vec{E}$ depends only on r and is directed radially outward (or inward if $Q < 0$).

SOLUTION (*a*) For our gaussian surface we choose a sphere of radius r $(r > r_0)$, labeled A_1 in Fig. 22–12. Since E depends only on r, Gauss's law gives, with $Q_{\text{encl}} = Q$,

$$\oint \vec{E} \cdot d\vec{A} = E(4\pi r^2) = \frac{Q}{\epsilon_0}$$

or

$$E = \frac{1}{4\pi\epsilon_0} \frac{Q}{r^2}.$$

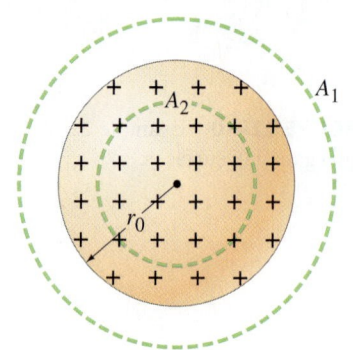

FIGURE 22–12 A solid sphere of uniform charge density. Example 22–4.

Again, the field outside a spherically symmetric distribution of charge is the same as that for a point charge of the same magnitude located at the center of the sphere.

(*b*) Inside the sphere, we choose for our gaussian surface a concentric sphere of radius r $(r < r_0)$, labeled A_2 in Fig. 22–12. From symmetry, the magnitude of $\vec{E}$ is the same at all points on A_2, and $\vec{E}$ is perpendicular to the surface, so

$$\oint \vec{E} \cdot d\vec{A} = E(4\pi r^2).$$

We must equate this to $Q_{\text{encl}}/\epsilon_0$ where Q_{encl} is the charge enclosed by A_2. Q_{encl} is not the total charge Q but only a portion of it. We define the **charge density**, ρ_E, as the charge per unit volume $(\rho_E = dQ/dV)$, and here we are given that $\rho_E = $ constant. So the charge enclosed by the gaussian surface A_2, a sphere of radius r, is

$$Q_{\text{encl}} = \left(\frac{\frac{4}{3}\pi r^3 \rho_E}{\frac{4}{3}\pi r_0^3 \rho_E} \right) Q = \frac{r^3}{r_0^3} Q.$$

Hence, from Gauss's law,

$$E(4\pi r^2) = \frac{Q_{\text{encl}}}{\epsilon_0} = \frac{r^3}{r_0^3} \frac{Q}{\epsilon_0}$$

or

$$E = \frac{1}{4\pi\epsilon_0} \frac{Q}{r_0^3} r. \qquad\qquad [r < r_0]$$

FIGURE 22–13 Magnitude of the electric field as a function of the distance r from the center of a uniformly charged solid sphere.

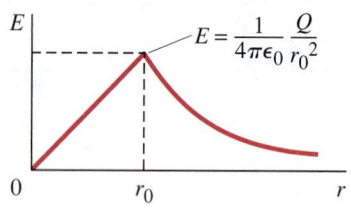

Thus the field increases linearly with r, until $r = r_0$. It then decreases as $1/r^2$, as plotted in Fig. 22–13.

EXERCISE E Return to the Chapter-Opening Question, page 591, and answer it again now. Try to explain why you may have answered differently the first time.

The results in Example 22–4 would have been difficult to obtain from Coulomb's law by integrating over the sphere. Using Gauss's law and the *symmetry* of the situation, this result is obtained rather easily, and shows the great power of Gauss's law. However, its use in this way is limited mainly to cases where the charge distribution has a high degree of symmetry. In such cases, we *choose* a simple surface on which $E = $ constant, so the integration is simple. Gauss's law holds, of course, for any surface.

EXAMPLE 22–5 **Nonuniformly charged solid sphere.** Suppose the charge density of the solid sphere in Fig. 22–12, Example 22–4, is given by $\rho_E = \alpha r^2$, where α is a constant. (a) Find α in terms of the total charge Q on the sphere and its radius r_0. (b) Find the electric field as a function of r inside the sphere.

APPROACH We divide the sphere up into concentric thin shells of thickness dr as shown in Fig. 22–14, and integrate (a) setting $Q = \int \rho_E \, dV$ and (b) using Gauss's law.

SOLUTION (a) A thin shell of radius r and thickness dr (Fig. 22–14) has volume $dV = 4\pi r^2 \, dr$. The total charge is given by

$$Q = \int \rho_E \, dV = \int_0^{r_0} (\alpha r^2)(4\pi r^2 \, dr) = 4\pi\alpha \int_0^{r_0} r^4 \, dr = \frac{4\pi\alpha}{5} r_0^5.$$

Thus $\alpha = 5Q/4\pi r_0^5$.

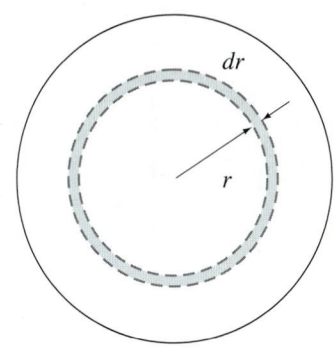

FIGURE 22–14 Example 22–5.

(b) To find E inside the sphere at distance r from its center, we apply Gauss's law to an imaginary sphere of radius r which will enclose a charge

$$Q_{\text{encl}} = \int_0^r \rho_E \, dV = \int_0^r (\alpha r^2) \, 4\pi r^2 \, dr = \int_0^r \left(\frac{5Q}{4\pi r_0^5} r^2 \right) 4\pi r^2 \, dr = Q \frac{r^5}{r_0^5}.$$

By *symmetry*, E will be the same at all points on the surface of a sphere of radius r, so Gauss's law gives

$$\oint \vec{\mathbf{E}} \cdot d\vec{\mathbf{A}} = \frac{Q_{\text{encl}}}{\epsilon_0}$$

$$(E)(4\pi r^2) = Q \frac{r^5}{\epsilon_0 r_0^5},$$

so

$$E = \frac{Qr^3}{4\pi\epsilon_0 r_0^5}.$$

EXAMPLE 22–6 **Long uniform line of charge.** A very long straight wire possesses a uniform positive charge per unit length, λ. Calculate the electric field at points near (but outside) the wire, far from the ends.

APPROACH Because of the *symmetry*, we expect the field to be directed radially outward and to depend only on the perpendicular distance, R, from the wire. Because of the cylindrical symmetry, the field will be the same at all points on a gaussian surface that is a cylinder with the wire along its axis, Fig. 22–15. $\vec{\mathbf{E}}$ is perpendicular to this surface at all points. For Gauss's law, we need a closed surface, so we include the flat ends of the cylinder. Since $\vec{\mathbf{E}}$ is parallel to the ends, there is no flux through the ends (the cosine of the angle between $\vec{\mathbf{E}}$ and $d\vec{\mathbf{A}}$ on the ends is $\cos 90° = 0$).

FIGURE 22–15 Calculation of $\vec{\mathbf{E}}$ due to a very long line of charge. Example 22–6.

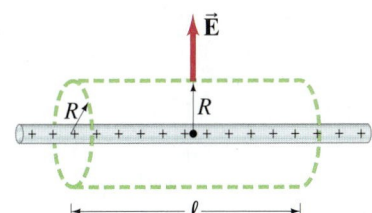

SOLUTION For our chosen gaussian surface Gauss's law gives

$$\oint \vec{\mathbf{E}} \cdot d\vec{\mathbf{A}} = E(2\pi R\ell) = \frac{Q_{\text{encl}}}{\epsilon_0} = \frac{\lambda\ell}{\epsilon_0},$$

where ℓ is the length of our chosen gaussian surface ($\ell \ll$ length of wire), and $2\pi R$ is its circumference. Hence

$$E = \frac{1}{2\pi\epsilon_0} \frac{\lambda}{R}.$$

NOTE This is the same result we found in Example 21–11 using Coulomb's law (we used x there instead of R), but here it took much less effort. Again we see the great power of Gauss's law.[†]

NOTE Recall from Chapter 10, Fig. 10–2, that we use R for the distance of a particle from an axis (cylindrical symmetry), but lower case r for the distance from a point (usually the origin 0).

[†]But note that the method of Example 21–11 allows calculation of E also for a short line of charge by using the appropriate limits for the integral, whereas Gauss's law is not readily adapted due to lack of symmetry.

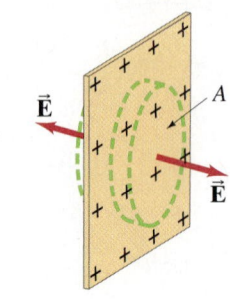

FIGURE 22–16 Calculation of the electric field outside a large uniformly charged nonconducting plane surface. Example 22–7.

EXAMPLE 22–7 **Infinite plane of charge.** Charge is distributed uniformly, with a surface charge density σ (σ = charge per unit area = dQ/dA), over a very large but very thin nonconducting flat plane surface. Determine the electric field at points near the plane.

APPROACH We choose as our gaussian surface a small closed cylinder whose axis is perpendicular to the plane and which extends through the plane as shown in Fig. 22–16. Because of the symmetry, we expect $\vec{E}$ to be directed perpendicular to the plane on both sides as shown, and to be uniform over the end caps of the cylinder, each of whose area is A.

SOLUTION Since no flux passes through the curved sides of our chosen cylindrical surface, all the flux is through the two end caps. So Gauss's law gives

$$\oint \vec{E} \cdot d\vec{A} = 2EA = \frac{Q_{encl}}{\epsilon_0} = \frac{\sigma A}{\epsilon_0},$$

where $Q_{encl} = \sigma A$ is the charge enclosed by our gaussian cylinder. The electric field is then

$$E = \frac{\sigma}{2\epsilon_0}.$$

NOTE This is the same result we obtained much more laboriously in Chapter 21, Eq. 21–7. The field is uniform for points far from the ends of the plane, and close to its surface.

EXAMPLE 22–8 **Electric field near any conducting surface.** Show that the electric field just outside the surface of any good conductor of arbitrary shape is given by

$$E = \frac{\sigma}{\epsilon_0},$$

where σ is the surface charge density on the conductor's surface at that point.

APPROACH We choose as our gaussian surface a small cylindrical box, as we did in the previous Example. We choose the cylinder to be very small in height, so that one of its circular ends is just above the conductor (Fig. 22–17). The other end is just below the conductor's surface, and the sides are perpendicular to it.

SOLUTION The electric field is zero inside a conductor and is perpendicular to the surface just outside it (Section 21–9), so electric flux passes only through the outside end of our cylindrical box; no flux passes through the short sides or inside end. We choose the area A (of the flat cylinder end) small enough so that E is essentially uniform over it. Then Gauss's law gives

$$\oint \vec{E} \cdot d\vec{A} = EA = \frac{Q_{encl}}{\epsilon_0} = \frac{\sigma A}{\epsilon_0},$$

so that

$$E = \frac{\sigma}{\epsilon_0}. \qquad \text{[at surface of conductor]} \quad \textbf{(22–5)}$$

NOTE This useful result applies for a conductor of any shape.

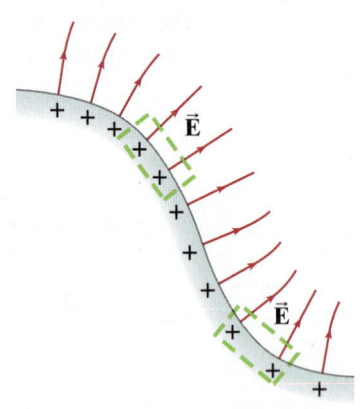

FIGURE 22–17 Electric field near surface of a conductor. Example 22–8.

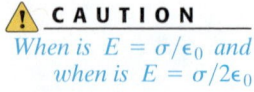

⚠ **CAUTION**
When is $E = \sigma/\epsilon_0$ and when is $E = \sigma/2\epsilon_0$

Why is it that the field outside a large plane nonconductor is $E = \sigma/2\epsilon_0$ (Example 22–7) whereas outside a conductor it is $E = \sigma/\epsilon_0$ (Example 22–8)? The reason for the factor of 2 comes not from conductor verses nonconductor. It comes instead from how we define charge per unit area σ. For a thin flat nonconductor, Fig. 22–16, the charge may be distributed throughout the volume (not only on the surface, as for a conductor). The charge per unit area σ represents all the charge throughout the thickness of the thin nonconductor. Also our gaussian surface has its ends outside the nonconductor on each side, so as to include all this charge.

For a conductor, on the other hand, the charge accumulates on the outer surfaces only. For a thin flat conductor, as shown in Fig. 22–18, the charge accumulates on both surfaces, and using the same small gaussian surface we did in Fig. 22–17, with one end inside and the other end outside the conductor, we came up with the result, $E = \sigma/\epsilon_0$. If we defined σ for a conductor, as we did for a nonconductor, σ would represent the charge per area for the entire conductor. Then Fig. 22–18 would show $\sigma/2$ as the surface charge on each surface, and Gauss's law would give $\int \vec{E} \cdot d\vec{A} = EA = (\sigma/2)A/\epsilon_0 = \sigma A/2\epsilon_0$ so $E = \sigma/2\epsilon_0$, just as for a nonconductor. We need to be careful about how we define charge per unit area σ.

We saw in Section 21–9 that in the static situation, the electric field inside any conductor must be zero even if it has a net charge. (Otherwise, the free charges in the conductor would move—until the net force on each, and hence $\vec{E}$, were zero.) We also mentioned there that any net electric charge on a conductor must all reside on its outer surface. This is readily shown using Gauss's law. Consider any charged conductor of any shape, such as that shown in Fig. 22–19, which carries a net charge Q. We choose the gaussian surface, shown dashed in the diagram, so that it all lies just below the surface of the conductor and encloses essentially the whole volume of the conductor. Our gaussian surface can be arbitrarily close to the surface, but still *inside* the conductor. The electric field is zero at all points on this gaussian surface since it is inside the conductor. Hence, from Gauss's law, Eq. 22–4, the net charge within the surface must be zero. Thus, there can be no net charge within the conductor. Any net charge must lie on the surface of the conductor.

If there is an empty cavity inside a conductor, can charge accumulate on that (inner) surface too? As shown in Fig. 22–20, if we imagine a gaussian surface (shown dashed) just inside the conductor above the cavity, we know that $\vec{E}$ must be zero everywhere on this surface since it is inside the conductor. Hence, by Gauss's law, *there can be no net charge at the surface of the cavity*.

But what if the cavity is not empty and there is a charge inside it?

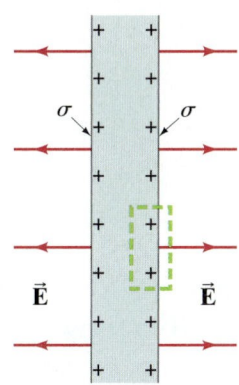

FIGURE 22–18 Thin flat charged conductor with surface charge density σ at each surface. For the conductor as a whole, the charge density is $\sigma' = 2\sigma$.

FIGURE 22–19 An insulated charged conductor of arbitrary shape, showing a gaussian surface (dashed) just below the surface of the conductor.

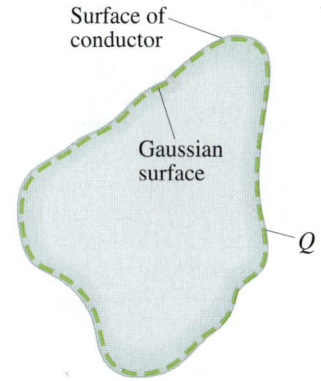

Surface of conductor

Gaussian surface

Q

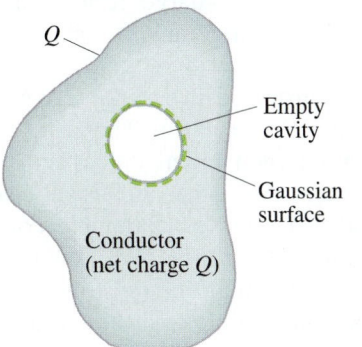

Q

Empty cavity

Gaussian surface

Conductor (net charge Q)

FIGURE 22–20 An empty cavity inside a charged conductor carries zero net charge.

FIGURE 22–21 Example 22–9.

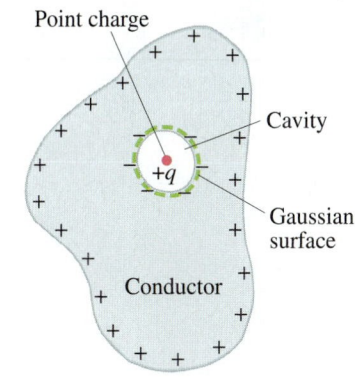

Point charge

Cavity

$+q$

Gaussian surface

Conductor

CONCEPTUAL EXAMPLE 22–9 **Conductor with charge inside a cavity.**
Suppose a conductor carries a net charge $+Q$ and contains a cavity, inside of which resides a point charge $+q$. What can you say about the charges on the inner and outer surfaces of the conductor?

RESPONSE As shown in Fig. 22–21, a gaussian surface just inside the conductor surrounding the cavity must contain zero net charge ($E = 0$ in a conductor). Thus a net charge of $-q$ must exist on the cavity surface. The conductor itself carries a net charge $+Q$, so its outer surface must carry a charge equal to $Q + q$. These results apply to a cavity of any shape.

EXERCISE F Which of the following statements about Gauss's law is correct? (*a*) If we know the charge enclosed by a surface, we always know the electric field everywhere at the surface. (*b*) When finding the electric field with Gauss's law, we always use a sphere for the gaussian surface. (*c*) If we know the total flux through a surface, we also know the total charge inside the surface. (*d*) We can only use Gauss's law if the electric field is constant in space.

Gauss's Law for Symmetric Charge Distributions

1. First identify the **symmetry** of the charge distribution: spherical, cylindrical, planar. This identification should suggest a gaussian surface for which $\vec{E}$ will be constant and/or zero on all or on parts of the surface: a sphere for spherical symmetry, a cylinder for cylindrical symmetry and a small cylinder or "pillbox" for planar symmetry.

2. Draw the appropriate gaussian surface making sure it passes through the point where you want to know the electric field.

3. Use the symmetry of the charge distribution to determine the direction of $\vec{E}$ at points on the gaussian surface.

4. Evaluate the flux, $\oint \vec{E} \cdot d\vec{A}$. With an appropriate gaussian surface, the dot product $\vec{E} \cdot d\vec{A}$ should be zero or equal to $\pm E\,dA$, with the magnitude of E being constant over all or parts of the surface.

5. Calculate the charge *enclosed* by the gaussian surface. Remember it's the enclosed charge that matters. Ignore all the charge outside the gaussian surface.

6. Equate the flux to the enclosed charge and solve for E.

FIGURE 22–22 (a) A charged conductor (metal ball) is lowered into an insulated metal can (a good conductor) carrying zero net charge. (b) The charged ball is touched to the can and all of its charge quickly flows to the outer surface of the can. (c) When the ball is then removed, it is found to carry zero net charge.

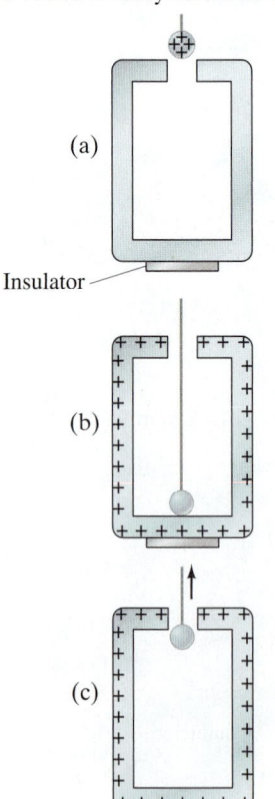

*22–4 Experimental Basis of Gauss's and Coulomb's Laws

Gauss's law predicts that any net charge on a conductor must lie only on its surface. But is this true in real life? Let us see how it can be verified experimentally. And in confirming this prediction of Gauss's law, Coulomb's law is also confirmed since the latter follows from Gauss's law, as we saw in Section 22–2. Indeed, the earliest observation that charge resides only on the outside of a conductor was recorded by Benjamin Franklin some 30 years before Coulomb stated his law.

A simple experiment is illustrated in Fig. 22–22. A metal can with a small opening at the top rests on an insulator. The can, a conductor, is initially uncharged (Fig. 22–22a). A charged metal ball (also a conductor) is lowered by an insulating thread into the can, and is allowed to touch the can (Fig. 22–22b). The ball and can now form a single conductor. Gauss's law, as discussed above, predicts that all the charge will flow to the outer surface of the can. (The flow of charge in such situations does not occur instantaneously, but the time involved is usually negligible). These predictions are confirmed in experiments by (1) connecting an electroscope to the can, which will show that the can is charged, and (2) connecting an electroscope to the ball after it has been withdrawn from the can (Fig. 22–22c), which will show that the ball carries zero charge.

The precision with which Coulomb's and Gauss's laws hold can be stated quantitatively by writing Coulomb's law as

$$F = k\frac{Q_1 Q_2}{r^{2+\delta}}.$$

For a perfect inverse-square law, $\delta = 0$. The most recent and precise experiments (1971) give $\delta = (2.7 \pm 3.1) \times 10^{-16}$. Thus Coulomb's and Gauss's laws are found to be valid to an extremely high precision!

Summary

The **electric flux** passing through a flat area A for a uniform electric field $\vec{E}$ is

$$\Phi_E = \vec{E} \cdot \vec{A}. \qquad (22\text{-}1b)$$

If the field is not uniform, the flux is determined from the integral

$$\Phi_E = \int \vec{E} \cdot d\vec{A}. \qquad (22\text{-}2)$$

The direction of the vector $\vec{A}$ or $d\vec{A}$ is chosen to be perpendicular to the surface whose area is A or dA, and points outward from an enclosed surface. The flux through a surface is proportional to the number of field lines passing through it.

Gauss's law states that the net flux passing through any closed surface is equal to the net charge Q_{encl} enclosed by the surface divided by ϵ_0:

$$\oint \vec{E} \cdot d\vec{A} = \frac{Q_{encl}}{\epsilon_0}. \qquad (22\text{-}4)$$

Gauss's law can in principle be used to determine the electric field due to a given charge distribution, but its usefulness is mainly limited to a small number of cases, usually where the charge distribution displays much symmetry. The real importance of Gauss's law is that it is a more general and elegant statement (than Coulomb's law) for the relation between electric charge and electric field. It is one of the basic equations of electromagnetism.

Questions

1. If the electric flux through a closed surface is zero, is the electric field necessarily zero at all points on the surface? Explain. What about the converse: If $\vec{E} = 0$ at all points on the surface is the flux through the surface zero?

2. Is the electric field $\vec{E}$ in Gauss's law, $\oint \vec{E} \cdot d\vec{A} = Q_{encl}/\epsilon_0$, created only by the charge Q_{encl}?

3. A point charge is surrounded by a spherical gaussian surface of radius r. If the sphere is replaced by a cube of side r, will Φ_E be larger, smaller, or the same? Explain.

4. What can you say about the flux through a closed surface that encloses an electric dipole?

5. The electric field $\vec{E}$ is zero at all points on a closed surface; is there necessarily no net charge within the surface? If a surface encloses zero net charge, is the electric field necessarily zero at all points on the surface?

6. Define gravitational flux in analogy to electric flux. Are there "sources" and "sinks" for the gravitational field as there are for the electric field? Discuss.

7. Would Gauss's law be helpful in determining the electric field due to an electric dipole?

8. A spherical basketball (a nonconductor) is given a charge Q distributed uniformly over its surface. What can you say about the electric field inside the ball? A person now steps on the ball, collapsing it, and forcing most of the air out without altering the charge. What can you say about the field inside now?

9. In Example 22–6, it may seem that the electric field calculated is due only to the charge on the wire that is enclosed by the cylinder chosen as our gaussian surface. In fact, the entire charge along the whole length of the wire contributes to the field. Explain how the charge outside the cylindrical gaussian surface of Fig. 22–15 contributes to E at the gaussian surface. [*Hint:* Compare to what the field would be due to a short wire.]

10. Suppose the line of charge in Example 22–6 extended only a short way beyond the ends of the cylinder shown in Fig. 22–15. How would the result of Example 22–6 be altered?

11. A point charge Q is surrounded by a spherical surface of radius r_0, whose center is at C. Later, the charge is moved to the right a distance $\frac{1}{2}r_0$, but the sphere remains where it was, Fig. 22–23. How is the electric flux Φ_E through the sphere changed? Is the electric field at the surface of the sphere changed? For each "yes" answer, describe the change.

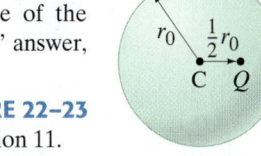

FIGURE 22–23
Question 11.

12. A solid conductor carries a net positive charge Q. There is a hollow cavity within the conductor, at whose center is a negative point charge $-q$ (Fig. 22–24). What is the charge on (a) the outer surface of the conductor and (b) the inner surface of the conductor's cavity?

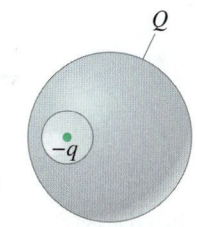

FIGURE 22–24
Question 12.

13. A point charge q is placed at the center of the cavity of a thin metal shell which is neutral. Will a charge Q placed outside the shell feel an electric force? Explain.

14. A small charged ball is inserted into a balloon. The balloon is then blown up slowly. Describe how the flux through the balloon's surface changes as the balloon is blown up. Consider both the total flux and the flux per unit surface area of the balloon.

Problems

22–1 Electric Flux

1. (I) A uniform electric field of magnitude 5.8×10^2 N/C passes through a circle of radius 13 cm. What is the electric flux through the circle when its face is (a) perpendicular to the field lines, (b) at 45° to the field lines, and (c) parallel to the field lines?

2. (I) The Earth possesses an electric field of (average) magnitude 150 N/C near its surface. The field points radially inward. Calculate the net electric flux outward through a spherical surface surrounding, and just beyond, the Earth's surface.

3. (II) A cube of side ℓ is placed in a uniform field E_0 with edges parallel to the field lines. (a) What is the net flux through the cube? (b) What is the flux through each of its six faces?

4. (II) A uniform field $\vec{E}$ is parallel to the axis of a hollow hemisphere of radius r, Fig. 22–25. (a) What is the electric flux through the hemispherical surface? (b) What is the result if $\vec{E}$ is instead perpendicular to the axis?

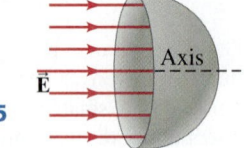

FIGURE 22–25
Problem 4.

22–2 Gauss's Law

5. (I) The total electric flux from a cubical box 28.0 cm on a side is $1.84 \times 10^3 \, \text{N} \cdot \text{m}^2/\text{C}$. What charge is enclosed by the box?

6. (I) Figure 22–26 shows five closed surfaces that surround various charges in a plane, as indicated. Determine the electric flux through each surface, S_1, S_2, S_3, S_4, and S_5. The surfaces are flat "pillbox" surfaces that extend only slightly above and below the plane in which the charges lie.

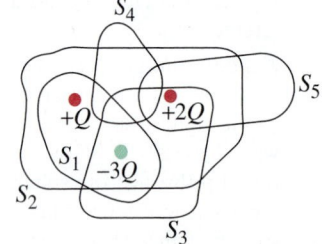

FIGURE 22–26
Problem 6.

7. (II) In Fig. 22–27, two objects, O_1 and O_2, have charges $+1.0 \, \mu\text{C}$ and $-2.0 \, \mu\text{C}$ respectively, and a third object, O_3, is electrically neutral. (a) What is the electric flux through the surface A_1 that encloses all the three objects? (b) What is the electric flux through the surface A_2 that encloses the third object only?

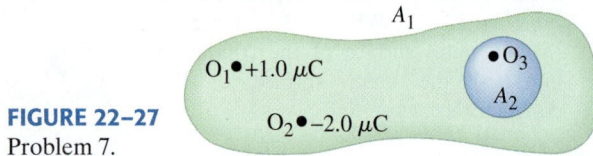

FIGURE 22–27
Problem 7.

8. (II) A ring of charge with uniform charge density is completely enclosed in a hollow donut shape. An exact copy of the ring is completely enclosed in a hollow sphere. What is the ratio of the flux out of the donut shape to that out of the sphere?

9. (II) In a certain region of space, the electric field is constant in direction (say horizontal, in the x direction), but its magnitude decreases from $E = 560 \, \text{N/C}$ at $x = 0$ to $E = 410 \, \text{N/C}$ at $x = 25 \, \text{m}$. Determine the charge within a cubical box of side $\ell = 25 \, \text{m}$, where the box is oriented so that four of its sides are parallel to the field lines (Fig. 22–28).

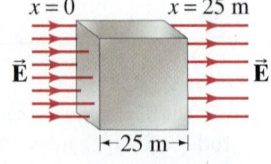

FIGURE 22–28
Problem 9.

10. (II) A point charge Q is placed at the center of a cube of side ℓ. What is the flux through one face of the cube?

11. (II) A 15.0-cm-long uniformly charged plastic rod is sealed inside a plastic bag. The total electric flux leaving the bag is $7.3 \times 10^5 \, \text{N} \cdot \text{m}^2/\text{C}$. What is the linear charge density on the rod?

22–3 Applications of Gauss's Law

12. (I) Draw the electric field lines around a negatively charged metal egg.

13. (I) The field just outside a 3.50-cm-radius metal ball is $6.25 \times 10^2 \, \text{N/C}$ and points toward the ball. What charge resides on the ball?

14. (I) Starting from the result of Example 22–3, show that the electric field just outside a uniformly charged spherical conductor is $E = \sigma/\epsilon_0$, consistent with Example 22–8.

15. (I) A long thin wire, hundreds of meters long, carries a uniformly distributed charge of $-7.2 \, \mu\text{C}$ per meter of length. Estimate the magnitude and direction of the electric field at points (a) 5.0 m and (b) 1.5 m perpendicular from the center of the wire.

16. (I) A metal globe has 1.50 mC of charge put on it at the north pole. Then -3.00 mC of charge is applied to the south pole. Draw the field lines for this system after it has come to equilibrium.

17. (II) A nonconducting sphere is made of two layers. The innermost section has a radius of 6.0 cm and a uniform charge density of $-5.0 \, \text{C/m}^3$. The outer layer has a uniform charge density of $+8.0 \, \text{C/m}^3$ and extends from an inner radius of 6.0 cm to an outer radius of 12.0 cm. Determine the electric field for (a) $0 < r < 6.0$ cm, (b) $6.0 \, \text{cm} < r < 12.0$ cm, and (c) $12.0 \, \text{cm} < r < 50.0$ cm. (d) Plot the magnitude of the electric field for $0 < r < 50.0$ cm. Is the field continuous at the edges of the layers?

18. (II) A solid metal sphere of radius 3.00 m carries a total charge of $-5.50 \, \mu\text{C}$. What is the magnitude of the electric field at a distance from the sphere's center of (a) 0.250 m, (b) 2.90 m, (c) 3.10 m, and (d) 8.00 m? How would the answers differ if the sphere were (e) a thin shell, or (f) a solid nonconductor uniformly charged throughout?

19. (II) A 15.0-cm-diameter nonconducting sphere carries a total charge of $2.25 \, \mu\text{C}$ distributed uniformly throughout its volume. Graph the electric field E as a function of the distance r from the center of the sphere from $r = 0$ to $r = 30.0$ cm.

20. (II) A flat square sheet of thin aluminum foil, 25 cm on a side, carries a uniformly distributed 275 nC charge. What, approximately, is the electric field (a) 1.0 cm above the center of the sheet and (b) 15 m above the center of the sheet?

21. (II) A spherical cavity of radius 4.50 cm is at the center of a metal sphere of radius 18.0 cm. A point charge $Q = 5.50 \, \mu\text{C}$ rests at the very center of the cavity, whereas the metal conductor carries no net charge. Determine the electric field at a point (a) 3.00 cm from the center of the cavity, (b) 6.00 cm from the center of the cavity, (c) 30.0 cm from the center.

22. (II) A point charge Q rests at the center of an uncharged thin spherical conducting shell. What is the electric field E as a function of r (a) for r less than the radius of the shell, (b) inside the shell, and (c) beyond the shell? (d) Does the shell affect the field due to Q alone? Does the charge Q affect the shell?

23. (II) A solid metal cube has a spherical cavity at its center as shown in Fig. 22–29. At the center of the cavity there is a point charge $Q = +8.00\,\mu C$. The metal cube carries a net charge $q = -6.10\,\mu C$ (not including Q). Determine (a) the total charge on the surface of the spherical cavity and (b) the total charge on the outer surface of the cube.

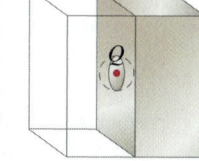

FIGURE 22–29
Problem 23.

24. (II) Two large, flat metal plates are separated by a distance that is very small compared to their height and width. The conductors are given equal but opposite uniform surface charge densities $\pm\sigma$. Ignore edge effects and use Gauss's law to show (a) that for points far from the edges, the electric field between the plates is $E = \sigma/\epsilon_0$ and (b) that outside the plates on either side the field is zero. (c) How would your results be altered if the two plates were nonconductors? (See Fig. 22–30).

FIGURE 22–30
Problems 24, 25, and 26.
$+\sigma$ $-\sigma$

25. (II) Suppose the two conducting plates in Problem 24 have the *same* sign and magnitude of charge. What then will be the electric field (a) between them and (b) outside them on either side? (c) What if the plates are nonconducting?

26. (II) The electric field between two square metal plates is $160\,N/C$. The plates are $1.0\,m$ on a side and are separated by $3.0\,cm$, as in Fig. 22–30. What is the charge on each plate? Neglect edge effects.

27. (II) Two thin concentric spherical shells of radii r_1 and r_2 $(r_1 < r_2)$ contain uniform surface charge densities σ_1 and σ_2, respectively (see Fig. 22–31). Determine the electric field for (a) $0 < r < r_1$, (b) $r_1 < r < r_2$, and (c) $r > r_2$. (d) Under what conditions will $E = 0$ for $r > r_2$? (e) Under what conditions will $E = 0$ for $r_1 < r < r_2$? Neglect the thickness of the shells.

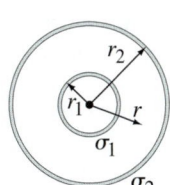

FIGURE 22–31 Two spherical shells (Problem 27).

28. (II) A spherical rubber balloon carries a total charge Q uniformly distributed on its surface. At $t = 0$ the nonconducting balloon has radius r_0 and the balloon is then slowly blown up so that r increases linearly to $2r_0$ in a time T. Determine the electric field as a function of time (a) just outside the balloon surface and (b) at $r = 3.2r_0$.

29. (II) Suppose the nonconducting sphere of Example 22–4 has a spherical cavity of radius r_1 centered at the sphere's center (Fig. 22–32). Assuming the charge Q is distributed uniformly in the "shell" (between $r = r_1$ and $r = r_0$), determine the electric field as a function of r for (a) $0 < r < r_1$, (b) $r_1 < r < r_0$, and (c) $r > r_0$.

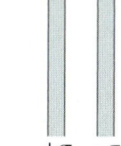

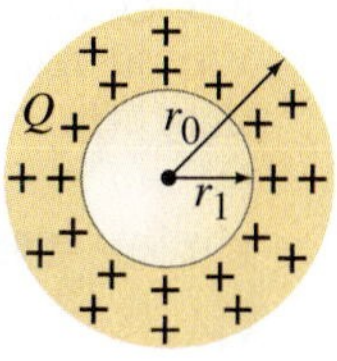

FIGURE 22–32
Problems 29, 30, 31, and 44.

30. (II) Suppose in Fig. 22–32, Problem 29, there is also a charge q at the center of the cavity. Determine the electric field for (a) $0 < r < r_1$, (b) $r_1 < r < r_0$, and (c) $r > r_0$.

31. (II) Suppose the thick spherical shell of Problem 29 is a conductor. It carries a total net charge Q and at its center there is a point charge q. What total charge is found on (a) the inner surface of the shell and (b) the outer surface of the shell? Determine the electric field for (c) $0 < r < r_1$, (d) $r_1 < r < r_0$, and (e) $r > r_0$.

32. (II) Suppose that at the center of the cavity inside the shell (charge Q) of Fig. 22–11 (and Example 22–3), there is a point charge q $(\neq \pm Q)$. Determine the electric field for (a) $0 < r < r_0$, and for (b) $r > r_0$. What are your answers if (c) $q = Q$ and (d) $q = -Q$?

33. (II) A long cylindrical shell of radius R_0 and length ℓ $(R_0 \ll \ell)$ possesses a uniform surface charge density (charge per unit area) σ (Fig. 22–33). Determine the electric field at points (a) outside the cylinder $(R > R_0)$ and (b) inside the cylinder $(0 < R < R_0)$; assume the points are far from the ends and not too far from the shell $(R \ll \ell)$. (c) Compare to the result for a long line of charge, Example 22–6. Neglect the thickness of shell.

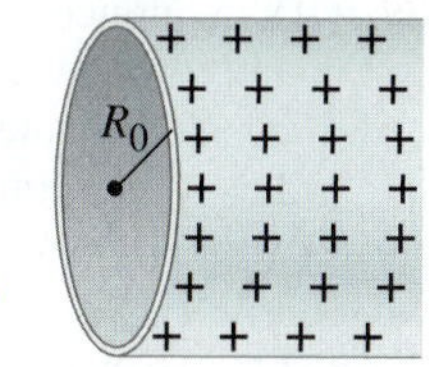

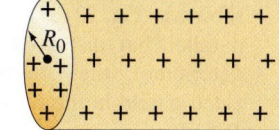

FIGURE 22–33
Problem 33.

34. (II) A very long solid nonconducting cylinder of radius R_0 and length ℓ $(R_0 \ll \ell)$ possesses a uniform volume charge density $\rho_E\,(C/m^3)$, Fig. 22–34. Determine the electric field at points (a) outside the cylinder $(R > R_0)$ and (b) inside the cylinder $(R < R_0)$. Do only for points far from the ends and for which $R \ll \ell$.

FIGURE 22–34
Problem 34.

35. (II) A thin cylindrical shell of radius R_1 is surrounded by a second concentric cylindrical shell of radius R_2 (Fig. 22–35). The inner shell has a total charge $+Q$ and the outer shell $-Q$. Assuming the length ℓ of the shells is much greater than R_1 or R_2, determine the electric field as a function of R (the perpendicular distance from the common axis of the cylinders) for (a) $0 < R < R_1$, (b) $R_1 < R < R_2$, and (c) $R > R_2$. (d) What is the kinetic energy of an electron if it moves between (and concentric with) the shells in a circular orbit of radius $(R_1 + R_2)/2$? Neglect thickness of shells.

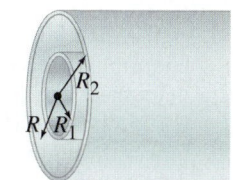

FIGURE 22–35
Problems 35, 36, and 37.

36. (II) A thin cylindrical shell of radius $R_1 = 6.5\,cm$ is surrounded by a second cylindrical shell of radius $R_2 = 9.0\,cm$, as in Fig. 22–35. Both cylinders are $5.0\,m$ long and the inner one carries a total charge $Q_1 = -0.88\,\mu C$ and the outer one $Q_2 = +1.56\,\mu C$. For points far from the ends of the cylinders, determine the electric field at a radial distance R from the central axis of (a) $3.0\,cm$, (b) $7.0\,cm$, and (c) $12.0\,cm$.

37. (II) (a) If an electron $(m = 9.1 \times 10^{-31}\,kg)$ escaped from the surface of the inner cylinder in Problem 36 (Fig. 22–35) with negligible speed, what would be its speed when it reached the outer cylinder? (b) If a proton $(m = 1.67 \times 10^{-27}\,kg)$ revolves in a circular orbit of radius $R = 7.0\,cm$ about the axis (i.e., between the cylinders), what must be its speed?

38. (II) A very long solid nonconducting cylinder of radius R_1 is uniformly charged with a charge density ρ_E. It is surrounded by a concentric cylindrical tube of inner radius R_2 and outer radius R_3 as shown in Fig. 22–36, and it too carries a uniform charge density ρ_E. Determine the electric field as a function of the distance R from the center of the cylinders for (a) $0 < R < R_1$, (b) $R_1 < R < R_2$, (c) $R_2 < R < R_3$, and (d) $R > R_3$. (e) If $\rho_E = 15\,\mu\text{C/m}^3$ and $R_1 = \frac{1}{2}R_2 = \frac{1}{3}R_3 = 5.0\,\text{cm}$, plot E as a function of R from $R = 0$ to $R = 20.0\,\text{cm}$. Assume the cylinders are very long compared to R_3.

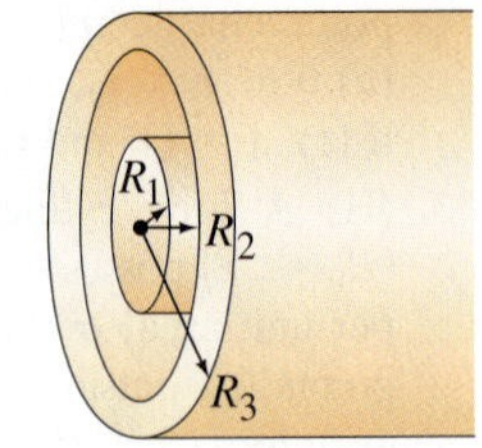

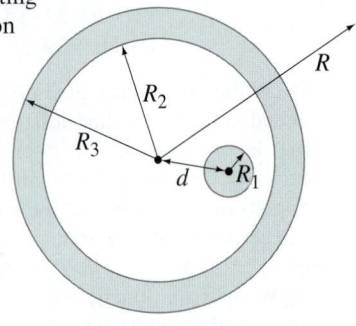

FIGURE 22–36
Problem 38.

39. (II) A nonconducting sphere of radius r_0 is uniformly charged with volume charge density ρ_E. It is surrounded by a concentric metal (conducting) spherical shell of inner radius r_1 and outer radius r_2, which carries a net charge $+Q$. Determine the resulting electric field in the regions (a) $0 < r < r_0$, (b) $r_0 < r < r_1$, (c) $r_1 < r < r_2$, and (d) $r > r_2$ where the radial distance r is measured from the center of the nonconducting sphere.

40. (II) A very long solid nonconducting cylinder of radius R_1 is uniformly charged with charge density ρ_E. It is surrounded by a cylindrical metal (conducting) tube of inner radius R_2 and outer radius R_3, which has no net charge (cross-sectional view shown in Fig. 22–37). If the axes of the two cylinders are parallel, but displaced from each other by a distance d, determine the resulting electric field in the region $R > R_3$, where the radial distance R is measured from the metal cylinder's axis. Assume $d < (R_2 - R_1)$.

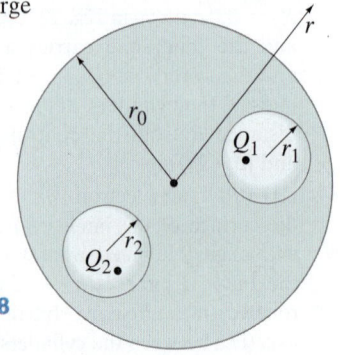

FIGURE 22–37
Problem 40.

41. (II) A flat ring (inner radius R_0, outer radius $4R_0$) is uniformly charged. In terms of the total charge Q, determine the electric field on the axis at points (a) $0.25R_0$ and (b) $75R_0$ from the center of the ring. [*Hint:* The ring can be replaced with two oppositely charged superposed disks.]

42. (II) An uncharged solid conducting sphere of radius r_0 contains two spherical cavities of radii r_1 and r_2, respectively. Point charge Q_1 is then placed within the cavity of radius r_1 and point charge Q_2 is placed within the cavity of radius r_2 (Fig. 22–38). Determine the resulting electric field (magnitude and direction) at locations outside the solid sphere ($r > r_0$), where r is the distance from its center.

FIGURE 22–38
Problem 42.

43. (III) A very large (i.e., assume infinite) flat slab of nonconducting material has thickness d and a uniform volume charge density $+\rho_E$. (a) Show that a uniform electric field exists outside of this slab. Determine its magnitude E and its direction (relative to the slab's surface). (b) As shown in Fig. 22–39, the slab is now aligned so that one of its surfaces lies on the line $y = x$. At time $t = 0$, a pointlike particle (mass m, charge $+q$) is located at position $\vec{r} = +y_0\hat{j}$ and has velocity $\vec{v} = v_0\hat{i}$. Show that the particle will collide with the slab if $v_0 \geq \sqrt{\sqrt{2}qy_0\rho_E d/\epsilon_0 m}$. Ignore gravity.

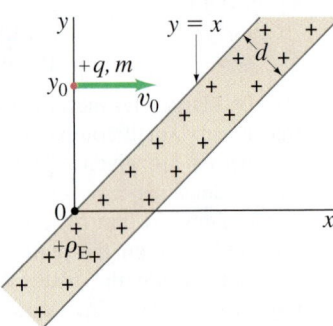

FIGURE 22–39
Problem 43.

44. (III) Suppose the density of charge between r_1 and r_0 of the hollow sphere of Problem 29 (Fig. 22–32) varies as $\rho_E = \rho_0 r_1/r$. Determine the electric field as a function of r for (a) $0 < r < r_1$, (b) $r_1 < r < r_0$, and (c) $r > r_0$. (d) Plot E versus r from $r = 0$ to $r = 2r_0$.

45. (III) Suppose two thin flat plates measure $1.0\,\text{m} \times 1.0\,\text{m}$ and are separated by $5.0\,\text{mm}$. They are oppositely charged with $\pm 15\,\mu\text{C}$. (a) Estimate the total force exerted by one plate on the other (ignore edge effects). (b) How much work would be required to move the plates from $5.0\,\text{mm}$ apart to $1.00\,\text{cm}$ apart?

46. (III) A flat slab of nonconducting material (Fig. 22–40) carries a uniform charge per unit volume, ρ_E. The slab has thickness d which is small compared to the height and breadth of the slab. Determine the electric field as a function of x (a) inside the slab and (b) outside the slab (at distances much less than the slab's height or breadth). Take the origin at the center of the slab.

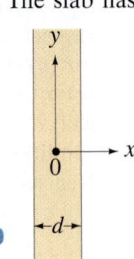

FIGURE 22–40
Problem 46.

47. (III) A flat slab of nonconducting material has thickness $2d$, which is small compared to its height and breadth. Define the x axis to be along the direction of the slab's thickness with the origin at the center of the slab (Fig. 22–41). If the slab carries a volume charge density $\rho_E(x) = -\rho_0$ in the region $-d \leq x < 0$, and $\rho_E(x) = +\rho_0$ in the region $0 < x \leq +d$, determine the electric field $\vec{E}$ as a function of x in the regions (a) outside the slab, (b) $0 < x \leq +d$, and (c) $-d \leq x < 0$. Let ρ_0 be a positive constant.

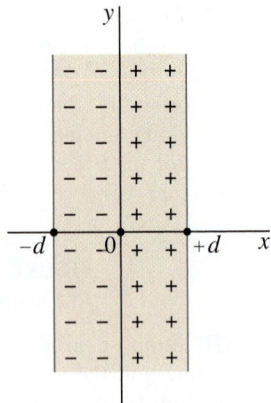

FIGURE 22–41
Problem 47.

48. (III) An extremely long, solid nonconducting cylinder has a radius R_0. The charge density within the cylinder is a function of the distance R from the axis, given by $\rho_E(R) = \rho_0(R/R_0)^2$. What is the electric field everywhere inside and outside the cylinder (far away from the ends) in terms of ρ_0 and R_0?

49. (III) Charge is distributed within a solid sphere of radius r_0 in such a way that the charge density is a function of the radial position within the sphere of the form: $\rho_E(r) = \rho_0(r/r_0)$. If the total charge within the sphere is Q (and positive), what is the electric field everywhere within the sphere in terms of Q, r_0, and the radial position r?

General Problems

50. A point charge Q is on the axis of a short cylinder at its center. The diameter of the cylinder is equal to its length ℓ (Fig. 22–42). What is the total flux through the curved sides of the cylinder? [*Hint*: First calculate the flux through the ends.]

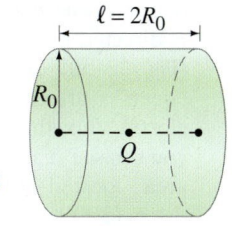

$\ell = 2R_0$

FIGURE 22–42
Problem 50.

51. Write Gauss's law for the gravitational field $\vec{g}$ (see Section 6–6).

52. The Earth is surrounded by an electric field, pointing inward at every point, of magnitude $E \approx 150\,\text{N/C}$ near the surface. (a) What is the net charge on the Earth? (b) How many excess electrons per square meter on the Earth's surface does this correspond to?

53. A cube of side ℓ has one corner at the origin of coordinates, and extends along the positive x, y, and z axes. Suppose the electric field in this region is given by $\vec{E} = (ay + b)\hat{j}$. Determine the charge inside the cube.

54. A solid nonconducting sphere of radius r_0 has a total charge Q which is distributed according to $\rho_E = br$, where ρ_E is the charge per unit volume, or charge density (C/m^3), and b is a constant. Determine (a) b in terms of Q, (b) the electric field at points inside the sphere, and (c) the electric field at points outside the sphere.

55. A point charge of $9.20\,\text{nC}$ is located at the origin and a second charge of $-5.00\,\text{nC}$ is located on the x axis at $x = 2.75\,\text{cm}$. Calculate the electric flux through a sphere centered at the origin with radius $1.00\,\text{m}$. Repeat the calculation for a sphere of radius $2.00\,\text{m}$.

56. A point charge produces an electric flux of $+235\,\text{N}\cdot\text{m}^2/\text{C}$ through a gaussian sphere of radius $15.0\,\text{cm}$ centered on the charge. (a) What is the flux through a gaussian sphere with a radius $27.5\,\text{cm}$? (b) What is the magnitude and sign of the charge?

57. A point charge Q is placed a distance $r_0/2$ above the surface of an imaginary spherical surface of radius r_0 (Fig. 22–43). (a) What is the electric flux through the sphere? (b) What range of values does E have at the surface of the sphere? (c) Is $\vec{E}$ perpendicular to the sphere at all points? (d) Is Gauss's law useful for obtaining E at the surface of the sphere?

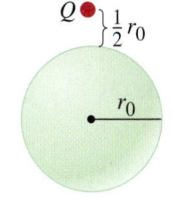

Q $\left.\right\} \frac{1}{2}r_0$

r_0

FIGURE 22–43
Problem 57.

58. Three large but thin charged sheets are parallel to each other as shown in Fig. 22–44. Sheet I has a total surface charge density of $6.5\,\text{nC/m}^2$, sheet II a charge of $-2.0\,\text{nC/m}^2$, and sheet III a charge of $5.0\,\text{nC/m}^2$. Estimate the force per unit area on each sheet, in N/m^2.

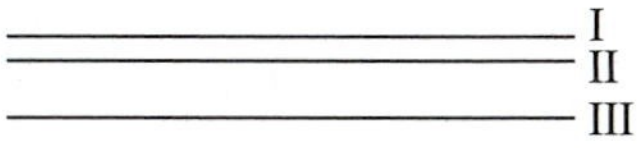

FIGURE 22–44
Problem 58.

I
II
III

59. Neutral hydrogen can be modeled as a positive point charge $+1.6 \times 10^{-19}\,\text{C}$ surrounded by a distribution of negative charge with volume density given by $\rho_E(r) = -Ae^{-2r/a_0}$ where $a_0 = 0.53 \times 10^{-10}\,\text{m}$ is called the *Bohr radius*, A is a constant such that the total amount of negative charge is $-1.6 \times 10^{-19}\,\text{C}$, and $e = 2.718\cdots$ is the base of the natural log. (a) What is the net charge inside a sphere of radius a_0? (b) What is the strength of the electric field at a distance a_0 from the nucleus? [*Hint*: Do not confuse the exponential number e with the elementary charge e which uses the same symbol but has a completely different meaning and value $(e = 1.6 \times 10^{-19}\,\text{C})$.]

60. A very large thin plane has uniform surface charge density σ. Touching it on the right (Fig. 22–45) is a long wide slab of thickness d with uniform volume charge density ρ_E. Determine the electric field (a) to the left of the plane, (b) to the right of the slab, and (c) everywhere inside the slab.

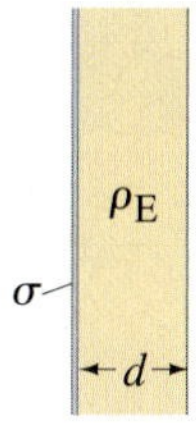

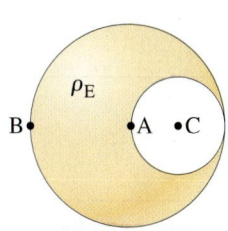

ρ_E

σ

$\leftarrow d \rightarrow$

FIGURE 22–45
Problem 60.

61. A sphere of radius r_0 carries a volume charge density ρ_E (Fig. 22–46). A spherical cavity of radius $r_0/2$ is then scooped out and left empty, as shown. (a) What is the magnitude and direction of the electric field at point A? (b) What is the direction and magnitude of the electric field at point B? Points A and C are at the centers of the respective spheres.

ρ_E

B• •A •C

FIGURE 22–46
Problem 61.

62. Dry air will break down and generate a spark if the electric field exceeds about $3 \times 10^6\,\text{N/C}$. How much charge could be packed onto the surface of a green pea (diameter $0.75\,\text{cm}$) before the pea spontaneously discharges?

63. Three very large sheets are separated by equal distances of 15.0 cm (Fig. 22–47). The first and third sheets are very thin and nonconducting and have charge per unit area σ of $+5.00\,\mu\text{C/m}^2$ and $-5.00\,\mu\text{C/m}^2$ respectively. The middle sheet is conducting but has no net charge. (a) What is the electric field inside the middle sheet? What is the electric field (b) between the left and middle sheets, and (c) between the middle and right sheets? (d) What is the charge density on the surface of the middle sheet facing the left sheet, and (e) on the surface facing the right sheet?

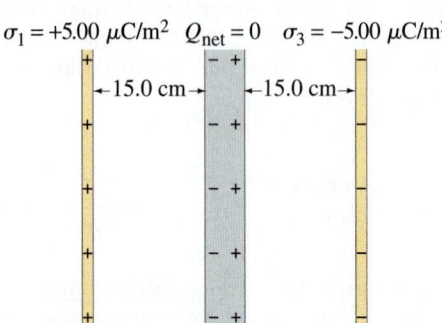

FIGURE 22–47   Problem 63.

64. In a cubical volume, 0.70 m on a side, the electric field is

$$\vec{\mathbf{E}} \;=\; E_0\!\left(1 + \frac{z}{a}\right)\hat{\mathbf{i}} \;+\; E_0\!\left(\frac{z}{a}\right)\hat{\mathbf{j}}$$

where $E_0 = 0.125\,\text{N/C}$ and $a = 0.70\,\text{m}$. The cube has its sides parallel to the coordinate axes, Fig. 22–48. Determine the net charge within the cube.

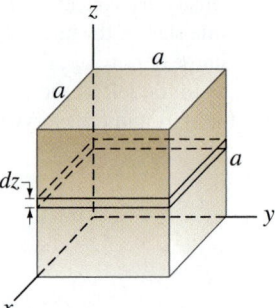

FIGURE 22–48
Problem 64.

65. A conducting spherical shell (Fig. 22–49) has inner radius = 10.0 cm, outer radius = 15.0 cm, and has a $+3.0\,\mu\text{C}$ point charge at the center. A charge of $-3.0\,\mu\text{C}$ is put on the conductor. (a) Where on the conductor does the $-3.0\,\mu\text{C}$ end up? (b) What is the electric field both inside and outside the shell?

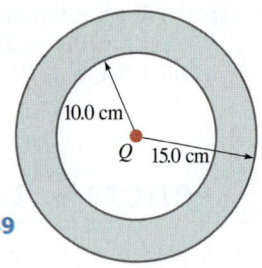

FIGURE 22–49
Problem 65.

66. A hemisphere of radius R is placed in a charge-free region of space where a uniform electric field exists of magnitude E directed perpendicular to the hemisphere's circular base (Fig. 22–50). (a) Using the definition of Φ_E through an "open" surface, calculate (via explicit integration) the electric flux through the hemisphere. [*Hint*: In Fig. 22–50 you can see that, on the surface of a sphere, the infinitesimal area located between the angles θ and $\theta + d\theta$ is $dA = (2\pi R \sin\theta)(R\,d\theta) = 2\pi R^2 \sin\theta\,d\theta$.] (b) Choose an appropriate gaussian surface and use Gauss's law to much more easily obtain the same result for the electric flux through the hemisphere.

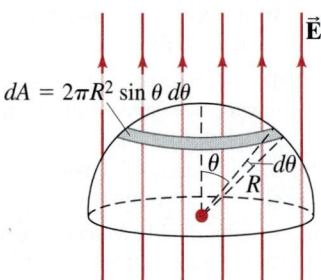

FIGURE 22–50
Problem 66.

*Numerical/Computer

*67. (III) An electric field is given by

$$\vec{\mathbf{E}} \;=\; E_{x0}\,e^{-\left(\frac{x+y}{a}\right)^2}\hat{\mathbf{i}} \;+\; E_{y0}\,e^{-\left(\frac{x+y}{a}\right)^2}\hat{\mathbf{j}},$$

where $E_{x0} = 50\,\text{N/C}$, $E_{y0} = 25\,\text{N/C}$, and $a = 1.0\,\text{m}$. Given a cube with sides parallel to the coordinate axes, with one corner at the origin (as in Fig. 22–48), and with sides of length 1.0 m, estimate the flux out of the cube using a spreadsheet or other numerical method. How much total charge is enclosed by the cube?

Answers to Exercises

A: (d).

B: (a).

C: (d).

D: (e).

E: (c).

F: (c).

We are used to voltage in our lives—a 12-volt car battery, 110 V or 220 V at home, 1.5 volt flashlight batteries, and so on. Here we see a Van de Graaff generator, whose voltage may reach 50,000 V or more. Voltage is the same as electric potential difference between two points. Electric potential is defined as the potential energy per unit charge.

The children here, whose hair stands on end because each hair has received the same sign of charge, are not harmed by the voltage because the Van de Graaff cannot provide much current before the voltage drops. (It is current through the body that is harmful, as we will see later.)

Electric Potential

C H A P T E R

23

CHAPTER-OPENING QUESTION—Guess now!

Consider a pair of parallel plates with equal and opposite charge densities, σ. Which of the following actions will increase the voltage between the plates (assuming fixed charge density)?

(a) Moving the plates closer together.
(b) Moving the plates apart.
(c) Doubling the area of the plates.
(d) Halving the area of the plates.

W e saw in Chapters 7 and 8 that the concept of energy was extremely useful in dealing with the subject of mechanics. The energy point of view is especially useful for electricity. It not only extends the law of conservation of energy, but it gives us another way to view electrical phenomena. Energy is also a powerful tool for solving Problems more easily in many cases than by using forces and electric fields.

CONTENTS

23–1 Electric Potential Energy and Potential Difference

23–2 Relation between Electric Potential and Electric Field

23–3 Electric Potential Due to Point Charges

23–4 Potential Due to Any Charge Distribution

23–5 Equipotential Surfaces

23–6 Electric Dipole Potential

23–7 $\vec{E}$ Determined from V

23–8 Electrostatic Potential Energy; the Electron Volt

*23–9 Cathode Ray Tube: TV and Computer Monitors, Oscilloscope

23–1 Electric Potential Energy and Potential Difference

Electric Potential Energy

To apply conservation of energy, we need to define electric potential energy as we did for other types of potential energy. As we saw in Chapter 8, potential energy can be defined only for a conservative force. The work done by a conservative force in moving an object between any two positions is independent of the path taken. The electrostatic force between any two charges (Eq. 21–1, $F = kQ_1Q_2/r^2$) is conservative since the dependence on position is just like the gravitational force, $1/r^2$, which we saw in Section 8–7 is conservative. Hence we can define potential energy U for the electrostatic force.

607

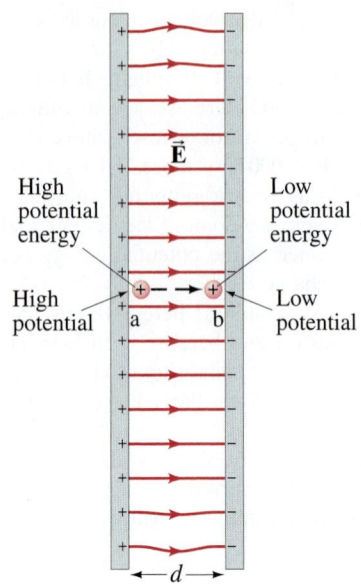

FIGURE 23–1 Work is done by the electric field in moving the positive charge q from position a to position b.

We saw in Chapter 8 that the change in potential energy between two points, a and b, equals the negative of the work done by the conservative force as an object moves from a to b: $\Delta U = -W$.

Thus we define the change in electric potential energy, $U_b - U_a$, when a point charge q moves from some point a to another point b, as the negative of the work done by the electric force as the charge moves from a to b. For example, consider the electric field between two equally but oppositely charged parallel plates; we assume their separation is small compared to their width and height, so the field $\vec{E}$ will be uniform over most of the region, Fig. 23–1. Now consider a tiny positive point charge q placed at point a very near the positive plate as shown. This charge q is so small it has no effect on $\vec{E}$. If this charge q at point a is released, the electric force will do work on the charge and accelerate it toward the negative plate. The work W done by the electric field E to move the charge a distance d is

$$W = Fd = qEd$$

where we used Eq. 21–5, $F = qE$. The change in electric potential energy equals the negative of the work done by the electric force:

$$U_b - U_a = -W = -qEd \qquad \text{[uniform } \vec{E}\text{]} \quad \textbf{(23–1)}$$

for this case of uniform electric field $\vec{E}$. In the case illustrated, the potential energy decreases (ΔU is negative); and as the charged particle accelerates from point a to point b in Fig. 23–1, the particle's kinetic energy K increases—by an equal amount. In accord with the conservation of energy, electric potential energy is transformed into kinetic energy, and the total energy is conserved. Note that the positive charge q has its greatest potential energy at point a, near the positive plate.[†] The reverse is true for a negative charge: its potential energy is greatest near the negative plate.

Electric Potential and Potential Difference

In Chapter 21, we found it useful to define the electric field as the force per unit charge. Similarly, it is useful to define the **electric potential** (or simply the **potential** when "electric" is understood) as the *electric potential energy per unit charge*. Electric potential is given the symbol V. If a positive test charge q in an electric field has electric potential energy U_a at some point a (relative to some zero potential energy), the electric potential V_a at this point is

$$V_a = \frac{U_a}{q}. \qquad \textbf{(23–2a)}$$

As we discussed in Chapter 8, only differences in potential energy are physically meaningful. Hence only the **difference in potential**, or the **potential difference**, between two points a and b (such as between a and b in Fig. 23–1) is measurable. When the electric force does positive work on a charge, the kinetic energy increases and the potential energy decreases. The difference in potential energy, $U_b - U_a$, is equal to the negative of the work, W_{ba}, done by the electric field as the charge moves from a to b; so the potential difference V_{ba} is

$$V_{ba} = \Delta V = V_b - V_a = \frac{U_b - U_a}{q} = -\frac{W_{ba}}{q}. \qquad \textbf{(23–2b)}$$

Note that electric potential, like electric field, does not depend on our test charge q. V depends on the other charges that create the field, not on q; q acquires potential energy by being in the potential V due to the other charges.

We can see from our definition that the positive plate in Fig. 23–1 is at a higher potential than the negative plate. Thus a positively charged object moves naturally from a high potential to a low potential. A negative charge does the reverse.

The unit of electric potential, and of potential difference, is joules/coulomb and is given a special name, the **volt**, in honor of Alessandro Volta (1745–1827) who is best known for inventing the electric battery. The volt is abbreviated V, so $1\,\text{V} = 1\,\text{J/C}$. Potential difference, since it is measured in volts, is often referred to as **voltage**.

[†] At this point the charge has its greatest ability to do work (on some other object or system).

If we wish to speak of the potential V_a at some point a, we must be aware that V_a depends on where the potential is chosen to be zero. The zero for electric potential in a given situation can be chosen arbitrarily, just as for potential energy, because only differences in potential energy can be measured. Often the ground, or a conductor connected directly to the ground (the Earth), is taken as zero potential, and other potentials are given with respect to ground. (Thus, a point where the voltage is 50 V is one where the difference of potential between it and ground is 50 V.) In other cases, as we shall see, we may choose the potential to be zero at an infinite distance ($r = \infty$).

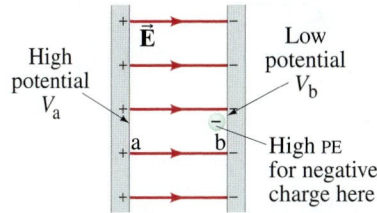

FIGURE 23–2 Central part of Fig. 23–1, showing a negative point charge near the negative plate, where its potential energy (PE) is high. Example 23–1.

CONCEPTUAL EXAMPLE 23–1 **A negative charge.** Suppose a negative charge, such as an electron, is placed near the negative plate in Fig. 23–1, at point b, shown here in Fig. 23–2. If the electron is free to move, will its electric potential energy increase or decrease? How will the electric potential change?

RESPONSE An electron released at point b will move toward the positive plate. As the electron moves toward the positive plate, its potential energy *decreases* as its kinetic energy gets larger, so $U_a < U_b$ and $\Delta U = U_a - U_b < 0$. But note that the electron moves from point b at low potential to point a at higher potential: $V_{ab} = V_a - V_b > 0$. (Potentials V_a and V_b are due to the charges on the plates, not due to the electron.) The sign of ΔU and ΔV are opposite because of the negative charge.

 **CAUTION**

A negative charge has high potential energy when potential V is low

Because the electric potential difference is defined as the potential energy difference per unit charge, then the change in potential energy of a charge q when moved between two points a and b is

$$\Delta U = U_b - U_a = q(V_b - V_a) = qV_{ba}. \qquad \text{(23–3)}$$

That is, if an object with charge q moves through a potential difference V_{ba}, its potential energy changes by an amount qV_{ba}. For example, if the potential difference between the two plates in Fig. 23–1 is 6 V, then a +1 C charge moved (say by an external force) from point b to point a will gain $(1\,\text{C})(6\,\text{V}) = 6\,\text{J}$ of electric potential energy. (And it will lose 6 J of electric potential energy if it moves from a to b.) Similarly, a +2 C charge will gain 12 J, and so on. Thus, electric potential difference is a measure of how much energy an electric charge can acquire in a given situation. And, since energy is the ability to do work, the electric potential difference is also a measure of how much work a given charge can do. The exact amount depends both on the potential difference and on the charge.

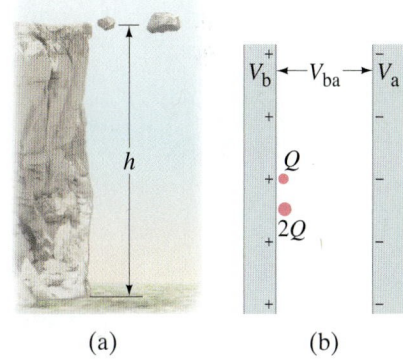

FIGURE 23–3 (a) Two rocks are at the same height. The larger rock has more potential energy. (b) Two charges have the same electric potential. The 2Q charge has more potential energy.

To better understand electric potential, let's make a comparison to the gravitational case when a rock falls from the top of a cliff. The greater the height, h, of a cliff, the more potential energy ($= mgh$) the rock has at the top of the cliff, relative to the bottom, and the more kinetic energy it will have when it reaches the bottom. The actual amount of kinetic energy it will acquire, and the amount of work it can do, depends both on the height of the cliff and the mass m of the rock. A large rock and a small rock can be at the same height h (Fig. 23–3a) and thus have the same "gravitational potential," but the larger rock has the greater potential energy (it has more mass). The electrical case is similar (Fig. 23–3b): the potential energy change, or the work that can be done, depends both on the potential difference (corresponding to the height of the cliff) and on the charge (corresponding to mass), Eq. 23–3. But note a significant difference: electric charge comes in two types, + and −, whereas gravitational mass is always +.

Sources of electrical energy such as batteries and electric generators are meant to maintain a potential difference. The actual amount of energy transformed by such a device depends on how much charge flows, as well as the potential difference (Eq. 23–3). For example, consider an automobile headlight connected to a 12.0-V battery. The amount of energy transformed (into light and thermal energy) is proportional to how much charge flows, which depends on how long the light is on. If over a given period of time 5.0 C of charge flows through the light, the total energy transformed is $(5.0\,\text{C})(12.0\,\text{V}) = 60\,\text{J}$. If the headlight is left on twice as long, 10.0 C of charge will flow and the energy transformed is $(10.0\,\text{C})(12.0\,\text{V}) = 120\,\text{J}$. Table 23–1 presents some typical voltages.

TABLE 23–1 Some Typical Potential Differences (Voltages)

Source	Voltage (approx.)
Thundercloud to ground	10^8 V
High-voltage power line	10^5–10^6 V
Power supply for TV tube	10^4 V
Automobile ignition	10^4 V
Household outlet	10^2 V
Automobile battery	12 V
Flashlight battery	1.5 V
Resting potential across nerve membrane	10^{-1} V
Potential changes on skin (EKG and EEG)	10^{-4} V

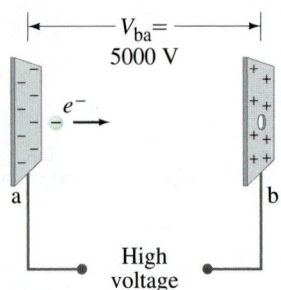

FIGURE 23–4 Electron accelerated in CRT. Example 23–2.

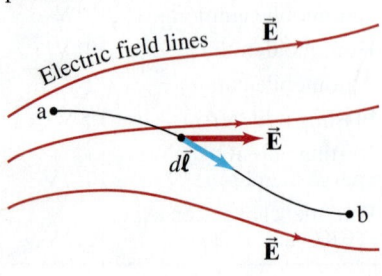

FIGURE 23–5 To find V_{ba} in a nonuniform electric field $\vec{E}$, we integrate $\vec{E} \cdot d\vec{\ell}$ from point a to point b.

EXAMPLE 23–2 **Electron in CRT.** Suppose an electron in a cathode ray tube (Section 23–9) is accelerated from rest through a potential difference $V_b - V_a = V_{ba} = +5000\,\text{V}$ (Fig. 23–4). (a) What is the change in electric potential energy of the electron? (b) What is the speed of the electron $(m = 9.1 \times 10^{-31}\,\text{kg})$ as a result of this acceleration?

APPROACH The electron, accelerated toward the positive plate, will decrease in potential energy by an amount $\Delta U = qV_{ba}$ (Eq. 23–3). The loss in potential energy will equal its gain in kinetic energy (energy conservation).

SOLUTION (a) The charge on an electron is $q = -e = -1.6 \times 10^{-19}\,\text{C}$. Therefore its change in potential energy is

$$\Delta U = qV_{ba} = (-1.6 \times 10^{-19}\,\text{C})(+5000\,\text{V}) = -8.0 \times 10^{-16}\,\text{J}.$$

The minus sign indicates that the potential energy decreases. The potential difference V_{ba} has a positive sign since the final potential V_b is higher than the initial potential V_a. Negative electrons are attracted toward a positive electrode and repelled away from a negative electrode.

(b) The potential energy lost by the electron becomes kinetic energy K. From conservation of energy (Eq. 8–9a), $\Delta K + \Delta U = 0$, so

$$\Delta K = -\Delta U$$
$$\tfrac{1}{2}mv^2 - 0 = -q(V_b - V_a) = -qV_{ba},$$

where the initial kinetic energy is zero since we are given that the electron started from rest. We solve for v:

$$v = \sqrt{-\frac{2qV_{ba}}{m}} = \sqrt{-\frac{2(-1.6 \times 10^{-19}\,\text{C})(5000\,\text{V})}{9.1 \times 10^{-31}\,\text{kg}}} = 4.2 \times 10^7\,\text{m/s}.$$

NOTE The electric potential energy does not depend on the mass, only on the charge and voltage. The speed *does* depend on m.

23–2 Relation between Electric Potential and Electric Field

The effects of any charge distribution can be described either in terms of electric field or in terms of electric potential. Electric potential is often easier to use because it is a scalar, as compared to electric field which is a vector. There is a crucial connection between the electric potential produced by a given arrangement of charges and the electric field due to those charges, which we now examine.

We start by recalling the relation between a conservative force $\vec{F}$ and the potential energy U associated with that force. As discussed in Section 8–2, the difference in potential energy between any two points in space, a and b, is given by Eq. 8–4:

$$U_b - U_a = -\int_a^b \vec{F} \cdot d\vec{\ell},$$

where $d\vec{\ell}$ is an infinitesimal increment of displacement, and the integral is taken along any path in space from point a to point b. For the electrical case, we are more interested in the potential difference, given by Eq. 23–2b, $V_{ba} = V_b - V_a = (U_b - U_a)/q$, rather than in the potential energy itself. Also, the electric field $\vec{E}$ at any point in space is defined as the force per unit charge (Eq. 21–3): $\vec{E} = \vec{F}/q$. Putting these two relations in the above equation gives us

$$V_{ba} = V_b - V_a = -\int_a^b \vec{E} \cdot d\vec{\ell}. \qquad \textbf{(23–4a)}$$

This is the general relation between electric field and potential difference. See Fig. 23–5. If we are given the electric field due to some arrangement of electric charge, we can use Eq. 23–4a to determine V_{ba}.

A simple special case is a uniform field. In Fig. 23–1, for example, a path parallel to the electric field lines from point a at the positive plate to point b at the negative plate gives (since $\vec{E}$ and $d\vec{\ell}$ are in the same direction at each point),

$$V_{ba} = V_b - V_a = -\int_a^b \vec{E} \cdot d\vec{\ell} = -E \int_a^b d\ell = -Ed$$

or

$$V_{ba} = -Ed \qquad \text{[only if } E \text{ is uniform] } \textbf{(23–4b)}$$

where d is the distance, parallel to the field lines, between points a and b. Be careful not to use Eq. 23–4b unless you are sure the electric field is uniform.

From either of Eqs. 23–4 we can see that the units for electric field intensity can be written as volts per meter (V/m) as well as newtons per coulomb (N/C). These are equivalent in general, since $1\,\text{N/C} = 1\,\text{N·m/C·m} = 1\,\text{J/C·m} = 1\,\text{V/m}$.

EXERCISE A Return to the Chapter-Opening Question, page 607, and answer it again now. Try to explain why you may have answered differently the first time.

EXAMPLE 23–3 **Electric field obtained from voltage.** Two parallel plates are charged to produce a potential difference of 50 V. If the separation between the plates is 0.050 m, calculate the magnitude of the electric field in the space between the plates (Fig. 23–6).

APPROACH We apply Eq. 23–4b to obtain the magnitude of E, assumed uniform.

SOLUTION The electric field magnitude is $E = V_{ba}/d = (50\,\text{V}/0.050\,\text{m}) = 1000\,\text{V/m}$.

EXAMPLE 23–4 **Charged conducting sphere.** Determine the potential at a distance r from the center of a charged conducting sphere of radius r_0 for (a) $r > r_0$, (b) $r = r_0$, (c) $r < r_0$. The total charge on the sphere is Q.

APPROACH The charge Q is distributed over the surface of the sphere since it is a conductor. We saw in Example 22–3 that the electric field outside a conducting sphere is

$$E = \frac{1}{4\pi\epsilon_0} \frac{Q}{r^2} \qquad [r > r_0]$$

and points radially outward (inward if $Q < 0$). Since we know $\vec{E}$, we can start by using Eq. 23–4a.

SOLUTION (a) We use Eq. 23–4a and integrate along a radial line with $d\vec{\ell}$ parallel to $\vec{E}$ (Fig. 23–7) between two points which are distances r_a and r_b from the sphere's center:

$$V_b - V_a = -\int_{r_a}^{r_b} \vec{E} \cdot d\vec{\ell} = -\frac{Q}{4\pi\epsilon_0} \int_{r_a}^{r_b} \frac{dr}{r^2} = \frac{Q}{4\pi\epsilon_0} \left(\frac{1}{r_b} - \frac{1}{r_a} \right)$$

and we set $d\ell = dr$. If we let $V = 0$ for $r = \infty$ (let's choose $V_b = 0$ at $r_b = \infty$), then at any other point r (for $r > r_0$) we have

$$V = \frac{1}{4\pi\epsilon_0} \frac{Q}{r}. \qquad [r > r_0]$$

We will see in the next Section that this same equation applies for the potential a distance r from a single point charge. Thus the electric potential outside a spherical conductor with a uniformly distributed charge is the same as if all the charge were at its center.

(b) As r approaches r_0, we see that

$$V = \frac{1}{4\pi\epsilon_0} \frac{Q}{r_0} \qquad [r = r_0]$$

at the surface of the conductor.

(c) For points within the conductor, $E = 0$. Thus the integral, $\int \vec{E} \cdot d\vec{\ell}$, between $r = r_0$ and any point within the conductor gives zero change in V. Hence V is constant within the conductor:

$$V = \frac{1}{4\pi\epsilon_0} \frac{Q}{r_0}. \qquad [r \leq r_0]$$

The whole conductor, not just its surface, is at this same potential. Plots of both E and V as a function of r are shown in Fig. 23–8 for a positively charged conducting sphere.

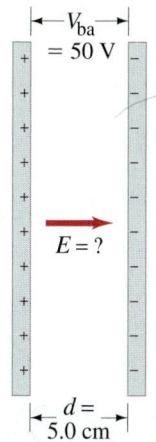

FIGURE 23–6 Example 23–3.

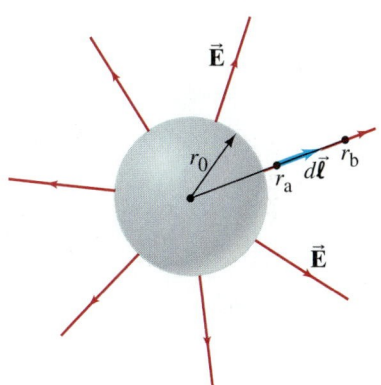

FIGURE 23–7 Example 23–4. Integrating $\vec{E} \cdot d\vec{\ell}$ for the field outside a spherical conductor.

FIGURE 23–8 (a) E versus r, and (b) V versus r, for a positively charged solid conducting sphere of radius r_0 (the charge distributes itself evenly on the surface); r is the distance from the center of the sphere.

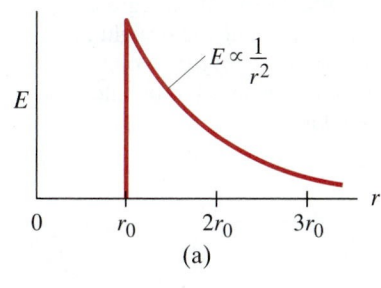

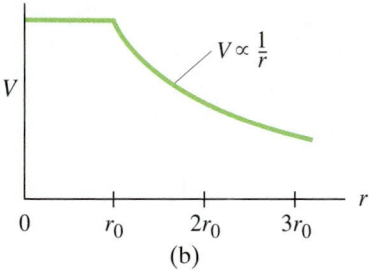

EXAMPLE 23–5 **Breakdown voltage.** In many kinds of equipment, very high voltages are used. A problem with high voltage is that the air can become ionized due to the high electric fields: free electrons in the air (produced by cosmic rays, for example) can be accelerated by such high fields to speeds sufficient to ionize O_2 and N_2 molecules by collision, knocking out one or more of their electrons. The air then becomes conducting and the high voltage cannot be maintained as charge flows. The breakdown of air occurs for electric fields of about 3×10^6 V/m. (a) Show that the breakdown voltage for a spherical conductor in air is proportional to the radius of the sphere, and (b) estimate the breakdown voltage in air for a sphere of diameter 1.0 cm.

APPROACH The electric potential at the surface of a spherical conductor of radius r_0 (Example 23–4), and the electric field just outside its surface, are

$$V = \frac{1}{4\pi\epsilon_0}\frac{Q}{r_0} \quad \text{and} \quad E = \frac{1}{4\pi\epsilon_0}\frac{Q}{r_0^2}.$$

SOLUTION (a) We combine these two equations and obtain

$$V = r_0 E. \qquad \text{[at surface of spherical conductor]}$$

(b) For $r_0 = 5 \times 10^{-3}$ m, the breakdown voltage in air is

$$V = (5 \times 10^{-3}\,\text{m})(3 \times 10^6\,\text{V/m}) \approx 15{,}000\,\text{V}.$$

When high voltages are present, a glow may be seen around sharp points, known as a **corona discharge**, due to the high electric fields at these points which ionize air molecules. The light we see is due to electrons jumping down to empty lower states. **Lightning rods**, with their sharp tips, are intended to ionize the surrounding air when a storm cloud is near, and to provide a conduction path to discharge a dangerous high-voltage cloud slowly, over a period of time. Thus lightning rods, connected to the ground, are intended to draw electric charge off threatening clouds before a large buildup of charge results in a swift destructive lightning bolt.

EXERCISE B On a dry day, a person can become electrically charged by rubbing against rugs and other ordinary objects. Suppose you notice a small shock as you reach for a metal doorknob, noting that the shock occurs along with a tiny spark when your hand is about 3.0 mm from the doorknob. As a rough estimate, use Eq. 23–4b to estimate the potential difference between your hand and the doorknob. (a) 9 V; (b) 90 V; (c) 900 V; (d) 9000 V; (e) none of these.

23–3 Electric Potential Due to Point Charges

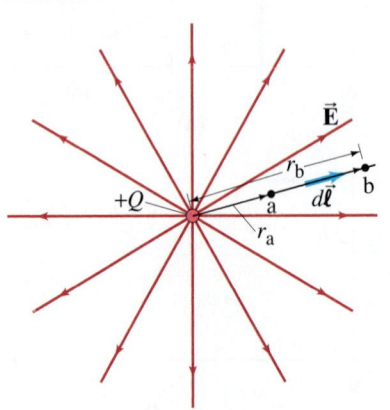

FIGURE 23–9 We integrate Eq. 23–4a along the straight line (shown in black) from point a to point b. The line ab is parallel to a field line.

The electric potential at a distance r from a single point charge Q can be derived directly from Eq. 23–4a, $V_b - V_a = -\int \vec{E} \cdot d\vec{\ell}$. The electric field due to a single point charge has magnitude (Eq. 21–4)

$$E = \frac{1}{4\pi\epsilon_0}\frac{Q}{r^2} \quad \text{or} \quad E = k\frac{Q}{r^2}$$

(where $k = 1/4\pi\epsilon_0 = 8.99 \times 10^9$ N·m²/C²), and is directed radially outward from a positive charge (inward if $Q < 0$). We take the integral in Eq. 23–4a along a (straight) field line (Fig. 23–9) from point a, a distance r_a from Q, to point b, a distance r_b from Q. Then $d\vec{\ell}$ will be parallel to $\vec{E}$ and $d\ell = dr$. Thus

$$V_b - V_a = -\int_{r_a}^{r_b}\vec{E}\cdot d\vec{\ell} = -\frac{Q}{4\pi\epsilon_0}\int_{r_a}^{r_b}\frac{1}{r^2}\,dr = \frac{1}{4\pi\epsilon_0}\left(\frac{Q}{r_b} - \frac{Q}{r_a}\right).$$

As mentioned earlier, only differences in potential have physical meaning. We are free, therefore, to choose the value of the potential at some one point to be whatever we please. It is common to choose the potential to be zero at infinity (let $V_b = 0$ at $r_b = \infty$). Then the electric potential V at a distance r from a single point charge is

$$V = \frac{1}{4\pi\epsilon_0}\frac{Q}{r}. \qquad \begin{bmatrix} \text{single point charge;} \\ V = 0 \text{ at } r = \infty \end{bmatrix} \qquad \textbf{(23–5)}$$

We can think of V here as representing the absolute potential, where $V = 0$ at $r = \infty$, or we can think of V as the potential difference between r and infinity. Notice that the potential V decreases with the first power of the distance, whereas the electric field (Eq. 21–4) decreases as the *square* of the distance. The potential near a positive charge is large, and it decreases toward zero at very large distances (Fig. 23–10). For a negative charge, the potential is negative and increases toward zero at large distances (Fig. 23–11). Equation 23–5 is sometimes called the **Coulomb potential** (it has its origin in Coulomb's law).

In Example 23–4 we found that the potential due to a uniformly charged sphere is given by the same relation, Eq. 23–5, for points outside the sphere. Thus we see that the potential outside a uniformly charged sphere is the same as if all the charge were concentrated at its center.

EXERCISE C What is the potential at a distance of 3.0 cm from a point charge $Q = -2.0 \times 10^{-9}$ C? (a) 600 V; (b) 60 V; (c) 6 V; (d) −600 V; (e) −60 V; (f) −6 V.

EXAMPLE 23–6 **Work required to bring two positive charges close together.** What minimum work must be done by an external force to bring a charge $q = 3.00\,\mu\text{C}$ from a great distance away (take $r = \infty$) to a point 0.500 m from a charge $Q = 20.0\,\mu\text{C}$?

APPROACH To find the work we cannot simply multiply the force times distance because the force is not constant. Instead we can set the change in potential energy equal to the (positive of the) work required of an *external* force (Chapter 8), and Eq. 23–3: $W = \Delta U = q(V_b - V_a)$. We get the potentials V_b and V_a using Eq. 23–5.

SOLUTION The work required is equal to the change in potential energy:

$$W = q(V_b - V_a)$$

$$= q\left(\frac{kQ}{r_b} - \frac{kQ}{r_a}\right),$$

where $r_b = 0.500$ m and $r_a = \infty$. The right-hand term within the parentheses is zero ($1/\infty = 0$) so

$$W = (3.00 \times 10^{-6}\,\text{C})\frac{(8.99 \times 10^9\,\text{N}\cdot\text{m}^2/\text{C}^2)(2.00 \times 10^{-5}\,\text{C})}{(0.500\,\text{m})} = 1.08\,\text{J}.$$

NOTE We could not use Eq. 23–4b here because it applies *only* to uniform fields. But we did use Eq. 23–3 because it is always valid.

To determine the electric field at points near a collection of two or more point charges requires adding up the electric fields due to each charge. Since the electric field is a vector, this can be time consuming or complicated. To find the electric potential at a point due to a collection of point charges is far easier, since the electric potential is a scalar, and hence you only need to add numbers (with appropriate signs) without concern for direction. This is a major advantage in using electric potential for solving Problems.

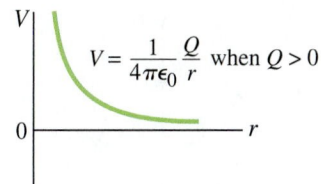

FIGURE 23–10 Potential V as a function of distance r from a single point charge Q when the charge is positive.

FIGURE 23–11 Potential V as a function of distance r from a single point charge Q when the charge is negative.

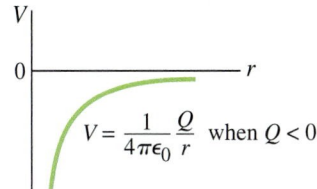

CAUTION

We cannot use $W = Fd$ when F is not constant

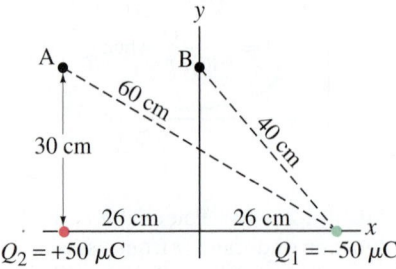

FIGURE 23-12 Example 23-7.
(See also Example 21-8, Fig. 21-27.)

⚠ **C A U T I O N**

*Potential is a scalar and
has no components*

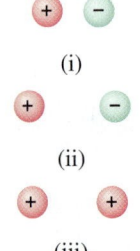

(i)

(ii)

(iii)

FIGURE 23-13 Exercise D.

EXAMPLE 23-7 **Potential above two charges.** Calculate the electric potential (a) at point A in Fig. 23-12 due to the two charges shown, and (b) at point B. [This is the same situation as Example 21-8, Fig. 21-27, where we calculated the electric field at these points.]

APPROACH The total potential at point A (or at point B) is the sum of the potentials at that point due to each of the two charges Q_1 and Q_2. The potential due to each single charge is given by Eq. 23-5. We do not have to worry about directions because electric potential is a scalar quantity. But we do have to keep track of the signs of charges.

SOLUTION (a) We add the potentials at point A due to each charge Q_1 and Q_2, and we use Eq. 23-5 for each:

$$V_A = V_{A2} + V_{A1}$$
$$= k\frac{Q_2}{r_{2A}} + k\frac{Q_1}{r_{1A}}$$

where $r_{1A} = 60$ cm and $r_{2A} = 30$ cm. Then

$$V_A = \frac{(9.0 \times 10^9\,\mathrm{N \cdot m^2/C^2})(5.0 \times 10^{-5}\,\mathrm{C})}{0.30\,\mathrm{m}}$$
$$+ \frac{(9.0 \times 10^9\,\mathrm{N \cdot m^2/C^2})(-5.0 \times 10^{-5}\,\mathrm{C})}{0.60\,\mathrm{m}}$$
$$= 1.50 \times 10^6\,\mathrm{V} - 0.75 \times 10^6\,\mathrm{V}$$
$$= 7.5 \times 10^5\,\mathrm{V}.$$

(b) At point B, $r_{1B} = r_{2B} = 0.40$ m, so

$$V_B = V_{B2} + V_{B1}$$
$$= \frac{(9.0 \times 10^9\,\mathrm{N \cdot m^2/C^2})(5.0 \times 10^{-5}\,\mathrm{C})}{0.40\,\mathrm{m}}$$
$$+ \frac{(9.0 \times 10^9\,\mathrm{N \cdot m^2/C^2})(-5.0 \times 10^{-5}\,\mathrm{C})}{0.40\,\mathrm{m}}$$
$$= 0\,\mathrm{V}.$$

NOTE The two terms in the sum in (b) cancel for any point equidistant from Q_1 and Q_2 $(r_{1B} = r_{2B})$. Thus the potential will be zero everywhere on the plane equidistant between the two opposite charges. This plane where V is constant is called an equipotential surface.

Simple summations like these can easily be performed for any number of point charges.

EXERCISE D Consider the three pairs of charges, Q_1 and Q_2, in Fig. 23-13. (a) Which set has a positive potential energy? (b) Which set has the most negative potential energy? (c) Which set requires the most work to separate the charges to infinity? Assume the charges all have the same magnitude.

23-4 Potential Due to Any Charge Distribution

If we know the electric field in a region of space due to any distribution of electric charge, we can determine the difference in potential between two points in the region using Eq. 23-4a, $V_{ba} = -\int_a^b \vec{E} \cdot d\vec{\ell}$. In many cases we don't know $\vec{E}$ as a function of position, and it may be difficult to calculate. We can calculate the potential V due to a given charge distribution in another way, using the potential due to a single point charge, Eq. 23-5:

$$V = \frac{1}{4\pi\epsilon_0}\frac{Q}{r},$$

where we choose $V = 0$ at $r = \infty$. Then we can sum over all the charges.

If we have n individual point charges, the potential at some point a (relative to $V = 0$ at $r = \infty$) is

$$V_\text{a} = \sum_{i=1}^{n} V_i = \frac{1}{4\pi\epsilon_0} \sum_{i=1}^{n} \frac{Q_i}{r_{ia}}, \qquad \textbf{(23–6a)}$$

where r_{ia} is the distance from the ith charge (Q_i) to the point a. (We already used this approach in Example 23–7.) If the charge distribution can be considered continuous, then

$$V = \frac{1}{4\pi\epsilon_0} \int \frac{dq}{r}, \qquad \textbf{(23–6b)}$$

where r is the distance from a tiny element of charge, dq, to the point where V is being determined.

EXAMPLE 23–8 **Potential due to a ring of charge.** A thin circular ring of radius R has a uniformly distributed charge Q. Determine the electric potential at a point P on the axis of the ring a distance x from its center, Fig. 23–14.

APPROACH We integrate over the ring using Eq. 23–6b.

SOLUTION Each point on the ring is equidistant from point P, and this distance is $(x^2 + R^2)^\frac{1}{2}$. So the potential at P is:

$$V = \frac{1}{4\pi\epsilon_0} \int \frac{dq}{r} = \frac{1}{4\pi\epsilon_0} \frac{1}{(x^2 + R^2)^\frac{1}{2}} \int dq = \frac{1}{4\pi\epsilon_0} \frac{Q}{(x^2 + R^2)^\frac{1}{2}}.$$

NOTE For points very far away from the ring, $x \gg R$, this result reduces to $(1/4\pi\epsilon_0)(Q/x)$, the potential of a point charge, as we should expect.

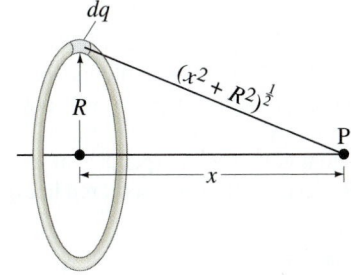

FIGURE 23–14 Example 23–8. Calculating the potential at point P, a distance x from the center of a uniform ring of charge.

EXAMPLE 23–9 **Potential due to a charged disk.** A thin flat disk, of radius R_0, has a uniformly distributed charge Q, Fig. 23–15. Determine the potential at a point P on the axis of the disk, a distance x from its center.

FIGURE 23–15 Example 23–9. Calculating the electric potential at point P on the axis of a uniformly charged thin disk.

APPROACH We divide the disk into thin rings of radius R and thickness dR and use the result of Example 23–8 to sum over the disk.

SOLUTION The charge Q is distributed uniformly, so the charge contained in each ring is proportional to its area. The disk has area πR_0^2 and each thin ring has area $dA = (2\pi R)(dR)$. Hence

$$\frac{dq}{Q} = \frac{2\pi R\, dR}{\pi R_0^2}$$

so

$$dq = Q \frac{(2\pi R)(dR)}{\pi R_0^2} = \frac{2QR\, dR}{R_0^2}.$$

Then the potential at P, using Eq. 23–6b in which r is replaced by $(x^2 + R^2)^\frac{1}{2}$, is

$$V = \frac{1}{4\pi\epsilon_0} \int \frac{dq}{(x^2 + R^2)^\frac{1}{2}} = \frac{2Q}{4\pi\epsilon_0 R_0^2} \int_0^{R_0} \frac{R\, dR}{(x^2 + R^2)^\frac{1}{2}} = \frac{Q}{2\pi\epsilon_0 R_0^2} (x^2 + R^2)^\frac{1}{2} \Big|_{R=0}^{R=R_0}$$

$$= \frac{Q}{2\pi\epsilon_0 R_0^2} \left[(x^2 + R_0^2)^\frac{1}{2} - x\right].$$

NOTE For $x \gg R_0$, this formula reduces to

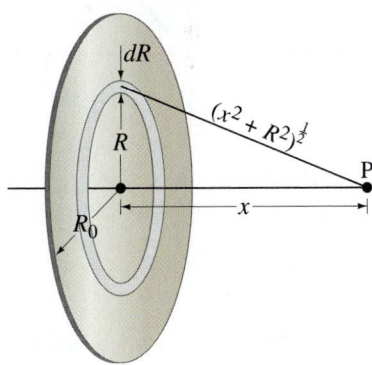

$$V \approx \frac{Q}{2\pi\epsilon_0 R_0^2} \left[x\left(1 + \frac{1}{2}\frac{R_0^2}{x^2}\right) - x\right] = \frac{Q}{4\pi\epsilon_0 x}.$$

This is the formula for a point charge, as we expect.

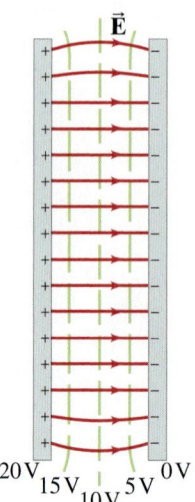

FIGURE 23–16 Equipotential lines (the green dashed lines) between two oppositely charged parallel plates. Note that they are perpendicular to the electric field lines (solid red lines).

23–5 Equipotential Surfaces

The electric potential can be represented graphically by drawing **equipotential lines** or, in three dimensions, **equipotential surfaces**. An equipotential surface is one on which all points are at the same potential. That is, the potential difference between any two points on the surface is zero, and no work is required to move a charge from one point to the other. An *equipotential surface must be perpendicular to the electric field* at any point. If this were not so—that is, if there were a component of $\vec{E}$ parallel to the surface—it would require work to move the charge along the surface against this component of $\vec{E}$; and this would contradict the idea that it is an equipotential surface. This can also be seen from Eq. 23–4a, $\Delta V = -\int \vec{E} \cdot d\vec{\ell}$. On a surface where V is constant, $\Delta V = 0$, so we must have either $\vec{E} = 0$, $d\vec{\ell} = 0$, or $\cos \theta = 0$ where θ is the angle between $\vec{E}$ and $d\vec{\ell}$. Thus in a region where $\vec{E}$ is not zero, the path $d\vec{\ell}$ along an equipotential must have $\cos \theta = 0$, meaning $\theta = 90°$ and $\vec{E}$ is perpendicular to the equipotential.

The fact that the electric field lines and equipotential surfaces are mutually perpendicular helps us locate the equipotentials when the electric field lines are known. In a normal two-dimensional drawing, we show equipotential *lines*, which are the intersections of equipotential surfaces with the plane of the drawing. In Fig. 23–16, a few of the equipotential lines are drawn (dashed green lines) for the electric field (red lines) between two parallel plates at a potential difference of 20 V. The negative plate is arbitrarily chosen to be zero volts and the potential of each equipotential line is indicated. Note that $\vec{E}$ points toward lower values of V.

FIGURE 23–17 Example 23–10. Electric field lines and equipotential surfaces for a point charge.

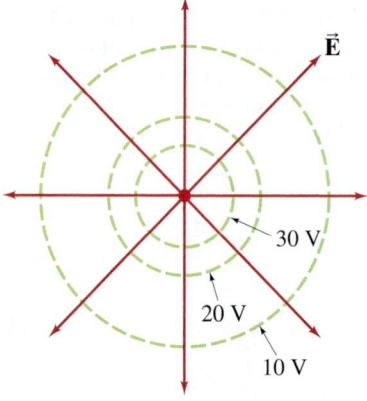

EXAMPLE 23–10 Point charge equipotential surfaces. For a single point charge with $Q = 4.0 \times 10^{-9}\,\text{C}$, sketch the equipotential surfaces (or lines in a plane containing the charge) corresponding to $V_1 = 10\,\text{V}$, $V_2 = 20\,\text{V}$, and $V_3 = 30\,\text{V}$.

APPROACH The electric potential V depends on the distance r from the charge (Eq. 23–5).

SOLUTION The electric field for a positive point charge is directed radially outward. Since the equipotential surfaces must be perpendicular to the lines of electric field, they will be spherical in shape, centered on the point charge, Fig. 23–17. From Eq. 23–5 we have $r = (1/4\pi\epsilon_0)(Q/V)$, so that for $V_1 = 10\,\text{V}$, $r_1 = (9.0 \times 10^9\,\text{N·m}^2/\text{C}^2)(4.0 \times 10^{-9}\,\text{C})/(10\,\text{V}) = 3.6\,\text{m}$, for $V_2 = 20\,\text{V}$, $r_2 = 1.8\,\text{m}$, and for $V_3 = 30\,\text{V}$, $r_3 = 1.2\,\text{m}$, as shown.

NOTE The equipotential surface with the largest potential is closest to the positive charge. How would this change if Q were negative?

FIGURE 23–18 Equipotential lines (green, dashed) are always perpendicular to the electric field lines (solid red) shown here for two equal but oppositely charged particles.

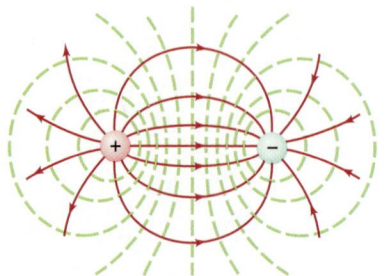

The equipotential lines for the case of two equal but oppositely charged particles are shown in Fig. 23–18 as green dashed lines. Equipotential lines and surfaces, unlike field lines, are always continuous and never end, and so continue beyond the borders of Figs. 23–16 and 23–18.

We saw in Section 21–9 that there can be no electric field within a conductor in the static case, for otherwise the free electrons would feel a force and would move. Indeed, the entire volume of *a conductor must be entirely at the same potential in the static case*, and the surface of a conductor is then an equipotential surface. (If it weren't, the free electrons at the surface would move, since whenever there is a potential difference between two points, free charges will move.) This is fully consistent with our result, discussed earlier, that the electric field at the surface of a conductor must be perpendicular to the surface.

A useful analogy for equipotential lines is a topographic map: the contour lines are essentially gravitational equipotential lines (Fig. 23–19).

FIGURE 23–19 A topographic map (here, a portion of the Sierra Nevada in California) shows continuous contour lines, each of which is at a fixed height above sea level. Here they are at 80 ft (25 m) intervals. If you walk along one contour line, you neither climb nor descend. If you cross lines, and especially if you climb perpendicular to the lines, you will be changing your gravitational potential (rapidly, if the lines are close together).

23–6 Electric Dipole Potential

Two equal point charges Q, of opposite sign, separated by a distance ℓ, are called an **electric dipole**, as we saw in Section 21–11. Also, the two charges we saw in Figs. 23–12 and 23–18 constitute an electric dipole, and the latter shows the electric field lines and equipotential surfaces for a dipole. Because electric dipoles occur often in physics, as well as in other fields, it is useful to examine them more closely.

The electric potential at an arbitrary point P due to a dipole, Fig. 23–20, is the sum of the potentials due to each of the two charges (we take $V = 0$ at $r = \infty$):

$$V = \frac{1}{4\pi\epsilon_0}\frac{Q}{r} + \frac{1}{4\pi\epsilon_0}\frac{(-Q)}{(r + \Delta r)} = \frac{1}{4\pi\epsilon_0}Q\left(\frac{1}{r} - \frac{1}{r + \Delta r}\right) = \frac{Q}{4\pi\epsilon_0}\frac{\Delta r}{r(r + \Delta r)},$$

where r is the distance from P to the positive charge and $r + \Delta r$ is the distance to the negative charge. This equation becomes simpler if we consider points P whose distance from the dipole is much larger than the separation of the two charges— that is, for $r \gg \ell$. From Fig. 23–20 we see that $\Delta r \approx \ell \cos\theta$; since $r \gg \Delta r = \ell \cos\theta$, we can neglect Δr in the denominator as compared to r. Therefore, we obtain

$$V = \frac{1}{4\pi\epsilon_0}\frac{Q\ell\cos\theta}{r^2} = \frac{1}{4\pi\epsilon_0}\frac{p\cos\theta}{r^2} \qquad \text{[dipole; } r \gg \ell\text{]} \quad \textbf{(23–7)}$$

where $p = Q\ell$ is called the **dipole moment**. We see that the potential decreases as the *square* of the distance from the dipole, whereas for a single point charge the potential decreases with the first power of the distance (Eq. 23–5). It is not surprising that the potential should fall off faster for a dipole; for when you are far from a dipole, the two equal but opposite charges appear so close together as to tend to neutralize each other.

Table 23–2 gives the dipole moments for several molecules. The + and − signs indicate on which atoms these charges lie. The last two entries are a part of many organic molecules and play an important role in molecular biology. A dipole moment has units of coulomb-meters (C·m), although for molecules a smaller unit called a *debye* is sometimes used: 1 debye = 3.33×10^{-30} C·m.

23–7 $\vec{E}$ Determined from V

We can use Eq. 23–4a, $V_b - V_a = -\int_a^b \vec{E} \cdot d\vec{\ell}$, to determine the difference in potential between two points if the electric field is known in the region between those two points. By inverting Eq. 23–4a, we can write the electric field in terms of the potential. Then the electric field can be determined from a knowledge of V. Let us see how to do this.

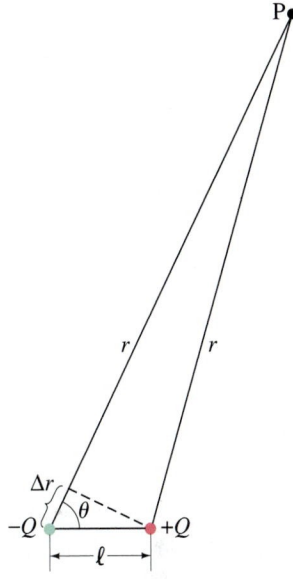

FIGURE 23–20 Electric dipole. Calculation of potential V at point P.

TABLE 23–2 Dipole Moments of Selected Molecules

Molecule	Dipole Moment (C · m)
$H_2^{(+)}O^{(-)}$	6.1×10^{-30}
$H^{(+)}Cl^{(-)}$	3.4×10^{-30}
$N^{(-)}H_3^{(+)}$	5.0×10^{-30}
$>N^{(-)}-H^{(+)}$	$\approx 3.0^\dagger \times 10^{-30}$
$>C^{(+)}=O^{(-)}$	$\approx 8.0^\dagger \times 10^{-30}$

†These groups often appear on larger molecules; hence the value for the dipole moment will vary somewhat, depending on the rest of the molecule.

We write Eq. 23–4a in differential form as

$$dV = -\vec{E} \cdot d\vec{\ell} = -E_\ell \, d\ell,$$

where dV is the infinitesimal difference in potential between two points a distance $d\ell$ apart, and E_ℓ is the component of the electric field in the direction of the infinitesimal displacement $d\vec{\ell}$. We can then write

$$E_\ell = -\frac{dV}{d\ell}. \qquad (23\text{–}8)$$

Thus *the component of the electric field in any direction is equal to the negative of the rate of change of the electric potential with distance in that direction.* The quantity $dV/d\ell$ is called the gradient of V in a particular direction. If the direction is not specified, the term *gradient* refers to that direction in which V changes most rapidly; this would be the direction of $\vec{E}$ at that point, so we can write

$$E = -\frac{dV}{d\ell}. \qquad [\text{if } d\vec{\ell} \| \vec{E}]$$

If $\vec{E}$ is written as a function of x, y, and z, and we let ℓ refer to the x, y, and z axes, then Eq. 23–8 becomes

$$E_x = -\frac{\partial V}{\partial x}, \qquad E_y = -\frac{\partial V}{\partial y}, \qquad E_z = -\frac{\partial V}{\partial z}. \qquad (23\text{–}9)$$

Here, $\partial V/\partial x$ is the "partial derivative" of V with respect to x, with y and z held constant.[†] For example, if $V(x, y, z) = (2 \, \text{V/m}^2)x^2 + (8 \, \text{V/m}^3)y^2z + (2 \, \text{V/m}^2)z^2$, then

$$E_x = -\partial V/\partial x = -(4 \, \text{V/m}^2)x,$$
$$E_y = -\partial V/\partial y = -(16 \, \text{V/m}^3)yz,$$

and

$$E_z = -\partial V/\partial z = -(8 \, \text{V/m}^3)y^2 - (4 \, \text{V/m}^2)z.$$

EXAMPLE 23–11 $\vec{E}$ **for ring and disk.** Use electric potential to determine the electric field at point P on the axis of (*a*) a circular ring of charge (Fig. 23–14) and (*b*) a uniformly charged disk (Fig. 23–15).

APPROACH We obtained V as a function of x in Examples 23–8 and 23–9, so we find E by taking derivatives (Eqs. 23–9).

SOLUTION (*a*) From Example 23–8,

$$V = \frac{1}{4\pi\epsilon_0} \frac{Q}{(x^2 + R^2)^{\frac{1}{2}}}.$$

Then

$$E_x = -\frac{\partial V}{\partial x} = \frac{1}{4\pi\epsilon_0} \frac{Qx}{(x^2 + R^2)^{\frac{3}{2}}}.$$

This is the same result we obtained in Example 21–9.

(*b*) From Example 23–9,

$$V = \frac{Q}{2\pi\epsilon_0 R_0^2}\left[(x^2 + R_0^2)^{\frac{1}{2}} - x\right],$$

so

$$E_x = -\frac{\partial V}{\partial x} = \frac{Q}{2\pi\epsilon_0 R_0^2}\left[1 - \frac{x}{(x^2 + R_0^2)^{\frac{1}{2}}}\right].$$

For points very close to the disk, $x \ll R_0$, this can be approximated by

$$E_x \approx \frac{Q}{2\pi\epsilon_0 R_0^2} = \frac{\sigma}{2\epsilon_0}$$

where $\sigma = Q/\pi R_0^2$ is the surface charge density. We also obtained these results in Chapter 21, Example 21–12 and Eq. 21–7.

[†]Equation 23–9 can be written as a vector equation,

$$\vec{E} = -\text{grad } V = -\vec{\nabla}V = -\left(\hat{i}\frac{\partial}{\partial x} + \hat{j}\frac{\partial}{\partial y} + \hat{k}\frac{\partial}{\partial z}\right)V$$

where the symbol $\vec{\nabla}$ is called the *del* or *gradient operator:* $\vec{\nabla} = \hat{i}\frac{\partial}{\partial x} + \hat{j}\frac{\partial}{\partial y} + \hat{k}\frac{\partial}{\partial z}.$

If we compare this last Example with Examples 21–9 and 21–12, we see that here, as for many charge distributions, it is easier to calculate V first, and then $\vec{E}$ from Eq. 23–9, rather than to calculate $\vec{E}$ due to each charge from Coulomb's law. This is because V due to many charges is a scalar sum, whereas $\vec{E}$ is a vector sum.

23–8 Electrostatic Potential Energy; the Electron Volt

Suppose a point charge q is moved between two points in space, a and b, where the electric potential due to other charges is V_a and V_b, respectively. The change in electrostatic potential energy of q in the field of these other charges is, according to Eq. 23–2b,

$$\Delta U = U_b - U_a = q(V_b - V_a).$$

Now suppose we have a system of several point charges. What is the electrostatic potential energy of the system? It is most convenient to choose the electric potential energy to be zero when the charges are very far (ideally infinitely far) apart. A single point charge, Q_1, in isolation, has no potential energy, because if there are no other charges around, no electric force can be exerted on it. If a second point charge Q_2 is brought close to Q_1, the potential due to Q_1 at the position of this second charge is

$$V = \frac{1}{4\pi\epsilon_0} \frac{Q_1}{r_{12}},$$

where r_{12} is the distance between the two. The potential energy of the two charges, relative to $V = 0$ at $r = \infty$, is

$$U = Q_2 V = \frac{1}{4\pi\epsilon_0} \frac{Q_1 Q_2}{r_{12}}. \tag{23–10}$$

This represents the work that needs to be done by an external force to bring Q_2 from infinity $(V = 0)$ to a distance r_{12} from Q_1. It is also the negative of the work needed to separate them to infinity.

If the system consists of three charges, the total potential energy will be the work needed to bring all three together. Equation 23–10 represents the work needed to bring Q_2 close to Q_1; to bring a third charge Q_3 so that it is a distance r_{13} from Q_1 and r_{23} from Q_2 requires work equal to

$$\frac{1}{4\pi\epsilon_0} \frac{Q_1 Q_3}{r_{13}} + \frac{1}{4\pi\epsilon_0} \frac{Q_2 Q_3}{r_{23}}.$$

So the potential energy of a system of three point charges is

$$U = \frac{1}{4\pi\epsilon_0} \left(\frac{Q_1 Q_2}{r_{12}} + \frac{Q_1 Q_3}{r_{13}} + \frac{Q_2 Q_3}{r_{23}} \right). \qquad [V = 0 \text{ at } r = \infty]$$

For a system of four charges, the potential energy would contain six such terms, and so on. (Caution must be used when making such sums to avoid double counting of the different pairs.)

The Electron Volt Unit

The joule is a very large unit for dealing with energies of electrons, atoms, or molecules (see Example 23–2), and for this purpose, the unit **electron volt** (eV) is used. One electron volt is defined as the energy acquired by a particle carrying a charge whose magnitude equals that on the electron $(q = e)$ as a result of moving through a potential difference of 1 V. Since $e = 1.6 \times 10^{-19}$ C, and since the change in potential energy equals qV, 1 eV is equal to $(1.6 \times 10^{-19} \text{ C})(1.0 \text{ V}) = 1.6 \times 10^{-19}$ J:

$$1 \text{ eV} = 1.6 \times 10^{-19} \text{ J}.$$

An electron that accelerates through a potential difference of 1000 V will lose 1000 eV of potential energy and will thus gain 1000 eV or 1 keV (kiloelectron volt) of kinetic energy. On the other hand, if a particle with a charge equal to twice the magnitude of the charge on the electron $(= 2e = 3.2 \times 10^{-19} \text{ C})$ moves through a potential difference of 1000 V, its energy will change by 2000 eV $= 2$ keV.

Although the electron volt is handy for *stating* the energies of molecules and elementary particles, it is not a proper SI unit. For calculations it should be converted to joules using the conversion factor given above. In Example 23–2, for example, the electron acquired a kinetic energy of 8.0×10^{-16} J. We normally would quote this energy as 5000 eV $(= 8.0 \times 10^{-16}$ J/1.6 $\times 10^{-19}$ J/eV$)$. But when determining the speed of a particle in SI units, we must use the kinetic energy in J.

EXERCISE E What is the kinetic energy of a He^{2+} ion released from rest and accelerated through a potential difference of 1.0 kV? (a) 1000 eV, (b) 500 eV, (c) 2000 eV, (d) 4000 eV, (e) 250 eV.

EXAMPLE 23–12 **Disassembling a hydrogen atom.** Calculate the work needed to "disassemble" a hydrogen atom. Assume that the proton and electron are initially separated by a distance equal to the "average" radius of the hydrogen atom in its ground state, 0.529×10^{-10} m, and that they end up an infinite distance apart from each other.

APPROACH The work necessary will be equal to the total energy, kinetic plus potential, of the electron and proton as an atom, compared to their total energy when infinitely far apart.

SOLUTION From Eq. 23–10 we have initially

$$U = \frac{1}{4\pi\epsilon_0}\frac{Q_1 Q_2}{r} = \frac{1}{4\pi\epsilon_0}\frac{(e)(-e)}{r} = \frac{-(8.99 \times 10^9\, \text{N} \cdot \text{m}^2/\text{C}^2)(1.60 \times 10^{-19}\,\text{C})^2}{(0.529 \times 10^{-10}\,\text{m})}$$
$$= -27.2(1.60 \times 10^{-19})\,\text{J} = -27.2\,\text{eV}.$$

This represents the potential energy. The total energy must include also the kinetic energy of the electron moving in an orbit of radius $r = 0.529 \times 10^{-10}$ m. From $F = ma$ for centripetal acceleration, we have

$$\frac{1}{4\pi\epsilon_0}\left(\frac{e^2}{r^2}\right) = \frac{mv^2}{r}.$$

Then

$$K = \tfrac{1}{2}mv^2 = \tfrac{1}{2}\left(\frac{1}{4\pi\epsilon_0}\right)\frac{e^2}{r}$$

which equals $-\tfrac{1}{2}U$ (as calculated above), so $K = +13.6$ eV. The total energy initially is $E = K + U = 13.6\,\text{eV} - 27.2\,\text{eV} = -13.6$ eV. To separate a stable hydrogen atom into a proton and an electron at rest very far apart ($U = 0$ at $r = \infty$, $K = 0$ because $v = 0$) requires $+13.6$ eV. This is, in fact, the measured ionization energy for hydrogen.

NOTE To treat atoms properly, we need to use quantum theory (Chapters 37 to 39). But our "classical" calculation does give the correct answer here.

EXERCISE F The kinetic energy of a 1000-kg automobile traveling 20 m/s (70 km/h) would be about (a) 100 GeV, (b) 1000 TeV, (c) 10^6 TeV, (d) 10^{12} TeV, (e) 10^{18} TeV.

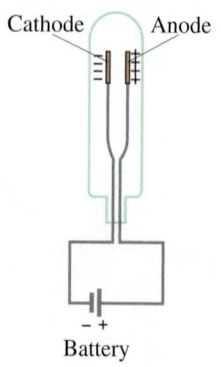

FIGURE 23–21 If the cathode inside the evacuated glass tube is heated to glowing, negatively charged "cathode rays" (electrons) are "boiled off" and flow across to the anode ($+$) to which they are attracted.

*23–9 Cathode Ray Tube: TV and Computer Monitors, Oscilloscope

An important device that makes use of voltage, and that allows us to "visualize" how a voltage changes in time, is the *cathode ray tube* (CRT). A CRT used in this way is an *oscilloscope*. The CRT has also been used for many years as the picture tube of television sets and computer monitors, but LCD (Chapter 35) and other screens are now common.

The operation of a CRT depends on the phenomenon of **thermionic emission** discovered by Thomas Edison (1847–1931). Consider two small plates (electrodes) inside an evacuated "bulb" or "tube" as shown in Fig. 23–21, to which is applied a potential difference. The negative electrode is called the **cathode**, the positive one the **anode**. If the negative cathode is heated (usually by an electric current, as in a lightbulb) so that it becomes hot and glowing, it is found that negative charge leaves the cathode and flows to the positive anode. These negative charges are now called electrons, but originally they were called **cathode rays** since they seemed to come from the cathode (see Section 27–7 on the discovery of the electron).

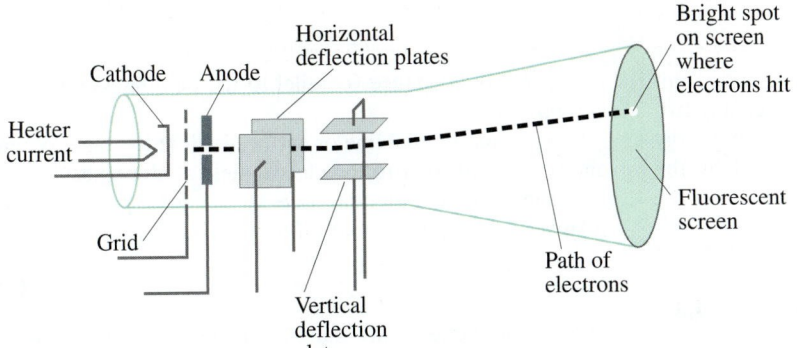

FIGURE 23–22 A cathode ray tube. Magnetic deflection coils are often used in place of the electric deflection plates shown here. The relative positions of the elements have been exaggerated for clarity.

The **cathode ray tube** (CRT) derives its name from the fact that inside an evacuated glass tube, a beam of cathode rays (electrons) is directed to various parts of a screen to produce a "picture." A simple CRT is diagrammed in Fig. 23–22. Electrons emitted by the heated cathode are accelerated by a high voltage (5000–50,000 V) applied between the anode and cathode. The electrons pass out of this "electron gun" through a small hole in the anode. The inside of the tube face is coated with a fluorescent material that glows when struck by electrons. A tiny bright spot is thus visible where the electron beam strikes the screen. Two horizontal and two vertical plates can deflect the beam of electrons when a voltage is applied to them. The electrons are deflected toward whichever plate is positive. By varying the voltage on the deflection plates, the bright spot can be placed at any point on the screen. Many CRTs use magnetic deflection coils (Chapter 27) instead of electric plates.

In the picture tube or monitor for a computer or television set, the electron beam is made to sweep over the screen in the manner shown in Fig. 23–23 by changing voltages applied to the deflection plates. For standard television in the United States, 525 lines constitutes a complete sweep in $\frac{1}{30}$ s, over the entire screen. High-definition TV provides more than double this number of lines (1080), giving greater picture sharpness. We see a picture because the image is retained by the fluorescent screen and by our eyes for about $\frac{1}{20}$ s. The picture we see consists of the varied brightness of the spots on the screen, controlled by the grid (a "porous" electrode, such as a wire grid, that allows passage of electrons). The grid limits the flow of electrons by means of the voltage (the "video signal") applied to it: the more negative this voltage, the more electrons are repelled and the fewer pass through. This video signal sent out by the TV station, and received by the TV set, is accompanied by signals that synchronize the grid voltage to the horizontal and vertical sweeps. (More in Chapter 31.)

An **oscilloscope** is a device for amplifying, measuring, and visually observing an electrical signal as a function of time on the screen of a CRT (a "signal" is usually a time-varying voltage). The electron beam is swept horizontally at a uniform rate in time by the horizontal deflection plates. The signal to be displayed is applied (after amplification) to the vertical deflection plates. The visible "trace" on the screen, which could be an electrocardiogram (Fig. 23–24), or a signal from an experiment on nerve conduction, is a plot of the signal voltage (vertically) versus time (horizontally).

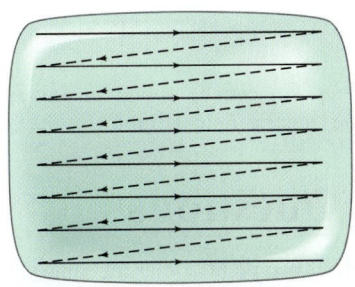

FIGURE 23–23 Electron beam sweeps across a television screen in a succession of horizontal lines. Each horizontal sweep is made by varying the voltage on the horizontal deflection plates. Then the electron beam is moved down a short distance by a change in voltage on the vertical deflection plates, and the process is repeated.

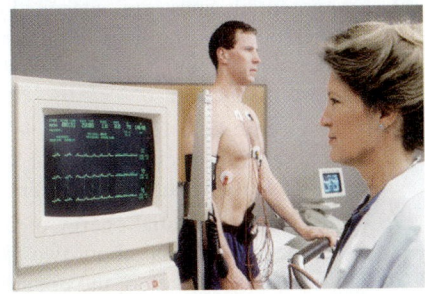

FIGURE 23–24 An electrocardiogram (ECG) trace displayed on a CRT.

Summary

Electric potential is defined as electric potential energy per unit charge. That is, the **electric potential difference** between any two points in space is defined as the difference in potential energy of a test charge q placed at those two points, divided by the charge q:

$$V_{ba} = \frac{U_b - U_a}{q}. \tag{23-2b}$$

Potential difference is measured in volts $(1\,\text{V} = 1\,\text{J/C})$ and is sometimes referred to as **voltage**.

The change in potential energy of a charge q when it moves through a potential difference V_{ba} is

$$\Delta U = qV_{ba}. \tag{23-3}$$

The potential difference V_{ba} between two points, a and b, is given by the relation

$$V_{ba} = V_b - V_a = -\int_a^b \vec{E} \cdot d\vec{\ell}. \tag{23-4a}$$

Thus V_{ba} can be found in any region where $\vec{E}$ is known. If the electric field is uniform, the integral is easy: $V_{ba} = -Ed$,

where d is the distance (parallel to the field lines) between the two points.

An **equipotential line** or **surface** is all at the same potential, and is perpendicular to the electric field at all points.

The electric potential due to a single point charge Q, relative to zero potential at infinity, is given by

$$V = \frac{1}{4\pi\epsilon_0}\frac{Q}{r}. \tag{23-5}$$

The potential due to any charge distribution can be obtained by summing (or integrating) over the potentials for all the charges.

The potential due to an **electric dipole** drops off as $1/r^2$. The **dipole moment** is $p = Q\ell$, where ℓ is the distance between the two equal but opposite charges of magnitude Q.

When V is known, the components of $\vec{E}$ can be found from the inverse of Eq. 23–4a, namely

$$E_x = -\frac{\partial V}{\partial x}, \qquad E_y = -\frac{\partial V}{\partial y}, \qquad E_z = -\frac{\partial V}{\partial z}. \tag{23-9}$$

[*Television and computer monitors traditionally use a **cathode ray tube** (CRT) that accelerates electrons by high voltage, and sweeps them across the screen in a regular way using deflection plates.]

Questions

1. If two points are at the same potential, does this mean that no work is done in moving a test charge from one point to the other? Does this imply that no force must be exerted? Explain.

2. If a negative charge is initially at rest in an electric field, will it move toward a region of higher potential or lower potential? What about a positive charge? How does the potential energy of the charge change in each instance?

3. State clearly the difference (a) between electric potential and electric field, (b) between electric potential and electric potential energy.

4. An electron is accelerated from rest by a potential difference of, say, 0.10 V. How much greater would its final speed be if it is accelerated with four times as much voltage? Explain.

5. Can a particle ever move from a region of low electric potential to one of high potential and yet have its electric potential energy decrease? Explain.

6. If $V = 0$ at a point in space, must $\vec{E} = 0$? If $\vec{E} = 0$ at some point, must $V = 0$ at that point? Explain. Give examples for each.

7. When dealing with practical devices, we often take the ground (the Earth) to be 0 V. (a) If instead we said the ground was -10 V, how would this affect V and E at other points? (b) Does the fact that the Earth carries a net charge affect the choice of V at its surface?

8. Can two equipotential lines cross? Explain.

9. Draw in a few equipotential lines in Fig. 21–34b and c.

10. What can you say about the electric field in a region of space that has the same potential throughout?

11. A satellite orbits the Earth along a gravitational equipotential line. What shape must the orbit be?

12. Suppose the charged ring of Example 23–8 was not uniformly charged, so that the density of charge was twice as great near the top as near the bottom. Assuming the total charge Q is unchanged, would this affect the potential at point P on the axis (Fig. 23–14)? Would it affect the value of $\vec{E}$ at that point? Is there a discrepancy here? Explain.

13. Consider a metal conductor in the shape of a football. If it carries a total charge Q, where would you expect the charge density σ to be greatest, at the ends or along the flatter sides? Explain. [*Hint*: Near the surface of a conductor, $E = \sigma/\epsilon_0$.]

14. If you know V at a point in space, can you calculate $\vec{E}$ at that point? If you know $\vec{E}$ at a point can you calculate V at that point? If not, what else must be known in each case?

15. A conducting sphere carries a charge Q and a second identical conducting sphere is neutral. The two are initially isolated, but then they are placed in contact. (a) What can you say about the potential of each when they are in contact? (b) Will charge flow from one to the other? If so, how much? (c) If the spheres do not have the same radius, how are your answers to parts (a) and (b) altered?

16. At a particular location, the electric field points due north. In what direction(s) will the rate of change of potential be (a) greatest, (b) least, and (c) zero?

17. Equipotential lines are spaced 1.00 V apart. Does the distance between the lines in different regions of space tell you anything about the relative strengths of $\vec{E}$ in those regions? If so, what?

18. If the electric field $\vec{E}$ is uniform in a region, what can you infer about the electric potential V? If V is uniform in a region of space, what can you infer about $\vec{E}$?

19. Is the electric potential energy of two unlike charges positive or negative? What about two like charges? What is the significance of the sign of the potential energy in each case?

Problems

23–1 Electric Potential

1. (I) What potential difference is needed to stop an electron that has an initial velocity $v = 5.0 \times 10^5$ m/s?

2. (I) How much work does the electric field do in moving a proton from a point with a potential of $+185$ V to a point where it is -55 V?

3. (I) An electron acquires 5.25×10^{-16} J of kinetic energy when it is accelerated by an electric field from plate A to plate B. What is the potential difference between the plates, and which plate is at the higher potential?

4. (II) The work done by an external force to move a $-9.10\,\mu$C charge from point a to point b is 7.00×10^{-4} J. If the charge was started from rest and had 2.10×10^{-4} J of kinetic energy when it reached point b, what must be the potential difference between a and b?

23–2 Potential Related to Electric Field

5. (I) Thunderclouds typically develop voltage differences of about 1×10^8 V. Given that an electric field of 3×10^6 V/m is required to produce an electrical spark within a volume of air, estimate the length of a thundercloud lightning bolt. [Can you see why, when lightning strikes from a cloud to the ground, the bolt actually has to propagate as a sequence of steps?]

6. (I) The electric field between two parallel plates connected to a 45-V battery is 1300 V/m. How far apart are the plates?

7. (I) What is the maximum amount of charge that a spherical conductor of radius 6.5 cm can hold in air?

8. (I) What is the magnitude of the electric field between two parallel plates 4.0 mm apart if the potential difference between them is 110 V?

9. (I) What minimum radius must a large conducting sphere of an electrostatic generating machine have if it is to be at 35,000 V without discharge into the air? How much charge will it carry?

10. (II) A manufacturer claims that a carpet will not generate more than 5.0 kV of static electricity. What magnitude of charge would have to be transferred between a carpet and a shoe for there to be a 5.0-kV potential difference between the shoe and the carpet? Approximate the shoe and the carpet as large sheets of charge separated by a distance $d = 1.0$ mm.

11. (II) A uniform electric field $\vec{E} = -4.20$ N/C $\hat{i}$ points in the negative x direction as shown in Fig. 23–25. The x and y coordinates of points A, B, and C are given on the diagram (in meters). Determine the differences in potential (a) V_{BA}, (b) V_{CB}, and (c) V_{CA}.

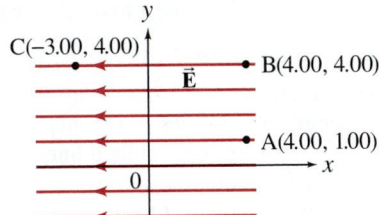

FIGURE 23–25
Problem 11.

12. (II) The electric potential of a very large isolated flat metal plate is V_0. It carries a uniform distribution of charge of surface density σ (C/m²), or $\sigma/2$ on each surface. Determine V at a distance x from the plate. Consider the point x to be far from the edges and assume x is much smaller than the plate dimensions.

13. (II) The Earth produces an inwardly directed electric field of magnitude 150 V/m near its surface. (a) What is the potential of the Earth's surface relative to $V = 0$ at $r = \infty$? (b) If the potential of the Earth is chosen to be zero, what is the potential at infinity? (Ignore the fact that positive charge in the ionosphere approximately cancels the Earth's net charge; how would this affect your answer?)

14. (II) A 32-cm-diameter conducting sphere is charged to 680 V relative to $V = 0$ at $r = \infty$. (a) What is the surface charge density σ? (b) At what distance will the potential due to the sphere be only 25 V?

15. (II) An insulated spherical conductor of radius r_1 carries a charge Q. A second conducting sphere of radius r_2 and initially uncharged is then connected to the first by a long conducting wire. (a) After the connection, what can you say about the electric potential of each sphere? (b) How much charge is transferred to the second sphere? Assume the connected spheres are far apart compared to their radii. (Why make this assumption?)

16. (II) Determine the difference in potential between two points that are distances R_a and R_b from a very long ($\gg R_a$ or R_b) straight wire carrying a uniform charge per unit length λ.

17. (II) Suppose the end of your finger is charged. (a) Estimate the breakdown voltage in air for your finger. (b) About what surface charge density would have to be on your finger at this voltage?

18. (II) Estimate the electric field in the membrane wall of a living cell. Assume the wall is 10 nm thick and has a potential of 0.10 V across it.

19. (II) A nonconducting sphere of radius r_0 carries a total charge Q distributed uniformly throughout its volume. Determine the electric potential as a function of the distance r from the center of the sphere for (a) $r > r_0$ and (b) $r < r_0$. Take $V = 0$ at $r = \infty$. (c) Plot V versus r and E versus r.

20. (III) Repeat Problem 19 assuming the charge density ρ_E increases as the square of the distance from the center of the sphere, and $\rho_E = 0$ at the center.

21. (III) The volume charge density ρ_E within a sphere of radius r_0 is distributed in accordance with the following spherically symmetric relation

$$\rho_E(r) = \rho_0\left[1 - \frac{r^2}{r_0^2}\right]$$

where r is measured from the center of the sphere and ρ_0 is a constant. For a point P inside the sphere ($r < r_0$), determine the electric potential V. Let $V = 0$ at infinity.

22. (III) A hollow spherical conductor, carrying a net charge $+Q$, has inner radius r_1 and outer radius $r_2 = 2r_1$ (Fig. 23–26). At the center of the sphere is a point charge $+Q/2$. (a) Write the electric field strength E in all three regions as a function of r. Then determine the potential as a function of r, the distance from the center, for (b) $r > r_2$, (c) $r_1 < r < r_2$, and (d) $0 < r < r_1$. (e) Plot both V and E as a function of r from $r = 0$ to $r = 2r_2$.

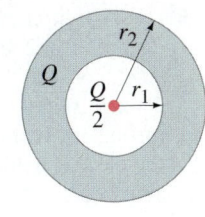

FIGURE 23–26
Problem 22.

23. (III) A very long conducting cylinder (length ℓ) of radius R_0 ($R_0 \ll \ell$) carries a uniform surface charge density σ (C/m^2). The cylinder is at an electric potential V_0. What is the potential, at points far from the end, at a distance R from the center of the cylinder? Determine for (a) $R > R_0$ and (b) $R < R_0$. (c) Is $V = 0$ at $R = \infty$ (assume $\ell = \infty$)? Explain.

23–3 Potential Due to Point Charges

24. (I) A point charge Q creates an electric potential of $+185$ V at a distance of 15 cm. What is Q (let $V = 0$ at $r = \infty$)?

25. (I) (a) What is the electric potential 0.50×10^{-10} m from a proton (charge $+e$)? Let $V = 0$ at $r = \infty$. (b) What is the potential energy of an electron at this point?

26. (II) Two point charges, $3.4\,\mu C$ and $-2.0\,\mu C$, are placed 5.0 cm apart on the x axis. At what points along the x axis is (a) the electric field zero and (b) the potential zero? Let $V = 0$ at $r = \infty$.

27. (II) A $+25\,\mu C$ point charge is placed 6.0 cm from an identical $+25\,\mu C$ point charge. How much work would be required by an external force to move a $+0.18\,\mu C$ test charge from a point midway between them to a point 1.0 cm closer to either of the charges?

28. (II) Point a is 26 cm north of a $-3.8\,\mu C$ point charge, and point b is 36 cm west of the charge (Fig. 23–27). Determine (a) $V_b - V_a$, and (b) $\vec{E}_b - \vec{E}_a$ (magnitude and direction).

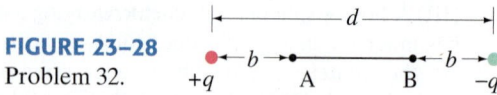

FIGURE 23–27
Problem 28.

26 cm, b, a, 36 cm, $Q = -3.8\,\mu C$

29. (II) How much voltage must be used to accelerate a proton (radius 1.2×10^{-15} m) so that it has sufficient energy to just "touch" a silicon nucleus? A silicon nucleus has a charge of $+14e$ and its radius is about 3.6×10^{-15} m. Assume the potential is that for point charges.

30. (II) Two identical $+5.5\,\mu C$ point charges are initially spaced 6.5 cm from each other. If they are released at the same instant from rest, how fast will they be moving when they are very far away from each other? Assume they have identical masses of 1.0 mg.

31. (II) An electron starts from rest 42.5 cm from a fixed point charge with $Q = -0.125$ nC. How fast will the electron be moving when it is very far away?

32. (II) Two equal but opposite charges are separated by a distance d, as shown in Fig. 23–28. Determine a formula for $V_{BA} = V_B - V_A$ for points B and A on the line between the charges situated as shown.

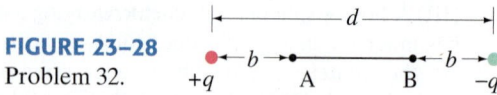

FIGURE 23–28
Problem 32.

d, $+q$, b, A, B, b, $-q$

23–4 Potential Due to Charge Distribution

33. (II) A thin circular ring of radius R (as in Fig. 23–14) has charge $+Q/2$ uniformly distributed on the top half, and $-Q/2$ on the bottom half. (a) What is the value of the electric potential at a point a distance x along the axis through the center of the circle? (b) What can you say about the electric field $\vec{E}$ at a distance x along the axis? Let $V = 0$ at $r = \infty$.

34. (II) Three point charges are arranged at the corners of a square of side ℓ as shown in Fig. 23–29. What is the potential at the fourth corner (point A), taking $V = 0$ at a great distance?

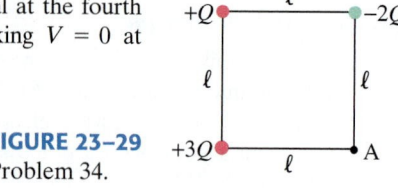

FIGURE 23–29
Problem 34.

ℓ, $+Q$, $-2Q$, ℓ, ℓ, $+3Q$, ℓ, A

35. (II) A flat ring of inner radius R_1 and outer radius R_2, Fig. 23–30, carries a uniform surface charge density σ. Determine the electric potential at points along the axis (the x axis). [*Hint*: Try substituting variables.]

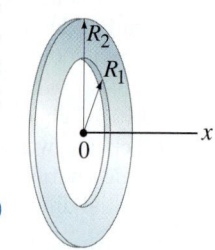

FIGURE 23–30
Problem 35.

R_2, R_1, 0, x

36. (II) A total charge Q is uniformly distributed on a thread of length ℓ. The thread forms a semicircle. What is the potential at the center? (Assume $V = 0$ at large distances.)

37. (II) A 12.0-cm-radius thin ring carries a uniformly distributed $15.0\,\mu C$ charge. A small 7.5-g sphere with a charge of $3.0\,\mu C$ is placed exactly at the center of the ring and given a very small push so it moves along the ring axis ($+x$ axis). How fast will the sphere be moving when it is 2.0 m from the center of the ring (ignore gravity)?

38. (II) A thin rod of length 2ℓ is centered on the x axis as shown in Fig. 23–31. The rod carries a uniformly distributed charge Q. Determine the potential V as a function of y for points along the y axis. Let $V = 0$ at infinity.

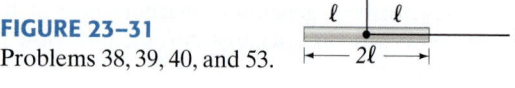

FIGURE 23–31
Problems 38, 39, 40, and 53.

y, ℓ, ℓ, 2ℓ, x

39. (II) Determine the potential $V(x)$ for points along the x axis outside the rod of Fig. 23–31 (Problem 38).

40. (III) The charge on the rod of Fig. 23–31 has a nonuniform linear charge distribution, $\lambda = ax$. Determine the potential V for (a) points along the y axis and (b) points along the x axis outside the rod.

41. (III) Suppose the flat circular disk of Fig. 23–15 (Example 23–9) has a nonuniform surface charge density $\sigma = aR^2$, where R is measured from the center of the disk. Find the potential $V(x)$ at points along the x axis, relative to $V = 0$ at $x = \infty$.

23–5 Equipotentials

42. (I) Draw a conductor in the shape of a football. This conductor carries a net negative charge, $-Q$. Draw in a dozen or so electric field lines and equipotential lines.

43. (II) Equipotential surfaces are to be drawn 100 V apart near a very large uniformly charged metal plate carrying a surface charge density $\sigma = 0.75\,\mu C/m^2$. How far apart (in space) are the equipotential surfaces?

44. (II) A metal sphere of radius $r_0 = 0.44$ m carries a charge $Q = 0.50\,\mu C$. Equipotential surfaces are to be drawn for 100-V intervals outside the sphere. Determine the radius r of (a) the first, (b) the tenth, and (c) the 100th equipotential from the surface.

23–6 Dipoles

45. (II) Calculate the electric potential due to a tiny dipole whose dipole moment is $4.8 \times 10^{-30}\ \text{C·m}$ at a point $4.1 \times 10^{-9}\ \text{m}$ away if this point is (a) along the axis of the dipole nearer the positive charge; (b) $45°$ above the axis but nearer the positive charge; (c) $45°$ above the axis but nearer the negative charge. Let $V = 0$ at $r = \infty$.

46. (III) The dipole moment, considered as a vector, points from the negative to the positive charge. The water molecule, Fig. 23–32, has a dipole moment $\vec{p}$ which can be considered as the vector sum of the two dipole moments $\vec{p}_1$ and $\vec{p}_2$ as shown. The distance between each H and the O is about $0.96 \times 10^{-10}\ \text{m}$; the lines joining the center of the O atom with each H atom make an angle of $104°$ as shown, and the net dipole moment has been measured to be $p = 6.1 \times 10^{-30}\ \text{C·m}$. (a) Determine the effective charge q on each H atom. (b) Determine the electric potential, far from the molecule, due to each dipole, $\vec{p}_1$ and $\vec{p}_2$, and show that

$$V = \frac{1}{4\pi\epsilon_0} \frac{p\cos\theta}{r^2},$$

where p is the magnitude of the net dipole moment, $\vec{p} = \vec{p}_1 + \vec{p}_2$, and V is the total potential due to both $\vec{p}_1$ and $\vec{p}_2$. Take $V = 0$ at $r = \infty$.

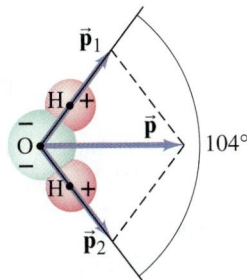

FIGURE 23–32
Problem 46.

23–7 $\vec{E}$ Determined from V

47. (I) Show that the electric field of a single point charge (Eq. 21–4) follows from Eq. 23–5, $V = (1/4\pi\epsilon_0)(Q/r)$.

48. (I) What is the potential gradient just outside the surface of a uranium nucleus $(Q = +92e)$ whose diameter is about $15 \times 10^{-15}\ \text{m}$?

49. (II) The electric potential between two parallel plates is given by $V(x) = (8.0\ \text{V/m})x + 5.0\ \text{V}$, with $x = 0$ taken at one of the plates and x positive in the direction toward the other plate. What is the charge density on the plates?

50. (II) The electric potential in a region of space varies as $V = by/(a^2 + y^2)$. Determine $\vec{E}$.

51. (II) In a certain region of space, the electric potential is given by $V = y^2 + 2.5xy - 3.5xyz$. Determine the electric field vector, $\vec{E}$, in this region.

52. (II) A dust particle with mass of $0.050\ \text{g}$ and a charge of $2.0 \times 10^{-6}\ \text{C}$ is in a region of space where the potential is given by $V(x) = (2.0\ \text{V/m}^2)x^2 - (3.0\ \text{V/m}^3)x^3$. If the particle starts at $x = 2.0\ \text{m}$, what is the initial acceleration of the charge?

53. (III) Use the results of Problems 38 and 39 to determine the electric field due to the uniformly charged rod of Fig. 23–31 for points (a) along the y axis and (b) along the x axis.

23–8 Electrostatic Potential Energy; Electron Volt

54. (I) How much work must be done to bring three electrons from a great distance apart to within $1.0 \times 10^{-10}\ \text{m}$ from one another (at the corners of an equilateral triangle)?

55. (I) What potential difference is needed to give a helium nucleus $(Q = 3.2 \times 10^{-19}\ \text{C})$ $125\ \text{keV}$ of kinetic energy?

56. (I) What is the speed of (a) a 1.5-keV (kinetic energy) electron and (b) a 1.5-keV proton?

57. (II) Many chemical reactions release energy. Suppose that at the beginning of a reaction, an electron and proton are separated by $0.110\ \text{nm}$, and their final separation is $0.100\ \text{nm}$. How much electric potential energy was lost in this reaction (in units of eV)?

58. (II) An alpha particle (which is a helium nucleus, $Q = +2e$, $m = 6.64 \times 10^{-27}\ \text{kg}$) is emitted in a radioactive decay with kinetic energy $5.53\ \text{MeV}$. What is its speed?

59. (II) Write the total electrostatic potential energy, U, for (a) four point charges and (b) five point charges. Draw a diagram defining all quantities.

60. (II) Four equal point charges, Q, are fixed at the corners of a square of side b. (a) What is their total electrostatic potential energy? (b) How much potential energy will a fifth charge, Q, have at the center of the square (relative to $V = 0$ at $r = \infty$)? (c) If constrained to remain in that plane, is the fifth charge in stable or unstable equilibrium? (d) If a negative $(-Q)$ charge is at the center, is it in stable equilibrium?

61. (II) An electron starting from rest acquires $1.33\ \text{keV}$ of kinetic energy in moving from point A to point B. (a) How much kinetic energy would a proton acquire, starting from rest at B and moving to point A? (b) Determine the ratio of their speeds at the end of their respective trajectories.

62. (II) Determine the total electrostatic potential energy of a conducting sphere of radius r_0 that carries a total charge Q distributed uniformly on its surface.

63. (II) The **liquid-drop model** of the nucleus suggests that high-energy oscillations of certain nuclei can split ("fission") a large nucleus into two unequal fragments plus a few neutrons. Using this model, consider the case of a uranium nucleus fissioning into two spherical fragments, one with a charge $q_1 = +38e$ and radius $r_1 = 5.5 \times 10^{-15}\ \text{m}$, the other with $q_2 = +54e$ and $r_2 = 6.2 \times 10^{-15}\ \text{m}$. Calculate the electric potential energy (MeV) of these fragments, assuming that the charge is uniformly distributed throughout the volume of each spherical nucleus and that their surfaces are initially in contact at rest. The electrons surrounding the nuclei can be neglected. This electric potential energy will then be entirely converted to kinetic energy as the fragments repel each other. How does your predicted kinetic energy of the fragments agree with the observed value associated with uranium fission (approximately $200\ \text{MeV}$ total)? $[1\ \text{MeV} = 10^6\ \text{eV}.]$

64. (III) Determine the total electrostatic potential energy of a nonconducting sphere of radius r_0 carrying a total charge Q distributed uniformly throughout its volume.

*23–9 CRT

***65.** (I) Use the ideal gas as a model to estimate the rms speed of a free electron in a metal at $273\ \text{K}$, and at $2700\ \text{K}$ (a typical temperature of the cathode in a CRT).

***66.** (III) Electrons are accelerated by $6.0\ \text{kV}$ in a CRT. The screen is $28\ \text{cm}$ wide and is $34\ \text{cm}$ from the 2.6-cm-long deflection plates. Over what range must the horizontally deflecting electric field vary to sweep the beam fully across the screen?

***67.** (III) In a given CRT, electrons are accelerated horizontally by $7.2\ \text{kV}$. They then pass through a uniform electric field E for a distance of $2.8\ \text{cm}$ which deflects them upward so they reach the top of the screen $22\ \text{cm}$ away, $11\ \text{cm}$ above the center. Estimate the value of E.

General Problems

68. If the electrons in a single raindrop, 3.5 mm in diameter, could be removed from the Earth (without removing the atomic nuclei), by how much would the potential of the Earth increase?

69. By rubbing a nonconducting material, a charge of 10^{-8} C can readily be produced. If this is done to a sphere of radius 15 cm, estimate the potential produced at the surface. Let $V = 0$ at $r = \infty$.

70. Sketch the electric field and equipotential lines for two charges of the same sign and magnitude separated by a distance d.

71. A $+33 \mu C$ point charge is placed 36 cm from an identical $+33 \mu C$ charge. A $-1.5 \mu C$ charge is moved from point a to point b, Fig. 23–33. What is the change in potential energy?

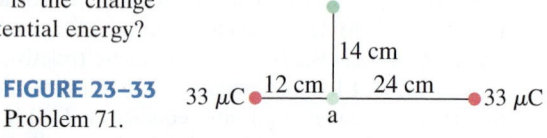

FIGURE 23–33
Problem 71.

72. At each corner of a cube of side ℓ there is a point charge Q, Fig. 23–34. (a) What is the potential at the center of the cube ($V = 0$ at $r = \infty$)? (b) What is the potential at each corner due to the other seven charges? (c) What is the total potential energy of this system?

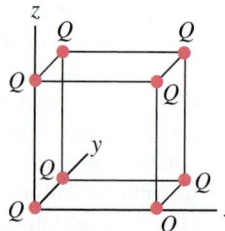

FIGURE 23–34
Problem 72.

73. In a television picture tube (CRT), electrons are accelerated by thousands of volts through a vacuum. If a television set is laid on its back, would electrons be able to move upward against the force of gravity? What potential difference, acting over a distance of 3.5 cm, would be needed to balance the downward force of gravity so that an electron would remain stationary? Assume that the electric field is uniform.

74. Four point charges are located at the corners of a square that is 8.0 cm on a side. The charges, going in rotation around the square, are Q, $2Q$, $-3Q$, and $2Q$, where $Q = 3.1 \mu C$ (Fig. 23–35). What is the total electric potential energy stored in the system, relative to $U = 0$ at infinite separation?

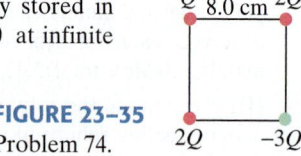

FIGURE 23–35
Problem 74.

75. In a **photocell**, ultraviolet (UV) light provides enough energy to some electrons in barium metal to eject them from the surface at high speed. See Fig. 23–36. To measure the maximum energy of the electrons, another plate above the barium surface is kept at a negative enough potential that the emitted electrons are slowed down and stopped, and return to the barium surface. If the plate voltage is -3.02 V (compared to the barium) when the fastest electrons are stopped, what was the speed of these electrons when they were emitted?

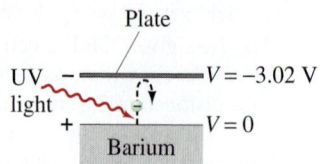

FIGURE 23–36
Problem 75.

76. An electron is accelerated horizontally from rest in a television picture tube by a potential difference of 5500 V. It then passes between two horizontal plates 6.5 cm long and 1.3 cm apart that have a potential difference of 250 V (Fig. 23–37). At what angle θ will the electron be traveling after it passes between the plates?

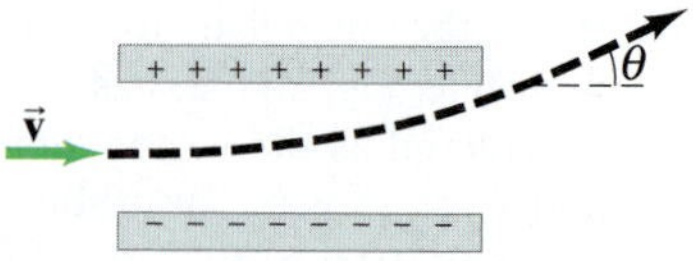

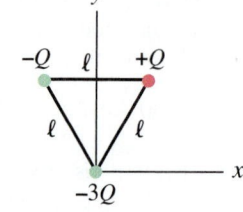

FIGURE 23–37
Problem 76.

77. Three charges are at the corners of an equilateral triangle (side ℓ) as shown in Fig. 23–38. Determine the potential at the midpoint of each of the sides. Let $V = 0$ at $r = \infty$.

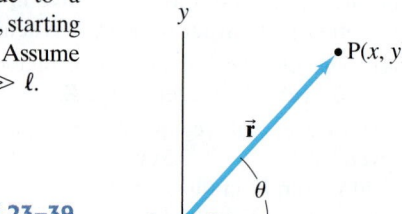

FIGURE 23–38
Problem 77.

78. Near the surface of the Earth there is an electric field of about 150 V/m which points downward. Two identical balls with mass $m = 0.340$ kg are dropped from a height of 2.00 m, but one of the balls is positively charged with $q_1 = 450 \mu C$, and the second is negatively charged with $q_2 = -450 \mu C$. Use conservation of energy to determine the difference in the speeds of the two balls when they hit the ground. (Neglect air resistance.)

79. A lightning flash transfers 4.0 C of charge and 4.8 MJ of energy to the Earth. (a) Between what potential difference did it travel? (b) How much water could this energy boil, starting from room temperature? [*Hint*: See Chapter 19.]

80. Determine the components of the electric field, E_x and E_y, as a function of x and y in the xy plane due to a dipole, Fig. 23–39, starting with Eq. 23–7. Assume $r = (x^2 + y^2)^{\frac{1}{2}} \gg \ell$.

FIGURE 23–39
Problem 80.

81. A nonconducting sphere of radius r_2 contains a concentric spherical cavity of radius r_1. The material between r_1 and r_2 carries a uniform charge density ρ_E (C/m^3). Determine the electric potential V, relative to $V = 0$ at $r = \infty$, as a function of the distance r from the center for (a) $r > r_2$, (b) $r_1 < r < r_2$, and (c) $0 < r < r_1$. Is V continuous at r_1 and r_2?

82. A thin flat nonconducting disk, with radius R_0 and charge Q, has a hole with a radius $R_0/2$ in its center. Find the electric potential $V(x)$ at points along the symmetry (x) axis of the disk (a line perpendicular to the disk, passing through its center). Let $V = 0$ at $x = \infty$.

83. A **Geiger counter** is used to detect charged particles emitted by radioactive nuclei. It consists of a thin, positively charged central wire of radius R_a surrounded by a concentric conducting cylinder of radius R_b with an equal negative charge (Fig. 23–40). The charge per unit length on the inner wire is λ (units C/m). The interior space between wire and cylinder is filled with low-pressure inert gas. Charged particles ionize some of these gas atoms; the resulting free electrons are attracted toward the positive central wire. If the radial electric field is strong enough, the freed electrons gain enough energy to ionize other atoms, causing an "avalanche" of electrons to strike the central wire, generating an electric "signal." Find the expression for the electric field between the wire and the cylinder, and show that the potential difference between R_a and R_b is

$$V_a - V_b = \left(\frac{\lambda}{2\pi\epsilon_0}\right) \ln\left(\frac{R_b}{R_a}\right).$$

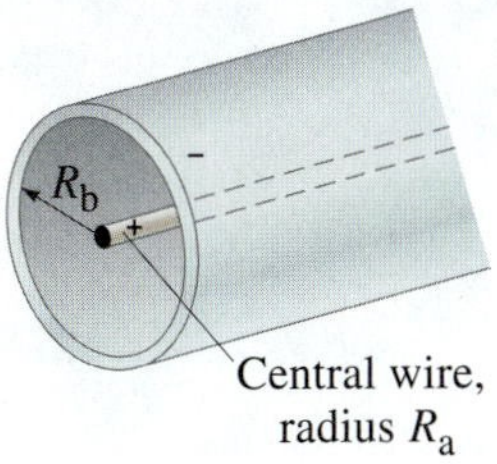

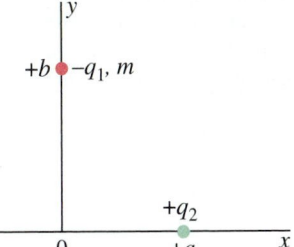

FIGURE 23–40
Problem 83.

Central wire, radius R_a

84. A **Van de Graaff generator** (Fig. 23–41) can develop a very large potential difference, even millions of volts. Electrons are pulled off the belt by the high voltage pointed electrode at A, leaving the belt positively charged. (Recall Example 23–5 where we saw that near sharp points the electric field is high and ionization can occur.) The belt carries the positive charge up inside the spherical shell where electrons from the large conducting sphere are attracted over to the pointed conductor at B, leaving the outer surface of the conducting sphere positively charged. As more charge is brought up, the sphere reaches extremely high voltage. Consider a Van de Graaff generator with a sphere of radius 0.20 m. (a) What is the electric potential on the surface of the sphere when electrical breakdown occurs? (Assume $V = 0$ at $r = \infty$.) (b) What is the charge on the sphere for the potential found in part (a)?

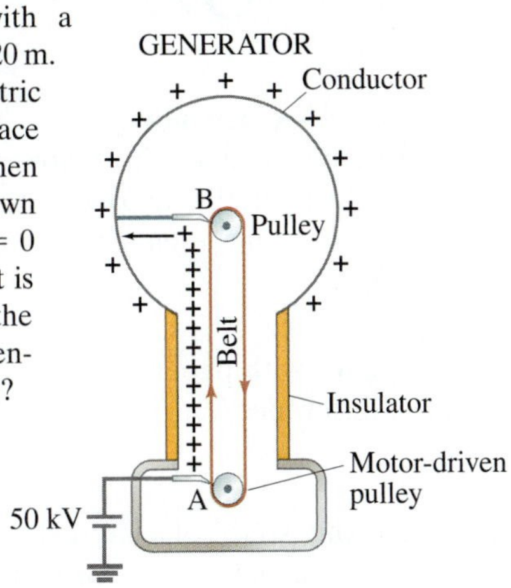

GENERATOR

+ + + Conductor

B Pulley

Belt

Insulator

Motor-driven pulley

A

FIGURE 23–41 50 kV
Problem 84.

85. The potential in a region of space is given by $V = B/(x^2 + R^2)^2$ where $B = 150 \text{ V} \cdot \text{m}^4$ and $R = 0.20$ m. (a) Find V at $x = 0.20$ m. (b) Find $\vec{E}$ as a function of x. (c) Find $\vec{E}$ at $x = 0.20$ m.

86. A charge $-q_1$ of mass m rests on the y axis at a distance b above the x axis. Two positive charges of magnitude $+q_2$ are fixed on the x axis at $x = +a$ and $x = -a$, respectively (Fig. 23–42). If the $-q_1$ charge is given an initial velocity v_0 in the positive y direction, what is the minimum value of v_0 such that the charge escapes to a point infinitely far away from the two positive charges?

y

$+b$ ● $-q_1, m$

$+q_2$ $+q_2$

$-a$ 0 $+a$ x

FIGURE 23–42
Problem 86.

Numerical/Computer

*87. (II) A dipole is composed of a -1.0 nC charge at $x = -1.0$ cm and a $+1.0$ nC charge at $x = +1.0$ cm. (a) Make a plot of V along the x axis from $x = 2.0$ cm to $x = 15$ cm. (b) On the same graph, plot the approximate V using Eq. 23–7 from $x = 2.0$ cm to $x = 15$ cm. Let $V = 0$ at $x = \infty$.

*88. (II) A thin flat disk of radius R_0 carries a total charge Q that is distributed uniformly over its surface. The electric potential at a distance x on the x axis is given by

$$V(x) = \frac{Q}{2\pi\epsilon_0 R_0^2} \left[(x^2 + R_0^2)^{\frac{1}{2}} - x\right].$$

(See Example 23–9.) Show that the electric field at a distance x on the x axis is given by

$$E(x) = \frac{Q}{2\pi\epsilon_0 R_0^2} \left(1 - \frac{x}{(x^2 + R_0^2)^{\frac{1}{2}}}\right).$$

Make graphs of $V(x)$ and $E(x)$ as a function of x/R_0 for $x/R_0 = 0$ to 4. (Do the calculations in steps of 0.1.) Use $Q = 5.0 \mu$C and $R_0 = 10$ cm for the calculation and graphs.

*89. (III) You are trying to determine an unknown amount of charge using only a voltmeter and a ruler, knowing that it is either a single sheet of charge or a point charge that is creating it. You determine the direction of greatest change of potential, and then measure potentials along a line in that direction. The potential versus position (note that the zero of position is arbitrary, and the potential is measured relative to ground) is measured as follows:

x (cm)	0.0	1.0	2.0	3.0	4.0	5.0	6.0	7.0	8.0	9.0
V (volts)	3.9	3.0	2.5	2.0	1.7	1.5	1.4	1.4	1.2	1.1

(a) Graph V versus position. Do you think the field is caused by a sheet or a point charge? (b) Graph the data in such a way that you can determine the magnitude of the charge and determine that value. (c) Is it possible to determine where the charge is from this data? If so, give the position of the charge.

Answers to Exercises

A: (b).

B: (d).

C: (d).

D: (a) iii, (b) i, (c) i.

E: (c).

F: (d).

Capacitors come in a wide range of sizes and shapes, only a few of which are shown here. A capacitor is basically two conductors that do not touch, and which therefore can store charge of opposite sign on its two conductors. Capacitors are used in a wide variety of circuits, as we shall see in this and later Chapters.

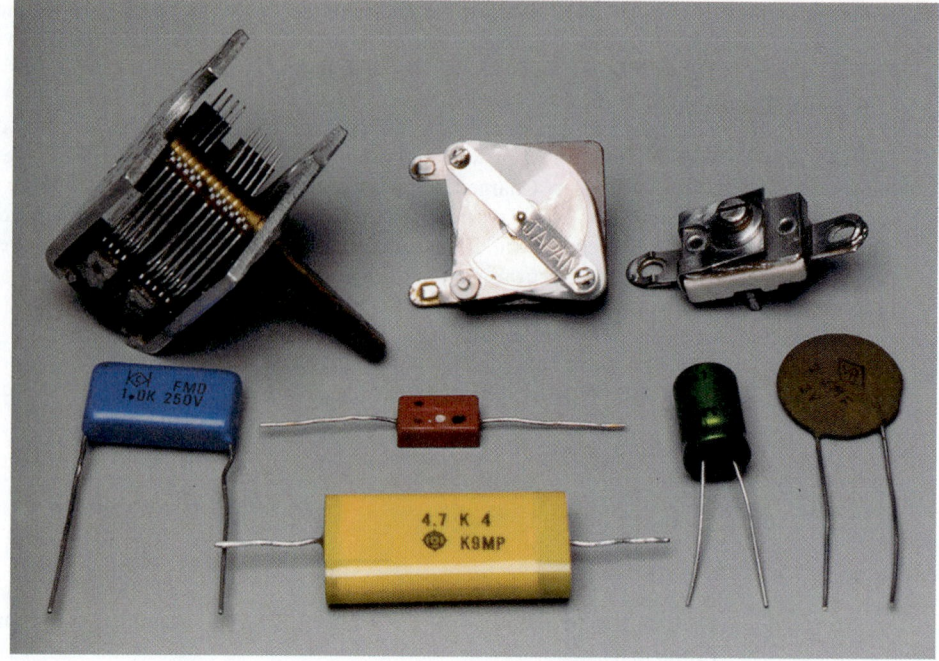

24

Capacitance, Dielectrics, Electric Energy Storage

CHAPTER-OPENING QUESTION—Guess now!

A fixed potential difference V exists between a pair of close parallel plates carrying opposite charges $+Q$ and $-Q$. Which of the following would not increase the magnitude of charge that you could put on the plates?

(a) Increase the size of the plates.
(b) Move the plates farther apart.
(c) Fill the space between the plates with paper.
(d) Increase the fixed potential difference V.
(e) None of the above.

This Chapter will complete our study of electrostatics. It deals first of all with an important device, the capacitor, which is used in many electronic circuits. We will also discuss electric energy storage and the effects of an insulator, or dielectric, on electric fields and potential differences.

CONTENTS

24–1 Capacitors
24–2 Determination of Capacitance
24–3 Capacitors in Series and Parallel
24–4 Electric Energy Storage
24–5 Dielectrics
*24–6 Molecular Description of Dielectrics

PHYSICS APPLIED
Uses of capacitors

24–1 Capacitors

A **capacitor** is a device that can store electric charge, and normally consists of two conducting objects (usually plates or sheets) placed near each other but not touching. Capacitors are widely used in electronic circuits. They store charge for later use, such as in a camera flash, and as energy backup in computers if the power fails. Capacitors also block surges of charge and energy to protect circuits.

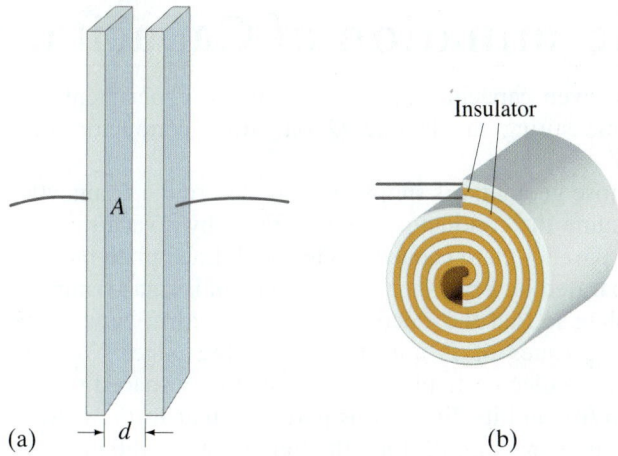

Insulator

FIGURE 24–1 Capacitors: diagrams of (a) parallel plate, (b) cylindrical (rolled up parallel plate).

(a) d (b)

Very tiny capacitors serve as memory for the "ones" and "zeros" of the binary code in the random access memory (RAM) of computers. Capacitors serve many other applications, some of which we will discuss.

A simple capacitor consists of a pair of parallel plates of area A separated by a small distance d (Fig. 24–1a). Often the two plates are rolled into the form of a cylinder with plastic, paper, or other insulator separating the plates, Fig. 24–1b. In a diagram, the symbol

⊣⊢   or   ⊣⊦   [capacitor symbol]

represents a capacitor. A battery, which is a source of voltage, is indicated by the symbol:

⊣⊦   [battery symbol]

with unequal arms.

If a voltage is applied across a capacitor by connecting the capacitor to a battery with conducting wires as in Fig. 24–2, the two plates quickly become charged: one plate acquires a negative charge, the other an equal amount of positive charge. Each battery terminal and the plate of the capacitor connected to it are at the same potential; hence the full battery voltage appears across the capacitor. For a given capacitor, it is found that the amount of charge Q acquired by each plate is proportional to the magnitude of the potential difference V between them:

$$Q = CV. \qquad (24–1)$$

The constant of proportionality, C, in the above relation is called the **capacitance** of the capacitor. The unit of capacitance is coulombs per volt and this unit is called a **farad** (F). Common capacitors have capacitance in the range of $1\,\text{pF}$ (picofarad $= 10^{-12}\,\text{F}$) to $10^3\,\mu\text{F}$ (microfarad $= 10^{-6}\,\text{F}$). The relation, Eq. 24–1, was first suggested by Volta in the late eighteenth century. The capacitance C does not in general depend on Q or V. Its value depends only on the size, shape, and relative position of the two conductors, and also on the material that separates them.

In Eq. 24–1, and from now on, we use simply V (in italics) to represent a potential difference, rather than V_{ba}, ΔV, or $V_b - V_a$, as previously. (Be sure not to confuse italic V and C which stand for voltage and capacitance, with non-italic V and C which stand for the units volts and coulombs).

EXERCISE A Graphs for charge versus voltage are shown in Fig. 24–3 for three capacitors, A, B, and C. Which has the greatest capacitance?

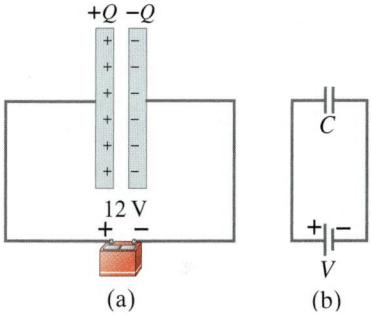

$+Q$ $-Q$

12 V

(a)   (b)

FIGURE 24–2 (a) Parallel-plate capacitor connected to a battery. (b) Same circuit shown using symbols.

FIGURE 24–3 Exercise A.

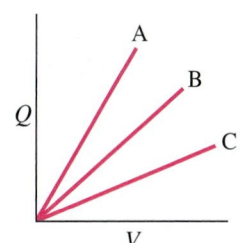

⚠ **CAUTION**

V = potential difference from here on

24–2 Determination of Capacitance

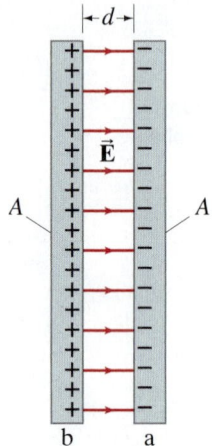

FIGURE 24–4 Parallel-plate capacitor, each of whose plates has area A. Fringing of the field is ignored.

The capacitance of a given capacitor can be determined experimentally directly from Eq. 24–1, by measuring the charge Q on either conductor for a given potential difference V.

For capacitors whose geometry is simple, we can determine C analytically, and in this Section we assume the conductors are separated by a vacuum or air. First, we determine C for a parallel-plate capacitor, Fig. 24–4. Each plate has area A and the two plates are separated by a distance d. We assume d is small compared to the dimensions of each plate so that the electric field $\vec{E}$ is uniform between them and we can ignore fringing (lines of $\vec{E}$ not straight) at the edges. We saw earlier (Example 21–13) that the electric field between two closely spaced parallel plates has magnitude $E = \sigma/\epsilon_0$ and its direction is perpendicular to the plates. Since σ is the charge per unit area, $\sigma = Q/A$, then the field between the plates is

$$E = \frac{Q}{\epsilon_0 A}.$$

The relation between electric field and electric potential, as given by Eq. 23–4a, is

$$V = V_{ba} = V_b - V_a = -\int_a^b \vec{E} \cdot d\vec{\ell}.$$

We can take the line integral along a path antiparallel to the field lines, from plate a to plate b; then $\theta = 180°$ and $\cos 180° = -1$, so

$$V = V_b - V_a = -\int_a^b E \, d\ell \cos 180° = +\int_a^b E \, d\ell = \frac{Q}{\epsilon_0 A} \int_a^b d\ell = \frac{Qd}{\epsilon_0 A}.$$

This relates Q to V, and from it we can get the capacitance C in terms of the geometry of the plates:

$$C = \frac{Q}{V} = \epsilon_0 \frac{A}{d}. \qquad \text{[parallel-plate capacitor]} \qquad \textbf{(24–2)}$$

Note from Eq. 24–2 that the value of C does not depend on Q or V, so Q is predicted to be proportional to V as is found experimentally.

EXAMPLE 24–1 **Capacitor calculations.** (a) Calculate the capacitance of a parallel-plate capacitor whose plates are 20 cm × 3.0 cm and are separated by a 1.0-mm air gap. (b) What is the charge on each plate if a 12-V battery is connected across the two plates? (c) What is the electric field between the plates? (d) Estimate the area of the plates needed to achieve a capacitance of 1 F, given the same air gap d.

APPROACH The capacitance is found by using Eq. 24–2, $C = \epsilon_0 A/d$. The charge on each plate is obtained from the definition of capacitance, Eq. 24–1, $Q = CV$. The electric field is uniform, so we can use Eq. 23–4b for the magnitude $E = V/d$. In (d) we use Eq. 24–2 again.

SOLUTION (a) The area $A = (20 \times 10^{-2}\,\text{m})(3.0 \times 10^{-2}\,\text{m}) = 6.0 \times 10^{-3}\,\text{m}^2$. The capacitance C is then

$$C = \epsilon_0 \frac{A}{d} = (8.85 \times 10^{-12}\,\text{C}^2/\text{N} \cdot \text{m}^2) \frac{6.0 \times 10^{-3}\,\text{m}^2}{1.0 \times 10^{-3}\,\text{m}} = 53\,\text{pF}.$$

(b) The charge on each plate is

$$Q = CV = (53 \times 10^{-12}\,\text{F})(12\,\text{V}) = 6.4 \times 10^{-10}\,\text{C}.$$

(c) From Eq. 23–4b for a uniform electric field, the magnitude of E is

$$E = \frac{V}{d} = \frac{12\,\text{V}}{1.0 \times 10^{-3}\,\text{m}} = 1.2 \times 10^4\,\text{V/m}.$$

(d) We solve for A in Eq. 24–2 and substitute $C = 1.0\,\text{F}$ and $d = 1.0\,\text{mm}$ to find that we need plates with an area

$$A = \frac{Cd}{\epsilon_0} \approx \frac{(1\,\text{F})(1.0 \times 10^{-3}\,\text{m})}{(9 \times 10^{-12}\,\text{C}^2/\text{N} \cdot \text{m}^2)} \approx 10^8\,\text{m}^2.$$

NOTE This is the area of a square 10^4 m or 10 km on a side. That is the size of a city like San Francisco or Boston! Large-capacitance capacitors will not be simple parallel plates.

EXERCISE B Two circular plates of radius 5.0 cm are separated by a 0.10-mm air gap. What is the magnitude of the charge on each plate when connected to a 12-V battery?

Not long ago, a capacitance greater than a few mF was unusual. Today capacitors are available that are 1 or 2 F, yet they are just a few cm on a side. Such capacitors are used as power backups, for example, in computer memory and electronics where the time and date can be maintained through tiny charge flow. [Capacitors are superior to rechargable batteries for this purpose because they can be recharged more than 10^5 times with no degradation.] Such high-capacitance capacitors can be made of "activated" carbon which has very high porosity, so that the surface area is very large; one tenth of a gram of activated carbon can have a surface area of 100 m^2. Furthermore, the equal and opposite charges exist in an electric "double layer" about 10^{-9} m thick. Thus, the capacitance of 0.1 g of activated carbon, whose internal area can be 10^2 m^2, is equivalent to a parallel-plate capacitor with $C \approx \epsilon_0 A/d = (8.85 \times 10^{-12} \text{ C}^2/\text{N}\cdot\text{m}^2)(10^2 \text{ m}^2)/(10^{-9} \text{ m}) \approx 1 \text{ F}$.

One type of computer keyboard operates by capacitance. As shown in Fig. 24–5, each key is connected to the upper plate of a capacitor. The upper plate moves down when the key is pressed, reducing the spacing between the capacitor plates, and increasing the capacitance (Eq. 24–2: smaller d, larger C). The *change* in capacitance results in an electric signal that is detected by an electronic circuit.

The proportionality, $C \propto A/d$ in Eq. 24–2, is valid also for a parallel-plate capacitor that is rolled up into a spiral cylinder, as in Fig. 24–1b. However, the constant factor, ϵ_0, must be replaced if an insulator such as paper separates the plates, as is usual, and this is discussed in Section 24–5. For a true cylindrical capacitor—consisting of two long coaxial cylinders—the result is somewhat different as the next Example shows.

PHYSICS APPLIED
Very high capacitance

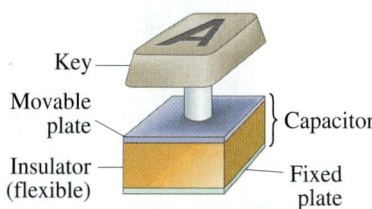

FIGURE 24–5 Key on a computer keyboard. Pressing the key reduces the capacitor spacing thus increasing the capacitance which can be detected electronically.

PHYSICS APPLIED
Computer key

EXAMPLE 24–2 **Cylindrical capacitor.** A cylindrical capacitor consists of a cylinder (or wire) of radius R_b surrounded by a coaxial cylindrical shell of inner radius R_a, Fig. 24–6a. Both cylinders have length ℓ which we assume is much greater than the separation of the cylinders, $R_a - R_b$, so we can neglect end effects. The capacitor is charged (by connecting it to a battery) so that one cylinder has a charge $+Q$ (say, the inner one) and the other one a charge $-Q$. Determine a formula for the capacitance.

APPROACH To obtain $C = Q/V$, we need to determine the potential difference V between the cylinders in terms of Q. We can use our earlier result (Example 21–11 or 22–6) that the electric field outside a long wire is directed radially outward and has magnitude $E = (1/2\pi\epsilon_0)(\lambda/R)$, where R is the distance from the axis and λ is the charge per unit length, Q/ℓ. Then $E = (1/2\pi\epsilon_0)(Q/\ell R)$ for points between the cylinders.

SOLUTION To obtain the potential difference V in terms of Q, we use this result for E in Eq. 23–4a, $V = V_b - V_a = -\int_a^b \vec{E}\cdot d\vec{\ell}$, and write the line integral from the outer cylinder to the inner one (so $V > 0$) along a radial line:[†]

$$V = V_b - V_a = -\int_a^b \vec{E}\cdot d\vec{\ell} = -\frac{Q}{2\pi\epsilon_0\ell}\int_{R_a}^{R_b}\frac{dR}{R}$$
$$= -\frac{Q}{2\pi\epsilon_0\ell}\ln\frac{R_b}{R_a} = \frac{Q}{2\pi\epsilon_0\ell}\ln\frac{R_a}{R_b}.$$

Q and V are proportional, and the capacitance C is

$$C = \frac{Q}{V} = \frac{2\pi\epsilon_0\ell}{\ln(R_a/R_b)}. \qquad \text{[cylindrical capacitor]}$$

NOTE If the space between cylinders, $R_a - R_b = \Delta R$ is small, we have $\ln(R_a/R_b) = \ln[(R_b + \Delta R)/R_b] = \ln[1 + \Delta R/R_b] \approx \Delta R/R_b$ (see Appendix A–3) so $C \approx 2\pi\epsilon_0\ell R_b/\Delta R = \epsilon_0 A/\Delta R$ because the area of cylinder b is $A = 2\pi R_b\ell$. This is just Eq. 24–2 ($d = \Delta R$), a nice check.

EXERCISE C What is the capacitance per unit length of a cylindrical capacitor with radii $R_a = 2.5$ mm and $R_b = 0.40$ mm? (a) 30 pF/m; (b) −30 pF/m; (c) 56 pF/m; (d) −56 pF/m; (e) 100 pF/m; (f) −100 pF/m.

FIGURE 24–6 (a) Cylindrical capacitor consists of two coaxial cylindrical conductors. (b) The electric field lines are shown in cross-sectional view.

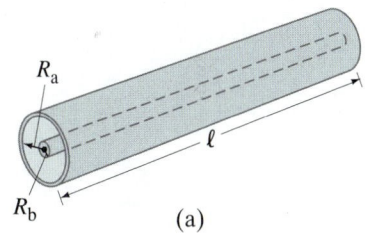

(a)

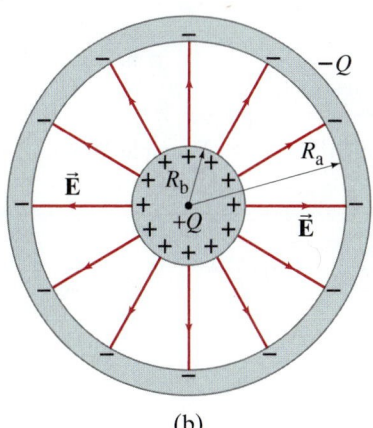

(b)

[†]Note that $\vec{E}$ points outward in Fig. 24–6b, but $d\vec{\ell}$ points inward for our chosen direction of integration; the angle between $\vec{E}$ and $d\vec{\ell}$ is 180° and $\cos 180° = -1$. Also, $d\ell = -dr$ because dr increases outward. These two minus signs cancel.

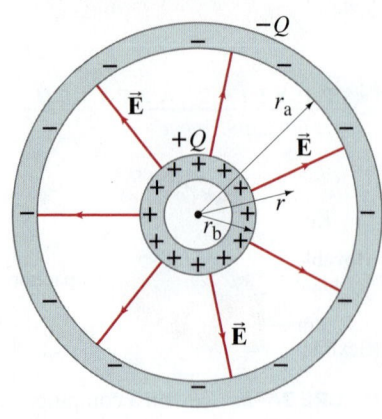

FIGURE 24–7 Cross section through the center of a spherical capacitor. The thin inner shell has radius r_b and the thin outer shell has radius r_a.

PROBLEM SOLVING

Checking with a limiting case

EXAMPLE 24–3 Spherical capacitor. A spherical capacitor consists of two thin concentric spherical conducting shells, of radius r_a and r_b as shown in Fig. 24–7. The inner shell carries a uniformly distributed charge Q on its surface, and the outer shell an equal but opposite charge $-Q$. Determine the capacitance of the two shells.

APPROACH In Example 22–3 we used Gauss's law to show that the electric field outside a uniformly charged conducting sphere is $E = Q/4\pi\epsilon_0 r^2$ as if all the charge were concentrated at the center. Now we use Eq. 23–4a, $V = -\int_a^b \vec{E} \cdot d\vec{\ell}$.

SOLUTION We integrate Eq. 23–4a along a radial line to obtain the potential difference between the two conducting shells:

$$V_{ba} = -\int_a^b \vec{E} \cdot d\vec{\ell} = -\frac{Q}{4\pi\epsilon_0} \int_{r_a}^{r_b} \frac{1}{r^2} dr$$

$$= \frac{Q}{4\pi\epsilon_0}\left(\frac{1}{r_b} - \frac{1}{r_a}\right) = \frac{Q}{4\pi\epsilon_0}\left(\frac{r_a - r_b}{r_a r_b}\right).$$

Finally,

$$C = \frac{Q}{V_{ba}} = 4\pi\epsilon_0\left(\frac{r_a r_b}{r_a - r_b}\right).$$

NOTE If the separation $\Delta r = r_a - r_b$ is very small, then $C = 4\pi\epsilon_0 r^2/\Delta r \approx \epsilon_0 A/\Delta r$ (since $A = 4\pi r^2$), which is the parallel-plate formula, Eq. 24–2.

A single isolated conductor can also be said to have a capacitance, C. In this case, C can still be defined as the ratio of the charge to absolute potential V on the conductor (relative to $V = 0$ at $r = \infty$), so that the relation

$$Q = CV$$

remains valid. For example, the potential of a single conducting sphere of radius r_b can be obtained from our results in Example 24–3 by letting r_a become infinitely large. As $r_a \to \infty$, then

$$V = \frac{Q}{4\pi\epsilon_0}\left(\frac{1}{r_b} - \frac{1}{r_a}\right) = \frac{1}{4\pi\epsilon_0}\frac{Q}{r_b};$$

so its capacitance is

$$C = \frac{Q}{V} = 4\pi\epsilon_0 r_b.$$

In practical cases, a single conductor may be near other conductors or the Earth (which can be thought of as the other "plate" of a capacitor), and these will affect the value of the capacitance.

FIGURE 24–8 Example 24–4.

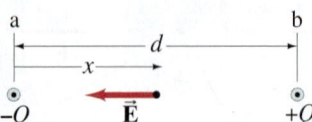

EXAMPLE 24–4 Capacitance of two long parallel wires. Estimate the capacitance per unit length of two very long straight parallel wires, each of radius R, carrying uniform charges $+Q$ and $-Q$, and separated by a distance d which is large compared to R ($d \gg R$), Fig. 24–8.

APPROACH We calculate the potential difference between the wires by treating the electric field at any point between them as the superposition of the two fields created by each wire. (The electric field inside each wire conductor is zero.)

SOLUTION The electric field outside of a long straight conductor was found in Examples 21–11 and 22–6 to be radial and given by $E = \lambda/(2\pi\epsilon_0 x)$ where λ is the charge per unit length, $\lambda = Q/\ell$. The total electric field at distance x from the left-hand wire in Fig. 24–8 has magnitude

$$E = \frac{\lambda}{2\pi\epsilon_0 x} + \frac{\lambda}{2\pi\epsilon_0(d - x)},$$

and points to the left (from $+$ to $-$). Now we find the potential difference between the two wires using Eq. 23–4a and integrating along the straight line from the surface of the negative wire to the surface of the positive wire, noting that $\vec{E}$ and $d\vec{\ell}$ point in opposite directions ($\vec{E} \cdot d\vec{\ell} < 0$):

$$V = V_b - V_a = -\int_a^b \vec{E} \cdot d\vec{\ell} = \left(\frac{\lambda}{2\pi\epsilon_0}\right) \int_R^{d-R} \left[\frac{1}{x} + \frac{1}{(d-x)}\right] dx$$

$$= \left(\frac{\lambda}{2\pi\epsilon_0}\right) \Big[\ln(x) - \ln(d - x)\Big]\Big|_R^{d-R}$$

$$= \left(\frac{\lambda}{2\pi\epsilon_0}\right) \Big[\ln(d - R) - \ln R - \ln R + \ln(d - R)\Big]$$

$$= \left(\frac{\lambda}{\pi\epsilon_0}\right) \Big[\ln(d - R) - \ln(R)\Big] \approx \left(\frac{\lambda}{\pi\epsilon_0}\right) \Big[\ln(d) - \ln(R)\Big].$$

We are given that $d \gg R$, so

$$V \approx \left(\frac{Q}{\pi\epsilon_0 \ell}\right) \left[\ln\left(\frac{d}{R}\right)\right].$$

The capacitance from Eq. 24–1 is $C = Q/V \approx (\pi\epsilon_0 \ell)/\ln(d/R)$, so the capacitance per unit length is given approximately by

$$\frac{C}{\ell} \approx \frac{\pi\epsilon_0}{\ln(d/R)}.$$

24–3 Capacitors in Series and Parallel

Capacitors are found in many electric circuits. By electric circuit we mean a closed path of conductors, usually wires connecting capacitors and/or other devices, in which charge can flow and which includes a source of voltage such as a battery. The battery voltage is usually given the symbol V, which means that V represents a potential *difference*. Capacitors can be connected together in various ways. Two common ways are in *series*, or in *parallel*, and we now discuss both.

A circuit containing three capacitors connected in **parallel** is shown in Fig. 24–9. They are in "parallel" because when a battery of voltage V is connected to points a and b, this voltage $V = V_{ab}$ exists across each of the capacitors. That is, since the left-hand plates of all the capacitors are connected by conductors, they all reach the same potential V_a when connected to the battery; and the right-hand plates each reach potential V_b. Each capacitor plate acquires a charge given by $Q_1 = C_1 V$, $Q_2 = C_2 V$, and $Q_3 = C_3 V$. The total charge Q that must leave the battery is then

$$Q = Q_1 + Q_2 + Q_3 = C_1 V + C_2 V + C_3 V.$$

Let us try to find a single equivalent capacitor that will hold the same charge Q at the same voltage $V = V_{ab}$. It will have a capacitance C_{eq} given by

$$Q = C_{eq} V.$$

Combining the two previous equations, we have

$$C_{eq} V = C_1 V + C_2 V + C_3 V = (C_1 + C_2 + C_3)V$$

or

$$C_{eq} = C_1 + C_2 + C_3. \qquad \text{[parallel]} \quad \textbf{(24–3)}$$

The net effect of connecting capacitors in parallel is thus to *increase* the capacitance. This makes sense because we are essentially increasing the area of the plates where charge can accumulate (see, for example, Eq. 24–2).

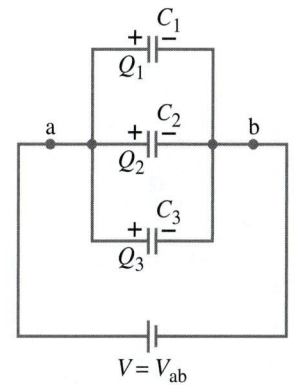

FIGURE 24–9 Capacitors in parallel:
$$C_{eq} = C_1 + C_2 + C_3.$$

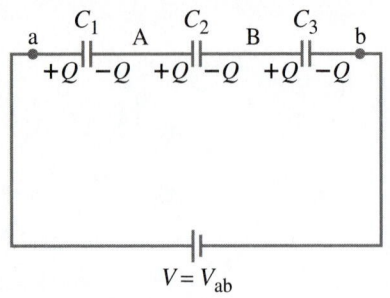

FIGURE 24-10 Capacitors in series:

$$\frac{1}{C_{eq}} = \frac{1}{C_1} + \frac{1}{C_2} + \frac{1}{C_3}.$$

Capacitors can also be connected in **series**: that is, end to end as shown in Fig. 24–10. A charge $+Q$ flows from the battery to one plate of C_1, and $-Q$ flows to one plate of C_3. The regions A and B between the capacitors were originally neutral; so the net charge there must still be zero. The $+Q$ on the left plate of C_1 attracts a charge of $-Q$ on the opposite plate. Because region A must have a zero net charge, there is thus $+Q$ on the left plate of C_2. The same considerations apply to the other capacitors, so we see the charge on each capacitor is the same value Q. A single capacitor that could replace these three in series without affecting the circuit (that is, Q and V the same) would have a capacitance C_{eq} where

$$Q = C_{eq}V.$$

Now the total voltage V across the three capacitors in series must equal the sum of the voltages across each capacitor:

$$V = V_1 + V_2 + V_3.$$

We also have for each capacitor $Q = C_1V_1$, $Q = C_2V_2$, and $Q = C_3V_3$, so we substitute for $V, V_1, V_2,$ and V_3 into the last equation and get

$$\frac{Q}{C_{eq}} = \frac{Q}{C_1} + \frac{Q}{C_2} + \frac{Q}{C_3} = Q\left(\frac{1}{C_1} + \frac{1}{C_2} + \frac{1}{C_3}\right)$$

or

$$\frac{1}{C_{eq}} = \frac{1}{C_1} + \frac{1}{C_2} + \frac{1}{C_3}. \qquad \text{[series]} \quad \textbf{(24-4)}$$

Notice that the equivalent capacitance C_{eq} is smaller than the smallest contributing capacitance.

EXERCISE D Consider two identical capacitors $C_1 = C_2 = 10\,\mu F$. What are the minimum and maximum capacitances that can be obtained by connecting these in series or parallel combinations? (*a*) $0.2\,\mu F, 5\,\mu F$; (*b*) $0.2\,\mu F, 10\,\mu F$; (*c*) $0.2\,\mu F, 20\,\mu F$; (*d*) $5\,\mu F, 10\,\mu F$; (*e*) $5\,\mu F, 20\,\mu F$; (*f*) $10\,\mu F, 20\,\mu F$.

Other connections of capacitors can be analyzed similarly using charge conservation, and often simply in terms of series and parallel connections.

FIGURE 24-11 Examples 24–5 and 24–6.

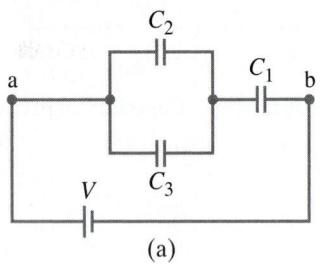

(a)

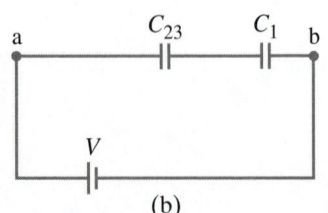

(b)

EXAMPLE 24–5 **Equivalent capacitance.** Determine the capacitance of a single capacitor that will have the same effect as the combination shown in Fig. 24–11a. Take $C_1 = C_2 = C_3 = C$.

APPROACH First we find the equivalent capacitance of C_2 and C_3 in parallel, and then consider that capacitance in series with C_1.

SOLUTION Capacitors C_2 and C_3 are connected in parallel, so they are equivalent to a single capacitor having capacitance

$$C_{23} = C_2 + C_3 = 2C.$$

This C_{23} is in series with C_1, Fig. 24–11b, so the equivalent capacitance of the entire circuit, C_{eq}, is given by

$$\frac{1}{C_{eq}} = \frac{1}{C_1} + \frac{1}{C_{23}} = \frac{1}{C} + \frac{1}{2C} = \frac{3}{2C}.$$

Hence the equivalent capacitance of the entire combination is $C_{eq} = \frac{2}{3}C$, and it is smaller than any of the contributing capacitors, $C_1 = C_2 = C_3 = C$.

EXAMPLE 24-6 **Charge and voltage on capacitors.** Determine the charge on each capacitor in Fig. 24–11a of Example 24–5 and the voltage across each, assuming $C = 3.0 \, \mu\text{F}$ and the battery voltage is $V = 4.0 \, \text{V}$.

APPROACH We have to work "backward" through Example 24–5. That is, we find the charge Q that leaves the battery, using the equivalent capacitance. Then we find the charge on each separate capacitor and the voltage across each. Each step uses Eq. 24–1, $Q = CV$.

SOLUTION The 4.0-V battery behaves as if it is connected to a capacitance $C_{\text{eq}} = \frac{2}{3}C = \frac{2}{3}(3.0 \, \mu\text{F}) = 2.0 \, \mu\text{F}$. Therefore the charge Q that leaves the battery, by Eq. 24–1, is

$$Q = CV = (2.0 \, \mu\text{F})(4.0 \, \text{V}) = 8.0 \, \mu\text{C}.$$

From Fig. 24–11a, this charge arrives at the negative plate of C_1, so $Q_1 = 8.0 \, \mu\text{C}$. The charge Q that leaves the positive plate of the battery is split evenly between C_2 and C_3 (symmetry: $C_2 = C_3$) and is $Q_2 = Q_3 = \frac{1}{2}Q = 4.0 \, \mu\text{C}$. Next, the voltages across C_2 and C_3 have to be the same. The voltage across each capacitor is obtained using $V = Q/C$. So

$$V_1 = Q_1/C_1 = (8.0 \, \mu\text{C})/(3.0 \, \mu\text{F}) = 2.7 \, \text{V}$$
$$V_2 = Q_2/C_2 = (4.0 \, \mu\text{C})/(3.0 \, \mu\text{F}) = 1.3 \, \text{V}$$
$$V_3 = Q_3/C_3 = (4.0 \, \mu\text{C})/(3.0 \, \mu\text{F}) = 1.3 \, \text{V}.$$

EXAMPLE 24-7 **Capacitors reconnected.** Two capacitors, $C_1 = 2.2 \, \mu\text{F}$ and $C_2 = 1.2 \, \mu\text{F}$, are connected in parallel to a 24-V source as shown in Fig. 24–12a. After they are charged they are disconnected from the source and from each other, and then reconnected directly to each other with plates of opposite sign connected together (see Fig. 24–12b). Find the charge on each capacitor and the potential across each after equilibrium is established.

APPROACH We find the charge $Q = CV$ on each capacitor initially. Charge is conserved, although rearranged after the switch. The two new voltages will have to be equal.

SOLUTION First we calculate how much charge has been placed on each capacitor after the power source has charged them fully, using Eq. 24–1:

$$Q_1 = C_1 V = (2.2 \, \mu\text{F})(24 \, \text{V}) = 52.8 \, \mu\text{C},$$
$$Q_2 = C_2 V = (1.2 \, \mu\text{F})(24 \, \text{V}) = 28.8 \, \mu\text{C}.$$

Next the capacitors are connected in parallel, Fig. 24–12b, and the potential difference across each must quickly equalize. Thus, the charge cannot remain as shown in Fig. 24–12b, but the charge must rearrange itself so that the upper plates at least have the same sign of charge, with the lower plates having the opposite charge as shown in Fig. 24–12c. Equation 24–1 applies for each:

$$q_1 = C_1 V' \quad \text{and} \quad q_2 = C_2 V',$$

where V' is the voltage across each capacitor after the charges have rearranged themselves. We don't know q_1, q_2, or V', so we need a third equation. This is provided by charge conservation. The charges have rearranged themselves between Figs. 24–12b and c. The total charge on the upper plates in those two Figures must be the same, so we have

$$q_1 + q_2 = Q_1 - Q_2 = 24.0 \, \mu\text{C}.$$

Combining the last three equations we find:

$$V' = (q_1 + q_2)/(C_1 + C_2) = 24.0 \, \mu\text{C}/3.4 \, \mu\text{F} = 7.06 \, \text{V} \approx 7.1 \, \text{V}$$
$$q_1 = C_1 V' = (2.2 \, \mu\text{F})(7.06 \, \text{V}) = 15.5 \, \mu\text{C} \approx 16 \, \mu\text{C}$$
$$q_2 = C_2 V' = (1.2 \, \mu\text{F})(7.06 \, \text{V}) = 8.5 \, \mu\text{C}$$

where we have kept only two significant figures in our final answers.

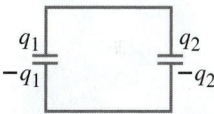

(a) Initial configuration.

(b) At the instant of reconnection only.

(c) A short time later.

FIGURE 24-12 Example 24–7.

24–4 Electric Energy Storage

A charged capacitor stores electrical energy. The energy stored in a capacitor will be equal to the work done to charge it. The net effect of charging a capacitor is to remove charge from one plate and add it to the other plate. This is what a battery does when it is connected to a capacitor. A capacitor does not become charged instantly. It takes time (Section 26–4). Initially, when the capacitor is uncharged, it requires no work to move the first bit of charge over. When some charge is on each plate, it requires work to add more charge of the same sign because of the electric repulsion. The more charge already on a plate, the more work required to add additional charge. The work needed to add a small amount of charge dq, when a potential difference V is across the plates, is $dW = V\,dq$. Since $V = q/C$ at any moment (Eq. 24–1), where C is the capacitance, the work needed to store a total charge Q is

$$ W = \int_0^Q V\,dq = \frac{1}{C}\int_0^Q q\,dq = \frac{1}{2}\frac{Q^2}{C}. $$

Thus we can say that the energy "stored" in a capacitor is

$$ U = \frac{1}{2}\frac{Q^2}{C} $$

when the capacitor C carries charges $+Q$ and $-Q$ on its two conductors. Since $Q = CV$, where V is the potential difference across the capacitor, we can also write

$$ U = \frac{1}{2}\frac{Q^2}{C} = \frac{1}{2}CV^2 = \frac{1}{2}QV. \tag{24–5} $$

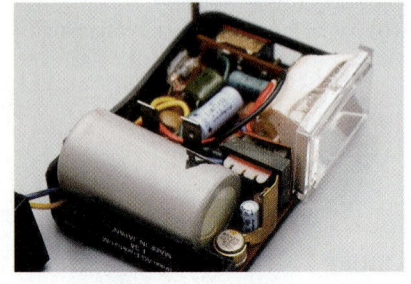

PHYSICS APPLIED

Camera flash

FIGURE 24–13 A camera flash unit.

EXAMPLE 24–8 **Energy stored in a capacitor.** A camera flash unit (Fig. 24–13) stores energy in a 150-μF capacitor at 200 V. (a) How much electric energy can be stored? (b) What is the power output if nearly all this energy is released in 1.0 ms?

APPROACH We use Eq. 24–5 in the form $U = \frac{1}{2}CV^2$ because we are given C and V.

SOLUTION The energy stored is

$$ U = \frac{1}{2}CV^2 = \frac{1}{2}(150 \times 10^{-6}\,\text{F})(200\,\text{V})^2 = 3.0\,\text{J}. $$

If this energy is released in $\frac{1}{1000}$ of a second, the power output is $P = U/t = (3.0\,\text{J})/(1.0 \times 10^{-3}\,\text{s}) = 3000\,\text{W}.$

CONCEPTUAL EXAMPLE 24–9 **Capacitor plate separation increased.** A parallel-plate capacitor carries charge Q and is then disconnected from a battery. The two plates are initially separated by a distance d. Suppose the plates are pulled apart until the separation is $2d$. How has the energy stored in this capacitor changed?

RESPONSE If we increase the plate separation d, we decrease the capacitance according to Eq. 24–2, $C = \epsilon_0 A/d$, by a factor of 2. The charge Q hasn't changed. So according to Eq. 24–5, where we choose the form $U = \frac{1}{2}Q^2/C$ because we know Q is the same and C has been halved, the reduced C means the potential energy stored increases by a factor of 2.

NOTE We can see why the energy stored increases from a physical point of view: the two plates are charged equal and opposite, so they attract each other. If we pull them apart, we must do work, so we raise their potential energy.

EXAMPLE 24–10 **Moving parallel capacitor plates.** The plates of a parallel-plate capacitor have area A, separation x, and are connected to a battery with voltage V. While connected to the battery, the plates are pulled apart until they are separated by $3x$. (a) What are the initial and final energies stored in the capacitor? (b) How much work is required to pull the plates apart (assume constant speed)? (c) How much energy is exchanged with the battery?

APPROACH The stored energy is given by Eq. 24–5: $U = \frac{1}{2}CV^2$, where $C = \epsilon_0 A/x$. Unlike Example 24–9, here the capacitor remains connected to the battery. Hence charge and energy can flow to or from the battery, and we can not set the work $W = \Delta U$. Instead, the work can be calculated from Eq. 7–7, $W = \int \vec{F} \cdot d\vec{\ell}$.

SOLUTION (a) When the separation is x, the capacitance is $C_1 = \epsilon_0 A/x$ and the energy stored is

$$U_1 = \frac{1}{2}C_1 V^2 = \frac{1}{2}\frac{\epsilon_0 A}{x}V^2.$$

When the separation is $3x$, $C_2 = \epsilon_0 A/3x$ and

$$U_2 = \frac{1}{2}\frac{\epsilon_0 A}{3x}V^2.$$

Then

$$\Delta U_{cap} = U_2 - U_1 = -\frac{\epsilon_0 AV^2}{3x}.$$

The potential energy decreases as the oppositely charged plates are pulled apart, which makes sense. The plates remain connected to the battery, so V does not change and C decreases; hence some charge leaves each plate $(Q = CV)$, causing U to decrease.

(b) The work done in pulling the plates apart is $W = \int_x^{3x} F \, d\ell = \int_x^{3x} QE \, d\ell$, where Q is the charge on one plate at a given moment when the plates are a distance ℓ apart, and E is the field due to the other plate at that instant. You might think we could use $E = V/\ell$ where ℓ is the separation of the plates (Eq. 23–4b). But we want the force on one plate (of charge Q) due to the electric field of the other plate only—which is half by *symmetry*: so we take $E = V/2\ell$. The charge at any separation ℓ is given by $Q = CV$, where $C = \epsilon_0 A/\ell$. Substituting, the work is

$$W = \int_{\ell=x}^{\ell=3x} QE\,d\ell = \frac{\epsilon_0 AV^2}{2}\int_x^{3x}\frac{d\ell}{\ell^2} = -\frac{\epsilon_0 AV^2}{2\ell}\Big|_{\ell=x}^{\ell=3x} = \frac{\epsilon_0 AV^2}{2}\left(\frac{-1}{3x} + \frac{1}{x}\right) = \frac{\epsilon_0 AV^2}{3x}.$$

As you might expect, the work required to pull these oppositely charged plates apart is positive.

(c) Even though the work done is positive, the potential energy decreased, which tells us that energy must have gone into the battery (as if charging it). Conservation of energy tells us that the work W done on the system must equal the change in potential energy of the capacitor plus that of the battery (kinetic energy can be assumed to be essentially zero):

$$W = \Delta U_{cap} + \Delta U_{batt}.$$

Thus the battery experiences a change in energy of

$$\Delta U_{batt} = W - \Delta U_{cap} = \frac{\epsilon_0 AV^2}{3x} + \frac{\epsilon_0 AV^2}{3x} = \frac{2\epsilon_0 AV^2}{3x}.$$

Thus charge flows back into the battery, raising its stored energy. In fact, the battery energy increase is double the work we do.

It is useful to think of the energy stored in a capacitor as being stored in the electric field between the plates. As an example let us calculate the energy stored in a parallel-plate capacitor in terms of the electric field.

We have seen (Eq. 23–4b) that the electric field $\vec{E}$ between two close parallel plates is (approximately) uniform and its magnitude is related to the potential difference by $V = Ed$ where d is the plate separation. Also, Eq. 24–2 tells us $C = \epsilon_0 A/d$ for a parallel-plate capacitor. Thus

$$U = \tfrac{1}{2}CV^2 = \tfrac{1}{2}\left(\frac{\epsilon_0 A}{d}\right)(E^2 d^2)$$
$$= \tfrac{1}{2}\epsilon_0 E^2 Ad.$$

The quantity Ad is the volume between the plates in which the electric field E exists. If we divide both sides by the volume, we obtain an expression for the energy per unit volume or **energy density**, u:

$$u = \text{energy density} = \tfrac{1}{2}\epsilon_0 E^2. \tag{24–6}$$

The *electric energy stored per unit volume in any region of space is proportional to the square of the electric field* in that region. We derived Eq. 24–6 for the special case of a parallel-plate capacitor. But it can be shown to be true for any region of space where there is an electric field. Note that the units check: for $(\epsilon_0 E^2)$ we have $(\text{C}^2/\text{N}\cdot\text{m}^2)(\text{N}/\text{C})^2 = \text{N}/\text{m}^2 = (\text{N}\cdot\text{m})/\text{m}^3 = \text{J}/\text{m}^3$.

Health Effects

The energy stored in a large capacitance can do harm, giving you a burn or a shock. One reason you are warned not to touch a circuit, or the inside of electronic devices, is because capacitors may still be carrying charge even if the external power has been turned off.

On the other hand, the basis of a **heart defibrillator** is a capacitor charged to a high voltage. A heart attack can be characterized by fast irregular beating of the heart, known as *ventricular* (or *cardiac*) *fibrillation*. The heart then does not pump blood to the rest of the body properly, and if it lasts for long, death results. A sudden, brief jolt of charge through the heart from a defibrillator can cause complete heart stoppage, sometimes followed by a resumption of normal beating. The defibrillator capacitor is charged to a high voltage, typically a few thousand volts, and is allowed to discharge very rapidly through the heart via a pair of wide contacts known as "paddles" that spread out the current over the chest (Fig. 24–14).

24–5 Dielectrics

In most capacitors there is an insulating sheet of material, such as paper or plastic, called a **dielectric** between the plates. This serves several purposes. First of all, dielectrics break down (allowing electric charge to flow) less readily than air, so higher voltages can be applied without charge passing across the gap. Furthermore, a dielectric allows the plates to be placed closer together without touching, thus allowing an increased capacitance because d is smaller in Eq. 24–2. Finally, it is found experimentally that if the dielectric fills the space between the two conductors, it increases the capacitance by a factor K which is known as the **dielectric constant**. Thus

$$C = KC_0, \tag{24–7}$$

where C_0 is the capacitance when the space between the two conductors of the capacitor is a vacuum, and C is the capacitance when the space is filled with a material whose dielectric constant is K.

The values of the dielectric constant for various materials are given in Table 24–1. Also shown in Table 24–1 is the **dielectric strength**, the maximum electric field before breakdown (charge flow) occurs.

For a parallel-plate capacitor (see Eq. 24–2),

$$C = K\epsilon_0 \frac{A}{d} \qquad \text{[parallel-plate capacitor]} \tag{24–8}$$

when the space between the plates is completely filled with a dielectric whose dielectric constant is K. (The situation when the dielectric only partially fills the space will be discussed shortly in Example 24–11.) The quantity $K\epsilon_0$ appears so

PHYSICS APPLIED
Shocks, burns, defibrillators

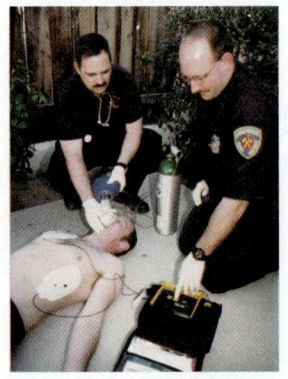

FIGURE 24–14 Heart defibrillator.

TABLE 24–1
Dielectric Constants (at 20°C)

Material	Dielectric constant K	Dielectric strength (V/m)
Vacuum	1.0000	
Air (1 atm)	1.0006	3×10^6
Paraffin	2.2	10×10^6
Polystyrene	2.6	24×10^6
Vinyl (plastic)	2–4	50×10^6
Paper	3.7	15×10^6
Quartz	4.3	8×10^6
Oil	4	12×10^6
Glass, Pyrex	5	14×10^6
Porcelain	6–8	5×10^6
Mica	7	150×10^6
Water (liquid)	80	
Strontium titanate	300	8×10^6

often in formulas that we define a new quantity

$$\epsilon = K\epsilon_0 \qquad (24\text{-}9)$$

called the **permittivity** of a material. Then the capacitance of a parallel-plate capacitor becomes

$$C = \epsilon \frac{A}{d}.$$

Note that ϵ_0 represents the permittivity of free space (a vacuum), as in Section 21–5.

The energy density stored in an electric field E (Section 24–4) in a dielectric is given by (see Eq. 24–6)

$$u = \tfrac{1}{2}K\epsilon_0 E^2 = \tfrac{1}{2}\epsilon E^2. \qquad [E \text{ in a dielectric}]$$

EXERCISE E Return to the Chapter-Opening Question, page 628, and answer it again now. Try to explain why you may have answered differently the first time.

Two simple experiments illustrate the effect of a dielectric. In the first, Fig. 24–15a, a battery of voltage V_0 is kept connected to a capacitor as a dielectric is inserted between the plates. If the charge on the plates without dielectric is Q_0, then when the dielectric is inserted, it is found experimentally (first by Faraday) that the charge Q on the plates is increased by a factor K,

$$Q = KQ_0. \qquad [\text{voltage constant}]$$

The capacitance has increased to $C = Q/V_0 = KQ_0/V_0 = KC_0$, which is Eq. 24–7. In a second experiment, Fig. 24–15b, a battery V_0 is connected to a capacitor C_0 which then holds a charge $Q_0 = C_0 V_0$. The battery is then disconnected, leaving the capacitor isolated with charge Q_0 and still at voltage V_0. Next a dielectric is inserted between the plates of the capacitor. The charge remains Q_0 (there is nowhere for the charge to go) but the voltage is found experimentally to drop by a factor K:

$$V = \frac{V_0}{K}. \qquad [\text{charge constant}]$$

Note that the capacitance changes to $C = Q_0/V = Q_0/(V_0/K) = KQ_0/V_0 = KC_0$, so this experiment too confirms Eq. 24–7.

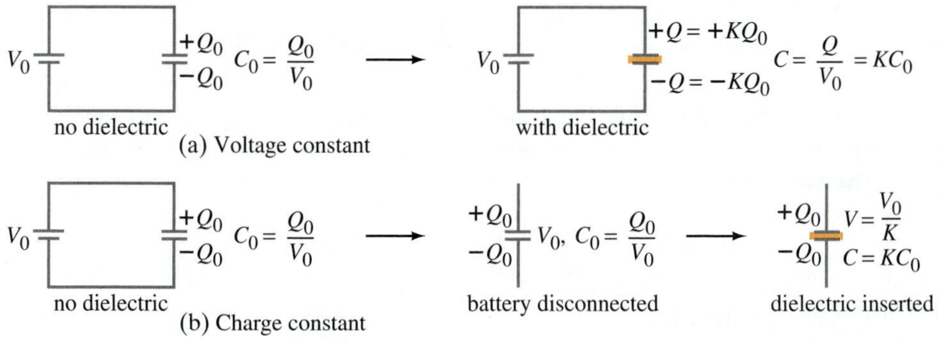

no dielectric with dielectric
(a) Voltage constant

no dielectric battery disconnected dielectric inserted
(b) Charge constant

FIGURE 24–15 Two experiments with a capacitor. Dielectric inserted with (a) voltage held constant, (b) charge held constant.

The electric field when a dielectric is inserted is also altered. When no dielectric is present, the electric field between the plates of a parallel-plate capacitor is given by Eq. 23–4b:

$$E_0 = \frac{V_0}{d},$$

where V_0 is the potential difference between the plates and d is their separation. If the capacitor is isolated so that the charge remains fixed on the plates when a dielectric is inserted, filling the space between the plates, the potential difference drops to $V = V_0/K$. So the electric field in the dielectric is

$$E = E_D = \frac{V}{d} = \frac{V_0}{Kd}$$

or

$$E_D = \frac{E_0}{K}. \qquad [\text{in a dielectric}] \quad (24\text{-}10)$$

The electric field in a dielectric is reduced by a factor equal to the dielectric constant. The field in a dielectric (or insulator) is not reduced all the way to zero as in a conductor. Equation 24–10 is valid even if the dielectric's width is smaller than the gap between the capacitor plates.

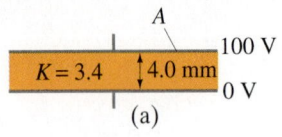

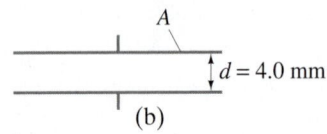

FIGURE 24–16 Example 24–11.

EXAMPLE 24–11 **Dielectric removal.** A parallel-plate capacitor, filled with a dielectric with $K = 3.4$, is connected to a 100-V battery (Fig. 24–16a). After the capacitor is fully charged, the battery is disconnected. The plates have area $A = 4.0\,\text{m}^2$, and are separated by $d = 4.0\,\text{mm}$. (a) Find the capacitance, the charge on the capacitor, the electric field strength, and the energy stored in the capacitor. (b) The dielectric is carefully removed, without changing the plate separation nor does any charge leave the capacitor (Fig. 24–16b). Find the new values of capacitance, electric field strength, voltage between the plates, and the energy stored in the capacitor.

APPROACH We use the formulas for parallel-plate capacitance and electric field with and without a dielectric.

SOLUTION (a) First we find the capacitance, with dielectric:

$$C = \frac{K\epsilon_0 A}{d} = \frac{3.4(8.85 \times 10^{-12}\,\text{C}^2/\text{N}\cdot\text{m}^2)(4.0\,\text{m}^2)}{4.0 \times 10^{-3}\,\text{m}}$$
$$= 3.0 \times 10^{-8}\,\text{F}.$$

The charge Q on the plates is

$$Q = CV = (3.0 \times 10^{-8}\,\text{F})(100\,\text{V}) = 3.0 \times 10^{-6}\,\text{C}.$$

The electric field between the plates is

$$E = \frac{V}{d} = \frac{100\,\text{V}}{4.0 \times 10^{-3}\,\text{m}} = 25\,\text{kV/m}.$$

Finally, the total energy stored in the capacitor is

$$U = \tfrac{1}{2}CV^2 = \tfrac{1}{2}(3.0 \times 10^{-8}\,\text{F})(100\,\text{V})^2 = 1.5 \times 10^{-4}\,\text{J}.$$

(b) The capacitance without dielectric decreases by a factor $K = 3.4$:

$$C_0 = \frac{C}{K} = \frac{(3.0 \times 10^{-8}\,\text{F})}{3.4} = 8.8 \times 10^{-9}\,\text{F}.$$

Because the battery has been disconnected, the charge Q can not change; when the dielectric is removed, $V = Q/C$ increases by a factor $K = 3.4$ to 340 V. The electric field is

$$E = \frac{V}{d} = \frac{340\,\text{V}}{4.0 \times 10^{-3}\,\text{m}} = 85\,\text{kV/m}.$$

The energy stored is

$$U = \tfrac{1}{2}CV^2 = \tfrac{1}{2}(8.8 \times 10^{-9}\,\text{F})(340\,\text{V})^2$$
$$= 5.1 \times 10^{-4}\,\text{J}.$$

NOTE Where did all this extra energy come from? The energy increased because work had to be done to remove the dielectric. The work required was $W = (5.1 \times 10^{-4}\,\text{J}) - (1.5 \times 10^{-4}\,\text{J}) = 3.6 \times 10^{-4}\,\text{J}$. (We will see in the next Section that work is required because of the force of attraction between induced charge on the dielectric and the charges on the plates, Fig. 24–17c.)

*24–6 Molecular Description of Dielectrics

Let us examine, from the molecular point of view, why the capacitance of a capacitor should be larger when a dielectric is between the plates. A capacitor whose plates are separated by an air gap has a charge $+Q$ on one plate and $-Q$ on the other (Fig. 24–17a). Assume it is isolated (not connected to a battery) so charge cannot flow to or from the plates. The potential difference between the plates, V_0, is given by Eq. 24–1:

$$Q = C_0 V_0,$$

where the subscripts refer to air between the plates. Now we insert a dielectric between the plates (Fig. 24–17b). Because of the electric field between the capacitor plates, the dielectric molecules will tend to become oriented as shown in Fig. 24–17b. If the dielectric molecules are *polar*, the positive end is attracted to the negative plate and vice versa. Even if the dielectric molecules are not polar, electrons within them will tend to move slightly toward the positive capacitor plate, so the effect is the same. The net effect of the aligned dipoles is a net negative charge on the outer edge of the dielectric facing the positive plate, and a net positive charge on the opposite side, as shown in Fig. 24–17c.

Some of the electric field lines, then, do not pass through the dielectric but instead end on charges induced on the surface of the dielectric as shown in Fig. 24–17c. Hence the electric field within the dielectric is less than in air. That is, the electric field between the capacitor plates, assumed filled by the dielectric, has been reduced by some factor K. The voltage across the capacitor is reduced by the same factor K because $V = Ed$ (Eq. 23–4b) and hence, by Eq. 24–1, $Q = CV$, the capacitance C must increase by that same factor K to keep Q constant.

As shown in Fig. 24–17d, the electric field within the dielectric E_D can be considered as the vector sum of the electric field $\vec{E}_0$ due to the "free" charges on the conducting plates, and the field $\vec{E}_{ind}$ due to the induced charge on the surfaces of the dielectric. Since these two fields are in opposite directions, the net field within the dielectric, $E_0 - E_{ind}$, is less than E_0. The precise relationship is given by Eq. 24–10, even if the dielectric does not fill the gap between the plates:

$$E_D = E_0 - E_{ind} = \frac{E_0}{K},$$

so

$$E_{ind} = E_0\left(1 - \frac{1}{K}\right).$$

The electric field between two parallel plates is related to the surface charge density, σ, by $E = \sigma/\epsilon_0$ (Example 21–13 or 22–8). Thus

$$E_0 = \sigma/\epsilon_0$$

where $\sigma = Q/A$ is the surface charge density on the conductor; Q is the net charge on the conductor and is often called the **free charge** (since charge is free to move in a conductor). Similarly, we define an equivalent induced surface charge density σ_{ind} on the dielectric; then

$$E_{ind} = \sigma_{ind}/\epsilon_0$$

where E_{ind} is the electric field due to the induced charge $Q_{ind} = \sigma_{ind} A$ on the surface of the dielectric, Fig. 24–17d. Q_{ind} is often called the **bound charge**, since it is on an insulator and is not free to move. Since $E_{ind} = E_0(1 - 1/K)$ as shown above, we now have

$$\sigma_{ind} = \sigma\left(1 - \frac{1}{K}\right) \qquad \textbf{(24–11a)}$$

and

$$Q_{ind} = Q\left(1 - \frac{1}{K}\right). \qquad \textbf{(24–11b)}$$

Since K is always greater than 1, we see that the charge induced on the dielectric is always less than the free charge on each of the capacitor plates.

+Q   −Q

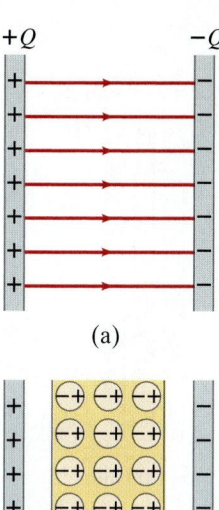

(a)

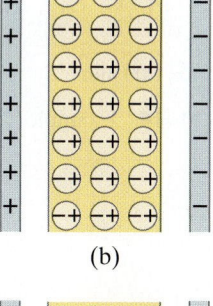

(b)

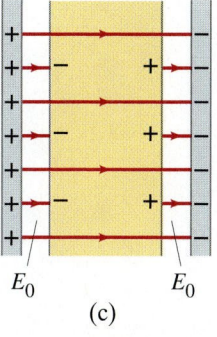

E_0 E_0

(c)

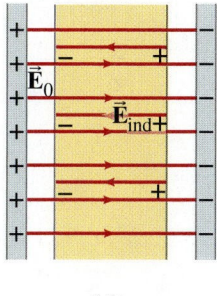

(d)

FIGURE 24–17 Molecular view of the effects of a dielectric.

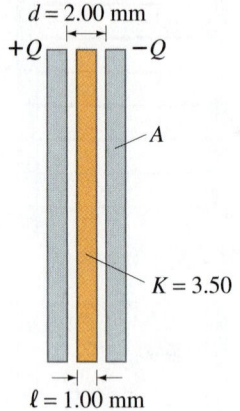

$d = 2.00$ mm

$+Q$ $-Q$

A

$K = 3.50$

$\ell = 1.00$ mm

FIGURE 24–18 Example 24–12.

EXAMPLE 24–12 **Dielectric partially fills capacitor.** A parallel-plate capacitor has plates of area $A = 250\ \text{cm}^2$ and separation $d = 2.00$ mm. The capacitor is charged to a potential difference $V_0 = 150$ V. Then the battery is disconnected (the charge Q on the plates then won't change), and a dielectric sheet ($K = 3.50$) of the same area A but thickness $\ell = 1.00$ mm is placed between the plates as shown in Fig. 24–18. Determine (*a*) the initial capacitance of the air-filled capacitor, (*b*) the charge on each plate before the dielectric is inserted, (*c*) the charge induced on each face of the dielectric after it is inserted, (*d*) the electric field in the space between each plate and the dielectric, (*e*) the electric field in the dielectric, (*f*) the potential difference between the plates after the dielectric is added, and (*g*) the capacitance after the dielectric is in place.

APPROACH We use the expressions for capacitance and charge developed in this Section plus (part *e*), Eq. 23–4a, $V = -\int \vec{E} \cdot d\vec{\ell}$.

SOLUTION (*a*) Before the dielectric is in place, the capacitance is

$$C_0 = \epsilon_0 \frac{A}{d} = (8.85 \times 10^{-12}\ \text{C}^2/\text{N} \cdot \text{m}^2)\left(\frac{2.50 \times 10^{-2}\ \text{m}^2}{2.00 \times 10^{-3}\ \text{m}}\right) = 111\ \text{pF}.$$

(*b*) The charge on each plate is

$$Q = C_0 V_0 = (1.11 \times 10^{-10}\ \text{F})(150\ \text{V}) = 1.66 \times 10^{-8}\ \text{C}.$$

(*c*) Equations 24–10 and 24–11 are valid even when the dielectric does not fill the gap, so (Eq. 24–11b)

$$Q_{\text{ind}} = Q\left(1 - \frac{1}{K}\right) = (1.66 \times 10^{-8}\ \text{C})\left(1 - \frac{1}{3.50}\right) = 1.19 \times 10^{-8}\ \text{C}.$$

(*d*) The electric field in the gaps between the plates and the dielectric (see Fig. 24–17c) is the same as in the absence of the dielectric since the charge on the plates has not been altered. The result of Example 21–13 can be used here, which gives $E_0 = \sigma/\epsilon_0$. [Or we can note that, in the absence of the dielectric, $E_0 = V_0/d = Q/C_0 d$ (since $V_0 = Q/C_0$) $= Q/\epsilon_0 A$ (since $C_0 = \epsilon_0 A/d$) which is the same result.] Thus

$$E_0 = \frac{Q}{\epsilon_0 A} = \frac{1.66 \times 10^{-8}\ \text{C}}{(8.85 \times 10^{-12}\ \text{C}^2/\text{N} \cdot \text{m}^2)(2.50 \times 10^{-2}\ \text{m}^2)} = 7.50 \times 10^4\ \text{V/m}.$$

(*e*) In the dielectric the electric field is (Eq. 24–10)

$$E_D = \frac{E_0}{K} = \frac{7.50 \times 10^4\ \text{V/m}}{3.50} = 2.14 \times 10^4\ \text{V/m}.$$

(*f*) To obtain the potential difference in the presence of the dielectric we use Eq. 23–4a, and integrate from the surface of one plate to the other along a straight line parallel to the field lines:

$$V = -\int \vec{E} \cdot d\vec{\ell} = E_0(d - \ell) + E_D \ell,$$

which can be simplified to

$$V = E_0\left(d - \ell + \frac{\ell}{K}\right)$$

$$= (7.50 \times 10^4\ \text{V/m})\left(1.00 \times 10^{-3}\ \text{m} + \frac{1.00 \times 10^{-3}\ \text{m}}{3.50}\right)$$

$$= 96.4\ \text{V}.$$

(*g*) In the presence of the dielectric, the capacitance is

$$C = \frac{Q}{V} = \frac{1.66 \times 10^{-8}\ \text{C}}{96.4\ \text{V}} = 172\ \text{pF}.$$

NOTE If the dielectric filled the space between the plates, the answers to (*f*) and (*g*) would be 42.9 V and 387 pF, respectively.

Summary

A **capacitor** is a device used to store charge (and electric energy), and consists of two nontouching conductors. The two conductors generally hold equal and opposite charges of magnitude Q. The ratio of this charge Q to the potential difference V between the conductors is called the **capacitance**, C:

$$C = \frac{Q}{V} \quad \text{or} \quad Q = CV. \tag{24-1}$$

The capacitance of a parallel-plate capacitor is proportional to the area A of each plate and inversely proportional to their separation d:

$$C = \epsilon_0 \frac{A}{d}. \tag{24-2}$$

When capacitors are connected in **parallel**, the equivalent capacitance is the sum of the individual capacitances:

$$C_{eq} = C_1 + C_2 + \cdots. \tag{24-3}$$

When capacitors are connected in **series**, the reciprocal of the equivalent capacitance equals the sum of the reciprocals of the individual capacitances:

$$\frac{1}{C_{eq}} = \frac{1}{C_1} + \frac{1}{C_2} + \cdots. \tag{24-4}$$

A charged capacitor stores an amount of electric energy given by

$$U = \tfrac{1}{2}QV = \tfrac{1}{2}CV^2 = \tfrac{1}{2}\frac{Q^2}{C}. \tag{24-5}$$

This energy can be thought of as stored in the electric field between the plates. In any electric field $\vec{E}$ in free space the **energy density** u (energy per unit volume) is

$$u = \tfrac{1}{2}\epsilon_0 E^2. \tag{24-6}$$

The space between the conductors contains a nonconducting material such as air, paper, or plastic. These materials are referred to as **dielectrics**, and the capacitance is proportional to a property of dielectrics called the **dielectric constant**, K (nearly equal to 1 for air). For a parallel-plate capacitor

$$C = K\epsilon_0 \frac{A}{d} = \epsilon \frac{A}{d} \tag{24-8}$$

where $\epsilon = K\epsilon_0$ is called the **permittivity** of the dielectric material. When a dielectric is present, the energy density is

$$u = \tfrac{1}{2}K\epsilon_0 E^2 = \tfrac{1}{2}\epsilon E^2.$$

Questions

1. Suppose two nearby conductors carry the same negative charge. Can there be a potential difference between them? If so, can the definition of capacitance, $C = Q/V$, be used here?

2. Suppose the separation of plates d in a parallel-plate capacitor is not very small compared to the dimensions of the plates. Would you expect Eq. 24–2 to give an overestimate or underestimate of the true capacitance? Explain.

3. Suppose one of the plates of a parallel-plate capacitor was moved so that the area of overlap was reduced by half, but they are still parallel. How would this affect the capacitance?

4. When a battery is connected to a capacitor, why do the two plates acquire charges of the same magnitude? Will this be true if the two conductors are different sizes or shapes?

5. Describe a simple method of measuring ϵ_0 using a capacitor.

6. Suppose three identical capacitors are connected to a battery. Will they store more energy if connected in series or in parallel?

7. A large copper sheet of thickness ℓ is placed between the parallel plates of a capacitor, but does not touch the plates. How will this affect the capacitance?

8. The parallel plates of an isolated capacitor carry opposite charges, Q. If the separation of the plates is increased, is a force required to do so? Is the potential difference changed? What happens to the work done in the pulling process?

9. How does the energy in a capacitor change if (a) the potential difference is doubled, (b) the charge on each plate is doubled, and (c) the separation of the plates is doubled, as the capacitor remains connected to a battery in each case?

10. If the voltage across a capacitor is doubled, the amount of energy it can store (a) doubles; (b) is halved; (c) is quadrupled; (d) is unaffected; (e) none of these.

11. An isolated charged capacitor has horizontal plates. If a thin dielectric is inserted a short way between the plates, Fig. 24–19, will it move left or right when it is released?

Dielectric

FIGURE 24–19
Question 11.

12. Suppose a battery remains connected to the capacitor in Question 11. What then will happen when the dielectric is released?

13. How does the energy stored in a capacitor change when a dielectric is inserted if (a) the capacitor is isolated so Q does not change; (b) the capacitor remains connected to a battery so V does not change?

14. For dielectrics consisting of polar molecules, how would you expect the dielectric constant to change with temperature?

15. A dielectric is pulled out from between the plates of a capacitor which remains connected to a battery. What changes occur to the capacitance, charge on the plates, potential difference, energy stored in the capacitor, and electric field?

16. We have seen that the capacitance C depends on the size, shape, and position of the two conductors, as well as on the dielectric constant K. What then did we mean when we said that C is a constant in Eq. 24–1?

17. What value might we assign to the dielectric constant for a good conductor? Explain.

Problems

24–1 Capacitors

1. (I) The two plates of a capacitor hold $+2800\,\mu C$ and $-2800\,\mu C$ of charge, respectively, when the potential difference is 930 V. What is the capacitance?

2. (I) How much charge flows from a 12.0-V battery when it is connected to a 12.6-μF capacitor?

3. (I) The potential difference between two short sections of parallel wire in air is 24.0 V. They carry equal and opposite charge of magnitude 75 pC. What is the capacitance of the two wires?

4. (I) The charge on a capacitor increases by $26\,\mu C$ when the voltage across it increases from 28 V to 78 V. What is the capacitance of the capacitor?

5. (II) A 7.7-μF capacitor is charged by a 125-V battery (Fig. 24–20a) and then is disconnected from the battery. When this capacitor (C_1) is then connected (Fig. 24–20b) to a second (initially uncharged) capacitor, C_2, the final voltage on each capacitor is 15 V. What is the value of C_2? [*Hint*: Charge is conserved.]

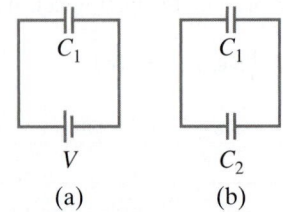

FIGURE 24–20 Problems 5 and 48.

6. (II) An isolated capacitor C_1 carries a charge Q_0. Its wires are then connected to those of a second capacitor C_2, previously uncharged. What charge will each carry now? What will be the potential difference across each?

7. (II) It takes 15 J of energy to move a 0.20-mC charge from one plate of a 15-μF capacitor to the other. How much charge is on each plate?

8. (II) A 2.70-μF capacitor is charged to 475 V and a 4.00-μF capacitor is charged to 525 V. (a) These capacitors are then disconnected from their batteries, and the positive plates are now connected to each other and the negative plates are connected to each other. What will be the potential difference across each capacitor and the charge on each? (b) What is the voltage and charge for each capacitor if plates of opposite sign are connected?

9. (II) Compact "ultracapacitors" with capacitance values up to several thousand farads are now commercially available. One application for ultracapacitors is in providing power for electrical circuits when other sources (such as a battery) are turned off. To get an idea of how much charge can be stored in such a component, assume a 1200-F ultracapacitor is initially charged to 12.0 V by a battery and is then disconnected from the battery. If charge is then drawn off the plates of this capacitor at a rate of 1.0 mC/s, say, to power the backup memory of some electrical gadget, how long (in days) will it take for the potential difference across this capacitor to drop to 6.0 V?

10. (II) In a **dynamic random access memory (DRAM)** computer chip, each memory cell chiefly consists of a capacitor for charge storage. Each of these cells represents a single binary-bit value of 1 when its 35-fF capacitor $\left(1\,\text{fF} = 10^{-15}\,\text{F}\right)$ is charged at 1.5 V, or 0 when uncharged at 0 V. (a) When it is fully charged, how many excess electrons are on a cell capacitor's negative plate? (b) After charge has been placed on a cell capacitor's plate, it slowly "leaks" off (through a variety of mechanisms) at a constant rate of 0.30 fC/s. How long does it take for the potential difference across this capacitor to decrease by 1.0% from its fully charged value? (Because of this leakage effect, the charge on a DRAM capacitor is "refreshed" many times per second.)

24–2 Determination of Capacitance

11. (I) To make a 0.40-μF capacitor, what area must the plates have if they are to be separated by a 2.8-mm air gap?

12. (I) What is the capacitance per unit length (F/m) of a coaxial cable whose inner conductor has a 1.0-mm diameter and the outer cylindrical sheath has a 5.0-mm diameter? Assume the space between is filled with air.

13. (I) Determine the capacitance of the Earth, assuming it to be a spherical conductor.

14. (II) Use Gauss's law to show that $\vec{\mathbf{E}} = 0$ inside the inner conductor of a cylindrical capacitor (see Fig. 24–6 and Example 24–2) as well as outside the outer cylinder.

15. (II) Dry air will break down if the electric field exceeds about 3.0×10^6 V/m. What amount of charge can be placed on a capacitor if the area of each plate is 6.8 cm²?

16. (II) An electric field of 4.80×10^5 V/m is desired between two parallel plates, each of area 21.0 cm² and separated by 0.250 cm of air. What charge must be on each plate?

17. (II) How strong is the electric field between the plates of a 0.80-μF air-gap capacitor if they are 2.0 mm apart and each has a charge of 92 μC?

18. (II) A large metal sheet of thickness ℓ is placed between, and parallel to, the plates of the parallel-plate capacitor of Fig. 24–4. It does not touch the plates, and extends beyond their edges. (a) What is now the net capacitance in terms of A, d, and ℓ? (b) If $\ell = 0.40\,d$, by what factor does the capacitance change when the sheet is inserted?

19. (III) Small distances are commonly measured capacitively. Consider an air-filled parallel-plate capacitor with fixed plate area $A = 25$ mm² and a variable plate-separation distance x. Assume this capacitor is attached to a capacitance-measuring instrument which can measure capacitance C in the range 1.0 pF to 1000.0 pF with an accuracy of $\Delta C = 0.1$ pF. (a) If C is measured while x is varied, over what range $\left(x_{\min} \leq x \leq x_{\max}\right)$ can the plate-separation distance (in μm) be determined by this setup? (b) Define Δx to be the accuracy (magnitude) to which x can be determined, and determine a formula for Δx. (c) Determine the percent accuracy to which $x_{\min}$ and $x_{\max}$ can be measured.

20. (III) In an **electrostatic air cleaner ("precipitator")**, the strong nonuniform electric field in the central region of a cylindrical capacitor (with outer and inner cylindrical radii R_a and R_b) is used to create ionized air molecules for use in charging dust and soot particles (Fig. 24–21). Under standard atmospheric conditions, if air is subjected to an electric field magnitude that exceeds its dielectric strength $E_S = 2.7 \times 10^6$ N/C, air molecules will dissociate into positively charged ions and free electrons. In a precipitator, the region within which air is ionized (the *corona discharge* region) occupies a cylindrical volume of radius R that is typically five times that of the inner cylinder. Assume a particular precipitator is constructed with $R_b = 0.10$ mm and $R_a = 10.0$ cm. In order to create a corona discharge region with radius $R = 5.0\ R_b$, what potential difference V should be applied between the precipitator's inner and outer conducting cylinders? [Besides dissociating air, the charged inner cylinder repels the resulting positive ions from the corona discharge region, where they are put to use in charging dust particles, which are then "collected" on the negatively charged outer cylinder.]

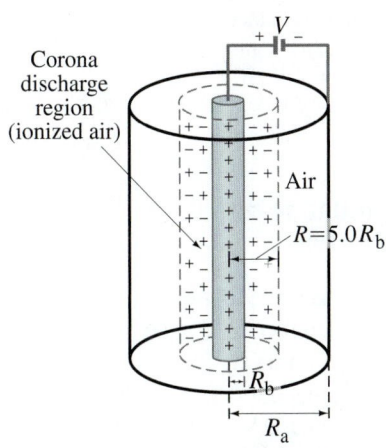

FIGURE 24–21 Problem 20.

24–3 Capacitors in Series and Parallel

21. (I) The capacitance of a portion of a circuit is to be reduced from 2900 pF to 1600 pF. What capacitance can be added to the circuit to produce this effect without removing existing circuit elements? Must any existing connections be broken to accomplish this?

22. (I) (a) Six 3.8-μF capacitors are connected in parallel. What is the equivalent capacitance? (b) What is their equivalent capacitance if connected in series?

23. (II) Given three capacitors, $C_1 = 2.0\ \mu$F, $C_2 = 1.5\ \mu$F, and $C_3 = 3.0\ \mu$F, what arrangement of parallel and series connections with a 12-V battery will give the minimum voltage drop across the 2.0-μF capacitor? What is the minimum voltage drop?

24. (II) Suppose three parallel-plate capacitors, whose plates have areas A_1, A_2, and A_3 and separations d_1, d_2, and d_3, are connected in parallel. Show, using only Eq. 24–2, that Eq. 24–3 is valid.

25. (II) An electric circuit was accidentally constructed using a 5.0-μF capacitor instead of the required 16-μF value. Without removing the 5.0-μF capacitor, what can a technician add to correct this circuit?

26. (II) Three conducting plates, each of area A, are connected as shown in Fig. 24–22. (a) Are the two capacitors thus formed connected in series or in parallel? (b) Determine C as a function of d_1, d_2, and A. Assume $d_1 + d_2$ is much less than the dimensions of the plates. (c) The middle plate can be moved (changing the values of d_1 and d_2), so as to vary the capacitance. What are the minimum and maximum values of the net capacitance?

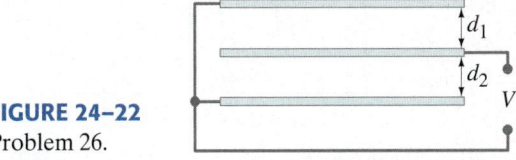

FIGURE 24–22
Problem 26.

27. (II) Consider three capacitors, of capacitance 3600 pF, 5800 pF, and 0.0100 μF. What maximum and minimum capacitance can you form from these? How do you make the connection in each case?

28. (II) A 0.50-μF and a 0.80-μF capacitor are connected in series to a 9.0-V battery. Calculate (a) the potential difference across each capacitor and (b) the charge on each. (c) Repeat parts (a) and (b) assuming the two capacitors are in parallel.

29. (II) In Fig. 24–23, suppose $C_1 = C_2 = C_3 = C_4 = C$. (a) Determine the equivalent capacitance between points a and b. (b) Determine the charge on each capacitor and the potential difference across each in terms of V.

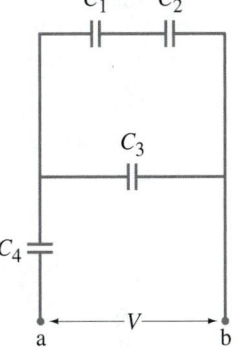

FIGURE 24–23
Problems 29 and 30.

30. (II) Suppose in Fig. 24–23 that $C_1 = C_2 = C_3 = 16.0\ \mu$F and $C_4 = 28.5\ \mu$F. If the charge on C_2 is $Q_2 = 12.4\ \mu$C, determine the charge on each of the other capacitors, the voltage across each capacitor, and the voltage V_{ab} across the entire combination.

31. (II) The switch S in Fig. 24–24 is connected downward so that capacitor C_2 becomes fully charged by the battery of voltage V_0. If the switch is then connected upward, determine the charge on each capacitor after the switching.

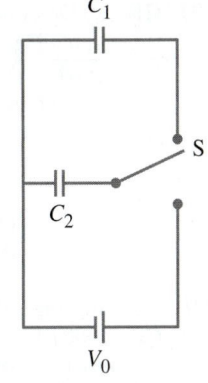

FIGURE 24–24
Problem 31.

32. (II) (*a*) Determine the equivalent capacitance between points a and b for the combination of capacitors shown in Fig. 24–25. (*b*) Determine the charge on each capacitor and the voltage across each if $V_{ba} = V$.

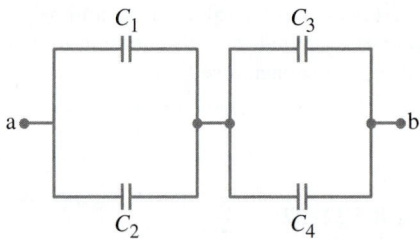

FIGURE 24–25 Problems 32 and 33.

33. (II) Suppose in Problem 32, Fig. 24–25, that $C_1 = C_3 = 8.0\,\mu F$, $C_2 = C_4 = 16\,\mu F$, and $Q_3 = 23\,\mu C$. Determine (*a*) the charge on each of the other capacitors, (*b*) the voltage across each capacitor, and (*c*) the voltage V_{ba} across the combination.

34. (II) Two capacitors connected in parallel produce an equivalent capacitance of $35.0\,\mu F$ but when connected in series the equivalent capacitance is only $5.5\,\mu F$. What is the individual capacitance of each capacitor?

35. (II) In the **capacitance bridge** shown in Fig. 24–26, a voltage V_0 is applied and the variable capacitor C_1 is adjusted until there is zero voltage between points a and b as measured on the voltmeter ($\bullet\!-\!\widehat{V}\!-\!\bullet$). Determine the unknown capacitance C_x if $C_1 = 8.9\,\mu F$ and the fixed capacitors have $C_2 = 18.0\,\mu F$ and $C_3 = 4.8\,\mu F$. Assume no charge flows through the voltmeter.

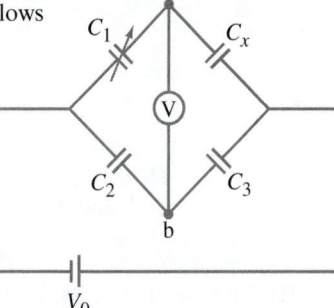

FIGURE 24–26
Problem 35.

36. (II) Two capacitors, $C_1 = 3200\,pF$ and $C_2 = 1800\,pF$, are connected in series to a 12.0-V battery. The capacitors are later disconnected from the battery and connected directly to each other, positive plate to positive plate, and negative plate to negative plate. What then will be the charge on each capacitor?

37. (II) (*a*) Determine the equivalent capacitance of the circuit shown in Fig. 24–27. (*b*) If $C_1 = C_2 = 2C_3 = 24.0\,\mu F$, how much charge is stored on each capacitor when $V = 35.0\,V$?

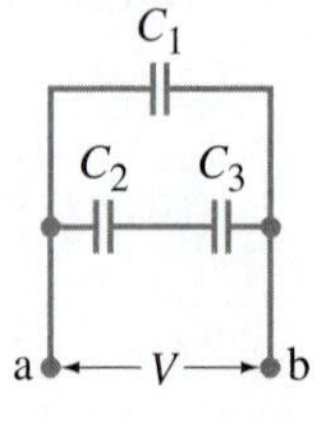

FIGURE 24–27
Problems 37, 38, and 45.

38. (II) In Fig. 24–27, let $C_1 = 2.00\,\mu F$, $C_2 = 3.00\,\mu F$, $C_3 = 4.00\,\mu F$, and $V = 24.0\,V$. What is the potential difference across each capacitor?

39. (III) Suppose one plate of a parallel-plate capacitor is tilted so it makes a small angle θ with the other plate, as shown in Fig. 24–28. Determine a formula for the capacitance C in terms of A, d, and θ, where A is the area of each plate and θ is small. Assume the plates are square. [*Hint*: Imagine the capacitor as many infinitesimal capacitors in parallel.]

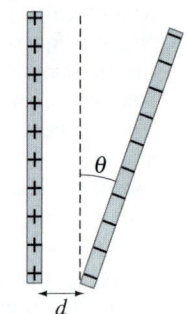

FIGURE 24–28
Problem 39.

40. (III) A voltage V is applied to the capacitor network shown in Fig. 24–29. (*a*) What is the equivalent capacitance? [*Hint*: Assume a potential difference V_{ab} exists across the network as shown; write potential differences for various pathways through the network from a to b in terms of the charges on the capacitors and the capacitances.] (*b*) Determine the equivalent capacitance if $C_2 = C_4 = 8.0\,\mu F$ and $C_1 = C_3 = C_5 = 4.5\,\mu F$.

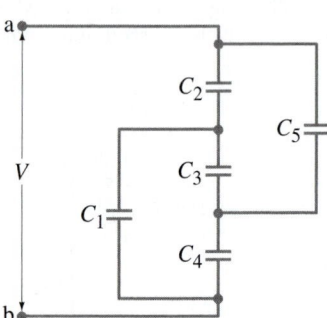

FIGURE 24–29
Problem 40.

24–4 Electric Energy Storage

41. (I) 2200 V is applied to a 2800-pF capacitor. How much electric energy is stored?

42. (I) There is an electric field near the Earth's surface whose intensity is about 150 V/m. How much energy is stored per cubic meter in this field?

43. (I) How much energy is stored by the electric field between two square plates, 8.0 cm on a side, separated by a 1.3-mm air gap? The charges on the plates are equal and opposite and of magnitude $420\,\mu C$.

44. (II) A parallel-plate capacitor has fixed charges $+Q$ and $-Q$. The separation of the plates is then tripled. (*a*) By what factor does the energy stored in the electric field change? (*b*) How much work must be done to increase the separation of the plates from d to $3.0d$? The area of each plate is A.

45. (II) In Fig. 24–27, let $V = 10.0\,V$ and $C_1 = C_2 = C_3 = 22.6\,\mu F$. How much energy is stored in the capacitor network?

46. (II) How much energy must a 28-V battery expend to charge a 0.45-μF and a 0.20-μF capacitor fully when they are placed (*a*) in parallel, (*b*) in series? (*c*) How much charge flowed from the battery in each case?

47. (II) (*a*) Suppose the outer radius R_a of a cylindrical capacitor was tripled, but the charge was kept constant. By what factor would the stored energy change? Where would the energy come from? (*b*) Repeat part (*a*), assuming the voltage remains constant.

48. (II) A 2.20-μF capacitor is charged by a 12.0-V battery. It is disconnected from the battery and then connected to an uncharged 3.50-μF capacitor (Fig. 24–20). Determine the total stored energy (*a*) before the two capacitors are connected, and (*b*) after they are connected. (*c*) What is the change in energy?

49. (II) How much work would be required to remove a metal sheet from between the plates of a capacitor (as in Problem 18a), assuming: (*a*) the battery remains connected so the voltage remains constant; (*b*) the battery is disconnected so the charge remains constant?

50. (II) (*a*) Show that each plate of a parallel-plate capacitor exerts a force

$$F = \frac{1}{2}\frac{Q^2}{\epsilon_0 A}$$

on the other, by calculating dW/dx where dW is the work needed to increase the separation by dx. (*b*) Why does using $F = QE$, with E being the electric field between the plates, give the wrong answer?

51. (II) Show that the electrostatic energy stored in the electric field outside an isolated spherical conductor of radius r_0 carrying a net charge Q is

$$U = \frac{1}{8\pi\epsilon_0}\frac{Q^2}{r_0}.$$

Do this in three ways: (*a*) Use Eq. 24–6 for the energy density in an electric field [*Hint*: Consider spherical shells of thickness dr]; (*b*) use Eq. 24–5 together with the capacitance of an isolated sphere (Section 24–2); (*c*) by calculating the work needed to bring all the charge Q up from infinity in infinitesimal bits dq.

52. (II) When two capacitors are connected in parallel and then connected to a battery, the total stored energy is 5.0 times greater than when they are connected in series and then connected to the same battery. What is the ratio of the two capacitances? (Before the battery is connected in each case, the capacitors are fully discharged.)

53. (II) For commonly used **CMOS** (complementary metal oxide semiconductor) digital circuits, the charging of the component capacitors C to their working potential difference V accounts for the major contribution of its energy input requirements. Thus, if a given logical operation requires such circuitry to charge its capacitors N times, we can assume that the operation requires an energy of $N(\frac{1}{2}CV^2)$. In the past 20 years, the capacitance in digital circuits has been reduced by a factor of about 20 and the voltage to which these capacitors are charged has been reduced from 5.0 V to 1.5 V. Also, present-day alkaline batteries hold about five times the energy of older batteries. Two present-day AA alkaline cells, each of which measures 1 cm diameter by 4 cm long, can power the logic circuitry of a hand-held **personal digital assistant** (PDA) with its display turned off for about two months. If an attempt was made to construct a similar PDA (i.e., same digital capabilities so N remains constant) 20 years ago, how many (older) AA batteries would have been required to power its digital circuitry for two months? Would this PDA fit in a pocket or purse?

24–5 Dielectrics

54. (I) What is the capacitance of two square parallel plates 4.2 cm on a side that are separated by 1.8 mm of paraffin?

55. (II) Suppose the capacitor in Example 24–11 remains connected to the battery as the dielectric is removed. What will be the work required to remove the dielectric in this case?

56. (II) How much energy would be stored in the capacitor of Problem 43 if a mica dielectric is placed between the plates? Assume the mica is 1.3 mm thick (and therefore fills the space between the plates).

57. (II) In the DRAM computer chip of Problem 10, the cell capacitor's two conducting parallel plates are separated by a 2.0-nm thick insulating material with dielectric constant $K = 25$. (*a*) Determine the area A (in μm^2) of the cell capacitor's plates. (*b*) In (older) "planar" designs, the capacitor was mounted on a silicon-wafer surface with its plates parallel to the plane of the wafer. Assuming the plate area A accounts for half of the area of each cell, estimate how many megabytes of memory can be placed on a 3.0-cm² silicon wafer with the planar design? (1 byte = 8 bits.)

58. (II) A 3500-pF air-gap capacitor is connected to a 32-V battery. If a piece of mica fills the space between the plates, how much charge will flow from the battery?

59. (II) Two different dielectrics each fill half the space between the plates of a parallel-plate capacitor as shown in Fig. 24–30. Determine a formula for the capacitance in terms of K_1, K_2, the area A of the plates, and the separation d. [*Hint*: Can you consider this capacitor as two capacitors in series or in parallel?]

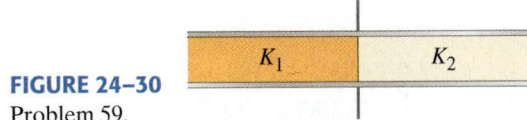

FIGURE 24–30
Problem 59.

60. (II) Two different dielectrics fill the space between the plates of a parallel-plate capacitor as shown in Fig. 24–31. Determine a formula for the capacitance in terms of K_1, K_2, the area A, of the plates, and the separation $d_1 = d_2 = d/2$. [*Hint*: Can you consider this capacitor as two capacitors in series or in parallel?]

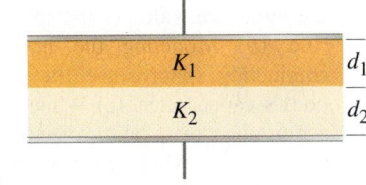

FIGURE 24–31
Problems 60 and 61.

61. (II) Repeat Problem 60 (Fig. 24–31) but assume the separation $d_1 \neq d_2$.

62. (II) Two identical capacitors are connected in parallel and each acquires a charge Q_0 when connected to a source of voltage V_0. The voltage source is disconnected and then a dielectric ($K = 3.2$) is inserted to fill the space between the plates of one of the capacitors. Determine (*a*) the charge now on each capacitor, and (*b*) the voltage now across each capacitor.

63. (III) A slab of width d and dielectric constant K is inserted a distance x into the space between the square parallel plates (of side ℓ) of a capacitor as shown in Fig. 24–32. Determine, as a function of x, (*a*) the capacitance, (*b*) the energy stored if the potential difference is V_0, and (*c*) the magnitude and direction of the force exerted on the slab (assume V_0 is constant).

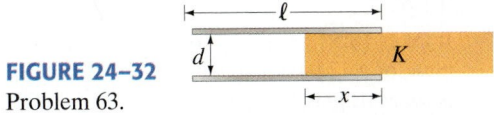

FIGURE 24–32
Problem 63.

64. (III) The quantity of liquid (such as cryogenic liquid nitrogen) available in its storage tank is often monitored by a capacitive level sensor. This sensor is a vertically aligned cylindrical capacitor with outer and inner conductor radii R_a and R_b, whose length ℓ spans the height of the tank. When a nonconducting liquid fills the tank to a height h ($\leq \ell$) from the tank's bottom, the dielectric in the lower and upper region between the cylindrical conductors is the liquid (K_{liq}) and its vapor (K_V), respectively (Fig. 24–33). (a) Determine a formula for the fraction F of the tank filled by liquid in terms of the level-sensor capacitance C. [Hint: Consider the sensor as a combination of two capacitors.] (b) By connecting a capacitance-measuring instrument to the level sensor, F can be monitored. Assume the sensor dimensions are $\ell = 2.0$ m, $R_a = 5.0$ mm, and $R_b = 4.5$ mm. For liquid nitrogen $(K_{\text{liq}} = 1.4, K_V = 1.0)$, what values of C (in pF) will correspond to the tank being completely full and completely empty?

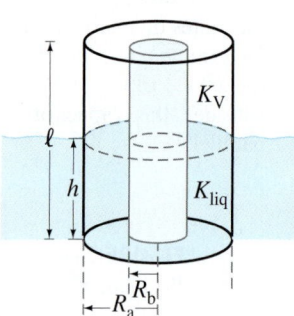

FIGURE 24–33
Problem 64.

*24–6 Molecular Description of Dielectrics

***65. (II)** Show that the capacitor in Example 24–12 with dielectric inserted can be considered as equivalent to three capacitors in series, and using this assumption show that the same value for the capacitance is obtained as was obtained in part (g) of the Example.

***66. (II)** Repeat Example 24–12 assuming the battery remains connected when the dielectric is inserted. Also, what is the free charge on the plates after the dielectric is added (let this be part (h) of this Problem)?

***67. (II)** Using Example 24–12 as a model, derive a formula for the capacitance of a parallel-plate capacitor whose plates have area A, separation d, with a dielectric of dielectric constant K and thickness ℓ ($\ell < d$) placed between the plates.

***68. (II)** In Example 24–12 what percent of the stored energy is stored in the electric field in the dielectric?

***69. (III)** The capacitor shown in Fig. 24–34 is connected to a 90.0-V battery. Calculate (and sketch) the electric field everywhere between the capacitor plates. Find both the free charge on the capacitor plate and the induced charge on the faces of the glass dielectric plate.

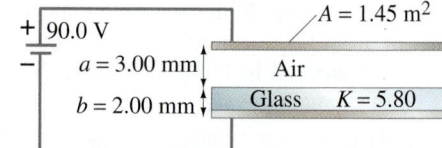

FIGURE 24–34
Problem 69.

General Problems

70. (a) A general rule for estimating the capacitance C of an isolated conducting sphere with radius r is C (in pF) $\approx r$ (in cm). That is, the numerical value of C in pF is about the same as the numerical value of the sphere's radius in cm. Justify this rule. (b) Modeling the human body as a 1-m-radius conducting sphere, use the given rule to estimate your body's capacitance. (c) While walking across a carpet, you acquire an excess "static electricity" charge Q and produce a 0.5-cm spark when reaching out to touch a metallic doorknob. The dielectric strength of air is 30 kV/cm. Use this information to estimate Q (in μC).

71. A *cardiac defibrillator* is used to shock a heart that is beating erratically. A capacitor in this device is charged to 7.5 kV and stores 1200 J of energy. What is its capacitance?

72. A homemade capacitor is assembled by placing two 9-in. pie pans 5.0 cm apart and connecting them to the opposite terminals of a 9-V battery. Estimate (a) the capacitance, (b) the charge on each plate, (c) the electric field halfway between the plates, and (d) the work done by the battery to charge the plates. (e) Which of the above values change if a dielectric is inserted?

73. An uncharged capacitor is connected to a 34.0-V battery until it is fully charged, after which it is disconnected from the battery. A slab of paraffin is then inserted between the plates. What will now be the voltage between the plates?

74. It takes 18.5 J of energy to move a 13.0-mC charge from one plate of a 17.0-μF capacitor to the other. How much charge is on each plate?

75. A huge 3.0-F capacitor has enough stored energy to heat 3.5 kg of water from 22°C to 95°C. What is the potential difference across the plates?

76. A coaxial cable, Fig. 24–35, consists of an inner cylindrical conducting wire of radius R_b surrounded by a dielectric insulator. Surrounding the dielectric insulator is an outer conducting sheath of radius R_a, which is usually "grounded." (a) Determine an expression for the capacitance per unit length of a cable whose insulator has dielectric constant K. (b) For a given cable, $R_b = 2.5$ mm and $R_a = 9.0$ mm. The dielectric constant of the dielectric insulator is $K = 2.6$. Suppose that there is a potential of 1.0 kV between the inner conducting wire and the outer conducting sheath. Find the capacitance per meter of the cable.

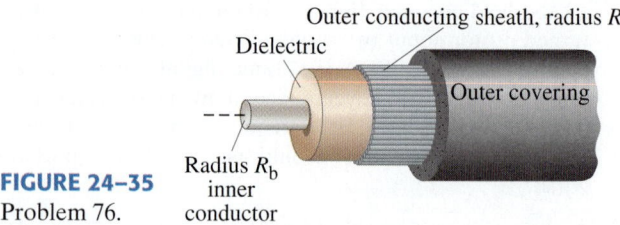

FIGURE 24–35
Problem 76.

77. The electric field between the plates of a paper-separated ($K = 3.75$) capacitor is 9.21×10^4 V/m. The plates are 1.95 mm apart and the charge on each plate is 0.675 μC. Determine the capacitance of this capacitor and the area of each plate.

78. Capacitors can be used as "electric charge counters." Consider an initially uncharged capacitor of capacitance C with its bottom plate grounded and its top plate connected to a source of electrons. (a) If N electrons flow onto the capacitor's top plate, show that the resulting potential difference V across the capacitor is directly proportional to N. (b) Assume the voltage-measuring device can accurately resolve voltage changes of about 1 mV. What value of C would be necessary to detect each new collected electron? (c) Using modern semiconductor technology, a micron-size capacitor can be constructed with parallel conducting plates separated by an insulating oxide of dielectric constant $K = 3$ and thickness $d = 100$ nm. To resolve the arrival of an individual electron on the plate of such a capacitor, determine the required value of ℓ (in μm) assuming square plates of side length ℓ.

79. A parallel-plate capacitor is isolated with a charge $\pm Q$ on each plate. If the separation of the plates is halved and a dielectric (constant K) is inserted in place of air, by what factor does the energy storage change? To what do you attribute the change in stored potential energy? How does the new value of the electric field between the plates compare with the original value?

80. In lightning storms, the potential difference between the Earth and the bottom of the thunderclouds can be as high as 35,000,000 V. The bottoms of thunderclouds are typically 1500 m above the Earth, and may have an area of 120 km². Modeling the Earth–cloud system as a huge capacitor, calculate (a) the capacitance of the Earth–cloud system, (b) the charge stored in the "capacitor," and (c) the energy stored in the "capacitor."

81. A multilayer film capacitor has a maximum voltage rating of 100 V and a capacitance of 1.0 μF. It is made from alternating sheets of metal foil connected together, separated by films of polyester dielectric. The sheets are 12.0 mm by 14.0 mm and the total thickness of the capacitor is 6.0 mm (not counting the thickness of the insulator on the outside). The metal foil is actually a very thin layer of metal deposited directly on the dielectric, so most of the thickness of the capacitor is due to the dielectric. The dielectric strength of the polyester is about 30×10^6 V/m. Estimate the dielectric constant of the polyester material in the capacitor.

82. A 3.5-μF capacitor is charged by a 12.4-V battery and then is disconnected from the battery. When this capacitor (C_1) is then connected to a second (initially uncharged) capacitor, C_2, the voltage on the first drops to 5.9 V. What is the value of C_2?

83. The power supply for a pulsed nitrogen laser has a 0.080-μF capacitor with a maximum voltage rating of 25 kV. (a) Estimate how much energy could be stored in this capacitor. (b) If 15% of this stored electrical energy is converted to light energy in a pulse that is 4.0-μs long, what is the power of the laser pulse?

84. A parallel-plate capacitor has square plates 12 cm on a side separated by 0.10 mm of plastic with a dielectric constant of $K = 3.1$. The plates are connected to a battery, causing them to become oppositely charged. Since the oppositely charged plates attract each other, they exert a pressure on the dielectric. If this pressure is 40.0 Pa, what is the battery voltage?

85. The variable capacitance of an old radio tuner consists of four plates connected together placed alternately between four other plates, also connected together (Fig. 24–36). Each plate is separated from its neighbor by 1.6 mm of air. One set of plates can move so that the area of overlap of each plate varies from 2.0 cm² to 9.0 cm². (a) Are these seven capacitors connected in series or in parallel? (b) Determine the range of capacitance values.

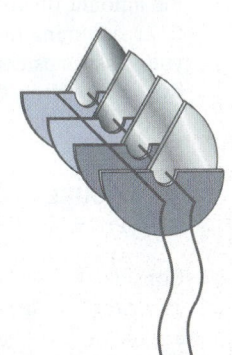

FIGURE 24–36
Problems 85 and 86.

86. A high-voltage supply can be constructed from a variable capacitor with interleaving plates which can be rotated as in Fig. 24–36. A version of this type of capacitor with more plates has a capacitance which can be varied from 10 pF to 1 pF. (a) Initially, this capacitor is charged by a 7500-V power supply when the capacitance is 8.0 pF. It is then disconnected from the power supply and the capacitance reduced to 1.0 pF by rotating the plates. What is the voltage across the capacitor now? (b) What is a major disadvantage of this as a high-voltage power supply?

87. A 175-pF capacitor is connected in series with an unknown capacitor, and as a series combination they are connected to a 25.0-V battery. If the 175-pF capacitor stores 125 pC of charge on its plates, what is the unknown capacitance?

88. A parallel-plate capacitor with plate area 2.0 cm² and air-gap separation 0.50 mm is connected to a 12-V battery, and fully charged. The battery is then disconnected. (a) What is the charge on the capacitor? (b) The plates are now pulled to a separation of 0.75 mm. What is the charge on the capacitor now? (c) What is the potential difference across the plates now? (d) How much work was required to pull the plates to their new separation?

89. In the circuit shown in Fig. 24–37, $C_1 = 1.0$ μF, $C_2 = 2.0$ μF, $C_3 = 2.4$ μF, and a voltage $V_{ab} = 24$ V is applied across points a and b. After C_1 is fully charged the switch is thrown to the right. What is the final charge and potential difference on each capacitor?

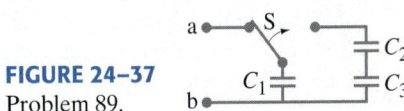

FIGURE 24–37
Problem 89.

90. The long cylindrical capacitor shown in Fig. 24–38 consists of four concentric cylinders, with respective radii R_a, R_b, R_c, and R_d. The cylinders b and c are joined by metal strips. Determine the capacitance per unit length of this arrangement. (Assume equal and opposite charges are placed on the innermost and outermost cylinders.)

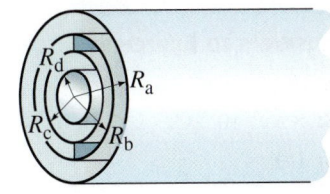

FIGURE 24–38
Problem 90.

91. A parallel-plate capacitor has plate area A, plate separation x, and has a charge Q stored on its plates (Fig. 24–39). Find the amount of work required to double the plate separation to $2x$, assuming the charge remains constant at Q. Show that your answer is consistent with the change in energy stored by the capacitor. (*Hint*: See Example 24–10.)

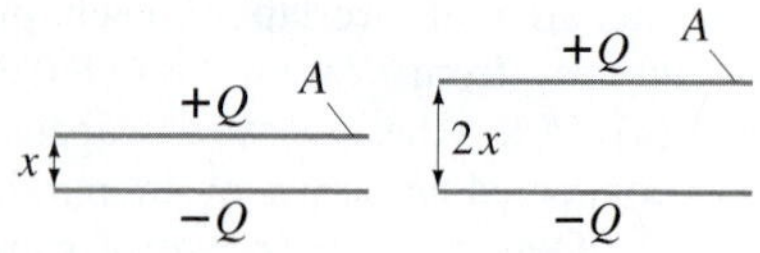

FIGURE 24–39
Problem 91.

92. Consider the use of capacitors as memory cells. A charged capacitor would represent a one and an uncharged capacitor a zero. Suppose these capacitors were fabricated on a silicon chip and each has a capacitance of 30 femto-farads $(1\ \text{fF} = 10^{-15}\ \text{F}.)$ The dielectric filling the space between the parallel plates has dielectric constant $K = 25$ and a dielectric strength of $1.0 \times 10^9\ \text{V/m}$. (*a*) If the operating voltage is 1.5 V, how many electrons would be stored on one of these capacitors when charged? (*b*) If no safety factor is allowed, how thin a dielectric layer could we use for operation at 1.5 V? (*c*) Using the layer thickness from your answer to part (*b*), what would be the area of the capacitor plates?

93. To get an idea how big a farad is, suppose you want to make a 1-F air-filled parallel-plate capacitor for a circuit you are building. To make it a reasonable size, suppose you limit the plate area to $1.0\ \text{cm}^2$. What would the gap have to be between the plates? Is this practically achievable?

94. A student wearing shoes with thin insulating soles is standing on a grounded metal floor when he puts his hand flat against the screen of a CRT computer monitor. The voltage inside the monitor screen, 6.3 mm from his hand, is 25,000 V. The student's hand and the monitor form a capacitor; the student is a conductor, and there is another capacitor between the floor and his feet. Using reasonable numbers for hand and foot areas, estimate the student's voltage relative to the floor. Assume vinyl-soled shoes 1 cm thick.

95. A parallel-plate capacitor with plate area $A = 2.0\ \text{m}^2$ and plate separation $d = 3.0\ \text{mm}$ is connected to a 45-V battery (Fig. 24–40a). (*a*) Determine the charge on the capacitor, the electric field, the capacitance, and the energy stored in the capacitor. (*b*) With the capacitor still connected to the battery, a slab of plastic with dielectric strength $K = 3.2$ is placed between the plates of the capacitor, so that the gap is completely filled with the dielectric. What are the new values of charge, electric field, capacitance, and the energy U stored in the capacitor?

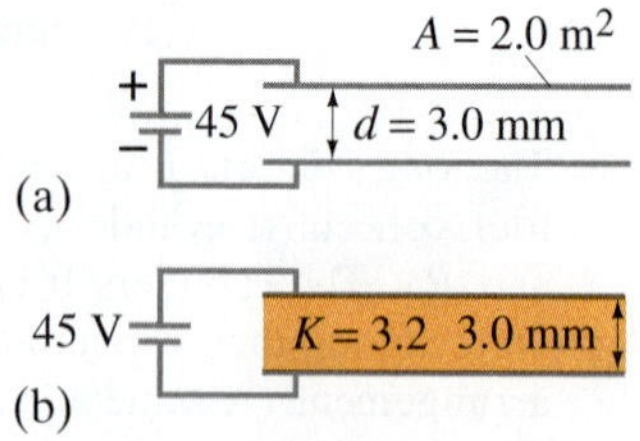

FIGURE 24–40
Problem 95.

96. Let us try to estimate the maximum "static electricity" charge that might result during each walking step across an insulating floor. Assume the sole of a person's shoe has area $A \approx 150\ \text{cm}^2$, and when the foot is lifted from the ground during each step, the sole acquires an excess charge Q from rubbing contact with the floor. (*a*) Model the sole as a plane conducting surface with Q uniformly distributed across it as the foot is lifted from the ground. If the dielectric strength of the air between the sole and floor as the foot is lifted is $E_S = 3 \times 10^6\ \text{N/C}$, determine Q_{max}, the maximum possible excess charge that can be transferred to the sole during each step. (*b*) Modeling a person as an isolated conducting sphere of radius $r \approx 1\ \text{m}$, estimate a person's capacitance. (*c*) After lifting the foot from the floor, assume the excess charge Q quickly redistributes itself over the entire surface area of the person. Estimate the maximum potential difference that the person can develop with respect to the floor.

97. Paper has a dielectric constant $K = 3.7$ and a dielectric strength of $15 \times 10^6\ \text{V/m}$. Suppose that a typical sheet of paper has a thickness of 0.030 mm. You make a "homemade" capacitor by placing a sheet of $21 \times 14\ \text{cm}$ paper between two aluminum foil sheets (Fig. 24–41). The thickness of the aluminum foil is 0.040 mm. (*a*) What is the capacitance C_0 of your device? (*b*) About how much charge could you store on your capacitor before it would break down? (*c*) Show in a sketch how you could overlay sheets of paper and aluminum for a parallel combination. If you made 100 such capacitors, and connected the edges of the sheets in parallel so that you have a single large capacitor of capacitance $100\ C_0$, how thick would your new large capacitor be? (*d*) What is the maximum voltage you can apply to this $100\ C_0$ capacitor without breakdown?

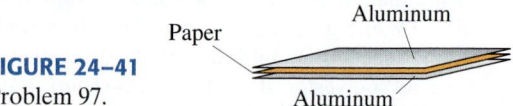

FIGURE 24–41
Problem 97.

*Numerical/Computer

***98.** (II) Six physics students were each given an air filled capacitor. Although the areas were different, the spacing between the plates, d, was the same for all six capacitors, but was unknown. Each student made a measurement of the area A and capacitance C of their capacitor. Below is a Table for their data. Using the combined data and a graphing program or spreadsheet, determine the spacing d between the plates.

Area (m²)	Capacitance (pF)
0.01	90
0.03	250
0.04	340
0.06	450
0.09	800
0.12	1050

Answers to Exercises

A: A.

B: $8.3 \times 10^{-9}\ \text{C}$.

C: (*a*).

D: (*e*).

E: (*b*).

The glow of the thin wire filament of a lightbulb is caused by the electric current passing through it. Electric energy is transformed to thermal energy (via collisions between moving electrons and atoms of the wire), which causes the wire's temperature to become so high that it glows. Electric current and electric power in electric circuits are of basic importance in everyday life. We examine both dc and ac in this Chapter, and include the microscopic analysis of electric current.

Electric Currents and Resistance

C H A P T E R

25

CHAPTER-OPENING QUESTION—Guess now!

The conductors shown are all made of copper and are at the same temperature. Which conductor would have the greatest resistance to the flow of charge entering from the left? Which would offer the least resistance?

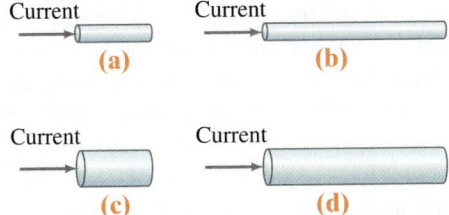

(a)　　　(b)

(c)　　　(d)

I n the previous four Chapters we have been studying static electricity: electric charges at rest. In this Chapter we begin our study of charges in motion, and we call a flow of charge an electric current.

In everyday life we are familiar with electric currents in wires and other conductors. Indeed, most practical electrical devices depend on electric current: current through a lightbulb, current in the heating element of a stove or electric heater, and currents in electronic devices. Electric currents can exist in conductors such as wires, and also in other devices such as the CRT of a television or computer monitor whose charged electrons flow through space (Section 23–9).

CONTENTS

25–1　The Electric Battery
25–2　Electric Current
25–3　Ohm's Law: Resistance and Resistors
25–4　Resistivity
25–5　Electric Power
25–6　Power in Household Circuits
25–7　Alternating Current
25–8　Microscopic View of Electric Current: Current Density and Drift Velocity
*25–9　Superconductivity
*25–10　Electrical Conduction in the Nervous System

FIGURE 25–1 Alessandro Volta. In this portrait, Volta exhibits his battery to Napoleon in 1801.

In electrostatic situations, we saw in Sections 21–9 and 22–3 that the electric field must be zero inside a conductor (if it weren't, the charges would move). But when charges are *moving* in a conductor, there usually *is* an electric field in the conductor. Indeed, an electric field is needed to set charges into motion, and to keep them in motion in any normal conductor. We can control the flow of charge using electric fields and electric potential (voltage), concepts we have just been discussing. In order to have a current in a wire, a potential difference is needed, which can be provided by a battery.

We first look at electric current from a macroscopic point of view: that is, current as measured in a laboratory. Later in the Chapter we look at currents from a microscopic (theoretical) point of view as a flow of electrons in a wire.

Until the year 1800, the technical development of electricity consisted mainly of producing a static charge by friction. It all changed in 1800 when Alessandro Volta (1745–1827; Fig. 25–1) invented the electric battery, and with it produced the first steady flow of electric charge—that is, a steady electric current.

25–1 The Electric Battery

The events that led to the discovery of the battery are interesting. For not only was this an important discovery, but it also gave rise to a famous scientific debate.

In the 1780s, Luigi Galvani (1737–1798), professor at the University of Bologna, carried out a series of experiments on the contraction of a frog's leg muscle through electricity produced by static electricity. Galvani found that the muscle also contracted when dissimilar metals were inserted into the frog. Galvani believed that the source of the electric charge was in the frog muscle or nerve itself, and that the metal merely transmitted the charge to the proper points. When he published his work in 1791, he termed this charge "animal electricity." Many wondered, including Galvani himself, if he had discovered the long-sought "life-force."

Volta, at the University of Pavia 200 km away, was skeptical of Galvani's results, and came to believe that the source of the electricity was not in the animal itself, but rather in the *contact between the dissimilar metals*. Volta realized that a moist conductor, such as a frog muscle or moisture at the contact point of two dissimilar metals, was necessary in the circuit if it was to be effective. He also saw that the contracting frog muscle was a sensitive instrument for detecting electric "tension" or "electromotive force" (his words for what we now call potential), in fact more sensitive than the best available electroscopes (Section 21–4) that he and others had developed.[†]

FIGURE 25–2 A voltaic battery, from Volta's original publication.

Volta's research found that certain combinations of metals produced a greater effect than others, and, using his measurements, he listed them in order of effectiveness. (This "electrochemical series" is still used by chemists today.) He also found that carbon could be used in place of one of the metals.

Volta then conceived his greatest contribution to science. Between a disc of zinc and one of silver, he placed a piece of cloth or paper soaked in salt solution or dilute acid and piled a "battery" of such couplings, one on top of another, as shown in Fig. 25–2. This "pile" or "battery" produced a much increased potential difference. Indeed, when strips of metal connected to the two ends of the pile were brought close, a spark was produced. Volta had designed and built the first electric battery; he published his discovery in 1800.

[†]Volta's most sensitive electroscope measured about 40 V per degree (angle of leaf separation). Nonetheless, he was able to estimate the potential differences produced by dissimilar metals in contact: for a silver–zinc contact he got about 0.7 V, remarkably close to today's value of 0.78 V.

Electric Cells and Batteries

A battery produces electricity by transforming chemical energy into electrical energy. Today a great variety of electric cells and batteries are available, from flashlight batteries to the storage battery of a car. The simplest batteries contain two plates or rods made of dissimilar metals (one can be carbon) called **electrodes**. The electrodes are immersed in a solution, such as a dilute acid, called the **electrolyte**. Such a device is properly called an **electric cell**, and several cells connected together is a **battery**, although today even a single cell is called a battery. The chemical reactions involved in most electric cells are quite complicated. Here we describe how one very simple cell works, emphasizing the physical aspects.

The cell shown in Fig. 25–3 uses dilute sulfuric acid as the electrolyte. One of the electrodes is made of carbon, the other of zinc. That part of each electrode outside the solution is called the **terminal**, and connections to wires and circuits are made here. The acid tends to dissolve the zinc electrode. Each zinc atom leaves two electrons behind on the electrode and enters the solution as a positive ion. The zinc electrode thus acquires a negative charge. As the electrolyte becomes positively charged, electrons are pulled off the carbon electrode by the electrolyte. Thus the carbon electrode becomes positively charged. Because there is an opposite charge on the two electrodes, there is a potential difference between the two terminals.

In a cell whose terminals are not connected, only a small amount of the zinc is dissolved, for as the zinc electrode becomes increasingly negative, any new positive zinc ions produced are attracted back to the electrode. Thus, a particular potential difference (or voltage) is maintained between the two terminals. If charge is allowed to flow between the terminals, say, through a wire (or a lightbulb), then more zinc can be dissolved. After a time, one or the other electrode is used up and the cell becomes "dead."

The voltage that exists between the terminals of a battery depends on what the electrodes are made of and their relative ability to be dissolved or give up electrons.

When two or more cells are connected so that the positive terminal of one is connected to the negative terminal of the next, they are said to be connected in *series* and their voltages add up. Thus, the voltage between the ends of two 1.5-V flashlight batteries connected in series is 3.0 V, whereas the six 2-V cells of an automobile storage battery give 12 V. Figure 25–4a shows a diagram of a common "dry cell" or "flashlight battery" used in portable radios and CD players, flashlights, etc., and Fig. 25–4b shows two smaller ones in series, connected to a flashlight bulb. A lightbulb consists of a thin, coiled wire (filament) inside an evacuated glass bulb, as shown in Fig. 25–5 and in the large photo opening this Chapter, page 651. The filament gets very hot (3000 K) and glows when charge passes through it.

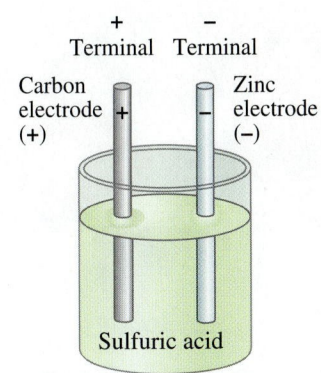

FIGURE 25–3 Simple electric cell.

FIGURE 25–4 (a) Diagram of an ordinary dry cell (like a D-cell or AA). The cylindrical zinc cup is covered on the sides; its flat bottom is the negative terminal. (b) Two dry cells (AA type) connected in series. Note that the positive terminal of one cell pushes against the negative terminal of the other.

FIGURE 25–5 A lightbulb: the fine wire of the filament becomes so hot that it glows. This type of lightbulb is called an incandescent bulb (as compared, say, to a fluorescent bulb).

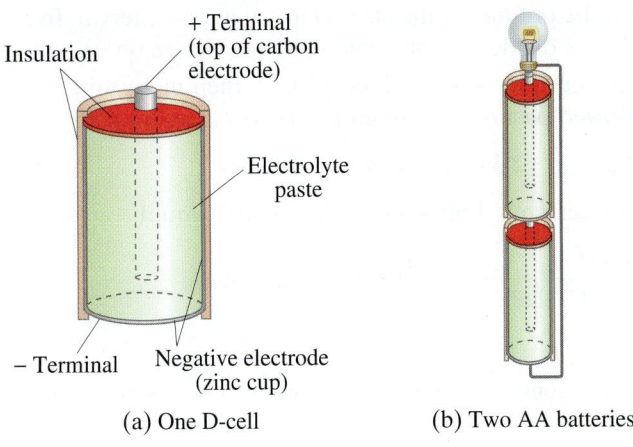

(a) One D-cell   (b) Two AA batteries

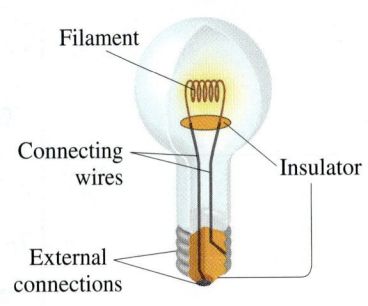

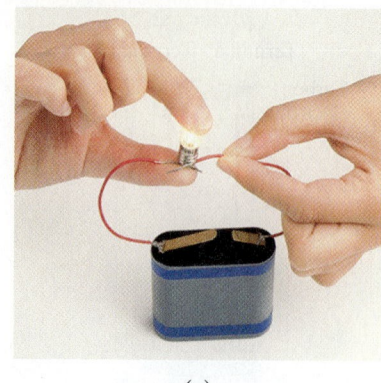

(a)

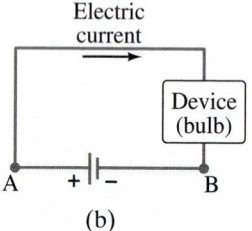

(b)

FIGURE 25–6 (a) A simple electric circuit. (b) Schematic drawing of the same circuit, consisting of a battery, connecting wires (thick gray lines), and a lightbulb or other device.

CAUTION

A battery does not create charge; a lightbulb does not destroy charge

25–2 Electric Current

The purpose of a battery is to produce a potential difference, which can then make charges move. When a continuous conducting path is connected between the terminals of a battery, we have an electric **circuit**, Fig. 25–6a. On any diagram of a circuit, as in Fig. 25–6b, we use the symbol

$$ \dashv \vert\vdash $$ [battery symbol]

to represent a battery. The device connected to the battery could be a lightbulb, a heater, a radio, or whatever. When such a circuit is formed, charge can flow through the wires of the circuit, from one terminal of the battery to the other, as long as the conducting path is continuous. Any flow of charge such as this is called an **electric current**.

More precisely, the electric current in a wire is defined as the net amount of charge that passes through the wire's full cross section at any point per unit time. Thus, the average current $\bar{I}$ is defined as

$$ \bar{I} = \frac{\Delta Q}{\Delta t}, \qquad (25\text{–}1a) $$

where ΔQ is the amount of charge that passes through the conductor at any location during the time interval Δt. The instantaneous current is defined by the derivative limit

$$ I = \frac{dQ}{dt}. \qquad (25\text{–}1b) $$

Electric current is measured in coulombs per second; this is given a special name, the **ampere** (abbreviated amp or A), after the French physicist André Ampère (1775–1836). Thus, $1\,\text{A} = 1\,\text{C/s}$. Smaller units of current are often used, such as the milliampere $(1\,\text{mA} = 10^{-3}\,\text{A})$ and microampere $(1\,\mu\text{A} = 10^{-6}\,\text{A})$.

A current can flow in a circuit only if there is a *continuous* conducting path. We then have a **complete circuit**. If there is a break in the circuit, say, a cut wire, we call it an **open circuit** and no current flows. In any single circuit, with only a single path for current to follow such as in Fig. 25–6b, a steady current at any instant is the same at one point (say, point A) as at any other point (such as B). This follows from the conservation of electric charge: charge doesn't disappear. A battery does not create (or destroy) any net charge, nor does a lightbulb absorb or destroy charge.

EXAMPLE 25–1 **Current is flow of charge.** A steady current of 2.5 A exists in a wire for 4.0 min. (*a*) How much total charge passed by a given point in the circuit during those 4.0 min? (*b*) How many electrons would this be?

APPROACH Current is flow of charge per unit time, Eqs. 25–1, so the amount of charge passing a point is the product of the current and the time interval. To get the number of electrons (*b*), we divide the total charge by the charge on one electron.

SOLUTION (*a*) Since the current was 2.5 A, or 2.5 C/s, then in 4.0 min (= 240 s) the total charge that flowed past a given point in the wire was, from Eq. 25–1a,

$$ \Delta Q = I\,\Delta t = (2.5\,\text{C/s})(240\,\text{s}) = 600\,\text{C}. $$

(*b*) The charge on one electron is $1.60 \times 10^{-19}\,\text{C}$, so 600 C would consist of

$$ \frac{600\,\text{C}}{1.60 \times 10^{-19}\,\text{C/electron}} = 3.8 \times 10^{21}\,\text{electrons}. $$

EXERCISE A If 1 million electrons per second pass a point in a wire, what is the current in amps?

CONCEPTUAL EXAMPLE 25–2 **How to connect a battery.** What is wrong with each of the schemes shown in Fig. 25–7 for lighting a flashlight bulb with a flashlight battery and a single wire?

RESPONSE (*a*) There is no closed path for charge to flow around. Charges might briefly start to flow from the battery toward the lightbulb, but there they run into a "dead end," and the flow would immediately come to a stop.

(*b*) Now there is a closed path passing to and from the lightbulb; but the wire touches only one battery terminal, so there is no potential difference in the circuit to make the charge move.

(*c*) Nothing is wrong here. This is a complete circuit: charge can flow out from one terminal of the battery, through the wire and the bulb, and into the other terminal. This scheme will light the bulb.

In many real circuits, wires are connected to a common conductor that provides continuity. This common conductor is called **ground**, usually represented as ⏚ or ⏚, and really is connected to the ground in a building or house. In a car, one terminal of the battery is called "ground," but is not connected to the ground—it is connected to the frame of the car, as is one connection to each lightbulb and other devices. Thus the car frame is a conductor in each circuit, ensuring a continuous path for charge flow.

We saw in Chapter 21 that conductors contain many free electrons. Thus, if a continuous conducting wire is connected to the terminals of a battery, negatively charged electrons flow in the wire. When the wire is first connected, the potential difference between the terminals of the battery sets up an electric field inside the wire[†] and parallel to it. Free electrons at one end of the wire are attracted into the positive terminal, and at the same time other electrons leave the negative terminal of the battery and enter the wire at the other end. There is a continuous flow of electrons throughout the wire that begins as soon as the wire is connected to *both* terminals. However, when the conventions of positive and negative charge were invented two centuries ago, it was assumed that positive charge flowed in a wire. For nearly all purposes, positive charge flowing in one direction is exactly equivalent to negative charge flowing in the opposite direction,[‡] as shown in Fig. 25–8. Today, we still use the historical convention of positive charge flow when discussing the direction of a current. So when we speak of the current direction in a circuit, we mean the direction positive charge would flow. This is sometimes referred to as **conventional current**. When we want to speak of the direction of electron flow, we will specifically state it is the electron current. In liquids and gases, both positive and negative charges (ions) can move.

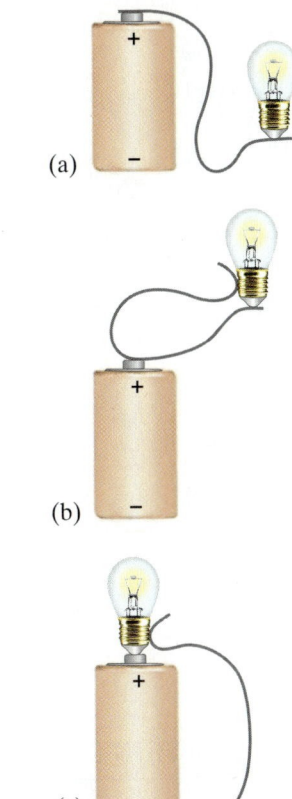

(a)

(b)

(c)

FIGURE 25–7 Example 25–2.

FIGURE 25–8 Conventional current from + to − is equivalent to a negative electron flow from − to +.

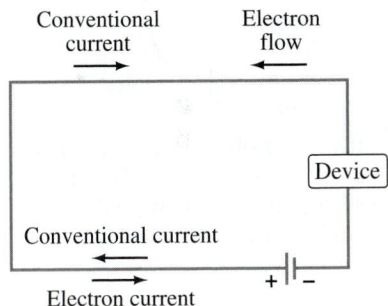

25–3 Ohm's Law: Resistance and Resistors

To produce an electric current in a circuit, a difference in potential is required. One way of producing a potential difference along a wire is to connect its ends to the opposite terminals of a battery. It was Georg Simon Ohm (1787–1854) who established experimentally that the current in a metal wire is proportional to the potential difference V applied to its two ends:

$$I \propto V.$$

If, for example, we connect a wire to the two terminals of a 6-V battery, the current flow will be twice what it would be if the wire were connected to a 3-V battery. It is also found that reversing the sign of the voltage does not affect the magnitude of the current.

[†]This does not contradict what was said in Section 21–9 that in the *static* case, there can be no electric field within a conductor since otherwise the charges would move. Indeed, when there is an electric field in a conductor, charges do move, and we get an electric current.

[‡]An exception is discussed in Section 27–8.

A useful analogy compares the flow of electric charge in a wire to the flow of water in a river, or in a pipe, acted on by gravity. If the river or pipe is nearly level, the flow rate is small. But if one end is somewhat higher than the other, the flow rate—or current—is greater. The greater the difference in height, the swifter the current. We saw in Chapter 23 that electric potential is analogous, in the gravitational case, to the height of a cliff. This applies in the present case to the height through which the fluid flows. Just as an increase in height can cause a greater flow of water, so a greater electric potential difference, or voltage, causes a greater electric current.

Exactly how large the current is in a wire depends not only on the voltage but also on the resistance the wire offers to the flow of electrons. The walls of a pipe, or the banks of a river and rocks in the middle, offer resistance to the water current. Similarly, electron flow is impeded because of interactions with the atoms of the wire. The higher this resistance, the less the current for a given voltage V. We then define electrical *resistance* so that the current is inversely proportional to the resistance: that is,

$$I = \frac{V}{R} \qquad\qquad \textbf{(25-2a)}$$

where R is the **resistance** of a wire or other device, V is the potential difference applied across the wire or device, and I is the current through it. Equation 25–2a is often written as

$$V = IR. \qquad\qquad \textbf{(25-2b)}$$

As mentioned above, Ohm found experimentally that in metal conductors R is a constant independent of V, a result known as **Ohm's law**. Equation 25–2b, $V = IR$, is itself sometimes called Ohm's law, but only when referring to materials or devices for which R is a constant independent of V. But R is not a constant for many substances other than metals, nor for devices such as diodes, vacuum tubes, transistors, and so on. Even for metals, R is not constant if the temperature changes much: for a lightbulb filament the measured resistance is low for small currents, but is much higher at its normal large operating current that puts it at the high temperature needed to make it glow (3000 K). Thus Ohm's "law" is not a fundamental law, but rather a description of a certain class of materials: metal conductors, whose temperature does not change much. Materials or devices that do not follow Ohm's law are said to be *nonohmic*. See Fig. 25–9.

The unit for resistance is called the **ohm** and is abbreviated Ω (Greek capital letter omega). Because $R = V/I$, we see that $1.0\,\Omega$ is equivalent to $1.0\,\text{V/A}$.

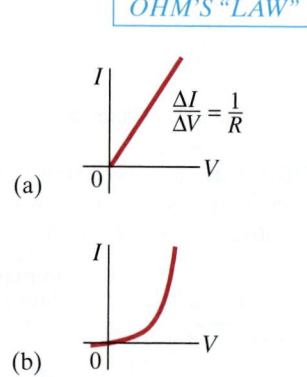

OHM'S "LAW"

(a)

(b)

FIGURE 25–9 Graphs of current vs. voltage for (a) a metal conductor which obeys Ohm's law, and (b) for a nonohmic device, in this case a semiconductor diode.

FIGURE 25–10 Example 25–3.

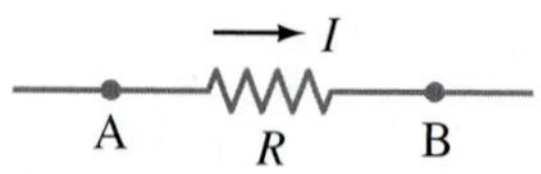

CONCEPTUAL EXAMPLE 25–3 **Current and potential.** Current I enters a resistor R as shown in Fig. 25–10. (a) Is the potential higher at point A or at point B? (b) Is the current greater at point A or at point B?

RESPONSE (a) Positive charge always flows from + to −, from high potential to low potential. Think again of the gravitational analogy: a mass will fall down from high gravitational potential to low. So for positive current I, point A is at a higher potential than point B.

(b) Conservation of charge requires that whatever charge flows into the resistor at point A, an equal amount of charge emerges at point B. Charge or current does not get "used up" by a resistor, just as an object that falls through a gravitational potential difference does not gain or lose mass. So the current is the same at A and B.

An electric potential decrease, as from point A to point B in Example 25–3, is often called a **potential drop** or a **voltage drop**.

EXAMPLE 25–4 **Flashlight bulb resistance.** A small flashlight bulb (Fig. 25–11) draws 300 mA from its 1.5-V battery. (*a*) What is the resistance of the bulb? (*b*) If the battery becomes weak and the voltage drops to 1.2 V, how would the current change?

APPROACH We can apply Ohm's law to the bulb, where the voltage applied across it is the battery voltage.

SOLUTION (*a*) We change 300 mA to 0.30 A and use Eq. 25–2a or b:

$$R = \frac{V}{I} = \frac{1.5 \text{ V}}{0.30 \text{ A}} = 5.0 \, \Omega.$$

(*b*) If the resistance stays the same, the current would be

$$I = \frac{V}{R} = \frac{1.2 \text{ V}}{5.0 \, \Omega} = 0.24 \text{ A} = 240 \text{ mA},$$

or a decrease of 60 mA.

NOTE With the smaller current in part (*b*), the bulb filament's temperature would be lower and the bulb less bright. Also, resistance does depend on temperature (Section 25–4), so our calculation is only a rough approximation.

EXERCISE B What resistance should be connected across a 9.0-V battery to make a 10-mA current? (*a*) 9 Ω, (*b*) 0.9 Ω, (*c*) 900 Ω, (*d*) 1.1 Ω, (*e*) 0.11 Ω.

All electric devices, from heaters to lightbulbs to stereo amplifiers, offer resistance to the flow of current. The filaments of lightbulbs (Fig. 25–5) and electric heaters are special types of wires whose resistance results in their becoming very hot. Generally, the connecting wires have very low resistance in comparison to the resistance of the wire filaments or coils, so the connecting wires usually have a minimal effect on the magnitude of the current. In many circuits, particularly in electronic devices, **resistors** are used to control the amount of current. Resistors have resistances ranging from less than an ohm to millions of ohms (see Figs. 25–12 and 25–13). The main types are "wire-wound" resistors which consist of a coil of fine wire, "composition" resistors which are usually made of carbon, and thin carbon or metal films.

When we draw a diagram of a circuit, we use the symbol

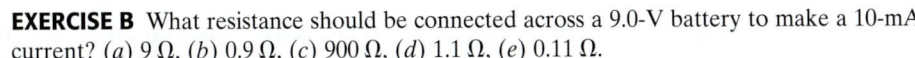
[resistor symbol]

to indicate a resistance. Wires whose resistance is negligible, however, are shown simply as straight lines.

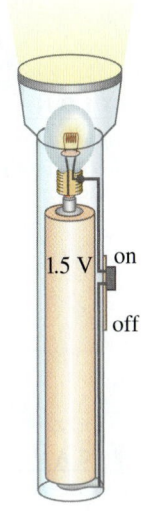

FIGURE 25–11 Flashlight (Example 25–4). Note how the circuit is completed along the side strip.

FIGURE 25–12 Photo of resistors (striped), plus other devices on a circuit board.

Resistor Color Code

Color	Number	Multiplier	Tolerance
Black	0	1	
Brown	1	10^1	1%
Red	2	10^2	2%
Orange	3	10^3	
Yellow	4	10^4	
Green	5	10^5	
Blue	6	10^6	
Violet	7	10^7	
Gray	8	10^8	
White	9	10^9	
Gold		10^{-1}	5%
Silver		10^{-2}	10%
No color			20%

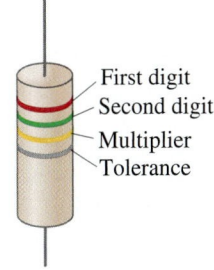

First digit
Second digit
Multiplier
Tolerance

FIGURE 25–13 The resistance value of a given resistor is written on the exterior, or may be given as a color code as shown above and in the Table: the first two colors represent the first two digits in the value of the resistance, the third color represents the power of ten that it must be multiplied by, and the fourth is the manufactured tolerance. For example, a resistor whose four colors are red, green, yellow, and silver has a resistance of $25 \times 10^4 \, \Omega = 250{,}000 \, \Omega = 250 \text{ k}\Omega$, plus or minus 10%. An alternate example of a simple code is a number such as 104, which means $R = 1.0 \times 10^4 \, \Omega$.

Some Helpful Clarifications

Here we briefly summarize some possible misunderstandings and clarifications. Batteries do not put out a constant current. Instead, batteries are intended to maintain a constant potential difference, or very nearly so. (Details in the next Chapter.) Thus a battery should be considered a source of voltage. The voltage is applied *across* a wire or device.

Electric current passes *through* a wire or device (connected to a battery), and its magnitude depends on that device's resistance. The resistance is a *property* of the wire or device. The voltage, on the other hand, is external to the wire or device, and is applied across the two ends of the wire or device. The current through the device might be called the "response": the current increases if the voltage increases or the resistance decreases, as $I = V/R$.

In a wire, the direction of the current is always parallel to the wire, no matter how the wire curves, just like water in a pipe. The direction of conventional (positive) current is from high potential $(+)$ toward lower potential $(-)$.

Current and charge do not increase or decrease or get "used up" when going through a wire or other device. The amount of charge that goes in at one end comes out at the other end.

25–4 Resistivity

It is found experimentally that the resistance R of any wire is directly proportional to its length ℓ and inversely proportional to its cross-sectional area A. That is,

$$R = \rho \frac{\ell}{A}, \qquad\qquad (25\text{–}3)$$

where ρ, the constant of proportionality, is called the **resistivity** and depends on the material used. Typical values of ρ, whose units are $\Omega \cdot m$ (see Eq. 25–3), are given for various materials in the middle column of Table 25–1, which is divided into the categories *conductors*, *insulators*, and *semiconductors* (see Section 21–3). The values depend somewhat on purity, heat treatment, temperature, and other factors. Notice that silver has the lowest resistivity and is thus the best conductor (although it is expensive). Copper is close, and much less expensive, which is why most wires are made of copper. Aluminum, although it has a higher resistivity, is much less dense than copper; it is thus preferable to copper in some situations, such as for transmission lines, because its resistance for the same weight is less than that for copper.

TABLE 25–1 Resistivity and Temperature Coefficients (at 20°C)

Material	Resistivity, $\rho \ (\Omega \cdot m)$	Temperature Coefficient, $\alpha \ (C°)^{-1}$
Conductors		
Silver	1.59×10^{-8}	0.0061
Copper	1.68×10^{-8}	0.0068
Gold	2.44×10^{-8}	0.0034
Aluminum	2.65×10^{-8}	0.00429
Tungsten	$5.6 \ \times 10^{-8}$	0.0045
Iron	9.71×10^{-8}	0.00651
Platinum	$10.6 \ \times 10^{-8}$	0.003927
Mercury	$98 \ \ \times 10^{-8}$	0.0009
Nichrome (Ni, Fe, Cr alloy)	$100 \ \ \times 10^{-8}$	0.0004
Semiconductors[†]		
Carbon (graphite)	$(3-60) \times 10^{-5}$	-0.0005
Germanium	$(1-500) \times 10^{-3}$	-0.05
Silicon	$0.1-60$	-0.07
Insulators		
Glass	$10^9 - 10^{12}$	
Hard rubber	$10^{13} - 10^{15}$	

[†] Values depend strongly on the presence of even slight amounts of impurities.

The reciprocal of the resistivity, called the **conductivity** σ, is

$$\sigma = \frac{1}{\rho} \qquad\qquad (25\text{–}4)$$

and has units of $(\Omega \cdot m)^{-1}$.

EXERCISE C Return to the Chapter-Opening Question, page 651, and answer it again now. Try to explain why you may have answered differently the first time.

EXERCISE D A copper wire has a resistance of $10\,\Omega$. What will its resistance be if it is only half as long? (a) $20\,\Omega$, (b) $10\,\Omega$, (c) $5\,\Omega$, (d) $1\,\Omega$, (e) none of these.

EXAMPLE 25–5 Speaker wires. Suppose you want to connect your stereo to remote speakers (Fig. 25–14). (a) If each wire must be 20 m long, what diameter copper wire should you use to keep the resistance less than $0.10\,\Omega$ per wire? (b) If the current to each speaker is 4.0 A, what is the potential difference, or voltage drop, across each wire?

APPROACH We solve Eq. 25–3 to get the area A, from which we can calculate the wire's radius using $A = \pi r^2$. The diameter is $2r$. In (b) we can use Ohm's law, $V = IR$.

SOLUTION (a) We solve Eq. 25–3 for the area A and find ρ for copper in Table 25–1:

$$A = \rho\frac{\ell}{R} = \frac{(1.68 \times 10^{-8}\,\Omega \cdot m)(20\,m)}{(0.10\,\Omega)} = 3.4 \times 10^{-6}\,m^2.$$

The cross-sectional area A of a circular wire is $A = \pi r^2$. The radius must then be at least

$$r = \sqrt{\frac{A}{\pi}} = 1.04 \times 10^{-3}\,m = 1.04\,mm.$$

The diameter is twice the radius and so must be at least $2r = 2.1$ mm.

(b) From $V = IR$ we find that the voltage drop across each wire is

$$V = IR = (4.0\,A)(0.10\,\Omega) = 0.40\,V.$$

NOTE The voltage drop across the wires reduces the voltage that reaches the speakers from the stereo amplifier, thus reducing the sound level a bit.

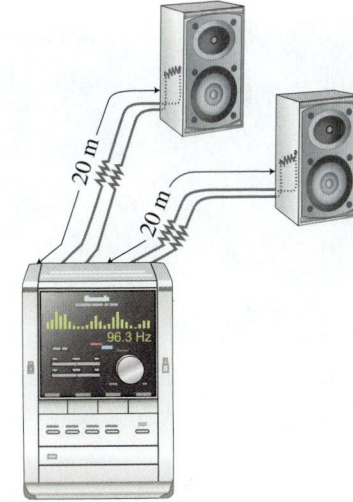

FIGURE 25–14 Example 25–5.

CONCEPTUAL EXAMPLE 25–6 Stretching changes resistance. Suppose a wire of resistance R could be stretched uniformly until it was twice its original length. What would happen to its resistance?

RESPONSE If the length ℓ doubles, then the cross-sectional area A is halved, because the volume $(V = A\ell)$ of the wire remains the same. From Eq. 25–3 we see that the resistance would increase by a factor of four $\left(2/\tfrac{1}{2} = 4\right)$.

EXERCISE E Copper wires in houses typically have a diameter of about 1.5 mm. How long a wire would have a 1.0-Ω resistance?

Temperature Dependence of Resistivity

The resistivity of a material depends somewhat on temperature. The resistance of metals generally increases with temperature. This is not surprising, for at higher temperatures, the atoms are moving more rapidly and are arranged in a less orderly fashion. So they might be expected to interfere more with the flow of electrons. If the temperature change is not too great, the resistivity of metals usually increases nearly linearly with temperature. That is,

$$\rho_T = \rho_0\big[1 + \alpha(T - T_0)\big] \qquad\qquad (25\text{–}5)$$

where ρ_0 is the resistivity at some reference temperature T_0 (such as $0°C$ or $20°C$), ρ_T is the resistivity at a temperature T, and α is the *temperature coefficient of resistivity*. Values for α are given in Table 25–1. Note that the temperature coefficient for semiconductors can be negative. Why? It seems that at higher temperatures, some of the electrons that are normally not free in a semiconductor become free and can contribute to the current. Thus, the resistance of a semiconductor can decrease with an increase in temperature.

EXAMPLE 25–7 **Resistance thermometer.** The variation in electrical resistance with temperature can be used to make precise temperature measurements. Platinum is commonly used since it is relatively free from corrosive effects and has a high melting point. Suppose at 20.0°C the resistance of a platinum resistance thermometer is 164.2 Ω. When placed in a particular solution, the resistance is 187.4 Ω. What is the temperature of this solution?

APPROACH Since the resistance R is directly proportional to the resistivity ρ, we can combine Eq. 25–3 with Eq. 25–5 to find R as a function of temperature T, and then solve that equation for T.

SOLUTION We multiply Eq. 25–5 by (ℓ/A) to obtain (see also Eq. 25–3)

$$R = R_0[1 + \alpha(T - T_0)].$$

Here $R_0 = \rho_0 \ell/A$ is the resistance of the wire at $T_0 = 20.0°C$. We solve this equation for T and find (see Table 25–1 for α)

$$T = T_0 + \frac{R - R_0}{\alpha R_0} = 20.0°C + \frac{187.4\ \Omega - 164.2\ \Omega}{(3.927 \times 10^{-3}(C°)^{-1})(164.2\ \Omega)} = 56.0°C.$$

NOTE Resistance thermometers have the advantage that they can be used at very high or low temperatures where gas or liquid thermometers would be useless.

NOTE More convenient for some applications is a *thermistor* (Fig. 25–15), which consists of a metal oxide or semiconductor whose resistance also varies in a repeatable way with temperature. Thermistors can be made quite small and respond very quickly to temperature changes.

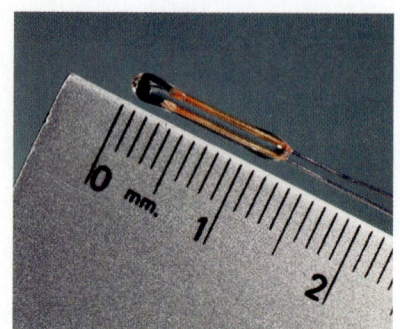

FIGURE 25–15 A thermistor shown next to a millimeter ruler for scale.

EXERCISE F The resistance of the tungsten filament of a common incandescent lightbulb is how many times greater at its operating temperature of 3000 K than its resistance at room temperature? (*a*) Less than 1% greater; (*b*) roughly 10% greater; (*c*) about 2 times greater; (*d*) roughly 10 times greater; (*e*) more than 100 times greater.

The value of α in Eq. 25–5 itself can depend on temperature, so it is important to check the temperature range of validity of any value (say, in a handbook of physical data). If the temperature range is wide, Eq. 25–5 is not adequate and terms proportional to the square and cube of the temperature are needed, but they are generally very small except when $T - T_0$ is large.

25–5 Electric Power

Electric energy is useful to us because it can be easily transformed into other forms of energy. Motors transform electric energy into mechanical energy, and are examined in Chapter 27.

In other devices such as electric heaters, stoves, toasters, and hair dryers, electric energy is transformed into thermal energy in a wire resistance known as a "heating element." And in an ordinary lightbulb, the tiny wire filament (Fig. 25–5 and Chapter-opening photo) becomes so hot it glows; only a few percent of the energy is transformed into visible light, and the rest, over 90%, into thermal energy. Lightbulb filaments and heating elements (Fig. 25–16) in household appliances have resistances typically of a few ohms to a few hundred ohms.

Electric energy is transformed into thermal energy or light in such devices, and there are many collisions between the moving electrons and the atoms of the wire. In each collision, part of the electron's kinetic energy is transferred to the atom with which it collides. As a result, the kinetic energy of the wire's atoms increases and hence the temperature of the wire element increases. The increased thermal energy can be transferred as heat by conduction and convection to the air in a heater or to food in a pan, by radiation to bread in a toaster, or radiated as light.

To find the power transformed by an electric device, recall that the energy transformed when an infinitesimal charge dq moves through a potential difference V is $dU = V\,dq$ (Eq. 23–3). Let dt be the time required for an amount of charge dq

FIGURE 25–16 Hot electric stove burner glows because of energy transformed by electric current.

to move through a potential difference V. Then the power P, which is the rate energy is transformed, is

$$P = \frac{dU}{dt} = \frac{dq}{dt}V.$$

The charge that flows per second, dq/dt, is the electric current I. Thus we have

$$P = IV. \tag{25-6}$$

This general relation gives us the power transformed by any device, where I is the current passing through it and V is the potential difference across it. It also gives the power delivered by a source such as a battery. The SI unit of electric power is the same as for any kind of power, the **watt** $(1\text{ W} = 1\text{ J/s})$.

The rate of energy transformation in a resistance R can be written in two other ways, starting with the general relation $P = IV$ and substituting in $V = IR$:

$$P = IV = I(IR) = I^2 R \tag{25-7a}$$

$$P = IV = \left(\frac{V}{R}\right)V = \frac{V^2}{R}. \tag{25-7b}$$

Equations 25–7a and b apply only to resistors, whereas Eq. 25–6, $P = IV$, is more general and applies to any device, including a resistor.

EXAMPLE 25–8 **Headlights.** Calculate the resistance of a 40-W automobile headlight designed for 12 V (Fig. 25–17).

APPROACH We solve Eq. 25–7b for R.

SOLUTION From Eq. 25–7b,

$$R = \frac{V^2}{P} = \frac{(12\text{ V})^2}{(40\text{ W})} = 3.6\ \Omega.$$

NOTE This is the resistance when the bulb is burning brightly at 40 W. When the bulb is cold, the resistance is much lower, as we saw in Eq. 25–5. Since the current is high when the resistance is low, lightbulbs burn out most often when first turned on.

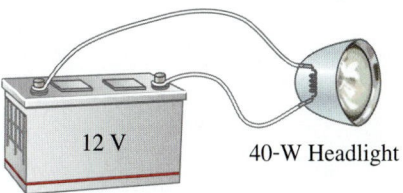

FIGURE 25–17 Example 25–8.

🏃 **PHYSICS APPLIED**

Why lightbulbs burn out when first turned on

⚠ **CAUTION**

You pay for energy, which is power × time, not for power

It is energy, not power, that you pay for on your electric bill. Since power is the *rate* energy is transformed, the total energy used by any device is simply its power consumption multiplied by the time it is on. If the power is in watts and the time is in seconds, the energy will be in joules since $1\text{ W} = 1\text{ J/s}$. Electric companies usually specify the energy with a much larger unit, the **kilowatt-hour** (kWh). One $\text{kWh} = (1000\text{ W})(3600\text{ s}) = 3.60 \times 10^6\text{ J}$.

EXAMPLE 25–9 **Electric heater.** An electric heater draws a steady 15.0 A on a 120-V line. How much power does it require and how much does it cost per month (30 days) if it operates 3.0 h per day and the electric company charges 9.2 cents per kWh?

APPROACH We use Eq. 25–6, $P = IV$, to find the power. We multiply the power (in kW) by the time (h) used in a month and by the cost per energy unit, $0.092 per kWh, to get the cost per month.

SOLUTION The power is

$$P = IV = (15.0\text{ A})(120\text{ V}) = 1800\text{ W}$$

or 1.80 kW. The time (in hours) the heater is used per month is $(3.0\text{ h/d})(30\text{ d}) = 90\text{ h}$, which at 9.2¢/kWh would cost $(1.80\text{ kW})(90\text{ h})(\$0.092/\text{kWh}) = \$15$.

NOTE Household current is actually alternating (ac), but our solution is still valid assuming the given values for V and I are the proper averages (rms) as we discuss in Section 25–7.

FIGURE 25–18 Example 25–10. A lightning bolt.

EXAMPLE 25–10 **ESTIMATE** **Lightning bolt.** Lightning is a spectacular example of electric current in a natural phenomenon (Fig. 25–18). There is much variability to lightning bolts, but a typical event can transfer 10^9 J of energy across a potential difference of perhaps 5×10^7 V during a time interval of about 0.2 s. Use this information to estimate (*a*) the total amount of charge transferred between cloud and ground, (*b*) the current in the lightning bolt, and (*c*) the average power delivered over the 0.2 s.

APPROACH We estimate the charge Q, recalling that potential energy change equals the potential difference ΔV times the charge Q, Eq. 23–3. We equate ΔU with the energy transferred, $\Delta U \approx 10^9$ J. Next, the current I is Q/t (Eq. 25–1a) and the power P is energy/time.

SOLUTION (*a*) From Eq. 23–3, the energy transformed is $\Delta U = Q \, \Delta V$. We solve for Q:

$$Q = \frac{\Delta U}{\Delta V} \approx \frac{10^9 \text{ J}}{5 \times 10^7 \text{ V}} = 20 \text{ coulombs.}$$

(*b*) The current during the 0.2 s is about

$$I = \frac{Q}{t} \approx \frac{20 \text{ C}}{0.2 \text{ s}} = 100 \text{ A.}$$

(*c*) The average power delivered is

$$P = \frac{\text{energy}}{\text{time}} = \frac{10^9 \text{ J}}{0.2 \text{ s}} = 5 \times 10^9 \text{ W} = 5 \text{ GW.}$$

We can also use Eq. 25–6:

$$P = IV = (100 \text{ A})(5 \times 10^7 \text{ V}) = 5 \text{ GW.}$$

NOTE Since most lightning bolts consist of several stages, it is possible that individual parts could carry currents much higher than the 100 A calculated above.

25–6 Power in Household Circuits

FIGURE 25–19 (a) Fuses. When the current exceeds a certain value, the metallic ribbon melts and the circuit opens. Then the fuse must be replaced. (b) One type of circuit breaker. The electric current passes through a bimetallic strip. When the current exceeds a safe level, the heating of the bimetallic strip causes the strip to bend so far to the left that the notch in the spring-loaded metal strip drops down over the end of the bimetallic strip; (c) the circuit then opens at the contact points (one is attached to the metal strip) and the outside switch is also flipped. As soon as the bimetallic strip cools down, it can be reset using the outside switch. Magnetic-type circuit breakers are discussed in Chapters 28 and 29.

The electric wires that carry electricity to lights and other electric appliances have some resistance, although usually it is quite small. Nonetheless, if the current is large enough, the wires will heat up and produce thermal energy at a rate equal to $I^2 R$, where R is the wire's resistance. One possible hazard is that the current-carrying wires in the wall of a building may become so hot as to start a fire. Thicker wires have less resistance (see Eq. 25–3) and thus can carry more current without becoming too hot. When a wire carries more current than is safe, it is said to be "overloaded." To prevent overloading, *fuses* or *circuit breakers* are installed in circuits. They are basically switches (Fig. 25–19)

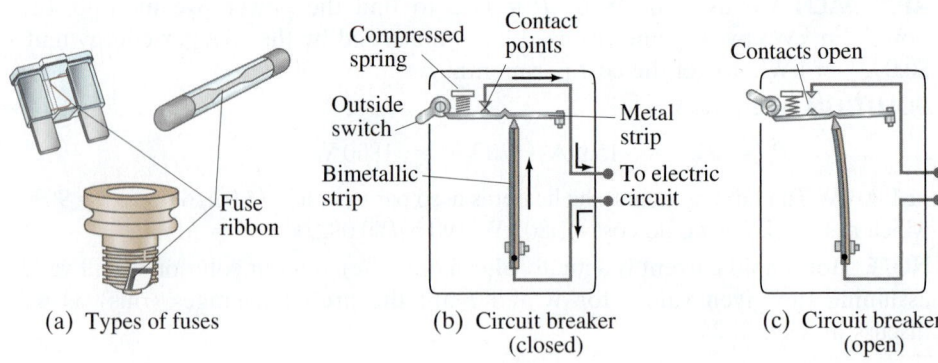

(a) Types of fuses

(b) Circuit breaker (closed)

(c) Circuit breaker (open)

that open the circuit when the current exceeds some particular value. A 20-A fuse or circuit breaker, for example, opens when the current passing through it exceeds 20 A. If a circuit repeatedly burns out a fuse or opens a circuit breaker, there are two possibilities: there may be too many devices drawing current in that circuit; or there is a fault somewhere, such as a "short." A short, or "short circuit," means that two wires have touched that should not have (perhaps because the insulation has worn through) so the resistance is much reduced and the current becomes very large. Short circuits should be remedied immediately.

Household circuits are designed with the various devices connected so that each receives the standard voltage (usually 120 V in the United States) from the electric company (Fig. 25–20). Circuits with the devices arranged as in Fig. 25–20 are called *parallel circuits*, as we will discuss in the next Chapter. When a fuse blows or circuit breaker opens, it is important to check the total current being drawn on that circuit, which is the sum of the currents in each device.

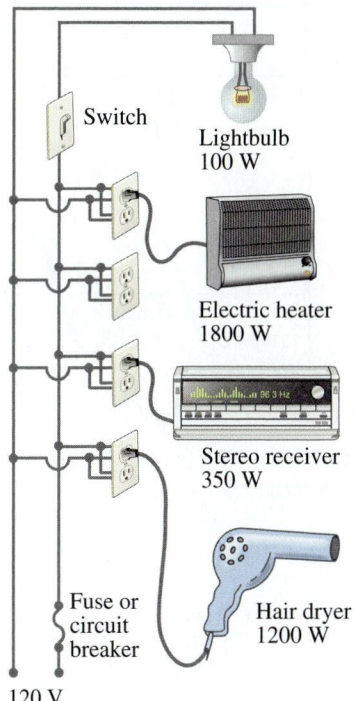

EXAMPLE 25–11 **Will a fuse blow?** Determine the total current drawn by all the devices in the circuit of Fig. 25–20.

APPROACH Each device has the same 120-V voltage across it. The current each draws from the source is found from $I = P/V$, Eq. 25–6.

SOLUTION The circuit in Fig. 25–20 draws the following currents: the lightbulb draws $I = P/V = 100\,\text{W}/120\,\text{V} = 0.8\,\text{A}$; the heater draws $1800\,\text{W}/120\,\text{V} = 15.0\,\text{A}$; the stereo draws a maximum of $350\,\text{W}/120\,\text{V} = 2.9\,\text{A}$; and the hair dryer draws $1200\,\text{W}/120\,\text{V} = 10.0\,\text{A}$. The total current drawn, if all devices are used at the same time, is

$$0.8\,\text{A} + 15.0\,\text{A} + 2.9\,\text{A} + 10.0\,\text{A} = 28.7\,\text{A}.$$

NOTE The heater draws as much current as 18 100-W lightbulbs. For safety, the heater should probably be on a circuit by itself.

FIGURE 25–20 Connection of household appliances.

If the circuit in Fig. 25–20 is designed for a 20-A fuse, the fuse should blow, and we hope it will, to prevent overloaded wires from getting hot enough to start a fire. Something will have to be turned off to get this circuit below 20 A. (Houses and apartments usually have several circuits, each with its own fuse or circuit breaker; try moving one of the devices to another circuit.) If the circuit is designed with heavier wire and a 30-A fuse, the fuse shouldn't blow—if it does, a short may be the problem. (The most likely place for a short is in the cord of one of the devices.) Proper fuse size is selected according to the wire used to supply the current. A properly rated fuse should *never* be replaced by a higher-rated one. A fuse blowing or a circuit breaker opening is acting like a switch, making an "open circuit." By an open circuit, we mean that there is no longer a complete conducting path, so no current can flow; it is as if $R = \infty$.

CONCEPTUAL EXAMPLE 25–12 **A dangerous extension cord.** Your 1800-W portable electric heater is too far from your desk to warm your feet. Its cord is too short, so you plug it into an extension cord rated at 11 A. Why is this dangerous?

RESPONSE 1800 W at 120 V draws a 15-A current. The wires in the extension cord rated at 11 A could become hot enough to melt the insulation and cause a fire.

EXERCISE G How many 60-W 120-V lightbulbs can operate on a 20-A line? (*a*) 2; (*b*) 3; (*c*) 6; (*d*) 20; (*e*) 40.

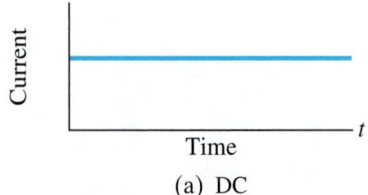

(a) DC

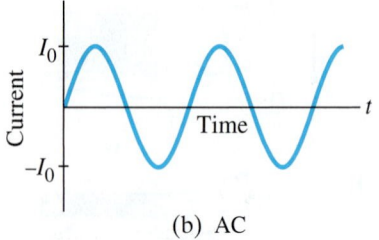

(b) AC

FIGURE 25–21 (a) Direct current. (b) Alternating current.

25–7 Alternating Current

When a battery is connected to a circuit, the current moves steadily in one direction. This is called a **direct current**, or **dc**. Electric generators at electric power plants, however, produce **alternating current**, or **ac**. (Sometimes capital letters are used, DC and AC.) An alternating current reverses direction many times per second and is commonly sinusoidal, as shown in Fig. 25–21. The electrons in a wire first move in one direction and then in the other. The current supplied to homes and businesses by electric companies is ac throughout virtually the entire world. We will discuss and analyze ac circuits in detail in Chapter 30. But because ac circuits are so common in real life, we will discuss some of their basic aspects here.

The voltage produced by an ac electric generator is sinusoidal, as we shall see later. The current it produces is thus sinusoidal (Fig. 25–21b). We can write the voltage as a function of time as

$$V = V_0 \sin 2\pi f t = V_0 \sin \omega t.$$

The potential V oscillates between $+V_0$ and $-V_0$, and V_0 is referred to as the **peak voltage**. The frequency f is the number of complete oscillations made per second, and $\omega = 2\pi f$. In most areas of the United States and Canada, f is 60 Hz (the unit "hertz," as we saw in Chapters 10 and 14, means cycles per second). In many other countries, 50 Hz is used.

Equation 25–2, $V = IR$, works also for ac: if a voltage V exists across a resistance R, then the current I through the resistance is

$$I = \frac{V}{R} = \frac{V_0}{R} \sin \omega t = I_0 \sin \omega t. \tag{25–8}$$

The quantity $I_0 = V_0/R$ is the **peak current**. The current is considered positive when the electrons flow in one direction and negative when they flow in the opposite direction. It is clear from Fig. 25–21b that an alternating current is as often positive as it is negative. Thus, the average current is zero. This does not mean, however, that no power is needed or that no heat is produced in a resistor. Electrons do move back and forth, and do produce heat. Indeed, the power transformed in a resistance R at any instant is

$$P = I^2 R = I_0^2 R \sin^2 \omega t.$$

Because the current is squared, we see that the power is always positive, as graphed in Fig. 25–22. The quantity $\sin^2 \omega t$ varies between 0 and 1; and it is not too difficult to show† that its average value is $\frac{1}{2}$, as indicated in Fig. 25–22. Thus, the *average power* transformed, $\overline{P}$, is

$$\overline{P} = \tfrac{1}{2} I_0^2 R.$$

Since power can also be written $P = V^2/R = (V_0^2/R)\sin^2 \omega t$, we also have that the average power is

$$\overline{P} = \tfrac{1}{2}\frac{V_0^2}{R}.$$

The average or mean value of the *square* of the current or voltage is thus what is important for calculating average power: $\overline{I^2} = \tfrac{1}{2} I_0^2$ and $\overline{V^2} = \tfrac{1}{2} V_0^2$. The square root of each of these is the **rms** (root-mean-square) value of the current or voltage:

$$I_{\mathrm{rms}} = \sqrt{\overline{I^2}} = \frac{I_0}{\sqrt{2}} = 0.707 I_0, \tag{25–9a}$$

$$V_{\mathrm{rms}} = \sqrt{\overline{V^2}} = \frac{V_0}{\sqrt{2}} = 0.707 V_0. \tag{25–9b}$$

The rms values of V and I are sometimes called the *effective values*.

FIGURE 25–22 Power transformed in a resistor in an ac circuit.

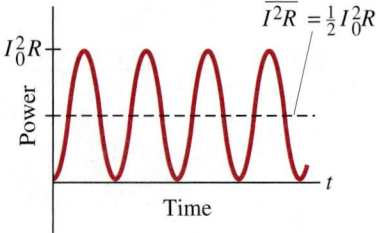

---

†A graph of $\cos^2 \omega t$ versus t is identical to that for $\sin^2 \omega t$ in Fig. 25–22, except that the points are shifted (by $\frac{1}{4}$ cycle) on the time axis. Hence the average value of $\sin^2$ and $\cos^2$, averaged over one or more full cycles, will be the same: $\overline{\sin^2 \omega t} = \overline{\cos^2 \omega t}$. From the trigonometric identity $\sin^2 \theta + \cos^2 \theta = 1$, we can write

$$\overline{(\sin^2 \omega t)} + \overline{(\cos^2 \omega t)} = 2\overline{(\sin^2 \omega t)} = 1.$$

Hence the average value of $\sin^2 \omega t$ is $\frac{1}{2}$.

They are useful because they can be substituted directly into the power formulas, Eqs. 25–6 and 25–7, to get the average power:

$$\overline{P} = I_{rms} V_{rms} \qquad\qquad \textbf{(25–10a)}$$

$$\overline{P} = \tfrac{1}{2} I_0^2 R = I_{rms}^2 R \qquad\qquad \textbf{(25–10b)}$$

$$\overline{P} = \tfrac{1}{2} \frac{V_0^2}{R} = \frac{V_{rms}^2}{R}. \qquad\qquad \textbf{(25–10c)}$$

Thus, a direct current whose values of I and V equal the rms values of I and V for an alternating current will produce the same power. Hence it is usually the rms value of current and voltage that is specified or measured. For example, in the United States and Canada, standard line voltage[†] is 120-V ac. The 120 V is V_{rms}; the peak voltage V_0 is

$$V_0 = \sqrt{2}\, V_{rms} = 170\ \text{V}.$$

In much of the world (Europe, Australia, Asia) the rms voltage is 240 V, so the peak voltage is 340 V.

EXAMPLE 25–13 Hair dryer. (*a*) Calculate the resistance and the peak current in a 1000-W hair dryer (Fig. 25–23) connected to a 120-V line. (*b*) What happens if it is connected to a 240-V line in Britain?

APPROACH We are given $\overline{P}$ and V_{rms}, so $I_{rms} = \overline{P}/V_{rms}$ (Eq. 25–10a or 25–6), and $I_0 = \sqrt{2}\, I_{rms}$. Then we find R from $V = IR$.

SOLUTION (*a*) We solve Eq. 25–10a for the rms current:

$$I_{rms} = \frac{\overline{P}}{V_{rms}} = \frac{1000\ \text{W}}{120\ \text{V}} = 8.33\ \text{A}.$$

Then

$$I_0 = \sqrt{2}\, I_{rms} = 11.8\ \text{A}.$$

The resistance is

$$R = \frac{V_{rms}}{I_{rms}} = \frac{120\ \text{V}}{8.33\ \text{A}} = 14.4\ \Omega.$$

The resistance could equally well be calculated using peak values:

$$R = \frac{V_0}{I_0} = \frac{170\ \text{V}}{11.8\ \text{A}} = 14.4\ \Omega.$$

(*b*) When connected to a 240-V line, more current would flow and the resistance would change with the increased temperature (Section 25–4). But let us make an estimate of the power transformed based on the same 14.4-Ω resistance. The average power would be

$$\overline{P} = \frac{V_{rms}^2}{R} = \frac{(240\ \text{V})^2}{(14.4\ \Omega)} = 4000\ \text{W}.$$

This is four times the dryer's power rating and would undoubtedly melt the heating element or the wire coils of the motor.

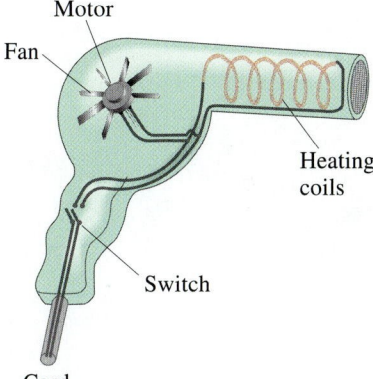

FIGURE 25–23 A hair dryer. Most of the current goes through the heating coils, a pure resistance; a small part goes to the motor to turn the fan. Example 25–13.

EXERCISE H Each channel of a stereo receiver is capable of an average power output of 100 W into an 8-Ω loudspeaker (see Fig. 25–14). What are the rms voltage and the rms current fed to the speaker (*a*) at the maximum power of 100 W, and (*b*) at 1.0 W when the volume is turned down?

[†]The line voltage can vary, depending on the total load; the frequency of 60 Hz or 50 Hz, however, remains extremely steady.

25–8 Microscopic View of Electric Current: Current Density and Drift Velocity

Up to now in this Chapter we have dealt mainly with a macroscopic view of electric current. We saw, however, that according to atomic theory, the electric current in metal wires is carried by negatively charged electrons, and that in liquid solutions current can also be carried by positive and/or negative ions. Let us now look at this microscopic picture in more detail.

When a potential difference is applied to the two ends of a wire of uniform cross section, the direction of the electric field $\vec{\mathbf{E}}$ is parallel to the walls of the wire (Fig. 25–24). The existence of $\vec{\mathbf{E}}$ within the conducting wire does not contradict our earlier result that $\vec{\mathbf{E}} = 0$ inside a conductor in the electrostatic case, as we are no longer dealing with the static case. Charges are free to move in a conductor, and hence can move under the action of the electric field. If all the charges are at rest, then $\vec{\mathbf{E}}$ must be zero (electrostatics).

We now define a new microscopic quantity, the **current density**, $\vec{\mathbf{j}}$. It is defined as the *electric current per unit cross-sectional area* at any point in space. If the current density $\vec{\mathbf{j}}$ in a wire of cross-sectional area A is uniform over the cross section, then j is related to the electric current by

$$ j = \frac{I}{A} \quad \text{or} \quad I = jA. \tag{25–11} $$

If the current density is not uniform, then the general relation is

$$ I = \int \vec{\mathbf{j}} \cdot d\vec{\mathbf{A}}, \tag{25–12} $$

where $d\vec{\mathbf{A}}$ is an element of surface and I is the current through the surface over which the integration is taken. The direction of the current density at any point is the direction that a positive charge would move when placed at that point—that is, the direction of $\vec{\mathbf{j}}$ at any point is generally the same as the direction of $\vec{\mathbf{E}}$, Fig. 25–24. The current density exists for any *point* in space. The current I, on the other hand, refers to a conductor as a whole, and hence is a macroscopic quantity.

The direction of $\vec{\mathbf{j}}$ is chosen to represent the direction of net flow of positive charge. In a conductor, it is negatively charged electrons that move, so they move in the direction of $-\vec{\mathbf{j}}$, or $-\vec{\mathbf{E}}$ (to the left in Fig. 25–24). We can imagine the free electrons as moving about randomly at high speeds, bouncing off the atoms of the wire (somewhat like the molecules of a gas—Chapter 18). When an electric field exists in the wire, Fig. 25–25, the electrons feel a force and initially begin to accelerate. But they soon reach a more or less steady average velocity in the direction of $\vec{\mathbf{E}}$, known as their **drift velocity**, $\vec{\boldsymbol{v}}_d$ (collisions with atoms in the wire keep them from accelerating further). The drift velocity is normally very much smaller than the electrons' average random speed.

We can relate the drift velocity v_d to the macroscopic current I in the wire. In a time Δt, the electrons will travel a distance $\ell = v_d \Delta t$ on average. Suppose the wire has cross-sectional area A. Then in time Δt, electrons in a volume $V = A\ell = Av_d \Delta t$ will pass through the cross section A of wire, as shown in Fig. 25–26. If there are n free electrons (each of charge $-e$) per unit volume ($n = N/V$), then the total charge ΔQ that passes through the area A in a time Δt is

$$ \Delta Q = (\text{no. of charges, } N) \times (\text{charge per particle}) $$
$$ = (nV)(-e) = -(nAv_d \Delta t)(e). $$

The current I in the wire is thus

$$ I = \frac{\Delta Q}{\Delta t} = -neAv_d. \tag{25–13} $$

The current density, $j = I/A$, is

$$ j = -nev_d. \tag{25–14} $$

In vector form, this is written

$$ \vec{\mathbf{j}} = -ne\vec{\mathbf{v}}_d, \tag{25–15} $$

where the minus sign indicates that the direction of (positive) current flow is opposite to the drift velocity of electrons.

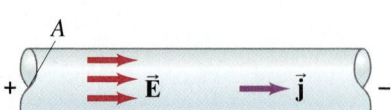

FIGURE 25–24 Electric field $\vec{\mathbf{E}}$ in a uniform wire of cross-sectional area A carrying a current I. The current density $j = I/A$.

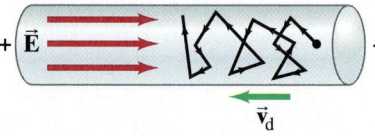

FIGURE 25–25 Electric field $\vec{\mathbf{E}}$ in a wire gives electrons in random motion a drift velocity v_d.

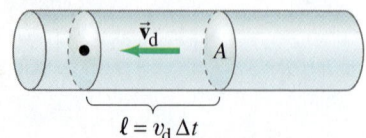

FIGURE 25–26 Electrons in the volume $A\ell$ will all pass through the cross section indicated in a time Δt, where $\ell = v_d \Delta t$.

We can generalize Eq. 25–15 to any type of charge flow, such as flow of ions in an electrolyte. If there are several types of ions (which can include free electrons), each of density n_i (number per unit volume), charge q_i ($q_i = -e$ for electrons) and drift velocity $\vec{v}_{di}$, then the net current density at any point is

$$\vec{j} = \sum_i n_i q_i \vec{v}_{di}. \qquad (25\text{–}16)$$

The total current I passing through an area A perpendicular to a uniform $\vec{j}$ is then

$$I = \sum_i n_i q_i v_{di} A.$$

EXAMPLE 25–14 **Electron speeds in a wire.** A copper wire 3.2 mm in diameter carries a 5.0-A current. Determine (a) the current density in the wire, and (b) the drift velocity of the free electrons. (c) Estimate the rms speed of electrons assuming they behave like an ideal gas at 20°C. Assume that one electron per Cu atom is free to move (the others remain bound to the atom).

APPROACH For (a) $j = I/A = I/\pi r^2$. For (b) we can apply Eq. 25–14 to find v_d if we can determine the number n of free electrons per unit volume. Since we assume there is one free electron per atom, the density of free electrons, n, is the same as the density of Cu atoms. The atomic mass of Cu is 63.5 u (see Periodic Table inside the back cover), so 63.5 g of Cu contains one mole or 6.02×10^{23} free electrons. The mass density of copper (Table 13–1) is $\rho_D = 8.9 \times 10^3 \text{ kg/m}^3$, where $\rho_D = m/V$. (We use ρ_D to distinguish it here from ρ for resistivity.) In (c) we use $K = \frac{3}{2}kT$, Eq. 18–4. (Do not confuse V for volume with V for voltage.)

SOLUTION (a) The current density is (with $r = \frac{1}{2}(3.2 \text{ mm}) = 1.6 \times 10^{-3} \text{ m}$)

$$j = \frac{I}{A} = \frac{I}{\pi r^2} = \frac{5.0 \text{ A}}{\pi (1.6 \times 10^{-3} \text{ m})^2} = 6.2 \times 10^5 \text{ A/m}^2.$$

(b) The number of free electrons per unit volume, $n = N/V$ (where $V = m/\rho_D$), is

$$n = \frac{N}{V} = \frac{N}{m/\rho_D} = \frac{N(1 \text{ mole})}{m(1 \text{ mole})}\rho_D$$

$$n = \left(\frac{6.02 \times 10^{23} \text{ electrons}}{63.5 \times 10^{-3} \text{ kg}}\right)(8.9 \times 10^3 \text{ kg/m}^3) = 8.4 \times 10^{28} \text{ m}^{-3}.$$

Then, by Eq. 25–14, the drift velocity has magnitude

$$v_d = \frac{j}{ne} = \frac{6.2 \times 10^5 \text{ A/m}^2}{(8.4 \times 10^{28} \text{ m}^{-3})(1.6 \times 10^{-19} \text{ C})} = 4.6 \times 10^{-5} \text{ m/s} \approx 0.05 \text{ mm/s}.$$

(c) If we model the free electrons as an ideal gas (a rather rough approximation), we use Eq. 18–5 to estimate the random rms speed of an electron as it darts around:

$$v_{rms} = \sqrt{\frac{3kT}{m}} = \sqrt{\frac{3(1.38 \times 10^{-23} \text{ J/K})(293 \text{ K})}{9.11 \times 10^{-31} \text{ kg}}} = 1.2 \times 10^5 \text{ m/s}.$$

The drift velocity (average speed in the direction of the current) is very much less than the rms thermal speed of the electrons, by a factor of about 10^9.

NOTE The result in (c) is an underestimate. Quantum theory calculations, and experiments, give the rms speed in copper to be about 1.6×10^6 m/s.

The drift velocity of electrons in a wire is very slow, only about 0.05 mm/s (Example 25–14 above), which means it takes an electron 20×10^3 s, or $5\frac{1}{2}$ h, to travel only 1 m. This is not, of course, how fast "electricity travels": when you flip a light switch, the light—even if many meters away—goes on nearly instantaneously. Why? Because electric fields travel essentially at the speed of light (3×10^8 m/s). We can think of electrons in a wire as being like a pipe full of water: when a little water enters one end of the pipe, almost immediately some water comes out at the other end.

*Electric Field Inside a Wire

Equation 25–2b, $V = IR$, can be written in terms of microscopic quantities as follows. We write the resistance R in terms of the resistivity ρ:

$$R = \rho \frac{\ell}{A};$$

and we write V and I as

$$I = jA \qquad \text{and} \qquad V = E\ell.$$

The last relation follows from Eqs. 23–4, where we assume the electric field is uniform within the wire and ℓ is the length of the wire (or a portion of the wire) between whose ends the potential difference is V. Thus, from $V = IR$, we have

$$E\ell = (jA)\left(\rho \frac{\ell}{A}\right) = j\rho\ell$$

so

$$j = \frac{1}{\rho} E = \sigma E, \tag{25–17}$$

where $\sigma = 1/\rho$ is the *conductivity* (Eq. 25–4). For a metal conductor, ρ and σ do not depend on V (and hence not on E). Therefore the current density $\vec{\mathbf{j}}$ is proportional to the electrical field $\vec{\mathbf{E}}$ in the conductor. This is the "microscopic" statement of Ohm's law. Equation 25–17, which can be written in vector form as

$$\vec{\mathbf{j}} = \sigma\vec{\mathbf{E}} = \frac{1}{\rho}\vec{\mathbf{E}},$$

is sometimes taken as the definition of conductivity σ and resistivity ρ.

EXAMPLE 25–15 **Electric field inside a wire.** What is the electric field inside the wire of Example 25–14?

APPROACH We use Eq. 25–17 and $\rho = 1.68 \times 10^{-8} \, \Omega \cdot \text{m}$ for copper.

SOLUTION Example 25–14 gives $j = 6.2 \times 10^5 \, \text{A/m}^2$, so

$$E = \rho j = (1.68 \times 10^{-8} \, \Omega \cdot \text{m})(6.2 \times 10^5 \, \text{A/m}^2) = 1.0 \times 10^{-2} \, \text{V/m}.$$

NOTE For comparison, the electric field between the plates of a capacitor is often much larger; in Example 24–1, for example, E is on the order of $10^4 \, \text{V/m}$. Thus we see that only a modest electric field is needed for current flow in practical cases.

*25–9 Superconductivity

At very low temperatures, well below 0°C, the resistivity (Section 25–4) of certain metals and certain compounds or alloys becomes zero as measured by the highest-precision techniques. Materials in such a state are said to be **superconducting**. It was first observed by H. K. Onnes (1853–1926) in 1911 when he cooled mercury below 4.2 K (-269°C) and found that the resistance of mercury suddenly dropped to zero. In general, superconductors become superconducting only below a certain *transition temperature* or *critical temperature*, T_C, which is usually within a few degrees of absolute zero. Current in a ring-shaped superconducting material has been observed to flow for years in the absence of a potential difference, with no measurable decrease. Measurements show that the resistivity ρ of superconductors is less than $4 \times 10^{-25} \, \Omega \cdot \text{m}$, which is over 10^{16} times smaller than that for copper, and is considered to be zero in practice. See Fig. 25–27.

Before 1986 the highest temperature at which a material was found to super-conduct was 23 K, which required liquid helium to keep the material cold. In 1987, a compound of yttrium, barium, copper, and oxygen (YBCO) was developed that can be superconducting at 90 K. This was an important breakthrough since liquid nitrogen, which boils at 77 K (sufficiently cold to keep the material superconducting), is more easily and cheaply obtained than the liquid helium needed for conventional superconductors. Superconductivity at temperatures as high as 160 K has been reported, though in fragile compounds.

FIGURE 25–27 A superconducting material has zero resistivity when its temperature is below T_C, its "critical temperature." At T_C, the resistivity jumps to a "normal" nonzero value and increases with temperature as most materials do (Eq. 25–5).

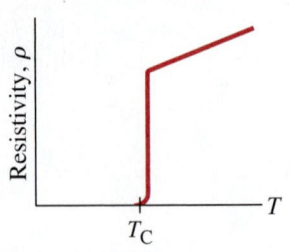

Most applications today use a bismuth-strontium-calcium-copper oxide, known (for short) as BSCCO. A major challenge is how to make a useable, bendable wire out of the BSCCO, which is very brittle. (One solution is to embed tiny filaments of the high-T_C superconductor in a metal alloy, which is not resistanceless, but the resistance is much less than that of a conventional copper cable.)

*25–10 Electrical Conduction in the Nervous System

The flow of electric charge in the human nervous system provides us the means for being aware of the world. Although the detailed functioning is not well understood, we do have a reasonable understanding of how messages are transmitted within the nervous system: they are electrical signals passing along the basic element of the nervous system, the *neuron*.

Neurons are living cells of unusual shape (Fig. 25–28). Attached to the main cell body are several small appendages known as *dendrites* and a long tail called the *axon*. Signals are received by the dendrites and are propagated along the axon. When a signal reaches the nerve endings, it is transmitted to the next neuron or to a muscle at a connection called a *synapse*.

A neuron, before transmitting an electrical signal, is in the so-called "resting state." Like nearly all living cells, neurons have a net positive charge on the outer surface of the cell membrane and a negative charge on the inner surface. This difference in charge, or "dipole layer," means that a potential difference exists across the cell membrane. When a neuron is not transmitting a signal, this "resting potential," normally stated as

$$V_{inside} - V_{outside},$$

is typically $-60\,\text{mV}$ to $-90\,\text{mV}$, depending on the type of organism. The most common ions in a cell are K^+, Na^+, and Cl^-. There are large differences in the concentrations of these ions inside and outside a cell, as indicated by the typical values given in Table 25–2. Other ions are also present, so the fluids both inside and outside the axon are electrically neutral. Because of the differences in concentration, there is a tendency for ions to diffuse across the membrane (see Section 18–7 on diffusion). However, in the resting state the cell membrane prevents any net flow of Na^+ (through a mechanism of "active pumping" of Na^+ out of the cell). But it does allow the flow of Cl^- ions, and less so of K^+ ions, and it is these two ions that produce the dipole charge layer on the membrane. Because there is a greater concentration of K^+ inside the cell than outside, more K^+ ions tend to diffuse outward across the membrane than diffuse inward. A K^+ ion that passes through the membrane becomes attached to the outer surface of the membrane, and leaves behind an equal negative charge that lies on the inner surface of the membrane (Fig. 25–29). Independently, Cl^- ions tend to diffuse *into* the cell since their concentration outside is higher. Both K^+ and Cl^- diffusion tends to charge the interior surface of the membrane negatively and the outside positively. As charge accumulates on the membrane surface, it becomes increasingly difficult for more ions to diffuse: K^+ ions trying to move outward, for example, are repelled by the positive charge already there. Equilibrium is reached when the tendency to diffuse because of the concentration difference is just balanced by the electrical potential difference across the membrane. The greater the concentration difference, the greater the potential difference across the membrane ($-60\,\text{mV}$ to $-90\,\text{mV}$).

The most important aspect of a neuron is not that it has a resting potential (most cells do), but rather that it can respond to a stimulus and conduct an electrical signal along its length. The stimulus could be thermal (when you touch a hot stove) or chemical (as in taste buds); it could be pressure (as on the skin or at the eardrum), or light (as in the eye); or it could be the electric stimulus of a signal coming from the brain or another neuron. In the laboratory, the stimulus is usually electrical and is applied by a tiny probe at some point on the neuron. If the stimulus exceeds some threshold, a voltage pulse will travel down the axon. This voltage pulse can be detected at a point on the axon using a voltmeter or an oscilloscope connected as in Fig. 25–30.

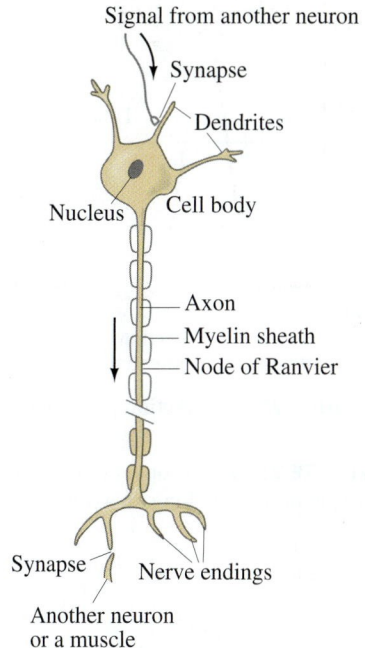

FIGURE 25–28 A simplified sketch of a typical neuron.

TABLE 25–2
Concentrations of Ions Inside and Outside a Typical Axon

	Concentration inside axon (mol/m³)	Concentration outside axon (mol/m³)
K^+	140	5
Na^+	15	140
Cl^-	9	125

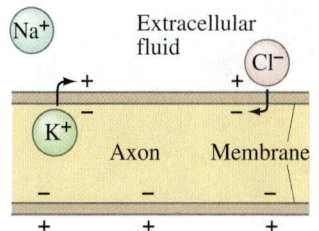

FIGURE 25–29 How a dipole layer of charge forms on a cell membrane.

FIGURE 25–30 Measuring the potential difference between the inside and outside of a nerve cell.

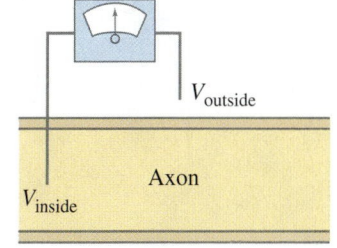

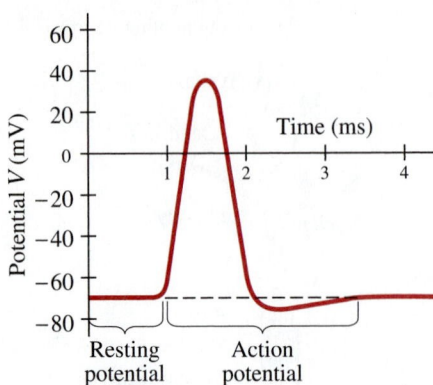

FIGURE 25–31 Action potential.

FIGURE 25–32 Propagation of an action potential along an axon membrane.

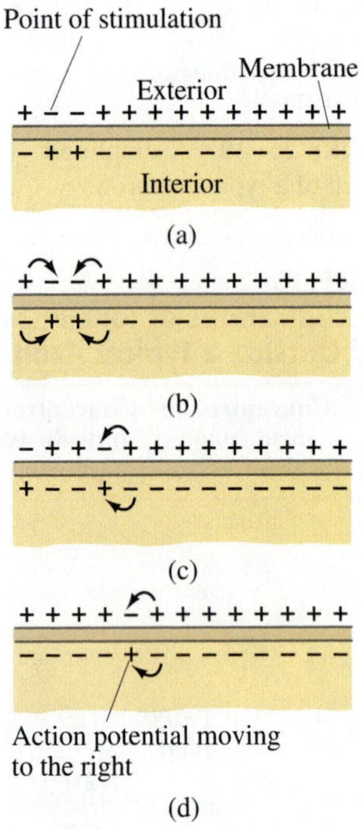

Point of stimulation

Exterior / Membrane

(a)

(b)

(c)

Action potential moving to the right

(d)

This voltage pulse has the shape shown in Fig. 25–31, and is called an **action potential**. As can be seen, the potential increases from a resting potential of about -70 mV and becomes a positive 30 mV or 40 mV. The action potential lasts for about 1 ms and travels down an axon with a speed of 30 m/s to 150 m/s. When an action potential is stimulated, the nerve is said to have "fired."

What causes the action potential? Apparently, the cell membrane has the ability to alter its permeability properties. At the point where the stimulus occurs, the membrane suddenly becomes much more permeable to Na^+ than to K^+ and Cl^- ions. Thus, Na^+ ions rush into the cell and the inner surface of the wall becomes positively charged, and the potential difference quickly swings positive ($\approx +35$ mV in Fig. 25–31). Just as suddenly, the membrane returns to its original characteristics: it becomes impermeable to Na^+ and in fact pumps out Na^+ ions. The diffusion of Cl^- and K^+ ions again predominates and the original resting potential is restored (-70 mV in Fig. 25–31).

What causes the action potential to travel along the axon? The action potential occurs at the point of stimulation, as shown in Fig. 25–32a. The membrane momentarily is positive on the inside and negative on the outside at this point. Nearby charges are attracted toward this region, as shown in Fig. 25–32b. The potential in these adjacent regions then drops, causing an action potential there. Thus, as the membrane returns to normal at the original point, nearby it experiences an action potential, so the action potential moves down the axon (Figs. 25–32c and d).

You may wonder if the number of ions that pass through the membrane would significantly alter the concentrations. The answer is no; and we can show why by treating the axon as a capacitor in the following Example.

EXAMPLE 25–16 ESTIMATE Capacitance of an axon. (a) Do an order-of-magnitude estimate for the capacitance of an axon 10 cm long of radius 10 μm. The thickness of the membrane is about 10^{-8} m, and the dielectric constant is about 3. (b) By what factor does the concentration (number of ions per volume) of Na^+ ions in the cell change as a result of one action potential?

APPROACH We model the membrane of an axon as a cylindrically shaped parallel-plate capacitor, with opposite charges on each side. The separation of the "plates" is the thickness of the membrane, $d \approx 10^{-8}$ m. We first calculate the area of the cylinder and then can use Eq. 24–8, $C = K\epsilon_0 A/d$, to find the capacitance. In (b), we use the voltage change during one action potential to find the amount of charge moved across the membrane.

SOLUTION (a) The area A is the area of a cylinder of radius r and length ℓ:

$$A = 2\pi r\ell \approx (6.28)(10^{-5}\,\text{m})(0.1\,\text{m}) \approx 6 \times 10^{-6}\,\text{m}^2.$$

From Eq. 24–8, we have

$$C = K\epsilon_0\frac{A}{d} \approx (3)(8.85 \times 10^{-12}\,\text{C}^2/\text{N}\cdot\text{m}^2)\frac{6 \times 10^{-6}\,\text{m}^2}{10^{-8}\,\text{m}} \approx 10^{-8}\,\text{F}.$$

(b) Since the voltage changes from -70 mV to about $+30$ mV, the total change is about 100 mV. The amount of charge that moves is then

$$Q = CV \approx (10^{-8}\,\text{F})(0.1\,\text{V}) = 10^{-9}\,\text{C}.$$

Each ion carries a charge $e = 1.6 \times 10^{-19}$ C, so the number of ions that flow per action potential is $Q/e = (10^{-9}\,\text{C})/(1.6 \times 10^{-19}\,\text{C}) \approx 10^{10}$. The volume of our cylindrical axon is

$$V = \pi r^2\ell \approx (3)(10^{-5}\,\text{m})^2(0.1\,\text{m}) = 3 \times 10^{-11}\,\text{m}^3,$$

and the concentration of Na^+ ions inside the cell (Table 25–2) is 15 mol/m^3 = $15 \times 6.02 \times 10^{23}$ ions/m^3 $\approx 10^{25}$ ions/m^3. Thus, the cell contains $(10^{25}$ ions/m$^3) \times (3 \times 10^{-11}\,\text{m}^3) \approx 3 \times 10^{14}$ Na^+ ions. One action potential, then, will change the concentration of Na^+ ions by about $10^{10}/(3 \times 10^{14}) = \frac{1}{3} \times 10^{-4}$, or 1 part in 30,000. This tiny change would not be measurable.

Thus, even 1000 action potentials will not alter the concentration significantly. The sodium pump does not, therefore, have to remove Na^+ ions quickly after an action potential, but can operate slowly over time to maintain a relatively constant concentration.

Summary

An electric **battery** serves as a source of nearly constant potential difference by transforming chemical energy into electric energy. A simple battery consists of two electrodes made of different metals immersed in a solution or paste known as an **electrolyte**.

Electric current, I, refers to the rate of flow of electric charge and is measured in **amperes** (A): 1 A equals a flow of 1 C/s past a given point.

The direction of **conventional current** is that of positive charge flow. In a wire, it is actually negatively charged electrons that move, so they flow in a direction opposite to the conventional current. A positive charge flow in one direction is almost always equivalent to a negative charge flow in the opposite direction. Positive conventional current always flows from a high potential to a low potential.

The **resistance** R of a device is defined by the relation

$$V = IR, \qquad (25\text{--}2)$$

where I is the current in the device when a potential difference V is applied across it. For materials such as metals, R is a constant independent of V (thus $I \propto V$), a result known as **Ohm's law**. Thus, the current I coming from a battery of voltage V depends on the resistance R of the circuit connected to it.

Voltage is applied *across* a device or between the ends of a wire. Current passes *through* a wire or device. Resistance is a property *of* the wire or device.

The unit of resistance is the **ohm** (Ω), where $1\,\Omega = 1\,\text{V/A}$. See Table 25–3.

TABLE 25–3 Summary of Units

Current	$1\,\text{A} = 1\,\text{C/s}$
Potential difference	$1\,\text{V} = 1\,\text{J/C}$
Power	$1\,\text{W} = 1\,\text{J/s}$
Resistance	$1\,\Omega = 1\,\text{V/A}$

The resistance R of a wire is inversely proportional to its cross-sectional area A, and directly proportional to its length ℓ and to a property of the material called its resistivity:

$$R = \frac{\rho \ell}{A}. \qquad (25\text{--}3)$$

The **resistivity**, ρ, increases with temperature for metals, but for semiconductors it may decrease.

The rate at which energy is transformed in a resistance R from electric to other forms of energy (such as heat and light) is equal to the product of current and voltage. That is, the **power** transformed, measured in watts, is given by

$$P = IV, \qquad (25\text{--}6)$$

which for resistors can be written as

$$P = I^2 R = \frac{V^2}{R}. \qquad (25\text{--}7)$$

The SI unit of power is the **watt** $(1\,\text{W} = 1\,\text{J/s})$.

The total electric energy transformed in any device equals the product of the power and the time during which the device is operated. In SI units, energy is given in joules $(1\,\text{J} = 1\,\text{W} \cdot \text{s})$, but electric companies use a larger unit, the **kilowatt-hour** $(1\,\text{kWh} = 3.6 \times 10^6\,\text{J})$.

Electric current can be **direct current** (**dc**), in which the current is steady in one direction; or it can be **alternating current** (**ac**), in which the current reverses direction at a particular frequency f, typically 60 Hz. Alternating currents are typically sinusoidal in time,

$$I = I_0 \sin \omega t, \qquad (25\text{--}8)$$

where $\omega = 2\pi f$, and are produced by an alternating voltage.

The **rms** values of sinusoidally alternating currents and voltages are given by

$$I_{\text{rms}} = \frac{I_0}{\sqrt{2}} \qquad \text{and} \qquad V_{\text{rms}} = \frac{V_0}{\sqrt{2}}, \qquad (25\text{--}9)$$

respectively, where I_0 and V_0 are the **peak** values. The power relationship, $P = IV = I^2 R = V^2/R$, is valid for the average power in alternating currents when the rms values of V and I are used.

Current density $\vec{\mathbf{j}}$ is the current per cross-sectional area. From a microscopic point of view, the current density is related to the number of charge carriers per unit volume, n, their charge, q, and their **drift velocity**, $\vec{\mathbf{v}}_{\text{d}}$, by

$$\vec{\mathbf{j}} = nq\vec{\mathbf{v}}_{\text{d}}. \qquad (25\text{--}16)$$

The electric field within a wire is related to $\vec{\mathbf{j}}$ by $\vec{\mathbf{j}} = \sigma \vec{\mathbf{E}}$ where $\sigma = 1/\rho$ is the **conductivity**.

[*At very low temperatures certain materials become **superconducting**, which means their electrical resistance becomes zero.]

[*The human nervous system operates via electrical conduction: when a nerve "fires," an electrical signal travels as a voltage pulse known as an **action potential**.]

Questions

1. What quantity is measured by a battery rating given in ampere-hours $(\text{A} \cdot \text{h})$?

2. When an electric cell is connected to a circuit, electrons flow away from the negative terminal in the circuit. But within the cell, electrons flow *to* the negative terminal. Explain.

3. When a flashlight is operated, what is being used up: battery current, battery voltage, battery energy, battery power, or battery resistance? Explain.

4. One terminal of a car battery is said to be connected to "ground." Since it is not really connected to the ground, what is meant by this expression?

5. When you turn on a water faucet, the water usually flows immediately. You don't have to wait for water to flow from the faucet valve to the spout. Why not? Is the same thing true when you connect a wire to the terminals of a battery?

6. Can a copper wire and an aluminum wire of the same length have the same resistance? Explain.

7. The equation $P = V^2/R$ indicates that the power dissipated in a resistor decreases if the resistance is increased, whereas the equation $P = I^2 R$ implies the opposite. Is there a contradiction here? Explain.

8. What happens when a lightbulb burns out?

9. If the resistance of a small immersion heater (to heat water for tea or soup, Fig. 25–33) was increased, would it speed up or slow down the heating process? Explain.

FIGURE 25–33
Question 9.

10. If a rectangular solid made of carbon has sides of lengths a, $2a$, and $3a$, how would you connect the wires from a battery so as to obtain (a) the least resistance, (b) the greatest resistance?

11. Explain why lightbulbs almost always burn out just as they are turned on and not after they have been on for some time.

12. Which draws more current, a 100-W lightbulb or a 75-W bulb? Which has the higher resistance?

13. Electric power is transferred over large distances at very high voltages. Explain how the high voltage reduces power losses in the transmission lines.

14. A 15-A fuse blows repeatedly. Why is it dangerous to replace this fuse with a 25-A fuse?

15. When electric lights are operated on low-frequency ac (say, 5 Hz), they flicker noticeably. Why?

16. Driven by ac power, the same electrons pass back and forth through your reading lamp over and over again. Explain why the light stays lit instead of going out after the first pass of electrons.

17. The heating element in a toaster is made of Nichrome wire. Immediately after the toaster is turned on, is the current (I_{rms}) in the wire increasing, decreasing, or staying constant? Explain.

18. Is current used up in a resistor? Explain.

19. Compare the drift velocities and electric currents in two wires that are geometrically identical and the density of atoms is similar, but the number of free electrons per atom in the material of one wire is twice that in the other.

20. A voltage V is connected across a wire of length ℓ and radius r. How is the electron drift velocity affected if (a) ℓ is doubled, (b) r is doubled, (c) V is doubled?

21. Why is it more dangerous to turn on an electric appliance when you are standing outside in bare feet than when you are inside wearing shoes with thick soles?

Problems

25–2 and 25–3 Electric Current, Resistance, Ohm's Law

(*Note*: The charge on one electron is 1.60×10^{-19} C.)

1. (I) A current of 1.30 A flows in a wire. How many electrons are flowing past any point in the wire per second?

2. (I) A service station charges a battery using a current of 6.7-A for 5.0 h. How much charge passes through the battery?

3. (I) What is the current in amperes if 1200 Na$^+$ ions flow across a cell membrane in 3.5 μs? The charge on the sodium is the same as on an electron, but positive.

4. (I) What is the resistance of a toaster if 120 V produces a current of 4.2 A?

5. (II) An electric clothes dryer has a heating element with a resistance of 8.6 Ω. (a) What is the current in the element when it is connected to 240 V? (b) How much charge passes through the element in 50 min? (Assume direct current.)

6. (II) A hair dryer draws 9.5 A when plugged into a 120-V line. (a) What is its resistance? (b) How much charge passes through it in 15 min? (Assume direct current.)

7. (II) A 4.5-V battery is connected to a bulb whose resistance is 1.6 Ω. How many electrons leave the battery per minute?

8. (II) A bird stands on a dc electric transmission line carrying 3100 A (Fig. 25–34). The line has 2.5×10^{-5} Ω resistance per meter, and the bird's feet are 4.0 cm apart. What is the potential difference between the bird's feet?

FIGURE 25–34
Problem 8.

9. (II) A 12-V battery causes a current of 0.60 A through a resistor. (a) What is its resistance, and (b) how many joules of energy does the battery lose in a minute?

10. (II) An electric device draws 6.50 A at 240 V. (a) If the voltage drops by 15%, what will be the current, assuming nothing else changes? (b) If the resistance of the device were reduced by 15%, what current would be drawn at 240 V?

25–4 Resistivity

11. (I) What is the diameter of a 1.00-m length of tungsten wire whose resistance is 0.32 Ω?

12. (I) What is the resistance of a 4.5-m length of copper wire 1.5 mm in diameter?

13. (II) Calculate the ratio of the resistance of 10.0 m of aluminum wire 2.0 mm in diameter, to 20.0 m of copper wire 1.8 mm in diameter.

14. (II) Can a 2.2-mm-diameter copper wire have the same resistance as a tungsten wire of the same length? Give numerical details.

15. (II) A sequence of potential differences V is applied across a wire (diameter = 0.32 mm, length = 11 cm) and the resulting currents I are measured as follows:

V (V)	0.100	0.200	0.300	0.400	0.500
I (mA)	72	144	216	288	360

(a) If this wire obeys Ohm's law, graphing I vs. V will result in a straight-line plot. Explain why this is so and determine the theoretical predictions for the straight line's slope and y-intercept. (b) Plot I vs. V. Based on this plot, can you conclude that the wire obeys Ohm's law (i.e., did you obtain a straight line with the expected y-intercept)? If so, determine the wire's resistance R. (c) Calculate the wire's resistivity and use Table 25–1 to identify the solid material from which it is composed.

16. (II) How much would you have to raise the temperature of a copper wire (originally at 20°C) to increase its resistance by 15%?

17. (II) A certain copper wire has a resistance of $10.0\,\Omega$. At what point along its length must the wire be cut so that the resistance of one piece is 4.0 times the resistance of the other? What is the resistance of each piece?

18. (II) Determine at what temperature aluminum will have the same resistivity as tungsten does at 20°C.

19. (II) A 100-W lightbulb has a resistance of about $12\,\Omega$ when cold (20°C) and $140\,\Omega$ when on (hot). Estimate the temperature of the filament when hot assuming an average temperature coefficient of resistivity $\alpha = 0.0045\,(\text{C}°)^{-1}$.

20. (II) Compute the voltage drop along a 26-m length of household no. 14 copper wire (used in 15-A circuits). The wire has diameter 1.628 mm and carries a 12-A current.

21. (II) Two aluminum wires have the same resistance. If one has twice the length of the other, what is the ratio of the diameter of the longer wire to the diameter of the shorter wire?

22. (II) A rectangular solid made of carbon has sides of lengths 1.0 cm, 2.0 cm, and 4.0 cm, lying along the x, y, and z axes, respectively (Fig. 25–35). Determine the resistance for current that passes through the solid in (a) the x direction, (b) the y direction, and (c) the z direction. Assume the resistivity is $\rho = 3.0 \times 10^{-5}\,\Omega\cdot\text{m}$.

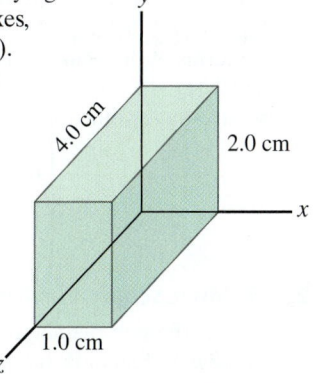

FIGURE 25–35
Problem 22.

23. (II) A length of aluminum wire is connected to a precision 10.00-V power supply, and a current of 0.4212 A is precisely measured at 20.0°C. The wire is placed in a new environment of unknown temperature where the measured current is 0.3818 A. What is the unknown temperature?

24. (II) Small changes in the length of an object can be measured using a **strain gauge** sensor, which is a wire with undeformed length ℓ_0, cross-sectional area A_0, and resistance R_0. This sensor is rigidly affixed to the object's surface, aligning its length in the direction in which length changes are to be measured. As the object deforms, the length of the wire sensor changes by $\Delta\ell$, and the resulting change ΔR in the sensor's resistance is measured. Assuming that as the solid wire is deformed to a length ℓ, its density (and volume) remains constant (only approximately valid), show that the strain $(= \Delta\ell/\ell_0)$ of the wire sensor, and thus of the object to which it is attached, is $\Delta R/2R_0$.

25. (II) A length of wire is cut in half and the two lengths are wrapped together side by side to make a thicker wire. How does the resistance of this new combination compare to the resistance of the original wire?

26. (III) For some applications, it is important that the value of a resistance not change with temperature. For example, suppose you made a 3.70-kΩ resistor from a carbon resistor and a Nichrome wire-wound resistor connected together so the total resistance is the sum of their separate resistances. What value should each of these resistors have (at 0°C) so that the combination is temperature independent?

27. (III) Determine a formula for the total resistance of a spherical shell made of material whose conductivity is σ and whose inner and outer radii are r_1 and r_2. Assume the current flows radially outward.

28. (III) The filament of a lightbulb has a resistance of $12\,\Omega$ at 20°C and $140\,\Omega$ when hot (as in Problem 19). (a) Calculate the temperature of the filament when it is hot, and take into account the change in length and area of the filament due to thermal expansion (assume tungsten for which the thermal expansion coefficient is $\approx 5.5 \times 10^{-6}\,\text{C}°^{-1}$). (b) In this temperature range, what is the percentage change in resistance due to thermal expansion, and what is the percentage change in resistance due solely to the change in ρ? Use Eq. 25–5.

29. (III) A 10.0-m length of wire consists of 5.0 m of copper followed by 5.0 m of aluminum, both of diameter 1.4 mm. A voltage difference of 85 mV is placed across the composite wire. (a) What is the total resistance (sum) of the two wires? (b) What is the current through the wire? (c) What are the voltages across the aluminum part and across the copper part?

30. (III) A hollow cylindrical resistor with inner radius r_1 and outer radius r_2, and length ℓ, is made of a material whose resistivity is ρ (Fig. 25–36). (a) Show that the resistance is given by

$$R = \frac{\rho}{2\pi\ell}\ln\frac{r_2}{r_1}$$

for current that flows radially outward. [*Hint*: Divide the resistor into concentric cylindrical shells and integrate.] (b) Evaluate the resistance R for such a resistor made of carbon whose inner and outer radii are 1.0 mm and 1.8 mm and whose length is 2.4 cm. (Choose $\rho = 15 \times 10^{-5}\,\Omega\cdot\text{m}$.) (c) What is the resistance in part (b) for current flowing *parallel* to the axis?

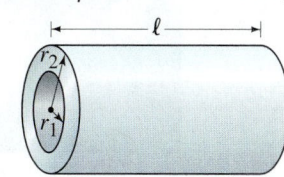

FIGURE 25–36
Problem 30.

25–5 and 25–6 Electric Power

31. (I) What is the maximum power consumption of a 3.0-V portable CD player that draws a maximum of 270 mA of current?

32. (I) The heating element of an electric oven is designed to produce 3.3 kW of heat when connected to a 240-V source. What must be the resistance of the element?

33. (I) What is the maximum voltage that can be applied across a 3.3-kΩ resistor rated at $\frac{1}{4}$ watt?

34. (I) (a) Determine the resistance of, and current through, a 75-W lightbulb connected to its proper source voltage of 110 V. (b) Repeat for a 440-W bulb.

35. (II) An electric power plant can produce electricity at a fixed power P, but the plant operator is free to choose the voltage V at which it is produced. This electricity is carried as an electric current I through a transmission line (resistance R) from the plant to the user, where it provides the user with electric power P'. (a) Show that the reduction in power $\Delta P = P - P'$ due to transmission losses is given by $\Delta P = P^2 R/V^2$. (b) In order to reduce power losses during transmission, should the operator choose V to be as large or as small as possible?

36. (II) A 120-V hair dryer has two settings: 850 W and 1250 W. (a) At which setting do you expect the resistance to be higher? After making a guess, determine the resistance at (b) the lower setting; and (c) the higher setting.

37. (II) A 115-V fish-tank heater is rated at 95 W. Calculate (a) the current through the heater when it is operating, and (b) its resistance.

38. (II) You buy a 75-W lightbulb in Europe, where electricity is delivered to homes at 240 V. If you use the lightbulb in the United States at 120 V (assume its resistance does not change), how bright will it be relative to 75-W 120-V bulbs? [*Hint*: Assume roughly that brightness is proportional to power consumed.]

39. (II) How many kWh of energy does a 550-W toaster use in the morning if it is in operation for a total of 6.0 min? At a cost of 9.0 cents/kWh, estimate how much this would add to your monthly electric energy bill if you made toast four mornings per week.

40. (II) At \$0.095/kWh, what does it cost to leave a 25-W porch light on day and night for a year?

41. (II) What is the total amount of energy stored in a 12-V, 75-A·h car battery when it is fully charged?

42. (II) An ordinary flashlight uses two D-cell 1.5-V batteries connected in series as in Fig. 25–4b (Fig. 25–37). The bulb draws 380 mA when turned on. (a) Calculate the resistance of the bulb and the power dissipated. (b) By what factor would the power increase if four D-cells in series were used with the same bulb? (Neglect heating effects of the filament.) Why shouldn't you try this?

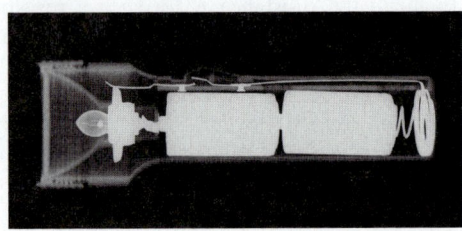

FIGURE 25–37 Problem 42.

43. (II) How many 75-W lightbulbs, connected to 120 V as in Fig. 25–20, can be used without blowing a 15-A fuse?

44. (II) An extension cord made of two wires of diameter 0.129 cm (no. 16 copper wire) and of length 2.7 m (9 ft) is connected to an electric heater which draws 15.0 A on a 120-V line. How much power is dissipated in the cord?

45. (II) A power station delivers 750 kW of power at 12,000 V to a factory through wires with total resistance 3.0 Ω. How much less power is wasted if the electricity is delivered at 50,000 V rather than 12,000 V?

46. (III) A small immersion heater can be used in a car to heat a cup of water for coffee or tea. If the heater can heat 120 mL of water from 25°C to 95°C in 8.0 min, (a) approximately how much current does it draw from the car's 12-V battery, and (b) what is its resistance? Assume the manufacturer's claim of 75% efficiency.

47. (III) The current in an electromagnet connected to a 240-V line is 17.5 A. At what rate must cooling water pass over the coils if the water temperature is to rise by no more than 6.50 C°?

48. (III) A 1.0-m-long round tungsten wire is to reach a temperature of 3100 K when a current of 15.0 A flows through it. What diameter should the wire be? Assume the wire loses energy only by radiation (emissivity ε = 1.0, Section 19–10) and the surrounding temperature is 20°C.

25–7 Alternating Current

49. (I) Calculate the peak current in a 2.7-kΩ resistor connected to a 220-V rms ac source.

50. (I) An ac voltage, whose peak value is 180 V, is across a 380-Ω resistor. What are the rms and peak currents in the resistor?

51. (II) Estimate the resistance of the 120-V_{rms} circuits in your house as seen by the power company, when (a) everything electrical is unplugged, and (b) there are two 75-W lightbulbs burning.

52. (II) The peak value of an alternating current in a 1500-W device is 5.4 A. What is the rms voltage across it?

53. (II) An 1800-W arc welder is connected to a 660-V_{rms} ac line. Calculate (a) the peak voltage and (b) the peak current.

54. (II) (a) What is the maximum instantaneous power dissipated by a 2.5-hp pump connected to a 240-V_{rms} ac power source? (b) What is the maximum current passing through the pump?

55. (II) A heater coil connected to a 240-V_{rms} ac line has a resistance of 44 Ω. (a) What is the average power used? (b) What are the maximum and minimum values of the instantaneous power?

56. (II) For a time-dependent voltage $V(t)$, which is periodic with period T, the rms voltage is defined to be $V_{rms} = \left[\frac{1}{T}\int_0^T V^2\,dt\right]^{\frac{1}{2}}$. Use this definition to determine V_{rms} (in terms of the peak voltage V_0) for (a) a sinusoidal voltage, i.e., $V(t) = V_0\sin(2\pi t/T)$ for $0 \le t \le T$; and (b) a positive square-wave voltage, i.e.,

$$V(t) = \begin{cases} V_0 & 0 \le t \le \dfrac{T}{2} \\ 0 & \dfrac{T}{2} \le t \le T \end{cases}.$$

25–8 Microscopic View of Electric Current

57. (II) A 0.65-mm-diameter copper wire carries a tiny current of 2.3 μA. Estimate (a) the electron drift velocity, (b) the current density, and (c) the electric field in the wire.

58. (II) A 5.80-m length of 2.0-mm-diameter wire carries a 750-mA current when 22.0 mV is applied to its ends. If the drift velocity is 1.7×10^{-5} m/s, determine (a) the resistance R of the wire, (b) the resistivity ρ, (c) the current density j, (d) the electric field inside the wire, and (e) the number n of free electrons per unit volume.

59. (II) At a point high in the Earth's atmosphere, He^{2+} ions in a concentration of $2.8 \times 10^{12}/m^3$ are moving due north at a speed of 2.0×10^6 m/s. Also, a $7.0 \times 10^{11}/m^3$ concentration of O_2^- ions is moving due south at a speed of 6.2×10^6 m/s. Determine the magnitude and direction of the current density $\vec{\mathbf{j}}$ at this point.

*25–10 Nerve Conduction

*60. (I) What is the magnitude of the electric field across an axon membrane 1.0×10^{-8} m thick if the resting potential is −70 mV?

*61. (II) A neuron is stimulated with an electric pulse. The action potential is detected at a point 3.40 cm down the axon 0.0052 s later. When the action potential is detected 7.20 cm from the point of stimulation, the time required is 0.0063 s. What is the speed of the electric pulse along the axon? (Why are two measurements needed instead of only one?)

*62. (III) During an action potential, Na^+ ions move into the cell at a rate of about 3×10^{-7} mol/m²·s. How much power must be produced by the "active Na^+ pumping" system to produce this flow against a +30-mV potential difference? Assume that the axon is 10 cm long and 20 μm in diameter.

General Problems

63. A person accidentally leaves a car with the lights on. If each of the two headlights uses 40 W and each of the two taillights 6 W, for a total of 92 W, how long will a fresh 12-V battery last if it is rated at 85 A·h? Assume the full 12 V appears across each bulb.

64. How many coulombs are there in 1.00 ampere-hour?

65. You want to design a portable electric blanket that runs on a 1.5-V battery. If you use copper wire with a 0.50-mm diameter as the heating element, how long should the wire be if you want to generate 15 W of heating power? What happens if you accidentally connect the blanket to a 9.0-V battery?

66. What is the average current drawn by a 1.0-hp 120-V motor? (1 hp = 746 W.)

67. The *conductance G* of an object is defined as the reciprocal of the resistance R; that is, $G = 1/R$. The unit of conductance is a *mho* (= ohm^{-1}), which is also called the *siemens* (S). What is the conductance (in siemens) of an object that draws 480 mA of current at 3.0 V?

68. The heating element of a 110-V, 1500-W heater is 3.5 m long. If it is made of iron, what must its diameter be?

69. (*a*) A particular household uses a 1.8-kW heater 2.0 h/day ("on" time), four 100-W lightbulbs 6.0 h/day, a 3.0-kW electric stove element for a total of 1.0 h/day, and miscellaneous power amounting to 2.0 kWh/day. If electricity costs $0.105 per kWh, what will be their monthly bill (30 d)? (*b*) How much coal (which produces 7500 kcal/kg) must be burned by a 35%-efficient power plant to provide the yearly needs of this household?

70. A small city requires about 15 MW of power. Suppose that instead of using high-voltage lines to supply the power, the power is delivered at 120 V. Assuming a two-wire line of 0.50-cm-diameter copper wire, estimate the cost of the energy lost to heat per hour per meter. Assume the cost of electricity is about 9.0 cents per kWh.

71. A 1400-W hair dryer is designed for 117 V. (*a*) What will be the percentage change in power output if the voltage drops to 105 V? Assume no change in resistance. (*b*) How would the actual change in resistivity with temperature affect your answer?

72. The wiring in a house must be thick enough so it does not become so hot as to start a fire. What diameter must a copper wire be if it is to carry a maximum current of 35 A and produce no more than 1.5 W of heat per meter of length?

73. Determine the resistance of the tungsten filament in a 75-W 120-V incandescent lightbulb (*a*) at its operating temperature of about 3000 K, (*b*) at room temperature.

74. Suppose a current is given by the equation $I = 1.80 \sin 210t$, where I is in amperes and t in seconds. (*a*) What is the frequency? (*b*) What is the rms value of the current? (*c*) If this is the current through a 24.0-Ω resistor, write the equation that describes the voltage as a function of time.

75. A microwave oven running at 65% efficiency delivers 950 W to the interior. Find (*a*) the power drawn from the source, and (*b*) the current drawn. Assume a source voltage of 120 V.

76. A 1.00-Ω wire is stretched uniformly to 1.20 times its original length. What is its resistance now?

77. 220 V is applied to two different conductors made of the same material. One conductor is twice as long and twice the diameter of the second. What is the ratio of the power transformed in the first relative to the second?

78. An electric heater is used to heat a room of volume 54 m³. Air is brought into the room at 5°C and is completely replaced twice per hour. Heat loss through the walls amounts to approximately 850 kcal/h. If the air is to be maintained at 20°C, what minimum wattage must the heater have? (The specific heat of air is about 0.17 kcal/kg·C°.)

79. A 2800-W oven is connected to a 240-V source. (*a*) What is the resistance of the oven? (*b*) How long will it take to bring 120 mL of 15°C water to 100°C assuming 75% efficiency? (*c*) How much will this cost at 11 cents/kWh?

80. A proposed electric vehicle makes use of storage batteries as its source of energy. Its mass is 1560 kg and it is powered by 24 batteries, each 12 V, 95 A·h. Assume that the car is driven on level roads at an average speed of 45 km/h, and the average friction force is 240 N. Assume 100% efficiency and neglect energy used for acceleration. No energy is consumed when the vehicle is stopped, since the engine doesn't need to idle. (*a*) Determine the horsepower required. (*b*) After approximately how many kilometers must the batteries be recharged?

81. A 12.5-Ω resistor is made from a coil of copper wire whose total mass is 15.5 g. What is the diameter of the wire, and how long is it?

82. A fish-tank heater is rated at 95 W when connected to 120 V. The heating element is a coil of Nichrome wire. When uncoiled, the wire has a total length of 3.8 m. What is the diameter of the wire?

83. A 100-W, 120-V lightbulb has a resistance of 12 Ω when cold (20°C) and 140 Ω when on (hot). Calculate its power consumption (*a*) at the instant it is turned on, and (*b*) after a few moments when it is hot.

84. In an automobile, the system voltage varies from about 12 V when the car is off to about 13.8 V when the car is on and the charging system is in operation, a difference of 15%. By what percentage does the power delivered to the headlights vary as the voltage changes from 12 V to 13.8 V? Assume the headlight resistance remains constant.

85. The Tevatron accelerator at Fermilab (Illinois) is designed to carry an 11-mA beam of protons traveling at very nearly the speed of light $(3.0 \times 10^8 \text{ m/s})$ around a ring 6300 m in circumference. How many protons are in the beam?

86. Lightbulb A is rated at 120 V and 40 W for household applications. Lightbulb B is rated at 12 V and 40 W for automotive applications. (*a*) What is the current through each bulb? (*b*) What is the resistance of each bulb? (*c*) In one hour, how much charge passes through each bulb? (*d*) In one hour, how much energy does each bulb use? (*e*) Which bulb requires larger diameter wires to connect its power source and the bulb?

87. An air conditioner draws 14 A at 220-V ac. The connecting cord is copper wire with a diameter of 1.628 mm. (*a*) How much power does the air conditioner draw? (*b*) If the total length of wire is 15 m, how much power is dissipated in the wiring? (*c*) If no. 12 wire, with a diameter of 2.053 mm, was used instead, how much power would be dissipated in the wiring? (*d*) Assuming that the air conditioner is run 12 h per day, how much money per month (30 days) would be saved by using no. 12 wire? Assume that the cost of electricity is 12 cents per kWh.

88. Copper wire of diameter 0.259 cm is used to connect a set of appliances at 120 V, which draw 1750 W of power total. (a) What power is wasted in 25.0 m of this wire? (b) What is your answer if wire of diameter 0.412 cm is used?

89. Battery-powered electricity is very expensive compared with that available from a wall receptacle. Estimate the cost per kWh of (a) an alkaline D-cell (cost $1.70) and (b) an alkaline AA-cell (cost $1.25). These batteries can provide a continuous current of 25 mA for 820 h and 120 h, respectively, at 1.5 V. Compare to normal 120-V ac house current at $0.10/kWh.

90. How far does an average electron move along the wires of a 550-W toaster during an alternating current cycle? The power cord has copper wires of diameter 1.7 mm and is plugged into a standard 60-Hz 120-V ac outlet. [Hint: The maximum current in the cycle is related to the maximum drift velocity. The maximum velocity in an oscillation is related to the maximum displacement; see Chapter 14.]

91. A copper pipe has an inside diameter of 3.00 cm and an outside diameter of 5.00 cm (Fig. 25–38). What is the resistance of a 10.0-m length of this pipe?

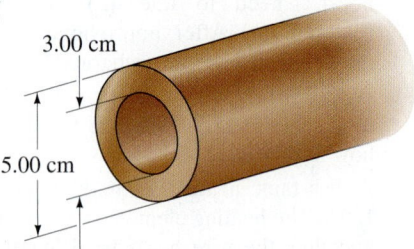

3.00 cm

5.00 cm

FIGURE 25–38
Problem 91.

92. For the wire in Fig. 25–39, whose diameter varies uniformly from a to b as shown, suppose a current $I = 2.0$ A enters at a. If $a = 2.5$ mm and $b = 4.0$ mm, what is the current density (assume uniform) at each end?

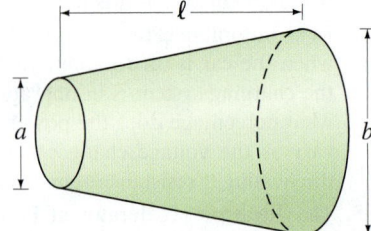

ℓ

a

b

FIGURE 25–39
Problems 92 and 93.

93. The cross section of a portion of wire increases uniformly as shown in Fig. 25–39 so it has the shape of a truncated cone. The diameter at one end is a and at the other it is b, and the total length along the axis is ℓ. If the material has resistivity ρ, determine the resistance R between the two ends in terms of $a, b, \ell,$ and ρ. Assume that the current flows uniformly through each section, and that the taper is small, i.e., $(b - a) \ll \ell$.

94. A tungsten filament used in a flashlight bulb operates at 0.20 A and 3.2 V. If its resistance at 20°C is 1.5 Ω, what is the temperature of the filament when the flashlight is on?

95. The level of liquid helium (temperature ≤ 4 K) in its storage tank can be monitored using a vertically aligned niobium–titanium (NbTi) wire, whose length ℓ spans the height of the tank. In this level-sensing setup, an electronic circuit maintains a constant electrical current I at all times in the NbTi wire and a voltmeter monitors the voltage difference V across this wire. Since the superconducting transition temperature for NbTi is 10 K, the portion of the wire immersed in the liquid helium is in the superconducting state, while the portion above the liquid (in helium vapor with temperature above 10 K) is in the normal state. Define $f = x/\ell$ to be the fraction of the tank filled with liquid helium (Fig. 25–40) and V_0 to be the value of V when the tank is empty ($f = 0$). Determine the relation between f and V (in terms of V_0).

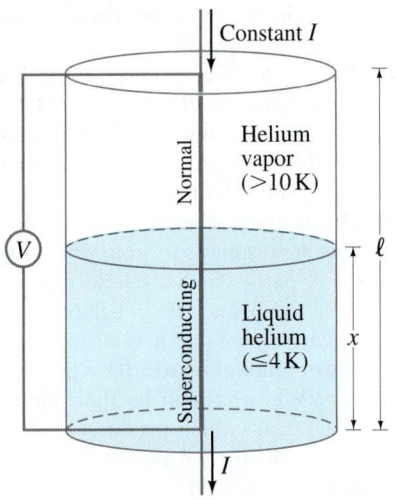

Constant I

Helium vapor (>10 K)

Liquid helium (≤4 K)

Normal

Superconducting

ℓ

x

I

FIGURE 25–40
Problem 95.

*Numerical/Computer

*96. (II) The resistance, R, of a particular thermistor as a function of temperature T is shown in this Table:

T (°C)	R (Ω)	T (°C)	R (Ω)
20	126,740	36	60,743
22	115,190	38	55,658
24	104,800	40	51,048
26	95,447	42	46,863
28	87,022	44	43,602
30	79,422	46	39,605
32	72,560	48	36,458
34	66,356	50	33,591

Determine what type of best-fit equation (linear, quadratic, exponential, other) describes the variation of R with T. The resistance of the thermistor is 57,641 Ω when embedded in a substance whose temperature is unknown. Based on your equation, what is the unknown temperature?

Answers to Exercises

A: 1.6×10^{-13} A.

B: (c).

C: (b), (c).

D: (c).

E: 110 m.

F: (d).

G: (e).

H: (a) 28 V, 3.5 A; (b) 2.8 V, 0.35 A.

These MP3 players contain circuits that are dc, at least in part. (The audio signal is ac.) The circuit diagram below shows a possible amplifier circuit for each stereo channel. We have already met two of the circuit elements shown: resistors and capacitors, and we discuss them in circuits in this Chapter. (The large triangle is an amplifier chip containing transistors.) We also discuss voltmeters and ammeters, and how they are built and used to make measurements.

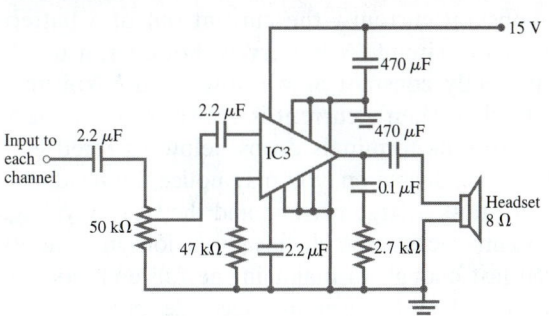

DC Circuits

CONTENTS

26–1 EMF and Terminal Voltage

26–2 Resistors in Series and in Parallel

26–3 Kirchhoff's Rules

26–4 Series and Parallel EMFs; Battery Charging

26–5 Circuits Containing Resistor and Capacitor (*RC* Circuits)

26–6 Electric Hazards

*26–7 Ammeters and Voltmeters

CHAPTER-OPENING QUESTION—Guess now!

The automobile headlight bulbs shown in the circuits here are identical. The connection which produces more light is

(a) circuit 1.

(b) circuit 2.

(c) both the same.

(d) not enough information.

Circuit 1   Circuit 2

Electric circuits are basic parts of all electronic devices from radio and TV sets to computers and automobiles. Scientific measurements, from physics to biology and medicine, make use of electric circuits. In Chapter 25, we discussed the basic principles of electric current. Now we will apply these principles to analyze dc circuits involving combinations of batteries, resistors, and capacitors. We also study the operation of some useful instruments.[†]

[†] AC circuits that contain only a voltage source and resistors can be analyzed like the dc circuits in this Chapter. However, ac circuits that contain capacitors and other circuit elements are more complicated, and we discuss them in Chapter 30.

TABLE 26–1 Symbols for Circuit Elements

Symbol	Device
⊣⊢	Battery
⊣⊢ or ⊣⊦	Capacitor
-⋀⋀-	Resistor
——	Wire with negligible resistance
⌒—	Switch
⏚ or ⏚	Ground

When we draw a diagram for a circuit, we represent batteries, capacitors, and resistors by the symbols shown in Table 26–1. Wires whose resistance is negligible compared with other resistance in the circuit are drawn simply as straight lines. Some circuit diagrams show a ground symbol (⏚ or ⏚) which may mean a real connection to the ground, perhaps via a metal pipe, or it may simply mean a common connection, such as the frame of a car.

For the most part in this Chapter, except in Section 26–5 on *RC* circuits, we will be interested in circuits operating in their steady state. That is, we won't be looking at a circuit at the moment a change is made in it, such as when a battery or resistor is connected or disconnected, but rather later when the currents have reached their steady values.

26–1 EMF and Terminal Voltage

To have current in an electric circuit, we need a device such as a battery or an electric generator that transforms one type of energy (chemical, mechanical, or light, for example) into electric energy. Such a device is called a **source** of **electromotive force** or of **emf**. (The term "electromotive force" is a misnomer since it does not refer to a "force" that is measured in newtons. Hence, to avoid confusion, we prefer to use the abbreviation, emf.) The *potential difference* between the terminals of such a source, when no current flows to an external circuit, is called the **emf** of the source. The symbol $\mathscr{E}$ is usually used for emf (don't confuse it with E for electric field), and its unit is volts.

A battery is not a source of constant current—the current out of a battery varies according to the resistance in the circuit. A battery *is*, however, a nearly constant voltage source, but not perfectly constant as we now discuss. You may have noticed in your own experience that when a current is drawn from a battery, the potential difference (voltage) across its terminals drops below its rated emf. For example, if you start a car with the headlights on, you may notice the headlights dim. This happens because the starter draws a large current, and the battery voltage drops as a result. The voltage drop occurs because the chemical reactions in a battery (Section 25–1) cannot supply charge fast enough to maintain the full emf. For one thing, charge must move (within the electrolyte) between the electrodes of the battery, and there is always some hindrance to completely free flow. Thus, a battery itself has some resistance, which is called its **internal resistance**; it is usually designated r.

A real battery is modeled as if it were a perfect emf $\mathscr{E}$ in series with a resistor r, as shown in Fig. 26–1. Since this resistance r is inside the battery, we can never separate it from the battery. The two points a and b in the diagram represent the two terminals of the battery. What we measure is the **terminal voltage** $V_{ab} = V_a - V_b$. When no current is drawn from the battery, the terminal voltage equals the emf, which is determined by the chemical reactions in the battery: $V_{ab} = \mathscr{E}$. However, when a current I flows naturally from the battery there is an internal drop in voltage equal to Ir. Thus the terminal voltage (the actual voltage) is[†]

$$V_{ab} = \mathscr{E} - Ir. \qquad (26-1)$$

For example, if a 12-V battery has an internal resistance of 0.1 Ω, then when 10 A flows from the battery, the terminal voltage is 12 V − (10 A)(0.1 Ω) = 11 V. The internal resistance of a battery is usually small. For example, an ordinary flashlight battery when fresh may have an internal resistance of perhaps 0.05 Ω. (However, as it ages and the electrolyte dries out, the internal resistance increases to many ohms.) Car batteries have lower internal resistance.

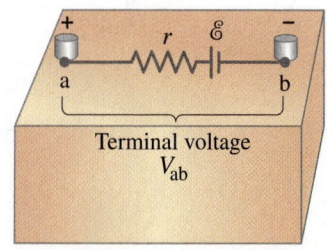

FIGURE 26–1 Diagram for an electric cell or battery.

FIGURE 26–2 Example 26–1.

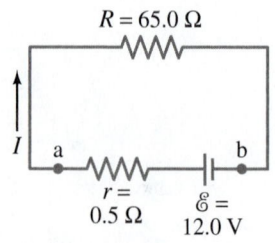

EXAMPLE 26–1 **Battery with internal resistance.** A 65.0-Ω resistor is connected to the terminals of a battery whose emf is 12.0 V and whose internal resistance is 0.5 Ω, Fig. 26–2. Calculate (*a*) the current in the circuit, (*b*) the terminal voltage of the battery, V_{ab}, and (*c*) the power dissipated in the resistor R and in the battery's internal resistance r.

APPROACH We first consider the battery as a whole, which is shown in Fig. 26–2 as an emf $\mathscr{E}$ and internal resistance r between points a and b. Then we apply $V = IR$ to the circuit itself.

[†]When a battery is being charged, a current is forced to pass through it; we then have to write
$$V_{ab} = \mathscr{E} + Ir.$$
See Section 26–4 or Problem 28 and Fig. 26–46.

SOLUTION (a) From Eq. 26–1, we have
$$V_{ab} = \mathscr{E} - Ir.$$
We apply Ohm's law (Eqs. 25–2) to this battery and the resistance R of the circuit: $V_{ab} = IR$. Hence $IR = \mathscr{E} - Ir$ or $\mathscr{E} = I(R + r)$, and so
$$I = \frac{\mathscr{E}}{R + r} = \frac{12.0 \text{ V}}{65.0\,\Omega + 0.5\,\Omega} = \frac{12.0 \text{ V}}{65.5\,\Omega} = 0.183 \text{ A}.$$
(b) The terminal voltage is
$$V_{ab} = \mathscr{E} - Ir = 12.0 \text{ V} - (0.183 \text{ A})(0.5\,\Omega) = 11.9 \text{ V}.$$
(c) The power dissipated (Eq. 25–7) in R is
$$P_R = I^2R = (0.183 \text{ A})^2(65.0\,\Omega) = 2.18 \text{ W},$$
and in r is
$$P_r = I^2r = (0.183 \text{ A})^2(0.5\,\Omega) = 0.02 \text{ W}.$$

EXERCISE A Repeat Example 26–1 assuming now that the resistance $R = 10.0\,\Omega$, whereas $\mathscr{E}$ and r remain as before.

In much of what follows, unless stated otherwise, we assume that the battery's internal resistance is negligible, and that the battery voltage given is its terminal voltage, which we will usually write simply as V rather than V_{ab}. Be careful not to confuse V (italic) for voltage and V (not italic) for the volt unit.

26–2 Resistors in Series and in Parallel

When two or more resistors are connected end to end along a single path as shown in Fig. 26–3a, they are said to be connected in **series**. The resistors could be simple resistors as were pictured in Fig. 25–12, or they could be lightbulbs (Fig. 26–3b), or heating elements, or other resistive devices. Any charge that passes through R_1 in Fig. 26–3a will also pass through R_2 and then R_3. Hence the same current I passes through each resistor. (If it did not, this would imply that either charge was not conserved, or that charge was accumulating at some point in the circuit, which does not happen in the steady state.)

We let V represent the potential difference (voltage) across all three resistors in Fig. 26–3a. We assume all other resistance in the circuit can be ignored, so V equals the terminal voltage supplied by the battery. We let V_1, V_2, and V_3 be the potential differences across each of the resistors, R_1, R_2, and R_3, respectively. From Ohm's law, $V = IR$, we can write $V_1 = IR_1$, $V_2 = IR_2$, and $V_3 = IR_3$. Because the resistors are connected end to end, energy conservation tells us that the total voltage V is equal to the sum of the voltages[†] across each resistor:
$$V = V_1 + V_2 + V_3 = IR_1 + IR_2 + IR_3. \quad \text{[series]} \quad (26\text{–}2)$$

Now let us determine the equivalent single resistance R_{eq} that would draw the same current I as our combination of three resistors in series; see Fig. 26–3c. Such a single resistance R_{eq} would be related to V by
$$V = IR_{eq}.$$
We equate this expression with Eq. 26–2, $V = I(R_1 + R_2 + R_3)$, and find
$$R_{eq} = R_1 + R_2 + R_3. \quad \text{[series]} \quad (26\text{–}3)$$

This is, in fact, what we expect. When we put several resistances in series, the total or equivalent resistance is the sum of the separate resistances. (Sometimes we may also call it the "net resistance.") This sum applies to any number of resistances in series. Note that when you add more resistance to the circuit, the current through the circuit will decrease. For example, if a 12-V battery is connected to a 4-Ω resistor, the current will be 3 A. But if the 12-V battery is connected to three 4-Ω resistors in series, the total resistance is 12 Ω and the current through the entire circuit will be only 1 A.

[†]To see in more detail why this is true, note that an electric charge q passing through R_1 loses an amount of potential energy equal to qV_1. In passing through R_2 and R_3, the potential energy U decreases by qV_2 and qV_3, for a total $\Delta U = qV_1 + qV_2 + qV_3$; this sum must equal the energy given to q by the battery, qV, so that energy is conserved. Hence $qV = q(V_1 + V_2 + V_3)$, and so $V = V_1 + V_2 + V_3$, which is Eq. 26–2.

FIGURE 26–3 (a) Resistances connected in series. (b) Resistances could be lightbulbs, or any other type of resistance. (c) Equivalent single resistance R_{eq} that draws the same current: $R_{eq} = R_1 + R_2 + R_3$.

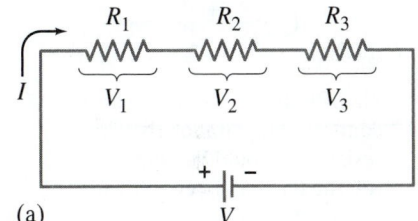

(a)

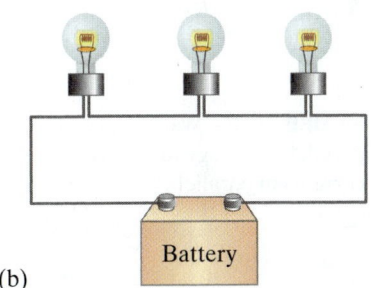

(b)

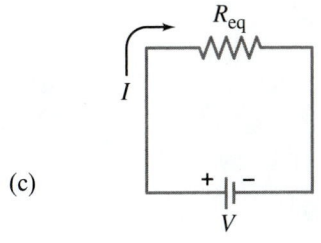

(c)

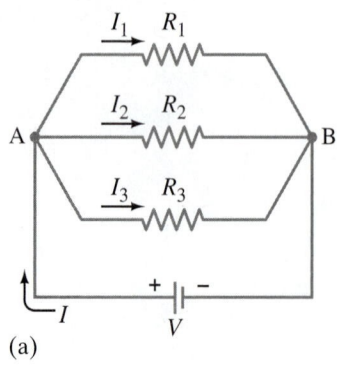

(a)

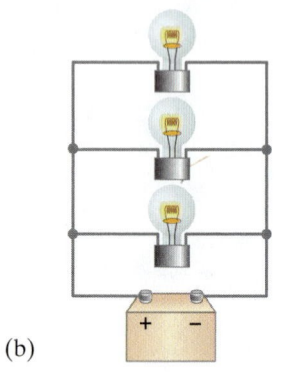

(b)

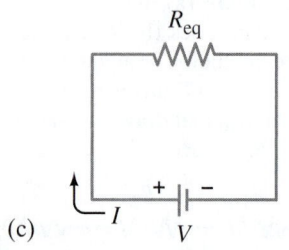

(c)

FIGURE 26–4 (a) Resistances connected in parallel. (b) The resistances could be lightbulbs. (c) The equivalent circuit with R_{eq} obtained from Eq. 26–4:
$$\frac{1}{R_{eq}} = \frac{1}{R_1} + \frac{1}{R_2} + \frac{1}{R_3}.$$

FIGURE 26–5 Water pipes in parallel—analogy to electric currents in parallel.

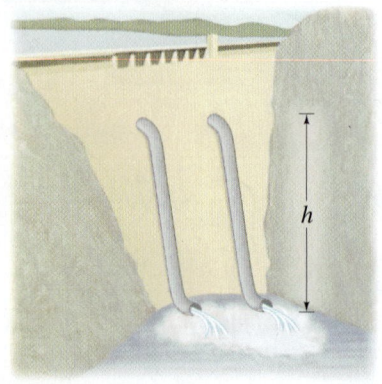

Another simple way to connect resistors is in **parallel** so that the current from the source splits into separate branches or paths, as shown in Fig. 26–4a and b. The wiring in houses and buildings is arranged so all electric devices are in parallel, as we already saw in Chapter 25, Fig. 25–20. With parallel wiring, if you disconnect one device (say, R_1 in Fig. 26–4a), the current to the other devices is not interrupted. Compare to a series circuit, where if one device (say, R_1 in Fig. 26–3a) is disconnected, the current *is* stopped to all the others.

In a parallel circuit, Fig. 26–4a, the total current I that leaves the battery splits into three separate paths. We let I_1, I_2, and I_3 be the currents through each of the resistors, R_1, R_2, and R_3, respectively. Because *electric charge is conserved*, the current I flowing into junction A (where the different wires or conductors meet, Fig. 26–4a) must equal the current flowing out of the junction. Thus

$$I = I_1 + I_2 + I_3. \qquad \text{[parallel]}$$

When resistors are connected in parallel, each has the same voltage across it. (Indeed, any two points in a circuit connected by a wire of negligible resistance are at the same potential.) Hence the full voltage of the battery is applied to each resistor in Fig. 26–4a. Applying Ohm's law to each resistor, we have

$$I_1 = \frac{V}{R_1}, \qquad I_2 = \frac{V}{R_2}, \qquad \text{and} \qquad I_3 = \frac{V}{R_3}.$$

Let us now determine what single resistor R_{eq} (Fig. 26–4c) will draw the same current I as these three resistances in parallel. This equivalent resistance R_{eq} must satisfy Ohm's law too:

$$I = \frac{V}{R_{eq}}.$$

We now combine the equations above:

$$I = I_1 + I_2 + I_3,$$
$$\frac{V}{R_{eq}} = \frac{V}{R_1} + \frac{V}{R_2} + \frac{V}{R_3}.$$

When we divide out the V from each term, we have

$$\frac{1}{R_{eq}} = \frac{1}{R_1} + \frac{1}{R_2} + \frac{1}{R_3}. \qquad \text{[parallel]} \quad \textbf{(26–4)}$$

For example, suppose you connect two 4-Ω loudspeakers to a single set of output terminals of your stereo amplifier or receiver. (Ignore the other channel for a moment—our two speakers are both connected to the left channel, say.) The equivalent resistance of the two 4-Ω "resistors" in parallel is

$$\frac{1}{R_{eq}} = \frac{1}{4\,\Omega} + \frac{1}{4\,\Omega} = \frac{2}{4\,\Omega} = \frac{1}{2\,\Omega},$$

and so $R_{eq} = 2\,\Omega$. Thus the net (or equivalent) resistance is *less* than each single resistance. This may at first seem surprising. But remember that when you connect resistors in parallel, you are giving the current additional paths to follow. Hence the net resistance will be less.

Equations 26–3 and 26–4 make good sense. Recalling Eq. 25–3 for resistivity, $R = \rho\ell/A$, we see that placing resistors in series increases the length and therefore the resistance; putting resistors in parallel increases the area through which current flows, thus reducing the overall resistance.

An analogy may help here. Consider two identical pipes taking in water near the top of a dam and releasing it below as shown in Fig. 26–5. The gravitational potential difference, proportional to the height h, is the same for both pipes, just as the voltage is the same for parallel resistors. If both pipes are open, rather than only one, twice as much water will flow through. That is, with two equal pipes open, the net resistance to the flow of water will be reduced, by half, just as for electrical resistors in parallel. Note that if both pipes are closed, the dam offers infinite resistance to the flow of water. This corresponds in the electrical case to an open circuit—when the path is not continuous and no current flows—so the electrical resistance is infinite.

EXERCISE B You have a 10-Ω and a 15-Ω resistor. What is the smallest and largest equivalent resistance that you can make with these two resistors?

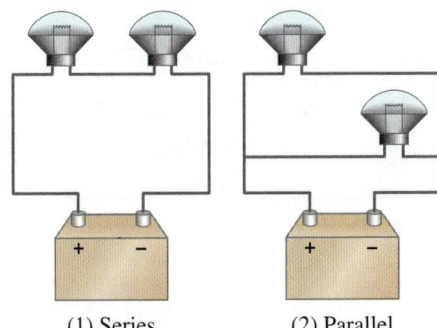

CONCEPTUAL EXAMPLE 26–2 **Series or parallel?** (*a*) The lightbulbs in Fig. 26–6 are identical. Which configuration produces more light? (*b*) Which way do you think the headlights of a car are wired? Ignore change of filament resistance *R* with current.

RESPONSE (*a*) The equivalent resistance of the parallel circuit is found from Eq. 26–4, $1/R_{eq} = 1/R + 1/R = 2/R$. Thus $R_{eq} = R/2$. The parallel combination then has lower resistance ($= R/2$) than the series combination $\left(R_{eq} = R + R = 2R\right)$. There will be more total current in the parallel configuration (2), since $I = V/R_{eq}$ and *V* is the same for both circuits. The total power transformed, which is related to the light produced, is $P = IV$, so the greater current in (2) means more light produced.
(*b*) Headlights are wired in parallel (2), because if one bulb goes out, the other bulb can stay lit. If they were in series (1), when one bulb burned out (the filament broke), the circuit would be open and no current would flow, so neither bulb would light.

NOTE When you answered the Chapter-Opening Question on page 677, was your answer circuit 2? Can you express any misconceptions you might have had?

(1) Series   (2) Parallel

FIGURE 26–6 Example 26–2.

FIGURE 26–7 Example 26–3.

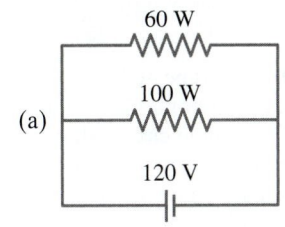

(a)

CONCEPTUAL EXAMPLE 26–3 **An illuminating surprise.** A 100-W, 120-V lightbulb and a 60-W, 120-V lightbulb are connected in two different ways as shown in Fig. 26–7. In each case, which bulb glows more brightly? Ignore change of filament resistance with current (and temperature).

RESPONSE (*a*) These are normal lightbulbs with their power rating given for 120 V. They both receive 120 V, so the 100-W bulb is naturally brighter.
(*b*) The resistance of the 100-W bulb is less than that of the 60-W bulb (calculated from $P = V^2/R$ at constant 120 V). Here they are connected in series and receive the same current. Hence, from $P = I^2R$ ($I =$ constant) the higher-resistance "60-W" bulb will transform more power and thus be brighter.

NOTE When connected in series as in (b), the two bulbs do *not* dissipate 60 W and 100 W because neither bulb receives 120 V.

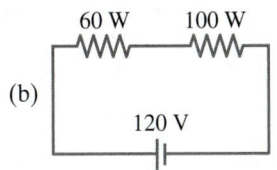

(b)

Note that whenever a group of resistors is replaced by the equivalent resistance, current and voltage and power in the rest of the circuit are unaffected.

EXAMPLE 26–4 **Circuit with series and parallel resistors.** How much current is drawn from the battery shown in Fig. 26–8a?

APPROACH The current *I* that flows out of the battery all passes through the 400-Ω resistor, but then it splits into I_1 and I_2 passing through the 500-Ω and 700-Ω resistors. The latter two resistors are in parallel with each other. We look for something that we already know how to treat. So let's start by finding the equivalent resistance, R_p, of the parallel resistors, 500 Ω and 700 Ω. Then we can consider this R_p to be in series with the 400-Ω resistor.

SOLUTION The equivalent resistance, R_p, of the 500-Ω and 700-Ω resistors in parallel is given by

$$\frac{1}{R_P} = \frac{1}{500 \ \Omega} + \frac{1}{700 \ \Omega} = 0.0020 \ \Omega^{-1} + 0.0014 \ \Omega^{-1} = 0.0034 \ \Omega^{-1}.$$

This is $1/R_P$, so we take the reciprocal to find R_P. It is a common mistake to forget to take this reciprocal. Notice that the units of reciprocal ohms, Ω^{-1}, are a reminder. Thus

$$R_P = \frac{1}{0.0034 \ \Omega^{-1}} = 290 \ \Omega.$$

This 290 Ω is the equivalent resistance of the two parallel resistors, and is in series with the 400-Ω resistor as shown in the equivalent circuit of Fig. 26–8b. To find the total equivalent resistance R_{eq}, we add the 400-Ω and 290-Ω resistances together, since they are in series, and find

$$R_{eq} = 400 \ \Omega + 290 \ \Omega = 690 \ \Omega.$$

The total current flowing from the battery is then

$$I = \frac{V}{R_{eq}} = \frac{12.0 \ V}{690 \ \Omega} = 0.0174 \ A \approx 17 \ mA.$$

NOTE This *I* is also the current flowing through the 400-Ω resistor, but not through the 500-Ω and 700-Ω resistors (both currents are less—see the next Example).

NOTE Complex resistor circuits can often be analyzed in this way, considering the circuit as a combination of series and parallel resistances.

FIGURE 26–8 (a) Circuit for Examples 26–4 and 26–5. (b) Equivalent circuit, showing the equivalent resistance of 290 Ω for the two parallel resistors in (a).

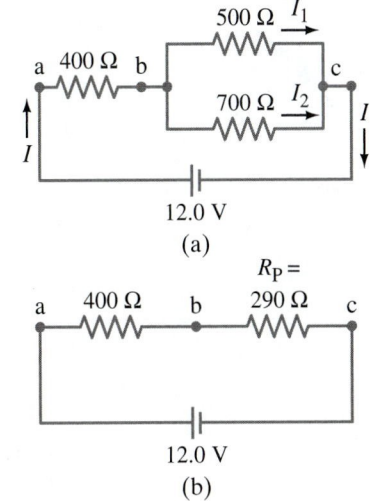

(a)

(b)

⚠ **CAUTION**

Remember to take the reciprocal

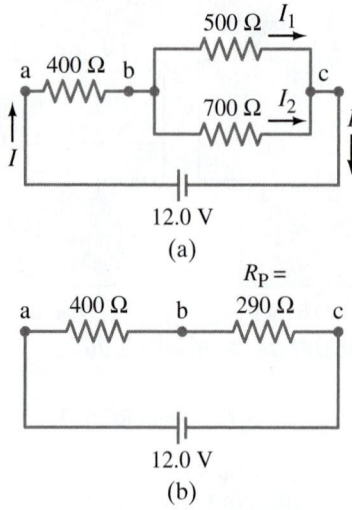

FIGURE 26–8 (repeated)
(a) Circuit for Examples 26–4 and 26–5. (b) Equivalent circuit, showing the equivalent resistance of 290 Ω for the two parallel resistors in (a).

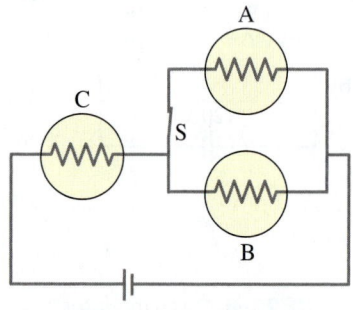

FIGURE 26–9 Example 26–6, three identical lightbulbs. Each yellow circle with -WW- inside represents a lightbulb and its resistance.

EXAMPLE 26–5 Current in one branch. What is the current through the 500-Ω resistor in Fig. 26–8a?

APPROACH We need to find the voltage across the 500-Ω resistor, which is the voltage between points b and c in Fig. 26–8a, and we call it V_{bc}. Once V_{bc} is known, we can apply Ohm's law, $V = IR$, to get the current. First we find the voltage across the 400-Ω resistor, V_{ab}, since we know that 17.4 mA passes through it (Example 26–4).

SOLUTION V_{ab} can be found using $V = IR$:
$$V_{ab} = (0.0174 \text{ A})(400 \, \Omega) = 7.0 \text{ V}.$$
Since the total voltage across the network of resistors is $V_{ac} = 12.0$ V, then V_{bc} must be $12.0 \text{ V} - 7.0 \text{ V} = 5.0 \text{ V}$. Then Ohm's law applied to the 500-Ω resistor tells us that the current I_1 through that resistor is
$$I_1 = \frac{5.0 \text{ V}}{500 \, \Omega} = 1.0 \times 10^{-2} \text{ A} = 10 \text{ mA}.$$
This is the answer we wanted. We can also calculate the current I_2 through the 700-Ω resistor since the voltage across it is also 5.0 V:
$$I_2 = \frac{5.0 \text{ V}}{700 \, \Omega} = 7 \text{ mA}.$$

NOTE When I_1 combines with I_2 to form the total current I (at point c in Fig. 26–8a), their sum is $10 \text{ mA} + 7 \text{ mA} = 17 \text{ mA}$. This equals the total current I as calculated in Example 26–4, as it should.

CONCEPTUAL EXAMPLE 26–6 Bulb brightness in a circuit. The circuit shown in Fig. 26–9 has three identical lightbulbs, each of resistance R. (a) When switch S is closed, how will the brightness of bulbs A and B compare with that of bulb C? (b) What happens when switch S is opened? Use a minimum of mathematics in your answers.

RESPONSE (a) With switch S closed, the current that passes through bulb C must split into two equal parts when it reaches the junction leading to bulbs A and B. It splits into equal parts because the resistance of bulb A equals that of B. Thus, bulbs A and B each receive half of C's current; A and B will be equally bright, but they will be less bright than bulb C ($P = I^2R$). (b) When the switch S is open, no current can flow through bulb A, so it will be dark. We now have a simple one-loop series circuit, and we expect bulbs B and C to be equally bright. However, the equivalent resistance of this circuit ($= R + R$) is greater than that of the circuit with the switch closed. When we open the switch, we increase the resistance and reduce the current leaving the battery. Thus, bulb C will be dimmer when we open the switch. Bulb B gets more current when the switch is open (you may have to use some mathematics here), and so it will be brighter than with the switch closed; and B will be as bright as C.

EXAMPLE 26–7 ESTIMATE A two-speed fan. One way a multiple-speed ventilation fan for a car can be designed is to put resistors in series with the fan motor. The resistors reduce the current through the motor and make it run more slowly. Suppose the current in the motor is 5.0 A when it is connected directly across a 12-V battery. (a) What series resistor should be used to reduce the current to 2.0 A for low-speed operation? (b) What power rating should the resistor have?

APPROACH An electric motor in series with a resistor can be treated as two resistors in series. The power comes from $P = IV$.

SOLUTION (a) When the motor is connected to 12 V and drawing 5.0 A, its resistance is $R = V/I = (12 \text{ V})/(5.0 \text{ A}) = 2.4 \, \Omega$. We will assume that this is the motor's resistance for all speeds. (This is an approximation because the current through the motor depends on its speed.) Then, when a current of 2.0 A is flowing, the voltage across the motor is $(2.0 \text{ A})(2.4 \, \Omega) = 4.8 \text{ V}$. The remaining $12.0 \text{ V} - 4.8 \text{ V} = 7.2 \text{ V}$ must appear across the series resistor. When 2.0 A flows through the resistor, its resistance must be $R = (7.2 \text{ V})/(2.0 \text{ A}) = 3.6 \, \Omega$. (b) The power dissipated by the resistor is $P = (7.2 \text{ V})(2.0 \text{ A}) = 14.4 \text{ W}$. To be safe, a power rating of 20 W would be appropriate.

EXAMPLE 26–8 **Analyzing a circuit.** A 9.0-V battery whose internal resistance r is $0.50\,\Omega$ is connected in the circuit shown in Fig. 26–10a. (a) How much current is drawn from the battery? (b) What is the terminal voltage of the battery? (c) What is the current in the 6.0-Ω resistor?

APPROACH To find the current out of the battery, we first need to determine the equivalent resistance R_{eq} of the entire circuit, including r, which we do by identifying and isolating simple series or parallel combinations of resistors. Once we find I from Ohm's law, $I = \mathscr{E}/R_{eq}$, we get the terminal voltage using $V_{ab} = \mathscr{E} - Ir$. For (c) we apply Ohm's law to the 6.0-Ω resistor.

SOLUTION (a) We want to determine the equivalent resistance of the circuit. But where do we start? We note that the 4.0-Ω and 8.0-Ω resistors are in parallel, and so have an equivalent resistance R_{eq1} given by

$$\frac{1}{R_{eq1}} = \frac{1}{8.0\,\Omega} + \frac{1}{4.0\,\Omega} = \frac{3}{8.0\,\Omega};$$

so $R_{eq1} = 2.7\,\Omega$. This $2.7\,\Omega$ is in series with the 6.0-Ω resistor, as shown in the equivalent circuit of Fig. 26–10b. The net resistance of the lower arm of the circuit is then

$$R_{eq2} = 6.0\,\Omega + 2.7\,\Omega = 8.7\,\Omega,$$

as shown in Fig. 26–10c. The equivalent resistance R_{eq3} of the 8.7-Ω and 10.0-Ω resistances in parallel is given by

$$\frac{1}{R_{eq3}} = \frac{1}{10.0\,\Omega} + \frac{1}{8.7\,\Omega} = 0.21\,\Omega^{-1},$$

so $R_{eq3} = (1/0.21\,\Omega^{-1}) = 4.8\,\Omega$. This $4.8\,\Omega$ is in series with the 5.0-Ω resistor and the 0.50-Ω internal resistance of the battery (Fig. 26–10d), so the total equivalent resistance R_{eq} of the circuit is $R_{eq} = 4.8\,\Omega + 5.0\,\Omega + 0.50\,\Omega = 10.3\,\Omega$. Hence the current drawn is

$$I = \frac{\mathscr{E}}{R_{eq}} = \frac{9.0\,\text{V}}{10.3\,\Omega} = 0.87\,\text{A}.$$

(b) The terminal voltage of the battery is

$$V_{ab} = \mathscr{E} - Ir = 9.0\,\text{V} - (0.87\,\text{A})(0.50\,\Omega) = 8.6\,\text{V}.$$

(c) Now we can work back and get the current in the 6.0-Ω resistor. It must be the same as the current through the $8.7\,\Omega$ shown in Fig. 26–10c (why?). The voltage across that $8.7\,\Omega$ will be the emf of the battery minus the voltage drops across r and the 5.0-Ω resistor: $V_{8.7} = 9.0\,\text{V} - (0.87\,\text{A})(0.50\,\Omega + 5.0\,\Omega)$. Applying Ohm's law, we get the current (call it I')

$$I' = \frac{9.0\,\text{V} - (0.87\,\text{A})(0.50\,\Omega + 5.0\,\Omega)}{8.7\,\Omega} = 0.48\,\text{A}.$$

This is the current through the 6.0-Ω resistor.

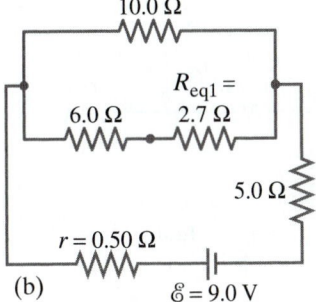

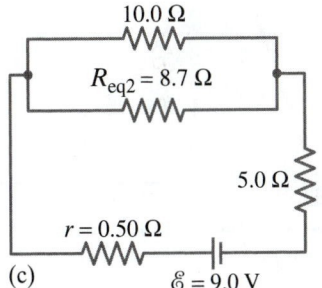

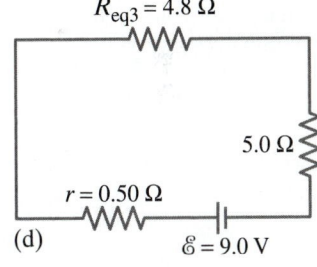

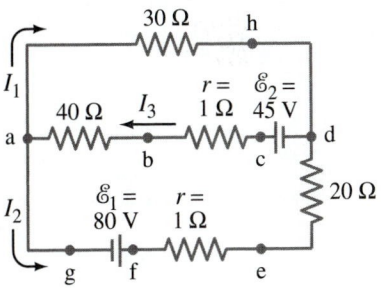

FIGURE 26–10 Circuit for Example 26–8, where r is the internal resistance of the battery.

26–3 Kirchhoff's Rules

In the last few Examples we have been able to find the currents in circuits by combining resistances in series and parallel, and using Ohm's law. This technique can be used for many circuits. However, some circuits are too complicated for that analysis. For example, we cannot find the currents in each part of the circuit shown in Fig. 26–11 simply by combining resistances as we did before.

To deal with such complicated circuits, we use Kirchhoff's rules, devised by G. R. Kirchhoff (1824–1887) in the mid-nineteenth century. There are two rules, and they are simply convenient applications of the laws of conservation of charge and energy.

FIGURE 26–11 Currents can be calculated using Kirchhoff's rules.

Kirchhoff's first rule or **junction rule** is based on the conservation of electric charge that we already used to derive the rule for parallel resistors. It states that

at any junction point, the sum of all currents entering the junction must equal the sum of all currents leaving the junction.

That is, whatever charge goes in must come out. We already saw an instance of this in the NOTE at the end of Example 26–5.

Kirchhoff's second rule or **loop rule** is based on the conservation of energy. It states that

the sum of the changes in potential around any closed loop of a circuit must be zero.

To see why this rule should hold, consider a rough analogy with the potential energy of a roller coaster on its track. When the roller coaster starts from the station, it has a particular potential energy. As it climbs the first hill, its potential energy increases and reaches a maximum at the top. As it descends the other side, its potential energy decreases and reaches a local minimum at the bottom of the hill. As the roller coaster continues on its path, its potential energy goes through more changes. But when it arrives back at the starting point, it has exactly as much potential energy as it had when it started at this point. Another way of saying this is that there was as much uphill as there was downhill.

Similar reasoning can be applied to an electric circuit. We will analyze the circuit of Fig. 26–11 shortly but first we consider the simpler circuit in Fig. 26–12. We have chosen it to be the same as the equivalent circuit of Fig. 26–8b already discussed. The current in this circuit is $I = (12.0\,\text{V})/(690\,\Omega) = 0.0174\,\text{A}$, as we calculated in Example 26–4. (We keep an extra digit in I to reduce rounding errors.) The positive side of the battery, point e in Fig. 26–12a, is at a high potential compared to point d at the negative side of the battery. That is, point e is like the top of a hill for a roller coaster. We follow the current around the circuit starting at any point. We choose to start at point d and follow a positive test charge completely around this circuit. As we go, we note all changes in potential. When the test charge returns to point d, the potential will be the same as when we started (total change in potential around the circuit is zero). We plot the changes in potential around the circuit in Fig. 26–12b; point d is arbitrarily taken as zero.

As our positive test charge goes from point d, which is the negative or low potential side of the battery, to point e, which is the positive terminal (high potential side) of the battery, the potential increases by 12.0 V. (This is like the roller coaster being pulled up the first hill.) That is,

$$V_{\text{ed}} = +12.0\,\text{V}.$$

When our test charge moves from point e to point a, there is no change in potential since there is no source of emf and we assume negligible resistance in the connecting wires. Next, as the charge passes through the 400-Ω resistor to get to point b, there is a decrease in potential of $V = IR = (0.0174\,\text{A})(400\,\Omega) = 7.0\,\text{V}$. The positive test charge is flowing "downhill" since it is heading toward the negative terminal of the battery, as indicated in the graph of Fig. 26–12b. Because this is a *decrease* in potential, we use a *negative* sign:

$$V_{\text{ba}} = V_{\text{b}} - V_{\text{a}} = -7.0\,\text{V}.$$

As the charge proceeds from b to c there is another potential decrease (a "voltage drop") of $(0.0174\,\text{A}) \times (290\,\Omega) = 5.0\,\text{V}$, and this too is a decrease in potential:

$$V_{\text{cb}} = -5.0\,\text{V}.$$

There is no change in potential as our test charge moves from c to d as we assume negligible resistance in the wires.

The sum of all the changes in potential around the circuit of Fig. 26–12 is

$$+12.0\,\text{V} - 7.0\,\text{V} - 5.0\,\text{V} = 0.$$

This is exactly what Kirchhoff's loop rule said it would be.

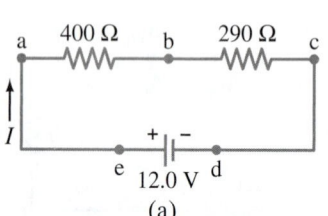

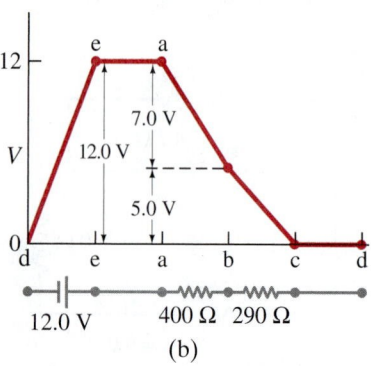

FIGURE 26–12 Changes in potential around the circuit in (a) are plotted in (b).

Kirchhoff's Rules

1. **Label the current** in each separate branch of the given circuit with a different subscript, such as I_1, I_2, I_3 (see Fig. 26–11 or 26–13). Each current refers to a segment between two junctions. Choose the direction of each current, using an arrow. The direction can be chosen arbitrarily: if the current is actually in the opposite direction, it will come out with a minus sign in the solution.

2. **Identify the unknowns.** You will need as many independent equations as there are unknowns. You may write down more equations than this, but you will find that some of the equations will be redundant (that is, not be independent in the sense of providing new information). You may use $V = IR$ for each resistor, which sometimes will reduce the number of unknowns.

3. **Apply Kirchhoff's junction rule** at one or more junctions.

4. **Apply Kirchhoff's loop rule** for one or more loops: follow each loop in one direction only. Pay careful attention to subscripts, and to signs:
   (a) For a resistor, apply Ohm's law; the potential difference is negative (a decrease) if your chosen loop direction is the same as the chosen current direction through that resistor; the potential difference is positive (an increase) if your chosen loop direction is opposite to the chosen current direction.
   (b) For a battery, the potential difference is positive if your chosen loop direction is from the negative terminal toward the positive terminal; the potential difference is negative if the loop direction is from the positive terminal toward the negative terminal.

5. **Solve the equations** algebraically for the unknowns. Be careful when manipulating equations not to err with signs. At the end, check your answers by plugging them into the original equations, or even by using any additional loop or junction rule equations not used previously.

EXAMPLE 26–9 **Using Kirchhoff's rules.** Calculate the currents I_1, I_2, and I_3 in the three branches of the circuit in Fig. 26–13 (which is the same as Fig. 26–11).

APPROACH AND SOLUTION

1. **Label the currents** and their directions. Figure 26–13 uses the labels I_1, I_2, and I_3 for the current in the three separate branches. Since (positive) current tends to move away from the positive terminal of a battery, we choose I_2 and I_3 to have the directions shown in Fig. 26–13. The direction of I_1 is not obvious in advance, so we arbitrarily chose the direction indicated. If the current actually flows in the opposite direction, our answer will have a negative sign.

2. **Identify the unknowns.** We have three unknowns and therefore we need three equations, which we get by applying Kirchhoff's junction and loop rules.

3. **Junction rule:** We apply Kirchhoff's junction rule to the currents at point a, where I_3 enters and I_2 and I_1 leave:

$$I_3 = I_1 + I_2. \qquad (a)$$

This same equation holds at point d, so we get no new information by writing an equation for point d.

4. **Loop rule:** We apply Kirchhoff's loop rule to two different closed loops. First we apply it to the upper loop ahdcba. We start (and end) at point a. From a to h we have a potential decrease $V_{ha} = -(I_1)(30\,\Omega)$. From h to d there is no change, but from d to c the potential increases by 45 V: that is, $V_{cd} = +45\,\text{V}$. From c to a the potential decreases through the two resistances by an amount $V_{ac} = -(I_3)(40\,\Omega + 1\,\Omega) = -(41\,\Omega)I_3$. Thus we have $V_{ha} + V_{cd} + V_{ac} = 0$, or

$$-30I_1 + 45 - 41I_3 = 0, \qquad (b)$$

where we have omitted the units (volts and amps) so we can more easily do the algebra. For our second loop, we take the outer loop ahdefga. (We could have chosen the lower loop abcdefga instead.) Again we start at point a and have $V_{ha} = -(I_1)(30\,\Omega)$, and $V_{dh} = 0$. But when we take our positive test charge from d to e, it actually is going uphill, against the current—or at least against the *assumed* direction of the current, which is what counts in this calculation. Thus $V_{ed} = I_2(20\,\Omega)$ has a *positive* sign. Similarly, $V_{fe} = I_2(1\,\Omega)$. From f to g there is a decrease in potential of 80 V since we go from the high potential terminal of the battery to the low. Thus $V_{gf} = -80\,\text{V}$. Finally, $V_{ag} = 0$, and the sum of the potential changes around this loop is

$$-30I_1 + (20 + 1)I_2 - 80 = 0. \qquad (c)$$

Our major work is done. The rest is algebra.

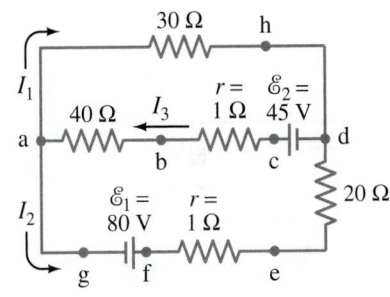

FIGURE 26–13 Currents can be calculated using Kirchhoff's rules. See Example 26–9.

5. Solve the equations. We have three equations—labeled (a), (b), and (c)—and three unknowns. From Eq. (c) we have

$$I_2 = \frac{80 + 30I_1}{21} = 3.8 + 1.4I_1. \tag{d}$$

From Eq. (b) we have

$$I_3 = \frac{45 - 30I_1}{41} = 1.1 - 0.73I_1. \tag{e}$$

We substitute Eqs. (d) and (e) into Eq. (a):

$$I_1 = I_3 - I_2 = 1.1 - 0.73I_1 - 3.8 - 1.4I_1.$$

We solve for I_1, collecting terms:

$$3.1I_1 = -2.7$$
$$I_1 = -0.87 \text{ A}.$$

The negative sign indicates that the direction of I_1 is actually opposite to that initially assumed and shown in Fig. 26–13. The answer automatically comes out in amperes because all values were in volts and ohms. From Eq. (d) we have

$$I_2 = 3.8 + 1.4I_1 = 3.8 + 1.4(-0.87) = 2.6 \text{ A},$$

and from Eq. (e)

$$I_3 = 1.1 - 0.73I_1 = 1.1 - 0.73(-0.87) = 1.7 \text{ A}.$$

This completes the solution.

NOTE The unknowns in different situations are not necessarily currents. It might be that the currents are given and we have to solve for unknown resistance or voltage. The variables are then different, but the technique is the same.

EXERCISE C Write the equation for the lower loop abcdefga of Example 26–9 and show, assuming the currents calculated in this Example, that the potentials add to zero for this lower loop.

26–4 Series and Parallel EMFs; Battery Charging

When two or more sources of emf, such as batteries, are arranged in series as in Fig. 26–14a, the total voltage is the algebraic sum of their respective voltages. On the other hand, when a 20-V and a 12-V battery are connected oppositely, as shown in Fig. 26–14b, the net voltage V_{ca} is 8 V (ignoring voltage drop across internal resistances). That is, a positive test charge moved from a to b gains in potential by 20 V, but when it passes from b to c it drops by 12 V. So the net change is $20 \text{ V} - 12 \text{ V} = 8 \text{ V}$. You might think that connecting batteries in reverse like this would be wasteful. For most purposes that would be true. But such a reverse arrangement is precisely how a battery charger works. In Fig. 26–14b, the 20-V source is charging up the 12-V battery. Because of its greater voltage, the 20-V source is forcing charge back into the 12-V battery: electrons are being forced into its negative terminal and removed from its positive terminal.

An automobile alternator keeps the car battery charged in the same way. A voltmeter placed across the terminals of a (12-V) car battery with the engine running fairly fast can tell you whether or not the alternator is charging the battery. If it is, the voltmeter reads 13 or 14 V. If the battery is not being charged, the voltage will be 12 V, or less if the battery is discharging. Car batteries can be recharged, but other batteries may not be rechargeable, since the chemical reactions in many cannot be reversed. In such cases, the arrangement of Fig. 26–14b would simply waste energy.

Sources of emf can also be arranged in parallel, Fig. 26–14c. With equal emfs, a parallel arrangement can provide more energy when large currents are needed. Each of the cells in parallel has to produce only a fraction of the total current, so the energy loss due to internal resistance is less than for a single cell; and the batteries will go dead less quickly.

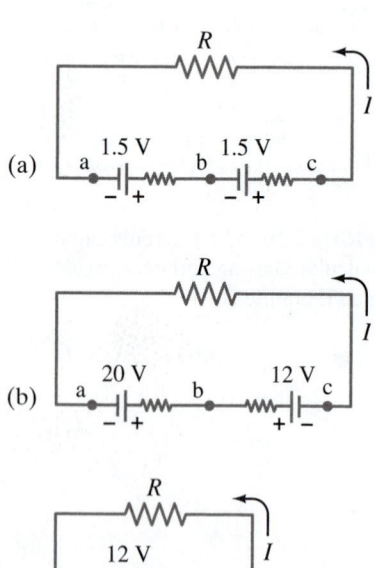

FIGURE 26–14 Batteries in series (a) and (b), and in parallel (c).

EXAMPLE 26–10 Jump starting a car. A good car battery is being used to jump start a car with a weak battery. The good battery has an emf of 12.5 V and internal resistance 0.020 Ω. Suppose the weak battery has an emf of 10.1 V and internal resistance 0.10 Ω. Each copper jumper cable is 3.0 m long and 0.50 cm in diameter, and can be attached as shown in Fig. 26–15. Assume the starter motor can be represented as a resistor $R_s = 0.15\ \Omega$. Determine the current through the starter motor (a) if only the weak battery is connected to it, and (b) if the good battery is also connected, as shown in Fig. 26–15.

APPROACH We apply Kirchhoff's rules, but in (b) we will first need to determine the resistance of the jumper cables using their dimensions and the resistivity ($\rho = 1.68 \times 10^{-8}\ \Omega\cdot$m for copper) as discussed in Section 25–4.

SOLUTION (a) The circuit with only the weak battery and no jumper cables is simple: an emf of 10.1 V connected to two resistances in series, $0.10\ \Omega + 0.15\ \Omega = 0.25\ \Omega$. Hence the current is $I = V/R = (10.1\ \text{V})/(0.25\ \Omega) = 40\ \text{A}$.

(b) We need to find the resistance of the jumper cables that connect the good battery. From Eq. 25–3, each has resistance $R_J = \rho\ell/A = (1.68 \times 10^{-8}\ \Omega\cdot\text{m})(3.0\ \text{m})/(\pi)(0.25 \times 10^{-2}\ \text{m})^2 = 0.0026\ \Omega$. Kirchhoff's loop rule for the full outside loop gives

$$12.5\ \text{V} - I_1(2R_J + r_1) - I_3 R_s = 0$$
$$12.5\ \text{V} - I_1(0.025\ \Omega) - I_3(0.15\ \Omega) = 0 \qquad (a)$$

since $(2R_J + r) = (0.0052\ \Omega + 0.020\ \Omega) = 0.025\ \Omega$.
The loop rule for the lower loop, including the weak battery and the starter, gives

$$10.1\ \text{V} - I_3(0.15\ \Omega) - I_2(0.10\ \Omega) = 0. \qquad (b)$$

The junction rule at point B gives

$$I_1 + I_2 = I_3. \qquad (c)$$

We have three equations in three unknowns. From Eq. (c), $I_1 = I_3 - I_2$ and we substitute this into Eq. (a):

$$12.5\ \text{V} - (I_3 - I_2)(0.025\ \Omega) - I_3(0.15\ \Omega) = 0,$$
$$12.5\ \text{V} - I_3(0.175\ \Omega) + I_2(0.025\ \Omega) = 0.$$

Combining this last equation with (b) gives $I_3 = 71\ \text{A}$, quite a bit better than in (a). The other currents are $I_2 = -5\ \text{A}$ and $I_1 = 76\ \text{A}$. Note that $I_2 = -5\ \text{A}$ is in the opposite direction from that assumed in Fig. 26–15. The terminal voltage of the weak 10.1-V battery is now $V_{BA} = 10.1\ \text{V} - (-5\ \text{A})(0.10\ \Omega) = 10.6\ \text{V}$.

NOTE The circuit shown in Fig. 26–15, without the starter motor, is how a battery can be charged. The stronger battery pushes charge back into the weaker battery.

EXERCISE D If the jumper cables of Example 26–10 were mistakenly connected in reverse, the positive terminal of each battery would be connected to the negative terminal of the other battery (Fig. 26–16). What would be the current I even before the starter motor is engaged (the switch S in Fig. 26–16 is open)? Why could this cause the batteries to explode?

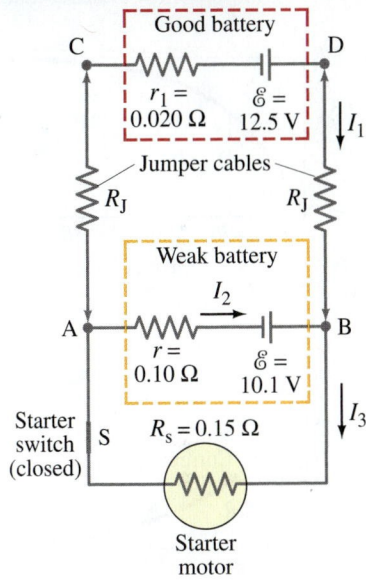

FIGURE 26–15 Example 26–10, a jump start.

FIGURE 26–16 Exercise D.

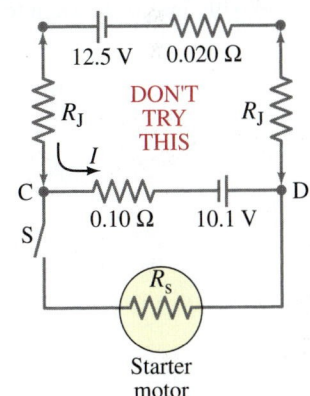

26–5 Circuits Containing Resistor and Capacitor (RC Circuits)

Our study of circuits in this Chapter has, until now, dealt with steady currents that do not change in time. Now we examine circuits that contain both resistance and capacitance. Such a circuit is called an **RC circuit**. RC circuits are common in everyday life: they are used to control the speed of a car's windshield wiper, and the timing of the change of traffic lights. They are used in camera flashes, in heart pacemakers, and in many other electronic devices. In RC circuits, we are not so interested in the final "steady state" voltage and charge on the capacitor, but rather in how these variables change in time.

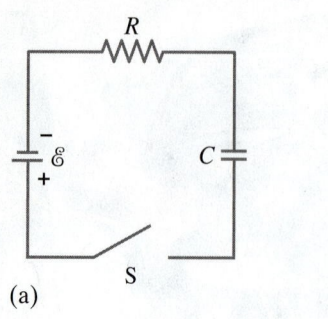

(a)

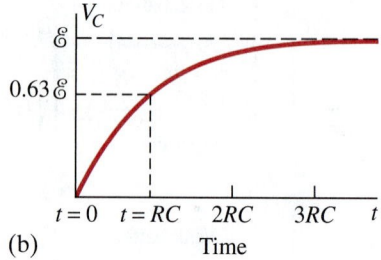

$t = 0$ $t = RC$ $2RC$ $3RC$ t

(b)   Time

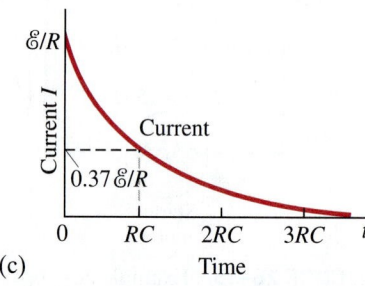

0 RC $2RC$ $3RC$ t

(c)   Time

FIGURE 26–17 After the switch S closes in the RC circuit shown in (a), the voltage across the capacitor increases with time as shown in (b), and the current through the resistor decreases with time as shown in (c).

Let us now examine the *RC* circuit shown in Fig. 26–17a. When the switch S is closed, current immediately begins to flow through the circuit. Electrons will flow out from the negative terminal of the battery, through the resistor *R*, and accumulate on the upper plate of the capacitor. And electrons will flow into the positive terminal of the battery, leaving a positive charge on the other plate of the capacitor. As charge accumulates on the capacitor, the potential difference across it increases $(V_C = Q/C)$, and the current is reduced until eventually the voltage across the capacitor equals the emf of the battery, $\mathscr{E}$. There is then no potential difference across the resistor, and no further current flows. Potential difference V_C across the capacitor thus increases in time as shown in Fig. 26–17b. The mathematical form of this curve—that is, V_C as a function of time—can be derived using conservation of energy (or Kirchhoff's loop rule). The emf $\mathscr{E}$ of the battery will equal the sum of the voltage drops across the resistor (IR) and the capacitor (Q/C):

$$\mathscr{E} = IR + \frac{Q}{C}.$$ (26–5)

The resistance *R* includes all resistance in the circuit, including the internal resistance of the battery; *I* is the current in the circuit at any instant, and *Q* is the charge on the capacitor at that same instant. Although $\mathscr{E}$, *R*, and *C* are constants, both *Q* and *I* are functions of time. The rate at which charge flows through the resistor $(I = dQ/dt)$ is equal to the rate at which charge accumulates on the capacitor. Thus we can write

$$\mathscr{E} = R\frac{dQ}{dt} + \frac{1}{C}Q.$$

This equation can be solved by rearranging it:

$$\frac{dQ}{C\mathscr{E} - Q} = \frac{dt}{RC}.$$

We now integrate from $t = 0$, when $Q = 0$, to time t when a charge Q is on the capacitor:

$$\int_0^Q \frac{dQ}{C\mathscr{E} - Q} = \frac{1}{RC}\int_0^t dt$$

$$-\ln(C\mathscr{E} - Q)\Big|_0^Q = \frac{t}{RC}\Big|_0^t$$

$$-\ln(C\mathscr{E} - Q) - (-\ln C\mathscr{E}) = \frac{t}{RC}$$

$$\ln(C\mathscr{E} - Q) - \ln(C\mathscr{E}) = -\frac{t}{RC}$$

$$\ln\left(1 - \frac{Q}{C\mathscr{E}}\right) = -\frac{t}{RC}.$$

We take the exponential[†] of both sides

$$1 - \frac{Q}{C\mathscr{E}} = e^{-t/RC}$$

or

$$Q = C\mathscr{E}(1 - e^{-t/RC}).$$ (26–6a)

The potential difference across the capacitor is $V_C = Q/C$, so

$$V_C = \mathscr{E}(1 - e^{-t/RC}).$$ (26–6b)

From Eqs. 26–6 we see that the charge *Q* on the capacitor, and the voltage V_C across it, increase from zero at $t = 0$ to maximum values $Q_{max} = C\mathscr{E}$ and $V_C = \mathscr{E}$ after a very long time. The quantity *RC* that appears in the exponent is called the **time constant** τ of the circuit:

$$\tau = RC.$$ (26–7)

It represents the time[‡] required for the capacitor to reach $(1 - e^{-1}) = 0.63$ or 63% of its full charge and voltage. Thus the product *RC* is a measure of how quickly the

---

[†]The constant *e*, known as the base for natural logarithms, has the value $e = 2.718\cdots$. Do not confuse this *e* with *e* for the charge on the electron.

[‡]The units of *RC* are $\Omega \cdot F = (V/A)(C/V) = C/(C/s) = s$.

capacitor gets charged. In a circuit, for example, where $R = 200 \, \text{k}\Omega$ and $C = 3.0 \, \mu\text{F}$, the time constant is $(2.0 \times 10^5 \, \Omega)(3.0 \times 10^{-6} \, \text{F}) = 0.60 \, \text{s}$. If the resistance is much lower, the time constant is much smaller. This makes sense, since a lower resistance will retard the flow of charge less. All circuits contain some resistance (if only in the connecting wires), so a capacitor never can be charged instantaneously when connected to a battery.

From Eqs. 26–6, it appears that Q and V_C never quite reach their maximum values within a finite time. However, they reach 86% of maximum in $2RC$, 95% in $3RC$, 98% in $4RC$, and so on. Q and V_C approach their maximum values asymptotically. For example, if $R = 20 \, \text{k}\Omega$ and $C = 0.30 \, \mu\text{F}$, the time constant is $(2.0 \times 10^4 \, \Omega)(3.0 \times 10^{-7} \, \text{F}) = 6.0 \times 10^{-3} \, \text{s}$. So the capacitor is more than 98% charged in less than $\frac{1}{40}$ of a second.

The current I through the circuit of Fig. 26–17a at any time t can be obtained by differentiating Eq. 26–6a:

$$I = \frac{dQ}{dt} = \frac{\mathscr{E}}{R} e^{-t/RC}. \qquad (26\text{–}8)$$

Thus, at $t = 0$, the current is $I = \mathscr{E}/R$, as expected for a circuit containing only a resistor (there is not yet a potential difference across the capacitor). The current then drops exponentially in time with a time constant equal to RC, as the voltage across the capacitor increases. This is shown in Fig. 26–17c. The time constant RC represents the time required for the current to drop to $1/e \approx 0.37$ of its initial value.

EXAMPLE 26–11 *RC circuit, with emf.* The capacitance in the circuit of Fig. 26–17a is $C = 0.30 \, \mu\text{F}$, the total resistance is $20 \, \text{k}\Omega$, and the battery emf is 12 V. Determine (*a*) the time constant, (*b*) the maximum charge the capacitor could acquire, (*c*) the time it takes for the charge to reach 99% of this value, (*d*) the current I when the charge Q is half its maximum value, (*e*) the maximum current, and (*f*) the charge Q when the current I is 0.20 its maximum value.

APPROACH We use Fig. 26–17 and Eqs. 26–5, 6, 7, and 8.

SOLUTION (*a*) The time constant is $RC = (2.0 \times 10^4 \, \Omega)(3.0 \times 10^{-7} \, \text{F}) = 6.0 \times 10^{-3} \, \text{s}$.

(*b*) The maximum charge would be $Q = C\mathscr{E} = (3.0 \times 10^{-7} \, \text{F})(12 \, \text{V}) = 3.6 \, \mu\text{C}$.

(*c*) In Eq. 26–6a, we set $Q = 0.99C\mathscr{E}$:

$$0.99C\mathscr{E} = C\mathscr{E}(1 - e^{-t/RC}),$$

or

$$e^{-t/RC} = 1 - 0.99 = 0.01.$$

Then

$$\frac{t}{RC} = -\ln(0.01) = 4.6$$

so

$$t = 4.6RC = 28 \times 10^{-3} \, \text{s}$$

or 28 ms (less than $\frac{1}{30}$ s).

(*d*) From part (*b*) the maximum charge is $3.6 \, \mu\text{C}$. When the charge is half this value, $1.8 \, \mu\text{C}$, the current I in the circuit can be found using the original differential equation, or Eq. 26–5:

$$I = \frac{1}{R}\left(\mathscr{E} - \frac{Q}{C}\right) = \frac{1}{2.0 \times 10^4 \, \Omega}\left(12 \, \text{V} - \frac{1.8 \times 10^{-6} \, \text{C}}{0.30 \times 10^{-6} \, \text{F}}\right) = 300 \, \mu\text{A}.$$

(*e*) The current is a maximum when there is no charge on the capacitor ($Q = 0$):

$$I_{\text{max}} = \frac{\mathscr{E}}{R} = \frac{12 \, \text{V}}{2.0 \times 10^4 \, \Omega} = 600 \, \mu\text{A}.$$

(*f*) Again using Eq. 26–5, now with $I = 0.20I_{\text{max}} = 120 \, \mu\text{A}$, we have

$$Q = C(\mathscr{E} - IR)$$
$$= (3.0 \times 10^{-7} \, \text{F})[12 \, \text{V} - (1.2 \times 10^{-4} \, \text{A})(2.0 \times 10^4 \, \Omega)] = 2.9 \, \mu\text{C}.$$

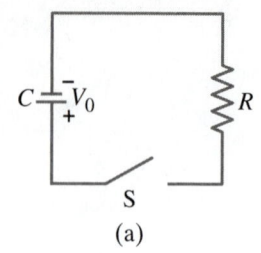

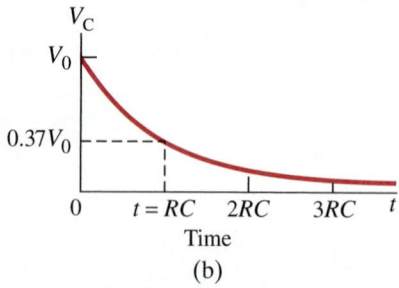

FIGURE 26–18 For the RC circuit shown in (a), the voltage V_C across the capacitor decreases with time, as shown in (b), after the switch S is closed at $t = 0$. The charge on the capacitor follows the same curve since $V_C \propto Q$.

The circuit just discussed involved the *charging* of a capacitor by a battery through a resistance. Now let us look at another situation: when a capacitor is already charged (say to a voltage V_0), and it is then allowed to *discharge* through a resistance R as shown in Fig. 26–18a. (In this case there is no battery.) When the switch S is closed, charge begins to flow through resistor R from one side of the capacitor toward the other side, until the capacitor is fully discharged. The voltage across the resistor at any instant equals that across the capacitor:

$$IR = \frac{Q}{C}.$$

The rate at which charge leaves the capacitor equals the negative of the current in the resistor, $I = -dQ/dt$, because the capacitor is discharging (Q is decreasing). So we write the above equation as

$$-\frac{dQ}{dt}R = \frac{Q}{C}.$$

We rearrange this to

$$\frac{dQ}{Q} = -\frac{dt}{RC}$$

and integrate it from $t = 0$ when the charge on the capacitor is Q_0, to some time t later when the charge is Q:

$$\ln\frac{Q}{Q_0} = -\frac{t}{RC}$$

or

$$Q = Q_0 e^{-t/RC}. \tag{26–9a}$$

The voltage across the capacitor $(V_C = Q/C)$ as a function of time is

$$V_C = V_0 e^{-t/RC}, \tag{26–9b}$$

where the initial voltage $V_0 = Q_0/C$. Thus the charge on the capacitor, and the voltage across it, decrease exponentially in time with a time constant RC. This is shown in Fig. 26–18b. The current is

$$I = -\frac{dQ}{dt} = \frac{Q_0}{RC}e^{-t/RC} = I_0 e^{-t/RC}, \tag{26–10}$$

and it too is seen to decrease exponentially in time with the same time constant RC. The charge on the capacitor, the voltage across it, and the current in the resistor all decrease to 37% of their original value in one time constant $t = \tau = RC$.

EXERCISE E In 10 time constants, the charge on the capacitor in Fig. 26–18 will be about (*a*) $Q_0/20{,}000$, (*b*) $Q_0/5000$, (*c*) $Q_0/1000$, (*d*) $Q_0/10$, (*e*) $Q_0/3$?

EXAMPLE 26–12 Discharging RC circuit. In the RC circuit shown in Fig. 26–19, the battery has fully charged the capacitor, so $Q_0 = C\mathscr{E}$. Then at $t = 0$ the switch is thrown from position a to b. The battery emf is 20.0 V, and the capacitance $C = 1.02 \,\mu\text{F}$. The current I is observed to decrease to 0.50 of its initial value in 40 μs. (*a*) What is the value of Q, the charge on the capacitor, at $t = 0$? (*b*) What is the value of R? (*c*) What is Q at $t = 60 \,\mu$s?

APPROACH At $t = 0$, the capacitor has charge $Q_0 = C\mathscr{E}$, and then the battery is removed from the circuit and the capacitor begins discharging through the resistor, as in Fig. 26–18. At any time t later (Eq. 26–9a) we have

$$Q = Q_0 e^{-t/RC} = C\mathscr{E}e^{-t/RC}.$$

FIGURE 26–19 Example 26–12.

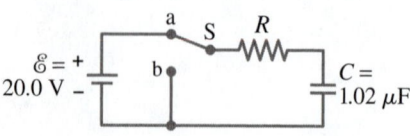

SOLUTION (*a*) At $t = 0$,

$$Q = Q_0 = C\mathscr{E} = (1.02 \times 10^{-6}\,\text{F})(20.0\,\text{V}) = 2.04 \times 10^{-5}\,\text{C} = 20.4\,\mu\text{C}.$$

(*b*) To find R, we are given that at $t = 40\,\mu\text{s}$, $I = 0.50 I_0$. Hence

$$0.50 I_0 = I_0 e^{-t/RC}.$$

Taking natural logs on both sides ($\ln 0.50 = -0.693$):

$$0.693 = \frac{t}{RC}$$

so

$$R = \frac{t}{(0.693)C} = \frac{(40 \times 10^{-6}\,\text{s})}{(0.693)(1.02 \times 10^{-6}\,\text{F})} = 57\,\Omega.$$

(*c*) At $t = 60\,\mu\text{s}$,

$$Q = Q_0 e^{-t/RC} = (20.4 \times 10^{-6}\,\text{C})e^{-\frac{60 \times 10^{-6}\,\text{s}}{(57\,\Omega)(1.02 \times 10^{-6}\,\text{F})}} = 7.3\,\mu\text{C}.$$

CONCEPTUAL EXAMPLE 26–13 **Bulb in *RC* circuit.** In the circuit of Fig. 26–20, the capacitor is originally uncharged. Describe the behavior of the lightbulb from the instant switch S is closed until a long time later.

RESPONSE When the switch is first closed, the current in the circuit is high and the lightbulb burns brightly. As the capacitor charges, the voltage across the capacitor increases causing the current to be reduced, and the lightbulb dims. As the potential difference across the capacitor approaches the same voltage as the battery, the current decreases toward zero and the lightbulb goes out.

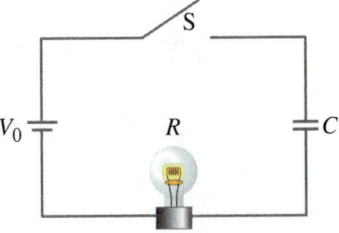

FIGURE 26–20 Example 26–13.

---

*Applications of *RC* Circuits

The charging and discharging in an *RC* circuit can be used to produce voltage pulses at a regular frequency. The charge on the capacitor increases to a particular voltage, and then discharges. One way of initiating the discharge of the capacitor is by the use of a gas-filled tube which has an electrical breakdown when the voltage across it reaches a certain value V_0. After the discharge is finished, the tube no longer conducts current and the recharging process repeats itself, starting at a lower voltage V_0'. Figure 26–21 shows a possible circuit, and the "sawtooth" voltage it produces.

A simple blinking light can be an application of a sawtooth oscillator circuit. Here the emf is supplied by a battery; the neon bulb flashes on at a rate of perhaps 1 cycle per second. The main component of a "flasher unit" is a moderately large capacitor.

The intermittent windshield wipers of a car can also use an *RC* circuit. The *RC* time constant, which can be changed using a multi-positioned switch for different values of R with fixed C, determines the rate at which the wipers come on.

FIGURE 26–21 (a) An *RC* circuit, coupled with a gas-filled tube as a switch, can produce a repeating "sawtooth" voltage, as shown in (b).

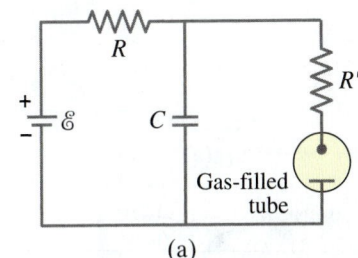

EXAMPLE 26–14 **ESTIMATE** **Resistor in a turn signal.** Estimate the order of magnitude of the resistor in a turn-signal circuit.

APPROACH A typical turn signal flashes perhaps twice per second, so the time constant is on the order of 0.5 s. A moderate capacitor might have $C = 1\,\mu\text{F}$.

SOLUTION Setting $\tau = RC = 0.5\,\text{s}$, we find

$$R = \frac{\tau}{C} = \frac{0.5\,\text{s}}{1 \times 10^{-6}\,\text{F}} \approx 500\,\text{k}\Omega.$$

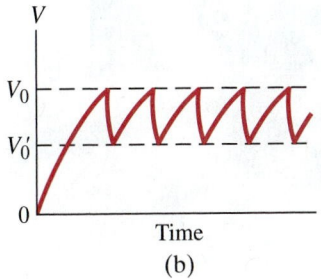

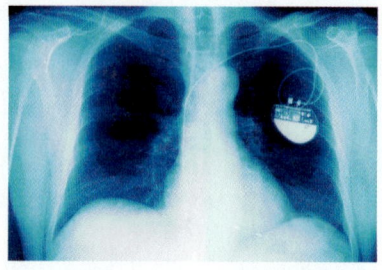

FIGURE 26–22 Electronic battery-powered pacemaker can be seen on the rib cage in this X-ray.

An interesting medical use of an *RC* circuit is the electronic heart pacemaker, which can make a stopped heart start beating again by applying an electric stimulus through electrodes attached to the chest. The stimulus can be repeated at the normal heartbeat rate if necessary. The heart itself contains *pacemaker* cells, which send out tiny electric pulses at a rate of 60 to 80 per minute. These signals induce the start of each heartbeat. In some forms of heart disease, the natural pacemaker fails to function properly, and the heart loses its beat. Such patients use *electronic pacemakers* which produce a regular voltage pulse that starts and controls the frequency of the heartbeat. The electrodes are implanted in or near the heart (Fig. 26–22), and the circuit contains a capacitor and a resistor. The charge on the capacitor increases to a certain point and then discharges a pulse to the heart. Then it starts charging again. The pulsing rate depends on the values of *R* and *C*.

26–6 Electric Hazards

Excess electric current can heat wires in buildings and cause fires, as discussed in Section 25–6. Electric current can also damage the human body or even be fatal. Electric current through the human body can cause damage in two ways: (1) Electric current heats tissue and can cause burns; (2) electric current stimulates nerves and muscles, and we feel a "shock." The severity of a shock depends on the magnitude of the current, how long it acts, and through what part of the body it passes. A current passing through vital organs such as the heart or brain is especially serious for it can interfere with their operation.

Most people can "feel" a current of about 1 mA. Currents of a few mA cause pain but rarely cause much damage in a healthy person. Currents above 10 mA cause severe contraction of the muscles, and a person may not be able to let go of the source of the current (say, a faulty appliance or wire). Death from paralysis of the respiratory system can occur. Artificial respiration, however, can sometimes revive a victim. If a current above about 80 to 100 mA passes across the torso, so that a portion passes through the heart for more than a second or two, the heart muscles will begin to contract irregularly and blood will not be properly pumped. This condition is called **ventricular fibrillation**. If it lasts for long, death results. Strangely enough, if the current is much larger, on the order of 1 A, death by heart failure may be less likely,[†] but such currents can cause serious burns, especially if concentrated through a small area of the body.

The seriousness of a shock depends on the applied voltage and on the effective resistance of the body. Living tissue has low resistance since the fluid of cells contains ions that can conduct quite well. However, the outer layer of skin, when dry, offers high resistance and is thus protective. The effective resistance between two points on opposite sides of the body when the skin is dry is in the range of 10^4 to $10^6\ \Omega$. But when the skin is wet, the resistance may be $10^3\ \Omega$ or less. A person who is barefoot or wearing thin-soled shoes will be in good contact with the ground, and touching a 120-V line with a wet hand can result in a current

FIGURE 26–23 A person receives an electric shock when the circuit is completed.

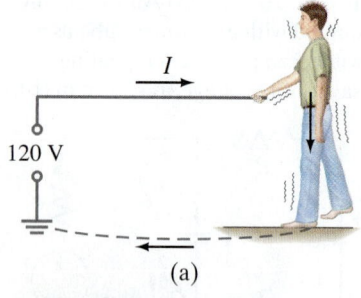

(a)

$$I = \frac{120\ \text{V}}{1000\ \Omega} = 120\ \text{mA}.$$

As we saw, this could be lethal.

A person who has received a shock has become part of a complete circuit. Figure 26–23 shows two ways the circuit might be completed when a person

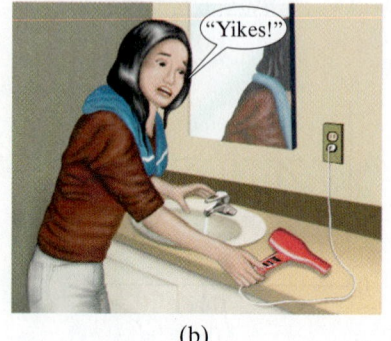

(b)

[†]Larger currents apparently bring the entire heart to a standstill. Upon release of the current, the heart returns to its normal rhythm. This may not happen when fibrillation occurs because, once started, it can be hard to stop. Fibrillation may also occur as a result of a heart attack or during heart surgery. A device known as a *defibrillator* (described in Section 24–4) can apply a brief high current to the heart, causing complete heart stoppage which is often followed by resumption of normal beating.

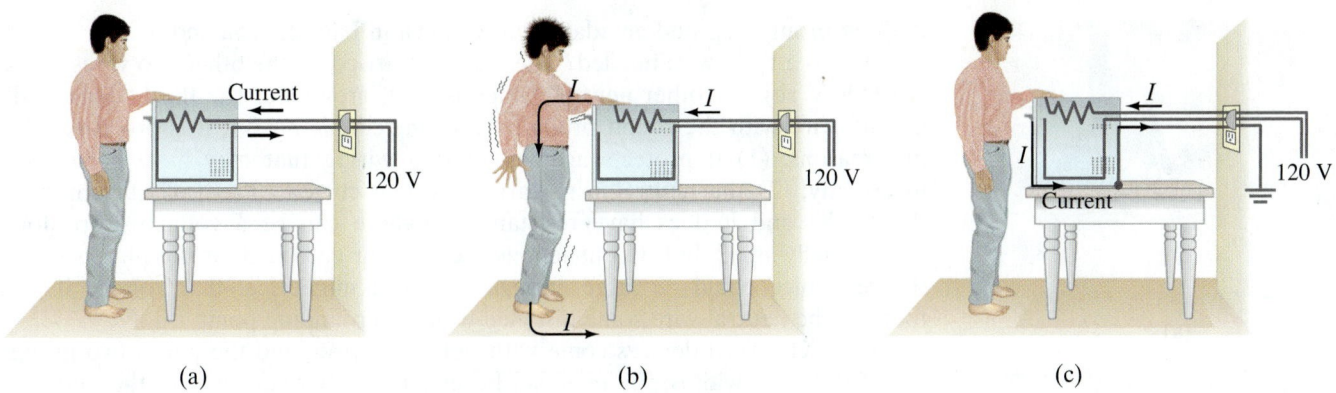

(a)   (b)   (c)

FIGURE 26–24 (a) An electric oven operating normally with a 2-prong plug. (b) Short to the case with ungrounded case: shock. (c) Short to the case with the case grounded by a 3-prong plug.

accidentally touches a "hot" electric wire—"hot" meaning a high potential such as 120 V (normal U.S. household voltage) relative to ground. The other wire of building wiring is connected to ground—either by a wire connected to a buried conductor, or via a metal water pipe into the ground. In Fig. 26–23a, the current passes from the high-voltage wire, through the person, to the ground through his bare feet, and back along the ground (a fair conductor) to the ground terminal of the source. If the person stands on a good insulator—thick rubber-soled shoes or a dry wood floor—there will be much more resistance in the circuit and consequently much less current through the person. If the person stands with bare feet on the ground, or is in a bathtub, there is lethal danger because the resistance is much less and the current greater. In a bathtub (or swimming pool), not only are you wet, which reduces your resistance, but the water is in contact with the drain pipe (typically metal) that leads to the ground. It is strongly recommended that you not touch anything electrical when wet or in bare feet. Building codes that require the use of non-metal pipes would be protective.

In Fig. 26–23b, a person touches a faulty "hot" wire with one hand, and the other hand touches a sink faucet (connected to ground via the pipe). The current is particularly dangerous because it passes across the chest, through the heart and lungs. A useful rule: if one hand is touching something electrical, keep your other hand in your pocket (don't use it!), and wear thick rubber-soled shoes. It is also a good idea to remove metal jewelry, especially rings (your finger is usually moist under a ring).

You can come into contact with a hot wire by touching a bare wire whose insulation has worn off, or from a bare wire inside an appliance when you're tinkering with it. (Always unplug an electrical device before investigating[†] its insides!) Another possibility is that a wire inside a device may break or lose its insulation and come in contact with the case. If the case is metal, it will conduct electricity. A person could then suffer a severe shock merely by touching the case, as shown in Fig. 26–24b. To prevent an accident, metal cases are supposed to be connected directly to ground by a separate ground wire. Then if a "hot" wire touches the grounded case, a short circuit to ground immediately occurs internally, as shown in Fig. 26–24c, and most of the current passes through the low-resistance ground wire rather than through the person. Furthermore, the high current should open the fuse or circuit breaker. Grounding a metal case is done by a separate ground wire connected to the third (round) prong of a 3-prong plug. Never cut off the third prong of a plug—it could save your life.

⚠ **CAUTION**

Keep one hand in your pocket when other touches electricity

🚶 **PHYSICS APPLIED**

Grounding and shocks

[†]Even then you can get a bad shock from a capacitor that hasn't been discharged until you touch it.

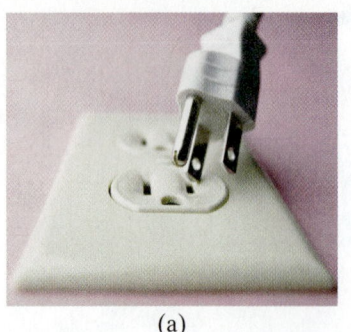

(a)

(b)

(c)

FIGURE 26–25 (a) A 3-prong plug, and (b) an adapter (gray) for old-fashioned 2-prong outlets—be sure to screw down the ground tab. (c) A polarized 2-prong plug.

FIGURE 26–26 Four wires entering a typical house. The color codes for wires are not always as shown here—be careful!

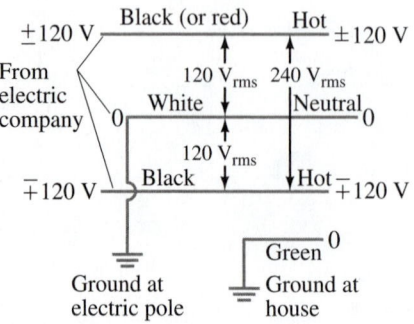

A three-prong plug, and an adapter, are shown in Figs. 26–25a and b.

Why is a third wire needed? The 120 V is carried by the other two wires—one **hot** (120 V ac), the other **neutral**, which is itself grounded. The third "dedicated" ground wire with the round prong may seem redundant. But it is protection for two reasons: (1) it protects against internal wiring that may have been done incorrectly; (2) the neutral wire carries normal current ("return" current from the 120 V) and it does have resistance; so there can be a voltage drop along it—normally small, but if connections are poor or corroded, or the plug is loose, the resistance could be large enough that you might feel that voltage if you touched the neutral wire some distance from its grounding point.

Some electrical devices come with only two wires, and the plug's two prongs are of different widths; the plug can be inserted only one way into the outlet so that the intended neutral (wider prong) in the device is connected to neutral in the wiring (Fig. 26–25c). For example, the screw threads of a lightbulb are meant to be connected to neutral (and the base contact to hot), to avoid shocks when changing a bulb in a possibly protruding socket. Devices with 2-prong plugs do *not* have their cases grounded; they are supposed to have double electric insulation. Take extra care anyway.

The insulation on a wire may be color coded. Hand-held meters may have red (hot) and black (ground) lead wires. But in a house, black is usually hot (or it may be red), whereas white is neutral and green is the dedicated ground, Fig. 26–26. But beware: these color codes cannot always be trusted. [In the U.S., three wires normally enter a house: two *hot* wires at 120 V each (which add together to 240 V for appliances or devices that run on 240 V) plus the grounded *neutral* (carrying return current for the two hots). See Fig. 26–26. The "dedicated" *ground* wire (non-current carrying) is a fourth wire that does not come from the electric company but enters the house from a nearby heavy stake in the ground or a buried metal pipe. The two hot wires can feed separate 120-V circuits in the house, so each 120-V circuit inside the house has only three wires, including ground.]

Normal circuit breakers (Sections 25–6 and 28–8) protect equipment and buildings from overload and fires. They protect humans only in some circumstances, such as the very high currents that result from a short, if they respond quickly enough. *Ground fault circuit interrupters* (GFCI), described in Section 29–8, are designed to protect people from the much lower currents (10 mA to 100 mA) that are lethal but would not throw a 15-A circuit breaker or blow a 20-A fuse.

It is current that harms, but it is voltage that drives the current. 30 volts is sometimes said to be the threshhold for danger. But even a 12-V car battery (which can supply large currents) can cause nasty burns and shock.

Another danger is **leakage current**, by which we mean a current along an unintended path. Leakage currents are often "capacitively coupled." For example, a wire in a lamp forms a capacitor with the metal case; charges moving in one conductor attract or repel charge in the other, so there is a current. Typical electrical codes limit leakage currents to 1 mA for any device. A 1-mA leakage current is usually harmless. It can be very dangerous, however, to a hospital patient with implanted electrodes connected to ground through the apparatus. This is due to the absence of the protective skin layer and because the current can pass directly through the heart as compared to the usual situation where the current enters at the hands and spreads out through the body. Although 100 mA may be needed to cause heart fibrillation when entering through the hands (very little of it actually passes through the heart), as little as 0.02 mA has been known to cause fibrillation when passing directly to the heart. Thus, a "wired" patient is in considerable danger from leakage current even from as simple an act as touching a lamp.

Finally, don't touch a downed power line (lethal!) or even get near it. A hot power line is at thousands of volts. A huge current can flow along the ground or pavement, from where the high-voltage wire touches the ground along its path to the grounding point of the neutral line, enough that the voltage between your two feet could be large. Tip: stand on one foot or run (only one foot touching the ground at a time).

*26–7 Ammeters and Voltmeters

An **ammeter** is used to measure current, and a **voltmeter** measures potential difference or voltage. Measurements of current and voltage are made with meters that are of two types: (1) *analog* meters, which display numerical values by the position of a pointer that can move across a scale (Fig. 26–27a); and (2) *digital* meters, which display the numerical value in numbers (Fig. 26–27b). We now discuss the meters themselves and how they work, then how they are connected to circuits to make measurements. Finally we will discuss how using meters affects the circuit being measured, possibly causing erroneous results—and what to do about it.

*Analog Ammeters and Voltmeters

The crucial part of an analog ammeter or voltmeter, in which the reading is by a pointer on a scale (Fig. 26–27a), is a *galvanometer*. The galvanometer works on the principle of the force between a magnetic field and a current-carrying coil of wire, and will be discussed in Chapter 27. For now, we merely need to know that the deflection of the needle of a galvanometer is proportional to the current flowing through it. The *full-scale current sensitivity* of a galvanometer, I_m, is the electric current needed to make the needle deflect full scale.

A galvanometer can be used directly to measure small dc currents. For example, a galvanometer whose sensitivity I_m is 50 μA can measure currents from about 1 μA (currents smaller than this would be hard to read on the scale) up to 50 μA. To measure larger currents, a resistor is placed in parallel with the galvanometer. Thus, an analog **ammeter**, represented by the symbol •–Ⓐ–•, consists of a galvanometer (•–Ⓖ–•) in parallel with a resistor called the **shunt resistor**, as shown in Fig. 26–28. ("Shunt" is a synonym for "in parallel.") The shunt resistance is R_{sh}, and the resistance of the galvanometer coil, through which current passes, is r. The value of R_{sh} is chosen according to the full-scale deflection desired; R_{sh} is normally very small—giving an ammeter a very small net resistance—so most of the current passes through R_{sh} and very little ($\lesssim 50\,\mu$A) passes through the galvanometer to deflect the needle.

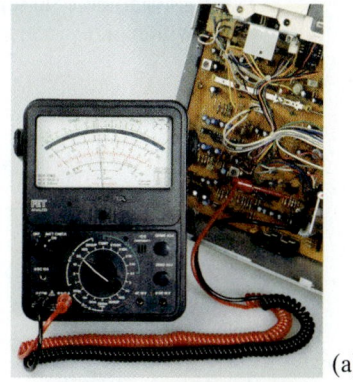

(a)

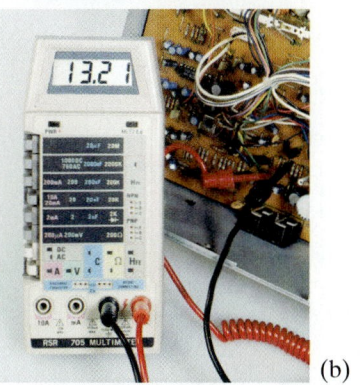

(b)

FIGURE 26–27 (a) An analog multimeter being used as a voltmeter. (b) An electronic digital meter.

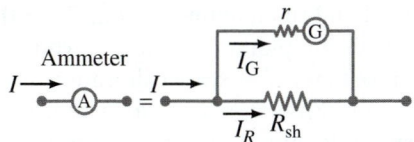

FIGURE 26–28 An ammeter is a galvanometer in parallel with a (shunt) resistor with low resistance, R_{sh}.

EXAMPLE 26–15 **Ammeter design.** Design an ammeter to read 1.0 A at full scale using a galvanometer with a full-scale sensitivity of 50 μA and a resistance $r = 30\,\Omega$. Check if the scale is linear.

APPROACH Only 50 μA $(= I_G = 0.000050\,\text{A})$ of the 1.0-A current must pass through the galvanometer to give full-scale deflection. The rest of the current $(I_R = 0.999950\,\text{A})$ passes through the small shunt resistor, R_{sh}, Fig. 26–28. The potential difference across the galvanometer equals that across the shunt resistor (they are in parallel). We apply Ohm's law to find R_{sh}.

SOLUTION Because $I = I_G + I_R$, when $I = 1.0$ A flows into the meter, we want I_R through the shunt resistor to be $I_R = 0.999950$ A. The potential difference across the shunt is the same as across the galvanometer, so Ohm's law tells us

$$I_R R_{sh} = I_G r;$$

then

$$R_{sh} = \frac{I_G r}{I_R} = \frac{(5.0 \times 10^{-5}\,\text{A})(30\,\Omega)}{(0.999950\,\text{A})} = 1.5 \times 10^{-3}\,\Omega,$$

or 0.0015 Ω. The shunt resistor must thus have a very low resistance and most of the current passes through it.

Because $I_G = I_R(R_{sh}/r)$ and (R_{sh}/r) is constant, we see that the scale is linear.

An analog **voltmeter** (•-Ⓥ-•) also consists of a galvanometer and a resistor. But the resistor R_{ser} is connected in series, Fig. 26–29, and it is usually large, giving a voltmeter a high internal resistance.

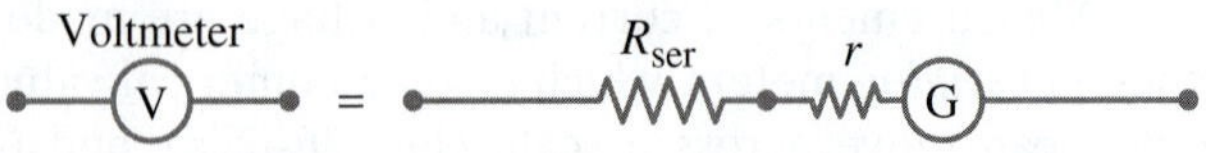

FIGURE 26–29 A voltmeter is a galvanometer in series with a resistor with high resistance, R_{ser}.

EXAMPLE 26–16 **Voltmeter design.** Using a galvanometer with internal resistance $r = 30\,\Omega$ and full-scale current sensitivity of $50\,\mu A$, design a voltmeter that reads from 0 to 15 V. Is the scale linear?

APPROACH When a potential difference of 15 V exists across the terminals of our voltmeter, we want $50\,\mu A$ to be passing through it so as to give a full-scale deflection.

SOLUTION From Ohm's law, $V = IR$, we have (see Fig. 26–29)

$$15\,V = (50\,\mu A)(r + R_{ser}),$$

so

$$R_{ser} = \frac{15\,V}{5.0 \times 10^{-5}\,A} - r = 300\,k\Omega - 30\,\Omega = 300\,k\Omega.$$

Notice that $r = 30\,\Omega$ is so small compared to the value of R_{ser} that it doesn't influence the calculation significantly. The scale will again be linear: if the voltage to be measured is 6.0 V, the current passing through the voltmeter will be $(6.0\,V)/(3.0 \times 10^5\,\Omega) = 2.0 \times 10^{-5}\,A$, or $20\,\mu A$. This will produce two-fifths of full-scale deflection, as required $(6.0\,V/15.0\,V = 2/5)$.

The meters just described are for direct current. A dc meter can be modified to measure ac (alternating current, Section 25–7) with the addition of diodes (Chapter 40), which allow current to flow in one direction only. An ac meter can be calibrated to read rms or peak values.

Voltmeters and ammeters can have several series or shunt resistors to offer a choice of range. **Multimeters** can measure voltage, current, and resistance. Sometimes a multimeter is called a VOM (Volt-Ohm-Meter or Volt-Ohm-Milliammeter).

An **ohmmeter** measures resistance, and must contain a battery of known voltage connected in series to a resistor (R_{ser}) and to an ammeter (Fig. 26–30). The resistor whose resistance is to be measured completes the circuit. The needle deflection is inversely proportional to the resistance. The scale calibration depends on the value of the series resistor. Because an ohmmeter sends a current through the device whose resistance is to be measured, it should not be used on very delicate devices that could be damaged by the current.

The **sensitivity** of a meter is generally specified on the face. It may be given as so many ohms per volt, which indicates how many ohms of resistance there are in the meter per volt of full-scale reading. For example, if the sensitivity is 30,000 Ω/V, this means that on the 10-V scale the meter has a resistance of 300,000 Ω, whereas on a 100-V scale the meter resistance is 3 $M\Omega$. The full-scale current sensitivity, I_m, discussed earlier, is just the reciprocal of the sensitivity in Ω/V.

*How to Connect Meters

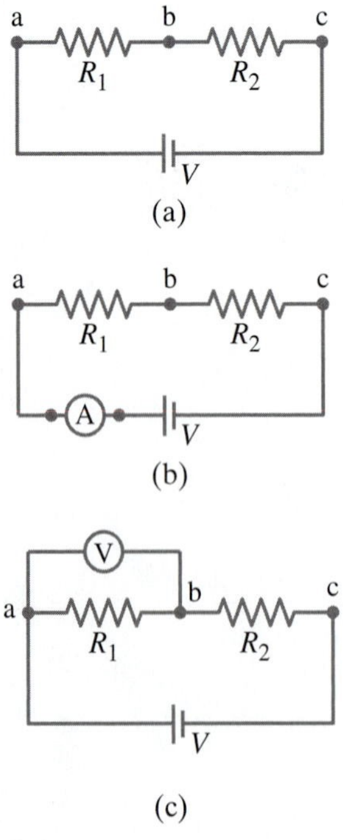

Suppose you wish to determine the current I in the circuit shown in Fig. 26–31a, and the voltage V across the resistor R_1. How exactly are ammeters and voltmeters connected to the circuit being measured?

Because an ammeter is used to measure the current flowing in the circuit, it must be inserted directly into the circuit, in series with the other elements, as shown in Fig. 26–31b. The smaller its internal resistance, the less it affects the circuit.

A voltmeter, on the other hand, is connected "externally," in parallel with the circuit element across which the voltage is to be measured. It is used to measure the potential difference between two points. Its two wire leads (connecting wires) are connected to the two points, as shown in Fig. 26–31c, where the voltage across R_1 is being measured. The larger its internal resistance, $(R_{ser} + r)$ in Fig. 26–29, the less it affects the circuit being measured.

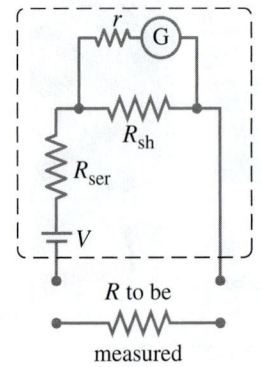

FIGURE 26–30 An ohmmeter.

FIGURE 26–31 Measuring current and voltage.

(a)

(b)

(c)

*Effects of Meter Resistance

It is important to know the sensitivity of a meter, for in many cases the resistance of the meter can seriously affect your results. Take the following Example.

EXAMPLE 26–17 **Voltage reading versus true voltage.** Suppose you are testing an electronic circuit which has two resistors, R_1 and R_2, each 15 kΩ, connected in series as shown in Fig. 26–32a. The battery maintains 8.0 V across them and has negligible internal resistance. A voltmeter whose sensitivity is 10,000 Ω/V is put on the 5.0-V scale. What voltage does the meter read when connected across R_1, Fig. 26–32b, and what error is caused by the finite resistance of the meter?

APPROACH The meter acts as a resistor in parallel with R_1. We use parallel and series resistor analyses and Ohm's law to find currents and voltages.

SOLUTION On the 5.0-V scale, the voltmeter has an internal resistance of $(5.0\,\text{V})(10,000\,\Omega/\text{V}) = 50,000\,\Omega$. When connected across R_1, as in Fig. 26–32b, we have this 50 kΩ in parallel with $R_1 = 15$ kΩ. The net resistance R_{eq} of these two is given by

$$\frac{1}{R_{eq}} = \frac{1}{50\,\text{k}\Omega} + \frac{1}{15\,\text{k}\Omega} = \frac{13}{150\,\text{k}\Omega};$$

so $R_{eq} = 11.5$ kΩ. This $R_{eq} = 11.5$ kΩ is in series with $R_2 = 15$ kΩ, so the total resistance of the circuit is now 26.5 kΩ (instead of the original 30 kΩ). Hence the current from the battery is

$$I = \frac{8.0\,\text{V}}{26.5\,\text{k}\Omega} = 3.0 \times 10^{-4}\,\text{A} = 0.30\,\text{mA}.$$

Then the voltage drop across R_1, which is the same as that across the voltmeter, is $(3.0 \times 10^{-4}\,\text{A})(11.5 \times 10^3\,\Omega) = 3.5$ V. [The voltage drop across R_2 is $(3.0 \times 10^{-4}\,\text{A})(15 \times 10^3\,\Omega) = 4.5$ V, for a total of 8.0 V.] If we assume the meter is precise, it will read 3.5 V. In the original circuit, without the meter, $R_1 = R_2$ so the voltage across R_1 is half that of the battery, or 4.0 V. Thus the voltmeter, because of its internal resistance, gives a low reading. In this case it is off by 0.5 V, or more than 10%.

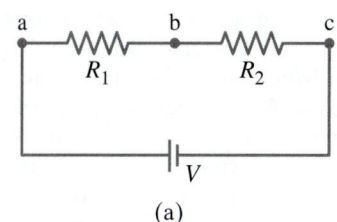

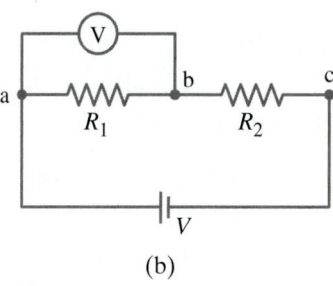

FIGURE 26–32 Example 26–17.

Example 26–17 illustrates how seriously a meter can affect a circuit and give a misleading reading. If the resistance of a voltmeter is much higher than the resistance of the circuit, however, it will have little effect and its readings can be trusted, at least to the manufactured precision of the meter, which for ordinary analog meters is typically 3% to 4% of full-scale deflection. An ammeter also can interfere with a circuit, but the effect is minimal if its resistance is much less than that of the circuit as a whole. For both voltmeters and ammeters, the more sensitive the galvanometer, the less effect it will have. A 50,000-Ω/V meter is far better than a 1000-Ω/V meter.

*Digital Meters

Digital meters (see Fig. 26–27b) are used in the same way as analog meters: they are inserted directly into the circuit, in series, to measure current (Fig. 26–31b), and connected "outside," in parallel with the circuit, to measure voltage (Fig. 26–31c).

The internal construction of digital meters, however, is different from that of analog meters in that digital meters do not use a galvanometer. The electronic circuitry and digital readout are more sensitive than a galvanometer, and have less effect on the circuit to be measured. When we measure dc voltages, a digital meter's resistance is very high, commonly on the order of 10 to 100 MΩ $(10^7 – 10^8\,\Omega)$, and doesn't change significantly when different voltage scales are selected. A 100-MΩ digital meter draws off very little current when connected across even a 1-MΩ resistance.

The precision of digital meters is exceptional, often one part in 10^4 $(=0.01\%)$ or better. This precision is not the same as accuracy, however. A precise meter of internal resistance $10^8\,\Omega$ will not give accurate results if used to measure a voltage across a 10^8-Ω resistor—in which case it is necessary to do a calculation like that in Example 26–17.

Whenever we make a measurement on a circuit, to some degree we affect that circuit (Example 26–17). This is true for other types of measurement as well: when we make a measurement on a system, we affect that system in some way. On a temperature measurement, for example, the thermometer can exchange heat with the system, thus altering its temperature. It is important to be able to make needed corrections, as we saw in Example 26–17.

Summary

A device that transforms another type of energy into electrical energy is called a **source** of **emf**. A battery behaves like a source of emf in series with an **internal resistance**. The emf is the potential difference determined by the chemical reactions in the battery and equals the terminal voltage when no current is drawn. When a current is drawn, the voltage at the battery's terminals is less than its emf by an amount equal to the potential decrease Ir across the internal resistance.

When resistances are connected in **series** (end to end in a single linear path), the equivalent resistance is the sum of the individual resistances:

$$R_{eq} = R_1 + R_2 + \cdots. \tag{26-3}$$

In a series combination, R_{eq} is greater than any component resistance.

When resistors are connected in **parallel**, the reciprocal of the equivalent resistance equals the sum of the reciprocals of the individual resistances:

$$\frac{1}{R_{eq}} = \frac{1}{R_1} + \frac{1}{R_2} + \cdots. \tag{26-4}$$

In a parallel connection, the net resistance is less than any of the individual resistances.

Kirchhoff's rules are helpful in determining the currents and voltages in circuits. Kirchhoff's **junction rule** is based on conservation of electric charge and states that the sum of all currents entering any junction equals the sum of all currents leaving that junction. The second, or **loop rule**, is based on conservation of energy and states that the algebraic sum of the changes in potential around any closed path of the circuit must be zero.

When an **RC circuit** containing a resistor R in series with a capacitance C is connected to a dc source of emf, the voltage across the capacitor rises gradually in time characterized by an exponential of the form $\left(1 - e^{-t/RC}\right)$, where the **time constant**,

$$\tau = RC, \tag{26-7}$$

is the time it takes for the voltage to reach 63 percent of its maximum value. The current through the resistor decreases as $e^{-t/RC}$.

A capacitor discharging through a resistor is characterized by the same time constant: in a time $\tau = RC$, the voltage across the capacitor drops to 37 percent of its initial value. The charge on the capacitor, and voltage across it, decreases as $e^{-t/RC}$, as does the current.

Electric shocks are caused by current passing through the body. To avoid shocks, the body must not become part of a complete circuit by allowing different parts of the body to touch objects at different potentials. Commonly, shocks are caused by one part of the body touching ground and another part touching a high electric potential.

[*An **ammeter** measures current. An analog ammeter consists of a galvanometer and a parallel **shunt resistor** that carries most of the current. An analog **voltmeter** consists of a galvanometer and a series resistor. An ammeter is inserted *into* the circuit whose current is to be measured. A voltmeter is external, being connected in parallel to the element whose voltage is to be measured. Digital voltmeters have greater internal resistance and affect the circuit to be measured less than do analog meters.]

Questions

1. Explain why birds can sit on power lines safely, whereas leaning a metal ladder up against a power line to fetch a stuck kite is extremely dangerous.

2. Discuss the advantages and disadvantages of Christmas tree lights connected in parallel versus those connected in series.

3. If all you have is a 120-V line, would it be possible to light several 6-V lamps without burning them out? How?

4. Two lightbulbs of resistance R_1 and R_2 $(R_2 > R_1)$ and a battery are all connected in series. Which bulb is brighter? What if they are connected in parallel? Explain.

5. Household outlets are often double outlets. Are these connected in series or parallel? How do you know?

6. With two identical lightbulbs and two identical batteries, how would you arrange the bulbs and batteries in a circuit to get the maximum possible total power to the lightbulbs? (Assume the batteries have negligible internal resistance.)

7. If two identical resistors are connected in series to a battery, does the battery have to supply more power or less power than when only one of the resistors is connected? Explain.

8. You have a single 60-W bulb on in your room. How does the overall resistance of your room's electric circuit change when you turn on an additional 100-W bulb?

9. When applying Kirchhoff's loop rule (such as in Fig. 26–33), does the sign (or direction) of a battery's emf depend on the direction of current through the battery? What about the terminal voltage?

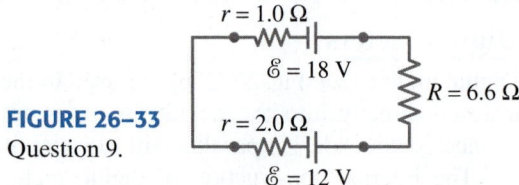

FIGURE 26–33
Question 9.

10. Compare and discuss the formulas for resistors and for capacitors when connected in series and in parallel.

11. For what use are batteries connected in series? For what use are they connected in parallel? Does it matter if the batteries are nearly identical or not in either case?

12. Can the terminal voltage of a battery ever exceed its emf? Explain.

13. Explain in detail how you could measure the internal resistance of a battery.

14. In an *RC* circuit, current flows from the battery until the capacitor is completely charged. Is the total energy supplied by the battery equal to the total energy stored by the capacitor? If not, where does the extra energy go?

15. Given the circuit shown in Fig. 26–34, use the words "increases," "decreases," or "stays the same" to complete the following statements:

(a) If R_7 increases, the potential difference between A and E _____. Assume no resistance in Ⓐ and ℰ.

(b) If R_7 increases, the potential difference between A and E _____. Assume Ⓐ and ℰ have resistance.

(c) If R_7 increases, the voltage drop across R_4 _____.

(d) If R_2 decreases, the current through R_1 _____.

(e) If R_2 decreases, the current through R_6 _____.

(f) If R_2 decreases, the current through R_3 _____.

(g) If R_5 increases, the voltage drop across R_2 _____.

(h) If R_5 increases, the voltage drop across R_4 _____.

(i) If R_2, R_5, and R_7 increase, ℰ $(r = 0)$ _____.

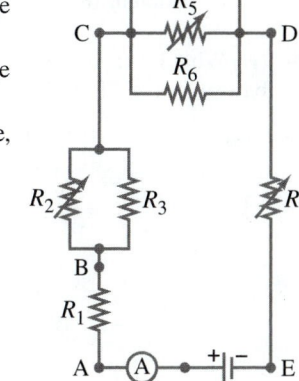

FIGURE 26–34
Question 15. R_2, R_5, and R_7 are *variable* resistors (you can change their resistance), given the symbol 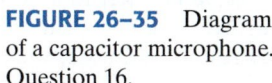.

16. Figure 26–35 is a diagram of a capacitor (or condenser) **microphone**. The changing air pressure in a sound wave causes one plate of the capacitor C to move back and forth. Explain how a current of the same frequency as the sound wave is produced.

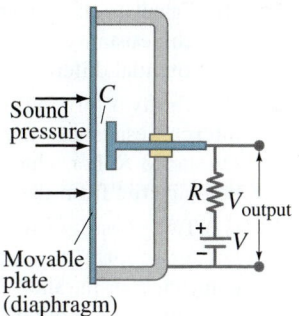

FIGURE 26–35 Diagram of a capacitor microphone. Question 16.

17. Design a circuit in which two different switches of the type shown in Fig. 26–36 can be used to operate the same lightbulb from opposite sides of a room.

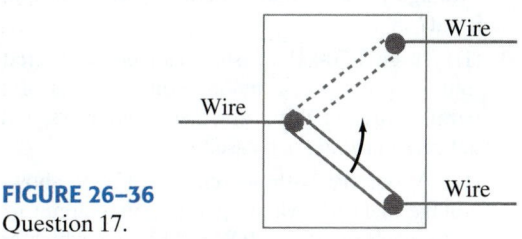

FIGURE 26–36
Question 17.

*18. What is the main difference between an analog voltmeter and an analog ammeter?

*19. What would happen if you mistakenly used an ammeter where you needed to use a voltmeter?

*20. Explain why an ideal ammeter would have zero resistance and an ideal voltmeter infinite resistance.

*21. A voltmeter connected across a resistor always reads *less* than the actual voltage across the resistor when the meter is not present. Explain.

*22. A small battery-operated flashlight requires a single 1.5-V battery. The bulb is barely glowing, but when you take the battery out and check it with a voltmeter, it registers 1.5 V. How would you explain this?

23. Different lamps might have batteries connected in either of the two arrangements shown in Fig. 26–37. What would be the advantages of each scheme?

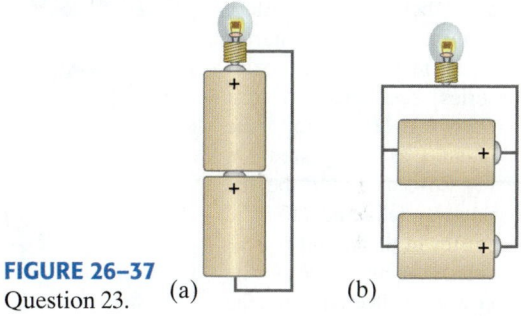

FIGURE 26–37
Question 23.   (a)   (b)

Problems

26–1 Emf and Terminal Voltage

1. (I) Calculate the terminal voltage for a battery with an internal resistance of 0.900 Ω and an emf of 6.00 V when the battery is connected in series with (a) an 81.0-Ω resistor, and (b) an 810-Ω resistor.

2. (I) Four 1.50-V cells are connected in series to a 12-Ω lightbulb. If the resulting current is 0.45 A, what is the internal resistance of each cell, assuming they are identical and neglecting the resistance of the wires?

3. (II) A 1.5-V dry cell can be tested by connecting it to a low-resistance ammeter. It should be able to supply at least 25 A. What is the internal resistance of the cell in this case, assuming it is much greater than that of the ammeter?

4. (II) What is the internal resistance of a 12.0-V car battery whose terminal voltage drops to 8.4 V when the starter draws 95 A? What is the resistance of the starter?

26–2 Resistors in Series and Parallel

In these Problems neglect the internal resistance of a battery unless the Problem refers to it.

5. (I) A 650-Ω and a 2200-Ω resistor are connected in series with a 12-V battery. What is the voltage across the 2200-Ω resistor?

6. (I) Three 45-Ω lightbulbs and three 65-Ω lightbulbs are connected in series. (a) What is the total resistance of the circuit? (b) What is the total resistance if all six are wired in parallel?

7. (I) Suppose that you have a 680-Ω, a 720-Ω, and a 1.20-kΩ resistor. What is (a) the maximum, and (b) the minimum resistance you can obtain by combining these?

8. (I) How many 10-Ω resistors must be connected in series to give an equivalent resistance to five 100-Ω resistors connected in parallel?

9. (II) Suppose that you have a 9.0-V battery and you wish to apply a voltage of only 4.0 V. Given an unlimited supply of 1.0-Ω resistors, how could you connect them so as to make a "voltage divider" that produces a 4.0-V output for a 9.0-V input?

10. (II) Three 1.70-kΩ resistors can be connected together in four different ways, making combinations of series and/or parallel circuits. What are these four ways, and what is the net resistance in each case?

11. (II) A battery with an emf of 12.0 V shows a terminal voltage of 11.8 V when operating in a circuit with two light-bulbs, each rated at 4.0 W (at 12.0 V), which are connected in parallel. What is the battery's internal resistance?

12. (II) Eight identical bulbs are connected in series across a 110-V line. (a) What is the voltage across each bulb? (b) If the current is 0.42 A, what is the resistance of each bulb, and what is the power dissipated in each?

13. (II) Eight bulbs are connected in parallel to a 110-V source by two long leads of total resistance 1.4 Ω. If 240 mA flows through each bulb, what is the resistance of each, and what fraction of the total power is wasted in the leads?

14. (II) The performance of the starter circuit in an automobile can be significantly degraded by a small amount of corrosion on a battery terminal. Figure 26–38a depicts a properly functioning circuit with a battery (12.5-V emf, 0.02-Ω internal resistance) attached via corrosion-free cables to a starter motor of resistance $R_S = 0.15\ \Omega$. Suppose that later, corrosion between a battery terminal and a starter cable introduces an extra series resistance of just $R_C = 0.10\ \Omega$ into the circuit as suggested in Fig. 26–38b. Let P_0 be the power delivered to the starter in the circuit free of corrosion, and let P be the power delivered to the circuit with corrosion. Determine the ratio P/P_0.

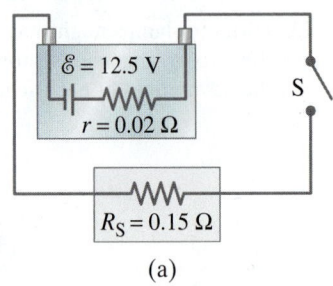

(a)

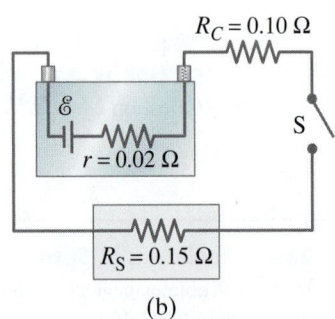

(b)

FIGURE 26–38
Problem 14.

15. (II) A close inspection of an electric circuit reveals that a 480-Ω resistor was inadvertently soldered in the place where a 370-Ω resistor is needed. How can this be fixed without removing anything from the existing circuit?

16. (II) Determine (a) the equivalent resistance of the circuit shown in Fig. 26–39, and (b) the voltage across each resistor.

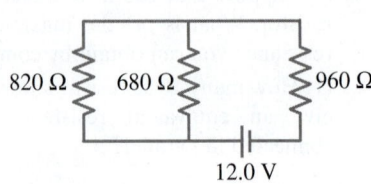

FIGURE 26–39
Problem 16.

820 Ω 680 Ω 960 Ω

12.0 V

17. (II) A 75-W, 110-V bulb is connected in parallel with a 25-W, 110-V bulb. What is the net resistance?

18. (II) (a) Determine the equivalent resistance of the "ladder" of equal 125-Ω resistors shown in Fig. 26–40. In other words, what resistance would an ohmmeter read if connected between points A and B? (b) What is the current through each of the three resistors on the left if a 50.0-V battery is connected between points A and B?

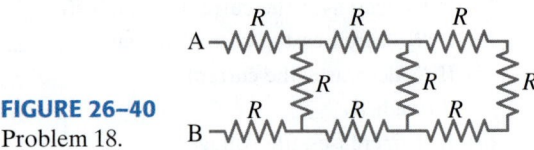

FIGURE 26–40
Problem 18.

19. (II) What is the net resistance of the circuit connected to the battery in Fig. 26–41?

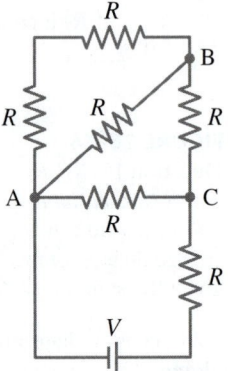

FIGURE 26–41
Problems 19 and 20.

20. (II) Calculate the current through each resistor in Fig. 26–41 if each resistance $R = 1.20\ k\Omega$ and $V = 12.0\ V$. What is the potential difference between points A and B?

21. (II) The two terminals of a voltage source with emf $\mathscr{E}$ and internal resistance r are connected to the two sides of a load resistance R. For what value of R will the maximum power be delivered from the source to the load?

22. (II) Two resistors when connected in series to a 110-V line use one-fourth the power that is used when they are connected in parallel. If one resistor is 3.8 kΩ, what is the resistance of the other?

23. (III) Three equal resistors (R) are connected to a battery as shown in Fig. 26–42. Qualitatively, what happens to (a) the voltage drop across each of these resistors, (b) the current flow through each, and (c) the terminal voltage of the battery, when the switch S is opened, after having been closed for a long time? (d) If the emf of the battery is 9.0 V, what is its terminal voltage when the switch is closed if the internal resistance r is 0.50 Ω and $R = 5.50\ \Omega$? (e) What is the terminal voltage when the switch is open?

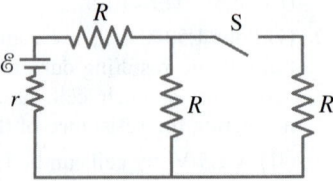

FIGURE 26–42
Problem 23.

24. (III) A 2.8-kΩ and a 3.7-kΩ resistor are connected in parallel; this combination is connected in series with a 1.8-kΩ resistor. If each resistor is rated at $\frac{1}{2}$ W (maximum without overheating), what is the maximum voltage that can be applied across the whole network?

25. (III) Consider the network of resistors shown in Fig. 26–43. Answer qualitatively: (*a*) What happens to the voltage across each resistor when the switch S is closed? (*b*) What happens to the current through each when the switch is closed? (*c*) What happens to the power output of the battery when the switch is closed? (*d*) Let $R_1 = R_2 = R_3 = R_4 = 125\,\Omega$ and $V = 22.0\,\text{V}$. Determine the current through each resistor before and after closing the switch. Are your qualitative predictions confirmed?

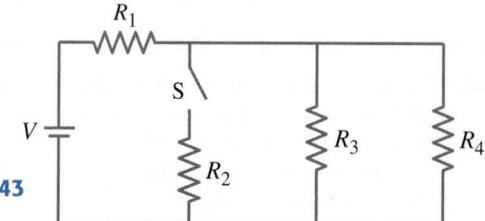

FIGURE 26–43
Problem 25.

26. (III) You are designing a wire resistance heater to heat an enclosed volume of gas. For the apparatus to function properly, this heater must transfer heat to the gas at a very constant rate. While in operation, the resistance of the heater will always be close to the value $R = R_0$, but may fluctuate slightly causing its resistance to vary a small amount $\Delta R\ (\ll R_0)$. To maintain the heater at constant power, you design the circuit shown in Fig. 26–44, which includes two resistors, each of resistance r. Determine the value for r so that the heater power will remain constant even if its resistance R fluctuates by a small amount. [*Hint:* If $\Delta R \ll R_0$, then $\Delta P \approx \Delta R\,\dfrac{dP}{dR}\Big|_{R=R_0}$.]

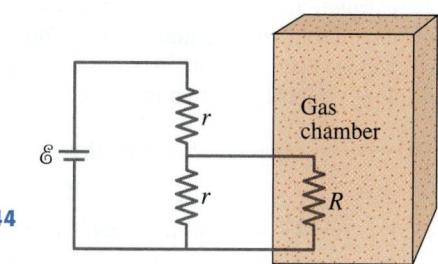

FIGURE 26–44
Problem 26.

26–3 Kirchhoff's Rules

27. (I) Calculate the current in the circuit of Fig. 26–45, and show that the sum of all the voltage changes around the circuit is zero.

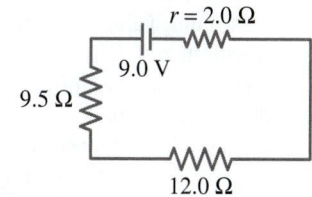

FIGURE 26–45
Problem 27.

28. (II) Determine the terminal voltage of each battery in Fig. 26–46.

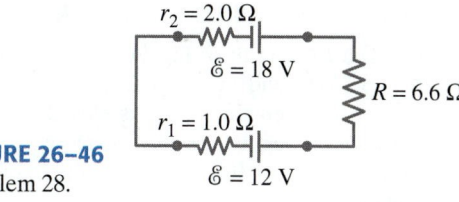

FIGURE 26–46
Problem 28.

29. (II) For the circuit shown in Fig. 26–47, find the potential difference between points a and b. Each resistor has $R = 130\,\Omega$ and each battery is 1.5 V.

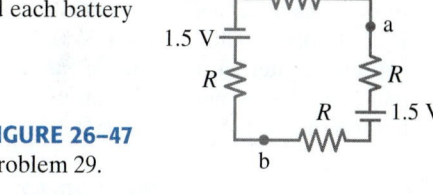

FIGURE 26–47
Problem 29.

30. (II) (*a*) A network of five equal resistors R is connected to a battery $\mathscr{E}$ as shown in Fig. 26–48. Determine the current I that flows out of the battery. (*b*) Use the value determined for I to find the single resistor R_{eq} that is equivalent to the five-resistor network.

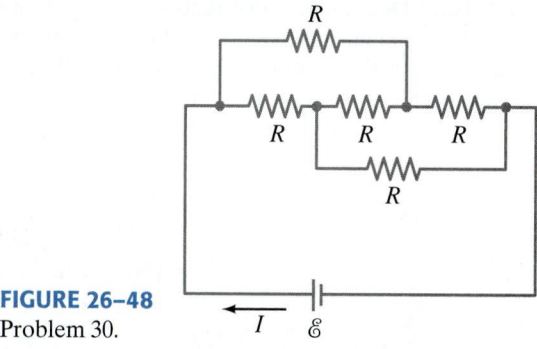

FIGURE 26–48
Problem 30.

31. (II) (*a*) What is the potential difference between points a and d in Fig. 26–49 (similar to Fig. 26–13, Example 26–9), and (*b*) what is the terminal voltage of each battery?

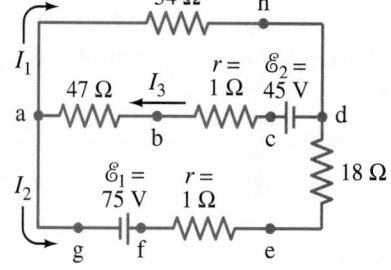

FIGURE 26–49
Problem 31.

32. (II) Calculate the currents in each resistor of Fig. 26–50.

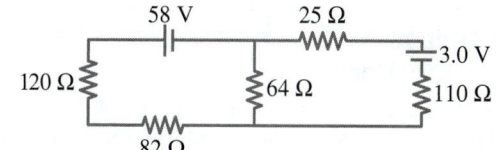

FIGURE 26–50 Problem 32.

33. (II) Determine the magnitudes and directions of the currents through R_1 and R_2 in Fig. 26–51.

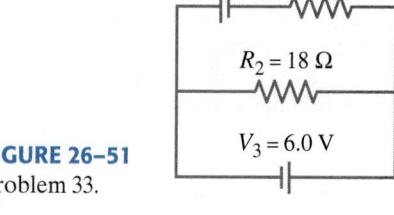

FIGURE 26–51
Problem 33.

34. (II) Determine the magnitudes and directions of the currents in each resistor shown in Fig. 26–52. The batteries have emfs of $\mathscr{E}_1 = 9.0$ V and $\mathscr{E}_2 = 12.0$ V and the resistors have values of $R_1 = 25\,\Omega$, $R_2 = 48\,\Omega$, and $R_3 = 35\,\Omega$. (a) Ignore internal resistance of the batteries. (b) Assume each battery has internal resistance $r = 1.0\,\Omega$.

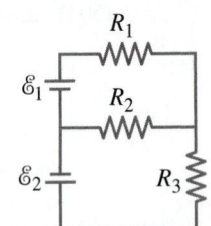

FIGURE 26–52
Problem 34.

35. (II) A voltage V is applied to n identical resistors connected in parallel. If the resistors are instead all connected in series with the applied voltage, show that the power transformed is decreased by a factor n^2.

36. (III) (a) Determine the currents I_1, I_2, and I_3 in Fig. 26–53. Assume the internal resistance of each battery is $r = 1.0\,\Omega$. (b) What is the terminal voltage of the 6.0-V battery?

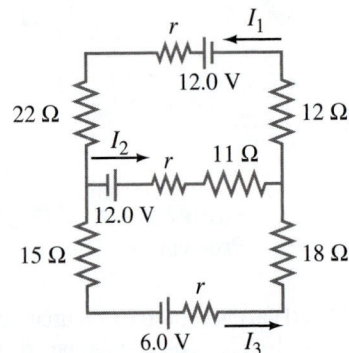

FIGURE 26–53
Problems 36 and 37.

37. (III) What would the current I_1 be in Fig. 26–53 if the 12-Ω resistor is shorted out (resistance = 0)? Let $r = 1.0\,\Omega$.

38. (III) Determine the current through each of the resistors in Fig. 26–54.

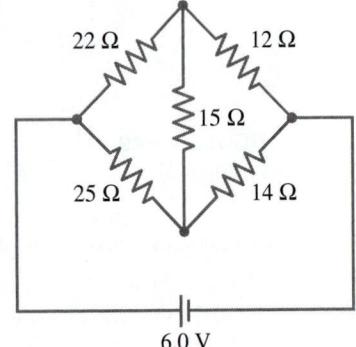

FIGURE 26–54
Problems 38 and 39.

39. (III) If the 25-Ω resistor in Fig. 26–54 is shorted out (resistance = 0), what then would be the current through the 15-Ω resistor?

40. (III) Twelve resistors, each of resistance R, are connected as the edges of a cube as shown in Fig. 26–55. Determine the equivalent resistance (a) between points a and b, the ends of a side; (b) between points a and c, the ends of a face diagonal; (c) between points a and d, the ends of the volume diagonal. [Hint: Apply an emf and determine currents; use symmetry at junctions.]

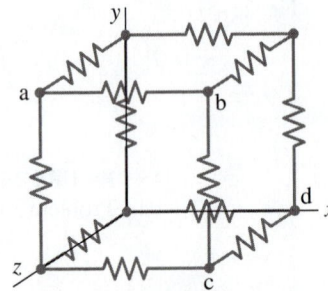

FIGURE 26–55
Problem 40.

41. (III) Determine the net resistance in Fig. 26–56 (a) between points a and c, and (b) between points a and b. Assume $R' \neq R$. [Hint: Apply an emf and determine currents; use symmetry at junctions.]

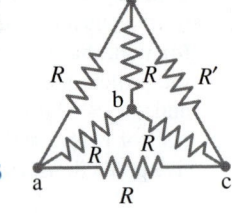

FIGURE 26–56
Problem 41.

26–4 Emfs Combined, Battery Charging

42. (II) Suppose two batteries, with unequal emfs of 2.00 V and 3.00 V, are connected as shown in Fig. 26–57. If each internal resistance is $r = 0.450\,\Omega$, and $R = 4.00\,\Omega$, what is the voltage across the resistor R?

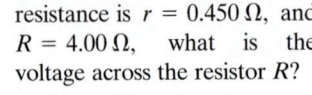

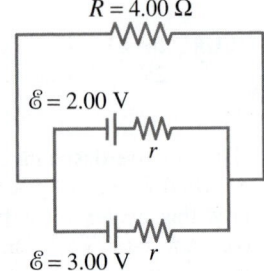

FIGURE 26–57
Problem 42.

26–5 *RC* Circuits

43. (I) Estimate the range of resistance needed to make a variable timer for typical intermittent windshield wipers if the capacitor used is on the order of 1 μF.

44. (II) In Fig. 26–58 (same as Fig. 26–17a), the total resistance is 15.0 kΩ, and the battery's emf is 24.0 V. If the time constant is measured to be 24.0 μs, calculate (a) the total capacitance of the circuit and (b) the time it takes for the voltage across the resistor to reach 16.0 V after the switch is closed.

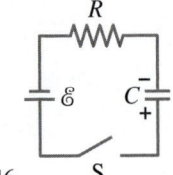

FIGURE 26–58
Problems 44 and 46.

45. (II) Two 3.8-μF capacitors, two 2.2-kΩ resistors, and a 12.0-V source are connected in series. Starting from the uncharged state, how long does it take for the current to drop from its initial value to 1.50 mA?

46. (II) How long does it take for the energy stored in a capacitor in a series *RC* circuit (Fig. 26–58) to reach 75% of its maximum value? Express answer in terms of the time constant $\tau = RC$.

47. (II) A parallel-plate capacitor is filled with a dielectric of dielectric constant K and high resistivity ρ (it conducts very slightly). This capacitor can be modeled as a pure capacitance C in parallel with a resistance R. Assume a battery places a charge $+Q$ and $-Q$ on the capacitor's opposing plates and is then disconnected. Show that the capacitor discharges with a time constant $\tau = K\varepsilon_0\rho$ (known as the *dielectric relaxation time*). Evaluate τ if the dielectric is glass with $\rho = 1.0 \times 10^{12}\,\Omega\cdot$m and $K = 5.0$.

48. (II) The *RC* circuit of Fig. 26–59 (same as Fig. 26–18a) has $R = 8.7$ kΩ and $C = 3.0\,\mu$F. The capacitor is at voltage V_0 at $t = 0$, when the switch is closed. How long does it take the capacitor to discharge to 0.10% of its initial voltage?

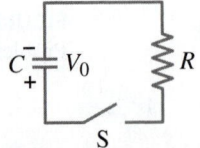

FIGURE 26–59
Problem 48.

49. (II) Consider the circuit shown in Fig. 26–60, where all resistors have the same resistance R. At $t = 0$, with the capacitor C uncharged, the switch is closed. (a) At $t = 0$, the three currents can be determined by analyzing a simpler, but equivalent, circuit. Identify this simpler circuit and use it to find the values of I_1, I_2, and I_3 at $t = 0$. (b) At $t = \infty$, the currents can be determined by analyzing a simpler, equivalent circuit. Identify this simpler circuit and implement it in finding the values of I_1, I_2, and I_3 at $t = \infty$. (c) At $t = \infty$, what is the potential difference across the capacitor?

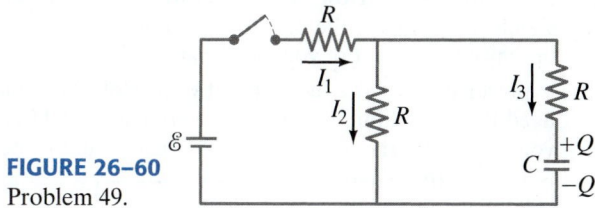

FIGURE 26–60
Problem 49.

50. (III) Determine the time constant for charging the capacitor in the circuit of Fig. 26–61. [*Hint:* Use Kirchhoff's rules.] (b) What is the maximum charge on the capacitor?

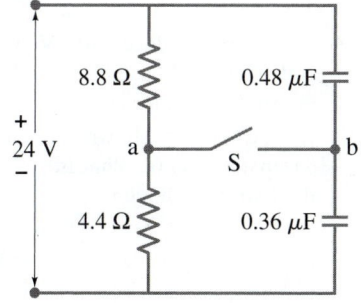

FIGURE 26–61
Problem 50.

51. (III) Two resistors and two uncharged capacitors are arranged as shown in Fig. 26–62. Then a potential difference of 24 V is applied across the combination as shown. (a) What is the potential at point a with switch S open? (Let $V = 0$ at the negative terminal of the source.) (b) What is the potential at point b with the switch open? (c) When the switch is closed, what is the final potential of point b? (d) How much charge flows through the switch S after it is closed?

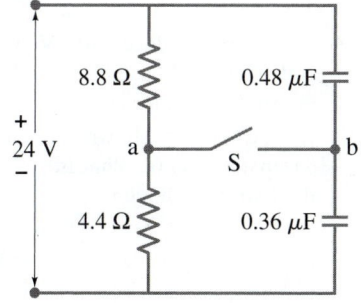

FIGURE 26–62
Problems 51 and 52.

52. (III) Suppose the switch S in Fig. 26–62 is closed. What is the time constant (or time constants) for charging the capacitors after the 24 V is applied?

*26–7 Ammeters and Voltmeters

*53. (I) An ammeter has a sensitivity of 35,000 Ω/V. What current in the galvanometer produces full-scale deflection?

*54. (I) What is the resistance of a voltmeter on the 250-V scale if the meter sensitivity is 35,000 Ω/V?

*55. (II) A galvanometer has a sensitivity of 45 kΩ/V and internal resistance 20.0 Ω. How could you make this into (a) an ammeter that reads 2.0 A full scale, or (b) a voltmeter reading 1.00 V full scale?

*56. (II) A galvanometer has an internal resistance of 32 Ω and deflects full scale for a 55-μA current. Describe how to use this galvanometer to make (a) an ammeter to read currents up to 25 A, and (b) a voltmeter to give a full scale deflection of 250 V.

*57. (II) A particular digital meter is based on an electronic module that has an internal resistance of 100 MΩ and a full-scale sensitivity of 400 mV. Two resistors connected as shown in Fig. 26–63 can be used to change the voltage range. Assume $R_1 = 10$ MΩ. Find the value of R_2 that will result in a voltmeter with a full-scale range of 40 V.

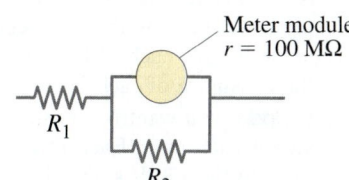

FIGURE 26–63
Problem 57.

*58. (II) A milliammeter reads 25 mA full scale. It consists of a 0.20-Ω resistor in parallel with a 33-Ω galvanometer. How can you change this ammeter to a voltmeter giving a full scale reading of 25 V without taking the ammeter apart? What will be the sensitivity (Ω/V) of your voltmeter?

*59. (II) A 45-V battery of negligible internal resistance is connected to a 44-kΩ and a 27-kΩ resistor in series. What reading will a voltmeter, of internal resistance 95 kΩ, give when used to measure the voltage across each resistor? What is the percent inaccuracy due to meter resistance for each case?

*60. (II) An ammeter whose internal resistance is 53 Ω reads 5.25 mA when connected in a circuit containing a battery and two resistors in series whose values are 650 Ω and 480 Ω. What is the actual current when the ammeter is absent?

*61. (II) A battery with $\mathscr{E} = 12.0$ V and internal resistance $r = 1.0$ Ω is connected to two 7.5-kΩ resistors in series. An ammeter of internal resistance 0.50 Ω measures the current, and at the same time a voltmeter with internal resistance 15 kΩ measures the voltage across one of the 7.5-kΩ resistors in the circuit. What do the ammeter and voltmeter read?

*62. (II) A 12.0-V battery (assume the internal resistance $= 0$) is connected to two resistors in series. A voltmeter whose internal resistance is 18.0 kΩ measures 5.5 V and 4.0 V, respectively, when connected across each of the resistors. What is the resistance of each resistor?

*63. (III) Two 9.4-kΩ resistors are placed in series and connected to a battery. A voltmeter of sensitivity 1000 Ω/V is on the 3.0-V scale and reads 2.3 V when placed across either resistor. What is the emf of the battery? (Ignore its internal resistance.)

*64. (III) When the resistor R in Fig. 26–64 is 35 Ω, the high-resistance voltmeter reads 9.7 V. When R is replaced by a 14.0-Ω resistor, the voltmeter reading drops to 8.1 V. What are the emf and internal resistance of the battery?

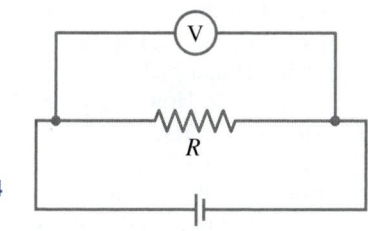

FIGURE 26–64
Problem 64.

General Problems

65. Suppose that you wish to apply a 0.25-V potential difference between two points on the human body. The resistance is about 1800 Ω, and you only have a 1.5-V battery. How can you connect up one or more resistors to produce the desired voltage?

66. A **three-way lightbulb** can produce 50 W, 100 W, or 150 W, at 120 V. Such a bulb contains two filaments that can be connected to the 120 V individually or in parallel. (a) Describe how the connections to the two filaments are made to give each of the three wattages. (b) What must be the resistance of each filament?

67. Suppose you want to run some apparatus that is 65 m from an electric outlet. Each of the wires connecting your apparatus to the 120-V source has a resistance per unit length of 0.0065 Ω/m. If your apparatus draws 3.0 A, what will be the voltage drop across the connecting wires and what voltage will be applied to your apparatus?

68. For the circuit shown in Fig. 26–18a, show that the decrease in energy stored in the capacitor from $t = 0$ until one time constant has elapsed equals the energy dissipated as heat in the resistor.

69. A heart pacemaker is designed to operate at 72 beats/min using a 6.5-μF capacitor in a simple RC circuit. What value of resistance should be used if the pacemaker is to fire (capacitor discharge) when the voltage reaches 75% of maximum?

70. Suppose that a person's body resistance is 950 Ω. (a) What current passes through the body when the person accidentally is connected to 110 V? (b) If there is an alternative path to ground whose resistance is 35 Ω, what current passes through the person? (c) If the voltage source can produce at most 1.5 A, how much current passes through the person in case (b)?

71. A **Wheatstone bridge** is a type of "bridge circuit" used to make measurements of resistance. The unknown resistance to be measured, R_x, is placed in the circuit with accurately known resistances R_1, R_2, and R_3 (Fig. 26–65). One of these, R_3, is a variable resistor which is adjusted so that when the switch is closed momentarily, the ammeter Ⓐ shows zero current flow. (a) Determine R_x in terms of R_1, R_2, and R_3. (b) If a Wheatstone bridge is "balanced" when $R_1 = 630\,\Omega$, $R_2 = 972\,\Omega$, and $R_3 = 78.6\,\Omega$, what is the value of the unknown resistance?

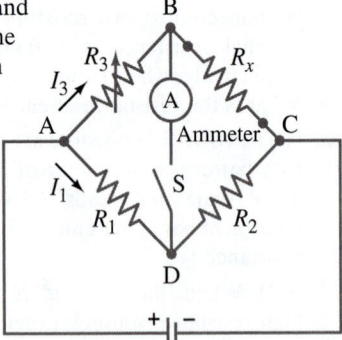

FIGURE 26–65
Problems 71 and 72.
Wheatstone bridge.

72. An unknown length of platinum wire 1.22 mm in diameter is placed as the unknown resistance in a Wheatstone bridge (see Problem 71, Fig. 26–65). Arms 1 and 2 have resistance of 38.0 Ω and 29.2 Ω, respectively. Balance is achieved when R_3 is 3.48 Ω. How long is the platinum wire?

73. The internal resistance of a 1.35-V mercury cell is 0.030 Ω, whereas that of a 1.5-V dry cell is 0.35 Ω. Explain why three mercury cells can more effectively power a 2.5-W hearing aid that requires 4.0 V than can three dry cells.

74. How many $\frac{1}{2}$-W resistors, each of the same resistance, must be used to produce an equivalent 3.2-kΩ, 3.5-W resistor? What is the resistance of each, and how must they be connected? Do not exceed $P = \frac{1}{2}$ W in each resistor.

75. A solar cell, 3.0 cm square, has an output of 350 mA at 0.80 V when exposed to full sunlight. A solar panel that delivers close to 1.3 A of current at an emf of 120 V to an external load is needed. How many cells will you need to create the panel? How big a panel will you need, and how should you connect the cells to one another? How can you optimize the output of your solar panel?

76. A power supply has a fixed output voltage of 12.0 V, but you need $V_T = 3.0$ V output for an experiment. (a) Using the voltage divider shown in Fig. 26–66, what should R_2 be if R_1 is 14.5 Ω? (b) What will the terminal voltage V_T be if you connect a load to the 3.0-V output, assuming the load has a resistance of 7.0 Ω?

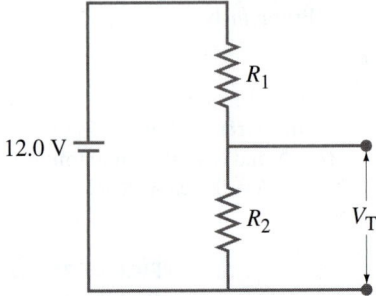

FIGURE 26–66
Problem 76.

77. The current through the 4.0-kΩ resistor in Fig. 26–67 is 3.10 mA. What is the terminal voltage V_{ba} of the "unknown" battery? (There are two answers. Why?)

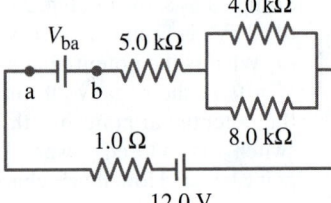

FIGURE 26–67
Problem 77.

78. A battery produces 40.8 V when 7.40 A is drawn from it, and 47.3 V when 2.80 A is drawn. What are the emf and internal resistance of the battery?

79. In the circuit shown in Fig. 26–68, the 33-Ω resistor dissipates 0.80 W. What is the battery voltage?

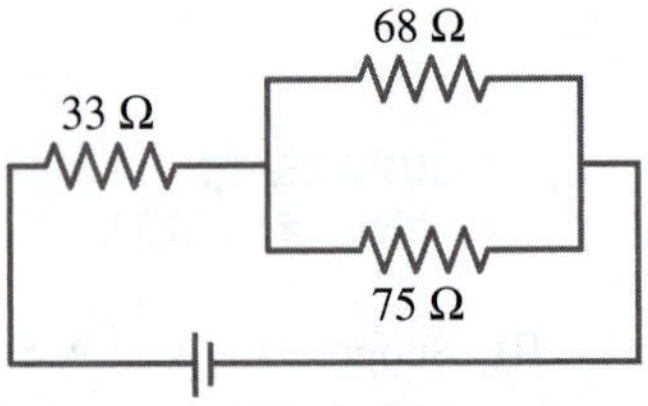

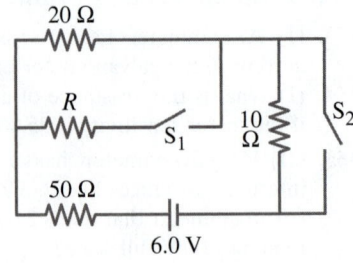

FIGURE 26–68
Problem 79.

80. The current through the 20-Ω resistor in Fig. 26–69 does not change whether the two switches S_1 and S_2 are both open or both closed. Use this clue to determine the value of the unknown resistance R.

FIGURE 26–69
Problem 80.

*81. (a) A voltmeter and an ammeter can be connected as shown in Fig. 26–70a to measure a resistance R. If V is the voltmeter reading, and I is the ammeter reading, the value of R will not quite be V/I (as in Ohm's law) because some of the current actually goes through the voltmeter. Show that the actual value of R is given by

$$\frac{1}{R} = \frac{I}{V} - \frac{1}{R_V},$$

where R_V is the voltmeter resistance. Note that $R \approx V/I$ if $R_V \gg R$. (b) A voltmeter and an ammeter can also be connected as shown in Fig. 26–70b to measure a resistance R. Show in this case that

$$R = \frac{V}{I} - R_A,$$

where V and I are the voltmeter and ammeter readings and R_A is the resistance of the ammeter. Note that $R \approx V/I$ if $R_A \ll R$.

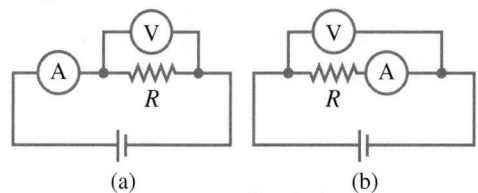

(a)   (b)

FIGURE 26–70 Problem 81.

82. (a) What is the equivalent resistance of the circuit shown in Fig. 26–71? (b) What is the current in the 18-Ω resistor? (c) What is the current in the 12-Ω resistor? (d) What is the power dissipation in the 4.5-Ω resistor?

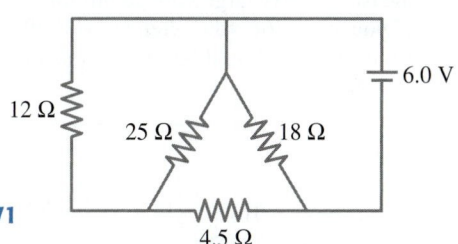

FIGURE 26–71
Problem 82.

83. A flashlight bulb rated at 2.0 W and 3.0 V is operated by a 9.0-V battery. To light the bulb at its rated voltage and power, a resistor R is connected in series as shown in Fig. 26–72. What value should the resistor have?

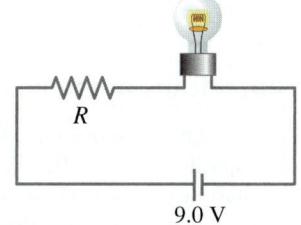

FIGURE 26–72
Problem 83.

84. Some light-dimmer switches use a variable resistor as shown in Fig. 26–73. The slide moves from position $x = 0$ to $x = 1$, and the resistance up to slide position x is proportional to x (the total resistance is $R_{var} = 150\,\Omega$ at $x = 1$). What is the power expended in the lightbulb if (a) $x = 1.00$, (b) $x = 0.65$, (c) $x = 0.35$?

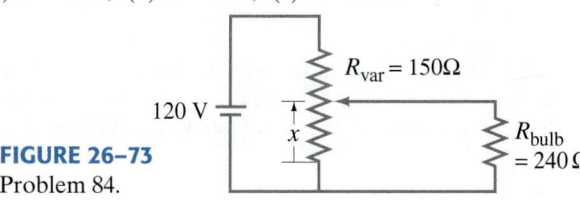

FIGURE 26–73
Problem 84.

85. A **potentiometer** is a device to precisely measure potential differences or emf, using a "null" technique. In the simple potentiometer circuit shown in Fig. 26–74, R' represents the total resistance of the resistor from A to B (which could be a long uniform "slide" wire), whereas R represents the resistance of only the part from A to the movable contact at C. When the unknown emf to be measured, $\mathscr{E}_x$, is placed into the circuit as shown, the movable contact C is moved until the galvanometer G gives a null reading (i.e., zero) when the switch S is closed. The resistance between A and C for this situation we call R_x. Next, a standard emf, $\mathscr{E}_s$, which is known precisely, is inserted into the circuit in place of $\mathscr{E}_x$ and again the contact C is moved until zero current flows through the galvanometer when the switch S is closed. The resistance between A and C now is called R_s. (a) Show that the unknown emf is given by

$$\mathscr{E}_x = \left(\frac{R_x}{R_s}\right)\mathscr{E}_s$$

where R_x, R_s, and $\mathscr{E}_s$ are all precisely known. The working battery is assumed to be fresh and to give a constant voltage. (b) A slide-wire potentiometer is balanced against a 1.0182-V standard cell when the slide wire is set at 33.6 cm out of a total length of 100.0 cm. For an unknown source, the setting is 45.8 cm. What is the emf of the unknown? (c) The galvanometer of a potentiometer has an internal resistance of 35 Ω and can detect a current as small as 0.012 mA. What is the minimum uncertainty possible in measuring an unknown voltage? (d) Explain the advantage of using this "null" method of measuring emf.

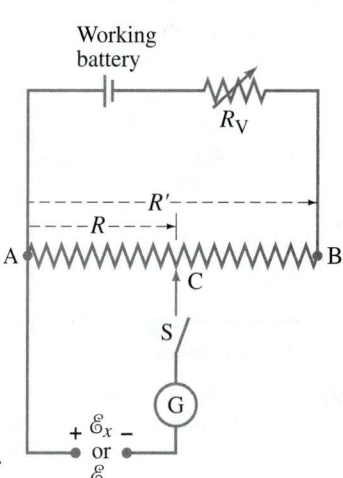

FIGURE 26–74
Potentiometer circuit.
Problem 85.

86. Electronic devices often use an RC circuit to protect against power outages as shown in Fig. 26–75. (a) If the protector circuit is supposed to keep the supply voltage at least 75% of full voltage for as long as 0.20 s, how big a resistance R is needed? The capacitor is 8.5 μF. Assume the attached "electronics" draws negligible current. (b) Between which two terminals should the device be connected, a and b, b and c, or a and c?

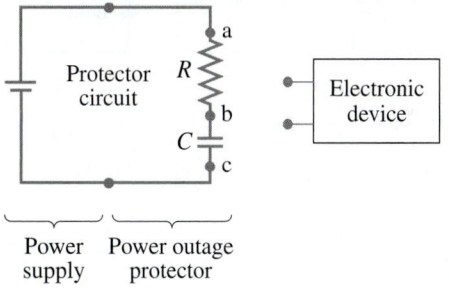

FIGURE 26–75
Problem 86.

87. The circuit shown in Fig. 26–76 is a primitive 4-bit **digital-to-analog converter** (DAC). In this circuit, to represent each digit (2^n) of a binary number, a "1" has the nth switch closed whereas zero ("0") has the switch open. For example, 0010 is represented by closing switch $n = 1$, while all other switches are open. Show that the voltage V across the 1.0-Ω resistor for the binary numbers 0001, 0010, 0100, and 1010 (which represent 1, 2, 4, 10) follows the pattern that you expect for a 4-bit DAC.

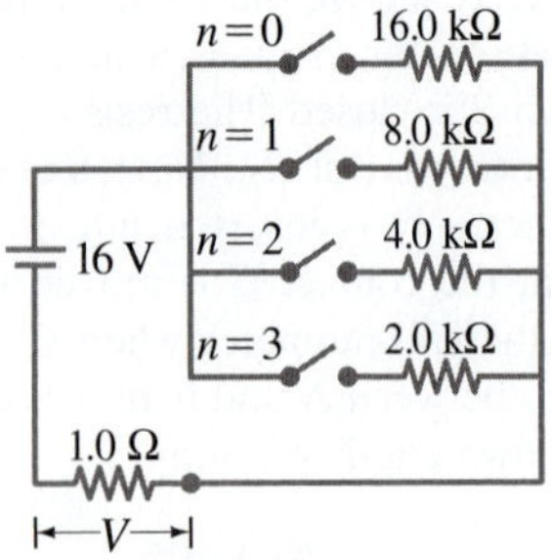

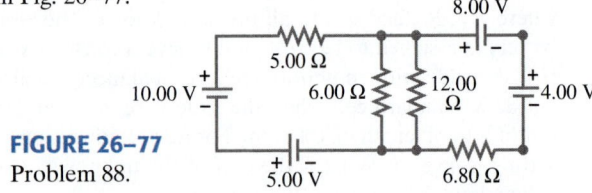

FIGURE 26–76
Problem 87.

88. Determine the current in each resistor of the circuit shown in Fig. 26–77.

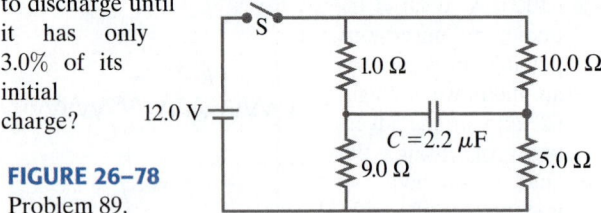

FIGURE 26–77
Problem 88.

89. In the circuit shown in Fig. 26–78, switch S is closed at time $t = 0$. (a) After the capacitor is fully charged, what is the voltage across it? How much charge is on it? (b) Switch S is now opened. How long does it now take for the capacitor to discharge until it has only 3.0% of its initial charge?

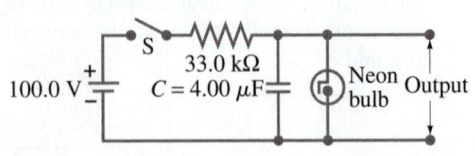

FIGURE 26–78
Problem 89.

90. Figure 26–79 shows the circuit for a simple **sawtooth oscillator**. At time $t = 0$, its switch S is closed. The neon bulb has initially infinite resistance until the voltage across it reaches 90.0 V, and then it begins to conduct with very little resistance (essentially zero). It stops conducting (its resistance becomes essentially infinite) when the voltage drops down to 65.0 V. (a) At what time t_1 does the neon bulb reach 90.0 V and start conducting? (b) At what time t_2 does the bulb reach 90.0 V for a second time and again become conducting? (c) Sketch the sawtooth waveform between $t = 0$ and $t = 0.70$ s.

FIGURE 26–79
Problem 90.

*91. Measurements made on circuits that contain large resistances can be confusing. Consider a circuit powered by a battery $\mathscr{E} = 15.000$ V with a 10.00-MΩ resistor in series with an unknown resistor R. As shown in Fig. 26–80, a particular voltmeter reads $V_1 = 366$ mV when connected across the 10.00-MΩ resistor, and this meter reads $V_2 = 7.317$ V when connected across R. Determine the value of R. [*Hint:* Define R_V as the voltmeter's internal resistance.]

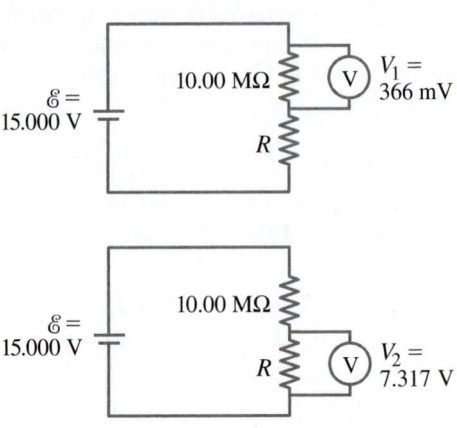

FIGURE 26–80 Problem 91.

*92. A typical voltmeter has an internal resistance of 10 MΩ and can only measure voltage differences of up to several hundred volts. Figure 26–81 shows the design of a probe to measure a very large voltage difference V using a voltmeter. If you want the voltmeter to read 50 V when $V = 50$ kV, what value R should be used in this probe?

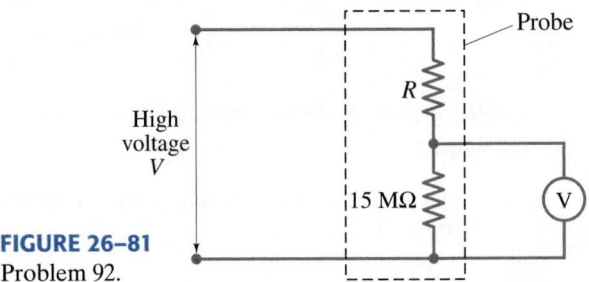

FIGURE 26–81
Problem 92.

* Numerical/Computer

*93. (II) An RC series circuit contains a resistor $R = 15$ kΩ, a capacitor $C = 0.30\,\mu$F, and a battery of emf $\mathscr{E} = 9.0$ V. Starting at $t = 0$, when the battery is connected, determine the charge Q on the capacitor and the current I in the circuit from $t = 0$ to $t = 10.0$ ms (at 0.1-ms intervals). Make graphs showing how the charge Q and the current I change with time within this time interval. From the graphs find the time at which the charge attains 63% of its final value, $C\mathscr{E}$, and the current drops to 37% of its initial value, $\mathscr{E}/R$.

Answers to Exercises

A: (a) 1.14 A; (b) 11.4 V; (c) $P_R = 13.1$ W, $P_r = 0.65$ W.

B: 6 Ω and 25 Ω.

C: $41I_3 - 45 + 21I_2 - 80 = 0$.

D: 180 A; this high current through the batteries could cause them to become very hot: the power dissipated in the weak battery would be $P = I^2r = (180\text{ A})^2(0.10\,\Omega) = 3200$ W!

E: (a).

Magnets produce magnetic fields, but so do electric currents. An electric current flowing in this straight wire produces a magnetic field which causes the tiny pieces of iron (iron "filings") to align in the field. We shall see in this Chapter how magnetic field is defined, and that the magnetic field direction is along the iron filings. The magnetic field lines due to the electric current in this long wire are in the shape of circles around the wire.

We also discuss how magnetic fields exert forces on electric currents and on charged particles, as well as useful applications of the interaction between magnetic fields and electric currents and moving electric charges.

I

Magnetism

C H A P T E R

27

CHAPTER-OPENING QUESTION—Guess now!

Which of the following can experience a force when placed in the magnetic field of a magnet?

(a) An electric charge at rest.
(b) An electric charge moving.
(c) An electric current in a wire.
(d) Another magnet.

The history of magnetism begins thousands of years ago. In a region of Asia Minor known as Magnesia, rocks were found that could attract each other. These rocks were called "magnets" after their place of discovery. Not until the nineteenth century, however, was it seen that magnetism and electricity are closely related. A crucial discovery was that electric currents produce magnetic effects (we will say "magnetic fields") like magnets do. All kinds of practical devices depend on magnetism, from compasses to motors, loudspeakers, computer memory, and electric generators.

CONTENTS

27–1 Magnets and Magnetic Fields
27–2 Electric Currents Produce Magnetic Fields
27–3 Force on an Electric Current in a Magnetic Field; Definition of $\vec{B}$
27–4 Force on an Electric Charge Moving in a Magnetic Field
27–5 Torque on a Current Loop; Magnetic Dipole Moment
*27–6 Applications: Motors, Loudspeakers, Galvanometers
27–7 Discovery and Properties of the Electron
27–8 The Hall Effect
*27–9 Mass Spectrometer

27–1 Magnets and Magnetic Fields

We have all observed a magnet attract paper clips, nails, and other objects made of iron, Fig. 27–1. Any magnet, whether it is in the shape of a bar or a horseshoe, has two ends or faces, called **poles**, which is where the magnetic effect is strongest. If a bar magnet is suspended from a fine thread, it is found that one pole of the magnet will always point toward the north. It is not known for sure when this fact was discovered, but it is known that the Chinese were making use of it as an aid to navigation by the eleventh century and perhaps earlier. This is the principle of a compass.

FIGURE 27–1 A horseshoe magnet attracts iron tacks and paper clips.

707

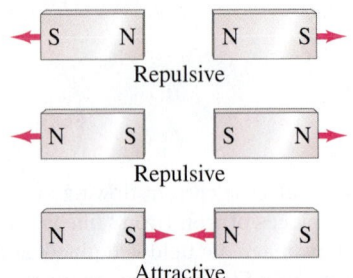

FIGURE 27–2 Like poles of a magnet repel; unlike poles attract. Red arrows indicate force direction.

FIGURE 27–3 If you split a magnet, you won't get isolated north and south poles; instead, two new magnets are produced, each with a north and a south pole.

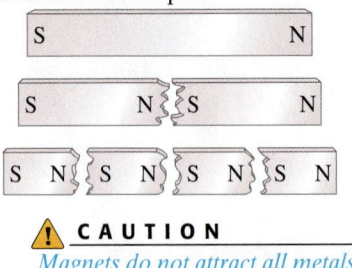

⚠ **CAUTION**

Magnets do not attract all metals

FIGURE 27–4 (a) Visualizing magnetic field lines around a bar magnet, using iron filings and compass needles. The red end of the bar magnet is its north pole. The N pole of a nearby compass needle points away from the north pole of the magnet. (b) Magnetic field lines for a bar magnet.

⚠ **CAUTION**

Magnetic field lines form closed loops, unlike electric field lines

A compass needle is simply a bar magnet which is supported at its center of gravity so that it can rotate freely. The pole of a freely suspended magnet that points toward geographic north is called the **north pole** of the magnet. The other pole points toward the south and is called the **south pole**.

It is a familiar observation that when two magnets are brought near one another, each exerts a force on the other. The force can be either attractive or repulsive and can be felt even when the magnets don't touch. If the north pole of one bar magnet is brought near the north pole of a second magnet, the force is repulsive. Similarly, if the south poles of two magnets are brought close, the force is repulsive. But when a north pole is brought near the south pole of another magnet, the force is attractive. These results are shown in Fig. 27–2, and are reminiscent of the forces between electric charges: like poles repel, and unlike poles attract. *But do not confuse magnetic poles with electric charge.* They are very different. One important difference is that a positive or negative electric charge can easily be isolated. But an isolated single magnetic pole has never been observed. If a bar magnet is cut in half, you do not obtain isolated north and south poles. Instead, two new magnets are produced, Fig. 27–3, each with north (N) and south (S) poles. If the cutting operation is repeated, more magnets are produced, each with a north and a south pole. Physicists have searched for isolated single magnetic poles (monopoles), but no **magnetic monopole** has ever been observed.

Only iron and a few other materials, such as cobalt, nickel, gadolinium, and some of their oxides and alloys, show strong magnetic effects. They are said to be **ferromagnetic** (from the Latin word *ferrum* for iron). Other materials show some slight magnetic effect, but it is very weak and can be detected only with delicate instruments. We will look in more detail at ferromagnetism in Section 28–7.

In Chapter 21, we used the concept of an electric field surrounding an electric charge. In a similar way, we can picture a **magnetic field** surrounding a magnet. The force one magnet exerts on another can then be described as the interaction between one magnet and the magnetic field of the other. Just as we drew electric field lines, we can also draw **magnetic field lines**. They can be drawn, as for electric field lines, so that (1) the direction of the magnetic field is tangent to a field line at any point, and (2) the number of lines per unit area is proportional to the strength of the magnetic field.

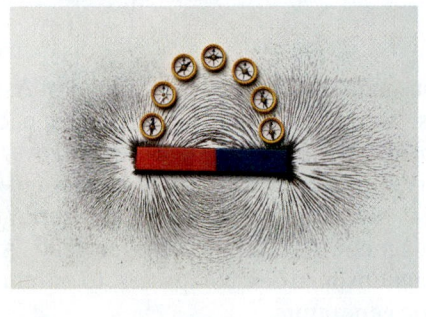

(a)

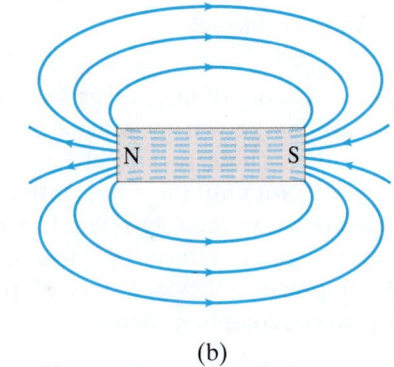

(b)

The *direction* of the magnetic field at a given point can be defined as the direction that the north pole of a compass needle would point if placed at that point. (A more precise definition will be given in Section 27–3.) Figure 27–4a shows how thin iron filings (acting like tiny magnets) reveal the magnetic field lines by lining up like the compass needles. The magnetic field determined in this way for the field surrounding a bar magnet is shown in Fig. 27–4b. Notice that because of our definition, the lines always point out from the north pole and in toward the south pole of a magnet (the north pole of a magnetic compass needle is attracted to the south pole of the magnet).

Magnetic field lines continue inside a magnet, as indicated in Fig. 27–4b. Indeed, given the lack of single magnetic poles, magnetic field lines always form closed loops, unlike electric field lines that begin on positive charges and end on negative charges.

Earth's Magnetic Field

The Earth's magnetic field is shown in Fig. 27–5. The pattern of field lines is as if there were an imaginary bar magnet inside the Earth. Since the north pole (N) of a compass needle points north, the Earth's magnetic pole which is in the geographic north is magnetically a south pole, as indicated in Fig. 27–5 by the S on the schematic bar magnet inside the Earth. Remember that the north pole of one magnet is attracted to the south pole of another magnet. Nonetheless, Earth's pole in the north is still often called the "north magnetic pole," or "geomagnetic north," simply because it is in the north. Similarly, the Earth's southern magnetic pole, which is near the geographic south pole, is magnetically a north pole (N). The Earth's magnetic poles do not coincide with the *geographic* poles, which are on the Earth's axis of rotation. The north magnetic pole, for example, is in the Canadian Arctic,[†] about 900 km from the geographic north pole, or "true north." This difference must be taken into account when you use a compass (Fig. 27–6). The angular difference between magnetic north and true (geographical) north is called the **magnetic declination**. In the U.S. it varies from 0° to about 20°, depending on location.

Notice in Fig. 27–5 that the Earth's magnetic field at most locations is not tangent to the Earth's surface. The angle that the Earth's magnetic field makes with the horizontal at any point is referred to as the **angle of dip**.

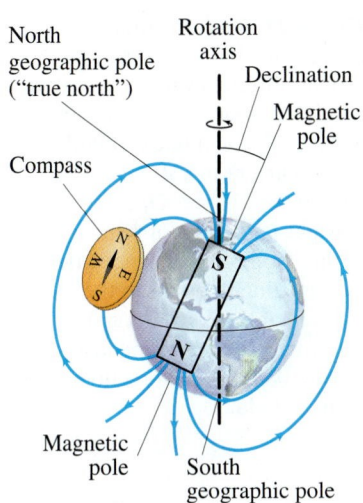

FIGURE 27–5 The Earth acts like a huge magnet; but its magnetic poles are not at the geographic poles, which are on the Earth's rotation axis.

| **EXERCISE A** Does the Earth's magnetic field have a greater magnitude near the poles or near the equator? [*Hint*: Note the field lines in Fig. 27–5.]

FIGURE 27–6 Using a map and compass in the wilderness. First you align the compass case so the needle points away from true north (N) exactly the number of degrees of declination as stated on the map (15° for the place shown on this topographic map of a part of California). Then align the map with true north, as shown, *not* with the compass needle.

Uniform Magnetic Field

The simplest magnetic field is one that is uniform—it doesn't change in magnitude or direction from one point to another. A perfectly uniform field over a large area is not easy to produce. But the field between two flat parallel pole pieces of a magnet is nearly uniform if the area of the pole faces is large compared to their separation, as shown in Fig. 27–7. At the edges, the field "fringes" out somewhat: the magnetic field lines are no longer quite parallel and uniform. The parallel evenly spaced field lines in the central region of the gap indicate that the field is uniform at points not too near the edges, much like the electric field between two parallel plates (Fig. 23–16).

FIGURE 27–7 Magnetic field between two wide poles of a magnet is nearly uniform, except near the edges.

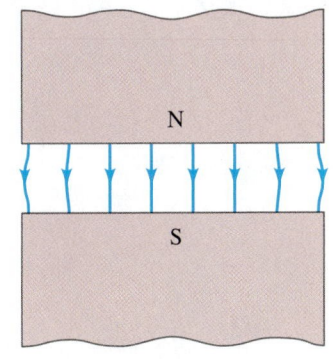

[†]Magnetic north is moving many kilometers a year at present. Magnetism in rocks suggests that the Earth's poles have not only moved significantly over geologic time, but have also reversed direction 400 times over the last 330 million years.

FIGURE 27–8 (a) Deflection of compass needles near a current-carrying wire, showing the presence and direction of the magnetic field. (b) Magnetic field lines around an electric current in a straight wire. See also the Chapter-Opening photo. (c) Right-hand rule for remembering the direction of the magnetic field: when the thumb points in the direction of the conventional current, the fingers wrapped around the wire point in the direction of the magnetic field.

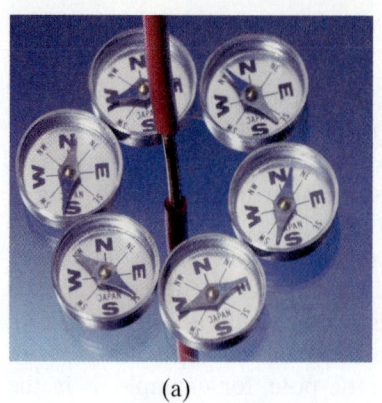

(a)

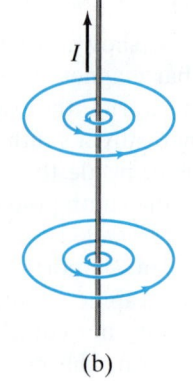

(b)

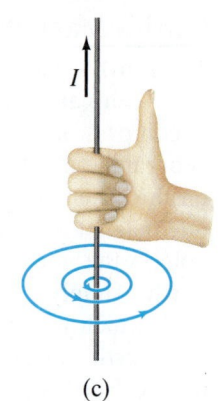

(c)

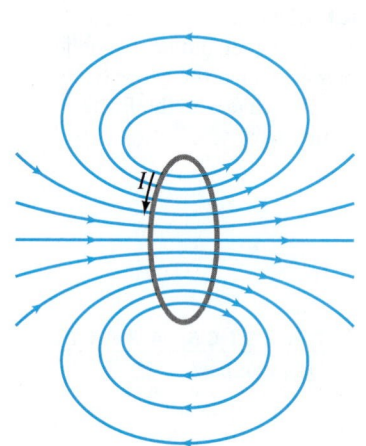

FIGURE 27–9 Magnetic field lines due to a circular loop of wire.

Right-hand-rule 1: magnetic field direction produced by electric current

FIGURE 27–10 Right-hand rule for determining the direction of the magnetic field relative to the current.

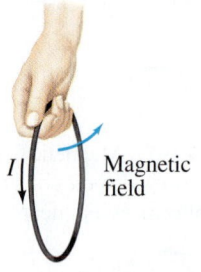

I Magnetic field

27–2 Electric Currents Produce Magnetic Fields

During the eighteenth century, many scientists sought to find a connection between electricity and magnetism. A stationary electric charge and a magnet were shown to have no influence on each other. But in 1820, Hans Christian Oersted (1777–1851) found that when a compass needle is placed near a wire, the needle deflects as soon as the two ends of the wire are connected to the terminals of a battery and the wire carries an electric current. As we have seen, a compass needle is deflected by a magnetic field. So Oersted's experiment showed that **an electric current produces a magnetic field**. He had found a connection between electricity and magnetism.

A compass needle placed near a straight section of current-carrying wire experiences a force, causing the needle to align tangent to a circle around the wire, Fig. 27–8a. Thus, the magnetic field lines produced by a current in a straight wire are in the form of circles with the wire at their center, Fig. 27–8b. The direction of these lines is indicated by the north pole of the compasses in Fig. 27–8a. There is a simple way to remember the direction of the magnetic field lines in this case. It is called a **right-hand rule**: grasp the wire with your right hand so that your thumb points in the direction of the conventional (positive) current; then your fingers will encircle the wire in the direction of the magnetic field, Fig. 27–8c.

The magnetic field lines due to a circular loop of current-carrying wire can be determined in a similar way using a compass. The result is shown in Fig. 27–9. Again the right-hand rule can be used, as shown in Fig. 27–10. Unlike the uniform field shown in Fig. 27–7, the magnetic fields shown in Figs. 27–8 and 27–9 are *not* uniform—the fields are different in magnitude and direction at different points.

EXERCISE B A straight wire carries a current directly toward you. In what direction are the magnetic field lines surrounding the wire?

27–3 Force on an Electric Current in a Magnetic Field; Definition of $\vec{B}$

In Section 27–2 we saw that an electric current exerts a force on a magnet, such as a compass needle. By Newton's third law, we might expect the reverse to be true as well: we should expect that *a magnet exerts a force on a current-carrying wire*. Experiments indeed confirm this effect, and it too was first observed by Oersted.

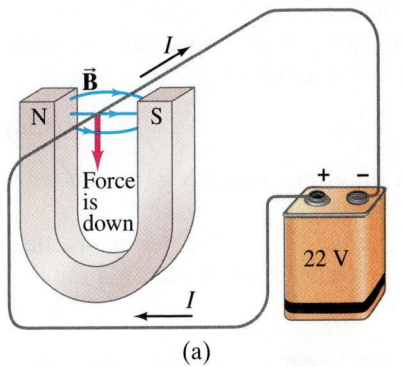

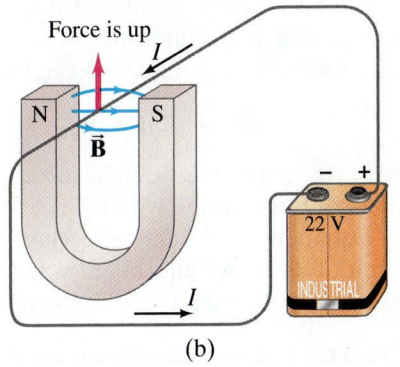

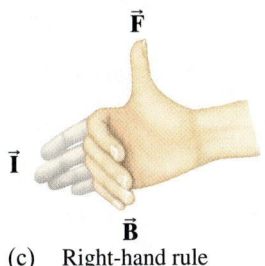

(a)   (b)   (c)   Right-hand rule

Suppose a straight wire is placed in the magnetic field between the poles of a horseshoe magnet as shown in Fig. 27–11. When a current flows in the wire, experiment shows that a force is exerted on the wire. But this force is *not* toward one or the other pole of the magnet. Instead, the force is directed at right angles to the magnetic field direction, downward in Fig. 27–11a. If the current is reversed in direction, the force is in the opposite direction, upward as shown in Fig. 27–11b. Experiments show that *the direction of the force is always perpendicular to the direction of the current and also perpendicular to the direction of the magnetic field,* $\vec{\mathbf{B}}$.

The direction of the force is given by another **right-hand rule**, as illustrated in Fig. 27–11c. Orient your right hand until your outstretched fingers can point in the direction of the conventional current I, and when you bend your fingers they point in the direction of the magnetic field lines, $\vec{\mathbf{B}}$. Then your outstretched thumb will point in the direction of the force $\vec{\mathbf{F}}$ on the wire.

This right-hand rule describes the direction of the force. What about the magnitude of the force on the wire? It is found experimentally that the magnitude of the force is directly proportional to the current I in the wire, and to the length ℓ of wire exposed to the magnetic field (assumed uniform). Furthermore, if the magnetic field B is made stronger, the force is found to be proportionally greater. The force also depends on the angle θ between the current direction and the magnetic field (Fig. 27–12), being proportional to $\sin \theta$. Thus, the force on a wire carrying a current I with length ℓ in a uniform magnetic field B is given by

$$F \ \propto \ I\ell B \sin \theta.$$

When the current is perpendicular to the field lines $(\theta = 90°)$, the force is strongest. When the wire is parallel to the magnetic field lines $(\theta = 0°)$, there is no force at all.

Up to now we have not defined the magnetic field strength precisely. In fact, the magnetic field B can be conveniently defined in terms of the above proportion so that the proportionality constant is precisely 1. Thus we have

$$F \ = \ I\ell B \sin \theta. \tag{27–1}$$

If the direction of the current is perpendicular to the field $\vec{\mathbf{B}}$ $(\theta = 90°)$, then the force is

$$F_{\max} \ = \ I\ell B. \qquad \left[\text{current} \perp \vec{\mathbf{B}}\right] \tag{27–2}$$

If the current is parallel to the field $(\theta = 0°)$, the force is zero. The magnitude of $\vec{\mathbf{B}}$ can be defined using Eq. 27–2 as $B = F_{\max}/I\ell$, where $F_{\max}$ is the magnitude of the force on a straight length ℓ of wire carrying a current I when the wire is perpendicular to $\vec{\mathbf{B}}$.

The relation between the force $\vec{\mathbf{F}}$ on a wire carrying current I, and the magnetic field $\vec{\mathbf{B}}$ that causes the force, can be written as a vector equation. To do so, we recall that the direction of $\vec{\mathbf{F}}$ is given by the right-hand rule (Fig. 27–11c), and the magnitude by Eq. 27–1. This is consistent with the definition of the vector cross product (see Section 11–2), so we can write

$$\vec{\mathbf{F}} \ = \ I\vec{\boldsymbol{\ell}} \times \vec{\mathbf{B}}; \tag{27–3}$$

here, $\vec{\boldsymbol{\ell}}$ is a vector whose magnitude is the length of the wire and its direction is along the wire (assumed straight) in the direction of the conventional (positive) current.

FIGURE 27–11 (a) Force on a current-carrying wire placed in a magnetic field $\vec{\mathbf{B}}$; (b) same, but current reversed; (c) right-hand rule for setup in (b).

Right-hand-rule 2:
force on current exerted by $\vec{\mathbf{B}}$

FIGURE 27–12 Current-carrying wire in a magnetic field. Force on the wire is directed into the page.

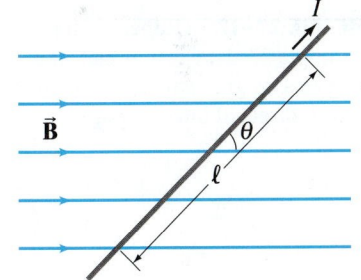

Equation 27–3 applies if the magnetic field is uniform and the wire is straight. If $\vec{\mathbf{B}}$ is not uniform, or if the wire does not everywhere make the same angle θ with $\vec{\mathbf{B}}$, then Eq. 27–3 can be written more generally as

$$d\vec{\mathbf{F}} = I\,d\vec{\boldsymbol{\ell}} \times \vec{\mathbf{B}}, \tag{27-4}$$

where $d\vec{\mathbf{F}}$ is the infinitesimal force acting on a differential length $d\vec{\boldsymbol{\ell}}$ of the wire. The total force on the wire is then found by integrating.

Equation 27–4 can serve (just as well as Eq. 27–2 or 27–3) as a practical definition of $\vec{\mathbf{B}}$. An equivalent way to define $\vec{\mathbf{B}}$, in terms of the force on a moving electric charge, is discussed in the next Section.

> **EXERCISE C** A wire carrying current I is perpendicular to a magnetic field of strength B. Assuming a fixed length of wire, which of the following changes will result in decreasing the force on the wire by a factor of 2? (a) Decrease the angle from 90° to 45°; (b) decrease the angle from 90° to 30°; (c) decrease the current in the wire to $I/2$; (d) decrease the magnetic field strength to $B/2$; (e) none of these will do it.

The SI unit for magnetic field B is the **tesla** (T). From Eqs. 27–1, 2, 3, or 4, we see that $1\,\text{T} = 1\,\text{N/A·m}$. An older name for the tesla is the "weber per meter squared" $(1\,\text{Wb/m}^2 = 1\,\text{T})$. Another unit sometimes used to specify magnetic field is a cgs unit, the **gauss** (G): $1\,\text{G} = 10^{-4}\,\text{T}$. A field given in gauss should always be changed to teslas before using with other SI units. To get a "feel" for these units, we note that the magnetic field of the Earth at its surface is about $\frac{1}{2}$ G or $0.5 \times 10^{-4}\,\text{T}$. On the other hand, the field near a small magnet attached to your refrigerator may be $100\,\text{G}$ $(0.01\,\text{T})$ whereas strong electromagnets can produce fields on the order of $2\,\text{T}$ and superconducting magnets can produce over $10\,\text{T}$.

EXAMPLE 27–1 **Magnetic force on a current-carrying wire.** A wire carrying a 30-A current has a length $\ell = 12\,\text{cm}$ between the pole faces of a magnet at an angle $\theta = 60°$ (Fig. 27–12). The magnetic field is approximately uniform at $0.90\,\text{T}$. We ignore the field beyond the pole pieces. What is the magnitude of the force on the wire?

APPROACH We use Eq. 27–1, $F = I\ell B \sin\theta$.

SOLUTION The force F on the 12-cm length of wire within the uniform field B is

$$F = I\ell B \sin\theta = (30\,\text{A})(0.12\,\text{m})(0.90\,\text{T})(0.866) = 2.8\,\text{N}.$$

> **EXERCISE D** A straight power line carries 30 A and is perpendicular to the Earth's magnetic field of $0.50 \times 10^{-4}\,\text{T}$. What magnitude force is exerted on 100 m of this power line?

On a diagram, when we want to represent an electric current or a magnetic field that is pointing out of the page (toward us) or into the page, we use $\odot$ or $\times$, respectively. The $\odot$ is meant to resemble the tip of an arrow pointing directly toward the reader, whereas the $\times$ or $\otimes$ resembles the tail of an arrow moving away. (See Figs. 27–13 and 27–14.)

EXAMPLE 27–2 **Measuring a magnetic field.** A rectangular loop of wire hangs vertically as shown in Fig. 27–13. A magnetic field $\vec{\mathbf{B}}$ is directed horizontally, perpendicular to the wire, and points out of the page at all points as represented by the symbol $\odot$. The magnetic field $\vec{\mathbf{B}}$ is very nearly uniform along the horizontal portion of wire ab (length $\ell = 10.0\,\text{cm}$) which is near the center of the gap of a large magnet producing the field. The top portion of the wire loop is free of the field. The loop hangs from a balance which measures a downward magnetic force (in addition to the gravitational force) of $F = 3.48 \times 10^{-2}\,\text{N}$ when the wire carries a current $I = 0.245\,\text{A}$. What is the magnitude of the magnetic field B?

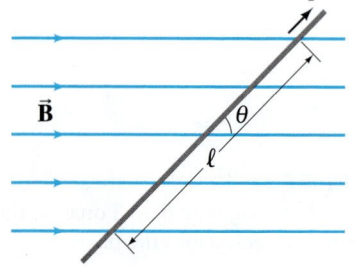

FIGURE 27–12 (Repeated for Example 27–1.) Current-carrying wire in a magnetic field. Force on the wire is directed into the page.

APPROACH Three straight sections of the wire loop are in the magnetic field: a horizontal section and two vertical sections. We apply Eq. 27–1 to each section and use the right-hand rule.

SOLUTION The magnetic force on the left vertical section of wire points to the left; the force on the vertical section on the right points to the right. These two forces are equal and in opposite directions and so add up to zero. Hence, the net magnetic force on the loop is that on the horizontal section ab, whose length is $\ell = 0.100$ m. The angle θ between $\vec{B}$ and the wire is $\theta = 90°$, so $\sin\theta = 1$. Thus Eq. 27–1 gives

$$B = \frac{F}{I\ell} = \frac{3.48 \times 10^{-2}\,\text{N}}{(0.245\,\text{A})(0.100\,\text{m})} = 1.42\,\text{T}.$$

NOTE This technique can be a precise means of determining magnetic field strength.

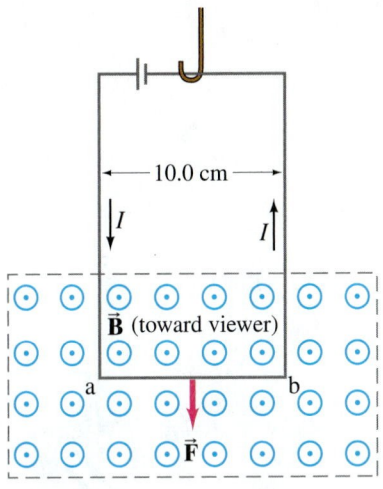

FIGURE 27–13 Measuring a magnetic field $\vec{B}$. Example 27–2.

EXAMPLE 27–3 Magnetic force on a semicircular wire. A rigid wire, carrying a current I, consists of a semicircle of radius R and two straight portions as shown in Fig. 27–14. The wire lies in a plane perpendicular to a uniform magnetic field $\vec{B}_0$. Note choice of x and y axis. The straight portions each have length ℓ within the field. Determine the net force on the wire due to the magnetic field $\vec{B}_0$.

APPROACH The forces on the two straight sections are equal $(= I\ell B_0)$ and in opposite directions, so they cancel. Hence the net force is that on the semicircular portion.

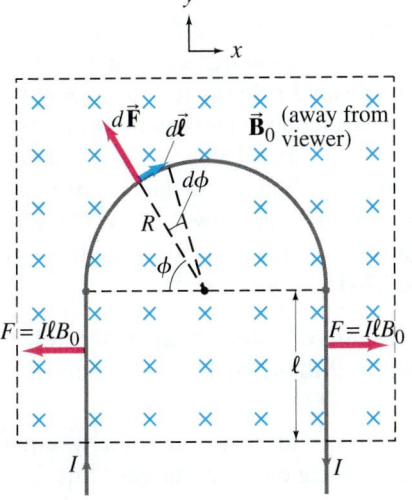

FIGURE 27–14 Example 27–3.

SOLUTION We divide the semicircle into short lengths $d\ell = R\,d\phi$ as indicated in Fig. 27–14, and use Eq. 27–4, $d\vec{F} = I\,d\vec{\ell} \times \vec{B}$, to find

$$dF = IB_0 R\,d\phi,$$

where dF is the force on the length $d\ell = R\,d\phi$, and the angle between $d\vec{\ell}$ and $\vec{B}_0$ is $90°$ (so $\sin\theta = 1$ in the cross product). The x component of the force $d\vec{F}$ on the segment $d\vec{\ell}$ shown, and the x component of $d\vec{F}$ for a symmetrically located $d\vec{\ell}$ on the other side of the semicircle, will cancel each other. Thus for the entire semicircle there will be no x component of force. Hence we need be concerned only with the y components, each equal to $dF\sin\phi$, and the total force will have magnitude

$$F = \int_0^\pi dF\sin\phi = IB_0 R\int_0^\pi \sin\phi\,d\phi = -IB_0 R\cos\phi\Big|_0^\pi = 2IB_0 R,$$

with direction vertically upward along the y axis in Fig. 27–14.

27–4 Force on an Electric Charge Moving in a Magnetic Field

We have seen that a current-carrying wire experiences a force when placed in a magnetic field. Since a current in a wire consists of moving electric charges, we might expect that freely moving charged particles (not in a wire) would also experience a force when passing through a magnetic field. Indeed, this is the case.

From what we already know we can predict the force on a single moving electric charge. If N such particles of charge q pass by a given point in time t, they constitute a current $I = Nq/t$. We let t be the time for a charge q to travel a distance ℓ in a magnetic field $\vec{B}$; then $\vec{\ell} = \vec{v}t$ where $\vec{v}$ is the velocity of the particle. Thus, the force on these N particles is, by Eq. 27–3, $\vec{F} = I\vec{\ell} \times \vec{B} = (Nq/t)(\vec{v}t) \times \vec{B} = Nq\vec{v} \times \vec{B}$. The force on *one* of the N particles is then

$$\vec{F} = q\vec{v} \times \vec{B}. \qquad (27\text{–}5a)$$

This basic and important result can be considered as an alternative way of defining the magnetic field $\vec{B}$, in place of Eq. 27–4 or 27–3. The magnitude of the force in Eq. 27–5a is

$$F = qvB \sin\theta. \qquad (27\text{–}5b)$$

This gives the magnitude of the force on a particle of charge q moving with velocity $\vec{v}$ at a point where the magnetic field has magnitude B. The angle between $\vec{v}$ and $\vec{B}$ is θ. The force is greatest when the particle moves perpendicular to $\vec{B}$ ($\theta = 90°$):

$$F_{max} = qvB. \qquad [\vec{v} \perp \vec{B}]$$

The force is *zero* if the particle moves *parallel* to the field lines ($\theta = 0°$). The *direction* of the force is perpendicular to the magnetic field $\vec{B}$ and to the velocity $\vec{v}$ of the particle. It is given again by a **right-hand rule** (as for any cross product): you orient your right hand so that your outstretched fingers point along the direction of the particle's velocity ($\vec{v}$), and when you bend your fingers they must point along the direction of $\vec{B}$. Then your thumb will point in the direction of the force. This is true only for *positively* charged particles, and will be "up" for the positive particle shown in Fig. 27–15. For negatively charged particles, the force is in exactly the opposite direction, "down" in Fig. 27–15.

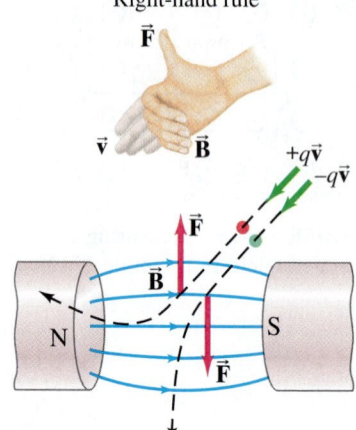

Right-hand rule

FIGURE 27–15 Force on charged particles due to a magnetic field is perpendicular to the magnetic field direction. If $\vec{v}$ is horizontal, then $\vec{F}$ is vertical.

Right-hand-rule 3:
force on moving charge exerted by $\vec{B}$

CONCEPTUAL EXAMPLE 27–4 | **Negative charge near a magnet.** A negative charge $-Q$ is placed at rest near a magnet. Will the charge begin to move? Will it feel a force? What if the charge were positive, $+Q$?

RESPONSE No to all questions. A charge at rest has velocity equal to zero. Magnetic fields exert a force only on moving electric charges (Eqs. 27–5).

EXERCISE E Return to the Chapter-Opening Question, page 707, and answer it again now. Try to explain why you may have answered differently the first time.

EXAMPLE 27–5 **Magnetic force on a proton.** A magnetic field exerts a force of 8.0×10^{-14} N toward the west on a proton moving vertically upward at a speed of 5.0×10^6 m/s (Fig. 27–16a). When moving horizontally in a northerly direction, the force on the proton is zero (Fig. 27–16b). Determine the magnitude and direction of the magnetic field in this region. (The charge on a proton is $q = +e = 1.6 \times 10^{-19}$ C.)

APPROACH Since the force on the proton is zero when moving north, the field must be in a north–south direction. In order to produce a force to the west when the proton moves upward, the right-hand rule tells us that $\vec{B}$ must point toward the north. (Your thumb points west and the outstretched fingers of your right hand point upward only when your bent fingers point north.) The magnitude of $\vec{B}$ is found using Eq. 27–5b.

SOLUTION Equation 27–5b with $\theta = 90°$ gives

$$B = \frac{F}{qv} = \frac{8.0 \times 10^{-14}\,\text{N}}{(1.6 \times 10^{-19}\,\text{C})(5.0 \times 10^6\,\text{m/s})} = 0.10\,\text{T}.$$

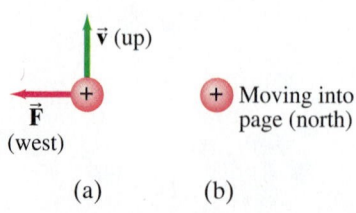

FIGURE 27–16 Example 27–5.

$\vec{v}$ (up)

$\vec{F}$
(west)

$+$ Moving into page (north)

(a)   (b)

EXAMPLE 27–6 ESTIMATE Magnetic force on ions during a nerve pulse.

Estimate the magnetic force due to the Earth's magnetic field on ions crossing a cell membrane during an action potential. Assume the speed of the ions is 10^{-2} m/s (Section 25–10).

APPROACH Using $F = qvB$, set the magnetic field of the Earth to be roughly $B \approx 10^{-4}$ T, and the charge $q \approx e \approx 10^{-19}$ C.

SOLUTION $F \approx (10^{-19}\,\text{C})(10^{-2}\,\text{m/s})(10^{-4}\,\text{T}) = 10^{-25}$ N.

NOTE This is an extremely small force. Yet it is thought migrating animals do somehow detect the Earth's magnetic field, and this is an area of active research.

The path of a charged particle moving in a plane perpendicular to a uniform magnetic field is a circle as we shall now show. In Fig. 27–17 the magnetic field is directed *into* the paper, as represented by ×'s. An electron at point P is moving to the right, and the force on it at this point is downward as shown (use the right-hand rule and reverse the direction for negative charge). The electron is thus deflected toward the page bottom. A moment later, say, when it reaches point Q, the force is still perpendicular to the velocity and is in the direction shown. Because the force is always perpendicular to $\vec{v}$, the magnitude of $\vec{v}$ does not change—the electron moves at constant speed. We saw in Chapter 5 that if the force on a particle is always perpendicular to its velocity $\vec{v}$, the particle moves in a circle and has a centripetal acceleration $a = v^2/r$ (Eq. 5–1). Thus a charged particle moves in a circular path with constant centripetal acceleration in a uniform magnetic field (see Example 27–7). The electron moves clockwise in Fig. 27–17. A positive particle in this field would feel a force in the opposite direction and would thus move counterclockwise.

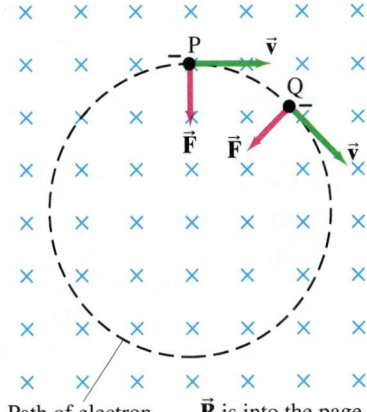

Path of electron $\vec{B}$ is into the page

FIGURE 27–17 Force exerted by a uniform magnetic field on a moving charged particle (in this case, an electron) produces a circular path.

EXAMPLE 27–7 Electron's path in a uniform magnetic field.

An electron travels at 2.0×10^7 m/s in a plane perpendicular to a uniform 0.010-T magnetic field. Describe its path quantitatively.

APPROACH The electron moves at speed v in a curved path and so must have a centripetal acceleration $a = v^2/r$ (Eq. 5–1). We find the radius of curvature using Newton's second law. The force is given by Eq. 27–5b with $\sin\theta = 1$: $F = qvB$.

SOLUTION We insert F and a into Newton's second law:

$$\Sigma F = ma$$
$$qvB = \frac{mv^2}{r}.$$

We solve for r and find

$$r = \frac{mv}{qB}.$$

Since $\vec{F}$ is perpendicular to $\vec{v}$, the magnitude of $\vec{v}$ doesn't change. From this equation we see that if $\vec{B} = $ constant, then $r = $ constant, and the curve must be a circle as we claimed above. To get r we put in the numbers:

$$r = \frac{(9.1 \times 10^{-31}\,\text{kg})(2.0 \times 10^7\,\text{m/s})}{(1.6 \times 10^{-19}\,\text{C})(0.010\,\text{T})} = 1.1 \times 10^{-2}\,\text{m} = 1.1\,\text{cm}.$$

NOTE See Fig. 27–18.

FIGURE 27–18 The blue ring inside the glass tube is the glow of a beam of electrons that ionize the gas molecules. The red coils of current-carrying wire produce a nearly uniform magnetic field, illustrating the circular path of charged particles in a uniform magnetic field.

The time T required for a particle of charge q moving with constant speed v to make one circular revolution in a uniform magnetic field $\vec{B}$ ($\perp \vec{v}$) is $T = 2\pi r/v$, where $2\pi r$ is the circumference of its circular path. From Example 27–7, $r = mv/qB$, so

$$T = \frac{2\pi m}{qB}.$$

Since T is the period of rotation, the frequency of rotation is

$$f = \frac{1}{T} = \frac{qB}{2\pi m}. \qquad \textbf{(27–6)}$$

This is often called the **cyclotron frequency** of a particle in a field because this is the frequency at which particles revolve in a cyclotron (see Problem 66).

CONCEPTUAL EXAMPLE 27–8 **Stopping charged particles.** Can a magnetic field be used to stop a single charged particle, as an electric field can?

RESPONSE No, because the force is always *perpendicular* to the velocity of the particle and thus cannot change the magnitude of its velocity. It also means the magnetic force cannot do work on the particle and so cannot change the kinetic energy of the particle.

Magnetic Fields

Magnetic fields are somewhat analogous to the electric fields of Chapter 21, but there are several important differences to recall:

1. The force experienced by a charged particle moving in a magnetic field is *perpendicular* to the direction of the magnetic field (and to the direction of the velocity of the particle), whereas the force exerted by an electric field is *parallel* to the direction of the field (and unaffected by the velocity of the particle).

2. The *right-hand rule*, in its different forms, is intended to help you determine the directions of magnetic field, and the forces they exert, and/or the directions of electric current or charged particle velocity. The right-hand rules (Table 27–1) are designed to deal with the "perpendicular" nature of these quantities.

TABLE 27–1 Summary of Right-hand Rules (= RHR)

Physical Situation	Example	How to Orient Right Hand	Result
1. Magnetic field produced by current (RHR-1)	I $\vec{\mathbf{B}}$ Fig. 27–8c	Wrap fingers around wire with thumb pointing in direction of current I	Fingers point in direction of $\vec{\mathbf{B}}$
2. Force on electric current I due to magnetic field (RHR-2)	$\vec{\mathbf{F}}$ $\vec{\mathbf{I}}$ $\vec{\mathbf{B}}$ Fig. 27–11c	Fingers point straight along current I, then bend along magnetic field $\vec{\mathbf{B}}$	Thumb points in direction of the force $\vec{\mathbf{F}}$
3. Force on electric charge $+q$ due to magnetic field (RHR-3)	$\vec{\mathbf{F}}$ $\vec{\mathbf{v}}$ $\vec{\mathbf{B}}$ Fig. 27–15	Fingers point along particle's velocity $\vec{\mathbf{v}}$, then along $\vec{\mathbf{B}}$	Thumb points in direction of the force $\vec{\mathbf{F}}$

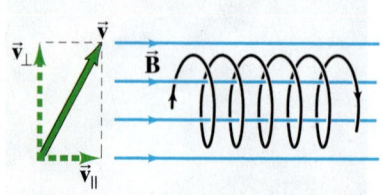

FIGURE 27–19 Example 27–9.

CONCEPTUAL EXAMPLE 27–9 **A helical path.** What is the path of a charged particle in a uniform magnetic field if its velocity is *not* perpendicular to the magnetic field?

RESPONSE The velocity vector can be broken down into components parallel and perpendicular to the field. The velocity component parallel to the field lines experiences no force $(\theta = 0)$, so this component remains constant. The velocity component perpendicular to the field results in circular motion about the field lines. Putting these two motions together produces a helical (spiral) motion around the field lines as shown in Fig. 27–19.

EXERCISE F What is the sign of the charge in Fig. 27–19? How would you modify the drawing if the sign were reversed?

*Aurora Borealis

Charged ions approach the Earth from the Sun (the "solar wind") and enter the atmosphere mainly near the poles, sometimes causing a phenomenon called the **aurora borealis** or "northern lights" in northern latitudes. To see why, consider Example 27–9 and Fig. 27–20 (see also Fig. 27–19). In Fig. 27–20 we imagine a stream of charged particles approaching the Earth. The velocity component *perpendicular* to the field for each particle becomes a circular orbit around the field lines, whereas the velocity component *parallel* to the field carries the particle along the field lines toward the poles. As a particle approaches the N pole, the magnetic field is stronger and the radius of the helical path becomes smaller.

A high concentration of charged particles ionizes the air, and as the electrons recombine with atoms, light is emitted (Chapter 37) which is the aurora. Auroras are especially spectacular during periods of high sunspot activity when the solar wind brings more charged particles toward Earth.

Lorentz Equation

If a particle of charge q moves with velocity $\vec{\mathbf{v}}$ in the presence of both a magnetic field $\vec{\mathbf{B}}$ and an electric field $\vec{\mathbf{E}}$, it will feel a force

$$\vec{\mathbf{F}} = q(\vec{\mathbf{E}} + \vec{\mathbf{v}} \times \vec{\mathbf{B}}) \qquad (27\text{–}7)$$

where we have made use of Eqs. 21–3 and 27–5a. Equation 27–7 is often called the **Lorentz equation** and is considered one of the basic equations in physics.

CONCEPTUAL EXAMPLE 27–10 | **Velocity selector, or filter: Crossed $\vec{\mathbf{E}}$ and $\vec{\mathbf{B}}$ fields.** Some electronic devices and experiments need a beam of charged particles all moving at nearly the same velocity. This can be achieved using both a uniform electric field and a uniform magnetic field, arranged so they are at right angles to each other. As shown in Fig. 27–21a, particles of charge q pass through slit S_1 and enter the region where $\vec{\mathbf{B}}$ points into the page and $\vec{\mathbf{E}}$ points down from the positive plate toward the negative plate. If the particles enter with different velocities, show how this device "selects" a particular velocity, and determine what this velocity is.

RESPONSE After passing through slit S_1, each particle is subject to two forces as shown in Fig. 27–21b. If q is positive, the magnetic force is upwards and the electric force downwards. (Vice versa if q is negative.) The exit slit, S_2, is assumed to be directly in line with S_1 and the particles' velocity $\vec{\mathbf{v}}$. Depending on the magnitude of $\vec{\mathbf{v}}$, some particles will be bent upwards and some downwards. The only ones to make it through the slit S_2 will be those for which the net force is zero: $\Sigma F = qvB - qE = 0$. Hence this device selects particles whose velocity is

$$v = \frac{E}{B}. \qquad (27\text{–}8)$$

This result does not depend on the sign of the charge q.

EXERCISE G A particle in a velocity selector as diagrammed in Fig. 27–21 hits below the exit hole, S_2. This means that the particle (*a*) is going faster than the selected speed; (*b*) is going slower than the selected speed; (*c*) answer *a* is true if $q > 0$, *b* is true if $q < 0$; (*d*) answer *a* is true if $q < 0$, *b* is true if $q > 0$.

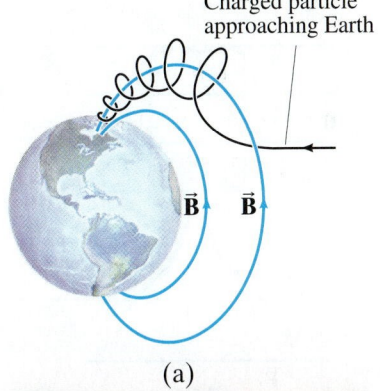

Charged particle approaching Earth

$\vec{\mathbf{B}}$ $\vec{\mathbf{B}}$

(a)

(b)

FIGURE 27–20 (a) Diagram showing a negatively charged particle that approaches the Earth and is "captured" by the magnetic field of the Earth. Such particles follow the field lines toward the poles as shown. (b) Photo of aurora borealis (here, in Kansas, where it is a rare sight).

FIGURE 27–21 A velocity selector: if $v = E/B$, the particles passing through S_1 make it through S_2.

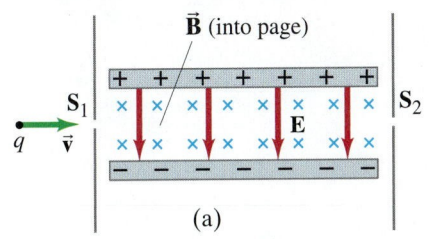

$\vec{\mathbf{B}}$ (into page)

S_1 S_2

q $\vec{\mathbf{v}}$ E

(a)

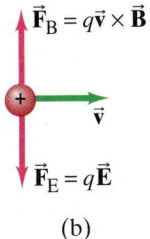

$\vec{\mathbf{F}}_B = q\vec{\mathbf{v}} \times \vec{\mathbf{B}}$

$\vec{\mathbf{v}}$

$\vec{\mathbf{F}}_E = q\vec{\mathbf{E}}$

(b)

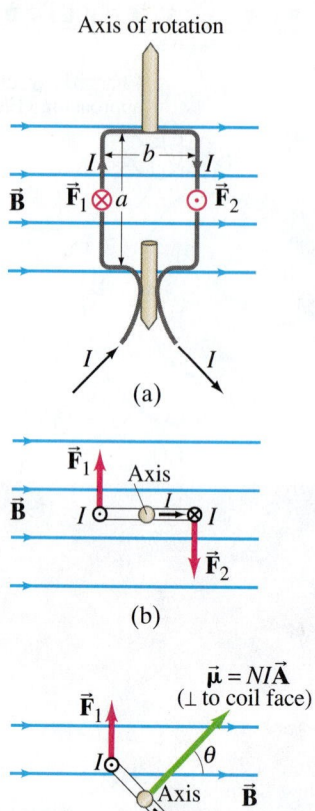

Axis of rotation

(a)

(b)

$\vec{\boldsymbol{\mu}} = NI\vec{\mathbf{A}}$
($\perp$ to coil face)

θ

Axis

(c)

FIGURE 27–22 Calculating the torque on a current loop in a magnetic field $\vec{\mathbf{B}}$. (a) Loop face parallel to $\vec{\mathbf{B}}$ field lines; (b) top view; (c) loop makes an angle to $\vec{\mathbf{B}}$, reducing the torque since the lever arm is reduced.

27–5 Torque on a Current Loop; Magnetic Dipole Moment

When an electric current flows in a closed loop of wire placed in an external magnetic field, as shown in Fig. 27–22, the magnetic force on the current can produce a torque. This is the principle behind a number of important practical devices, including motors and analog voltmeters and ammeters, which we discuss in the next Section. The interaction between a current and a magnetic field is important in other areas as well, including atomic physics.

Current flows through the rectangular loop in Fig. 27–22a, whose face we assume is parallel to $\vec{\mathbf{B}}$. $\vec{\mathbf{B}}$ exerts no force and no torque on the horizontal segments of wire because they are parallel to the field and $\sin\theta = 0$ in Eq. 27–1. But the magnetic field does exert a force on each of the vertical sections of wire as shown, $\vec{\mathbf{F}}_1$ and $\vec{\mathbf{F}}_2$ (see also top view, Fig. 27–22b). By right-hand-rule 2 (Fig. 27–11c or Table 27–1) the direction of the force on the upward current on the left is in the opposite direction from the equal magnitude force $\vec{\mathbf{F}}_2$ on the downward current on the right. These forces give rise to a net torque that acts to rotate the coil about its vertical axis.

Let us calculate the magnitude of this torque. From Eq. 27–2 (current $\perp \vec{\mathbf{B}}$), the force $F = IaB$, where a is the length of the vertical arm of the coil. The lever arm for each force is $b/2$, where b is the width of the coil and the "axis" is at the midpoint. The torques produced by $\vec{\mathbf{F}}_1$ and $\vec{\mathbf{F}}_2$ act in the same direction, so the total torque is the sum of the two torques:

$$\tau = IaB\frac{b}{2} + IaB\frac{b}{2} = IabB = IAB,$$

where $A = ab$ is the area of the coil. If the coil consists of N loops of wire, the current is then NI, so the torque becomes

$$\tau = NIAB.$$

If the coil makes an angle θ with the magnetic field, as shown in Fig. 27–22c, the forces are unchanged, but each lever arm is reduced from $\frac{1}{2}b$ to $\frac{1}{2}b\sin\theta$. Note that the angle θ is taken to be the angle between $\vec{\mathbf{B}}$ and the perpendicular to the face of the coil, Fig. 27–22c. So the torque becomes

$$\tau = NIAB\sin\theta. \qquad (27\text{–}9)$$

This formula, derived here for a rectangular coil, is valid for any shape of flat coil.

The quantity NIA is called the **magnetic dipole moment** of the coil and is considered a vector:

$$\vec{\boldsymbol{\mu}} = NI\vec{\mathbf{A}}, \qquad (27\text{–}10)$$

where the direction of $\vec{\mathbf{A}}$ (and therefore of $\vec{\boldsymbol{\mu}}$) is *perpendicular* to the plane of the coil (the green arrow in Fig. 27–22c) consistent with the right-hand rule (cup your right hand so your fingers wrap around the loop in the direction of current flow, then your thumb points in the direction of $\vec{\boldsymbol{\mu}}$ and $\vec{\mathbf{A}}$). With this definition of $\vec{\boldsymbol{\mu}}$, we can rewrite Eq. 27–9 in vector form:

$$\vec{\boldsymbol{\tau}} = NI\vec{\mathbf{A}} \times \vec{\mathbf{B}}$$

or

$$\vec{\boldsymbol{\tau}} = \vec{\boldsymbol{\mu}} \times \vec{\mathbf{B}}, \qquad (27\text{–}11)$$

which gives the correct magnitude and direction for the torque $\vec{\boldsymbol{\tau}}$.

Equation 27–11 has the same form as Eq. 21–9b for an electric dipole (with electric dipole moment $\vec{\mathbf{p}}$) in an electric field $\vec{\mathbf{E}}$, which is $\vec{\boldsymbol{\tau}} = \vec{\mathbf{p}} \times \vec{\mathbf{E}}$. And just as an electric dipole has potential energy given by $U = -\vec{\mathbf{p}} \cdot \vec{\mathbf{E}}$ when in an electric field, we expect a similar form for a magnetic dipole in a magnetic field. In order to rotate a current loop (Fig. 27–22) so as to increase θ, we must do work against the torque due to the magnetic field.

Hence the potential energy depends on angle (see Eq. 10–22, the work-energy principle for rotational motion) as

$$U = \int \tau \, d\theta = \int NIAB \sin\theta \, d\theta = -\mu B \cos\theta + C.$$

If we choose $U = 0$ at $\theta = \pi/2$, then the arbitrary constant C is zero and the potential energy is

$$U = -\mu B \cos\theta = -\vec{\mu} \cdot \vec{B}, \qquad (27\text{–}12)$$

as expected (compare Eq. 21–10). Bar magnets and compass needles, as well as current loops, can be considered as magnetic dipoles. Note the striking similarities of the fields produced by a bar magnet and a current loop, Figs. 27–4b and 27–9.

EXAMPLE 27–11 **Torque on a coil.** A circular coil of wire has a diameter of 20.0 cm and contains 10 loops. The current in each loop is 3.00 A, and the coil is placed in a 2.00-T external magnetic field. Determine the maximum and minimum torque exerted on the coil by the field.

APPROACH Equation 27–9 is valid for any shape of coil, including circular loops. Maximum and minimum torque are determined by the angle θ the coil makes with the magnetic field.

SOLUTION The area of one loop of the coil is

$$A = \pi r^2 = \pi(0.100 \, \text{m})^2 = 3.14 \times 10^{-2} \, \text{m}^2.$$

The maximum torque occurs when the coil's face is parallel to the magnetic field, so $\theta = 90°$ in Fig. 27–22c, and $\sin\theta = 1$ in Eq. 27–9:

$$\tau = NIAB \sin\theta = (10)(3.00 \, \text{A})(3.14 \times 10^{-2} \, \text{m}^2)(2.00 \, \text{T})(1) = 1.88 \, \text{N} \cdot \text{m}.$$

The minimum torque occurs if $\sin\theta = 0$, for which $\theta = 0°$, and then $\tau = 0$ from Eq. 27–9.

NOTE If the coil is free to turn, it will rotate toward the orientation with $\theta = 0°$.

EXAMPLE 27–12 **Magnetic moment of a hydrogen atom.** Determine the magnetic dipole moment of the electron orbiting the proton of a hydrogen atom at a given instant, assuming (in the Bohr model) it is in its ground state with a circular orbit of radius $0.529 \times 10^{-10} \, \text{m}$. [This is a very rough picture of atomic structure, but nonetheless gives an accurate result.]

APPROACH We start by setting the electrostatic force on the electron due to the proton equal to $ma = mv^2/r$ since the electron's acceleration is centripetal.

SOLUTION The electron is held in its orbit by the coulomb force, so Newton's second law, $F = ma$, gives

$$\frac{e^2}{4\pi\epsilon_0 r^2} = \frac{mv^2}{r};$$

so

$$v = \sqrt{\frac{e^2}{4\pi\epsilon_0 mr}}$$

$$= \sqrt{\frac{(8.99 \times 10^9 \, \text{N} \cdot \text{m}^2/\text{C}^2)(1.60 \times 10^{-19} \, \text{C})^2}{(9.11 \times 10^{-31} \, \text{kg})(0.529 \times 10^{-10} \, \text{m})}} = 2.19 \times 10^6 \, \text{m/s}.$$

Since current is the electric charge that passes a given point per unit time, the revolving electron is equivalent to a current

$$I = \frac{e}{T} = \frac{ev}{2\pi r},$$

where $T = 2\pi r/v$ is the time required for one orbit. Since the area of the orbit is $A = \pi r^2$, the magnetic dipole moment is

$$\mu = IA = \frac{ev}{2\pi r}(\pi r^2) = \tfrac{1}{2} evr$$

$$= \tfrac{1}{2}(1.60 \times 10^{-19} \, \text{C})(2.19 \times 10^6 \, \text{m/s})(0.529 \times 10^{-10} \, \text{m}) = 9.27 \times 10^{-24} \, \text{A} \cdot \text{m}^2,$$

or $9.27 \times 10^{-24} \, \text{J/T}$.

*27–6 Applications: Motors, Loudspeakers, Galvanometers

*Electric Motors

An **electric motor** changes electric energy into (rotational) mechanical energy. A motor works on the principle that a torque is exerted on a coil of current-carrying wire suspended in the magnetic field of a magnet, described in Section 27–5. The coil is mounted on a large cylinder called the **rotor** or **armature**, Fig. 27–23, so that it can rotate continuously in one direction. Actually, there are several coils, although only one is indicated in the Figure. The armature is mounted on a shaft or axle. When the armature is in the position shown in Fig. 27–23, the magnetic field exerts forces on the current in the loop as shown (perpendicular to $\vec{B}$ and to the current direction). However, when the coil, which is rotating clockwise in Fig. 27–23, passes beyond the vertical position, the forces would then act to return the coil back to vertical if the current remained the same. But if the current could somehow be reversed at that critical moment, the forces would reverse, and the coil would continue rotating in the same direction. Thus, alternation of the current is necessary if a motor is to turn continuously in one direction. This can be achieved in a **dc motor** with the use of **commutators** and **brushes**: as shown in Fig. 27–24, input current passes through stationary brushes that rub against the conducting commutators mounted on the motor shaft. At every half revolution, each commutator changes its connection over to the other brush. Thus the current in the coil reverses every half revolution as required for continuous rotation.

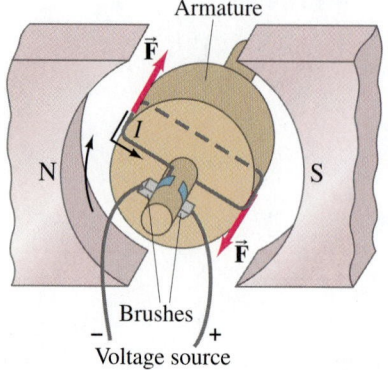

FIGURE 27–23 Diagram of a simple dc motor.

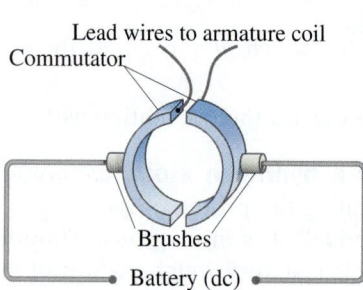

FIGURE 27–24 The commutator-brush arrangement in a dc motor ensures alternation of the current in the armature to keep rotation continuous. The commutators are attached to the motor shaft and turn with it, whereas the brushes remain stationary.

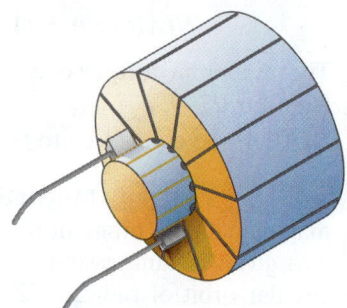

FIGURE 27–25 Motor with many windings.

Most motors contain several coils, called *windings*, each located in a different place on the armature, Fig. 27–25. Current flows through each coil only during a small part of a revolution, at the time when its orientation results in the maximum torque. In this way, a motor produces a much steadier torque than can be obtained from a single coil.

An **ac motor**, with ac current as input, can work without commutators since the current itself alternates. Many motors use wire coils to produce the magnetic field (electromagnets) instead of a permanent magnet. Indeed the design of most motors is more complex than described here, but the general principles remain the same.

*Loudspeakers

A **loudspeaker** also works on the principle that a magnet exerts a force on a current-carrying wire. The electrical output of a stereo or TV set is connected to the wire leads of the speaker. The speaker leads are connected internally to a coil of wire, which is itself attached to the speaker cone, Fig. 27–26. The speaker cone is usually made of stiffened cardboard and is mounted so that it can move back and forth freely. A permanent magnet is mounted directly in line with the coil of wire. When the alternating current of an audio signal flows through the wire coil, which is free to move within the magnet, the coil experiences a force due to the magnetic field of the magnet. (The force is to the right at the instant shown in Fig. 27–26.)

FIGURE 27–26 Loudspeaker.

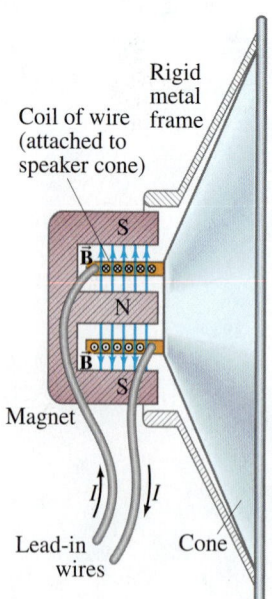

As the current alternates at the frequency of the audio signal, the coil and attached speaker cone move back and forth at the same frequency, causing alternate compressions and rarefactions of the adjacent air, and sound waves are produced. A speaker thus changes electrical energy into sound energy, and the frequencies and intensities of the emitted sound waves can be an accurate reproduction of the electrical input.

*Galvanometer

The basic component of analog meters (those with pointer and dial), including analog ammeters, voltmeters, and ohmmeters, is a galvanometer. We have already seen how these meters are designed (Section 26–7), and now we can examine how the crucial element, a galvanometer, works. As shown in Fig. 27–27, a **galvanometer** consists of a coil of wire (with attached pointer) suspended in the magnetic field of a permanent magnet. When current flows through the loop of wire, the magnetic field exerts a torque on the loop, as given by Eq. 27–9,

$$\tau = NIAB \sin \theta.$$

This torque is opposed by a spring which exerts a torque τ_s approximately proportional to the angle ϕ through which it is turned (Hooke's law). That is,

$$\tau_s = k\phi,$$

where k is the stiffness constant of the spring. The coil and attached pointer rotate to the angle where the torques balance. When the needle is in equilibrium at rest, the torques are equal: $k\phi = NIAB \sin \theta$, or

$$\phi = \frac{NIAB \sin \theta}{k}.$$

The deflection of the pointer, ϕ, is directly proportional to the current I flowing in the coil, but also depends on the angle θ the coil makes with $\vec{\mathbf{B}}$. For a useful meter we need ϕ to depend only on the current I, independent of θ. To solve this problem, magnets with curved pole pieces are used and the galvanometer coil is wrapped around a cylindrical iron core as shown in Fig. 27–28. The iron tends to concentrate the magnetic field lines so that $\vec{\mathbf{B}}$ always points parallel to the face of the coil at the wire outside the core. The force is then always perpendicular to the face of the coil, and the torque will not vary with angle. Thus ϕ will be proportional to I as required.

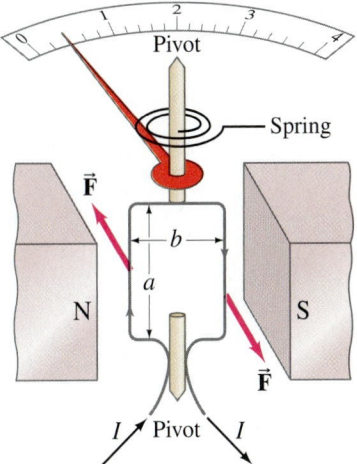

FIGURE 27–27 Galvanometer.

FIGURE 27–28 Galvanometer coil wrapped on an iron core.

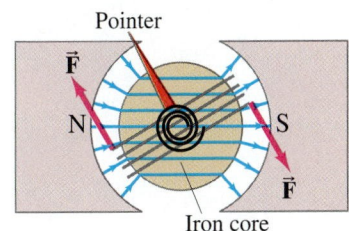

27–7 Discovery and Properties of the Electron

The electron plays a basic role in our understanding of electricity and magnetism today. But its existence was not suggested until the 1890s. We discuss it here because magnetic fields were crucial for measuring its properties.

Toward the end of the nineteenth century, studies were being done on the discharge of electricity through rarefied gases. One apparatus, diagrammed in Fig. 27–29, was a glass tube fitted with electrodes and evacuated so only a small amount of gas remained inside. When a very high voltage was applied to the electrodes, a dark space seemed to extend outward from the cathode (negative electrode) toward the opposite end of the tube; and that far end of the tube would glow. If one or more screens containing a small hole was inserted as shown, the glow was restricted to a tiny spot on the end of the tube. It seemed as though something being emitted by the cathode traveled to the opposite end of the tube. These "somethings" were named **cathode rays**.

There was much discussion at the time about what these rays might be. Some scientists thought they might resemble light. But the observation that the bright spot at the end of the tube could be deflected to one side by an electric or magnetic field suggested that cathode rays could be charged particles; and the direction of the deflection was consistent with a negative charge. Furthermore, if the tube contained certain types of rarefied gas, the path of the cathode rays was made visible by a slight glow.

FIGURE 27–29 Discharge tube. In some models, one of the screens is the anode (positive plate).

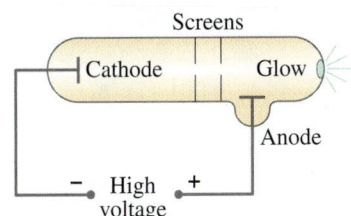

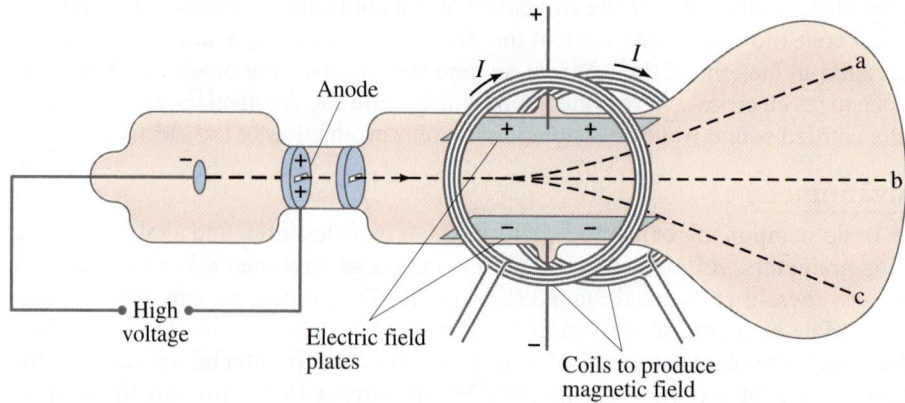

FIGURE 27–30 Cathode rays deflected by electric and magnetic fields.

Estimates of the charge e of the (assumed) cathode-ray particles, as well as of their charge-to-mass ratio e/m, had been made by 1897. But in that year, J. J. Thomson (1856–1940) was able to measure e/m directly, using the apparatus shown in Fig. 27–30. Cathode rays are accelerated by a high voltage and then pass between a pair of parallel plates built into the tube. The voltage applied to the plates produces an electric field, and a pair of coils produces a magnetic field. When only the electric field is present, say with the upper plate positive, the cathode rays are deflected upward as in path a in Fig. 27–30. If only a magnetic field exists, say inward, the rays are deflected downward along path c. These observations are just what is expected for a negatively charged particle. The force on the rays due to the magnetic field is $F = evB$, where e is the charge and v is the velocity of the cathode rays. In the absence of an electric field, the rays are bent into a curved path, so we have, from $F = ma$,

$$evB = m\frac{v^2}{r},$$

and thus

$$\frac{e}{m} = \frac{v}{Br}.$$

The radius of curvature r can be measured and so can B. The velocity v can be found by applying an electric field in addition to the magnetic field. The electric field E is adjusted so that the cathode rays are undeflected and follow path b in Fig. 27–30. This is just like the velocity selector of Example 27–10 where the force due to the electric field, $F = eE$, is balanced by the force due to the magnetic field, $F = evB$. Thus $eE = evB$ and $v = E/B$. Combining this with the above equation we have

$$\frac{e}{m} = \frac{E}{B^2 r}. \tag{27–13}$$

The quantities on the right side can all be measured so that although e and m could not be determined separately, the ratio e/m could be determined. The accepted value today is $e/m = 1.76 \times 10^{11}$ C/kg. Cathode rays soon came to be called **electrons**.

It is worth noting that the "discovery" of the electron, like many others in science, is not quite so obvious as discovering gold or oil. Should the discovery of the electron be credited to the person who first saw a glow in the tube? Or to the person who first called them cathode rays? Perhaps neither one, for they had no conception of the electron as we know it today. In fact, the credit for the discovery is generally given to Thomson, but not because he was the first to see the glow in the tube. Rather it is because he believed that this phenomenon was due to tiny negatively charged particles and made careful measurements on them. Furthermore he argued that these particles were constituents of atoms, and not ions or atoms themselves as many thought, and he developed an electron theory of matter. His view is close to what we accept today, and this is why Thomson is credited with the "discovery." Note, however, that neither he nor anyone else ever actually saw an electron itself. We discuss this briefly, for it illustrates the fact that discovery in science is not always a clear-cut matter. In fact some philosophers of science think the word "discovery" is often not appropriate, such as in this case.

Thomson believed that an electron was not an atom, but rather a constituent, or part, of an atom. Convincing evidence for this came soon with the determination of the charge and the mass of the cathode rays. Thomson's student J. S. Townsend made the first direct (but rough) measurements of e in 1897. But it was the more refined **oil-drop experiment** of Robert A. Millikan (1868–1953) that yielded a precise value for the charge on the electron and showed that charge comes in discrete amounts. In this experiment, tiny droplets of mineral oil carrying an electric charge were allowed to fall under gravity between two parallel plates, Fig. 27–31. The electric field E between the plates was adjusted until the drop was suspended in midair. The downward pull of gravity, mg, was then just balanced by the upward force due to the electric field. Thus $qE = mg$, so the charge $q = mg/E$. The mass of the droplet was determined by measuring its terminal velocity in the absence of the electric field. Sometimes the drop was charged negatively, and sometimes positively, suggesting that the drop had acquired or lost electrons (by friction, leaving the atomizer). Millikan's painstaking observations and analysis presented convincing evidence that any charge was an integral multiple of a smallest charge, e, that was ascribed to the electron, and that the value of e was 1.6×10^{-19} C. This value of e, combined with the measurement of e/m, gives the mass of the electron to be $(1.6 \times 10^{-19}\ \text{C})/(1.76 \times 10^{11}\ \text{C/kg}) = 9.1 \times 10^{-31}$ kg. This mass is less than a thousandth the mass of the smallest atom, and thus confirmed the idea that the electron is only a part of an atom. The accepted value today for the mass of the electron is $m_e = 9.11 \times 10^{-31}$ kg.

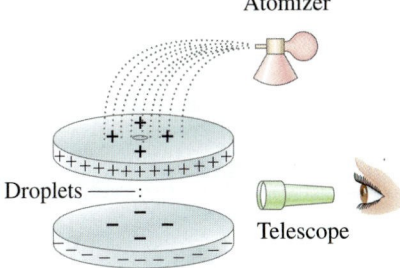

FIGURE 27–31 Millikan's oil-drop experiment.

CRT, Revisited

The cathode ray tube (CRT), which can serve as the picture tube of TV sets, oscilloscopes, and computer monitors, was discussed in Chapter 23. There, in Fig. 23–22, we saw a design using electric deflection plates to maneuver the electron beam. Many CRTs, however, make use of the magnetic field produced by coils to maneuver the electron beam. They operate much like the coils shown in Fig. 27–30.

27–8 The Hall Effect

When a current-carrying conductor is held fixed in a magnetic field, the field exerts a sideways force on the charges moving in the conductor. For example, if electrons move to the right in the rectangular conductor shown in Fig. 27–32a, the inward magnetic field will exert a downward force on the electrons $\vec{\mathbf{F}}_B = -e\vec{\mathbf{v}}_d \times \vec{\mathbf{B}}$, where $\vec{\mathbf{v}}_d$ is the drift velocity of the electrons (Section 25–8). Thus the electrons will tend to move nearer to face D than face C. There will thus be a potential difference between faces C and D of the conductor. This potential difference builds up until the electric field $\vec{\mathbf{E}}_H$ that it produces exerts a force, $e\vec{\mathbf{E}}_H$, on the moving charges that is equal and opposite to the magnetic force. This effect is called the **Hall effect** after E. H. Hall, who discovered it in 1879. The difference of potential produced is called the **Hall emf**.

The electric field due to the separation of charge is called the *Hall field*, $\vec{\mathbf{E}}_H$, and points downward in Fig. 27–32a, as shown. In equilibrium, the force due to this electric field is balanced by the magnetic force $ev_d B$, so

$$eE_H = ev_d B.$$

Hence $E_H = v_d B$. The Hall emf is then (Eq. 23–4b, assuming the conductor is long and thin so E_H is uniform)

$$\mathscr{E}_H = E_H d = v_d Bd, \tag{27–14}$$

where d is the width of the conductor.

A current of negative charges moving to the right is equivalent to positive charges moving to the left, at least for most purposes. But the Hall effect can distinguish these two. As can be seen in Fig. 27–32b, positive particles moving to the left are deflected downward, so that the bottom surface is positive relative to the top surface. This is the reverse of part (a). Indeed, the direction of the emf in the Hall effect first revealed that it is negative particles that move in metal conductors.

FIGURE 27–32 The Hall effect. (a) Negative charges moving to the right as the current. (b) Positive charges moving to the left as the current.

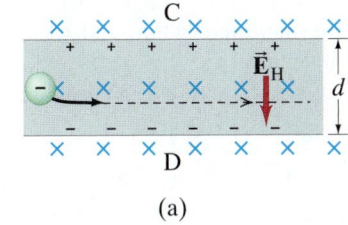

(a)

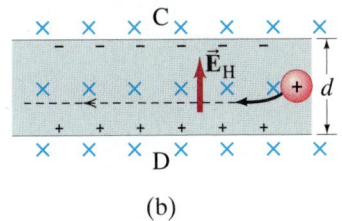

(b)

The magnitude of the Hall emf is proportional to the strength of the magnetic field. The Hall effect can thus be used to measure magnetic field strengths. First the conductor, called a *Hall probe*, is calibrated with known magnetic fields. Then, for the same current, its emf output will be a measure of B. Hall probes can be made very small and are convenient and accurate to use.

The Hall effect can also be used to measure the drift velocity of charge carriers when the external magnetic field B is known. Such a measurement also allows us to determine the density of charge carriers in the material.

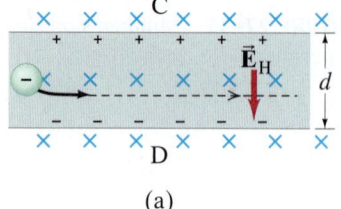

(a)

FIGURE 27–32a (Repeated here for Example 27–13.)

EXAMPLE 27–13 **Drift velocity using the Hall effect.** A long copper strip 1.8 cm wide and 1.0 mm thick is placed in a 1.2-T magnetic field as in Fig. 27–32a. When a steady current of 15 A passes through it, the Hall emf is measured to be $1.02 \, \mu V$. Determine the drift velocity of the electrons and the density of free (conducting) electrons in the copper.

APPROACH We use Eq. 27–14 to obtain the drift velocity, and Eq. 25–13 of Chapter 25 to find the density of conducting electrons.

SOLUTION The drift velocity (Eq. 27–14) is

$$v_d = \frac{\mathcal{E}_H}{Bd} = \frac{1.02 \times 10^{-6} \, \text{V}}{(1.2 \, \text{T})(1.8 \times 10^{-2} \, \text{m})} = 4.7 \times 10^{-5} \, \text{m/s}.$$

The density of charge carriers n is obtained from Eq. 25–13, $I = nev_d A$, where A is the cross-sectional area through which the current I flows. Then

$$n = \frac{I}{ev_d A} = \frac{15 \, \text{A}}{(1.6 \times 10^{-19} \, \text{C})(4.7 \times 10^{-5} \, \text{m/s})(1.8 \times 10^{-2} \, \text{m})(1.0 \times 10^{-3} \, \text{m})}$$

$$= 11 \times 10^{28} \, \text{m}^{-3}.$$

This value for the density of free electrons in copper, $n = 11 \times 10^{28}$ per m^3, is the experimentally measured value. It represents *more* than one free electron per atom, which as we saw in Example 25–14 is $8.4 \times 10^{28} \, \text{m}^{-3}$.

FIGURE 27–33 Bainbridge-type mass spectrometer. The magnetic fields B and B′ point out of the paper (indicated by the dots), for positive ions.

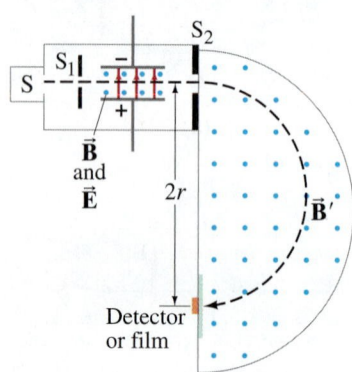

*27–9 Mass Spectrometer

A **mass spectrometer** is a device to measure masses of atoms. It is used today not only in physics but also in chemistry, geology, and medicine, often to identify atoms (and their concentration) in given samples. As shown in Fig. 27–33, ions are produced by heating, or by an electric current, in the source or sample S. The particles, of mass m and electric charge q, pass through slit S_1 and enter crossed electric and magnetic fields. Ions follow a straight-line path in this "velocity selector" (as in Example 27–10) if the electric force qE is balanced by the magnetic force qvB: that is, if qE = qvB, or v = E/B. Thus only those ions whose speed is v = E/B will pass through undeflected and emerge through slit S_2. In the semicircular region, after S_2, there is only a magnetic field, B′, so the ions follow a circular path. The radius of the circular path is found from their mark on film (or detectors) if B′ is fixed; or else r is fixed by the position of a detector and B′ is varied until detection occurs. Newton's second law, $\Sigma F = ma$, applied to an ion moving in a circle under the influence only of the magnetic field B′ gives $qvB' = mv^2/r$. Since v = E/B, we have

$$m = \frac{qB'r}{v} = \frac{qBB'r}{E}.$$

All the quantities on the right side are known or can be measured, and thus m can be determined.

Historically, the masses of many atoms were measured this way. When a pure substance was used, it was sometimes found that two or more closely spaced marks would appear on the film. For example, neon produced two marks whose radii corresponded to atoms of mass 20 and 22 atomic mass units (u). Impurities were ruled out and it was concluded that there must be two types of neon with different masses. These different forms were called **isotopes**. It was soon found that most elements are mixtures of isotopes, and the difference in mass is due to different numbers of neutrons (discussed in Chapter 41).

EXAMPLE 27–14 **Mass spectrometry.** Carbon atoms of atomic mass 12.0 u are found to be mixed with another, unknown, element. In a mass spectrometer with fixed B', the carbon traverses a path of radius 22.4 cm and the unknown's path has a 26.2-cm radius. What is the unknown element? Assume the ions of both elements have the same charge.

APPROACH The carbon and unknown atoms pass through the same electric and magnetic fields. Hence their masses are proportional to the radius of their respective paths (see equation on previous page).

SOLUTION We write a ratio for the masses, using the equation at the bottom of the previous page:

$$\frac{m_x}{m_C} = \frac{qBB'r_x/E}{qBB'r_C/E} = \frac{26.2 \text{ cm}}{22.4 \text{ cm}} = 1.17.$$

Thus $m_x = 1.17 \times 12.0 \text{ u} = 14.0 \text{ u}$. The other element is probably nitrogen (see the Periodic Table, inside the back cover).

NOTE The unknown could also be an isotope such as carbon-14 $\left(^{14}_{6}\text{C}\right)$. See Appendix F. Further physical or chemical analysis would be needed.

Summary

A magnet has two **poles**, north and south. The north pole is that end which points toward geographic north when the magnet is freely suspended. Like poles of two magnets repel each other, whereas unlike poles attract.

We can picture that a **magnetic field** surrounds every magnet. The SI unit for magnetic field is the **tesla** (T).

Electric currents produce magnetic fields. For example, the lines of magnetic field due to a current in a straight wire form circles around the wire, and the field exerts a force on magnets (or currents) near it.

A magnetic field exerts a force on an electric current. The force on an infinitesimal length of wire $d\vec{\boldsymbol{\ell}}$ carrying a current I in a magnetic field $\vec{\mathbf{B}}$ is

$$d\vec{\mathbf{F}} = I\,d\vec{\boldsymbol{\ell}} \times \vec{\mathbf{B}}. \quad (27\text{–}4)$$

If the field $\vec{\mathbf{B}}$ is uniform over a straight length $\vec{\boldsymbol{\ell}}$ of wire, then the force is

$$\vec{\mathbf{F}} = I\vec{\boldsymbol{\ell}} \times \vec{\mathbf{B}} \quad (27\text{–}3)$$

which has magnitude

$$F = I\ell B \sin\theta \quad (27\text{–}1)$$

where θ is the angle between magnetic field $\vec{\mathbf{B}}$ and the wire. The direction of the force is perpendicular to the wire and to the magnetic field, and is given by the right-hand rule. This relation serves as the definition of magnetic field $\vec{\mathbf{B}}$.

Similarly, a magnetic field $\vec{\mathbf{B}}$ exerts a force on a charge q moving with velocity $\vec{\mathbf{v}}$ given by

$$\vec{\mathbf{F}} = q\vec{\mathbf{v}} \times \vec{\mathbf{B}}. \quad (27\text{–}5a)$$

The magnitude of the force is

$$F = qvB \sin\theta, \quad (27\text{–}5b)$$

where θ is the angle between $\vec{\mathbf{v}}$ and $\vec{\mathbf{B}}$.

The path of a charged particle moving perpendicular to a uniform magnetic field is a circle.

If both electric and magnetic fields ($\vec{\mathbf{E}}$ and $\vec{\mathbf{B}}$) are present, the force on a charge q moving with velocity $\vec{\mathbf{v}}$ is

$$\vec{\mathbf{F}} = q\vec{\mathbf{E}} + q\vec{\mathbf{v}} \times \vec{\mathbf{B}}. \quad (27\text{–}7)$$

The torque on a current loop in a magnetic field $\vec{\mathbf{B}}$ is

$$\vec{\boldsymbol{\tau}} = \vec{\boldsymbol{\mu}} \times \vec{\mathbf{B}}, \quad (27\text{–}11)$$

where $\vec{\boldsymbol{\mu}}$ is the **magnetic dipole moment** of the loop:

$$\vec{\boldsymbol{\mu}} = NI\vec{\mathbf{A}}. \quad (27\text{–}10)$$

Here N is the number of coils carrying current I in the loop and $\vec{\mathbf{A}}$ is a vector perpendicular to the plane of the loop (use right-hand rule, fingers along current in loop) and has magnitude equal to the area of the loop.

The measurement of the charge-to-mass ratio (e/m) of the electron was done using magnetic and electric fields. The charge e on the electron was first measured in the Millikan oil-drop experiment and then its mass was obtained from the measured value of the e/m ratio.

[*In the **Hall effect**, moving charges in a conductor placed in a magnetic field are forced to one side, producing an emf between the two sides of the conductor.]

[*A **mass spectrometer** uses magnetic and electric fields to measure the mass of ions.]

Questions

1. A compass needle is not always balanced parallel to the Earth's surface, but one end may dip downward. Explain.

2. Draw the magnetic field lines around a straight section of wire carrying a current horizontally to the left.

3. A horseshoe magnet is held vertically with the north pole on the left and south pole on the right. A wire passing between the poles, equidistant from them, carries a current directly away from you. In what direction is the force on the wire?

4. In the relation $\vec{F} = I\vec{\ell} \times \vec{B}$, which pairs of the vectors $(\vec{F}, \vec{\ell}, \vec{B})$ are always at 90°? Which can be at other angles?

5. The magnetic field due to current in wires in your home can affect a compass. Discuss the effect in terms of currents, including if they are ac or dc.

6. If a negatively charged particle enters a region of uniform magnetic field which is perpendicular to the particle's velocity, will the kinetic energy of the particle increase, decrease, or stay the same? Explain your answer. (Neglect gravity and assume there is no electric field.)

7. In Fig. 27–34, charged particles move in the vicinity of a current-carrying wire. For each charged particle, the arrow indicates the direction of motion of the particle, and the + or − indicates the sign of the charge. For each of the particles, indicate the direction of the magnetic force due to the magnetic field produced by the wire.

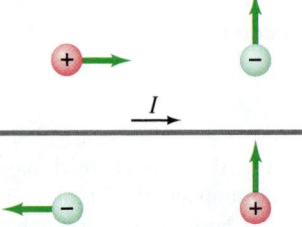

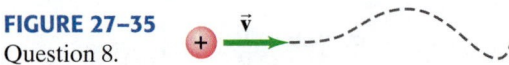

FIGURE 27–34
Question 7.

8. A positively charged particle in a nonuniform magnetic field follows the trajectory shown in Fig. 27–35. Indicate the direction of the magnetic field at points near the path, assuming the path is always in the plane of the page, and indicate the relative magnitudes of the field in each region.

FIGURE 27–35
Question 8.

9. Note that the pattern of magnetic field lines surrounding a bar magnet is similar to that of the electric field around an electric dipole. From this fact, predict how the magnetic field will change with distance (*a*) when near one pole of a very long bar magnet, and (*b*) when far from a magnet as a whole.

10. Explain why a strong magnet held near a CRT television screen causes the picture to become distorted. Also, explain why the picture sometimes goes completely black where the field is the strongest. [But don't risk damage to your TV by trying this.]

11. Describe the trajectory of a negatively charged particle in the velocity selector of Fig. 27–21 if its speed exceeds E/B. What is its trajectory if $v < E/B$? Would it make any difference if the particle were positively charged?

12. Can you set a resting electron into motion with a steady magnetic field? With an electric field? Explain.

13. A charged particle is moving in a circle under the influence of a uniform magnetic field. If an electric field that points in the same direction as the magnetic field is turned on, describe the path the charged particle will take.

14. The force on a particle in a magnetic field is the idea behind **electromagnetic pumping**. It is used to pump metallic fluids (such as sodium) and to pump blood in artificial heart machines. The basic design is shown in Fig. 27–36. An electric field is applied perpendicular to a blood vessel and to a magnetic field. Explain how ions are caused to move. Do positive and negative ions feel a force in the same direction?

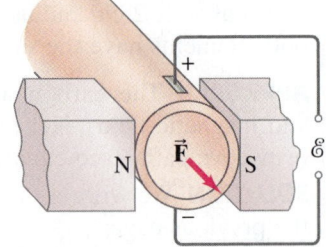

FIGURE 27–36
Electromagnetic pumping in a blood vessel.
Question 14.

15. A beam of electrons is directed toward a horizontal wire carrying a current from left to right (Fig. 27–37). In what direction is the beam deflected?

FIGURE 27–37
Question 15.

16. A charged particle moves in a straight line through a particular region of space. Could there be a nonzero magnetic field in this region? If so, give two possible situations.

17. If a moving charged particle is deflected sideways in some region of space, can we conclude, for certain, that $\vec{B} \neq 0$ in that region? Explain.

18. How could you tell whether moving electrons in a certain region of space are being deflected by an electric field or by a magnetic field (or by both)?

19. How can you make a compass without using iron or other ferromagnetic material?

20. Describe how you could determine the dipole moment of a bar magnet or compass needle.

21. In what positions (if any) will a current loop placed in a uniform magnetic field be in (*a*) stable equilibrium, and (*b*) unstable equilibrium?

*22. A rectangular piece of semiconductor is inserted in a magnetic field and a battery is connected to its ends as shown in Fig. 27–38. When a sensitive voltmeter is connected between points a and b, it is found that point a is at a higher potential than b. What is the sign of the charge carriers in this semiconductor material?

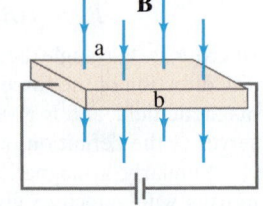

FIGURE 27–38
Question 22.

*23. Two ions have the same mass, but one is singly ionized and the other is doubly ionized. How will their positions on the film of the mass spectrometer of Fig. 27–33 differ?

Problems

27–3 Force on Electric Current in Magnetic Field

1. (I) (a) What is the force per meter of length on a straight wire carrying a 9.40-A current when perpendicular to a 0.90-T uniform magnetic field? (b) What if the angle between the wire and field is 35.0°?

2. (I) Calculate the magnitude of the magnetic force on a 240-m length of wire stretched between two towers and carrying a 150-A current. The Earth's magnetic field of 5.0×10^{-5} T makes an angle of 68° with the wire.

3. (I) A 1.6-m length of wire carrying 4.5 A of current toward the south is oriented horizontally. At that point on the Earth's surface, the dip angle of the Earth's magnetic field makes an angle of 41° to the wire. Estimate the magnitude of the magnetic force on the wire due to the Earth's magnetic field of 5.5×10^{-5} T at this point.

4. (II) The magnetic force per meter on a wire is measured to be only 25 percent of its maximum possible value. Sketch the relationship of the wire and the field if the force had been a maximum, and sketch the relationship as it actually is, calculating the angle between the wire and the magnetic field.

5. (II) The force on a wire is a maximum of 7.50×10^{-2} N when placed between the pole faces of a magnet. The current flows horizontally to the right and the magnetic field is vertical. The wire is observed to "jump" toward the observer when the current is turned on. (a) What type of magnetic pole is the top pole face? (b) If the pole faces have a diameter of 10.0 cm, estimate the current in the wire if the field is 0.220 T. (c) If the wire is tipped so that it makes an angle of 10.0° with the horizontal, what force will it now feel?

6. (II) Suppose a straight 1.00-mm-diameter copper wire could just "float" horizontally in air because of the force due to the Earth's magnetic field $\vec{\mathbf{B}}$, which is horizontal, perpendicular to the wire, and of magnitude 5.0×10^{-5} T. What current would the wire carry? Does the answer seem feasible? Explain briefly.

7. (II) A stiff wire 50.0 cm long is bent at a right angle in the middle. One section lies along the z axis and the other is along the line $y = 2x$ in the xy plane. A current of 20.0 A flows in the wire—down the z axis and out the line in the xy plane. The wire passes through a uniform magnetic field given by $\vec{\mathbf{B}} = (0.318\hat{\mathbf{i}})$T. Determine the magnitude and direction of the total force on the wire.

8. (II) A long wire stretches along the x axis and carries a 3.0-A current to the right $(+x)$. The wire is in a uniform magnetic field $\vec{\mathbf{B}} = (0.20\hat{\mathbf{i}} - 0.36\hat{\mathbf{j}} + 0.25\hat{\mathbf{k}})$ T. Determine the components of the force on the wire per cm of length.

9. (II) A current-carrying circular loop of wire (radius r, current I) is partially immersed in a magnetic field of constant magnitude B_0 directed out of the page as shown in Fig. 27–39. Determine the net force on the loop due to the field in terms of θ_0. (Note that θ_0 points to the dashed line, above which $B = 0$.)

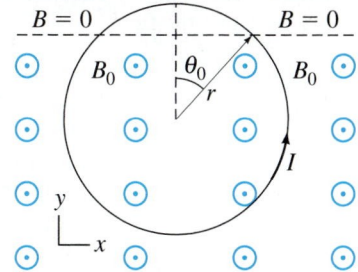

FIGURE 27–39
Problem 9.

10. (II) A 2.0-m-long wire carries a current of 8.2 A and is immersed within a uniform magnetic field $\vec{\mathbf{B}}$. When this wire lies along the $+x$ axis, a magnetic force $\vec{\mathbf{F}} = (-2.5\hat{\mathbf{j}})$ N acts on the wire, and when it lies on the $+y$ axis, the force is $\vec{\mathbf{F}} = (2.5\hat{\mathbf{i}} - 5.0\hat{\mathbf{k}})$ N. Find $\vec{\mathbf{B}}$.

11. (III) A curved wire, connecting two points a and b, lies in a plane perpendicular to a uniform magnetic field $\vec{\mathbf{B}}$ and carries a current I. Show that the resultant magnetic force on the wire, no matter what its shape, is the same as that on a straight wire connecting the two points carrying the same current I. See Fig. 27–40.

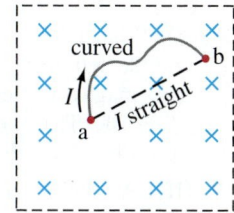

FIGURE 27–40
Problem 11.

12. (III) A circular loop of wire, of radius r, carries current I. It is placed in a magnetic field whose straight lines seem to diverge from a point a distance d below the loop on its axis. (That is, the field makes an angle θ with the loop at all points, Fig. 27–41, where $\tan \theta = r/d$.) Determine the force on the loop.

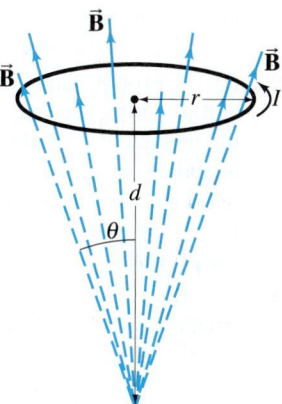

FIGURE 27–41
Problem 12.

27–4 Force on Charge Moving in Magnetic Field

13. (I) Determine the magnitude and direction of the force on an electron traveling 8.75×10^5 m/s horizontally to the east in a vertically upward magnetic field of strength 0.45 T.

14. (I) An electron is projected vertically upward with a speed of 1.70×10^6 m/s into a uniform magnetic field of 0.480 T that is directed horizontally away from the observer. Describe the electron's path in this field.

15. (I) Alpha particles of charge $q = +2e$ and mass $m = 6.6 \times 10^{-27}$ kg are emitted from a radioactive source at a speed of 1.6×10^7 m/s. What magnetic field strength would be required to bend them into a circular path of radius $r = 0.18$ m?

16. (I) Find the direction of the force on a negative charge for each diagram shown in Fig. 27–42, where $\vec{\mathbf{v}}$ (green) is the velocity of the charge and $\vec{\mathbf{B}}$ (blue) is the direction of the magnetic field. ($\otimes$ means the vector points inward. $\odot$ means it points outward, toward you.)

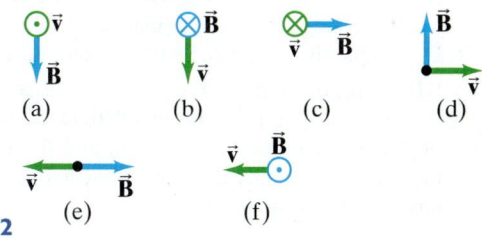

FIGURE 27–42
Problem 16.

17. (I) Determine the direction of $\vec{\mathbf{B}}$ for each case in Fig. 27–43, where $\vec{\mathbf{F}}$ represents the maximum magnetic force on a positively charged particle moving with velocity $\vec{\mathbf{v}}$.

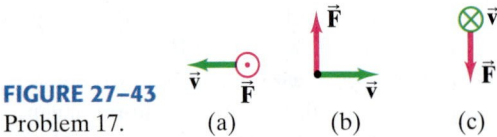

FIGURE 27–43
Problem 17.   (a)   (b)   (c)

18. (II) What is the velocity of a beam of electrons that goes undeflected when passing through perpendicular electric and magnetic fields of magnitude 8.8×10^3 V/m and 7.5×10^{-3} T, respectively? What is the radius of the electron orbit if the electric field is turned off?

19. (II) A doubly charged helium atom whose mass is 6.6×10^{-27} kg is accelerated by a voltage of 2700 V. (a) What will be its radius of curvature if it moves in a plane perpendicular to a uniform 0.340-T field? (b) What is its period of revolution?

20. (II) A proton (mass m_p), a deuteron $(m = 2m_\mathrm{p}, Q = e)$, and an alpha particle $(m = 4m_\mathrm{p}, Q = 2e)$ are accelerated by the same potential difference V and then enter a uniform magnetic field $\vec{\mathbf{B}}$, where they move in circular paths perpendicular to $\vec{\mathbf{B}}$. Determine the radius of the paths for the deuteron and alpha particle in terms of that for the proton.

21. (II) For a particle of mass m and charge q moving in a circular path in a magnetic field B, (a) show that its kinetic energy is proportional to r^2, the square of the radius of curvature of its path, and (b) show that its angular momentum is $L = qBr^2$, about the center of the circle.

22. (II) An electron moves with velocity $\vec{\mathbf{v}} = (7.0\hat{\mathbf{i}} - 6.0\hat{\mathbf{j}}) \times 10^4$ m/s in a magnetic field $\vec{\mathbf{B}} = (-0.80\hat{\mathbf{i}} + 0.60\hat{\mathbf{j}})$ T. Determine the magnitude and direction of the force on the electron.

23. (II) A 6.0-MeV (kinetic energy) proton enters a 0.20-T field, in a plane perpendicular to the field. What is the radius of its path? See Section 23–8.

24. (II) An electron experiences the greatest force as it travels 2.8×10^6 m/s in a magnetic field when it is moving northward. The force is vertically upward and of magnitude 8.2×10^{-13} N. What is the magnitude and direction of the magnetic field?

25. (II) A proton moves through a region of space where there is a magnetic field $\vec{\mathbf{B}} = (0.45\hat{\mathbf{i}} + 0.38\hat{\mathbf{j}})$ T and an electric field $\vec{\mathbf{E}} = (3.0\hat{\mathbf{i}} - 4.2\hat{\mathbf{j}}) \times 10^3$ V/m. At a given instant, the proton's velocity is $\vec{\mathbf{v}} = (6.0\hat{\mathbf{i}} + 3.0\hat{\mathbf{j}} - 5.0\hat{\mathbf{k}}) \times 10^3$ m/s. Determine the components of the total force on the proton.

26. (II) An electron experiences a force $\vec{\mathbf{F}} = (3.8\hat{\mathbf{i}} - 2.7\hat{\mathbf{j}}) \times 10^{-13}$ N when passing through a magnetic field $\vec{\mathbf{B}} = (0.85\text{ T})\hat{\mathbf{k}}$. Determine the components of the electron's velocity.

27. (II) A particle of charge q moves in a circular path of radius r in a uniform magnetic field $\vec{\mathbf{B}}$. If the magnitude of the magnetic field is doubled, and the kinetic energy of the particle remains constant, what happens to the angular momentum of the particle?

28. (II) An electron enters a uniform magnetic field $B = 0.28$ T at a 45° angle to $\vec{\mathbf{B}}$. Determine the radius r and pitch p (distance between loops) of the electron's helical path assuming its speed is 3.0×10^6 m/s. See Fig. 27–44.

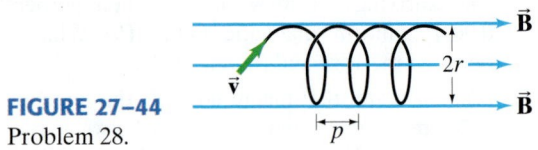

FIGURE 27–44
Problem 28.

29. (II) A particle with charge q and momentum p, initially moving along the x axis, enters a region where a uniform magnetic field $\vec{\mathbf{B}} = B_0\hat{\mathbf{k}}$ extends over a width $x = \ell$ as shown in Fig. 27–45. The particle is deflected a distance d in the $+y$ direction as it traverses the field. Determine (a) whether q is positive or negative, and (b) the magnitude of its momentum p.

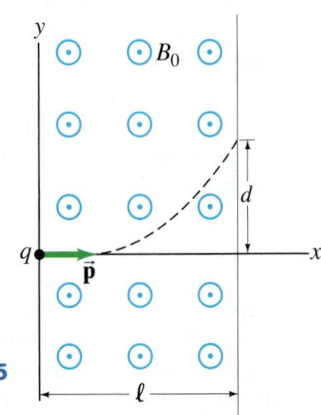

FIGURE 27–45
Problem 29.

30. (II) The path of protons emerging from an accelerator must be bent by 90° by a "bending magnet" so as not to strike a barrier in their path a distance d from their exit hole in the accelerator. Show that the field $\vec{\mathbf{B}}$ in the bending magnet, which we assume is uniform and can extend over an area $d \times d$, must have magnitude $B \geq (2mK/e^2d^2)^{\frac{1}{2}}$, where m is the mass of a proton and K is its kinetic energy.

31. (III) Suppose the Earth's magnetic field at the equator has magnitude 0.50×10^{-4} T and a northerly direction at all points. Estimate the speed a singly ionized uranium ion $(m = 238\text{ u}, q = e)$ would need to circle the Earth 5.0 km above the equator. Can you ignore gravity? [Ignore relativity.]

32. (III) A 3.40-g bullet moves with a speed of 155 m/s perpendicular to the Earth's magnetic field of 5.00×10^{-5} T. If the bullet possesses a net charge of 18.5×10^{-9} C, by what distance will it be deflected from its path due to the Earth's magnetic field after it has traveled 1.00 km?

33. (III) A proton moving with speed $v = 1.3 \times 10^5$ m/s in a field-free region abruptly enters an essentially uniform magnetic field $B = 0.850$ T $(\vec{\mathbf{B}} \perp \vec{\mathbf{v}})$. If the proton enters the magnetic field region at a 45° angle as shown in Fig. 27–46, (a) at what angle does it leave, and (b) at what distance x does it exit the field?

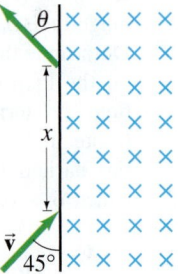

FIGURE 27–46
Problem 33.

34. (III) A particle with charge $+q$ and mass m travels in a uniform magnetic field $\vec{\mathbf{B}} = B_0\hat{\mathbf{k}}$. At time $t = 0$, the particle's speed is v_0 and its velocity vector lies in the xy plane directed at an angle of 30° with respect to the y axis as shown in Fig. 27–47. At a later time $t = t_\alpha$, the particle will cross the x axis at $x = \alpha$. In terms of q, m, v_0, and B_0, determine (*a*) α, and (*b*) t_α.

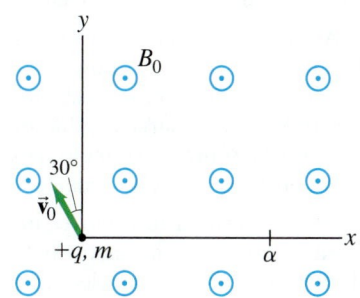

FIGURE 27–47
Problem 34.

27–5 Torque on a Current Loop; Magnetic Moment

35. (I) How much work is required to rotate the current loop (Fig. 27–22) in a uniform magnetic field $\vec{\mathbf{B}}$ from (*a*) $\theta = 0° \ (\vec{\boldsymbol{\mu}} \| \vec{\mathbf{B}})$ to $\theta = 180°$, (*b*) $\theta = 90°$ to $\theta = -90°$?

36. (I) A 13.0-cm-diameter circular loop of wire is placed with the plane of the loop parallel to the uniform magnetic field between the pole pieces of a large magnet. When 4.20 A flows in the coil, the torque on it is 0.185 m·N. What is the magnetic field strength?

37. (II) A circular coil 18.0 cm in diameter and containing twelve loops lies flat on the ground. The Earth's magnetic field at this location has magnitude 5.50×10^{-5} T and points into the Earth at an angle of 66.0° below a line pointing due north. If a 7.10-A clockwise current passes through the coil, determine (*a*) the torque on the coil, and (*b*) which edge of the coil rises up, north, east, south, or west.

38. (II) Show that the magnetic dipole moment μ of an electron orbiting the proton nucleus of a hydrogen atom is related to the orbital momentum L of the electron by

$$\mu = \frac{e}{2m} L.$$

39. (II) A 15-loop circular coil 22 cm in diameter lies in the xy plane. The current in each loop of the coil is 7.6 A clockwise, and an external magnetic field $\vec{\mathbf{B}} = (0.55\hat{\mathbf{i}} + 0.60\hat{\mathbf{j}} - 0.65\hat{\mathbf{k}})$ T passes through the coil. Determine (*a*) the magnetic moment of the coil, $\vec{\boldsymbol{\mu}}$; (*b*) the torque on the coil due to the external magnetic field; (*c*) the potential energy U of the coil in the field (take the same zero for U as we did in our discussion of Fig. 27–22).

40. (III) Suppose a nonconducting rod of length d carries a uniformly distributed charge Q. It is rotated with angular velocity ω about an axis perpendicular to the rod at one end, Fig. 27–48. Show that the magnetic dipole moment of this rod is $\frac{1}{6}Q\omega d^2$. [*Hint*: Consider the motion of each infinitesimal length of the rod.]

FIGURE 27–48
Problem 40.
Axis — d

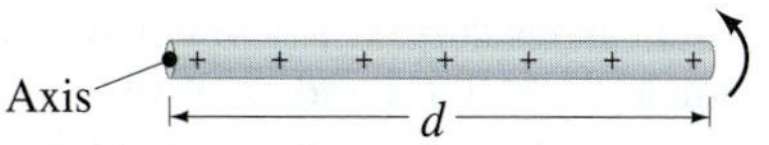

*27–6 Motors, Galvanometers

***41.** (I) If the current to a motor drops by 12%, by what factor does the output torque change?

***42.** (I) A galvanometer needle deflects full scale for a 63.0-μA current. What current will give full-scale deflection if the magnetic field weakens to 0.800 of its original value?

***43.** (I) If the restoring spring of a galvanometer weakens by 15% over the years, what current will give full-scale deflection if it originally required 46 μA?

27–7 Discovery of Electron

44. (I) What is the value of q/m for a particle that moves in a circle of radius 8.0 mm in a 0.46-T magnetic field if a crossed 260-V/m electric field will make the path straight?

45. (II) An oil drop whose mass is determined to be 3.3×10^{-15} kg is held at rest between two large plates separated by 1.0 cm as in Fig. 27–31. If the potential difference between the plates is 340 V, how many excess electrons does this drop have?

27–8 Hall Effect

46. (II) A Hall probe, consisting of a rectangular slab of current-carrying material, is calibrated by placing it in a known magnetic field of magnitude 0.10 T. When the field is oriented normal to the slab's rectangular face, a Hall emf of 12 mV is measured across the slab's width. The probe is then placed in a magnetic field of unknown magnitude B, and a Hall emf of 63 mV is measured. Determine B assuming that the angle θ between the unknown field and the plane of the slab's rectangular face is (*a*) $\theta = 90°$, and (*b*) $\theta = 60°$.

47. (II) A Hall probe used to measure magnetic field strengths consists of a rectangular slab of material (free-electron density n) with width d and thickness t, carrying a current I along its length ℓ. The slab is immersed in a magnetic field of magnitude B oriented perpendicular to its rectangular face (of area ℓd), so that a Hall emf $\mathscr{E}_H$ is produced across its width d. The probe's magnetic sensitivity, defined as $K_H = \mathscr{E}_H/IB$, indicates the magnitude of the Hall emf achieved for a given applied magnetic field and current. A slab with a large K_H is a good candidate for use as a Hall probe. (*a*) Show that $K_H = 1/ent$. Thus, a good Hall probe has small values for both n and t. (*b*) As possible candidates for the material used in a Hall probe, consider (i) a typical metal $(n \approx 1 \times 10^{29}/\text{m}^3)$ and (ii) a (doped) semiconductor $(n \approx 3 \times 10^{22}/\text{m}^3)$. Given that a semiconductor slab can be manufactured with a thickness of 0.15 mm, how thin (nm) should a metal slab be to yield a K_H value equal to that of the semiconductor slab? Compare this metal slab thickness with the 0.3-nm size of a typical metal atom. (*c*) For the typical semiconductor slab described in part (*b*), what is the expected value for $\mathscr{E}_H$ when $I = 100$ mA and $B = 0.1$ T?

48. (II) A rectangular sample of a metal is 3.0 cm wide and 680 μm thick. When it carries a 42-A current and is placed in a 0.80-T magnetic field it produces a 6.5-μV Hall emf. Determine: (*a*) the Hall field in the conductor; (*b*) the drift speed of the conduction electrons; (*c*) the density of free electrons in the metal.

49. (II) In a probe that uses the Hall effect to measure magnetic fields, a 12.0-A current passes through a 1.50-cm-wide 1.30-mm-thick strip of sodium metal. If the Hall emf is 1.86 μV, what is the magnitude of the magnetic field (take it perpendicular to the flat face of the strip)? Assume one free electron per atom of Na, and take its specific gravity to be 0.971.

50. (II) The Hall effect can be used to measure blood flow rate because the blood contains ions that constitute an electric current. (*a*) Does the sign of the ions influence the emf? (*b*) Determine the flow velocity in an artery 3.3 mm in diameter if the measured emf is 0.13 mV and *B* is 0.070 T. (In actual practice, an alternating magnetic field is used.)

*27–9 Mass Spectrometer

*51. (I) In a mass spectrometer, germanium atoms have radii of curvature equal to 21.0, 21.6, 21.9, 22.2, and 22.8 cm. The largest radius corresponds to an atomic mass of 76 u. What are the atomic masses of the other isotopes?

*52. (II) One form of mass spectrometer accelerates ions by a voltage *V* before they enter a magnetic field *B*. The ions are assumed to start from rest. Show that the mass of an ion is $m = qB^2R^2/2V$, where *R* is the radius of the ions' path in the magnetic field and *q* is their charge.

*53. (II) Suppose the electric field between the electric plates in the mass spectrometer of Fig. 27–33 is 2.48×10^4 V/m and the magnetic fields are $B = B' = 0.58$ T. The source contains carbon isotopes of mass numbers 12, 13, and 14 from a long dead piece of a tree. (To estimate atomic masses, multiply by 1.66×10^{-27} kg.) How far apart are the lines formed by the singly charged ions of each type on the photographic film? What if the ions were doubly charged?

*54. (II) A mass spectrometer is being used to monitor air pollutants. It is difficult, however, to separate molecules with nearly equal mass such as CO (28.0106 u) and N_2 (28.0134 u). How large a radius of curvature must a spectrometer have if these two molecules are to be separated at the film or detectors by 0.65 mm?

*55. (II) An unknown particle moves in a straight line through crossed electric and magnetic fields with $E = 1.5$ kV/m and $B = 0.034$ T. If the electric field is turned off, the particle moves in a circular path of radius $r = 2.7$ cm. What might the particle be?

General Problems

56. Protons move in a circle of radius 5.10 cm in a 0.625-T magnetic field. What value of electric field could make their paths straight? In what direction must the electric field point?

57. Protons with momentum 3.8×10^{-16} kg·m/s are magnetically steered clockwise in a circular path 2.0 km in diameter at Fermi National Accelerator Laboratory in Illinois. Determine the magnitude and direction of the field in the magnets surrounding the beam pipe.

58. A proton and an electron have the same kinetic energy upon entering a region of constant magnetic field. What is the ratio of the radii of their circular paths?

59. Two stiff parallel wires a distance *d* apart in a horizontal plane act as rails to support a light metal rod of mass *m* (perpendicular to each rail), Fig. 27–49. A magnetic field $\vec{B}$, directed vertically upward (outward in diagram), acts throughout. At $t = 0$, a constant current *I* begins to flow through the system. Determine the speed of the rod, which starts from rest at $t = 0$, as a function of time (*a*) assuming no friction between the rod and the rails, and (*b*) if the coefficient of friction is μ_k. (*c*) In which direction does the rod move, east or west, if the current through it heads north?

60. Suppose the rod in Fig. 27–49 (Problem 59) has mass $m = 0.40$ kg and length 22 cm and the current through it is $I = 36$ A. If the coefficient of static friction is $\mu_s = 0.50$, determine the minimum magnetic field $\vec{B}$ (not necessarily vertical) that will just cause the rod to slide. Give the magnitude of $\vec{B}$ and its direction relative to the vertical (outwards towards us).

61. Near the equator, the Earth's magnetic field points almost horizontally to the north and has magnitude $B = 0.50 \times 10^{-4}$ T. What should be the magnitude and direction for the velocity of an electron if its weight is to be exactly balanced by the magnetic force?

62. Calculate the magnetic force on an airplane which has acquired a net charge of 1850 μC and moves with a speed of 120 m/s perpendicular to the Earth's magnetic field of 5.0×10^{-5} T.

63. A motor run by a 9.0-V battery has a 20 turn square coil with sides of length 5.0 cm and total resistance 24 Ω. When spinning, the magnetic field felt by the wire in the coil is 0.020 T. What is the maximum torque on the motor?

64. Estimate the approximate maximum deflection of the electron beam near the center of a CRT television screen due to the Earth's 5.0×10^{-5} T field. Assume the screen is 18 cm from the electron gun, where the electrons are accelerated (*a*) by 2.0 kV, or (*b*) by 28 kV. Note that in color TV sets, the beam must be directed accurately to within less than 1 mm in order to strike the correct phosphor. Because the Earth's field is significant here, mu-metal shields are used to reduce the Earth's field in the CRT. (See Section 23–9.)

65. The rectangular loop of wire shown in Fig. 27–22 has mass *m* and carries current *I*. Show that if the loop is oriented at an angle $\theta \ll 1$ (in radians), then when it is released it will execute simple harmonic motion about $\theta = 0$. Calculate the period of the motion.

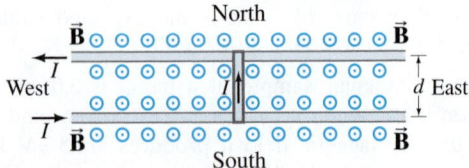

FIGURE 27–49 Looking down on a rod sliding on rails. Problems 59 and 60.

66. The **cyclotron** (Fig. 27–50) is a device used to accelerate elementary particles such as protons to high speeds. Particles starting at point A with some initial velocity travel in circular orbits in the magnetic field B. The particles are accelerated to higher speeds each time they pass in the gap between the metal "dees," where there is an electric field E. (There is no electric field within the hollow metal dees.) The electric field changes direction each half-cycle, due to an ac voltage $V = V_0 \sin 2\pi ft$, so that the particles are increased in speed at each passage through the gap. (a) Show that the frequency f of the voltage must be $f = Bq/2\pi m$, where q is the charge on the particles and m their mass. (b) Show that the kinetic energy of the particles increases by $2qV_0$ each revolution, assuming that the gap is small. (c) If the radius of the cyclotron is 0.50 m and the magnetic field strength is 0.60 T, what will be the maximum kinetic energy of accelerated protons in MeV?

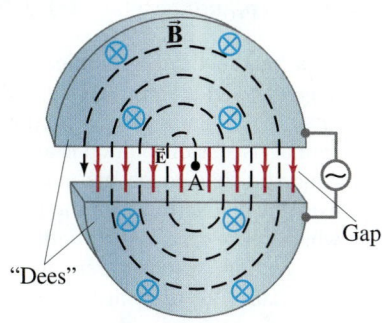

FIGURE 27–50
A cyclotron.
Problem 66.

"Dees"

67. Magnetic fields are very useful in particle accelerators for "beam steering"; that is, magnetic fields can be used to change the beam's direction without altering its speed (Fig. 27–51). Show how this could work with a beam of protons. What happens to protons that are not moving with the speed that the magnetic field is designed for? If the field extends over a region 5.0 cm wide and has a magnitude of 0.38 T, by approximately what angle will a beam of protons traveling at 0.85×10^7 m/s be bent?

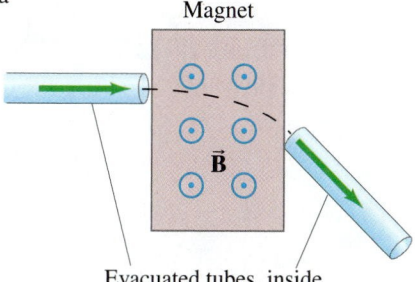

Magnet

$\vec{B}$

FIGURE 27–51
Problem 67.

Evacuated tubes, inside of which the protons move with velocity indicated by the green arrows

68. A square loop of aluminum wire is 20.0 cm on a side. It is to carry 15.0 A and rotate in a uniform 1.35-T magnetic field as shown in Fig. 27–52. (a) Determine the minimum diameter of the wire so that it will not fracture from tension or shear. Assume a safety factor of 10. (See Table 12–2.) (b) What is the resistance of a single loop of this wire?

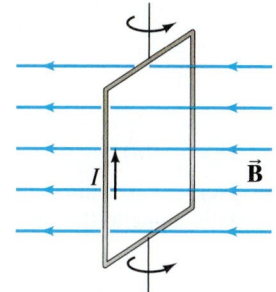

I $\vec{B}$

FIGURE 27–52
Problem 68.

69. A sort of "projectile launcher" is shown in Fig. 27–53. A large current moves in a closed loop composed of fixed rails, a power supply, and a very light, almost frictionless bar touching the rails. A 1.8 T magnetic field is perpendicular to the plane of the circuit. If the rails are a distance $d = 24$ cm apart, and the bar has a mass of 1.5 g, what constant current flow is needed to accelerate the bar from rest to 25 m/s in a distance of 1.0 m? In what direction must the field point?

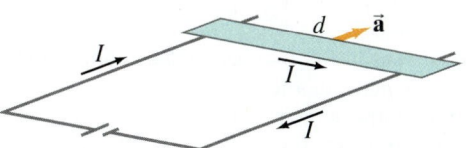

FIGURE 27–53   Problem 69.

70. (a) What value of magnetic field would make a beam of electrons, traveling to the right at a speed of 4.8×10^6 m/s, go undeflected through a region where there is a uniform electric field of 8400 V/m pointing vertically up? (b) What is the direction of the magnetic field if it is known to be perpendicular to the electric field? (c) What is the frequency of the circular orbit of the electrons if the electric field is turned off?

71. In a certain cathode ray tube, electrons are accelerated horizontally by 25 kV. They then pass through a uniform magnetic field B for a distance of 3.5 cm, which deflects them upward so they reach the top of the screen 22 cm away, 11 cm above the center. Estimate the value of B.

72. **Zeeman effect.** In the Bohr model of the hydrogen atom, the electron is held in its circular orbit of radius r about its proton nucleus by electrostatic attraction. If the atoms are placed in a weak magnetic field $\vec{B}$, the rotation frequency of electrons rotating in a plane perpendicular to $\vec{B}$ is changed by an amount

$$\Delta f = \pm \frac{eB}{4\pi m}$$

where e and m are the charge and mass of an electron. (a) Derive this result, assuming the force due to $\vec{B}$ is much less than that due to electrostatic attraction of the nucleus. (b) What does the $\pm$ sign indicate?

73. A proton follows a spiral path through a gas in a magnetic field of 0.018 T, perpendicular to the plane of the spiral, as shown in Fig. 27–54. In two successive loops, at points P and Q, the radii are 10.0 mm and 8.5 mm, respectively. Calculate the change in the kinetic energy of the proton as it travels from P to Q.

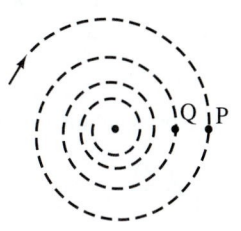

FIGURE 27–54   Problem 73.

74. The net force on a current loop whose face is perpendicular to a uniform magnetic field is zero, since contributions to the net force from opposite sides of the loop cancel. However, if the field varies in magnitude from one side of the loop to the other, then there can be a net force on the loop. Consider a square loop with sides whose length is a, located with one side at $x = b$ in the xy plane (Fig. 27–55). A magnetic field is directed along z, with a magnitude that varies with x according to

$$B = B_0 \left(1 - \frac{x}{b} \right).$$

If the current in the loop circulates counterclockwise (that is, the magnetic dipole moment of the loop is along the z axis), find an expression for the net force on the loop.

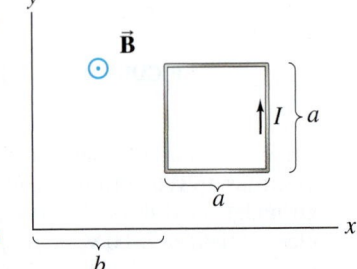

FIGURE 27–55
Problem 74.

75. The power cable for an electric trolley (Fig. 27–56) carries a horizontal current of 330 A toward the east. The Earth's magnetic field has a strength 5.0×10^{-5} T and makes an angle of dip of 22° at this location. Calculate the magnitude and direction of the magnetic force on a 5.0-m length of this cable.

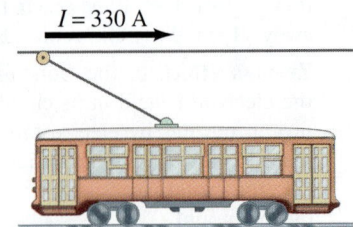

FIGURE 27–56
Problem 75.

76. A uniform conducting rod of length d and mass m sits atop a fulcrum, which is placed a distance $d/4$ from the rod's left-hand end and is immersed in a uniform magnetic field of magnitude B directed into the page (Fig. 27–57). An object whose mass M is 8.0 times greater than the rod's mass is hung from the rod's left-hand end. What current (direction and magnitude) should flow through the rod in order for it to be "balanced" (i.e., be at rest horizontally) on the fulcrum? (Flexible connecting wires which exert negligible force on the rod are not shown.)

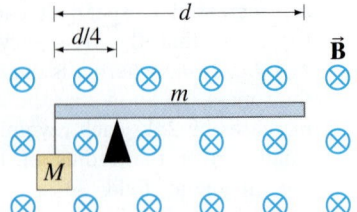

FIGURE 27–57
Problem 76.

77. In a simple device for measuring the magnitude B of a magnetic field, a conducting rod (length $d = 1.0$ m, mass $m = 150$ g) hangs from a friction-free pivot and is oriented so that its axis of rotation is aligned with the direction of the magnetic field to be measured. Thin flexible wires (which exert negligible force on the rod) carry a current $I = 12$ A, which causes the rod to deflect an angle θ with respect to the vertical, where it remains at rest (Fig. 27–58). (a) Is the current flowing upward (toward the pivot) or downward in Fig. 27–58? (b) If $\theta = 13°$, determine B. (c) What is the largest magnetic field magnitude that can be measured using this device?

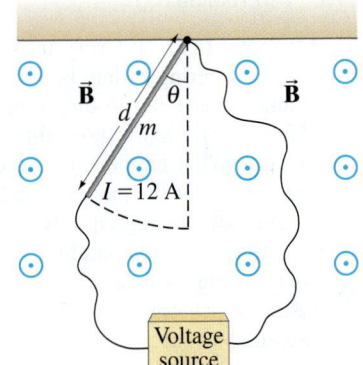

FIGURE 27–58
Problem 77.

Answers to Exercises

A: Near the poles, where the field lines are closer together.

B: Counterclockwise.

C: $(b), (c), (d)$.

D: 0.15 N.

E: $(b), (c), (d)$.

F: Negative; the direction of the helical path would be reversed (still going to the right).

G: (d).

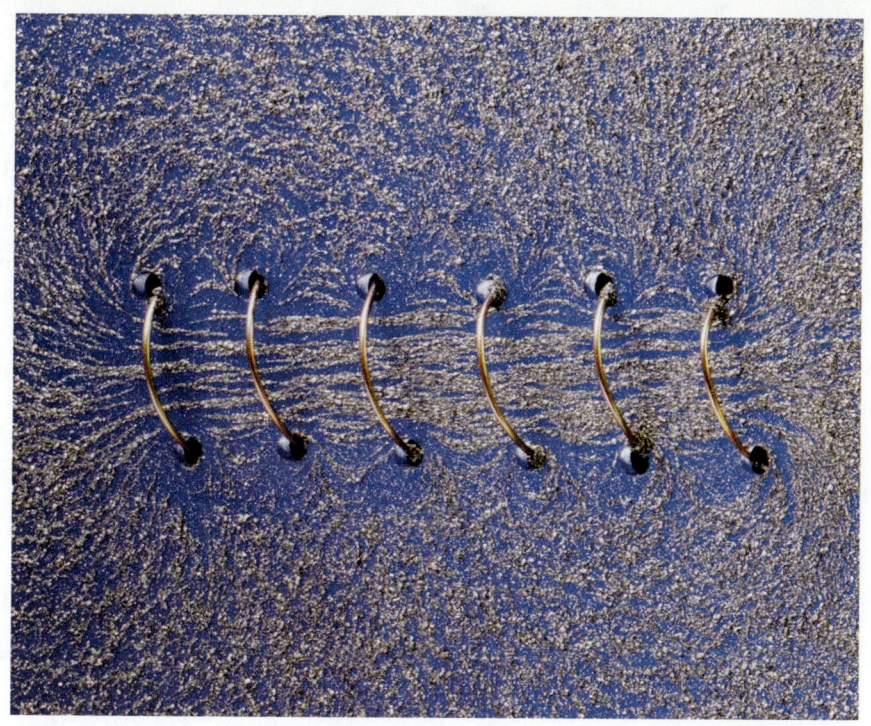

A long coil of wire with many closely spaced loops is called a solenoid. When a long solenoid carries an electric current, a nearly uniform magnetic field is produced within the loops as suggested by the alignment of the iron filings in this photo. The magnitude of the field inside a solenoid is readily found using Ampère's law, one of the great general laws of electromagnetism, relating magnetic fields and electric currents. We examine these connections in detail in this Chapter, as well as other means for producing magnetic fields.

28

Sources of Magnetic Field

CHAPTER-OPENING QUESTION—Guess now!

Which of the following will produce a magnetic field?

- **(a)** An electric charge at rest.
- **(b)** A moving electric charge.
- **(c)** An electric current.
- **(d)** The voltage of a battery not connected to anything.
- **(e)** Any piece of iron.
- **(f)** A piece of any metal.

I n the previous Chapter, we discussed the effects (forces and torques) that a magnetic field has on electric currents and on moving electric charges. We also saw that magnetic fields are produced not only by magnets but also by electric currents (Oersted's great discovery). It is this aspect of magnetism, the production of magnetic fields, that we discuss in this Chapter. We will now see how magnetic field strengths are determined for some simple situations, and discuss some general relations between magnetic fields and their sources, electric current. Most elegant is Ampère's law. We also study the Biot-Savart Law, which can be very helpful for solving practical problems.

CONTENTS

28–1 Magnetic Field Due to a Straight Wire

28–2 Force between Two Parallel Wires

28–3 Definitions of the Ampere and the Coulomb

28–4 Ampère's Law

28–5 Magnetic Field of a Solenoid and a Toroid

28–6 Biot-Savart Law

28–7 Magnetic Materials— Ferromagnetism

*28–8 Electromagnets and Solenoids—Applications

*28–9 Magnetic Fields in Magnetic Materials; Hysteresis

*28–10 Paramagnetism and Diamagnetism

28–1 Magnetic Field Due to a Straight Wire

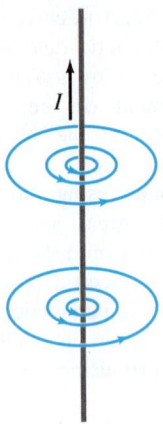

FIGURE 28–1 Same as Fig. 27–8b. Magnetic field lines around a long straight wire carrying an electric current I.

FIGURE 28–2 Example 28–1.

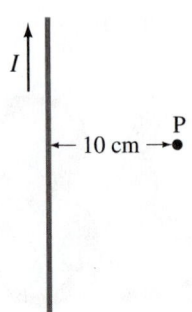

We saw in Section 27–2 that the magnetic field due to the electric current in a long straight wire is such that the field lines are circles with the wire at the center (Fig. 28–1). You might expect that the field strength at a given point would be greater if the current flowing in the wire were greater; and that the field would be less at points farther from the wire. This is indeed the case. Careful experiments show that the magnetic field B due to a long straight wire at a point near it is directly proportional to the current I in the wire and inversely proportional to the distance r from the wire:

$$B \propto \frac{I}{r}.$$

This relation $B \propto I/r$ is valid as long as r, the perpendicular distance to the wire, is much less than the distance to the ends of the wire (i.e., the wire is long).

The proportionality constant is written[†] as $\mu_0/2\pi$; thus,

$$B = \frac{\mu_0}{2\pi}\frac{I}{r}. \qquad \text{[near a long straight wire]} \quad \textbf{(28–1)}$$

The value of the constant μ_0, which is called the **permeability of free space**, is $\mu_0 = 4\pi \times 10^{-7}\,\text{T·m/A}$.

EXAMPLE 28–1 **Calculation of $\vec{\mathbf{B}}$ near a wire.** An electric wire in the wall of a building carries a dc current of 25 A vertically upward. What is the magnetic field due to this current at a point P, 10 cm due north of the wire (Fig. 28–2)?

APPROACH We assume the wire is much longer than the 10-cm distance to the point P so we can apply Eq. 28–1.

SOLUTION According to Eq. 28–1:

$$B = \frac{\mu_0 I}{2\pi r} = \frac{(4\pi \times 10^{-7}\,\text{T·m/A})(25\,\text{A})}{(2\pi)(0.10\,\text{m})} = 5.0 \times 10^{-5}\,\text{T},$$

or 0.50 G. By the right-hand rule (Table 27–1, page 716), the field due to the current points to the west (into the page in Fig. 28–2) at point P.

NOTE The wire's field has about the same magnitude as Earth's magnetic field, so a compass at P would not point north but in a northwesterly direction.

NOTE Most electrical wiring in buildings consists of cables with two wires in each cable. Since the two wires carry current in opposite directions, their magnetic fields cancel to a large extent, but may still affect sensitive electronic devices.

> ⚠️ **CAUTION**
> *A compass, near a current, may not point north*

EXERCISE A In Example 25–10 we saw that a typical lightning bolt produces a 100-A current for 0.2 s. Estimate the magnetic field 10 m from a lightning bolt. Would it have a significant effect on a compass?

EXAMPLE 28–2 **Magnetic field midway between two currents.** Two parallel straight wires 10.0 cm apart carry currents in opposite directions (Fig. 28–3). Current $I_1 = 5.0\,\text{A}$ is out of the page, and $I_2 = 7.0\,\text{A}$ is into the page. Determine the magnitude and direction of the magnetic field halfway between the two wires.

APPROACH The magnitude of the field produced by each wire is calculated from Eq. 28–1. The direction of *each* wire's field is determined with the right-hand rule. The total field is the vector sum of the two fields at the midway point.

SOLUTION The magnetic field lines due to current I_1 form circles around the wire of I_1, and right-hand-rule-1 (Fig. 27–8c) tells us they point counterclockwise around the wire. The field lines due to I_2 form circles around the wire of I_2 and point clockwise, Fig. 28–3. At the midpoint, both fields point upward as shown, and so add together.

FIGURE 28–3 Example 28–2. Wire 1 carrying current I_1 out towards us, and wire 2 carrying current I_2 into the page, produce magnetic fields whose lines are circles around their respective wires.

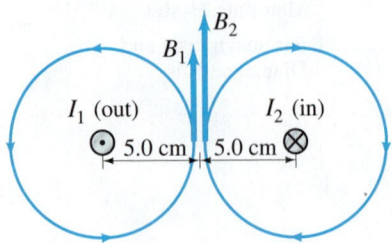

[†]The constant is chosen in this complicated way so that Ampère's law (Section 28–4), which is considered more fundamental, will have a simple and elegant form.

The midpoint is 0.050 m from each wire, and from Eq. 28–1 the magnitudes of B_1 and B_2 are

$$B_1 = \frac{\mu_0 I_1}{2\pi r} = \frac{(4\pi \times 10^{-7}\,\text{T}\cdot\text{m/A})(5.0\,\text{A})}{2\pi(0.050\,\text{m})} = 2.0 \times 10^{-5}\,\text{T};$$

$$B_2 = \frac{\mu_0 I_2}{2\pi r} = \frac{(4\pi \times 10^{-7}\,\text{T}\cdot\text{m/A})(7.0\,\text{A})}{2\pi(0.050\,\text{m})} = 2.8 \times 10^{-5}\,\text{T}.$$

The total field is *up* with a magnitude of

$$B = B_1 + B_2 = 4.8 \times 10^{-5}\,\text{T}.$$

EXERCISE B Suppose both I_1 and I_2 point into the page in Fig. 28–3. What then is the field midway between the two wires?

CONCEPTUAL EXAMPLE 28–3 **Magnetic field due to four wires.** Figure 28–4 shows four long parallel wires which carry equal currents into or out of the page as shown. In which configuration, (a) or (b), is the magnetic field greater at the center of the square?

RESPONSE It is greater in (a). The arrows illustrate the directions of the field produced by each wire; check it out, using the right-hand rule to confirm these results. The net field at the center is the superposition of the four fields, which will point to the left in (a) and is zero in (b).

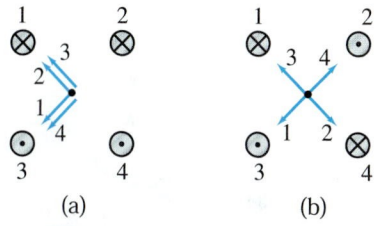

FIGURE 28–4 Example 28–3.

28–2 Force between Two Parallel Wires

We have seen that a wire carrying a current produces a magnetic field (magnitude given by Eq. 28–1 for a long straight wire). Also, a current-carrying wire feels a force when placed in a magnetic field (Section 27–3, Eq. 27–1). Thus, we expect that two current-carrying wires will exert a force on each other.

Consider two long parallel wires separated by a distance d, as in Fig. 28–5a. They carry currents I_1 and I_2, respectively. Each current produces a magnetic field that is "felt" by the other, so each must exert a force on the other. For example, the magnetic field B_1 produced by I_1 in Fig. 28–5 is given by Eq. 28–1, which at the location of wire 2 is

$$B_1 = \frac{\mu_0}{2\pi}\frac{I_1}{d}.$$

See Fig. 28–5b, where the field due *only* to I_1 is shown. According to Eq. 27–2, the force F_2 exerted by B_1 on a length ℓ_2 of wire 2, carrying current I_2, is

$$F_2 = I_2 B_1 \ell_2.$$

Note that the force on I_2 is due only to the field produced by I_1. Of course, I_2 also produces a field, but it does not exert a force on itself. We substitute B_1 into the formula for F_2 and find that the force on a length ℓ_2 of wire 2 is

$$F_2 = \frac{\mu_0}{2\pi}\frac{I_1 I_2}{d}\ell_2. \qquad \text{[parallel wires]} \quad \textbf{(28–2)}$$

If we use right-hand-rule-1 of Fig. 27–8c, we see that the lines of B_1 are as shown in Fig. 28–5b. Then using right-hand-rule-2 of Fig. 27–11c, we see that the force exerted on I_2 will be to the left in Fig. 28–5b. That is, I_1 exerts an attractive force on I_2 (Fig. 28–6a). This is true as long as the currents are in the same direction. If I_2 is in the opposite direction, the right-hand rule indicates that the force is in the opposite direction. That is, I_1 exerts a repulsive force on I_2 (Fig. 28–6b).

Reasoning similar to that above shows that the magnetic field produced by I_2 exerts an equal but opposite force on I_1. We expect this to be true also from Newton's third law. Thus, as shown in Fig. 28–6, parallel currents in the same direction attract each other, whereas parallel currents in opposite directions repel.

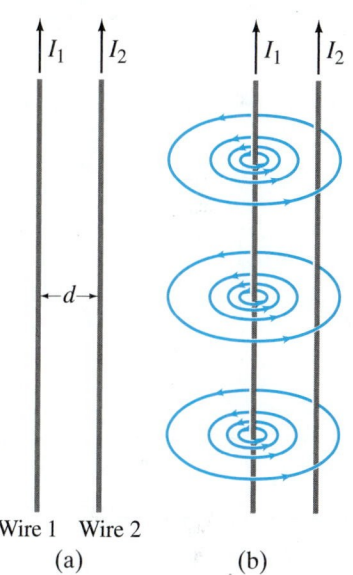

FIGURE 28–5 (a) Two parallel conductors carrying currents I_1 and I_2. (b) Magnetic field $\vec{B}_1$ produced by I_1. (Field produced by I_2 is not shown.) $\vec{B}_1$ points into page at position of I_2.

FIGURE 28–6 (a) Parallel currents in the same direction exert an attractive force on each other. (b) Antiparallel currents (in opposite directions) exert a repulsive force on each other.

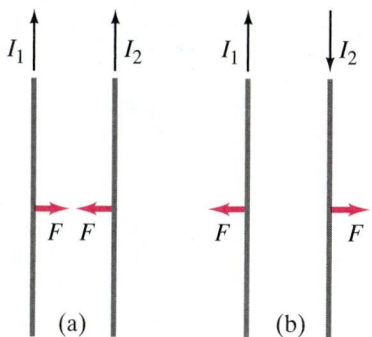

EXAMPLE 28–4 **Force between two current-carrying wires.** The two wires of a 2.0-m-long appliance cord are 3.0 mm apart and carry a current of 8.0 A dc. Calculate the force one wire exerts on the other.

APPROACH Each wire is in the magnetic field of the other when the current is on, so we can apply Eq. 28–2.

SOLUTION Equation 28–2 gives

$$F = \frac{(4\pi \times 10^{-7}\,\text{T·m/A})(8.0\,\text{A})^2(2.0\,\text{m})}{(2\pi)(3.0 \times 10^{-3}\,\text{m})} = 8.5 \times 10^{-3}\,\text{N}.$$

The currents are in opposite directions (one toward the appliance, the other away from it), so the force would be repulsive and tend to spread the wires apart.

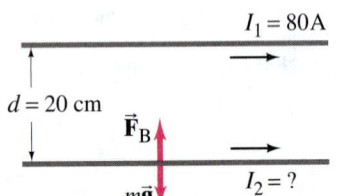

$I_1 = 80\,\text{A}$

$d = 20$ cm

$\vec{\mathbf{F}}_\text{B}$

$m\vec{\mathbf{g}}$

$I_2 = ?$

FIGURE 28–7 Example 28–5.

EXAMPLE 28–5 **Suspending a wire with a current.** A horizontal wire carries a current $I_1 = 80$ A dc. A second parallel wire 20 cm below it (Fig. 28–7) must carry how much current I_2 so that it doesn't fall due to gravity? The lower wire has a mass of 0.12 g per meter of length.

APPROACH If wire 2 is not to fall under gravity, which acts downward, the magnetic force on it must be upward. This means that the current in the two wires must be in the same direction (Fig. 28–6). We can find the current I_2 by equating the magnitudes of the magnetic force and the gravitational force on the wire.

SOLUTION The force of gravity on wire 2 is downward. For each 1.0 m of wire length, the gravitational force has magnitude

$$F = mg = (0.12 \times 10^{-3}\,\text{kg/m})(1.0\,\text{m})(9.8\,\text{m/s}^2) = 1.18 \times 10^{-3}\,\text{N}.$$

The magnetic force on wire 2 must be upward, and Eq. 28–2 gives

$$F = \frac{\mu_0}{2\pi} \frac{I_1 I_2}{d} \ell$$

where $d = 0.20$ m and $I_1 = 80$ A. We solve this for I_2 and set the two force magnitudes equal (letting $\ell = 1.0$ m):

$$I_2 = \frac{2\pi d}{\mu_0 I_1}\left(\frac{F}{\ell}\right) = \frac{2\pi(0.20\,\text{m})}{(4\pi \times 10^{-7}\,\text{T·m/A})(80\,\text{A})}\frac{(1.18 \times 10^{-3}\,\text{N/m})}{(1.0\,\text{m})} = 15\,\text{A}.$$

28–3 Definitions of the Ampere and the Coulomb

You may have wondered how the constant μ_0 in Eq. 28–1 could be exactly $4\pi \times 10^{-7}$ T·m/A. Here is how it happened. With an older definition of the ampere, μ_0 was measured experimentally to be very close to this value. Today, μ_0 is *defined* to be exactly $4\pi \times 10^{-7}$ T·m/A. This could not be done if the ampere were defined independently. The ampere, the unit of current, is now defined in terms of the magnetic field B it produces using the defined value of μ_0.

In particular, we use the force between two parallel current-carrying wires, Eq. 28–2, to define the ampere precisely. If $I_1 = I_2 = 1$ A exactly, and the two wires are exactly 1 m apart, then

$$\frac{F}{\ell} = \frac{\mu_0}{2\pi} \frac{I_1 I_2}{d} = \frac{(4\pi \times 10^{-7}\,\text{T·m/A})}{(2\pi)}\frac{(1\,\text{A})(1\,\text{A})}{(1\,\text{m})} = 2 \times 10^{-7}\,\text{N/m}.$$

Thus, *one* **ampere** *is defined as that current flowing in each of two long parallel wires 1 m apart, which results in a force of exactly 2×10^{-7} N per meter of length of each wire.*

This is the precise definition of the ampere. The **coulomb** is then defined as being *exactly* one ampere-second: $1 \, \text{C} = 1 \, \text{A} \cdot \text{s}$. The value of k or ϵ_0 in Coulomb's law (Section 21–5) is obtained from experiment.

This may seem a rather roundabout way of defining quantities. The reason behind it is the desire for **operational definitions** of quantities—that is, definitions of quantities that can actually be measured given a definite set of operations to carry out. For example, the unit of charge, the coulomb, could be defined in terms of the force between two equal charges after defining a value for ϵ_0 or k in Eqs. 21–1 or 21–2. However, to carry out an actual experiment to measure the force between two charges is very difficult. For one thing, any desired amount of charge is not easily obtained precisely; and charge tends to leak from objects into the air. The amount of current in a wire, on the other hand, can be varied accurately and continuously (by putting a variable resistor in a circuit). Thus the force between two current-carrying conductors is far easier to measure precisely. This is why we first define the ampere, and then define the coulomb in terms of the ampere. At the National Institute of Standards and Technology in Maryland, precise measurement of current is made using circular coils of wire rather than straight lengths because it is more convenient and accurate.

Electric and magnetic field strengths are also defined operationally: the electric field in terms of the measurable force on a charge, via Eq. 21–3; and the magnetic field in terms of the force per unit length on a current-carrying wire, via Eq. 27–2.

28–4 Ampère's Law

In Section 28–1 we saw that Eq. 28–1 gives the relation between the current in a long straight wire and the magnetic field it produces. This equation is valid *only* for a long straight wire. Is there a general relation between a current in a wire of any shape and the magnetic field around it? The answer is yes: the French scientist André Marie Ampère (1775–1836) proposed such a relation shortly after Oersted's discovery. Consider an arbitrary closed path around a current as shown in Fig. 28–8, and imagine this path as being made up of short segments each of length $\Delta \ell$. First, we take the product of the length of each segment times the component of $\vec{\mathbf{B}}$ parallel to that segment (call this component $B_\parallel$). If we now sum all these terms, according to Ampère, the result will be equal to μ_0 times the net current I_{encl} that passes through the surface enclosed by the path:

$$\sum B_\parallel \, \Delta \ell \ = \ \mu_0 \, I_{\text{encl}}.$$

The lengths $\Delta \ell$ are chosen so that $B_\parallel$ is essentially constant along each length. The sum must be made over a *closed path*; and I_{encl} is the net current passing through the surface bounded by this closed path (orange in Fig. 28–8). In the limit $\Delta \ell \to 0$, this relation becomes

$$\oint \vec{\mathbf{B}} \cdot d\vec{\boldsymbol{\ell}} \ = \ \mu_0 \, I_{\text{encl}}, \qquad \qquad \textbf{(28–3)}$$

where $d\vec{\boldsymbol{\ell}}$ is an infinitesimal length vector and the vector dot product assures that the parallel component of $\vec{\mathbf{B}}$ is taken. Equation 28–3 is known as **Ampère's law**. The integrand in Eq. 28–3 is taken around a closed path, and I_{encl} is the current passing through the space enclosed by the chosen path or loop.

FIGURE 28–8 Arbitrary path enclosing a current, for Ampère's law. The path is broken down into segments of equal length $\Delta \ell$.

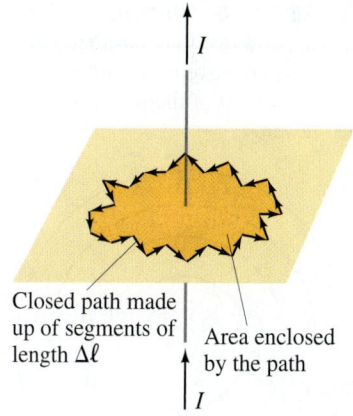

I

Closed path made up of segments of length $\Delta \ell$

Area enclosed by the path

I

AMPÈRE'S LAW

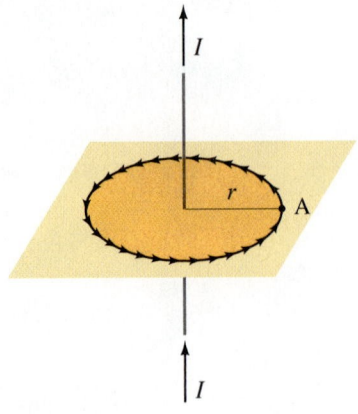

FIGURE 28–9 Circular path of radius r.

To understand Ampère's law better, let us apply it to the simple case of a single long straight wire carrying a current I which we've already examined, and which served as an inspiration for Ampère himself. Suppose we want to find the magnitude of $\vec{B}$ at some point A which is a distance r from the wire (Fig. 28–9). We know the magnetic field lines are circles with the wire at their center. So to apply Eq. 28–3 we choose as our path of integration a circle of radius r. The choice of path is ours, so we choose one that will be convenient: at any point on this circular path, $\vec{B}$ will be tangent to the circle. Furthermore, since all points on the path are the same distance from the wire, by symmetry we expect B to have the same magnitude at each point. Thus for any short segment of the circle (Fig. 28–9), $\vec{B}$ will be parallel to that segment, and (setting $I_{\text{encl}} = I$)

$$\mu_0 I = \oint \vec{B} \cdot d\vec{\ell}$$

$$= \oint B \, d\ell = B \oint d\ell = B(2\pi r),$$

where $\oint d\ell = 2\pi r$, the circumference of the circle. We solve for B and obtain

$$B = \frac{\mu_0 I}{2\pi r}.$$

This is just Eq. 28–1 for the field near a long straight wire as discussed earlier.

Ampère's law thus works for this simple case. A great many experiments indicate that Ampère's law is valid in general. However, as with Gauss's law for the electric field, its practical value as a means to calculate the magnetic field is limited mainly to simple or symmetric situations. Its importance is that it relates the magnetic field to the current in a direct and mathematically elegant way. Ampère's law is thus considered one of the basic laws of electricity and magnetism. It is valid for any situation where the currents and fields are steady and not changing in time, and no magnetic materials are present.

We now can see why the constant in Eq. 28–1 is written $\mu_0/2\pi$. This is done so that only μ_0 appears in Eq. 28–3, rather than, say, $2\pi k$ if we had used k in Eq. 28–1. In this way, the more fundamental equation, Ampère's law, has the simpler form.

It should be noted that the $\vec{B}$ in Ampère's law is not necessarily due only to the current I_{encl}. Ampère's law, like Gauss's law for the electric field, is valid in general. $\vec{B}$ is the field at each point in space along the chosen path due to all sources—including the current I enclosed by the path, but also due to any other sources. For example, the field surrounding two parallel current-carrying wires is the vector sum of the fields produced by each, and the field lines are shown in Fig. 28–10. If the path chosen for the integral (Eq. 28–3) is a circle centered on one of the wires with radius less than the distance between the wires (the dashed line in Fig. 28–10), only the current (I_1) in the encircled wire is included on the right side of Eq. 28–3. $\vec{B}$ on the left side of the equation must be the total $\vec{B}$ at each point due to both wires. Note also that $\oint \vec{B} \cdot d\vec{\ell}$ for the path shown in Fig. 28–10 is the same whether the second wire is present or not (in both cases, it equals $\mu_0 I_1$ according to Ampère's law). How can this be? It can be so because the fields due to the two wires partially cancel one another at some points between them, such as point C in the diagram ($\vec{B} = 0$ at a point midway between the wires if $I_1 = I_2$); at other points, such as D in Fig. 28–10, the fields add together to produce a larger field. In the *sum*, $\oint \vec{B} \cdot d\vec{\ell}$, these effects just balance so that $\oint \vec{B} \cdot d\vec{\ell} = \mu_0 I_1$, whether the second wire is there or not. The integral $\oint \vec{B} \cdot d\vec{\ell}$ will be the same in each case, even though $\vec{B}$ will not be the same at every point for each of the two cases.

FIGURE 28–10 Magnetic field lines around two long parallel wires whose equal currents, I_1 and I_2, are coming out of the paper toward the viewer.

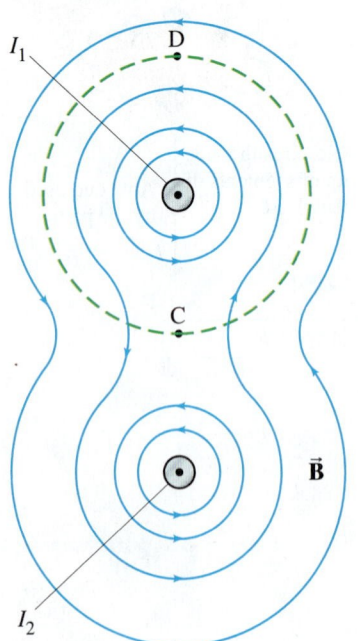

EXAMPLE 28–6 **Field inside and outside a wire.** A long straight cylindrical wire conductor of radius R carries a current I of uniform current density in the conductor. Determine the magnetic field due to this current at (a) points outside the conductor ($r > R$), and (b) points inside the conductor ($r < R$). See Fig. 28–11. Assume that r, the radial distance from the axis, is much less than the length of the wire. (c) If $R = 2.0$ mm and $I = 60$ A, what is B at $r = 1.0$ mm, $r = 2.0$ mm, and $r = 3.0$ mm?

APPROACH We can use *symmetry*: Because the wire is long, straight, and cylindrical, we expect from symmetry that the magnetic field must be the same at all points that are the same distance from the center of the conductor. There is no reason why any such point should have preference over others at the same distance from the wire (they are physically equivalent). So B must have the same value at all points the same distance from the center. We also expect $\vec{B}$ to be tangent to circles around the wire (Fig. 28–1), so we choose a circular path of integration as we did in Fig. 28–9.

SOLUTION (a) We apply Ampère's law, integrating around a circle ($r > R$) centered on the wire (Fig. 28–11a), and then $I_{encl} = I$:

$$\oint \vec{B} \cdot d\vec{\ell} = B(2\pi r) = \mu_0 I_{encl}$$

or

$$B = \frac{\mu_0 I}{2\pi r}, \qquad\qquad [r > R]$$

which is the same result as for a thin wire.

(b) Inside the wire ($r < R$), we again choose a circular path concentric with the cylinder; we expect $\vec{B}$ to be tangential to this path, and again, because of the symmetry, it will have the same magnitude at all points on the circle. The current enclosed in this case is less than I by a factor of the ratio of the areas:

$$I_{encl} = I \frac{\pi r^2}{\pi R^2}.$$

So Ampère's law gives

$$\oint \vec{B} \cdot d\vec{\ell} = \mu_0 I_{encl}$$

$$B(2\pi r) = \mu_0 I \left(\frac{\pi r^2}{\pi R^2} \right)$$

so

$$B = \frac{\mu_0 I r}{2\pi R^2}. \qquad\qquad [r < R]$$

The field is zero at the center of the conductor and increases linearly with r until $r = R$; beyond $r = R$, B decreases as $1/r$. This is shown in Fig. 28–11b. Note that these results are valid only for points close to the center of the conductor as compared to its length. For a current to flow, there must be connecting wires (to a battery, say), and the field due to these conducting wires, if not very far away, will destroy the assumed symmetry.

(c) At $r = 2.0$ mm, the surface of the wire, $r = R$, so

$$B = \frac{\mu_0 I}{2\pi R} = \frac{(4\pi \times 10^{-7}\,\text{T·m/A})(60\,\text{A})}{(2\pi)(2.0 \times 10^{-3}\,\text{m})} = 6.0 \times 10^{-3}\,\text{T}.$$

We saw in (b) that inside the wire B is linear in r. So at $r = 1.0$ mm, B will be half what it is at $r = 2.0$ mm or 3.0×10^{-3} T. Outside the wire, B falls off as $1/r$, so at $r = 3.0$ mm it will be two-thirds as great as at $r = 2.0$ mm, or $B = 4.0 \times 10^{-3}$ T. To check, we use our result in (a), $B = \mu_0 I/2\pi r$, which gives the same result.

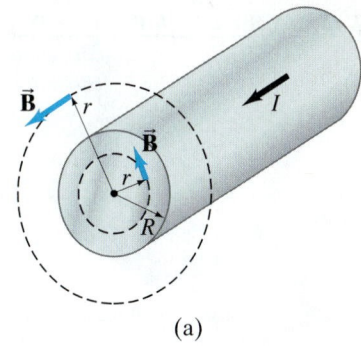

(a)

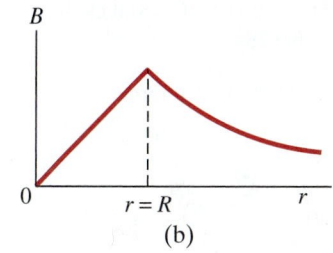

(b)

FIGURE 28–11 Magnetic field inside and outside a cylindrical conductor (Example 28–6).

 **CAUTION**

Connecting wires can destroy assumed symmetry

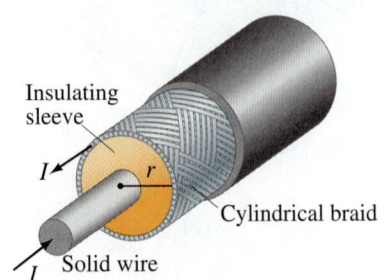

FIGURE 28–12 Coaxial cable. Example 28–7.

FIGURE 28–13 Exercise C.

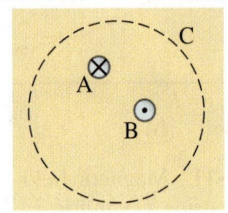

FIGURE 28–14 Example 28–8.

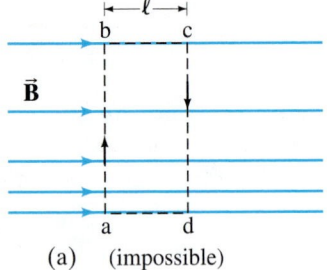

(a) (impossible)

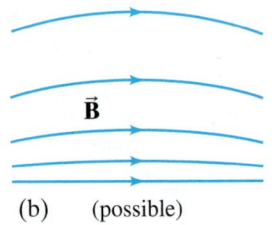

(b) (possible)

CONCEPTUAL EXAMPLE 28–7 | **Coaxial cable.** A *coaxial cable* is a single wire surrounded by a cylindrical metallic braid, as shown in Fig. 28–12. The two conductors are separated by an insulator. The central wire carries current to the other end of the cable, and the outer braid carries the return current and is usually considered ground. Describe the magnetic field (*a*) in the space between the conductors, and (*b*) outside the cable.

RESPONSE (*a*) In the space between the conductors, we can apply Ampère's law for a circular path around the center wire, just as we did for the case shown in Figs. 28–9 and 28–11. The magnetic field lines will be concentric circles centered on the center of the wire, and the magnitude is given by Eq. 28–1. The current in the outer conductor has no bearing on this result. (Ampère's law uses only the current enclosed *inside* the path; as long as the currents outside the path don't affect the *symmetry* of the field, they do not contribute to the field along the path at all). (*b*) Outside the cable, we can draw a similar circular path, for we expect the field to have the same cylindrical symmetry. Now, however, there are two currents enclosed by the path, and they add up to zero. The field outside the cable is zero.

The nice feature of coaxial cables is that they are self-shielding: no stray magnetic fields exist outside the cable. The outer cylindrical conductor also shields external electric fields from coming in (see also Example 21–14). This makes them ideal for carrying signals near sensitive equipment. Audiophiles use coaxial cables between stereo equipment components and even to the loudspeakers.

EXERCISE C In Fig. 28–13, A and B are wires each carrying a 3.0-A current but in opposite directions. On the circle C, which statement is true? (*a*) $B = 0$; (*b*) $\oint \vec{B} \cdot d\vec{\ell} = 0$; (*c*) $B = 3\mu_0$; (*d*) $B = -3\mu_0$; (*e*) $\oint \vec{B} \cdot d\vec{\ell} = 6\mu_0$.

EXAMPLE 28–8 **A nice use for Ampère's law.** Use Ampère's law to show that in any region of space where there are no currents the magnetic field cannot be both unidirectional and nonuniform as shown in Fig. 28–14a.

APPROACH The wider spacing of lines near the top of Fig. 28–14a indicates the field $\vec{B}$ has a smaller magnitude at the top than it does lower down. We apply Ampère's law to the rectangular path abcd shown dashed in Fig. 28–14a.

SOLUTION Because no current is enclosed by the chosen path, Ampère's law gives

$$\oint \vec{B} \cdot d\vec{\ell} = 0.$$

The integral along sections ab and cd is zero, since $\vec{B} \perp d\vec{\ell}$. Thus

$$\oint \vec{B} \cdot d\vec{\ell} = B_{bc}\ell - B_{da}\ell = (B_{bc} - B_{da})\ell,$$

which is not zero since the field B_{bc} along the path bc is less than the field B_{da} along path da. Hence we have a contradiction: $\oint \vec{B} \cdot d\vec{\ell}$ cannot be both zero (since $I = 0$) and nonzero. Thus we have shown that a nonuniform unidirectional field is not consistent with Ampère's law. A nonuniform field whose direction also changes, as in Fig. 28–14b, is consistent with Ampère's law (convince yourself this is so), and possible. The fringing of a permanent magnet's field (Fig. 27–7) has this shape.

Ampère's Law

1. Ampère's law, like Gauss's law, is always a valid statement. But as a calculation tool it is limited mainly to systems with a high degree of symmetry. The first step in applying Ampère's law is to identify useful **symmetry**.

2. Choose an integration path that reflects the symmetry (see the Examples). Search for paths where B has constant magnitude along the entire path or along segments of the path. Make sure your integration path passes through the point where you wish to evaluate the magnetic field.

3. Use symmetry to determine the direction of $\vec{B}$ along the integration path. With a smart choice of path, $\vec{B}$ will be either parallel or perpendicular to the path.

4. Determine the enclosed current, I_{encl}. Be careful with signs. Let the fingers of your right hand curl along the direction of $\vec{B}$ so that your thumb shows the direction of positive current. If you have a solid conductor and your integration path does not enclose the full current, you can use the current density (current per unit area) multiplied by the enclosed area (as in Example 28–6).

28-5 Magnetic Field of a Solenoid and a Toroid

A long coil of wire consisting of many loops is called a **solenoid**. Each loop produces a magnetic field as was shown in Fig. 27–9. In Fig. 28–15a, we see the field due to a solenoid when the coils are far apart. Near each wire, the field lines are very nearly circles as for a straight wire (that is, at distances that are small compared to the curvature of the wire). Between any two wires, the fields due to each loop tend to cancel. Toward the center of the solenoid, the fields add up to give a field that can be fairly large and fairly uniform. For a long solenoid with closely packed coils, the field is nearly uniform and parallel to the solenoid axis within the entire cross section, as shown in Fig. 28–15b. The field outside the solenoid is very small compared to the field inside, except near the ends. Note that the same number of field lines that are concentrated inside the solenoid, spread out into the vast open space outside.

We now use Ampère's law to determine the magnetic field inside a very long (ideally, infinitely long) closely packed solenoid. We choose the path abcd shown in Fig. 28–16, far from either end, for applying Ampère's law. We will consider this path as made up of four segments, the sides of the rectangle: ab, bc, cd, da. Then the left side of Eq. 28–3, Ampère's law, becomes

$$\oint \vec{\mathbf{B}} \cdot d\vec{\boldsymbol{\ell}} = \int_a^b \vec{\mathbf{B}} \cdot d\vec{\boldsymbol{\ell}} + \int_b^c \vec{\mathbf{B}} \cdot d\vec{\boldsymbol{\ell}} + \int_c^d \vec{\mathbf{B}} \cdot d\vec{\boldsymbol{\ell}} + \int_d^a \vec{\mathbf{B}} \cdot d\vec{\boldsymbol{\ell}}.$$

The field outside the solenoid is so small as to be negligible compared to the field inside. Thus the first term in this sum will be zero. Furthermore, $\vec{\mathbf{B}}$ is perpendicular to the segments bc and da inside the solenoid, and is nearly zero between and outside the coils,

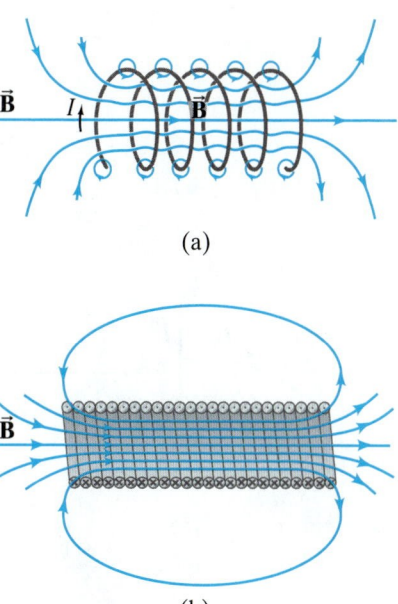

FIGURE 28–15 Magnetic field due to a solenoid: (a) loosely spaced turns, (b) closely spaced turns.

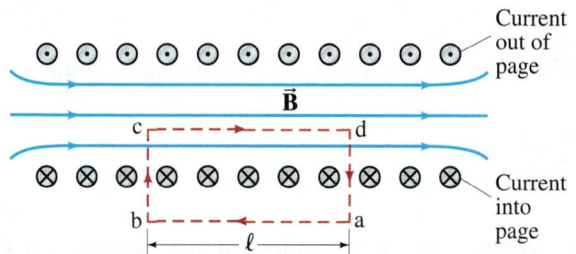

FIGURE 28–16 Cross-sectional view into a solenoid. The magnetic field inside is straight except at the ends. Red dashed lines indicate the path chosen for use in Ampère's law. $\odot$ and $\otimes$ are electric current direction (in the wire loops) out of the page and into the page.

so these terms too are zero. Therefore we have reduced the integral to the segment cd where $\vec{\mathbf{B}}$ is the nearly uniform field inside the solenoid, and is parallel to $d\vec{\boldsymbol{\ell}}$, so

$$\oint \vec{\mathbf{B}} \cdot d\vec{\boldsymbol{\ell}} = \int_c^d \vec{\mathbf{B}} \cdot d\vec{\boldsymbol{\ell}} = B\ell,$$

where ℓ is the length cd. Now we determine the current enclosed by this loop for the right side of Ampère's law, Eq. 28–3. If a current I flows in the wire of the solenoid, the total current enclosed by our path abcd is NI where N is the number of loops our path encircles (five in Fig. 28–16). Thus Ampère's law gives us

$$B\ell = \mu_0 NI.$$

If we let $n = N/\ell$ be the *number of loops per unit length*, then

$$B = \mu_0 nI. \qquad \text{[solenoid]} \quad \textbf{(28–4)}$$

This is the magnitude of the magnetic field within a solenoid. Note that B depends only on the number of loops per unit length, n, and the current I. The field does not depend on position within the solenoid, so B is uniform. This is strictly true only for an infinite solenoid, but it is a good approximation for real ones for points not close to the ends.

EXAMPLE 28–9 **Field inside a solenoid.** A thin 10-cm-long solenoid used for fast electromechanical switching has a total of 400 turns of wire and carries a current of 2.0 A. Calculate the field inside near the center.

APPROACH We use Eq. 28–4, where the number of turns per unit length is $n = 400/0.10\,\text{m} = 4.0 \times 10^3\,\text{m}^{-1}$.

SOLUTION $B = \mu_0 nI = (4\pi \times 10^{-7}\,\text{T}\cdot\text{m/A})(4.0 \times 10^3\,\text{m}^{-1})(2.0\,\text{A}) = 1.0 \times 10^{-2}\,\text{T}.$

A close look at Fig. 28–15 shows that the field outside of a solenoid is much like that of a bar magnet (Fig. 27–4). Indeed, a solenoid acts like a magnet, with one end acting as a north pole and the other as south pole, depending on the direction of the current in the loops. Since magnetic field lines leave the north pole of a magnet, the north poles of the solenoids in Fig. 28–15 are on the right.

Solenoids have many practical applications, and we discuss some of them later in the Chapter, in Section 28–8.

EXAMPLE 28–10 **Toroid.** Use Ampère's law to determine the magnetic field (a) inside and (b) outside a toroid, which is like a solenoid bent into the shape of a circle as shown in Fig. 28–17a.

APPROACH The magnetic field lines inside the toroid will be circles concentric with the toroid. (If you think of the toroid as a solenoid bent into a circle, the field lines bend along with the solenoid.) The direction of $\vec{\mathbf{B}}$ is clockwise. We choose as our path of integration one of these field lines of radius r inside the toroid as shown by the dashed line labeled "path 1" in Fig. 28–17a. We make this choice to use the *symmetry* of the situation, so B will be tangent to the path and will have the same magnitude at all points along the path (although it is not necessarily the same across the whole cross section of the toroid). This chosen path encloses *all* the coils; if there are N coils, each carrying current I, then $I_{encl} = NI$.

SOLUTION (a) Ampère's law applied along this path gives

$$\oint \vec{\mathbf{B}} \cdot d\vec{\boldsymbol{\ell}} = \mu_0 I_{encl}$$

$$B(2\pi r) = \mu_0 N I,$$

where N is the total number of coils and I is the current in each of the coils. Thus

$$B = \frac{\mu_0 N I}{2\pi r}.$$

The magnetic field B is not uniform within the toroid: it is largest along the inner edge (where r is smallest) and smallest at the outer edge. However, if the toroid is large, but thin (so that the difference between the inner and outer radii is small compared to the average radius), the field will be essentially uniform within the toroid. In this case, the formula for B reduces to that for a straight solenoid $B = \mu_0 n I$ where $n = N/(2\pi r)$ is the number of coils per unit length. (b) Outside the toroid, we choose as our path of integration a circle concentric with the toroid, "path 2" in Fig. 28–17a. This path encloses N loops carrying current I in one direction and N loops carrying the same current in the opposite direction. (Figure 28–17b shows the directions of the current for the parts of the loop on the inside and outside of the toroid.) Thus the net current enclosed by path 2 is zero. For a very tightly packed toroid, all points on path 2 are equidistant from the toroid and equivalent, so we expect B to be the same at all points along the path. Hence, Ampère's law gives

$$\oint \vec{\mathbf{B}} \cdot d\vec{\boldsymbol{\ell}} = \mu_0 I_{encl}$$

$$B(2\pi r) = 0$$

or

$$B = 0.$$

The same is true for a path taken at a radius smaller than that of the toroid. So there is no field exterior to a very tightly wound toroid. It is all inside the loops.

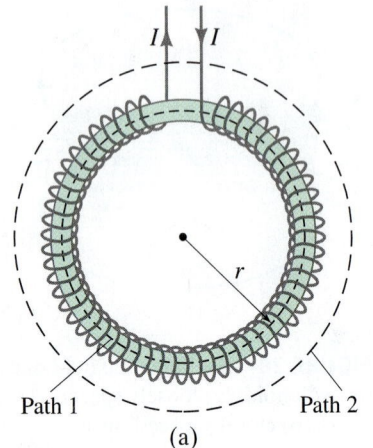

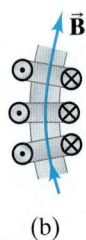

FIGURE 28–17 (a) A toroid. (b) A section of the toroid showing direction of the current for three loops: ⊙ means current toward you, ⊗ means current away from you.

28–6 Biot-Savart Law

The usefulness of Ampère's law for determining the magnetic field $\vec{B}$ due to particular electric currents is restricted to situations where the symmetry of the given currents allows us to evaluate $\oint \vec{B} \cdot d\vec{\ell}$ readily. This does not, of course, invalidate Ampère's law nor does it reduce its fundamental importance. Recall the electric case, where Gauss's law is considered fundamental but is limited in its use for actually calculating $\vec{E}$. We must often determine the electric field $\vec{E}$ by another method summing over contributions due to infinitesimal charge elements dq via Coulomb's law: $dE = (1/4\pi\epsilon_0)(dq/r^2)$. A magnetic equivalent to this infinitesimal form of Coulomb's law would be helpful for currents that do not have great symmetry. Such a law was developed by Jean Baptiste Biot (1774–1862) and Felix Savart (1791–1841) shortly after Oersted's discovery in 1820 that a current produces a magnetic field.

According to Biot and Savart, a current I flowing in any path can be considered as many tiny (infinitesimal) current elements, such as in the wire of Fig. 28–18. If $d\vec{\ell}$ represents any infinitesimal length along which the current is flowing, then the magnetic field, $d\vec{B}$, at any point P in space, due to this element of current, is given by

$$d\vec{B} = \frac{\mu_0 I}{4\pi} \frac{d\vec{\ell} \times \hat{\mathbf{r}}}{r^2}, \qquad (28\text{–}5)$$

Biot-Savart law

where $\vec{r}$ is the displacement vector from the element $d\vec{\ell}$ to the point P, and $\hat{\mathbf{r}} = \vec{r}/r$ is the unit vector (magnitude = 1) in the direction of $\vec{r}$ (see Fig. 28–18).

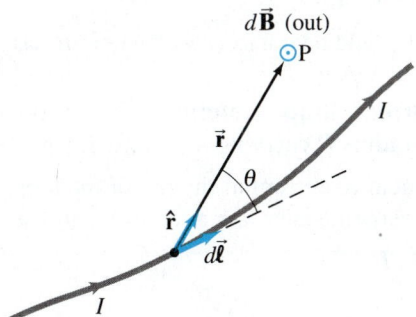

FIGURE 28–18 Biot-Savart law: the field at P due to current element $I d\vec{\ell}$ is $d\vec{B} = (\mu_0 I/4\pi)(d\vec{\ell} \times \hat{\mathbf{r}}/r^2)$.

Equation 28–5 is known as the **Biot-Savart law**. The magnitude of $d\vec{B}$ is

$$dB = \frac{\mu_0 I d\ell \sin \theta}{4\pi r^2}, \qquad (28\text{–}6)$$

where θ is the angle between $d\vec{\ell}$ and $\vec{r}$ (Fig. 28–18). The total magnetic field at point P is then found by summing (integrating) over all current elements:

$$\vec{B} = \int d\vec{B} = \frac{\mu_0 I}{4\pi} \int \frac{d\vec{\ell} \times \hat{\mathbf{r}}}{r^2}.$$

Note that this is a *vector* sum. The Biot-Savart law is the magnetic equivalent of Coulomb's law in its infinitesimal form. It is even an inverse square law, like Coulomb's law.

An important difference between the Biot-Savart law and Ampère's law (Eq. 28–3) is that in Ampère's law $\left[\oint \vec{B} \cdot d\vec{\ell} = \mu_0 I_{\text{encl}} \right]$, $\vec{B}$ is not necessarily due only to the current enclosed by the path of integration. But in the Biot-Savart law the field $d\vec{B}$ in Eq. 28–5 is due only, and entirely, to the current element $I d\vec{\ell}$. To find the total $\vec{B}$ at any point in space, it is necessary to include *all* currents.

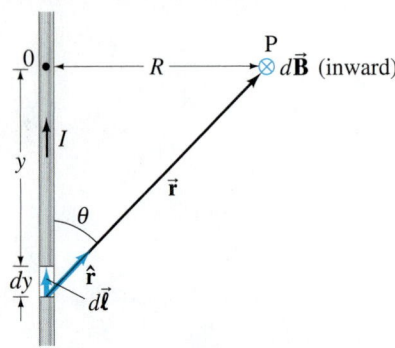

FIGURE 28–19 Determining $\vec{B}$ due to a long straight wire using the Biot-Savart law.

EXAMPLE 28–11 $\vec{B}$ **due to current I in straight wire.** For the field near a long straight wire carrying a current I, show that the Biot-Savart law gives the same result as Eq. 28–1, $B = \mu_0 I / 2\pi r$.

APPROACH We calculate the magnetic field in Fig. 28–19 at point P, which is a perpendicular distance R from an infinitely long wire. The current is moving upwards, and both $d\vec{\ell}$ and $\hat{r}$, which appear in the cross product of Eq. 28–5, are in the plane of the page. Hence the direction of the field $d\vec{B}$ due to each element of current must be directed into the plane of the page as shown (right-hand rule for the cross product $d\vec{\ell} \times \hat{r}$). Thus all the $d\vec{B}$ have the same direction at point P, and add up to give $\vec{B}$ the same direction consistent with our previous results (Figs. 28–1 and 28–11).

SOLUTION The magnitude of $\vec{B}$ will be

$$B = \frac{\mu_0 I}{4\pi} \int_{y=-\infty}^{+\infty} \frac{dy \sin \theta}{r^2},$$

where $dy = d\ell$ and $r^2 = R^2 + y^2$. Note that we are integrating over y (the length of the wire) so R is considered constant. Both y and θ are variables, but they are not independent. In fact, $y = -R/\tan \theta$. Note that we measure y as positive upward from point 0, so for the current element we are considering $y < 0$. Then

$$dy = +R \csc^2 \theta \, d\theta = \frac{R \, d\theta}{\sin^2 \theta} = \frac{R \, d\theta}{(R/r)^2} = \frac{r^2 \, d\theta}{R}.$$

From Fig. 28–19 we can see that $y = -\infty$ corresponds to $\theta = 0$ and that $y = +\infty$ corresponds to $\theta = \pi$ radians. So our integral becomes

$$B = \frac{\mu_0 I}{4\pi} \frac{1}{R} \int_{\theta=0}^{\pi} \sin \theta \, d\theta = -\frac{\mu_0 I}{4\pi R} \cos \theta \Big|_0^\pi = \frac{\mu_0 I}{2\pi R}.$$

This is just Eq. 28–1 for the field near a long wire, where R has been used instead of r.

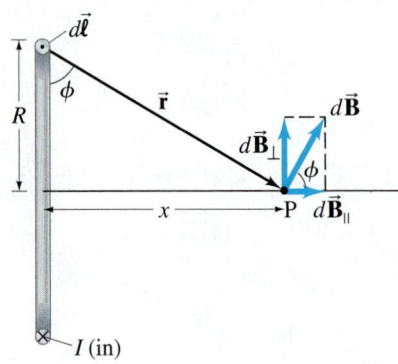

FIGURE 28–20 Determining $\vec{B}$ due to a current loop.

EXAMPLE 28–12 **Current loop.** Determine $\vec{B}$ for points on the axis of a circular loop of wire of radius R carrying a current I, Fig. 28–20.

APPROACH For an element of current at the top of the loop, the magnetic field $d\vec{B}$ at point P on the axis is perpendicular to $\vec{r}$ as shown, and has magnitude (Eq. 28–5)

$$dB = \frac{\mu_0 I \, d\ell}{4\pi r^2}$$

since $d\vec{\ell}$ is perpendicular to $\vec{r}$ so $|d\vec{\ell} \times \hat{r}| = d\ell$. We can break $d\vec{B}$ down into components $dB_\parallel$ and $dB_\perp$, which are parallel and perpendicular to the axis as shown.

SOLUTION When we sum over all the elements of the loop, *symmetry* tells us that the perpendicular components will cancel on opposite sides, so $B_\perp = 0$. Hence, the total $\vec{B}$ will point along the axis, and will have magnitude

$$B = B_\parallel = \int dB \cos \phi = \int dB \frac{R}{r} = \int dB \frac{R}{(R^2 + x^2)^{\frac{1}{2}}},$$

where x is the distance of P from the center of the ring, and $r^2 = R^2 + x^2$. Now we put in dB from the equation above and integrate around the current loop, noting that all segments $d\vec{\ell}$ of current are the same distance, $(R^2 + x^2)^{\frac{1}{2}}$, from point P:

$$B = \frac{\mu_0 I}{4\pi} \frac{R}{(R^2 + x^2)^{\frac{3}{2}}} \int d\ell = \frac{\mu_0 I R^2}{2(R^2 + x^2)^{\frac{3}{2}}}$$

since $\int d\ell = 2\pi R$, the circumference of the loop.

NOTE At the very center of the loop (where $x = 0$) the field has its maximum value

$$B = \frac{\mu_0 I}{2R}. \qquad \text{[at center of current loop]}$$

Recall from Section 27–5 that a current loop, such as that just discussed (Fig. 28–20), is considered a **magnetic dipole**. We saw there that a current loop has a magnetic dipole moment

$$\mu = NIA,$$

where A is the area of the loop and N is the number of coils in the loop, each carrying current I. We also saw in Chapter 27 that a magnetic dipole placed in an external magnetic field experiences a torque and possesses potential energy, just like an electric dipole. In Example 28–12, we looked at another aspect of a magnetic dipole: the magnetic field *produced by* a magnetic dipole has magnitude, along the dipole axis, of

$$B = \frac{\mu_0 I R^2}{2\left(R^2 + x^2\right)^{\frac{3}{2}}}.$$

We can write this in terms of the magnetic dipole moment $\mu = IA = I\pi R^2$ (for a single loop $N = 1$):

$$B = \frac{\mu_0}{2\pi} \frac{\mu}{\left(R^2 + x^2\right)^{\frac{3}{2}}}. \qquad \text{[magnetic dipole]} \quad \textbf{(28–7a)}$$

(Be careful to distinguish μ for dipole moment from μ_0, the magnetic permeability constant.) For distances far from the loop, $x \gg R$, this becomes

$$B \approx \frac{\mu_0}{2\pi} \frac{\mu}{x^3}. \qquad \left[\begin{matrix} \text{on axis,} \\ \text{magnetic dipole, } x \gg R \end{matrix}\right] \quad \textbf{(28–7b)}$$

The magnetic field on the axis of a magnetic dipole decreases with the cube of the distance, just as the electric field does for an electric dipole. B decreases as the cube of the distance also for points not on the axis, although the multiplying factor is not the same. The magnetic field due to a current loop can be determined at various points using the Biot-Savart law and the results are in accord with experiment. The field lines around a current loop are shown in Fig. 28–21.

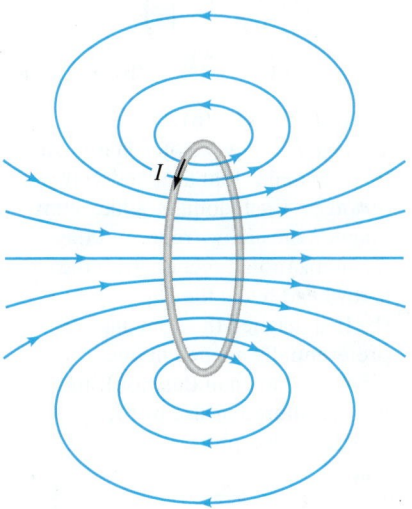

FIGURE 28–21 Magnetic field due to a circular loop of wire. (Same as Fig. 27–9.)

EXAMPLE 28–13 $\vec{B}$ **due to a wire segment.** One quarter of a circular loop of wire carries a current I as shown in Fig. 28–22. The current I enters and leaves on straight segments of wire, as shown; the straight wires are along the radial direction from the center C of the circular portion. Find the magnetic field at point C.

APPROACH The current in the straight sections produces no magnetic field at point C because $d\vec{\ell}$ and $\hat{r}$ in the Biot-Savart law (Eq. 28–5) are parallel and therefore $d\vec{\ell} \times \hat{r} = 0$. Each piece $d\vec{\ell}$ of the curved section of wire produces a field $d\vec{B}$ that points into the page at C (right-hand rule).

SOLUTION The magnitude of each $d\vec{B}$ due to each $d\vec{\ell}$ of the circular portion of wire is (Eq. 28–6)

$$dB = \frac{\mu_0 I\, d\ell}{4\pi R^2}$$

where $r = R$ is the radius of the curved section, and $\sin\theta$ in Eq. 28–6 is $\sin 90° = 1$. With $r = R$ for all pieces $d\vec{\ell}$, we integrate over a quarter of a circle.

$$B = \int dB = \frac{\mu_0 I}{4\pi R^2} \int d\ell = \frac{\mu_0 I}{4\pi R^2}\left(\frac{1}{4} 2\pi R\right) = \frac{\mu_0 I}{8R}.$$

FIGURE 28–22 Example 28–13.

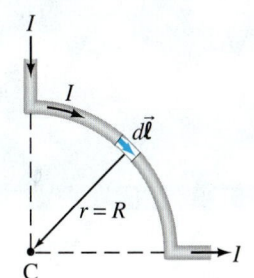

Magnetic Materials—Ferromagnetism

Magnetic fields can be produced (1) by magnetic materials (magnets) and (2) by electric currents. Common magnetic materials include ordinary magnets, iron cores in motors and electromagnets, recording tape, computer hard drives and magnetic stripes on credit cards. We saw in Section 27–1 that iron (and a few other materials) can be made into strong magnets. These materials are said to be **ferromagnetic**. We now look into the sources of ferromagnetism.

A bar magnet, with its two opposite poles near either end, resembles an electric dipole (equal-magnitude positive and negative charges separated by a distance). Indeed, a bar magnet is sometimes referred to as a "magnetic dipole." There are opposite "poles" separated by a distance. And the magnetic field lines of a bar magnet form a pattern much like that for the electric field of an electric dipole: compare Fig. 21–34a with Fig. 27–4 (or 28–24).

Microscopic examination reveals that a piece of iron is made up of tiny regions known as **domains**, less than 1 mm in length or width. Each domain behaves like a tiny magnet with a north and a south pole. In an unmagnetized piece of iron, these domains are arranged randomly, as shown in Fig. 28–23a. The magnetic effects of the domains cancel each other out, so this piece of iron is not a magnet. In a magnet, the domains are preferentially aligned in one direction as shown in Fig. 28–23b (downward in this case). A magnet can be made from an unmagnetized piece of iron by placing it in a strong magnetic field. (You can make a needle magnetic, for example, by stroking it with one pole of a strong magnet.) The magnetization direction of domains may actually rotate slightly to be more nearly parallel to the external field, and the borders of domains may move so domains with magnetic orientation parallel to the external field grow larger (compare Figs. 28–23a and b).

We can now explain how a magnet can pick up unmagnetized pieces of iron like paper clips. The field of the magnet's south pole (say) causes a slight realignment of the domains in the unmagnetized object, which then becomes a temporary magnet with its north pole facing the south pole of the permanent magnet; thus, attraction results. Similarly, elongated iron filings in a magnetic field acquire aligned domains and align themselves to reveal the shape of the magnetic field, Fig. 28–24.

An iron magnet can remain magnetized for a long time, and is referred to as a "permanent magnet." But if you drop a magnet on the floor or strike it with a hammer, you can jar the domains into randomness and the magnet loses some or all of its magnetism. Heating a permanent magnet can also cause loss of magnetism, for raising the temperature increases the random thermal motion of atoms, which tends to randomize the domains. Above a certain temperature known as the **Curie temperature** (1043 K for iron), a magnet cannot be made at all. Iron, nickel, cobalt, gadolinium, and certain alloys are ferromagnetic at room temperature; several other elements and alloys have low Curie temperature and thus are ferromagnetic only at low temperatures. Most other metals, such as aluminum and copper, do not show any noticeable magnetic effect (but see Section 28–10).

The striking similarity between the fields produced by a bar magnet and by a loop of electric current (Figs. 27–4b and 28–21) offers a clue that perhaps magnetic fields produced by electric currents may have something to do with ferromagnetism. According to modern atomic theory, atoms can be visualized as having electrons that orbit around a central nucleus. The electrons are charged, and so constitute an electric current and therefore produce a magnetic field; but the fields due to orbiting electrons generally all add up to zero. Electrons themselves produce an additional magnetic field, as if they and their electric charge were spinning about their own axes. It is the magnetic field due to electron **spin**[†] that is believed to produce ferromagnetism in most ferromagnetic materials.

It is believed today that *all* magnetic fields are caused by electric currents. This means that magnetic field lines always form closed loops, unlike electric field lines which begin on positive charges and end on negative charges.

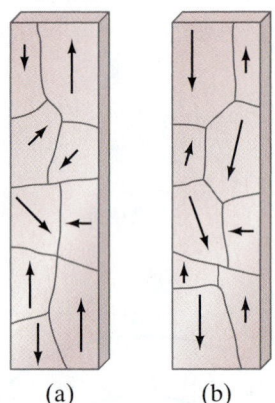

FIGURE 28–23 (a) An unmagnetized piece of iron is made up of domains that are randomly arranged. Each domain is like a tiny magnet; the arrows represent the magnetization direction, with the arrowhead being the N pole. (b) In a magnet, the domains are preferentially aligned in one direction (down in this case), and may be altered in size by the magnetization process.

(a)   (b)

FIGURE 28–24 Iron filings line up along magnetic field lines due to a permanent magnet.

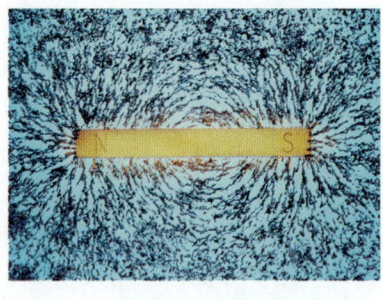

⚠ **CAUTION**
$\vec{B}$ lines form closed loops, $\vec{E}$ start on ⊕ and end on ⊖

[†]The name "spin" comes from an early suggestion that this intrinsic magnetic moment arises from the electron "spinning" on its axis (as well as "orbiting" the nucleus) to produce the extra field. However this view of a spinning electron is oversimplified and not valid.

EXERCISE D Return to the Chapter-Opening Question, page 733, and answer it again now. Try to explain why you may have answered differently the first time.

*28–8 Electromagnets and Solenoids—Applications

A long coil of wire consisting of many loops of wire, as discussed in Section 28–5, is called a solenoid. The magnetic field within a solenoid can be fairly large since it will be the sum of the fields due to the current in each loop (see Fig. 28–25). The solenoid acts like a magnet; one end can be considered the north pole and the other the south pole, depending on the direction of the current in the loops (use the right-hand rule). Since the magnetic field lines leave the north pole of a magnet, the north pole of the solenoid in Fig. 28–25 is on the right.

If a piece of iron is placed inside a solenoid, the magnetic field is increased greatly because the domains of the iron are aligned by the magnetic field produced by the current. The resulting magnetic field is the sum of that due to the current and that due to the iron, and can be hundreds or thousands of times larger than that due to the current alone (see Section 28–9). This arrangement is called an **electromagnet**. The alloys of iron used in electromagnets acquire and lose their magnetism quite readily when the current is turned on or off, and so are referred to as "soft iron." (It is "soft" only in a magnetic sense.) Iron that holds its magnetism even when there is no externally applied field is called "hard iron." Hard iron is used in permanent magnets. Soft iron is usually used in electromagnets so that the field can be turned on and off readily. Whether iron is hard or soft depends on heat treatment, type of alloy, and other factors.

Electromagnets have many practical applications, from use in motors and generators to producing large magnetic fields for research. Sometimes an iron core is not present—the magnetic field comes only from the current in the wire coils. When the current flows continuously in a normal electromagnet, a great deal of waste heat (I^2R power) can be produced. Cooling coils, which are tubes carrying water, are needed to absorb the heat in larger installations.

For some applications, the current-carrying wires are made of superconducting material kept below the transition temperature (Section 25–9). Very high fields can be produced with superconducting wire without an iron core. No electric power is needed to maintain large current in the superconducting coils, which means large savings of electricity; nor must huge amounts of heat be dissipated. It is not a free ride, though, because energy is needed to keep the superconducting coils at the necessary low temperature.

Another useful device consists of a solenoid into which a rod of iron is partially inserted. This combination is also referred to as a solenoid. One simple use is as a doorbell (Fig. 28–26). When the circuit is closed by pushing the button, the coil effectively becomes a magnet and exerts a force on the iron rod. The rod is pulled into the coil and strikes the bell. A large solenoid is used in the starters of cars; when you engage the starter, you are closing a circuit that not only turns the starter motor, but activates a solenoid that first moves the starter into direct contact with the gears on the engine's flywheel. Solenoids are used as switches in many devices. They have the advantage of moving mechanical parts quickly and accurately.

*Magnetic Circuit Breakers

Modern circuit breakers that protect houses and buildings from overload and fire contain not only a "thermal" part (bimetallic strip as described in Section 25–6, Fig. 25–19) but also a magnetic sensor. If the current is above a certain level, the magnetic field it produces pulls an iron plate that breaks the same contact points as in Fig. 25–19b and c. In more sophisticated circuit breakers, including ground fault circuit interrupters (GFCIs—discussed in Section 29–8), a solenoid is used. The iron rod of Fig. 28–26, instead of striking a bell, strikes one side of a pair of points, opening them and opening the circuit. Magnetic circuit breakers react quickly (<10 msec), and for buildings are designed to react to the high currents of shorts (but not shut off for the start-up surges of motors).

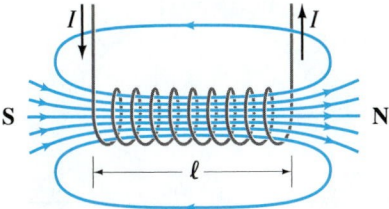

FIGURE 28–25 Magnetic field of a solenoid. The north pole of this solenoid, thought of as a magnet, is on the right, and the south pole is on the left.

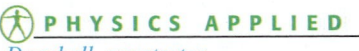

PHYSICS APPLIED
Electromagnets and solenoids

PHYSICS APPLIED
Doorbell, car starter

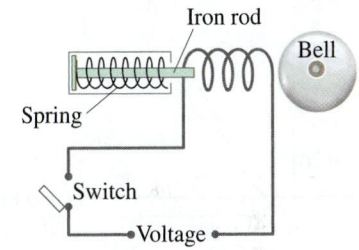

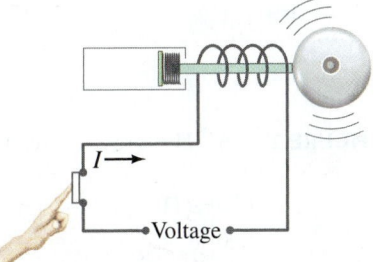

FIGURE 28–26 Solenoid used as a doorbell.

PHYSICS APPLIED
Magnetic circuit breakers

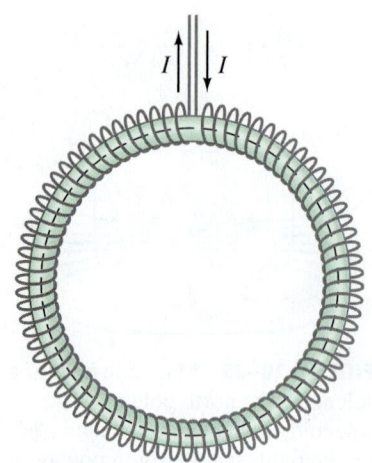

I↑ ↓I

FIGURE 28–27 Iron-core toroid.

FIGURE 28–28 Total magnetic field B in an iron-core toroid as a function of the external field B_0 (B_0 is caused by the current I in the coil).

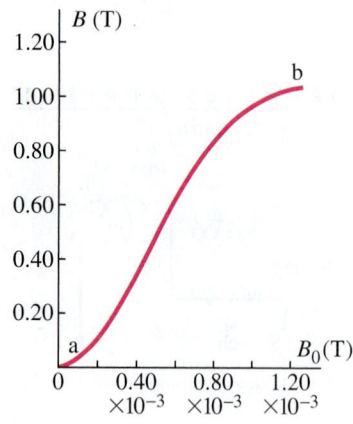

FIGURE 28–29 Hysteresis curve.

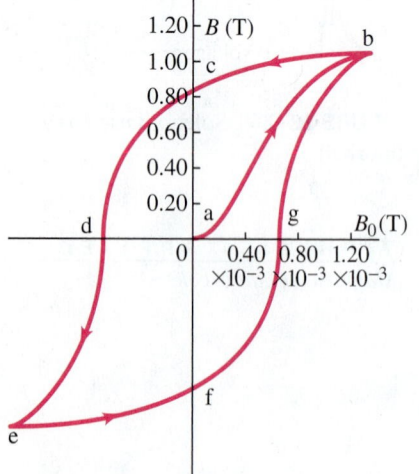

*28–9 Magnetic Fields in Magnetic Materials; Hysteresis

The field of a long solenoid is directly proportional to the current. Indeed, Eq. 28–4 tells us that the field B_0 inside a solenoid is given by

$$B_0 = \mu_0 nI.$$

This is valid if there is only air inside the coil. If we put a piece of iron or other ferromagnetic material inside the solenoid, the field will be greatly increased, often by hundreds or thousands of times. This occurs because the domains in the iron become preferentially aligned by the external field. The resulting magnetic field is the sum of that due to the current and that due to the iron. It is sometimes convenient to write the total field in this case as a sum of two terms:

$$\vec{\mathbf{B}} = \vec{\mathbf{B}}_0 + \vec{\mathbf{B}}_M. \qquad (28\text{–}8)$$

Here, $\vec{\mathbf{B}}_0$ refers to the field due only to the current in the wire (the "external field"). It is the field that would be present in the absence of a ferromagnetic material. Then $\vec{\mathbf{B}}_M$ represents the additional field due to the ferromagnetic material itself; often $\vec{\mathbf{B}}_M \gg \vec{\mathbf{B}}_0$.

The total field inside a solenoid in such a case can also be written by replacing the constant μ_0 in Eq. 28–4 by another constant, μ, characteristic of the material inside the coil:

$$B = \mu nI; \qquad (28\text{–}9)$$

μ is called the **magnetic permeability** of the material (do not confuse it with $\vec{\boldsymbol{\mu}}$ for magnetic dipole moment). For ferromagnetic materials, μ is much greater than μ_0. For all other materials, its value is very close to μ_0 (Section 28–10). The value of μ, however, is not constant for ferromagnetic materials; it depends on the value of the external field B_0, as the following experiment shows.

Measurements on magnetic materials are generally done using a toroid, which is essentially a long solenoid bent into the shape of a circle (Fig. 28–27), so that practically all the lines of $\vec{\mathbf{B}}$ remain within the toroid. Suppose the toroid has an iron core that is initially unmagnetized and there is no current in the windings of the toroid. Then the current I is slowly increased, and B_0 increases linearly with I. The total field B also increases, but follows the curved line shown in the graph of Fig. 28–28. (Note the different scales: $B \gg B_0$.) Initially, point a, the domains (Section 28–7) are randomly oriented. As B_0 increases, the domains become more and more aligned until at point b, nearly all are aligned. The iron is said to be approaching **saturation**. Point b is typically 70% of full saturation. (If B_0 is increased further, the curve continues to rise very slowly, and reaches 98% saturation only when B_0 reaches a value about a thousandfold above that at point b; the last few domains are very difficult to align.) Next, suppose the external field B_0 is reduced by decreasing the current in the toroid coils. As the current is reduced to zero, shown as point c in Fig. 28–29, the domains do not become completely random. Some permanent magnetism remains. If the current is then reversed in direction, enough domains can be turned around so $B = 0$ (point d). As the reverse current is increased further, the iron approaches saturation in the opposite direction (point e). Finally, if the current is again reduced to zero and then increased in the original direction, the total field follows the path efgb, again approaching saturation at point b.

Notice that the field did not pass through the origin (point a) in this cycle. The fact that the curves do not retrace themselves on the same path is called **hysteresis**. The curve bcdefgb is called a **hysteresis loop**. In such a cycle, much energy is transformed to thermal energy (friction) due to realigning of the domains. It can be shown that the energy dissipated in this way is proportional to the area of the hysteresis loop.

At points c and f, the iron core is magnetized even though there is no current in the coils. These points correspond to a permanent magnet. For a permanent magnet, it is desired that ac and af be as large as possible. Materials for which this is true are said to have high **retentivity**.

Materials with a broad hysteresis curve as in Fig. 28–29 are said to be magnetically "hard" and make good permanent magnets. On the other hand, a hysteresis curve such as that in Fig. 28–30 occurs for "soft" iron, which is preferred for electromagnets and transformers (Section 29–6) since the field can be more readily switched off, and the field can be reversed with less loss of energy.

A ferromagnetic material can be demagnetized—that is, made unmagnetized. This can be done by reversing the magnetizing current repeatedly while decreasing its magnitude. This results in the curve of Fig. 28–31. The heads of a tape recorder are demagnetized in this way. The alternating magnetic field acting at the heads due to a handheld demagnetizer is strong when the demagnetizer is placed near the heads and decreases as it is moved slowly away. Video and audio tapes themselves can be erased and ruined by a magnetic field, as can computer hard disks, other magnetic storage devices, and the magnetic stripes on credit cards.

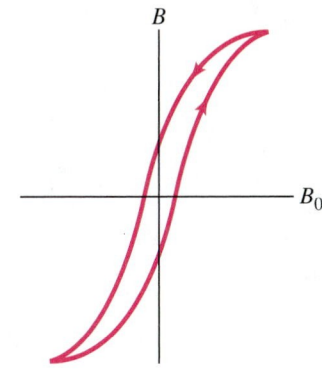

FIGURE 28–30 Hysteresis curve for soft iron.

FIGURE 28–31 Successive hysteresis loops during demagnetization.

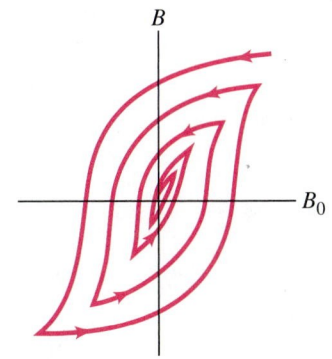

*28–10 Paramagnetism and Diamagnetism

All materials are magnetic to at least a tiny extent. Nonferromagnetic materials fall into two principal classes: *paramagnetic*, in which the magnetic permeability μ is very slightly greater than μ_0; and *diamagnetic*, in which μ is very slightly less than μ_0. The ratio of μ to μ_0 for any material is called the **relative permeability** K_m:

$$K_m = \frac{\mu}{\mu_0}.$$

Another useful parameter is the **magnetic susceptibility** χ_m defined as

$$\chi_m = K_m - 1.$$

Paramagnetic substances have $K_m > 1$ and $\chi_m > 0$, whereas diamagnetic substances have $K_m < 1$ and $\chi_m < 0$. See Table 28–1, and note how small the effect is.

TABLE 28–1 Paramagnetism and Diamagnetism: Magnetic Susceptibilities

Paramagnetic substance	χ_m	Diamagnetic substance	χ_m
Aluminum	2.3×10^{-5}	Copper	-9.8×10^{-6}
Calcium	1.9×10^{-5}	Diamond	-2.2×10^{-5}
Magnesium	1.2×10^{-5}	Gold	-3.6×10^{-5}
Oxygen (STP)	2.1×10^{-6}	Lead	-1.7×10^{-5}
Platinum	2.9×10^{-4}	Nitrogen (STP)	-5.0×10^{-9}
Tungsten	6.8×10^{-5}	Silicon	-4.2×10^{-6}

The difference between paramagnetic and diamagnetic materials can be understood theoretically at the molecular level on the basis of whether or not the molecules have a permanent magnetic dipole moment. One type of **paramagnetism** occurs in materials whose molecules (or ions) have a permanent magnetic dipole moment.[†] In the absence of an external field, the molecules are randomly oriented and no magnetic effects are observed. However, when an external magnetic field is applied, say, by putting the material in a solenoid, the applied field exerts a torque on the magnetic dipoles (Section 27–5), tending to align them parallel to the field. The total magnetic field (external plus that due to aligned magnetic dipoles) will be slightly greater than B_0. The thermal motion of the molecules reduces the alignment, however.

[†] Other types of paramagnetism also occur whose origin is different from that described here, such as in metals where free electrons can contribute.

A useful quantity is the **magnetization vector**, $\vec{\mathbf{M}}$, defined as the magnetic dipole moment per unit volume,

$$\vec{\mathbf{M}} = \frac{\vec{\boldsymbol{\mu}}}{V},$$

where $\vec{\boldsymbol{\mu}}$ is the magnetic dipole moment of the sample and V its volume. It is found experimentally that M is directly proportional to the external magnetic field (tending to align the dipoles) and inversely proportional to the kelvin temperature T (tending to randomize dipole directions). This is called *Curie's law*, after Pierre Curie (1859–1906), who first noted it:

$$M = C\frac{B}{T},$$

where C is a constant. If the ratio B/T is very large (B very large or T very small) Curie's law is no longer accurate; as B is increased (or T decreased), the magnetization approaches some maximum value, M_{max}. This makes sense, of course, since M_{max} corresponds to complete alignment of all the permanent magnetic dipoles. However, even for very large magnetic fields, $\approx 2.0\,\text{T}$, deviations from Curie's law are normally noted only at very low temperatures, on the order of a few kelvins.

Ferromagnetic materials, as mentioned in Section 28–7, are no longer ferromagnetic above a characteristic temperature called the Curie temperature (1043 K for iron). Above this Curie temperature, they generally are paramagnetic.

Diamagnetic materials (for which μ_m is slightly less than μ_0) are made up of molecules that have no permanent magnetic dipole moment. When an external magnetic field is applied, magnetic dipoles are induced, but the induced magnetic dipole moment is in the direction opposite to that of the field. Hence the total field will be slightly less than the external field. The effect of the external field—in the crude model of electrons orbiting nuclei—is to increase the "orbital" speed of electrons revolving in one direction, and to decrease the speed of electrons revolving in the other direction; the net result is a net dipole moment opposing the external field. Diamagnetism is present in all materials, but is weaker even than paramagnetism and so is overwhelmed by paramagnetic and ferromagnetic effects in materials that display these other forms of magnetism.

Summary

The magnetic field B at a distance r from a long straight wire is directly proportional to the current I in the wire and inversely proportional to r:

$$B = \frac{\mu_0}{2\pi}\frac{I}{r}. \qquad \text{(28–1)}$$

The magnetic field lines are circles centered at the wire.

The force that one long current-carrying wire exerts on a second parallel current-carrying wire 1 m away serves as the definition of the ampere unit, and ultimately of the coulomb as well.

Ampère's law states that the line integral of the magnetic field $\vec{\mathbf{B}}$ around any closed loop is equal to μ_0 times the total net current I_{encl} enclosed by the loop:

$$\oint \vec{\mathbf{B}} \cdot d\vec{\boldsymbol{\ell}} = \mu_0 I_{\text{encl}}. \qquad \text{(28–3)}$$

The magnetic field inside a long tightly wound solenoid is

$$B = \mu_0 n I \qquad \text{(28–4)}$$

where n is the number of coils per unit length and I is the current in each coil.

The **Biot-Savart law** is useful for determining the magnetic field due to a known arrangement of currents. It states that

$$d\vec{\mathbf{B}} = \frac{\mu_0 I}{4\pi}\frac{d\vec{\boldsymbol{\ell}} \times \hat{\mathbf{r}}}{r^2}, \qquad \text{(28–5)}$$

where $d\vec{\mathbf{B}}$ is the contribution to the total field at some point P due to a current I along an infinitesimal length $d\vec{\boldsymbol{\ell}}$ of its path, and $\hat{\mathbf{r}}$ is the unit vector along the direction of the displacement vector $\vec{\mathbf{r}}$ from $d\vec{\boldsymbol{\ell}}$ to P. The total field $\vec{\mathbf{B}}$ will be the integral over all $d\vec{\mathbf{B}}$.

Iron and a few other materials can be made into strong permanent magnets. They are said to be **ferromagnetic**. Ferromagnetic materials are made up of tiny **domains**—each a tiny magnet—which are preferentially aligned in a permanent magnet, but randomly aligned in a nonmagnetized sample.

[*When a ferromagnetic material is placed in a magnetic field B_0 due to a current, say inside a solenoid or toroid, the material becomes magnetized. When the current is turned off, however, the material remains magnetized, and when the current is increased in the opposite direction (and then again reversed), a graph of the total field B versus B_0 is a **hysteresis loop**, and the fact that the curves do not retrace themselves is called **hysteresis**.]

[*All materials exhibit some magnetic effects. Nonferromagnetic materials have much smaller paramagnetic or diamagnetic properties.]

Questions

1. The magnetic field due to current in wires in your home can affect a compass. Discuss the problem in terms of currents, depending on whether they are ac or dc, and their distance away.

2. Compare and contrast the magnetic field due to a long straight current and the electric field due to a long straight line of electric charge at rest (Section 21–7).

3. Two insulated long wires carrying equal currents I cross at right angles to each other. Describe the magnetic force one exerts on the other.

4. A horizontal wire carries a large current. A second wire carrying a current in the same direction is suspended below it. Can the current in the upper wire hold the lower wire in suspension against gravity? Under what conditions will the lower wire be in equilibrium?

5. A horizontal current-carrying wire, free to move in Earth's gravitational field, is suspended directly above a second, parallel, current-carrying wire. (a) In what direction is the current in the lower wire? (b) Can the upper wire be held in stable equilibrium due to the magnetic force of the lower wire? Explain.

6. (a) Write Ampère's law for a path that surrounds both conductors in Fig. 28–10. (b) Repeat, assuming the lower current, I_2, is in the opposite direction $(I_2 = -I_1)$.

7. Suppose the cylindrical conductor of Fig. 28–11a has a concentric cylindrical hollow cavity inside it (so it looks like a pipe). What can you say about $\vec{B}$ in the cavity?

8. Explain why a field such as that shown in Fig. 28–14b is consistent with Ampère's law. Could the lines curve upward instead of downward?

9. What would be the effect on B inside a long solenoid if (a) the diameter of all the loops was doubled, or (b) the spacing between loops was doubled, or (c) the solenoid's length was doubled along with a doubling in the total number of loops.

10. Use the Biot-Savart law to show that the field of the current loop in Fig. 28–21 is correct as shown for points off the axis.

11. Do you think $\vec{B}$ will be the same for all points in the plane of the current loop of Fig. 28–21? Explain.

12. Why does twisting the lead-in wires to electrical devices reduce the magnetic effects of the leads?

13. Compare the Biot-Savart law with Coulomb's law. What are the similarities and differences?

14. How might you define or determine the magnetic pole strength (the magnetic equivalent of a single electric charge) for (a) a bar magnet, (b) a current loop?

15. How might you measure the magnetic dipole moment of the Earth?

16. A type of magnetic switch similar to a solenoid is a **relay** (Fig. 28–32). A relay is an electromagnet (the iron rod inside the coil does not move) which, when activated, attracts a piece of iron on a pivot. Design a relay to close an electrical switch. A relay is used when you need to switch on a circuit carrying a very large current but you do not want that large current flowing through the main switch. For example, the starter switch of a car is connected to a relay so that the large current needed for the starter doesn't pass to the dashboard switch.

FIGURE 28–32
Question 16.

17. A heavy magnet attracts, from rest, a heavy block of iron. Before striking the magnet the block has acquired considerable kinetic energy. (a) What is the source of this kinetic energy? (b) When the block strikes the magnet, some of the latter's domains may be jarred into randomness; describe the energy transformations.

18. Will a magnet attract any metallic object, such as those made of aluminum, or only those made of iron? (Try it and see.) Why is this so?

19. An unmagnetized nail will not attract an unmagnetized paper clip. However, if one end of the nail is in contact with a magnet, the other end *will* attract a paper clip. Explain.

20. Can an iron rod attract a magnet? Can a magnet attract an iron rod? What must you consider to answer these questions?

21. How do you suppose the first magnets found in Magnesia were formed?

22. Why will either pole of a magnet attract an unmagnetized piece of iron?

23. Suppose you have three iron rods, two of which are magnetized but the third is not. How would you determine which two are the magnets without using any additional objects?

24. Two iron bars attract each other no matter which ends are placed close together. Are both magnets? Explain.

*25. Describe the magnetization curve for (a) a paramagnetic substance and (b) a diamagnetic substance, and compare to that for a ferromagnetic substance (Fig. 28–29).

*26. Can all materials be considered (a) diamagnetic, (b) paramagnetic, (c) ferromagnetic? Explain.

Problems

28–1 and 28–2 Straight Wires, Magnetic Field, and Force

1. (I) Jumper cables used to start a stalled vehicle often carry a 65-A current. How strong is the magnetic field 3.5 cm from one cable? Compare to the Earth's magnetic field $(5.0 \times 10^{-5}\,\text{T})$.

2. (I) If an electric wire is allowed to produce a magnetic field no larger than that of the Earth $(0.50 \times 10^{-4}\,\text{T})$ at a distance of 15 cm from the wire, what is the maximum current the wire can carry?

3. (I) Determine the magnitude and direction of the force between two parallel wires 25 m long and 4.0 cm apart, each carrying 35 A in the same direction.

4. (I) A vertical straight wire carrying an upward 28-A current exerts an attractive force per unit length of 7.8×10^{-4} N/m on a second parallel wire 7.0 cm away. What current (magnitude and direction) flows in the second wire?

5. (I) In Fig. 28–33, a long straight wire carries current I out of the page toward the viewer. Indicate, with appropriate arrows, the direction of $\vec{B}$ at each of the points C, D, and E in the plane of the page.

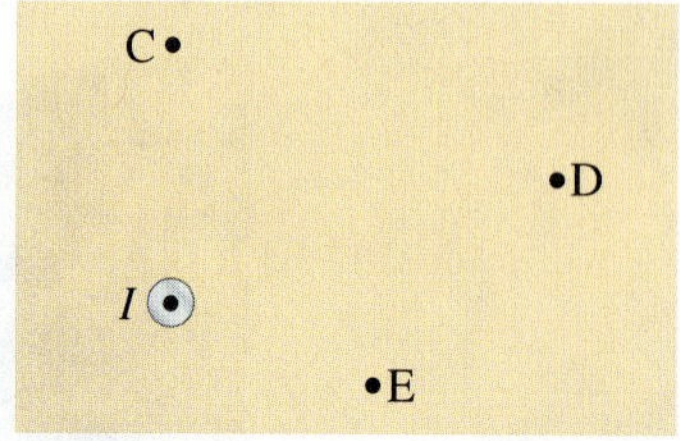

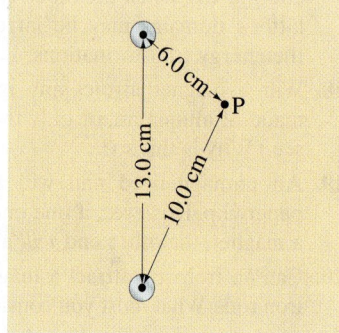

FIGURE 28–33
Problem 5.

6. (II) An experiment on the Earth's magnetic field is being carried out 1.00 m from an electric cable. What is the maximum allowable current in the cable if the experiment is to be accurate to ± 2.0%?

7. (II) Two long thin parallel wires 13.0 cm apart carry 35-A currents in the same direction. Determine the magnetic field vector at a point 10.0 cm from one wire and 6.0 cm from the other (Fig. 28–34).

FIGURE 28–34
Problem 7.

8. (II) A horizontal compass is placed 18 cm due south from a straight vertical wire carrying a 43-A current downward. In what direction does the compass needle point at this location? Assume the horizontal component of the Earth's field at this point is 0.45×10^{-4} T and the magnetic declination is 0°.

9. (II) A long horizontal wire carries 24.0 A of current due north. What is the net magnetic field 20.0 cm due west of the wire if the Earth's field there points downward, 44° below the horizontal, and has magnitude 5.0×10^{-5} T?

10. (II) A straight stream of protons passes a given point in space at a rate of 2.5×10^9 protons/s. What magnetic field do they produce 2.0 m from the beam?

11. (II) Determine the magnetic field midway between two long straight wires 2.0 cm apart in terms of the current I in one when the other carries 25 A. Assume these currents are (a) in the same direction, and (b) in opposite directions.

12. (II) Two straight parallel wires are separated by 6.0 cm. There is a 2.0-A current flowing in the first wire. If the magnetic field strength is found to be zero between the two wires at a distance of 2.2 cm from the first wire, what is the magnitude and direction of the current in the second wire?

13. (II) Two long straight wires each carry a current I out of the page toward the viewer, Fig. 28–35. Indicate, with appropriate arrows, the direction of $\vec{B}$ at each of the points 1 to 6 in the plane of the page. State if the field is zero at any of the points.

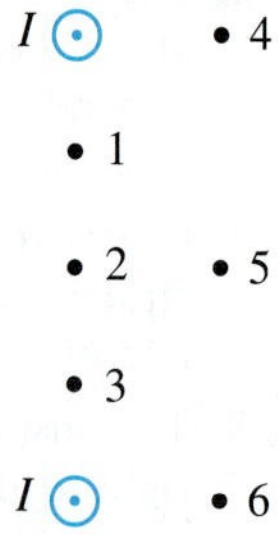

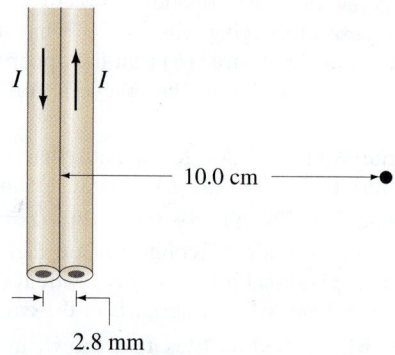

FIGURE 28–35
Problem 13.

14. (II) A long pair of insulated wires serves to conduct 28.0 A of dc current to and from an instrument. If the wires are of negligible diameter but are 2.8 mm apart, what is the magnetic field 10.0 cm from their midpoint, in their plane (Fig. 28–36)? Compare to the magnetic field of the Earth.

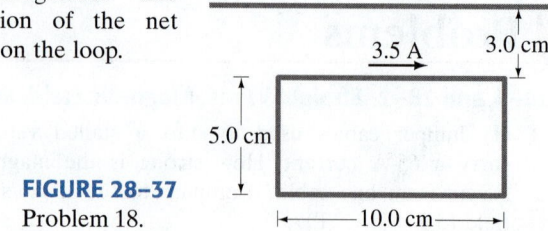

FIGURE 28–36 Problems 14 and 15.

15. (II) A third wire is placed in the plane of the two wires shown in Fig. 28–36 parallel and just to the right. If it carries 25.0 A upward, what force per meter of length does it exert on each of the other two wires? Assume it is 2.8 mm from the nearest wire, center to center.

16. (II) A power line carries a current of 95 A west along the tops of 8.5-m-high poles. (a) What is the magnitude and direction of the magnetic field produced by this wire at the ground directly below? How does this compare with the Earth's field of about $\frac{1}{2}$ G? (b) Where would the line's field cancel the Earth's?

17. (II) A compass needle points 28° E of N outdoors. However, when it is placed 12.0 cm to the east of a vertical wire inside a building, it points 55° E of N. What is the magnitude and direction of the current in the wire? The Earth's field there is 0.50×10^{-4} T and is horizontal.

18. (II) A rectangular loop of wire is placed next to a straight wire, as shown in Fig. 28–37. There is a current of 3.5 A in both wires. Determine the magnitude and direction of the net force on the loop.

FIGURE 28–37
Problem 18.

19. (II) Let two long parallel wires, a distance d apart, carry equal currents I in the same direction. One wire is at $x = 0$, the other at $x = d$, Fig. 28–38. Determine $\vec{\mathbf{B}}$ along the x axis between the wires as a function of x.

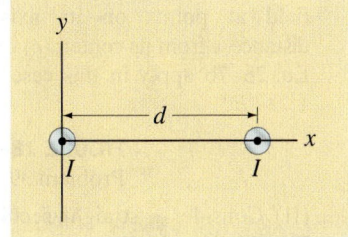

FIGURE 28–38
Problems 19 and 20.

20. (II) Repeat Problem 19 if the wire at $x = 0$ carries twice the current $(2I)$ as the other wire, and in the opposite direction.

21. (II) Two long wires are oriented so that they are perpendicular to each other. At their closest, they are 20.0 cm apart (Fig. 28–39). What is the magnitude of the magnetic field at a point midway between them if the top one carries a current of 20.0 A and the bottom one carries 12.0 A?

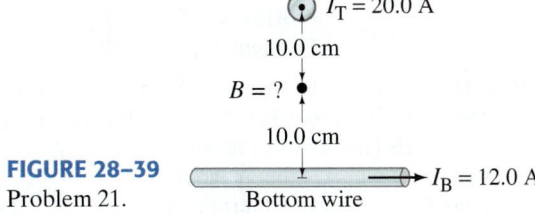

FIGURE 28–39
Problem 21.

22. (II) Two long parallel wires 8.20 cm apart carry 16.5-A currents in the same direction. Determine the magnetic field vector at a point P, 12.0 cm from one wire and 13.0 cm from the other. See Fig. 28–40. [*Hint*: Use the law of cosines.]

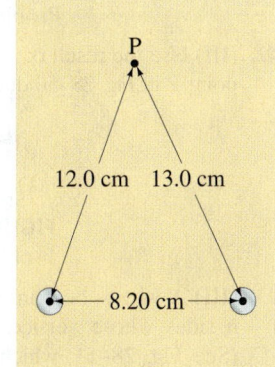

FIGURE 28–40
Problem 22.

23. (III) A very long flat conducting strip of width d and negligible thickness lies in a horizontal plane and carries a uniform current I across its cross section. (*a*) Show that at points a distance y directly above its center, the field is given by

$$B = \frac{\mu_0 I}{\pi d} \tan^{-1} \frac{d}{2y},$$

assuming the strip is infinitely long. [*Hint*: Divide the strip into many thin "wires," and sum (integrate) over these.] (*b*) What value does B approach for $y \gg d$? Does this make sense? Explain.

24. (III) A triangular loop of side length a carries a current I (Fig. 28–41). If this loop is placed a distance d away from a very long straight wire carrying a current I', determine the force on the loop.

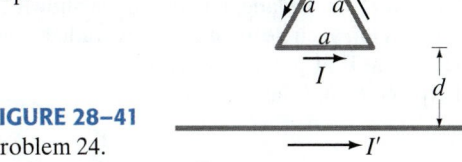

FIGURE 28–41
Problem 24.

28–4 and 28–5 Ampère's Law, Solenoids and Toroids

25. (I) A 40.0-cm-long solenoid 1.35 cm in diameter is to produce a field of 0.385 mT at its center. How much current should the solenoid carry if it has 765 turns of wire?

26. (I) A 32-cm-long solenoid, 1.8 cm in diameter, is to produce a 0.30-T magnetic field at its center. If the maximum current is 4.5 A, how many turns must the solenoid have?

27. (I) A 2.5-mm-diameter copper wire carries a 33-A current (uniform across its cross section). Determine the magnetic field: (*a*) at the surface of the wire; (*b*) inside the wire, 0.50 mm below the surface; (*c*) outside the wire 2.5 mm from the surface.

28. (II) A toroid (Fig. 28–17) has a 50.0-cm inner diameter and a 54.0-cm outer diameter. It carries a 25.0 A current in its 687 coils. Determine the range of values for B inside the toroid.

29. (II) A 20.0-m-long copper wire, 2.00 mm in diameter including insulation, is tightly wrapped in a single layer with adjacent coils touching, to form a solenoid of diameter 2.50 cm (outer edge). What is (*a*) the length of the solenoid and (*b*) the field at the center when the current in the wire is 16.7 A?

30. (II) (*a*) Use Eq. 28–1, and the vector nature of $\vec{\mathbf{B}}$, to show that the magnetic field lines around two long parallel wires carrying equal currents $I_1 = I_2$ are as shown in Fig. 28–10. (*b*) Draw the equipotential lines around two stationary positive electric charges. (*c*) Are these two diagrams similar? Identical? Why or why not?

31. (II) A coaxial cable consists of a solid inner conductor of radius R_1, surrounded by a concentric cylindrical tube of inner radius R_2 and outer radius R_3 (Fig. 28–42). The conductors carry equal and opposite currents I_0 distributed uniformly across their cross sections. Determine the magnetic field at a distance R from the axis for: (*a*) $R < R_1$; (*b*) $R_1 < R < R_2$; (*c*) $R_2 < R < R_3$; (*d*) $R > R_3$. (*e*) Let $I_0 = 1.50$ A, $R_1 = 1.00$ cm, $R_2 = 2.00$ cm, and $R_3 = 2.50$ cm. Graph B from $R = 0$ to $R = 3.00$ cm.

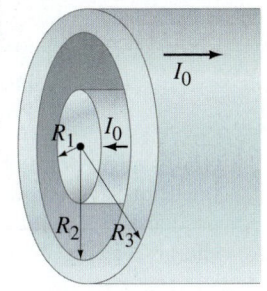

FIGURE 28–42
Problems 31 and 32.

32. (III) Suppose the current in the coaxial cable of Problem 31, Fig. 28–42, is not uniformly distributed, but instead the current density j varies linearly with distance from the center: $j_1 = C_1 R$ for the inner conductor and $j_2 = C_2 R$ for the outer conductor. Each conductor still carries the same total current I_0, in opposite directions. Determine the magnetic field in terms of I_0 in the same four regions of space as in Problem 31.

28–6 Biot-Savart Law

33. (I) The Earth's magnetic field is essentially that of a magnetic dipole. If the field near the North Pole is about 1.0×10^{-4} T, what will it be (approximately) 13,000 km above the surface at the North Pole?

34. (II) A wire, in a plane, has the shape shown in Fig. 28–43, two arcs of a circle connected by radial lengths of wire. Determine $\vec{\mathbf{B}}$ at point C in terms of R_1, R_2, θ, and the current I.

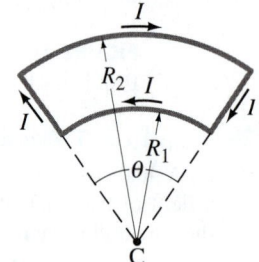

FIGURE 28–43
Problem 34.

35. (II) A circular conducting ring of radius R is connected to two exterior straight wires at two ends of a diameter (Fig. 28–44). The current I splits into unequal portions (as shown) while passing through the ring. What is $\vec{\mathbf{B}}$ at the center of the ring?

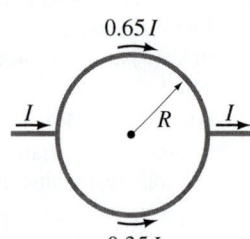

FIGURE 28–44
Problem 35.

36. (II) A small loop of wire of radius 1.8 cm is placed at the center of a wire loop with radius 25.0 cm. The planes of the loops are perpendicular to each other, and a 7.0-A current flows in each. Estimate the torque the large loop exerts on the smaller one. What simplifying assumption did you make?

37. (II) A wire is formed into the shape of two half circles connected by equal-length straight sections as shown in Fig. 28–45. A current I flows in the circuit clockwise as shown. Determine (a) the magnitude and direction of the magnetic field at the center, C, and (b) the magnetic dipole moment of the circuit.

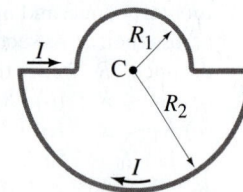

FIGURE 28–45
Problem 37.

38. (II) A single point charge q is moving with velocity $\vec{\mathbf{v}}$. Use the Biot-Savart law to show that the magnetic field $\vec{\mathbf{B}}$ it produces at a point P, whose position vector relative to the charge is $\vec{\mathbf{r}}$ (Fig. 28–46), is given by

$$\vec{\mathbf{B}} = \frac{\mu_0}{4\pi} \frac{q\vec{\mathbf{v}} \times \vec{\mathbf{r}}}{r^3}.$$

(Assume v is much less than the speed of light.)

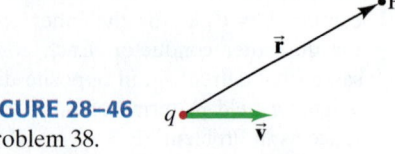

FIGURE 28–46
Problem 38.

39. (II) A nonconducting circular disk, of radius R, carries a uniformly distributed electric charge Q. The plate is set spinning with angular velocity ω about an axis perpendicular to the plate through its center (Fig. 28–47). Determine (a) its magnetic dipole moment and (b) the magnetic field at points on its axis a distance x from its center; (c) does Eq. 28–7b apply in this case for $x \gg R$?

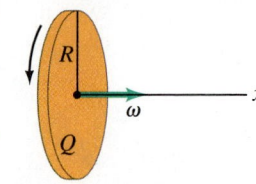

FIGURE 28–47
Problem 39.

40. (II) Consider a straight section of wire of length d, as in Fig. 28–48, which carries a current I. (a) Show that the magnetic field at a point P a distance R from the wire along its perpendicular bisector is

$$B = \frac{\mu_0 I}{2\pi R} \frac{d}{(d^2 + 4R^2)^{\frac{1}{2}}}.$$

(b) Show that this is consistent with Example 28–11 for an infinite wire.

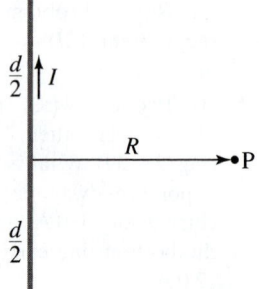

FIGURE 28–48
Problem 40.

41. (II) A segment of wire of length d carries a current I as shown in Fig. 28–49. (a) Show that for points along the positive x axis (the axis of the wire), such as point Q, the magnetic field $\vec{\mathbf{B}}$ is zero. (b) Determine a formula for the field at points along the y axis, such as point P.

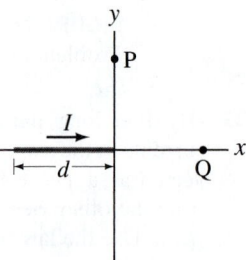

FIGURE 28–49
Problem 41.

42. (III) Use the result of Problem 41 to find the magnetic field at point P in Fig. 28–50 due to the current in the square loop.

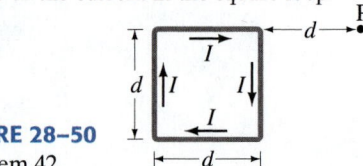

FIGURE 28–50
Problem 42.

43. (III) A wire is bent into the shape of a regular polygon with n sides whose vertices are a distance R from the center. (See Fig. 28–51, which shows the special case of $n = 6$.) If the wire carries a current I_0, (a) determine the magnetic field at the center; (b) if n is allowed to become very large ($n \to \infty$), show that the formula in part (a) reduces to that for a circular loop (Example 28–12).

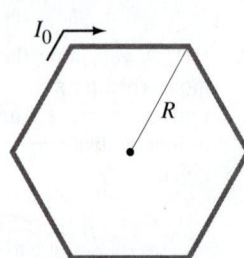

FIGURE 28–51
Problem 43.

44. (III) Start with the result of Example 28–12 for the magnetic field along the axis of a single loop to obtain the field inside a very long solenoid with n turns per meter (Eq. 28–4) that stretches from $+\infty$ to $-\infty$.

45. (III) A single rectangular loop of wire, with sides a and b, carries a current I. An xy coordinate system has its origin at the lower left corner of the rectangle with the x axis parallel to side b (Fig. 28–52) and the y axis parallel to side a. Determine the magnetic field B at all points (x, y) within the loop.

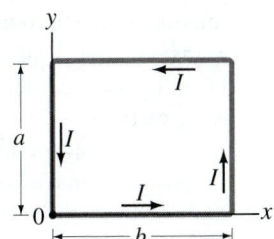

FIGURE 28–52
Problem 45.

46. (III) A square loop of wire, of side d, carries a current I. (a) Determine the magnetic field B at points on a line perpendicular to the plane of the square which passes through the center of the square (Fig. 28–53). Express B as a function of x, the distance along the line from the center of the square. (b) For $x \gg d$, does the square appear to be a magnetic dipole? If so, what is its dipole moment?

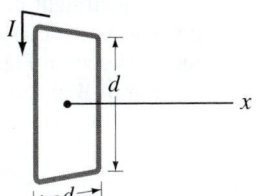

FIGURE 28–53
Problem 46.

28–7 Magnetic Materials—Ferromagnetism

47. (II) An iron atom has a magnetic dipole moment of about $1.8 \times 10^{-23}\,\text{A}\cdot\text{m}^2$. (a) Determine the dipole moment of an iron bar 9.0 cm long, 1.2 cm wide, and 1.0 cm thick, if it is 100 percent saturated. (b) What torque would be exerted on this bar when placed in a 0.80-T field acting at right angles to the bar?

*28–9 Magnetic Materials; Hysteresis

***48. (I)** The following are some values of B and B_0 for a piece of annealed iron as it is being magnetized:

$B_0(10^{-4}\,\text{T})$	0.0	0.13	0.25	0.50	0.63	0.78	1.0	1.3
$B(\text{T})$	0.0	0.0042	0.010	0.028	0.043	0.095	0.45	0.67

$B_0(10^{-4}\,\text{T})$	1.9	2.5	6.3	13.0	130	1300	10,000	
$B(\text{T})$	1.01	1.18	1.44	1.58	1.72	2.26	3.15	

Determine the magnetic permeability μ for each value and plot a graph of μ versus B_0.

***49. (I)** A large thin toroid has 285 loops of wire per meter, and a 3.0-A current flows through the wire. If the relative permeability of the iron is $\mu/\mu_0 = 2200$, what is the total field B inside the toroid?

***50. (II)** An iron-core solenoid is 38 cm long and 1.8 cm in diameter, and has 640 turns of wire. The magnetic field inside the solenoid is 2.2 T when 48 A flows in the wire. What is the permeability μ at this high field strength?

General Problems

51. Three long parallel wires are 3.5 cm from one another. (Looking along them, they are at three corners of an equilateral triangle.) The current in each wire is 8.00 A, but its direction in wire M is opposite to that in wires N and P (Fig. 28–54). Determine the magnetic force per unit length on each wire due to the other two.

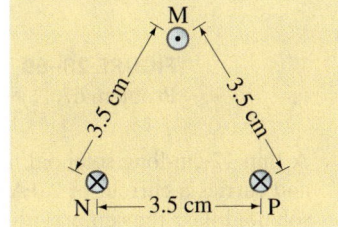

FIGURE 28–54
Problems 51, 52, and 53.

52. In Fig. 28–54, determine the magnitude and direction of the magnetic field midway between points M and N.

53. In Fig. 28–54 the top wire is 1.00-mm-diameter copper wire and is suspended in air due to the two magnetic forces from the bottom two wires. The current is 40.0 A in each of the two bottom wires. Calculate the required current flow in the suspended wire.

54. An electron enters a large solenoid at a 7.0° angle to the axis. If the field is a uniform $3.3 \times 10^{-2}\,\text{T}$, determine the radius and pitch (distance between loops) of the electron's helical path if its speed is $1.3 \times 10^7\,\text{m/s}$.

55. Two long straight parallel wires are 15 cm apart. Wire A carries 2.0-A current. Wire B's current is 4.0 A in the same direction. (a) Determine the magnetic field due to wire A at the position of wire B. (b) Determine the magnetic field due to wire B at the position of wire A. (c) Are these two magnetic fields equal and opposite? Why or why not? (d) Determine the force per unit length on wire A due to wire B, and that on wire B due to wire A. Are these two forces equal and opposite? Why or why not?

56. A rectangular loop of wire carries a 2.0-A current and lies in a plane which also contains a very long straight wire carrying a 10.0-A current as shown in Fig. 28–55. Determine (a) the net force and (b) the net torque on the loop due to the straight wire.

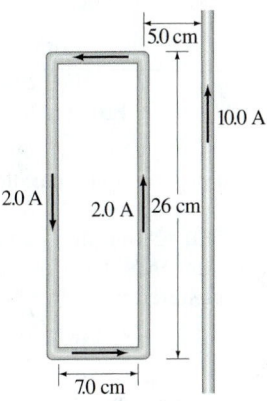

FIGURE 28–55
Problem 56.

57. A very large flat conducting sheet of thickness t carries a uniform current density $\vec{\jmath}$ throughout (Fig. 28–56). Determine the magnetic field (magnitude and direction) at a distance y above the plane. (Assume the plane is infinitely long and wide.)

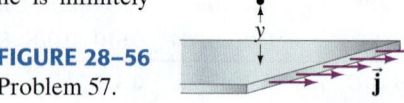

FIGURE 28–56
Problem 57.

58. A long horizontal wire carries a current of 48 A. A second wire, made of 1.00-mm-diameter copper wire and parallel to the first, is kept in suspension magnetically 5.0 cm below (Fig. 28–57). (a) Determine the magnitude and direction of the current in the lower wire. (b) Is the lower wire in stable equilibrium? (c) Repeat parts (a) and (b) if the second wire is suspended 5.0 cm *above* the first due to the first's magnetic field.

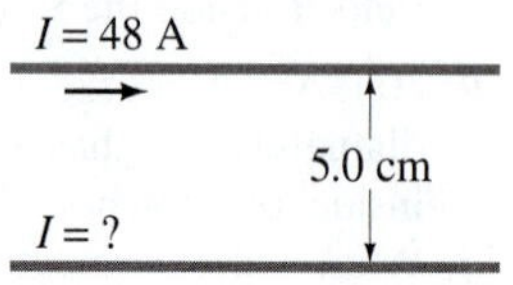

$I = 48$ A

5.0 cm

FIGURE 28–57
Problem 58.

$I = ?$

59. A square loop of wire, of side d, carries a current I. Show that the magnetic field at the center of the square is

$$B = \frac{2\sqrt{2}\,\mu_0 I}{\pi d}.$$

[*Hint*: Determine $\vec{B}$ for each segment of length d.]

60. In Problem 59, if you reshaped the square wire into a circle, would B increase or decrease at the center? Explain.

61. Helmholtz coils are two identical circular coils having the same radius R and the same number of turns N, separated by a distance equal to the radius R and carrying the same current I in the same direction. (See Fig. 28–58.) They are used in scientific instruments to generate nearly uniform magnetic fields.(They can be seen in the photo, Fig. 27–18.) (a) Determine the magnetic field B at points x along the line joining their centers. Let $x = 0$ at the center of one coil, and $x = R$ at the center of the other. (b) Show that the field midway between the coils is particularly uniform by showing that $\dfrac{dB}{dx} = 0$ and $\dfrac{d^2B}{dx^2} = 0$ at the midpoint between the coils. (c) If $R = 10.0$ cm, $N = 250$ turns and $I = 2.0$ A, what is the field at the midpoint between the coils, $x = R/2$?

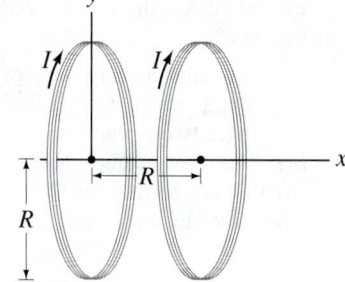

FIGURE 28–58
Problem 61.

62. For two long parallel wires separated by a distance d, carrying currents I_1 and I_2 as in Fig. 28–10, show directly (Eq. 28–1) that Ampère's law is valid (but do not use Ampère's law) for a circular path of radius r ($r < d$) centered on I_1:

$$\oint \vec{B} \cdot d\vec{\ell} = \mu_0 I_1.$$

63. Near the Earth's poles the magnetic field is about 1 G $(1 \times 10^{-4}\,\text{T})$. Imagine a simple model in which the Earth's field is produced by a single current loop around the equator. Estimate roughly the current this loop would carry.

64. A 175-g model airplane charged to 18.0 mC and traveling at 2.8 m/s passes within 8.6 cm of a wire, nearly parallel to its path, carrying a 25-A current. What acceleration (in g's) does this interaction give the airplane?

65. Suppose that an electromagnet uses a coil 2.0 m in diameter made from square copper wire 2.0 mm on a side; the power supply produces 35 V at a maximum power output of 1.0 kW. (a) How many turns are needed to run the power supply at maximum power? (b) What is the magnetic field strength at the center of the coil? (c) If you use a greater number of turns and this same power supply, will a greater magnetic field result? Explain.

66. Four long straight parallel wires located at the corners of a square of side d carry equal currents I_0 perpendicular to the page as shown in Fig. 28–59. Determine the magnitude and direction of $\vec{B}$ at the center C of the square.

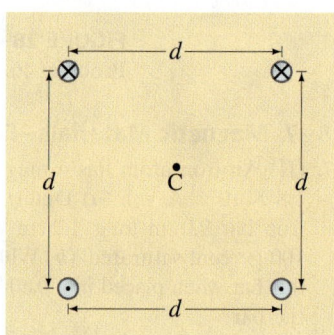

FIGURE 28–59
Problem 66.

67. Determine the magnetic field at the point P due to a very long wire with a square bend as shown in Fig. 28–60. The point P is halfway between the two corners. [*Hint*: You can use the results of Problems 40 and 41.]

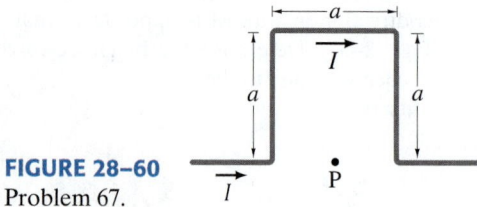

FIGURE 28–60
Problem 67.

68. A thin 12-cm-long solenoid has a total of 420 turns of wire and carries a current of 2.0 A. Calculate the field inside the solenoid near the center.

69. A 550-turn solenoid is 15 cm long. The current into it is 33 A. A 3.0-cm-long straight wire cuts through the center of the solenoid, along a diameter. This wire carries a 22-A current downward (and is connected by other wires that don't concern us). What is the force on this wire assuming the solenoid's field points due east?

70. You have 1.0 kg of copper and want to make a practical solenoid that produces the greatest possible magnetic field for a given voltage. Should you make your copper wire long and thin, short and fat, or something else? Consider other variables, such as solenoid diameter, length, and so on.

71. A small solenoid (radius r_a) is inside a larger solenoid (radius $r_b > r_a$). They are coaxial with n_a and n_b turns per unit length, respectively. The solenoids carry the same current, but in opposite directions. Let r be the radial distance from the common axis of the solenoids. If the magnetic field inside the inner solenoid $(r < r_a)$ is to be in the opposite direction as the field between the solenoids $(r_a < r < r_b)$, but have half the magnitude, determine the required ratio n_b/n_a.

72. Find B at the center of the 4.0-cm-radius semicircle in Fig. 28–61. The straight wires extend a great distance outward to the left and carry a current $I = 6.0A$.

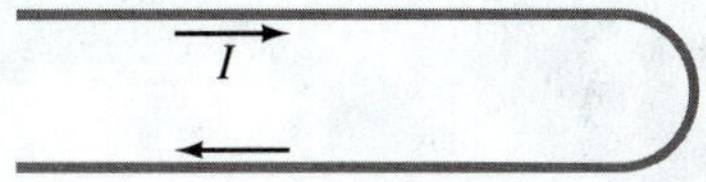

FIGURE 28–61 Problem 72.

73. The design of a magneto-optical atom trap requires a magnetic field B that is directly proportional to position x along an axis. Such a field perturbs the absorption of laser light by atoms in the manner needed to spatially confine atoms in the trap. Let us demonstrate that "anti-Helmholtz" coils will provide the required field $B = Cx$, where C is a constant. Anti-Helmholtz coils consist of two identical circular wire coils, each with radius R and N turns, carrying current I in opposite directions (Fig. 28–62). The coils share a common axis (defined as the x axis with $x = 0$ at the midpoint (0) between the coils). Assume that the centers of the coils are separated by a distance equal to the radius R of the coils. (a) Show that the magnetic field at position x along the x axis is given by

$$B(x) = \frac{4\mu_0 NI}{R}\left\{\left[4 + \left(1 - \frac{2x}{R}\right)^2\right]^{-\frac{3}{2}} - \left[4 + \left(1 + \frac{2x}{R}\right)^2\right]^{-\frac{3}{2}}\right\}.$$

(b) For small excursions from the origin where $|x| \ll R$, show that the magnetic field is given by $B \approx Cx$, where the constant $C = 48\mu_0 NI/25\sqrt{5}\,R^2$. ($c$) For optimal atom trapping, dB/dx should be about 0.15 T/m. Assume an atom trap uses anti-Helmholtz coils with $R = 4.0$ cm and $N = 150$. What current should flow through the coils? [Coil separation equal to coil radius, as assumed in this problem, is not a strict requirement for anti-Helmholtz coils.]

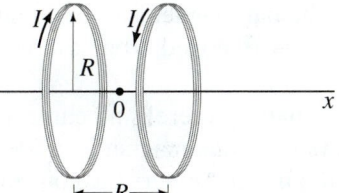

FIGURE 28–62 Problem 75.

74. You want to get an idea of the magnitude of magnetic fields produced by overhead power lines. You estimate that a transmission wire is about 12 m above the ground. The local power company tells you that the line operates at 15 kV and provide a maximum of 45 MW to the local area. Estimate the maximum magnetic field you might experience walking under such a power line, and compare to the Earth's field. [For an ac current, values are rms, and the magnetic field will be changing.]

* Numerical/Computer

* **75.** (II) A circular current loop of radius 15 cm containing 250 turns carries a current of 2.0 A. Its center is at the origin and its axis lies along the x axis. Calculate the magnetic field B at a point x on the x axis for $x = -40$ cm to $+40$ cm in steps of 2 cm and make a graph of B as a function of x.

* **76.** (III) A set of Helmholtz coils (see Problem 61, Fig. 28–58) have a radius $R = 10.0$ cm and are separated by a distance $R = 10.0$ cm. Each coil has 250 loops carrying a current $I = 2.0$ A. (a) Determine the total magnetic field B along the x axis (the center line for the two coils) in steps of 0.2 cm from the center of one coil ($x = 0$) to the center of the other ($x = R$). (b) Graph B as a function of x. (c) By what % does B vary from $x = 5.0$ cm to $x = 6.0$ cm?

Answers to Exercises

A: 2×10^{-6} T; not at this distance, and then only briefly.

B: 0.8×10^{-5} T, up.

C: (b).

D: (b), (c).

One of the great laws of physics is Faraday's law of induction, which says that a changing magnetic flux produces an induced emf. This photo shows a bar magnet moving inside a coil of wire, and the galvanometer registers an induced current. This phenomenon of electromagnetic induction is the basis for many practical devices, including generators, alternators, transformers, tape recording, and computer memory.

Electromagnetic Induction and Faraday's Law

CONTENTS

29–1 Induced EMF

29–2 Faraday's Law of Induction; Lenz's Law

29–3 EMF Induced in a Moving Conductor

*29–4 Electric Generators

29–5 Back EMF and Counter Torque; Eddy Currents

29–6 Transformers and Transmission of Power

29–7 A Changing Magnetic Flux Produces an Electric Field

*29–8 Applications of Induction: Sound Systems, Computer Memory, Seismograph, GFCI

CHAPTER-OPENING QUESTION—Guess now!

In the photograph above, the bar magnet is inserted into the coil of wire, and is left there for 1 minute; then it is removed from the coil. What would an observer watching the galvanometer see?

(a) No change; without a battery there is no current to detect.

(b) A small current flows while the magnet is inside the coil of wire.

(c) A current spike as the magnet enters the coil, and then nothing.

(d) A current spike as the magnet enters the coil, and then a steady small current.

(e) A current spike as the magnet enters the coil, then nothing, and then a current spike in the opposite direction as the magnet leaves the coil.

I n Chapter 27, we discussed two ways in which electricity and magnetism are related: (1) an electric current produces a magnetic field; and (2) a magnetic field exerts a force on an electric current or moving electric charge. These discoveries were made in 1820–1821. Scientists then began to wonder: if electric currents produce a magnetic field, is it possible that a magnetic field can produce an electric current? Ten years later the American Joseph Henry (1797–1878) and the Englishman Michael Faraday (1791–1867) independently found that it was possible. Henry actually made the discovery first. But Faraday published his results earlier and investigated the subject in more detail. We now discuss this phenomenon and some of its world-changing applications including the electric generator.

29–1 Induced EMF

In his attempt to produce an electric current from a magnetic field, Faraday used an apparatus like that shown in Fig. 29–1. A coil of wire, X, was connected to a battery. The current that flowed through X produced a magnetic field that was intensified by the ring-shaped iron core around which the wire was wrapped. Faraday hoped that a strong steady current in X would produce a great enough magnetic field to produce a current in a second coil Y wrapped on the same iron ring. This second circuit, Y, contained a galvanometer to detect any current but contained no battery.

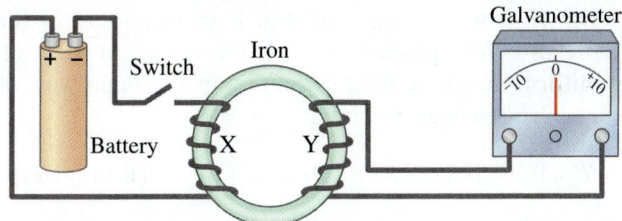

FIGURE 29–1 Faraday's experiment to induce an emf.

He met no success with constant currents. But the long-sought effect was finally observed when Faraday noticed the galvanometer in circuit Y deflect strongly at the moment he closed the switch in circuit X. And the galvanometer deflected strongly in the opposite direction when he opened the switch in X. A constant current in X produced a constant magnetic field which produced *no* current in Y. Only when the current in X was starting or stopping was a current produced in Y.

Faraday concluded that although a constant magnetic field produces no current in a conductor, a *changing* magnetic field can produce an electric current. Such a current is called an **induced current**. When the magnetic field through coil Y changes, a current occurs in Y as if there were a source of emf in circuit Y. We therefore say that

a changing magnetic field induces an emf.

> ⚠ **CAUTION**
> *Changing $\vec{B}$, not $\vec{B}$ itself, induces current*

Faraday did further experiments on **electromagnetic induction**, as this phenomenon is called. For example, Fig. 29–2 shows that if a magnet is moved quickly into a coil of wire, a current is induced in the wire. If the magnet is quickly removed, a current is induced in the opposite direction ($\vec{B}$ through the coil decreases). Furthermore, if the magnet is held steady and the coil of wire is moved toward or away from the magnet, again an emf is induced and a current flows. Motion or change is required to induce an emf. It doesn't matter whether the magnet or the coil moves. It is their *relative motion* that counts.

> ⚠ **CAUTION**
> *Relative motion—magnet or coil moving induces current*

FIGURE 29–2   (a) A current is induced when a magnet is moved toward a coil, momentarily increasing the magnetic field through the coil. (b) The induced current is opposite when the magnet is moved away from the coil ($\vec{B}$ decreases). Note that the galvanometer zero is at the center of the scale and the needle deflects left or right, depending on the direction of the current. In (c), no current is induced if the magnet does not move relative to the coil. It is the relative motion that counts here: the magnet can be held steady and the coil moved, which also induces an emf.

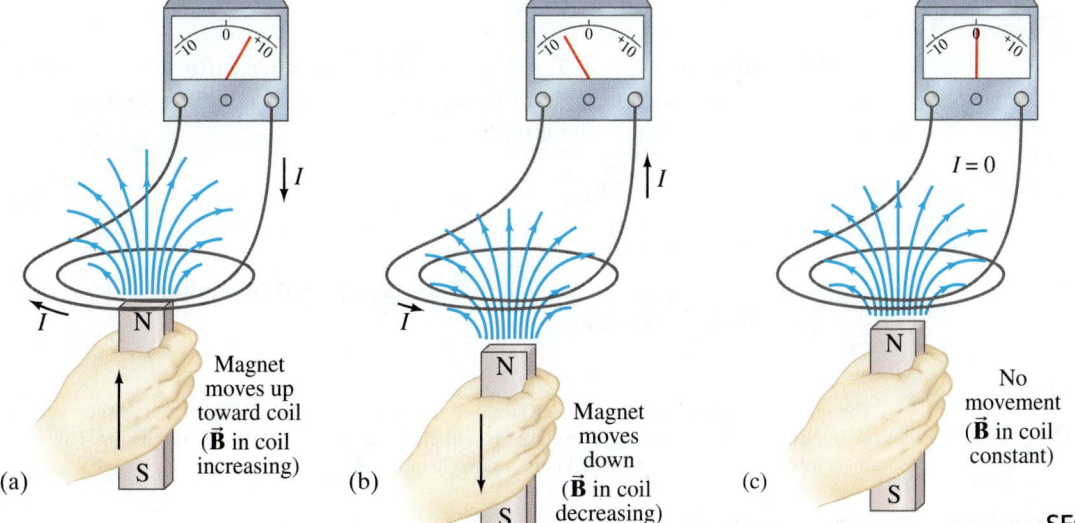

(a) Magnet moves up toward coil ($\vec{B}$ in coil increasing)

(b) Magnet moves down ($\vec{B}$ in coil decreasing)

(c) No movement ($\vec{B}$ in coil constant)

EXERCISE A Return to the Chapter-Opening Question, page 758, and answer it again now. Try to explain why you may have answered differently the first time.

29–2 Faraday's Law of Induction; Lenz's Law

Faraday investigated quantitatively what factors influence the magnitude of the emf induced. He found first of all that the more rapidly the magnetic field changes, the greater the induced emf. He also found that the induced emf depends on the area of the circuit loop. Thus we say that the emf is proportional to the rate of change of the **magnetic flux**, Φ_B, passing through the circuit or loop of area A. Magnetic flux for a uniform magnetic field is defined in the same way we did for electric flux in Chapter 22, namely as

$$\Phi_B = B_\perp A = BA \cos\theta = \vec{\mathbf{B}} \cdot \vec{\mathbf{A}}. \qquad [\vec{\mathbf{B}} \text{ uniform}] \quad (29\text{–}1a)$$

Here $B_\perp$ is the component of the magnetic field $\vec{\mathbf{B}}$ perpendicular to the face of the loop, and θ is the angle between $\vec{\mathbf{B}}$ and the vector $\vec{\mathbf{A}}$ (representing the area) whose direction is perpendicular to the face of the loop. These quantities are shown in Fig. 29–3 for a square loop of side ℓ whose area is $A = \ell^2$. If the area is of some other shape, or $\vec{\mathbf{B}}$ is not uniform, the magnetic flux can be written[†]

$$\Phi_B = \int \vec{\mathbf{B}} \cdot d\vec{\mathbf{A}}. \qquad (29\text{–}1b)$$

As we saw in Chapter 27, the lines of $\vec{\mathbf{B}}$ (like lines of $\vec{\mathbf{E}}$) can be drawn such that the number of lines per unit area is proportional to the field strength. Then the flux Φ_B can be thought of as being proportional to the *total number of lines passing through the area enclosed by the loop*. This is illustrated in Fig. 29–4, where the loop is viewed from the side (on edge). For $\theta = 90°$, no magnetic field lines pass through the loop and $\Phi_B = 0$, whereas Φ_B is a maximum when $\theta = 0°$. The unit of magnetic flux is the tesla-meter²; this is called a **weber**: $1 \text{ Wb} = 1 \text{ T} \cdot \text{m}^2$.

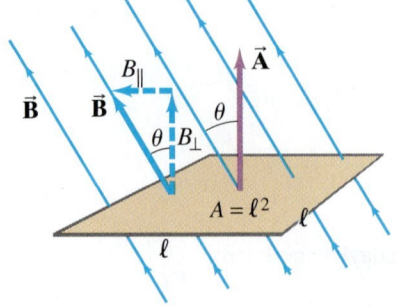

FIGURE 29–3 Determining the flux through a flat loop of wire. This loop is square, of side ℓ and area $A = \ell^2$.

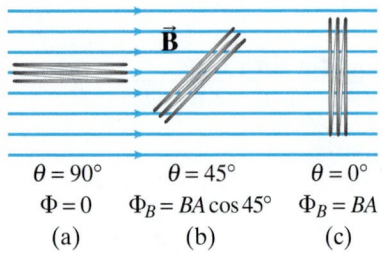

FIGURE 29–4 Magnetic flux Φ_B is proportional to the number of lines of $\vec{\mathbf{B}}$ that pass through the loop.

$\theta = 90°$ $\theta = 45°$ $\theta = 0°$
$\Phi = 0$ $\Phi_B = BA \cos 45°$ $\Phi_B = BA$
(a)   (b)   (c)

FIGURE 29–5 Example 29–1.

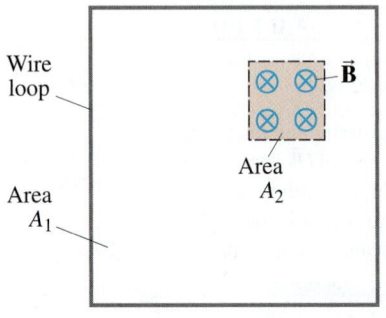

Wire loop

$\vec{\mathbf{B}}$

Area A_2

Area A_1

CONCEPTUAL EXAMPLE 29–1 **Determining flux.** A square loop of wire encloses area A_1 as shown in Fig. 29–5. A uniform magnetic field $\vec{\mathbf{B}}$ perpendicular to the loop extends over the area A_2. What is the magnetic flux through the loop A_1?

RESPONSE We assume that the magnetic field is zero outside the area A_2. The total magnetic flux through area A_1 is the flux through area A_2, which by Eq. 29–1a for a uniform field is BA_2, plus the flux through the remaining area $(= A_1 - A_2)$, which is zero because $B = 0$. So the total flux is $\Phi_B = BA_2 + 0(A_1 - A_2) = BA_2$. It is *not* equal to BA_1 because $\vec{\mathbf{B}}$ is not uniform over A_1.

With our definition of flux, Eqs. 29–1, we can now write down the results of Faraday's investigations: The emf induced in a circuit is equal to the rate of change of magnetic flux through the circuit:

FARADAY'S LAW OF INDUCTION

$$\mathscr{E} = -\frac{d\Phi_B}{dt}. \qquad (29\text{–}2a)$$

This fundamental result is known as **Faraday's law of induction**, and is one of the basic laws of electromagnetism.

[†]The integral is taken over an open surface—that is, one bounded by a closed curve such as a circle or square. In the present discussion, the area is that enclosed by the loop under discussion. The area is not an enclosed surface as we used in Gauss's law, Chapter 22.

If the circuit contains N loops that are closely wrapped so the same flux passes through each, the emfs induced in each loop add together, so

$$\mathcal{E} = -N\frac{d\Phi_B}{dt}.$$

[N loops] **(29–2b)**

EXAMPLE 29–2 **A loop of wire in a magnetic field.** A square loop of wire of side $\ell = 5.0\,\text{cm}$ is in a uniform magnetic field $B = 0.16\,\text{T}$. What is the magnetic flux in the loop (a) when $\vec{\mathbf{B}}$ is perpendicular to the face of the loop and (b) when $\vec{\mathbf{B}}$ is at an angle of 30° to the area $\vec{\mathbf{A}}$ of the loop? (c) What is the magnitude of the average current in the loop if it has a resistance of 0.012 Ω and it is rotated from position (b) to position (a) in 0.14 s?

APPROACH We use the definition $\Phi_B = \vec{\mathbf{B}} \cdot \vec{\mathbf{A}}$ to calculate the magnetic flux. Then we use Faraday's law of induction to find the induced emf in the coil, and from that the induced current ($I = \mathcal{E}/R$).

SOLUTION The area of the coil is $A = \ell^2 = (5.0 \times 10^{-2}\,\text{m})^2 = 2.5 \times 10^{-3}\,\text{m}^2$, and the direction of $\vec{\mathbf{A}}$ is perpendicular to the face of the loop (Fig. 29–3).

(a) $\vec{\mathbf{B}}$ is perpendicular to the coil's face, and thus parallel to $\vec{\mathbf{A}}$ (Fig. 29–3), so

$$\Phi_B = \vec{\mathbf{B}} \cdot \vec{\mathbf{A}}$$
$$= BA\cos 0° = (0.16\,\text{T})(2.5 \times 10^{-3}\,\text{m}^2)(1) = 4.0 \times 10^{-4}\,\text{Wb}.$$

(b) The angle between $\vec{\mathbf{B}}$ and $\vec{\mathbf{A}}$ is 30°, so

$$\Phi_B = \vec{\mathbf{B}} \cdot \vec{\mathbf{A}}$$
$$= BA\cos\theta = (0.16\,\text{T})(2.5 \times 10^{-3}\,\text{m}^2)\cos 30° = 3.5 \times 10^{-4}\,\text{Wb}.$$

(c) The magnitude of the induced emf is

$$\mathcal{E} = \frac{\Delta\Phi_B}{\Delta t} = \frac{(4.0 \times 10^{-4}\,\text{Wb}) - (3.5 \times 10^{-4}\,\text{Wb})}{0.14\,\text{s}} = 3.6 \times 10^{-4}\,\text{V}.$$

The current is then

$$I = \frac{\mathcal{E}}{R} = \frac{3.6 \times 10^{-4}\,\text{V}}{0.012\,\Omega} = 0.030\,\text{A} = 30\,\text{mA}.$$

The minus signs in Eqs. 29–2a and b are there to remind us in which direction the induced emf acts. Experiments show that

a current produced by an induced emf moves in a direction so that the magnetic field created by that current opposes the original change in flux.

This is known as **Lenz's law.** Be aware that we are now discussing two distinct magnetic fields: (1) the changing magnetic field or flux that induces the current, and (2) the magnetic field produced by the induced current (all currents produce a field). The second field opposes the change in the first.

Lenz's law can be said another way, valid even if no current can flow (as when a circuit is not complete):

An induced emf is always in a direction that opposes the original change in flux that caused it.

⚠ **CAUTION**

Distinguish two different magnetic fields

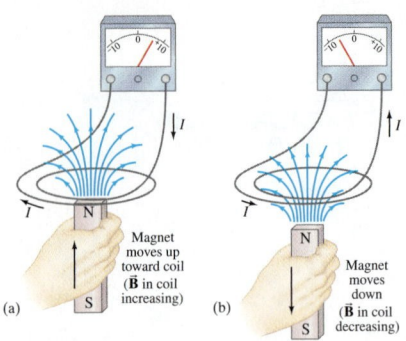

FIGURE 29–2 (repeated).

Let us apply Lenz's law to the relative motion between a magnet and a coil, Fig. 29–2. The changing flux through the coil induces an emf in the coil, producing a current. This induced current produces its own magnetic field. In Fig. 29–2a the distance between the coil and the magnet decreases. The magnet's magnetic field (and number of field lines) through the coil increases, and therefore the flux increases. The magnetic field of the magnet points upward. To oppose the upward increase, the magnetic field inside the coil produced by the induced current needs to point *downward*. Thus, Lenz's law tells us that the current moves as shown (use the right-hand rule). In Fig. 29–2b, the flux *decreases* (because the magnet is moved away and B decreases), so the induced current in the coil produces an *upward* magnetic field through the coil that is "trying" to maintain the status quo. Thus the current in Fig. 29–2b is in the opposite direction from Fig. 29–2a.

It is important to note that an emf is induced whenever there is a change in *flux* through the coil, and we now consider some more possibilities.

FIGURE 29–6 A current can be induced by changing the area of the coil, even though B doesn't change. Here the area is reduced by pulling on its sides: the *flux* through the coil is reduced as we go from (a) to (b). Here the brief induced current acts in the direction shown so as to try to maintain the original flux ($\Phi = BA$) by producing its own magnetic field into the page. That is, as the area A decreases, the current acts to increase B in the original (inward) direction.

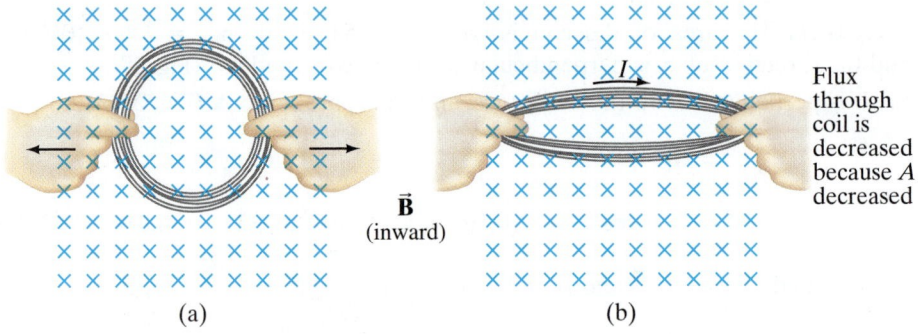

Since magnetic flux $\Phi_B = \int \vec{\mathbf{B}} \cdot d\vec{\mathbf{A}} = \int B \cos \theta \, dA$, we see that an emf can be induced in three ways: (1) by a changing magnetic field B; (2) by changing the area A of the loop in the field; or (3) by changing the loop's orientation θ with respect to the field. Figures 29–1 and 29–2 illustrated case 1. Examples of cases 2 and 3 are illustrated in Figs. 29–6 and 29–7, respectively.

FIGURE 29–7 A current can be induced by rotating a coil in a magnetic field. The flux through the coil changes from (a) to (b) because θ (in Eq. 29–1a, $\Phi = BA \cos \theta$) went from $0°$ ($\cos \theta = 1$) to $90°$ ($\cos \theta = 0$).

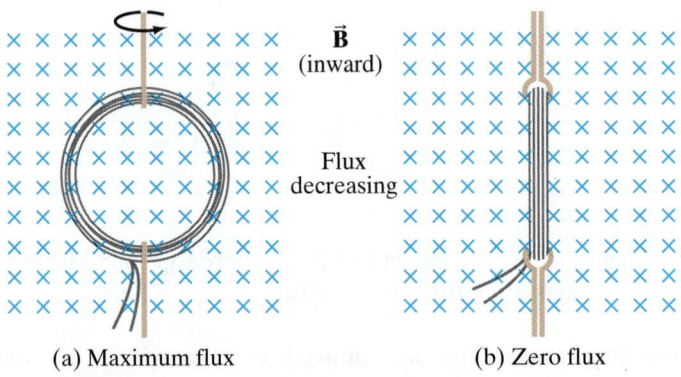

(a) Maximum flux   (b) Zero flux

FIGURE 29–8 Example 29–3: An induction stove.

CONCEPTUAL EXAMPLE 29–3 | **Induction stove.** In an induction stove (Fig. 29–8), an ac current exists in a coil that is the "burner" (a burner that never gets hot). Why will it heat a metal pan but not a glass container?

RESPONSE The ac current sets up a changing magnetic field that passes through the pan bottom. This changing magnetic field induces a current in the pan bottom, and since the pan offers resistance, electric energy is transformed to thermal energy which heats the pot and its contents. A glass container offers such high resistance that little current is induced and little energy is transferred $\left(P = V^2/R\right)$.

Lenz's Law

Lenz's law is used to determine the direction of the (conventional) electric current induced in a loop due to a change in magnetic flux inside the loop. To produce an induced current you need

(a) a closed conducting loop, and

(b) an external magnetic flux through the loop that is changing in time.

1. Determine whether the magnetic flux ($\Phi_B = BA \cos \theta$) inside the loop is decreasing, increasing, or unchanged.

2. The magnetic field due to the induced current: (a) points in the same direction as the external field if the flux is decreasing; (b) points in the opposite direction from the external field if the flux is increasing; or (c) is zero if the flux is not changing.

3. Once you know the direction of the induced magnetic field, use the right-hand rule to find the direction of the induced current.

4. Always keep in mind that there are two magnetic fields: (1) an external field whose flux must be changing if it is to induce an electric current, and (2) a magnetic field produced by the induced current.

FIGURE 29–9 Example 29–4.

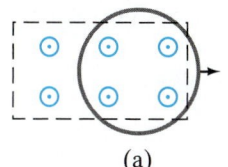 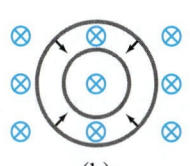 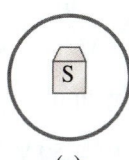 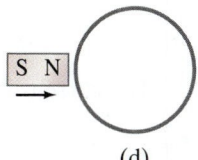 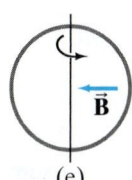

(a)	(b)	(c)	(d)	(e)
Pulling a round loop to the right out of a magnetic field which points out of the page	Shrinking a loop in a magnetic field pointing into the page	S magnetic pole moving from below, up toward the loop	N magnetic pole moving toward loop in the plane of the page	Rotating the loop by pulling the left side toward us and pushing the right side in; the magnetic field points from right to left

CONCEPTUAL EXAMPLE 29–4 **Practice with Lenz's law.** In which direction is the current induced in the circular loop for each situation in Fig. 29–9?

RESPONSE (*a*) Initially, the magnetic field pointing out of the page passes through the loop. If you pull the loop out of the field, magnetic flux through the loop decreases; so the induced current will be in a direction to maintain the decreasing flux through the loop: the current will be counterclockwise to produce a magnetic field outward (toward the reader).

(*b*) The external field is into the page. The coil area gets smaller, so the flux will decrease; hence the induced current will be clockwise, producing its own field into the page to make up for the flux decrease.

(*c*) Magnetic field lines point into the S pole of a magnet, so as the magnet moves toward us and the loop, the magnet's field points into the page and is getting stronger. The current in the loop will be induced in the counterclockwise direction in order to produce a field $\vec{\mathbf{B}}$ *out* of the page.

(*d*) The field is in the plane of the loop, so no magnetic field lines pass through the loop and the flux through the loop is zero throughout the process; hence there is no change in external magnetic flux with time, and there will be no induced emf or current in the loop.

(*e*) Initially there is no flux through the loop. When you start to rotate the loop, the external field through the loop begins increasing to the left. To counteract this change in flux, the loop will have current induced in a counterclockwise direction so as to produce its own field to the right.

⚠ **C A U T I O N**

Magnetic field created by induced current opposes change in external flux, not necessarily opposing the external field

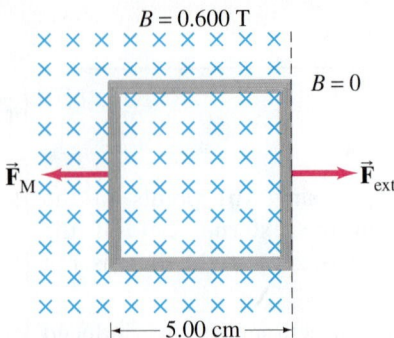

$B = 0.600\ \text{T}$

$B = 0$

$\vec{F}_M$

$\vec{F}_{ext}$

5.00 cm

FIGURE 29–10 Example 29–5. The square coil in a magnetic field $B = 0.600\ \text{T}$ is pulled abruptly to the right to a region where $B = 0$.

FIGURE 29–11 Exercise B.

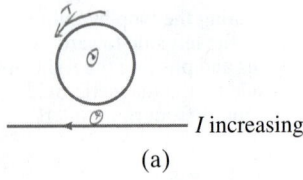

I increasing

(a)

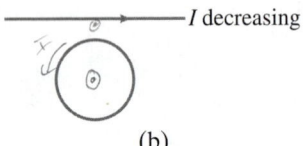

I decreasing

(b)

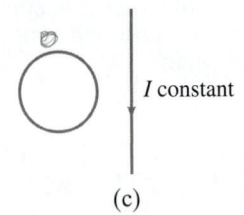

I constant

(c)

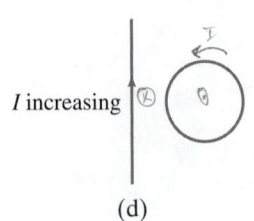

I increasing

(d)

EXAMPLE 29–5 **Pulling a coil from a magnetic field.** A 100-loop square coil of wire, with side $\ell = 5.00\ \text{cm}$ and total resistance $100\ \Omega$, is positioned perpendicular to a uniform 0.600-T magnetic field, as shown in Fig. 29–10. It is quickly pulled from the field at constant speed (moving perpendicular to $\vec{B}$) to a region where B drops abruptly to zero. At $t = 0$, the right edge of the coil is at the edge of the field. It takes 0.100 s for the whole coil to reach the field-free region. Find (a) the rate of change in flux through the coil, and (b) the emf and current induced. (c) How much energy is dissipated in the coil? (d) What was the average force required (F_{ext})?

APPROACH We start by finding how the magnetic flux, $\Phi_B = BA$, changes during the time interval $\Delta t = 0.100\ \text{s}$. Faraday's law then gives the induced emf and Ohm's law gives the current.

SOLUTION (a) The area of the coil is $A = \ell^2 = (5.00 \times 10^{-2}\ \text{m})^2 = 2.50 \times 10^{-3}\ \text{m}^2$. The flux through one loop is initially $\Phi_B = BA = (0.600\ \text{T})(2.50 \times 10^{-3}\ \text{m}^2) = 1.50 \times 10^{-3}\ \text{Wb}$. After 0.100 s, the flux is zero. The rate of change in flux is constant (because the coil is square), equal to

$$\frac{\Delta \Phi_B}{\Delta t} = \frac{0 - (1.50 \times 10^{-3}\ \text{Wb})}{0.100\ \text{s}} = -1.50 \times 10^{-2}\ \text{Wb/s}.$$

(b) The emf induced (Eq. 29–2) in the 100-loop coil during this 0.100-s interval is

$$\mathcal{E} = -N\frac{\Delta \Phi_B}{\Delta t} = -(100)(-1.50 \times 10^{-2}\ \text{Wb/s}) = 1.50\ \text{V}.$$

The current is found by applying Ohm's law to the $100\text{-}\Omega$ coil:

$$I = \frac{\mathcal{E}}{R} = \frac{1.50\ \text{V}}{100\ \Omega} = 1.50 \times 10^{-2}\ \text{A} = 15.0\ \text{mA}.$$

By Lenz's law, the current must be clockwise to produce more $\vec{B}$ into the page and thus oppose the decreasing flux into the page.

(c) The total energy dissipated in the coil is the product of the power $(= I^2R)$ and the time:

$$E = Pt = I^2Rt = (1.50 \times 10^{-2}\ \text{A})^2(100\ \Omega)(0.100\ \text{s}) = 2.25 \times 10^{-3}\ \text{J}.$$

(d) We can use the result of part (c) and apply the work-energy principle: the energy dissipated E is equal to the work W needed to pull the coil out of the field (Chapters 7 and 8). Because $W = \bar{F}d$ where $d = 5.00\ \text{cm}$, then

$$\bar{F} = \frac{W}{d} = \frac{2.25 \times 10^{-3}\ \text{J}}{5.00 \times 10^{-2}\ \text{m}} = 0.0450\ \text{N}.$$

Alternate Solution (d) We can also calculate the force directly using $\vec{F} = I\vec{\ell} \times \vec{B}$, Eq. 27–3, which here for constant $\vec{B}$ is $F = I\ell B$. The force the magnetic field exerts on the top and bottom sections of the square coil of Fig. 29–10 are in opposite directions and cancel each other. The magnetic force $\vec{F}_M$ exerted on the left vertical section of the square coil acts to the left as shown because the current is up (clockwise). The right side of the loop is in the region where $\vec{B} = 0$. Hence the external force, to the right, needed to just overcome the magnetic force to the left (on $N = 100$ loops) is

$$F_{ext} = NI\ell B = (100)(0.0150\ \text{A})(0.0500\ \text{m})(0.600\ \text{T}) = 0.0450\ \text{N},$$

which is the same answer, confirming our use of energy conservation above.

EXERCISE B What is the direction of the induced current in the circular loop due to the current shown in each part of Fig. 29–11?

29–3 EMF Induced in a Moving Conductor

Another way to induce an emf is shown in Fig. 29–12a, and this situation helps illuminate the nature of the induced emf. Assume that a uniform magnetic field $\vec{B}$ is perpendicular to the area bounded by the U-shaped conductor and the movable rod resting on it. If the rod is made to move at a speed v, it travels a distance $dx = v\,dt$ in a time dt. Therefore, the area of the loop increases by an amount $dA = \ell\,dx = \ell v\,dt$ in a time dt. By Faraday's law there is an induced emf $\mathscr{E}$ whose magnitude is given by

$$\mathscr{E} = \frac{d\Phi_B}{dt} = \frac{B\,dA}{dt} = \frac{B\ell v\,dt}{dt} = B\ell v. \qquad (29\text{–}3)$$

Equation 29–3 is valid as long as B, ℓ, and v are mutually perpendicular. (If they are not, we use only the components of each that are mutually perpendicular.) An emf induced on a conductor moving in a magnetic field is sometimes called *motional emf*.

We can also obtain Eq. 29–3 without using Faraday's law. We saw in Chapter 27 that a charged particle moving perpendicular to a magnetic field B with speed v experiences a force $\vec{F} = q\vec{v} \times \vec{B}$ (Eq. 27–5a). When the rod of Fig. 29–12a moves to the right with speed v, the electrons in the rod also move with this speed. Therefore, since $\vec{v} \perp \vec{B}$, each electron feels a force $F = qvB$, which acts up the page as shown in Fig. 29–12b. If the rod was not in contact with the U-shaped conductor, electrons would collect at the upper end of the rod, leaving the lower end positive (see signs in Fig. 29–12b). There must thus be an induced emf. If the rod is in contact with the U-shaped conductor (Fig. 29–12a), the electrons will flow into the U. There will then be a clockwise (conventional) current in the loop. To calculate the emf, we determine the work W needed to move a charge q from one end of the rod to the other against this potential difference: $W = \text{force} \times \text{distance} = (qvB)(\ell)$. The emf equals the work done per unit charge, so $\mathscr{E} = W/q = qvB\ell/q = B\ell v$, the same result[†] as from Faraday's law above, Eq. 29–3.

EXERCISE C In what direction will the electrons flow in Fig. 29–12 if the rod moves to the left, decreasing the area of the current loop?

EXAMPLE 29–6 ESTIMATE **Does a moving airplane develop a large emf?** An airplane travels 1000 km/h in a region where the Earth's magnetic field is about $5 \times 10^{-5}\,\text{T}$ and is nearly vertical (Fig. 29–13). What is the potential difference induced between the wing tips that are 70 m apart?

APPROACH We consider the wings to be a 70-m-long conductor moving through the Earth's magnetic field. We use Eq. 29–3 to get the emf.

SOLUTION Since $v = 1000\,\text{km/h} = 280\,\text{m/s}$, and $\vec{v} \perp \vec{B}$, we have

$$\mathscr{E} = B\ell v = (5 \times 10^{-5}\,\text{T})(70\,\text{m})(280\,\text{m/s}) \approx 1\,\text{V}.$$

NOTE Not much to worry about.

EXAMPLE 29–7 **Electromagnetic blood-flow measurement.** The rate of blood flow in our body's vessels can be measured using the apparatus shown in Fig. 29–14, since blood contains charged ions. Suppose that the blood vessel is 2.0 mm in diameter, the magnetic field is 0.080 T, and the measured emf is 0.10 mV. What is the flow velocity v of the blood?

APPROACH The magnetic field $\vec{B}$ points horizontally from left to right (N pole toward S pole). The induced emf acts over the width $\ell = 2.0\,\text{mm}$ of the blood vessel, perpendicular to $\vec{B}$ and $\vec{v}$ (Fig. 29–14), just as in Fig. 29–12. We can then use Eq. 29–3 to get v. ($\vec{v}$ in Fig. 29–14 corresponds to $\vec{v}$ in Fig. 29–12.)

SOLUTION We solve for v in Eq. 29–3:

$$v = \frac{\mathscr{E}}{B\ell} = \frac{(1.0 \times 10^{-4}\,\text{V})}{(0.080\,\text{T})(2.0 \times 10^{-3}\,\text{m})} = 0.63\,\text{m/s}.$$

NOTE In actual practice, an alternating current is used to produce an alternating magnetic field. The induced emf is then alternating.

[†]This force argument, which is basically the same as for the Hall effect (Section 27–8), explains this one way of inducing an emf. It does not explain the general case of electromagnetic induction.

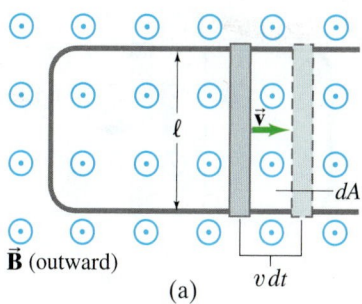

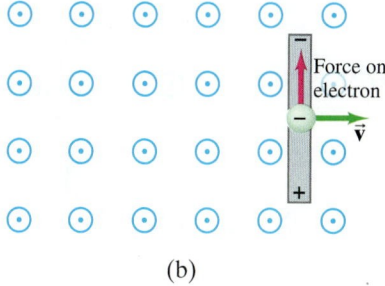

FIGURE 29–12 (a) A conducting rod is moved to the right on a U-shaped conductor in a uniform magnetic field $\vec{B}$ that points out of the page. The induced current is clockwise. (b) Upward force on an electron in the metal rod (moving to the right) due to $\vec{B}$ pointing out of the page; hence electrons can collect at top of rod, leaving + charge at bottom.

FIGURE 29–13 Example 29–6.

PHYSICS APPLIED
Blood-flow measurement

FIGURE 29–14 Measurement of blood velocity from the induced emf. Example 29–7.

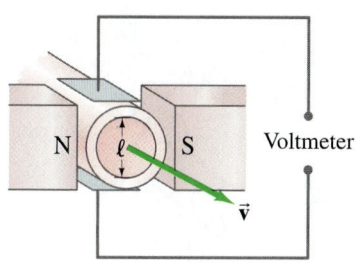

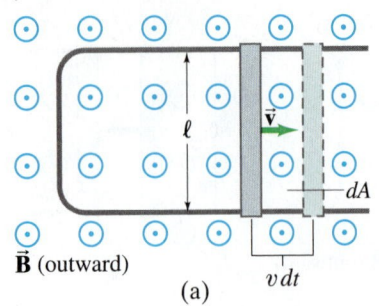

$\vec{B}$ (outward)

(a)

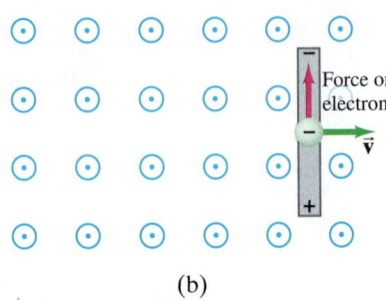

(b)

FIGURE 29–12 (repeated)
(a) A conducting rod is moved to the right on a U-shaped conductor in a uniform magnetic field $\vec{B}$ that points out of the page. The induced current is clockwise. (b) Upward force on an electron in the metal rod (moving to the right) due to $\vec{B}$ pointing out of the page; hence electrons can collect at top of rod, leaving + charge at bottom.

EXAMPLE 29–8 **Force on the rod.** To make the rod of Fig. 29–12a move to the right at constant speed v, you need to apply an external force on the rod to the right. (a) Explain and determine the magnitude of the required force. (b) What external power is needed to move the rod? (Do not confuse this external force on the rod with the upward force on the electrons shown in Fig. 29–12b.)

APPROACH When the rod moves to the right, electrons flow upward in the rod according to the right-hand rule. So the conventional current is downward in the rod. We can see this also from Lenz's law: the outward magnetic flux through the loop is increasing, so the induced current must oppose the increase. Thus the current is clockwise so as to produce a magnetic field into the page (right-hand rule). The magnetic force on the moving rod is $\vec{F} = I\vec{l} \times \vec{B}$ for a constant $\vec{B}$ (Eq. 27–3). The right-hand rule tells us this magnetic force is to the left, and is thus a "drag force" opposing our effort to move the rod to the right.

SOLUTION (a) The magnitude of the external force, to the right, needs to balance the magnetic force $F = I\ell B$, to the left. The current $I = \mathscr{E}/R = B\ell v/R$ (see Eq. 29–3), and the resistance R is that of the whole circuit: the rod and the U-shaped conductor. The force F required to move the rod is thus

$$F = I\ell B = \left(\frac{B\ell v}{R}\right)\ell B = \frac{B^2\ell^2}{R}v.$$

If B, ℓ, and R are constant, then a constant speed v is produced by a constant external force. (Constant R implies that the parallel rails have negligible resistance.)

(b) The external power needed to move the rod for constant R is

$$P_{\text{ext}} = Fv = \frac{B^2\ell^2v^2}{R}.$$

The power dissipated in the resistance is $P = I^2R$. With $I = \mathscr{E}/R = B\ell v/R$,

$$P_{\text{R}} = I^2R = \frac{B^2\ell^2v^2}{R},$$

so the power input equals the power dissipated in the resistance at any moment.

29–4 Electric Generators

We discussed alternating currents (ac) briefly in Section 25–7. Now we examine how ac is generated: by an **electric generator** or **dynamo**, one of the most important practical results of Faraday's great discovery. A generator transforms mechanical energy into electric energy, just the opposite of what a motor does. A simplified diagram of an **ac generator** is shown in Fig. 29–15. A generator consists of many loops of wire (only one is shown) wound on an *armature* that can rotate in a magnetic field. The axle is turned by some mechanical means (falling water, steam turbine, car motor belt), and an emf is induced in the rotating coil. An electric current is thus the *output* of a generator. Suppose in Fig. 29–15 that the armature is rotating clockwise; then $\vec{F} = q\vec{v} \times \vec{B}$ applied to charged particles in the wire (or Lenz's law) tells us that the (conventional) current in the wire labeled b on the armature is outward, toward us; therefore the current is outward from brush b. (Each brush is fixed and presses against a continuous slip ring that rotates with the armature.) After one-half revolution, wire b will be where wire a is now in the drawing, and the current then at brush b will be inward. Thus the current produced is alternating.

FIGURE 29–15 An ac generator.

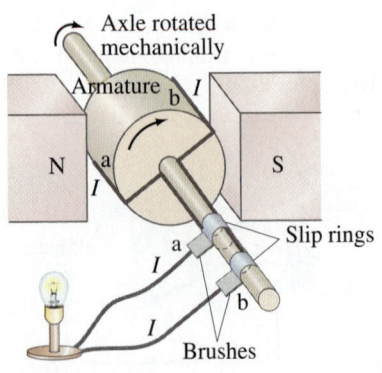

Let us assume the loop is being made to rotate in a uniform magnetic field $\vec{B}$ with constant angular velocity ω. From Faraday's law (Eq. 29–2a), the induced emf is

$$\mathscr{E} = -\frac{d\Phi_B}{dt} = -\frac{d}{dt}\int \vec{B}\cdot d\vec{A} = -\frac{d}{dt}[BA\cos\theta]$$

where A is the area of the loop and θ is the angle between $\vec{B}$ and $\vec{A}$. Since $\omega = d\theta/dt$, then $\theta = \theta_0 + \omega t$. We arbitrarily take $\theta_0 = 0$, so

$$\mathscr{E} = -BA\frac{d}{dt}(\cos\omega t) = BA\omega\sin\omega t.$$

If the rotating coil contains N loops,

$$\mathscr{E} = NBA\omega\sin\omega t$$
$$= \mathscr{E}_0\sin\omega t. \qquad \textbf{(29–4)}$$

Thus the output emf is sinusoidal (Fig. 29–16) with amplitude $\mathscr{E}_0 = NBA\omega$. Such a rotating coil in a magnetic field is the basic operating principle of an ac generator.

The frequency $f\ (=\omega/2\pi)$ is 60 Hz for general use in the United States and Canada, whereas 50 Hz is used in many countries. Most of the power generated in the United States is done at steam plants, where the burning of fossil fuels (coal, oil, natural gas) boils water to produce high-pressure steam that turns a turbine connected to the generator axle. Falling water from the top of a dam (hydroelectric) is also common (Fig. 29–17). At nuclear power plants, the nuclear energy released is used to produce steam to turn turbines. Indeed, a heat engine (Chapter 20) connected to a generator is the principal means of generating electric power. The frequency of 60 Hz or 50 Hz is maintained very precisely by power companies, and in doing Problems, we will assume it is at least as precise as other numbers given.

EXAMPLE 29–9 An ac generator. The armature of a 60-Hz ac generator rotates in a 0.15-T magnetic field. If the area of the coil is $2.0 \times 10^{-2}\,\mathrm{m^2}$, how many loops must the coil contain if the peak output is to be $\mathscr{E}_0 = 170\,\mathrm{V}$?

APPROACH From Eq. 29–4 we see that the maximum emf is $\mathscr{E}_0 = NBA\omega$.

SOLUTION We solve Eq. 29–4 for N with $\omega = 2\pi f = (6.28)(60\,\mathrm{s^{-1}}) = 377\,\mathrm{s^{-1}}$:

$$N = \frac{\mathscr{E}_0}{BA\omega} = \frac{170\,\mathrm{V}}{(0.15\,\mathrm{T})(2.0\times10^{-2}\,\mathrm{m^2})(377\,\mathrm{s^{-1}})} = 150\ \text{turns}.$$

A **dc generator** is much like an ac generator, except the slip rings are replaced by split-ring commutators, Fig. 29–18a, just as in a dc motor (Section 27–6). The output of such a generator is as shown and can be smoothed out by placing a capacitor in parallel with the output (Section 26–5). More common is the use of many armature windings, as in Fig. 29–18b, which produces a smoother output.

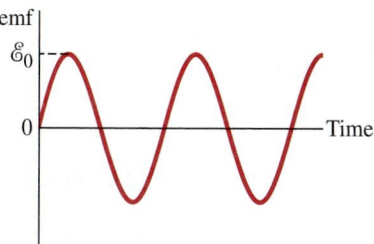
FIGURE 29–16 An ac generator produces an alternating current. The output emf $\mathscr{E} = \mathscr{E}_0\sin\omega t$, where $\mathscr{E}_0 = NAB\omega$ (Eq. 29–4).

PHYSICS APPLIED
Power plants

FIGURE 29–17 Water-driven generators at the base of Bonneville Dam, Oregon.

PHYSICS APPLIED
DC generator

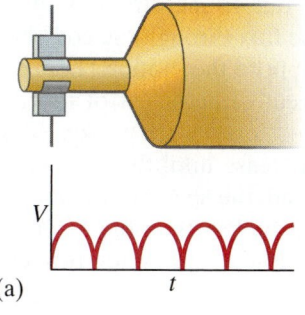
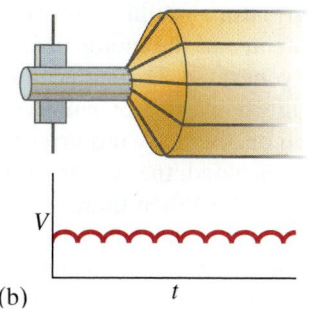

FIGURE 29–18 (a) A dc generator with one set of commutators, and (b) a dc generator with many sets of commutators and windings.

FIGURE 29–19 (a) Simplified schematic diagram of an alternator. The input current to the rotor from the battery is connected through continuous slip rings. Sometimes the rotor electromagnet is replaced by a permanent magnet. (b) Actual shape of an alternator. The rotor is made to turn by a belt from the engine. The current in the wire coil of the rotor produces a magnetic field inside it on its axis that points horizontally from left to right, thus making north and south poles of the plates attached at either end. These end plates are made with triangular fingers that are bent over the coil—hence there are alternating N and S poles quite close to one another, with magnetic field lines between them as shown by the blue lines. As the rotor turns, these field lines pass through the fixed stator coils (shown on the right for clarity, but in operation the rotor rotates within the stator), inducing a current in them, which is the output.

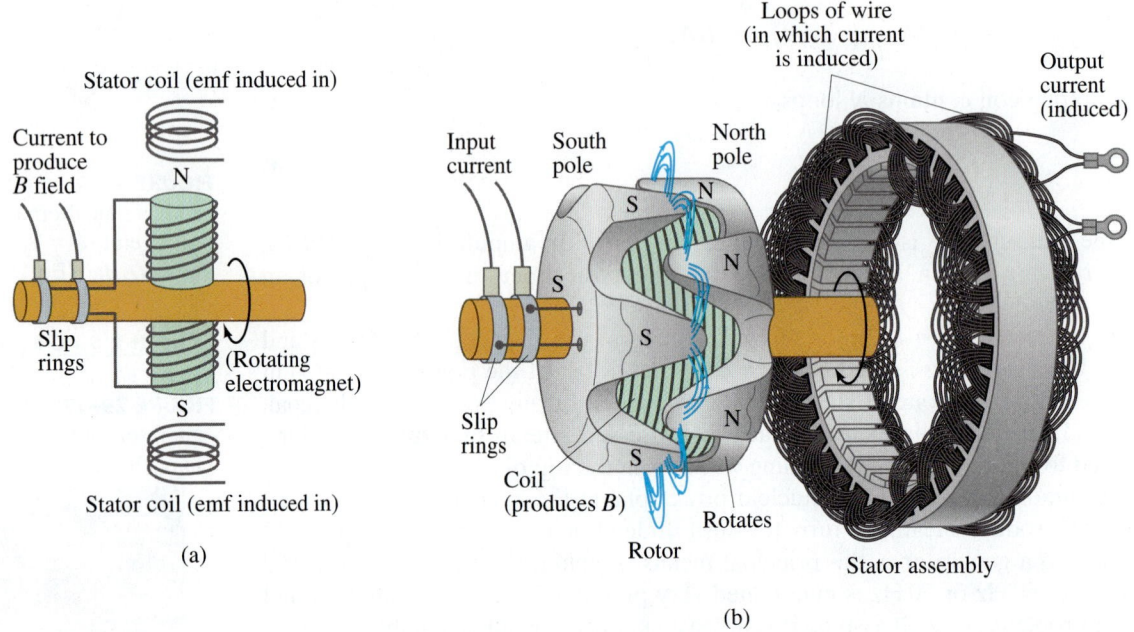

(a)

(b)

Automobiles used to use dc generators. Today they mainly use **alternators**, which avoid the problems of wear and electrical arcing (sparks) across the split-ring commutators of dc generators. Alternators differ from generators in that an electromagnet, called the *rotor*, is fed by current from the battery and is made to rotate by a belt from the engine. The magnetic field of the turning rotor passes through a surrounding set of stationary coils called the *stator* (Fig. 29–19), inducing an alternating current in the stator coils, which is the output. This ac output is changed to dc for charging the battery by the use of semiconductor diodes, which allow current flow in one direction only.

*29–5 Back EMF and Counter Torque; Eddy Currents

*Back EMF, in a Motor

A motor turns and produces mechanical energy when a current is made to flow in it. From our description in Section 27–6 of a simple dc motor, you might expect that the armature would accelerate indefinitely due to the torque on it. However, as the armature of the motor turns, the magnetic flux through the coil changes and an emf is generated. This induced emf acts to oppose the motion (Lenz's law) and is called the **back emf** or **counter emf**. The greater the speed of the motor, the greater the back emf. A motor normally turns and does work on something, but if there were no load, the motor's speed would increase until the back emf equaled the input voltage. When there is a mechanical load, the speed of the motor may be limited also by the load. The back emf will then be less than the external applied voltage. The greater the mechanical load, the slower the motor rotates and the lower is the back emf ($\mathscr{E} \propto \omega$, Eq. 29–4).

EXAMPLE 29–10 **Back emf in a motor.** The armature windings of a dc motor have a resistance of 5.0 Ω. The motor is connected to a 120-V line, and when the motor reaches full speed against its normal load, the back emf is 108 V. Calculate (a) the current into the motor when it is just starting up, and (b) the current when the motor reaches full speed.

APPROACH As the motor is just starting up, it is turning very slowly, so there is no induced back emf. The only voltage is the 120-V line. The current is given by Ohm's law with $R = 5.0\ \Omega$. At full speed, we must include as emfs both the 120-V applied emf and the opposing back emf.

SOLUTION (a) At start up, the current is controlled by the 120 V applied to the coil's 5.0-Ω resistance. By Ohm's law,

$$I = \frac{V}{R} = \frac{120\ \text{V}}{5.0\ \Omega} = 24\ \text{A}.$$

(b) When the motor is at full speed, the back emf must be included in the equivalent circuit shown in Fig. 29–20. In this case, Ohm's law (or Kirchhoff's rule) gives

$$120\ \text{V} - 108\ \text{V} = I(5.0\ \Omega).$$

Therefore

$$I = \frac{12\ \text{V}}{5.0\ \Omega} = 2.4\ \text{A}.$$

NOTE This result shows that the current can be very high when a motor first starts up. This is why the lights in your house may dim when the motor of the refrigerator (or other large motor) starts up. The large initial current causes the voltage to the lights and at the outlets to drop, since the house wiring has resistance and there is some voltage drop across it when large currents are drawn.

CONCEPTUAL EXAMPLE 29–11 **Motor overload.** When using an appliance such as a blender, electric drill, or sewing machine, if the appliance is overloaded or jammed so that the motor slows appreciably or stops while the power is still connected, the device can burn out and be ruined. Explain why this happens.

RESPONSE The motors are designed to run at a certain speed for a given applied voltage, and the designer must take the expected back emf into account. If the rotation speed is reduced, the back emf will not be as high as expected ($\mathcal{E} \propto \omega$, Eq. 29–4), and the current will increase, and may become large enough that the windings of the motor heat up to the point of ruining the motor.

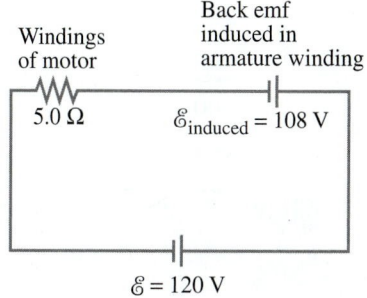

Back emf induced in armature winding

Windings of motor

5.0 Ω $\mathcal{E}_{\text{induced}} = 108\ \text{V}$

$\mathcal{E} = 120\ \text{V}$

FIGURE 29–20 Circuit of a motor showing induced back emf. Example 29–10.

PHYSICS APPLIED
Burning out a motor

*Counter Torque

In a generator, the situation is the reverse of that for a motor. As we saw, the mechanical turning of the armature induces an emf in the loops, which is the output. If the generator is not connected to an external circuit, the emf exists at the terminals but there is no current. In this case, it takes little effort to turn the armature. But if the generator *is* connected to a device that draws current, then a current flows in the coils of the armature. Because this current-carrying coil is in an external magnetic field, there will be a torque exerted on it (as in a motor), and this torque opposes the motion (use the right-hand rule for the force on a wire in Fig. 29–15). This is called a **counter torque**. The greater the electrical load—that is, the more current that is drawn—the greater will be the counter torque. Hence the external applied torque will have to be greater to keep the generator turning. This makes sense from the conservation of energy principle. More mechanical-energy input is needed to produce more electrical-energy output.

EXERCISE D A bicycle headlight is powered by a generator that is turned by the bicycle wheel. (a) If you pedal faster, how does the power to the light change? (b) Does the generator resist being turned as the bicycle's speed increases, and if so how?

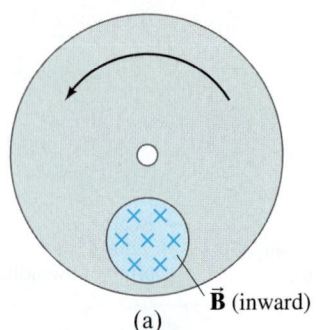

(a)

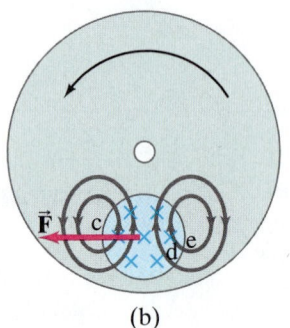

(b)

FIGURE 29–21 Production of eddy currents in a rotating wheel. The grey lines in (b) indicate induced current.

FIGURE 29–22 Airport metal detector.

🚶 **PHYSICS APPLIED**
Airport metal detector

FIGURE 29–23 Repairing a step-down transformer on a utility pole.

*Eddy Currents

Induced currents are not always confined to well-defined paths such as in wires. Consider, for example, the rotating metal wheel in Fig. 29–21a. An external magnetic field is applied to a limited area of the wheel as shown and points into the page. The section of wheel in the magnetic field has an emf induced in it because the conductor is moving, carrying electrons with it. The flow of induced (conventional) current in the wheel is upward in the region of the magnetic field (Fig. 29–21b), and the current follows a downward return path outside that region. Why? According to Lenz's law, the induced currents oppose the change that causes them. Consider the part of the wheel labeled c in Fig. 29–21b, where the magnetic field is zero but is just about to enter a region where $\vec{\mathbf{B}}$ points into the page. To oppose this inward increase in magnetic field, the induced current is counterclockwise to produce a field pointing out of the page (right-hand-rule 1). Similarly, region d is about to move to e, where $\vec{\mathbf{B}}$ is zero; hence the current is clockwise to produce an inward field opposed to this decreasing flux inward. These currents are referred to as **eddy currents**. They can be present in any conductor that is moving across a magnetic field or through which the magnetic flux is changing.

In Fig. 29–21b, the magnetic field exerts a force $\vec{\mathbf{F}}$ on the induced currents it has created, and that force opposes the rotational motion. Eddy currents can be used in this way as a smooth braking device on, say, a rapid-transit car. In order to stop the car, an electromagnet can be turned on that applies its field either to the wheels or to the moving steel rail below. Eddy currents can also be used to dampen (reduce) the oscillation of a vibrating system. Eddy currents, however, can be a problem. For example, eddy currents induced in the armature of a motor or generator produce heat ($P = I\mathscr{E}$) and waste energy. To reduce the eddy currents, the armatures are *laminated*; that is, they are made of very thin sheets of iron that are well insulated from one another. The total path length of the eddy currents is confined to each slab, which increases the total resistance; hence the current is less and there is less wasted energy.

Walk-through metal detectors at airports (Fig. 29–22) detect metal objects using electromagnetic induction and eddy currents. Several coils are situated in the walls of the walk-through at different heights. In a technique called "pulse induction," the coils are given repeated brief pulses of current (on the order of microseconds), hundreds or thousands of times a second. Each pulse in a coil produces a magnetic field for a very brief period of time. When a passenger passes through the walk-through, any metal object being carried will have eddy currents induced in it. The eddy currents persist briefly after each input pulse, and the small magnetic field produced by the persisting eddy current (before the next external pulse) can be detected, setting off an alert or alarm. Stores and libraries sometimes use similar systems to discourage theft.

29–6 Transformers and Transmission of Power

A transformer is a device for increasing or decreasing an ac voltage. Transformers are found everywhere: on utility poles (Fig. 29–23) to reduce the high voltage from the electric company to a usable voltage in houses (120 V or 240 V), in chargers for cell phones, laptops, and other electronic devices, in CRT monitors and in your car to give the needed high voltage (to the spark plugs), and in many other applications. A **transformer** consists of two coils of wire known as the **primary** and **secondary** coils. The two coils can be interwoven (with insulated wire); or they can be linked by an iron core which is laminated to minimize eddy-current losses (Section 29–5), as shown in Fig. 29–24. Transformers are designed so that (nearly) all the magnetic flux produced by the current in the primary coil also passes through the secondary coil, and we assume this is true in what follows. We also assume that energy losses (in resistance and hysteresis) can be ignored—a good approximation for real transformers, which are often better than 99% efficient.

When an ac voltage is applied to the primary coil, the changing magnetic field it produces will induce an ac voltage of the same frequency in the secondary coil. However, the voltage will be different according to the number of loops in each coil. From Faraday's law, the voltage or emf induced in the secondary coil is

$$V_S = N_S \frac{d\Phi_B}{dt},$$

where N_S is the number of turns in the secondary coil, and $d\Phi_B/dt$ is the rate at which the magnetic flux changes.

The input primary voltage, V_P, is related to the rate at which the flux changes through it,

$$V_P = N_P \frac{d\Phi_B}{dt},$$

where N_P is the number of turns in the primary coil. This follows because the changing flux produces a back emf, $N_P \, d\Phi_B/dt$, in the primary that exactly balances the applied voltage V_P if the resistance of the primary can be ignored (Kirchhoff's rules). We divide these two equations, assuming little or no flux is lost, to find

$$\frac{V_S}{V_P} = \frac{N_S}{N_P}. \qquad \textbf{(29–5)}$$

This **transformer equation** tells how the secondary (output) voltage is related to the primary (input) voltage; V_S and V_P in Eq. 29–5 can be the rms values (Section 25–7) for both, or peak values for both. DC voltages don't work in a transformer because there would be no changing magnetic flux.

If the secondary coil contains more loops than the primary coil $(N_S > N_P)$, we have a **step-up transformer**. The secondary voltage is greater than the primary voltage. For example, if the secondary coil has twice as many turns as the primary coil, then the secondary voltage will be twice that of the primary voltage. If N_S is less than N_P, we have a **step-down transformer**.

Although ac voltage can be increased (or decreased) with a transformer, we don't get something for nothing. Energy conservation tells us that the power output can be no greater than the power input. A well-designed transformer can be greater than 99% efficient, so little energy is lost to heat. The power output thus essentially equals the power input. Since power $P = IV$ (Eq. 25–6), we have

$$I_P V_P = I_S V_S,$$

or

$$\frac{I_S}{I_P} = \frac{N_P}{N_S}. \qquad \textbf{(29–6)}$$

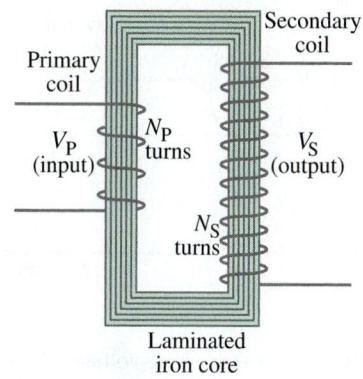

FIGURE 29–24 Step-up transformer $(N_P = 4, N_S = 12)$.

EXAMPLE 29–12 **Cell phone charger.** The charger for a cell phone contains a transformer that reduces 120-V (or 240-V) ac to 5.0-V ac to charge the 3.7-V battery (Section 26–4). (It also contains diodes to change the 5.0-V ac to 5.0-V dc.) Suppose the secondary coil contains 30 turns and the charger supplies 700 mA. Calculate (a) the number of turns in the primary coil, (b) the current in the primary, and (c) the power transformed.

APPROACH We assume the transformer is ideal, with no flux loss, so we can use Eq. 29–5 and then Eq. 29–6.

SOLUTION (a) This is a step-down transformer, and from Eq. 29–5 we have

$$N_P = N_S \frac{V_P}{V_S} = \frac{(30)(120 \text{ V})}{(5.0 \text{ V})} = 720 \text{ turns.}$$

(b) From Eq. 29–6

$$I_P = I_S \frac{N_S}{N_P} = (0.70 \text{ A})\left(\frac{30}{720}\right) = 29 \text{ mA.}$$

(c) The power transformed is

$$P = I_S V_S = (0.70 \text{ A})(5.0 \text{ V}) = 3.5 \text{ W.}$$

NOTE The power in the primary coil, $P = (0.029 \text{ A})(120 \text{ V}) = 3.5 \text{ W}$, is the same as the power in the secondary coil. There is 100% efficiency in power transfer for our ideal transformer.

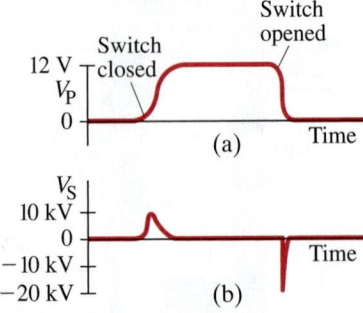

FIGURE 29–25 A dc voltage turned on and off as shown in (a) produces voltage pulses in the secondary (b). Voltage scales in (a) and (b) are not the same.

🏃 **PHYSICS APPLIED**

Transformers help power transmission

A transformer operates only on ac. A dc current in the primary coil does not produce a changing flux and therefore induces no emf in the secondary. However, if a dc voltage is applied to the primary through a switch, at the instant the switch is opened or closed there will be an induced voltage in the secondary. For example, if the dc is turned on and off as shown in Fig. 29–25a, the voltage induced in the secondary is as shown in Fig. 29–25b. Notice that the secondary voltage drops to zero when the dc voltage is steady. This is basically how, in the **ignition system** of an automobile, the high voltage is created to produce the spark across the gap of a spark plug that ignites the gas-air mixture. The transformer is referred to simply as an "ignition coil," and transforms the 12 V of the battery (when switched off in the primary) into a spike of as much as 30 kV in the secondary.

Transformers play an important role in the transmission of electricity. Power plants are often situated some distance from metropolitan areas, so electricity must then be transmitted over long distances (Fig. 29–26). There is always some power loss in the transmission lines, and this loss can be minimized if the power is transmitted at high voltage, using transformers, as the following Example shows.

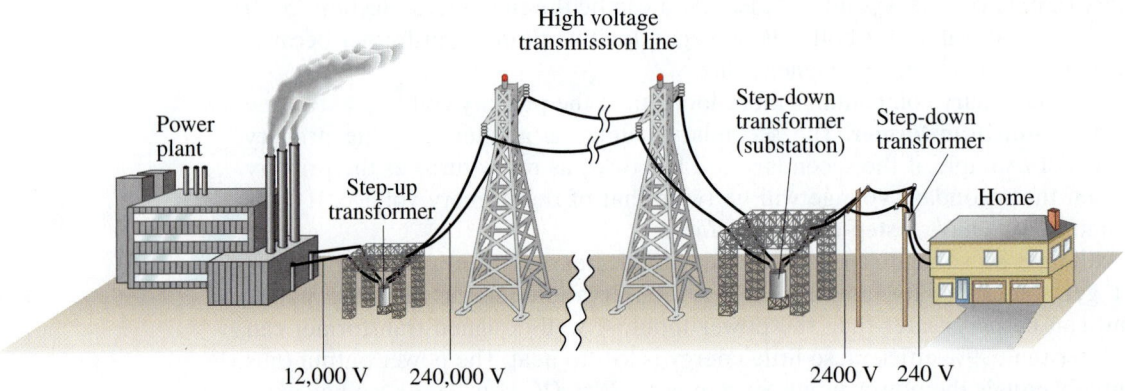

FIGURE 29–26 The transmission of electric power from power plants to homes makes use of transformers at various stages.

EXAMPLE 29–13 **Transmission lines.** An average of 120 kW of electric power is sent to a small town from a power plant 10 km away. The transmission lines have a total resistance of 0.40 Ω. Calculate the power loss if the power is transmitted at (*a*) 240 V and (*b*) 24,000 V.

APPROACH We cannot use $P = V^2/R$ because if R is the resistance of the transmission lines, we don't know the voltage drop along them; the given voltages are applied across the lines plus the load (the town). But we can determine the current I in the lines $(= P/V)$, and then find the power loss from $P_L = I^2R$, for both cases (*a*) and (*b*).

SOLUTION (*a*) If 120 kW is sent at 240 V, the total current will be

$$I = \frac{P}{V} = \frac{1.2 \times 10^5 \text{ W}}{2.4 \times 10^2 \text{ V}} = 500 \text{ A}.$$

The power loss in the lines, P_L, is then

$$P_L = I^2R = (500 \text{ A})^2(0.40 \text{ Ω}) = 100 \text{ kW}.$$

Thus, over 80% of all the power would be wasted as heat in the power lines!

(b) If 120 kW is sent at 24,000 V, the total current will be

$$I = \frac{P}{V} = \frac{1.2 \times 10^5 \, \text{W}}{2.4 \times 10^4 \, \text{V}} = 5.0 \, \text{A}.$$

The power loss in the lines is then

$$P_L = I^2 R = (5.0 \, \text{A})^2 (0.40 \, \Omega) = 10 \, \text{W},$$

which is less than $\frac{1}{100}$ of 1%: a far better efficiency!

NOTE We see that the higher voltage results in less current, and thus less power is wasted as heat in the transmission lines. It is for this reason that power is usually transmitted at very high voltages, as high as 700 kV.

The great advantage of ac, and a major reason it is in nearly universal use, is that the voltage can easily be stepped up or down by a transformer. The output voltage of an electric generating plant is stepped up prior to transmission. Upon arrival in a city, it is stepped down in stages at electric substations prior to distribution. The voltage in lines along city streets is typically 2400 V or 7200 V (but sometimes less), and is stepped down to 240 V or 120 V for home use by transformers (Figs. 29–23 and 29–26).

Fluorescent lights require a very high voltage initially to ionize the gas inside the bulb. The high voltage is obtained using a step-up transformer, called a ballast, and can be replaced independently of the bulb in many fluorescent light fixtures. When the ballast starts to fail, the tube is slow to light. Replacing the bulb will not solve the problem. In newer compact fluorescent bulbs designed to replace incandescent bulbs, the ballast (transformer) is part of the bulb, and is very small.

29–7 A Changing Magnetic Flux Produces an Electric Field

We have seen in earlier Chapters (especially Chapter 25, Section 25–8) that when an electric current flows in a wire, there is an electric field in the wire that does the work of moving the electrons in the wire. In this Chapter we have seen that a changing magnetic flux induces a current in the wire, which implies that there is an electric field in the wire induced by the changing magnetic flux. Thus we come to the important conclusion that

a changing magnetic flux produces an electric field.

This result applies not only to wires and other conductors, but is actually a general result that applies to any region in space. Indeed, an electric field will be produced at any point in space where there is a changing magnetic field.

Faraday's Law—General Form

We can put these ideas into mathematical form by generalizing our relation between an electric field and the potential difference between two points a and b: $V_{ab} = \int_a^b \vec{\mathbf{E}} \cdot d\vec{\boldsymbol{\ell}}$ (Eq. 23–4a) where $d\vec{\boldsymbol{\ell}}$ is an element of displacement along the path of integration. The emf $\mathscr{E}$ induced in a circuit is equal to the work done per unit charge by the electric field, which equals the integral of $\vec{\mathbf{E}} \cdot d\vec{\boldsymbol{\ell}}$ around the closed path:

$$\mathscr{E} = \oint \vec{\mathbf{E}} \cdot d\vec{\boldsymbol{\ell}}. \tag{29–7}$$

We combine this with Eq. 29–2a, to obtain a more elegant and general form of Faraday's law

$$\oint \vec{\mathbf{E}} \cdot d\vec{\boldsymbol{\ell}} = -\frac{d\Phi_B}{dt} \tag{29–8}$$

FARADAY'S LAW
(general form)

which relates the changing magnetic flux to the electric field it produces. The integral on the left is taken around a path enclosing the area through which the magnetic flux Φ_B is changing. This more elegant statement of Faraday's law (Eq. 29–8) is valid not only in conductors, but in any region of space. To illustrate this, let us take an Example.

FIGURE 29–27 Example 29–14.
(a) Side view of nearly constant $\vec{\mathbf{B}}$.
(b) Top view, for determining the
electric field $\vec{\mathbf{E}}$ at point P. (c) Lines of
$\vec{\mathbf{E}}$ produced by increasing $\vec{\mathbf{B}}$ (pointing
outward). (d) Graph of E vs. r.

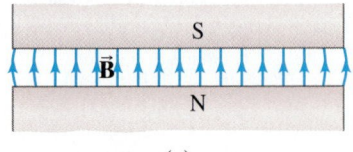

(a)

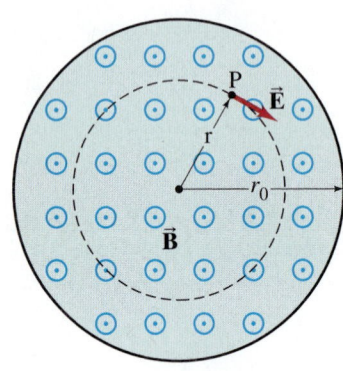

(b)

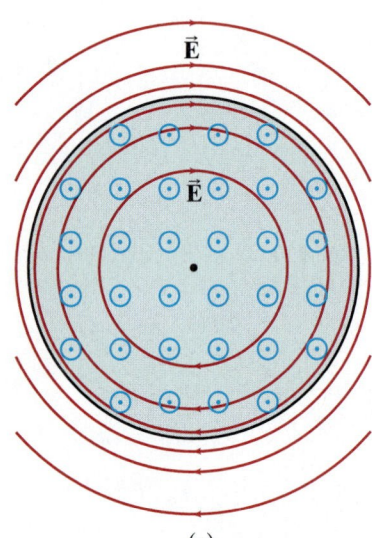

(c)

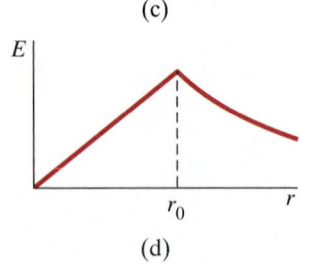

(d)

EXAMPLE 29–14 $\vec{\mathbf{E}}$ **produced by changing** $\vec{\mathbf{B}}$**.** A magnetic field $\vec{\mathbf{B}}$ between the pole faces of an electromagnet is nearly uniform at any instant over a circular area of radius r_0 as shown in Figs. 29–27a and b. The current in the windings of the electromagnet is increasing in time so that $\vec{\mathbf{B}}$ changes in time at a constant rate $d\vec{\mathbf{B}}/dt$ at each point. Beyond the circular region $(r > r_0)$, we assume $\vec{\mathbf{B}} = 0$ at all times. Determine the electric field $\vec{\mathbf{E}}$ at any point P a distance r from the center of the circular area due to the changing $\vec{\mathbf{B}}$.

APPROACH The changing magnetic flux through a circle of radius r, shown dashed in Fig. 29–27b, will produce an emf around this circle. Because all points on the dashed circle are equivalent physically, the electric field too will show this symmetry and will be in the plane perpendicular to $\vec{\mathbf{B}}$. Thus we can expect $\vec{\mathbf{E}}$ to be perpendicular to $\vec{\mathbf{B}}$ and to be tangent to the circle of radius r. The direction of $\vec{\mathbf{E}}$ will be as shown in Fig. 29–27b and c, since by Lenz's law the induced $\vec{\mathbf{E}}$ needs to be capable of producing a current that generates a magnetic field opposing the original change in $\vec{\mathbf{B}}$. By *symmetry*, we also expect $\vec{\mathbf{E}}$ to have the same magnitude at all points on the circle of radius r.

SOLUTION We take the circle shown in Fig 29–27b as our path of integration in Eq. 29–8. We ignore the minus sign so we can concentrate on magnitude since we already found the direction of $\vec{\mathbf{E}}$ from Lenz's law, and obtain

$$E(2\pi r) = (\pi r^2)\frac{dB}{dt}, \qquad [r < r_0]$$

since $\Phi_B = BA = B(\pi r^2)$ at any instant. We solve for E and obtain

$$E = \frac{r}{2}\frac{dB}{dt}. \qquad [r < r_0]$$

This expression is valid up to the edge of the circle $(r \le r_0)$, beyond which $\vec{\mathbf{B}} = 0$. If we now consider a point where $r > r_0$, the flux through a circle of radius r is $\Phi_B = \pi r_0^2 B$. Then Eq. 29–8 gives

$$E(2\pi r) = \pi r_0^2 \frac{dB}{dt} \qquad [r > r_0]$$

or

$$E = \frac{r_0^2}{2r}\frac{dB}{dt}. \qquad [r > r_0]$$

Thus the magnitude of the induced electric field increases linearly from zero at the center of the magnet to $E = (dB/dt)(r_0/2)$ at the edge, and then decreases inversely with distance in the region beyond the edge of the magnetic field. The electric field lines are circles as shown in Fig. 29–27c. A graph of E vs. r is shown in Fig. 29–27d.

EXERCISE E Consider the magnet shown in Fig. 29–27 with a radius $r_0 = 6.0$ cm. If the magnetic field changes uniformly from 0.040 T to 0.090 T in 0.18 s, what is the magnitude of the resulting electric field at (a) $r = 3.0$ cm and (b) $r = 9.0$ cm?

*Forces Due to Changing $\vec{\mathbf{B}}$ are Nonconservative

Example 29–14 illustrates an important difference between electric fields produced by changing magnetic fields and electric fields produced by electric charges at rest (electrostatic fields). Electric field lines produced in the electrostatic case (Chapters 21 to 24) start and stop on electric charges. But the electric field lines produced by a changing magnetic field are continuous; they form closed loops. This distinction goes even further and is an important one. In the electrostatic case, the potential difference between two points is given by (Eq. 23–4a)

$$V_{ba} = V_b - V_a = -\int_a^b \vec{\mathbf{E}} \cdot d\vec{\boldsymbol{\ell}}.$$

If the integral is around a closed loop, so points a and b are the same, then $V_{ba} = 0$.

Hence the integral of $\vec{\mathbf{E}} \cdot d\vec{\boldsymbol{\ell}}$ around a closed path is zero:

$$\oint \vec{\mathbf{E}} \cdot d\vec{\boldsymbol{\ell}} = 0. \qquad\qquad \text{[electrostatic field]}$$

This followed from the fact that the electrostatic force (Coulomb's law) is a conservative force, and so a potential energy function could be defined. Indeed, the relation above, $\oint \vec{\mathbf{E}} \cdot d\vec{\boldsymbol{\ell}} = 0$, tells us that the work done per unit charge around any closed path is zero (or the work done between any two points is independent of path—see Chapter 8), which is a property only of a conservative force. But in the nonelectrostatic case, when the electric field is produced by a changing magnetic field, the integral around a closed path is *not* zero, but is given by Eq. 29–8:

$$\oint \vec{\mathbf{E}} \cdot d\vec{\boldsymbol{\ell}} = -\frac{d\Phi_B}{dt}.$$

We thus come to the conclusion that the forces due to changing magnetic fields are *nonconservative*. We are not able therefore to define a potential energy, or potential function, at a given point in space for the nonelectrostatic case. Although static electric fields are *conservative fields*, the electric field produced by a changing magnetic field is a **nonconservative field**.

*29–8 Applications of Induction: Sound Systems, Computer Memory, Seismograph, GFCI

*Microphone

There are various types of *microphones*, and many operate on the principle of induction. In one form, a microphone is just the inverse of a loudspeaker (Section 27–6). A small coil connected to a membrane is suspended close to a small permanent magnet, as shown in Fig. 29–28. The coil moves in the magnetic field when sound waves strike the membrane and this motion induces an emf. The frequency of the induced emf will be just that of the impinging sound waves, and this emf is the "signal" that can be amplified and sent to loudspeakers, or sent to a recorder.

*Read/Write on Tape and Disks

Recording and playback on tape or disks is done by magnetic *heads*. Recording tapes for use in audio and video tape recorders contain a thin layer of magnetic oxide on a thin plastic tape. During recording, the audio and/or video signal voltage is sent to the recording head, which acts as a tiny electromagnet (Fig. 29–29) that magnetizes the tiny section of tape passing over the narrow gap in the head at each instant. In playback, the changing magnetism of the moving tape at the gap causes corresponding changes in the magnetic field within the soft-iron head, which in turn induces an emf in the coil (Faraday's law). This induced emf is the output signal that can be amplified and sent to a loudspeaker (audio) or to the picture tube (video). In audio and video recorders, the signals may be *analog*—they vary continuously in amplitude over time. The variation in degree of magnetization of the tape at any point reflects the variation in amplitude and frequency of the audio or video signal.

Digital information, such as used on computer hard drives or on magnetic computer tape and some types of digital tape recorders, is read and written using heads that are basically the same as just described (Fig. 29–29). The essential difference is in the signals, which are not analog, but are digital, and in particular binary, meaning that only two values are possible for each of the extremely high number of predetermined spaces on the tape or disk. The two possible values are usually referred to as 1 and 0. The signal voltage does not vary continuously but rather takes on only two values, +5 V and 0 V, for example, corresponding to the 1 or 0. Thus, information is carried as a series of **bits**, each of which can have only one of two values, 1 or 0.

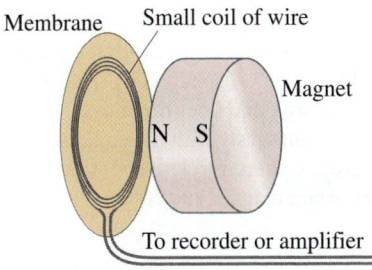

FIGURE 29–28 Diagram of a microphone that works by induction.

FIGURE 29–29 (a) Read/Write (playback/recording) head for tape or disk. In writing or recording, the electric input signal to the head, which acts as an electromagnet, magnetizes the passing tape or disk. In reading or playback, the changing magnetic field of the passing tape or disk induces a changing magnetic field in the head, which in turn induces in the coil an emf that is the output signal.
(b) Photo of a hard drive showing several platters and read/write heads that can quickly move from the edge of the disk to the center.

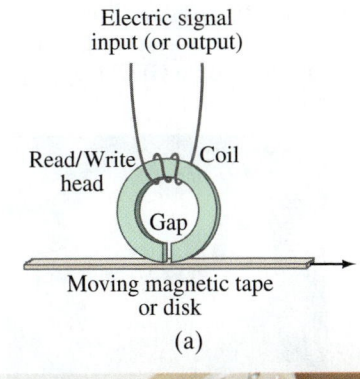

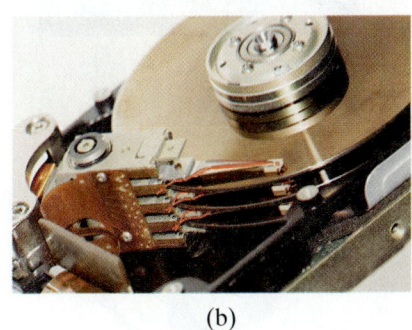

(b)

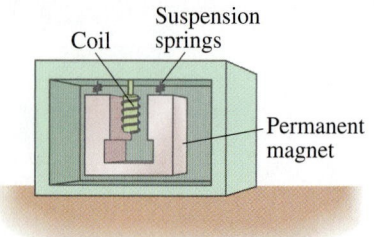

Permanent
magnet

FIGURE 29–30 One type of seismograph, in which the coil is fixed to the case and moves with the Earth. The magnet, suspended by springs, has inertia and does not move instantaneously with the coil (and case), so there is relative motion between magnet and coil.

*Credit Card Swipe

When you swipe your credit card at a store or gas station, the magnetic stripe on the back of the card passes over a read head just as in a tape recorder or computer. The magnetic stripe contains personal information about your account and connects by telephone line for approval if your account is in order.

*Seismograph

In geophysics, a **seismograph** measures the intensity of earthquake waves using a magnet and a coil of wire. Either the magnet or the coil is fixed to the case, and the other is inertial (suspended by a spring; Fig. 29–30). The relative motion of magnet and coil when the Earth shakes induces an emf output.

*Ground Fault Circuit Interrupter (GFCI)

Fuses and circuit breakers (Sections 25–6 and 28–8) protect buildings from fire, and apparatus from damage, due to undesired high currents. But they do not turn off the current until it is very much greater than that which causes permanent damage to humans or death ($\approx 100\,\text{mA}$). If fast enough, they may protect in case of a short. A *ground fault circuit interrupter* (GFCI) is meant to protect humans; GFCIs can react to currents as small as 5 mA.

FIGURE 29–31 A ground fault circuit interrupter (GFCI).

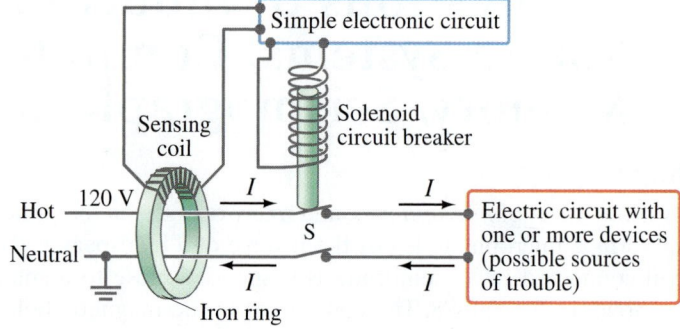

FIGURE 29–32 (a) A GFCI wall outlet. GFCIs can be recognized because they have "test" and "reset" buttons. (b) Add-on GFCI that plugs into outlet.

(a)

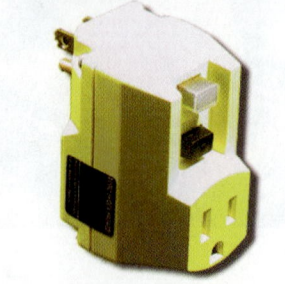

(b)

Electromagnetic induction is the physical basis of a GFCI. As shown in Fig. 29–31, the two conductors of a power line leading to an electrical circuit or device (red) pass through a small iron ring. Around the ring are many loops of thin wire that serve as a sensing coil. Under normal conditions (no ground fault), the current moving in the hot wire is exactly balanced by the returning current in the neutral wire. If something goes wrong and the hot wire touches the ungrounded metal case of the device or appliance, some of the entering current can pass through a person who touches the case and then to ground (a *ground fault*). Then the return current in the neutral wire will be less than the entering current in the hot wire, so there is a *net current* passing through the GFCI's iron ring. Because the current is ac, it is changing and the current difference produces a changing magnetic field in the iron, thus inducing an emf in the sensing coil wrapped around the iron. For example, if a device draws 8.0 A, and there is a ground fault through a person of 100 mA ($= 0.1$ A), then 7.9 A will appear in the neutral wire. The emf induced in the sensing coil by this 100-mA difference is amplified by a simple transistor circuit and sent to its own solenoid circuit breaker that opens the circuit at the switch S, thus protecting your life.

If the case of the faulty device is grounded, the current difference is even higher when there is a fault, and the GFCI trips immediately.

GFCIs can sense currents as low as 5 mA and react in 1 msec, saving lives. They can be small enough to fit as a wall outlet (Fig. 29–32a), or as a plug-in unit into which you plug a hair dryer or toaster (Fig. 29–32b). It is especially important to have GFCIs installed in kitchens, bathrooms, outdoors, and near swimming pools, where people are most in danger of touching ground. GFCIs always have a "test" button (to be sure it works) and a "reset" button (after it goes off).

Summary

The **magnetic flux** passing through a loop is equal to the product of the area of the loop times the perpendicular component of the (uniform) magnetic field: $\Phi_B = B_\perp A = BA \cos\theta$. If $\vec{B}$ is not uniform, then

$$\Phi_B = \int \vec{B} \cdot d\vec{A}. \tag{29-1b}$$

If the magnetic flux through a coil of wire changes in time, an emf is induced in the coil. The magnitude of the induced emf equals the time rate of change of the magnetic flux through the loop times the number N of loops in the coil:

$$\mathscr{E} = -N\frac{d\Phi_B}{dt}. \tag{29-2b}$$

This is **Faraday's law of induction**.

The induced emf can produce a current whose magnetic field opposes the original change in flux (**Lenz's law**).

We can also see from Faraday's law that a straight wire of length ℓ moving with speed v perpendicular to a magnetic field of strength B has an emf induced between its ends equal to:

$$\mathscr{E} = B\ell v. \tag{29-3}$$

Faraday's law also tells us that *a changing magnetic field produces an electric field.* The mathematical relation is

$$\oint \vec{E} \cdot d\vec{\ell} = -\frac{d\Phi_B}{dt} \tag{29-8}$$

and is the general form of Faraday's law. The integral on the left is taken around the loop through which the magnetic flux Φ_B is changing.

An electric **generator** changes mechanical energy into electrical energy. Its operation is based on Faraday's law: a coil of wire is made to rotate uniformly by mechanical means in a magnetic field, and the changing flux through the coil induces a sinusoidal current, which is the output of the generator.

[*A motor, which operates in the reverse of a generator, acts like a generator in that a **back emf** is induced in its rotating coil; since this counter emf opposes the input voltage, it can act to limit the current in a motor coil. Similarly, a generator acts somewhat like a motor in that a **counter torque** acts on its rotating coil.]

A **transformer**, which is a device to change the magnitude of an ac voltage, consists of a primary coil and a secondary coil. The changing flux due to an ac voltage in the primary coil induces an ac voltage in the secondary coil. In a 100% efficient transformer, the ratio of output to input voltages (V_S/V_P) equals the ratio of the number of turns N_S in the secondary to the number N_P in the primary:

$$\frac{V_S}{V_P} = \frac{N_S}{N_P}. \tag{29-5}$$

The ratio of secondary to primary current is in the inverse ratio of turns:

$$\frac{I_S}{I_P} = \frac{N_P}{N_S}. \tag{29-6}$$

[*Microphones, ground fault circuit interrupters, seismographs, and read/write heads for computer drives and tape recorders are applications of electromagnetic induction.]

Questions

1. What would be the advantage, in Faraday's experiments (Fig. 29–1), of using coils with many turns?

2. What is the difference between magnetic flux and magnetic field?

3. Suppose you are holding a circular ring of wire and suddenly thrust a magnet, south pole first, away from you toward the center of the circle. Is a current induced in the wire? Is a current induced when the magnet is held steady within the ring? Is a current induced when you withdraw the magnet? In each case, if your answer is yes, specify the direction.

4. Two loops of wire are moving in the vicinity of a very long straight wire carrying a steady current as shown in Fig. 29–33. Find the direction of the induced current in each loop.

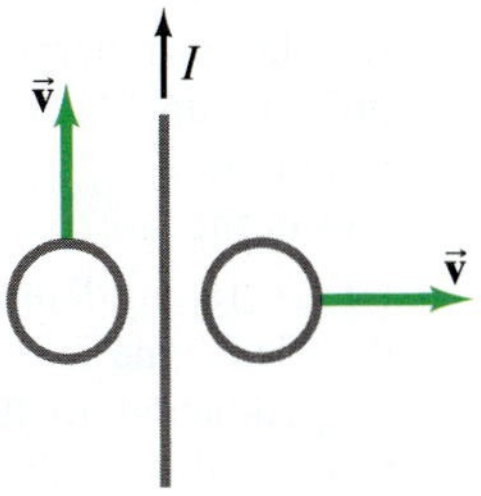

FIGURE 29–33
Questions 4 and 5.

5. Is there a force between the two loops discussed in Question 4? If so, in what direction?

6. Suppose you are looking along a line through the centers of two circular (but separate) wire loops, one behind the other. A battery is suddenly connected to the front loop, establishing a clockwise current. (a) Will a current be induced in the second loop? (b) If so, when does this current start? (c) When does it stop? (d) In what direction is this current? (e) Is there a force between the two loops? (f) If so, in what direction?

7. The battery mentioned in Question 6 is disconnected. Will a current be induced in the second loop? If so, when does it start and stop? In what direction is this current?

8. In what direction will the current flow in Fig. 29–12a if the rod moves to the left, which decreases the area of the loop to the left?

9. In Fig. 29–34, determine the direction of the induced current in resistor R_A (a) when coil B is moved toward coil A, (b) when coil B is moved away from A, (c) when the resistance R_B is increased.

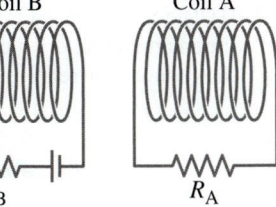

FIGURE 29–34
Question 9.

10. In situations where a small signal must travel over a distance, a *shielded cable* is used in which the signal wire is surrounded by an insulator and then enclosed by a cylindrical conductor carrying the return current (Fig. 28–12). Why is a "shield" necessary?

11. What is the advantage of placing the two insulated electric wires carrying ac close together or even twisted about each other?

12. Which object will fall faster in a nonuniform magnetic field, a conducting loop with radius ℓ or a straight wire of length $\ell/2$?

13. A region where no magnetic field is desired is surrounded by a sheet of low-resistivity metal. (*a*) Will this sheet shield the interior from a rapidly changing magnetic field outside? Explain. (*b*) Will it act as a shield to a static magnetic field? (*c*) What if the sheet is superconducting (resistivity = 0)?

14. A cell phone charger contains a transformer. Why can't you just buy one universal charger to charge your old cell phone, your new cell phone, your drill, and your toy electric train?

15. An enclosed transformer has four wire leads coming from it. How could you determine the ratio of turns on the two coils without taking the transformer apart? How would you know which wires paired with which?

16. The use of higher-voltage lines in homes—say, 600 V or 1200 V—would reduce energy waste. Why are they not used?

17. A transformer designed for a 120-V ac input will often "burn out" if connected to a 120-V dc source. Explain. [*Hint*: The resistance of the primary coil is usually very low.]

*18. Explain why, exactly, the lights may dim briefly when a refrigerator motor starts up. When an electric heater is turned on, the lights may stay dimmed as long as the heater is on. Explain the difference.

*19. Use Fig. 29–15 plus the right-hand rules to show why the counter torque in a generator *opposes* the motion.

*20. Will an eddy current brake (Fig. 29–21) work on a copper or aluminum wheel, or must the wheel be ferromagnetic? Explain.

*21. It has been proposed that eddy currents be used to help sort solid waste for recycling. The waste is first ground into tiny pieces and iron removed with a dc magnet. The waste then is allowed to slide down an incline over permanent magnets. How will this aid in the separation of nonferrous metals (Al, Cu, Pb, brass) from nonmetallic materials?

*22. The pivoted metal bar with slots in Fig. 29–35 falls much more quickly through a magnetic field than does a solid bar. Explain.

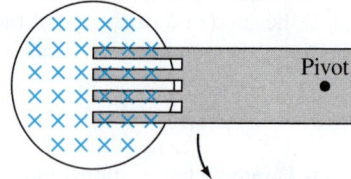

FIGURE 29–35
Question 22.

*23. If an aluminum sheet is held between the poles of a large bar magnet, it requires some force to pull it out of the magnetic field even though the sheet is not ferromagnetic and does not touch the pole faces. Explain.

*24. A bar magnet falling inside a vertical metal tube reaches a terminal velocity even if the tube is evacuated so that there is no air resistance. Explain.

*25. A metal bar, pivoted at one end, oscillates freely in the absence of a magnetic field; but in a magnetic field, its oscillations are quickly damped out. Explain. (This *magnetic damping* is used in a number of practical devices.)

*26. Since a magnetic microphone is basically like a loudspeaker, could a loudspeaker (Section 27–6) actually serve as a microphone? That is, could you speak into a loudspeaker and obtain an output signal that could be amplified? Explain. Discuss, in light of your response, how a microphone and loudspeaker differ in construction.

Problems

29–1 and 29–2 Faraday's Law of Induction

1. (I) The magnetic flux through a coil of wire containing two loops changes at a constant rate from −58 Wb to +38 Wb in 0.42 s. What is the emf induced in the coil?

2. (I) The north pole of the magnet in Fig. 29–36 is being inserted into the coil. In which direction is the induced current flowing through the resistor *R*?

FIGURE 29–36
Problem 2.

3. (I) The rectangular loop shown in Fig. 29–37 is pushed into the magnetic field which points inward. In what direction is the induced current?

FIGURE 29–37
Problem 3.

4. (I) A 22.0-cm-diameter loop of wire is initially oriented perpendicular to a 1.5-T magnetic field. The loop is rotated so that its plane is parallel to the field direction in 0.20 s. What is the average induced emf in the loop?

5. (II) A circular wire loop of radius $r = 12$ cm is immersed in a uniform magnetic field $B = 0.500$ T with its plane normal to the direction of the field. If the field magnitude then decreases at a constant rate of -0.010 T/s, at what rate should r increase so that the induced emf within the loop is zero?

6. (II) A 10.8-cm-diameter wire coil is initially oriented so that its plane is perpendicular to a magnetic field of 0.68 T pointing up. During the course of 0.16 s, the field is changed to one of 0.25 T pointing down. What is the average induced emf in the coil?

7. (II) A 16-cm-diameter circular loop of wire is placed in a 0.50-T magnetic field. (*a*) When the plane of the loop is perpendicular to the field lines, what is the magnetic flux through the loop? (*b*) The plane of the loop is rotated until it makes a 35° angle with the field lines. What is the angle θ in Eq. 29–1a for this situation? (*c*) What is the magnetic flux through the loop at this angle?

8. (II) (*a*) If the resistance of the resistor in Fig. 29–38 is slowly increased, what is the direction of the current induced in the small circular loop inside the larger loop? (*b*) What would it be if the small loop were placed outside the larger one, to the left?

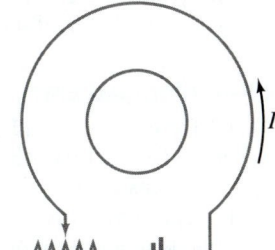

FIGURE 29–38
Problem 8.

9. (II) If the solenoid in Fig. 29–39 is being pulled away from the loop shown, in what direction is the induced current in the loop?

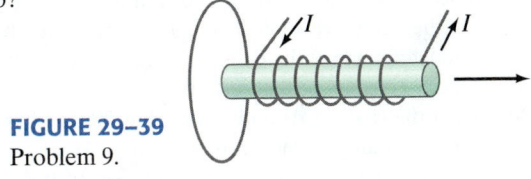

FIGURE 29–39
Problem 9.

10. (II) The magnetic field perpendicular to a circular wire loop 8.0 cm in diameter is changed from +0.52 T to −0.45 T in 180 ms, where + means the field points away from an observer and − toward the observer. (*a*) Calculate the induced emf. (*b*) In what direction does the induced current flow?

11. (II) A circular loop in the plane of the paper lies in a 0.75-T magnetic field pointing into the paper. If the loop's diameter changes from 20.0 cm to 6.0 cm in 0.50 s, (*a*) what is the direction of the induced current, (*b*) what is the magnitude of the average induced emf, and (*c*) if the coil resistance is 2.5 Ω, what is the average induced current?

12. (II) Part of a single rectangular loop of wire with dimensions shown in Fig. 29–40 is situated inside a region of uniform magnetic field of 0.650 T. The total resistance of the loop is 0.280 Ω. Calculate the force required to pull the loop from the field (to the right) at a constant velocity of 3.40 m/s. Neglect gravity.

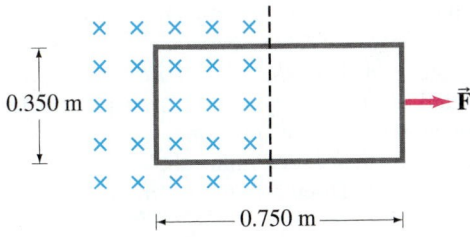

FIGURE 29–40   Problem 12.

13. (II) While demonstrating Faraday's law to her class, a physics professor inadvertently moves the gold ring on her finger from a location where a 0.80-T magnetic field points along her finger to a zero-field location in 45 ms. The 1.5-cm-diameter ring has a resistance and mass of 55 $\mu\Omega$ and 15 g, respectively. (*a*) Estimate the thermal energy produced in the ring due to the flow of induced current. (*b*) Find the temperature rise of the ring, assuming all of the thermal energy produced goes into increasing the ring's temperature. The specific heat of gold is 129 J/kg·C°.

14. (II) A 420-turn solenoid, 25 cm long, has a diameter of 2.5 cm. A 15-turn coil is wound tightly around the center of the solenoid. If the current in the solenoid increases uniformly from 0 to 5.0 A in 0.60 s, what will be the induced emf in the short coil during this time?

15. (II) A 22.0-cm-diameter coil consists of 28 turns of circular copper wire 2.6 mm in diameter. A uniform magnetic field, perpendicular to the plane of the coil, changes at a rate of 8.65×10^{-3} T/s. Determine (*a*) the current in the loop, and (*b*) the rate at which thermal energy is produced.

16. (II) A power line carrying a sinusoidally varying current with frequency $f = 60$ Hz and peak value $I_0 = 55$ kA runs at a height of 7.0 m across a farmer's land (Fig. 29–41). The farmer constructs a vertically oriented 2.0-m-high 10-turn rectangular wire coil below the power line. The farmer hopes to use the induced voltage in this coil to power 120-Volt electrical equipment, which requires a sinusoidally varying voltage with frequency $f = 60$ Hz and peak value $V_0 = 170$ V. What should the length ℓ of the coil be? Would this be unethical?

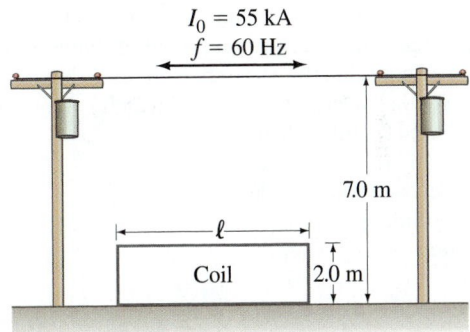

FIGURE 29–41   Problem 16.

17. (II) The magnetic field perpendicular to a single 18.2-cm-diameter circular loop of copper wire decreases uniformly from 0.750 T to zero. If the wire is 2.35 mm in diameter, how much charge moves past a point in the coil during this operation?

18. (II) The magnetic flux through each loop of a 75-loop coil is given by $(8.8t - 0.51t^3) \times 10^{-2}$ T·m², where the time t is in seconds. (*a*) Determine the emf $\mathcal{E}$ as a function of time. (*b*) What is $\mathcal{E}$ at $t = 1.0$ s and $t = 4.0$ s?

19. (II) A 25-cm-diameter circular loop of wire has a resistance of 150 Ω. It is initially in a 0.40-T magnetic field, with its plane perpendicular to $\vec{\mathbf{B}}$, but is removed from the field in 120 ms. Calculate the electric energy dissipated in this process.

20. (II) The area of an elastic circular loop decreases at a constant rate, $dA/dt = -3.50 \times 10^{-2}$ m²/s. The loop is in a magnetic field $B = 0.28$ T whose direction is perpendicular to the plane of the loop. At $t = 0$, the loop has area $A = 0.285$ m². Determine the induced emf at $t = 0$, and at $t = 2.00$ s.

21. (II) Suppose the radius of the elastic loop in Problem 20 increases at a constant rate, $dr/dt = 4.30$ cm/s. Determine the emf induced in the loop at $t = 0$ and at $t = 1.00$ s.

22. (II) A single circular loop of wire is placed inside a long solenoid with its plane perpendicular to the axis of the solenoid. The area of the loop is A_1 and that of the solenoid, which has n turns per unit length, is A_2. A current $I = I_0 \cos \omega t$ flows in the solenoid turns. What is the induced emf in the small loop?

23. (II) We are looking down on an elastic conducting loop with resistance $R = 2.0 \, \Omega$, immersed in a magnetic field. The field's magnitude is uniform spatially, but varies with time t according to $B(t) = \alpha t$, where $\alpha = 0.60 \, \text{T/s}$. The area A of the loop also increases at a constant rate, according to $A(t) = A_0 + \beta t$, where $A_0 = 0.50 \, \text{m}^2$ and $\beta = 0.70 \, \text{m}^2/\text{s}$. Find the magnitude and direction (clockwise or counterclockwise, when viewed from above the page) of the induced current within the loop at time $t = 2.0 \, \text{s}$ if the magnetic field (a) is parallel to the plane of the loop to the right; (b) is perpendicular to the plane of the loop, down.

24. (II) Inductive battery chargers, which allow transfer of electrical power without the need for exposed electrical contacts, are commonly used in appliances that need to be safely immersed in water, such as electric toothbrushes. Consider the following simple model for the power transfer in an inductive charger (Fig. 29–42). Within the charger's plastic base, a primary coil of diameter d with n_P turns per unit length is connected to a home's ac wall outlet so that a current $I = I_0 \sin(2\pi f t)$ flows within it. When the toothbrush is seated on the base, an N-turn secondary coil inside the toothbrush has a diameter only slightly greater than d and is centered on the primary. Find an expression for the emf induced in the secondary coil. [This induced emf recharges the battery.]

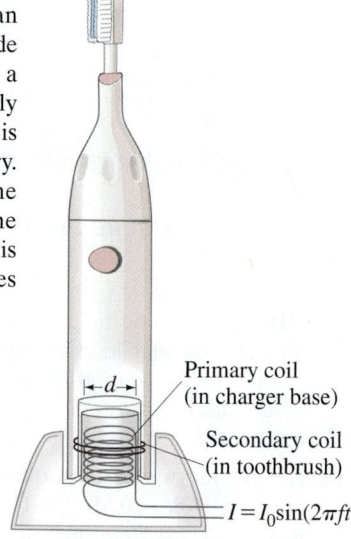

FIGURE 29–42
Problem 24.

25. (III) (a) Determine the magnetic flux through a square loop of side a (Fig. 29–43) if one side is parallel to, and a distance b from, a straight wire that carries a current I. (b) If the loop is pulled away from the wire at speed v, what emf is induced in it? (c) Does the induced current flow clockwise or counterclockwise? (d) Determine the force F required to pull the loop away.

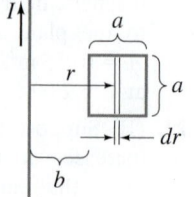

FIGURE 29–43
Problems 25 and 26.

26. (III) Determine the emf induced in the square loop in Fig. 29–43 if the loop stays at rest and the current in the straight wire is given by $I(t) = (15.0 \, \text{A}) \sin(2500 t)$ where t is in seconds. The distance a is 12.0 cm, and b is 15.0 cm.

29–3 Motional EMF

27. (I) The moving rod in Fig. 29–12b is 13.2 cm long and generates an emf of 120 mV while moving in a 0.90-T magnetic field. What is its speed?

28. (I) The moving rod in Fig. 29–12b is 12.0 cm long and is pulled at a speed of 15.0 cm/s. If the magnetic field is 0.800 T, calculate the emf developed.

29. (II) In Fig. 29–12a, the rod moves to the right with a speed of 1.3 m/s and has a resistance of 2.5 Ω. The rail separation is $\ell = 25.0 \, \text{cm}$. The magnetic field is 0.35 T, and the resistance of the U-shaped conductor is 25.0 Ω at a given instant. Calculate (a) the induced emf, (b) the current in the U-shaped conductor, and (c) the external force needed to keep the rod's velocity constant at that instant.

30. (II) If the U-shaped conductor in Fig. 29–12a has resistivity ρ, whereas that of the moving rod is negligible, derive a formula for the current I as a function of time. Assume the rod starts at the bottom of the U at $t = 0$, and moves with uniform speed v in the magnetic field B. The cross-sectional area of the rod and all parts of the U is A.

31. (II) Suppose that the U-shaped conductor and connecting rod in Fig. 29–12a are oriented vertically (but still in contact) so that the rod is falling due to the gravitational force. Find the terminal speed of the rod if it has mass $m = 3.6$ grams, length $\ell = 18$ cm, and resistance $R = 0.0013 \, \Omega$. It is falling in a uniform horizontal field $B = 0.060 \, \text{T}$. Neglect the resistance of the U-shaped conductor.

32. (II) When a car drives through the Earth's magnetic field, an emf is induced in its vertical 75.0-cm-long radio antenna. If the Earth's field $(5.0 \times 10^{-5} \, \text{T})$ points north with a dip angle of 45°, what is the maximum emf induced in the antenna and which direction(s) will the car be moving to produce this maximum value? The car's speed is 30.0 m/s on a horizontal road.

33. (II) A conducting rod rests on two long frictionless parallel rails in a magnetic field $\vec{B}$ ($\perp$ to the rails and rod) as in Fig. 29–44. (a) If the rails are horizontal and the rod is given an initial push, will the rod travel at constant speed even though a magnetic field is present? (b) Suppose at $t = 0$, when the rod has speed $v = v_0$, the two rails are connected electrically by a wire from point a to point b. Assuming the rod has resistance R and the rails have negligible resistance, determine the speed of the rod as a function of time. Discuss your answer.

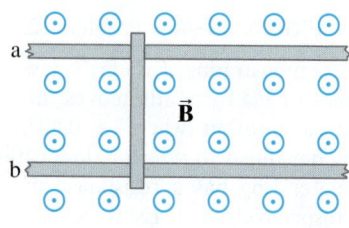

FIGURE 29–44 Problems 33 and 34.

34. (III) Suppose a conducting rod (mass m, resistance R) rests on two frictionless and resistanceless parallel rails a distance ℓ apart in a uniform magnetic field $\vec{\mathbf{B}}$ ($\perp$ to the rails and to the rod) as in Fig. 29–44. At $t = 0$, the rod is at rest and a source of emf is connected to the points a and b. Determine the speed of the rod as a function of time if (a) the source puts out a constant current I, (b) the source puts out a constant emf $\mathscr{E}_0$. (c) Does the rod reach a terminal speed in either case? If so, what is it?

35. (III) A short section of wire, of length a, is moving with velocity $\vec{\mathbf{v}}$, parallel to a very long wire carrying a current I as shown in Fig. 29–45. The near end of the wire section is a distance b from the long wire. Assuming the vertical wire is very long compared to $a + b$, determine the emf between the ends of the short section. Assume $\vec{\mathbf{v}}$ is (a) in the same direction as I, (b) in the opposite direction to I.

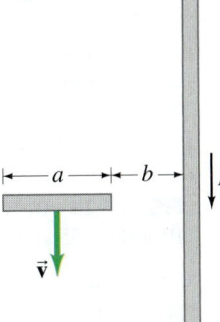

FIGURE 29–45
Problem 35.

29–4 Generators

36. (I) The generator of a car idling at 875-rpm produces 12.4 V. What will the output be at a rotation speed of 1550 rpm assuming nothing else changes?

37. (I) A simple generator is used to generate a peak output voltage of 24.0 V. The square armature consists of windings that are 5.15 cm on a side and rotates in a field of 0.420 T at a rate of 60.0 rev/s. How many loops of wire should be wound on the square armature?

38. (II) A simple generator has a 480-loop square coil 22.0 cm on a side. How fast must it turn in a 0.550-T field to produce a 120-V peak output?

39. (II) Show that the rms output of an ac generator is $V_{rms} = NAB\omega/\sqrt{2}$ where $\omega = 2\pi f$.

40. (II) A 250-loop circular armature coil with a diameter of 10.0 cm rotates at 120 rev/s in a uniform magnetic field of strength 0.45 T. What is the rms voltage output of the generator? What would you do to the rotation frequency in order to double the rms voltage output?

*29–5 Back EMF, Counter Torque; Eddy Current

*41. (I) The back emf in a motor is 72 V when operating at 1200 rpm. What would be the back emf at 2500 rpm if the magnetic field is unchanged?

*42. (I) A motor has an armature resistance of 3.05 Ω. If it draws 7.20 A when running at full speed and connected to a 120-V line, how large is the back emf?

*43. (II) What will be the current in the motor of Example 29–10 if the load causes it to run at half speed?

*44. (II) The back emf in a motor is 85 V when the motor is operating at 1100 rpm. How would you change the motor's magnetic field if you wanted to reduce the back emf to 75 V when the motor was running at 2300 rpm?

*45. (II) A dc generator is rated at 16 kW, 250 V, and 64 A when it rotates at 1000 rpm. The resistance of the armature windings is 0.40 Ω. (a) Calculate the "no-load" voltage at 1000 rpm (when there is no circuit hooked up to the generator). (b) Calculate the full-load voltage (i.e. at 64 A) when the generator is run at 750 rpm. Assume that the magnitude of the magnetic field remains constant.

29–6 Transformers

[Assume 100% efficiency, unless stated otherwise.]

46. (I) A transformer has 620 turns in the primary coil and 85 in the secondary coil. What kind of transformer is this, and by what factor does it change the voltage? By what factor does it change the current?

47. (I) Neon signs require 12 kV for their operation. To operate from a 240-V line, what must be the ratio of secondary to primary turns of the transformer? What would the voltage output be if the transformer were connected backward?

48. (II) A model-train transformer plugs into 120-V ac and draws 0.35 A while supplying 7.5 A to the train. (a) What voltage is present across the tracks? (b) Is the transformer step-up or step-down?

49. (II) The output voltage of a 75-W transformer is 12 V, and the input current is 22 A. (a) Is this a step-up or a step-down transformer? (b) By what factor is the voltage multiplied?

50. (II) If 65 MW of power at 45 kV (rms) arrives at a town from a generator via 3.0-Ω transmission lines, calculate (a) the emf at the generator end of the lines, and (b) the fraction of the power generated that is wasted in the lines.

51. (II) Assume a voltage source supplies an ac voltage of amplitude V_0 between its output terminals. If the output terminals are connected to an external circuit, and an ac current of amplitude I_0 flows out of the terminals, then the equivalent resistance of the external circuit is $R_{eq} = V_0/I_0$. (a) If a resistor R is connected directly to the output terminals, what is R_{eq}? (b) If a transformer with N_P and N_S turns in its primary and secondary, respectively, is placed between the source and the resistor as shown in Fig. 29–46, what is R_{eq}? [Transformers can be used in ac circuits to alter the apparent resistance of circuit elements, such as loud speakers, in order to maximize transfer of power.]

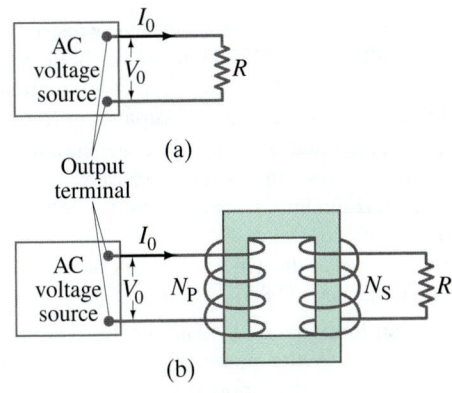

FIGURE 29–46 Problem 51.

52. (III) Design a dc transmission line that can transmit 225 MW of electricity 185 km with only a 2.0% loss. The wires are to be made of aluminum and the voltage is 660 kV.

53. (III) Suppose 85 kW is to be transmitted over two 0.100-Ω lines. Estimate how much power is saved if the voltage is stepped up from 120 V to 1200 V and then down again, rather than simply transmitting at 120 V. Assume the transformers are each 99% efficient.

29–7 Changing Φ_B Produces $\vec{\mathbf{E}}$

54. (II) In a circular region, there is a uniform magnetic field $\vec{\mathbf{B}}$ pointing into the page (Fig. 29–47). An xy coordinate system has its origin at the circular region's center. A free positive point charge $+Q = 1.0\,\mu\text{C}$ is initially at rest at a position $x = +10$ cm on the x axis. If the magnitude of the magnetic field is now decreased at a rate of -0.10 T/s, what force (magnitude and direction) will act on $+Q$?

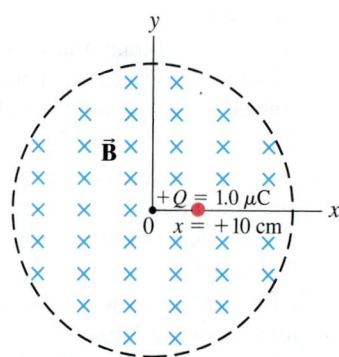

FIGURE 29–47
Problem 54.

55. (II) The **betatron**, a device used to accelerate electrons to high energy, consists of a circular vacuum tube placed in a magnetic field (Fig. 29–48), into which electrons are injected. The electromagnet produces a field that (1) keeps the electrons in their circular orbit inside the tube, and (2) increases the speed of the electrons when B changes. (a) Explain how the electrons are accelerated. (See Fig. 29–48.) (b) In what direction are the electrons moving in Fig. 29–48 (give directions as if looking down from above)? (c) Should B increase or decrease to accelerate the electrons? (d) The magnetic field is actually 60 Hz ac; show that the electrons can be accelerated only during $\frac{1}{4}$ of a cycle $\left(\frac{1}{240}\,\text{s}\right)$. (During this time they make hundreds of thousands of revolutions and acquire very high energy.)

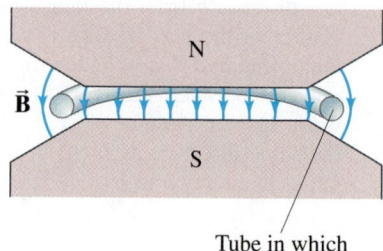

FIGURE 29–48
Problems 55 and 56.

Tube in which electrons orbit

56. (III) Show that the electrons in a betatron, Problem 55 and Fig. 29–48, are accelerated at constant radius if the magnetic field B_0 at the position of the electron orbit in the tube is equal to half the average value of the magnetic field (B_{avg}) over the area of the circular orbit at each moment: $B_0 = \frac{1}{2}B_{\text{avg}}$. (This is the reason the pole faces have a rather odd shape, as indicated in Fig. 29–48.)

57. (III) Find a formula for the net electric field in the moving rod of Problem 34 as a function of time for each case, (a) and (b).

General Problems

58. Suppose you are looking at two current loops in the plane of the page as shown in Fig. 29–49. When the switch S is closed in the left-hand coil, (a) what is the direction of the induced current in the other loop? (b) What is the situation after a "long" time? (c) What is the direction of the induced current in the right-hand loop if that loop is quickly pulled horizontally to the right (S having been closed for a long time)?

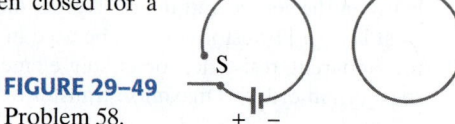

FIGURE 29–49
Problem 58.

59. A square loop 27.0 cm on a side has a resistance of 7.50 Ω. It is initially in a 0.755-T magnetic field, with its plane perpendicular to $\vec{\mathbf{B}}$, but is removed from the field in 40.0 ms. Calculate the electric energy dissipated in this process.

60. Power is generated at 24 kV at a generating plant located 85 km from a town that requires 65 MW of power at 12 kV. Two transmission lines from the plant to the town each have a resistance of 0.10 Ω/km. What should the output voltage of the transformer at the generating plant be for an overall transmission efficiency of 98.5%, assuming a perfect transformer?

61. A circular loop of area 12 m² encloses a magnetic field perpendicular to the plane of the loop; its magnitude is $B(t) = (10\,\text{T/s})t$. The loop is connected to a 7.5-Ω resistor and a 5.0-pF capacitor in series. When fully charged, how much charge is stored on the capacitor?

62. The primary windings of a transformer which has an 85% efficiency are connected to 110-V ac. The secondary windings are connected across a 2.4-Ω, 75-W lightbulb. (a) Calculate the current through the primary windings of the transformer. (b) Calculate the ratio of the number of primary windings of the transformer to the number of secondary windings of the transformer.

63. A pair of power transmission lines each have a 0.80-Ω resistance and carry 740 A over 9.0 km. If the rms input voltage is 42 kV, calculate (a) the voltage at the other end, (b) the power input, (c) power loss in the lines, and (d) the power output.

64. Show that the power loss in transmission lines, P_L, is given by $P_L = (P_T)^2 R_L/V^2$, where P_T is the power transmitted to the user, V is the delivered voltage, and R_L is the resistance of the power lines.

65. A high-intensity desk lamp is rated at 35 W but requires only 12 V. It contains a transformer that converts 120-V household voltage. (a) Is the transformer step-up or step-down? (b) What is the current in the secondary coil when the lamp is on? (c) What is the current in the primary coil? (d) What is the resistance of the bulb when on?

66. Two resistanceless rails rest 32 cm apart on a 6.0° ramp. They are joined at the bottom by a 0.60-Ω resistor. At the top a copper bar of mass 0.040 kg (ignore its resistance) is laid across the rails. The whole apparatus is immersed in a vertical 0.55-T field. What is the terminal (steady) velocity of the bar as it slides frictionlessly down the rails?

67. A coil with 150 turns, a radius of 5.0 cm, and a resistance of 12 Ω surrounds a solenoid with 230 turns/cm and a radius of 4.5 cm; see Fig. 29–50. The current in the solenoid changes at a constant rate from 0 to 2.0 A in 0.10 s. Calculate the magnitude and direction of the induced current in the 150-turn coil.

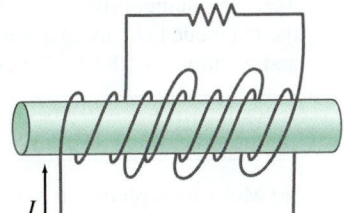

FIGURE 29–50
Problem 67.

68. A **search coil** for measuring B (also called a **flip coil**) is a small coil with N turns, each of cross-sectional area A. It is connected to a so-called **ballistic galvanometer**, which is a device to measure the total charge Q that passes through it in a short time. The flip coil is placed in the magnetic field to be measured with its face perpendicular to the field. It is then quickly rotated 180° about a diameter. Show that the total charge Q that flows in the induced current during this short "flip" time is proportional to the magnetic field B. In particular, show that B is given by

$$B = \frac{QR}{2NA}$$

where R is the total resistance of the circuit, including that of the coil and that of the ballistic galvanometer which measures the charge Q.

69. A ring with a radius of 3.0 cm and a resistance of 0.025 Ω is rotated about an axis through its diameter by 90° in a magnetic field of 0.23 T perpendicular to that axis. What is the largest number of electrons that would flow past a fixed point in the ring as this process is accomplished?

70. A flashlight can be made that is powered by the induced current from a magnet moving through a coil of wire. The coil and magnet are inside a plastic tube that can be shaken causing the magnet to move back and forth through the coil. Assume the magnet has a maximum field strength of 0.05 T. Make reasonable assumptions and specify the size of the coil and the number of turns necessary to light a standard 1-watt, 3-V flashlight bulb.

*71. A small electric car overcomes a 250-N friction force when traveling 35 km/h. The electric motor is powered by ten 12-V batteries connected in series and is coupled directly to the wheels whose diameters are 58 cm. The 270 armature coils are rectangular, 12 cm by 15 cm, and rotate in a 0.60-T magnetic field. (a) How much current does the motor draw to produce the required torque? (b) What is the back emf? (c) How much power is dissipated in the coils? (d) What percent of the input power is used to drive the car?

72. What is the energy dissipated as a function of time in a circular loop of 18 turns of wire having a radius of 10.0 cm and a resistance of 2.0 Ω if the plane of the loop is perpendicular to a magnetic field given by

$$B(t) = B_0 e^{-t/\tau}$$

with $B_0 = 0.50$ T and $\tau = 0.10$ s?

73. A thin metal rod of length ℓ rotates with angular velocity ω about an axis through one end (Fig. 29–51). The rotation axis is perpendicular to the rod and is parallel to a uniform magnetic field $\vec{B}$. Determine the emf developed between the ends of the rod.

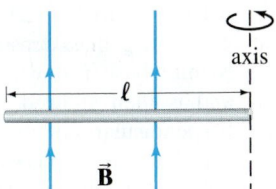

FIGURE 29–51
Problem 73.

*74. The magnetic field of a "shunt-wound" dc motor is produced by field coils placed in parallel with the armature coils. Suppose that the field coils have a resistance of 36.0 Ω and the armature coils 3.00 Ω. The back emf at full speed is 105 V when the motor is connected to 115 V dc. (a) Draw the equivalent circuit for the situations when the motor is just starting and when it is running full speed. (b) What is the total current drawn by the motor at start up? (c) What is the total current drawn when the motor runs at full speed?

75. Apply Faraday's law, in the form of Eq. 29–8, to show that the static electric field between the plates of a parallel-plate capacitor cannot drop abruptly to zero at the edges, but must, in fact, fringe. Use the path shown dashed in Fig. 29–52.

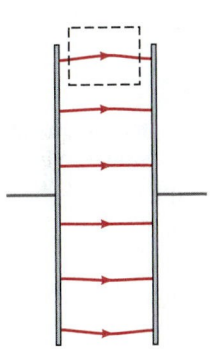

FIGURE 29–52
Problem 75.

76. A circular metal disk of radius R rotates with angular velocity ω about an axis through its center perpendicular to its face. The disk rotates in a uniform magnetic field B whose direction is parallel to the rotation axis. Determine the emf induced between the center and the edges.

77. What is the magnitude and direction of the electric field at each point in the rotating disk of Problem 76?

78. A circular-shaped circuit of radius r, containing a resistance R and capacitance C, is situated with its plane perpendicular to a spatially uniform magnetic field $\vec{B}$ directed into the page (Fig. 29–53). Starting at time $t = 0$, the voltage difference $V_{ba} = V_b - V_a$ across the capacitor plates is observed to increase with time t according to $V_{ba} = V_0(1 - e^{-t/\tau})$, where V_0 and τ are positive constants. Determine dB/dt, the rate at which the magnetic field magnitude changes with time. Is B becoming larger or smaller as time increases?

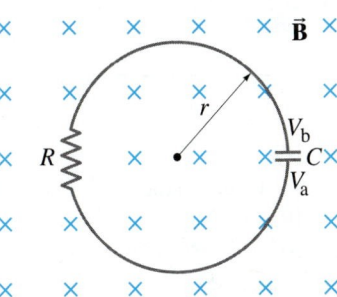

FIGURE 29–53
Problem 78.

79. In a certain region of space near Earth's surface, a uniform horizontal magnetic field of magnitude B exists above a level defined to be $y = 0$. Below $y = 0$, the field abruptly becomes zero (Fig. 29–54). A vertical square wire loop has resistivity ρ, mass density ρ_m, diameter d, and side length ℓ. It is initially at rest with its lower horizontal side at $y = 0$ and is then allowed to fall under gravity, with its plane perpendicular to the direction of the magnetic field. (a) While the loop is still partially immersed in the magnetic field (as it falls into the zero-field region), determine the magnetic "drag" force that acts on it at the moment when its speed is v. (b) Assume that the loop achieves a terminal velocity v_T before its upper horizontal side exits the field. Determine a formula for v_T. (c) If the loop is made of copper and $B = 0.80$ T, find v_T.

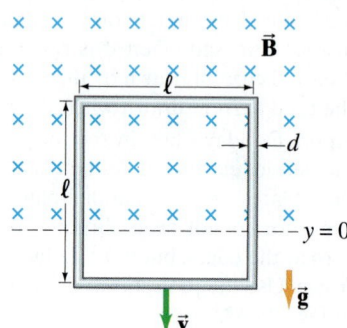

FIGURE 29–54
Problem 79.

***80.** (III) In an experiment, a coil was mounted on a low-friction cart that moved through the magnetic field B of a permanent magnet. The speed of the cart v and the induced voltage V were simultaneously measured, as the cart moved through the magnetic field, using a computer-interfaced motion sensor and a voltmeter. The Table below shows the collected data:

Speed, v (m/s)	0.367	0.379	0.465	0.623	0.630
Induced voltage, V (V)	0.128	0.135	0.164	0.221	0.222

(a) Make a graph of the induced voltage, V, vs. the speed, v. Determine a best-fit linear equation for the data. Theoretically, the relationship between V and v is given by $V = BN\ell v$ where N is the number of turns of the coil, B is the magnetic field, and ℓ is the average of the inside and outside widths of the coil. In the experiment, $B = 0.126$ T, $N = 50$, and $\ell = 0.0561$ m. (b) Find the % error between the slope of the experimental graph and the theoretical value for the slope. (c) For each of the measured speeds v, determine the theoretical value of V and find the % error of each.

Answers to Exercises

A: (e).

B: (a) Counterclockwise; (b) clockwise; (c) zero; (d) counterclockwise.

C: Electrons flow clockwise (conventional current counterclockwise).

D: (a) increases; (b) yes; increases (counter torque).

E: (a) 4.2×10^{-3} V/m; (b) 5.6×10^{-3} V/m.

A spark plug in a car receives a high voltage, which produces a high enough electric field in the air across its gap to pull electrons off the atoms in the air–gasoline mixture and form a spark. The high voltage is produced, from the basic 12 V of the car battery, by an induction coil which is basically a transformer or mutual inductance. Any coil of wire has a self-inductance, and a changing current in it causes an emf to be induced. Such inductors are useful in many circuits.

C H A P T E R

30

Inductance, Electromagnetic Oscillations, and AC Circuits

CHAPTER-OPENING QUESTION—Guess now!

Consider a circuit with only a capacitor C and a coil of many loops of wire (called an inductor, L) as shown. If the capacitor is initially charged $(Q = Q_0)$, what will happen when the switch S is closed?

(a) Nothing will happen—the capacitor will remain charged with charge $Q = Q_0$.

(b) The capacitor will quickly discharge and remain discharged $(Q = 0)$.

(c) Current will flow until the positive charge is on the opposite plate of the capacitor, and then will reverse—back and forth.

(d) The energy initially in the capacitor $(U_E = \frac{1}{2}Q_0^2/C)$ will all transfer to the coil and then remain that way.

(e) The system will quickly transfer half of the capacitor energy to the coil and then remain that way.

CONTENTS

30–1 Mutual Inductance

30–2 Self-Inductance

30–3 Energy Stored in a Magnetic Field

30–4 *LR* Circuits

30–5 *LC* Circuits and Electromagnetic Oscillations

30–6 *LC* Oscillations with Resistance (*LRC* Circuit)

30–7 AC Circuits with AC Source

30–8 *LRC* Series AC Circuit

30–9 Resonance in AC Circuits

*30–10 Impedance Matching

*30–11 Three-Phase AC

W e discussed in the last Chapter how a changing magnetic flux through a circuit induces an emf in that circuit. Before that we saw that an electric current produces a magnetic field. Combining these two ideas, we could predict that a changing current in one circuit ought to induce an emf and a current in a second nearby circuit, and even induce an emf in itself. We already saw an example in the previous Chapter (transformers), but now we will treat this effect in a more general way in terms of what we will call mutual inductance and self-inductance. The concept of inductance also gives us a springboard to treat energy storage in a magnetic field. This Chapter concludes with an analysis of circuits that contain inductance as well as resistance and/or capacitance.

30–1 Mutual Inductance

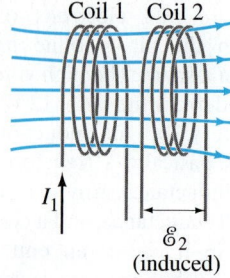

FIGURE 30–1 A changing current in one coil will induce a current in the second coil.

If two coils of wire are placed near each other, as in Fig. 30–1, a changing current in one will induce an emf in the other. According to Faraday's law, the emf $\mathcal{E}_2$ induced in coil 2 is proportional to the rate of change of magnetic flux passing through it. This flux is due to the current I_1 in coil 1, and it is often convenient to express the emf in coil 2 in terms of the current in coil 1.

We let Φ_{21} be the magnetic flux in each loop of coil 2 created by the current in coil 1. If coil 2 contains N_2 closely wrapped loops, then $N_2 \Phi_{21}$ is the total flux passing through coil 2. If the two coils are fixed in space, $N_2 \Phi_{21}$ is proportional to the current I_1 in coil 1; the proportionality constant is called the **mutual inductance**, M_{21}, defined by

$$M_{21} = \frac{N_2 \Phi_{21}}{I_1}. \tag{30–1}$$

The emf $\mathcal{E}_2$ induced in coil 2 due to a changing current in coil 1 is, by Faraday's law,

$$\mathcal{E}_2 = -N_2 \frac{d\Phi_{21}}{dt}.$$

We combine this with Eq. 30–1 rewritten as $\Phi_{21} = M_{21} I_1 / N_2$ (and take its derivative) and obtain

$$\mathcal{E}_2 = -M_{21} \frac{dI_1}{dt}. \tag{30–2}$$

This relates the change in current in coil 1 to the emf it induces in coil 2. The mutual inductance of coil 2 with respect to coil 1, M_{21}, is a "constant" in that it does not depend on I_1; M_{21} depends on "geometric" factors such as the size, shape, number of turns, and relative positions of the two coils, and also on whether iron (or some other ferromagnetic material) is present. For example, if the two coils in Fig. 30–1 are farther apart, fewer lines of flux can pass through coil 2, so M_{21} will be less. For some arrangements, the mutual inductance can be calculated (see Example 30–1). More often it is determined experimentally.

Suppose, now, we consider the reverse situation: when a changing current in coil 2 induces an emf in coil 1. In this case,

$$\mathcal{E}_1 = -M_{12} \frac{dI_2}{dt}$$

where M_{12} is the mutual inductance of coil 1 with respect to coil 2. It is possible to show, although we will not prove it here, that $M_{12} = M_{21}$. Hence, for a given arrangement we do not need the subscripts and we can let

$$M = M_{12} = M_{21},$$

so that

$$\mathcal{E}_1 = -M \frac{dI_2}{dt} \tag{30–3a}$$

and

$$\mathcal{E}_2 = -M \frac{dI_1}{dt}. \tag{30–3b}$$

The SI unit for mutual inductance is the henry (H), where $1\,\text{H} = 1\,\text{V}\cdot\text{s/A} = 1\,\Omega\cdot\text{s}$.

EXERCISE A Two coils which are close together have a mutual inductance of 330 mH. (*a*) If the emf in coil 1 is 120 V, what is the rate of change of the current in coil 2? (*b*) If the rate of change of current in coil 1 is 36 A/s, what is the emf in coil 2?

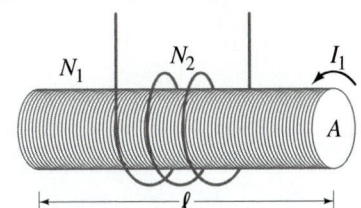

EXAMPLE 30–1 **Solenoid and coil.** A long thin solenoid of length ℓ and cross-sectional area A contains N_1 closely packed turns of wire. Wrapped around it is an insulated coil of N_2 turns, Fig. 30–2. Assume all the flux from coil 1 (the solenoid) passes through coil 2, and calculate the mutual inductance.

APPROACH We first determine the flux produced by the solenoid, all of which passes uniformly through coil N_2, using Eq. 28–4 for the magnetic field inside the solenoid:

$$B = \mu_0 \frac{N_1}{\ell} I_1,$$

where $n = N_1/\ell$ is the number of loops in the solenoid per unit length, and I_1 is the current in the solenoid.

SOLUTION The solenoid is closely packed, so we assume that all the flux in the solenoid stays inside the secondary coil. Then the flux Φ_{21} through coil 2 is

$$\Phi_{21} = BA = \mu_0 \frac{N_1}{\ell} I_1 A.$$

Then the mutual inductance is

$$M = \frac{N_2 \Phi_{21}}{I_1} = \frac{\mu_0 N_1 N_2 A}{\ell}.$$

NOTE We calculated M_{21}; if we had tried to calculate M_{12}, it would have been difficult. Given $M_{12} = M_{21} = M$, we did the simpler calculation to obtain M. Note again that M depends only on geometric factors, and not on the currents.

FIGURE 30–2 Example 30–1.

CONCEPTUAL EXAMPLE 30–2 | **Reversing the coils.** How would Example 30–1 change if the coil with N_2 turns was inside the solenoid rather than outside the solenoid?

RESPONSE The magnetic field inside the solenoid would be unchanged. The flux through the coil would be BA where A is the area of the coil, not of the solenoid as in Example 30–1. Solving for M would give the same formula except that A would refer to the coil, and would be smaller.

EXERCISE B Which solenoid and coil combination shown in Fig. 30–3 has the largest mutual inductance? Assume each solenoid is the same.

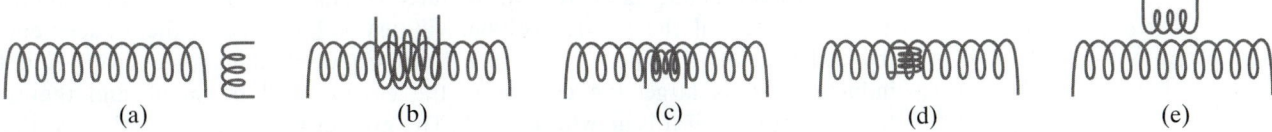

(a) (b) (c) (d) (e)

FIGURE 30–3 Exercise B.

A transformer is an example of mutual inductance in which the coupling is maximized so that nearly all flux lines pass through both coils. Mutual inductance has other uses as well, including some types of *pacemakers* used to maintain blood flow in heart patients (Section 26–5). Power in an external coil is transmitted via mutual inductance to a second coil in the pacemaker at the heart. This type has the advantage over battery-powered pacemakers in that surgery is not needed to replace a battery when it wears out.

Mutual inductance can sometimes be a problem, however. Any changing current in a circuit can induce an emf in another part of the same circuit or in a different circuit even though the conductors are not in the shape of a coil. The mutual inductance M is usually small unless coils with many turns and/or iron cores are involved. However, in situations where small voltages are being used, problems due to mutual inductance often arise. Shielded cable, in which an inner conductor is surrounded by a cylindrical grounded conductor (Fig. 28–12), is often used to reduce the problem.

PHYSICS APPLIED
Pacemaker

30–2 Self-Inductance

The concept of inductance applies also to a single isolated coil of N turns. When a changing current passes through a coil (or solenoid), a changing magnetic flux is produced inside the coil, and this in turn induces an emf in that same coil. This induced emf opposes the change in flux (Lenz's law). For example, if the current through the coil is increasing, the increasing magnetic flux induces an emf that opposes the original current and tends to retard its increase. If the current is decreasing in the coil, the decreasing flux induces an emf in the same direction as the current, thus tending to maintain the original current.

The magnetic flux Φ_B passing through the N turns of a coil is proportional to the current I in the coil, so we define the **self-inductance** L (in analogy to mutual inductance, Eq. 30–1) as

$$L = \frac{N\Phi_B}{I}. \tag{30–4}$$

Then the emf $\mathscr{E}$ induced in a coil of self-inductance L is, from Faraday's law,

$$\mathscr{E} = -N\frac{d\Phi_B}{dt} = -L\frac{dI}{dt}. \tag{30–5}$$

Like mutual inductance, self-inductance is measured in henrys. The magnitude of L depends on the geometry and on the presence of a ferromagnetic material. Self-inductance (inductance, for short) can be defined, as above, for any circuit or part of a circuit.

Circuits always contain some inductance, but often it is quite small unless the circuit contains a coil of many turns. A coil that has significant self-inductance L is called an **inductor**. Inductance is shown on circuit diagrams by the symbol

$$\text{—\small{oooo}—};\qquad\qquad\text{[inductor symbol]}$$

any resistance an inductor has should also be shown separately. Inductance can serve a useful purpose in certain circuits. Often, however, inductance is to be avoided in a circuit. Precision resistors are normally wire wound and thus would have inductance as well as resistance. The inductance can be minimized by winding the insulated wire back on itself in the opposite sense so that the current going in opposite directions produces little net magnetic flux; this is called a **noninductive winding**.

If an inductor has negligible resistance, it is the inductance (or induced emf) that controls a changing current. If a source of changing or alternating voltage is applied to the coil, this applied voltage will just be balanced by the induced emf of the coil (Eq. 30–5). Thus we can see from Eq. 30–5 that, for a given $\mathscr{E}$, if the inductance L is large, the change in the current will be small, and therefore the current itself if it is ac will be small. The greater the inductance, the less the ac current. An inductance thus acts something like a resistance to impede the flow of alternating current. We use the term *reactance* or *impedance* for this quality of an inductor. We will discuss reactance and impedance more fully in Sections 30–7 and 30–8. We shall see that reactance depends not only on the inductance L, but also on the frequency. Here we mention one example of its importance. The resistance of the primary in a transformer is usually quite small, perhaps less than $1\,\Omega$. If resistance alone limited the current in a transformer, tremendous currents would flow when a high voltage was applied. Indeed, a dc voltage applied to a transformer can burn it out. It is the induced emf (or reactance) of the coil that limits the current to a reasonable value.

Common inductors have inductances in the range from about $1\,\mu\text{H}$ to about $1\,\text{H}$ (where $1\,\text{H} = 1\,\text{henry} = 1\,\Omega\cdot\text{s}$).

EXAMPLE 30–3 **Solenoid inductance.** (*a*) Determine a formula for the self-inductance L of a tightly wrapped and long solenoid containing N turns of wire in its length ℓ and whose cross-sectional area is A. (*b*) Calculate the value of L if $N = 100$, $\ell = 5.0\,\text{cm}$, $A = 0.30\,\text{cm}^2$ and the solenoid is air filled.

APPROACH To determine the inductance L, it is usually simplest to start with Eq. 30–4, so we need to first determine the flux.

SOLUTION (a) According to Eq. 28–4, the magnetic field inside a solenoid (ignoring end effects) is constant: $B = \mu_0 nI$ where $n = N/\ell$. The flux is $\Phi_B = BA = \mu_0 NIA/\ell$, so

$$L = \frac{N\Phi_B}{I} = \frac{\mu_0 N^2 A}{\ell}.$$

(b) Since $\mu_0 = 4\pi \times 10^{-7}\,\text{T}\cdot\text{m/A}$, then

$$L = \frac{(4\pi \times 10^{-7}\,\text{T}\cdot\text{m/A})(100)^2(3.0 \times 10^{-5}\,\text{m}^2)}{(5.0 \times 10^{-2}\,\text{m})} = 7.5\,\mu\text{H}.$$

NOTE Magnetic field lines "stray" out of the solenoid (see Fig. 28–15), especially near the ends, so our formula is only an approximation.

CONCEPTUAL EXAMPLE 30–4 | **Direction of emf in inductor.** Current passes through the coil in Fig. 30–4 from left to right as shown. (a) If the current is increasing with time, in which direction is the induced emf? (b) If the current is decreasing in time, what then is the direction of the induced emf?

RESPONSE (a) From Lenz's law we know that the induced emf must oppose the change in magnetic flux. If the current is increasing, so is the magnetic flux. The induced emf acts to oppose the increasing flux, which means it acts like a source of emf that opposes the outside source of emf driving the current. So the induced emf in the coil acts to oppose I in Fig. 30–4a. In other words, the inductor might be thought of as a battery with a positive terminal at point A (tending to block the current entering at A), and negative at point B. (b) If the current is decreasing, then by Lenz's law the induced emf acts to bolster the flux—like a source of emf reinforcing the external emf. The induced emf acts to increase I in Fig. 30–4b, so in this situation you can think of the induced emf as a battery with its negative terminal at point A to attract more $(+)$ current to move to the right.

EXAMPLE 30–5 **Coaxial cable inductance.** Determine the inductance per unit length of a coaxial cable whose inner conductor has a radius r_1 and the outer conductor has a radius r_2, Fig. 30–5. Assume the conductors are thin hollow tubes so there is no magnetic field within the inner conductor, and the magnetic field inside both thin conductors can be ignored. The conductors carry equal currents I in opposite directions.

APPROACH We need to find the magnetic flux, $\Phi_B = \int \vec{\mathbf{B}}\cdot d\vec{\mathbf{A}}$, between the conductors. The lines of $\vec{\mathbf{B}}$ are circles surrounding the inner conductor (only one is shown in Fig. 30–5a). From Ampère's law, $\oint \vec{\mathbf{B}}\cdot d\vec{\boldsymbol{\ell}} = \mu_0 I$, the magnitude of the field along the circle at a distance r from the center, when the inner conductor carries a current I, is (Example 28–6):

$$B = \frac{\mu_0 I}{2\pi r}.$$

The magnetic flux through a rectangle of width dr and length ℓ (along the cable, Fig. 30–5b), a distance r from the center, is

$$d\Phi_B = B(\ell\, dr) = \frac{\mu_0 I}{2\pi r}\ell\, dr.$$

SOLUTION The total flux in a length ℓ of cable is

$$\Phi_B = \int d\Phi_B = \frac{\mu_0 I\ell}{2\pi}\int_{r_1}^{r_2}\frac{dr}{r} = \frac{\mu_0 I\ell}{2\pi}\ln\frac{r_2}{r_1}.$$

Since the current I all flows in one direction in the inner conductor, and the same current I all flows in the opposite direction in the outer conductor, we have only one turn, so $N = 1$ in Eq. 30–4. Hence the self-inductance for a length ℓ is

$$L = \frac{\Phi_B}{I} = \frac{\mu_0 \ell}{2\pi}\ln\frac{r_2}{r_1}.$$

The inductance per unit length is

$$\frac{L}{\ell} = \frac{\mu_0}{2\pi}\ln\frac{r_2}{r_1}.$$

Note that L depends only on geometric factors and not on the current I.

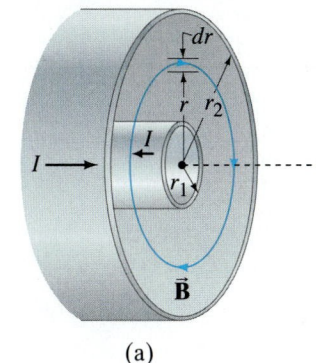

FIGURE 30–4 Example 30–4. The $+$ and $-$ signs refer to the induced emf due to the changing current, as if points A and B were the terminals of a battery (and the coiled loops were the inside of the battery).

FIGURE 30–5 Example 30–5. Coaxial cable: (a) end view, (b) side view (cross section).

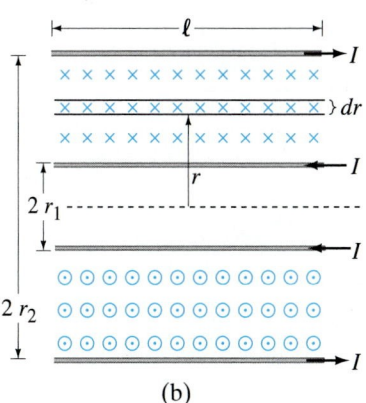

30–3 Energy Stored in a Magnetic Field

When an inductor of inductance L is carrying a current I which is changing at a rate dI/dt, energy is being supplied to the inductor at a rate

$$P = I\mathscr{E} = LI\frac{dI}{dt}$$

where P stands for power and we used[†] Eq. 30–5. Let us calculate the work needed to increase the current in an inductor from zero to some value I. Using this last equation, the work dW done in a time dt is

$$dW = P\,dt = LI\,dI.$$

Then the total work done to increase the current from zero to I is

$$W = \int dW = \int_0^I LI\,dI = \tfrac{1}{2}LI^2.$$

This work done is equal to the energy U stored in the inductor when it is carrying a current I (and we take $U = 0$ when $I = 0$):

$$U = \tfrac{1}{2}LI^2. \tag{30–6}$$

This can be compared to the energy stored in a capacitor, C, when the potential difference across it is V (see Section 24–4):

$$U = \tfrac{1}{2}CV^2.$$

EXERCISE C What is the inductance of an inductor if it has a stored energy of 1.5 J when there is a current of 2.5 A in it? (a) 0.48 H, (b) 1.2 H, (c) 2.1 H, (d) 4.7 H, (e) 19 H.

Just as the energy stored in a capacitor can be considered to reside in the electric field between its plates, so the energy in an inductor can be considered to be stored in its magnetic field. To write the energy in terms of the magnetic field, let us use the result of Example 30–3, that the inductance of an ideal solenoid (end effects ignored) is $L = \mu_0 N^2 A/\ell$. Because the magnetic field B in a solenoid is related to the current I by $B = \mu_0 NI/\ell$, we have

$$U = \tfrac{1}{2}LI^2 = \frac{1}{2}\left(\frac{\mu_0 N^2 A}{\ell}\right)\left(\frac{B\ell}{\mu_0 N}\right)^2$$

$$= \frac{1}{2}\frac{B^2}{\mu_0}A\ell.$$

We can think of this energy as residing in the volume enclosed by the windings, which is $A\ell$. Then the energy per unit volume or **energy density** is

$$u = \text{energy density} = \frac{1}{2}\frac{B^2}{\mu_0}. \tag{30–7}$$

This formula, which was derived for the special case of a solenoid, can be shown to be valid for any region of space where a magnetic field exists. If a ferromagnetic material is present, μ_0 is replaced by μ. This equation is analogous to that for electric field, $\tfrac{1}{2}\epsilon_0 E^2$, Eq. 24–6.

30–4 LR Circuits

Any inductor will have some resistance. We represent this situation by drawing its inductance L and its resistance R separately, as in Fig. 30–6a. The resistance R could also include any other resistance present in the circuit. Now we ask, what happens when a battery or other source of dc voltage V_0 is connected in series to such an LR circuit?

[†]No minus sign here because we are supplying power to oppose the emf of the inductor.

At the instant the switch connecting the battery is closed, the current starts to flow. It is opposed by the induced emf in the inductor which means point B in Fig. 30–6a is positive relative to point C. However, as soon as current starts to flow, there is also a voltage drop of magnitude IR across the resistance. Hence the voltage applied across the inductance is reduced and the current increases less rapidly. The current thus rises gradually as shown in Fig. 30–6b, and approaches the steady value $I_{max} = V_0/R_0$, for which all the voltage drop is across the resistance.

We can show this analytically by applying Kirchhoff's loop rule to the circuit of Fig. 30–6a. The emfs in the circuit are the battery voltage V_0 and the emf $\mathscr{E} = -L(dI/dt)$ in the inductor opposing the increasing current. Hence the sum of the potential changes around the loop is

$$V_0 - IR - L\frac{dI}{dt} = 0,$$

where I is the current in the circuit at any instant. We rearrange this to obtain

$$L\frac{dI}{dt} + RI = V_0. \tag{30–8}$$

This is a linear differential equation and can be integrated in the same way we did in Section 26–5 for an RC circuit. We rewrite Eq. 30–8 and then integrate:

$$\int_{I=0}^{I} \frac{dI}{V_0 - IR} = \int_0^t \frac{dt}{L}.$$

Then

$$-\frac{1}{R}\ln\left(\frac{V_0 - IR}{V_0}\right) = \frac{t}{L}$$

or

$$I = \frac{V_0}{R}\left(1 - e^{-t/\tau}\right) \tag{30–9}$$

where

$$\tau = \frac{L}{R} \tag{30–10}$$

is the **time constant** of the LR circuit. The symbol τ represents the time required for the current I to reach $(1 - 1/e) = 0.63$ or 63% of its maximum value (V_0/R). Equation 30–9 is plotted in Fig. 30–6b. (Compare to the RC circuit, Section 26–5.)

| **EXERCISE D** Show that L/R does have dimensions of time. (See Section 1–7.)

Now let us flip the switch in Fig. 30–6a so that the battery is taken out of the circuit, and points A and C are connected together as shown in Fig. 30–7 at the moment when the switching occurs (call it $t = 0$) and the current is I_0. Then the differential equation (Eq. 30–8) becomes (since $V_0 = 0$):

$$L\frac{dI}{dt} + RI = 0.$$

We rearrange this equation and integrate:

$$\int_{I_0}^{I} \frac{dI}{I} = -\int_0^t \frac{R}{L}\,dt$$

where $I = I_0$ at $t = 0$, and $I = I$ at time t.

We integrate this last equation to obtain

$$\ln\frac{I}{I_0} = -\frac{R}{L}t$$

or

$$I = I_0 e^{-t/\tau} \tag{30–11}$$

where again the time constant is $\tau = L/R$. The current thus decays exponentially to zero as shown in Fig. 30–8.

This analysis shows that there is always some "reaction time" when an electromagnet, for example, is turned on or off. We also see that an LR circuit has properties similar to an RC circuit (Section 26–5). Unlike the capacitor case, however, the time constant here is *inversely* proportional to R.

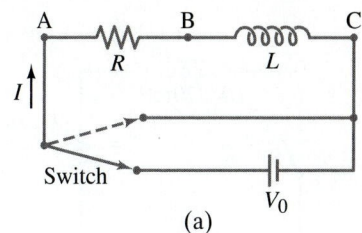

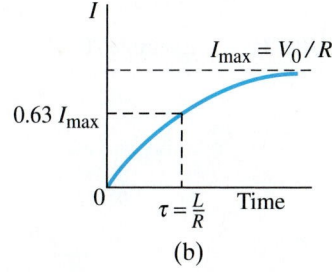

FIGURE 30–6 (a) LR circuit; (b) growth of current when connected to battery.

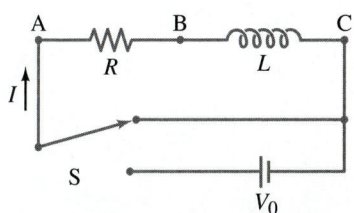

FIGURE 30–7 The switch is flipped quickly so the battery is removed but we still have a circuit. The current at this moment (call it $t = 0$) is I_0.

FIGURE 30–8 Decay of the current in Fig. 30–7 in time after the battery is switched out of the circuit.

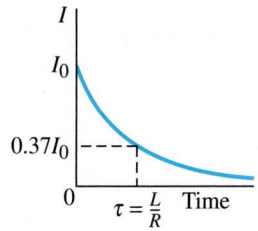

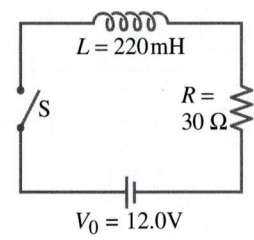

FIGURE 30–9 Example 30–6.

EXAMPLE 30–6 **An *LR* circuit.** At $t = 0$, a 12.0-V battery is connected in series with a 220-mH inductor and a total of 30-Ω resistance, as shown in Fig. 30–9. (*a*) What is the current at $t = 0$? (*b*) What is the time constant? (*c*) What is the maximum current? (*d*) How long will it take the current to reach half its maximum possible value? (*e*) At this instant, at what rate is energy being delivered by the battery, and (*f*) at what rate is energy being stored in the inductor's magnetic field?

APPROACH We have the situation shown in Figs. 30–6a and b, and we can apply the equations we just developed.

SOLUTION (*a*) The current cannot instantaneously jump from zero to some other value when the switch is closed because the inductor opposes the change $\left(\mathcal{E}_L = -L(dI/dt)\right)$. Hence just after the switch is closed, I is still zero at $t = 0$ and then begins to increase.

(*b*) The time constant is, from Eq. 30–10, $\tau = L/R = (0.22\,\text{H})/(30\,\Omega) = 7.3\,\text{ms}$.

(*c*) The current reaches its maximum steady value after a long time, when $dI/dt = 0$ so $I_{max} = V_0/R = 12.0\,\text{V}/30\,\Omega = 0.40\,\text{A}$.

(*d*) We set $I = \frac{1}{2}I_{max} = V_0/2R$ in Eq. 30–9, which gives us

$$1 - e^{-t/\tau} = \tfrac{1}{2}$$

or

$$e^{-t/\tau} = \tfrac{1}{2}.$$

We solve for t:

$$t = \tau \ln 2 = (7.3 \times 10^{-3}\,\text{s})(0.69) = 5.0\,\text{ms}.$$

(*e*) At this instant, $I = I_{max}/2 = 200\,\text{mA}$, so the power being delivered by the battery is

$$P = IV = (0.20\,\text{A})(12\,\text{V}) = 2.4\,\text{W}.$$

(*f*) From Eq. 30–6, the energy stored in an inductor L at any instant is

$$U = \tfrac{1}{2}LI^2$$

where I is the current in the inductor at that instant. The *rate* at which the energy changes is

$$\frac{dU}{dt} = LI\frac{dI}{dt}.$$

We can differentiate Eq. 30–9 to obtain dI/dt, or use the differential equation, Eq. 30–8, directly:

$$\frac{dU}{dt} = I\left(L\frac{dI}{dt}\right) = I(V_0 - RI)$$
$$= (0.20\,\text{A})[12\,\text{V} - (30\,\Omega)(0.20\,\text{A})] = 1.2\,\text{W}.$$

Since only part of the battery's power is feeding the inductor at this instant, where is the rest going?

EXERCISE E A resistor in series with an inductor has a time constant of 10 ms. When the same resistor is placed in series with a 5-μF capacitor, the time constant is 5×10^{-6} s. What is the value of the inductor? (*a*) 5 μH; (*b*) 10 μH; (*c*) 5 mH; (*d*) 10 mH; (*e*) not enough information to determine it.

Surge protection

An inductor can act as a "surge protector" for sensitive electronic equipment that can be damaged by high currents. If equipment is plugged into a standard wall plug, a sudden "surge," or increase, in voltage will normally cause a corresponding large change in current and damage the electronics. However, if there is an inductor in series with the voltage to the device, the sudden change in current produces an opposing emf preventing the current from reaching dangerous levels.

30–5 LC Circuits and Electromagnetic Oscillations

In any electric circuit, there can be three basic components: resistance, capacitance, and inductance, in addition to a source of emf. (There can also be more complex components, such as diodes or transistors.) We have previously discussed both RC and LR circuits. Now we look at an LC circuit, one that contains only a capacitance C and an inductance, L, Fig. 30–10. This is an idealized circuit in which we assume there is no resistance; in the next Section we will include resistance. Let us suppose the capacitor in Fig. 30–10 is initially charged so that one plate has charge Q_0 and the other plate has charge $-Q_0$, and the potential difference across it is $V = Q/C$ (Eq. 24–1). Suppose that at $t = 0$, the switch is closed. The capacitor immediately begins to discharge. As it does so, the current I through the inductor increases. We now apply Kirchhoff's loop rule (sum of potential changes around a loop is zero):

$$-L\frac{dI}{dt} + \frac{Q}{C} = 0.$$

Because charge leaves the positive plate on the capacitor to produce the current I as shown in Fig. 30–10, the charge Q on the (positive) plate of the capacitor is decreasing, so $I = -dQ/dt$. We can then rewrite the above equation as

$$\frac{d^2Q}{dt^2} + \frac{Q}{LC} = 0. \tag{30–12}$$

This is a familiar differential equation. It has the same form as the equation for simple harmonic motion (Chapter 14, Eq. 14–3). The solution of Eq. 30–12 can be written as

$$Q = Q_0 \cos(\omega t + \phi) \tag{30–13}$$

where Q_0 and ϕ are constants that depend on the initial conditions. We insert Eq. 30–13 into Eq. 30–12, noting that $d^2Q/dt^2 = -\omega^2 Q_0 \cos(\omega t + \phi)$; thus

$$-\omega^2 Q_0 \cos(\omega t + \phi) + \frac{1}{LC} Q_0 \cos(\omega t + \phi) = 0$$

or

$$\left(-\omega^2 + \frac{1}{LC}\right) \cos(\omega t + \phi) = 0.$$

This relation can be true for all times t only if $\left(-\omega^2 + 1/LC\right) = 0$, which tells us that

$$\omega = 2\pi f = \sqrt{\frac{1}{LC}}. \tag{30–14}$$

Equation 30–13 shows that the charge on the capacitor in an LC circuit oscillates sinusoidally. The current in the inductor is

$$I = -\frac{dQ}{dt} = \omega Q_0 \sin(\omega t + \phi)$$

$$= I_0 \sin(\omega t + \phi); \tag{30–15}$$

so the current too is sinusoidal. The maximum value of I is $I_0 = \omega Q_0 = Q_0/\sqrt{LC}$. Equations 30–13 and 30–15 for Q and I when $\phi = 0$ are plotted in Fig. 30–11.

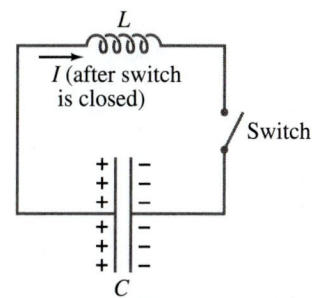

FIGURE 30–10 An LC circuit.

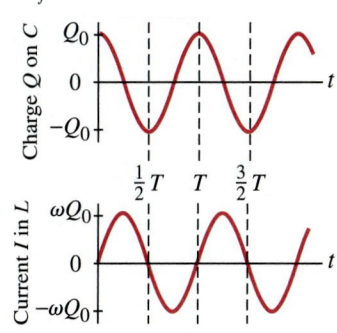

FIGURE 30–11 Charge Q and current I in an LC circuit. The period $T = \frac{1}{f} = \frac{2\pi}{\omega} = 2\pi\sqrt{LC}$.

Now let us look at LC oscillations from the point of view of energy. The energy stored in the electric field of the capacitor at any time t is (see Eq. 24–5):

$$U_E = \frac{1}{2}\frac{Q^2}{C} = \frac{Q_0^2}{2C}\cos^2(\omega t + \phi).$$

The energy stored in the magnetic field of the inductor at the same instant is (Eq. 30–6)

$$U_B = \frac{1}{2}LI^2 = \frac{L\omega^2 Q_0^2}{2}\sin^2(\omega t + \phi) = \frac{Q_0^2}{2C}\sin^2(\omega t + \phi)$$

where we used Eq. 30–14. If we let $\phi = 0$, then at times $t = 0$, $t = \frac{1}{2}T$, $t = T$, and so on (where T is the period $= 1/f = 2\pi/\omega$), we have $U_E = Q_0^2/2C$ and $U_B = 0$. That is, all the energy is stored in the electric field of the capacitor. But at $t = \frac{1}{4}T, \frac{3}{4}T$, and so on, $U_E = 0$ and $U_B = Q_0^2/2C$, and so all the energy is stored in the magnetic field of the inductor. At any time t, the total energy is

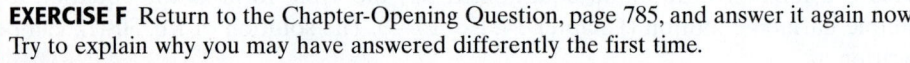

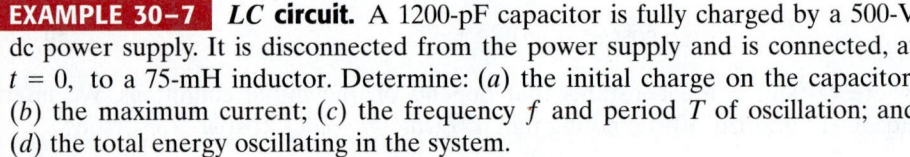

$$
\begin{aligned}
U = U_E + U_B &= \frac{1}{2}\frac{Q^2}{C} + \frac{1}{2}LI^2 \\
&= \frac{Q_0^2}{2C}\left[\cos^2(\omega t + \phi) + \sin^2(\omega t + \phi)\right] = \frac{Q_0^2}{2C}. \quad \textbf{(30–16)}
\end{aligned}
$$

Hence the total energy is constant, and energy is conserved.

What we have in this LC circuit is an **LC oscillator** or **electromagnetic oscillation**. The charge Q oscillates back and forth, from one plate of the capacitor to the other, and repeats this continuously. Likewise, the current oscillates back and forth as well. They are also energy oscillations: when Q is a maximum, the energy is all stored in the electric field of the capacitor; but when Q reaches zero, the current I is a maximum and all the energy is stored in the magnetic field of the inductor. Thus the energy oscillates between being stored in the electric field of the capacitor and in the magnetic field of the inductor. See Fig. 30–12.

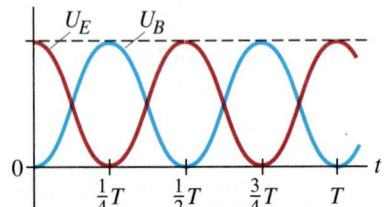

FIGURE 30–12 Energy U_E (red line) and U_B (blue line) stored in the capacitor and the inductor as a function of time. Note how the energy oscillates between electric and magnetic. The dashed line at the top is the (constant) total energy $U = U_E + U_B$.

EXERCISE F Return to the Chapter-Opening Question, page 785, and answer it again now. Try to explain why you may have answered differently the first time.

EXAMPLE 30–7 **LC circuit.** A 1200-pF capacitor is fully charged by a 500-V dc power supply. It is disconnected from the power supply and is connected, at $t = 0$, to a 75-mH inductor. Determine: (a) the initial charge on the capacitor; (b) the maximum current; (c) the frequency f and period T of oscillation; and (d) the total energy oscillating in the system.

APPROACH We use the analysis above, and the definition of capacitance $Q = CV$ (Chapter 24).

SOLUTION (a) The 500-V power supply, before being disconnected, charged the capacitor to a charge of

$$Q_0 = CV = (1.2 \times 10^{-9}\,\text{F})(500\,\text{V}) = 6.0 \times 10^{-7}\,\text{C}.$$

(b) The maximum current, I_{max}, is (see Eqs. 30-14 and 30–15)

$$I_{max} = \omega Q_0 = \frac{Q_0}{\sqrt{LC}} = \frac{(6.0 \times 10^{-7}\,\text{C})}{\sqrt{(0.075\,\text{H})(1.2 \times 10^{-9}\,\text{F})}} = 63\,\text{mA}.$$

(c) Equation 30–14 gives us the frequency:

$$f = \frac{\omega}{2\pi} = \frac{1}{(2\pi\sqrt{LC})} = 17\,\text{kHz},$$

and the period T is

$$T = \frac{1}{f} = 6.0 \times 10^{-5}\,\text{s}.$$

(d) Finally the total energy (Eq. 30–16) is

$$U = \frac{Q_0^2}{2C} = \frac{(6.0 \times 10^{-7}\,\text{C})^2}{2(1.2 \times 10^{-9}\,\text{F})} = 1.5 \times 10^{-4}\,\text{J}.$$

30-6 LC Oscillations with Resistance (LRC Circuit)

The *LC* circuit discussed in the previous Section is an idealization. There is always some resistance *R* in any circuit, and so we now discuss such a simple *LRC* circuit, Fig. 30–13.

Suppose again that the capacitor is initially given a charge Q_0 and the battery or other source is then removed from the circuit. The switch is closed at $t = 0$. Since there is now a resistance in the circuit, we expect some of the energy to be converted to thermal energy, and so we don't expect undamped oscillations as in a pure *LC* circuit. Indeed, if we use Kirchhoff's loop rule around this circuit, we obtain

$$-L\frac{dI}{dt} - IR + \frac{Q}{C} = 0,$$

which is the same equation we had in Section 30–5 with the addition of the voltage drop *IR* across the resistor. Since $I = -dQ/dt$, as we saw in Section 30–5, this equation becomes

$$L\frac{d^2Q}{dt^2} + R\frac{dQ}{dt} + \frac{1}{C}Q = 0. \qquad \textbf{(30–17)}$$

This second-order differential equation in the variable *Q* has precisely the same form as that for the damped harmonic oscillator, Eq. 14–15:

$$m\frac{d^2x}{dt^2} + b\frac{dx}{dt} + kx = 0.$$

Hence we can analyze our *LRC* circuit in the same way as for damped harmonic motion, Section 14–7. Our system may undergo damped oscillations, curve A in Fig. 30–14 (underdamped system), or it may be critically damped (curve B), or overdamped (curve C), depending on the relative values of *R*, *L*, and *C*. Using the results of Section 14–7, with *m* replaced by *L*, *b* by *R*, and *k* by C^{-1}, we find that the system will be underdamped when

$$R^2 < \frac{4L}{C},$$

and overdamped for $R^2 > 4L/C$. Critical damping (curve B in Fig. 30–14) occurs when $R^2 = 4L/C$. If *R* is smaller than $\sqrt{4L/C}$, the angular frequency, ω', will be

$$\omega' = \sqrt{\frac{1}{LC} - \frac{R^2}{4L^2}} \qquad \textbf{(30–18)}$$

(compare to Eq. 14–18). And the charge *Q* as a function of time will be

$$Q = Q_0 e^{-\frac{R}{2L}t}\cos(\omega't + \phi) \qquad \textbf{(30–19)}$$

where ϕ is a phase constant (compare to Eq. 14–19).

Oscillators are an important element in many electronic devices: radios and television sets use them for tuning, tape recorders use them (the "bias frequency") when recording, and so on. Because some resistance is always present, electrical oscillators generally need a periodic input of power to compensate for the energy converted to thermal energy in the resistance.

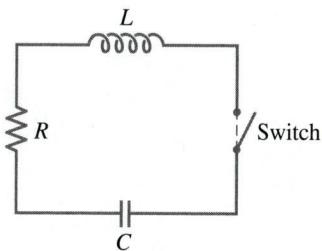

FIGURE 30–13 An *LRC* circuit.

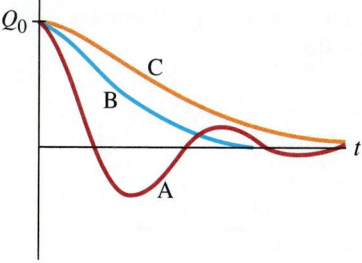

FIGURE 30–14 Charge *Q* on the capacitor in an *LRC* circuit as a function of time: curve A is for underdamped oscillation $(R^2 < 4L/C)$, curve B is for critically damped $(R^2 = 4L/C)$, and curve C is for overdamped $(R^2 > 4L/C)$.

EXAMPLE 30–8 **Damped oscillations.** At $t = 0$, a 40-mH inductor is placed in series with a resistance $R = 3.0\,\Omega$ and a charged capacitor $C = 4.8\,\mu\text{F}$. (a) Show that this circuit will oscillate. (b) Determine the frequency. (c) What is the time required for the charge amplitude to drop to half its starting value? (d) What value of R will make the circuit nonoscillating?

APPROACH We first check R^2 vs. $4L/C$; then use Eqs. 30–18 and 30–19.

SOLUTION (a) In order to oscillate, the circuit must be underdamped, so we must have $R^2 < 4L/C$. Since $R^2 = 9.0\,\Omega^2$ and $4L/C = 4(0.040\,\text{H})/(4.8 \times 10^{-6}\,\text{F}) = 3.3 \times 10^4\,\Omega^2$, this relation is satisfied, so the circuit will oscillate.

(b) We use Eq. 30–18:

$$f' = \frac{\omega'}{2\pi} = \frac{1}{2\pi}\sqrt{\frac{1}{LC} - \frac{R^2}{4L^2}} = 3.6 \times 10^2\,\text{Hz}.$$

(c) From Eq. 30–19, the amplitude will be half when

$$e^{-\frac{R}{2L}t} = \tfrac{1}{2}$$

or

$$t = \frac{2L}{R}\ln 2 = 18\,\text{ms}.$$

(d) To make the circuit critically damped or overdamped, we must use the criterion $R^2 \geq 4L/C = 3.3 \times 10^4\,\Omega^2$. Hence we must have $R \geq 180\,\Omega$.

30–7 AC Circuits with AC Source

We have previously discussed circuits that contain combinations of resistor, capacitor, and inductor, but only when they are connected to a dc source of emf or to no source. Now we discuss these circuit elements when they are connected to a source of alternating voltage that produces an alternating current (ac).

First we examine, one at a time, how a resistor, a capacitor, and an inductor behave when connected to a source of alternating voltage, represented by the symbol

[alternating voltage]

which produces a sinusoidal voltage of frequency f. We assume in each case that the emf gives rise to a current

$$I = I_0 \cos 2\pi ft = I_0 \cos \omega t \tag{30–20}$$

where t is time and I_0 is the peak current. Remember (Section 25–7) that $V_{\text{rms}} = V_0/\sqrt{2}$ and $I_{\text{rms}} = I_0/\sqrt{2}$ (Eqs. 25–9).

Resistor

When an ac source is connected to a resistor as in Fig. 30–15a, the current increases and decreases with the alternating voltage according to Ohm's law

$$V = IR = I_0 R \cos \omega t = V_0 \cos \omega t$$

where $V_0 = I_0 R$ is the peak voltage as a function of time. Figure 30–15b shows the voltage (red curve) and the current (blue curve). Because the current is zero when the voltage is zero and the current reaches a peak when the voltage does, we say that the current and voltage are **in phase**. Energy is transformed into heat (Section 25–7), at an average rate

$$\overline{P} = \overline{IV} = I_{\text{rms}}^2 R = V_{\text{rms}}^2/R.$$

Inductor

In Fig. 30–16a an inductor of inductance L (symbol —◠◠◠—) is connected to the ac source. We ignore any resistance it might have (it is usually small). The voltage applied to the inductor will be equal to the "back" emf generated in the inductor by the changing current as given by Eq. 30–5. This is because the sum of the electric potential changes around any closed circuit must add up to zero, according to Kirchhoff's rule.

FIGURE 30–15 (a) Resistor connected to an ac source. (b) Current (blue curve) is in phase with the voltage (red) across a resistor.

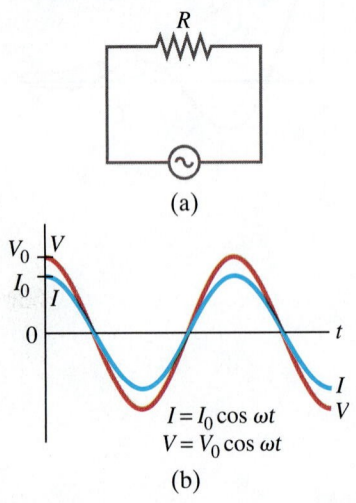

Thus

$$V - L\frac{dI}{dt} = 0$$

or (inserting Eq. 30–20)

$$V = L\frac{dI}{dt} = -\omega L I_0 \sin\omega t. \tag{30–21}$$

Using the identity $\sin\theta = -\cos(\theta + 90°)$ we can write

$$V = \omega L I_0 \cos(\omega t + 90°) = V_0 \cos(\omega t + 90°) \tag{30–22a}$$

where

$$V_0 = I_0 \omega L \tag{30–22b}$$

is the peak voltage. The current I and voltage V as a function of time are graphed for the inductor in Fig. 30–16b. It is clear from this graph, as well as from Eqs. 30–22, that the current and voltage are out of phase by a quarter cycle, which is equivalent to $\pi/2$ radians or 90°. We see from the graph that

the current lags the voltage by 90° in an inductor.

That is, the current in an inductor reaches its peaks a quarter cycle later than the voltage does. Alternatively, we can say that the voltage leads the current by 90°.

Because the current and voltage in an inductor are out of phase by 90°, the product IV (= power) is as often positive as it is negative (Fig. 30–16b). So no energy is transformed in an inductor on the average; and no energy is dissipated as thermal energy.

Just as a resistor impedes the flow of charge, so too an inductor impedes the flow of charge in an alternating current due to the back emf produced. For a resistor R, the peak current and peak voltage are related by $V_0 = I_0 R$. We can write a similar relation for an inductor:

$$V_0 = I_0 X_L \qquad \left[\begin{array}{c}\text{maximum or rms values,}\\\text{not at any instant}\end{array}\right] \tag{30–23a}$$

where, from Eq. 30–22b (and using $\omega = 2\pi f$ where f is the frequency of the ac),

$$X_L = \omega L = 2\pi f L. \tag{30–23b}$$

The term X_L is called the **inductive reactance** of the inductor, and has units of ohms. The greater X_L is, the more it impedes the flow of charge and the smaller the current. X_L is larger for higher frequencies f and larger inductance L.

Equation 30–23a is valid for peak values I_0 and V_0; it is also valid for rms values, $V_{\text{rms}} = I_{\text{rms}} X_L$. Because the peak values of current and voltage are not reached at the same time, Eq. 30–23a is *not valid at a particular instant*, as is the case for a resistor $(V = IR)$.

Note from Eq. 30–23b that if $\omega = 2\pi f = 0$ (so the current is dc), there is no back emf and no impedance to the flow of charge.

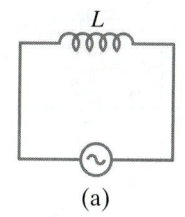

(a)

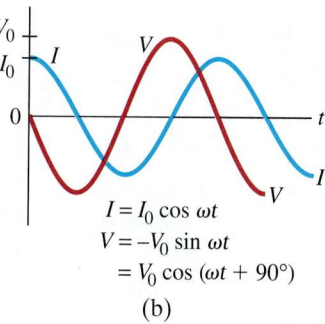

$$I = I_0 \cos\omega t$$
$$V = -V_0 \sin\omega t$$
$$= V_0 \cos(\omega t + 90°)$$

(b)

FIGURE 30–16 (a) Inductor connected to an ac source. (b) Current (blue curve) lags voltage (red curve) by a quarter cycle or 90°.

EXAMPLE 30–9 **Reactance of a coil.** A coil has a resistance $R = 1.00\,\Omega$ and an inductance of 0.300 H. Determine the current in the coil if (*a*) 120-V dc is applied to it, (*b*) 120-V ac (rms) at 60.0 Hz is applied.

APPROACH When the voltage is dc, there is no inductive reactance $(X_L = 2\pi f L = 0$ since $f = 0)$, so we apply Ohm's law for the resistance. When the voltage is ac, we calculate the reactance X_L and then use Eq. 30–23a.

SOLUTION (*a*) With dc, we have no X_L so we simply apply Ohm's law:

$$I = \frac{V}{R} = \frac{120\,\text{V}}{1.00\,\Omega} = 120\,\text{A}.$$

(*b*) The inductive reactance is

$$X_L = 2\pi f L = (6.283)(60.0\,\text{s}^{-1})(0.300\,\text{H}) = 113\,\Omega.$$

In comparison to this, the resistance can be ignored. Thus,

$$I_{\text{rms}} = \frac{V_{\text{rms}}}{X_L} = \frac{120\,\text{V}}{113\,\Omega} = 1.06\,\text{A}.$$

NOTE It might be tempting to say that the total impedance is $113\,\Omega + 1\,\Omega = 114\,\Omega$. This might imply that about 1% of the voltage drop is across the resistor, or about 1 V; and that across the inductance is 119 V. Although the 1 V across the resistor is correct, the other statements are not true because of the alteration in phase in an inductor. This will be discussed in the next Section.

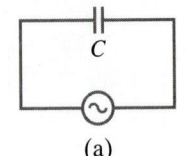

(a)

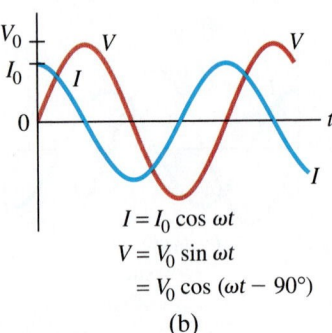

$I = I_0 \cos \omega t$
$V = V_0 \sin \omega t$
$\quad = V_0 \cos (\omega t - 90°)$

(b)

FIGURE 30–17 (a) Capacitor connected to an ac source. (b) Current leads voltage by a quarter cycle, or 90°.

Capacitor

When a capacitor is connected to a battery, the capacitor plates quickly acquire equal and opposite charges; but no steady current flows in the circuit. A capacitor prevents the flow of a dc current. But if a capacitor is connected to an alternating source of voltage, as in Fig. 30–17a, an alternating current will flow continuously. This can happen because when the ac voltage is first turned on, charge begins to flow and one plate acquires a negative charge and the other a positive charge. But when the voltage reverses itself, the charges flow in the opposite direction. Thus, for an alternating applied voltage, an ac current is present in the circuit continuously.

Let us look at this in more detail. By Kirchhoff's loop rule, the applied source voltage must equal the voltage V across the capacitor at any moment:

$$V = \frac{Q}{C}$$

where C is the capacitance and Q is the charge on the capacitor plates. The current I at any instant (given as $I = I_0 \cos \omega t$, Eq. 30–20) is

$$I = \frac{dQ}{dt} = I_0 \cos \omega t.$$

Hence the charge Q on the plates at any instant is given by

$$Q = \int_0^t dQ = \int_0^t I_0 \cos \omega t \, dt = \frac{I_0}{\omega} \sin \omega t.$$

Then the voltage across the capacitor is

$$V = \frac{Q}{C} = I_0 \left(\frac{1}{\omega C} \right) \sin \omega t.$$

Using the trigonometric identity $\sin \theta = \cos(90° - \theta) = \cos(\theta - 90°)$, we can rewrite this as

$$V = I_0 \left(\frac{1}{\omega C} \right) \cos(\omega t - 90°) = V_0 \cos(\omega t - 90°) \quad \textbf{(30–24a)}$$

where

$$V_0 = I_0 \left(\frac{1}{\omega C} \right) \quad \textbf{(30–24b)}$$

is the peak voltage. The current $I \left(= I_0 \cos \omega t \right)$ and voltage V (Eq. 30–24a) across the capacitor are graphed in Fig. 30–17b. It is clear from this graph, as well as a comparison of Eq. 30–24a with Eq. 30–20, that the current and voltage are out of phase by a quarter cycle or 90° ($\pi/2$ radians):

The current leads the voltage across a capacitor by 90°.

Alternatively we can say that the voltage lags the current by 90°. This is the opposite of what happens for an inductor.

Because the current and voltage are out of phase by 90°, the average power dissipated is zero, just as for an inductor. Energy from the source is fed to the capacitor, where it is stored in the electric field between its plates. As the field decreases, the energy returns to the source. Thus *only a resistance will dissipate energy* as thermal energy in an ac circuit.

A relationship between the applied voltage and the current in a capacitor can be written just as for an inductance:

$$V_0 = I_0 X_C \qquad \left[\begin{array}{c} \text{maximum or rms values,} \\ \text{not at any instant} \end{array} \right] \quad \textbf{(30–25a)}$$

where X_C is the **capacitive reactance** of the capacitor, and has units of ohms; X_C is given by (see Eq. 30–24b):

$$X_C = \frac{1}{\omega C} = \frac{1}{2\pi f C}. \quad \textbf{(30–25b)}$$

When frequency f and/or capacitance C are smaller, X_C is larger and thus impedes the flow of charge more. That is, when X_C is larger, the current is smaller (Eq. 30–25a). In the next Section we use the term **impedance** to represent reactances and resistance.

Equation 30–25a relates the peak values of V and I, or the rms values $\left(V_{rms} = I_{rms} X_C \right)$. But it is not valid at a particular instant because I and V are not in phase.

Note from Eq. 30–25b that for dc conditions, $\omega = 2\pi f = 0$ and X_C becomes infinite. This is as it should be, since a pure capacitor does not pass dc current. Also, note that the reactance of an inductor increases with frequency, but that of a capacitor decreases with frequency.

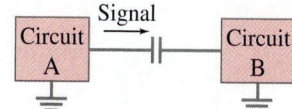

(a) High-pass filter

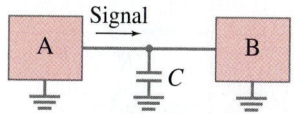

(b) Low-pass filter

EXAMPLE 30–10 **Capacitor reactance.** What is the rms current in the circuit of Fig. 30–17a if $C = 1.0\,\mu\text{F}$ and $V_{\text{rms}} = 120\,\text{V}$? Calculate (a) for $f = 60\,\text{Hz}$, and then (b) for $f = 6.0 \times 10^5\,\text{Hz}$.

APPROACH We find the reactance using Eq. 30–25b, and solve for current in the equivalent form of Ohm's law, Eq. 30–25a.

SOLUTION (a) $X_C = 1/2\pi fC = 1/(6.28)(60\,\text{s}^{-1})(1.0 \times 10^{-6}\,\text{F}) = 2.7\,\text{k}\Omega$. The rms current is (Eq. 30–25a):

$$I_{\text{rms}} = \frac{V_{\text{rms}}}{X_C} = \frac{120\,\text{V}}{2.7 \times 10^3\,\Omega} = 44\,\text{mA}.$$

(b) For $f = 6.0 \times 10^5\,\text{Hz}$, X_C will be $0.27\,\Omega$ and $I_{\text{rms}} = 440\,\text{A}$, vastly larger!

NOTE The dependence on f is dramatic. For high frequencies, the capacitive reactance is very small, and the current can be large.

Two common applications of capacitors are illustrated in Fig. 30–18a and b. In Fig. 30–18a, circuit A is said to be capacitively coupled to circuit B. The purpose of the capacitor is to prevent a dc voltage from passing from A to B but allowing an ac signal to pass relatively unimpeded (if C is sufficiently large, Eq. 30–25b). The capacitor in Fig. 30–18a is called a **high-pass filter** because it allows high-frequency ac to pass easily, but not dc.

In Fig. 30–18b, the capacitor passes ac to ground. In this case, a dc voltage can be maintained between circuits A and B, but an ac signal leaving A passes to ground instead of into B. Thus the capacitor in Fig. 30–18b acts like a **low-pass filter** when a constant dc voltage is required; any high-frequency variation in voltage will pass to ground instead of into circuit B. (Very low-frequency ac will also be able to reach circuit B, at least in part.)

Loudspeakers having separate "woofer" (low-frequency speaker) and "tweeter" (high-frequency speaker) may use a simple "cross-over" that consists of a capacitor in the tweeter circuit to impede low-frequency signals, and an inductor in the woofer circuit to impede high-frequency signals $(X_L = 2\pi fL)$. Hence mainly low-frequency sounds reach and are emitted by the woofer. See Fig. 30–18c.

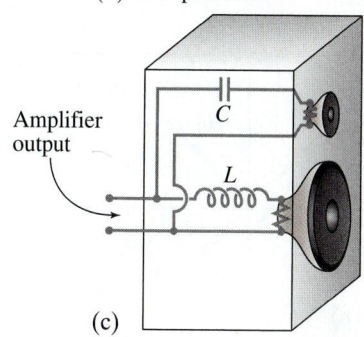

FIGURE 30–18 (a) and (b) Two common uses for a capacitor as a filter. (c) Simple loudspeaker cross-over.

🅐 **PHYSICS APPLIED**
Capacitors as filters

🅐 **PHYSICS APPLIED**
Loudspeaker cross-over

EXERCISE G At what frequency is the reactance of a 1.0-μF capacitor equal to 500 Ω? (a) 320 Hz, (b) 500 Hz, (c) 640 Hz, (d) 2000 Hz, (e) 4000 Hz.

EXERCISE H At what frequency is the reactance of a 1.0-μH inductor equal to 500 Ω? (a) 80 Hz, (b) 500 Hz, (c) 80 MHz, (d) 160 MHz, (e) 500 MHz.

30–8 LRC Series AC Circuit

Let us examine a circuit containing all three elements in series: a resistor R, an inductor L, and a capacitor C, Fig. 30–19. If a given circuit contains only two of these elements, we can still use the results of this Section by setting $R = 0$, $X_L = 0$, or $X_C = 0$, as needed. We let V_R, V_L, and V_C represent the voltage across each element at a *given instant* in time; and V_{R0}, V_{L0}, and V_{C0} represent the *maximum* (peak) values of these voltages. The voltage across each of the elements will follow the phase relations we discussed in the previous Section. At any instant the voltage V supplied by the source will be, by Kirchhoff's loop rule,

$$V = V_R + V_L + V_C. \tag{30–26}$$

Because the various voltages are not in phase, they do not reach their peak values at the same time, so the peak voltage of the source V_0 will *not* equal $V_{R0} + V_{L0} + V_{C0}$.

FIGURE 30–19 An LRC circuit.

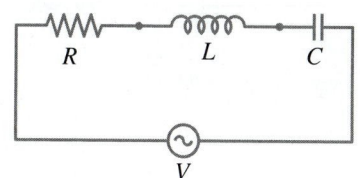

⚠️ **CAUTION**
Peak voltages do not add to yield source voltage

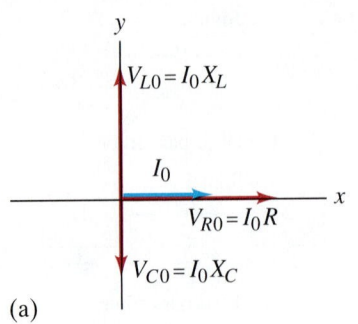

(a)

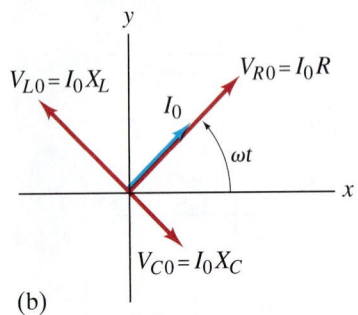

(b)

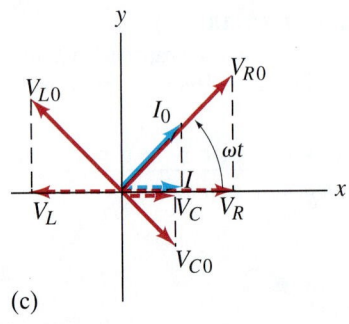

(c)

FIGURE 30–20 Phasor diagram for a series *LRC* circuit at (a) $t = 0$, (b) a time t later. (c) Projections on x axis reflect Eqs. 30–20, 30–22a, and 30–24a.

FIGURE 30–21 Phasor diagram for a series *LRC* circuit showing the sum vector, V_0.

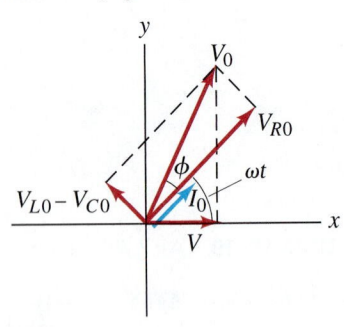

Let us now find the impedance of an *LRC* circuit as a whole (the effect of R, X_C, and X_L), as well as the peak current I_0, and the phase relation between V and I. The current at any instant must be the same at all points in the circuit. Thus the *currents in each element are in phase with each other, even though the voltages are not.* We choose our origin in time $(t = 0)$ so that the current I at any time t is (as in Eq. 30–20)

$$I = I_0 \cos \omega t.$$

We analyze an *LRC* circuit using[†] a **phasor diagram**. Arrows (treated like vectors) are drawn in an xy coordinate system to represent each voltage. The *length of each arrow represents the magnitude of the peak voltage across each element*:

$$V_{R0} = I_0 R, \qquad V_{L0} = I_0 X_L, \qquad \text{and} \qquad V_{C0} = I_0 X_C.$$

V_{R0} is in phase with the current and is initially $(t = 0)$ drawn along the positive x axis, as is the current (Fig. 30–20a). V_{L0} leads the current by 90°, so it leads V_{R0} by 90° and is initially drawn along the positive y axis. V_{C0} lags the current by 90°, so V_{C0} is drawn initially along the negative y axis, Fig. 30–20a.

If we let the vector diagram rotate counterclockwise at frequency f, we get the diagram shown in Fig. 30–20b; after a time, t, each arrow has rotated through an angle ωt. Then the *projections of each arrow on the x axis represent the voltages across each element* at the instant t (Fig. 30–20c). For example $I = I_0 \cos \omega t$. Compare Eqs. 30–22a and 30–24a with Fig. 30–20c to confirm the validity of the phasor diagram.

The sum of the projections of the three voltage vectors represents the instantaneous voltage across the whole circuit, V. Therefore, the vector sum of these vectors will be the vector that represents the peak source voltage, V_0, as shown in Fig. 30–21 where it is seen that V_0 makes an angle ϕ with I_0 and V_{R0}. As time passes, V_0 rotates with the other vectors, so the instantaneous voltage V (projection of V_0 on the x axis) is (see Fig. 30–21)

$$V = V_0 \cos(\omega t + \phi).$$

The voltage V across the whole circuit must equal the source voltage (Fig. 30–19). Thus the voltage from the source is out of phase[‡] with the current by an angle ϕ.

From this analysis we can now determine the total **impedance** Z of the circuit, which is defined in analogy to resistance and reactance as

$$V_{\text{rms}} = I_{\text{rms}} Z, \qquad \text{or} \qquad V_0 = I_0 Z. \qquad \textbf{(30–27)}$$

From Fig. 30–21 we see, using the Pythagorean theorem (V_0 is the hypotenuse of a right triangle), that

$$V_0 = \sqrt{V_{R0}^2 + (V_{L0} - V_{C0})^2}$$

$$= I_0 \sqrt{R^2 + (X_L - X_C)^2}.$$

Thus, from Eq. 30–27, the total impedance Z is

$$Z = \sqrt{R^2 + (X_L - X_C)^2} \qquad \textbf{(30–28a)}$$

$$= \sqrt{R^2 + \left(\omega L - \frac{1}{\omega C}\right)^2}. \qquad \textbf{(30–28b)}$$

Also from Fig. 30–21, we can find the phase angle ϕ between voltage and current:

$$\tan \phi = \frac{V_{L0} - V_{C0}}{V_{R0}} = \frac{I_0(X_L - X_C)}{I_0 R} = \frac{X_L - X_C}{R}. \qquad \textbf{(30–29a)}$$

[†]We could instead do our analysis by rewriting Eq. 30–26 as a differential equation (setting $V_C = Q/C$, $V_R = IR = (dQ/dt)R$, and $V_L = L\, dI/dt$) and trying to solve the differential equation. The differential equation we would get would look like Eq. 14–21 in Section 14–8 (on forced vibrations), and would be solved in the same way. Phasor diagrams are easier, and at the same time give us some physical insight.

[‡]As a check, note that if $R = X_C = 0$, then $\phi = 90°$, and V_0 would lead the current by 90°, as it must for an inductor alone. Similarly, if $R = L = 0$, $\phi = -90°$ and V_0 would lag the current by 90°, as it must for a capacitor alone.

We can also write

$$\cos \phi = \frac{V_{R0}}{V_0} = \frac{I_0 R}{I_0 Z} = \frac{R}{Z}. \qquad \text{(30–29b)}$$

Figure 30–21 was drawn for the case $X_L > X_C$, and the current lags the source voltage by ϕ. When the reverse is true, $X_L < X_C$, then ϕ in Eqs. 30–29 is less than zero, and the current leads the source voltage.

We saw earlier that power is dissipated only by a resistance; none is dissipated by inductance or capacitance. Therefore, the average power $\overline{P} = I_{rms}^2 R$. But from Eq. 30–29b, $R = Z \cos \phi$. Therefore

$$\overline{P} = I_{rms}^2 Z \cos \phi = I_{rms} V_{rms} \cos \phi. \qquad \text{(30–30)}$$

The factor $\cos \phi$ is referred to as the **power factor** of the circuit. For a pure resistor, $\cos \phi = 1$ and $\overline{P} = I_{rms} V_{rms}$. For a capacitor or inductor alone, $\phi = -90°$ or $+90°$, respectively, so $\cos \phi = 0$ and no power is dissipated.

EXAMPLE 30–11 *LRC* **circuit.** Suppose $R = 25.0 \, \Omega$, $L = 30.0 \, \text{mH}$, and $C = 12.0 \, \mu\text{F}$ in Fig. 30–19, and they are connected in series to a 90.0-V ac (rms) 500-Hz source. Calculate (*a*) the current in the circuit, (*b*) the voltmeter readings (rms) across each element, (*c*) the phase angle ϕ, and (*d*) the power dissipated in the circuit.

APPROACH To obtain the current we need to determine the impedance (Eq. 30–28 plus Eqs. 30–23b and 30–25b), and then use $I_{rms} = V_{rms}/Z$. Voltage drops across each element are found using Ohm's law or equivalent for each element: $V_R = IR$, $V_L = IX_L$, and $V_C = IX_C$.

SOLUTION (*a*) First, we find the reactance of the inductor and capacitor at $f = 500 \, \text{Hz} = 500 \, \text{s}^{-1}$:

$$X_L = 2\pi f L = 94.2 \, \Omega, \qquad X_C = \frac{1}{2\pi f C} = 26.5 \, \Omega.$$

Then the total impedance is

$$Z = \sqrt{R^2 + (X_L - X_C)^2} = \sqrt{(25.0 \, \Omega)^2 + (94.2 \, \Omega - 26.5 \, \Omega)^2} = 72.2 \, \Omega.$$

From the impedance version of Ohm's law, Eq. 30–27,

$$I_{rms} = \frac{V_{rms}}{Z} = \frac{90.0 \, \text{V}}{72.2 \, \Omega} = 1.25 \, \text{A}.$$

(*b*) The rms voltage across each element is

$$(V_R)_{rms} = I_{rms} R = (1.25 \, \text{A})(25.0 \, \Omega) = 31.2 \, \text{V}$$
$$(V_L)_{rms} = I_{rms} X_L = (1.25 \, \text{A})(94.2 \, \Omega) = 118 \, \text{V}$$
$$(V_C)_{rms} = I_{rms} X_C = (1.25 \, \text{A})(26.5 \, \Omega) = 33.1 \, \text{V}.$$

NOTE These voltages do *not* add up to the source voltage, 90.0 V (rms). Indeed, the rms voltage across the inductance *exceeds* the source voltage. This can happen because the different voltages are out of phase with each other, and at any instant one voltage can be negative, to compensate for a large positive voltage of another. The rms voltages, however, are always positive by definition. Although the rms voltages need not add up to the source voltage, the instantaneous voltages at any time must add up to the source voltage at that instant.

(*c*) The phase angle ϕ is given by Eq. 30–29b,

$$\cos \phi = \frac{R}{Z} = \frac{25.0 \, \Omega}{72.2 \, \Omega} = 0.346,$$

so $\phi = 69.7°$. Note that ϕ is positive because $X_L > X_C$ in this case, so $V_{L0} > V_{C0}$ in Fig. 30–21.

(*d*) $\overline{P} = I_{rms} V_{rms} \cos \phi = (1.25 \, \text{A})(90.0 \, \text{V})(25.0 \, \Omega/72.2 \, \Omega) = 39.0 \, \text{W}.$

⚠ **C A U T I O N**

Individual peak or rms voltages do NOT add up to source voltage (due to phase differences)

30–9 Resonance in AC Circuits

The rms current in an LRC series circuit is given by (see Eqs. 30–27 and 30–28b):

$$I_{rms} = \frac{V_{rms}}{Z} = \frac{V_{rms}}{\sqrt{R^2 + \left(\omega L - \frac{1}{\omega C}\right)^2}}. \qquad (30\text{–}31)$$

Because the reactance of inductors and capacitors depends on the frequency f ($= \omega/2\pi$) of the source, the current in an LRC circuit will depend on frequency. From Eq. 30–31 we can see that the current will be maximum at a frequency that satisfies

$$\left(\omega L - \frac{1}{\omega C}\right) = 0.$$

We solve this for ω and call the solution ω_0:

$$\omega_0 = \sqrt{\frac{1}{LC}}. \qquad \text{[resonance]} \quad (30\text{–}32)$$

When $\omega = \omega_0$, the circuit is in **resonance**, and $f_0 = \omega_0/2\pi$ is the **resonant frequency** of the circuit. At this frequency, $X_C = X_L$, so the impedance is purely resistive and $\cos\phi = 1$. A graph of I_{rms} versus ω is shown in Fig. 30–22 for particular values of R, L, and C. For small R compared to X_L and X_C, the resonance peak will be higher and sharper. When R is very small, the circuit approaches the pure LC circuit we discussed in Section 30–5. When R is large compared to X_L and X_C, the resonance curve is relatively flat—there is little frequency dependence.

This electrical resonance is analogous to mechanical resonance, which we discussed in Chapter 14. The energy transferred to the system by the source is a maximum at resonance whether it is electrical resonance, the oscillation of a spring, or pushing a child on a swing (Section 14–8). That this is true in the electrical case can be seen from Eq. 30–30; at resonance, $\cos\phi = 1$, and power $\overline{P}$ is a maximum. A graph of power versus frequency peaks like that for the current, Fig. 30–22.

Electric resonance is used in many circuits. Radio and TV sets, for example, use resonant circuits for tuning in a station. Many frequencies reach the circuit, but a significant current flows only for those at or near the resonant frequency. Either L or C is variable so that different stations can be tuned in.

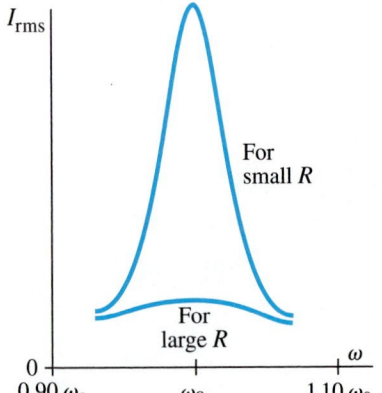

FIGURE 30–22 Current in LRC circuit as a function of angular frequency, ω, showing resonance peak at $\omega = \omega_0 = \sqrt{1/LC}$.

*30–10 Impedance Matching

It is common to connect one electric circuit to a second circuit. For example, a TV antenna is connected to a TV receiver, an amplifier is connected to a loudspeaker; electrodes for an electrocardiogram are connected to a recorder. Maximum power is transferred from one to the other, with a minimum of loss, when the output impedance of the one device matches the input impedance of the second.

To show why, we consider simple circuits that contain only resistance. In Fig. 30–23 the source in circuit 1 could represent the signal from an antenna or a laboratory probe, and R_1 represents its resistance including internal resistance of the source. R_1 is called the output impedance (or resistance) of circuit 1. The output of circuit 1 is across the terminals a and b which are connected to the input of circuit 2 which may be very complicated. We let R_2 be the equivalent "input resistance" of circuit 2.

FIGURE 30–23 Output of the circuit on the left is input to the circuit on the right.

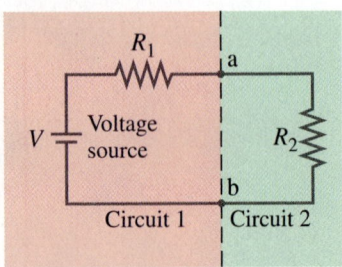

The power delivered to circuit 2 is $P = I^2 R_2$ where $I = V/(R_1 + R_2)$. Thus

$$P = I^2 R_2 = \frac{V^2 R_2}{(R_1 + R_2)^2}.$$

If the resistance of the source is R_1, what value should R_2 have so that the maximum power is transferred to circuit 2? To determine this, we take the derivative of P with respect to R_2 and set it equal to zero, which gives

$$V^2 \left[\frac{1}{(R_1 + R_2)^2} - \frac{2R_2}{(R_1 + R_2)^3} \right] = 0$$

or

$$R_2 = R_1.$$

Thus, the maximum power is transmitted when the *output impedance* of one device *equals the input impedance* of the second. This is called **impedance matching**.

In an ac circuit that contains capacitors and inductors, the different phases are important and the analysis is more complicated. However, the same result holds: to maximize power transfer it is important to match impedances $(Z_2 = Z_1)$.

In addition, it is possible to seriously distort a signal if impedances do not match, and this can lead to meaningless or erroneous experimental results.

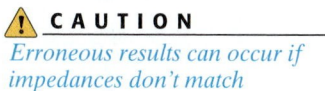

⚠ CAUTION

Erroneous results can occur if impedances don't match

*30–11 Three-Phase AC

Transmission lines typically consist of four wires, rather than two. One of these wires is the ground; the remaining three are used to transmit three-phase ac power which is a superposition of three ac voltages 120° out of phase with each other:

$$V_1 = V_0 \sin \omega t$$
$$V_2 = V_0 \sin(\omega t + 2\pi/3)$$
$$V_3 = V_0 \sin(\omega t + 4\pi/3).$$

(See Fig. 30–24.) Why is three-phase power used? We saw in Fig. 25–22 that single-phase ac (i.e., the voltage V_1 by itself) delivers power to the load in pulses. A much smoother flow of power can be delivered using three-phase power. Suppose that each of the three voltages making up the three-phase source is hooked up to a resistor R. Then the power delivered is:

$$P = \frac{1}{R}(V_1^2 + V_2^2 + V_3^2).$$

You can show that this power is a constant equal to $3V_0^2/2R$, which is three times the rms power delivered by a single-phase source. This smooth flow of power makes electrical equipment run smoothly. Although houses use single-phase ac power, most industrial-grade machinery is wired for three-phase power.

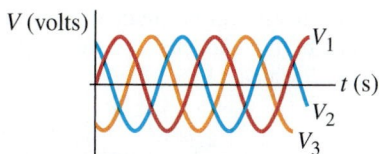

FIGURE 30–24 The three voltages, out of phase by 120° ($=\frac{2}{3}\pi$ radians), in a three-phase power line.

EXAMPLE 30–12 **Three-phase circuit.** In a three-phase circuit, 266 V rms exists between line 1 and ground. What is the rms voltage between lines 2 and 3?

SOLUTION We are given $V_{rms} = V_0/\sqrt{2} = 266$ V. Hence $V_0 = 376$ V. Now $V_3 - V_2 = V_0[\sin(\omega t + 4\pi/3) - \sin(\omega t + 2\pi/3)] = 2V_0 \sin\frac{1}{2}(\frac{2\pi}{3}) \cos\frac{1}{2}(2\omega t)$ where we used the identity: $\sin A - \sin B = 2\sin\frac{1}{2}(A - B)\cos\frac{1}{2}(A + B)$. The rms voltage is

$$(V_3 - V_2)_{rms} = \frac{1}{\sqrt{2}} 2V_0 \sin\frac{\pi}{3} = \sqrt{2}(376\text{ V})(0.866) = 460\text{ V (rms)}.$$

Summary

A changing current in a coil of wire will induce an emf in a second coil placed nearby. The **mutual inductance**, M, is defined as the proportionality constant between the induced emf $\mathcal{E}_2$ in the second coil and the time rate of change of current in the first:

$$\mathcal{E}_2 = -M \, dI_1/dt. \qquad \text{(30–3b)}$$

We can also write M as

$$M = \frac{N_2 \Phi_{21}}{I_1} \qquad \text{(30–1)}$$

where Φ_{21} is the magnetic flux through coil 2 with N_2 loops, produced by the current I_1 in another coil (coil 1).

Within a single coil, a changing current induces an opposing emf, $\mathcal{E}$, so a coil has a **self-inductance** L defined by

$$\mathcal{E} = -L \, dI/dt. \qquad \text{(30–5)}$$

This induced emf acts as an *impedance* to the flow of an alternating current. We can also write L as

$$L = N \frac{\Phi_B}{I} \qquad \text{(30–4)}$$

where Φ_B is the flux through the inductance when a current I flows in its N loops.

When the current in an inductance L is I, the energy stored in the inductance is given by

$$U = \tfrac{1}{2} L I^2. \qquad \text{(30–6)}$$

This energy can be thought of as being stored in the magnetic field of the inductor. The energy density u in any magnetic field B is given by

$$u = \frac{1}{2} \frac{B^2}{\mu_0}, \qquad \text{(30–7)}$$

where μ_0 is replaced by μ if a ferromagnetic material is present.

When an inductance L and resistor R are connected in series to a constant source of emf, V_0, the current rises according to an exponential of the form

$$I = \frac{V_0}{R}\left(1 - e^{-t/\tau}\right), \qquad \text{(30–9)}$$

where

$$\tau = L/R \qquad \text{(30–10)}$$

is the **time constant**. The current eventually levels out at $I = V_0/R$. If the battery is suddenly switched out of the **LR circuit**, and the circuit remains complete, the current drops exponentially, $I = I_0 e^{-t/\tau}$, with the same time constant τ.

The current in a pure **LC circuit** (or charge on the capacitor) would oscillate sinusoidally. The energy too would oscillate back and forth between electric and magnetic, from the capacitor to the inductor, and back again. If such a circuit has resistance (*LRC*), and the capacitor at some instant is charged, it can undergo damped oscillations or exhibit critically damped or overdamped behavior.

Capacitance and inductance offer *impedance* to the flow of alternating current just as resistance does. This impedance is referred to as **reactance**, X, and is defined (as for resistors) as the proportionality constant between voltage and current (either the rms or peak values). Across an inductor,

$$V_0 = I_0 X_L, \qquad \text{(30–23a)}$$

and across a capacitor,

$$V_0 = I_0 X_C. \qquad \text{(30–25a)}$$

The reactance of an inductor increases with frequency:

$$X_L = \omega L. \qquad \text{(30–23b)}$$

where $\omega = 2\pi f$ and f is the frequency of the ac. The reactance of a capacitor decreases with frequency:

$$X_C = \frac{1}{\omega C}. \qquad \text{(30–25b)}$$

Whereas the current through a resistor is always in phase with the voltage across it, this is not true for inductors and capacitors: in an inductor, the current lags the voltage by 90°, and in a capacitor the current leads the voltage by 90°.

In an ac **LRC series circuit**, the total **impedance** Z is defined by the equivalent of $V = IR$ for resistance: namely $V_0 = I_0 Z$ or $V_{\rm rms} = I_{\rm rms} Z$. The impedance Z is related to R, C, and L by

$$Z = \sqrt{R^2 + (X_L - X_C)^2}. \qquad \text{(30–28a)}$$

The current in the circuit lags (or leads) the source voltage by an angle ϕ given by $\cos\phi = R/Z$. Only the resistor in an ac *LRC* circuit dissipates energy, and at a rate

$$\overline{P} = I_{\rm rms}^2 Z \cos\phi \qquad \text{(30–30)}$$

where the factor $\cos\phi$ is referred to as the **power factor**.

An *LRC* series circuit **resonates** at a frequency given by

$$\omega_0 = \sqrt{\frac{1}{LC}}. \qquad \text{(30–32)}$$

The rms current in the circuit is largest when the applied voltage has a frequency equal to $f_0 (= \omega_0/2\pi)$. The lower the resistance R, the higher and sharper the resonance peak.

Questions

1. How would you arrange two flat circular coils so that their mutual inductance was (*a*) greatest, (*b*) least (without separating them by a great distance)?

2. Suppose the second coil of N_2 turns in Fig. 30–2 were moved so it was near the end of the solenoid. How would this affect the mutual inductance?

3. Would two coils with mutual inductance also have self-inductance? Explain.

4. Is the energy density inside a solenoid greatest near the ends of the solenoid or near its center?

5. If you are given a fixed length of wire, how would you shape it to obtain the greatest self-inductance? The least?

6. Does the emf of the battery in Fig. 30–6a affect the time needed for the *LR* circuit to reach (*a*) a given fraction of its maximum possible current, (*b*) a given value of current?

7. A circuit with large inductive time constant carries a steady current. If a switch is opened, there can be a very large (and sometimes dangerous) spark or "arcing over." Explain.

8. At the instant the battery is connected into the LR circuit of Fig. 30–6a, the emf in the inductor has its maximum value even though the current is zero. Explain.

9. What keeps an LC circuit oscillating even after the capacitor has discharged completely?

10. Is the ac current in the inductor always the same as the current in the resistor of the LRC circuit of Fig. 30–13?

11. When an ac generator is connected to an LRC circuit, where does the energy come from ultimately? Where does it go? How do the values of L, C, and R affect the energy supplied by the generator?

12. In an ac LRC circuit, if $X_L > X_C$, the circuit is said to be predominantly "inductive." And if $X_C > X_L$, the circuit is said to be predominantly "capacitive." Discuss the reasons for these terms. In particular, do they say anything about the relative values of L and C at a given frequency?

13. Do the results of Section 30–8 approach the proper expected results when ω approaches zero? What are the expected results?

14. Under what conditions is the impedance in an LRC circuit a minimum?

15. Is it possible for the instantaneous power output of an ac generator connected to an LRC circuit ever to be negative? Explain.

16. In an ac LRC circuit, does the power factor, $\cos \phi$, depend on frequency? Does the power dissipated depend on frequency?

17. Describe briefly how the frequency of the source emf affects the impedance of (a) a pure resistance, (b) a pure capacitance, (c) a pure inductance, (d) an LRC circuit near resonance (R small), (e) an LRC circuit far from resonance (R small).

18. Discuss the response of an LRC circuit as $R \rightarrow 0$ when the frequency is (a) at resonance, (b) near resonance, (c) far from resonance. Is there energy dissipation in each case? Discuss the transformations of energy that occur in each case.

19. An LRC resonant circuit is often called an *oscillator* circuit. What is it that oscillates?

20. Compare the oscillations of an LRC circuit to the vibration of a mass m on a spring. What do L and C correspond to in the mechanical system?

Problems

30–1 Mutual Inductance

1. (II) A 2.44-m-long coil containing 225 loops is wound on an iron core (average $\mu = 1850\mu_0$) along with a second coil of 115 loops. The loops of each coil have a radius of 2.00 cm. If the current in the first coil drops uniformly from 12.0 A to zero in 98.0 ms, determine: (a) the mutual inductance M; (b) the emf induced in the second coil.

2. (II) Determine the mutual inductance per unit length between two long solenoids, one inside the other, whose radii are r_1 and r_2 $(r_2 < r_1)$ and whose turns per unit length are n_1 and n_2.

3. (II) A small thin coil with N_2 loops, each of area A_2, is placed inside a long solenoid, near its center. The solenoid has N_1 loops in its length ℓ and has area A_1. Determine the mutual inductance as a function of θ, the angle between the plane of the small coil and the axis of the solenoid.

4. (III) A long straight wire and a small rectangular wire loop lie in the same plane, Fig. 30–25. Determine the mutual inductance in terms of ℓ_1, ℓ_2, and w. Assume the wire is very long compared to ℓ_1, ℓ_2, and w, and that the rest of its circuit is very far away compared to ℓ_1, ℓ_2, and w.

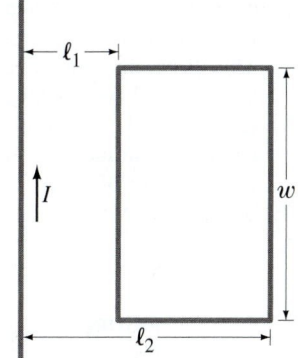

FIGURE 30–25
Problem 4.

30–2 Self-Inductance

5. (I) If the current in a 280-mH coil changes steadily from 25.0 A to 10.0 A in 360 ms, what is the magnitude of the induced emf?

6. (I) How many turns of wire would be required to make a 130-mH inductance out of a 30.0-cm-long air-filled coil with a diameter of 4.2 cm?

7. (I) What is the inductance of a coil if the coil produces an emf of 2.50 V when the current in it changes from -28.0 mA to $+25.0$ mA in 12.0 ms?

8. (II) An air-filled cylindrical inductor has 2800 turns, and it is 2.5 cm in diameter and 21.7 cm long. (a) What is its inductance? (b) How many turns would you need to generate the same inductance if the core were filled with iron of magnetic permeability 1200 times that of free space?

9. (II) A coil has 3.25-Ω resistance and 440-mH inductance. If the current is 3.00 A and is increasing at a rate of 3.60 A/s, what is the potential difference across the coil at this moment?

10. (II) If the outer conductor of a coaxial cable has radius 3.0 mm, what should be the radius of the inner conductor so that the inductance per unit length does not exceed 55 nH per meter?

11. (II) To demonstrate the large size of the henry unit, a physics professor wants to wind an air-filled solenoid with self-inductance of 1.0 H on the outside of a 12-cm diameter plastic hollow tube using copper wire with a 0.81-mm diameter. The solenoid is to be tightly wound with each turn touching its neighbor (the wire has a thin insulating layer on its surface so the neighboring turns are not in electrical contact). How long will the plastic tube need to be and how many kilometers of copper wire will be required? What will be the resistance of this solenoid?

12. (II) The wire of a tightly wound solenoid is unwound and used to make another tightly wound solenoid of 2.5 times the diameter. By what factor does the inductance change?

13. (II) A toroid has a rectangular cross section as shown in Fig. 30–26. Show that the self-inductance is

$$L = \frac{\mu_0 N^2 h}{2\pi} \ln \frac{r_2}{r_1}$$

where N is the total number of turns and r_1, r_2, and h are the dimensions shown in Fig. 30–26. [*Hint*: Use Ampère's law to get B as a function of r inside the toroid, and integrate.]

FIGURE 30–26
Problems 13 and 19.
A toroid of rectangular cross section, with N turns carrying a current I.

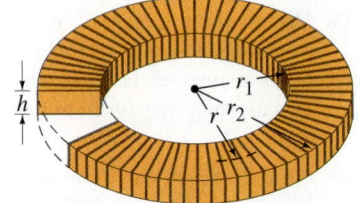

14. (II) Ignoring any mutual inductance, what is the equivalent inductance of two inductors connected (*a*) in series, (*b*) in parallel?

30–3 Magnetic Energy Storage

15. (I) The magnetic field inside an air-filled solenoid 38.0 cm long and 2.10 cm in diameter is 0.600 T. Approximately how much energy is stored in this field?

16. (I) Typical large values for electric and magnetic fields attained in laboratories are about 1.0×10^4 V/m and 2.0 T. (*a*) Determine the energy density for each field and compare. (*b*) What magnitude electric field would be needed to produce the same energy density as the 2.0-T magnetic field?

17. (II) What is the energy density at the center of a circular loop of wire carrying a 23.0-A current if the radius of the loop is 28.0 cm?

18. (II) Calculate the magnetic and electric energy densities at the surface of a 3.0-mm-diameter copper wire carrying a 15-A current.

19. (II) For the toroid of Fig. 30–26, determine the energy density in the magnetic field as a function of r ($r_1 < r < r_2$) and integrate this over the volume to obtain the total energy stored in the toroid, which carries a current I in each of its N loops.

20. (II) Determine the total energy stored per unit length in the magnetic field between the coaxial cylinders of a coaxial cable (Fig. 30–5) by using Eq. 30–7 for the energy density and integrating over the volume.

21. (II) A long straight wire of radius R carries current I uniformly distributed across its cross-sectional area. Find the magnetic energy stored per unit length in the interior of this wire.

30–4 *LR* Circuits

22. (II) After how many time constants does the current in Fig. 30–6 reach within (*a*) 5.0%, (*b*) 1.0%, and (*c*) 0.10% of its maximum value?

23. (II) How many time constants does it take for the potential difference across the resistor in an *LR* circuit like that in Fig. 30–7 to drop to 3.0% of its original value?

24. (II) It takes 2.56 ms for the current in an *LR* circuit to increase from zero to 0.75 its maximum value. Determine (*a*) the time constant of the circuit, (*b*) the resistance of the circuit if $L = 31.0$ mH.

25. (II) (*a*) Determine the energy stored in the inductor L as a function of time for the *LR* circuit of Fig. 30–6a. (*b*) After how many time constants does the stored energy reach 99.9% of its maximum value?

26. (II) In the circuit of Fig. 30–27, determine the current in each resistor (I_1, I_2, I_3) at the moment (*a*) the switch is closed, (*b*) a long time after the switch is closed. After the switch has been closed for a long time, and then reopened, what is each current (*c*) just after it is opened, (*d*) after a long time?

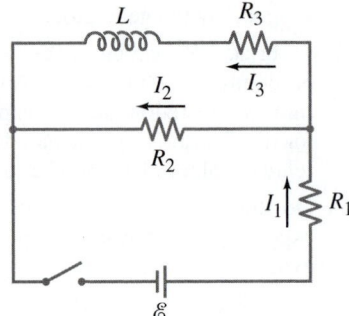

FIGURE 30–27
Problem 26.

27. (II) (*a*) In Fig. 30–28a, assume that the switch S has been in position A for sufficient time so that a steady current $I_0 = V_0/R$ flows through the resistor R. At time $t = 0$, the switch is quickly switched to position B and the current through R decays according to $I = I_0 e^{-t/\tau}$. Show that the maximum emf $\mathscr{E}_{max}$ induced in the inductor during this time period equals the battery voltage V_0. (*b*) In Fig. 30–28b, assume that the switch has been in position A for sufficient time so that a steady current $I_0 = V_0/R$ flows through the resistor R. At time $t = 0$, the switch is quickly switched to position B and the current decays through resistor R' (which is much greater than R) according to $I = I_0 e^{-t/\tau'}$. Show that the maximum emf $\mathscr{E}_{max}$ induced in the inductor during this time period is $(R'/R)V_0$. If $R' = 55R$ and $V_0 = 120$ V, determine $\mathscr{E}_{max}$. [When a mechanical switch is opened, a high-resistance air gap is created, which is modeled as R' here. This Problem illustrates why high-voltage sparking can occur if a current-carrying inductor is suddenly cut off from its power source.]

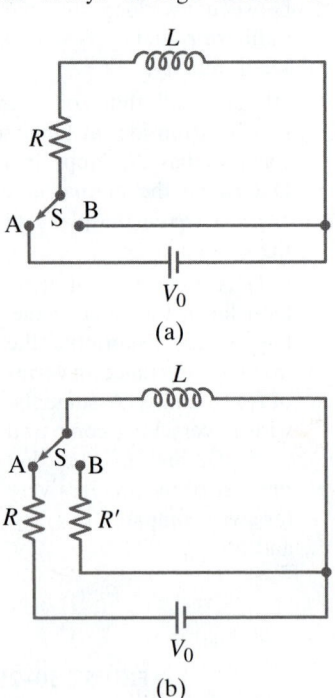

FIGURE 30–28
Problem 27.

(a)

(b)

28. (II) You want to turn on the current through a coil of self-inductance L in a controlled manner, so you place it in series with a resistor $R = 2200\,\Omega$, a switch, and a dc voltage source $V_0 = 240\,\text{V}$. After closing the switch, you find that the current through the coil builds up to its steady-state value with a time constant τ. You are pleased with the current's steady-state value, but want τ to be half as long. What new values should you use for R and V_0?

29. (II) A 12-V battery has been connected to an LR circuit for sufficient time so that a steady current flows through the resistor $R = 2.2\,\text{k}\Omega$ and inductor $L = 18\,\text{mH}$. At $t = 0$, the battery is removed from the circuit and the current decays exponentially through R. Determine the emf $\mathscr{E}$ across the inductor as time t increases. At what time is $\mathscr{E}$ greatest and what is this maximum value (V)?

30. (III) Two tightly wound solenoids have the same length and circular cross-sectional area. But solenoid 1 uses wire that is 1.5 times as thick as solenoid 2. (a) What is the ratio of their inductances? (b) What is the ratio of their inductive time constants (assuming no other resistance in the circuits)?

30–5 LC Circuits and Oscillations

31. (I) The variable capacitor in the tuner of an AM radio has a capacitance of 1350 pF when the radio is tuned to a station at 550 kHz. (a) What must be the capacitance for a station at 1600 kHz? (b) What is the inductance (assumed constant)? Ignore resistance.

32. (I) (a) If the initial conditions of an LC circuit were $I = I_0$ and $Q = 0$ at $t = 0$, write Q as a function of time. (b) Practically, how could you set up these initial conditions?

33. (II) In some experiments, short distances are measured by using capacitance. Consider forming an LC circuit using a parallel-plate capacitor with plate area A, and a known inductance L. (a) If charge is found to oscillate in this circuit at frequency $f = \omega/2\pi$ when the capacitor plates are separated by distance x, show that $x = 4\pi^2 A\epsilon_0 f^2 L$. (b) When the plate separation is changed by Δx, the circuit's oscillation frequency will change by Δf. Show that $\Delta x/x \approx 2(\Delta f/f)$. (c) If f is on the order of 1 MHz and can be measured to a precision of $\Delta f = 1\,\text{Hz}$, with what percent accuracy can x be determined? Assume fringing effects at the capacitor's edges can be neglected.

34. (II) A 425-pF capacitor is charged to 135 V and then quickly connected to a 175-mH inductor. Determine (a) the frequency of oscillation, (b) the peak value of the current, and (c) the maximum energy stored in the magnetic field of the inductor.

35. (II) At $t = 0$, let $Q = Q_0$, and $I = 0$ in an LC circuit. (a) At the first moment when the energy is shared equally by the inductor and the capacitor, what is the charge on the capacitor? (b) How much time has elapsed (in terms of the period T)?

30–6 LC Oscillations with Resistance

36. (II) A damped LC circuit loses 3.5% of its electromagnetic energy per cycle to thermal energy. If $L = 65\,\text{mH}$ and $C = 1.00\,\mu\text{F}$, what is the value of R?

37. (II) In an oscillating LRC circuit, how much time does it take for the energy stored in the fields of the capacitor and inductor to fall to 75% of the initial value? (See Fig. 30–13; assume $R \ll \sqrt{4L/C}$.)

38. (III) How much resistance must be added to a pure LC circuit $(L = 350\,\text{mH}, \quad C = 1800\,\text{pF})$ to change the oscillator's frequency by 0.25%? Will it be increased or decreased?

30–7 AC Circuits; Reactance

39. (I) At what frequency will a 32.0-mH inductor have a reactance of $660\,\Omega$?

40. (I) What is the reactance of a 9.2-μF capacitor at a frequency of (a) 60.0 Hz, (b) 1.00 MHz?

41. (I) Plot a graph of the reactance of a 1.0-μF capacitor as a function of frequency from 10 Hz to 1000 Hz.

42. (I) Calculate the reactance of, and rms current in, a 36.0-mH radio coil connected to a 250-V (rms) 33.3-kHz ac line. Ignore resistance.

43. (II) A resistor R is in parallel with a capacitor C, and this parallel combination is in series with a resistor R'. If connected to an ac voltage source of frequency ω, what is the equivalent impedance of this circuit at the two extremes in frequency (a) $\omega = 0$, and (b) $\omega = \infty$?

44. (II) What is the inductance L of the primary of a transformer whose input is 110 V at 60 Hz and the current drawn is 3.1 A? Assume no current in the secondary.

45. (II) (a) What is the reactance of a 0.086-μF capacitor connected to a 22-kV (rms), 660-Hz line? (b) Determine the frequency and the peak value of the current.

46. (II) A capacitor is placed in parallel with some device, B, as in Fig. 30–18b, to filter out stray high-frequency signals, but to allow ordinary 60-Hz ac to pass through with little loss. Suppose that circuit B in Fig. 30–18b is a resistance $R = 490\,\Omega$ connected to ground, and that $C = 0.35\,\mu\text{F}$. What percent of the incoming current will pass through C rather than R if it is (a) 60 Hz; (b) 60,000 Hz?

47. (II) A current $I = 1.80 \cos 377t$ (I in amps, t in seconds, and the "angle" is in radians) flows in a series LR circuit in which $L = 3.85\,\text{mH}$ and $R = 1.35\,\text{k}\Omega$. What is the average power dissipation?

30–8 LRC Series AC Circuit

48. (I) A 10.0-kΩ resistor is in series with a 26.0-mH inductor and an ac source. Calculate the impedance of the circuit if the source frequency is (a) 55.0 Hz; (b) 55,000 Hz.

49. (I) A 75-Ω resistor and a 6.8-μF capacitor are connected in series to an ac source. Calculate the impedance of the circuit if the source frequency is (a) 60 Hz; (b) 6.0 MHz.

50. (I) For a 120-V, 60-Hz voltage, a current of 70 mA passing through the body for 1.0 s could be lethal. What must be the impedance of the body for this to occur?

51. (II) A 2.5-kΩ resistor in series with a 420-mH inductor is driven by an ac power supply. At what frequency is the impedance double that of the impedance at 60 Hz?

52. (II) (a) What is the rms current in a series RC circuit if $R = 3.8\,\text{k}\Omega$, $C = 0.80\,\mu\text{F}$, and the rms applied voltage is 120 V at 60.0 Hz? (b) What is the phase angle between voltage and current? (c) What is the power dissipated by the circuit? (d) What are the voltmeter readings across R and C?

53. (II) An ac voltage source is connected in series with a 1.0-μF capacitor and a 750-Ω resistor. Using a digital ac voltmeter, the amplitude of the voltage source is measured to be 4.0 V rms, while the voltages across the resistor and across the capacitor are found to be 3.0 V rms and 2.7 V rms, respectively. Determine the frequency of the ac voltage source. Why is the voltage measured across the voltage source not equal to the sum of the voltages measured across the resistor and across the capacitor?

54. (II) Determine the total impedance, phase angle, and rms current in an LRC circuit connected to a 10.0-kHz, 725-V (rms) source if $L = 32.0$ mH, $R = 8.70$ kΩ, and $C = 6250$ pF.

55. (II) (a) What is the rms current in a series LR circuit when a 60.0-Hz, 120-V rms ac voltage is applied, where $R = 965$ Ω and $L = 225$ mH? (b) What is the phase angle between voltage and current? (c) How much power is dissipated? (d) What are the rms voltage readings across R and L?

56. (II) A 35-mH inductor with 2.0-Ω resistance is connected in series to a 26-μF capacitor and a 60-Hz, 45-V (rms) source. Calculate (a) the rms current, (b) the phase angle, and (c) the power dissipated in this circuit.

57. (II) A 25-mH coil whose resistance is 0.80 Ω is connected to a capacitor C and a 360-Hz source voltage. If the current and voltage are to be in phase, what value must C have?

58. (II) A 75-W lightbulb is designed to operate with an applied ac voltage of 120 V rms. The bulb is placed in series with an inductor L, and this series combination is then connected to a 60-Hz 240-V rms voltage source. For the bulb to operate properly, determine the required value for L. Assume the bulb has resistance R and negligible inductance.

59. (II) In the LRC circuit of Fig. 30–19, suppose $I = I_0 \sin \omega t$ and $V = V_0 \sin(\omega t + \phi)$. Determine the instantaneous power dissipated in the circuit from $P = IV$ using these equations and show that on the average, $\overline{P} = \frac{1}{2} V_0 I_0 \cos \phi$, which confirms Eq. 30–30.

60. (II) An LRC series circuit with $R = 150$ Ω, $L = 25$ mH, and $C = 2.0$ μF is powered by an ac voltage source of peak voltage $V_0 = 340$ V and frequency $f = 660$ Hz. (a) Determine the peak current that flows in this circuit. (b) Determine the phase angle of the source voltage relative to the current. (c) Determine the peak voltage across R and its phase angle relative to the source voltage. (d) Determine the peak voltage across L and its phase angle relative to the source voltage. (e) Determine the peak voltage across C and its phase angle relative to the source voltage.

61. (II) An LR circuit can be used as a "phase shifter." Assume that an "input" source voltage $V = V_0 \sin(2\pi ft + \phi)$ is connected across a series combination of an inductor $L = 55$ mH and resistor R. The "output" of this circuit is taken across the resistor. If $V_0 = 24$ V and $f = 175$ Hz, determine the value of R so that the output voltage V_R lags the input voltage V by 25°. Compare (as a ratio) the peak output voltage with V_0.

30–9 Resonance in AC Circuits

62. (I) A 3800-pF capacitor is connected in series to a 26.0-μH coil of resistance 2.00 Ω. What is the resonant frequency of this circuit?

63. (I) What is the resonant frequency of the LRC circuit of Example 30–11? At what rate is energy taken from the generator, on the average, at this frequency?

64. (II) An LRC circuit has $L = 4.15$ mH and $R = 3.80$ kΩ. (a) What value must C have to produce resonance at 33.0 kHz? (b) What will be the maximum current at resonance if the peak external voltage is 136 V?

65. (II) The frequency of the ac voltage source (peak voltage V_0) in an LRC circuit is tuned to the circuit's resonant frequency $f_0 = 1/(2\pi\sqrt{LC})$. (a) Show that the peak voltage across the capacitor is $V_{C0} = V_0 T_0/2\pi \tau$), where $T_0 (= 1/f_0)$ is the period of the resonant frequency and $\tau = RC$ is the time constant for charging the capacitor C through a resistor R. (b) Define $\beta = T_0/(2\pi\tau)$ so that $V_{C0} = \beta V_0$. Then β is the "amplification" of the source voltage across the capacitor. If a particular LRC circuit contains a 2.0-nF capacitor and has a resonant frequency of 5.0 kHz, what value of R will yield $\beta = 125$?

66. (II) Capacitors made from piezoelectric materials are commonly used as sound transducers ("speakers"). They often require a large operating voltage. One method for providing the required voltage is to include the speaker as part of an LRC circuit as shown in Fig. 30–29, where the speaker is modeled electrically as the capacitance $C = 1.0$ nF. Take $R = 35$ Ω and $L = 55$ mH. (a) What is the resonant frequency f_0 for this circuit? (b) If the voltage source has peak amplitude $V_0 = 2.0$ V at frequency $f = f_0$, find the peak voltage V_{C0} across the speaker (i.e., the capacitor C). (c) Determine the ratio V_{C0}/V_0.

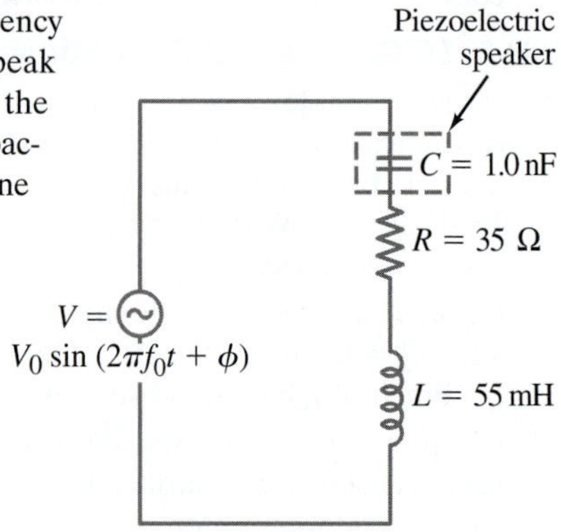

FIGURE 30–29
Problem 66.

67. (II) (a) Determine a formula for the average power $\overline{P}$ dissipated in an LRC circuit in terms of L, R, C, ω, and V_0. (b) At what frequency is the power a maximum? (c) Find an approximate formula for the width of the resonance peak in average power, $\Delta\omega$, which is the difference in the two (angular) frequencies where $\overline{P}$ has half its maximum value. Assume a sharp peak.

68. (II) (a) Show that oscillation of charge Q on the capacitor of an LRC circuit has amplitude

$$Q_0 = \frac{V_0}{\sqrt{(\omega R)^2 + \left(\omega^2 L - \dfrac{1}{C}\right)^2}}.$$

(b) At what angular frequency, ω', will Q_0 be a maximum? (c) Compare to a forced damped harmonic oscillator (Chapter 14), and discuss. (See also Question 20 in this Chapter.)

69. (II) A resonant circuit using a 220-nF capacitor is to resonate at 18.0 kHz. The air-core inductor is to be a solenoid with closely packed coils made from 12.0 m of insulated wire 1.1 mm in diameter. How many loops will the inductor contain?

*30–10 Impedance Matching

*70. (II) The output of an electrocardiogram amplifier has an impedance of 45 kΩ. It is to be connected to an 8.0-Ω loudspeaker through a transformer. What should be the turns ratio of the transformer?

General Problems

71. A 2200-pF capacitor is charged to 120 V and then quickly connected to an inductor. The frequency of oscillation is observed to be 17 kHz. Determine (a) the inductance, (b) the peak value of the current, and (c) the maximum energy stored in the magnetic field of the inductor.

72. At $t = 0$, the current through a 60.0-mH inductor is 50.0 mA and is increasing at the rate of 78.0 mA/s. What is the initial energy stored in the inductor, and how long does it take for the energy to increase by a factor of 5.0 from the initial value?

73. At time $t = 0$, the switch in the circuit shown in Fig. 30–30 is closed. After a sufficiently long time, steady currents I_1, I_2, and I_3 flow through resistors R_1, R_2, and R_3, respectively. Determine these three currents.

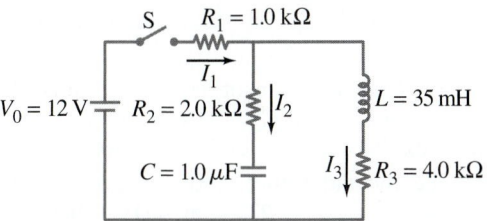

FIGURE 30–30 Problem 73.

74. (a) Show that the self-inductance L of a toroid (Fig. 30–31) of radius r_0 containing N loops each of diameter d is

$$L \approx \frac{\mu_0 N^2 d^2}{8r_0}$$

if $r_0 \gg d$. Assume the field is uniform inside the toroid; is this actually true? Is this result consistent with L for a solenoid? Should it be? (b) Calculate the inductance L of a large toroid if the diameter of the coils is 2.0 cm and the diameter of the whole ring is 66 cm. Assume the field inside the toroid is uniform. There are a total of 550 loops of wire.

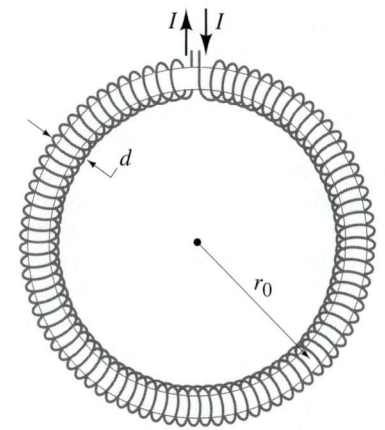

FIGURE 30–31
A toroid.
Problem 74.

75. A pair of straight parallel thin wires, such as a lamp cord, each of radius r, are a distance ℓ apart and carry current to a circuit some distance away. Ignoring the field within each wire, show that the inductance per unit length is $(\mu_0/\pi) \ln[(\ell - r)/r]$.

76. Assuming the Earth's magnetic field averages about 0.50×10^{-4} T near the surface of the Earth, estimate the total energy stored in this field in the first 5.0 km above the Earth's surface.

77. (a) For an underdamped LRC circuit, determine a formula for the energy $U = U_E + U_B$ stored in the electric and magnetic fields as a function of time. Give answer in terms of the initial charge Q_0 on the capacitor. (b) Show how dU/dt is related to the rate energy is transformed in the resistor, $I^2 R$.

78. An electronic device needs to be protected against sudden surges in current. In particular, after the power is turned on the current should rise to no more than 7.5 mA in the first 75 μs. The device has resistance 150 Ω and is designed to operate at 33 mA. How would you protect this device?

79. The circuit shown in Fig. 30–32a can integrate (in the calculus sense) the input voltage V_{in}, if the time constant L/R is large compared with the time during which V_{in} varies. Explain how this integrator works and sketch its output for the square wave signal input shown in Fig. 30–32b. [*Hint*: Write Kirchhoff's loop rule for the circuit. Multiply each term in this differential equation (in I) by a factor $e^{Rt/L}$ to make it easier to integrate.]

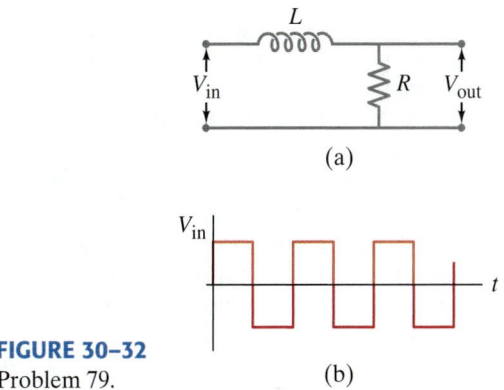

FIGURE 30–32
Problem 79.

80. Suppose circuit B in Fig. 30–18a consists of a resistance $R = 550 \,\Omega$. The filter capacitor has capacitance $C = 1.2 \,\mu$F. Will this capacitor act to eliminate 60-Hz ac but pass a high-frequency signal of frequency 6.0 kHz? To check this, determine the voltage drop across R for a 130-mV signal of frequency (a) 60 Hz; (b) 6.0 kHz.

81. An ac voltage source $V = V_0 \sin(\omega t + 90°)$ is connected across an inductor L and current $I = I_0 \sin(\omega t)$ flows in this circuit. Note that the current and source voltage are 90° out of phase. (a) Directly calculate the average power delivered by the source over one period T of its sinusoidal cycle via the integral $\overline{P} = \int_0^T VI \, dt/T$. (b) Apply the relation $\overline{P} = I_{rms} V_{rms} \cos \phi$ to this circuit and show that the answer you obtain is consistent with that found in part (a). Comment on your results.

82. A circuit contains two elements, but it is not known if they are L, R, or C. The current in this circuit when connected to a 120-V 60-Hz source is 5.6 A and lags the voltage by 65°. What are the two elements and what are their values?

83. A 3.5-kΩ resistor in series with a 440-mH inductor is driven by an ac power supply. At what frequency is the impedance double that of the impedance at 60 Hz?

84. (a) What is the rms current in an RC circuit if $R = 5.70\,\text{k}\Omega$, $C = 1.80\,\mu\text{F}$, and the rms applied voltage is 120 V at 60.0 Hz? (b) What is the phase angle between voltage and current? (c) What is the power dissipated by the circuit? (d) What are the voltmeter readings across R and C?

85. An inductance coil draws 2.5 A dc when connected to a 45-V battery. When connected to a 60-Hz 120-V (rms) source, the current drawn is 3.8 A (rms). Determine the inductance and resistance of the coil.

86. The **Q-value** of a resonance circuit can be defined as the ratio of the voltage across the capacitor (or inductor) to the voltage across the resistor, at resonance. The larger the Q factor, the sharper the resonance curve will be and the sharper the tuning. (a) Show that the Q factor is given by the equation $Q = (1/R)\sqrt{L/C}$. (b) At a resonant frequency $f_0 = 1.0\,\text{MHz}$, what must be the value of L and R to produce a Q factor of 350? Assume that $C = 0.010\,\mu\text{F}$.

87. Show that the fraction of electromagnetic energy lost (to thermal energy) per cycle in a lightly damped $(R^2 \ll 4L/C)$ LRC circuit is approximately

$$\frac{\Delta U}{U} = \frac{2\pi R}{L\omega} = \frac{2\pi}{Q}.$$

The quantity Q can be defined as $Q = L\omega/R$, and is called the Q-value, or *quality factor*, of the circuit and is a measure of the damping present. A high Q-value means smaller damping and less energy input required to maintain oscillations.

88. In a series LRC circuit, the inductance is 33 mH, the capacitance is 55 nF, and the resistance is $1.50\,\text{k}\Omega$. At what frequencies is the power factor equal to 0.17?

89. In our analysis of a series LRC circuit, Fig. 30–19, suppose we chose $V = V_0 \sin \omega t$. (a) Construct a phasor diagram, like that of Fig. 30–21, for this case. (b) Write a formula for the current I, defining all terms.

90. A voltage $V = 0.95 \sin 754t$ is applied to an LRC circuit (I is in amperes, t is in seconds, V is in volts, and the "angle" is in radians) which has $L = 22.0\,\text{mH}$, $R = 23.2\,\text{k}\Omega$, and $C = 0.42\,\mu\text{F}$. (a) What is the impedance and phase angle? (b) How much power is dissipated in the circuit? (c) What is the rms current and voltage across each element?

91. *Filter circuit.* Figure 30–33 shows a simple filter circuit designed to pass dc voltages with minimal attenuation and to remove, as much as possible, any ac components (such as 60-Hz line voltage that could cause hum in a stereo receiver, for example). Assume $V_{in} = V_1 + V_2$ where V_1 is dc and $V_2 = V_{20} \sin \omega t$, and that any resistance is very small. (a) Determine the current through the capacitor: give amplitude and phase (assume $R = 0$ and $X_L > X_C$). (b) Show that the ac component of the output voltage, $V_{2\,out}$, equals $(Q/C) - V_1$, where Q is the charge on the capacitor at any instant, and determine the amplitude and phase of $V_{2\,out}$. (c) Show that the attenuation of the ac voltage is greatest when $X_C \ll X_L$, and calculate the ratio of the output to input ac voltage in this case. (d) Compare the dc output voltage to input voltage.

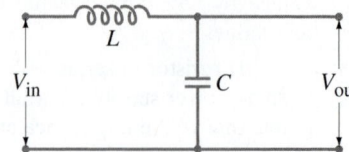

FIGURE 30–33
Problems 91 and 92.

92. Show that if the inductor L in the filter circuit of Fig. 30–33 (Problem 91) is replaced by a large resistor R, there will still be significant attenuation of the ac voltage and little attenuation of the dc voltage if the input dc voltage is high and the current (and power) are low.

93. A resistor R, capacitor C, and inductor L are connected in parallel across an ac generator as shown in Fig. 30–34. The source emf is $V = V_0 \sin \omega t$. Determine the current as a function of time (including amplitude and phase): (a) in the resistor, (b) in the inductor, (c) in the capacitor. (d) What is the total current leaving the source? (Give amplitude I_0 and phase.) (e) Determine the impedance Z defined as $Z = V_0/I_0$. (f) What is the power factor?

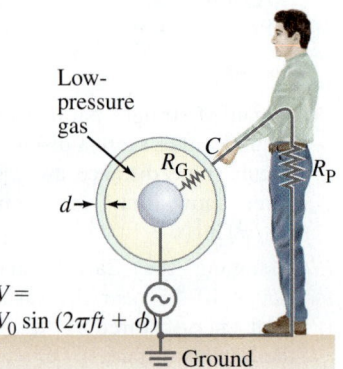

FIGURE 30–34
Problem 93.

94. Suppose a series LRC circuit has two resistors, R_1 and R_2, two capacitors, C_1 and C_2, and two inductors, L_1 and L_2, all in series. Calculate the total impedance of the circuit.

95. Determine the inductance L of the primary of a transformer whose input is 220 V at 60 Hz when the current drawn is 4.3 A. Assume no current in the secondary.

96. In a *plasma globe*, a hollow glass sphere is filled with low-pressure gas and a small spherical metal electrode is located at its center. Assume an ac voltage source of peak voltage V_0 and frequency f is applied between the metal sphere and the ground, and that a person is touching the outer surface of the globe with a fingertip, whose approximate area is $1.0\,\text{cm}^2$. The equivalent circuit for this situation is shown in Fig. 30–35, where R_G and R_P are the resistances of the gas and the person, respectively, and C is the capacitance formed by the gas, glass, and finger. (a) Determine C assuming it is a parallel-plate capacitor. The conductive gas and the person's fingertip form the opposing plates of area $A = 1.0\,\text{cm}^2$. The plates are separated by glass (dielectric constant $K = 5.0$) of thickness $d = 2.0\,\text{mm}$. (b) In a typical plasma globe, $f = 12\,\text{kHz}$. Determine the reactance X_C of C at this frequency in $\text{M}\Omega$. (c) The voltage may be $V_0 = 2500\,\text{V}$. With this high voltage, the dielectric strength of the gas is exceeded and the gas becomes ionized. In this "plasma" state, the gas emits light ("sparks") and is highly conductive so that $R_G \ll X_C$. Assuming also that $R_P \ll X_C$, estimate the peak current that flows in the given circuit. Is this level of current dangerous? (d) If the plasma globe operated at $f = 1.0\,\text{MHz}$, estimate the peak current that would flow in the given circuit. Is this level of current dangerous?

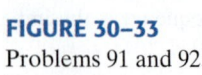

FIGURE 30–35
Problem 96.

97. You have a small electromagnet that consumes 350 W from a residential circuit operating at 120 V at 60 Hz. Using your ac multimeter, you determine that the unit draws 4.0 A rms. What are the values of the inductance and the internal resistance?

98. An inductor L in series with a resistor R, driven by a sinusoidal voltage source, responds as described by the following differential equation:

$$V_0 \sin \omega t = L\frac{dI}{dt} + RI.$$

Show that a current of the form $I = I_0 \sin(\omega t - \phi)$ flows through the circuit by direct substitution into the differential equation. Determine the amplitude of the current (I_0) and the phase difference ϕ between the current and the voltage source.

99. In a certain LRC series circuit, when the ac voltage source has a particular frequency f, the peak voltage across the inductor is 6.0 times greater than the peak voltage across the capacitor. Determine f in terms of the resonant frequency f_0 of this circuit.

100. For the circuit shown in Fig. 30–36, $V = V_0 \sin \omega t$. Calculate the current in each element of the circuit, as well as the total impedance. [*Hint:* Try a trial solution of the form $I = I_0 \sin(\omega t + \phi)$ for the current leaving the source.]

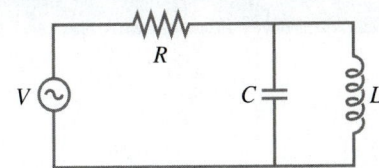

FIGURE 30–36
Problem 100.

101. To detect vehicles at traffic lights, wire loops with dimensions on the order of 2 m are often buried horizontally under roadways. Assume the self-inductance of such a loop is $L = 5.0$ mH and that it is part of an LRC circuit as shown in Fig. 30–37 with $C = 0.10 \, \mu\text{F}$ and $R = 45 \, \Omega$. The ac voltage has frequency f and rms voltage V_{rms}. (*a*) The frequency f is chosen to match the resonant frequency f_0 of the circuit. Find f_0 and determine what the rms voltage $(V_R)_{\text{rms}}$ across the resistor will be when $f = f_0$. (*b*) Assume that f, C, and R never change, but that, when a car is located above the buried loop, the loop's self-inductance decreases by 10% (due to induced eddy currents in the car's metal parts). Determine by what factor the voltage $(V_R)_{\text{rms}}$ decreases in this situation in comparison to no car above the loop. [Monitoring $(V_R)_{\text{rms}}$ detects the presence of a car.]

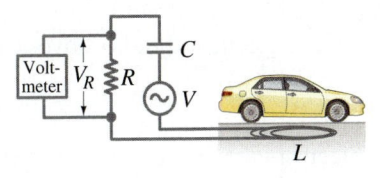

FIGURE 30–37
Problem 101.

102. For the circuit shown in Fig. 30–38, show that if the condition $R_1 R_2 = L/C$ is satisfied then the potential difference between points a and b is zero for all frequencies.

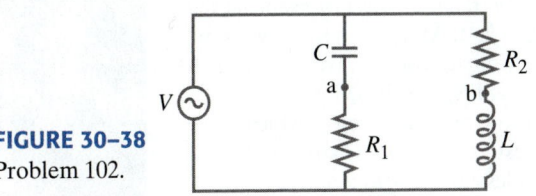

FIGURE 30–38
Problem 102.

*Numerical/Computer

* **103.** (II) The RC circuit shown in Fig. 30–39 is called a **low-pass filter** because it passes low-frequency ac signals with less attenuation than high-frequency ac signals. (*a*) Show that the voltage gain is $A = V_{\text{out}}/V_{\text{in}} = 1/(4\pi^2 f^2 R^2 C^2 + 1)^{\frac{1}{2}}$. (*b*) Discuss the behavior of the gain A for $f \to 0$ and $f \to \infty$. (*c*) Choose $R = 850 \, \Omega$ and $C = 1.0 \times 10^{-6}$ F, and graph $\log A$ versus $\log f$ with suitable scales to show the behavior of the circuit at low and high frequencies.

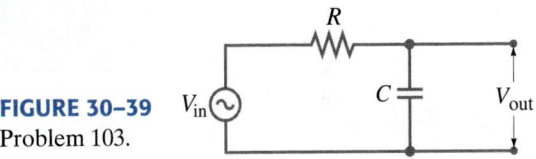

FIGURE 30–39
Problem 103.

* **104.** (II) The RC circuit shown in Fig. 30–40 is called a **high-pass filter** because it passes high-frequency ac signals with less attenuation than low-frequency ac signals. (*a*) Show that the voltage gain is $A = V_{\text{out}}/V_{\text{in}} = 2\pi f RC/(4\pi^2 f^2 R^2 C^2 + 1)^{\frac{1}{2}}$. (*b*) Discuss the behavior of the gain A for $f \to 0$ and $f \to \infty$. (*c*) Choose $R = 850 \, \Omega$ and $C = 1.0 \times 10^{-6}$ F, and then graph $\log A$ versus $\log f$ with suitable scales to show the behavior of the circuit at high and low frequencies.

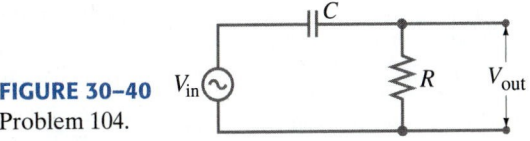

FIGURE 30–40
Problem 104.

* **105.** (III) Write a computer program or use a spreadsheet program to plot I_{rms} for an ac LRC circuit with a sinusoidal voltage source (Fig. 30–19) with $V_{\text{rms}} = 0.100$ V. For $L = 50 \, \mu\text{H}$ and $C = 50 \, \mu\text{F}$, plot the I_{rms} graph for (*a*) $R = 0.10 \, \Omega$, and (*b*) $R = 1.0 \, \Omega$ from $\omega = 0.1\omega_0$ to $\omega = 3.0\omega_0$ on the same graph.

Answers to Exercises

A: (*a*) 360 A/s; (*b*) 12 V.

B: (*b*).

C: (*a*).

D: From Eq. 30–5, L has dimensions VT/A so (L/R) has dimensions $(VT/A)/(V/A) = T$.

E: (*d*).

F: (*c*)

G: (*a*).

H: (*c*).

Wireless technology is all around us: in this photo we see a Bluetooth earpiece for wireless telephone communication and a wi-fi computer. The wi-fi antenna is just visible at the lower left. All these devices work by electromagnetic waves traveling through space, based on the great work of Maxwell which we investigate in this Chapter. Modern wireless devices are applications of Marconi's development of long distance transmission of information a century ago.

We will see in this Chapter that Maxwell predicted the existence of EM waves from his famous equations. Maxwell's equations themselves are a magnificent summary of electromagnetism. We will also examine how EM waves carry energy and momentum.

CHAPTER

31

CONTENTS

31–1 Changing Electric Fields Produce Magnetic Fields; Ampère's Law and Displacement Current
31–2 Gauss's Law for Magnetism
31–3 Maxwell's Equations
31–4 Production of Electromagnetic Waves
31–5 Electromagnetic Waves, and Their Speed, Derived from Maxwell's Equations
31–6 Light as an Electromagnetic Wave and the Electromagnetic Spectrum
31–7 Measuring the Speed of Light
31–8 Energy in EM Waves; the Poynting Vector
31–9 Radiation Pressure
31–10 Radio and Television; Wireless Communication

Maxwell's Equations and Electromagnetic Waves

CHAPTER-OPENING QUESTION—Guess now!

Which of the following best describes the difference between radio waves and X-rays?

- **(a)** X-rays are radiation while radio waves are electromagnetic waves.
- **(b)** Both can be thought of as electromagnetic waves. They differ only in wavelength and frequency.
- **(c)** X-rays are pure energy. Radio waves are made of fields, not energy.
- **(d)** Radio waves come from electric currents in an antenna. X-rays are not related to electric charge.
- **(e)** The fact that X-rays can expose film, and radio waves cannot, means they are fundamentally different.

The culmination of electromagnetic theory in the nineteenth century was the prediction, and the experimental verification, that waves of electromagnetic fields could travel through space. This achievement opened a whole new world of communication: first the wireless telegraph, then radio and television, and more recently cell phones, remote-control devices, wi-fi, and Bluetooth. Most important was the spectacular prediction that visible light is an electromagnetic wave.

The theoretical prediction of electromagnetic waves was the work of the Scottish physicist James Clerk Maxwell (1831–1879; Fig. 31–1), who unified, in one magnificent theory, all the phenomena of electricity and magnetism.

The development of electromagnetic theory in the early part of the nineteenth century by Oersted, Ampère, and others was not actually done in terms of electric and magnetic fields. The idea of the field was introduced somewhat later by Faraday, and was not generally used until Maxwell showed that all electric and magnetic phenomena could be described using only four equations involving electric and magnetic fields. These equations, known as **Maxwell's equations**, are the basic equations for all electromagnetism. They are fundamental in the same sense that Newton's three laws of motion and the law of universal gravitation are for mechanics. In a sense, they are even more fundamental, since they are consistent with the theory of relativity (Chapter 36), whereas Newton's laws are not. Because all of electromagnetism is contained in this set of four equations, Maxwell's equations are considered one of the great triumphs of human intellect.

Before we discuss Maxwell's equations and electromagnetic waves, we first need to discuss a major new prediction of Maxwell's, and, in addition, Gauss's law for magnetism.

FIGURE 31–1 James Clerk Maxwell (1831–1879).

31–1 Changing Electric Fields Produce Magnetic Fields; Ampère's Law and Displacement Current

Ampère's Law

That a magnetic field is produced by an electric current was discovered by Oersted, and the mathematic relation is given by Ampère's law (Eq. 28–3):

$$\oint \vec{\mathbf{B}} \cdot d\vec{\boldsymbol{\ell}} = \mu_0 I_{\text{encl}}.$$

Is it possible that magnetic fields could be produced in another way as well? For if a changing magnetic field produces an electric field, as discussed in Section 29–7, then perhaps the reverse might be true as well: that *a changing electric field will produce a magnetic field*. If this were true, it would signify a beautiful *symmetry* in nature.

To back up this idea that a changing electric field might produce a magnetic field, we use an indirect argument that goes something like this. According to Ampère's law, we divide any chosen closed path into short segments $d\vec{\boldsymbol{\ell}}$, take the dot product of each $d\vec{\boldsymbol{\ell}}$ with the magnetic field $\vec{\mathbf{B}}$ at that segment, and sum (integrate) all these products over the chosen closed path. That sum will equal μ_0 times the total current I that passes through a surface bounded by the path of the line integral. When we applied Ampère's law to the field around a straight wire (Section 28–4), we imagined the current as passing through the circular area enclosed by our circular loop, and that area is the flat surface 1 shown in Fig. 31–2. However, we could just as well use the sackshaped surface 2 in Fig. 31–2 as the surface for Ampère's law, since the same current I passes through it.

Now consider the closed circular path for the situation of Fig. 31–3, where a capacitor is being discharged. Ampère's law works for surface 1 (current I passes through surface 1), but it does not work for surface 2, since no current passes through surface 2. There is a magnetic field around the wire, so the left side of Ampère's law ($\int \vec{\mathbf{B}} \cdot d\vec{\boldsymbol{\ell}}$) is not zero; yet no current flows through surface 2, so the right side of Ampère's law *is* zero. We seem to have a contradiction of Ampère's law.

There is a magnetic field present in Fig. 31–3, however, only if charge is flowing to or away from the capacitor plates. The changing charge on the plates means that the electric field between the plates is changing in time. Maxwell resolved the problem of no current through surface 2 in Fig. 31–3 by proposing that there needs to be an extra term on the right in Ampère's law involving the changing electric field.

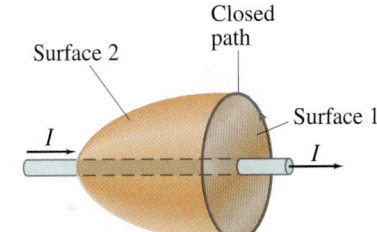

FIGURE 31–2 Ampère's law applied to two different surfaces bounded by the same closed path.

FIGURE 31–3 A capacitor discharging. A conduction current passes through surface 1, but no conduction current passes through surface 2. An extra term is needed in Ampère's law.

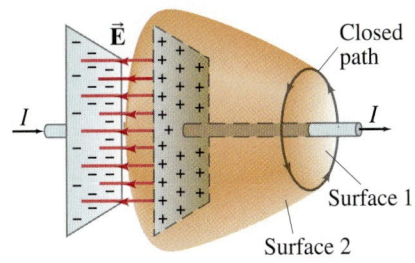

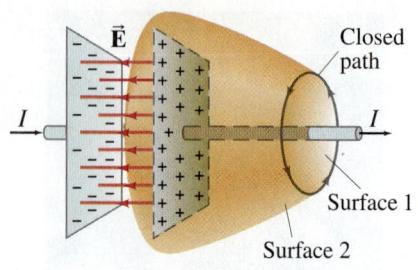

FIGURE 31-3 (repeated) See text.

*Ampère's law
(general form)*

Let us see what this term should be by determining it for the changing electric field between the capacitor plates in Fig. 31–3. The charge Q on a capacitor of capacitance C is $Q = CV$ where V is the potential difference between the plates (Eq. 24–1). Also recall that $V = Ed$ (Eq. 23–4) where d is the (small) separation of the plates and E is the (uniform) electric field strength between them, if we ignore any fringing of the field. Also, for a parallel-plate capacitor, $C = \epsilon_0 A/d$, where A is the area of each plate (Eq. 24–2). We combine these to obtain:

$$Q = CV = \left(\epsilon_0 \frac{A}{d}\right)(Ed) = \epsilon_0 AE.$$

If the charge on each plate changes at a rate dQ/dt, the electric field changes at a proportional rate. That is, by differentiating this expression for Q, we have:

$$\frac{dQ}{dt} = \epsilon_0 A \frac{dE}{dt}.$$

Now dQ/dt is also the current I flowing into or out of the capacitor:

$$I = \frac{dQ}{dt} = \epsilon_0 A \frac{dE}{dt} = \epsilon_0 \frac{d\Phi_E}{dt}$$

where $\Phi_E = EA$ is the **electric flux** through the closed path (surface 2 in Fig. 31–3). In order to make Ampère's law work for surface 2 in Fig. 31–3, as well as for surface 1 (where current I flows), we therefore write:

$$\oint \vec{B} \cdot d\vec{\ell} = \mu_0 I_{\text{encl}} + \mu_0 \epsilon_0 \frac{d\Phi_E}{dt}. \qquad (31\text{–}1)$$

This equation represents the general form of **Ampère's law**,[†] and embodies Maxwell's idea that a magnetic field can be caused not only by an ordinary electric current, but also by a changing electric field or changing electric flux. Although we arrived at it for a special case, Eq. 31–1 has proved valid in general. The last term on the right in Eq. 31–1 is usually very small, and not easy to measure experimentally.

EXAMPLE 31–1 **Charging capacitor.** A 30-pF air-gap capacitor has circular plates of area $A = 100\,\text{cm}^2$. It is charged by a 70-V battery through a 2.0-Ω resistor. At the instant the battery is connected, the electric field between the plates is changing most rapidly. At this instant, calculate (a) the current into the plates, and (b) the rate of change of electric field between the plates. (c) Determine the magnetic field induced between the plates. Assume $\vec{E}$ is uniform between the plates at any instant and is zero at all points beyond the edges of the plates.

APPROACH In Section 26–5 we discussed RC circuits and saw that the charge on a capacitor being charged, as a function of time, is

$$Q = CV_0(1 - e^{-t/RC}),$$

where V_0 is the voltage of the battery. To find the current at $t = 0$, we differentiate this and substitute the values $V_0 = 70\,\text{V}$, $C = 30\,\text{pF}$, $R = 2.0\,\Omega$.

SOLUTION (a) We take the derivative of Q and evaluate it at $t = 0$:

$$\left.\frac{dQ}{dt}\right|_{t=0} = \left.\frac{CV_0}{RC} e^{-t/RC}\right|_{t=0} = \frac{V_0}{R} = \frac{70\,\text{V}}{2.0\,\Omega} = 35\,\text{A}.$$

This is the rate at which charge accumulates on the capacitor and equals the current flowing in the circuit at $t = 0$.

(b) The electric field between two closely spaced conductors is given by (Eq. 21–8)

$$E = \frac{\sigma}{\epsilon_0} = \frac{Q/A}{\epsilon_0}$$

as we saw in Chapter 21 (see Example 21–13).

[†]Actually, there is a third term on the right for the case when a magnetic field is produced by magnetized materials. This can be accounted for by changing μ_0 to μ, but we will mainly be interested in cases where no magnetic material is present. In the presence of a dielectric, ϵ_0 is replaced by $\epsilon = K\epsilon_0$ (see Section 24–5).

Hence

$$\frac{dE}{dt} = \frac{dQ/dt}{\epsilon_0 A} = \frac{35\,\text{A}}{(8.85 \times 10^{-12}\,\text{C}^2/\text{N}\cdot\text{m}^2)(1.0 \times 10^{-2}\,\text{m}^2)} = 4.0 \times 10^{14}\,\text{V/m}\cdot\text{s}.$$

(c) Although we will not prove it, we might expect the lines of $\vec{\mathbf{B}}$, because of *symmetry*, to be circles, and to be perpendicular to $\vec{\mathbf{E}}$, as shown in Fig. 31–4; this is the same symmetry we saw for the inverse situation of a changing magnetic field producing an electric field (Section 29–7, see Fig. 29–27). To determine the magnitude of B between the plates we apply Ampère's law, Eq. 31–1, with the current $I_{\text{encl}} = 0$:

$$\oint \vec{\mathbf{B}} \cdot d\vec{\boldsymbol{\ell}} = \mu_0 \epsilon_0 \frac{d\Phi_E}{dt}.$$

We choose our path to be a circle of radius r, centered at the center of the plate, and thus following a magnetic field line such as the one shown in Fig. 31–4. For $r \leq r_0$ (the radius of plate) the flux through a circle of radius r is $E(\pi r^2)$ since E is assumed uniform between the plates at any moment. So from Ampère's law we have

$$B(2\pi r) = \mu_0 \epsilon_0 \frac{d}{dt}(\pi r^2 E)$$

$$= \mu_0 \epsilon_0 \pi r^2 \frac{dE}{dt}.$$

Hence

$$B = \frac{\mu_0 \epsilon_0}{2} r \frac{dE}{dt}. \qquad\qquad [r \leq r_0]$$

We assume $\vec{\mathbf{E}} = 0$ for $r > r_0$, so for points beyond the edge of the plates all the flux is contained within the plates $(\text{area} = \pi r_0^2)$ and $\Phi_E = E\pi r_0^2$. Thus Ampère's law gives

$$B(2\pi r) = \mu_0 \epsilon_0 \frac{d}{dt}(\pi r_0^2 E)$$

$$= \mu_0 \epsilon_0 \pi r_0^2 \frac{dE}{dt}$$

or

$$B = \frac{\mu_0 \epsilon_0 r_0^2}{2r} \frac{dE}{dt}. \qquad\qquad [r \geq r_0]$$

B has its maximum value at $r = r_0$ which, from either relation above (using $r_0 = \sqrt{A/\pi} = 5.6\,\text{cm}$), is

$$B_{\text{max}} = \frac{\mu_0 \epsilon_0 r_0}{2} \frac{dE}{dt}$$

$$= \tfrac{1}{2}(4\pi \times 10^{-7}\,\text{T}\cdot\text{m/A})(8.85 \times 10^{-12}\,\text{C}^2/\text{N}\cdot\text{m}^2)(5.6 \times 10^{-2}\,\text{m})(4.0 \times 10^{14}\,\text{V/m}\cdot\text{s})$$

$$= 1.2 \times 10^{-4}\,\text{T}.$$

This is a very small field and lasts only briefly (the time constant $RC = 6.0 \times 10^{-11}\,\text{s}$) and so would be very difficult to measure.

Let us write the magnetic field B outside the capacitor plates of Example 31–1 in terms of the current I that leaves the plates. The electric field between the plates is $E = \sigma/\epsilon_0 = Q/(\epsilon_0 A)$, as we saw in part b, so $dE/dt = I/(\epsilon_0 A)$. Hence B for $r > r_0$ is,

$$B = \frac{\mu_0 \epsilon_0 r_0^2}{2r} \frac{dE}{dt} = \frac{\mu_0 \epsilon_0 r_0^2}{2r} \frac{I}{\epsilon_0 \pi r_0^2} = \frac{\mu_0 I}{2\pi r}.$$

This is the same formula for the field that surrounds a wire (Eq. 28–1). Thus the B field outside the capacitor is the same as that outside the wire. In other words, the magnetic field produced by the changing electric field between the plates is the same as that produced by the current in the wire.

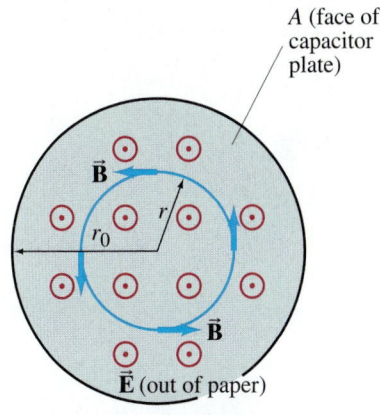

FIGURE 31–4 Frontal view of a circular plate of a parallel-plate capacitor. $\vec{\mathbf{E}}$ between plates points out toward viewer; lines of $\vec{\mathbf{B}}$ are circles. (Example 31–1.)

Displacement Current

Maxwell interpreted the second term on the right in Eq. 31–1 as being *equivalent* to an electric current. He called it a **displacement current**, I_D. An ordinary current I is then called a **conduction current**. Ampère's law can then be written

$$\oint \vec{\mathbf{B}} \cdot d\vec{\boldsymbol{\ell}} = \mu_0 \left(I + I_D \right)_{\text{encl}} \tag{31–2}$$

where

Displacement current

$$I_D = \epsilon_0 \frac{d\Phi_E}{dt}. \tag{31–3}$$

The term "displacement current" was based on an old discarded theory. Don't let it confuse you: I_D does not represent a flow of electric charge[†], nor is there a displacement.

31–2 Gauss's Law for Magnetism

We are almost in a position to state Maxwell's equations, but first we need to discuss the magnetic equivalent of Gauss's law. As we saw in Chapter 29, for a magnetic field $\vec{\mathbf{B}}$ the *magnetic flux* Φ_B through a surface is defined as

$$\Phi_B = \int \vec{\mathbf{B}} \cdot d\vec{\mathbf{A}}$$

where the integral is over the area of either an open or a closed surface. The magnetic flux through a closed surface—that is, a surface which completely encloses a volume—is written

$$\Phi_B = \oint \vec{\mathbf{B}} \cdot d\vec{\mathbf{A}}.$$

In the electric case, we saw in Section 22–2 that the electric flux Φ_E through a closed surface is equal to the total net charge Q enclosed by the surface, divided by ϵ_0 (Eq. 22–4):

$$\oint \vec{\mathbf{E}} \cdot d\vec{\mathbf{A}} = \frac{Q}{\epsilon_0}.$$

This relation is Gauss's law for electricity.

We can write a similar relation for the magnetic flux. We have seen, however, that in spite of intense searches, no isolated magnetic poles (monopoles)—the magnetic equivalent of single electric charges—have ever been observed. Hence, **Gauss's law for magnetism** is

$$\oint \vec{\mathbf{B}} \cdot d\vec{\mathbf{A}} = 0. \tag{31–4}$$

FIGURE 31–5 Magnetic field lines for a bar magnet.

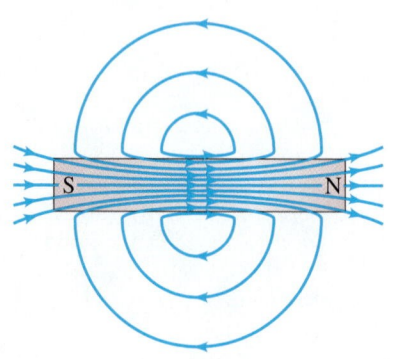

In terms of magnetic field lines, this relation tells us that as many lines enter the enclosed volume as leave it. If, indeed, magnetic monopoles do not exist, then there are no "sources" or "sinks" for magnetic field lines to start or stop on, corresponding to electric field lines starting on positive charges and ending on negative charges. Magnetic field lines must then be continuous. Even for a bar magnet, a magnetic field $\vec{\mathbf{B}}$ exists inside as well as outside the magnetic material, and the lines of $\vec{\mathbf{B}}$ are closed loops as shown in Fig. 31–5.

[†]The interpretation of the changing electric field as a current does fit in well with our discussion in Chapter 30 where we saw that an alternating current can be said to pass through a capacitor (although charge doesn't). It also means that Kirchhoff's junction rule will be valid even at a capacitor plate: conduction current flows into the plate, but no conduction current flows out of the plate—instead a "displacement current" flows out of one plate (toward the other plate).

31–3 Maxwell's Equations

With the extension of Ampère's law given by Eq. 31–1, plus Gauss's law for magnetism (Eq. 31–4), we are now ready to state all four of Maxwell's equations. We have seen them all before in the past ten Chapters. In the absence of dielectric or magnetic materials, **Maxwell's equations** are:

$$\oint \vec{E} \cdot d\vec{A} = \frac{Q}{\epsilon_0} \tag{31–5a}$$

$$\oint \vec{B} \cdot d\vec{A} = 0 \tag{31–5b}$$

$$\oint \vec{E} \cdot d\vec{\ell} = -\frac{d\Phi_B}{dt} \tag{31–5c}$$

$$\oint \vec{B} \cdot d\vec{\ell} = \mu_0 I + \mu_0 \epsilon_0 \frac{d\Phi_E}{dt}. \tag{31–5d}$$

MAXWELL'S EQUATIONS

The first two of Maxwell's equations are the same as Gauss's law for electricity (Chapter 22, Eq. 22–4) and Gauss's law for magnetism (Section 31–2, Eq. 31–4). The third is Faraday's law (Chapter 29, Eq. 29–8) and the fourth is Ampère's law as modified by Maxwell (Eq. 31–1). (We dropped the subscripts on Q_{encl} and I_{encl} for simplicity.)

They can be summarized in words: (1) a generalized form of Coulomb's law relating electric field to its sources, electric charges; (2) the same for the magnetic field, except that if there are no magnetic monopoles, magnetic field lines are continuous—they do not begin or end (as electric field lines do on charges); (3) an electric field is produced by a changing magnetic field; (4) a magnetic field is produced by an electric current or by a changing electric field.

Maxwell's equations are the basic equations for all electromagnetism, and are as fundamental as Newton's three laws of motion and the law of universal gravitation. Maxwell's equations can also be written in differential form; see Appendix E.

In earlier Chapters, we have seen that we can treat electric and magnetic fields separately if they do not vary in time. But we cannot treat them independently if they do change in time. For a changing magnetic field produces an electric field; and a changing electric field produces a magnetic field. An important outcome of these relations is the production of electromagnetic waves.

31–4 Production of Electromagnetic Waves

A magnetic field will be produced in empty space if there is a changing electric field. A changing magnetic field produces an electric field that is itself changing. This changing electric field will, in turn, produce a magnetic field, which will be changing, and so it too will produce a changing electric field; and so on. Maxwell found that the net result of these interacting changing fields was a *wave* of electric and magnetic fields that can propagate (travel) through space! We now examine, in a simplified way, how such **electromagnetic waves** can be produced.

Consider two conducting rods that will serve as an "antenna" (Fig. 31–6a). Suppose these two rods are connected by a switch to the opposite terminals of a battery. When the switch is closed, the upper rod quickly becomes positively charged and the lower one negatively charged. Electric field lines are formed as indicated by the lines in Fig. 31–6b. While the charges are flowing, a current exists whose direction is indicated by the black arrows. A magnetic field is therefore produced near the antenna. The magnetic field lines encircle the rod-like antenna and therefore, in Fig. 31–6, $\vec{B}$ points into the page ($\otimes$) on the right and out of the page ($\odot$) on the left. How far out do these electric and magnetic fields extend? In the static case, the fields extend outward indefinitely far. However, when the switch in Fig. 31–6 is closed, the fields quickly appear nearby, but it takes time for them to reach distant points. Both electric and magnetic fields store energy, and this energy cannot be transferred to distant points at infinite speed.

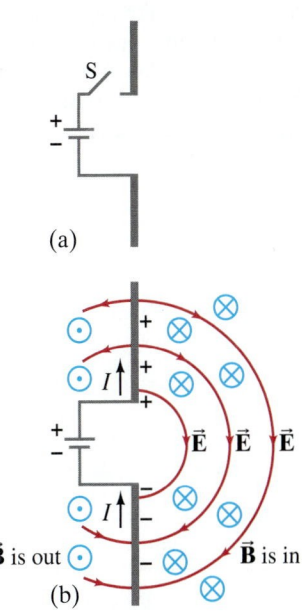

FIGURE 31–6 Fields produced by charge flowing into conductors. It takes time for the $\vec{E}$ and $\vec{B}$ fields to travel outward to distant points. The fields are shown to the right of the antenna, but they move out in all directions, symmetrically about the (vertical) antenna.

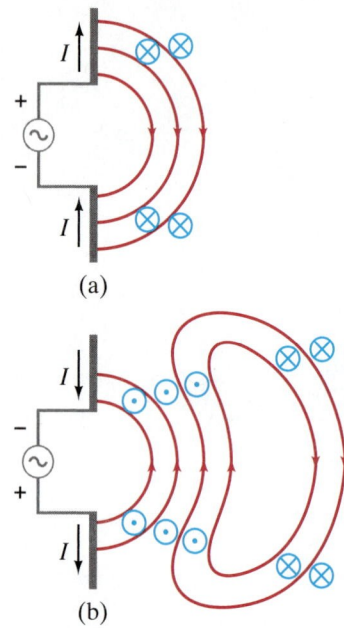

(a)

(b)

FIGURE 31–7 Sequence showing electric and magnetic fields that spread outward from oscillating charges on two conductors (the antenna) connected to an ac source (see the text).

FIGURE 31–8 (a) The radiation fields (far from the antenna) produced by a sinusoidal signal on the antenna. The red closed loops represent electric field lines. The magnetic field lines, perpendicular to the page and represented by blue ⊗ and ⊙, also form closed loops. (b) Very far from the antenna the wave fronts (field lines) are essentially flat over a fairly large area, and are referred to as *plane waves*.

Now we look at the situation of Fig. 31–7 where our antenna is connected to an ac generator. In Fig. 31–7a, the connection has just been completed. Charge starts building up and fields form just as in Fig. 31–6. The + and − signs in Fig. 31–7a indicate the net charge on each rod at a given instant. The black arrows indicate the direction of the current. The electric field is represented by the red lines in the plane of the page; and the magnetic field, according to the right-hand rule, is into (⊗) or out of (⊙) the page, in blue. In Fig. 31–7b, the voltage of the ac generator has reversed in direction; the current is reversed and the new magnetic field is in the opposite direction. Because the new fields have changed direction, the old lines fold back to connect up to some of the new lines and form closed loops as shown.[†] The old fields, however, don't suddenly disappear; they are on their way to distant points. Indeed, because a changing magnetic field produces an electric field, and a changing electric field produces a magnetic field, this combination of changing electric and magnetic fields moving outward is self-supporting, no longer depending on the antenna charges.

The fields not far from the antenna, referred to as the *near field*, become quite complicated, but we are not so interested in them. We are instead mainly interested in the fields far from the antenna (they are generally what we detect), which we refer to as the **radiation field**, or *far field*. The electric field lines form loops, as shown in Fig. 31–8, and continue moving outward. The magnetic field lines also form closed loops, but are not shown since they are perpendicular to the page. Although the lines are shown only on the right of the source, fields also travel in other directions. The field strengths are greatest in directions perpendicular to the oscillating charges; and they drop to zero along the direction of oscillation—above and below the antenna in Fig. 31–8.

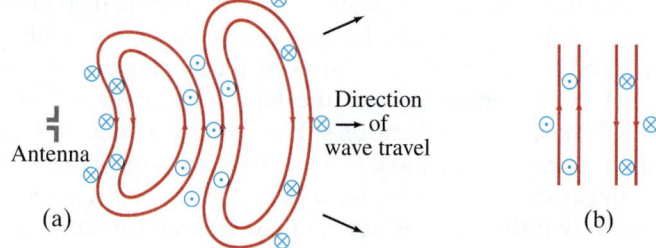

(a)

Antenna

Direction of wave travel

(b)

The magnitudes of both $\vec{E}$ and $\vec{B}$ in the radiation field are found to decrease with distance as $1/r$. (Compare this to the static electric field given by Coulomb's law where $\vec{E}$ decreases as $1/r^2$.) The energy carried by the electromagnetic wave is proportional (as for any wave, Chapter 15) to the square of the amplitude, E^2 or B^2, as will be discussed further in Section 31–8, so the intensity of the wave decreases as $1/r^2$.

Several things about the radiation field can be noted from Fig. 31–8. First, *the electric and magnetic fields at any point are perpendicular to each other, and to the direction of wave travel.* Second, we can see that the fields alternate in direction ($\vec{B}$ is into the page at some points and out of the page at others; $\vec{E}$ points up at some points and down at others). Thus, the field strengths vary from a maximum in one direction, to zero, to a maximum in the other direction. The electric and magnetic fields are "in phase": that is, they each are zero at the same points and reach their maxima at the same points in space. Finally, very far from the antenna (Fig. 31–8b) the field lines are quite flat over a reasonably large area, and the waves are referred to as **plane waves**.

If the source voltage varies sinusoidally, then the electric and magnetic field strengths in the radiation field will also vary sinusoidally. The sinusoidal character of the waves is diagrammed in Fig. 31–9, which shows the field directions and magnitudes plotted as a function of position. Notice that $\vec{B}$ and $\vec{E}$ are perpendicular to each other and to the direction of travel (= the direction of the wave velocity $\vec{v}$). The direction of $\vec{v}$ can be had from a right-hand rule using $\vec{E} \times \vec{B}$: fingers along $\vec{E}$, then along $\vec{B}$, gives $\vec{v}$ along thumb.

[†]We are considering waves traveling through empty space. There are no charges for lines of $\vec{E}$ to start or stop on, so they form closed loops. Magnetic field lines always form closed loops.

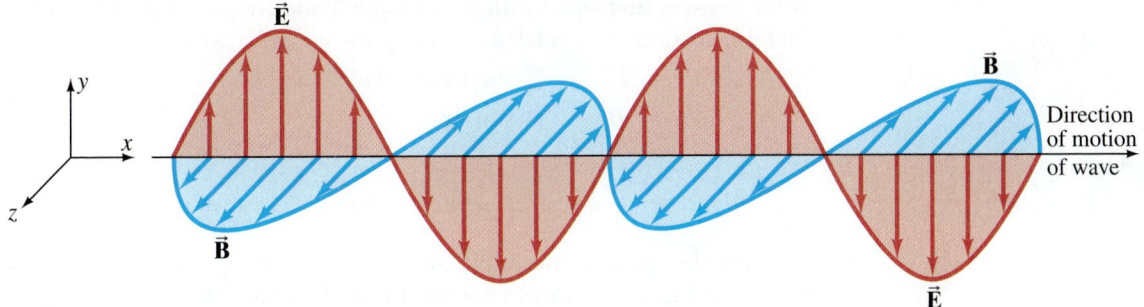

We call these waves electromagnetic (EM) waves. They are *transverse* waves because the amplitude is perpendicular to the direction of wave travel. However, EM waves are always waves of *fields*, not of matter (like waves on water or a rope). Because they are fields, EM waves can propagate in empty space.

As we have seen, EM waves are produced by electric charges that are oscillating, and hence are undergoing acceleration. In fact, we can say in general that

accelerating electric charges give rise to electromagnetic waves.

Electromagnetic waves can be produced in other ways as well, requiring description at the atomic and nuclear levels, as we will discuss later.

EXERCISE A At a particular instant in time, a wave has its electric field pointing north and its magnetic field pointing up. In which direction is the wave traveling? (*a*) South, (*b*) west, (*c*) east, (*d*) down, (*e*) not enough information.

31–5 Electromagnetic Waves, and Their Speed, Derived from Maxwell's Equations

Let us now examine how the existence of EM waves follows from Maxwell's equations. We will see that Maxwell's prediction of the existence of EM waves was startling. Equally startling was the speed at which they were predicted to travel.

We begin by considering a region of free space, where there are *no charges or conduction currents*—that is, far from the source so that the wave fronts (the field lines in Fig. 31–8) are essentially flat over a reasonable area. We call them **plane waves**, as we saw, because at any instant $\vec{\mathbf{E}}$ and $\vec{\mathbf{B}}$ are uniform over a reasonably large plane perpendicular to the direction of propagation. We choose a coordinate system, so that the wave is traveling in the x direction with velocity $\vec{\mathbf{v}} = v\hat{\mathbf{i}}$, with $\vec{\mathbf{E}}$ parallel to the y axis and $\vec{\mathbf{B}}$ parallel to the z axis, as in Fig. 31–9.

Maxwell's equations, with $Q = I = 0$, become

$$\oint \vec{\mathbf{E}} \cdot d\vec{\mathbf{A}} = 0 \tag{31–6a}$$

$$\oint \vec{\mathbf{B}} \cdot d\vec{\mathbf{A}} = 0 \tag{31–6b}$$

$$\oint \vec{\mathbf{E}} \cdot d\vec{\boldsymbol{\ell}} = -\frac{d\Phi_B}{dt} \tag{31–6c}$$

$$\oint \vec{\mathbf{B}} \cdot d\vec{\boldsymbol{\ell}} = \mu_0 \epsilon_0 \frac{d\Phi_E}{dt}. \tag{31–6d}$$

Notice the beautiful *symmetry* of these equations. The term on the right in the last equation, conceived by Maxwell, is essential for this symmetry. It is also essential if electromagnetic waves are to be produced, as we will now see.

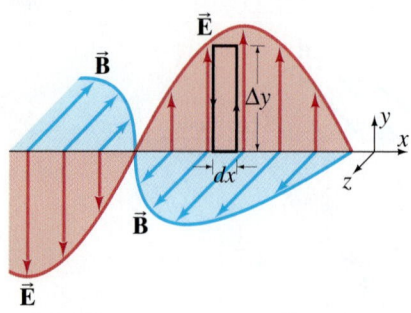

FIGURE 31–10 Applying Faraday's law to the rectangle $(\Delta y)(dx)$.

If the wave is sinusoidal with wavelength λ and frequency f, then, as we saw in Chapter 15, Section 15–4, such a traveling wave can be written as

$$E = E_y = E_0 \sin(kx - \omega t)$$
$$B = B_z = B_0 \sin(kx - \omega t)$$

(31–7)

where

$$k = \frac{2\pi}{\lambda}, \qquad \omega = 2\pi f, \qquad \text{and} \qquad f\lambda = \frac{\omega}{k} = v,$$

(31–8)

with v being the speed of the wave. Although visualizing the wave as sinusoidal is helpful, we will not have to assume so in most of what follows.

Consider now a small rectangle in the plane of the electric field as shown in Fig. 31–10. This rectangle has a finite height Δy, and a very thin width which we take to be the infinitesimal distance dx. First we show that $\vec{\mathbf{E}}$, $\vec{\mathbf{B}}$, and $\vec{\mathbf{v}}$ are in the orientation shown by applying Lenz's law to this rectangular loop. The changing magnetic flux through this loop is related to the electric field around the loop by Faraday's law (Maxwell's third equation, Eq. 31–6c). For the case shown, B through the loop is decreasing in time (the wave is moving to the right). So the electric field must be in a direction to oppose this change, meaning E must be greater on the right side of the loop than on the left, as shown (so it could produce a counterclockwise current whose magnetic field would act to oppose the change in Φ_B—but of course there is no current). This brief argument shows that the orientation of $\vec{\mathbf{E}}$, $\vec{\mathbf{B}}$, and $\vec{\mathbf{v}}$ are in the correct relation as shown. That is, $\vec{\mathbf{v}}$ is in the direction of $\vec{\mathbf{E}} \times \vec{\mathbf{B}}$. Now let us apply Faraday's law, which is Maxwell's third equation (Eq. 31–6c),

$$\oint \vec{\mathbf{E}} \cdot d\vec{\boldsymbol{\ell}} = -\frac{d\Phi_B}{dt}$$

to the rectangle of height Δy and width dx shown in Fig. 31–10. First we consider $\oint \vec{\mathbf{E}} \cdot d\vec{\boldsymbol{\ell}}$. Along the short top and bottom sections of length dx, $\vec{\mathbf{E}}$ is perpendicular to $d\vec{\boldsymbol{\ell}}$, so $\vec{\mathbf{E}} \cdot d\vec{\boldsymbol{\ell}} = 0$. Along the vertical sides, we let E be the electric field along the left side, and on the right side where it will be slightly larger, it is $E + dE$. Thus, if we take our loop counterclockwise,

$$\oint \vec{\mathbf{E}} \cdot d\vec{\boldsymbol{\ell}} = (E + dE)\,\Delta y - E\,\Delta y = dE\,\Delta y.$$

For the right side of Faraday's law, the magnetic flux through the loop changes as

$$\frac{d\Phi_B}{dt} = \frac{dB}{dt}\,dx\,\Delta y,$$

since the area of the loop, $(dx)(\Delta y)$, is not changing. Thus, Faraday's law gives us

$$dE\,\Delta y = -\frac{dB}{dt}\,dx\,\Delta y$$

or

$$\frac{dE}{dx} = -\frac{dB}{dt}.$$

Actually, both E and B are functions of position x and time t. We should therefore use partial derivatives:

$$\frac{\partial E}{\partial x} = -\frac{\partial B}{\partial t}$$

(31–9)

where $\partial E/\partial x$ means the derivative of E with respect to x while t is held fixed, and $\partial B/\partial t$ is the derivative of B with respect to t while x is kept fixed.

We can obtain another important relation between E and B in addition to Eq. 31–9. To do so, we consider now a small rectangle in the plane of $\vec{\mathbf{B}}$, whose length and width are Δz and dx as shown in Fig. 31–11. To this rectangular loop we apply Maxwell's fourth equation (the extension of Ampère's law), Eq. 31–6d:

$$\oint \vec{\mathbf{B}} \cdot d\vec{\boldsymbol{\ell}} = \mu_0 \epsilon_0 \frac{d\Phi_E}{dt}$$

where we have taken $I = 0$ since we assume the absence of conduction currents.

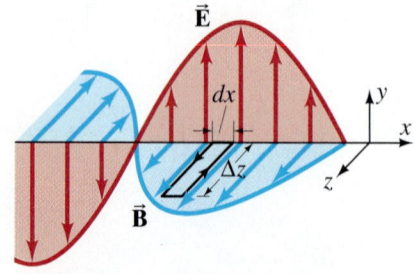

FIGURE 31–11 Applying Maxwell's fourth equation to the rectangle $(\Delta z)(dx)$.

Along the short sides (dx), $\vec{B} \cdot d\vec{\ell}$ is zero since $\vec{B}$ is perpendicular to $d\vec{\ell}$. Along the longer sides (Δz), we let B be the magnetic field along the left side of length Δz, and $B + dB$ be the field along the right side. We again integrate counterclockwise, so

$$\oint \vec{B} \cdot d\vec{\ell} = B \Delta z - (B + dB) \Delta z = -dB \Delta z.$$

The right side of Maxwell's fourth equation is

$$\mu_0 \epsilon_0 \frac{d\Phi_E}{dt} = \mu_0 \epsilon_0 \frac{dE}{dt} dx \Delta z.$$

Equating the two expressions, we obtain

$$-dB \Delta z = \mu_0 \epsilon_0 \frac{dE}{dt} dx \Delta z$$

or

$$\frac{\partial B}{\partial x} = -\mu_0 \epsilon_0 \frac{\partial E}{\partial t} \qquad\qquad \textbf{(31–10)}$$

where we have replaced dB/dx and dE/dt by the proper partial derivatives as before.

We can use Eqs. 31–9 and 31–10 to obtain a relation between the magnitudes of $\vec{E}$ and $\vec{B}$, and the speed v. Let E and B be given by Eqs. 31–7 as a function of x and t. When we apply Eq. 31–9, taking the derivatives of E and B as given by Eqs. 31–7, we obtain

$$kE_0 \cos(kx - \omega t) = \omega B_0 \cos(kx - \omega t)$$

or

$$\frac{E_0}{B_0} = \frac{\omega}{k} = v,$$

since $v = \omega/k$ (see Eq. 31–8 or 15–12). Since E and B are in phase, we see that E and B are related by

$$\frac{E}{B} = v \qquad\qquad \textbf{(31–11)}$$

at any point in space, where v is the velocity of the wave.

Now we apply Eq. 31–10 to the sinusoidal fields (Eqs. 31–7) and we obtain

$$kB_0 \cos(kx - \omega t) = \mu_0 \epsilon_0 \omega E_0 \cos(kx - \omega t)$$

or

$$\frac{B_0}{E_0} = \frac{\mu_0 \epsilon_0 \omega}{k} = \mu_0 \epsilon_0 v.$$

We just saw that $B_0/E_0 = 1/v$, so

$$\mu_0 \epsilon_0 v = \frac{1}{v}.$$

Solving for v we find

$$v = c = \frac{1}{\sqrt{\epsilon_0 \mu_0}}, \qquad\qquad \textbf{(31–12)}$$

where c is the special symbol for the speed of electromagnetic waves in free space. We see that c is a constant, independent of the wavelength or frequency. If we put in values for ϵ_0 and μ_0 we find

$$c = \frac{1}{\sqrt{\epsilon_0 \mu_0}} = \frac{1}{\sqrt{(8.85 \times 10^{-12}\,\text{C}^2/\text{N}\cdot\text{m}^2)(4\pi \times 10^{-7}\,\text{T}\cdot\text{m/A})}}$$

$$= 3.00 \times 10^8\,\text{m/s}.$$

This is a remarkable result. For this is precisely equal to the measured speed of light!

EXAMPLE 31–2 Determining $\vec{E}$ and $\vec{B}$ in EM waves. Assume a 60.0-Hz EM wave is a sinusoidal wave propagating in the z direction with $\vec{E}$ pointing in the x direction, and $E_0 = 2.00\text{ V/m}$. Write vector expressions for $\vec{E}$ and $\vec{B}$ as functions of position and time.

APPROACH We find λ from $\lambda f = v = c$. Then we use Fig. 31–9 and Eqs. 31–7 and 31–8 for the mathematical form of traveling electric and magnetic fields of an EM wave.

SOLUTION The wavelength is

$$\lambda = \frac{c}{f} = \frac{3.00 \times 10^8\text{ m/s}}{60.0\text{ s}^{-1}} = 5.00 \times 10^6\text{ m}.$$

From Eq. 31–8 we have

$$k = \frac{2\pi}{\lambda} = \frac{2\pi}{5.00 \times 10^6\text{ m}} = 1.26 \times 10^{-6}\text{ m}^{-1}$$

$$\omega = 2\pi f = 2\pi(60.0\text{ Hz}) = 377\text{ rad/s}.$$

From Eq. 31–11 with $v = c$, we find that

$$B_0 = \frac{E_0}{c} = \frac{2.00\text{ V/m}}{3.00 \times 10^8\text{ m/s}} = 6.67 \times 10^{-9}\text{ T}.$$

The direction of propagation is that of $\vec{E} \times \vec{B}$, as in Fig. 31–9. With $\vec{E}$ pointing in the x direction, and the wave propagating in the z direction, $\vec{B}$ must point in the y direction. Using Eqs. 31–7 we find:

$$\vec{E} = \hat{i}(2.00\text{ V/m})\sin\left[(1.26 \times 10^{-6}\text{ m}^{-1})z - (377\text{ rad/s})t\right]$$

$$\vec{B} = \hat{j}(6.67 \times 10^{-9}\text{ T})\sin\left[(1.26 \times 10^{-6}\text{ m}^{-1})z - (377\text{ rad/s})t\right]$$

*Derivation of Speed of Light (General)

We can derive the speed of EM waves without having to assume sinusoidal waves by combining Eqs. 31–9 and 31–10 as follows. We take the derivative, with respect to t of Eq. 31–10

$$\frac{\partial^2 B}{\partial t\, \partial x} = -\mu_0 \epsilon_0 \frac{\partial^2 E}{\partial t^2}.$$

We next take the derivative of Eq. 31–9 with respect to x:

$$\frac{\partial^2 E}{\partial x^2} = -\frac{\partial^2 B}{\partial t\, \partial x}.$$

Since $\partial^2 B / \partial t\, \partial x$ appears in both relations, we obtain

$$\frac{\partial^2 E}{\partial t^2} = \frac{1}{\mu_0 \epsilon_0} \frac{\partial^2 E}{\partial x^2}. \tag{31–13a}$$

By taking other derivatives of Eqs. 31–9 and 31–10 we obtain the same relation for B:

$$\frac{\partial^2 B}{\partial t^2} = \frac{1}{\mu_0 \epsilon_0} \frac{\partial^2 B}{\partial x^2}. \tag{31–13b}$$

Both of Eqs. 31–13 have the form of the **wave equation** for a plane wave traveling in the x direction, as discussed in Section 15–5 (Eq. 15–16):

$$\frac{\partial^2 D}{\partial t^2} = v^2 \frac{\partial^2 D}{\partial x^2},$$

where D stands for any type of displacement. We see that the velocity v for Eqs. 31–13 is given by

$$v^2 = \frac{1}{\mu_0 \epsilon_0}$$

in agreement with Eq. 31–12. Thus we see that a natural outcome of Maxwell's equations is that E and B obey the wave equation for waves traveling with speed $v = 1/\sqrt{\mu_0 \epsilon_0}$. It was on this basis that Maxwell predicted the existence of electromagnetic waves and predicted their speed.

31–6 Light as an Electromagnetic Wave and the Electromagnetic Spectrum

The calculations in Section 31–5 gave the result that Maxwell himself determined: that the speed of EM waves in empty space is given by

$$c = \frac{E}{B} = \frac{1}{\sqrt{\epsilon_0 \mu_0}} = 3.00 \times 10^8 \, \text{m/s},$$

the same as the measured speed of light in vacuum.

Light had been shown some 60 years previously to behave like a wave (we'll discuss this in Chapter 34). But nobody knew what kind of wave it was. What is it that is oscillating in a light wave? Maxwell, on the basis of the calculated speed of EM waves, argued that light must be an electromagnetic wave. This idea soon came to be generally accepted by scientists, but not fully until after EM waves were experimentally detected. EM waves were first generated and detected experimentally by Heinrich Hertz (1857–1894) in 1887, eight years after Maxwell's death. Hertz used a spark-gap apparatus in which charge was made to rush back and forth for a short time, generating waves whose frequency was about 10^9 Hz. He detected them some distance away using a loop of wire in which an emf was produced when a changing magnetic field passed through. These waves were later shown to travel at the speed of light, 3.00×10^8 m/s, and to exhibit all the characteristics of light such as reflection, refraction, and interference. The only difference was that they were not visible. Hertz's experiment was a strong confirmation of Maxwell's theory.

The wavelengths of visible light were measured in the first decade of the nineteenth century, long before anyone imagined that light was an electromagnetic wave. The wavelengths were found to lie between 4.0×10^{-7} m and 7.5×10^{-7} m, or 400 nm to 750 nm $(1 \, \text{nm} = 10^9 \, \text{m})$. The frequencies of visible light can be found using Eq. 15–1 or 31–8, which we rewrite here:

$$c = \lambda f, \tag{31–14}$$

where f and λ are the frequency and wavelength, respectively, of the wave. Here, c is the speed of light, 3.00×10^8 m/s; it gets the special symbol c because of its universality for all EM waves in free space. Equation 31–14 tells us that the frequencies of visible light are between 4.0×10^{14} Hz and 7.5×10^{14} Hz. (Recall that $1 \, \text{Hz} = 1$ cycle per second $= 1 \, \text{s}^{-1}$.)

But visible light is only one kind of EM wave. As we have seen, Hertz produced EM waves of much lower frequency, about 10^9 Hz. These are now called **radio waves**, because frequencies in this range are used to transmit radio and TV signals. Electromagnetic waves, or EM radiation as we sometimes call it, have been produced or detected over a wide range of frequencies. They are usually categorized as shown in Fig. 31–12, which is known as the **electromagnetic spectrum**.

FIGURE 31–12
Electromagnetic spectrum.

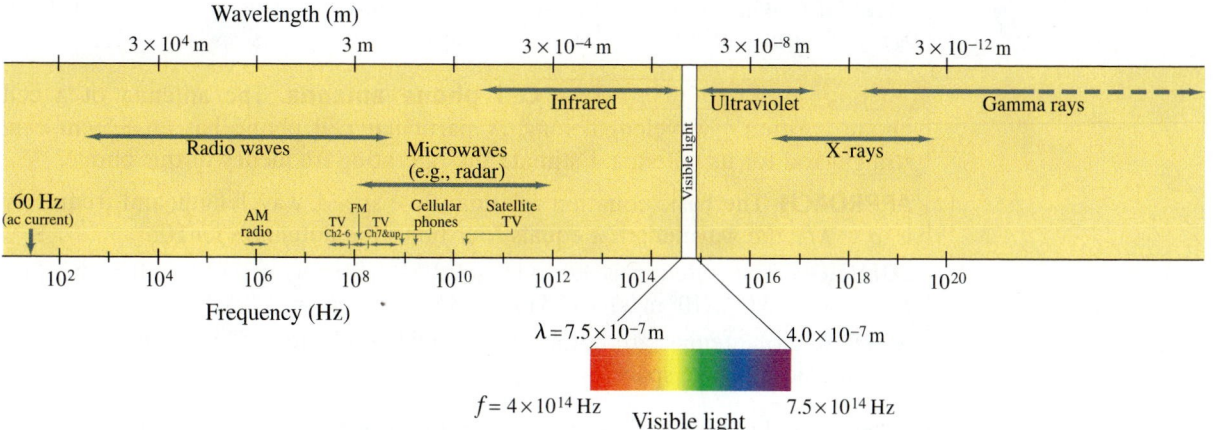

Radio waves and microwaves can be produced in the laboratory using electronic equipment (Fig. 31–7). Higher-frequency waves are very difficult to produce electronically. These and other types of EM waves are produced in natural processes, as emission from atoms, molecules, and nuclei (more on this later). EM waves can be produced by the acceleration of electrons or other charged particles, such as electrons in the antenna of Fig. 31–7. Another example is X-rays, which are produced (Chapter 35) when fast-moving electrons are rapidly decelerated upon striking a metal target. Even the visible light emitted by an ordinary incandescent light is due to electrons undergoing acceleration within the hot filament.

We will meet various types of EM waves later. However, it is worth mentioning here that infrared (IR) radiation (EM waves whose frequency is just less than that of visible light) is mainly responsible for the heating effect of the Sun. The Sun emits not only visible light but substantial amounts of IR and UV (ultraviolet) as well. The molecules of our skin tend to "resonate" at infrared frequencies, so it is these that are preferentially absorbed and thus warm us up. We humans experience EM waves differently, depending on their wavelengths: Our eyes detect wavelengths between about 4×10^{-7} m and 7.5×10^{-7} m (visible light), whereas our skin detects longer wavelengths (IR). Many EM wavelengths we don't detect directly at all.

EXERCISE B Return to the Chapter-Opening Question, page 812, and answer it again now. Try to explain why you may have answered differently the first time.

CAUTION

Sound and EM waves are different

Light and other electromagnetic waves travel at a speed of 3×10^8 m/s. Compare this to sound, which travels (see Chapter 16) at a speed of about 300 m/s in air, a million times slower; or to typical freeway speeds of a car, 30 m/s (100 km/h, or 60 mi/h), 10 million times slower than light. EM waves differ from sound waves in another big way: sound waves travel in a medium such as air, and involve motion of air molecules; EM waves do not involve any material—only fields, and they can travel in empty space.

EXAMPLE 31–3 **Wavelengths of EM waves.** Calculate the wavelength (a) of a 60-Hz EM wave, (b) of a 93.3-MHz FM radio wave, and (c) of a beam of visible red light from a laser at frequency 4.74×10^{14} Hz.

APPROACH All of these waves are electromagnetic waves, so their speed is $c = 3.00 \times 10^8$ m/s. We solve for λ in Eq. 31–14: $\lambda = c/f$.

SOLUTION (a) $\lambda = \dfrac{c}{f} = \dfrac{3.00 \times 10^8 \text{ m/s}}{60 \text{ s}^{-1}} = 5.0 \times 10^6$ m,

or 5000 km. 60 Hz is the frequency of ac current in the United States, and, as we see here, one wavelength stretches all the way across the continental USA.

(b) $\lambda = \dfrac{3.00 \times 10^8 \text{ m/s}}{93.3 \times 10^6 \text{ s}^{-1}} = 3.22$ m.

The length of an FM antenna is about half this $(\frac{1}{2}\lambda)$, or $1\frac{1}{2}$ m.

(c) $\lambda = \dfrac{3.00 \times 10^8 \text{ m/s}}{4.74 \times 10^{14} \text{ s}^{-1}} = 6.33 \times 10^{-7}$ m ($= 633$ nm).

EXERCISE C What are the frequencies of (a) an 80-m-wavelength radio wave, and (b) an X-ray of wavelength 5.5×10^{-11} m?

EXAMPLE 31–4 **ESTIMATE** **Cell phone antenna.** The antenna of a cell phone is often $\frac{1}{4}$ wavelength long. A particular cell phone has an 8.5-cm-long straight rod for its antenna. Estimate the operating frequency of this phone.

APPROACH The basic equation relating wave speed, wavelength, and frequency is $c = \lambda f$; the wavelength λ equals four times the antenna's length.

SOLUTION The antenna is $\frac{1}{4}\lambda$ long, so $\lambda = 4(8.5 \text{ cm}) = 34$ cm $= 0.34$ m. Then $f = c/\lambda = (3.0 \times 10^8 \text{ m/s})/(0.34 \text{ m}) = 8.8 \times 10^8$ Hz $= 880$ MHz.

NOTE Radio antennas are not always straight conductors. The conductor may be a round loop to save space. See Fig. 31–21b.

EXERCISE D How long should a $\frac{1}{4}\lambda$ antenna be for an aircraft radio operating at 165 MHz?

Electromagnetic waves can travel along transmission lines as well as in empty space. When a source of emf is connected to a transmission line—be it two parallel wires or a coaxial cable (Fig. 31–13)—the electric field within the wire is not set up immediately at all points along the wires. This is based on the same argument we used in Section 31–4 with reference to Fig. 31–7. Indeed, it can be shown that if the wires are separated by empty space or air, the electrical signal travels along the wires at the speed $c = 3.0 \times 10^8$ m/s. For example, when you flip a light switch, the light actually goes on a tiny fraction of a second later. If the wires are in a medium whose electric permittivity is ϵ and magnetic permeability is μ (Sections 24–5 and 28–9, respectively), the speed is not given by Eq. 31–12, but by

$$v = \frac{1}{\sqrt{\epsilon\mu}}.$$

FIGURE 31–13   Coaxial cable.

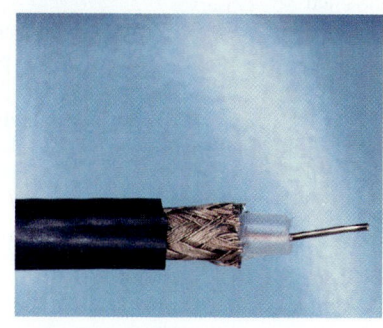

EXAMPLE 31–5 **ESTIMATE** **Phone call time lag.** You make a telephone call from New York to a friend in London. Estimate how long it will take the electrical signal generated by your voice to reach London, assuming the signal is (*a*) carried on a telephone cable under the Atlantic Ocean, and (*b*) sent via satellite 36,000 km above the ocean. Would this cause a noticeable delay in either case?

APPROACH The signal is carried on a telephone wire or in the air via satellite. In either case it is an electromagnetic wave. Electronics as well as the wire or cable slow things down, but as a rough estimate we take the speed to be $c = 3.0 \times 10^8$ m/s.

SOLUTION The distance from New York to London is about 5000 km.
(*a*) The time delay via the cable is $t = d/c \approx (5 \times 10^6$ m$)/(3.0 \times 10^8$ m/s$) = 0.017$ s.
(*b*) Via satellite the time would be longer because communications satellites, which are usually geosynchronous (Example 6–6), move at a height of 36,000 km. The signal would have to go up to the satellite and back down, or about 72,000 km. The actual distance the signal would travel would be a little more than this as the signal would go up and down on a diagonal. Thus $t = d/c \approx 7.2 \times 10^7$ m$/(3 \times 10^8$ m/s$) = 0.24$ s.

NOTE When the signal travels via the underwater cable, there is only a hint of a delay and conversations are fairly normal. When the signal is sent via satellite, the delay *is* noticeable. The length of time between the end of when you speak and your friend receives it and replies, and then you hear the reply, is about a half second beyond the normal time in a conversation. This is enough to be noticeable, and you have to adjust for it so you don't start talking again while your friend's reply is on the way back to you.

EXERCISE E If you are on the phone via satellite to someone only 100 km away, would you hear the same effect?

EXERCISE F If your voice traveled as a sound wave, how long would it take to go from New York to London?

31–7 Measuring the Speed of Light

Galileo attempted to measure the speed of light by trying to measure the time required for light to travel a known distance between two hilltops. He stationed an assistant on one hilltop and himself on another, and ordered the assistant to lift the cover from a lamp the instant he saw a flash from Galileo's lamp. Galileo measured the time between the flash of his lamp and when he received the light from his assistant's lamp. The time was so short that Galileo concluded it merely represented human reaction time, and that the speed of light must be extremely high.

The first successful determination that the speed of light is finite was made by the Danish astronomer Ole Roemer (1644–1710). Roemer had noted that the carefully measured orbital period of Io, a moon of Jupiter with an average period of 42.5 h, varied slightly, depending on the relative position of Earth and Jupiter. He attributed this variation in the apparent period to the change in distance between the Earth and Jupiter during one of Io's periods, and the time it took light to travel the extra distance. Roemer concluded that the speed of light—though great—is finite.

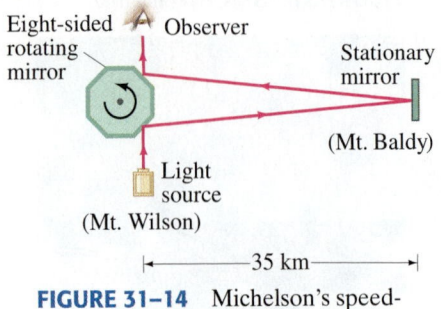

FIGURE 31–14 Michelson's speed-of-light apparatus (not to scale).

Eight-sided rotating mirror
Observer
Stationary mirror
(Mt. Baldy)
Light source
(Mt. Wilson)
|←——————35 km——————→|

Since then a number of techniques have been used to measure the speed of light. Among the most important were those carried out by the American Albert A. Michelson (1852–1931). Michelson used the rotating mirror apparatus diagrammed in Fig. 31–14 for a series of high-precision experiments carried out from 1880 to the 1920s. Light from a source would hit one face of a rotating eight-sided mirror. The reflected light traveled to a stationary mirror a large distance away and back again as shown. If the rotating mirror was turning at just the right rate, the returning beam of light would reflect from one of the eight mirrors into a small telescope through which the observer looked. If the speed of rotation was only slightly different, the beam would be deflected to one side and would not be seen by the observer. From the required speed of the rotating mirror and the known distance to the stationary mirror, the speed of light could be calculated. In the 1920s, Michelson set up the rotating mirror on the top of Mt. Wilson in southern California and the stationary mirror on Mt. Baldy (Mt. San Antonio) 35 km away. He later measured the speed of light in vacuum using a long evacuated tube.

Today the speed of light, c, in vacuum is taken as

$$c = 2.99792458 \times 10^8 \text{ m/s},$$

and is defined to be this value. This means that the standard for length, the meter, is no longer defined separately. Instead, as we noted in Section 1–4, the meter is now formally defined as the distance light travels in vacuum in 1/299,792,458 of a second. We usually round off c to

$$c = 3.00 \times 10^8 \text{ m/s}$$

when extremely precise results are not required. In air, the speed is only slightly less.

31–8 Energy in EM Waves; the Poynting Vector

Electromagnetic waves carry energy from one region of space to another. This energy is associated with the moving electric and magnetic fields. In Section 24–4, we saw that the energy density u_E (J/m^3) stored in an electric field E is $u_E = \frac{1}{2}\epsilon_0 E^2$ (Eq. 24–6). The energy density stored in a magnetic field B, as we discussed in Section 30–3, is given by $u_B = \frac{1}{2}B^2/\mu_0$ (Eq. 30–7). Thus, the total energy stored per unit volume in a region of space where there is an electromagnetic wave is

$$u = u_E + u_B = \frac{1}{2}\epsilon_0 E^2 + \frac{1}{2}\frac{B^2}{\mu_0}. \quad \textbf{(31–15)}$$

In this equation, E and B represent the electric and magnetic field strengths of the wave at any instant in a small region of space. We can write Eq. 31–15 in terms of the E field alone, using Eqs. 31–11 $(B = E/c)$ and 31–12 $(c = 1/\sqrt{\epsilon_0\mu_0})$ to obtain

$$u = \frac{1}{2}\epsilon_0 E^2 + \frac{1}{2}\frac{\epsilon_0\mu_0 E^2}{\mu_0} = \epsilon_0 E^2. \quad \textbf{(31–16a)}$$

Note here that the energy density associated with the B field equals that due to the E field, and each contributes half to the total energy. We can also write the energy density in terms of the B field only:

$$u = \epsilon_0 E^2 = \epsilon_0 c^2 B^2 = \frac{B^2}{\mu_0}, \quad \textbf{(31–16b)}$$

or in one term containing both E and B,

$$u = \epsilon_0 E^2 = \epsilon_0 EcB = \frac{\epsilon_0 EB}{\sqrt{\epsilon_0\mu_0}} = \sqrt{\frac{\epsilon_0}{\mu_0}}\, EB. \quad \textbf{(31–16c)}$$

Equations 31–16 give the energy density in any region of space at any instant.

Now let us determine the energy the wave transports per unit time per unit area. This is given by a vector $\vec{S}$, which is called the **Poynting vector**.[†] The units of $\vec{S}$ are W/m^2. The direction of $\vec{S}$ is the direction in which the energy is transported, which is the direction in which the wave is moving.

[†]After J. H. Poynting (1852–1914).

Let us imagine the wave is passing through an area A perpendicular to the x axis as shown in Fig. 31–15. In a short time dt, the wave moves to the right a distance $dx = c\,dt$ where c is the wave speed. The energy that passes through A in the time dt is the energy that occupies the volume $dV = A\,dx = Ac\,dt$. The energy density u is $u = \epsilon_0 E^2$ where E is the electric field in this volume at the given instant. So the total energy dU contained in this volume dV is the energy density u times the volume: $dU = u\,dV = (\epsilon_0 E^2)(Ac\,dt)$. Therefore the energy crossing the area A per time dt is

$$S = \frac{1}{A}\frac{dU}{dt} = \epsilon_0 c E^2. \qquad (31\text{–}17)$$

Since $E = cB$ and $c = 1/\sqrt{\epsilon_0 \mu_0}$, this can also be written:

$$S = \epsilon_0 c E^2 = \frac{cB^2}{\mu_0} = \frac{EB}{\mu_0}.$$

The direction of $\vec{S}$ is along $\vec{v}$, perpendicular to $\vec{E}$ and $\vec{B}$, so the Poynting vector $\vec{S}$ can be written

$$\vec{S} = \frac{1}{\mu_0}(\vec{E} \times \vec{B}). \qquad (31\text{–}18)$$

Equation 31–17 or 31–18 gives the energy transported per unit area per unit time at any *instant*. We often want to know the *average* over an extended period of time since the frequencies are usually so high we don't detect the rapid time variation. If E and B are sinusoidal, then $\overline{E^2} = E_0^2/2$, just as for electric currents and voltages (Section 25–7), where E_0 is the *maximum* value of E. Thus we can write for the magnitude of the Poynting vector, on the average,

$$\overline{S} = \frac{1}{2}\epsilon_0 c E_0^2 = \frac{1}{2}\frac{c}{\mu_0}B_0^2 = \frac{E_0 B_0}{2\mu_0}, \qquad (31\text{–}19a)$$

where B_0 is the maximum value of B. This time averaged value of $\vec{S}$ is the **intensity**, defined as the average power transferred across unit area (Section 15–3). We can also write for the average value of S:

$$\overline{S} = \frac{E_{rms}B_{rms}}{\mu_0} \qquad (31\text{–}19b)$$

where E_{rms} and B_{rms} are the rms values $\left(E_{rms} = \sqrt{\overline{E^2}},\ B_{rms} = \sqrt{\overline{B^2}}\right)$.

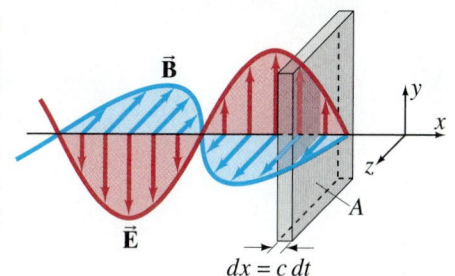

FIGURE 31–15 Electromagnetic wave carrying energy through area A.

EXAMPLE 31–6 *E and B from the Sun.* Radiation from the Sun reaches the Earth (above the atmosphere) at a rate of about $1350\,\text{J/s}\cdot\text{m}^2\ (= 1350\,\text{W/m}^2)$. Assume that this is a single EM wave, and calculate the maximum values of E and B.

APPROACH We solve Eq. 31–19a $\left(\overline{S} = \frac{1}{2}\epsilon_0 c E_0^2\right)$ for E_0 in terms of $\overline{S}$ using $\overline{S} = 1350\,\text{J/s}\cdot\text{m}^2$.

SOLUTION
$$E_0 = \sqrt{\frac{2\overline{S}}{\epsilon_0 c}} = \sqrt{\frac{2(1350\,\text{J/s}\cdot\text{m}^2)}{(8.85 \times 10^{-12}\,\text{C}^2/\text{N}\cdot\text{m}^2)(3.00 \times 10^8\,\text{m/s})}}$$
$$= 1.01 \times 10^3\,\text{V/m}.$$

From Eq. 31–11, $B = E/c$, so

$$B_0 = \frac{E_0}{c} = \frac{1.01 \times 10^3\,\text{V/m}}{3.00 \times 10^8\,\text{m/s}} = 3.37 \times 10^{-6}\,\text{T}.$$

NOTE Although B has a small numerical value compared to E (because of the way the different units for E and B are defined), B contributes the same energy to the wave as E does, as we saw earlier (Eqs. 31–15 and 16).

31–9 Radiation Pressure

If electromagnetic waves carry energy, then we might expect them to also carry linear momentum. When an electromagnetic wave encounters the surface of an object, a force will be exerted on the surface as a result of the momentum transfer ($F = dp/dt$), just as when a moving object strikes a surface. The force per unit area exerted by the waves is called **radiation pressure**, and its existence was predicted by Maxwell. He showed that if a beam of EM radiation (light, for example) is completely absorbed by an object, then the momentum transferred is

$$\Delta p = \frac{\Delta U}{c} \qquad \begin{bmatrix} \text{radiation} \\ \text{fully} \\ \text{absorbed} \end{bmatrix} \quad \textbf{(31–20a)}$$

where ΔU is the energy absorbed by the object in a time Δt, and c is the speed of light.[†] If instead, the radiation is fully reflected (suppose the object is a mirror), then the momentum transferred is twice as great, just as when a ball bounces elastically off a surface:

$$\Delta p = \frac{2\,\Delta U}{c}. \qquad \begin{bmatrix} \text{radiation} \\ \text{fully} \\ \text{reflected} \end{bmatrix} \quad \textbf{(31–20b)}$$

If a surface absorbs some of the energy, and reflects some of it, then $\Delta p = a\,\Delta U/c$, where a is a factor between 1 and 2.

Using Newton's second law we can calculate the force and the pressure exerted by radiation on the object. The force F is given by

$$F = \frac{dp}{dt}.$$

The average rate that energy is delivered to the object is related to the Poynting vector by

$$\frac{dU}{dt} = \overline{S}A,$$

where A is the cross-sectional area of the object which intercepts the radiation. The radiation pressure P (assuming full absorption) is given by (see Eq. 31–20a)

$$P = \frac{F}{A} = \frac{1}{A}\frac{dp}{dt} = \frac{1}{Ac}\frac{dU}{dt} = \frac{\overline{S}}{c}. \qquad \begin{bmatrix} \text{fully} \\ \text{absorbed} \end{bmatrix} \quad \textbf{(31–21a)}$$

If the light is fully reflected, the pressure is twice as great (Eq. 31–20b):

$$P = \frac{2\overline{S}}{c}. \qquad \begin{bmatrix} \text{fully} \\ \text{reflected} \end{bmatrix} \quad \textbf{(31–21b)}$$

EXAMPLE 31–7 **ESTIMATE** **Solar pressure.** Radiation from the Sun that reaches the Earth's surface (after passing through the atmosphere) transports energy at a rate of about $1000\ \text{W/m}^2$. Estimate the pressure and force exerted by the Sun on your outstretched hand.

APPROACH The radiation is partially reflected and partially absorbed, so let us estimate simply $P = \overline{S}/c$.

SOLUTION $P \approx \dfrac{\overline{S}}{c} = \dfrac{1000\ \text{W/m}^2}{3 \times 10^8\ \text{m/s}} \approx 3 \times 10^{-6}\ \text{N/m}^2.$

An estimate of the area of your outstretched hand might be about 10 cm by 20 cm, so $A = 0.02\ \text{m}^2$. Then the force is

$$F = PA \approx (3 \times 10^{-6}\ \text{N/m}^2)(0.02\ \text{m}^2) \approx 6 \times 10^{-8}\ \text{N}.$$

NOTE These numbers are tiny. The force of gravity on your hand, for comparison, is maybe a half pound, or with $m = 0.2\ \text{kg}$, $mg \approx (0.2\ \text{kg})(9.8\ \text{m/s}^2) \approx 2\ \text{N}$. The radiation pressure on your hand is imperceptible compared to gravity.

[†]Very roughly, if we think of light as particles (and we do—see Chapter 37), the force that would be needed to bring such a particle moving at speed c to "rest" (i.e. absorption) is $F = \Delta p/\Delta t$. But F is also related to energy by Eq. 8–7, $F = \Delta U/\Delta x$, so $\Delta p = F\,\Delta t = \Delta U/(\Delta x/\Delta t) = \Delta U/c$ where we identify $(\Delta x/\Delta t)$ with the speed of light c.

EXAMPLE 31–8 **ESTIMATE** **A solar sail.** Proposals have been made to use the radiation pressure from the Sun to help propel spacecraft around the solar system. (*a*) About how much force would be applied on a 1 km × 1 km highly reflective sail, and (*b*) by how much would this increase the speed of a 5000-kg spacecraft in one year? (*c*) If the spacecraft started from rest, about how far would it travel in a year?

APPROACH Pressure P is force per unit area, so $F = PA$. We use the estimate of Example 31–7, doubling it for a reflecting surface $P = 2\overline{S}/c$. We find the acceleration from Newton's second law, and assume it is constant, and then find the speed from $v = v_0 + at$. The distance traveled is given by $x = \frac{1}{2}at^2$.

SOLUTION (*a*) Doubling the result of Example 31–7, the solar pressure is $2\overline{S}/c = 6 \times 10^{-6}\,\text{N/m}^2$. Then the force is $F \approx PA = (6 \times 10^{-6}\,\text{N/m}^2)(10^6\,\text{m}^2) \approx 6\,\text{N}$. (*b*) The acceleration is $a \approx F/m \approx (6\,\text{N})/(5000\,\text{kg}) \approx 1.2 \times 10^{-3}\,\text{m/s}^2$. The speed increase is $v - v_0 = at = (1.2 \times 10^{-3}\,\text{m/s}^2)(365\,\text{days})(24\,\text{hr/day})(3600\,\text{s/hr}) \approx 4 \times 10^4\,\text{m/s}$ ($\approx 150{,}000\,\text{km/h!}$). (*c*) Starting from rest, this acceleration would result in a distance of about $\frac{1}{2}at^2 \approx 6 \times 10^{11}\,\text{m}$ in a year, about four times the Sun-Earth distance. The starting point should be far from the Earth so the Earth's gravitational force is small compared to 6 N.

NOTE A large sail providing a small force over a long time can result in a lot of motion.

Although you cannot directly feel the effects of radiation pressure, the phenomenon is quite dramatic when applied to atoms irradiated by a finely focused laser beam. An atom has a mass on the order of $10^{-27}\,\text{kg}$, and a laser beam can deliver energy at a rate of $1000\,\text{W/m}^2$. This is the same intensity used in Example 31–7, but here a radiation pressure of $10^{-6}\,\text{N/m}^2$ would be very significant on a molecule whose mass might be 10^{-23} to $10^{-26}\,\text{kg}$. It is possible to move atoms and molecules around by steering them with a laser beam, in a device called "optical tweezers." Optical tweezers have some remarkable applications. They are of great interest to biologists, especially since optical tweezers can manipulate live microorganisms, and components within a cell, without damaging them. Optical tweezers have been used to measure the elastic properties of DNA by pulling each end of the molecule with such a laser "tweezers."

PHYSICS APPLIED
Optical tweezers

31–10 Radio and Television; Wireless Communication

PHYSICS APPLIED
Wireless transmission

Electromagnetic waves offer the possibility of transmitting information over long distances. Among the first to realize this and put it into practice was Guglielmo Marconi (1874–1937) who, in the 1890s, invented and developed wireless communication. With it, messages could be sent at the speed of light without the use of wires. The first signals were merely long and short pulses that could be translated into words by a code, such as the "dots" and "dashes" of the Morse code: they were digital wireless, believe it or not. In 1895 Marconi sent wireless signals a kilometer or two in Italy. By 1901 he had sent test signals 3000 km across the ocean from Newfoundland, Canada, to Cornwall, England. In 1903 he sent the first practical commercial messages from Cape Cod, Massachusetts, to England: the London *Times* printed news items sent from its New York correspondent. 1903 was also the year of the first powered airplane flight by the Wright brothers. The hallmarks of the modern age—wireless communication and flight—date from the same year. Our modern world of wireless communication, including radio, television, cordless phones, cell phones, Bluetooth, wi-fi, and satellite communication, are simply modern applications of Marconi's pioneering work.

The next decade saw the development of vacuum tubes. Out of this early work radio and television were born. We now discuss briefly (1) how radio and TV signals are transmitted, and (2) how they are received at home.

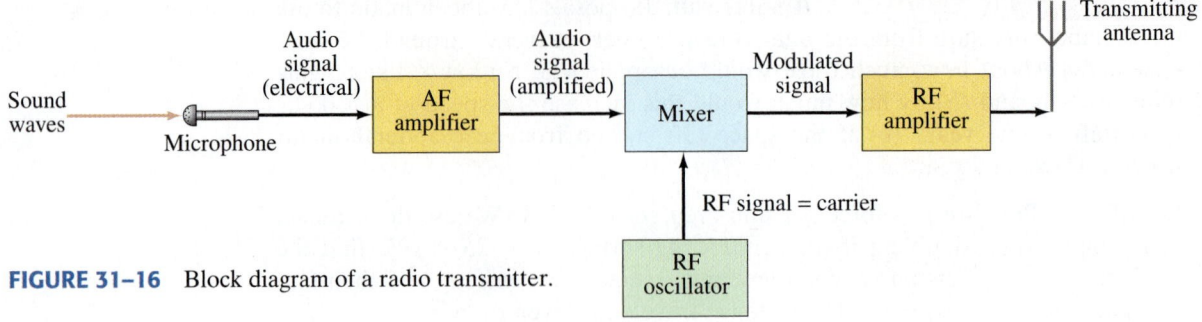

FIGURE 31–16 Block diagram of a radio transmitter.

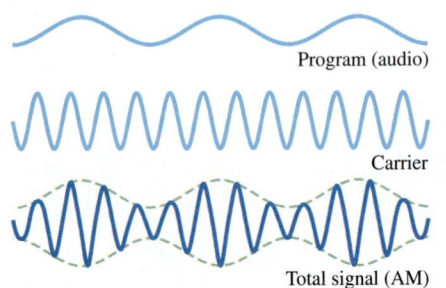

Program (audio)

Carrier

Total signal (AM)

FIGURE 31–17 In amplitude modulation (AM), the amplitude of the carrier signal is made to vary in proportion to the audio signal's amplitude.

PHYSICS APPLIED
AM and FM

FIGURE 31–18 In frequency modulation (FM), the frequency of the carrier signal is made to change in proportion to the audio signal's amplitude. This method is used by FM radio and television.

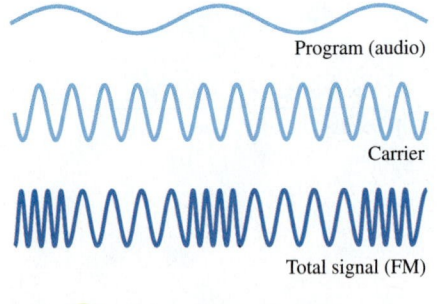

Program (audio)

Carrier

Total signal (FM)

PHYSICS APPLIED
Radio and TV receivers

The process by which a radio station transmits information (words and music) is outlined in Fig. 31–16. The audio (sound) information is changed into an electrical signal of the same frequencies by, say, a microphone or magnetic read/write head. This electrical signal is called an audiofrequency (AF) signal, since the frequencies are in the audio range (20 to 20,000 Hz). The signal is amplified electronically and is then mixed with a radio-frequency (RF) signal called its **carrier frequency**, which represents that station. AM radio stations have carrier frequencies from about 530 kHz to 1700 kHz. For example, "710 on your dial" means a station whose carrier frequency is 710 kHz. FM radio stations have much higher carrier frequencies, between 88 MHz and 108 MHz. The carrier frequencies for broadcast TV stations in the United States lie between 54 MHz and 72 MHz, between 76 MHz and 88 MHz, between 174 MHz and 216 MHz, and between 470 MHz and 698 MHz.

The mixing of the audio and carrier frequencies is done in two ways. In **amplitude modulation** (AM), the amplitude of the high-frequency carrier wave is made to vary in proportion to the amplitude of the audio signal, as shown in Fig. 31–17. It is called "amplitude modulation" because the *amplitude* of the carrier is altered ("modulate" means to change or alter). In **frequency modulation** (FM), the *frequency* of the carrier wave is made to change in proportion to the audio signal's amplitude, as shown in Fig. 31–18. The mixed signal is amplified further and sent to the transmitting antenna, where the complex mixture of frequencies is sent out in the form of EM waves. In digital communication, the signal is put into a digital form (Section 29–8) which modulates the carrier.

A television transmitter works in a similar way, using FM for audio and AM for video; both audio and video signals (see Section 23–9) are mixed with carrier frequencies.

Now let us look at the other end of the process, the reception of radio and TV programs at home. A simple radio receiver is diagrammed in Fig. 31–19. The EM waves sent out by all stations are received by the antenna. The signals the antenna detects and sends to the receiver are very small and contain frequencies from many different stations. The receiver selects out a particular RF frequency (actually a narrow range of frequencies) corresponding to a particular station using a resonant *LC* circuit (Sections 30–6 and 30–9).

FIGURE 31–19 Block diagram of a simple radio receiver.

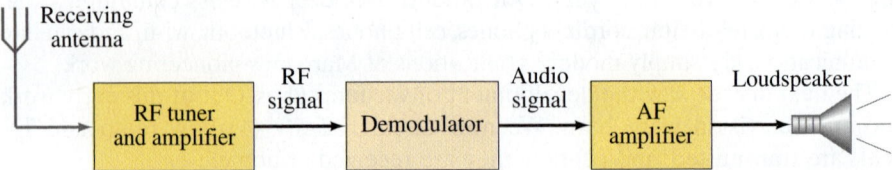

A simple way of tuning a station is shown in Fig. 31–20. A particular station is "tuned in" by adjusting C and/or L so that the resonant frequency of the circuit equals that of the station's carrier frequency. The signal, containing both audio and carrier frequencies, next goes to the *demodulator*, or *detector* (Fig. 31–19), where "demodulation" takes place—that is, the RF carrier frequency is separated from the audio signal. The audio signal is amplified and sent to a loudspeaker or headphones.

Modern receivers have more stages than those shown. Various means are used to increase the sensitivity and selectivity (ability to detect weak signals and distinguish them from other stations), and to minimize distortion of the original signal.[†]

A television receiver does similar things to both the audio and the video signals. The audio signal goes finally to the loudspeaker, and the video signal to the monitor, such as a *cathode ray tube* (CRT) or LCD screen (Sections 23–9 and 35–12).

One kind of antenna consists of one or more conducting rods; the electric field in the EM waves exerts a force on the electrons in the conductor, causing them to move back and forth at the frequencies of the waves (Fig. 31–21a). A second type of antenna consists of a tubular coil of wire which detects the magnetic field of the wave: the changing B field induces an emf in the coil (Fig. 31–21b).

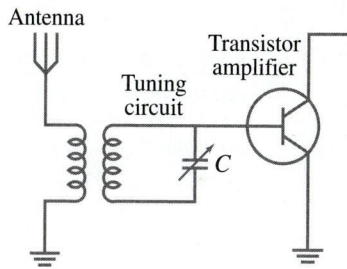

FIGURE 31–20 Simple tuning stage of a radio.

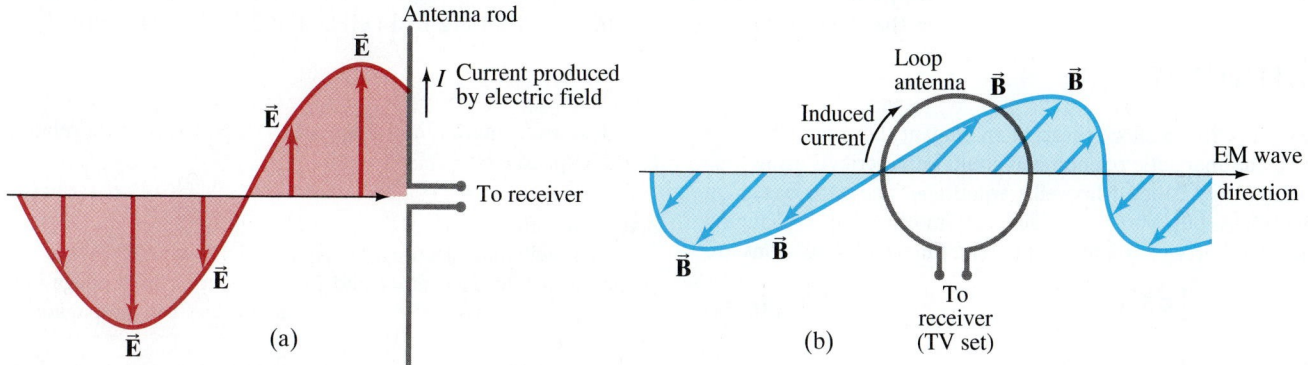

FIGURE 31–21 Antennas. (a) Electric field of EM wave produces a current in an antenna consisting of straight wire or rods. (b) Changing magnetic field induces an emf and current in a loop antenna.

A satellite dish (Fig. 31–22) consists of a parabolic reflector that focuses the EM waves onto a "horn," similar to a concave mirror telescope (Fig. 33–38).

FIGURE 31–22 A satellite dish.

EXAMPLE 31–9 Tuning a station. Calculate the transmitting wavelength of an FM radio station that transmits at 100 MHz.

APPROACH Radio is transmitted as an EM wave, so the speed is $c = 3.0 \times 10^8 \, \text{m/s}$. The wavelength is found from Eq. 31–14, $\lambda = c/f$.

SOLUTION The carrier frequency is $f = 100 \, \text{MHz} = 1.0 \times 10^8 \, \text{s}^{-1}$, so

$$\lambda = \frac{c}{f} = \frac{(3.0 \times 10^8 \, \text{m/s})}{(1.0 \times 10^8 \, \text{s}^{-1})} = 3.0 \, \text{m}.$$

NOTE The wavelengths of other FM signals (88 MHz to 108 MHz) are close to the 3.0-m wavelength of this station. FM antennas are typically 1.5 m long, or about a half wavelength. This length is chosen so that the antenna reacts in a resonant fashion and thus is more sensitive to FM frequencies. AM radio antennas would have to be very long to be either $\frac{1}{2}\lambda$ or $\frac{1}{4}\lambda$.

[†]For *FM stereo broadcasting*, two signals are carried by the carrier wave. One signal contains frequencies up to about 15 kHz, which includes most audio frequencies. The other signal includes the same range of frequencies, but 19 kHz is added to it. A stereo receiver subtracts this 19,000-Hz signal and distributes the two signals to the left and right channels. The first signal consists of the sum of left and right channels (L + R), so mono radios detect all the sound. The second signal is the difference between left and right (L − R). Hence the receiver must add and subtract the two signals to get pure left and right signals for each channel.

Other EM Wave Communications

PHYSICS APPLIED

Cell phones, radio control, remote control, cable TV, and satellite TV and radio

The various regions of the radio-wave spectrum are assigned by governmental agencies for various purposes. Besides those mentioned above, there are "bands" assigned for use by ships, airplanes, police, military, amateurs, satellites and space, and radar. Cell phones, for example, are complete radio transmitters and receivers. In the U.S., CDMA cell phones function on two different bands: 800 MHz and 1900 MHz (= 1.9 GHz). Europe, Asia, and much of the rest of the world use a different system: the international standard called GSM (Global System for Mobile Communication), on 900-MHz and 1800-MHz bands. The U.S. now also has the GSM option (at 850 MHz and 1.9 GHz), as does much of the rest of the Americas. A 700-MHz band is now being made available for cell phones (it used to carry TV broadcast channels 52–69, now no longer used). Radio-controlled toys (cars, sailboats, robotic animals, etc.) can use various frequencies from 27 MHz to 75 MHz. Automobile remote (keyless) entry may operate around 300 MHz or 400 MHz.

Cable TV channels are carried as electromagnetic waves along a coaxial cable (see Fig. 31–13) rather than being broadcast and received through the "air." The channels are in the same part of the EM spectrum, hundreds of MHz, but some are at frequencies not available for TV broadcast. Digital satellite TV and radio are carried in the microwave portion of the spectrum (12 to 14 GHz and 2.3 GHz, respectively).

Summary

James Clerk Maxwell synthesized an elegant theory in which all electric and magnetic phenomena could be described using four equations, now called **Maxwell's equations**. They are based on earlier ideas, but Maxwell added one more—that a changing electric field produces a magnetic field. Maxwell's equations are

$$\oint \vec{E} \cdot d\vec{A} = \frac{Q}{\epsilon_0} \tag{31–5a}$$

$$\oint \vec{B} \cdot d\vec{A} = 0 \tag{31–5b}$$

$$\oint \vec{E} \cdot d\vec{\ell} = -\frac{d\Phi_B}{dt} \tag{31–5c}$$

$$\oint \vec{B} \cdot d\vec{\ell} = \mu_0 I + \mu_0 \epsilon_0 \frac{d\Phi_E}{dt}. \tag{31–5d}$$

The first two are Gauss's laws for electricity and for magnetism; the other two are Faraday's law and Ampère's law (as extended by Maxwell), respectively.

Maxwell's theory predicted that transverse **electromagnetic (EM) waves** would be produced by accelerating electric charges, and these waves would propagate through space at the speed of light c, given by

$$c = \frac{1}{\sqrt{\epsilon_0 \mu_0}} = 3.00 \times 10^8 \, \text{m/s}. \tag{31–12}$$

The wavelength λ and frequency f of EM waves are related to their speed c by

$$c = \lambda f, \tag{31–14}$$

just as for other waves.

The oscillating electric and magnetic fields in an EM wave are perpendicular to each other and to the direction of propagation. EM waves are waves of fields, not matter, and can propagate in empty space.

After EM waves were experimentally detected in the late 1800s, the idea that light is an EM wave (although of much higher frequency than those detected directly) became generally accepted. The **electromagnetic spectrum** includes EM waves of a wide variety of wavelengths, from microwaves and radio waves to visible light to X-rays and gamma rays, all of which travel through space at a speed $c = 3.00 \times 10^8 \, \text{m/s}$.

The energy carried by EM waves can be described by the **Poynting vector**

$$\vec{S} = \frac{1}{\mu_0} \vec{E} \times \vec{B} \tag{31–18}$$

which gives the rate energy is carried across unit area per unit time when the electric and magnetic fields in an EM wave in free space are $\vec{E}$ and $\vec{B}$.

EM waves carry momentum and exert a **radiation pressure** proportional to the intensity S of the wave.

Questions

1. An electric field $\vec{E}$ points away from you, and its magnitude is increasing. Will the induced magnetic field be clockwise or counterclockwise? What if $\vec{E}$ points toward you and is decreasing?

2. What is the direction of the displacement current in Fig. 31–3? (*Note:* The capacitor is discharging.)

3. Why is it that the magnetic field of a displacement current in a capacitor is so much harder to detect than the magnetic field of a conduction current?

4. Are there any good reasons for calling the term $\mu_0 \epsilon_0 \, d\Phi_E/dt$ in Eq. 31–1 as being due to a "current"? Explain.

5. The electric field in an EM wave traveling north oscillates in an east–west plane. Describe the direction of the magnetic field vector in this wave.

6. Is sound an electromagnetic wave? If not, what kind of wave is it?

7. Can EM waves travel through a perfect vacuum? Can sound waves?

8. When you flip a light switch, does the overhead light go on immediately? Explain.

9. Are the wavelengths of radio and television signals longer or shorter than those detectable by the human eye?

10. What does the wavelength calculated in Example 31–2 tell you about the phase of a 60-Hz ac current that starts at a power plant as compared to its phase at a house 200 km away?

11. When you connect two loudspeakers to the output of a stereo amplifier, should you be sure the lead wires are equal in length so that there will not be a time lag between speakers? Explain.

12. In the electromagnetic spectrum, what type of EM wave would have a wavelength of 10^3 km; 1 km; 1 m; 1 cm; 1 mm; 1 μm?

13. Can radio waves have the same frequencies as sound waves (20 Hz–20,000 Hz)?

14. Discuss how cordless telephones make use of EM waves. What about cellular telephones?

15. Can two radio or TV stations broadcast on the same carrier frequency? Explain.

16. If a radio transmitter has a vertical antenna, should a receiver's antenna (rod type) be vertical or horizontal to obtain best reception?

17. The carrier frequencies of FM broadcasts are much higher than for AM broadcasts. On the basis of what you learned about diffraction in Chapter 15, explain why AM signals can be detected more readily than FM signals behind low hills or buildings.

18. A lost person may signal by flashing a flashlight on and off using Morse code. This is actually a modulated EM wave. Is it AM or FM? What is the frequency of the carrier, approximately?

Problems

31–1 $\vec{B}$ Produced by Changing $\vec{E}$

1. (I) Determine the rate at which the electric field changes between the round plates of a capacitor, 6.0 cm in diameter, if the plates are spaced 1.1 mm apart and the voltage across them is changing at a rate of 120 V/s.

2. (I) Calculate the displacement current I_D between the square plates, 5.8 cm on a side, of a capacitor if the electric field is changing at a rate of 2.0×10^6 V/m·s.

3. (II) At a given instant, a 2.8-A current flows in the wires connected to a parallel-plate capacitor. What is the rate at which the electric field is changing between the plates if the square plates are 1.60 cm on a side?

4. (II) A 1500-nF capacitor with circular parallel plates 2.0 cm in diameter is accumulating charge at the rate of 38.0 mC/s at some instant in time. What will be the induced magnetic field strength 10.0 cm radially outward from the center of the plates? What will be the value of the field strength after the capacitor is fully charged?

5. (II) Show that the displacement current through a parallel-plate capacitor can be written $I_D = C\, dV/dt$, where V is the voltage across the capacitor at any instant.

6. (II) Suppose an air-gap capacitor has circular plates of radius $R = 2.5$ cm and separation $d = 1.6$ mm. A 76.0-Hz emf, $\mathscr{E} = \mathscr{E}_0 \cos \omega t$, is applied to the capacitor. The maximum displacement current is 35 μA. Determine (a) the maximum conduction current I, (b) the value of $\mathscr{E}_0$, (c) the maximum value of $d\Phi_E/dt$ between the plates. Neglect fringing.

7. (III) Suppose that a circular parallel-plate capacitor has radius $R_0 = 3.0$ cm and plate separation $d = 5.0$ mm. A sinusoidal potential difference $V = V_0 \sin(2\pi ft)$ is applied across the plates, where $V_0 = 150$ V and $f = 60$ Hz. (a) In the region between the plates, show that the magnitude of the induced magnetic field is given by $B = B_0(R) \cos(2\pi ft)$, where R is the radial distance from the capacitor's central axis. (b) Determine the expression for the amplitude $B_0(R)$ of this time-dependent (sinusoidal) field when $R \le R_0$, and when $R > R_0$. (c) Plot $B_0(R)$ in tesla for the range $0 \le R \le 10$ cm.

31–5 EM Waves

8. (I) If the electric field in an EM wave has a peak magnitude of 0.57×10^{-4} V/m, what is the peak magnitude of the magnetic field strength?

9. (I) If the magnetic field in a traveling EM wave has a peak magnitude of 12.5 nT, what is the peak magnitude of the electric field?

10. (I) In an EM wave traveling west, the B field oscillates vertically and has a frequency of 80.0 kHz and an rms strength of 7.75×10^{-9} T. Determine the frequency and rms strength of the electric field. What is its direction?

11. (II) The electric field of a plane EM wave is given by $E_x = E_0 \cos(kz + \omega t)$, $E_y = E_z = 0$. Determine (a) the direction of propagation and (b) the magnitude and direction of $\vec{B}$.

12. (III) Consider two possible candidates $E(x, t)$ as solutions of the wave equation for an EM wave's electric field. Let A and α be constants. Show that (a) $E(x, t) = Ae^{-\alpha(x-vt)^2}$ satisfies the wave equation, and that (b) $E(x, t) = Ae^{-(\alpha x^2 - vt)}$ does not satisfy the wave equation.

31–6 Electromagnetic Spectrum

13. (I) What is the frequency of a microwave whose wavelength is 1.50 cm?

14. (I) (a) What is the wavelength of a 25.75×10^9 Hz radar signal? (b) What is the frequency of an X-ray with wavelength 0.12 nm?

15. (I) How long does it take light to reach us from the Sun, 1.50×10^8 km away?

16. (I) An EM wave has frequency 8.56×10^{14} Hz. What is its wavelength, and how would we classify it?

17. (I) Electromagnetic waves and sound waves can have the same frequency. (a) What is the wavelength of a 1.00-kHz electromagnetic wave? (b) What is the wavelength of a 1.00-kHz sound wave? (The speed of sound in air is 341 m/s.) (c) Can you hear a 1.00-kHz electromagnetic wave?

18. (II) Pulsed lasers used for science and medicine produce very brief bursts of electromagnetic energy. If the laser light wavelength is 1062 nm (Neodymium–YAG laser), and the pulse lasts for 38 picoseconds, how many wavelengths are found within the laser pulse? How brief would the pulse need to be to fit only one wavelength?

19. (II) How long would it take a message sent as radio waves from Earth to reach Mars (a) when nearest Earth, (b) when farthest from Earth?

20. (II) An electromagnetic wave has an electric field given by
$$\vec{E} = \hat{i}(225 \text{ V/m}) \sin[(0.077 \text{ m}^{-1})z - (2.3 \times 10^7 \text{ rad/s})t].$$
(a) What are the wavelength and frequency of the wave? (b) Write down an expression for the magnetic field.

31-7 Speed of Light

21. (II) What is the minimum angular speed at which Michelson's eight-sided mirror would have had to rotate to reflect light into an observer's eye by succeeding mirror faces (1/8 of a revolution, Fig. 31–14)?

31-8 EM Wave Energy; Poynting Vector

22. (I) The $\vec{E}$ field in an EM wave has a peak of $26.5 \, \text{mV/m}$. What is the average rate at which this wave carries energy across unit area per unit time?

23. (II) The magnetic field in a traveling EM wave has an rms strength of $22.5 \, \text{nT}$. How long does it take to deliver $335 \, \text{J}$ of energy to $1.00 \, \text{cm}^2$ of a wall that it hits perpendicularly?

24. (II) How much energy is transported across a $1.00 \, \text{cm}^2$ area per hour by an EM wave whose E field has an rms strength of $32.8 \, \text{mV/m}$?

25. (II) A spherically spreading EM wave comes from a 1500-W source. At a distance of $5.0 \, \text{m}$, what is the intensity, and what is the rms value of the electric field?

26. (II) If the amplitude of the B field of an EM wave is $2.5 \times 10^{-7} \, \text{T}$, (a) what is the amplitude of the E field? (b) What is the average power per unit area of the EM wave?

27. (II) What is the energy contained in a 1.00-m^3 volume near the Earth's surface due to radiant energy from the Sun? See Example 31–6.

28. (II) A 15.8-mW laser puts out a narrow beam $2.00 \, \text{mm}$ in diameter. What are the rms values of E and B in the beam?

29. (II) Estimate the average power output of the Sun, given that about $1350 \, \text{W/m}^2$ reaches the upper atmosphere of the Earth.

30. (II) A high-energy pulsed laser emits a 1.0-ns-long pulse of average power $1.8 \times 10^{11} \, \text{W}$. The beam is $2.2 \times 10^{-3} \, \text{m}$ in radius. Determine (a) the energy delivered in each pulse, and (b) the rms value of the electric field.

31. (II) How practical is solar power for various devices? Assume that on a sunny day, sunlight has an intensity of $1000 \, \text{W/m}^2$ at the surface of Earth and that, when illuminated by that sunlight, a solar-cell panel can convert 10% of the sunlight's energy into electric power. For each device given below, calculate the area A of solar panel needed to power it. (a) A calculator consumes $50 \, \text{mW}$. Find A in cm^2. Is A small enough so that the solar panel can be mounted directly on the calculator that it is powering? (b) A hair dryer consumes $1500 \, \text{W}$. Find A in m^2. Assuming no other electronic devices are operating within a house at the same time, is A small enough so that the hair dryer can be powered by a solar panel mounted on the house's roof? (c) A car requires 20 hp for highway driving at constant velocity (this car would perform poorly in situations requiring acceleration). Find A in m^2. Is A small enough so that this solar panel can be mounted directly on the car and power it in "real time"?

32. (III) (a) Show that the Poynting vector $\vec{S}$ points radially inward toward the center of a circular parallel-plate capacitor when it is being charged as in Example 31–1. (b) Integrate $\vec{S}$ over the cylindrical boundary of the capacitor gap to show that the rate at which energy enters the capacitor is equal to the rate at which electrostatic energy is being stored in the electric field of the capacitor (Section 24–4). Ignore fringing of $\vec{E}$.

33. (III) The Arecibo radio telescope in Puerto Rico can detect a radio wave with an intensity as low as $1 \times 10^{-23} \, \text{W/m}^2$. As a "best-case" scenario for communication with extraterrestrials, consider the following: suppose an advanced civilization located at point A, a distance x away from Earth, is somehow able to harness the entire power output of a Sun-like star, converting that power completely into a radio-wave signal which is transmitted uniformly in all directions from A. (a) In order for Arecibo to detect this radio signal, what is the maximum value for x in light-years $(1 \, \text{ly} \approx 10^{16} \, \text{m})$? (b) How does this maximum value compare with the 100,000-ly size of our Milky Way galaxy? The intensity of sunlight at Earth's orbital distance from the Sun is $1350 \, \text{W/m}^2$.

31-9 Radiation Pressure

34. (II) Estimate the radiation pressure due to a 75-W bulb at a distance of $8.0 \, \text{cm}$ from the center of the bulb. Estimate the force exerted on your fingertip if you place it at this point.

35. (II) Laser light can be focused (at best) to a spot with a radius r equal to its wavelength λ. Suppose that a 1.0-W beam of green laser light $(\lambda = 5 \times 10^{-7} \, \text{m})$ is used to form such a spot and that a cylindrical particle of about that size (let the radius and height equal r) is illuminated by the laser as shown in Fig. 31–23. Estimate the acceleration of the particle, if its density equals that of water and it absorbs the radiation. [This order-of-magnitude calculation convinced researchers of the feasibility of "optical tweezers," p. 829.]

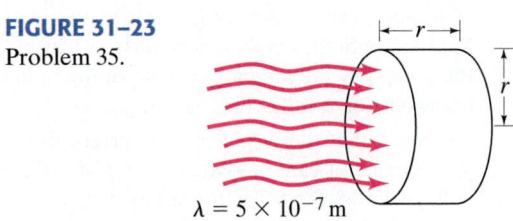

FIGURE 31–23
Problem 35.

$\lambda = 5 \times 10^{-7} \, \text{m}$

36. (II) The powerful laser used in a laser light show provides a 3-mm diameter beam of green light with a power of 3 W. When a space-walking astronaut is outside the Space Shuttle, her colleague inside the Shuttle playfully aims such a laser beam at the astronaut's space suit. The masses of the suited astronaut and the Space Shuttle are $120 \, \text{kg}$ and $103,000 \, \text{kg}$, respectively. (a) Assuming the suit is perfectly reflecting, determine the "radiation-pressure" force exerted on the astronaut by the laser beam. (b) Assuming the astronaut is separated from the Shuttle's center of mass by 20 m, model the Shuttle as a sphere in order to estimate the gravitation force it exerts on the astronaut. (c) Which of the two forces is larger, and by what factor?

37. (II) What size should the solar panel on a satellite orbiting Jupiter be if it is to collect the same amount of radiation from the Sun as a 1.0-m^2 solar panel on a satellite orbiting Earth?

38. (I) What is the range of wavelengths for (*a*) FM radio (88 MHz to 108 MHz) and (*b*) AM radio (535 kHz to 1700 kHz)?

39. (I) Estimate the wavelength for 1.9-GHz cell phone reception.

40. (I) The variable capacitor in the tuner of an AM radio has a capacitance of 2200 pF when the radio is tuned to a station at 550 kHz. What must the capacitance be for a station near the other end of the dial, 1610 kHz?

41. (II) A certain FM radio tuning circuit has a fixed capacitor $C = 620$ pF. Tuning is done by a variable inductance. What range of values must the inductance have to tune stations from 88 MHz to 108 MHz?

42. (II) A satellite beams microwave radiation with a power of 12 kW toward the Earth's surface, 550 km away. When the beam strikes Earth, its circular diameter is about 1500 m. Find the rms electric field strength of the beam at the surface of the Earth.

General Problems

43. A 1.60-m-long FM antenna is oriented parallel to the electric field of an EM wave. How large must the electric field be to produce a 1.00-mV (rms) voltage between the ends of the antenna? What is the rate of energy transport per square meter?

44. Who will hear the voice of a singer first: a person in the balcony 50.0 m away from the stage (see Fig. 31–24), or a person 1500 km away at home whose ear is next to the radio listening to a live broadcast? Roughly how much sooner? Assume the microphone is a few centimeters from the singer and the temperature is 20°C.

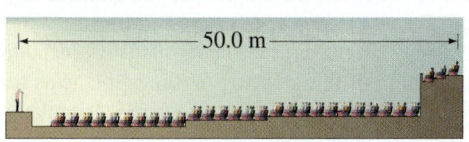

FIGURE 31–24 Problem 44.

45. Light is emitted from an ordinary lightbulb filament in wave-train bursts about 10^{-8} s in duration. What is the length in space of such wave trains?

46. Radio-controlled clocks throughout the United States receive a radio signal from a transmitter in Fort Collins, Colorado, that accurately (within a microsecond) marks the beginning of each minute. A slight delay, however, is introduced because this signal must travel from the transmitter to the clocks. Assuming Fort Collins is no more than 3000 km from any point in the U.S., what is the longest travel-time delay?

47. A radio voice signal from the *Apollo* crew on the Moon (Fig. 31–25) was beamed to a listening crowd from a radio speaker. If you were standing 25 m from the loudspeaker, what was the total time lag between when you heard the sound and when the sound entered a microphone on the Moon and traveled to Earth?

FIGURE 31–25
Problem 47.

48. Cosmic microwave background radiation fills all space with an average energy density of 4×10^{-14} J/m³. (*a*) Find the rms value of the electric field associated with this radiation. (*b*) How far from a 7.5-kW radio transmitter emitting uniformly in all directions would you find a comparable value?

49. What are E_0 and B_0 2.00 m from a 75-W light source? Assume the bulb emits radiation of a single frequency uniformly in all directions.

50. Estimate the rms electric field in the sunlight that hits Mars, knowing that the Earth receives about 1350 W/m² and that Mars is 1.52 times farther from the Sun (on average) than is the Earth.

51. At a given instant in time, a traveling EM wave is noted to have its maximum magnetic field pointing west and its maximum electric field pointing south. In which direction is the wave traveling? If the rate of energy flow is 560 W/m², what are the maximum values for the two fields?

52. How large an emf (rms) will be generated in an antenna that consists of a circular coil 2.2 cm in diameter having 320 turns of wire, when an EM wave of frequency 810 kHz transporting energy at an average rate of 1.0×10^{-4} W/m² passes through it? [*Hint*: you can use Eq. 29–4 for a generator, since it could be applied to an observer moving with the coil so that the magnetic field is oscillating with the frequency $f = \omega/2\pi$.]

53. The average intensity of a particular TV station's signal is 1.0×10^{-13} W/m² when it arrives at a 33-cm-diameter satellite TV antenna. (*a*) Calculate the total energy received by the antenna during 6.0 hours of viewing this station's programs. (*b*) What are the amplitudes of the E and B fields of the EM wave?

54. A radio station is allowed to broadcast at an average power not to exceed 25 kW. If an electric field amplitude of 0.020 V/m is considered to be acceptable for receiving the radio transmission, estimate how many kilometers away you might be able to hear this station.

55. A point source emits light energy uniformly in all directions at an average rate P_0 with a single frequency f. Show that the peak electric field in the wave is given by

$$E_0 = \sqrt{\frac{\mu_0 c P_0}{2\pi r^2}}.$$

56. Suppose a 35-kW radio station emits EM waves uniformly in all directions. (*a*) How much energy per second crosses a 1.0-m² area 1.0 km from the transmitting antenna? (*b*) What is the rms magnitude of the $\vec{E}$ field at this point, assuming the station is operating at full power? What is the rms voltage induced in a 1.0-m-long vertical car antenna (*c*) 1.0 km away, (*d*) 50 km away?

57. What is the maximum power level of a radio station so as to avoid electrical breakdown of air at a distance of 0.50 m from the transmitting antenna? Assume the antenna is a point source. Air breaks down in an electric field of about 3×10^6 V/m.

58. In free space ("vacuum"), where the net charge and current flow is zero, the speed of an EM wave is given by $v = 1/\sqrt{\epsilon_0 \mu_0}$. If, instead, an EM wave travels in a nonconducting ("dielectric") material with dielectric constant K, then $v = 1/\sqrt{K\epsilon_0\mu_0}$. For frequencies corresponding to the visible spectrum (near 5×10^{14} Hz), the dielectric constant of water is 1.77. Predict the speed of light in water and compare this value (as a percentage) with the speed of light in a vacuum.

59. The metal walls of a microwave oven form a cavity of dimensions 37 cm $\times$ 37 cm $\times$ 20 cm. When 2.45-GHz microwaves are continuously introduced into this cavity, reflection of incident waves from the walls set up standing waves with nodes at the walls. Along the 37-cm dimension of the oven, how many nodes exist (excluding the nodes at the wall) and what is the distance between adjacent nodes? [Because no heating occurs at these nodes, most microwaves rotate food while operating.]

60. Imagine that a steady current I flows in a straight cylindrical wire of radius R_0 and resistivity ρ. (a) If the current is then changed at a rate dI/dt, show that a displacement current I_D exists in the wire of magnitude $\epsilon_0\rho(dI/dt)$. (b) If the current in a copper wire is changed at the rate of 1.0 A/ms, determine the magnitude of I_D. (c) Determine the magnitude of the magnetic field B_D (T) created by I_D at the surface of a copper wire with $R_0 = 1.0$ mm. Compare (as a ratio) B_D with the field created at the surface of the wire by a steady current of 1.0 A.

61. The electric field of an EM wave pulse traveling along the x axis in free space is given by $E_y = E_0 \exp[-\alpha^2x^2 - \beta^2t^2 + 2\alpha\beta xt]$, where E_0, α, and β are positive constants. (a) Is the pulse moving in the $+x$ or $-x$ direction? (b) Express β in terms of α and c (speed of light in free space). (c) Determine the expression for the magnetic field of this EM wave.

62. Suppose that a right-moving EM wave overlaps with a left-moving EM wave so that, in a certain region of space, the total electric field in the y direction and magnetic field in the z direction are given by $E_y = E_0\sin(kx - \omega t) + E_0\sin(kx + \omega t)$ and $B_z = B_0\sin(kx - \omega t) - B_0\sin(kx + \omega t)$. (a) Find the mathematical expression that represents the standing electric and magnetic waves in the y and z directions, respectively. (b) Determine the Poynting vector and find the x locations at which it is zero at all times.

63. The electric and magnetic fields of a certain EM wave in free space are given by $\vec{E} = E_0\sin(kx - \omega t)\hat{j} + E_0\cos(kx - \omega t)\hat{k}$ and $\vec{B} = B_0\cos(kx - \omega t)\hat{j} - B_0\sin(kx - \omega t)\hat{k}$. (a) Show that $\vec{E}$ and $\vec{B}$ are perpendicular to each other at all times. (b) For this wave, $\vec{E}$ and $\vec{B}$ are in a plane parallel to the yz plane. Show that the wave moves in a direction perpendicular to both $\vec{E}$ and $\vec{B}$. (c) At any arbitrary choice of position x and time t, show that the magnitudes of $\vec{E}$ and $\vec{B}$ always equal E_0 and B_0, respectively. (d) At $x = 0$, draw the orientation of $\vec{E}$ and $\vec{B}$ in the yz plane at $t = 0$. Then qualitatively describe the motion of these vectors in the yz plane as time increases. [Note: The EM wave in this Problem is "circularly polarized."]

Answers to Exercises

A: (c).

B: (b).

C: (a) 3.8×10^6 Hz; (b) 5.5×10^{18} Hz.

D: 45 cm.

E: Yes; the signal still travels 72,000 km.

F: Over 4 hours.

Reflection from still water, as from a glass mirror, can be analyzed using the ray model of light.

Is this picture right side up? How can you tell? What are the clues? Notice the people and position of the Sun. Ray diagrams, which we will learn to draw in this Chapter, can provide the answer. See Example 32–3.

In this first Chapter on light and optics, we use the ray model of light to understand the formation of images by mirrors, both plane and curved (spherical). We also begin our study of refraction—how light rays bend when they go from one medium to another—which prepares us for our study in the next Chapter of lenses, which are the crucial part of so many optical instruments.

Light: Reflection and Refraction

CHAPTER-OPENING QUESTION—Guess now!

A 2.0-m-tall person is standing 2.0 m from a flat vertical mirror staring at her image. What minimum height must the mirror have if the person is to see her entire body, from the top of her head to her feet?

(a) 0.50 m.
(b) 1.0 m.
(c) 1.5 m.
(d) 2.0 m.
(e) 2.5 m.

The sense of sight is extremely important to us, for it provides us with a large part of our information about the world. How do we see? What is the something called *light* that enters our eyes and causes the sensation of sight? How does light behave so that we can see everything that we do? We saw in Chapter 31 that light can be considered a form of electromagnetic radiation. We now examine the subject of light in detail in the next four Chapters.

We see an object in one of two ways: (1) the object may be a *source* of light, such as a lightbulb, a flame, or a star, in which case we see the light emitted directly from the source; or, more commonly, (2) we see an object by light *reflected* from it. In the latter case, the light may have originated from the Sun, artificial lights, or a campfire. An understanding of how objects *emit* light was not achieved until the 1920s, and will be discussed in Chapter 37. How light is *reflected* from objects was understood earlier, and will be discussed here, in Section 32–2.

CONTENTS

32–1 The Ray Model of Light

32–2 Reflection; Image Formation by a Plane Mirror

32–3 Formation of Images by Spherical Mirrors

32–4 Index of Refraction

32–5 Refraction: Snell's Law

32–6 Visible Spectrum and Dispersion

32–7 Total Internal Reflection; Fiber Optics

*32–8 Refraction at a Spherical Surface

32–1 The Ray Model of Light

A great deal of evidence suggests that *light travels in straight lines* under a wide variety of circumstances. For example, a source of light like the Sun casts distinct shadows, and the light from a laser pointer appears to be a straight line. In fact, we infer the positions of objects in our environment by assuming that light moves from the object to our eyes in straight-line paths. Our orientation to the physical world is based on this assumption.

This reasonable assumption is the basis of the **ray model** of light. This model assumes that light travels in straight-line paths called **light rays**. Actually, a ray is an idealization; it is meant to represent an extremely narrow beam of light. When we see an object, according to the ray model, light reaches our eyes from each point on the object. Although light rays leave each point in many different directions, normally only a small bundle of these rays can enter an observer's eye, as shown in Fig. 32–1. If the person's head moves to one side, a different bundle of rays will enter the eye from each point.

We saw in Chapter 31 that light can be considered as an electromagnetic wave. Although the ray model of light does not deal with this aspect of light (we discuss the wave nature of light in Chapters 34 and 35), the ray model has been very successful in describing many aspects of light such as reflection, refraction, and the formation of images by mirrors and lenses.[†] Because these explanations involve straight-line rays at various angles, this subject is referred to as **geometric optics**.

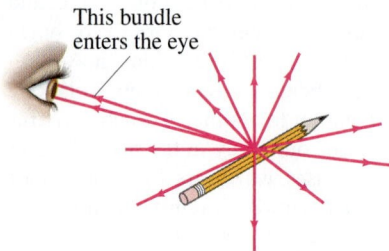

FIGURE 32–1 Light rays come from each single point on an object. A small bundle of rays leaving one point is shown entering a person's eye.

32–2 Reflection; Image Formation by a Plane Mirror

When light strikes the surface of an object, some of the light is reflected. The rest can be absorbed by the object (and transformed to thermal energy) or, if the object is transparent like glass or water, part can be transmitted through. For a very smooth shiny object such as a silvered mirror, over 95% of the light may be reflected.

When a narrow beam of light strikes a flat surface (Fig. 32–2), we define the **angle of incidence**, θ_i, to be the angle an incident ray makes with the normal (perpendicular) to the surface, and the **angle of reflection**, θ_r, to be the angle the reflected ray makes with the normal. It is found that the *incident and reflected rays lie in the same plane with the normal to the surface*, and that

the angle of reflection equals the angle of incidence, $\theta_r = \theta_i$.

This is the **law of reflection**, and it is depicted in Fig. 32–2. It was known to the ancient Greeks, and you can confirm it yourself by shining a narrow flashlight beam or a laser pointer at a mirror in a darkened room.

FIGURE 32–2 Law of reflection: (a) Shows a 3-D view of an incident ray being reflected at the top of a flat surface; (b) shows a side or "end-on" view, which we will usually use because of its clarity.

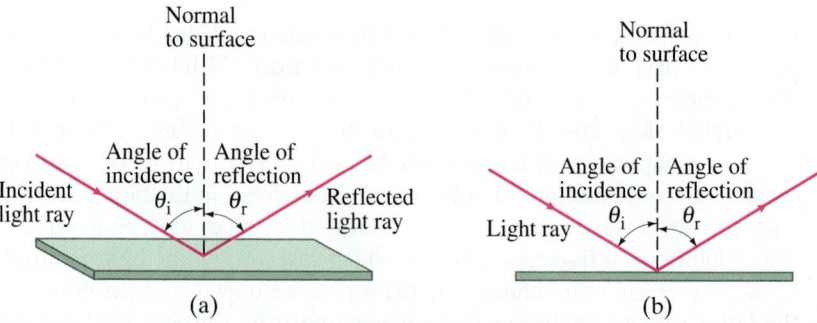

[†]In ignoring the wave properties of light we must be careful that when the light rays pass by objects or through apertures, these must be large compared to the wavelength of the light (so the wave phenomena of interference and diffraction, as discussed in Chapter 15, can be ignored), and we ignore what happens to the light at the edges of objects until we get to Chapters 34 and 35.

When light is incident upon a rough surface, even microscopically rough such as this page, it is reflected in many directions, as shown in Fig. 32–3. This is called **diffuse reflection**. The law of reflection still holds, however, at each small section of the surface. Because of diffuse reflection in all directions, an ordinary object can be seen at many different angles by the light reflected from it. When you move your head to the side, different reflected rays reach your eye from each point on the object (such as this page), Fig. 32–4a. Let us compare diffuse reflection to reflection from a mirror, which is known as **specular reflection**. ("Speculum" is Latin for mirror.) When a narrow beam of light shines on a mirror, the light will not reach your eye unless your eye is positioned at just the right place where the law of reflection is satisfied, as shown in Fig. 32–4b. This is what gives rise to the special image-forming properties of mirrors.

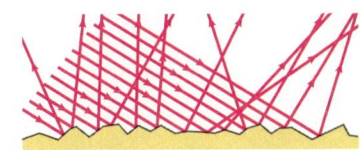

FIGURE 32–3 Diffuse reflection from a rough surface.

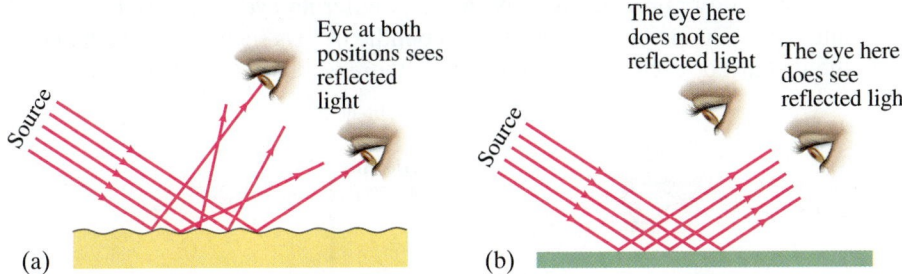

(a)   (b)

FIGURE 32–4 A narrow beam of light shines on (a) white paper, and (b) a mirror. In part (a), you can see with your eye the white light reflected at various positions because of diffuse reflection. But in part (b), you see the reflected light only when your eye is placed correctly $(\theta_r = \theta_i)$; mirror reflection is also known as specular reflection. (Galileo, using similar arguments, showed that the Moon must have a rough surface rather than a highly polished surface like a mirror, as some people thought.)

EXAMPLE 32–1 **Reflection from flat mirrors.** Two flat mirrors are perpendicular to each other. An incoming beam of light makes an angle of 15° with the first mirror as shown in Fig. 32–5a. What angle will the outgoing beam make with the second mirror?

APPROACH We sketch the path of the beam as it reflects off the two mirrors, and draw the two normals to the mirrors for the two reflections. We use geometry and the law of reflection to find the various angles.

FIGURE 32–5 Example 32–1.

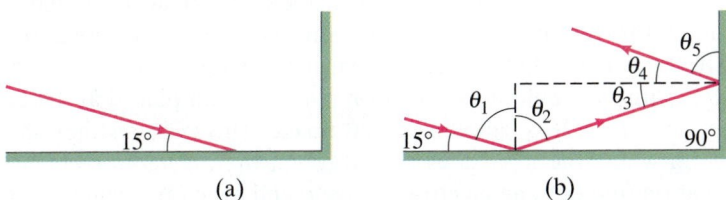

(a)   (b)

SOLUTION In Fig. 32–5b, $\theta_1 + 15° = 90°$, so $\theta_1 = 75°$; by the law of reflection $\theta_2 = \theta_1 = 75°$ too. The two normals to the two mirrors are perpendicular to each other, so $\theta_2 + \theta_3 + 90° = 180°$ as for any triangle. Thus $\theta_3 = 180° - 90° - 75° = 15°$. By the law of reflection, $\theta_4 = \theta_3 = 15°$, so $\theta_5 = 75°$ is the angle the reflected ray makes with the second mirror surface.

NOTE The outgoing ray is parallel to the incoming ray. Red reflectors on bicycles and cars use this principle.

When you look straight into a mirror, you see what appears to be yourself as well as various objects around and behind you, Fig. 32–6. Your face and the other objects look as if they are in front of you, beyond the mirror. But what you see in the mirror is an **image** of the objects, including yourself, that are in front of the mirror.

FIGURE 32–6 When you look in a mirror, you see an image of yourself and objects around you. You don't see yourself as others see you, because left and right appear reversed in the image.

A "plane" mirror is one with a smooth flat reflecting surface. Figure 32–7 shows how an image is formed by a plane mirror according to the ray model. We are viewing the mirror, on edge, in the diagram of Fig. 32–7, and the rays are shown reflecting from the front surface. (Good mirrors are generally made by putting a highly reflective metallic coating on one surface of a very flat piece of glass.) Rays from two different points on an object (the bottle on the left in Fig. 32–7) are shown: two rays are shown leaving from a point on the top of the bottle, and two more from a point on the bottom. Rays leave each point on the object going in many directions, but only those that enclose the bundle of rays that enter the eye from each of the two points are shown. Each set of diverging rays that reflect from the mirror and enter the eye *appear* to come from a single point (called the image point) behind the mirror, as shown by the dashed lines. That is, our eyes and brain interpret any rays that enter an eye as having traveled straight-line paths. The point from which each bundle of rays seems to come is one point on the image. For each point on the object, there is a corresponding image point.

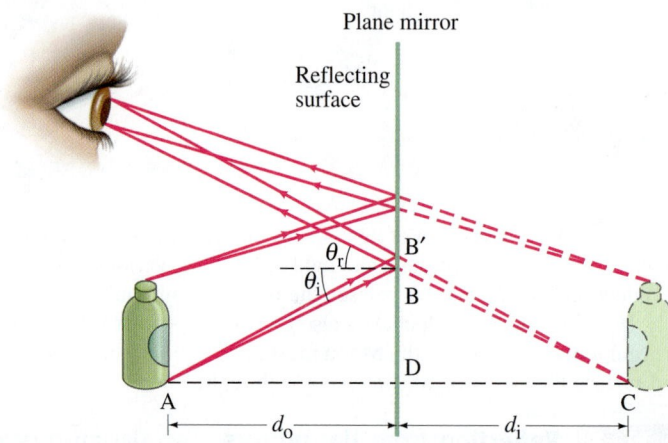

FIGURE 32–7 Formation of a virtual image by a plane mirror.

Let us concentrate on the two rays that leave point A on the object in Fig. 32–7, and strike the mirror at points B and B'. We use geometry for the rays at B. The angles ADB and CDB are right angles; and because of the law of reflection, $\theta_i = \theta_r$ at point B. Therefore, angles ABD and CBD are also equal. The two triangles ABD and CBD are thus congruent, and the length AD = CD. That is, the image appears as far behind the mirror as the object is in front. The **image distance**, d_i (perpendicular distance from mirror to image, Fig. 32–7), equals the **object distance**, d_o (perpendicular distance from object to mirror). From the geometry, we can also see that the height of the image is the same as that of the object.

The light rays do not actually pass through the image location itself in Fig. 32–7. (Note where the red lines are dashed to show they are our projections, not rays.) The image would not appear on paper or film placed at the location of the image. Therefore, it is called a **virtual image**. This is to distinguish it from a **real image** in which the light does pass through the image and which therefore could appear on film or in an electronic sensor, and even on a white sheet of paper or screen placed at the position of the image. Our eyes can see both real and virtual images, as long as the diverging rays enter our pupils. We will see that curved mirrors and lenses can form real images, as well as virtual. A movie projector lens, for example, produces a real image that is visible on the screen.

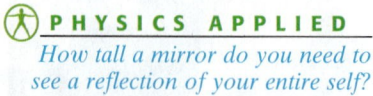

PHYSICS APPLIED

How tall a mirror do you need to see a reflection of your entire self?

EXAMPLE 32–2   How tall must a full-length mirror be? A woman 1.60 m tall stands in front of a vertical plane mirror. What is the minimum height of the mirror, and how close must its lower edge be to the floor, if she is to be able to see her whole body? Assume her eyes are 10 cm below the top of her head.

APPROACH For her to see her whole body, light rays from the top of her head and from the bottom of her foot must reflect from the mirror and enter her eye: see Fig. 32–8. We don't show two rays diverging from each point as we did in Fig. 32–7, where we wanted to find where the image is. Now that we know the image is the same distance behind a plane mirror as the object is in front, we only need to show one ray leaving point G (top of head) and one ray leaving point A (her toe), and then use simple geometry.

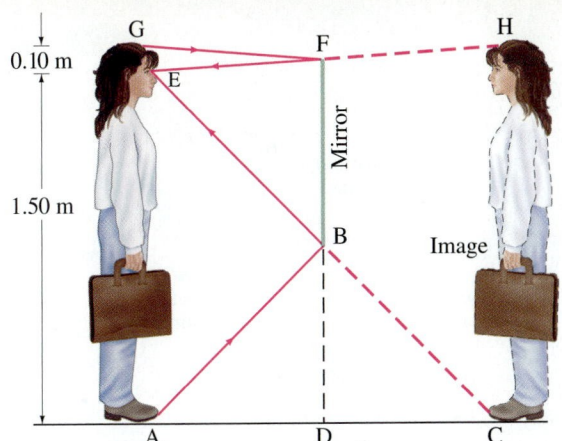

G

F

H

0.10 m

E

Mirror

1.50 m

B

Image

A

D

C

FIGURE 32–8 Seeing oneself in a mirror. Example 32–2.

SOLUTION First consider the ray that leaves her foot at A, reflects at B, and enters the eye at E. The mirror needs to extend no lower than B. The angle of reflection equals the angle of incidence, so the height BD is half of the height AE. Because AE = 1.60 m − 0.10 m = 1.50 m, then BD = 0.75 m. Similarly, if the woman is to see the top of her head, the top edge of the mirror only needs to reach point F, which is 5 cm below the top of her head (half of GE = 10 cm). Thus, DF = 1.55 m, and the mirror needs to have a vertical height of only (1.55 m − 0.75 m) = 0.80 m. The mirror's bottom edge must be 0.75 m above the floor.

NOTE We see that a mirror, if positioned well, need be only half as tall as a person for that person to see all of himself or herself.

EXERCISE A Does the result of Example 32–2 depend on your distance from the mirror? (Try it.)

EXERCISE B Return to the Chapter-Opening Question, page 837, and answer it again now. Try to explain why you may have answered differently the first time.

EXERCISE C Suppose you are standing about 3 m in front of a mirror in a hair salon. You can see yourself from your head to your waist, where the end of the mirror cuts off the rest of your image. If you walk closer to the mirror (*a*) you will not be able to see any more of your image; (*b*) you will be able to see more of your image, below your waist; (*c*) you will see less of your image, with the cutoff rising to be above your waist.

CONCEPTUAL EXAMPLE 32–3 | **Is the photo upside down?** Close examination of the photograph on the first page of this Chapter reveals that in the top portion, the image of the Sun is seen clearly, whereas in the lower portion, the image of the Sun is partially blocked by the tree branches. Show why the reflection is not the same as the real scene by drawing a sketch of this situation, showing the Sun, the camera, the branch, and two rays going from the Sun to the camera (one direct and one reflected). Is the photograph right side up?

RESPONSE We need to draw two diagrams, one assuming the photo on p. 837 is right side up, and another assuming it is upside down. Figure 32–9 is drawn assuming the photo is upside down. In this case, the Sun blocked by the tree would be the direct view, and the full view of the Sun the reflection: the ray which reflects off the water and into the camera travels at an angle below the branch, whereas the ray that travels directly to the camera passes through the branches. This works. Try to draw a diagram assuming the photo is right side up (thus assuming that the image of the Sun in the reflection is higher above the horizon than it is as viewed directly). It won't work. The photo on p. 837 is upside down.

Also, what about the people in the photo? Try to draw a diagram showing why they don't appear in the reflection. [*Hint*: Assume they are not sitting on the edge of poolside, but back from the edge a bit.] Then try to draw a diagram of the reverse (i.e., assume the photo is right side up so the people are visible only in the reflection). Reflected images are not perfect replicas when different planes (distances) are involved.

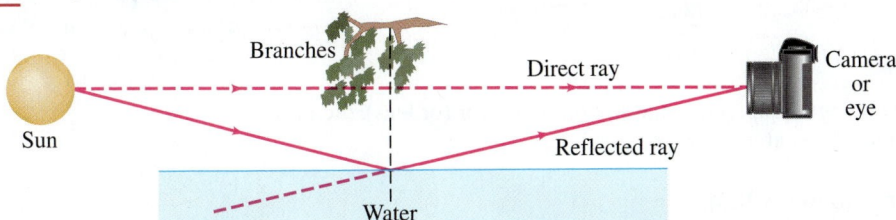

Branches

Direct ray

Camera or eye

Sun

Reflected ray

Water

FIGURE 32–9 Example 32–3.

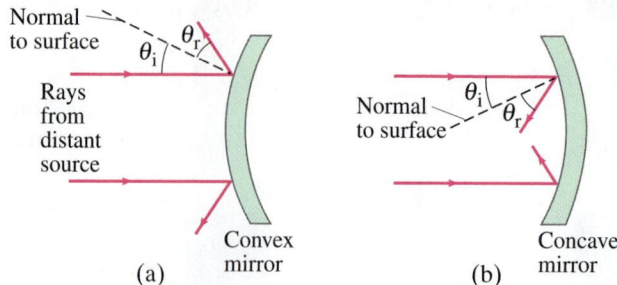

FIGURE 32–10 Mirrors with convex and concave spherical surfaces. Note that $\theta_r = \theta_i$ for each ray.

(a) Convex mirror

(b) Concave mirror

32–3 Formation of Images by Spherical Mirrors

Reflecting surfaces do not have to be flat. The most common *curved* mirrors are *spherical*, which means they form a section of a sphere. A spherical mirror is called **convex** if the reflection takes place on the outer surface of the spherical shape so that the center of the mirror surface bulges out toward the viewer (Fig. 32–10a). A mirror is called **concave** if the reflecting surface is on the inner surface of the sphere so that the mirror surface sinks away from the viewer (like a "cave"), Fig. 32–10b. Concave mirrors are used as shaving or cosmetic mirrors (Fig. 32–11a) because they magnify, and convex mirrors are sometimes used on cars and trucks (rearview mirrors) and in shops (to watch for thieves), because they take in a wide field of view (Fig. 32–11b).

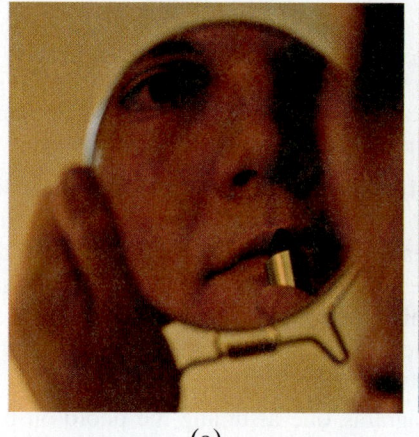

FIGURE 32–11 (a) A concave cosmetic mirror gives a magnified image. (b) A convex mirror in a store reduces image size and so includes a wide field of view.

(a)

(b)

Focal Point and Focal Length

To see how spherical mirrors form images, we first consider an object that is very far from a concave mirror. For a distant object, as shown in Fig. 32–12, the rays from each point on the object that strike the mirror will be nearly parallel. *For an object infinitely far away* (the Sun and stars approach this), *the rays would be precisely parallel.*

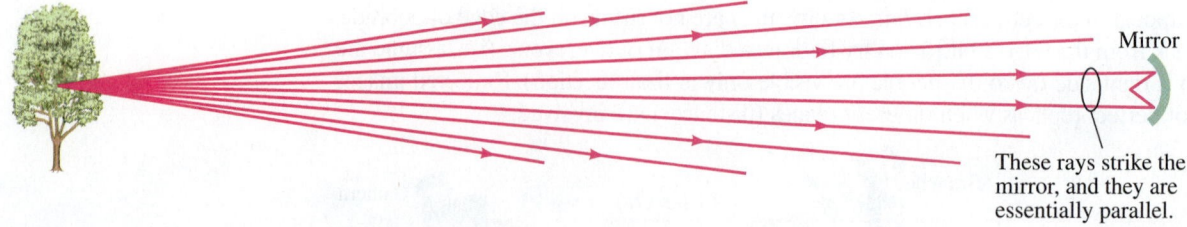

Mirror

These rays strike the mirror, and they are essentially parallel.

FIGURE 32–12 If the object's distance is large compared to the size of the mirror (or lens), the rays are nearly parallel. They are parallel for an object at infinity (∞).

Now consider such parallel rays falling on a concave mirror as in Fig. 32–13. The law of reflection holds for each of these rays at the point each strikes the mirror. As can be seen, they are not all brought to a single point. In order to form a sharp image, the rays must come to a point. Thus a spherical mirror will not make as sharp an image as a plane mirror will. However, as we show below, if the mirror is small compared to its radius of curvature, so that a reflected ray makes only a *small angle* with the incident ray (2θ in Fig. 32–14), then the rays will cross each other at very nearly a single point, or **focus**. In the case shown in Fig. 32–14, the incoming rays are parallel to the **principal axis**, which is defined as the straight line perpendicular to the curved surface at its center (line CA in Fig. 32–14). The point F, where incident parallel rays come to a focus after reflection, is called the **focal point** of the mirror. The distance between F and the center of the mirror, length FA, is called the **focal length**, f, of the mirror. The focal point is also the *image point for an object infinitely far away* along the principal axis. The image of the Sun, for example, would be at F.

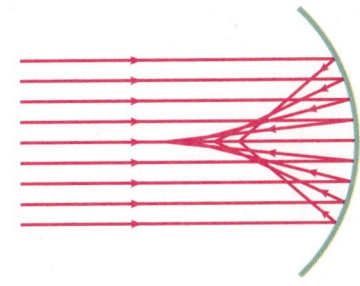

FIGURE 32–13 Parallel rays striking a concave spherical mirror do not intersect (or focus) at precisely a single point. (This "defect" is referred to as "spherical aberration.")

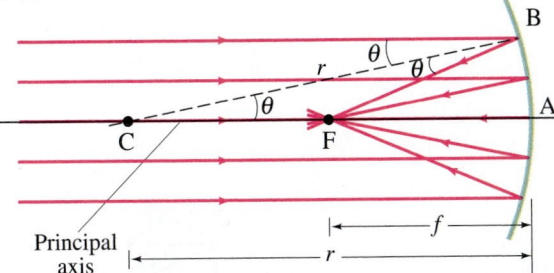

FIGURE 32–14 Rays parallel to the principal axis of a concave spherical mirror come to a focus at F, the focal point, as long as the mirror is small in width as compared to its radius of curvature, r, so that the rays are "paraxial"— that is, make only small angles with the horizontal axis.

Now we will show, for a mirror whose reflecting surface is small compared to its radius of curvature, that the rays very nearly meet at a common point, F, and we will also calculate the focal length f. In this approximation, we consider only rays that make a small angle with the principal axis; such rays are called **paraxial rays**, and their angles are exaggerated in Fig. 32–14 to make the labels clear. First we consider a ray that strikes the mirror at B in Fig. 32–14. The point C is the center of curvature of the mirror (the center of the sphere of which the mirror is a part). So the dashed line CB is equal to r, the radius of curvature, and CB is normal to the mirror's surface at B. The incoming ray that hits the mirror at B makes an angle θ with this normal, and hence the reflected ray, BF, also makes an angle θ with the normal (law of reflection). Note that angle BCF is also θ as shown. The triangle CBF is isosceles because two of its angles are equal. Thus we have length CF = BF. We assume the mirror surface is small compared to the mirror's radius of curvature, so the angles are small, and the length FB is nearly equal to length FA. In this approximation, FA = FC. But FA = f, the focal length, and CA = $2 \times$ FA = r. Thus the focal length is half the radius of curvature:

$$f = \frac{r}{2}. \qquad\qquad \text{[spherical mirror]} \quad \textbf{(32–1)}$$

We assumed only that the angle θ was small, so this result applies for all other incident paraxial rays. Thus all paraxial rays pass through the same point F, the focal point.

Since it is only approximately true that the rays come to a perfect focus at F, the more curved the mirror, the worse the approximation (Fig. 32–13) and the more blurred the image. This "defect" of spherical mirrors is called **spherical aberration**; we will discuss it more with regard to lenses in Chapter 33. A **parabolic reflector**, on the other hand, will reflect the rays to a perfect focus. However, because parabolic shapes are much harder to make and thus much more expensive, spherical mirrors are used for most purposes. (Many astronomical telescopes use parabolic reflectors.) We consider here only spherical mirrors and we will assume that they are small compared to their radius of curvature so that the image is sharp and Eq. 32–1 holds.

Image Formation—Ray Diagrams

We saw that for an object at infinity, the image is located at the focal point of a concave spherical mirror, where $f = r/2$. But where does the image lie for an object not at infinity? First consider the object shown as an arrow in Fig. 32–15a, which is placed between F and C at point O (O for object). Let us determine where the image will be for a given point O′ at the top of the object.

(a) Ray 1 goes out from O′ parallel to the axis and reflects through F.

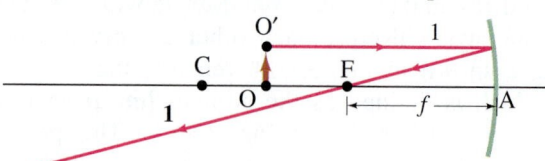

(b) Ray 2 goes through F and then reflects back parallel to the axis.

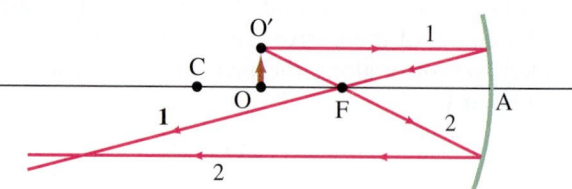

FIGURE 32–15 Rays leave point O′ on the object (an arrow). Shown are the three most useful rays for determining where the image I′ is formed. [Note that our mirror is not small compared to f, so our diagram will not give the precise position of the image.]

(c) Ray 3 is chosen perpendicular to mirror, and so must reflect back on itself and go through C (center of curvature).

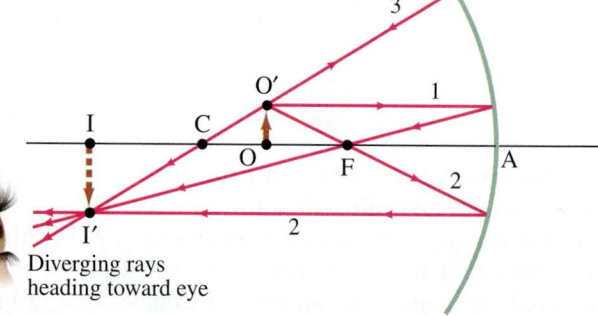

Diverging rays heading toward eye

To do this we can draw several rays and make sure these reflect from the mirror such that the angle of reflection equals the angle of incidence. Many rays could be drawn leaving any point on an object, but determining the image position is simplified if we deal with three particularly simple rays. These are the rays labeled 1, 2, and 3 in Fig. 32–15 and we draw them leaving object point O′ as follows:

Ray 1 is drawn parallel to the axis; therefore after reflection it must pass along a line through F (Fig. 32–15a).

Ray 2 leaves O′ and is made to pass through F (Fig. 32–15b); therefore it must reflect so it is parallel to the axis.

Ray 3 passes through C, the center of curvature (Fig. 32–15c); it is along a radius of the spherical surface and is perpendicular to the mirror, so it is reflected back on itself.

All three rays leave a single point O′ on the object. After reflection from a (small) mirror, the point at which these rays cross is the image point I′. All other rays from the same object point will also pass through this image point. To find the image point for any object point, only these three types of rays need to be drawn. Only two of these rays are needed, but the third serves as a check.

We have shown the image point in Fig. 32–15 only for a single point on the object. Other points on the object are imaged nearby, so a complete image of the object is formed, as shown by the dashed arrow in Fig. 32–15c. Because the light actually passes through the image itself, this is a **real image** that will appear on a piece of paper or film placed there. This can be compared to the virtual image formed by a plane mirror (the light does not actually pass through that image, Fig. 32–7).

The image in Fig. 32–15 can be seen by the eye when the eye is placed to the left of the image, so that some of the rays diverging from each point on the image (as point I′) can enter the eye as shown in Fig. 32–15c. (See also Figs. 32–1 and 32–7.)

Mirror Equation and Magnification

Image points can be determined, roughly, by drawing the three rays as just described, Fig. 32–15. But it is difficult to draw small angles for the "paraxial" rays as we assumed. For more accurate results, we now derive an equation that gives the image distance if the object distance and radius of curvature of the mirror are known. To do this, we refer to Fig. 32–16. The **object distance**, d_o, is the distance of the object (point O) from the center of the mirror. The **image distance**, d_i, is the distance of the image (point I) from the center of the mirror. The height of the object OO' is called h_o and the height of the image, I'I, is h_i.

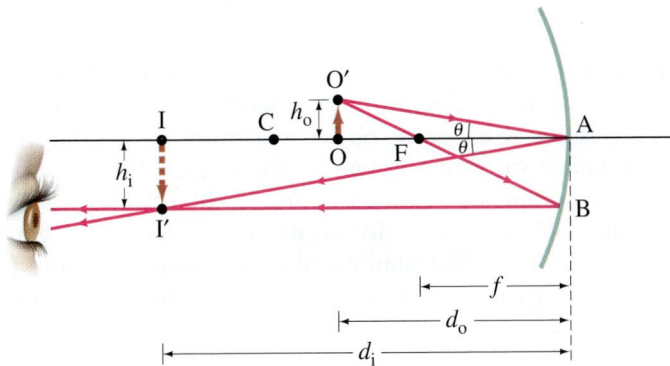

FIGURE 32–16 Diagram for deriving the mirror equation. For the derivation, we assume the mirror size is small compared to its radius of curvature.

Two rays leaving O' are shown: O'FBI' (same as ray 2 in Fig. 32–15) and O'AI', which is a fourth type of ray that reflects at the center of the mirror and can also be used to find an image point. The ray O'AI' obeys the law of reflection, so the two right triangles O'AO and I'AI are similar. Therefore, we have

$$\frac{h_o}{h_i} = \frac{d_o}{d_i}.$$

For the other ray shown, O'FBI', the triangles O'FO and AFB are also similar because the angles are equal and we use the approximation AB = h_i (mirror small compared to its radius). Furthermore FA = f, the focal length of the mirror, so

$$\frac{h_o}{h_i} = \frac{\text{OF}}{\text{FA}} = \frac{d_o - f}{f}.$$

The left sides of the two preceding expressions are the same, so we can equate the right sides:

$$\frac{d_o}{d_i} = \frac{d_o - f}{f}.$$

We now divide both sides by d_o and rearrange to obtain

$$\frac{1}{d_o} + \frac{1}{d_i} = \frac{1}{f}. \qquad \text{(32–2)} \qquad \textit{Mirror equation}$$

This is the equation we were seeking. It is called the **mirror equation** and relates the object and image distances to the focal length f (where $f = r/2$).

The **lateral magnification**, m, of a mirror is defined as the height of the image divided by the height of the object. From our first set of similar triangles above, or the first equation on this page, we can write:

$$m = \frac{h_i}{h_o} = -\frac{d_i}{d_o}. \qquad \text{(32–3)}$$

The minus sign in Eq. 32–4 is inserted as a convention. Indeed, we must be careful about the signs of all quantities in Eqs. 32–2 and 32–3. Sign conventions are chosen so as to give the correct locations and orientations of images, as predicted by ray diagrams.

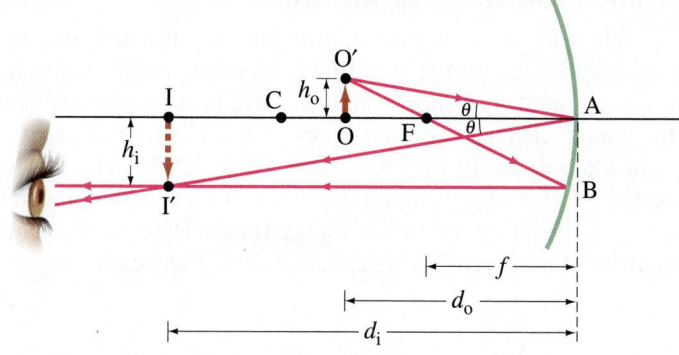

FIGURE 32–16 (Repeated from previous page.)

PROBLEM SOLVING

Sign conventions

The sign conventions we use are: the image height h_i is positive if the image is upright, and negative if inverted, relative to the object (assuming h_o is taken as positive); d_i or d_o is positive if image or object is in front of the mirror (as in Fig. 32–16); if either image or object is behind the mirror, the corresponding distance is negative (an example can be seen in Fig. 32–18, Example 32–6). Thus the magnification (Eq. 32–3) is positive for an upright image and negative for an inverted image (upside down). We summarize sign conventions more fully in the Problem Solving Strategy following our discussion of convex mirrors later in this Section.

Concave Mirror Examples

EXAMPLE 32–4 **Image in a concave mirror.** A 1.50-cm-high diamond ring is placed 20.0 cm from a concave mirror with radius of curvature 30.0 cm. Determine (a) the position of the image, and (b) its size.

APPROACH We determine the focal length from the radius of curvature (Eq. 32–1), $f = r/2 = 15.0$ cm. The ray diagram is basically like that shown in Fig. 32–16 (repeated here on this page), since the object is between F and C. The position and size of the image are found from Eqs. 32–2 and 32–3.

SOLUTION Referring to Fig. 32–16, we have CA $= r = 30.0$ cm, FA $= f = 15.0$ cm, and OA $= d_o = 20.0$ cm.

(a) From Eq. 32–2,

$$\frac{1}{d_i} = \frac{1}{f} - \frac{1}{d}$$

$$= \frac{1}{15.0 \text{ cm}} - \frac{1}{20.0 \text{ cm}} = 0.0167 \text{ cm}^{-1}.$$

⚠ CAUTION

Remember to take the reciprocal

So $d_i = 1/(0.0167 \text{ cm}^{-1}) = 60.0$ cm. Because d_i is positive, the image is 60.0 cm in front of the mirror, on the same side as the object.

(b) From Eq. 32–3, the magnification is

$$m = -\frac{d_i}{d_o}$$

$$= -\frac{60.0 \text{ cm}}{20.0 \text{ cm}} = -3.00.$$

The image is 3.0 times larger than the object, and its height is

$$h_i = mh_o = (-3.00)(1.5 \text{ cm}) = -4.5 \text{ cm}.$$

The minus sign reminds us that the image is inverted, as in Fig. 32–16.

NOTE When an object is further from a concave mirror than the focal point, we can see from Fig. 32–15 or 32–16 that the image is always inverted and real.

CONCEPTUAL EXAMPLE 32–5 **Reversible rays.** If the object in Example 32–4 is placed instead where the image is (see Fig. 32–16), where will the new image be?

RESPONSE The mirror equation is *symmetric* in d_o and d_i. Thus the new image will be where the old object was. Indeed, in Fig. 32–16 we need only reverse the direction of the rays to get our new situation.

EXAMPLE 32–6 **Object closer to concave mirror.** A 1.00-cm-high object is placed 10.0 cm from a concave mirror whose radius of curvature is 30.0 cm. (*a*) Draw a ray diagram to locate (approximately) the position of the image. (*b*) Determine the position of the image and the magnification analytically.

APPROACH We draw the ray diagram using the rays as in Fig. 32–15, page 844. An analytic solution uses Eqs. 32–1, 32–2, and 32–3.

SOLUTION (*a*) Since $f = r/2 = 15.0$ cm, the object is between the mirror and the focal point. We draw the three rays as described earlier (Fig. 32–15); they are shown leaving the tip of the object in Fig. 32–17. Ray 1 leaves the tip of our object heading toward the mirror parallel to the axis, and reflects through F. Ray 2 cannot head toward F because it would not strike the mirror; so ray 2 must point as if it started at F (dashed line) and heads to the mirror, and then is reflected parallel to the principal axis. Ray 3 is perpendicular to the mirror, as before. The rays reflected from the mirror diverge and so never meet at a point. They appear, however, to be coming from a point behind the mirror. This point locates the image of the tip of the arrow. The image is thus behind the mirror and is *virtual*. (Why?)

(*b*) We use Eq. 32–2 to find d_i when $d_o = 10.0$ cm:

$$\frac{1}{d_i} = \frac{1}{f} - \frac{1}{d_o} = \frac{1}{15.0 \text{ cm}} - \frac{1}{10.0 \text{ cm}} = \frac{2-3}{30.0 \text{ cm}} = -\frac{1}{30.0 \text{ cm}}.$$

Therefore, $d_i = -30.0$ cm. The minus sign means the image is behind the mirror, which our diagram also told us. The magnification is $m = -d_i/d_o = -(-30.0 \text{ cm})/(10.0 \text{ cm}) = +3.00$. So the image is 3.00 times larger than the object. The plus sign indicates that the image is upright (same as object), which is consistent with the ray diagram, Fig. 32–17.

NOTE The image distance cannot be obtained accurately by measuring on Fig. 32–17, because our diagram violates the paraxial ray assumption (we draw rays at steeper angles to make them clearly visible).

NOTE When the object is located inside the focal point of a concave mirror $(d_o < f)$, the image is always upright and vertical. And if the object O in Fig. 32–17 is you, you see yourself clearly, because the reflected rays at point O are diverging. Your image is upright and enlarged.

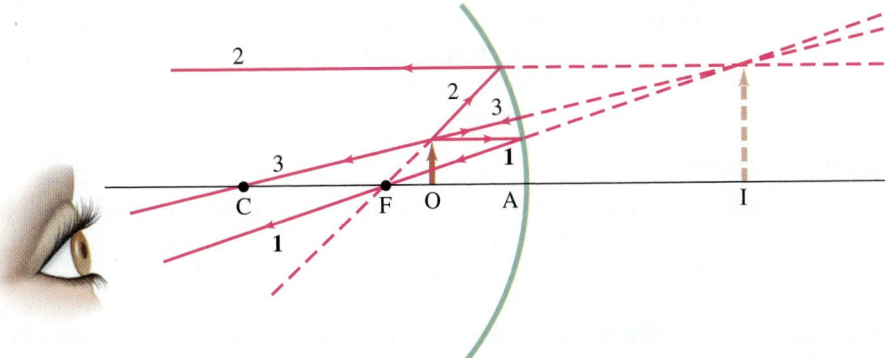

FIGURE 32–17 Object placed within the focal point F. The image is *behind* the mirror and is *virtual*, Example 32–6. [Note that the vertical scale (height of object = 1.0 cm) is different from the horizontal (OA = 10.0 cm) for ease of drawing, and reduces the precision of the drawing.]

It is useful to compare Figs. 32–15 and 32–17. We can see that if the object is within the focal point $(d_o < f)$, as in Fig. 32–17, the image is virtual, upright, and magnified. This is how a shaving or cosmetic mirror is used—you must place your head closer to the mirror than the focal point if you are to see yourself right-side up (see the photograph of Fig. 32–11a). If the object is *beyond* the focal point, as in Fig. 32–15, the image is real and inverted (upside down—and hard to use!). Whether the magnification has magnitude greater or less than 1.0 in the latter case depends on the position of the object relative to the center of curvature, point C. Practice making ray diagrams with various object distances.

The mirror equation also holds for a plane mirror: the focal length is $f = r/2 = \infty$, and Eq. 32–2 gives $d_i = -d_o$.

Seeing the Image

For a person's eye to see a sharp image, the eye must be at a place where it intercepts diverging rays from points on the image, as is the case for the eye's position in Figs. 32–15 and 32–16. Our eyes are made to see normal objects, which always means the rays are diverging toward the eye as shown in Fig. 32–1. (Or, for very distant objects like stars, the rays become essentially parallel, as in Fig. 32–12.) If you placed your eye between points O and I in Fig. 32–16, for example, *converging* rays from the object OO′ would enter your eye and the lens of your eye could not bring them to a focus; you would see a blurry image. [We will discuss the eye more in Chapter 33.]

FIGURE 32–18 You can see a clear inverted image of your face when you are beyond C $(d_o > 2f)$, because the rays that arrive at your eye are diverging. Standard rays 2 and 3 are shown leaving point O on your nose. Ray 2 (and other nearby rays) enters your eye. Notice that rays are diverging as they move to the left of image point I.

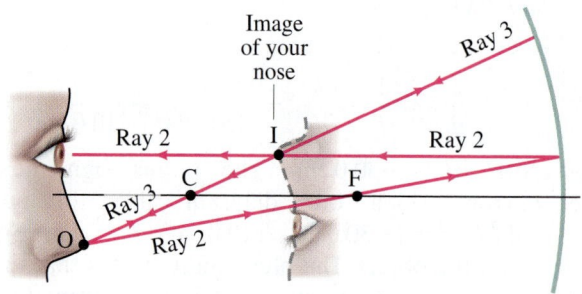

FIGURE 32–19 Convex mirror: (a) the focal point is at F, behind the mirror; (b) the image I of the object at O is virtual, upright, and smaller than the object. [Not to scale for Example 32–7.]

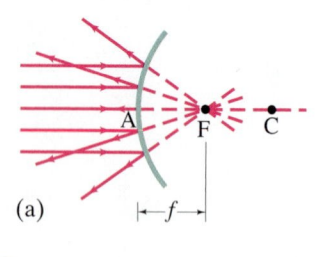

(a)

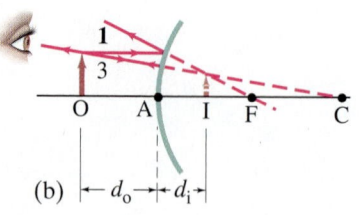

(b)

If *you* are the object OO′ in Fig. 32–16, situated between F and C, and are trying to see yourself in the mirror, you would see a blur; but the person whose eye is shown in Fig. 32–16 can see you clearly. You can see yourself clearly, but upside down, if you are to the left of C in Fig. 32–16, where $d_o > 2f$. Why? Because then the rays reflected from the image will be *diverging* at your position as demonstrated in Fig. 32–18, and your eye can then focus them. You can also see yourself clearly, and right-side up, if you are closer to the mirror than its focal point $(d_o < f)$, as we saw in Example 32–6, Fig. 32–17.

Convex Mirrors

The analysis used for concave mirrors can be applied to **convex** mirrors. Even the mirror equation (Eq. 32–2) holds for a convex mirror, although the quantities involved must be carefully defined. Figure 32–19a shows parallel rays falling on a convex mirror. Again spherical aberration is significant (Fig. 32–13), unless we assume the mirror's size is very small compared to its radius of curvature. The reflected rays diverge, but seem to come from point F behind the mirror. This is the **focal point**, and its distance from the center of the mirror (point A) is the **focal length**, f. It is easy to show that again $f = r/2$. We see that an object at infinity produces a virtual image in a convex mirror. Indeed, no matter where the object is placed on the reflecting side of a convex mirror, the image will be virtual and upright, as indicated in Fig. 32–19b. To find the image we draw rays 1 and 3 according to the rules used before on the concave mirror, as shown in Fig. 32–19b. Note that although rays 1 and 3 don't actually pass through points F and C, the line along which each is drawn does (shown dashed).

The mirror equation, Eq. 32–2, holds for convex mirrors but the focal length f must be considered negative, as must the radius of curvature. The proof is left as a Problem. It is also left as a Problem to show that Eq. 32–3 for the magnification is also valid.

Spherical Mirrors

1. Always **draw a ray diagram** even though you are going to make an analytic calculation—the diagram serves as a check, even if not precise. From one point on the object, draw at least two, preferably three, of the easy-to-draw rays using the rules described in Fig. 32–15. The image point is where the reflected rays intersect or appear to intersect.

2. Apply the **mirror equation**, Eq. 32–2, and the **magnification equation**, Eq. 32–3. It is crucially important to follow the sign conventions—see the next point.

3. **Sign Conventions**
   (a) When the object, image, or focal point is on the reflecting side of the mirror (on the left in our drawings), the corresponding distance is positive. If any of these points is behind the mirror (on the right) the corresponding distance is negative.[†]
   (b) The image height h_i is positive if the image is upright, and negative if inverted, relative to the object (h_o is always taken as positive).

4. **Check** that the analytical solution is consistent with the ray diagram.

[†]Object distances are positive for material objects, but can be negative in systems with more than one mirror or lens—see Section 33–3.

EXAMPLE 32–7 **Convex rearview mirror.** An external rearview car mirror is convex with a radius of curvature of 16.0 m (Fig. 32–20). Determine the location of the image and its magnification for an object 10.0 m from the mirror.

Convex rearview mirror

APPROACH We follow the steps of the Problem Solving Strategy explicitly.

SOLUTION

1. **Draw a ray diagram.** The ray diagram will be like Fig. 32–19b, but the large object distance $(d_o = 10.0\,\text{m})$ makes a precise drawing difficult. We have a convex mirror, so r is negative by convention.

2. **Mirror and magnification equations.** The center of curvature of a convex mirror is behind the mirror, as is its focal point, so we set $r = -16.0\,\text{m}$ so that the focal length is $f = r/2 = -8.0\,\text{m}$. The object is in front of the mirror, $d_o = 10.0\,\text{m}$. Solving the mirror equation, Eq. 32–2, for $1/d_i$ gives

$$\frac{1}{d_i} = \frac{1}{f} - \frac{1}{d_o} = \frac{1}{-8.0\,\text{m}} - \frac{1}{10.0\,\text{m}} = \frac{-10.0 - 8.0}{80.0\,\text{m}} = -\frac{18}{80.0\,\text{m}}.$$

FIGURE 32–20 Example 32–7.

Thus $d_i = -80.0\,\text{m}/18 = -4.4\,\text{m}$. Equation 32–3 gives the magnification

$$m = -\frac{d_i}{d_o} = -\frac{(-4.4\,\text{m})}{(10.0\,\text{m})} = +0.44.$$

3. **Sign conventions.** The image distance is negative, $-4.4\,\text{m}$, so the image is *behind* the mirror. The magnification is $m = +0.44$, so the image is *upright* (same orientation as object) and less than half as tall as the object.

4. **Check.** Our results are consistent with Fig. 32–19b.

Convex rearview mirrors on vehicles sometimes come with a warning that objects are closer than they appear in the mirror. The fact that d_i may be smaller than d_o (as in Example 32–7) seems to contradict this observation. The real reason the object seems farther away is that its image in the convex mirror is *smaller* than it would be in a plane mirror, and we judge distance of ordinary objects such as other cars mostly by their size.

TABLE 32–1 Indices of Refraction[†]

Material	$n = \dfrac{c}{v}$
Vacuum	1.0000
Air (at STP)	1.0003
Water	1.33
Ethyl alcohol	1.36
Glass	
Fused quartz	1.46
Crown glass	1.52
Light flint	1.58
Lucite or Plexiglas	1.51
Sodium chloride	1.53
Diamond	2.42

[†] $\lambda = 589$ nm.

32–4 Index of Refraction

We saw in Chapter 31 that the speed of light in vacuum is

$$c = 2.99792458 \times 10^8 \, \text{m/s},$$

which is usually rounded off to

$$3.00 \times 10^8 \, \text{m/s}$$

when extremely precise results are not required.

In air, the speed is only slightly less. In other transparent materials such as glass and water, the speed is always less than that in vacuum. For example, in water light travels at about $\frac{3}{4}c$. The ratio of the speed of light in vacuum to the speed v in a given material is called the **index of refraction**, n, of that material:

$$n = \frac{c}{v}. \qquad \textbf{(32–4)}$$

The index of refraction is never less than 1, and values for various materials are given in Table 32–1. For example, since $n = 2.42$ for diamond, the speed of light in diamond is

$$v = \frac{c}{n} = \frac{(3.00 \times 10^8 \, \text{m/s})}{2.42} = 1.24 \times 10^8 \, \text{m/s}.$$

As we shall see later, n varies somewhat with the wavelength of the light—except in vacuum—so a particular wavelength is specified in Table 32–1, that of yellow light with wavelength $\lambda = 589$ nm.

That light travels more slowly in matter than in vacuum can be explained at the atomic level as being due to the absorption and reemission of light by atoms and molecules of the material.

32–5 Refraction: Snell's Law

When light passes from one transparent medium into another with a different index of refraction, part of the incident light is reflected at the boundary. The remainder passes into the new medium. If a ray of light is incident at an angle to the surface (other than perpendicular), the ray changes direction as it enters the new medium. This change in direction, or bending, is called **refraction**.

Figure 32–21a shows a ray passing from air into water. Angle θ_1 is the angle the incident ray makes with the normal (perpendicular) to the surface and is called the **angle of incidence**. Angle θ_2 is the **angle of refraction**, the angle the refracted ray makes with the normal to the surface. Notice that the ray bends toward the normal when entering the water. This is always the case when the ray enters a medium where the speed of light is *less* (and the index of refraction greater, Eq. 32–4). If light travels from one medium into a second where its speed is *greater*, the ray bends away from the normal; this is shown in Fig. 32–21b for a ray traveling from water to air.

⚠ **CAUTION**

Angles θ_1 and θ_2 are measured from the perpendicular, not from surface

FIGURE 32–21 Refraction.
(a) Light refracted when passing from air (n_1) into water (n_2): $n_2 > n_1$.
(b) Light refracted when passing from water (n_1) into air (n_2): $n_1 > n_2$.

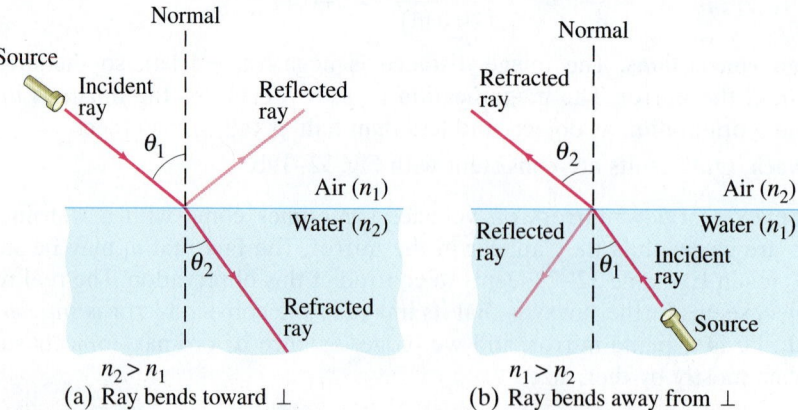

$n_2 > n_1$
(a) Ray bends toward ⊥

$n_1 > n_2$
(b) Ray bends away from ⊥

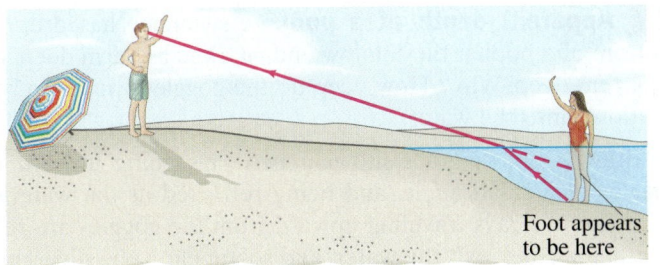

Foot appears to be here

FIGURE 32–22 Ray diagram showing why a person's legs look shorter when standing in waist-deep water: the path of light traveling from the bather's foot to the observer's eye bends at the water's surface, and our brain interprets the light as having traveled in a straight line, from higher up (dashed line).

Refraction is responsible for a number of common optical illusions. For example, a person standing in waist-deep water appears to have shortened legs. As shown in Fig. 32–22, the rays leaving the person's foot are bent at the surface. The observer's brain assumes the rays to have traveled a straight-line path (dashed red line), and so the feet appear to be higher than they really are. Similarly, when you put a straw in water, it appears to be bent (Fig. 32–23).

Snell's Law

The angle of refraction depends on the speed of light in the two media and on the incident angle. An analytical relation between θ_1 and θ_2 in Fig. 32–21 was arrived at experimentally about 1621 by Willebrord Snell (1591–1626). It is known as **Snell's law** and is written:

$$n_1 \sin \theta_1 = n_2 \sin \theta_2. \qquad \textbf{(32–5)}$$

θ_1 is the angle of incidence and θ_2 is the angle of refraction; n_1 and n_2 are the respective indices of refraction in the materials. See Fig. 32–21. The incident and refracted rays lie in the same plane, which also includes the perpendicular to the surface. Snell's law is the **law of refraction**. (Snell's law was derived in Section 15–10 where Eq. 15–19 is just a combination of Eqs. 32–5 and 32–4. We also derive it in Chapter 34 using the wave theory of light.)

It is clear from Snell's law that if $n_2 > n_1$, then $\theta_2 < \theta_1$. That is, if light enters a medium where n is greater (and its speed is less), then the ray is bent toward the normal. And if $n_2 < n_1$, then $\theta_2 > \theta_1$, so the ray bends away from the normal. This is what we saw in Fig. 32–21.

EXERCISE D Light passes from a medium with $n = 1.3$ into a medium with $n = 1.5$. Is the light bent toward or away from the perpendicular to the interface?

EXAMPLE 32–8 **Refraction through flat glass.** Light traveling in air strikes a flat piece of uniformly thick glass at an incident angle of 60°, as shown in Fig. 32–24. If the index of refraction of the glass is 1.50, (*a*) what is the angle of refraction θ_A in the glass; (*b*) what is the angle θ_B at which the ray emerges from the glass?

APPROACH We apply Snell's law at the first surface, where the light enters the glass, and again at the second surface where it leaves the glass and enters the air.

SOLUTION (*a*) The incident ray is in air, so $n_1 = 1.00$ and $n_2 = 1.50$. Applying Snell's law where the light enters the glass $(\theta_1 = 60°)$ gives

$$\sin \theta_A = \frac{1.00}{1.50} \sin 60° = 0.5774,$$

so $\theta_A = 35.3°$.

(*b*) Since the faces of the glass are parallel, the incident angle at the second surface is just θ_A (simple geometry), so $\sin \theta_A = 0.5774$. At this second interface, $n_1 = 1.50$ and $n_2 = 1.00$. Thus the ray re-enters the air at an angle $\theta_B (= \theta_2)$ given by

$$\sin \theta_B = \frac{1.50}{1.00} \sin \theta_A = 0.866,$$

and $\theta_B = 60°$. The direction of a light ray is thus unchanged by passing through a flat piece of glass of uniform thickness.

NOTE This result is valid for any angle of incidence. The ray is displaced slightly to one side, however. You can observe this by looking through a piece of glass (near its edge) at some object and then moving your head to the side slightly so that you see the object directly. It "jumps."

FIGURE 32–23 A straw in water looks bent even when it isn't.

SNELL'S LAW (LAW OF REFRACTION)

FIGURE 32–24 Light passing through a piece of glass (Example 32–8).

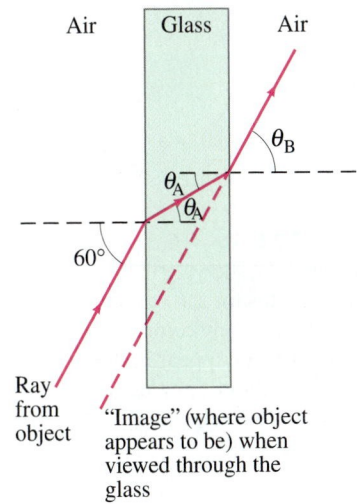

Air   Glass   Air

θ_B

θ_A

θ_A

60°

Ray from object

"Image" (where object appears to be) when viewed through the glass

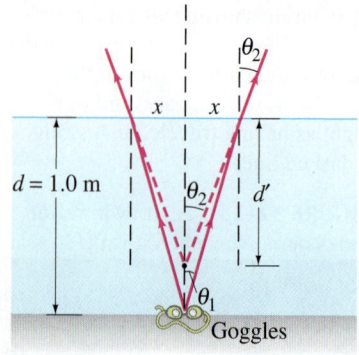

FIGURE 32–25 Example 32–9.

EXAMPLE 32–9 **Apparent depth of a pool.** A swimmer has dropped her goggles to the bottom of a pool at the shallow end, marked as 1.0 m deep. But the goggles don't look that deep. Why? How deep do the goggles appear to be when you look straight down into the water?

APPROACH We draw a ray diagram showing two rays going upward from a point on the goggles at a small angle, and being refracted at the water's (flat) surface, Fig. 32–25. The two rays traveling upward from the goggles are refracted *away* from the normal as they exit the water, and so appear to be diverging from a point above the goggles (dashed lines), which is why the water seems less deep than it actually is.

SOLUTION To calculate the apparent depth d' (Fig. 32–25), given a real depth $d = 1.0$ m, we use Snell's law with $n_1 = 1.33$ for water and $n_2 = 1.0$ for air:

$$\sin \theta_2 = n_1 \sin \theta_1.$$

We are considering only small angles, so $\sin \theta \approx \tan \theta \approx \theta$, with θ in radians. So Snell's law becomes

$$\theta_2 \approx n_1 \theta_1.$$

From Fig. 32–25, we see that

$$\theta_2 \approx \tan \theta_2 = \frac{x}{d'} \quad \text{and} \quad \theta_1 \approx \tan \theta_1 = \frac{x}{d}.$$

Putting these into Snell's law, $\theta_2 \approx n_1 \theta_1$, we get

$$\frac{x}{d'} \approx n_1 \frac{x}{d}$$

or

$$d' \approx \frac{d}{n_1} = \frac{1.0 \text{ m}}{1.33} = 0.75 \text{ m}.$$

The pool seems only three-fourths as deep as it actually is.

32–6 Visible Spectrum and Dispersion

An obvious property of visible light is its color. Color is related to the wavelengths or frequencies of the light. (How this was discovered will be discussed in Chapter 34.) Visible light—that to which our eyes are sensitive—has wavelengths in air in the range of about 400 nm to 750 nm.[†] This is known as the **visible spectrum**, and within it lie the different colors from violet to red, as shown in Fig. 32–26. Light with wavelength shorter than 400 nm (= violet) is called **ultraviolet** (UV), and light with wavelength greater than 750 nm (= red) is called **infrared** (IR).[‡] Although human eyes are not sensitive to UV or IR, some types of photographic film and digital cameras do respond to them.

A prism can separate white light into a rainbow of colors, as shown in Fig. 32–27. This happens because the index of refraction of a material depends on the wavelength, as shown for several materials in Fig. 32–28. White light is a

FIGURE 32–26 The spectrum of visible light, showing the range of wavelengths for the various colors as seen in air. Many colors, such as brown, do not appear in the spectrum; they are made from a mixture of wavelengths.

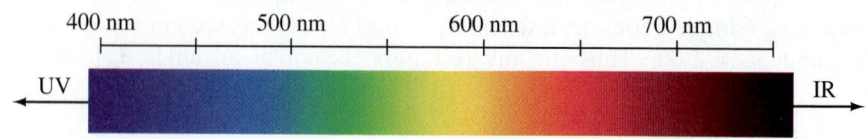

[†]Sometimes the angstrom (Å) unit is used when referring to light: $1 \text{ Å} = 1 \times 10^{-10}$ m. Then visible light falls in the wavelength range of 4000 Å to 7500 Å.

[‡]The complete electromagnetic spectrum is illustrated in Fig. 31–12.

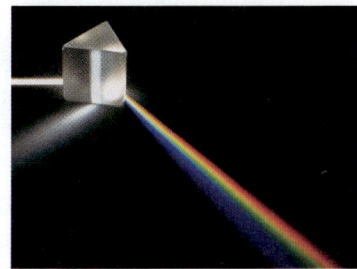

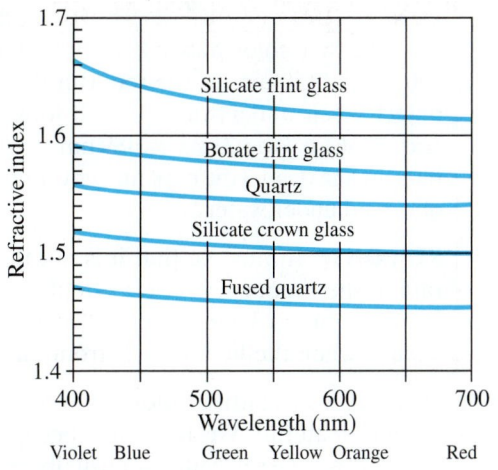

FIGURE 32–27 White light passing through a prism is broken down into its constituent colors.

FIGURE 32–28 Index of refraction as a function of wavelength for various transparent solids.

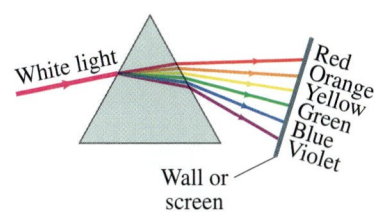

FIGURE 32–29 White light dispersed by a prism into the visible spectrum.

mixture of all visible wavelengths, and when incident on a prism, as in Fig. 32–29, the different wavelengths are bent to varying degrees. Because the index of refraction is greater for the shorter wavelengths, violet light is bent the most and red the least, as indicated. This spreading of white light into the full spectrum is called **dispersion**.

Rainbows are a spectacular example of dispersion—by drops of water. You can see rainbows when you look at falling water droplets with the Sun behind you. Figure 32–30 shows how red and violet rays are bent by spherical water droplets and are reflected off the back surface of the droplet. Red is bent the least and so reaches the observer's eyes from droplets higher in the sky, as shown in the diagram. Thus the top of the rainbow is red.

PHYSICS APPLIED

Rainbows

FIGURE 32–30 (a) Ray diagram explaining how a rainbow (b) is formed.

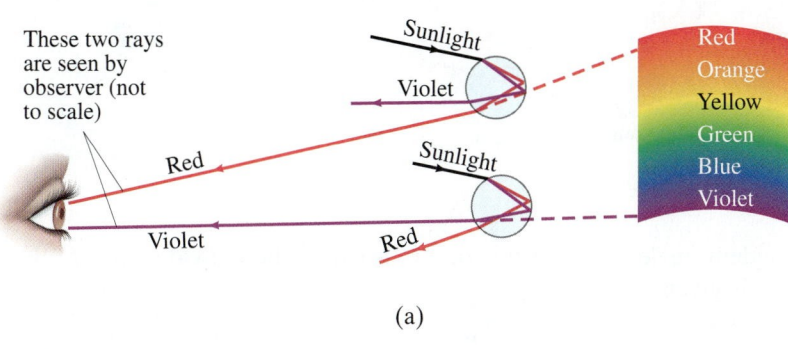

(a)

(b)

The visible spectrum, Fig. 32–26, does not show all the colors seen in nature. For example, there is no brown in Fig. 32–26. Many of the colors we see are a mixture of wavelengths. For practical purposes, most natural colors can be reproduced using three primary colors. They are red, green, and blue for direct source viewing such as TV and computer monitors. For inks used in printing, the primary colors are cyan (the color of the margin notes in this book), yellow, and magenta (the color we use for light rays in diagrams).

For any wave, its velocity v is related to its wavelength λ and frequency f by $v = f\lambda$ (Eq. 15–1 or 31–14). When a wave travels from one material into another, the frequency of the wave does not change across the boundary since a point (an atom) at the boundary oscillates at that frequency. Thus if light goes from air into a material with index of refraction n, the wavelength becomes (recall Eq. 32–4):

$$\lambda_n = \frac{v}{f} = \frac{c}{nf} = \frac{\lambda}{n} \qquad (32–6)$$

where λ is the wavelength in vacuum or air and λ_n is the wavelength in the material with index of refraction n.

Observed color of light under water.
We said that color depends on wavelength. For example, for an object emitting 650 nm light in air, we see red. But this is true only in air. If we observe this same object when under water, it still looks red. But the wavelength in water λ_n is (Eq. 32–6) 650 nm/1.33 = 489 nm. Light with wavelength 489 nm would appear blue in air. Can you explain why the light appears red rather than blue when observed under water?

RESPONSE It must be that it is not the wavelength that the eye responds to, but rather the frequency. For example, the frequency of 650 nm red light in air is $f = c/\lambda = (3.0 \times 10^8 \, \text{m/s})/(650 \times 10^{-9} \, \text{m}) = 4.6 \times 10^{14} \, \text{Hz}$, and does not change when the light travels from one medium to another. Only λ changes.

NOTE If we classified colors by frequency, the color assignments would be valid for any material. We typically specify colors by *wavelength* in air (even if less general) not just because we usually see objects in air, but because wavelength is what is commonly measured (it is easier to measure than frequency).

32–7 Total Internal Reflection; Fiber Optics

When light passes from one material into a second material where the index of refraction is less (say, from water into air), the light bends away from the normal, as for rays I and J in Fig. 32–31. At a particular incident angle, the angle of refraction will be 90°, and the refracted ray would skim the surface (ray K) in this

FIGURE 32–31 Since $n_2 < n_1$, light rays are totally internally reflected if the incident angle $\theta_1 > \theta_C$, as for ray L. If $\theta_1 < \theta_C$, as for rays I and J, only a part of the light is reflected, and the rest is refracted.

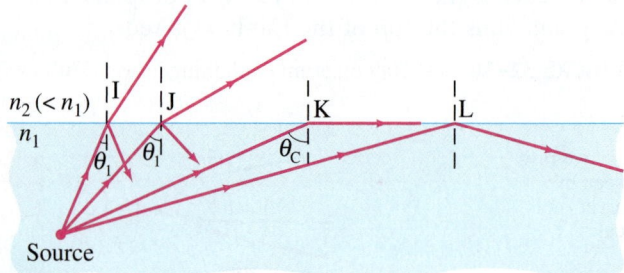

case. The incident angle at which this occurs is called the **critical angle**, θ_C. From Snell's law, θ_C is given by

$$\sin \theta_C = \frac{n_2}{n_1} \sin 90° = \frac{n_2}{n_1}. \tag{32–7}$$

For any incident angle less than θ_C, there will be a refracted ray, although part of the light will also be reflected at the boundary. However, for incident angles greater than θ_C, Snell's law would tell us that $\sin \theta_2$ is greater than 1.00. Yet the sine of an angle can never be greater than 1.00. In this case there is no refracted ray at all, and *all of the light is reflected*, as for ray L in Fig. 32–31. This effect is called **total internal reflection**. Total internal reflection can occur only when light strikes a boundary where the medium beyond has a lower index of refraction.

⚠ **CAUTION**

Total internal reflection (occurs only if refractive index is smaller beyond boundary)

EXERCISE E Fill a sink with water. Place a waterproof watch just below the surface with the watch's flat crystal parallel to the water surface. From above you can still see the watch reading. As you move your head to one side far enough what do you see, and why?

CONCEPTUAL EXAMPLE 32–11 **View up from under water.** Describe what a person would see who looked up at the world from beneath the perfectly smooth surface of a lake or swimming pool.

RESPONSE For an air–water interface, the critical angle is given by

$$\sin \theta_C = \frac{1.00}{1.33} = 0.750.$$

Therefore, $\theta_C = 49°$. Thus the person would see the outside world compressed into a circle whose edge makes a 49° angle with the vertical. Beyond this angle, the person would see reflections from the sides and bottom of the lake or pool (Fig. 32–32).

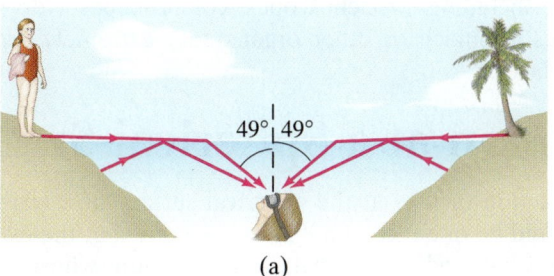

(a)

(b)

FIGURE 32–32 (a) Light rays, and (b) view looking upward from beneath the water (the surface of the water must be very smooth). Example 32–11.

Diamonds achieve their brilliance from a combination of dispersion and total internal reflection. Because diamonds have a very high index of refraction of about 2.4, the critical angle for total internal reflection is only 25°. The light dispersed into a spectrum inside the diamond therefore strikes many of the internal surfaces of the diamond before it strikes one at less than 25° and emerges.

After many such reflections, the light has traveled far enough that the colors have become sufficiently separated to be seen individually and brilliantly by the eye after leaving the diamond.

Many optical instruments, such as binoculars, use total internal reflection within a prism to reflect light. The advantage is that very nearly 100% of the light is reflected, whereas even the best mirrors reflect somewhat less than 100%. Thus the image is brighter, especially after several reflections. For glass with $n = 1.50$, $\theta_C = 41.8°$. Therefore, 45° prisms will reflect all the light internally, if oriented as shown in the binoculars of Fig. 32–33.

EXERCISE F If 45.0° plastic lenses are used in binoculars, what minimum index of refraction must the plastic have?

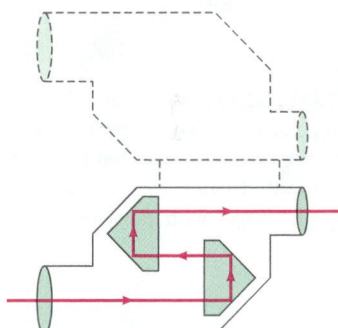

FIGURE 32–33 Total internal reflection of light by prisms in binoculars.

Fiber Optics

Total internal reflection is the principle behind **fiber optics**. Glass and plastic fibers as thin as a few micrometers in diameter are common. A bundle of such tiny fibers is called a **light pipe** or cable, and light[†] can be transmitted along it with almost no loss because of total internal reflection. Figure 32–34 shows how light traveling down a thin fiber makes only glancing collisions with the walls so that total internal reflection occurs. Even if the light pipe is bent into a complicated shape, the critical angle still won't be exceeded, so light is transmitted practically undiminished to the other end. Very small losses do occur, mainly by reflection at the ends and absorption within the fiber.

Important applications of fiber-optic cables are in communications and medicine. They are used in place of wire to carry telephone calls, video signals, and computer data. The signal is a modulated light beam (a light beam whose intensity can be varied) and data is transmitted at a much higher rate and with less loss and less interference than an electrical signal in a copper wire. Fibers have been developed that can support over one hundred separate wavelengths, each modulated to carry up to 10 gigabits $(10^{10}$ bits$)$ of information per second. That amounts to a terabit $(10^{12}$ bits$)$ per second for the full one hundred wavelengths.

FIGURE 32–34 Light reflected totally at the interior surface of a glass or transparent plastic fiber.

⊛ PHYSICS APPLIED
Fiber optics in communications

[†]Fiber optic devices can use not only visible light but also infrared light, ultraviolet light, and microwaves.

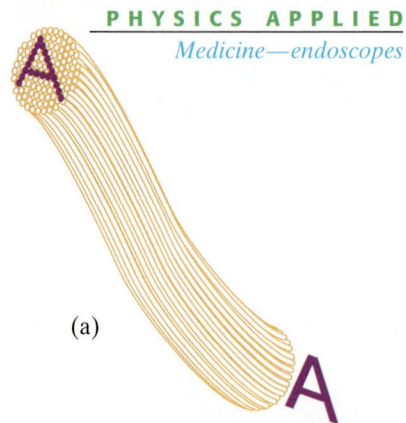

(a)

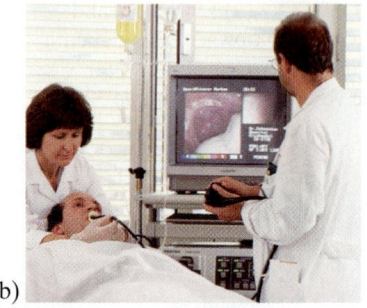

(b)

FIGURE 32–35 (a) How a fiber-optic image is made. (b) Example of a fiber-optic device inserted through the mouth to view the gastro-intestinal tract, with image on screen.

The sophisticated use of fiber optics to transmit a clear picture is particularly useful in medicine, Fig. 32–35. For example, a patient's lungs can be examined by inserting a light pipe known as a bronchoscope through the mouth and down the bronchial tube. Light is sent down an outer set of fibers to illuminate the lungs. The reflected light returns up a central core set of fibers. Light directly in front of each fiber travels up that fiber. At the opposite end, a viewer sees a series of bright and dark spots, much like a TV screen—that is, a picture of what lies at the opposite end. Lenses are used at each end. The image may be viewed directly or on a monitor screen or film. The fibers must be optically insulated from one another, usually by a thin coating of material with index of refraction less than that of the fiber. The more fibers there are, and the smaller they are, the more detailed the picture. Such instruments, including bronchoscopes, colonoscopes (for viewing the colon), and endoscopes (stomach or other organs), are extremely useful for examining hard-to-reach places.

*32–8 Refraction at a Spherical Surface

We now examine the refraction of rays at the spherical surface of a transparent material. Such a surface could be one face of a lens or the cornea of the eye. To be general, let us consider an object which is located in a medium whose index of refraction is n_1, and rays from each point on the object can enter a medium whose index of refraction is n_2. The radius of curvature of the spherical boundary is R, and its center of curvature is at point C, Fig. 32–36. We now show that all rays leaving a point O on the object will be focused at a single point I, the image point, if we consider only paraxial rays: rays that make a small angle with the axis.

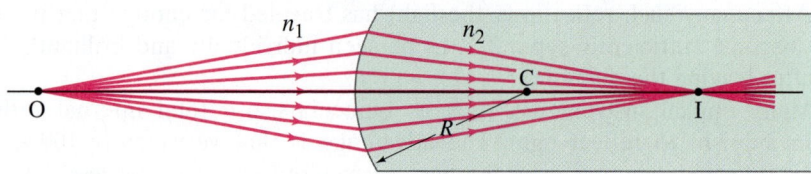

FIGURE 32–36 Rays from a point O on an object will be focused at a single image point I by a spherical boundary between two transparent materials $(n_2 > n_1)$, as long as the rays make small angles with the axis.

To do so, we consider a single ray that leaves point O as shown in Fig. 32–37. From Snell's law, Eq. 32–5, we have

$$n_1 \sin \theta_1 = n_2 \sin \theta_2.$$

We are assuming that angles $\theta_1, \theta_2, \alpha, \beta,$ and γ are small, so $\sin \theta \approx \theta$ (in radians), and Snell's law becomes, approximately,

$$n_1 \theta_1 = n_2 \theta_2.$$

Also, $\beta + \phi = 180°$ and $\theta_2 + \gamma + \phi = 180°$, so

$$\beta = \gamma + \theta_2.$$

Similarly for triangle OPC,

$$\theta_1 = \alpha + \beta.$$

FIGURE 32–37 Diagram for showing that all paraxial rays from O focus at the same point I $(n_2 > n_1)$.

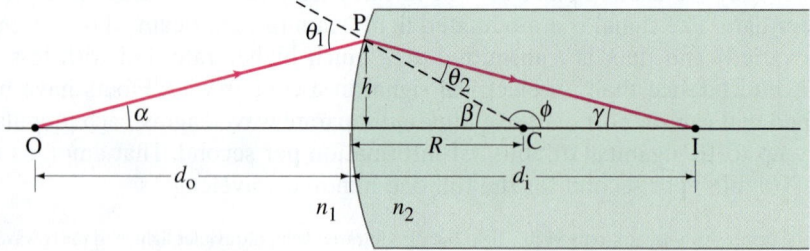

These three relations can be combined to yield

$$n_1 \alpha + n_2 \gamma = (n_2 - n_1)\beta.$$

Since we are considering only the case of small angles, we can write, approximately,

$$\alpha = \frac{h}{d_o}, \qquad \beta = \frac{h}{R}, \qquad \gamma = \frac{h}{d_i},$$

where d_o and d_i are the object and image distances and h is the height as shown in Fig. 32–37. We substitute these into the previous equation, divide through by h, and obtain

$$\frac{n_1}{d_o} + \frac{n_2}{d_i} = \frac{n_2 - n_1}{R}. \qquad (32\text{–}8)$$

For a given object distance d_o, this equation tells us d_i, the image distance, does not depend on the angle of a ray. Hence all paraxial rays meet at the same point I. This is true only for rays that make small angles with the axis and with each other, and is equivalent to assuming that the width of the refracting spherical surface is small compared to its radius of curvature. If this assumption is not true, the rays will not converge to a point; there will be spherical aberration, just as for a mirror (see Fig. 32–13), and the image will be blurry. (Spherical aberration will be discussed further in Section 33–10.)

We derived Eq. 32–8 using Fig. 32–37 for which the spherical surface is convex (as viewed by the incoming ray). It is also valid for a concave surface—as can be seen using Fig. 32–38—if we use the following conventions:

1. If the surface is convex (so the center of curvature C is on the side of the surface opposite to that from which the light comes), R is positive; if the surface is concave (C on the same side from which the light comes) R is negative.

2. The image distance, d_i, follows the same convention: positive if on the opposite side from where the light comes, negative if on the same side.

3. The object distance is positive if on the same side from which the light comes (this is the normal case, although when several surfaces bend the light it may not be so), otherwise it is negative.

For the case shown in Fig. 32–38 with a concave surface, both R and d_i are negative when used in Eq. 32–8. Note, in this case, that the image is virtual.

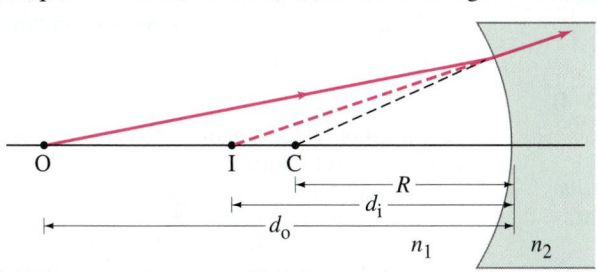

FIGURE 32–38 Rays from O refracted by a concave surface form a virtual image ($n_2 > n_1$). Per our conventions, $R < 0$, $d_i < 0$, $d_o > 0$.

EXAMPLE 32–12 **Apparent depth II.** A person looks vertically down into a 1.0-m-deep pool. How deep does the water appear to be?

APPROACH Example 32–9 solved this problem using Snell's law. Here we use Eq. 32–8.

SOLUTION A ray diagram is shown in Fig. 32–39. Point O represents a point on the pool's bottom. The rays diverge and appear to come from point I, the image. We have $d_o = 1.0$ m and, for a flat surface, $R = \infty$. Then Eq. 32–8 becomes

$$\frac{1.33}{1.0\,\text{m}} + \frac{1.00}{d_i} = \frac{(1.00 - 1.33)}{\infty} = 0.$$

Hence $d_i = -(1.0\,\text{m})/(1.33) = -0.75$ m. So the pool appears to be only three-fourths as deep as it actually is, the same result we found in Example 32–9. The minus sign tells us the image point I is on the same side of the surface as O, and the image is virtual. At angles other than vertical, this conclusion must be modified.

FIGURE 32–39 Example 32–12.

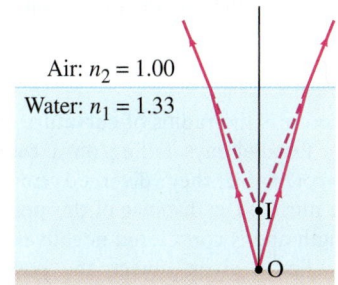

Air: $n_2 = 1.00$

Water: $n_1 = 1.33$

EXAMPLE 32–13 **A spherical "lens."** A point source of light is placed at a distance of 25.0 cm from the center of a glass sphere $(n = 1.5)$ of radius 10.0 cm, Fig. 32–40. Find the image of the source.

APPROACH As shown in Fig. 32–40, there are two refractions, and we treat them successively, one at a time. The light rays from the source first refract from the convex glass surface facing the source. We analyze this first refraction, treating it as in Fig. 32–36, ignoring the back side of the sphere.

SOLUTION Using Eq. 32–8 (assuming paraxial rays) with $n_1 = 1.0$, $n_2 = 1.5$, $R = 10.0$ cm, and $d_o = 25.0$ cm $-$ 10.0 cm $=$ 15.0 cm, we solve for the image distance as formed at surface 1, d_{i1}:

$$\frac{1}{d_{i1}} = \frac{1}{n_2}\left(\frac{(n_2 - n_1)}{R} - \frac{n_1}{d_o}\right) = \frac{1}{1.5}\left(\frac{1.5 - 1.0}{10.0\ \text{cm}} - \frac{1.0}{15.0\ \text{cm}}\right) = -\frac{1}{90.0\ \text{cm}}.$$

Thus, the image of the first refraction is located 90.0 cm *to the left* of the front surface. This image (I_1) now serves as the object for the refraction occurring at the back surface (surface 2) of the sphere. This surface is concave so $R = -10.0$ cm, and we consider a ray close to the axis. Then the object distance is $d_{o2} = 90.0$ cm $+$ 2(10.0 cm) $=$ 110.0 cm, and Eq. 32–8 yields, with $n_1 = 1.5$, $n_2 = 1.0$,

$$\frac{1}{d_{i2}} = \frac{1}{1.0}\left(\frac{1.0 - 1.5}{-10.0\ \text{cm}} - \frac{1.5}{110.0\ \text{cm}}\right) = \frac{4.0}{110.0\ \text{cm}}$$

so $d_{i2} = 28$ cm. Thus, the final image is located a distance 28 cm from the back side of the sphere.

FIGURE 32–40 Example 32–13.

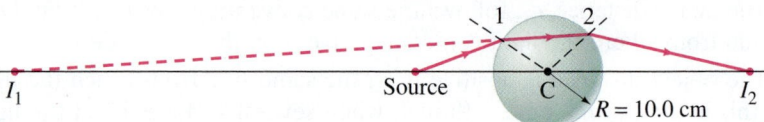

Summary

Light appears to travel along straight-line paths, called **rays**, at a speed v that depends on the **index of refraction**, n, of the material; that is

$$v = \frac{c}{n}, \qquad (32\text{–}4)$$

where c is the speed of light in vacuum.

When light reflects from a flat surface, *the angle of reflection equals the angle of incidence*. This **law of reflection** explains why mirrors can form **images**.

In a **plane mirror**, the image is virtual, upright, the same size as the object, and is as far behind the mirror as the object is in front.

A **spherical mirror** can be concave or convex. A **concave** spherical mirror focuses parallel rays of light (light from a very distant object) to a point called the **focal point**. The distance of this point from the mirror is the **focal length** f of the mirror and

$$f = \frac{r}{2} \qquad (32\text{–}1)$$

where r is the radius of curvature of the mirror.

Parallel rays falling on a **convex mirror** reflect from the mirror as if they diverged from a common point behind the mirror. The distance of this point from the mirror is the focal length and is considered negative for a convex mirror.

For a given object, the position and size of the image formed by a mirror can be found by ray tracing. Algebraically,

the relation between image and object distances, d_i and d_o, and the focal length f, is given by the **mirror equation**:

$$\frac{1}{d_o} + \frac{1}{d_i} = \frac{1}{f}. \qquad (32\text{–}2)$$

The ratio of image height to object height, which equals the magnification m of a mirror, is

$$m = \frac{h_i}{h_o} = -\frac{d_i}{d_o}. \qquad (32\text{–}3)$$

If the rays that converge to form an image actually pass through the image, so the image would appear on film or a screen placed there, the image is said to be a **real image**. If the rays do not actually pass through the image, the image is a **virtual image**.

When light passes from one transparent medium into another, the rays bend or refract. The **law of refraction (Snell's law)** states that

$$n_1 \sin\theta_1 = n_2 \sin\theta_2, \qquad (32\text{–}5)$$

where n_1 and θ_1 are the index of refraction and angle with the normal to the surface for the incident ray, and n_2 and θ_2 are for the refracted ray.

When light of wavelength λ enters a medium with index of refraction n, the wavelength is reduced to

$$\lambda_n = \frac{\lambda}{n}. \qquad (32\text{–}6)$$

The frequency does not change.

The frequency or wavelength of light determines its color. The **visible spectrum** in air extends from about 400 nm (violet) to about 750 nm (red).

Glass prisms break white light down into its constituent colors because the index of refraction varies with wavelength, a phenomenon known as **dispersion**.

When light rays reach the boundary of a material where the index of refraction decreases, the rays will be **totally internally reflected** if the incident angle, θ_1, is such that Snell's law would predict $\sin \theta_2 > 1$. This occurs if θ_1 exceeds the critical angle θ_C given by

$$\sin \theta_C = \frac{n_2}{n_1}. \tag{32-7}$$

Questions

1. What would be the appearance of the Moon if it had (a) a rough surface; (b) a polished mirrorlike surface?

2. Archimedes is said to have burned the whole Roman fleet in the harbor of Syracuse by focusing the rays of the Sun with a huge spherical mirror. Is this[†] reasonable?

3. What is the focal length of a plane mirror? What is the magnification of a plane mirror?

4. An object is placed along the principal axis of a spherical mirror. The magnification of the object is −3.0. Is the image real or virtual, inverted or upright? Is the mirror concave or convex? On which side of the mirror is the image located?

5. Using the rules for the three rays discussed with reference to Fig. 32–15, draw ray 2 for Fig. 32–19b.

6. Does the mirror equation, Eq. 32–2, hold for a plane mirror? Explain.

7. If a concave mirror produces a real image, is the image necessarily inverted?

8. How might you determine the speed of light in a solid, rectangular, transparent object?

9. When you look at the Moon's reflection from a ripply sea, it appears elongated (Fig. 32–41). Explain.

FIGURE 32–41
Question 9.

10. How can a spherical mirror have a negative object distance?

11. What is the angle of refraction when a light ray is incident perpendicular to the boundary between two transparent materials?

12. When you look down into a swimming pool or a lake, are you likely to overestimate or underestimate its depth? Explain. How does the apparent depth vary with the viewing angle? (Use ray diagrams.)

13. Draw a ray diagram to show why a stick or straw looks bent when part of it is under water (Fig. 32–23).

[†]Students at MIT did a feasibility study. See www.mit.edu.

14. When a wide beam of parallel light enters water at an angle, the beam broadens. Explain.

15. You look into an aquarium and view a fish inside. One ray of light from the fish as it emerges from the tank is shown in Fig. 32–42. The apparent position of the fish is also shown. In the drawing, indicate the approximate position of the actual fish. Briefly justify your answer.

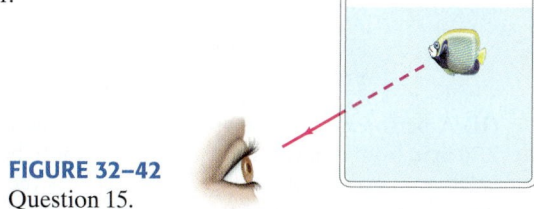

FIGURE 32–42
Question 15.

16. How can you "see" a round drop of water on a table even though the water is transparent and colorless?

17. A ray of light is refracted through three different materials (Fig. 32–43). Rank the materials according to their index of refraction, least to greatest.

FIGURE 32–43
Question 17.

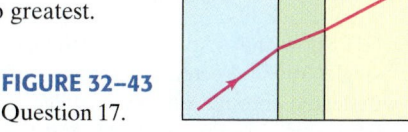

18. Can a light ray traveling in air be totally reflected when it strikes a smooth water surface if the incident angle is chosen correctly? Explain.

19. When you look up at an object in air from beneath the surface in a swimming pool, does the object appear to be the same size as when you see it directly in air? Explain.

20. What type of mirror is shown in Fig. 32–44?

FIGURE 32–44
Question 20.

21. Light rays from stars (including our Sun) always bend toward the vertical direction as they pass through the Earth's atmosphere. (a) Why does this make sense? (b) What can you conclude about the apparent positions of stars as viewed from Earth?

Problems

32–2 Reflection; Plane Mirrors

1. (I) When you look at yourself in a 60-cm-tall plane mirror, you see the same amount of your body whether you are close to the mirror or far away. (Try it and see.) Use ray diagrams to show why this should be true.

2. (I) Suppose that you want to take a photograph of yourself as you look at your image in a mirror 2.8 m away. For what distance should the camera lens be focused?

3. (II) Two plane mirrors meet at a 135° angle, Fig. 32–45. If light rays strike one mirror at 38° as shown, at what angle ϕ do they leave the second mirror?

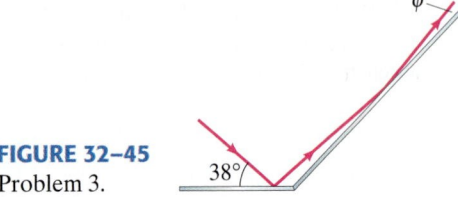

FIGURE 32–45
Problem 3.

38°

4. (II) A person whose eyes are 1.64 m above the floor stands 2.30 m in front of a vertical plane mirror whose bottom edge is 38 cm above the floor, Fig. 32–46. What is the horizontal distance x to the base of the wall supporting the mirror of the nearest point on the floor that can be seen reflected in the mirror?

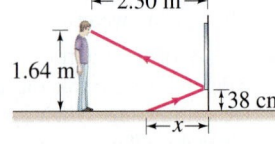

|←2.30 m→|

1.64 m

FIGURE 32–46
Problem 4.

‡38 cm

←x→

5. (II) Show that if two plane mirrors meet at an angle ϕ, a single ray reflected successively from both mirrors is deflected through an angle of 2ϕ independent of the incident angle. Assume $\phi < 90°$ and that only two reflections, one from each mirror, take place.

6. (II) Suppose you are 88 cm from a plane mirror. What area of the mirror is used to reflect the rays entering one eye from a point on the tip of your nose if your pupil diameter is 4.5 mm?

7. (II) Stand up two plane mirrors so they form a 90.0° angle as in Fig. 32–47. When you look into this double mirror, you see yourself as others see you, instead of reversed as in a single mirror. Make a ray diagram to show how this occurs.

FIGURE 32–47
Problems 7 and 8.

8. (III) Suppose a third mirror is placed beneath the two shown in Fig. 32–47, so that all three are perpendicular to each other. (a) Show that for such a "corner reflector," any incident ray will return in its original direction after three reflections. (b) What happens if it makes only two reflections?

32–3 Spherical Mirrors

9. (I) A solar cooker, really a concave mirror pointed at the Sun, focuses the Sun's rays 18.8 cm in front of the mirror. What is the radius of the spherical surface from which the mirror was made?

10. (I) How far from a concave mirror (radius 24.0 cm) must an object be placed if its image is to be at infinity?

11. (I) When walking toward a concave mirror you notice that the image flips at a distance of 0.50 m. What is the radius of curvature of the mirror?

12. (II) A small candle is 35 cm from a concave mirror having a radius of curvature of 24 cm. (a) What is the focal length of the mirror? (b) Where will the image of the candle be located? (c) Will the image be upright or inverted?

13. (II) You look at yourself in a shiny 9.2-cm-diameter Christmas tree ball. If your face is 25.0 cm away from the ball's front surface, where is your image? Is it real or virtual? Is it upright or inverted?

14. (II) A mirror at an amusement park shows an upright image of any person who stands 1.7 m in front of it. If the image is three times the person's height, what is the radius of curvature of the mirror? (See Fig. 32–44.)

15. (II) A dentist wants a small mirror that, when 2.00 cm from a tooth, will produce a 4.0× upright image. What kind of mirror must be used and what must its radius of curvature be?

16. (II) Some rearview mirrors produce images of cars to your rear that are smaller than they would be if the mirror were flat. Are the mirrors concave or convex? What is a mirror's radius of curvature if cars 18.0 m away appear 0.33 their normal size?

17. (II) You are standing 3.0 m from a convex security mirror in a store. You estimate the height of your image to be half of your actual height. Estimate the radius of curvature of the mirror.

18. (II) An object 3.0 mm high is placed 18 cm from a convex mirror of radius of curvature 18 cm. (a) Show by ray tracing that the image is virtual, and estimate the image distance. (b) Show that the (negative) image distance can be computed from Eq. 32–2 using a focal length of -9.0 cm. (c) Compute the image size, using Eq. 32–3.

19. (II) The image of a distant tree is virtual and very small when viewed in a curved mirror. The image appears to be 16.0 cm behind the mirror. What kind of mirror is it, and what is its radius of curvature?

20. (II) Use two techniques, (a) a ray diagram, and (b) the mirror equation, to show that the magnitude of the magnification of a concave mirror is less than 1 if the object is beyond the center of curvature C ($d_o > r$), and is greater than 1 if the object is within C ($d_o < r$).

21. (II) Show, using a ray diagram, that the magnification m of a convex mirror is $m = -d_i/d_o$, just as for a concave mirror. [*Hint*: Consider a ray from the top of the object that reflects at the center of the mirror.]

22. (II) Use ray diagrams to show that the mirror equation, Eq. 32–2, is valid for a convex mirror as long as f is considered negative.

23. (II) The magnification of a convex mirror is $+0.55×$ for objects 3.2 m from the mirror. What is the focal length of this mirror?

24. (II) (a) Where should an object be placed in front of a concave mirror so that it produces an image at the same location as the object? (b) Is the image real or virtual? (c) Is the image inverted or upright? (d) What is the magnification of the image?

25. (II) A 4.5-cm tall object is placed 26 cm in front of a spherical mirror. It is desired to produce a virtual image that is upright and 3.5 cm tall. (a) What type of mirror should be used? (b) Where is the image located? (c) What is the focal length of the mirror? (d) What is the radius of curvature of the mirror?

26. (II) A shaving or makeup mirror is designed to magnify your face by a factor of 1.35 when your face is placed 20.0 cm in front of it. (a) What type of mirror is it? (b) Describe the type of image that it makes of your face. (c) Calculate the required radius of curvature for the mirror.

27. (II) A concave mirror has focal length f. When an object is placed a distance $d_o > f$ from this mirror, a real image with magnification m is formed. (a) Show that $m = f/(f - d_o)$. (b) Sketch m vs. d_o over the range $f < d_o < +\infty$ where $f = 0.45$ m. (c) For what value of d_o will the real image have the same (lateral) size as the object? (d) To obtain a real image that is much larger than the object, in what general region should the object be placed relative to the mirror?

28. (II) Let the focal length of a convex mirror be written as $f = -|f|$. Show that the magnification m of an object a distance d_o from this mirror is given by $m = |f|/(d_o + |f|)$. Based on this relation, explain why your nose looks bigger than the rest of your face when looking into a convex mirror.

29. (II) A spherical mirror of focal length f produces an image of an object with magnification m. (a) Show that the object is a distance $d_o = f\left(1 - \dfrac{1}{m}\right)$ from the reflecting side of the mirror. (b) Use the relation in part (a) to show that, no matter where an object is placed in front of a convex mirror, its image will have a magnification in the range $0 \leq m \leq +1$.

30. (III) An object is placed a distance r in front of a wall, where r exactly equals the radius of curvature of a certain concave mirror. At what distance from the wall should this mirror be placed so that a real image of the object is formed on the wall? What is the magnification of the image?

31. (III) A short thin object (like a short length of wire) of length ℓ is placed along the axis of a spherical mirror (perpendicular to the glass surface). Show that its image has length $\ell' = m^2\ell$ so the *longitudinal magnification* is equal to $-m^2$ where m is the normal "lateral" magnification, Eq. 32–3. Why the minus sign? [*Hint*: Find the image positions for both ends of the wire, and assume ℓ is very small.]

32–4 Index of Refraction

32. (I) The speed of light in ice is 2.29×10^8 m/s. What is the index of refraction of ice?

33. (I) What is the speed of light in (a) ethyl alcohol, (b) lucite, (c) crown glass?

34. (I) Our nearest star (other than the Sun) is 4.2 light years away. That is, it takes 4.2 years for the light to reach Earth. How far away is it in meters?

35. (I) How long does it take light to reach us from the Sun, 1.50×10^8 km away?

36. (II) The speed of light in a certain substance is 88% of its value in water. What is the index of refraction of that substance?

37. (II) Light is emitted from an ordinary lightbulb filament in wave-train bursts of about 10^{-8} s in duration. What is the length in space of such wave trains?

32–5 Snell's Law

38. (I) A diver shines a flashlight upward from beneath the water at a 38.5° angle to the vertical. At what angle does the light leave the water?

39. (I) A flashlight beam strikes the surface of a pane of glass ($n = 1.56$) at a 63° angle to the normal. What is the angle of refraction?

40. (I) Rays of the Sun are seen to make a 33.0° angle to the vertical beneath the water. At what angle above the horizon is the Sun?

41. (I) A light beam coming from an underwater spotlight exits the water at an angle of 56.0°. At what angle of incidence did it hit the air–water interface from below the surface?

42. (II) A beam of light in air strikes a slab of glass ($n = 1.56$) and is partially reflected and partially refracted. Determine the angle of incidence if the angle of reflection is twice the angle of refraction.

43. (II) A light beam strikes a 2.0-cm-thick piece of plastic with a refractive index of 1.62 at a 45° angle. The plastic is on top of a 3.0-cm-thick piece of glass for which $n = 1.47$. What is the distance D in Fig. 32–48?

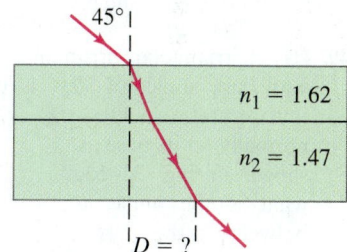

FIGURE 32–48
Problem 43.

44. (II) An aquarium filled with water has flat glass sides whose index of refraction is 1.56. A beam of light from outside the aquarium strikes the glass at a 43.5° angle to the perpendicular (Fig. 32–49). What is the angle of this light ray when it enters (a) the glass, and then (b) the water? (c) What would be the refracted angle if the ray entered the water directly?

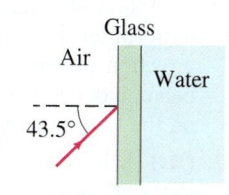

FIGURE 32–49
Problem 44.

45. (II) In searching the bottom of a pool at night, a watchman shines a narrow beam of light from his flashlight, 1.3 m above the water level, onto the surface of the water at a point 2.5 m from his foot at the edge of the pool (Fig. 32–50). Where does the spot of light hit the bottom of the pool, measured from the bottom of the wall beneath his foot, if the pool is 2.1 m deep?

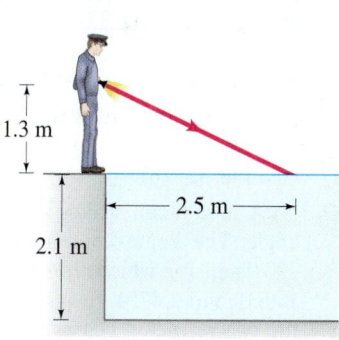

FIGURE 32–50
Problem 45.

46. (II) The block of glass $(n = 1.5)$ shown in cross section in Fig. 32–51 is surrounded by air. A ray of light enters the block at its left-hand face with incident angle θ_1 and reemerges into the air from the right-hand face directed parallel to the block's base. Determine θ_1.

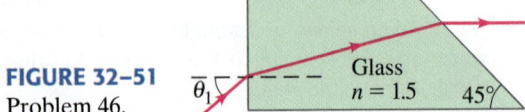

FIGURE 32–51
Problem 46.

Glass
$n = 1.5$
45°

47. (II) A laser beam of diameter $d_1 = 3.0$ mm in air has an incident angle $\theta_1 = 25°$ at a flat air–glass surface. If the index of refraction of the glass is $n = 1.5$, determine the diameter d_2 of the beam after it enters the glass.

48. (II) Light is incident on an equilateral glass prism at a 45.0° angle to one face, Fig. 32–52. Calculate the angle at which light emerges from the opposite face. Assume that $n = 1.54$.

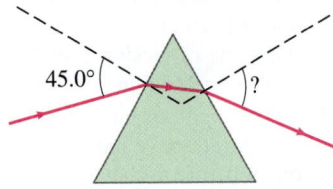

45.0°
?

FIGURE 32–52
Problems 48 and 65.

49. (II) A triangular prism made of crown glass $(n = 1.52)$ with base angles of 30.0° is surrounded by air. If parallel rays are incident normally on its base as shown in Fig. 32–53, what is the angle ϕ between the two emerging rays?

ϕ

30.0° $n = 1.52$ 30.0°

FIGURE 32–53
Problem 49.

50. (II) Show in general that for a light beam incident on a uniform layer of transparent material, as in Fig. 32–24, the direction of the emerging beam is parallel to the incident beam, independent of the incident angle θ. Assume the air on the two sides of the transparent material is the same.

51. (III) A light ray is incident on a flat piece of glass with index of refraction n as in Fig. 32–24. Show that if the incident angle θ is small, the emerging ray is displaced a distance $d = t\theta(n - 1)/n$, where t is the thickness of the glass, θ is in radians, and d is the perpendicular distance between the incident ray and the (dashed) line of the emerging ray (Fig. 32–24).

32–6 Visible Spectrum; Dispersion

52. (I) By what percent is the speed of blue light (450 nm) less than the speed of red light (680 nm), in silicate flint glass (see Fig. 32–28)?

53. (I) A light beam strikes a piece of glass at a 60.00° incident angle. The beam contains two wavelengths, 450.0 nm and 700.0 nm, for which the index of refraction of the glass is 1.4831 and 1.4754, respectively. What is the angle between the two refracted beams?

54. (II) A parallel beam of light containing two wavelengths, $\lambda_1 = 465$ nm and $\lambda_2 = 652$ nm, enters the silicate flint glass of an equilateral prism as shown in Fig. 32–54. At what angle does each beam leave the prism (give angle with normal to the face)? See Fig. 32–28.

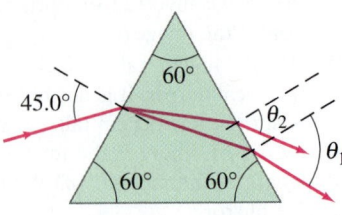

FIGURE 32–54
Problem 54.

55. (III) A ray of light with wavelength λ is incident from air at precisely 60° $(= \theta)$ on a spherical water drop of radius r and index of refraction n (which depends on λ). When the ray reemerges into the air from the far side of the drop, it has been deflected an angle ϕ from its original direction as shown in Fig. 32–55. By how much does the value of ϕ for violet light $(n = 1.341)$ differ from the value for red light $(n = 1.330)$?

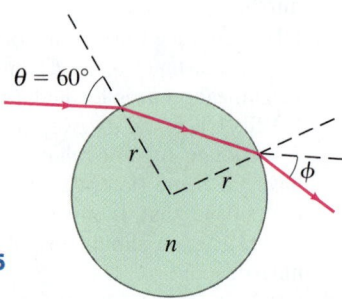

FIGURE 32–55
Problem 55.

56. (III) For visible light, the index of refraction n of glass is roughly 1.5, although this value varies by about 1% across the visible range. Consider a ray of white light incident from air at angle θ_1 onto a flat piece of glass. (a) Show that, upon entering the glass, the visible colors contained in this incident ray will be dispersed over a range $\Delta\theta_2$ of refracted angles given approximately by

$$\Delta\theta_2 \approx \frac{\sin\theta_1}{\sqrt{n^2 - \sin^2\theta_1}} \frac{\Delta n}{n}.$$

[*Hint*: For x in radians, $(d/dx)(\sin^{-1} x) = 1/\sqrt{1 - x^2}$.] (b) If $\theta_1 = 0°$, what is $\Delta\theta_2$ in degrees? (c) If $\theta_1 = 90°$, what is $\Delta\theta_2$ in degrees?

32–7 Total Internal Reflection

57. (I) What is the critical angle for the interface between water and diamond? To be internally reflected, the light must start in which material?

58. (I) The critical angle for a certain liquid–air surface is 49.6°. What is the index of refraction of the liquid?

59. (II) A beam of light is emitted in a pool of water from a depth of 72.0 cm. Where must it strike the air–water interface, relative to the spot directly above it, in order that the light does *not* exit the water?

60. (II) A ray of light, after entering a light fiber, reflects at an angle of 14.5° with the long axis of the fiber, as in Fig. 32–56. Calculate the distance along the axis of the fiber that the light ray travels between successive reflections off the sides of the fiber. Assume that the fiber has an index of refraction of 1.55 and is 1.40×10^{-4} m in diameter.

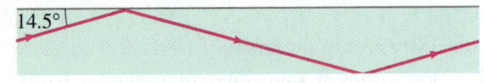

14.5°

FIGURE 32–56 Problem 60.

61. (II) A beam of light is emitted 8.0 cm beneath the surface of a liquid and strikes the surface 7.6 cm from the point directly above the source. If total internal reflection occurs, what can you say about the index of refraction of the liquid?

62. (II) Figure 32–57 shows a liquid-detecting prism device that might be used inside a washing machine or other liquid-containing appliance. If no liquid covers the prism's hypotenuse, total internal reflection of the beam from the light source produces a large signal in the light sensor. If liquid covers the hypotenuse, some light escapes from the prism into the liquid and the light sensor's signal decreases. Thus a large signal from the light sensor indicates the absence of liquid in the reservoir. If this device is designed to detect the presence of water, determine the allowable range for the prism's index of refraction n. Will the device work properly if the prism is constructed from (inexpensive) lucite? For lucite, $n = 1.5$.

FIGURE 32–57 Problem 62.

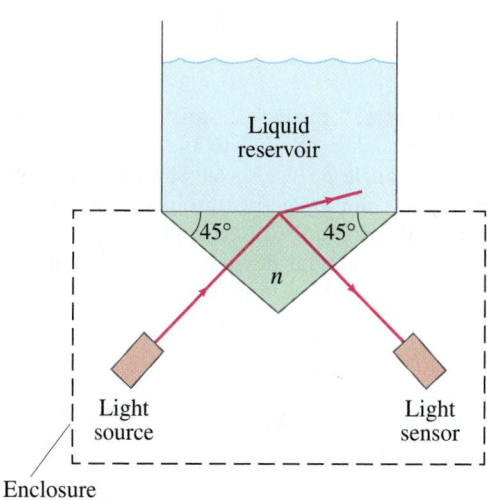

63. (II) Two rays A and B travel down a cylindrical optical fiber of diameter $d = 75.0 \,\mu m$, length $\ell = 1.0 \,km$, and index of refraction $n_1 = 1.465$. Ray A travels a straight path down the fiber's axis, whereas ray B propagates down the fiber by repeated reflections at the critical angle each time it impinges on the fiber's boundary. Determine the extra time Δt it takes for ray B to travel down the entire fiber in comparison with ray A (Fig. 32–58), assuming (a) the fiber is surrounded by air, (b) the fiber is surrounded by a cylindrical glass "cladding" with index of refraction $n_2 = 1.460$.

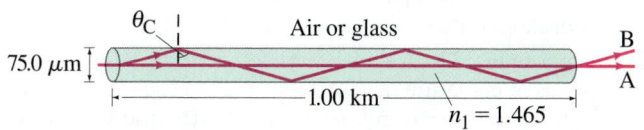

FIGURE 32–58 Problem 63.

64. (II) (a) What is the minimum index of refraction for a glass or plastic prism to be used in binoculars (Fig. 32–33) so that total internal reflection occurs at 45°? (b) Will binoculars work if their prisms (assume $n = 1.58$) are immersed in water? (c) What minimum n is needed if the prisms are immersed in water?

65. (III) Suppose a ray strikes the left face of the prism in Fig. 32–52 at 45.0° as shown, but is totally internally reflected at the opposite side. If the apex angle (at the top) is $\theta = 60.0°$, what can you say about the index of refraction of the prism?

66. (III) A beam of light enters the end of an optic fiber as shown in Fig. 32–59. (a) Show that we can guarantee total internal reflection at the side surface of the material (at point A), if the index of refraction is greater than about 1.42. In other words, regardless of the angle α, the light beam reflects back into the material at point A, assuming air outside.

FIGURE 32–59 Problem 66.

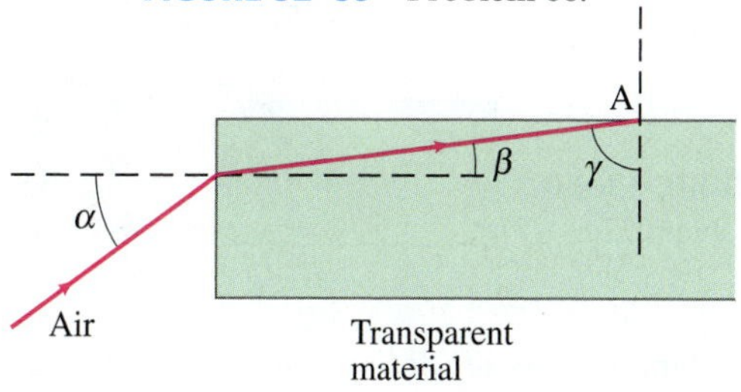

32–8 Refraction at Spherical Surface

*67. (II) A 13.0-cm-thick plane piece of glass ($n = 1.58$) lies on the surface of a 12.0-cm-deep pool of water. How far below the top of the glass does the bottom of the pool seem, as viewed from directly above?

*68. (II) A fish is swimming in water inside a thin spherical glass bowl of uniform thickness. Assuming the radius of curvature of the bowl is 28.0 cm, locate the image of the fish if the fish is located: (a) at the center of the bowl; (b) 20.0 cm from the side of the bowl between the observer and the center of the bowl. Assume the fish is small.

*69. (III) In Section 32–8, we derived Eq. 32–8 for a convex spherical surface with $n_2 > n_1$. Using the same conventions and using diagrams similar to Fig. 32–37, show that Eq. 32–8 is valid also for (a) a convex spherical surface with $n_2 < n_1$, (b) a concave spherical surface with $n_2 > n_1$, and (c) a concave spherical surface with $n_2 < n_1$.

*70. (III) A coin lies at the bottom of a 0.75-m-deep pool. If a viewer sees it at a 45° angle, where is the image of the coin, relative to the coin? [*Hint*: The image is found by tracing back to the intersection of two rays.]

General Problems

71. Two identical concave mirrors are set facing each other 1.0 m apart. A small lightbulb is placed halfway between the mirrors. A small piece of paper placed just to the left of the bulb prevents light from the bulb from directly shining on the left mirror, but light reflected from the right mirror still reaches the left mirror. A good image of the bulb appears on the left side of the piece of paper. What is the focal length of the mirrors?

72. A slab of thickness D, whose two faces are parallel, has index of refraction n. A ray of light incident from air onto one face of the slab at incident angle θ_1 splits into two rays A and B. Ray A reflects directly back into the air, while B travels a total distance ℓ within the slab before reemerging from the slab's face a distance d from its point of entry (Fig. 32–60). (a) Derive expressions for ℓ and d in terms of D, n, and θ_1. (b) For normal incidence (i.e., $\theta_1 = 0°$) show that your expressions yield the expected values for ℓ and d.

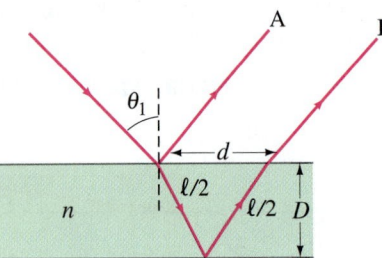

FIGURE 32–60
Problem 72.

73. Two plane mirrors are facing each other 2.2 m apart as in Fig. 32–61. You stand 1.5 m away from one of these mirrors and look into it. You will see multiple images of yourself. (a) How far away from you are the first three images of yourself in the mirror in front of you? (b) Are these first three images facing toward you or away from you?

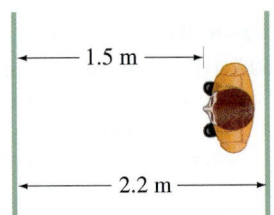

FIGURE 32–61
Problem 73.

74. We wish to determine the depth of a swimming pool filled with water by measuring the width ($x = 5.50$ m) and then noting that the bottom edge of the pool is just visible at an angle of 13.0° above the horizontal as shown in Fig. 32–62. Calculate the depth of the pool.

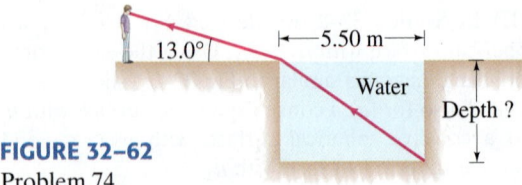

FIGURE 32–62
Problem 74.

75. A 1.80-m-tall person stands 3.80 m from a convex mirror and notices that he looks precisely half as tall as he does in a plane mirror placed at the same distance. What is the radius of curvature of the convex mirror? (Assume that $\sin\theta \approx \theta$.) [*Hint:* The viewing *angle* is half.]

76. The critical angle of a certain piece of plastic in air is $\theta_C = 39.3°$. What is the critical angle of the same plastic if it is immersed in water?

77. Each student in a physics lab is assigned to find the location where a bright object may be placed in order that a concave mirror, with radius of curvature $r = 46$ cm, will produce an image three times the size of the object. Two students complete the assignment at different times using identical equipment, but when they compare notes later, they discover that their answers for the object distance are not the same. Explain why they do not necessarily need to repeat the lab, and justify your response with a calculation.

78. A kaleidoscope makes symmetric patterns with two plane mirrors having a 60° angle between them as shown in Fig. 32–63. Draw the location of the images (some of them images of images) of the object placed between the mirrors.

FIGURE 32–63
Problem 78.

79. When light passes through a prism, the angle that the refracted ray makes relative to the incident ray is called the deviation angle δ, Fig. 32–64. Show that this angle is a minimum when the ray passes through the prism symmetrically, perpendicular to the bisector of the apex angle ϕ, and show that the minimum deviation angle, δ_m, is related to the prism's index of refraction n by

$$n = \frac{\sin\frac{1}{2}(\phi + \delta_m)}{\sin\phi/2}.$$

[*Hint:* For θ in radians, $(d/d\theta)(\sin^{-1}\theta) = 1/\sqrt{1-\theta^2}$.]

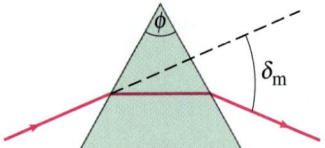

FIGURE 32–64
Problems 79 and 80.

80. If the apex angle of a prism is $\phi = 72°$ (see Fig. 32–64), what is the minimum incident angle for a ray if it is to emerge from the opposite side (i.e., not be totally internally reflected), given $n = 1.58$?

81. **Fermat's principle** states that "light travels between two points along the path that requires the least time, as compared to other nearby paths." From Fermat's principle derive (a) the law of reflection ($\theta_i = \theta_r$) and (b) the law of refraction (Snell's law). [*Hint:* Choose two appropriate points so that a ray between them can undergo reflection or refraction. Draw a rough path for a ray between these points, and write down an expression of the time required for light to travel the arbitrary path chosen. Then take the derivative to find the minimum.]

*82. Suppose Fig. 32–36 shows a cylindrical rod whose end has a radius of curvature $R = 2.0$ cm, and the rod is immersed in water with index of refraction of 1.33. The rod has index of refraction 1.53. Find the location and height of the image of an object 2.0 mm high located 23 cm away from the rod.

83. An optical fiber is a long transparent cylinder of diameter d and index of refraction n. If this fiber is bent sharply, some light hitting the side of the cylinder may escape rather than reflect back into the fiber (Fig. 32–65). What is the smallest radius r at a short bent section for which total internal reflection will be assured for light initially travelling parallel to the axis of the fiber?

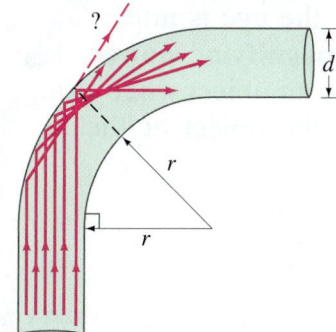

FIGURE 32–65
Problem 83.

84. An object is placed 15 cm from a certain mirror. The image is half the height of the object, inverted, and real. How far is the image from the mirror, and what is the radius of curvature of the mirror?

85. The end faces of a cylindrical glass rod $(n = 1.51)$ are perpendicular to the sides. Show that a light ray entering an end face at any angle will be totally internally reflected inside the rod when it strikes the sides. Assume the rod is in air. What if it were in water?

86. The paint used on highway signs often contains small transparent spheres which provide nighttime illumination of the sign's lettering by retro-reflecting vehicle headlight beams. Consider a light ray from air incident on one such sphere of radius r and index of refraction n. Let θ be its incident angle, and let the ray follow the path shown in Fig. 32–66, so that the ray exits the sphere in the direction exactly antiparallel to its incoming direction. Considering only rays for which $\sin \theta$ can be approximated as θ, determine the required value for n.

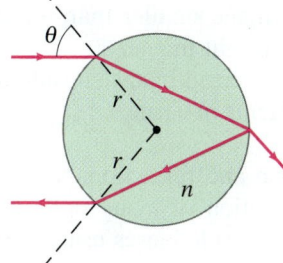

FIGURE 32–66
Problem 86.

*87. (II) The index of refraction, n, of crown flint glass at different wavelengths (λ) of light are given in the Table below.

λ (nm)	1060	546.1	365.0	312.5
n	1.50586	1.51978	1.54251	1.5600

Make a graph of n versus λ. The variation in index of refraction with wavelength is given by the Cauchy equation $n = A + B/\lambda^2$. Make another graph of n versus $1/\lambda^2$ and determine the constants A and B for the glass by fitting the data with a straight line.

*88. (III) Consider a ray of sunlight incident from air on a spherical raindrop of radius r and index of refraction n. Defining θ to be its incident angle, the ray then follows the path shown in Fig. 32–67, exiting the drop at a "scattering angle" ϕ compared with its original incoming direction. (a) Show that $\phi = 180° + 2\theta - 4\sin^{-1}(\sin\theta/n)$. (b) The parallel rays of sunlight illuminate a raindrop with rays of all possible incident angles from 0° to 90°. Plot ϕ vs. θ in the range $0° \le \theta \le 90°$, in 0.5° steps, assuming $n = 1.33$ as is appropriate for water at visible-light wavelengths. (c) From your plot, you should find that a fairly large fraction of the incident angles have nearly the same scattering angle. Approximately what fraction of the possible incident angles is within roughly 1° of $\phi = 139°$? [This subset of incident rays is what creates the rainbow. Wavelength-dependent variations in n cause the rainbow to form at slightly different ϕ for the various visible colors.]

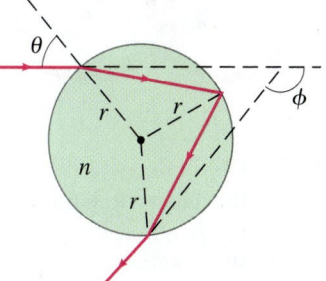

FIGURE 32–67
Problem 88.

Answers to Exercises

A: No.

B: (b).

C: (a).

D: Toward.

E: The face becomes shiny: total internal reflection.

F: 1.414.

Of the many optical devices we discuss in this Chapter, the magnifying glass is the simplest. Here it is magnifying part of page 886 of this Chapter, which describes how the magnifying glass works according to the ray model. In this Chapter we examine thin lenses in detail, seeing how to determine image position as a function of object position and the focal length of the lens, based on the ray model of light. We then examine optical devices including film and digital cameras, the human eye, eyeglasses, telescopes, and microscopes.

subtends a _____. As shown in Fig. 33–33a, the object is placed at the focal point _____ **magnifying glass** _____ converging lens produces a virtual image, which _____ of the eye is to focus on it. If the eye is relaxed, _____ comparison of part (a) or _____ se the object is exactly at the focal point. viewed at the near point with _____ when you "focus" on the object _____ object subtends at the eye is much la _____ part (b), in which **magnification** or **magnifying power**, _____ eye, re _____ angle subtended by an object when using _____ _____ gular unaided eye, with the object at the nea _____ ratio of the normal eye): _____ _____ subtended using the _____ the eye ($N = 25$ cm for a

$$M = \frac{\theta'}{\theta},$$ **(33–5)**

where θ and θ' are shown in Fig. 33–33 _____ $\theta' = h/d_o$ (Fig. 33–33a), where length by noting that $\theta = h/N$ (Fig. 33– _____ angles are small so θ and θ' equal h is the height of the object and we ass _____ r least eye strain), the image will _____ heir sines and tangents. If the eye is _____ at the focal point; see Fig. 33–34. _____ at infinity and the object will b _____ _____ $d_o = f$ and $\theta' = h/f$. Th _____

33 Lenses and Optical Instruments

CONTENTS

33–1 Thin Lenses; Ray Tracing

33–2 The Thin Lens Equation; Magnification

33–3 Combinations of Lenses

*33–4 Lensmaker's Equation

33–5 Cameras: Film and Digital

33–6 The Human Eye; Corrective Lenses

33–7 Magnifying Glass

33–8 Telescopes

*33–9 Compound Microscope

*33–10 Aberrations of Lenses and Mirrors

CHAPTER-OPENING QUESTION—Guess now!

A converging lens, like the type used in a magnifying glass,

(a) always produces a magnified image (image taller than the object).

(b) can also produce an image smaller than the object.

(c) always produces an upright image.

(d) can also produce an inverted image (upside down).

(e) None of these statements are true.

The laws of reflection and refraction, particularly the latter, are the basis for explaining the operation of many optical instruments. In this Chapter we discuss and analyze simple lenses using the model of ray optics discussed in the previous Chapter. We then analyze a number of optical instruments, from the magnifying glass and the human eye to telescopes and microscopes. The importance of lenses is that they form images of objects, as shown in Fig. 33–1.

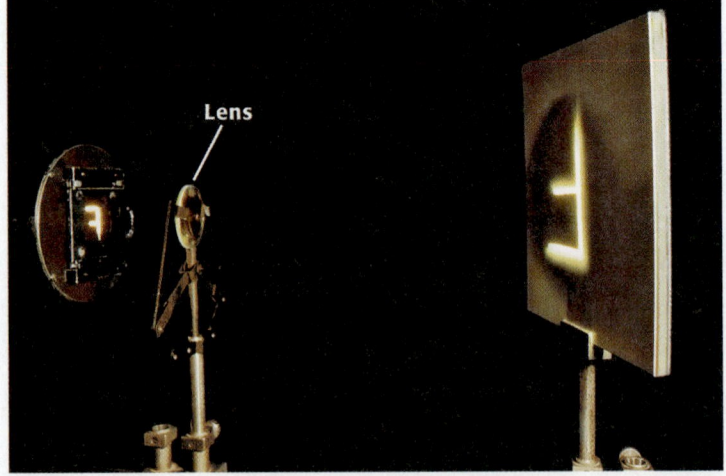

FIGURE 33–1 Converging lens (in holder) forms an image (large "F" on screen at right) of a bright object (illuminated "F" at the left).

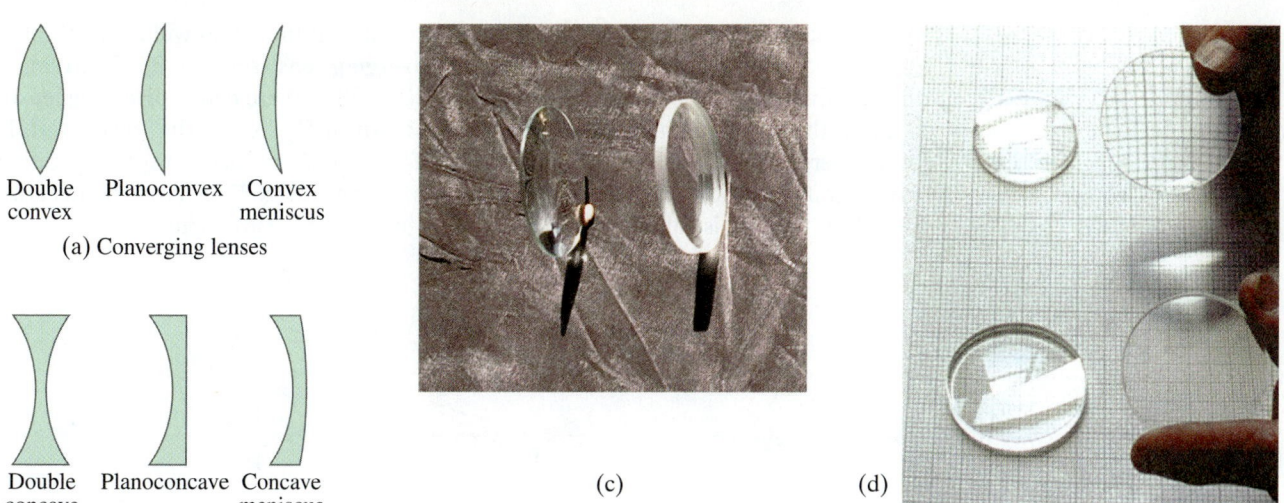

Double convex   Planoconvex   Convex meniscus

(a) Converging lenses

Double concave   Planoconcave   Concave meniscus

(b) Diverging lenses

(c)

(d)

FIGURE 33–2 (a) Converging lenses and (b) diverging lenses, shown in cross section. Converging lenses are thicker in the center whereas diverging lenses are thinner in the center. (c) Photo of a converging lens (on the left) and a diverging lens (right). (d) Converging lenses (above), and diverging lenses (below), lying flat, and raised off the paper to form images.

33–1 Thin Lenses; Ray Tracing

The most important simple optical device is no doubt the thin lens. The development of optical devices using lenses dates to the sixteenth and seventeenth centuries, although the earliest record of eyeglasses dates from the late thirteenth century. Today we find lenses in eyeglasses, cameras, magnifying glasses, telescopes, binoculars, microscopes, and medical instruments. A thin lens is usually circular, and its two faces are portions of a sphere. (Cylindrical faces are also possible, but we will concentrate on spherical.) The two faces can be concave, convex, or plane. Several types are shown in Fig. 33–2 a and b in cross section.

Consider parallel rays striking the double convex lens shown in cross section in Fig. 33–3a. We assume the lens is made of material such as glass or transparent plastic, with index of refraction greater than that of the air outside. The **axis** of a lens is a straight line passing through the center of the lens and perpendicular to its two surfaces (Fig. 33–3). From Snell's law, we can see that each ray in Fig. 33–3a is bent toward the axis when the ray enters the lens and again when it leaves the lens at the back surface. (Note the dashed lines indicating the normals to each surface for the top ray.) If rays parallel to the axis fall on a thin lens, they will be focused to a point called the **focal point**, F. This will not be precisely true for a lens with spherical surfaces. But it will be very nearly true—that is, parallel rays will be focused to a tiny region that is nearly a point—if the diameter of the lens is small compared to the radii of curvature of the two lens surfaces. This criterion is satisfied by a **thin lens**, one that is very thin compared to its diameter, and we consider only thin lenses here.

The rays from a point on a distant object are essentially parallel—see Fig. 32–12. Therefore we can say that *the focal point is the image point for an object at infinity on the lens axis*. Thus, the focal point of a lens can be found by locating the point where the Sun's rays (or those of some other distant object) are brought to a sharp image, Fig. 33–4. The distance of the focal point from the center of the lens is called the **focal length**, f. A lens can be turned around so that light can pass through it from the opposite side. The focal length is *the same* on both sides, as we shall see later, even if the curvatures of the two lens surfaces are different. If parallel rays fall on a lens at an angle, as in Fig. 33–3b, they focus at a point F_a. The plane containing all focus points, such as F and F_a in Fig. 33–3b, is called the **focal plane** of the lens.

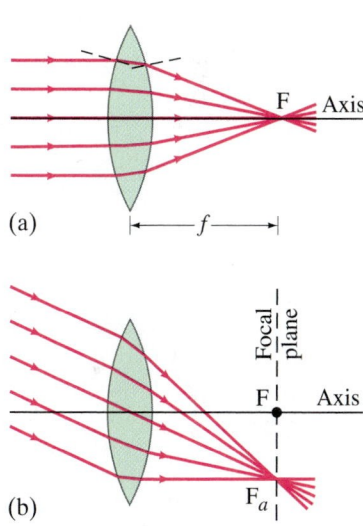

(a)

(b)

FIGURE 33–3 Parallel rays are brought to a focus by a converging thin lens.

FIGURE 33–4 Image of the Sun burning wood.

Any lens (in air) that is thicker in the center than at the edges will make parallel rays converge to a point, and is called a **converging lens** (see Fig. 33–2a). Lenses that are thinner in the center than at the edges (Fig. 33–2b) are called **diverging lenses** because they make parallel light diverge, as shown in Fig. 33–5. The focal point, F, of a diverging lens is defined as that point from which refracted rays, originating from parallel incident rays, seem to emerge as shown in Fig. 33–5. And the distance from F to the lens is called the **focal length**, f, just as for a converging lens.

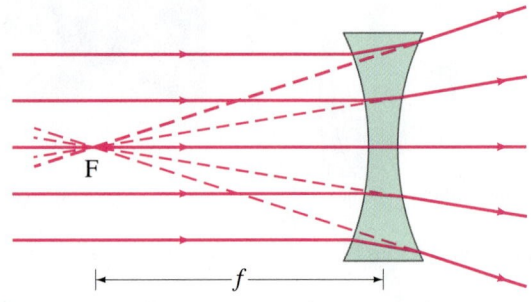

FIGURE 33–5 Diverging lens.

Optometrists and ophthalmologists, instead of using the focal length, use the reciprocal of the focal length to specify the strength of eyeglass (or contact) lenses. This is called the **power**, P, of a lens:

$$P = \frac{1}{f}. \tag{33-1}$$

The unit for lens power is the **diopter** (D), which is an inverse meter: $1\,D = 1\,m^{-1}$. For example, a 20-cm-focal-length lens has a power $P = 1/(0.20\,m) = 5.0\,D$. We will mainly use the focal length, but we will refer again to the power of a lens when we discuss eyeglass lenses in Section 33–6.

The most important parameter of a lens is its focal length f. For a converging lens, f is easily measured by finding the image point for the Sun or other distant objects. Once f is known, the image position can be calculated for any object. To find the image point by drawing rays would be difficult if we had to determine the refractive angles at the front surface of the lens and again at the back surface where the ray exits. We can save ourselves a lot of effort by making use of certain facts we already know, such as that a ray parallel to the axis of the lens passes (after refraction) through the focal point. To determine an image point, we can consider only the three rays indicated in Fig. 33–6, which uses an arrow (on the left) as the object, and a converging lens forming an image (dashed arrow) to the right. These rays, emanating from a single point on the object, are drawn as if the lens were infinitely thin, and we show only a single sharp bend at the center line of the lens instead of the refractions at each surface. These three rays are drawn as follows:

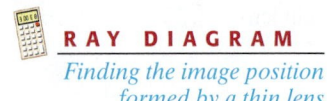

RAY DIAGRAM
Finding the image position formed by a thin lens

Ray 1 is drawn parallel to the axis, Fig. 33–6a; therefore it is refracted by the lens so that it passes along a line through the focal point F behind the lens. (See also Fig. 33–3a.)

Ray 2 is drawn on a line passing through the other focal point F′ (front side of lens in Fig. 33–6) and emerges from the lens parallel to the axis, Fig. 33–6b.

Ray 3 is directed toward the very center of the lens, where the two surfaces are essentially parallel to each other, Fig. 33–6c; this ray therefore emerges from the lens at the same angle as it entered; the ray would be displaced slightly to one side, as we saw in Example 32–8, but since we assume the lens is thin, we draw ray 3 straight through as shown.

The point where these three rays cross is the image point for that object point. Actually, any two of these rays will suffice to locate the image point, but drawing the third ray can serve as a check.

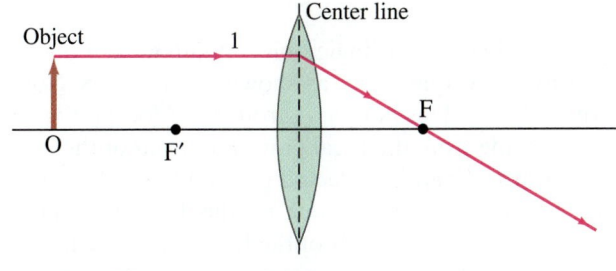

Center line

Object

O   F'   F

(a) Ray 1 leaves one point on object going parallel to the axis, then refracts through focal point behind the lens.

FIGURE 33–6 Finding the image by ray tracing for a converging lens. Rays are shown leaving one point on the object (an arrow). Shown are the three most useful rays, leaving the tip of the object, for determining where the image of that point is formed.

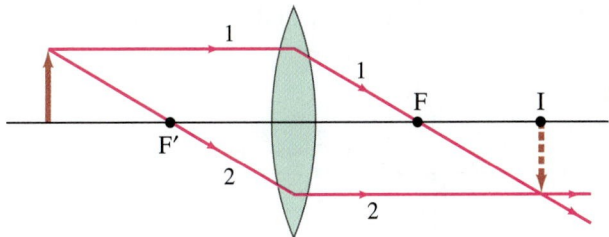

F'   F   I

(b) Ray 2 passes through F' in front of the lens; therefore it is parallel to the axis behind the lens.

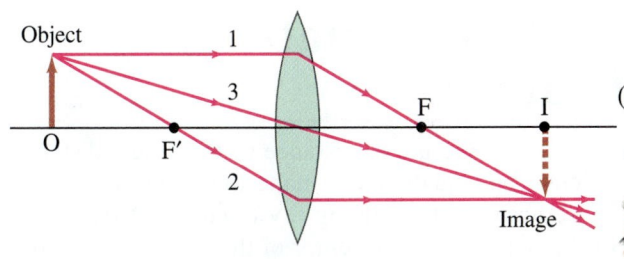

Object

O   F'   F   I

Image

(c) Ray 3 passes straight through the center of the lens (assumed very thin).

Using these three rays for one object point, we can find the image point for that point of the object (the top of the arrow in Fig. 33–6). The image points for all other points on the object can be found similarly to determine the complete image of the object. Because the rays actually pass through the image for the case shown in Fig. 33–6, it is a **real image** (see page 840). The image could be detected by film or electronic sensor, or actually seen on a white surface or screen placed at the position of the image (Fig. 33–7a).

FIGURE 33–7 (a) A converging lens can form a real image (here of a distant building, upside down) on a screen. (b) That same real image is also directly visible to the eye. [Figure 33–2d shows images (graph paper) seen by the eye made by both diverging and converging lenses.]

CONCEPTUAL EXAMPLE 33–1 **Half-blocked lens.** What happens to the image of an object if the top half of a lens is covered by a piece of cardboard?

RESPONSE Let us look at the rays in Fig. 33–6. If the top half (or any half of the lens) is blocked, you might think that half the image is blocked. But in Fig. 33–6c, we see how the rays used to create the "top" of the image pass through both the top and the bottom of the lens. Only three of many rays are shown—many more rays pass through the lens, and they can form the image. You don't lose the image, but covering part of the lens cuts down on the total light received and reduces the brightness of the image.

NOTE If the lens is partially blocked by your thumb, you may notice an out of focus image of part of that thumb.

(a)

Seeing the Image

The image can also be seen directly by the eye when the eye is placed behind the image, as shown in Fig. 33–6c, so that some of the rays diverging from each point on the image can enter the eye. We can see a sharp image only for rays *diverging* from each point on the image, because we see normal objects when diverging rays from each point enter the eye as was shown in Fig. 32–1. Your eye cannot focus rays converging on it; if your eye was positioned between points F and I in Fig. 33–6c, it would not see a clear image. (More about our eyes in Section 33–6.) Figure 33–7 shows an image seen (a) on a screen and (b) directly by the eye (and a camera) placed behind the image. The eye can see both real and virtual images (see next page) as long as the eye is positioned so rays diverging from the image enter it.

(b)

Diverging Lens

By drawing the same three rays emerging from a single object point, we can determine the image position formed by a diverging lens, as shown in Fig. 33–8. Note that ray 1 is drawn parallel to the axis, but does not pass through the focal point F′ behind the lens. Instead it seems to come from the focal point F in front of the lens (dashed line). Ray 2 is directed toward F′ and is refracted parallel to the lens axis by the lens. Ray 3 passes directly through the center of the lens. The three refracted rays seem to emerge from a point on the left of the lens. This is the image point, I. Because the rays do not pass through the image, it is a **virtual image**. Note that the eye does not distinguish between real and virtual images—both are visible.

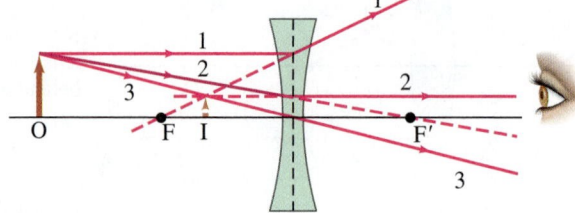

FIGURE 33–8 Finding the image by ray tracing for a diverging lens.

33–2 The Thin Lens Equation; Magnification

We now derive an equation that relates the image distance to the object distance and the focal length of a thin lens. This equation will make the determination of image position quicker and more accurate than doing ray tracing. Let d_o be the object distance, the distance of the object from the center of the lens, and d_i be the image distance, the distance of the image from the center of the lens.

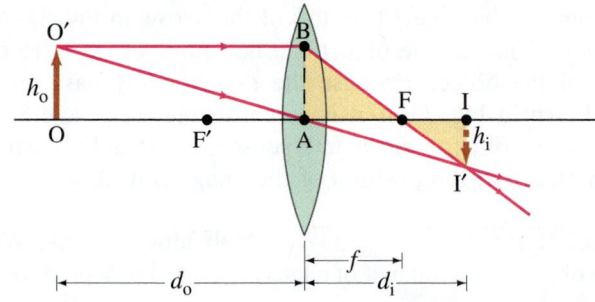

FIGURE 33–9 Deriving the lens equation for a converging lens.

Let h_o and h_i refer to the heights of the object and image. Consider the two rays shown in Fig. 33–9 for a converging lens, assumed to be very thin. The right triangles FI′I and FBA (highlighted in yellow) are similar because angle AFB equals angle IFI′; so

$$\frac{h_i}{h_o} = \frac{d_i - f}{f},$$

since length AB = h_o. Triangles OAO′ and IAI′ are similar as well. Therefore,

$$\frac{h_i}{h_o} = \frac{d_i}{d_o}.$$

We equate the right sides of these two equations (the left sides are the same), and divide by d_i to obtain

$$\frac{1}{f} - \frac{1}{d_i} = \frac{1}{d_o}$$

or

THIN LENS EQUATION

$$\frac{1}{d_o} + \frac{1}{d_i} = \frac{1}{f}.$$

(33–2)

This is called the **thin lens equation**. It relates the image distance d_i to the object distance d_o and the focal length f. It is the most useful equation in geometric optics.

(Interestingly, it is exactly the same as the mirror equation, Eq. 32–2). If the object is at infinity, then $1/d_o = 0$, so $d_i = f$. Thus the focal length is the image distance for an object at infinity, as mentioned earlier.

We can derive the lens equation for a diverging lens using Fig. 33–10. Triangles IAI′ and OAO′ are similar; and triangles IFI′ and AFB are similar. Thus (noting that length $AB = h_o$)

$$\frac{h_i}{h_o} = \frac{d_i}{d_o} \quad \text{and} \quad \frac{h_i}{h_o} = \frac{f - d_i}{f}.$$

When we equate the right sides of these two equations and simplify, we obtain

$$\frac{1}{d_o} - \frac{1}{d_i} = -\frac{1}{f}.$$

This equation becomes the same as Eq. 33–2 if we make f and d_i negative. That is, we take f to be *negative for a diverging lens*, and d_i negative when the image is on the same side of the lens as the light comes from. Thus Eq. 33–2 will be valid for both converging and diverging lenses, and for *all* situations, if we use the following **sign conventions**:

1. The focal length is positive for converging lenses and negative for diverging lenses.
2. The object distance is positive if the object is on the side of the lens from which the light is coming (this is usually the case, although when lenses are used in combination, it might not be so); otherwise, it is negative.
3. The image distance is positive if the image is on the opposite side of the lens from where the light is coming; if it is on the same side, d_i is negative. Equivalently, the image distance is positive for a real image and negative for a virtual image.
4. The height of the image, h_i, is positive if the image is upright, and negative if the image is inverted relative to the object. (h_o is always taken as upright and positive.)

The **lateral magnification**, m, of a lens is defined as the ratio of the image height to object height, $m = h_i/h_o$. From Figs. 33–9 and 33–10 and the conventions just stated (for which we'll need a minus sign below), we have

$$m = \frac{h_i}{h_o} = -\frac{d_i}{d_o}. \tag{33–3}$$

For an upright image the magnification is positive ($h_i > 0$ and $d_i < 0$), and for an inverted image the magnification is negative ($h_i < 0$ and $d_i > 0$).

From sign convention 1, it follows that the power (Eq. 33–1) of a converging lens, in diopters, is positive, whereas the power of a diverging lens is negative. A converging lens is sometimes referred to as a **positive lens**, and a diverging lens as a **negative lens**.

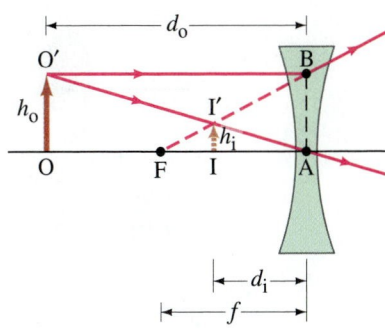

FIGURE 33–10 Deriving the lens equation for a diverging lens.

⚠ **CAUTION**

Focal length is negative for diverging lens

 **PROBLEM SOLVING**

SIGN CONVENTIONS for lenses

PROBLEM SOLVING

Thin Lenses

1. Draw a **ray diagram**, as precise as possible, but even a rough one can serve as confirmation of analytic results. Choose one point on the object and draw at least two, or preferably three, of the easy-to-draw rays described in Figs. 33–6 and 33–8. The image point is where the rays intersect.

2. For analytic solutions, solve for unknowns in the **thin lens equation** (Eq. 33–2) and the **magnification equation** (Eq. 33–3). The thin lens equation involves reciprocals—don't forget to take the reciprocal.

3. Follow the **sign conventions** listed just above.

4. Check that your analytic answers are **consistent** with your ray diagram.

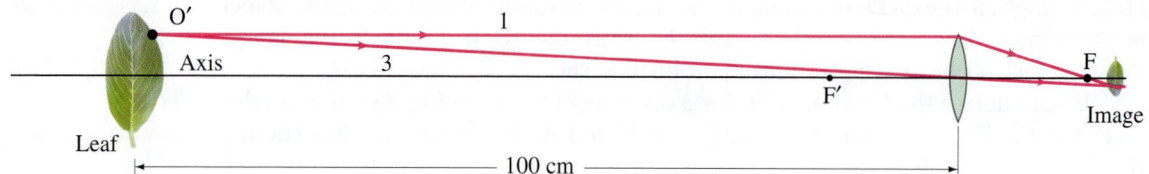

FIGURE 33–11 Example 33–2. (Not to scale.)

EXAMPLE 33–2 **Image formed by converging lens.** What is (*a*) the position, and (*b*) the size, of the image of a 7.6-cm-high leaf placed 1.00 m from a +50.0-mm-focal-length camera lens?

APPROACH We follow the steps of the Problem Solving Strategy explicitly.

SOLUTION

1. **Ray diagram.** Figure 33–11 is an approximate ray diagram, showing only rays 1 and 3 for a single point on the leaf. We see that the image ought to be a little behind the focal point F, to the right of the lens.

2. **Thin lens and magnification equations.** (*a*) We find the image position analytically using the thin lens equation, Eq. 33–2. The camera lens is converging, with $f = +5.00$ cm, and $d_o = 100$ cm, and so the thin lens equation gives

$$\frac{1}{d_i} = \frac{1}{f} - \frac{1}{d_o} = \frac{1}{5.00 \text{ cm}} - \frac{1}{100 \text{ cm}}$$

$$= \frac{20.0 - 1.0}{100 \text{ cm}} = \frac{19.0}{100 \text{ cm}}.$$

Then, taking the reciprocal,

$$d_i = \frac{100 \text{ cm}}{19.0} = 5.26 \text{ cm},$$

or 52.6 mm behind the lens.

(*b*) The magnification is

$$m = -\frac{d_i}{d_o} = -\frac{5.26 \text{ cm}}{100 \text{ cm}} = -0.0526,$$

so

$$h_i = mh_o = (-0.0526)(7.6 \text{ cm}) = -0.40 \text{ cm}.$$

The image is 4.0 mm high.

3. **Sign conventions.** The image distance d_i came out positive, so the image is behind the lens. The image height is $h_i = -0.40$ cm; the minus sign means the image is inverted.

4. **Consistency.** The analytic results of steps (2) and (3) are consistent with the ray diagram, Fig. 33–11: the image is behind the lens and inverted.

NOTE Part (*a*) tells us that the image is 2.6 mm farther from the lens than the image for an object at infinity, which would equal the focal length, 50.0 mm. Indeed, when focusing a camera lens, the closer the object is to the camera, the farther the lens must be from the sensor or film.

EXERCISE A If the leaf (object) of Example 33–2 is moved farther from the lens, does the image move closer to or farther from the lens? (Don't calculate!)

EXAMPLE 33–3 Object close to converging lens. An object is placed 10 cm from a 15-cm-focal-length converging lens. Determine the image position and size (*a*) analytically, and (*b*) using a ray diagram.

APPROACH We first use Eqs. 33–2 and 33–3 to obtain an analytic solution, and then confirm with a ray diagram using the special rays 1, 2, and 3 for a single object point.

SOLUTION (*a*) Given $f = 15$ cm and $d_o = 10$ cm, then

$$\frac{1}{d_i} = \frac{1}{15 \text{ cm}} - \frac{1}{10 \text{ cm}} = -\frac{1}{30 \text{ cm}},$$

and $d_i = -30$ cm. (Remember to take the reciprocal!) Because d_i is negative, the image must be virtual and on the same side of the lens as the object. The magnification

$$m = -\frac{d_i}{d_o} = -\frac{-30 \text{ cm}}{10 \text{ cm}} = 3.0.$$

⚠️ **CAUTION**

Don't forget to take the reciprocal

The image is three times as large as the object and is upright. This lens is being used as a simple magnifying glass, which we discuss in more detail in Section 33–7.

(*b*) The ray diagram is shown in Fig. 33–12 and confirms the result in part (*a*). We choose point O′ on the top of the object and draw ray 1, which is easy. But ray 2 may take some thought: if we draw it heading toward F′, it is going the wrong way—so we have to draw it as if coming from F′ (and so dashed), striking the lens, and then going out parallel to the lens axis. We project it back parallel, with a dashed line, as we must do also for ray 1, in order to find where they cross. Ray 3 is drawn through the lens center, and it crosses the other two rays at the image point, I′.

NOTE From Fig. 33–12 we can see that, whenever an object is placed between a converging lens and its focal point, the image is virtual.

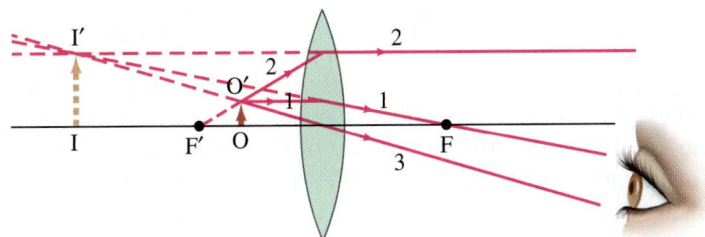

FIGURE 33–12 An object placed within the focal point of a converging lens produces a virtual image. Example 33–3.

EXAMPLE 33–4 Diverging lens. Where must a small insect be placed if a 25-cm-focal-length diverging lens is to form a virtual image 20 cm from the lens, on the same side as the object?

APPROACH The ray diagram is basically that of Fig. 33–10 because our lens here is diverging and our image is in front of the lens within the focal distance. (It would be a valuable exercise to draw the ray diagram to scale, precisely, now.) The insect's distance, d_o, can be calculated using the thin lens equation.

SOLUTION The lens is diverging, so f is negative: $f = -25$ cm. The image distance must be negative too because the image is in front of the lens (sign conventions), so $d_i = -20$ cm. Equation 33–2 gives

$$\frac{1}{d_o} = \frac{1}{f} - \frac{1}{d_i} = -\frac{1}{25 \text{ cm}} + \frac{1}{20 \text{ cm}} = \frac{-4 + 5}{100 \text{ cm}} = \frac{1}{100 \text{ cm}}.$$

So the object must be 100 cm in front of the lens.

FIGURE 33–13 Exercise C.

(a)

(b)

EXERCISE B Return to the Chapter-Opening Question, page 866, and answer it again now. Try to explain why you may have answered differently the first time.

EXERCISE C Figure 33–13 shows a converging lens held above three equal-sized letters A. In (a) the lens is 5 cm from the paper, and in (b) the lens is 15 cm from the paper. Estimate the focal length of the lens. What is the image position for each case?

33–3 Combinations of Lenses

Optical instruments typically use lenses in combination. When light passes through more than one lens, we find the image formed by the first lens as if it were alone. This image becomes the *object* for the second lens, and we find the image then formed by this second lens, which is the final image if there are only two lenses. The total magnification will be the product of the separate magnifications of each lens, as we shall see. Even if the second lens intercepts the light from the first lens before it forms an image, this technique still works.

EXAMPLE 33–5 A two-lens system. Two converging lenses, A and B, with focal lengths $f_A = 20.0$ cm and $f_B = 25.0$ cm, are placed 80.0 cm apart, as shown in Fig. 33–14a. An object is placed 60.0 cm in front of the first lens as shown in Fig. 33–14b. Determine (*a*) the position, and (*b*) the magnification, of the final image formed by the combination of the two lenses.

APPROACH Starting at the tip of our object O, we draw rays 1, 2, and 3 for the first lens, A, and also a ray 4 which, after passing through lens A, acts as "ray 3" (through the center) for the second lens, B. Ray 2 for lens A exits parallel, and so is ray 1 for lens B. To determine the position of the image I_A formed by lens A, we use Eq. 33–2 with $f_A = 20.0$ cm and $d_{oA} = 60.0$ cm. The distance of I_A (lens A's image) from lens B is the object distance d_{oB} for lens B. The final image is found using the thin lens equation, this time with all distances relative to lens B. For (*b*) the magnifications are found from Eq. 33–3 for each lens in turn.

SOLUTION (*a*) The object is a distance $d_{oA} = +60.0$ cm from the first lens, A, and this lens forms an image whose position can be calculated using the thin lens equation:

$$\frac{1}{d_{iA}} = \frac{1}{f_A} - \frac{1}{d_{oA}} = \frac{1}{20.0 \text{ cm}} - \frac{1}{60.0 \text{ cm}} = \frac{3-1}{60.0 \text{ cm}} = \frac{1}{30.0 \text{ cm}}.$$

So the first image I_A is at $d_{iA} = 30.0$ cm behind the first lens. This image becomes the object for the second lens, B. It is a distance $d_{oB} = 80.0 \text{ cm} - 30.0 \text{ cm} = 50.0$ cm in front of lens B, as shown in Fig. 33–14b. The image formed by lens B, again using the thin lens equation, is at a distance d_{iB} from the lens B:

$$\frac{1}{d_{iB}} = \frac{1}{f_B} - \frac{1}{d_{oB}} = \frac{1}{25.0 \text{ cm}} - \frac{1}{50.0 \text{ cm}} = \frac{2-1}{50.0 \text{ cm}} = \frac{1}{50.0 \text{ cm}}.$$

Hence $d_{iB} = 50.0$ cm behind lens B. This is the final image—see Fig. 33–14b.

FIGURE 33–14 Two lenses, A and B, used in combination, Example 33–5. The small numbers refer to the easily drawn rays.

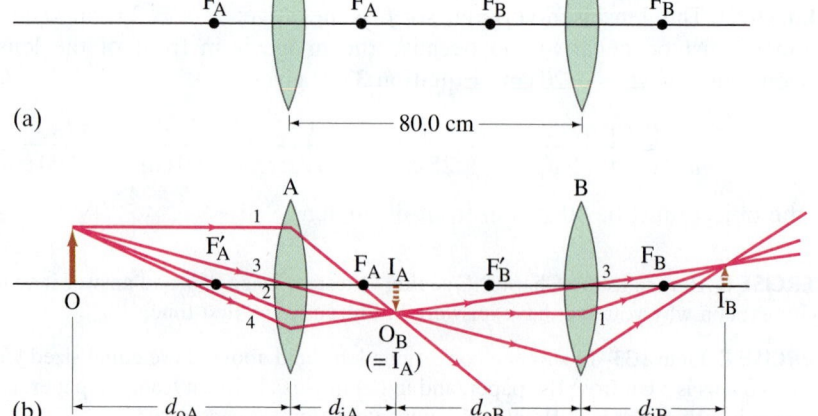

(b) Lens A has a magnification (Eq. 33–3)

$$m_A = -\frac{d_{iA}}{d_{oA}} = -\frac{30.0 \text{ cm}}{60.0 \text{ cm}} = -0.500.$$

Thus, the first image is inverted and is half as high as the object (again Eq. 33–3):

$$h_{iA} = m_A h_{oA} = -0.500 h_{oA}.$$

Lens B takes this image as object and changes its height by a factor

$$m_B = -\frac{d_{iB}}{d_{oB}} = -\frac{50.0 \text{ cm}}{50.0 \text{ cm}} = -1.000.$$

The second lens reinverts the image (the minus sign) but doesn't change its size. The final image height is (remember h_{oB} is the same as h_{iA})

$$h_{iB} = m_B h_{oB} = m_B h_{iA} = m_B m_A h_{oA} = (m_{total}) h_{oA}.$$

The total magnification is the product of m_A and m_B, which here equals $m_{total} = m_A m_B = (-1.000)(-0.500) = +0.500$, or half the original height, and the final image is upright.

EXAMPLE 33–6 **Measuring f for a diverging lens.** To measure the focal length of a diverging lens, a converging lens is placed in contact with it, as shown in Fig. 33–15. The Sun's rays are focused by this combination at a point 28.5 cm, behind the lenses as shown. If the converging lens has a focal length f_C of 16.0 cm, what is the focal length f_D of the diverging lens? Assume both lenses are thin and the space between them is negligible.

APPROACH The image distance for the first lens equals its focal length (16.0 cm) since the object distance is infinity (∞). The position of this image, even though it is never actually formed, acts as the object for the second (diverging) lens. We apply the thin lens equation to the diverging lens to find its focal length, given that the final image is at $d_i = 28.5$ cm.

SOLUTION If the diverging lens was absent, the converging lens would form the image at its focal point—that is, at a distance $f_C = 16.0$ cm behind it (dashed lines in Fig. 33–15). When the diverging lens is placed next to the converging lens, we treat the image formed by the first lens as the *object* for the second lens. Since this object lies to the right of the diverging lens, this is a situation where d_o is negative (see the sign conventions, page 871). Thus, for the diverging lens, the object is virtual and $d_o = -16.0$ cm. The diverging lens forms the image of this virtual object at a distance $d_i = 28.5$ cm away (given). Thus,

$$\frac{1}{f_D} = \frac{1}{d_o} + \frac{1}{d_i} = \frac{1}{-16.0 \text{ cm}} + \frac{1}{28.5 \text{ cm}} = -0.0274 \text{ cm}^{-1}.$$

We take the reciprocal to find $f_D = -1/(0.0274 \text{ cm}^{-1}) = -36.5$ cm.

NOTE If this technique is to work, the converging lens must be "stronger" than the diverging lens—that is, it must have a focal length whose magnitude is less than that of the diverging lens. (Rays from the Sun are focused 28.5 cm behind the combination, so the focal length of the total combination is $f_T = 28.5$ cm.)

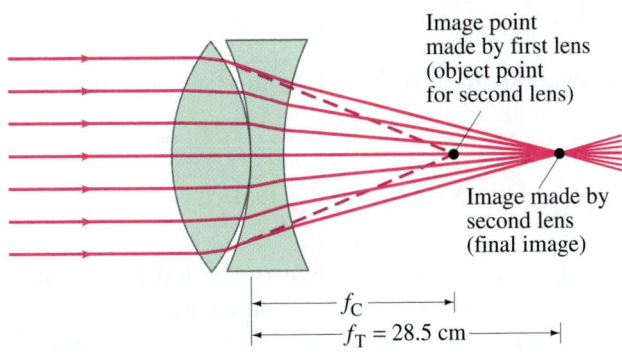

Image point made by first lens (object point for second lens)

Image made by second lens (final image)

FIGURE 33–15 Determining the focal length of a diverging lens. Example 33–6.

f_C

$f_T = 28.5$ cm

*33–4 Lensmaker's Equation

In this Section, we will show that parallel rays are brought to a focus at a single point for a thin lens. At the same time, we will also derive an equation that relates the focal length of a lens to the radii of curvature of its two surfaces and its index of refraction, which is known as the lensmaker's equation.

In Fig. 33–16, a ray parallel to the axis of a lens is refracted at the front surface of the lens at point A_1 and is refracted at the back surface at point A_2. This ray then passes through point F, which we call the focal point for this ray. Point A_1 is a height h_1 above the axis, and point A_2 is height h_2 above the axis. C_1 and C_2 are the centers of curvature of the two lens surfaces; so the length $C_1 A_1 = R_1$, the radius of curvature of the front surface, and $C_2 A_2 = R_2$ is the radius of the second surface. We consider a double convex lens and by convention choose the radii of both lens surfaces as positive. The thickness of the lens has been grossly exaggerated so that the various angles would be clear. But we will assume that the lens is actually very thin and that angles between the rays and the axis are small. In this approximation, the sines and tangents of all the angles will be equal to the angles themselves in radians. For example, $\sin \theta_1 \approx \tan \theta_1 \approx \theta_1$ (radians).

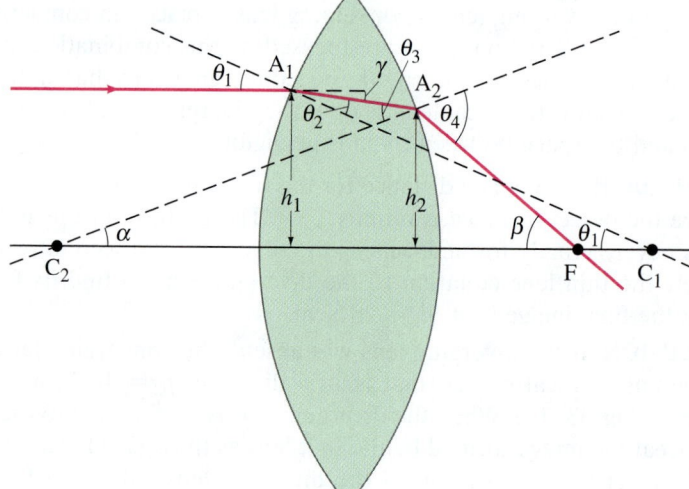

FIGURE 33–16 Diagram of a ray passing through a lens for derivation of the lensmaker's equation.

To this approximation, then, Snell's law tells us that

$$\theta_1 = n\theta_2$$
$$\theta_4 = n\theta_3$$

where n is the index of refraction of the glass, and we assume that the lens is surrounded by air $(n = 1)$. Notice also in Fig. 33–16 that

$$\theta_1 \approx \sin \theta_1 = \frac{h_1}{R_1}$$

$$\alpha \approx \frac{h_2}{R_2}$$

$$\beta \approx \frac{h_2}{f}.$$

The last follows because the distance from F to the lens (assumed very thin) is f. From the diagram, the angle γ is

$$\gamma = \theta_1 - \theta_2.$$

A careful examination of Fig. 33–16 shows also that

$$\alpha = \theta_3 - \gamma.$$

This can be seen by drawing a horizontal line to the left from point A_2, which divides the angle θ_3 into two parts. The upper part equals γ and the lower part equals α. (The opposite angles between an oblique line and two parallel lines are equal.) Thus, $\theta_3 = \gamma + \alpha$. Furthermore, by drawing a horizontal line to the right from point A_2, we divide θ_4 into two parts. The upper part is α and the lower is β.

Thus

$$\theta_4 = \alpha + \beta.$$

We now combine all these equations:

$$\alpha = \theta_3 - \gamma = \frac{\theta_4}{n} - (\theta_1 - \theta_2) = \frac{\alpha}{n} + \frac{\beta}{n} - \theta_1 + \theta_2,$$

or

$$\frac{h_2}{R_2} = \frac{h_2}{nR_2} + \frac{h_2}{nf} - \frac{h_1}{R_1} + \frac{h_1}{nR_1}.$$

Because the lens is thin, $h_1 \approx h_2$ and all h's can be canceled from all the numerators. We then multiply through by n and rearrange to find that

$$\frac{1}{f} = (n - 1)\left(\frac{1}{R_1} + \frac{1}{R_2}\right). \qquad \text{(33–4)}$$

This is called the **lensmaker's equation**. It relates the focal length of a thin lens to the radii of curvature of its two surfaces and its index of refraction. Notice that f for a thin lens does not depend on h_1 or h_2. Thus the position of the point F does not depend on where the ray strikes the lens. Hence, all rays parallel to the axis of a thin lens will pass through the same point F, which we wished to prove.

In our derivation, both surfaces are convex and R_1 and R_2 are considered positive.[†] Equation 33–4 also works for lenses with one or both surfaces concave; but for a concave surface, the radius must be considered *negative*.

Notice in Eq. 33–4 that the equation is *symmetrical* in R_1 and R_2. Thus, if a lens is turned around so that light impinges on the other surface, the focal length is the same even if the two lens surfaces are different.

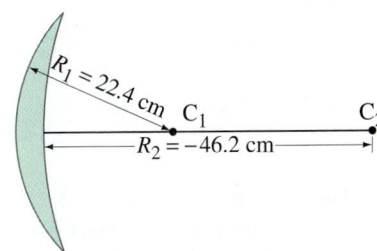

! CAUTION
Sign conventions

EXAMPLE 33–7 **Calculating f for a converging lens.** A convex meniscus lens (Figs. 33–2a and 33–17) is made from glass with $n = 1.50$. The radius of curvature of the convex surface is 22.4 cm and that of the concave surface is 46.2 cm. (*a*) What is the focal length? (*b*) Where will the image be for an object 2.00 m away?

APPROACH We use Eq. 33–4, noting that R_2 is negative because it refers to the concave surface.

SOLUTION (*a*) $R_1 = 22.4$ cm and $R_2 = -46.2$ cm.
Then

$$\frac{1}{f} = (1.50 - 1.00)\left(\frac{1}{22.4\text{ cm}} - \frac{1}{46.2\text{ cm}}\right)$$

$$= 0.0115\text{ cm}^{-1}.$$

So

$$f = \frac{1}{0.0115\text{ cm}^{-1}} = 87.0\text{ cm}$$

and the lens is converging. Notice that if we turn the lens around so that $R_1 = -46.2$ cm and $R_2 = +22.4$ cm, we get the same result.
(*b*) From the lens equation, with $f = 0.870$ m and $d_o = 2.00$ m, we have

$$\frac{1}{d_i} = \frac{1}{f} - \frac{1}{d_o} = \frac{1}{0.870\text{ m}} - \frac{1}{2.00\text{ m}}$$

$$= 0.649\text{ m}^{-1},$$

so $d_i = 1/0.649\text{ m}^{-1} = 1.54$ m.

FIGURE 33–17 Example 33–7.

EXERCISE D A Lucite planoconcave lens (see Fig. 33–2b) has one flat surface and the other has $R = -18.4$ cm. What is the focal length? Is the lens converging or diverging?

[†]Some books use a different convention—for example, R_1 and R_2 are considered positive if their centers of curvature are to the right of the lens, in which case a minus sign appears in their equivalent of Eq. 33–4.

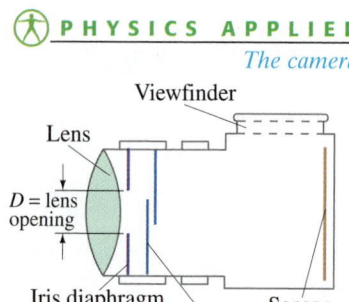

FIGURE 33–18 A simple camera.

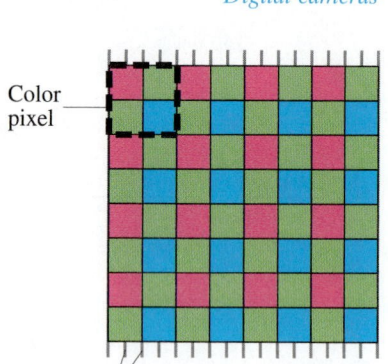

FIGURE 33–19 Portion of a typical CCD sensor. A square group of four pixels $^{RG}_{GB}$ is sometimes called a "color pixel."

FIGURE 33–20 Suppose we take a picture that includes a thin black line (our object) on a white background. The *image* of this black line has a colored halo (red above, blue below) due to the mosaic arrangement of color filter pixels, as shown by the colors transmitted. Computer averaging can minimize color problems such as this (the green at top and bottom of image can be averaged with nearby pixels to give white or nearly so) but the image is consequently "softened" or blurred. The layered color pixel described in the text would avoid this artifact.

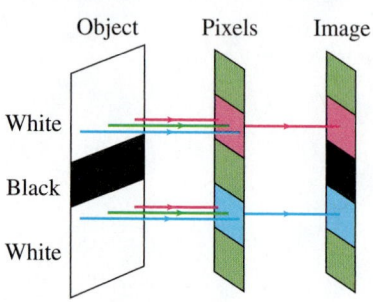

33–5 Cameras: Film and Digital

The basic elements of a **camera** are a lens, a light-tight box, a shutter to let light pass through the lens only briefly, and in a digital camera an electronic sensor or in a traditional camera a piece of film (Fig. 33–18). When the shutter is opened for a brief "exposure," light from external objects in the field of view is focused by the lens as an image on the film or sensor. Film contains light-sensitive chemicals that change when light strikes them. In the development process, chemical reactions cause the changed areas to turn opaque, so the image is recorded on the film.[†]

You can see the image yourself if you remove the back of a conventional camera and view through a piece of tissue paper (on which an image can form) placed where the film should be with the shutter open.

Digital Cameras, Electronic Sensors (CCD, CMOS)

In a **digital camera**, the film is replaced by a semiconductor sensor. Two types are in common use: **CCD** (*charge-coupled device*) and **CMOS** (*complementary metal oxide semiconductor*). A CCD sensor is made up of millions of tiny **pixels** ("picture elements")—see Figs. 35–42 and 33–19. A 6-MP (6-megapixel) sensor[‡] would contain about 2000 pixels vertically by 3000 pixels horizontally over an area of perhaps 4×6 mm or even 24×36 mm. Light reaching any pixel liberates electrons from the semiconductor. The more intense the light, the more charge accumulates during the brief exposure time. Conducting electrodes carry each pixel's charge (serially in time, row by row—hence the name "charge-coupled") to a central processor that stores the relative brightness of pixels, and allows reformation of the image later onto a computer screen or printer. A CCD is fully reusable. Once the pixel charges are transferred to memory, a new picture can be taken.

A CMOS sensor also uses a silicon semiconductor, and incorporates some electronics within each pixel, allowing parallel readout.

Color is achieved by red, green, and blue filters over alternating pixels as shown in Fig. 33–19, similar to a color CRT or LCD screen. The sensor type shown in Fig. 33–19 contains twice as many green pixels as red or blue (because green is claimed to have a stronger influence on the sensation of sharpness). The computer-analyzed color at each pixel is that pixel's intensity averaged with the intensities of the nearest-neighbor colors.

To reduce the amount of memory for each picture, compression programs average over pixels, but with a consequent loss of sharpness, or "resolution."

*Digital Artifacts

Digital cameras can produce image artifacts (artificial effects in the image not present in the original) resulting from the imaging process. One example using the "mosaic" of pixels (Fig. 33–19) is described in Fig. 33–20. An alternative technology uses a semitransparent silicon semiconductor layer system wherein different wavelengths of light penetrate silicon to different depths: each pixel is a sandwich of partly transparent layers, one for each color. The top layer can absorb blue light, allowing green and red light to pass through. The second layer absorbs green and the bottom layer detects the red. All three colors are detected by each pixel, resulting in better color resolution and fewer artifacts.

[†]This is called a *negative*, because the black areas correspond to bright objects and vice versa. The same process occurs during printing to produce a black-and-white "positive" picture from the negative. Color film has three emulsion layers (or dyes) corresponding to the three primary colors.

[‡]Each different color of pixel in a CCD is counted as a separate pixel. In contrast, in an LCD screen (Section 35–12), a group of three subpixels is counted as one pixel, a more conservative count.

Camera Adjustments

There are three main adjustments on good-quality cameras: shutter speed, f-stop, and focusing. Although most cameras today make these adjustments automatically, it is valuable to understand these adjustments to use a camera effectively. For special or top-quality work, a manual camera is indispensable (Fig. 33–21).

Exposure time or shutter speed   This refers to how quickly the digital sensor can make an accurate reading, or how long the shutter of a camera is open and the film or sensor exposed. It could vary from a second or more ("time exposures"), to $\frac{1}{1000}$ s or less. To avoid blurring from camera movement, exposure times shorter than $\frac{1}{100}$ s are normally needed. If the object is moving, even shorter exposure times are needed to "stop" the action. If the exposure (or sampling) time is not fast enough, the image will be blurred by camera shake no matter how many pixels a digital camera claims. Blurring in low light conditions is more of a problem with cell-phone cameras whose sensors are not usually the most sophisticated. Digital still cameras or cell phones that take short videos must have a fast enough "sampling" time and fast "clearing" (of the charge) time so as to take pictures at least 15 frames per second, and preferably 30 fps.

f-stop   The amount of light reaching the film must be carefully controlled to avoid **underexposure** (too little light so the picture is dark and only the brightest objects show up) or **overexposure** (too much light, so that all bright objects look the same, with a consequent lack of contrast and a "washed-out" appearance). Most cameras these days make f-stop and shutter speed adjustments automatically. A high quality camera controls the exposure with a "stop" or iris diaphragm, whose opening is of variable diameter, placed behind the lens (Fig. 33–18). The size of the opening is controlled (automatically or manually) to compensate for bright or dark lighting conditions, the sensitivity of the sensor or film,[†] and for different shutter speeds. The size of the opening is specified by the **f-number** or **f-stop**, defined as

$$f\text{-stop} = \frac{f}{D},$$

where f is the focal length of the lens and D is the diameter of the lens opening (Fig. 33–18). For example, when a 50-mm-focal-length lens has an opening $D = 25$ mm, we say it is set at $f/2$. When this lens is set at $f/8$, the opening is only $6\frac{1}{4}$ mm $\left(50/6\frac{1}{4} = 8\right)$. For faster shutter speeds, or low light conditions, a wider lens opening must be used to get a proper exposure, which corresponds to a smaller f-stop number. The smaller the f-stop number, the larger the opening and the more light passes through the lens to the sensor or film. The smallest f-number of a lens (largest opening) is referred to as the *speed* of the lens. The best lenses may have a speed of $f/2.0$, or even faster. The advantage of a fast lens is that it allows pictures to be taken under poor lighting conditions. Good quality lenses consist of several elements to reduce the defects present in simple thin lenses (Section 33–10). Standard f-stops are

$$1.0, \quad 1.4, \quad 2.0, \quad 2.8, \quad 4.0, \quad 5.6, \quad 8, \quad 11, \quad 16, \quad 22, \quad \text{and} \quad 32$$

(Fig. 33–21). Each of these stops corresponds to a diameter reduction by a factor of about $\sqrt{2} \approx 1.4$. Because the amount of light reaching the film is proportional to the *area* of the opening, and therefore proportional to the diameter squared, each standard f-stop corresponds to a factor of 2 in light intensity reaching the film.

Focusing   Focusing is the operation of placing the lens at the correct position relative to the sensor or film for the sharpest image. The image distance is smallest for objects at infinity (the symbol ∞ is used for infinity) and is equal to the focal length. For closer objects, the image distance is greater than the focal length, as can be seen from the lens equation, $1/f = 1/d_\mathrm{o} + 1/d_\mathrm{i}$ (Eq. 33–2). To focus on nearby objects, the lens must therefore be moved away from the sensor or film, and this is usually done on a manual camera by turning a ring on the lens.

FIGURE 33–21   On this camera, the f-stops and the focusing ring are on the camera lens. Shutter speeds are selected on the small wheel on top of the camera body.

[†]Different films have different sensitivities to light, referred to as the "film speed" and specified as an "ISO (or ASA) number." A "faster" film is more sensitive and needs less light to produce a good image. Faster films are grainier so offer less sharpness (resolution) when enlarged. Digital cameras may have a "gain" or "ISO" adjustment for sensitivity. Adjusting a CCD to be "faster" for low light conditions results in "noise," the digital equivalent of graininess.

FIGURE 33–22 Photos taken with a camera lens (a) focused on a nearby object with distant object blurry, and (b) focused on a more distant object with nearby object blurry.

(a)

(b)

If the lens is focused on a nearby object, a sharp image of it will be formed, but the image of distant objects may be blurry (Fig. 33–22). The rays from a point on the distant object will be out of focus—they will form a circle on the sensor or film as shown (exaggerated) in Fig. 33–23. The distant object will thus produce an image consisting of overlapping circles and will be blurred. These circles are called **circles of confusion**. To include near and distant objects in the same photo, you (or the camera) can try setting the lens focus at an intermediate position. For a given distance setting, there is a range of distances over which the circles of confusion will be small enough that the images will be reasonably sharp. This is called the **depth of field**. The depth of field varies with the lens opening. If the lens opening is smaller, only rays through the central part of the lens are accepted, and these form smaller circles of confusion for a given object distance. Hence, at smaller lens openings, a greater range of object distances will fit within the circle of confusion criterion, so the depth of field is greater.[†] For a sensor or film width of 36 mm (including 35-mm film cameras), the depth of field is usually based on a maximum circle of confusion diameter of 0.003 mm.

FIGURE 33–23 When the lens is positioned to focus on a nearby object, points on a distant object produce circles and are therefore blurred. (The effect is shown greatly exaggerated.)

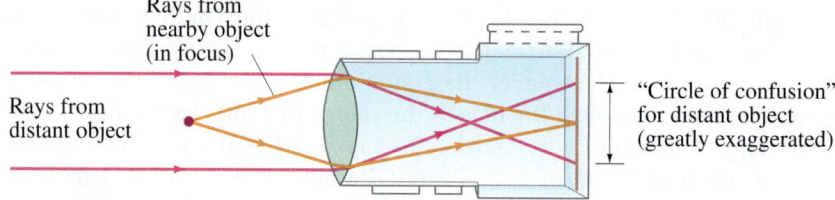

Rays from nearby object (in focus)

Rays from distant object

"Circle of confusion" for distant object (greatly exaggerated)

EXAMPLE 33–8 **Camera focus.** How far must a 50.0-mm-focal-length camera lens be moved from its infinity setting to sharply focus an object 3.00 m away?

APPROACH For an object at infinity, the image is at the focal point, by definition. For an object distance of 3.00 m, we use the thin lens equation, Eq. 33–2, to find the image distance (distance of lens to film or sensor).

SOLUTION When focused at infinity, the lens is 50.0 mm from the film. When focused at $d_o = 3.00$ m, the image distance is given by the lens equation,

$$\frac{1}{d_i} = \frac{1}{f} - \frac{1}{d_o} = \frac{1}{50.0 \text{ mm}} - \frac{1}{3000 \text{ mm}} = \frac{3000 - 50}{(3000)(50.0) \text{ mm}} = \frac{2950}{150,000 \text{ mm}}.$$

We solve for d_i and find $d_i = 50.8$ mm, so the lens needs to move 0.8 mm away from the film or digital sensor.

[†]Smaller lens openings, however, result in reduced resolution due to diffraction (discussed in Chapter 35). Best resolution is typically found around $f/8$.

CONCEPTUAL EXAMPLE 33–9 **Shutter speed.** To improve the depth of field, you "stop down" your camera lens by two f-stops from $f/4$ to $f/8$. What should you do to the shutter speed to maintain the same exposure?

RESPONSE The amount of light admitted by the lens is proportional to the area of the lens opening. Reducing the lens opening by two f-stops reduces the diameter by a factor of 2, and the area by a factor of 4. To maintain the same exposure, the shutter must be open four times as long. If the shutter speed had been $\frac{1}{500}$ s, you would have to increase the exposure time to $\frac{1}{125}$ s.

*Picture Sharpness

The sharpness of a picture depends not only on accurate focusing and short exposure times, but also on the graininess of the film, or the number of pixels for a digital camera. Fine-grained films are "slower," meaning they require longer exposures for a given light level. Digital cameras have averaging (or "compression") programs, such as JPEG, which reduce memory size by averaging over pixels where little contrast is detected. Hence it is unusual to use all pixels available. They also average over pixels in low light conditions, resulting in a less sharp photo.

The quality of the lens strongly affects the image quality, and we discuss lens resolution and diffraction effects in Sections 33–10 and 35–4. The sharpness, or *resolution*, of a lens is often given as so many lines per millimeter, measured by photographing a standard set of parallel lines on fine-grain film or high quality sensor, or as so many dots per inch (dpi). The minimum spacing of distinguishable lines or dots gives the resolution; 50 lines/mm is reasonable, 100 lines/mm is quite good ($= 100$ dots/mm ≈ 2500 dpi on the sensor).

EXAMPLE 33–10 **Pixels and resolution.** A 6-MP (6-megapixel) digital camera offers a maximum resolution of 2000 × 3000 pixels on a 16-mm × 24-mm CCD sensor. How sharp should the lens be to make use of this resolution?

APPROACH We find the number of pixels per millimeter and require the lens to be at least that good.

SOLUTION We can either take the image height (2000 pixels in 16 mm) or the width (3000 pixels in 24 mm):

$$\frac{3000 \text{ pixels}}{24 \text{ mm}} = 125 \text{ pixels/mm.}$$

We would want the lens to be able to resolve at least 125 lines or dots per mm as well, which would be a very good lens. If the lens is not this good, fewer pixels and less memory could be used.

NOTE Increasing lens resolution is a tougher problem today than is squeezing more pixels on a CCD or CMOS. The sensor for high MP cameras must also be physically larger for greater light sensitivity (low light conditions).

EXAMPLE 33–11 **Blown-up photograph.** An enlarged photograph looks sharp at normal viewing distances if the dots or lines are resolved to about 10 dots/mm. Would an 8 × 10-inch enlargement of a photo taken by the camera in Example 33–10 seem sharp? To what maximum size could you enlarge this 2000 × 3000-pixel image?

APPROACH We assume the image is 2000 × 3000 pixels on a 16 × 24-mm CCD as in Example 33–10, or 125 pixels/mm. We make an enlarged photo 8 × 10 in. = 20 cm × 25 cm.

SOLUTION The short side of the CCD is 16 mm = 1.6 cm long, and that side of the photograph is 8 inches or 20 cm. Thus the size is increased by a factor of 20 cm/1.6 cm = 12.5× (or 25 cm/2.4 cm ≈ 10×). To fill the 8 × 10-in. paper, we assume the enlargement is 12.5×. The pixels are thus enlarged 12.5×; so the pixel count of 125/mm on the CCD becomes 10 per mm on the print. Hence an 8 × 10-inch print is just about the maximum possible for a sharp photograph with 6 megapixels. If you feel 7 dots per mm is good enough, you can enlarge to maybe 11 × 14 inches.

*Telephotos and Wide-angles

Camera lenses are categorized into normal, telephoto, and wide angle, according to focal length and film size. A **normal lens** covers the sensor or film with a field of view that corresponds approximately to that of normal vision. A normal lens for 35-mm film has a focal length in the vicinity of 50 mm. The best digital cameras aim for a sensor of the same size[†] (24 mm × 36mm). (If the sensor is smaller, digital cameras sometimes specify focal lengths to correspond with classic 35-mm cameras.) **Telephoto lenses** act like telescopes to magnify images. They have longer focal lengths than a normal lens: as we saw in Section 33–2 (Eq. 33–3), the height of the image for a given object distance is proportional to the image distance, and the image distance will be greater for a lens with longer focal length. For distant objects, the image height is very nearly proportional to the focal length. Thus a 200-mm telephoto lens for use with a 35-mm camera gives a 4× magnification over the normal 50-mm lens. A **wide-angle lens** has a shorter focal length than normal: a wider field of view is included, and objects appear smaller. A **zoom lens** is one whose focal length can be changed so that you seem to zoom up to, or away from, the subject as you change the focal length.

Digital cameras may have an **optical zoom** meaning the lens can change focal length and maintain resolution. But an "electronic" or **digital zoom** just enlarges the dots (pixels) with loss of sharpness.

Different types of viewing systems are used in cameras. In some cameras, you view through a small window just above the lens as in Fig. 33–18. In a **single-lens reflex** camera (SLR), you actually view through the lens with the use of prisms and mirrors (Fig. 33–24). A mirror hangs at a 45° angle behind the lens and flips up out of the way just before the shutter opens. SLRs have the advantage that you can see almost exactly what you will get. Digital cameras use an LCD display, and it too can show what you will get on the photo if it is carefully designed.

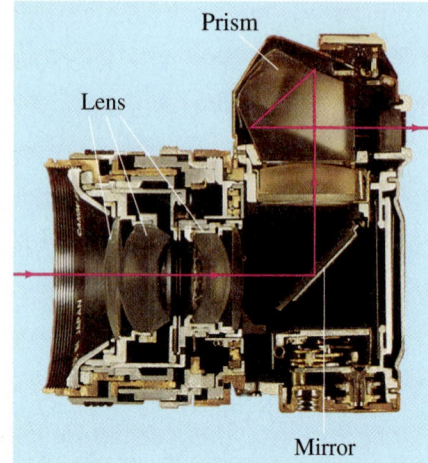

FIGURE 33–24 Single-lens reflex (SLR) camera, showing how the image is viewed through the lens with the help of a movable mirror and prism.

PHYSICS APPLIED
The eye

FIGURE 33–25 Diagram of a human eye.

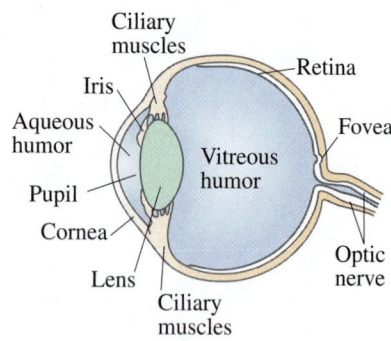

33–6 The Human Eye; Corrective Lenses

The human eye resembles a camera in its basic structure (Fig. 33–25), but is far more sophisticated. The interior of the eye is filled with a transparent gel-like substance called the *vitreous humor* with index of refraction $n = 1.337$. Light enters this enclosed volume through the cornea and lens. Between the cornea and lens is a watery fluid, the aqueous humor (*aqua* is "water" in Latin) with $n = 1.336$. A diaphragm, called the **iris** (the colored part of your eye), adjusts automatically to control the amount of light entering the eye, similar to a camera. The hole in the iris through which light passes (the **pupil**) is black because no light is reflected from it (it's a hole), and very little light is reflected back out from the interior of the eye. The **retina**, which plays the role of the film or sensor in a camera, is on the curved rear surface of the eye. The retina consists of a complex array of nerves and receptors known as *rods* and *cones* which act to change light energy into electrical signals that travel along the nerves. The reconstruction of the image from all these tiny receptors is done mainly in the brain, although some analysis may also be done in the complex interconnected nerve network at the retina itself. At the center of the retina is a small area called the **fovea**, about 0.25 mm in diameter, where the cones are very closely packed and the sharpest image and best color discrimination are found.

Unlike a camera, the eye contains no shutter. The equivalent operation is carried out by the nervous system, which analyzes the signals to form images at the rate of about 30 per second. This can be compared to motion picture or television cameras, which operate by taking a series of still pictures at a rate of 24 (movies) or 30 (U.S. television) per second; their rapid projection on the screen gives the appearance of motion.

[†]A "35-mm camera" uses film that is physically 35 mm wide; that 35 mm is not to be confused with a focal length. 35-mm film has sprocket holes, so only 24 mm of its height is used for the photo; the width is usually 36 mm for stills. Thus one frame is 24 mm × 36 mm. Movie frames are 18 mm × 24 mm.

The lens of the eye ($n = 1.386$ to 1.406) does little of the bending of the light rays. Most of the refraction is done at the front surface of the **cornea** ($n = 1.376$) at its interface with air ($n = 1.0$). The lens acts as a fine adjustment for focusing at different distances. This is accomplished by the ciliary muscles (Fig. 33–25), which change the curvature of the lens so that its focal length is changed. To focus on a distant object, the ciliary muscles of the eye are relaxed and the lens is thin, as shown in Fig. 33–26a, and parallel rays focus at the focal point (on the retina). To focus on a nearby object, the muscles contract, causing the center of the lens to thicken, Fig. 33–26b, thus shortening the focal length so that images of nearby objects can be focused on the retina, behind the new focal point. This focusing adjustment is called **accommodation**.

The closest distance at which the eye can focus clearly is called the **near point** of the eye. For young adults it is typically 25 cm, although younger children can often focus on objects as close as 10 cm. As people grow older, the ability to accommodate is reduced and the near point increases. A given person's **far point** is the farthest distance at which an object can be seen clearly. For some purposes it is useful to speak of a **normal eye** (a sort of average over the population), defined as an eye having a near point of 25 cm and a far point of infinity. To check your own near point, place this book close to your eye and slowly move it away until the type is sharp.

The "normal" eye is sort of an ideal. Many people have eyes that do not accommodate within the "normal" range of 25 cm to infinity, or have some other defect. Two common defects are nearsightedness and farsightedness. Both can be corrected to a large extent with lenses—either eyeglasses or contact lenses.

In **nearsightedness**, or *myopia*, the eye can focus only on nearby objects. The far point is not infinity but some shorter distance, so distant objects are not seen clearly. It is usually caused by an eyeball that is too long, although sometimes it is the curvature of the cornea that is too great. In either case, images of distant objects are focused in front of the retina. A diverging lens, because it causes parallel rays to diverge, allows the rays to be focused at the retina (Fig. 33–27a) and thus corrects this defect.

In **farsightedness**, or *hyperopia*, the eye cannot focus on nearby objects. Although distant objects are usually seen clearly, the near point is somewhat greater than the "normal" 25 cm, which makes reading difficult. This defect is caused by an eyeball that is too short or (less often) by a cornea that is not sufficiently curved. It is corrected by a converging lens, Fig. 33–27b. Similar to hyperopia is *presbyopia*, which refers to the lessening ability of the eye to accommodate as a person ages, and the near point moves out. Converging lenses also compensate for this.

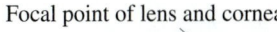
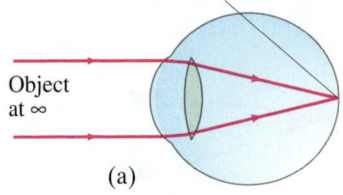

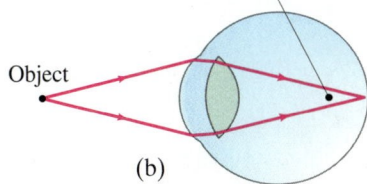

FIGURE 33–26 Accommodation by a normal eye: (a) lens relaxed, focused at infinity; (b) lens thickened, focused on a nearby object.

PHYSICS APPLIED
Corrective lenses

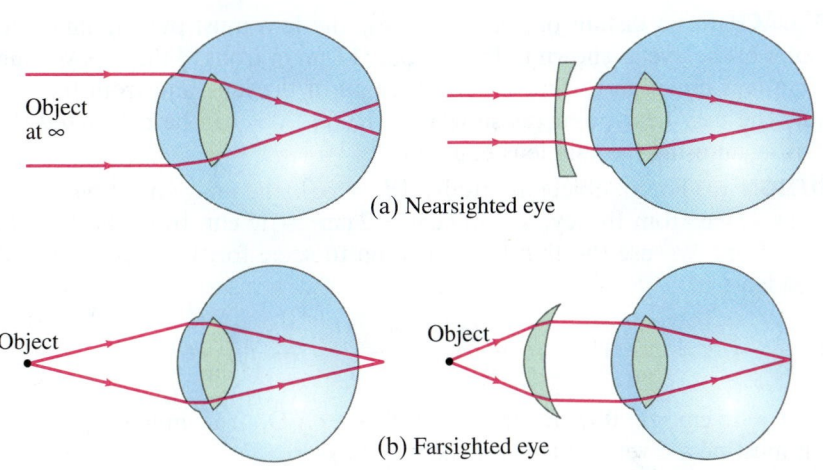

(a) Nearsighted eye

(b) Farsighted eye

FIGURE 33–27 Correcting eye defects with lenses: (a) a nearsighted eye, which cannot focus clearly on distant objects, can be corrected by use of a diverging lens; (b) a farsighted eye, which cannot focus clearly on nearby objects, can be corrected by use of a converging lens.

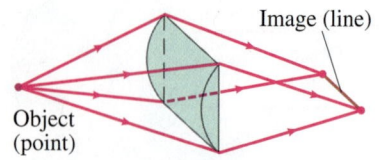

FIGURE 33–28 A cylindrical lens forms a line image of a point object because it is converging in one plane only.

Astigmatism is usually caused by an out-of-round cornea or lens so that point objects are focused as short lines, which blurs the image. It is as if the cornea were spherical with a cylindrical section superimposed. As shown in Fig. 33–28, a cylindrical lens focuses a point into a line parallel to its axis. An astigmatic eye may focus rays in one plane, such as the vertical plane, at a shorter distance than it does for rays in a horizontal plane. Astigmatism is corrected with the use of a compensating cylindrical lens. Lenses for eyes that are nearsighted or farsighted as well as astigmatic are ground with superimposed spherical and cylindrical surfaces, so that the radius of curvature of the correcting lens is different in different planes.

EXAMPLE 33–12 Farsighted eye. Sue is farsighted with a near point of 100 cm. Reading glasses must have what lens power so that she can read a newspaper at a distance of 25 cm? Assume the lens is very close to the eye.

APPROACH When the object is placed 25 cm from the lens, we want the image to be 100 cm away on the *same* side of the lens (so the eye can focus it), and so the image is virtual, as shown in Fig. 33–29, and $d_i = -100$ cm will be negative. We use the thin lens equation (Eq. 33–2) to determine the needed focal length. Optometrists' prescriptions specify the power ($P = 1/f$, Eq. 33–1) given in diopters $(1\,\text{D} = 1\,\text{m}^{-1})$.

SOLUTION Given that $d_o = 25$ cm and $d_i = -100$ cm, the thin lens equation gives

$$\frac{1}{f} = \frac{1}{d_o} + \frac{1}{d_i} = \frac{1}{25\,\text{cm}} + \frac{1}{-100\,\text{cm}} = \frac{4-1}{100\,\text{cm}} = \frac{1}{33\,\text{cm}}.$$

So $f = 33$ cm $= 0.33$ m. The power P of the lens is $P = 1/f = +3.0$ D. The plus sign indicates that it is a converging lens.

NOTE We chose the image position to be where the eye can actually focus. The lens needs to put the image there, given the desired placement of the object (newspaper).

FIGURE 33–29 Lens of reading glasses (Example 33–12).

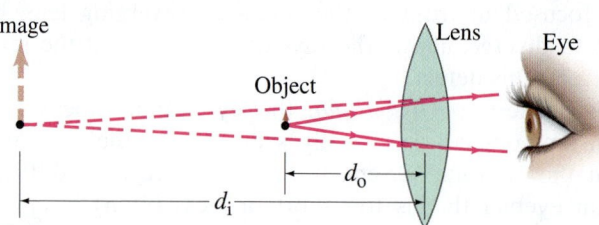

FIGURE 33–30 Example 33–13.

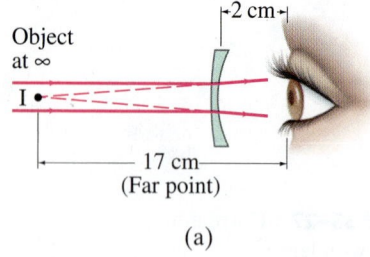

(a)

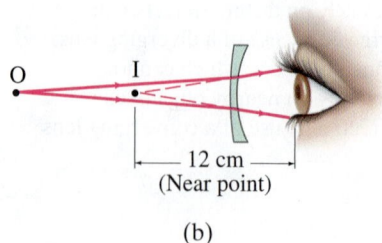

(b)

EXAMPLE 33–13 Nearsighted eye. A nearsighted eye has near and far points of 12 cm and 17 cm, respectively. (*a*) What lens power is needed for this person to see distant objects clearly, and (*b*) what then will be the near point? Assume that the lens is 2.0 cm from the eye (typical for eyeglasses).

APPROACH For a distant object $(d_o = \infty)$, the lens must put the image at the far point of the eye as shown in Fig. 33–30a, 17 cm in front of the eye. We can use the thin lens equation to find the focal length of the lens, and from this its lens power. The new near point (as shown in Fig. 33–30b) can be calculated for the lens by again using the thin lens equation.

SOLUTION (*a*) For an object at infinity $(d_o = \infty)$, the image must be in front of the lens 17 cm from the eye or $(17\,\text{cm} - 2\,\text{cm}) = 15\,\text{cm}$ from the lens; hence $d_i = -15$ cm. We use the thin lens equation to solve for the focal length of the needed lens:

$$\frac{1}{f} = \frac{1}{d_o} + \frac{1}{d_i} = \frac{1}{\infty} + \frac{1}{-15\,\text{cm}} = -\frac{1}{15\,\text{cm}}.$$

So $f = -15$ cm $= -0.15$ m or $P = 1/f = -6.7$ D. The minus sign indicates that it must be a diverging lens for the myopic eye.

(b) The near point when glasses are worn is where an object is placed (d_o) so that the lens forms an image at the "near point of the naked eye," namely 12 cm from the eye. That image point is $(12\,\text{cm} - 2\,\text{cm}) = 10\,\text{cm}$ in front of the lens, so $d_i = -0.10\,\text{m}$ and the thin lens equation gives

$$\frac{1}{d_o} = \frac{1}{f} - \frac{1}{d_i} = -\frac{1}{0.15\,\text{m}} + \frac{1}{0.10\,\text{m}} = \frac{-2 + 3}{0.30\,\text{m}} = \frac{1}{0.30\,\text{m}}.$$

So $d_o = 30\,\text{cm}$, which means the near point when the person is wearing glasses is 30 cm in front of the lens, or 32 cm from the eye.

Suppose contact lenses are used to correct the eye in Example 33–13. Since contacts are placed directly on the cornea, we would not subtract out the 2.0 cm for the image distances. That is, for distant objects $d_i = f = -17\,\text{cm}$, so $P = 1/f = -5.9\,\text{D}$. The new near point would be 41 cm. Thus we see that a contact lens and an eyeglass lens will require slightly different powers, or focal lengths, for the same eye because of their different placements relative to the eye. We also see that glasses in this case give a better near point than contacts.

PHYSICS APPLIED
Contact lenses

EXERCISE E What power contact lens is needed for an eye to see distant objects if its far point is 25 cm?

When your eyes are under water, distant underwater objects look blurry because at the water–cornea interface, the difference in indices of refraction is very small: $n = 1.33$ for water, 1.376 for the cornea. Hence light rays are bent very little and are focused far behind the retina, Fig. 33–31a. If you wear goggles or a face mask, you restore an air–cornea interface ($n = 1.0$ and 1.376, respectively) and the rays can be focused, Fig. 33–31b.

PHYSICS APPLIED
Underwater vision

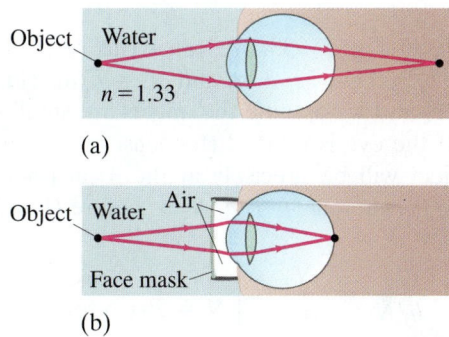

(a)

(b)

FIGURE 33–31 (a) Under water, we see a blurry image because light rays are bent much less than in air. (b) If we wear goggles, we again have an air–cornea interface and can see clearly.

33–7 Magnifying Glass

Much of the remainder of this Chapter will deal with optical devices that are used to produce magnified images of objects. We first discuss the **simple magnifier**, or **magnifying glass**, which is simply a converging lens (see Chapter-Opening Photo).

How large an object appears, and how much detail we can see on it, depends on the size of the image it makes on the retina. This, in turn, depends on the angle subtended by the object at the eye. For example, a penny held 30 cm from the eye looks twice as tall as one held 60 cm away because the angle it subtends is twice as great (Fig. 33–32). When we want to examine detail on an object, we bring it up close to our eyes so that it subtends a greater angle. However, our eyes can accommodate only up to a point (the near point), and we will assume a standard distance of $N = 25\,\text{cm}$ as the near point in what follows.

FIGURE 33–32 When the same object is viewed at a shorter distance, the image on the retina is greater, so the object appears larger and more detail can be seen. The angle θ that the object subtends in (a) is greater than in (b). *Note:* This is not a normal ray diagram because we are showing only one ray from each point.

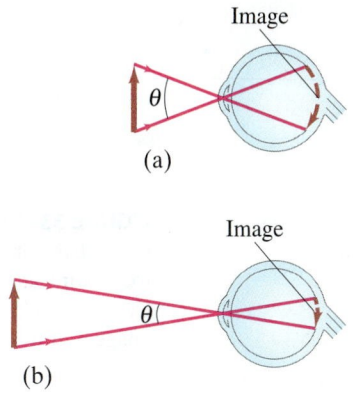

(a)

(b)

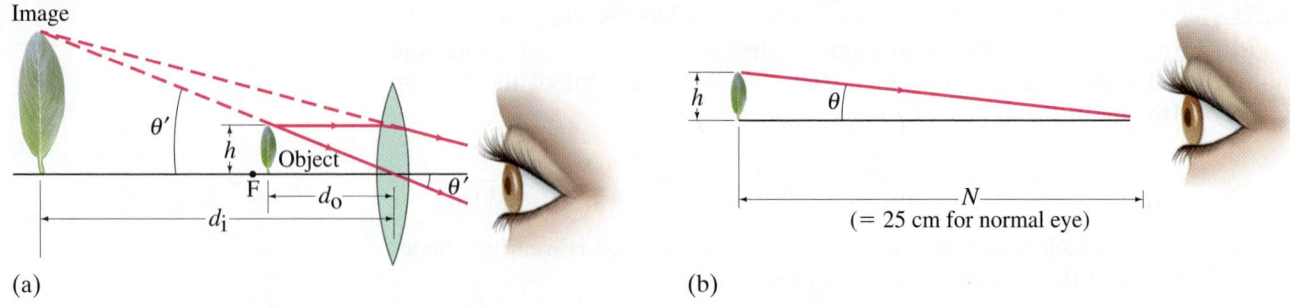

(a)

(b)

FIGURE 33–33 Leaf viewed (a) through a magnifying glass, and (b) with the unaided eye. The eye is focused at its near point in both cases.

A magnifying glass allows us to place the object closer to our eye so that it subtends a greater angle. As shown in Fig. 33–33a, the object is placed at the focal point or just within it. Then the converging lens produces a virtual image, which must be at least 25 cm from the eye if the eye is to focus on it. If the eye is relaxed, the image will be at infinity, and in this case the object is exactly at the focal point. (You make this slight adjustment yourself when you "focus" on the object by moving the magnifying glass.)

A comparison of part (a) of Fig. 33–33 with part (b), in which the same object is viewed at the near point with the unaided eye, reveals that the angle the object subtends at the eye is much larger when the magnifier is used. The **angular magnification** or **magnifying power**, M, of the lens is defined as the ratio of the angle subtended by an object when using the lens, to the angle subtended using the unaided eye, with the object at the near point N of the eye ($N = 25$ cm for a normal eye):

$$M = \frac{\theta'}{\theta}, \tag{33–5}$$

where θ and θ' are shown in Fig. 33–33. We can write M in terms of the focal length by noting that $\theta = h/N$ (Fig. 33–33b) and $\theta' = h/d_o$ (Fig. 33–33a), where h is the height of the object and we assume the angles are small so θ and θ' equal their sines and tangents. If the eye is relaxed (for least eye strain), the image will be at infinity and the object will be precisely at the focal point; see Fig. 33–34. Then $d_o = f$ and $\theta' = h/f$, whereas $\theta = h/N$ as before (Fig. 33–33b). Thus

$$M = \frac{\theta'}{\theta} = \frac{h/f}{h/N} = \frac{N}{f}. \quad \left[\begin{array}{l} \text{eye focused at } \infty; \\ N = 25 \text{ cm for normal eye} \end{array} \right] \tag{33–6a}$$

We see that the shorter the focal length of the lens, the greater the magnification.[†]

The magnification of a given lens can be increased a bit by moving the lens and adjusting your eye so it focuses on the image at the eye's near point. In this case, $d_i = -N$ (see Fig. 33–33a) if your eye is very near the magnifier.

[†]Simple single-lens magnifiers are limited to about 2 or 3× because of blurring due to spherical aberration (Section 33–10).

FIGURE 33–34 With the eye relaxed, the object is placed at the focal point, and the image is at infinity. Compare to Fig. 33–33a where the image is at the eye's near point.

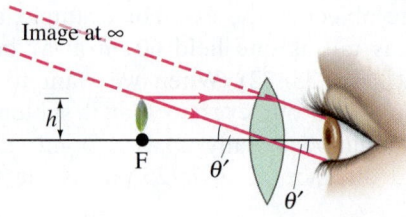

Then the object distance d_o is given by

$$\frac{1}{d_o} = \frac{1}{f} - \frac{1}{d_i} = \frac{1}{f} + \frac{1}{N}.$$

We see from this equation that $d_o = fN/(f + N) < f$, as shown in Fig. 33–33a, since $N/(f + N)$ must be less than 1. With $\theta' = h/d_o$ the magnification is

$$M = \frac{\theta'}{\theta} = \frac{h/d_o}{h/N} = \frac{N}{d_o} = N\left(\frac{1}{f} + \frac{1}{N}\right)$$

or

$$M = \frac{N}{f} + 1. \qquad \left[\begin{array}{c}\text{eye focused at near point, } N; \\ N = 25 \text{ cm for normal eye}\end{array}\right] \quad \textbf{(33–6b)}$$

We see that the magnification is slightly greater when the eye is focused at its near point, as compared to when it is relaxed.

EXAMPLE 33–14 **ESTIMATE** **A jeweler's "loupe."** An 8-cm-focal-length converging lens is used as a "jeweler's loupe," which is a magnifying glass. Estimate (a) the magnification when the eye is relaxed, and (b) the magnification if the eye is focused at its near point $N = 25$ cm.

APPROACH The magnification when the eye is relaxed is given by Eq. 33–6a. When the eye is focused at its near point, we use Eq. 33–6b and we assume the lens is near the eye.

SOLUTION (a) With the relaxed eye focused at infinity,

$$M = \frac{N}{f} = \frac{25 \text{ cm}}{8 \text{ cm}} \approx 3\times.$$

(b) The magnification when the eye is focused at its near point $(N = 25 \text{ cm})$, and the lens is near the eye, is

$$M = 1 + \frac{N}{f} = 1 + \frac{25}{8} \approx 4\times.$$

33–8 Telescopes

A telescope is used to magnify objects that are very far away. In most cases, the object can be considered to be at infinity.

Galileo, although he did not invent it,[†] developed the telescope into a usable and important instrument. He was the first to examine the heavens with the telescope (Fig. 33–35), and he made world-shaking discoveries, including the moons of Jupiter, the phases of Venus, sunspots, the structure of the Moon's surface, and that the Milky Way is made up of a huge number of individual stars.

[†]Galileo built his first telescope in 1609 after having heard of such an instrument existing in Holland. The first telescopes magnified only three to four times, but Galileo soon made a 30-power instrument. The first Dutch telescope seems to date from about 1604, but there is a reference suggesting it may have been copied from an Italian telescope built as early as 1590. Kepler (see Chapter 6) gave a ray description (in 1611) of the Keplerian telescope, which is named for him because he first described it, although he did not build it.

FIGURE 33–35 (a) Objective lens (mounted now in an ivory frame) from the telescope with which Galileo made his world-shaking discoveries, including the moons of Jupiter. (b) Later telescopes made by Galileo.

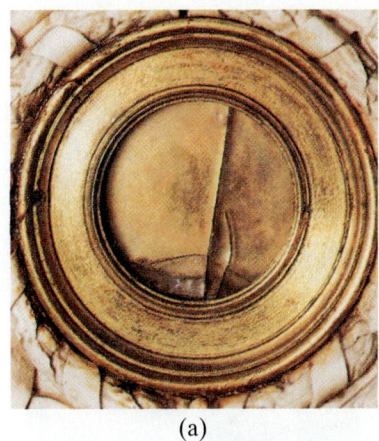

(a)

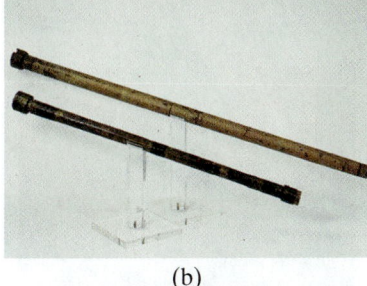

(b)

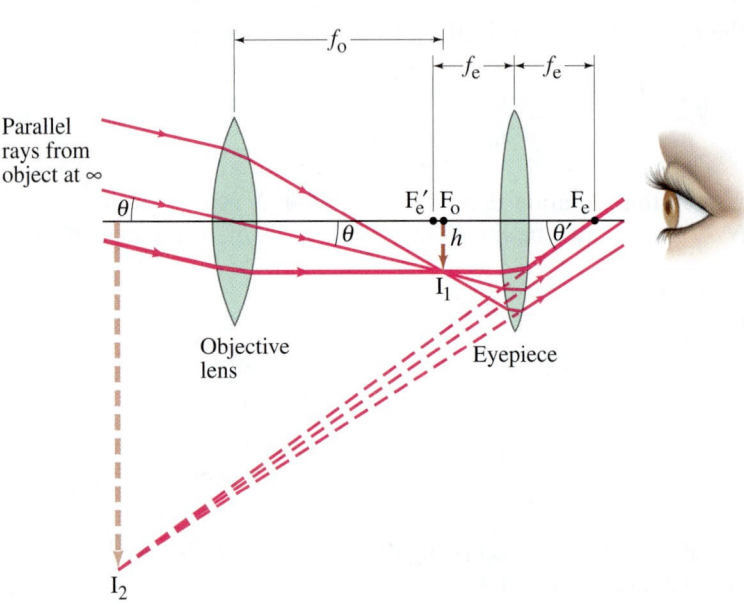

FIGURE 33–36 Astronomical telescope (refracting). Parallel light from one point on a distant object $(d_o = \infty)$ is brought to a focus by the objective lens in its focal plane. This image (I_1) is magnified by the eyepiece to form the final image I_2. Only two of the rays shown entering the objective are standard rays (2 and 3) as described in Fig. 33–6.

Several types of **astronomical telescope** exist. The common **refracting** type, sometimes called **Keplerian**, contains two converging lenses located at opposite ends of a long tube, as illustrated in Fig. 33–36. The lens closest to the object is called the **objective lens** (focal length f_o) and forms a real image I_1 of the distant object in the plane of its focal point F_o (or near it if the object is not at infinity). The second lens, called the **eyepiece** (focal length f_e), acts as a magnifier. That is, the eyepiece magnifies the image I_1 formed by the objective lens to produce a second, greatly magnified image, I_2, which is virtual and inverted. If the viewing eye is relaxed, the eyepiece is adjusted so the image I_2 is at infinity. Then the real image I_1 is at the focal point F_e' of the eyepiece, and the distance between the lenses is $f_o + f_e$ for an object at infinity.

To find the total angular magnification of this telescope, we note that the angle an object subtends as viewed by the unaided eye is just the angle θ subtended at the telescope objective. From Fig. 33–36 we can see that $\theta \approx h/f_o$, where h is the height of the image I_1 and we assume θ is small so that $\tan\theta \approx \theta$. Note, too, that the thickest of the three rays drawn in Fig. 33–36 is parallel to the axis before it strikes the eyepiece and therefore is refracted through the eyepiece focal point F_e on the far side. Thus, $\theta' \approx h/f_e$ and the **total magnifying power** (that is, angular magnification, which is what is always quoted) of this telescope is

$$M = \cdot\frac{\theta'}{\theta} = \frac{(h/f_e)}{(h/f_o)} = -\frac{f_o}{f_e},$$ [telescope] **(33–7)**

where we have inserted a minus sign to indicate that the image is inverted. To achieve a large magnification, the objective lens should have a long focal length and the eyepiece a short focal length.

FIGURE 33–37 This large refracting telescope was built in 1897 and is housed at Yerkes Observatory in Wisconsin. The objective lens is 102 cm (40 inches) in diameter, and the telescope tube is about 19 m long. Example 33–15.

EXAMPLE 33–15 **Telescope magnification.** The largest optical refracting telescope in the world is located at the Yerkes Observatory in Wisconsin, Fig. 33–37. It is referred to as a "40-inch" telescope, meaning that the diameter of the objective is 40 in., or 102 cm. The objective lens has a focal length of 19 m, and the eyepiece has a focal length of 10 cm. (a) Calculate the total magnifying power of this telescope. (b) Estimate the length of the telescope.

APPROACH Equation 33–7 gives the magnification. The length of the telescope is the distance between the two lenses.

SOLUTION (a) From Eq. 33–7 we find

$$M = -\frac{f_o}{f_e} = -\frac{19\text{ m}}{0.10\text{ m}} = -190\times.$$

(b) For a relaxed eye, the image I_1 is at the focal point of both the eyepiece and the objective lenses. The distance between the two lenses is thus $f_o + f_e \approx 19$ m, which is essentially the length of the telescope.

EXERCISE F A 40× telescope has a 1.2-cm focal length eyepiece. What is the focal length of the objective lens?

For an astronomical telescope to produce bright images of faint stars, the objective lens must be large to allow in as much light as possible. Indeed, the diameter of the objective lens (and hence its "light-gathering power") is an important parameter for an astronomical telescope, which is why the largest ones are specified by giving the objective diameter (such as the 10-meter Keck telescope in Hawaii). The construction and grinding of large lenses is very difficult. Therefore, the largest telescopes are **reflecting telescopes** which use a curved mirror as the objective, Fig. 33–38. A mirror has only one surface to be ground and can be supported along its entire surface[†] (a large lens, supported at its edges, would sag under its own weight). Often, the eyepiece lens or mirror (see Fig. 33–38) is removed so that the real image formed by the objective mirror can be recorded directly on film or on an electronic sensor (CCD or CMOS, Section 33–5).

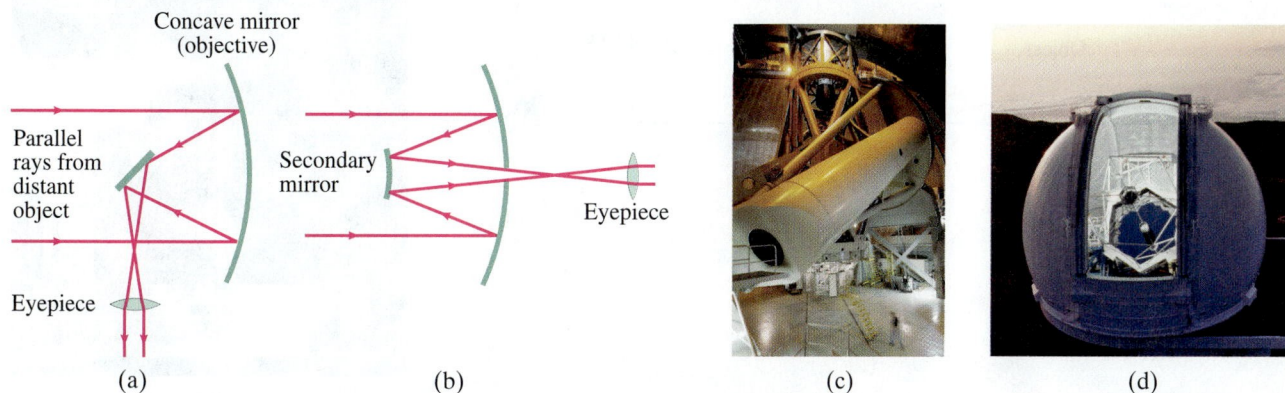

FIGURE 33–38 A concave mirror can be used as the objective of an astronomical telescope. Arrangement (a) is called the Newtonian focus, and (b) the Cassegrainian focus. Other arrangements are also possible. (c) The 200-inch (mirror diameter) Hale telescope on Palomar Mountain in California. (d) The 10-meter Keck telescope on Mauna Kea, Hawaii. The Keck combines thirty-six 1.8-meter six-sided mirrors into the equivalent of a very large single reflector, 10 m in diameter.

A **terrestrial telescope**, for viewing objects on Earth, must provide an upright image—seeing normal objects upside down would be difficult (much less important for viewing stars). Two designs are shown in Fig. 33–39. The **Galilean** type, which Galileo used for his great astronomical discoveries, has a diverging lens as eyepiece which intercepts the converging rays from the objective lens before they reach a focus, and acts to form a virtual upright image, Fig. 33–39a. This design is still used in opera glasses. The tube is reasonably short, but the field of view is small. The second type, shown in Fig. 33–39b, is often called a **spyglass** and makes use of a third convex lens that acts to make the image upright as shown. A spyglass must be quite long. The most practical design today is the **prism binocular** which was shown in Fig. 32–33. The objective and eyepiece are converging lenses. The prisms reflect the rays by total internal reflection and shorten the physical size of the device, and they also act to produce an upright image. One prism reinverts the image in the vertical plane, the other in the horizontal plane.

FIGURE 33–39 Terrestrial telescopes that produce an upright image: (a) Galilean; (b) spyglass, or erector type.

[†]Another advantage of mirrors is that they exhibit no chromatic aberration because the light doesn't pass through them; and they can be ground in a parabolic shape to correct for spherical aberration (Section 33–10). The reflecting telescope was first proposed by Newton.

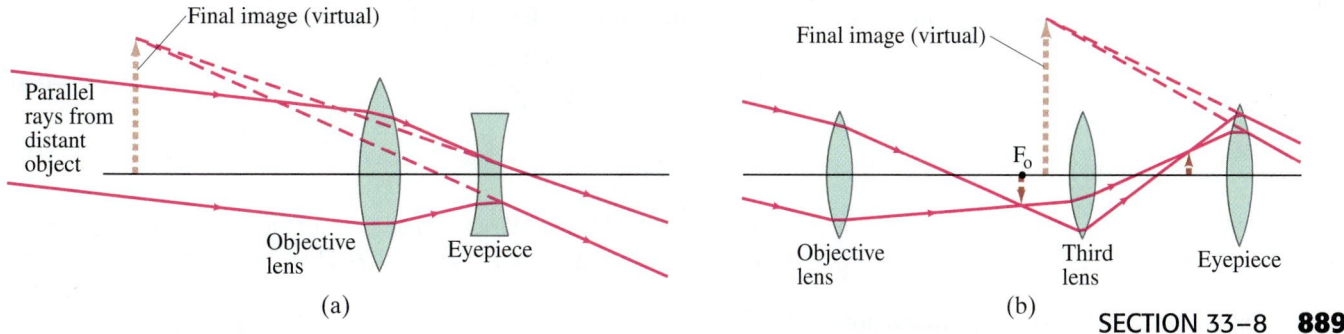

*33–9 Compound Microscope

The compound **microscope**, like the telescope, has both objective and eyepiece (or ocular) lenses, Fig. 33–40. The design is different from that for a telescope because a microscope is used to view objects that are very close, so the object distance is very small. The object is placed just beyond the objective's focal point as shown in Fig. 33–40a. The image I_1 formed by the objective lens is real, quite far from the objective lens, and much enlarged. The eyepiece is positioned so that this image is near the eyepiece focal point F_e. The image I_1 is magnified by the eyepiece into a very large virtual image, I_2, which is seen by the eye and is inverted. Modern microscopes use a third "tube" lens behind the objective, but we will analyze the simpler arrangement shown in Fig. 33–40a.

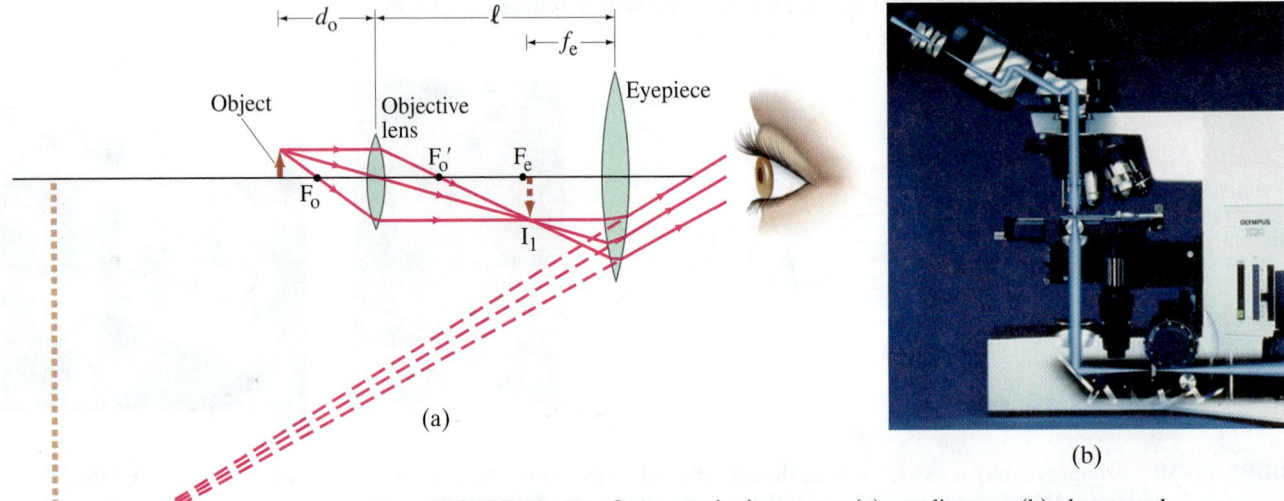

FIGURE 33–40 Compound microscope: (a) ray diagram, (b) photograph (illumination comes from the lower right, then up through the slide holding the object).

The overall magnification of a microscope is the product of the magnifications produced by the two lenses. The image I_1 formed by the objective lens is a factor m_o greater than the object itself. From Fig. 33–40a and Eq. 33–3 for the lateral magnification of a simple lens, we have

$$m_o = \frac{h_i}{h_o} = \frac{d_i}{d_o} = \frac{\ell - f_e}{d_o}, \tag{33–8}$$

where d_o and d_i are the object and image distances for the objective lens, ℓ is the distance between the lenses (equal to the length of the barrel), and we ignored the minus sign in Eq. 33–3 which only tells us that the image is inverted. We set $d_i = \ell - f_e$, which is true only if the eye is relaxed, so that the image I_1 is at the eyepiece focal point F_e. The eyepiece acts like a simple magnifier. If we assume that the eye is relaxed, the eyepiece angular magnification M_e is (from Eq. 33–6a)

$$M_e = \frac{N}{f_e}, \tag{33–9}$$

where the near point $N = 25 \text{ cm}$ for the normal eye. Since the eyepiece enlarges the image formed by the objective, the overall angular magnification M is the product of the lateral magnification of the objective lens, m_o, times the angular magnification, M_e, of the eyepiece lens (Eqs. 33–8 and 33–9):

$$M = M_e m_o = \left(\frac{N}{f_e}\right)\left(\frac{\ell - f_e}{d_o}\right) \qquad \text{[microscope]} \tag{33–10a}$$

$$\approx \frac{N\ell}{f_e f_o}. \qquad \text{[}f_o \text{ and } f_e \ll \ell\text{]} \tag{33–10b}$$

The approximation, Eq. 33–10b, is accurate when f_e and f_o are small compared to ℓ, so $\ell - f_e \approx \ell$, and the object is near F_o so $d_o \approx f_o$ (Fig. 33–40a). This is a good approximation for large magnifications, which are obtained when f_o and f_e are very small (they are in the denominator of Eq. 33–10b). To make lenses of very short focal length, compound lenses involving several elements must be used to avoid serious aberrations, as discussed in the next Section.

EXAMPLE 33–16 **Microscope.** A compound microscope consists of a $10\times$ eyepiece and a $50\times$ objective $17.0\,\text{cm}$ apart. Determine (a) the overall magnification, (b) the focal length of each lens, and (c) the position of the object when the final image is in focus with the eye relaxed. Assume a normal eye, so $N = 25\,\text{cm}$.

APPROACH The overall magnification is the product of the eyepiece magnification and the objective magnification. The focal length of the eyepiece is found from Eq. 33–6a or 33–9 for the magnification of a simple magnifier. For the objective lens, it is easier to next find d_o (part c) using Eq. 33–8 before we find f_o.

SOLUTION (a) The overall magnification is $(10\times)(50\times) = 500\times$.
(b) The eyepiece focal length is (Eq. 33–9) $f_e = N/M_e = 25\,\text{cm}/10 = 2.5\,\text{cm}$. Next we solve Eq. 33–8 for d_o, and find

$$d_o = \frac{\ell - f_e}{m_o} = \frac{(17.0\,\text{cm} - 2.5\,\text{cm})}{50} = 0.29\,\text{cm}.$$

Then, from the thin lens equation for the objective with $d_i = \ell - f_e = 14.5\,\text{cm}$ (see Fig. 33–40a),

$$\frac{1}{f_o} = \frac{1}{d_o} + \frac{1}{d_i} = \frac{1}{0.29\,\text{cm}} + \frac{1}{14.5\,\text{cm}} = 3.52\,\text{cm}^{-1};$$

so $f_o = 1/(3.52\,\text{cm}^{-1}) = 0.28\,\text{cm}$.
(c) We just calculated $d_o = 0.29\,\text{cm}$, which is very close to f_o.

*33–10 Aberrations of Lenses and Mirrors

Earlier in this Chapter we developed a theory of image formation by a thin lens. We found, for example, that all rays from each point on an object are brought to a single point as the image point. This result, and others, were based on approximations for a thin lens, mainly that all rays make small angles with the axis and that we can use $\sin\theta \approx \theta$. Because of these approximations, we expect deviations from the simple theory, which are referred to as **lens aberrations**. There are several types of aberration; we will briefly discuss each of them separately, but all may be present at one time.

Consider an object at any point (even at infinity) on the axis of a lens with spherical surfaces. Rays from this point that pass through the outer regions of the lens are brought to a focus at a different point from those that pass through the center of the lens. This is called **spherical aberration**, and is shown exaggerated in Fig. 33–41. Consequently, the image seen on a screen or film will not be a point but a tiny circular patch of light. If the sensor or film is placed at the point C, as indicated, the circle will have its smallest diameter, which is referred to as the **circle of least confusion**. Spherical aberration is present whenever spherical surfaces are used. It can be reduced by using nonspherical (= aspherical) lens surfaces, but grinding such lenses is difficult and expensive. Spherical aberration can be reduced by the use of several lenses in combination, and by using primarily the central part of lenses.

PHYSICS APPLIED
Lens aberrations

FIGURE 33–41 Spherical aberration (exaggerated). Circle of least confusion is at C.

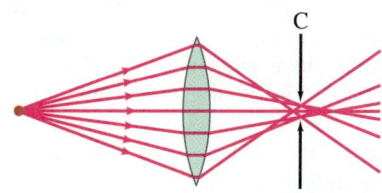

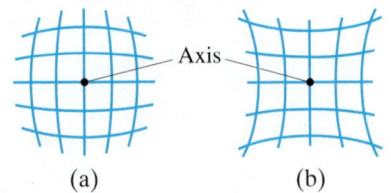

FIGURE 33–42 Distortion: lenses may image a square grid of perpendicular lines to produce (a) barrel distortion or (b) pincushion distortion. These distortions can be seen in the photograph of Fig. 33–2d.

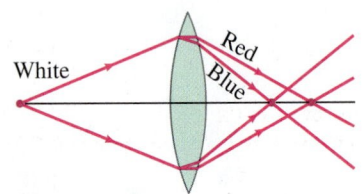

FIGURE 33–43 Chromatic aberration. Different colors are focused at different points.

FIGURE 33–44 Achromatic doublet.

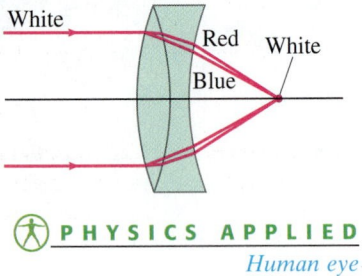

⊕ **PHYSICS APPLIED**

Human eye

For object points off the lens axis, additional aberrations occur. Rays passing through the different parts of the lens cause spreading of the image that is noncircular. There are two effects: **coma** (because the image of a point is comet-shaped rather than a circle) and **off-axis astigmatism**.† Furthermore, the image points for objects off the axis but at the same distance from the lens do not fall on a flat plane but on a curved surface—that is, the focal plane is not flat. (We expect this because the points on a flat plane, such as the film in a camera, are not equidistant from the lens.) This aberration is known as **curvature of field** and is a problem in cameras and other devices where the film is placed in a flat plane. In the eye, however, the retina is curved, which compensates for this effect.

Another aberration, **distortion**, is a result of variation of magnification at different distances from the lens axis. Thus a straight-line object some distance from the axis may form a curved image. A square grid of lines may be distorted to produce "barrel distortion," or "pincushion distortion," Fig. 33–42. The former is common in extreme wide-angle lenses.

All the above aberrations occur for monochromatic light and hence are referred to as *monochromatic aberrations*. Normal light is not monochromatic, and there will also be **chromatic aberration**. This aberration arises because of dispersion—the variation of index of refraction of transparent materials with wavelength (Section 32–6). For example, blue light is bent more than red light by glass. So if white light is incident on a lens, the different colors are focused at different points, Fig. 33–43, and have slightly different magnifications resulting in colored fringes in the image. Chromatic aberration can be eliminated for any two colors (and reduced greatly for all others) by the use of two lenses made of different materials with different indices of refraction and dispersion. Normally one lens is converging and the other diverging, and they are often cemented together (Fig. 33–44). Such a lens combination is called an **achromatic doublet** (or "color-corrected" lens).

To reduce aberrations, high-quality lenses are **compound lenses** consisting of many simple lenses, referred to as **elements**. A typical high-quality camera lens may contain six to eight (or more) elements. For simplicity we will usually indicate lenses in diagrams as if they were simple lenses.

The human eye is also subject to aberrations, but they are minimal. Spherical aberration, for example, is minimized because (1) the cornea is less curved at the edges than at the center, and (2) the lens is less dense at the edges than at the center. Both effects cause rays at the outer edges to be bent less strongly, and thus help to reduce spherical aberration. Chromatic aberration is partially compensated for because the lens absorbs the shorter wavelengths appreciably and the retina is less sensitive to the blue and violet wavelengths. This is just the region of the spectrum where dispersion—and thus chromatic aberration—is greatest (Fig. 32–28).

Spherical mirrors (Section 32–3) also suffer aberrations including spherical aberration (see Fig. 32–13). Mirrors can be ground in a parabolic shape to correct for aberrations, but they are much harder to make and therefore very expensive. Spherical mirrors do not, however, exhibit chromatic aberration because the light does not pass through them (no refraction, no dispersion).

---

†Although the effect is the same as for astigmatism in the eye (Section 33–6), the cause is different. Off-axis astigmatism is no problem in the eye because objects are clearly seen only at the fovea, on the lens axis.

Summary

A lens uses refraction to produce a real or virtual image. Parallel rays of light are focused to a point, called the **focal point**, by a **converging** lens. The distance of the focal point from the lens is called the **focal length** f of the lens.

After parallel rays pass through a **diverging** lens, they appear to diverge from a point, its focal point; and the corresponding focal length is considered negative.

The **power** P of a lens, which is

$$P = \frac{1}{f} \tag{33–1}$$

is given in diopters, which are units of inverse meters (m^{-1}).

For a given object, the position and size of the image formed by a lens can be found approximately by ray tracing.

Algebraically, the relation between image and object distances, d_i and d_o, and the focal length f, is given by the **thin lens equation**:

$$\frac{1}{d_o} + \frac{1}{d_i} = \frac{1}{f}. \qquad (33\text{--}2)$$

The ratio of image height to object height, which equals the **lateral magnification** m, is

$$m = \frac{h_i}{h_o} = -\frac{d_i}{d_o}. \qquad (33\text{--}3)$$

When using the various equations of geometric optics, it is important to remember the **sign conventions** for all quantities involved: carefully review them (page 871) when doing Problems.

When two (or more) thin lenses are used in combination to produce an image, the thin lens equation can be used for each lens in sequence. The image produced by the first lens acts as the object for the second lens.

[*The **lensmaker's equation** relates the radii of curvature of the lens surfaces and the len's index of refraction to the focal length of the lens.]

A **camera** lens forms an image on film, or on an electronic sensor (CCD or CMOS) in a digital camera, by allowing light in through a shutter. The image is focused by moving the lens relative to the film, and the **f-stop** (or lens opening) must be adjusted for the brightness of the scene and the chosen shutter speed. The f-stop is defined as the ratio of the focal length to the diameter of the lens opening.

The human **eye** also adjusts for the available light—by opening and closing the iris. It focuses not by moving the lens, but by adjusting the shape of the lens to vary its focal length. The image is formed on the retina, which contains an array of receptors known as rods and cones.

Diverging eyeglass or contact lenses are used to correct the defect of a nearsighted eye, which cannot focus well on distant objects. Converging lenses are used to correct for defects in which the eye cannot focus on close objects.

A **simple magnifier** is a converging lens that forms a virtual image of an object placed at (or within) the focal point. The **angular magnification**, when viewed by a relaxed normal eye, is

$$M = \frac{N}{f}, \qquad (33\text{--}6a)$$

where f is the focal length of the lens and N is the near point of the eye (25 cm for a "normal" eye).

An **astronomical telescope** consists of an **objective** lens or mirror, and an **eyepiece** that magnifies the real image formed by the objective. The **magnification** is equal to the ratio of the objective and eyepiece focal lengths, and the image is inverted:

$$M = -\frac{f_o}{f_e}. \qquad (33\text{--}7)$$

[*A compound **microscope** also uses objective and eyepiece lenses, and the final image is inverted. The total magnification is the product of the magnifications of the two lenses and is approximately

$$M \approx \frac{N\ell}{f_e f_o}, \qquad (33\text{--}10b)$$

where ℓ is the distance between the lenses, N is the near point of the eye, and f_o and f_e are the focal lengths of objective and eyepiece, respectively.]

[*Microscopes, telescopes, and other optical instruments are limited in the formation of sharp images by **lens aberrations**. These include **spherical aberration**, in which rays passing through the edge of a lens are not focused at the same point as those that pass near the center; and **chromatic aberration**, in which different colors are focused at different points. Compound lenses, consisting of several elements, can largely correct for aberrations.]

Questions

1. Where must the film be placed if a camera lens is to make a sharp image of an object far away?

2. A photographer moves closer to his subject and then refocuses. Does the camera lens move farther away from or closer to the sensor? Explain.

3. Can a diverging lens form a real image under any circumstances? Explain.

4. Use ray diagrams to show that a real image formed by a thin lens is always inverted, whereas a virtual image is always upright if the object is real.

5. Light rays are said to be "reversible." Is this consistent with the thin lens equation? Explain.

6. Can real images be projected on a screen? Can virtual images? Can either be photographed? Discuss carefully.

7. A thin converging lens is moved closer to a nearby object. Does the real image formed change (a) in position, (b) in size? If yes, describe how.

8. Compare the mirror equation with the thin lens equation. Discuss similarities and differences, especially the sign conventions for the quantities involved.

9. A lens is made of a material with an index of refraction $n = 1.30$. In air, it is a converging lens. Will it still be a converging lens if placed in water? Explain, using a ray diagram.

10. Explain how you could have a virtual object.

11. A dog with its tail in the air stands facing a converging lens. If the nose and the tail are each focused on a screen in turn, which will have the greater magnification?

12. A cat with its tail in the air stands facing a converging lens. Under what circumstances (if any) would the image of the nose be virtual and the image of the tail be real? Where would the image of the rest of the cat be?

13. Why, in Example 33–6, must the converging lens have a shorter focal length than the diverging lens if the latter's focal length is to be determined by combining them?

14. The thicker a double convex lens is in the center as compared to its edges, the shorter its focal length for a given lens diameter. Explain.

15. Does the focal length of a lens depend on the fluid in which it is immersed? What about the focal length of a spherical mirror? Explain.

16. An underwater lens consists of a carefully shaped thin-walled plastic container filled with air. What shape should it have in order to be (a) converging (b) diverging? Use ray diagrams to support your answer.

17. Consider two converging lenses separated by some distance. An object is placed so that the image from the first lens lies exactly at the focal point of the second lens. Will this combination produce an image? If so, where? If not, why not?

18. Will a nearsighted person who wears corrective lenses in her glasses be able to see clearly underwater when wearing those glasses? Use a diagram to show why or why not.

19. You can tell whether people are nearsighted or farsighted by looking at the width of their face through their glasses. If a person's face appears narrower through the glasses, (Fig. 33–45), is the person farsighted or nearsighted?

FIGURE 33–45
Question 19.

20. The human eye is much like a camera—yet, when a camera shutter is left open and the camera is moved, the image will be blurred. But when you move your head with your eyes open, you still see clearly. Explain.

21. In attempting to discern distant details, people will sometimes squint. Why does this help?

22. Is the image formed on the retina of the human eye upright or inverted? Discuss the implications of this for our perception of objects.

23. Reading glasses use converging lenses. A simple magnifier is also a converging lens. Are reading glasses therefore magnifiers? Discuss the similarities and differences between converging lenses as used for these two different purposes.

24. Why must a camera lens be moved farther from the film to focus on a closer object?

*25. Spherical aberration in a thin lens is minimized if rays are bent equally by the two surfaces. If a planoconvex lens is used to form a real image of an object at infinity, which surface should face the object? Use ray diagrams to show why.

*26. For both converging and diverging lenses, discuss how the focal length for red light differs from that for violet light.

Problems

33–1 and 33–2 Thin Lenses

1. (I) A sharp image is located 373 mm behind a 215-mm-focal-length converging lens. Find the object distance (a) using a ray diagram, (b) by calculation.

2. (I) Sunlight is observed to focus at a point 18.5 cm behind a lens. (a) What kind of lens is it? (b) What is its power in diopters?

3. (I) (a) What is the power of a 23.5-cm-focal-length lens? (b) What is the focal length of a -6.75-D lens? Are these lenses converging or diverging?

4. (II) A certain lens focuses an object 1.85 m away as an image 48.3 cm on the other side of the lens. What type of lens is it and what is its focal length? Is the image real or virtual?

5. (II) A 105-mm-focal-length lens is used to focus an image on the sensor of a camera. The maximum distance allowed between the lens and the sensor plane is 132 mm. (a) How far ahead of the sensor should the lens be if the object to be photographed is 10.0 m away? (b) 3.0 m away? (c) 1.0 m away? (d) What is the closest object this lens could photograph sharply?

6. (II) A stamp collector uses a converging lens with focal length 28 cm to view a stamp 18 cm in front of the lens. (a) Where is the image located? (b) What is the magnification?

7. (II) It is desired to magnify reading material by a factor of 2.5× when a book is placed 9.0 cm behind a lens. (a) Draw a ray diagram and describe the type of image this would be. (b) What type of lens is needed? (c) What is the power of the lens in diopters?

8. (II) A -8.00-D lens is held 12.5 cm from an ant 1.00 mm high. Describe the position, type, and height of the image.

9. (II) An object is located 1.50 m from an 8.0-D lens. By how much does the image move if the object is moved (a) 0.90 m closer to the lens, and (b) 0.90 m farther from the lens?

10. (II) (a) How far from a 50.0-mm-focal-length lens must an object be placed if its image is to be magnified 2.50× and be real? (b) What if the image is to be virtual and magnified 2.50×?

11. (II) How far from a converging lens with a focal length of 25 cm should an object be placed to produce a real image which is the same size as the object?

12. (II) (a) A 2.80-cm-high insect is 1.30 m from a 135-mm-focal-length lens. Where is the image, how high is it, and what type is it? (b) What if $f = -135$ mm?

13. (II) A bright object and a viewing screen are separated by a distance of 86.0 cm. At what location(s) between the object and the screen should a lens of focal length 16.0 cm be placed in order to produce a sharp image on the screen? [Hint: first draw a diagram.]

14. (II) How far apart are an object and an image formed by an 85-cm-focal-length converging lens if the image is 2.95× larger than the object and is real?

15. (II) Show analytically that the image formed by a converging lens (a) is real and inverted if the object is beyond the focal point $(d_o > f)$, and (b) is virtual and upright if the object is within the focal point $(d_o < f)$. Next, describe the image if the object is itself an image (formed by another lens), and its position is on the opposite side of the lens from the incoming light, (c) for $-d_o > f$, and (d) for $0 < -d_o < f$.

16. (II) A converging lens has focal length f. When an object is placed a distance $d_o > f$ from this lens, a real image with magnification m is formed. (a) Show that $m = f/(f - d_o)$. (b) Sketch m vs. d_o over the range $f < d_o < \infty$ where $f = 0.45$ cm. (c) For what value of d_o will the real image have the same (lateral) size as the object? (d) To obtain a real image that is much larger than the object, in what general region should the object be placed relative to the lens?

17. (II) In a slide or movie projector, the film acts as the object whose image is projected on a screen (Fig. 33–46). If a 105-mm-focal-length lens is to project an image on a screen 6.50 m away, how far from the lens should the slide be? If the slide is 36 mm wide, how wide will the picture be on the screen?

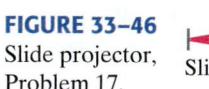 

FIGURE 33–46
Slide projector, Problem 17.
Slide
Lens
Screen

18. (III) A bright object is placed on one side of a converging lens of focal length f, and a white screen for viewing the image is on the opposite side. The distance $d_T = d_i + d_o$ between the object and the screen is kept fixed, but the lens can be moved. (a) Show that if $d_T > 4f$, there will be two positions where the lens can be placed and a sharp image will be produced on the screen. (b) If $d_T < 4f$, show that there will be no lens position where a sharp image is formed. (c) Determine a formula for the distance between the two lens positions in part (a), and the ratio of the image sizes.

19. (III) (a) Show that the lens equation can be written in the *Newtonian form*:

$$xx' = f^2,$$

where x is the distance of the object from the focal point on the front side of the lens, and x' is the distance of the image to the focal point on the other side of the lens. Calculate the location of an image if the object is placed 48.0 cm in front of a convex lens with a focal length of 38.0 cm using (b) the standard form of the thin lens formula, and (c) the Newtonian form, derived above.

33–3 Lens Combinations

20. (II) A diverging lens with $f = -33.5$ cm is placed 14.0 cm behind a converging lens with $f = 20.0$ cm. Where will an object at infinity be focused?

21. (II) Two 25.0-cm-focal-length converging lenses are placed 16.5 cm apart. An object is placed 35.0 cm in front of one lens. Where will the final image formed by the second lens be located? What is the total magnification?

22. (II) A 34.0-cm-focal-length converging lens is 24.0 cm behind a diverging lens. Parallel light strikes the diverging lens. After passing through the converging lens, the light is again parallel. What is the focal length of the diverging lens? [*Hint*: first draw a ray diagram.]

23. (II) The two converging lenses of Example 33–5 are now placed only 20.0 cm apart. The object is still 60.0 cm in front of the first lens as in Fig. 33–14. In this case, determine (a) the position of the final image, and (b) the overall magnification. (c) Sketch the ray diagram for this system.

24. (II) A diverging lens with a focal length of -14 cm is placed 12 cm to the right of a converging lens with a focal length of 18 cm. An object is placed 33 cm to the left of the converging lens. (a) Where will the final image be located? (b) Where will the image be if the diverging lens is 38 cm from the converging lens?

25. (II) Two lenses, one converging with focal length 20.0 cm and one diverging with focal length -10.0 cm, are placed 25.0 cm apart. An object is placed 60.0 cm in front of the converging lens. Determine (a) the position and (b) the magnification of the final image formed. (c) Sketch a ray diagram for this system.

26. (II) A diverging lens is placed next to a converging lens of focal length f_C, as in Fig. 33–15. If f_T represents the focal length of the combination, show that the focal length of the diverging lens, f_D, is given by

$$\frac{1}{f_D} = \frac{1}{f_T} - \frac{1}{f_C}.$$

27. (II) A lighted candle is placed 36 cm in front of a converging lens of focal length $f_1 = 13$ cm, which in turn is 56 cm in front of another converging lens of focal length $f_2 = 16$ cm (see Fig. 33–47). (a) Draw a ray diagram and estimate the location and the relative size of the final image. (b) Calculate the position and relative size of the final image.

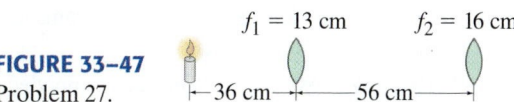

FIGURE 33–47
Problem 27.
$f_1 = 13$ cm $f_2 = 16$ cm
$\leftarrow$ 36 cm $\rightarrow$ $\leftarrow$ 56 cm $\rightarrow$

*33–4 Lensmaker's Equation

*28. (I) A double concave lens has surface radii of 33.4 cm and 28.8 cm. What is the focal length if $n = 1.58$?

*29. (I) Both surfaces of a double convex lens have radii of 31.4 cm. If the focal length is 28.9 cm, what is the index of refraction of the lens material?

*30. (I) Show that if the lens of Example 33–7 is reversed, the focal length is unchanged.

*31. (I) A planoconvex lens (Fig. 33–2a) is to have a focal length of 18.7 cm. If made from fused quartz, what must be the radius of curvature of the convex surface?

*32. (II) An object is placed 90.0 cm from a glass lens ($n = 1.52$) with one concave surface of radius 22.0 cm and one convex surface of radius 18.5 cm. Where is the final image? What is the magnification?

*33. (II) A prescription for a corrective lens calls for $+3.50$ diopters. The lensmaker grinds the lens from a "blank" with $n = 1.56$ and convex front surface of radius of curvature of 30.0 cm. What should be the radius of curvature of the other surface?

33–5 Camera

34. (I) A properly exposed photograph is taken at $f/16$ and $\frac{1}{120}$ s. What lens opening is required if the shutter speed is $\frac{1}{1000}$ s?

35. (I) A television camera lens has a 17-cm focal length and a lens diameter of 6.0 cm. What is its f-number?

36. (II) A "pinhole" camera uses a tiny pinhole instead of a lens. Show, using ray diagrams, how reasonably sharp images can be formed using such a pinhole camera. In particular, consider two point objects 2.0 cm apart that are 1.0 m from a 1.0-mm-diameter pinhole. Show that on a piece of film 7.0 cm behind the pinhole the two objects produce two separate circles that do not overlap.

37. (II) Suppose that a correct exposure is $\frac{1}{250}$ s at $f/11$. Under the same conditions, what exposure time would be needed for a pinhole camera (Problem 36) if the pinhole diameter is 1.0 mm and the film is 7.0 cm from the hole?

38. (II) Human vision normally covers an angle of about 40° horizontally. A "normal" camera lens then is defined as follows: When focused on a distant horizontal object which subtends an angle of 40°, the lens produces an image that extends across the full horizontal extent of the camera's light-recording medium (film or electronic sensor). Determine the focal length f of the "normal" lens for the following types of cameras: (a) a 35-mm camera that records images on film 36 mm wide; (b) a digital camera that records images on a charge-coupled device (CCD) 1.00 cm wide.

39. (II) A nature photographer wishes to photograph a 38-m tall tree from a distance of 65 m. What focal-length lens should be used if the image is to fill the 24-mm height of the sensor?

33–6 Eye and Corrective Lenses

40. (I) A human eyeball is about 2.0 cm long and the pupil has a maximum diameter of about 8.0 mm. What is the "speed" of this lens?

41. (II) A person struggles to read by holding a book at arm's length, a distance of 55 cm away. What power of reading glasses should be prescribed for her, assuming they will be placed 2.0 cm from the eye and she wants to read at the "normal" near point of 25 cm?

42. (II) Reading glasses of what power are needed for a person whose near point is 105 cm, so that he can read a computer screen at 55 cm? Assume a lens–eye distance of 1.8 cm.

43. (II) If the nearsighted person in Example 33–13 wore contact lenses corrected for the far point ($= \infty$), what would be the near point? Would glasses be better in this case?

44. (II) An eye is corrected by a -4.50-D lens, 2.0 cm from the eye. (a) Is this eye near- or farsighted? (b) What is this eye's far point without glasses?

45. (II) A person's right eye can see objects clearly only if they are between 25 cm and 78 cm away. (a) What power of contact lens is required so that objects far away are sharp? (b) What will be the near point with the lens in place?

46. (II) A person has a far point of 14 cm. What power glasses would correct this vision if the glasses were placed 2.0 cm from the eye? What power contact lenses, placed on the eye, would the person need?

47. (II) One lens of a nearsighted person's eyeglasses has a focal length of -23.0 cm and the lens is 1.8 cm from the eye. If the person switches to contact lenses placed directly on the eye, what should be the focal length of the corresponding contact lens?

48. (II) What is the focal length of the eye lens system when viewing an object (a) at infinity, and (b) 38 cm from the eye? Assume that the lens–retina distance is 2.0 cm.

49. (II) A nearsighted person has near and far points of 10.6 and 20.0 cm respectively. If she puts on contact lenses with power $P = -4.00$ D, what are her new near and far points?

50. (II) The closely packed cones in the fovea of the eye have a diameter of about 2 μm. For the eye to discern two images on the fovea as distinct, assume that the images must be separated by at least one cone that is not excited. If these images are of two point-like objects at the eye's 25-cm near point, how far apart are these barely resolvable objects? Assume the diameter of the eye (cornea-to-fovea distance) is 2.0 cm.

33–7 Magnifying Glass

51. (I) What is the focal length of a magnifying glass of 3.8× magnification for a relaxed normal eye?

52. (I) What is the magnification of a lens used with a relaxed eye if its focal length is 13 cm?

53. (I) A magnifier is rated at 3.0× for a normal eye focusing on an image at the near point. (a) What is its focal length? (b) What is its focal length if the 3.0× refers to a relaxed eye?

54. (II) Sherlock Holmes is using an 8.80-cm-focal-length lens as his magnifying glass. To obtain maximum magnification, where must the object be placed (assume a normal eye), and what will be the magnification?

55. (II) A small insect is placed 5.85 cm from a $+6.00$-cm-focal-length lens. Calculate (a) the position of the image, and (b) the angular magnification.

56. (II) A 3.40-mm-wide bolt is viewed with a 9.60-cm-focal-length lens. A normal eye views the image at its near point. Calculate (a) the angular magnification, (b) the width of the image, and (c) the object distance from the lens.

57. (II) A magnifying glass with a focal length of 9.5 cm is used to read print placed at a distance of 8.3 cm. Calculate (a) the position of the image; (b) the angular magnification.

58. (II) A magnifying glass is rated at 3.0× for a normal eye that is relaxed. What would be the magnification for a relaxed eye whose near point is (a) 65 cm, and (b) 17 cm? Explain the differences.

59. (II) A converging lens of focal length $f = 12$ cm is being used by a writer as a magnifying glass to read some fine print on his book contract. Initially, the writer holds the lens above the fine print so that its image is at infinity. To get a better look, he then moves the lens so that the image is at his 25-cm near point. How far, and in what direction (toward or away from the fine print) did the writer move the lens? Assume the writer's eye is adjusted to remain always very near the magnifying glass.

33–8 Telescopes

60. (I) What is the magnification of an astronomical telescope whose objective lens has a focal length of 78 cm, and whose eyepiece has a focal length of 2.8 cm? What is the overall length of the telescope when adjusted for a relaxed eye?

61. (I) The overall magnification of an astronomical telescope is desired to be 35×. If an objective of 88 cm focal length is used, what must be the focal length of the eyepiece? What is the overall length of the telescope when adjusted for use by the relaxed eye?

62. (II) A 7.0× binocular has 3.0-cm-focal-length eyepieces. What is the focal length of the objective lenses?

63. (II) An astronomical telescope has an objective with focal length 75 cm and a $+35$ D eyepiece. What is the total magnification?

64. (II) An astronomical telescope has its two lenses spaced 78.0 cm apart. If the objective lens has a focal length of 75.5 cm, what is the magnification of this telescope? Assume a relaxed eye.

65. (II) A Galilean telescope adjusted for a relaxed eye is 33.8 cm long. If the objective lens has a focal length of 36.0 cm, what is the magnification?

66. (II) What is the magnifying power of an astronomical telescope using a reflecting mirror whose radius of curvature is 6.4 m and an eyepiece whose focal length is 2.8 cm?

67. (II) The Moon's image appears to be magnified 120× by a reflecting astronomical telescope with an eyepiece having a focal length of 3.1 cm. What are the focal length and radius of curvature of the main (objective) mirror?

68. (II) A 120× astronomical telescope is adjusted for a relaxed eye when the two lenses are 1.25 m apart. What is the focal length of each lens?

69. (II) An astronomical telescope longer than about 50 cm is not easy to hold by hand. Based on this fact, estimate the maximum angular magnification achievable for a telescope designed to be handheld. Assume its eyepiece lens, if used as a magnifying glass, provides a magnification of 5× for a relaxed eye with near point $N = 25$ cm.

70. (III) A reflecting telescope (Fig. 33–38b) has a radius of curvature of 3.00 m for its objective mirror and a radius of curvature of -1.50 m for its eyepiece mirror. If the distance between the two mirrors is 0.90 m, how far in front of the eyepiece should you place the electronic sensor to record the image of a star?

71. (III) A 7.5× pair of binoculars has an objective focal length of 26 cm. If the binoculars are focused on an object 4.0 m away (from the objective), what is the magnification? (The 7.5× refers to objects at infinity; Eq. 33–7 holds only for objects at infinity and not for nearby ones.)

*33–9 Microscopes

*72. (I) A microscope uses an eyepiece with a focal length of 1.50 cm. Using a normal eye with a final image at infinity, the barrel length is 17.5 cm and the focal length of the objective lens is 0.65 cm. What is the magnification of the microscope?

*73. (I) A 680× microscope uses a 0.40-cm-focal-length objective lens. If the barrel length is 17.5 cm, what is the focal length of the eyepiece? Assume a normal eye and that the final image is at infinity.

*74. (I) A 17-cm-long microscope has an eyepiece with a focal length of 2.5 cm and an objective with a focal length of 0.28 cm. What is the approximate magnification?

*75. (II) A microscope has a 13.0× eyepiece and a 58.0× objective lens 20.0 cm apart. Calculate (a) the total magnification, (b) the focal length of each lens, and (c) where the object must be for a normal relaxed eye to see it in focus.

*76. (II) Repeat Problem 75 assuming that the final image is located 25 cm from the eyepiece (near point of a normal eye).

*77. (II) A microscope has a 1.8-cm-focal-length eyepiece and a 0.80-cm objective. Assuming a relaxed normal eye, calculate (a) the position of the object if the distance between the lenses is 16.8 cm, and (b) the total magnification.

*78. (II) The eyepiece of a compound microscope has a focal length of 2.80 cm and the objective lens has $f = 0.740$ cm. If an object is placed 0.790 cm from the objective lens, calculate (a) the distance between the lenses when the microscope is adjusted for a relaxed eye, and (b) the total magnification.

*79. (II) An inexpensive instructional lab microscope allows the user to select its objective lens to have a focal length of 32 mm, 15 mm, or 3.9 mm. It also has two possible eyepieces with magnifications 5× and 10×. Each objective forms a real image 160 mm beyond its focal point. What are the largest and smallest overall magnifications obtainable with this instrument?

*80. (III) Given two 12-cm-focal-length lenses, you attempt to make a crude microscope using them. While holding these lenses a distance 55 cm apart, you position your microscope so that its objective lens is distance d_o from a small object. Assume your eye's near point $N = 25$ cm. (a) For your microscope to function properly, what should d_o be? (b) Assuming your eye is relaxed when using it, what magnification M does your microscope achieve? (c) Since the length of your microscope is not much greater than the focal lengths of its lenses, the approximation $M \approx N\ell/f_e f_o$ is not valid. If you apply this approximation to your microscope, what % error do you make in your microscope's true magnification?

*33–10 Lens Aberrations

*81. (II) A planoconvex lens (Fig. 33–2a) has one flat surface and the other has $R = 15.3$ cm. This lens is used to view a red and yellow object which is 66.0 cm away from the lens. The index of refraction of the glass is 1.5106 for red light and 1.5226 for yellow light. What are the locations of the red and yellow images formed by the lens?

*82. (II) An achromatic lens is made of two very thin lenses, placed in contact, that have focal lengths $f_1 = -28$ cm and $f_2 = +25$ cm. (a) Is the combination converging or diverging? (b) What is the net focal length?

General Problems

83. A 200-mm-focal-length lens can be adjusted so that it is 200.0 mm to 206.4 mm from the film. For what range of object distances can it be adjusted?

84. If a 135-mm telephoto lens is designed to cover object distances from 1.30 m to ∞, over what distance must the lens move relative to the plane of the sensor or film?

85. For a camera equipped with a 58-mm-focal-length lens, what is the object distance if the image height equals the object height? How far is the object from the image on the film?

86. Show that for objects very far away (assume infinity), the magnification of any camera lens is proportional to its focal length.

87. A small object is 25.0 cm from a diverging lens as shown in Fig. 33–48. A converging lens with a focal length of 12.0 cm is 30.0 cm to the right of the diverging lens. The two-lens system forms a real inverted image 17.0 cm to the right of the converging lens. What is the focal length of the diverging lens?

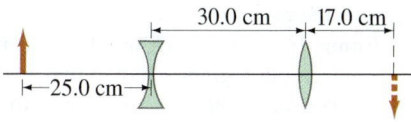

FIGURE 33–48 Problem 87.

88. A converging lens with focal length of 13.0 cm is placed in contact with a diverging lens with a focal length of −20.0 cm. What is the focal length of the combination, and is the combination converging or diverging?

89. An astronomical telescope has a magnification of 8.0×. If the two lenses are 28 cm apart, determine the focal length of each lens.

90. (a) Show that if two thin lenses of focal lengths f_1 and f_2 are placed in contact with each other, the focal length of the combination is given by $f_T = f_1 f_2 / (f_1 + f_2)$. (b) Show that the power P of the combination of two lenses is the sum of their separate powers, $P = P_1 + P_2$.

91. How large is the image of the Sun on film used in a camera with (a) a 28-mm-focal-length lens, (b) a 50-mm-focal-length lens, and (c) a 135-mm-focal-length lens? (d) If the 50-mm lens is considered normal for this camera, what relative magnification does each of the other two lenses provide? The Sun has diameter 1.4×10^6 km, and it is 1.5×10^8 km away.

92. Two converging lenses are placed 30.0 cm apart. The focal length of the lens on the right is 20.0 cm, and the focal length of the lens on the left is 15.0 cm. An object is placed to the left of the 15.0-cm-focal-length lens. A final image from both lenses is inverted and located halfway between the two lenses. How far to the left of the 15.0-cm-focal-length lens is the original object?

93. When an object is placed 60.0 cm from a certain converging lens, it forms a real image. When the object is moved to 40.0 cm from the lens, the image moves 10.0 cm farther from the lens. Find the focal length of this lens.

94. Figure 33–49 was taken from the NIST Laboratory (National Institute of Standards and Technology) in Boulder, CO, 2 km from the hiker in the photo. The Sun's image was 15 mm across on the film. Estimate the focal length of the camera lens (actually a telescope). The Sun has diameter 1.4×10^6 km, and it is 1.5×10^8 km away.

FIGURE 33–49 Problem 94.

95. A movie star catches a reporter shooting pictures of her at home. She claims the reporter was trespassing. To prove her point, she gives as evidence the film she seized. Her 1.75-m height is 8.25 mm high on the film, and the focal length of the camera lens was 220 mm. How far away from the subject was the reporter standing?

96. As early morning passed toward midday, and the sunlight got more intense, a photographer noted that, if she kept her shutter speed constant, she had to change the f-number from $f/5.6$ to $f/16$. By what factor had the sunlight intensity increased during that time?

97. A child has a near point of 15 cm. What is the maximum magnification the child can obtain using an 8.5-cm-focal-length magnifier? What magnification can a normal eye obtain with the same lens? Which person sees more detail?

98. A woman can see clearly with her right eye only when objects are between 45 cm and 155 cm away. Prescription bifocals should have what powers so that she can see distant objects clearly (upper part) and be able to read a book 25 cm away (lower part) with her right eye? Assume that the glasses will be 2.0 cm from the eye.

99. What is the magnifying power of a +4.0-D lens used as a magnifier? Assume a relaxed normal eye.

100. A physicist lost in the mountains tries to make a telescope using the lenses from his reading glasses. They have powers of +2.0 D and +4.5 D, respectively. (a) What maximum magnification telescope is possible? (b) Which lens should be used as the eyepiece?

101. A 50-year-old man uses +2.5-D lenses to read a newspaper 25 cm away. Ten years later, he must hold the paper 32 cm away to see clearly with the same lenses. What power lenses does he need now in order to hold the paper 25 cm away? (Distances are measured from the lens.)

102. An object is moving toward a converging lens of focal length f with constant speed v_o such that its distance d_o from the lens is always greater than f. (a) Determine the velocity v_i of the image as a function of d_o. (b) Which direction (toward or away from the lens) does the image move? (c) For what d_o does the image's speed equal the object's speed?

103. The objective lens and the eyepiece of a telescope are spaced 85 cm apart. If the eyepiece is +23 D, what is the total magnification of the telescope?

*104. Two converging lenses, one with $f = 4.0$ cm and the other with $f = 44$ cm, are made into a telescope. (a) What are the length and magnification? Which lens should be the eyepiece? (b) Assume these lenses are now combined to make a microscope; if the magnification needs to be 25×, how long would the microscope be?

105. Sam purchases +3.50-D eyeglasses which correct his faulty vision to put his near point at 25 cm. (Assume he wears the lenses 2.0 cm from his eyes.) (a) Calculate the focal length of Sam's glasses. (b) Calculate Sam's near point without glasses. (c) Pam, who has normal eyes with near point at 25 cm, puts on Sam's glasses. Calculate Pam's near point with Sam's glasses on.

106. The proper functioning of certain optical devices (e.g., optical fibers and spectrometers) requires that the input light be a collection of diverging rays within a cone of half-angle θ (Fig. 33–50). If the light initially exists as a collimated beam (i.e., parallel rays), show that a single lens of focal length f and diameter D can be used to create the required input light if $D/f = 2 \tan \theta$. If $\theta = 3.5°$ for a certain spectrometer, what focal length lens should be used if the lens diameter is 5.0 cm?

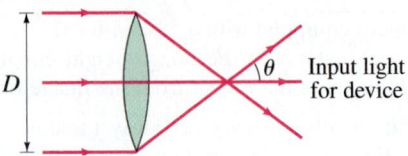

FIGURE 33–50 Problem 106.

107. In a science-fiction novel, an intelligent ocean-dwelling creature's eye functions underwater with a near point of 25 cm. This creature would like to create an underwater magnifier out of a thin plastic container filled with air. What shape should the air-filled plastic container have (i.e., determine radii of curvature of its surfaces) in order for it to be used by the creature as a 3.0× magnifier? Assume the eye is focused at its near point.

108. A telephoto lens system obtains a large magnification in a compact package. A simple such system can be constructed out of two lenses, one converging and one diverging, of focal lengths f_1 and $f_2 = -\frac{1}{2}f_1$, respectively, separated by a distance $\ell = \frac{3}{4}f_1$ as shown in Fig. 33–51. (a) For a distant object located at distance d_o from the first lens, show that the first lens forms an image with magnification $m_1 \approx -f_1/d_o$ located very close to its focal point. Go on to show that the total magnification for the two-lens system is $m \approx -2f_1/d_o$. (b) For an object located at infinity, show that the two-lens system forms an image that is a distance $\frac{5}{4}f_1$ behind the first lens. (c) A single 250-mm-focal-length lens would have to be mounted about 250 mm from a camera's film in order to produce an image of a distant object at d_o with magnification $-(250\text{ mm})/d_o$. To produce an image of this object with the same magnification using the two-lens system, what value of f_1 should be used and how far in front of the film should the first lens be placed? How much smaller is the "focusing length" (i.e., first lens-to-final image distance) of this two-lens system in comparison with the 250-mm "focusing length" of the equivalent single lens?

*109. (III) In the "magnification" method, the focal length f of a converging lens is found by placing an object of known size at various locations in front of the lens and measuring the resulting real-image distances d_i and their associated magnifications m (minus sign indicates that image is inverted). The data taken in such an experiment are given here:

d_i (cm)	20	25	30	35	40
m	−0.43	−0.79	−1.14	−1.50	−1.89

(a) Show analytically that a graph of m vs. d_i should produce a straight line. What are the theoretically expected values for the slope and the y-intercept of this line? [*Hint*: d_o is not constant.] (b) Using the data above, graph m vs. d_i and show that a straight line does indeed result. Use the slope of this line to determine the focal length of the lens. Does the y-intercept of your plot have the expected value? (c) In performing such an experiment, one has the practical problem of locating the exact center of the lens since d_i must be measured from this point. Imagine, instead, that one measures the image distance d_i' from the back surface of the lens, which is a distance ℓ from the lens's center. Then, $d_i = d_i' + \ell$. Show that, when implementing the magnification method in this fashion, a plot of m vs. d_i' will still result in a straight line. How can f be determined from this straight line?

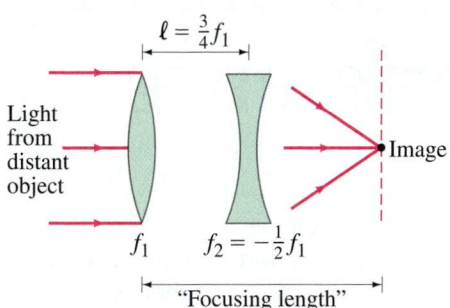

$$\ell = \tfrac{3}{4}f_1$$

Light from distant object

Image

f_1 $f_2 = -\tfrac{1}{2}f_1$

"Focusing length"

FIGURE 33–51 Problem 108.

Answers to Exercises

A: Closer to it.

B: (b) and (d) are true.

C: $f = 10$ cm for both cases; $d_i = -10$ cm in (a) and 30 cm in (b).

D: −36 cm; diverging.

E: $P = -4.0$ D.

F: 48 cm.

The beautiful colors from the surface of this soap bubble can be nicely explained by the wave theory of light. A soap bubble is a very thin spherical film filled with air. Light reflected from the outer and inner surfaces of this thin film of soapy water interferes constructively to produce the bright colors. Which color we see at any point depends on the thickness of the soapy water film at that point and also on the viewing angle. Near the top of the bubble, we see a small black area surrounded by a silver or white area. The bubble's thickness is smallest at that black spot, perhaps only about 30 nm thick, and is fully transparent (we see the black background).

We cover fundamental aspects of the wave nature of light, including two-slit interference and interference in thin films.

The Wave Nature of Light; Interference

CONTENTS

34–1 Waves versus Particles; Huygens' Principle and Diffraction

34–2 Huygens' Principle and the Law of Refraction

34–3 Interference—Young's Double-Slit Experiment

*34–4 Intensity in the Double-Slit Interference Pattern

34–5 Interference in Thin Films

*34–6 Michelson Interferometer

*34–7 Luminous Intensity

CHAPTER-OPENING QUESTION—Guess now!

When a thin layer of oil lies on top of water or wet pavement, you can often see swirls of color. We also see swirls of color on the soap bubble shown above. What causes these colors?

(a) Additives in the oil or soap reflect various colors.

(b) Chemicals in the oil or soap absorb various colors.

(c) Dispersion due to differences in index of refraction in the oil or soap.

(d) The interactions of the light with a thin boundary layer where the oil (or soap) and the water have mixed irregularly.

(e) Light waves reflected from the top and bottom surfaces of the thin oil or soap film can add up constructively for particular wavelengths.

That light carries energy is obvious to anyone who has focused the Sun's rays with a magnifying glass on a piece of paper and burned a hole in it. But how does light travel, and in what form is this energy carried? In our discussion of waves in Chapter 15, we noted that energy can be carried from place to place in basically two ways: by particles or by waves. In the first case, material objects or particles can carry energy, such as an avalanche of rocks or rushing water. In the second case, water waves and sound waves, for example, can carry energy over long distances even though the oscillating particles of the medium do not travel these distances. In view of this, what can we say about the nature of light: does light travel as a stream of particles away from its source, or does light travel in the form of waves that spread outward from the source?

Historically, this question has turned out to be a difficult one. For one thing, light does not reveal itself in any obvious way as being made up of tiny particles; nor do we see tiny light waves passing by as we do water waves. The evidence seemed to favor first one side and then the other until about 1830, when most physicists had accepted the wave theory. By the end of the nineteenth century, light was considered to be an *electromagnetic wave* (Chapter 31). In the early twentieth century, light was shown to have a particle nature as well, as we shall discuss in Chapter 37. We now speak of the wave–particle duality of light. The wave theory of light remains valid and has proved very successful. We now investigate the evidence for the wave theory and how it has been used to explain a wide range of phenomena.

34–1 Waves versus Particles; Huygens' Principle and Diffraction

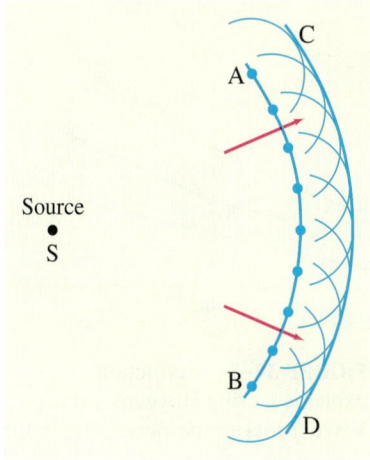

FIGURE 34–1 Huygens' principle, used to determine wave front CD when wave front AB is given.

FIGURE 34–2 Huygens' principle is consistent with diffraction (a) around the edge of an obstacle, (b) through a large hole, (c) through a small hole whose size is on the order of the wavelength of the wave.

The Dutch scientist Christian Huygens (1629–1695), a contemporary of Newton, proposed a wave theory of light that had much merit. Still useful today is a technique Huygens developed for predicting the future position of a wave front when an earlier position is known. By a wave front, we mean all the points along a two- or three-dimensional wave that form a wave crest—what we simply call a "wave" as seen on the ocean. Wave fronts are perpendicular to rays as we already discussed in Chapter 15 (Fig. 15–20). **Huygens' principle** can be stated as follows: *Every point on a wave front can be considered as a source of tiny wavelets that spread out in the forward direction at the speed of the wave itself. The new wave front is the envelope of all the wavelets*—that is, *the tangent to all of them*.

As a simple example of the use of Huygens' principle, consider the wave front AB in Fig. 34–1, which is traveling away from a source S. We assume the medium is *isotropic*—that is, the speed v of the waves is the same in all directions. To find the wave front a short time t after it is at AB, tiny circles are drawn with radius $r = vt$. The centers of these tiny circles are blue dots on the original wave front AB, and the circles represent Huygens' (imaginary) wavelets. The tangent to all these wavelets, the curved line CD, is the new position of the wave front.

Huygens' principle is particularly useful for analyzing what happens when waves impinge on an obstacle and the wave fronts are partially interrupted. Huygens' principle predicts that waves bend in behind an obstacle, as shown in Fig. 34–2. This is just what water waves do, as we saw in Chapter 15 (Figs. 15–31 and 15–32). The bending of waves behind obstacles into the "shadow region" is known as **diffraction**. Since diffraction occurs for waves, but not for particles, it can serve as one means for distinguishing the nature of light.

Note, as shown in Fig. 34–2, that diffraction is most prominent when the size of the opening is on the order of the wavelength of the wave. If the opening is much larger than the wavelength, diffraction goes unnoticed.

Does light exhibit diffraction? In the mid-seventeenth century, the Jesuit priest Francesco Grimaldi (1618–1663) had observed that when sunlight entered a darkened room through a tiny hole in a screen, the spot on the opposite wall was larger than would be expected from geometric rays. He also observed that the border of the image was not clear but was surrounded by colored fringes. Grimaldi attributed this to the diffraction of light.

The wave model of light nicely accounts for diffraction, and we discuss diffraction in detail in the next Chapter. But the ray model (Chapter 32) cannot account for diffraction, and it is important to be aware of such limitations to the ray model. Geometric optics using rays is successful in a wide range of situations only because normal openings and obstacles are much larger than the wavelength of the light, and so relatively little diffraction or bending occurs.

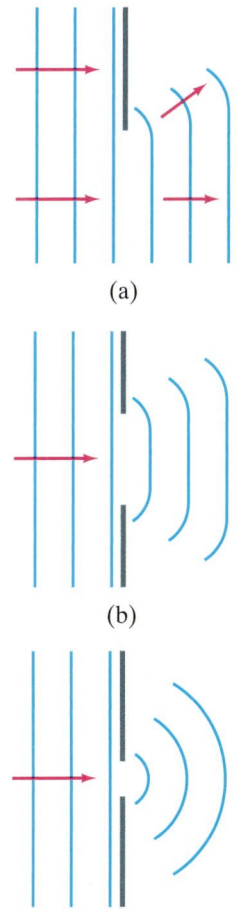

(a)

(b)

(c)

34–2 Huygens' Principle and the Law of Refraction

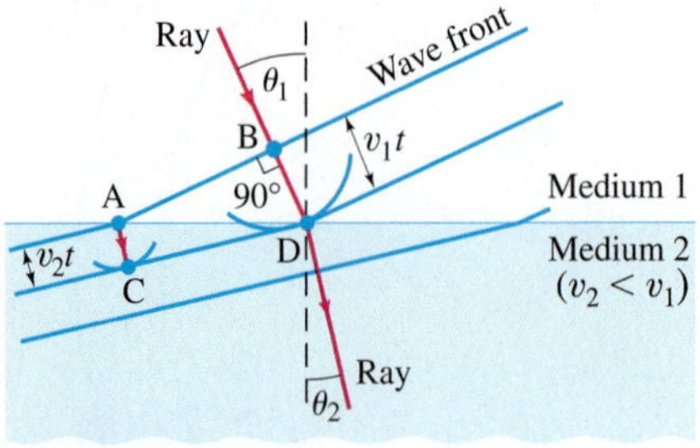

FIGURE 34–3 Refraction explained, using Huygens' principle. Wave fronts are perpendicular to the rays.

The laws of reflection and refraction were well known in Newton's time. The law of reflection could not distinguish between the two theories we just discussed: waves versus particles. For when waves reflect from an obstacle, the angle of incidence equals the angle of reflection (Fig. 15–21). The same is true of particles—think of a tennis ball without spin striking a flat surface.

The law of refraction is another matter. Consider a ray of light entering a medium where it is bent toward the normal, as when traveling from air into water. As shown in Fig. 34–3, this bending can be constructed using Huygens' principle if we assume the speed of light is less in the second medium $(v_2 < v_1)$. In time t, point B on wave front AB (perpendicular to the incoming ray) travels a distance $v_1 t$ to reach point D. Point A on the wave front, traveling in the second medium, goes a distance $v_2 t$ to reach point C, and $v_2 t < v_1 t$. Huygens' principle is applied to points A and B to obtain the curved wavelets shown at C and D. The wave front is tangent to these two wavelets, so the new wave front is the line CD. Hence the rays, which are perpendicular to the wave fronts, bend toward the normal if $v_2 < v_1$, as drawn.

Newton favored a particle theory of light which predicted the opposite result, that the speed of light would be greater in the second medium $(v_2 > v_1)$. Thus the wave theory predicts that the speed of light in water, for example, is less than in air; and Newton's particle theory predicts the reverse. An experiment to actually measure the speed of light in water was performed in 1850 by the French physicist Jean Foucault, and it confirmed the wave-theory prediction. By then, however, the wave theory was already fully accepted, as we shall see in the next Section.

Snell's law of refraction follows directly from Huygens' principle, given that the speed of light v in any medium is related to the speed in a vacuum, c, and the index of refraction, n, by Eq. 32–4: that is, $v = c/n$. From the Huygens' construction of Fig. 34–3, angle ADC is equal to θ_2 and angle BAD is equal to θ_1. Then for the two triangles that have the common side AD, we have

$$\sin \theta_1 = \frac{v_1 t}{\text{AD}}, \qquad \sin \theta_2 = \frac{v_2 t}{\text{AD}}.$$

We divide these two equations and obtain

$$\frac{\sin \theta_1}{\sin \theta_2} = \frac{v_1}{v_2}.$$

Then, by Eq. 32–4 $v_1 = c/n_1$ and $v_2 = c/n_2$, so we have

$$n_1 \sin \theta_1 = n_2 \sin \theta_2,$$

which is Snell's law of refraction, Eq. 32–5. (The law of reflection can be derived from Huygens' principle in a similar way: see Problem 1 at the end of this Chapter.)

When a light wave travels from one medium to another, its frequency does not change, but its wavelength does. This can be seen from Fig. 34–3, where each of the blue lines representing a wave front corresponds to a crest (peak) of the wave. Then

$$\frac{\lambda_2}{\lambda_1} = \frac{v_2 t}{v_1 t} = \frac{v_2}{v_1} = \frac{n_1}{n_2},$$

where, in the last step, we used Eq. 32–4, $v = c/n$. If medium 1 is a vacuum (or air), so $n_1 = 1, v_1 = c$, and we call λ_1 simply λ, then the wavelength in another medium of index of refraction $n (= n_2)$ will be

$$\lambda_n = \frac{\lambda}{n}. \tag{34–1}$$

This result is consistent with the frequency f being unchanged no matter what medium the wave is traveling in, since $c = f\lambda$.

EXERCISE A A light beam in air with wavelength = 500 nm, frequency = 6.0×10^{14} Hz, and speed = 3.0×10^8 m/s goes into glass which has an index of refraction = 1.5. What are the wavelength, frequency, and speed of the light in the glass?

(a)

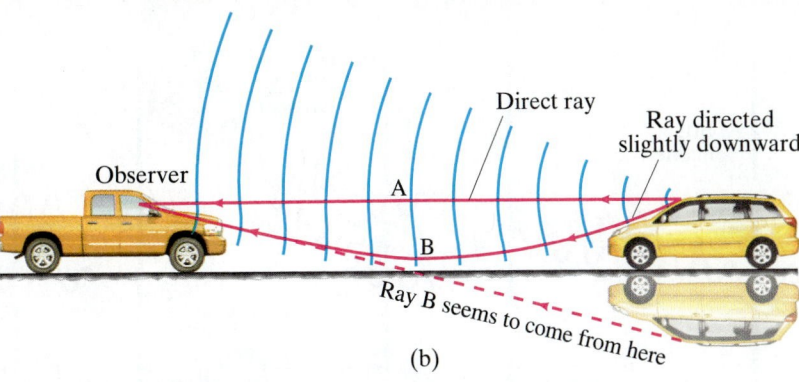

Direct ray

Ray directed slightly downward

Observer

A

B

Ray B seems to come from here

(b)

FIGURE 34–4 (a) A highway mirage. (b) Drawing (greatly exaggerated) showing wave fronts and rays to explain highway mirages. Note how sections of the wave fronts near the ground move faster and so are farther apart.

Wave fronts can be used to explain how mirages are produced by refraction of light. For example, on a hot day motorists sometimes see a mirage of water on the highway ahead of them, with distant vehicles seemingly reflected in it (Fig. 34–4a). On a hot day, there can be a layer of very hot air next to the roadway (made hot by the Sun beating on the road). Hot air is less dense than cooler air, so the index of refraction is slightly lower in the hot air. In Fig. 34–4b, we see a diagram of light coming from one point on a distant car (on the right) heading left toward the observer. Wave fronts and two rays (perpendicular to the wave fronts) are shown. Ray A heads directly at the observer and follows a straight-line path, and represents the normal view of the distant car. Ray B is a ray initially directed slightly downward, but it bends slightly as it moves through layers of air of different index of refraction. The wave fronts, shown in blue in Fig. 34–4b, move slightly faster in the layers of air nearer the ground. Thus ray B is bent as shown, and seems to the observer to be coming from below (dashed line) as if reflected off the road. Hence the mirage.

PHYSICS APPLIED

Highway mirages

34–3 Interference—Young's Double-Slit Experiment

In 1801, the Englishman Thomas Young (1773–1829) obtained convincing evidence for the wave nature of light and was even able to measure wavelengths for visible light. Figure 34–5a shows a schematic diagram of Young's famous double-slit experiment.

FIGURE 34–5 (a) Young's double-slit experiment. (b) If light consists of particles, we would expect to see two bright lines on the screen behind the slits. (c) In fact, many lines are observed. The slits and their separation need to be very thin.

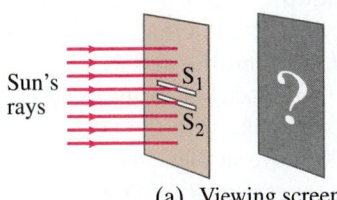

Sun's rays

S_1
S_2

(a) Viewing screen

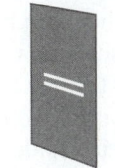

(b) Viewing screen (particle theory prediction)

(c) Viewing screen (actual)

To have light from a single source, Young used the Sun passing through a very narrow slit in a window covering. This beam of parallel rays falls on a screen containing two closely spaced slits, S_1 and S_2. (The slits and their separation are very narrow, not much larger than the wavelength of the light.) If light consists of tiny particles, we might expect to see two bright lines on a screen placed behind the slits as in (b). But instead a series of bright lines are seen, as in (c). Young was able to explain this result as a **wave-interference** phenomenon.

To understand why, we consider the simple situation of plane waves of light of a single wavelength—called **monochromatic**, meaning "one color"—falling on the two slits as shown in Fig. 34–6. Because of diffraction, the waves leaving the two small slits spread out as shown. This is equivalent to the interference pattern produced when two rocks are thrown into a lake (Fig. 15–23), or when sound from two loudspeakers interferes (Fig. 16–15). Recall Section 15–8 on wave interference.

FIGURE 34–6 If light is a wave, light passing through one of two slits should interfere with light passing through the other slit.

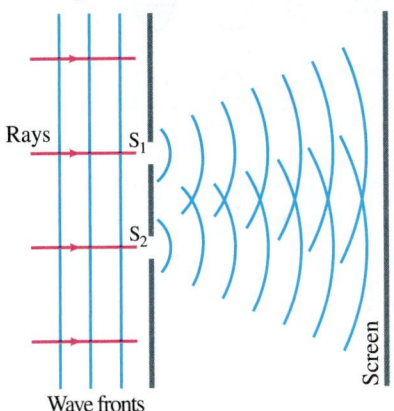

Rays

S_1

S_2

Wave fronts

Screen

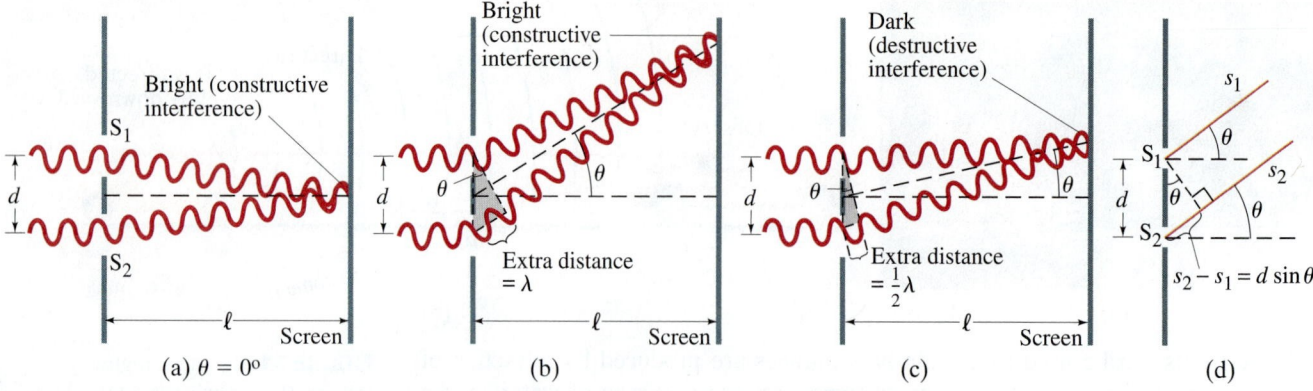

FIGURE 34–7 How the wave theory explains the pattern of lines seen in the double-slit experiment. (a) At the center of the screen the waves from each slit travel the same distance and are in phase. (b) At this angle θ, the lower wave travels an extra distance of one whole wavelength, and the waves are in phase; note from the shaded triangle that the path difference equals $d \sin \theta$. (c) For this angle θ, the lower wave travels an extra distance equal to one-half wavelength, so the two waves arrive at the screen fully out of phase. (d) A more detailed diagram showing the geometry for parts (b) and (c).

FIGURE 34–8 Two traveling waves are shown undergoing (a) constructive interference, (b) destructive interference. (See also Section 15–8.)

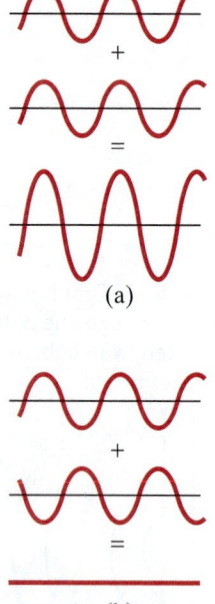

To see how an interference pattern is produced on the screen, we make use of Fig. 34–7. Waves of wavelength λ are shown entering the slits S_1 and S_2, which are a distance d apart. The waves spread out in all directions after passing through the slits (Fig. 34–6), but they are shown only for three different angles θ. In Fig. 34–7a, the waves reaching the center of the screen are shown ($\theta = 0°$). The waves from the two slits travel the same distance, so they are in phase: a crest of one wave arrives at the same time as a crest of the other wave. Hence the amplitudes of the two waves add to form a larger amplitude as shown in Fig. 34–8a. This is **constructive interference**, and there is a bright area at the center of the screen. Constructive interference also occurs when the paths of the two rays differ by one wavelength (or any whole number of wavelengths), as shown in Fig. 34–7b; also here there will be brightness on the screen. But if one ray travels an extra distance of one-half wavelength (or $\frac{3}{2}\lambda$, $\frac{5}{2}\lambda$, and so on), the two waves are exactly out of phase when they reach the screen: the crests of one wave arrive at the same time as the troughs of the other wave, and so they add to produce zero amplitude (Fig. 34–8b). This is **destructive interference**, and the screen is dark, Fig. 34–7c. Thus, there will be a series of bright and dark lines (or **fringes**) on the viewing screen.

To determine exactly where the bright lines fall, first note that Fig. 34–7 is somewhat exaggerated; in real situations, the distance d between the slits is very small compared to the distance ℓ to the screen. The rays from each slit for each case will therefore be essentially parallel, and θ is the angle they make with the horizontal as shown in Fig. 34–7d. From the shaded right triangles shown in Figs. 34–7b and c, we can see that the extra distance traveled by the lower ray is $d \sin \theta$ (seen more clearly in Fig. 34–7d). Constructive interference will occur, and a bright fringe will appear on the screen, when the *path difference*, $d \sin \theta$, equals a whole number of wavelengths:

$$d \sin \theta = m\lambda, \qquad m = 0, 1, 2, \cdots. \qquad \begin{bmatrix} \text{constructive} \\ \text{interference} \\ \text{(bright)} \end{bmatrix} \quad \textbf{(34–2a)}$$

The value of m is called the **order** of the interference fringe. The first order ($m = 1$), for example, is the first fringe on each side of the central fringe (which is at $\theta = 0$, $m = 0$). Destructive interference occurs when the path difference $d \sin \theta$ is $\frac{1}{2}\lambda$, $\frac{3}{2}\lambda$, and so on:

$$d \sin \theta = (m + \tfrac{1}{2})\lambda, \qquad m = 0, 1, 2, \cdots. \qquad \begin{bmatrix} \text{destructive} \\ \text{interference} \\ \text{(dark)} \end{bmatrix} \quad \textbf{(34–2b)}$$

The bright fringes are peaks or maxima of light intensity, the dark fringes are minima.

The intensity of the bright fringes is greatest for the central fringe $(m = 0)$ and decreases for higher orders, as shown in Fig. 34–9. How much the intensity decreases with increasing order depends on the width of the two slits.

Constructive interference

(a)

CONCEPTUAL EXAMPLE 34–1 **Interference pattern lines.** (a) Will there be an infinite number of points on the viewing screen where constructive and destructive interference occur, or only a finite number of points? (b) Are neighboring points of constructive interference uniformly spaced, or is the spacing between neighboring points of constructive interference not uniform?

RESPONSE (a) When you look at Eqs. 34–2a and b you might be tempted to say, given the statement $m = 0, 1, 2, \cdots$ beside the equations, that there are an infinite number of points of constructive and destructive interference. However, recall that $\sin \theta$ cannot exceed 1. Thus, there is an upper limit to the values of m that can be used in these equations. For Eq. 34–2a, the maximum value of m is the integer closest in value but smaller than d/λ. So there are a *finite* number of points of constructive and destructive interference no matter how large the screen. (b) The spacing between neighboring points of constructive or destructive interference is not uniform: The spacing gets larger as θ gets larger, and you can verify this statement mathematically. For small values of θ the spacing is nearly uniform as you will see in Example 34–2.

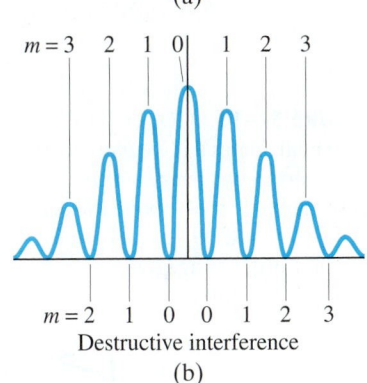

FIGURE 34–9 (a) Interference fringes produced by a double-slit experiment and detected by photographic film placed on the viewing screen. The arrow marks the central fringe. (b) Graph of the intensity of light in the interference pattern. Also shown are values of m for Eq. 34–2a (constructive interference) and Eq. 34–2b (destructive interference).

EXAMPLE 34–2 **Line spacing for double-slit interference.** A screen containing two slits 0.100 mm apart is 1.20 m from the viewing screen. Light of wavelength $\lambda = 500$ nm falls on the slits from a distant source. Approximately how far apart will adjacent bright interference fringes be on the screen?

APPROACH The angular position of bright (constructive interference) fringes is found using Eq. 34–2a. The distance between the first two fringes (say) can be found using right triangles as shown in Fig. 34–10.

SOLUTION Given $d = 0.100$ mm $= 1.00 \times 10^{-4}$ m, $\lambda = 500 \times 10^{-9}$ m, and $\ell = 1.20$ m, the first-order fringe $(m = 1)$ occurs at an angle θ given by

$$\sin \theta_1 = \frac{m\lambda}{d} = \frac{(1)(500 \times 10^{-9}\,\text{m})}{1.00 \times 10^{-4}\,\text{m}} = 5.00 \times 10^{-3}.$$

This is a very small angle, so we can take $\sin \theta \approx \theta$, with θ in radians. The first-order fringe will occur a distance x_1 above the center of the screen (see Fig. 34–10), given by $x_1/\ell = \tan \theta_1 \approx \theta_1$, so

$$x_1 \approx \ell\theta_1 = (1.20\,\text{m})(5.00 \times 10^{-3}) = 6.00\,\text{mm}.$$

The second-order fringe $(m = 2)$ will occur at

$$x_2 \approx \ell\theta_2 = \ell\frac{2\lambda}{d} = 12.0\,\text{mm}$$

above the center, and so on. Thus the lower order fringes are 6.00 mm apart.

NOTE The spacing between fringes is essentially uniform until the approximation $\sin \theta \approx \theta$ is no longer valid.

FIGURE 34–10 Examples 34–2 and 34–3. For small angles θ (give θ in radians), the interference fringes occur at distance $x = \theta\ell$ above the center fringe $(m = 0)$; θ_1 and x_1 are for the first-order fringe $(m = 1)$; θ_2 and x_2 are for $m = 2$.

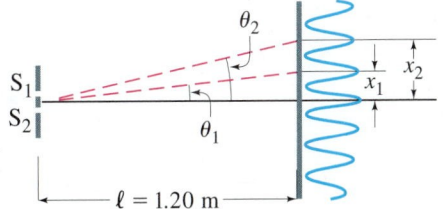

CONCEPTUAL EXAMPLE 34–3 **Changing the wavelength.** (a) What happens to the interference pattern shown in Fig. 34–10, Example 34–2, if the incident light (500 nm) is replaced by light of wavelength 700 nm? (b) What happens instead if the wavelength stays at 500 nm but the slits are moved farther apart?

RESPONSE (a) When λ increases in Eq. 34–2a but d stays the same, then the angle θ for bright fringes increases and the interference pattern spreads out. (b) Increasing the slit spacing d reduces θ for each order, so the lines are closer together.

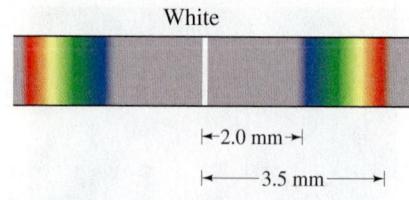

White

|←2.0 mm→|

|←——3.5 mm——→|

FIGURE 34–11 First-order fringes are a full spectrum, like a rainbow. Also Example 34–4.

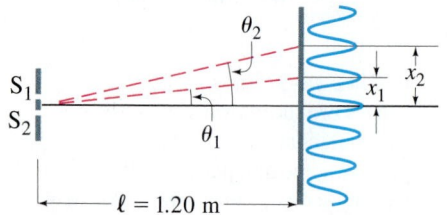

FIGURE 34–10 (Repeated.) For small angles θ (give θ in radians), the interference fringes occur at distance $x = \theta\ell$ above the center fringe $(m = 0)$; θ_1 and x_1 are for the first-order fringe $(m = 1)$, θ_2 and x_2 are for $m = 2$.

$\ell = 1.20$ m

From Eqs. 34–2 we can see that, except for the zeroth-order fringe at the center, the position of the fringes depends on wavelength. Consequently, when white light falls on the two slits, as Young found in his experiments, the central fringe is white, but the first- (and higher-) order fringes contain a spectrum of colors like a rainbow; θ was found to be smallest for violet light and largest for red (Fig. 34–11). By measuring the position of these fringes, Young was the first to determine the wavelengths of visible light (using Eqs. 34–2). In doing so, he showed that what distinguishes different colors physically is their wavelength (or frequency), an idea put forward earlier by Grimaldi in 1665.

EXAMPLE 34–4 **Wavelengths from double-slit interference.** White light passes through two slits 0.50 mm apart, and an interference pattern is observed on a screen 2.5 m away. The first-order fringe resembles a rainbow with violet and red light at opposite ends. The violet light is about 2.0 mm and the red 3.5 mm from the center of the central white fringe (Fig. 34–11). Estimate the wavelengths for the violet and red light.

APPROACH We find the angles for violet and red light from the distances given and the diagram of Fig. 34–10. Then we use Eq. 34–2a to obtain the wavelengths. Because 3.5 mm is much less than 2.5 m, we can use the small-angle approximation.

SOLUTION We use Eq. 34–2a with $m = 1$ and $\sin\theta \approx \tan\theta \approx \theta$. Then for violet light, $x = 2.0$ mm, so (see also Fig. 34–10)

$$\lambda = \frac{d\sin\theta}{m} \approx \frac{d\theta}{m} \approx \frac{d}{m}\frac{x}{\ell} = \left(\frac{5.0 \times 10^{-4}\,\text{m}}{1}\right)\left(\frac{2.0 \times 10^{-3}\,\text{m}}{2.5\,\text{m}}\right) = 4.0 \times 10^{-7}\,\text{m},$$

or 400 nm. For red light, $x = 3.5$ mm, so

$$\lambda = \frac{d}{m}\frac{x}{\ell} = \left(\frac{5.0 \times 10^{-4}\,\text{m}}{1}\right)\left(\frac{3.5 \times 10^{-3}\,\text{m}}{2.5\,\text{m}}\right) = 7.0 \times 10^{-7}\,\text{m} = 700\,\text{nm}.$$

Coherence

The two slits in Fig. 34–7 act as if they were two sources of radiation. They are called **coherent sources** because the waves leaving them have the same wavelength and frequency, and bear the same phase relationship to each other at all times. This happens because the waves come from a single source to the left of the two slits in Fig. 34–7, splitting the original beam into two. An interference pattern is observed only when the sources are coherent. If two tiny lightbulbs replaced the two slits, an interference pattern would not be seen. The light emitted by one lightbulb would have a random phase with respect to the second bulb, and the screen would be more or less uniformly illuminated. Two such sources, whose output waves have phases that bear no fixed relationship to each other over time, are called **incoherent sources**.

*34–4 Intensity in the Double-Slit Interference Pattern

We saw in Section 34–3 that the interference pattern produced by the coherent light from two slits, S_1 and S_2 (Figs. 34–7 and 34–9), produces a series of bright and dark fringes. If the two monochromatic waves of wavelength λ are in phase at the slits, the maxima (brightest points) occur at angles θ given by (Eqs. 34–2)

$$d\sin\theta = m\lambda,$$

and the minima (darkest points) when

$$d\sin\theta = \left(m + \tfrac{1}{2}\right)\lambda,$$

where m is an integer $(m = 0, 1, 2, \cdots)$.

We now determine the intensity of the light at all points in the pattern, assuming that if either slit were covered, the light passing through the other would diffract sufficiently to illuminate a large portion of the screen uniformly. The intensity I of the light at any point is proportional to the square of its wave amplitude (Section 15–3). Treating light as an electromagnetic wave, I is proportional to the square of the electric field E (or to the magnetic field B, Section 31–8): $I \propto E^2$. The electric field $\vec{E}$ at any point P (see Fig. 34–12) will be the sum of the electric field vectors of the waves coming from each of the two slits, $\vec{E}_1$ and $\vec{E}_2$. Since $\vec{E}_1$ and $\vec{E}_2$ are essentially parallel (on a screen far away compared to the slit separation), the magnitude of the electric field at angle θ (that is, at point P) will be

$$E_\theta = E_1 + E_2 .$$

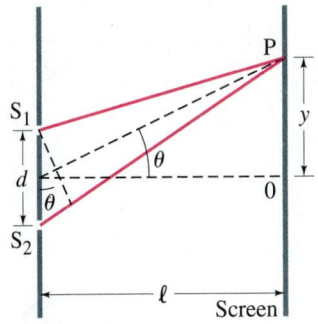

FIGURE 34–12 Determining the intensity in a double-slit interference pattern. Not to scale: in fact $\ell \gg d$, and the two rays become essentially parallel.

Both E_1 and E_2 vary sinusoidally with frequency $f = c/\lambda$, but they differ in phase, depending on their different travel distances from the slits. The electric field at P can then be written for the light from each of the two slits, using $\omega = 2\pi f$, as

$$\begin{aligned} E_1 &= E_{10} \sin \omega t \\ E_2 &= E_{20} \sin(\omega t + \delta) \end{aligned} \qquad \textbf{(34–3)}$$

where E_{10} and E_{20} are their respective amplitudes and δ is the phase difference. The value of δ depends on the angle θ, so let us now determine δ as a function of θ.

At the center of the screen (point 0), $\delta = 0$. If the difference in path length from P to S_1 and S_2 is $d \sin \theta = \lambda/2$, the two waves are exactly out of phase so $\delta = \pi$ (or 180°). If $d \sin \theta = \lambda$, the two waves differ in phase by $\delta = 2\pi$. In general, then, δ is related to θ by

$$\frac{\delta}{2\pi} = \frac{d \sin \theta}{\lambda}$$

or

$$\delta = \frac{2\pi}{\lambda} d \sin \theta . \qquad \textbf{(34–4)}$$

To determine $E_\theta = E_1 + E_2$, we add the two scalars E_1 and E_2 which are sine functions differing by the phase δ. One way to determine the sum of E_1 and E_2 is to use a **phasor diagram**. (We used this technique before, in Chapter 30.) As shown in Fig. 34–13, we draw an arrow of length E_{10} to represent the amplitude of E_1 (Eq. 34–3); and the arrow of length E_{20}, which we draw to make a fixed angle δ with E_{10}, represents the amplitude of E_2. When the diagram rotates at angular frequency ω about the origin, the projections of E_{10} and E_{20} on the vertical axis represent E_1 and E_2 as a function of time (see Eq. 34–3). We let $E_{\theta 0}$ be the "vector" sum[†] of E_{10} and E_{20}; $E_{\theta 0}$ is the amplitude of the sum $E_\theta = E_1 + E_2$, and the projection of $E_{\theta 0}$ on the vertical axis is just E_θ. If the two slits provide equal illumination, so that $E_{10} = E_{20} = E_0$, then from *symmetry* in Fig. 34–13, the angle $\phi = \delta/2$, and we can write

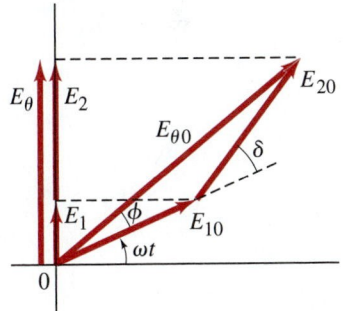

FIGURE 34–13 Phasor diagram for double-slit interference pattern.

$$E_\theta = E_{\theta 0} \sin \left(\omega t + \frac{\delta}{2} \right) . \qquad \textbf{(34–5a)}$$

From Fig. 34–13 we can also see that

$$E_{\theta 0} = 2 E_0 \cos \phi = 2 E_0 \cos \frac{\delta}{2} . \qquad \textbf{(34–5b)}$$

Combining Eqs. 34–5a and b, we obtain

$$E_\theta = 2 E_0 \cos \frac{\delta}{2} \sin \left(\omega t + \frac{\delta}{2} \right) , \qquad \textbf{(34–5c)}$$

where δ is given by Eq. 34–4.

[†]We are not adding the actual electric field vectors; instead we are using the "phasor" technique to add the amplitudes, taking into account the phase difference of the two waves.

We are not really interested in E_θ as a function of time, since for visible light the frequency (10^{14} to 10^{15} Hz) is much too high to be noticeable. We are interested in the average intensity, which is proportional to the amplitude squared, $E_{\theta 0}^2$. We now drop the word "average," and we let I_θ ($I_\theta \propto E_\theta^2$) be the intensity at any point P at an angle θ to the horizontal. We let I_0 be the intensity at point O, the center of the screen, where $\theta = \delta = 0$, so $I_0 \propto (E_{10} + E_{20})^2 = (2E_0)^2$. Then the ratio I_θ/I_0 is equal to the ratio of the squares of the electric-field amplitudes at these two points, so

$$\frac{I_\theta}{I_0} = \frac{E_{\theta 0}^2}{(2E_0)^2} = \cos^2\frac{\delta}{2}$$

where we used Eq. 34–5b. Thus the intensity I_θ at any point is related to the maximum intensity at the center of the screen by

$$I_\theta = I_0 \cos^2\frac{\delta}{2}$$

$$= I_0 \cos^2\left(\frac{\pi d \sin\theta}{\lambda}\right)$$

(34–6)

where δ was given by Eq. 34–4. This is the relation we sought.

From Eq. 34–6 we see that maxima occur where $\cos\delta/2 = \pm 1$, which corresponds to $\delta = 0, 2\pi, 4\pi, \cdots$. From Eq. 34–4, δ has these values when

$$d \sin\theta = m\lambda, \qquad m = 0, 1, 2, \cdots.$$

Minima occur where $\delta = \pi, 3\pi, 5\pi, \cdots$, which corresponds to

$$d \sin\theta = (m + \tfrac{1}{2})\lambda, \qquad m = 0, 1, 2, \cdots.$$

These are the same results we obtained in Section 34–3. But now we know not only the position of maxima and minima, but from Eq. 34–6 we can determine the intensity at all points.

In the usual situation where the distance ℓ to the screen from the slits is large compared to the slit separation d ($\ell \gg d$), if we consider only points P whose distance y from the center (point O) is small compared to ℓ ($y \ll \ell$)—see Fig. 34–12—then

$$\sin\theta = \frac{y}{\ell}.$$

From this it follows (see Eq. 34–4) that

$$\delta = \frac{2\pi}{\lambda}\frac{d}{\ell}y.$$

Equation 34–6 then becomes

$$I_\theta = I_0\left[\cos\left(\frac{\pi d}{\lambda\ell}y\right)\right]^2. \qquad\qquad [y \ll \ell, d \ll \ell] \quad (34\text{–}7)$$

| **EXERCISE B** What are the values for the intensity I_θ when (a) $y = 0$, (b) $y = \lambda\ell/4d$, and (c) $y = \lambda\ell/2d$?

The intensity I_θ as a function of the phase difference δ is plotted in Fig. 34–14. In the approximation of Eq. 34–7, the horizontal axis could as well be y, the position on the screen.

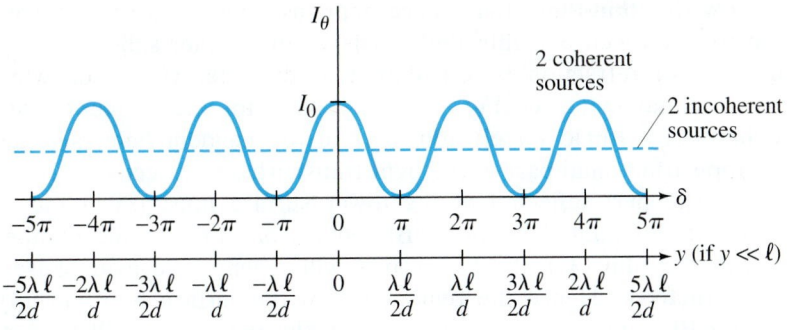

FIGURE 34–14 Intensity I as a function of phase difference δ and position on screen y (assuming $y \ll \ell$).

The intensity pattern expressed in Eqs. 34–6 and 34–7, and plotted in Fig. 34–14, shows a series of maxima of equal height, and is based on the assumption that each slit (alone) would illuminate the screen uniformly. This is never quite true, as we shall see when we discuss diffraction in the next Chapter. We will see that the center maximum is strongest and each succeeding maximum to each side is a little less strong.

EXAMPLE 34–5 **Antenna intensity.** Two radio antennas are located close to each other as shown in Fig. 34–15, separated by a distance d. The antennas radiate in phase with each other, emitting waves of intensity I_0 at wavelength λ. (a) Calculate the net intensity as a function of θ for points very far from the antennas. (b) For $d = \lambda$, determine I and find in which directions I is a maximum and a minimum. (c) Repeat part (b) when $d = \lambda/2$.

APPROACH This setup is similar to Young's double-slit experiment.

SOLUTION (a) Points of constructive and destructive interference are still given by Eqs. 34–2a and b, and the net intensity as a function of θ is given by Eq. 34–6. (b) We let $d = \lambda$ in Eq. 34–6, and find for the intensity,

$$I = I_0 \cos^2(\pi \sin \theta).$$

I is a maximum, equal to I_0, when $\sin \theta = 0$, 1, or -1, meaning $\theta = 0, 90°, 180°$, and $270°$. I is zero when $\sin \theta = \frac{1}{2}$ and $-\frac{1}{2}$, for which $\theta = 30°, 150°, 210°$, and $330°$. (c) For $d = \lambda/2$, I is maximized for $\theta = 0$ and $180°$, and minimized for $90°$ and $270°$.

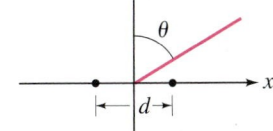

FIGURE 34–15 Example 34–5. The two dots represent the antennas.

34–5 Interference in Thin Films

Interference of light gives rise to many everyday phenomena such as the bright colors reflected from soap bubbles and from thin oil or gasoline films on water, Fig. 34–16. In these and other cases, the colors are a result of constructive interference between light reflected from the two surfaces of the thin film.

FIGURE 34–16 Thin film interference patterns seen in (a) a soap bubble, (b) a thin film of soapy water, and (c) a thin layer of oil on wet pavement.

(a)   (b)   (c)

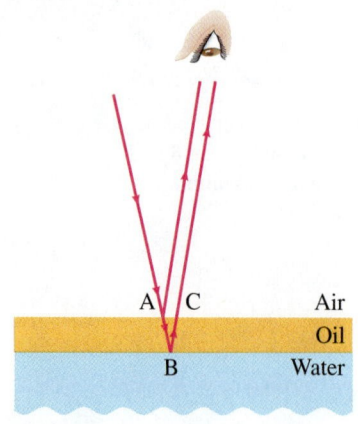

FIGURE 34–17 Light reflected from the upper and lower surfaces of a thin film of oil lying on water. This analysis assumes the light strikes the surface nearly perpendicularly, but is shown here at an angle so we can display each ray.

To see how this **thin-film interference** happens, consider a smooth surface of water on top of which is a thin uniform layer of another substance, say an oil whose index of refraction is less than that of water (we'll see why we assume this in a moment); see Fig. 34–17. Assume for the moment that the incident light is of a single wavelength. Part of the incident light is reflected at A on the top surface, and part of the light transmitted is reflected at B on the lower surface. The part reflected at the lower surface must travel the extra distance ABC. If this *path difference* ABC equals one or a whole number of wavelengths in the film (λ_n), the two waves will reach the eye in phase and interfere constructively. Hence the region AC on the surface film will appear bright. But if ABC equals $\frac{1}{2}\lambda_n, \frac{3}{2}\lambda_n$, and so on, the two waves will be exactly out of phase and destructive interference occurs: the area AC on the film will show no reflection—it will be dark (or better, transparent to the dark material below). The wavelength λ_n is *the wavelength in the film:* $\lambda_n = \lambda/n$, where n is the index of refraction in the film and λ is the wavelength in vacuum. See Eq. 34–1.

When white light falls on such a film, the path difference ABC will equal λ_n (or $m\lambda_n$, with $m =$ an integer) for only one wavelength at a given viewing angle. The color corresponding to λ (λ in air) will be seen as very bright. For light viewed at a slightly different angle, the path difference ABC will be longer or shorter and a different color will undergo constructive interference. Thus, for an extended (nonpoint) source emitting white light, a series of bright colors will be seen next to one another. Variations in thickness of the film will also alter the path difference ABC and therefore affect the color of light that is most strongly reflected.

EXERCISE C Return to the Chapter-Opening Question, page 900, and answer it again now. Try to explain why you may have answered differently the first time.

When a curved glass surface is placed in contact with a flat glass surface, Fig. 34–18, a series of concentric rings is seen when illuminated from above by either white light (as shown) or by monochromatic light. These are called **Newton's rings**[†] and they are due to interference between waves reflected by the top and bottom surfaces of the very thin *air gap* between the two pieces of glass. Because this gap (which is equivalent to a thin film) increases in width from the central contact point out to the edges, the extra path length for the lower ray (equal to BCD) varies; where it equals $0, \frac{1}{2}\lambda, \lambda, \frac{3}{2}\lambda, 2\lambda$, and so on, it corresponds to constructive and destructive interference; and this gives rise to the series of bright colored circles seen in Fig. 34–18b. The color you see at a given radius corresponds to constructive interference; at that radius, other colors partially or fully destructively interfere. (If monochromatic light is used, the rings are alternately bright and dark.)

[†]Although Newton gave an elaborate description of them, they had been first observed and described by his contemporary, Robert Hooke.

FIGURE 34–18 Newton's rings. (a) Light rays reflected from upper and lower surfaces of the thin air gap can interfere. (b) Photograph of interference patterns using white light.

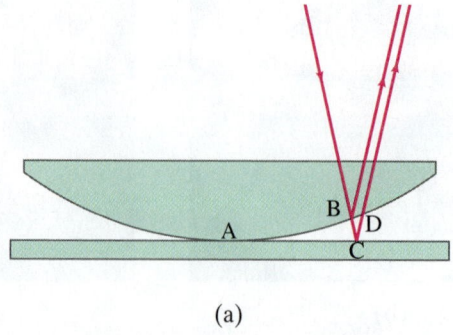

(a)

(b)

The point of contact of the two glass surfaces (A in Fig. 34–18a) is dark in Fig. 34–18b. Since the path difference is zero here, our previous analysis would suggest that the waves reflected from each surface are in phase and so this central area ought to be bright. But it is dark, which tells us something else is happening here: the two waves must be completely out of phase. This can happen if one of the waves, upon reflection, flips over—a crest becomes a trough—see Fig. 34–19. We say that the reflected wave has undergone a phase change of 180°, or of half a wave cycle. Indeed, this and other experiments reveal that, at normal incidence,

a beam of light reflected by a material with index of refraction greater than that of the material in which it is traveling, changes phase by 180° or $\frac{1}{2}$ cycle;

see Fig. 34–19. This phase change acts just like a path difference of $\frac{1}{2}\lambda$. If the index of refraction of the reflecting material is less than that of the material in which the light is traveling, no phase change occurs.[†]

Thus the wave reflected at the curved surface above the air gap in Fig. 34–18a undergoes no change in phase. But the wave reflected at the lower surface, where the beam in air strikes the glass, undergoes a $\frac{1}{2}$-cycle phase change, equivalent to a $\frac{1}{2}\lambda$ path difference. Thus the two waves reflected near the point of contact A of the two glass surfaces (where the air gap approaches zero thickness) will be a half cycle (or 180°) out of phase, and a dark spot occurs. Bright colored rings will occur when the path difference is $\frac{1}{2}\lambda, \frac{3}{2}\lambda$, and so on, because the phase change at one surface effectively adds a path difference of $\frac{1}{2}\lambda \left(=\frac{1}{2}\text{ cycle}\right)$. (If monochromatic light is used, the bright Newton's rings will be separated by dark bands which occur when the path difference BCD in Fig. 34–18a is equal to an integral number of wavelengths.)

Returning for a moment to Fig. 34–17, the light reflecting at both interfaces, air–oil and oil–water, underwent a phase change of 180° equivalent to a path difference of $\frac{1}{2}\lambda$, since we assumed $n_{\text{water}} > n_{\text{oil}} > n_{\text{air}}$; since the phase changes were equal, they didn't affect our analysis.

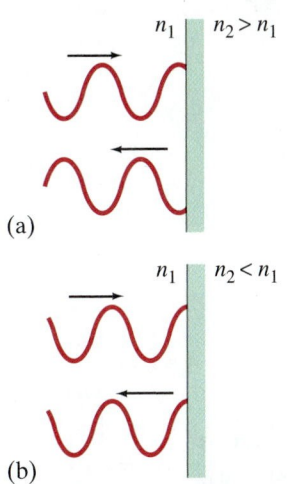

FIGURE 34–19 (a) Reflected ray changes phase by 180° or $\frac{1}{2}$ cycle if $n_2 > n_1$, but (b) does not if $n_2 < n_1$.

FIGURE 34–20 (a) Light rays reflected from the upper and lower surfaces of a thin wedge of air interfere to produce bright and dark bands. (b) Pattern observed when glass plates are optically flat; (c) pattern when plates are not so flat. See Example 34–6.

EXAMPLE 34–6 **Thin film of air, wedge-shaped.** A very fine wire 7.35×10^{-3} mm in diameter is placed between two flat glass plates as in Fig. 34–20a. Light whose wavelength in air is 600 nm falls (and is viewed) perpendicular to the plates and a series of bright and dark bands is seen, Fig. 34–20b. How many light and dark bands will there be in this case? Will the area next to the wire be bright or dark?

APPROACH We need to consider two effects: (1) path differences for rays reflecting from the two close surfaces (thin wedge of air between the two glass plates), and (2) the $\frac{1}{2}$-cycle phase change at the lower surface (point E in Fig. 34–20a), where rays in air can enter glass. Because of the phase change at the lower surface, there will be a dark band (no reflection) when the path difference is $0, \lambda, 2\lambda, 3\lambda$, and so on. Since the light rays are perpendicular to the plates, the extra path length equals $2t$, where t is the thickness of the air gap at any point.

SOLUTION Dark bands will occur where

$$2t = m\lambda, \qquad m = 0, 1, 2, \cdots.$$

Bright bands occur when $2t = \left(m + \frac{1}{2}\right)\lambda$, where m is an integer. At the position of the wire, $t = 7.35 \times 10^{-6}$ m. At this point there will be $2t/\lambda = (2)(7.35 \times 10^{-6}\text{ m})/(6.00 \times 10^{-7}\text{ m}) = 24.5$ wavelengths. This is a "half integer," so the area next to the wire will be bright. There will be a total of 25 dark lines along the plates, corresponding to path lengths of $0\lambda, 1\lambda, 2\lambda, 3\lambda, \ldots, 24\lambda$, including the one at the point of contact A ($m = 0$). Between them, there will be 24 bright lines plus the one at the end, or 25.

NOTE The bright and dark bands will be straight only if the glass plates are extremely flat. If they are not, the pattern is uneven, as in Fig. 34–20c. Thus we see a very precise way of testing a glass surface for flatness. Spherical lens surfaces can be tested for precision by placing the lens on a flat glass surface and observing Newton's rings (Fig. 34–18b) for perfect circularity.

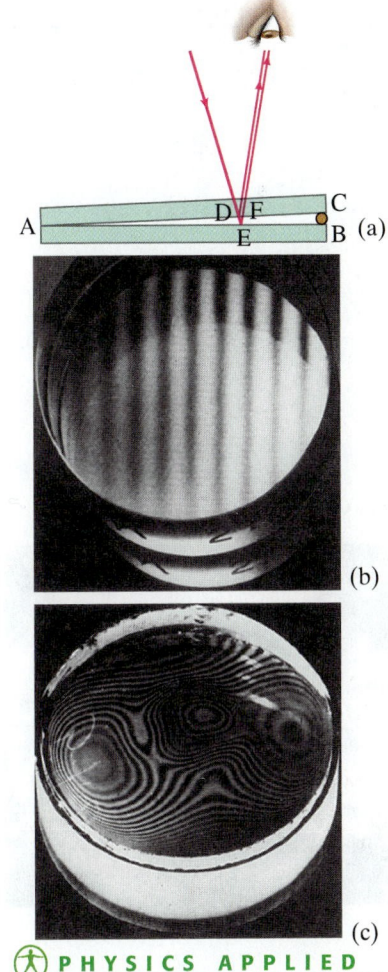

PHYSICS APPLIED
Testing glass for flatness

[†]This result can be derived from Maxwell's equations. It corresponds to the reflection of a wave traveling along a cord when it reaches the end; as we saw in Fig. 15–18, if the end is tied down, the wave changes phase and the pulse flips over, but if the end is free, no phase change occurs.

If the wedge between the two glass plates of Example 34–6 is filled with some transparent substance other than air—say, water—the pattern shifts because the wavelength of the light changes. In a material where the index of refraction is n, the wavelength is $\lambda_n = \lambda/n$, where λ is the wavelength in vacuum (Eq. 34–1). For instance, if the thin wedge of Example 34–6 were filled with water, then $\lambda_n = 600\,\text{nm}/1.33 = 450\,\text{nm}$; instead of 25 dark lines, there would be 33.

When white light (rather than monochromatic light) is incident on the thin wedge of air in Figs. 34–18a or 34–20a, a colorful series of fringes is seen because constructive interference occurs for different wavelengths in the reflected light at different thicknesses along the wedge.

A soap bubble (Fig. 34–16a and Chapter-Opening Photo) is a thin spherical shell (or film) with air inside. The variations in thickness of a soap bubble film gives rise to bright colors reflected from the soap bubble. (There is air on both sides of the bubble film.) Similar variations in film thickness produce the bright colors seen reflecting from a thin layer of oil or gasoline on a puddle or lake (Fig. 34–16c). Which wavelengths appear brightest also depends on the viewing angle.

EXAMPLE 34–7 **Thickness of soap bubble skin.** A soap bubble appears green ($\lambda = 540\,\text{nm}$) at the point on its front surface nearest the viewer. What is the smallest thickness the soap bubble film could have? Assume $n = 1.35$.

APPROACH Assume the light is reflected perpendicularly from the point on a spherical surface nearest the viewer, Fig. 34–21. The light rays also reflect from the inner surface of the soap bubble film as shown. The path difference of these two reflected rays is $2t$, where t is the thickness of the soap film. Light reflected from the first (outer) surface undergoes a 180° phase change (index of refraction of soap is greater than that of air), whereas reflection at the second (inner) surface does not. To determine the thickness t for an interference maximum, we must use the wavelength of light in the soap ($n = 1.35$).

SOLUTION The 180° phase change at only one surface is equivalent to a $\frac{1}{2}\lambda$ path difference. Therefore, green light is bright when the minimum path difference equals $\frac{1}{2}\lambda_n$. Thus, $2t = \lambda/2n$, so

$$t = \frac{\lambda}{4n} = \frac{(540\,\text{nm})}{(4)(1.35)} = 100\,\text{nm}.$$

This is the smallest thickness; but the green color is more likely to be seen at the *next* thickness that gives constructive interference, $2t = 3\lambda/2n$, because other colors would be more fully cancelled by destructive interference. The more likely thickness is $3\lambda/4n = 300\,\text{nm}$, or even $5\lambda/4n = 500\,\text{nm}$. Note that green is seen in air, so $\lambda = 540\,\text{nm}$ (not λ/n).

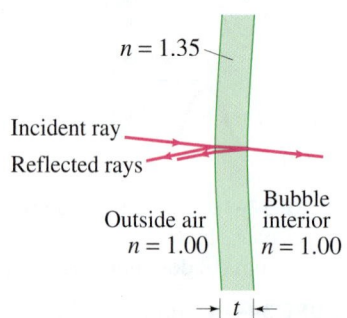

FIGURE 34–21 Example 34–7. The incident and reflected rays are assumed to be perpendicular to the bubble's surface. They are shown at a slight angle so we can distinguish them.

$n = 1.35$

Incident ray

Reflected rays

Outside air $n = 1.00$

Bubble interior $n = 1.00$

t

*Colors in a Thin Soap Film

The thin film of soapy water shown in Fig. 34–16b (repeated here) has stood vertically for a long time so that gravity has pulled much of the soapy water toward the bottom. The top section is so thin (perhaps 30 nm thick) that light reflected from the front and back surfaces have almost no path difference. Thus the 180° phase change at the front surface assures that the two reflected waves are 180° out of phase for all wavelengths of visible light. The white light incident on this thin film does not reflect at the top part of the film. Thus the top **is** transparent, and we see the background which is black.

Below the black area at the top, there is a thin blue line, and then a white band. The film thickness is perhaps 75 to 100 nm, so the shortest wavelength (blue) light begins to partially interfere constructively; but just below, where the thickness is slightly greater (100 nm), the path length is reasonably close to $\lambda/2$ for much of the spectrum and we see white or silver. (Why? Recall that red starts at 600 nm in air; so most colors in the spectrum lie between 450 nm and 600 nm in air; but in water the wavelengths are $n = 1.33$ times smaller, 340 nm to 450 nm, so a 100 nm thickness is a 200 nm path length, not far from $\lambda/2$ for most colors.) Immediately below the white band we see a brown band (around 200 nm in thickness) where selected wavelengths (not all) are close to exactly λ and those colors destructively interfere, leaving only a few colors to partially interfere constructively, giving us murky brown.

FIGURE 34–16b (Repeated.)

Farther down, with increasing thickness t, a path length $2t = 510$ nm corresponds nicely to $\frac{3}{2}\lambda$ for blue, but not for other colors, so we see blue ($\frac{3}{2}\lambda$ path difference plus $\frac{1}{2}\lambda$ phase change = constructive interference). Other colors experience constructive interference (at $\frac{3}{2}\lambda$ and then at $\frac{5}{2}\lambda$) at still greater thicknesses, so we see a series of separated colors something like a rainbow.

In the soap bubble of our Chapter-Opening Photo (p. 900), similar things happen: at the top (where the film is thinnest) we see black and then silver-white, just as within the loop shown in Fig. 34–16b. And examine the oil film on wet pavement shown in Fig. 34–16c (repeated here); the oil film is thickest at the center and thins out toward the edges. Notice the whitish outer ring where most colors constructively interfere, which would suggest a thickness on the order of 100 nm as discussed above for the white band in the soap film. Beyond the outer white band of the oil film, Fig. 34–16c, there is still some oil, but the film is so thin that reflected light from upper and lower surfaces destructively interfere and you can see right through this very thin oil film.

FIGURE 34–16c (Repeated.)

Lens Coatings

An important application of thin-film interference is in the coating of glass to make it "nonreflecting," particularly for lenses. A glass surface reflects about 4% of the light incident upon it. Good-quality cameras, microscopes, and other optical devices may contain six to ten thin lenses. Reflection from all these surfaces can reduce the light level considerably, and multiple reflections produce a background haze that reduces the quality of the image. By reducing reflection, transmission is increased. A very thin coating on the lens surfaces can reduce reflections considerably. The thickness of the film is chosen so that light (at least for one wavelength) reflecting from the front and rear surfaces of the film destructively interferes. The amount of reflection at a boundary depends on the difference in index of refraction between the two materials. Ideally, the coating material should have an index of refraction which is the geometric mean $(= \sqrt{n_1 n_2})$ of those for air and glass, so that the amount of reflection at each surface is about equal. Then destructive interference can occur nearly completely for one particular wavelength depending on the thickness of the coating. Nearby wavelengths will at least partially destructively interfere, but a single coating cannot eliminate reflections for all wavelengths. Nonetheless, a single coating can reduce total reflection from 4% to 1% of the incident light. Often the coating is designed to eliminate the center of the reflected spectrum (around 550 nm). The extremes of the spectrum—red and violet—will not be reduced as much. Since a mixture of red and violet produces purple, the light seen reflected from such coated lenses is purple (Fig. 34–22). Lenses containing two or three separate coatings can more effectively reduce a wider range of reflecting wavelengths.

PHYSICS APPLIED
Lens coatings

FIGURE 34–22 A coated lens. Note color of light reflected from the front lens surface.

Interference

1. **Interference effects** depend on the simultaneous arrival of two or more waves at the same point in space.

2. **Constructive interference** occurs when the waves arrive in phase with each other: a crest of one wave arrives at the same time as a crest of the other wave. The amplitudes of the waves then add to form a larger amplitude. Constructive interference also occurs when the path difference is exactly one full wavelength or any integer multiple of a full wavelength: 1λ, 2λ, 3λ, $\cdots$.

3. **Destructive interference** occurs when a crest of one wave arrives at the same time as a trough of the other wave. The amplitudes add, but they are of opposite sign, so the total amplitude is reduced to zero if the two amplitudes are equal. Destructive interference occurs whenever the phase difference is half a wave cycle, or the path difference is a half-integral number of wavelengths. Thus, the total amplitude will be zero if two identical waves arrive one-half wavelength out of phase, or $(m + \frac{1}{2})\lambda$ out of phase, where m is an integer.

4. For thin-film interference, an extra half-wavelength **phase shift** occurs when light **reflects** from an optically more dense medium (going from a material of lesser toward greater index of refraction).

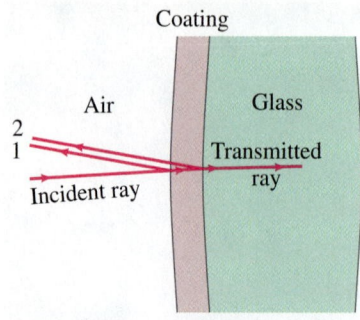

FIGURE 34–23 Example 34–8. Incident ray of light is partially reflected at the front surface of a lens coating (ray 1) and again partially reflected at the rear surface of the coating (ray 2), with most of the energy passing as the transmitted ray into the glass.

EXAMPLE 34–8 **Nonreflective coating.** What is the thickness of an optical coating of MgF_2 whose index of refraction is $n = 1.38$ and which is designed to eliminate reflected light at wavelengths (in air) around 550 nm when incident normally on glass for which $n = 1.50$?

APPROACH We explicitly follow the procedure outlined in the Problem Solving Strategy on page 913.

SOLUTION

1. **Interference effects.** Consider two rays reflected from the front and rear surfaces of the coating on the lens as shown in Fig. 34–23. The rays are drawn not quite perpendicular to the lens so we can see each of them. These two reflected rays will interfere with each other.

2. **Constructive interference.** We want to eliminate reflected light, so we do not consider constructive interference.

3. **Destructive interference.** To eliminate reflection, we want reflected rays 1 and 2 to be $\frac{1}{2}$ cycle out of phase with each other so that they destructively interfere. The phase difference is due to the path difference $2t$ traveled by ray 2, as well as any phase change in either ray due to reflection.

4. **Reflection phase shift.** Rays 1 and 2 *both* undergo a change of phase by $\frac{1}{2}$ cycle when they reflect from the coating's front and rear surfaces, respectively (at both surfaces the index of refraction increases). Thus there is no net change in phase due to the reflections. The net phase difference will be due to the extra path $2t$ taken by ray 2 in the coating, where $n = 1.38$. We want $2t$ to equal $\frac{1}{2}\lambda_n$ so that destructive interference occurs, where $\lambda_n = \lambda/n$ is the wavelength in the coating. With $2t = \lambda_n/2 = \lambda/2n$, then

$$t = \frac{\lambda_n}{4} = \frac{\lambda}{4n} = \frac{(550 \text{ nm})}{(4)(1.38)} = 99.6 \text{ nm}.$$

NOTE We could have set $2t = \left(m + \frac{1}{2}\right)\lambda_n$, where m is an integer. The smallest thickness ($m = 0$) is usually chosen because destructive interference will occur over the widest angle.

NOTE Complete destructive interference occurs only for the given wavelength of visible light. Longer and shorter wavelengths will have only partial cancellation.

*34–6 Michelson Interferometer

A useful instrument involving wave interference is the **Michelson interferometer** (Fig. 34–24),[†] invented by the American Albert A. Michelson (Section 31–7). Monochromatic light from a single point on an extended source is shown striking a half-silvered mirror M_S. This **beam splitter** mirror M_S has a thin layer of silver that reflects only half the light that hits it, so that half of the beam passes through to a fixed mirror M_2, where it is reflected back. The other half is reflected by M_S to a mirror M_1 that is movable (by a fine-thread screw), where it is also reflected back. Upon its return, part of beam 1 passes through M_S and reaches the eye; and part of beam 2, on its return, is reflected by M_S into the eye. If the two path lengths are identical, the two coherent beams entering the eye constructively interfere and brightness will be seen. If the movable mirror is moved a distance $\lambda/4$, one beam will travel an extra distance equal to $\lambda/2$ (because it travels back and forth over the distance $\lambda/4$). In this case, the two beams will destructively interfere and darkness will be seen. As M_1 is moved farther, brightness will recur (when the path difference is λ), then darkness, and so on.

FIGURE 34–24 Michelson interferometer.

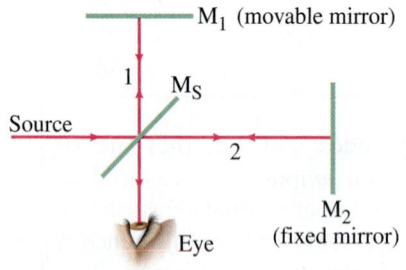

Very precise length measurements can be made with an interferometer. The motion of mirror M_1 by only $\frac{1}{4}\lambda$ produces a clear difference between brightness and darkness. For $\lambda = 400$ nm, this means a precision of 100 nm or 10^{-4} mm! If mirror M_1 is tilted very slightly, the bright or dark spots are seen instead as a series of bright and dark lines or "fringes." By counting the number of fringes (or fractions thereof) that pass a reference line, extremely precise length measurements can be made.

[†]There are other types of interferometer, but Michelson's is the best known.

*34–7 Luminous Intensity

The *intensity* of light, as for any electromagnetic wave, is measured by the Poynting vector in W/m^2, and the total power output of a source can be measured in watts (the *radiant flux*). But for measuring the visual sensation we call brightness, we must consider *only* the visible spectrum as well as the eye's sensitivity to different wavelengths—the eye is most sensitive in the central, 550-nm (green), portion of the spectrum.

These factors are taken into account in the quantity **luminous flux**, F_ℓ, whose unit is the **lumen** (lm). One lumen is equivalent to $\frac{1}{683}$ watts of 555-nm light.

Since the luminous flux from a source may not be uniform over all directions, we define the **luminous intensity** I_ℓ as the luminous flux per unit solid angle[†] (steradian). Its unit is the **candela** (cd) where $1\,\text{cd} = 1\,\text{lm/sr}$, and it is one of the seven basic quantities in the SI. (See Section 1–4 and Table 1–5.)

The **illuminance**, E_ℓ, is the luminous flux incident on a surface per unit area of the surface: $E_\ell = F_\ell/A$. Its unit is the lumen per square meter (lm/m^2) and is a measure of the illumination falling on a surface.[‡]

> **EXAMPLE 34–9** **Lightbulb illuminance.** The brightness of a particular type of 100-W lightbulb is rated at 1700 lm. Determine (*a*) the luminous intensity and (*b*) the illuminance at a distance of 2.0 m.
>
> **APPROACH** Assume the light output is uniform in all directions.
>
> **SOLUTION** (*a*) A full sphere corresponds to 4π sr. Hence, $I_\ell = 1700\,\text{lm}/4\pi\,\text{sr} = 135\,\text{cd}$. It does not depend on distance. (*b*) At $d = 2.0\,\text{m}$ from the source, the luminous flux per unit area is
>
> $$E_\ell = \frac{F_\ell}{4\pi d^2} = \frac{1700\,\text{lm}}{(4\pi)(2.0\,\text{m})^2} = 34\,\text{lm/m}^2.$$
>
> The illuminance decreases as the square of the distance.

[†]A solid angle is a sort of two-dimensional angle and is measured in steradians. Think of a solid angle starting at a point and intercepting an area ΔA on a sphere of radius r surrounding that point. The solid angle has magnitude $\Delta A/r^2$ steradians. A solid angle including all of space intercepts the full surface area of the sphere, $4\pi r^2$, and so has magnitude $4\pi r^2/r^2 = 4\pi$ steradians. (Compare to a normal angle, for which a full circle subtends 2π radians.)

[‡]The British unit is the foot-candle, or lumen per square foot.

▌Summary

The wave theory of light is strongly supported by the observations that light exhibits **interference** and **diffraction**. Wave theory also explains the refraction of light and the fact that light travels more slowly in transparent solids and liquids than it does in air.

An aid to predicting wave behavior is **Huygens' principle**, which states that every point on a wave front can be considered as a source of tiny wavelets that spread out in the forward direction at the speed of the wave itself. The new wave front is the envelope (the common tangent) of all the wavelets.

The wavelength of light in a medium with index of refraction n is

$$\lambda_n = \frac{\lambda}{n}, \tag{34–1}$$

where λ is the wavelength in vacuum; the frequency is not changed.

Young's double-slit experiment clearly demonstrated the interference of light. The observed bright spots of the interference pattern are explained as constructive interference between the beams coming through the two slits, where the beams differ in path length by an integral number of wavelengths. The dark areas in between are due to destructive interference when the path lengths differ by $\frac{1}{2}\lambda, \frac{3}{2}\lambda$, and so on. The angles θ at which **constructive interference** occurs are given by

$$\sin\theta = m\frac{\lambda}{d}, \tag{34–2a}$$

where λ is the wavelength of the light, d is the separation of the slits, and m is an integer $(0, 1, 2, \cdots)$. **Destructive interference** occurs at angles θ given by

$$\sin\theta = \left(m + \tfrac{1}{2}\right)\frac{\lambda}{d}, \tag{34–2b}$$

where m is an integer $(0, 1, 2, \cdots)$.

The light intensity I_θ at any point in a double-slit interference pattern can be calculated using a phasor diagram, which predicts that

$$I_\theta = I_0 \cos^2 \frac{\delta}{2} \qquad (34\text{–}6)$$

where I_0 is the intensity at $\theta = 0$ and the phase angle δ is

$$\delta = \frac{2\pi d}{\lambda} \sin \theta. \qquad (34\text{–}4)$$

Two sources of light are perfectly **coherent** if the waves leaving them are of the same single frequency and maintain the same phase relationship at all times. If the light waves from the two sources have a random phase with respect to each other over time (as for two incandescent lightbulbs) the two sources are **incoherent**.

Light reflected from the front and rear surfaces of a thin film of transparent material can interfere constructively or destructively, depending on the path difference. A phase change of 180° or $\frac{1}{2}\lambda$ occurs when the light reflects at a surface where the index of refraction increases. Such **thin-film interference** has many practical applications, such as lens coatings and using Newton's rings to check the uniformity of glass surfaces.

Questions

1. Does Huygens' principle apply to sound waves? To water waves?
2. What is the evidence that light is energy?
3. Why is light sometimes described as rays and sometimes as waves?
4. We can hear sounds around corners but we cannot see around corners; yet both sound and light are waves. Explain the difference.
5. Can the wavelength of light be determined from reflection or refraction measurements?
6. Two rays of light from the same source destructively interfere if their path lengths differ by how much?
7. Monochromatic red light is incident on a double slit and the interference pattern is viewed on a screen some distance away. Explain how the fringe pattern would change if the red light source is replaced by a blue light source.
8. If Young's double-slit experiment were submerged in water, how would the fringe pattern be changed?

9. Compare a double-slit experiment for sound waves to that for light waves. Discuss the similarities and differences.
10. Suppose white light falls on the two slits of Fig. 34–7, but one slit is covered by a red filter (700 nm) and the other by a blue filter (450 nm). Describe the pattern on the screen.
11. Why doesn't the light from the two headlights of a distant car produce an interference pattern?
12. Why are interference fringes noticeable only for a *thin* film like a soap bubble and not for a thick piece of glass, say?
13. Why are Newton's rings (Fig. 34–18) closer together farther from the center?
14. Some coated lenses appear greenish yellow when seen by reflected light. What wavelengths do you suppose the coating is designed to eliminate completely?
15. A drop of oil on a pond appears bright at its edges where its thickness is much less than the wavelengths of visible light. What can you say about the index of refraction of the oil compared to that of water?

Problems

34–2 Huygens' Principle

1. (II) Derive the law of reflection—namely, that the angle of incidence equals the angle of reflection from a flat surface—using Huygens' principle for waves.

34–3 Double-Slit Interference

2. (I) Monochromatic light falling on two slits 0.018 mm apart produces the fifth-order bright fringe at a 9.8° angle. What is the wavelength of the light used?
3. (I) The third-order bright fringe of 610 nm light is observed at an angle of 28° when the light falls on two narrow slits. How far apart are the slits?
4. (II) Monochromatic light falls on two very narrow slits 0.048 mm apart. Successive fringes on a screen 6.00 m away are 8.5 cm apart near the center of the pattern. Determine the wavelength and frequency of the light.
5. (II) If 720-nm and 660-nm light passes through two slits 0.68 mm apart, how far apart are the second-order fringes for these two wavelengths on a screen 1.0 m away?
6. (II) A red laser from the physics lab is marked as producing 632.8-nm light. When light from this laser falls on two closely spaced slits, an interference pattern formed on a wall several meters away has bright fringes spaced 5.00 mm apart near the center of the pattern. When the laser is replaced by a small laser pointer, the fringes are 5.14 mm apart. What is the wavelength of light produced by the pointer?

7. (II) Light of wavelength λ passes through a pair of slits separated by 0.17 mm, forming a double-slit interference pattern on a screen located a distance 35 cm away. Suppose that the image in Fig. 34–9a is an actual-size reproduction of this interference pattern. Use a ruler to measure a pertinent distance on this image; then utilize this measured value to determine λ (nm).
8. (II) Light of wavelength 680 nm falls on two slits and produces an interference pattern in which the third-order bright fringe is 38 mm from the central fringe on a screen 2.6 m away. What is the separation of the two slits?
9. (II) A parallel beam of light from a He–Ne laser, with a wavelength 633 nm, falls on two very narrow slits 0.068 mm apart. How far apart are the fringes in the center of the pattern on a screen 3.8 m away?
10. (II) A physics professor wants to perform a lecture demonstration of Young's double-slit experiment for her class using the 633-nm light from a He–Ne laser. Because the lecture hall is very large, the interference pattern will be projected on a wall that is 5.0 m from the slits. For easy viewing by all students in the class, the professor wants the distance between the $m = 0$ and $m = 1$ maxima to be 25 cm. What slit separation is required in order to produce the desired interference pattern?

11. (II) Suppose a thin piece of glass is placed in front of the lower slit in Fig. 34–7 so that the two waves enter the slits 180° out of phase (Fig. 34–25). Describe in detail the interference pattern on the screen.

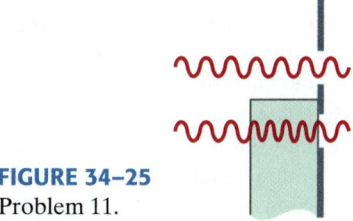

FIGURE 34–25
Problem 11.

12. (II) In a double-slit experiment it is found that blue light of wavelength 480 nm gives a second-order maximum at a certain location on the screen. What wavelength of visible light would have a minimum at the same location?

13. (II) Two narrow slits separated by 1.0 mm are illuminated by 544 nm light. Find the distance between adjacent bright fringes on a screen 5.0 m from the slits.

14. (II) In a double-slit experiment, the third-order maximum for light of wavelength 500 nm is located 12 mm from the central bright spot on a screen 1.6 m from the slits. Light of wavelength 650 nm is then projected through the same slits. How far from the central bright spot will the second-order maximum of this light be located?

15. (II) Light of wavelength 470 nm in air falls on two slits 6.00×10^{-2} mm apart. The slits are immersed in water, as is a viewing screen 50.0 cm away. How far apart are the fringes on the screen?

16. (II) A very thin sheet of plastic $(n = 1.60)$ covers one slit of a double-slit apparatus illuminated by 680-nm light. The center point on the screen, instead of being a maximum, is dark. What is the (minimum) thickness of the plastic?

*34–4 Intensity in Two-Slit Interference

***17. (I)** If one slit in Fig. 34–12 is covered, by what factor does the intensity at the center of the screen change?

***18. (II)** Derive an expression similar to Eq. 34–2 which gives the angles for which the double-slit intensity is one-half its maximum value, $I_\theta = \frac{1}{2} I_0$.

***19. (II)** Show that the angular full width at half maximum of the central peak in a double-slit interference pattern is given by $\Delta\theta = \lambda/2d$ if $\lambda \ll d$.

***20. (II)** In a two-slit interference experiment, the path length to a certain point P on the screen differs for one slit in comparison with the other by 1.25λ. (a) What is the phase difference between the two waves arriving at point P? (b) Determine the intensity at P, expressed as a fraction of the maximum intensity I_0 on the screen.

***21. (III)** Suppose that one slit of a double-slit apparatus is wider than the other so that the intensity of light passing through it is twice as great. Determine the intensity I as a function of position (θ) on the screen for coherent light.

***22. (III)** (a) Consider three equally spaced and equal-intensity coherent sources of light (such as adding a third slit to the two slits of Fig. 34–12). Use the phasor method to obtain the intensity as a function of the phase difference δ (Eq. 34–4). (b) Determine the positions of maxima and minima.

34–5 Thin-Film Interference

23. (I) If a soap bubble is 120 nm thick, what wavelength is most strongly reflected at the center of the outer surface when illuminated normally by white light? Assume that $n = 1.32$.

24. (I) How far apart are the dark fringes in Example 34–6 if the glass plates are each 28.5 cm long?

25. (II) (a) What is the smallest thickness of a soap film $(n = 1.33)$ that would appear black if illuminated with 480-nm light? Assume there is air on both sides of the soap film. (b) What are two other possible thicknesses for the film to appear black? (c) If the thickness t was much less than λ, why would the film also appear black?

26. (II) A lens appears greenish yellow $(\lambda = 570$ nm is strongest) when white light reflects from it. What minimum thickness of coating $(n = 1.25)$ do you think is used on such a glass $(n = 1.52)$ lens, and why?

27. (II) A thin film of oil $(n_o = 1.50)$ with varying thickness floats on water $(n_w = 1.33)$. When it is illuminated from above by white light, the reflected colors are as shown in Fig. 34–26. In air, the wavelength of yellow light is 580 nm. (a) Why are there no reflected colors at point A? (b) What is the oil's thickness t at point B?

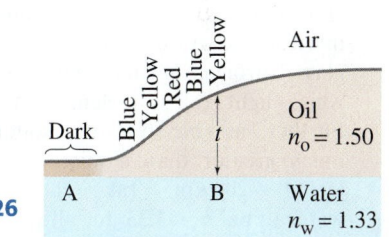

FIGURE 34–26
Problem 27.

28. (II) A thin oil slick $(n_o = 1.50)$ floats on water $(n_w = 1.33)$. When a beam of white light strikes this film at normal incidence from air, the only enhanced reflected colors are red (650 nm) and violet (390 nm). From this information, deduce the (minimum) thickness t of the oil slick.

29. (II) A total of 31 bright and 31 dark Newton's rings (not counting the dark spot at the center) are observed when 560-nm light falls normally on a planoconvex lens resting on a flat glass surface (Fig. 34–18). How much thicker is the center than the edges?

30. (II) A fine metal foil separates one end of two pieces of optically flat glass, as in Fig. 34–20. When light of wavelength 670 nm is incident normally, 28 dark lines are observed (with one at each end). How thick is the foil?

31. (II) How thick (minimum) should the air layer be between two flat glass surfaces if the glass is to appear bright when 450-nm light is incident normally? What if the glass is to appear dark?

32. (II) A uniform thin film of alcohol $(n = 1.36)$ lies on a flat glass plate $(n = 1.56)$. When monochromatic light, whose wavelength can be changed, is incident normally, the reflected light is a minimum for $\lambda = 512$ nm and a maximum for $\lambda = 635$ nm. What is the minimum thickness of the film?

33. (II) Show that the radius r of the mth dark Newton's ring, as viewed from directly above (Fig. 34–18), is given by $r = \sqrt{m\lambda R}$ where R is the radius of curvature of the curved glass surface and λ is the wavelength of light used. Assume that the thickness of the air gap is much less than R at all points and that $r \ll R$. [Hint: Use the binomial expansion.]

34. (II) Use the result of Problem 33 to show that the distance between adjacent dark Newton's rings is

$$\Delta r \approx \sqrt{\frac{\lambda R}{4m}}$$

for the mth ring, assuming $m \gg 1$.

35. (II) When a Newton's ring apparatus (Fig. 34–18) is immersed in a liquid, the diameter of the eighth dark ring decreases from 2.92 cm to 2.54 cm. What is the refractive index of the liquid? [Hint: see Problem 33.]

36. (II) A planoconvex lucite lens 3.4 cm in diameter is placed on a flat piece of glass as in Fig. 34–18. When 580-nm light is incident normally, 44 bright rings are observed, the last one right at the edge. What is the radius of curvature of the lens surface, and the focal length of the lens? [Hint: see Problem 33.]

37. (II) Let's explore why only "thin" layers exhibit thin-film interference. Assume a layer of water, sitting atop a flat glass surface, is illuminated from the air above by white light (all wavelengths from 400 nm to 700 nm). Further, assume that the water layer's thickness t is much greater than a micron ($=1000$ nm); in particular, let $t = 200\,\mu$m. Take the index of refraction for water to be $n = 1.33$ for all visible wavelengths. (a) Show that a visible color will be reflected from the water layer if its wavelength is $\lambda = 2nt/m$, where m is an integer. (b) Show that the two extremes in wavelengths (400 nm and 700 nm) of the incident light are both reflected from the water layer and determine the m-value associated with each. (c) How many other visible wavelengths, besides $\lambda = 400$ nm and 700 nm, are reflected from the "thick" layer of water? (d) How does this explain why such a thick layer does not reflect colorfully, but is white or grey?

38. (III) A single optical coating reduces reflection to zero for $\lambda = 550$ nm. By what factor is the intensity reduced by the coating for $\lambda = 430$ nm and $\lambda = 670$ nm as compared to no coating? Assume normal incidence.

*34–6 Michelson Interferometer

*39. (II) How far must the mirror M_1 in a Michelson interferometer be moved if 650 fringes of 589-nm light are to pass by a reference line?

*40. (II) What is the wavelength of the light entering an interferometer if 384 bright fringes are counted when the movable mirror moves 0.125 mm?

*41. (II) A micrometer is connected to the movable mirror of an interferometer. When the micrometer is tightened down on a thin metal foil, the net number of bright fringes that move, compared to the empty micrometer, is 272. What is the thickness of the foil? The wavelength of light used is 589 nm.

*42. (III) One of the beams of an interferometer (Fig. 34–27) passes through a small evacuated glass container 1.155 cm deep. When a gas is allowed to slowly fill the container, a total of 176 dark fringes are counted to move past a reference line. The light used has a wavelength of 632.8 nm. Calculate the index of refraction of the gas at its final density, assuming that the interferometer is in vacuum.

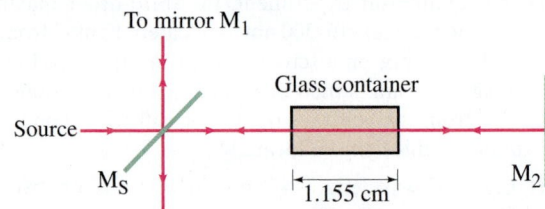

FIGURE 34–27 Problem 42.

*43. (III) The yellow sodium D lines have wavelengths of 589.0 and 589.6 nm. When they are used to illuminate a Michelson interferometer, it is noted that the interference fringes disappear and reappear periodically as the mirror M_1 is moved. Why does this happen? How far must the mirror move between one disappearance and the next?

*34–7 Luminous Intensity

*44. (II) The illuminance of direct sunlight on Earth is about 10^5 lm/m^2. Estimate the luminous flux and luminous intensity of the Sun.

*45. (II) The luminous efficiency of a lightbulb is the ratio of luminous flux to electric power input. (a) What is the luminous efficiency of a 100-W, 1700-lm bulb? (b) How many 40-W, 60-lm/W fluorescent lamps would be needed to provide an illuminance of 250 lm/m^2 on a factory floor of area 25 m × 30 m? Assume the lights are 10 m above the floor and that half their flux reaches the floor.

General Problems

46. Light of wavelength 5.0×10^{-7} m passes through two parallel slits and falls on a screen 4.0 m away. Adjacent bright bands of the interference pattern are 2.0 cm apart. (a) Find the distance between the slits. (b) The same two slits are next illuminated by light of a different wavelength, and the fifth-order minimum for this light occurs at the same point on the screen as the fourth-order minimum for the previous light. What is the wavelength of the second source of light?

47. Television and radio waves reflecting from mountains or airplanes can interfere with the direct signal from the station. (a) What kind of interference will occur when 75-MHz television signals arrive at a receiver directly from a distant station, and are reflected from a nearby airplane 122 m directly above the receiver? Assume $\frac{1}{2}\lambda$ change in phase of the signal upon reflection. (b) What kind of interference will occur if the plane is 22 m closer to the receiver?

48. A radio station operating at 88.5 MHz broadcasts from two identical antennas at the same elevation but separated by a 9.0-m horizontal distance d, Fig. 34–28. A maximum signal is found along the midline, perpendicular to d at its midpoint and extending horizontally in both directions. If the midline is taken as $0°$, at what other angle(s) θ is a maximum signal detected? A minimum signal? Assume all measurements are made much farther than 9.0 m from the antenna towers.

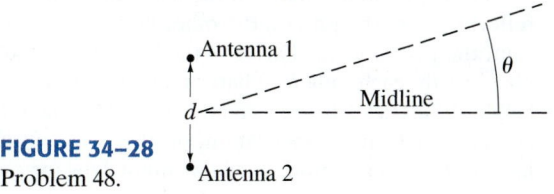

FIGURE 34–28
Problem 48.

49. Light of wavelength 690 nm passes through two narrow slits 0.66 mm apart. The screen is 1.60 m away. A second source of unknown wavelength produces its second-order fringe 1.23 mm closer to the central maximum than the 690-nm light. What is the wavelength of the unknown light?

50. Monochromatic light of variable wavelength is incident normally on a thin sheet of plastic film in air. The reflected light is a maximum only for $\lambda = 491.4$ nm and $\lambda = 688.0$ nm in the visible spectrum. What is the thickness of the film $(n = 1.58)$? [*Hint*: Assume successive values of m.]

*51. Suppose the mirrors in a Michelson interferometer are perfectly aligned and the path lengths to mirrors M_1 and M_2 are identical. With these initial conditions, an observer sees a bright maximum at the center of the viewing area. Now one of the mirrors is moved a distance x. Determine a formula for the intensity at the center of the viewing area as a function of x, the distance the movable mirror is moved from the initial position.

52. A highly reflective mirror can be made for a particular wavelength at normal incidence by using two thin layers of transparent materials of indices of refraction n_1 and n_2 $(1 < n_1 < n_2)$ on the surface of the glass $(n > n_2)$. What should be the minimum thicknesses d_1 and d_2 in Fig. 34–29 in terms of the incident wavelength λ, to maximize reflection?

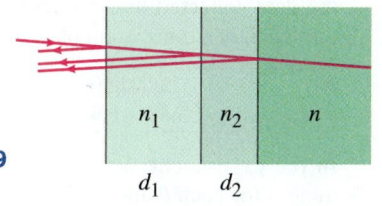

FIGURE 34–29
Problem 52.

53. Calculate the minimum thickness needed for an antireflective coating $(n = 1.38)$ applied to a glass lens in order to eliminate (*a*) blue (450 nm), or (*b*) red (720 nm) reflections for light at normal incidence.

54. Stealth aircraft are designed to not reflect radar, whose wavelength is typically 2 cm, by using an antireflecting coating. Ignoring any change in wavelength in the coating, estimate its thickness.

55. Light of wavelength λ strikes a screen containing two slits a distance d apart at an angle θ_i to the normal. Determine the angle θ_m at which the mth-order maximum occurs.

56. Consider two antennas radiating 6.0-MHz radio waves in phase with each other. They are located at points S_1 and S_2, separated by a distance $d = 175$ m, Fig. 34–30. Determine the points on the y axis where the signals from the two sources will be out of phase (crests of one meet troughs of the other).

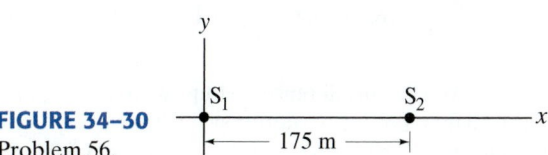

FIGURE 34–30
Problem 56.

57. What is the minimum (non-zero) thickness for the air layer between two flat glass surfaces if the glass is to appear dark when 680-nm light is incident normally? What if the glass is to appear bright?

58. *Lloyd's mirror* provides one way of obtaining a double-slit interference pattern from a single source so the light is coherent. As shown in Fig. 34–31, the light that reflects from the plane mirror appears to come from the virtual image of the slit. Describe in detail the interference pattern on the screen.

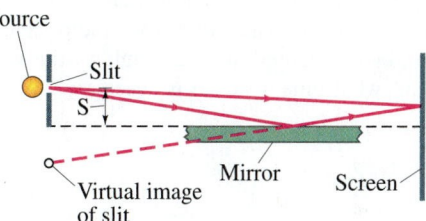

FIGURE 34–31 Problem 58.

59. Consider the antenna array of Example 34–5, Fig. 34–15. Let $d = \lambda/2$, and suppose that the two antennas are now $180°$ out of phase with each other. Find the directions for constructive and destructive interference, and compare with the case when the sources are in phase. (These results illustrate the basis for directional antennas.)

60. Suppose you viewed the light *transmitted* through a thin film layered on a flat piece of glass. Draw a diagram, similar to Fig. 34–17 or 34–23, and describe the conditions required for maxima and minima. Consider all possible values of index of refraction. Discuss the relative size of the minima compared to the maxima and to zero.

61. A thin film of soap $(n = 1.34)$ coats a piece of flat glass $(n = 1.52)$. How thick is the film if it reflects 643-nm red light most strongly when illuminated normally by white light?

62. Two identical sources S_1 and S_2, separated by distance d, coherently emit light of wavelength λ uniformly in all directions. Defining the x axis with its origin at S_1 as shown in Fig. 34–32, find the locations (expressed as multiples of λ) where the signals from the two sources are out of phase along this axis for $x > 0$, if $d = 3\lambda$.

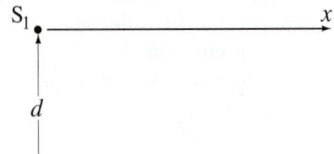

FIGURE 34–32
Problem 62.

63. A two-slit interference set-up with slit separation $d = 0.10\ \text{mm}$ produces interference fringes at a particular set of angles θ_m (where $m = 0, 1, 2, \ldots$) for red light of frequency $f = 4.6 \times 10^{14}\ \text{Hz}$. If one wishes to construct an analogous two-slit interference set-up that produces interference fringes at the same set of angles θ_m for room-temperature sound of middle-C frequency $f_S = 262\ \text{Hz}$, what should the slit separation d_S be for this analogous set-up?

64. A radio telescope, whose two antennas are separated by 55 m, is designed to receive 3.0-MHz radio waves produced by astronomical objects. The received radio waves create 3.0-MHz electronic signals in the telescope's left and right antennas. These signals then travel by equal-length cables to a centrally located amplifier, where they are added together. The telescope can be "pointed" to a certain region of the sky by adding the instantaneous signal from the right antenna to a "time-delayed" signal received by the left antenna a time Δt ago. (This time delay of the left signal can be easily accomplished with the proper electronic circuit.) If a radio astronomer wishes to "view" radio signals arriving from an object oriented at a 12° angle to the vertical as in Fig. 34–33, what time delay Δt is necessary?

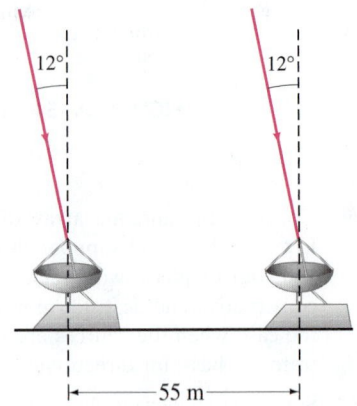

FIGURE 34–33
Problem 64.

65. In a compact disc (CD), digital information is stored as a sequence of raised surfaces called "pits" and recessed surfaces called "lands." Both pits and lands are highly reflective and are embedded in a thick plastic material with index of refraction $n = 1.55$ (Fig. 34–34). As a 780-nm wavelength (in air) laser scans across the pit–land sequence, the transition between a neighboring pit and land is sensed by monitoring the intensity of reflected laser light from the CD. At the moment when half the width of the laser beam is reflected from the pit and the other half from the land, we want the two reflected halves of the beam to be 180° out of phase with each other. What should be the (minimum) height difference t between a pit and land? [When this light enters a detector, cancellation of the two out-of-phase halves of the beam produces a minimum detector output.]

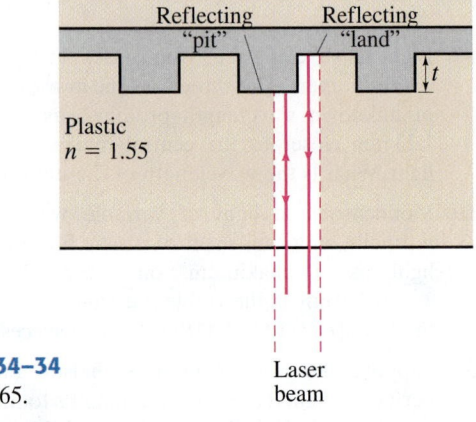

FIGURE 34–34
Problem 65.

***Numerical/Computer**

*66. (II) A Michelson interferometer can be used to determine the index of refraction of a glass plate. A glass plate (thickness t) is placed on a platform that can rotate. The plate is placed in the light's path between the beam splitter and either the fixed or movable mirror, so that its thickness is in the direction of the laser beam. The platform is rotated to various angles, and the number of fringes shifted is counted. It can be shown that if N is the number of fringes shifted when the angle of rotation changes by θ, the index of refraction is $n = (2t - N\lambda)(1 - \cos\theta)/[2t(1 - \cos\theta) - N\lambda]$ where t is the thickness of the plate. The accompanying Table shows the data collected by a student in determining the index of refraction of a transparent plate by a Michelson interferometer.

N	25	50	75	100	125	150
θ (degree)	5.5	6.9	8.6	10.0	11.3	12.5

In the experiment $\lambda = 632.8\ \text{nm}$ and $t = 4.0\ \text{mm}$. Determine n for each θ and find the average n.

Answers to Exercises

A: 333 nm; 6.0×10^{14} Hz; 2.0×10^8 m/s.

B: (a) I_0; (b) $0.50I_0$; (c) 0.

C: (e).

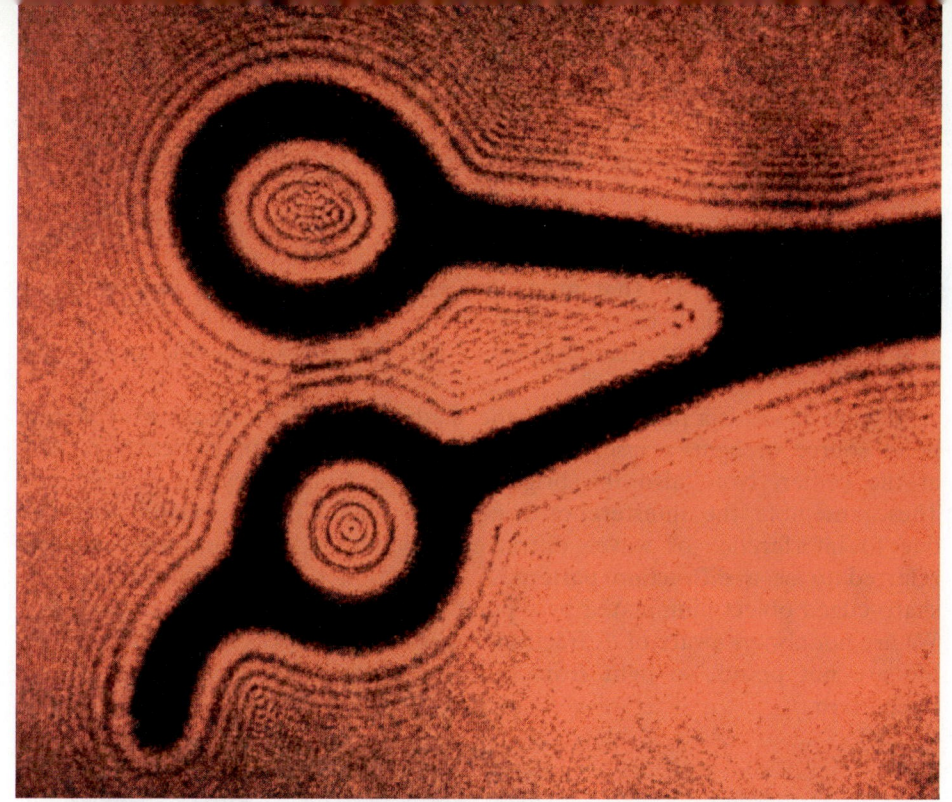

Parallel coherent light from a laser, which acts as nearly a point source, illuminates these shears. Instead of a clean shadow, there is a dramatic diffraction pattern, which is a strong confirmation of the wave theory of light. Diffraction patterns are washed out when typical extended sources of light are used, and hence are not seen, although a careful examination of shadows will reveal fuzziness. We will examine diffraction by a single slit, and how it affects the double-slit pattern. We also discuss diffraction gratings and diffraction of X-rays by crystals. We will see how diffraction affects the resolution of optical instruments, and that the ultimate resolution can never be greater than the wavelength of the radiation used. Finally we study the polarization of light.

Diffraction and Polarization

CHAPTER-OPENING QUESTION—Guess now!

Because of diffraction, a light microscope has a maximum useful magnification of about

- **(a)** $50\times$;
- **(b)** $100\times$;
- **(c)** $500\times$;
- **(d)** $2000\times$;
- **(e)** $5000\times$;

and the smallest objects it can resolve have a size of about

- **(a)** 10 nm;
- **(b)** 100 nm;
- **(c)** 500 nm;
- **(d)** 2500 nm;
- **(e)** 5500 nm.

Young's double-slit experiment put the wave theory of light on a firm footing. But full acceptance came only with studies on diffraction more than a decade later, in the 1810s and 1820s.

We have already discussed diffraction briefly with regard to water waves (Section 15–11) as well as for light (Section 34–1). We have seen that it refers to the spreading or bending of waves around edges. Now we look at diffraction in more detail, including its important practical effects of limiting the amount of detail, or *resolution*, that can be obtained with any optical instrument such as telescopes, cameras, and the eye.

CONTENTS

35–1   Diffraction by a Single Slit or Disk

*35–2   Intensity in Single-Slit Diffraction Pattern

*35–3   Diffraction in the Double-Slit Experiment

35–4   Limits of Resolution; Circular Apertures

35–5   Resolution of Telescopes and Microscopes; the λ Limit

*35–6   Resolution of the Human Eye and Useful Magnification

35–7   Diffraction Grating

35–8   The Spectrometer and Spectroscopy

*35–9   Peak Widths and Resolving Power for a Diffraction Grating

35–10 X-Rays and X-Ray Diffraction

35–11 Polarization

*35–12 Liquid Crystal Displays (LCD)

*35–13 Scattering of Light by the Atmosphere

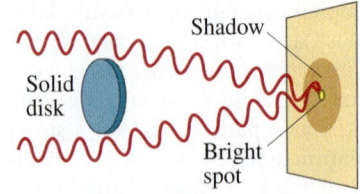

FIGURE 35–1 If light is a wave, a bright spot will appear at the center of the shadow of a solid disk illuminated by a point source of monochromatic light.

In 1819, Augustin Fresnel (1788–1827) presented to the French Academy a wave theory of light that predicted and explained interference and diffraction effects. Almost immediately Siméon Poisson (1781–1840) pointed out a counter-intuitive inference: according to Fresnel's wave theory, if light from a point source were to fall on a solid disk, part of the incident light would be diffracted around the edges and would constructively interfere at the center of the shadow (Fig. 35–1). That prediction seemed very unlikely. But when the experiment was actually carried out by François Arago, the bright spot was seen at the very center of the shadow (Fig. 35–2a). This was strong evidence for the wave theory.

Figure 35–2a is a photograph of the shadow cast by a coin using a coherent point source of light, a laser in this case. The bright spot is clearly present at the center. Note also the bright and dark fringes beyond the shadow. These resemble the interference fringes of a double slit. Indeed, they are due to interference of waves diffracted around the disk, and the whole is referred to as a **diffraction pattern**. A diffraction pattern exists around any sharp-edged object illuminated by a point source, as shown in Fig. 35–2b and c. We are not always aware of diffraction because most sources of light in everyday life are not points, so light from different parts of the source washes out the pattern.

FIGURE 35–2 Diffraction pattern of (a) a circular disk (a coin), (b) razorblade, (c) a single slit, each illuminated by a coherent point source of monochromatic light, such as a laser.

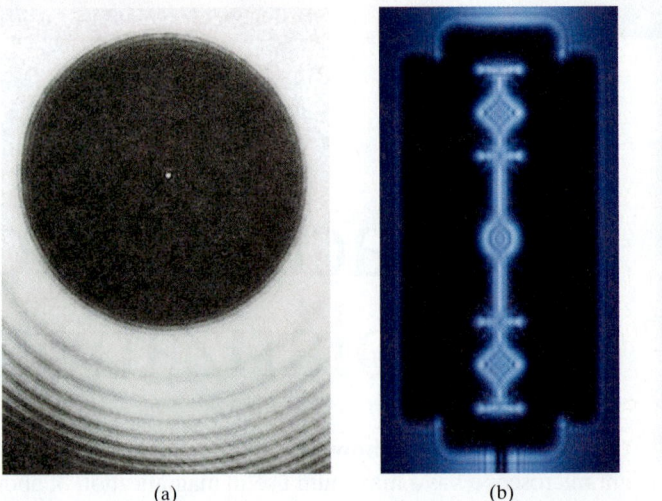

(a)   (b)   (c)

35–1 Diffraction by a Single Slit or Disk

To see how a diffraction pattern arises, we will analyze the important case of monochromatic light passing through a narrow slit. We will assume that parallel rays (plane waves) of light fall on the slit of width D, and pass through to a viewing screen very far away. If the viewing screen is not far away, lenses can be used to make the rays parallel.[†] As we know from studying water waves and from Huygens' principle, the waves passing through the slit spread out in all directions. We will now examine how the waves passing through different parts of the slit interfere with each other.

Parallel rays of monochromatic light pass through the narrow slit as shown in Fig. 35–3a. The slit width D is on the order of the wavelength λ of the light, but the slit's length (in and out of page) is large compared to λ. The light falls on a screen which is assumed to be very far away, so the rays heading for any point are very nearly parallel before they meet at the screen.

[†]Such a diffraction pattern, involving parallel rays, is called *Fraunhofer diffraction*. If the screen is close and no lenses are used, it is called *Fresnel diffraction*. The analysis in the latter case is rather involved, so we consider only the limiting case of Fraunhofer diffraction.

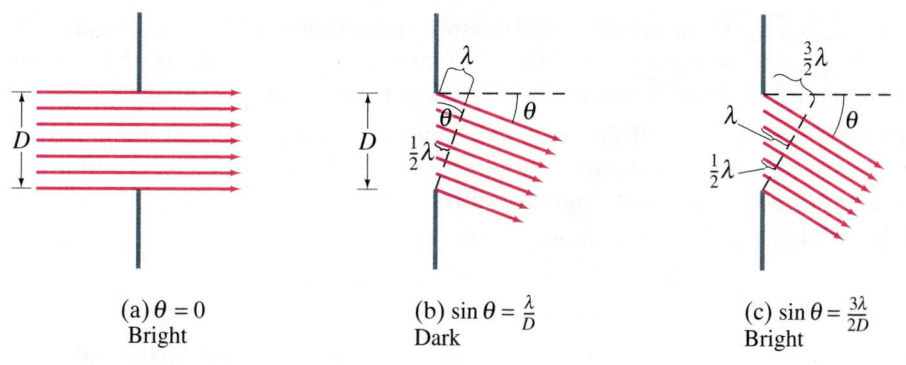
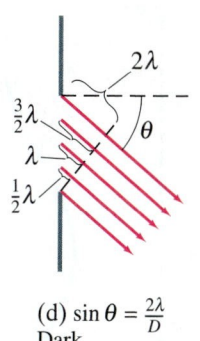

(a) $\theta = 0$ Bright	(b) $\sin\theta = \frac{\lambda}{D}$ Dark	(c) $\sin\theta = \frac{3\lambda}{2D}$ Bright	(d) $\sin\theta = \frac{2\lambda}{D}$ Dark

FIGURE 35–3 Analysis of diffraction pattern formed by light passing through a narrow slit of width D.

First we consider rays that pass straight through as in Fig. 35–3a. They are all in phase, so there will be a central bright spot on the screen (see Fig. 35–2c). In Fig. 35–3b, we consider rays moving at an angle θ such that the ray from the top of the slit travels exactly one wavelength farther than the ray from the bottom edge to reach the screen. The ray passing through the very center of the slit will travel one-half wavelength farther than the ray at the bottom of the slit. These two rays will be exactly out of phase with one another and so will destructively interfere when they overlap at the screen. Similarly, a ray slightly above the bottom one will cancel a ray that is the same distance above the central one. Indeed, each ray passing through the lower half of the slit will cancel with a corresponding ray passing through the upper half. Thus, all the rays destructively interfere in pairs, and so the light intensity will be zero on the viewing screen at this angle. The angle θ at which this takes place can be seen from Fig. 35–3b to occur when $\lambda = D\sin\theta$, so

$$\sin\theta = \frac{\lambda}{D}. \qquad\qquad \text{[first minimum]} \quad \textbf{(35–1)}$$

The light intensity is a maximum at $\theta = 0°$ and decreases to a minimum (intensity $=$ zero) at the angle θ given by Eq. 35–1.

Now consider a larger angle θ such that the top ray travels $\frac{3}{2}\lambda$ farther than the bottom ray, as in Fig. 35–3c. In this case, the rays from the bottom third of the slit will cancel in pairs with those in the middle third because they will be $\lambda/2$ out of phase. However, light from the top third of the slit will still reach the screen, so there will be a bright spot centered near $\sin\theta \approx 3\lambda/2D$, but it will not be nearly as bright as the central spot at $\theta = 0°$. For an even larger angle θ such that the top ray travels 2λ farther than the bottom ray, Fig. 35–3d, rays from the bottom quarter of the slit will cancel with those in the quarter just above it because the path lengths differ by $\lambda/2$. And the rays through the quarter of the slit just above center will cancel with those through the top quarter. At this angle there will again be a minimum of zero intensity in the diffraction pattern. A plot of the intensity as a function of angle is shown in Fig. 35–4. This corresponds well with the photo of Fig. 35–2c. Notice that minima (zero intensity) occur on both sides of center at

$$D\sin\theta = m\lambda, \qquad m = \pm 1, \pm 2, \pm 3, \cdots, \qquad \text{[minima]} \quad \textbf{(35–2)}$$

but *not* at $m = 0$ where there is the strongest maximum. Between the minima, smaller intensity maxima occur at approximately (not exactly) $m \approx \frac{3}{2}, \frac{5}{2}, \cdots$.

Note that the *minima* for a diffraction pattern, Eq. 35–2, satisfy a criterion that looks very similar to that for the *maxima* (bright fringes) for double-slit interference, Eq. 34–2a. Also note that D is a single slit width, whereas d in Eq. 34–2 is the distance between two slits.

FIGURE 35–4 Intensity in the diffraction pattern of a single slit as a function of $\sin\theta$. Note that the central maximum is not only much higher than the maxima to each side, but it is also twice as wide ($2\lambda/D$ wide) as any of the others (only λ/D wide each).

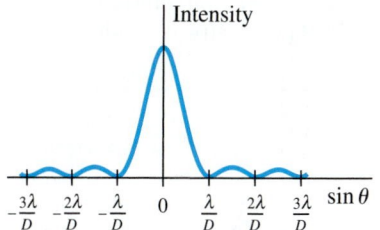

⚠ **CAUTION**

Don't confuse Eqs. 34–2 for interference with Eq. 35–1 for diffraction: note the differences

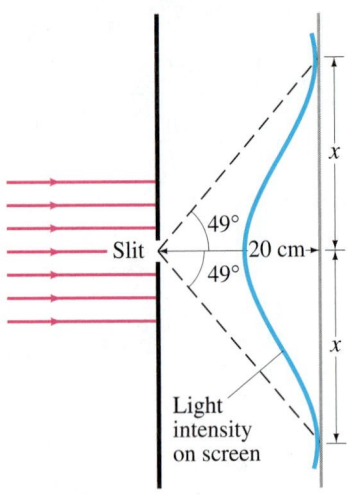

FIGURE 35–5 Example 35–1.

EXAMPLE 35–1 **Single-slit diffraction maximum.** Light of wavelength 750 nm passes through a slit 1.0×10^{-3} mm wide. How wide is the central maximum (a) in degrees, and (b) in centimeters, on a screen 20 cm away?

APPROACH The width of the central maximum goes from the first minimum on one side to the first minimum on the other side. We use Eq. 35–1 to find the angular position of the first single-slit diffraction minimum.

SOLUTION (a) The first minimum occurs at

$$\sin \theta = \frac{\lambda}{D} = \frac{7.5 \times 10^{-7} \, \text{m}}{1.0 \times 10^{-6} \, \text{m}} = 0.75.$$

So $\theta = 49°$. This is the angle between the center and the first minimum, Fig. 35–5. The angle subtended by the whole central maximum, between the minima above and below the center, is twice this, or 98°.

(b) The width of the central maximum is $2x$, where $\tan \theta = x/20$ cm. So $2x = 2(20 \, \text{cm})(\tan 49°) = 46$ cm.

NOTE A large width of the screen will be illuminated, but it will not normally be very bright since the amount of light that passes through such a small slit will be small and it is spread over a large area. Note also that we *cannot* use the small-angle approximation here ($\theta \approx \sin \theta \approx \tan \theta$) because θ is large.

EXERCISE A In Example 35–1, red light ($\lambda = 750$ nm) was used. If instead yellow light at 575 nm had been used, would the central maximum be wider or narrower?

CONCEPTUAL EXAMPLE 35–2 **Diffraction spreads.** Light shines through a rectangular hole that is narrower in the vertical direction than the horizontal, Fig. 35–6. (a) Would you expect the diffraction pattern to be more spread out in the vertical direction or in the horizontal direction? (b) Should a rectangular loudspeaker horn at a stadium be high and narrow, or wide and flat?

RESPONSE (a) From Eq. 35–1 we can see that if we make the slit (width D) narrower, the pattern spreads out more. This is consistent with our study of waves in Chapter 15. The diffraction through the rectangular hole will be wider vertically, since the opening is smaller in that direction. (b) For a loudspeaker, the sound pattern desired is one spread out horizontally, so the horn should be tall and narrow (rotate Fig. 35–6 by 90°).

FIGURE 35–6 Example 35–2.

*35–2 Intensity in Single-Slit Diffraction Pattern

We have determined the positions of the minima in the diffraction pattern produced by light passing through a single slit, Eq. 35–2. We now discuss a method for predicting the amplitude and intensity at any point in the pattern using the phasor technique already discussed in Section 34–4.

Let us consider the slit divided into N very thin strips of width Δy as indicated in Fig. 35–7. Each strip sends light in all directions toward a screen on the right. Again we take the rays heading for any particular point on the distant screen to be parallel, all making an angle θ with the horizontal as shown. We choose the strip width Δy to be much smaller than the wavelength λ of the monochromatic light falling on the slit, so all the light from a given strip is in phase. The strips are of equal size, and if the whole slit is uniformly illuminated, we can take the electric field wave amplitudes ΔE_0 from each thin strip to be equal as long as θ is not too large. However, the separate amplitudes from the different strips will differ in phase. The phase difference in the light coming from adjacent strips will be (see Section 34–4, Eq. 34–4)

$$\Delta \beta = \frac{2\pi}{\lambda} \Delta y \sin \theta \qquad (35–3)$$

since the difference in path length is $\Delta y \sin \theta$.

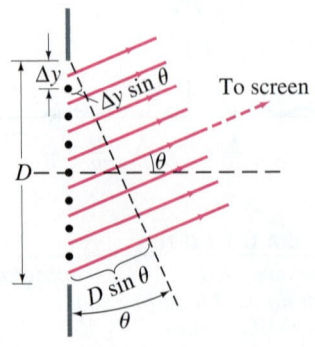

FIGURE 35–7 Slit of width D divided into N strips of width Δy.

The total amplitude on the screen at any angle θ will be the sum of the separate wave amplitudes due to each strip. These wavelets have the same amplitude ΔE_0 but differ in phase. To obtain the total amplitude, we can use a phasor diagram as we did in Section 34–4 (Fig. 34–13). The phasor diagrams for four different angles θ are shown in Fig. 35–8. At the center of the screen, $\theta = 0$, the waves from each strip are all in phase ($\Delta \beta = 0$, Eq. 35–3), so the arrows representing each ΔE_0 line up as shown in Fig. 35–8a. The total amplitude of the light arriving at the center of the screen is then $E_0 = N \Delta E_0$.

FIGURE 35–8 Phasor diagram for single-slit diffraction, giving the total amplitude E_θ at four different angles θ.

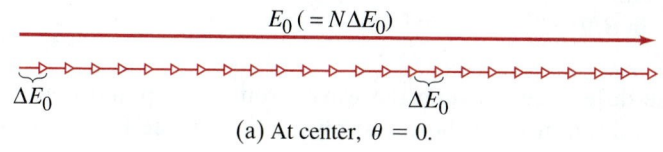

(a) At center, $\theta = 0$.

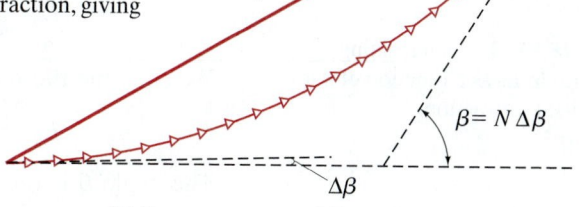

(b) Between center and first minimum.

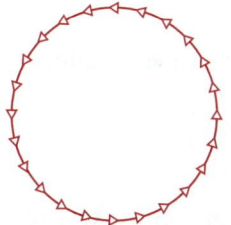

(c) First minimum, $E_\theta = 0$ ($\beta = 2\pi = 360°$).

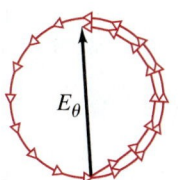

(d) Near secondary maximum.

At a small angle θ, for a point on the distant screen not far from the center, Fig. 35–8b shows how the wavelets of amplitude ΔE_0 add up to give E_θ, the total amplitude on the screen at this angle θ. Note that each wavelet differs in phase from the adjacent one by $\Delta \beta$. The phase difference between the wavelets from the top and bottom edges of the slit is

$$\beta = N \Delta \beta = \frac{2\pi}{\lambda} N \Delta y \sin \theta = \frac{2\pi}{\lambda} D \sin \theta \qquad (35–4)$$

where $D = N \Delta y$ is the total width of the slit. Although the "arc" in Fig. 35–8b has length $N \Delta E_0$, and so would equal E_0 (total amplitude at $\theta = 0$), the amplitude of the total wave E_θ at angle θ is the *vector* sum of each wavelet amplitude and so is equal to the length of the chord as shown. The chord is shorter than the arc, so $E_\theta < E_0$.

For greater θ, we eventually come to the case, illustrated in Fig. 35–8c, where the chain of arrows closes on itself. In this case the vector sum is zero, so $E_\theta = 0$ for this angle θ. This corresponds to the first minimum. Since $\beta = N \Delta \beta$ is 360° or 2π in this case, we have from Eq. 35–3,

$$2\pi = N \Delta \beta = N \left(\frac{2\pi}{\lambda} \Delta y \sin \theta \right)$$

or, since the slit width $D = N \Delta y$,

$$\sin \theta = \frac{\lambda}{D}.$$

Thus the first minimum $(E_\theta = 0)$ occurs where $\sin \theta = \lambda/D$, which is the same result we obtained in the previous Section, Eq. 35–1.

For even greater values of θ, the chain of arrows spirals beyond 360°. Figure 35–8d shows the case near the secondary maximum next to the first minimum. Here $\beta = N \Delta \beta \approx 360° + 180° = 540°$ or 3π. When greater angles θ are considered, new maxima and minima occur. But since the total length of the coil remains constant, equal to $N \Delta E_0 (= E_0)$, each succeeding maximum is smaller and smaller as the coil winds in on itself.

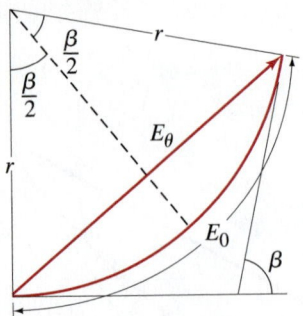

FIGURE 35–9 Determining amplitude E_θ as a function of θ for single-slit diffraction.

To obtain a quantitative expression for the amplitude (and intensity) for any point on the screen (that is, for any angle θ), we now consider the limit $N \to \infty$ so Δy becomes the infinitesimal width dy. In this case, the diagrams of Fig. 35–8 become smooth curves, one of which is shown in Fig. 35–9. For any angle θ, the wave amplitude on the screen is E_θ, equal to the chord in Fig. 35–9. The length of the arc is E_0, as before. If r is the radius of curvature of the arc, then

$$\frac{E_\theta}{2} = r \sin \frac{\beta}{2}.$$

Using radian measure for $\beta/2$, we also have

$$\frac{E_0}{2} = r \frac{\beta}{2}.$$

We combine these to obtain

$$E_\theta = E_0 \frac{\sin \beta/2}{\beta/2}. \tag{35–5}$$

The angle β is the phase difference between the waves from the top and bottom edges of the slit. The path difference for these two rays is $D \sin \theta$ (see Fig. 35–7 as well as Eq. 35–4), so

$$\beta = \frac{2\pi}{\lambda} D \sin \theta. \tag{35–6}$$

Intensity is proportional to the square of the wave amplitude, so the intensity I_θ at any angle θ is, from Eq. 35–5,

$$I_\theta = I_0 \left(\frac{\sin \beta/2}{\beta/2} \right)^2 \tag{35–7}$$

where $I_0 (\propto E_0^2)$ is the intensity at $\theta = 0$ (the central maximum). We can combine Eqs. 35–7 and 35–6 (although it is often simpler to leave them as separate equations) to obtain

$$I_\theta = I_0 \left[\frac{\sin\left(\dfrac{\pi D \sin \theta}{\lambda} \right)}{\left(\dfrac{\pi D \sin \theta}{\lambda} \right)} \right]^2. \tag{35–8}$$

According to Eq. 35–8, minima $(I_\theta = 0)$ occur where $\sin(\pi D \sin \theta / \lambda) = 0$, which means $\pi D \sin \theta / \lambda$ must be $\pi, 2\pi, 3\pi$, and so on, or

$$D \sin \theta = m\lambda, \qquad m = 1, 2, 3, \cdots \qquad \text{[minima]}$$

which is what we have obtained previously, Eq. 35–2. Notice that m cannot be zero: when $\beta/2 = \pi D \sin \theta / \lambda = 0$, the denominator as well as the numerator in Eqs. 35–7 or 35–8 vanishes. We can evaluate the intensity in this case by taking the limit as $\theta \to 0$ (or $\beta \to 0$); for very small angles, $\sin \beta/2 \approx \beta/2$, so $(\sin \beta/2)/(\beta/2) \to 1$ and $I_\theta = I_0$, the *maximum* at the center of the pattern.

The intensity I_θ as a function of θ, as given by Eq. 35–8, corresponds to the diagram of Fig. 35–4.

EXAMPLE 35–3 ESTIMATE Intensity at secondary maxima. Estimate the intensities of the first two secondary maxima to either side of the central maximum.

APPROACH The secondary maxima occur close to halfway between the minima, at about

$$\frac{\beta}{2} = \frac{\pi D \sin \theta}{\lambda} \approx (m + \tfrac{1}{2})\pi. \qquad m = 1, 2, 3, \cdots$$

The actual maxima are not quite at these points—their positions can be determined by differentiating Eq. 35–7 (see Problem 14)—but we are only seeking an estimate.

SOLUTION Using these values for β in Eq. 35–7 or 35–8, with $\sin(m + \tfrac{1}{2})\pi = 1$, gives

$$I_\theta = \frac{I_0}{(m + \tfrac{1}{2})^2 \pi^2}. \qquad m = 1, 2, 3, \cdots$$

For $m = 1$ and 2, we get

$$I_\theta = \frac{I_0}{22.2} = 0.045I_0 \qquad\qquad [m = 1]$$

$$I_\theta = \frac{I_0}{61.7} = 0.016I_0. \qquad\qquad [m = 2]$$

The first maximum to the side of the central peak has only 1/22, or 4.5%, the intensity of the central peak, and succeeding ones are smaller still, just as we can see in Fig. 35–4 and the photo of Fig. 35–2c.

Diffraction by a circular opening produces a similar pattern (though circular rather than rectangular) and is of great practical importance, since lenses are essentially circular apertures through which light passes. We will discuss this in Section 35–4 and see how diffraction limits the resolution (or sharpness) of images.

*35–3 Diffraction in the Double-Slit Experiment

When we analyzed Young's double-slit experiment in Section 34–4, we assumed that the central portion of the screen was uniformly illuminated. This is equivalent to assuming the slits are infinitesimally narrow, so that the central diffraction peak is spread out over the whole screen. This can never be the case for real slits; diffraction reduces the intensity of the bright interference fringes to the side of center so they are not all of the same height as they were shown in Fig. 34–14. (They were shown more correctly in Fig. 34–9b.)

To calculate the intensity in a double-slit interference pattern, including diffraction, let us assume the slits have equal widths D and their centers are separated by a distance d. Since the distance to the screen is large compared to the slit separation d, the wave amplitude due to each slit is essentially the same at each point on the screen. Then the total wave amplitude at any angle θ will no longer be

$$E_{\theta 0} = 2E_0 \cos\frac{\delta}{2},$$

as was given by Eq. 34–5b. Rather, it must be modified, because of diffraction, by Eq. 35–5, so that

$$E_{\theta 0} = 2E_0\left(\frac{\sin\beta/2}{\beta/2}\right)\cos\frac{\delta}{2}.$$

Thus the intensity will be given by

$$I_\theta = I_0\left(\frac{\sin\beta/2}{\beta/2}\right)^2\left(\cos\frac{\delta}{2}\right)^2 \qquad\qquad (35\text{–}9)$$

where $I_0 = 4E_0^2$, and from Eqs. 35–6 and 34–4 we have

$$\frac{\beta}{2} = \frac{\pi}{\lambda}D\sin\theta \qquad \text{and} \qquad \frac{\delta}{2} = \frac{\pi}{\lambda}d\sin\theta.$$

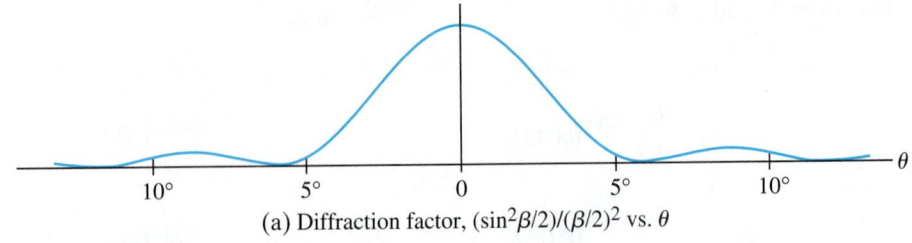

(a) Diffraction factor, $(\sin^2\beta/2)/(\beta/2)^2$ vs. θ

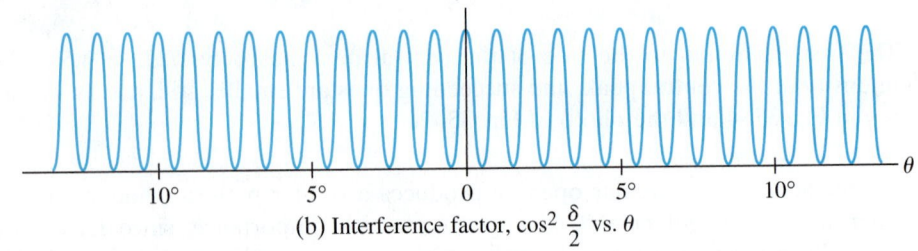

(b) Interference factor, $\cos^2\dfrac{\delta}{2}$ vs. θ

FIGURE 35–10
(a) Diffraction factor,
(b) interference factor, and
(c) the resultant intensity I_θ,
plotted as a function of θ
for $d = 6D = 60\lambda$.

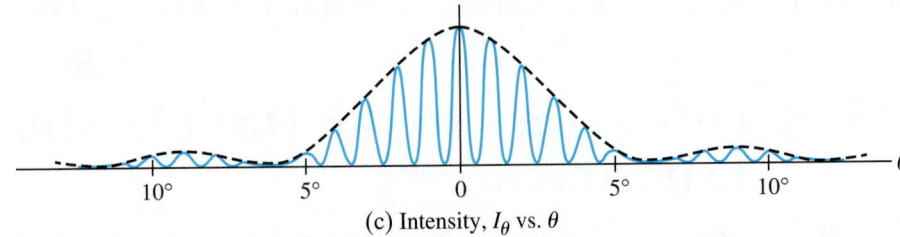

(c) Intensity, I_θ vs. θ

FIGURE 35–11 Photographs of a double-slit interference pattern using a laser beam showing effects of diffraction. In both cases $d = 0.50$ mm, whereas $D = 0.040$ mm in (a) and 0.080 mm in (b).

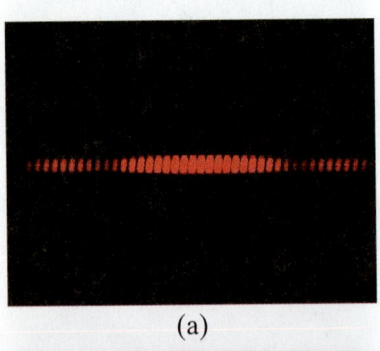

(a)

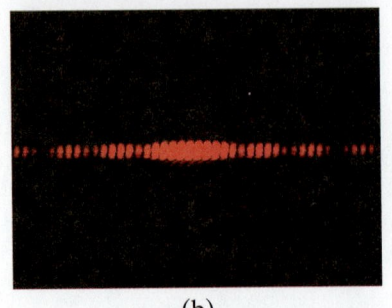

(b)

Equation 35–9 for the intensity in a double-slit pattern, as we just saw, is

$$I_\theta = I_0\left(\frac{\sin\beta/2}{\beta/2}\right)^2\left(\cos\frac{\delta}{2}\right)^2. \tag{35–9}$$

The first term in parentheses is sometimes called the "diffraction factor" and the second one the "interference factor." These two factors are plotted in Fig. 35–10a and b for the case when $d = 6D$, and $D = 10\lambda$. (Figure 35–10b is essentially the same as Fig. 34–14.) Figure 35–10c shows the product of these two curves (times I_0) which is the actual intensity as a function of θ (or as a function of position on the screen for θ not too large) as given by Eq. 35–9. As indicated by the dashed lines in Fig. 35–10c, the diffraction factor acts as a sort of envelope that limits the interference peaks.

EXAMPLE 35–4 **Diffraction plus interference.** Show why the central diffraction peak in Fig. 35–10c contains 11 interference fringes.

APPROACH The first minimum in the diffraction pattern occurs where

$$\sin\theta = \frac{\lambda}{D}.$$

Since $d = 6D$,

$$d\sin\theta = 6D\left(\frac{\lambda}{D}\right) = 6\lambda.$$

SOLUTION From Eq. 34–2a, interference peaks (maxima) occur for $d\sin\theta = m\lambda$ where m can be $0, 1, \cdots$ or any integer. Thus the diffraction minimum ($d\sin\theta = 6\lambda$) coincides with $m = 6$ in the interference pattern, so the $m = 6$ peak won't appear. Hence the central diffraction peak encloses the central interference peak ($m = 0$) and five peaks ($m = 1$ to 5) on each side for a total of 11. Since the sixth order doesn't appear, it is said to be a "missing order."

Notice from Example 35–4 that the number of interference fringes in the central diffraction peak depends only on the ratio d/D. It does not depend on wavelength λ. The actual spacing (in angle, or in position on the screen) does depend on λ. For the case illustrated, $D = 10\lambda$, and so the first diffraction minimum occurs at $\sin\theta = \lambda/D = 0.10$ or about $6°$.

The decrease in intensity of the interference fringes away from the center, as graphed in Fig. 35–10, is shown in Fig. 35–11.

Interference vs. Diffraction

The patterns due to interference and diffraction arise from the same phenomenon—the superposition of coherent waves of different phase. The distinction between them is thus not so much physical as for convenience of description, as in this Section where we analyzed the two-slit pattern in terms of interference and diffraction separately. In general, we use the word "diffraction" when referring to an analysis by superposition of many infinitesimal and usually contiguous sources, such as when we subdivide a source into infinitesimal parts. We use the term "interference" when we superpose the wave from a finite (and usually small) number of coherent sources.

35–4 Limits of Resolution; Circular Apertures

The ability of a lens to produce distinct images of two point objects very close together is called the **resolution** of the lens. The closer the two images can be and still be seen as distinct (rather than overlapping blobs), the higher the resolution. The resolution of a camera lens, for example, is often specified as so many dots or lines per millimeter, as mentioned in Section 33–5.

Two principal factors limit the resolution of a lens. The first is lens aberrations. As we saw in Chapter 33, because of spherical and other aberrations, a point object is not a point on the image but a tiny blob. Careful design of compound lenses can reduce aberrations significantly, but they cannot be eliminated entirely. The second factor that limits resolution is *diffraction*, which cannot be corrected for because it is a natural result of the wave nature of light. We discuss it now.

In Section 35–1, we saw that because light travels as a wave, light from a point source passing through a slit is spread out into a diffraction pattern (Figs. 35–2 and 35–4). A lens, because it has edges, acts like a round slit. When a lens forms the image of a point object, the image is actually a tiny diffraction pattern. Thus *an image would be blurred even if aberrations were absent*.

In the analysis that follows, we assume that the lens is free of aberrations, so we can concentrate on diffraction effects and how much they limit the resolution of a lens. In Fig. 35–4 we saw that the diffraction pattern produced by light passing through a rectangular slit has a central maximum in which most of the light falls. This central peak falls to a minimum on either side of its center at an angle θ given by $\sin\theta = \lambda/D$ (this is Eq. 35–1), where D is the slit width and λ the wavelength of light used. θ is the angular half-width of the central maximum, and for small angles can be written

$$\theta \approx \sin\theta = \frac{\lambda}{D}.$$

There are also low-intensity fringes beyond.

For a lens, or any circular hole, the image of a point object will consist of a *circular* central peak (called the *diffraction spot* or *Airy disk*) surrounded by faint circular fringes, as shown in Fig. 35–12a. The central maximum has an angular half width given by

$$\theta = \frac{1.22\lambda}{D},$$

where D is the diameter of the circular opening. This is a theoretical result for a perfect circle or lens. For real lenses or circles, the factor is on the order of 1 to 2. This formula differs from that for a slit (Eq. 35–1) by the factor 1.22. This factor appears because the width of a circular hole is not uniform (like a rectangular slit) but varies from its diameter D to zero. A mathematical analysis shows that the "average" width is $D/1.22$. Hence we get the equation above rather than Eq. 35–1. The intensity of light in the diffraction pattern from a point source of light passing through a circular opening is shown in Fig. 35–13. The image for a non-point source is a superposition of such patterns. For most purposes we need consider only the central spot, since the concentric rings are so much dimmer.

If two point objects are very close, the diffraction patterns of their images will overlap as shown in Fig. 35–12b. As the objects are moved closer, a separation is

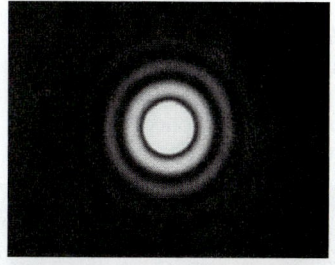
(a)

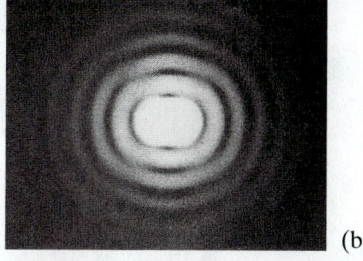
(b)

FIGURE 35–12 Photographs of images (greatly magnified) formed by a lens, showing the diffraction pattern of an image for: (a) a single point object; (b) two point objects whose images are barely resolved.

FIGURE 35–13 Intensity of light across the diffraction pattern of a circular hole.

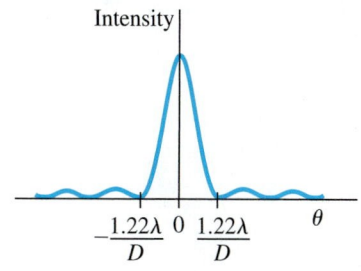

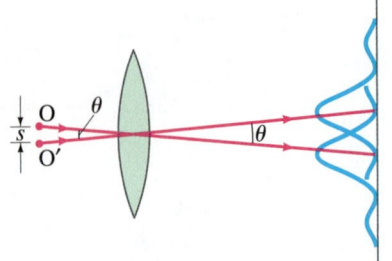

FIGURE 35–14 The *Rayleigh criterion.* Two images are just resolvable when the center of the diffraction peak of one is directly over the first minimum in the diffraction pattern of the other. The two point objects O and O′ subtend an angle θ at the lens; only one ray (it passes through the center of the lens) is drawn for each object, to indicate the center of the diffraction pattern of its image.

FIGURE 35–15 Hubble Space Telescope, with Earth in the background. The flat orange panels are solar cells that collect energy from the Sun.

PHYSICS APPLIED

How well the eye can see

reached where you can't tell if there are two overlapping images or a single image. The separation at which this happens may be judged differently by different observers. However, a generally accepted criterion is that proposed by Lord Rayleigh (1842–1919). This **Rayleigh criterion** states that *two images are just resolvable when the center of the diffraction disk of one image is directly over the first minimum in the diffraction pattern of the other.* This is shown in Fig. 35–14. Since the first minimum is at an angle $\theta = 1.22\lambda/D$ from the central maximum, Fig. 35–14 shows that two objects can be considered *just resolvable* if they are separated by at least an angle θ given by

$$\theta = \frac{1.22\lambda}{D}.$$
[θ in radians] **(35–10)**

In this equation, D is the diameter of the lens, and applies also to a mirror diameter. This is the limit on resolution set by the wave nature of light due to diffraction. A smaller angle means better resolution: you can make out closer objects. We see from Eq. 35–10 that using a shorter wavelength λ can reduce θ and thus increase resolution.

EXERCISE B Green light (550 nm) passes through a 25-mm-diameter camera lens. What is the angular half-width of the resulting diffraction pattern? (a) 2.7×10^{-5} degrees, (b) 1.5×10^{-3} degrees, (c) 3.2°, (d) 27°, (e) 1.5×10^{3} degrees.

EXAMPLE 35–5 **Hubble Space Telescope.** The Hubble Space Telescope (HST) is a reflecting telescope that was placed in orbit above the Earth's atmosphere, so its resolution would not be limited by turbulence in the atmosphere (Fig. 35–15). Its objective diameter is 2.4 m. For visible light, say $\lambda = 550\,\text{nm}$, estimate the improvement in resolution the Hubble offers over Earth-bound telescopes, which are limited in resolution by movement of the Earth's atmosphere to about half an arc second. (Each degree is divided into 60 minutes each containing 60 seconds, so $1° = 3600$ arc seconds.)

APPROACH Angular resolution for the Hubble is given (in radians) by Eq. 35–10. The resolution for Earth telescopes is given, and we first convert it to radians so we can compare.

SOLUTION Earth-bound telescopes are limited to an angular resolution of

$$\theta = \tfrac{1}{2}\left(\frac{1}{3600}\right)^{°}\left(\frac{2\pi\,\text{rad}}{360°}\right) = 2.4 \times 10^{-6}\,\text{rad}.$$

The Hubble, on the other hand, is limited by diffraction (Eq. 35–10) which for $\lambda = 550\,\text{nm}$ is

$$\theta = \frac{1.22\lambda}{D} = \frac{1.22(550 \times 10^{-9}\,\text{m})}{2.4\,\text{m}} = 2.8 \times 10^{-7}\,\text{rad},$$

thus giving almost ten times better resolution $(2.4 \times 10^{-6}\,\text{rad}/2.8 \times 10^{-7}\,\text{rad} \approx 9\times)$.

EXAMPLE 35–6 **ESTIMATE** **Eye resolution.** You are in an airplane at an altitude of 10,000 m. If you look down at the ground, estimate the minimum separation s between objects that you could distinguish. Could you count cars in a parking lot? Consider only diffraction, and assume your pupil is about 3.0 mm in diameter and $\lambda = 550\,\text{nm}$.

APPROACH We use the Rayleigh criterion, Eq. 35–10, to estimate θ. The separation s of objects is $s = \ell\theta$, where $\ell = 10^4\,\text{m}$ and θ is in radians.

SOLUTION In Eq. 35–10, we set $D = 3.0\,\text{mm}$ for the opening of the eye:

$$s = \ell\theta = \ell\frac{1.22\lambda}{D} = \frac{(10^4\,\text{m})(1.22)(550 \times 10^{-9}\,\text{m})}{3.0 \times 10^{-3}\,\text{m}} = 2.2\,\text{m}.$$

Yes, you could just resolve a car (roughly 2 m wide by 3 or 4 m long) and count them.

35–5 Resolution of Telescopes and Microscopes; the λ Limit

You might think that a microscope or telescope could be designed to produce any desired magnification, depending on the choice of focal lengths and quality of the lenses. But this is not possible, because of diffraction. An increase in magnification above a certain point merely results in magnification of the diffraction patterns. This can be highly misleading since we might think we are seeing details of an object when we are really seeing details of the diffraction pattern. To examine this problem, we apply the Rayleigh criterion: two objects (or two nearby points on one object) are just resolvable if they are separated by an angle θ (Fig. 35–14) given by Eq. 35–10:

$$\theta = \frac{1.22\lambda}{D}.$$

This formula is valid for either a microscope or a telescope, where D is the diameter of the objective lens or mirror. For a telescope, the resolution is specified by stating θ as given by this equation.[†]

FIGURE 35–16 The 300-meter radiotelescope in Arecibo, Puerto Rico, uses radio waves (Fig. 31–12) instead of visible light.

EXAMPLE 35–7 Telescope resolution (radio wave vs. visible light). What is the theoretical minimum angular separation of two stars that can just be resolved by (a) the 200-inch telescope on Palomar Mountain (Fig. 33–38c); and (b) the Arecibo radiotelescope (Fig. 35–16), whose diameter is 300 m and whose radius of curvature is also 300 m. Assume $\lambda = 550$ nm for the visible-light telescope in part (a), and $\lambda = 4$ cm (the shortest wavelength at which the radiotelescope has been operated) in part (b).

APPROACH We apply the Rayleigh criterion (Eq. 35–10) for each telescope.

SOLUTION (a) Since $D = 200$ in. $= 5.1$ m, we have from Eq. 35–10 that

$$\theta = \frac{1.22\lambda}{D} = \frac{(1.22)(5.50 \times 10^{-7}\,\text{m})}{(5.1\,\text{m})} = 1.3 \times 10^{-7}\,\text{rad},$$

or 0.75×10^{-5} deg. (Note that this is equivalent to resolving two points less than 1 cm apart from a distance of 100 km!)

(b) For radio waves with $\lambda = 0.04$ m emitted by stars, the resolution is

$$\theta = \frac{(1.22)(0.04\,\text{m})}{(300\,\text{m})} = 1.6 \times 10^{-4}\,\text{rad}.$$

The resolution is less because the wavelength is so much larger, but the larger objective collects more radiation and thus detects fainter objects.

NOTE In both cases, we determined the limit set by diffraction. The resolution for a visible-light Earth-bound telescope is not this good because of aberrations and, more importantly, turbulence in the atmosphere. In fact, large-diameter objectives are not justified by increased resolution, but by their greater light-gathering ability—they allow more light in, so fainter objects can be seen. Radiotelescopes are not hindered by atmospheric turbulence, and the resolution found in (b) is a good estimate.

[†]Earth-bound telescopes with large-diameter objectives are usually limited not by diffraction but by other effects such as turbulence in the atmosphere. The resolution of a high-quality microscope, on the other hand, normally *is* limited by diffraction; microscope objectives are complex compound lenses containing many elements of small diameter (since f is small), thus reducing aberrations.

For a microscope, it is more convenient to specify the actual distance, s, between two points that are just barely resolvable: see Fig. 35–14. Since objects are normally placed near the focal point of the microscope objective, the angle subtended by two objects is $\theta = s/f$, so $s = f\theta$. If we combine this with Eq. 35–10, we obtain the **resolving power (RP)** of a microscope

$$RP = s = f\theta = \frac{1.22\lambda f}{D},\qquad (35\text{--}11)$$

where f is the objective lens' focal length (not frequency). This distance s is called the resolving power of the lens because it is the minimum separation of two object points that can just be resolved—assuming the highest quality lens since this limit is imposed by the wave nature of light. A smaller RP means better resolution, better detail.

EXERCISE C What is the resolving power of a microscope with a 5-mm-diameter objective which has $f = 9\,\text{mm}$? (a) 550 nm, (b) 750 nm, (c) 1200 nm, (d) 0.05 nm, (e) 0.005 nm.

Diffraction sets an ultimate limit on the detail that can be seen on any object. In Eq. 35–11 for resolving power of a microscope, the focal length of the lens cannot practically be made less than (approximately) the radius of the lens, and even that is very difficult (see the lensmaker's equation, Eq. 33–4). In this best case, Eq. 35–11 gives, with $f \approx D/2$.

$$RP \approx \frac{\lambda}{2}.\qquad (35\text{--}12)$$

Thus we can say, to within a factor of 2 or so, that

it is not possible to resolve detail of objects smaller than the wavelength of the radiation being used.

This is an important and useful rule of thumb.

Compound lenses in microscopes are now designed so well that the actual limit on resolution is often set by diffraction—that is, by the wavelength of the light used. To obtain greater detail, one must use radiation of shorter wavelength. The use of UV radiation can increase the resolution by a factor of perhaps 2. Far more important, however, was the discovery in the early twentieth century that electrons have wave properties (Chapter 37) and that their wavelengths can be very small. The wave nature of electrons is utilized in the electron microscope (Section 37–8), which can magnify 100 to 1000 times more than a visible-light microscope because of the much shorter wavelengths. X-rays, too, have very short wavelengths and are often used to study objects in great detail (Section 35–10).

λ limits resolution

*35–6 Resolution of the Human Eye and Useful Magnification

The resolution of the human eye is limited by several factors, all of roughly the same order of magnitude. The resolution is best at the fovea, where the cone spacing is smallest, about $3\,\mu\text{m}\,(=3000\,\text{nm})$. The diameter of the pupil varies from about 0.1 cm to about 0.8 cm. So for $\lambda = 550\,\text{nm}$ (where the eye's sensitivity is greatest), the diffraction limit is about $\theta \approx 1.22\lambda/D \approx 8 \times 10^{-5}\,\text{rad}$ to $6 \times 10^{-4}\,\text{rad}$. The eye is about 2 cm long, giving a resolving power (Eq. 35–11) of $s \approx (2 \times 10^{-2}\,\text{m})(8 \times 10^{-5}\,\text{rad}) \approx 2\,\mu\text{m}$ at best, to about $10\,\mu\text{m}$ at worst (pupil small). Spherical and chromatic aberration also limit the resolution to about $10\,\mu\text{m}$. The net result is that the eye can just resolve objects whose angular separation is around

$$5 \times 10^{-4}\,\text{rad.}\qquad \begin{bmatrix} \text{best eye} \\ \text{resolution} \end{bmatrix}$$

This corresponds to objects separated by 1 cm at a distance of about 20 m.

The typical near point of a human eye is about 25 cm. At this distance, the eye can just resolve objects that are $(25\,\text{cm})(5 \times 10^{-4}\,\text{rad}) \approx 10^{-4}\,\text{m} = \frac{1}{10}\,\text{mm}$ apart. Since the best light microscopes can resolve objects no smaller than about 200 nm at best (Eq. 35–12 for violet light, $\lambda = 400\,\text{nm}$), the useful magnification

[= (resolution by naked eye)/(resolution by microscope)] is limited to about

$$\frac{10^{-4}\,\text{m}}{200 \times 10^{-9}\,\text{m}} \approx 500\times. \qquad \begin{bmatrix} \text{maximum useful} \\ \text{microscope magnification} \end{bmatrix}$$

In practice, magnifications of about 1000× are often used to minimize eyestrain. Any greater magnification would simply make visible the diffraction pattern produced by the microscope objective lens.

Now you have the answers to the Chapter-Opening Question: (c), by the equation above, and (c) by the λ rule.

35–7 Diffraction Grating

A large number of equally spaced parallel slits is called a **diffraction grating**, although the term "interference grating" might be as appropriate. Gratings can be made by precision machining of very fine parallel lines on a glass plate. The untouched spaces between the lines serve as the slits. Photographic transparencies of an original grating serve as inexpensive gratings. Gratings containing 10,000 lines per centimeter are common, and are very useful for precise measurements of wavelengths. A diffraction grating containing slits is called a **transmission grating**. Another type of diffraction grating is the **reflection grating**, made by ruling fine lines on a metallic or glass surface from which light is reflected and analyzed. The analysis is basically the same as for a transmission grating, which we now discuss.

The analysis of a diffraction grating is much like that of Young's double-slit experiment. We assume parallel rays of light are incident on the grating as shown in Fig. 35–17. We also assume that the slits are narrow enough so that diffraction by each of them spreads light over a very wide angle on a distant screen beyond the grating, and interference can occur with light from all the other slits. Light rays that pass through each slit without deviation $(\theta = 0°)$ interfere constructively to produce a bright line at the center of the screen. Constructive interference also occurs at an angle θ such that rays from adjacent slits travel an extra distance of $\Delta\ell = m\lambda$, where m is an integer. If d is the distance between slits, then we see from Fig. 35–17 that $\Delta\ell = d \sin\theta$, and

$$\sin\theta = \frac{m\lambda}{d}, \qquad m = 0, 1, 2, \cdots \qquad \begin{bmatrix} \text{diffraction grating,} \\ \text{principal maxima} \end{bmatrix} \quad \textbf{(35–13)}$$

is the criterion to have a brightness maximum. This is the same equation as for the double-slit situation, and again m is called the order of the pattern.

There is an important difference between a double-slit and a multiple-slit pattern. The bright maxima are much *sharper* and *narrower* for a grating. Why? Suppose that the angle θ is increased just slightly beyond that required for a maximum. In the case of only two slits, the two waves will be only slightly out of phase, so nearly full constructive interference occurs. This means the maxima are wide (see Fig. 34–9). For a grating, the waves from two adjacent slits will also not be significantly out of phase. But waves from one slit and those from a second one a few hundred slits away may be exactly out of phase; all or nearly all the light can cancel in pairs in this way. For example, suppose the angle θ is very slightly different from its first-order maximum, so that the extra path length for a pair of adjacent slits is not exactly λ but rather 1.0010λ. The wave through one slit and another one 500 slits below will have a path difference of $1\lambda + (500)(0.0010\lambda) = 1.5000\lambda$, or $1\frac{1}{2}$ wavelengths, so the two will cancel. A pair of slits, one below each of these, will also cancel. That is, the light from slit 1 cancels with that from slit 501; light from slit 2 cancels with that from slit 502, and so on. Thus even for a tiny angle[†] corresponding to an extra path length of $\frac{1}{1000}\lambda$, there is much destructive interference, and so the maxima are very narrow. The more lines there are in a grating, the sharper will be the peaks (see Fig. 35–18). Because a grating produces much sharper lines than two slits alone can (and much brighter lines because there are many more slits), a grating is a far more precise device for measuring wavelengths.

[†]Depending on the total number of slits, there may or may not be complete cancellation for such an angle, so there will be very tiny peaks between the main maxima (see Fig. 35–18b), but they are usually much too small to be seen.

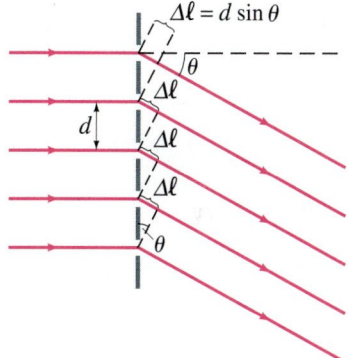

FIGURE 35–17 Diffraction grating.

> ⚠ **CAUTION**
>
> *Diffraction grating is analyzed using interference formulas, not diffraction formulas*

FIGURE 35–18 Intensity as a function of viewing angle θ (or position on the screen) for (a) two slits, (b) six slits. For a diffraction grating, the number of slits is very large $(\approx 10^4)$ and the peaks are narrower still.

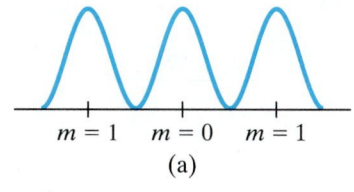

$m = 1 \qquad m = 0 \qquad m = 1$
(a)

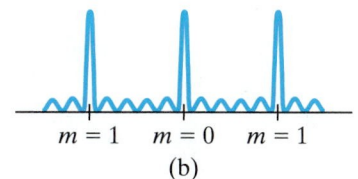

$m = 1 \qquad m = 0 \qquad m = 1$
(b)

Suppose the light striking a diffraction grating is not monochromatic, but consists of two or more distinct wavelengths. Then for all orders other than $m = 0$, each wavelength will produce a maximum at a different angle (Fig. 35–19a), just as for a double slit. If white light strikes a grating, the central ($m = 0$) maximum will be a sharp white peak. But for all other orders, there will be a distinct spectrum of colors spread out over a certain angular width, Fig. 35–19b. Because a diffraction grating spreads out light into its component wavelengths, the resulting pattern is called a **spectrum**.

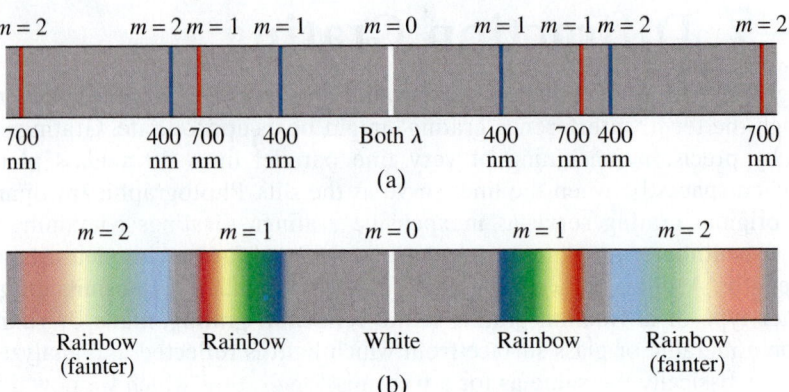

FIGURE 35–19 Spectra produced by a grating: (a) two wavelengths, 400 nm and 700 nm; (b) white light. The second order will normally be dimmer than the first order. (Higher orders are not shown.) If grating spacing is small enough, the second and higher orders will be missing.

EXAMPLE 35–8 **Diffraction grating: lines.** Determine the angular positions of the first- and second-order maxima for light of wavelength 400 nm and 700 nm incident on a grating containing 10,000 lines/cm.

APPROACH First we find the distance d between grating lines: if the grating has N lines in 1 m, then the distance between lines must be $d = 1/N$ meters. Then we use Eq. 35–13, $\sin \theta = m\lambda/d$, to find the angles for the two wavelengths for $m = 1$ and 2.

SOLUTION The grating contains 1.00×10^4 lines/cm $= 1.00 \times 10^6$ lines/m, which means the distance between lines is $d = (1/1.00 \times 10^6)$ m $= 1.00 \times 10^{-6}$ m $= 1.00\ \mu$m. In first order ($m = 1$), the angles are

$$\sin \theta_{400} = \frac{m\lambda}{d} = \frac{(1)(4.00 \times 10^{-7}\,\text{m})}{1.00 \times 10^{-6}\,\text{m}} = 0.400$$

$$\sin \theta_{700} = \frac{(1)(7.00 \times 10^{-7}\,\text{m})}{1.00 \times 10^{-6}\,\text{m}} = 0.700$$

so $\theta_{400} = 23.6°$ and $\theta_{700} = 44.4°$. In second order,

$$\sin \theta_{400} = \frac{2\lambda}{d} = \frac{(2)(4.00 \times 10^{-7}\,\text{m})}{1.00 \times 10^{-6}\,\text{m}} = 0.800$$

$$\sin \theta_{700} = \frac{(2)(7.00 \times 10^{-7}\,\text{m})}{1.00 \times 10^{-6}\,\text{m}} = 1.40$$

so $\theta_{400} = 53.1°$. But the second order does not exist for $\lambda = 700$ nm because $\sin \theta$ cannot exceed 1. No higher orders will appear.

EXAMPLE 35–9 **Spectra overlap.** White light containing wavelengths from 400 nm to 750 nm strikes a grating containing 4000 lines/cm. Show that the blue at $\lambda = 450$ nm of the third-order spectrum overlaps the red at 700 nm of the second order.

APPROACH We use $\sin \theta = m\lambda/d$ to calculate the angular positions of the $m = 3$ blue maximum and the $m = 2$ red one.

SOLUTION The grating spacing is $d = (1/4000)$ cm $= 2.50 \times 10^{-6}$ m. The blue of the third order occurs at an angle θ given by

$$\sin \theta = \frac{m\lambda}{d} = \frac{(3)(4.50 \times 10^{-7}\,\text{m})}{(2.50 \times 10^{-6}\,\text{m})} = 0.540.$$

Red in second order occurs at

$$\sin \theta = \frac{(2)(7.00 \times 10^{-7}\,\text{m})}{(2.50 \times 10^{-6}\,\text{m})} = 0.560,$$

which is a greater angle; so the second order overlaps into the beginning of the third-order spectrum.

CONCEPTUAL EXAMPLE 35–10 | **Compact disk.** When you look at the surface of a music CD (Fig. 35–20), you see the colors of a rainbow. (*a*) Estimate the distance between the curved lines (they are read by a laser). (*b*) Estimate the distance between lines, noting that a CD contains at most 80 min of music, that it rotates at speeds from 200 to 500 rev/min, and that $\frac{2}{3}$ of its 6 cm radius contains the lines.

RESPONSE (*a*) The CD acts like a reflection diffraction grating. To satisfy Eq. 35–13, we might estimate the line spacing as one or a few (2 or 3) wavelengths ($\lambda \approx 550$ nm) or 0.5 to 1.5 μm. (*b*) Average rotation speed of 350 rev/min times 80 min gives 28,000 total rotations or 28,000 lines, which are spread over $\left(\frac{2}{3}\right)(6\,\text{cm}) = 4$ cm. So we have a sort of reflection diffraction grating with about $(28,000\,\text{lines})/(4\,\text{cm}) = 7000$ lines/cm. The distance d between lines is roughly 1 cm/7000 lines $\approx 1.4 \times 10^{-6}$ m $= 1.4\,\mu$m. Our results in (*a*) and (*b*) agree.

FIGURE 35–20 A compact disk, Example 35–10.

35–8 The Spectrometer and Spectroscopy

A **spectrometer** or **spectroscope**, Fig. 35–21, is a device to measure wavelengths accurately using a diffraction grating (or a prism) to separate different wavelengths of light. Light from a source passes through a narrow slit S in the "collimator." The slit is at the focal point of the lens L, so parallel light falls on the grating. The movable telescope can bring the rays to a focus. Nothing will be seen in the viewing telescope unless it is positioned at an angle θ that corresponds to a diffraction peak (first order is usually used) of a wavelength emitted by the source. The angle θ can be measured to very high accuracy, so the wavelength of a line can be determined to high accuracy using Eq. 35–13:

$$\lambda = \frac{d}{m}\sin\theta,$$

where m is an integer representing the order, and d is the distance between grating lines. The line you see in a spectrometer corresponding to each wavelength is actually an image of the slit S. A narrower slit results in dimmer light but we can measure the angular positions more precisely. If the light contains a continuous range of wavelengths, then a continuous spectrum is seen in the spectroscope.

The spectrometer in Fig. 35–21 uses a transmission grating. Others may use a reflection grating, or sometimes a prism. A prism works because of dispersion (Section 32–6), bending light of different wavelengths into different angles. (A prism is not a linear device and must be calibrated.)

An important use of a spectrometer is for the identification of atoms or molecules. When a gas is heated or an electric current is passed through it, the gas emits a characteristic **line spectrum**. That is, only certain discrete wavelengths of light are emitted, and these are different for different elements and compounds.[†] Figure 35–22 shows the line spectra for a number of elements in the gas state. Line spectra occur only for gases at high temperatures and low pressure and density. The light from heated solids, such as a lightbulb filament, and even from a dense gaseous object such as the Sun, produces a **continuous spectrum** including a wide range of wavelengths.

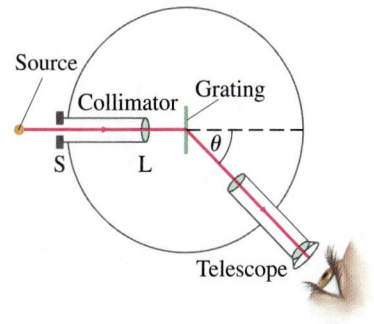

FIGURE 35–21 Spectrometer or spectroscope.

Source

Collimator   Grating

S   L θ

Telescope

Eye

[†]Why atoms and molecules emit line spectra was a great mystery for many years and played a central role in the development of modern quantum theory, as we shall see in Chapter 37.

Atomic hydrogen

Mercury

Sodium

Solar absorption spectrum

FIGURE 35–22 Line spectra for the gases indicated, and the spectrum from the Sun showing absorption lines.

Figure 35–22 also shows the Sun's "continuous spectrum," which contains a number of *dark* lines (only the most prominent are shown), called **absorption lines**. Atoms and molecules can absorb light at the same wavelengths at which they emit light. The Sun's absorption lines are due to absorption by atoms and molecules in the cooler outer atmosphere of the Sun, as well as by atoms and molecules in the Earth's atmosphere. A careful analysis of all these thousands of lines reveals that at least two-thirds of all elements are present in the Sun's atmosphere. The presence of elements in the atmosphere of other planets, in interstellar space, and in stars, is also determined by spectroscopy.

Spectroscopy is useful for determining the presence of certain types of molecules in laboratory specimens where chemical analysis would be difficult. For example, biological DNA and different types of protein absorb light in particular regions of the spectrum (such as in the UV). The material to be examined, which is often in solution, is placed in a monochromatic light beam whose wavelength is selected by the placement angle of a diffraction grating or prism. The amount of absorption, as compared to a standard solution without the specimen, can reveal not only the presence of a particular type of molecule, but also its concentration.

Light emission and absorption also occur outside the visible part of the spectrum, such as in the UV and IR regions. Glass absorbs light in these regions, so reflection gratings and mirrors (in place of lenses) are used. Special types of film or sensors are used for detection.

PHYSICS APPLIED
Chemical and biochemical analysis by spectroscopy

EXAMPLE 35–11 **Hydrogen spectrum.** Light emitted by hot hydrogen gas is observed with a spectroscope using a diffraction grating having 1.00×10^4 lines/cm. The spectral lines nearest to the center $(0°)$ are a violet line at $24.2°$, a blue line at $25.7°$, a blue-green line at $29.1°$, and a red line at $41.0°$ from the center. What are the wavelengths of these spectral lines of hydrogen?

APPROACH The wavelengths can be determined from the angles by using $\lambda = (d/m) \sin \theta$ where d is the spacing between slits, and m is the order of the spectrum (Eq. 35–13).

SOLUTION Since these are the closest lines to $\theta = 0°$, this is the first-order spectrum $(m = 1)$. The slit spacing is $d = 1/(1.00 \times 10^4 \, \text{cm}^{-1}) = 1.00 \times 10^{-6} \, \text{m}$. The wavelength of the violet line is

$$\lambda = \left(\frac{d}{m}\right) \sin \theta = \left(\frac{1.00 \times 10^{-6} \, \text{m}}{1}\right) \sin 24.2° = 4.10 \times 10^{-7} \, \text{m} = 410 \, \text{nm}.$$

The other wavelengths are:

blue: $\qquad \lambda = (1.00 \times 10^{-6} \, \text{m}) \sin 25.7° = 434 \, \text{nm},$

blue-green: $\quad \lambda = (1.00 \times 10^{-6} \, \text{m}) \sin 29.1° = 486 \, \text{nm},$

red: $\qquad \lambda = (1.00 \times 10^{-6} \, \text{m}) \sin 41.0° = 656 \, \text{nm}.$

NOTE In an unknown mixture of gases, these four spectral lines need to be seen to identify that the mixture contains hydrogen.

936 CHAPTER 35

*35–9 Peak Widths and Resolving Power for a Diffraction Grating

We now look at the pattern of maxima produced by a multiple-slit grating using phasor diagrams. We can determine a formula for the width of each peak, and we will see why there are tiny maxima between the principal maxima, as indicated in Fig. 35–18b. First of all, it should be noted that the two-slit and six-slit patterns shown in Fig. 35–18 were drawn assuming very narrow slits so that diffraction does not limit the height of the peaks. For real diffraction gratings, this is not normally the case: the slit width D is often not much smaller than the slit separation d, and diffraction thus limits the intensity of the peaks so the central peak $(m = 0)$ is brighter than the side peaks. We won't worry about this effect on intensity except to note that if a diffraction minimum coincides with a particular order of the interference pattern, that order will not appear. (For example, if $d = 2D$, all the even orders, $m = 2, 4, \cdots$, will be missing. Can you see why? Hint: See Example 35–4.)

Figures 35–23 and 35–24 show phasor diagrams for a two-slit and a six-slit grating, respectively. Each short arrow represents the amplitude of a wave from a single slit, and their vector sum (as phasors) represents the total amplitude for a given viewing angle θ. Part (a) of each Figure shows the phasor diagram at $\theta = 0°$, at the center of the pattern, which is the central maximum $(m = 0)$. Part (b) of each Figure shows the condition for the adjacent minimum: where the arrows first close on themselves (add to zero) so the amplitude E_θ is zero. For two slits, this occurs when the two separate amplitudes are 180° out of phase. For six slits, it occurs when each amplitude makes a 60° angle with its neighbor. For two slits, the minimum occurs when the phase between slits is $\delta = 2\pi/2$ (in radians); for six slits it occurs when the phase δ is $2\pi/6$; and in the general case of N slits, the minimum occurs for a phase difference between adjacent slits of

$$\delta = \frac{2\pi}{N}. \tag{35–14}$$

What does this correspond to in θ? First note that δ is related to θ by (Eq. 34–4)

$$\frac{\delta}{2\pi} = \frac{d \sin \theta}{\lambda} \qquad \text{or} \qquad \delta = \frac{2\pi}{\lambda} d \sin \theta. \tag{35–15}$$

Let us call $\Delta\theta_0$ the angular position of the minimum next to the peak at $\theta = 0$. Then

$$\frac{\delta}{2\pi} = \frac{d \sin \Delta\theta_0}{\lambda}.$$

We insert Eq. 35–14 for δ and find

$$\sin \Delta\theta_0 = \frac{\lambda}{Nd}. \tag{35–16a}$$

Since $\Delta\theta_0$ is usually small (N is usually very large for a grating), $\sin \Delta\theta_0 \approx \Delta\theta_0$, so in the small angle limit we can write

$$\Delta\theta_0 = \frac{\lambda}{Nd}. \tag{35–16b}$$

It is clear from either of the last two relations that the larger N is, the narrower will be the central peak. (For $N = 2, \sin \Delta\theta_0 = \lambda/2d$, which is what we obtained earlier for the double slit, Eq. 34–2b with $m = 0$.)

Either of Eqs. 35–16 shows why the peaks become narrower for larger N. The origin of the small secondary maxima between the principal peaks (see Fig. 35–18b) can be deduced from the diagram of Fig. 35–25. This is just a continuation of Fig. 35–24b (where $\delta = 60°$); but now the phase δ has been increased to almost 90°, where E_θ is a relative maximum. Note that E_θ is much less than E_0 (Fig. 35–24a), so the intensity in this secondary maximum is much smaller than in a principal peak. As δ (and θ) is increased further, E_θ again decreases to zero (a "double circle"), then reaches another tiny maximum, and so on. Eventually the diagram unfolds again and when $\delta = 360°$, all the amplitudes again lie in a straight line (as in Fig. 35–24a) corresponding to the next principal maximum $(m = 1$ in Eq. 35–13).

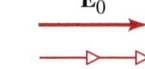

(a) Central maximum: $\theta = 0, \delta = 0$

$E = 0$

$\delta = 180°$

(b) Minimum: $\delta = 180°$

FIGURE 35–23 Phasor diagram for two slits (a) at the central maximum, (b) at the nearest minimum.

FIGURE 35–24 Phasor diagram for six slits (a) at the central maximum, (b) at the nearest minimum.

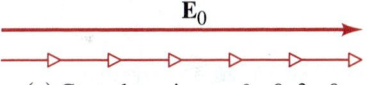

(a) Central maximum: $\theta = 0, \delta = 0$

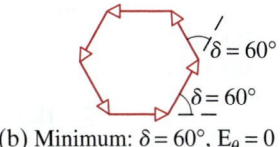

(b) Minimum: $\delta = 60°$, $E_\theta = 0$

FIGURE 35–25 Phasor diagram for the secondary peak.

Equation 35–16b gives the half-width of the central $(m = 0)$ peak. To determine the half-width of higher order peaks, $\Delta\theta_m$ for order m, we differentiate Eq. 35–15 so as to relate the change $\Delta\delta$ in δ, to the change $\Delta\theta$ in the angle θ:

$$\Delta\delta \approx \frac{d\delta}{d\theta}\Delta\theta = \frac{2\pi d}{\lambda}\cos\theta\,\Delta\theta.$$

If $\Delta\theta_m$ represents the half-width of a peak of order m $(m = 1, 2, \cdots)$—that is, the angle between the peak maximum and the minimum to either side—then $\Delta\delta = 2\pi/N$ as given by Eq. 35–14. We insert this into the above relation and find

$$\Delta\theta_m = \frac{\lambda}{Nd\cos\theta_m}, \qquad (35\text{–}17)$$

where θ_m is the angular position of the mth peak as given by Eq. 35–13. This derivation is valid, of course, only for small $\Delta\delta$ $(= 2\pi/N)$ which is indeed the case for real gratings since N is on the order of 10^4 or more.

An important property of any diffraction grating used in a spectrometer is its ability to resolve two very closely spaced wavelengths (wavelength difference $= \Delta\lambda$). The **resolving power** R of a grating is defined as

$$R = \frac{\lambda}{\Delta\lambda}. \qquad (35\text{–}18)$$

With a little work, using Eq. 35–17, we can show that $\Delta\lambda = \lambda/Nm$ where N is the total number of grating lines and m is the order. Then we have

$$R = \frac{\lambda}{\Delta\lambda} = Nm. \qquad (35\text{–}19)$$

The larger the value of R, the closer two wavelengths can be resolvable. If R is given, the minimum separation $\Delta\lambda$ between two wavelengths near λ, is (by Eq. 35–18)

$$\Delta\lambda = \frac{\lambda}{R}.$$

EXAMPLE 35–12 **Resolving two close lines.** Yellow sodium light, which consists of two wavelengths, $\lambda_1 = 589.00$ nm and $\lambda_2 = 589.59$ nm, falls on a 7500-line/cm diffraction grating. Determine (a) the maximum order m that will be present for sodium light, (b) the width of grating necessary to resolve the two sodium lines.

APPROACH We first find $d = 1\,\text{cm}/7500 = 1.33 \times 10^{-6}\,\text{m}$, and then use Eq. 35–13 to find m. For (b) we use Eqs. 35–18 and 35–19.

SOLUTION (a) The maximum value of m at $\lambda = 589$ nm, using Eq. 35–13 with $\sin\theta \leq 1$, is

$$m = \frac{d}{\lambda}\sin\theta \leq \frac{d}{\lambda} = \frac{1.33 \times 10^{-6}\,\text{m}}{5.89 \times 10^{-7}\,\text{m}} = 2.26,$$

so $m = 2$ is the maximum order present.
(b) The resolving power needed is

$$R = \frac{\lambda}{\Delta\lambda} = \frac{589\,\text{nm}}{0.59\,\text{nm}} = 1000.$$

From Eq. 35–19, the total number N of lines needed for the $m = 2$ order is $N = R/m = 1000/2 = 500$, so the grating need only be $500/7500\,\text{cm}^{-1} = 0.0667\,\text{cm}$ wide. A typical grating is a few centimeters wide, and so will easily resolve the two lines.

35–10 X-Rays and X-Ray Diffraction

In 1895, W. C. Roentgen (1845–1923) discovered that when electrons were accelerated by a high voltage in a vacuum tube and allowed to strike a glass or metal surface inside the tube, fluorescent minerals some distance away would glow, and photographic film would become exposed. Roentgen attributed these effects to a new type of radiation (different from cathode rays). They were given the name **X-rays** after the algebraic symbol x, meaning an unknown quantity. He soon found that X-rays penetrated through some materials better than through others, and within a few weeks he presented the first X-ray photograph (of his wife's hand). The production of X-rays today is usually done in a tube (Fig. 35–26) similar to Roentgen's, using voltages of typically 30 kV to 150 kV.

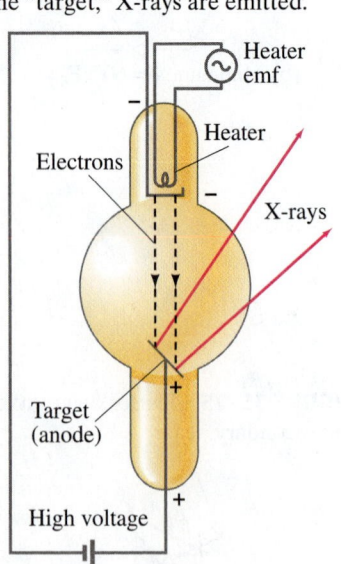

FIGURE 35–26 X-ray tube. Electrons emitted by a heated filament in a vacuum tube are accelerated by a high voltage. When they strike the surface of the anode, the "target," X-rays are emitted.

Heater emf

Heater

Electrons

X-rays

Target (anode)

High voltage

Investigations into the nature of X-rays indicated they were not charged particles (such as electrons) since they could not be deflected by electric or magnetic fields. It was suggested that they might be a form of invisible light. However, they showed no diffraction or interference effects using ordinary gratings. Indeed, if their wavelengths were much smaller than the typical grating spacing of 10^{-6} m $(= 10^3$ nm), no effects would be expected. Around 1912, Max von Laue (1879–1960) suggested that if the atoms in a crystal were arranged in a regular array (see Fig. 17–2a), such a crystal might serve as a diffraction grating for very short wavelengths on the order of the spacing between atoms, estimated to be about 10^{-10} m $(= 10^{-1}$ nm). Experiments soon showed that X-rays scattered from a crystal did indeed show the peaks and valleys of a diffraction pattern (Fig. 35–27). Thus it was shown, in a single blow, that X-rays have a wave nature and that atoms are arranged in a regular way in crystals. Today, X-rays are recognized as electromagnetic radiation with wavelengths in the range of about 10^{-2} nm to 10 nm, the range readily produced in an X-ray tube.

We saw in Section 35–5 that light of shorter wavelength provides greater resolution when we are examining an object microscopically. Since X-rays have much shorter wavelengths than visible light, they should in principle offer much greater resolution. However, there seems to be no effective material to use as lenses for the very short wavelengths of X-rays. Instead, the clever but complicated technique of **X-ray diffraction** (or **crystallography**) has proved very effective for examining the microscopic world of atoms and molecules. In a simple crystal such as NaCl, the atoms are arranged in an orderly cubical fashion, Fig. 35–28, with atoms spaced a distance d apart. Suppose that a beam of X-rays is incident on the crystal at an angle ϕ to the surface, and that the two rays shown are reflected from two subsequent planes of atoms as shown. The two rays will constructively interfere if the extra distance ray I travels is a whole number of wavelengths farther than the distance ray II travels. This extra distance is $2d \sin \phi$. Therefore, constructive interference will occur when

$$m\lambda = 2d \sin \phi, \qquad m = 1, 2, 3, \cdots, \qquad \textbf{(35–20)}$$

where m can be any integer. (Notice that ϕ is *not* the angle with respect to the normal to the surface.) This is called the **Bragg equation** after W. L. Bragg (1890–1971), who derived it and who, together with his father W. H. Bragg (1862–1942), developed the theory and technique of X-ray diffraction by crystals in 1912–1913. If the X-ray wavelength is known and the angle ϕ is measured, the distance d between atoms can be obtained. This is the basis for X-ray crystallography.

EXERCISE D When X-rays of wavelength 0.10×10^{-9} m are scattered from a sodium chloride crystal, a second-order diffraction peak is observed at 21°. What is the spacing between the planes of atoms for this scattering?

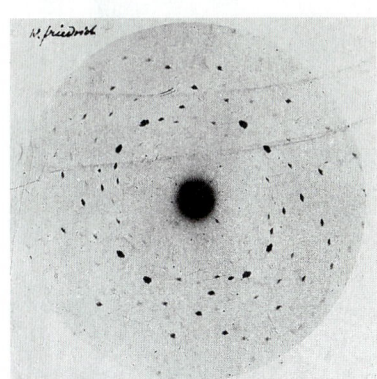

FIGURE 35–27 This X-ray diffraction pattern is one of the first observed by Max von Laue in 1912 when he aimed a beam of X-rays at a zinc sulfide crystal. The diffraction pattern was detected directly on a photographic plate.

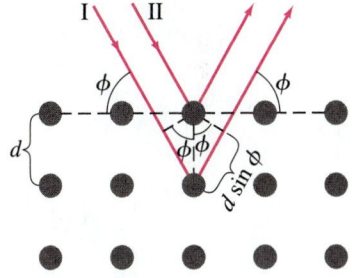

FIGURE 35–28 X-ray diffraction by a crystal.

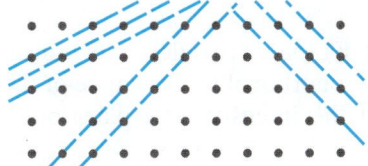

FIGURE 35–29 X-rays can be diffracted from many possible planes within a crystal.

Actual X-ray diffraction patterns are quite complicated. First of all, a crystal is a three-dimensional object, and X-rays can be diffracted from different planes at different angles within the crystal, as shown in Fig. 35–29. Although the analysis is complex, a great deal can be learned about any substance that can be put in crystalline form.

X-ray diffraction has also been very useful in determining the structure of biologically important molecules, such as the double helix structure of DNA, worked out by James Watson and Francis Crick in 1953. See Fig. 35–30, and for models of the double helix, Figs. 21–47a and 21–48. Around 1960, the first detailed structure of a protein molecule, myoglobin, was elucidated with the aid of X-ray diffraction. Soon the structure of an important constituent of blood, hemoglobin, was worked out, and since then the structures of a great many molecules have been determined with the help of X-rays.

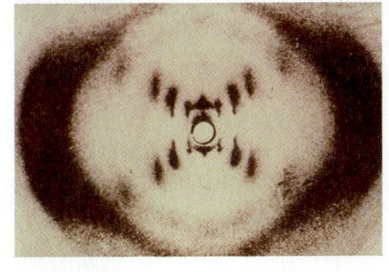

FIGURE 35–30 X-ray diffraction photo of DNA molecules taken by Rosalind Franklin in the early 1950s. The cross of spots suggested that DNA is a helix.

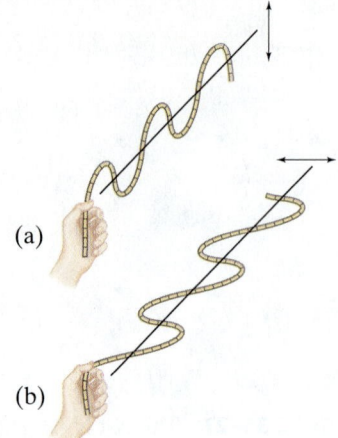

FIGURE 35–31 Transverse waves on a rope polarized (a) in a vertical plane and (b) in a horizontal plane.

FIGURE 35–32 (a) Vertically polarized wave passes through a vertical slit, but (b) a horizontally polarized wave will not.

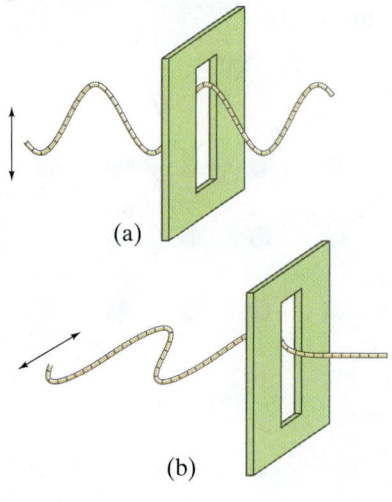

FIGURE 35–33 (below) (a) Oscillation of the electric field vectors in unpolarized light. The light is traveling into or out of the page. (b) Electric field in linear polarized light.

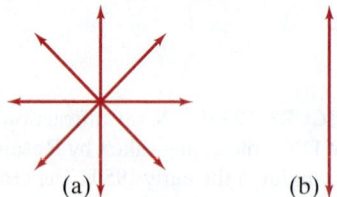

35–11 Polarization

An important and useful property of light is that it can be *polarized*. To see what this means, let us examine waves traveling on a rope. A rope can be set into oscillation in a vertical plane as in Fig. 35–31a, or in a horizontal plane as in Fig. 35–31b. In either case, the wave is said to be **linearly polarized** or **plane-polarized**—that is, the oscillations are in a plane.

If we now place an obstacle containing a vertical slit in the path of the wave, Fig. 35–32, a vertically polarized wave passes through the vertical slit, but a horizontally polarized wave will not. If a horizontal slit were used, the vertically polarized wave would be stopped. If both types of slit were used, both types of wave would be stopped by one slit or the other. Note that polarization can exist *only* for *transverse waves*, and not for longitudinal waves such as sound. The latter oscillate only along the direction of motion, and neither orientation of slit would stop them.

Light is not necessarily polarized. It can also be **unpolarized**, which means that the source has oscillations in many planes at once, as shown in Fig. 35–33. An ordinary incandescent lightbulb emits unpolarized light, as does the Sun.

Polaroids (Polarization by Absorption)

Plane-polarized light can be obtained from unpolarized light using certain crystals such as tourmaline. Or, more commonly, we use a **Polaroid sheet**. (Polaroid materials were invented in 1929 by Edwin Land.) A Polaroid sheet consists of long complex molecules arranged parallel to one another. Such a Polaroid acts like a series of parallel slits to allow one orientation of polarization to pass through nearly undiminished. This direction is called the *transmission axis* of the Polaroid. Polarization perpendicular to this direction is absorbed almost completely by the Polaroid.

Absorption by a Polaroid can be explained at the molecular level. An electric field $\vec{E}$ that oscillates parallel to the long molecules can set electrons into motion along the molecules, thus doing work on them and transferring energy. Hence, if $\vec{E}$ is parallel to the molecules, it gets absorbed. An electric field $\vec{E}$ perpendicular to the long molecules does not have this possibility of doing work and transferring its energy, and so passes through freely. When we speak of the *transmission axis* of a Polaroid, we mean the direction for which $\vec{E}$ is passed, so a Polaroid axis is *perpendicular* to the long molecules. If we want to think of there being slits between the parallel molecules in the sense of Fig. 35–32, then Fig. 35–32 would apply for the $\vec{B}$ field in the EM wave, not the $\vec{E}$ field.

If a beam of plane-polarized light strikes a Polaroid whose transmission axis is at an angle θ to the incident polarization direction, the beam will emerge plane-polarized parallel to the Polaroid transmission axis, and the amplitude of E will be reduced to $E \cos \theta$, Fig. 35–34. Thus, a Polaroid passes only that component of polarization (the electric field vector, $\vec{E}$) that is parallel to its transmission axis. Because the intensity of a light beam is proportional to the square of the amplitude (Sections 15–3 and 31–8), we see that the intensity of a plane-polarized beam transmitted by a polarizer is

$$I = I_0 \cos^2 \theta, \qquad \left[\begin{array}{c}\text{intensity of plane polarized}\\\text{wave reduced by polarizer}\end{array}\right] \quad \textbf{(35–21)}$$

where I_0 is the incoming intensity and θ is the angle between the polarizer transmission axis and the plane of polarization of the incoming wave.

FIGURE 35–34 (right) Vertical Polaroid transmits only the vertical component of a wave (electric field) incident upon it.

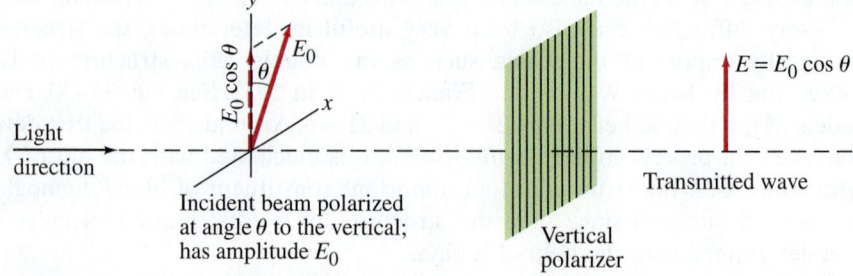

A Polaroid can be used as a **polarizer** to produce plane-polarized light from unpolarized light, since only the component of light parallel to the axis is transmitted. A Polaroid can also be used as an **analyzer** to determine (1) if light is polarized and (2) the plane of polarization. A Polaroid acting as an analyzer will pass the same amount of light independent of the orientation of its axis if the light is unpolarized; try rotating one lens of a pair of Polaroid sunglasses while looking through it at a lightbulb. If the light is polarized, however, when you rotate the Polaroid the transmitted light will be a maximum when the plane of polarization is parallel to the Polaroid's axis, and a minimum when perpendicular to it. If you do this while looking at the sky, preferably at right angles to the Sun's direction, you will see that skylight is polarized. (Direct sunlight is unpolarized, but don't look directly at the Sun, even through a polarizer, for damage to the eye may occur.) If the light transmitted by an analyzer Polaroid falls to zero at one orientation, then the light is 100% plane-polarized. If it merely reaches a minimum, the light is *partially polarized*.

Unpolarized light consists of light with random directions of polarization. Each of these polarization directions can be resolved into components along two mutually perpendicular directions. On average, an unpolarized beam can be thought of as two plane-polarized beams of equal magnitude perpendicular to one another. When unpolarized light passes through a polarizer, one component is eliminated. So the intensity of the light passing through is reduced by half since half the light is eliminated: $I = \frac{1}{2}I_0$ (Fig. 35–35).

When two Polaroids are *crossed*—that is, their polarizing axes are perpendicular to one another—unpolarized light can be entirely stopped. As shown in Fig. 35–36, unpolarized light is made plane-polarized by the first Polaroid (the polarizer).

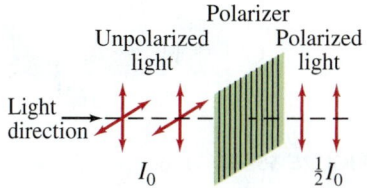

FIGURE 35–35 Unpolarized light has equal intensity vertical and horizontal components. After passing through a polarizer, one of these components is eliminated. The intensity of the light is reduced to half.

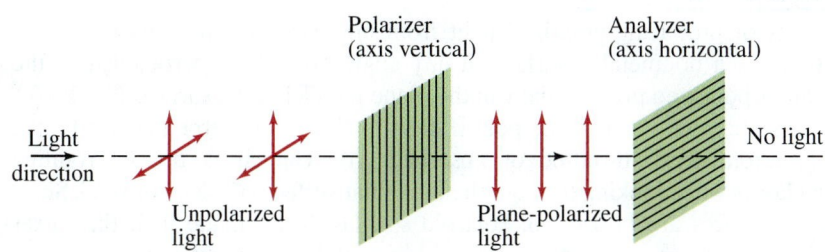

FIGURE 35–36 Crossed Polaroids completely eliminate light.

The second Polaroid, the analyzer, then eliminates this component since its transmission axis is perpendicular to the first. You can try this with Polaroid sunglasses (Fig. 35–37). Note that Polaroid sunglasses eliminate 50% of unpolarized light because of their polarizing property; they absorb even more because they are colored.

EXAMPLE 35–13 **Two Polaroids at 60°.** Unpolarized light passes through two Polaroids; the axis of one is vertical and that of the other is at 60° to the vertical. Describe the orientation and intensity of the transmitted light.

APPROACH Half of the unpolarized light is absorbed by the first Polaroid, and the remaining light emerges plane polarized. When that light passes through the second Polaroid, the intensity is further reduced according to Eq. 35–21, and the plane of polarization is then along the axis of the second Polaroid.

SOLUTION The first Polaroid eliminates half the light, so the intensity is reduced by half: $I_1 = \frac{1}{2}I_0$. The light reaching the second polarizer is vertically polarized and so is reduced in intensity (Eq. 35–21) to

$$I_2 = I_1(\cos 60°)^2 = \frac{1}{4}I_1.$$

Thus, $I_2 = \frac{1}{8}I_0$. The transmitted light has an intensity one-eighth that of the original and is plane-polarized at a 60° angle to the vertical.

FIGURE 35–37 Crossed Polaroids. When the two polarized sunglass lenses overlap, with axes perpendicular, almost no light passes through.

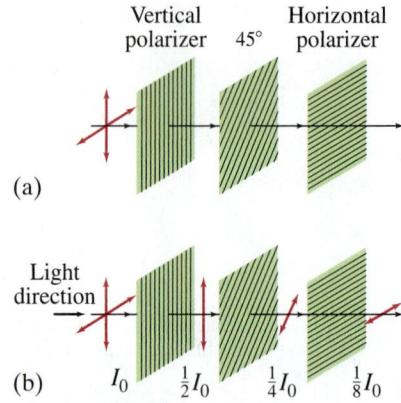

Vertical polarizer   45°   Horizontal polarizer

(a)

Light direction

(b) I_0 $\frac{1}{2}I_0$ $\frac{1}{4}I_0$ $\frac{1}{8}I_0$

FIGURE 35–38   Example 35–14.

CONCEPTUAL EXAMPLE 35–14 **Three Polaroids.** We saw in Fig. 35–36 that when unpolarized light falls on two crossed Polaroids (axes at 90°), no light passes through. What happens if a third Polaroid, with axis at 45° to each of the other two, is placed between them (Fig. 35–38a)?

RESPONSE We start just as in Example 35–13 and recall again that light emerging from each Polaroid is polarized parallel to that Polaroid's axis. Thus the angle in Eq. 35–21 is that between the transmission axes of each pair of Polaroids taken in turn. The first Polaroid changes the unpolarized light to plane-polarized and reduces the intensity from I_0 to $I_1 = \frac{1}{2}I_0$. The second polarizer further reduces the intensity by $(\cos 45°)^2$, Eq. 35–21:

$$I_2 = I_1(\cos 45°)^2 = \frac{1}{2}I_1 = \frac{1}{4}I_0.$$

The light leaving the second polarizer is plane polarized at 45° (Fig. 35–38b) relative to the third polarizer, so the third one reduces the intensity to

$$I_3 = I_2(\cos 45°)^2 = \frac{1}{2}I_2,$$

or $I_3 = \frac{1}{8}I_0$. Thus $\frac{1}{8}$ of the original intensity gets transmitted.

NOTE If we don't insert the 45° Polaroid, zero intensity results (Fig. 35–36).

EXERCISE E How much light would pass through if the 45° polarizer in Example 35–14 was placed not between the other two polarizers but (a) before the vertical (first) polarizer, or (b) after the horizontal polarizer?

Polarization by Reflection

Another means of producing polarized light from unpolarized light is by reflection. When light strikes a nonmetallic surface at any angle other than perpendicular, the reflected beam is polarized preferentially in the plane parallel to the surface, Fig. 35–39. In other words, the component with polarization in the plane perpendicular to the surface is preferentially transmitted or absorbed. You can check this by rotating Polaroid sunglasses while looking through them at a flat surface of a lake or road. Since most outdoor surfaces are horizontal, Polaroid sunglasses are made with their axes vertical to eliminate the more strongly reflected horizontal component, and thus reduce glare. People who go fishing wear Polaroids to eliminate reflected glare from the surface of a lake or stream and thus see beneath the water more clearly (Fig. 35–40).

FIGURE 35–39 Light reflected from a nonmetallic surface, such as the smooth surface of water in a lake, is partially polarized parallel to the surface.

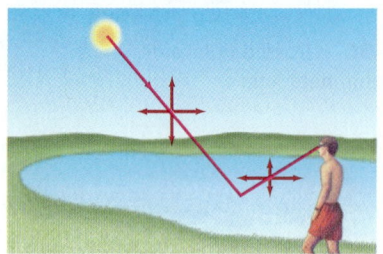

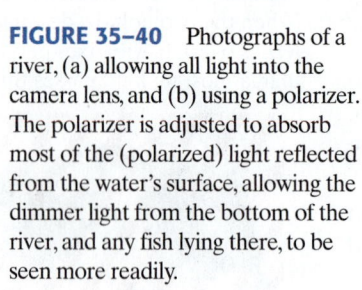

FIGURE 35–40 Photographs of a river, (a) allowing all light into the camera lens, and (b) using a polarizer. The polarizer is adjusted to absorb most of the (polarized) light reflected from the water's surface, allowing the dimmer light from the bottom of the river, and any fish lying there, to be seen more readily.

(a)

(b)

The amount of polarization in the reflected beam depends on the angle, varying from no polarization at normal incidence to 100% polarization at an angle known as the **polarizing angle** θ_p.[†] This angle is related to the index of refraction of the two materials on either side of the boundary by the equation

$$\tan \theta_p = \frac{n_2}{n_1},\qquad\qquad\text{(35–22a)}$$

where n_1 is the index of refraction of the material in which the beam is traveling, and n_2 is that of the medium beyond the reflecting boundary. If the beam is traveling in air, $n_1 = 1$, and Eq. 35–22a becomes

$$\tan \theta_p = n.\qquad\qquad\text{(35–22b)}$$

The polarizing angle θ_p is also called **Brewster's angle**, and Eqs. 35–22 *Brewster's law*, after the Scottish physicist David Brewster (1781–1868), who worked it out experimentally in 1812. Equations 35–22 can be derived from the electromagnetic wave theory of light. It is interesting that at Brewster's angle, the reflected ray and the transmitted (refracted) ray make a 90° angle to each other; that is, $\theta_p + \theta_r = 90°$, where θ_r is the refraction angle (Fig. 35–41). This can be seen by substituting Eq. 35–22a, $n_2 = n_1 \tan \theta_p = n_1 \sin \theta_p / \cos \theta_p$, into Snell's law, $n_1 \sin \theta_p = n_2 \sin \theta_r$, which gives $\cos \theta_p = \sin \theta_r$ which can only hold if $\theta_p = 90° - \theta_r$.

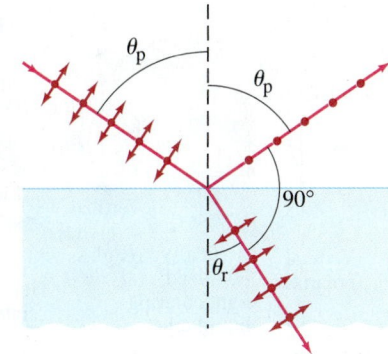

FIGURE 35–41 At θ_p the reflected light is plane-polarized parallel to the surface, and $\theta_p + \theta_r = 90°$, where θ_r is the refraction angle. (The large dots represent vibrations perpendicular to the page.)

EXAMPLE 35–15 **Polarizing angle.** (a) At what incident angle is sunlight reflected from a lake plane-polarized? (b) What is the refraction angle?

APPROACH The polarizing angle at the surface is Brewster's angle, Eq. 35–22b. We find the angle of refraction from Snell's law.

SOLUTION (a) We use Eq. 35–22b with $n = 1.33$, so $\tan \theta_p = 1.33$ giving $\theta_p = 53.1°$. (b) From Snell's law, $\sin \theta_r = \sin \theta_p / n = \sin 53.1° / 1.33 = 0.601$ giving $\theta_r = 36.9°$.

NOTE $\theta_p + \theta_r = 53.1° + 36.9° = 90.0°$, as expected.

*35–12 Liquid Crystal Displays (LCD)

A wonderful use of polarization is in a **liquid crystal display** (LCD). LCDs are used as the display in hand-held calculators, digital wrist watches, cell phones, and in beautiful color flat-panel computer and television screens.

A liquid crystal display is made up of many tiny rectangles called **pixels**, or "picture elements." The picture you see depends on which pixels are dark or light and of what color, as suggested in Fig. 35–42 for a simple black and white picture.

Liquid crystals are organic materials that at room temperature exist in a phase that is neither fully solid nor fully liquid. They are sort of gooey, and their molecules display a randomness of position characteristic of liquids, as we discussed in Section 17–1 and Fig. 17–2. They also show some of the orderliness of a solid crystal (Fig. 17–2a), but only in one dimension. The liquid crystals we find useful are made up of relatively rigid rod-like molecules that interact weakly with each other and tend to align parallel to each other, as shown in Fig. 35–43.

FIGURE 35–42 Example of an image made up of many small squares or *pixels* (picture elements). This one has rather poor resolution.

FIGURE 35–43 Liquid crystal molecules tend to align in one dimension (parallel to each other) but have random positions (left-right, up-down).

[†]Only a fraction of the incident light is reflected at the surface of a transparent medium. Although this reflected light is 100% polarized (if $\theta = \theta_p$), the remainder of the light, which is transmitted into the new medium, is only partially polarized.

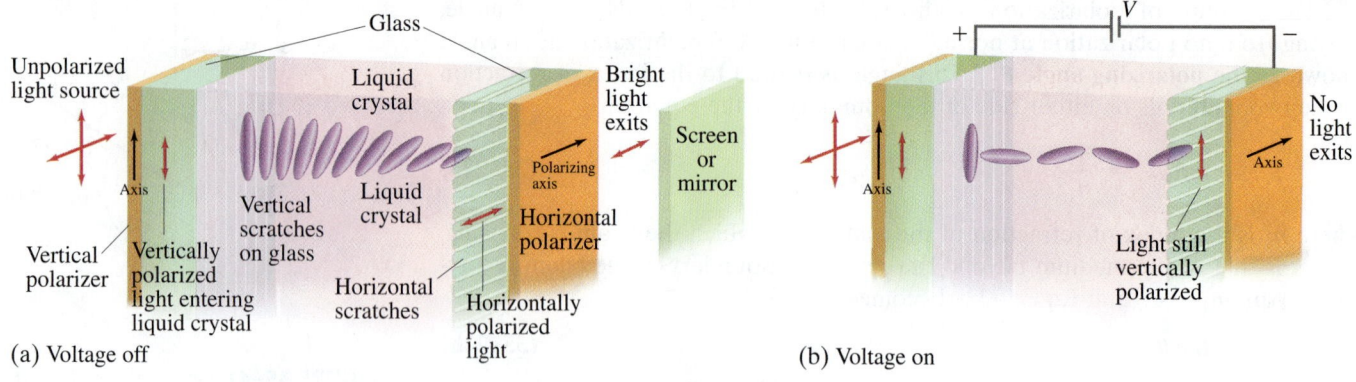

(a) Voltage off

(b) Voltage on

FIGURE 35–44 (a) "Twisted" form of liquid crystal. Light polarization plane is rotated 90°, and so is transmitted by the horizontal polarizer. Only one line of molecules is shown. (b) Molecules disoriented by electric field. Plane of polarization is not changed, so light does not pass through the horizontal polarizer. (The transparent electrodes are not shown.)

FIGURE 35–45 Calculator LCD display. The black segments or pixels have a voltage applied to them. Note that the 8 uses all seven segments (pixels), whereas other numbers use fewer.

In a simple LCD, each pixel (picture element) contains a liquid crystal sandwiched between two glass plates whose inner surfaces have been brushed to form nanometer-wide parallel scratches. The rod-like liquid crystal molecules in contact with the scratches tend to line up along the scratches. The two plates typically have their scratches at 90° to each other, and the weak forces between the rod-like molecules tend to keep them nearly aligned with their nearest neighbors, resulting in the twisted pattern shown in Fig. 35–44a.

The outer surfaces of the glass plates each have a thin film polarizer, they too oriented at 90° to each other. Unpolarized light incident from the left becomes plane-polarized and the liquid crystal molecules keep this polarization aligned with their rod-like shape. That is, the plane of polarization of the light rotates with the molecules as the light passes through the liquid crystal. The light emerges with its plane of polarization rotated by 90°, and passes through the second polarizer readily (Fig. 35–44a). A tiny LCD pixel in this situation will appear bright.

Now suppose a voltage is applied to transparent electrodes on each glass plate of the pixel. The rod-like molecules are polar (or can acquire an internal separation of charge due to the applied electric field). The applied voltage tends to align the molecules and they no longer follow the twisted pattern shown in Fig. 35–44a, with the end molecules always lying in a plane parallel to the glass plates. Instead the applied electric field tends to align the molecules flat, left to right (perpendicular to the glass plates), and they don't affect the light polarization significantly. The entering plane-polarized light no longer has its plane of polarization rotated as it passes through, and no light can exit through the second (horizontal) polarizer (Fig. 35–44b). With the voltage on, the pixel appears dark.[†]

The simple display screens of watches and calculators use ambient light as the source (you can't see the display in the dark), and a mirror behind the LCD to reflect the light back. There are only a few pixels, corresponding to the elongated segments needed to form the numbers from 0 to 9 (and letters in some displays), as seen in Fig. 35–45. Any pixels to which a voltage is applied appear dark and form part of a number. With no voltage, pixels pass light through the polarizers to the mirror and back out, which forms a bright background to the dark numbers on the display.

Color television and computer LCDs are more sophisticated. A color pixel consists of three cells, or subpixels, each covered with a red, green, or blue filter. Varying brightnesses of these three primary colors can yield almost any natural color. A good-quality screen consists of a million or more pixels. Behind this array of pixels is a light source, often thin fluorescent tubes the diameter of a straw. The light passes through the pixels, or not, depending on the voltage applied to each subpixel, as in Fig. 35–44a and b.

[†]In some displays, the polarizers are parallel to each other (the scratches remain at 90° to maintain the twist). Then voltage off results in black (no light), and voltage on results in bright light.

*35–13 Scattering of Light by the Atmosphere

Sunsets are red, the sky is blue, and skylight is polarized (at least partially). These phenomena can be explained on the basis of the *scattering* of light by the molecules of the atmosphere. In Fig. 35–46 we see unpolarized light from the Sun impinging on a molecule of the Earth's atmosphere. The electric field of the EM wave sets the electric charges within the molecule into oscillation, and the molecule absorbs some of the incident radiation. But the molecule quickly reemits this light since the charges are oscillating. As discussed in Section 31–4, oscillating electric charges produce EM waves. The intensity is strongest along the direction perpendicular to the oscillation, and drops to zero along the line of oscillation (Section 31–4). In Fig. 35–46 the motion of the charges is resolved into two components. An observer at right angles to the direction of the sunlight, as shown, will see plane-polarized light because no light is emitted along the line of the other component of the oscillation. (When viewing along the line of an oscillation, you don't see that oscillation, and hence see no waves made by it.) At other viewing angles, both components will be present; one will be stronger, however, so the light appears partially polarized. Thus, the process of scattering explains the polarization of skylight.

Scattering of light by the Earth's atmosphere depends on wavelength λ. For particles much smaller than the wavelength of light (such as molecules of air), the particles will be less of an obstruction to long wavelengths than to short ones. The scattering decreases, in fact, as $1/\lambda^4$. Blue and violet light are thus scattered much more than red and orange, which is why the sky looks blue. At sunset, the Sun's rays pass through a maximum length of atmosphere. Much of the blue has been taken out by scattering. The light that reaches the surface of the Earth, and reflects off clouds and haze, is thus lacking in blue. That is why sunsets appear reddish.

The dependence of scattering on $1/\lambda^4$ is valid only if the scattering objects are much smaller than the wavelength of the light. This is valid for oxygen and nitrogen molecules whose diameters are about 0.2 nm. Clouds, however, contain water droplets or crystals that are much larger than λ. They scatter all frequencies of light nearly uniformly. Hence clouds appear white (or gray, if shadowed).

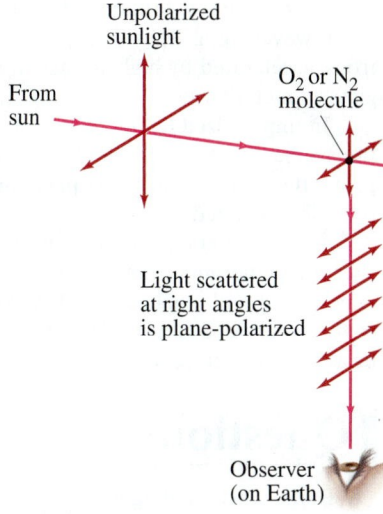

FIGURE 35–46 Unpolarized sunlight scattered by molecules of the air. An observer at right angles sees plane-polarized light, since the component of oscillation along the line of sight emits no light along that line.

 **PHYSICS APPLIED**
Why the sky is blue
Why sunsets are red

 **PHYSICS APPLIED**
Why clouds are white

Summary

Diffraction refers to the fact that light, like other waves, bends around objects it passes, and spreads out after passing through narrow slits. This bending gives rise to a **diffraction pattern** due to interference between rays of light that travel different distances.

Light passing through a very narrow slit of width D (on the order of the wavelength λ) will produce a pattern with a bright central maximum of half-width θ given by

$$\sin\theta = \frac{\lambda}{D},\qquad(35\text{–}1)$$

flanked by fainter lines to either side.

The minima in the diffraction pattern occur at

$$D\sin\theta = m\lambda\qquad(35\text{–}2)$$

where $m = 1, 2, 3, \cdots$, but not $m = 0$ (for which the pattern has its strongest maximum).

The **intensity** at any point in the single-slit diffraction pattern can be calculated using **phasor** diagrams. The same technique can be used to determine the intensity of the pattern produced by two slits.

The pattern for two-slit interference can be described as a series of maxima due to interference of light from the two slits, modified by an "envelope" due to diffraction at each slit.

The wave nature of light limits the sharpness or **resolution** of images. Because of diffraction, it is not possible to *discern details smaller than the wavelength* of the radiation being used. The useful magnification of a light microscope is limited by diffraction to about 500×.

A **diffraction grating** consists of many parallel slits or lines, each separated from its neighbors by a distance d. The peaks of constructive interference occur at angles θ given by

$$\sin\theta = \frac{m\lambda}{d},\qquad(35\text{–}13)$$

where $m = 0, 1, 2, \cdots$. The peaks of constructive interference are much brighter and sharper for a diffraction grating than for a simple two-slit apparatus. [*Peak width is inversely proportional to the total number of lines in the grating.]

[*A diffraction grating (or a prism) is used in a **spectrometer** to separate different colors or to observe **line spectra**. For a given order m, θ depends on λ. Precise determination of wavelength can be done with a spectrometer by careful measurement of θ.]

X-rays are a form of electromagnetic radiation of very short wavelength. They are produced when high-speed electrons, accelerated by high voltage in an evacuated tube, strike a glass or metal target.

In **unpolarized light**, the electric field vectors oscillate in all transverse directions. If the electric vector oscillates only in one plane, the light is said to be **plane-polarized**. Light can also be partially polarized.

When an unpolarized light beam passes through a **Polaroid** sheet, the emerging beam is plane-polarized. When a light beam is polarized and passes through a Polaroid, the intensity varies as the Polaroid is rotated. Thus a Polaroid can act as a **polarizer** or as an **analyzer**.

The intensity I of a plane-polarized light beam incident on a Polaroid is reduced by the factor

$$I = I_0 \cos^2 \theta \qquad (35\text{-}21)$$

where θ is the angle between the axis of the Polaroid and the initial plane of polarization.

Light can also be partially or fully **polarized by reflection**. If light traveling in air is reflected from a medium of index of refraction n, the reflected beam will be *completely* plane-polarized if the incident angle θ_p is given by

$$\tan \theta_p = n. \qquad (35\text{-}22b)$$

The fact that light can be polarized shows that it must be a transverse wave.

Questions

1. Radio waves and light are both electromagnetic waves. Why can a radio receive a signal behind a hill when we cannot see the transmitting antenna?

2. Hold one hand close to your eye and focus on a distant light source through a narrow slit between two fingers. (Adjust your fingers to obtain the best pattern.) Describe the pattern that you see.

3. Explain why diffraction patterns are more difficult to observe with an extended light source than for a point source. Compare also a monochromatic source to white light.

4. For diffraction by a single slit, what is the effect of increasing (a) the slit width, (b) the wavelength?

5. Describe the single-slit diffraction pattern produced when white light falls on a slit having a width of (a) 50 nm, (b) 50,000 nm.

6. What happens to the diffraction pattern of a single slit if the whole apparatus is immersed in (a) water, (b) a vacuum, instead of in air.

7. In the single-slit diffraction pattern, why does the first off-center maximum not occur at exactly $\sin \theta = \frac{3}{2} \lambda / D$?

8. Discuss the similarities, and differences, of double-slit interference and single-slit diffraction.

*9. Figure 35–10 shows a two-slit interference pattern for the case when d is larger than D. Can the reverse case occur, when d is less than D?

*10. When both diffraction and interference are taken into account in the double-slit experiment, discuss the effect of increasing (a) the wavelength, (b) the slit separation, (c) the slit width.

11. Does diffraction limit the resolution of images formed by (a) spherical mirrors, (b) plane mirrors?

12. Do diffraction effects occur for virtual as well as real images?

13. Give at least two advantages for the use of large reflecting mirrors in astronomical telescopes.

14. Atoms have diameters of about 10^{-8} cm. Can visible light be used to "see" an atom? Explain.

15. Which color of visible light would give the best resolution in a microscope? Explain.

16. Could a diffraction grating just as well be called an interference grating? Discuss.

17. Suppose light consisting of wavelengths between 400 nm and 700 nm is incident normally on a diffraction grating. For what orders (if any) would there be overlap in the observed spectrum? Does your answer depend on the slit spacing?

18. What is the difference in the interference patterns formed by two slits 10^{-4} cm apart as compared to a diffraction grating containing 10^4 lines/cm?

19. White light strikes (a) a diffraction grating and (b) a prism. A rainbow appears on a wall just below the direction of the horizontal incident beam in each case. What is the color of the top of the rainbow in each case? Explain.

20. Explain why there are tiny peaks between the main peaks produced by a diffraction grating illuminated with monochromatic light. Why are the peaks so tiny?

21. What does polarization tell us about the nature of light?

22. How can you tell if a pair of sunglasses is polarizing or not?

*23. What would be the color of the sky if the Earth had no atmosphere?

Problems

[Note: Assume light passing through slits is in phase, unless stated otherwise.]

35–1 Single-Slit Diffraction

1. (I) If 680-nm light falls on a slit 0.0365 mm wide, what is the angular width of the central diffraction peak?

2. (I) Monochromatic light falls on a slit that is 2.60×10^{-3} mm wide. If the angle between the first dark fringes on either side of the central maximum is 32.0° (dark fringe to dark fringe), what is the wavelength of the light used?

3. (II) Light of wavelength 580 nm falls on a slit that is 3.80×10^{-3} mm wide. Estimate how far the first brightest diffraction fringe is from the strong central maximum if the screen is 10.0 m away.

4. (II) Consider microwaves which are incident perpendicular to a metal plate which has a 1.6-cm slit in it. Discuss the angles at which there are diffraction minima for wavelengths of (a) 0.50 cm, (b) 1.0 cm, and (c) 3.0 cm.

5. (II) If parallel light falls on a single slit of width D at a 23.0° angle to the normal, describe the diffraction pattern.

6. (II) Monochromatic light of wavelength 633 nm falls on a slit. If the angle between the first bright fringes on either side of the central maximum is 35°, estimate the slit width.

7. (II) If a slit diffracts 580-nm light so that the diffraction maximum is 6.0 cm wide on a screen 2.20 m away, what will be the width of the diffraction maximum for light with a wavelength of 460 nm?

8. (II) (a) For a given wavelength λ, what is the minimum slit width for which there will be no diffraction minima? (b) What is the minimum slit width so that no visible light exhibits a diffraction minimum?

9. (II) When blue light of wavelength 440 nm falls on a single slit, the first dark bands on either side of center are separated by 55.0°. Determine the width of the slit.

10. (II) A single slit 1.0 mm wide is illuminated by 450-nm light. What is the width of the central maximum (in cm) in the diffraction pattern on a screen 5.0 m away?

11. (II) Coherent light from a laser diode is emitted through a rectangular area $3.0\,\mu\text{m} \times 1.5\,\mu\text{m}$ (horizontal-by-vertical). If the laser light has a wavelength of 780 nm, determine the angle between the first diffraction minima (a) above and below the central maximum, (b) to the left and right of the central maximum.

*35–2 Intensity, Single-Slit Diffraction Pattern

*12. (II) If you double the width of a single slit, the intensity of the light passing through the slit is doubled. (a) Show, however, that the intensity at the center of the screen increases by a factor of 4. (b) Explain why this does not violate conservation of energy.

*13. (II) Light of wavelength 750 nm passes through a slit 1.0 μm wide and a single-slit diffraction pattern is formed vertically on a screen 25 cm away. Determine the light intensity I 15 cm above the central maximum, expressed as a fraction of the central maximum's intensity I_0.

*14. (III) (a) Explain why the secondary maxima in the single-slit diffraction pattern do not occur precisely at $\beta/2 = (m + \frac{1}{2})\pi$ where $m = 1, 2, 3, \cdots$. (b) By differentiating Eq. 35–7 with respect to β show that the secondary maxima occur when $\beta/2$ satisfies the relation $\tan(\beta/2) = \beta/2$. (c) Carefully and precisely plot the curves $y = \beta/2$ and $y = \tan \beta/2$. From their intersections, determine the values of β for the first and second secondary maxima. What is the percent difference from $\beta/2 = (m + \frac{1}{2})\pi$?

*35–3 Intensity, Double-Slit Diffraction

*15. (II) If a double-slit pattern contains exactly nine fringes in the central diffraction peak, what can you say about the slit width and separation? Assume the first diffraction minimum occurs at an interference minimum.

*16. (II) Design a double-slit apparatus so that the central diffraction peak contains precisely seventeen fringes. Assume the first diffraction minimum occurs at (a) an interference mimimum, (b) an interference maximum.

*17. (II) 605-nm light passes through a pair of slits and creates an interference pattern on a screen 2.0 m behind the slits. The slits are separated by 0.120 mm and each slit is 0.040 mm wide. How many constructive interference fringes are formed on the screen? [Many of these fringes will be of very low intensity.]

*18. (II) In a double-slit experiment, if the central diffraction peak contains 13 interference fringes, how many fringes are contained within each secondary diffraction peak (between $m = +1$ and $+2$ in Eq. 35–2). Assume the first diffraction minimum occurs at an interference minimum.

*19. (II) Two 0.010-mm-wide slits are 0.030 mm apart (center to center). Determine (a) the spacing between interference fringes for 580 nm light on a screen 1.0 m away and (b) the distance between the two diffraction minima on either side of the central maximum of the envelope.

*20. (II) Suppose $d = D$ in a double-slit apparatus, so that the two slits merge into one slit of width $2D$. Show that Eq. 35–9 reduces to the correct equation for single-slit diffraction.

*21. (II) In a double-slit experiment, let $d = 5.00\,D = 40.0\,\lambda$. Compare (as a ratio) the intensity of the third-order interference maximum with that of the zero-order maximum.

*22. (II) How many fringes are contained in the central diffraction peak for a double-slit pattern if (a) $d = 2.00D$, (b) $d = 12.0D$, (c) $d = 4.50D$, (d) $d = 7.20D$.

*23. (III) (a) Derive an expression for the intensity in the interference pattern for three equally spaced slits. Express in terms of $\delta = 2\pi d \sin \theta / \lambda$ where d is the distance between adjacent slits and assume the slit width $D \approx \lambda$. (b) Show that there is only one secondary maximum between principal peaks.

35–4 and 35–5 Resolution Limits

24. (I) What is the angular resolution limit (degrees) set by diffraction for the 100-inch (254-cm mirror diameter) Mt. Wilson telescope ($\lambda = 560$ nm)?

25. (II) Two stars 16 light-years away are barely resolved by a 66-cm (mirror diameter) telescope. How far apart are the stars? Assume $\lambda = 550$ nm and that the resolution is limited by diffraction.

26. (II) The nearest neighboring star to the Sun is about 4 light-years away. If a planet happened to be orbiting this star at an orbital radius equal to that of the Earth–Sun distance, what minimum diameter would an Earth-based telescope's aperture have to be in order to obtain an image that resolved this star–planet system? Assume the light emitted by the star and planet has a wavelength of 550 nm.

27. (II) If you shine a flashlight beam toward the Moon, estimate the diameter of the beam when it reaches the Moon. Assume that the beam leaves the flashlight through a 5.0-cm aperture, that its white light has an average wavelength of 550 nm, and that the beam spreads due to diffraction only.

28. (II) Suppose that you wish to construct a telescope that can resolve features 7.5 km across on the moon, 384,000 km away. You have a 2.0-m-focal-length objective lens whose diameter is 11.0 cm. What focal-length eyepiece is needed if your eye can resolve objects 0.10 mm apart at a distance of 25 cm? What is the resolution limit set by the size of the objective lens (that is, by diffraction)? Use $\lambda = 560$ nm.

29. (II) The normal lens on a 35-mm camera has a focal length of 50.0 mm. Its aperture diameter varies from a maximum of 25 mm ($f/2$) to a minimum of 3.0 mm ($f/16$). Determine the resolution limit set by diffraction for ($f/2$) and ($f/16$). Specify as the number of lines per millimeter resolved on the detector or film. Take $\lambda = 560$ nm.

35–7 and 35–8 Diffraction Grating, Spectroscopy

30. (I) At what angle will 480-nm light produce a second-order maximum when falling on a grating whose slits are 1.35×10^{-3} cm apart?

31. (I) A source produces first-order lines when incident normally on a 12,000-line/cm diffraction grating at angles 28.8°, 36.7°, 38.6°, and 47.9°. What are the wavelengths?

32. (I) A 3500-line/cm grating produces a third-order fringe at a 26.0° angle. What wavelength of light is being used?

33. (I) A grating has 6800 lines/cm. How many spectral orders can be seen (400 to 700 nm) when it is illuminated by white light?

34. (II) How many lines per centimeter does a grating have if the third order occurs at a 15.0° angle for 650-nm light?

35. (II) Red laser light from a He–Ne laser ($\lambda = 632.8$ nm) is used to calibrate a diffraction grating. If this light creates a second-order fringe at 53.2° after passing through the grating, and light of an unknown wavelength λ creates a first-order fringe at 20.6°, find λ.

36. (II) White light containing wavelengths from 410 nm to 750 nm falls on a grating with 7800 lines/cm. How wide is the first-order spectrum on a screen 2.80 m away?

37. (II) A diffraction grating has 6.0×10^5 lines/m. Find the angular spread in the second-order spectrum between red light of wavelength 7.0×10^{-7} m and blue light of wavelength 4.5×10^{-7} m.

38. (II) A tungsten–halogen bulb emits a continuous spectrum of ultraviolet, visible, and infrared light in the wavelength range 360 nm to 2000 nm. Assume that the light from a tungsten–halogen bulb is incident on a diffraction grating with slit spacing d and that the first-order brightness maximum for the wavelength of 1200 nm occurs at angle θ. What other wavelengths within the spectrum of incident light will produce a brightness maximum at this same angle θ? [Optical filters are used to deal with this bothersome effect when a continuous spectrum of light is measured by a spectrometer.]

39. (II) Show that the second- and third-order spectra of white light produced by a diffraction grating always overlap. What wavelengths overlap?

40. (II) Two first-order spectrum lines are measured by a 9650 line/cm spectroscope at angles, on each side of center, of $+26°38'$, $+41°02'$ and $-26°18'$, $-40°27'$. Calculate the wavelengths based on these data.

41. (II) Suppose the angles measured in Problem 40 were produced when the spectrometer (but not the source) was submerged in water. What then would be the wavelengths (in air)?

42. (II) The first-order line of 589-nm light falling on a diffraction grating is observed at a 16.5° angle. How far apart are the slits? At what angle will the third order be observed?

43. (II) White light passes through a 610-line/mm diffraction grating. First-order and second-order visible spectra ("rainbows") appear on the wall 32 cm away as shown in Fig. 35–47. Determine the widths ℓ_1 and ℓ_2 of the two "rainbows" (400 nm to 700 nm). In which order is the "rainbow" dispersed over a larger distance?

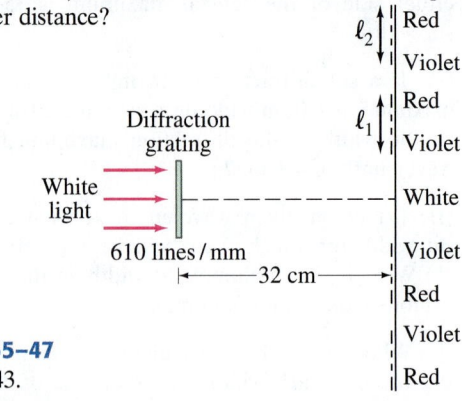

FIGURE 35–47
Problem 43.

44. (II) **Missing orders** occur for a diffraction grating when a diffraction minimum coincides with an interference maximum. Let D be the width of each slit and d the separation of slits. (a) Show that if $d = 2D$, all even orders ($m = 2, 4, 6, \cdots$) are missing. (b) Show there will be missing orders whenever

$$\frac{d}{D} = \frac{m_1}{m_2}$$

where m_1 and m_2 are integers. (c) Discusss the case $d = D$, the limit in which the space between slits becomes negligible.

45. (II) Monochromatic light falls on a transmission diffraction grating at an angle ϕ to the normal. (a) Show that Eq. 35–13 for diffraction maxima must be replaced by

$$d(\sin \phi + \sin \theta) = \pm m\lambda. \qquad m = 0, 1, 2, \cdots.$$

(b) Explain the $\pm$ sign. (c) Green light with a wavelength of 550 nm is incident at an angle of 15° to the normal on a diffraction grating with 5000 lines/cm. Find the angles at which the first-order maxima occur.

*35–9 Grating, Peak Widths, Resolving Power

***46.** (II) A 6500-line/cm diffraction grating is 3.18 cm wide. If light with wavelengths near 624 nm falls on the grating, how close can two wavelengths be if they are to be resolved in any order? What order gives the best resolution?

***47.** (II) A diffraction grating has 16,000 rulings in its 1.9 cm width. Determine (a) its resolving power in first and second orders, and (b) the minimum wavelength resolution ($\Delta\lambda$) it can yield for $\lambda = 410$ nm.

***48.** (II) Let 580-nm light be incident normally on a diffraction grating for which $d = 3.00D = 1050$ nm. (a) How many orders (principal maxima) are present? (b) If the grating is 1.80 cm wide, what is the full angular width of each principal maximum?

35–10 X-Ray Diffraction

49. (II) X-rays of wavelength 0.138 nm fall on a crystal whose atoms, lying in planes, are spaced 0.285 nm apart. At what angle ϕ (relative to the surface, Fig. 35–28) must the X-rays be directed if the first diffraction maximum is to be observed?

50. (II) First-order Bragg diffraction is observed at 26.8° relative to the crystal surface, with spacing between atoms of 0.24 nm. (a) At what angle will second order be observed? (b) What is the wavelength of the X-rays?

51. (II) If X-ray diffraction peaks corresponding to the first three orders ($m = 1, 2,$ and 3) are measured, can both the X-ray wavelength λ and lattice spacing d be determined? Prove your answer.

35–11 Polarization

52. (I) Two polarizers are oriented at 65° to one another. Unpolarized light falls on them. What fraction of the light intensity is transmitted?

53. (I) Two Polaroids are aligned so that the light passing through them is a maximum. At what angle should one of them be placed so the intensity is subsequently reduced by half?

54. (I) What is Brewster's angle for an air–glass $(n = 1.58)$ surface?

55. (I) What is Brewster's angle for a diamond submerged in water if the light is hitting the diamond $(n = 2.42)$ while traveling in the water?

56. (II) The critical angle for total internal reflection at a boundary between two materials is 55°. What is Brewster's angle at this boundary? Give two answers, one for each material.

57. (II) At what angle should the axes of two Polaroids be placed so as to reduce the intensity of the incident unpolarized light to (a) $\frac{1}{3}$, (b) $\frac{1}{10}$?

58. (II) Two polarizers are oriented at 36.0° to one another. Light polarized at an 18.0° angle to each polarizer passes through both. What is the transmitted intensity (%)?

59. (II) What would Brewster's angle be for reflections off the surface of water for light coming from beneath the surface? Compare to the angle for total internal reflection, and to Brewster's angle from above the surface.

60. (II) Unpolarized light passes through six successive Polaroid sheets each of whose axis makes a 45° angle with the previous one. What is the intensity of the transmitted beam?

61. (II) Two polarizers A and B are aligned so that their transmission axes are vertical and horizontal, respectively. A third polarizer is placed between these two with its axis aligned at angle θ with respect to the vertical. Assuming vertically polarized light of intensity I_0 is incident upon polarizer A, find an expression for the light intensity I transmitted through this three-polarizer sequence. Calculate the derivative $dI/d\theta$; then use it to find the angle θ that maximizes I.

62. (III) The percent polarization P of a partially polarized beam of light is defined as

$$ P = \frac{I_{max} - I_{min}}{I_{max} + I_{min}} \times 100 $$

where I_{max} and I_{min} are the maximum and minimum intensities that are obtained when the light passes through a polarizer that is slowly rotated. Such light can be considered as the sum of two unequal plane-polarized beams of intensities I_{max} and I_{min} perpendicular to each other. Show that the light transmitted by a polarizer, whose axis makes an angle ϕ to the direction in which I_{max} is obtained, has intensity

$$ \frac{1 + p \cos 2\phi}{1 + p} I_{max} $$

where $p = P/100$ is the "fractional polarization."

General Problems

63. When violet light of wavelength 415 nm falls on a single slit, it creates a central diffraction peak that is 8.20 cm wide on a screen that is 2.85 m away. How wide is the slit?

64. A series of polarizers are each placed at a 10° interval from the previous polarizer. Unpolarized light is incident on this series of polarizers. How many polarizers does the light have to go through before it is $\frac{1}{4}$ of its original intensity?

65. The wings of a certain beetle have a series of parallel lines across them. When normally incident 480-nm light is reflected from the wing, the wing appears bright when viewed at an angle of 56°. How far apart are the lines?

66. A teacher stands well back from an outside doorway 0.88 m wide, and blows a whistle of frequency 850 Hz. Ignoring reflections, estimate at what angle(s) it is *not* possible to hear the whistle clearly on the playground outside the doorway. Assume 340 m/s for the speed of sound.

67. Light is incident on a diffraction grating with 7600 lines/cm and the pattern is viewed on a screen located 2.5 m from the grating. The incident light beam consists of two wavelengths, $\lambda_1 = 4.4 \times 10^{-7}\,\text{m}$ and $\lambda_2 = 6.8 \times 10^{-7}\,\text{m}$. Calculate the linear distance between the first-order bright fringes of these two wavelengths on the screen.

68. How many lines per centimeter must a grating have if there is to be no second-order spectrum for any visible wavelength?

69. When yellow sodium light, $\lambda = 589\,\text{nm}$, falls on a diffraction grating, its first-order peak on a screen 66.0 cm away falls 3.32 cm from the central peak. Another source produces a line 3.71 cm from the central peak. What is its wavelength? How many lines/cm are on the grating?

70. Two of the lines of the atomic hydrogen spectrum have wavelengths of 656 nm and 410 nm. If these fall at normal incidence on a grating with 8100 lines/cm, what will be the angular separation of the two wavelengths in the first-order spectrum?

71. (a) How far away can a human eye distinguish two car headlights 2.0 m apart? Consider only diffraction effects and assume an eye diameter of 6.0 mm and a wavelength of 560 nm. (b) What is the minimum angular separation an eye could resolve when viewing two stars, considering only diffraction effects? In reality, it is about 1′ of arc. Why is it not equal to your answer in (b)?

72. A laser beam passes through a slit of width 1.0 cm and is pointed at the Moon, which is approximately 380,000 km from the Earth. Assume the laser emits waves of wavelength 633 nm (the red light of a He–Ne laser). Estimate the width of the beam when it reaches the Moon.

73. A He–Ne gas laser which produces monochromatic light of wavelength $\lambda = 6.328 \times 10^{-7}\,\text{m}$ is used to calibrate a reflection grating in a spectroscope. The first-order diffraction line is found at an angle of 21.5° to the incident beam. How many lines per meter are there on the grating?

74. The entrance to a boy's bedroom consists of two doorways, each 1.0 m wide, which are separated by a distance of 3.0 m. The boy's mother yells at him through the two doors as shown in Fig. 35–48, telling him to clean up his room. Her voice has a frequency of 400 Hz. Later, when the mother discovers the room is still a mess, the boy says he never heard her telling him to clean his room. The velocity of sound is 340 m/s. (a) Find all of the angles θ (Fig. 35–48) at which no sound will be heard within the bedroom when the mother yells. Assume sound is completely absorbed when it strikes a bedroom wall. (b) If the boy was at the position shown when his mother yelled, does he have a good explanation for not having heard her? Explain.

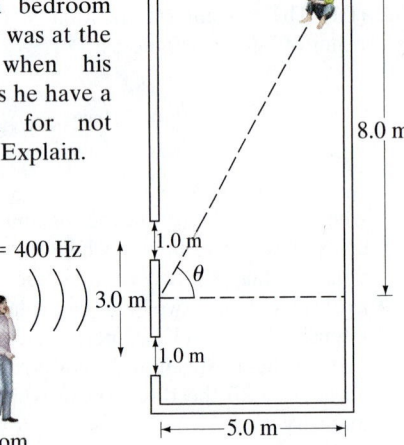

FIGURE 35–48
Problem 74.

75. At what angle above the horizon is the Sun when light reflecting off a smooth lake is polarized most strongly?

76. Unpolarized light falls on two polarizer sheets whose axes are at right angles. (a) What fraction of the incident light intensity is transmitted? (b) What fraction is transmitted if a third polarizer is placed between the first two so that its axis makes a 66° angle with the axis of the first polarizer? (c) What if the third polarizer is in front of the other two?

77. At what angle should the axes of two Polaroids be placed so as to reduce the intensity of the incident unpolarized light by an additional factor (after the first Polaroid cuts it in half) of (a) 4, (b) 10, (c) 100?

78. Four polarizers are placed in succession with their axes vertical, at 30.0° to the vertical, at 60.0° to the vertical, and at 90.0° to the vertical. (a) Calculate what fraction of the incident unpolarized light is transmitted by the four polarizers. (b) Can the transmitted light be *decreased* by removing one of the polarizers? If so, which one? (c) Can the transmitted light intensity be extinguished by removing polarizers? If so, which one(s)?

79. Spy planes fly at extremely high altitudes (25 km) to avoid interception. Their cameras are reportedly able to discern features as small as 5 cm. What must be the minimum aperture of the camera lens to afford this resolution? (Use λ = 580 nm.)

80. Two polarizers are oriented at 48° to each other and plane-polarized light is incident on them. If only 25% of the light gets through both of them, what was the initial polarization direction of the incident light?

81. X-rays of wavelength 0.0973 nm are directed at an unknown crystal. The second diffraction maximum is recorded when the X-rays are directed at an angle of 23.4° relative to the crystal surface. What is the spacing between crystal planes?

82. X-rays of wavelength 0.10 nm fall on a microcrystalline powder sample. The sample is located 12 cm from the photographic film. The crystal structure of the sample has an atomic spacing of 0.22 nm. Calculate the radii of the diffraction rings corresponding to first- and second-order scattering. Note in Fig. 35–28 that the X-ray beam is deflected through an angle 2φ.

83. The Hubble Space Telescope with an objective diameter of 2.4 m, is viewing the Moon. Estimate the minimum distance between two objects on the Moon that the Hubble can distinguish. Consider diffraction of light with wavelength 550 nm. Assume the Hubble is near the Earth.

84. The Earth and Moon are separated by about 400×10^6 m. When Mars is 8×10^{10} m from Earth, could a person standing on Mars resolve the Earth and its Moon as two separate objects without a telescope? Assume a pupil diameter of 5 mm and λ = 550 nm.

85. A slit of width $D = 22\ \mu m$ is cut through a thin aluminum plate. Light with wavelength λ = 650 nm passes through this slit and forms a single-slit diffraction pattern on a screen a distance ℓ = 2.0 m away. Defining x to be the distance between the first minima ($m = +1$ and $m = -1$) in this diffraction pattern, find the change Δx in this distance when the temperature T of the metal plate is changed by amount $\Delta T = 55$ C°. [*Hint*: Since λ ≪ D, the first minima occur at a small angle.]

*Numerical/Computer

*86. (II) A student shined a laser light onto a single slit of width 0.04000 mm. He placed a screen at a distance of 1.490 m from the slit to observe the diffraction pattern of the laser light. The accompanying Table shows the distances of the dark fringes from the center of the central bright fringe for different orders.

Order number, m:	1	2	3	4	5	6	7	8
Distance (m)	0.0225	0.0445	0.0655	0.0870	0.1105	0.1320	0.1540	0.1775

Determine the angle of diffraction, θ, and sin θ for each order. Make a graph of sin θ vs. order number, m, and find the wavelength, λ, of the laser from the best-fit straight line.

*87. (III) Describe how to rotate the plane of polarization of a plane-polarized beam of light by 90° and produce only a 10% loss in intensity, using polarizers. Let N be the number of polarizers and θ be the (same) angle between successive polarizers.

88. (III) The "full-width at half-maximum" (FWHM) of the central peak for single-slit diffraction is defined as the angle $\Delta\theta$ between the two points on either side of center where the intensity is $\frac{1}{2}I_0$. (a) Determine $\Delta\theta$ in terms of (λ/D). Use graphs or a spreadsheet to solve $\sin \alpha = \alpha/\sqrt{2}$. (b) Determine $\Delta\theta$ (in degrees) for $D = \lambda$ and for $D = 100\lambda$.

Answers to Exercises

A: Narrower.

B: (b).

C: (c).

D: 0.28 nm.

E: Zero for both (a) and (b), because the two successive polarizers at 90° cancel all light. The 45° Polaroid must be inserted *between* the other two if any transmission is to occur.

An early science fantasy book (1940), called *Mr Tompkins in Wonderland* by physicist George Gamow, imagined a world in which the speed of light was only 10 m/s (20 mi/h). Mr Tompkins had studied relativity and when he began "speeding" on a bicycle, he "expected that he would be immediately shortened, and was very happy about it as his increasing figure had lately caused him some anxiety. To his great surprise, however, nothing happened to him or to his cycle. On the other hand, the picture around him completely changed. The streets grew shorter, the windows of the shops began to look like narrow slits, and the policeman on the corner became the thinnest man he had ever seen. 'By Jove!' exclaimed Mr Tompkins excitedly, 'I see the trick now. This is where the word *relativity* comes in.'"

Relativity does indeed predict that objects moving relative to us at high speed, close to the speed of light c, are shortened in length. We don't notice it as Mr Tompkins did, because $c = 3 \times 10^8$ m/s is incredibly fast. We will study length contraction, time dilation, simultaneity non-agreement, and how energy and mass are equivalent ($E = mc^2$).

The Special Theory of Relativity

36

CHAPTER-OPENING QUESTION—Guess now!

A rocket is headed away from Earth at a speed of 0.80c. The rocket fires a missile at a speed of 0.70c (the missile is aimed away from Earth and leaves the rocket at 0.70c relative to the rocket). How fast is the missile moving relative to Earth?

(a) 1.50c;
(b) a little less than 1.50c;
(c) a little over c;
(d) a little under c;
(e) 0.75c.

Physics at the end of the nineteenth century looked back on a period of great progress. The theories developed over the preceding three centuries had been very successful in explaining a wide range of natural phenomena. Newtonian mechanics beautifully explained the motion of objects on Earth and in the heavens. Furthermore, it formed the basis for successful treatments of fluids, wave motion, and sound. Kinetic theory explained the behavior of gases and other materials. Maxwell's theory of electromagnetism not only brought together and explained electric and magnetic phenomena, but it predicted the existence of electromagnetic waves that would behave in every way just like light—so light came to be thought of as an electromagnetic wave. Indeed, it seemed that the natural world, as seen through the eyes of physicists, was very well explained. A few puzzles remained, but it was felt that these would soon be explained using already known principles.

CONTENTS

36–1 Galilean–Newtonian Relativity

*36–2 The Michelson–Morley Experiment

36–3 Postulates of the Special Theory of Relativity

36–4 Simultaneity

36–5 Time Dilation and the Twin Paradox

36–6 Length Contraction

36–7 Four-Dimensional Space–Time

36–8 Galilean and Lorentz Transformations

36–9 Relativistic Momentum

36–10 The Ultimate Speed

36–11 $E = mc^2$; Mass and Energy

36–12 Doppler Shift for Light

36–13 The Impact of Special Relativity

951

FIGURE 36–1 Albert Einstein (1879–1955), one of the great minds of the twentieth century, was the creator of the special and general theories of relativity.

It did not turn out so simply. Instead, these puzzles were to be solved only by the introduction, in the early part of the twentieth century, of two revolutionary new theories that changed our whole conception of nature: the *theory of relativity* and *quantum theory*.

Physics as it was known at the end of the nineteenth century (what we've covered up to now in this book) is referred to as **classical physics**. The new physics that grew out of the great revolution at the turn of the twentieth century is now called **modern physics**. In this Chapter, we present the special theory of relativity, which was first proposed by Albert Einstein (1879–1955; Fig. 36–1) in 1905. In Chapter 37, we introduce the equally momentous quantum theory.

36–1 Galilean–Newtonian Relativity

Einstein's special theory of relativity deals with how we observe events, particularly how objects and events are observed from different frames of reference. This subject had, of course, already been explored by Galileo and Newton.

The special theory of relativity deals with events that are observed and measured from so-called **inertial reference frames** (Sections 4–2 and 11–8), which are reference frames in which Newton's first law is valid: if an object experiences no net force, the object either remains at rest or continues in motion with constant speed in a straight line. It is usually easiest to analyze events when they are observed and measured by observers at rest in an inertial frame. The Earth, though not quite an inertial frame (it rotates), is close enough that for most purposes we can consider it an inertial frame. Rotating or otherwise accelerating frames of reference are noninertial frames,[†] and won't concern us in this Chapter (they are dealt with in Einstein's general theory of relativity).

A reference frame that moves with constant velocity with respect to an inertial frame is itself also an inertial frame, since Newton's laws hold in it as well. When we say that we observe or make measurements from a certain reference frame, it means that we are at rest in that reference frame.

Both Galileo and Newton were aware of what we now call the **relativity principle** applied to mechanics: that *the basic laws of physics are the same in all inertial reference frames*. You may have recognized its validity in everyday life. For example, objects move in the same way in a smoothly moving (constant-velocity) train or airplane as they do on Earth. (This assumes no vibrations or rocking which would make the reference frame noninertial.) When you walk, drink a cup of soup, play pool, or drop a pencil on the floor while traveling in a train, airplane, or ship moving at constant velocity, the objects move just as they do when you are at rest on Earth. Suppose you are in a car traveling rapidly at constant velocity. If you drop a coin from above your head inside the car, how will it fall? It falls straight downward with respect to the car, and hits the floor directly below the point of release, Fig. 36–2a.

FIGURE 36–2 A coin is dropped by a person in a moving car. The upper views show the moment of the coin's release, the lower views are a short time later. (a) In the reference frame of the car, the coin falls straight down (and the tree moves to the left). (b) In a reference frame fixed on the Earth, the coin has an initial velocity (= to car's) and follows a curved (parabolic) path.

[†]On a rotating platform (say a merry-go-round), for example, an object at rest starts moving outward even though no object exerts a force on it. This is therefore not an inertial frame. See Section 11–8.

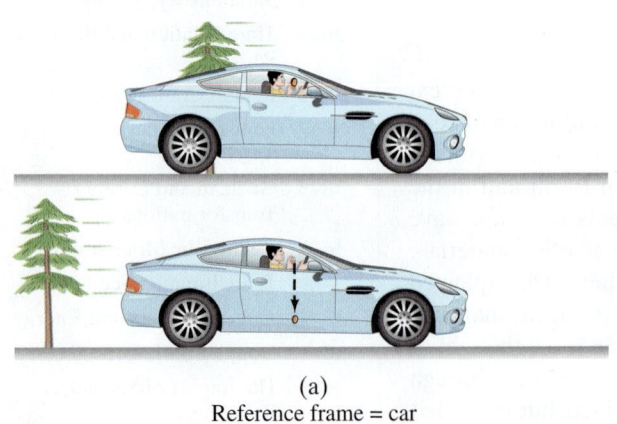

(a)
Reference frame = car

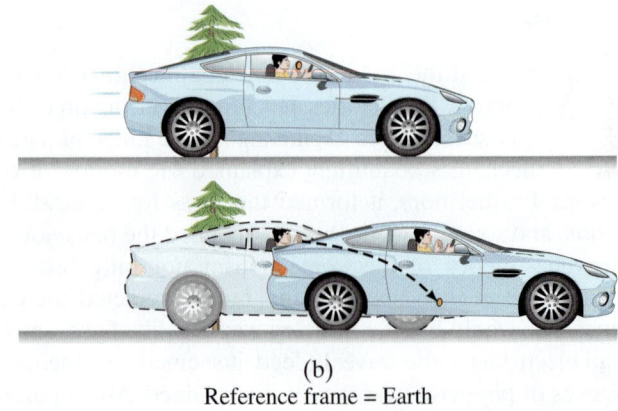

(b)
Reference frame = Earth

This is just how objects fall on the Earth—straight down—and thus our experiment in the moving car is in accord with the relativity principle. (If you drop the coin out the car's window, this won't happen because the moving air drags the coin backward relative to the car.)

Note in this example, however, that to an observer on the Earth, the coin follows a curved path, Fig. 36–2b. The actual path followed by the coin is different as viewed from different frames of reference. This does not violate the relativity principle because this principle states that the *laws* of physics are the same in all inertial frames. The same law of gravity, and the same laws of motion, apply in both reference frames. The acceleration of the coin is the same in both reference frames. The difference in Figs. 36–2a and b is that in the Earth's frame of reference, the coin has an initial velocity (equal to that of the car). The laws of physics therefore predict it will follow a parabolic path like any projectile (Chapter 3). In the car's reference frame, there is no initial velocity, and the laws of physics predict that the coin will fall straight down. The laws are the same in both reference frames, although the specific paths are different.

Galilean–Newtonian relativity involves certain unprovable assumptions that make sense from everyday experience. It is assumed that the lengths of objects are the same in one reference frame as in another, and that time passes at the same rate in different reference frames. In classical mechanics, then, space and time intervals are considered to be **absolute**: their measurement does not change from one reference frame to another. The mass of an object, as well as all forces, are assumed to be unchanged by a change in inertial reference frame.

The position of an object, however, is different when specified in different reference frames, and so is velocity. For example, a person may walk inside a bus toward the front with a speed of 2 m/s. But if the bus moves 10 m/s with respect to the Earth, the person is then moving with a speed of 12 m/s with respect to the Earth. The acceleration of an object, however, is the same in any inertial reference frame according to classical mechanics. This is because the change in velocity, and the time interval, will be the same. For example, the person in the bus may accelerate from 0 to 2 m/s in 1.0 seconds, so $a = 2 \text{ m/s}^2$ in the reference frame of the bus. With respect to the Earth, the acceleration is $(12 \text{ m/s} - 10 \text{ m/s})/(1.0 \text{ s}) = 2 \text{ m/s}^2$, which is the same.

Since neither F, m, nor a changes from one inertial frame to another, then Newton's second law, $F = ma$, does not change. Thus Newton's second law satisfies the relativity principle. It is easily shown that the other laws of mechanics also satisfy the relativity principle.

That the laws of mechanics are the same in all inertial reference frames implies that no one inertial frame is special in any sense. We express this important conclusion by saying that **all inertial reference frames are equivalent** for the description of mechanical phenomena. No one inertial reference frame is any better than another. A reference frame fixed to a car or an aircraft traveling at constant velocity is as good as one fixed on the Earth. When you travel smoothly at constant velocity in a car or airplane, it is just as valid to say you are at rest and the Earth is moving as it is to say the reverse.[†] There is no experiment you can do to tell which frame is "really" at rest and which is moving. Thus, there is no way to single out one particular reference frame as being at absolute rest.

A complication arose, however, in the last half of the nineteenth century. Maxwell's comprehensive and successful theory of electromagnetism (Chapter 31) predicted that light is an electromagnetic wave. Maxwell's equations gave the velocity of light c as 3.00×10^8 m/s; and this is just what is measured. The question then arose: in what reference frame does light have precisely the value predicted by Maxwell's theory? It was assumed that light would have a different speed in different frames of reference. For example, if observers were traveling on a rocket ship at a speed of 1.0×10^8 m/s away from a source of light, we might expect them to measure the speed of the light reaching them to be $(3.0 \times 10^8 \text{ m/s}) - (1.0 \times 10^8 \text{ m/s}) = 2.0 \times 10^8 \text{ m/s}$. But Maxwell's equations have no provision for relative velocity. They predicted the speed of light to be $c = 3.0 \times 10^8$ m/s, which seemed to imply that there must be some preferred reference frame where c would have this value.

⚠ CAUTION

Laws are the same, but paths may be different in different reference frames

⚠ CAUTION

Position and velocity are different in different reference frames, but length is the same (classical)

[†]We are ignoring the rotation and curvature of the Earth.

We discussed in Chapters 15 and 16 that waves can travel on water and along ropes or strings, and sound waves travel in air and other materials. Nineteenth-century physicists viewed the material world in terms of the laws of mechanics, so it was natural for them to assume that light too must travel in some *medium*. They called this transparent medium the **ether** and assumed it permeated all space.[†] It was therefore assumed that the velocity of light given by Maxwell's equations must be with respect to the ether.

At first it appeared that Maxwell's equations did *not* satisfy the relativity principle. They were simplest in the frame where $c = 3.00 \times 10^8$ m/s; that is, in a reference frame at rest in the ether. In any other reference frame, extra terms would have to be added to take into account the relative velocity. Thus, although most of the laws of physics obeyed the relativity principle, the laws of electricity and magnetism apparently did not. Einstein's second postulate (Section 36–3) resolved this problem: Maxwell's equations do satisfy relativity.

Scientists soon set out to determine the speed of the Earth relative to this absolute frame, whatever it might be. A number of clever experiments were designed. The most direct were performed by A. A. Michelson and E. W. Morley in the 1880s. They measured the difference in the speed of light in different directions using Michelson's interferometer (Section 34–6). They expected to find a difference depending on the orientation of their apparatus with respect to the ether. For just as a boat has different speeds relative to the land when it moves upstream, downstream, or across the stream, so too light would be expected to have different speeds depending on the velocity of the ether past the Earth.

Strange as it may seem, they detected no difference at all. This was a great puzzle. A number of explanations were put forth over a period of years, but they led to contradictions or were otherwise not generally accepted. This **null result** was one of the great puzzles at the end of the nineteenth century.

Then in 1905, Albert Einstein proposed a radical new theory that reconciled these many problems in a simple way. But at the same time, as we shall see, it completely changed our ideas of space and time.

*36–2 The Michelson–Morley Experiment

The Michelson–Morley experiment was designed to measure the speed of the *ether*—the medium in which light was assumed to travel—with respect to the Earth. The experimenters thus hoped to find an absolute reference frame, one that could be considered to be at rest.

One of the possibilities nineteenth-century scientists considered was that the ether is fixed relative to the Sun, for even Newton had taken the Sun as the center of the universe. If this were the case (there was no guarantee, of course), the Earth's speed of about 3×10^4 m/s in its orbit around the Sun could produce a change of 1 part in 10^4 in the speed of light $(3.0 \times 10^8$ m/s$)$. Direct measurement of the speed of light to this precision was not possible. But A. A. Michelson, later with the help of E. W. Morley, was able to use his interferometer (Section 34–6) to measure the difference in the speed of light in different directions to this precision.

This famous experiment is based on the principle shown in Fig. 36–3. Part (a) is a diagram of the Michelson interferometer, and it is assumed that the "ether wind" is moving with speed v to the right. (Alternatively, the Earth is assumed to move to the left with respect to the ether at speed v.) The light from a source is split into two beams by a half-silvered mirror M_S. One beam travels to mirror M_1 and the other to mirror M_2. The beams are reflected by M_1 and M_2 and are joined again after passing through M_S. The now superposed beams interfere with each other and the resultant is viewed by the observer's eye as an interference pattern (discussed in Section 34–6).

Whether constructive or destructive interference occurs at the center of the interference pattern depends on the relative phases of the two beams after they have traveled their separate paths. Let us consider an analogy of a boat traveling up and

[†]The medium for light waves could not be air, since light travels from the Sun to Earth through nearly empty space. Therefore, another medium was postulated, the ether. The ether was not only transparent but, because of difficulty in detecting it, was assumed to have zero density.

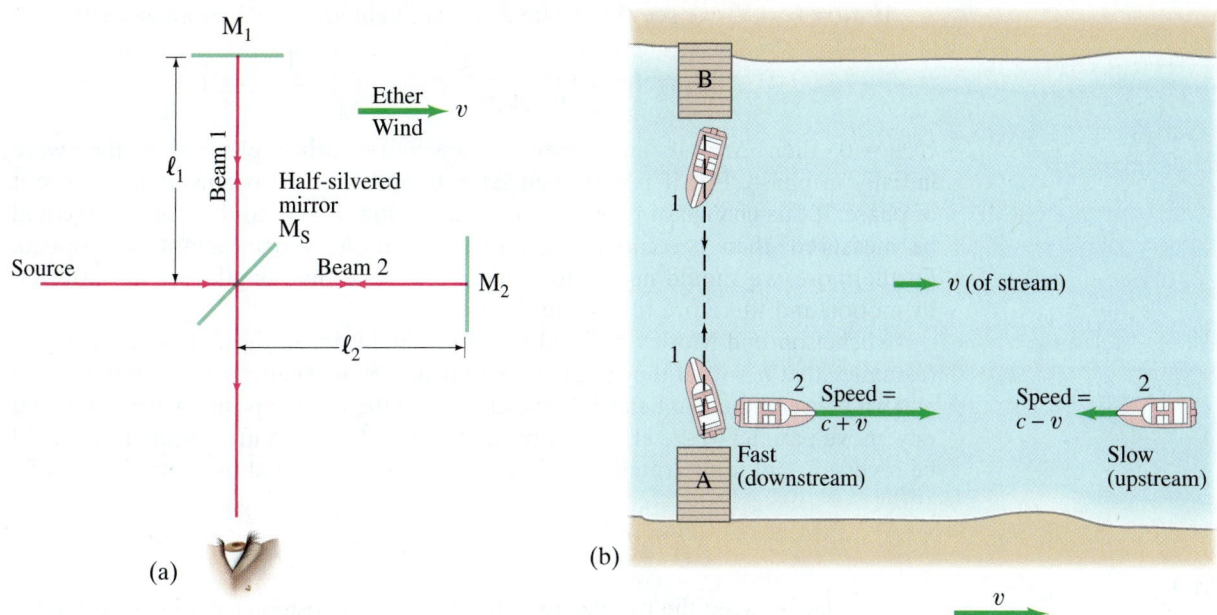

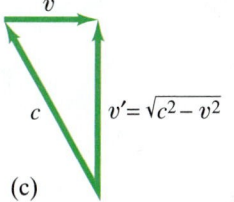

FIGURE 36–3 The Michelson–Morley experiment. (a) Michelson interferometer. (b) Boat analogy: boat 1 goes across the stream and back; boat 2 goes downstream and back upstream (boat has speed c relative to the water). (c) Calculation of the velocity of boat (or light beam) traveling perpendicular to the current (or ether wind).

down, and across, a river whose current moves with speed v, as shown in Fig. 36–3b. In still water, the boat can travel with speed c (not the speed of light in this case).

First we consider beam 2 in Fig. 36–3a, which travels parallel to the "ether wind." In its journey from M_S to M_2, the light would travel with speed $c + v$, according to classical physics, just as for a boat traveling downstream (see Fig. 36–3b) we add the speed of the river water to the boat's own speed (relative to the water) to get the boat's speed relative to the shore. Since the beam travels a distance ℓ_2, the time it takes to go from M_S to M_2 would be $t = \ell_2/(c + v)$. To make the return trip from M_2 to M_S, the light moves against the ether wind (like the boat going upstream), so its relative speed is expected to be $c - v$. The time for the return trip would be $\ell_2/(c - v)$. The total time for beam 2 to travel from M_S to M_2 and back to M_S is

$$t_2 = \frac{\ell_2}{c + v} + \frac{\ell_2}{c - v} = \frac{2\ell_2}{c\left(1 - v^2/c^2\right)}.$$

Now let us consider beam 1, which travels crosswise to the ether wind. Here the boat analogy (Fig. 36–3b) is especially helpful. The boat is to go from wharf A to wharf B directly across the stream. If it heads directly across, the stream's current will drag it downstream. To reach wharf B, the boat must head at an angle upstream. The precise angle depends on the magnitudes of c and v, but is of no interest to us in itself. Part (c) of Fig. 36–3 shows how to calculate the velocity v' of the boat relative to Earth as it crosses the stream. Since c, v, and v' form a right triangle, we have that $v' = \sqrt{c^2 - v^2}$. The boat has the same speed when it returns. If we now apply these principles to light beam 1 in Fig. 36–3a, we expect the beam to travel with speed $\sqrt{c^2 - v^2}$ in going from M_S to M_1 and back again. The total distance traveled is $2\ell_1$, so the time required for beam 1 to make the round trip would be $2\ell_1/\sqrt{c^2 - v^2}$, or

$$t_1 = \frac{2\ell_1}{c\sqrt{1 - v^2/c^2}}.$$

Notice that the denominator in this equation for t_1 involves a square root, whereas that for t_2 does not.

If $\ell_1 = \ell_2 = \ell$, we see that beam 2 will lag behind beam 1 by an amount

$$\Delta t = t_2 - t_1 = \frac{2\ell}{c}\left(\frac{1}{1 - v^2/c^2} - \frac{1}{\sqrt{1 - v^2/c^2}}\right).$$

If $v = 0$, then $\Delta t = 0$, and the two beams will return in phase since they were initially in phase. But if $v \neq 0$, then $\Delta t \neq 0$, and the two beams will return out of phase. If this change of phase from the condition $v = 0$ to that for $v \neq 0$ could be measured, then v could be determined. But the Earth cannot be stopped. Furthermore, we should not be too quick to assume that lengths are not affected by motion and therefore to assume $\ell_1 = \ell_2$.

Michelson and Morley realized that they could detect the difference in phase (assuming that $v \neq 0$) if they rotated their apparatus by $90°$, for then the interference pattern between the two beams should change. In the rotated position, beam 1 would now move parallel to the ether and beam 2 perpendicular to it. Thus the roles could be reversed, and in the rotated position the times (designated by primes) would be

$$t_1' = \frac{2\ell_1}{c\left(1 - v^2/c^2\right)} \qquad \text{and} \qquad t_2' = \frac{2\ell_2}{c\sqrt{1 - v^2/c^2}}.$$

The time lag between the two beams in the nonrotated position (unprimed) would be

$$\Delta t = t_2 - t_1 = \frac{2\ell_2}{c\left(1 - v^2/c^2\right)} - \frac{2\ell_1}{c\sqrt{1 - v^2/c^2}}.$$

In the rotated position, the time difference would be

$$\Delta t' = t_2' - t_1' = \frac{2\ell_2}{c\sqrt{1 - v^2/c^2}} - \frac{2\ell_1}{c\left(1 - v^2/c^2\right)}.$$

When the rotation is made, the fringes of the interference pattern (Section 34–6) will shift an amount determined by the difference:

$$\Delta t - \Delta t' = \frac{2}{c}\left(\ell_1 + \ell_2\right)\left(\frac{1}{1 - v^2/c^2} - \frac{1}{\sqrt{1 - v^2/c^2}}\right).$$

This expression can be considerably simplified if we assume that $v/c \ll 1$. In this case we can use the binomial expansion (Appendix A), so

$$\frac{1}{1 - v^2/c^2} \approx 1 + \frac{v^2}{c^2} \qquad \text{and} \qquad \frac{1}{\sqrt{1 - v^2/c^2}} \approx 1 + \frac{1}{2}\frac{v^2}{c^2}.$$

Then

$$\Delta t - \Delta t' \approx \frac{2}{c}\left(\ell_1 + \ell_2\right)\left(1 + \frac{v^2}{c^2} - 1 - \frac{1}{2}\frac{v^2}{c^2}\right)$$

$$\approx \left(\ell_1 + \ell_2\right)\frac{v^2}{c^3}.$$

Now we assume $v = 3.0 \times 10^4 \, \text{m/s}$, the speed of the Earth in its orbit around the Sun. In Michelson and Morley's experiments, the arms ℓ_1 and ℓ_2 were about 11 m long. The time difference would then be about

$$\frac{(22 \, \text{m})(3.0 \times 10^4 \, \text{m/s})^2}{(3.0 \times 10^8 \, \text{m/s})^3} \approx 7.3 \times 10^{-16} \, \text{s}.$$

For visible light of wavelength $\lambda = 5.5 \times 10^{-7} \, \text{m}$, say, the frequency would be $f = c/\lambda = (3.0 \times 10^8 \, \text{m/s})/(5.5 \times 10^{-7} \, \text{m}) = 5.5 \times 10^{14} \, \text{Hz}$, which means that wave crests pass by a point every $1/(5.5 \times 10^{14} \, \text{Hz}) = 1.8 \times 10^{-15} \, \text{s}$. Thus, with a time difference of $7.3 \times 10^{-16} \, \text{s}$, Michelson and Morley should have noted a movement in the interference pattern of $(7.3 \times 10^{-16} \, \text{s})/(1.8 \times 10^{-15} \, \text{s}) = 0.4$ fringe. They could easily have detected this, since their apparatus was capable of observing a fringe shift as small as 0.01 fringe.

But they found *no significant fringe shift whatever!* They set their apparatus at various orientations. They made observations day and night so that they would be at various orientations with respect to the Sun (due to the Earth's rotation).

They tried at different seasons of the year (the Earth at different locations due to its orbit around the Sun). Never did they observe a significant fringe shift.

This **null result** was one of the great puzzles of physics at the end of the nineteenth century. To explain it was a difficult challenge. One possibility to explain the null result was put forth independently by G. F. Fitzgerald and H. A. Lorentz (in the 1890s) in which they proposed that any length (including the arm of an interferometer) contracts by a factor $\sqrt{1 - v^2/c^2}$ in the direction of motion through the ether. According to Lorentz, this could be due to the ether affecting the forces between the molecules of a substance, which were assumed to be electrical in nature. This theory was eventually replaced by the far more comprehensive theory proposed by Albert Einstein in 1905—the special theory of relativity.

36–3 Postulates of the Special Theory of Relativity

The problems that existed at the start of the twentieth century with regard to electromagnetic theory and Newtonian mechanics were beautifully resolved by Einstein's introduction of the theory of relativity in 1905. Unaware of the Michelson–Morley null result, Einstein was motivated by certain questions regarding electromagnetic theory and light waves. For example, he asked himself: "What would I see if I rode a light beam?" The answer was that instead of a traveling electromagnetic wave, he would see alternating electric and magnetic fields at rest whose magnitude changed in space, but did not change in time. Such fields, he realized, had never been detected and indeed were not consistent with Maxwell's electromagnetic theory. He argued, therefore, that it was unreasonable to think that the speed of light relative to any observer could be reduced to zero, or in fact reduced at all. This idea became the second postulate of his theory of relativity.

In his famous 1905 paper, Einstein proposed doing away completely with the idea of the ether and the accompanying assumption of a preferred or absolute reference frame at rest. This proposal was embodied in two postulates. The first postulate was an extension of the Galilean–Newtonian relativity principle to include not only the laws of mechanics but also those of the rest of physics, including electricity and magnetism:

> *First postulate (the relativity principle)*: **The laws of physics have the same form in all inertial reference frames.**

The first postulate can also be stated as: *There is no experiment you can do in an inertial reference frame to tell if you are at rest or moving uniformly at constant velocity.*

The second postulate is consistent with the first:

> *Second postulate (constancy of the speed of light)*: **Light propagates through empty space with a definite speed c independent of the speed of the source or observer.**

These two postulates form the foundation of Einstein's **special theory of relativity**. It is called "special" to distinguish it from his later "general theory of relativity," which deals with noninertial (accelerating) reference frames (Chapter 44). The special theory, which is what we discuss here, deals only with inertial frames.

The second postulate may seem hard to accept, for it seems to violate common sense. First of all, we have to think of light traveling through empty space. Giving up the ether is not too hard, however, since it had never been detected. But the second postulate also tells us that the speed of light in vacuum is always the same, 3.00×10^8 m/s, no matter what the speed of the observer or the source. Thus, a person traveling toward or away from a source of light will measure the same speed for that light as someone at rest with respect to the source. This conflicts with our everyday experience: we would expect to have to add in the velocity of the observer. On the other hand, perhaps we can't expect our everyday experience to be helpful when dealing with the high velocity of light. Furthermore, the null result of the Michelson–Morley experiment is fully consistent with the second postulate.[†]

[†]The Michelson–Morley experiment can also be considered as evidence for the first postulate, since it was intended to measure the motion of the Earth relative to an absolute reference frame. Its failure to do so implies the absence of any such preferred frame.

Einstein's proposal has a certain beauty. By doing away with the idea of an absolute reference frame, it was possible to reconcile classical mechanics with Maxwell's electromagnetic theory. The speed of light predicted by Maxwell's equations *is* the speed of light in vacuum in *any* reference frame.

Einstein's theory required us to give up common sense notions of space and time, and in the following Sections we will examine some strange but interesting consequences of special relativity. Our arguments for the most part will be simple ones. We will use a technique that Einstein himself did: we will imagine very simple experimental situations in which little mathematics is needed. In this way, we can see many of the consequences of relativity theory without getting involved in detailed calculations. Einstein called these "thought" experiments.

36–4 Simultaneity

An important consequence of the theory of relativity is that we can no longer regard time as an absolute quantity. No one doubts that time flows onward and never turns back. But the time interval between two events, and even whether or not two events are simultaneous, depends on the observer's reference frame. By an **event**, which we use a lot here, we mean something that happens at a particular place and at a particular time.

Two events are said to occur simultaneously if they occur at exactly the same time. But how do we know if two events occur precisely at the same time? If they occur at the same point in space—such as two apples falling on your head at the same time—it is easy. But if the two events occur at widely separated places, it is more difficult to know whether the events are simultaneous since we have to take into account the time it takes for the light from them to reach us. Because light travels at finite speed, a person who sees two events must calculate back to find out when they actually occurred. For example, if two events are *observed* to occur at the same time, but one actually took place farther from the observer than the other, then the more distant one must have occurred earlier, and the two events were not simultaneous.

We now imagine a simple thought experiment. Assume an observer, called O, is located exactly halfway between points A and B where two events occur, Fig. 36–4. Suppose the two events are lightning that strikes the points A and B, as shown. For brief events like lightning, only short pulses of light (blue in Fig. 36–4) will travel outward from A and B and reach O. Observer O "sees" the events when the pulses of light reach point O. If the two pulses reach O at the same time, then the two events had to be simultaneous. This is because the two light pulses travel at the same speed (postulate 2), and since the distance OA equals OB, the time for the light to travel from A to O and B to O must be the same. Observer O can then definitely state that the two events occurred simultaneously. On the other hand, if O sees the light from one event before that from the other, then the former event occurred first.

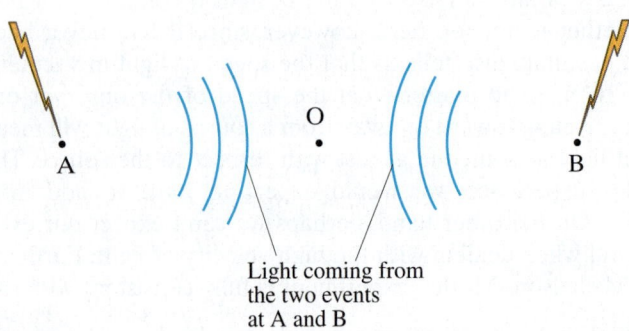

FIGURE 36–4 A moment after lightning strikes at points A and B, the pulses of light (shown as blue waves) are traveling toward the observer O, but O "sees" the lightning only when the light reaches O.

Light coming from the two events at A and B

The question we really want to examine is this: if two events are simultaneous to an observer in one reference frame, are they also simultaneous to another observer moving with respect to the first? Let us call the observers O_1 and O_2 and assume they are fixed in reference frames 1 and 2 that move with speed v relative to one another. These two reference frames can be thought of as two rockets or two trains (Fig. 36–5). O_2 says that O_1 is moving to the right with speed v, as in Fig. 36–5a; and O_1 says O_2 is moving to the left with speed v, as in Fig. 36–5b. Both viewpoints are legitimate according to the relativity principle. [There is no third point of view which will tell us which one is "really" moving.]

Now suppose that observers O_1 and O_2 observe and measure two lightning strikes. The lightning bolts mark both trains where they strike: at A_1 and B_1 on O_1's train, and at A_2 and B_2 on O_2's train, Fig. 36–6a. For simplicity, we assume that O_1 is exactly halfway between A_1 and B_1, and that O_2 is halfway between A_2 and B_2. Let us first put ourselves in O_2's reference frame, so we observe O_1 moving to the right with speed v. Let us also assume that the two events occur *simultaneously* in O_2's frame, and just at the instant when O_1 and O_2 are opposite each other, Fig. 36–6a. A short time later, Fig. 36–6b, the light from A_2 and B_2 reaches O_2 at the same time (we assumed this). Since O_2 knows (or measures) the distances O_2A_2 and O_2B_2 as equal, O_2 knows the two events are simultaneous in the O_2 reference frame.

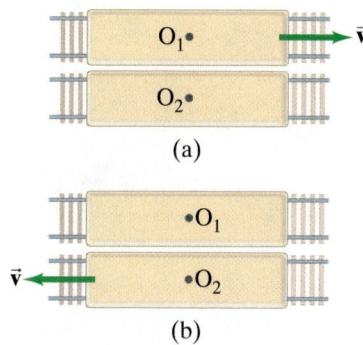

FIGURE 36–5 Observers O_1 and O_2, on two different trains (two different reference frames), are moving with relative speed v. O_2 says that O_1 is moving to the right (a); O_1 says that O_2 is moving to the left (b). Both viewpoints are legitimate: it all depends on your reference frame.

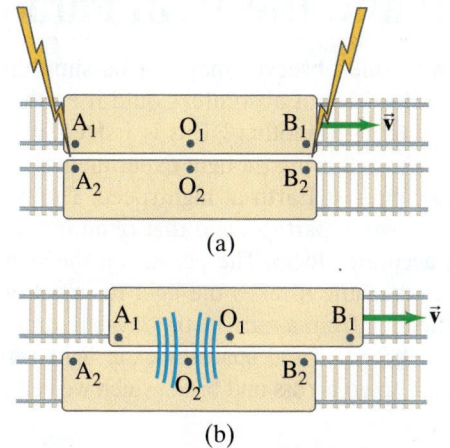

FIGURE 36–6 Thought experiment on simultaneity. In both (a) and (b) we are in the reference frame of observer O_2, who sees the reference frame of O_1 moving to the right. In (a), one lightning bolt strikes the two reference frames at A_1 and A_2, and a second lightning bolt strikes at B_1 and B_2. (b) A moment later, the light (shown in blue) from the two events reaches O_2 at the same time. So according to observer O_2, the two bolts of lightning struck simultaneously. But in O_1's reference frame, the light from B_1 has already reached O_1, whereas the light from A_1 has not yet reached O_1. So in O_1's reference frame, the event at B_1 must have preceded the event at A_1. Simultaneity in time is not absolute.

But what does observer O_1 observe and measure? From our (O_2) reference frame, we can predict what O_1 will observe. We see that O_1 moves to the right during the time the light is traveling to O_1 from A_1 and B_1. As shown in Fig. 36–6b, we can see from our O_2 reference frame that the light from B_1 has already passed O_1, whereas the light from A_1 has not yet reached O_1. That is, O_1 observes the light coming from B_1 before observing the light coming from A_1. Given (1) that light travels at the same speed c in any direction and in any reference frame, and (2) that the distance O_1A_1 equals O_1B_1, then observer O_1 can only conclude that the event at B_1 occurred before the event at A_1. The two events are *not* simultaneous for O_1, even though they are for O_2.

We thus find that two events which take place at different locations and are simultaneous to one observer, are actually not simultaneous to a second observer who moves relative to the first.

It may be tempting to ask: "Which observer is right, O_1 or O_2?" The answer, according to relativity, is that they are *both* right. There is no "best" reference frame we can choose to determine which observer is right. Both frames are equally good. We can only conclude that *simultaneity is not an absolute concept*, but is relative. We are not aware of this lack of agreement on simultaneity in everyday life because the effect is noticeable only when the relative speed of the two reference frames is very large (near c), or the distances involved are very large.

EXERCISE A Examine the experiment of Fig. 36–6 from O_1's reference frame. In this case, O_1 will be at rest and will see event B_1 occur before A_1. Will O_1 recognize that O_2, who is moving with speed v to the left, will see the two events as simultaneous? [*Hint:* Draw a diagram equivalent to Fig. 36–6.]

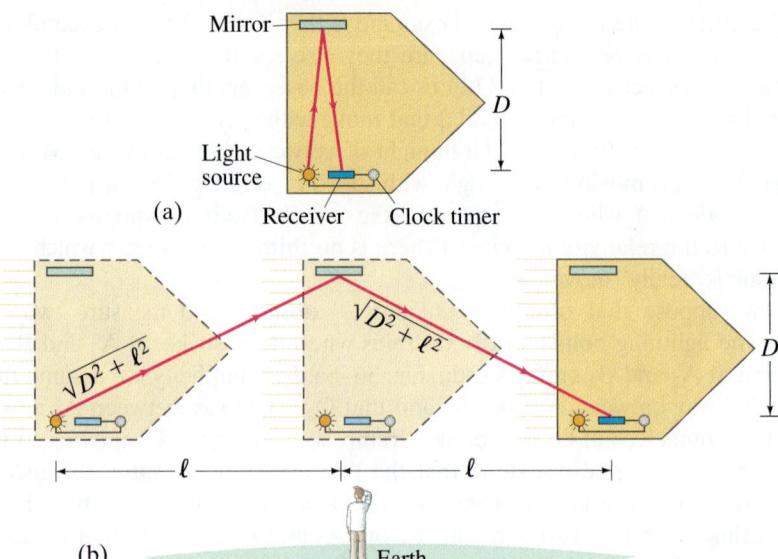

FIGURE 36–7 Time dilation can be shown by a thought experiment: the time it takes for light to travel across a spaceship and back is longer for the observer on Earth (b) than for the observer on the spaceship (a).

(a) Light source / Receiver / Clock timer / Mirror

(b) Earth

36–5 Time Dilation and the Twin Paradox

The fact that two events simultaneous to one observer may not be simultaneous to a second observer suggests that time itself is not absolute. Could it be that time passes differently in one reference frame than in another? This is, indeed, just what Einstein's theory of relativity predicts, as the following thought experiment shows.

Figure 36–7 shows a spaceship traveling past Earth at high speed. The point of view of an observer on the spaceship is shown in part (a), and that of an observer on Earth in part (b). Both observers have accurate clocks. The person on the spaceship (Fig. 36–7a) flashes a light and measures the time it takes the light to travel directly across the spaceship and return after reflecting from a mirror (the rays are drawn at a slight angle for clarity). In the reference frame of the spaceship, the light travels a distance $2D$ at speed c; so the time required to go across and back, which we call Δt_0, is

$$\Delta t_0 = 2D/c.$$

The observer on Earth, Fig. 36–7b, observes the same process. But to this observer, the spaceship is moving. So the light travels the diagonal path shown going across the spaceship, reflecting off the mirror, and returning to the sender. Although the light travels at the same speed to this observer (the second postulate), it travels a greater distance. Hence the time required, as measured by the observer on Earth, will be *greater* than that measured by the observer on the spaceship.

Let us determine the time interval Δt measured by the observer on Earth between sending and receiving the light. In time Δt, the spaceship travels a distance $2\ell = v\,\Delta t$ where v is the speed of the spaceship (Fig. 36–7b). The light travels a total distance on its diagonal path (Pythagorean theorem) of $2\sqrt{D^2 + \ell^2}$, where $\ell = v\,\Delta t/2$. Therefore

$$c = \frac{2\sqrt{D^2 + \ell^2}}{\Delta t} = \frac{2\sqrt{D^2 + v^2(\Delta t)^2/4}}{\Delta t}.$$

We square both sides,

$$c^2 = \frac{4D^2}{(\Delta t)^2} + v^2,$$

and solve for Δt, to find

$$\Delta t = \frac{2D}{c\sqrt{1 - v^2/c^2}}.$$

We combine this equation for Δt with the formula above, $\Delta t_0 = 2D/c$, and find

TIME DILATION

$$\Delta t = \frac{\Delta t_0}{\sqrt{1 - v^2/c^2}}. \tag{36–1a}$$

Since $\sqrt{1 - v^2/c^2}$ is always less than 1, we see that $\Delta t > \Delta t_0$. That is, the time interval between the two events (the sending of the light, and its reception on the spaceship) is *greater* for the observer on Earth than for the observer on the spaceship. This is a general result of the theory of relativity, and is known as **time dilation**. Stated simply, the time dilation effect says that

> **clocks moving relative to an observer are measured to run more slowly (as compared to clocks at rest relative to that observer).**

However, we should not think that the clocks are somehow at fault. Time is actually measured to pass more slowly in any moving reference frame as compared to your own. This remarkable result is an inevitable outcome of the two postulates of the theory of relativity.

The factor $1/\sqrt{1 - v^2/c^2}$ occurs so often in relativity that we often give it the shorthand symbol γ (the Greek letter "gamma"), and write Eq. 36–1a as

$$\Delta t = \gamma \, \Delta t_0 \tag{36-1b}$$

where

$$\gamma = \frac{1}{\sqrt{1 - v^2/c^2}}. \tag{36-2}$$

Note that γ is never less than one, and has no units. At normal speeds, $\gamma = 1$ to a few decimal places; in general, $\gamma \geq 1$.

The concept of time dilation may be hard to accept, for it contradicts our experience. We can see from Eqs. 36–1 that the time dilation effect is indeed negligible unless v is reasonably close to c. If v is much less than c, then the term v^2/c^2 is much smaller than the 1 in the denominator of Eq. 36–1a, and then $\Delta t \approx \Delta t_0$ (see Example 36–2). The speeds we experience in everyday life are much smaller than c, so it is little wonder we don't ordinarily notice time dilation. Experiments have tested the time dilation effect, and have confirmed Einstein's predictions. In 1971, for example, extremely precise atomic clocks were flown around the Earth in jet planes. The speed of the planes $(10^3 \, \text{km/h})$ was much less than c, so the clocks had to be accurate to nanoseconds $(10^{-9} \, \text{s})$ in order to detect any time dilation. They were this accurate, and they confirmed Eqs. 36–1 to within experimental error. Time dilation had been confirmed decades earlier, however, by observations on "elementary particles" which have very small masses (typically 10^{-30} to $10^{-27} \, \text{kg}$) and so require little energy to be accelerated to speeds close to the speed of light, c. Many of these elementary particles are not stable and decay after a time into lighter particles. One example is the muon, whose mean lifetime is $2.2 \, \mu\text{s}$ when at rest. Careful experiments showed that when a muon is traveling at high speeds, its lifetime is measured to be longer than when it is at rest, just as predicted by the time dilation formula.

EXAMPLE 36–1 **Lifetime of a moving muon.** (*a*) What will be the mean lifetime of a muon as measured in the laboratory if it is traveling at $v = 0.60c = 1.80 \times 10^8 \, \text{m/s}$ with respect to the laboratory? Its mean lifetime at rest is $2.20 \, \mu\text{s} = 2.20 \times 10^{-6} \, \text{s}$. (*b*) How far does a muon travel in the laboratory, on average, before decaying?

APPROACH If an observer were to move along with the muon (the muon would be at rest to this observer), the muon would have a mean life of $2.20 \times 10^{-6} \, \text{s}$. To an observer in the lab, the muon lives longer because of time dilation. We find the mean lifetime using Eq. 36–1a and the average distance using $d = v \, \Delta t$.

SOLUTION (*a*) From Eq. 36–1a with $v = 0.60c$, we have

$$\Delta t = \frac{\Delta t_0}{\sqrt{1 - v^2/c^2}} = \frac{2.20 \times 10^{-6} \, \text{s}}{\sqrt{1 - 0.36c^2/c^2}} = \frac{2.20 \times 10^{-6} \, \text{s}}{\sqrt{0.64}} = 2.8 \times 10^{-6} \, \text{s}.$$

(*b*) Relativity predicts that a muon with speed $1.80 \times 10^8 \, \text{m/s}$ would travel an average distance $d = v \, \Delta t = (1.80 \times 10^8 \, \text{m/s})(2.8 \times 10^{-6} \, \text{s}) = 500 \, \text{m}$, and this is the distance that is measured experimentally in the laboratory.

NOTE At a speed of $1.8 \times 10^8 \, \text{m/s}$, classical physics would tell us that with a mean life of $2.2 \, \mu\text{s}$, an average muon would travel $d = vt = (1.8 \times 10^8 \, \text{m/s})(2.2 \times 10^{-6} \, \text{s}) = 400 \, \text{m}$. This is shorter than the distance measured.

EXERCISE B What is the muon's mean lifetime (Example 36–1) if it is traveling at $v = 0.90c$? (*a*) $0.42 \, \mu\text{s}$; (*b*) $2.3 \, \mu\text{s}$; (*c*) $5.0 \, \mu\text{s}$; (*d*) $5.3 \, \mu\text{s}$; (*e*) $12.0 \, \mu\text{s}$.

We need to clarify how to use Eqs. 36–1, and the meaning of Δt and Δt_0. The equation is true only when Δt_0 represents the time interval between the two events in a reference frame where the two events occur at *the same point in space* (as in Fig. 36–7a where the two events are the light flash being sent and being received). This time interval, Δt_0, is called the **proper time**. Then Δt in Eqs. 36–1 represents the time interval between the two events as measured in a reference frame moving with speed v with respect to the first. In Example 36–1 above, Δt_0 (and not Δt) was set equal to 2.2×10^{-6} s because it is only in the rest frame of the muon that the two events ("birth" and "decay") occur at the same point in space. The proper time Δt_0 is the shortest time between the events any observer can measure. In any other moving reference frame, the time Δt is greater.

EXAMPLE 36–2 **Time dilation at 100 km/h.** Let us check time dilation for everyday speeds. A car traveling 100 km/h covers a certain distance in 10.00 s according to the driver's watch. What does an observer at rest on Earth measure for the time interval?

APPROACH The car's speed relative to Earth is 100 km/h = $(1.00 \times 10^5 \, \text{m})/(3600 \, \text{s})$ = 27.8 m/s. The driver is at rest in the reference frame of the car, so we set $\Delta t_0 = 10.00$ s in the time dilation formula.

SOLUTION We use Eq. 36–1a:

$$\Delta t = \frac{\Delta t_0}{\sqrt{1 - \frac{v^2}{c^2}}} = \frac{10.00 \, \text{s}}{\sqrt{1 - \left(\frac{27.8 \, \text{m/s}}{3.00 \times 10^8 \, \text{m/s}}\right)^2}} = \frac{10.00 \, \text{s}}{\sqrt{1 - \left(8.59 \times 10^{-15}\right)}}.$$

If you put these numbers into a calculator, you will obtain $\Delta t = 10.00$ s, since the denominator differs from 1 by such a tiny amount. Indeed, the time measured by an observer on Earth would show no difference from that measured by the driver, even with the best instruments. A computer that could calculate to a large number of decimal places would reveal a difference between Δt and Δt_0. We can estimate the difference using the binomial expansion (Appendix A),

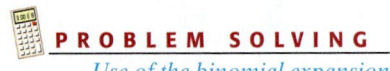 **PROBLEM SOLVING**

Use of the binomial expansion

$$(1 \pm x)^n \approx 1 \pm nx. \qquad\qquad [\text{for } x \ll 1]$$

In our time dilation formula, we have the factor $\gamma = \left(1 - v^2/c^2\right)^{-\frac{1}{2}}$. Thus

$$\Delta t = \gamma \, \Delta t_0 = \Delta t_0 \left(1 - \frac{v^2}{c^2}\right)^{-\frac{1}{2}} \approx \Delta t_0 \left(1 + \frac{1}{2} \frac{v^2}{c^2}\right)$$

$$\approx 10.00 \, \text{s} \left[1 + \frac{1}{2} \left(\frac{27.8 \, \text{m/s}}{3.00 \times 10^8 \, \text{m/s}}\right)^2\right] \approx 10.00 \, \text{s} + 4 \times 10^{-14} \, \text{s}.$$

So the difference between Δt and Δt_0 is predicted to be 4×10^{-14} s, an extremely small amount.

EXERCISE C A certain atomic clock keeps perfect time on Earth. If the clock is taken on a spaceship traveling at a speed $v = 0.60c$, does this clock now run slow according to the people (*a*) on the spaceship, (*b*) on Earth?

EXAMPLE 36–3 **Reading a magazine on a spaceship.** A passenger on a high-speed spaceship traveling between Earth and Jupiter at a steady speed of $0.75c$ reads a magazine which takes 10.0 min according to her watch. (*a*) How long does this take as measured by Earth-based clocks? (*b*) How much farther is the spaceship from Earth at the end of reading the article than it was at the beginning?

APPROACH (*a*) The time interval in one reference frame is related to the time interval in the other by Eq. 36–1a or b. (*b*) At constant speed, distance is speed × time. Since there are two times (a Δt and a Δt_0) we will get two distances, one for each reference frame. [This surprising result is explored in the next Section (36–6).]

SOLUTION (a) The given 10.0-min time interval is the proper time—starting and finishing the magazine happen at the same place on the spaceship. Earth clocks measure

$$\Delta t = \frac{\Delta t_0}{\sqrt{1 - \dfrac{v^2}{c^2}}} = \frac{10.00 \text{ min}}{\sqrt{1 - (0.75)^2}} = 15.1 \text{ min}.$$

(b) In the Earth frame, the rocket travels a distance $D = v\,\Delta t = (0.75c)(15.1 \text{ min}) = (0.75)(3.0 \times 10^8 \text{ m/s})(15.1 \text{ min} \times 60 \text{ s/min}) = 2.04 \times 10^{11}$ m. In the spaceship's frame, the Earth is moving away from the spaceship at $0.75c$, but the time is only 10.0 min, so the distance is measured to be $D_0 = v\,\Delta t_0 = (2.25 \times 10^8 \text{ m/s})(600 \text{ s}) = 1.35 \times 10^{11}$ m.

Values for $\gamma = 1/\sqrt{1 - v^2/c^2}$ at a few speeds v are given in Table 36–1.

Space Travel?

Time dilation has aroused interesting speculation about space travel. According to classical (Newtonian) physics, to reach a star 100 light-years away would not be possible for ordinary mortals (1 light-year is the distance light can travel in 1 year $= 3.0 \times 10^8 \text{ m/s} \times 3.16 \times 10^7 \text{ s} = 9.5 \times 10^{15}$ m). Even if a spaceship could travel at close to the speed of light, it would take over 100 years to reach such a star. But time dilation tells us that the time involved could be less. In a spaceship traveling at $v = 0.999c$, the time for such a trip would be only about $\Delta t_0 = \Delta t \sqrt{1 - v^2/c^2} = (100 \text{ yr})\sqrt{1 - (0.999)^2} = 4.5$ yr. Thus time dilation allows such a trip, but the enormous practical problems of achieving such speeds may not be possible to overcome, certainly not in the near future.

In this example, 100 years would pass on Earth, whereas only 4.5 years would pass for the astronaut on the trip. Is it just the clocks that would slow down for the astronaut? No. All processes, including aging and other life processes, run more slowly for the astronaut according to the Earth observer. But to the astronaut, time would pass in a normal way. The astronaut would experience 4.5 years of normal sleeping, eating, reading, and so on. And people on Earth would experience 100 years of ordinary activity.

Twin Paradox

Not long after Einstein proposed the special theory of relativity, an apparent paradox was pointed out. According to this **twin paradox**, suppose one of a pair of 20-year-old twins takes off in a spaceship traveling at very high speed to a distant star and back again, while the other twin remains on Earth. According to the Earth twin, the astronaut twin will age less. Whereas 20 years might pass for the Earth twin, perhaps only 1 year (depending on the spacecraft's speed) would pass for the traveler. Thus, when the traveler returns, the earthbound twin could expect to be 40 years old whereas the traveling twin would be only 21.

This is the viewpoint of the twin on the Earth. But what about the traveling twin? If all inertial reference frames are equally good, won't the traveling twin make all the claims the Earth twin does, only in reverse? Can't the astronaut twin claim that since the Earth is moving away at high speed, time passes more slowly on Earth and the twin on Earth will age less? This is the opposite of what the Earth twin predicts. They cannot both be right, for after all the spacecraft returns to Earth and a direct comparison of ages and clocks can be made.

There is, however, no contradiction here. One of the viewpoints is indeed incorrect. The consequences of the special theory of relativity—in this case, time dilation—can be applied only by observers in an inertial reference frame. The Earth is such a frame (or nearly so), whereas the spacecraft is not. The spacecraft accelerates at the start and end of its trip and when it turns around at the far point of its journey. During the acceleration, the twin on the spacecraft is not in an inertial frame. In between, the astronaut twin may be in an inertial frame (and is justified in saying the Earth twin's clocks run slow), but it is not always the same frame. So she cannot use special relativity to predict their relative ages when she returns to Earth. The Earth twin stays in the same inertial frame, and we can thus trust her predictions based on special relativity. Thus, there is no paradox. The prediction of the Earth twin that the traveling twin ages less is the proper one.

TABLE 36–1 Values of γ

v	γ
0	1.000
$0.01c$	1.000
$0.10c$	1.005
$0.50c$	1.15
$0.90c$	2.3
$0.99c$	7.1

*Global Positioning System (GPS)

Airplanes, cars, boats, and hikers use **global positioning system** (**GPS**) receivers to tell them quite accurately where they are, at a given moment. The 24 global positioning system satellites send out precise time signals using atomic clocks. Your receiver compares the times received from at least four satellites, all of whose times are carefully synchronized to within 1 part in 10^{13}. By comparing the time differences with the known satellite positions and the fixed speed of light, the receiver can determine how far it is from each satellite and thus where it is on the Earth. It can do this to a typical accuracy of 15 m, if it has been constructed to make corrections such as the one below due to special relativity.

CONCEPTUAL EXAMPLE 36–4 | **A relativity correction to GPS.** GPS satellites move at about $4 \text{ km/s} = 4000 \text{ m/s}$. Show that a good GPS receiver needs to correct for time dilation if it is to produce results consistent with atomic clocks accurate to 1 part in 10^{13}.

RESPONSE Let us calculate the magnitude of the time dilation effect by inserting $v = 4000 \text{ m/s}$ into Eq. 36–1a:

$$\Delta t = \frac{1}{\sqrt{1 - \dfrac{v^2}{c^2}}} \Delta t_0$$

$$= \frac{1}{\sqrt{1 - \left(\dfrac{4 \times 10^3 \text{ m/s}}{3 \times 10^8 \text{ m/s}}\right)^2}} \Delta t_0$$

$$= \frac{1}{\sqrt{1 - 1.8 \times 10^{-10}}} \Delta t_0.$$

We use the binomial expansion: $(1 \pm x)^n \approx 1 \pm nx$ for $x \ll 1$ (see Appendix A) which here is $(1 - x)^{-\frac{1}{2}} \approx 1 + \frac{1}{2}x$. That is

$$\Delta t = \left(1 + \tfrac{1}{2}(1.8 \times 10^{-10})\right) \Delta t_0 = \left(1 + 9 \times 10^{-11}\right) \Delta t_0.$$

The time "error" divided by the time interval is

$$\frac{(\Delta t - \Delta t_0)}{\Delta t_0} = 1 + 9 \times 10^{-11} - 1 = 9 \times 10^{-11} \approx 1 \times 10^{-10}.$$

Time dilation, if not accounted for, would introduce an error of about 1 part in 10^{10}, which is 1000 times greater than the precision of the atomic clocks. Not correcting for time dilation means a receiver could give much poorer position accuracy.

NOTE GPS devices must make other corrections as well, including effects associated with general relativity.

36–6 Length Contraction

Time intervals are not the only things different in different reference frames. Space intervals—lengths and distances—are different as well, according to the special theory of relativity, and we illustrate this with a thought experiment.

Observers on Earth watch a spacecraft traveling at speed v from Earth to, say, Neptune, Fig. 36–8a. The distance between the planets, as measured by the Earth observers, is ℓ_0. The time required for the trip, measured from Earth, is

$$\Delta t = \frac{\ell_0}{v}. \qquad \text{[Earth observer]}$$

In Fig. 36–8b we see the point of view of observers on the spacecraft. In this frame of reference, the spaceship is at rest; Earth and Neptune move[†] with speed v. The time between departure of Earth and arrival of Neptune (observed from the spacecraft) is the "proper time," since the two events occur at the same point in space (i.e., on the spacecraft). Therefore the time interval is less for the spacecraft

[†]We assume v is much greater than the relative speed of Neptune and Earth, so the latter can be ignored.

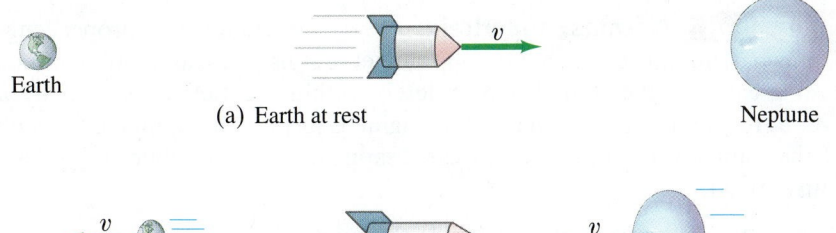

(a) Earth at rest

Earth

Neptune

(b) Spacecraft at rest

Earth

Neptune

FIGURE 36–8 (a) A spaceship traveling at very high speed from Earth to the planet Neptune, as seen from Earth's frame of reference. (b) According to an observer on the spaceship, Earth and Neptune are moving at the very high speed v: Earth leaves the spaceship, and a time Δt_0 later Neptune arrives at the spaceship.

observers than for the Earth observers. That is, because of time dilation (Eq. 36–1a), the time for the trip as viewed by the spacecraft is

$$\Delta t_0 = \Delta t \sqrt{1 - v^2/c^2} = \Delta t/\gamma. \qquad \text{[spacecraft observer]}$$

Because the spacecraft observers measure the same speed but less time between these two events, they also measure the distance as less. If we let ℓ be the distance between the planets as viewed by the spacecraft observers, then $\ell = v \Delta t_0$, which we can rewrite as $\ell = v \Delta t_0 = v \Delta t \sqrt{1 - v^2/c^2} = \ell_0\sqrt{1 - v^2/c^2}$. Thus we have the important result that

$$\ell = \ell_0\sqrt{1 - v^2/c^2} \qquad \textbf{(36–3a)}$$

LENGTH CONTRACTION

or, using γ (Eq. 36–2),

$$\ell = \frac{\ell_0}{\gamma}. \qquad \textbf{(36–3b)}$$

This is a general result of the special theory of relativity and applies to lengths of objects as well as to distance between objects. The result can be stated most simply in words as:

> **the length of an object moving relative to an observer is measured to be shorter along its direction of motion than when it is at rest.**

This is called **length contraction**. The length ℓ_0 in Eqs. 36–3 is called the **proper length**. It is the length of the object (or distance between two points whose positions are measured at the same time) as determined by *observers at rest* with respect to the object. Equations 36–3 give the length ℓ that will be measured by observers when the object travels past them at speed v.

It is important to note that length contraction occurs *only along the direction of motion*. For example, the moving spaceship in Fig. 36–8a is shortened in length, but its height is the same as when it is at rest.

Length contraction, like time dilation, is not noticeable in everyday life because the factor $\sqrt{1 - v^2/c^2}$ in Eq. 36–3a differs from 1.00 significantly only when v is very large.

⚠ **CAUTION**
Proper length is measured in reference frame where the two positions are at rest

FIGURE 36–9 Example 36–5.

EXAMPLE 36–5 **Painting's contraction.** A rectangular painting measures 1.00 m tall and 1.50 m wide. It is hung on the side wall of a spaceship which is moving past the Earth at a speed of 0.90c. See Fig. 36–9a. (a) What are the dimensions of the picture according to the captain of the spaceship? (b) What are the dimensions as seen by an observer on the Earth?

APPROACH We apply the length contraction formula, Eq. 36–3a, to the dimension parallel to the motion; v is the speed of the painting relative to the observer.

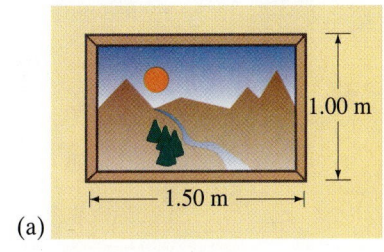
(a)

1.00 m

1.50 m

SOLUTION (a) The painting is at rest ($v = 0$) on the spaceship so it (as well as everything else in the spaceship) looks perfectly normal to everyone on the spaceship. The captain sees a 1.00-m by 1.50-m painting.
(b) Only the dimension in the direction of motion is shortened, so the height is unchanged at 1.00 m, Fig. 36–9b. The length, however, is contracted to

$$\ell = \ell_0\sqrt{1 - \frac{v^2}{c^2}}$$
$$= (1.50\,\text{m})\sqrt{1 - (0.90)^2} = 0.65\,\text{m}.$$

So the picture has dimensions 1.00 m × 0.65 m.

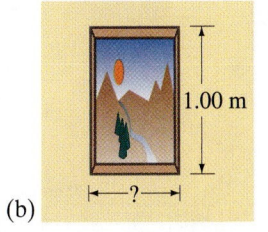
(b)

1.00 m

?

EXAMPLE 36-6 **A fantasy supertrain.** A very fast train with a proper length of 500 m is passing through a 200-m-long tunnel. Let us imagine the train's speed to be so great that the train fits completely within the tunnel as seen by an observer at rest on the Earth. That is, the engine is just about to emerge from one end of the tunnel at the time the last car disappears into the other end. What is the train's speed?

APPROACH Since the train just fits inside the tunnel, its length measured by the person on the ground is 200 m. The length contraction formula, Eq. 36–3a or b, can thus be used to solve for v.

SOLUTION Substituting $\ell = 200$ m and $\ell_0 = 500$ m into Eq. 36–3a gives

$$200 \text{ m} = 500 \text{ m} \sqrt{1 - \frac{v^2}{c^2}};$$

dividing both sides by 500 m and squaring, we get

$$(0.40)^2 = 1 - \frac{v^2}{c^2}$$

or

$$\frac{v}{c} = \sqrt{1 - (0.40)^2}$$

and

$$v = 0.92c.$$

NOTE No real train could go this fast. But it is fun to think about.

NOTE An observer on the *train* would *not* see the two ends of the train inside the tunnel at the same time. Recall that observers moving relative to each other do not agree about simultaneity.

EXERCISE D What is the length of the tunnel as measured by observers on the train in Example 36–6?

CONCEPTUAL EXAMPLE 36-7 **Resolving the train and tunnel length.** Observers at rest on the Earth see a very fast 200-m-long train pass through a 200-m-long tunnel (as in Example 36–6) so that the train momentarily disappears from view inside the tunnel. Observers on the train measure the train's length to be 500 m and the tunnel's length to be only 80 m (Exercise D, using Eq. 36–3a). Clearly a 500-m-long train cannot fit inside an 80-m-long tunnel. How is this apparent inconsistency explained?

RESPONSE Events simultaneous in one reference frame may not be simultaneous in another. Let the engine emerging from one end of the tunnel be "event A," and the last car disappearing into the other end of the tunnel "event B." To observers in the Earth frame, events A and B are simultaneous. To observers on the train, however, the events are not simultaneous. In the train's frame, event A occurs before event B. As the engine emerges from the tunnel, observers on the train observe the last car as still $500 \text{ m} - 80 \text{ m} = 420 \text{ m}$ from the entrance to the tunnel.

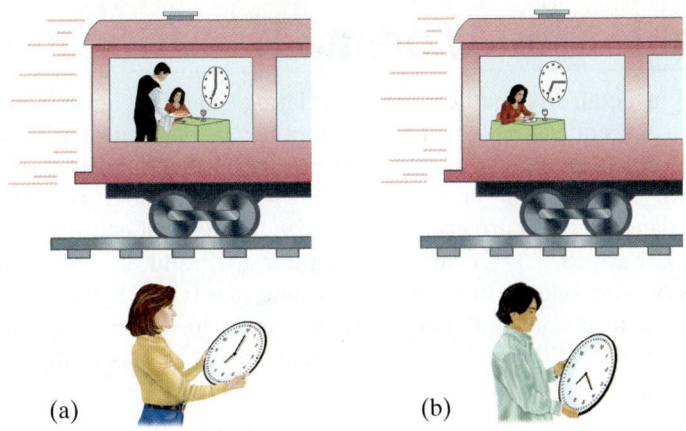

FIGURE 36–10 According to an accurate clock on a fast-moving train, a person (a) begins dinner at 7:00 and (b) finishes at 7:15. At the beginning of the meal, two observers on Earth set their watches to correspond with the clock on the train. These observers measure the eating time as 20 minutes.

(a)   (b)

36–7 Four-Dimensional Space–Time

Let us imagine a person is on a train moving at a very high speed, say $0.65c$, Fig. 36–10. This person begins a meal at 7:00 and finishes at 7:15, according to a clock on the train. The two events, beginning and ending the meal, take place at the same point on the train. So the proper time between these two events is 15 min. To observers on Earth, the meal will take longer—20 min according to Eqs. 36–1. Let us assume that the meal was served on a 20-cm-diameter plate. To observers on the Earth, the plate is only 15 cm wide (length contraction). Thus, to observers on the Earth, the meal looks smaller but lasts longer.

In a sense the two effects, time dilation and length contraction, balance each other. When viewed from the Earth, what an object seems to lose in size it gains in length of time it lasts. Space, or length, is exchanged for time.

Considerations like this led to the idea of **four-dimensional space–time**: space takes up three dimensions and time is a fourth dimension. Space and time are intimately connected. Just as when we squeeze a balloon we make one dimension larger and another smaller, so when we examine objects and events from different reference frames, a certain amount of space is exchanged for time, or vice versa.

Although the idea of four dimensions may seem strange, it refers to the idea that any object or event is specified by four quantities—three to describe where in space, and one to describe when in time. The really unusual aspect of four-dimensional space–time is that space and time can intermix: a little of one can be exchanged for a little of the other when the reference frame is changed.

It is difficult for most of us to understand the idea of four-dimensional space–time. Somehow we feel, just as physicists did before the advent of relativity, that space and time are completely separate entities. Yet we have found in our thought experiments that they are not completely separate. And think about Galileo and Newton. Before Galileo, the vertical direction, that in which objects fall, was considered to be distinctly different from the two horizontal dimensions. Galileo showed that the vertical dimension differs only in that it happens to be the direction in which gravity acts. Otherwise, all three dimensions are equivalent, a viewpoint we all accept today. Now we are asked to accept one more dimension, time, which we had previously thought of as being somehow different. This is not to say that there is no distinction between space and time. What relativity has shown is that space and time determinations are not independent of one another.

In Galilean–Newtonian relativity, the time interval between two events, Δt, and the distance between two events or points, Δx, are invariant quantities no matter what inertial reference frame they are viewed from. Neither of these quantities is invariant according to Einstein's relativity. But there is an invariant quantity in four-dimensional space–time, called the **space–time interval**, which is $(\Delta s)^2 = (c\,\Delta t)^2 - (\Delta x)^2$. We leave it as a Problem (97) to show that this quantity is indeed invariant under a Lorentz transformation (Section 36–8).

36–8 Galilean and Lorentz Transformations

We now examine in detail the mathematics of relating quantities in one inertial reference frame to the equivalent quantities in another. In particular, we will see how positions and velocities *transform* (that is, change) from one frame to the other.

We begin with the classical or Galilean viewpoint. Consider two inertial reference frames S and S' which are each characterized by a set of coordinate axes, Fig. 36–11. The axes x and y (z is not shown) refer to S and x' and y' to S'. The x' and x axes overlap one another, and we assume that frame S' moves to the right in the x direction at constant speed v with respect to S. For simplicity let us assume the origins 0 and 0' of the two reference frames are superimposed at time $t = 0$.

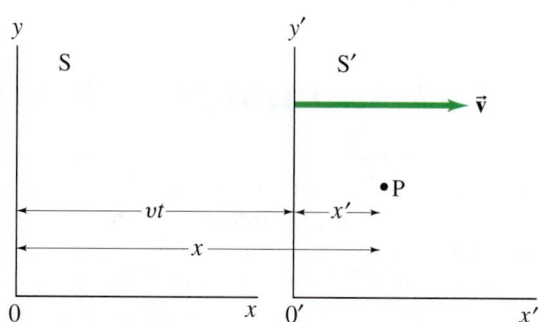

FIGURE 36–11 Inertial reference frame S' moves to the right at constant speed v with respect to frame S.

Now consider an event that occurs at some point P (Fig. 36–11) represented by the coordinates x', y', z' in reference frame S' at the time t'. What will be the coordinates of P in S? Since S and S' initially overlap precisely, after a time t', S' will have moved a distance vt'. Therefore, at time t', $x = x' + vt'$. The y and z coordinates, on the other hand, are not altered by motion along the x axis; thus $y = y'$ and $z = z'$. Finally, since time is assumed to be absolute in Galilean–Newtonian physics, clocks in the two frames will agree with each other; so $t = t'$. We summarize these in the following **Galilean transformation equations**:

$$
\begin{aligned}
x &= x' + vt' \\
y &= y' \\
z &= z' \\
t &= t'.
\end{aligned}
\qquad \text{[Galilean]} \quad \textbf{(36–4)}
$$

These equations give the coordinates of an event in the S frame when those in the S' frame are known. If those in the S frame are known, then the S' coordinates are obtained from

$$
x' = x - vt, \qquad y' = y, \qquad z' = z, \qquad t' = t. \quad \text{[Galilean]}
$$

These four equations are the "inverse" transformation and are very easily obtained from Eqs. 36–4. Notice that the effect is merely to exchange primed and unprimed quantities and replace v by $-v$. This makes sense because from the S' frame, S moves to the left (negative x direction) with speed v.

Now suppose the point P in Fig. 36–11 represents a particle that is moving. Let the components of its velocity vector in S' be u'_x, u'_y, u'_z. (We use u to distinguish it from the relative velocity of the two frames, v.) Now $u'_x = dx'/dt'$, $u'_y = dy'/dt'$ and $u'_z = dz'/dt'$. The velocity of P as seen from S will have components u_x, u_y, and u_z. We can show how these are related to the velocity components in S' by differentiating Eqs. 36–4. For u_x we get

$$
u_x = \frac{dx}{dt} = \frac{d(x' + vt')}{dt'} = u'_x + v
$$

since v is assumed constant. For the other components, $u'_y = u_y$ and $u'_z = u_z$, so we have

$$u_x = u'_x + v$$
$$u_y = u'_y \qquad\qquad \text{[Galilean]} \quad \textbf{(36–5)}$$
$$u_z = u'_z.$$

These are known as the **Galilean velocity transformation equations**. We see that the y and z components of velocity are unchanged, but the x components differ by v: $u_x = u'_x + v$. This is just what we have used before (see Chapter 3, Section 3–9) when dealing with relative velocity.

The Galilean transformations, Eqs. 36–4 and 36–5, are valid only when the velocities involved are much less than c. We can see, for example, that the first of Eqs. 36–5 will not work for the speed of light: light traveling in S′ with speed $u'_x = c$ would have speed $c + v$ in S, whereas the theory of relativity insists it must be c in S. Clearly, then, a new set of transformation equations is needed to deal with relativistic velocities.

We derive the required equation, looking again at Fig. 36–11. We will try the simple assumption that the transformation is linear and of the form

$$x = \gamma(x' + vt'), \qquad y = y', \qquad z = z'. \qquad \textbf{(i)}$$

That is, we modify the first of Eqs. 36–4 by multiplying by a constant γ which is yet to be determined[†] ($\gamma = 1$ non-relativistically). But we assume the y and z equations are unchanged since there is no length contraction in these directions. We will not assume a form for t, but will derive it. The inverse equations must have the same form with v replaced by $-v$. (The principle of relativity demands it, since S′ moving to the right with respect to S is equivalent to S moving to the left with respect to S′.) Therefore

$$x' = \gamma(x - vt). \qquad \textbf{(ii)}$$

Now if a light pulse leaves the common origin of S and S′ at time $t = t' = 0$, after a time t it will have traveled a distance $x = ct$ or $x' = ct'$ along the x axis. Therefore, from Eqs. (i) and (ii) above,

$$ct = \gamma(ct' + vt') = \gamma(c + v)t', \qquad \textbf{(iii)}$$

$$ct' = \gamma(ct - vt) = \gamma(c - v)t. \qquad \textbf{(iv)}$$

We substitute t' from Eq. (iv) into Eq. (iii) and find $ct = \gamma(c + v)\gamma(c - v)(t/c) = \gamma^2(c^2 - v^2)t/c$. We cancel out the t on each side and solve for γ to find

$$\gamma = \frac{1}{\sqrt{1 - v^2/c^2}}.$$

The constant γ here has the same value as the γ we used before, Eq. 36–2. Now that we have found γ, we need only find the relation between t and t'. To do so, we combine $x' = \gamma(x - vt)$ with $x = \gamma(x' + vt')$:

$$x' = \gamma(x - vt) = \gamma(\gamma[x' + vt'] - vt).$$

We solve for t and find $t = \gamma(t' + vx'/c^2)$. In summary,

$$x = \gamma(x' + vt')$$
$$y = y'$$
$$z = z' \qquad\qquad \textbf{(36–6)}$$
$$t = \gamma\left(t' + \frac{vx'}{c^2}\right)$$

LORENTZ

TRANSFORMATIONS

These are called the **Lorentz transformation equations**. They were first proposed, in a slightly different form, by Lorentz in 1904 to explain the null result of the Michelson–Morley experiment and to make Maxwell's equations take the same form in all inertial reference frames. A year later Einstein derived them independently based on his theory of relativity. Notice that not only is the x equation modified as compared to the Galilean transformation, but so is the t equation; indeed, we see directly in this last equation how the space and time coordinates mix.

[†]γ here is not assumed to be given by Eq. 36–2.

Deriving Length Contraction

We now derive the length contraction formula, Eq. 36–3, from the Lorentz transformation equations. We consider two reference frames S and S' as in Fig. 36–11.

Let an object of length ℓ_0 be at rest on the x axis in S. The coordinates of its two end points are x_1 and x_2, so that $x_2 - x_1 = \ell_0$. At any instant in S', the end points will be at x_1' and x_2' as given by the Lorentz transformation equations. The length measured in S' is $\ell = x_2' - x_1'$. An observer in S' measures this length by measuring x_2' and x_1' at the same time (in the S' reference frame), so $t_2' = t_1'$. Then, from the first of Eqs. 36–6,

$$\ell_0 = x_2 - x_1 = \frac{1}{\sqrt{1 - v^2/c^2}}(x_2' + vt_2' - x_1' - vt_1').$$

Since $t_2' = t_1'$, we have

$$\ell_0 = \frac{1}{\sqrt{1 - v^2/c^2}}(x_2' - x_1') = \frac{\ell}{\sqrt{1 - v^2/c^2}},$$

or

$$\ell = \ell_0\sqrt{1 - v^2/c^2},$$

which is Eq. 36–3.

Deriving Time Dilation

We now derive the time-dilation formula, Eq. 36–1a, using the Lorentz transformation equations.

The time Δt_0 between two events that occur at the same place $(x_2' = x_1')$ in S' is measured to be $\Delta t_0 = t_2' - t_1'$. Since $x_2' = x_1'$, then from the last of Eqs. 36–6, the time Δt between the events as measured in S is

$$\Delta t = t_2 - t_1 = \frac{1}{\sqrt{1 - v^2/c^2}}\left(t_2' + \frac{vx_2'}{c^2} - t_1' - \frac{vx_1'}{c^2}\right)$$

$$= \frac{1}{\sqrt{1 - v^2/c^2}}(t_2' - t_1') = \frac{\Delta t_0}{\sqrt{1 - v^2/c^2}},$$

which is Eq. 36–1a. Note that we chose S' to be the frame in which the two events occur at the same place, so that $x_1' = x_2'$ and the terms containing x_1' and x_2' cancel out.

Relativistic Addition of Velocities

The relativistically correct velocity equations are readily obtained by differentiating Eqs. 36–6 with respect to time. For example (using $\gamma = 1/\sqrt{1 - v^2/c^2}$ and the chain rule for derivatives):

$$u_x = \frac{dx}{dt} = \frac{d}{dt}\left[\gamma(x' + vt')\right]$$

$$= \frac{d}{dt'}\left[\gamma(x' + vt')\right]\frac{dt'}{dt} = \gamma\left[\frac{dx'}{dt'} + v\right]\frac{dt'}{dt}.$$

But $dx'/dt' = u_x'$ and $dt'/dt = 1/(dt/dt') = 1/\left[\gamma(1 + vu_x'/c^2)\right]$ where we have differentiated the last of Eqs. 36–6 with respect to time. Therefore

$$u_x = \frac{\left[\gamma(u_x' + v)\right]}{\left[\gamma(1 + vu_x'/c^2)\right]} = \frac{u_x' + v}{1 + vu_x'/c^2}.$$

The others are obtained in the same way and we collect them here:

LORENTZ

VELOCITY

TRANSFORMATIONS

$$u_x = \frac{u_x' + v}{1 + vu_x'/c^2} \tag{36–7a}$$

$$u_y = \frac{u_y'\sqrt{1 - v^2/c^2}}{1 + vu_x'/c^2} \tag{36–7b}$$

$$u_z = \frac{u_z'\sqrt{1 - v^2/c^2}}{1 + vu_x'/c^2}. \tag{36–7c}$$

Note that even though the relative velocity $\vec{v}$ is in the x direction, if the object has y or z components of velocity, they too are affected by v and the x component of the object's velocity. This was not true for the Galilean transformation, Eqs. 36–5.

EXAMPLE 36–8 **Adding velocities.** Calculate the speed of rocket 2 in Fig. 36–12 with respect to Earth.

APPROACH Consider Earth as reference frame S, and rocket 1 as reference frame S′. Rocket 2 moves with speed $u' = 0.60c$ with respect to rocket 1. Rocket 1 has speed $v = 0.60c$ with respect to Earth. The velocities are along the same straight line which we take to be the x (and x') axis. We need use only the first of Eqs. 36–7.

SOLUTION The speed of rocket 2 with respect to Earth is

$$u = \frac{u' + v}{1 + \frac{vu'}{c^2}} = \frac{0.60c + 0.60c}{1 + \frac{(0.60c)(0.60c)}{c^2}} = \frac{1.20c}{1.36} = 0.88c.$$

NOTE The Galilean transformation would have given $u = 1.20c$.

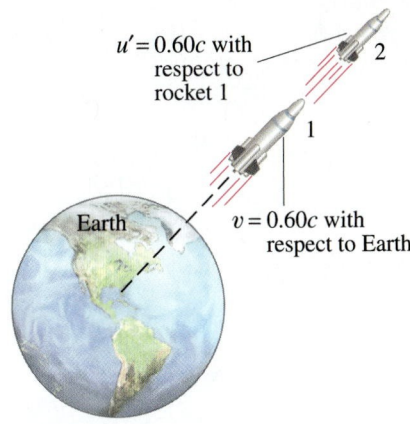

FIGURE 36–12 Rocket 1 moves away at speed $v = 0.60c$. Rocket 2 is fired from rocket 1 with speed $u' = 0.60c$. What is the speed of rocket 2 with respect to the Earth?

EXERCISE E Use Eqs. 36–7 to calculate the speed of rocket 2 in Fig. 36–12 relative to Earth if it was shot from rocket 1 at a speed $u' = 3000 \, \text{km/s} = 0.010c$. Assume rocket 1 had a speed $v = 6000 \, \text{km/s} = 0.020c$.

EXERCISE F Return to the Chapter-Opening Question, page 951, and answer it again now. Try to explain why you may have answered differently the first time.

Notice that Eqs. 36–7 reduce to the classical (Galilean) forms for velocities small compared to the speed of light, since $1 + vu'/c^2 \approx 1$ for v and $u' \ll c$. At the other extreme, let rocket 1 in Fig. 36–12 send out a beam of light, so that $u' = c$. Then Eq. 36–7a tells us the speed of light relative to Earth is

$$u = \frac{0.60c + c}{1 + \frac{(0.60c)(c)}{c^2}} = c,$$

which is consistent with the second postulate of relativity.

36–9 Relativistic Momentum

So far in this Chapter, we have seen that two basic mechanical quantities, length and time intervals, need modification because they are relative—their value depends on the reference frame from which they are measured. We might expect that other physical quantities might need some modification according to the theory of relativity, such as momentum, energy, and mass.

The analysis of collisions between two particles shows that if we want to preserve the law of conservation of momentum in relativity, we must redefine momentum as

$$p = \frac{mv}{\sqrt{1 - v^2/c^2}} = \gamma mv. \tag{36–8}$$

Here γ is shorthand for $1/\sqrt{1 - v^2/c^2}$ as before (Eq. 36–2). For speeds much less than the speed of light, Eq. 36–8 gives the classical momentum, $p = mv$.

Relativistic momentum has been tested many times on tiny elementary particles (such as muons), and it has been found to behave in accord with Eq. 36–8. We derive Eq. 36–8 in the optional subsection on the next page.

EXAMPLE 36–9 Momentum of moving electron. Compare the momentum of an electron when it has a speed of (a) 4.00×10^7 m/s in the CRT of a television set, and (b) $0.98c$ in an accelerator used for cancer therapy.

APPROACH We use Eq. 36–8 for the momentum of a moving electron.

SOLUTION (a) At $v = 4.00 \times 10^7$ m/s, the electron's momentum is

$$p = \frac{mv}{\sqrt{1 - \dfrac{v^2}{c^2}}} = \frac{mv}{\sqrt{1 - \dfrac{(4.00 \times 10^7 \text{ m/s})^2}{(3.00 \times 10^8 \text{ m/s})^2}}} = 1.01mv.$$

The factor $\gamma = 1/\sqrt{1 - v^2/c^2} \approx 1.01$, so the momentum is only about 1% greater than the classical value. (If we put in the mass of an electron, $m = 9.11 \times 10^{-31}$ kg, the momentum is $p = 1.01mv = 3.68 \times 10^{-23}$ kg·m/s.)
(b) With $v = 0.98c$, the momentum is

$$p = \frac{mv}{\sqrt{1 - \dfrac{v^2}{c^2}}} = \frac{mv}{\sqrt{1 - \dfrac{(0.98c)^2}{c^2}}} = \frac{mv}{\sqrt{1 - (0.98)^2}} = 5.0mv.$$

An electron traveling at 98% the speed of light has $\gamma = 5.0$ and a momentum 5.0 times its classical value.

Newton's second law, stated in its most general form, is

$$\vec{\mathbf{F}} = \frac{d\vec{\mathbf{p}}}{dt} = \frac{d}{dt}(\gamma m\vec{\mathbf{v}}) = \frac{d}{dt}\left(\frac{m\vec{\mathbf{v}}}{\sqrt{1 - v^2/c^2}}\right) \tag{36–9}$$

and is valid relativistically.

*Derivation of Relativistic Momentum

Classically, momentum is a conserved quantity. We hope to find a formula for momentum that will also be valid relativistically. To do so, let us assume it has the general form given by $p = fmv$ where f is some function of v: $f(v)$. We consider a hypothetical collision between two objects—a thought experiment—and see what form $f(v)$ must take if momentum is to be conserved.

Our thought experiment involves the elastic collision of two identical balls, A and B. We consider two inertial reference frames, S and S′, moving along the x axis with a speed v with respect to each other, Fig. 36–13. That is, frame S′ moves to the right with velocity $\vec{\mathbf{v}}$ as seen by observers on frame S; and frame S moves to the left with $-\vec{\mathbf{v}}$ as seen by observers on S′. In reference frame S, ball A is thrown with speed u in the $+y$ direction. In reference frame S′, ball B is thrown with speed u in the negative $y′$ direction. The two balls are thrown at just the right time so that they collide. We assume that they rebound elastically and, from *symmetry*, that each moves with the same speed u back in the opposite direction in its thrower's reference frame. Figure 36–13a shows the collision as seen by an observer in reference frame S; and Fig. 36–13b shows the collision as seen from reference frame S′. In reference frame S, ball A has $v_x = 0$ both before and after the collision; it has $v_y = +u$ before the collision and $-u$ after the collision. In frame S′, ball A has x component of velocity $u'_x = -v$ both before and after the collision, and a $y′$ component (Eq. 36–7b with $u'_x = -v$) of magnitude

$$u'_y = u \sqrt{1 - v^2/c^2}.$$

The same holds true for ball B, except in reverse. The velocity components are indicated in Fig. 36–13.

We now apply the law of conservation of momentum, which we hope remains valid in relativity, even if momentum has to be redefined. That is, we assume that the total momentum before the collision is equal to the total momentum after the

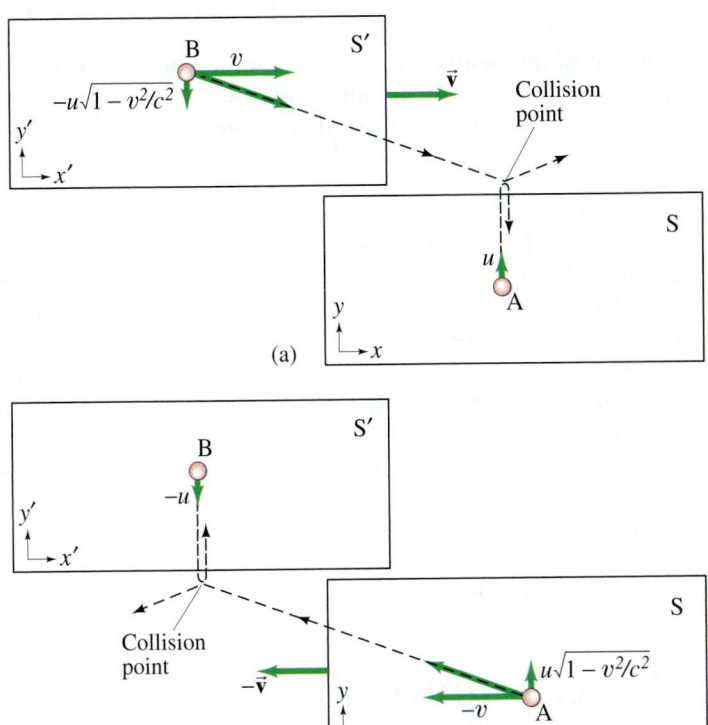

(a)

(b)

FIGURE 36–13 Deriving the momentum formula. Collision as seen by observers (a) in reference frame S, (b) in reference frame S′.

collision. We apply conservation of momentum to the y component of momentum in reference frame S (Fig. 36–13a). To make our task easier, let us assume $u \ll v$ so that the speed of ball B as seen in reference frame S is essentially v. Then B's y component of momentum in S before collision is $-f(v)mu\sqrt{1 - v^2/c^2}$ and after the collision is $+f(v)mu\sqrt{1 - v^2/c^2}$. Ball A in S has y component $f(u)mu$ before and $-f(u)mu$ after the collision. (We use $f(u)$ for A because its speed in S is only u.) Conservation of momentum in S for the y component is

$$(p_A + p_B)_{\text{before}} = (p_A + p_B)_{\text{after}}$$

$$f(u)mu - f(v)mu\sqrt{1 - v^2/c^2} = -f(u)mu + f(v)mu\sqrt{1 - v^2/c^2}.$$

We solve this for $f(v)$ and obtain

$$f(v) = \frac{f(u)}{\sqrt{1 - v^2/c^2}}.$$

To simplify this relation so we can solve for f, let us allow u to become very small so that it approaches zero (this corresponds to a glancing collision with one of the balls essentially at rest and the other moving with speed v). Then the momentum terms $f(u)mu$ are in the nonrelativistic realm and take on the classical form, simply mu, meaning that $f(u) = 1$. So the previous equation becomes

$$f(v) = \frac{1}{\sqrt{1 - v^2/c^2}}.$$

We see that $f(v)$ comes out to be the factor we used before and called γ, and here has been shown to be valid for ball A. Using Fig. 36–13b we can derive the same relation for ball B. Thus we can conclude that we need to define the relativistic momentum of a particle moving with velocity $\vec{v}$ as

$$\vec{p} = \frac{m\vec{v}}{\sqrt{1 - v^2/c^2}} = \gamma m\vec{v}.$$

With this definition the law of conservation of momentum will remain valid even in the relativistic realm. This relativistic momentum formula (Eq. 36–8) has been tested countless times on tiny elementary particles and been found valid.

*Relativistic Mass

The relativistic definition of momentum, Eq. 36–8, is sometimes interpreted as an increase in the mass of an object. In this interpretation, a particle can have a **relativistic mass**, m_{rel}, which increases with speed according to

$$m_{\text{rel}} = \frac{m}{\sqrt{1 - v^2/c^2}}.$$

In this "mass-increase" formula, m is referred to as the **rest mass** of the object. With this interpretation, *the mass of an object appears to increase as its speed increases.* But we must be careful in the use of relativistic mass. We cannot just plug it into formulas like $F = ma$ or $K = \frac{1}{2}mv^2$. For example, if we substitute it into $F = ma$, we obtain a formula that does not agree with experiment. If however, we write Newton's second law in its more general form, $\vec{F} = d\vec{p}/dt$, we do get a correct result (Eq. 36–9).

Also, be careful *not* to think a mass acquires more particles or more molecules as its speed becomes very large. It doesn't. In fact, many physicists believe an object has only one mass (its rest mass), and that it is only the momentum that increases with speed.

Whenever we talk about the mass of an object, we will always mean its rest mass (a fixed value).

36–10 The Ultimate Speed

A basic result of the special theory of relativity is that the speed of an object cannot equal or exceed the speed of light. That the speed of light is a natural speed limit in the universe can be seen from any of Eqs. 36–1, 36–2, 36–8, or the addition of velocities formula. It is perhaps easiest to see from Eq. 36–8. As an object is accelerated to greater and greater speeds, its momentum becomes larger and larger. Indeed, if v were to equal c, the denominator in this equation would be zero, and the momentum would be infinite. To accelerate an object up to $v = c$ would thus require infinite energy, and so is not possible.

36–11 $E = mc^2$; Mass and Energy

If momentum needs to be modified to fit with relativity as we just saw in Eq. 36–8, then we might expect energy too would need to be rethought. Indeed, Einstein not only developed a new formula for kinetic energy, but also found a new relation between mass and energy, and the startling idea that mass is a form of energy.

We start with the work-energy principle (Chapter 7), hoping it is still valid in relativity and will give verifiable results. That is, we assume the net work done on a particle is equal to its change in kinetic energy (K). Using this principle, Einstein showed that at high speeds the formula $K = \frac{1}{2}mv^2$ is not correct. Instead, as we show in the optional Subsection on page 978, the kinetic energy of a particle of mass m traveling at speed v is given by

$$K = \frac{mc^2}{\sqrt{1 - v^2/c^2}} - mc^2. \qquad \textbf{(36–10a)}$$

In terms of $\gamma = 1/\sqrt{1 - v^2/c^2}$ we can rewrite Eq. 36–10a as

$$K = \gamma mc^2 - mc^2 = (\gamma - 1)mc^2. \qquad \textbf{(36–10b)}$$

Equations 36–10 require some interpretation. The first term increases with the speed v of the particle. The second term, mc^2, is constant; it is called the **rest energy** of the particle, and represents a form of energy that a particle has even when at rest. Note that if a particle is at rest $(v = 0)$ the first term in Eq. 36–10a becomes mc^2, so $K = 0$ as it should.

We can rearrange Eq. 36–10b to get

$$\gamma mc^2 = mc^2 + K.$$

We call γmc^2 the *total energy* E of the particle (assuming no potential energy), because it equals the rest energy plus the kinetic energy:

$$E = K + mc^2. \tag{36–11a}$$

The total energy can also be written, using Eqs. 36–10, as

$$E = \gamma mc^2 = \frac{mc^2}{\sqrt{1 - v^2/c^2}}. \tag{36–11b}$$

For a particle at rest in a given reference frame, K is zero in Eq. 36–11a, so the total energy is its rest energy:

$$E = mc^2. \tag{36–12}$$

MASS RELATED TO ENERGY

Here we have Einstein's famous formula, $E = mc^2$. This formula mathematically relates the concepts of energy and mass. But if this idea is to have any physical meaning, then mass ought to be convertible to other forms of energy and vice versa. Einstein suggested that this might be possible, and indeed changes of mass to other forms of energy, and vice versa, have been experimentally confirmed countless times in nuclear and elementary particle physics. For example, an electron and a positron (= a positive electron, Section 37–5) have often been observed to collide and disappear, producing pure electromagnetic radiation. The amount of electromagnetic energy produced is found to be exactly equal to that predicted by Einstein's formula, $E = mc^2$. The reverse process is also commonly observed in the laboratory: electromagnetic radiation under certain conditions can be converted into material particles such as electrons (see Section 37–5 on pair production). On a larger scale, the energy produced in nuclear power plants is a result of the loss in mass of the uranium fuel as it undergoes the process called fission. Even the radiant energy we receive from the Sun is an example of $E = mc^2$; the Sun's mass is continually decreasing as it radiates electromagnetic energy outward.

The relation $E = mc^2$ is now believed to apply to all processes, although the changes are often too small to measure. That is, when the energy of a system changes by an amount ΔE, the mass of the system changes by an amount Δm given by

$$\Delta E = (\Delta m)(c^2).$$

In a nuclear reaction where an energy E is required or released, the masses of the reactants and the products will be different by $\Delta m = \Delta E/c^2$.

EXAMPLE 36–10 **Pion's kinetic energy.** A π^0 meson $(m = 2.4 \times 10^{-28}\,\text{kg})$ travels at a speed $v = 0.80c = 2.4 \times 10^8\,\text{m/s}$. What is its kinetic energy? Compare to a classical calculation.

APPROACH We use Eq. 36–10 and compare to $\frac{1}{2}mv^2$.

SOLUTION We substitute values into Eq. 36–10a or b

$$K = (\gamma - 1)mc^2$$

where

$$\gamma = \frac{1}{\sqrt{1 - v^2/c^2}} = \frac{1}{\sqrt{1 - (0.80)^2}} = 1.67.$$

Then

$$K = (1.67 - 1)(2.4 \times 10^{-28}\,\text{kg})(3.0 \times 10^8\,\text{m/s})^2$$

$$= 1.4 \times 10^{-11}\,\text{J}.$$

PROBLEM SOLVING
Relativistic kinetic energy

Notice that the units of mc^2 are $\text{kg}\cdot\text{m}^2/\text{s}^2$, which is the joule.

NOTE If we were to do a classical calculation we would obtain $K = \frac{1}{2}mv^2 = \frac{1}{2}(2.4 \times 10^{-28}\,\text{kg})(2.4 \times 10^8\,\text{m/s})^2 = 6.9 \times 10^{-12}\,\text{J}$, about half as much, but this is not a correct result. Note that $\frac{1}{2}\gamma mv^2$ also does not work.

EXERCISE G A proton is traveling in an accelerator with a speed of $1.0 \times 10^8\,\text{m/s}$. By what factor does the proton's kinetic energy increase if its speed is doubled? (a) 1.3, (b) 2.0, (c) 4.0, (d) 5.6.

EXAMPLE 36–11 **Energy from nuclear decay.** The energy required or released in nuclear reactions and decays comes from a change in mass between the initial and final particles. In one type of radioactive decay, an atom of uranium ($m = 232.03714\,u$) decays to an atom of thorium ($m = 228.02873\,u$) plus an atom of helium ($m = 4.00260\,u$) where the masses given are in atomic mass units ($1\,u = 1.6605 \times 10^{-27}\,kg$). Calculate the energy released in this decay.

APPROACH The initial mass minus the total final mass gives the mass loss in atomic mass units (u); we convert that to kg, and multiply by c^2 to find the energy released, $\Delta E = \Delta m c^2$.

SOLUTION The initial mass is 232.03714 u, and after the decay the mass is 228.02873 u + 4.00260 u = 232.03133 u, so there is a decrease in mass of 0.00581 u. This mass, which equals $(0.00581\,u)(1.66 \times 10^{-27}\,kg) = 9.64 \times 10^{-30}\,kg$, is changed into energy. By $\Delta E = \Delta m c^2$, we have

$$\Delta E = (9.64 \times 10^{-30}\,kg)(3.0 \times 10^8\,m/s)^2 = 8.68 \times 10^{-13}\,J.$$

Since $1\,MeV = 1.60 \times 10^{-13}\,J$ (Section 23–8), the energy released is 5.4 MeV.

In the tiny world of atoms and nuclei, it is common to quote energies in eV (electron volts) or multiples such as MeV ($10^6\,eV$). Momentum (see Eq. 36–8) can be quoted in units of eV/c (or MeV/c). And mass can be quoted (from $E = mc^2$) in units of eV/c^2 (or MeV/c^2). Note the use of c to keep the units correct. The rest masses of the electron and the proton are readily shown to be $0.511\,MeV/c^2$ and $938\,MeV/c^2$, respectively. See also the Table inside the front cover.

EXAMPLE 36–12 **A 1-TeV proton.** The Tevatron accelerator at Fermilab in Illinois can accelerate protons to a kinetic energy of 1.0 TeV ($10^{12}\,eV$). What is the speed of such a proton?

APPROACH We solve the kinetic energy formula, Eq. 36–10a, for v.

SOLUTION The rest energy of a proton is $mc^2 = 938\,MeV$ or $9.38 \times 10^8\,eV$. Compared to the kinetic energy of $10^{12}\,eV$, the rest energy can be neglected, so we simplify Eq. 36–10a to

$$K \approx \frac{mc^2}{\sqrt{1 - v^2/c^2}}.$$

We solve this for v in the following steps:

$$\sqrt{1 - \frac{v^2}{c^2}} = \frac{mc^2}{K};$$

$$1 - \frac{v^2}{c^2} = \left(\frac{mc^2}{K}\right)^2;$$

$$\frac{v^2}{c^2} = 1 - \left(\frac{mc^2}{K}\right)^2 = 1 - \left(\frac{9.38 \times 10^8\,eV}{1.0 \times 10^{12}\,eV}\right)^2;$$

$$v = \sqrt{1 - (9.38 \times 10^{-4})^2}\,c = 0.99999956\,c.$$

So the proton is traveling at a speed very nearly equal to c.

At low speeds, $v \ll c$, the relativistic formula for kinetic energy reduces to the classical one, as we now show by using the binomial expansion, $(1 \pm x)^n = 1 \pm nx + n(n-1)x^2/2! + \cdots$. With $n = -\frac{1}{2}$, we expand the square root in Eq. 36–10a

$$K = mc^2\left(\frac{1}{\sqrt{1 - v^2/c^2}} - 1\right)$$

so that

$$K \approx mc^2\left(1 + \frac{1}{2}\frac{v^2}{c^2} + \cdots - 1\right)$$

$$\approx \tfrac{1}{2}mv^2.$$

The dots in the first expression represent very small terms in the expansion which we neglect since we assumed that $v \ll c$. Thus at low speeds, the relativistic form for kinetic energy reduces to the classical form, $K = \frac{1}{2}mv^2$. This makes relativity a viable theory in that it can predict accurate results at low speed as well as at high. Indeed, the other equations of special relativity also reduce to their classical equivalents at ordinary speeds: length contraction, time dilation, and modifications to momentum as well as kinetic energy, all disappear for $v \ll c$ since $\sqrt{1 - v^2/c^2} \approx 1$.

A useful relation between the total energy E of a particle and its momentum p can also be derived. The momentum of a particle of mass m and speed v is given by Eq. 36–8

$$p = \gamma mv = \frac{mv}{\sqrt{1 - v^2/c^2}}.$$

The total energy is

$$E = K + mc^2$$

or

$$E = \gamma mc^2 = \frac{mc^2}{\sqrt{1 - v^2/c^2}}.$$

We square this equation (and we insert "$v^2 - v^2$" which is zero, but will help us):

$$E^2 = \frac{m^2c^2(v^2 - v^2 + c^2)}{1 - v^2/c^2}$$

$$= p^2c^2 + \frac{m^2c^4(1 - v^2/c^2)}{1 - v^2/c^2}$$

or

$$E^2 = p^2c^2 + m^2c^4. \tag{36–13}$$

Thus, the total energy can be written in terms of the momentum p, or in terms of the kinetic energy (Eqs. 36–11), where we have assumed there is no potential energy.

We can rewrite Eq. 36–13 as $E^2 - p^2c^2 = m^2c^4$. Since the mass m of a given particle is the same in any reference frame, we see that the quantity $E^2 - p^2c^2$ must also be the same in any reference frame. Thus, at any given moment the total energy E and momentum p of a particle will be different in different reference frames, but the quantity $E^2 - p^2c^2$ will have the same value in all inertial reference frames. We say that the quantity $E^2 - p^2c^2$ is **invariant** under a Lorentz transformation.

*When Do We Use Relativistic Formulas?

From a practical point of view, we do not have much opportunity in our daily lives to use the mathematics of relativity. For example, the γ factor, $\gamma = 1/\sqrt{1 - v^2/c^2}$, which appears in many relativistic formulas, has a value of 1.005 when $v = 0.10c$. Thus, for speeds even as high as $0.10c = 3.0 \times 10^7$ m/s, the factor $\sqrt{1 - v^2/c^2}$ in relativistic formulas gives a numerical correction of less than 1%. For speeds less than $0.10c$, or unless mass and energy are interchanged, we don't usually need to use the more complicated relativistic formulas, and can use the simpler classical formulas.

If you are given a particle's mass m and its kinetic energy K, you can do a quick calculation to determine if you need to use relativistic formulas or if classical ones are good enough. You simply compute the ratio K/mc^2 because (Eq. 36–10b)

$$\frac{K}{mc^2} = \gamma - 1 = \frac{1}{\sqrt{1 - v^2/c^2}} - 1.$$

If this ratio comes out to be less than, say, 0.01, then $\gamma \leq 1.01$ and relativistic equations will correct the classical ones by about 1%. If your expected precision is no better than 1%, classical formulas are good enough. But if your precision is 1 part in 1000 (0.1%) then you would want to use relativistic formulas. If your expected precision is only 10%, you need relativity if $(K/mc^2) \gtrsim 0.1$.

EXERCISE H For 1% accuracy, does an electron with $K = 100$ eV need to be treated relativistically? [*Hint*: The mass of an electron is 0.511 MeV.]

*Deriving Relativistic Energy

To find the mathematical relationship between mass and energy, we assume that the work-energy theorem is still valid in relativity for a particle, and we take the motion of the particle to be along the x axis. The work done to increase a particle's speed from zero to v is

$$W = \int_i^f F\, dx = \int_i^f \frac{dp}{dt}\, dx = \int_i^f \frac{dp}{dt}\, v\, dt = \int_i^f v\, dp$$

where i and f refer to the initial $(v = 0)$ and final $(v = v)$ states. Since $d(pv) = p\, dv + v\, dp$ we can write

$$v\, dp = d(pv) - p\, dv$$

so

$$W = \int_i^f d(pv) - \int_i^f p\, dv.$$

The first term on the right of the equal sign is

$$\int_i^f d(pv) = pv \Big|_i^f = (\gamma mv)v = \frac{mv^2}{\sqrt{1 - v^2/c^2}}.$$

The second term in our equation for W above is easily integrated since

$$\frac{d}{dv}\left(\sqrt{1 - v^2/c^2}\right) = -(v/c^2)/\sqrt{1 - v^2/c^2},$$

and so becomes

$$-\int_i^f p\, dv = -\int_0^v \frac{mv}{\sqrt{1 - v^2/c^2}}\, dv = mc^2\sqrt{1 - v^2/c^2}\,\Big|_0^v$$

$$= mc^2\sqrt{1 - v^2/c^2} - mc^2.$$

Finally, we have for W:

$$W = \frac{mv^2}{\sqrt{1 - v^2/c^2}} + mc^2\sqrt{1 - v^2/c^2} - mc^2.$$

We multiply the second term on the right by $\sqrt{1 - v^2/c^2}/\sqrt{1 - v^2/c^2} = 1$, and obtain

$$W = \frac{mc^2}{\sqrt{1 - v^2/c^2}} - mc^2.$$

By the work-energy theorem, the work done on the particle must equal its final kinetic energy K since the particle started from rest. Therefore

$$K = \frac{mc^2}{\sqrt{1 - v^2/c^2}} - mc^2$$

$$= \gamma mc^2 - mc^2 = (\gamma - 1)mc^2,$$

which are Eqs. 36–10.

*36–12 Doppler Shift for Light

In Section 16–7 we discussed how the frequency and wavelength of sound are altered if the source of the sound and the observer are moving toward or away from each other. When a source is moving toward us, the frequency is higher than when the source is at rest. If the source moves away from us, the frequency is lower. We obtained four different equations for the Doppler shift (Eqs. 16–9a and b, Eqs. 16–10a and b), depending on the direction of the relative motion and whether the source or the observer is moving. The Doppler effect occurs also for light; but the shifted frequency or wavelength is given by slightly different equations, and there are only two of them, because for light—according to special

relativity—we can make no distinction between motion of the source and motion of the observer. (Recall that sound travels in a medium such as air, whereas light does not—there is no evidence for an ether.)

To derive the Doppler shift for light, let us consider a light source and an observer that move toward each other, and let their relative velocity be v as measured in the reference frame of either the source or the observer. Figure 36–14a shows a source at rest emitting light waves of frequency f_0 and wavelength $\lambda_0 = c/f_0$. Two wavecrests are shown, a distance λ_0 apart, the second crest just having been emitted. In Fig. 36–14b, the source is shown moving at speed v toward a stationary observer who will see the wavelength λ being somewhat less than λ_0. (This is much like Fig. 16–19 for sound.) Let Δt represent the time between crests as detected by the observer, whose reference frame is shown in Fig. 36–14b. From Fig. 36–14b we see that

$$\lambda = c\,\Delta t - v\,\Delta t,$$

where $c\,\Delta t$ is the distance crest 1 has moved in the time Δt after it was emitted, and $v\,\Delta t$ is the distance the source has moved in time Δt. So far our derivation has not differed from that for sound (Section 16–7). Now we invoke the theory of relativity. The time between emission of wavecrests has undergone time dilation:

$$\Delta t = \Delta t_0 / \sqrt{1 - v^2/c^2}$$

where Δt_0 is the time between emissions of wavecrests in the reference frame where the source is at rest (the "proper" time). In the source's reference frame (Fig. 36–14a), we have

$$\Delta t_0 = \frac{1}{f_0} = \frac{\lambda_0}{c}$$

(Eqs. 5–2 and 31–14). Thus

$$\lambda = (c - v)\,\Delta t = (c - v)\frac{\Delta t_0}{\sqrt{1 - v^2/c^2}} = \frac{(c - v)}{\sqrt{c^2 - v^2}}\lambda_0$$

or

$$\lambda = \lambda_0 \sqrt{\frac{c - v}{c + v}}. \qquad \begin{bmatrix} \text{source and observer} \\ \text{moving toward} \\ \text{each other} \end{bmatrix} \quad \textbf{(36–14a)}$$

The frequency f is (recall $\lambda_0 = c/f_0$)

$$f = \frac{c}{\lambda} = f_0 \sqrt{\frac{c + v}{c - v}}. \qquad \begin{bmatrix} \text{source and observer} \\ \text{moving toward} \\ \text{each other} \end{bmatrix} \quad \textbf{(36–14b)}$$

Here f_0 is the frequency of the light as seen in the source's reference frame, and f is the frequency as measured by an observer moving toward the source or toward whom the source is moving. Equations 36–14 depend only on the relative velocity v. For relative motion *away* from each other we set $v < 0$ in Eqs. 36–14, and obtain

$$\lambda = \lambda_0 \sqrt{\frac{c + v}{c - v}} \qquad \textbf{(36–15a)}$$

$$\begin{bmatrix} \text{source and observer} \\ \text{moving away from} \\ \text{each other} \end{bmatrix}$$

$$f = f_0 \sqrt{\frac{c - v}{c + v}}. \qquad \textbf{(36–15b)}$$

From Eqs. 36–14 and 36–15 we see that light from a source moving toward us will have a higher frequency and shorter wavelength, whereas if a light source moves away from us, we will see a lower frequency and a longer wavelength. In the latter case, visible light will have its wavelength lengthened toward the red end of the visible spectrum (Fig. 32–26), an effect called a **redshift**. As we will see in the next Chapter, all atoms have their own distinctive signature in terms of the frequencies of the light they emit. In 1929 the American astronomer Edwin Hubble (1889–1953) found that radiation from atoms in many galaxies is redshifted. That is, the frequencies of light emitted are lower than those emitted by stationary atoms on Earth, suggesting that the galaxies are receding from us. This is the origin of the idea that the universe is expanding.

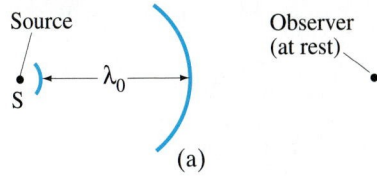

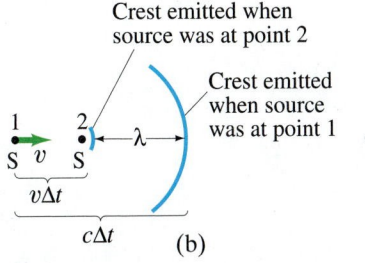

FIGURE 36–14 Doppler shift for light. (a) Source and observer at rest. (b) Source moving toward stationary observer.

EXAMPLE 36–13 **Speeding through a red light.** A driver claims that he did not go through a red light because the light was Doppler shifted and appeared green. Calculate the speed of a driver in order for a red light to appear green.

APPROACH We apply the Doppler shift equation for red light ($\lambda_0 \approx 650$ nm) and green light ($\lambda \approx 500$ nm).

SOLUTION Equation 36–14a holds for the source and the object moving toward each other:

$$\lambda = \lambda_0 \sqrt{\frac{c - v}{c + v}}.$$

We square this equation:

$$\frac{c - v}{c + v} = \left(\frac{\lambda}{\lambda_0}\right)^2$$

where $(\lambda/\lambda_0)^2 = (500 \text{ nm}/650 \text{ nm})^2 = 0.59$. We solve for v:

$$v = c\left[\frac{1 - (\lambda/\lambda_0)^2}{1 + (\lambda/\lambda_0)^2}\right] = 0.26c.$$

With this defense, the driver might not be guilty of running a red light, but he would clearly be guilty of speeding.

36–13 The Impact of Special Relativity

A great many experiments have been performed to test the predictions of the special theory of relativity. Within experimental error, no contradictions have been found. Scientists have therefore accepted relativity as an accurate description of nature.

At speeds much less than the speed of light, the relativistic formulas reduce to the old classical ones, as we have discussed. We would, of course, hope—or rather, insist—that this be true since Newtonian mechanics works so well for objects moving with speeds $v \ll c$. This insistence that a more general theory (such as relativity) give the same results as a more restricted theory (such as classical mechanics which works for $v \ll c$) is called the **correspondence principle**. The two theories must correspond where their realms of validity overlap. Relativity thus does not contradict classical mechanics. Rather, it is a more general theory, of which classical mechanics is now considered to be a limiting case.

The importance of relativity is not simply that it gives more accurate results, especially at very high speeds. Much more than that, it has changed the way we view the world. The concepts of space and time are now seen to be relative, and intertwined with one another, whereas before they were considered absolute and separate. Even our concepts of matter and energy have changed: either can be converted to the other. The impact of relativity extends far beyond physics. It has influenced the other sciences, and even the world of art and literature; it has, indeed, entered the general culture.

From a practical point of view, we do not have much opportunity in our daily lives to use the mathematics of relativity. For example, the γ factor $1/\sqrt{1 - v^2/c^2}$, which appears in relativistic formulas, has a value of only 1.005 even for a speed as high as $0.10c = 3.0 \times 10^7$ m/s, giving a correction of less than 1%. For speeds less than $0.10c$, or unless mass and energy are interchanged, we don't usually need to use the more complicated relativistic formulas, and can use the simpler classical formulas.

relativity—we can make no distinction between motion of the source and motion of the observer. (Recall that sound travels in a medium such as air, whereas light does not—there is no evidence for an ether.)

To derive the Doppler shift for light, let us consider a light source and an observer that move toward each other, and let their relative velocity be v as measured in the reference frame of either the source or the observer. Figure 36–14a shows a source at rest emitting light waves of frequency f_0 and wavelength $\lambda_0 = c/f_0$. Two wavecrests are shown, a distance λ_0 apart, the second crest just having been emitted. In Fig. 36–14b, the source is shown moving at speed v toward a stationary observer who will see the wavelength λ being somewhat less than λ_0. (This is much like Fig. 16–19 for sound.) Let Δt represent the time between crests as detected by the observer, whose reference frame is shown in Fig. 36–14b. From Fig. 36–14b we see that

$$\lambda = c\,\Delta t - v\,\Delta t,$$

where $c\,\Delta t$ is the distance crest 1 has moved in the time Δt after it was emitted, and $v\,\Delta t$ is the distance the source has moved in time Δt. So far our derivation has not differed from that for sound (Section 16–7). Now we invoke the theory of relativity. The time between emission of wavecrests has undergone time dilation:

$$\Delta t = \Delta t_0/\sqrt{1 - v^2/c^2}$$

where Δt_0 is the time between emissions of wavecrests in the reference frame where the source is at rest (the "proper" time). In the source's reference frame (Fig. 36–14a), we have

$$\Delta t_0 = \frac{1}{f_0} = \frac{\lambda_0}{c}$$

(Eqs. 5–2 and 31–14). Thus

$$\lambda = (c - v)\,\Delta t = (c - v)\frac{\Delta t_0}{\sqrt{1 - v^2/c^2}} = \frac{(c - v)}{\sqrt{c^2 - v^2}}\lambda_0$$

or

$$\lambda = \lambda_0\sqrt{\frac{c - v}{c + v}}. \qquad \begin{bmatrix} \text{source and observer} \\ \text{moving toward} \\ \text{each other} \end{bmatrix} \quad \textbf{(36–14a)}$$

The frequency f is (recall $\lambda_0 = c/f_0$)

$$f = \frac{c}{\lambda} = f_0\sqrt{\frac{c + v}{c - v}}. \qquad \begin{bmatrix} \text{source and observer} \\ \text{moving toward} \\ \text{each other} \end{bmatrix} \quad \textbf{(36–14b)}$$

Here f_0 is the frequency of the light as seen in the source's reference frame, and f is the frequency as measured by an observer moving toward the source or toward whom the source is moving. Equations 36–14 depend only on the relative velocity v. For relative motion *away* from each other we set $v < 0$ in Eqs. 36–14, and obtain

$$\lambda = \lambda_0\sqrt{\frac{c + v}{c - v}} \qquad \textbf{(36–15a)}$$

$$\begin{bmatrix} \text{source and observer} \\ \text{moving away from} \\ \text{each other} \end{bmatrix}$$

$$f = f_0\sqrt{\frac{c - v}{c + v}}. \qquad \textbf{(36–15b)}$$

From Eqs. 36–14 and 36–15 we see that light from a source moving toward us will have a higher frequency and shorter wavelength, whereas if a light source moves away from us, we will see a lower frequency and a longer wavelength. In the latter case, visible light will have its wavelength lengthened toward the red end of the visible spectrum (Fig. 32–26), an effect called a **redshift**. As we will see in the next Chapter, all atoms have their own distinctive signature in terms of the frequencies of the light they emit. In 1929 the American astronomer Edwin Hubble (1889–1953) found that radiation from atoms in many galaxies is redshifted. That is, the frequencies of light emitted are lower than those emitted by stationary atoms on Earth, suggesting that the galaxies are receding from us. This is the origin of the idea that the universe is expanding.

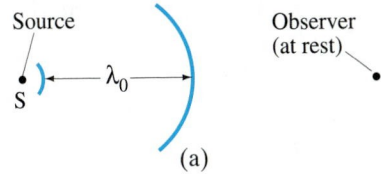

FIGURE 36–14 Doppler shift for light. (a) Source and observer at rest. (b) Source moving toward stationary observer.

EXAMPLE 36–13 **Speeding through a red light.** A driver claims that he did not go through a red light because the light was Doppler shifted and appeared green. Calculate the speed of a driver in order for a red light to appear green.

APPROACH We apply the Doppler shift equation for red light ($\lambda_0 \approx 650$ nm) and green light ($\lambda \approx 500$ nm).

SOLUTION Equation 36–14a holds for the source and the object moving toward each other:

$$\lambda = \lambda_0 \sqrt{\frac{c - v}{c + v}}.$$

We square this equation:

$$\frac{c - v}{c + v} = \left(\frac{\lambda}{\lambda_0}\right)^2$$

where $(\lambda/\lambda_0)^2 = (500\,\text{nm}/650\,\text{nm})^2 = 0.59$. We solve for v:

$$v = c\left[\frac{1 - (\lambda/\lambda_0)^2}{1 + (\lambda/\lambda_0)^2}\right] = 0.26c.$$

With this defense, the driver might not be guilty of running a red light, but he would clearly be guilty of speeding.

36–13 The Impact of Special Relativity

A great many experiments have been performed to test the predictions of the special theory of relativity. Within experimental error, no contradictions have been found. Scientists have therefore accepted relativity as an accurate description of nature.

At speeds much less than the speed of light, the relativistic formulas reduce to the old classical ones, as we have discussed. We would, of course, hope—or rather, insist—that this be true since Newtonian mechanics works so well for objects moving with speeds $v \ll c$. This insistence that a more general theory (such as relativity) give the same results as a more restricted theory (such as classical mechanics which works for $v \ll c$) is called the **correspondence principle**. The two theories must correspond where their realms of validity overlap. Relativity thus does not contradict classical mechanics. Rather, it is a more general theory, of which classical mechanics is now considered to be a limiting case.

The importance of relativity is not simply that it gives more accurate results, especially at very high speeds. Much more than that, it has changed the way we view the world. The concepts of space and time are now seen to be relative, and intertwined with one another, whereas before they were considered absolute and separate. Even our concepts of matter and energy have changed: either can be converted to the other. The impact of relativity extends far beyond physics. It has influenced the other sciences, and even the world of art and literature; it has, indeed, entered the general culture.

From a practical point of view, we do not have much opportunity in our daily lives to use the mathematics of relativity. For example, the γ factor $1/\sqrt{1 - v^2/c^2}$, which appears in relativistic formulas, has a value of only 1.005 even for a speed as high as $0.10c = 3.0 \times 10^7$ m/s, giving a correction of less than 1%. For speeds less than $0.10c$, or unless mass and energy are interchanged, we don't usually need to use the more complicated relativistic formulas, and can use the simpler classical formulas.

relativity—we can make no distinction between motion of the source and motion of the observer. (Recall that sound travels in a medium such as air, whereas light does not—there is no evidence for an ether.)

To derive the Doppler shift for light, let us consider a light source and an observer that move toward each other, and let their relative velocity be v as measured in the reference frame of either the source or the observer. Figure 36–14a shows a source at rest emitting light waves of frequency f_0 and wavelength $\lambda_0 = c/f_0$. Two wavecrests are shown, a distance λ_0 apart, the second crest just having been emitted. In Fig. 36–14b, the source is shown moving at speed v toward a stationary observer who will see the wavelength λ being somewhat less than λ_0. (This is much like Fig. 16–19 for sound.) Let Δt represent the time between crests as detected by the observer, whose reference frame is shown in Fig. 36–14b. From Fig. 36–14b we see that

$$\lambda = c\,\Delta t - v\,\Delta t,$$

where $c\,\Delta t$ is the distance crest 1 has moved in the time Δt after it was emitted, and $v\,\Delta t$ is the distance the source has moved in time Δt. So far our derivation has not differed from that for sound (Section 16–7). Now we invoke the theory of relativity. The time between emission of wavecrests has undergone time dilation:

$$\Delta t = \Delta t_0 / \sqrt{1 - v^2/c^2}$$

where Δt_0 is the time between emissions of wavecrests in the reference frame where the source is at rest (the "proper" time). In the source's reference frame (Fig. 36–14a), we have

$$\Delta t_0 = \frac{1}{f_0} = \frac{\lambda_0}{c}$$

(Eqs. 5–2 and 31–14). Thus

$$\lambda = (c - v)\,\Delta t = (c - v)\frac{\Delta t_0}{\sqrt{1 - v^2/c^2}} = \frac{(c - v)}{\sqrt{c^2 - v^2}}\lambda_0$$

or

$$\lambda = \lambda_0 \sqrt{\frac{c - v}{c + v}}. \qquad \begin{bmatrix} \text{source and observer} \\ \text{moving toward} \\ \text{each other} \end{bmatrix} \quad \textbf{(36–14a)}$$

The frequency f is (recall $\lambda_0 = c/f_0$)

$$f = \frac{c}{\lambda} = f_0 \sqrt{\frac{c + v}{c - v}}. \qquad \begin{bmatrix} \text{source and observer} \\ \text{moving toward} \\ \text{each other} \end{bmatrix} \quad \textbf{(36–14b)}$$

Here f_0 is the frequency of the light as seen in the source's reference frame, and f is the frequency as measured by an observer moving toward the source or toward whom the source is moving. Equations 36–14 depend only on the relative velocity v. For relative motion *away* from each other we set $v < 0$ in Eqs. 36–14, and obtain

$$\lambda = \lambda_0 \sqrt{\frac{c + v}{c - v}} \qquad \textbf{(36–15a)}$$

$$\begin{bmatrix} \text{source and observer} \\ \text{moving away from} \\ \text{each other} \end{bmatrix}$$

$$f = f_0 \sqrt{\frac{c - v}{c + v}}. \qquad \textbf{(36–15b)}$$

From Eqs. 36–14 and 36–15 we see that light from a source moving toward us will have a higher frequency and shorter wavelength, whereas if a light source moves away from us, we will see a lower frequency and a longer wavelength. In the latter case, visible light will have its wavelength lengthened toward the red end of the visible spectrum (Fig. 32–26), an effect called a **redshift**. As we will see in the next Chapter, all atoms have their own distinctive signature in terms of the frequencies of the light they emit. In 1929 the American astronomer Edwin Hubble (1889–1953) found that radiation from atoms in many galaxies is redshifted. That is, the frequencies of light emitted are lower than those emitted by stationary atoms on Earth, suggesting that the galaxies are receding from us. This is the origin of the idea that the universe is expanding.

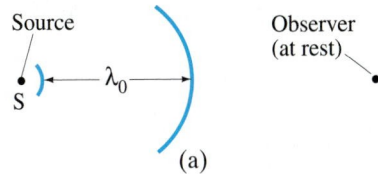

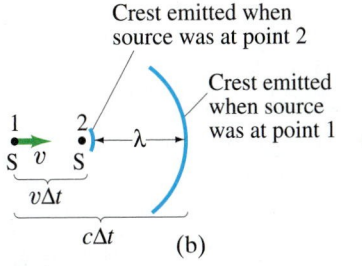

FIGURE 36–14 Doppler shift for light. (a) Source and observer at rest. (b) Source moving toward stationary observer.

EXAMPLE 36–13 **Speeding through a red light.** A driver claims that he did not go through a red light because the light was Doppler shifted and appeared green. Calculate the speed of a driver in order for a red light to appear green.

APPROACH We apply the Doppler shift equation for red light ($\lambda_0 \approx 650 \, \text{nm}$) and green light ($\lambda \approx 500 \, \text{nm}$).

SOLUTION Equation 36–14a holds for the source and the object moving toward each other:

$$\lambda = \lambda_0 \sqrt{\frac{c - v}{c + v}}.$$

We square this equation:

$$\frac{c - v}{c + v} = \left(\frac{\lambda}{\lambda_0}\right)^2$$

where $(\lambda/\lambda_0)^2 = (500 \, \text{nm}/650 \, \text{nm})^2 = 0.59$. We solve for v:

$$v = c\left[\frac{1 - (\lambda/\lambda_0)^2}{1 + (\lambda/\lambda_0)^2}\right] = 0.26c.$$

With this defense, the driver might not be guilty of running a red light, but he would clearly be guilty of speeding.

36–13 The Impact of Special Relativity

A great many experiments have been performed to test the predictions of the special theory of relativity. Within experimental error, no contradictions have been found. Scientists have therefore accepted relativity as an accurate description of nature.

At speeds much less than the speed of light, the relativistic formulas reduce to the old classical ones, as we have discussed. We would, of course, hope—or rather, insist—that this be true since Newtonian mechanics works so well for objects moving with speeds $v \ll c$. This insistence that a more general theory (such as relativity) give the same results as a more restricted theory (such as classical mechanics which works for $v \ll c$) is called the **correspondence principle**. The two theories must correspond where their realms of validity overlap. Relativity thus does not contradict classical mechanics. Rather, it is a more general theory, of which classical mechanics is now considered to be a limiting case.

The importance of relativity is not simply that it gives more accurate results, especially at very high speeds. Much more than that, it has changed the way we view the world. The concepts of space and time are now seen to be relative, and intertwined with one another, whereas before they were considered absolute and separate. Even our concepts of matter and energy have changed: either can be converted to the other. The impact of relativity extends far beyond physics. It has influenced the other sciences, and even the world of art and literature; it has, indeed, entered the general culture.

From a practical point of view, we do not have much opportunity in our daily lives to use the mathematics of relativity. For example, the γ factor $1/\sqrt{1 - v^2/c^2}$, which appears in relativistic formulas, has a value of only 1.005 even for a speed as high as $0.10c = 3.0 \times 10^7 \, \text{m/s}$, giving a correction of less than 1%. For speeds less than $0.10c$, or unless mass and energy are interchanged, we don't usually need to use the more complicated relativistic formulas, and can use the simpler classical formulas.

Summary

An **inertial reference frame** is one in which Newton's law of inertia holds. Inertial reference frames can move at constant velocity relative to one another; accelerating reference frames are **noninertial**.

The **special theory of relativity** is based on two principles: the **relativity principle**, which states that the laws of physics are the same in all inertial reference frames, and the principle of the **constancy of the speed of light**, which states that the speed of light in empty space has the same value in all inertial reference frames.

One consequence of relativity theory is that two events that are simultaneous in one reference frame may not be simultaneous in another. Other effects are **time dilation**: moving clocks are measured to run slow; and **length contraction**: the length of a moving object is measured to be shorter (in its direction of motion) than when it is at rest. Quantitatively,

$$\Delta t = \frac{\Delta t_0}{\sqrt{1 - v^2/c^2}} = \gamma \, \Delta t_0 \qquad (36\text{--}1)$$

$$\ell = \ell_0 \sqrt{1 - v^2/c^2} = \frac{\ell_0}{\gamma} \qquad (36\text{--}3)$$

where ℓ and Δt are the length and time interval of objects (or events) observed as they move by at the speed v; ℓ_0 and Δt_0 are the **proper length** and **proper time**—that is, the same quantities as measured in the rest frame of the objects or events. The quantity γ is shorthand for

$$\gamma = \frac{1}{\sqrt{1 - v^2/c^2}}. \qquad (36\text{--}2)$$

The theory of relativity has changed our notions of space and time, and of momentum, energy, and mass. Space and time are seen to be intimately connected, with time being the fourth dimension in addition to the three dimensions of space.

The **Lorentz transformations** relate the positions and times of events in one inertial reference frame to their positions and times in a second inertial reference frame.

$$\begin{aligned}
x &= \gamma(x' + vt') \\
y &= y' \\
z &= z' \qquad\qquad\qquad (36\text{--}6) \\
t &= \gamma\left(t' + \frac{vx'}{c^2}\right)
\end{aligned}$$

where $\gamma = 1/\sqrt{1 - v^2/c^2}$.

Velocity addition also must be done in a special way. All these relativistic effects are significant only at high speeds, close to the speed of light, which itself is the ultimate speed in the universe.

The **momentum** of an object is given by

$$p = \gamma m v = \frac{mv}{\sqrt{1 - v^2/c^2}}. \qquad (36\text{--}8)$$

Mass and energy are interconvertible. The equation

$$E = mc^2 \qquad (36\text{--}12)$$

tells how much energy E is needed to create a mass m, or vice versa. Said another way, $E = mc^2$ is the amount of energy an object has because of its mass m. The law of conservation of energy must include mass as a form of energy.

The kinetic energy K of an object moving at speed v is given by

$$K = (\gamma - 1)mc^2 = \frac{mc^2}{\sqrt{1 - v^2/c^2}} - mc^2 \qquad (36\text{--}10)$$

where m is the mass of the object. The total energy E, if there is no potential energy, is

$$\begin{aligned}
E &= K + mc^2 \\
&= \gamma mc^2.
\end{aligned} \qquad (36\text{--}11)$$

The momentum p of an object is related to its total energy E (assuming no potential energy) by

$$E^2 = p^2c^2 + m^2c^4. \qquad (36\text{--}13)$$

Questions

1. You are in a windowless car in an exceptionally smooth train moving at constant velocity. Is there any physical experiment you can do in the train car to determine whether you are moving? Explain.

2. You might have had the experience of being at a red light when, out of the corner of your eye, you see the car beside you creep forward. Instinctively you stomp on the brake pedal, thinking that you are rolling backward. What does this say about absolute and relative motion?

3. A worker stands on top of a moving railroad car, and throws a heavy ball straight up (from his point of view). Ignoring air resistance, will the ball land back in his hand or behind him?

4. Does the Earth really go around the Sun? Or is it also valid to say that the Sun goes around the Earth? Discuss in view of the relativity principle (that there is no best reference frame). Explain.

5. If you were on a spaceship traveling at $0.5c$ away from a star, at what speed would the starlight pass you?

6. The time dilation effect is sometimes expressed as "moving clocks run slowly." Actually, this effect has nothing to do with motion affecting the functioning of clocks. What then does it deal with?

7. Does time dilation mean that time actually passes more slowly in moving reference frames or that it only *seems* to pass more slowly?

8. A young-looking woman astronaut has just arrived home from a long trip. She rushes up to an old gray-haired man and in the ensuing conversation refers to him as her son. How might this be possible?

9. If you were traveling away from Earth at speed $0.5c$, would you notice a change in your heartbeat? Would your mass, height, or waistline change? What would observers on Earth using telescopes say about you?

10. Do time dilation and length contraction occur at ordinary speeds, say 90 km/h?

11. Suppose the speed of light were infinite. What would happen to the relativistic predictions of length contraction and time dilation?

12. Discuss how our everyday lives would be different if the speed of light were only 25 m/s.

13. Explain how the length contraction and time dilation formulas might be used to indicate that c is the limiting speed in the universe.

14. The drawing at the start of this Chapter shows the street as seen by Mr Tompkins, where the speed of light is $c = 20$ mi/h. What does Mr Tompkins look like to the people standing on the street (Fig. 36–15)? Explain.

FIGURE 36–15
Question 14.
Mr Tompkins as seen by people on the sidewalk. See also Chapter-Opening figure on page 951.

15. An electron is limited to travel at speeds less than c. Does this put an upper limit on the momentum of an electron? If so, what is this upper limit? If not, explain.

16. Can a particle of nonzero mass attain the speed of light?

17. Does the equation $E = mc^2$ conflict with the conservation of energy principle? Explain.

18. If mass is a form of energy, does this mean that a spring has more mass when compressed than when relaxed?

19. It is not correct to say that "matter can neither be created nor destroyed." What must we say instead?

20. Is our intuitive notion that velocities simply add, as in Section 3–9, completely wrong?

Problems

36–5 and 36–6 Time Dilation, Length Contraction

1. (I) A spaceship passes you at a speed of $0.850c$. You measure its length to be 38.2 m. How long would it be when at rest?

2. (I) A certain type of elementary particle travels at a speed of 2.70×10^8 m/s. At this speed, the average lifetime is measured to be 4.76×10^{-6} s. What is the particle's lifetime at rest?

3. (II) According to the special theory of relativity, the factor γ that determines the length contraction and the time dilation is given by $\gamma = 1/\sqrt{1 - v^2/c^2}$. Determine the numerical values of γ for an object moving at speed $v = 0.01c, 0.05c, 0.10c, 0.20c, 0.30c, 0.40c, 0.50c, 0.60c, 0.70c, 0.80c, 0.90c$, and $0.99c$. Make a graph of γ versus v.

4. (II) If you were to travel to a star 135 light-years from Earth at a speed of 2.80×10^8 m/s, what would you measure this distance to be?

5. (II) What is the speed of a pion if its average lifetime is measured to be 4.40×10^{-8} s? At rest, its average lifetime is 2.60×10^{-8} s.

6. (II) In an Earth reference frame, a star is 56 light-years away. How fast would you have to travel so that to you the distance would be only 35 light-years?

7. (II) Suppose you decide to travel to a star 65 light-years away at a speed that tells you the distance is only 25 light-years. How many years would it take you to make the trip?

8. (II) At what speed v will the length of a 1.00-m stick look 10.0% shorter (90.0 cm)?

9. (II) Escape velocity from the Earth is 11.2 km/s. What would be the percent decrease in length of a 65.2-m-long spacecraft traveling at that speed as seen from Earth?

10. (II) A friend speeds by you in her spacecraft at a speed of $0.760c$. It is measured in your frame to be 4.80 m long and 1.35 m high. (a) What will be its length and height at rest? (b) How many seconds elapsed on your friend's watch when 20.0 s passed on yours? (c) How fast did you appear to be traveling according to your friend? (d) How many seconds elapsed on your watch when she saw 20.0 s pass on hers?

11. (II) At what speed do the relativistic formulas for (a) length and (b) time intervals differ from classical values by 1.00%? (This is a reasonable way to estimate when to do relativistic calculations rather than classical.)

12. (II) A certain star is 18.6 light-years away. How long would it take a spacecraft traveling $0.950c$ to reach that star from Earth, as measured by observers: (a) on Earth, (b) on the spacecraft? (c) What is the distance traveled according to observers on the spacecraft? (d) What will the spacecraft occupants compute their speed to be from the results of (b) and (c)?

13. (II) Suppose a news report stated that starship *Enterprise* had just returned from a 5-year voyage while traveling at $0.74c$. (a) If the report meant 5.0 years of *Earth time*, how much time elapsed on the ship? (b) If the report meant 5.0 years of *ship time*, how much time passed on Earth?

14. (II) An unstable particle produced in an accelerator experiment travels at constant velocity, covering 1.00 m in 3.40 ns in the lab frame before changing ("decaying") into other particles. In the rest frame of the particle, determine (a) how long it lived before decaying, (b) how far it moved before decaying.

15. (II) When it is stationary, the half-life of a certain subatomic particle is T_0. That is, if N of these particles are present at a certain time, then a time T_0 later only $N/2$ particles will be present, assuming the particles are at rest. A beam carrying N such particles per second is created at position $x = 0$ in a high-energy physics laboratory. This beam travels along the x axis at speed v in the laboratory reference frame and it is found that only $N/2$ particles per second travel in the beam at $x = 2cT_0$, where c is the speed of light. Find the speed v of the particles within the beam.

16. (II) In its own reference frame, a box has the shape of a cube 2.0 m on a side. This box is loaded onto the flat floor of a spaceship and the spaceship then flies past us with a horizontal speed of $0.80c$. What is the volume of the box as we observe it?

17. (II) When at rest, a spaceship has the form of an isosceles triangle whose two equal sides have length 2ℓ and whose base has length ℓ. If this ship flies past an observer with a relative velocity of $v = 0.95c$ directed along its base, what are the lengths of the ship's three sides according to the observer?

18. (II) How fast must a pion be moving on average to travel 25 m before it decays? The average lifetime, at rest, is 2.6×10^{-8} s.

36–8 Lorentz Transformations

19. (I) An observer on Earth sees an alien vessel approach at a speed of 0.60c. The *Enterprise* comes to the rescue (Fig. 36–16), overtaking the aliens while moving directly toward Earth at a speed of 0.90c relative to Earth. What is the relative speed of one vessel as seen by the other?

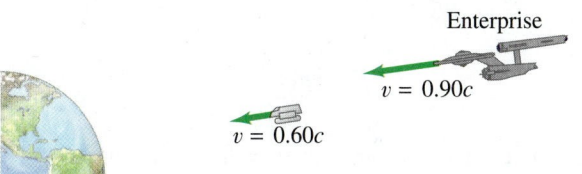

Enterprise

$v = 0.90c$

$v = 0.60c$

FIGURE 36–16 Problem 19.

20. (I) Suppose in Fig. 36–11 that the origins of S and S′ overlap at $t = t′ = 0$ and that S′ moves at speed $v = 30 \, \text{m/s}$ with respect to S. In S′, a person is resting at a point whose coordinates are $x′ = 25 \, \text{m}$, $y′ = 20 \, \text{m}$, and $z′ = 0$. Calculate this person's coordinates in S (x, y, z) at (a) $t = 3.5 \, \text{s}$, (b) $t = 10.0 \, \text{s}$. Use the Galilean transformation.

21. (I) Repeat Problem 20 using the Lorentz transformation and a relative speed $v = 1.80 \times 10^8 \, \text{m/s}$, but choose the time t to be (a) $3.5 \, \mu\text{s}$ and (b) $10.0 \, \mu\text{s}$.

22. (II) In Problem 21, suppose that the person moves with a velocity whose components are $u′_x = u′_y = 1.10 \times 10^8 \, \text{m/s}$, What will be her velocity with respect to S? (Give magnitude and direction.)

23. (II) Two spaceships leave Earth in opposite directions, each with a speed of 0.60c with respect to Earth. (a) What is the velocity of spaceship 1 relative to spaceship 2? (b) What is the velocity of spaceship 2 relative to spaceship 1?

24. (II) Reference frame S′ moves at speed $v = 0.92c$ in the $+x$ direction with respect to reference frame S. The origins of S and S′ overlap at $t = t′ = 0$. An object is stationary in S′ at position $x′ = 100 \, \text{m}$. What is the position of the object in S when the clock in S reads $1.00 \, \mu\text{s}$ according to the (a) Galilean and (b) Lorentz transformation equations?

25. (II) A spaceship leaves Earth traveling at 0.61c. A second spaceship leaves the first at a speed of 0.87c with respect to the first. Calculate the speed of the second ship with respect to Earth if it is fired (a) in the same direction the first spaceship is already moving, (b) directly backward toward Earth.

26. (II) Your spaceship, traveling at 0.90c, needs to launch a probe out the forward hatch so that its speed relative to the planet that you are approaching is 0.95c. With what speed must it leave your ship?

27. (II) A spaceship traveling at 0.76c away from Earth fires a module with a speed of 0.82c at right angles to its own direction of travel (as seen by the spaceship). What is the speed of the module, and its direction of travel (relative to the spaceship's direction), as seen by an observer on Earth?

28. (II) If a particle moves in the xy plane of system S (Fig. 36–11) with speed u in a direction that makes an angle θ with the x axis, show that it makes an angle $\theta′$ in S′ given by $\tan \theta′ = (\sin \theta)\sqrt{1 - v^2/c^2}/(\cos \theta - v/u)$.

29. (II) A stick of length ℓ_0, at rest in reference frame S, makes an angle θ with the x axis. In reference frame S′, which moves to the right with velocity $\vec{v} = v\hat{i}$ with respect to S, determine (a) the length ℓ of the stick, and (b) the angle $\theta′$ it makes with the $x′$ axis.

30. (III) In the old West, a marshal riding on a train traveling 35.0 m/s sees a duel between two men standing on the Earth 55.0 m apart parallel to the train. The marshal's instruments indicate that in his reference frame the two men fired simultaneously. (a) Which of the two men, the first one the train passes (A) or the second one (B) should be arrested for firing the first shot? That is, in the gunfighter's frame of reference, who fired first? (b) How much earlier did he fire? (c) Who was struck first?

31. (III) Two lightbulbs, A and B, are placed at rest on the x axis at positions $x_A = 0$ and $x_B = +\ell$. In this reference frame, the bulbs are turned on simultaneously. Use the Lorentz transformations to find an expression for the time interval between when the bulbs are turned on as measured by an observer moving at velocity v in the $+x$ direction. According to this observer, which bulb is turned on first?

32. (III) An observer in reference frame S notes that two events are separated in space by 220 m and in time by $0.80 \, \mu\text{s}$. How fast must reference frame S′ be moving relative to S in order for an observer in S′ to detect the two events as occurring at the same location in space?

33. (III) A farm boy studying physics believes that he can fit a 12.0-m long pole into a 10.0-m long barn if he runs fast enough, carrying the pole. Can he do it? Explain in detail. How does this fit with the idea that when he is running the barn looks even shorter than 10.0 m?

36–9 Relativistic Momentum

34. (I) What is the momentum of a proton traveling at $v = 0.75c$?

35. (II) (a) A particle travels at $v = 0.10c$. By what percentage will a calculation of its momentum be wrong if you use the classical formula? (b) Repeat for $v = 0.60c$.

36. (II) A particle of mass m travels at a speed $v = 0.26c$. At what speed will its momentum be doubled?

37. (II) An unstable particle is at rest and suddenly decays into two fragments. No external forces act on the particle or its fragments. One of the fragments has a speed of 0.60c and a mass of $6.68 \times 10^{-27} \, \text{kg}$, while the other has a mass of $1.67 \times 10^{-27} \, \text{kg}$. What is the speed of the less massive fragment?

38. (II) What is the percent change in momentum of a proton that accelerates (a) from 0.45c to 0.80c, (b) from 0.80c to 0.98c?

36–11 Relativistic Energy

39. (I) Calculate the rest energy of an electron in joules and in MeV $(1 \, \text{MeV} = 1.60 \times 10^{-13} \, \text{J})$.

40. (I) When a uranium nucleus at rest breaks apart in the process known as fission in a nuclear reactor, the resulting fragments have a total kinetic energy of about 200 MeV. How much mass was lost in the process?

41. (I) The total annual energy consumption in the United States is about $8 \times 10^{19} \, \text{J}$. How much mass would have to be converted to energy to fuel this need?

42. (I) Calculate the mass of a proton in MeV/c^2.

43. (II) Suppose there was a process by which two photons, each with momentum $0.50 \, \text{MeV}/c$, could collide and make a single particle. What is the maximum mass that the particle could possess?

44. (II) (a) How much work is required to accelerate a proton from rest up to a speed of 0.998c? (b) What would be the momentum of this proton?

45. (II) How much energy can be obtained from conversion of 1.0 gram of mass? How much mass could this energy raise to a height of 1.0 km above the Earth's surface?

46. (II) To accelerate a particle of mass m from rest to speed $0.90c$ requires work W_1. To accelerate the particle from speed $0.90c$ to $0.99c$, requires work W_2. Determine the ratio W_2/W_1.

47. (II) What is the speed of a particle when its kinetic energy equals its rest energy?

48. (II) What is the momentum of a 950-MeV proton (that is, its kinetic energy is 950 MeV)?

49. (II) Calculate the kinetic energy and momentum of a proton traveling 2.80×10^8 m/s.

50. (II) What is the speed of an electron whose kinetic energy is 1.25 MeV?

51. (II) What is the speed of a proton accelerated by a potential difference of 125 MV?

52. (II) Two identical particles of mass m approach each other at equal and opposite speeds, v. The collision is completely inelastic and results in a single particle at rest. What is the mass of the new particle? How much energy was lost in the collision? How much kinetic energy was lost in this collision?

53. (II) What is the speed of an electron just before it hits a television screen after being accelerated from rest by the 28,000 V of the picture tube?

54. (II) The kinetic energy of a particle is 45 MeV. If the momentum is 121 MeV/c, what is the particle's mass?

55. (II) Calculate the speed of a proton $(m = 1.67 \times 10^{-27}$ kg$)$ whose kinetic energy is exactly half (a) its total energy, (b) its rest energy.

56. (II) Calculate the kinetic energy and momentum of a proton $(m = 1.67 \times 10^{-27}$ kg$)$ traveling 8.15×10^7 m/s. By what percentages would your calculations have been in error if you had used classical formulas?

57. (II) Suppose a spacecraft of mass 17,000 kg is accelerated to $0.18c$. (a) How much kinetic energy would it have? (b) If you used the classical formula for kinetic energy, by what percentage would you be in error?

*58. (II) What magnetic field B is needed to keep 998-GeV protons revolving in a circle of radius 1.0 km (at, say, the Fermilab synchrotron)? Use the relativistic mass. The proton's rest mass is 0.938 GeV/c^2. $(1$ GeV $= 10^9$ eV.) [Hint: In relativity, $m_{\text{rel}} v^2/r = qvB$ is still valid in a magnetic field, where $m_{\text{rel}} = \gamma m$.]

59. (II) The americium nucleus, $^{241}_{95}$Am, decays to a neptunium nucleus, $^{237}_{93}$Np, by emitting an alpha particle of mass 4.00260 u and kinetic energy 5.5 MeV. Estimate the mass of the neptunium nucleus, ignoring its recoil, given that the americium mass is 241.05682 u.

60. (II) Make a graph of the kinetic energy versus momentum for (a) a particle of nonzero mass, and (b) a particle with zero mass.

61. (II) A negative muon traveling at 43% the speed of light collides head on with a positive muon traveling at 55% the speed of light. The two muons (each of mass 105.7 MeV/c^2) annihilate, and produce how much electromagnetic energy?

62. (II) Show that the kinetic energy K of a particle of mass m is related to its momentum p by the equation

$$p = \sqrt{K^2 + 2Kmc^2}/c.$$

63. (III) (a) In reference frame S, a particle has momentum $\vec{p} = p_x \hat{i}$ along the positive x axis. Show that in frame S′, which moves with speed v as in Fig. 36–11, the momentum has components

$$p'_x = \frac{p_x - vE/c^2}{\sqrt{1 - v^2/c^2}}$$
$$p'_y = p_y$$
$$p'_z = p_z$$
$$E' = \frac{E - p_x v}{\sqrt{1 - v^2/c^2}}.$$

(These transformation equations hold, actually, for any direction of $\vec{p}$, as long as the motion of S′ is along the x axis.) (b) Show that $p_x, p_y, p_z, E/c$ transform according to the Lorentz transformation in the same way as x, y, z, ct.

36–12 Doppler Shift for Light

64. (II) A certain galaxy has a Doppler shift given by $f_0 - f = 0.0987 f_0$. Estimate how fast it is moving away from us.

65. (II) A spaceship moving toward Earth at $0.70c$ transmits radio signals at 95.0 MHz. At what frequency should Earth receivers be tuned?

66. (II) Starting from Eq. 36–15a, show that the Doppler shift in wavelength is

$$\frac{\Delta\lambda}{\lambda} = \frac{v}{c}$$

if $v \ll c$.

67. (III) A radar "speed gun" emits microwaves of frequency $f_0 = 36.0$ GHz. When the gun is pointed at an object moving toward it at speed v, the object senses the microwaves at the Doppler-shifted frequency f. The moving object reflects these microwaves at this same frequency f. The stationary radar apparatus detects these reflected waves at a Doppler-shifted frequency f'. The gun combines its emitted wave at f_0 and its detected wave at f'. These waves interfere, creating a beat pattern whose beat frequency is $f_{\text{beat}} = f' - f_0$. (a) Show that

$$v \approx \frac{c f_{\text{beat}}}{2 f_0},$$

if $f_{\text{beat}} \ll f_0$. If $f_{\text{beat}} = 6670$ Hz, what is v (km/h)? (b) If the object's speed is different by Δv, show that the difference in beat frequency Δf_{beat} is given by

$$\Delta f_{\text{beat}} = \frac{2 f_0 \Delta v}{c}.$$

If the accuracy of the speed gun is to be 1 km/h, to what accuracy must the beat frequency be measured?

68. (III) A certain atom emits light of frequency f_0 when at rest. A monatomic gas composed of these atoms is at temperature T. Some of the gas atoms move toward and others away from an observer due to their random thermal motion. Using the rms speed of thermal motion, show that the fractional difference between the Doppler-shifted frequencies for atoms moving directly toward the observer and directly away from the observer is $\Delta f/f_0 \approx 2\sqrt{3kT/mc^2}$; assume $mc^2 \gg 3kT$. Evaluate $\Delta f/f_0$ for a gas of hydrogen atoms at 550 K. [This "Doppler-broadening" effect is commonly used to measure gas temperature, such as in astronomy.]

General Problems

69. An atomic clock is taken to the North Pole, while another stays at the Equator. How far will they be out of synchronization after 2.0 years has elapsed? [*Hint*: Use the binomial expansion, Appendix A.]

70. A spaceship in distress sends out two escape pods in opposite directions. One travels at a speed $v_1 = -0.60c$ in one direction, and the other travels at a speed $v_2 = +0.50c$ in the other direction, as observed from the spaceship. What speed does the first escape pod measure for the second escape pod?

71. An airplane travels 1300 km/h around the Earth in a circle of radius essentially equal to that of the Earth, returning to the same place. Using special relativity, estimate the difference in time to make the trip as seen by Earth and airplane observers. [*Hint*: Use the binomial expansion, Appendix A.]

72. The nearest star to Earth is Proxima Centauri, 4.3 light-years away. (*a*) At what constant velocity must a spacecraft travel from Earth if it is to reach the star in 4.6 years, as measured by travelers on the spacecraft? (*b*) How long does the trip take according to Earth observers?

73. A quasar emits familiar hydrogen lines whose wave-lengths are 7.0% longer than what we measure in the laboratory. (*a*) Using the Doppler formula for light, estimate the speed of this quasar. (*b*) What result would you obtain if you used the "classical" Doppler shift discussed in Chapter 16?

74. A healthy astronaut's heart rate is 60 beats/min. Flight doctors on Earth can monitor an astronaut's vital signs remotely while in flight. How fast would an astronaut have to be flying away from Earth in order for the doctor to measure her having a heart rate of 30 beats/min?

75. A spacecraft (reference frame S′) moves past Earth (reference frame S) at velocity $\vec{v}$, which points along the x and $x′$ axes. The spacecraft emits a light beam (speed c) along its $y′$ axis as shown in Fig. 36–17. (*a*) What angle θ does this light beam make with the x axis in the Earth's reference frame? (*b*) Use velocity transformations to show that the light moves with speed c also in the Earth's reference frame. (*c*) Compare these relativistic results to what you would have obtained classically (Galilean transformations).

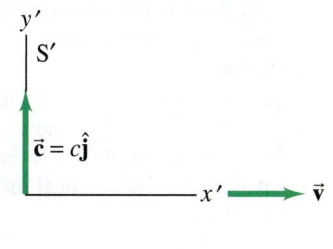

FIGURE 36–17
Problem 75.

76. Rocket A passes Earth at a speed of $0.65c$. At the same time, rocket B passes Earth moving $0.85c$ relative to Earth in the same direction. How fast is B moving relative to A when it passes A?

77. (*a*) What is the speed v of an electron whose kinetic energy is 14,000 times its rest energy? You can state the answer as the difference $c - v$. Such speeds are reached in the Stanford Linear Accelerator, SLAC. (*b*) If the electrons travel in the lab through a tube 3.0 km long (as at SLAC), how long is this tube in the electrons' reference frame? [*Hint*: Use the binomial expansion.]

78. As a rough rule, anything traveling faster than about $0.1c$ is called *relativistic*—that is, special relativity is a significant effect. Determine the speed of an electron in a hydrogen atom (radius 0.53×10^{-10} m) and state whether or not it is relativistic. (Treat the electron as though it were in a circular orbit around the proton.)

79. What minimum amount of electromagnetic energy is needed to produce an electron and a positron together? A positron is a particle with the same mass as an electron, but has the opposite charge. (Note that electric charge is conserved in this process. See Section 37–5.)

80. How many grams of matter would have to be totally destroyed to run a 75-W lightbulb for 1.0 year?

81. If E is the total energy of a particle with zero potential energy, show that $dE/dp = v$, where p and v are the momentum and velocity of the particle, respectively.

82. A free neutron can decay into a proton, an electron, and a neutrino. Assume the neutrino's mass is zero; the other masses can be found in the Table inside the front cover. Determine the total kinetic energy shared among the three particles when a neutron decays at rest.

83. The Sun radiates energy at a rate of about 4×10^{26} W. (*a*) At what rate is the Sun's mass decreasing? (*b*) How long does it take for the Sun to lose a mass equal to that of Earth? (*c*) Estimate how long the Sun could last if it radiated constantly at this rate.

84. An unknown particle is measured to have a negative charge and a speed of 2.24×10^8 m/s. Its momentum is determined to be 3.07×10^{-22} kg·m/s. Identify the particle by finding its mass.

85. How much energy would be required to break a helium nucleus into its constituents, two protons and two neutrons? The rest masses of a proton (including an electron), a neutron, and neutral helium are, respectively, 1.00783 u, 1.00867 u, and 4.00260 u. (This energy difference is called the *total binding energy* of the $^4_2$He nucleus.)

86. Show analytically that a particle with momentum p and energy E has a speed given by

$$v = \frac{pc^2}{E} = \frac{pc}{\sqrt{m^2c^2 + p^2}}.$$

87. Two protons, each having a speed of $0.985c$ in the laboratory, are moving toward each other. Determine (*a*) the momentum of each proton in the laboratory, (*b*) the total momentum of the two protons in the laboratory, and (*c*) the momentum of one proton as seen by the other proton.

88. When two moles of hydrogen molecules (H_2) and one mole of oxygen molecules (O_2) react to form two moles of water (H_2O), the energy released is 484 kJ. How much does the mass decrease in this reaction? What % of the total original mass of the system does this mass change represent?

89. The fictional starship *Enterprise* obtains its power by combining matter and antimatter, achieving complete conversion of mass into energy. If the mass of the *Enterprise* is approximately 6×10^9 kg, how much mass must be converted into kinetic energy to accelerate it from rest to one-tenth the speed of light?

90. A spaceship and its occupants have a total mass of 180,000 kg. The occupants would like to travel to a star that is 35 light-years away at a speed of 0.70c. To accelerate, the engine of the spaceship changes mass directly to energy. How much mass will be converted to energy to accelerate the spaceship to this speed? Assume the acceleration is rapid, so the speed for the entire trip can be taken to be 0.70c, and ignore decrease in total mass for the calculation. How long will the trip take according to the astronauts on board?

91. In a nuclear reaction two identical particles are created, traveling in opposite directions. If the speed of each particle is 0.85c, relative to the laboratory frame of reference, what is one particle's speed relative to the other particle?

92. A 32,000-kg spaceship is to travel to the vicinity of a star 6.6 light-years from Earth. Passengers on the ship want the (one-way) trip to take no more than 1.0 year. How much work must be done on the spaceship to bring it to the speed necessary for this trip?

93. Suppose a 14,500-kg spaceship left Earth at a speed of 0.98c. What is the spaceship's kinetic energy? Compare with the total U.S. annual energy consumption (about 10^{20} J).

94. A pi meson of mass m_π decays at rest into a muon (mass m_μ) and a neutrino of negligible or zero mass. Show that the kinetic energy of the muon is $K_\mu = (m_\pi - m_\mu)^2 c^2 / (2m_\pi)$.

95. Astronomers measure the distance to a particular star to be 6.0 light-years (1 ly = distance light travels in 1 year). A spaceship travels from Earth to the vicinity of this star at steady speed, arriving in 2.50 years as measured by clocks on the spaceship. (a) How long does the trip take as measured by clocks in Earth's reference frame (assumed inertial)? (b) What distance does the spaceship travel as measured in its own reference frame?

96. A 1.88-kg mass oscillates on the end of a spring whose spring stiffness constant is $k = 84.2$ N/m. If this system is in a spaceship moving past Earth at 0.900c, what is its period of oscillation according to (a) observers on the ship, and (b) observers on Earth?

97. Show that the space–time interval, $(c \Delta t)^2 - (\Delta x)^2$, is invariant, meaning that all observers in all inertial reference frames calculate the same number for this quantity for any pair of events.

98. A slab of glass with index of refraction n moves to the right with speed v. A flash of light is emitted at point A (Fig. 36–18) and passes through the glass arriving at point B a distance ℓ away. The glass has thickness d in the reference frame where it is at rest, and the speed of light in the glass is c/n. How long does it take the light to go from point A to point B according to an observer at rest with respect to points A and B? Check your answer for the cases $v = c$, $v = 0$, and $n = 1$.

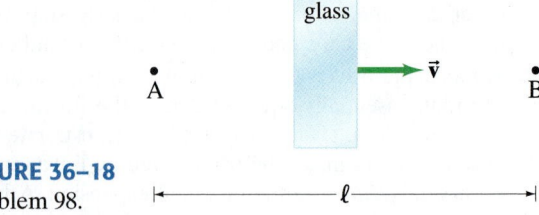

FIGURE 36–18
Problem 98.

*Numerical/Computer

* **99.** (II) For a 1.0-kg mass, make a plot of the kinetic energy as a function of speed for speeds from 0 to 0.9c, using both the classical formula $\left(K = \frac{1}{2}mv^2\right)$ and the correct relativistic formula $\left(K = (\gamma - 1)mc^2\right)$.

* **100.** (III) A particle of mass m is projected horizontally at a relativistic speed v_0 in the $+x$ direction. There is a constant downward force F acting on the particle. Using the definition of relativistic momentum $\vec{p} = \gamma m \vec{v}$ and Newton's second law $\vec{F} = d\vec{p}/dt$, (a) show that the x and y components of the velocity of the particle at time t are given by

$$v_x(t) = p_0 c / (m^2 c^2 + p_0^2 + F^2 t^2)^{\frac{1}{2}}$$

$$v_y(t) = -Fct / (m^2 c^2 + p_0^2 + F^2 t^2)^{\frac{1}{2}}$$

where p_0 is the initial momentum of the particle. (b) Assume the particle is an electron $(m = 9.11 \times 10^{-31}$ kg), with $v_0 = 0.50c$ and $F = 1.00 \times 10^{-15}$ N. Calculate the values of v_x and v_y of the electron as a function of time t from $t = 0$ to $t = 5.00 \, \mu$s in intervals of $0.05 \, \mu$s. Graph the values to show how the velocity components change with time during this interval. (c) Is the path parabolic, as it would be in classical mechanics (Sections 3–7 and 3–8)? Explain.

Answers to Exercises

A: Yes.

B: (c).

C: (a) No; (b) yes.

D: 80 m.

E: 0.030c, same as classical, to an accuracy of better than 0.1%.

F: (d).

G: (d).

H: No.

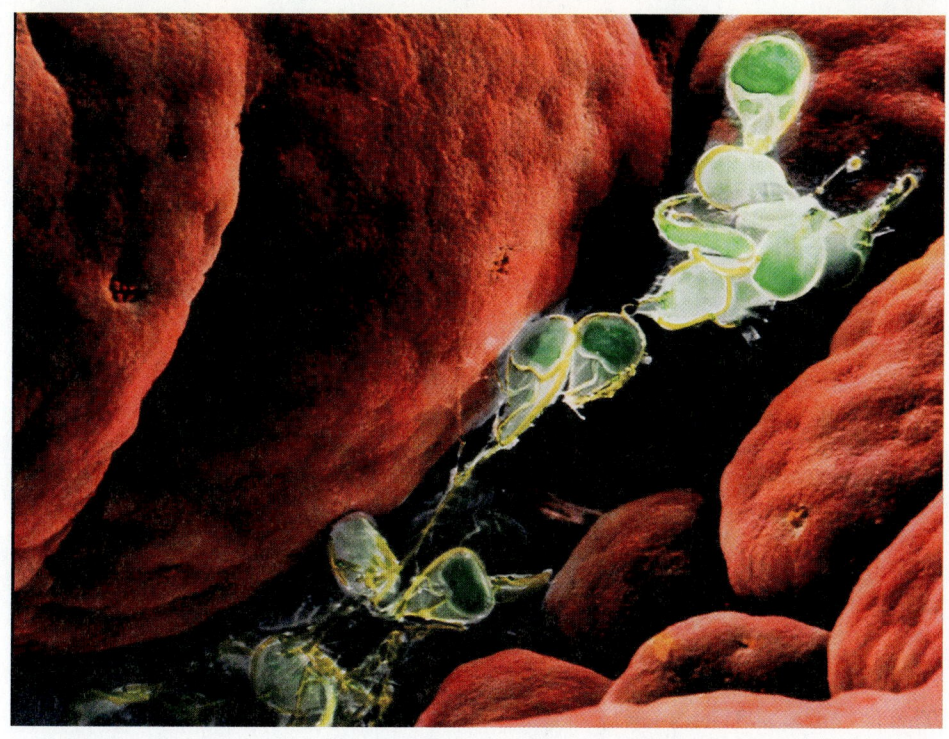

Electron microscopes produce images using electrons which have wave properties just as light does. Since the wavelength of electrons can be much smaller than that of visible light, much greater resolution and magnification can be obtained. A scanning electron microscope (SEM) can produce images with a three-dimensional quality, as for these *Giardia* cells inside a human small intestine. Magnification here is about 2000×. *Giardia* is on the minds of backpackers because it has become too common in untreated water, even in the high mountains, and causes an unpleasant intestinal infection not easy to get rid of.

C H A P T E R

37

Early Quantum Theory and Models of the Atom

CHAPTER-OPENING QUESTION—Guess now!

It has been found experimentally that

(a) light behaves as a wave.
(b) light behaves as a particle.
(c) electrons behave as particles.
(d) electrons behave as waves.
(e) all of the above are true.
(f) none of the above are true.

The second aspect of the revolution that shook the world of physics in the early part of the twentieth century was the quantum theory (the other was Einstein's theory of relativity). Unlike the special theory of relativity, the revolution of quantum theory required almost three decades to unfold, and many scientists contributed to its development. It began in 1900 with Planck's quantum hypothesis, and culminated in the mid-1920s with the theory of quantum mechanics of Schrödinger and Heisenberg which has been so effective in explaining the structure of matter.

CONTENTS

37–1 Planck's Quantum Hypothesis; Blackbody Radiation

37–2 Photon Theory of Light and the Photoelectric Effect

37–3 Energy, Mass, and Momentum of a Photon

37–4 Compton Effect

37–5 Photon Interactions; Pair Production

37–6 Wave–Particle Duality; the Principle of Complementarity

37–7 Wave Nature of Matter

*37–8 Electron Microscopes

37–9 Early Models of the Atom

37–10 Atomic Spectra: Key to the Structure of the Atom

37–11 The Bohr Model

37–12 de Broglie's Hypothesis Applied to Atoms

37–1 Planck's Quantum Hypothesis; Blackbody Radiation

Blackbody Radiation

One of the observations that was unexplained at the end of the nineteenth century was the spectrum of light emitted by hot objects. We saw in Section 19–10 that all objects emit radiation whose total intensity is proportional to the fourth power of the Kelvin (absolute) temperature (T^4). At normal temperatures $(\approx 300\,\text{K})$, we are not aware of this electromagnetic radiation because of its low intensity.

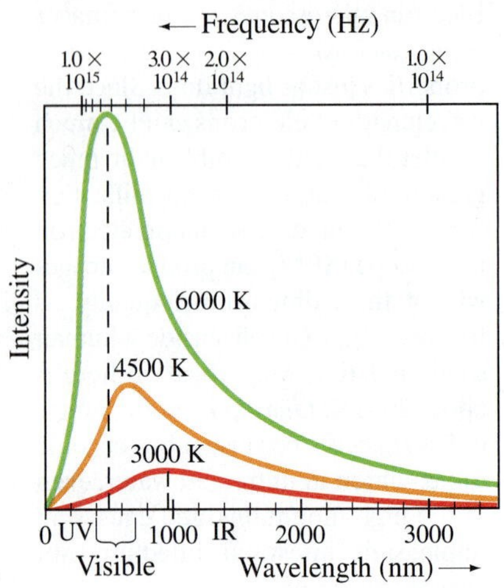

FIGURE 37–1 Measured spectra of wavelengths and frequencies emitted by a blackbody at three different temperatures.

At higher temperatures, there is sufficient infrared radiation that we can feel heat if we are close to the object. At still higher temperatures (on the order of 1000 K), objects actually glow, such as a red-hot electric stove burner or the heating element in a toaster. At temperatures above 2000 K, objects glow with a yellow or whitish color, such as white-hot iron and the filament of a lightbulb. The light emitted is of a continuous range of wavelengths or frequencies, and the *spectrum* is a plot of intensity vs. wavelength or frequency. As the temperature increases, the electromagnetic radiation emitted by objects not only increases in total intensity but reaches a peak at higher and higher frequencies.

The spectrum of light emitted by a hot dense object is shown in Fig. 37–1 for an idealized **blackbody**. A blackbody is a body that would absorb all the radiation falling on it (and so would appear black under reflection when illuminated by other sources). The radiation such an idealized blackbody would emit when hot and luminous, called **blackbody radiation** (though not necessarily black in color), approximates that from many real objects. The 6000-K curve in Fig. 37–1, corresponding to the temperature of the surface of the Sun, peaks in the visible part of the spectrum. For lower temperatures, the total radiation drops considerably and the peak occurs at longer wavelengths (or lower frequencies). (This is why objects glow with a red color at around 1000 K.) It is found experimentally that the wavelength at the peak of the spectrum, λ_P, is related to the Kelvin temperature T by

$$\lambda_P T = 2.90 \times 10^{-3}\,\mathrm{m\cdot K}. \qquad \textbf{(37–1)}$$

This is known as **Wien's law**.

EXAMPLE 37–1 **The Sun's surface temperature.** Estimate the temperature of the surface of our Sun, given that the Sun emits light whose peak intensity occurs in the visible spectrum at around 500 nm.

APPROACH We assume the Sun acts as a blackbody, and use $\lambda_P = 500\,\mathrm{nm}$ in Wien's law (Eq. 37–1).

SOLUTION Wien's law gives

$$T = \frac{2.90 \times 10^{-3}\,\mathrm{m\cdot K}}{\lambda_P} = \frac{2.90 \times 10^{-3}\,\mathrm{m\cdot K}}{500 \times 10^{-9}\,\mathrm{m}} \approx 6000\,\mathrm{K}.$$

EXAMPLE 37–2 **Star color.** Suppose a star has a surface temperature of 32,500 K. What color would this star appear?

APPROACH We assume the star emits radiation as a blackbody, and solve for λ_P in Wien's law, Eq. 37–1.

SOLUTION From Wien's law we have

$$\lambda_P = \frac{2.90 \times 10^{-3}\,\mathrm{m\cdot K}}{T} = \frac{2.90 \times 10^{-3}\,\mathrm{m\cdot K}}{3.25 \times 10^{4}\,\mathrm{K}} = 89.2\,\mathrm{nm}.$$

The peak is in the UV range of the spectrum, and will be way to the left in Fig. 37–1. In the visible region, the curve will be descending, so the shortest visible wavelengths will be strongest. Hence the star will appear bluish (or blue-white).

NOTE This example helps us to understand why stars have different colors (reddish for the coolest stars, orangish, yellow, white, bluish for "hotter" stars.)

FIGURE 37–2 Comparison of the Wien and the Rayleigh–Jeans theories to that of Planck, which closely follows experiment. The dashed lines show lack of agreement of older theories.

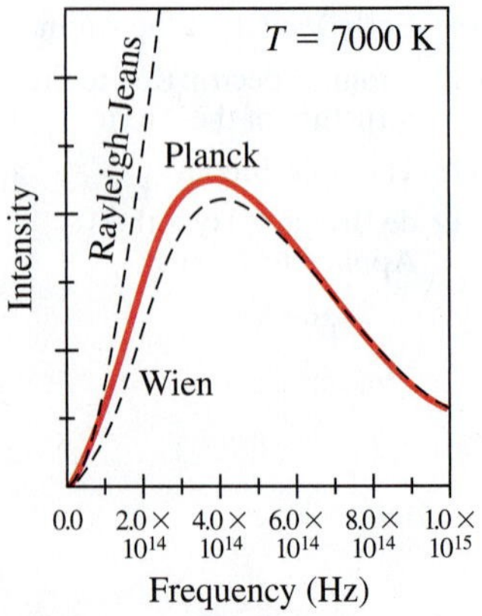

Planck's Quantum Hypothesis

A major problem facing scientists in the 1890s was to explain blackbody radiation. Maxwell's electromagnetic theory had predicted that oscillating electric charges produce electromagnetic waves, and the radiation emitted by a hot object could be due to the oscillations of electric charges in the molecules of the material. Although this would explain where the radiation came from, it did not correctly predict the observed spectrum of emitted light. Two important theoretical curves based on classical ideas were those proposed by W. Wien (in 1896) and by Lord Rayleigh (in 1900). The latter was modified later by J. Jeans and since then has been known as the Rayleigh–Jeans theory. As experimental data came in, it became clear that neither Wien's nor the Rayleigh–Jeans formulations were in accord with experiment (see Fig. 37–2).

In the year 1900 Max Planck (1858–1947) proposed an empirical formula that nicely fit the data (now often called *Planck's radiation formula*):

$$I(\lambda, T) = \frac{2\pi hc^2 \lambda^{-5}}{e^{hc/\lambda kT} - 1}.$$

$I(\lambda, T)$ is the radiation intensity as a function of wavelength λ at the temperature T; k is Boltzman's constant, c is the speed of light, and h is a new constant, now called **Planck's constant**. The value of h was estimated by Planck by fitting his formula for the blackbody radiation curve to experiment. The value accepted today is

$$h = 6.626 \times 10^{-34} \, \text{J} \cdot \text{s}.$$

To provide a theoretical basis for his formula, Planck made a new and radical assumption: that the energy of the oscillations of atoms within molecules cannot have just any value; instead each has energy which is a multiple of a minimum value related to the frequency of oscillation by

$$E = hf.$$

Planck's assumption suggests that the energy of any molecular vibration could be only a whole number multiple of the minimum energy hf:

$$E = nhf, \qquad n = 1, 2, 3, \cdots, \qquad \textbf{(37–2)}$$

where n is called a **quantum number** ("quantum" means "discrete amount" as opposed to "continuous"). This idea is often called **Planck's quantum hypothesis**, although little attention was brought to this point at the time. In fact, it appears that Planck considered it more as a mathematical device to get the "right answer" rather than as an important discovery in its own right. Planck himself continued to seek a classical explanation for the introduction of h. The recognition that this was an important and radical innovation did not come until later, after about 1905 when others, particularly Einstein, entered the field.

The quantum hypothesis, Eq. 37–2, states that the energy of an oscillator can be $E = hf$, or $2hf$, or $3hf$, and so on, but there cannot be vibrations with energies between these values. That is, energy would not be a continuous quantity as had been believed for centuries; rather it is **quantized**—it exists only in discrete amounts. The smallest amount of energy possible (hf) is called the **quantum of energy**. Recall from Chapter 14 that the energy of an oscillation is proportional to the amplitude squared. Another way of expressing the quantum hypothesis is that not just any amplitude of vibration is possible. The possible values for the amplitude are related to the frequency f.

A simple analogy may help. Compare a ramp, on which a box can be placed at any height, to a flight of stairs on which the box can have only certain discrete amounts of potential energy, as shown in Fig. 37–3.

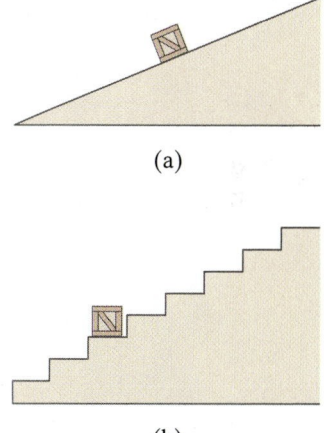

(a)

(b)

FIGURE 37–3 Ramp versus stair analogy. (a) On a ramp, a box can have continuous values of potential energy. (b) But on stairs, the box can have only discrete (quantized) values of energy.

37–2 Photon Theory of Light and the Photoelectric Effect

In 1905, the same year that he introduced the special theory of relativity, Einstein made a bold extension of the quantum idea by proposing a new theory of light. Planck's work had suggested that the vibrational energy of molecules in a radiating object is quantized with energy $E = nhf$, where n is an integer and f is the frequency of molecular vibration. Einstein argued that when light is emitted by a molecular oscillator, the molecule's vibrational energy of nhf must decrease by an amount hf (or by $2hf$, etc.) to another integer times hf, such as $(n - 1)hf$. Then to conserve energy, the light ought to be emitted in packets, or *quanta*, each with an energy

$$E = hf, \qquad \textbf{(37–3)}$$

where f is here the frequency of the emitted light.

Again h is Planck's constant. Since all light ultimately comes from a radiating source, this suggests that perhaps *light is transmitted as tiny particles*, or **photons**, as they are now called, as well as via waves predicted by Maxwell's electromagnetic theory. The photon theory of light was a radical departure from classical ideas. Einstein proposed a test of the quantum theory of light: quantitative measurements on the photoelectric effect.

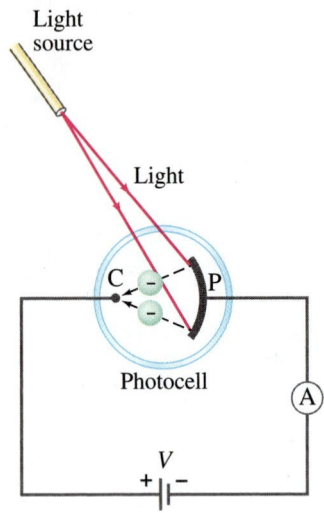

Light source

Light

C P

Photocell

Ⓐ

V
+ −

FIGURE 37–4 The photoelectric effect.

When light shines on a metal surface, electrons are found to be emitted from the surface. This effect is called the **photoelectric effect** and it occurs in many materials, but is most easily observed with metals. It can be observed using the apparatus shown in Fig. 37–4. A metal plate P and a smaller electrode C are placed inside an evacuated glass tube, called a **photocell**. The two electrodes are connected to an ammeter and a source of emf, as shown. When the photocell is in the dark, the ammeter reads zero. But when light of sufficiently high frequency illuminates the plate, the ammeter indicates a current flowing in the circuit. We explain completion of the circuit by imagining that electrons, ejected from the plate by the impinging radiation, flow across the tube from the plate to the "collector" C as indicated in Fig. 37–4.

That electrons should be emitted when light shines on a metal is consistent with the electromagnetic (EM) wave theory of light: the electric field of an EM wave could exert a force on electrons in the metal and eject some of them. Einstein pointed out, however, that the wave theory and the photon theory of light give very different predictions on the details of the photoelectric effect. For example, one thing that can be measured with the apparatus of Fig. 37–4 is the maximum kinetic energy (K_{max}) of the emitted electrons. This can be done by using a variable voltage source and reversing the terminals so that electrode C is negative and P is positive. The electrons emitted from P will be repelled by the negative electrode, but if this reverse voltage is small enough, the fastest electrons will still reach C and there will be a current in the circuit. If the reversed voltage is increased, a point is reached where the current reaches zero—no electrons have sufficient kinetic energy to reach C. This is called the *stopping potential*, or *stopping voltage*, V_0, and from its measurement, K_{max} can be determined using conservation of energy (loss of kinetic energy = gain in potential energy):

$$K_{max} = eV_0.$$

Now let us examine the details of the photoelectric effect from the point of view of the wave theory versus Einstein's particle theory.

First the wave theory, assuming monochromatic light. The two important properties of a light wave are its intensity and its frequency (or wavelength). When these two quantities are varied, the wave theory makes the following predictions:

Wave

theory

predictions

1. If the light intensity is increased, the number of electrons ejected and their maximum kinetic energy should be increased because the higher intensity means a greater electric field amplitude, and the greater electric field should eject electrons with higher speed.

2. The frequency of the light should not affect the kinetic energy of the ejected electrons. Only the intensity should affect K_{max}.

The photon theory makes completely different predictions. First we note that in a monochromatic beam, all photons have the same energy $(= hf)$. Increasing the intensity of the light beam means increasing the number of photons in the beam, but does not affect the energy of each photon as long as the frequency is not changed. According to Einstein's theory, an electron is ejected from the metal by a collision with a single photon. In the process, all the photon energy is transferred to the electron and the photon ceases to exist. Since electrons are held in the metal by attractive forces, some minimum energy W_0 is required just to get an electron out through the surface. W_0 is called the **work function**, and is a few electron volts $(1\,eV = 1.6 \times 10^{-19}\,J)$ for most metals. If the frequency f of the incoming light is so low that hf is less than W_0, then the photons will not have enough energy to eject any electrons at all. If $hf > W_0$, then electrons will be ejected and energy will be conserved in the process. That is, the input energy (of the photon), hf, will equal the outgoing kinetic energy K of the electron plus the energy required to get it out of the metal, W:

$$hf = K + W. \tag{37–4a}$$

The least tightly held electrons will be emitted with the most kinetic energy (K_{max}), in which case W in this equation becomes the work function W_0, and K becomes K_{max}:

$$hf = K_{\text{max}} + W_0. \qquad \text{[least bound electrons]} \quad \textbf{(37–4b)}$$

Many electrons will require more energy than the bare minimum (W_0) to get out of the metal, and thus the kinetic energy of such electrons will be less than the maximum.

From these considerations, the photon theory makes the following predictions:

1. An increase in intensity of the light beam means more photons are incident, so more electrons will be ejected; but since the energy of each photon is not changed, the maximum kinetic energy of electrons is not changed by an increase in intensity.

2. If the frequency of the light is increased, the maximum kinetic energy of the electrons increases linearly, according to Eq. 37–4b. That is,

$$K_{\text{max}} = hf - W_0.$$

This relationship is plotted in Fig. 37–5.

3. If the frequency f is less than the "cutoff" frequency f_0, where $hf_0 = W_0$, no electrons will be ejected, no matter how great the intensity of the light.

Photon

theory

predictions

These predictions of the photon theory are clearly very different from the predictions of the wave theory. In 1913–1914, careful experiments were carried out by R. A. Millikan. The results were fully in agreement with Einstein's photon theory.

One other aspect of the photoelectric effect also confirmed the photon theory. If extremely low light intensity is used, the wave theory predicts a time delay before electron emission so that an electron can absorb enough energy to exceed the work function. The photon theory predicts no such delay—it only takes one photon (if its frequency is high enough) to eject an electron—and experiments showed no delay. This too confirmed Einstein's photon theory.

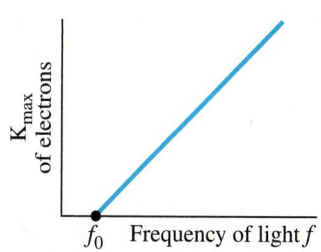

FIGURE 37–5 Photoelectric effect: the maximum kinetic energy of ejected electrons increases linearly with the frequency of incident light. No electrons are emitted if $f < f_0$.

EXAMPLE 37–3 **Photon energy.** Calculate the energy of a photon of blue light, $\lambda = 450\,\text{nm}$ in air (or vacuum).

APPROACH The photon has energy $E = hf$ (Eq. 37–3) where $f = c/\lambda$.

SOLUTION Since $f = c/\lambda$, we have

$$E = hf = \frac{hc}{\lambda} = \frac{(6.63 \times 10^{-34}\,\text{J} \cdot \text{s})(3.0 \times 10^8\,\text{m/s})}{(4.5 \times 10^{-7}\,\text{m})} = 4.4 \times 10^{-19}\,\text{J},$$

or $(4.4 \times 10^{-19}\,\text{J})/(1.60 \times 10^{-19}\,\text{J/eV}) = 2.8\,\text{eV}.$ (See definition of eV in Section 23–8, $1\,\text{eV} = 1.60 \times 10^{-19}\,\text{J}.$)

EXAMPLE 37–4 **ESTIMATE** **Photons from a lightbulb.** Estimate how many visible light photons a 100-W lightbulb emits per second. Assume the bulb has a typical efficiency of about 3% (that is, 97% of the energy goes to heat).

APPROACH Let's assume an average wavelength in the middle of the visible spectrum, $\lambda \approx 500\,\text{nm}$. The energy of each photon is $E = hf = hc/\lambda$. Only 3% of the 100-W power is emitted as light, or $3\,\text{W} = 3\,\text{J/s}$. The number of photons emitted per second equals the light output of $3\,\text{J/s}$ divided by the energy of each photon.

SOLUTION The energy emitted in one second ($= 3\,\text{J}$) is $E = Nhf$ where N is the number of photons emitted per second and $f = c/\lambda$. Hence

$$N = \frac{E}{hf} = \frac{E\lambda}{hc} = \frac{(3\,\text{J})(500 \times 10^{-9}\,\text{m})}{(6.63 \times 10^{-34}\,\text{J} \cdot \text{s})(3.0 \times 10^8\,\text{m/s})} \approx 8 \times 10^{18}$$

per second, or almost 10^{19} photons emitted per second, an enormous number.

EXERCISE A Compare a light beam that contains infrared light of a single wavelength, 1000 nm, with a beam of monochromatic UV at 100 nm, both of the same intensity. Are there more 100-nm photons or more 1000-nm photons?

EXAMPLE 37-5 **Photoelectron speed and energy.** What is the kinetic energy and the speed of an electron ejected from a sodium surface whose work function is $W_0 = 2.28\,\text{eV}$ when illuminated by light of wavelength (a) 410 nm, (b) 550 nm?

APPROACH We first find the energy of the photons ($E = hf = hc/\lambda$). If the energy is greater than W_0, then electrons will be ejected with varying amounts of kinetic energy, with a maximum of $K_{max} = hf - W_0$.

SOLUTION (a) For $\lambda = 410\,\text{nm}$,

$$hf = \frac{hc}{\lambda} = 4.85 \times 10^{-19}\,\text{J} \quad \text{or} \quad 3.03\,\text{eV}.$$

The maximum kinetic energy an electron can have is given by Eq. 37–4b, $K_{max} = 3.03\,\text{eV} - 2.28\,\text{eV} = 0.75\,\text{eV}$, or $(0.75\,\text{eV})(1.60 \times 10^{-19}\,\text{J/eV}) = 1.2 \times 10^{-19}\,\text{J}$. Since $K = \frac{1}{2}mv^2$ where $m = 9.11 \times 10^{-31}\,\text{kg}$,

$$v_{max} = \sqrt{\frac{2K}{m}} = 5.1 \times 10^5\,\text{m/s}.$$

Most ejected electrons will have less kinetic energy and less speed than these maximum values.

(b) For $\lambda = 550\,\text{nm}$, $hf = hc/\lambda = 3.61 \times 10^{-19}\,\text{J} = 2.26\,\text{eV}$. Since this photon energy is less than the work function, no electrons are ejected.

NOTE In (a) we used the nonrelativistic equation for kinetic energy. If v had turned out to be more than about $0.1c$, our calculation would have been inaccurate by at least a percent or so, and we would probably prefer to redo it using the relativistic form (Eq. 36–10).

EXERCISE B Determine the lowest frequency and the longest wavelength needed to emit electrons from sodium.

It is easy to show that the energy of a photon in electron volts, when given the wavelength λ in nm, is

$$E\,(\text{eV}) = \frac{1.240 \times 10^3\,\text{eV}\cdot\text{nm}}{\lambda\,(\text{nm})}. \qquad \text{[photon energy in eV]}$$

Applications of the Photoelectric Effect

The photoelectric effect, besides playing an important historical role in confirming the photon theory of light, also has many practical applications. Burglar alarms and automatic door openers often make use of the photocell circuit of Fig. 37–4. When a person interrupts the beam of light, the sudden drop in current in the circuit activates a switch—often a solenoid—which operates a bell or opens the door. UV or IR light is sometimes used in burglar alarms because of its invisibility. Many smoke detectors use the photoelectric effect to detect tiny amounts of smoke that interrupt the flow of light and so alter the electric current. Photographic light meters use this circuit as well. Photocells are used in many other devices, such as absorption spectrophotometers, to measure light intensity. One type of film sound track is a variably shaded narrow section at the side of the film. Light passing through the film is thus "modulated," and the output electrical signal of the photocell detector follows the frequencies on the sound track. See Fig. 37–6. For many applications today, the vacuum-tube photocell of Fig. 37–4 has been replaced by a semiconductor device known as a **photodiode** (Section 40–9). In these semiconductors, the absorption of a photon liberates a bound electron, which changes the conductivity of the material, so the current through a photodiode is altered.

FIGURE 37–6 Optical sound track on movie film. In the projector, light from a small source (different from that for the picture) passes through the sound track on the moving film. The light and dark areas on the sound track vary the intensity of the transmitted light which reaches the photocell, whose output current is then a replica of the original sound. This output is amplified and sent to the loudspeakers. High-quality projectors can show movies containing several parallel sound tracks to go to different speakers around the theater.

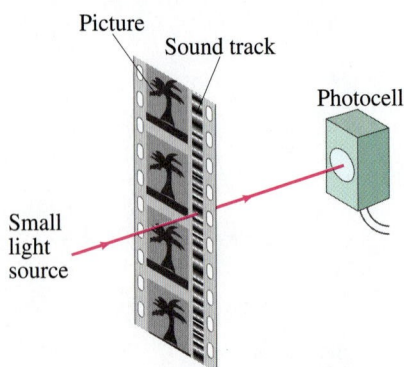

37–3 Energy, Mass, and Momentum of a Photon

We have just seen (Eq. 37–3) that the total energy of a single photon is given by $E = hf$. Because a photon always travels at the speed of light, it is truly a relativistic particle. Thus we must use relativistic formulas for dealing with its energy and momentum. The momentum of any particle of mass m is given by $p = mv/\sqrt{1 - v^2/c^2}$. Since $v = c$ for a photon, the denominator is zero. To avoid having an infinite momentum, we conclude that the photon's mass must be zero: $m = 0$. This makes sense too because a photon can never be at rest (it always moves at the speed of light). A photon's kinetic energy is its total energy:

$$K = E = hf. \qquad \text{[photon]}$$

The momentum of a photon can be obtained from the relativistic formula (Eq. 36–13) $E^2 = p^2c^2 + m^2c^4$ where we set $m = 0$, so $E^2 = p^2c^2$ or

$$p = \frac{E}{c}. \qquad \text{[photon]}$$

> ⚠ **CAUTION**
>
> *Momentum of photon is not mv*

Since $E = hf$ for a photon, its momentum is related to its wavelength by

$$p = \frac{E}{c} = \frac{hf}{c} = \frac{h}{\lambda}. \qquad \textbf{(37–5)}$$

EXAMPLE 37–6 **ESTIMATE** **Photon momentum and force.** Suppose the 10^{19} photons emitted per second from the 100-W lightbulb in Example 37–4 were all focused onto a piece of black paper and absorbed. (*a*) Calculate the momentum of one photon and (*b*) estimate the force all these photons could exert on the paper.

APPROACH Each photon's momentum is obtained from Eq. 37–5, $p = h/\lambda$. Next, each absorbed photon's momentum changes from $p = h/\lambda$ to zero. We use Newton's second law, $F = \Delta p/\Delta t$, to get the force. Let $\lambda = 500\,\text{nm}$.

SOLUTION (*a*) Each photon has a momentum

$$p = \frac{h}{\lambda} = \frac{6.63 \times 10^{-34}\,\text{J} \cdot \text{s}}{500 \times 10^{-9}\,\text{m}} = 1.3 \times 10^{-27}\,\text{kg} \cdot \text{m/s}.$$

(*b*) Using Newton's second law for $N = 10^{19}$ photons (Example 37–4) whose momentum changes from h/λ to 0, we obtain

$$F = \frac{\Delta p}{\Delta t} = \frac{Nh/\lambda - 0}{1\,\text{s}} = N\frac{h}{\lambda} \approx (10^{19}\,\text{s}^{-1})(10^{-27}\,\text{kg} \cdot \text{m/s}) \approx 10^{-8}\,\text{N}.$$

This is a tiny force, but we can see that a very strong light source could exert a measurable force, and near the Sun or a star the force due to photons in electromagnetic radiation could be considerable. See Section 31–9.

EXAMPLE 37–7 **Photosynthesis.** In *photosynthesis*, pigments such as chlorophyll in plants capture the energy of sunlight to change CO_2 to useful carbohydrate. About nine photons are needed to transform one molecule of CO_2 to carbohydrate and O_2. Assuming light of wavelength $\lambda = 670\,\text{nm}$ (chlorophyll absorbs most strongly in the range 650 nm to 700 nm), how efficient is the photosynthetic process? The reverse chemical reaction releases an energy of 4.9 eV/molecule of CO_2.

APPROACH The efficiency is the minimum energy required (4.9 eV) divided by the actual energy absorbed, nine times the energy (hf) of one photon.

SOLUTION The energy of nine photons, each of energy $hf = hc/\lambda$ is $(9)(6.63 \times 10^{-34}\,\text{J} \cdot \text{s})(3.0 \times 10^8\,\text{m/s})/(6.7 \times 10^{-7}\,\text{m}) = 2.7 \times 10^{-18}\,\text{J}$ or 17 eV. Thus the process is (4.9 eV/17 eV) = 29% efficient.

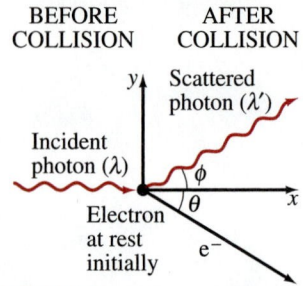

BEFORE COLLISION   **AFTER COLLISION**

FIGURE 37–7 The Compton effect. A single photon of wavelength λ strikes an electron in some material, knocking it out of its atom. The scattered photon has less energy (some energy is given to the electron) and hence has a longer wavelength λ'.

FIGURE 37–8 Plots of intensity of radiation scattered from a target such as graphite (carbon), for three different angles. The values for λ' match Eq. 37–6. For (a) $\phi = 0°$, $\lambda' = \lambda_0$. In (b) and (c) a peak is found not only at λ' due to photons scattered from free electrons (or very nearly free), but also a peak at almost precisely λ_0. The latter is due to scattering from electrons very tightly bound to their atoms so the mass in Eq. 37–6 becomes very large (mass of the atom) and $\Delta\lambda$ becomes very small.

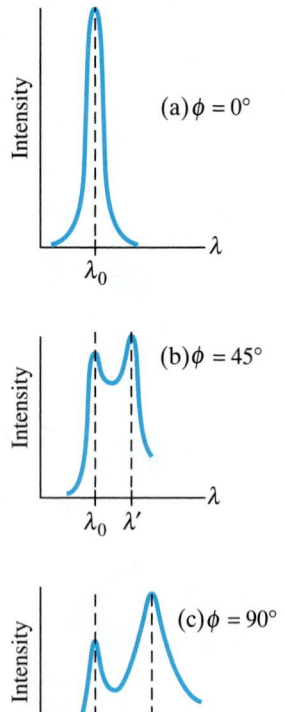

37–4 Compton Effect

Besides the photoelectric effect, a number of other experiments were carried out in the early twentieth century which also supported the photon theory. One of these was the **Compton effect** (1923) named after its discoverer, A. H. Compton (1892–1962). Compton scattered short-wavelength light (actually X-rays) from various materials. He found that the scattered light had a slightly longer wavelength than did the incident light, and therefore a slightly lower frequency indicating a loss of energy. He explained this result on the basis of the photon theory as incident photons colliding with electrons of the material, Fig. 37–7. Using Eq. 37–5 for momentum of a photon, Compton applied the laws of conservation of momentum and energy to the collision of Fig. 37–7 and derived the following equation for the wavelength of the scattered photons:

$$\lambda' = \lambda + \frac{h}{m_e c}(1 - \cos\phi), \qquad \text{(37–6a)}$$

where m_e is the mass of the electron. For $\phi = 0$, the wavelength is unchanged (there is no collision for this case of the photon passing straight through). At any other angle, λ' is longer than λ. The difference in wavelength,

$$\Delta\lambda = \lambda' - \lambda = \frac{h}{m_e c}(1 - \cos\phi), \qquad \text{(37–6b)}$$

is called the **Compton shift**. The quantity $h/m_e c$, which has the dimensions of length, is called the **Compton wavelength** λ_C of a free electron,

$$\lambda_C = \frac{h}{m_e c} = 2.43 \times 10^{-3}\,\text{nm} = 2.43\,\text{pm}. \qquad \text{[electron]}$$

Equations 37–6 predict that λ' depends on the angle ϕ at which the photons are detected. Compton's measurements of 1923 were consistent with this formula, confirming the value of λ_C and the dependence of λ' on ϕ. See Fig. 37–8. The wave theory of light predicts no wavelength shift: an incoming electromagnetic wave of frequency f should set electrons into oscillation at the same frequency f, and such oscillating electrons should reemit EM waves of this same frequency f (Chapter 31), and would not change with the angle ϕ. Hence the Compton effect adds to the firm experimental foundation for the photon theory of light.

EXERCISE C When a photon scatters off an electron by the Compton effect, which of the following increase: its energy, frequency, or wavelength?

EXAMPLE 37–8 **X-ray scattering.** X-rays of wavelength 0.140 nm are scattered from a very thin slice of carbon. What will be the wavelengths of X-rays scattered at (a) 0°, (b) 90°, (c) 180°?

APPROACH This is an example of the Compton effect, and we use Eq. 37–6a to find the wavelengths.

SOLUTION (a) For $\phi = 0°$, $\cos\phi = 1$ and $1 - \cos\phi = 0$. Then Eq. 37–6 gives $\lambda' = \lambda = 0.140\,\text{nm}$. This makes sense since for $\phi = 0°$, there really isn't any collision as the photon goes straight through without interacting.

(b) For $\phi = 90°$, $\cos\phi = 0$, and $1 - \cos\phi = 1$. So

$$\lambda' = \lambda + \frac{h}{m_e c} = 0.140\,\text{nm} + \frac{6.63 \times 10^{-34}\,\text{J·s}}{(9.11 \times 10^{-31}\,\text{kg})(3.00 \times 10^8\,\text{m/s})}$$

$$= 0.140\,\text{nm} + 2.4 \times 10^{-12}\,\text{m} = 0.142\,\text{nm};$$

that is, the wavelength is longer by one Compton wavelength (= 0.0024 nm for an electron).

(c) For $\phi = 180°$, which means the photon is scattered backward, returning in the direction from which it came (a direct "head-on" collision), $\cos \phi = -1$, and $1 - \cos \phi = 2$. So

$$\lambda' = \lambda + 2 \frac{h}{m_e c} = 0.140 \text{ nm} + 2(0.0024 \text{ nm}) = 0.145 \text{ nm}.$$

NOTE The maximum shift in wavelength occurs for backward scattering, and it is twice the Compton wavelength.

The Compton effect has been used to diagnose bone disease such as osteoporosis. Gamma rays, which are photons of even shorter wavelength than X-rays, coming from a radioactive source are scattered off bone material. The total intensity of the scattered radiation is proportional to the density of electrons, which is in turn proportional to the bone density. Changes in the density of bone can indicate the onset of osteoporosis.

PHYSICS APPLIED
Measuring bone density

*Derivation of Compton Shift

If the incoming photon in Fig. 37–7 has wavelength λ, then its total energy and momentum are

$$E = hf = \frac{hc}{\lambda} \quad \text{and} \quad p = \frac{h}{\lambda}.$$

After the collision of Fig. 37–7, the photon scattered at the angle ϕ has a wavelength which we call λ'. Its energy and momentum are

$$E' = \frac{hc}{\lambda'} \quad \text{and} \quad p' = \frac{h}{\lambda'}.$$

The electron, assumed at rest before the collision but free to move when struck, is scattered at an angle θ as shown in Fig. 37–8. The electron's kinetic energy is (see Eq. 36–10):

$$K_e = \left(\frac{1}{\sqrt{1 - v^2/c^2}} - 1 \right) m_e c^2$$

where m_e is the mass of the electron and v is its speed. The electron's momentum is

$$p_e = \frac{1}{\sqrt{1 - v^2/c^2}} m_e v.$$

We apply conservation of energy to the collision (see Fig. 37–7):

$$\text{incoming photon} \longrightarrow \text{scattered photon} + \text{electron}$$

$$\frac{hc}{\lambda} = \frac{hc}{\lambda'} + \left(\frac{1}{\sqrt{1 - v^2/c^2}} - 1 \right) m_e c^2.$$

We apply conservation of momentum to the x and y components of momentum:

$$\frac{h}{\lambda} = \frac{h}{\lambda'} \cos \phi + \frac{m_e v \cos \theta}{\sqrt{1 - v^2/c^2}}$$

$$0 = \frac{h}{\lambda'} \sin \phi - \frac{m_e v \sin \theta}{\sqrt{1 - v^2/c^2}}.$$

We can combine these three equations to eliminate v and θ, and we obtain, as Compton did, an equation for the wavelength of the scattered photon in terms of its scattering angle ϕ:

$$\lambda' = \lambda + \frac{h}{m_e c} (1 - \cos \phi),$$

which is Eq. 37–6a.

37–5 Photon Interactions; Pair Production

When a photon passes through matter, it interacts with the atoms and electrons. There are four important types of interactions that a photon can undergo:

1. The *photoelectric effect*: A photon may knock an electron out of an atom and in the process the photon disappears.

2. The photon may knock an atomic electron to a higher energy state in the atom if its energy is not sufficient to knock the electron out altogether. In this process the photon also disappears, and all its energy is given to the atom. Such an atom is then said to be in an *excited state*, and we shall discuss it more later.

3. The photon can be scattered from an electron (or a nucleus) and in the process lose some energy; this is the *Compton effect* (Fig. 37–7). But notice that the photon is not slowed down. It still travels with speed c, but its frequency will be lower because it has lost some energy.

4. *Pair production*: A photon can actually create matter, such as the production of an electron and a positron, Fig. 37–9. (A positron has the same mass as an electron, but the opposite charge, $+e$.)

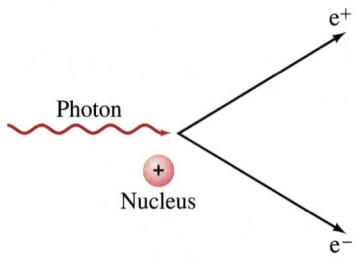

FIGURE 37–9 Pair production: a photon disappears and produces an electron and a positron.

In process 4, **pair production**, the photon disappears in the process of creating the electron–positron pair. This is an example of mass being created from pure energy, and it occurs in accord with Einstein's equation $E = mc^2$. Notice that a photon cannot create an electron alone since electric charge would not then be conserved. The inverse of pair production also occurs: if an electron collides with a positron, the two **annihilate** each other and their energy, including their mass, appears as electromagnetic energy of photons. Because of this process, positrons usually do not last long in nature.

EXAMPLE 37–9 **Pair production.** (*a*) What is the minimum energy of a photon that can produce an electron–positron pair? (*b*) What is this photon's wavelength?

APPROACH The minimum photon energy E equals the rest energy (mc^2) of the two particles created, via Einstein's famous equation $E = m_0 c^2$ (Eq. 36–12). There is no energy left over, so the particles produced will have zero kinetic energy. The wavelength is $\lambda = c/f$ where $E = hf$ for the original photon.

SOLUTION (*a*) Because $E = mc^2$, and the mass created is equal to two electron masses, the photon must have energy

$$E = 2(9.11 \times 10^{-31}\,\text{kg})(3.0 \times 10^8\,\text{m/s})^2 = 1.64 \times 10^{-13}\,\text{J} = 1.02\,\text{MeV}$$

$(1\,\text{MeV} = 10^6\,\text{eV} = 1.60 \times 10^{-13}\,\text{J})$. A photon with less energy cannot undergo pair production.

(*b*) Since $E = hf = hc/\lambda$, the wavelength of a 1.02-MeV photon is

$$\lambda = \frac{hc}{E} = \frac{(6.63 \times 10^{-34}\,\text{J·s})(3.0 \times 10^8\,\text{m/s})}{(1.64 \times 10^{-13}\,\text{J})} = 1.2 \times 10^{-12}\,\text{m},$$

which is 0.0012 nm. Such photons are in the gamma-ray (or very short X-ray) region of the electromagnetic spectrum (Fig. 31–12).

NOTE Photons of higher energy (shorter wavelength) can also create an electron–positron pair, with the excess energy becoming kinetic energy of the particles.

Pair production cannot occur in empty space, for momentum could not be conserved. In Example 37–9, for instance, energy is conserved, but only enough energy was provided to create the electron–positron pair at rest, and thus with zero momentum, which could not equal the initial momentum of the photon. Indeed, it can be shown that at any energy, an additional massive object, such as an atomic nucleus, must take part in the interaction to carry off some of the momentum.

37–6 Wave–Particle Duality; the Principle of Complementarity

The photoelectric effect, the Compton effect, and other experiments have placed the particle theory of light on a firm experimental basis. But what about the classic experiments of Young and others (Chapters 34 and 35) on interference and diffraction which showed that the wave theory of light also rests on a firm experimental basis?

We seem to be in a dilemma. Some experiments indicate that light behaves like a wave; others indicate that it behaves like a stream of particles. These two theories seem to be incompatible, but both have been shown to have validity. Physicists finally came to the conclusion that this duality of light must be accepted as a fact of life. It is referred to as the **wave–particle duality**. Apparently, light is a more complex phenomenon than just a simple wave or a simple beam of particles.

To clarify the situation, the great Danish physicist Niels Bohr (1885–1962, Fig. 37–10) proposed his famous **principle of complementarity**. It states that to understand an experiment, sometimes we find an explanation using wave theory and sometimes using particle theory. Yet we must be aware of both the wave and particle aspects of light if we are to have a full understanding of light. Therefore these two aspects of light complement one another.

It is not easy to "visualize" this duality. We cannot readily picture a combination of wave and particle. Instead, we must recognize that the two aspects of light are different "faces" that light shows to experimenters.

Part of the difficulty stems from how we think. Visual pictures (or models) in our minds are based on what we see in the everyday world. We apply the concepts of waves and particles to light because in the macroscopic world we see that energy is transferred from place to place by these two methods. We cannot see directly whether light is a wave or particle, so we do indirect experiments. To explain the experiments, we apply the models of waves or of particles to the nature of light. But these are abstractions of the human mind. When we try to conceive of what light really "is," we insist on a visual picture. Yet there is no reason why light should conform to these models (or visual images) taken from the macroscopic world. The "true" nature of light—if that means anything—is not possible to visualize. The best we can do is recognize that our knowledge is limited to the indirect experiments, and that in terms of everyday language and images, light reveals both wave and particle properties.

It is worth noting that Einstein's equation $E = hf$ itself links the particle and wave properties of a light beam. In this equation, E refers to the energy of a particle; and on the other side of the equation, we have the frequency f of the corresponding wave.

FIGURE 37–10 Niels Bohr (right), walking with Enrico Fermi along the Appian Way outside Rome. This photo shows one important way physics is done.

⚠ **CAUTION**

*Not correct to say light is a wave and/or a particle. Light can **act** like a wave or like a particle.*

37–7 Wave Nature of Matter

In 1923, Louis de Broglie (1892–1987) extended the idea of the wave–particle duality. He appreciated the *symmetry* in nature, and argued that if light sometimes behaves like a wave and sometimes like a particle, then perhaps those things in nature thought to be particles—such as electrons and other material objects—might also have wave properties. De Broglie proposed that the wavelength of a material particle would be related to its momentum in the same way as for a photon, Eq. 37–5, $p = h/\lambda$. That is, for a particle having linear momentum $p = mv$, the wavelength λ is given by

$$\lambda = \frac{h}{p}, \tag{37–7}$$

and is valid classically ($p = mv$ for $v \ll c$) and relativistically ($p = \gamma mv = mv/\sqrt{1 - v^2/c^2}$). This is sometimes called the **de Broglie wavelength** of a particle.

EXAMPLE 37–10 **Wavelength of a ball.** Calculate the de Broglie wavelength of a 0.20-kg ball moving with a speed of 15 m/s.

APPROACH We use Eq. 37–7.

SOLUTION $\lambda = \dfrac{h}{p} = \dfrac{h}{mv} = \dfrac{(6.6 \times 10^{-34}\,\text{J·s})}{(0.20\,\text{kg})(15\,\text{m/s})} = 2.2 \times 10^{-34}\,\text{m}.$

Ordinary size objects, such as the ball of Example 37–10, have unimaginably small wavelengths. Even if the speed is extremely small, say 10^{-4} m/s, the wavelength would be about 10^{-29} m. Indeed, the wavelength of any ordinary object is much too small to be measured and detected. The problem is that the properties of waves, such as interference and diffraction, are significant only when the size of objects or slits is not much larger than the wavelength. And there are no known objects or slits to diffract waves only 10^{-30} m long, so the wave properties of ordinary objects go undetected.

But tiny elementary particles, such as electrons, are another matter. Since the mass m appears in the denominator of Eq. 37–7, a very small mass should have a much larger wavelength.

EXAMPLE 37–11 **Wavelength of an electron.** Determine the wavelength of an electron that has been accelerated through a potential difference of 100 V.

APPROACH If the kinetic energy is much less than the rest energy, we can use the classical formula, $K = \frac{1}{2}mv^2$ (see Section 36–11). For an electron, $mc^2 = 0.511$ MeV. We then apply conservation of energy: the kinetic energy acquired by the electron equals its loss in potential energy. After solving for v, we use Eq. 37–7 to find the de Broglie wavelength.

SOLUTION Gain in kinetic energy equals loss in potential energy: $\Delta U = eV - 0$. Thus $K = eV$, so $K = 100$ eV. The ratio $K/mc^2 = 100\,\text{eV}/(0.511 \times 10^6\,\text{eV}) \approx 10^{-4}$, so relativity is not needed. Thus

$$\frac{1}{2}mv^2 = eV$$

and

$$v = \sqrt{\frac{2eV}{m}} = \sqrt{\frac{(2)(1.6 \times 10^{-19}\,\text{C})(100\,\text{V})}{(9.1 \times 10^{-31}\,\text{kg})}} = 5.9 \times 10^6\,\text{m/s}.$$

Then

$$\lambda = \frac{h}{mv} = \frac{(6.63 \times 10^{-34}\,\text{J·s})}{(9.1 \times 10^{-31}\,\text{kg})(5.9 \times 10^6\,\text{m/s})} = 1.2 \times 10^{-10}\,\text{m},$$

or 0.12 nm.

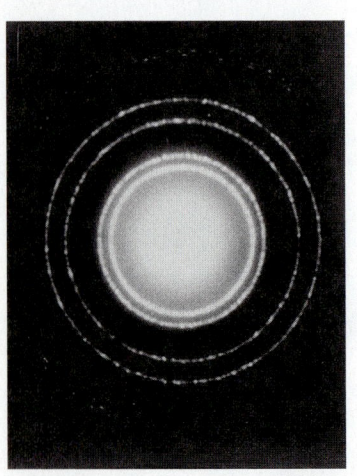

FIGURE 37–11 Diffraction pattern of electrons scattered from aluminum foil, as recorded on film.

EXERCISE D As a particle travels faster, does its de Broglie wavelength decrease, increase, or remain the same?

From Example 37–11, we see that electrons can have wavelengths on the order of 10^{-10} m, and even smaller. Although small, this wavelength can be detected: the spacing of atoms in a crystal is on the order of 10^{-10} m and the orderly array of atoms in a crystal could be used as a type of diffraction grating, as was done earlier for X-rays (see Section 35–10). C. J. Davisson and L. H. Germer performed the crucial experiment; they scattered electrons from the surface of a metal crystal and, in early 1927, observed that the electrons were scattered into a pattern of regular peaks. When they interpreted these peaks as a diffraction pattern, the wavelength of the diffracted electron wave was found to be just that predicted by de Broglie, Eq. 37–7. In the same year, G. P. Thomson (son of J. J. Thomson) used a different experimental arrangement and also detected diffraction of electrons. (See Fig. 37–11. Compare it to X-ray diffraction, Section 35–10.) Later experiments showed that protons, neutrons, and other particles also have wave properties.

Thus the wave–particle duality applies to material objects as well as to light. The principle of complementarity applies to matter as well. That is, we must be aware of both the particle and wave aspects in order to have an understanding of matter, including electrons. But again we must recognize that a visual picture of a "wave–particle" is not possible.

EXAMPLE 37–12 **Electron diffraction.** The wave nature of electrons is manifested in experiments where an electron beam interacts with the atoms on the surface of a solid. By studying the angular distribution of the diffracted electrons, one can indirectly measure the geometrical arrangement of atoms. Assume that the electrons strike perpendicular to the surface of a solid (see Fig. 37–12), and that their energy is low, $K = 100\,\text{eV}$, so that they interact only with the surface layer of atoms. If the smallest angle at which a diffraction maximum occurs is at $24°$, what is the separation d between the atoms on the surface?

SOLUTION Treating the electrons as waves, we need to determine the condition where the difference in path traveled by the wave diffracted from adjacent atoms is an integer multiple of the de Broglie wavelength, so that constructive interference occurs. The path length difference is $d \sin\theta$; so for the smallest value of θ we must have

$$d \sin\theta = \lambda.$$

However, λ is related to the (non-relativistic) kinetic energy K by

$$K = \frac{p^2}{2m_e} = \frac{h^2}{2m_e\lambda^2}.$$

Thus

$$\lambda = \frac{h}{\sqrt{2m_e K}} = \frac{(6.63 \times 10^{-34}\,\text{J·s})}{\sqrt{2(9.11 \times 10^{-31}\,\text{kg})(100\,\text{eV})(1.6 \times 10^{-19}\,\text{J/eV})}} = 0.123\,\text{nm}.$$

The surface inter-atomic spacing is

$$d = \frac{\lambda}{\sin\theta} = \frac{0.123\,\text{nm}}{\sin 24°} = 0.30\,\text{nm}.$$

EXERCISE E Return to the Chapter-Opening Question, page 987, and answer it again now. Try to explain why you may have answered differently the first time.

PHYSICS APPLIED
Electron diffraction

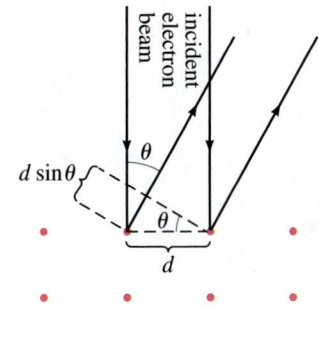

FIGURE 37–12 Example 37–12. The red dots represent atoms in an orderly array in a solid.

What Is an Electron?

We might ask ourselves: "What is an electron?" The early experiments of J. J. Thomson (Section 27–7) indicated a glow in a tube, and that glow moved when a magnetic field was applied. The results of these and other experiments were best interpreted as being caused by tiny negatively charged particles which we now call electrons. No one, however, has actually seen an electron directly. The drawings we sometimes make of electrons as tiny spheres with a negative charge on them are merely convenient pictures (now recognized to be inaccurate). Again we must rely on experimental results, some of which are best interpreted using the particle model and others using the wave model. These models are mere pictures that we use to extrapolate from the macroscopic world to the tiny microscopic world of the atom. And there is no reason to expect that these models somehow reflect the reality of an electron. We thus use a wave or a particle model (whichever works best in a situation) so that we can talk about what is happening. But we should not be led to believe that an electron *is* a wave or a particle. Instead we could say that an electron is the set of its properties that we can measure. Bertrand Russell said it well when he wrote that an electron is "a logical construction."

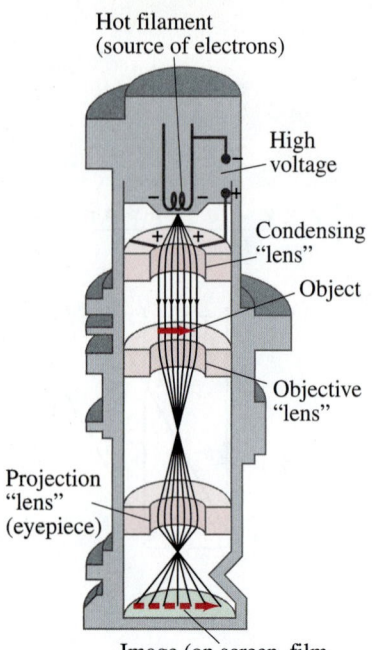

Hot filament
(source of electrons)

High
voltage

Condensing
"lens"

Object

Objective
"lens"

Projection
"lens"
(eyepiece)

Image (on screen, film,
or semiconductor detector)

FIGURE 37–13 Transmission electron microscope. The magnetic field coils are designed to be "magnetic lenses," which bend the electron paths and bring them to a focus, as shown.

PHYSICS APPLIED

Electron microscope

FIGURE 37–14 Scanning electron microscope. Scanning coils move an electron beam back and forth across the specimen. Secondary electrons produced when the beam strikes the specimen are collected and modulate the intensity of the beam in the CRT to produce a picture.

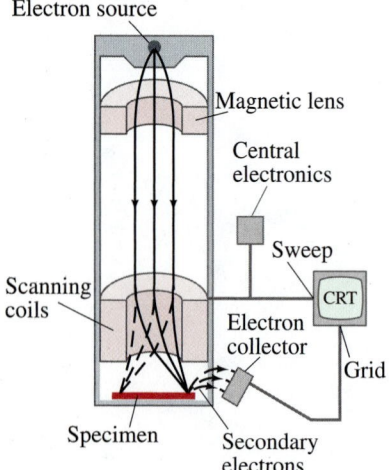

Electron source

Magnetic lens

Central
electronics

Sweep

CRT

Scanning
coils

Electron
collector

Grid

Specimen

Secondary
electrons

*37–8 Electron Microscopes

The idea that electrons have wave properties led to the development of the **electron microscope**, which can produce images of much greater magnification than does a light microscope. Figures 37–13 and 37–14 are diagrams of two types, developed around the middle of the twentieth century: the **transmission electron microscope**, which produces a two-dimensional image, and the **scanning electron microscope** (SEM), which produces images with a three-dimensional quality. In both types, the objective and eyepiece lenses are actually magnetic fields that exert forces on the electrons to bring them to a focus. The fields are produced by carefully designed current-carrying coils of wire. Photographs using each type are shown in Fig. 37–15.

As discussed in Section 35–5, the maximum resolution of details on an object is about the size of the wavelength of the radiation used to view it. Electrons accelerated by voltages on the order of 10^5 V have wavelengths of about 0.004 nm. The maximum resolution obtainable would be on this order, but in practice, aberrations in the magnetic lenses limit the resolution in transmission electron microscopes to at best about 0.1 to 0.5 nm. This is still 10^3 times better than that attainable with a visible-light microscope, and corresponds to a useful magnification of about a million. Such magnifications are difficult to attain, and more common magnifications are 10^4 to 10^5. The maximum resolution attainable with a scanning electron microscope is somewhat less, typically 5 to 10 nm although new high-resolution SEMs approach 1 nm.

We discuss other sophisticated electron microscopes in the next Chapter, Section 38–10.

FIGURE 37–15 Electron micrographs (in false color) of viruses attacking a cell of the bacterium *Escherichia coli*: (a) transmission electron micrograph ($\approx$ 50,000$\times$); (b) scanning electron micrograph ($\approx$ 35,000$\times$).

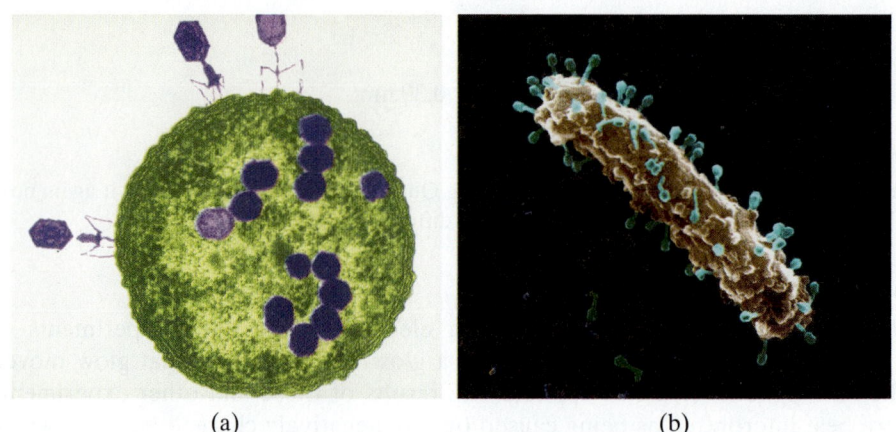

(a)

(b)

37–9 Early Models of the Atom

The idea that matter is made up of atoms was accepted by most scientists by 1900. With the discovery of the electron in the 1890s, scientists began to think of the atom itself as having a structure with electrons as part of that structure. We now introduce our modern approach to the atom and the quantum theory with which it is intertwined.[†]

[†]Some readers may say: "Tell us the facts as we know them today, and don't bother us with the historical background and its outmoded theories." Such an approach would ignore the creative aspect of science and thus give a false impression of how science develops. Moreover, it is not really possible to understand today's view of the atom without insight into the concepts that led to it.

A typical model of the atom in the 1890s visualized the atom as a homogeneous sphere of positive charge inside of which there were tiny negatively charged electrons, a little like plums in a pudding, Fig. 37–16.

Around 1911, Ernest Rutherford (1871–1937) and his colleagues performed experiments whose results contradicted the plum-pudding model of the atom. In these experiments a beam of positively charged alpha (α) particles was directed at a thin sheet of metal foil such as gold, Fig. 37–17a. (These newly discovered α particles were emitted by certain radioactive materials and were soon shown to be doubly ionized helium atoms—that is, having a charge of $+2e$.) It was expected from the plum-pudding model that the alpha particles would not be deflected significantly because electrons are so much lighter than alpha particles, and the alpha particles should not have encountered any massive concentration of positive charge to strongly repel them. The experimental results completely contradicted these predictions. It was found that most of the alpha particles passed through the foil unaffected, as if the foil were mostly empty space. And of those deflected, a few were deflected at very large angles—some even backward, nearly in the direction from which they had come. This could happen, Rutherford reasoned, only if the positively charged alpha particles were being repelled by a massive positive charge concentrated in a very small region of space (see Fig. 37–17b).

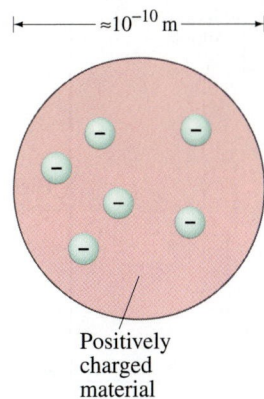

FIGURE 37–16 Plum-pudding model of the atom.

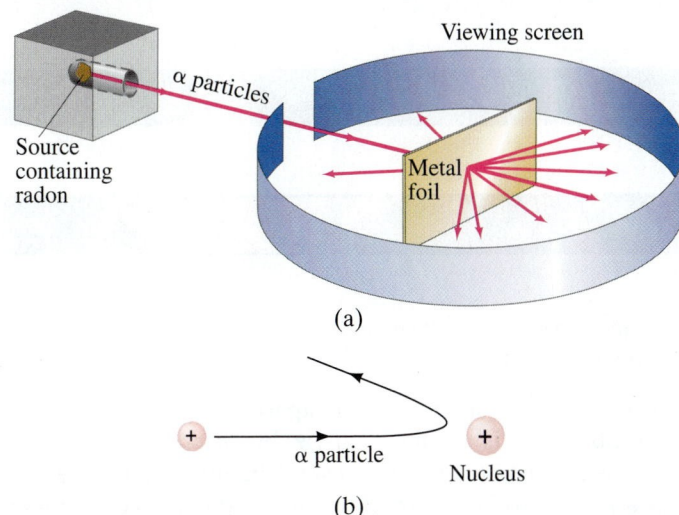

(a)

(b)

FIGURE 37–17 (a) Experimental setup for Rutherford's experiment: α particles emitted by radon are deflected by a thin metallic foil and a few rebound backward; (b) backward rebound of α particles explained as the repulsion from a heavy positively charged nucleus.

He hypothesized that the atom must consist of a tiny but massive positively charged nucleus, containing over 99.9% of the mass of the atom, surrounded by electrons some distance away. The electrons would be moving in orbits about the nucleus—much as the planets move around the Sun—because if they were at rest, they would fall into the nucleus due to electrical attraction, Fig. 37–18. Rutherford's experiments suggested that the nucleus must have a radius of about 10^{-15} to 10^{-14} m. From kinetic theory, and especially Einstein's analysis of Brownian motion (see Section 17–1), the radius of atoms was estimated to be about 10^{-10} m. Thus the electrons would seem to be at a distance from the nucleus of about 10,000 to 100,000 times the radius of the nucleus itself. (If the nucleus were the size of a baseball, the atom would have the diameter of a big city several kilometers across.) So an atom would be mostly empty space.

Rutherford's "planetary" model of the atom (also called the "nuclear model of the atom") was a major step toward how we view the atom today. It was not, however, a complete model and presented some major problems, as we shall see.

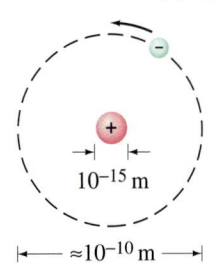

FIGURE 37–18 Rutherford's model of the atom, in which electrons orbit a tiny positive nucleus (not to scale). The atom is visualized as mostly empty space.

37–10 Atomic Spectra: Key to the Structure of the Atom

Earlier in this Chapter we saw that heated solids (as well as liquids and dense gases) emit light with a continuous spectrum of wavelengths. This radiation is assumed to be due to oscillations of atoms and molecules, which are largely governed by the interaction of each atom or molecule with its neighbors.

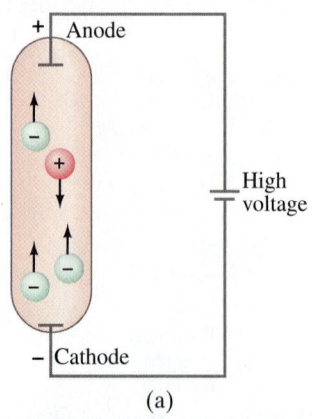

(a)

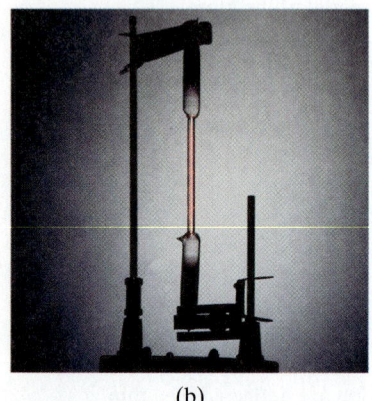

(b)

FIGURE 37–19 Gas-discharge tube: (a) diagram; (b) photo of an actual discharge tube for hydrogen.

FIGURE 37–21 Balmer series of lines for hydrogen.

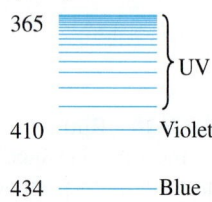

365

UV

410 —————— Violet

434 —————— Blue

486 —————— Blue-green

λ
(nm)

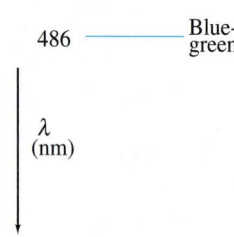

656 —————— Red

Rarefied gases can also be excited to emit light. This is done by intense heating, or more commonly by applying a high voltage to a "discharge tube" containing the gas at low pressure, Fig. 37–19. The radiation from excited gases had been observed early in the nineteenth century, and it was found that the spectrum was not continuous, but *discrete*. Since excited gases emit light of only certain wavelengths, when this light is analyzed through the slit of a spectroscope or spectrometer, a **line spectrum** is seen rather than a continuous spectrum. The line spectra in the visible region emitted by a number of elements are shown below in Fig. 37–20, and in Chapter 35, Fig. 35–22. The **emission spectrum** is characteristic of the material and can serve as a type of "fingerprint" for identification of the gas.

We also saw (Chapter 35) that if a continuous spectrum passes through a rarefied gas, dark lines are observed in the emerging spectrum, at wavelengths corresponding to lines normally emitted by the gas. This is called an **absorption spectrum** (Fig. 37–20c), and it became clear that gases can absorb light at the same frequencies at which they emit. Using film sensitive to ultraviolet and to infrared light, it was found that gases emit and absorb discrete frequencies in these regions as well as in the visible.

(a)

(b)

(c)

FIGURE 37–20 Emission spectra of the gases (a) atomic hydrogen, (b) helium, and (c) the *solar absorption* spectrum.

In low-density gases, the atoms are far apart on the average and hence the light emitted or absorbed is assumed to be by *individual atoms* rather than through interactions between atoms, as in a solid, liquid, or dense gas. Thus the line spectra serve as a key to the structure of the atom: any theory of atomic structure must be able to explain why atoms emit light only of discrete wavelengths, and it should be able to predict what these wavelengths are.

Hydrogen is the simplest atom—it has only one electron. It also has the simplest spectrum. The spectrum of most atoms shows little apparent regularity. But the spacing between lines in the hydrogen spectrum decreases in a regular way, Fig. 37–21. Indeed, in 1885, J. J. Balmer (1825–1898) showed that the four lines in the visible portion of the hydrogen spectrum (with measured wavelengths 656 nm, 486 nm, 434 nm, and 410 nm) have wavelengths that fit the formula

$$\frac{1}{\lambda} = R\left(\frac{1}{2^2} - \frac{1}{n^2}\right), \qquad n = 3, 4, \cdots. \qquad \textbf{(37–8)}$$

Here n takes on the values 3, 4, 5, 6 for the four visible lines, and R, called the **Rydberg constant**, has the value $R = 1.0974 \times 10^7 \, \text{m}^{-1}$. Later it was found that this **Balmer series** of lines extended into the UV region, ending at $\lambda = 365$ nm, as shown in Fig. 37–21. Balmer's formula, Eq. 37–8, also worked for these lines with higher integer values of n. The lines near 365 nm became too close together to distinguish, but the limit of the series at 365 nm corresponds to $n = \infty$ (so $1/n^2 = 0$ in Eq. 37–8).

Later experiments on hydrogen showed that there were similar series of lines in the UV and IR regions, and each series had a pattern just like the Balmer series, but at different wavelengths, Fig. 37–22. Each of these series was found to fit a formula with the same form as Eq. 37–8 but with the $1/2^2$ replaced by $1/1^2$, $1/3^2$, $1/4^2$, and so on. For example, the so-called **Lyman series** contains lines

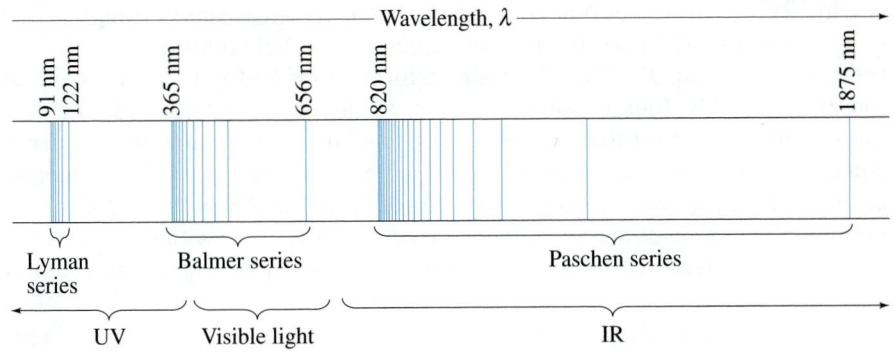

with wavelengths from 91 nm to 122 nm (in the UV region) and fits the formula

$$\frac{1}{\lambda} = R\left(\frac{1}{1^2} - \frac{1}{n^2}\right), \qquad n = 2, 3, \cdots.$$

The wavelengths of the **Paschen series** (in the IR region) fit

$$\frac{1}{\lambda} = R\left(\frac{1}{3^2} - \frac{1}{n^2}\right), \qquad n = 4, 5, \cdots.$$

The Rutherford model was unable to explain why atoms emit line spectra. It had other difficulties as well. According to the Rutherford model, electrons orbit the nucleus, and since their paths are curved the electrons are accelerating. Hence they should give off light like any other accelerating electric charge (Chapter 31), with a frequency equal to its orbital frequency. Since light carries off energy and energy is conserved, the electron's own energy must decrease to compensate. Hence electrons would be expected to spiral into the nucleus. As they spiraled inward, their frequency would increase in a short time and so too would the frequency of the light emitted. Thus the two main difficulties with the Rutherford model are these: (1) it predicts that light of a continuous range of frequencies will be emitted, whereas experiment shows line spectra; (2) it predicts that atoms are unstable—electrons would quickly spiral into the nucleus—but we know that atoms in general are stable, because there is stable matter all around us.

Clearly Rutherford's model was not sufficient. Some sort of modification was needed, and Niels Bohr provided it in a model that included the quantum hypothesis. Although the Bohr model has been superceded, it did provide a crucial stepping stone to our present understanding. And some aspects of the Bohr model are still useful today, so we examine it in detail in the next Section.

37–11 The Bohr Model

Bohr had studied in Rutherford's laboratory for several months in 1912 and was convinced that Rutherford's planetary model of the atom had validity. But in order to make it work, he felt that the newly developing quantum theory would somehow have to be incorporated in it. The work of Planck and Einstein had shown that in heated solids, the energy of oscillating electric charges must change discontinuously—from one discrete energy state to another, with the emission of a quantum of light. Perhaps, Bohr argued, the electrons in an atom also cannot lose energy continuously, but must do so in quantum "jumps." In working out his model during the next year, Bohr postulated that electrons move about the nucleus in circular orbits, but that only certain orbits are allowed. He further postulated that an electron in each orbit would have a definite energy and would move in the orbit *without radiating energy* (even though this violated classical ideas since accelerating electric charges are supposed to emit EM waves; see Chapter 31). He thus called the possible orbits **stationary states**. Light is emitted, he hypothesized, only when an electron jumps from a higher (upper) stationary state to another of lower energy, Fig. 37–23. When such a transition occurs, a single photon of light is emitted whose energy, by energy conservation, is given by

$$hf = E_U - E_L, \tag{37–9}$$

where E_U refers to the energy of the upper state and E_L the energy of the lower state.

FIGURE 37–23 An atom emits a photon (energy $= hf$) when its energy changes from E_U to a lower energy E_L.

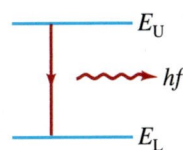

In 1912–13, Bohr set out to determine what energies these orbits would have in the simplest atom, hydrogen; the spectrum of light emitted could then be predicted from Eq. 37–9. In the Balmer formula he had the key he was looking for. Bohr quickly found that his theory would be in accord with the Balmer formula if he assumed that the electron's angular momentum L is quantized and equal to an integer n times $h/2\pi$. As we saw in Chapter 11 angular momentum is given by $L = I\omega$, where I is the moment of inertia and ω is the angular velocity. For a single particle of mass m moving in a circle of radius r with speed v, $I = mr^2$ and $\omega = v/r$; hence, $L = I\omega = (mr^2)(v/r) = mvr$. Bohr's **quantum condition** is

$$L = mvr_n = n\frac{h}{2\pi}, \qquad n = 1, 2, 3, \cdots, \qquad \textbf{(37–10)}$$

where n is an integer and r_n is the radius of the nth possible orbit. The allowed orbits are numbered $1, 2, 3, \cdots$, according to the value of n, which is called the **principal quantum number** of the orbit.

Equation 37–10 did not have a firm theoretical foundation. Bohr had searched for some "quantum condition," and such tries as $E = hf$ (where E represents the energy of the electron in an orbit) did not give results in accord with experiment. Bohr's reason for using Eq. 37–10 was simply that it worked; and we now look at how. In particular, let us determine what the Bohr theory predicts for the measurable wavelengths of emitted light.

An electron in a circular orbit of radius r_n (Fig. 37–24) would have a centripetal acceleration v^2/r_n produced by the electrical force of attraction between the negative electron and the positive nucleus. This force is given by Coulomb's law,

$$F = \frac{1}{4\pi\epsilon_0}\frac{(Ze)(e)}{r_n^2}.$$

The charge on the nucleus is $+Ze$, where Z is the number of positive charges[†] (i.e., protons). For the hydrogen atom, $Z = +1$.

In Newton's second law, $F = ma$, we substitute Coulomb's law for F, and $a = v^2/r_n$ for a particular allowed orbit of radius r_n, and obtain

$$F = ma$$
$$\frac{1}{4\pi\epsilon_0}\frac{Ze^2}{r_n^2} = \frac{mv^2}{r_n}.$$

We solve this for r_n, and then substitute for v from Eq. 37–10 (which says $v = nh/2\pi m r_n$):

$$r_n = \frac{Ze^2}{4\pi\epsilon_0 mv^2} = \frac{Ze^2 4\pi^2 mr_n^2}{4\pi\epsilon_0 n^2 h^2}.$$

We solve for r_n (it appears on both sides, so we cancel one of them) and find

$$r_n = \frac{n^2 h^2 \epsilon_0}{\pi m Z e^2} = \frac{n^2}{Z}r_1 \qquad \textbf{(37–11)}$$

where

$$r_1 = \frac{h^2 \epsilon_0}{\pi m e^2}.$$

Equation 37–11 gives the radii of all possible orbits. The smallest orbit is for $n = 1$, and for hydrogen $(Z = 1)$ has the value

$$r_1 = \frac{(1)^2(6.626 \times 10^{-34}\,\text{J}\cdot\text{s})^2(8.85 \times 10^{-12}\,\text{C}^2/\text{N}\cdot\text{m}^2)}{(3.14)(9.11 \times 10^{-31}\,\text{kg})(1.602 \times 10^{-19}\,\text{C})^2}$$

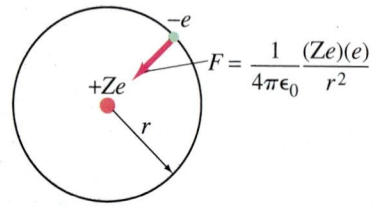

FIGURE 37–24 Electric force (Coulomb's law) keeps the negative electron in orbit around the positively charged nucleus.

[†]We include Z in our derivation so that we can treat other single-electron ("hydrogenlike") atoms such as the ions He$^+$ ($Z = 2$) and Li^{2+} ($Z = 3$). Helium in the neutral state has two electrons: if one electron is missing, the remaining He$^+$ ion consists of one electron revolving around a nucleus of charge $+2e$. Similarly, doubly ionized lithium, Li^{2+}, also has a single electron, and in this case $Z = 3$.

or
$$r_1 = 0.529 \times 10^{-10}\,\text{m}. \qquad (37\text{–}12)$$

The radius of the smallest orbit in hydrogen, r_1, is sometimes called the **Bohr radius**. From Eq. 37–11, we see that the radii of the larger orbits[†] increase as n^2, so

$$r_2 = 4r_1 = 2.12 \times 10^{-10}\,\text{m},$$
$$r_3 = 9r_1 = 4.76 \times 10^{-10}\,\text{m},$$
$$\vdots$$
$$r_n = n^2 r_1.$$

The first four orbits are shown in Fig. 37–25. Notice that, according to Bohr's model, an electron can exist only in the orbits given by Eq. 37–11. There are no allowable orbits in between.

For an atom with $Z \neq 1$, we can write the orbital radii, r_n, using Eq. 37–11:

$$r_n = \frac{n^2}{Z}(0.529 \times 10^{-10}\,\text{m}), \qquad n = 1, 2, 3, \cdots. \qquad (37\text{–}13)$$

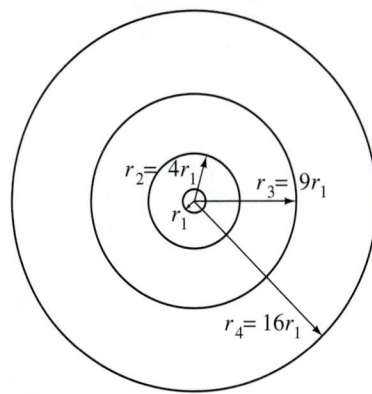

FIGURE 37–25 The four smallest orbits in the Bohr model of hydrogen; $r_1 = 0.529 \times 10^{-10}$ m.

In each of its possible orbits, the electron would have a definite energy, as the following calculation shows. The total energy equals the sum of the kinetic and potential energies. The potential energy of the electron is given by $U = qV = -eV$, where V is the potential due to a point charge $+Ze$ as given by Eq. 23–5: $V = (1/4\pi\epsilon_0)(Q/r) = (1/4\pi\epsilon_0)(Ze/r)$. So

$$U = -eV = -\frac{1}{4\pi\epsilon_0}\frac{Ze^2}{r}.$$

The total energy E_n for an electron in the nth orbit of radius r_n is the sum of the kinetic and potential energies:

$$E_n = \tfrac{1}{2}mv^2 - \frac{1}{4\pi\epsilon_0}\frac{Ze^2}{r_n}.$$

When we substitute v from Eq. 37–10 and r_n from Eq. 37–11 into this equation, we obtain

$$E_n = -\left(\frac{Z^2 e^4 m}{8\epsilon_0^2 h^2}\right)\left(\frac{1}{n^2}\right), \qquad n = 1, 2, 3, \cdots. \qquad (37\text{–}14a)$$

If we evaluate the constant term in Eq. 37–14a and convert it to electron volts, as is customary in atomic physics, we obtain

$$E_n = -(13.6\,\text{eV})\frac{Z^2}{n^2}, \qquad n = 1, 2, 3, \cdots. \qquad (37\text{–}14b)$$

The lowest energy level $(n = 1)$ for hydrogen $(Z = 1)$ is

$$E_1 = -13.6\,\text{eV}.$$

Since n^2 appears in the denominator of Eq. 37–14b, the energies of the larger orbits in hydrogen $(Z = 1)$ are given by

$$E_n = \frac{-13.6\,\text{eV}}{n^2}.$$

For example,

$$E_2 = \frac{-13.6\,\text{eV}}{4} = -3.40\,\text{eV},$$

$$E_3 = \frac{-13.6\,\text{eV}}{9} = -1.51\,\text{eV}.$$

We see that not only are the orbit radii quantized, but from Eqs. 37–14 so is the energy. The quantum number n that labels the orbit radii also labels the energy levels. The lowest **energy level** or **energy state** has energy E_1, and is called the **ground state**. The higher states, E_2, E_3, and so on, are called **excited states**. The fixed energy levels are also called **stationary states**.

[†]Be careful not to believe that these well-defined orbits actually exist. Today electrons are better thought of as forming "clouds," as discussed in Chapter 39.

Notice that although the energy for the larger orbits has a smaller numerical value, all the energies are less than zero. Thus, $-3.4\,\text{eV}$ is a higher energy than $-13.6\,\text{eV}$. Hence the orbit closest to the nucleus (r_1) has the lowest energy. The reason the energies have negative values has to do with the way we defined the zero for potential energy (U). For two point charges, $U = (1/4\pi\epsilon_0)(q_1 q_2/r)$ corresponds to zero potential energy when the two charges are infinitely far apart. Thus, an electron that can just barely be free from the atom by reaching $r = \infty$ (or, at least, far from the nucleus) with zero kinetic energy will have $E = 0$, corresponding to $n = \infty$ in Eqs. 37–14. If an electron is free and has kinetic energy, then $E > 0$. To remove an electron that is part of an atom requires an energy input (otherwise atoms would not be stable). Since $E \geq 0$ for a free electron, then an electron bound to an atom needs to have $E < 0$. That is, energy must be added to bring its energy up, from a negative value, to at least zero in order to free it.

The minimum energy required to remove an electron from an atom initially in the ground state is called the **binding energy** or **ionization energy**. The ionization energy for hydrogen has been measured to be $13.6\,\text{eV}$, and this corresponds precisely to removing an electron from the lowest state, $E_1 = -13.6\,\text{eV}$, up to $E = 0$ where it can be free.

Spectra Lines Explained

It is useful to show the various possible energy values as horizontal lines on an energy-level diagram. This is shown for hydrogen in Fig. 37–26.[†] The electron in a hydrogen atom can be in any one of these levels according to Bohr's theory. But it could never be in between, say at $-9.0\,\text{eV}$. At room temperature, nearly all H atoms will be in the ground state ($n = 1$). At higher temperatures, or during an electric discharge when there are many collisions between free electrons and atoms, many atoms can be in excited states ($n > 1$). Once in an excited state, an atom's electron can jump down to a lower state, and give off a photon in the process. This is, according to the Bohr model, the origin of the emission spectra of excited gases.

[†]Note that above $E = 0$, an electron is free and can have any energy (E is not quantized). Thus there is a continuum of energy states above $E = 0$, as indicated in the energy-level diagram of Fig. 37–26.

FIGURE 37–26 Energy-level diagram for the hydrogen atom, showing the transitions for the spectral lines of the Lyman, Balmer, and Paschen series (Fig. 37–22). Each vertical arrow represents an atomic transition that gives rise to the photons of one spectral line (a single wavelength or frequency).

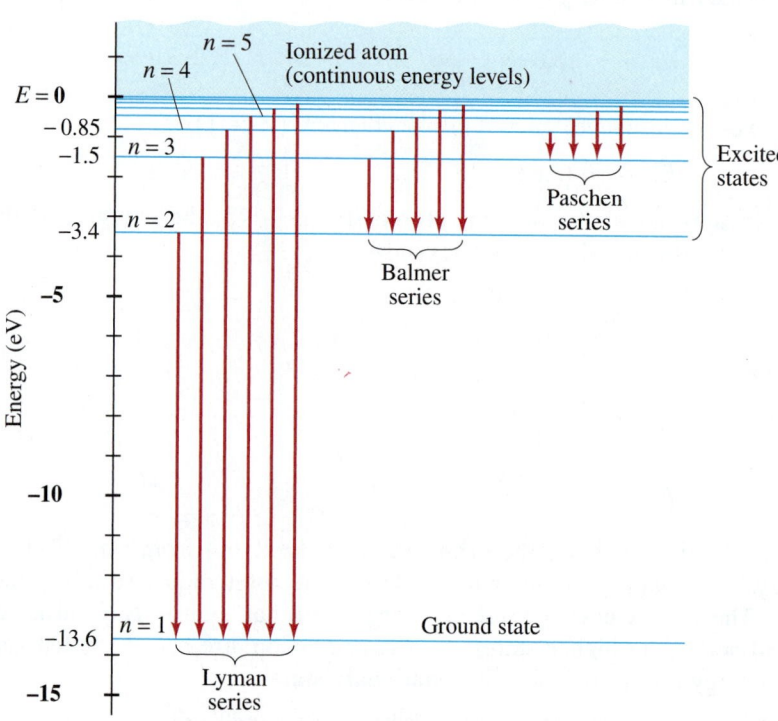

The vertical arrows in Fig. 37–26 represent the transitions or jumps that correspond to the various observed spectral lines. For example, an electron jumping from the level $n = 3$ to $n = 2$ would give rise to the 656-nm line in the Balmer series, and the jump from $n = 4$ to $n = 2$ would give rise to the 486-nm line (see Fig. 37–21). We can predict wavelengths of the spectral lines emitted by combining Eq. 37–9 with Eq. 37–14a. Since $hf = hc/\lambda$, we have from Eq. 37–9

$$\frac{1}{\lambda} = \frac{hf}{hc} = \frac{1}{hc}(E_n - E_{n'}),$$

where n refers to the upper state and n' to the lower state. Then using Eq. 37–14a,

$$\frac{1}{\lambda} = \frac{Z^2 e^4 m}{8\epsilon_0^2 h^3 c}\left(\frac{1}{(n')^2} - \frac{1}{(n)^2}\right). \qquad (37\text{–}15)$$

This theoretical formula has the same form as the experimental Balmer formula, Eq. 37–8, with $n' = 2$. Thus we see that the Balmer series of lines corresponds to transitions or "jumps" that bring the electron down to the second energy level. Similarly, $n' = 1$ corresponds to the Lyman series and $n' = 3$ to the Paschen series (see Fig. 37–26).

When the constant in Eq. 37–15 is evaluated with $Z = 1$, it is found to have the measured value of the Rydberg constant, $R = 1.0974 \times 10^7\,\text{m}^{-1}$ in Eq. 37–8, in accord with experiment (see Problem 58).

The great success of Bohr's model is that it gives an explanation for why atoms emit line spectra, and accurately predicts the wavelengths of emitted light for hydrogen. The Bohr model also explains absorption spectra: photons of just the right wavelength can knock an electron from one energy level to a higher one. To conserve energy, only photons that have just the right energy will be absorbed. This explains why a continuous spectrum of light entering a gas will emerge with dark (absorption) lines at frequencies that correspond to emission lines (Fig. 37–20c).

The Bohr theory also ensures the stability of atoms. It establishes stability by decree: the ground state is the lowest state for an electron and there is no lower energy level to which it can go and emit more energy. Finally, as we saw above, the Bohr theory accurately predicts the ionization energy of 13.6 eV for hydrogen. However, the Bohr model was not so successful for other atoms, and has been superseded as we shall discuss in the next Chapter. We discuss the Bohr model because it *was* an important start, and because we still use the concept of stationary states, the ground state, and transitions between states. Also, the terminology used in the Bohr model is still used by chemists and spectroscopists.

EXAMPLE 37–13 **Wavelength of a Lyman line.** Use Fig. 37–26 to determine the wavelength of the first Lyman line, the transition from $n = 2$ to $n = 1$. In what region of the electromagnetic spectrum does this lie?

APPROACH We use Eq. 37–9, $hf = E_U - E_L$, with the energies obtained from Fig. 37–26 to find the energy and the wavelength of the transition. The region of the electromagnetic spectrum is found using the EM spectrum in Fig. 31–12.

SOLUTION In this case, $hf = E_2 - E_1 = \{-3.4\,\text{eV} - (-13.6\,\text{eV})\} = 10.2\,\text{eV}$ $= (10.2\,\text{eV})(1.60 \times 10^{-19}\,\text{J/eV}) = 1.63 \times 10^{-18}\,\text{J}$. Since $\lambda = c/f$, we have

$$\lambda = \frac{c}{f} = \frac{hc}{E_2 - E_1} = \frac{(6.63 \times 10^{-34}\,\text{J}\cdot\text{s})(3.00 \times 10^8\,\text{m/s})}{1.63 \times 10^{-18}\,\text{J}} = 1.22 \times 10^{-7}\,\text{m},$$

or 122 nm, which is in the UV region of the EM spectrum, Fig. 31–12. See also Fig. 37–22.

NOTE An alternate approach would be to use Eq. 37–15 to find λ, and it gives the same result.

EXAMPLE 37–14 **Wavelength of a Balmer line.** Determine the wavelength of light emitted when a hydrogen atom makes a transition from the $n = 6$ to the $n = 2$ energy level according to the Bohr model.

APPROACH We can use Eq. 37–15 or its equivalent, Eq. 37–8, with $R = 1.097 \times 10^7 \, \text{m}^{-1}$.

SOLUTION We find

$$\frac{1}{\lambda} = \left(1.097 \times 10^7 \, \text{m}^{-1}\right)\left(\frac{1}{4} - \frac{1}{36}\right) = 2.44 \times 10^6 \, \text{m}^{-1}.$$

So $\lambda = 1/(2.44 \times 10^6 \, \text{m}^{-1}) = 4.10 \times 10^{-7} \, \text{m}$ or 410 nm. This is the fourth line in the Balmer series, Fig. 37–21, and is violet in color.

EXERCISE F The energy of the photon emitted when a hydrogen atom goes from the $n = 6$ state to the $n = 3$ state is (a) 0.378 eV; (b) 0.503 eV; (c) 1.13 eV; (d) 3.06 eV; (e) 13.6 eV.

EXAMPLE 37–15 **Absorption wavelength.** Use Fig. 37–26 to determine the maximum wavelength that hydrogen in its ground state can absorb. What would be the next smaller wavelength that would work?

APPROACH Maximum wavelength corresponds to minimum energy, and this would be the jump from the ground state up to the first excited state (Fig. 37–26). The next smaller wavelength occurs for the jump from the ground state to the second excited state. In each case, the energy difference can be used to find the wavelength.

SOLUTION The energy needed to jump from the ground state to the first excited state is $13.6 \, \text{eV} - 3.4 \, \text{eV} = 10.2 \, \text{eV}$; the required wavelength, as we saw in Example 37–13, is 122 nm. The energy to jump from the ground state to the second excited state is $13.6 \, \text{eV} - 1.5 \, \text{eV} = 12.1 \, \text{eV}$, which corresponds to a wavelength

$$\lambda = \frac{c}{f} = \frac{hc}{hf} = \frac{hc}{E_3 - E_1}$$

$$= \frac{(6.63 \times 10^{-34} \, \text{J·s})(3.00 \times 10^8 \, \text{m/s})}{(12.1 \, \text{eV})(1.60 \times 10^{-19} \, \text{J/eV})} = 103 \, \text{nm}.$$

EXAMPLE 37–16 **He$^+$ ionization energy.** (a) Use the Bohr model to determine the ionization energy of the He$^+$ ion, which has a single electron. (b) Also calculate the maximum wavelength a photon can have to cause ionization.

APPROACH We want to determine the minimum energy required to lift the electron from its ground state and to barely reach the free state at $E = 0$. The ground state energy of He$^+$ is given by Eq. 37–14b with $n = 1$ and $Z = 2$.

SOLUTION (a) Since all the symbols in Eq. 37–14b are the same as for the calculation for hydrogen, except that Z is 2 instead of 1, we see that E_1 will be $Z^2 = 2^2 = 4$ times the E_1 for hydrogen:

$$E_1 = 4(-13.6 \, \text{eV}) = -54.4 \, \text{eV}.$$

Thus, to ionize the He$^+$ ion should require 54.4 eV, and this value agrees with experiment.

(b) The maximum wavelength photon that can cause ionization will have energy $hf = 54.4 \, \text{eV}$ and wavelength

$$\lambda = \frac{c}{f} = \frac{hc}{hf} = \frac{(6.63 \times 10^{-34} \, \text{J·s})(3.00 \times 10^8 \, \text{m/s})}{(54.4 \, \text{eV})(1.60 \times 10^{-19} \, \text{J/eV})} = 22.8 \, \text{nm}.$$

NOTE If the atom absorbed a photon of greater energy (wavelength shorter than 22.8 nm), the atom could still be ionized and the freed electron would have kinetic energy of its own. If $\lambda > 22.8$ nm, the photon has too little energy to cause ionization.

In this last Example, we saw that E_1 for the He$^+$ ion is four times more negative than that for hydrogen. Indeed, the energy-level diagram for He$^+$ looks just like that for hydrogen, Fig. 37–26, except that the numerical values for each energy level are four times larger. Note, however, that we are talking here about the He$^+$ *ion*. Normal (neutral) helium has two electrons and its energy level diagram is entirely different.

Hydrogen at 20°C. Estimate the average kinetic energy of whole hydrogen atoms (not just the electrons) at room temperature, and use the result to explain why nearly all H atoms are in the ground state at room temperature, and hence emit no light.

RESPONSE According to kinetic theory (Chapter 18), the average kinetic energy of atoms or molecules in a gas is given by Eq. 18–4:

$$\bar{K} = \tfrac{3}{2}kT,$$

where $k = 1.38 \times 10^{-23}$ J/K is Boltzmann's constant, and T is the kelvin (absolute) temperature. Room temperature is about $T = 300$ K, so

$$\bar{K} = \tfrac{3}{2}(1.38 \times 10^{-23}\,\text{J/K})(300\,\text{K}) = 6.2 \times 10^{-21}\,\text{J},$$

or, in electron volts:

$$\bar{K} = \frac{6.2 \times 10^{-21}\,\text{J}}{1.6 \times 10^{-19}\,\text{J/eV}} = 0.04\,\text{eV}.$$

The average kinetic energy of an atom as a whole is thus very small compared to the energy between the ground state and the next higher energy state ($13.6\,\text{eV} - 3.4\,\text{eV} = 10.2\,\text{eV}$). Any atoms in excited states quickly fall to the ground state and emit light. Once in the ground state, collisions with other atoms can transfer energy of only 0.04 eV on the average. A small fraction of atoms can have much more energy (see Section 18–2 on the distribution of molecular speeds), but even a kinetic energy that is 10 times the average is not nearly enough to excite atoms into states above the ground state. Thus, at room temperature, nearly all atoms are in the ground state. Atoms can be excited to upper states by very high temperatures, or by passing a current of high energy electrons through the gas, as in a discharge tube (Fig. 37–19).

Correspondence Principle

We should note that Bohr made some radical assumptions that were at variance with classical ideas. He assumed that electrons in fixed orbits do not radiate light even though they are accelerating (moving in a circle), and he assumed that angular momentum is quantized. Furthermore, he was not able to say how an electron moved when it made a transition from one energy level to another. On the other hand, there is no real reason to expect that in the tiny world of the atom electrons would behave as ordinary-sized objects do. Nonetheless, he felt that where quantum theory overlaps with the macroscopic world, it should predict classical results. This is the **correspondence principle**, already mentioned in regard to relativity (Section 36–13). This principle does work for Bohr's theory of the hydrogen atom. The orbit sizes and energies are quite different for $n = 1$ and $n = 2$, say. But orbits with $n = 100{,}000{,}000$ and $100{,}000{,}001$ would be very close in radius and energy (see Fig. 37–26). Indeed, jumps between such large orbits (which would approach macroscopic sizes), would be imperceptible. Such orbits would thus appear to be continuously spaced, which is what we expect in the everyday world.

Finally, it must be emphasized that the well-defined orbits of the Bohr model do not actually exist. The Bohr model is only a model, not reality. The idea of electron orbits was rejected a few years later, and today electrons are thought of (Chapter 39) as forming "probability clouds."

37–12 de Broglie's Hypothesis Applied to Atoms

Bohr's theory was largely of an *ad hoc* nature. Assumptions were made so that theory would agree with experiment. But Bohr could give no reason why the orbits were quantized, nor why there should be a stable ground state. Finally, ten years later, a reason was proposed by Louis de Broglie. We saw in Section 37–7 that in 1923, de Broglie proposed that material particles, such as electrons, have a wave nature; and that this hypothesis was confirmed by experiment several years later.

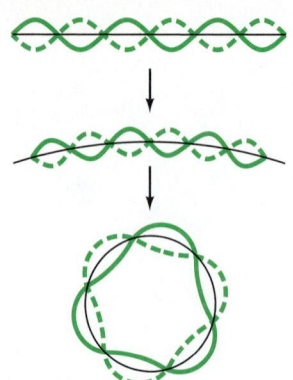

FIGURE 37–27 An ordinary standing wave compared to a circular standing wave.

FIGURE 37–28 When a wave does not close (and hence interferes destructively with itself), it rapidly dies out.

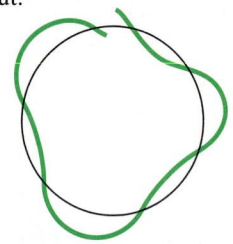

FIGURE 37–29 Standing circular waves for two, three, and five wavelengths on the circumference; n, the number of wavelengths, is also the quantum number.

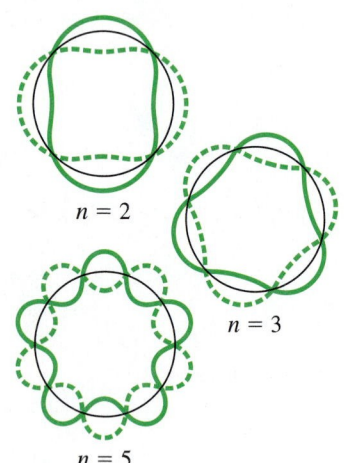

$n = 2$

$n = 3$

$n = 5$

One of de Broglie's original arguments in favor of the wave nature of electrons was that it provided an explanation for Bohr's theory of the hydrogen atom. According to de Broglie, a particle of mass m moving with a nonrelativistic speed v would have a wavelength (Eq. 37–7) of

$$\lambda = \frac{h}{mv}.$$

Each electron orbit in an atom, he proposed, is actually a standing wave. As we saw in Chapter 15, when a violin or guitar string is plucked, a vast number of wavelengths are excited. But only certain ones—those that have nodes at the ends—are sustained. These are the *resonant* modes of the string. Waves with other wavelengths interfere with themselves upon reflection and their amplitudes quickly drop to zero. With electrons moving in circles, according to Bohr's theory, de Broglie argued that the electron wave was a *circular* standing wave that closes on itself, Fig. 37–27. If the wavelength of a wave does not close on itself, as in Fig. 37–28, destructive interference takes place as the wave travels around the loop, and the wave quickly dies out. Thus, the only waves that persist are those for which the circumference of the circular orbit contains a whole number of wavelengths, Fig. 37–29. The circumference of a Bohr orbit of radius r_n is $2\pi r_n$, so to have constructive interference, we need

$$2\pi r_n = n\lambda, \qquad n = 1, 2, 3, \cdots.$$

When we substitute $\lambda = h/mv$, we get $2\pi r_n = nh/mv$, or

$$mvr_n = \frac{nh}{2\pi}.$$

This is just the *quantum condition* proposed by Bohr on an *ad hoc* basis, Eq. 37–10. It is from this equation that the discrete orbits and energy levels were derived. Thus we have a first explanation for the quantized orbits and energy states in the Bohr model: they are due to the wave nature of the electron, and only resonant "standing" waves can persist.[†] This implies that the *wave–particle duality* is at the root of atomic structure.

In viewing the circular electron waves of Fig. 37–29, the electron is not to be thought of as following the oscillating wave pattern. In the Bohr model of hydrogen, the electron moves in a circle. The circular wave, on the other hand, represents the *amplitude* of the electron "matter wave," and in Fig. 37–29 the wave amplitude is shown superimposed on the circular path of the particle orbit for convenience.

Bohr's theory worked well for hydrogen and for one-electron ions. But it did not prove successful for multi-electron atoms. Bohr's theory could not predict line spectra even for the next simplest atom, helium. It could not explain why some emission lines are brighter than others, nor why some lines are split into two or more closely spaced lines ("fine structure"). A new theory was needed and was indeed developed in the 1920s. This new and radical theory is called *quantum mechanics*. It finally solved the problem of atomic structure, but it gives us a very different view of the atom: the idea of electrons in well-defined orbits was replaced with the idea of electron "clouds." This new theory of quantum mechanics has given us a wholly different view of the basic mechanisms underlying physical processes.

[†]We note, however, that Eq. 37–10 is no longer considered valid, as discussed in Chapter 39.

Summary

Quantum theory has its origins in **Planck's quantum hypothesis** that molecular oscillations are **quantized**: their energy E can only be integer (n) multiples of hf, where h is Planck's constant and f is the natural frequency of oscillation:

$$E = nhf. \qquad (37\text{–}2)$$

This hypothesis explained the spectrum of radiation emitted by a **blackbody** at high temperature.

Einstein proposed that for some experiments, light could be pictured as being emitted and absorbed as **quanta** (particles), which we now call **photons**, each with energy

$$E = hf \qquad (37\text{–}3)$$

and momentum

$$p = \frac{E}{c} = \frac{hf}{c} = \frac{h}{\lambda}. \qquad (37\text{–}5)$$

He proposed the photoelectric effect as a test for the photon theory of light. In the **photoelectric effect**, the photon theory says that each incident photon can strike an electron in a material and eject it if the photon has sufficient energy. The maximum energy of ejected electrons is then linearly related to the frequency of the incident light.

The photon theory is also supported by the **Compton effect** and the observation of electron–positron **pair production**.

The **wave–particle duality** refers to the idea that light and matter (such as electrons) have both wave and particle properties. The wavelength of an object is given by

$$\lambda = \frac{h}{p}, \qquad (37\text{–}7)$$

where p is the momentum of the object ($p = mv$ for a particle of mass m and speed v).

The **principle of complementarity** states that we must be aware of both the particle and wave properties of light and of matter for a complete understanding of them.

Early models of the atom include the plum-pudding model, and Rutherford's planetary (or nuclear) model of an atom which consists of a tiny but massive positively charged nucleus surrounded (at a relatively great distance) by electrons.

To explain the **line spectra** emitted by atoms, as well as the stability of atoms, **Bohr's theory** postulated that: (1) electrons bound in an atom can only occupy orbits for which the angular momentum is quantized, which results in discrete values for the radius and energy; (2) an electron in such a **stationary state** emits no radiation; (3) if an electron jumps to a lower state, it emits a photon whose energy equals the difference in energy between the two states; (4) the angular momentum L of atomic electrons is quantized by the rule

$$L = \frac{nh}{2\pi}, \qquad (37\text{–}10)$$

where n is an integer called the **quantum number**. The $n = 1$ state is the **ground state**, which in hydrogen has an energy $E_1 = -13.6\,\text{eV}$. Higher values of n correspond to **excited states**, and their energies are

$$E_n = -(13.6\,\text{eV})\frac{Z^2}{n^2}. \qquad (37\text{–}14b)$$

Atoms are excited to these higher states by collisions with other atoms or electrons, or by absorption of a photon of just the right frequency.

De Broglie's hypothesis that electrons (and other matter) have a wavelength $\lambda = h/mv$ gave an explanation for Bohr's quantized orbits by bringing in the wave–particle duality: the orbits correspond to circular standing waves in which the circumference of the orbit equals a whole number of wavelengths.

Questions

1. What can be said about the relative temperatures of whitish-yellow, reddish, and bluish stars? Explain.

2. If energy is radiated by all objects, why can we not see most of them in the dark?

3. Does a lightbulb at a temperature of 2500 K produce as white a light as the Sun at 6000 K? Explain.

4. Darkrooms for developing black-and-white film were sometimes lit by a red bulb. Why red? Would such a bulb work in a darkroom for developing color photographs?

5. If the threshold wavelength in the photoelectric effect increases when the emitting metal is changed to a different metal, what can you say about the work functions of the two metals?

6. Explain why the existence of a cutoff frequency in the photoelectric effect more strongly favors a particle theory rather than a wave theory of light.

7. UV light causes sunburn, whereas visible light does not. Suggest a reason.

8. The work functions for sodium and cesium are 2.28 eV and 2.14 eV, respectively. For incident photons of a given frequency, which metal will give a higher maximum kinetic energy for the electrons?

9. (a) Does a beam of infrared photons always have less energy than a beam of ultraviolet photons? Explain. (b) Does a single photon of infrared light always have less energy than a single photon of ultraviolet light?

10. Light of 450-nm wavelength strikes a metal surface, and a stream of electrons emerges from the metal. If light of the same intensity but of wavelength 400 nm strikes the surface, are more electrons emitted? Does the energy of the emitted electrons change? Explain.

11. Explain how the photoelectric circuit of Fig. 37–4 could be used in (a) a burglar alarm, (b) a smoke detector, (c) a photographic light meter.

12. If an X-ray photon is scattered by an electron, does the photon's wavelength change? If so, does it increase or decrease?

13. In both the photoelectric effect and in the Compton effect, a photon collides with an electron causing the electron to fly off. What then, is the difference between the two processes?

14. Consider a point source of light. How would the intensity of light vary with distance from the source according to (a) wave theory, (b) particle (photon) theory? Would this help to distinguish the two theories?

15. If an electron and a proton travel at the same speed, which has the shorter de Broglie wavelength? Explain.

16. Why do we say that light has wave properties? Why do we say that light has particle properties?

17. Why do we say that electrons have wave properties? Why do we say that electrons have particle properties?

18. What are the differences between a photon and an electron? Be specific: make a list.

19. In Rutherford's planetary model of the atom, what keeps the electrons from flying off into space?

20. How can you tell if there is oxygen near the surface of the Sun?

21. When a wide spectrum of light passes through hydrogen gas at room temperature, absorption lines are observed that correspond only to the Lyman series. Why don't we observe the other series?

22. Explain how the closely spaced energy levels for hydrogen near the top of Fig. 37–26 correspond to the closely spaced spectral lines at the top of Fig. 37–21.

23. Is it possible for the de Broglie wavelength of a "particle" to be greater than the dimensions of the particle? To be smaller? Is there any direct connection?

24. In a helium atom, which contains two electrons, do you think that on average the electrons are closer to the nucleus or farther away than in a hydrogen atom? Why?

25. How can the spectrum of hydrogen contain so many lines when hydrogen contains only one electron?

26. The Lyman series is brighter than the Balmer series because this series of transitions ends up in the most common state for hydrogen, the ground state. Why then was the Balmer series discovered first?

27. Use conservation of momentum to explain why photons emitted by hydrogen atoms have slightly less energy than that predicted by Eq. 37–9.

28. Suppose we obtain an emission spectrum for hydrogen at very high temperature (when some of the atoms are in excited states), and an absorption spectrum at room temperature, when all atoms are in the ground state. Will the two spectra contain identical lines?

Problems

37–1 Planck's Quantum Hypothesis

1. (I) Estimate the peak wavelength for radiation from (a) ice at 273 K, (b) a floodlamp at 3500 K, (c) helium at 4.2 K, (d) for the universe at $T = 2.725$ K, assuming blackbody emission. In what region of the EM spectrum is each?

2. (I) How hot is metal being welded if it radiates most strongly at 460 nm?

3. (I) An HCl molecule vibrates with a natural frequency of 8.1×10^{13} Hz. What is the difference in energy (in joules and electron volts) between successive values of the oscillation energy?

4. (II) Estimate the peak wavelength of light issuing from the pupil of the human eye (which approximates a blackbody) assuming normal body temperature.

5. (III) Planck's radiation law is given by:

$$I(\lambda, T) = \frac{2\pi h c^2 \lambda^{-5}}{e^{hc/\lambda kT} - 1}$$

where $I(\lambda, T)$ is the rate energy is radiated per unit surface area per unit wavelength interval at wavelength λ and Kelvin temperature T. (a) Show that Wien's displacement law follows from this relationship. (b) Determine the value of h from the experimental value of $\lambda_P T$ given in the text. [You may want to use graphing techniques.] (c) Derive the T^4 dependence of the rate at which energy is radiated (as in the Stefan-Boltzmann law, Eq. 19–17), by integrating Planck's formula over all wavelengths; that is, show that

$$\int I(\lambda, T) \, d\lambda \propto T^4.$$

37–2 and 37–3 Photons and the Photoelectric Effect

6. (I) What is the energy of photons (in joules) emitted by a 104.1-MHz FM radio station?

7. (I) What is the energy range (in joules and eV) of photons in the visible spectrum, of wavelength 410 nm to 750 nm?

8. (I) A typical gamma ray emitted from a nucleus during radioactive decay may have an energy of 380 keV. What is its wavelength? Would we expect significant diffraction of this type of light when it passes through an everyday opening, such as a door?

9. (I) About 0.1 eV is required to break a "hydrogen bond" in a protein molecule. Calculate the minimum frequency and maximum wavelength of a photon that can accomplish this.

10. (I) Calculate the momentum of a photon of yellow light of wavelength 6.20×10^{-7} m.

11. (I) What minimum frequency of light is needed to eject electrons from a metal whose work function is 4.8×10^{-19} J?

12. (I) What is the longest wavelength of light that will emit electrons from a metal whose work function is 3.70 eV?

13. (II) What wavelength photon would have the same energy as a 145-gram baseball moving 30.0 m/s?

14. (II) The human eye can respond to as little as 10^{-18} J of light energy. For a wavelength at the peak of visual sensitivity, 550 nm, how many photons lead to an observable flash?

15. (II) The work functions for sodium, cesium, copper, and iron are 2.3, 2.1, 4.7, and 4.5 eV, respectively. Which of these metals will not emit electrons when visible light shines on it?

16. (II) In a photoelectric-effect experiment it is observed that no current flows unless the wavelength is less than 520 nm. (a) What is the work function of this material? (b) What is the stopping voltage required if light of wavelength 470 nm is used?

17. (II) What is the maximum kinetic energy of electrons ejected from barium $(W_0 = 2.48 \text{ eV})$ when illuminated by white light, $\lambda = 410$ to 750 nm?

18. (II) Barium has a work function of 2.48 eV. What is the maximum kinetic energy of electrons if the metal is illuminated by UV light of wavelength 365 nm? What is their speed?

19. (II) When UV light of wavelength 285 nm falls on a metal surface, the maximum kinetic energy of emitted electrons is 1.70 eV. What is the work function of the metal?

20. (II) The threshold wavelength for emission of electrons from a given surface is 320 nm. What will be the maximum kinetic energy of ejected electrons when the wavelength is changed to (a) 280 nm, (b) 360 nm?

21. (II) When 230-nm light falls on a metal, the current through a photoelectric circuit (Fig. 37–4) is brought to zero at a stopping voltage of 1.84 V. What is the work function of the metal?

22. (II) A certain type of film is sensitive only to light whose wavelength is less than 630 nm. What is the energy (eV and kcal/mol) needed for the chemical reaction to occur which causes the film to change?

23. (II) The range of visible light wavelengths extends from about 410 nm to 750 nm. (a) Estimate the minimum energy (eV) necessary to initiate the chemical process on the retina that is responsible for vision. (b) Speculate as to why, at the other end of the visible range, there is a threshold photon energy beyond which the eye registers no sensation of sight. Determine this threshold photon energy (eV).

24. (II) In a photoelectric experiment using a clean sodium surface, the maximum energy of the emitted electrons was measured for a number of different incident frequencies, with the following results.

Frequency ($\times 10^{14}$ Hz)	Energy (eV)
11.8	2.60
10.6	2.11
9.9	1.81
9.1	1.47
8.2	1.10
6.9	0.57

Plot the graph of these results and find: (a) Planck's constant; (b) the cutoff frequency of sodium; (c) the work function.

25. (II) A **photomultiplier tube** (a very sensitive light sensor), is based on the photoelectric effect: incident photons strike a metal surface and the resulting ejected electrons are collected. By counting the number of collected electrons, the number of incident photons (i.e., the incident light intensity) can be determined. (a) If a photomultiplier tube is to respond properly for incident wavelengths throughout the visible range (410 nm to 750 nm), what is the maximum value for the work function W_0 (eV) of its metal surface? (b) If W_0 for its metal surface is above a certain threshold value, the photomultiplier will only function for incident ultraviolet wavelengths and be unresponsive to visible light. Determine this threshold value (eV).

26. (III) A group of atoms is confined to a very small (point-like) volume in a laser-based **atom trap**. The incident laser light causes each atom to emit 1.0×10^6 photons of wavelength 780 nm every second. A sensor of area 1.0 cm² measures the light intensity emanating from the trap to be 1.6 nW when placed 25 cm away from the trapped atoms. Assuming each atom emits photons with equal probability in all directions, determine the number of trapped atoms.

27. (III) Assume light of wavelength λ is incident on a metal surface, whose work function is known precisely (i.e., its uncertainty is better than 0.1% and can be ignored). Show that if the stopping voltage can be determined to an accuracy of ΔV_0, the fractional uncertainty (magnitude) in wavelength is

$$\frac{\Delta \lambda}{\lambda} = \frac{\lambda e}{hc} \Delta V_0.$$

Determine this fractional uncertainty if $\Delta V_0 = 0.01$ V and $\lambda = 550$ nm.

37–4 Compton Effect

28. (I) A high-frequency photon is scattered off of an electron and experiences a change of wavelength of 1.5×10^{-4} nm. At what angle must a detector be placed to detect the scattered photon (relative to the direction of the incoming photon)?

29. (II) Determine the Compton wavelength for (a) an electron, (b) a proton. (c) Show that if a photon has wavelength equal to the Compton wavelength of a particle, the photon's energy is equal to the rest energy of the particle.

30. (II) X-rays of wavelength $\lambda = 0.120$ nm are scattered from carbon. What is the expected Compton wavelength shift for photons detected at angles (relative to the incident beam) of exactly (a) 60°, (b) 90°, (c) 180°?

31. (II) In the Compton effect, determine the ratio ($\Delta\lambda/\lambda$) of the maximum change $\Delta\lambda$ in a photon's wavelength to the photon's initial wavelength λ, if the photon is (a) a visible-light photon with $\lambda = 550$ nm, (b) an X-ray photon with $\lambda = 0.10$ nm.

32. (II) A 1.0-MeV gamma-ray photon undergoes a sequence of Compton-scattering events. If the photon is scattered at an angle of 0.50° in each event, estimate the number of events required to convert the photon into a visible-light photon with wavelength 555 nm. You can use an expansion for small θ; see Appendix A. [Gamma rays created near the center of the Sun are transformed to visible wavelengths as they travel to the Sun's surface through a sequence of small-angle Compton scattering events.]

33. (III) In the Compton effect, a 0.160-nm photon strikes a free electron in a head-on collision and knocks it into the forward direction. The rebounding photon recoils directly backward. Use conservation of (relativistic) energy and momentum to determine (a) the kinetic energy of the electron, and (b) the wavelength of the recoiling photon. Use Eq. 37–5, but not Eq. 37–6.

34. (III) In the Compton effect (see Fig. 37–7), use the relativistic equations for conservation of energy and of linear momentum to show that the Compton shift in wavelength is given by Eq. 37–6.

37–5 Pair Production

35. (I) How much total kinetic energy will an electron–positron pair have if produced by a 2.67-MeV photon?

36. (II) What is the longest wavelength photon that could produce a proton–antiproton pair? (Each has a mass of 1.67×10^{-27} kg.)

37. (II) What is the minimum photon energy needed to produce a $\mu^+ - \mu^-$ pair? The mass of each μ (muon) is 207 times the mass of an electron. What is the wavelength of such a photon?

38. (II) An electron and a positron, each moving at 2.0×10^5 m/s, collide head on, disappear, and produce two photons moving in opposite directions, each with the same energy and momentum. Determine the energy and momentum of each photon.

39. (II) A gamma-ray photon produces an electron and a positron, each with a kinetic energy of 375 keV. Determine the energy and wavelength of the photon.

37–7 Wave Nature of Matter

40. (I) Calculate the wavelength of a 0.23-kg ball traveling at 0.10 m/s.

41. (I) What is the wavelength of a neutron ($m = 1.67 \times 10^{-27}$ kg) traveling at 8.5×10^4 m/s?

42. (I) Through how many volts of potential difference must an electron be accelerated to achieve a wavelength of 0.21 nm?

43. (II) What is the theoretical limit of resolution for an electron microscope whose electrons are accelerated through 85 kV? (Relativistic formulas should be used.)

44. (II) The speed of an electron in a particle accelerator is 0.98c. Find its de Broglie wavelength. (Use relativistic momentum.)

45. (II) Calculate the ratio of the kinetic energy of an electron to that of a proton if their wavelengths are equal. Assume that the speeds are nonrelativistic.

46. (II) Neutrons can be used in diffraction experiments to probe the lattice structure of crystalline solids. Since the neutron's wavelength needs to be on the order of the spacing between atoms in the lattice, about 0.3 nm, what should the speed of the neutrons be?

47. (II) An electron has a de Broglie wavelength $\lambda = 6.0 \times 10^{-10}$ m. (a) What is its momentum? (b) What is its speed? (c) What voltage was needed to accelerate it to this speed?

48. (II) What is the wavelength of an electron of energy (a) 20 eV, (b) 200 eV, (c) 2.0 keV?

49. (II) Show that if an electron and a proton have the same nonrelativistic kinetic energy, the proton has the shorter wavelength.

50. (II) Calculate the de Broglie wavelength of an electron in a TV picture tube if it is accelerated by 33,000 V. Is it relativistic? How does its wavelength compare to the size of the "neck" of the tube, typically 5 cm? Do we have to worry about diffraction problems blurring our picture on the screen?

51. (II) After passing through two slits separated by a distance of 3.0 μm, a beam of electrons creates an interference pattern with its second-order maximum at an angle of 55°. Find the speed of the electrons in this beam.

*37–8 Electron Microscope

***52. (II)** What voltage is needed to produce electron wavelengths of 0.28 nm? (Assume that the electrons are nonrelativistic.)

***53. (II)** Electrons are accelerated by 3450 V in an electron microscope. Estimate the maximum possible resolution of the microscope.

37–10 and 37–11 Bohr Model

54. (I) For the three hydrogen transitions indicated below, with n being the initial state and n' being the final state, is the transition an absorption or an emission? Which is higher, the initial state energy or the final state energy of the atom? Finally, which of these transitions involves the largest energy photon? (a) $n = 1$, $n' = 3$; (b) $n = 6$, $n' = 2$; (c) $n = 4$, $n' = 5$.

55. (I) How much energy is needed to ionize a hydrogen atom in the $n = 3$ state?

56. (I) (a) Determine the wavelength of the second Balmer line ($n = 4$ to $n = 2$ transition) using Fig. 37–26. Determine likewise (b) the wavelength of the third Lyman line and (c) the wavelength of the first Balmer line.

57. (I) Calculate the ionization energy of doubly ionized lithium, Li^{2+}, which has $Z = 3$.

58. (I) Evaluate the Rydberg constant R using the Bohr model (compare Eqs. 37–8 and 37–15) and show that its value is $R = 1.0974 \times 10^7$ m^{-1}.

59. (II) What is the longest wavelength light capable of ionizing a hydrogen atom in the ground state?

60. (II) In the Sun, an ionized helium (He^+) atom makes a transition from the $n = 5$ state to the $n = 2$ state, emitting a photon. Can that photon be absorbed by hydrogen atoms present in the Sun? If so, between what energy states will the hydrogen atom transition occur?

61. (II) What wavelength photon would be required to ionize a hydrogen atom in the ground state and give the ejected electron a kinetic energy of 20.0 eV?

62. (II) For what maximum kinetic energy is a collision between an electron and a hydrogen atom in its ground state definitely elastic?

63. (II) Construct the energy-level diagram for the He^+ ion (like Fig. 37–26).

64. (II) Construct the energy-level diagram (like Fig. 37–26) for doubly ionized lithium, Li^{2+}.

65. (II) Determine the electrostatic potential energy and the kinetic energy of an electron in the ground state of the hydrogen atom.

66. (II) An excited hydrogen atom could, in principle, have a diameter of 0.10 mm. What would be the value of n for a Bohr orbit of this size? What would its energy be?

67. (II) Is the use of nonrelativistic formulas justified in the Bohr atom? To check, calculate the electron's velocity, v, in terms of c, for the ground state of hydrogen, and then calculate $\sqrt{1 - v^2/c^2}$.

68. (II) A hydrogen atom has an angular momentum of 5.273×10^{-34} kg·m^2/s. According to the Bohr model, what is the energy (eV) associated with this state?

69. (II) Assume hydrogen atoms in a gas are initially in their ground state. If free electrons with kinetic energy 12.75 eV collide with these atoms, what photon wavelengths will be emitted by the gas?

70. (II) Suppose an electron was bound to a proton, as in the hydrogen atom, but by the gravitational force rather than by the electric force. What would be the radius, and energy, of the first Bohr orbit?

71. (II) *Correspondence principle*: Show that for large values of n, the difference in radius Δr between two adjacent orbits (with quantum numbers n and $n - 1$) is given by

$$\Delta r = r_n - r_{n-1} \approx \frac{2r_n}{n},$$

so $\Delta r/r_n \to 0$ as $n \to \infty$ in accordance with the correspondence principle. [Note that we can check the correspondence principle by either considering large values of n ($n \to \infty$) or by letting $h \to 0$. Are these equivalent?]

General Problems

72. If a 75-W lightbulb emits 3.0% of the input energy as visible light (average wavelength 550 nm) uniformly in all directions, estimate how many photons per second of visible light will strike the pupil (4.0 mm diameter) of the eye of an observer 250 m away.

73. At low temperatures, nearly all the atoms in hydrogen gas will be in the ground state. What minimum frequency photon is needed if the photoelectric effect is to be observed?

74. A beam of 125-eV electrons is scattered from a crystal, as in X-ray diffraction, and a first-order peak is observed at $\theta = 38°$. What is the spacing between planes in the diffracting crystal? (See Section 35–10.)

75. A microwave oven produces electromagnetic radiation at $\lambda = 12.2$ cm and produces a power of 860 W. Calculate the number of microwave photons produced by the microwave oven each second.

76. Sunlight reaching the Earth has an intensity of about 1350 W/m². Estimate how many photons per square meter per second this represents. Take the average wavelength to be 550 nm.

77. A beam of red laser light $(\lambda = 633\ \text{nm})$ hits a black wall and is fully absorbed. If this light exerts a total force $F = 6.5\ \text{nN}$ on the wall, how many photons per second are hitting the wall?

78. The Big Bang theory states that the beginning of the universe was accompanied by a huge burst of photons. Those photons are still present today and make up the so-called cosmic microwave background radiation. The universe radiates like a blackbody with a temperature of about 2.7 K. Calculate the peak wavelength of this radiation.

79. An electron and a positron collide head on, annihilate, and create two 0.755-MeV photons traveling in opposite directions. What were the initial kinetic energies of electron and positron?

80. By what potential difference must (a) a proton $(m = 1.67 \times 10^{-27}\ \text{kg})$, and (b) an electron $(m = 9.11 \times 10^{-31}\ \text{kg})$, be accelerated to have a wavelength $\lambda = 6.0 \times 10^{-12}\ \text{m}$?

81. In some of Rutherford's experiments (Fig. 37–17) the α particles $(\text{mass} = 6.64 \times 10^{-27}\ \text{kg})$ had a kinetic energy of 4.8 MeV. How close could they get to the center of a silver nucleus $(\text{charge} = +47e)$? Ignore the recoil motion of the nucleus.

82. Show that the magnitude of the electrostatic potential energy of an electron in any Bohr orbit of a hydrogen atom is twice the magnitude of its kinetic energy in that orbit.

83. Calculate the ratio of the gravitational force to the electric force for the electron in a hydrogen atom. Can the gravitational force be safely ignored?

84. Electrons accelerated by a potential difference of 12.3 V pass through a gas of hydrogen atoms at room temperature. What wavelengths of light will be emitted?

85. In a particular photoelectric experiment, a stopping potential of 2.70 V is measured when ultraviolet light of wavelength 380 nm is incident on the metal. If blue light of wavelength 440 nm is used, what is the new stopping potential?

86. In an X-ray tube (see Fig. 35–26 and discussion in Section 35–10), the high voltage between filament and target is V. After being accelerated through this voltage, an electron strikes the target where it is decelerated (by positively charged nuclei) and in the process one or more X-ray photons are emitted. (a) Show that the photon of shortest wavelength will have

$$\lambda_0 = \frac{hc}{eV}.$$

(b) What is the shortest wavelength of X-ray emitted when accelerated electrons strike the face of a 33-kV television picture tube?

87. The intensity of the Sun's light in the vicinity of Earth is about 1350 W/m². Imagine a spacecraft with a mirrored square sail of dimension 1.0 km. Estimate how much thrust (in newtons) this craft will experience due to collisions with the Sun's photons. [*Hint:* Assume the photons bounce perpendicularly off the sail with no change in the magnitude of their momentum.]

88. Photons of energy 9.0 eV are incident on a metal. It is found that current flows from the metal until a stopping potential of 4.0 V is applied. If the wavelength of the incident photons is doubled, what is the maximum kinetic energy of the ejected electrons? What would happen if the wavelength of the incident photons was tripled?

89. Light of wavelength 360 nm strikes a metal whose work function is 2.4 eV. What is the shortest de Broglie wavelength for the electrons that are produced as photoelectrons?

90. Visible light incident on a diffraction grating with slit spacing of 0.012 mm has the first maximum at an angle of 3.5° from the central peak. If electrons could be diffracted by the same grating, what electron velocity would produce the same diffraction pattern as the visible light?

91. (a) Suppose an unknown element has an absorption spectrum with lines corresponding to 2.5, 4.7, and 5.1 eV above its ground state, and an ionization energy of 11.5 eV. Draw an energy level diagram for this element. (b) If a 5.1-eV photon is absorbed by an atom of this substance, in which state was the atom before absorbing the photon? What will be the energies of the photons that can subsequently be emitted by this atom?

92. Light of wavelength 424 nm falls on a metal which has a work function of 2.28 eV. (a) How much voltage should be applied to bring the current to zero? (b) What is the maximum speed of the emitted electrons? (c) What is the de Broglie wavelength of these electrons?

93. (a) Apply Bohr's assumptions to the Earth–Moon system to calculate the allowed energies and radii of motion. (b) Given the known distance between Earth and the Moon, is the quantization of the energy and radius apparent?

94. Show that the wavelength of a particle of mass m with kinetic energy K is given by the relativistic formula $\lambda = hc/\sqrt{K^2 + 2mc^2 K}$.

95. A small flashlight is rated at 3.0 W. As the light leaves the flashlight in one direction, a reaction force is exerted on the flashlight in the opposite direction. Estimate the size of this reaction force.

96. At the atomic-scale, the electron volt and nanometer are well-suited units for energy and distance, respectively. (a) Show that the energy E in eV of a photon, whose wavelength λ is in nm, is given by

$$E = \frac{1240\ \text{eV} \cdot \text{nm}}{\lambda\ (\text{nm})}.$$

(b) How much energy (eV) does a 650-nm photon have?

*97. Three fundamental constants of nature—the gravitational constant G, Planck's constant h, and the speed of light c—have the dimensions of $[L^3/MT^2]$, $[ML^2/T]$, and $[L/T]$, respectively. (a) Find the mathematical combination of these fundamental constants that has the dimension of time. This combination is called the "Planck time" t_P and is thought to be the earliest time, after the creation of the universe, at which the currently known laws of physics can be applied. (b) Determine the numerical value of t_P. (c) Find the mathematical combination of these fundamental constants that has the dimension of length. This combination is called the "Planck length" λ_P and is thought to be the smallest length over which the currently known laws of physics can be applied. (d) Determine the numerical value of λ_P.

98. Imagine a free particle of mass m bouncing back and forth between two perfectly reflecting walls, separated by distance ℓ. Imagine that the two oppositely directed matter waves associated with this particle interfere to create a standing wave with a node at each of the walls. Show that the ground state (first harmonic) and first excited state (second harmonic) have (non-relativistic) kinetic energies $h^2/8m\ell^2$ and $h^2/2m\ell^2$, respectively.

99. (a) A rubidium atom $(m = 85 \, u)$ is at rest with one electron in an excited energy level. When the electron jumps to the ground state, the atom emits a photon of wavelength $\lambda = 780$ nm. Determine the resulting (nonrelativistic) recoil speed v of the atom. (b) The recoil speed sets the lower limit on the temperature to which an ideal gas of rubidium atoms can be cooled in a laser-based **atom trap**. Using the kinetic theory of gases (Chapter 18), estimate this "lowest achievable" temperature.

100. A rubidium atom (atomic mass 85) is initially at room temperature and has a velocity $v = 290$ m/s due to its thermal motion. Consider the absorption of photons by this atom from a laser beam of wavelength $\lambda = 780$ nm. Assume the rubidium atom's initial velocity v is directed into the laser beam (the photons are moving right and the atom is moving left) and that the atom absorbs a new photon every 25 ns. How long will it take for this process to completely stop ("cool") the rubidium atom? [*Note*: a more detailed analysis predicts that the atom can be slowed to about 1 cm/s by this light absorption process, but it cannot be completely stopped.]

*Numerical/Computer

101. (III) (a) Graph Planck's radiation formula (top of page 989) as a function of wavelength from $\lambda = 20$ nm to 2000 nm in 20 nm steps for two lightbulb filaments, one at 2700 K and the other at 3300 K. Plot both curves on the same set of axes. (b) Approximately how much more intense is the visible light from the hotter bulb? Use numerical integration.

102. (III) Estimate what % of emitted sunlight energy is in the visible range. Use Planck's radiation formula (top of page 989) and numerical integration.

103. (III) Potassium has one of the lowest work functions of all metals and so is useful in photoelectric devices using visible light. Light from a source is incident on a potassium surface. Data for the stopping voltage V_0 as a function of wavelength λ is shown below. (a) Explain why a graph of V_0 vs. $1/\lambda$ is expected to yield a straight line. What are the theoretical expectations for the slope and the y-intercept of this line? (b) Using the data below, graph V_0 vs. $1/\lambda$ and show that a straight-line plot does indeed result. (c) Determine the slope a and y-intercept b of this line. Using your values for a and b, determine (d) potassium's work function (eV) and (e) Planck's constant h (J·s).

$\lambda \, (\mu m)$	0.400	0.430	0.460	0.490	0.520
$V_0 \, (V)$	0.803	0.578	0.402	0.229	0.083

Answers to Exercises

A: More 1000-nm photons (lower frequency).

B: 5.50×10^{14} Hz, 545 nm.

C: Only λ.

D: Decrease.

E: (e).

F: (c).

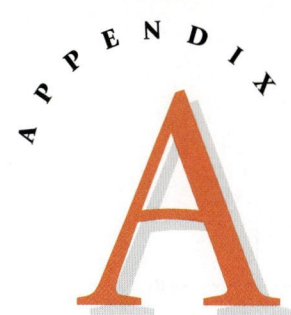

A Mathematical Formulas

A–1 Quadratic Formula

If
$$ax^2 + bx + c = 0$$

then
$$x = \frac{-b \pm \sqrt{b^2 - 4ac}}{2a}$$

A–2 Binomial Expansion

$$(1 \pm x)^n = 1 \pm nx + \frac{n(n-1)}{2!} x^2 \pm \frac{n(n-1)(n-2)}{3!} x^3 + \cdots$$

$$(x + y)^n = x^n \left(1 + \frac{y}{x}\right)^n = x^n \left(1 + n\frac{y}{x} + \frac{n(n-1)}{2!} \frac{y^2}{x^2} + \cdots\right)$$

A–3 Other Expansions

$$e^x = 1 + x + \frac{x^2}{2!} + \frac{x^3}{3!} + \cdots$$

$$\ln(1 + x) = x - \frac{x^2}{2} + \frac{x^3}{3} - \frac{x^4}{4} + \cdots$$

$$\sin\theta = \theta - \frac{\theta^3}{3!} + \frac{\theta^5}{5!} - \cdots$$

$$\cos\theta = 1 - \frac{\theta^2}{2!} + \frac{\theta^4}{4!} - \cdots$$

$$\tan\theta = \theta + \frac{\theta^3}{3} + \frac{2}{15}\theta^5 + \cdots \qquad |\theta| < \frac{\pi}{2}$$

In general:
$$f(x) = f(0) + \left(\frac{df}{dx}\right)_0 x + \left(\frac{d^2 f}{dx^2}\right)_0 \frac{x^2}{2!} + \cdots$$

A–4 Exponents

$$(a^n)(a^m) = a^{n+m}$$
$$(a^n)(b^n) = (ab)^n$$
$$(a^n)^m = a^{nm}$$

$$\frac{1}{a^n} = a^{-n}$$
$$a^n a^{-n} = a^0 = 1$$
$$a^{\frac{1}{2}} = \sqrt{a}$$

A–5 Areas and Volumes

Object	Surface area	Volume
Circle, radius r	πr^2	—
Sphere, radius r	$4\pi r^2$	$\frac{4}{3}\pi r^3$
Right circular cylinder, radius r, height h	$2\pi r^2 + 2\pi rh$	$\pi r^2 h$
Right circular cone, radius r, height h	$\pi r^2 + \pi r\sqrt{r^2 + h^2}$	$\frac{1}{3}\pi r^2 h$

A–6 Plane Geometry

1. *Equal angles:*

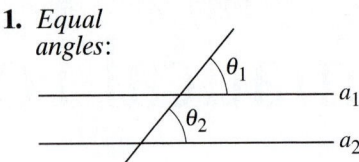

FIGURE A–1 If line a_1 is parallel to line a_2, then $\theta_1 = \theta_2$.

2. *Equal angles:*

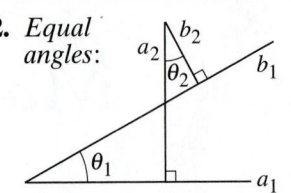

FIGURE A–2 If $a_1 \perp a_2$ and $b_1 \perp b_2$, then $\theta_1 = \theta_2$.

3. The sum of the angles in any plane triangle is $180°$.

4. *Pythagorean theorem:*

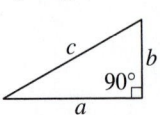

FIGURE A–3

In any right triangle (one angle $= 90°$) of sides a, b, and c:

$$a^2 + b^2 = c^2$$

where c is the length of the hypotenuse (opposite the $90°$ angle).

5. *Similar triangles:* Two triangles are said to be similar if all three of their angles are equal (in Fig. A–4, $\theta_1 = \phi_1$, $\theta_2 = \phi_2$, and $\theta_3 = \phi_3$). Similar triangles can have different sizes and different orientations.

(*a*) Two triangles are similar if any two of their angles are equal. (This follows because the third angles must also be equal since the sum of the angles of a triangle is $180°$.)

(*b*) The ratios of corresponding sides of two similar triangles are equal (Fig. A–4):

$$\frac{a_1}{b_1} = \frac{a_2}{b_2} = \frac{a_3}{b_3}.$$

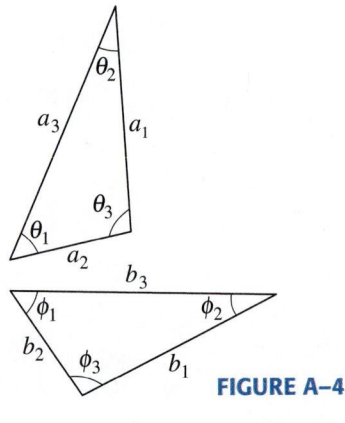

FIGURE A–4

6. *Congruent triangles:* Two triangles are congruent if one can be placed precisely on top of the other. That is, they are similar triangles and they have the same size. Two triangles are congruent if any of the following holds:

(*a*) The three corresponding sides are equal.

(*b*) Two sides and the enclosed angle are equal ("side-angle-side").

(*c*) Two angles and the enclosed side are equal ("angle-side-angle").

A–7 Logarithms

Logarithms are defined in the following way:

$$\text{if} \quad y = A^x, \quad \text{then} \quad x = \log_A y.$$

That is, the logarithm of a number y to the base A is that number which, as the exponent of A, gives back the number y. For **common logarithms**, the base is 10, so

$$\text{if} \quad y = 10^x, \quad \text{then} \quad x = \log y.$$

The subscript 10 on $\log_{10}$ is usually omitted when dealing with common logs. Another important base is the exponential base $e = 2.718\ldots$, a natural number. Such logarithms are called **natural logarithms** and are written ln. Thus,

$$\text{if} \quad y = e^x, \quad \text{then} \quad x = \ln y.$$

For any number y, the two types of logarithm are related by

$$\ln y = 2.3026 \log y.$$

Some simple rules for logarithms are as follows:

$$\log(ab) = \log a + \log b, \tag{i}$$

which is true because if $a = 10^n$ and $b = 10^m$, then $ab = 10^{n+m}$. From the definition of logarithm, $\log a = n$, $\log b = m$, and $\log(ab) = n + m$; hence, $\log(ab) = n + m = \log a + \log b$. In a similar way, we can show that

$$\log\left(\frac{a}{b}\right) = \log a - \log b \qquad \textbf{(ii)}$$

and

$$\log a^n = n \log a. \qquad \textbf{(iii)}$$

These three rules apply to any kind of logarithm.

If you do not have a calculator that calculates logs, you can easily use a **log table**, such as the small one shown here (Table A–1): the number N whose log we want is given to two digits. The first digit is in the vertical column to the left, the second digit is in the horizontal row across the top. For example, Table A–1 tells us that $\log 1.0 = 0.000$, $\log 1.1 = 0.041$, and $\log 4.1 = 0.613$. Table A–1 does not include the decimal point. The Table gives logs for numbers between 1.0 and 9.9. For larger or smaller numbers, we use rule (i) above, $\log(ab) = \log a + \log b$. For example, $\log(380) = \log(3.8 \times 10^2) = \log(3.8) + \log(10^2)$. From the Table, $\log 3.8 = 0.580$; and from rule (iii) above $\log(10^2) = 2 \log(10) = 2$, since $\log(10) = 1$. [This follows from the definition of the logarithm: if $10 = 10^1$, then $1 = \log(10)$.] Thus,

$$\begin{aligned} \log(380) &= \log(3.8) + \log(10^2) \\ &= 0.580 + 2 \\ &= 2.580. \end{aligned}$$

Similarly,

$$\begin{aligned} \log(0.081) &= \log(8.1) + \log(10^{-2}) \\ &= 0.908 - 2 = -1.092. \end{aligned}$$

The reverse process of finding the number N whose log is, say, 2.670, is called "taking the **antilogarithm**." To do so, we separate our number 2.670 into two parts, making the separation at the decimal point:

$$\begin{aligned} \log N = 2.670 &= 2 + 0.670 \\ &= \log 10^2 + 0.670. \end{aligned}$$

We now look at Table A–1 to see what number has its log equal to 0.670; none does, so we must **interpolate**: we see that $\log 4.6 = 0.663$ and $\log 4.7 = 0.672$. So the number we want is between 4.6 and 4.7, and closer to the latter by $\frac{7}{9}$. Approximately we can say that $\log 4.68 = 0.670$. Thus

$$\begin{aligned} \log N &= 2 + 0.670 \\ &= \log(10^2) + \log(4.68) = \log(4.68 \times 10^2), \end{aligned}$$

so $N = 4.68 \times 10^2 = 468$.

If the given logarithm is negative, say, -2.180, we proceed as follows:

$$\begin{aligned} \log N = -2.180 &= -3 + 0.820 \\ &= \log 10^{-3} + \log 6.6 = \log 6.6 \times 10^{-3}, \end{aligned}$$

so $N = 6.6 \times 10^{-3}$. Notice that we added to our given logarithm the next largest integer (3 in this case) so that we have an integer, plus a decimal number between 0 and 1.0 whose antilogarithm can be looked up in the Table.

TABLE A–1 Short Table of Common Logarithms

N	0.0	0.1	0.2	0.3	0.4	0.5	0.6	0.7	0.8	0.9
1	000	041	079	114	146	176	204	230	255	279
2	301	322	342	362	380	398	415	431	447	462
3	477	491	505	519	531	544	556	568	580	591
4	602	613	623	633	643	653	663	672	681	690
5	699	708	716	724	732	740	748	756	763	771
6	778	785	792	799	806	813	820	826	833	839
7	845	851	857	863	869	875	881	886	892	898
8	903	908	914	919	924	929	935	940	944	949
9	954	959	964	968	973	978	982	987	991	996

A–8 Vectors

Vector addition is covered in Sections 3–2 to 3–5.
Vector multiplication is covered in Sections 3–3, 7–2, and 11–2.

A–9 Trigonometric Functions and Identities

The trigonometric functions are defined as follows (see Fig. A–5, o = side opposite, a = side adjacent, h = hypotenuse. Values are given in Table A–2):

$$\sin \theta = \frac{o}{h} \qquad\qquad \csc \theta = \frac{1}{\sin \theta} = \frac{h}{o}$$

$$\cos \theta = \frac{a}{h} \qquad\qquad \sec \theta = \frac{1}{\cos \theta} = \frac{h}{a}$$

$$\tan \theta = \frac{o}{a} = \frac{\sin \theta}{\cos \theta} \qquad\qquad \cot \theta = \frac{1}{\tan \theta} = \frac{a}{o}$$

and recall that

$$a^2 + o^2 = h^2 \qquad\qquad \text{[Pythagorean theorem]}.$$

Figure A–6 shows the signs (+ or −) that cosine, sine, and tangent take on for angles θ in the four quadrants (0° to 360°). Note that angles are measured counterclockwise from the x axis as shown; negative angles are measured from *below* the x axis, clockwise: for example, −30° = +330°, and so on.

The following are some useful identities among the trigonometric functions:

$$\sin^2 \theta + \cos^2 \theta = 1$$

$$\sec^2 \theta - \tan^2 \theta = 1, \quad \csc^2 \theta - \cot^2 \theta = 1$$

$$\sin 2\theta = 2 \sin \theta \cos \theta$$

$$\cos 2\theta = \cos^2 \theta - \sin^2 \theta = 2 \cos^2 \theta - 1 = 1 - 2 \sin^2 \theta$$

$$\tan 2\theta = \frac{2 \tan \theta}{1 - \tan^2 \theta}$$

$$\sin(A \pm B) = \sin A \cos B \pm \cos A \sin B$$

$$\cos(A \pm B) = \cos A \cos B \mp \sin A \sin B$$

$$\tan(A \pm B) = \frac{\tan A \pm \tan B}{1 \mp \tan A \tan B}$$

$$\sin(180° - \theta) = \sin \theta$$

$$\cos(180° - \theta) = -\cos \theta$$

$$\sin(90° - \theta) = \cos \theta$$

$$\cos(90° - \theta) = \sin \theta$$

$$\sin(-\theta) = -\sin \theta$$

$$\cos(-\theta) = \cos \theta$$

$$\tan(-\theta) = -\tan \theta$$

$$\sin\tfrac{1}{2}\theta = \sqrt{\frac{1 - \cos \theta}{2}}, \quad \cos\tfrac{1}{2}\theta = \sqrt{\frac{1 + \cos \theta}{2}}, \quad \tan\tfrac{1}{2}\theta = \sqrt{\frac{1 - \cos \theta}{1 + \cos \theta}}$$

$$\sin A \pm \sin B = 2 \sin\left(\frac{A \pm B}{2}\right) \cos\left(\frac{A \mp B}{2}\right).$$

For any triangle (see Fig. A–7):

$$\frac{\sin \alpha}{a} = \frac{\sin \beta}{b} = \frac{\sin \gamma}{c} \qquad\qquad \text{[Law of sines]}$$

$$c^2 = a^2 + b^2 - 2ab \cos \gamma. \qquad\qquad \text{[Law of cosines]}$$

Values of sine, cosine, tangent are given in Table A–2.

FIGURE A–5

FIGURE A–6

First Quadrant
(0° to 90°)
$x > 0$
$y > 0$

$\sin \theta = y/r > 0$
$\cos \theta = x/r > 0$
$\tan \theta = y/x > 0$

Second Quadrant
(90° to 180°)
$x < 0$
$y > 0$

$\sin \theta > 0$
$\cos \theta < 0$
$\tan \theta < 0$

Third Quadrant
(180° to 270°)
$x < 0$
$y < 0$

$\sin \theta < 0$
$\cos \theta < 0$
$\tan \theta > 0$

Fourth Quadrant
(270° to 360°)
$x > 0$
$y < 0$

$\sin \theta < 0$
$\cos \theta > 0$
$\tan \theta < 0$

FIGURE A–7

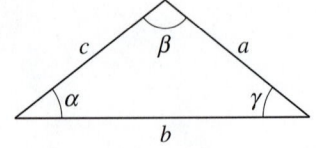

TABLE A–2 Trigonometric Table: Numerical Values of Sin, Cos, Tan

Angle in Degrees	Angle in Radians	Sine	Cosine	Tangent	Angle in Degrees	Angle in Radians	Sine	Cosine	Tangent
0°	0.000	0.000	1.000	0.000					
1°	0.017	0.017	1.000	0.017	46°	0.803	0.719	0.695	1.036
2°	0.035	0.035	0.999	0.035	47°	0.820	0.731	0.682	1.072
3°	0.052	0.052	0.999	0.052	48°	0.838	0.743	0.669	1.111
4°	0.070	0.070	0.998	0.070	49°	0.855	0.755	0.656	1.150
5°	0.087	0.087	0.996	0.087	50°	0.873	0.766	0.643	1.192
6°	0.105	0.105	0.995	0.105	51°	0.890	0.777	0.629	1.235
7°	0.122	0.122	0.993	0.123	52°	0.908	0.788	0.616	1.280
8°	0.140	0.139	0.990	0.141	53°	0.925	0.799	0.602	1.327
9°	0.157	0.156	0.988	0.158	54°	0.942	0.809	0.588	1.376
10°	0.175	0.174	0.985	0.176	55°	0.960	0.819	0.574	1.428
11°	0.192	0.191	0.982	0.194	56°	0.977	0.829	0.559	1.483
12°	0.209	0.208	0.978	0.213	57°	0.995	0.839	0.545	1.540
13°	0.227	0.225	0.974	0.231	58°	1.012	0.848	0.530	1.600
14°	0.244	0.242	0.970	0.249	59°	1.030	0.857	0.515	1.664
15°	0.262	0.259	0.966	0.268	60°	1.047	0.866	0.500	1.732
16°	0.279	0.276	0.961	0.287	61°	1.065	0.875	0.485	1.804
17°	0.297	0.292	0.956	0.306	62°	1.082	0.883	0.469	1.881
18°	0.314	0.309	0.951	0.325	63°	1.100	0.891	0.454	1.963
19°	0.332	0.326	0.946	0.344	64°	1.117	0.899	0.438	2.050
20°	0.349	0.342	0.940	0.364	65°	1.134	0.906	0.423	2.145
21°	0.367	0.358	0.934	0.384	66°	1.152	0.914	0.407	2.246
22°	0.384	0.375	0.927	0.404	67°	1.169	0.921	0.391	2.356
23°	0.401	0.391	0.921	0.424	68°	1.187	0.927	0.375	2.475
24°	0.419	0.407	0.914	0.445	69°	1.204	0.934	0.358	2.605
25°	0.436	0.423	0.906	0.466	70°	1.222	0.940	0.342	2.747
26°	0.454	0.438	0.899	0.488	71°	1.239	0.946	0.326	2.904
27°	0.471	0.454	0.891	0.510	72°	1.257	0.951	0.309	3.078
28°	0.489	0.469	0.883	0.532	73°	1.274	0.956	0.292	3.271
29°	0.506	0.485	0.875	0.554	74°	1.292	0.961	0.276	3.487
30°	0.524	0.500	0.866	0.577	75°	1.309	0.966	0.259	3.732
31°	0.541	0.515	0.857	0.601	76°	1.326	0.970	0.242	4.011
32°	0.559	0.530	0.848	0.625	77°	1.344	0.974	0.225	4.331
33°	0.576	0.545	0.839	0.649	78°	1.361	0.978	0.208	4.705
34°	0.593	0.559	0.829	0.675	79°	1.379	0.982	0.191	5.145
35°	0.611	0.574	0.819	0.700	80°	1.396	0.985	0.174	5.671
36°	0.628	0.588	0.809	0.727	81°	1.414	0.988	0.156	6.314
37°	0.646	0.602	0.799	0.754	82°	1.431	0.990	0.139	7.115
38°	0.663	0.616	0.788	0.781	83°	1.449	0.993	0.122	8.144
39°	0.681	0.629	0.777	0.810	84°	1.466	0.995	0.105	9.514
40°	0.698	0.643	0.766	0.839	85°	1.484	0.996	0.087	11.43
41°	0.716	0.656	0.755	0.869	86°	1.501	0.998	0.070	14.301
42°	0.733	0.669	0.743	0.900	87°	1.518	0.999	0.052	19.081
43°	0.750	0.682	0.731	0.933	88°	1.536	0.999	0.035	28.636
44°	0.768	0.695	0.719	0.966	89°	1.553	1.000	0.017	57.290
45°	0.785	0.707	0.707	1.000	90°	1.571	1.000	0.000	∞

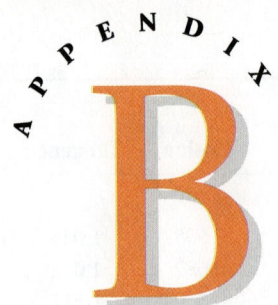

Derivatives and Integrals

B–1 Derivatives: General Rules

(See also Section 2–3.)

$$\frac{dx}{dx} = 1$$

$$\frac{d}{dx}[af(x)] = a\frac{df}{dx} \qquad (a = \text{constant})$$

$$\frac{d}{dx}[f(x) + g(x)] = \frac{df}{dx} + \frac{dg}{dx}$$

$$\frac{d}{dx}[f(x)g(x)] = \frac{df}{dx}g + f\frac{dg}{dx}$$

$$\frac{d}{dx}[f(y)] = \frac{df}{dy}\frac{dy}{dx} \qquad \text{[chain rule]}$$

$$\frac{dx}{dy} = \frac{1}{\left(\dfrac{dy}{dx}\right)} \qquad \text{if } \frac{dy}{dx} \neq 0.$$

B–2 Derivatives: Particular Functions

$$\frac{da}{dx} = 0 \qquad (a = \text{constant})$$

$$\frac{d}{dx}x^n = nx^{n-1}$$

$$\frac{d}{dx}\sin ax = a\cos ax$$

$$\frac{d}{dx}\cos ax = -a\sin ax$$

$$\frac{d}{dx}\tan ax = a\sec^2 ax$$

$$\frac{d}{dx}\ln ax = \frac{1}{x}$$

$$\frac{d}{dx}e^{ax} = ae^{ax}$$

B–3 Indefinite Integrals: General Rules

(See also Section 7–3.)

$$\int dx = x$$

$$\int a\,f(x)\,dx = a\int f(x)\,dx \qquad (a = \text{constant})$$

$$\int [f(x) + g(x)]\,dx = \int f(x)\,dx + \int g(x)\,dx$$

$$\int u\,dv = uv - \int v\,du \qquad \text{(integration by parts)}$$

B–4 Indefinite Integrals: Particular Functions

(An arbitrary constant can be added to the right side of each equation.)

$$\int a\, dx = ax \qquad (a = \text{constant})$$

$$\int x^m\, dx = \frac{1}{m+1} x^{m+1} \qquad (m \neq -1)$$

$$\int \sin ax\, dx = -\frac{1}{a} \cos ax$$

$$\int \cos ax\, dx = \frac{1}{a} \sin ax$$

$$\int \tan ax\, dx = \frac{1}{a} \ln|\sec ax|$$

$$\int \frac{1}{x}\, dx = \ln x$$

$$\int e^{ax}\, dx = \frac{1}{a} e^{ax}$$

$$\int \frac{dx}{x^2 + a^2} = \frac{1}{a} \tan^{-1} \frac{x}{a}$$

$$\int \frac{dx}{x^2 - a^2} = \frac{1}{2a} \ln\left(\frac{x-a}{x+a}\right) \qquad (x^2 > a^2)$$

$$= -\frac{1}{2a} \ln\left(\frac{a+x}{a-x}\right) \qquad (x^2 < a^2)$$

$$\int \frac{dx}{\sqrt{x^2 \pm a^2}} = \ln(x + \sqrt{x^2 \pm a^2})$$

$$\int \frac{dx}{\sqrt{a^2 - x^2}} = \sin^{-1}\left(\frac{x}{a}\right) = -\cos^{-1}\left(\frac{x}{a}\right) \qquad \text{if } x^2 \leq a^2$$

$$\int \frac{dx}{(x^2 \pm a^2)^{\frac{3}{2}}} = \frac{\pm x}{a^2 \sqrt{x^2 \pm a^2}}$$

$$\int \frac{x\, dx}{(x^2 \pm a^2)^{\frac{3}{2}}} = \frac{-1}{\sqrt{x^2 \pm a^2}}$$

$$\int \sin^2 ax\, dx = \frac{x}{2} - \frac{\sin 2ax}{4a}$$

$$\int x e^{-ax}\, dx = -\frac{e^{-ax}}{a^2}(ax + 1)$$

$$\int x^2 e^{-ax}\, dx = -\frac{e^{-ax}}{a^3}(a^2 x^2 + 2ax + 2)$$

B–5 A Few Definite Integrals

$$\int_0^\infty x^n e^{-ax}\, dx = \frac{n!}{a^{n+1}}$$

$$\int_0^\infty x^2 e^{-ax^2}\, dx = \sqrt{\frac{\pi}{16a^3}}$$

$$\int_0^\infty e^{-ax^2}\, dx = \sqrt{\frac{\pi}{4a}}$$

$$\int_0^\infty x^3 e^{-ax^2}\, dx = \frac{1}{2a^2}$$

$$\int_0^\infty x e^{-ax^2}\, dx = \frac{1}{2a}$$

$$\int_0^\infty x^{2n} e^{-ax^2}\, dx = \frac{1 \cdot 3 \cdot 5 \cdots (2n-1)}{2^{n+1} a^n} \sqrt{\frac{\pi}{a}}$$

C More on Dimensional Analysis

An important use of dimensional analysis (Section 1–7) is to obtain the *form* of an equation: how one quantity depends on others. To take a concrete example, let us try to find an expression for the period T of a simple pendulum. First, we try to figure out what T could depend on, and make a list of these variables. It might depend on its length ℓ, on the mass m of the bob, on the angle of swing θ, and on the acceleration due to gravity, g. It might also depend on air resistance (we would use the viscosity of air), the gravitational pull of the Moon, and so on; but everyday experience suggests that the Earth's gravity is the major force involved, so we ignore the other possible forces. So let us assume that T is a function of ℓ, m, θ, and g, and that each of these factors is present to some power:

$$T = C\ell^w m^x \theta^y g^z.$$

C is a dimensionless constant, and w, x, y, and z are exponents we want to solve for. We now write down the dimensional equation (Section 1–7) for this relationship:

$$[T] = [L]^w[M]^x[L/T^2]^z.$$

Because θ has no dimensions (a radian is a length divided by a length—see Eq. 10–1a), it does not appear. We simplify and obtain

$$[T] = [L]^{w+z}[M]^x[T]^{-2z}$$

To have dimensional consistency, we must have

$$1 = -2z$$
$$0 = w + z$$
$$0 = x.$$

We solve these equations and find that $z = -\frac{1}{2}, w = \frac{1}{2}$, and $x = 0$. Thus our desired equation must be

$$T = C\sqrt{\ell/g}\,f(\theta), \tag{C–1}$$

where $f(\theta)$ is some function of θ that we cannot determine using this technique. Nor can we determine in this way the dimensionless constant C. (To obtain C and f, we would have to do an analysis such as that in Chapter 14 using Newton's laws, which reveals that $C = 2\pi$ and $f \approx 1$ for small θ). But look what we *have* found, using only dimensional consistency. We obtained the form of the expression that relates the period of a simple pendulum to the major variables of the situation, ℓ and g (see Eq. 14–12c), and saw that it does not depend on the mass m.

How did we do it? And how useful is this technique? Basically, we had to use our intuition as to which variables were important and which were not. This is not always easy, and often requires a lot of insight. As to usefulness, the final result in our example could have been obtained from Newton's laws, as in Chapter 14. But in many physical situations, such a derivation from other laws cannot be done. In those situations, dimensional analysis can be a powerful tool.

In the end, any expression derived by the use of dimensional analysis (or by any other means, for that matter) must be checked against experiment. For example, in our derivation of Eq. C–1, we can compare the periods of two pendulums of different lengths, ℓ_1 and ℓ_2, whose amplitudes (θ) are the same. For, using Eq. C–1, we would have

$$\frac{T_1}{T_2} = \frac{C\sqrt{\ell_1/g}\,f(\theta)}{C\sqrt{\ell_2/g}\,f(\theta)} = \sqrt{\frac{\ell_1}{\ell_2}}.$$

Because C and $f(\theta)$ are the same for both pendula, they cancel out, so we can experimentally determine if the ratio of the periods varies as the ratio of the square roots of the lengths. This comparison to experiment checks our derivation, at least in part; C and $f(\theta)$ could be determined by further experiments.

D Gravitational Force due to a Spherical Mass Distribution

In Chapter 6 we stated that the gravitational force exerted by or on a uniform sphere acts as if all the mass of the sphere were concentrated at its center, if the other object (exerting or feeling the force) is outside the sphere. In other words, the gravitational force that a uniform sphere exerts on a particle outside it is

$$F = G\frac{mM}{r^2}, \qquad \text{[m outside sphere of mass M]}$$

where m is the mass of the particle, M the mass of the sphere, and r the distance of m from the center of the sphere. Now we will derive this result. We will use the concepts of infinitesimally small quantities and integration.

First we consider a very thin, uniform spherical shell (like a thin-walled basketball) of mass M whose thickness t is small compared to its radius R (Fig. D–1). The force on a particle of mass m at a distance r from the center of the shell can be calculated as the vector sum of the forces due to all the particles of the shell. We imagine the shell divided up into thin (infinitesimal) circular strips so that all points on a strip are equidistant from our particle m. One of these circular strips, labeled AB, is shown in Fig. D–1. It is $R\,d\theta$ wide, t thick, and has a radius $R\sin\theta$. The force on our particle m due to a tiny piece of the strip at point A is represented by the vector $\vec{F}_A$ shown. The force due to a tiny piece of the strip at point B, which is diametrically opposite A, is the force $\vec{F}_B$. We take the two pieces at A and B to be of equal mass, so $F_A = F_B$. The horizontal components of $\vec{F}_A$ and $\vec{F}_B$ are each equal to

$$F_A \cos\phi$$

and point toward the center of the shell. The vertical components of $\vec{F}_A$ and $\vec{F}_B$ are of equal magnitude and point in opposite directions, and so cancel. Since for every point on the strip there is a corresponding point diametrically opposite (as with A and B), we see that the net force due to the entire strip points toward the center of the shell. Its magnitude will be

$$dF = G\frac{m\,dM}{\ell^2}\cos\phi,$$

where dM is the mass of the entire circular strip and ℓ is the distance from all points on the strip to m, as shown. We write dM in terms of the density ρ; by density we mean the mass per unit volume (Section 13–2). Hence, $dM = \rho\,dV$, where dV is the volume of the strip and equals $(2\pi R\sin\theta)(t)(R\,d\theta)$. Then the force dF due to the circular strip shown is

$$dF = G\frac{m\rho 2\pi R^2 t \sin\theta\,d\theta}{\ell^2}\cos\phi. \qquad \textbf{(D–1)}$$

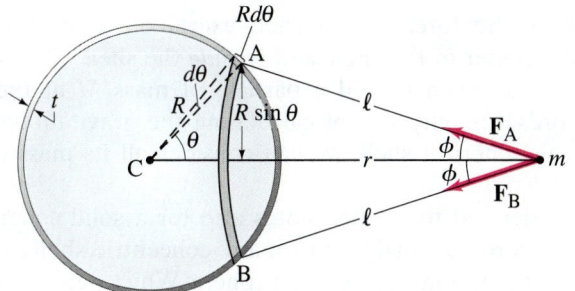

FIGURE D–1 Calculating the gravitational force on a particle mass m due to a uniform sphere shell of radius R and mass '

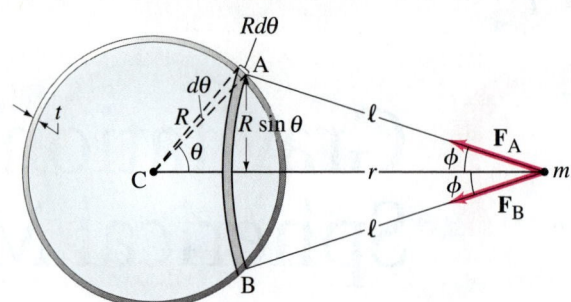

FIGURE D–1 (repeated)
Calculating the gravitational force on a particle of mass m due to a uniform spherical shell of radius R and mass M.

To get the total force F that the entire shell exerts on the particle m, we must integrate over all the circular strips: that is, we integrate

$$dF = G \frac{m\rho 2\pi R^2 t \sin\theta \, d\theta}{\ell^2} \cos\phi \qquad \text{(D–1)}$$

from $\theta = 0°$ to $\theta = 180°$. But our expression for dF contains ℓ and ϕ, which are functions of θ. From Fig. D–1 we can see that

$$\ell \cos\phi = r - R\cos\theta.$$

Furthermore, we can write the law of cosines for triangle CmA:

$$\cos\theta = \frac{r^2 + R^2 - \ell^2}{2rR}. \qquad \text{(D–2)}$$

With these two expressions we can reduce our three variables (ℓ, θ, ϕ) to only one, which we take to be ℓ. We do two things with Eq. D–2: (1) We put it into the equation for $\ell\cos\phi$ above:

$$\cos\phi = \frac{1}{\ell}(r - R\cos\theta) = \frac{r^2 + \ell^2 - R^2}{2r\ell}.$$

and (2) we take the differential of both sides of Eq. D–2 (because $\sin\theta \, d\theta$ appears in the expression for dF, Eq. D–1), considering r and R to be constants when summing over the strips:

$$-\sin\theta \, d\theta = -\frac{2\ell \, d\ell}{2rR} \qquad \text{or} \qquad \sin\theta \, d\theta = \frac{\ell \, d\ell}{rR}.$$

We insert these into Eq. D–1 for dF and find

$$dF = Gm\rho\pi t \frac{R}{r^2}\left(1 + \frac{r^2 - R^2}{\ell^2}\right) d\ell.$$

Now we integrate to get the net force on our thin shell of radius R. To integrate over all the strips ($\theta = 0°$ to $180°$), we must go from $\ell = r - R$ to $\ell = r + R$ (see Fig. D–1). Thus,

$$F = Gm\rho\pi t \frac{R}{r^2}\left[\ell - \frac{r^2 - R^2}{\ell}\right]_{\ell=r-R}^{\ell=r+R}$$

$$= Gm\rho\pi t \frac{R}{r^2}(4R).$$

The volume V of the spherical shell is its area $(4\pi R^2)$ times the thickness t. Hence the mass $M = \rho V = \rho 4\pi R^2 t$, and finally

$$F = G\frac{mM}{r^2}. \qquad \left[\begin{array}{c}\text{particle of mass } m \text{ outside a} \\ \text{thin uniform spherical shell of mass } M\end{array}\right]$$

This result gives us the force a thin shell exerts on a particle of mass m a distance r from the center of the shell, and *outside* the shell. We see that the force is the same as that between m and a particle of mass M at the center of the shell. In other words, for purposes of calculating the gravitational force exerted on or by a uniform spherical shell, we can consider all its mass concentrated at its center.

What we have derived for a shell holds also for a solid sphere, since a solid sphere can be considered as made up of many concentric shells, from $R = 0$ to $R = R_0$, where R_0 is the radius of the solid sphere. Why? Because if each shell has mass dM, we write for each shell, $dF = Gm\,dM/r^2$, where r is the distance from the center C to mass m and is the same for all shells. Then the total force equals the sum or integral over dM, which gives the total mass M. Thus the result

$$F = G\frac{mM}{r^2} \qquad \begin{bmatrix} \text{particle of mass } m \text{ outside} \\ \text{solid sphere of mass } M \end{bmatrix} \quad \textbf{(D–3)}$$

is valid for a solid sphere of mass M even if the density varies with distance from the center. (It is not valid if the density varies within each shell—that is, depends not only on R.) Thus the gravitational force exerted on or by spherical objects, including nearly spherical objects like the Earth, Sun, and Moon, can be considered to act as if the objects were point particles.

This result, Eq. D–3, is true only if the mass m is outside the sphere. Let us next consider a point mass m that is located inside the spherical shell of Fig. D–1. Here, r would be less than R, and the integration over ℓ would be from $\ell = R - r$ to $\ell = R + r$, so

$$\left[\ell - \frac{r^2 - R^2}{\ell} \right]_{R-r}^{R+r} = 0.$$

Thus the force on any mass inside the shell would be zero. This result has particular importance for the electrostatic force, which is also an inverse square law. For the gravitational situation, we see that at points within a solid sphere, say 1000 km below the Earth's surface, only the mass up to that radius contributes to the net force. The outer shells beyond the point in question contribute zero net gravitational effect.

The results we have obtained here can also be reached using the gravitational analog of Gauss's law for electrostatics (Chapter 22).

Differential Form of Maxwell's Equations

Maxwell's equations can be written in another form that is often more convenient than Eqs. 31–5. This material is usually covered in more advanced courses, and is included here simply for completeness.

We quote here two theorems, without proof, that are derived in vector analysis textbooks. The first is called **Gauss's theorem** or the **divergence theorem**. It relates the integral over a surface of any vector function $\vec{\mathbf{F}}$ to a volume integral over the volume enclosed by the surface:

$$\oint_{\text{Area } A} \vec{\mathbf{F}} \cdot d\vec{\mathbf{A}} = \int_{\text{Volume } V} \vec{\boldsymbol{\nabla}} \cdot \vec{\mathbf{F}} \, dV.$$

The operator $\vec{\boldsymbol{\nabla}}$ is the **del operator**, defined in Cartesian coordinates as

$$\vec{\boldsymbol{\nabla}} = \hat{\mathbf{i}} \frac{\partial}{\partial x} + \hat{\mathbf{j}} \frac{\partial}{\partial y} + \hat{\mathbf{k}} \frac{\partial}{\partial z}.$$

The quantity

$$\vec{\boldsymbol{\nabla}} \cdot \vec{\mathbf{F}} = \frac{\partial F_x}{\partial x} + \frac{\partial F_y}{\partial y} + \frac{\partial F_z}{\partial z}$$

is called the **divergence** of $\vec{\mathbf{F}}$. The second theorem is **Stokes's theorem**, and relates a line integral around a closed path to a surface integral over any surface enclosed by that path:

$$\oint_{\text{Line}} \vec{\mathbf{F}} \cdot d\vec{\boldsymbol{\ell}} = \int_{\text{Area } A} \vec{\boldsymbol{\nabla}} \times \vec{\mathbf{F}} \cdot d\vec{\mathbf{A}}.$$

The quantity $\vec{\boldsymbol{\nabla}} \times \vec{\mathbf{F}}$ is called the **curl** of $\vec{\mathbf{F}}$. (See Section 11–2 on the vector product.)

We now use these two theorems to obtain the differential form of Maxwell's equations in free space. We apply Gauss's theorem to Eq. 31–5a (Gauss's law):

$$\oint_A \vec{\mathbf{E}} \cdot d\vec{\mathbf{A}} = \int \vec{\boldsymbol{\nabla}} \cdot \vec{\mathbf{E}} \, dV = \frac{Q}{\epsilon_0}.$$

Now the charge Q can be written as a volume integral over the charge density ρ: $Q = \int \rho \, dV.$ Then

$$\int \vec{\boldsymbol{\nabla}} \cdot \vec{\mathbf{E}} \, dV = \frac{1}{\epsilon_0} \int \rho \, dV.$$

Both sides contain volume integrals over the same volume, and for this to be true over *any* volume, whatever its size or shape, the integrands must be equal:

$$\vec{\boldsymbol{\nabla}} \cdot \vec{\mathbf{E}} = \frac{\rho}{\epsilon_0}. \tag{E–1}$$

This is the differential form of Gauss's law. The second of Maxwell's equations, $\oint \vec{\mathbf{B}} \cdot d\vec{\mathbf{A}} = 0,$ is treated in the same way, and we obtain

$$\vec{\boldsymbol{\nabla}} \cdot \vec{\mathbf{B}} = 0. \tag{E–2}$$

Next, we apply Stokes's theorem to the third of Maxwell's equations,

$$\oint \vec{\mathbf{E}} \cdot d\vec{\boldsymbol{\ell}} = \int \vec{\boldsymbol{\nabla}} \times \vec{\mathbf{E}} \cdot d\vec{\mathbf{A}} = -\frac{d\Phi_B}{dt}.$$

Since the magnetic flux $\Phi_B = \int \vec{\mathbf{B}} \cdot d\vec{\mathbf{A}}$, we have

$$\int \vec{\boldsymbol{\nabla}} \times \vec{\mathbf{E}} \cdot d\vec{\mathbf{A}} = -\frac{\partial}{\partial t} \int \vec{\mathbf{B}} \cdot d\vec{\mathbf{A}}$$

where we use the partial derivative, $\partial \vec{\mathbf{B}}/\partial t$, since B may also depend on position. These are surface integrals over the same area, and to be true over any area, even a very small one, we must have

$$\vec{\boldsymbol{\nabla}} \times \vec{\mathbf{E}} = -\frac{\partial \vec{\mathbf{B}}}{\partial t}. \qquad \text{(E–3)}$$

This is the third of Maxwell's equations in differential form. Finally, to the last of Maxwell's equations,

$$\oint \vec{\mathbf{B}} \cdot d\vec{\boldsymbol{\ell}} = \mu_0 I + \mu_0 \epsilon_0 \frac{d\Phi_E}{dt},$$

we apply Stokes's theorem and write $\Phi_E = \int \vec{\mathbf{E}} \cdot d\vec{\mathbf{A}}$:

$$\int \vec{\boldsymbol{\nabla}} \times \vec{\mathbf{B}} \cdot d\vec{\mathbf{A}} = \mu_0 I + \mu_0 \epsilon_0 \frac{\partial}{\partial t} \int \vec{\mathbf{E}} \cdot d\vec{\mathbf{A}}.$$

The conduction current I can be written in terms of the current density $\vec{\mathbf{j}}$, using Eq. 25–12:

$$I = \int \vec{\mathbf{j}} \cdot d\vec{\mathbf{A}}.$$

Then Maxwell's fourth equation becomes:

$$\int \vec{\boldsymbol{\nabla}} \times \vec{\mathbf{B}} \cdot d\vec{\mathbf{A}} = \mu_0 \int \vec{\mathbf{j}} \cdot d\vec{\mathbf{A}} + \mu_0 \epsilon_0 \frac{\partial}{\partial t} \int \vec{\mathbf{E}} \cdot d\vec{\mathbf{A}}.$$

For this to be true over any area A, whatever its size or shape, the integrands on each side of the equation must be equal:

$$\vec{\boldsymbol{\nabla}} \times \vec{\mathbf{B}} = \mu_0 \vec{\mathbf{j}} + \mu_0 \epsilon_0 \frac{\partial \vec{\mathbf{E}}}{\partial t}. \qquad \text{(E–4)}$$

Equations E–1, 2, 3, and 4 are Maxwell's equations in differential form for free space. They are summarized in Table E–1.

TABLE E–1 Maxwell's Equations in Free Space[†]

Integral form	Differential form
$\oint \vec{\mathbf{E}} \cdot d\vec{\mathbf{A}} = \dfrac{Q}{\epsilon_0}$	$\vec{\boldsymbol{\nabla}} \cdot \vec{\mathbf{E}} = \dfrac{\rho}{\epsilon_0}$
$\oint \vec{\mathbf{B}} \cdot d\vec{\mathbf{A}} = 0$	$\vec{\boldsymbol{\nabla}} \cdot \vec{\mathbf{B}} = 0$
$\oint \vec{\mathbf{E}} \cdot d\vec{\boldsymbol{\ell}} = -\dfrac{d\Phi_B}{dt}$	$\vec{\boldsymbol{\nabla}} \times \vec{\mathbf{E}} = -\dfrac{\partial \vec{\mathbf{B}}}{\partial t}$
$\oint \vec{\mathbf{B}} \cdot d\vec{\boldsymbol{\ell}} = \mu_0 I + \mu_0 \epsilon_0 \dfrac{d\Phi_E}{dt}$	$\vec{\boldsymbol{\nabla}} \times \vec{\mathbf{B}} = \mu_0 \vec{\mathbf{j}} + \mu_0 \epsilon_0 \dfrac{\partial \vec{\mathbf{E}}}{\partial t}$

[†]$\vec{\boldsymbol{\nabla}}$ stands for the *del operator* $\vec{\boldsymbol{\nabla}} = \hat{\mathbf{i}} \dfrac{\partial}{\partial x} + \hat{\mathbf{j}} \dfrac{\partial}{\partial y} + \hat{\mathbf{k}} \dfrac{\partial}{\partial z}$ in Cartesian coordinates.

Selected Isotopes

(1) Atomic Number Z	(2) Element	(3) Symbol	(4) Mass Number A	(5) Atomic Mass†	(6) % Abundance (or Radioactive Decay‡ Mode)	(7) Half-life (if radioactive)
0	(Neutron)	n	1	1.008665	β^-	10.23 min
1	Hydrogen	H	1	1.007825	99.9885%	
	Deuterium	d or D	2	2.014082	0.0115%	
	Tritium	t or T	3	3.016049	β^-	12.33 yr
2	Helium	He	3	3.016029	0.000137%	
			4	4.002603	99.999863%	
3	Lithium	Li	6	6.015123	7.59%	
			7	7.016005	92.41%	
4	Beryllium	Be	7	7.016930	EC, γ	53.22 days
			9	9.012182	100%	
5	Boron	B	10	10.012937	19.9%	
			11	11.009305	80.1%	
6	Carbon	C	11	11.011434	β^+, EC	20.39 min
			12	12.000000	98.93%	
			13	13.003355	1.07%	
			14	14.003242	β^-	5730 yr
7	Nitrogen	N	13	13.005739	β^+, EC	9.965 min
			14	14.003074	99.632%	
			15	15.000109	0.368%	
8	Oxygen	O	15	15.003066	β^+, EC	122.24 s
			16	15.994915	99.757%	
			18	17.999161	0.205%	
9	Fluorine	F	19	18.998403	100%	
10	Neon	Ne	20	19.992440	90.48%	
			22	21.991385	9.25%	
11	Sodium	Na	22	21.994436	β^+, EC, γ	2.6027 yr
			23	22.989769	100%	
			24	23.990963	β^-, γ	14.959 h
12	Magnesium	Mg	24	23.985042	78.99%	
13	Aluminum	Al	27	26.981539	100%	
14	Silicon	Si	28	27.976927	92.2297%	
			31	30.975363	β^-, γ	157.3 min
15	Phosphorus	P	31	30.973762	100%	
			32	31.973907	β^-	14.262 days

† The masses given in column (5) are those for the neutral atom, including the Z electrons.

‡ Chapter 41; EC = electron capture.

(1) Atomic Number Z	(2) Element	(3) Symbol	(4) Mass Number A	(5) Atomic Mass	(6) % Abundance (or Radioactive Decay Mode)	(7) Half-life (if radioactive)
16	Sulfur	S	32	31.972071	94.9%	
			35	34.969032	β^-	87.51 days
17	Chlorine	Cl	35	34.968853	75.78%	
			37	36.965903	24.22%	
18	Argon	Ar	40	39.962383	99.600%	
19	Potassium	K	39	38.963707	93.258%	
			40	39.963998	0.0117%	
					β^-, EC, γ, β^+	1.248×10^9 yr
20	Calcium	Ca	40	39.962591	96.94%	
21	Scandium	Sc	45	44.955912	100%	
22	Titanium	Ti	48	47.947946	73.72%	
23	Vanadium	V	51	50.943960	99.750%	
24	Chromium	Cr	52	51.940508	83.789%	
25	Manganese	Mn	55	54.938045	100%	
26	Iron	Fe	56	55.934938	91.75%	
27	Cobalt	Co	59	58.933195	100%	
			60	59.933817	β^-, γ	5.2711 yr
28	Nickel	Ni	58	57.935343	68.077%	
			60	59.930786	26.223%	
29	Copper	Cu	63	62.929598	69.17%	
			65	64.927790	30.83%	
30	Zinc	Zn	64	63.929142	48.6%	
			66	65.926033	27.9%	
31	Gallium	Ga	69	68.925574	60.108%	
32	Germanium	Ge	72	71.922076	27.5%	
			74	73.921178	36.3%	
33	Arsenic	As	75	74.921596	100%	
34	Selenium	Se	80	79.916521	49.6%	
35	Bromine	Br	79	78.918337	50.69%	
36	Krypton	Kr	84	83.911507	57.00%	
37	Rubidium	Rb	85	84.911790	72.17%	
38	Strontium	Sr	86	85.909260	9.86%	
			88	87.905612	82.58%	
			90	89.907738	β^-	28.79 yr
39	Yttrium	Y	89	88.905848	100%	
40	Zirconium	Zr	90	89.904704	51.4%	
41	Niobium	Nb	93	92.906378	100%	
42	Molybdenum	Mo	98	97.905408	24.1%	
43	Technetium	Tc	98	97.907216	β^-, γ	4.2×10^6 yr
44	Ruthenium	Ru	102	101.904349	31.55%	
45	Rhodium	Rh	103	102.905504	100%	
46	Palladium	Pd	106	105.903486	27.33%	
47	Silver	Ag	107	106.905097	51.839%	
			109	108.904752	48.161%	
48	Cadmium	Cd	114	113.903359	28.7%	
49	Indium	In	115	114.903878	95.71%; β^-	4.41×10^{14} yr
50	Tin	Sn	120	119.902195	32.58%	
51	Antimony	Sb	121	120.903816	57.21%	

(1) Atomic Number Z	(2) Element	(3) Symbol	(4) Mass Number A	(5) Atomic Mass	(6) % Abundance (or Radioactive Decay Mode)	(7) Half-life (if radioactive)
52	Tellurium	Te	130	129.906224	34.1%; $\beta^-\beta^-$	$>7.9 \times 10^{22}$ yr
53	Iodine	I	127	126.904473	100%	
			131	130.906125	β^-, γ	8.0207 days
54	Xenon	Xe	132	131.904154	26.89%	
			136	135.907219	8.87%; $\beta^-\beta^-$	$>3.6 \times 10^{20}$ yr
55	Cesium	Cs	133	132.905452	100%	
56	Barium	Ba	137	136.905827	11.232%	
			138	137.905247	71.70%	
57	Lanthanum	La	139	138.906353	99.910%	
58	Cerium	Ce	140	139.905439	88.45%	
59	Praseodymium	Pr	141	140.907653	100%	
60	Neodymium	Nd	142	141.907723	27.2%	
61	Promethium	Pm	145	144.912749	EC, α	17.7 yr
62	Samarium	Sm	152	151.919732	26.75%	
63	Europium	Eu	153	152.921230	52.19%	
64	Gadolinium	Gd	158	157.924104	24.84%	
65	Terbium	Tb	159	158.925347	100%	
66	Dysprosium	Dy	164	163.929175	28.2%	
67	Holmium	Ho	165	164.930322	100%	
68	Erbium	Er	166	165.930293	33.6%	
69	Thulium	Tm	169	168.934213	100%	
70	Ytterbium	Yb	174	173.938862	31.8%	
71	Lutetium	Lu	175	174.940772	97.41%	
72	Hafnium	Hf	180	179.946550	35.08%	
73	Tantalum	Ta	181	180.947996	99.988%	
74	Tungsten (wolfram)	W	184	183.950931	30.64%; α	$>3 \times 10^{17}$ yr
75	Rhenium	Re	187	186.955753	62.60%; β^-	4.35×10^{10} yr
76	Osmium	Os	191	190.960930	β^-, γ	15.4 days
			192	191.961481	40.78%	
77	Iridium	Ir	191	190.960594	37.3%	
			193	192.962926	62.7%	
78	Platinum	Pt	195	194.964791	33.832%	
79	Gold	Au	197	196.966569	100%	
80	Mercury	Hg	199	198.968280	16.87%	
			202	201.970643	29.9%	
81	Thallium	Tl	205	204.974428	70.476%	
82	Lead	Pb	206	205.974465	24.1%	
			207	206.975897	22.1%	
			208	207.976652	52.4%	
			210	209.984188	β^-, γ, α	22.20 yr
			211	210.988737	β^-, γ	36.1 min
			212	211.991898	β^-, γ	10.64 h
			214	213.999805	β^-, γ	26.8 min
83	Bismuth	Bi	209	208.980399	100%	
			211	210.987269	α, γ, β^-	2.14 min
84	Polonium	Po	210	209.982874	α, γ, EC	138.376 days
			214	213.995201	α, γ	164.3 μs
85	Astatine	At	218	218.008694	α, β^-	1.5 s

(1) Atomic Number Z	(2) Element	(3) Symbol	(4) Mass Number A	(5) Atomic Mass	(6) % Abundance (or Radioactive Decay Mode)	(7) Half-life (if radioactive)
86	Radon	Rn	222	222.017578	α, γ	3.8235 days
87	Francium	Fr	223	223.019736	β^-, γ, α	22.00 min
88	Radium	Ra	226	226.025410	α, γ	1600 yr
89	Actinium	Ac	227	227.027752	β^-, γ, α	21.772 yr
90	Thorium	Th	228	228.028741	α, γ	1.9116 yr
			232	232.038055	100%; α, γ	1.405×10^{10} yr
91	Protactinium	Pa	231	231.035884	α, γ	3.276×10^4 yr
92	Uranium	U	232	232.037156	α, γ	68.9 yr
			233	233.039635	α, γ	1.592×10^5 yr
			235	235.043930	0.720%; α, γ	7.04×10^8 yr
			236	236.045568	α, γ	2.342×10^7 yr
			238	238.050788	99.274%; α, γ	4.468×10^9 yr
			239	239.054293	β^-, γ	23.45 min
93	Neptunium	Np	237	237.048173	α, γ	2.144×10^6 yr
			239	239.052939	β^-, γ	2.356 days
94	Plutonium	Pu	239	239.052163	α, γ	24,110 yr
			244	244.064204	α	8.00×10^7 yr
95	Americium	Am	243	243.061381	α, γ	7370 yr
96	Curium	Cm	247	247.070354	α, γ	1.56×10^7 yr
97	Berkelium	Bk	247	247.070307	α, γ	1380 yr
98	Californium	Cf	251	251.079587	α, γ	898 yr
99	Einsteinium	Es	252	252.082980	α, EC, γ	471.7 days
100	Fermium	Fm	257	257.095105	α, γ	100.5 days
101	Mendelevium	Md	258	258.098431	α, γ	51.5 days
102	Nobelium	No	259	259.10103	α, EC	58 min
103	Lawrencium	Lr	262	262.10963	α, EC, fission	≈ 4 h
104	Rutherfordium	Rf	263	263.11255	fission	10 min
105	Dubnium	Db	262	262.11408	α, fission, EC	35 s
106	Seaborgium	Sg	266	266.12210	α, fission	≈ 21 s
107	Bohrium	Bh	264	264.12460	α	≈ 0.44 s
108	Hassium	Hs	269	269.13406	α	≈ 10 s
109	Meitnerium	Mt	268	268.13870	α	21 ms
110	Darmstadtium	Ds	271	271.14606	α	≈ 70 ms
111	Roentgenium	Rg	272	272.15360	α	3.8 ms
112		Uub	277	277.16394	α	≈ 0.7 ms

Preliminary evidence (unconfirmed) has been reported for elements 113, 114, 115, 116 and 118.

Answers to Odd-Numbered Problems

CHAPTER 1

1. (a) 1.4×10^{10} y;

(b) 4.4×10^{17} s.

3. (a) 1.156×10^{0};

(b) 2.18×10^{1};

(c) 6.8×10^{-3};

(d) 3.2865×10^{2};

(e) 2.19×10^{-1};

(f) 4.44×10^{2}.

5. 4.6%.

7. 1.00×10^{5} s.

9. 0.24 rad.

11. (a) 0.2866 m;

(b) 0.000085 V;

(c) 0.00076 kg;

(d) 0.0000000000600 s;

(e) 0.0000000000000225 m;

(f) 2,500,000,000 V.

13. $5'10'' = 1.8$ m, 165 lbs = 75.2 kg.

15. (a) 1 ft^2 = 0.111 yd^2;

(b) 1 m^2 = 10.8 ft^2.

17. (a) 3.9×10^{-9} in.;

(b) 1.0×10^{8} atoms.

19. (a) 1 km/h = 0.621 mi/h;

(b) 1 m/s = 3.28 ft/s;

(c) 1 km/h = 0.278 m/s.

21. (a) 9.46×10^{15} m;

(b) 6.31×10^{4} AU;

(c) 7.20 AU/h.

23. (a) 3.80×10^{13} m^2;

(b) 13.4.

25. 6×10^{5} books.

27. 5×10^{4} L.

29. (a) 1800.

31. 5×10^{4} m.

33. 6.5×10^{6} m.

35. $[M/L^3]$.

37. (a) Cannot;

(b) can;

(c) can.

39. (1×10^{-5})%, 8 significant figures.

41. (a) 3.16×10^{7} s;

(b) 3.16×10^{16} ns;

(c) 3.17×10^{-8} y.

43. 2×10^{-4} m.

45. 1×10^{11} gal/y.

47. 9 cm/y.

49. 2×10^{9} kg/y.

51. 75 min.

53. 4×10^{5} metric tons, 1×10^{8} gal.

55. 1×10^{3} days

57. 210 yd, 190 m.

59. (a) 0.10 nm;

(b) 1.0×10^{5} fm;

(c) 1.0×10^{10} Å;

(d) 9.5×10^{25} Å.

61. (a) 3%, 3%;

(b) 0.7%, 0.2%.

63. 8×10^{-2} m^3.

65. L/m, L/y, L.

67. (a) 13.4;

(b) 49.3.

69. 4×10^{51} kg.

CHAPTER 2

1. 61 m.

3. 0.65 cm/s, no.

5. 300 m/s, 1 km every 3 sec.

7. (a) 9.26 m/s;

(b) 3.1 m/s.

9. (a) 0.3 m/s;

(b) 1.2 m/s;

(c) 0.30 m/s;

(d) 1.4 m/s;

(e) -0.95 m/s.

11. 2.0×10^{1} s.

13. (a) 5.4×10^{3} m;

(b) 72 min.

15. (a) 61 km/h;

(b) 0.

17. (a) 16 m/s;

(b) +5 m/s.

19. 6.73 m/s.

21. 5 s.

23. (a) 48 s;

(b) 90 s to 108 s;

(c) 0 to 42 s, 65 s to 83 s, 90 s to 108 s;

(d) 65 s to 83 s.

25. (a) 21.2 m/s;

(b) 2.00 m/s^2.

27. 17.0 m/s^2.

29. (a) m/s, m/s^2;

(b) $2B$ m/s^2;

(c) $(A + 10B)$ m/s, $2B$ m/s^2;

(d) $A - 3Bt^{-4}$.

31. 1.5 m/s^2, 99 m.

33. 240 m/s^2.

35. 4.41 m/s^2, 2.61 s.

37. 45.0 m.

39. (a) 560 m;

(b) 47 s;

(c) 23 m, 21 m.

41. (a) 96 m;

(b) 76 m.

43. 27 m/s.

45. 117 km/h.

47. 0.49 m/s^2.

49. 1.6 s.

51. (a) 20 m;

(b) 4 s.

53. 1.16 s.

55. 5.18 s.

57. (a) 25 m/s;

(b) 33 m;

(c) 1.2 s;

(d) 5.2 s.

59. (a) 14 m/s;

(b) fifth floor.

61. 1.3 m.

63. 18.8 m/s, 18.1 m.

65. 52 m.

67. 106 m.

69. (a) $\dfrac{g}{k}\left(1 - e^{-kt}\right)$;

(b) $\dfrac{g}{k}$.

71. 6.

73. 1.3 m.

75. (b) 10 m;

(c) 40 m.

77. 5.2×10^{-2} m/s^2.

79. 4.6 m/s to 5.4 m/s, 5.8 m/s to 6.7 m/s, smaller range of velocities.

81. (a) 5.39 s;

(b) 40.3 m/s;

(c) 90.9 m.

83. (*a*) 8.7 min;
(*b*) 7.3 min.

85. 2.3.

87. Stop.

89. 1.5 poles.

91. 0.44 m/min, 2.9 burgers/min.

93. (*a*) Where the slopes are the same;
(*b*) bicycle A;
(*c*) when the two graphs cross; first crossing, B passing A; second crossing, A passing B;
(*d*) B until the slopes are equal, A after that;
(*e*) same.

95. (*c*)

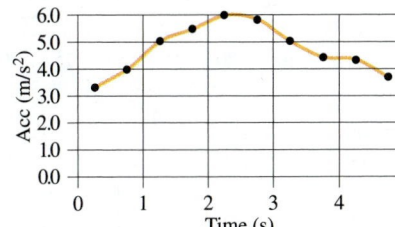

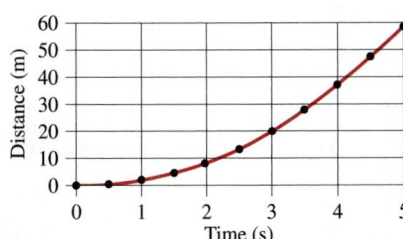

97. (*b*) 6.8 m.

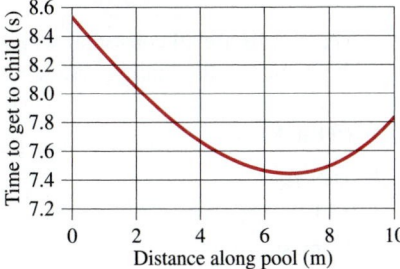

CHAPTER 3

1. 286 km, 11° south of west.

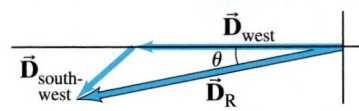

3. 10.1, −39.4°.

5. (*a*)

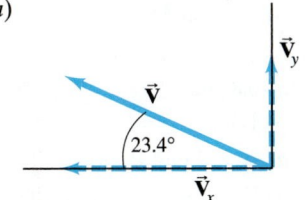

(*b*) −22.8, 9.85;
(*c*) 24.8, 23.4° above the −*x* axis.

7. (*a*) 625 km/h, 553 km/h;
(*b*) 1560 km, 1380 km.

9. (*a*) 4.2 at 315°;
(*b*) $1.0\hat{\mathbf{i}} - 5.0\hat{\mathbf{j}}$ or 5.1 at 280°.

11. (*a*) $-53.7\hat{\mathbf{i}} + 1.31\hat{\mathbf{j}}$ or 53.7 at 1.4° above −*x* axis;
(*b*) $53.7\hat{\mathbf{i}} - 1.31\hat{\mathbf{j}}$ or 53.7 at 1.4° below +*x* axis, they are opposite.

13. (*a*) $-92.5\hat{\mathbf{i}} - 19.4\hat{\mathbf{j}}$ or 94.5 at 11.8° below −*x* axis;
(*b*) $122\hat{\mathbf{i}} - 86.6\hat{\mathbf{j}}$ or 150 at 35.3° below +*x* axis.

15. $(-2450\text{ m})\hat{\mathbf{i}} + (3870\text{ m})\hat{\mathbf{j}}$
$+ (2450\text{ m})\hat{\mathbf{k}}$, 5190 m.

17. $(9.60\hat{\mathbf{i}} - 2.00t\hat{\mathbf{k}})$ m/s,
$(-2.00\hat{\mathbf{k}})$ m/s².

19. Parabola.

21. (*a*) 4.0*t* m/s, 3.0*t* m/s;
(*b*) 5.0*t* m/s;
(*c*) $(2.0t^2\hat{\mathbf{i}} + 1.5t^2\hat{\mathbf{j}})$ m;
(*d*) $v_x = 8.0$ m/s, $v_y = 6.0$ m/s, $v = 10.0$ m/s, $\vec{\mathbf{r}} = (8.0\hat{\mathbf{i}} + 6.0\hat{\mathbf{j}})$ m.

23. (*a*) $(3.16\hat{\mathbf{i}} + 2.78\hat{\mathbf{j}})$ cm/s;
(*b*) 4.21 cm/s at 41.3°.

25. (*a*) $(6.0t\hat{\mathbf{i}} - 18.0t^2\hat{\mathbf{j}})$ m/s, $(6.0\hat{\mathbf{i}} - 36.0t\hat{\mathbf{j}})$ m/s²;
(*b*) $(19\hat{\mathbf{i}} - 94\hat{\mathbf{j}})$ m, $(15\hat{\mathbf{i}} - 110\hat{\mathbf{j}})$ m/s.

27. 414 m at −65.0°.

29. 44 m, 6.9 m.

31. 18°, 72°.

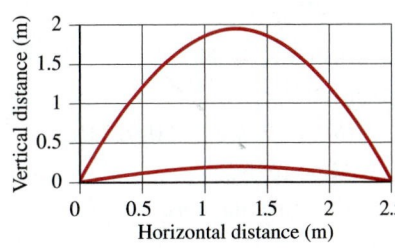

33. 2.26 s.

35. 22.3 m.

37. 39 m.

41. (*a*) 12 s;
(*b*) 62 m.

43. 5.5 s.

45. (*a*) $(2.3\hat{\mathbf{i}} + 2.5\hat{\mathbf{j}})$ m/s;
(*b*) 5.3 m;
(*c*) $(2.3\hat{\mathbf{i}} - 10.2\hat{\mathbf{j}})$ m/s.

47. No, 0.76 m too low; 4.5 m to 34.7 m.

51. $\tan^{-1} gt/v_0$.

53. (*a*) 50.0 m;
(*b*) 6.39 s;
(*c*) 221 m;
(*d*) 38.3 m/s at 25.7°.

55. $\frac{1}{2}\tan^{-1}\left(-\frac{1}{\tan\phi}\right) = \frac{\phi}{2} + \frac{\pi}{4}$.

57. $(10.5\text{ m/s})\hat{\mathbf{i}}, (6.5\text{ m/s})\hat{\mathbf{i}}$.

59. 1.41 m/s.

61. 23 s, 23 m.

63. (*a*) 11.2 m/s, 27° above the horizontal;
(*b*) 11.2 m/s, 27° below the horizontal.

65. 6.3°, west of south.

67. (*a*) 46 m;
(*b*) 92 s.

69. (*a*) 1.13 m/s;
(*b*) 3.20 m/s.

71. 43.6° north of east.

73. $(66\text{ m})\hat{\mathbf{i}} - (35\text{ m})\hat{\mathbf{j}} - (12\text{ m})\hat{\mathbf{k}}$, 76 m, 28° south of east, 9° below the horizontal.

75. 131 km/h, 43.1° north of east.

77. 7.0 m/s.

79. 1.8 m/s².

81. 1.9 m/s, 2.7 s.

83. (*a*) $\dfrac{Dv}{(v^2 - u^2)}$;
(*b*) $\dfrac{D}{\sqrt{v^2 - u^2}}$.

85. 54°.

87. $[(1.5\text{ m})\hat{\mathbf{i}} - (2.0t\text{ m})\hat{\mathbf{i}}]$
$+ [(-3.1\text{ m})\hat{\mathbf{j}} + (1.75t^2\text{ m})\hat{\mathbf{j}}$,
$(3.5\text{ m/s}^2)\hat{\mathbf{j}}$, parabolic.

89. Row at an angle of 24.9° upstream and run 104 m along the bank in a total time of 862 seconds.

91. 69.9° north of east.

93. (*a*) 13 m;
(*b*) 31° below the horizontal.

95. 5.1 s.

97. (*a*) 13 m/s, 12 m/s;
(*b*) 33 m.

99. (*a*) $x = (3.03t - 0.0265)$ m, 3.03 m/s;
(*b*) $y = (0.158 - 0.855t + 6.09t^2)$ m, 12.2 m/s².

CHAPTER 4

1. 77 N.

3. (a) 6.7×10^2 N;

    (b) 1.2×10^2 N;

    (c) 2.5×10^2 N;

    (d) 0.

5. 1.3×10^6 N, 39%, 1.3×10^6 N.

7. 2.1×10^2 N.

9. $m > 1.5$ kg.

11. 89.8 N.

13. 1.8 m/s², up.

15. Descend with $a \geq 2.2$ m/s².

17. -2800 m/s², 280 g's, 1.9×10^5 N.

19. (a) 7.5 s, 13 s, 7.5 s;

    (b) 12%, 0%, -12%;

    (c) 55%.

21. (a) 3.1 m/s²;

    (b) 25 m/s;

    (c) 78 s.

23. 3.3×10^3 N.

25. (a) 150 N;

    (b) 14.5 m/s.

27. (a) 47.0 N;

    (b) 17.0 N;

    (c) 0.

29. (a)   (b)

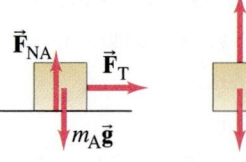

31. (a) 1.5 m;

    (b) 11.5 kN, no.

33. (a) 31 N, 63 N;

    (b) 35 N, 71 N.

35. 6.3×10^3 N, 8.4×10^3 N.

37. (a) 19.0 N at 237.5°, 1.03 m/s² at 237.5°;

    (b) 14.0 N at 51.0°, 0.758 m/s² at 51.0°.

39. $\dfrac{5}{2}\dfrac{F_0}{m}t_0^2$.

41. 4.0×10^2 m.

43. 12°.

45. (a) 9.9 N;

    (b) 260 N.

47. (a) $m_E g - F_T = m_E a$;

    $F_T - m_C g = m_C a$;

    (b) 0.68 m/s², 10,500 N.

49. (a) 2.8 m;

    (b) 2.5 s.

51. (a)

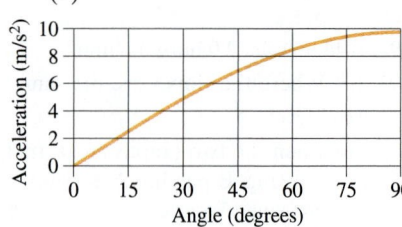

    (b) $g\dfrac{m_B}{m_A + m_B}$, $g\dfrac{m_A\,m_B}{m_A + m_B}$.

53. $g\,\dfrac{m_B + \dfrac{\ell_B}{\ell_A + \ell_B}m_C}{m_A + m_B + m_C}$.

55. $(m + M)g\tan\theta$.

57. 1.52 m/s², 18.3 N, 19.8 N.

59. $\dfrac{(m_A + m_B + m_C)m_B}{\sqrt{(m_A^2 - m_B^2)}}\,g$.

61. (a) $\left(\dfrac{2y}{\ell} - 1\right)g$;

    (b) $\sqrt{2gy_0\left(1 - \dfrac{y_0}{\ell}\right)}$;

    (c) $\dfrac{2}{3}\sqrt{g\ell}$.

63. 6.3 N.

65. 2.0 s, no change.

67. (a) $g\dfrac{(m_A\sin\theta - m_B)}{(m_A + m_B)}$;

    (b) $m_A\sin\theta > m_B$

        (m_A down the plane),

    $m_A\sin\theta < m_B$

        (m_A up the plane).

69. (a) $\dfrac{m_B\sin\theta_B - m_A\sin\theta_A}{m_A + m_B}\,g$;

    (b) 6.8 kg, 26 N;

    (c) 0.74.

71. 9.9°.

73. (a) $41\,\dfrac{N}{m/s}$;

    (b) 1.4×10^2 N.

75. (a) $Mg/2$;

    (b) $Mg/2$, $Mg/2$, $3Mg/2$, Mg.

77. 8.7×10^2 N,

    72° above the horizontal.

79. (a) 0.6 m/s²;

    (b) 1.5×10^5 N.

81. 1.76×10^4 N.

83. 3.8×10^2 N, 7.6×10^2 N.

85. 3.4 m/s.

87. (a) 23 N;

    (b) 3.8 N.

89. (a) $g\sin\theta$, $\sqrt{\dfrac{2\ell}{g\sin\theta}}$,

    $\sqrt{2\ell g\sin\theta}$, $mg\cos\theta$;

    (b)

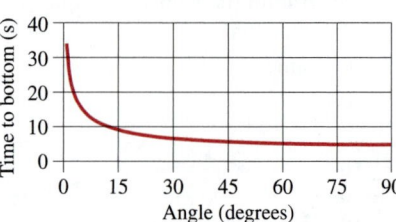

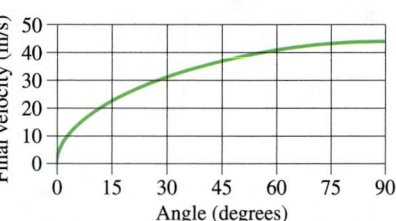

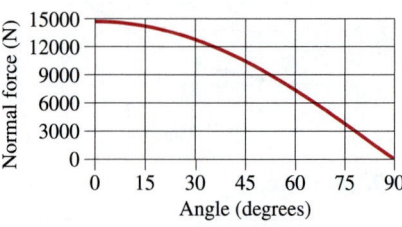

The graphs are all consistent with the results of the limiting cases.

CHAPTER 5

1. 65 N, 0.

3. 0.20.

5. 8.8 m/s².

7. 1.0×10^2 N, 0.48.

9. 0.51.

11. 4.2 m.

13. 1.2×10^3 N.

15. (a) 0.67;

    (b) 6.8 m/s;

    (c) 16 m/s.

17. (a) 1.7 m/s²;

    (b) 4.3×10^2 N;

    (c) 1.7 m/s², 2.2×10^2 N.

19. (a) 0.80 m;

    (b) 1.3 s.

21. (a) A will pull B along;

    (b) B will eventually catch up to A;

(c) $\mu_A < \mu_B$: $a =$
$$g\left[\frac{(m_A + m_B)\sin\theta - (\mu_A m_A + \mu_B m_B)\cos\theta}{(m_A + m_B)}\right],$$
$$F_T = g\frac{m_A m_B}{(m_A + m_B)}(\mu_B - \mu_A)\cos\theta,$$
$$\mu_A > \mu_B: a_A = g(\sin\theta - \mu_A\cos\theta),$$
$$a_B = g(\sin\theta - \mu_B\cos\theta), F_T = 0.$$

23. (a) 5.0 kg;
(b) 6.7 kg.

25. (a) $\dfrac{v_0^2}{2dg\cos\theta} - \tan\theta$;
(b) $\mu_s \geq \tan\theta$.

27. (a) 0.22 s;
(b) 0.16 m.

29. 0.51.

31. (a) 82 N;
(b) 4.5 m/s^2.

33. $(M + m)g\dfrac{(\sin\theta + \mu\cos\theta)}{(\cos\theta - \mu\sin\theta)}$.

35. (a) 1.41 m/s^2;
(b) 31.7 N.

37. $\sqrt{rg}$.

39. 30 m.

41. 31 m/s.

43. 0.9 g's.

45. 9.0 rev/min.

47. (a) 1.9×10^3 m;
(b) 5.4×10^3 N;
(c) 3.8×10^3 N.

49. 3.0×10^2 N.

51. 0.164.

53. (a) 7960 N;
(b) 588 N;
(c) 29.4 m/s.

55. 6.2 m/s.

57. (b) $\vec{v} = (-6.0\text{ m/s})\sin(3.0\text{ rad/s }t)\hat{i}$
$+ (6.0\text{ m/s})\cos(3.0\text{ rad/s }t)\hat{j}$,
$\vec{a} = (-18\text{ m/s}^2)\cos(3.0\text{ rad/s }t)\hat{i}$
$+ (-18\text{ m/s}^2)\sin(3.0\text{ rad/s }t)\hat{j}$;
(c) $v = 6.0$ m/s, $a = 18$ m/s^2.

59. 17 m/s $\leq v \leq$ 32 m/s.

61. (a) $a_t = (\pi/2)$ m/s^2, $a_c = 0$;
(b) $a_t = (\pi/2)$ m/s^2,
$a_c = (\pi^2/8)$ m/s^2;
(c) $a_t = (\pi/2)$ m/s^2,
$a_c = (\pi^2/2)$ m/s^2.

63. (a) 1.64 m/s;
(b) 3.45 m/s.

65. m/b.

67. (a) $\dfrac{mg}{b} + \left(v_0 - \dfrac{mg}{b}\right)e^{-\frac{b}{m}t}$;
(b) $-\dfrac{mg}{b} + \left(v_0 + \dfrac{mg}{b}\right)e^{-\frac{b}{m}t}$.

69. (a) 14 kg/m;
(b) 570 N.

71. $\dfrac{mg}{b}\left[t + \dfrac{m}{b}\left(e^{-\frac{b}{m}t} - 1\right)\right], ge^{-\frac{b}{m}t}$.

75. 10 m.

77. 0.46.

79. 102 N, 0.725.

81. Yes, 14 m/s.

83. 28.3 m/s, 0.410 rev/s.

85. 3500 N, 1900 N.

87. 35°.

89. 132 m.

91. (a) 55 s;
(b) centripetal component of the normal force.

93. (a) $\theta = \cos^{-1}\dfrac{g}{4\pi^2 rf^2}$;
(b) 73.6°;
(c) no.

95. 82°.

97. (a) 16 m/s;
(b) 13 m/s.

99. (a) 0.88 m/s^2;
(b) 0.98 m/s^2.

101. (a) 42.2 m/s;
(b) 35.6 m, 52.6 m.

103. (a)

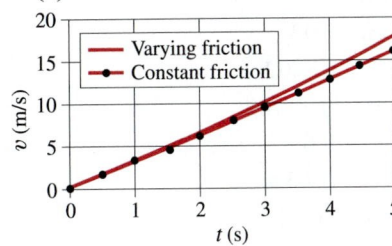

(b)

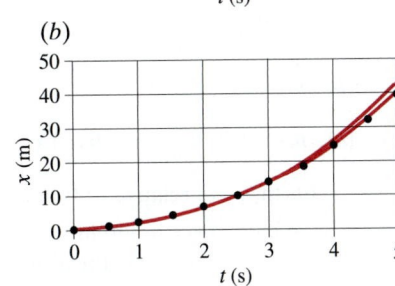

(c) speed: −12%, position: −6.6%.

CHAPTER 6

1. 1610 N.

3. 1.9 m/s^2.

5. $\frac{2}{9}$.

7. 0.91 g's.

9. 1.4×10^{-8} N at 45°.

11. $Gm^2\left\{\left[\dfrac{2}{x_0^2} + \dfrac{3x_0}{(x_0^2 + y_0^2)^{3/2}}\right]\hat{i} \right.$
$\left. + \left[\dfrac{4}{y_0^2} + \dfrac{3y_0}{(x_0^2 + y_0^2)^{3/2}}\right]\hat{j}\right\}$.

13. $2^{1/3} \approx 1.26$ times larger.

15. 3.46×10^8 m from the center of the Earth.

19. (b) g decreases as r increases;
(c) 9.42 m/s^2 approximate,
9.43 m/s^2 exact.

21. 9.78 m/s^2, 0.099° south of radially inward.

23. 7.52×10^3 m/s.

25. 1.7 m/s^2 upward.

27. 7.20×10^3 s.

29. (a) 520 N;
(b) 520 N;
(c) 690 N;
(d) 350 N;
(e) 0.

31. (a) 59 N, toward the Moon;
(b) 110 N, away from the Moon.

33. (a) They are executing centripetal motion;
(b) 9.6×10^{29} kg.

35. $\sqrt{\dfrac{GM}{\ell}}$.

37. 5070 s, or 84.5 min.

39. 160 y.

41. 2×10^8 y.

43. Europa: 671×10^3 km;
Ganymede: 1070×10^3 km;
Callisto: 1880×10^3 km.

45. (a) 180 AU;
(b) 360 AU;
(c) 360/1.

47. (a) $\log T = \frac{3}{2}\log r + \frac{1}{2}\log\left(\dfrac{4\pi^2}{Gm_J}\right)$,
slope $= \frac{3}{2}$,
y-intercept $= \frac{1}{2}\log\left(\dfrac{4\pi^2}{Gm_J}\right)$;
(b)

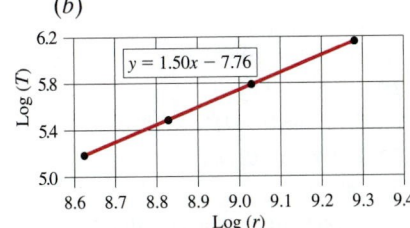

slope = 1.50 as predicted,
$m_J = 1.97 \times 10^{27}$ kg.

49. (a) 5.95×10^{-3} m/s^2;
(b) no, only by about 0.06%.

51. 2.64×10^6 m.

53. (a) 4.38×10^7 m/s^2;
(b) 2.8×10^9 N;
(c) 9.4×10^3 m/s.

55. $T_{inner} = 2.0 \times 10^4$ s,
$T_{outer} = 7.1 \times 10^4$ s.

57. 5.4×10^{12} m, it is still in the solar system, nearest to Pluto's orbit.

59. 2.3 g's.

61. 7.4×10^{36} kg, 3.7×10^6 M_{Sun}.

65. 1.21×10^6 m.

67. $V_{\text{deposit}} = 5 \times 10^7$ m³,
$r_{\text{deposit}} = 200$ m;
$m_{\text{deposit}} = 4 \times 10^{10}$ kg.

69. 8.99 days.

71. $0.44r$.

73. (a) 53 N;
(b) 3.1×10^{26} kg.

77. 1×10^{-10} m³/kg·s².

79. (a)

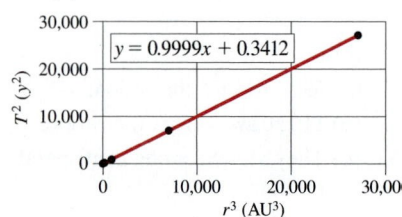
$y = 0.9999x + 0.3412$

(b) 39.44 AU.

CHAPTER 7

1. 7.7×10^3 J.

3. 1.47×10^4 J.

5. 6000 J.

7. 4.5×10^5 J.

9. 590 J.

11. (a) 1700 N;
(b) −6600 J;
(c) 6600 J;
(d) 0.

13. (a) 1.1×10^7 J;
(b) 5.0×10^7 J.

15. −490 J, 0, 490 J.

21. $1.5\hat{\mathbf{i}} - 3.0\hat{\mathbf{j}}$.

23. (a) 7.1;
(b) −250;
(c) 2.0×10^1.

25. $-1.4\hat{\mathbf{i}} + 2.0\hat{\mathbf{j}}$.

27. 52.5°, 48.0°, 115°.

29. 113.4° or 301.4°.

31. (a) 130°;
(b) negative sign says that the angle is obtuse.

35. 0.11 J.

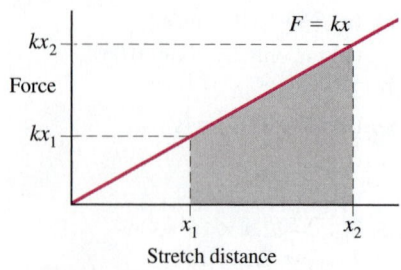

37. 3.0×10^3 J.

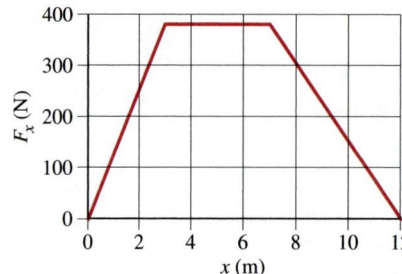

39. 2800 J.

41. 670 J.

43. $\frac{1}{2}kX^2 + \frac{1}{4}aX^4 + \frac{1}{5}bX^5$.

45. 4.0 J.

47. $\dfrac{\sqrt{3}\pi RF}{2}$.

49. 72 J.

51. (a) $\sqrt{3}$;
(b) $\frac{1}{4}$.

53. -4.5×10^5 J.

55. 3.0×10^2 N.

57. (a) $\sqrt{\dfrac{Fx}{m}}$;
(b) $\sqrt{\dfrac{3Fx}{4m}}$.

59. 8.3×10^4 N/m.

61. 1400 J.

63. (a) 640 J;
(b) −470 J;
(c) 0;
(d) 4.3 m/s.

65. 27 m/s.

67. (a) $\frac{1}{2}mv_2^2\left(1 + 2\dfrac{v_1}{v_2}\right)$;
(b) $\frac{1}{2}mv_2^2$;
(c) $\frac{1}{2}mv_2^2\left(1 + 2\dfrac{v_1}{v_2}\right)$ relative to Earth, $\frac{1}{2}mv_2^2$ relative to train;
(d) the ball moves different distances during the throwing process in the two frames of reference.

69. (a) 2.04×10^5 J;
(b) 21.0 m/s;
(c) 2.37 m.

71. 1710 J.

73. (a) 32.2 J;
(b) 554 J;
(c) −333 J;
(d) 0;
(e) 253 J.

75. 12.3 J.

77. $\dfrac{A}{k}e^{-0.10k}$.

79. 86 kJ, 42°.

81. 1.5 N.

83. 2×10^7 N/m.

85. 6.7°, 10°.

87. (a) 130 N, yes (≈ 29 lbs);
(b) 470 N, perhaps not (≈ 110 lbs).

89. (a) 1.5×10^4 J;
(b) 18 m/s.

93. (a) $F = 10.0x$;
(b) 10.0 N/m;
(c) 2.00 N.

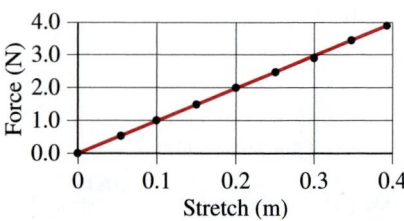

CHAPTER 8

1. 0.924 m.

3. 54 cm.

5. (a) 42.0 J;
(b) 11 J;
(c) same as part (a), unrelated to part (b).

7. (a) Yes, the expression for the work depends only on the endpoints;
(b) $U(x) = \frac{1}{2}kx^2 - \frac{1}{4}ax^4 - \frac{1}{5}bx^5 + C$.

9. $U(x) = -\dfrac{k}{2x^2} + \dfrac{k}{8\,\text{m}^2}$.

11. 49 m/s.

13. 6.5 m/s.

15. (a) 93 N/m;
(b) 22 m/s².

19. (a) 7.47 m/s;
(b) 3.01 m.

21. No, $D = 2d$.

23. (a) $\sqrt{v_0^2 + \dfrac{k}{m}x_0^2}$;
(b) $\sqrt{x_0^2 + \dfrac{m}{k}v_0^2}$.

25. (a) 2.29 m/s;
(b) 1.98 m/s;
(c) 1.98 m/s;
(d) 0.870 N, 0.800 N, 0.800 N;
(e) 2.59 m/s, 2.31 m/s, 2.31 m/s.

27. $k = \dfrac{12Mg}{h}$.

29. 3.9×10^7 J.

31. (a) 25 m/s;
(b) 370 m.

33. 12 m/s.

35. 0.020.

37. 0.40.

39. (a) 25%;
(b) 6.3 m/s, 5.4 m/s;
(c) primarily into heat energy.

41. For a mass of 75 kg, the energy change is 740 J.

43. (a) 0.13 m;
(b) 0.77;
(c) 0.5 m/s.

45. (a) $\dfrac{GM_E m_s}{2r_s}$;
(b) $-\dfrac{GM_E m_s}{r_s}$;
(c) $-\frac{1}{2}$.

47. $\frac{1}{4}$.

49. (a) 6.2×10^5 m/s;
(b) 4.2×10^4 m/s,
$v_{esc \, at \, Earth \, orbit} = \sqrt{2}v_{Earth \, orbit}$.

53. (a) 1.07×10^4 m/s;
(b) 1.16×10^4 m/s;
(c) 1.12×10^4 m/s.

55. (a) $-\sqrt{\dfrac{GM_E}{2r^3}}$;
(b) 1.09×10^4 m/s.

57. $\dfrac{GMm}{12r_E}$.

59. 1.12×10^4 m/s.

63. 510 N.

65. 2.9×10^4 W or 38 hp.

67. 4.2×10^3 N, opposing the velocity.

69. 510 W.

71. 2×10^6 W.

73. (a) -2.0×10^2 W;
(b) 3800 W;
(c) -120 W;
(d) 1200 W.

75. The mass oscillates between $+x_0$ and $-x_0$, with a maximum speed at $x = 0$.

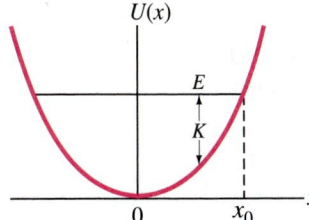

77. (a) $r_{U\text{min}} = \left(\dfrac{2b}{a}\right)^{\frac{1}{6}}$, $r_{U\text{max}} = 0$;

(b) $r_{U=0} = \left(\dfrac{b}{a}\right)^{\frac{1}{6}}$;

(c)

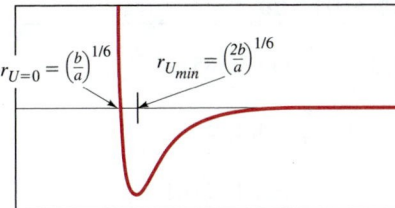

(d) $E < 0$: bound oscillatory motion between two turning points, $E > 0$: unbounded;

(e) $r_{F>0} < \left(\dfrac{2b}{a}\right)^{\frac{1}{6}}$,

$r_{F<0} > \left(\dfrac{2b}{a}\right)^{\frac{1}{6}}$,

$r_{F=0} = \left(\dfrac{2b}{a}\right)^{\frac{1}{6}}$;

(f) $F(r) = \dfrac{12b}{r^{13}} - \dfrac{6a}{r^7}$.

79. 2.52×10^4 W.

81. (a) 42 m/s;
(b) 2.6×10^5 W.

83. (a) 28.2 m/s;
(b) 116 m.

85. (a) $\sqrt{2g\ell}$;
(b) $\sqrt{1.2g\ell}$.

89. (a) 8.9×10^5 J;
(b) 5.0×10^1 W, 6.6×10^{-2} hp;
(c) 330 W, 0.44 hp.

91. (a) 29°;
(b) 480 N;
(c) 690 N.

93. 5800 W or 7.8 hp.

95. (a) 2.8 m;
(b) 1.5 m;
(c) 1.5 m.

97. 1.7×10^5 m³.

99. (a) 5220 m/s;
(b) 3190 m/s.

101. (a) 1500 m;
(b) 170 m/s.

103. 60 m.

105. (a) 79 m/s;
(b) 2.4×10^7 W.

107. (a) 2.2×10^5 J;
(b) 22 m/s;
(c) -1.4 m.

109. $x = \sqrt{\dfrac{a}{b}}$.

CHAPTER 9

1. 5.9×10^7 N.

3. $(9.6t\hat{\mathbf{i}} - 8.9\hat{\mathbf{k}})$ N.

5. $4.35 \text{ kg·m/s} (\hat{\mathbf{j}} - \hat{\mathbf{i}})$.

7. 1.40×10^2 kg.

9. 2.0×10^4 kg.

11. 4.9×10^3 m/s.

13. -0.966 m/s.

15. 1:2.

17. $\frac{3}{2}v_0\hat{\mathbf{i}} - v_0\hat{\mathbf{j}}$.

19. $(4.0\hat{\mathbf{i}} + 3.3\hat{\mathbf{j}} - 3.3\hat{\mathbf{k}})$ m/s.

21. (a) $(116\hat{\mathbf{i}} + 58.0\hat{\mathbf{j}})$ m/s;
(b) 5.02×10^5 J.

23. (a) 2.0 kg·m/s, forward;
(b) 5.8×10^2 N, forward.

25. 2.1 kg·m/s, to the left.

27. 0.11 N.

29. 1.5 kg·m/s.

31. (a) $\dfrac{2mv}{\Delta t}$;
(b) $\dfrac{2mv}{t}$.

33. (a) $0.98 \text{ N} + (1.4 \text{ N/s})t$;
(b) 13.3 N;
(c) $\big[(0.62 \text{ N/m}^2) \times$
$\overline{\sqrt{2.5 \text{ m}} - (0.070 \text{ m/s})t}\big]$
$+ (1.4 \text{ N/s})t$, 13.2 N.

35. 1.60 m/s (west), 3.20 m/s (east).

37. (a) 3.7 m/s;
(b) 0.67 kg.

39. (a) 1.00;
(b) 0.890;
(c) 0.286;
(d) 0.0192.

41. (a) 0.37 m;
(b) -1.6 m/s, 6.4 m/s;
(c) yes.

43. (a) $\dfrac{-M}{m + M}$;
(b) -0.96.

45. 3.0×10^3 J, 4.5×10^3 J.

47. 0.11 kg·m/s, upward.

49. (b) $e = \sqrt{\dfrac{h'}{h}}$.

51. (a) 890 m/s;
(b) 0.999 of initial kinetic energy lost.

53. (a) 7.1×10^{-2} m/s;
(b) -5.4 m/s, 4.1 m/s;
(c) 0, 0.13 m/s, reasonable;
(d) 0.17 m/s, 0, not reasonable;
(e) in this case, -4.0 m/s, 3.1 m/s, reasonable.

55. 1.14×10^{-22} kg·m/s, $147°$ from the electron's momentum, $123°$ from the neutrino's momentum.

57. (a) $30°$;
(b) $v'_A = v'_B = \dfrac{v}{\sqrt{3}}$;
(c) $\frac{2}{3}$.

59. 39.9 u.

63. 6.5×10^{-11} m.

65. $(1.2\,\text{m})\hat{\mathbf{i}} - (1.2\,\text{m})\hat{\mathbf{j}}$.

67. $0\hat{\mathbf{i}} + \dfrac{2r}{\pi}\hat{\mathbf{j}}$.

69. $0\hat{\mathbf{i}} + 0\hat{\mathbf{j}} + \frac{3}{4}h\hat{\mathbf{k}}$.

71. $0\hat{\mathbf{i}} + \dfrac{4R}{3\pi}\hat{\mathbf{j}}$.

73. (a) 4.66×10^6 m from the center of the Earth.

75. (a) 5.7 m;
(b) 4.2 m;
(c) 4.3 m.

77. 0.41 m toward the initial position of the 85-kg person.

79. $v\,\dfrac{m}{m + M}$, upward, balloon also stops.

81. 0.93 hp.

83. -76 m/s.

85. Good possibility of a "scratch" shot.

87. 11 bounces.

89. 1.4 m.

91. 50%.

93. (a) $v = \dfrac{M_0 v_0}{M_0 + \dfrac{dM}{dt}t}$;
(b) 8.2 m/s, yes.

95. 112 km/h or 70 mi/h.

97. 21 m.

99. (a) 1.9 m/s;
(b) -0.3 m/s, 1.5 m/s;
(c) 0.6 cm, 12 cm.

101. $m < \frac{1}{3}M$ or $m < 2.33$ kg.

103. (a) 8.6 m;
(b) 40 m.

105. 29.6 km/s.

107. 0.38 m, 1.5 m.

109. (a) 1.3×10^5 N;
(b) -83 m/s².

111. 12 kg.

113. 0.2 km/s, in the original direction of m_A.

CHAPTER 10

1. (a) $\dfrac{\pi}{4}$ rad, 0.785 rad;
(b) $\dfrac{\pi}{3}$ rad, 1.05 rad;
(c) $\dfrac{\pi}{2}$ rad, 1.57 rad;
(d) 2π rad, 6.283 rad;
(e) $\dfrac{89\pi}{36}$ rad, 7.77 rad.

3. 5.3×10^3 m.

5. (a) 260 rad/s;
(b) 46 m/s, 1.2×10^4 m/s².

7. (a) 1.05×10^{-1} rad/s;
(b) 1.75×10^{-3} rad/s;
(c) 1.45×10^{-4} rad/s;
(d) 0.

9. (a) 464 m/s;
(b) 185 m/s;
(c) 328 m/s.

11. $36{,}000$ rev/min.

13. (a) 1.5×10^{-4} rad/s²;
(b) 1.6×10^{-2} m/s², 6.2×10^{-4} m/s².

15. (a) $-\hat{\mathbf{i}}$, $\hat{\mathbf{k}}$;
(b) 56.2 rad/s, $38.5°$ from $-x$ axis towards $+z$ axis;
(c) 1540 rad/s², $-\hat{\mathbf{j}}$.

17. $28{,}000$ rev.

19. (a) -0.47 rad/s²;
(b) 190 s.

21. (a) 0.69 rad/s²;
(b) 9.9 s.

23. (a) $\omega = \frac{1}{3}5.0t^3 - \frac{1}{2}8.5t^2$;
(b) $\theta = \frac{1}{12}5.0t^4 - \frac{1}{6}8.5t^3$;
(c) $\omega(2.0\,\text{s}) = -4$ rad/s, $\theta(2.0\,\text{s}) = -5$ rad.

25. 1.4 m·N, clockwise.

27. $mg(\ell_2 - \ell_1)$, clockwise.

29. 270 N, 1700 N.

31. 1.81 kg·m².

33. (a) 9.0×10^{-2} m·N;
(b) 12 s.

35. 56 m·N.

37. (a) 0.94 kg·m²;
(b) 2.4×10^{-2} m·N.

39. (a) 78 rad/s²;
(b) 670 N.

41. 2.2×10^4 m·N.

43. 17.5 m/s.

45. (a) $14M\ell^2$;
(b) $\frac{14}{3}M\ell\alpha$;
(c) perpendicular to the rod and the axis.

47. (a) 1.90×10^3 kg·m²;
(b) 7.5×10^3 m·N.

49. (a) R_0;
(b) $\sqrt{\frac{1}{2}R_0^2 + \frac{1}{12}w^2}$;
(c) $\sqrt{\frac{1}{2}}R_0$;
(d) $\sqrt{\frac{1}{2}(R_1^2 + R_2^2)}$;
(e) $\sqrt{\frac{2}{5}}r_0$;
(f) $\sqrt{\frac{1}{12}}\ell$;
(g) $\sqrt{\frac{1}{3}}\ell$;
(h) $\sqrt{\frac{1}{12}(\ell^2 + w^2)}$.

51. $a = \dfrac{(m_B - m_A)}{(m_A + m_B + I/R^2)}g$,
compared to
$a_{I=0} = \dfrac{(m_B - m_A)}{(m_A + m_B)}g$.

53. (a) 9.70 rad/s²;
(b) 11.6 m/s²;
(c) 585 m/s²;
(d) 4.27×10^3 N;
(e) $1.14°$.

57. (a) $5.3Mr_0^2$; (b) -15%.

59. (a) 3.9 cm from center along line connecting the small weight and the center;
(b) 0.42 kg·m².

61. (b) $\frac{1}{12}M\ell^2$, $\frac{1}{12}Mw^2$.

63. $22{,}200$ J.

65. $14{,}200$ J.

67. 1.4 m/s.

69. 8.22 m/s.

71. 7.0×10^1 J.

73. (a) 8.37 m/s, 32.9 rad/sec.
(b) $\frac{5}{2}$;
(c) the translational speed and the energy relationship are independent of both mass and radius, but the rotational speed depends on the radius.

75. $\sqrt{\frac{10}{7}g(R_0 - r_0)}$.

77. (a) 4.06 m/s;
(b) 8.99 J;
(c) 0.158.

79. (a) 4.1×10^5 J;
(b) 18%;
(c) 1.3 m/s²;
(d) 6%.

81. (a) 1.6 m/s;
(b) 0.48 m.

83. $\frac{\ell}{2}, \frac{\ell}{2}$.

85. (a) 0.84 m/s;
    (b) 96%.

87. 2.0 m·N, from the arm swinging the sling.

89. (a) $\frac{\omega_R}{\omega_F} = \frac{N_F}{N_R}$;
    (b) 4.0;
    (c) 1.5.

91. (a) 1.7×10^8 J;
    (b) 2.2×10^3 rad/s;
    (c) 25 min.

93. $\dfrac{Mg\sqrt{2Rh - h^2}}{R - h}$.

95. $\dfrac{\lambda_0 \ell^3}{6}$.

97. 5.0×10^2 m·N.

99. (a) 1.6 m;
    (b) 1.1 m.

101. (a) $\dfrac{x}{y} g$;
    (b) x should be as small as possible, y should be as large as possible, and the rider should move upward and toward the rear of the bicycle;
    (c) 3.6 m/s².

103. $\sqrt{\dfrac{3g\ell}{4}}$.

105. $\tau = $
    $[(0.300 \text{ m}) \cos\theta + 0.200 \text{ m}](500 \text{ N})$

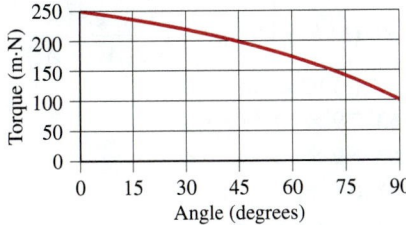

CHAPTER 11

1. 3.98 kg·m²/s.

3. (a) L is conserved: If I increases, ω must decrease;
    (b) increased by a factor of 1.3.

5. 0.38 rev/s.

7. (a) 7.1×10^{33} kg·m²/s;
    (b) 2.7×10^{40} kg·m²/s.

9. (a) $-\dfrac{I_W}{I_P} \omega_W$;
    (b) $-\dfrac{I_W}{2I_P} \omega_W$;
    (c) $\omega_W \dfrac{I_W}{I_P}$;
    (d) 0.

11. (a) 0.55 rad/s;
    (b) 420 J, 240 J.

13. 0.48 rad/s, 0.80 rad/s.

15. $\frac{1}{2}\omega$.

17. (a) 3.7×10^{16} J;
    (b) 1.9×10^{20} kg·m²/s.

19. −0.32 rad/s.

23. 45°.

27. $(25\hat{\mathbf{i}} \pm 14\hat{\mathbf{j}} \mp 19\hat{\mathbf{k}})$ m·kN.

29. (a) $-7.0\hat{\mathbf{i}} - 11\hat{\mathbf{j}} + 0.5\hat{\mathbf{k}}$;
    (b) 170°.

37. $(-55\hat{\mathbf{i}} - 45\hat{\mathbf{j}} + 49\hat{\mathbf{k}})$ kg·m²/s.

39. (a) $(\frac{1}{6}M + \frac{7}{9}m)\ell^2\omega^2$;
    (b) $(\frac{1}{3}M + \frac{14}{9}m)\ell^2\omega$.

41. (a) $\left[(M_A + M_B)R_0 + \dfrac{I}{R_0}\right]v$;
    (b) $\dfrac{M_B g}{M_A + M_B + \dfrac{I}{R_0^2}}$.

45. $F_A = \dfrac{(d + r_A \cos\phi)m_A r_A \omega^2 \sin\phi}{2d}$,
    $F_B = \dfrac{(d - r_A \cos\phi)m_A r_A \omega^2 \sin\phi}{2d}$.

47. $\dfrac{m^2 v^2}{g(m + M)(m + \frac{4}{3}M)}$.

49. $\Delta\omega/\omega_0 = -8.4 \times 10^{-13}$.

51. $v_{CM} = \dfrac{m}{M + m} v$,
    ω (about CM) $= \left(\dfrac{12m}{4M + 7m}\right)\dfrac{v}{\ell}$.

53. 8.3×10^{-4} kg·m².

55. 8.0 rad/s.

57. 14 rev/min, CCW when viewed from above.

59. (a) 9.80 m/s², along a radial line;
    (b) 9.78 m/s², 0.0988° south from a radial line;
    (c) 9.77 m/s², along a radial line.

61. Due north or due south.

63. $(mr\omega^2 - F_{fr})\hat{\mathbf{i}}$
    $+ (F_{spoke} - 2m\omega v)\hat{\mathbf{j}}$
    $+ (F_N - mg)\hat{\mathbf{k}}$.

65. (a) $(-24\hat{\mathbf{i}} + 28\hat{\mathbf{j}} - 14\hat{\mathbf{k}})$ kg·m²/s;
    (b) $(16\hat{\mathbf{j}} - 8.0\hat{\mathbf{k}})$ m·N.

67. (b) 0.750.

69. $v[-\sin(\omega t)\hat{\mathbf{i}} + \cos(\omega t)\hat{\mathbf{j}}]$,
    $\vec{\boldsymbol{\omega}} = \left(\dfrac{v}{R}\right)\hat{\mathbf{k}}$.

71. (a) The wheel will turn to the right;
    (b) $\Delta L/L_0 = 0.19$.

73. (a) 820 kg·m²/s²;
    (b) 820 m·N;
    (c) 930 W.

75. $\vec{\mathbf{a}}_{tan} = -R\alpha \sin\theta\hat{\mathbf{i}} + R\alpha \cos\theta\hat{\mathbf{j}}$;
    (a) $mR^2\alpha\hat{\mathbf{k}}$;
    (b) $mR^2\alpha\hat{\mathbf{k}}$.

77. 0.965.

79. (a) There is zero net torque exerted about any axis through the skater's center of mass;
    (b) $f_{single\ axel} = 2.5$ rad/s, $f_{triple\ axel} = 6.5$ rad/s.

81. (a) 17,000 rev/s;
    (b) 4300 rev/s.

83. (a) $\omega = \left(12 \dfrac{\text{rad/s}}{\text{m}}\right)x$;
    (b)

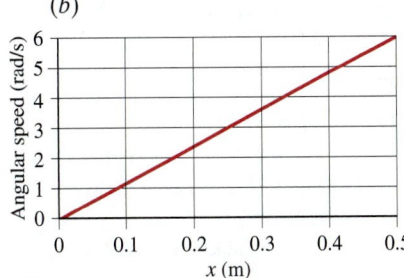

CHAPTER 12

1. 528 N, $(1.20 \times 10^2)°$ clockwise from $\vec{\mathbf{F}}_A$.

3. 6.73 kg.

5. (a) $F_A = 1.5 \times 10^3$ N down, $F_B = 2.0 \times 10^3$ N up;
    (b) $F_A = 1.8 \times 10^3$ N down, $F_B = 2.6 \times 10^3$ N up.

7. (a) 230 N;
    (b) 2100 N.

9. -2.9×10^3 N, 1.5×10^4 N.

11. 3400 N, 2900 N.

13. 0.28 m.

15. 6300 N, 6100 N.

17. 1600 N.

19. 1400 N, 2100 N.

21. (a) 410 N;
    (b) 410 N, 328 N.

23. 120 N.

25. 550 N.

27. (a)

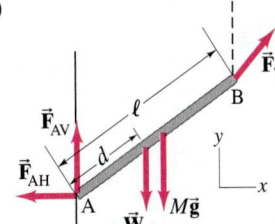

    (b) $F_{AH} = 51$ N, $F_{AV} = -9$ N;
    (c) 2.4 m.

29. $F_{top} = 55.2$ N right, 63.7 N up,
$F_{bottom} = 55.2$ N left, 63.7 N up.

31. 5.2 m/s^2.

33. 2.5 m at the top.

35. (a) 1.8×10^5 N/m^2;
(b) 3.5×10^{-6}.

37. (a) 1.4×10^6 N/m^2;
(b) 6.9×10^{-6};
(c) 6.6×10^{-5} m.

39. 9.6×10^6 N/m^2.

41. (a) 1.3×10^2 m·N, clockwise;
(b) the wall;
(c) all three are present.

43. (a) 393 N;
(b) thicker.

45. (a) 3.7×10^{-5} m^2;
(b) 2.7×10^{-3} m.

47. 1.3 cm.

49. (a) $F_T = 150$ kN;
$\mathbf{F}_A = 170$ kN, 23° above AC;
(b) $F_{DE} = F_{DB} = F_{BC} = 76$ kN,
tension;
$F_{CE} = 38$ kN, compression;
$F_{DC} = F_{AB} = 76$ kN, compression;
$F_{CA} = 114$ kN, compression.

51. (a) 5.5×10^{-2} m^2;
(b) 8.6×10^{-2} m^2.

53. $F_{AB} = F_{BD} = F_{DE} = 7.5 \times 10^4$ N,
compression;
$F_{BC} = F_{CD} = 7.5 \times 10^4$ N, tension;
$F_{CE} = F_{AC} = 3.7 \times 10^4$ N, tension.

55. $F_{AB} = F_{JG} = \dfrac{3\sqrt{2}}{2} F$, compression;
$F_{AC} = F_{JH} = F_{CE} = F_{HE} = \dfrac{3}{2}F$,
tension;
$F_{BC} = F_{GH} = F$, tension;
$F_{BE} = F_{GE} = \dfrac{\sqrt{2}}{2} F$, tension;
$F_{BD} = F_{GD} = 2F$, compression;
$F_{DE} = 0$.

57. 0.249 kg, 0.194 kg, 0.0554 kg.

59. (a) $Mg\sqrt{\dfrac{h}{2R - h}}$;
(b) $Mg\dfrac{\sqrt{h(2R - h)}}{R - h}$.

61. (a)

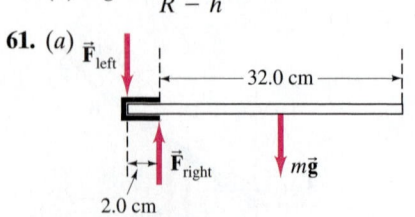

(b) $mg = 65$ N, $F_{right} = 550$ N,
$F_{left} = 490$ N;
(c) 11 m·N.

63. 29°.

65. 3.8.

67. 5.0×10^5 N, 3.2 m.

69. (a) 650 N;
(b) $F_A = 0$, $F_B = 1300$ N;
(c) $F_A = 160$ N, $F_B = 1140$ N;
(d) $F_A = 810$ N, $F_B = 490$ N.

71. He can walk only 0.95 m to the right
of the right support, and 0.83 m to
the left of the left support.

73. $F_{left} = 120$ N, $F_{right} = 210$ N.

75. $F/A =$
3.8×10^5 N/m$^2 <$ tissue strength.

77. $F_A = 1.7 \times 10^4$ N,
$F_B = 7.7 \times 10^3$ N.

79. 2.5 m.

81. (a) 6500 m;
(b) 6400 m.

83. 570 N.

85. 45°.

87. (a) $2.4w$;
(b) $2.6w$, 32° above the horizontal.

89. (a) $(4.5 \times 10^{-6})\%$;
(b) 9.0×10^{-18} m.

91. 150 N, 0.83 m.

93. (a) $mg\left(1 - \dfrac{r_0}{h}\cot\theta\right)$;
(b) $\dfrac{h}{r_0} - \cot\theta$.

95. (b) 46°, 51°, 11%.

97. (a)

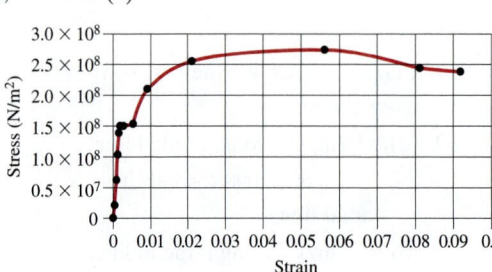

(b)

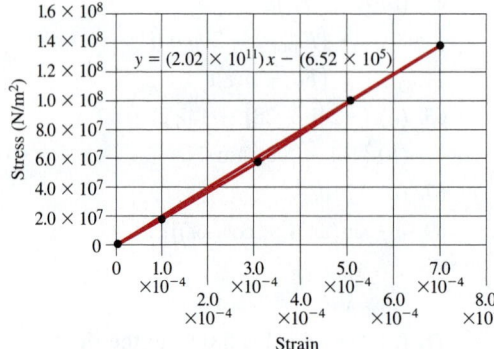

Elastic Modulus $= 2.02 \times 10^{11}$ N/m^2.

1. 3×10^{11} kg.

3. 6.7×10^2 kg.

5. 0.8547.

7. (a) 5510 kg/m^3;
(b) 5520 kg/m^3, 0.3%.

9. (a) 8.1×10^7 N/m^2;
(b) 2×10^5 N/m^2.

11. 13 m.

13. 6990 kg.

15. (a) 2.8×10^7 N, 1.2×10^5 N/m^2;
(b) 1.2×10^5 N/m^2.

17. 683 kg/m^3.

19. 3.35×10^4 N/m^2.

21. (a) 1.32×10^5 Pa;
(b) 9.7×10^4 Pa.

23. (c) $0.38h$, no.

27. 2990 kg/m^3.

29. 920 kg.

31. Iron or steel.

33. 1.1×10^{-2} m^3.

35. 10.5%.

37. (b) Above.

39. 3600 balloons.

43. 2.8 m/s.

45. 1.0×10^1 m/s.

47. 1.8×10^5 N/m^2.

49. 1.2×10^5 N.

51. 9.7×10^4 Pa.

57. $\frac{1}{2}$.

59. (b) $h = \left[\sqrt{h_0} - t\sqrt{\dfrac{gA_1^2}{2(A_2^2 - A_1^2)}}\right]^2$
(c) 92 s.

63. 7.9×10^{-2} Pa·s.

65. 6.9×10^3 Pa.

67. 0.10 m.

69. (a) Laminar;
(b) turbulent.

71. 1.0 m.

73. 0.012 N.

75. 1.5 mm.

79. (a) 0.75 m;
(b) 0.65 m;
(c) 1.1 m.

81. 0.047 atm.

83. 0.24 N.

85. 1.0 m.

87. 5.3 km.

89. (a) 88 Pa/s;
(b) 5.0×10^1 s.

91. 5×10^{18} kg.

93. (a) 8.5 m/s;
   (b) 0.24 L/s;
   (c) 0.85 m/s.

95. $d\left(\dfrac{v_0^2}{v_0^2 + 2gy}\right)^{\frac{1}{4}}$.

97. 170 m/s.

99. 1.2×10^4 N.

101. 4.9 s.

CHAPTER 14

1. 0.72 m.

3. 1.5 Hz.

5. 350 N/m.

7. 0.13 m/s, 0.12 m/s², 1.2%.

9. (a) 0.16 N/m;
   (b) 2.8 Hz.

11. $\dfrac{\sqrt{3k/M}}{2\pi}$.

13. (a) 2.5 m, 3.5 m;
   (b) 0.25 Hz, 0.50 Hz;
   (c) 4.0 s, 2.0 s;
   (d) $x_A = (2.5\text{ m})\sin(\tfrac{1}{2}\pi t)$,
      $x_B = (3.5\text{ m})\cos(\pi t)$.

15. (a) $y(t) =$
      $(0.280\text{ m})\sin[(34.3\text{ rad/s})t]$;
   (b) $t_{\text{longest}} =$
      $4.59 \times 10^{-2}\text{ s} + n(0.183\text{ s})$,
      $n = 0, 1, 2, \cdots$;
      $t_{\text{shortest}} =$
      $1.38 \times 10^{-1}\text{ s} + n(0.183\text{ s})$,
      $n = 0, 1, 2, \cdots$.

17. (a) 1.6 s, $\tfrac{5}{8}$ Hz;
   (b) 3.3 m, -7.5 m/s;
   (c) -13 m/s, 29 m/s².

19. 0.75 s.

21. 3.1 s, 6.3 s, 9.4 s.

23. 88.8 N/m, 17.8 m.

27. (a) 0.650 m;
   (b) 1.18 Hz;
   (c) 13.3 J;
   (d) 11.2 J, 2.1 J.

29.

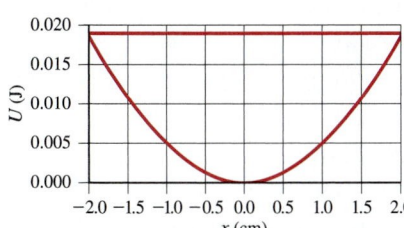

   (a) 0.011 J;
   (b) 0.008 J;
   (c) 0.5 m/s.

31. 10.2 m/s.

33. $A_{\text{high energy}} = \sqrt{5}A_{\text{low energy}}$.

35. (a) 430 N/m;
   (b) 3.7 kg.

37. 309.8 m/s.

39. (a) 0.410 s, 2.44 Hz;
   (b) 0.148 m;
   (c) 34.6 m/s²;
   (d) $x = (0.148\text{ m})\sin(4.87\pi t)$;
   (e) 2.00 J;
   (f) 1.68 J.

41. 2.2 s.

43. (a) $-5.4°$;
   (b) 8.4°;
   (c) $-13°$.

45. $\tfrac{1}{3}$.

47. $\sqrt{2g\ell(1 - \cos\theta)}$.

49. 0.41 g.

51. (a) $\theta = \theta_0\cos(\omega t + \phi)$, $\omega = \sqrt{\dfrac{K}{I}}$.

53. 2.9 s.

55. 1.08 s.

57. Decreased by a factor of 6.

59. (a) (-1.21×10^{-3})%;
   (b) 32.3 periods.

63. (a) 0°;
   (b) 0, $\pm A$;
   (c) $\tfrac{1}{2}\pi$ or 90°.

65. 3.1 m/s.

67. 23.7.

69. (a) 170 s;
   (b) 1.3×10^{-5} W;
   (c) 1.0×10^{-3} Hz on either side.

71. 0.11 m.

73. (a) $1.22\,f$;
   (b) $0.71\,f$.

75. (a) 0.41 s;
   (b) 9 mm.

77. 0.9922 m, 1.6 mm, 0.164 m.

79. $x = \pm\dfrac{\sqrt{3}A}{2} \approx \pm 0.866A$.

81. $\rho_{\text{water}}\,g(\text{area}_{\text{bottom side}})$.

83. (a) 130 N/m;
   (b) 0.096 m.

85. (a) $x = \pm\dfrac{\sqrt{3}x_0}{2} \approx \pm 0.866x_0$;
   (b) $x = \pm\tfrac{1}{2}x_0$.

87. 84.5 min.

89. 1.25 Hz.

91. ~ 3000 N/m.

93. (a) $k = \dfrac{4\pi^2}{\text{slope}}$, y-intercept $= 0$;
   (b) slope $= 0.13$ s²/kg,
      y-intercept $= 0.14$ s²

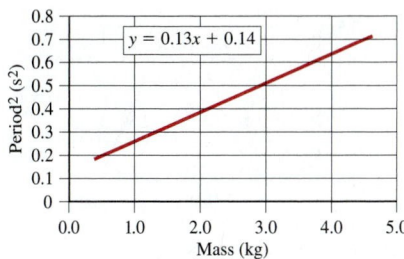

   (c) $k = \dfrac{4\pi^2}{\text{slope}} = 310$ N/m,
      y-intercept $= \dfrac{4\pi^2 m_0}{k}$,
      $m_0 = 1.1$ kg;
   (d) portion of spring's mass that is
      effectively oscillating.

CHAPTER 15

1. 2.7 m/s.

3. (a) 1400 m/s;
   (b) 4100 m/s;
   (c) 5100 m/s.

5. 0.62 m.

7. 4.3 N.

9. (a) 78 m/s;
   (b) 8300 N.

11. (a)

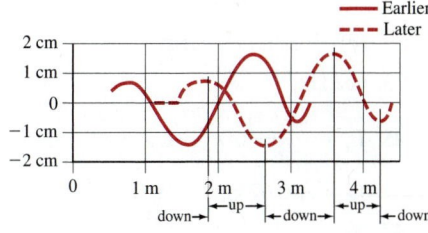

   (b) -4 cm/s.

13. 18 m.

15. $A_{\text{more energy}}/A_{\text{less energy}} = \sqrt{3}$.

19. (a) 0.38 W;
   (b) 0.25 cm.

21. (b) 420 W.

23. $D = A\sin\left[2\pi\left(\dfrac{x}{\lambda} + \dfrac{t}{T}\right) + \phi\right]$.

25. (a) 41 m/s;
   (b) 6.4×10^4 m/s²;
   (c) 35 m/s, 3.2×10^4 m/s².

27. (b) $D =$
$(0.45 \text{ m}) \cos[2.6(x - 2.0t) + 1.2]$;
(d) $D =$
$(0.45 \text{ m}) \cos[2.6(x + 2.0t) + 1.2]$.

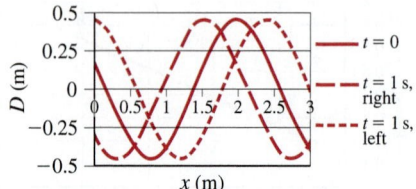

29. $D = (0.020 \text{ cm}) \times$
$\sin\left[(9.54 \text{ m}^{-1})x - (3290 \text{ rad/s})t + \frac{3}{2}\pi\right]$

31. Yes, it is a solution.

35. Yes, it is a solution.

37. (a) 0.84 m;
(b) 0.26 N;
(c) 0.59 m.

39. (a) $t = \frac{2}{v}\sqrt{D^2 + \left(\frac{x}{2}\right)^2}$;

(b) slope $= \frac{1}{v^2}$,

y-intercept $= \frac{4}{v^2}D^2$.

41. (a)

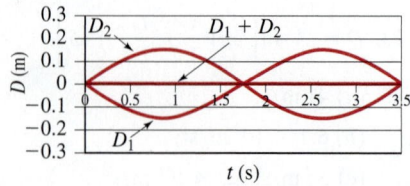

(b)

(c) all kinetic energy.

43. 662 Hz.

45. $T_n = \frac{(1.5 \text{ s})}{n}$, $n = 1, 2, 3, \cdots$,

$f_n = n(0.67 \text{ Hz})$, $n = 1, 2, 3, \cdots$.

47. $f_{0.50}/f_{1.00} = \sqrt{2}$.

49. 80 Hz.

53. 11.

55. (a) $D_2 = 4.2 \sin(0.84x + 47t + 2.1)$;
(b) $8.4 \sin(0.84x + 2.1) \cos(47t)$.

57. 315 Hz.

59. (a)

(b)

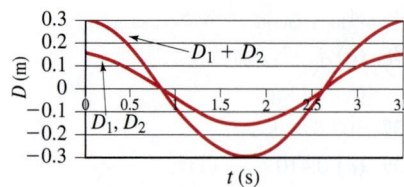

61. $n = 4$, $n = 8$, and $n = 12$.

63. $x = \pm\left(n + \frac{1}{2}\right)\frac{\pi}{2}$ m, $n = 0, 1, 2, \cdots$.

65. 5.2 km/s.

67. $(3.0 \times 10^1)°$.

69. 44°.

71. (a) 0.042 m;
(b) 0.55 radians.

73. The speed is greater in the less dense rod, by a factor of $\sqrt{2.5} = 1.6$.

75. (a) 0.05 m;
(b) 2.25.

77. 0.69 m.

79. (a) $t = 0$ s;

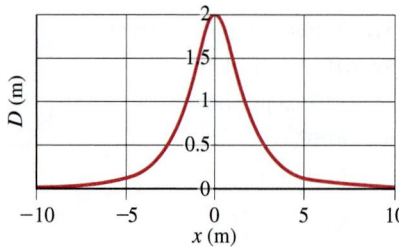

(b) $D = \frac{4.0 \text{ m}^3}{(x - 2.4t)^2 + 2.0 \text{ m}^2}$

(c) $t = 1.0$ s, moving right;

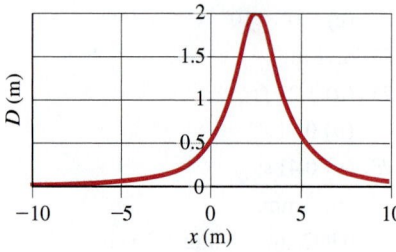

(d) $D = \frac{4.0 \text{ m}^3}{(x + 2.4t)^2 + 2.0 \text{ m}^2}$,

$t = 1.0$ s, moving left.

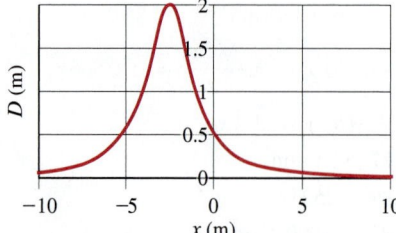

81. (a) G: 784 Hz, 1180 Hz, B: 988 Hz, 1480 Hz;
(b) 1.59;
(c) 1.26;
(d) 0.630.

83. 6.3 m from the end where the first pulse originated.

85. $\lambda = \frac{4\ell}{2n - 1}$, $n = 1, 2, 3, \cdots$.

87. $D(x, t) =$
$(3.5 \text{ cm}) \cos(0.10\pi x - 1.5\pi t)$, with x in cm and t in s.

89. 12 min.

93. speed $= 0.50$ m/s; direction of motion $= +x$, period $= 2\pi$ s, wavelength $= \pi$ m.

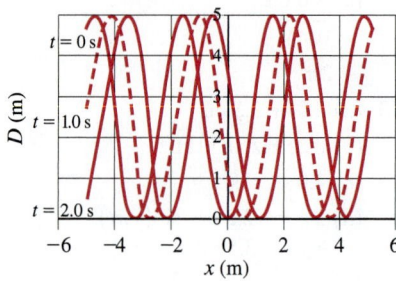

CHAPTER 16

1. 340 m.

3. (a) 1.7 cm to 17 m;
(b) 2.3×10^{-5} m.

5. (a) 0.17 m;
(b) 11 m;
(c) 0.5%.

7. 41 m.

9. (a) 8%;
(b) 4%.

11. (a) 4.4×10^{-5} Pa;
(b) 4.4×10^{-3} Pa.

13. (a) 5.3 m;
(b) 675 Hz;
(c) 3600 m/s;
(d) 1.0×10^{-13} m.

15. 63 dB.

17. (a) 10^9;
(b) 10^{12}.

19. 2.9×10^{-9} J.

21. 124 dB.

23. (a) 9.4×10^{-6} W;
(b) 8.0×10^6 people.

25. (a) 122 dB, 114 dB;
(b) no.

27. 7 dB.

29. (a) The higher frequency wave, 2.6;
(b) 6.8.

31. (a) 3.2×10^{-5} m;
(b) 3.0×10^1 Pa.

33. 1.24 m.

35. (a) 69.2 Hz, 207 Hz, 346 Hz, 484 Hz;
(b) 138 Hz, 277 Hz, 415 Hz, 553 Hz.

37. 8.6 mm to 8.6 m.

39. (a) 0.18 m;
(b) 1.1 m;
(c) 440 Hz, 0.78 m.

41. −3.0%.

43. (a) 1.31 m;
(b) 3, 4, 5, 6.

45. 3.65 cm, 7.09 cm, 10.3 cm, 13.4 cm, 16.3 cm, 19.0 cm.

47. 4.3 m, open.

49. 21.4 Hz, 42.8 Hz.

51. 3430 Hz, 10,300 Hz, 17,200 Hz, relatively sensitive frequencies.

53. ± 0.50 Hz.

55. 346 Hz.

57. 10 beats/s.

59. (a) 221.5 Hz or 218.5 Hz;
(b) 1.4% increase, 1.3% decrease.

61. (a) 1470 Hz;
(b) 1230 Hz.

63. (a) 2430 Hz, 2420 Hz, difference of 10 Hz;
(b) 4310 Hz, 3370 Hz, difference of 940 Hz;
(c) 34,300 Hz, 4450 Hz, difference of 29,900 Hz;
(d) $f'_{\text{source moving}} \approx f'_{\text{observer moving}}$
$= f\left(1 + \dfrac{v_{\text{object}}}{v_{\text{sound}}}\right)$.

65. (a) 1420 Hz, 1170 Hz;
(b) 1520 Hz, 1080 Hz;
(c) 1330 Hz, 1240 Hz.

67. 3 Hz.

69. (a) Every 1.3 s;
(b) every 15 s.

71. 8.9 cm/s.

73. (a) 93;
(b) 0.62°.

77. 19 km.

79. (a) 57 Hz, 69 Hz, 86 Hz, 110 Hz, 170 Hz.

81. 90 dB.

83. 11 W.

85. 51 dB.

87. 1.07.

89. (a) 280 m/s, 57 N;
(b) 0.19 m;
(c) 880 Hz, 1320 Hz.

91. 3 Hz.

93. 141 Hz, 422 Hz, 703 Hz, 984 Hz.

95. 22 m/s.

97. (a) No beats;
(b) 20 Hz;
(c) no beats.

99. 55.2 kHz.

101. 11.5 m.

103. 2.3 Hz.

105. 17 km/h.

107. (a) 3400 Hz;
(b) 1.50 m;
(c) 0.10 m.

109. (a)

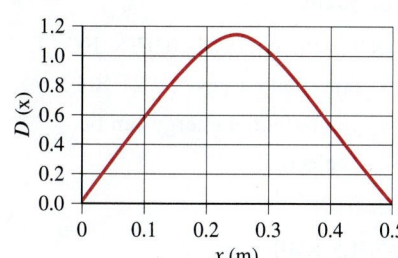

(b)

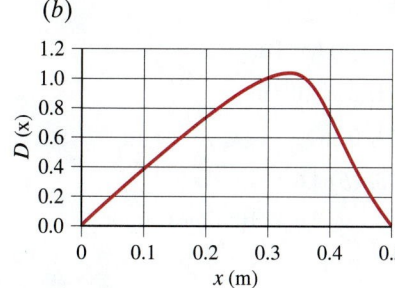

CHAPTER 17

1. $N_{\text{Au}} = 0.548 N_{\text{Ag}}$.

3. (a) 20°C;
(b) 3500°F.

5. 102.9°F.

7. 0.08 m.

9. 1.6×10^{-6} m for Super Invar™, 9.6×10^{-5} m for steel, steel is 60× as much.

11. 981 kg/m³.

13. −69°C.

15. 3.9 cm³.

17. (a) 5.0×10^{-5}/C°;
(b) copper.

21. (a) 2.7 cm;
(b) 0.3 cm.

23. 55 min.

25. 3.0×10^7 N/m².

27. (a) 27°C;
(b) 5500 N.

29. −459.67°F.

31. 1.35 m³.

33. 1.25 kg/m³.

35. 181°C.

37. (a) 22.8 m³;
(b) 1.88 atm.

39. 1660 atm.

41. 313°C.

43. 3.49 atm.

45. −130°C.

47. 7.0 min.

49. Ideal = 0.588 m³, actual = 0.598 m³ (nonideal behavior).

51. 2.69×10^{25} molecules/m³.

53. 4×10^{-17} Pa.

55. 300 molecules/cm³.

57. 19 molecules/breath.

59. (a) 71.2 torr;
(b) 180°C.

61. 223 K.

63. (a) Low;
(b) 0.025%.

65. 20%.

67. 9.9 L, not advisable.

69. (a) 1100 kg;
(b) 100 kg.

71. (a) Lower;
(b) 0.36%.

73. 1.1×10^{44} molecules.

75. 3.34 nm.

77. 13 h.

79. (a) 0.66×10^3 kg/m³;
(b) −3%.

81. ± 0.11 C°.

83. 3.6 m.

85. 3% increase.

87.

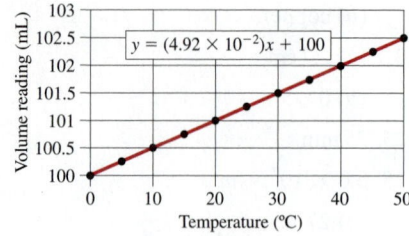

Slope of the line: 4.92×10^{-2} ml/°C,
relative β: 492×10^{-6}/°C,
β for the liquid: 501×10^{-6}/°C,
which liquid: glycerin.

CHAPTER 18

1. (a) 5.65×10^{-21} J;

   (b) 3.7×10^3 J.

3. 1.29.

5. 3.5×10^{-9} m/s.

7. (a) 4.5;

   (b) 5.2.

9. $\sqrt{3}$.

13. (b) 5.6%.

15. 1.004.

17. (a) 461 m/s;

   (b) 26 round trips/s.

19. Double the temperature.

21. (a) 710 m/s;

   (b) 240 K;

   (c) 650 m/s, 240 K, yes.

23. Vapor.

25. (a) Vapor;

   (b) solid.

27. 3600 Pa.

29. 355 torr or 4.73×10^4 Pa or
0.466 atm.

31. 92°C.

33. 1.99×10^5 Pa or 1.97 atm.

35. 70 g.

37. 16.6°C.

39. (a) Slope $= -5.00 \times 10^3$ K,
y intercept $= 24.9$.
Let $P_0 = 1$ Pa in this graph:

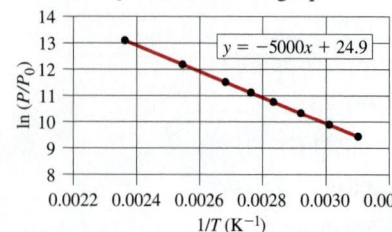

41. (a) 3.1×10^6 Pa;

   (b) 3.2×10^6 Pa.

43. (b) $a = 0.365$ N·m⁴/mol²,
$b = 4.28 \times 10^{-5}$ m³/mol.

45. (a) 0.10 Pa;

   (b) 3×10^7 Pa.

47. 2.1×10^{-7} m, stationary targets,
effective radius of $r_{H_2} + r_{air}$.

49. (b) 4.7×10^7 s⁻¹.

51. $\frac{1}{40}$.

53. 3.5 h, convection is much more
important than diffusion.

55. (b) 4×10^{-11} mol/s;

   (c) 0.6 s.

57. 260 m/s, 3.7×10^{-22} atm.

59. (a) 290 m/s;

   (b) 9.5 m/s.

61. 50 cm.

63. Kinetic energy $= 6.07 \times 10^{-21}$ J,
potential energy $= 5.21 \times 10^{-25}$ J,
yes, potential energy can be
neglected.

65. 0.07%.

67. 1.5×10^5 K.

69. (a) 2800 Pa;

   (b) 650 Pa.

71. 2×10^{13} m.

73. 0.36 kg.

75. (b) 4.6×10^9 Hz,
2.3×10^5 times larger.

77. 0.21.

CHAPTER 19

1. 10.7°C.

3. (a) 1.0×10^7 J;

   (b) 2.9 kWh;

   (c) $0.29 per day, no.

5. 4.2×10^5 J, 1.0×10^2 kcal.

7. 6.0×10^6 J.

9. (a) 3.3×10^5 J;

   (b) 56 min.

11. 6.9 min.

13. 39.9°C.

15. 2.3×10^3 J/kg·C°.

17. 54 C°.

19. 0.31 kg.

21. (a) 5.1×10^5 J;

   (b) 1.5×10^5 J.

23. 4700 kcal.

25. 360 m/s.

27.

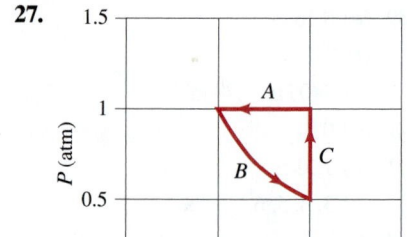

29. (a) 0;

   (b) -365 kJ.

31. (a) 480 J;

   (b) 0;

   (c) 480 J into gas.

33. (a) 4350 J;

   (b) 4350 J;

   (c) 0.

35. -4.0×10^2 K.

37. 236 J.

39. (a) 3.0×10^1 J;

   (b) 68 J;

   (c) -84 J;

   (d) -114 J;

   (e) -15 J.

41. $RT \ln \dfrac{(V_2 - b)}{(V_1 - b)} + a\left(\dfrac{1}{V_2} - \dfrac{1}{V_1}\right)$.

43. 43 C°.

45. 83.7 g/mol, krypton.

47. 48 C°.

49. (a) 6230 J;

   (b) 2490 J;

   (c) 8720 J.

51. 0.457 atm, -39°C.

53. (a) 404 K, 195 K;

   (b) -1.59×10^4 J;

   (c) 0;

   (d) -1.59×10^4 J.

55. (a)

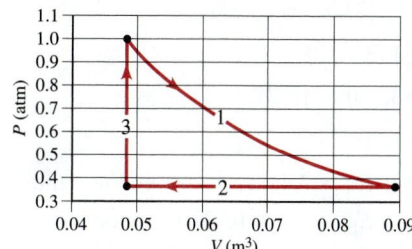

(b) 209 K;

(c) $Q_{1\to2} = 0$,
$\Delta E_{1\to2} = -2480$ J,
$W_{1\to2} = 2480$ J,
$Q_{2\to3} = -3740$ J,
$\Delta E_{2\to3} = -2240$ J,
$W_{2\to3} = -1490$ J,
$Q_{3\to1} = 4720$ J,
$\Delta E_{3\to1} = 4720$ J,
$W_{3\to1} = 0$;

(d) $Q_{\text{cycle}} = 990$ J,
$\Delta E_{\text{cycle}} = 0$,
$W_{\text{cycle}} = 990$ J.

57. (a) 5.0×10^1 W;
(b) 17 W.

59. 21 h.

61. (a) Ceramic: 14 W, shiny: 2.0 W;
(b) ceramic: 11 C°, shiny: 1.6 C°.

63. (a) 1.73×10^{17} W;
(b) 278 K or 5°C.

65. 28%.

67. (b) 4.8 C°/s;
(c) 0.60 C°/cm.

69. 6.4 Cal.

71. 4×10^{15} J.

73. 1 C°.

75. 3.6 kg.

77. 0.14 C°.

79. (a) 800 W;
(b) 5.3 g.

81. 1.1 days.

83. (a) 4.79 cm;
(b)

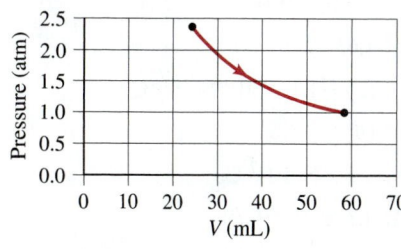

(c) $Q = 4.99$ J, $\Delta E = 0$, $W = 4.99$ J.

85. 110°C.

87. 305 J.

89. (a) 1.9×10^5 J;
(b) 4.4×10^5 J;
(c)

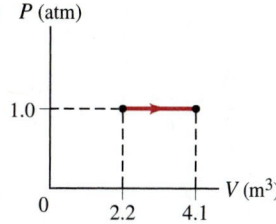

91. 2200 J.

CHAPTER 20

1. 0.25.

3. 0.16.

5. 0.21.

7. (b) 0.55.

9. 0.74.

13. 1.4×10^{13} J/h.

15. 1400 m.

17. 660°C.

19. (a) 4.1×10^5 Pa, 2.1×10^5 Pa;
(b) 34 L, 17 L;
(c) 2100 J;
(d) -1500 J;
(e) 600 J;
(f) 0.3.

21. 8.55.

23. 5.4.

25. (a) -4°C;
(b) 29%.

27. (a) 230 J;
(b) 390 J.

29. (a) 3.1×10^4 J;
(b) 2.7 min.

31. 91 L.

33. 0.20 J/K.

35. 5×10^4 J/K.

37. $5.49 \times 10^{-2} \dfrac{\text{J/K}}{\text{s}}$.

39. 9.3 J/K.

41. (a) 93 m J/K, yes;
(b) -93 m J/K, no; m in kg (SI).

43. (a) 1010 J/K;
(b) 1020 J/K;
(c) -9.0×10^2 J/K.

45. (a) Adiabatic;
(b) $\Delta S_{\text{adiabatic}} = 0$,
$\Delta S_{\text{isothermal}} = -nR \ln 2$;
(c) $\Delta S_{\text{environment adiabatic}} = 0$,
$\Delta S_{\text{environment isothermal}} = nR \ln 2$.

47. (a) All processes are reversible.

49. $\dfrac{T}{nC_V}$.

53. 2.1×10^5 J.

55. (a) $\frac{5}{16}$;
(b) $\frac{1}{64}$.

57. (a) 2.47×10^{-23} J/K;
(b) -9.2×10^{-22} J/K;
(c) these are many orders of magnitude smaller, due to the relatively small number of microstates for the coins.

59. (a) 1.79×10^6 kWh;
(b) 9.6×10^4 kW.

61. 12 MW.

63. (a) 0.41 mol;
(b) 396 K;
(c) 810 J;
(d) -700 J;
(e) 810 J;
(f) 0.13;
(g) 0.24.

65. (a) 110 kg/s;
(b) 9.3×10^7 gal/h.

67. (a) 18 km³/days;
(b) 120 km².

69. (a) 0.19;
(b) 0.23.

71. (a) 5.0 C°;
(b) 72.8 J/kg·K.

73. 1700 J/K.

75. 57 W or 0.076 hp.

77. $e_{\text{Sterling}} =$
$$\left(\frac{T_H - T_L}{T_H}\right)\left[\frac{\ln\left(\frac{V_b}{V_a}\right)}{\ln\left(\frac{V_b}{V_a}\right) + \frac{3}{2}\left(\frac{T_H - T_L}{T_H}\right)}\right],$$
$e_{\text{Sterling}} < e_{\text{Carnot}}$.

79. (a)

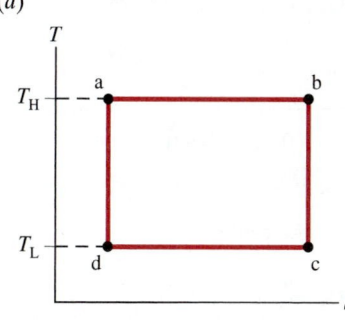

(b) W_{net}.

81. 16 kg.

83. 3.61×10^{-2} J/K.

CHAPTER 21

1. 2.7×10^{-3} N.

3. 7200 N.

5. $(4.9 \times 10^{-14})\%$.

7. 4.88 cm.

9. -5.8×10^8 C, 0.

11. (a) $q_1 = q_2 = \frac{1}{2}Q_T$;

     (b) $q_1 = 0, q_2 = Q_T$.

13. $F_1 = 0.53$ N at $265°$,

     $F_2 = 0.33$ N at $112°$,

     $F_3 = 0.26$ N at $53°$.

15. $F = 2.96 \times 10^7$ N, away from

     center of square.

17. 1.0×10^{12} electrons.

19. (a) $\pi\sqrt{\dfrac{md^3}{kQq}}$;

     (b) 0.2 ps.

21. 3.08×10^{-16} N west.

23. 1.10×10^7 N/C up.

25. $(172\,\hat{\mathbf{j}})$ N/C.

27. 1.01×10^{14} m/s^2, opposite to the

     field.

29.

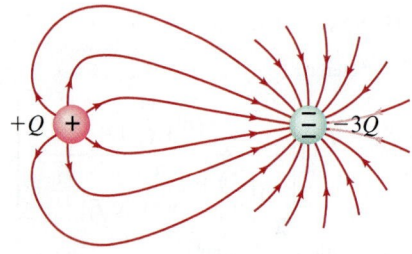

31. $(-4.7 \times 10^{11}\,\hat{\mathbf{i}})$ N/C

     $- (1.6 \times 10^{11}\,\hat{\mathbf{j}})$ N/C;

     or

     5.0×10^{11} N/C at $199°$.

33. $E = 2.60 \times 10^4$ N/C, away from

     the center.

35. $\dfrac{4kQxa}{(x^2 - a^2)^2}$, left.

37. $\dfrac{\lambda}{2\pi\varepsilon_0}\sqrt{\dfrac{1}{x^2} + \dfrac{1}{y^2}}\cdot\tan^{-1}\dfrac{x}{y}$.

39.

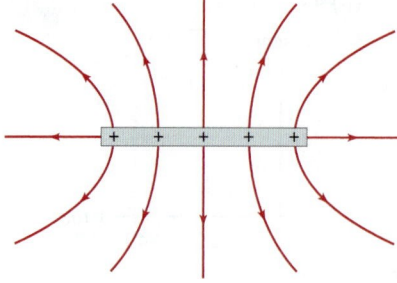

41. $\frac{1}{4}$.

43. (a) $\dfrac{Qy}{2\pi\varepsilon_0(y^2 + \ell^2)^{3/2}}$.

45. 1.8×10^6 N/C, away from the wire.

47. $\dfrac{8\lambda\ell z}{\pi\varepsilon_0(\ell^2 + 4z^2)\sqrt{4z^2 + 2\ell^2}}$, vertical.

49. $-\dfrac{2\lambda\sin\theta_0}{4\pi\varepsilon_0 R}\,\hat{\mathbf{i}}$.

51. (a) $\dfrac{\lambda}{4\pi\varepsilon_0 x(x^2 + \ell^2)^{1/2}}$

     $\times\,(\ell\hat{\mathbf{i}} + [x - (x^2 + \ell^2)^{1/2}]\hat{\mathbf{j}})$.

53. $\dfrac{Q}{4\pi\varepsilon_0 x(x + \ell)}$.

55. $\dfrac{Q(x\hat{\mathbf{i}} - \frac{2a}{\pi}\hat{\mathbf{j}})}{4\pi\varepsilon_0(x^2 + a^2)^{3/2}}$.

57. (a) $(-3.5 \times 10^{15}$ m/s$^2)\,\hat{\mathbf{i}}$

     $- (1.41 \times 10^{16}$ m/s$^2)\,\hat{\mathbf{j}}$;

     (b) $166°$ counterclockwise from the

     initial direction.

59. $-23°$.

61. (b) $2\pi\sqrt{\dfrac{4\pi\varepsilon_0 mR^3}{qQ}}$.

63. (a) 3.4×10^{-20} C;

     (b) no;

     (c) 8.5×10^{-26} m·N;

     (d) 2.5×10^{-26} J.

65. (a) θ very small;

     (b) $\dfrac{1}{2\pi}\sqrt{\dfrac{pE}{I}}$.

67. (a) In the direction of the dipole.

69. 3.5×10^9 C.

71. 6.8×10^5 C, negative.

73. 1.0×10^7 electrons.

75. 5.71×10^{13} C.

77. 1.6 m from Q_2, 3.6 m from Q_1.

79. $\dfrac{1.08 \times 10^7}{[3.00 - \cos(13.9t)]^2}$ N/C (upwards).

81. 5×10^{-9} C.

83. 8.0×10^{-9} C.

85. $18°$.

87. $E_A = 3.4 \times 10^4$ N/C, to the right;

     $E_B = 2.3 \times 10^4$ N/C, to the left;

     $E_C = 5.6 \times 10^3$ N/C, to the right;

     $E_D = 3.4 \times 10^3$ N/C, to the left.

89. -7.66×10^{-6} C, unstable.

91. (a) 9.18×10^6 N/C, down;

     (b) 1.63×10^{-4} C/m^2.

93. (a) $\dfrac{a}{\sqrt{2}} = 7.07$ cm;

     (b) yes;

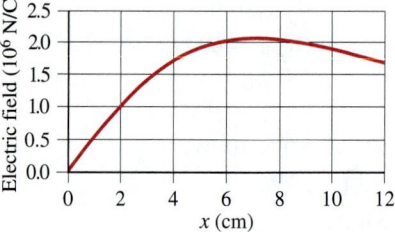

     (c) and (d)

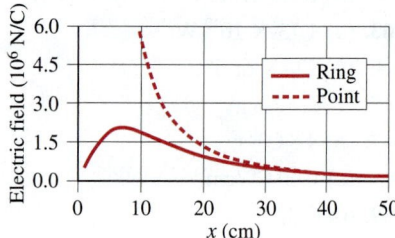

     (e) 37 cm.

CHAPTER 22

1. (a) 31 N·m^2/C;

     (b) 22 N·m^2/C;

     (c) 0.

3. (a) 0;

     (b) 0, 0, 0, 0, $E_0\ell^2$, $-E_0\ell^2$.

5. 1.63×10^{-8} C.

7. (a) -1.1×10^5 N·m^2/C;

     (b) 0.

9. -8.3×10^{-7} C.

11. 4.3×10^{-5} C/m.

13. -8.52×10^{-11} C.

15. (a) -2.6×10^4 N/C (toward wire);

     (b) -8.6×10^4 N/C (toward wire).

17. (a) $-(1.9 \times 10^{11}\ \text{N/C·m})r$;
   (b) $-(1.1 \times 10^{8}\ \text{N·m}^2/\text{C})/r^2$
     $+ (3.0 \times 10^{11}\ \text{N/C·m})r$;
   (c) $(4.1 \times 10^{8}\ \text{N·m}^2/\text{C})/r^2$;
   (d) yes.

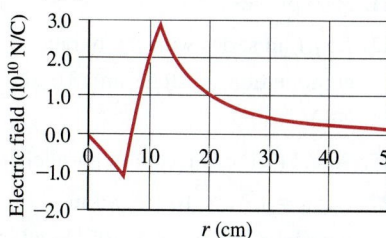

19.

21. (a) $5.5 \times 10^{7}\ \text{N/C}$ (outward);
   (b) 0;
   (c) $5.5 \times 10^{5}\ \text{N/C}$ (outward).

23. (a) $-8.00\ \mu\text{C}$;
   (b) $+1.90\ \mu\text{C}$.

25. (a) 0;
   (b) $\dfrac{\sigma}{\varepsilon_0}$ (outward, if both plates are positive);
   (c) same.

27. (a) 0;
   (b) $\dfrac{r_1^2 \sigma_1}{\varepsilon_0 r^2}$;
   (c) $\dfrac{(r_1^2 \sigma_1 + r_2^2 \sigma_2)}{\varepsilon_0 r^2}$;
   (d) $\sigma_1 = -\left(\dfrac{r_2}{r_1}\right)^2 \sigma_2$;
   (e) $\sigma_1 = 0$, or place $Q = -4\pi\sigma_1 r_1^2$ inside r_1.

29. (a) 0;
   (b) $\dfrac{Q}{4\pi\varepsilon_0}\left(\dfrac{1}{r_0^3 - r_1^3}\right)\left(\dfrac{r^3 - r_1^3}{r^2}\right)$;
   (c) $\dfrac{kQ}{r^2}$.

31. (a) $-q$;
   (b) $Q + q$;
   (c) $\dfrac{kq}{r^2}$;
   (d) 0;
   (e) $\dfrac{k(q + Q)}{r^2}$.

33. (a) $\dfrac{\sigma R_0}{\varepsilon_0 R}$, radially outward;
   (b) 0;
   (c) same for $R > R_0$ if $\lambda = 2\pi R_0 \sigma$.

35. (a) 0;
   (b) $\dfrac{1}{2\pi\varepsilon_0}\dfrac{(Q/\ell)}{r}$;
   (c) 0;
   (d) $\dfrac{e}{4\pi\varepsilon_0}\left(\dfrac{Q}{\ell}\right)$.

37. (a) $1.9 \times 10^{7}\ \text{m/s}$;
   (b) $5.5 \times 10^{5}\ \text{m/s}$.

39. (a) $\dfrac{\rho_E r}{3\varepsilon_0}$;
   (b) $\dfrac{\rho_E r_0^3}{3\varepsilon_0 r^2}$;
   (c) 0;
   (d) $\left(\dfrac{\rho_E r_0^3}{3\varepsilon_0} + \dfrac{Q}{4\pi\varepsilon_0}\right)\dfrac{1}{r^2}$.

41. (a) 0;
   (b) $\dfrac{Q}{2500\pi\varepsilon_0 R_0^2}$.

43. (a) $\dfrac{\rho_E d}{2\varepsilon_0}$ away from surface.

45. (a) 13 N (attractive);
   (b) 0.064 J.

47. (a) 0;
   (b) $-\dfrac{\rho_0(d - x)}{\varepsilon_0}\hat{\mathbf{i}}$;
   (c) $-\dfrac{\rho_0(d + x)}{\varepsilon_0}\hat{\mathbf{i}}$.

49. $\dfrac{Q}{4\pi\varepsilon_0}\dfrac{r^2}{r_0^4}$, radially outward.

51. $\Phi = \oint \vec{g} \cdot d\vec{A} = -4\pi G M_{\text{enc}}$.

53. $a\ell^3\varepsilon_0$.

55. $475\ \text{N·m}^2/\text{C}$, $475\ \text{N·m}^2/\text{C}$.

57. (a) 0;
   (b) $E_{\max} = \dfrac{Q}{\pi\varepsilon_0 r_0^2}$, $E_{\min} = \dfrac{Q}{25\pi\varepsilon_0 r_0^2}$;
   (c) no;
   (d) no.

59. (a) $1.1 \times 10^{-19}\ \text{C}$;
   (b) $3.5 \times 10^{11}\ \text{N/C}$.

61. (a) $\dfrac{\rho_E r_0}{6\varepsilon_0}$, right;
   (b) $\dfrac{17}{54}\dfrac{\rho_E r_0}{\varepsilon_0}$, left.

63. (a) 0;
   (b) $5.65 \times 10^{5}\ \text{N/C}$, right;
   (c) $5.65 \times 10^{5}\ \text{N/C}$, right;
   (d) $-5.00 \times 10^{-6}\ \text{C/m}^3$;
   (e) $+5.00 \times 10^{-6}\ \text{C/m}^3$.

65. (a) On inside surface of shell.
   (b) $r < 0.10\ \text{m}$,
     $$E = \left(\dfrac{2.7 \times 10^4}{r^2}\right)\ \text{N/C};$$
     $r > 0.10\ \text{m}$, $E = 0$.

67. $-46\ \text{N·m}^2/\text{C}$, $-4.0 \times 10^{-10}\ \text{C}$.

CHAPTER 23

1. $-0.71\ \text{V}$.

3. 3280 V, plate B has a higher potential.

5. 30 m.

7. $1.4\ \mu\text{C}$.

9. 1.2 cm, 46 nC.

11. (a) 0;
   (b) $-29.4\ \text{V}$;
   (c) $-29.4\ \text{V}$.

13. (a) $-9.6 \times 10^{8}\ \text{V}$;
   (b) $9.6 \times 10^{8}\ \text{V}$.

15. (a) They are equal;
   (b) $Q\left(\dfrac{r_2}{r_1 + r_2}\right)$.

17. (a) 10–20 kV;
   (b) $30\ \mu\text{C/m}^2$.

19. (a) $\dfrac{Q}{4\pi\varepsilon_0 r}$;
   (b) $\dfrac{Q}{8\pi\varepsilon_0 r_0}\left(3 - \dfrac{r^2}{r_0^2}\right)$;
   (c) Let $V_0 = V$ at $r = r_0$, and $E_0 = E$ at $r = r_0$:

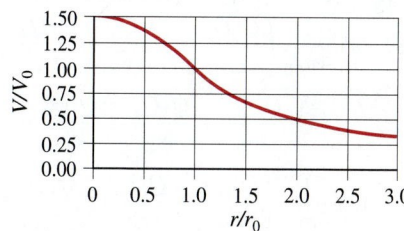

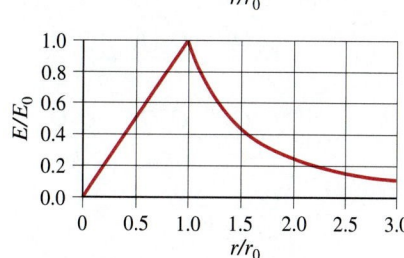

21. $\dfrac{\rho_0}{\varepsilon_0}\left(\dfrac{r_0^2}{4} - \dfrac{r^2}{6} + \dfrac{r^4}{20 r_0^2}\right)$.

23. (a) $\dfrac{R_0 \sigma}{\varepsilon_0}\ln\left(\dfrac{R_0}{R}\right) + V_0$;
   (b) V_0;
   (c) no, from part (a) $V \to -\infty$ due to length of wire.

25. (a) 29 V;
   (b) $-4.6 \times 10^{-18}\ \text{J}$.

27. 0.34 J.

29. 4.2 MV.

31. 9.64×10^5 m/s.

33. (a) 0;

(b) $E_x = 0$,

$$E_y = \frac{Q}{4\pi\varepsilon_0} \frac{R}{(x^2 + R^2)^{3/2}}, \text{ looks}$$

like a dipole.

35. $\frac{\sigma}{2\varepsilon_0}\left(\sqrt{R_2^2 + x^2} - \sqrt{R_1^2 + x^2}\right)$.

37. 29 m/s.

39. $\frac{Q}{8\pi\varepsilon_0 \ell} \ln\left(\frac{x + \ell}{x - \ell}\right)$.

41. $\frac{a}{6\varepsilon_0}(R^2 - 2x^2)\sqrt{R^2 + x^2} + \frac{a|x|^3}{3\varepsilon_0}$.

43. 2 mm.

45. (a) 2.6 mV;

(b) 1.8 mV;

(c) -1.8 mV.

49. -7.1×10^{-11} C/m^2 on $x = 0$ plate, 7.1×10^{-11} C/m^2 on other plate.

51. $(-2.5y + 3.5yz)\hat{\mathbf{i}}$
$+ (-2y - 2.5x + 3.5xz)\hat{\mathbf{j}}$
$+ (3.5xy)\hat{\mathbf{k}}$.

53. (a) $\frac{Q}{4\pi\varepsilon_0}\left(\frac{1}{y\sqrt{\ell^2 + y^2}}\right)\hat{\mathbf{j}}$;

(b) $\frac{Q}{4\pi\varepsilon_0}\left(\frac{1}{x^2 - \ell^2}\right)\hat{\mathbf{i}}$.

55. -62.5 kV.

57. 1.3 eV.

59. (a) $\frac{1}{4\pi\varepsilon_0}\left(\frac{Q_1 Q_2}{r_{12}} + \frac{Q_1 Q_3}{r_{13}} + \frac{Q_1 Q_4}{r_{14}}\right.$

$\left. + \frac{Q_2 Q_3}{r_{23}} + \frac{Q_2 Q_4}{r_{24}} + \frac{Q_3 Q_4}{r_{34}}\right)$;

(b) $\frac{1}{4\pi\varepsilon_0}\left(\frac{Q_1 Q_2}{r_{12}} + \frac{Q_1 Q_3}{r_{13}} + \frac{Q_1 Q_4}{r_{14}}\right.$

$+ \frac{Q_1 Q_5}{r_{15}} + \frac{Q_2 Q_3}{r_{23}} + \frac{Q_2 Q_4}{r_{24}}$

$+ \frac{Q_2 Q_5}{r_{25}} + \frac{Q_3 Q_4}{r_{34}} + \frac{Q_3 Q_5}{r_{35}}$

$\left. + \frac{Q_4 Q_5}{r_{45}}\right)$.

61. (a) 1.33 keV;

(b) $v_e/v_p = 42.8$.

63. 250 MeV, same order of magnitude as observed values.

65. 1.11×10^5 m/s, 3.5×10^5 m/s.

67. 0.26 MV/m.

69. 600 V.

71. 1.5 J.

73. Yes, 2.0 pV.

75. 1.03×10^6 m/s.

77. $-\frac{\sqrt{3}Q}{2\pi\varepsilon_0 \ell}, \frac{Q}{\pi\varepsilon_0 \ell}\left(\frac{\sqrt{3}}{6} - 2\right),$

$-\frac{Q}{\pi\varepsilon_0 \ell}\left(1 + \frac{\sqrt{3}}{6}\right)$.

79. (a) 1.2 MV;

(b) 1.8 kg.

81. (a) $\frac{\rho_E(r_2^3 - r_1^3)}{3\varepsilon_0 r}$;

(b) $\frac{\rho_E}{\varepsilon_0}\left(\frac{r_2^2}{2} - \frac{r^2}{6} - \frac{r_1^3}{3r}\right)$;

(c) $\frac{\rho_E}{2\varepsilon_0}(r_2^2 - r_1^2)$; yes.

83. $\vec{\mathbf{E}} = \frac{\lambda}{2\pi\varepsilon_0 R}$, radially outward.

85. (a) 23 kV;

(b) $\frac{4Bx\hat{\mathbf{i}}}{(x^2 + R^2)^3}$;

(c) $(2.3 \times 10^5 \text{ N/C})\hat{\mathbf{i}}$.

87. (a) and (b):

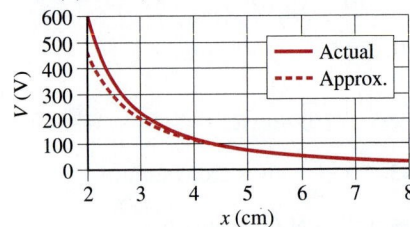

89. (a) Point charge;

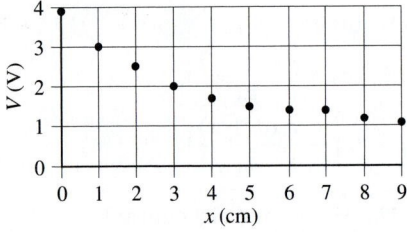

(b) 1.5×10^{-11} C;

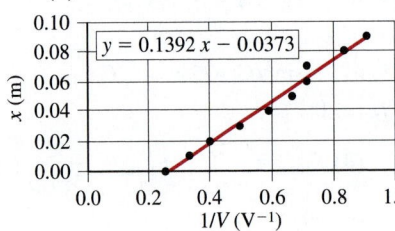

(c) $x = -3.7$ cm.

CHAPTER 24

1. 3.0 μF.

3. 3.1 pF.

5. 56 μF.

7. 1.1 C.

9. 83 days.

11. 130 m^2.

13. 7.10×10^{-4} F.

15. 18 nC.

17. 5.8×10^4 V/m.

19. (a) 0.22 μm $\leq x \leq$ 220 μm;

(b) $\frac{x^2 \Delta C}{\varepsilon_0 A}$;

(c) 0.01%, 10%.

21. 3600 pF, yes.

23. 1.5 μF in series with the parallel combination of 2.0 μF and 3.0 μF, 2.8 V.

25. Add 11 μF connected in parallel.

27. $C_{max} = 1.94 \times 10^{-8}$ F, all in parallel, $C_{min} = 1.8 \times 10^{-9}$ F, all in series.

29. (a) $\frac{3}{5}C$;

(b) $Q_1 = Q_2 = \frac{1}{5}CV, Q_3 = \frac{2}{5}CV$,

$Q_4 = \frac{3}{5}CV, V_1 = V_2 = \frac{1}{5}V$,

$V_3 = \frac{2}{5}V, V_4 = \frac{3}{5}V$.

31. $Q_1 = \frac{C_1 C_2}{C_1 + C_2}V_0, Q_2 = \frac{C_2^2}{C_1 + C_2}V_0$.

33. (a) $Q_1 = 23$ μC, $Q_2 = Q_4 = 46$ μC;

(b) $V_1 = V_2 = V_3 = V_4 = 2.9$ V;

(c) 5.8 V.

35. 2.4 μF.

37. (a) $C_1 + \frac{C_2 C_3}{C_2 + C_3}$;

(b) $Q_1 = 8.40 \times 10^{-4}$ C,

$Q_2 = Q_3 = 2.80 \times 10^{-4}$ C.

39. $C = \frac{\varepsilon_0 A}{d}\left(1 - \frac{\theta\sqrt{A}}{2d}\right)$.

41. 6.8×10^{-3} J.

43. 2.0×10^3 J.

45. 1.70×10^{-3} J.

47. (a) $\frac{U_f}{U_i} = \frac{\ln\left(\dfrac{3R_a}{R_b}\right)}{\ln\left(\dfrac{R_a}{R_b}\right)} > 1$,

work done to enlarge cylinder;

(b) $\frac{U_f}{U_i} = \frac{\ln\left(\dfrac{R_a}{R_b}\right)}{\ln\left(\dfrac{3R_a}{R_b}\right)} < 1$,

charge moved to battery.

49. (a) $-\frac{\varepsilon_0 A\ell V_0^2}{2d(d - \ell)}$;

(b) $\frac{\varepsilon_0 A\ell V_0^2}{2(d - \ell)^2}$.

53. 2200 batteries, no.

55. 1.1×10^{-4} J.

57. (a) $0.32 \, \mu\text{m}^2$;

(b) 59 megabytes.

59. $\dfrac{\varepsilon_0 A}{2d}(K_1 + K_2)$.

61. $\dfrac{\varepsilon_0 A K_1 K_2}{(d_1 K_2 + d_2 K_1)}$.

63. (a) $\dfrac{\varepsilon_0 \ell^2}{d}\left[1 + (K - 1)\dfrac{x}{\ell}\right]$;

(b) $\dfrac{V_0^2 \varepsilon_0 \ell^2}{2d}\left[1 + (K - 1)\dfrac{x}{\ell}\right]$;

(c) $\dfrac{V_0^2 \varepsilon_0 \ell}{2d}(K - 1)$, left.

67. $\dfrac{\varepsilon_0 A}{d - \ell + \dfrac{\ell}{K}}$.

69. $E_{\text{air}} = 2.69 \times 10^4 \, \text{V/m}$,

$E_{\text{glass}} = 4.64 \times 10^3 \, \text{V/m}$,

$Q_{\text{free}} = 0.345 \, \mu\text{C}, Q_{\text{ind}} = 0.286 \, \mu\text{C}$.

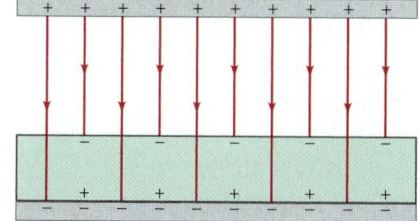

71. $43 \, \mu\text{F}$.

73. 15 V.

75. 840 V.

77. $3.76 \times 10^{-9} \, \text{F}, 0.221 \, \text{m}^2$.

79. $\dfrac{1}{2K}$, work done by the electric

field, $\dfrac{1}{K}$.

81. 1.2.

83. (a) 25 J;

(b) 940 kW.

85. (a) Parallel;

(b) 7.7 pF to 35 pF.

87. 5.15 pF.

89. $Q_1 = 11 \, \mu\text{C}, Q_2 = 13 \, \mu\text{C}$,

$Q_3 = 13 \, \mu\text{C}, V_1 = 11 \, \text{V}$,

$V_2 = 6.3 \, \text{V}, V_3 = 5.2 \, \text{V}$.

91. $\dfrac{Q^2 x}{2\varepsilon_0 A}$.

93. $9 \times 10^{-16} \, \text{m}$, no.

95. (a) $0.27 \, \mu\text{C}, 15 \, \text{kV/m}, 5.9 \, \text{nF}$,

$6.0 \, \mu\text{J}$;

(b) $0.85 \, \mu\text{C}, 15 \, \text{kV/m}, 19 \, \text{nF}, 19 \, \mu\text{J}$.

97. (a) 32 nF;

(b) $14 \, \mu\text{C}$;

(c) 7.0 mm;

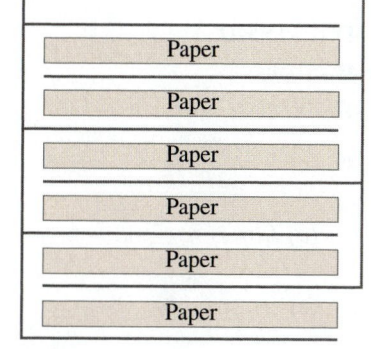

(d) 450 V.

CHAPTER 25

1. 8.13×10^{18} electrons/s.

3. 5.5×10^{-11} A.

5. (a) 28 A;

(b) 8.4×10^4 C.

7. 1.1×10^{21} electrons/min.

9. (a) $2.0 \times 10^1 \, \Omega$;

(b) 430 J.

11. 0.47 mm.

13. 0.64.

15. (a) Slope $= 1/R$, y-intercept $= 0$;

(b) yes, $R = 1.39 \, \Omega$;

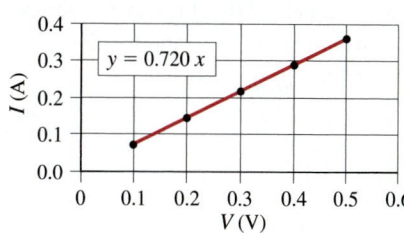

(c) $1.0 \times 10^{-6} \, \Omega \cdot \text{m}$, nichrome.

17. At 1/5.0 of its length, $2.0 \, \Omega$, $8.0 \, \Omega$.

19. 2400°C.

21. $\sqrt{2}$.

23. 44.1°C.

25. One-quarter of the original.

27. $\dfrac{1}{4\pi\sigma}\left(\dfrac{1}{r_1} - \dfrac{1}{r_2}\right)$.

29. (a) $0.14 \, \Omega$;

(b) 0.60 A;

(c) $V_{\text{Al}} = 52 \, \text{mV}, V_{\text{Cu}} = 33 \, \text{mV}$.

31. 0.81 W.

33. 29 V.

35. (b) As large as possible.

37. (a) 0.83 A;

(b) $140 \, \Omega$.

39. 0.055 kWh, 7.9 cents/month.

41. $0.90 \, \text{kWh} = 3.2 \times 10^6 \, \text{J}$.

43. 24 lightbulbs.

45. 11 kW.

47. $0.15 \, \text{kg/s} = 150 \, \text{mL/s}$.

49. 0.12 A.

51. (a) ∞;

(b) $96 \, \Omega$.

53. (a) 930 V;

(b) 3.9 A.

55. (a) 1.3 kW;

(b) max $= 2.6 \, \text{kW}$, min $= 0$.

57. (a) $5.1 \times 10^{-10} \, \text{m/s}$;

(b) $6.9 \, \text{A/m}^2$;

(c) $1.2 \times 10^{-7} \, \text{V/m}$.

59. $2.5 \, \text{A/m}^2$, north.

61. 35 m/s, delay time from stimulus to action.

63. 11 hr.

65. 1.8 m, it would generate 540 W of heat and could start a fire.

67. 0.16 S.

69. (a) \$35/month;

(b) 1300 kg/year.

71. (a) -19% change;

(b) % change would be slightly less.

73. (a) $190 \, \Omega$;

(b) $15 \, \Omega$.

75. (a) 1500 W;

(b) 12 A.

77. 2:1.

79. (a) $21 \, \Omega$;

(b) 2.0×10^1 s;

(c) 0.17 cents.

81. 36.0 m, 0.248 mm.

83. (a) 1200 W;

(b) 100 W.

85. 1.4×10^{12} protons.

87. (a) 3.1 kW;

(b) 24 W;

(c) 15 W;

(d) 38 cents/month.

89. (a) \$55/kWh;

(b) \$280/kWh, D-cells and AA-cells are $550\times$ and $2800\times$, respectively, more expensive.

91. $1.34 \times 10^{-4} \, \Omega$.

93. $\dfrac{4\ell\rho}{ab\pi}$.

95. $f = 1 - \dfrac{V}{V_0}$.

CHAPTER 26

1. (a) 5.93 V;

   (b) 5.99 V.

3. 0.060 Ω.

5. 9.3 V.

7. (a) 2.60 kΩ;

   (b) 270 Ω.

9. Connect nine 1.0-Ω resistors in series with battery; then connect output voltage circuit across four consecutive resistors.

11. 0.3 Ω.

13. 450 Ω, 0.024.

15. Solder a 1.6-kΩ resistor in parallel with 480-Ω resistor.

17. 120 Ω.

19. $\frac{13}{8} R$.

21. $R = r$.

23. (a) V_{left} decreases,

     V_{middle} increases,

     $V_{\text{right}} = 0$;

   (b) I_{left} decreases,

     I_{middle} increases,

     $I_{\text{right}} = 0$;

   (c) terminal voltage increases;

   (d) 8.5 V;

   (e) 8.6 V.

25. (a) V_1 and V_2 increase, V_3 and V_4 decrease;

   (b) I_1 and I_2 increase, I_3 and I_4 decrease;

   (c) increases;

   (d) before: $I_1 = 117\,\text{mA}$, $I_2 = 0$,

     $I_3 = I_4 = 59\,\text{mA}$;

     after: $I_1 = 132\,\text{mA}$,

     $I_2 = I_3 = I_4 = 44\,\text{mA}$, yes.

27. 0.38 A.

29. 0.

31. (a) 29 V;

   (b) 43 V, 73 V.

33. $I_1 = 0.68$ A left, $I_2 = 0.33$ A left.

37. 0.70 A.

39. 0.17 A.

41. (a) $\dfrac{R(5R' + 3R)}{8(R' + R)}$;

   (b) $\dfrac{R}{2}$.

43. 1 – 15 MΩ.

45. 5.0 ms.

47. 44 s.

49. (a) $I_1 = \dfrac{2\mathscr{E}}{3R}$, $I_2 = I_3 = \dfrac{\mathscr{E}}{3R}$;

   (b) $I_1 = I_2 = \dfrac{\mathscr{E}}{2R}$, $I_3 = 0$;

   (c) $\dfrac{\mathscr{E}}{2}$.

51. (a) 8.0 V;

   (b) 14 V;

   (c) 8.0 V;

   (d) 4.8 μC.

53. 29 μA.

55. (a) Place in parallel with 0.22-mΩ shunt resistor;

   (b) place in series with 45-kΩ resistor.

57. 100 kΩ.

59. $V_{44} = 24$ V, $V_{27} = 15$ V;

   -15%, -15%.

61. 0.960 mA, 4.8 V.

63. 12 V.

65. Connect a 9.0-kΩ resistor in series with human body and battery.

67. 2.5 V, 117 V.

69. 92 kΩ.

71. (a) $\dfrac{R_2 R_3}{R_1}$;

   (b) 121 Ω.

73. Terminal voltage of mercury cell (3.99 V) is closer to 4.0 V than terminal voltage of dry cell (3.84 V).

75. 150 cells, 0.54 m², connect in series; connect four such sets in parallel to total 600 cells and deliver 120 V.

77. Counterclockwise current: -24 V, clockwise current: $+48$ V.

79. 10.7 V.

83. 9.0 Ω.

85. (b) 1.39 V;

   (c) 0.42 mV;

   (d) no current from "working" battery is needed to "power" galvanometer.

87. 1.0 mV, 2.0 mV, 4.0 mV, 10.0 mV.

89. (a) 6.8 V, 15 μC;

   (b) 48 μs.

91. 200 MΩ.

93. 4.5 ms.

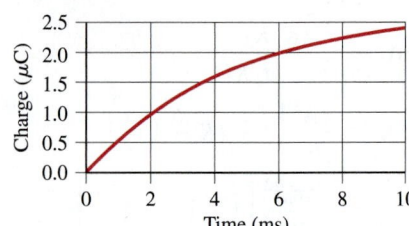

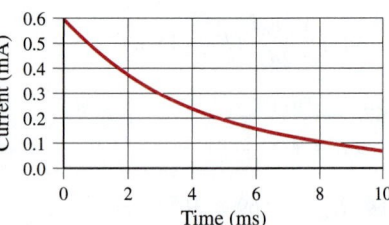

CHAPTER 27

1. (a) 8.5 N/m;

   (b) 4.9 N/m.

3. 2.6×10^{-4} N.

5. (a) South pole;

   (b) 3.41 A;

   (c) 7.39×10^{-2} N.

7. 2.13 N, 41.8° below negative y axis.

9. $\left(-2IrB_0 \sin\theta_0\right)\hat{\mathbf{j}}$.

13. 6.3×10^{-14} N, north.

15. 1.8 T.

17. (a) Downward;

   (b) into page;

   (c) right.

19. (a) 0.031 m;

   (b) 3.8×10^{-7} s.

23. 1.8 m.

25. $(0.78\hat{\mathbf{i}} - 1.0\hat{\mathbf{j}} + 0.1\hat{\mathbf{k}}) \times 10^{-15}$ N.

27. $L_{\text{final}} = \frac{1}{2} L_{\text{initial}}$.

29. (a) Negative;

   (b) $qB_0 \left(\dfrac{\ell^2 + d^2}{2d}\right)$.

31. 1.3×10^8 m/s, yes.

33. (a) 45°;

   (b) 2.3×10^{-3} m.

35. (a) $2NIAB$;

   (b) 0.

37. (a) 4.85×10^{-5} m·N;

   (b) north.

39. (a) $\left(-4.3\,\hat{\mathbf{k}}\right)$ A·m²;

   (b) $(2.6\hat{\mathbf{i}} - 2.4\hat{\mathbf{j}})$ m·N;

   (c) -2.8 J.

41. 12%.

43. 39 μA.

45. 6 electrons.

47. (b) 0.05 nm, about $\frac{1}{6}$ the size of a typical metal atom;

    (c) 10 mV.

49. 0.820 T.

51. 70 u, 72 u, 73 u, and 74 u.

53. 1.5 mm, 1.5 mm, 0.77 mm, 0.77 mm.

55. $^2_1$H, $^4_2$He.

57. 2.4 T, upwards.

59. (a) $\dfrac{IBd}{m}t$;

    (b) $\left(\dfrac{IBd}{m} - \mu_k g\right)t$;

    (c) east.

61. 1.1×10^{-6} m/s, west.

63. 3.8×10^{-4} m·N.

65. $\pi\left[\dfrac{mb(3a + b)}{3NIBa(a + b)}\right]^{1/2}$.

67. They do not enter second tube, 12°.

69. 1.1 A, down.

71. 7.3×10^{-3} T.

73. -6.9×10^{-20} J.

75. 0.083 N, northerly and 68° above the horizontal.

77. (a) Downward;

    (b) 28 mT;

    (c) 0.12 T.

CHAPTER 28

1. 0.37 mT, 7.4 times larger.

3. 0.15 N, toward other wire.

7. 0.12 mT, 82° above directly right.

9. 3.8×10^{-5} T, 17° below the horizontal to north.

11. (a) $(2.0 \times 10^{-5})(25 - I)$ T;

    (b) $(2.0 \times 10^{-5})(25 + I)$ T.

15. Closer wire: 0.050 N/m, attractive, farther wire: 0.025 N/m, repulsive.

17. 17 A, downward.

19. $\dfrac{\mu_0 I}{2\pi}\left(\dfrac{d - 2x}{x(d - x)}\right)\hat{\mathbf{j}}$.

21. 46.6 μT.

23. (b) $\dfrac{\mu_0 I}{2\pi y}$, yes, looks like B from long straight wire.

25. 0.160 A.

27. (a) 5.3 mT;

    (b) 3.2 mT;

    (c) 1.8 mT.

29. (a) 0.554 m;

    (b) 10.5 mT.

31. (a) $\dfrac{\mu_0 I_0 R}{2\pi R_1^2}$;

    (b) $\dfrac{\mu_0 I_0}{2\pi R}$;

    (c) $\dfrac{\mu_0 I_0}{2\pi R}\left(\dfrac{R_3^2 - R^2}{R_3^2 - R_2^2}\right)$;

    (d) 0;

    (e)

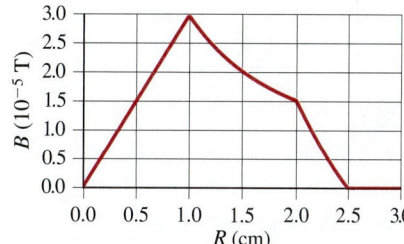

33. 3.6×10^{-6} T.

35. $0.075 \mu_0 I/R$.

37. (a) $\dfrac{\mu_0 I}{4}\left(\dfrac{1}{R_1} + \dfrac{1}{R_2}\right)$, into the page;

    (b) $\dfrac{\pi I(R_1^2 + R_2^2)}{2}$, into the page.

39. (a) $\dfrac{Q\omega R^2}{4}\hat{\mathbf{i}}$;

    (b) $\dfrac{\mu_0 Q\omega}{2\pi R^2}\left(\dfrac{R^2 + 2x^2}{\sqrt{R^2 + x^2}} - 2x\right)\hat{\mathbf{i}}$;

    (c) yes.

41. (b) $\dfrac{\mu_0 I}{4\pi y}\left(\dfrac{d}{\sqrt{d^2 + y^2}}\right)\hat{\mathbf{k}}$.

43. (a) $\dfrac{n\mu_0 I \tan(\pi/n)}{2\pi R}$, into the page.

45. $\dfrac{\mu_0 I}{4\pi}\left[\dfrac{\sqrt{x^2 + y^2}}{xy} + \dfrac{\sqrt{y^2 + (b - x)^2}}{(b - x)y}\right.$

    $+ \dfrac{\sqrt{(a - y)^2 + (b - x)^2}}{(a - y)(b - x)}$

    $\left. + \dfrac{\sqrt{(a - y)^2 + x^2}}{x(a - y)}\right]$,

    out of page.

47. (a) 16 A·m²;

    (b) 13 m·N.

49. 2.4 T.

51. $(\vec{F}/\ell)_M = 6.3 \times 10^{-4}$ N/m at 90°,

    $(\vec{F}/\ell)_N = 3.7 \times 10^{-4}$ N/m at 300°,

    $(\vec{F}/\ell)_P = 3.7 \times 10^{-4}$ N/m at 240°.

53. 170 A.

55. (a) 2.7×10^{-6} T;

    (b) 5.3×10^{-6} T;

    (c) no, no Newton's third-law-type of relationship;

    (d) both 1.1×10^{-5} N/m, yes, Newton's third law holds.

57. $\dfrac{\mu_0 tj}{2}$, to the left above sheet (with current coming toward you).

61. (a) $\dfrac{N\mu_0 IR^2}{2}$

    $\times \left(\dfrac{1}{(R^2 + x^2)^{3/2}} + \dfrac{1}{(R^2 + (x - R)^2)^{3/2}}\right)$;

    (b) 4.5 mT.

63. 3×10^9 A.

65. (a) 46 turns;

    (b) 0.83 mT;

    (c) no.

67. $\dfrac{\mu_0 I\sqrt{5}}{2\pi a}$, into the page.

69. 0.10 N, south.

71. $\frac{2}{3}$.

73. (c) 1.5 A.

75.

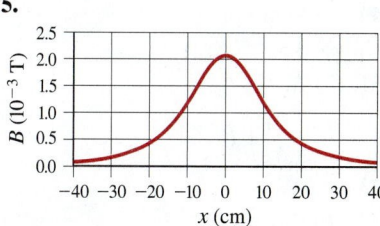

CHAPTER 29

1. -460 V.

3. Counterclockwise.

5. 1.2 mm/s.

7. (a) 0.010 Wb;

    (b) 55°;

    (c) 5.8 mWb.

9. Counterclockwise.

11. (a) Clockwise;

    (b) 43 mV;

    (c) 17 mA.

13. (a) 8.1 mJ;

    (b) 4.2×10^{-3} C°.

15. (a) 0.15 A;

    (b) 1.4 mW.

17. 8.81 C.

19. 21 μJ.

21. 23 mV, 26 mV.

23. (a) 0;

    (b) 0.99 A, counterclockwise.

25. (a) $\dfrac{\mu_0 Ia}{2\pi}\ln\!\left(1 + \dfrac{a}{b}\right);$

   (b) $\dfrac{\mu_0 Ia^2v}{2\pi b(a + b)};$

   (c) clockwise;

   (d) $\dfrac{\mu_0^2 I^2 a^4 v}{4\pi^2 b^2 (a + b)^2 R}.$

27. 1.0 m/s.

29. (a) 0.11 V;

   (b) 4.1 mA;

   (c) 0.36 mN.

31. 0.39 m/s.

33. (a) Yes;

   (b) $v_0\, e^{-B^2\ell^2 t/mR}.$

35. (a) $\dfrac{v\mu_0 I}{2\pi}\ln\!\left(1 + \dfrac{a}{b}\right);$

   (b) $-\dfrac{v\mu_0 I}{2\pi}\ln\!\left(1 + \dfrac{a}{b}\right).$

37. 57.2 loops.

41. 150 V.

43. 13 A.

45. (a) 2.4 kV;

   (b) 190 V.

47. 50, 4.8 V.

49. (a) Step-up;

   (b) 3.5.

51. (a) R;

   (b) $\left(\dfrac{N_P}{N_S}\right)^2 R.$

53. 98 kW.

55. (b) Clockwise;

   (c) increase.

57. (a) $\dfrac{IR}{\ell};$

   (b) $\dfrac{\mathscr{E}_0}{\ell}\, e^{-B^2\ell^2 t/mR}.$

59. 10.1 mJ.

61. 0.6 nC.

63. (a) 41 kV;

   (b) 31 MW;

   (c) 0.88 MW;

   (d) 3.0×10^7 W.

65. (a) Step-down;

   (b) 2.9 A;

   (c) 0.29 A;

   (d) 4.1 Ω.

67. 57 mA, left to right through resistor.

69. 2.3×10^{17} electrons.

71. (a) 25 A;

   (b) 98 V;

   (c) 600 W;

   (d) 81%.

73. $\frac{1}{2}B\omega\ell^2.$

77. $B\omega R$, radially in toward axis.

79. (a) $\dfrac{\pi d^2 B^2 \ell v}{16\rho};$

   (b) $16\rho\rho_m g/B^2;$

   (c) 3.7 cm/s.

CHAPTER 30

1. (a) 31.0 mH;

   (b) 3.79 V.

3. $\dfrac{\mu_0 N_1 N_2 A_2 \sin\theta}{\ell}.$

5. 12 V.

7. 0.566 H.

9. 11.3 V.

11. 46 m, 21 km, 0.70 kΩ.

15. 18.9 J.

17. 1.06×10^{-3} J/m³.

19. $\dfrac{\mu_0 N^2 I^2}{8\pi^2 r^2}, \dfrac{\mu_0 N^2 I^2 h}{4\pi}\ln\!\left(\dfrac{r_2}{r_1}\right).$

21. $\dfrac{\mu_0 I^2}{16\pi}.$

23. 3.5 time constants.

25. (a) $\dfrac{LV_0^2}{2R^2}\left(1 - e^{-t/\tau}\right)^2;$

   (b) 7.6 time constants.

27. (b) 6600 V.

29. $(12\text{ V})e^{-t/8.2\,\mu s}, 0, 12$ V.

31. (a) 0.16 nF;

   (b) 62 μH.

33. (c) $\left(2 \times 10^{-4}\right)\%.$

35. (a) $\dfrac{Q_0}{\sqrt{2}};$

   (b) $\frac{1}{8}T.$

37. $\dfrac{L}{R}\ln\!\left(\frac{4}{3}\right) = (0.29)\dfrac{L}{R}.$

39. 3300 Hz.

41.

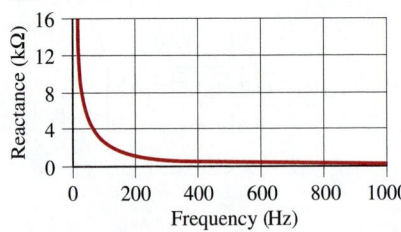

43. (a) $R + R'$;

   (b) R'.

45. (a) 2800 Ω;

   (b) 660 Hz, 11 A.

47. 2190 W.

49. (a) 0.40 kΩ;

   (b) 75 Ω.

51. 1600 Hz.

53. 240 Hz, voltages are out of phase.

55. (a) 0.124 A;

   (b) 5.02°;

   (c) 14.8 W;

   (d) 0.120 kV, 10.5 V.

57. 7.8 μF.

59. $I_0 V_0 \sin\omega t \sin(\omega t + \phi).$

61. 130 Ω, 0.91.

63. 265 Hz, 324 W.

65. (b) 130 Ω.

67. (a) $\dfrac{V_0^2 R}{2\left[R^2 + \left(\omega L - \dfrac{1}{\omega C}\right)^2\right]};$

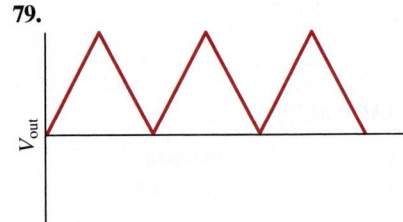

   (b) $\dfrac{1}{2\pi}\sqrt{\dfrac{1}{LC}};$

   (c) $\dfrac{R}{L}.$

69. 37 loops.

71. (a) 0.040 H;

   (b) 28 mA;

   (c) 16 μJ.

73. 2.4 mA, 0, 2.4 mA.

77. (a) $\dfrac{Q_0^2}{2C}\, e^{-Rt/L};$

   (b) $\dfrac{dU}{dt} = -I^2 R.$

79.

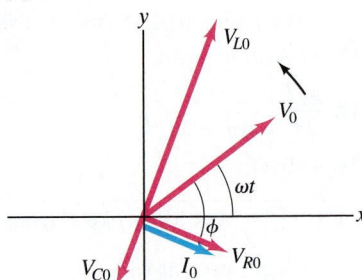

81. (a) 0;

   (b) 0, 90° out of phase.

83. 2.2 kHz.

85. 69 mH, 18 Ω.

89. (a)

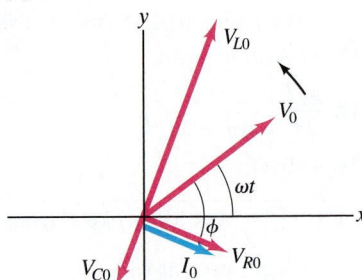

   (b) $\dfrac{V_0}{\sqrt{R^2 + \left(\omega L - \dfrac{1}{\omega C}\right)^2}}\sin(\omega t - \phi),$

   $\phi = \tan^{-1}\dfrac{\omega L - \dfrac{1}{\omega C}}{R}.$

91. (a) $\left(\dfrac{V_{20}}{\omega L - \dfrac{1}{\omega C}}\right)\sin\left(\omega t - \tfrac{1}{2}\pi\right)$;

(b) $\left(\dfrac{V_{20}}{\omega^2 LC - 1}\right)\sin(\omega t - \pi)$;

(c) $\dfrac{1}{\omega^2 LC}$;

(d) $V_{1\,\text{out}} = V_1$.

93. (a) $\dfrac{V_0}{R}\sin\omega t$;

(b) $\dfrac{V_0}{X_L}\sin\left(\omega t - \tfrac{1}{2}\pi\right)$;

(c) $\dfrac{V_0}{X_C}\sin\left(\omega t + \tfrac{1}{2}\pi\right)$

(d) $\dfrac{V_0}{R}\sqrt{1+\left(R\omega C - \dfrac{R}{\omega L}\right)^2}\sin(\omega t+\phi)$,

$\phi = \tan^{-1}\left(R\omega C - \dfrac{R}{\omega L}\right)$;

(e) $\dfrac{R}{\sqrt{1+\left(R\omega C - \dfrac{R}{\omega L}\right)^2}}$;

(f) $\dfrac{1}{\sqrt{1+\left(R\omega C - \dfrac{R}{\omega L}\right)^2}}$.

95. 0.14 H.

97. 54 mH, 22 Ω.

99. $\sqrt{6.0}\,f_0 = 2.4 f_0$.

101. (a) 7.1 kHz, V_{rms};

(b) 0.90.

103. (b) For $f \to 0\ A \to 1$; for $f \to \infty,\ A \to 0$;

(c) f is in s^{-1}:

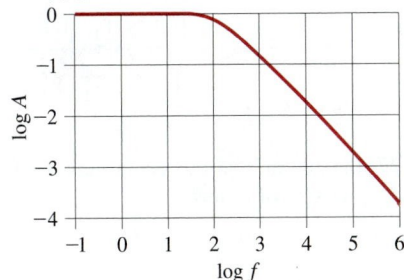

105.

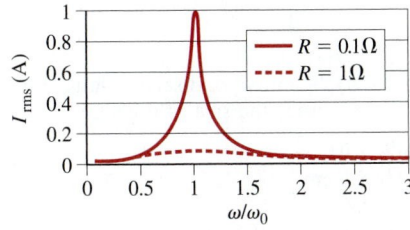

CHAPTER 31

1. 110 V/m·s.

3. 1.2×10^{15} V/m·s.

7. (b) $R \le R_0$: $\left(6.3 \times 10^{-11}\text{ T/m}\right)R$,

$R > R_0$: $\dfrac{5.7 \times 10^{-14}\text{ T·m}}{R}$,

R in meters;

(c)

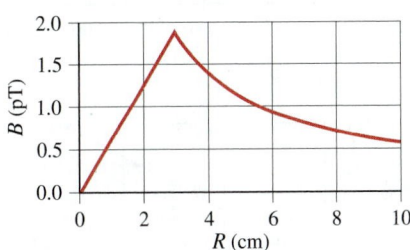

9. 0.14 pT.

11. (a) $-\hat{\mathbf{k}}$;

(b) $\dfrac{E_0}{c},\ -\hat{\mathbf{j}}$.

13. 2.00×10^{10} Hz.

15. 5.00×10^2 s = 8.33 min.

17. (a) 3.00×10^5 m;

(b) 34.1 cm;

(c) no.

19. (a) 261 s;

(b) 1260 s.

21. 3.4 krad/s.

23. 2.77×10^7 s.

25. 4.8 W/m^2, 42 V/m.

27. 4.50 μJ.

29. 3.80×10^{26} W.

31. (a) 5 cm^2, yes;

(b) 20 m^2, yes;

(c) 100 m^2, no.

33. (a) 2×10^8 ly;

(b) 2000 times larger.

35. 8×10^6 m/s^2.

37. 27 m^2.

39. 16 cm.

41. 3.5 nH to 5.3 nH.

43. $E_{\text{rms}} = 6.25 \times 10^{-4}$ V/m, 1.04×10^{-9} W/m^2.

45. 3 m.

47. 1.5 s.

49. 34 V/m, 0.11 μT.

51. Down, 2.2 μT, 650 V/m.

53. (a) 0.18 nJ;

(b) 8.7 μV/m, 2.9×10^{-14} T.

57. 4×10^{10} W.

59. 5 nodes, 6.1 cm.

61. (a) $+x$;

(b) $\beta = \alpha c$;

(c) $\dfrac{E_0}{c}\,e^{-(\alpha x - \beta t)^2}$.

63. (d) Both $\vec{\mathbf{E}}$ and $\vec{\mathbf{B}}$ rotate counterclockwise.

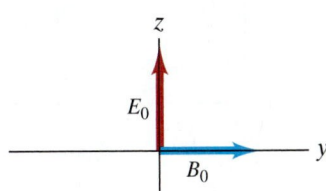

CHAPTER 32

1.

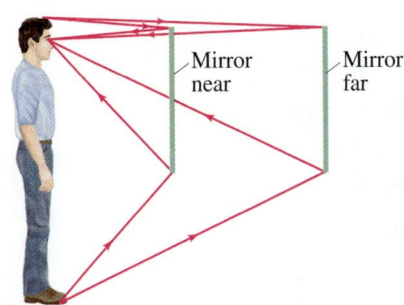

3. 7°.

7.

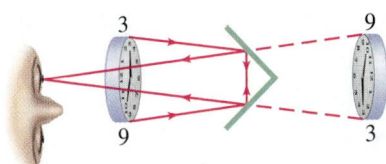

9. 37.6 cm.

11. 1.0 m.

13. 2.1 cm behind front of bulb, virtual, upright.

15. Concave, 5.3 cm.

17. 6.0 m.

19. Convex, 32.0 cm.

21.

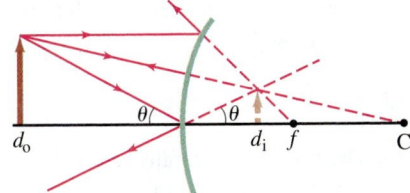

23. -3.9 m.

25. (a) Convex;

    (b) 20 cm behind mirror;

    (c) -91 cm;

    (d) 1.8 m.

27. (b)

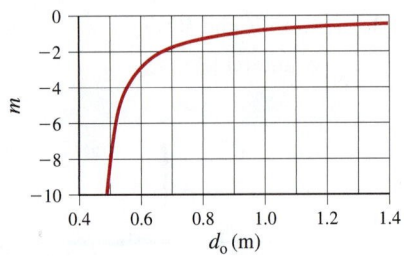

    (c) 0.90 m;

    (d) just beyond focal point.

31. Because the image is inverted.

33. (a) 2.21×10^8 m/s;

    (b) 1.99×10^8 m/s;

    (c) 1.97×10^8 m/s.

35. 8.33 min.

37. 3 m.

39. 35°.

41. 38.6°.

43. 2.6 cm.

45. 4.4 m.

47. 3.2 mm.

49. 38.9°.

53. 0.22°.

55. 0.80°.

57. 61.7°, lucite.

59. 82.1 cm.

61. $n \geq 1.5$.

63. (a) 2.3 μs;

    (b) 17 ns.

65. $n = 1.72$.

67. 17.25 cm.

71. 0.25 m, 0.50 m.

73. (a) 3.0 m, 4.4 m, 7.4 m;

    (b) toward, away, toward.

75. 3.80 m.

77. 31 cm for real image, 15 cm for virtual image.

83. $\dfrac{d}{n-1}$.

85. The light would totally internally reflect only if $\theta_i \leq 32.5°$.

87. $A = 1.5005$, $B = 5740$ nm^2.

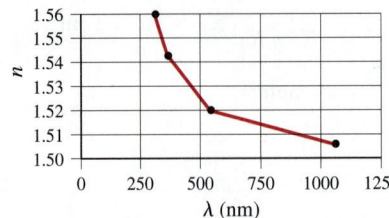

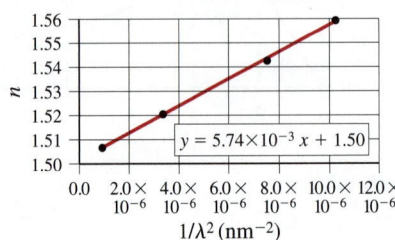

CHAPTER 33

1. (a)

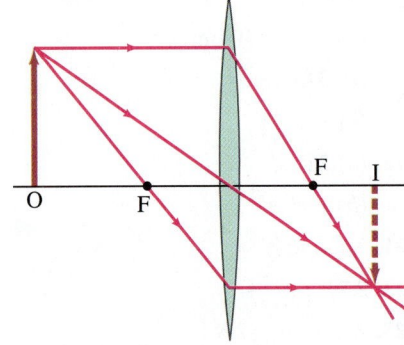

    (b) 508 mm.

3. (a) 4.26 D, converging;

    (b) -14.8 cm, diverging.

5. (a) 106 mm;

    (b) 109 mm;

    (c) 117 mm;

    (d) an object 0.513 m away.

7. (a) Virtual, upright, magnified;

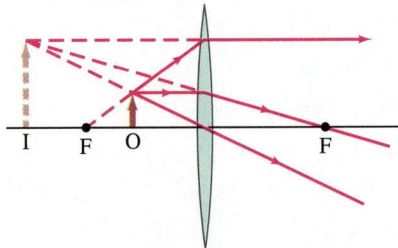

    (b) converging;

    (c) 6.7 D.

9. (a) 0.02 m;

    (b) 0.004 m.

11. 50 cm.

13. 21.3 cm, 64.7 cm.

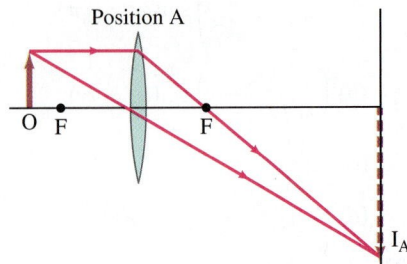

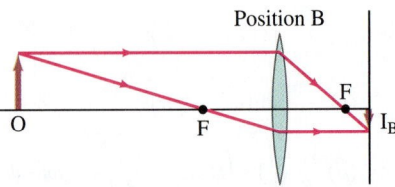

15. (c) Real, upright; (d) real, upright.

17. 0.107 m, 2.2 m.

19. (c) 182 cm; (d) 182 cm.

21. 18.5 cm beyond second lens, $-0.651\times$.

23. (a) 7.14 cm beyond second lens;

    (b) $-0.357\times$; (c)

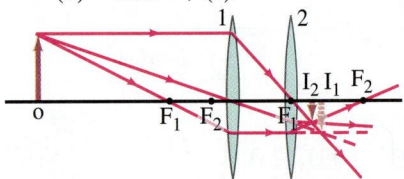

25. (a) 0.10 m to right of diverging lens; (b) $-1.0\times$;

    (c)

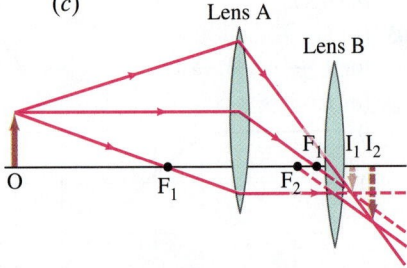

27. (a) 30 cm beyond second lens, half the size of object;

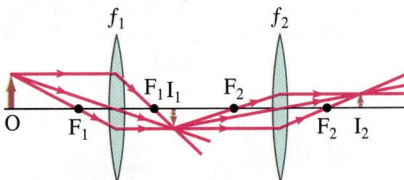

    (b) 29 cm beyond second lens, 0.46 times the size of object.

29. 1.54.

31. 8.6 cm.

33. 34 cm.

35. $f/2.8$.

37. $\frac{1}{6}$ s.

39. 41 mm.

41. $+2.5$ D.

43. 41 cm, yes.

45. (a) −1.3 D;
  (b) 37 cm.

47. −24.8 cm.

49. 18.4 cm, 1.00 m.

51. 6.6 cm.

53. (a) 13 cm;
  (b) 8 cm.

55. (a) −234 cm;
  (b) 4.27×.

57. (a) −66 cm;
  (b) 3.0×.

59. 4 cm, toward.

61. 2.5 cm, 91 cm.

63. −26×.

65. 16×.

67. 3.7 m, 7.4 m.

69. −9×.

71. 8.0×.

73. 1.6 cm.

75. (a) 754×;
  (b) 0.307 cm;
  (c) 0.312 cm.

77. (a) 0.85 cm;
  (b) 250×.

79. 410×, 25×.

81. 79.4 cm, 75.5 cm.

83. 6.450 m ≤ d_0 ≤ ∞.

85. 116 mm, 232 mm.

87. −19.0 cm.

89. 3.1 cm, 25 cm.

91. (a) 0.26 mm;
  (b) 0.47 mm;
  (c) 1.3 mm;
  (d) 0.56×, 2.7×.

93. 20.0 cm.

95. 47 m.

97. 2.8×, 3.9×, person with normal eye.

99. 1.0×.

101. +3.4 D.

103. −19×.

105. (a) 28.6 cm;
  (b) 120 cm;
  (c) 14.7 cm.

107. −6.2 cm.

109. (a) −1/f, 1;
  (b) 14 cm, yes,
    y-intercept = 1.03;

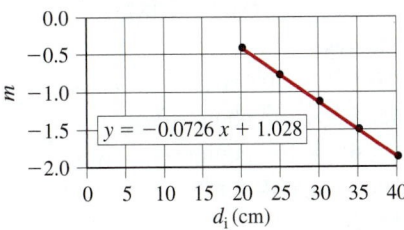

  (c) f = −1/slope.

CHAPTER 34

3. 3.9 μm.

5. 0.2 mm.

7. 390 nm.

9. 3.5 cm.

11. Inverted, starts with central dark line, and every place there was bright fringe before is now dark fringe and vice versa.

13. 2.7 mm.

15. 2.94 mm.

17. $\frac{1}{4}$.

21. $I_0 \left[\dfrac{3 + 2\sqrt{2} \cos\left(\dfrac{2\pi d \sin\theta}{\lambda}\right)}{3 + 2\sqrt{2}} \right]$.

23. 634 nm.

25. (a) 180 nm;
  (b) 361 nm, 541 nm.

27. (b) 290 nm.

29. 8.68 μm.

31. 113 nm, 225 nm.

35. 1.32.

37. (c) 570.

39. 0.191 mm.

41. 80.1 μm.

43. 0.3 mm.

45. (a) 2.5%;
  (b) 160 lamps.

47. (a) Constructive;
  (b) destructive.

49. 440 nm.

51. $I_0 \cos^2\left(\dfrac{2\pi x}{\lambda}\right)$.

53. (a) 81.5 nm;
  (b) 0.130 μm.

55. $\theta = \sin^{-1}\left(\sin\theta_i \pm \dfrac{m\lambda}{d}\right)$.

57. 340 nm, 170 nm.

59. Constructive: 90°, 270°; destructive: 0°, 180°; exactly switched.

61. 240 nm.

63. 0.20 km.

65. 126 nm.

CHAPTER 35

1. 37.3 mrad = 2.13°.

3. 2.35 m.

5. Entire pattern is shifted, with central maximum at 23° to the normal.

7. 4.8 cm.

9. 953 nm.

11. (a) 63°;
  (b) 30°.

---

13. 0.15.

15. $d = 5D$.

17. 265 fringes.

19. (a) 1.9 cm;
  (b) 12 cm.

21. 0.255.

23. (a) $I_\theta = I_0 \left(\dfrac{1 + 2\cos\delta}{3}\right)^2$.

25. 1.5×10^{11} m.

27. 1.0×10^4 m.

29. 730 lines/mm, 88 lines/mm.

31. 0.40 μm, 0.50 μm, 0.52 μm, 0.62 μm.

33. Two full orders, plus part of a third order.

35. 556 nm.

37. 24°.

39. $\lambda_2 > 600$ nm overlap with $\lambda_3 < 467$ nm.

41. $\lambda_1 = 614$ nm, $\lambda_2 = 9.00 \times 10^2$ nm.

43. 7 cm, 35 cm, second order.

45. (c) −32°, 0.9°.

47. (a) 16,000 and 32,000;
  (b) 26 pm, 13 pm.

49. 14.0°.

51. No.

53. 45°.

55. 61.2°.

57. (a) 35.3°;
  (b) 63.4°.

59. 36.9°, smaller than both angles.

61. $I = \dfrac{I_0}{4} \sin^2(2\theta)$, 45°.

63. 28.9 μm.

65. 580 nm.

67. 0.6 m.

69. 658 nm, 853 lines/cm.

71. (a) 18 km;
  (b) 23″, atmospheric distortions make it worse.

73. 5.79×10^5 lines/m.

75. 36.9°.

77. (a) 60°;
  (b) 71.6°;
  (c) 84.3°.

79. 0.4 m.

81. 0.245 nm.

83. 110 m.

85. −0.2 mm.

87. Use 24 polarizers, each rotated 3.75° from previous axis.

CHAPTER 36

1. 72.5 m.

3. 1.00, 1.00, 1.01, 1.02, 1.05, 1.09, 1.15, 1.25, 1.40, 1.67, 2.29, 7.09.

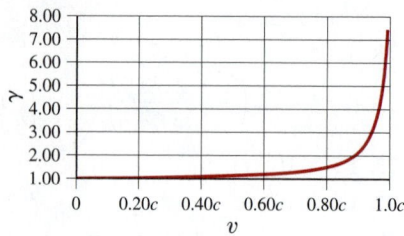

5. 2.42×10^8 m/s.

7. 27 y.

9. $(6.97 \times 10^{-8})\%$.

11. (a) $0.141c$;
    (b) $0.140c$.

13. (a) 3.4 y;
    (b) 7.4 y.

15. $0.894c$.

17. Base: 0.30ℓ, sides: 1.94ℓ.

19. $0.65c$.

21. (a) (650 m, 20 m, 0);
    (b) (1820 m, 20 m, 0).

23. (a) $0.88c$;
    (b) $0.88c$.

25. (a) $0.97c$;
    (b) $0.55c$.

27. $0.93c$ at $35°$.

29. (a) $\ell_0\sqrt{1 - \dfrac{v^2}{c^2}\cos^2\theta}$;

    (b) $\tan^{-1}\left[\dfrac{\tan\theta}{\sqrt{1 - \dfrac{v^2}{c^2}}}\right]$.

31. $t'_B - t'_A = -\dfrac{v\ell}{c^2\sqrt{1 - \dfrac{v^2}{c^2}}}$,

    B is turned on first.

33. Not possible in boy's frame of reference.

35. (a) -0.5%;
    (b) -20%.

37. $0.95c$.

39. 8.20×10^{-14} J, 0.511 MeV.

41. 900 kg.

43. $1.00\,\text{MeV}/c^2$, or 1.78×10^{-30} kg.

45. 9.0×10^{13} J, 9.2×10^9 kg.

47. $0.866c$.

49. 1670 MeV, $2440\,\text{MeV}/c$.

51. $0.470c$.

53. $0.32c$.

55. $0.866c$, $0.745c$.

57. (a) 2.5×10^{19} J;
    (b) -2.4%.

59. 237.04832 u.

61. 244 MeV.

65. 226 MHz.

67. (a) 1.00×10^2 km/h;
    (b) 67 Hz.

69. 75 μs.

71. 8.0×10^{-8} s.

73. (a) $0.72c$;
    (b) $1.5c$.

75. (a) $\tan^{-1}\sqrt{\dfrac{c^2}{v^2} - 1}$;

    (c) $\tan^{-1}\dfrac{c}{v}$, $u = \sqrt{c^2 + v^2}$.

77. (a) 0.77 m/s;
    (b) 0.21 m.

79. 1.022 MeV.

83. (a) 4×10^9 kg/s;
    (b) 4×10^7 y;
    (c) 1×10^{13} y.

85. 28.32 MeV.

87. (a) 2.86×10^{-18} kg·m/s;
    (b) 0;
    (c) 3.31×10^{-17} kg·m/s.

89. 3×10^7 kg.

91. $0.987c$.

93. 5.3×10^{21} J, 53 times as great.

95. (a) 6.5y;
    (b) 2.3 ly.

99.

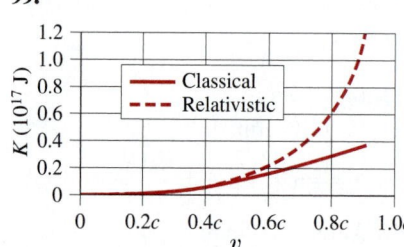

CHAPTER 37

1. (a) 10.6 μm, far infrared;
   (b) 829 nm, infrared;
   (c) 0.73 mm, microwave;
   (d) 1.06 mm, microwave.

3. 5.4×10^{-20} J, 0.34 eV.

5. (b) 6.62×10^{-34} J·s.

7. 2.7×10^{-19} J $< E < 4.5 \times 10^{-19}$ J,
   1.7 eV $< E <$ 2.8 eV.

9. 2×10^{13} Hz, 1×10^{-5} m.

11. 7.2×10^{14} Hz.

13. 3.05×10^{-27} m.

15. Copper and iron.

17. 0.55 eV.

19. 2.66 eV.

21. 3.56 eV.

23. (a) 1.66 eV;
    (b) 3.03 eV.

25. (a) 1.66 eV;
    (b) 3.03 eV.

27. 0.004, or 0.4%.

29. (a) 2.43 pm;
    (b) 1.32 fm.

31. (a) 8.8×10^{-6};
    (b) 0.049.

33. (a) 229 eV;
    (b) 0.165 nm.

35. 1.65 MeV.

37. 212 MeV, 5.88 fm.

39. 1.772 MeV, 702 fm.

41. 4.7 pm.

43. 4.0 pm.

45. 1840.

47. (a) 1.1×10^{-24} kg·m/s;
    (b) 1.2×10^6 m/s;
    (c) 4.2 V.

51. 590 m/s.

53. 20.9 pm.

55. 1.51 eV

57. 122 eV

59. 91.4 nm.

61. 37.0 nm.

63.

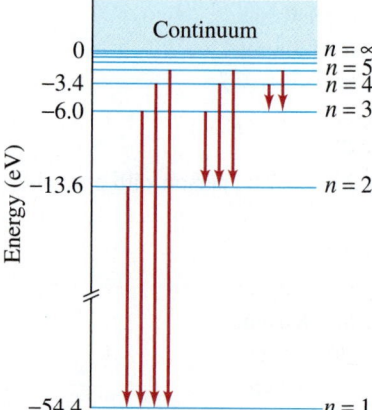

65. -27.2 eV, 13.6 eV.

67. Yes, $(7.30 \times 10^{-3})\,c$, 0.9999.

69. 97.23 nm, 102.6 nm, 121.5 nm, 486.2 nm, 656.3 nm, 1875 nm.

71. Yes.

73. 3.28×10^{15} Hz.

75. 5.3×10^{26} photons/s.

77. 6.2×10^{18} photons/s.

79. 0.244 MeV for both.

81. 28 fm.

83. 4.4×10^{-40}, yes.

85. 2.25 V.

87. 9.0 N.

89. 1.2 nm.

91. (*a*)

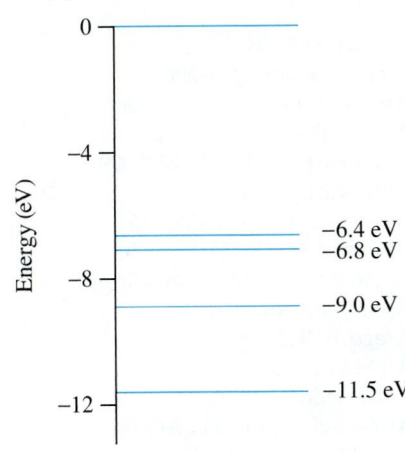

(*b*) Ground state, 0.4 eV, 2.2 eV, 2.5 eV, 2.6 eV, 4.7 eV, 5.1 eV.

93. $E_n = -\dfrac{2.84 \times 10^{165} \text{ J}}{n^2}$,

$r_n = n^2(5.17 \times 10^{-129} \text{ m})$, no.

95. 1.0×10^{-8} N.

97. (*a*) $\sqrt{\dfrac{Gh}{c^5}}$;

(*b*) 1.34×10^{-43} s;

(*c*) $\sqrt{\dfrac{Gh}{c^3}}$;

(*d*) 4.05×10^{-35} m.

99. (*a*) 6.0×10^{-3} m/s;

(*b*) 1.2×10^{-7} K.

101. (*a*)

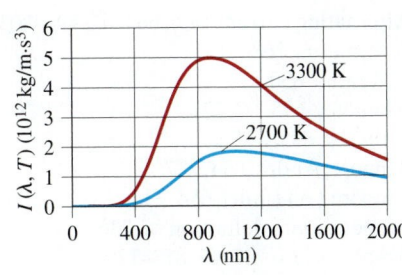

(*b*) 4.7 times more intense.

103. (*a*) $\dfrac{hc}{e}$, $-\dfrac{W_0}{e}$;

(*b*)

1.2×10^{-6} V·m, -2.31 V;

(*c*) 2.31 eV;

(*d*) 6.63×10^{-34} J·s.

Index

Note: The abbreviation *defn* means the page cited gives the definition of the term; *fn* means the reference is in a footnote; *pr* means it is found in a Problem or Question; *ff* means "also the following pages."

Aberration:
 chromatic, 889 *fn*, 892, 932
 of lenses, 891–92, 929, 931
 spherical, 843, 857, 891, 892, 932
Absolute pressure, 345
Absolute space, 953, 957
Absolute temperature scale, 457, 464
Absolute time, 953
Absolute zero, 464, 549
Absorption lines, 936, 1002
Absorption spectra, 936, 1002
Absorption wavelength, 1008
Ac circuits, 677 *fn*, 790–803
Ac generator, 766–67
Ac motor, 720
Accelerating reference frames, 85, 88, 155–56
Acceleration, 24–42, 60–62
 angular, 251–6, 258–63
 average, 24–26
 centripetal, 120 *ff*
 constant, 28–29, 62
 Coriolis, 301–2
 in g's, 37
 due to gravity, 34–35, 143
 instantaneous, 27–28, 60–61
 of the Moon, 140
 motion at constant, 28–39, 62–71
 radial, 120 *ff*, 128
 related to force, 86–88
 tangential, 128–29, 251–52
 uniform, 28–39, 62–71
 variable, 39–43
Accelerometer, 100
Accommodation of eye, 883
Accuracy, 3–5
 precision vs., 5
Achromatic doublet, 892
Action at a distance, 154, 568
Action potential, 670
Action-reaction (Newton's third law), 89–91
Activation energy, 481
Active solar heating, 550
Addition of vectors, 52–58
Addition of velocities:
 classical, 71–74
 relativistic, 970–71
Adhesion (*defn*), 360

Adiabatic lapse rate, 525 *pr*
Adiabatic processes, 508, 514–15
Air bags, 31
Air columns, vibrations of, 431 *ff*
Air conditioners, 537–38
Air parcel, 525 *pr*
Air pollution, 551
Air resistance, 34–35, 129–30
Airplane wing, 356–57
Airy disk, 929
Alternating current (ac), 664–65, 796–803
Alternators, 768
AM radio, 830
Ammeter, 695–97, 721
 digital, 695, 697
Ampère, André, 654, 737
Ampère (A) (unit), 654, 736
 operational definition of, 736
Ampère's law, 737–43, 813–17
Amplitude, 371, 397, 404
 intensity related to, 430
 pressure, 427
 of vibration, 371
 of wave, 371, 402, 404, 426, 430
Amplitude modulation (AM), 830
Analog information, 775
Analog meters, 695–97, 721
Analyzer (of polarized light), 941
Aneroid barometer, 347
Aneroid gauge, 347
Angle, 7 *fn*, 249
 Brewster's, 943
 critical, 854
 of dip, 709
 of incidence, 410, 415, 838, 850
 phase, 373, 405, 800
 polarizing, 943–44
 radian measure of, 249
 of reflection, 410, 838
 of refraction, 415, 850
 solid, 7 *fn*, 915 *fn*
Angstrom (unit), 17 *pr*, 852 *fn*
Angular acceleration, 251–56, 258–63
Angular displacement, 250, 381
Angular frequency, 373
Angular magnification, 886
Angular momentum, 285–89, 291–300
 in atoms, 1004
 conservation, law of, 285–89, 297–98
 directional nature of, 288–89, 291 *ff*
 of a particle, 291–92
 relation between torque and, 292–97
 vector, 288, 291
Angular position, 249
Angular quantities, 249 *ff*
 vector nature, 254

Angular velocity, 250–55
 of precession, 299–300
Annihilation of e⁻ and e⁺, 996
Anode, 620
Antenna, 812, 817, 824, 831, 909
Antilogarithm, A-3
Antinodes, 412, 433, 434, 435
Apparent weight, 148–49, 350
Apparent weightlessness, 148–49
Approximations, 9–12
Arago, F., 922
Arches, 327–28
Archimedes, 349–50
Archimedes' principle, 348–52
 and geology, 351
Area, 9, A-1, inside back cover
 under a curve or graph, 169–71
Arecibo, 931
Aristotle, 2, 84
Armature, 720, 766
Arteriosclerosis, 359
ASA number, 879 *fn*
Associative property, 54
Asteroids, 159 *pr*, 162 *pr*, 210 *pr*, 247 *pr*, 308 *pr*
Astigmatism, 884, 892, 892 *fn*
Astronomical telescope, 888–89
Atmosphere, scattering of light by, 945
Atmosphere (unit), 345
Atmospheric pressure, 344–48
 decrease with altitude, 344
Atom trap, 1013 *pr*, 1016 *pr*
Atomic emission spectra, 936, 1002
Atomic mass, 455
Atomic mass unit, 7, 455
Atomic spectra, 1001–3, 1006–8
Atomic structure:
 Bohr model of, 1003–9
 early models of, 1000–1
 nuclear model of, 1001
 planetary model of, 1001
Atomic theory of matter, 455–56, 559
Atomic weight, 455 *fn*
Atoms, 455–6, 468–9, 476–82, 486–90, 1000–10
 angular momentum in, 1004
 binding energy in, 1006
 Bohr model of, 1003–9
 and de Broglie's hypothesis, 1009–10
 distance between, 456
 electric charge in, 561
 energy levels in, 1003–9
 hydrogen, 1002–10
 ionization energy in, 1006–8
 (*see also* Atomic structure; Kinetic theory)

Attack angle, 356
Atwood's machine, 99, 279 *pr*, 294
Audible range, 425
Audiofrequency signal, 830
Aurora borealis, 717
Autofocusing camera, 426
Average acceleration, 24–26
Average speed, 20, 480–82
Average velocity, 20–22
Average velocity vector, 60
Avogadro, Amedeo, 468
Avogadro's hypothesis, 468
Avogadro's number, 468–69
Axial vector, 254 *fn*
Axis, instantaneous, 268
Axis of rotation (*defn*), 249
Axis of lens, 867
Axon, 669–70

Back, forces in, 337 *pr*
Back emf, 768–70
Bainbridge-type mass spectrometer, 724
Balance, 317–18
Balance a car wheel, 296
Ballistic galvanometer, 783 *pr*
Ballistic pendulum, 226
Balloons:
    helium, 467
    hot air, 454
Balmer, J. J., 1002
Balmer formula, 1002, 1007
Balmer series, 1002, 1007–8
Banking of curves, 126–27
Bar (unit), 345
Barometer, 347
Barrel distortion, 892
Base quantities, 7
Base units (*defn*), 7
Baseball, 82 *pr*, 163, 303 *pr*, 310 *pr*, 357
Baseball curve, and Bernoulli's principle, 357
Basketball, 82 *pr*, 105 *pr*
Battery, 609, 652–3, 655, 658, 678
    automobile, charging, 678 *fn*, 686–87
    chargers, inductive, 780 *pr*
Beam splitter, 914
Beams, 322, 323–6
Beat frequency, 438–39
Beats, 438–39
Bel (unit), 428
Bell, Alexander Graham, 428
Bernoulli, Daniel, 354
Bernoulli's equation, 354–58
Bernoulli's principle, 354–57
Betatron, 782 *pr*
Bicycle, 181 *pr*, 281 *pr*, 283 *pr*, 289, 295, 309 *pr*
Bimetallic-strip thermometer, 457
Binding energy:
    in atoms, 1006
    in molecules, 211 *pr*
Binoculars, 855, 889
Binomial expansion, A-1, inside back cover
Biological evolution, and entropy, 545

Biot, Jean Baptiste, 743
Biot-Savart law, 743–45
Bismuth-strontium-calcium-copper oxide
    (BSCCO), 669
Bits, 775
Blackbody, 988
Blackbody radiation, 987–88
Black holes, 155–56, 160 *pr*, 161 *pr*
Blood flow, 353, 357, 359, 361, 366 *pr*, 453 *pr*
Blood-flow measurement, electromagnetic,
    453 *pr*, 765
Blue sky, 945
Body fat, 368 *pr*
Bohr, Niels, 997, 1003–4, 1009
Bohr model of atom, 1003–9
Bohr radius, 1005
Boiling, 485 (*see also* Phase, changes of)
Boiling point, 457, 485, 503
Boltzmann, Ludwig, 546
Boltzmann constant, 468, 547
Bound charge, 641
Bow wave, 443–44
Boyle, Robert, 464
Boyle's law, 464, 477
Bragg, W. H., 939
Bragg, W. L., 939
Bragg equation, 939
Brahe, Tycho, 149
Brake, hydraulic, 346
Braking a car, 32, 174, 272–73
Brayton cycle, 557 *pr*
Breakdown voltage, 612
Breaking point, 319
Breaking the sound barrier, 444
Breath, molecules in, 469
Brewster, D., 944, 949 *pr*
Brewster's angle and law, 944, 949 *pr*
Bridge circuit, 704 *pr*
Bridges, 324–27, 335 *pr*, 386
British engineering system of units, 7
Broglie, Louis de, 997, 1009
Bronchoscope, 856
Brown, Robert, 455
Brownian motion, 455
Brunelleschi, Filippo, 328
Brushes, 720, 766
BSCCO, 669
Btu (unit), 497
Bulk modulus, 319, 321
Buoyancy, 348–52
    center of, 364 *pr*
Buoyant force, 348–49
Burglar alarms, 992

Cable television, 832
Calculator errors, 4
Calculator LCD display, 944
Caloric, 497
Calories (unit), 497
    relation to joule, 497
Calorimeter, 501
Calorimetry, 500–5

Camera, 878–82
    autofocusing, 426
Camera flash unit, 636
Candela (cd) (unit), 915
Cantilever, 315
Capacitance, 629–42
    of axon, 670
Capacitance bridge, 646 *pr*
Capacitive reactance, 798–99
Capacitor discharge, 690–91
Capacitor microphone, 699 *pr*
Capacitors, 628–42
    charging of, 813–15
    in circuits, 633–35, 687–92, 798–99
    energy stored in, 636–38
    as filters, 798–99
    reactance of, 798–99
    with *R* or *L*, 687–92, 793 *ff*
    in series and parallel, 633–35
    uses of, 799
Capacity, 629–42, 670
Capillaries, 353, 360
Capillarity, 359–60
Car:
    battery charging, 686–7
    power needs, 203
    stopping of, 32, 174, 272–73
Carnot, N. L. Sadi, 533
Carnot cycle, 533
Carnot efficiency, 534
    and second law of thermodynamics,
    534–35
Carnot engine, 533–35
Carnot's theorem, 535
Carrier frequency, 830
Caruso, Enrico, 386
Cassegrainian focus, 889
Cathedrals, 327
Cathode, 620
Cathode ray tube (CRT), 620–21, 723, 831
Cathode rays, 620, 721–22 (*see also* Electron)
Causal laws, 152
Causality, 152
Cavendish, Henry, 141, 144
CCD, 878
CDs, 44 *pr*, 45 *pr*, 920 *pr*, 935
CDMA cell phone, 832
Cell, electric, 653, 678
Cell phone, 771, 812, 824, 832
Celsius temperature scale, 457–58
Center of buoyancy, 364 *pr*
Center of gravity (CG), 232
Center of mass (CM), 230–36
    and angular momentum, 293
    and moment of inertia, 259, 264, 268–71
    and sport, 192, 193
    and statics, 313
    and translational motion, 234–36, 268–9
Centi- (prefix), 7
Centigrade scale, 457–58
Centiliter (cL) (unit), 7
Centimeter (cm) (unit), 7

Centipoise (cP) (unit), 358
Centrifugal (pseudo) force, 123, 300
Centrifugal pump, 361
Centrifugation, 122
Centripetal acceleration, 120
Centripetal force, 122–24
Cgs system of units, 7
Change of phase (or state), 482–86, 502–5
Charge, 506 *ff* (*see also* Electric charge)
Charge density, 596
Charge-coupled device (CCD), 878
Charging a battery, 678 *fn*, 686–87
Charles, Jacques, 464
Charles's law, 464
Chemical reactions, rate of, 481
Chimney, and Bernoulli effect, 357
Cholesterol, 359
Chord, 23, 250
Chromatic aberration, 889 *fn*, 892
Chromatography, 490
Circle of confusion, 880, 891
Circuit, electric (*see* Electric circuits)
Circuit breaker, 662–63, 694, 747, 776
Circular apertures, 929–31
Circular motion, 119–29
   nonuniform, 128–29
   uniform, 119–25
Circulating pump, 361
Classical physics (*defn*), 2, 952
Clausius, R. J. E., 529, 539
Clausius equation of state, 487
Clausius statement of second law of
      thermodynamics, 529, 537
Closed system (*defn*), 500
Closed tube, 434
Cloud color, 945
CM, 230–36 (*see* Center of mass)
CMOS, 647 *pr*, 878
Coating of lenses, optical, 913–14
Coaxial cable, 740, 789, 825
Coefficient:
   of kinetic friction, 113–14
   of linear expansion, 459–63
   of performance (COP), 537, 538
   of restitution, 243 *pr*
   of static friction, 113–14
   of viscosity, 358
   of volume expansion, 460, 461
Coherence, 906
Coherent light, 906
Cohesion (*defn*), 360
Coil (*see* Inductor)
Collision:
   completely inelastic, 225
   conservation of energy and
         momentum in, 217–19, 222–29
   elastic, 222–25
   and impulse, 220–21
   inelastic, 222, 225–27, 238
   nuclear, 225, 228–29
Colloids, 340
Colonoscope, 856

Color:
   in digital camera, 878
   of light related to frequency and
         wavelength, 852–4, 903, 906, 912
   of star, 988
Coma, 892
Common logarithms, A-2–A-3
Commutative property, 53, 167, 290
Commutator, 720
Compact disks, 44 *pr*, 45 *pr*, 920 *pr*, 935
Compass, magnetic, 707–8, 709
Complementarity, principle of, 997
Complementary metal oxide
      semiconductor (CMOS), 647 *pr*, 878
Complete circuit, 654
Completely inelastic collisions, 225
Complex wave, 408, 436
Components of vector, 55–59
Composite wave, 408, 436
Composition resistors, 657
Compound lenses, 892
Compound microscope, 890–91
Compounds, 455 *fn*
Compression (longitudinal wave), 398, 401
Compressive stress, 321
Compton, A. H., 994
Compton effect, 994–95, 996
Compton shift, 994
   derivation of, 995
Compton wavelength, 994
Computer:
   and digital information, 775
   disks, 775
   hard drive, 253, 775
   keyboard, 631
   memory, 644 *pr*
   monitor, 621, 943
   printers, 582–83
Computer chips, 16 *pr*
Concave mirror, 842, 846–48, 889
Concentration gradient, 489, 516 *fn*
Concrete, prestressed and reinforced, 323
Condensation, 484
Condenser microphone, 699 *pr*
Conduction:
   charging by, 562–63
   electrical, 561, 651–97
   of heat, 515–17
   in nervous system, 669–70
Conduction current (*defn*), 816
Conduction electrons, 561
Conductivity:
   electrical, 659, 668
   thermal, 515
Conductors:
   electric, 561, 577, 654 *ff*
   heat, 516
Cones, 882
Conical pendulum, 125
Conservation of energy, 189–201, 506–7
   in collisions, 222

Conservation laws, 163, 190
   of angular momentum, 285–89, 297–98
   of electric charge, 560
   of energy, 189–201, 506–7
   of mechanical energy, 189–95
   of momentum, 217–21
Conservative field, 775
Conservative forces, 184–85
Conserved quantity, 163, 190
Constants, values of: inside front cover
Constant-volume gas thermometer, 458
Constructive interference, 410–11, 437,
      904, 913, 914
Contact force, 84, 92, 95
Contact lens, 885
Continental drift, 351
Continuity, equation of, 353
Continuous wave, 397
Convection, 517
Conventional current (*defn*), 655
Conventions, sign (geometric optics),
      845–46, 849, 871
Converging lens, 866 *ff*
Conversion factors, 8, inside front cover
Converting units, 8–9
Convex mirror, 842, 848–49
Conveyor belt, 236–37, 244 *pr*
Coordinate axes, 19
Copier, electrostatic, 569, 582–83
Cord, tension in, 97
Coriolis acceleration, 301–2
Cornea, 883
Corona discharge, 612, 645 *pr*
Corrective lenses, 883–85
Correspondence principle, 980, 1009
Coulomb, Charles, 563
Coulomb (C) (unit), 564, 737
   operational definition of, 737
Coulomb's law, 563–67, 593–94, 600, 817
   vector form of, 567
Counter emf, 768–70
Counter torque, 769
Creativity in science, 2–3
Credit card swipe, 776
Crick, F., 939
Critical angle, 854
Critical damping, 383
Critical point, 483
Critical temperature, 483, 668
Cross product, vector, 289–91
Crossed Polaroids, 941–42
CRT, 620–21, 723, 831
Crystal lattice, 456
Crystallography, 939
Curie, P., 750
Curie temperature, 746, 750
Curie's law, 750
Curl, A-12
Current, electric (*see* Electric current)
Current density, 666–68
Current sensitivity, 695
Curvature of field, 892

Curvature of space, 155–56
Cycle (*defn*), 371
Cyclotron, 731 *pr*
Cyclotron frequency, 715

DAC, 706 *pr*
Damping and damped harmonic motion,
    382–85
Davisson, C. J., 998
dB (unit), 428–31
Dc (*defn*), 664
Dc circuits, 677–97
Dc generator, 767, 768
Dc motor, 720
de Broglie, Louis, 997, 1009
de Broglie wavelength, 997–98, 1009–10
    applied to atoms, 1009–10
Debye (unit), 617
Debye equation, 527 *pr*, 558 *pr*
Deceleration, 26
Decibels (dB) (unit), 428–31
Declination, magnetic, 709
Defects of the eye, 883–85, 892
Defibrillator, heart, 638, 692 *fn*
Definite integrals, 41, A-7
Degradation of energy, 545–46
Degrees of freedom, 512–13
Dehumidifier, 558 *pr*
Del operator, 618 *fn*, A-12
Demagnetization, 749
Demodulator, 831
Dendrites, 669
Density, 340–41
    and floating, 351
Depth of field, 880
Derivatives, 22–23, 27, A-6, inside back
        cover
    partial, 189, 406
Derived quantities, 7
Destructive interference, 410, 437, 904,
        913, 914
Detergents and surface tension, 360
Determinism, 152
Dew point, 486
Diamagnetism, 749–50
Diamond, 855
Dielectric constant, 638
Dielectric strength, 638
Dielectrics, 638–40
    molecular description of, 640–42
Diesel engine, 508, 527 *pr*, 553 *pr*
Differential equation, 372
Diffraction, 901, 921–39
    by circular opening, 929–30
    as distinguished from interference, 929
    in double-slit experiment, 921, 927–28,
        933, 947 *pr*
    of electrons, 998–99
    Fraunhofer, 922 *fn*
    Fresnel, 922 *fn*
    of light, 901, 921–39
    as limit to resolution, 929–33

by single slit, 922–27
    X-ray, 938–39
    of water waves, 416
Diffraction factor, 928
Diffraction grating, 933–35
    resolving power of, 937–38
Diffraction limit of lens resolution, 929–30
Diffraction patterns, 922
    of circular opening, 929
    of single slit, 922–27
    X-ray, 938–39
Diffraction spot or disk, 929–30
Diffuse reflection, 839
Diffusion, 489–90
    Fick's law of, 489
Diffusion constant, 489
Diffusion equation, 489
Diffusion time, 490
Digital artifact, 878
Digital camera, 878–82
Digital information, 775
Digital voltmeter, 695, 697
Digital-to-analog converter (DAC),
        706 *pr*
Dimensional analysis, 12–13, 16 *pr*, 134 *pr*,
        135 *pr*, 418 *pr*, 1015 *pr*, A-8
Dimensions, 12–13
Diopter (D) (unit), 868
Dip, angle of, 709
Dipole antenna, 817–18
Dipole layer, 669
Dipoles and dipole moments:
    electric, 576, 579–80, 617, 641
    magnetic, 718–19, 745
Direct current (dc), 664 (*see also* Electric
        current)
Discharge tube, 1002
Discovery in science, 721–23
Disorder and order, 544–45
Dispersion, 409, 853
Displacement, 20–21, 371, 380, 404
    angular, 250, 381
    resultant, 52–53
    vector, 20, 52–54, 59–60
    in vibrational motion, 371
    of wave, 404
Displacement current, 816
Dissipative forces, 196–98
    energy conservation with, 197–99
Distance:
    image, 840, 845
    object, 840, 845
    relativity of, 964–70
Distortion, by lenses, 892
Distributive property, 167, 290
Diver, 286
Divergence, A-12
Divergence theorem, A-12
Diverging lens, 867 *ff*
DNA, 581–82, 936, 939
Domains, magnetic, 746
Domes, 328

Door opener, automatic, 992
Doorbell, 747
Doppler, J. C., 439 *fn*
Doppler effect:
    for light, 443, 978–80
    for sound, 439–43
Doppler flow meter, 442, 453 *pr*
Dot (scalar) product, 167–68
Double-slit experiment (light), 903–6
    intensity in pattern, 906–9, 927–29
Drag force, 129–30, 356, 368 *pr*
DRAM, 644 *pr*, 647 *pr*
Drift velocity, 666–68, 723, 724
Dry cell, 653
Dry ice, 483
Duality, wave-particle, 997–99, 1009–10
Dulong and Petit value, 513
Dynamic lift, 356–57
Dynamic random access memory
        (DRAM), 644 *pr*, 647 *pr*
Dynamics, 19, 84 *ff*
    of rotational motion, 258 *ff*
Dynamo, 766–68
Dyne (unit), 87

Ear:
    discomfort, altitude, 367 *pr*
    response of, 431
Earth:
    as concentric shells, 142–43, A-9–A-11
    electric field of, 589 *pr*, 605 *pr*
    estimating radius of, 11, 15 *pr*
    as inertial frame, 85, 137 *pr*, 145–46
    magnetic field and magnetic poles of,
        709
    mass, radius, etc.: inside front cover
    mass determination, 144
    precession of axis, 303 *pr*
Earthquake waves, 401, 403
Eccentricity, 150
ECG, 609, 621
Echolocation, 400
Eddy currents (electric), 770
Eddy currents (fluids), 352
Edison, Thomas, 620
Effective values, 664–65
Efficiency, 203, 531, 534
    Carnot, 534
    and Otto cycle, 536
Einstein, Albert, 155, 455, 513, 952, 954,
        957–58, 961, 969, 989
EKG, 609, 621
Elapsed time, 20–21
Elastic collisions, 222–25
Elastic limit, 319
Elastic moduli, 319
    and speed of sound waves, 400
Elastic potential energy, 188 *ff*
Elastic region, 319
Elasticity, 318–22
El Capitan, 77 *pr*, 363 *pr*
Electric battery, 609, 652–53, 655, 658, 678
Electric car, 675 *pr*

Electric cell, 653, 678
Electric charge, 560 *ff*
  in atom, 561
  bound and free, 641
  conservation of, 560
  continuous charge distributions, 572–75
  and Coulomb's law, 563–67
  of electron, 564
  elementary, 564
  free, 641
  induced, 562–63, 641
  motion of, in electric field, 578–79
  motion of, in magnetic field, 714–17
  point (*defn*), 565
  quantization of, 564
  test, 568
  types of, 560
Electric circuits, 654–55, 662–65, 677–97, 790–803
  ac, 664–65, 677 *fn*, 796–803
  complete, 654
  containing capacitors, 633–35, 687–92, 798 *ff*
  dc, 677–97
  impedance matching of, 802–3
  and Kirchhoff's rules, 683–86
  LC, 793–96
  LR, 790–92
  LRC, 795–803
  open, 654
  parallel, 633, 663, 680
  RC, 687–92
  resonant, 802
  series, 634, 659
  time constants of, 688, 791
Electric conductivity, 659, 668
  in nervous system, 669–70
Electric current, 651, 654–58, 662–69, 683 *ff*
  alternating (ac), 664–65, 677 *fn*, 796–803
  conduction (*defn*), 816
  conventional, 655
  density, 666–68
  direct (dc) (*defn*), 664
  displacement, 816
  hazards of, 692–94
  leakage, 694
  magnetic force on, 710–19
  microscopic view of, 666–68
  and Ohm's law, 655–58
  peak, 664
  produced by magnetic field, 759–60
  produces magnetic field, 710–13, 746
  rms, 664–65
  (*see also* Electric circuits)
Electric dipole, 576, 579–80, 617, 641
Electric energy, 607–9, 619–20, 636–38, 660–62
  stored in capacitor, 636–38
  stored in electric field, 637–38
Electric energy resources, 550

Electric field, 568–83, 591–600, 610–12, 617–19, 775
  calculation of, 568–75, 595–600, 610–11, 617–19
  and conductors, 577, 655 *fn*
  continuous charge distributions, 572–75
  in dielectric, 639–40
  of and by dipole, 579–80
  of Earth, 589 *pr,* 605 *pr*
  in EM wave, 817–18
  energy stored in, 637–38
  and Gauss's law, 591–600
  inside a wire, 668
  motion of charged particle in, 578–79
  produced by changing magnetic field, 759–60, 773–75
  produces magnetic field, 813–16
  relation to electric potential, 610–12, 617–19
Electric field lines, 575–76
Electric flux, 592–93, 814
Electric force, 559, 563–67, 717
  Coulomb's law for, 563–67
  in molecular biology, 581–82
Electric generator, 766–68
Electric hazards, 692–94
Electric motor, 720
  counter emf in, 768–69
Electric plug, 693–94
Electric potential, 607–18
  of dipole, 617
  due to point charges, 612–15
  equipotential surfaces, 616–17
  relation to electric field, 610–12, 617–19
  (*see also* Potential difference)
Electric potential energy, 607–10, 619–20, 636–38
Electric power, 660–63
  in ac circuits, 665, 790, 792, 797, 798, 801, 802, 803
  generation, 766–8
  in household circuits, 662–63
  and impedance matching, 802–3
  transmission of, 770–73
Electric quadrupole, 589 *pr*
Electric shielding, 577, 740
Electric shock, 692–94
Electric stove burner, 660
Electric vehicle, 675 *pr*
Electricity, 559–836
Electricity, static, 559 *ff*
Electrocardiogram (ECG, EKG), 609, 621
Electrochemical series, 652
Electrode, 653
Electrolyte, 653
Electromagnet, 747
Electromagnetic force, 155, 717
Electromagnetic induction, 758 *ff*
Electromagnetic oscillations, 793–96, 802
Electromagnetic pumping, 726 *pr*
Electromagnetic spectrum, 823, 852–54
Electromagnetic (EM) waves, 817–32
  (*see also* Light)

Electrometer, 563
Electromotive force (emf), 678–79, 758–67, 768 (*see also* Emf)
Electron:
  as cathode rays, 620, 721
  charge on, 564, 722–23
  conduction, 561
  defined, 999
  discovery of, 721–23
  free, 561
  mass of, 723
  measurement of charge on, 723
  measurement of *e/m,* 722–23
  momentum of, 972
  motion of, in electric field, 578–79
  in pair production, 996
  path in magnetic field, 715
  photoelectron, 992
  speed of, 666–68
  spin, 746
  wavelength of, 998
Electron diffraction, 999
Electron gun, 621
Electron microscope, 987, 1000
Electron spin, 746
Electron volt (eV) (unit), 619–20
Electronic pacemakers, 692
Electroscope, 562–63, 652 *fn*
Electrostatic air cleaner, 645 *pr*
Electrostatic copier, 569, 582–83
Electrostatic force, 565, 581–82
  potential energy for, 607–8
Electrostatic potential energy, 619–20
Electrostatic unit (esu), 564 *fn*
Electrostatics, 560–642
Electroweak force, 155, 559 *fn*
Elementary charge, 564
Elements, 455 *fn*
  in compound lenses, 892
  periodic table of: inside back cover
Elevator and counterweight, 99
Ellipse, 150
EM waves, 817–32 (*see also* Light)
Emf, 678–79, 758–66, 767, 768
  back, 768–69
  counter, 768–69
  of generator, 766–69
  Hall, 723–24
  induced, 758–69, 789
  motional, 765–66
  RC circuit with, 689
  series and parallel, 686–87
  sources of, 678, 758–68
Emission spectra, 987–88, 1001–3, 1005–8
Emissivity, 518
Endoscopes, 856
Energy, 163, 172, 183–200, 505–7, 607 *ff*
  activation, 481
  binding, 985 *pr,* 1006
  conservation of, 189–201, 506–7
  degradation of, 545–46
  electric, 607–9, 619–20, 636–38, 660–63
  in EM waves, 826–27

Energy (*continued*)
  equipartition of, 512–13
  and first law of thermodynamics, 505–7
  geothermal, 550
  gravitational potential, 186–88, 191, 194–95, 199–201
  internal, 196, 498–99
  ionization, 1006, 1008
  kinetic, 172–73, 265–69, 974–6
  and mass, 974–78
  mechanical, 189–95
  molecular kinetic, 478–79
  nuclear, 530 *fn*, 550
  photon, 989–93
  potential, 186–89, 607–10, 619–20, 636–38 (*see also* Electric potential; Potential energy)
  quantization of, 989, 1003–9
  relation to work, 172–76, 186, 197–99, 265–67, 978
  relativistic, 974–8
  rest, 974–76
  rotational, 265–67 and *ff*, 499
  in simple harmonic motion, 377–78
  solar, 550
  thermal, 196, 498
  total binding, 985 *pr*
  transformation of, 196, 201
  translational kinetic, 172–74
  unavailability of, 545–46
  units of, 164, 173, 256
  vibrational, 377–8, 499
Energy density:
  in electric field, 638, 639
  in magnetic field, 790, 826
Energy levels, in atoms, 1003–9
Energy states, in atoms, 1003–9
Energy transfer, heat as, 497
Engine:
  diesel, 508, 527 *pr*, 553 *pr*
  internal combustion, 530–2, 535–6
  power, 202–3
  steam, 530
Entropy, 539–48
  and biological evolution, 545
  as order to disorder, 544–45
  and second law of thermodynamics, 541–48
  as a state variable, 540
  statistical interpretation, 546–48
  and time's arrow, 544
Equally tempered chromatic scale, 431
Equation of continuity, 353
Equation of motion, 372
Equation of state, 463
  Clausius, 487
  ideal gas, 466
  van der Waals, 486–87
Equilibrium (*defn*), 204–5, 311, 312–13, 317
  first condition for, 312
  neutral, 205, 317

  second condition for, 313
  stable, 205, 317
  thermal, 459
  unstable, 205, 317
Equilibrium position (vibrational motion), 370
Equilibrium state, 463
Equipartition of energy, 512–13
Equipotential lines, 616–17
Equipotential surface, 616–17
Equivalence, principle of, 155–56
Erg (unit), 164
Escape velocity, 201
Escher drawing, 206 *pr*
Estimated uncertainty, 3
Estimating, 9–12
Ether, 954–57
Evaporation, 484
  and latent heat, 505
Event, 958 *ff*
Everest, Mt., 6, 8, 144, 485
Evolution and entropy, 545
Exact differential, 506 *fn*
Excited state, of atom, 996, 1005 *ff*
Expansion, linear and volume, 318–21
Expansion, thermal, 459–62
Expansions, mathematical, A-1
Expansions, in waves, 398
Exponents, A-1, inside back cover
Exposure time, 879
Extension cord, 663
External force, 218, 234
Extraterrestrials, possible communication with, 834 *pr*
Eye:
  aberrations of, 892
  accommodation, 883
  defects of, 883–85, 892
  far and near points of, 883
  normal (*defn*), 883
  resolution of, 930, 932–33
  structure and function of, 882–85
Eyeglass lenses, 883–85
Eyepiece, 888

Fahrenheit temperature scale, 457–58
Falling objects, 34–39
Far field, 818
Far point of eye, 883
Farad (F) (unit of capacitance), 629
Faraday, Michael, 154, 568, 758–60
Faraday cage, 577
Faraday's law, 760–61, 773–74, 817
Farsightedness, 883, 884
Fermat's principle, 864 *pr*
Fermi, Enrico, 12, 997
Ferromagnetism and ferromagnetic materials, 708, 746–49
Fiber optics, 855–56
Fick's law of diffusion, 489
Fictitious (inertial) forces, 300–1

Field, 154
  conservative and nonconservative, 775
  electric, 568–83, 591–600, 610–12, 617–19, 775 (*see also* Electric field)
  gravitational, 154, 156, 576
  magnetic, 707–17, 733–50 (*see also* Magnetic field)
  vector, 575
Film speed, 879 *fn*
Filter circuit, 799, 810 *pr*, 811 *pr*
First law of motion, 84–85
First law of thermodynamics, 505–7
  applications, 507–11
  extended, 507
Fission, 550
Fitzgerald, G. F., 957
Flasher unit, 691
Flashlight, 659
Flip coil, 783 *pr*
Floating, 351
Flow of fluids, 352–61
Flow rate, 353
Fluid dynamics, 352–61
Fluids, 339–61 (*see also* Flow of fluids; Gases; Liquids; Pressure)
Fluorescent lightbulb ballast, 773
Flux:
  electric, 592–93, 814
  magnetic, 760, 773–75, 816
Flying buttresses, 327
Flywheel, 266, 281 *pr*
FM radio, 830–31, 831 *fn*
f-number, 879
Focal length:
  of lens, 867–8, 875, 876–7, 882
  of spherical mirror, 842–43, 848
Focal plane, 867
Focal point, 842–43, 848, 867–68
Focus, 843
Focusing, of camera, 879–80
Football kicks, 66, 69
Foot-candle (*defn*), 915 *fn*
Foot-pounds (unit), 164
Force, 83–102, 155, 184–85, 215, 234–5
  addition of, 95, 143
  buoyant, 348–49
  centrifugal (pseudo), 123, 300
  centripetal, 122–24
  conservative, 184–85
  contact, 84, 92, 95
  Coriolis, 301
  definition of, 87
  diagram, 95
  dissipative, 196, 197–98
  drag, 129–30
  electromagnetic, 155, 717
  electrostatic, 563–7
  electroweak, 155
  in equilibrium, 312–13
  exerted by inanimate object, 90
  external, 218, 234
  fictitious, 300–1

ce (continued)
  of friction, 85–87, 113–19
  of gravity, 84, 92–94, 140 ff, 155
  inertial, 300–1
  magnetic, 707, 710–19
  measurement of, 84
  in muscles and joints, 278 pr, 315,
    330 pr, 331 pr, 332 pr, 336 pr, 337 pr
  net, 85–88, 95 ff
  in Newton's laws, 83–102, 215, 218, 234–5
  nonconservative, 185
  normal, 92–94
  pseudoforce, 300–1
  relation of momentum to, 215, 235
  resistive, 129–30
  restoring, 170, 370
  strong nuclear, 155
  types of, in nature, 155, 559 fn
  units of, 87
  velocity-dependent, 129–30
  viscous, 358–59
  weak nuclear, 155
  (see also Electric force; Magnetic force)
Force diagrams, 95
Force pumps, 348, 361
Forced oscillations, 385–87
Fossil-fuel power plants, 550
Foucault, J., 902
Four-dimensional space-time, 967
Fourier analysis, 436
Fourier integral, 408
Fourier's theorem, 408
Fovea, 882
Fracture, 322–23
Frame of reference, 19, 85, 300–2, 952 ff
  inertial, 85, 88, 300, 952 ff
  noninertial, 85, 88, 156, 300–2, 952
Franklin, Benjamin, 560, 600
Franklin, Rosalind, 939
Fraunhofer diffraction, 922 fn
Free-body diagrams, 95–96, 102
Free charge, 641
Free electrons, 561
Free expansion, 510–11, 542, 548
Free fall, 34–39, 148
Freezing (see Phase, change of)
Freezing point, 457 fn, 503
Frequency, 121, 253, 371, 397
  angular, 373
  of audible sound, 425, 431
  beat, 438–39
  of circular motion, 121
  collision, 494 pr
  fundamental, 413, 432, 433–35
  infrasonic, 426
  of light, 823, 853, 854
  natural, 374, 385, 412
  resonant, 385, 412–13
  of rotation, 253
  ultrasonic, 426, 445
  of vibration, 371, 382, 412
  of wave, 397

Frequency modulation (FM), 830, 831 fn
Fresnel, A., 922
Fresnel diffraction, 922 fn
Friction, 85, 113–19
  coefficients of, 113–14
  helping us to walk, 90
  kinetic, 113 ff
  rolling, 113, 273–74
  static, 114, 270
Fringe shift, 956
Fringes, interference, 904–6
f-stop (defn), 879
Fulcrum, 313
Full-scale current sensitivity, 695
Fundamental constants: inside front cover
Fundamental frequency, 413, 432, 433–35
Fuse, 662–63

Galaxy, black hole, 160 pr
Galilean telescope, 887 fn, 889
Galilean transformation, 968–69
Galilean-Newtonian relativity, 952–54,
  968–69
Galileo, 2, 18, 34, 51, 62, 84–85, 346, 348,
  380, 457, 825, 839, 887, 887 fn, 952, 968
Galvani, Luigi, 652
Galvanometer, 695–96, 721, 783 pr
Gamow, George, 951
Gas constant, 466
Gas laws, 463–65
Gas vs. vapor, 483
Gas-discharge tube, 1002
Gases, 340, 463–90
  adiabatic expansion of, 514–15
  ideal, 463–70, 476–81
  kinetic theory of, 476–94
  molar specific heats for, 511–12
  real, 482–87
Gauge pressure, 345
Gauges, pressure, 347
Gauss, K. F., 591
Gauss (G) (unit), 712
Gauss's law, 591–600
  for magnetism, 816, 817
Gauss's theorem, A-12
Gay-Lussac, Joseph, 464
Gay-Lussac's law, 464, 468, 469
Geiger counter, 627 pr
General motion, 230, 267–74, 292–93
General theory of relativity, 155–56
Generator, electric, 766–68
  Van de Graaff, 607, 627 pr
Geometric optics, 838–91
Geometry, A-2
Geosynchronous satellite, 147
Geothermal energy, 550
Germer, L. H., 998
GFCI, 694, 776
Glasses, eye, 883–85
Global positioning satellite (GPS), 16 pr,
  160 pr, 964

Global System for Mobile
  Communication (GSM), 832
Global warming, 551
Golf putt, 48 pr
GPS, 16 pr, 160 pr, 964
Gradient:
  concentration, 489, 516 fn
  of electric potential, 618
  pressure, 359, 516 fn
  temperature, 516
  velocity, 358
Gradient operator (del), 618 fn
Gram (unit), 7
Grand unified theories (GUT), 155
Graphical analysis, 40–43
Grating, 933–38
Gravitation, universal law of, 139–43,
  199–201, 564
Gravitational constant (G), 141
Gravitational field, 154, 156, 576
Gravitational force, 84, 92–94, 140–43
  and ff, 155
  due to spherical mass distribution,
    142–43, A-9–A-11
Gravitational mass, 155–56
Gravitational potential, 609, 617
Gravitational potential energy, 186–88,
  199–201
  and escape velocity, 201
Gravitational slingshot effect, 246 pr
Gravity, 34–39, 92, 139 ff
  acceleration of, 34–39, 87 fn, 92, 143–45
  center of, 232
  force of, 84, 92–94, 139 ff
  free fall under, 34–39, 148
  specific, 341
Gravity anomalies, 144
Greek alphabet: inside front cover
Grimaldi, F., 901, 906
Ground fault, 776
Ground fault circuit interrupter (GFCI),
  694, 776
Ground state, of atom, 1005
Ground wire, 693, 694
Grounding, electrical, 562, 655
GSM, 832
GUT, 155
Gyration, radius of, 279 pr
Gyroscope, 299–300

Hair dryer, 665
Hale telescope, 889
Hall, E. H., 723
Hall effect, Hall emf, Hall field, Hall
  probe, 723–24
Halley's comet, 160 pr
Hard drive, 253
Harmonic motion:
  damped, 382–85
  forced, 386
  simple, 372–79
Harmonic wave, 405

Harmonics, 413, 432–35
Hazards, electric, 692–94
Headlights, 609, 661, 677
Hearing, 431
Heart, 361
    defibrillator, 638, 648 *pr*, 692
    pacemaker, 692, 787
Heartbeats, number of, 12
Heat, 196, 496–528
    calorimetry, 500–5
    compared to work, 505
    conduction, 515–17
    convection, 517
    distinguished from internal energy and
        temperature, 498
    as energy transfer, 497
    in first law of thermodynamics, 505–7
    of fusion, 502
    latent, 502–5
    mechanical equivalent of, 497
    radiation, 517–20
    of vaporization, 502
Heat capacity, 522 *pr* (*see also* Specific heat)
Heat conduction to skin, 525 *pr*
Heat death, 546
Heat engine, 529, 530–32
    Carnot, 533–35
    efficiency of, 531–32
    internal combustion, 530–32
    operating temperatures, 530
    steam, 530–31
    temperature difference, 531
Heat of fusion, 502
Heat of vaporization, 502
Heat pump, 536, 538–39
Heat reservoir, 508
Heat transfer, 515–20
    conduction, 515–17
    convection, 517
    radiation, 517–20
Heating element, 665
Heisenberg, W., 987
Helicopter drop, 51, 70
Helium I and II, 483
Helmholtz coils, 756 *pr*
Henry, Joseph, 758
Henry (H) (unit), 786
Hertz, Heinrich, 823
Hertz (Hz) (unit of frequency), 253, 371
High-pass filter, 799, 811 *pr*
Highway curves, banked and unbanked,
    126–27
Hooke, Robert, 318, 910 *fn*
Hooke's law, 170, 188, 318, 370
Horizontal (*defn*), 92 *fn*
Horizontal range (*defn*), 68
Horsepower, 202–3
Hot air balloons, 454
Hot wire, 693, 694
Household circuits, 662–63
HST, 930
Hubble, Edwin, 979

Hubble Space Telescope (HST), 930
Humidity, 485–86
Huygens, C., 901
Huygens' principle, 901–3
Hydraulic brake, 346
Hydraulic lift, 346
Hydraulic press, 364 *pr*
Hydrodynamics, 352
Hydroelectric power, 550
Hydrogen atom:
    Bohr theory of, 1003–9
    magnetic moment of, 719
    spectrum of, 936, 1002–3
Hydrogen bond, 581
Hydrogen-like atoms, 1004 *fn*, 1008, 1010
Hydrometer, 351
Hyperopia, 883
Hysteresis, 748–49
    hysteresis loop, 748

Ice skater, 284, 286, 309 *pr*
Ideal gas, 466, 476 *ff*
    kinetic theory of, 476–89
Ideal gas law, 465–66, 482
    internal energy of, 498–99
    in terms of molecules, 468–69
Ideal gas temperature scale, 469–70, 534
Ignition, automobile, 609, 772
Illuminance, 915
Image:
    formed by lens, 867 *ff*
    formed by plane mirror, 838–41
    formed by spherical mirror, 842–49, 889
    real, 840, 844, 869
    seeing, 847, 848, 869
    as tiny diffraction pattern, 929–30
    ultrasound, 445–46
    virtual, 840, 870
Image artifact, 878
Image distance, 840, 845, 857, 870–1
Imaging, medical, 445–46
Imbalance, rotational, 296–97
Impedance, 788, 800–3
Impedance matching, 802–3
Impulse, 220–21
Impulsive forces, 221
Inch (in.) (unit), 6
Incidence, angle of, 410, 415, 838, 850
Incident waves, 410, 415
Inclines, motion on, 101
Incoherent source of light, 906
Indefinite integrals, A-6–A-7
Index of refraction, 850
    dependence on wavelength
        (dispersion), 853
    in Snell's law, 851
Induced current, 759
Induced electric charge, 562–63, 641
Induced emf, 758–66, 789
    counter, 768–69
    in electric generator, 766–68
    in transformer, 770–73

Inductance, 786–89
    in ac circuits, 790–803
    of coaxial cable, 789
    mutual, 786–87
    self-, 788–89
Induction:
    charging by, 562–63
    electromagnetic, 758, 759–60
    Faraday's law of, 760–61, 773–74
Induction stove, 762
Inductive battery charger, 780 *pr*
Inductive reactance, 797
Inductor, 785
    in circuits, 790–803
    energy stored in, 790
    reactance of, 797
Inelastic collisions, 222, 225–29
Inertia, 85
    moment of, 258–60
Inertial forces, 300–1
Inertial mass, 155
Inertial reference frame, 85, 88, 137 *pr*,
        300, 952 *ff*
    Earth as, 85, 137 *pr*, 145–46
    equivalence of all, 952–53, 957
    transformations between, 968–71
Infrared (IR) radiation, 824, 852, 936
Infrasonic waves, 426
Initial conditions, 373
Inkjet printer, 583
In-phase waves, 411
Instantaneous axis, 268
Insulators:
    electrical, 561, 658
    thermal, 516
Integrals, 39–43, 169–70, A-6, A-7, A-12,
        A-13, inside back cover
    definite, A-7
    indefinite, A-6, A-7
    surface, A-13
    volume, A-12
Intensity, 402–3, 427 *ff*
    in interference and diffraction patterns,
        906–9, 924–28
    of light, 915
    of Poynting vector, 827
    of sound, 427–31
Interference, 410–11, 437–8, 903–14
    constructive, 410–11, 437, 904, 913, 914
    destructive, 410, 437, 904, 913, 914
    as distinguished from diffraction, 929
    of light waves, 903–14, 928–29
    of sound waves, 437–39
    by thin films, 909–14
    of water waves, 411
    of waves on a string, 410
Interference factor, 928
Interference fringes, 904–6, 956
Interference pattern:
    double-slit, 903–9
        including diffraction, 927–29
    multiple slit, 933–6

...rferometers, 914, 954–7
...ntermodulation distortion, 408 *fn*
Internal combustion engine, 530–31
Internal energy, 196, 498–99
    distinguished from heat and
        temperature, 498
    of an ideal gas, 498–99
Internal reflection, total, 854–56
Internal resistance, 678–79
Interpolation, A-3
Invariant quantity, 977
Inverse square law, 140 *ff*, 403, 429, 563–4
Ion (*defn*), 561
Ionization energy, 1006–8
IR radiation, 824, 852, 936
Irreversible process, 533
Iris, 882
ISO number, 879 *fn*
Isobaric processes, 508
Isochoric processes, 508
Isolated system, 218, 500
Isotherm, 507
Isothermal processes, 507–8
Isotopes, 725
    table of, A-14–A-17
Isovolumetric (isochoric) process, 508

Jars and lids, 461, 465
Jeans, J., 988
Jeweler's loupe, 887
Joints, 324
    method of, 325
Joule, James Prescott, 497
Joule (j) (unit), 164, 173, 256, 619, 620,
    661
    relation to calorie, 497
Jump start, 687
Junction rule, Kirchhoff's, 684 *ff*
Jupiter, moons of, 150, 151, 158 *pr*,
    159–60 *pr*, 825, 887

Karate blow, 221
Keck telescope, 889
Kelvin (K) (unit), 464
Kelvin temperature scale, 464, 548–49
Kelvin-Planck statement of the second
    law of thermodynamics, 532, 535
Kepler, Johannes, 149–50, 887 *fn*
Keplerian telescope, 887 *fn*, 888
Kepler's laws, 149–53, 298
Keyboard, computer, 631
Kilo- (prefix), 7
Kilocalorie (kcal) (unit), 497
Kilogram (kg) (unit), 6, 86, 87
Kilometer (km) (unit), 7
Kilowatt-hour (kWh) (unit), 661
Kinematics, 18–43, 51–74, 248–55
    for rotational motion, 248–55
    translational motion, 18–43, 51–74
    for uniform circular motion, 119–22
    vector kinematics, 59–74

Kinetic energy, 172–75, 189 *ff*, 265–69,
    974–6
    of CM, 268–9
    in collisions, 222–23, 225–26
    and electric potential energy, 608
    of gas atoms and molecules, 478–9,
        498–9, 512–13
    molecular, relation to temperature,
        478–9, 498–9, 512–13
    of photon, 993
    relativistic, 974–78
    rotational, 265–69
    translational, 172–73
Kinetic friction, 113 *ff*
    coefficient of, 113
Kinetic theory, 455, 476–90
    basic postulates, 477
    boiling, 485
    diffusion, 489–90
    evaporation, 484
    ideal gas law, 476–80
    kinetic energy near absolute zero, 480
    of latent heat, 505
    mean free path, 487–88
    molecular speeds, distribution of,
        480–82
    van der Waals equation of state,
        486–87
Kirchhoff, G. R., 683
Kirchhoff's rules, 683–86, 816 *fn*

Ladder, forces on, 317, 338 *pr*
Lagrange, Joseph-Louis, 153
Lagrange Point, 153
Laminar flow, 352
Land, Edwin, 940
Laser printer, 583
Latent heats, 502–5
Lateral magnification, 845–46, 871
Lattice structure, 456
Laue, Max von, 939
Law (*defn*), 3
LC circuit, 793–96
LC oscillation, 793–96
LCD, 831, 878 *fn*, 943–44
Leakage current, 694
Length:
    focal, 843, 867, 868, 875, 877
    proper, 965
    relativity of, 964–70
    standard of, 6, 914
Length contraction, 964–67, 970
Lens, 866–92
    achromatic, 892
    coating of, 913–14
    color-corrected, 892
    combination of, 874–75
    compound, 892
    contact, 885
    converging, 866 *ff*
    corrective, 883–85
    cylindrical, 884
    diverging, 867 *ff*

of eye, 883
eyeglass, 883–85
eyepiece, 888
focal length of, 867, 868, 875, 877
    magnetic, 1000
magnification of, 871
negative, 871
normal, 882
objective, 888, 890
ocular, 890
positive, 871
power of (diopters), 868
resolution of, 881, 929–32
spherical, 858
telephoto, 882
thin (*defn*), 867
wide-angle, 882, 892
zoom, 882
Lens aberrations, 891–92, 929, 931
Lens elements, 892
Lensmaker's equation, 876–77
Lenz's law, 761–64
Level, sound, 428–30
Level range formula, 68–69
Lever, 177 *pr*, 313
Lever arm, 256
Lift, dynamic, 356–57
Light, 823, 825–26, 837–946
    coherent sources of, 906
    color of, and wavelength, 852–54, 903,
        906, 912
    dispersion of, 853
    Doppler shift for, 978–80
    as electromagnetic wave, 823–26
    frequencies of, 823, 853, 854
    incoherent sources of, 906
    infrared (IR), 823, 824, 852, 936, 948 *pr*
    intensity of, 915
    monochromatic (*defn*), 903
    as particles, 902, 989–97
    photon (particle) theory of, 989–97
    polarized, 940–43, 949 *pr*
    ray model of, 838 *ff*, 867 *ff*
    scattering, 945
    from sky, 945
    speed of, 6, 822, 825–26, 850, 902, 953,
        957, 975
    ultraviolet (UV), 823, 824, 852
    unpolarized (*defn*), 940
    velocity of, 6, 822, 825–26, 850, 902,
        953, 957, 975
    visible, 823, 852–54
    wave theory of, 900–45
    wavelengths of, 823, 852–4, 903, 906,
        912
    wave-particle duality of, 997
    white, 852–53
    (*see also* Diffraction; Intensity;
        Interference; Reflection; Refraction)
Light meter (photographic), 992
Light pipe, 855
Light rays, 838 *ff*, 867 *ff*

Lightbulb, 651, 653, 656, 657, 660, 704 *pr*, 773, 915, 991
Light-gathering power, 889
Lightning, 425, 662
Lightning rod, 612
Light-year (unit), 15 *pr*
Line integral, 169
Line spectrum, 935, 936, 1002 *ff*
Line voltage, 665
Linear expansion (thermal), 459–61
    coefficient of, 459–60
Linear momentum, 214–35
Linear waves, 402
Linearly polarized light, 940 *ff*
Lines of force, 575–76, 708
Liquefaction, 463–66, 476, 482
Liquid crystal, 340, 483, 943–44
Liquid crystal display (LCD), 878 *fn*, 943–44
Liquid-drop model, 625 *pr*
Liquid-in-glass thermometer, 457
Liquids, 340 *ff*, 455–56 (*see also* Phase, changes of)
Lloyd's mirror, 919 *pr*
Logarithms, A-2–A-3, inside back cover
Log table, A-3
Longitudinal waves, 398 *ff*
    and earthquakes, 401
    velocity of, 400–1
    (*see also* Sound waves)
Loop rule, Kirchhoff's, 684 *ff*
Lorentz, H. A., 957
Lorentz equation, 717
Lorentz transformation, 969–71
Loudness, 425, 427, 429 (*see also* Intensity)
Loudness control, 431
Loudness level, 431
Loudspeakers, 375, 428–29, 720–21, 799
    concert time delay, 452 *pr*
Low-pass filter, 799, 811 *pr*
LR circuit, 790–92
LRC circuit, 795–96, 799–801
Lumen (lm) (unit), 915
Luminous flux, 915
Luminous intensity, 915
Lyman series, 1002–3, 1006, 1007

Mach, E., 443 *fn*
Mach number, 443
Macroscopic description of a system, 454, 456
Macroscopic properties, 454, 456
Macrostate of system, 546–47
Magnet, 707–9, 746–47
    domains of, 746
    electro-, 747
    permanent, 746
    superconducting, 747
Magnetic circuit breakers, 747
Magnetic damping, 778 *pr*
Magnetic declination, 709

Magnetic deflection coils, 621
Magnetic dipoles and magnetic dipole moments, 718–19, 745
Magnetic domains, 746
Magnetic field, 707–17, 733–50
    of circular loop, 744–45
    definition of, 708
    determination of, 712–13, 738–45
    direction of, 708, 710, 716
    of Earth, 709
    energy stored in, 790
    hysteresis, 748–49
    induces emf, 759–73
    motion of charged particle in, 714–17
    produced by changing electric field, 813–16
    produced by electric current, 710, 741–2, 743–6 (*see also* Ampère's law)
    produces electric field and current, 773–75
    of solenoid, 741–42
    sources of, 733–51
    of straight wire, 711–12, 734–35
    of toroid, 742
    uniform, 709
Magnetic field lines, 708
Magnetic flux, 760 *ff*, 773–75, 816, 820
Magnetic force, 707, 710–19
    on electric current, 710–14, 718–19
    on moving electric charges, 714–17
Magnetic induction, 710 (*see also* Magnetic field)
Magnetic lens, 1000
Magnetic moment, 718–19, 745
Magnetic monopole, 708
Magnetic permeability, 734, 748
Magnetic poles, 707–9
    of Earth, 709
Magnetic susceptibility (*defn*), 749
Magnetic tape and disks, 775
Magnetism, 707–90
Magnetization vector, 750
Magnification:
    angular, 886
    lateral, 845–46, 871
    of lens, 871
    of lens combination, 874–75
    of magnifying glass, 885–87
    of microscope, 890–91, 932, 933, 1000
    of mirror, 845
    sign conventions for, 845–46, 849, 871
    of telescope, 888, 931
    useful, 932–3, 1000
Magnifier, simple, 866, 885–87
Magnifying glass, 866, 885–87
Magnifying mirror, 848
Magnifying power, 886 (*see also* Magnification)
Manometer, 346
Marconi, Guglielmo, 829
Mars, 150, 151

Mass, 6, 86–88, 155
    atomic, 455
    center of, 230–33
    gravitational vs. inertial, 155
    molecular, 455, 465
    of photon, 993
    precise definition of, 88
    in relativity theory, 974
    rest, 974
    standard of, 6–7
    table of, 7
    units of, 6–7, 87
    variable, systems of, 236–38
Mass spectrometer (spectrograph), 724–25
Mass-energy transformation, 974–78
Mathematical expansions, A-1
Mathematical signs and symbols: inside front cover
Matter:
    states of, 340, 455–56
    wave nature of, 997–99
Maxwell distribution of molecular speeds, 480–82, 547
Maxwell, James Clerk, 480, 813, 817, 819–20, 822, 823, 953–4
Maxwell's equations, 813, 817, 819–22, 911 *fn*, 951, 953, 954, 958, 969
    differential form of, A-12–A-13
    in free space, A-13
Maxwell's preferred reference frame, 953–4
Mean free path, 487–88
Measurements, 3–5
Mechanical advantage, 100, 313, 346
Mechanical energy, 189–95
Mechanical equivalent of heat, 497
Mechanical oscillations, 369
Mechanical waves, 395
Mechanics, 18–445 (*see also* Motion)
    definition, 19
Medical imaging, 445–46
Melting point, 503–5 (*see also* Phase, changes of)
Mercury barometer, 347
Mercury-in-glass thermometer, 457–58
Metal detector, 770
Meter (m) (unit), 6
Meters, 695–97, 721
    correction for resistance of, 697
Metric (SI) multipliers: inside front cover
Metric (SI) system, 7
Mho (unit), 675 *pr*
Michelson, A. A., 826, 914, 954–57
Michelson interferometer, 914, 954–57
Michelson-Morley experiment, 954–57
Microampere (μA) (unit), 654
Micrometer, 10–11
Microphones:
    capacitor, 699 *pr*
    magnetic, 775

.oscope, 890–91, 931–33
compound, 890–91
electron, 1000
magnification of, 890–91, 932, 933, 1000
resolving power of, 932
useful magnification, 932–33, 1000
Microscopic description of a system, 454, 456, 476 *ff*
Microscopic properties, 454, 456, 476 *ff*
Microstate of a system, 546
Microwaves, 824
Milliampere (mA) (unit), 654
Millikan, R. A., 723, 991
Millikan oil-drop experiment, 723
Millimeter (mm) (unit), 7
Mirage, 903
Mirror equation, 845–49
Mirrors, 838–49
aberrations of, 889 *fn*, 891–92
concave, 842–49, 889
convex, 842, 848–49
focal length of, 842–43, 848
Lloyd's, 919 *pr*
plane, 838–42
used in telescope, 889
Missing orders, 948 *pr*
Mr Tompkins in Wonderland (Gamow), 951, 982
MKS (meter-kilogram-second) system (*defn*), 7
mm-Hg (unit), 346
Models, 2–3
Modern physics (*defn*), 2, 952
Modulation:
amplitude, 830
frequency, 830, 831 *fn*
Moduli of elasticity, 319
Molar specific heat, 511–13
Mole (mol) (unit), 465
volume of, for ideal gas, 465
Molecular biology, electric force in, 581–82
Molecular kinetic energy, 478–79, 498–99, 512–13
Molecular mass, 455, 465
Molecular speeds, 480–82
Molecular weight, 455 *fn*
Molecules, 455 *fn*
polar, 561, 579
Moment arm, 256
Moment of a force about an axis, 256
Moment of inertia, 258–60
determining, 263–65, 382
parallel-axis theorem, 264–65
perpendicular-axis theorem, 265
Momentum, 214–38
angular, 285–89, 291–300, 1003
center of mass (CM), 230–33
in collisions, 217–29
conservation of angular, 285–87, 297–98
conservation of linear, 217 *ff*
linear, 214–38
of photon, 993

relation of force to, 215–16, 218, 220–21, 235, 236, 972, 974
relativistic, 971–73, 977, 978
Monochromatic aberration, 892
Monochromatic light (*defn*), 903
Moon:
centripetal acceleration of, 121, 140
force on, 140, 142
work on, 167
Morley, E. W., 954–57
Motion, 18–300, 951–80
of charged particle in electric field, 578–79
circular, 119–29
at constant acceleration, 28–29
damped, 382–85
description of (kinematics), 18–43, 51–74
in free fall, 34–39, 148
harmonic, 372–77, 382–85
Kepler's laws of planetary, 149–53, 298
linear, 18–43
Newton's laws of, 84–91, 95–96, 112 *ff*, 215, 218, 234, 235, 259–63, 292–93, 972
nonuniform circular, 128–29
oscillatory, 369 *ff*
periodic (*defn*), 370
projectile, 51, 62–71
rectilinear, 18–43
and reference frames, 19
relative, 71–74, 951–80
rolling, 267–73
rotational, 248–302
simple harmonic (SHM), 372–77
translational, 18–239
uniform circular, 119–25
uniformly accelerated, 28–39
at variable acceleration, 39–43
vibrational, 369 *ff*
of waves, 395–416
Motion sensor, 448 *pr*
Motional emf, 765–66
Motor, electric, 720
back emf in, 768–69
Mountaineering, 106 *pr*, 110 *pr*, 137 *pr*, 182 *pr*
Mt. Everest, 6, 8, 144, 161 *pr*, 364 *pr*, 485
MP3 player, 677
Multimeter, 696
Multiplication of vectors, 55, 167–68, 289–91
Muscles and joints, forces in, 278 *pr*, 315, 330 *pr*, 331 *pr*, 332 *pr*, 336 *pr*, 337 *pr*
Musical instruments, 413, 422 *pr*, 424, 431–36
Musical scale, 431
Mutual inductance, 786–87
Myopia, 883

Natural frequency, 374, 385, 412 (*see also* Resonant frequency)
Natural logarithms, A-2
Near field, 818
Near point, of eye, 883

Nearsightedness, 883, 884–85
Negative, photographic, 878 *fn*
Negative electric charge (*defn*), 560
Negative lens, 871
Neptune, 150, 152
Nerve pulse, 669–70, 715
Nervous system, electrical conduction in, 669–70
Net force, 85–88, 95 *ff*
Net resistance, 679
Neuron, 669
Neutral equilibrium, 205, 317
Neutral wire, 694
Neutron, 561
Neutron star, 287
Newton, Isaac, 18, 85–86, 89, 139–40, 155, 568, 889 *fn*, 902, 910 *fn*, 952
Newton (N) (unit), 87
Newtonian focus, 889
Newtonian mechanics, 83–156
Newton's first law of motion, 84–85
Newton's law of universal gravitation, 139, 140–43, 199–201
Newton's laws of motion, 84–91, 95–96, 112 *ff*, 215, 218, 234–35, 259–63, 292–93, 972
Newton's rings, 910–11
Newton's second law, 86–88, 90, 95–96, 215, 218, 234–35, 953, 972
for rotation, 259–63, 292–93
for a system of particles, 234–35, 292–93
Newton's synthesis, 152
Newton's third law of motion, 89–91
Nodes, 412, 433, 434, 435
Nonconductors, 561, 638–42, 658
Nonconservative field, 775
Nonconservative forces, 185
Noninductive winding, 788
Noninertial reference frames, 85, 88, 156, 300–2
Nonohmic device, 656
Nonreflecting glass, 913–14
Nonuniform circular motion, 128–29
Normal eye (*defn*), 883
Normal force, 92–94
Normal lens, 882
North pole, Earth, 709
North pole, of magnet, 708
Nuclear collision, 225, 227–29
Nuclear decay, 976
Nuclear energy, 530 *fn*, 550
Nuclear force, 155, 212 *pr*
Nucleus, liquid-drop model of, 625 *pr*
Null result, 952–55, 957
Numerical integration, 40–43

Object distance, 840, 845, 857, 870–71
Objective lens, 890, 932
Observations, 2, 952
Oersted, H. C., 710
Off-axis astigmatism, 892

Ohm, G. S., 655
Ohm (Ω) (unit), 656
Ohmmeter, 696, 721
Ohm's law, 655–58, 668, 680, 685
Oil-drop experiment, 723
One-dimensional wave equation, 407
Onnes, H. K., 668
Open circuit, 654
Open system, 500
Open tube, 434
Open-tube manometer, 346–47
Operating temperatures, heat engines, 530
Operational definitions, 7, 737
Optical coating, 913–14
Optical illusion, 851, 903
Optical instruments, 878–92, 914, 929–38
Optical sound track, 992
Optical tweezers, 105 pr, 829
Optical zoom, 882
Optics:
    geometric, 838–91
    physical, 900–45
Order and disorder, 544–45
Order of interference or diffraction
        pattern, 904–6, 933–34, 936, 939,
        948 pr
Order-of-magnitude estimate, 9–12, 102
Organ pipe, 435
Oscillations, 369–89
    of air columns, 434–6
    damped harmonic motion, 382–85
    displacement, 371
    forced, 385–87
    mechanical, 369
    of molecules, 512–13
    of physical pendulum, 381–82
    simple harmonic motion (SHM), 372–77
    as source of waves, 397
    of a spring, 370–71
    on strings, 412–14, 431–3
    of torsion pendulum, 382
Oscilloscope, 620, 621
Osteoporosis, diagnosis of, 995
Otto cycle, 535–36
Out-of-phase waves, 411
Overdamped system, 383
Overexposure, 879
Overtones, 413, 432, 433

Pacemaker, heart, 692, 787
Page thickness, 10–11
Pair production, 996
Pantheon, dome of, 328
Parabola, 51, 71, 326
Parabolic mirror, 843
Parallel-axis theorem, 264–65
Parallel circuits, 633, 663, 680
Parallelogram method of adding vectors, 54
Paramagnetism, 749–50
Paraxial rays (defn), 843
Partial derivatives, 189, 406
Partial pressure, 485–86

Partially polarized, 945
Particle (defn), 19
Particulate pollution, 15 pr
Pascal, Blaise, 341, 346, 363 pr
Pascal (Pa) (unit of pressure), 341
Pascal's principle, 346
Paschen series, 1003, 1006, 1007
Passive solar heating, 550
PDA, 647 pr
Peak current, 664
Peak voltage, 664
Peak widths, of diffraction grating, 937–38
Peaks, tallest, 8
Pendulum:
    ballistic, 226
    conical, 125
    physical, 381–82
    simple, 13, 195, 379–81
    torsion, 382
Pendulum clock, 380
Percent uncertainty, 3–4, 5
    and significant figures, 5
Performance, coefficient of (COP), 537, 538
Perfume atomizer, 356
Period, 121, 253, 371, 397
    of circular motion, 121
    of pendulums, 13, 380, A-8
    of planets, 150–51
    of rotation, 253–54
    of vibration, 371
    of wave, 397
Periodic motion, 370 ff
Periodic table: inside back cover
Periodic wave, 397
Permeability, magnetic, 734, 748
Permittivity, 565, 639
Perpendicular-axis theorem, 265
Personal digital assistant (PDA), 647 pr
Perturbations, 152
Phase:
    in ac circuit, 796–802
    changes of, 482–83, 502–5
    of matter, 340, 456
    of waves, 404, 411
Phase angle, 373, 405, 800
Phase diagram, 483
Phase shift, 911, 913, 914
Phase transitions, 482–83, 502–5
Phase velocity, 404
Phasor diagram:
    ac circuits, 800
    interference and diffraction of light,
        907, 925, 937
Phon (unit), 431
Photocell, 626 pr, 990
Photocell circuit, 990, 992
Photoconductivity, 582
Photocopier, 569, 582–83
Photodiode, 992
Photoelectric effect, 989–92, 996
Photoelectron, 992
Photographic film, 878, 879

Photon, 989–97
    energy of, 993
    mass of, 993
    momentum of, 993
Photon interactions, 996
Photon theory of light, 989–97
Photosynthesis, 993
Photovoltaic (solar) cells, 550
Physical pendulum, 381–82
Physics:
    classical (defn), 2, 952
    modern (defn), 2, 952
Piano tuner, 12
Pin, structural, 323
Pincushion distortion, 892
Pipe, vibrating air columns in, 431 ff
Pitch of a sound, 425
Pixel, 878, 881, 943–4
Planck, Max, 989
Planck length, 13
Planck time, 16 pr, 1015 pr
Planck's constant, 989
Planck's quantum hypothesis, 988–89
Plane geometry, A-2
Plane waves, 410, 818, 819
Plane-polarized light, 940
Planetary motion, 149–53, 298
Planets, 149–53, 158 pr, 247 pr, 309 pr
Plasma, 340
Plasma globe, 810 pr
Plastic region, 319
Plate tectonics, 351
Plum-pudding model of atom, 1001
Pluto, 150, 152
Point charge (defn), 565, 569
    potential, 612–15
Point particle, 19, 96
Point rule, Kirchhoff's, 816 ff
Poise (P) (unit), 358
Poiseuille, J. L., 358
Poiseuille's equation, 358–59
Poisson, Siméon, 922
Polar molecules, 561, 579, 641
Polarization of light, 940–44, 949 pr
    by absorption, 940–42
    plane, 940–44
    by reflection, 942–43
    of skylight, 945
Polarizer, 941–44
Polarizing angle, 943
Polaroid, 940–42
Pole vault, 183, 192–93
Poles, magnetic, 707–9
    of Earth, 709
Pollution, 549–50
Pool depth, apparent, 852
Position vector, 59–60
Positive electric charge (defn), 560
Positive lens, 871
Positron, 996
Post-and-beam construction, 321
Potential difference, electric, 608 ff (see
        also Electric potential; Voltage)

...tial energy, 186–89 and *ff*
  diagrams, 204–5
  elastic, 188, 193, 194, 377–78
  electric, 607–10, 619–20, 636–38
  gravitational, 186–88, 199–201
  related to force, 188–89
Potentiometer, 705 *pr*
Pound (lb) (unit), 87
Power, 201–3, 660–65, 801
  rating of an engine, 202–3
  (*see also* Electric power)
Power factor (ac circuit), 801
Power generation, 767
Power of a lens, 868
Power plants, 767
Power transmission, 770–73
Powers of ten, 5
Poynting, J. H., 826 *fn*
Poynting vector, 826–27
Precession, 299
  of Earth, 303 *pr*
Precipitator, 645 *pr*
Precision, 5
Presbyopia, 883
Prescriptive laws, 3
Pressure, 341–45
  absolute, 345
  atmospheric, 344–48
  in fluids, 341–45
  in a gas, 345, 463–65, 478, 482–87
  gauge, 345
  head, 343
  hydraulic, 346
  measurement of, 346–48
  partial, 485
  and Pascal's principle, 346
  radiation, 828–29
  units for and conversions, 341, 345, 347
  vapor, 484–85, 491
Pressure amplitude, 427, 430–31
Pressure cooker, 485, 493 *pr*
Pressure gauges, 347
Pressure gradient, 359
Pressure head, 343
Pressure waves, 401, 426 *ff*
Prestressed concrete, 323
Primary coil, 770
Principal axis, 843
Principia (Newton), 85, 139
Principle, 3 (*see proper name*)
Principle of correspondence, 980, 1009
Principle of complementarity, 997
Principle of equipartition of energy, 512–13
Principle of equivalence, 155–56
Principle of superposition, 407–9, 436, 565, 569
Principal quantum number, 1004 *ff*
Printers, inkjet and laser, 583
Prism, 852–53
Prism binoculars, 855, 889
Probability:
  and entropy, 546–48
  in kinetic theory, 476–82

Problem-solving strategies, 30, 58, 64, 96, 102, 125, 166, 198, 229, 261, 314, 504, 551, 571, 685, 716, 741, 763, 849, 871, 913
Processes, reversible and irreversible (*defn*), 533
Projectile, horizontal range of, 68–69
Projectile motion, 51, 62, 71
  kinematic equations for (table), 64
  parabolic, 71
Proper length, 965
Proper time, 962
Proportional limit, 318–19
Proton-proton collision, 228–29
Pseudoforce, 300–1
Pseudovector, 254 *fn*
psi (unit), 341
PT diagram, 483
Pulley, 99–100
Pulse, wave, 396
Pulse induction, 770
Pulse-echo technique, 445–46
Pumps, 348, 361
  heat, 538–39
Pupil, 882
PV diagrams, 482–83, 487, 507
P waves, 401, 403, 416
Pythagorean theorem, A-2–A-4

Quadratic equation, 36
Quadratic formula, 38, A-1, inside back cover
Quadrupole, electric, 589 *pr*
Quality factor (*Q*-value) of a resonant system, 387, 392 *pr*, 810 *pr*
Quality of sound, 436
Quantities, base and derived, 7
Quantization:
  of angular momentum, 1004
  of electric charge, 564
  of energy, 989, 1003–9
Quantum condition, Bohr's, 1004, 1010
Quantum hypothesis, Planck's, 988–89
Quantum numbers, 989
  principal, 1004 *ff*
Quantum (quanta) of energy, 989
Quantum theory, 952, 987–1010
  of atoms, 1003–10
  of blackbody radiation, 987–88
  of light, 987–97
  of specific heat, 513
Quarks, 564 *fn*
Quartz oscillator, 450 *pr*
Quasistatic process (*defn*), 508
Q-value (quality factor) of a resonant system, 387, 392 *pr*, 810 *pr*

Radar, 446 *fn*, 823
Radial acceleration, 120 *ff*, 128
Radian (rad), measure for angles, 249–50

Radiant flux, 915
Radiation:
  blackbody, 987–88
  emissivity of, 518
  infrared (IR), 823–4, 852
  microwave, 823–4
  seasons and, 519
  solar constant and, 519
  thermal, 517–20
  ultraviolet (UV), 823–4, 852
  X-ray, 823–4, 938–39, 950 *pr*
Radiation field, 818
Radiation pressure, 828–29
Radio, 829–32
Radio waves, 823–24, 931
Radiotelescope, 931
Radius of curvature, 129
Radius of Earth estimate, 11, 15 *pr*
Radius of gyration, 279 *pr*
Rainbow, 853
RAM (Random access memory), 629, 644 *pr*
Raman effect, 1016
Ramp vs. stair analogy, 989
Random access memory (RAM), 629, 644 *pr*
Range of projectile, 68–69
Rapid estimating, 9–12
Rapid transit system, 49 *pr*
Rarefactions, in waves, 398
Ray, 410, 838
  paraxial (*defn*), 843
Ray diagram, 844, 849, 871
Ray model of light, 838 *ff*, 867 *ff*
Ray tracing, 838 *ff*, 867 *ff*
Rayleigh, Lord, 930, 988
Rayleigh criterion, 930
Rayleigh–Jeans theory, 988
RC circuit, 687–92
Reactance, 788, 797, 798
  capacitive, 798
  inductive, 797
  (*see also* Impedance)
Reaction time, 791
Read/Write head, 775
Real image, 840, 844, 869
Rearview mirror, 849
Receivers, radio and television, 830–31
Recoil, 220
Redshift, 443, 979
Reference frames, 19, 85, 300–2, 952 *ff*
  accelerating, 85, 88, 156, 300–2
  inertial, 85, 88, 300, 952 *ff*
  noninertial, 85, 88, 156, 300–2
  rotating, 300–2
  transformations between, 968–71
Reflecting telescope, 889
Reflection:
  angle of, 410, 838
  diffuse, 839
  law of, 409–10, 838
  and lens coating, 913
  of light, 837, 838–42

Reflection (*continued*)
  phase changes during, 909–14
  polarization by, 942–43
  specular, 839
  from thin films, 909–14
  total internal, 421 *pr*, 854–56
  of waves on a cord, 409
Reflection grating, 933
Reflectors, 865 *pr*
Refracting telescope, 888
Refraction, 415–16, 850 *ff*, 902–3
  angle of, 415, 850
  index of, 850
  law of, 415, 851, 902–3
  of light, 850–52, 902–3
  and Snell's law, 850–52
  at spherical surface, 856–58
  by thin lenses, 867–70
  of water waves, 415
Refrigerators, 536–38
  coefficient of performance (COP) of, 537
Regelation, 491 *pr*
Reinforced concrete, 323
Relative humidity, 485
Relative motion, 71–4, 951–80
Relative permeability, 749
Relative velocity, 71–74, 959 *ff*, 968 *ff*
Relativistic addition of velocities, 970–71
Relativistic energy, 974–8
Relativistic mass, 974
Relativistic momentum, 971–73, 977
  derivation of, 972–73
Relativity, Galilean-Newtonian, 952–54,
    968–69
Relativity, general theory of, 155–56
Relativity, special theory of, 951–80
  constancy of speed of light, 957
  four-dimensional space-time, 967
  impact of, 980
  and length, 964–67
  and Lorentz transformation, 968–71
  and mass, 974
  mass-energy relation in, 974–78
  postulates of, 957–58
  simultaneity in, 958–59
  and time, 959–64, 967
Relativity principle, 952–53, 957 *ff*
Relay, 751 *pr*
Resistance and resistors, 656–58, 661, 796
  in ac circuit, 796 *ff*
  with capacitor, 687–92, 795–802
  color code, 657
  and electric currents, 651 *ff*
  with inductor, 790–92, 795–802
  internal, 678–79
  in LRC circuit, 795–803
  of meter, 697
  in series and parallel, 679–83
  shunt, 695
  and superconductivity, 668–69
Resistance thermometer, 660
Resistive force, 129–30

Resistivity, 658–60
  temperature coefficient of, 659–60
Resolution:
  of diffraction grating, 937–39
  of electron microscope, 1000
  of eye, 930, 932–33
  of lens, 881, 929–32
  of light microscope, 932–3
  limits of, 929–32
  of telescope, 931
  of vectors, 55–58
Resolving power, 932, 938
Resonance, 385–87
  in ac circuit, 802
Resonant frequency, 385, 412–13, 432–35,
    802
Resonant oscillation, 385–86
Resonant peak, width of, 387
Rest energy, 974–76
Rest mass, 974
Resting potential, 669–70
Restitution, coefficient of, 243 *pr*
Restoring force, 170, 370
Resultant vector, 52–54, 57–58
Retentivity (magnetic), 749
Retina, 882
Reversible cycle, 533–35, 540
Reversible process, 533
Revolutions per second (rev/s), 253
Reynold's number, 366 *pr*
Rifle recoil, 220
Right-hand rule, 254, 710, 711, 714, 716,
    735, 763
Rigid object (*defn*), 249
  rotational motion of, 248–74, 294–97
  translational motion of, 234–6, 268–70
Rms (root-mean-square):
  current, 664–65
  speed, 479–82
  voltage, 664–65
Rock climbing, 106 *pr*, 110 *pr*, 137 *pr*,
    182 *pr*
Rocket propulsion, 83, 90, 219, 238
Roemer, Ole, 825
Roentgen, W. C., 938
Roller coaster, 191, 198
Rolling friction, 113, 273–74
Rolling motion, 267–73
  instantaneous axis of, 268
  total kinetic energy, 268
  without slipping, 267–71
Root-mean-square (rms) current, 664–65
Root-mean-square (rms) speed, 479–82
Root-mean-square (rms) voltage,
    664–65
Rotating reference frames, 300–2
Rotation, 248–302
  axis of (*defn*), 249
  frequency of (*defn*), 253
  of rigid body, 248–74, 294–97
Rotational imbalance, 296–97
Rotational inertia, 258, 259–60 (*see also*
    Moment of inertia)

Rotational kinetic energy, 265–67
  molecular, 499, 512–13
Rotational motion, 248–302
Rotational plus translational motion,
    267–68
Rotational work, 266
Rotor, 720, 768
Rough calculations, 9–12
Runway, 29
Russell, Bertrand, 999
Rutherford, Ernest, 1001
Rutherford's model of the atom, 1001
R-value, 517
Rydberg constant, 1002, 1007

S wave, 401
SAE, viscosity numbers, 358 *fn*
Safety factor, 322
Sailboats, and Bernoulli's principle, 357
Satellite dish, 831
Satellites, 139, 146–49
Saturated vapor pressure, 484
Saturation (magnetic), 748
Savart, Felix, 743
Sawtooth oscillator, 691, 706 *pr*
Sawtooth voltage, 691
Scalar (*defn*), 52
Scalar components, 55
Scalar (dot) product, 167–68
Scalar quantities, 52
Scale, musical, 431
Scanning electron microscope, 987, 1000
Scattering of light, 945
Schrödinger, Erwin, 987
Scientific notation, 5
Scuba diving, 473 *pr*, 475 *pr*, 495 *pr*,
    527 *pr*
Search coil, 783 *pr*
Seasons, 519
Second (s) (unit), 6
Second law of motion, 86–88, 90, 95–96,
    215, 218, 234–35, 953, 972
  for rotation, 259–63, 292–93
  for a system of particles, 234–35,
    292–93
Second law of thermodynamics, 529–48
  and Carnot efficiency, 534–35
  Clausius statement of, 529, 537
  and efficiency, 531–32
  and entropy, 539–48, 551
  general statement of, 543, 544, 548
  heat engine, 529, 530–32
  and irreversible processes, 533
  Kelvin-Planck statement of, 532, 535
  refrigerators, air conditioners, and heat
    pumps, 536–39
  reversible processes, 533
  and statistical interpretation of
    entropy, 546–48
  and time's arrow, 544
Secondary coil, 770
See saw, 314

nograph, 776
lf-inductance, 788–89
SEM, 987, 1000
Semiconductors, 561, 658
   resistivity of, 658
Sensitivity of meters, 696, 697
Series circuit, 634, 659
Shear modulus, 319, 321
Shear stress, 321
Shielded cable, 740, 789, 825
Shielding, electrical, 577, 740
SHM, 372–77
Shock absorbers, 369, 371, 383
Shock waves, 443–44
Short circuit, 663
Shunt resistor, 695
Shutter speed, 879, 881
SI (Système International) units, 7
   SI derived units: inside front cover
Siemens (S) (unit), 675 *pr*
Sign conventions (geometric optics),
   845–46, 849, 871
Significant figures, 4–5
   percent uncertainty vs., 5
Simple harmonic motion (SHM), 372–79
   applied to pendulums, 379–82
   related to uniform circular motion, 379
   sinusoidal nature of, 372
Simple harmonic oscillator (SHO), 372–79
   acceleration of, 374
   energy in, 377–78
   velocity and acceleration of, 374
Simple machines:
   lever, 177 *pr*, 313
   pulley, 99–100
Simple magnifier, 885–87
Simple pendulum, 13, 195, 379–81
   with damping, 384
Simultaneity, 958–60
Single-lens reflex (SLR) camera, 882
Single-slit diffraction, 922–27
Sinusoidal curve, 372 *ff*
Sinusoidal traveling wave, 404–6
Siphon, 362 *pr*, 368 *pr*
Skater, 284, 286, 309 *pr*
Skidding car, 126–27
Skier, 112, 117, 149, 183, 211 *pr*
Sky color, 945
Sky diver, 77 *pr*, 105 *pr*, 138 *pr*
Slope, of a curve, 23
SLR camera, 882
Slug (unit), 87
Snell, W., 851
Snell's law, 851–52, 856, 876, 902
Snowboarder, 51, 133 *pr*
Soap bubble, 900, 909, 912–13
Soaps, 360
Solar and Heliospheric Observatory
   (SOHO) satellite, 153
Solar (photovoltaic) cell, 550
Solar absorption spectrum, 936, 1002
Solar constant, 519

Solar energy, 550
Solar pressure, 828
Solar sail, 829
Solenoid, 733, 741–42, 747, 748–49, 788–89
Solid angle, 7 *fn*, 915 *fn*
Solids, 318 *ff*, 340, 455–56 (*see also* Phase,
   changes of)
Sonar, 444–45
Sonic boom, 444
Sonogram, 445
Sound, 424–46
   audible range of, 425
   and beats, 438–39
   dBs of, 428–31
   Doppler effect of, 439–43
   ear's response to, 431
   infrasonic, 426
   intensity of, 427–31
   interference of, 437–39
   level of, 428–31
   loudness of, 425, 427, 429
   loudness level of, 431
   mathematical representation of, 426–27
   pitch of, 425
   pressure amplitude of, 427, 430–31
   quality of, 436
   shock waves of, 443–44
   and sonic boom, 444
   sound level of, 428–31
   sources of, 431–36
   speed of, 425–6
   supersonic, 426, 443–44
   timbre of, 436
   tone color of, 436
   ultrasonic, 425, 445–46
Sound barrier, 444
Sound level, 428–31
Sound spectrum, 436
Sound track, optical, 992
Sound waves, 424–46 (*see also* Sound)
Sounding board, 433
Sounding box, 433
Soundings, 444
Source of emf, 678, 758–68
South pole, Earth, 709
South pole, of magnet, 708
Space:
   absolute, 953, 957
   curvature of, 155–56
   relativity of, 964–70
Space–time (4-D), 967
Space shuttle, 139
Space station, 131 *pr*, 149
Space travel, 963
Spark plug, 785
Speaker wires, 659
Special theory of relativity, 951–80 (*see
   also* Relativity, special theory of)
Specific gravity, 341, 351
Specific heat, 499–500
   for gases, 511–13
   for solids, 513

Spectrometer:
   light, 935–36
   mass, 724–25
Spectroscope and spectroscopy, 935–36,
   948 *pr*
Spectrum:
   absorption, 936, 1002
   atomic emission, 936, 1001–3, 1006–8
   continuous, 935, 988
   electromagnetic, 823, 852
   emitted by hot object, 987–88
   line, 935–36, 1002 *ff*
   visible light, 852–54
Specular reflection, 839
Speed, 20
   average, 20, 480–82
   of EM waves, 821–2, 825
   instantaneous, 22
   of light (*see entry below*)
   molecular, 480–82
   most probable, 480–82
   rms (root-mean-square), 479, 480, 482
   of sound (*see entry below*)
   (*see also* Velocity)
Speed of light, 6, 822, 825–26, 850, 902,
   953, 957, 975
   constancy of, 957
   measurement of, 825–26
   as ultimate speed, 974
Speed of sound, 425–26
   infrasonic, 426
   supersonic, 426, 443–44
Spherical aberration, 843, 891, 892, 929,
   932
Spherical mirrors, image formed by,
   842–49, 889, 892
Spherical shells, Earth, 142–43, A-9–A-11
Spherical wave, 403, 410
Spiderman, 179 *pr*
Spin, electron, 746
Spinning top, 299–300
Spring:
   potential energy of, 188, 193–94, 377–78
   vibration of, 370 *ff*
Spring constant, 170, 370
Spring equation, 170, 370
Spring stiffness constant, 170, 370
Spyglass, 889
Square wave, 409
Stable equilibrium, 204–5, 317
Standard conditions (STP), 466
Standard length, 6, 914
Standard mass, 6
Standard of time, 6
Standard temperature and pressure
   (STP), 466
Standards and units, 6–7
Standing waves, 412–15
   fundamental frequency of, 413
   mathematical representation of,
   414–15
   natural frequencies of, 412
   resonant frequencies of, 412–13

Star:
    color of, 988
    neutron, 287
    size of, 520
State:
    changes of, 482–83, 502–5
    energy, in atoms, 1003–9
    equation of, 463
        for an ideal gas, 466, 468
        van der Waals, 486–87
    of matter, 340, 456
    as physical condition of system, 454, 463
    of a system, 454
State variable, 455, 506, 539, 540
Static electricity, 559–642
Static equilibrium, 311–24
Static friction, 114, 270
    coefficient of, 113–14
Statics, 311–28
Stationary states in atom, 1003–10
Statistics and entropy, 546–8
Stator, 768
Steam engine, 528, 530–31
Stefan-Boltzmann constant, 518
Stefan-Boltzmann law (or equation), 518
Step-down transformer, 771
Step-up transformer, 771
Stirling cycle, 557 pr
Stokes's theorem, A-12–A-13
Stopping a car, 32, 174, 272–73
Stopping potential, 990
Stopping voltage, 990
STP, 466
Strain, 320–21
Strain gauge, 673
Streamline (defn), 352
Streamline flow, 352
Strength of materials, 319, 322
Stress, 320–21
    compressive, 321
    shear, 321
    tensile, 320–21
    thermal, 463
Stringed instruments, 413, 432–33
Strings, vibrating, 412–15, 431–33
Strong nuclear force, 155
Struts, 324
Sublimation, 483
Sublimation point, 483
Subtraction of vectors, 54–55
Suction, 348
Sun:
    mass determination, 152
    surface temperature of, 988
Sunglasses, polarized, 941, 942
Sunsets, 945
Superconducting magnets, 747
Superconductivity, 668–69
Superdome (New Orleans, LA), 328
Superfluidity, 483
Superposition, principle of, 407, 408–9, 436, 565, 569

Supersaturated air, 486
Supersonic speed, 426, 443
Surface area formulas, A-1, inside back cover
Surface charge density, 641
Surface tension, 359–60
Surface waves, 402, 410
Surfactants, 360
Surge protector, 792
Suspension bridge, 326
SUV rollover, 308 pr
S wave, 401
Symmetry, 10, 37, 140, 228, 233, 296, 313, 323, 325, 563 fn, 565, 571, 572, 573, 579, 580, 593, 595, 596, 597, 598, 600, 635, 637, 713, 738, 739, 740, 742, 743, 744, 774, 813, 815, 819, 847, 877, 907, 972, 997
Synapse, 669
Système International (SI), 7, inside front cover
Systems, 98, 454, 500
    closed, 500
    isolated, 218, 500
    open, 500
    as set of objects, 98, 454
    of units, 7
    of variable mass, 236–38

Tacoma Narrows Bridge, 386
Tail-to-tip method of adding vectors, 53–54
Tangential acceleration, 128–29, 251–52
Tape recorder, 749, 775
Telephone, cell, 771, 812, 824, 832
Telephoto lens, 882
Telescopes, 887–89, 930–31
    astronomical, 888–89
    Galilean, 887, 889
    Keplerian, 887 fn, 888
    magnification of, 888
    reflecting, 889
    refracting, 888–9
    resolution of, 930–31
    terrestrial, 889
Television, 621, 830–32, 943–4
Temperature, 456–59, 464, 469, 548–59
    absolute, 464, 469–70, 548–59
    Celsius (or centigrade), 457–58
    critical, 483
    Curie, 746, 750
    distinguished from heat and internal energy, 498
    Fahrenheit, 457–58
    human body, 458, 505
    Kelvin, 464, 469–70, 548–49
    molecular interpretation of, 476–80
    operating (of heat engine), 530
    relation to molecular velocities, 476–82
    scales of, 457–58, 464
    transition, 668
Temperature coefficient of resistivity, 658, 659–60

Tennis serve, 81 pr, 216, 220
Tensile strength, 322
Tensile stress, 320–21
Tension (stress), 320–21
Tension in flexible cord, 97
Terminal, of battery, 653, 655
Terminal velocity, 35 fn, 129–30
Terminal voltage, 678–79
Terrestrial telescope, 889
Tesla (T) (unit), 712
Test charge, 568
Testing, of ideas/theories, 2
Theories (general), 3
Thermal conductivity, 515
Thermal contact, 459
Thermal energy, 196, 498
    distinguished from heat and temperature, 498
    transformation of electric to, 660
    (see also Internal energy)
Thermal equilibrium, 459
Thermal expansion, 459–62
    anomalous behavior of water below 4°C, 462
    coefficients of, 460
    linear expansion, 459–61
    volume expansion, 461–62
Thermal pollution, 549–50
Thermal radiation, 519
Thermal resistance, 517
Thermal stress, 463
Thermionic emission, 620
Thermistor, 660
Thermodynamic probability, 547
Thermodynamic temperature scale, 548–49
Thermodynamics, 455, 496–520, 528–51
    first law of, 505–7
    second law of, 529–48
    third law of, 539 fn, 548–49
    zeroth law of, 459
Thermography, 519
Thermometers, 457–58
    bimetallic-strip, 457
    constant-volume gas, 458–59
    liquid-in-glass, 457
    mercury-in-glass thermometer, 457–58
    resistance, 660
Thermos bottle, 521 pr
Thermostat, 471 pr
Thin lens equation, 870–73
Thin lenses, 867–77 and ff
Thin-film interference, 909–14
Third law of motion, 89–91
Third law of thermodynamics, 539 fn, 548–49
Thomson, G. P., 998
Thomson, J. J., 722–3, 998, 999
Thought experiment, 958 and ff
    definition, 958
Three-dimensional waves, 402–3
Three-phase ac, 803
Three-way lightbulb, 704 pr
Threshold of hearing, 431

eshold of pain, 431

hrust, 237

TIA, 357

Tidal wave, 397

Timbre, 436

Time:
absolute, 953
proper, 962
relativity of, 958–64, 967, 968–71
standard of, 6

Time's arrow, 544

Time constant, 688, 791

Time dilation, 959–64, 970

Time intervals, 6, 21

Tire pressure, 468

Tire pressure gauge, 347

Tone color, 436

Toner, 583

Top, spinning, 299–300

Topographic map, 617

Toroid, 742, 748

Torque, 256–60 and ff, 290 ff
counter, 769
on current loop, 718–19

Torr (unit), 346–47

Torricelli, Evangelista, 346, 347–48, 356

Torricelli's theorem, 356

Torsion balance, 563

Torsion pendulum, 382

Total binding energy, 985 pr

Total internal reflection, 854–56

Townsend, J. S., 723

Transformation of energy, 196, 201

Transformations:
Galilean, 968–69
Lorentz, 969–71

Transformer, 770–73, 787

Transformer equation, 771

Transient ischemic attack (TIA), 357

Transition temperature, 668

Translational kinetic energy, 172–73

Translational motion, 18–239
and center of mass (CM), 234–36, 268–9

Transmission electron microscope, 1000

Transmission grating, 933 ff

Transmission lines, 772–73, 825

Transmission of electricity, 772–73

Transverse waves, 398 ff
and earthquakes, 401
velocity of, 399

Traveling sinusoidal wave, mathematical representation of, 404–6

Triangulation, 11

Trigonometric functions and identities, 56, 57, A-4–A-5, inside back cover

Trigonometric table, A-5

Triple point, 469, 483

Trough, 397

Trusses, 324–27

Tsunami, 397

Tubes:
flow in, 353–5, 357, 358–59
vibrating column of air in, 431 ff

Turbine, 549, 767

Turbulent flow, 352, 357

Turn signal, automobile, 691

Turning points, 204

Twin paradox, 963

Two-dimensional waves, 402

Tycho Brahe, 149

Tyrolean traverse, 106 pr, 338 pr

Ultimate speed, 974

Ultimate strength, 319, 322

Ultracapacitors, 644 pr

Ultrasonic frequencies, 426, 445

Ultrasonic waves, 426, 442, 445–46

Ultrasound, 445

Ultrasound imaging, 445–46

Ultraviolet (UV) light, 823, 824, 852

Unavailability of energy, 545–46

Uncertainty (in measurements), 3–5
estimated, 3
percent, 3–4, 5

Underdamped system, 383

Underexposure, 879

Underwater vision, 885

Unified atomic mass units (u), 7, 455

Uniform circular motion, 119–25
dynamics of, 122–25
kinematics of, 119–22

Uniformly accelerated motion, 28 ff, 62 ff

Uniformly accelerated rotational motion, 255

Unit conversion, 8–9, inside front cover

Unit vectors, 59

Units of measurement, 6
converting, 8–9, inside front cover
prefixes, 7
in problem solving, 9, 30, 102

Units and standards, 6–7

Universal gas constant, 466

Universal law of gravitation, 139, 140–43, 199–201, 564

Unpolarized light (defn), 941

Unstable equilibrium, 205, 317

Uranus, 150, 152

Useful magnification, 932–33

UV light, 823, 824, 852

Vacuum pump, 361

Van de Graaff generator, 607, 627 pr

van der Waals, J. D., 486

van der Waals equation of state, 486–87

van der Waals gas, 487

Vapor (defn), 483 (see also Gases)

Vapor pressure, 484–85

Vaporization, latent heat of, 502, 503, 505

Variable acceleration, 39–43

Variable mass systems, 236–38

Vector cross product, 289–90

Vector displacement, 20, 52–54, 59–60

Vector field, 575

Vector kinematics, 59–74

Vector product, 289–90

Vector sum, 52–58, 95, 143, 217

Vectors, 20, 52–62, 167–68, 289–90
addition of, 52–58
components of, 55–59
cross product, 289–90
kinematics, 59–74
magnetization, 750
multiplication of, 55, 167–68, 289–90
multiplication, by a scalar, 55
position, 59–60, 62
Poynting, 826–27
resolution of, 55–58
resultant, 52–54, 57–58
scalar (dot) product, 167–68
subtraction of, 54–55
sum, 52–58, 95, 143
unit, 59
vector (cross) product, 289–90

Velocity, 20–24, 60
addition of, 71–74, 970–71
angular, 250–55
average, 20–22, 60
drift, 666–68, 723, 724
escape, 201
of EM waves, 819–22
gradient, 358
instantaneous, 22–24, 60
of light, 6, 822, 825–26, 850, 902, 953, 957, 975
molecular, and relation to temperature, 479–82
phase, 404–5
relative, 71–74
relativistic addition of, 970–71
rms (root-mean-square velocity), 479–82
of sound, 425
supersonic, 426, 443
terminal, 35 fn, 129–30
of waves, 397, 399–401

Velocity selector, 717

Velocity-dependent forces, 129–30

Ventricular fibrillation, 638, 692

Venturi meter, 357

Venturi tube, 357

Venus, 150, 158 pr, 887

Vertical (defn), 92 fn

Vibrating strings, 412–15, 431–33

Vibration, 369–86
of air columns, 434–6
forced, 385–7
molecular, 499, 512–13
as source of waves, 397
of spring, 370 ff
on strings, 412–14, 431–3
(see also Oscillations)

Vibrational energy, 377–8
molecular, 499, 513

Virtual image, 840, 870

Viscosity, 352, 353 fn, 358–59
coefficient of, 358

Viscous force, 358–59

Visible light, wavelengths of, 823, 852–54

Visible spectrum, 852–54

Volt (V) (unit), 608

Volt-Ohm-Meter/Volt-Ohm-Milliammeter (VOM), 696
Volta, Alessandro, 608, 629, 652
Voltage, 607, 608 *ff*, 653 *ff*, 678 *ff*
 breakdown, 612
 electric field related to, 610–11, 617–19
 hazards of, 692–4
 measuring, 695–97
 peak, 664
 rms, 664
 terminal, 678–79
 (*see also* Electric potential)
Voltage drop, 684 (*see* Voltage)
Voltaic battery, 652
Voltmeter, 695–97, 721
 digital, 695, 697
Volume change under pressure, 321
Volume expansion (thermal), 460, 461–62
 coefficient of, 461
Volume formulas, A-1, inside back cover
Volume rate of flow, 353
VOM, 696
von Laue, Max, 939

Walking, 90
Water:
 anomalous behavior below 4° C, 462
 cohesion of, 360
 density of, 340–41, 351
 dipole moment of, 617
 and electric shock, 693
 expansion of, 462
 latent heats of, 503
 polar nature of, 561, 579, 617
 properties of: inside front cover
 saturated vapor pressure, 484
 specific gravity of, 341, 351
 thermal expansion of, 462
 triple point of, 469, 483
Watson, J., 939
Watt, James, 202 *fn*
Watt (W) (unit), 202, 661
Wave(s), 395–416, 817 *ff*, 823 *ff*, 900–45
 amplitude of, 397, 402, 404, 426, 430
 complex, 408, 436
 composite, 408, 436
 compression, 398, 401
 continuous (*defn*), 397
 diffraction of, 416, 901, 921–39
 dispersion, 409, 853
 displacement of, 404 *ff*
 earthquake, 401, 402, 403, 416
 electromagnetic, 817–32 (*see also* Light)
 energy in, 402–3
 frequency, 397
 front, 410
 harmonic (*defn*), 405
 incident, 410, 415
 infrasonic, 426
 in-phase, 411
 intensity, 402–3, 427–31, 826–7
 interference of, 410–11, 437–8, 903–14
 light, 821–6, 900–45 (*see* Light)
 linear, 402

longitudinal (*defn*), 398
 velocity of, 400–1
 mathematical representation of, 404–6, 426–27
 of matter, 997–9, 1009–10
 mechanical, 395–416
 number, 404
 one-dimensional, 402–3
 P, 401, 403, 416
 periodic (*defn*), 397
 phase of, 404
 plane, 410, 818
 power, 402
 pressure, 401, 426
 pulse, 396
 radio, 823–24, 931
 reflection of, 409–10
 refraction of, 415–16
 S, 401
 shock, 443–44
 sound, 424–46, 824
 spherical, 403, 410
 square, 409
 standing, 412–15, 431–35
 surface, 402, 410
 three-dimensional, 402–3
 tidal, 397
 transmission of, 409
 transverse, 398 *ff*, 819, 840
 velocity of, 399
 traveling, 404–6
 two-dimensional, 402
 types of, 398–99 (*see also* Light)
 ultrasonic, 426, 442
 velocity of, 397, 399–401
 water, 395 *ff*
 (*see also* Light)
Wave equation, 406–8, 822
Wave front, 410, 901
Wave intensity, 402–3, 427–31, 826–7
Wave motion (*see* Waves; Light; Sound)
Wave nature of matter, 997–9, 1009–10
Wave number (*defn*), 404
Wave theory of light, 900–45
Wave velocity, 397, 399–401 (*see also* Light; Sound)
Wave-particle duality:
 of light, 997
 of matter, 997–999, 1009–10
Wavelength:
 definition, 397
 Compton, 994
 de Broglie, 997–98, 1009
 depending on index of refraction, 853, 902
 as limit to resolution, 932
 of material particles, 997–9, 1009–10
Weak nuclear force, 155
Weather, 302, 525 *pr*
Weber (Wb) (unit), 760
Weight, 84, 86, 92–94, 143
 as a force, 86, 92
 force of gravity, 84, 92–94, 143
 mass compared to, 86, 92

Weightlessness, 148–49
Wheatstone bridge, 704 *pr*
Wheel balancing, 296
White light, 852–53
Wide-angle lens, 882, 892
Wien, W., 988
Wien's displacement law, 988
Wien's radiation theory, 988
Wind instruments, 433–36
Wind power, 550
Windings, 720
Windshield wipers, 691
Wing of an airplane, lift on, 356–57
Wireless communication, 812, 829–32
Wire-wound resistors, 657
Work, 163–76, 497, 505–10
 to bring positive charges together, 613
 compared to heat, 505
 defined, 164, 169, 505 *ff*
 done by a constant force, 164–66
 done by a gas, 508 *ff*
 done by a spring force, 170–71
 done by a varying force, 168–71
 in first law of thermodynamics, 505–7
 from heat engines, 530 *ff*
 and power, 201
 relation to energy, 172–74, 186–89, 197, 201, 266
 rotational, 266
 units of, 164
Work function, 990–91
Work-energy principle, 172–73, 176, 266, 974, 978
 energy conservation vs., 197
 general derivation of, 176
 as reformulation of Newton's laws, 173
Working substance (*defn*), 530

X-rays, 823, 824, 938–39
 in electromagnetic spectrum, 823
X-ray crystallography, 939
X-ray diffraction, 938–39
X-ray scattering, 994–95

YBCO superconductor, 668
Yerkes Observatory, 888
Young, Thomas, 903, 906
Young's double-slit experiment, 903–9, 927–9
Young's modulus, 319
Yo-Yo, 271, 281 *pr*
Yttrium, barium, copper, oxygen superconductor, (YBCO), 668

Zeeman effect, 731 *pr*
Zero, absolute, temperature of, 464, 549
Zeroth law of thermodynamics, 459
Zoom lens, 882

Photo Credits

20-1a-c Leonard Lessin/Peter Arnold, Inc. **20-15a** Corbis Digital Stock **20-15b** Warren Gretz/NREL/US DOE/Photo Researchers, Inc. **20-15c** Lionel Delevingne/Stock Boston **Table 20-2 top-bottom** © Royalty-Free/Corbis; Billy Hustace/Stone/Allstock/Getty Images; Michael Collier; Inga Spence/Visuals Limited **20-19** Geoff Tompkinson/Science Photo Library/Photo Researchers, Inc. **20-22** Inga Spence/Visuals Unlimited **20-23** Michael Collier **CO-21** Richard Megna/Fundamental Photographs, NYC **21-39** Michael J. Lutch/Boston Museum of Science **21-46** Dr. Gopal Murti/Science Photo Library/Photo Researchers, Inc. **CO-23** Lester V. Bergman/Corbis **23-19** Douglas C. Giancoli **23-24** Jon Feingersh **CO-24** Tom Pantages **24-13** Tom Pantages **24-14** J. Reid/Custom Medical Stock Photo, Inc. **CO-25** Mahaux Photography/Image Bank/Getty Images **25-1** J.-L. Charmet/ Photo Researchers, Inc. **25-6a** Dave King/Dorling Kindersley Media Library **25-12** Tom Pantages **25-15** Richard Megna/ Fundamental Photographs, NYC **25-16** Mark C. Burnett/Photo Researchers, Inc. **25-18** A&J Verkaik/Bettmann/Corbis **25-33** Alexandra Truitt & Jerry Marshall **25-34** Scott T. Smith/Bettmann/Corbis **25-37** Jim Wehtje/Photodisc/Getty Images **CO-26** Dino Vournas/Reuters Ltd. **26-15a** Alamy Images **26-22** Charles O'Rear/Corbis **26-25a** Photodisc/Getty Images **26-25b** William E. Ferguson **26-25c** Ed Degginger/Color-Pic, Inc. **26-27a** Paul Silverman/Fundamental Photographs, NYC **26-27b** Paul Silverman/Fundamental Photographs, NYC **CO-27** Richard Megna/Fundamental Photographs, NYC **27-1** Michael Newman/ PhotoEdit, Inc. **27-4a** Stephen Oliver/Dorling Kindersley Media Library **27-6** Mary Teresa Giancoli **27-8a** Richard Megna/ Fundamental Photographs, NYC **27-18** Richard Megna/Fundamental Photographs, NYC **27-2b** Steven Hausler/Hays Daily News/ AP Wide World Photos **CO-28** Richard Megna/Fundamental Photographs, NYC **28-24** Richard Megna/Fundamental Photographs, NYC **28-32** Clive Streeter/Dorling Kindersley Media Library **CO-29** Richard Megna/Fundamental Photographs, NYC **29-8** Diva de Provence/ DIVA Induction froid **29-13** Jeff Hunter/Image Bank/Getty Images **29-17** Rick Bowmer/AP Wide World Photos **29-22** Jack Hollingsworth/ Photodisc/Getty Images **29-23** Robert Houser **29-29b** Terence Kearey **29-32a** Richard Megna/ Fundamental Photographs, NYC **29-32b** Christian Botting **CO-30** Corbis Royalty Free **CO-31** Douglas C. Giancoli **31-1** American Institute of Physics **31-13** The Image Works **31-22** Spencer Grant/PhotoEdit, Inc. **31-25** World Perspectives/Stone/ Allstock/Getty Images **CO-32** Douglas C. Giancoli **32-6** Douglas C. Giancoli **32-11a** Mary Teresa Giancoli and Suzanne Saylor **32-11b** Francesco Campani **32-20** Travel Pix Ltd./Super Stock, Inc. **32-23** Giuseppe Molesini, Istituto Nazionale di Ottica Florence **32-27** David Parker/Science Photo Library/Photo Researchers, Inc. **32-30b** Lewis Kemper/Photolibrary.com **32-35b** Mitterer/ Mauritus, GMBH/ Phototake NYC **32-41** Douglas C. Giancoli **32-44** Mary Teresa Giancoli **CO-33** Richard Megna/Fundamental Photographers, NYC **33-1** Douglas C. Giancoli **33-2c** Douglas C. Giancoli **33-2d** Douglas C. Giancoli **33-4** Kari Erik Marttila/Kari Erik Marttila Photography **33-7a** Douglas C. Giancoli **33-7b** Douglas C. Giancoli **33-13a** Scott Dudley **33-13b** Scott Dudley **33-21** Mary Teresa Giancoli **33-22a** Mary Teresa Giancoli **33-22b** Mary Teresa Giancoli **33-35a** Franca Principe/Istituto e Museo di Storia della Scienza **33-35b** Franca Principe/Istituto e Museo di Storia della Scienza **33-37** Yerkes Observatory **33-38c** Sandy Huffaker/ Getty Images **33-38d** Roger Ressmeyer/Corbis **33-40b** Olympus America Inc. **33-45** Ron Chapple/Ron Chapple Photography **33-49** NOAA Space Environment Center **CO-34** Giuseppe Molesini, Istituto Nazionale di Ottica Florence **34-4a** John M. Duany IV/Fundamental Photographs, NYC **34-9a** Giuseppe Molesini, Istituto Nazionale di Ottica Florence **34-16a/b/c** Giuseppe Molesini, Istituto Nazionale di Ottica Florence **34-18b** Giuseppe Molesini, Istituto Nazionale di Ottica Florence **34-20b/c** Bausch & Lomb Inc. **34-22** Kristen Brochmann/Fundamental Photographs, NYC **CO-35** Richard Megna/Fundamental Photographs, NYC **35-2a** P. M. Rinard/American Journal of Physics **35-2b** Ken Kay/Fundamental Photographs, NYC **35-2c** Ken Kay/Fundamental Photographs, NYC **35-11a/b** Richard Megna/Fundamental Photographs, NYC **35-12a/b** Springer-Verlang GmbH & Co. KG **35-15** Space Telescope Science Institute **35-16** David Parker/Photo Researchers, Inc. **35-20** Spike Mafford/Photodisc/Getty Images **35-22** Wabash Instrument Corp./Fundamental Photographs, NYC **35-27** Burndy Library **35-30** Rosalind Franklin/Photo Researchers, Inc. **35-37** Diane Schiumo/Fundamental Photographs, NYC **35-40a/b** Douglas C. Giancoli **35-45** Texas Instruments Inc. **CO-36** Cambridge University Press; "The City Blocks Became Still Shorter" photo from page 4 of the book "Mr Tompkins in Paperback" by George Gamow. Reprinted with the permission of Cambridge University Press **36-1** Bettmann/Corbis **36-16** Cambridge University Press; "Unbelievably Shortened" photo from page 3 of the book "Mr Tompkins in Paperback" by George Gamow. Reprinted with permission of Cambridge University Press **CO-37** P. M. Motta & F. M. Magliocca/Science Photo Library/Photo Researchers, Inc. **37-10** S. A. Goudsmit/American Institute of Physics/Niels Bohr Library/AIP Emilio Segrè Visual Archives **37-11** Education Development Center, Inc. **37-15a** Lee D. Simon/ Science Source/Photo Researchers, Inc. **37-15b** Oliver Meckes/ Max Planck Institut Tubigen/Photo Researchers, Inc. **37-19b** Richard Megna/Fundamental Photographs, NYC **37-20** Wabash Instrument Corp./Fundamental Photographs, NYC.

Table of Contents Photos p. iii **left** © Reuters/Corbis; **right** Agence Zoom/Getty Images **p. iv** **left** Ben Margot/AP Wide World Photos; **right** Kai Pfaffenbach/Reuters Limited **p. v** Jerry Driendl/Taxi/Getty Images **p. vi** **left** Richard Price/Photographer's Choice/Getty Images; **right** Frank Herholdt/Stone/Getty Images **p. viii** Richard Megna/Fundamental Photographs, NYC **p. ix** **left** Richard Megna/Fundamental Photographs, NYC, **right** Giuseppe Molesini, Istituto Nazionale di Ottica Florence **p. x** © Richard Cummins/Corbis **p. xi** Lester Lefkowitz/Taxi/Getty Images **p. xvii** J. C. Soto.

Useful Geometry Formulas—Areas, Volumes

Circumference of circle $C = \pi d = 2\pi r$

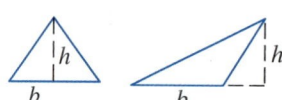

Area of circle $A = \pi r^2 = \dfrac{\pi d^2}{4}$

Area of rectangle $A = \ell w$

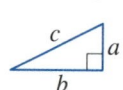

Area of parallelogram $A = bh$

Area of triangle $A = \frac{1}{2} hb$

Right triangle
(Pythagoras) $c^2 = a^2 + b^2$

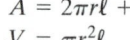

Sphere: surface area $A = 4\pi r^2$
volume $V = \frac{4}{3}\pi r^3$

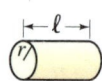

Rectangular solid:
volume $V = \ell wh$

Cylinder (right):
surface area $A = 2\pi r\ell + 2\pi r^2$
volume $V = \pi r^2 \ell$

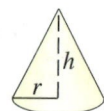

Right circular cone:
surface area $A = \pi r^2 + \pi r \sqrt{r^2 + h^2}$
volume $V = \frac{1}{3}\pi r^2 h$

Quadratic Formula

Equation with unknown x, in the form

$$ax^2 + bx + c = 0,$$

has solutions

$$x = \frac{-b \pm \sqrt{b^2 - 4ac}}{2a}.$$

Exponents

$(a^n)(a^m) = a^{n+m}$ [Example: $(a^3)(a^2) = a^5$]
$(a^n)(b^n) = (ab)^n$ [Example: $(a^3)(b^3) = (ab)^3$]
$(a^n)^m = a^{nm}$ $\begin{bmatrix}\text{Example: } (a^3)^2 = a^6 \\ \text{Example: } (a^{\frac{1}{4}})^4 = a\end{bmatrix}$

$a^{-1} = \dfrac{1}{a}$ $a^{-n} = \dfrac{1}{a^n}$ $a^0 = 1$

$a^{\frac{1}{2}} = \sqrt{a}$ $a^{\frac{1}{4}} = \sqrt{\sqrt{a}}$

$(a^n)(a^{-m}) = \dfrac{a^n}{a^m} = a^{n-m}$ [Ex.: $(a^5)(a^{-2}) = a^3$]

$\dfrac{a^n}{b^n} = \left(\dfrac{a}{b}\right)^n$

Logarithms [Appendix A–7; Table A–1]

If $y = 10^x$,   then $x = \log_{10} y = \log y$.
If $y = e^x$,   then $x = \log_e y = \ln y$.

$\log(ab) = \log a + \log b$

$\log\left(\dfrac{a}{b}\right) = \log a - \log b$

$\log a^n = n \log a$

Some Derivatives and Integrals†

$$\frac{d}{dx}x^n = nx^{n-1} \qquad \int \sin ax\, dx = -\frac{1}{a}\cos ax$$

$$\frac{d}{dx}\sin ax = a\cos ax \qquad \int \cos ax\, dx = \frac{1}{a}\sin ax$$

$$\frac{d}{dx}\cos ax = -a\sin ax \qquad \int \frac{1}{x}\, dx = \ln x$$

$$\int x^m\, dx = \frac{1}{m+1}x^{m+1} \qquad \int e^{ax}\, dx = \frac{1}{a}e^{ax}$$

† See Appendix B for more.

Binomial Expansion

$$(1 \pm x)^n = 1 \pm nx + \frac{n(n-1)}{2\cdot 1}x^2 \pm \frac{n(n-1)(n-2)}{3\cdot 2\cdot 1}x^3 + \cdots \quad [\text{for } x^2 < 1]$$

$$\approx 1 \pm nx \quad [\text{for } x \ll 1]$$

Trigonometric Formulas [Appendix A–9]

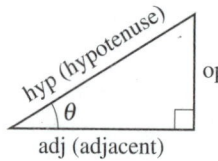

$\sin\theta = \dfrac{\text{opp}}{\text{hyp}}$

$\cos\theta = \dfrac{\text{adj}}{\text{hyp}}$

$\tan\theta = \dfrac{\text{opp}}{\text{adj}}$

$\text{adj}^2 + \text{opp}^2 = \text{hyp}^2$ (Pythagorean theorem)

$\tan\theta = \dfrac{\sin\theta}{\cos\theta}$

$\sin^2\theta + \cos^2\theta = 1$

$\sin 2\theta = 2\sin\theta\cos\theta$

$\cos 2\theta = (\cos^2\theta - \sin^2\theta) = (1 - 2\sin^2\theta) = (2\cos^2\theta - 1)$

$\sin(180° - \theta) = \sin\theta$ $\cos(180° - \theta) = -\cos\theta$
$\sin(90° - \theta) = \cos\theta$
$\cos(90° - \theta) = \sin\theta$
$\sin\frac{1}{2}\theta = \sqrt{(1 - \cos\theta)/2}$ $\cos\frac{1}{2}\theta = \sqrt{(1 + \cos\theta)/2}$
$\sin\theta \approx \theta$ [for small $\theta \lesssim 0.2$ rad]
$\cos\theta \approx 1 - \dfrac{\theta^2}{2}$ [for small $\theta \lesssim 0.2$ rad]
$\sin(A \pm B) = \sin A \cos B \pm \cos A \sin B$
$\cos(A \pm B) = \cos A \cos B \mp \sin A \sin B$

For any triangle:
$c^2 = a^2 + b^2 - 2ab\cos\gamma$ (law of cosines)
$\dfrac{\sin\alpha}{a} = \dfrac{\sin\beta}{b} = \dfrac{\sin\gamma}{c}$ (law of sines)

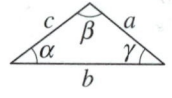